SIXTH EDITION

EARTH

PORTRAIT OF A PLANET

SIXTH EDITION

EARTH

PORTRAIT OF A PLANET

Stephen Marshak

UNIVERSITY OF ILLINOIS

W. W. NORTON & COMPANY
NEW YORK • LONDON

W. W. Norton & Company has been independent since its founding in 1923, when William Warder Norton and Mary D. Herter Norton first published lectures delivered at the People's Institute, the adult education division of New York City's Cooper Union. The firm soon expanded its program beyond the Institute, publishing books by celebrated academics from America and abroad. By midcentury, the two major pillars of Norton's publishing program—trade books and college texts—were firmly established. In the 1950s, the Norton family transferred control of the company to its employees, and today—with a staff of four hundred and a comparable number of trade, college, and professional titles published each year—W. W. Norton & Company stands as the largest and oldest publishing house owned wholly by its employees.

Editor: Jake Schindel
Senior Project Editor: Thomas Foley
Associate Production Director: Benjamin Reynolds
Assistant Editor: Rachel Goodman
Copy Editor: Norma Sims Roche
Managing Editor, College: Marian Johnson
Managing Editor, College Digital Media: Kim Yi
Digital Media Editor: Robert Bellinger
Associate Media Editors: Arielle Holstein and Gina Forsythe
Assistant Media Editor: Liz Vogt
Media Project Editor: Marcus Van Harpen
Editorial Assistant, Digital Media: Kelly Smith
Marketing Manager, Geology: Katie Sweeney
Design Director: Rubina Yeh
Designer: Lissi Sigillo
Director of College Permissions: Megan Schindel
Photography Editor: Trish Marx
Developmental Editor for the First Edition: Susan Gaustad
Composition and page layout by MPS North America LLC
MPS Project Manager: Jackie Strohl
Illustrations for the Second, Third, Fourth, and Fifth Editions by Precision Graphics / Lachina
Illustrations for the Sixth Edition: Stan Maddock and Joanne Brummett
Manufacturing by Transcontinental Interglobe—Beauceville, Quebec

Permission to use copyrighted material is included in the backmatter of this book.

Library of Congress Cataloging-in-Publication Data

Name: Marshak, Stephen, 1955– author.
Title: Earth : portrait of a planet / Stephen Marshak
  (University of Illinois).
Description: Sixth edition. | New York : W.W. Norton & Company, [2019] |
  Includes bibliographical references and index.
Identifiers: LCCN 2018010310 | ISBN 9780393617511 (pbk.)
Subjects: LCSH: Geology—Textbooks.
Classification: LCC QE26.3 .M36 2019 | DDC 550--dc23 LC record available at https://lccn.loc.gov/2018010310

W. W. Norton & Company, Inc., 500 Fifth Avenue, New York, NY 10110
wwnorton.com

W. W. Norton & Company Ltd., 15 Carlisle Street, London W1D 3BS

1 2 3 4 5 6 7 8 9 0

To Kathy, David, Emma, and Michelle

Brief Contents

Special Features

Contents

PART I   OUR ISLAND IN SPACE

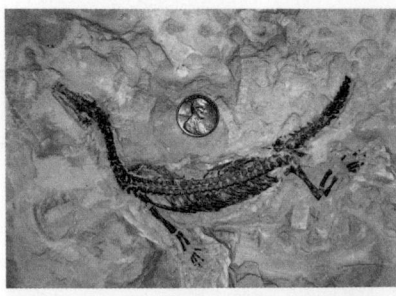

CHAPTER 3

Drifting Continents and Spreading Seas   66

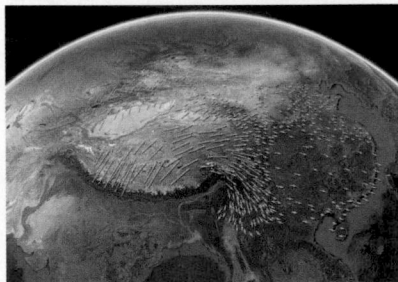

CHAPTER 4

The Way the Earth Works: Plate Tectonics   90

PART II EARTH MATERIALS

CHAPTER 5

Patterns in Nature: Minerals   120

INTERLUDE A
Introducing Rocks   148

CHAPTER 6
Up from the Inferno: Magma and Igneous Rocks   160

INTERLUDE B
A Surface Veneer: Sediments and Soils   190

CHAPTER 7
Pages of the Earth's Past: Sedimentary Rocks   210

CHAPTER 8
Metamorphism: A Process of Change 242

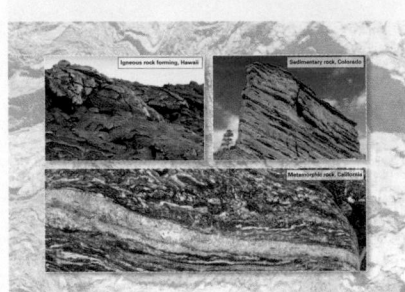

INTERLUDE C
The Rock Cycle in the Earth System 270

PART III TECTONIC ACTIVITY OF A DYNAMIC PLANET

CHAPTER 9
The Wrath of Vulcan: Volcanic Eruptions 280

CHAPTER 10
A Violent Pulse: Earthquakes   322

INTERLUDE D
The Earth's Interior, Revisited: Seismic Layering, Gravity, and the Magnetic Field   370

CHAPTER 11
Crags, Cracks, and Crumples: Crustal Deformation and Mountain Building   390

PART IV HISTORY BEFORE HISTORY

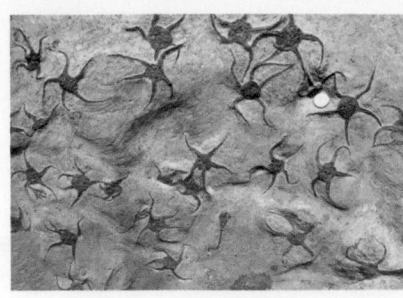

PART V EARTH RESOURCES

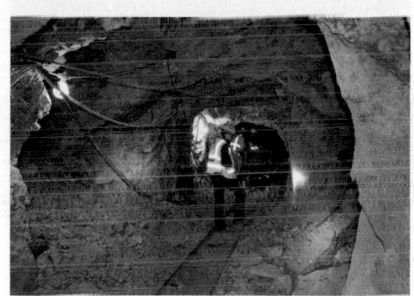

PART VI PROCESSES AND PROBLEMS AT THE EARTH'S SURFACE

CHAPTER 18
Restless Realm: Oceans and Coasts   676

CHAPTER 19
A Hidden Reserve: Groundwater   720

CHAPTER 20
An Envelope of Gas: The Earth's Atmosphere and Climate   756

CHAPTER 21

Dry Regions: The Geology of Deserts 802

CHAPTER 22

Amazing Ice: Glaciers and Ice Ages 832

CHAPTER 23

Global Change in the Earth System 876

Metric Conversion Chart   A-1

The Periodic Table of Elements   A-2

Glossary   G-1

Credits   C-1

Index   I-1

Preface

Narrative Themes

Why do earthquakes, volcanoes, floods, and landslides happen? What causes mountains to rise? How do beautiful landscapes develop? How have climate and life changed through time? When did the Earth form, and by what process? Where do we dig to find valuable metals, and where do we drill to find oil? Does sea level change? Do continents move? The study of geology addresses these important questions and many more. But from the birth of the discipline, in the late 18th century, until the mid-20th century, geologists considered each question largely in isolation, without pondering its relation to the others. This approach changed, beginning in the 1960s, in response to the formulation of two paradigm-shifting ideas that have unified thinking about the Earth and its features. The first idea, called the *theory of plate tectonics*, states that the Earth's outer shell, rather than being static, consists of discrete plates that slowly move, relative to each other, so that the map of our planet continuously changes. Plate interactions cause earthquakes and volcanoes, build mountains, provide gases that make up the atmosphere, and affect the distribution of life on our planet. The second idea, the *Earth System concept*, emphasizes that the Earth's water, land, atmosphere, and living inhabitants are dynamically interconnected, so that materials constantly cycle among various living and nonliving reservoirs on, above, and within the planet. In the context of this idea, we have come to realize that the history of life is intimately linked to the history of the physical Earth, and vice versa.

Earth: Portrait of a Planet, Sixth Edition, is an introduction to the study of our planet that uses the theory of plate tectonics as well as the Earth System perspective throughout, to weave together a number of narrative themes, including:

1. The solid Earth, the oceans, the atmosphere, and life interact in complex ways.
2. Many important geologic processes involve the interactions of plates—pieces of the Earth's outer, relatively rigid shell.
3. The Earth is a planet formed, like other planets, from dust and gas. But, in contrast to other planets, the Earth is a dynamic place where new geologic features continue to form, and old ones continue to be destroyed.
4. The Earth is very old—indeed, about 4.56 billion years have passed since its birth. During this time, the map of the planet and its surface features have changed, and life has evolved.
5. Internal processes (driven by the Earth's interior heat) and external processes (driven by heat from the Sun) interact at the Earth's surface to produce complex landscapes.
6. Geologic knowledge can help society understand, and perhaps reduce, the danger of natural hazards, such as earthquakes, volcanoes, landslides, and floods.
7. Energy and mineral resources come from the Earth and are formed by geologic phenomena. Geologic study can help locate these resources and mitigate the consequences of their use.
8. Geology is a science, and the ideas of science come from observation, calculation, and experiment by researchers—it is a human endeavor. Furthermore, geology utilizes ideas from physics, chemistry, and biology, so the study of geology provides an excellent opportunity for students to improve their science literacy.

These narrative themes serve as the book's take-home message, a message that students hopefully will remember long after they finish their introductory geology course. In effect, the themes provide a mental framework on which students can organize and connect ideas, and can develop a modern, coherent image of our planet.

Pedagogical Approach

Educational research demonstrates that students learn best when they actively engage with a combination of narrative text and narrative art. Some students respond more to the words of a textbook, which help to organize information, provide answers to questions, fill in the essential steps that link ideas together, and help a student develop a context for understanding ideas. Some students respond more to figures and photos, for images help students comprehend, visualize, and remember the narrative. And some respond best to active learning, an approach where students can practice their knowledge by putting ideas to work. *Earth: Portrait of a Planet*, Sixth Edition, provides all three of these learning tools. The text has been crafted to be engaging, the art has been configured to tell a story, the chapters are laid out to help students internalize key principles, and the online activities have been designed both to engage students and to provide active feedback. This book's narrative doesn't merely provide a dry statement of facts. Rather, it provides the story behind the story—the reasoning and observation that led to our current understanding, as well as an explanation of the processes that cause a particular geologic phenomenon.

Each chapter starts with a list of *Learning Objectives* that frame key pedagogical goals for each chapter. These objectives are revisited in the end-of-chapter *Review Questions* and in the *Smartwork5 online activities. Take-Home Message* panels, which include both a brief summary and a key question, appear at the end of each section to help students solidify key themes before proceeding to the next section. Throughout the chapter, brief *Did You Ever Wonder?* questions prompt students with real-life questions they may already have thought about—answers to these questions occur in the nearby text. *See for Yourself* panels guide students to visit spectacular examples of geologic features, using the power of Google Earth. They allow students to apply their newly acquired knowledge to the interpretation of real-world examples. In the ebook version of the text, these features are live links that "fly" students to the precise locations discussed. Each chapter concludes with a concise, two-page chapter summary that reinforces understanding and provides a concise study tool at the same time. *Review Questions* at the end of each chapter include two parts: the first addresses basic concepts; and the second, labeled as *On Further Thought*, stimulates critical thinking opportunities that invite students to think beyond the basics. Some of the questions use visuals from the chapter.

To enhance active-learning opportunities, the *Smartwork5 online activity system* has been developed specifically for *Earth: Portrait of a Planet*, Sixth Edition. *Smartwork5* offers a wide range of visual exercises, including ranking, labeling, and sorting questions. *Smartwork5* questions make the textbook art interactive, and they integrate the Narrative Art Videos, animations, and simulations that have been created to accompany the text. Questions are designed to give students answer-specific feedback when they are incorrect, coaching them toward developing a thorough understanding of the core concepts discussed in the book.

Organization

The topics covered in this book have been arranged so that students can build their knowledge of geology on a foundation of overarching principles. To set the stage, the book starts by describing processes that led to the formation of the Earth, in the context of scientific cosmology. It then introduces the architecture of our planet, from surface to center. With this basic background, students are prepared to delve into plate tectonics theory. Plate tectonics appears early in the book, so that students can relate the content of subsequent chapters to the theory. Knowledge of plate tectonics, for example, helps students understand the suite of chapters on minerals, rocks, and the rock cycle. Knowledge of plate tectonics and rocks together, in turn, provides a basis for studying volcanoes, earthquakes, and mountains. And with this background, students are prepared to see how the map of the Earth has changed through the vast expanse of geologic time, and how energy and mineral resources have developed. The book's final chapters address processes occurring at or near the Earth's surface, such as the flow of rivers, the evolution of coasts, and the carving of landscapes by glaciers. We also consider the problems that the Earth's surface processes can cause, such as landslides and floods. This part concludes with a topic of growing concern in society—global change, particularly climate change.

In addition to numbered chapters, the book contains several *Interludes*. These are, in effect, "mini-chapters" that focus on topics that are self-contained but are not broad enough to require an entire chapter. By placing selected topics in interludes, we can keep the numbered chapters reasonable in length, and can provide additional flexibility in sequencing topics within a course.

Although the sequence of chapters and interludes was chosen for a reason, this book is designed to be flexible enough for instructors to choose their own strategies for teaching geology. Therefore, each self-contained chapter reiterates relevant material where necessary. For example, if instructors prefer to introduce minerals and rocks before plate tectonics, they simply need to reorder the reading assignments. A low-cost, loose-leaf version of the book allows instructors to have students bring to class only the chapters that they need.

We have used a tiered approach in highlighting terminology in *Earth: Portrait of a Planet*, Sixth Edition. Terminology, the basic vocabulary of a subject, serves an important purpose in simplifying the discussion of topics. For example, once students understand the formal definition of a mineral, the term can be used again in subsequent discussion without further explanation or redundancy. Too much new vocabulary, however, can be overwhelming. So we have tried to keep the book's *Guide Terms* (set in boldface and referenced at the end of each chapter for studying purposes) to a minimum. Other terms, less significant but still useful, appear in italic when presented, to provide additional visual focus for students as they read the chapters. We take care not to use vocabulary until it has been completely introduced and defined.

Special Features of this Edition

Earth: Portrait of a Planet, Sixth Edition, contains a number of new or revised features that distinguish it from all competing texts.

Narrative Art, *What a Geologist Sees,* and *See for Yourself*

It's difficult to understand many features of the Earth System without being able to see them. To help students visualize these and other features, this book is lavishly illustrated with figures that try to give a realistic context for the particular feature, without overwhelming students with too much extraneous detail. The talented artists who worked on the book have used the latest computer graphics software, resulting in the most sophisticated pedagogical art ever provided by a geoscience text. Many figures have been updated with an eye toward improving students' 3-D visualization skills. The figures have also been reconfigured to be more friendly and intuitive. All of the plate tectonics figures have been revised in this Sixth Edition in order to provide students the clearest, most vibrant, and most accurate visual understanding of the Earth's interior dynamics.

In addition to the drawn art, the book also boasts over 1,000 stunning photographs from all around the world. Many of the photographs were taken by the author, in order to illustrate the exact concept under discussion. Where appropriate, photographs are accompanied by annotated sketches named *What a Geologist Sees.* These figures allow students to see how geologists perceive the world around them and to encourage students to start thinking like geologists.

Throughout the book, drawings and photographs have been integrated into *narrative art,* which has been laid out, labeled, and annotated to tell a story—the figures are drawn to teach! Subcaptions are positioned adjacent to relevant parts of each figure, labels point out key features, and balloons provide important annotation. Subparts are arranged to convey time progression, where relevant. The color schemes of drawings have been tied to those of relevant photos, so that students can easily relate features in the drawings to those in the photos. The author has also written and narrated over a dozen Narrative Art videos, which bring this art to life.

Google Earth provides an amazing opportunity for students to visit and tour important geologic sites wherever they occur. Throughout the book, we provide *See for Yourself* panels, which offer coordinates and descriptions of geologic features that students can visit at the touch of a finger, or the click of a mouse. The adjacent box provides a quick guide for using these panels.

Featured Paintings—*Geology at a Glance*

In addition to individual figures, artists Gary Hincks and Stan Maddock have created spectacular two-page annotated paintings for each chapter. These paintings, called *Geology at a Glance,* integrate key concepts introduced in the chapters, visually emphasize the relationships between components of the Earth System, and allow students a way to review a subject . . . at a glance. The Sixth Edition includes three new paintings, illustrating the formation of minerals (Chapter 5), the life cycle of a tornado (Chapter 20), and the consequences of sea-level change (Chapter 23).

Enhanced Coverage of Current Topics

To ensure that *Earth: Portrait of a Planet* continues to reflect the latest research discoveries and help students understand geologic events that have been featured in current news, we have updated many topics throughout the book. For example, the Sixth Edition discusses the causes of and lessons learned from recent natural disasters such as Hurricanes Harvey, Irma, and Maria (Chapter 20), and assesses the impact of recent earthquakes in Nepal, Japan, and Ecuador (Chapter 8). The Sixth Edition also includes updated coverage of the economics of oil and other energy resources, and the difference between conventional and unconventional reserves (Chapter 14). These topics, along with expanded discussion of climate change and its impacts (Chapter 23), highlight the relevance of physical geology concepts and phenomena to students' lives today.

Other notable new content in the Sixth Edition includes a revision of the paleomagnetism discussion to makes this topic more accessible (Chapter 3); new coverage of mantle modeling technology, and how it has changed our understanding of the appearance and behavior of subducted plates (Interlude D); new introductions to the concepts of phylogenetics, ecosystems, and paleoecology (Interlude E); and an intensive revision of the explanation of the Coriolis force and other atmospheric concepts, using text and figures developed in collaboration with atmospheric scientist Robert Rauber of the University of Illinois, coauthor of the First Edition of a separate book, *Earth Science.* This reworking ensures that students have access to the most contemporary and accurate explanations of these important processes (Chapter 18).

See for Yourself
Using *Google Earth*

Visiting the SFY Field Sites Identified in the Text

There's no better way to appreciate geology then to see it first-hand in the field. The challenge is that the great variety of geologic features that we discuss in this book can't be visited from any one locality. So, even if your class takes geology field trips during the semester, at best you'll see examples of just a few geologic settings. Fortunately, Google Earth makes it possible to fly to spectacular geologic field sites anywhere in the world in a matter of seconds—you can take a virtual field trip electronically. In each chapter in this book, *See for Yourself* panels identify geologic sites that you can explore on your own computer (Mac or PC) using Google Earth software, or on your Apple/Android smartphone or tablet with the appropriate Google Earth app.

To get started, follow these three simple steps:

1. Check to see if Google Earth is installed on your personal computer, smartphone, or tablet. If not, download the free software from https://www.google.com/earth, or access the desktop version at earth.google.com. You can also download the app from the Apple or Android app store.

2. Each See for Yourself panel in the margin of the chapter provides a thumbnail photo of a geologically interesting site, as well as a very brief description of the site. The panel also provides the latitude and longitude of the site.

3. Open Google Earth and enter the coordinates of the site in the search window. As an example, let's find Mt. Fuji, a beautiful volcano in Japan. We note that the coordinates in the See for Yourself panel are as follows:

Latitude	35°21'41.78"N
Longitude	138°43'50.74"E

Type these coordinates into the search window of Google Earth as:

35 21 41.78N, 138 43 50.74E

with the degree, minute, and second symbols left blank. When you click "Enter" or "Return," your device will bring you to the viewpoint right above Mt. Fuji, as illustrated by the left-hand thumbnail above.

Google Earth contains many built-in and easy-to-use tools that allow you to vary the elevation, tilt, orientation, and position of your viewpoint, so that you can tour around the feature, see it from many different perspectives, and thus develop a three-dimensional sense of the feature. In the case of Mt. Fuji, you'll be able to see its cone-like shape and the crater at its top. By zooming out to higher elevation, you can instantly perceive the context of the given geologic feature—for example, if you fly up into space above Mt. Fuji, you will see its position relative to the tectonic plate boundaries of the western Pacific. The thumbnail on the right below shows the view of the same location you'll see if you tilt your viewing direction and look north.

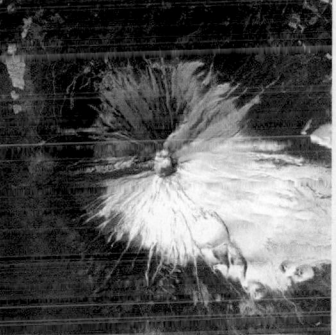 

Vertical view, looking down.   **Inclined view, looking north.**

Need More Help?

If you're having trouble, please visit digital.wwnorton.com/earth6. There you will find a video showing how to download and install Google Earth, additional instructions on how to find the *See for Yourself* sites, links to Google Earth videos describing basic functions, and links to any hardware and software requirements. Also, notes addressing Google Earth updates will be available at this site.

We also offer a separate book—the *Geotours Workbook*, Second Edition (ISBN 978-1-324-00096-9), by Scott Wilkerson, Beth Wilkerson, and Stephen Marshak—that identifies additional interesting geologic sites to visit, provides active-learning exercises linked to the sites, and explains how you can create your own virtual field trips.

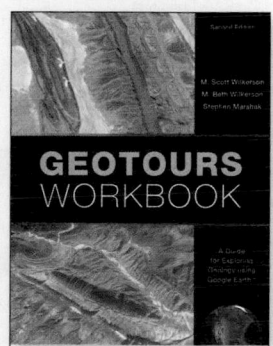

Smartwork5 Online Activities

Smartwork5

Smartwork5 is Norton's tablet-friendly, online activity platform. Both the system and its physical geology content were designed with the feedback of hundreds of instructors, resulting in unparalleled ease of use for students and instructors alike.

Smartwork5 features easy-to-deploy, highly visual assignments that provide students with answer-specific feedback. Students get the coaching they need to work through the assignments, while instructors get real-time assessment of student progress via automatic grading and item analysis. The question bank features a wide range of higher-order questions such as ranking, labeling, and sorting. All of the Narrative Art videos, animations, and interactive simulations are integrated directly into Smartwork5 questions—making them assignable. Smartwork5 also contains *What a Geologist Sees* questions that take students to sites not mentioned in the book, so they can apply their knowledge just as a geologist would. In addition, Smartwork5 offers reading quizzes for each chapter and *Geotours*-guided inquiry activities using Google Earth.

Based on instructor feedback, Smartwork5 offers three types of pre-made activity:

- Chapter **Reading Quizzes**, designed to help students prepare for lecture
- Chapter **Activities**, consisting of highly visual exercises covering all chapter Learning Objectives
- **Geotours Worksheets**—guided inquiry activities that use Google Earth

Smartwork5 is fully customizable, meaning that instructors can add or remove questions, create assignments, write their own questions, or modify ours. Easy and intuitive tools allow instructors to filter questions by chapter, section, question type, and learning objective.

SMARTWORK5 features a variety of question types to get students working hands-on with geologic concepts.

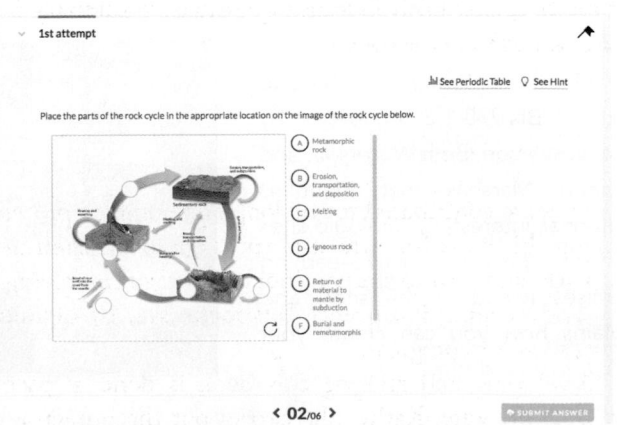

All Smartwork5 content is written by geology instructors. Our Sixth Edition Smartwork5 authors include Heather Lehto of Angelo State University, Tobin Hindle of Florida Atlantic University, Christine Clark of Eastern Michigan University, and Jacqueline Richard of Delgado Community College.

Media and Ancillaries

Animations, Simulations, and Videos

Earth: Portrait of a Planet, Sixth Edition, provides a rich collection of new animations, developed by Alex Glass of Duke University, working with Heather Cook of California State University, San Marcos. These illustrate geologic processes in a consistent style and with a 3-D perspective. Interactive simulations allow students to control variables and see the resulting output. The newest simulations are designed to help students understand basic terminology.

Narrative Art Videos, written and narrated by Stephen Marshak, bring both textbook art and supplementary field photos to life. And a robust suite of over 100 real-world video clips illustrate key processes, concepts, and natural phenomena.

ANIMATIONS illustrate geologic processes.

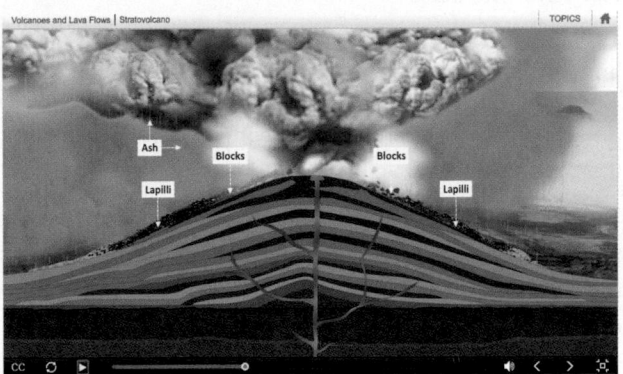

All the videos, animations, and interactive simulations are free, require no special software, and are available in a variety of settings to offer ultimate flexibility for instructors and students: on the Norton Digital Landing Page (digital.wwnorton.com/earth6); in LMS-compatible coursepacks; integrated into Smartwork5 questions; linked to the ebook; and as linked resources in the new Interactive Instructor's Guide that accompanies the text (iig.wwnorton.com/earth6/full).

Mobile-ready E-book

Earth Portrait of a Planet, Sixth Edition is available in a new format perfect for tablets and phones. Within the ebook, art expands for a closer look, links send you to geologic locations in Google Maps, animations and videos link out from each chapter, and pop-up key terms allow for quick review. It's also easy to highlight, take notes, and search the text.

The *Geotours Workbook*

Created by Scott Wilkerson, Beth Wilkerson, and Stephen Marshak, the *Geotours Workbook,* Second Edition, provides active-learning opportunities that take students on virtual field trips to see outstanding examples of geology at locations around the world, using Google Earth. Arranged by topic, the questions in the *Geotours Workbook* have been designed for auto-grading, and they are available as worksheets, both in print format (these come free with the book and include complete user instructions and advanced instruction) or electronically with automatic grading through *Smartwork5* or your campus LMS. The *Geotours Workbook* also provides instructions that will allow instructors or students to make their own geotours. Request a sample copy to preview each worksheet.

Lecture PowerPoints and Image Files

Norton provides a variety of electronic presentations of art and photographs in the book to enhance the classroom experience. These include:

- *Lecture Slides*—Designed for instant classroom use, these slides utilize photographs and line art from the book in a form that has been optimized for use in the PowerPoint environment. The art has been relabeled and resized for projection. Lecture Slides also include supplemental photographs. For the Sixth Edition, the lecture bullet PowerPoints were revised by Brian Zimmer of Appalachian State University, and accuracy-checked by Lalo Guerrero of Portland Community College.
- *Labeled and Unlabeled Image Files*—These include all art from the book, formatted as JPEGs, pre-pasted into PowerPoints. We offer one set in which all labeling has been stripped for use in quizzes and clicker questions, and one set in which the labeling has been retained. Individual JPEGs are also available for download.
- *Quarterly Update PowerPoints*—Norton offers a quarterly update service that provides new PowerPoint slides, with instructor support, covering recent geologic events. Monthly updates are authored by Paul Brandes.

Instructor's Manual and Test Bank

The *Instructor's Manual*, prepared by Robin Nagy of Houston Community College, and accuracy-checked by Kerry Workman-Ford of California State University, Fresno, is designed to help instructors prepare lectures and exams. It contains detailed Learning Objectives, Chapter Summaries, and complete answers to the end-of-chapter *Review Questions* and *On Further Thought Questions* for every chapter and interlude.

The *Test Bank*, written by Heather Cook of California State University, San Marcos, and Geoffrey Cook of University of California, San Diego, has been revised not only to correlate with this new Edition, but to provide greater, more rounded assessment than ever before. Expert accuracy checkers Angela Aranda of California State University, Fullerton, and Haraldur Karlsson of Texas Tech University have ensured that every question in the Test Bank is scientifically reliable and truly tests students' understanding of the most important topics in each chapter, so that the questions can be assigned with confidence.

Interactive Instructor's Guide— https://iig.wwnorton.com/earth6/full

New for the Sixth Edition, the Interactive Instructor's Guide is a dynamic, searchable, online resource that provides *all* instructor resources in one place. With content tagged by book chapter and section, learning objective, and keyword, instructors can find what they need, when they need it— whether a Real World Video, an in-class activity idea, or the Lecture Slides for the chapter.

LMS Coursepacks

Available at no cost to professors or students, *Norton Coursepacks* bring high-quality Norton digital media into a new or existing online course. *Coursepacks* contain ready-made content for your campus LMS. For *Earth: Portrait of a Planet*, Sixth Edition, content includes the full suite of animations, simulations, and videos keyed to core figures in each chapter; the Test Bank; reading quizzes authored by Cynthia Liutkus-Pierce of Appalachian State University and accuracy-checked by Karen Koy of Missouri Western State University; new European case studies; Geotour questions; vocabulary flashcards; and links to the ebook.

Acknowledgments

Many people contributed to the long and complex process of bringing this book from the concept stage to the shelf in the first place, and now to the continuous effort of improving the book and keeping it current. Textbooks are, by definition, always a work in progress.

Developing and revising this book is done in partnership with my wife, Kathy. She carries out the immense task of pulling together content and writing changes introduced in my other books, *Essentials of Geology*, Fifth Edition, and *Earth Science*, First Edition, including suggestions from users and reviewers, to produce *Earth: Portrait of a Planet*, Sixth Edition. Kathy edits new text, cross-checks many sets of proofs, and manages the never-ending inflow and outflow of proofs that perpetually occupy our dining-room table. Without her efforts, the updating of *Earth: Portrait of a Planet* through the years would not be possible. We are grateful to our daughter, Emma, and our son, David, who provided valuable feedback and several of the photos—and who also served as scale in some of the photos. They allowed "the book" to become a member of our household, for more than two decades, and tolerated the overabundance of geo-stops on family trips.

Kathy and I are very grateful to the staff of W. W. Norton & Company for their incredible, continuing efforts during the development of this book and its companions, over the past two decades. It has been a privilege to work with an employee-owned company that is willing to collaborate so closely with its authors. In particular, I would like to thank Jake Schindel, the geology editor at Norton, who has injected new enthusiasm and ideas into the project, working steadfastly to bring order to the chaos of juggling multiple titles at once. His skill in editing, ability to oversee many moving parts, and his friendly reminders of deadlines, led this book to completion on an accelerated production schedule. Thom Foley, the book's senior project editor, as always, does an amazing job of guiding the book through production. He somehow keeps track of all the drafts, all the changes, and all the figures for a lengthy and complicated project, while remaining incredibly calm and providing essential institutional memory for the geology team at Norton. It's thanks to Thom that everything somehow manages to get done, and that mistakes are few and far between. Rob Bellinger continues to keep the technology component of the book at the cutting edge, by introducing new web tools and overseeing the development of Smartwork5 for the book. Katie Sweeney has done wonders as the marketing manager for the book, by helping to determine how to meet the needs of adopters worldwide. I also wish to thank the previous editors for the book. Eric Svendsen ably oversaw the Fourth and Fifth Editions. And our dear friend, the late Jack Repcheck, served as the editor for the first three editions of the book. Jack suggested many of the innovations that strengthened the book, and his instincts about what works in textbook publishing brought the book to the attention of a wider geological community than I ever thought possible. He will always be remembered as an understanding friend and a fountain of sage advice.

Moving a new edition of *Earth: Portrait of a Planet* from concept to completion involves a large team of professionals. The artists, Joanne Brummett and Stan Maddock, in Champaign, Illinois, have created beauty and enhanced pedagogy with the new illustrations that they have rendered for this Edition. Their work, along with the past work of other artists once anchored at Precision Graphics, set the bar for the quality of art in geology textbooks. Stan established the initial style of the book's art and developed innovative ways of visualizing geologic phenomena. Trish Marx has done a fantastic job with the Herculean task of finding, organizing, and crediting photographs and bringing the photo selection process into the 21st century. Lissi Sigillo creatively developed a clean and friendly page design. I am also grateful to Rob Bellinger, Cailin Barrett Bressack, Liz Vogt, Kim Yi, Marcus Van Harpen, Leah Clark, Francesca Olivo, Arielle Holstein, Gina Forsythe, and Kelly Smith for their innovative approach to ancillary and e-media development. Thanks also go to Marcus Van Harpen, Lizz Thabet, and Mateus Teixeira for their work on the tablet and mobile e-books, and to associate production director Benjamin Reynolds, who coordinated the back-and-forth between the publisher and various vendors and suppliers. Norma Sims-Roche has done an excellent job as copy editor for this Sixth Edition, and Associate Editor Rachel Goodman provided consistent editorial support and trouble-shooting throughout the process of making this book.

The six editions of this book and its cousin, *Essentials of Geology*, have benefited greatly from input by expert reviewers for specific chapters, by general reviewers of the entire book, and by comments from faculty and students who have used the book and were kind enough to contact the author or the publisher with suggestions and corrections. We gratefully acknowledge the contributions of the individuals listed below, who have provided invaluable input into this and past editions either through comments or reviews. I apologize if I've inadvertently left anyone off the list.

Jack C. Allen, Bucknell University

David W. Anderson, San Jose State University

Martin Appold, University of Missouri, Columbia

Philip Astwood, University of South Carolina

Eric Baer, Highline University

Victor Baker, University of Arizona

Julie Baldwin, University of Montana

Miriam Barquero-Molina, University of Missouri

Sandra Barr, Acadia University

Keith Bell, Carleton University

Mary Lou Bevier, University of British Columbia

Jim Black, Tarrant County College

Daniel Blake, University of Illinois

Andy Bobyarchick, University of North Carolina, Charlotte

Ted Bornhorst, Michigan Technological University

Michael Bradley, Eastern Michigan University

Mike Branney, University of Leicester, UK

Sam Browning, Massachusetts Institute of Technology

Bill Buhay, University of Winnipeg

Rachel Burks, Towson University

Peter Burns, University of Notre Dame

Katherine Cashman, University of Oregon

Cinzia Cervato, Iowa State University

George S. Clark, University of Manitoba

Kevin Cole, Grand Valley State University

Patrick M. Colgan, Northeastern University

Peter Copeland, University of Houston

John W. Creasy, Bates College

Norbert Cygan, Chevron Oil, retired

Michael Dalman, Blinn College

Peter DeCelles, University of Arizona

Carlos Dengo, ExxonMobil Exploration Company

Meredith Denton-Hedrick, Austin Community College, Cypress Creek

John Dewey, University of California, Davis

Charles Dimmick, Central Connecticut State University

Robert T. Dodd, Stony Brook University

Missy Eppes, University of North Carolina, Charlotte

Eric Essene, University of Michigan

David Evans, Yale University

James E. Evans, Bowling Green State University

Susan Everett, University of Michigan, Dearborn

Dori Farthing, State University of New York, Geneseo

Mark Feigenson, Rutgers University

Grant Ferguson, St. Francis Xavier University

Eric Ferré, Southern Illinois University

Leon Follmer, Illinois Geological Survey

Nels Forman, University of North Dakota

Bruce Fouke, University of Illinois

David Furbish, Vanderbilt University

Steve Gao, University of Missouri

Grant Garvin, John Hopkins University

Christopher Geiss, Trinity College, Connecticut

Bryan Gibbs, Richland Community College

Gayle Gleason, State University of New York, Cortland

Patrick Gonsoulin-Getty, University of Connecticut

Cyrena Goodrich, Kingsborough Community College

William D. Gosnold, University of North Dakota

Lisa Greer, William & Mary College

Steve Guggenheim, University of Illinois, Chicago

Henry Halls, University of Toronto, Mississauga

Bryce M. Hand, Syracuse University

Anders Hellstrom, Stockholm University

Tom Henyey, University of South Carolina

Bruce Herbert, Texas A & M University

James Hinthorne, University of Texas, Pan American

Paul Hoffman, Harvard University

Curtis Hollabaugh, University of West Georgia

Bernie Housen, Western Washington University

Mary Hubbard, Kansas State University

Paul Hudak, University of North Texas

Melissa Hudley, University of North Carolina, Chapel Hill

Warren Huff, University of Cincinnati

Neal Iverson, Iowa State University

Charles Jones, University of Pittsburgh

Donna M. Jurdy, Northwestern University

Thomas Juster, University of Southern Florida

H. Karlsson, Texas Tech

Daniel Karner, Sonoma State University

Dennis Kent, Lamont Doherty/Rutgers

Charles Kerton, Iowa State University

Susan Kieffer, University of Illinois

Jeffrey Knott, California State University, Fullerton

Ulrich Kruse, University of Illinois

Robert S. Kuhlman, Montgomery County Community College

Lee Kump, Pennsylvania State University

David R. Lageson, Montana State University

Robert Lawrence, Oregon State University

Heather Lehto, Angelo State University

Scott Lockert, Bluefield Holdings

Leland Timothy Long, Georgia Tech

Craig Lundstrom, University of Illinois

John A. Madsen, University of Delaware

Jerry Magloughlin, Colorado State University

Scott Marshall, Appalachian State University

Kyle Mayborn, Western Illinois University

Jennifer McGuire, Texas A&M University

Judy McIlrath, University of South Florida

Paul Meijer, Utrecht University, Netherlands

Aric Mine, California State University, Fresno

Jamie Dustin Mitchem, California University of Pennsylvania

Alan Mix, Oregon State University

Marguerite Moloney, Nicholls State University

Otto Muller, Alfred University

Kristen Myshrall, University of Connecticut

Kathy Nagy, University of Illinois, Chicago

Pamela Nelson, Glendale Community College

Wendy Nelson, Towson University

Robert Nowack, Purdue University

Charlie Onasch, Bowling Green State University

David Osleger, University of California, Davis

Bill Patterson, University of Saskatchewan

Eric Peterson, Illinois State University

Ginny Peterson, Grand Valley State University

Stephen Piercey, Laurentian University

Adrian Pittari, University of Waikato, New Zealand

Lisa M. Pratt, Indiana University

Eriks Puris, Portland Community College

Mark Ragan, University of Iowa

Robert Rauber, University of Illinois

Bob Reynolds, Central Oregon Community College

Joshua J. Roering, University of Oregon

Eric Sandvol, University of Missouri

William E. Sanford, Colorado State University

Jeffrey Schaffer, Napa Valley Community College

Roy Schlische, Rutgers University

Sahlemedhin Sertsu, Bowie State University

Anne Sheehan, University of Colorado

Roger D. Shew, University of North Carolina,
  Wilmington

Doug Shakel, Pima Community College

Norma Small-Warren, Howard University

Donny Smoak, University of South Florida

David Sparks, Texas A&M University

Angela Speck, University of Missouri

Larry Standlee, University of Texas, Arlington

Tim Stark, University of Illinois

Seth Stein, Northwestern University

David Stetty, Jacksonville State University

Kevin G. Stewart, University of North Carolina,
  Chapel Hill

Michael Stewart, University of Illinois

Don Stierman, University of Toledo

Gina Marie Seegers Szablewski, University of Wisconsin, Milwaukee

Barbara Tewksbury, Hamilton College

Thomas M. Tharp, Purdue University

Kathryn Thornbjarnarson, San Diego State University

Robert Thorson, University of Connecticut

Basil Tikoff, University of Wisconsin

Spencer Titley, University of Arizona

Robert T. Todd, Stony Brook University

Torbjörn Törnqvist, University of Illinois, Chicago

Jon Tso, Radford University

James Tyburczy, Arizona State University

Stacey Verardo, George Mason University

Barry Weaver, University of Oklahoma

John Werner, Seminole State College of Florida

Alan Whittington, University of Missouri

John Wickham, University of Texas, Arlington

Lorraine Wolf, Auburn University

Christopher J. Woltemade, Shippensburg University

Jackie Wood, Delgado Community College, City Park

Kerry Workman-Ford, California State University,
  Fresno

Thanks!

I am very grateful to the faculty who have selected *Earth: Portrait of a Planet* for their classes, and to the students who engage so energetically with the book. I particularly appreciate receiving questions and corrections from readers that help to improve the book and keep it as accurate as possible. I continue to welcome comments and can be reached at smarshak@illinois.edu.

Stephen Marshak

About the Author

Stephen Marshak is a professor of geology at the University of Illinois, Urbana-Champaign, where he also serves as the Director of the School of Earth, Society, and Environment. He holds an A.B. from Cornell University, an M.S. from the University of Arizona, and a Ph.D. from Columbia University. Steve's research interests lie in structural geology and tectonics, and he has participated in field projects on a number of continents. Steve loves teaching and has won his college's and university's highest teaching awards. He also received the 2012 Neil Miner Award from the National Association of Geoscience Teachers (NAGT), for "exceptional contributions to the stimulation of interest in the Earth sciences." In addition to research papers and *Earth: Portrait of a Planet*, Steve has authored *Essentials of Geology*, and has co-authored *Earth Science; Laboratory Manual for Introductory Geology; Earth Structure: An Introduction to Structural Geology and Tectonics*, and *Basic Methods of Structural Geology*.

ANOTHER VIEW Geology students enjoying the view from outcrops in northern Scotland.

And Just What Is Geology?

By the end of this prelude, you should be able to . . .

A. describe the scope and applications of geology.

B. explain the foundational themes of modern geologic study.

C. demonstrate how geologists employ the scientific method.

D. provide a basic definition of the theory of plate tectonics.

E. explain what geologists mean by the Earth System concept.

F. name the main layers of the Earth's interior.

P.1 In Search of Ideas

We arrived in the late-night darkness at a campsite in western Arizona. Here in the desert, so little rain falls over the course of a year that hardly any plants can survive, and rocks form ledges on many hills. Under the dry sky, there's no need for tents, so we could rest under the stars with our sleeping bags on a bed of sand. At dawn, the red rays of the first sunlight made the slope of the steep-sided hill near our campsite start to glow, and we could see our target, a prominent ledge of rusty-brown rock near the top of the hill. To reach it, though, we'd have to climb a steep slope littered with jagged boulders.

After a quick breakfast, we loaded our day packs with water bottles and granola bars, slathered on a layer of sunscreen, and set off toward the slope. It was the breezeless morning of what was going to be a truly hot day, and we wanted to gain elevation before the Sun rose too high in the sky. After a tiring hour finding our way through the boulder obstacle course, we reached the base of the ledge and decided to take a rest before our final ascent. But just as we leaned back to rest our backs against a rock, we heard an unnerving vibration. Somewhere nearby, too close for comfort, a rattlesnake shook an urgent warning. Rest would have to wait, and we scrambled up the ledge. It was the right choice, for the view from the top of the surrounding landscape was amazing **(Fig. P.1a)**. But the rocks beneath our feet were even more amazing. Close up, we could see curving ribbons of light and dark layers, cut by stripes of white quartz. The ledge preserved the story of a distant age in our planet's past when the rock we now stood on was kilometers below ground level and was able to flow like soft plastic, but ever so slowly **(Fig. P.1b)**. We now set to the task of figuring out what it all meant.

FIGURE P.1 Geologic exploration provides beautiful views and mysteries to solve.

(a) A view of the western Arizona desert is not just beautiful—it holds clues to the Earth's past and to the changes taking place today.

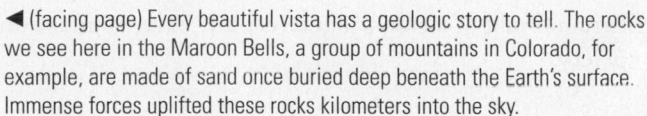

◀ (facing page) Every beautiful vista has a geologic story to tell. The rocks we see here in the Maroon Bells, a group of mountains in Colorado, for example, are made of sand once buried deep beneath the Earth's surface. Immense forces uplifted these rocks kilometers into the sky.

(b) The contortions of the rock layers speak of a time when the rock flowed like soft plastic.

(a) Cliff exposures in the desert of Utah.

(b) The shore in Massachusetts.

(c) A rainforest in Peru.

(d) Mountains in Alaska.

Geologists—scientists who study the Earth—do indeed explore remote deserts, such as the one pictured in Figure P.1. They also head to high mountains, damp rainforests, frigid glaciers, and deep canyons **(Fig. P.2)**. Such efforts can strike people in other professions as a strange way to make a living. This sentiment underlies a description by the Scottish poet Walter Scott (1771–1832) of geologists at work: "Some rin uphill and down dale, knapping the chucky stanes to pieces wi' hammers, like sae mony road-makers run daft—they say it is to see how the warld was made!" But Scott had it right—to see how the world was made, to see how it continues to evolve, to find its resources, to protect against its natural hazards, and to predict what its future may bring. These are the questions that have driven geologists to explore the Earth, on all continents and in all oceans, from the equator to the poles and everywhere in between.

Geologic discovery continues today. But, while some geologists continue to work in the field with hammers and hand lenses, others have moved into laboratories that employ techniques of chemistry or sophisticated electronic instruments to analyze microscopic quantities of Earth materials **(Fig. P.3a, b)**. Still others use satellites to detect the motions of continents or the stability of volcanoes, high-speed computers to locate earthquakes or to analyze the flow of underground water, and scale models to simulate flow in rivers **(Fig. P.3c, d)**. For over two centuries, geologists have pored over the Earth in search of ideas to explain the processes that form and change our planet. In this Prelude, we look more closely at the questions geologists ask and have tried to answer. You'll see that the results of this work are not just of academic interest but have implications for society.

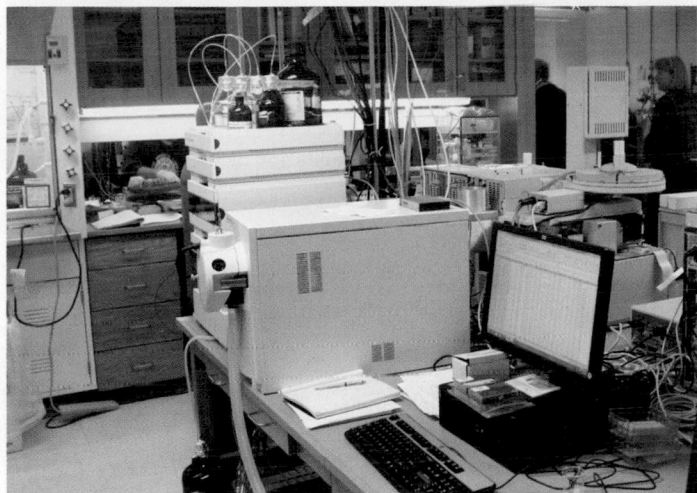

(a) Analytical equipment in a geology research laboratory.

(b) Laboratories provide an opportunity to carry out controlled experiments.

(c) A computer facility working with large amounts of data.

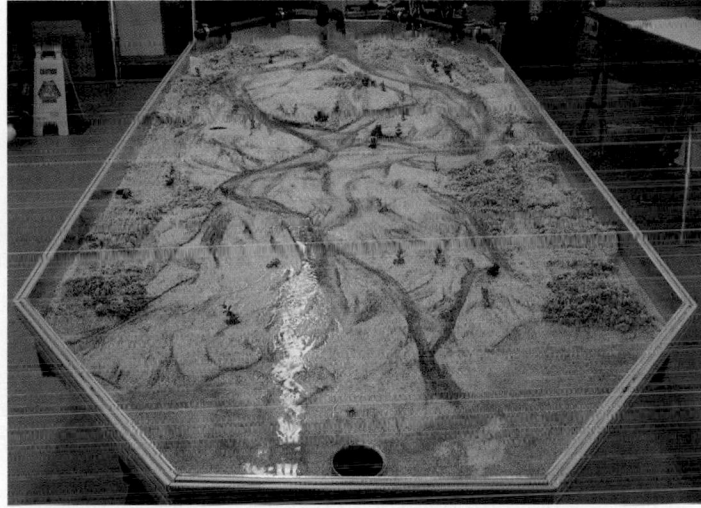

(d) A stream of water flowing over sand can simulate the evolution of a river.

P.2 Why Study Geology?

Geology (also called *geoscience*) is the study of the Earth. It encompasses studies that characterize the formation and composition of this planet, the causes of mountain building and ice ages, the record of life's evolution, and the history of climate change. Geology also addresses practical problems such as how to keep pollution out of groundwater, how to find oil and useful minerals, and how to avoid landslides. You can get a sense of the different kinds of problems that geologists work on by examining a list of the many subdisciplines of geology **(Table P.1)**.

Hundreds of thousands of people worldwide pursue careers in geology—mostly in energy, mining, water, engineering, and environmental companies. A smaller number work in universities or colleges, government-sponsored geological surveys, and research laboratories. Nevertheless, since most people reading this book will not become professional geologists, it's fair to ask the question, "Why should people, in general, study geology?"

First, geology may be one of the most practical subjects you can learn. When news reports begin with "Scientists say . . ." and then continue with "an earthquake occurred today off Japan," or "landslides will threaten the city," or "chemicals from a toxic waste dump will ruin the town's water supply," or "there's only a limited supply of oil left," or "the floods of

TABLE P.1 Principal Subdisciplines of Geology (Geoscience)

Name	Subject(s) of Study
Engineering geology	Aspects of geology relevant to understanding slope stability or to building tunnels, dams, mines, roads, or foundations
Environmental geology	Interactions between the environment and geologic materials, and the contamination of the near-surface realm of the Earth by pollutants
Geochemistry	Chemical composition and behavior of materials in the Earth, and chemical reactions in natural environments
Geochronology	The age (in years) of geologic materials, the Earth, and extraterrestrial objects
Geomorphology	Landscape formation and evolution
Geophysics	Physical characteristics of the Earth (such as the Earth's magnetic field and gravity field) and causes of forces that affect the Earth
Hydrogeology	Groundwater, its movement, and its reaction with rock and soil
Mineralogy	Physical properties, structure, and chemical behavior of minerals
Paleontology	Fossils and the evolution of life as preserved in the rock record
Petrology	Rocks and their formation
Sedimentology	Sediments and their deposition
Seismology	Earthquakes and the Earth's interior as revealed by earthquake waves
Stratigraphy	The succession of sedimentary rock layers and the record of Earth's history that they contain
Structural geology	Rock deformation (bending and breaking) in response to the application of force associated with mountain building
Tectonics	The origin and significance of regional-scale geologic features
Volcanology	Volcanic eruptions and their products, and volcanic hazards

the last few days are the worst on record," the scientists that the reports refer to are geologists. In fact, ask yourself the following questions, and you'll realize that geologic processes, phenomena, and materials play major roles in daily life:

- Do you live in a region threatened by landslides, volcanoes, earthquakes, or floods **(Fig. P.4a, b)**?

- Are you worried about the price of energy or about whether there will be a war in an oil-supplying country **(Fig. P.4c)**?

- Do you ever wonder where the copper in your home's wires comes from? Or the lithium in the battery of your cell phone?

- Have you seen fields of green crops surrounded by desert and wondered where the irrigation water comes from?

- Are you worried about the consequences of deforestation **(Fig. P.4d)**?

- Would you like to buy a dream house near a beach or a river?

- Are you concerned about how toxic waste can migrate underground into your town's well water?

Addressing these questions requires a basic understanding of geology. This knowledge may help you avoid building your home on a hazardous floodplain or fault zone, on an unstable slope, or along a rapidly eroding coast. With an understanding of groundwater, you may be able to find a good site for a well. With knowledge of the geologic controls on resource distribution, you may be able to invest more wisely in the resource industry or to understand the context of political choices regarding energy policy.

Second, the study of geology gives you an awareness of the planet that no other field can. As you will see, the Earth is a complicated world, where living organisms, oceans, atmosphere, and solid rock interact with one another in a great variety of ways. Geologic study reveals the Earth's antiquity (it's about 4.56 billion years old) and demonstrates how profoundly the planet has changed during its existence. What our ancestors considered to be the center of the Universe has become,

FIGURE P.4 Geology provides insight into natural hazards and resource exploration.

(a) Collapsed buildings in the aftermath of an earthquake.

(b) The natural power of a hurricane can level a beachside community.

(c) Coal is one of many resources that comes from the Earth.

(d) The human-generated power of a chain saw and a bulldozer can level a forest.

with the development of geologic perspective, our "island in space" today. And what was believed to be an unchanging orb originating at the same time as humanity has become a dynamic planet that not only existed long before people did, but continues to evolve.

Third, the study of geology puts the accomplishments and consequences of human civilization in a broader context. View the aftermath of a large earthquake, flood, or hurricane, and it's clear that the might of natural geologic phenomena greatly exceeds the strength of human-made structures. But watch a bulldozer clear a swath of forest, a dynamite explosion remove the top of a hill, or a prairie field turn into a housing development, and it's clear that people can also change the face of the Earth, often at rates exceeding those of natural geologic processes.

Finally, when you finish reading this book, your view of the world may be forever colored by geologic curiosity. If you walk in the mountains, you may remember that mountains rise and fall over time in response to forces that shape and reshape the Earth's surface. If you hear about a natural disaster, you may

think about the various phenomena that trigger such disasters. And as you drive past rock exposures along a highway, you won't just see featureless masses of gray but will pick out layers and structures providing a record of the Earth's amazing history.

P.3 Themes of This Book

A number of narrative themes appear—and reappear—throughout this text. These themes, listed below, can be viewed as the book's overall take-home message.

- *Geology is a synthesis of many sciences:* The study of geology can help you understand physical science in general, for geology applies many of the basic concepts of physics and chemistry to the interpretation of visible phenomena. As you learn about the Earth, you'll also be learning about the behavior of matter and energy, and about the nature of chemical reactions.

The Earth System

External energy

Mountain uplift

Sun

Thunderhead

Lightning

Rain and snow

Continental glacier

City

Ocean

Desert

Rocky coastline

Valley

Arid mountains

Mining

Lakes

Field pattern

Deciduous forest

Forested mountains

Beach

Tropical rainforest

Coral reef

Shark

When you stand on the surface of the Earth, you can see the wondrous ways in which components of the Earth System interact. The geosphere consists of the solid part of our planet. You see it wherever you see exposed rock, sediment, or soil. Most of it lies underground, in the internal layers of our planet. The hydrosphere consists of all liquid water at or near the surface of the Earth. It fills oceans, lakes, underground pores, and occurs as gas in the atmosphere. The cryosphere consists of frozen water, mostly in glaciers. The biosphere consists of living organisms, from bacteria to whales. The atmosphere is the envelope of gas that encircles the planet. Flow in the air and sea transfer heat and water around the planet.

Internal energy rising from the interior and external energy coming to the Earth from the Sun keep the Earth System dynamic, so that materials cycle from component to component over time. Human society is having a growing impact on the Earth System, by extracting resources, building and farming on its surface, and emitting waste.

Internal energy

Cirrus clouds

Jet stream

Moon

Aurora

Ice and snow

Wind system

Coniferous forest

Evaporation

Volcanic islands

Industrial pollution

Cold surface current

Surface waters

Delta

Swamps

Warm surface current

Twilight zone

Abyssal zone

Seafloor

Whale

Bacteria and plankton

Giant squid

Deep-sea current

Black smokers

- *The Earth has an internal structure:* The Earth does not have a homogeneous interior, but rather consists of concentric layers. From center to surface, our planet has a *core*, *mantle*, and *crust*. We live on the surface of the crust, where it meets the atmosphere and the oceans (**Fig. P.5a**).

- *The outer layer of the Earth consists of moving plates:* In the 1960s, geologists recognized that the crust, together with the uppermost part of the underlying mantle, forms a 100- to 150-km-thick semi-rigid shell called the **lithosphere**. Distinct boundaries separate this shell into discrete pieces, called **plates**, which move very slowly relative to one another over a softer part of the mantle called the **asthenosphere** (**Fig. P.5b, c**). The theory that describes this movement and its consequences is called the **theory of plate tectonics**, and it serves as the foundation for understanding most geologic phenomena. Plate movements and interactions produce earthquakes, volcanoes, and mountain ranges, and cause the map of the Earth's surface to change very slowly over time.

- *We can picture the Earth as a complex system:* The Earth is not static, but rather can be pictured as a dynamic entity whose components move and change. Our planet's interior, solid surface, oceans, atmosphere, and life all interact with one another in many ways to yield the land, oceans, and air in which we and other organisms can live. Geologists refer to this interconnected web of interacting materials and processes as the **Earth System** (**Geology at a Glance**, pp. 6–7). Within the Earth System, certain materials cycle among different types of rock, or among

FIGURE P.5 The Earth's interior and the theory of plate tectonics.

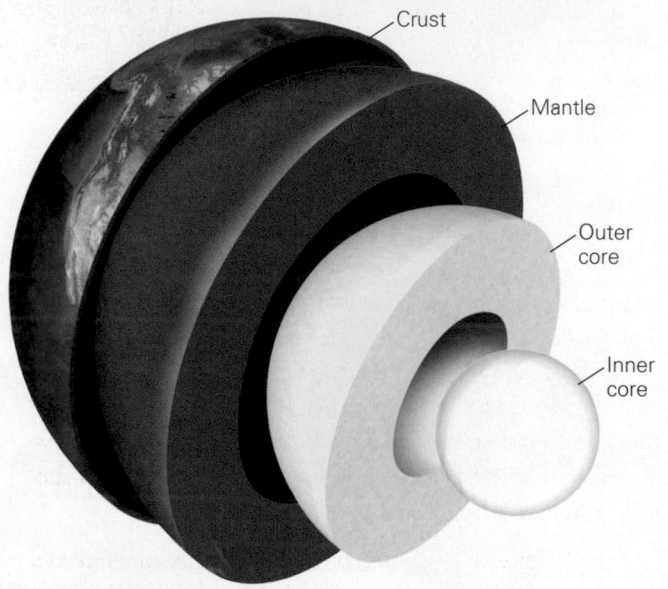

Crust
Mantle
Outer core
Inner core

(a) Simplistically, the interior of the Earth can be pictured as a set of concentric shells.

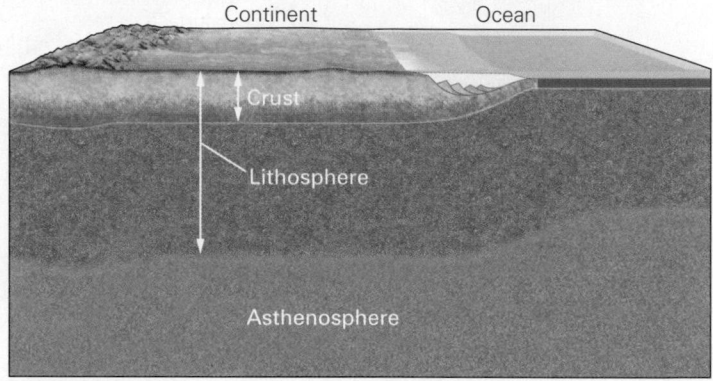

Continent   Ocean
Crust
Lithosphere
Asthenosphere

(b) The crust and the uppermost part of the mantle together constitute the relatively rigid lithosphere. The asthenosphere lies beneath the lithosphere.

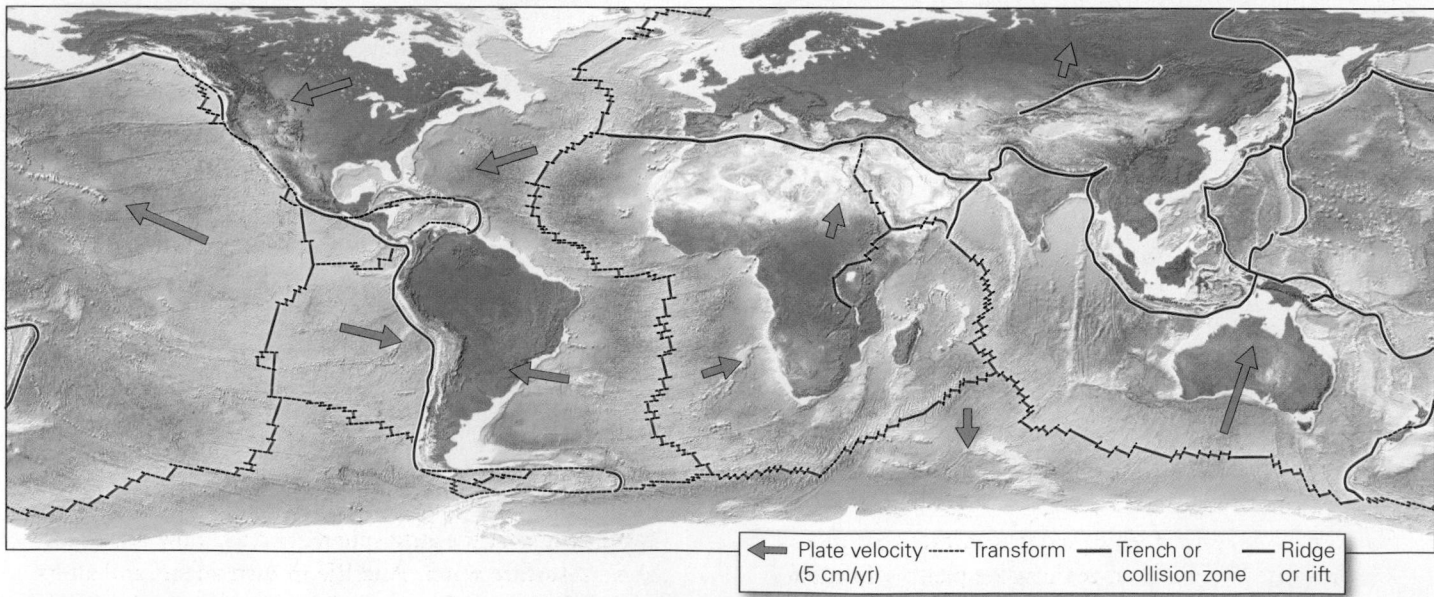

Plate velocity (5 cm/yr) ------ Transform —— Trench or collision zone —— Ridge or rift

(c) A simplified map of the Earth's plates. The arrows indicate the direction each plate is moving, and the length of the arrow indicates plate velocity (the longer the arrow, the faster the motion)

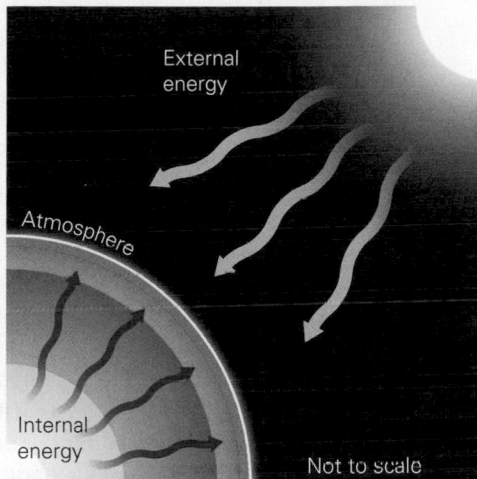

External energy

Atmosphere

Internal energy

Not to scale

(b) The difference between internal and external energy in the Earth System.

(a) In this scenic view in Switzerland, we see many aspects of the Earth System—air, water, ice, rock, life, and human activity.

rock, sea, air, and life **(Fig. P.6a)**. Over time, the distribution of these materials among various components of the Earth System can change, as can the characteristics of the components.

- *The Earth is a planet:* Despite the uniqueness of the Earth System, we can think of the Earth as a planet, formed like the other planets of the Solar System. But because of the way the Earth System operates, our planet differs from other planets by having plate tectonics, an oxygen-rich atmosphere and a liquid-water ocean, and abundant life.

- *Internal and external processes drive geologic phenomena:* **Internal processes** are those driven by heat from inside the Earth. Plate movement is an example. Because plate movements cause mountain building, earthquakes, and volcanoes, we consider all of these phenomena internal processes **(Fig. P.6b)**. **External processes** are those driven by energy coming to the Earth from the Sun. The heat produced by this energy drives the movement of air and water, which grinds and sculpts the Earth's surface and transports the debris to new locations, where it accumulates. As we'll see, **gravity**—the pull that one mass exerts on another—plays an important role in both internal and external processes. The interaction between internal and external processes forms the mountains, canyons, beaches, and plains of our planet.

- *The Earth is very old:* Geologic data indicate that the Earth formed 4.56 billion years ago—plenty of time for geologic processes to generate and destroy landscapes, for life forms to evolve, and for the map of the planet to

change. For example, plate movement at rates of only a few centimeters per year can move a continent thousands of kilometers if those movements continue for hundreds of millions of years. There's time enough to build mountains and time enough to grind them down, many times over. The Earth has a history, and it extends far into the past, long before human ancestors appeared.

- *The geologic time scale divides the Earth's history into intervals:* To refer to specific portions of geologic time, geologists developed the **geologic time scale (Fig. P.7)**. The last 541 million years comprise the *Phanerozoic Eon*, and all time before that falls into the *Precambrian*. The Precambrian can be further divided into three main intervals named, from oldest to youngest, the *Hadean*, the *Archean*, and the *Proterozoic Eons*. The Phanerozoic Eon, in turn, can be divided into three main intervals named, from oldest to youngest, the *Paleozoic*, the *Mesozoic*, and the *Cenozoic Eras*.

- *Geologic phenomena affect society:* Volcanoes, earthquakes, landslides, floods, groundwater, energy sources, and mineral reserves are of vital interest to every inhabitant of this planet. Therefore, throughout this book we emphasize the linkages among geology, the environment, and society.

- *Physical aspects of the Earth System interact with life processes:* All life on this planet depends on physical features such as the minerals in soil; the temperature, humidity, and composition of the atmosphere; and the flow of surface and subsurface water. And life in turn affects and alters physical features. For example, the oxygen in the Earth's atmosphere comes from photosynthesis, a life activity in

The Scientific Method

Sometime during the past 200 million years, a large block of rock or metal, which had been orbiting the Sun, slammed into our planet. It made contact at a site in what is now the central United States, a landscape of flat cornfields. The impact of this block—a *meteorite*—released more energy than a nuclear bomb. A cloud of shattered rock and dust blasted skyward, and once-horizontal layers of rock from deep below the ground sprang upward and tilted on end beneath the gaping hole left by the impact. When the dust had settled, a huge crater surrounded by debris marked the surface of the Earth at the impact site. Later in Earth history, running water and blowing wind wore down this jagged scar. Some 15,000 years ago, sand, gravel, and mud carried by a vast glacier buried what remained, hiding it entirely from view **(Fig. BxP.1)**. Wow! So much history beneath a cornfield. How do we know this? It takes scientific investigation.

The movies often portray science as a dangerous tool, capable of creating Frankenstein's monster, and scientists as nerdy characters with thick glasses and poor taste in clothes. In reality, **science** is simply the use of observation, experiment, and calculation to explain how nature operates, and scientists are simply people who study and try to understand natural phenomena. Scientists guide their work using the **scientific method**, a sequence of steps for systematically analyzing scientific problems in a way that leads to verifiable results. Let's see how geologists employed the steps of the scientific method to come up with the meteorite-impact story.

1. *Recognizing the problem.* Any scientific project, like any detective story, begins by identifying a mystery. The cornfield mystery came to light when water drillers discovered limestone, a rock typically made of shell fragments, just below the 15,000-year-old glacial sediment. In surrounding regions, the rock beneath the glacial sediment consists of sandstone, a rock made of cemented-together sand grains. Since limestone can be used to build roads, make cement, and produce the agricultural lime used in treating soil, workers stripped off the glacial sediment and dug a quarry to excavate the limestone. They were surprised to find that rock layers exposed in the quarry were tilted steeply and had been shattered by large cracks. In the surrounding regions, all rock layers are horizontal like the layers in a birthday cake, the limestone layer lies underneath the sandstone, and the rocks contain relatively few cracks. Curious geologists came to investigate, and they soon realized that the geologic features of the land just beneath the cornfield presented a problem to be explained: What phenomena had brought limestone up close to the Earth's surface, had tilted the layering in the rocks, and had shattered the rocks?

2. *Collecting data.* The scientific method proceeds with the collection of observations or clues that point to an answer. Geologists studied the quarry and determined the age of its rocks, measured the orientation of the rock layers, and documented (made a written or photographic record of) the fractures that broke up the rocks.

3. *Proposing hypotheses.* A scientific **hypothesis** is merely a possible explanation, involving only natural processes, that can explain a set of observations. Scientists propose hypotheses during or after their initial data collection. In this example, the geologists working in the quarry came up with two alternative hypotheses: either the features in this region resulted from a volcanic explosion, or they were caused by a meteorite impact.

4. *Testing hypotheses.* Because a hypothesis is just an idea that can be either right or wrong, scientists must put hypotheses through a series of tests to see if they work. The geologists at the quarry compared their field observations with published observations made at other sites of volcanic explosions and meteorite impacts, and they studied the results of experiments designed to simulate such events. If the geologic features visible in the quarry were the result of volcanism, the quarry should contain specific types of rock formed by the freezing of molten rock. But no such rocks were found. If, however, the features were produced by an impact, the rocks should contain **shatter cones**, tiny cracks that fan out from a point. Shatter cones can be overlooked, so the geologists returned to the quarry specifically to search for them and found them in abundance. The impact hypothesis passed the test!

Our description of the scientific method is somewhat idealized, however, because sometimes serendipity works its way into the process, and scientists make discoveries by chance. Furthermore, since we can't travel back in time, we can't always test all geologic hypotheses as thoroughly as we might like.

Theories are scientific ideas supported by an abundance of evidence; they have passed many tests and have failed none. Scientists are much more confident in the correctness of a theory than of a hypothesis. Continued study in the quarry

---

plants. This oxygen permits complex animals to survive and affects chemical reactions among air, water, and rock. Without the physical Earth, life could not exist; but without life, this planet's surface might have become a frozen wasteland like that of Mars, or a cloud-enshrouded oven like that of Venus.

- *The Earth has changed dramatically in many ways over geologic time and continues to change:* The landscape that you see outside your window today is not what you would have seen a thousand, a million, or a billion years ago. Over Earth history, the planet's surface, the atmosphere's composition, and sea level have all changed. Change continues today,

FIGURE BxP.1 An ancient meteorite impact excavates a crater and permanently changes rock beneath the surface.

Rock layers

The impact produces shatter cones that open in the direction away from the impact.

Impact direction

0   3 cm

(a) A meteorite strikes the surface of the ancient Earth.

(b) The force of the impact excavates a crater and fractures rock layers underground.

Glacial till layer

Faults

(c) Erosion removes the crater but leaves the underground disruption. Much later, the land is buried by glacial sediment.

Time

eventually yielded so much evidence for impact that the impact hypothesis came to be viewed as a theory. You may notice that in everyday conversation, people commonly use the word "theory" as a synonym for an untested or barely tested speculation. In scientific discussion, the word has a much more restricted meaning. Only after it has passed certain tests, without failing any, can an idea be called a theory.

Scientists continue to test theories over a long time. Successful theories withstand these tests and are supported by so many observations that they become part of a discipline's foundation. (As you will discover in Chapter 3, geologists consider the idea that continents have moved around the surface of the Earth to be a theory because so much evidence supports it.) However, some theories may eventually be disproven and replaced by better ones.

In a few cases, scientists have been able to devise concise statements that completely describe a specific relationship or phenomenon. Such statements, called **scientific laws**, apply without exception for a given range of conditions. Newton's law of gravitation serves as an example—it is a simple mathematical expression that always defines the invisible pull exerted by one mass on another. Note that scientific laws do not in themselves explain a phenomenon, and in this way they differ from theories. For example, the law of gravity does not explain why gravity exists, but the theory of evolution does provide an explanation of why evolution occurs.

and some aspects of the Earth System are changing faster than ever before because of human activity.

- *Most of the resources that we use come from geologic materials:* Modern society uses vast quantities of oil, gas, coal, metal, concrete, and other materials. All of these come from the Earth **(Fig. P.8)**.

- *Science comes from observation, and people make scientific discoveries:* Science does not consist of subjective guesses or arbitrary dogmas, but rather of a consistent set of objective statements resulting from the application of the scientific method **(Box P.1)**. Every scientific idea must be tested thoroughly and should be used only when

FIGURE P.7 The geologic time scale.

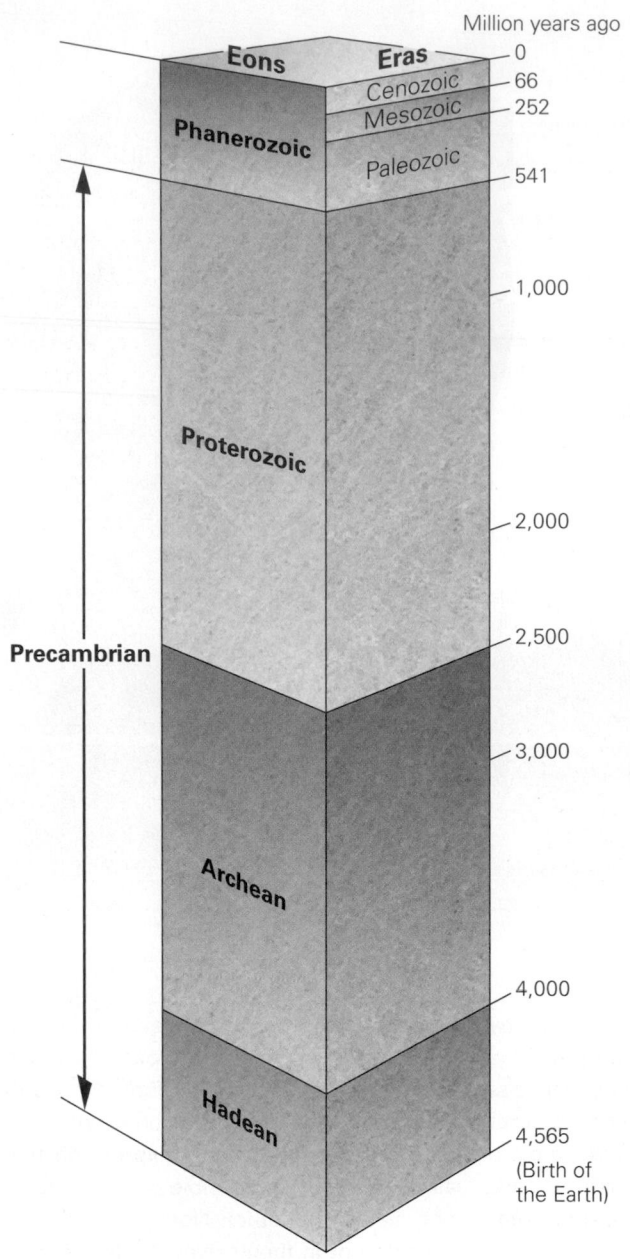

(a) The scale has been divided into eons and eras.

One thousand years ago = 1 **Ka**
(Ka stands for kilo-annum)

One million years ago = 1 **Ma**
(Ma stands for mega-annum)

One billion years ago = 1 **Ga**
(Ga stands for giga-annum)

(b) Abbreviations for time units.

supported by documented observations. Furthermore, scientific ideas do not appear out of nowhere; they are the result of human efforts. Wherever possible, this book shows where geologic ideas came from and tries to answer the question, "How do we know that?"

As you read this book, please keep these themes in mind. Don't view geology as a list of words to memorize, but rather as an interconnected set of concepts to digest. Most of all, enjoy yourself as you learn about the most fascinating planet in the Universe.

FIGURE P.8 Workers excavate limestone in a quarry near Chicago. This rock commonly consists of shells and shell fragments, and can be used in the production of concrete.

ANOTHER VIEW Students scramble across the rocky shores of Scotland's coast. . ."to see how the warld was made."

Prelude Review

SUMMARY

- Geologists are scientists who study the Earth. They search for the answers to the mysteries of our home planet, from why volcanoes explode to where we can find minerals.

- Geologic study can involve field exploration, laboratory experiments, high-tech measurements, and calculations with computers.

- Geologic research not only provides answers to academic questions such as how the Earth formed, but also addresses practical problems such as how to find groundwater or how to avoid landslides. Many people pursue careers as geologists.

- A set of themes underlies geologic thinking. Key concepts are that the Earth's outer shell consists of moving plates whose interactions produce earthquakes, volcanoes, and mountains; that the Earth is very old; and that interacting realms of material on the planet constitute the Earth System.

GUIDE TERMS

asthenosphere (p. 8)
Earth System (p. 8)
external processes (p. 9)
geologic time scale (p. 9)
geologist (p. 2)

geology (p. 3)
gravity (p. 9)
hypothesis (p. 10)
internal processes (p. 9)
lithosphere (p. 8)

plate (p. 8)
science (p. 10)
scientific laws (p. 11)
scientific method (p. 10)
shatter cones (p. 10)

theory (p. 10)
theory of plate tectonics
 (p. 8)

REVIEW QUESTIONS

The letters following each Review Question refer to the corresponding Learning Objective from the Chapter Opener.

1. What are some of the practical applications of geology? **(A)**

2. Explain the difference between internal processes and external processes. **(B)**

3. How would the Earth's atmosphere differ if life didn't exist? **(B)**

4. Explain the difference between a hypothesis and a theory, in the context of science. **(C)**

5. What is the basic premise of the theory of plate tectonics? **(D)**

6. What are the main layers of the Earth's interior? Label them in the diagram. **(F)**

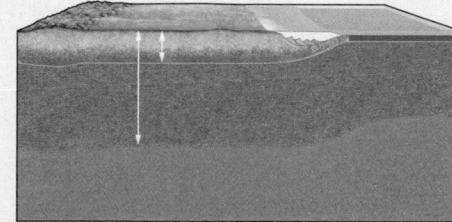

7. What do geologists mean by the statement, "the Earth is a complex system?" **(E)**

8. What are the sources of data that geologists can use to understand the Earth? **(C)**

9. What are the major subdivisions of geologic time? Which time unit is longer, the Precambrian or the Paleozoic? **(B)**

10. This mine truck carries 100 tons of coal. Where does this resource, and others like it, come from? **(B)**

PART I

OUR ISLAND IN SPACE

When you gaze out toward the horizon from a mountaintop, the Earth seems endless, and before the modern era, many people thought it was. But to astronauts flying to the Moon, the Earth looks like a small, shining globe—they can see half the planet in a single glance. From the astronauts' perspective, we are living on an island in space. The Earth may not be endless, but it is a very special planet: its temperature and composition, unlike those of the other planets in the Solar System, make it habitable. Part I of this book introduces the internal structure and dynamic character of our home planet.

1 **Cosmology and the Birth of the Earth**

2 **Journey to the Center of the Earth**

3 **Drifting Continents and Spreading Seas**

4 **The Way the Earth Works: Plate Tectonics**

In Chapter 1 we outline scientific ideas about how the Earth, and the Universe around it, came to be. Then, in Chapter 2, we take a quick tour of the planet to get a sense of its composition and its various layers. With this background, we're ready to move on to Chapter 3, where we'll encounter the 20th century revolution in geology that yielded the set of ideas we now call the theory of plate tectonics. In Chapter 4, we'll delve into our modern understanding of this theory, which proposes that the outer layer of the Earth consists of plates that move with respect to one another. We'll see that plate tectonics provides a scientific explanation for a great variety of geologic features—from the formation of continents to the distribution of fossils—and truly serves as the unifying concept for all of geology.

◄ As viewed by a NASA astronaut in orbit, the blue curve of the Earth's horizon stands in sharp contrast to the black vacuum of space beyond.

CHAPTER 1

Cosmology and the Birth of the Earth

By the end of this chapter, you should be able to . . .

A. assess how people's perceptions of the Earth's place in the Universe have changed over the centuries.

B. explain modern concepts concerning the basic architecture of our Universe and its components.

C. assess the evidence for the expanding Universe and the Big Bang theory.

D. describe where the elements that make up matter came from.

E. explain the nebula theory, a scientific model that explains how stars and planets form.

1.1 Introduction

Sometime in the distant past, humans developed the capacity for complex, conscious thought. This amazing ability, which distinguishes our species from all others, brought with it the gift of curiosity, an innate desire to understand and explain the workings and origin of the **Universe**, meaning space and everything it contains, including our home, the Earth. For most of human history, such musings have spawned legends in which heroes, gods, and goddesses used supernatural powers to ignite the Sun, to make the Moon glow, to mold the Earth from nothingness, to sculpt its surface into dramatic shapes, and to speckle the night with points of light. Only recently have people applied scientific principles (see Box P.1) to the systematic study of the overall structure and history of the Universe, thereby establishing the modern discipline of **scientific cosmology.**

In the context of scientific cosmology, the Universe contains of two basic entities: **matter**, the substance that makes up objects, and **energy**, the inherent ability of a region of space and the matter within it to do "work" (to change itself or its surroundings). We can refer to the amount of matter in an object as its **mass**, so an object with greater mass contains more matter.

Scientific cosmology provides a foundation from which we can begin to explore the composition, structure, and evolution of the Earth. So we begin this book on geology, the study of the Earth, with a chapter that outlines key principles of scientific cosmology. In effect, we must look outward in order to be able to understand what we'll see when we look inward at the Earth **(Fig. 1.1)**. We start by characterizing the architecture of the Universe overall, and then focus on our neighborhood, the Solar System, in particular. Next, we introduce the Big Bang theory, which researchers use to explain the formation of the Universe, and the production of the elements that eventually came together to form the Sun, the Earth, and other **celestial objects** (naturally occurring bodies in the Universe). We conclude by discussing the nebular theory for the birth of the Earth itself.

◄ (facing page) The Hubble Space Telescope can see into space from its perch in Earth orbit above the atmosphere. In this photo, taken through the telescope, we see gas and dust in Nebula S106 (3,300 light-years from the Earth), a birthplace of new stars.

1.2 An Image of Our Universe

What Is the Structure of the Universe?

Think about the mysterious spectacle of a clear night sky **(Fig. 1.2a)**. What objects sparkle up there? How big are they? How far away are they? How do they move? How are they arranged? Ancient cultures around the world thought long and hard about such questions, and by 3,000 years ago, keen observers—the first **astronomers** (people who study celestial bodies)—had realized that what they could see above had a recognizable order. Specifically, they noticed that most of the thousands of points of light visible to the naked eye move slowly across the sky nightly, as if revolving around a fixed point **(Fig. 1.2b)**, and that the positions of these points relative to one another remain fixed. These points became known as the *stars*. In contrast, a handful of the lights in the night sky etch seemingly complex paths, moving relative to one another and relative to the backdrop of stars **(Fig. 1.2c)**. These lights came to be known as the *planets* (from the Greek *planēs*, which means wanderer).

Early observers did not know what the stars and planets were, and they didn't understand the relationship among these points of light and the Earth, Moon, and Sun. It wasn't at all obvious, in fact, that the Earth might itself be a planet and the Sun a star. In the days of the Greek philosopher Homer (ca. 850 B.C.E.), for example, people in the Mediterranean

FIGURE 1.1 Astronauts in the Lunar Landing Module of Apollo 11 saw "Earthrise" over the Moon's horizon. Such images forever changed humanity's perception of our island in space.

FIGURE 1.2 The sky at night, showing a variety of celestial objects.

(a) Imagine what it would be like *not* to know what celestial objects, such as the Moon, are. Until the past few hundred years, we didn't.

(b) When viewed over a whole night, stars in the northern hemisphere appear to revolve around Polaris, the North Star.

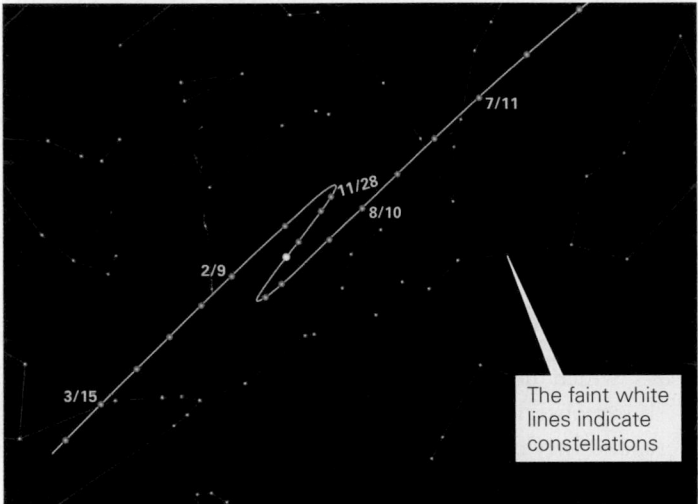

The faint white lines indicate constellations

(c) Certain objects in the night sky appear to move relative to the backdrop of stars. These "wanderers" are planets. The numbers give dates (month/day) when the planet shown was at a given location.

region considered the Earth to be a flat disk, with land toward the center and water around the margins, lying at the center of a celestial sphere, a dome to which the stars were attached. Philosophers of Homer's day argued about the nature of the Sun and why it produced heat and light: to some, the Sun was a burning bowl of oil, while to others it was a ball of red-hot iron. Many favored the notion that movements of celestial bodies represented the activities of gods and goddesses, and they named *constellations*—distinctive arrangements of stars—after these deities. Other societies developed quite different mythologies, full of symbolism, in which to interpret the heavens.

In the Western world, two distinct schools of thought developed concerning the arrangement of stars and planets and their relationship to the Earth, Sun, and Moon. The first school advocated a **geocentric model (Fig. 1.3a)**, in which the Earth sits without moving at the center of the Universe while the Moon and the planets whirl around it in circular orbits, all within a shell of stars. The second school advocated a **heliocentric model (Fig. 1.3b)**, in which the Sun lies at the center of the Universe, with the Earth and other planets orbiting around it. The geocentric image gained a widespread following due to the influence of an Egyptian mathematician, Ptolemy (100–170 c.e.), who developed equations that seemed to predict the wanderings of the planets, in the context of the geocentric model, with remarkable accuracy. During the Middle Ages (ca. 476–1400 c.e.), church leaders in Europe adopted Ptolemy's geocentric image as dogma because it seemed to justify the comforting thought that humanity's home occupies the most important place in the Universe. Eventually, anyone who disagreed with this view risked charges of heresy.

Then came the Renaissance. In 15th-century Europe, bold thinkers spawned a new age of exploration and scientific discovery. Thanks to the efforts of the Polish astronomer Nicolaus Copernicus (1473–1543) and the Italian astronomer Galileo Galilei (1564–1642), people eventually came to realize that the Earth and planets did indeed orbit the Sun, so the Earth could not possibly be at the center of the Universe. The idea solidified when the Dutch astronomer Johannes Kepler (1571–1630) showed that Ptolemy was wrong and that planets follow elliptical orbits. And when the British physicist Isaac Newton (1643–1727) explained gravity, the attractive force that one mass exerts on another (Box 1.1), it finally became possible to understand why celestial objects display the motions that they do.

Stars and Galaxies

After Galileo popularized use of the telescope for looking skyward, astronomers gained the ability to see and measure features progressively farther into space and gradually refined

(a) Ptolemy's geocentric image of the Universe puts the Earth at the center.

Earth

Sun

(b) The heliocentric image of the Universe puts the Sun at the center, as envisioned by Copernicus.

our understanding of the Universe's structure. We now realize that, although it looks like a point of light, a **star** is actually an immense sphere of incandescent gas that emits intense energy— stars resemble our Sun but lie farther away. Furthermore, stars do not fly randomly through the Universe. Rather, gravity holds them together in immense groups called **galaxies**. Our Sun is one of over 300 billion stars in the *Milky Way Galaxy*. From our vantage point on the Earth, the Milky Way looks like a hazy band **(Fig. 1.4a)**, but if we could view the Milky Way from a great distance, it would look like a flattened spiral with vast curving arms that slowly swirl around a glowing, disk-like center **(Fig. 1.4b)**. Presently, our Sun lies near the outer edge of one of these arms. Astronomers estimate that as many as a trillion galaxies sit within the visible Universe **(Fig. 1.4c)**.

Clearly, human understanding of the Earth's place in the Universe has evolved radically over the past few centuries. Neither the Earth nor the Sun, nor even the Milky Way, occupies the center of the Universe. While the Earth seems quite special to us because we live here, there's nothing particularly special about its position in the Universe.

The Nature of Our Solar System

Our Sun's gravitational pull holds on to many objects, which, together with the Sun, comprise the **Solar System (Fig. 1.5a)**. Most of the mass of the Solar System—99.8%, to be exact— resides in the Sun itself. The remaining 0.2% includes a great variety of objects, the largest of which are the planets. Astronomers define a **planet** as an object that orbits a star, is roughly spherical, and has "cleared its neighborhood of other objects." The last phrase in this definition sounds a bit strange at first, but it merely implies that a planet's gravity has pulled in all particles of matter in its orbit. According to this definition, formalized in 2005, our Solar System includes eight planets: Mercury, Venus, Earth, Mars, Jupiter, Saturn, Uranus, and Neptune. Until 2005, astronomers considered one more object, Pluto, to be a planet. But Pluto has not cleared its orbit, so it does not fit the modern definition of a planet and has been dropped from the roster. The eight planets orbit the Sun in the same direction and more or less in the same plane **(Fig. 1.5b)**.

Significantly, land-based instruments, as well as those of the Kepler Space Telescope (launched into orbit in 2009), have allowed astronomers to locate more than 3,500 *exoplanets* (planets that orbit stars other than our Sun) as of 2017. Some of these resemble the Earth in size. In fact, astronomers now estimate that the Milky Way hosts billions of Earth-sized planets.

Planets in our Solar System differ radically from one another in both size and composition. The *inner planets* (Mercury, Venus, Earth, and Mars), the ones closer to the Sun, are relatively small. We call them the **terrestrial planets** because, like the Earth, they consist of a shell of rock surrounding a ball of metal. The *outer planets* (Jupiter, Saturn, Uranus, and Neptune) are known as the **giant planets**, or Jovian planets. The largest, Jupiter, contains 318 times as much mass as the Earth and accounts for about 71% of the non-solar mass in the Solar System. The overall composition of giant planets differs markedly from that of terrestrial planets. Specifically, most of the matter in Jupiter and Saturn consists of hydrogen and helium as a gas, liquid, or as a strange liquid-metal state. Because so little solid material occurs in these planets, astronomers refer to them as *gas giants*. In contrast, most of the matter in Neptune and Uranus consists of water, carbon dioxide, and methane that has been frozen into solid ice. (Note that, in the context of discussing Solar System materials, the term *ice* refers to any compound, not just water, that can evaporate relatively easily at Earth-surface conditions.)

BOX 1.1 SCIENCE TOOLBOX . . .

Force and Energy

Types of Forces We use the word *force* frequently in everyday English—but what does it mean, scientifically? Isaac Newton defined a *force* as simply a push or pull that causes velocity of an object to change in magnitude (speed) and/or direction. In your everyday experience, you constantly see or feel the effects of forces. Forces can cause objects to slow down or go faster. Forces can also tear, stretch, squash, spin, and twist objects and can make them float or sink. Any *system* (a defined volume of space and everything in it) contains energy, meaning that it has an inherent ability or capacity to do work. In this context, *work* formally refers to the product of the force and the distance over which the force acts—you do work when you lift this book 10 cm, and you do more work when you lift it 20 cm. Doing work transfers energy from one part of a system to another, but the total amount of energy in the system cannot be changed (for energy cannot be created or destroyed).

Physicists (scientists who study matter, energy, and the interactions between them) distinguish between two general types of force. The first type, a *contact force*, or *mechanical force*, results when one *mass* moves and comes in contact with another. You apply a mechanical force to a boulder when you push on it **(Fig. Bx1.1a)**, and the wind applies a mechanical force to a sail when it blows. The second type, a non-contact force, or *field force*, applies across a distance; gravity and magnetism serve as examples.

Gravity is the force of attraction between two masses—it is what holds you to the surface of the Earth and pulls objects from higher elevations to lower ones **(Fig. Bx1.1b)**. The strength of gravity depends on the quantity of matter in the two masses and on the distance between them. For example, you feel a much stronger gravitational pull to the huge Earth than you do to a small baseball. *Weight* is the force that an object exerts due to gravitational pull, so the stronger the gravitational pull, the greater the weight of an object. For example, on the Moon, you weigh much less than on the Earth because the Moon is smaller, so it exerts less gravitational pull.

Magnetism, simplistically, is the force generated by electricity flowing in a wire, or by special materials called magnets. Unlike gravity, magnetic force can be attractive (pulling objects together) or repulsive (pushing them apart). Over short distances, the magnetic force of even a small magnet can be larger than the gravitational force produced by the Earth. Therefore, you can overcome gravity and lift objects with a magnet **(Fig. Bx1.1c)**.

Types of Energy We've defined energy as the inherent ability of a region of space or the matter within it to do work. This definition simply means that energy can make matter move or change the character of matter. For example, energy can cause a car to roll along the highway, it can turn an ice cube into a puddle of water, or it can make a light bulb glow. Physicists distinguish among several forms of energy: *kinetic energy*, the energy that an object has due to its movement; *potential energy*, the energy stored within a material; and *radiant energy*, the energy that a glowing object sends outward in the form of electromagnetic waves.

One type of energy can be converted into another. For example, as long as it remains stationary, a boulder sitting at the top of a hill stores potential energy. When the boulder starts to roll, this energy changes into kinetic energy. Similarly, radiant energy from the Sun, when absorbed by a solar panel on the roof of a house, can be changed into potential energy stored in a battery. This potential energy can later be turned into the kinetic energy of a moving car.

FIGURE Bx1.1 Examples of the forces of nature in everyday life. Mechanical and field forces are very familiar.

(a) A person applies a mechanical force to push a boulder.

(b) Gravity, a field force, pulls a person down a zip line.

(c) A magnet produces a field force sufficient to hold onto these clips.

FIGURE 1.4 A galaxy may contain about 300 billion stars.

(a) The Milky Way on a clear night. The "haze" actually consists of billions of faraway stars. A comet appears in the lower right.

(b) A spiral galaxy that looks like the Milky Way, as viewed from the top.

(c) A Hubble Space Telescope view of deep space, showing some of the billions of galaxies in the Universe.

FIGURE 1.5 The relative sizes and positions of planets in the Solar System.

(a) Relative sizes of the planets. All are much smaller than the Sun, but the gas-giant planets are much larger than the terrestrial planets. Jupiter has a diameter about 11.2 times greater than that of the Earth.

(b) Relative positions of the planets. This figure is not to scale. If the Sun in this figure was the size of a large orange, the Earth would be the size of a sesame seed 15 meters away. Note that all planetary orbits lie roughly in the same plane.

FIGURE 1.6 Examples of moons in our Solar System.

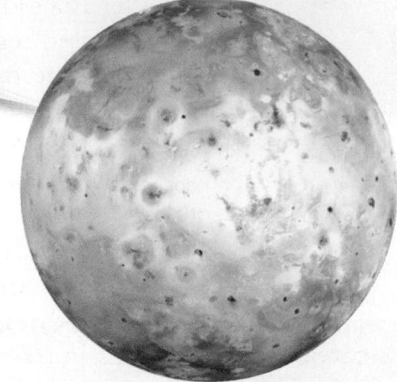

The Moon orbits the Earth. It's composed of rock and hosts craters and large mare.

Io orbits Jupiter. It's composed of rock and remains volcanically active.

Ganymede orbits Jupiter. It consists of both rock and ice, and may contain a liquid water mantle.

Enceladus orbits Saturn. It consists of ice and features blue stripes and erupting geysers.

Deimos orbits Mars. It's composed of rock and is non-spherical.

has at least 62. Moons vary greatly in size, composition, and surface characteristics—some consist mostly of rock, and others mostly of ice. Some have surfaces pockmarked by the impact of meteorites billions of years ago, while others have been resurfaced by materials erupted from volcanoes or geysers **(Fig. 1.6)**.

- *Asteroids:* An **asteroid** is a relatively small rocky or metallic object that orbits the Sun. Most lie in the region between the orbits of Mars and Jupiter. Known asteroids range in diameter from 1 cm to 930 km.

- *Kuiper Belt and Oort Cloud Objects:* About a trillion bodies of ice, with diameters ranging from less than 1 cm to 2,000 km, lie outside the orbit of Neptune. Those that occur in a donut-like ring closer in comprise the *Kuiper Belt*. More distant ones form the *Oort Cloud*, which has a somewhat spherical shape. Most of these icy objects are tiny (less than a few centimeters across), but some are much bigger.

- *Dwarf Planets:* Asteroids and Kuiper Belt objects with a diameter greater than about 900 km are known as **dwarf planets**. Though only five have yet been identified, there may be up to 200 dwarf planets. The largest known dwarf planets, Pluto and Eris, have diameters of about 2,300 km. The *New Horizons* space probe visited Pluto in 2015–2016, allowing people to see, for the first time, the amazing, complex texture of its surface **(Fig. 1.7)**.

- *Comets:* Kuiper Belt or Oort Cloud objects that follow elliptical orbits that bring them into the inner Solar

FIGURE 1.7 A composite photo in true color of Pluto, as seen by the *New Horizons* space probe in 2015.

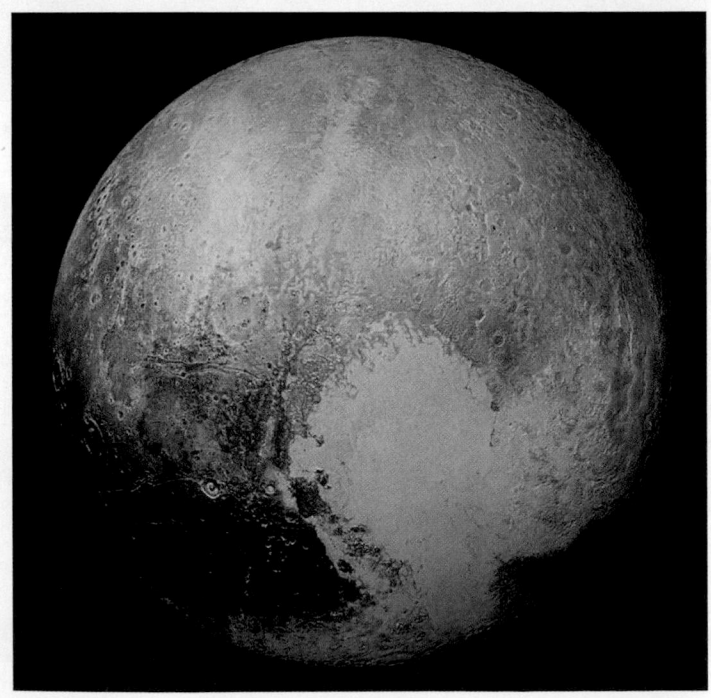

Because so much ice occurs in these planets, astronomers call them *ice giants*.

In addition to planets, the Solar System contains a great variety of other objects:

- *Moons:* A **moon** is a solid object, of detectable size, that orbits a planet. All planets except Mercury and Venus have moons. For example, our home planet has one (we call it *the Moon*), Mars has two, Jupiter has at least 63, and Saturn

FIGURE 1.8 Over 2,000 years ago, Eratosthenes calculated the circumference of the Earth using simple geometry.

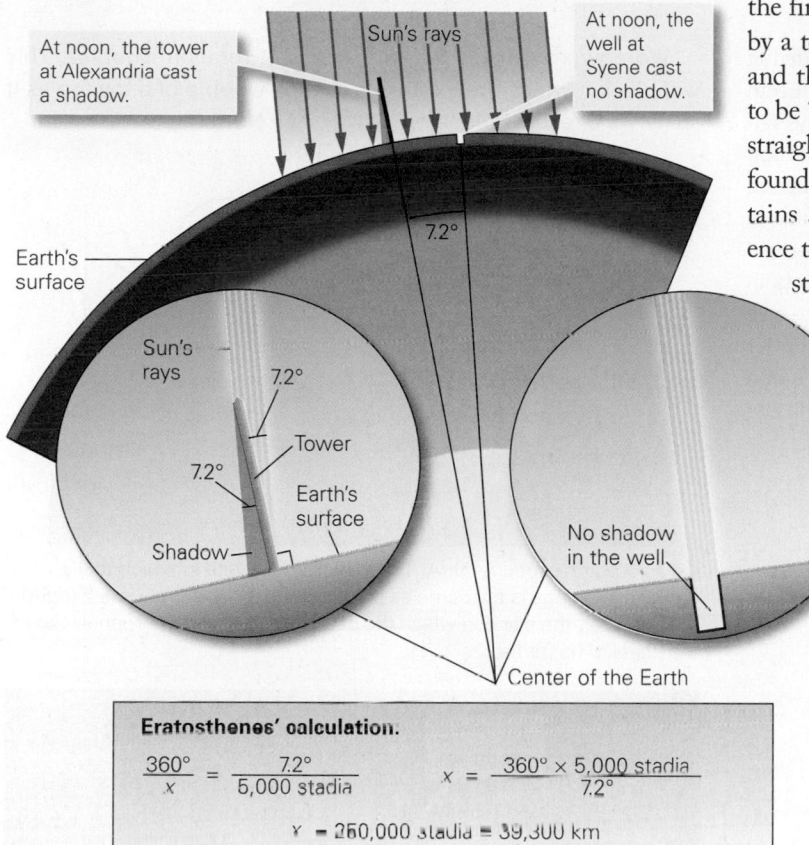

At noon, the tower at Alexandria cast a shadow.

Sun's rays

At noon, the well at Syene cast no shadow.

7.2°

Earth's surface

Sun's rays

7.2°

7.2°

Tower

7.2°

Earth's surface

Shadow

No shadow in the well.

Center of the Earth

Eratosthenes' calculation.

$$\frac{360°}{x} = \frac{7.2°}{5,000 \text{ stadia}} \qquad x = \frac{360° \times 5,000 \text{ stadia}}{7.2°}$$

$$x = 250,000 \text{ stadia} = 39,300 \text{ km}$$

System are called **comets**. When comets—which can be thought of as large, dirty snowballs (see Chapter 2)—fly relatively close to the Sun, they heat up and release long tails of gas and dust.

Developing a Sense of Shape and Scale

The concept that "space is vast" has become ingrained in modern culture—indeed, we use the term *astronomical distance* to mean "really, really far away." How did our sense of the scale of the Universe, and the objects in it, come to be?

Let's start by considering the dimensions of the Earth itself. The Greek astronomer Eratosthenes (ca. 276–194 B.C.E.), came up with the first good estimate of the Earth's circumference. Eratosthenes served as the chief librarian at the famous library of Alexandria in Egypt, one of the great centers of learning in the ancient Mediterranean region. One day, he came across a report noting that in the southern Egyptian city of Syene (modern-day Aswan), the Sun lit the base of a deep vertical well precisely at noon on the first day of summer. Eratosthenes deduced that the Sun's rays at noon on this day must be exactly perpendicular to the Earth's surface at Syene, and that if the Earth was spherical,

then the Sun's rays could not simultaneously be perpendicular to the Earth's surface at Alexandria, 800 km to the north. So on the first day of summer, Eratosthenes measured the shadow cast by a tower in Alexandria at noon. The angle between the tower and the Sun's rays, as indicated by the shadow's length, proved to be 7.2° (Fig. 1.8). He then commanded a servant to pace out a straight line from Alexandria to Syene. The sore-footed servant found the distance to be 5,000 stadia. Knowing that a circle contains 360°, Eratosthenes then calculated the Earth's circumference to be 250,000 stadia, using a simple equation. Given that 1 stadium = 0.1572 km, his answer came within 2% of today's accepted value of 40,008 km.

Not long after Eratosthenes' discovery, Greek mathematicians used ingenious geometric calculations to estimate that the distance to the Moon is about 30 times the Earth's diameter. This number comes close to the true distance of, on average, 381,555 km. But it wasn't until the 17th century that astronomers figured out that the mean distance between the Earth and the Sun is 149,600,000 km. As for the stars, the ancient Greeks realized that they must be much farther away than the Sun in order for them to appear as a fixed backdrop behind the Moon and planets, but the Greeks had no way of calculating the actual distance. Our modern documentation of the vastness of the Universe began in 1838, when astronomers determined that the nearest star to the Earth, Alpha Centauri, lies 40.85 trillion kilometers away.

The discovery that light travels at a constant speed of about 300,000 km per second in space provided astronomers with a convenient way to describe the huge distances between objects in space. They define a large distance by stating how long it takes for light to traverse that distance. For example, it takes light about 1.3 seconds to travel from the Earth to the Moon, so we can say that the Moon lies about 1.3 light-seconds away. Similarly, we can say that the Sun lies 8.3 light-minutes away. A **light-year**, the distance that light travels in one Earth year, is about 9.5 trillion kilometers. When you look up at Alpha Centauri, 4.37 light-years distant, you see light that started on its journey to the Earth 4.37 years ago.

Astronomers didn't develop techniques for measuring the distance to very distant stars and galaxies until the 20th century. With these techniques (described in astronomy books), they determined that the Milky Way itself has a diameter of about 100,000 light-years. Other galaxies reside so far away that, to the naked eye, they look like stars in the night sky. In fact, the nearest spiral galaxy to ours, Andromeda, lies 2.2 million light-years away. The farthest celestial objects that can be seen with the naked eye are

Did you ever wonder...
how far away the stars are?

BOX 1.2 CONSIDER THIS . . .

Foucault's Pendulum

It wasn't until the middle of the 19th century that Léon Foucault (1819–1868), a French physicist, actually proved that the Earth spins on its axis **(Fig. Bx1.2a)**. He made this discovery by setting a heavy pendulum, attached to a long cable, in motion **(Fig. Bx1.2b)**. As the pendulum continued to swing, Foucault noted that the plane in which it oscillated was perpendicular to the Earth's surface and that this plane rotated around a vertical axis. If Newton's *first law of motion*—which states that objects in motion remain in motion and objects at rest remain at rest—was correct, this phenomenon meant that the Earth was rotating under the pendulum while the pendulum continued to swing in the same plane **(Fig. Bx1.2c)**. Foucault displayed his discovery beneath the great dome of the Panthéon in Paris, to much acclaim.

We now know that the Earth's axis of rotation wobbles. This wobble, known as precession, is like the wobble of a toy top as it spins. As a result, Polaris isn't always the North Star.

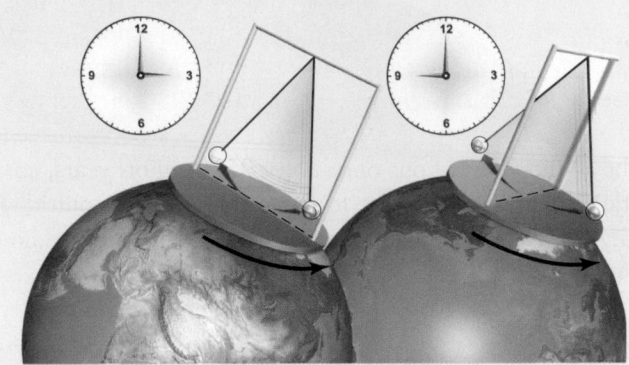

(b) Foucault's experiment. At Time 1 (left), the plane in which the pendulum swings is the same as the plane of its frame. At Time 2 (right), 6 hours later, the plane in which the pendulum swings is perpendicular to the plane of its frame.

FIGURE Bx 1.2 Proving the Earth rotates on its axis.

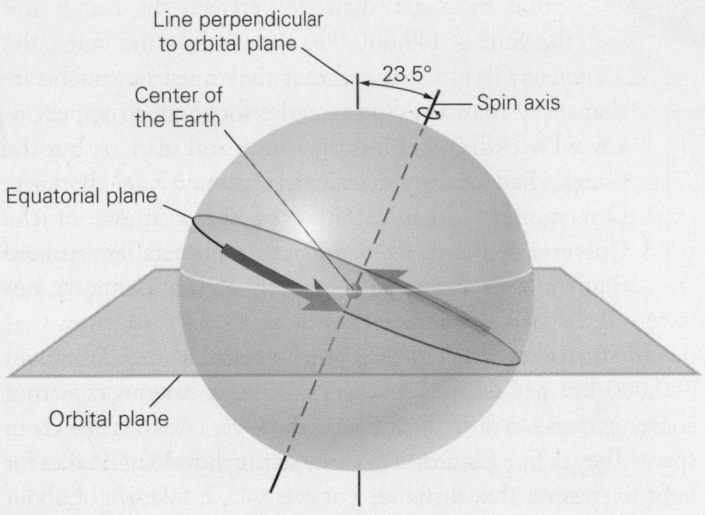

(a) The Earth rotates around an axis. The axis, an imaginary line that pierces the Earth at the poles and goes through the center of the planet, tilts relative to the plane of the Earth's orbit.

(c) An exact replica of Foucault's original pendulum on display in the Panthéon, Paris.

about 3 million light-years away. Light from these objects that we see today started on its path to the Earth a million years before the first *hominins* (direct human ancestors) walked the Earth. Powerful telescopes allow us to see much farther. The farthest object yet detected lies over 13 billion light-years away. When such numbers became known, people came to the realization that the dimensions of the Universe are truly staggering!

It's hard to fathom the distances to planets and stars without visualizing more familiar objects. Imagine a *scale model* (a model in which dimensions retain the proper proportion) in which an orange represents the Sun. At this scale, the Earth would be a sesame seed at a distance of 15 m from the orange, and Alpha Centauri would lie 2,000 km from the orange.

Motions of the Heavens

As you sit in your chair reading this book, you may think that you are motionless, but you aren't. It just seems that way because everything in the room around you moves at exactly the same velocity as you do. But relative to an observer in intergalactic space, in fact, you're moving quite fast!

Let's consider the components of this motion. First of all, the Earth, like all planets, rotates or "spins" on its *axis*, the imaginary line that passes through the center of the Earth and pierces the planet's surface at its two poles. The Earth's axis currently tilts at about 23.5° relative to the plane of its orbit. Because of this rotation, a person sitting on the **equator** (the circumference

Did you ever wonder...
how fast you are traveling
through space?

on the surface of the Earth, at a distance halfway between the poles) moves at about 1,674 km per hour, relative to a stationary observer hovering in space nearby—faster than the speed of sound! This spin makes the Sun and stars appear to cross the sky daily **(Box 1.2)**. The Earth also orbits the Sun, traveling counterclockwise along a 920-million-kilometer, slightly elliptical path that takes a full year to complete—it moves along this orbit at about 30 km per second (108,000 km per hour). And finally, the whole Solar System revolves around the center of the Milky Way about once every 250 million years, so we hurtle through space, relative to an observer standing outside the Milky Way, at about 200 km per second (720,000 km per hour).

TAKE-HOME MESSAGE

People once thought the Earth lay at the center of the Universe. Now it's clear that it is one of eight spinning planets orbiting our Sun, itself one of 300 billion stars of the revolving, spiral-shaped Milky Way Galaxy. Hundreds of billions of galaxies speckle the immense visible Universe.

QUICK QUESTION: Imagine that the distance between the Sun and Alpha Centauri is represented by the length of a city bus. At this scale, what is the diameter of the Milky Way?

1.3 Forming the Universe

We stand on a planet, in orbit around a star, speeding through space on the arm of a galaxy. Beyond our galaxy lie hundreds of billions of other galaxies. Where did all this "stuff"—the matter of the Universe—come from, and when did it first form? For most of human history, a scientific solution to these questions seemed intractable. But in the 1920s, unexpected observations about the nature of light from distant galaxies set astronomers on a path of discovery that ultimately led to a scientific model of Universe formation known as the *Big Bang theory*, and this idea has become the foundation of scientific cosmology. To explain these observations, we must first introduce an important phenomenon called the Doppler effect. Therefore, we begin this section by developing an understanding of how the Doppler effect modifies the light seen in telescopes. We then show how this understanding leads to the idea that the Universe expands, and then to the conclusion that this expansion began during the Big Bang.

Waves and the Doppler Effect

When a train whistle screams, the sound you hear moves through the air from the whistle to your ear in the form of sound waves. *Waves* are disturbances that transmit energy from one point to another in the form of periodic motions. You're probably most familiar with the passage of water waves, during which the surface of the water goes up and down in a direction perpendicular to the direction the wave moves. The ups form *crests*, and the downs form *troughs*. Sound waves are different. As a sound wave passes, air moves back and forth, alternately compressing and expanding, in the direction that the wave moves. The whistle sound is not just one wave, but rather is a succession of many waves. We refer to the distance between successive waves as the **wavelength** and to the number of waves that pass a point in a given time interval as the **frequency**. If the wavelength decreases, more waves pass a point in a given time interval, so the frequency increases. The "pitch" of a sound—its note on the musical scale—depends on the frequency of the sound waves.

Now imagine that you are standing next to the tracks, and a train moves toward you. The train whistle's sound gets louder as the train approaches, but its pitch remains the same. Then, the instant the train passes, the pitch abruptly changes—it sounds like a lower note on the musical scale. Why? When the train moves toward you, the sound has a higher frequency (the waves lie closer together, so the wavelength is smaller) because the sound's source, the whistle, moved slightly closer to you between the instant that it emits one wave and the instant that it emits the next **(Fig. 1.9a)**. When the train moves away from you, the sound has a lower frequency (the waves are farther apart) because the whistle has moved slightly farther from you between the instant it emits one wave and the instant it emits the next. An Austrian physicist, C. J. Doppler (1803–1853), first explained this phenomenon, so we now refer to the change in frequency that happens when a wave source moves as the **Doppler effect**.

Light energy also moves in the form of waves. Physicists consider light to be a form of **electromagnetic radiation**, energy that can be released by hot or glowing objects and can be transmitted through a vacuum. We can represent light waves symbolically by a periodic succession of crests and troughs. (Shape-wise these resemble water waves, but otherwise they are very different in character.) Visible light comes in many colors—the colors of the rainbow. The color you see depends on the frequency of the light waves, just as the pitch of a sound you hear depends on the frequency of sound waves. Specifically, red light has a longer wavelength (lower frequency) than does blue light.

The Doppler effect also applies to light. If a light source moves away from you, the light you see becomes redder as the light shifts to longer wavelengths or lower frequencies.

FIGURE 1.9 Manifestations of the Doppler effect for sound and for light.

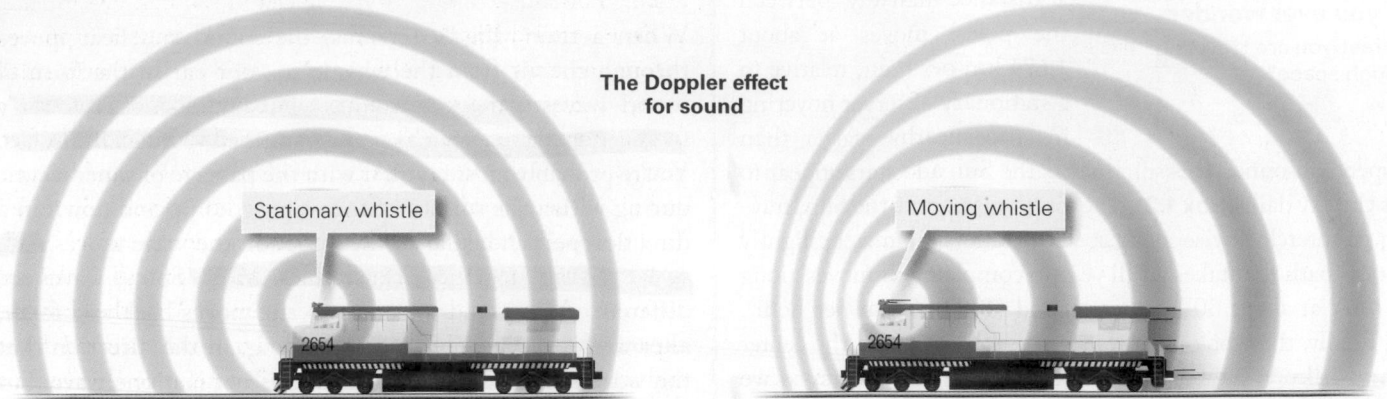

The Doppler effect for sound

Stationary whistle

Moving whistle

2654

2654

(a) The wavelength of sound waves emitted by a stationary train is the same in all directions. Waves behind a moving train have a longer wavelength than those in front.

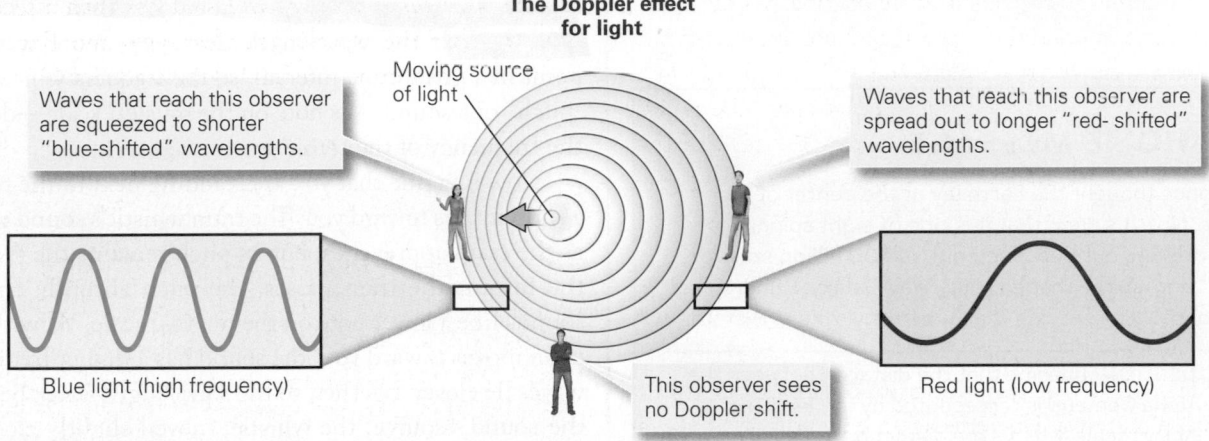

The Doppler effect for light

Moving source of light

Waves that reach this observer are squeezed to shorter "blue-shifted" wavelengths.

Waves that reach this observer are spread out to longer "red-shifted" wavelengths.

Blue light (high frequency)

This observer sees no Doppler shift.

Red light (low frequency)

(b) The wavelength of blue light is less than that of red light. If a light source moves very fast, the Doppler effect results in a shifting of the wavelengths. The observed shift depends on the position of the observer.

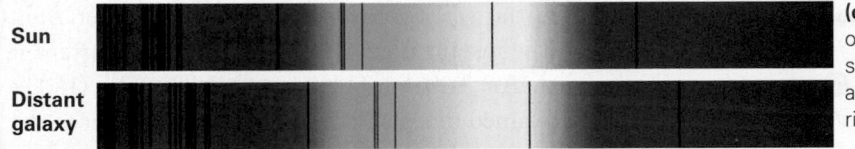

Sun

Distant galaxy

(c) The atoms in a star absorb certain specic wavelengths of light. We see these wavelengths as dark lines on a light spectrum. Note that the lines from a galaxy a billion light-years away are shifted toward the red end of the spectrum (i.e., to the right), relative to the lines from our own Sun.

If the source moves toward you, the light you see becomes more blue as the light shifts to higher frequencies. We call these changes the **red shift** and the **blue shift**, respectively (Fig. 1.9b).

Does the Size of the Universe Change?

In the 1920s, astronomers such as Edwin Hubble, after whom the Hubble Space Telescope was named, braved many a frosty night beneath the open dome of a mountaintop observatory in order to aim telescopes into deep space. These researchers were searching for distant galaxies. At first, they only wanted to document the location and shape of newly discovered galaxies. Eventually, however, they also began to study the wavelength of light produced by the distant galaxies. The results yielded

a surprise that would forever change humanity's perception of the Universe. To their amazement, astronomers found that the light coming to the Earth from all distant galaxies displayed a red shift, relative to the light coming from nearby stars (Fig. 1.9c).

Around 1929, Hubble concluded that the red shift must be a consequence of the Doppler effect, and so galaxies exhibiting a red shift must be moving away from the Earth at an immense velocity. How can all galaxies be moving away from us? Hubble puzzled over this question and finally recognized the solution: the whole Universe must be expanding. This idea came to be known as the **expanding Universe theory**. To picture the expanding

Did you ever wonder . . .
if galaxies move?

FIGURE 1.10 A raisin-bread analogy for the expanding Universe. As the dough expands, each raisin moves farther away from the others.

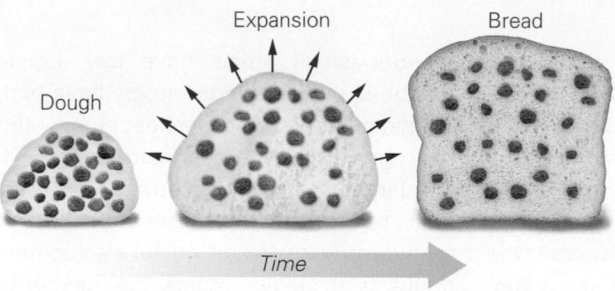

Dough   Expansion   Bread

Time

Universe, imagine a ball of bread dough with raisins scattered throughout. As the dough bakes and expands into a loaf, each raisin moves away from its neighbors, in every direction **(Fig. 1.10)**. Note that if two raisins were originally 1 cm apart, after a given time interval they spread to 2 cm apart, and that during the same time interval raisins that were originally 4 cm apart spread to 8 cm apart—so the farther apart the raisins are to start with, the faster they move apart. By this analogy, galaxies that lie farther away from the Earth are moving away from us faster than do galaxies closer to us, so farther galaxies exhibit a greater red shift than do nearer ones.

The Big Bang

Hubble's ideas started a revolution in cosmological thinking. Now we picture the Universe as an expanding bubble, in which galaxies race away from each other at incredible speeds. This image immediately triggers a key question of cosmology: Did the expansion begin at some specific time in the past? If it did, then that instant would mark the physical beginning of the Universe, the beginning of our space and time.

Most astronomers have concluded that expansion did indeed begin at a specific time, with a cataclysmic explosion called the Big Bang. According to the **Big Bang theory**, all matter and energy—everything that now constitutes the Universe—was initially packed into an infinitesimally small point, called a *singularity*. The point exploded and the Universe began, according to current estimates, 13.8 Ga (billion years ago).

Of course, no one was present at the instant of the Big Bang, so no one actually saw it happen. But by combining clever calculations with careful observations, researchers have developed a consistent model of how the Universe evolved, beginning an instant after the explosion. According to this model, the Universe consisted entirely of energy during the first instants of its existence. Atoms, or even the smallest subatomic particles that make up atoms, did not exist **(Box 1.3)**. Within a few seconds, however, the Universe had cooled enough that hydrogen atoms could begin to form **(Box 1.4)**. And by the time the Universe reached an age of 3 minutes, when its temperature had fallen below 1 billion degrees and its diameter had grown to about 53 million

kilometers, hydrogen atoms could fuse together to form helium atoms. Formation of new atomic nuclei in the first few minutes of time is called **Big Bang nucleosynthesis** because it happened before any stars existed. This process could produce only small atoms, meaning those containing a small number of protons (an atomic number less than 5), and it happened very rapidly. In fact, virtually all of the new atoms consisted of hydrogen and helium, and all of these atoms existed by the end of the first 5 minutes.

Eventually, the Universe cooled enough for chemical bonds to bind atoms together in molecules. Most notably, hydrogen atoms bonded to form molecular hydrogen (H_2). As the Universe expanded and cooled further, atoms and molecules slowed down and accumulated into patchy clouds called **nebulae**. The earliest nebulae of the Universe consisted almost entirely of hydrogen and helium gas. The initial expansion of the Universe took place very rapidly. This *inflationary epoch* lasted less than a second. After that, the Universe expanded at a less rapid rate. Recent research suggests that its expansion has begun to accelerate **(Fig. 1.11)**.

FIGURE 1.11 The concept of the expanding Universe and the Big Bang.

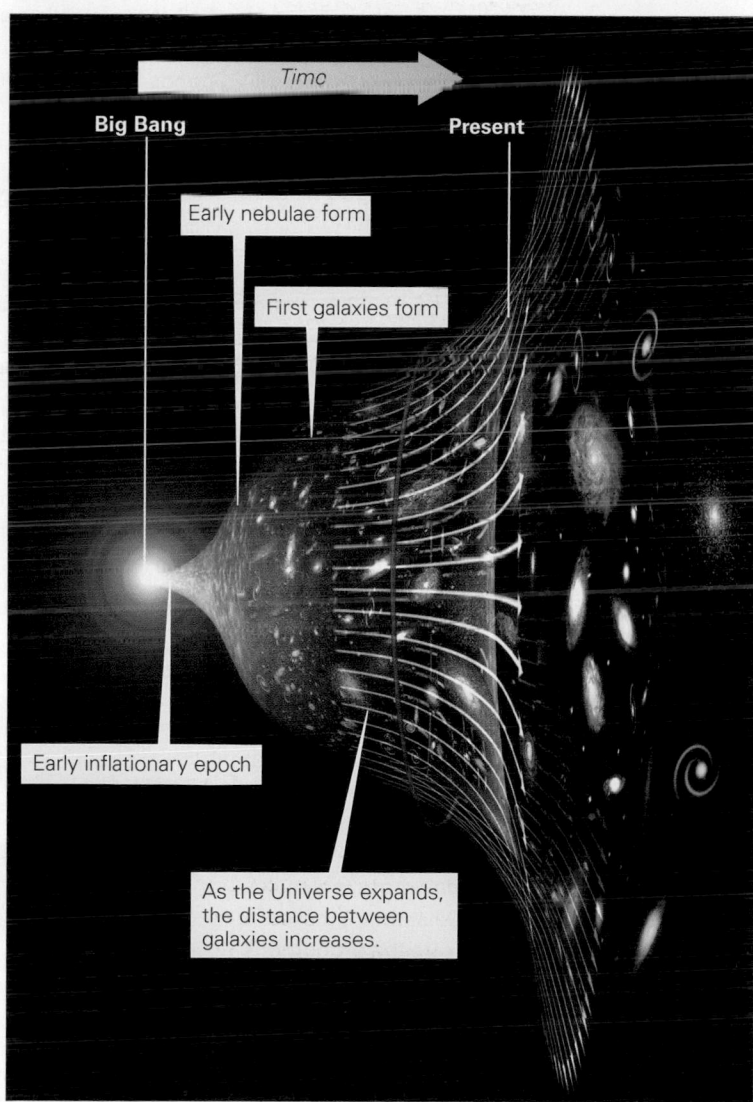

Time

Big Bang   Present

Early nebulae form

First galaxies form

Early inflationary epoch

As the Universe expands, the distance between galaxies increases.

BOX 1.3 SCIENCE TOOLBOX . . .

Matter and Energy

Matter, Atoms, and Molecules What does matter consist of? A Greek philosopher named Democritus (ca. 460–370 B.C.E.) reasoned that you cannot keep dividing a volume of matter into progressively smaller pieces, because if you did, you would eventually end up with nothing. Since it's not possible to make something out of nothing, there must be a smallest piece of matter that can't be subdivided further. Democritus proposed the name *atom* for these smallest pieces, from the Greek word *atomos*, which means indivisible.

Our modern understanding of matter developed in the 17th century, when *chemists* (scientists who study the properties, composition, and behavior of matter) recognized that some substances, such as hydrogen and oxygen, cannot be divided to form other substances, whereas others, such as water and salt, can be. The "undividable" substances came to be known as **elements**. John Dalton (1766–1844) adopted the word **atom** for the smallest piece of an element that has the property of the element. During the 18th and 19th centuries, chemists identified 92 naturally occurring elements on Earth, and during the 20th century, physicists produced more than a dozen more. Each element has a name and a symbol: for example, N = nitrogen, H = hydrogen, Fe = iron, and Ag = silver. In 1869, Dmitri Mendeleev (1834–1907) recognized that groups of elements share similar characteristics, and he organized the elements into a chart that we now call the *periodic table* (see Additional Maps and Charts at the back of this book).

Atoms don't always occur in isolation, but rather can attach to other atoms. Chemists refer to a combination of two or more atoms as a **molecule**, and to the "glue" that holds one atom to another as a **chemical bond**. Some molecules contain only one element, whereas others contain atoms of different elements; a material whose molecules contain more than one element is called a **compound**. We can indicate the proportions of different elements in a molecule by means of a **chemical formula**. A hydrogen molecule, for example, contains two hydrogen atoms, so its formula is H_2; note that hydrogen is not a compound. Water, in contrast, is a compound, for its molecules contain two hydrogen atoms and one oxygen atom. A molecule of water has the formula H_2O **(Fig. Bx1.3a)**.

What's Inside an Atom? Experiments in the early 20th century revealed that atoms consist of smaller pieces, known as *subatomic particles*, of which there are three types: **neutrons**, which have a neutral charge, **protons**, which have a positive charge, and **electrons**, which have a negative charge **(Fig. Bx1.3b)**. (*Charge*, simplistically, refers to the way in which a particle responds to a magnet or an electric current; unlike charges attract, whereas like charges repel). Protons and neutrons pack together in the **nucleus**, a dense ball at the atom's center. The bonds holding these particles together are known as **nuclear bonds**. Electrons swirl around the nucleus in an electron cloud. Protons and neutrons are about the same size, but an electron contains only 1/1,836 as much mass as a proton.

Electron clouds have a complex internal structure. Specifically, electrons group into intervals called **electron shells**. Electrons in inner shells concentrate closer to the nucleus, while those in outer shells lie farther away. Roughly speaking, the diameter of an electron cloud is 10,000 times greater than that of a nucleus, yet the cloud contains only 0.05% of an atom's mass. So, in effect, atoms consist mostly of empty space!

The number of protons in the nucleus of an atom defines the element's **atomic number**—small atoms have low atomic numbers, whereas large atoms have high atomic numbers. The smallest atom is the hydrogen atom, with an atomic number of 1. The largest naturally occurring atom, uranium, has an atomic number of 92. Except for hydrogen nuclei, all nuclei also contain neutrons. In smaller atoms, the number of neutrons roughly equals the number of protons, but in larger atoms, the number of neutrons exceeds the number of protons. We define the **atomic mass** of an atom as approximately the sum of the number of neutrons and the number of protons. For example, an oxygen nucleus contains 8 protons and 8 neutrons, so it has an atomic mass of 16.

States of Matter Matter can exist in different forms, known as **states of matter**, depending on the degree to which the atoms or molecules in the matter hold together **(Fig. Bx1.3c)**. A material in which all the atoms or molecules bond so tightly that the material can retain its shape for a long time is called a **solid**. In contrast, a material in which the atoms occur in disorganized clumps or chains that can flow past one another fairly easily, so that the material can conform to the shape of its container even though it retains the same volume, is called a **liquid**. Finally, in a **gas**, atoms or molecules are not bonded to one another, so they can move around easily in all directions. As a consequence, a gas expands to fill any container it's placed in. When environmental conditions change, a material may

FIGURE Bx1.3 The nature of atoms.

(a) Two ways of portraying a water molecule. The red ball is oxygen; the small balls are hydrogen.

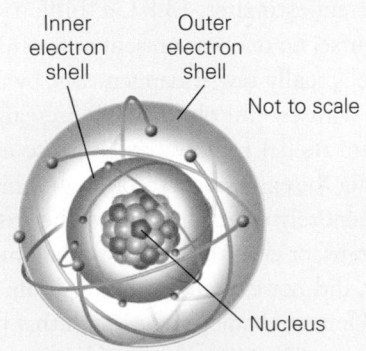

Inner electron shell

Outer electron shell

Not to scale

Nucleus

(b) An image of an atom with a nucleus orbited by electrons.

change from one state to another. For example, during **condensation**, a gas becomes a liquid and during **freezing**, a liquid becomes a solid.

The Energy Stored in Bonds In Box 1.1, we introduced the concept of *potential energy*, the energy stored in a material. Chemical bonds and nuclear bonds store potential energy, and this energy can be released when such bonds break or form. Chemical bonds break and form during a **chemical reaction**. During such reactions, some molecules break apart and others form. During a *nuclear reaction*, in which nuclear bonds break or form, a small amount of matter converts into energy following Einstein's famous equation, $E = mc^2$. (In this equation, E stands for energy, m stands for mass, and c stands for the speed of light.) Nuclear reactions during which nuclei break apart are called **fission reactions**, whereas those during which nuclei of two smaller atoms fuse together are called **fusion reactions** (**Fig. Bx1.3d, e**). Fission generates energy in nuclear power plants and in the explosion of an atomic bomb, whereas fusion reactions power the Sun and take place during the explosion of a hydrogen bomb. Fusion can occur only when nuclei collide at extremely high velocity, because the nuclear force that binds nuclei together operates only at subatomic distances.

FIGURE Bx1.3 *(continued)*

(c) The three common states of matter. Solids keep their shape, liquids conform to the shape of their container without changing density, and gases expand to fill their container.

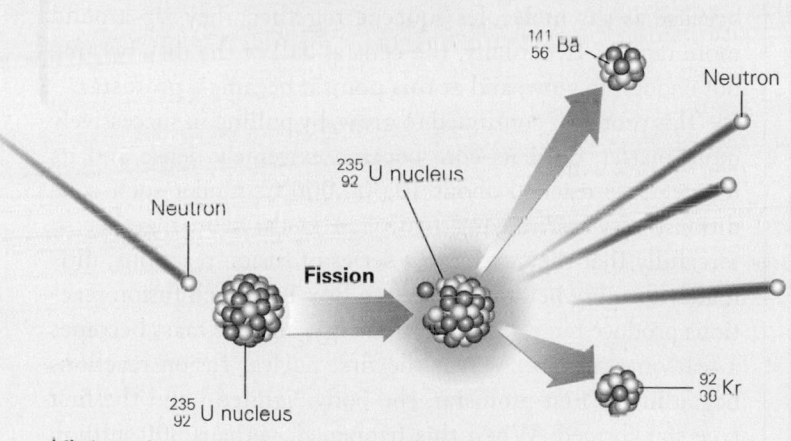

(d) A uranium atom splits during nuclear fission.

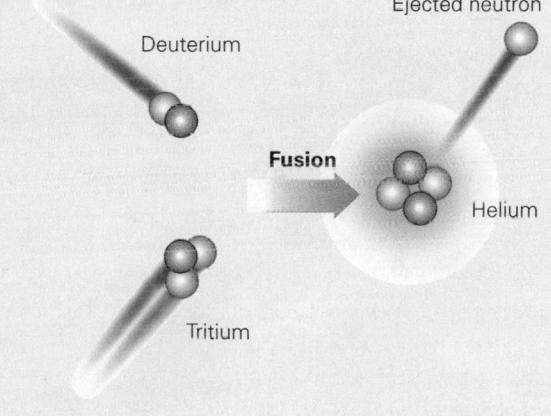

(e) Two atoms (versions of hydrogen) stick together to form one atom of helium during nuclear fusion.

Birth of the First Stars

When the Universe reached its 200 millionth birthday, it contained immense, slowly swirling, dark nebulae separated by vast voids of empty space (**Fig. 1.12**). The Universe could not remain this way forever, though, because of the invisible but persistent pull of gravity. Eventually, molecules began sticking together to form specks of ice, and gravity began to remold the Universe pervasively and permanently.

All matter exerts gravitational pull—a type of force—on its surroundings, and as Isaac Newton first pointed out, the amount of pull depends on the amount of mass. Somewhere in the young Universe, the gravitational pull of an initially more massive region of a nebula began to suck in surrounding ice and gas and, in a grand example of the rich getting richer, grew in mass and therefore density (mass per unit volume). As this denser region attracted progressively more matter, the matter compacted into a smaller region, and the initial swirling movement of a nebula transformed into a rotation around an axis. As matter continued to move inward, cramming into a progressively smaller volume, the rotation rate became faster and faster. (A similar phenomenon happens when a spinning

BOX 1.4 SCIENCE TOOLBOX . . .

Heat and Temperature

In everyday English, the words *heat* and *temperature* may seem interchangeable, but in scientific discussion, they have different, distinct definitions. Scientists refer to the total kinetic energy contained in matter due to the motion of its particles as **thermal energy**, or **heat**. A jar of gas that has been warmed in an oven contains more heat than the same-sized jar that has been sitting in a refrigerator.

The **temperature** of a material is a number that represents the average velocity of particles moving within the material. These movements can include the vibration of atoms or molecules in place and the physical movement of atoms or molecules from one location to another. The faster the particles vibrate or move, the higher the temperature. When heat is added to a material, its temperature rises. For example, when you place a jar of gas on a hot stove, heat moves into the jar so that the temperature of the gas rises because the gas molecules start to move faster. If, however, you place the jar in a cold refrigerator, heat moves out of the jar, and the gas molecules slow down.

With the above concepts in mind, consider the following experiment: Imagine that you have two jars, both containing gas at the same temperature. One jar contains 2 grams of gas, and the other jar contains 1 gram of gas. The jar containing 2 grams of gas holds more heat because it contains more atoms and, therefore, more total kinetic energy.

Scientists use the *centigrade scale* (also known as the Celsius scale) for calibrating changes in temperatures in the metric system of measurement. In the United States, weather reports typically give temperatures using the *Fahrenheit scale* of the English system of measurement. Water freezes at 0°C, or 32°F, and it boils at 100°C, or 212°F. Note that a single degree in the centigrade scale represents a larger temperature change than does a single degree in the Fahrenheit scale.

Matter becomes its coldest when its atoms or molecules stand still. We call this temperature absolute zero, or 0 K (pronounced "zero kay"), where K stands for Kelvin (after Lord Kelvin, 1824–1907, a British physicist), another unit of temperature. You simply can't get colder than absolute zero, meaning that you can't extract any thermal energy from a substance at 0 K (−273.15°C).

FIGURE 1.12 A representation of a starless nebula in the early Universe. In reality, the clouds would have been dark, for no stars existed. This rendition is modified from a Hubble Space Telescope photograph.

ice skater pulls her arms inward.) Because of its increased rotation, the nebula evolved into a disk shape. As more and more matter rained down onto the disk, it continued to grow, until eventually, gravity collapsed the inner portion of the disk into a dense ball. As the matter compressed into a smaller and smaller space, its temperature increased dramatically because as gas molecules squeeze together, they zip around more rapidly. Eventually, the central ball of the disk became hot enough to glow, and at this point it became a **protostar**.

This protostar continued to grow, by pulling in successively more matter, until its core became extremely dense and its temperature reached about 10,000,000°C. Under such conditions, very fast-moving hydrogen nuclei slam together so forcefully that they undergo a series of fusion reactions, ultimately forming helium nuclei (see Box 1.3). Such fusion reactions produce huge amounts of energy, and the mass becomes a fearsome furnace. When the first nuclear fusion reactions began in the first protostar, the body "ignited," and the first true star formed. When this happened, perhaps 800 million years after the Big Bang, the first starlight illuminated the newborn Universe. This process would soon happen again and again, and many *first-generation stars* came into existence.

First-generation stars tended to be massive, perhaps 100 times the mass of the Sun. Astronomers have shown that the larger the star, the hotter it burns, and the faster it runs out of fuel and dies. A huge star may survive only a few million to a few tens of millions of years before it explodes. The explosion of a giant star produces a **supernova** (from the Latin word *nova*, meaning new), so named because closer examples can be seen from the Earth, where they appear as "new" stars in the sky. As a result, not long after the first generation of stars formed, the Universe began to be peppered with the first generation of supernovae.

TAKE-HOME MESSAGE

Study of the red shift shows that all distant galaxies are moving away from us, an observation that requires the Universe to be expanding. According to the Big Bang theory, this expansion and, therefore, the beginning of our Universe, began with a cataclysmic explosion at 13.8 Ga. Atoms formed during the Big Bang collected into nebulae, which gravity collapsed into dense balls that began to produce energy by nuclear fusion reactions. These objects were the first stars.

QUICK QUESTION: The farther a galaxy lies from the Earth, the greater the red shift it exhibits. Why?

FIGURE 1.13 Element factories in space.

(a) A photo of solar (stellar) wind streaming into space.

1.4 We Are All Made of Stardust

Where Do Elements Come From?

The nebulae from which the first-generation stars formed consisted entirely of small atoms (atoms with atomic numbers less than 5) because only these small atoms were generated by Big Bang nucleosynthesis. In contrast, the Universe of today contains 92 naturally occurring elements. Where did the other 87 elements come from? In other words, how did elements such as carbon, sulfur, silicon, iron, gold, and uranium form? These elements, which are common on the Earth, have atomic numbers greater than 5. For example, carbon has an atomic number of 6, and iron has an atomic number of 26. Physicists have shown that these elements form during the life cycle of stars by the process of **stellar nucleosynthesis**. Because of stellar nucleosynthesis, we can consider stars to be "element factories," constantly fashioning larger atoms out of smaller atoms.

What happens to the atoms produced by stellar nucleosynthesis? Some escape into space during the star's lifetime, simply by moving fast enough to overcome the star's gravitational pull. The stream of atoms emitted from a star during its lifetime is its **stellar wind (Fig. 1.13a)**. Most atoms, however, escape only when a star dies. A relatively low-mass star (like our Sun) releases a large shell of gas as it dies, ballooning into a *red giant* during the process, whereas a relatively high-mass star blasts matter into space during a *supernova explosion* **(Fig. 1.13b)**. Most very large atoms (those with atomic numbers greater than that of iron) can form only at the ultrahigh temperatures that develop in a supernova explosion. The process of forming these larger, heavier atoms is, therefore, called *supernova nucleogenesis.*

(b) Very heavy elements form during supernova explosions. Here we see the rapidly expanding shell of gas ejected into space from an explosion whose light reached the Earth in 1054 c.e. This shell is called the Crab Nebula.

When the first generation of stars died, it left a legacy of new, heavier elements, which then mixed with residual gas from the Big Bang. A second generation of stars formed out of these compositionally more diverse nebulae. When these stars died, they contributed heavier elements to third-generation stars. Succeeding generations of stars contain a greater proportion of heavier elements. (Nevertheless, most of the visible mass of the Universe still consists of 74% hydrogen and 24% helium.) Because not all stars live for the same duration of time, at any given moment the Universe contains many different generations of stars. Our Sun may be a third-, fourth-, or fifth-generation star. The mix of elements we find on the Earth includes relics

TABLE 1.1 Top 10 Elements in the Milky Way (out of 1 Million Atoms)

Hydrogen	739,000
Helium	240,000
Oxygen	10,400
Carbon	4,600
Neon	1,340
Iron	1,090
Nitrogen	960
Silicon	650
Magnesium	580
Sulfur	440
Other (approx.)	940

Did you ever wonder... where the atoms in your body first formed?

one another by gravity. Eventually, as we've seen, gravity pulls a swirling nebula inward and it coalesces into a spinning disk with a bulbous center. In the case of our Solar System the central "bulb" of this disk became the Sun, whereas the remainder of the Solar System formed from the material in the flattened outer part of the disk, a region now known as the **protoplanetary disk**. Material in the disk does not fall to the center of the disk because it's moving fast enough to stay in orbit.

The protoplanetary disk of our Solar System contained all 92 elements, some as isolated atoms and some bonded to others in molecules, in a mixture of gases produced during the Big Bang as well as gases expelled from earlier generations of stars of primordial gas from the Big Bang as well as the disgorged guts of dead stars. Think of it—the elements that make up your body once resided inside a star, or even a supernova! The proportions of elements reflect the processes of element formation and the degree to which stellar gases and supernova gases mix **(Table 1.1)**.

FIGURE 1.14 Nebular theory for planet formation.

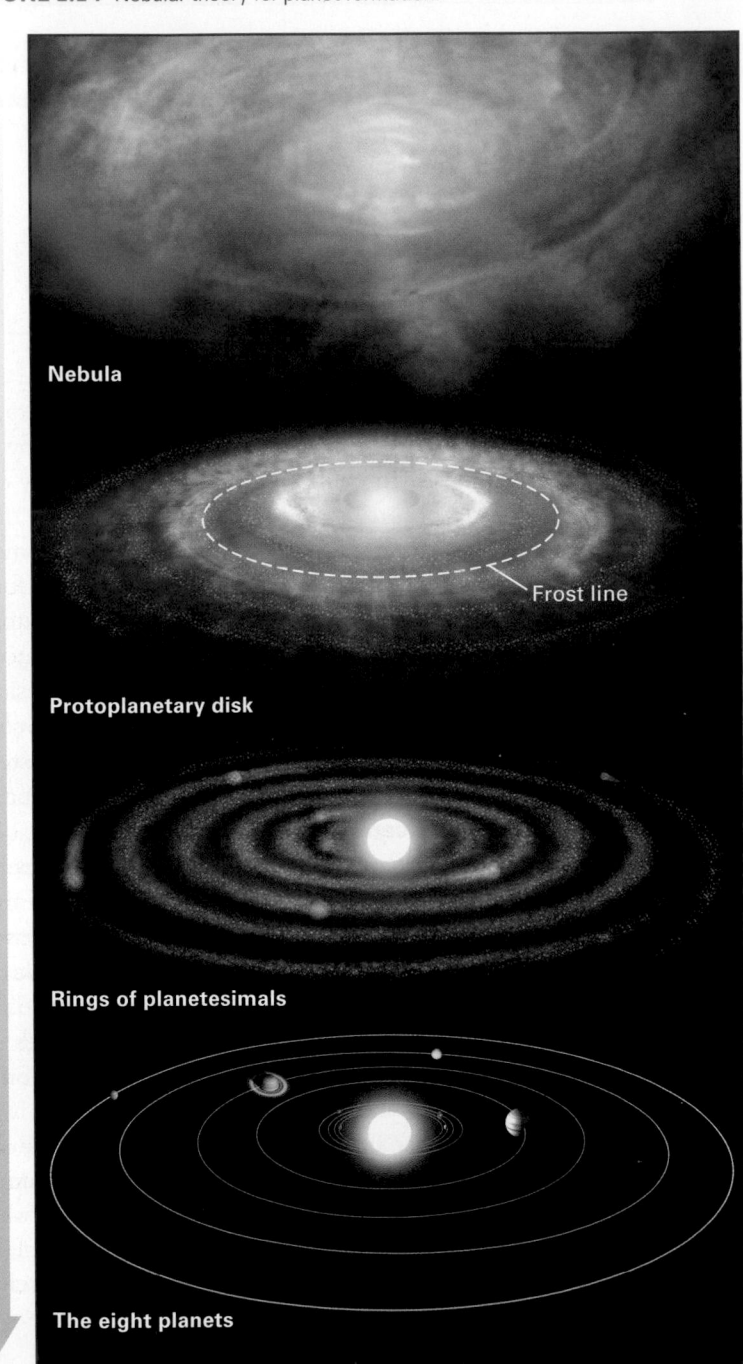

Nebula

Frost line

Protoplanetary disk

Rings of planetesimals

The eight planets

Time

The Nebular Theory

Earlier in this chapter, we described how stars formed from nebulae that consisted mostly of small atoms formed by Big Bang nucleosynthesis. But we delayed addressing the question of how the planets and other objects in our Solar System originated until we had discussed the production of heavier elements, such as oxygen, silicon, and iron, because planets such as the Earth consist predominantly of these elements. Now that we've shown how stars and supernovae can serve as element factories producing larger atoms, we return to the early history of the Solar System and add an explanation for the origin of planets, moons, asteroids, and comets.

In the context of scientific cosmology, the Sun and all other objects in the Solar System formed from material that had been swirling about in a nebula, an idea now known as the **nebular theory** or, more recently, as the *condensation theory* **(Fig. 1.14)**. According to the theory, the process of Solar System formation involved several stages **(Geology at a Glance, pp. 34–35)**. Such a process begins when tiny ice and *dust particles* (specks of solids made of materials that do not evaporate easily) condense in a nebula. Other atoms or molecules attach to these particles, building into masses that are large enough to be attracted to

FIGURE 1.15 Forming solids in the Solar System.

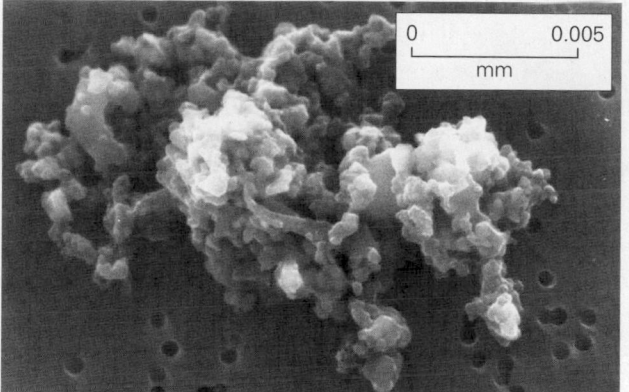

(a) At first, solids collected into dust-sized aggregates.

(c) The disk then evolved into a revolving ground of debris.

(b) Eventually, the dust collected into grainy masses, similar to this meteorite fragment.

and from supernova explosions. Geologists divide the materials formed from these atoms and molecules into two classes. Volatile materials—such as hydrogen, helium, methane, ammonia, water, and carbon dioxide—can exist as gas at the Earth's surface. Under the pressure and temperature conditions of regions in the protoplanetary disk close to the Sun, all volatile material remained in a gaseous state. But beyond a certain distance called the *frost line*, volatiles froze to form ice. (Note again that we don't limit use of the word *ice* for frozen water alone.) Refractory materials are those that melt only at high temperatures. These are the elements which, in the coldness of space, solidify to form solid soot-sized particles of dust. In the protoplanetary disk, some of this dust consisted mostly of metal, but a large portion of the dust formed in this way contained molecules of silicon and oxygen bonded to various metal atoms. We'll see in Chapter 5 that such materials are known as *silicate minerals* and that they form most of the rock of the Earth, so clumps of silicate-mineral dust were rock-like in character. Refractory materials could remain solid in the region of the protoplanetary disk close to the Sun.

Initially, the protoplanetary disk may have been fairly homogeneous, meaning that it had much the same composition throughout, and most volatiles were frozen. But when the

Sun became a nuclear inferno, the *solar wind* (the stellar wind produced by our Sun) evaporated volatile materials and blew them outward, to distances beyond the frost line. Consequently, the inner part of the protoplanetary disk ended up with higher concentrations of dust, whereas the outer portions ended up with higher concentrations of ice. As this was happening, gravity caused the gas, dust, and ice of the disk to separate into a series of concentric rings in which density exceeded that of the space between the rings.

How did these dusty, icy, and gassy rings transform into planets? Even before the proto-Sun ignited, the material of the surrounding rings began to clump and bind together due to gravity (Fig 1.15). First, soot-sized particles merged into sand-sized grains. Then these grains stuck together to form grainy basketball-sized blocks, which in turn collided. If the collision was slow, the blocks stuck together or simply bounced apart. If the collision was fast, one or both of the blocks shattered, producing smaller fragments that recombined later. Eventually, enough blocks coalesced to form **planetesimals**, solid bodies whose diameter exceeded about 1 km. Because of their mass, planetesimals exerted enough gravitational attraction to pull in other objects that were nearby. Figuratively, planetesimals acted like vacuum cleaners, sucking in small pieces of dust and ice as well as smaller planetesimals that lay in their orbit, and in the process they grew progressively larger—astronomers refer to this process of planetary growth as *accretion*. Eventually, victors in the competition to attract matter grew into **protoplanets**, bodies approaching the size of today's inner planets. Once a protoplanet succeeded in incorporating virtually all the debris within its orbit, it became a full-fledged planet.

Early stages in the planet-forming process probably occurred very quickly—some computer models suggest that it may have taken less than 1 million years to go from the dust-and-gas

Forming the Planets and the Earth-Moon System

2. Gravity pulls gas and dust inward to form a protoplanetary disk. Eventually, a glowing ball—the proto-Sun—forms at the center of the disk.

1. Forming the Solar System, according to the nebular theory: A nebula forms from hydrogen and helium left over from the Big Bang, as well as from heavier elements that were produced by fusion reactions in stars or during explosions of stars.

6. Gravity reshapes the proto-Earth into a sphere. The interior of the Earth differentiates into a core and mantle.

5. Forming the planets from planetesimals: Planetesimals grow by continuous collisions. Gradually, an irregularly shaped proto-Earth develops. The interior heats up and becomes soft.

7. Soon after the Earth forms, a protoplanet collides with it, blasting debris that forms a ring around the Earth.

8. The Moon forms from the ring of debris.

3. "Dust" (particles of refractory materials) concentrates in the inner rings, while "ice" (particles of volatile materials) concentrates in the outer rings. Eventually, the dense ball of gas at the center of the disk becomes hot enough for fusion reactions to begin. When it ignites, it becomes the Sun.

4. Dust and ice particles collide and stick together, forming planetesimals.

9. Eventually, the atmosphere develops from volcanic gases. When the Earth becomes cool enough, moisture condenses and rains to fill the oceans. Some gases may be added by passing comets.

stage to the large planetesimal stage. Planets may have grown from planetesimals over the course of the next few million years. In the inner orbits, where rings of the protoplanetary disk consisted mostly of dust (refractory materials), small terrestrial planets composed mostly of rock and metal formed. The outer rings of the protoplanetary disk, in contrast, contained huge volumes of volatile materials. Protoplanets formed in the outer rings gathered up these volatile materials to form the giant planets. Any refractory materials in these planets sank to the center, so these planets consist of a small refractory (metal and rock) ball surrounded by very thick shells of volatile materials (in the form of gas, liquid, or ice).

When did the planets form? As we'll see later in the book, rocks now exposed on the surface of the Earth are much younger than the Solar System. But certain *meteorites*, objects that have fallen to the Earth from space (see Chapter 2), appear to be leftover planetesimals. Using dating techniques introduced in Chapter 12, geologists have determined that some materials in these meteorites formed as long as 4.57 billion years ago, and consider that date to be the birth date of the Solar System. This date means that the Solar System formed about 9 billion years after the Big Bang and is only about a third as old as the Universe.

Differentiation of the Earth and Formation of the Moon

When planetesimals first formed, they had a fairly homogeneous distribution of material throughout because the smaller pieces from which they formed all had much the same composition and accreted in no particular order. But large planetesimals did not stay homogeneous for long because they began to heat up inside. The heat came primarily from three sources: the heat produced by collisions (similar to the heat produced when you bang on a nail with a hammer and they both get warm); the heat produced when matter squeezes into a smaller volume (similar to the heat produced when you compress air with a pump); and the heat produced by the decay of radioactive elements (similar to the heat produced by a nuclear power plant). In bodies whose temperature rose sufficiently to cause internal melting, denser metals (mostly iron) separated out and sank to the center of the body, whereas lighter rocky materials (mostly silicate minerals) remained in a shell surrounding the center. By this process, called **differentiation**, protoplanets and large planetesimals developed internal layering early in their history (**Fig. 1.16**). As we will see later, the central ball of metal constitutes the body's *core* and the outer, rocky shell constitutes its *mantle*. Recent studies suggest that the terrestrial planets had differentiated by about 4.56 Ga.

In the early days of the Solar System, planets continued to be bombarded by meteorites even after the Sun had ignited and differentiation had occurred. Bombardment also contributed to heating the planets. Most geologists favor a model in which a particularly large collision between the Earth and a large planetesimal or protoplanet at about 4.5 Ga produced our Moon. Moon formation happened because the collision was so cataclysmic that much of the colliding body disintegrated and evaporated, along with a large part of the Earth's mantle. A ring of debris formed around the remaining, now-molten Earth. This ring quickly coalesced by accretion to form the Moon. When first formed, the Moon orbited much closer to the Earth than it does today. Not all moons in the Solar System necessarily formed in this manner, however. Some may have been independent

> **Did you ever wonder...**
> if the Moon is as old as the Earth?

FIGURE 1.16 Differentiation of the Earth's interior.

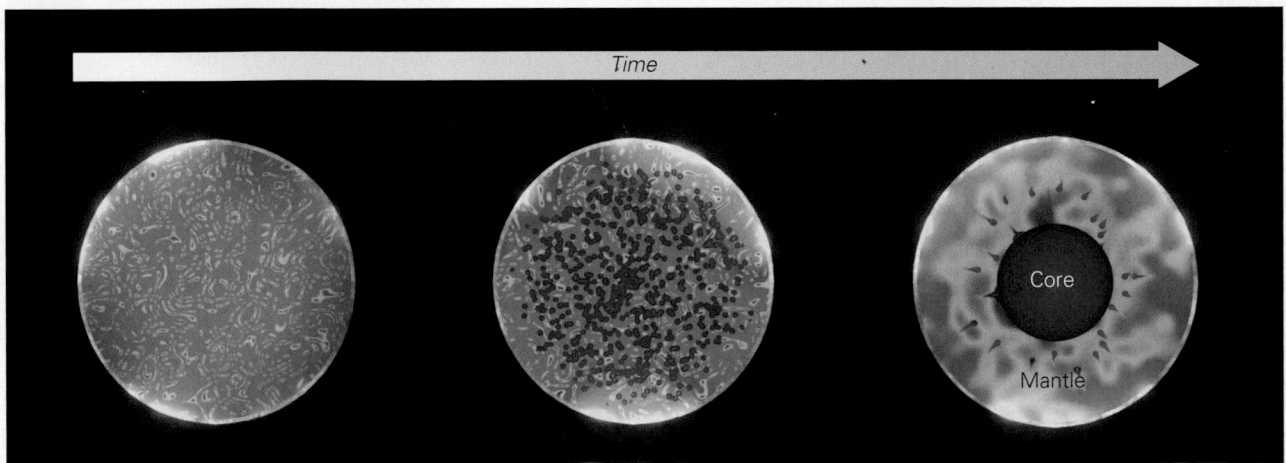

(a) Early on, the Earth was fairly homogeneous inside.

(b) When the temperature got hot enough, iron began to melt.

(c) The iron accumulated at the center of the planet to form a metallic core.

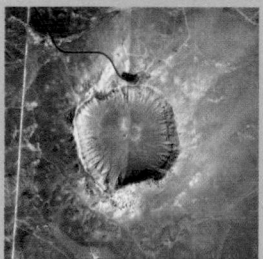

protoplanets or comets that were captured by a larger planet's gravity. Sporadic heavy bombardments of planets by meteorites continued until about 3.9 Ga. These bombardments may have pulverized and heated planetary surfaces. Rarer, but still significant, impacts left craters on planets.

Making the Earth Round

Small planetesimals were jagged or irregular in shape, and small asteroids today have irregular shapes. Planets, on the other hand, are more or less spherical. Why? Simply put, when a protoplanet gets big enough, and becomes warm enough inside, gravity can change its shape. To picture how, imagine a block of cheese warming in an oven. As the cheese gets softer and softer, gravity causes it to spread out in a pancake-like blob. This model shows that gravitational force alone can cause material to change shape if the material has become soft enough. Now let's apply this model to planetary growth.

The rock composing a small planetesimal remains cool and strong enough so that the force of gravity alone cannot cause the rock to flow. But once a planetesimal grows beyond a certain critical size (about 300 to 500 km in diameter), its interior becomes warm and soft enough to flow simply in response to gravity. As a consequence, protrusions are pulled inward toward the center so that the planetesimal re-forms into a special shape—a sphere—that permits the force of gravity to be nearly the same at all points on its surface, for in a sphere, mass is evenly distributed around the center.

Forming the Ocean and Atmosphere

Unlike the other terrestrial planets, the Earth today has an ocean of liquid water and an atmosphere that consists primarily of molecular nitrogen (N_2) and molecular oxygen (O_2). Where did the ocean and atmosphere come from? Let's take a brief look at how these entities, without which life as we know it could not exist, came to be.

When the Earth first formed, its atmosphere probably consisted mostly of molecular hydrogen and helium. But as the planet warmed, the temperature of these gases increased and the atoms were able to move so fast that, like rockets, they escaped the pull of gravity and were blown away from the Earth by the solar wind. Gradually, they were replaced by gases released from erupting volcanoes. These gases (volatile materials) were originally bonded to solid materials of the Earth's mantle. But when portions of the mantle melted to produce molten rock that rose to the surface, the volatiles separated from the solids and bubbled out. Comets bombarding the Earth may have brought in additional gases. The atmosphere evolved into one consisting mostly of water, carbon dioxide (CO_2), ammonia, and methane—all gases released by "outgassing" of the Earth's interior. You would suffocate instantly if you were to breathe this early atmosphere.

When the Earth cooled sufficiently for water to condense (change its state from gas to liquid), rain fell and the oceans accumulated. The concentration of water in the atmosphere decreased substantially as a consequence. The CO_2 from the atmosphere then dissolved in the water and precipitated into solids that settled out of the ocean and eventually became trapped in the crust as rock. As a result, the concentration of CO_2 diminished substantially. Because it does not react with other Earth materials, N_2 remained in the atmosphere and eventually became the dominant component of the atmosphere. Only later, after the appearance of organisms capable of carrying out photosynthesis, an O_2-generating reaction, did molecular oxygen appear in the atmosphere. As discussed in Chapter 13, the concentration of O_2 did not become significant until about 600 million years ago.

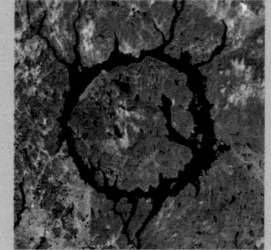

TAKE-HOME MESSAGE

Heavier elements formed in stars and supernovae added to gases in nebulae from which new generations of stars formed. Planets formed from rings of dust and ice orbiting the stars, so we are all formed of stardust. As they formed, planets differentiated, with denser materials sinking to the center.

QUICK QUESTION: Why do the gas-giant planets orbit farther from the Sun than the terrestrial planets?

Chapter 1 Review

SUMMARY

- The geocentric model of the Universe placed the Earth at the center of the Universe. The heliocentric model, which gained acceptance during the Renaissance, placed the Sun at the center. The geocentric model remained popular until the 17th century.

- Eratosthenes measured the size of the Earth in ancient times, but it was not until fairly recently that astronomers accurately determined the distances from the Earth to the Sun, planets, and stars. Distances in the Universe are so large that they must be described in light-years.

- The Earth is one of eight planets orbiting the Sun. Our Solar System lies on the outer edge of the Milky Way Galaxy, which contains about 300 billion stars. Perhaps a trillion galaxies populate the Universe.

- The red shift of light from distant galaxies, a manifestation of the Doppler effect, indicates that all distant galaxies are moving away from the Earth. This observation supports the expanding Universe theory. Most astronomers agree that this expansion began after the Big Bang, a cataclysmic explosion that occurred about 13.8 Ga.

- The first atoms of the Universe (hydrogen and helium) developed within minutes of the Big Bang. These atoms formed vast gas clouds called nebulae.

- Gravity caused clumps of gas and ice in the nebulae to coalesce into flattened disks with bulbous centers. The protostars at the centers of these disks eventually became dense and hot enough that fusion reactions began in them. When this happened, they became true stars.

- Heavier elements form during fusion reactions in stars; the heaviest are mostly made during supernova explosions. The Earth and the life forms on it contain elements that could only have been produced during the life cycle of stars, so we are all made of stardust.

- According to the nebular (condensation) theory of Solar System formation, planets developed from the rings of gas and dust surrounding the proto-Sun. The gas and dust condensed into planetesimals, which then clumped together to form protoplanets and finally true planets. Inner rings became the terrestrial planets. Outer rings grew into the giant planets.

- The Moon formed from debris ejected when a protoplanet collided with the Earth in the young Solar System.

- During differentiation, the interior of a planet separates into layers. A planet assumes a near-spherical shape when it becomes so soft that gravity can smooth out irregularities.

GUIDE TERMS

asteroid (p. 22)
astronomer (p. 17)
atom (p. 28)
atomic mass (p. 28)
atomic number (p. 28)
Big Bang nucleosynthesis (p. 27)
Big Bang theory (p. 27)
blue shift (p. 26)
celestial object (p. 17)
chemical bond (p. 28)
chemical formula (p. 28)
chemical reaction (p. 29)

comet (p. 23)
compound (p. 28)
condensation (p. 29)
density (p. 29)
differentiation (p. 36)
Doppler effect (p. 25)
dwarf planet (p. 22)
electromagnetic radiation (p. 25)
electron (p. 28)
electron shell (p. 28)
element (p. 28)
energy (pp. 17, 20)
equator (p. 24)

expanding Universe theory (p. 26)
fission reaction (p. 29)
freezing (p. 29)
frequency (p. 25)
fusion reaction (p. 29)
galaxy (p. 19)
gas (p. 28)
geocentric model (p. 18)
giant planet (p. 19)
gravity (p. 18)
heat (p. 30)
heliocentric model (p. 18)
light-year (p. 23)

liquid (p. 28)
magnetism (p. 20)
mass (p. 17)
matter (p. 17)
molecule (p. 28)
moon (p. 22)
nebulae (p. 27)
nebular theory (p. 32)
neutron (p. 28)
nuclear bond (p. 28)
nucleus (p. 28)
planet (p. 19)
planetesimal (p. 33)
proton (p. 28)

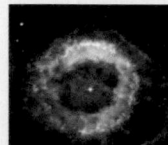
GEOTOURS *THIS CHAPTER'S GEOTOURS WORKSHEET (A) FEATURES QUESTIONS AND GOOGLE EARTH SITES ON:*

- Nebular supernovae
- Spiral galaxies
- Meteorite impacts
- The Moon
- Solar System scaling
- Stars

REVIEW QUESTIONS

The letters following each Review Question refer to the corresponding Learning Objective from the Chapter Opener.

1. Contrast the geocentric and heliocentric Universe concepts. **(A)**

2. Describe how Foucault's pendulum demonstrates that the Earth rotates on its axis. **(A)**

3. How did Eratosthenes calculate the Earth's circumference? **(A)**

4. Why do planets appear to move with respect to stars? **(B)**

5. Imagine you hear the main characters in a low-budget science-fiction movie say that they will "return 10 light-years from now." What's wrong with their use of the term? **(B)**

6. Describe how the Doppler effect works. If the light you see from a distant galaxy has undergone a blue shift, is the galaxy traveling toward or away from you? **(C)**

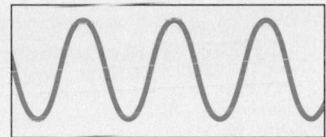

7. What does the red shift of the galaxies tell us about their motion with respect to the Earth? **(C)**

8. What is the Big Bang, and when do astronomers conclude it occured? **(C)**

9. When did hydrogen and helium atoms form? **(D)**

10. This image shows a nebula formed by a supernova explosion. Would it contain heavier elements? **(D)**

11. Describe the steps in the formation of the Solar System according to the nebular theory. **(E)**

12. Describe a process that may have produced the moon. **(E)**

13. Why did the Earth differentiate into layers? **(E)**

14. Why is the Earth round? **(E)**

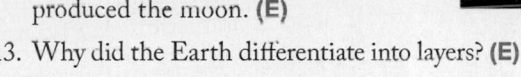

ON FURTHER THOUGHT

15. Could a planet with a composition similar to that of the Earth have formed around a first-generation star? Explain your answer. **(E)**

16. Astronomers discovered that distant galaxies move away from the Earth more rapidly than nearby ones. To see why, make a model of the problem by drawing three equally spaced dots along a cut rubber band. The dot at one end represents the Earth, and the other two dots represent galaxies. Stretch the rubber band to twice its length. This stretching represents Universe expansion. Pretend that it took 1 second to stretch the line (so Time = 1 second). Calculate the velocities of the two galaxies, using the equation: Velocity = Distance ÷ Time. **(C)**

17. Why do all planets orbit the Sun in roughly the same plane? **(E)**

ONLINE RESOURCES

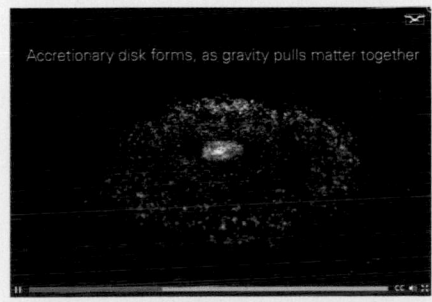

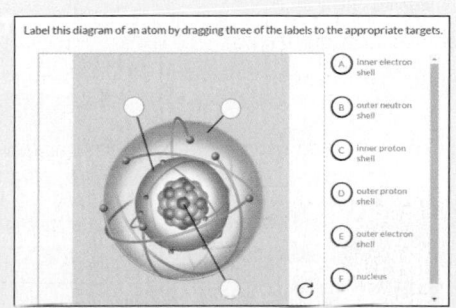

Videos
This chapter features videos showing how our Solar System and the Earth formed, including coverage of nebular theory, proto-Earth, the formation of the Moon, and the beginning of the atmosphere.

Smartwork5
This chapter features visual labeling exercises on topics such as atomic structures and our Solar System.

Journey to the Center of the Earth

By the end of this chapter, you should be able to . . .

A. describe the objects besides the Sun and planets that make up the Solar System.

B. describe the nature of the magnetic field and atmosphere that surround our planet.

C. list the distinct interacting realms within the Earth System.

D. distinguish the internal layers (crust, mantle, and core) of the Earth.

E. explain the relationship between the lithosphere and the asthenosphere.

2.1 Introduction

In 1961 a Russian cosmonaut, Yuri Gagarin, became the first human to orbit the Earth, and before the end of the decade, two Americans, Neil Armstrong and Buzz Aldrin, became the first to walk on the Moon. These exploits were truly amazing: for the first time, humans saw their home planet from a distance and gained an appreciation of its uniqueness and limits. Though people have not yet traveled farther than the Moon, we have sent spacecraft to other planets and beyond. In fact, in 2013, a spacecraft named *Voyager 1*, launched in 1977 to fly by Jupiter and Saturn, passed out of our Solar System. *Voyager 1* carries a message of greeting, inscribed on a copper disk, from humanity to whomever—or whatever—it might encounter tens of thousands of years in the future. Should *Voyager 1* ever reach an *exoplanet* (a planet orbiting another star), the craft's instruments might restart and begin investigating.

Let's turn the tables and speculate about what an imaginary spacecraft sent from an exoplanet toward the Earth would detect. In this chapter, we follow this "exoprobe" as it enters and traverses the Solar System and then explores the nature of the Earth's immediate surroundings and, finally, our planet's surface. Then we turn our attention downward to characterize the interior of the Earth, from its surface to its center, as detected by measurements of the Earth's shape and gravitational pull and confirmed by instruments placed on the surface. This high-speed fantasy tour provides a foundation from which we can develop geologic themes introduced throughout the remainder of this book.

2.2 Welcome to the Neighborhood

A Journey through the Solar System

For most of its journey toward our Solar System, the exoprobe travels through **interstellar space**, the region between stars.

◀ (facing page) This dramatic gorge cut by the Rio Grande in New Mexico barely scratches the surface of the Earth. Its height represents only the top 0.008% of our planet's radius.

This region hosts a **vacuum** (an absence of matter) so profound that it contains less than one atom per liter—by comparison, air at sea level contains 27,000,000,000,000,000,000,000,000 (or, in scientific notation, 2.7×10^{22}) atoms per liter. The atoms in interstellar space may be leftovers of the Big Bang or may be **cosmic rays**, high-energy atomic nuclei ejected into space at extreme velocity by supernova explosions.

Eventually, the exoprobe begins to feel the ever-so-weak pull of the Sun's gravity—this happens at a distance of about 50,000 AU from the Sun. (An **AU**, or **astronomical unit**, the distance between the Earth and the Sun, equals about 150 million kilometers.) As it continues to move toward the Sun, the exoprobe detects specks, flakes, and balls of ice (frozen volatile materials such as water, carbon dioxide, ammonia, and methane), either leftovers of the nebulae from which our Solar System formed or fragments scattered into space soon after its formation. Together, these objects make up the **Oort Cloud**, whose inner edge lies at a distance of about 3,500 AU from the Sun.

> **Did you ever wonder...**
> what defines the edge of our Solar System?

At a distance of about 200 AU, our exoprobe enters the **heliosphere**. The region within the heliosphere contains predominantly solar-wind particles, electrons and protons ejected into space from our Sun, whereas the region outside contains predominantly cosmic rays ejected from other objects in space toward our Sun. The heliosphere technically defines the "edge" of the Solar System that *Voyager 1* crossed. Then, at a distance of 30 to 55 AU, our spaceship traverses the **Kuiper Belt**, a diffuse ring of icy objects that once formed the outer reaches of the protoplanetary disk. About 100,000 of the objects in the Kuiper Belt have diameters over 100 km **(Fig. 2.1)**, and some, such as Pluto and Eris, have diameters of over 400 km and are known as *dwarf planets*. *Comets* originate from the Kuiper Belt, and to a lesser extent, from the Oort Cloud **(Box 2.1)**. All told, the Kuiper Belt and Oort Cloud could contain a trillion objects with a combined mass that may approach that of Jupiter.

The orbit of the ice giant Neptune, our Solar System's outermost true planet, defines the inner edge of the Kuiper Belt—once we've passed this orbit, we're traversing **interplanetary space**. In interplanetary space, the concentration of atoms increases to between 5,000 and 100,000 per liter. So, while

FIGURE 2.1 What a spacecraft would see if it traversed the Solar System and its surroundings.

(a) The Oort Cloud is a diffuse cloud of icy particles held in by the Sun's gravity.

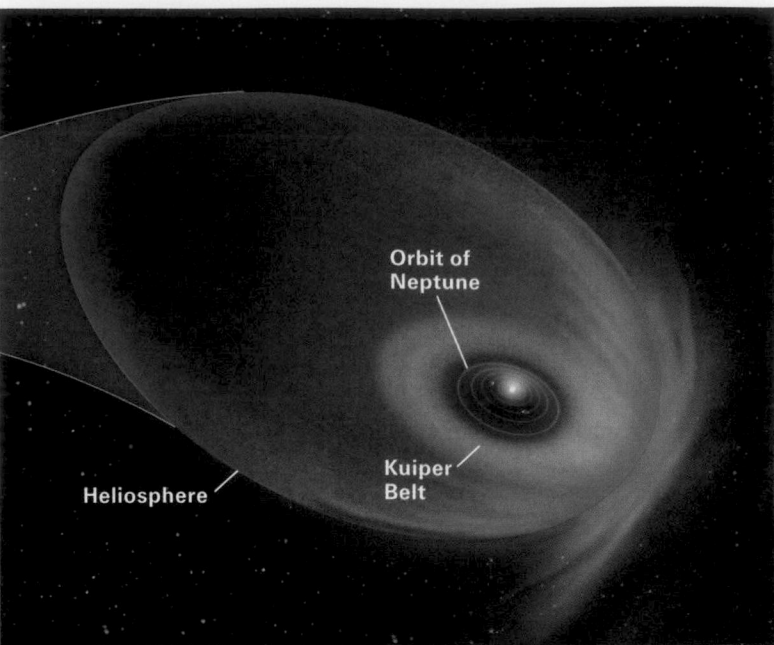

(b) The heliosphere, which is very tiny compared with the Oort Cloud, technically delimits the edge of the Solar System. The Kuiper Belt and planets lie within it.

(c) Most asteroids lie between the orbit of Mars and Jupiter. Some asteroids (known as the Trojans and the Greeks) border Jupiter's orbit, while the Hildas lie between the main asteroid belt and Jupiter.

this region still represents a profound vacuum, it's much denser than interstellar space. As our exoprobe zooms across the plane containing the orbits of all the planets, it passes Uranus, the second ice-giant planet; the two gas-giant planets (Saturn and Jupiter); and the *asteroid belt*, a diffuse band about 2.5 AU across that contains about 10 million small solid objects (see Box 2.1). In the inner part of the Solar System we find the four terrestrial planets—reddish Mars, bluish Earth (with its large Moon), greenish cloud-enshrouded Venus, and heavily cratered Mercury **(Fig. 2.2)**. Even from the distance of space, the Earth

looks special, so the exoprobe's computer chooses to approach and study it more closely.

The Earth's Magnetic Field

As the exoprobe nears the Earth, its instruments begin to detect the Earth's magnetic field, which acts like a signpost shouting, "Approaching Earth!" A magnetic field is the region measurably affected by the force emanating from a magnet. This force, which grows progressively stronger closer

FIGURE 2.2 An explorer from outer space would quickly realize that the four terrestrial planets look quite different from one another.

Mercury has a crater-pocked surface and icy poles.

Dense clouds hide the surface of Venus.

The surface of Mars varies in elevation and is host to craters and polar ice caps.

The Earth's surface has both land and sea. The atmosphere is largely transparent.

to the magnet, can cause another magnet or charged particles to move. The Earth's magnetic field, like the magnetic field around a familiar bar magnet, is a **dipole,** meaning that it has two ends: a north pole and a south pole. If you bring two magnets close to each other, the unlike poles attract (pull toward each other), and the like poles repel (push away from each other). By convention, physicists represent the orientation of a magnetic dipole by an arrow that points from the south pole to the north pole, and they depict the magnetic field of a magnet by a set of invisible **magnetic field lines** that curve through the space around the magnet **(Fig. 2.3a).** Arrowheads along these lines point in a direction to complete a loop. Magnetized materials, such as iron filings or compass needles, when placed in a magnetic field, align with the field lines.

Because it is dipolar, we can simplistically represent the Earth's magnetic field as emanating from an imaginary bar magnet in the planet's interior **(Fig. 2.3b).** The north pole of this bar, as defined by physicists, lies near the south geographic pole of the Earth, whereas the south pole of the bar lies near the north geographic pole. (The **geographic poles** are the places where the Earth's axis of rotation intersects the planet's surface. Presently, this axis tilts at an angle of 23.4° relative to the plane of the Earth's orbit.) Geologists and geographers, by convention, refer to the magnetic pole closer to the north geographic pole as the *north magnetic pole,* and to the magnetic pole closer to the south geographic pole as the *south magnetic pole.* This way, the north-seeking end of a compass needle points toward the north geographic pole.

The solar wind interacts with the Earth's magnetic field, distorting it into a huge teardrop pointing away from the Sun. Fortunately, the magnetic field deflects most (but not all)

solar-wind particles, so that they do not reach the Earth's surface. In other words, the magnetic field serves as a shield against solar-wind particles, which can be dangerous to life forms. Physicists refer to the region inside this magnetic shield as the **magnetosphere (Fig. 2.3c).**

Though it protects the Earth from most of the solar wind, the magnetic field does not stop our imaginary exoprobe. At distances of 3,000 km to 10,500 km out from the Earth, the spacecraft encounters the *Van Allen radiation belts,* named for the physicist who first recognized them in 1959. The Van Allen belts, where the magnetic field starts to strengthen, trap both cosmic rays and any solar-wind particles that were moving so fast that they could penetrate the weaker outer part of the magnetic field. Therefore, the Van Allen belts serve as a second line of defense that protects life on the Earth from dangerous radiation. But they don't trap all incoming particles. Some make it past the Van Allen belts and follow magnetic field lines to the polar regions of the Earth. When these particles interact with gas atoms nearer the Earth, they cause the gases to glow, like the gases in neon signs, producing spectacular **aurorae (Fig. 2.3d).** The *aurora borealis* occurs in high latitudes of the northern hemisphere, while the *aurora australis* occurs in high latitudes of the southern hemisphere.

A Plunge through the Atmosphere

The exoprobe descends to an elevation of 600 km and goes into orbit around the planet, circling the globe once every 96 minutes. At this elevation, the spacecraft skims through the outer reaches of the Earth's **atmosphere,** the gaseous cloak that envelops the planet. The atmosphere contains a mixture

BOX 2.1 CONSIDER THIS . . .

Comets and Asteroids: The Other Stuff of the Solar System

Comets and asteroids are of growing concern to humanity because some of them have orbits that may cross the orbit of the Earth and thus could collide with our planet. In fact, astronomers have, so far, identified over 6,500 *near-Earth objects* that could potentially strike the Earth. What are comets and asteroids, and how do they differ from each other?

A **comet** is a planetesimal composed of ice and dust. Its highly elliptical orbit brings it so close to the Sun that during part of its journey, some of its ice

evaporates. The evaporated ice becomes a glowing tail of gas that points away from the Sun. Dust particles bonded to the ice also head into space when the ice evaporates. These particles form a separate tail (**Fig. Bx2.1a, b**). Comets that take less than 200 years to orbit the Sun originate from the Kuiper Belt, whereas those with longer orbits originate from the Oort Cloud. Objects become comets when gravity from planets tugs on them and sends them on a trajectory to the inner Solar System.

In recent decades, researchers have sent spacecraft to observe comets close-up. Such studies confirm that comets consist primarily of frozen water (H_2O), carbon dioxide (CO_2), methane (CH_4), ammonia (NH_3), and other volatile compounds, along with a variety of organic chemicals and dust (tiny rocky or metallic particles). Considering these components, astronomers often refer to comets as "dirty snowballs." Some geologists have speculated that abundant impacts of comets with the Earth, early in

FIGURE Bx2.1 Images of asteroids and comets.

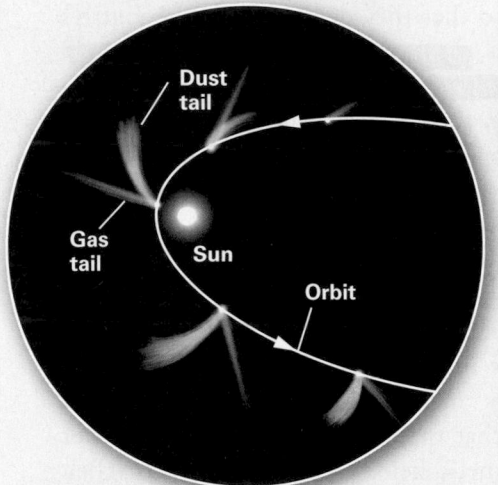

(a) Photograph of comet Hale-Bopp, which approached the Earth in 1997. The head of this comet is about 40 km across. The two tails are visible.

(b) Hartley 2, a comet viewed by the EPOXI spacecraft in 2010. The comet, which is about 1.5 km long, looks like a dirty snowball and emits jets of gas.

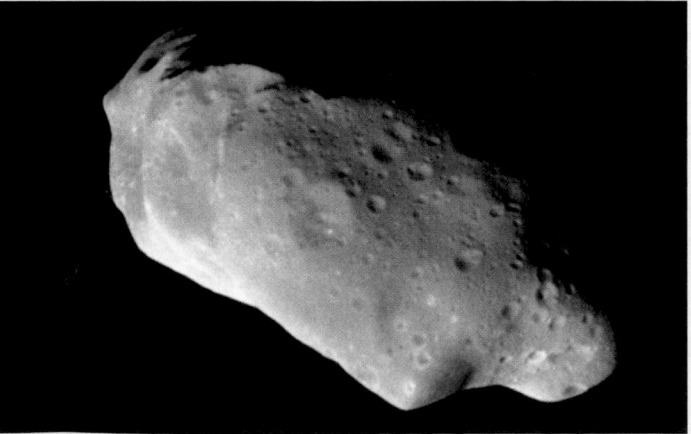

(c) The largest asteroid, Ceres, has a diameter of about 950 km, so it's a dwarf planet.

(d) Photograph of the asteroid Ida, a body that is about 56 km long.

Solar System history, might have brought water and even molecules from which life evolved.

An **asteroid** is a body of solid rock and/or metal that orbits the Sun. Most reside in the asteroid belt between the orbits of Mars and Jupiter. Some asteroids are small rocky planetesimals that were never incorporated into larger bodies, whereas others are fragments of once-large planetesimals that had differentiated into a metal core surrounded by a rocky mantle before being shattered into fragments by collisions early in the history of the Solar System. While most asteroids are very small—from dust to basketball sized—astronomers have found 1,000 asteroids with diameters greater than 30 km and estimate that there may be about 2 million more with diameters greater than 1 km.

Though asteroids are numerous, their combined mass equals only about 4% that of the Earth's Moon. Notably, half of the mass of the asteroid belt resides in the four largest asteroids. Of these, the most massive, Ceres, is spherical and has a diameter of 950 km **(Fig. Bx2.1c)**; water and ice surround its rocky/metallic center. Most other asteroids do not have differentiated interiors and are too small for their own gravity to reshape them into spheres, so they have irregular, crater-pocked surfaces **(Fig. Bx2.1d)**. The material in the asteroid belt never merged to form a planet. In recent decades, several space probes have visited asteroids.

FIGURE 2.3 A magnetic field permeates the space around the Earth. It can be represented by a bar magnet.

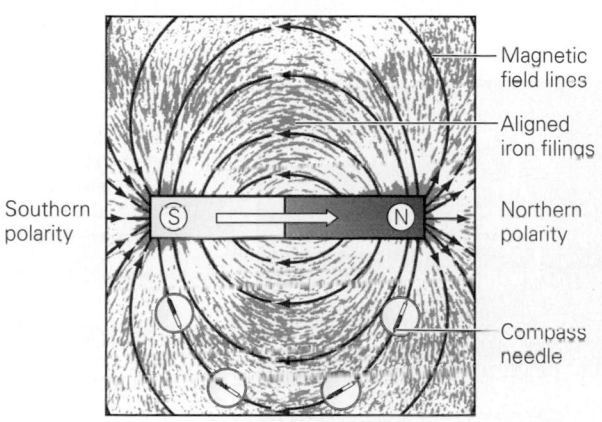

(a) A bar magnet produces a magnetic field. Magnetic field lines point into the "south pole" and out from the "north pole."

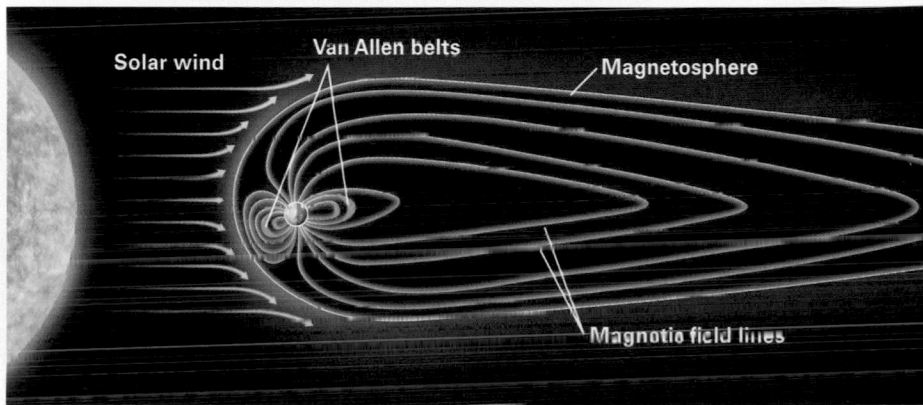

(c) The Earth behaves like a magnetic dipole, but the field lines are distorted by the solar wind. The Van Allen radiation belts trap charged particles.

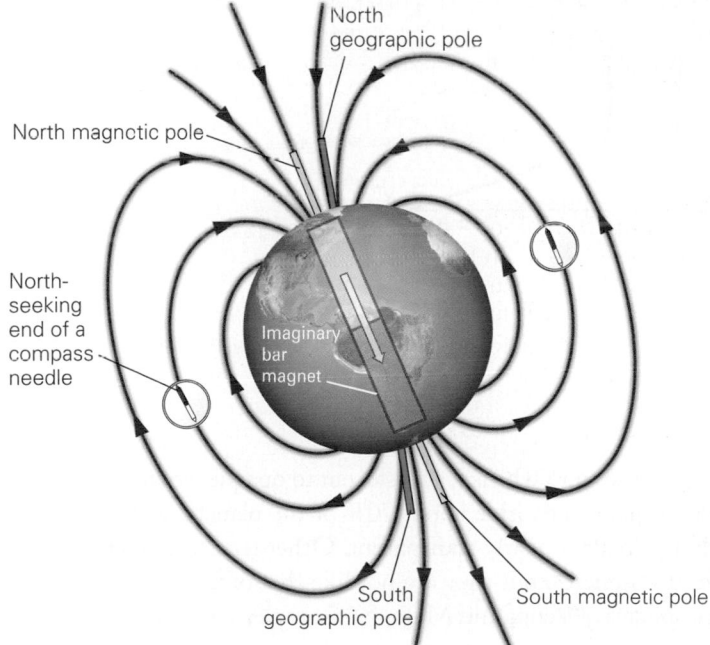

(b) We can represent the Earth's magnetic field by an imaginary bar magnet inside.

(d) Charged particles flow toward the Earth's magnetic poles and cause gases in the atmosphere to glow, forming colorful aurorae in polar skies.

FIGURE 2.4 Characteristics of the atmosphere that envelops the Earth.

Nitrogen (N$_2$)
78.08%

Other gases
(0.97%)

Oxygen (O$_2$)
20.95%

(b) Composition of the atmosphere—nitrogen and oxygen dominate.

(a) The haze of the atmosphere fades up into the blackness of space.

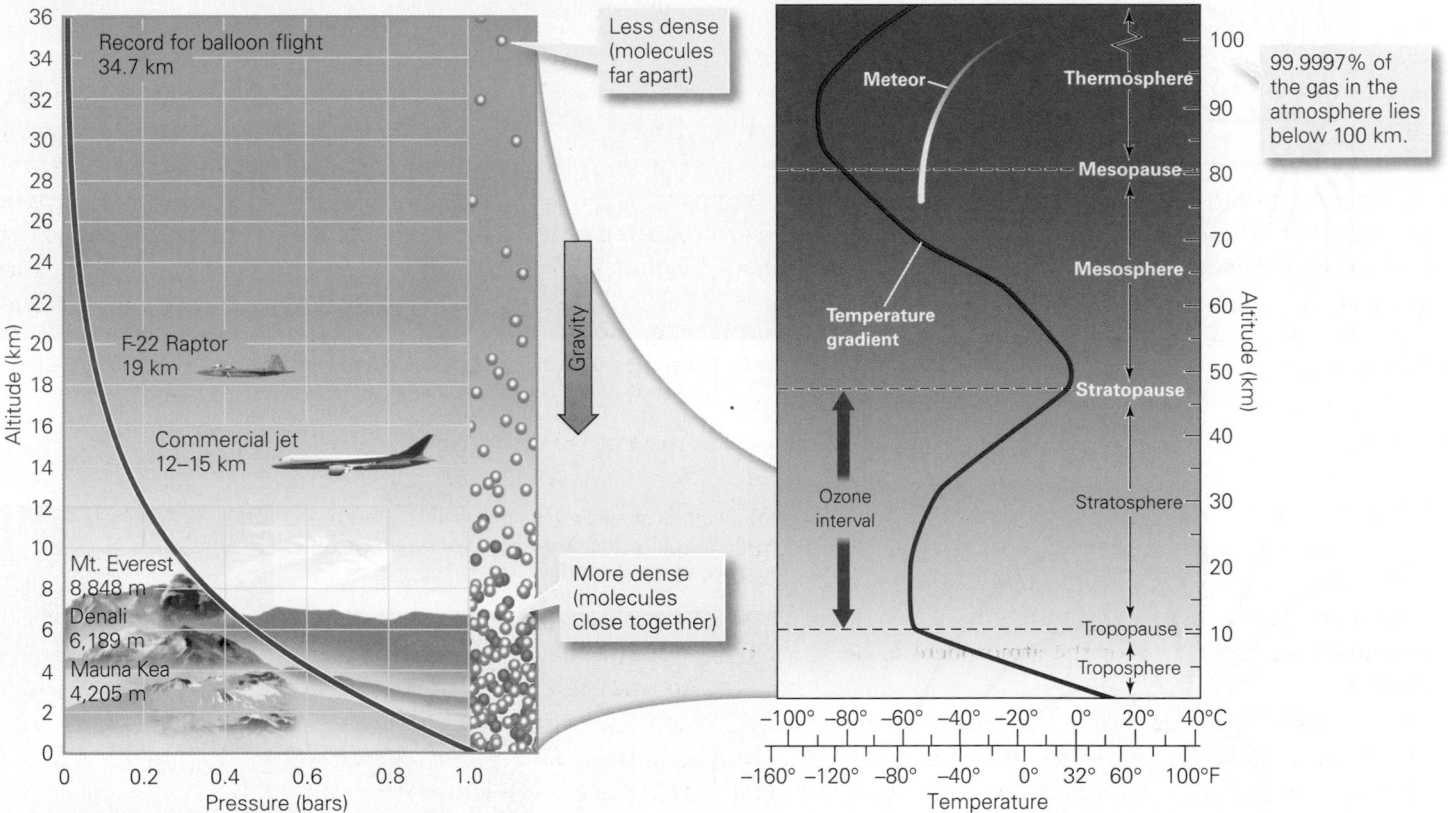

(c) Molecules pack together more tightly at the base of the atmosphere, so atmospheric pressure changes with elevation.

(d) The atmosphere can be divided into several distinct layers. We live in the troposphere.

of gases known as *air*. The probe's instruments determine that air consists of 78% nitrogen (N$_2$) and 21% oxygen (O$_2$), along with minor amounts (1% total) of several other gases, including argon, carbon dioxide, neon, methane, ozone, carbon monoxide, and sulfur dioxide **(Fig. 2.4a, b)**. Air also contains variable amounts of water (H$_2$O) vapor, which, at lower elevations, locally condenses into whitish, translucent to opaque clouds. At any given time, clouds hide about 70% of the planet's surface; air without clouds is nearly transparent. Other terrestrial planets have atmospheres, but they are not like that of the Earth—the atmospheres of Venus and Mars, for example, consist mostly of CO$_2$ gas.

The density of the atmosphere progressively increases closer to the Earth, for the weight of overlying air squeezes down on the air below, pushing gas molecules in the air closer together. At the Earth's surface, molecules are close enough together that the atmosphere has a density of 1.2 g/L. While vastly greater than that of interplanetary space, this density is only 12% that of water. *Air pressure*, the amount of push that air exerts on material surrounding it, also increases closer to the surface because of the weight of the overlying atmosphere **(Fig. 2.4c)**. Technically, we specify pressure in units of force (push) per unit area. For example, we could specify atmospheric pressure in pounds (a force) per square inch (an area), or kilograms per square centimeter. Air pressure at sea level averages 14.7 pounds per square inch, or 1.04 kg/cm². Researchers generally use other units, such as atmospheres (abbreviated atm) and bars. Air pressure at sea level is 1 atm. An atmosphere and a bar are almost the same: 1 atm = 1.01 bars. Notably, air pressure at the Earth's surface greatly exceeds that of Mars (0.006 bars) but not that of Venus (90 bars).

Since air pressure decreases with increasing elevation, the air pressure at the peak of Mt. Everest, 8.85 km above sea level, is only 0.3 bars. People can't survive for long at elevations above about 5.5 km, where the air pressure is about 0.5 bars (50% that at sea level), so since jet planes fly at altitudes above 5.5 km, their cabins must be pressurized by pumping air in.

The decrease in density with elevation means that 99% of atmospheric gas lies at elevations below 50 km, and the atmosphere can barely be detected at elevations above 120 km. While gravity still attracts air molecules to the Earth for half the distance to the Moon, there are so few molecules at elevations above about 600 km that the molecules no longer collide and interact like those of a gas. As a consequence, researchers generally consider the top of the atmosphere to lie at about 600 km.

The character of the atmosphere changes with increasing distance from the Earth's surface. Because of these changes, atmospheric scientists divide the atmosphere into layers. Most winds and clouds develop only in the lowest layer, the *troposphere*. The layers of the atmosphere that lie above the troposphere are named, in sequence from base to top: the stratosphere, the mesosphere, and the thermosphere **(Fig. 2.4d)**. As described further in Chapter 20, the boundaries between layers are defined as elevations where temperature stops decreasing and starts increasing, or vice versa. Boundaries are named for the underlying layer. For example, the boundary between the troposphere and the overlying stratosphere is called the *tropopause*.

TAKE-HOME MESSAGE

The Earth is one of eight planets and countless other, smaller objects that orbit the Sun. The Earth produces a magnetic field that deflects solar wind. An atmosphere composed mostly of N_2 and O_2 gas surrounds the planet; 99% of this atmosphere lies below an elevation of 50 km, so it is a thin envelope indeed.

QUICK QUESTION: What causes the aurorae?

2.3 Basic Characteristics of the Earth

The Earth System

Once in orbit, the exoprobe surveys the Earth's surface and detects several distinct components. Beneath the *atmosphere* (the gaseous envelope that we've just discussed) lie the *hydrosphere* (surface and near-surface liquid water), the *cryosphere* (surface and near-surface ice and snow), the *biosphere* (the great variety of living organisms), and the *solid Earth*. Geologists refer to the combination of these components, and the complex interactions among them, as the **Earth System**.

Of all the planets in the Solar System, only the Earth currently has liquid water, for our planet lies within the habitable zone, the distance from the Sun at which temperatures range between the boiling and freezing points of water **(Fig. 2.5)**. On planets closer to the Sun than the habitable zone, all water evaporates, and on planets farther away, all water exists only as solid ice. Astronomers estimate that the habitable zone in our Solar System extends between about 0.8 and 2.5 AU. This region includes the orbits of Earth and Mars and some of the asteroid belt. Venus orbits near the inner edge of the habitable zone. The fact that surface temperatures on Venus are way too hot for life (460°C) and those on

FIGURE 2.5 In our Solar System, the habitable zone lies between about 0.8 and 2.5 AU. The scale is a logrithmic, in astronomical units (AU).

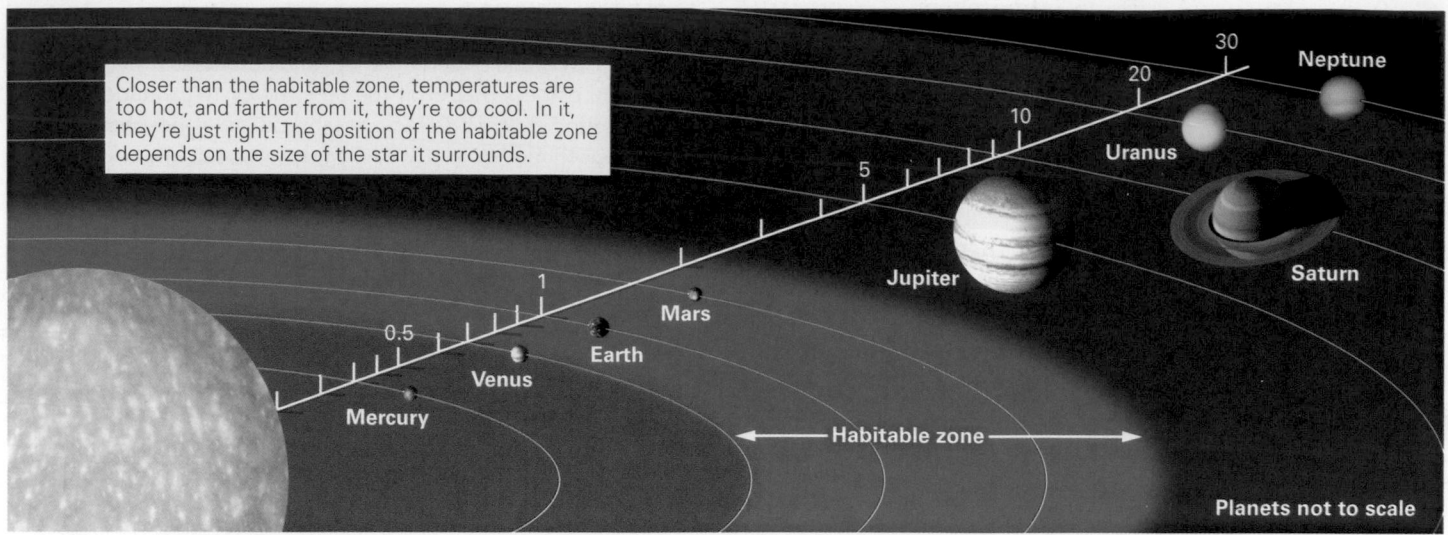

Closer than the habitable zone, temperatures are too hot, and farther from it, they're too cool. In it, they're just right! The position of the habitable zone depends on the size of the star it surrounds.

The Southern Alps

LATITUDE
44°9'31.68" S

LONGITUDE
169°46'9.60" E

Look down from 25 km and tilt to look north.

The rugged topography of the Southern Alps of New Zealand shows how the elevation of the land surface can vary.

Mars are too cold (−55°C) has to do with the respective abilities of their atmospheres to trap heat. Venus has a dense atmosphere that traps a lot of heat, and Mars has a very thin atmosphere that allows heat to escape.

As we'll see throughout this book, the Earth System is dynamic. Its surface and the objects on it move, and its atmosphere and oceans circulate; materials from its interior spill out on the surface, and materials from the surface sink into the interior. The energy driving all this activity ultimately comes from heat inside the Earth, from gravity, and from the Sun's heat and light.

Land and Sea

As the exoprobe continues to orbit the Earth, it makes a basic map of the surface by collecting information about spatial variations in the composition of the Earth's surface and determines that *dry land* (continents and islands) covers about 30% of the surface, whereas **surface water** covers the remaining 70% **(Fig. 2.6)**. Most surface water contains salt and resides in the oceans (or sea), but some is fresh (salt-free) and fills lakes and rivers. Our instruments also detect **ground-water**, the water that fills cracks and holes (pores) beneath the land surface. Finally, we find that ice covers significant areas of

land and sea in polar regions and at high elevations, and that living organisms populate the land, sea, air, and even the upper few kilometers of the subsurface.

To finish off its map of the Earth's surface, the spacecraft detects that the planet's land surface has **topography**, variations in its elevation, and identifies plains, mountains, and valleys. Notably, the highest point on land, the peak of Mt. Everest, lies about 8.9 km above sea level, while the lowest point, the shore of the Dead Sea, lies about 0.4 km below sea level. The exoprobe's instruments also measure **bathymetry**, the variations in depth of the ocean floor. The sea's surface looks much the same everywhere, because by definition it lies, on average, at sea level (changing temporarily by centimeters to tens of meters as waves or tides pass by). However, the seafloor hidden below displays distinct bathymetric realms. Most of the seafloor comprises broad **abyssal plains**, where the flat seafloor lies at a depth of 4 to 5 km below sea level. The seafloor rises to shallower depths along **mid-ocean ridges**, elongate submarine mountain chains that rise as much as 2.5 km above the abyssal plains, and descends to greater depths in **deep-sea trenches**, elongate troughs in which the seafloor lies as much as 10.9 km below sea level. Note that the Earth's total relief, from the floor of the

---

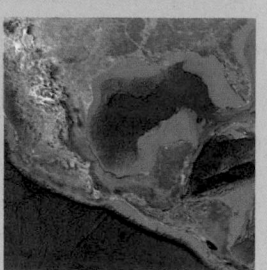

Bathymetry

LATITUDE
20°8'31.57" N

LONGITUDE
92°51'12.35" W

Look down from 5,000 km.

This view shows the bathymetry of the Gulf of Mexico, the eastern Pacific, and the Caribbean. Lighter blues are shallower.

---

deepest trench to the peak of the highest mountain, is 19.8 km, only 0.3% of the Earth's radius (6,371 km). In fact, if the Earth were the size of a billiard ball, it would actually feel smoother than a billiard ball.

A graph called a **hypsometric curve**, plotting surface elevation on the vertical axis and the percentage of the Earth's surface on the horizontal axis, shows that only a relatively small proportion of the Earth's surface occurs at very high elevations (mountains) or at great depths (deep trenches). In fact, most of the land surface lies within just a kilometer of sea level, and most of the seafloor lies between 4 and 5 km deep **(Fig. 2.7)**. Consequently, changes in sea level of tens to a couple of hundred meters would dramatically change the area of dry land.

Did you ever wonder . . .
what the average elevation of the land is?

What Is the Solid Earth Made of?

To complete its initial exploration of the Earth, the exoprobe sends robot landers down to the surface to sample and analyze **Earth materials** that make up the solid planet. Of the 92 naturally occurring elements (produced by fusion reactions in stars and supernova explosions) that make up the Earth, 91.2% of the Earth's mass consists of only 4: iron, oxygen, silicon, and magnesium **(Fig. 2.8)**. The elements of the Earth bond together to form a great variety of materials that can be classified into several basic categories, which we introduce here. (All will be discussed in more depth later in this book.)

- *Organic chemicals:* A carbon-containing compound that either occurs in living organisms or has characteristics that resemble those of compounds in living organisms is called an **organic chemical**.

- *Minerals:* A solid, natural substance in which atoms are arranged in an orderly pattern is a **mineral**. A coherent sample of a mineral that grew to its present shape is a **crystal**. Most minerals are *inorganic* chemicals, meaning that they are not organic.

- *Glass:* A solid in which atoms are not arranged in an orderly pattern is called **glass**.

- *Melts:* A **melt** forms when solid materials become hot and transform into liquid. *Molten rock* is a type of melt. Geologists distinguish between *magma*, molten rock beneath the Earth's surface, and *lava*, molten rock that has flowed out onto the Earth's surface.

- *Rocks:* A coherent aggregate of mineral crystals or grains, or a mass of natural glass, is called a **rock**. Geologists recognize three main groups of rocks:

FIGURE 2.6 This map of the Earth shows variations in elevation on both the land surface and the seafloor. Darker blues are deeper water in the oceans. Greens are lower elevation on land.

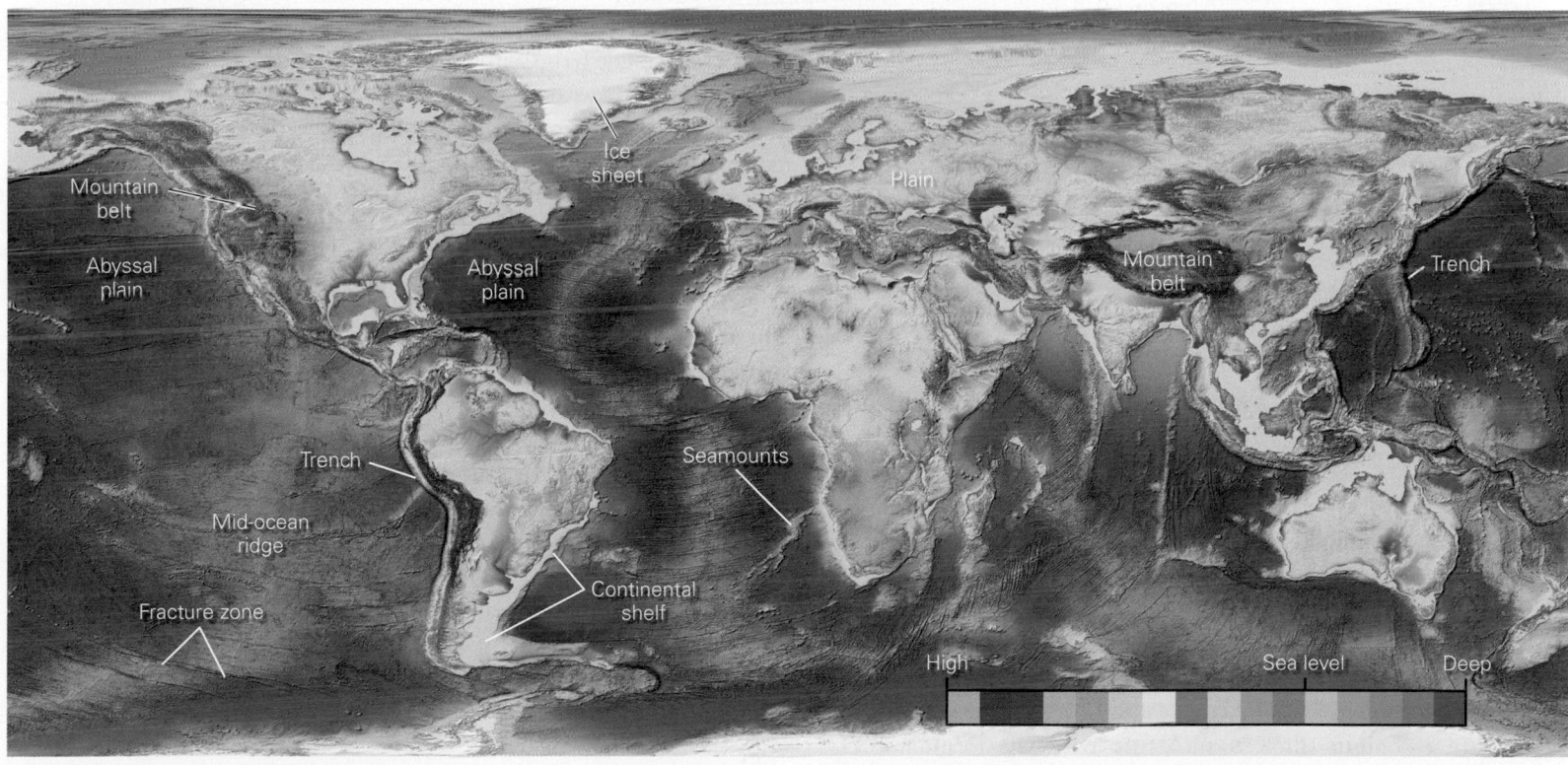

FIGURE 2.7 A hypsometric curve indicates the proportions of the Earth's solid surface at different elevations. Continental shelf areas are the submerged margins of continents. Mountains and deep trenches cover relatively little area.

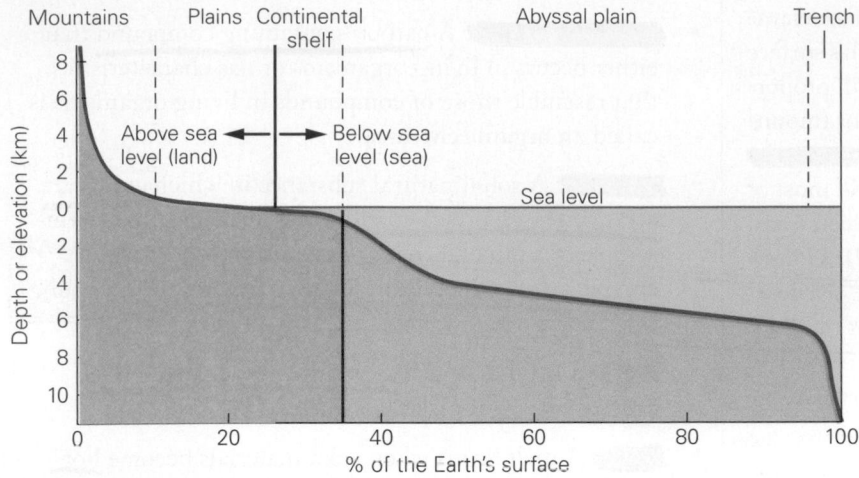

FIGURE 2.8 The proportions of major elements making up the mass of the whole Earth. Iron and oxygen dominate.

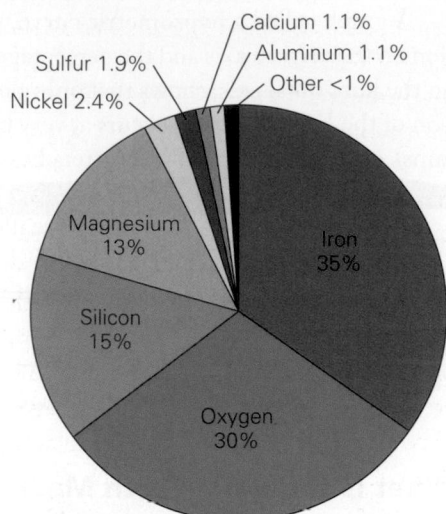

Igneous rock: A rock that forms when molten (liquid) rock cools and freezes solid, either underground or at the surface of the Earth.

Sedimentary rock: A rock that forms either from fragments that broke off pre-existing rock and were then cemented together, or from minerals that precipitate out of a water solution at or near the Earth's surface.

Metamorphic rock: A rock that forms when pre-existing rock undergoes changes in response to changes in temperature and pressure.

- *Grains:* We use the term **grain** either for an individual crystal within an igneous or metamorphic rock or for an individual fragment derived from a once-larger mineral sample or rock body.

- *Sediment:* An accumulation of loose (unconsolidated) grains, meaning grains that have not been cemented together, makes up **sediment**. Gravel and sand are types of sediment.

- *Metals:* A solid composed entirely of metal atoms (such as iron, aluminum, copper, and tin) is a **metal**. Metal can be stretched into wires or flattened into sheets; it tends to be shiny and can conduct electricity. An *alloy* is a mixture containing more than one type of metal.

- *Volatiles:* As noted in Chapter 1, a material that can transform into gas at the conditions found at the Earth's surface can be called a **volatile**.

Most of the Earth consists of *silicate minerals:* minerals that contain silica, a molecule with the formula SiO_2

(silicon plus oxygen), either alone or bonded to other atoms or molecules. Not surprisingly, geologists refer to rocks composed of silicate minerals as **silicate rocks**. Much of the Earth's volume consists of igneous silicate rocks.

We'll be discussing silicate minerals further in Chapter 5, and silicate rocks in Chapter 6, but in order to discuss the overall internal structure of the Earth, we need to introduce the fundamental classes of igneous silicate rocks here. These classes differ from one another in terms of the ratio between the silica that they contain and the sum of iron oxide (FeO) and magnesium oxide (MgO) that they contain. **Table 2.1** provides the names of these fundamental classes, listed in order from the greatest proportion of silica (at the top) to the smallest proportion of silica (at the bottom). The table also lists some of the more common rock names representative of each class. Some of these names may be familiar to you, but most are probably not. In cases where the table provides two rock names, the two have the same chemical composition, but differ in grain size. For example, a fine-grained mafic rock is called *basalt,* whereas a coarse-grained mafic rock is called *gabbro.* (In this context, fine means small and coarse means large.)

Pressure and Temperature inside the Earth

To keep underground tunnels from collapsing under the pressure created by the weight of overlying rock, mining engineers must design sturdy support structures. It's no surprise that deeper tunnels require stronger supports. The downward

> **Did you ever wonder . . .**
> how hot it gets at the center of this planet?

push from the weight of overlying rock increases with depth, simply because the mass of the overlying rock layer increases with depth. At the Earth's center, pressure probably reaches about 3,600,000 atm.

Temperature also increases with depth in the Earth. Even on a cool winter's day, miners who chisel away at gold veins exposed in tunnels 3.2 km below the surface swelter in temperatures of about 53°C. We refer to the rate of change in temperature with depth as the **geothermal gradient,** or *geotherm*. In the upper part of the crust, the geothermal gradient averages between 15°C and 30°C per kilometer. At greater depths, the rate decreases to 10°C per kilometer or less. So, 32 km below the surface of a continent, the temperature reaches 400°C to 700°C depending on location. No one has ever directly measured the temperature at the Earth's center, but calculations suggest it may exceed 4,700°C, close to the Sun's surface temperature of 5,500°C **(Fig. 2.9)**.

TABLE 2.1 Classes of Igneous Rocks, Distinguished by Composition*

Class Name	Relative Silica Content	Examples Fine-grained	Examples Coarse-grained	Relative Density
Felsic (silicic)	*More silica*	Rhyolite	Granite	*Less dense*
Intermediate		Andesite	Diorite	
Mafic		Basalt	Gabbro	
Ultramafic	*Less silica*	Komatiite	Peridotite	*More dense*

*Here, "composition" represents the ratio $SiO_2/(FeO + MgO)$.

FIGURE 2.9 The Earth's geothermal gradient, or *geotherm*, shows how temperature increases with depth.

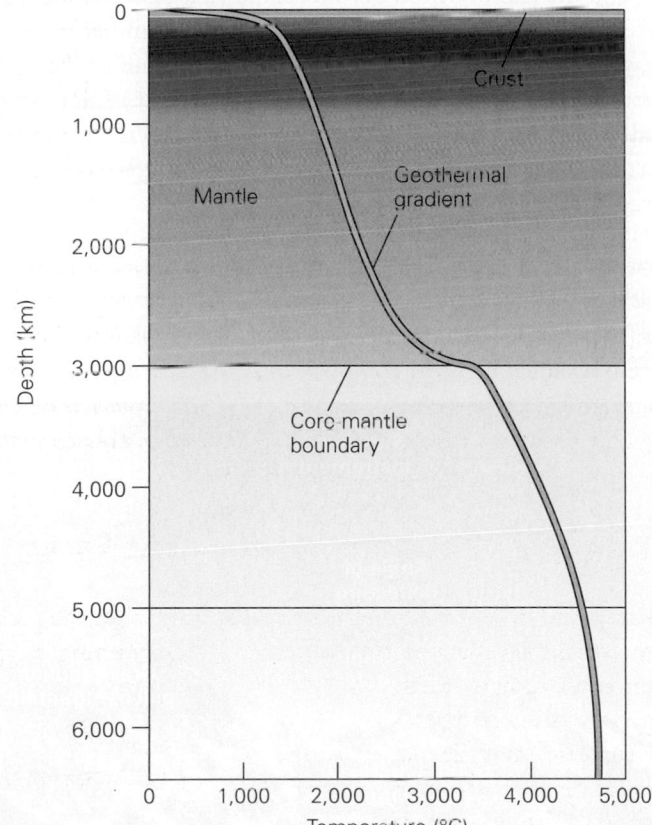

2.4 How Do We Know That the Earth Has Layers?

The deepest well ever drilled, a 12.3-km-deep hole in Russia, penetrates only the upper 0.2% of the Earth's 6,371-km radius, so it represents a mere pinprick. We literally live on the thin skin of our planet, its interior forever inaccessible to our direct observation.

Lacking the ability to observe the Earth's interior firsthand, pre-20th-century writers and artists dreamed up fanciful images of it. For example, to the ancient Greeks, the subsurface was Hades, the "underworld," which they pictured to be a realm of gloom populated by the dead. In Chinese lore, the underworld was a complex maze, and in Buddhist mythology, it was a succession of caverns. The English poet

FIGURE 2.10 An image of the Earth's interior, following a description from *Paradise Lost*, an epic poem published by John Milton in 1667, painted by John Martin (1789–1854).

John Milton (1608–1674) described the subsurface realm as "a dungeon horrible, on all sides round, as one great furnace flamed; yet from those flames, no light" **(Fig. 2.10)**. In the 18th and 19th centuries, some European writers thought the Earth's interior resembled a sponge, containing open caverns variously filled with molten rock, water, or air, an image that seemed to explain the source of volcanoes and water springs. The French science-fiction writer Jules Verne used this image as the basis of his popular 1864 novel *Journey to the Center of the Earth*, in which three explorers wander through interconnected caverns to reach the Earth's center.

Modern scientific studies give a very different picture of the Earth's interior. Except in the upper few kilometers, the interior contains no open spaces. Rather, it consists of distinct rock shells surrounding an iron ball. What clues led to this interpretation?

Early Clues to the Earth's Interior

The first key to understanding the Earth's interior came from studies that provided an estimate of our planet's average density. The volume of the Earth had been known since Eratosthenes first measured its dimensions, so all that was needed to determine the density was a measure of the planet's mass, because density = mass ÷ volume. In 1776, the British Royal Astronomer, Nevil Maskelyne, came up with the first realistic estimate of its mass. He obtained this

> **Did you ever wonder . . .**
> what's deep inside the Earth?

estimate by observing the deflection, due to the gravitational attraction of a nearby mountain, of a lead ball (called a plumb bob) suspended from a wire. Maskelyne reasoned that if the mountain weren't there, the lead ball would be pulled straight down by the Earth's gravity. The mass of the mountain pulled the ball sideways, toward it, so the wire made a slight angle relative to vertical **(Fig. 2.11)**. Since Newton's law of gravity states that the strength of gravitational pull depends on the amount of mass present, the amount of deflection represents the mass of the mountain relative to that of the whole Earth. From his measurements, Maskelyne calculated that the Earth had a density of 4.5 gm/cm³. In 1798, another British scientist used a different method for measurement and came up with a value of about 5.5 gm/cm³, the number we use today. Since average rocks at the Earth's surface have a density of only about 3.0 gm/cm³, researchers immediately realized that the interior of the Earth must contain denser material than its outermost layer and couldn't possibly be full of caverns or other open spaces.

What could the denser material of the Earth's interior be? Researchers eventually concluded that this material must be a metal, for only metals could have sufficiently high density. (At the Earth's surface, iron has a density of 7.9 gm/cm³). With this idea in mind, they then addressed the question of where the metal of the Earth's interior resides. They realized that since the Earth has the shape of a sphere, the metal must be concentrated near the center—otherwise, centrifugal force due to the spin of the Earth on its axis would pull the equator out, and the planet would become somewhat disk-shaped. (To picture why, consider that when you swing a hammer in a circle, your hand feels more force if you hold the end of the light wooden shaft rather than the heavy metal head.) Laboratory studies show that at the immense pressures occurring in the Earth's core, the metal that makes up the core would be

FIGURE 2.11 A surveyor noticed that a plumb bob was deflected by an angle β owing to the gravitational attraction of a nearby mountain. The angle represents the ratio between the mass of the mountain and the mass of the whole Earth.

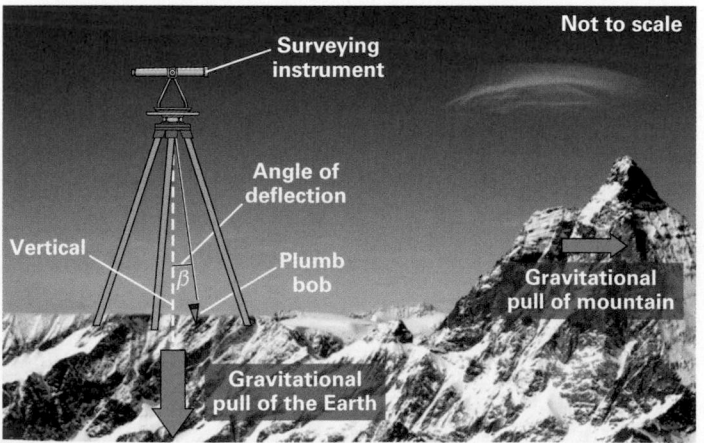

FIGURE 2.12 An early image of the Earth's internal layers.

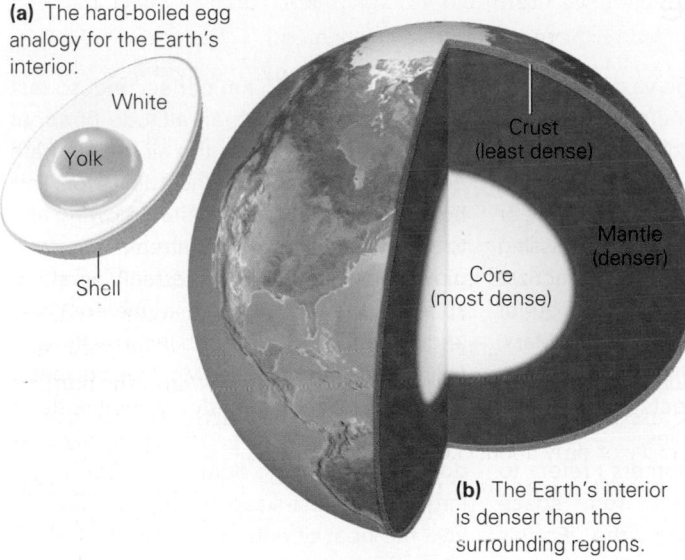

(a) The hard-boiled egg analogy for the Earth's interior.

White

Yolk

Shell

Crust (least dense)

Mantle (denser)

Core (most dense)

(b) The Earth's interior is denser than the surrounding regions.

almost twice as dense as it is on the Earth's surface because the pressure squeezes molecules closer together.

Researchers then wondered about the state of the materials in the Earth—are they solid or liquid (molten)? Eventually, they concluded that even though molten rock occasionally oozed out of the interior at volcanoes, and therefore must exist somewhere inside the Earth, most of the interior must be solid, because if it weren't, the land surface would rise and fall due to daily tides much more than it actually does. (The forces that produce tides, as discussed in Chapter 18, cause the liquid sea surface to rise and fall by as much as 14 m per day, but the land rises and falls by less than 0.5 m per day.)

Taking into account all the above observations on the density, shape, and tidal behavior of the Earth, researchers realized, by the end of the 19th century, that the Earth resembled a hard-boiled egg in that it had three principal layers: a not-so-dense *crust* (like an eggshell) composed of rocks such as granite, basalt, and gabbro (see Table 2.1); a denser solid *mantle* in the middle (like an egg white), composed of a then-unknown material; and a very dense *core* (like an egg yolk), composed of a then-unknown metal **(Fig. 2.12)**. Many questions remained: How thick are the layers? Are the boundaries between layers sharp or gradational? And what, exactly, do the layers consist of? Data available at the time could not provide answers, and it would take 20th-century studies of earthquakes to give us the detailed image of the interior that we have today.

Notably, studies by spacecraft sent from the Earth to other terrestrial planets in the Solar System reveal that all terrestrial planets share the same basic internal structure—with a crust, mantle, and core—though the relative thicknesses of the different layers on other planets are not the same as those of the Earth.

Clues from the Study of Earthquakes: Refining the Image of the Interior

To understand how earthquakes can provide information about the Earth's interior, we first need to understand what an earthquake is. Most earthquakes happen when rock within the outer portion of the Earth suddenly breaks along a *fault* (a fracture on which slip occurs), and generates energy that travels through the surrounding rock outward from the break. You can simulate this process, on a small scale, by snapping a stick between your hands—the "shock" that you feel is energy that propagated along the stick from the break to your hands **(Fig. 2.13)**. When energy reaches the Earth's surface and causes it to vibrate (move up and down or back and forth). Geologists use the term **earthquake** for both the sudden movement that generates vibrations, and the ground shaking that results when these vibrations reach the Earth's surface.

In 1889, a physicist in Germany noticed that a pendulum in his lab appeared to begin moving back and forth without having been touched. He reasoned that the pendulum was actually standing still while the Earth vibrated under it. A few days later, he read in a newspaper that a large earthquake had taken place in Japan about 12 minutes before the movement of his pendulum began. The physicist deduced that the energy generated by the earthquake had traveled all the way through the Earth from Japan and had slightly shaken his laboratory in Germany. Subsequent research revealed that earthquake energy moves through rock or along the Earth's surface in the form of

FIGURE 2.13 Faulting and earthquakes.

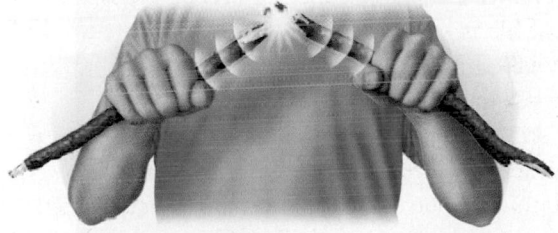

(a) Snapping a stick generates vibrations that pass through the stick.

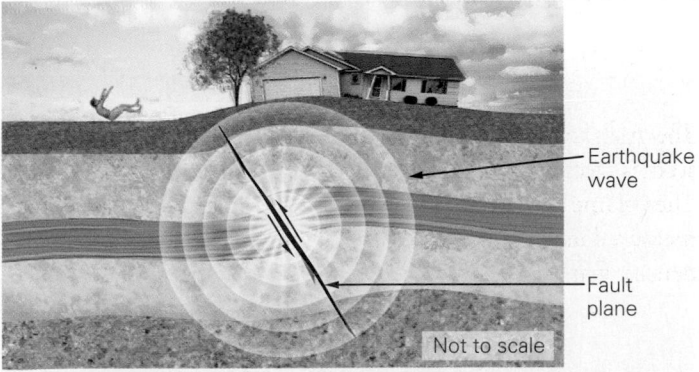

Earthquake wave

Fault plane

Not to scale

(b) Similarly, when the rock inside the Earth suddenly breaks and slips, forming a fracture called a fault, it generates shock waves that pass through the Earth and shake the surface.

BOX 2.2 CONSIDER THIS . . .

Meteorites: Clues to What's Inside

During the early days of the Solar System, the Earth collided with and incorporated countless planetesimals and smaller fragments of solid material lying in its path. Intense bombardment ceased about 3.9 Ga, but even today collisions with space objects continue, and nearly 1 million kg of material (rock, metal, dust, and ice) fall to the Earth, on average, every year. The vast majority of this material consists of fragments derived from comets and asteroids sent tumbling into the path of the Earth after billiard-ball-like collisions with each other out in space or because of the gravitational pull of a passing object that nudged it into a new direction. Some of the material, however, consists of chips of the Moon or Mars, ejected into space when large objects collided with those bodies.

Astronomers refer to any object from space that enters the Earth's atmosphere as a *meteoroid*. Meteoroids move at speeds of 20 to 75 km per second, so fast that when they reach an altitude of about 150 km, they compress the air in their path enough to make the air incredibly hot. This heat causes the object to start glowing and to evaporate, leaving a streak of bright, glowing gas. The glowing streak, an atmospheric phenomenon, is a *meteor*, also known colloquially, though incorrectly, as a falling star **(Fig. Bx2.2a)**. Most visible meteors evaporate completely by an altitude of about 30 km. But dust-sized ones may slow down sufficiently to float to the Earth, and larger ones (fist-sized or bigger) can survive the heat of entry to reach the surface of the planet. In some cases, meteoroids explode as brilliant fireballs in midair.

FIGURE Bx2.2 Meteors and meteorites.

(a) A shower of meteors over Hong Kong in 2001.

(c) The Barringer meteor crater in Arizona, formed about 50,000 years ago, is 1.1 km in diameter.

The meteorite that formed the crater was 50 m across.

(b) In February 2013, a meteor exploded to form a fireball above the Russian city of Chelyabinsk. The shock wave blew out windows and knocked people over.

(d) Examples of stony meteorites (left) and iron meteorites (right).

Extraterrestrial objects that reach the Earth's surface are called **meteorites**. Almost all meteorites that have struck the Earth are small and have not caused notable damage on the Earth. In fact, during human history, only a few have smashed through houses, dented cars, or bruised people. But two huge fireballs have caused significant damage. Specifically, a small comet exploded above Tunguska, Siberia, in 1908, flattening trees over an area of 2,150 square kilometers. More recently, in 2013, a small asteroid blew up 23 km above Chelyabinsk, Russia **(Fig. Bx2.2b)**. This explosion was about 25 times larger than that of the atomic bomb over Hiroshima, and its shock waves blasted out windows and knocked down walls, injuring about 1,500 people. Huge impacts have been observed elsewhere in the Solar System. For example, in 1994, astronomers observed four such impacts when a fragmented comet struck Jupiter. One of the impacts resulted in a 6-million-megaton explosion—this would be equivalent to blowing up 600 times the entire nuclear arsenal on the Earth all at once! Even larger impacts in the geologic past may have been responsible for disrupting life on Earth, as we will see later in this book. Some of these impacts have left craters that we can see today **(Fig. Bx2.2c)**.

Most meteorites are asteroidal or planetary fragments, for the icy material of small cometary bodies is too fragile to survive the fall. Researchers recognize three basic classes of meteorites: *iron* (made of iron-nickel alloy), *stony* (made of rock), and *stony iron* (rock embedded in a matrix of metal). Of all known meteorites, about 93% are stony and 6% are iron **(Fig. Bx2.2d)**. From their composition, researchers have concluded that some meteors (a special subcategory of stony meteorites called *chondrites*, because they contain small spherical nodules called chondrules) are asteroids derived from planetesimals that never underwent differentiation into a core and mantle. All other stony meteorites and all iron meteorites are asteroids derived from planetesimals that had differentiated into a metallic core and a rocky mantle early in Solar System history but later shattered into fragments. Stony meteorites are probably similar in composition to the Earth's mantle, and iron meteorites are probably similar in composition to the Earth's core. Most meteorites formed about 4.54 to 4.56 Ga. Some chondrites formed about 4.57 Ga, making them the oldest Solar System materials ever measured. Geologists consider the average composition of meteorites to be representative of the average composition of the whole Earth.

---

waves, called either *seismic waves* or earthquake waves. You can get a sense of what such waves look like by pushing suddenly on a spring or by jerking the end of a rope. (Chapter 10 provides further detail about earthquakes and seismic waves.)

Geoscientists immediately realized that the study of seismic waves traveling through the Earth could serve as a tool for exploring the Earth's interior, much as ultrasound today helps doctors study a patient's insides. That's because seismic waves travel at different velocities through different materials. So, by detecting depths at which seismic-wave velocities suddenly change, geoscientists can pinpoint the boundaries between layers and even recognize subtler boundaries within the main layers, as we now see. (Interlude D provides further detail about how the study of earthquake waves defines the Earth's layers.)

> ### TAKE-HOME MESSAGE
>
> Measurements of the Earth's mass, shape, and tidal behavior led to the conclusion that the Earth has three principal internal layers: the crust, the mantle, and the core. Study of earthquake waves passing through the interior has refined this image.
>
> **QUICK QUESTION:** What observation led 19th-century researchers to conclude that the Earth has a metal core?

2.5 What Are the Layers Made of?

We've already pointed out that the material composing the Earth's insides must be much denser than familiar surface rocks such as granite and basalt. How can we determine what this material consists of? Geoscientists have used several approaches to provide insight into the composition of the Earth's interior, including (1) examination of meteorite composition, for some meteorites are fragments that came from the interiors of planetesimals that had an Earth-like core and mantle **(Box 2.2)**; (2) studies to characterize materials that could be sources of the igneous rocks found in the crust; (3) analysis of mantle fragments that were carried up into the crust with igneous melts; and (4) laboratory measurement of the densities of known materials under high pressures and temperatures to see how they compare with estimated densities of the Earth's interior.

As a result of this work, we now have a pretty clear sense of what the layers inside the Earth are made of, though new findings continue to refine this image. Let's now look at the properties of individual layers, starting with the outermost layer, the crust **(Fig. 2.14)**.

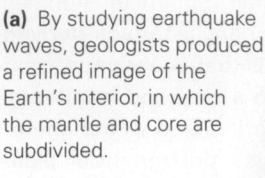

FIGURE 2.14 A modern view of the Earth's interior layers.

(a) By studying earthquake waves, geologists produced a refined image of the Earth's interior, in which the mantle and core are subdivided.

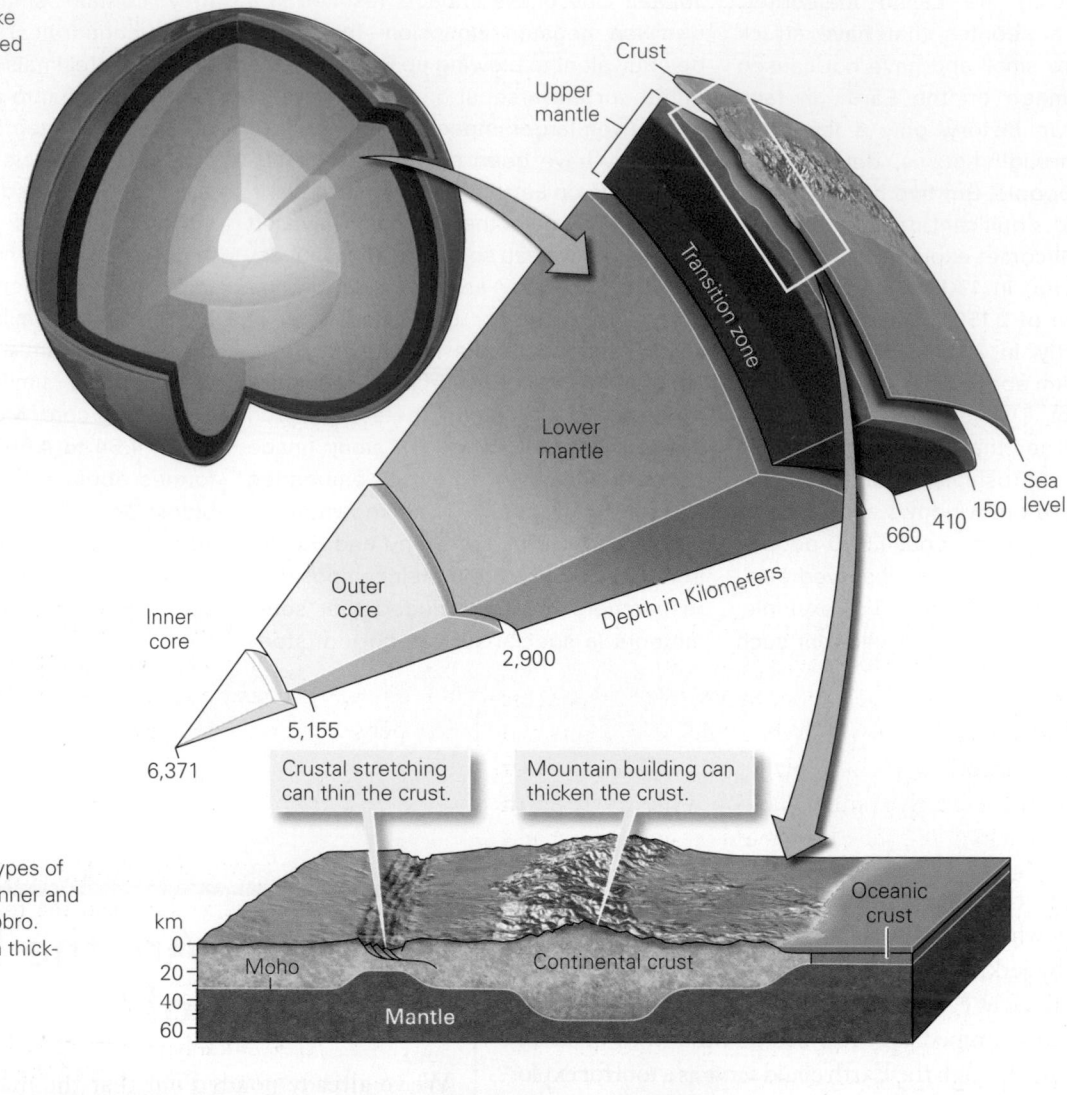

Crust
Upper mantle
Transition zone
Lower mantle
Sea level
660 410 150
Depth in Kilometers
Outer core
Inner core
2,900
5,155
6,371

Crustal stretching can thin the crust.

Mountain building can thicken the crust.

(b) There are two basic types of crust. Oceanic crust is thinner and consists of basalt and gabbro. Continental crust varies in thickness and rock type.

km
0
20
40
60
Moho
Continental crust
Oceanic crust
Mantle

Different Earth materials have different densities.

Increasing density

Granite	Basalt	Gabbro	Peridotite	Iron-nickel alloy
density = 2.7 g/cm³	density = 2.9 g/cm³	density = 3.0 g/cm³	density = 3.3 g/cm³	density = 7.5 g/cm³

(c) Examples of materials that characterize different layers inside the Earth. Continental crust has an average composition similar to granite, whereas oceanic crust consists of basalt and gabbro. The mantle consists of peridotite and the core of iron-nickel alloy.

The Crust

When you stand on the surface of the Earth, you are standing on top of its outermost layer, the **crust**. The crust is our home and the source of all our resources. Chemically, it's distinctly different from the whole Earth **(Fig. 2.15)**. How thick is this all-important layer? Or, in other words, at what depth does the crust-mantle boundary lie? An answer came from the work of Andrija Mohorovičić, a researcher working in Zagreb, Croatia. In 1909, Mohorovičić discovered that the velocity of seismic waves suddenly increased at a depth of a few tens of kilometers beneath the surface of continents, and he suggested that this increase was caused by an abrupt change in the properties of rock. He proposed that this change in seismic velocity

FIGURE 2.15 A table and a graph illustrating the abundance of elements in the Earth's crust.

Element	Symbol	% by weight	% by volume	% by atoms
Oxygen	O	46.3	93.8	60.5
Silicon	Si	28.0	0.9	20.5
Aluminum	Al	8.1	0.8	6.2
Iron	Fe	5.5	0.5	1.9
Calcium	Ca	3.4	1.0	1.9
Magnesium	Mg	2.8	0.3	1.4
Sodium	Na	2.4	1.2	2.5
Potassium	K	2.3	1.5	1.8
All others	—	1.2	>0.1	3.3

(a) A chart of relative abundances. Oxygen (O) and silicon (Si) account for almost three-quarters of the weight. By far, oxygen is the most abundant type of atom.

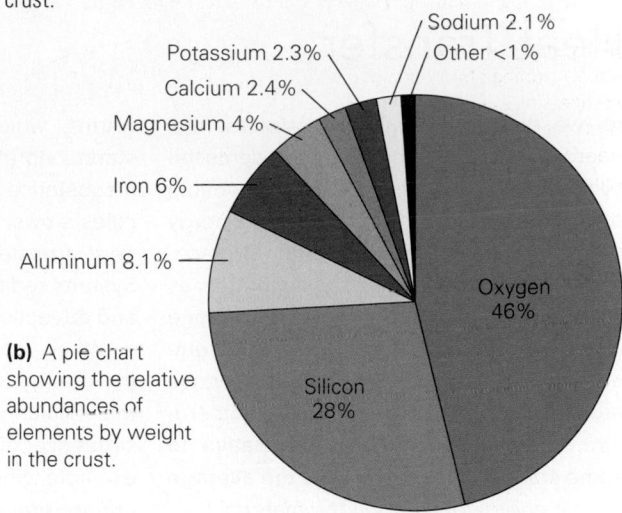

(b) A pie chart showing the relative abundances of elements by weight in the crust.

represents the crust-mantle boundary. Today we refer to this boundary as the **Moho** in Mohorovičić's honor.

Studies since Mohorovičić's time show that the thickness of the crust, defined as the depth to the Moho, varies between 7 and 70 km, depending on location. The crust is only about 0.1% to 1.0% of the Earth's radius (6,371 km), so if the Earth were the size of a balloon, the crust would be about the thickness of the balloon's skin. The crust does not consist simply of cooled mantle, like the skin on cooled chocolate pudding. Rather, it contains a variety of rocks that differ in composition (chemical makeup) from the underlying mantle rock. Geologists distinguish between two fundamentally different types of crust: **oceanic crust**, which underlies the seafloor, and **continental crust**, which underlies continents.

Oceanic crust is only 7 to 10 km thick, so at highway speeds (100 km per hour), you could drive a distance equal to the thickness of the oceanic crust in only 5 minutes. The top portion of oceanic crust is a blanket of sediment, generally less than 1 km thick, that consists of clay and tiny shells that settle like snow out of seawater. Beneath this blanket, the oceanic crust consists of a layer of basalt and, below that, a layer of gabbro.

Continental crust, in contrast to oceanic crust, varies in thickness from 25 to 70 km. The thinnest crust lies beneath regions called rifts, where the crust is being stretched apart and, therefore, has been thinned. Very thick continental crust occurs beneath mountain belts growing at locations where two continents are being squeezed together, causing the crust to shorten horizontally and thicken vertically. The broad plains found in the interiors of continents generally have a thickness of 35 to 50 km, about four to six times that of oceanic crust. Also, in contrast to oceanic crust, continental crust contains a great variety of rock types, ranging from mafic to felsic in composition. On average, upper continental crust is less mafic than oceanic crust—it has a felsic (granite-like) to intermediate (diorite-like) composition—so continental crust overall is less dense than oceanic crust.

The Mantle

The **mantle** is a 2,820- to 2,890-km-thick shell that surrounds the core. In contrast to the crust, the mantle consists entirely of peridotite, a dark and dense ultramafic rock that's quite rare at the Earth's surface. Notably, the mantle accounts for most of the Earth's volume, so—perhaps surprisingly—peridotite, a rock that most people have never seen, is actually the most abundant rock in our planet!

Did you ever wonder... what the most abundant rock of the Earth is?

Geoscientists have found that the velocity of seismic waves changes markedly, in a step-like manner, in the portion of the mantle that lies between 410 and 660 km deep. Based on this observation, they divide the mantle into two sublayers—the **upper mantle**, down to a depth of 660 km, and the **lower mantle**, from 660 km down to about 2,890 km. The portion of the upper mantle between 410 and 660 km, in which the steps in seismic velocity occur, is known as the **transition zone**. (Interlude D will explain why these steps exist.)

Almost all of the mantle is solid rock. But even though it's solid, mantle rock below a depth of about 100 km beneath the ocean floor, and of about 150 km beneath continents, stays so hot that it's soft enough to flow. This flow, however, takes place extremely slowly—at a rate of less than 15 cm a year.

BOX 2.3 SCIENCE TOOLBOX . . .

Heat Transfer

As we discussed in Box 1.4, the thermal energy or *heat* in a substance represents the sum of the kinetic energy (energy of motion) of all the substance's atoms. This energy comes from the back-and-forth displacements that an atom makes as it vibrates, as well as the movement of an atom from one place to another. When we say that one object is hotter or colder than another, we are describing its temperature. *Temperature* is a measure of warmth relative to some standard and represents the average kinetic energy of atoms in the material.

Heat can be transferred from one object, place, or material to another, and can be measured in calories: 1,000 calories can heat 1 kg of water by 1°C. When heat is added to a substance, the substance warms, which means that its molecules start to vibrate or move more rapidly. When a substance cools, the motion of its molecules slows. There are four ways in which *heat transfer* takes place in the Earth System: radiation, conduction, convection, and advection.

When electromagnetic waves transmit energy into a body or out of a body, the resulting heating or cooling is a consequence of *radiation* (Fig. Bx2.3a). For example, when sunlight strikes the ground during the day, radiative heating takes place. Similarly, when heat rises from the ground at night, the air warms and the ground cools.

Conduction takes place when you stick the end of an iron bar in a fire (Fig. Bx2.3b).

The iron atoms at the fire-licked end of the bar start to vibrate more energetically; they gradually incite atoms farther up the bar to start jiggling, and these atoms in turn set atoms even farther along in motion. In this way, heat slowly flows along the bar until you feel it with your hand.

Convection takes place when you set a pot of water on a stove (Fig. Bx2.3c). The heat from the stove warms the water at the base of the pot by making the molecules of that water vibrate faster and move around more. As a consequence, the density of the water at the base of the pot decreases, for as you heat a liquid, the atoms move away from one another, and the liquid expands. For a time, cold water remains at the top of the pot. But eventually, the warmer, less-dense water at depth becomes buoyant relative to the colder, denser water higher up. In a gravitational field, a buoyant material rises (like a Styrofoam ball in a pool of water) if the material above it is weak enough to flow out of the way. Since liquid water can flow easily, the hot water rises. When this happens, cold water sinks to take its place. Thus, during convection, the actual flow of the material itself carries heat. The trajectory of flow defines *convective cells*.

Advection, a less-familiar process, happens when a fluid that flows through cracks and pores within a solid material carries heat with it (Fig. Bx2.3d). The heat brought by the fluid conductively heats up the adjacent solid that the fluid passes through. Advection takes place, for example, if you pass hot water through a metal pipe and the pipe itself becomes hot. In the Earth, advection occurs where molten rock rises through the crust beneath a volcano and heats up the crust in the process.

FIGURE Bx2.3 The four processes of heat transfer.

(a) Radiation from sunlight warms the Earth.

40°C

Cool

Hot

90°C

(c) Convection takes place when moving fluid carries heat with it. Hot fluid rises while cool fluid sinks, setting up a convective cell.

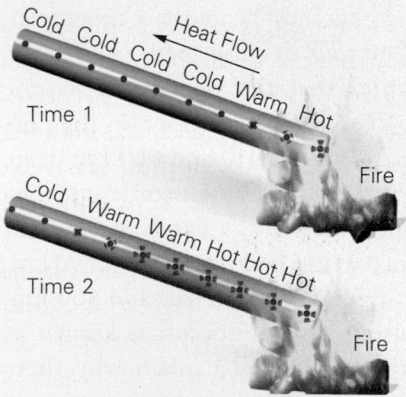

Cold Cold Cold Cold Warm Hot

Heat Flow

Time 1

Fire

Cold Warm Warm Hot Hot Hot

Time 2

Fire

(b) Conduction occurs when you heat the end of an iron bar in a flame. Heat flows from the hot region toward the cold region as vibrating atoms cause their neighbors to vibrate.

Cool   Warm   Hot   Hot   Warm   Cool

Molten rock

(d) During advection, a hot liquid (such as molten rock) rises into cooler material, and heat then conducts from the hot liquid into the cooler material.

Note, therefore, that "soft" in the context of describing the mantle does not mean liquid, and it does not mean that you could indent the mantle by pushing on it with your finger—it simply means that over long periods of time (thousands to millions of years), mantle rock can change shape significantly without breaking. Geologists often use the adjective *plastic* to indicate such behavior.

Earlier, we implied that whereas most mantle is solid, some is not. We made this distinction because, at a depth of 100 to 300 km beneath most ocean floor (and at a few other localities), a small percentage of the mantle has melted. This melt occurs in thin films or tiny drops between the solid grains in mantle rock.

Although, overall, the temperature of the mantle increases with depth, its temperature can also vary significantly with location even at the same depth. Warmer regions of mantle are less dense than adjacent cooler regions, so warmer regions are buoyant relative to cooler regions. As a result, warmer regions tend to flow upward, or *upwell*, while cooler regions flow downward, or *downwell*. In other words, the mantle undergoes very slow convection. The process resembles the flow of water in a simmering pot on a stove **(Box 2.3)**.

The Core

Early calculations suggested that the **core** had the same density as gold, so people once dreamed that vast riches lay at the heart of our planet. Alas, geologists eventually concluded that the core consists of a far less glamorous material, iron alloy (>80% iron mixed with nickel and lesser amounts of sulfur, oxygen, silicon, and other elements). Studies of seismic waves led geoscientists to divide the core into two parts, the *outer core* (between about 2,890 and 5,155 km deep) and the *inner core* (from 5,155 km down to the Earth's center at 6,371 km). The outer core consists of liquid iron alloy. It can exist as a liquid because the temperature in the outer core rises so high that even the great pressures squeezing the region cannot keep atoms locked into a solid framework. The iron alloy of the outer core can flow, and this flow generates the Earth's magnetic field (**Geology at a Glance**, pp. 60–61). Computer models suggest that because the Earth spins on its axis, the flow in the outer core follows spiral paths aligned with the axis.

The inner core, with a radius of about 1,220 km, consists of solid iron alloy that may reach a temperature of over 4,700°C. Even though it is hotter than the outer core, the inner core stays solid because it endures immense pressure. The pressure keeps atoms locked together tightly in very dense crystals. As the Earth slowly cools over geologic time, the base of the outer core solidifies and becomes, by definition, part of the inner core. Researchers estimate that the diameter of the inner core grows, as a result, by about 1 mm per year.

2.6 The Lithosphere and the Asthenosphere

So far, we have identified three major layers (crust, mantle, and core) of the Earth that differ compositionally from one another. Because of these differences, abrupt changes in seismic-wave velocity pinpoint the depths of these boundaries. An alternative way of thinking about Earth layers comes from studying the degree to which the material making up a layer can flow when subjected to geologic pushes or pulls over a relatively short time period. In this context, we distinguish between *rigid materials*, which can bend or break but do not flow easily, and *plastic materials*, which are relatively soft and can flow without breaking.

Let's apply this concept to the outer portion of the Earth. Geologists have determined that the outer 100 to 150 km of the Earth behaves as a relatively rigid solid. In other words, the Earth has an outer shell composed of rock that does not flow, overall, on a time scale of years to centuries. This outer layer is called the **lithosphere**, and it consists of the crust plus the uppermost, and therefore coolest, part of the mantle. We refer to the portion of the mantle within the lithosphere as the **lithospheric mantle**. Note that the terms *lithosphere* and *crust* are not synonymous—the crust makes up just the upper part of the lithosphere. Most of the lithosphere actually consists of mantle rock, specifically, mantle rock that has cooled enough to behave rigidly. (Cooler rock tends to be more rigid, while hotter rock tends to be more plastic. To picture this contrast, compare the behavior of a wax candle that you've just taken out of a freezer with that of one that has been warmed by the summer Sun on a hot day.)

The Earth from Surface to Center

If we could remove all the clouds and water that hide much of the solid surface of the Earth from view, we would see that both the land areas and the seafloor have plains and mountains. And if we could break open the Earth, we would see that its interior consists of a series of concentric layers (crust, mantle, and core) that differ from one another in terms of their composition and seismic velocity. Notably, oceanic crust differs from continental crust, both in thickness and in composition. Further, the mantle can be divided into two sublayers (upper mantle and lower mantle), and the core can be divided into an outer core of liquid iron alloy and an inner core of solid iron alloy. All terrestrial planets, as well as the Earth's Moon, have differentiated to form a crust, mantle, and core. But the relative thicknesses of the different layers are not the same for all planets. When discussing plate tectonics, it is convenient to call the outer part of the Earth, a relatively rigid shell composed of the crust and uppermost mantle, the lithosphere, and to refer to the underlying warmer, more plastic portion of the mantle as the asthenosphere.

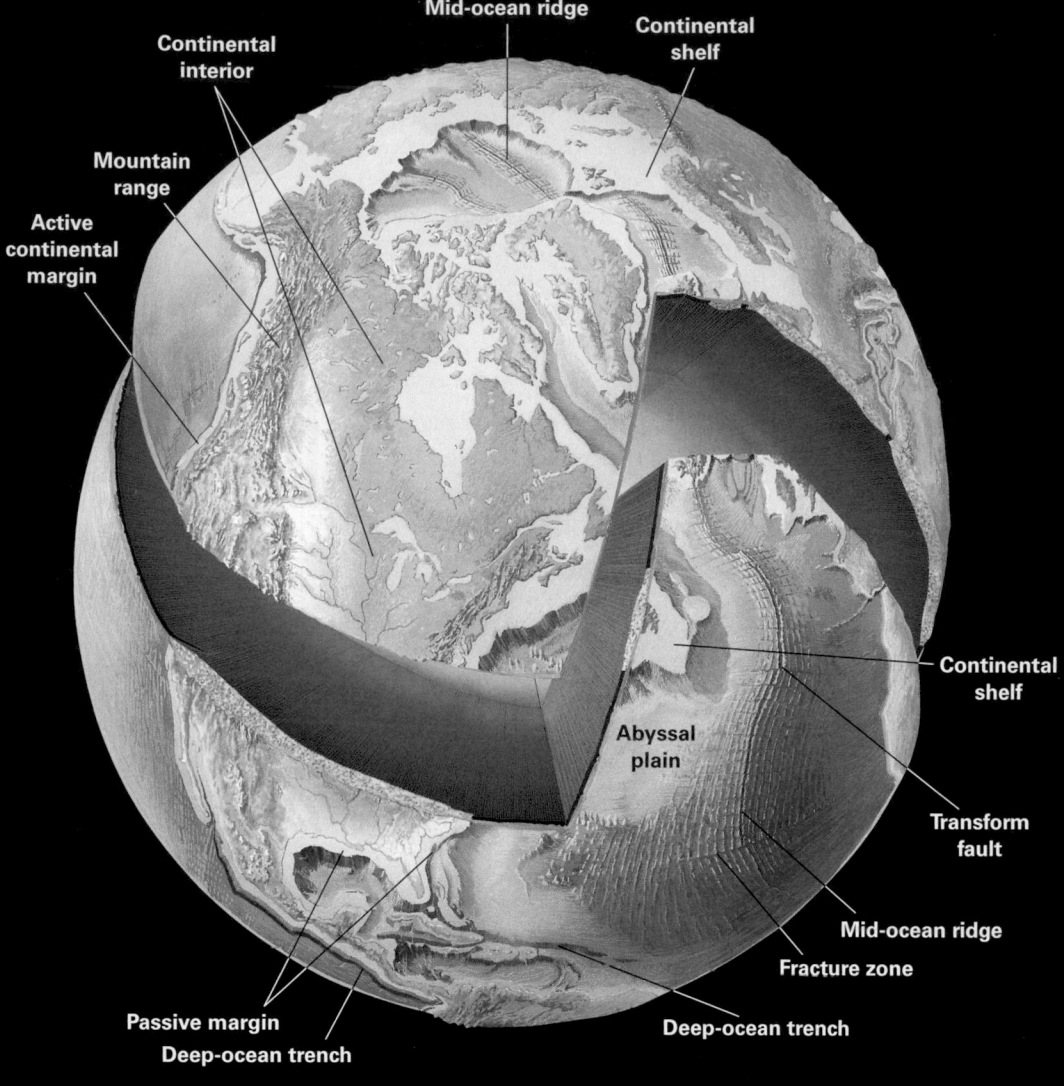

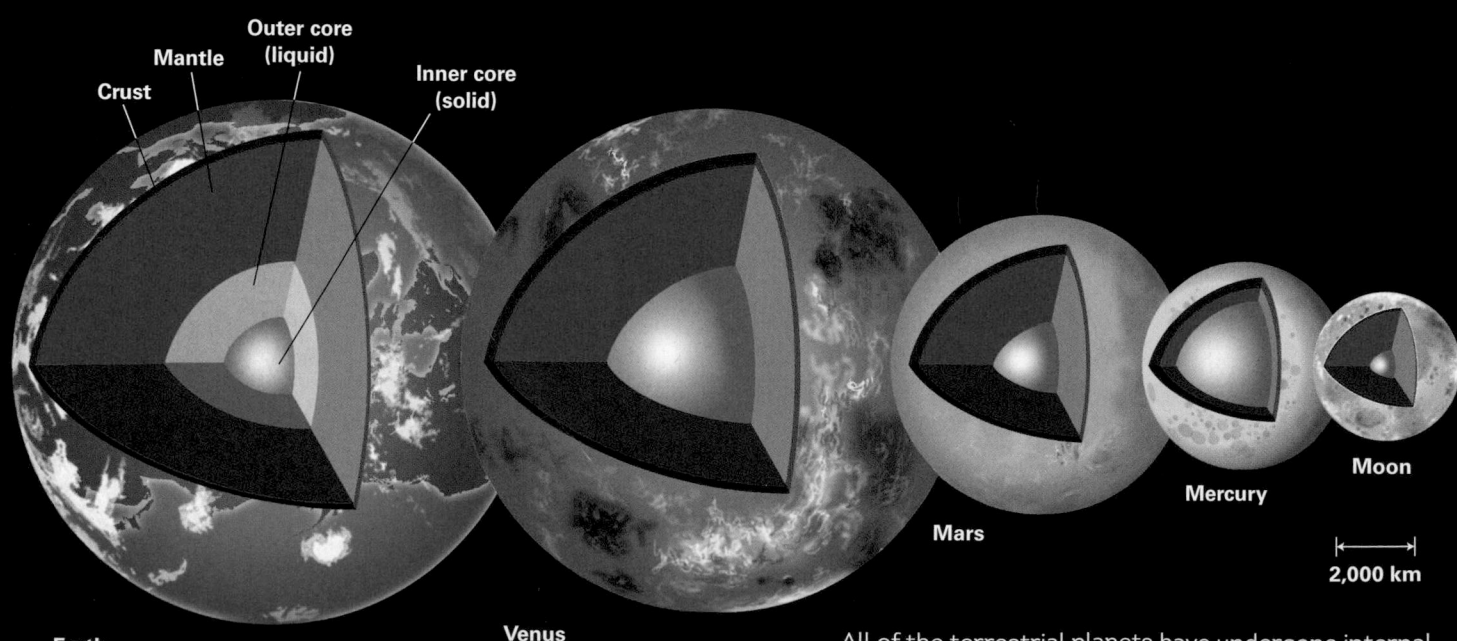

All of the terrestrial planets have undergone internal differentiation into a mantle and core. But the relative diameters of mantle and core differ from those of the Earth.

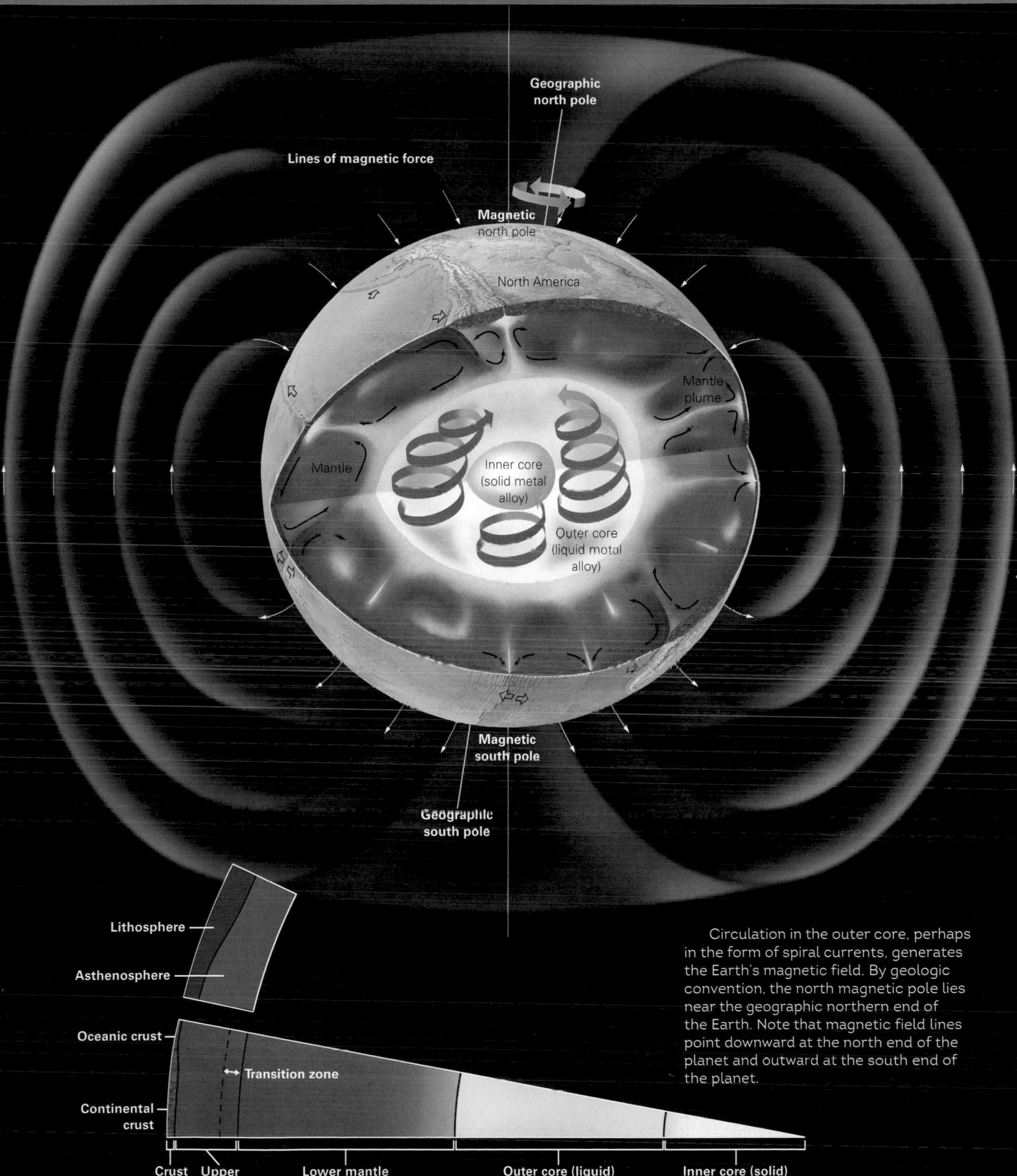

Geographic
north pole

Lines of magnetic force

Magnetic
north pole

North America

Mantle
plume

Mantle

Inner core
(solid metal
alloy)

Outer core
(liquid metal
alloy)

Magnetic
south pole

Geographic
south pole

Lithosphere

Asthenosphere

Oceanic crust

Transition zone

Continental
crust

Crust   Upper   Lower mantle   Outer core (liquid)   Inner core (solid)

Circulation in the outer core, perhaps in the form of spiral currents, generates the Earth's magnetic field. By geologic convention, the north magnetic pole lies near the geographic northern end of the Earth. Note that magnetic field lines point downward at the north end of the planet and outward at the south end of the planet.

FIGURE 2.16 A block diagram of the lithosphere emphasizing the difference between continental and oceanic lithosphere.

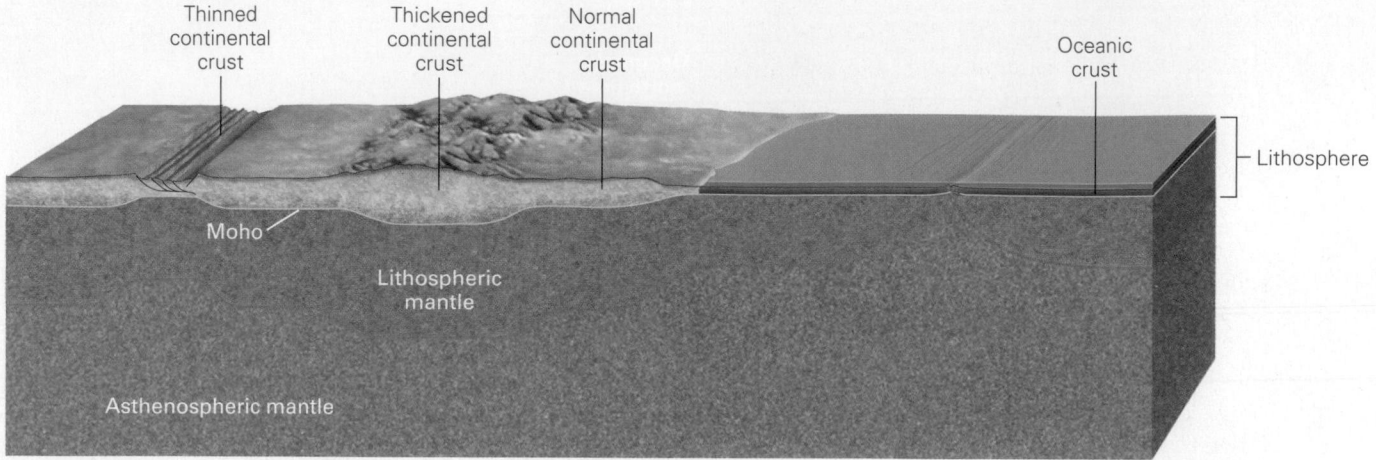

Geologists distinguish between two types of lithosphere **(Fig. 2.16)**. Most *oceanic lithosphere*, meaning lithosphere topped by oceanic crust, has a thickness of about 100 km. (As we'll see later, oceanic lithosphere is thinner along mid-ocean ridges.) In contrast, *continental lithosphere*, topped by continental crust, generally has a thickness of 150 to 200 km.

The lithosphere overlies a region called the asthenosphere, the plastic portion of the mantle that can flow and undergoes convection. The boundary between the lithosphere and asthenosphere occurs where the temperature reaches about 1,280°C, and tends to be gradational. At this temperature, mantle rock (peridotite) becomes soft enough to flow, for rock gets softer as it gets hotter because thermal energy causes chemical bonds to break more easily. Keep in mind that even though the asthenosphere flows, it is not molten overall. As we noted earlier, the mantle contains only tiny amounts of melt in films and drops between solid grains **(Fig. 2.17)**, and this zone of *partial melt* occurs only in the upper part of the asthenosphere (between depths of 100 and 300 km) beneath the ocean floor. While we can define the top of the asthenosphere as the base of the lithosphere (i.e., at a depth of about 100 km below oceanic abyssal plains and 150 to 200 km below the surfaces of continents), we can't really assign a specific depth to the base of the asthenosphere because all of the mantle has the ability to flow. For purposes of discussion, however, some geologists place the base of the asthenosphere at the top of the lower mantle.

> ## TAKE-HOME MESSAGE
>
> The crust and the outermost mantle together constitute the rigid lithosphere, a layer that overlies the softer, flowable asthenosphere. The behavior (rigid vs. plastic) of the mantle depends on its temperature.
>
> **QUICK QUESTION:** Is the asthenosphere entirely a liquid?

FIGURE 2.17 The igneous rock of the mantle consists of many grains of minerals that have grown together. Each grain has a surface, or "grain boundary." When partial melting takes place, films and pockets of melt (red) form along grain boundaries.

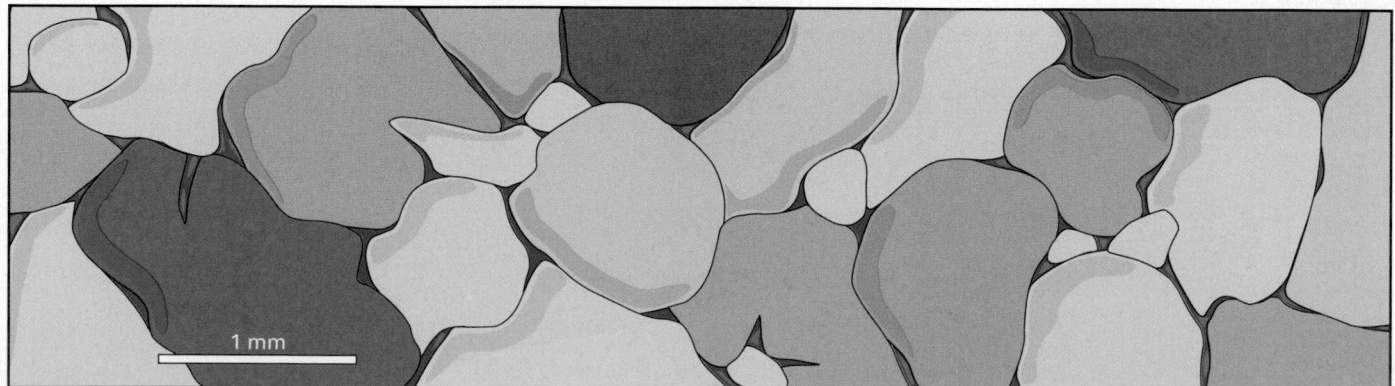

ANOTHER VIEW The green rock represents a piece of the mantle. Rising magma carried it upwards, through the crust, and out of the summit of a volcano in Antarctica.

Chapter 2 Review

SUMMARY

- A traverse from interstellar space into the Solar System passes through the Oort Cloud and crosses the edge of the heliosphere.

- Inside the heliosphere lies the Kuiper Belt, the outer planets, the asteroid belt, and the inner planets.

- A magnetic field surrounds the Earth. The field shields it from solar wind. Closer to the Earth, the field produces the Van Allen belts, which also trap cosmic rays.

- A layer of gas, the atmosphere, surrounds the Earth. Air in the atmosphere consists of 78% nitrogen, 21% oxygen, and 1% other gases. Air pressure decreases with elevation, so 99% of the gas in the atmosphere resides below 50 km.

- The surface of the Earth can be divided into land (30%) and ocean (70%). Most of the land surface lies within 1 km of sea level, and most of the seafloor is at a depth of 4 to 5 km. The land surface and the seafloor display great variation in elevation and depth, respectively.

- Earth materials include organic chemicals, minerals, glasses, rocks (igneous, metamorphic, and sedimentary), grains, sediment, metals, melts, and volatiles. Most rocks on the Earth contain silica (SiO_2), so they are called silicate rocks. Felsic, intermediate, mafic, and ultramafic igneous rocks have different densities.

- The Earth's interior can be divided into three compositionally distinct layers: the crust, the mantle, and the core. The first recognition of this division came from studying the density and shape of the Earth. The image has been refined by studying seismic waves.

- Pressure and temperature both increase with depth in the Earth. The rate of increase in temperature with depth is the geothermal gradient.

- Studies of seismic waves reveal the existence of the liquid outer core and the solid inner core. The mantle can be divided into the upper mantle and the lower mantle.

- The crust varies in thickness from 7–10 km (beneath oceans) to 25–70 km (beneath continents). Oceanic crust has a mafic composition, whereas average upper continental crust has a felsic to intermediate composition. The mantle consists of ultramafic rock, and the core of iron alloy. Flow in the outer core generates the magnetic field.

- The crust plus the upper part of the mantle constitute the lithosphere, a relatively rigid shell up to 150 km thick. The lithosphere lies over the asthenosphere, mantle that can flow plastically and, therefore, can convect.

GUIDE TERMS

abyssal plain (p. 48)
asteroid (p. 45)
asthenosphere (p. 62)
astronomical unit (AU) (p. 41)
atmosphere (p. 43)
aurorae (p. 43)
bathymetry (p. 48)
comet (p. 44)
continental crust (p. 57)
convection (p. 59)
core (p. 59)
cosmic rays (p. 41)
crust (p. 56)
crystal (p. 49)

deep-sea trench (p. 48)
dipole (p. 43)
Earth materials (p. 49)
earthquake (p. 53)
Earth System (p. 47)
geographic pole (p. 43)
geothermal gradient (p. 51)
glass (p. 49)
grain (p. 50)
groundwater (p. 48)
habitable zone (p. 47)
heliosphere (p. 41)
hypsometric curve (p. 49)
interplanetary space (p. 41)

interstellar space (p. 41)
Kuiper Belt (p. 41)
lithosphere (p. 59)
lithospheric mantle (p. 59)
lower mantle (p. 57)
magnetic field (p. 42)
magnetic field lines (p. 43)
magnetosphere (p. 43)
mantle (p. 57)
melt (p. 49)
metal (p. 50)
meteorite (p. 55)
mid-ocean ridge (p. 48)
mineral (p. 49)

Moho (p. 57)
oceanic crust (p. 57)
Oort Cloud (p. 41)
organic chemical (p. 49)
rock (p. 49)
sediment (p. 50)
silicate rock (p. 50)
surface water (p. 48)
topography (p. 48)
transition zone (p. 57)
upper mantle (p. 57)
vacuum (p. 41)
volatile (p. 50)

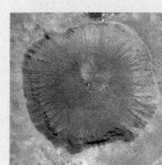 **GEOTOURS** *THIS CHAPTER'S GEOTOURS WORKSHEET (A) FEATURES QUESTIONS AND GOOGLE EARTH SITES ON:*

- Meteorite impacts
- Solar System scaling

REVIEW QUESTIONS

The letters following each Review Question refer to the corresponding Learning Objective from the Chapter Opener.

1. What occupies the space between planets? **(A)**

2. Name the features that a spacecraft traversing the Solar System and its surroundings would encounter. **(A)**

3. What is the Earth's magnetic field? Draw a representation of the field on a piece of paper. **(B)**

4. How does the magnetic field interact with solar wind and cosmic rays? **(B)**

5. What does the Earth's atmosphere consist of? Why would you die of suffocation if you were to parachute from an airplane at an elevation of 12 km? **(B)**

6. What is the proportion of land area to sea area? What is the average elevation of continents? Of seafloor? **(C)**

7. What are the two most abundant elements in the Earth? Describe the major categories of materials constituting the Earth. **(C)**

8. What are silicate rocks? Give examples of such rocks, and explain how they differ from one another. **(C)**

9. How did researchers first obtain a realistic estimate of the Earth's average density? What observations led to the realization that the Earth must be largely solid and that a particularly dense core lies at its center? **(C)**

10. What are seismic waves? Does their velocity change as they pass through the Earth? **(D)**

11. Identify the principal layers of the Earth in the figure. **(D)**

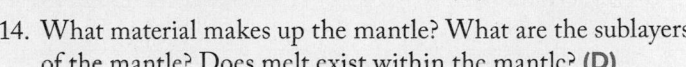

12. How do temperature and pressure change with depth in the Earth? **(D)**

13. What is the Moho? How was it first recognized? Describe the differences between continental and oceanic crust. **(D)**

14. What material makes up the mantle? What are the sublayers of the mantle? Does melt exist within the mantle? **(D)**

15. What is the core composed of? How do the inner core and outer core differ from each other? How can we obtain samples of materials that resemble those of the core? **(D)**

16. What is the difference between a meteor and a meteorite? Are all meteorites composed of the same material? **(A)**

17. What is the difference between lithosphere and asthenosphere? Identify the layers  and the Moho on the figure. At what depth does the lithosphere-asthenosphere boundary occur? **(E)**

ON FURTHER THOUGHT

18. Recent observations suggest that the Moon has a very small, solid core that is less than 3% of its mass. In comparison, the Earth's core is about 33% of its mass. Explain why this difference might exist. (Hint: Recall the model for Moon formation that we presented in Chapter 1.) **(D)**

19. Popular media sometimes imply that the crust floats on a "sea of magma." Is this a correct image. **(D)**

ONLINE RESOURCES

Videos
This chapter features videos on how solar storms helped make life on the Earth possible and how researchers discovered the Earth's inner core.

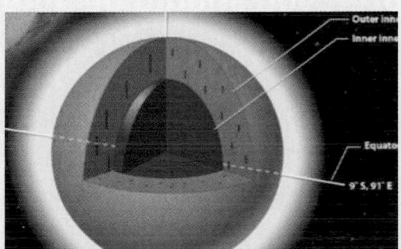

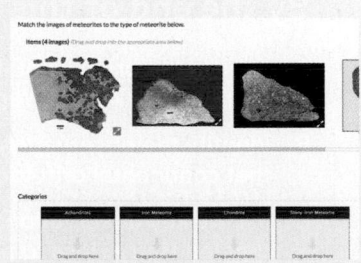

Smartwork5
This chapter features questions on topics such as the layers of the Earth and the composition of the Earth's interior.

CHAPTER 3
Drifting Continents and Spreading Seas

By the end of this chapter, you should be able to . . .

A. explain the premise of the continental-drift hypothesis proposed by Alfred Wegener.

B. identify the observations that Wegener used to show that continental drift took place.

C. list the key observations from the study of the seafloor that led Harry Hess to propose seafloor spreading.

D. explain how studies of paleomagnetism later proved that continents indeed move relative to one another.

E. describe some observations that can be used to prove that seafloor spreading happens.

> It is only by combing the information furnished by all the earth sciences that we can hope to determine "truth" here.
>
> —Alfred Wegener (German meteorologist, 1880–1930)

3.1 Introduction

In September 1930, 15 explorers led by a German meteorologist, Alfred Wegener, set out across the endless snowfields of Greenland to resupply two weather observers stranded at a remote camp. The observers were planning to spend the long polar night recording wind speeds and temperatures on Greenland's polar plateau. Wegener was a scientist well known not only to researchers studying climate but also to geologists. Some 15 years earlier, he had published a small book, *The Origin of the Continents and Oceans*, in which he had dared to challenge geologists' long-held assumption that the continents had remained fixed in position through all of Earth history. Wegener proposed instead that the continents had once fit together like pieces of a giant jigsaw puzzle, making one vast **supercontinent**. He suggested that this supercontinent, which he named **Pangaea** (pronounced pan-jee-ah), from the Greek *pan* (all) plus *gaia* (Earth), later fragmented into separate continents that drifted apart, moving slowly to their present positions **(Fig. 3.1)**. The phenomenon that Wegener proposed came to be known as *continental drift*.

Wegener presented many observations that he believed proved that continental drift had occurred, but he met strong resistance from his peers. At a widely publicized 1926 geology conference in New York City, a group of celebrated American professors posed the question, "What force could possibly be great enough to move the immense mass of a continent?" Wegener's writings didn't provide a good answer, so most of the group rejected continental drift. Four years later, Wegener faced an even greater challenge—survival itself. Sadly, he lost this battle too. On October 30, 1930, Wegener and a companion reached the observers, dropped off enough supplies to last the winter, and set out on the return trip the next day, but they never made it home.

Had Wegener survived to old age, he would have seen that his hypothesis, rather than being forgotten, would become the foundation of a scientific revolution. Today geologists accept Wegener's basic conclusion and take for granted the concept that the map of the Earth slowly changes as continents seemingly waltz around this planet's surface, variously combining and breaking apart, over geologic time. In fact, Pangaea wasn't the only supercontinent in Earth history—during the past few billion years, other supercontinents have formed and broken into pieces that later combined again.

The scientific revolution began in 1960 when an American geologist, Harry Hess, proposed that as continents move apart, new ocean floor forms between them by a process that his contemporary, Robert Dietz, named *seafloor spreading*. Hess also suggested that continents can move toward each other when the old ocean floor between them sinks back down into the Earth's interior, a process now called *subduction*. During the 1960s, geologists came to realize that continental movement, seafloor spreading, and subduction, along with a wide range of other geologic phenomena, happen because the Earth's outer, relatively rigid shell consists of about 20 distinct pieces—now called *plates*—that slowly move relative to one another. Because this idea has passed many tests, it has gained the status of a theory (see Box P.1), which we now call the theory of *plate tectonics*, from the Greek word *tekton*, which means builder. Plate movements effectively "build" regional geologic features. Geologists consider plate tectonics to be the grand unifying theory of geology because it so successfully explains a great many geologic processes and features, as we will see.

In this chapter, we introduce the observations that led Wegener to propose his continental-drift hypothesis. Next we learn how observations about the seafloor made by geologists during the mid-20th century led to the proposal of seafloor spreading and how the idea was tested and shown to be correct. Finally, we look at paleomagnetism, the record of the Earth's magnetic field in the past, which provides key proofs of continental drift and seafloor spreading. In Chapter 4 we will build on these concepts and describe the many facets of modern plate tectonics theory.

3.2 Wegener's Evidence for Continental Drift

Before Wegener, geologists viewed the continents and oceans as "immobile"—fixed in position throughout geologic time. According to Wegener, however, the positions of continents change over time. Specifically, Wegener suggested that a vast supercontinent, Pangaea, existed at the end of the Paleozoic Era

◀ (facing page) Fossils of *Mesosaurus*, a small coastal-marine reptile, occur in South America and Africa, continents now separated by the South Atlantic Ocean. How did these animals cross such a wide ocean? Alfred Wegener suggested that they didn't. Rather, the continents were once adjacent, and later drifted apart when the ocean basin grew.

FIGURE 3.1 Alfred Wegener and his model of continental drift.

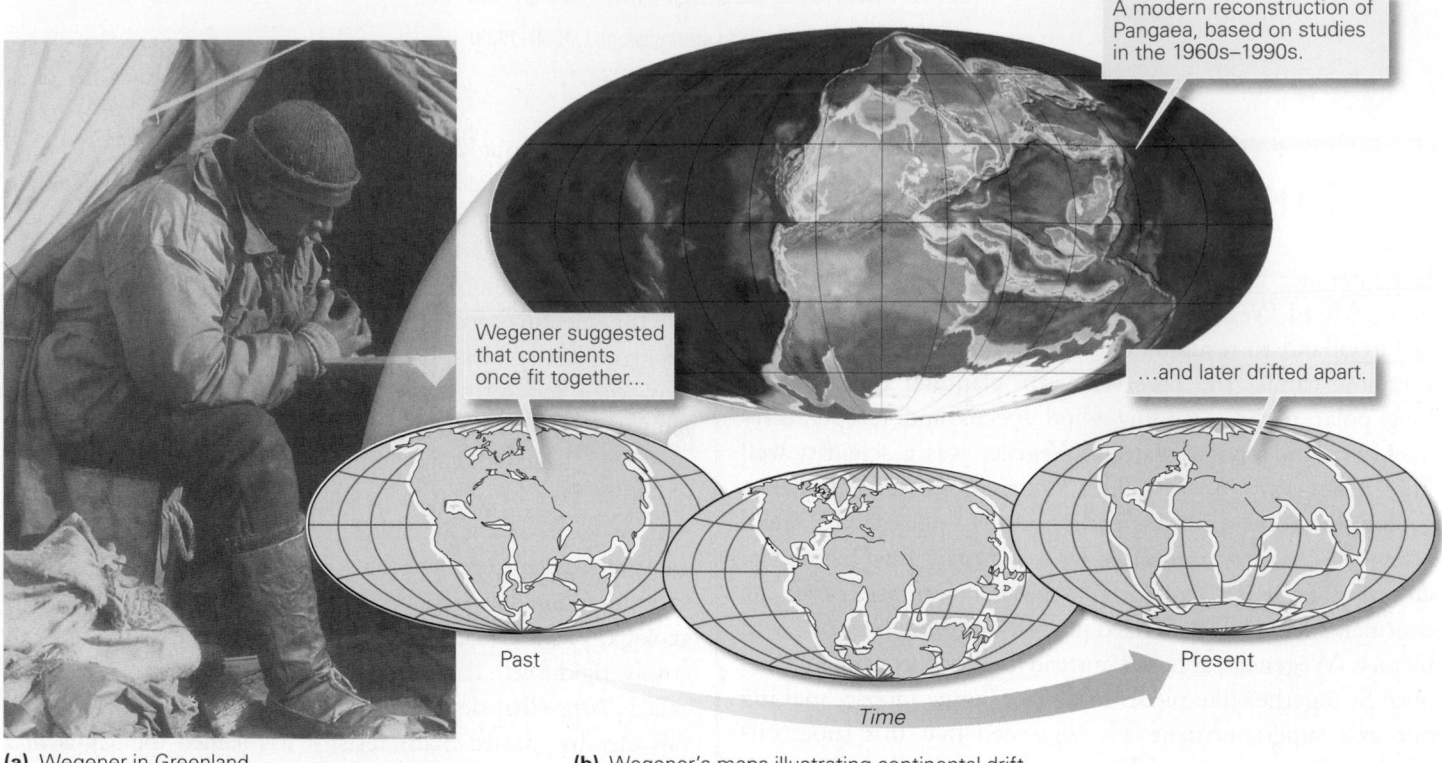

A modern reconstruction of Pangaea, based on studies in the 1960s–1990s.

Wegener suggested that continents once fit together...

...and later drifted apart.

Past

Present

Time

(a) Wegener in Greenland.

(b) Wegener's maps illustrating continental drift.

and that it broke apart during the Mesozoic (see Fig. P.7 for a simplified geologic time scale). The resulting smaller continents, the ones that exist today, then moved away from one another. Let's look at some of Wegener's arguments and see what led him to formulate this hypothesis of **continental drift**.

The Fit of the Continents

Almost as soon as maps of the Atlantic coastlines became available in the 1500s, scholars noticed the fit of the continents. Specifically, the northwestern coast of Africa looks like it could tuck in against the eastern coast of North America, and the bulge of eastern South America could nestle cozily into the indentation of southwestern Africa. Australia, Antarctica, and India could all connect to the southeast of Africa, while Greenland, Europe, and Asia could pack against the northeastern margin of North America. In fact, all the continents could be joined, with remarkably few overlaps or gaps, to form a single supercontinent, Pangaea **(Fig. 3.2)**. Wegener concluded that the fit was too good to be coincidence and, therefore, that the continents once did fit together.

> **Did you ever wonder . . .**
> why coasts of the Atlantic look like they fit together?

Locations of Past Glaciations

Glaciers are rivers or sheets of ice that flow across the land surface. As a glacier flows, it carries sediment grains of all sizes (clay, silt, sand, pebbles, and boulders; see Chapter 2 for a definition of *sediment*). Hard grains protruding from the base of the moving ice carve scratches, called *striations*, into the substrate. In some cases, it's possible to tell the direction of flow from striations. When the ice melts, it leaves the sediment in a deposit called *till*, which may bury the striations. As a consequence, the occurrence of till and striations at a location serves as evidence that the location was covered by a glacier in the past. By studying the age of glacial till deposits, geologists realized that large areas of land were

FIGURE 3.2 The "Bullard fit" of the continents. In 1965, Edward Bullard used a computer to fit the continents together and demonstrate how minor the gaps and overlaps are, although the match still isn't perfect.

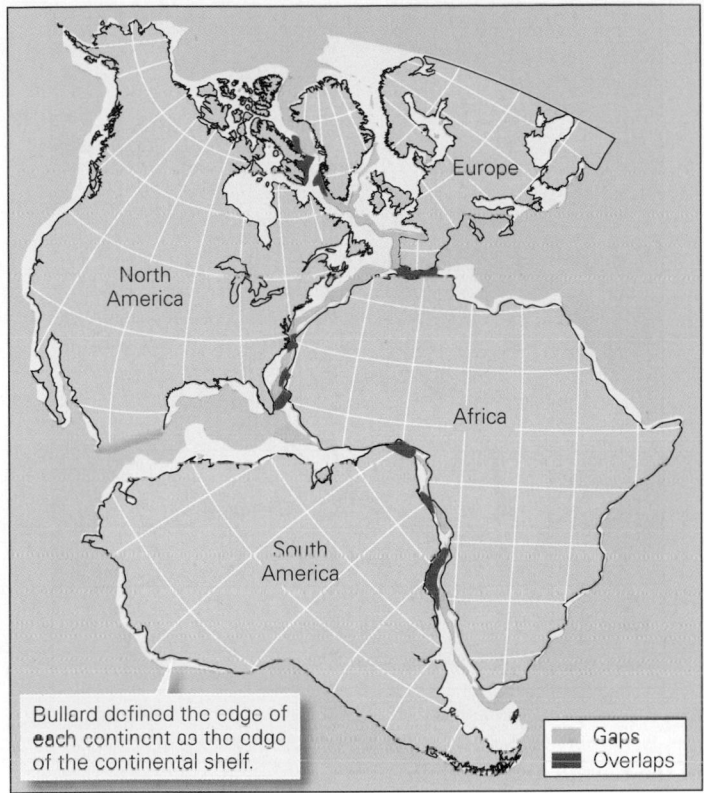

Bullard defined the edge of each continent as the edge of the continental shelf.

Gaps
Overlaps

FIGURE 3.3 Evidence for Pangaea, based on the distribution of glacial features.

(a) Glacial striations of late Paleozoic age on the surface of bedrock along the southern coast of Australia.

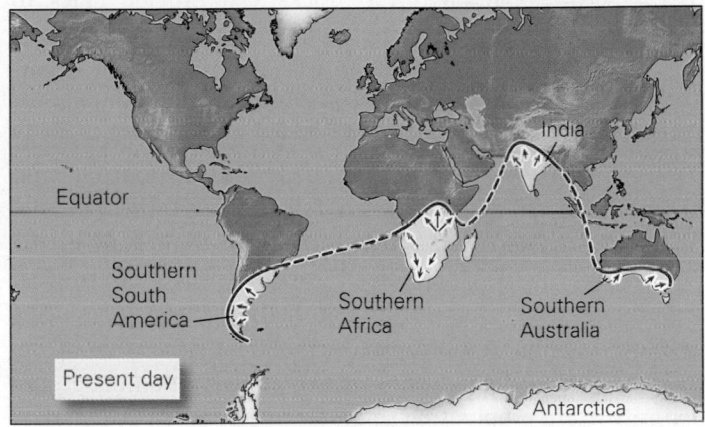

(b) A map showing the distribution of late Paleozoic glacial deposits and the orientation of associated striations.

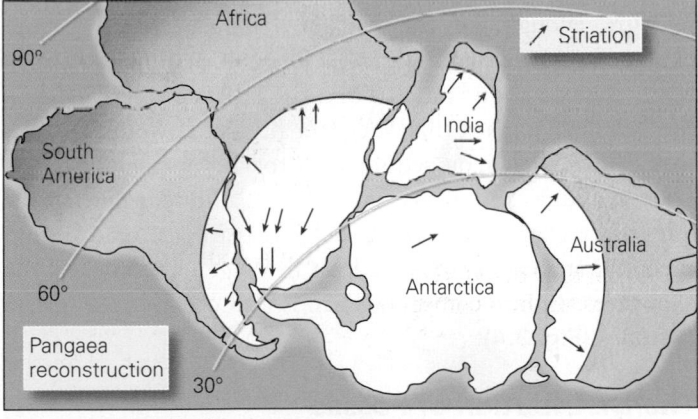

(c) In Wegener's reconstruction of Pangaea, the glaciated areas connect to outline a region of late Paleozoic southern polar ice caps.

covered by glaciers during discrete time intervals called *ice ages*. One of these ice ages occurred from 280 to 260 Ma (million years ago), near the end of the Paleozoic Era.

Wegener was a climate scientist by training, and he studied the Arctic, so it's no surprise that he had a strong interest in glaciers. He knew that glaciers form at high (polar) latitudes today, so he was bothered by the discovery of till and striations indicative of late Paleozoic glaciation in southern South America, southern Africa, southern India, Antarctica, and southern Australia, for, with the exception of Antarctica, these continents do not currently lie in high latitudes (Fig. 3.3). Wegener also noted that most striations associated with these late Paleozoic deposits seemed to point from the sea into the continents—this, too, was puzzling because glaciers today form on land and flow out to sea, so striations should point toward the coast. So Wegener plotted the distribution of the glacial deposits and the orientation of the striations on a map, then cut out the continents and fit them together to make Pangaea. To his amazement, all late Paleozoic glaciated areas lay adjacent to one another on his map of Pangaea, forming a single coherent ice sheet. Furthermore, when he determined the direction of movement, he found that it was roughly outward from the interior of this ice sheet to the sea. In other words, Wegener concluded that the distribution of glaciations at the end of the Paleozoic

Era could easily be explained if the continents had been united in Pangaea, with the southern part of Pangaea lying at polar latitudes, but could not be explained if continents had always been in their present positions.

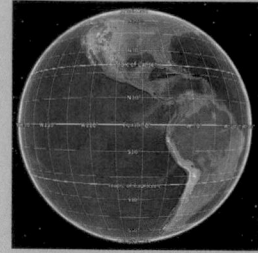
The Distribution of Climate Belts

If the southern part of Pangaea lay near the South Pole at the end of the Paleozoic Era, then during this same time interval, southern North America, southern Europe, and northwestern Africa would have straddled the equator and would have had tropical or subtropical climates. Wegener searched for evidence supporting this idea by studying characteristics of late Paleozoic sedimentary rocks from these regions, because the material making up sedimentary rocks can provide clues to the climate at the time the sediment was deposited. For example, in the swamps and jungles of tropical regions, thick deposits of plant material accumulate, and when deeply buried, this material transforms into coal. And in the clear, shallow seas of tropical regions, large reefs made from the shells of marine organisms develop. Finally, in subtropical regions, on either side of the tropical climate belt, where desert climates dominate, salt deposits (from evaporating seawater or salt lakes) accumulate and sand dunes grow. Wegener speculated that the distribution of late Paleozoic coal, reef, sand-dune, and salt deposits could define climate belts on Pangaea.

Sure enough, in the belt of Pangaea that Wegener expected to be equatorial, late Paleozoic sedimentary rock layers include abundant coal and the relics of reefs. And in the belts of Pangaea that Wegener predicted would be subtropical, late Paleozoic sedimentary rock layers include relics of desert dunes and deposits of salt. On a present-day map of our planet, exposures of these rock layers are scattered around the globe at a variety of latitudes. On Wegener's Pangaea, the exposures align in continuous bands that occupy appropriate latitudes (Fig. 3.4).

The Distribution of Fossils

Today, different continents provide homes for different species. Kangaroos, for example, live in the wild only in Australia. Similarly, many kinds of plants grow only on one continent and not on others. Why? Because land-dwelling or coastal species of animals and plants cannot swim across vast oceans, they evolved independently on different continents. During the period

FIGURE 3.4 Each symbol shows the location of an environment in which a distinctive type of sediment accumulates. The locations lie in belts, at appropriate latitudes, on a map of Pangaea.

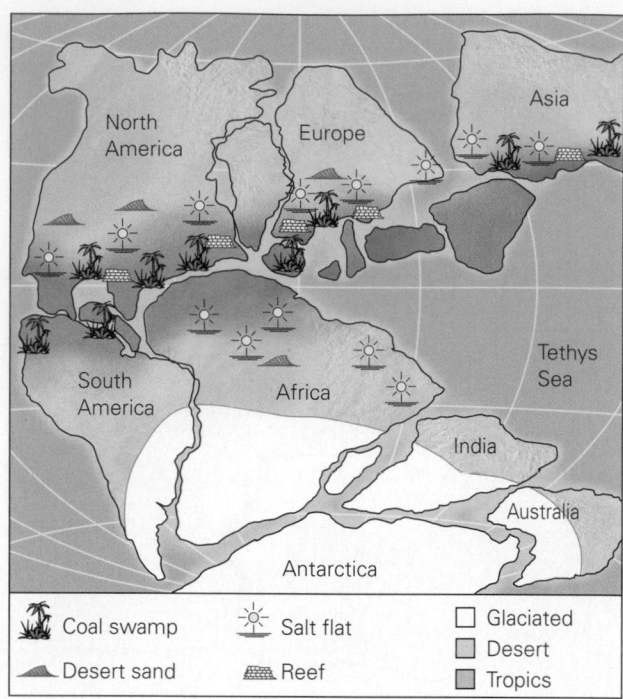

of the Earth's history when all continents were in contact, however, land animals and plants could have migrated relatively easily among many continents.

With this concept in mind, Wegener plotted occurrences of fossils representing land-dwelling species that lived during the late Paleozoic and early Mesozoic Eras (between 300 and 210 Ma). He found that such fossils exist on continents now separated by oceans (Fig. 3.5). Wegener argued, therefore, that the distribution of fossils requires that the continents were adjacent to one another in the late Paleozoic and early Mesozoic.

Did you ever wonder . . .
if you could have once walked from New York to Paris?

Matching Geologic Units

Art historians can recognize a Picasso painting, and architects know what makes a building's style Victorian. Similarly, geologists can identify distinctive assemblages of rocks. Wegener found that the same distinctive Precambrian (formed before 541 Ma) rock assemblages occurred on the eastern coast of South America and the western coast of Africa, regions now separated by an ocean (Fig. 3.6a). If the continents had been joined to create Pangaea in the past, then these matching rock groups would have been adjacent to each other and, therefore, could have composed

FIGURE 3.5 Fossil evidence for the existence of Pangaea.

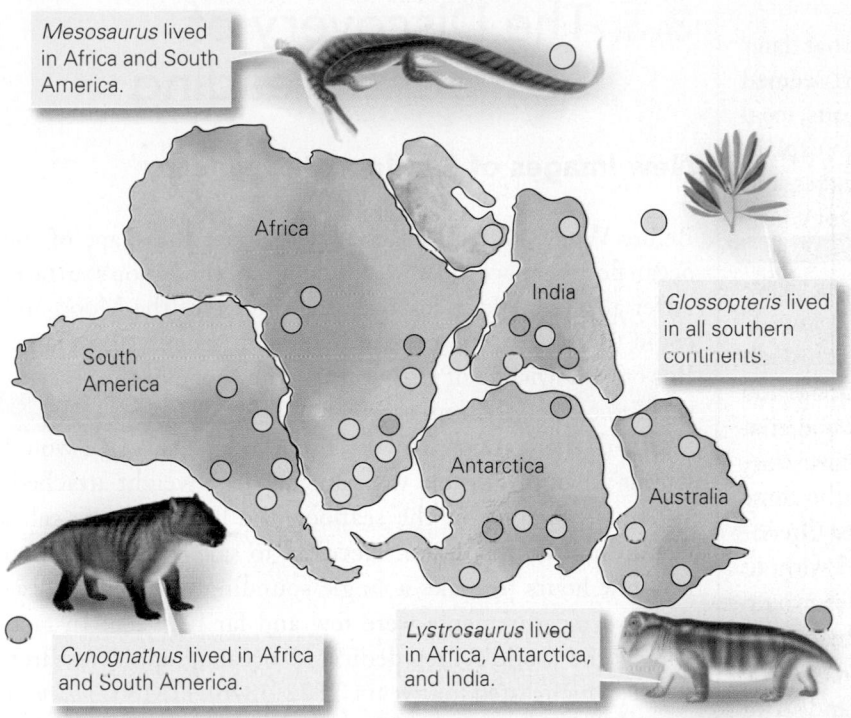

Mesosaurus lived in Africa and South America.

Africa

India

Glossopteris lived in all southern continents.

South America

Antarctica

Australia

Cynognathus lived in Africa and South America.

Lystrosaurus lived in Africa, Antarctica, and India.

(a) The distribution of fossil localities shows that Mesozoic land-dwelling or coastal organisms occur on continents that were adjacent in Pangaea.

(b) Fossils of *Glossopteris*, a land plant from Australia. Such fossils have been found on other continents.

FIGURE 3.6 Further evidence of continental drift: rocks on different sides of the ocean match.

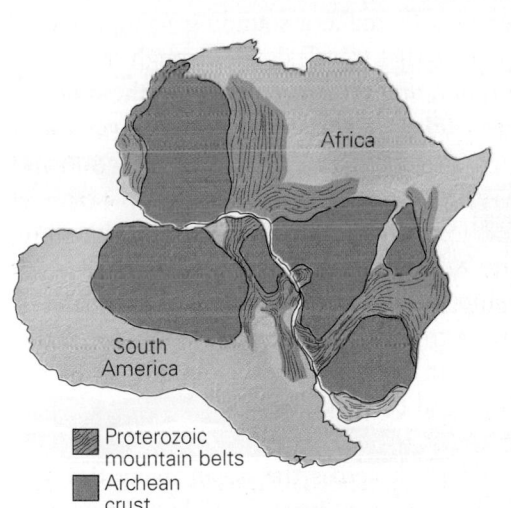

Africa

South America

Proterozoic mountain belts
Archean crust

(a) Distinctive belts of rock in South America would align with similar ones in Africa if the Atlantic Ocean didn't exist.

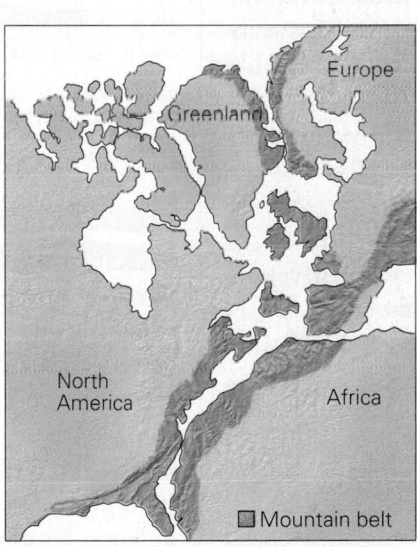

Europe

Greenland

North America

Africa

Mountain belt

(b) If the Atlantic didn't exist, Paleozoic mountain belts on both coasts would be adjacent.

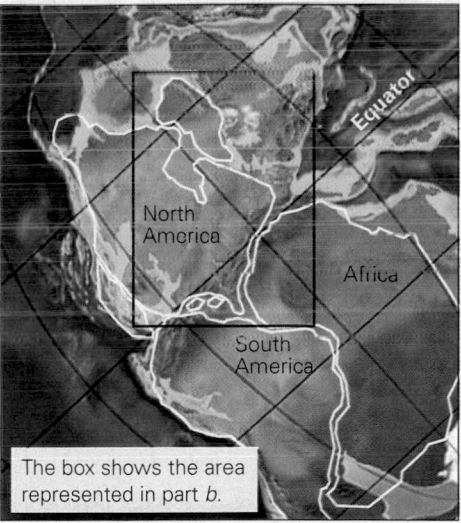

Equator

North America

Africa

South America

The box shows the area represented in part *b*.

(c) A modern reconstruction showing the positions of mountain belts in Pangaea. Modern continents are outlined in white.

continuous blocks or belts. Wegener also noted that features of the Appalachian mountain belt of the United States and Canada closely resemble those of mountain belts in southern Greenland, Great Britain, Scandinavia, and northwestern Africa **(Fig. 3.6b, c)**, regions that would have lain adjacent to North America in Pangaea. Wegener thus demonstrated that not only did the coastlines of continents match, but also the rocks adjacent to the coastlines.

Criticism of Wegener's Ideas

Wegener's model of a supercontinent (Pangaea) that later broke up to form smaller continents that moved apart seemed to explain the distribution of ancient glacial deposits, coal swamps, deserts, and reefs. The model also seemed to explain the distribution of certain distinctive rock assemblages and fossils. Clearly, Wegener had compiled a strong case for continental drift. But, as we noted earlier, he could not adequately explain how or why continents moved. Wegener's writings gave the impression that continents somehow "plowed" through the ocean floor as the keel of a ship plows through water, but that's not possible because ocean-floor rock is too strong to flow out of the way. Wegener also suggested that centrifugal force, due to the Earth's spin, could drive continental movement, but that's not possible because the force isn't strong enough. He left on his final expedition to Greenland having failed to convince his peers, and he died without knowing that his ideas, after lying dormant for decades, would be reborn as the basis of the broader theory of plate tectonics.

In effect, Wegener was simply ahead of his time. In the three decades that followed Wegener's death, a handful of iconoclasts continued to champion his notions. Among them was Arthur Holmes, a British geophysicist who argued that huge convection cells existed inside the Earth, slowly transporting hot rock from the deep mantle up to the base of the crust. Holmes speculated that continents might be forced apart in response to convective flow in the mantle and that the continents rode like rafts on the top of convective cells. But most geologists remained unconvinced and didn't realize that Wegener's bold proposal would one day change the whole discipline of geology forever. The door to the discovery of plate tectonics opened in the mid-20th century, when technologies became available to provide the data needed. Between 1930 and 1960, geologists learned how to determine the age of rocks, and how to "see" the ocean floor. In the next section, we describe some of the key results of this work.

TAKE-HOME MESSAGE

In the early 20th century, Alfred Wegener argued that the continents had once been connected in a supercontinent, Pangaea, that later broke up to produce smaller continents that "drifted" apart. He showed that the matching shapes of coastlines, as well as the distribution of ancient glaciers, climate belts, fossils, and rock units, make better sense if Pangaea existed. But Wegener couldn't convince his peers about continental drift.

QUICK QUESTION: Why were Wegener's peers skeptical of continental drift?

3.3 The Discovery of Seafloor Spreading

New Images of Seafloor Bathymetry

Before World War II, we knew less about the shape of the ocean floor than we did about the shape of the Moon's surface. After all, we could at least see the surface of the Moon and could use a telescope to map its craters, ridges, and plains. But our knowledge of **bathymetry** (the shape of the seafloor surface) came only from scattered *soundings* (depth measurements) of the seafloor. To take a sounding, a surveyor would let out a length of cable with a heavy lead weight attached. When the weight hit the seafloor, the length of the cable would indicate the depth. Needless to say, it could take up to a few hours to make a single sounding of the deep seafloor, so measurements were few and far between. In fact, during the world's first dedicated oceanographic research cruise, which lasted four years (1872–1876), HMS *Challenger* took only 360 soundings. Nevertheless, these measurements did hint at the existence of submarine mountain ranges and deep-sea troughs.

Military needs during World War II gave a huge boost to seafloor exploration, for as submarine fleets grew, navies required detailed information about bathymetry. The invention of *sonar* (echo sounding) permitted such information to be gathered quickly. To make a sounding using sonar, a ship emits a sound pulse that travels down through the water, bounces off the seafloor, and returns up through the water as an echo to a receiver **(Fig. 3.7)**. Because sound waves travel at a known velocity, the time between the sound's emission and the echo's detection indicates the distance between the ship and the seafloor (Velocity = Distance ÷ Time; therefore: Distance = Velocity × Time). Sound waves travel much faster than ships do: a ship moves only about 2 m in the time it takes for a sound wave to travel to the deep seafloor and back. Therefore, observers can obtain a continuous record of the depth to the seafloor and can produce a *bathymetric profile*, a graph showing how depth varies with location along a line. By cruising back and forth across the ocean many times at different locations, investigators eventually obtained enough bathymetric profiles to construct a *bathymetric map* of the seafloor. (Researchers now produce such maps much more rapidly and accurately using satellite data). Bathymetric maps reveal several important features **(Fig. 3.8)**.

- *Mid-ocean ridges:* The seafloor includes **abyssal plains,** broad flat regions that lie at a depth of 4 to 5 km below sea level, and **mid-ocean ridges,** elongate submarine mountain ranges whose peaks lie about 2 to 2.5 km below

FIGURE 3.7 How does sonar work?

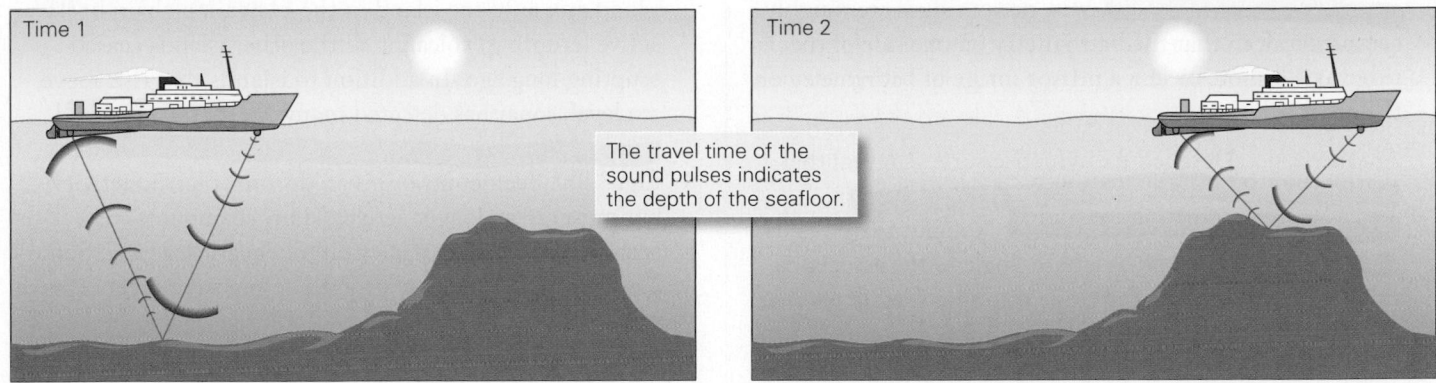

The travel time of the sound pulses indicates the depth of the seafloor.

(a) As the ship moves, it constantly sends out sound pulses. These pulses bounce off the seafloor and return to the ship as echoes.

(b) If the ship goes over a shallow area, the travel time of the pulses decreases.

FIGURE 3.8 Seafloor bathymetry.

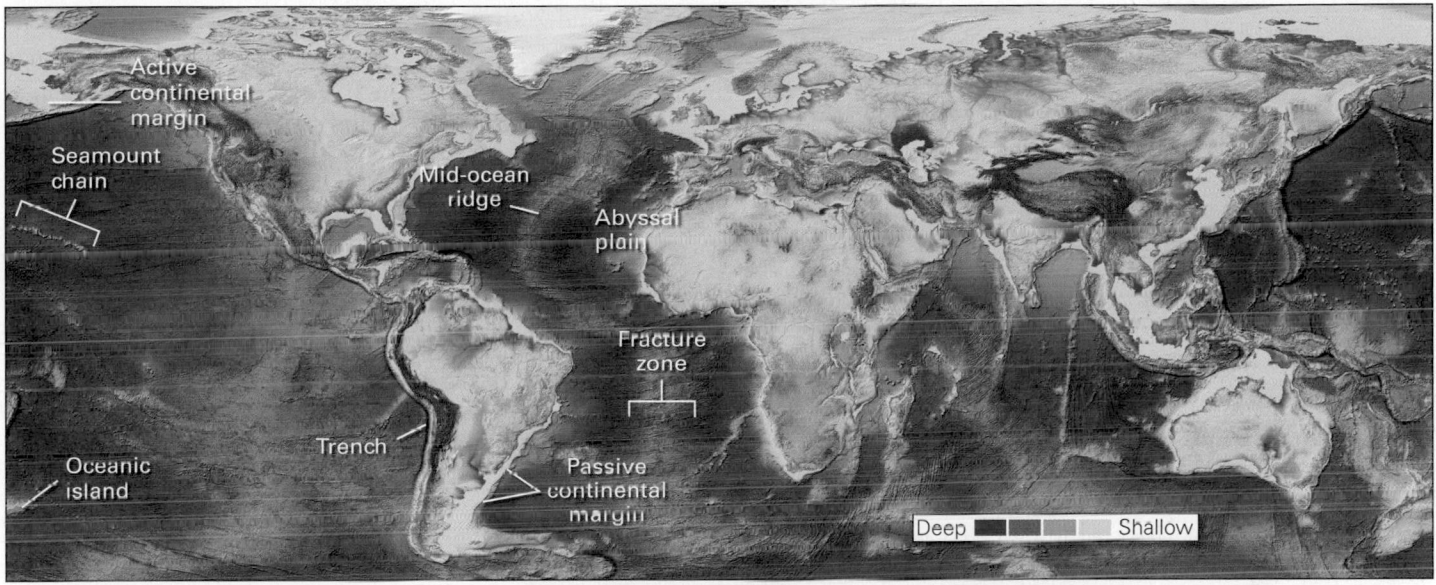

(a) A modern image of seafloor bathymetry. (Older versions were based on sonar studies; this one uses satellite data.) The colors indicate water depth.

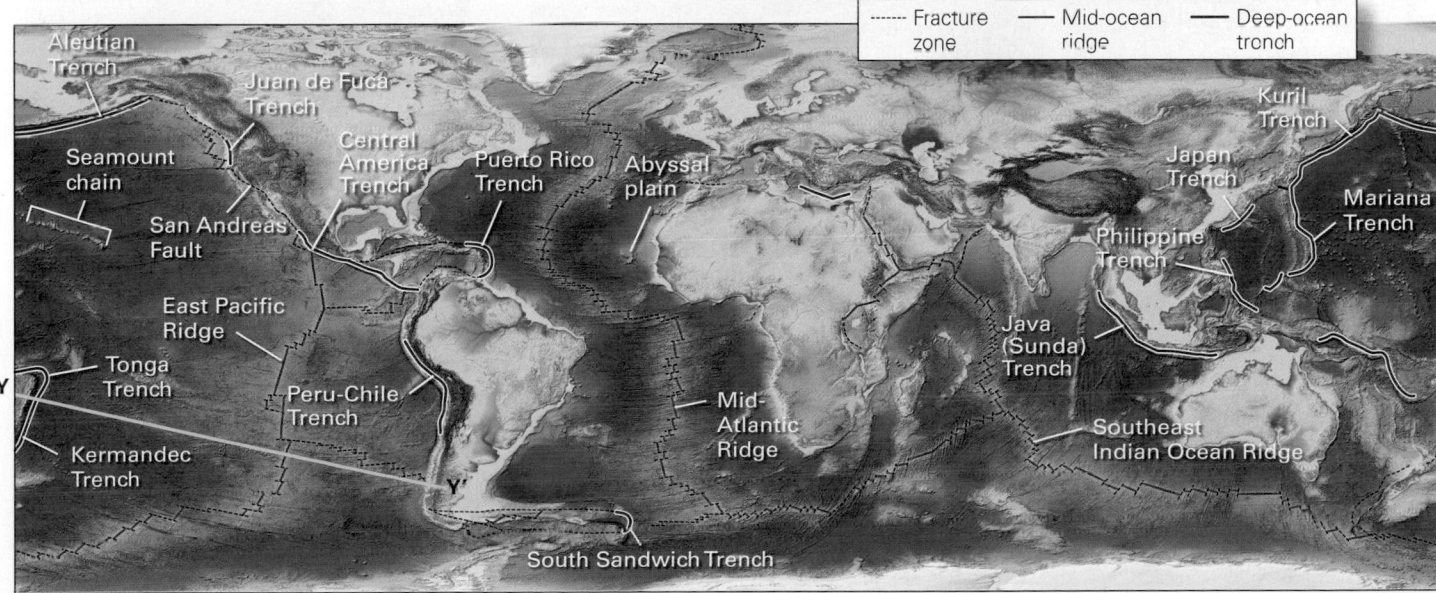

(b) On this map, major bathymetric features have been annotated with symbols.

sea level (Fig. 3.9a, b). Geologists call the crest of a mid-ocean ridge the **ridge axis**. Mid-ocean ridges are roughly symmetrical, in that the bathymetry on one side of the ridge axis is more or less a mirror image of bathymetry on the other side.

- *Fracture zones:* Detailed bathymetric surveys reveal that narrow bands of vertical cracks and broken-up rock locally cut across mid-ocean ridges. Notably, these bands, or **fracture zones**, trend at a high angle to the associated ridge axis and separate the ridge into small segments that do not align with one another (Fig. 3.9c). Fracture zones become less distinct away from the ridge axis and are not visible on the surface of abyssal plains.

- *Deep-sea trenches:* Along much of the perimeter of the Pacific Ocean, and at several other localities as well, the seafloor reaches depths greater than 5 km. These deep areas define elongate troughs that geologists now refer to as **trenches** (Fig. 3.10). Some trenches have depths in the range of 8 to 10 km, more than twice the depth of the abyssal plains. In fact, the deepest trench, the Mariana Trench of the western Pacific, reaches a depth of 10.9 km, deep enough to swallow Mt. Everest without a trace. All trenches border **volcanic arcs**, chains of active volcanoes. Some volcanic arcs form an archipelago of islands, whereas others fringe the edge of a continent.

- *Seamount chains:* Numerous volcanic islands poke up from the ocean floor, and not all of them lie within volcanic island arcs. The Hawaiian Island chain, for example, lies in the middle of the Pacific. In contrast to a volcanic island arc, only one island in the Hawaiian chain hosts active (erupting) volcano; all the other islands ceased erupting long ago. In addition to islands that rise above sea level, sonar has detected many **seamounts** (isolated submarine mountains), which also occur in chains (see Fig. 3.8b). Seamounts originated from volcanic activity, but most are no longer active. Many seamounts were islands at one time but later sank beneath sea level. Some have flat tops due to reef growth before submergence; such seamounts are called *guyots.*

New Observations on the Nature of Oceanic Crust

By the mid-20th century, geologists had discovered many important characteristics of the seafloor and had filled in huge blanks on the map of the Earth. These discoveries led them to realize that oceanic crust differs markedly from continental crust and, further, that bathymetric features of the ocean floor provide clues to the origin of the crust.

To start with, researchers found that a layer of sediment, composed of clay and the tiny shells of dead plankton, covers much of the ocean floor. Even at its thickest, it rarely exceeds 1 km in thickness. Given observed rates of sediment accumulation, the sediment layer is far too thin to have been accumulating for the entirety of Earth history. Furthermore, the sediment layer becomes progressively thicker away from the

FIGURE 3.9 Bathymetric characteristics of mid-ocean ridges.

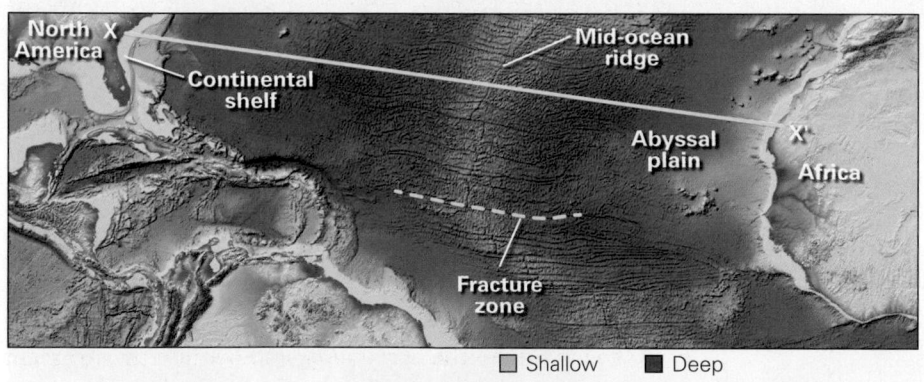

(a) The central Mid-Atlantic Ridge. The yellow line shows the position of the profile depicted in part (c).

☐ Shallow ■ Deep

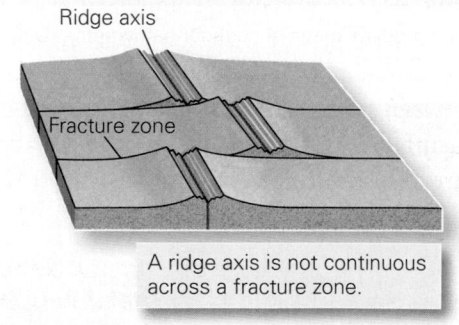

A ridge axis is not continuous across a fracture zone.

(c) Mid-ocean ridges are segmented. Segments link at fracture zones.

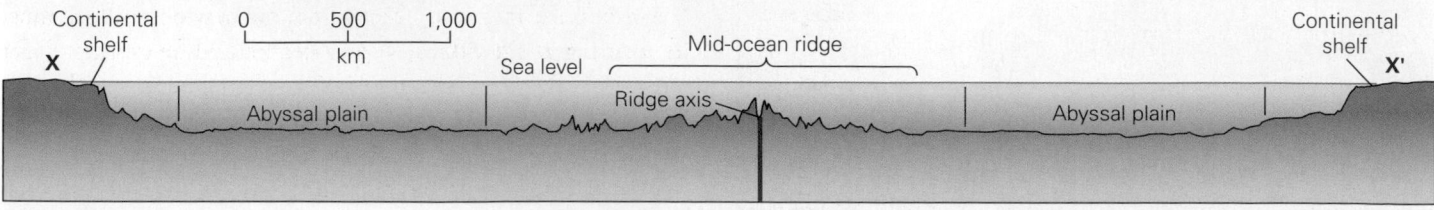

(b) A profile of the Mid-Atlantic Ridge, showing how the ridge rises above the abyssal plains.

FIGURE 3.10 Bathymetry of deep-sea trenches.

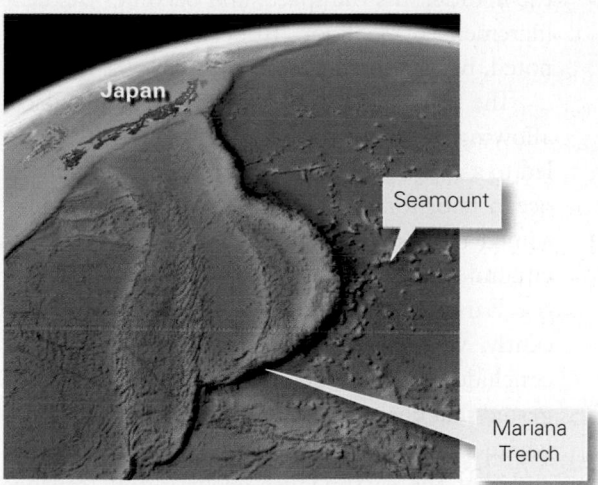

(a) The Mariana Trench lies on the western side of the Pacific. Water depths reach 11 km.

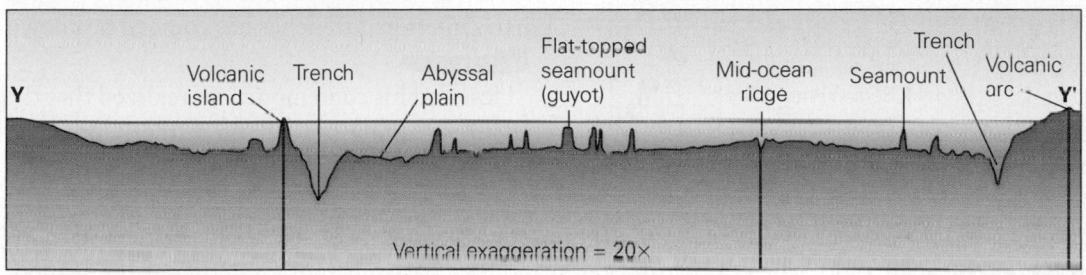

(b) A bathymetric profile of the Pacific (see Figure 3.8h for the location). Note that trenches can have ocean floor on either side, or can have a continent on one side. All trenches border volcanic arcs. Seamounts can rise to shallow depths.

mid-ocean ridge axis—in fact, there's almost no sediment at all near the ridge axis. This realization was initially based on surveys made by sending loud sound waves (resembling seismic waves) into the seafloor and recording the depth below the seafloor at which the waves reflected off the boundary between soft sediment above and harder basalt below. Subsequently, researchers drilled into the seafloor and could measure sediment thickness directly **(Fig. 3.11)**.

FIGURE 3.11 Drilling into the sediment layer of the ocean floor confirms that the basal sediment in contact with basalt gets older the farther away it is from the ridge. The sediment at location A is older than the sediment at location D.

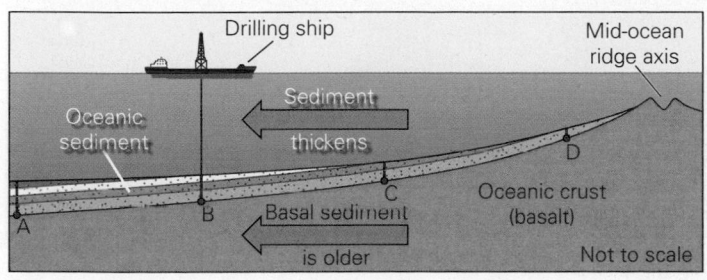

By dredging and drilling samples from the seafloor, geologists learned that oceanic crust does not have the same composition as continental crust. Beneath its sediment cover, oceanic-crust bedrock consists primarily of basalt—it does not display the great variety of rock types found on continents.

Heat flow, the rate at which heat rises from the Earth's interior up through the crust, is not the same everywhere in the ocean floor. Specifically, more heat rises beneath mid-ocean ridges than elsewhere. This observation led researchers to speculate that molten rock might be rising into the crust just below the mid-ocean ridge axis, for hot molten rock could bring heat into the crust (see Box 2.3).

When maps showing the distribution of earthquakes in oceanic regions became available in the years after World War II, it became clear that earthquakes do not take place randomly, but rather occur in distinct zones called **seismic belts (Fig. 3.12)**. Some seismic belts follow trenches, some follow mid-ocean ridge axes, and others lie along portions of fracture zones. Since earthquakes define locations where rocks break and move, geologists concluded that these bathymetric features coincide with places where movements of the crust are taking place.

One more key observation came from detailed studies of mid-ocean ridge bathymetry. Researchers found that some mid-ocean ridge axes occupy a narrow (a few kilometers wide), elongate trough that lies a few hundred meters deeper than its borders. In this regard, the bathymetry of a mid-ocean ridge resembles the topography of the East African Rift Valley, a place where the crust of Africa is stretching and breaking apart, and molten rock from below rises and erupts at volcanoes.

Hess's "Essay in Geopoetry"

In the late 1950s, Harry Hess, after studying the observations described above, concluded that because the sediment layer on the ocean floor is so thin overall, the ocean floor must be much younger than the continents, and because the sediment thickens progressively away from mid-ocean ridges, the ridges themselves are probably younger than the deeper parts of the ocean floor. If this is so, then somehow new ocean floor must be forming at the ridges, so an ocean could be getting wider with time. But how? The association of earthquakes with mid-ocean ridges suggested to Hess that the seafloor is breaking at the ridges. Furthermore, the discovery of high heat flow along mid-ocean ridge axes and the similarity of ridges to the East African rift indicated that molten rock is rising up beneath

FIGURE 3.12 A 1953 map showing the distribution of earthquake locations in the ocean basins. Note that earthquakes occur in belts.

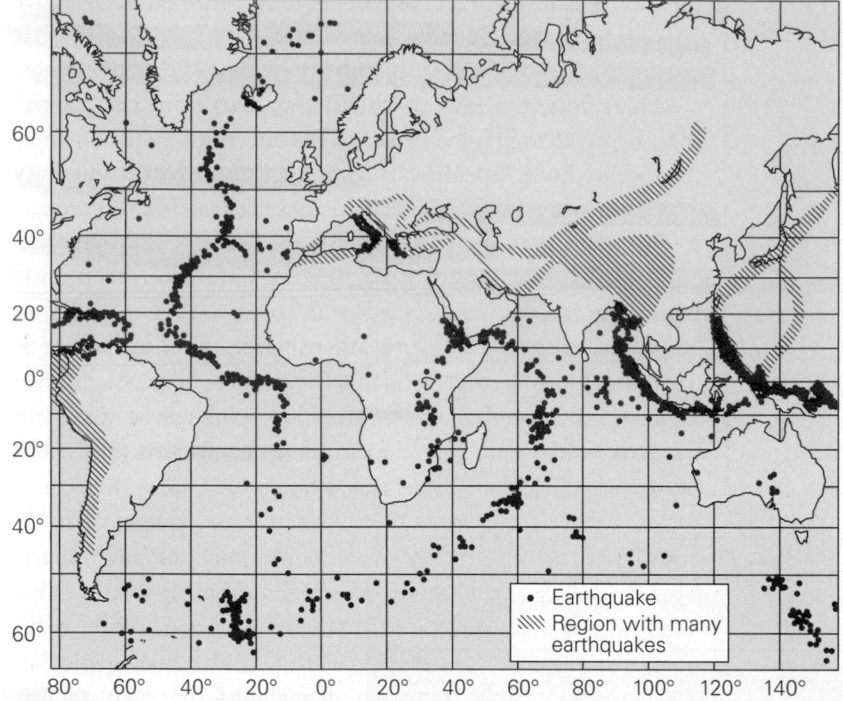

- • Earthquake
- ⫽⫽ Region with many earthquakes

away from the ridge axis, more melt rises from the mantle, fills the space, and becomes the next increment of seafloor. Robert Dietz, as we've noted, named the process **seafloor spreading**.

The idea that seafloor spreading takes place, allowing ocean basins to grow wider with time, led to a dilemma. Geologists realized that if new ocean floor forms, old ocean floor must be consumed or destroyed somewhere, or the Earth's circumference would have to increase, meaning the Earth would have to be expanding significantly, which the vast majority of studies had concluded wasn't possible. Researchers suggested that deep-sea trenches might be the places where the seafloor sinks back into the mantle and that the earthquakes occurring at trenches are evidence of this movement. The process by which ocean floor bends and sinks back into the Earth's interior at trenches has come to be known as **subduction**.

Hess and his contemporaries realized that the seafloor-spreading hypothesis instantly provided the long-sought explanation of how continental drift occurs. Continents passively move apart as the seafloor between them spreads at mid-

ridges and that the oceanic crust is stretching. In 1960, when Hess finally saw how these observations fit together, he wrote a manuscript in which he proposed that mantle rises beneath mid-ocean ridges and melts; that at the ridge axis, this melt solidifies to form oceanic crust; and that, once formed, the new crust cracks, splits apart, and moves away from the ridge **(Fig. 3.13)**. As each increment of seafloor forms and moves

ocean ridges, and they passively move together as the seafloor between them sinks back into the mantle at trenches. The idea seemed to be so good that Hess referred to his description of it as "geopoetry."

> **Did you ever wonder...**
> if the distance between New York and Paris changes?

FIGURE 3.13 Harry Hess's early concept of seafloor spreading (1962). Hess implied, incorrectly, that only the crust moved. We will see that this sketch is an oversimplification and contains errors.

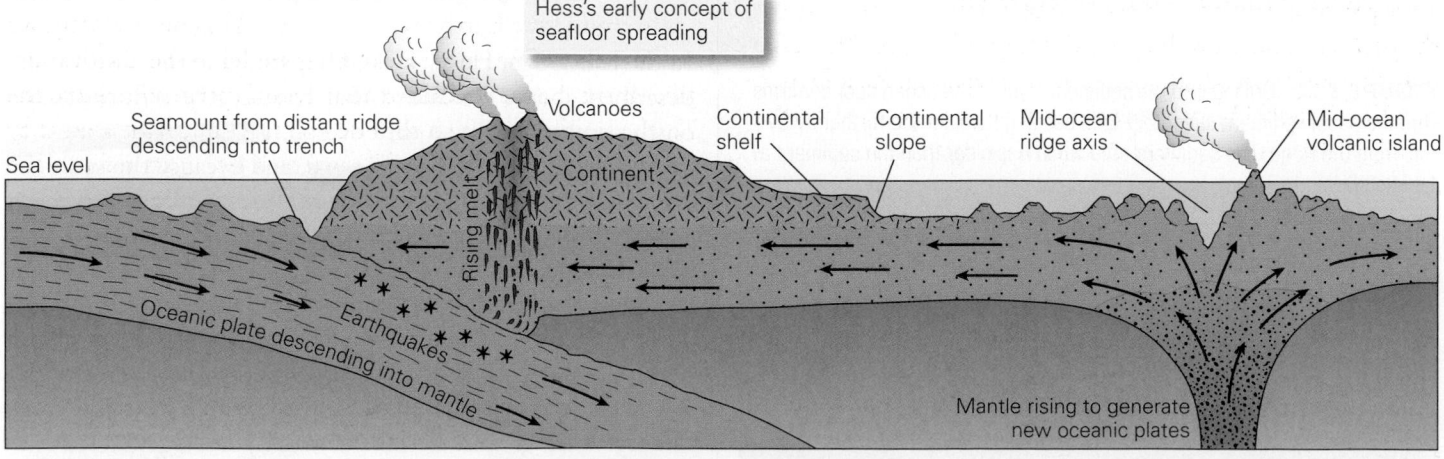

Hess's early concept of seafloor spreading

Seamount from distant ridge descending into trench — Volcanoes — Continental shelf — Continental slope — Mid-ocean ridge axis — Mid-ocean volcanic island

Sea level — Rising melt — Continent — Oceanic plate descending into mantle — Earthquakes — Mantle rising to generate new oceanic plates

3.4 Paleomagnetism—Proving Continental Drift and Seafloor Spreading

The publication of Hess's hypothesis that ocean basins grow by seafloor spreading suddenly made Wegener's hypothesis that continents drift seem like a real possibility. But, in order for both hypotheses to be accepted as theory, there had to be proof of movement. A method to directly measure continental movement on a short time scale did not exist in 1962. Fortunately, a discovery made by Chinese sailors over 1,500 years ago provided a basis for measuring such movement on a long time scale. The sailors had discovered that a piece of lodestone, when suspended from a thread, points in a northerly direction and can help guide a voyage. Lodestone exhibits this behavior because it consists of magnetite, an iron-rich mineral that, like a compass needle, aligns with the Earth's *magnetic field lines* (see Chapter 2). While not as magnetic as lodestone, several rock types contain trace amounts of magnetite or other magnetic minerals, so they behave overall like weak magnets. The study of such magnetic behavior led to the realization that rocks preserve **paleomagnetism**, a record of the Earth's magnetic field in the past. An understanding of paleomagnetism provided proof of both continental drift and sea-floor spreading. As a foundation for introducing paleomagnetism, we first provide further detail on the basic nature of the Earth's magnetic field.

The Earth's Magnetic Field: Some Further Details

As we learned in Chapter 2, we can represent the Earth's magnetic field by a *magnetic dipole*, an imaginary arrow that points from the north magnetic pole to the south magnetic pole, and passes through the center of the Earth. Magnetic field lines curve across space from one end of the dipole to the other. Why does the Earth produce such a magnetic field? Recall that the Earth's outer core consists of molten iron alloy, which constantly undergoes convective flow due to variations in density that develop in it (see Box 2.3). As physics textbooks explain, iron is an electrical conductor, and flow of a conductor in the presence of a magnetic field generates an electric current. This current, in turn, maintains the magnetic field. Put in the terminology of physics, the outer core behaves like a *self-sustaining dynamo*.

Convective flow in the outer core does not take place in random directions. According to one model, the rotation of the Earth around its axis causes the outer core's molten iron to flow in spirals whose center lines align, more or less, with the Earth's axis of rotation (**Fig. 3.14a**). As a result, the overall magnetic dipole representing the Earth's magnetic field also aligns with the axis of rotation, more or less, so our planet's magnetic poles lie near its geographic poles.

Over time, convection spirals within the outer core probably change their shape and orientation, causing the Earth's magnetic dipole to wobble and its magnetic poles to move a little. As a result, at any given time, the magnetic dipole and the Earth's axis of rotation are not exactly parallel. This situation means that at any given time, a compass needle doesn't point to geographic north and doesn't trend exactly parallel to a line of longitude. The angle between the direction in which a compass needle points and a line of longitude at a given location is called the **magnetic declination** (**Fig. 3.14b**). Today, the magnetic axis and the axis of rotation differ by several degrees, and the north magnetic pole lies a few hundred kilometers away from "true north" (geographic north). This distance changes measurably every year as the magnetic pole trundles across the Arctic Ocean toward Siberia at a rate of 50 to 60 km per year. Significantly, a map depicting the North Pole's changing position over the past 2,000 years reveals that the pole follows a fairly random path that stays north of the Arctic Circle (**Fig. 3.14c**). Because of this randomness, geologists assume that, averaged over thousands of years, the Earth's magnetic poles and geographic poles roughly coincide.

As we've seen, invisible magnetic field lines curve through space around the Earth (**Fig. 3.14d**). Looking more closely, we see that these lines are horizontal (parallel to the Earth's surface) at the magnetic equator and vertical (perpendicular to the Earth's surface) at the magnetic poles. At mid-latitudes, they tilt at an angle to the Earth's surface. Geologists refer to the angle between the magnetic field and the surface of the Earth at a given location as the **magnetic inclination**. The magnetic field has an inclination of 0° at the magnetic equator and 90° at the magnetic pole. If you make a *magnetic inclinometer* by placing a magnetic needle on a horizontal axis so it can pivot up and down, you can observe magnetic inclination directly. (Regular compasses don't show inclination because their needles have been balanced to compensate for a downward pull.)

FIGURE 3.14 Features of the Earth's magnetic field.

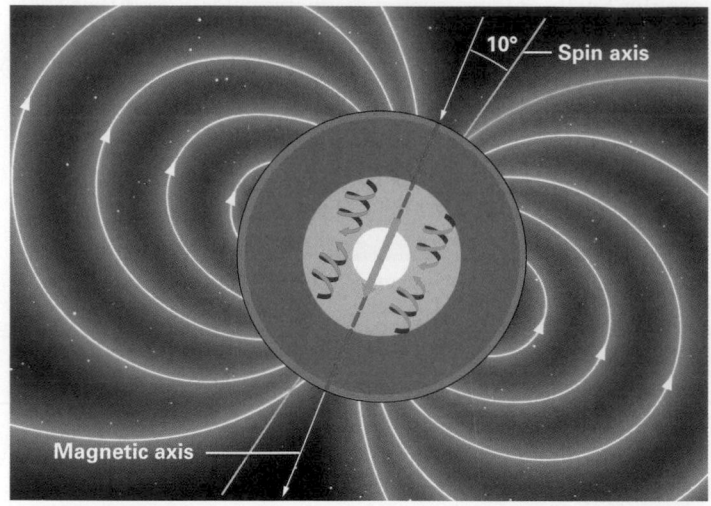

(a) The magnetic axis is not parallel to the axis of rotation. The field is generated by flow in the outer core.

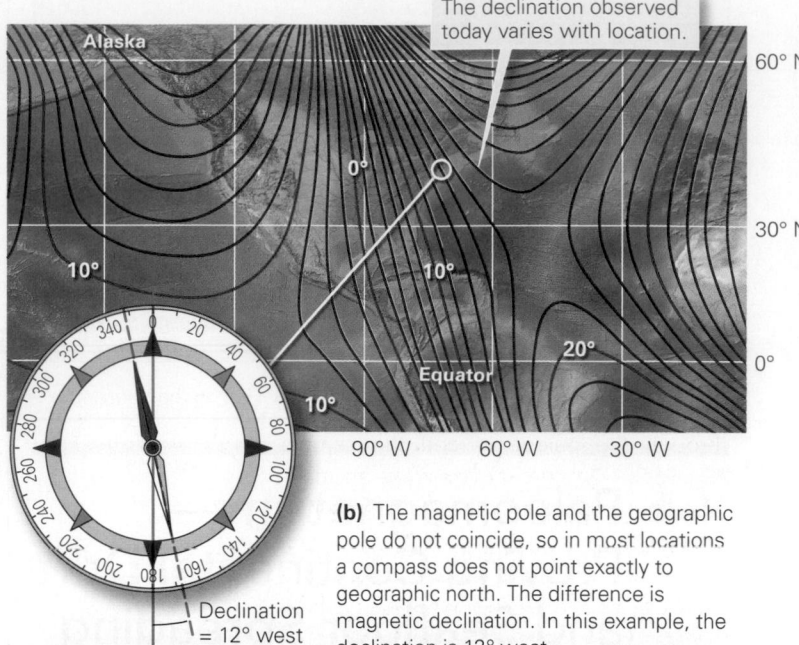

The declination observed today varies with location.

Declination = 12° west

(b) The magnetic pole and the geographic pole do not coincide, so in most locations a compass does not point exactly to geographic north. The difference is magnetic declination. In this example, the declination is 12° west.

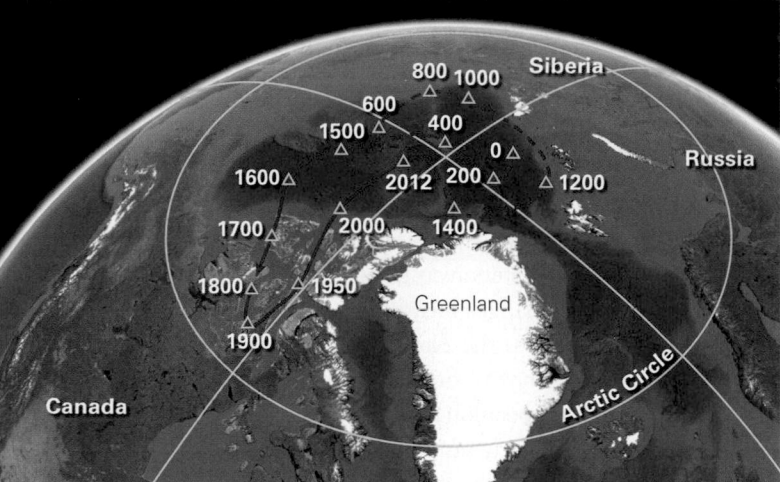

(c) A simplified map showing the changing position of the north magnetic pole over the past 2,000 years. Before about 1600, the position was not as well constrained, so the path is dashed.

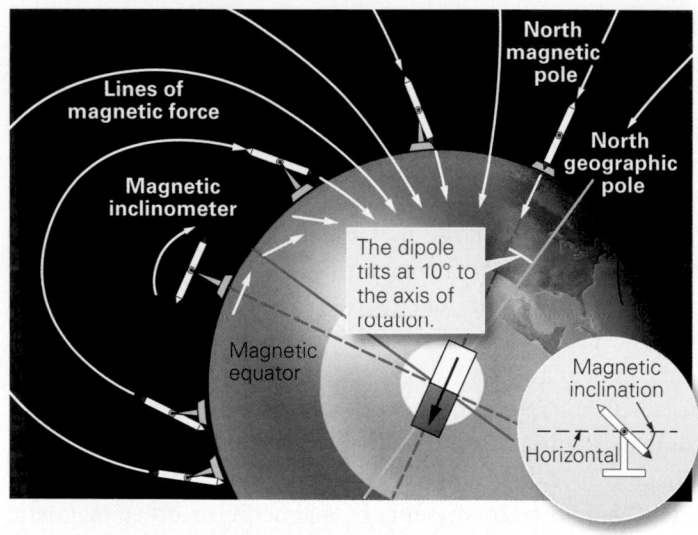

(d) The Earth's magnetic field lines curve, so the tilt of a magnetic needle changes with latitude. This tilt is the magnetic inclination.

What Is Paleomagnetism?

In the early 20th century, when researchers developed instruments that could measure the very weak magnetic field produced by some kinds of rock, they made a surprising discovery. The orientation of the dipole representing the magnetic field of an ancient rock does not necessarily point to the present-day magnetic pole, as a compass needle does **(Fig. 3.15a)**. Geologists interpreted this observation to mean that these rocks preserve *paleomagnetism*. They introduced the term **paleopole** to refer to the supposed position of the Earth's magnetic pole relative to the rock at the time in the past when the rock became magnetized **(Box 3.1)**.

Subsequent work revealed that paleomagnetism develops in rocks for different reasons. For example, in an igneous rock such as basalt (see Table 2.1), paleomagnetism forms when the rock solidifies and cools from melt. Why? In hot melt, thermal energy causes iron-bearing molecules to vibrate, spin, and move about randomly, so at any given instant, each iron-bearing molecule's dipole points in a different direction. The fields produced by these differently oriented dipoles cancel one another out. As the melt cools and solid rock starts to form, molecular motion slows down, so the dipoles, like tiny compass needles, align with the Earth's magnetic field. Finally, when crystals of solid magnetite grow and cool, the dipoles lock into alignment, yielding a measureable magnetic field that won't change as long as the rock remains sufficiently cool **(Fig. 3.15b)**. In a sedimentary rock, paleomagnetism can form when magnetic minerals grow in the spaces between

BOX 3.1 CONSIDER THIS . . .

Finding Paleopoles

How do you find a paleopole position from the orientation of a paleomagnetic dipole in a rock sample? The horizontal projection of the dipole arrow on the Earth's surface (picture the projection as the shadow cast on the Earth's surface by the arrow if the Sun were directly overhead) is like a compass needle that points to the paleopole; that is, the projection defines an imaginary great circle around the Earth that passes through the paleopole and the sample (Fig. Bx3.1). This great circle is like an imaginary "paleolongitude" line. Note that when drawing the circle, we assume that the declination at the time the sample was magnetized equals 0, because we assume that, averaged over time, the magnetic pole coincides with the geographic pole.

To find the specific position of the paleopole on this great circle, we must look at the inclination of the paleomagnetic dipole in the rock. Recall that inclination depends on latitude (see Fig. 3.14d). So, the inclination of the paleomagnetic dipole defines the paleolatitude of the sample with respect to the paleopole, and paleolatitude simply represents the distance (measured in degrees) from the pole along the great circle to where the sample formed.

FIGURE Bx3.1 How to find a paleopole from a paleomagnetic measurement.

(a) In a sample of rock (represented by the cube), the paleomagnetic dipole arrow has a declination, D, in the horizontal plane, and an inclination, I, in the vertical plane.

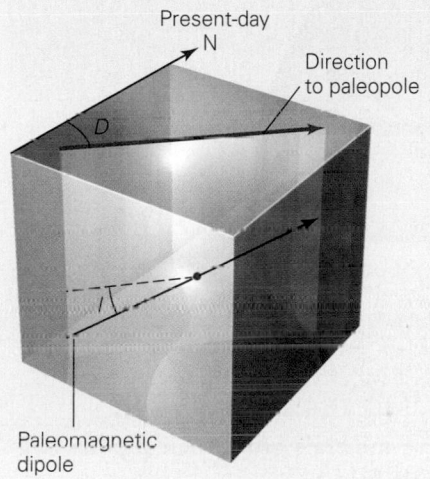

(b) The D of the sample points to the paleopole, P. So, the location of the rock outcrop and P lie on a circumference that represents a line of paleolongitude. The inclination indicates the paleolatitude of the sample and can be represented as a distance measured along the paleolongitude.

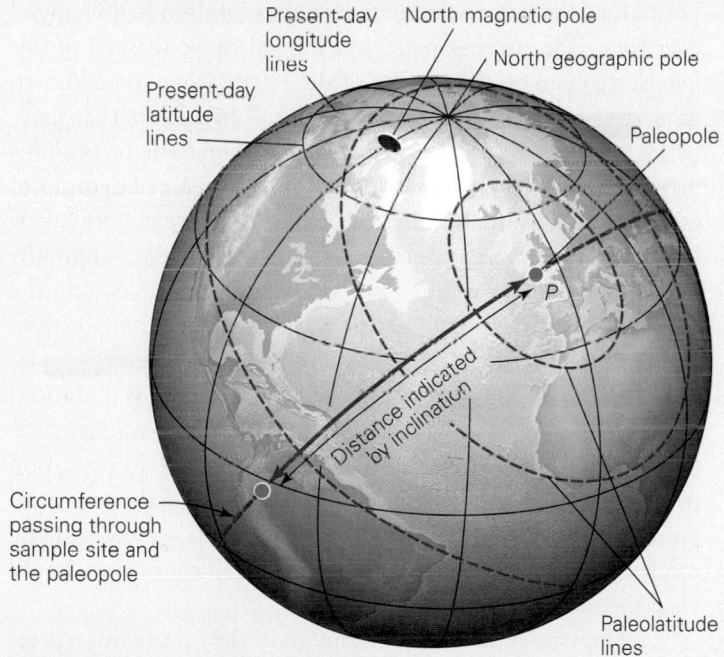

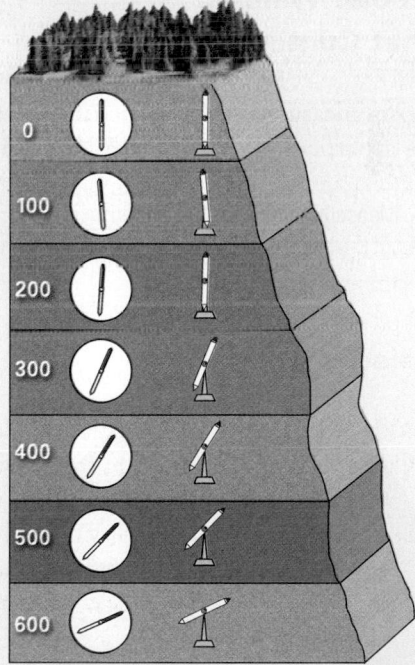

(c) Paleomagnetic inclination and declination change in a succession of rock layers at a given location. Numbers indicate the layer's age in millions of years.

FIGURE 3.15 Paleomagnetism and how it can form.

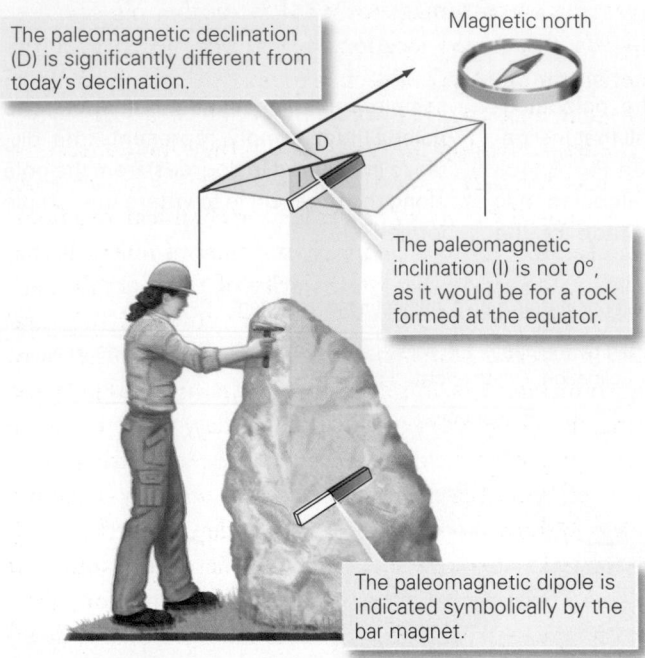

The paleomagnetic declination (D) is significantly different from today's declination.

Magnetic north

The paleomagnetic inclination (I) is not 0°, as it would be for a rock formed at the equator.

The paleomagnetic dipole is indicated symbolically by the bar magnet.

(a) A geologist finds an ancient rock sample at a location on the equator, where declination today is 0°. The orientation of the rock's paleomagnetism is different from that of today's field.

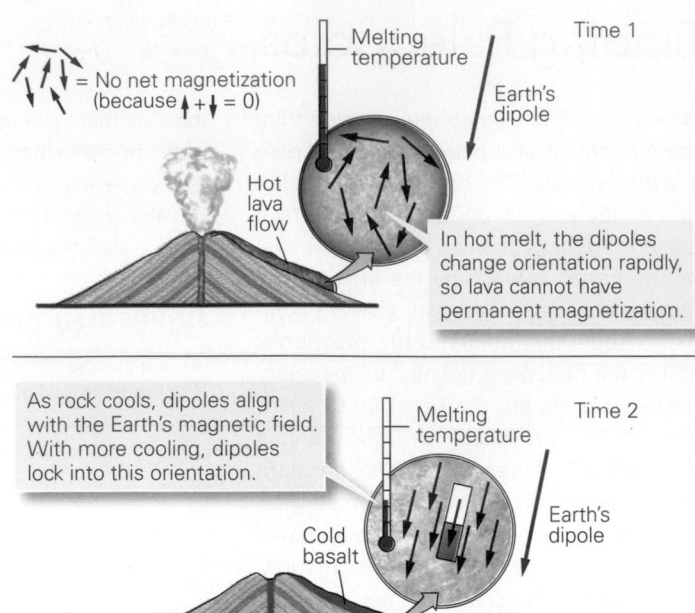

= No net magnetization (because ↑ + ↓ = 0)

Time 1

Melting temperature

Earth's dipole

Hot lava flow

In hot melt, the dipoles change orientation rapidly, so lava cannot have permanent magnetization.

As rock cools, dipoles align with the Earth's magnetic field. With more cooling, dipoles lock into this orientation.

Melting temperature

Time 2

Earth's dipole

Cold basalt

(b) Paleomagnetism can form when melt cools and becomes solid rock.

sediment grains; as these minerals grow, their dipoles align with the Earth's magnetic field.

Apparent Polar Wander: A Proof That Continents Move

When geologists measured paleomagnetism in rocks of different ages at a particular location, they discovered that the

FIGURE 3.16 Measurements of paleomagnetism in a succession of rock layers on a continent (x) define an apparent polar-wander path relative to the continent.

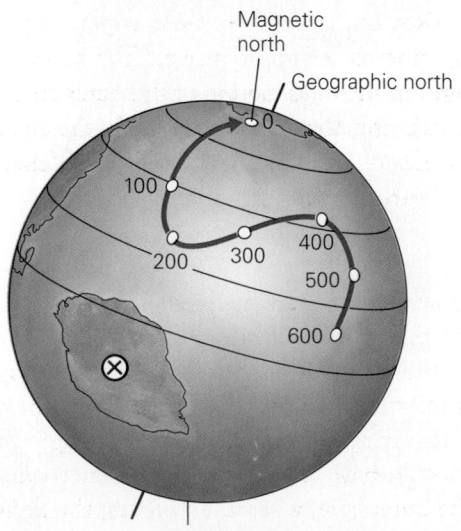

Magnetic north

Geographic north

0

100

200

300

400

500

600

position of the paleopole relative to that location had changed over time. The curving path that the paleopole seemed to follow, as mapped on the surface of the Earth, came to be known as an **apparent polar-wander path** (Fig. 3.16). (Note that these paths represent movements over tens to hundreds of millions of years, not the random movements on the scale of decades to centuries that we mentioned earlier.) When researchers calculated the first apparent polar-wander path, Wegener's hypothesis of continental drift had not yet been widely accepted. The researchers assumed that continents remained fixed in position, and therefore, that the Earth's magnetic poles actually moved long distances over time. Additional work, however, showed that each continent has a different apparent polar-wander path (Fig. 3.17). The concept that continents stay fixed in position over time cannot explain this observation, for if the magnetic poles really moved relative to fixed continents, paleomagnetic measurements in successions of rocks from all continents should yield the same apparent polar-wander path.

Geologists suddenly realized that they were interpreting apparent polar-wander paths backward. It's not the magnetic poles that move relative to fixed continents; rather it's the continents that move relative to fixed magnetic poles (Fig. 3.18). Because each continent has its own unique apparent polar-wander path, the continents must be moving relative to one another. Therefore, the discovery that different continents have different apparent polar-wander paths proved that Wegener was right all along—continents do drift!

FIGURE 3.17 Apparent polar-wander paths, as viewed looking down on the North Pole. If the continents were fixed in position and the poles really wandered, then all continents would display the same paths. But, they do not.

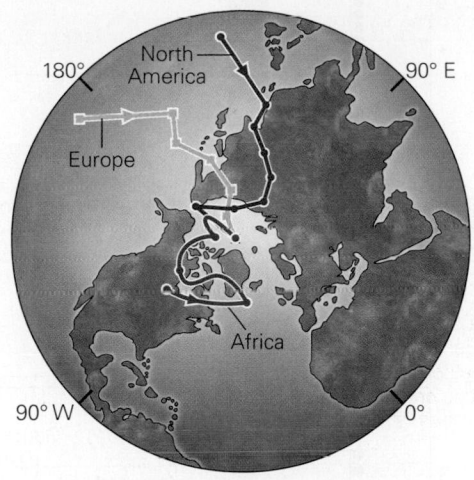

FIGURE 3.18 The alternate interpretations of apparent polar-wander paths.

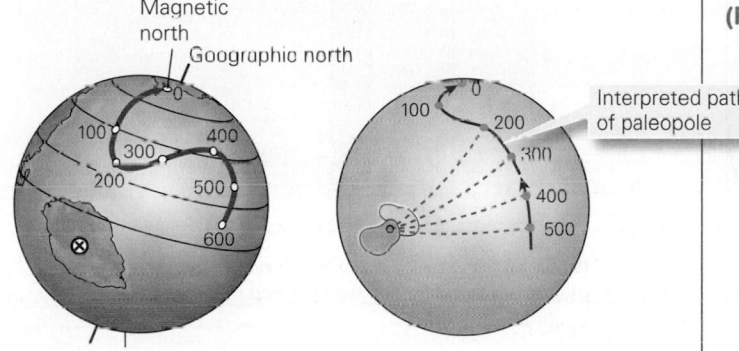

(a) If one assumes that the continent is fixed, then the paleopole follows a long path, ending up at the modern pole.

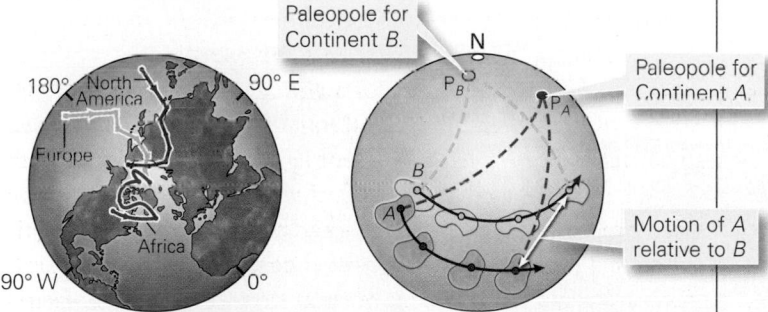

(b) If one assumes that the magnetic pole is fixed (at N), then the continents move. Each follows a path around a different fixed paleopole. Over time, the continents move relative to one another.

Marine Magnetic Anomalies: A Proof of Seafloor Spreading

Discovering Magnetic Reversals Through human history, the north-seeking end of a compass needle has always pointed toward the north magnetic pole, and geologists once assumed that it had done so for all of geologic time. But research on paleomagnetism led to another big surprise. Geologists discovered locations where the polarity of the magnetism in one layer of rock is opposite to the polarity of the adjacent layer. In other words, the dipole measured in one layer points in the opposite direction to the dipole measured in the next layer. At first, they thought that such *reversals* could be caused by lightning strikes or chemical reactions. But eventually, they found enough examples of reversals that they had to conclude that the polarity of the Earth's magnetic field itself really does reverse every now and then and that such reversals can happen quickly. Put another way, during some intervals of time, the Earth has **normal polarity**, meaning the same polarity that it has today, with the north magnetic pole near the north geographic pole. During other intervals of time, the Earth has **reversed polarity**, meaning that its polarity is opposite to that of today, with the north magnetic pole near the south geographic pole. Note that when a polarity reversal takes place, only the direction of the dipole changes—the Earth itself does not turn upside down **(Fig. 3.19a)**.

At first, researchers had no way of knowing when these reversals took place. This situation changed when geologists developed techniques for measuring the *numerical ages* of rocks, meaning the age of rocks in years (see Chapter 12). By recording the paleomagnetic polarity preserved in a succession of basalt layers, each of which had been dated numerically, a group of researchers established a **magnetic-reversal chronology**, a chart showing when reversals happened and, therefore, how much time occurred between reversals **(Fig. 3.19b, c)**. Their initial chart extended back 4.5 million years. The researchers referred to a major interval of a given polarity as a **chron**. Chrons typically include short-duration intervals of opposite polarity known as **subchrons**. They named chrons after famous researchers and subchrons after localities.

The question of why reversals take place still puzzles geologists, but researchers using supercomputer models are getting closer to an answer. The models show that changes in the fluid motion of the outer core can trigger reversals, and that during reversals the magnetic field first weakens and becomes complicated (chaotic) before reconfiguring with a different polarity **(Fig. 3.20)**.

Interpreting "Seafloor Stripes" During the years when some researchers were figuring out the magnetic-reversal chronology, others were measuring subtle variations in the strength of the magnetic field with location. Using a *magnetometer*, an instrument that measures magnetic field strength, they found that in some localities, the field has greater strength than they would have expected to be produced just by circulation in the outer core, whereas in others, the field is slightly

FIGURE 3.19 Magnetic polarity reversals and the chronology of reversals.

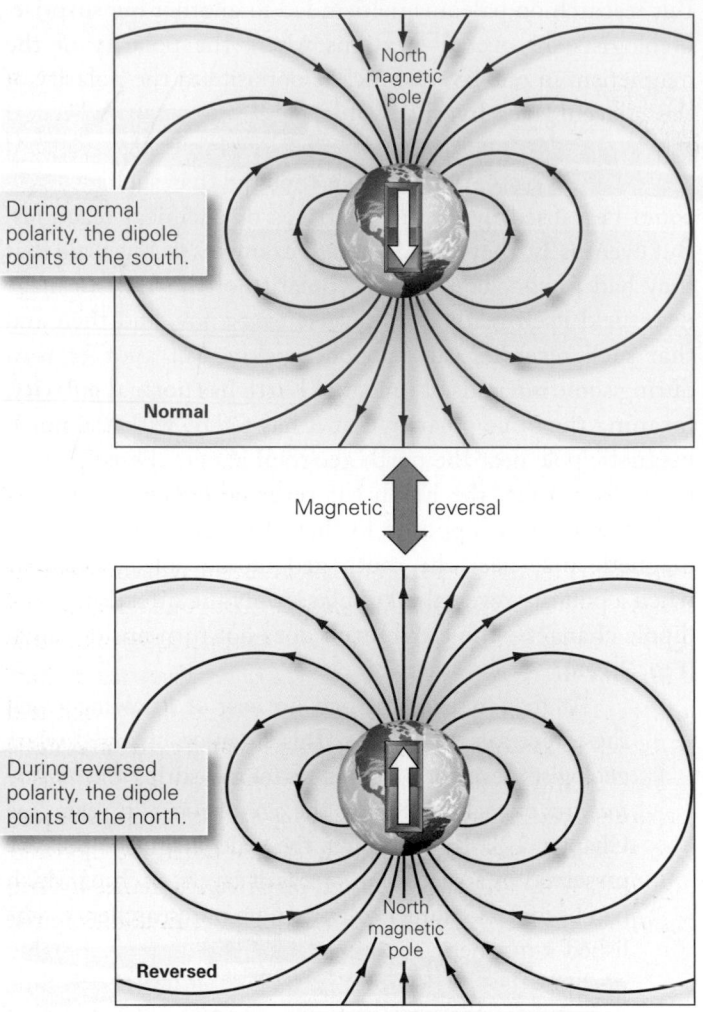

During normal polarity, the dipole points to the south.

Normal

Magnetic reversal

During reversed polarity, the dipole points to the north.

Reversed

(a) Geologists proposed that the Earth's magnetic field reverses polarity every now and then.

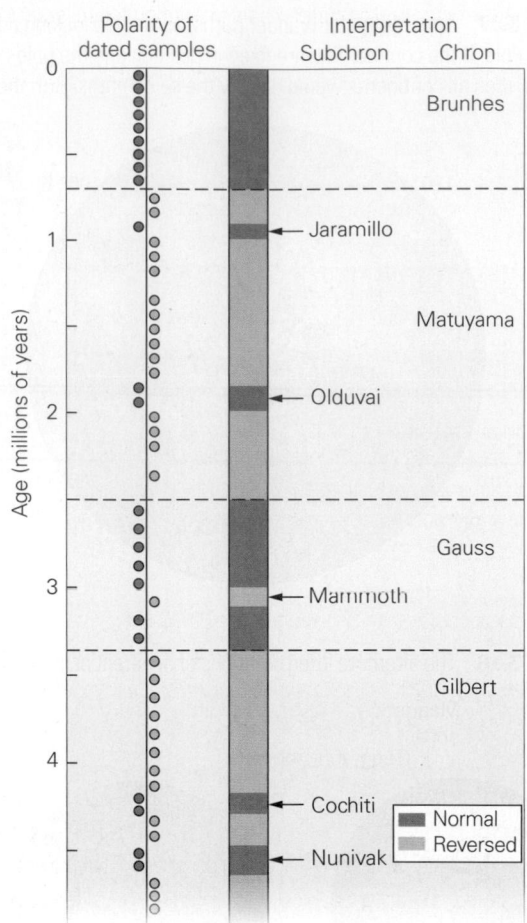

(b) Observations of such layers led to the production of a reversal chronology with named polarity intervals, or chrons.

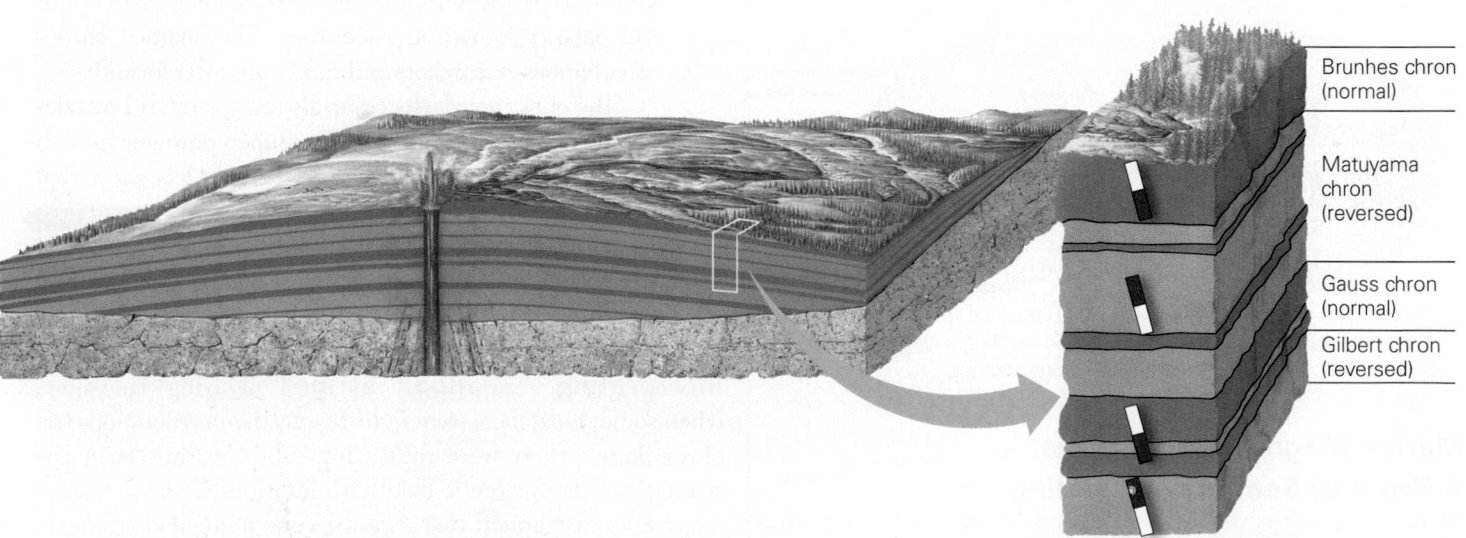

(c) Successive layers of lava build at a volcano over time. Some layers have normal polarity, whereas others have reversed polarity. The column identifies the chrons.

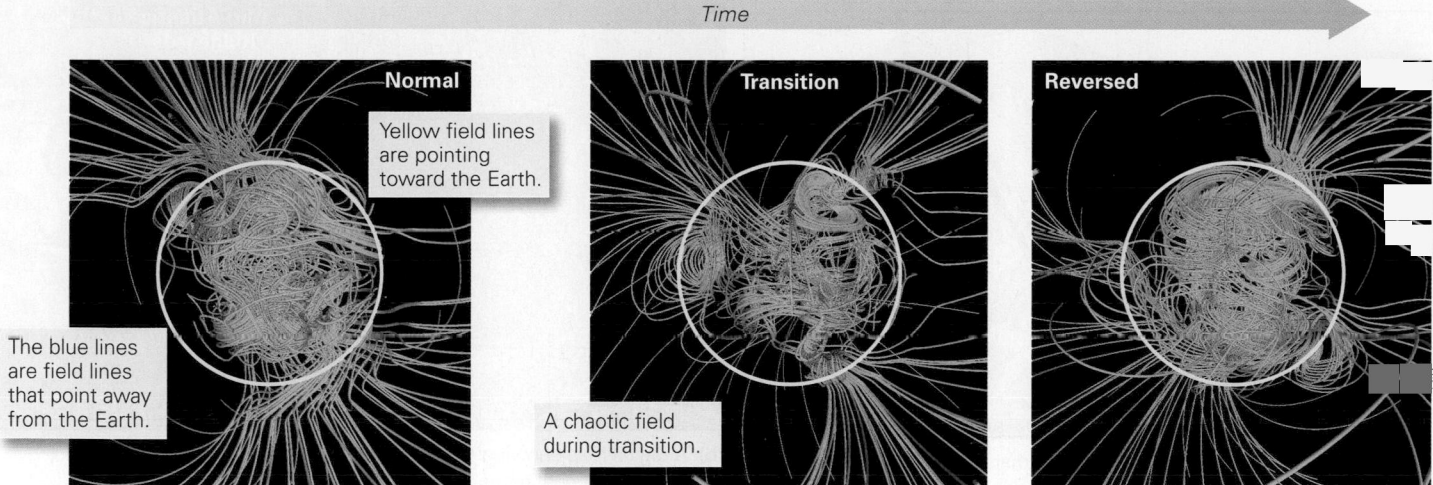

Time

weaker than expected. Researchers coined the term **magnetic anomaly** for the difference between the expected strength of the Earth's field at a location and its observed strength at that location. Field strengths greater than expected are *positive anomalies*, whereas field strengths less than expected are *negative anomalies*.

On land, the pattern of magnetic anomalies depends primarily on the composition of rock below the ground. For example, rocks that contain more iron tend to be more magnetic, so a magnetometer placed over them records a positive anomaly. As we'll discuss later in this book, continents contain

many different rock types, and rock bodies may have irregular shapes. As a result, magnetic anomalies recorded on land tend to have irregular shapes. In contrast, the upper layer of oceanic crust consists almost entirely of basalt, so when researchers attached magnetometers to ships and towed them back and forth across the oceans, they expected that **marine magnetic anomalies** might look different from those on land. The difference turned out to be astounding **(Fig. 3.21a, b)**. The pattern of marine magnetic anomalies, when portrayed by coloring positive anomalies darker and negative anomalies lighter, resembles the stripes on a zebra **(Fig. 3.21c)**. In addition, the

FIGURE 3.21 The discovery of marine magnetic anomalies.

(a) A ship towing a magnetometer detects changes in the strength of the magnetic field on the seafloor.

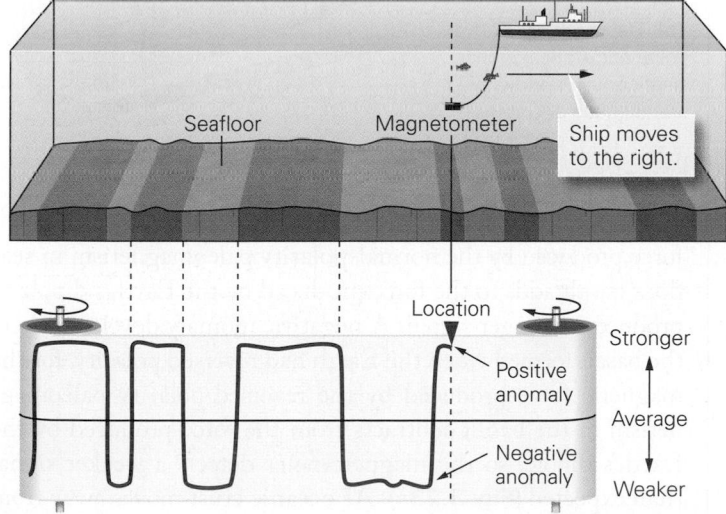

(b) On a paper record, intervals of stronger magnetism (positive anomalies) alternate with intervals of weaker magnetism (negative anomalies).

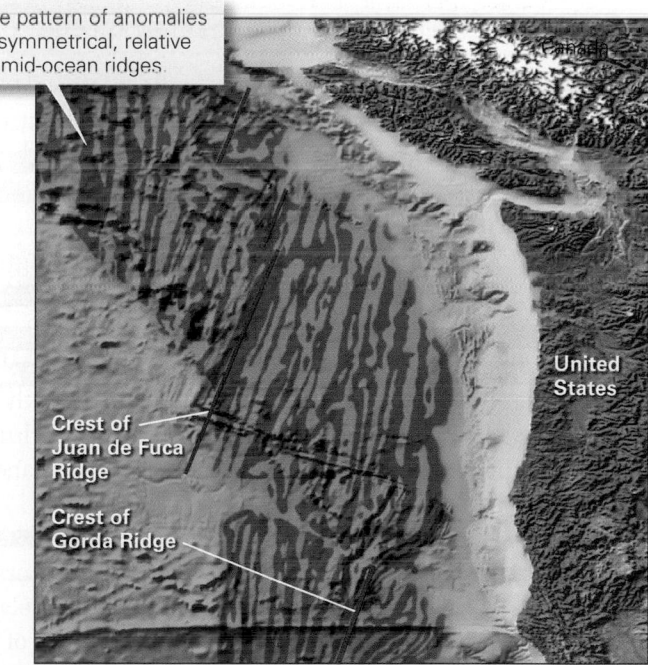

(c) A map showing areas of positive anomalies (dark) and negative anomalies (light) off the west coast of North America. The pattern of anomalies resembles candy-cane stripes.

FIGURE 3.22 Symmetry in the patterns of magnetic anomalies.

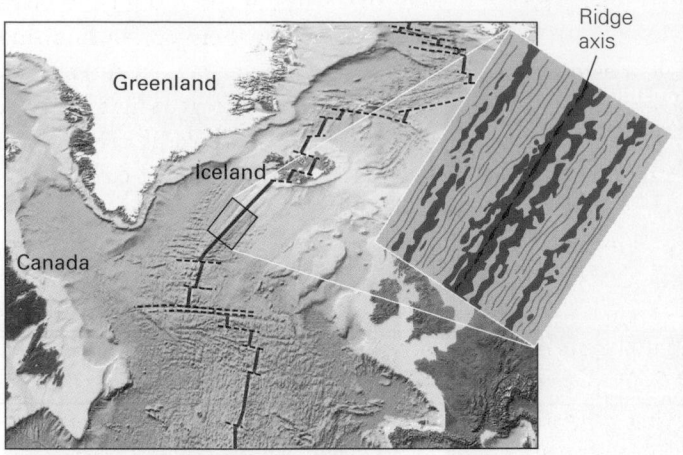

(a) The seafloor-spreading model predicts that magnetic anomalies are symmetrical relative to the mid-ocean ridge.

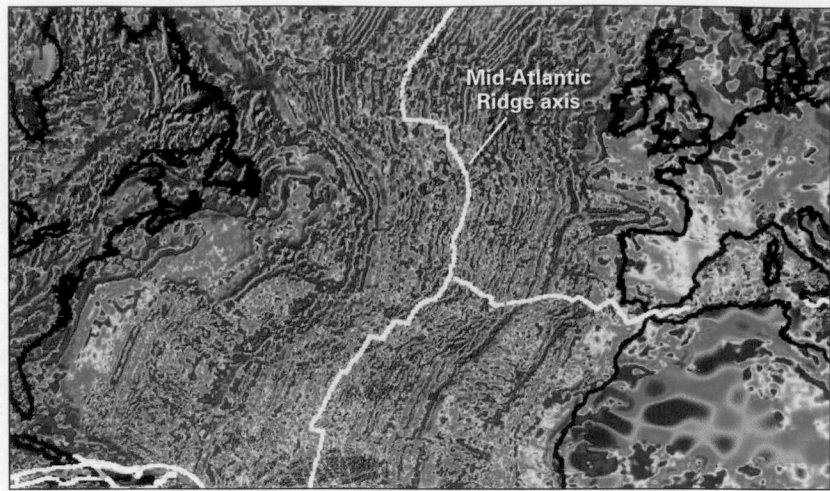

(b) A digital map showing the magnetic anomalies of the North Atlantic, and of adjacent continents. Note the striped pattern of the seafloor. (Anomalies on land don't show this pattern because they are controlled by the distribution of different rock types.)
Source: Korhonen, et al., 2007, © CCGM-CGMW.

FIGURE 3.23 The progressive development of magnetic anomalies and the long-term reversal chronology.

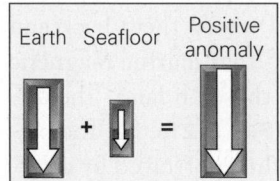 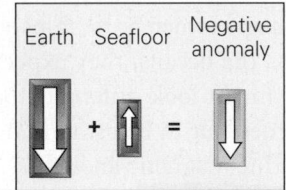

(a) Positive anomalies form when seafloor rock has the same polarity as the present magnetic field. Negative anomalies form when seafloor rock has polarity that is opposite to the present field.

The anomaly pattern represents alternating stripes of normal-polarity and reversed-polarity seafloor.

Positive

Negative

Earth's field

(b) The width of magnetic stripes on the seafloor is proportional to the duration of chrons.

Brunhes   Matuyama   Gauss   Gilbert

pattern isn't random. Not only do these "seafloor stripes" trend parallel to the axis of the nearest mid-ocean ridge, but they also vary in width, and the pattern of stripes on one side of a mid-ocean ridge looks like the mirror image of the pattern on the other side **(Fig. 3.22)**.

At first, researchers were baffled by marine magnetic anomalies. Eventually, however, they realized that the pattern of stripes close to a mid-ocean ridge resembles the pattern of chrons and subchrons in the magnetic-reversal chronology chart, in that the relative widths of the stripes are exactly the same as the relative durations of the chrons and subchrons. What does the stripe-like pattern of marine magnetic anomalies mean?

In 1963, researchers came to the realization that the striped pattern develops because magnetic reversals are occurring while seafloor spreading takes place. Specifically, the polarity of basalt in the oceanic crust depends on the polarity of the Earth's magnetic field at the time the crust forms. A positive anomaly develops where the basalt making up the seafloor formed when the Earth had normal polarity. The magnetic

force produced by the normal-polarity paleomagnetism in seafloor basalt adds to the force produced by the Earth's dipole to produce a stronger signal. A negative anomaly develops where the basalt formed when the Earth had reversed polarity, for the magnetic force produced by the reversed-polarity paleomagnetism in the basalt subtracts from the force produced by the Earth's dipole, so the magnetometer detects a weaker signal than expected **(Fig. 3.23a)**. As oceanic crust moves away from the ridge axis, it carries its paleomagnetic character with it. Seafloor spreading along a segment of ridge takes place at a fairly constant rate, so the relative widths of the stripes correspond to the relative durations of polarity intervals **(Fig. 3.23b)**.

FIGURE 3.24 The reversal chronology for the past 170 million years.

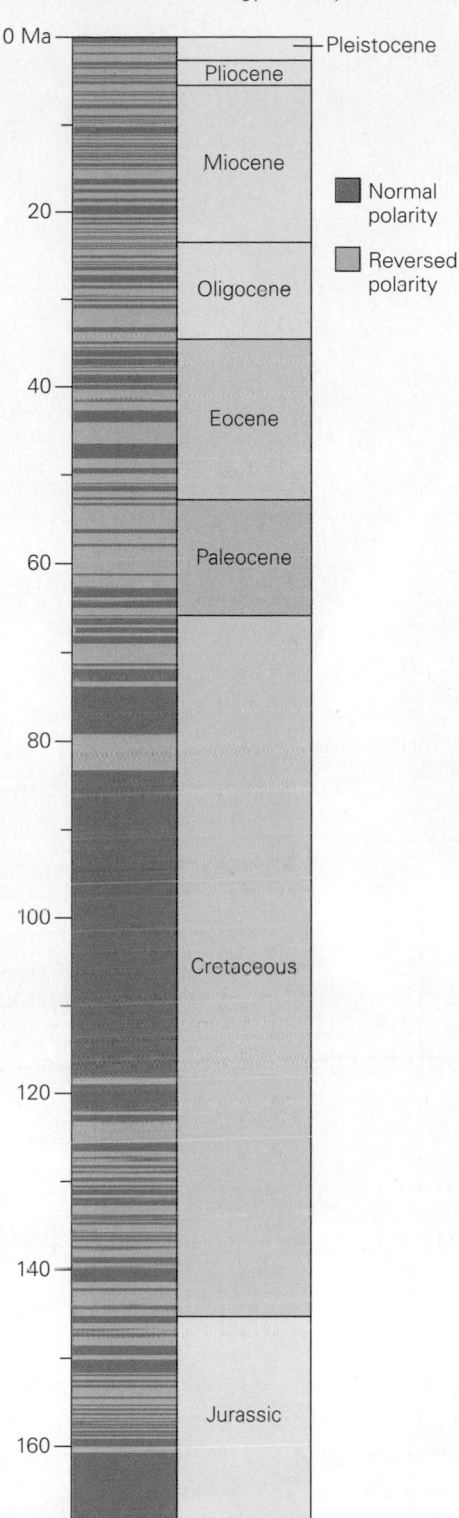

The discovery and explanation of marine magnetic anomalies now serves as a proof that seafloor spreading takes place (**Geology at a Glance**, pp. 86–87). Furthermore, by knowing the ages of the stripes produced during the last 4.5 million years, researchers can calculate the rate at which seafloor moves relative to the ridge axis. For example, if the distance from the ridge axis to the edge of the stripe that corresponds to the Gilbert chron is 80 km, then there has been 80 km (= 8 million cm) of movement in 3.6 million years, because the Gilbert chron began at 3.6 Ma. Remembering that Velocity = Distance ÷ Time, a quick calculation indicates that the seafloor has moved away from the ridge at a rate of 2.2 cm per year. By assuming that seafloor spreading along a given ridge takes place at a constant rate over a long period, geologists can define the timing of magnetic reversals for time intervals before 4.5 Ma (**Fig. 3.24**). In fact, by using this method, they eventually determined that the oldest seafloor currently on the Earth formed about 200 Ma.

So by the early 1960s, when the Beatles were topping the pop charts, it had become clear that Wegener had been right all along—continents do move. But, though the case for such movement had been greatly strengthened by the discovery of apparent polar-wander paths, it really took the proposal and proof of seafloor spreading to make believers of most geologists. Very quickly, as we will see in the next chapter, these ideas became the basis of the theory of plate tectonics.

TAKE-HOME MESSAGE

Certain rocks preserve paleomagnetism, a record of the past position of the Earth's magnetic poles, relative to a rock. By measuring paleomagnetism in rocks of different ages from various continents, geologists could show that over time, the movement of one continent, relative to the magnetic pole, is different from the movement of another continent. Therefore, continents must move relative to one another. This discovery proved continental drift. The Earth's magnetic field reverses polarity every now and then. As a result, alternating bands of normal and reversed polarity, causing positive and negative anomalies over the seafloor, develop as new oceanic crust forms at a mid-ocean ridge. The discovery of such marine magnetic anomalies proved that seafloor spreading is taking place.

QUICK QUESTION: Why do geologists insert the word "apparent" in front of "polar-wander path?"

Magnetic Reversals and Marine Magnetic Anomalies

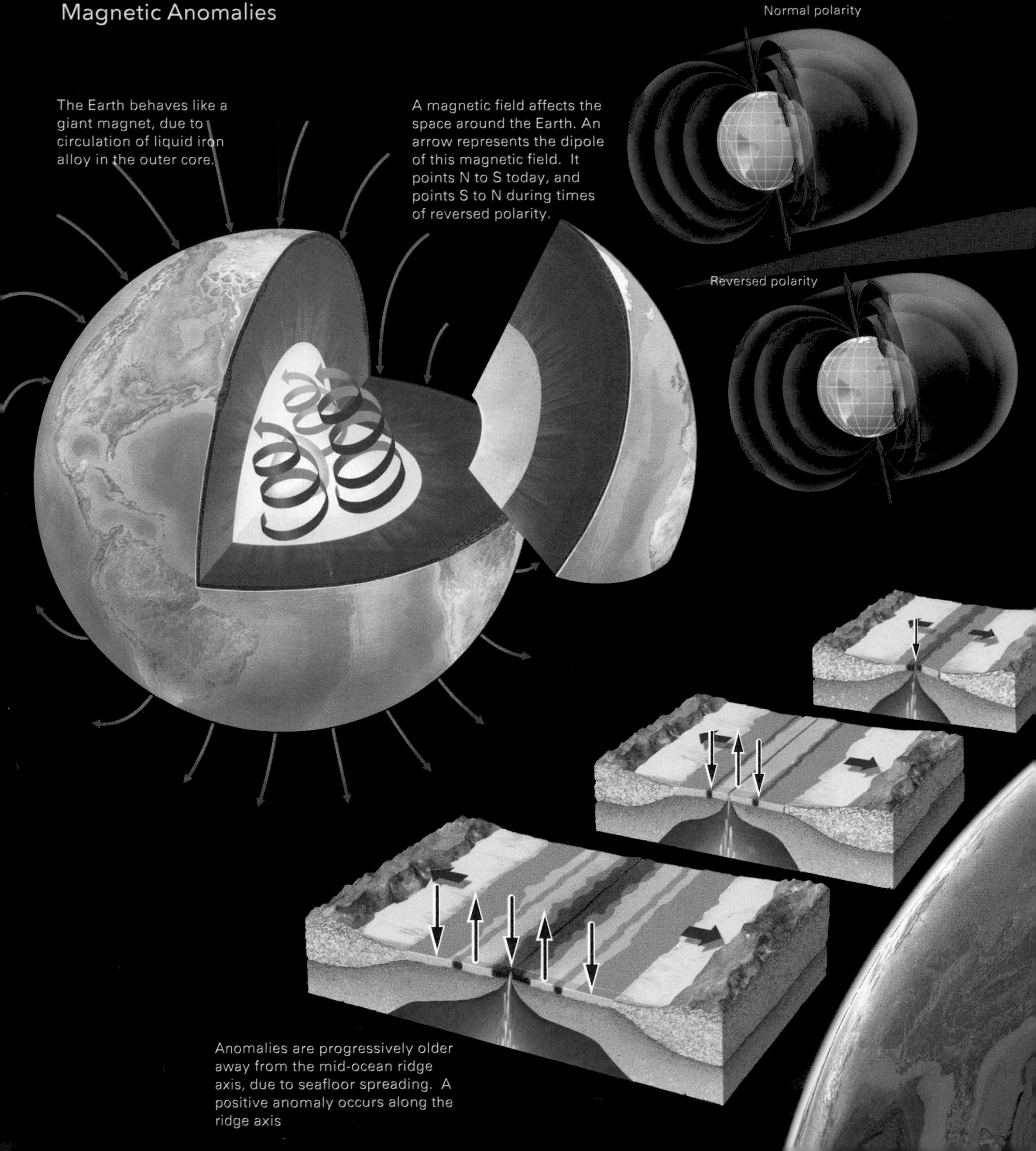

The Earth behaves like a giant magnet, due to circulation of liquid iron alloy in the outer core.

A magnetic field affects the space around the Earth. An arrow represents the dipole of this magnetic field. It points N to S today, and points S to N during times of reversed polarity.

Normal polarity

Reversed polarity

Anomalies are progressively older away from the mid-ocean ridge axis, due to seafloor spreading. A positive anomaly occurs along the ridge axis

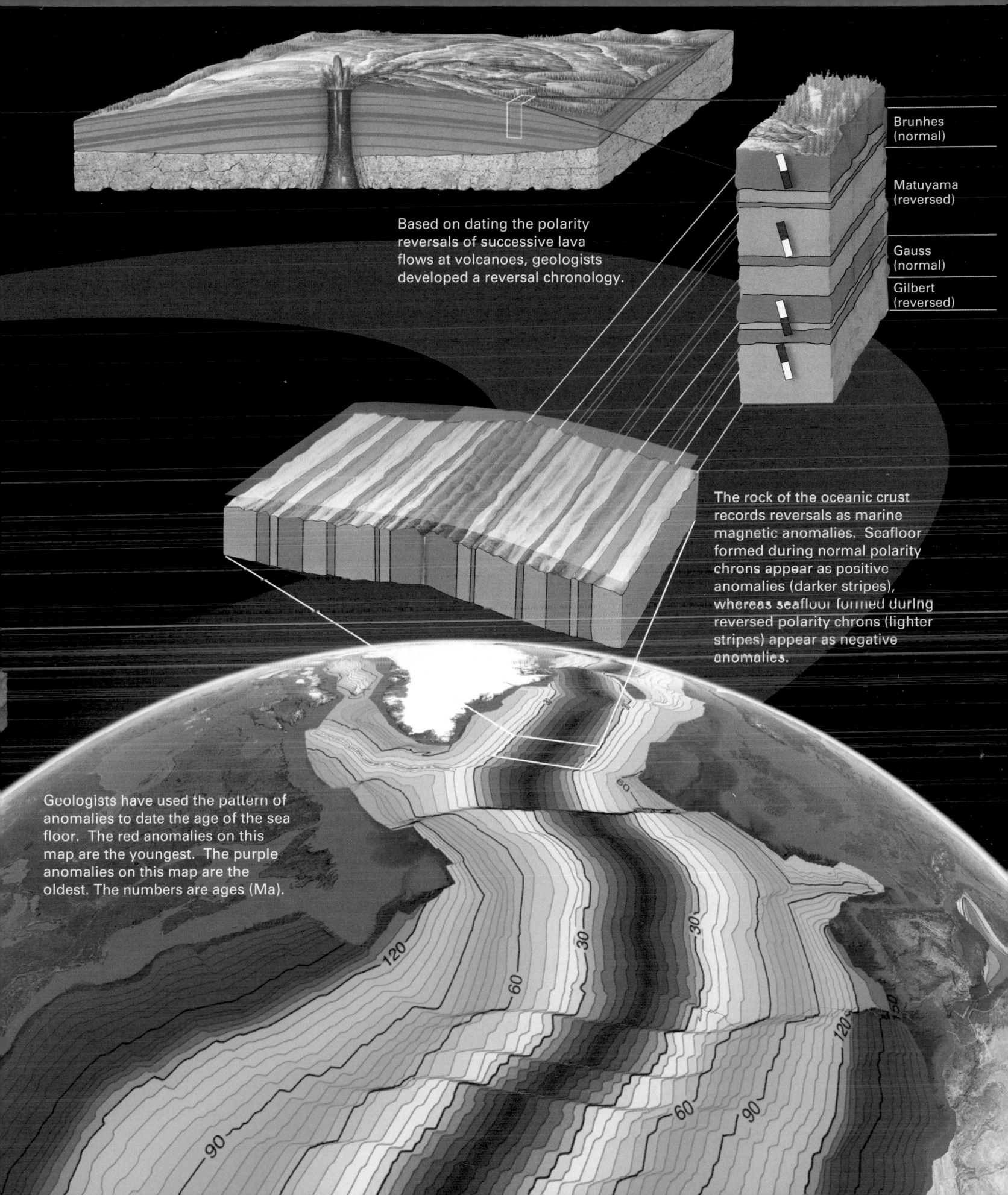

Based on dating the polarity reversals of successive lava flows at volcanoes, geologists developed a reversal chronology.

Brunhes (normal)

Matuyama (reversed)

Gauss (normal)

Gilbert (reversed)

The rock of the oceanic crust records reversals as marine magnetic anomalies. Seafloor formed during normal polarity chrons appear as positive anomalies (darker stripes), whereas seafloor formed during reversed polarity chrons (lighter stripes) appear as negative anomalies.

Geologists have used the pattern of anomalies to date the age of the sea floor. The red anomalies on this map are the youngest. The purple anomalies on this map are the oldest. The numbers are ages (Ma).

120

60

30

30

60

90

60

120

150

90

Chapter 3 Review

SUMMARY

- Alfred Wegener proposed that continents had once been joined together to form a single huge supercontinent (Pangaea) and had subsequently drifted apart. This idea is the continental-drift hypothesis.

- Wegener drew from several different sources of data to support his hypothesis: (1) coastlines on opposite sides of the ocean match up; (2) the distribution of late Paleozoic glaciers can be explained if the glaciers made up a polar ice cap over the southern end of Pangaea; (3) the distribution of late Paleozoic equatorial climatic belts is compatible with the concept of Pangaea; (4) the distribution of fossil species suggests the existence of a supercontinent; (5) distinctive rock assemblages that are now on opposite sides of the ocean were adjacent on Pangaea.

- Despite all the observations that supported continental drift, most geologists did not initially accept the idea because no one could explain how continents could move. It took decades of new data collection before the idea could be reconsidered.

- Rocks retain a record of the Earth's magnetic field as it existed at the time the rocks formed. This record is called paleomagnetism. By measuring paleomagnetism in successively older rocks, geologists found that the apparent position of the Earth's magnetic pole relative to the rocks changes through time. Successive positions of the pole define an apparent polar-wander path.

- Apparent polar-wander paths differ for different continents. This can be explained if continents move in respect to one another as the Earth's magnetic poles remain roughly fixed.

- The invention of sonar permitted explorers to make detailed maps of the seafloor. These maps revealed the existence of mid-ocean ridges, deep-ocean trenches, seamount chains, and fracture zones. Other measurements showed that heat flow is generally greater near the axis of a mid-ocean ridge.

- Hess's hypothesis of seafloor spreading states that new seafloor forms at mid-ocean ridges, above a band of upwelling mantle, then spreads symmetrically away from the ridge axis. Therefore, an ocean basin grows wider with time, and continents on either side drift apart. Eventually, the ocean floor sinks back into the mantle at deep-ocean trenches.

- Magnetometer surveys of the seafloor revealed marine magnetic anomalies. Positive anomalies (magnetic field strength is greater than expected) and negative anomalies (magnetic field strength is less than expected) are arranged in alternating stripes.

- During the 1950s, geologists documented that the Earth's magnetic field reverses polarity every now and then. The record of reversals, dated by isotopic techniques, is called the magnetic-reversal chronology.

- A proof of seafloor spreading came from the interpretation of marine magnetic anomalies. Seafloor that forms when the Earth has normal polarity results in positive anomalies, whereas seafloor that forms when the Earth has reversed polarity results in negative anomalies. Anomalies are symmetric with respect to a mid-ocean ridge axis, and their widths are proportional to the duration of polarity chrons. Study of anomalies allows us to calculate the rate of spreading.

- Drilling of the seafloor confirmed that its age increases away from the mid-ocean ridge axis, proving seafloor spreading.

GUIDE TERMS

abyssal plain (p. 72)

apparent polar-wander path (p. 80)

bathymetry (p. 72)

chron (p. 81)

continental drift (p. 68)

fracture zone (p. 74)

heat flow (p. 75)

magnetic anomaly (pp. 81–83)

magnetic declination (p. 77)

magnetic inclination (p. 77)

magnetic-reversal chronology (p. 81)

marine magnetic anomaly (p. 83)

mid-ocean ridge (p. 72)

normal polarity (p. 81)

paleomagnetism (p. 77)

paleopole (p. 78)

Pangaea (p. 67)

reversed polarity (p. 81)

ridge axis (p. 74)

seafloor spreading (p. 76)

seamount (p. 74)

seismic belt (p. 75)

subchron (p. 81)

subduction (p. 76)

supercontinent (p. 67)

trench (p. 74)

volcanic arc (p. 74)

GEOTOURS *THIS CHAPTER'S GEOTOURS WORKSHEET (B) FEATURES QUESTIONS AND GOOGLE EARTH SITES ON:*

- Seafloor spreading
- Tectonic plate boundaries

REVIEW QUESTIONS

The letters following each Review Question refer to the corresponding Learning Objective from the Chapter Opener.

1. What was Wegener's continental-drift hypothesis? **(A)**

2. How does the fit of the coastlines around the Atlantic support continental drift? **(B)**

3. Explain the distribution of late Paleozoic glaciation. **(B)**

4. How does the distribution of climatic belts support continental drift? **(B)**

5. Was it possible for a dinosaur to walk from New York to Paris when Pangaea existed? Explain your answer. **(A)**

6. Why were geologists initially skeptical of Wegener's continental-drift hypothesis? **(A)**

7. Describe the basic bathymetric characteristics of mid-ocean ridges, deep-ocean trenches, and seamount chains. **(C)**

8. Describe the hypothesis of seafloor spreading. **(C)**

9. What is paleomagnetism and how does it form? What does the red arrow in the figure represent? **(D)**

10. Describe how the angle of inclination of the Earth's magnetic field varies with latitude. How can paleomagnetic inclination be used to determine the ancient latitude of a continent? **(D)**

11. How did the observations of heat flow and seismicity support the hypothesis of seafloor spreading? **(E)**

12. What is a magnetic reversal? **(D)**

13. What is a marine magnetic anomaly? How is it detected? **(D)**

14. What do the pink steps in the figure represent? Why aren't they all the same width? **(E)**

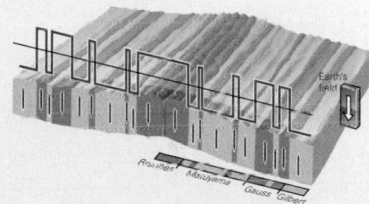

15. How do geologists calculate rates of seafloor spreading? **(C)**

16. Did drilling into the seafloor contribute further proof of seafloor spreading? If so, how? **(E)**

ON FURTHER THOUGHT

17. The geologic record suggests that when supercontinents break up, a pulse of rapid evolution, with many new species appearing and many existing species becoming extinct, takes place. Why might this be? (Hint: Consider how the environment, both global and local, might change as a result of breakup, and keep in mind the widely held idea that competition for resources drives evolution.) **(A)**

18. Why are the marine magnetic anomalies bordering the East Pacific Rise in the southeastern Pacific Ocean wider than those bordering the Mid-Atlantic Ridge in the South Atlantic Ocean? **(D)**

ONLINE RESOURCES

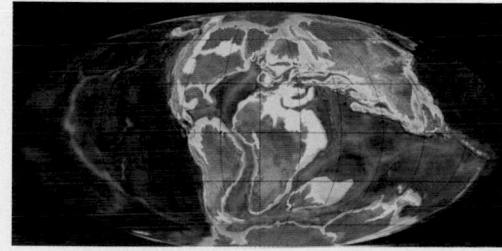

Videos
This chapter features videos on the breakup of Pangaea and subduction trenches.

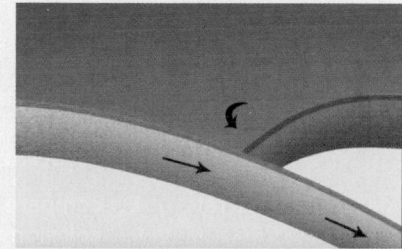

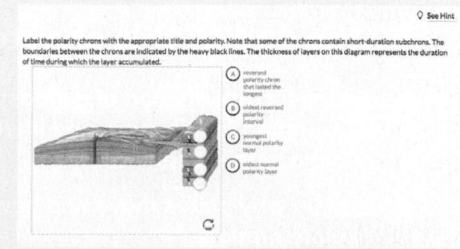

Smartwork5
This chapter features questions on topics including seafloor spreading and paleomagnetism.

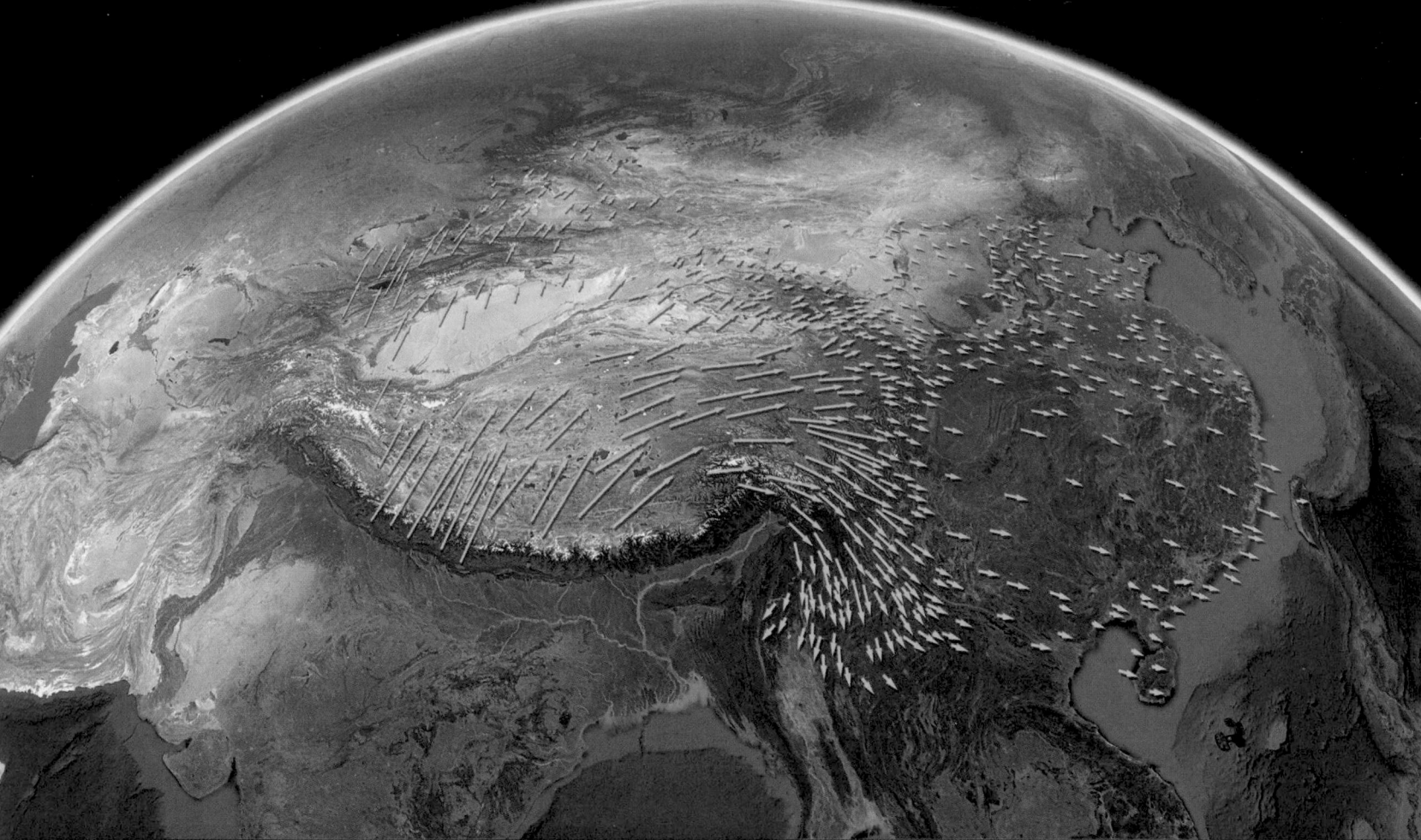

CHAPTER 4

The Way the Earth Works: Plate Tectonics

By the end of this chapter, you should be able to . . .

A. sketch a cross section of a lithosphere plate that shows its variation in thickness and its relationship to the crust and the asthenosphere.

B. explain why plates move and how to determine the rate of plate motion.

C. use a map showing the distribution of earthquakes to locate plate boundaries, and identify continental margins that are not plate boundaries.

D. distinguish among the three types of plate boundaries, and characterize the assemblage of geologic features associated with each type.

E. compare a rift with a mid-ocean ridge, and a continental collision zone with a subduction zone.

F. provide a model explaining the concept and consequences of a hot spot.

> If you start any large theory, such as quantum mechanics, plate tectonics, evolution,
> it takes about 40 years for mainstream science to come around.
>
> —James Lovelock (British scientist, born 1919)

4.1 Introduction

Thomas Kuhn, an influential historian of science, argued that scientific thinking advances in distinct steps. According to Kuhn, scientists mold their interpretations of the natural world to an established line of reasoning—a *paradigm*—that can remain unchanged for a long time until an inspired thinker proposes a radically new idea that works better and forms the basis of a new paradigm. Almost immediately, the broader scientific community scraps old ideas and formulates new ones consistent with the new idea. Kuhn called such abrupt changes in thinking a **scientific revolution**. Such a revolution happened in biology when Darwin proposed the theory of evolution by natural selection, and in physics, when Einstein proposed the theory of relativity.

In geology, a scientific revolution took place following the proposal of the theory of plate tectonics (often called simply **plate tectonics**), which states that the outer layer of the Earth, the lithosphere, consists of separate pieces, or plates, that move with respect to one another. This theory required geoscientists to cast aside interpretations rooted in the older paradigm of fixed continents and led to a complete restructuring of how geologists think about Earth history. Compare this book with a geology textbook from the 1950s, and you will instantly see the difference.

Alfred Wegener planted the seed of plate tectonics theory with his proposal of continental drift in 1915. But this seed lay dormant for 45 years, until discoveries about the ocean floor and about apparent polar wander set the stage for Harry Hess to propose seafloor spreading. The realization a few years later that the pattern of marine magnetic anomalies served as proof of seafloor spreading led many researchers to drop what they'd been doing and turn their attention to examining the broad implications of seafloor spreading. By 1968, thanks primarily to the work of more than two dozen different researchers, the seafloor-spreading hypothesis had bloomed into plate tectonics theory. Researchers clarified the concept

◄ (facing page) India collided with Asia over 40 million years ago. As the continents squeezed together, the Himalayas and the Tibetan Plateau rose, yielding some of the most dramatic landscapes of our planet. Modern GPS measurements demonstrate that movements continue to take place today. Each arrow represents the orientation and speed of the underlying land surface, relative to a point in the interior of Asia.

of a plate, described the types of plate boundaries, calculated plate motions, related plate tectonics to earthquakes and volcanoes, showed how plate motions could generate mountain belts and seamount chains, and defined the history of past plate motions. After these researchers presented their ideas to standing-room-only audiences at conferences between 1968 and 1970, the geoscience community, with few exceptions, embraced plate tectonics as the basis of a new paradigm and has been building on it ever since. In fact, plate tectonics has become geology's grand unifying theory.

To begin our explanation of the key elements of plate tectonics theory, we characterize lithosphere plates and distinguish among the three types of plate boundaries. Next, we describe the nature of geologic activity that occurs at each boundary. We then look at hot spots and other special locations on plates and examine the breakup and collision of continents. Finally, we consider models proposed to explain why plates move and measurements that define rates of movement.

4.2 What Do We Mean By Plate Tectonics?

The Concept of a Lithosphere Plate

As we saw at the end of Chapter 2, the lithosphere, which consists of the crust plus the uppermost part of the upper mantle, behaves as a relatively hard layer, meaning that when a force pushes or pulls on it, it does not flow, but rather bends or breaks (Fig. 4.1). The lithosphere lies over a relatively soft layer called the asthenosphere, composed of mantle that can flow when acted on by force. As a result, the asthenosphere convects like water in a pot, though much more slowly (see Box 2.3). To highlight this difference, geologists say that the lithosphere is *rigid*, whereas the asthenosphere is *plastic*.

The base of the lithosphere lies in the upper mantle. In fact, both the mantle part of the lithosphere (known as the **lithospheric mantle**) and the asthenosphere consist of the same very dense ultramafic rock. The boundary between the two is controlled by temperature. Lithospheric mantle consists of mantle rock cooler than 1,280°C, for such rock behaves rigidly, whereas asthenosphere consists of mantle rock warmer than 1,280°C, for such rock behaves plastically. Keep in

FIGURE 4.1 The lithosphere is fairly rigid, but when a heavy load, such as a glacier or volcano, builds on its surface, the surface bends down. This can happen because underlying plastic asthenosphere can flow out of the way.

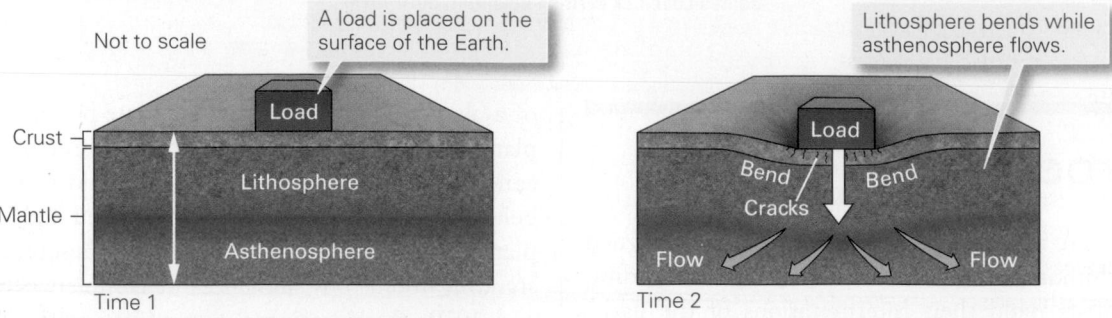

FIGURE 4.2 The differences between continental lithosphere and oceanic lithosphere.

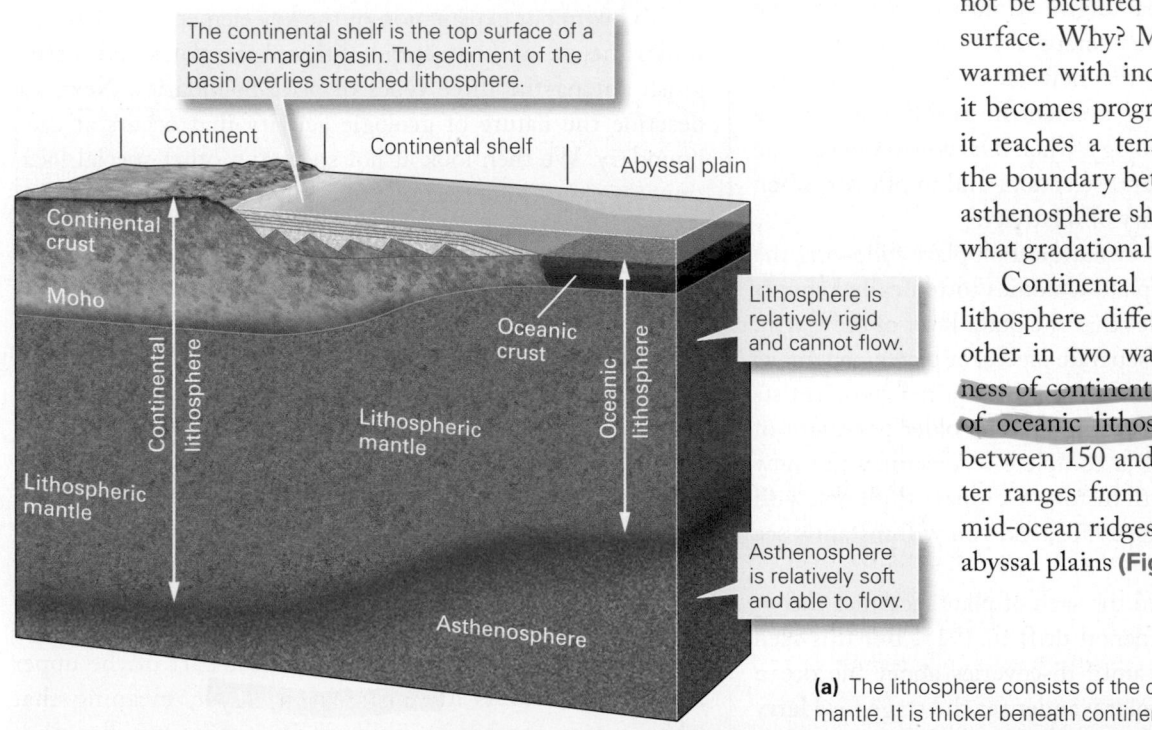

(a) The lithosphere consists of the crust plus the uppermost mantle. It is thicker beneath continents than beneath oceans.

mind, however, that the boundary should not be pictured as a smooth, hard planar surface. Why? Mantle rock gets gradually warmer with increasing depth. Therefore, it becomes progressively more plastic once it reaches a temperature of 1,280°C. So the boundary between the lithosphere and asthenosphere should be pictured as somewhat gradational.

Continental lithosphere and oceanic lithosphere differ significantly from each other in two ways. First, the total thickness of continental lithosphere exceeds that of oceanic lithosphere: the former ranges between 150 and 200 km, whereas the latter ranges from less than 10 km beneath mid-ocean ridges to about 100 km beneath abyssal plains **(Fig. 4.2)**. Second, the crustal

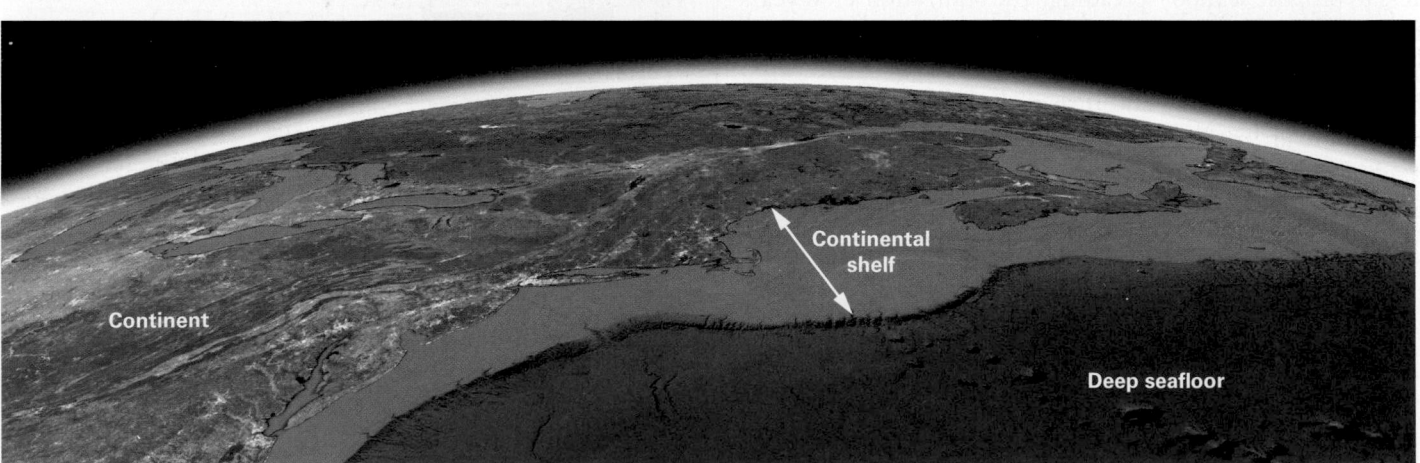

(b) An oblique view of the eastern coast of North America. Note the elevation difference between the land surface and the deep seafloor. This difference is due to contrasts between continental and oceanic lithosphere.

layer of continental lithosphere is thicker and less dense than that of oceanic lithosphere. Continental crust ranges from 25 to 70 km thick and consists of relatively low-density felsic and intermediate rock (see Table 2.1). In contrast, the crust of oceanic lithosphere is only 7 to 10 km thick and consists of dense mafic rock.

If you examine a cross section (a vertical slice) through the upper few hundred kilometers of the Earth, you'll see that the top of the lithosphere (the Earth's surface) varies in elevation, and that the base of the lithosphere varies in depth (see Fig. 4.2a). These differences reflect the thickness and composition of the crust as well as the thickness of the lithospheric mantle. Here, let's focus on how the nature of the crust affects the elevation of the lithosphere's top surface. Looking again at Figure 4.2a, note that the continental lithosphere, which includes a relatively thick layer of continental crust, has a higher surface elevation than does the oceanic lithosphere, which includes a relatively thin layer of oceanic crust (Fig. 4.3a). To picture why, let's make a simplistic model from rafts of wood (Fig. 4.3b). In this model, we use oak to represent very dense (ultramafic) lithospheric mantle, pine to represent dense (mafic) oceanic crust, and cork to represent less dense (felsic or intermediate) continental crust. Our model of continental lithosphere consists of a thick layer of cork over a thick layer of oak, and our model of oceanic lithosphere consists of a thin layer of pine over a thin layer of oak. When we place these rafts in a tub of water, the surface of the raft representing continental lithosphere sits higher than that of the raft representing oceanic lithosphere. (We discuss this phenomenon, which geologists refer to as *isostasy*, more completely in Interlude D and Chapter 11.)

Unlike the intact shell of a new chicken's egg, the lithosphere consists of about 20 discrete pieces. As noted earlier, we call the pieces **lithosphere plates**, or simply **plates**, and the contacts between them are **plate boundaries (Fig. 4.4a, b)**. Of the 20 or so plates, geoscientists consider about 12 of them to be *major plates* with relatively large areas, and the rest to be *microplates*. Some plates consist entirely of oceanic lithosphere, whereas others consist of both oceanic and continental lithosphere. Some plates have familiar names (the North American Plate, the African Plate), whereas others have less familiar names (the Cocos Plate, the Juan de Fuca Plate, the Nazca Plate).

While several plate boundaries coincide with a *continental margin*, the borderline between a continent and an ocean, many do not. For this reason, we distinguish between **active margins**, which are plate boundaries, and **passive margins**, which are not plate boundaries. Along a passive margin, older

FIGURE 4.3 Why is the land surface higher than the seafloor surface?

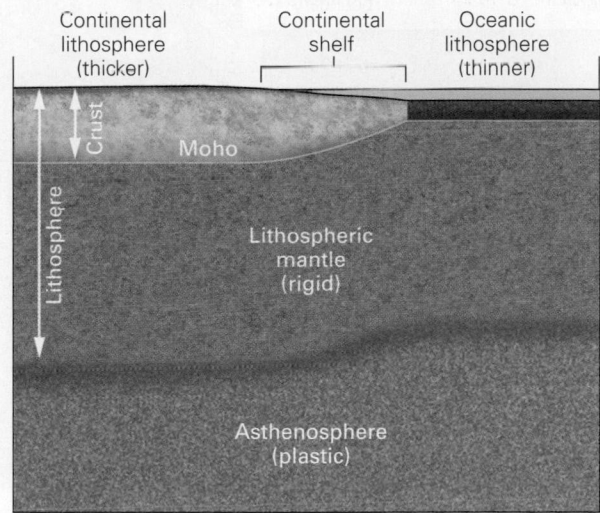

(a) The surface of continental lithosphere is higher than the surface of oceanic lithosphere because they have different crusts and, overall, different thicknesses.

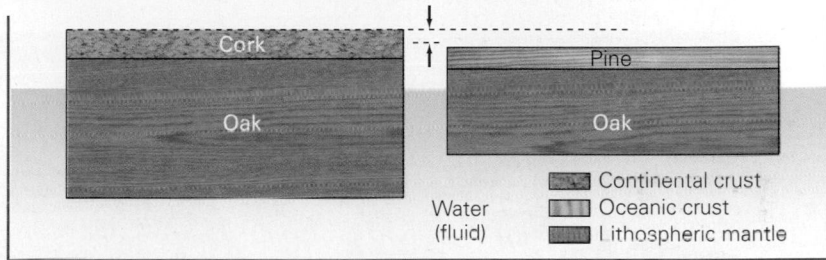

(b) A model provides insight. The model is not ideal because water is less dense than oak, but asthenosphere is less dense than lithospheric mantle.

continental crust is relatively thin and lies buried beneath a 10- to 15-km-thick accumulation of younger sediment. Geologists refer to this accumulation as a *passive-margin basin*. The surface of this sediment pile is a broad area of shallow seafloor (less than 500 m deep) called the *continental shelf* (see Fig. 4.2b), home to the major fisheries of the world.

The Basic Principles of Plate Tectonics Summarized

With the background provided above, we can summarize the basic principles of plate tectonics theory concisely as follows:

- The Earth's lithosphere is divided into plates that move relative to one another. As a plate moves, its internal area remains mostly, but not perfectly, rigid and intact.

- The motion of one plate relative to its neighbor takes place by slip along plate boundaries.

- Continents are parts of some plates, and as these plates move, the continents move with them. (This movement is the "drift" that Wegener recognized but couldn't explain.)

- Because of plate tectonics, the map of the Earth's surface constantly changes.

FIGURE 4.4 The locations of plate boundaries and the distribution of earthquakes.

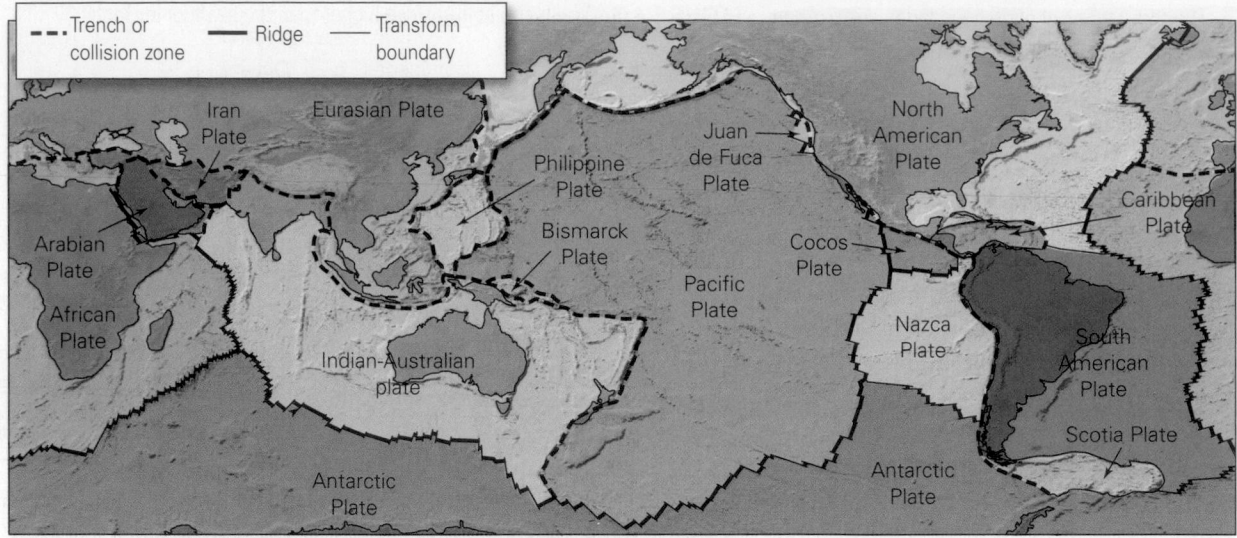

(a) A map of major plates shows that some consist entirely of oceanic lithosphere, whereas others consist of both continental and oceanic lithosphere. Active continental margins lie along plate boundaries; passive margins do not.

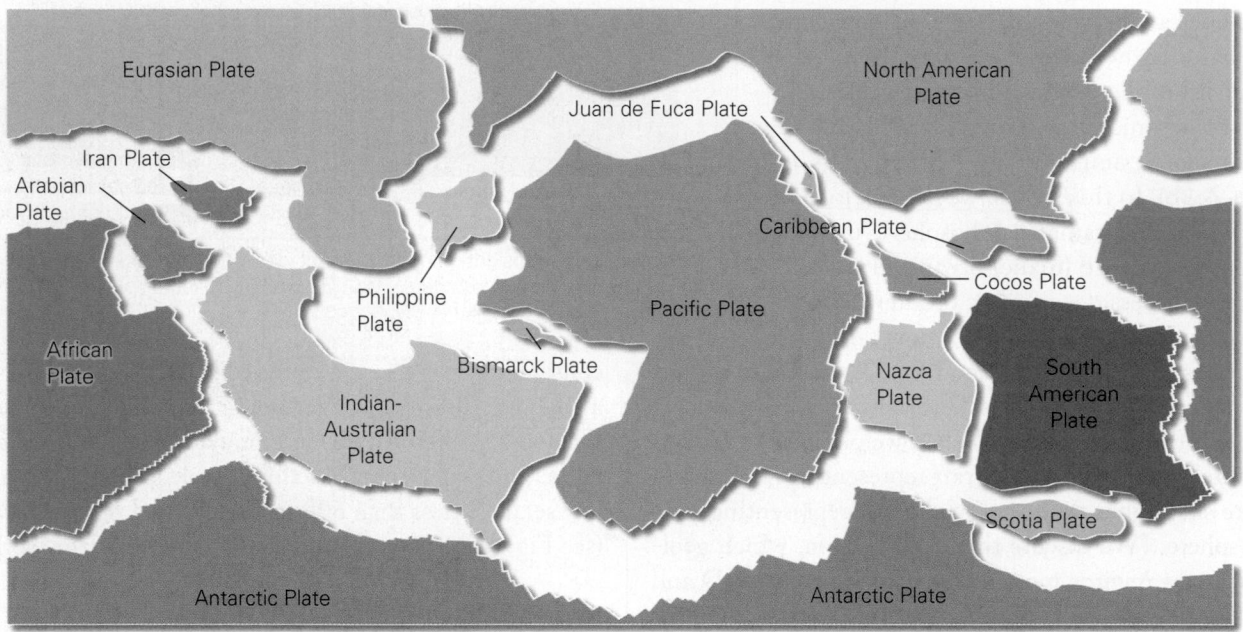

(b) An exploded view of the plates emphasizes the variation in their shape and size.

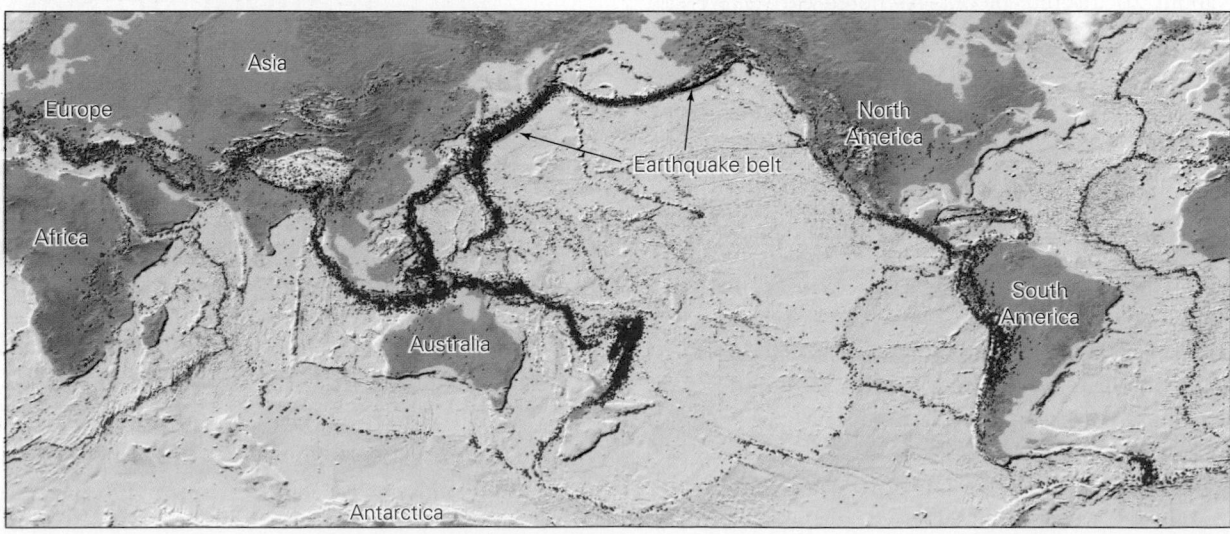

(c) The locations of most earthquakes (red dots) fall in distinct bands that correspond to plate boundaries. Relatively few earthquakes occur in the stabler plate interiors.

FIGURE 4.5 The three types of plate boundaries are distinguished by the nature of relative plate movement.

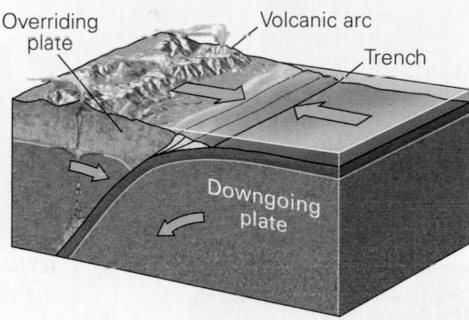

(a) At a *divergent boundary*, two plates move away from the axis of a mid-ocean ridge. New oceanic lithosphere forms.

(b) At a *convergent boundary*, two plates move toward each other; the downgoing plate sinks beneath the overriding plate.

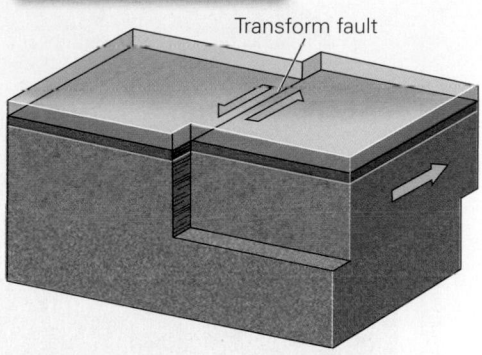

(c) At a *transform boundary*, two plates slide past each other on a vertical fault surface.

Identifying Plate Boundaries

How do we recognize the location of a plate boundary? The answer becomes clear from looking at a map showing the locations of earthquakes **(Fig. 4.4c)**. Recall from Chapter 2 that earthquakes are vibrations generated where rock breaks and suddenly slides along a fault (a fracture on which sliding occurs). The **focus** of the earthquake is the spot where the fault slips, and the **epicenter** marks the point on the surface of the Earth directly above the focus. Earthquake epicenters do not speckle the Earth's surface randomly, like buckshot on a target. Rather, the majority occur in relatively narrow, distinct belts. These **seismic belts** define the positions of plate boundaries, because the fracturing and sliding that take place along these boundaries as plates move generate earthquakes. *Plate interiors*, regions away from the plate boundaries, remain relatively earthquake free because very little movement takes place within them.

> **Did you ever wonder . . .**
> why earthquakes don't occur everywhere?

We can distinguish among three types of plate boundaries simply by the *relative motions* of the plates on either side of the boundary **(Fig. 4.5)**. Geologists refer to a boundary at which two plates move apart from each other as a *divergent boundary*, a boundary at which two plates move toward each other so that one plate sinks beneath the other as a *convergent boundary*, and a boundary at which two plates slide sideways past each other as a *transform boundary*. Each type of boundary looks and behaves differently from the others, as we will see in the next three sections of this chapter.

TAKE-HOME MESSAGE

The Earth's rigid shell, its lithosphere, is divided into about 20 plates that move relative to one another. The motion of one plate relative to its neighbor is accommodated by sliding along plate boundaries. Seismic belts, therefore, define plate boundaries.

QUICK QUESTION: Can a plate include both continental and oceanic lithosphere? Are all continental margins seismically active?

4.3 Divergent Boundaries and Seafloor Spreading

At a **divergent boundary**, or *spreading boundary*, two oceanic plates move apart (diverge) by the process of seafloor spreading. During the process, an open space does not develop between the diverging plates. Rather, as the plates move apart, new ocean floor forms along the divergent boundary **(Fig. 4.6a)**. This process takes place at mid-ocean ridges, so geologists commonly refer to a divergent boundary simply as a *mid-ocean ridge* or, even more simply, a *ridge*.

To characterize a divergent boundary more completely, let's look at one example, the Mid-Atlantic Ridge, in more detail **(Fig. 4.6b)**. The Mid-Atlantic Ridge extends from the waters between northern Greenland and northern Scandinavia southward across the equator to the latitude of South America's

FIGURE 4.6 Divergent boundaries are delineated by mid-ocean ridges, where seafloor spreading occurs.

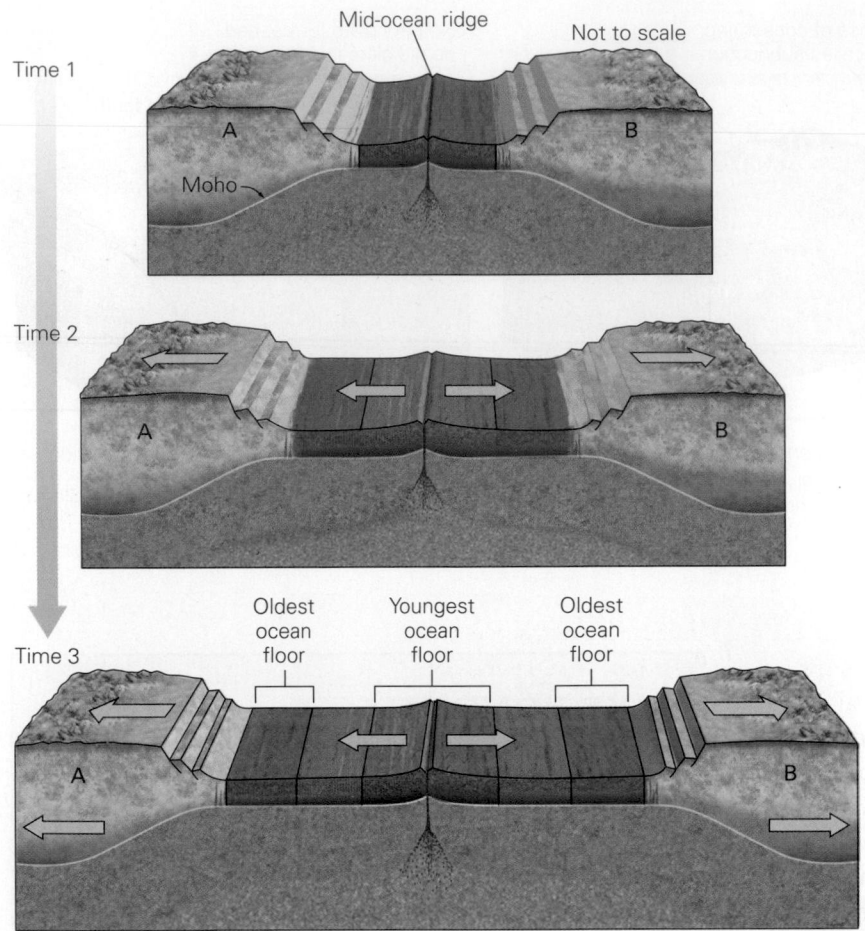

(a) During seafloor spreading, the ocean floor gets wider, and continents on either side move apart. New oceanic crust forms at the ridge axis. Rising asthenosphere melts beneath the axis.

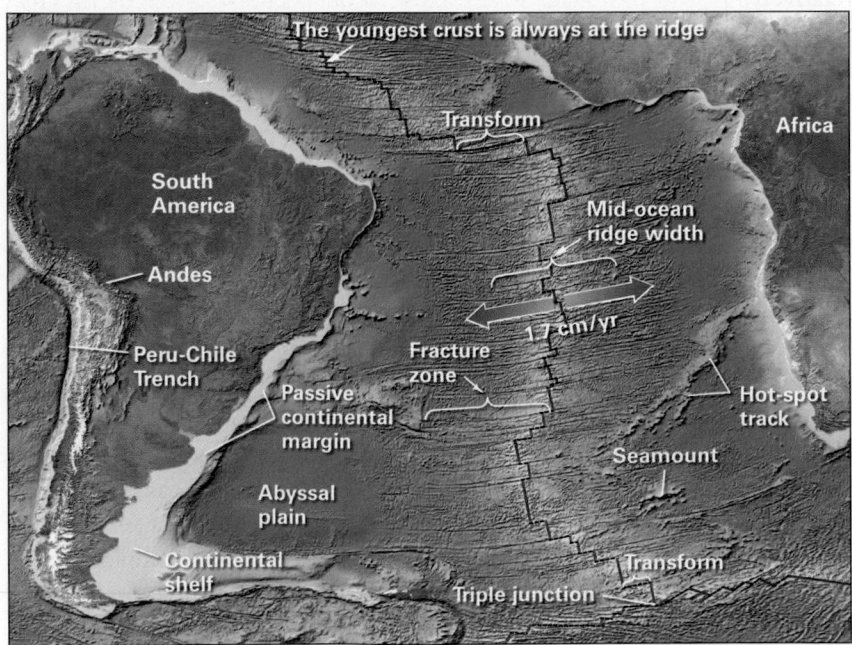

(b) The Mid-Atlantic Ridge in the South Atlantic Ocean. The lighter shades of blue are shallower water depths.

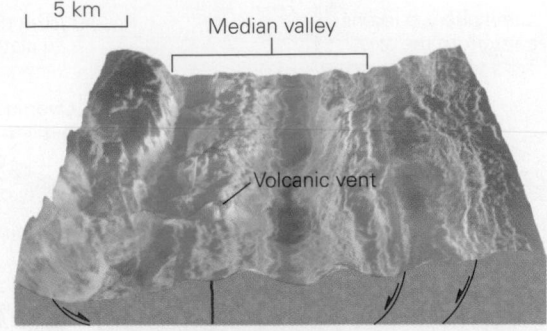

(c) A detailed digital map showing the surface of the median valley.

southern tip. It rises about 2 km above the depth of the Atlantic abyssal plains, so its crest lies at water depths of 2 to 2.5 km. The centerline of the mid-ocean ridge is called the *ridge axis*. Typically, a 1 km-deep elongate trough, the *median valley*, follows the trace of the ridge axis **(Fig. 4.6c)**. Geologists have found that the formation of new oceanic crust takes place on or beneath the floor of the median valley. A series of steep scarps form the walls of the trough. Outside of the trough, sea-floor slopes gently, overall from the ridge toward the abyssal plain, reaching abyssal-plain depth at a distance of about 500 to 800 km from the ridge axis (see Fig. 3.9). The Mid-Atlantic Ridge has a symmetrical shape, in that its eastern half looks like a mirror image of its western half.

How Does Oceanic Crust Form at a Mid-Ocean Ridge?

In the years since Hess's proposal of the seafloor-spreading model, geologists have studied mid-ocean ridges intensely and have developed a model of how oceanic crust forms at a ridge **(Fig. 4.7)**. According to this model, asthenosphere flows very slowly upward beneath the lithosphere, from deeper in the mantle to just beneath the ridge axis. In this context, "very slow" means that the rise takes place at rates of no more than several centimeters per year. When rising asthenosphere reaches a depth of 60–30 km beneath the seafloor, up to 15% of it melts. (Chapter 6 explains why.) Not all of the minerals in the ultramafic rock of the asthenosphere melt equally during this pro-cess, so the magma that forms has a composition different from that of its source rock. Specifically, the magma has a mafic composition: it contains relatively more silica than does the ultramafic rock

FIGURE 4.7 Formation of new oceanic crust occurs on and below the median valley.

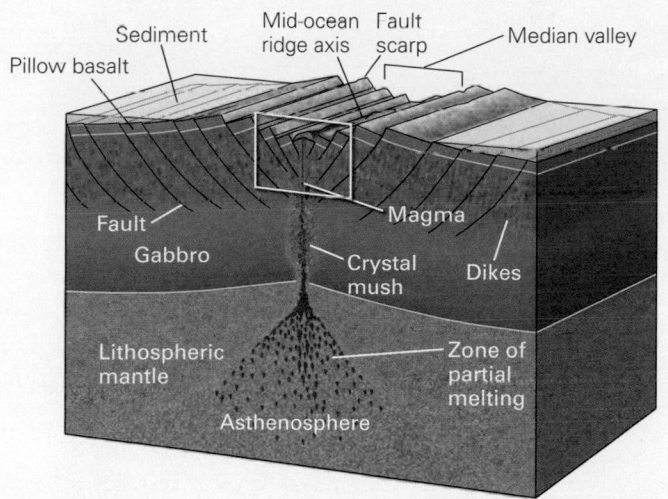

(a) Beneath a mid-ocean ridge lies a magma chamber. Gabbro forms on the sides of the magma chamber. Basalt dikes protrude upward.

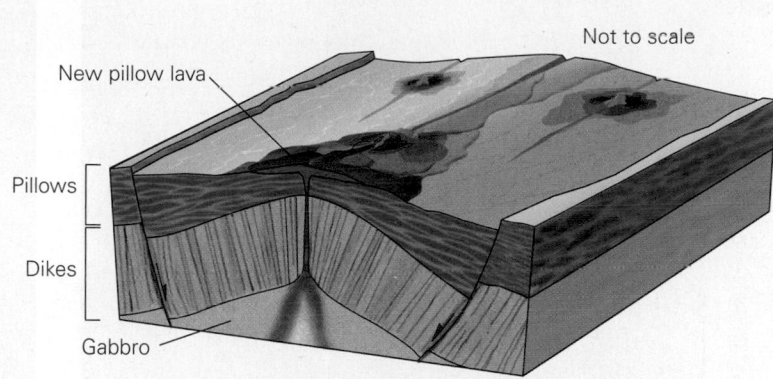

(b) An enlargement of the median valley illustrates that lenses of pillows spill out of distinct fissures where a dike gets close to the surface. The newly formed crust breaks up along faults.

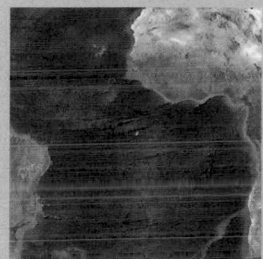
from which it formed. The density of the magma is less than that of the surrounding asthenosphere, so the magma behaves buoyantly and rises. Some of it accumulates in a region called a *magma chamber* that lies at a depth of 2 to 7 km beneath the seafloor. At any given time, the magma chamber contains a "mush" of solid mineral crystals and liquid melt. Some of this material cools along the sides of the magma chamber and solidifies to become a coarse-grained, mafic igneous rock called *gabbro*. Some rises still higher into vertical cracks, in which it solidifies to form wall-like sheets of basalt; these sheets are called *basalt dikes*. Magma that makes it all the way to the surface of the seafloor spills out of small submarine volcanoes as lava. This lava cools rapidly to form a layer of *pillow basalt*, consisting of piles of meter-wide basalt blobs, or "pillows" **(Fig. 4.8a)**.

Note that newly formed oceanic crust consists of three distinct layers: a layer of pillow basalt overlies a layer of basalt dikes, which in turn overlies a layer of gabbro. Clearly, oceanic crust does not form simply by cooling and solidification of the top of the mantle.

We can't easily see the submarine volcanoes of a mid-ocean ridge, which lie beneath 2 to 3 km of water, but they have been detected by observers in research submersibles. These observers have also found chimneys spewing hot, mineralized water along the ridge axis. These chimneys are known as **black smokers** because the water they emit looks like a cloud of dark smoke—the color comes from a suspension of tiny mineral grains that precipitate in cold seawater the instant that the rising mineralized water cools **(Fig. 4.8b)**. Black smokers form because, over time, seawater percolates down into the oceanic crust through a network of cracks. Heat from rising magma warms up the water, which then starts to rise. The hot water dissolves minerals and carries them along, in solution, as it spews back into the sea.

As soon as it forms, new oceanic crust moves away from the ridge axis, and when this happens, more magma rises from below, so still more crust forms. In other words, like a vast, moving conveyor belt, magma from the mantle rises to the Earth's surface at the ridge, solidifies to form oceanic crust, and then moves laterally away from the ridge. Because all seafloor forms at mid-ocean ridges, the youngest seafloor occurs at the ridge axis, and seafloor becomes progressively older away from the axis. In the Atlantic Ocean, the oldest seafloor lies adjacent to the passive continental margins on either side of the ocean **(Fig. 4.9)**. The oldest seafloor on our planet underlies the western Pacific Ocean; this crust formed about 200 Ma (million years ago).

The tension (stretching force) applied to newly formed solid crust as spreading takes place breaks this new crust, resulting in the formation of faults. Movement (slip) on these faults generates earthquakes and produces scarps that border the median valley of the mid-ocean ridge.

Did you ever wonder...
whether all ocean floor has the same age?

FIGURE 4.8 Eruption on mid-ocean ridges.

0
~1 m

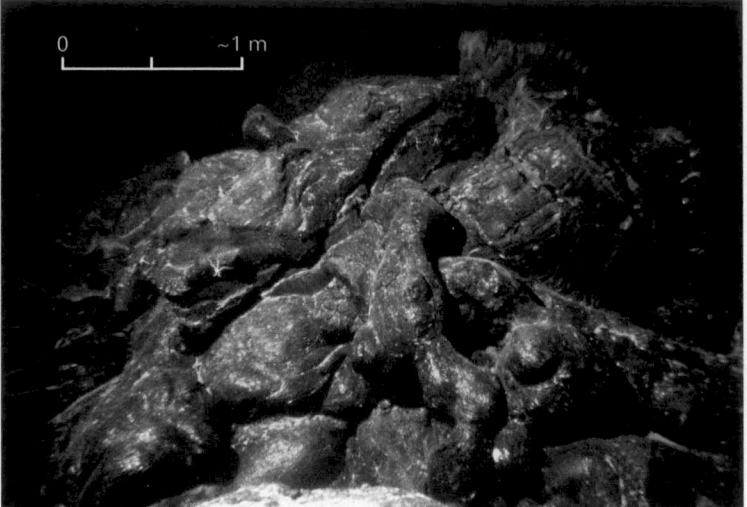

(a) Recently erupted pillow basalt from the Juan de Fuca Ridge.

Black smoker
Thermometer
Living organisms

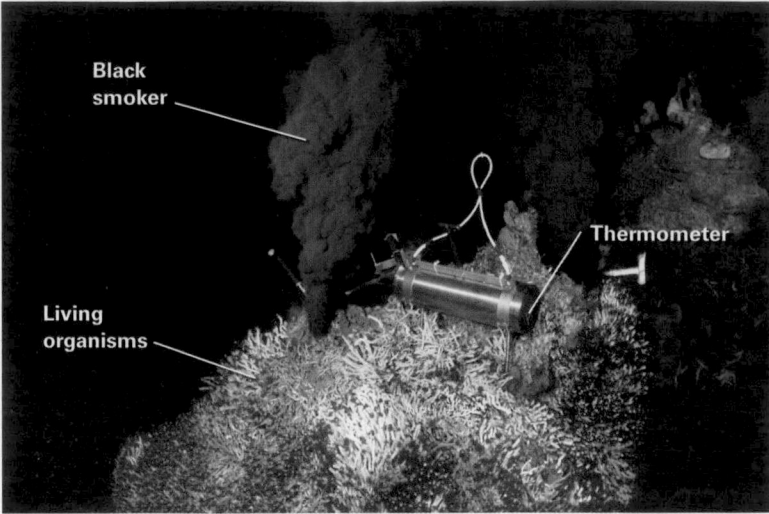

(b) A column of superheated water gushing from a vent (known as a black smoker) along the ridge. Bacteria, shrimp, and worms live around the vent.

How Does the Lithospheric Mantle Form at a Mid-Ocean Ridge?

We've seen that oceanic crust forms from magma that lies beneath mid-ocean ridges. What about the formation of the lithospheric-mantle portion of oceanic lithosphere? Recall that this part consists of the cooler uppermost layer of the mantle, in which temperatures are less than about 1,280°C. At the ridge axis, temperatures reach 1,280°C almost at the base of the crust because of the presence of rising hot asthenosphere and hot magma, so lithospheric mantle doesn't exist beneath the ridge axis. But as the newly formed oceanic crust moves away from the ridge axis, the crust and the uppermost mantle directly beneath it gradually cool by losing heat to the ocean above. As soon as mantle rock cools below 1,280°C, it becomes, by definition, part of the lithosphere because at this

FIGURE 4.9 This map of the world shows the age of the seafloor. Note that the seafloor grows older with increasing distance from the ridge axis.

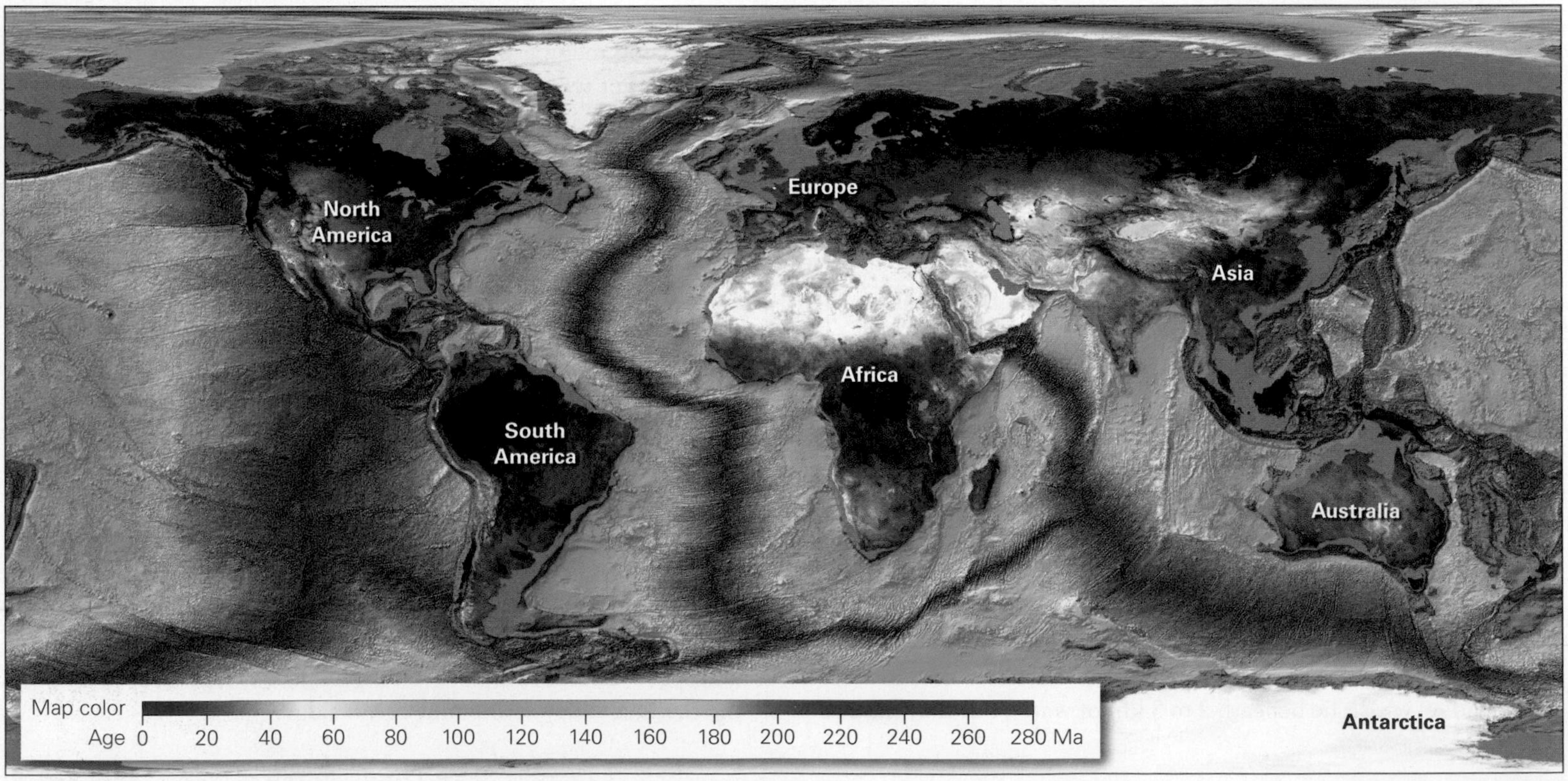

North America
Europe
Asia
Africa
South America
Australia
Antarctica
Map color
Age 0 20 40 60 80 100 120 140 160 180 200 220 240 260 280 Ma

temperature, it begins to behave relatively rigidly. As oceanic lithosphere continues to move away from the ridge axis, it continues to cool, so the lithospheric mantle, and therefore the oceanic lithosphere as a whole, grows progressively thicker (Fig. 4.10). Note that the process of forming the lithospheric mantle doesn't change the thickness of the overlying oceanic crust, for the crust is formed entirely at the ridge axis. The rate of cooling and lithospheric-mantle thickening decreases with distance from the ridge axis. In fact, by the time the lithosphere reaches about 80 million years old, it has just attained maximum thickness. As lithosphere thickens and gets cooler and denser, it sinks down into the asthenosphere, like a ship taking on ballast. So, the ocean becomes deeper over older ocean floor than over younger ocean floor. That's why abyssal plains are deeper than mid-ocean ridges.

FIGURE 4.10 Changes accompanying the aging of lithosphere.

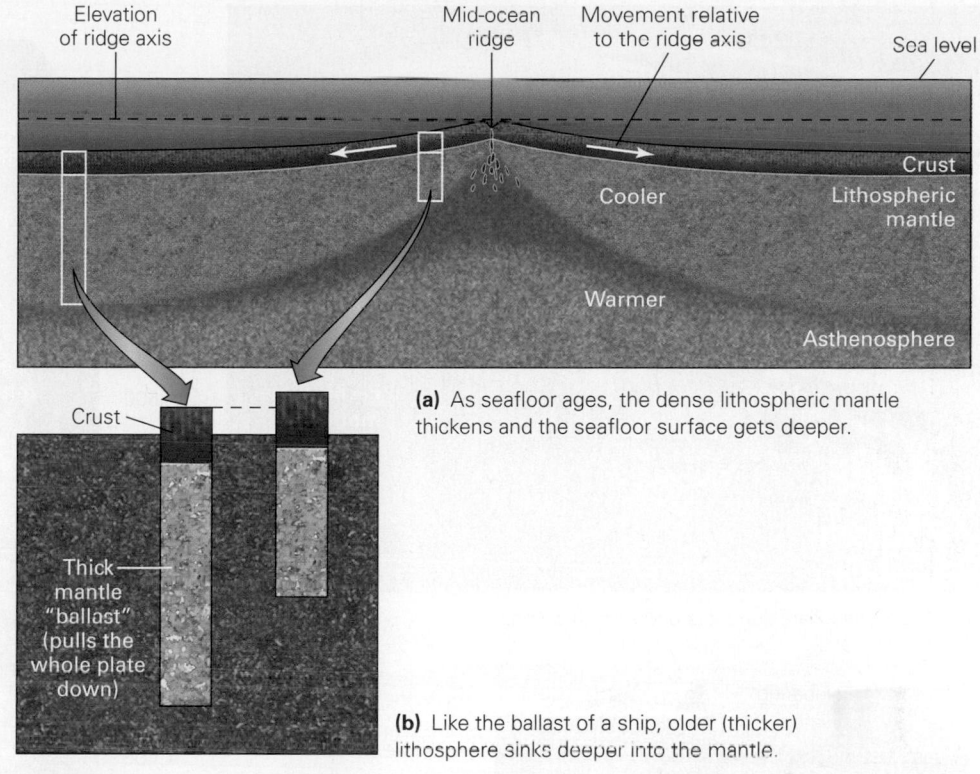

(a) As seafloor ages, the dense lithospheric mantle thickens and the seafloor surface gets deeper.

(b) Like the ballast of a ship, older (thicker) lithosphere sinks deeper into the mantle.

TAKE-HOME MESSAGE

Seafloor spreading occurs at divergent boundaries, defined by mid-ocean ridges. Hot asthenosphere rises beneath the mid-ocean ridge and starts to melt when it reaches shallower depths. The resulting magma rises still farther. Some solidifies underground as gabbro, some injects at a shallower level to produce vertical sheets (dikes) of basalt, and some erupts from small volcanoes along the ridge axis to form pillow basalt. The gabbro and basalt produced by these processes form new oceanic crust. As plates move away from the axis, they cool, and lithospheric mantle forms and thickens underneath this crust.

QUICK QUESTION: What are black smokers, and why do they form?

4.4 Convergent Boundaries and Subduction

At a **convergent boundary**, two plates, at least one of which is oceanic, move toward each other. But rather than butting against each other like angry rams, one oceanic plate bends and sinks down into the asthenosphere beneath the other plate. Geologists refer to the sinking process as **subduction**, so convergent boundaries are also known as *subduction zones*. Because subduction at a convergent boundary consumes old oceanic lithosphere, geologists also refer to convergent boundaries as *consuming boundaries*, and because deep-ocean trenches develop at convergent boundaries, they are sometimes simply called **trenches (Fig. 4.11a)**. The amount of oceanic-plate consumption worldwide, averaged over time, equals the amount of seafloor spreading worldwide, so the surface area of the Earth remains constant through time.

Subduction occurs for a simple reason: the overall density of an oceanic plate, once the plate has aged at least 10 million years, exceeds that of the underlying asthenosphere. Therefore, where a plate bends down and starts to slip into the mantle, it will continue to sink like an anchor falling to the bottom of a lake **(Fig. 4.11b)**. As the lithosphere sinks, asthenosphere flows out of its way, just as water flows out of the way of an anchor. But, the asthenosphere's viscosity, meaning its resistance to flow, allows it to flow only very slowly. Therefore, oceanic lithosphere can sink only very slowly—at a rate of less than about 15 cm per year. (To visualize the difference, imagine how much faster a coin can sink through water than it can through honey; honey's viscosity exceeds that of water.) Because of the asthenosphere's high viscosity, the broad horizontal interior of an oceanic plate can't just sink straight down because there's too much resistance to flow in the underlying asthenosphere.

At consuming boundaries, the *downgoing plate*, meaning the plate that has been subducted, must be composed of

FIGURE 4.11 During the process of subduction, oceanic lithosphere sinks back into the deeper mantle.

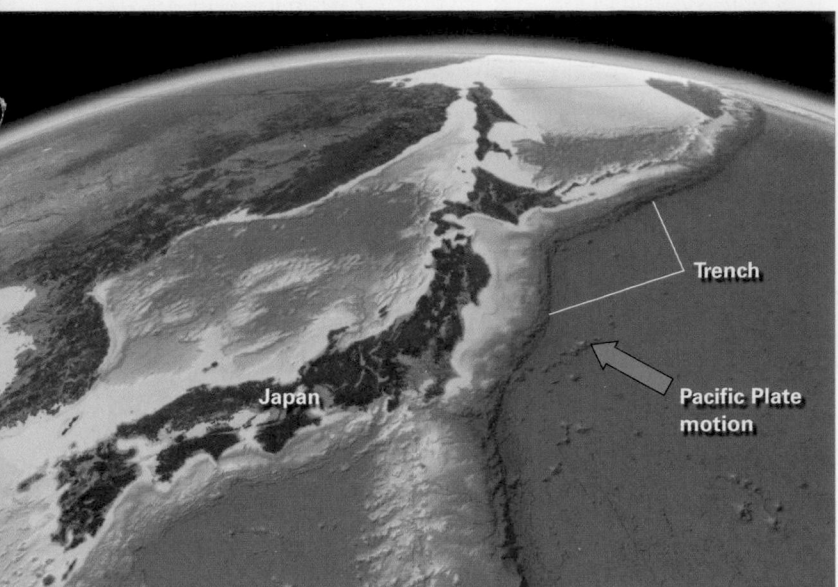

(a) The Pacific Plate subducts underneath Japan.

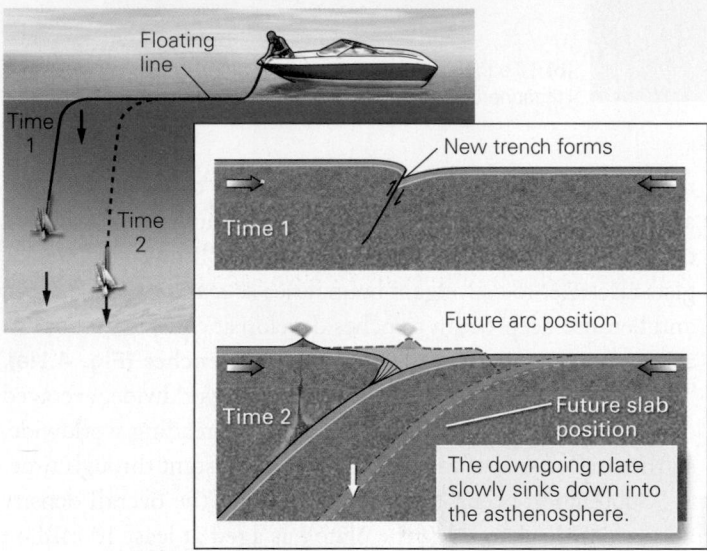

(b) Sinking of the downgoing plate resembles sinking of an anchor attached to a rope.

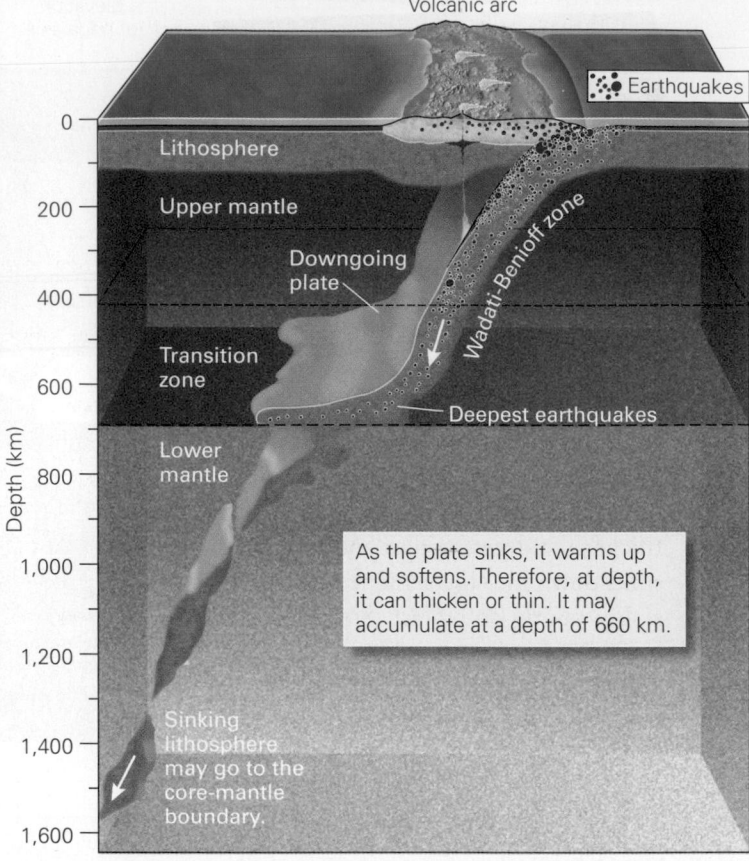

(c) A belt of earthquakes (dots) defines the position of the downgoing plate in the region above a depth of about 660 km. Sometimes plates "pile up" at a depth of 660 km.

oceanic lithosphere. The *overriding plate*, the one that is not sinking at a subduction zone, can consist of either oceanic or continental lithosphere. Continental lithosphere does not get subducted significantly because its thick crust of felsic and intermediate rocks is not dense enough to sink into the mantle, and thus acts like a life preserver, keeping the continent afloat. Because continental crust cannot be completely subducted, some continental crust has survived for over 3.8 billion years, much longer than the oldest oceanic lithosphere.

Earthquakes and the Fate of Subducted Plates

At convergent boundaries, the downgoing plate grinds along the base of the overriding plate, a process that generates large earthquakes. These earthquakes occur fairly close to the Earth's surface, so some of them cause massive destruction in coastal cities (see Chapter 10). But earthquakes also happen in downgoing plates at greater depths. In fact, geologists can detect earthquakes within subducted lithosphere down to a depth of 660 km. The band of earthquakes in a downgoing plate is called a **Wadati-Benioff zone**, after its discoverers **(Fig. 4.11c)**. At depths greater than 660 km, conditions change, and earthquakes no longer occur. Downgoing plates, however, do continue to sink below a depth of 660 km—they just do so without generating earthquakes. As a result, the lower mantle may be a graveyard for old subducted plates.

Geologic Features of a Convergent Boundary

To become familiar with the various geologic features that develop along a convergent boundary, let's look at an example more closely. Trenches form where a plate bends as it starts to sink into the asthenosphere. As the downgoing plate slides under the overriding plate, sediment (clay and plankton) that has settled on the surface of the downgoing plate, as well as sand that has fallen into the trench from the shores of nearby land, gets scraped off and incorporated in a wedge-shaped mass known as an **accretionary prism** (Fig. 4.12a, b). An accretionary prism forms in basically the same way as a pile of snow in front of a plow, and like the snow, the sediment becomes squashed and contorted during the process. Sediment eroded from the land accumulates in a forearc basin, behind the accretionary prism.

A chain of volcanoes known as a **volcanic arc** develops behind the accretionary prism. As we will see in Chapter 6, the magma that feeds these volcanoes forms above the surface of the downgoing plate where the plate reaches a depth of about 150 km below the Earth's surface. If the volcanic arc forms where an oceanic plate subducts beneath continental lithosphere, the resulting chain of volcanoes grows on the continent and forms a *continental volcanic arc*. In some cases, compression between the converging plates produces a belt of faults in the region behind the arc. If the volcanic arc forms where one oceanic plate subducts beneath another oceanic plate, the resulting volcanoes form a chain of islands known as a **volcanic island arc** (Fig. 4.12c). A *marginal sea*, or *back-arc basin*—a small ocean basin behind an island arc—forms either in cases where subduction happens to begin offshore, trapping oceanic lithosphere behind the arc, or where stretching of the lithosphere behind the arc leads to the formation of a small, spreading ridge behind the arc (Fig. 4.12d).

TAKE-HOME MESSAGE

At a convergent boundary, an oceanic plate sinks into the mantle beneath the edge of another plate. A volcanic arc and a trench delineate such plate boundaries, and earthquakes happen along the contact between the two plates as well as in the downgoing slab. Volcanic arcs can form on the edge of a continent or as a chain of islands in the sea.

QUICK QUESTION: Can continents be completely subducted?

FIGURE 4.12 The development of a convergent boundary varies with location.

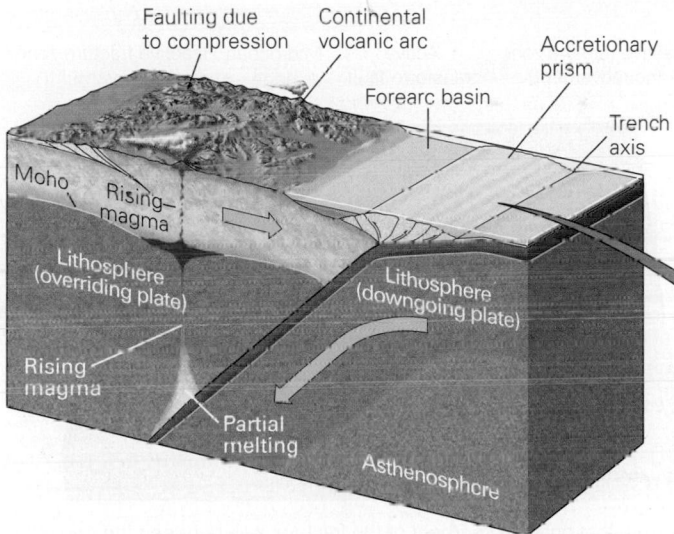

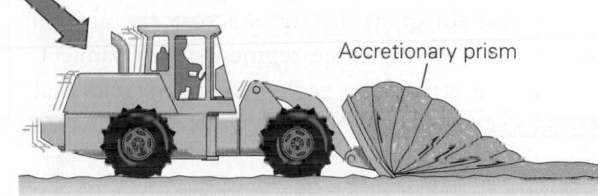

(a) A subduction zone along the edge of a continent. Here compression has caused faulting behind the arc.

(b) The overriding plate acts like a bulldozer, scraping sediment off the downgoing plate to build an accretionary prism.

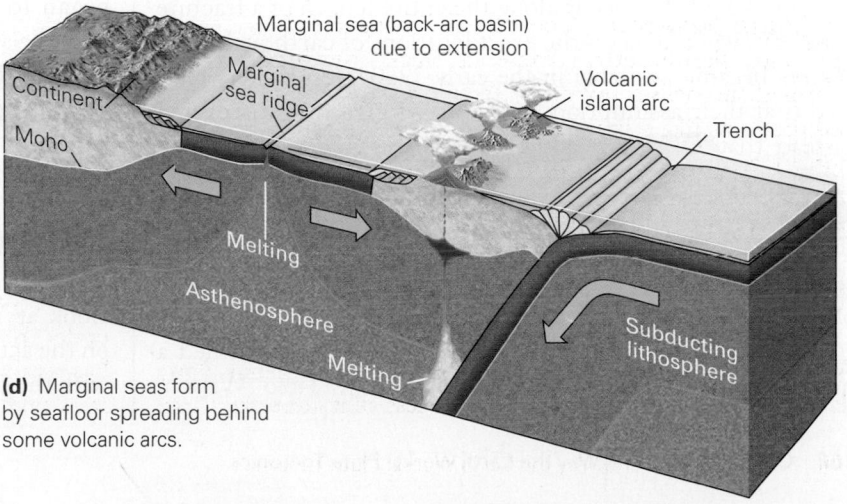

(c) Subduction beneath oceanic lithosphere produces an island arc.

(d) Marginal seas form by seafloor spreading behind some volcanic arcs.

4.5 Transform Boundaries

When researchers began to explore the bathymetry of mid-ocean ridges in detail, they discovered that mid-ocean ridges are not long, uninterrupted lines, but rather consist of short segments that appear to be offset from each other at their ends

FIGURE 4.13 The concept of transform faulting.

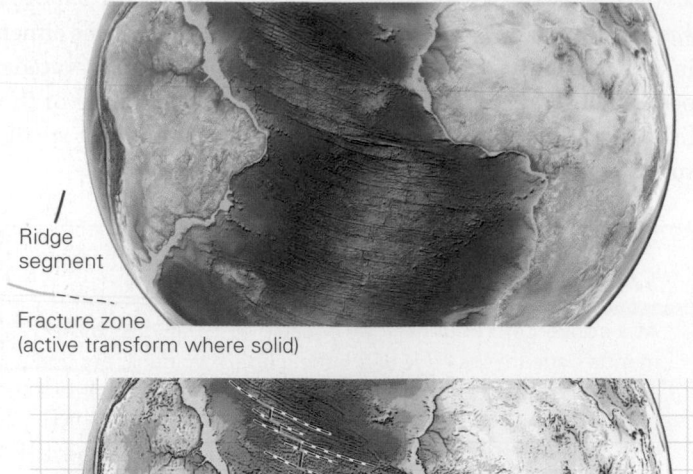

Ridge segment

Fracture zone (active transform where solid)

What a Geologist Sees

(a) Numerous transform faults segment the Mid-Atlantic Ridge.

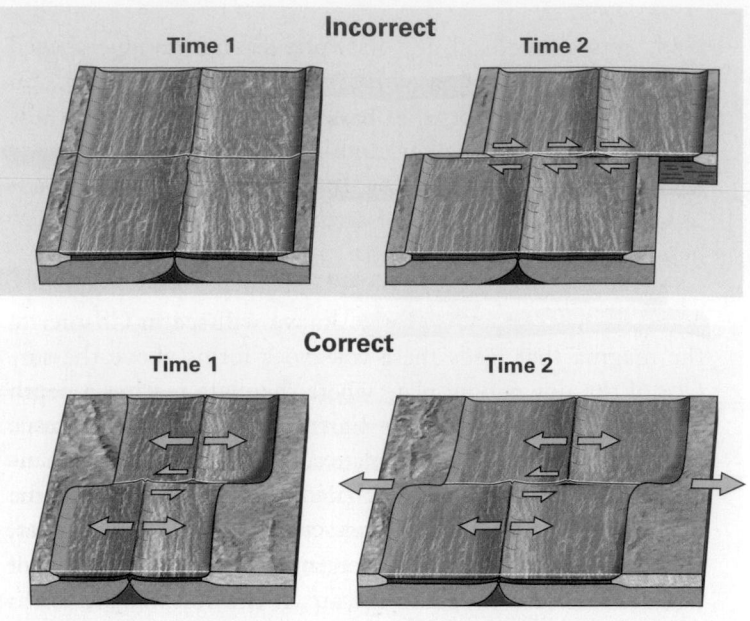

Incorrect

Time 1   Time 2

Correct

Time 1   Time 2

(b) A comparison of the old, incorrect model of transform faults with Wilson's new, correct model required by the seafloor-spreading hypothesis.

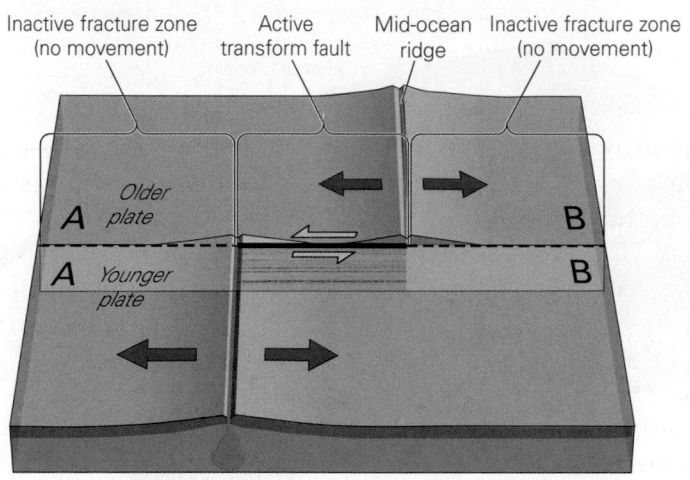

Inactive fracture zone (no movement)   Active transform fault   Mid-ocean ridge   Inactive fracture zone (no movement)

A   *Older plate*   B

A   *Younger plate*   B

(c) Note that only the segment of the fracture zone between the two ridge segments is active.

(Fig. 4.13a). A **fracture zone**, a narrow belt of rough and rubbly seafloor, intersects the end of each ridge segment at a right angle and links it to the next ridge segment. Notably, most fracture zones extend for great distances across the abyssal plains, beyond the ends of the ridge segments they connect. Using detailed bathymetric data, geologists discovered that fracture zones contain numerous steep, elongate cracks that trend parallel to the zone (i.e., perpendicular to the ridge segments) and appear to break up the oceanic crust.

Originally, researchers assumed that the entire length of a fracture zone was an actively slipping fault that displaced mid-ocean ridge segments. If this model were correct, then earthquakes should occur along the entire length of a fracture zone. But when data on the exact locations of earthquake epicenters became available in the early 1960s, researchers realized that their assumption had to be wrong. They discovered instead that earthquakes happen only along the segment of the fracture zone that links the end of a ridge segment to the end of the next ridge segment. Earthquakes don't occur along the portions of fracture zones that extend beyond the ends of ridge segments, out into the abyssal plains, so these portions of fracture zones are not actively slipping faults.

The nature of movement on fracture zones remained a mystery until a Canadian geophysicist, J. Tuzo Wilson,

began to think about fracture zones in the context of the seafloor-spreading concept. Wilson proposed that fracture zones formed at the same time as the ridge axis itself, so the ridge consisted of separate segments to start with. These segments were not offset by fracture zones. With this idea in mind, he drew a sketch map showing two ridge-axis segments linked by a fracture zone, and he drew arrows to indicate the direction in which oceanic crust was moving, relative to the ridge axis, due to seafloor spreading (Fig. 4.13b). Look at Wilson's arrows. Clearly, the movement direction on the active portion of the fracture zone must be compatible with the spreading direction, so it's opposite to the movement

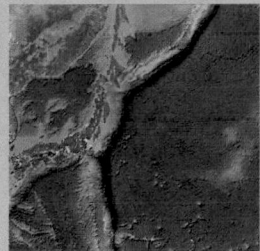
direction that researchers originally thought took place. Furthermore, in Wilson's model, slip occurs only along the segment of the fracture zone between the two ridge segments. Plates on opposite sides of inactive fracture zones move together as one plate.

Wilson introduced the term **transform boundary** (or *transform fault*) for the actively slipping segment of a fracture zone between two ridge segments, and he pointed out that these zones represent a third type of plate boundary. At a transform boundary, one plate slides sideways, relative to its neighbor, on a vertical fault. As a result, the slip direction on the fault is horizontal (parallel to the Earth's surface), so no new plate forms, and no old plate is consumed, at a transform boundary (Fig. 4.13c).

So far we've discussed only transform boundaries along mid-ocean ridges. Not all transform boundaries link ridge segments, however. Some, such as the Alpine fault of New Zealand, link trenches, while others link a trench to a ridge segment. Also, not all transform faults occur in oceanic lithosphere; a few cut across continental lithosphere. The San Andreas fault, for example, which cuts across western California, defines part of the boundary between the North American Plate and the Pacific Plate. The portion of California that lies to the west of the fault (including Los Angeles) is part of the Pacific Plate, while the portion that lies to the east of the fault is part of the North American Plate (Fig. 4.14).

FIGURE 4.14 The San Andreas fault is a continental transform boundary.

Fault
trace

(a) In Southern California, the San Andreas fault cuts a dry landscape. The fault trace is in the narrow valley. The land has been pushed up slightly along the fault.

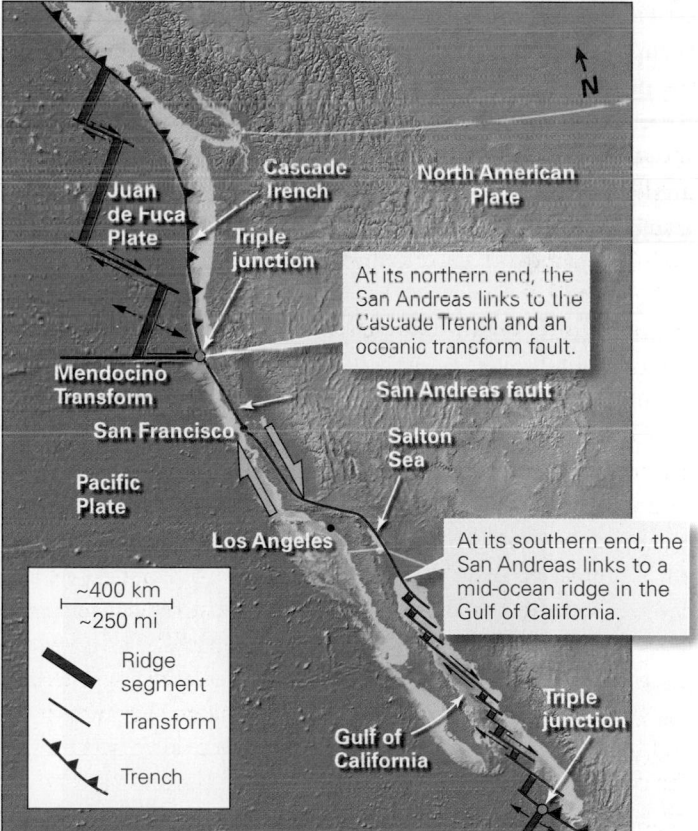

Juan de Fuca Plate

Cascade Trench

North American Plate

Triple junction

At its northern end, the San Andreas links to the Cascade Trench and an oceanic transform fault.

Mendocino Transform

San Francisco

San Andreas fault

Salton Sea

Pacific Plate

Los Angeles

At its southern end, the San Andreas links to a mid-ocean ridge in the Gulf of California.

~400 km
~250 mi

Ridge segment

Transform

Trench

Gulf of California

Triple junction

(b) The San Andreas fault is a transform plate boundary between the North American and Pacific Plates. The Pacific Plate is moving northwest, relative to the North American Plate.

TAKE-HOME MESSAGE

At transform boundaries, one plate slips sideways past another along a vertical fault. Thus, there is no production or destruction of lithosphere at a transform boundary. Most transform boundaries link segments of mid-ocean ridges, but some, such as the San Andreas fault, cut across continental crust.

QUICK QUESTION: Why do earthquakes occur only on the portion of an oceanic fracture zone that links mid-ocean ridge segments?

4.6 Special Locations in the Plate Mosaic

Triple Junctions

Geologists refer to a point where three plate boundaries intersect as a triple junction. We name triple junctions after the types of boundaries that intersect at them. For example, the triple junction formed where the Southwest Indian Ocean Ridge intersects two arms of the Mid–Indian Ocean Ridge is a ridge-ridge-ridge triple junction (**Fig. 4.15a**), and the triple junction north of San Francisco is a trench-transform-transform triple junction (**Fig. 4.15b**).

Hot Spots

Both the volcanoes of volcanic arcs and those of mid-ocean ridges are *plate-boundary volcanoes*, in that they formed as a consequence of movement along the boundary. Not all volcanoes on the Earth are plate-boundary volcanoes, however. For example, the currently erupting volcanoes of the Big Island of Hawaii lie over 4,000 kilometers from the nearest ridge or trench. And Yellowstone National Park, which occupies the remnants of a huge volcano that erupted during the past million years, lies well into the interior of the North American Plate. Worldwide, geoscientists have identified about 100 volcanoes that exist as isolated points, not as a consequence of movement at a plate boundary, and refer to these points as **hot spots** (**Fig. 4.16**). Many hot spots lie in the interiors of plates, away from plate boundaries, but some lie on mid-ocean ridges.

What causes hot spots? In the early 1960s, J. Tuzo Wilson noted that active hot-spot volcanoes (*active volcanoes* are those that have erupted recently and may erupt in the future) occur only at the end of a chain of inactive, or *extinct*, volcanic islands and seamounts (ones that will never erupt again). This configuration differs from that found in volcanic arcs along convergent boundaries: at volcanic arcs, all of the volcanoes remain active, because subduction happens along the length of the arc. Wilson proposed that hot-spot volcanoes form above a localized magma source whose position in the mantle stays fixed, more or less, relative to the moving plate above. In Wilson's model, the active volcano represents the present-day location of the magma source, whereas the associated chain of inactive volcanic islands represents locations on the plate that were once over the magma source, but progressively moved off the source as the plate moved. Once the moving plate has carried the volcano away from the magma source, the volcano becomes inactive, but as long as the source persists, melt continues to be produced, so a new, younger volcano grows. Eventually, the younger volcano moves off the hot spot and becomes inactive, and another still younger one forms. The chain of inactive volcanoes that formed in succession is now known as a **hot-spot track**.

Geologists don't all agree on why magma forms at hot spots, and the issue remains a topic of active research. Most researchers, however, favor a model in which magma forms at the top of a **mantle plume**, a column of very hot rock rising from deeper in the mantle, up through the asthenosphere, to the base of the lithosphere (**Fig. 4.17a, b**). Rock in the plume, though solid, is soft enough to flow, and rises buoyantly

FIGURE 4.15 Examples of triple junctions. The triple junctions are marked by dots.

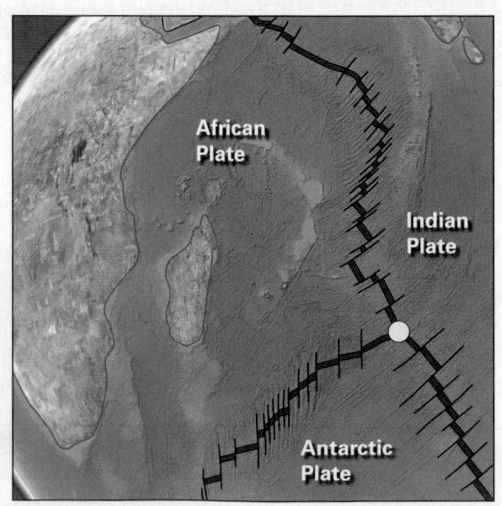
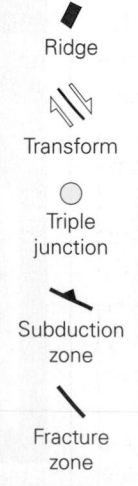

(a) A ridge-ridge-ridge triple junction occurs in the Indian Ocean.

(b) A trench-transform-transform triple junction occurs at the north end of the San Andreas fault.

FIGURE 4.16 The dots represent the locations of selected hot-spot volcanoes. The red lines represent hot-spot tracks. The most recent volcano (dot) is at one end of this track. Some of these volcanoes are extinct. Some hot spots are fairly recent and do not have tracks. Dashed tracks were broken by seafloor spreading.

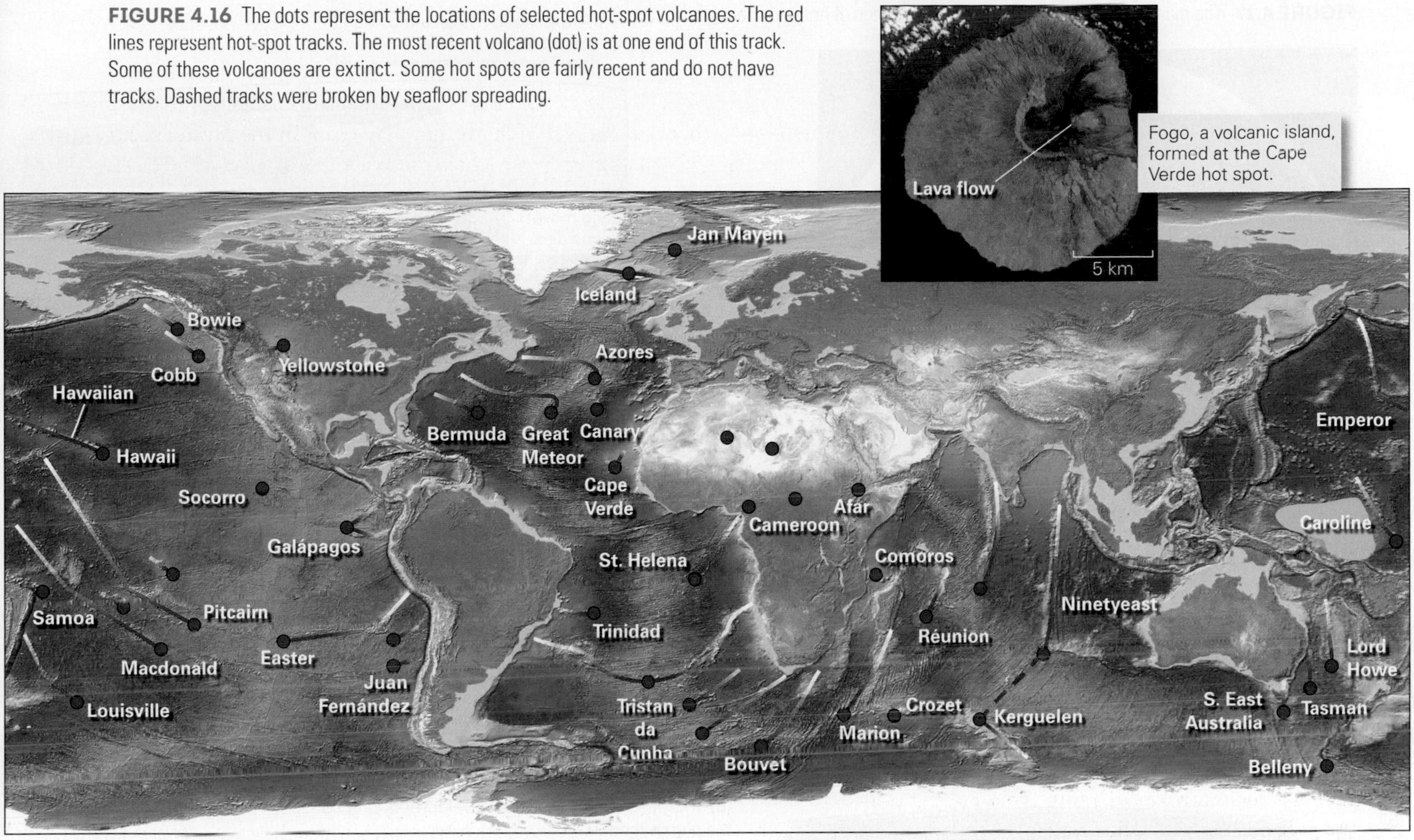

Fogo, a volcanic island, formed at the Cape Verde hot spot.

Lava flow

5 km

because it's less dense than the somewhat cooler rock of the surrounding mantle. According to the plume model, when the hot rock of the plume reaches the base of the lithosphere, it starts to melt and produces magma that seeps up through the lithosphere (see Chapter 6). Some of this magma reaches the Earth's surface and erupts to produce a hot-spot volcano.

We can apply Wilson's model of hot-spot track formation to the Hawaiian island chain and to its continuation, the Hawaiian-Emperor seamount chain **(Fig. 4.17c)**. Volcanic eruptions occur today on the island of Hawaii, at the southeastern end of the chain. All of the other islands and seamounts to the northwest are remnants of inactive volcanoes that have sunk below sea level **(Fig. 4.17d)**, and they get progressively older to the northwest.

As we've noted, some hot spots lie within continents. For example, several have been active in the interior of Africa, and one now underlies Yellowstone National Park. The famous geysers (natural steam and hot-water fountains) of Yellowstone exist because hot magma, formed at the Yellowstone hot spot, lies not far below the surface of the park. A few hot spots lie on mid-ocean ridges. Where this happens, a volcanic island may protrude above sea level because the hot spot produces more magma than does a normal mid-ocean ridge. Iceland, for example, formed where a hot spot underlies the Mid-Atlantic Ridge. The extra volcanism of the hot spot built up the island so that it rises almost 3 km above other places on the Mid-Atlantic Ridge.

TAKE-HOME MESSAGE

A triple junction marks the point where three plate boundaries join. A hot spot is a point where volcanism occurs independently of plate movement. Hot spots may be due to melting at the top of a mantle plume. As a plate moves over a plume, a hot-spot track develops.

QUICK QUESTION: Why is the volcano at the end of a hotspot track the only one to be active in the track?

FIGURE 4.17 The mantle plume hypothesis for the formation of hot-spot tracks.

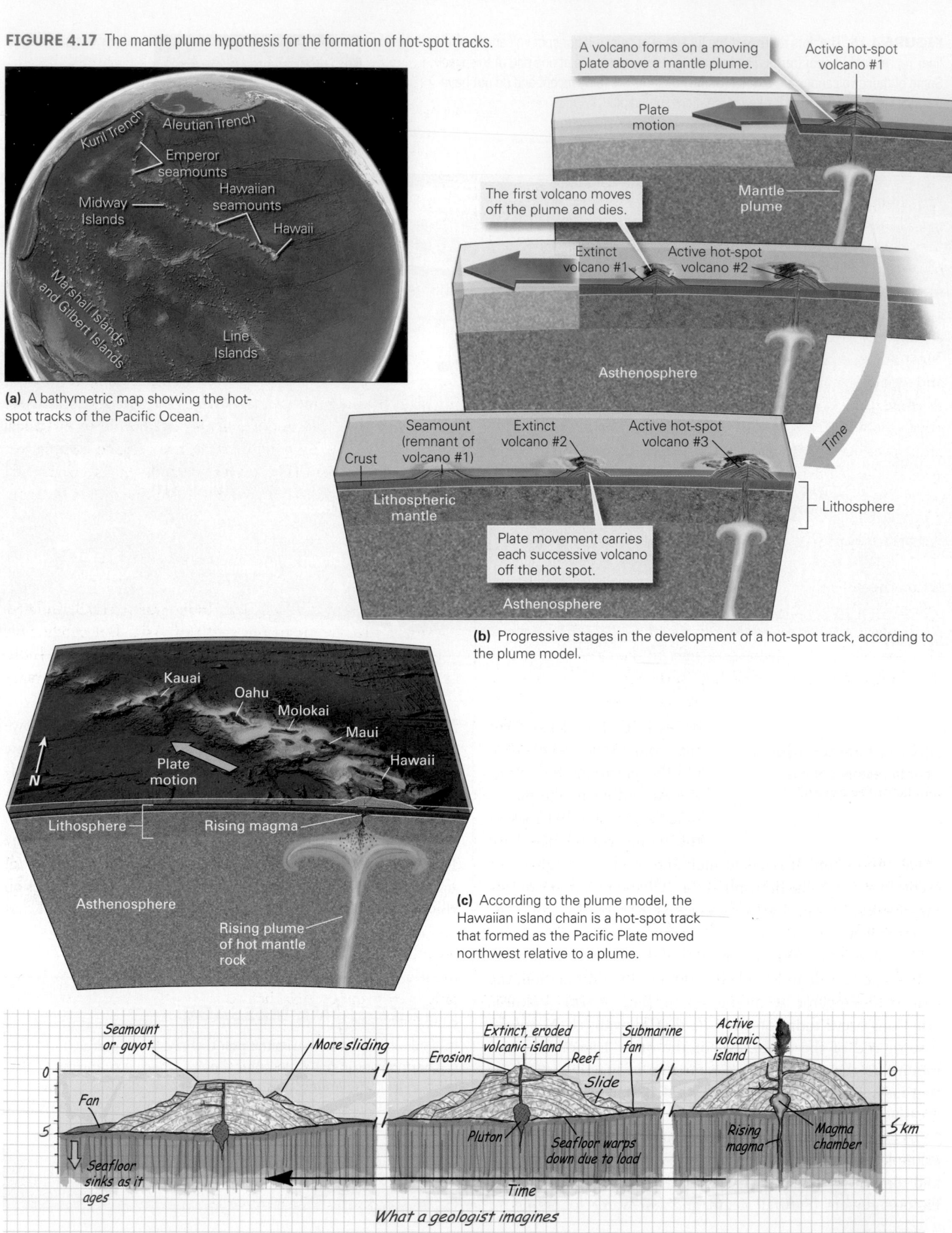

(a) A bathymetric map showing the hot-spot tracks of the Pacific Ocean.

A volcano forms on a moving plate above a mantle plume.

Active hot-spot volcano #1

Plate motion

Mantle plume

The first volcano moves off the plume and dies.

Extinct volcano #1

Active hot-spot volcano #2

Asthenosphere

Time

Seamount (remnant of volcano #1)

Crust

Extinct volcano #2

Active hot-spot volcano #3

Lithospheric mantle

Lithosphere

Plate movement carries each successive volcano off the hot spot.

Asthenosphere

(b) Progressive stages in the development of a hot-spot track, according to the plume model.

Kauai

Oahu

Molokai

Maui

Hawaii

N

Plate motion

Lithosphere

Rising magma

Asthenosphere

Rising plume of hot mantle rock

(c) According to the plume model, the Hawaiian island chain is a hot-spot track that formed as the Pacific Plate moved northwest relative to a plume.

Seamount or guyot

More sliding

Extinct, eroded volcanic island

Erosion

Reef

Submarine fan

Active volcanic island

0

Slide

0

Fan

Pluton

Seafloor warps down due to load

Rising magma

Magma chamber

5 km

5

Seafloor sinks as it ages

Time

What a geologist imagines

(d) As a volcano moves off the hot spot, it gradually sinks below sea level due to sinking of the plate, erosion, and submarine landslides.

4.7 How Do Plate Boundaries Form, and How Do They Die?

The configuration of plates and plate boundaries visible on our planet today has not existed for all of geologic history and will not exist indefinitely into the future. Because of plate motion, oceanic plates form and are later consumed, and continents merge into supercontinents, which later split apart. How does a new divergent boundary come into existence, and how does an existing convergent boundary eventually cease to exist? Most new divergent boundaries form when a continent splits and separates into two continents. We call this process *rifting.* A convergent boundary ceases to exist when a piece of relatively buoyant lithosphere, such as a continent or an island arc, moves into the subduction zone and, in effect, jams up the system. We call this process *collision.*

Continental Rifting

A **continental rift** is a linear belt along which continental lithosphere undergoes horizontal stretching in a direction perpendicular to the trend of the rift and thinning in the vertical direction (Fig. 4.18a). Geologists refer to the overall process of stretching and thinning as **rifting**. In the upper 15 km or so of the continental crust, which is relatively cold and brittle, stretching causes rock to break and faults to develop. During this faulting, blocks of the upper crust slip down along fault surfaces, leading to the formation of a low area, called a *rift valley*, that gradually becomes buried by sediment. At greater depth, continental rock is warmer, so stretching can take place plastically, without breaking the rock.

> **Did you ever wonder ...**
> if continents really split apart, and if so, how?

As continental lithosphere thins during rifting, the underlying hot asthenosphere rises, and the asthenosphere starts to melt, just as it does beneath mid-ocean ridges, producing magma that rises up into the crust. Some of the molten rock reaches the surface and erupts at volcanoes along the rift. If rifting continues for a long enough time, the continent will break in two, a new mid-ocean ridge will form, and seafloor spreading will begin. The relict of the rift, where the continental lithosphere had been stretched and thinned, develops into a passive margin. (The triangular wedges shown in Figure 4.2a are blocks that were displaced by faulting during rifting.) In some cases, however, rifting stops before the continent splits in two, and the low-lying rift valley eventually fills with sediment.

Then the rift remains as a permanent scar in the crust, defined by a belt of faults, volcanic rocks, and a thick layer of sediment.

Geologists refer to places where rifting takes place today as *active rifts*. Active rifting produced the *Basin and Range Province*, a region in the western United States in which numerous narrow mountain ridges are separated by flat valleys (Fig. 4.18b). The mountain ridges are blocks of crust that slipped down faults, whereas the valleys are sediment-filled basins. The Basin and Range Province is the widest rift on our planet. Another active rift, the *East African Rift*, extends in a northsouth direction for over 3,500 km (Fig. 4.18c). To astronauts in orbit, the rift looks like a giant gash in the crust. On the ground, it consists of a deep rift valley bordered on both sides by high cliffs formed by faulting. Along the length of the rift, several major volcanoes smoke and fume—these include snow-crested Mt. Kilimanjaro, towering over 6 km above the savannah. The East African Rift links to the Red Sea and the Gulf of Aden, both of which are narrow oceans in which rifting has gone to completion and new mid-ocean ridges have formed. At its north end, the Red Sea Rift dies out in the Gulf of Suez (Fig. 14.18d).

Collision

After the breakup of Pangaea, India was a small, isolated continent that lay far to the south of Asia. But subduction consumed the ocean between India and Asia, and India moved northward, finally slamming into the southern margin of Asia about 40 to 50 Ma. Continental crust is too buoyant to subduct completely. (It may seem strange to think of rock as being buoyant, but continental crustal rock has a density of about 2.8 g/cm^3, almost 18% less than that of mantle rock, which has a density of about 3.4 g/cm^3.) So when India collided with Asia, the attached oceanic plate broke off and sank down into the deep mantle, but the crust of India slid only partly under Asia before it couldn't go any farther. Its continued northward movement pushed it into Asia, squeezing the rocks and sediment that once lay between the two continents into the 8-km-high welt that we now know as the Himalayas. During this process, not only did the surface of the Earth rise, but the crust

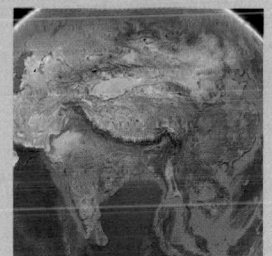

FIGURE 4.18 During the process of rifting, lithosphere stretches.

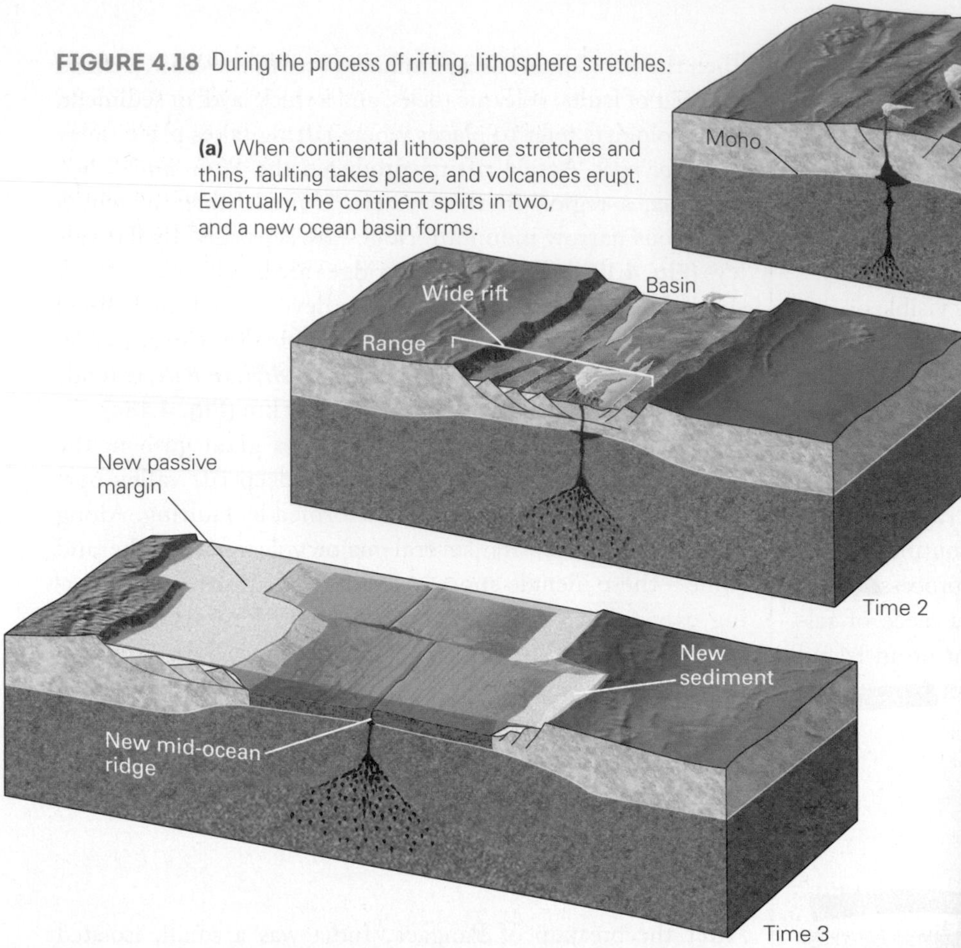

(a) When continental lithosphere stretches and thins, faulting takes place, and volcanoes erupt. Eventually, the continent splits in two, and a new ocean basin forms.

Moho

Time 1

Wide rift

Range

Basin

Time 2

New passive margin

New mid-ocean ridge

New sediment

Time 3

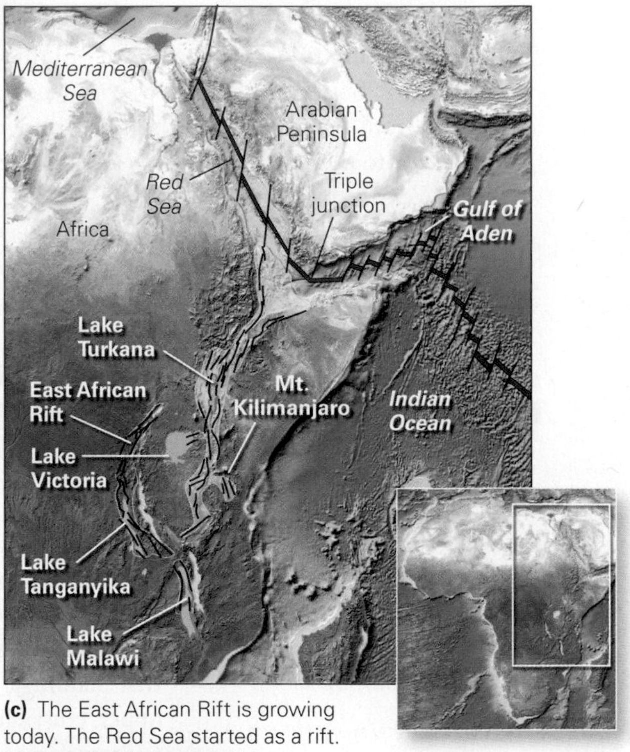

(c) The East African Rift is growing today. The Red Sea started as a rift. The inset shows map locations.

Mediterranean Sea

Arabian Peninsula

Red Sea

Triple junction

Africa

Gulf of Aden

Lake Turkana

East African Rift

Mt. Kilimanjaro

Indian Ocean

Lake Victoria

Lake Tanganyika

Lake Malawi

(b) The Basin and Range Province is a rift. Faulting bounds the narrow north-south-trending mountains, which are separated by basins. The arrows indicate the direction of stretching.

Snake River Plain

Reno

Salt Lake City

Basin and Range

Colorado Plateau

Sierra Nevada

San Andreas fault

Rio Grande Rift

N

250 km

(d) Astronauts can see how rifting has opened up gulfs on either side of the Sinai Peninsula.

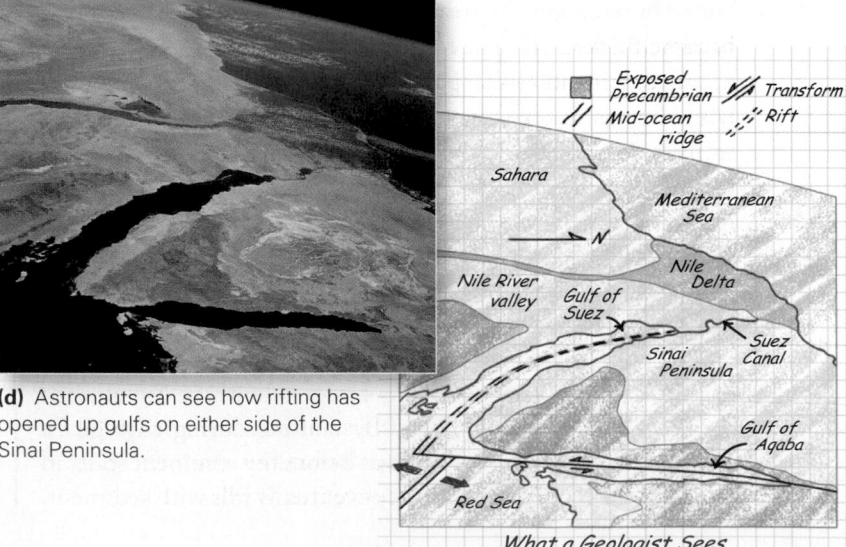

Exposed Precambrian

Transform

Mid-ocean ridge

Rift

Sahara

Mediterranean Sea

N

Nile River valley

Gulf of Suez

Nile Delta

Sinai Peninsula

Suez Canal

Gulf of Aqaba

Red Sea

What a Geologist Sees

FIGURE 4.19 Continental collision.

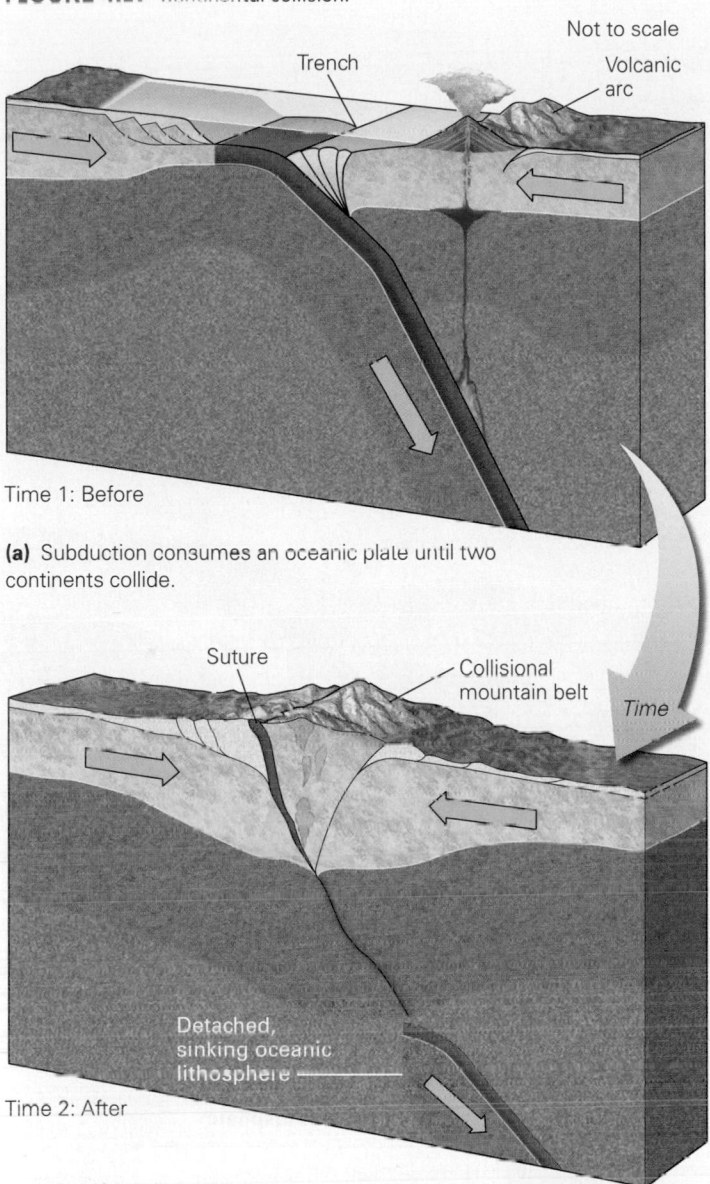

Not to scale

Trench

Volcanic arc

Time 1: Before

(a) Subduction consumes an oceanic plate until two continents collide.

Suture

Collisional mountain belt

Time

Detached, sinking oceanic lithosphere

Time 2: After

(b) After the collision, the oceanic plate detaches and sinks into the mantle. Rock caught in the collision zone gets broken, bent, and squashed and forms a mountain range.

became thicker. In fact, the crust beneath the Himalayas has attained a thickness of 60 to 70 km, nearly twice the thickness of normal continental crust. The boundary between what were once two separate continents is called a **suture**; slivers of ocean crust trapped between colliding continents may be found along the plane of a suture.

Geoscientists refer to the process during which two buoyant pieces of lithosphere converge and squeeze together as **collision (Fig. 4.19)**. Not all collisions involve two full-sized continents—some involve island arcs, continental slivers, hotspot volcanoes, or broad areas of unusually thick ocean crust called oceanic plateaus. Regardless of what collides, when a collision finishes, the convergent boundary that once existed between the two colliding pieces ceases to exist.

Collisions have yielded some of the Earth's highest mountain ranges, such as the Himalayas and the Alps. They have also produced major mountain ranges in the geologic past. These old ranges eventually eroded away, so today we see only their relics. For example, the Appalachian Mountains in the eastern United States formed as a consequence of three collisions. After the last one, a collision between Africa and North America that happened at about 280 Ma, North America became part of the Pangaea supercontinent. We'll discuss collisional mountain belts further in Chapter 11.

TAKE-HOME MESSAGE

Rifting can split a continent in two and can lead to the formation of a new divergent boundary. When two buoyant crustal blocks, such as continents and island arcs, collide, a mountain belt forms and subduction ceases.

QUICK QUESTION: Do rifting and collision affect crustal thickness? If so, how?

4.8 Moving Plates

Forces Acting on Plates

We've now discussed the many facets of plate tectonics theory (**Geology at a Glance**, pp. 110–111). But to complete the story, we need to address the major question Wegener couldn't answer: what drives plate motion? When geoscientists first proposed plate tectonics, they speculated that the plates moved simply because convective flow in the asthenosphere actively dragged them along, as if the plates were rafts on a flowing river. In fact, early images depicting plate motion showed simple elliptical *convection cells*—flow paths—in the asthenosphere. In these images, the cells were positioned such that asthenosphere flowed up beneath mid-ocean ridges and sank down at subduction zones (**Fig. 4.20a**). At first glance, this model looked plausible, but on closer examination, it failed. Among other reasons, it is impossible to draw a three-dimensional global configuration of convection cells that can explain the complex geometry of real plate boundaries on the Earth. Gradually, geoscientists came to the conclusion that while convective flow does occur within the asthenosphere, it does not drive plate motion alone. In other words, hot asthenosphere does rise in some places and sink in others because of temperature contrasts, and this convection does influence plate motion. But the local directions of convective flow do not necessarily control the local directions of plate motion.

The Theory of Plate Tectonics

Hot-spot volcano

Transform boundary

Volcanic arc

Trench

Continental rift

Convergent boundary

Subducting oceanic lithosphere

Continental crust

Continental lithosphere

Collisional mountain belt

Lithospheric mantle

Asthenosphere

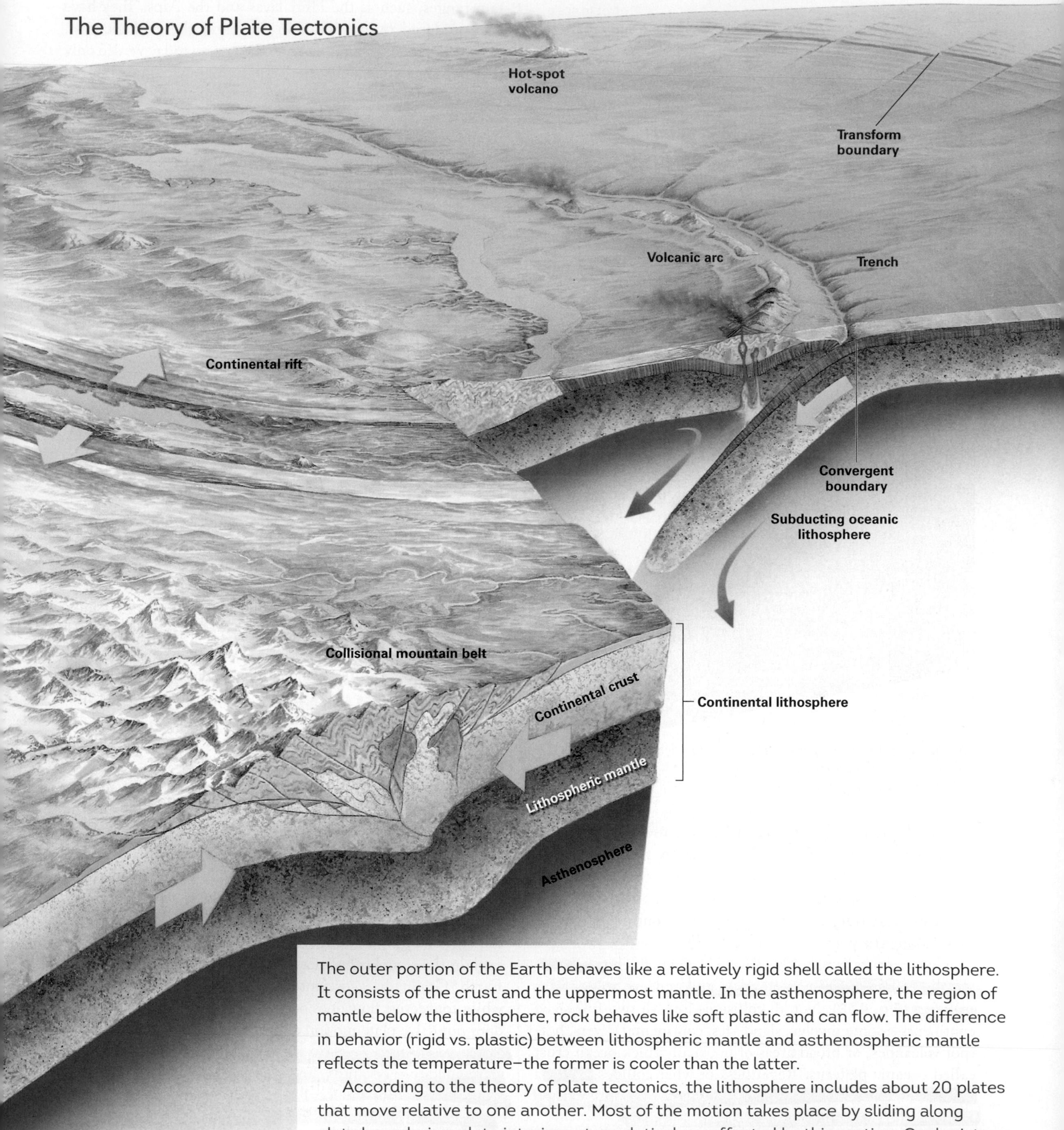

The outer portion of the Earth behaves like a relatively rigid shell called the lithosphere. It consists of the crust and the uppermost mantle. In the asthenosphere, the region of mantle below the lithosphere, rock behaves like soft plastic and can flow. The difference in behavior (rigid vs. plastic) between lithospheric mantle and asthenospheric mantle reflects the temperature—the former is cooler than the latter.

According to the theory of plate tectonics, the lithosphere includes about 20 plates that move relative to one another. Most of the motion takes place by sliding along plate boundaries; plate interiors stay relatively unaffected by this motion. Geologists recognize three kinds of plate boundaries. (1) *Convergent boundaries*: Here two plates

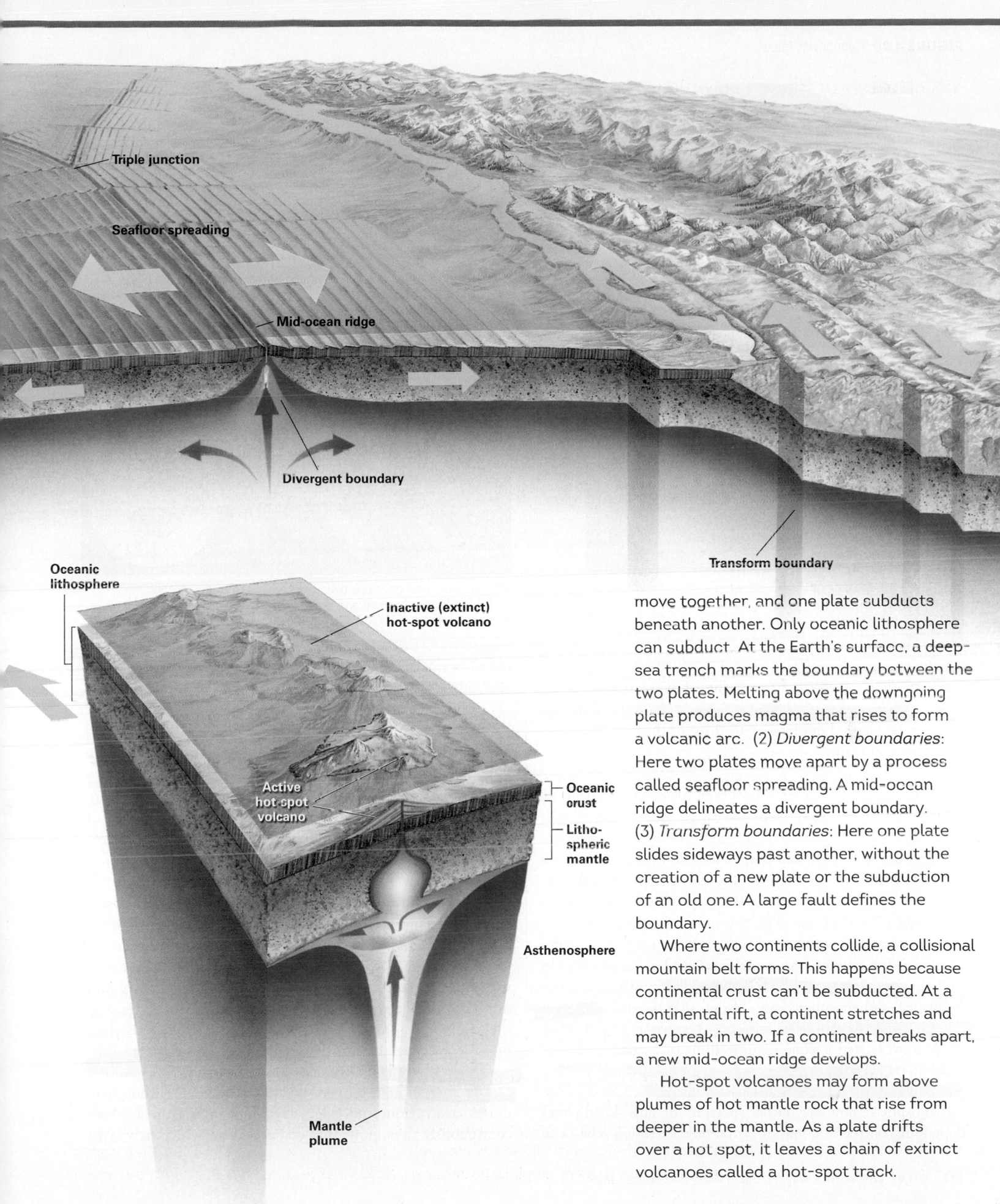

Triple junction

Seafloor spreading

Mid-ocean ridge

Divergent boundary

Transform boundary

Oceanic lithosphere

Inactive (extinct) hot-spot volcano

Active hot-spot volcano

Oceanic crust

Lithospheric mantle

Asthenosphere

Mantle plume

move together, and one plate subducts beneath another. Only oceanic lithosphere can subduct. At the Earth's surface, a deep-sea trench marks the boundary between the two plates. Melting above the downgoing plate produces magma that rises to form a volcanic arc. (2) *Divergent boundaries:* Here two plates move apart by a process called seafloor spreading. A mid-ocean ridge delineates a divergent boundary. (3) *Transform boundaries:* Here one plate slides sideways past another, without the creation of a new plate or the subduction of an old one. A large fault defines the boundary.

Where two continents collide, a collisional mountain belt forms. This happens because continental crust can't be subducted. At a continental rift, a continent stretches and may break in two. If a continent breaks apart, a new mid-ocean ridge develops.

Hot-spot volcanoes may form above plumes of hot mantle rock that rise from deeper in the mantle. As a plate drifts over a hot spot, it leaves a chain of extinct volcanoes called a hot-spot track.

FIGURE 4.20 Plate-driving forces.

Old idea

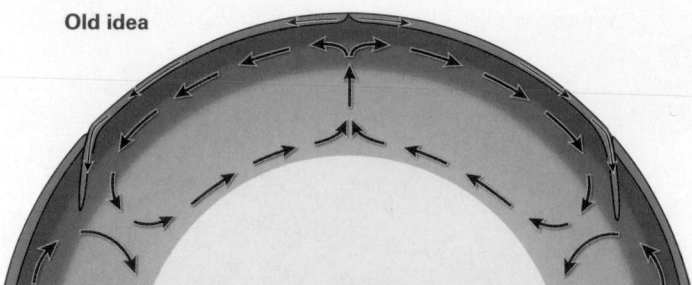

(a) The old, incorrect, image of simple convection cells in the asthenosphere.

Modern image

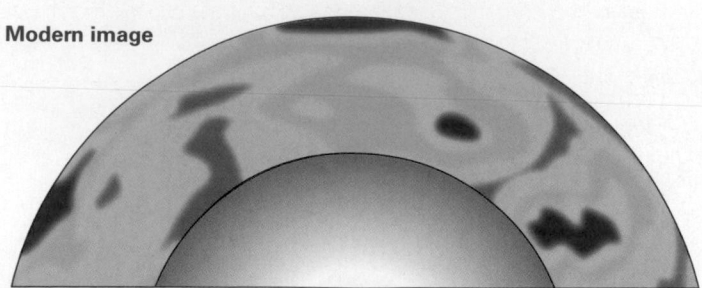

(b) The colors represent the difference between the observed velocity of seismic waves and the expected velocity. Geologists interpret red areas to be warmer, upwelling regions and blue areas to be cooler, downwelling regions.

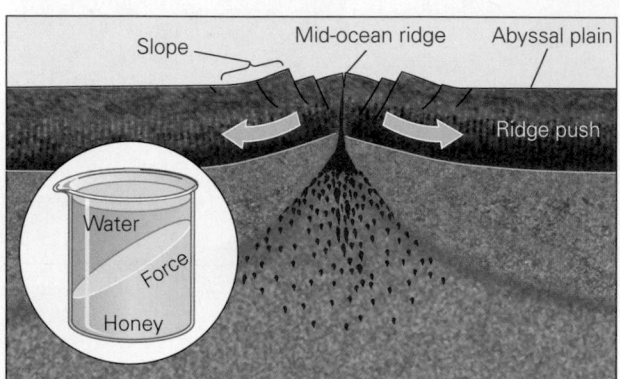

(c) Ridge-push force develops because the region of a rift is elevated. Like a wedge of honey with a sloping surface, the mass of the ridge pushes sideways.

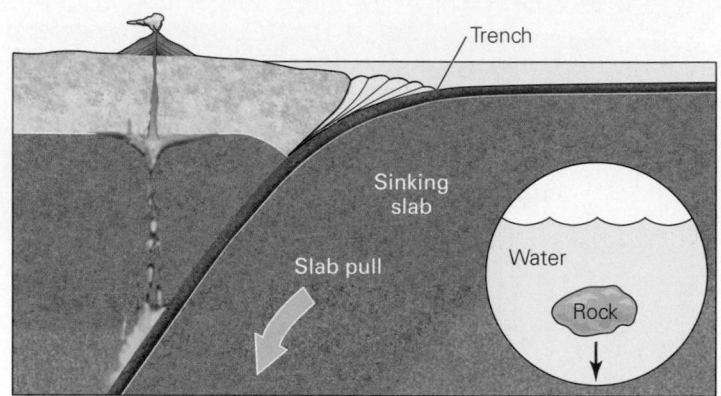

(d) Slab-pull force develops because lithosphere is denser than the underlying asthenosphere and sinks like a stone in water (though much more slowly).

Researchers now prefer a model in which convection plays a role in plate motion, but is not the whole story. Two other forces, ridge push and slab pull, contribute significantly to driving the plates. Let's look at all three of these *plate-driving mechanisms* in turn.

Convection Recall that lithosphere forms at a mid-ocean ridge, then moves away from the ridge until, eventually, it sinks back into the mantle at a trench. Because the material forming the plate starts out hot, cools, and then sinks, we can view the plate itself as the top of a convection cell and plate motion as a form of convection. But in this view, convection is effectively a consequence of plate motion, not the cause.

Can mantle convection beneath plates actually cause plates to move? The answer may come from studies demonstrating that the mantle contains zones of **upwelling**, where warmer (buoyant) asthenosphere rises from depth, and zones of **downwelling**, where cooler asthenosphere sinks (Fig. 4.20b). As it moves from a zone of upwelling to a zone of downwelling, the flowing asthenosphere probably does exert a force, or *shear*, on the bases of overlying plates. Conceivably, asthenospheric flow may either speed up or slow down plates depending on the orientation of the flow direction relative to

the movement direction of the overlying plate. New research also suggests that asthenospheric flow pushes on downgoing plates and affects the angle at which they sink.

Ridge-push force **Ridge-push force** is an outward-directed force that contributes to moving plates away from a mid-ocean ridge axis. It develops because the lithosphere of mid-ocean ridges lies at a higher elevation than that of the adjacent abyssal plains (Fig. 4.20c). To understand ridge-push force, imagine that you have a glass containing a layer of water over a layer of honey. By tilting the glass momentarily and then returning it to its upright position, you can create a temporary slope in the boundary between these substances. While the boundary has this slope, gravity causes the elevated honey to push against the glass adjacent to the side where the honey surface lies at lower elevation. The geometry of a mid-ocean ridge resembles this situation, for the surface of the seafloor lies higher along a mid-ocean ridge axis than in adjacent abyssal plains. Gravity causes the elevated lithosphere at the ridge axis to push on the lithosphere that lies farther from the axis, making it move away. As lithosphere moves away from the ridge axis, new hot asthenosphere continuously rises, providing material from which new crust

FIGURE 4.21 Relative and absolute plate velocities. The black arrows show the rate and direction at which the plate on one side of a boundary is moving with respect to the plate on the other side. Outward-pointing arrows indicate spreading (divergent boundaries), inward-pointing arrows indicate subduction (convergent boundaries), and parallel arrows show transform motion. The length of an arrow represents the velocity. The red arrows show the velocity of the plates with respect to a fixed point in the mantle.

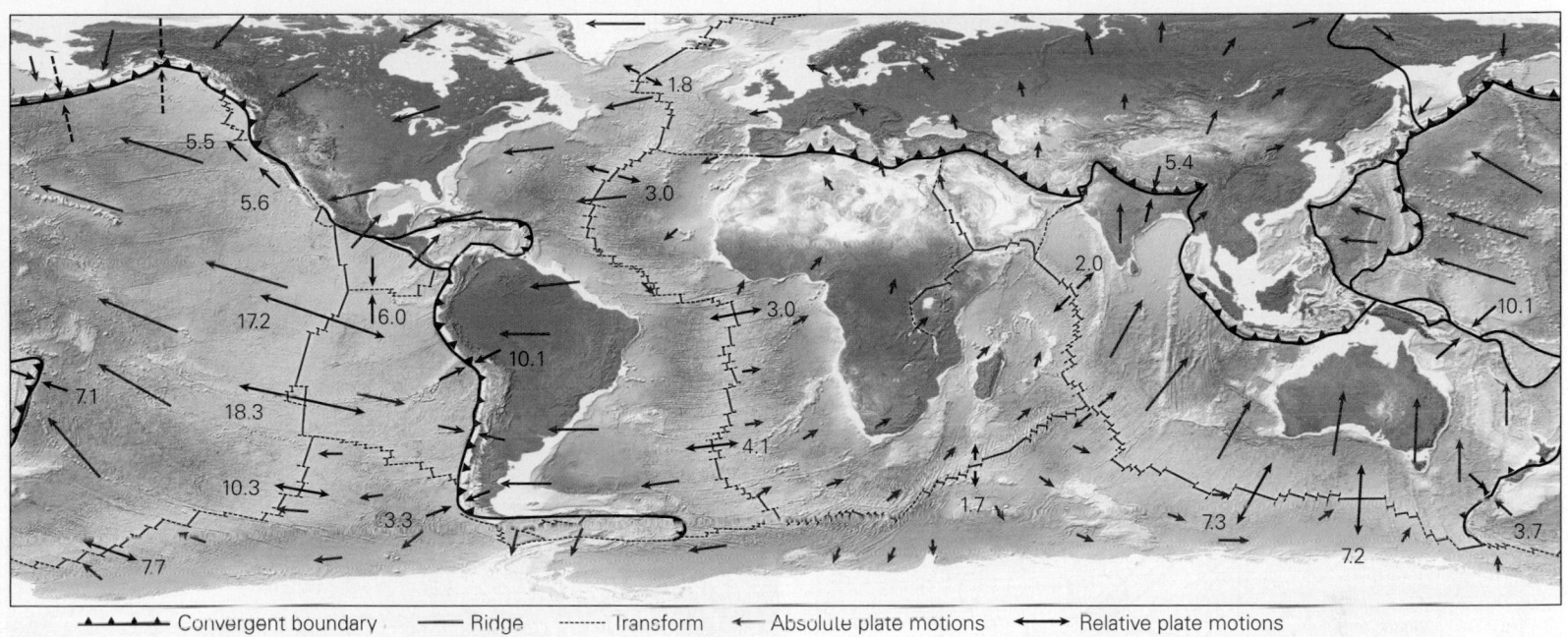

◄▲▲▲▲ Convergent boundary ——— Ridge ------- Transform ◄— Absolute plate motions ◄——► Relative plate motions
(5.5 cm per year)

and new lithospheric mantle forms. Note that the upward movement of asthenosphere beneath a mid-ocean ridge is a *consequence* of seafloor spreading, not the cause.

Slab-pull force **Slab-pull force** arises simply because lithosphere that was formed more than 10 million years ago has become denser than asthenosphere, so it can sink into the asthenosphere (Fig. 4.20d). Therefore, once any part of an oceanic plate starts to sink, it gradually pulls the rest of the plate along behind it, like an anchor pulling down the anchor line.

The Velocity of Plate Motions

How fast do plates move? It depends on your frame of reference. The movement of one plate with respect to another is called **relative plate velocity,** whereas the movement of a plate with respect to a fixed location inside the Earth is called **absolute plate velocity** (Fig. 4.21). To illustrate this concept, imagine two cars (A and B) speeding in the same direction down the highway. From the viewpoint of a tree along the side of the road, A zips by at 100 km an hour, while B moves at 80 km an hour. But relative to B, A moves at only 20 km an hour. The motion of a car relative to the tree represents its absolute motion, whereas the motion of one car with respect to another represents its relative motion.

One way of determining relative plate motions comes from the study of marine magnetic anomalies. If you measure the distance between a mid-ocean ridge axis and a magnetic

anomaly of known age on the seafloor formed at the ridge, then you can calculate the velocity of a plate relative to the ridge axis using the equation

$$\text{Velocity relative to the ridge} = \text{Distance from ridge} \div \text{Age of anomaly}$$

To determine the relative velocity of the plate on one side of the ridge with respect to the plate on the other side, simply multiply by 2, because spreading at ridges takes place symmetrically.

One way of estimating absolute plate motions comes from the assumption that the location of a hot spot does not change much over time, in which case the hot-spot track on a plate provides a record of the plate's absolute velocity. (In reality, hot spots are not perfectly fixed, so this method provides only an approximation of absolute plate velocity.) For example, by measuring the ages of volcanic rocks collected from islands and seamounts of the Hawaiian-Emperor chain that lie at various distances from the Big Island, geologists can estimate the absolute velocity of the Pacific Plate (see Fig. 4.17a). We obtain the rate of motion by the equation

$$\text{Rate} = \text{Distance of a seamount from Hawaii} \div \text{Age of the seamount}$$

The orientation of the chain gives the direction of motion. Note that the Hawaiian chain runs northwest, whereas the Emperor chain trends north-northwest (see Fig. 4.17a). Since rocks from the seamount at the bend are 47 million years old, geologists conclude that the direction of Pacific Plate motion changed at 47 Ma.

Working from the calculations described earlier, geologists have determined that relative plate motions on the Earth today occur at rates of 1 to 15 cm per year. Can we detect such slow motions? Yes, by using the *global positioning system (GPS)*, the same technology that automobile drivers can use to find their destinations. By setting up a fixed GPS receiver that collects data over many months, geologists can detect displacements as small as about 2 mm per year. Since plates move at 5 to 75 times this rate, we can indeed see the plates move—and this observation serves as the ultimate proof of plate tectonics (**Fig. 4.22**).

> **Did you ever wonder . . .**
> whether we can really "see" continents drift?

FIGURE 4.22 GPS is used to measure plate motions at many locations on the Earth.

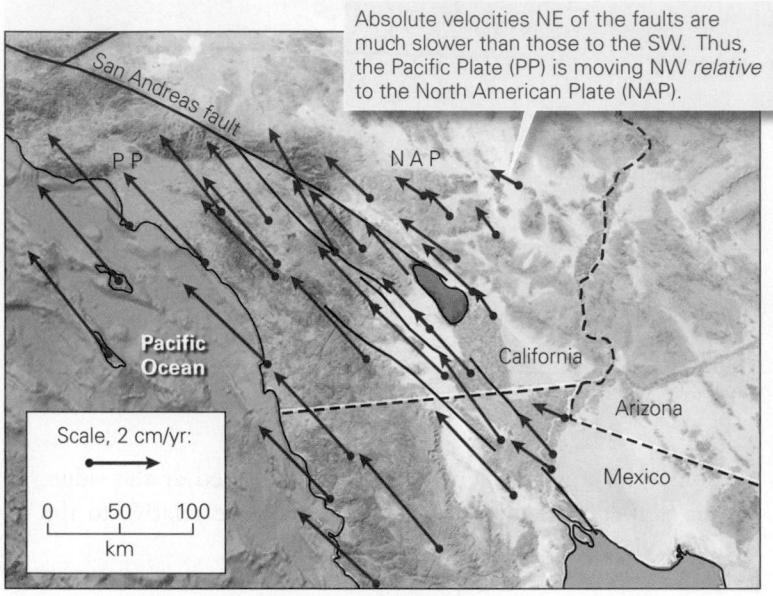

(a) GPS measurements in southern California show that the region west of the San Andreas fault system, a plate boundary, is moving northwest up to 6 cm per year. The length of an arrow represents the velocity.

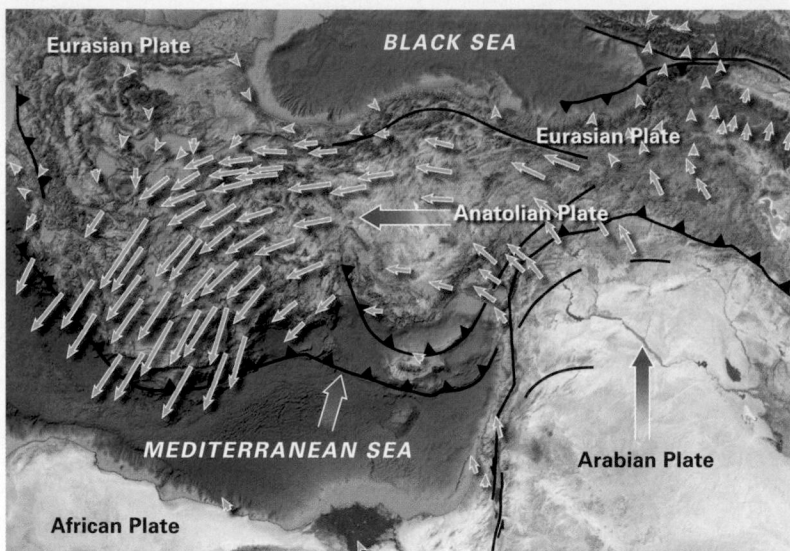

(b) GPS measurements indicate that Turkey is moving west relative to Asia. Each yellow arrow indicates the velocity of the point at the end of the arrow. Red arrows give overall plate-movement direction.

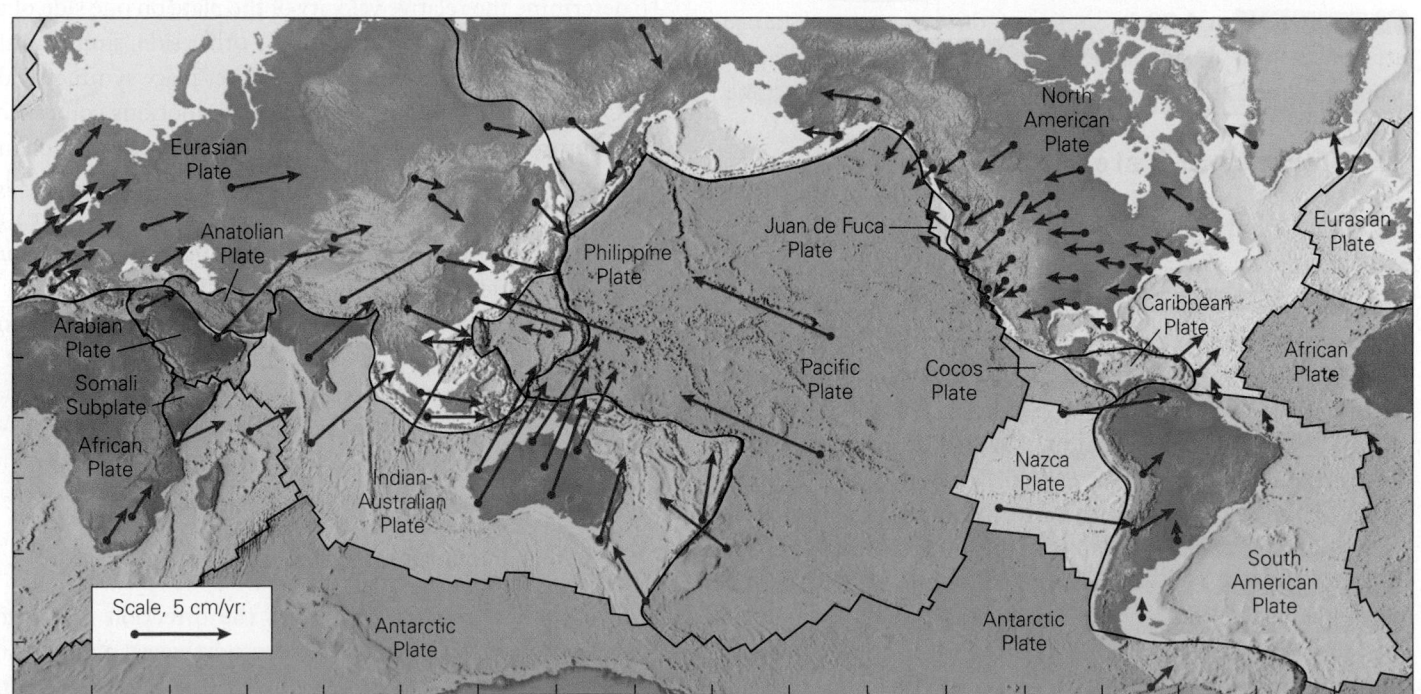

(c) A global map of plate velocities, determined by GPS.

The rates of plate motion are very slow in comparison to the rate at which we walk or drive. In fact, plates move at about the rate that your fingernails grow. But even at these slow rates, plate motions can yield large displacements, given the immensity of geologic time. For example, at a rate of 2.5 cm per year, a plate can move 25 km in a million years, and 2,500 km in 100 million years! At this rate, the Atlantic Ocean opened to its present width, 4,500 km, in the 180 million years since the breakup of Pangaea.

Taking into account many data sources that define the motion of plates, geologists have greatly refined the image of continental drift that Wegener tried so hard to prove nearly a century ago. We can now see that the map of our planet's surface has evolved radically during the past 400 million years and even before **(Fig. 4.23)**.

TAKE-HOME MESSAGE

Plate motion takes place in response to ridge push, slab pull, and convective shear. This motion takes place at rates of 1 to 15 cm per year. Relative motion specifies the rate that a plate moves relative to its neighbor, whereas absolute motion specifies the rate that a plate moves relative to a fixed point beneath the plate. GPS measurements can now detect relative plate motions directly.

QUICK QUESTION: What causes the map-view bend in the Hawaiian-Emperor seamount chain?

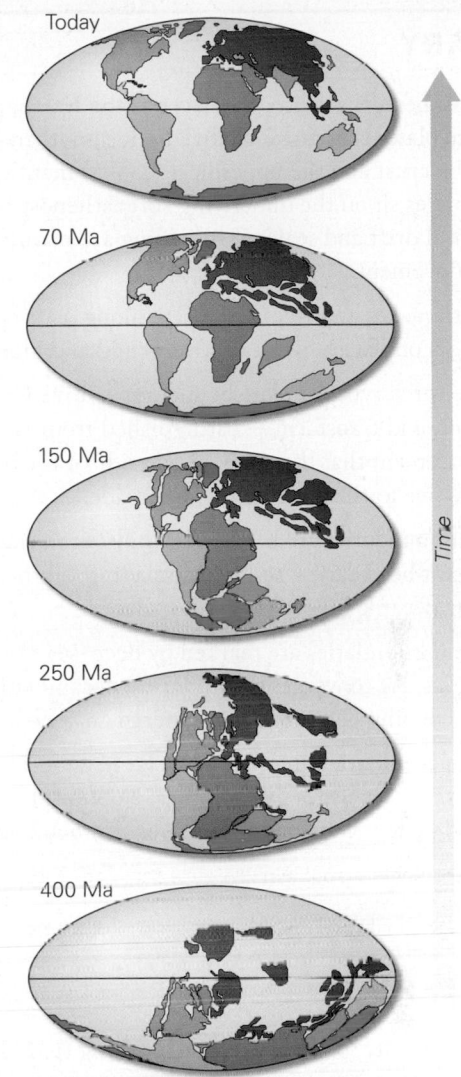

FIGURE 4.23 Due to plate tectonics, the map of the Earth's surface slowly changes. Here we see the assembly and the later breakup of Pangaea during the past 400 million years.

Today

70 Ma

150 Ma

250 Ma

400 Ma

Time

ANOTHER VIEW A GPS (Global Positioning System) station set up to monitor land movements in southern California. Instruments such as these can see plate tectonics in real time!

Solar panel

GPS receiver

Chapter 4 Review

SUMMARY

- The lithosphere, the rigid outer layer of the Earth, consists of discrete plates that move relative to one another. Plates include the crust and the uppermost (cooler) mantle. Lithosphere plates sit on the underlying soft asthenosphere. Continental drift and seafloor spreading are manifestations of plate movement.

- Most earthquakes and volcanoes occur along plate boundaries; the interiors of plates remain relatively rigid and intact.

- There are three types of plate boundaries—divergent, convergent, and transform—distinguished from one another by the movement that the plate on one side of the boundary makes relative to the plate on the other side.

- Divergent boundaries are marked by mid-ocean ridges. At divergent boundaries, seafloor spreading, a process that forms new oceanic lithosphere, takes place.

- Convergent boundaries are marked by deep-sea trenches and volcanic arcs. At convergent boundaries, oceanic lithosphere of the downgoing plate subducts beneath an overriding plate.

- Transform boundaries are marked by large faults at which one plate slides sideways past another. No new plate forms and no old plate is consumed at a transform boundary.

- Triple junctions are points where three plate boundaries intersect.

- At a hot spot, volcanism occurs at an isolated volcano. As a plate moves over the hot spot, the volcano moves off and becomes inactive, and a new volcano forms over the hot spot; the chain of volcanoes defines a hot-spot track. Hot spots may form above mantle plumes.

- A large continent can split into two smaller ones by the process of rifting. During rifting, continental lithosphere stretches and thins. If it finally breaks apart, a new mid-ocean ridge forms and seafloor spreading begins.

- Convergent boundaries cease to exist when a buoyant piece of crust moves into the subduction zone. When that happens, collision occurs. Collision can thicken crust and build large mountain belts.

- Ridge-push force and slab-pull force contribute to driving plate motions. Plates move at rates of about 1 to 15 cm per year. We can describe the motions of plates relative to one another or relative to a fixed point. Modern satellite measurements can detect these motions.

GUIDE TERMS

absolute plate velocity (p. 113)
accretionary prism (p. 101)
active margin (p. 93)
asthenosphere (p. 91)
black smoker (p. 97)
collision (p. 109)
continental rift (p. 107)
convergent (consuming) boundary (p. 99)
divergent boundary (p. 95)

downwelling (p. 112)
epicenter (p. 95)
focus (p. 95)
fracture zone (p. 102)
hot spot (p. 104)
hot-spot track (p. 104)
lithosphere (p. 91)
lithosphere plate (p. 93)
lithospheric mantle (p. 91)
mantle plume (p. 104)

passive margin (p. 93)
plate (p. 93)
plate boundary (p. 93)
plate tectonics (p. 91)
relative plate velocity (p. 113)
ridge-push force (p. 112)
rifting (p. 107)
scientific revolution (p. 91)
seismic belt (p. 95)
slab-pull force (p. 113)

subduction (p. 99)
suture (p. 109)
transform boundary (p. 103)
trench (p. 99)
triple junction (p. 104)
upwelling (p. 112)
volcanic arc (p. 101)
volcanic island arc (p. 101)
Wadati-Benioff zone (p. 100)

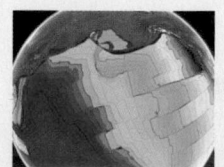 **GEOTOURS** *THIS CHAPTER'S GEOTOURS WORKSHEET (B) FEATURES QUESTIONS AND GOOGLE EARTH SITES ON:*

- Divergent boundaries
- Convergent boundaries

- Transform boundaries
- Hot spots

- Seafloor spreading

REVIEW QUESTIONS

The letters following each Review Question refer to the corresponding Learning Objective from the Chapter Opener.

1. What are the characteristics of a lithosphere plate? **(A)**

2. How does oceanic lithosphere differ from continental lithosphere? **(A)**

3. What are the basic premises of plate tectonics? **(B)**

4. How do we identify a plate boundary? **(C)**

5. Describe the three types of plate boundaries. **(D)**

6. How does oceanic crust form along a mid-ocean ridge? **(E)**

7. Describe the process of subduction at a consuming boundary. **(E)**

8. Describe the major features of a convergent boundary. Identify them in the drawing. **(D)**

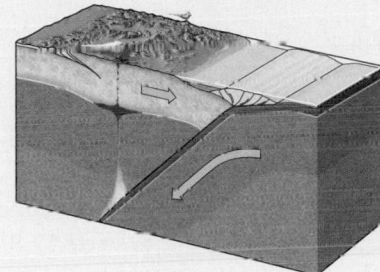

9. Describe the motion that takes place on a transform boundary. **(D)**

10. What phenomenon may produce a hot-spot track, and how can hot-spot tracks be used to track the past motions of a plate? What direction is the Pacific Plate moving in the figure? **(F)**

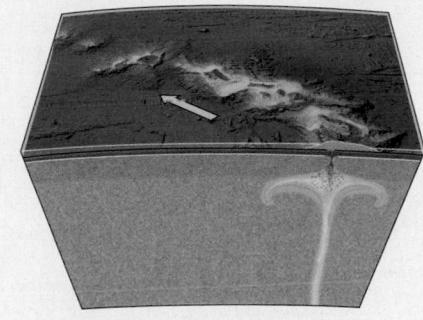

11. Describe the characteristics of a continental rift, and give examples of where rifting takes place today. **(E)**

12. Describe the process of continental collision, and give examples of where this process has occurred. **(E)**

13. Discuss the major forces that move lithosphere plates. **(B)**

14. Explain the difference between relative plate velocity and absolute plate velocity. **(B)**

ON FURTHER THOUGHT

15. The Pacific Plate moves north relative to the North American Plate at a rate of 6 cm per year. How long will it take Los Angeles (a city on the Pacific Plate) to move northward by 480 km, the present distance between Los Angeles and San Francisco? **(B)**

16. Look at a map of the western Pacific Ocean and examine the position of Japan with respect to mainland Asia. Japan's crust contains rocks similar to those of eastern Asia. Presently, many active volcanoes occur along the length of Japan. With these facts in mind, explain how the Japan Sea (the region between Japan and the mainland) formed. **(C)**

ONLINE RESOURCES

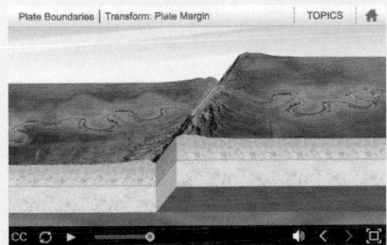

Animations
This chapter features animations on the different types of plate boundaries.

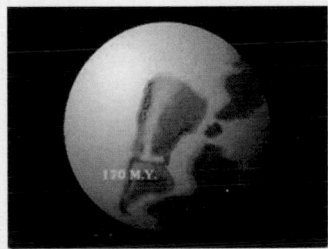

Videos
This chapter includes videos on plate motions over time, subduction trenches, and deep ocean volcanoes.

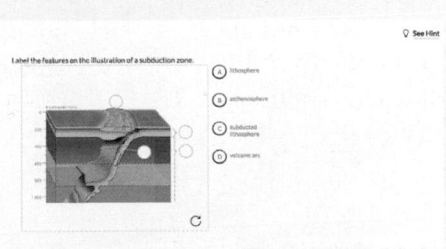

Smartwork5
This chapter features questions on topics including plate boundaries, plate formation, and plate movement.

PART II

EARTH MATERIALS

Imagine that you are standing on the prairie of the midwestern United States. The grass beneath your feet roots in a rich black soil, which grades downward into sediment (clay, sand, and gravel). The sediment covers countless layers of sedimentary rock (sandstone, shale, coal, and limestone), forming a veneer that in turn covers a "basement" of granite and other very ancient rocks. Descending still deeper takes you through 40 km of continental crust down to Moho. Below the Moho? Even more rock—almost 3,000 km of rock—down to the molten iron alloy of the outer core. How did these materials form? How do we describe and classify them? Are Earth materials permanent, or can they change over time?

In this part of the book, we learn about the great variety of materials that make up the crust and mantle of the solid Earth. We begin, in Chapter 5, with minerals, the building blocks of rock. Then, in Interlude A, we see how geologists distinguish among three categories of rock—igneous, sedimentary, and metamorphic based on how they form. In each succeeding chapter (6, 7, and 8), we examine one rock category in greater detail. Interlude B helps explain the origin of the sediments that develop into sedimentary rocks and also introduces soil. Finally, Interlude C shows us how materials in the dynamic Earth System can change over time as they pass through the rock cycle.

◄ A cliff face along the shore of Lake Superior, in Michigan, exposes bedrock consisting of Precambrian granite, overlain by a layer of loose sediment. The sediment has evolved into a soil that anchors the roots of a forest.

CHAPTER 5

Patterns in Nature: Minerals

By the end of this chapter, you should be able to . . .

A. explain why the term *mineral* has a very special meaning in a geologic context.

B. describe the processes by which minerals can form.

C. explain how geologists organize thousands of different minerals into just a few classes.

D. specify which minerals are the most common ones on the Earth, and describe how they are classified.

E. identify common mineral specimens based on their properties.

F. distinguish ordinary minerals from gems, and describe how to produce the shiny facets of gems.

> I died a mineral, and became a plant. I died as plant and rose to animal,
> I died as animal and I was Man. Why should I fear?
>
> —Jalal-Uddin Rumi (Persian mystic and poet, 1207–1273)

5.1 Introduction

In Greek legend, the god Dionysus, in a drunken rage, vowed to kill the next mortal that he saw. Just then, a beautiful young woman named Amethyst walked by, and Dionysus ordered two fearsome tigers to attack her. The goddess Artemis prevented a tragedy by changing Amethyst into a pure white statue made of quartz, which was much harder than the tigers' teeth. The statue was so beautiful that Dionysus sobered up and regretted his rashness. In remorse, he spilled his wine onto the statue as an offering, and the wine stained the quartz, turning it into purple amethyst (see chapter opening photo). The word *amethyst* comes from the Greek *amethustos*, meaning not intoxicated. For centuries afterward, amethyst was thought of as an antidote for drunkenness. In fact, revelers would wear amulets of amethyst, or drink from goblets made of it, or put fragments of it into their drinks, thinking that it somehow neutralized the alcohol in their wine. It doesn't—amethyst has no effect on alcohol and can't prevent inebriation. But legend aside, the beauty of amethyst has made it popular with jewelry makers for millennia (Fig. 5.1).

Amethyst, the maroon version of a common mineral, quartz, is one of about 4,000 minerals that **mineralogists**, people who specialize in the study of minerals, have identified so far. The list of known minerals continues to grow, for mineralogists discover about 50 new ones every year. Each mineral has a name. Some names come from Latin, Greek, German, or English words describing a certain characteristic of the mineral (for instance, albite comes from the Latin word for white, orthoclase comes from the German words meaning splits at right angles, and olivine has an olive color); some honor a person (sillimanite was named for Benjamin Silliman, a 19th-century mineralogist); some indicate the place where the mineral was first recognized (illite came from rocks in Illinois); and some refer to a particular element in the mineral (chromite contains chromium). The vast majority of mineral types are rare, forming only under special conditions. Fewer than 50 can be considered common rock-forming minerals of the Earth's crust.

Ancient Greek philosophers pondered minerals, and medieval alchemists puttered with them, but a true scientific description of minerals did not appear until 1556, when German physician Georgius Agricola's *De Re Metallica* (*On the Nature of Metals*) was published. In 1669, more than a century after Agricola's work, Nicholas Steno, a Danish scientist, discovered important geometric characteristics of minerals. Steno's work became the basis for the systematic study of minerals, a task that occupied many researchers during the next two centuries.

The examination of minerals with an optical microscope began in 1828, but although such work helped in mineral identification, it could not reveal the arrangement of atoms inside minerals. That understanding came in the early 20th century, when researchers developed methods for using X-rays to study mineral structure. By the middle of the century, new tools, such as electron microscopes and microprobes, made it possible to see details of mineral structure and to analyze mineral composition in a matter of seconds.

Why study minerals? Without exaggeration, we can say that minerals serve as the building blocks of our planet, for minerals make up most rocks and sediments of the solid Earth. Minerals are also important from a practical standpoint, in their use as resources by mankind (see Chapter 15). For example, *industrial minerals* provide the raw materials for

FIGURE 5.1 A royal crown containing a variety of valuable jewels.

◀ (facing page) This cluster of amethyst crystals, each about 1 cm across, precipitated from water passing through open spaces in a layer of basalt from southern Brazil.

FIGURE 5.2 Copper ore is a useful mineral that serves as a source of copper metal.

The malachite in this partially polished chunk grew in a succession of layers.

5 cm

(a) Malachite, a type of copper ore ($Cu_2[CO_3][OH]_2$), contains copper plus other chemicals.

(b) The copper used for pots comes from copper ore.

manufacturing chemicals, concrete, and wallboard; *ore minerals* are the source of valuable metals, like copper and gold **(Fig. 5.2)**, and yield energy resources, such as uranium; and *gems*—beautiful forms of certain minerals—delight the eye in jewelry. In recent decades, it has also become clear that certain minerals pose health and environmental hazards. No wonder **mineralogy**, the study of minerals, fascinates professionals and amateurs alike.

This chapter, an introduction to mineralogy, begins with the geologic definition of a mineral. We then look at how minerals form and at the main characteristics that enable us to identify minerals. Finally, we describe the basic scheme that mineralogists use to classify minerals. This chapter assumes that you understand fundamental concepts of matter and energy, especially the nature of atoms, molecules, and chemical bonds. If you are rusty on these topics, please study **Box 5.1** before going further.

5.2 What Is a Mineral?

In everyday English, the word *mineral* has many uses. When you play the game 20 Questions, a mineral is anything that's not animal or vegetable, and if you read food ingredients, the word refers to certain nutrients that people need in order to be healthy. Geologists use the term in a very specific way. To a geologist, a **mineral** is a naturally occurring, solid, crystalline material formed by geologic processes, which has a definable chemical composition. Almost all minerals are

inorganic. Let's pull apart this mouthful of a definition and examine its meaning in detail.

- *Naturally occurring:* True minerals grow in nature, not in factories. In recent decades, chemists have learned how to manufacture materials that have characteristics virtually identical to those of real minerals. Such materials can be referred to as *synthetic minerals*.

- *Formed by geologic processes:* Traditionally, this phrase implied that minerals were the result only of solidification of molten rock or direct precipitation from a water solution, processes that did not involve living organisms. Increasingly, however, geologists recognize life's involvement in the Earth System. So geologists now consider solid, crystalline materials produced by organisms to be minerals, too. To avoid confusion, the term **biogenic mineral** may be used when discussing such materials.

- *Solid:* Matter in the solid state can maintain its shape indefinitely, so it will not conform to the shape of its container. Minerals cannot be liquids (such as oil or water) or gases (such as air).

- *Crystalline material:* In a **crystalline material**, the atoms reside in an orderly, fixed pattern, locked in place by chemical bonds **(Fig. 5.3a)**. The three-dimensional geometric arrangement of atoms or ions that defines that pattern is called a **crystal structure**.

  > **Did you ever wonder . . .**
  > how a "cut crystal" glass differs from a clear quartz crystal?

- *Definable chemical composition:* This phrase simply means that it's possible to write a chemical formula for a mineral. Some minerals contain only one element—for example, diamond and graphite have the formula C because they consist entirely of carbon. Most minerals, however, are compounds of two or more elements. For example, quartz has the formula SiO_2—it contains the elements silicon and oxygen in the proportion of one silicon atom for every two oxygen atoms. Some mineral formulas are complicated: for example, the formula for biotite is $K(Mg,Fe)_3(AlSi_3O_{10})(OH)_2$. According to this formula, the proportion of magnesium to iron can vary in biotite.

- *Inorganic:* To understand the meaning of *inorganic*, we must first understand what we mean by *organic*. Organic chemicals consist of molecules that (1) include carbon-carbon and/or carbon-hydrogen bonds and (2) either form in living organisms or have structures similar to those of chemicals that form in living organisms. Sugar ($C_{12}H_{22}O_{11}$), fat, plastic, propane, and protein, for example, are organic chemicals. Some organic chemicals contain only carbon and hydrogen, while others include other elements, such as oxygen, nitrogen, or phosphorous. Almost all minerals are inorganic, in that they are not organic chemicals. But we have to add the qualifier "almost all" because mineralogists now consider a few dozen organic substances formed by "the action of geologic processes on organic materials" to be minerals. Examples include the crystals that grow in ancient deposits of bat guano.

With the geologic definition of a mineral in mind, we can distinguish between a mineral and a **glass**. Both minerals and glasses are solids, in that they can retain their shape indefinitely, but a mineral is crystalline, while a glass is not. This means that the atoms, ions, or molecules in a mineral are ordered into a geometric arrangement, like soldiers standing in formation, but those in a glass are arranged in a semi-chaotic way, in small clusters or chains that are neither oriented in the same way nor spaced at regular intervals, like guests at a party **(Fig. 5.3b)**.

Let's examine some everyday materials and see if they are minerals, geologically. Is motor oil a mineral? No—it's an organic liquid. Is table salt a mineral? Yes—it's a natural solid crystalline compound with the formula NaCl. Is the hard material making up the shell of an oyster considered to be a mineral? Yes, it's a biogenic mineral—it has the same composition and structure as an inorganic mineral. Is rock candy a mineral? No. Even though it is solid and crystalline, it's not made by geologic processes and it consists of sugar (an organic chemical).

FIGURE 5.3 The nature of crystalline and noncrystalline materials.

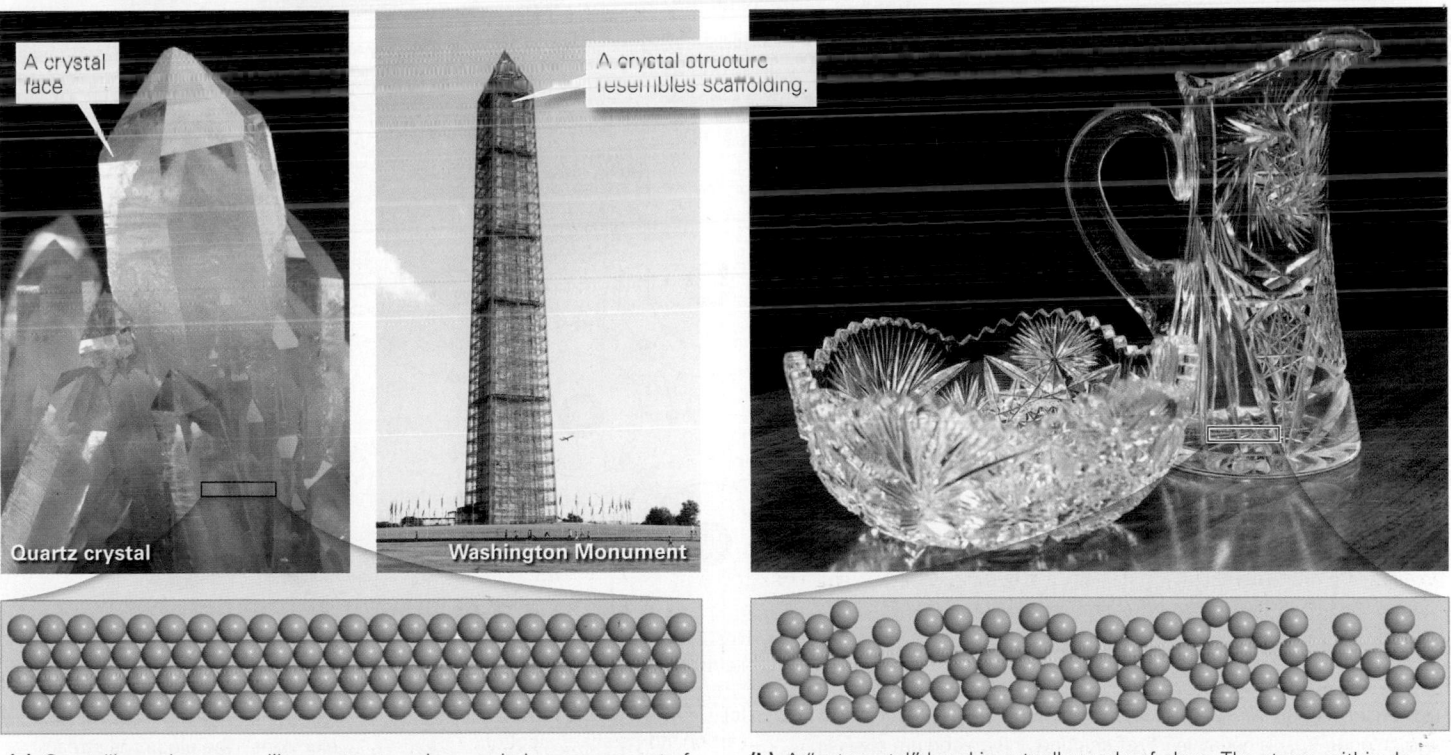

(a) Crystalline substances, like quartz, contain an orderly arrangement of atoms. The geometry of the arrangement defines the crystal structure.

(b) A "cut crystal" bowl is actually made of glass. The atoms within do not have an orderly arrangement.

BOX 5.1 SCIENCE TOOLBOX . . .

Some Basic Concepts from Chemistry: A Quick Review

To describe minerals, we must use the basic vocabulary of chemistry. We introduce this vocabulary below in an order that permits each successive term to use previous terms.

- *Element:* A pure substance that cannot be separated into other materials is an **element**. There are 92 naturally occurring elements. Each has a name (e.g., hydrogen, carbon, silicon, oxygen, uranium) and a corresponding symbol (e.g., respectively, H, C, Si, O, U).
- *Atoms and their components:* The smallest piece of an element retaining the characteristics of the element is an **atom**. Atoms are so small that over 5 trillion (5,000,000,000,000) could fit on the head of a pin. An atom consists of a nucleus surrounded by a cloud of orbiting electrons. A *nucleus* is a compact ball of protons and neutrons. (The only exception is the element hydrogen: the nucleus of a hydrogen atom contains only one proton and no neutrons.) The mass of an electron is only about 1/1,836 that of a proton.
- *Electron cloud:* The electron cloud consists of distinct orbitals, or electron shells, each of which contains a specific number of electrons **(Fig. Bx5.1a)**. The outermost shell defines the "surface" of an atom. The diameter of an electron cloud is about 100,000 times greater than that of

a nucleus, so an atom consists mostly of empty space.

- *Charge:* The charge of a particle characterizes the way the particle responds to an electrical current or to a magnet. Electrons have a negative charge, protons have a positive charge, and neutrons are neutral (they have no charge). Like charges repel (push away from each other) whereas unlike charges attract (pull toward each other).
- *Atomic number:* The number of protons in an atom of an element is the **atomic number** of the element. Hydrogen, the smallest atom, has an atomic number of 1, helium has an atomic number of 2, and uranium, the largest naturally occurring atom, has an atomic number of 92. The atomic number determines the elemental identity of an atom.
- *Atomic mass:* The **atomic mass** approximately equals the number of protons plus neutrons in an atom of an element. (Technically, this simple

sum is the "atomic mass number.") Hydrogen has an atomic mass of 1, helium has an atomic mass of 4, and the most common form of uranium has an atomic mass of 238.

- *Isotope:* Atoms that have the same atomic number, but a different atomic mass, are called *isotopes* of an element. For example, ^{238}U and ^{235}U are isotopes of uranium—both have atomic numbers of 92, but ^{238}U

FIGURE Bx5.1 Examples of states of matter and chemical bonds.

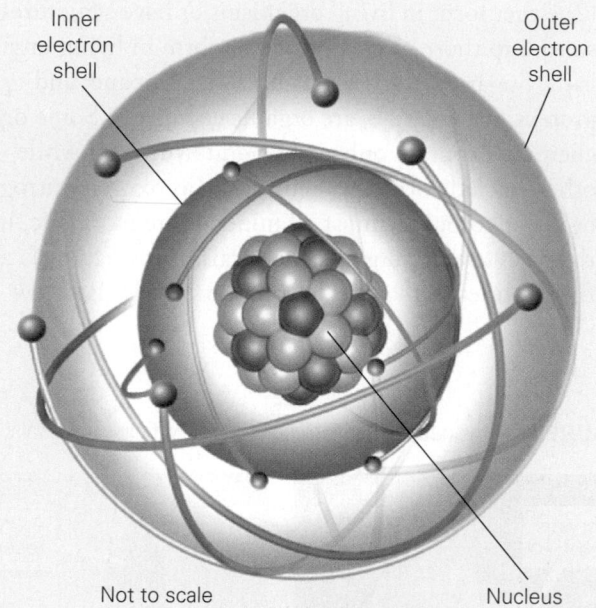

Inner electron shell

Outer electron shell

Not to scale

Nucleus

(a) A drawing of an atom.

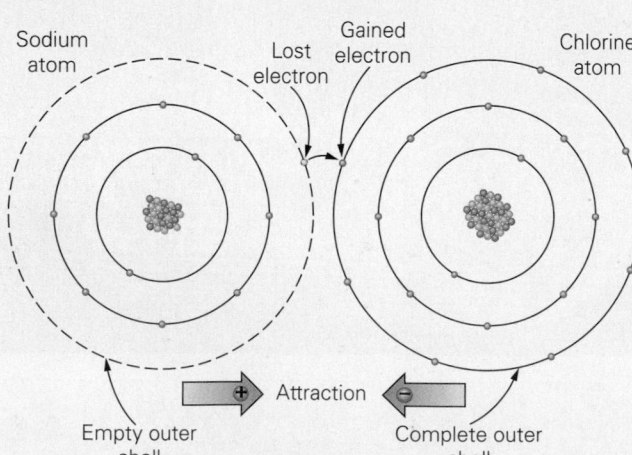

Sodium atom

Lost electron

Gained electron

Chlorine atom

Attraction

Empty outer shell

Complete outer shell

(b) An ionic bond forms between a positive ion of sodium (Na$^+$) and chloride (Cl$^-$), a negative ion of chlorine. When sodium gives up one electron to chlorine, so that both have filled shells, halite (NaCl) is produced.

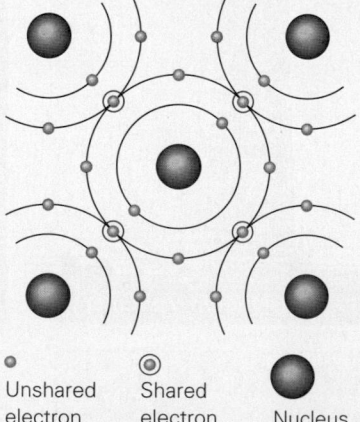

Unshared electron

Shared electron

Nucleus

(c) Covalent bonds form when carbon atoms share electrons so that all have filled electron shells.

(d) In metallically bonded material, nuclei and their inner shells of electrons float in a "sea" of free electrons. The electrons stream through the metal if there is an electrical current.

has 146 neutrons while ^{235}U has 143. (The atomic weight of an atom differs slightly from the atomic mass, because it reflects the proportions of different naturally occurring isotopes.)

- *Ion:* An atom that has the same number of electrons as protons is neutral; an atom that is not neutral is an **ion**. Chemists refer to an ion with excess negative charge (because it has more electrons than protons) as an *anion*, and to an ion with excess positive charge (because it has more protons than electrons) as a *cation*. We indicate the charge with a superscript. For example, Cl^- has a single excess electron; Fe^{2+} has lost two electrons.

- *Chemical bond:* An attractive force that holds two or more atoms together is a **chemical bond (Fig. Bx5.1b–d)**. Chemists recognize several different types of chemical bonds. *Covalent bonds* form when atoms share electrons. *Ionic bonds* form when a cation and anion (ions with opposite charges) get close together and attract each other. In materials with *metallic bonds*, electrons from the outer shell can move freely.

- *Molecule:* Two or more atoms held together by chemical bonds constitute a **molecule**. The atoms may be of the same element or of different elements.

- *Ionic molecule:* If a molecule gains or loses electrons, it becomes an *ionic molecule*. As an example, the carbonate ion (CO_3^{-2}) is an ionic molecule.

- *Compound:* A pure substance that can be subdivided into two or more elements is called a compound. The smallest piece of a compound that retains the characteristics of the compound is a molecule.

- *State of matter:* The form of a substance, which reflects the degree to which the atoms or molecules making up the matter are bonded together, is called the state of matter. The most common states at the Earth's surface are solid, liquid, and gas **(Fig. Bx5.1e)**. A solid can maintain its shape for a long time, a liquid will flow and conform to its container's shape, and a gas expands outward (when not confined). A fourth state, plasma, exists only at very high temperatures.

- *Change of state:* During a **change of state**, matter converts from one state to another. Such changes usually take place in response to a change in temperature, pressure, or both. During *condensation*, a gas becomes a liquid; during *evaporation*, a liquid becomes a gas; during *freezing*, a liquid becomes a solid; and during *sublimation*, a solid becomes a gas.

- *Chemical:* We can use the term **chemical** as a general name for a pure substance (either an element or a compound).

- *Chemical formula:* A shorthand recipe that itemizes the various elements in a chemical and specifies their relative proportions is called a **chemical formula**. For example, the formula for water, H_2O, indicates that water consists of molecules in which two hydrogens bond to one oxygen.

- *Chemical reaction:* During a **chemical reaction**, chemical bonds break or form. The process can cause molecules of a compound to separate or can lead to the formation of new molecules of a different compound.

- *Chemical equation:* Chemists represent chemical reactions in the form of an equation. *Reactants* are the starting elements or compounds that appear on the left side of the equation. *Products*, the new atoms or molecules formed during the reaction, appear on the right side of the equation. For example, the burning of methane gas can be represented by the formula: $CH_4 + O_2 \rightarrow CO_2 + H_2O$. This equation states that methane reacts with oxygen to form carbon dioxide and water.

- *Mixture:* A combination of two or more elements or compounds that can be separated without a chemical reaction is a mixture. For example, a cereal composed of bran flakes and raisins is a mixture—you can separate the raisins from the flakes without destroying either.

- *Solution:* A type of material in which one chemical (the *solute*) has dissolved in another (the *solvent*) is a solution. In solutions, a solute may separate into ions. For example, when salt (NaCl) dissolves in water, it separates into sodium (Na^+) and chloride (Cl^-) ions. In a solution, solute molecules fit between solvent molecules.

- *Concentration:* Chemists refer to the amount of solute per unit volume of a solution as the **concentration** of the solute. For example, the concentration of salt in seawater is 3.5%. This means that every 100 g of seawater contains 3.5 g of salt.

- *Precipitate:* When a solution becomes *oversaturated*, meaning that it contains more solute than it has room for, the excess solute forms solid particles that settle out of the solution. These particles are a **precipitate**. The word is also used as a verb in reference to the process of forming solid grains that settle from a solution. For example, we can say that when saltwater evaporates, it becomes oversaturated, and solid salt crystals start to precipitate.

FIGURE Bx5.1 (*continued*)

(e) Three states of water occur in this landscape in Iceland. Ice and snow consist of solid water, liquid water flows in the stream, and invisible water vapor rises in the air. The mist we see consists of suspended liquid droplets and ice crystals.

5.3 Beauty in Patterns: Crystals and Their Structure

What Is a Crystal?

The word *crystal* brings to mind sparkling chandeliers, elegant wine goblets, and shiny jewels. But, as is the case with the word *mineral*, geologists use a narrower definition in scientific discussion. A **crystal** is a single, continuous (uninterrupted) piece of a crystalline material bounded by flat surfaces, called **crystal faces**, that form naturally as the crystal grows. The word *crystal* comes from the Greek word *krystallos*, meaning ice.

Many crystals have beautiful shapes that look like they belong in the pages of a geometry book. Significantly, the angle between two adjacent crystal faces of any mineral specimen is identical to the angle between the corresponding faces of any other specimen of the same mineral. For example, a perfectly formed quartz crystal looks like an obelisk **(Fig. 5.4a)**. The angle between the faces of the columnar part of a quartz crystal is exactly 120°. This rule, discovered by Nicolas Steno, holds regardless of whether the whole crystal is big or small and regardless of whether all of the faces are the same size, for it depends on the geometry of the mineral's crystal structure. Because the crystal structure in one mineral isn't necessarily the same as the structure in another, the angle between crystal faces in one mineral isn't necessarily the same as that in another.

Crystals come in a great variety of shapes, including cubes, trapezoids, pyramids, octahedrons, blades, needles, columns, and obelisks **(Fig. 5.4b)**. Because crystals have a regular geometric form, people have always considered them to be special, perhaps even a source of magical powers. For example, shamans of some cultures made talismans of crystals, for these items supposedly brought power to their wearer or warded off evil spirits. Scientists have demonstrated, however, that the presence of a crystal has no physical effect on health or mood.

FIGURE 5.4 Some characteristics of crystals.

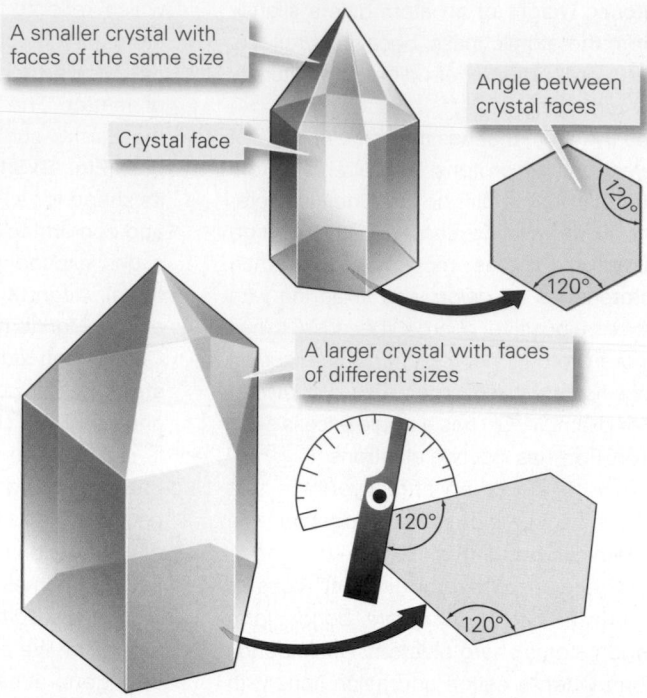

A smaller crystal with faces of the same size

Crystal face

Angle between crystal faces

A larger crystal with faces of different sizes

(a) Regardless of specimen size, the angle between two adjacent crystal faces is consistent in a particular mineral.

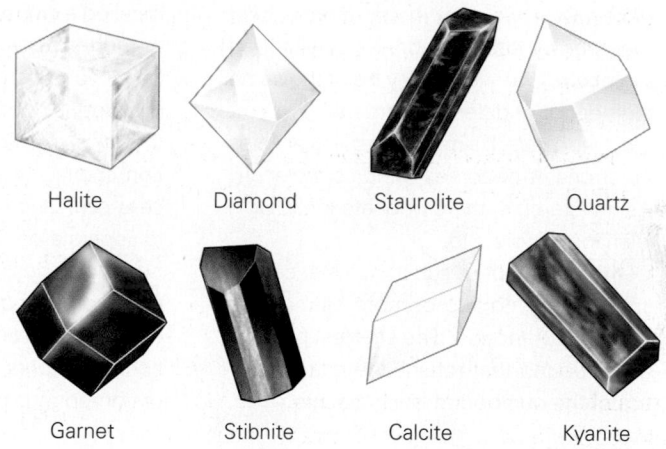

Halite   Diamond   Staurolite   Quartz

Garnet   Stibnite   Calcite   Kyanite

(b) Crystals come in a variety of shapes, including cubes, prisms, blades, and pyramids. Some terminate at a point and some terminate with flat surfaces.

For millennia, crystals have inspired awe because of the way they sparkle. But such behavior comes simply from the way that crystals interact with light.

Looking inside a Mineral

What do the insides of a mineral actually look like? This problem became a focus of study for centuries. An answer finally came from the work of a German physicist, Max von Laue,

FIGURE 5.5 Using X-rays and electron beams to characterize the internal structure of minerals.

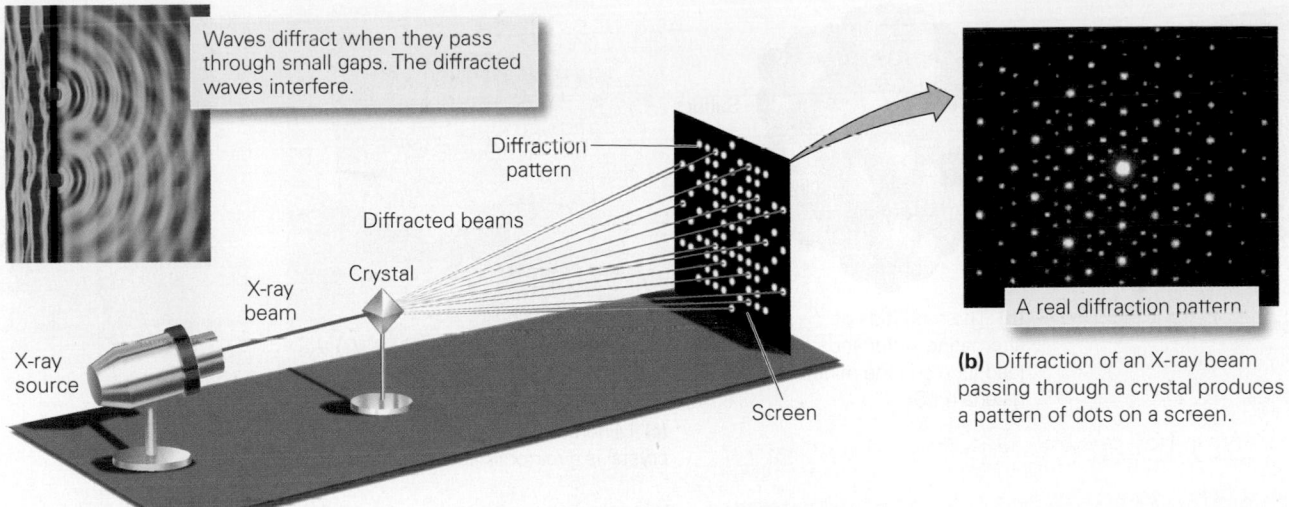

Waves diffract when they pass through small gaps. The diffracted waves interfere.

Diffraction pattern

Diffracted beams

Crystal

X-ray beam

X-ray source

Screen

A real diffraction pattern

(a) Water waves diffract when they pass through a gap. The waves produced by adjacent gaps interfere. Where crests overlap, the signal is stronger. X rays passing through a crystal behave the same way.

(b) Diffraction of an X-ray beam passing through a crystal produces a pattern of dots on a screen.

in 1912. He showed that an X-ray beam passing through a crystal breaks up into many tiny beams that produce a pattern of dots on a screen. Physicists refer to this phenomenon as diffraction; it occurs when waves interact with regularly spaced objects whose spacing approaches the wavelength of the waves—you can see diffraction of ocean waves when they pass through gaps in a seawall **(Fig. 5.5a)**. Von Laue concluded that for a crystal to cause diffraction, the atoms within it must be regularly spaced, and the spacing must be comparable to the wavelength of X rays. Eventually, Von Laue and others learned how to use *X-ray diffraction patterns* as a basis for defining the specific arrangement of atoms in crystals **(Fig. 5.5b)**. As we've noted, this arrangement defines the crystal structure of a mineral. Modern studies of minerals, using extremely powerful electron microscopes, now allow us to see the regular geometric arrangement of atoms in a crystal **(Fig. 5.6)**.

If you've ever looked at wallpaper, you've seen an example of a pattern **(Fig. 5.7a)**. Crystal structures contain one of nature's most spectacular examples of a pattern. In crystals, the pattern is defined by the regular spacing of atoms and, if the crystal contains more than one element, by the regular alternation of atoms **(Fig. 5.7b)**. (Mineralogists refer to a three-dimensional arrangement of points representing this pattern as a *crystal lattice*.) The pattern of atoms in a crystal may control the shape of the crystal. For example, if the crystal structure of a mineral has the shape of a cube, then a crystal of the mineral will have faces that intersect at 90° angles—galena

(PbS) and halite (NaCl) have such a cubic shape. Because of the pattern of atoms in a crystal structure, the structure has symmetry, meaning that the shape of one part of the structure represents the mirror image of the shape of a neighboring part. For example, if you were to cut a halite crystal or a water crystal (snowflake) in half and place the half against a mirror, it would look whole again **(Fig. 5.7c)**.

We can construct a model of a crystal's structure using a cluster of balls packed together in an orderly way **(Fig. 5.8)**. In such models, different colors or sizes represent different atoms, ions, or ionic molecules. In some models, sticks represent chemical bonds. In diamond, which consists entirely of covalently bonded carbon atoms, all of the balls represent carbon atoms, whereas in halite (rock salt), composed of ionically bonded sodium and chloride ions, balls representing cations of sodium (Na$^+$) alternate with balls representing anions of chloride (Cl$^-$). And in calcite (CaCO$_3$), balls representing

FIGURE 5.6 A transmission electron microscope (TEM) allows scientists to see the regular geometric arrangement of atoms in a crystal.

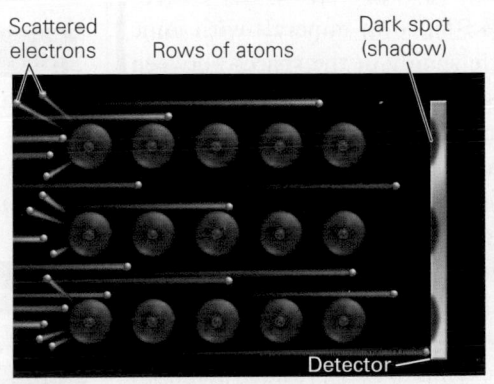

Scattered electrons

Rows of atoms

Dark spot (shadow)

Detector

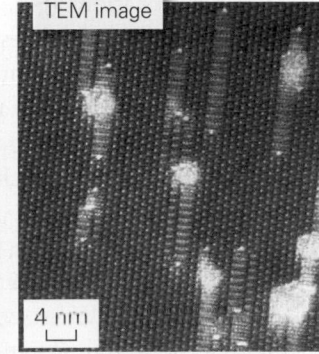

TEM image

4 nm

(a) When a TEM shoots beams of electrons at a material, some electrons scatter off atoms, but some pass between gaps and make a dark spot on a recorder.

(b) The result is an image showing the pattern of atoms in the material.

FIGURE 5.7 The concept of patterns and symmetry in minerals.

(a) The repetition of a flower motif on wallpaper.

(b) The repetition of alternating sulfur and lead atoms in the mineral galena (PbS).

Sulfur
Lead

Mirror
Halite

Mirror
Snowflake

(c) Minerals display symmetry. One-half of a halite crystal or a water crystal is a mirror image of the other.

FIGURE 5.8 A model of the crystal structure of halite (NaCl). The model on the left shows larger chloride ions (blue) interspersed with smaller sodium ions (white). The model on the right shows the same relation, but uses sticks to represent chemical bonds.

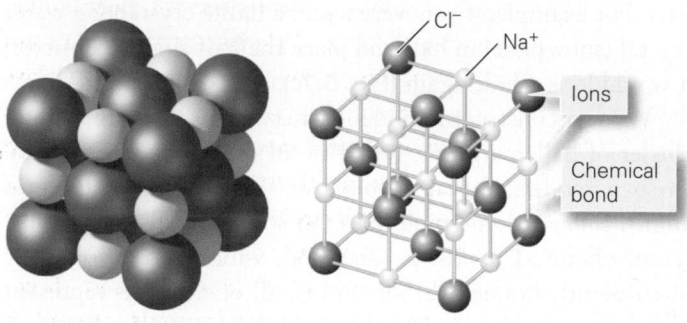

Cl⁻
Na⁺
Ions
Chemical bond

calcium (Ca^+) cations alternate with balls representing carbonate (CO^{3-}) anions.

The "packing" of atoms or ions, meaning the way that they fit together in a crystal structure, differs from mineral to mineral. Because anions have extra electrons, they tend to be bigger than cations **(Fig. 5.9a)**, so in minerals with ionic bonding, cations tend to nestle snugly in the spaces between anions. A variety of different geometries of packing can occur **(Fig. 5.9b)**. In halite, for example, six chloride ions surround each sodium ion, producing an overall arrangement of atoms that, as we've seen, defines the shape of a cube.

Some groups of atoms, ions, or ionic molecules can be arranged in more than one way, yielding different minerals. Two or more different minerals that have the same chemical composition but different crystal structures are called **polymorphs** of a material. As an example, let's consider diamond and graphite, polymorphs of carbon. In diamond, each atom packs together with four neighbors to form a tetrahedron. As

a result, some naturally formed diamond crystals have the shape of a double tetrahedron **(Fig. 5.10a)**. Graphite, another mineral composed entirely of carbon, behaves very differently from diamond. In contrast to diamond, graphite is so soft that we use it as the "lead" in a pencil, for when a pencil moves across paper, tiny flakes of graphite peel off the pencil point and adhere to the paper. This behavior occurs because the carbon atoms in graphite lie in sheets **(Fig. 5.10b)**. Only weak bonds hold the sheets together, so the sheets can separate from each other easily.

The Formation and Destruction of Minerals

A new mineral crystal can form in one of five ways (**Geology at a Glance**, pp. 130–131). First, it can form by the solidification of a melt, meaning the *freezing* of a liquid. For example, ice crystals grow when water freezes. Second, it can form by *precipitation* from a water solution, meaning that atoms, molecules, or ions dissolved in water bond together and separate out of the water. Salt crystals, for example, develop when saltwater evaporates. Third, it can form by solid-state *diffusion*, the movement of atoms or ions through a solid to arrange into a new crystal structure, a process that takes place very slowly. For example, garnets grow by diffusion in solid rock when the rock is subjected to high heat and pressure. Fourth, minerals can form at interfaces between the physical and biological components of the Earth System by *biomineralization*. This occurs when living organisms cause minerals to precipitate either within or on their bodies or immediately adjacent to their bodies. For example, clams and other shelled organisms extract ions from water to produce their mineral shells. Fifth, minerals can precipitate directly from a gas. This process typically occurs around volcanic vents or geysers, for at such locations volcanic gases or steam enter the atmosphere and

FIGURE 5.9 The various sizes of ions and the ways they pack together in minerals.

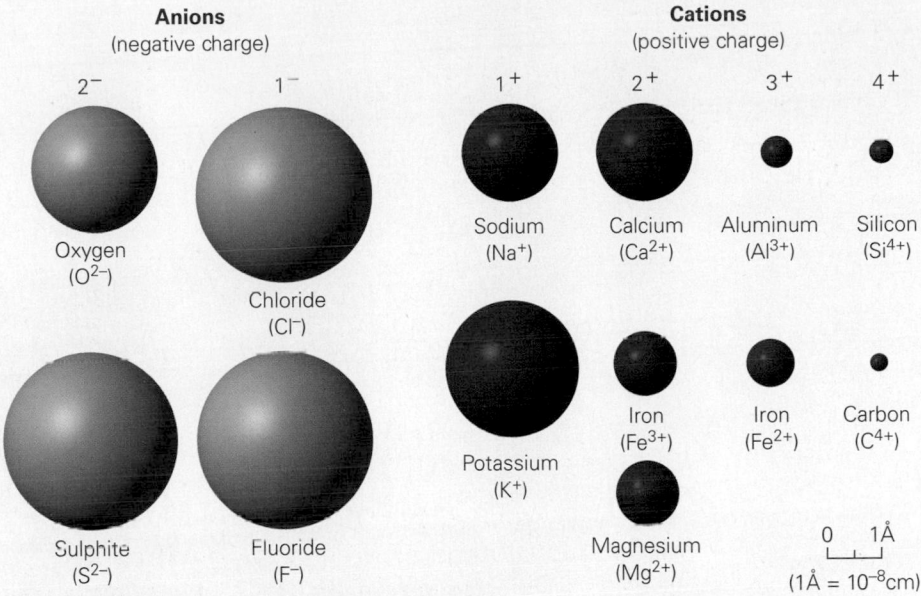

Anions
(negative charge)

Cations
(positive charge)

2^- Oxygen (O^{2-})

1^- Chloride (Cl^-)

1^+ Sodium (Na^+)

2^+ Calcium (Ca^{2+})

3^+ Aluminum (Al^{3+})

4^+ Silicon (Si^{4+})

Potassium (K^+)

Iron (Fe^{3+})

Iron (Fe^{2+})

Carbon (C^{4+})

Sulphite (S^{2-})

Fluoride (F^-)

Magnesium (Mg^{2+})

0   1Å
(1Å = 10^{-8}cm)

(a) Ions come in a wide range of sizes. The difference in size depends, in part, on the number of electrons they contain.

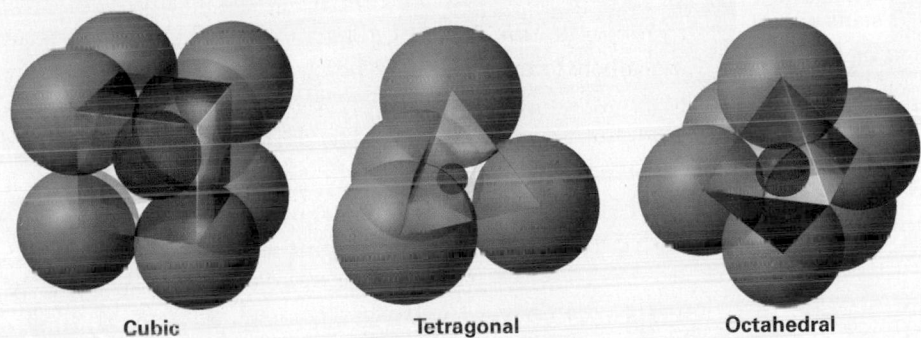

Cubic   **Tetragonal**   **Octahedral**

(b) Ions can pack together in different ways. Each conguration can be described by a geometric shape.

cool abruptly. Some of the bright yellow sulfur deposits found in volcanic regions form in this way.

The first step in forming a crystal happens when the chance assembly of a seed, an extremely small crystal, takes place **(Fig. 5.11a)**. Once the seed exists, other atoms in the surrounding material attach themselves to the faces of the seed and the seed grows into a crystal. As the crystal grows, its faces move outward but maintain the same orientation **(Fig. 5.11b)**, so the youngest part of the new crystal lies at its outer edge. A growing crystal develops its particular crystal shape based on the geometry of its internal crystal structure. The overall shape of a crystal also depends on the relative rates at which different crystal faces grow. Therefore, some crystals have roughly the same dimensions in all directions, but others grow to become needle-like, sheet-like, or blade-like. If a crystal grows uninhibited into an open space, it displays well-formed crystal faces, and geologists refer to it as a *euhedral crystal*. Crystals in a **geode**, a mineral-lined cavity in rock, are typically euhedral **(Fig. 5.11c)**. Commonly, however, the growth of a crystal may be restricted in one or more directions because other crystals around it

FIGURE 5.10 The nature of crystalline structure in minerals.

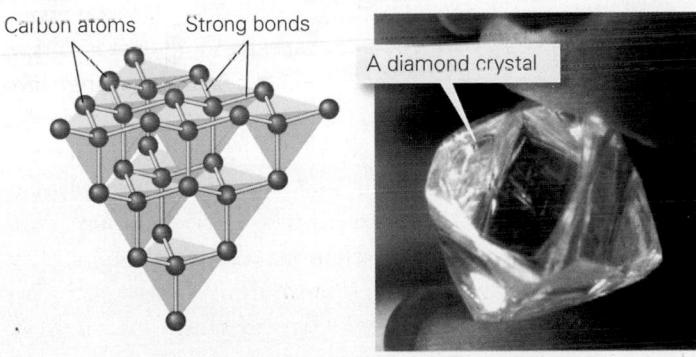

Carbon atoms   Strong bonds

A diamond crystal

(a) In a diamond, carbon atoms are arranged in tetrahedra. All of the bonds are strong.

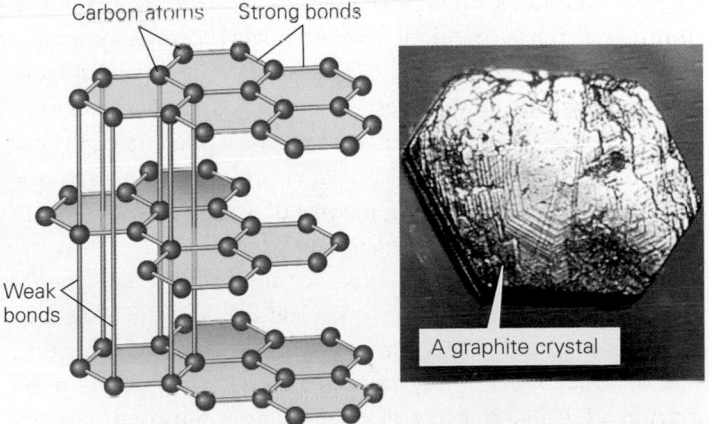

Carbon atoms   Strong bonds

Weak bonds

A graphite crystal

(b) Graphite consists of carbon atoms arranged in hexagonal sheets. The sheets are connected by weak bonds.

MINERAL FORMATION

Solidification from a melt happens when molten rock, such as lava erupted from a volcano, cools down. Different minerals grow in succession.

At first, a few crystals form. They remain surrounded by melt.

A mafic melt starts to cool.

Eventually, all the melt solidifies.

Temperature decreases over time

Diffusion can happen in a solid rock, though very slowly. During the process, atoms migrate through the crystal. New minerals, such as garnet, can grow in the rock.

Fumerole

Sulfur crystals

Precipitation of minerals from volcanic gas can also occur. Yellow sulfur crystals form this way.

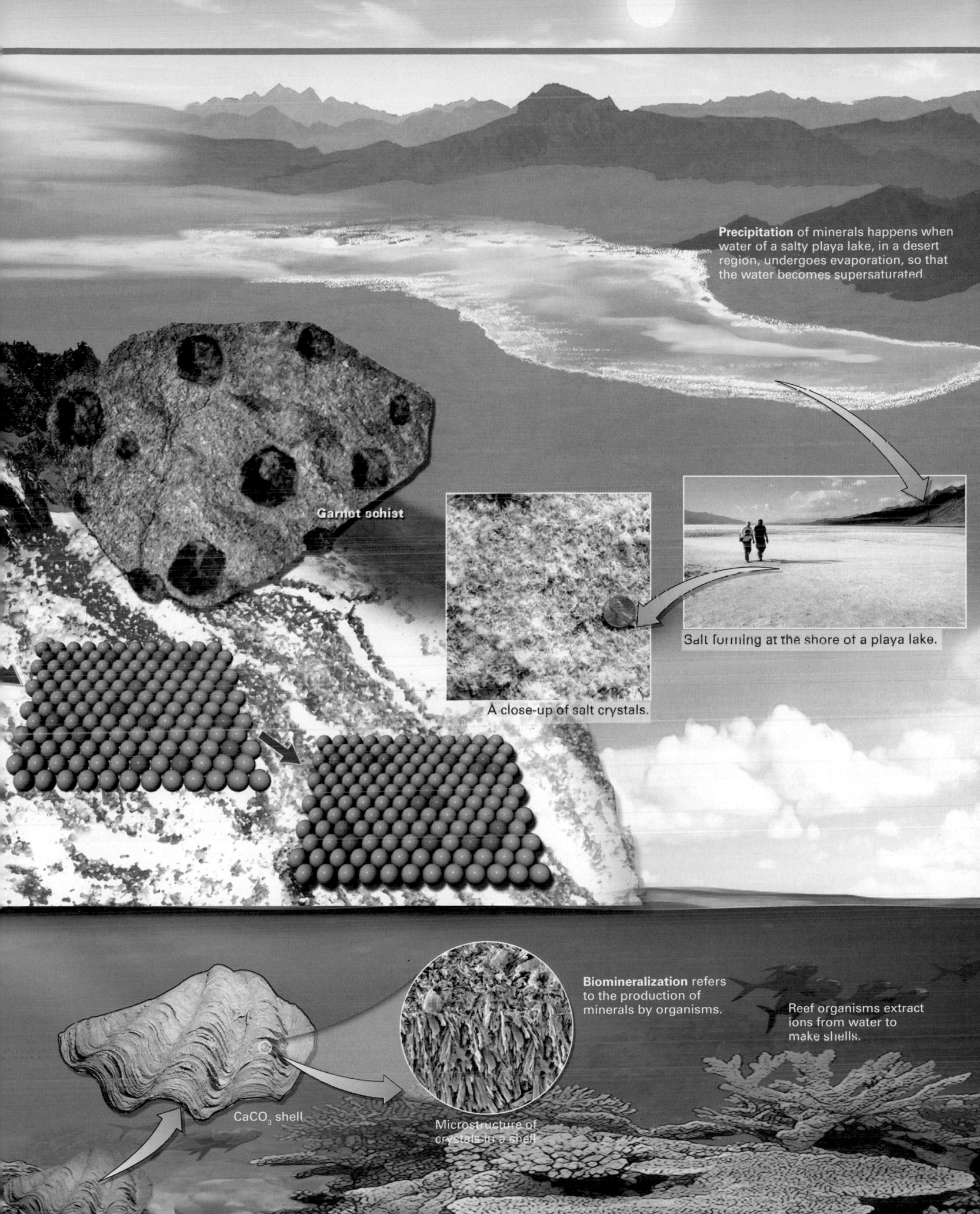

Precipitation of minerals happens when water of a salty playa lake, in a desert region, undergoes evaporation, so that the water becomes supersaturated.

Garnet schist

A close-up of salt crystals.

Salt forming at the shore of a playa lake.

Biomineralization refers to the production of minerals by organisms.

Reef organisms extract ions from water to make shells.

$CaCO_3$ shell

Microstructure of crystals in a shell

FIGURE 5.11 The growth of crystals.

Ions attach to the crystal face.

(a) New crystals nucleate and begin to precipitate out of a water solution. As time progresses, they grow into the open space.

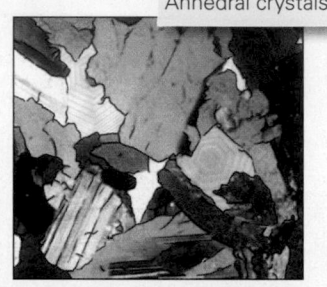

Anhedral crystals

(b) New crystals grow outward from the central seed. As time passes, they maintain their shape until they interfere with one another. A crystal growing in a confined space will be anhedral.

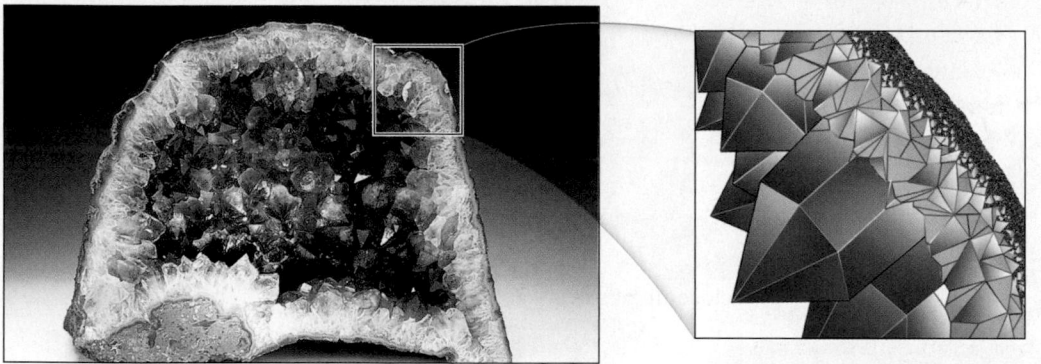

(c) A geode from Brazil consists of purple quartz crystals (amethyst) that grew from the wall into the center. The enlargement sketch indicates that the crystals are euhedral.

A mineral can be destroyed by melting, by dissolution, or by some other chemical reaction. Melting involves heating a mineral to a temperature at which thermal vibrations of the atoms or ions in the crystal structure break the chemical bonds holding them to the lattice. The atoms or ions then separate, either individually or in small groups, to move around again freely. Dissolution takes place when a mineral is immersed in a solvent, such as water. Atoms or ions then separate from the crystal face to be surrounded by solvent molecules. Chemical reactions can destroy a mineral when it comes in contact with reactive materials. For example, iron-bearing minerals react with air and water to form rust (iron oxide). Reactions that take place by diffusion in solid rock can effectively "digest" an assemblage of minerals and replace them with others. The action of microbes in the environment can also destroy minerals. In effect, some microbes can "eat" certain minerals; the microbes use the energy stored in the chemical bonds that hold the atoms of the mineral together as their source of energy for metabolism.

act as obstacles. In such cases, minerals grow to fill the space available, and their shapes are controlled by the shape of their surroundings. Minerals without well-formed crystal faces are known as *anhedral grains*.

Crystals formed by precipitation from a solution develop when the solution becomes oversaturated; that is, when the number of dissolved ions per unit volume of solution becomes so great that they can get close enough to one another to bond together. In an unsaturated solution, solvent molecules surround solute molecules, which prevents the latter from coming in contact. In the case of crystals formed by the solidification of a melt, atoms begin to attach to a seed when the melt becomes sufficiently cool that thermal vibrations can no longer break apart the attraction between the seed and the atoms in the melt.

TAKE-HOME MESSAGE

The regular geometric arrangement of atoms defines the crystal structure of a mineral. Minerals can form by solidification of a melt, by precipitation from a water solution or a gas, or by rearrangement of atoms in a solid, and can be destroyed by melting, dissolution, or chemical reaction.

QUICK QUESTION: Are the faces of a crystal randomly oriented?

5.4 How Can You Tell One Mineral from Another?

Amateur and professional mineralogists get a kick out of recognizing minerals. They might hover around a display case in a museum and name specimens without bothering to look at the labels. How do they do it? The trick lies in learning to recognize the basic *physical properties* (material characteristics) that distinguish one mineral from another. Some physical properties, such as shape and color, can be seen from a distance. Others, such as hardness and magnetization, can be determined only by handling the specimen or by performing an *identification test* on it. Such tests include scratching the mineral against another object, placing it near a magnet, weighing it, tasting it, or placing a drop of acid on it. Let's examine some of the physical properties most commonly used in mineral identification.

Color Color results from the way a mineral interacts with light. Sunlight contains the whole spectrum of colors, each with a different wavelength. A mineral absorbs certain wavelengths, so the color you see when looking at a specimen represents the wavelengths the mineral does not absorb. Certain minerals always have the same color, but many come in a range of colors **(Fig. 5.12a)**. Color variations in a mineral commonly reflect the presence of impurities. For example, trace amounts of iron may give quartz a reddish color.

Streak The streak of a mineral refers to the color of a powder produced by pulverizing the mineral. You can obtain a streak by scraping the mineral against an unglazed ceramic plate **(Fig. 5.12b)**. The color of a mineral powder tends to be less variable than the color of a whole crystal, so it provides a more reliable clue to a mineral's identity. Calcite, for example, always yields a white streak even though pieces of calcite may be white, pink, or clear.

Luster Luster refers to the way a mineral surface scatters light. Geoscientists describe luster simply by comparing the appearance of the mineral with the appearance of a familiar substance. For example, minerals that look like metal have a *metallic luster*, whereas those that do not have a *nonmetallic luster* **(Fig. 5.12c, d)**. Terms used for different versions of nonmetallic luster include silky, glassy, satiny, resinous, pearly, and earthy —the adjectives are self-explanatory.

Hardness Hardness indicates the relative ability of a mineral to resist scratching, and therefore represents the ability of bonds in the crystal structure to resist being broken. Hard minerals can scratch soft minerals, but soft minerals cannot scratch hard ones, because the atoms or ions in crystals of a hard mineral are more strongly bonded together than those in a soft mineral. Diamond, the hardest mineral known, can scratch anything, which is why it's used to cut glass. In the early 1800s, a mineralogist named Friedrich Mohs listed some minerals in sequence of relative hardness. This list, the **Mohs hardness scale**, helps in mineral identification; a mineral with a Mohs hardness of 5 can scratch all minerals with a hardness of 5 or less. When you use the scale **(Table 5.1)**, it helps to compare the hardness of a mineral with that of a common item such as your fingernail, a penny, or a glass plate. Note that the numbers on the Mohs hardness scale do not specify the true relative differences in hardness of minerals. For example, on the Mohs scale, talc has a hardness of 1 and quartz has a hardness of 7. But this does not mean that quartz is 7 times harder than talc. Careful tests show that quartz is actually about 100 times harder than talc.

Specific gravity Specific gravity represents the density of a mineral, as defined by the ratio between the weight of a volume of the mineral and the weight of an equal volume of water at 4°C. For example, one cubic centimeter of quartz has a weight of 2.65 grams, whereas one cubic centimeter

TABLE 5.1 Mohs Hardness Scale

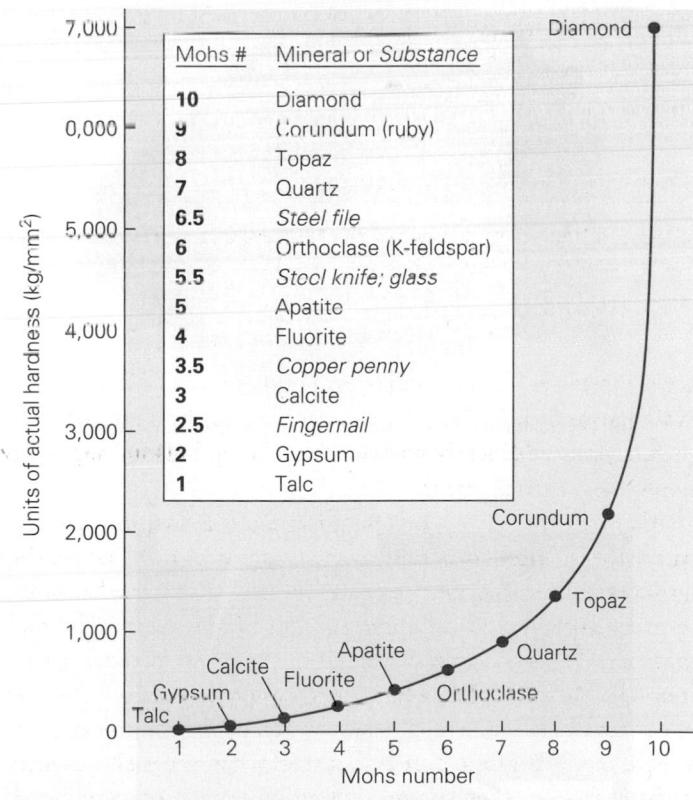

Mohs #	Mineral or *Substance*
10	Diamond
9	Corundum (ruby)
8	Topaz
7	Quartz
6.5	*Steel file*
6	Orthoclase (K-feldspar)
5.5	*Steel knife; glass*
5	Apatite
4	Fluorite
3.5	*Copper penny*
3	Calcite
2.5	*Fingernail*
2	Gypsum
1	Talc

Mohs's numbers are relative—in reality, diamond is 3.5 times harder than corundum, as the graph shows.

FIGURE 5.12 Physical characteristics of minerals.

(a) Color is diagnostic of some minerals, but not all. For example, quartz can come in many colors.

(b) To obtain the streak of a mineral, rub it against a porcelain plate. The streak consists of mineral powder.

(c) Pyrite has a metallic luster because it gleams like metal.

(d) Feldspar has a nonmetallic luster.

(e) Crystal habit refers to the shape of the crystal. Wulfenite crystals are very thin, tabular plates.

(f) Kyanite crystals are bladed.

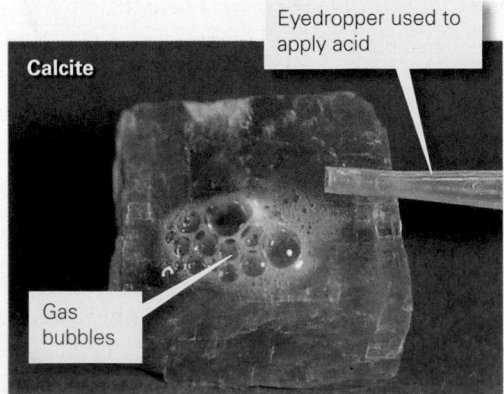

(g) Calcite reacts with hydrochloric acid to produce carbon dioxide gas.

(h) Magnetite is magnetic.

FIGURE 5.13 The nature of mineral cleavage and fracture.

(a) Mica has one strong plane of cleavage and splits into sheets.

(b) Pyroxene has two planes of cleavage that intersect at 90°.

(c) Amphibole has two planes of cleavage that intersect at 60°.

Halite breaks into cubes.

(d) Halite has three mutually perpendicular planes of cleavage.

Calcite breaks into rhombs.

(e) Calcite has three planes of cleavage, one of which is inclined.

(f) Diamond has four planes of cleavage, each inclined to the others.

of water has a weight of 1.00 gram, so the specific gravity of quartz is 2.65. In practice, you can develop a sense of specific gravity by hefting minerals in your hands. A piece of galena (lead ore) "feels" heavier than a similar-sized piece of quartz.

Crystal habit The **crystal habit** of a mineral refers to the shape of a single crystal with well-formed crystal faces or to the character of an aggregate of many well-formed crystals that grew together as a group **(Fig. 5.12e, f)**. The habit depends on the internal arrangement of atoms in the crystal. When describing habit, mineralogists commonly refer to common geometric shapes using adjectives such as cubic, prismatic, bladed, platy, or fibrous. The relative dimensions of the crystal depend on relative rates of crystal growth in different directions. For example, crystals that grow rapidly in one direction but slowly in the other two directions are needle-like. In some cases, the crystal habit of a mineral poses a potential health hazard **(Box 5.2)**.

Special properties Some minerals have distinctive properties that readily distinguish them from other minerals. For example, calcite ($CaCO_3$) reacts with dilute hydrochloric acid (HCl) to produce carbon dioxide (CO_2) gas **(Fig. 5.12g)**. Dolomite ($CaMg[CO_3]_2$) also reacts with acid but not as

strongly. Graphite makes a gray mark on paper, magnetite attracts a magnet **(Fig. 5.12h)**, halite tastes salty, and plagioclase has striations (thin parallel corrugations or stripes) on its crystal faces.

Fracture and cleavage Different minerals fracture (break) in different ways. These depend on the internal arrangement of atoms in the minerals. If a mineral breaks to form distinct planar surfaces that have a specific orientation in relation to the crystal structure, then we say that the mineral has **cleavage**, and we refer to each surface as a *cleavage plane* **(Fig. 5.13a–f)**.

Cleavage planes form in the directions where the bonds holding atoms together in the crystal are weakest. Some minerals have one direction of cleavage. For example, mica has very weak bonds in one direction but strong bonds in the other two directions, so it easily splits into parallel sheets; the surface of each sheet is a cleavage plane. Other minerals have two or three directions of cleavage that intersect at a specific angle. For example, halite has three sets of cleavage planes that intersect at right angles, so halite crystals break into little cubes. Cleavage planes may be hard to distinguish from crystal faces **(Fig. 5.14a)**. Materials that have no cleavage at all (because bonding is equally strong in all directions) break either by forming irregular fractures or by forming conchoidal fractures

FIGURE 5.14 The different ways that minerals can break.

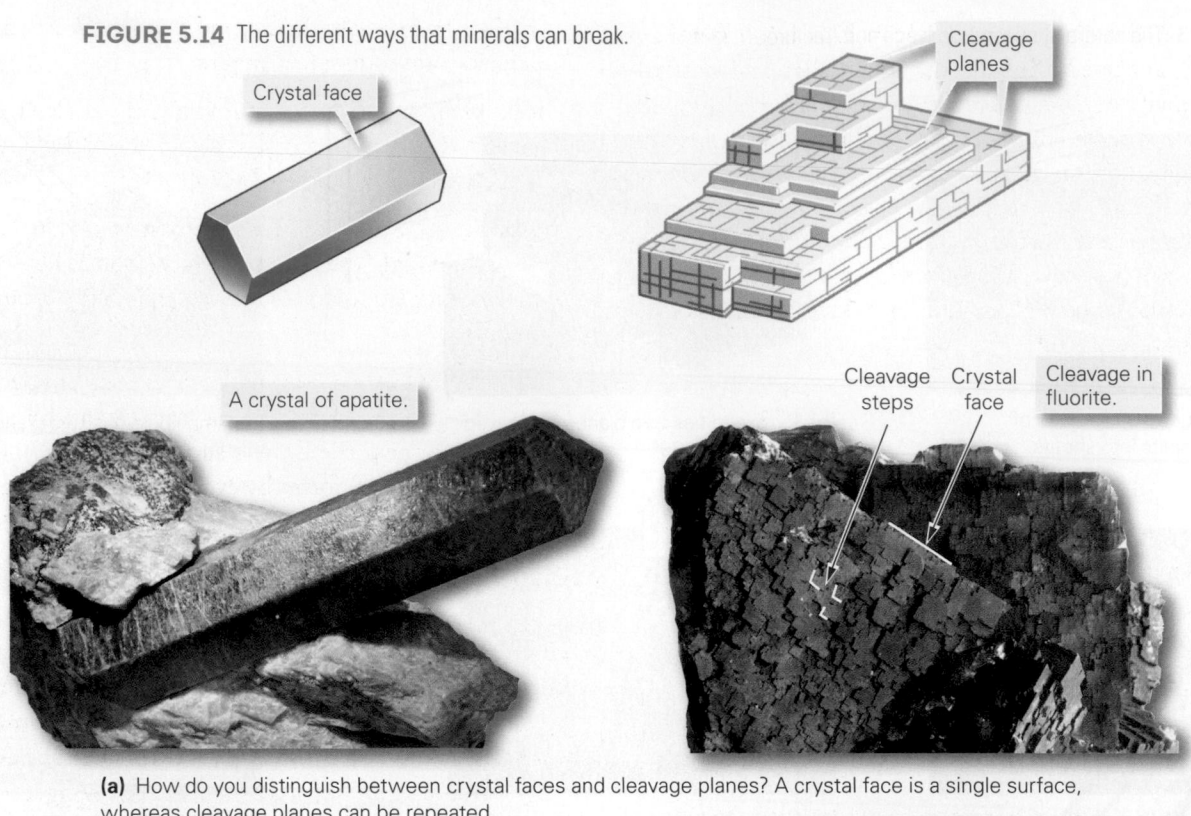

(a) How do you distinguish between crystal faces and cleavage planes? A crystal face is a single surface, whereas cleavage planes can be repeated.

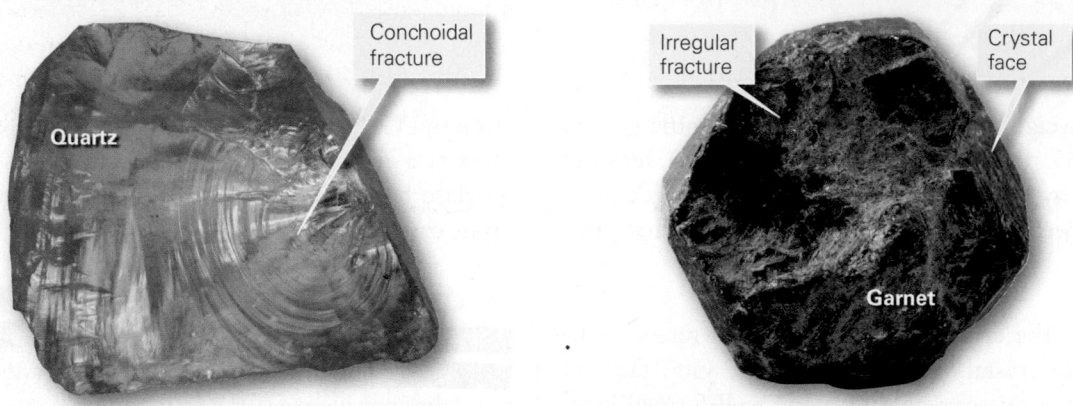

(b) Minerals without cleavage can develop irregular or conchoidal fractures.

(Fig. 5.14b). **Conchoidal fractures** are smoothly curving, clamshell-shaped surfaces; they typically form in glass.

TAKE-HOME MESSAGE

The properties of minerals (such as color, streak, luster, crystal habit, hardness, specific gravity, cleavage, magnetism, and reaction with acid) are a manifestation of the crystal structure and chemical composition of minerals and can be used for mineral identification.

QUICK QUESTION: What minerals react with acid to produce CO_2 bubbles?

5.5 Organizing Knowledge: Mineral Classification

Just about every object you come across in daily life has been classified in some way, because classification schemes help organize information and streamline discussion. Biologists classify animals based on how they reproduce, whether they have skeletons or not, and whether they breathe oxygen in air or oxygen in water. Botanists classify plants according to the way they reproduce and by the shape of their leaves. In the case of minerals, a good means of classification eluded researchers until it became possible to determine the chemical makeup of minerals. A Swedish chemist, Baron Jöns Jacob Berzelius (1779–1848), analyzed minerals and noted chemical similarities among many of them. Berzelius, along with his students, established that most minerals can be classified by specifying the principal anion or anionic group within the mineral (see Box 5.1). Using this approach, it's possible to divide the 4,000 known minerals into a small number of groups, or **mineral classes**. We now take a look at principal mineral classes, focusing especially on silicates, the class that constitutes most of the rock in the Earth.

Mineral Classes

Mineralogists distinguish several principal classes of minerals. Here are some of the major ones.

- *Silicates:* All **silicates** contain the SiO_4^{4-} anionic group. Examples of minerals in this class include quartz (SiO_2) and feldspar (such as $KAlSi_3O_8$). We will learn more about silicates in the next section.

- *Sulfides:* Sulfides consist of a metal cation bonded to a sulfide anion (S^{2-}). Examples include galena (PbS) and pyrite (FeS_2).

- *Oxides:* Oxides consist of metal cations bonded to oxygen anions. Typical oxide minerals include hematite (Fe_2O_3) and magnetite (Fe_3O_4).

- *Halides:* The anion in a halide is a halogen or salt producing ion (such as chloride, Cl^-, or fluoride, F^-).

Halite, or rock salt (NaCl), and fluorite (CaF_2), a source of fluoride, are common examples.

- *Carbonates:* In **carbonate minerals**, the molecule CO_3^{2-} serves as the anionic group. Examples include calcite ($CaCO_3$) and dolomite ($CaMg[CO_3]^2$).

- *Native metals:* Native metals consist of pure masses of a single metal. The metal atoms are bonded by metallic bonds. Copper and gold, for example, may occur as native metals.

- *Sulfates:* Sulfates consist of metal cations bonded to SO_4^{2-} anionic groups. Many sulfates form by precipitation out of water at or near the Earth's surface **(Fig. 5.15)**. Gypsum ($CaSO_4 \cdot 2H_2O$) forms by evaporation of saltwater.

Silicates: The Major Rock-Forming Minerals

Silicate minerals make up over 95% of the continental crust. Rocks of the oceanic crust and of the Earth's mantle consist almost entirely of silicates, so silicates are the most common minerals on the Earth. As noted earlier, silicates contain the SiO_4^{4-} anionic group. In this group, four oxygen atoms surround a single silicon atom, thereby defining the corners of a tetrahedron, a pyramid-like shape with four triangular faces **(Fig. 5.16a)**. We refer to this anionic group as the **silicon-oxygen tetrahedron** (or, informally, as the *silica tetrahedron*), and it acts, in effect, as the building block of silicate minerals.

FIGURE 5.15 This photo is real, not a computer collage! We're seeing the world's largest-known mineral crystals jutting from the walls of a cave in Chihuahua, Mexico. These crystals of the mineral gypsum ($CaSO_4 \cdot 2H_2O$) formed by precipitation from a water solution.

BOX 5.2 **CONSIDER THIS . . .**

Asbestos and Health: When Crystal Habit Matters!

Asbestos minerals have been in the news for decades because epidemiology studies show an association between asbestos inhalation and human health problems. Understanding the crystal habit of these minerals provides some insight into the origin of possible asbestos hazards.

The name asbestos doesn't refer to a single mineral, but rather to a group of six different minerals. These minerals share a key characteristic—all have a fibrous habit in that they contain clusters of needle-like crystals that are about 20 times longer than they are wide (Fig. Bx5.2a). "White asbestos," which accounts for more than 90% of the asbestos used, consists of the mineral chrysotile ($Mg_3[Si_2O_5][OH]_4$). It forms in serpentinite, a green rock produced when hot-water solutions interact with the olivine-rich rock that makes up oceanic lithosphere. "Brown asbestos" includes various types of amphibole minerals, such as grunerite ($Fe_7Si_8O_{22}[OH]_2$).

Between the mid-1800s and the 1990s, manufacturers used asbestos in a great variety of applications because its fibrous character allowed it to be woven into other materials to make strong, fire-resistant and chemical-resistant floor tiles, insulation, roof shingles, firefighters' clothing, gaskets, and brakes (Fig. Bx5.2b). When intact and incorporated into other materials, asbestos isn't dangerous because it is not poisonous and does not emit fumes. The problem arises when people inhale asbestos dust, as can happen during the mining of asbestos, the manufacturing of asbestos-containing materials, or the remodeling or demolition of rooms or buildings that contain asbestos. Because asbestos has a fibrous habit, it breaks into tiny (700 times smaller than human hair) needle-like shards that can move through human breathing passages. These shards get embedded in lungs, where they cause irritation and clogging and inhibit exchange of oxygen, resulting in asbestosis. For reasons that doctors don't fully understand, the fibers can also interact chemically and/or mechanically with DNA in cells, causing genetic mutations that trigger various kinds of lung cancer. Researchers still debate about the relative dangers of different kinds of asbestos, but

FIGURE Bx5.2 Asbestos has distinctive characteristics that make it useful, as well as potentially hazardous.

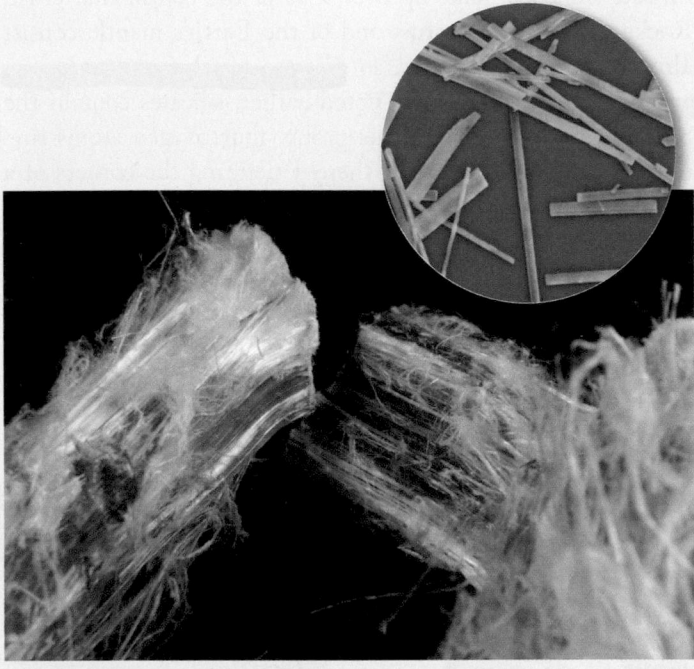

(a) Asbestos minerals have a fibrous habit. As the inset photomicrograph emphasizes, individual fibers are like tiny needles.

(b) Asbestos, mixed in with vinyl, makes stronger floor tiles. But when old tiles break up, the resulting dust can contain asbestos fibers.

(c) When removing asbestos, workers must wear protective gear and seal off the area.

it appears that white asbestos tends to be less dangerous than brown asbestos, perhaps because of its fiber shape.

When health concerns associated with asbestos became publicized in the 1970s, years of litigation followed, eventually resulting, in the mid-1980s, in bans on the use of asbestos. As a consequence, renovations of rooms containing asbestos tiles or insulation must begin with a very expensive asbestos abatement, during which workers in protective clothing seal off the asbestos-containing workspace from surrounding areas before scraping away and collecting the asbestos (Fig. Bx5.2c). In many cases, if asbestos is left alone and can be covered by a safer material, the hazard can be mitigated. It's only when workers sand or break up asbestos materials that the fibers can enter the air.

FIGURE 5.16 The structure of silicate minerals.

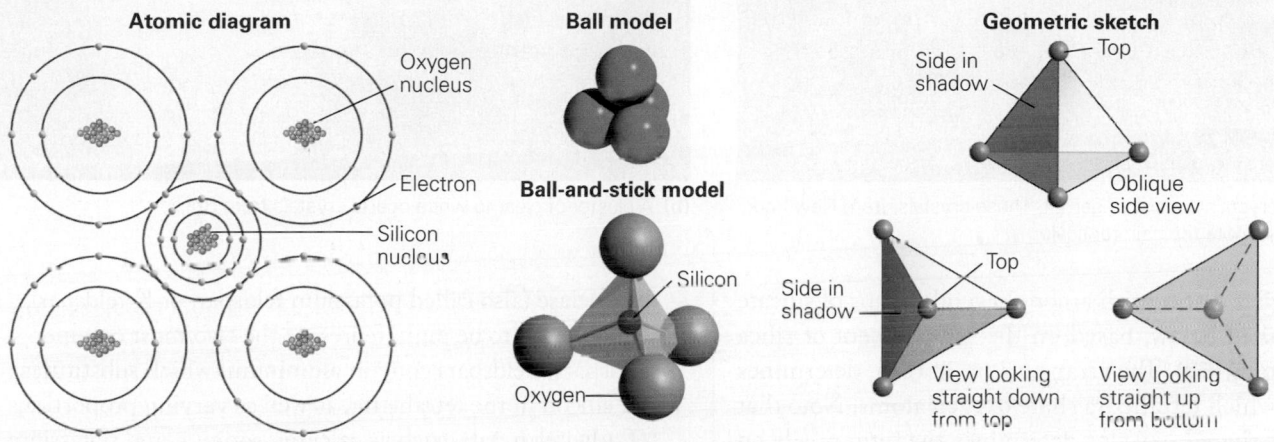

(a) The fundamental building block of a silicate mineral is the silicon-oxygen tetrahedron. Oxygen atoms occupy the corners of the tetrahedron, and silicon lies at the center. Geologists portray the tetrahedron in a number of different ways.

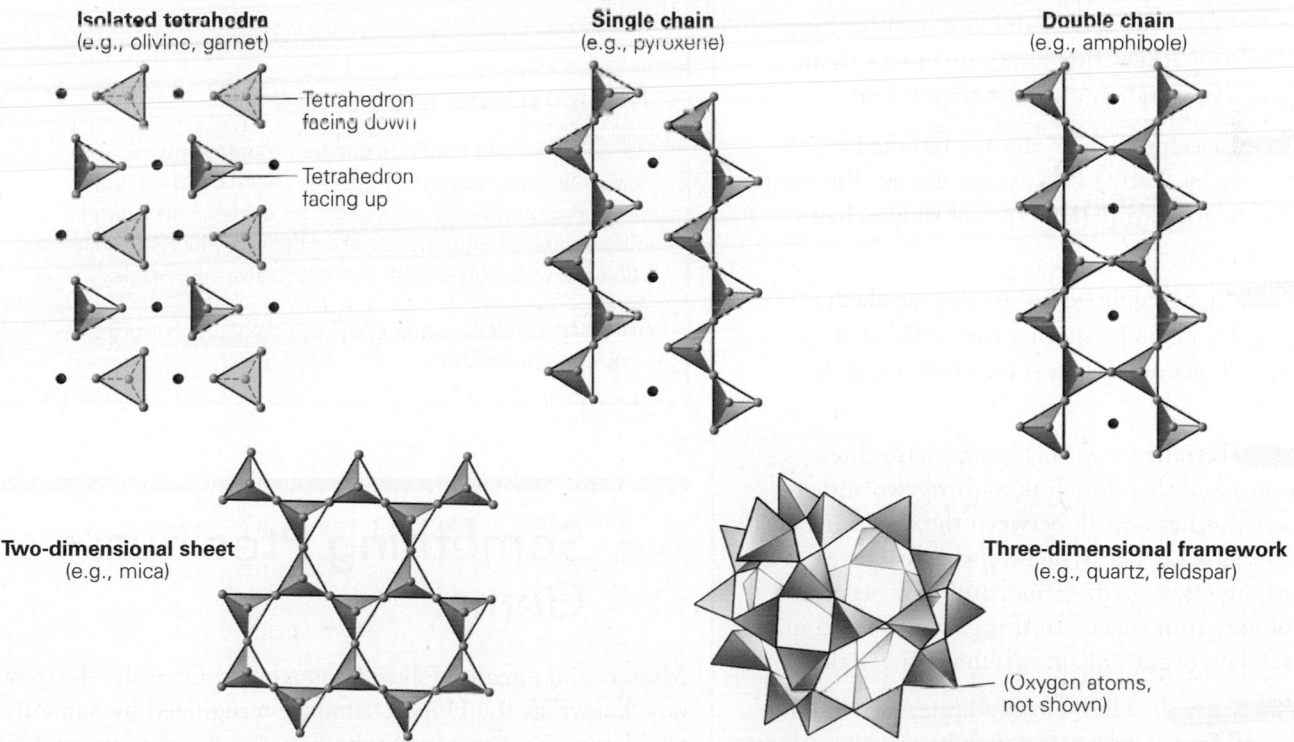

(b) The classes of silicate minerals differ from one another by the way in which the silicon-oxygen tetrahedra are linked. Where the tetrahedra link, they share an oxygen atom. Oxygen atoms are shown in blue. Positive ions (not shown) occupy spaces between tetrahedra.

FIGURE 5.17 Examples of silicate materials.

(a) The maroon crystals consist of garnet. These crystals, from New York, are unusually large; note the coin for scale.

(b) A cluster of clear to white quartz crystals from Brazil.

Mineralogists distinguish among several groups of silicate minerals, detailed below, based on the arrangement of silica tetrahedra **(Fig. 5.16b)**. The arrangement, in turn, determines the degree to which tetrahedra share oxygen atoms. Note that the number of shared oxygens determines the ratio of silicon (Si) to oxygen (O) in the mineral.

- *Independent tetrahedra:* In this group, tetrahedra are independent and do not share any oxygen atoms. The attraction between the tetrahedra and positive ions holds crystals together. This group includes olivine, a glassy green mineral, and garnet **(Fig. 5.17a)**.

- *Single chains:* In a single-chain silicate, tetrahedra link to form a chain by sharing two oxygen atoms. The most common of the many different types of single-chain silicates are pyroxenes.

- *Double chains:* In a double-chain silicate, tetrahedra link to form a double chain by sharing two or three oxygen atoms. Amphiboles are the most common minerals in this group.

- *Sheet silicates:* Tetrahedra in this group share three oxygen atoms and therefore link to form two-dimensional sheets. Other ions fit between the sheets in sheet silicates. Because of their structure, sheet silicates have a single strong cleavage direction, and they occur in "books" of very thin sheets. In this group, we find micas and clays. Clays occur only in extremely tiny flakes.

- *Framework silicates:* In a framework silicate, each tetrahedron shares all four oxygen atoms with its neighbors, forming a three-dimensional structure. Examples include feldspar and quartz. Plagioclase, which tends to be white, and

orthoclase (also called potassium feldspar, or K-feldspar), which tends to be pink, represent the two most common feldspars. Feldspars contain aluminum, which substitutes for silicon in the tetrahedra, as well as varying proportions of other elements, such as calcium, sodium, and potassium. Quartz, in contrast, contains only silicon and oxygen; the ratio of silicon to oxygen in quartz is 1:2, so the mineral has the familiar formula SiO_2 **(Fig. 5.17b)**.

TAKE-HOME MESSAGE

The 4,000 known minerals can be organized into a relatively small number of classes based on chemical makeup. Examples include silicates, oxides, carbonates, and sulfides. Most minerals are silicates, which contain silicon-oxygen tetrahedra arranged in various ways.

QUICK QUESTION: What is the principal anionic group in carbonate minerals?

5.6 Something Precious: Gems!

Mystery and romance follow famous gems. Consider the stone now known as the Hope Diamond, recognized by name the world over. No one knows who first dug it out of the ground **(Box 5.3)**. Was it mined in the 1600s, or was it stolen off an ancient religious monument? What we do know is that in

BOX 5.3 CONSIDER THIS . . .

Where Do Diamonds Come From?

Diamonds consist of carbon, which typically accumulates only at or near the Earth's surface. Experiments demonstrate that the temperatures and pressures needed to form diamond generally occur only at depths of about 150 km below the Earth's surface. Under these conditions, the carbon atoms in graphite, arranged in hexagonal sheets, rearrange to form the much stronger and more compact structure of diamond (see Fig. 5.10). (Because engineers can duplicate these conditions in the laboratory, corporations manufacture several tons of synthetic diamonds a year.)

How does carbon get down into the mantle, where it transforms into diamond? Geologists speculate that subduction or collision provides the means of carrying carbon-containing rocks and sediments from the Earth's surface into the mantle. This carbon transforms into diamond, some of which becomes trapped beneath continents. But if diamonds form at great depth, then how do they return to the surface? In some cases rifting of the continental crust causes a small part of the underlying lithospheric mantle to melt. Magma generated during this process rises to the surface, bringing the diamonds up with it. Near the surface, the magma cools and solidifies to form a special kind of igneous rock called *kimberlite* (named for Kimberley, South Africa, where it

was first found). Diamonds brought up with the magma lie embedded in the kimberlite (Fig. Bx5.3). Kimberlite magma tends to rise rapidly into the upper crust, where it solidifies to form rock in carrot-shaped bodies called kimberlite pipes that are 50 to 200 m across and at least 1 km deep.

In places where diamonds occur in solid kimberlite, they can be obtained only by digging up the kimberlite and crushing it to separate out the diamonds. But nature can also break diamonds free from the Earth on its own. In places where kimberlite has been exposed at the ground surface for a long time, the rock reacts chemically with water and air (a process called weathering; see Interlude B). These reactions cause most minerals in kimberlite to disintegrate, yielding sediment that washes away in rivers. Diamonds are so hard that they remain as solid grains in river gravel. So, many diamonds have been obtained simply by separating them from recent or ancient river gravel.

Diamond-bearing kimberlite pipes occur in many places around the world, particularly where very old continental lithosphere exists. Many of the diamonds we use today may have first formed over 3 billion years ago. Southern and central Africa, Siberia, northwestern Canada, India, Brazil, Borneo, Australia, and the U.S. Rocky Mountains all have pipes. Rivers and glaciers, however,

have transported diamond-bearing sediments great distances from their original sources. In fact, diamonds have even been found in farm fields of the midwestern United States.

Not all natural diamonds are valuable; value depends on color and clarity. Diamonds that contain imperfections (cracks, or specks of other material) or are dark gray in color cannot be used for jewelry. These stones, called *industrial diamonds*, make great abrasives, for diamond powder is so hard (10 on the Mohs hardness scale) that it can be used to grind away any other substance.

Gem-quality diamonds come in a range of sizes. Jewelers measure diamond size in carats, where 1 *carat* equals 200 mg (0.2 g). In English units of measurement, 1 ounce equals 142 carats. (Note that a carat measures gemstone weight, whereas a *karat* specifies the purity of gold.) The largest diamond ever found, a stone called the Cullinan Diamond, was discovered in South Africa in 1905. It weighed 3,106 carats (621 g) before being cut. By comparison, the diamond on a typical engagement ring weighs less than 1 carat. Diamonds are rare, but not as rare as their price perhaps suggests. A worldwide consortium of diamond producers stockpiles the stones so as not to flood the market and drive the price down.

FIGURE Bx5.3 Diamond occurrences.

(a) A diamond mine in northern Canada.

A diamond embedded in solid kimberlite.

(b) A kimberlite sample containing a diamond.

the 1600s, a French trader named Jean-Baptiste Tavernier obtained a large (112.5 carats, where 1 carat = 200 mg), rare blue diamond in India and carried it back to France. King Louis XIV bought the diamond and had it fashioned into a jewel of 68 carats. This jewel vanished in 1762 during a burglary. Perhaps it was lost forever—perhaps not. In 1830, a 44.5-carat blue diamond mysteriously appeared on the jewel market for sale. Henry Hope, a British banker, purchased the stone, which then became known as the Hope Diamond **(Fig. 5.18)**. It changed hands several times until 1958, when the famous New York jeweler Harry Winston donated it to the Smithsonian Institution in Washington, D.C., where it now sits behind bulletproof glass in a heavily guarded display.

What makes stones like the Hope Diamond so special that people risk life and fortune to obtain them? What is the difference between gemstones and other minerals? A gemstone is a mineral that has special value because it is rare and people consider it beautiful. A **gem** is a cut and finished gemstone ready to be

set in jewelry. Jewelers sometimes distinguish between precious stones (such as diamond, ruby, sapphire, and emerald), which are particularly rare and expensive, and semiprecious stones (such as topaz, tourmaline, aquamarine, and garnet), which are not as rare or expensive. But the distinction is rather arbitrary. All the stones mentioned so far are transparent crystals, though most have some color **(Table 5.2)**. The category of semiprecious stones also includes opaque or translucent minerals such as lapis, malachite (see Fig. 5.2a), and opal.

In everyday language, people also think of pearls and amber as gemstones. Unlike diamonds and garnets, which form inorganically in rocks, pearls form in living oysters when the oyster extracts calcium and carbonate ions from water and precipitates them around an impurity, such as a sand grain, embedded in its body. Pearls, therefore, result from biomineralization. Most pearls used in jewelry today are "cultured" pearls, made by artificially introducing round sand grains into oysters in order to stimulate pearl production.

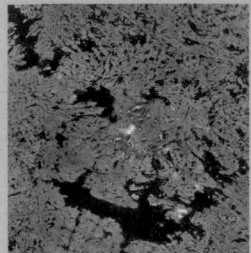

Organic processes yield amber—it consists of fossilized tree sap. But because amber consists of organic compounds that are not arranged in a crystal structure, it does not meet the definition of a mineral.

"Rare" means hard to find, and some gemstones are indeed hard to find. Many of the places where diamonds, for example, occur are in isolated regions of Congo, South Africa, Brazil, Canada, Russia, India, and Borneo (see Box 5.3). In some cases, it is not the mineral itself, but rather the "gem-quality" versions of the mineral, that are rare. For example, garnets occur in many rocks in such abundance that people use them as industrial abrasives. But most garnets are quite small and contain inclusions (specks of other minerals and/or bubbles) or fractures, so they are not particularly beautiful. Gem-quality garnets—clean, clear, large, unfractured

FIGURE 5.18 The Hope Diamond, now on display at the Smithsonian Institution in Washington, D.C.

TABLE 5.2 Precious and semiprecious materials used in jewelry

Gem Name	Material/Formula	Comments
Amber	Fossilized tree sap	Composed of organic chemicals; amber is not strictly a mineral.
Amethyst	Quartz/SiO_2	The best examples precipitate from water in openings in igneous rocks; a deep purple version of quartz.
Aquamarine	Beryl/$Be_3Al_2Si_6O_{18}$	A bluish version of emerald.
Diamond	Diamond/C	Brought to the surface from the mantle in igneous bodies called diamond pipes; may later be mixed in deposits of sediment.
Emerald	Beryl/$Be_3Al_2Si_6O_{18}$	Occurs in coarse igneous rocks (pegmatites; see Chapter 6).
Garnet	Garnet/(e.g., $Mg_3Al_2[SiO_4]_3$)	A variety of types differ in composition (Ca, Fe, Mg, and Mn versions); occurs in metamorphic rocks (see Chapter 8).
Jade	Jadeite/$NaAlSi_2O_6$ Nephrite/$Ca_2(Mg,Fe)_5Si_8O_{22}(OH)_2$	Jade can be one of two minerals, jadeite (a pyroxene) or nephrite (an amphibole); both occur in metamorphic rocks.
Opal	Composed of microscopic spheres of hydrated silica packed together	Most opal comes from a single mining district in central Australia; occurs in bedrock that has reacted with water near the surface.
Pearl	Aragonite/$CaCO_3$	Formed by oysters, which secrete coatings around sand grains that are accidentally embedded in the soft parts of the organism. Cultured pearls are formed the same way, but the impurity is a spherical bead that is intentionally introduced.
Ruby	Corundum/Al_2O_3	The red color is due to chromium impurities; found in coarse igneous rocks called pegmatites and as a result of contact metamorphism (see Chapters 6 and 8).
Sapphire	Corundum/Al_2O_3	A blue version of ruby.
Topaz	$Al_2SiO_4(F,OH)_2$	Found in igneous rocks, a result of the reaction of rock with hot water.
Tourmaline	$Na(Mg,Fe)_3Al_6(BO_3)_3(Si_6O_{18})(OH,F)_4$	Forms in igneous and metamorphic rocks.
Turquoise	$CuAl_6(PO_4)_4(OH)_8 \cdot 4H_2O$	Found in copper-bearing rocks; a popular jewelry gem in the American Southwest.

crystals—are unusual. In some cases, gemstones are merely pretty and rare versions of more common minerals. For example, ruby has the same composition and crystal structure as the common mineral corundum (Al_2O_3), emerald is a clear version of the common mineral beryl **(Fig. 5.19a)**, and peridot is the gemstone version of the common mineral olivine. As for the beauty of a gemstone, that quality lies basically in its color and, in the case of transparent gems, its "fire"—the way the mineral bends and internally reflects the light passing through it and disperses the light into a spectrum.

Gemstones form in many ways. Some solidify from a melt, some form by diffusion, some precipitate out of a water solution in cracks, and some result from the chemical interaction of rock with water near the Earth's surface. Many gems come from pegmatites, particularly coarse-grained igneous rocks formed by the solidification of "steamy" (water-rich) melt.

Most gems used in jewelry have been "cut," meaning that workers have made the smooth **facets** on the gem by using a faceting machine **(Fig. 5.19b)**. In other words, facets are not the natural crystal faces of the mineral, nor are they cleavage planes, though gem cutters sometimes orient facets parallel to cleavage directions and will try to break a large gemstone into smaller pieces by splitting it on a cleavage plane. A faceting machine consists of a doping arm, a device that holds a stone in a specific orientation, and a lap, a rotating disk covered with a wet paste of grinding powder and water. The gem cutter fixes a gemstone to the end of the doping arm and positions the arm so that it holds the stone against the moving lap. The movement of the lap grinds a facet. After, completing one facet, the gem cutter rotates the arm by a specific angle and grinds another facet. The geometry of the facets defines the cut of the stone. Different cuts have names, such

Did you ever wonder...
how jewelers make the facets on a jewel?

FIGURE 5.19 Cutting gemstones.

(a) Emerald is a green, transparent variety of the mineral beryl.

Non-gem-quality beryl

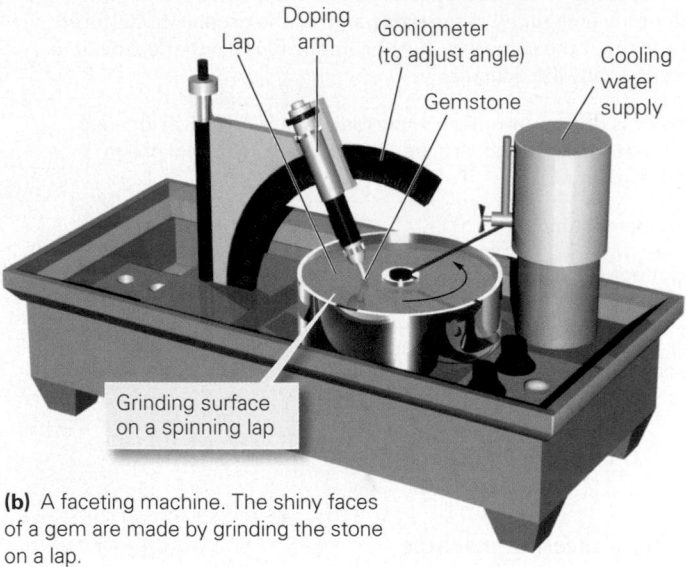

(b) A faceting machine. The shiny faces of a gem are made by grinding the stone on a lap.

Lap

Doping arm

Goniometer (to adjust angle)

Gemstone

Cooling water supply

Grinding surface on a spinning lap

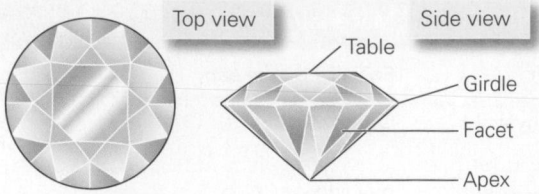

Top view

Side view

Table

Girdle

Facet

Apex

(c) There are many different "cuts" for a gem. Here we see the top and side views of a brilliant-cut diamond.

(d) Corundrum (Al_2O_3) comes in many colors. Gem-quality versions, including ruby and sapphire, can be cut into many shapes.

as "brilliant," "French," "star," and "pear." Grinding facets takes a lot of work—a typical engagement-ring diamond with a brilliant cut has 57 facets **(Fig. 5.19c)**!

Some mineral specimens have special value simply because their geometry and color before cutting are beautiful. Prize specimens exhibit shapes and colors reminiscent of fine art and may sell for tens of thousands of dollars or more **(Fig. 5.19d)**. It's no wonder that mineral "hounds" risk their necks looking for a cluster of crystals protruding from the dripping roof of a collapsing mine or hidden in a crack near the smoking summit of a volcano.

TAKE-HOME MESSAGE

Gemstones are particularly rare and beautiful minerals. The gems or jewels found in jewelry have been faceted using a lap. The facets are not natural crystal faces or cleavage surfaces. The fire of a jewel comes from the way it reflects light internally.

QUICK QUESTION: What's the difference between a facet on a gem and a crystal face?

ANOTHER VIEW A museum specimen of red quartz crystals, colored by trace amounts of iron oxide.

SUMMARY

- Minerals are naturally occurring, solid substances, formed by geologic processes, with a definable chemical composition, and are characterized by an orderly arrangement of atoms, ions, or molecules in a crystalline structure. Most minerals are inorganic.

- In the crystalline structure of minerals, atoms occur in a specific pattern—one of nature's finest examples of ordering.

- Minerals can form by the solidification of a melt, precipitation from a water solution, diffusion through a solid, biomineralization (precipitation by organisms), and precipitation from a gas.

- About 4,000 different types of minerals are known, each with distinctive physical properties (such as color, streak, luster, hardness, specific gravity, crystal habit, and cleavage).

- The unique physical properties of a mineral reflect its chemical composition and crystal structure. By observing these physical properties, you can identify minerals.

- Minerals can be classified according to their chemical composition. Mineral classes include silicates, sulfides, oxides, halides, carbonates, native metals, and sulfates.

- The silicate minerals are the most common on the Earth. The silicon-oxygen tetrahedron, a silicon atom surrounded by four oxygen atoms, provides the fundamental building block of silicate minerals.

- Groups of silicate minerals are distinguished from one another by the ways in which the silicon-oxygen tetrahedra that constitute them are linked.

- Gemstones are minerals known for their beauty and rarity. Gem cutters produce facets on cut gems used in jewelry by grinding and polishing the stones with a faceting machine.

GUIDE TERMS

atom (p. 124)
atomic mass (p. 124)
atomic number (p. 124)
biogenic mineral (p. 122)
carbonate mineral (p. 137)
change of state (p. 125)
chemical (p. 125)
chemical bond (p. 125)
chemical formula (p. 125)
chemical reaction (p. 125)
cleavage (p. 135)

color (p. 133)
concentration (p. 125)
conchoidal fracture (p. 136)
crystal (p. 126)
crystal face (p. 126)
crystal habit (p. 135)
crystalline material (p. 122)
crystal structure (p. 122)
diffraction (p. 127)
element (p. 124)
facet (p. 143)

gem (p. 142)
geode (p. 129)
glass (p. 123)
hardness (p. 133)
ion (p. 125)
luster (p. 133)
mineral (p. 122)
mineral class (p. 137)
mineralogist (p. 121)
mineralogy (p. 122)
Mohs hardness scale (p. 133)

molecule (p. 125)
polymorph (p. 128)
precipitate (p. 125)
silicates (p. 137)
silicon-oxygen tetrahedron (p. 137)
specific gravity (p. 133)
streak (p. 133)

GEOTOURS *THIS CHAPTER'S GEOTOURS WORKSHEET (C) FEATURES QUESTIONS AND GOOGLE EARTH SITES ON:*

- Rare Earth elements
- Physical properties of minerals
- Diamond mines and conflict diamonds
- Mineral reactions after coal mining

REVIEW QUESTIONS

The letters following each Review Question refer to the corresponding Learning Objective from the Chapter Opener.

1. What is a mineral, as geologists understand the term? How does this definition differ from the everyday usage of the word? **(A)**

2. Why isn't glass considered to be a mineral? **(A)**

3. Salt is a mineral, but the plastic making up an inexpensive pen is not. Why not? **(A)**

4. Describe the several ways that minerals can form. **(B)**

5. Why do some minerals occur as euhedral crystals, whereas others occur as anhedral grains? Which describes the crystals in the figure? **(B)**

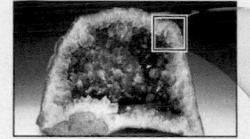 

6. List and define the principal physical properties used to identify a mineral. **(E)**

7. How can you determine the hardness of a mineral? What is the Mohs hardness scale? **(E)**

8. How do you distinguish cleavage planes from crystal faces on a mineral? How does each type of surface form? **(E)**

9. On what basis do geologists separate minerals into classes? **(C)**

10. On what basis do mineralogists organize silicate minerals into distinct groups? What type of silicate group does the diagram portray? **(D)**

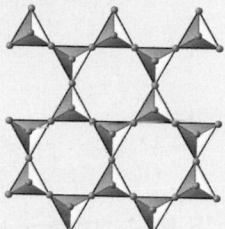

11. How does the bonding in mica determine cleavage in mica crystals? **(E)**

12. Why are some minerals considered gemstones? How do gem cutters make the facets on a gem? **(F)**

ON FURTHER THOUGHT

13. Compare the chemical formula of magnetite with that of biotite. Why do we mine magnetite to obtain iron supplies, but we don't mine biotite? **(C)**

14. Imagine that you are given two milky white crystals, each about 2 cm across. You are told that one of the crystals is plagioclase and the other is quartz. How can you determine which is which? **(E)**

15. Could you use crushed calcite to grind and form facets on a diamond? Why or why not? **(E)**

ONLINE RESOURCES

 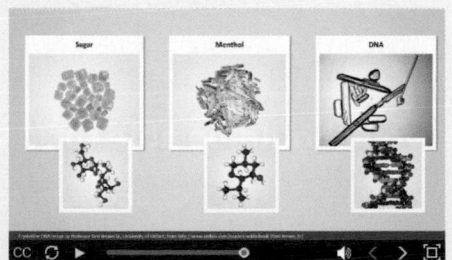 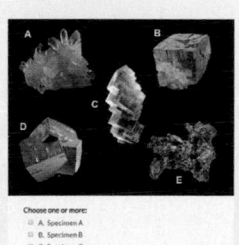

Animations
This chapter features animations on the formation, classification, and composition of various types of minerals.

Smartwork5
This chapter includes visual matching and labeling exercises designed to help students better understand crystal structure, how crystals grow, and the physical characteristics of minerals.

Introducing Rocks

By the end of this interlude, you should be able to . . .

A. provide a geologic definition of *rock*.

B. explain the basis that geologists use to classify rocks into three classes.

C. discuss the tools that can be used to study rocks.

A.1 Introduction

During the 1849 gold rush in the Sierra Nevada of California, only a few lucky individuals actually became rich. The rest of the "forty-niners" either slunk home in debt or took up less glamorous jobs in boom towns such as San Francisco. These towns grew rapidly, and soon the American west coast was demanding large quantities of manufactured goods from factories on the east coast. Making the goods was no problem, but getting them to California meant either a stormy ocean voyage or a trek with stubborn mule teams through the deserts of Nevada and Utah. The time was ripe to build a railroad linking the east and west coasts of North America, so with much fanfare, a consortium of companies set to work in 1863. The Union Pacific surveyed a route that crossed the Sierra Nevada, and as the Civil War raged, the company transported thousands of Chinese laborers across the Pacific in the squalor of unventilated cargo holds and set them to work chipping ledges around, and blasting tunnels through, the range's towering peaks. Sadly, untold numbers of laborers died of frostbite and exhaustion, and from mistimed blasts, landslides, and avalanches.

Through their efforts, workers building the transcontinental railroad certainly gained an intimate knowledge of how rock feels and behaves—it's solid, heavy, and mighty hard! They also found that some rocks break easily into layers, but others do not, and that some rocks are dark colored, while others are light colored. Like anyone who looks closely at rock exposures, they realized that rocks are not just gray, featureless masses, but rather come in a great variety of colors, textures, and configurations.

Why are there so many distinct types of rocks? The answer is simple: many different processes can produce rocks, and many different materials can make up rocks. Because of the relationship between rock type and the process of formation, rocks provide a record of geologic events over the Earth's history and give us insight into interactions among components of the Earth System. We devote the next few chapters to a discussion of rocks and a description of how rocks form. To provide a general introduction to these chapters, this interlude provides the geologic definition of the term *rock*, describes the basic components of rock, and characterizes the three principal classes of rocks. We also describe a few of the methods that geologists use to study rocks.

◀ (facing page) Wave-carved outcrops along the beach near Brisbane, Australia.

A.2 What Is Rock?

To a geologist, **rock** is a coherent, naturally occurring solid that consists of an aggregate of minerals or, less commonly, a body of glass. To understand this definition, let's look at its components more closely.

- *Coherent:* A rock holds together and must be broken in order to be separated into smaller pieces. As a result of its coherence, rock can form cliffs or be carved into construction blocks or sculptures **(Fig. A.1)**. A pile of unattached mineral grains does not constitute a rock.

FIGURE A.1 Rock is coherent, so it can be very strong and durable.

(a) The strength of rocks can hold up tall cliffs.

(b) Rock, because of its durability, was used to build this ancient castle in Portugal.

FIGURE A.2 Rocks that are aggregates of mineral grains and/or crystals can be clastic or crystalline.

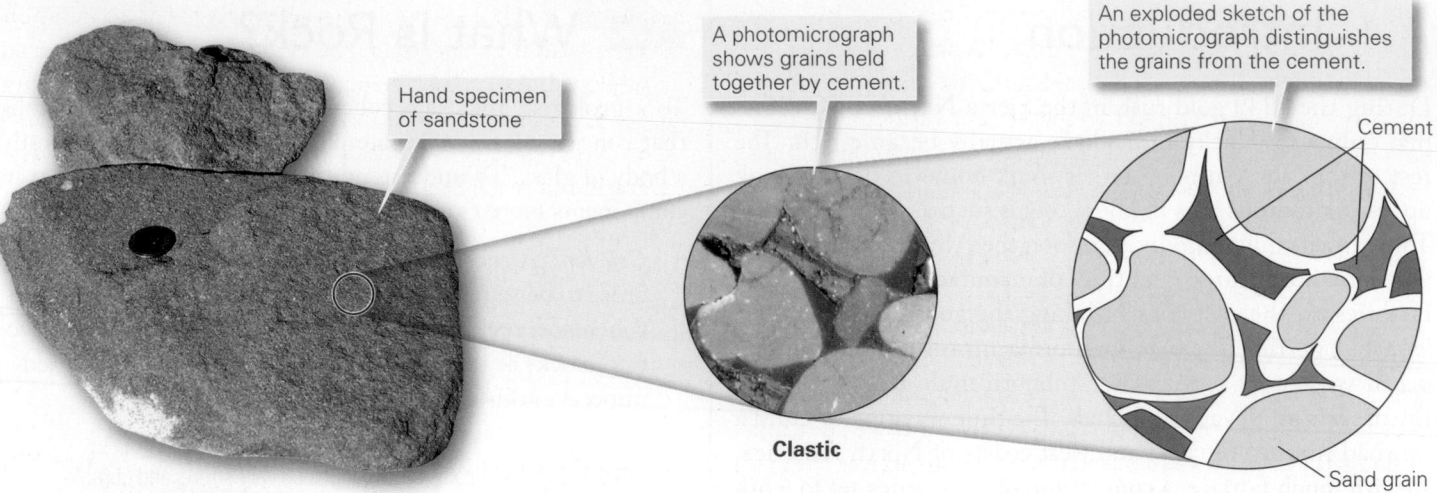

Hand specimen of sandstone

A photomicrograph shows grains held together by cement.

An exploded sketch of the photomicrograph distinguishes the grains from the cement.

Cement

Clastic

Sand grain

(a) Sandstone is a clastic rock, for it consists of separate grains that have been cemented together.

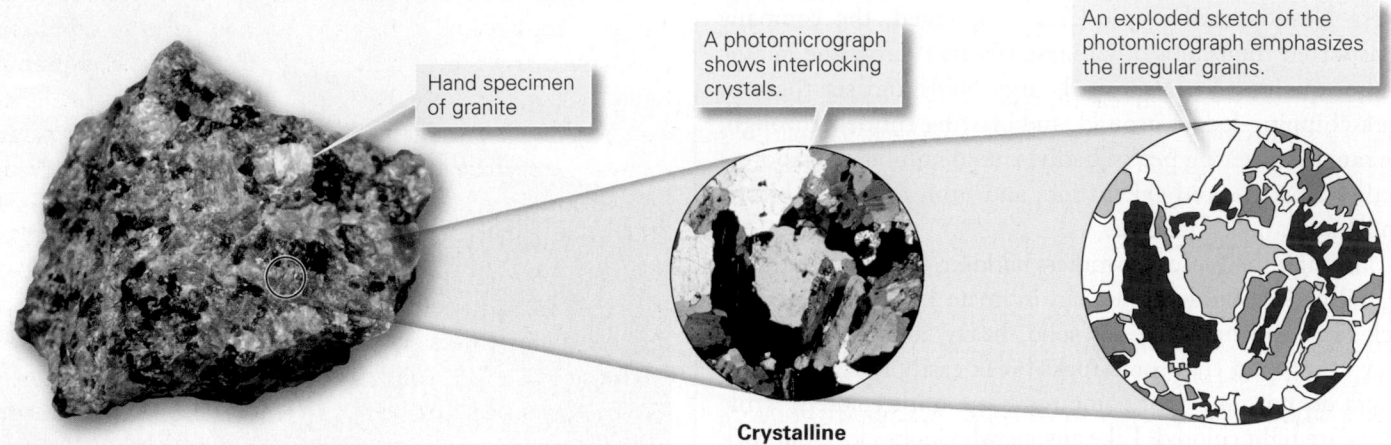

Hand specimen of granite

A photomicrograph shows interlocking crystals.

An exploded sketch of the photomicrograph emphasizes the irregular grains.

Crystalline

(b) Granite is a crystalline rock, for it consists of interlocking crystals that grew together.

- *Naturally occurring:* Geologists consider only materials formed by natural processes to be rocks. Manufactured products (concrete, bricks, or cinder blocks) are not rocks.

- *An aggregate of minerals or a body of glass:* Most rocks consist of an aggregate (a collection) of many mineral grains and/or crystals. Note that the terms *grain* and *crystal* are often used interchangeably. But to be picky, we use the word **grain** as a more general term that can refer to a piece of mineral, to a fragment broken off a once larger piece of a mineral or pre-existing rock, or to a fragment of glass. We restrict the term **crystal** to a continuous (uninterrupted) piece of a single mineral that grew to its present shape and may display crystal faces. (A large crystal, or aggregate of crystals, of a single mineral can be called a *mineral specimen,* even if it is meters long.) Some rocks contain only one kind of mineral, whereas others contain several different kinds. A few rock types that form at volcanoes consist of natural glass, which

may occur either as a homogeneous body or as an accumulation of tiny shards.

What holds rock together? Grains in rock connect to form a coherent mass either because natural **cement,** composed of minerals that precipitated from water in the space between grains, glues them together **(Fig. A.2a)**, or because they interlock with one another and fit together like pieces in a jigsaw puzzle **(Fig. A.2b)**. We refer to rocks whose grains are held in place by cement as **clastic rocks,** and to rocks whose crystals interlock with one another as **crystalline rocks**. A glassy rock can be coherent either because it originated as a continuous mass and therefore does not contain separate grains, or because it formed by welding together separate glass grains when the grains were still hot.

Regardless of whether a rock consists of glass or minerals, we can think of it, fundamentally, as a collection of chemicals. Significantly, not all rocks contain the same chemicals. For example, *granite*—a rock commonly used for gravestones, building facades, and kitchen counters—contains oxygen, silicon, aluminum, potassium, sodium, calcium, iron, and magnesium, whereas *marble*—a rock favored by sculptors—contains oxygen, carbon, and calcium. In both granite and marble, the elements bond together to form minerals. Granite contains quartz, feldspar, mica, pyroxene, and amphibole, whereas marble contains mainly calcite.

By averaging the chemical compositions of rocks, geologists have determined that oxygen and silicon are by far the most common elements in rocks of the Earth's crust and mantle (Table A.1). That's because most rock consists of silicate minerals. At or near the Earth's surface, however, living organisms play a role in the formation of some rock types, so a significant proportion of rock in the upper crust of continents consists of carbonate minerals extracted from water to form shells. Rocks made of oxides, sulfides, and sulfates serve as important resources for metals and industrial materials, but they constitute only a small percentage of our planet's rocks.

At the surface of the Earth, rock occurs either as broken chunks or fragments (such as pebbles, cobbles, or boulders) that fell down a slope or were transported in ice, water, or wind, or as **bedrock** that remains attached to the Earth's crust. Geologists refer to an exposure of bedrock as an **outcrop.** An outcrop may be a rounded knob out in a field, a ledge forming a cliff or ridge, or a stream cut (where running water has cut down into bedrock); or it may be the face of a human-made road cut, rail cut, or excavation (Fig. A.3). To people who live in cities or forests or on farmland, outcrops of bedrock may be unfamiliar, for vegetation, loose sediment

(sand, mud, gravel), soil, water, asphalt, concrete, or buildings may cover bedrock. Engineers prefer to set the foundations of large buildings and dams on strong bedrock, rather than on weak sediment. Therefore, planners may take into account the depth between the ground surface and the top of bedrock when siting large construction projects, to save on the cost of excavation to reach bedrock.

A.3 The Basis of Rock Classification

When the systematic study of geology began in the late 18th century, geologists grappled with the challenge of developing a meaningful classification scheme for rocks because classification helps organize information and helps to emphasize similarities and differences. One of the earliest classification schemes, popular in the late 18th century, distinguished three groups of Earth materials—so-called primary, secondary, and tertiary—based on the incorrect perception that the groups had formed in a time succession. According to this scheme, first formulated by Abraham Werner (1749–1817), a German mineralogist, a "universal ocean" containing dissolved minerals covered the Earth when the planet first formed. Precipitation of minerals from this solution produced early "primary rocks," which included hard crystalline granite. Werner proposed that later, as sea level dropped, the action of rivers, waves, and wind wore down exposed primary rocks and produced debris that settled in the ocean to form "secondary rocks," which occurred in layers over the primary material. Loose gravels, sands, and muds—materials that had not turned to rock—made up the most recent "tertiary materials." Because Werner's classification scheme viewed most rock formation as a process that took place only in water, his followers came to be known as the *Neptunists*, after the Roman god of the sea.

Meanwhile, a Scottish doctor and scientist named James Hutton (1726–1797) was exploring the outcrops in his home country. Hutton was a keen thinker who lived in Edinburgh, a hotbed of intellectual discourse during the Age of Enlightenment, a time when everything from political institutions to scientific paradigms became fodder for debate. Hutton sought insight into how rocks might form by looking for evidence of processes that could produce the characteristics he observed. For example, he watched sand settle on a beach and concluded that clastic rocks formed from cementation of grains. He examined outcrops in which bodies of crystalline rocks appeared to have pushed into other rocks and concluded that these crystalline rocks formed by solidification of magma that had been injected into older rock. He also noticed that

TABLE A.1 Chemical Composition of Rocks in the Earth

| Element | % by weight | |
	Mantle	Crust (continental)
O	44.8	46.6
Si	21.5	27.7
Mg	22.8	1.5
Fe	5.8	5.0
Al	2.2	8.1
Ca	2.3	3.6
Na	0.3	2.8
K	0.03	2.6
Other	0.3	2.1

FIGURE A.3 Types of outcrops. The left column illustrates natural outcrops; the right column shows exposures of bedrock that were produced by human activities such as mining or road and rail construction.

Outcrop in the woods, Illinois

Road cut, Maryland

Stream cut, New York

Quarry, Illinois

Mountain cliffs, Colorado

Railroad cut, Illinois

rocks adjacent to bodies of now-solid magma had somehow been altered, and he attributed this change to "subterranean heat." Hutton, like Werner, attracted followers—Hutton's group came to be known as the *Plutonists*, after the Roman god of the underworld, because they favored the idea that the

formation of certain rocks involved melts that had risen from deeper within the Earth.

In the last decades of the 18th century, as the armed rebellions that led to the formation of the United States and the Republic of France raged, a battle of ideas concerning the

FIGURE A.4 Examples of the three major classes of rocks.

(a) Lava (molten rock that has reached the Earth's surface) freezes quickly to form igneous rock. Here, the molten tip of a brand-new flow still glows red. Older flows are already solid.

(b) Sand, formed from grains eroded from these rock cliffs, collects on the beach. If buried and turned to rock, it becomes layers of sandstone, like that making up the cliffs.

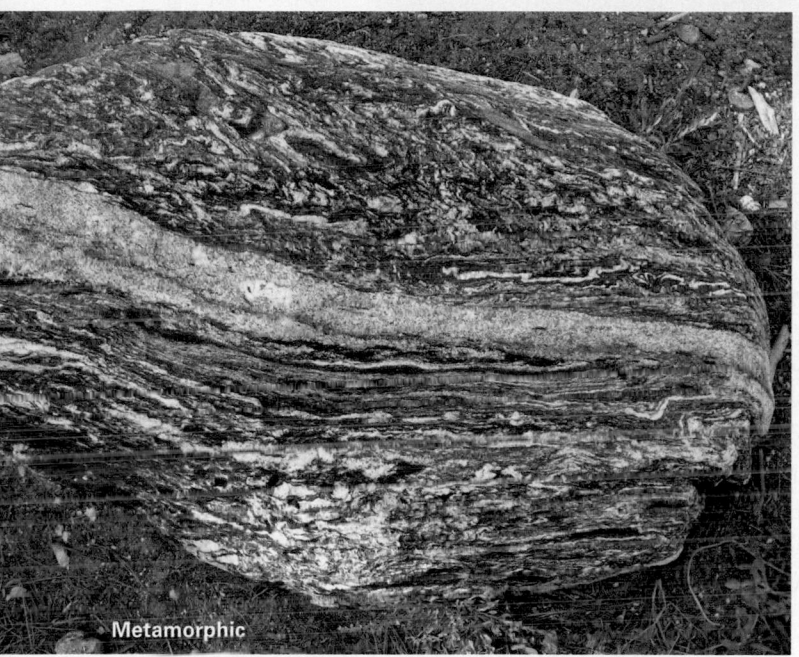

(c) Metamorphic rock forms when pre-existing rocks endure changes in temperature and pressure and/or are subjected to shearing, stretching, or squashing.

origin of rocks rattled the infant science of geology. This battle, pitting the Neptunists against the Plutonists, lasted for years. In the end, the Plutonists won, for they eventually demonstrated beyond a shadow of a doubt that certain crystalline rocks must have been in molten form when emplaced. Geologists came to understand that different rocks formed in different ways and concluded that rocks can best be classified on the basis of how they formed. This *genetic scheme* for classifying rocks—a scheme based on the origin (genesis) of

rocks—became a foundation for the modern classification of rocks. Because Hutton fostered this key idea, as well as many other ideas that we describe later in this book, modern geologists revere Hutton as the father of geology.

In the modern genetic scheme of rock classification, geologists recognize three basic rock classes: (1) **igneous rocks**, which form by the freezing (solidification) of molten rock **(Fig. A.4a)**; (2) **sedimentary rocks**, which form either by the cementing together of grains broken off pre-existing rocks or by the precipitation of mineral crystals out of water solutions at or near the Earth's surface **(Fig. A.4b)**; and (3) **metamorphic rocks**, which form when pre-existing rocks change character in response to a change in pressure and temperature conditions, and/or as a result of squashing, stretching, or shearing under conditions such that the rocks do not crack and break **(Fig. A.4c)**. Metamorphic change occurs in the solid state, which means that it does not involve melting. In the context of modern plate tectonics theory, different rock types form in different geologic settings **(Fig. A.5)**. In succeeding chapters, we will explore these settings in more detail.

Each of the three classes contains many different individual rock types, distinguished from one another by physical characteristics such as the following:

- *Grain size and shape:* The grains in rock come in a wide range of sizes. Some grains are so small that they can't be seen without a microscope, whereas others are as big as a car or larger. Rocks also differ in terms of the range of grain sizes that the rocks contain and in terms of grain shape **(Fig. A.6)**. Specifically, in some rocks, all grains have the same size, whereas other rocks contain grains of many different sizes. And in some rocks, all grains are **equant**, meaning that they have the same dimensions in all directions, whereas in others, the grains are **inequant**, meaning that the grains do not have the same dimensions in all directions.

- *Composition:* The term **rock composition** refers to the proportions of different chemicals that make up the rock. The proportions of chemicals, in turn, affect the proportions of different minerals constituting the rock. As you will see, however, chemical composition alone does not completely control the assemblage of minerals present in a rock. For example, two rocks with the same chemical composition can have totally different assemblages of minerals if each rock formed under different pressure and temperature conditions. That's because environmental factors such as pressure and temperature influence the growth of minerals.

- *Texture:* This term refers to the configuration of grains in a rock, that is, the way grains connect to one another and whether or not inequant grains align parallel to each other. The concept of rock texture will become easier to grasp as we look at different examples of rocks in the following chapters.

- *Layering:* Some rock bodies contain distinct layers, defined by bands of different compositions, grain sizes, or textures or by the alignment of inequant grains so that they parallel each other. Different types of layering occur in different kinds of rocks. For example, the layering in sedimentary rocks is called **bedding**, whereas the layering in metamorphic rocks is called **metamorphic foliation (Fig. A.7)**.

FIGURE A.6 Describing grains in rock.

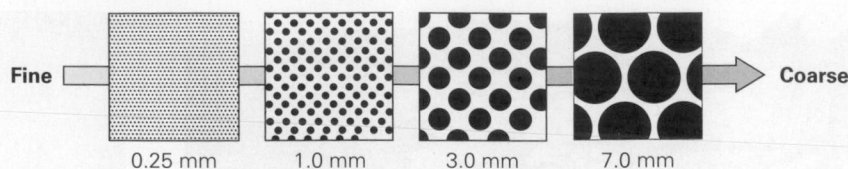

(a) Geologists define grain size by using this comparison chart.

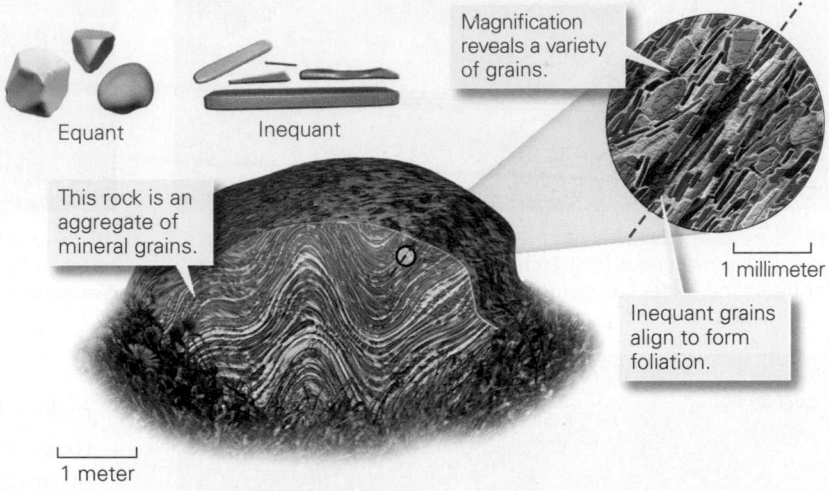

(b) Grains in rock come in a variety of shapes. Some are equant, whereas some are inequant. In this example of metamorphic rock, inequant grains align to define a foliation.

Each distinct rock type has a name. These names come from a variety of sources. Some names reflect the dominant mineral making up the rock, some represent the name of a region where the rock was first discovered or is particularly abundant, some come from ancient legends, some incorporate a root word of Latin or Greek origin, and some started as traditional names used by people in an area where the rock was found. Many rock names date back to antiquity, but some

FIGURE A.5 A cross section illustrating various geologic settings in which rocks form.

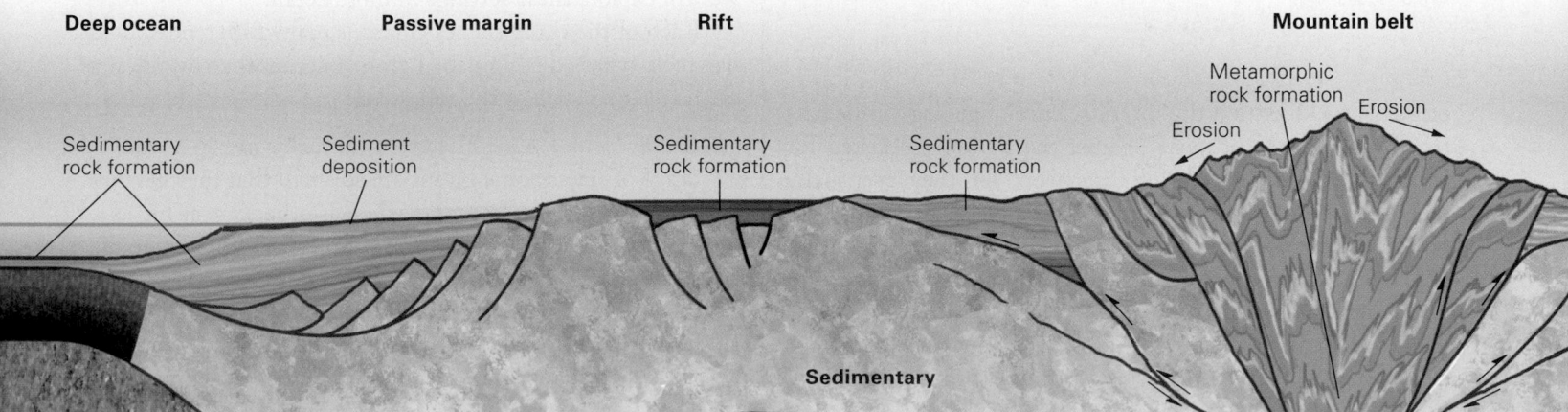

FIGURE A.7 Layering in rock.

(a) Bedding in sedimentary rock, here defined by alternating layers of coarser and finer grains, as exposed on a cliff in Utah.

(b) Foliation in this outcrop of metamorphic rock near Mecca, California, is defined by alternating light and dark layers. The color of a layer depends on the minerals that make up that layer.

were assigned only in recent decades. All told, geologists use hundreds of different rock names, though in this book we introduce only about 30.

A.4 Studying Rock

Outcrop Observations

The study of rocks begins by observing rock in an outcrop. If the outcrop is big enough, an examination will reveal relationships between the rock you're interested in and the rocks around it and will allow you to detect layering. Geologists carefully record observations about an outcrop, then may use a hammer to break off a **hand specimen (Fig. A.8a–c)**—a fist-sized piece of a rock—that they can examine more closely with a *hand lens*, a type of high-quality magnifying glass **(Fig. A.8d)**. (Breaking rocks with hammers can be dangerous and should be done only with appropriate eye protection.) Observation with a hand lens enables geologists to identify sand-sized or larger mineral grains and may allow them to characterize the rock's texture.

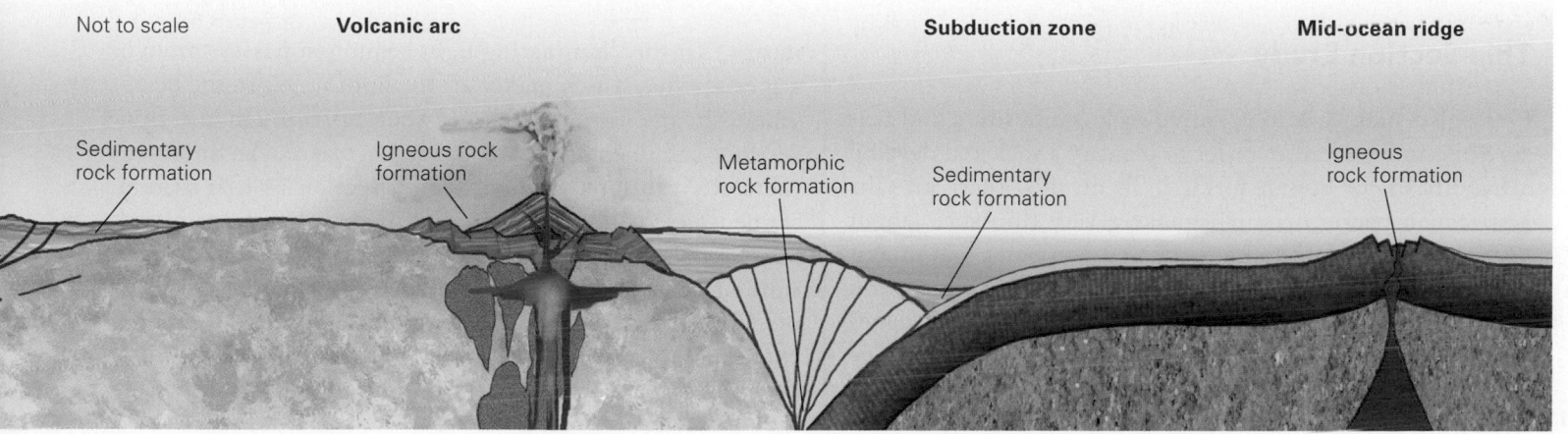

FIGURE A.8 Basic tools for studying rocks in the field.

(b) A rock hammer.

(c) A hand specimen.

(d) A hand lens.

(a) A field geologist examining a hand specimen (on a cold day).

Thin-Section Study

Geologists may have to examine rock composition and texture in minute detail in order to identify a rock and develop a hypothesis for how it formed. To do this, they can take a specimen back to the lab, make a very thin slice (about 0.03 mm thick, the thickness of a human hair), and mount it on a glass slide **(Fig. A.9a–c)**. The resulting **thin section** can be studied with a petrographic microscope (*petro* comes from the Greek word for rock). A **petrographic microscope** differs from an ordinary microscope in that it illuminates the thin section with transmitted polarized light. This means that the illuminating light beam first passes through a special filter, which makes all the light waves in the beam vibrate in the same plane, and then up through the thin section, before entering a polarized eyepiece. An observer, therefore, can look through a thin section as if it were a stained-glass window. When illuminated with transmitted polarized light, each type of mineral grain displays a unique suite of colors **(Fig. A.9d)**. The specific color the observer sees depends on both the identity of the grain and its orientation with respect to the waves of polarized light, for a crystal interferes with polarized light and allows only certain wavelengths to pass through.

FIGURE A.9 Studying rocks in thin section.

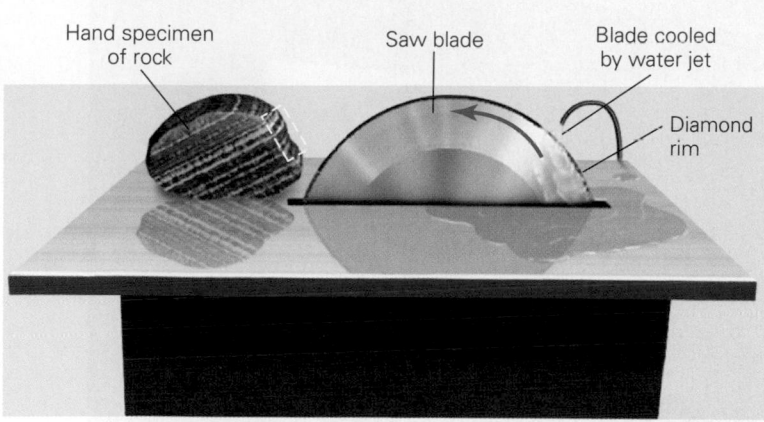

Hand specimen of rock · Saw blade · Blade cooled by water jet · Diamond rim

(a) Using a special saw, a geologist cuts a thin chip of a rock specimen.

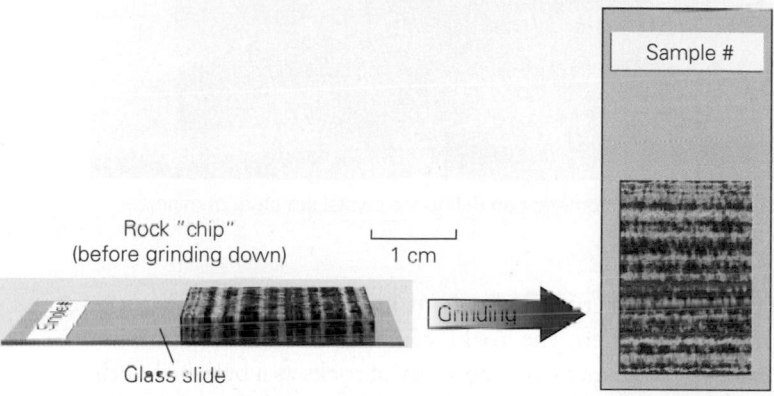

Rock "chip" (before grinding down) · 1 cm · Glass slide · Grinding · Sample #

(b) The geologist glues the chip to a glass slide and grinds it down until it is so thin that light can pass through it.

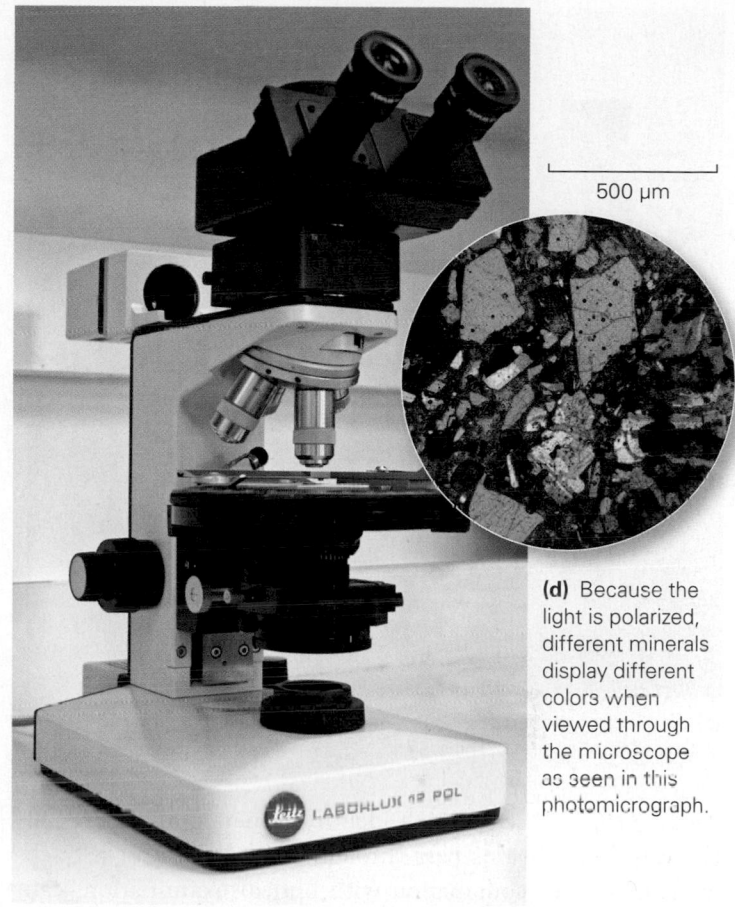

500 µm

(d) Because the light is polarized, different minerals display different colors when viewed through the microscope as seen in this photomicrograph.

(c) With a petrographic microscope, it's possible to view thin sections with polarized light that shines through the sample from below.

The brilliant colors and strange shapes visible in a thin section rival the beauty of an abstract painting. By examining a thin section with a petrographic microscope, geologists can identify most of the minerals constituting the rock and can describe the way in which the grains connect to one another. To convey the information visible in the thin section, they can attach a camera to the microscope eyepiece to take a **photomicrograph**.

How can you make a thin section? You start by using a special rock saw with a very thin, rapidly spinning, water-cooled diamond-studded blade (see Fig. A.9a). The blade slowly grinds a very thin groove into the rock—the hard diamonds embedded in the saw blade scratch and pulverize minerals as the saw blade rubs against the rock. By making four cuts, you can produce a small rectangular block, or "chip," of rock (see Fig. A.9b). Then, you cement the chip to a glass microscope slide with an epoxy adhesive (glue) and cut off the excess rock with the rock saw. By pressing the slide facedown against a *lap*, a spinning plate coated with abrasive, you can grind the

chip down until only a thin slice, still cemented to the glass slide, remains, ready for examination with a petrographic microscope.

High-Tech Analytical Equipment

Beginning in the 1950s, high-tech electronic instruments became available that have enabled geologists to examine rocks on an even finer scale than is possible with a petrographic microscope. Modern research laboratories typically obtain instruments such as *scanning electron microscopes* (SEMs), which can image the surface of a rock chip at extremely high magnification and can map the distribution of elements in the chip; *electron microprobes*, which can focus a beam of electrons on a small part of a grain to create a signal that defines the chemical composition of the mineral **(Fig. A.10)**; *mass spectrometers*, which analyze the proportions of different isotopes of elements contained in a rock; and

FIGURE A.10 High-tech equipment used to analyze rocks.

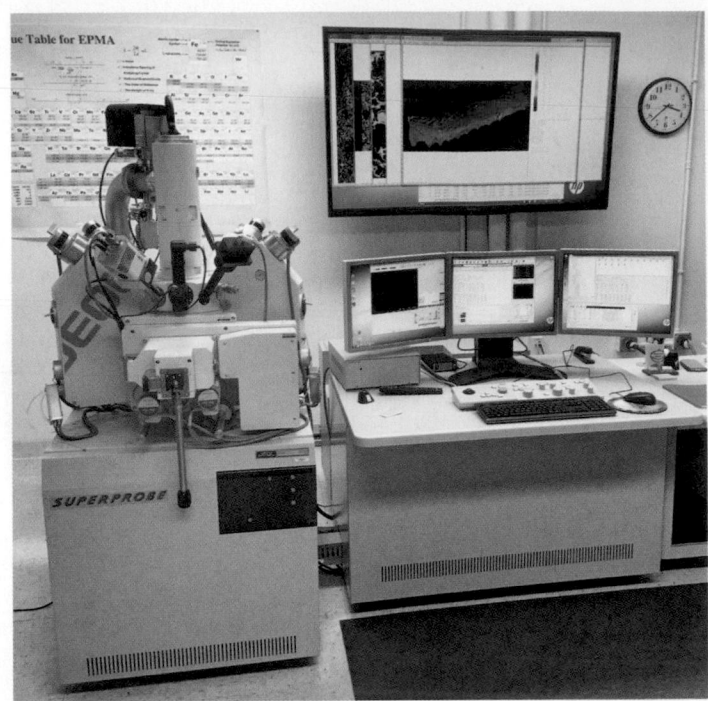

(a) An electron microprobe uses a beam of electrons to analyze the chemical composition of minerals.

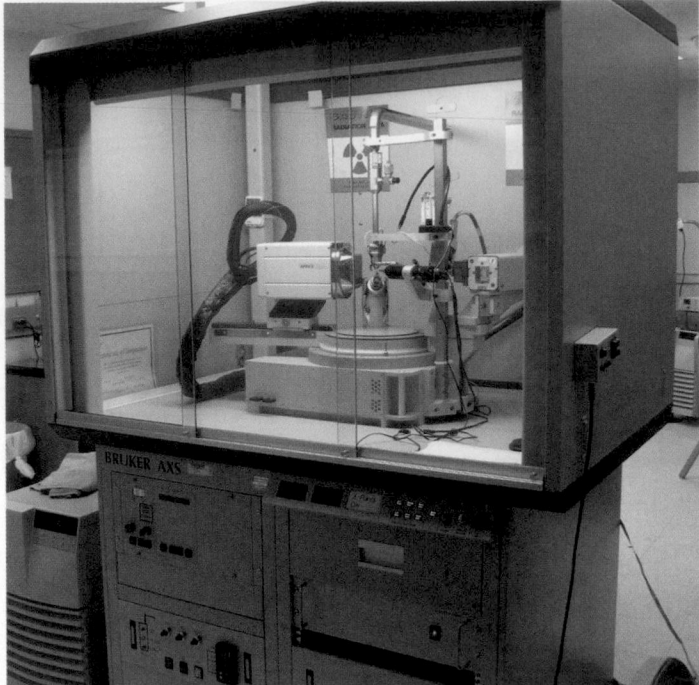

(b) An X-ray diffractometer can define the crystal structure of minerals in rocks.

X-ray diffractometers, which identify minerals by looking at the way X-ray beams pass through crystals in a rock. Such instruments, in conjunction with optical examination, can provide geologists with highly detailed characterizations of rocks, which in turn help them understand how the rocks formed and where the rocks came from. This information enables geologists to use the study of rocks as a basis for deciphering the Earth's history.

ANOTHER VIEW Students puzzle over the features visible in this outcrop of sandstone, in Utah, that might provide clues to its geologic history.

Interlude A Review

SUMMARY

- Rock is a coherent, naturally occurring solid, consisting of an aggregate of minerals or of a body of glass. Nonglassy rocks can be classified as crystalline or clastic.

- Bedrock consists of rock that remains connected to the underlying crust. Outcrops, exposures of bedrock at the Earth's surface, can be natural or human-made.

- Geologists classify a rock as igneous, sedimentary, or metamorphic based on how the rock formed.

- A variety of characteristics prove helpful in describing rocks. Examples include grain size and shape, composition, texture, and the nature of layering.

- Hand lenses, petrographic microscopes (after making thin sections), and sophisticated electronic equipment help geologists interpret the origin of rocks.

GUIDE TERMS

bedding (p. 154)
bedrock (p. 151)
cement (p. 150)
clastic rock (p. 150)
crystal (p. 150)
crystalline rock (p. 150)

equant (p. 154)
grain (p. 150)
hand specimen (p. 155)
igneous rock (p. 153)
inequant (p. 154)

metamorphic foliation (p. 154)
metamorphic rock (p. 153)
outcrop (p. 151)
petrographic microscope (p. 156)

photomicrograph (p. 157)
rock (p. 149)
rock composition (p. 154)
sedimentary rock (p. 153)
thin section (p. 156)

REVIEW QUESTIONS

The letters following each Review Question refer to the corresponding Learning Objective from the Chapter Opener.

1. How do geologists define the term *rock*? Can a brick be considered a rock? Explain your answer. **(A)**

2. Explain the difference between a clastic and a crystalline rock. **(A)**

3. Give examples of different kinds of rock outcrops. **(A)** Can you find outcrops everywhere? Explain your answer. **(A)**

4. On what basis do geologists define rocks into three classes? What are these classes? **(B)**

5. Distinguish between an equant and an inequant grain. **(B)**

6. What type of layering does the photograph show?

7. Give two examples of types of layering that can occur in rock. **(B)**

8. What are thin sections, how are they examined, and what do they allow you to see? **(C)**

9. Name examples of high-tech equipment that can be used to study rocks. What extra information can geologists learn by using such equipment? **(C)**

ONLINE RESOURCES

Animations
This interlude features animations on distinguishing rock groups and conducting rock analysis.

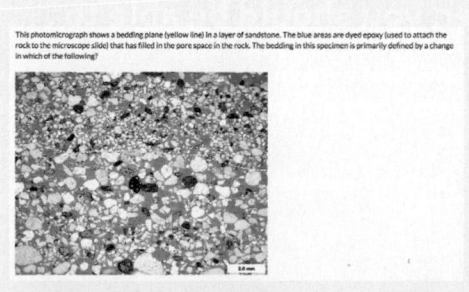

This photomicrograph shows a bedding plane (yellow line) in a layer of sandstone. The blue areas are dyed epoxy (used to attach the rock to the microscope slide) that has filled in the pore space in the rock. The bedding in this specimen is primarily defined by a change in which of the following?

Smartwork5
This interlude features visual identification questions on rock classifications and the three main rock groups.

CHAPTER 6

Up from the Inferno: Magma and Igneous Rocks

By the end of this chapter, you should be able to . . .

A. distinguish between magma and lava, and explain the chemical distinctions among different types of molten rock.

B. describe the special places in the Earth where magma forms.

C. explain why melt moves to locations where it solidifies and how solidification takes place.

D. describe and classify different kinds of igneous rocks.

E. discuss where and why igneous activity happens in the context of plate tectonics theory.

> Granite—it seems inevitable to begin with granite, even though so many people have ended with it, lying under those glossy pinkish slabs labeled in gold or black . . .
>
> —Jacquetta Hawkes (British archaeologist and writer; 1910–1996)

6.1 Introduction

Every now and then, you can see hot molten rock, or *melt*, fountaining from an opening on the Big Island of Hawaii. The incandescent liquid, which has a temperature of 1,100°C to 1,200°C when first disgorged from the Earth's interior, may accumulate in a seething pool around the vent or may spill downhill as a syrupy red-yellow stream **(Fig. 6.1a)**. As it flows, the melt burns its way through forests, submerges roads, and incinerates houses **(Fig. 6.1b)**. In places close to the vent, where the slope is steep and the melt follows a channel, it can move swiftly, cascading over escarpments at speeds of up to 60 km per hour, but farther from its source, where the melt has cooled or has spread onto gentler slopes, it slows to a few kilometers per hour. As it cools, the melt's surface darkens and crusts over, forming a carapace that insulates the still-hot, sticky mass oozing below. Finally, the melt stops moving entirely and, over the course of hours to weeks, freezes into a hard black solid through and through **(Fig. 6.1c)**. You've witnessed the formation of new **igneous rock**: rock made by the freezing of a melt. Considering the fiery heat of the melt from which igneous rock solidifies, the name *igneous*, from the Latin *ignis*, meaning fire, makes sense. Because we can see certain types of igneous rocks forming at the Earth's surface, it may seem like the process happens only at the surface. But that's not the case. In fact, the vast majority of igneous rock forms by cooling and solidification underground, hidden from our view.

It may seem strange to speak of "freezing" in the context of rock formation, for most people think of freezing as the transformation of liquid water to solid ice when the temperature drops below 0°C. Nevertheless, the freezing of liquid melt to form solid igneous rock represents the same phenomenon—the solidification of a liquid by forming crystals and/or glass. But unlike water, igneous rock freezes at high temperatures, between 650°C and 1,100°C. To put such temperatures in perspective, keep in mind that your home oven attains a maximum temperature of only 260°C (500°F).

◀ (facing page) The "Elephant Rocks," boulders that weathered in place to form unusual shapes, attract tourists to the St. Francis Mountains in Missouri. They consist of reddish granite that formed from magma that oozed into the crust almost 1.5 billion years ago.

A great variety of igneous rocks exist in the Earth's crust—they make up all of the ocean crust beneath the thin veneer of seafloor sediment, as well as vast volumes of the continental crust. To understand why and how these rocks form, and why there are so many different kinds, we begin by discussing why melt forms, why it rises toward the surface, how it flows, and how it freezes at or below the surface. We then examine the scheme that geologists use to classify igneous rocks. Finally, we relate the formation of igneous rocks to plate tectonics theory.

6.2 Why Do Melts Form?

Discussing Molten Rock and Its Occurrences

Geologists use special names for molten rock and for the settings in which it occurs. Specifically, we refer to melt that's underground as **magma** and to melt that has emerged at the Earth's surface as **lava**. A vent from which lava emerges is a **volcano,** and an event during which melt flows from or explodes out of a volcano is a **volcanic eruption**. We also use the word *volcano* in reference to a hill or mountain built from the products of eruption. Eruptions can produce a *lava fountain,* in which lava forcefully rises meters to hundreds of meters into the air; a *lava lake,* in which lava pools over the vent; or a **lava flow,** in which lava moves in a stream down a slope. We also refer to the rock sheet or mound formed when this lava cools to form rock as a lava flow. Explosive volcanic eruptions can send clouds of shattered pre-existing igneous rock as well as droplets or blobs of lava skyward. These fragmental materials are called **pyroclastic debris**, from the Greek word *pyro,* meaning fire, and *klastos,* meaning broken **(Fig. 6.2)**. There's a lot more to say about volcanoes and their eruptions, so we'll explore the topic further in Chapter 9.

Geologists distinguish between two main categories of igneous rock based on where it solidifies **(Fig. 6.3a)**. Rock that forms either by the freezing of lava above ground, after it spills out (*extrudes*) onto the surface of the Earth and comes into contact with air or water, or by the cementing or welding together of pyroclastic debris, is **extrusive igneous rock** **(Fig. 6.3b)**. As we've noted, vastly more igneous rock forms from magma that pushed its way, or *intruded*, into pre-existing rock, or **wall rock,** and solidified out of view underground. We refer to such rock as *intrusive igneous rock* and to a body of such rock as an **igneous intrusion**

FIGURE 6.1 Formation and evolution of lava flows.

Lava fountain

Active lava flow

As the lava cools, it darkens.

(a) Lava erupts as a fountain from a volcanic vent on Hawaii. A fast-moving river of lava then flows downslope.

Smoke comes from burning vegetation.

Time

(b) At a distance from the vent, the lava has completely crusted over with new rock, but the interior of the flow remains molten.

(c) Eventually, the flow cools completely and becomes a layer of new rock. This flow engulfed a road on Hawaii.

(Fig. 6.3c). In the intrusive realm, magma may accumulate in an irregularly shaped zone called a **magma chamber**, in a chimney-like column, along planar cracks, or in thin sheets between pre-existing layers. Each of these processes yields a different shape of intrusion.

Why Is It Hot inside the Earth?

Clearly, if the Earth were not hot inside, igneous processes could not take place. Where does our planet's internal heat come from? Some of this heat was left over from the Earth's early days. Recall that, according to the nebula theory, our planet formed from the collision and merging of countless planetesimals. Every time a collision occurred, its kinetic energy (energy of motion) transformed into heat energy. (You can simulate this phenomenon by banging a hammer repeatedly on a nail—the head of the nail becomes quite warm.) Then, as the Earth grew, gravity pulled matter inward until eventually, the weight of overlying material squeezed the matter inside tightly together. Such compression made the Earth's insides even hotter, just as compressing a gas with a piston makes it hotter. Eventually, the Earth became hot enough for iron inside it to melt, and the dense iron sank to the center

FIGURE 6.2 A volcanic eruption can produce both lava and pyroclastic debris.

Pyroclastic debris

Ash cloud

Lava flow

Solidified lava flow

Ash flow in motion

Ash flow deposits

FIGURE 6.3 The intrusive and extrusive realms.

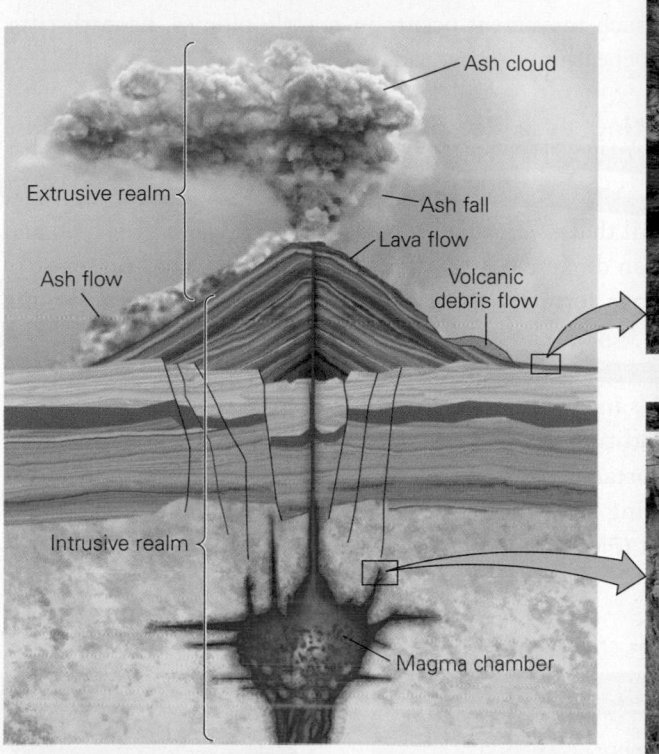

(b) Extrusive rocks include lava flows and pyroclastic layers.

(a) The intrusive realm lies underground and the extrusive realm lies above ground. Lava flows, as well as various types of explosive eruptions, all produce extrusive rocks.

(c) An intrusion of basalt (dark rock) cuts across an earlier intrusion of granite (light rock).

to form the core. Friction between sinking iron and its surroundings generated still more heat, just as rubbing your hands together generates heat. Soon after the Earth's formation, but after its differentiation into the mantle and core, a Mars-sized object collided with the Earth (see Chapter 1). This collision generated vast amounts of heat. And even after the Earth had grown to become a planet, intense bombardment continued to add heat energy—this bombardment didn't cease until about 3.9 billion years ago. Taken together, collisions and differentiation made the early Earth so hot that it was at least partially molten throughout, and its surface may, at times, have been an ocean of lava.

Ever since the heat-producing catastrophes of its early days, the Earth has radiated heat into space and, therefore, has slowly cooled. Eventually, the sea of lava on its surface solidified and formed igneous rock. Therefore, we can think of igneous rock as the first rock of this planet's crust to form. Significantly, if no heat had been added to the Earth after the era of intense bombardment, the Earth might have become too cold by now for igneous activity to take place today. Such cooling hasn't happened because of the presence of radioactive elements in the Earth (primarily in the crust). Decay of a single radioactive atom produces only a tiny amount of heat, but the cumulative heat of radioactive decay in the Earth has been sufficient to slow the cooling of this planet overall. Therefore, the Earth remains very hot today, with temperatures at the base of the lithosphere reaching almost 1,300°C, and temperatures at the planet's center exceeding 4,700°C. (By comparison, the surface of our Sun glows at about 5,700°C.)

Causes of Melting

Despite the very high temperatures inside the Earth, the common image that the solid crust floats on a sea of molten rock is not correct. In fact, under the very high pressures that exist inside the Earth, rock remains solid except in special places where local conditions allow some melt to form. Such conditions can develop both in

Did you ever wonder...
whether the Earth's crust floats on a magma sea?

the upper part of the asthenosphere and in the lower part of the crust. Let's now discuss each of the special physical conditions that lead to melting.

Melting Due to a Decrease in Pressure We can portray how temperature changes with increasing depth in the Earth by drawing a **geotherm**, a curving line on a graph that plots temperature on the horizontal axis and pressure on the vertical axis. The graph emphasizes that temperature always increases with depth, but that the rate of increase varies with depth. By reading the graph, we see that temperatures comparable to those of lava exist in the upper mantle **(Fig. 6.4a)**. But, as we've noted, even though the upper mantle has a very high temperature, its rock stays solid because it is also under great pressure from the weight of overlying rock. Simplistically, pressure squeezes atoms together so that they can't easily break free from solid mineral crystals.

Because pressure prevents melting, a decrease in pressure can permit melting. Specifically, if the pressure affecting hot mantle rock decreases while the rock's temperature remains nearly unchanged, the rock may melt. This kind of melting, called **decompression melting**, takes place in locations where

hot mantle rock rises to shallower depths in the Earth. As we'll see, such movement occurs in mantle plumes, beneath rifts, and beneath mid-ocean ridges **(Fig. 6.4b)**.

Melting as a Result of the Addition of Volatiles Magma also forms at locations where chemicals called volatiles have the opportunity to mix with hot mantle rock. Recall that *volatiles* are substances, such as water (H_2O) and carbon dioxide (CO_2), that evaporate easily and can exist in gaseous forms at the Earth's surface. When volatiles mix with hot, dry rock, they help break chemical bonds and the rock begins to melt. In effect, adding volatiles decreases a rock's melting temperature. Geologists refer to melting due to addition of volatiles as **flux melting**. Water plays the most important role in triggering melting. We'll see that flux melting happens in the mantle above subducting oceanic crust **(Fig. 6.5a)**.

FIGURE 6.4 Decompression melting.

Temperature (°C)

(a) Decompression takes place when the pressure acting on hot rock decreases. As this graph of pressure and temperature conditions in the Earth shows, when rock rises from point A to point B, the pressure decreases a lot, but the rock cools only a little, so the rock begins to melt.

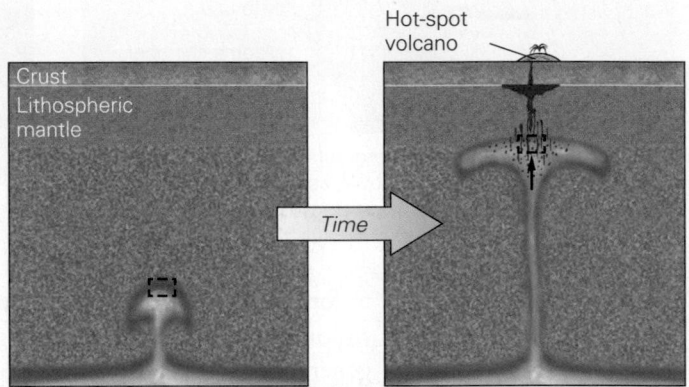

Decompression melting in a mantle plume

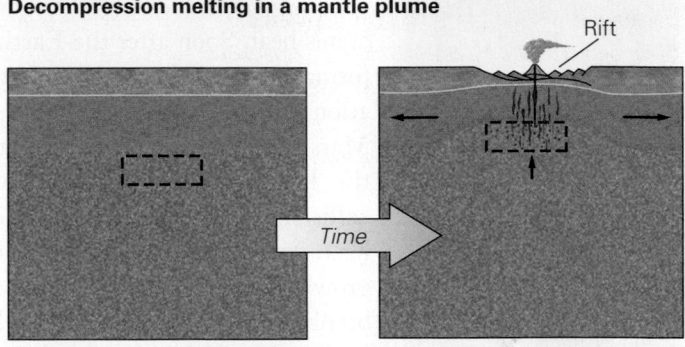

Decompression melting beneath a rift

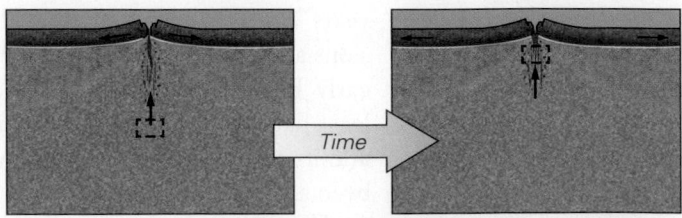

Decompression melting beneath a mid-ocean ridge

(b) The conditions leading to decompression melting occur in several different geologic environments. In each case, a volume of hot asthenosphere (outlined by dashed lines) rises to a shallower depth, and magma (red dots) forms.

FIGURE 6.5 Flux melting and heat-transfer melting.

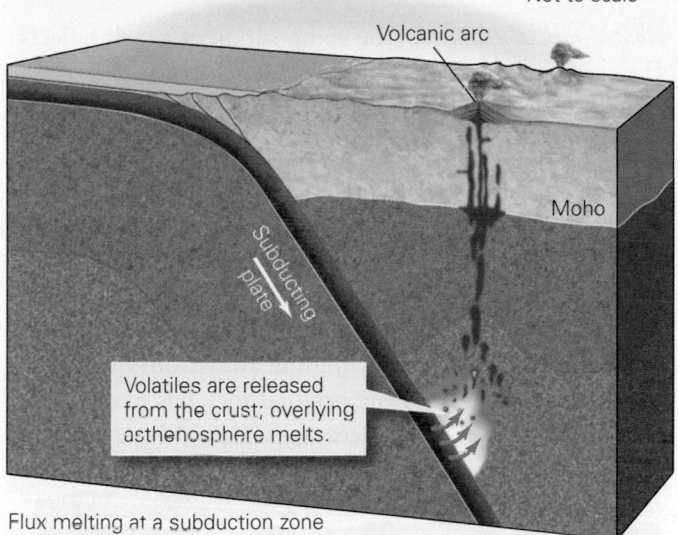

Flux melting at a subduction zone

(a) When volatiles enter hot mantle rock above a subducting plate, flux melting takes place.

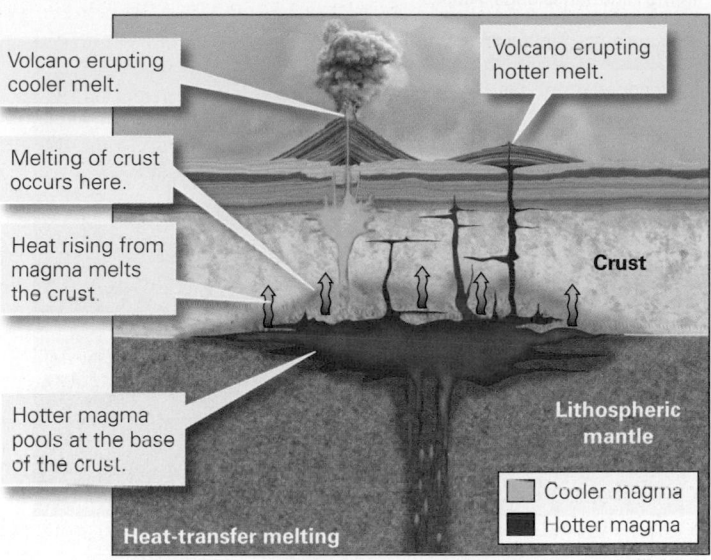

(b) When rising magma brings heat up with it, heat-transfer melting can take place in the overlying or surrounding rock. (We'll see that the cooler magma is felsic, and the hotter magma is mafic.)

Melting as a Result of Heat Transfer When very hot magma from the mantle rises up into the crust, it brings substantial amounts of heat with it. This heat can be conducted into the wall rock surrounding an intrusion and can raise the temperature of the wall rock. In some cases, the added heat may be sufficient to cause the wall rock to begin melting. To picture the process, imagine injecting hot fudge into ice cream—the fudge conveys heat to the ice cream, raises its temperature, and causes it to melt. We call this process heat-transfer melting because it results from the movement of thermal energy from a hotter material to a cooler one. As we'll see, heat-transfer melting can happen in rifts, along convergent boundaries, and at hot spots, for at all these locations, magma from the mantle can carry heat up into rocks that have a lower melting temperature **(Fig. 6.5b)**.

> ### TAKE-HOME MESSAGE
>
> Molten rock underground is magma, and molten rock that has come out of a vent at the Earth's surface is lava. Solidification of magma produces intrusive igneous rock; solidification of lava or pyroclastic debris produces extrusive igneous rock. Temperature increases with depth at a rate defined by the geotherm. Nevertheless, the mantle and crust are mostly solid, so production of magma occurs only in special locations where there is decompression, addition of volatiles, or heat transfer.
>
> **QUICK QUESTION:** Why hasn't the Earth cooled sufficiently, over geologic time, to become entirely solid?

6.3 What's in Molten Rock?

Molten Rock as a Chemical Soup

In a general sense, we can picture molten rock (magma or lava) to be just a hot liquid consisting of many chemicals. For convenience, geologists tend to describe the chemical composition of molten rock in terms of the proportions of oxides that it contains—an *oxide*, in this context, is simply a molecule consisting of an element bonded to oxygen. Common oxides in magma or lava include silica (SiO_2), aluminum oxide (Al_2O_3), iron oxide (FeO or Fe_2O_3), calcium oxide (CaO), magnesium oxide (MgO), sodium oxide (Na_2O), and potassium oxide (K_2O). In molten rock, oxide molecules do not lie in an orderly crystalline structure, but bond together instead in clusters or short chains that can move with respect to one another.

Geologists distinguish between "dry" melts, which contain no volatiles, and "wet" melts, which do. In fact, wet melts may include up to 15% dissolved volatiles such as water (H_2O), carbon dioxide (CO_2), nitrogen (N_2), hydrogen (H_2), and sulfur dioxide (SO_2). These volatiles come out of the Earth at volcanoes in the form of gas **(Fig. 6.6)**. Typically, water constitutes at least half of the gas erupting at a volcano. Note that because they can release volatiles, magma and lava provide not only the molecules that make up rocks, but also the molecules that comprise the Earth's water and air.

FIGURE 6.6 A cloud of gas, mixed with ash, rising above a volcano in the Aleutian Islands, Alaska.

Volcanoes erupt gas as well as liquid rock.

Chemical Variability of Molten Rock

Imagine four pots of molten chocolate simmering on a stove. Each pot contains a different type of chocolate. One pot contains white chocolate, one milk chocolate, one semisweet chocolate, and one dark chocolate. It's no surprise that different kinds of molten chocolate yield different kinds of solid chocolate; each type differs from the others in taste and color. Like molten chocolate, melts (magmas and lavas) differ from one another in terms of the proportions of chemicals that they contain. Geologists distinguish among four major compositional types of molten rock depending, overall, on the proportion of silica relative to the combination of magnesium oxide and iron oxide that a melt contains **(Table 6.1)**. *Mafic melts* contain a relatively high proportion of magnesium oxide and iron oxide relative to silica—the "ma-" in the word stands for magnesium, and the "-fic" comes from the Latin word for iron. *Ultramafic melts* have an even higher proportion of magnesium oxide and iron oxide relative to silica. *Felsic melts* have a relatively high proportion of silica relative to magnesium oxide and iron oxide. (Geologists may use the term *silicic* interchangeably with *felsic*.) *Intermediate melts* are so named because their composition lies partway between those of mafic and felsic melts.

TABLE 6.1 The Four Categories of Magma

Felsic (or silicic) magma	67–76% silica*
Intermediate magma	53–66% silica
Mafic magma	46–52% silica
Ultramafic magma	38–45% silica

*The numbers provided are "weight percent," meaning the proportion of the magma's weight that consists of silica.

Why do melts of so many compositions form in the Earth? Several factors play a role:

- *Source-rock composition:* The composition of a melt reflects the composition of the solid from which it was derived. Not all melts form from the same source rock, so not all melts have the same composition.

- *Partial melting:* Under the temperature and pressure conditions that occur in the Earth, only 2% to 30% of a source rock can melt to produce magma at a given location. The temperatures at sites of magma production simply never get high enough to melt the entire source rock, and magma tends to migrate away from the site of melting before all of the original rock has melted. Geologists refer to the process by which only part of an original rock melts to produce magma as **partial melting (Fig. 6.7a)**. Magmas formed by partial melting are more felsic than the source rock from which they were derived because more silica enters the liquid, as melting begins, than remains behind in the still-solid source. For example, partial melting of an ultramafic rock produces a mafic magma.

- *Assimilation:* As magma sits in a magma chamber before completely solidifying, it may incorporate chemicals dissolved from the wall rock of the chamber or from blocks that detach from the wall and sink into the magma **(Fig. 6.7b)**. This process, called contamination or **assimilation**, changes a melt's composition.

- *Magma mixing:* Different magmas formed in different locations from different source rock may enter the same magma chamber. In some cases, the originally distinct magmas mix, or mutually dissolve in each other, to produce a new, different magma. Thoroughly mixing a felsic magma with a mafic magma in equal proportions produces an intermediate magma.

TAKE-HOME MESSAGE

Molten rock contains many oxide chemicals, such as silica. Geologists distinguish among four types of molten rock based on composition, as defined by the proportion of silica to magnesium oxide and iron oxide. The composition of a given magma depends on many factors, both at the site where melting takes place and in the magma chamber where the melt resides.

QUICK QUESTION: Why doesn't a magma formed by the melting of a given source rock have the same composition as the source rock?

FIGURE 6.7 Phenomena that can affect the composition of magma.

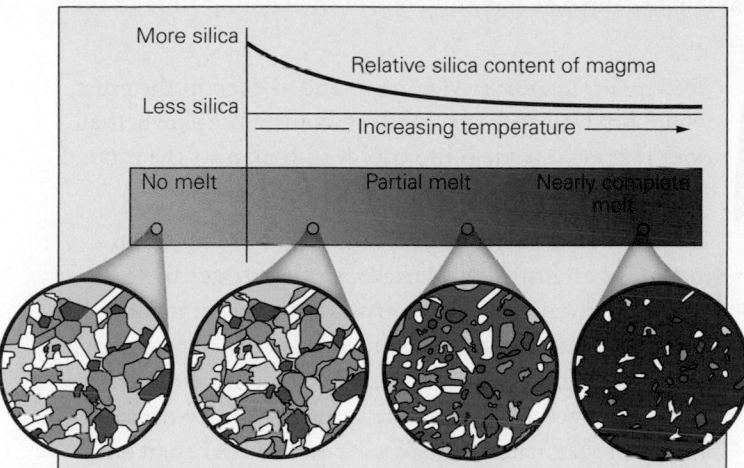

(a) Partial melting: The first-formed melt will be richer in silica than the original rock. As melting continues, magma becomes increasingly mafic.

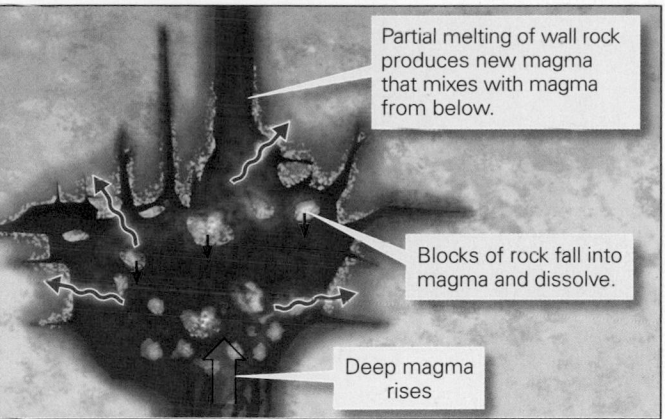

(b) Assimilation: Heat provided by magma partially melts wall rock; the new magma may then mix with the original magma. Also, blocks of wall rock can dissolve in the original magma, and that wall rock may chemically react with the magma.

6.4 Movement and Solidification of Molten Rock

If magma stayed put once it formed, new igneous rocks would not develop in or on the crust. But it doesn't stay put—magma tends to move upward, away from where it formed. It may eventually stop and freeze underground, but as we've seen, some molten rock reaches the Earth's surface to erupt as lava at volcanoes. The rise of molten rock serves an important role in the Earth System in that it transfers material from deeper parts of the Earth upward and provides the raw material from which new rocks, as well as the atmosphere and oceans, form.

Why Does Magma Rise?

Magma in the Earth rises for two reasons. First, because magma is less dense than surrounding rock, a buoyancy force acts on it to drive it upward. A similar process happens when you submerge a piece of Styrofoam in a bucket of water—because Styrofoam is less dense than water, it rises upward through the water. Second, the weight of rock produces pressure at depth, and this pressure squeezes magma upward. A similar process occurs when you step in a mud puddle while barefoot, and your weight squeezes mud up between your toes.

What Controls the Speed at Which Molten Rock Flows?

The resistance to flow, or **viscosity**, of a liquid affects the speed at which the liquid moves. We can say, for example, that molasses has greater viscosity than water because it flows more slowly than water. The viscosity of all molten rock (magma or lava) exceeds that of molasses, but not all molten rock has the same viscosity. The viscosity of molten rock depends primarily on temperature, volatile content, and silica content. Specifically, hotter melt tends to be less viscous than cooler melt because thermal energy breaks bonds and allows atoms or molecules to move more easily. A melt containing more volatiles has lower viscosity than does a dry (volatile-free) melt because volatiles also tend to break apart silicate molecules. Finally, mafic melt is less viscous than felsic melt because relatively more silicon-oxygen tetrahedra occur in felsic melt, and the tetrahedra tend to link together to create long chains that can't move past one another easily. With these relationships in mind, it's not surprising that a very hot mafic lava has relatively low viscosity and can flow in thin sheets over wide regions, but cool felsic lava has relatively high viscosity and clumps up at the volcanic vent to form a bulbous mound **(Fig. 6.8)**.

Transforming Melt into Rock

If a melt stayed at its point of origin, and nothing in its surroundings changed, it would remain molten. But melts don't last forever. Rather, they eventually solidify, or freeze. Most commonly, freezing takes place simply because a melt

FIGURE 6.8 Viscosity affects lava behavior.

(a) Mafic lava has relatively low viscosity. It can erupt in fountains, move long distances, and form thin lava flows.

(b) Felsic to intermediate lava is very viscous. When it erupts, it may form a mound-like lava dome around the volcano's vent.

cools below its freezing temperature. Cooling happens when magma rises toward the Earth's surface because the temperature of the crust decreases. If magma becomes trapped underground as an intrusion, it slowly loses heat to the surrounding wall rock, drops below its freezing temperature, and solidifies. If magma reaches the Earth's surface and extrudes as lava on the ground surface, it cools because it comes in contact with much cooler air or water. In some cases, magmas freeze, in part, due to loss of volatiles—we saw earlier that the addition of volatiles to hot rock can cause it to melt. Similarly, the subtraction of volatiles from magma can cause it to freeze. Volatiles bubble out of magma as the magma rises and the pressure acting on it decreases.

The time it takes for a magma to cool depends on how fast it can transfer heat to its surroundings. To see why, think about the process of cooling coffee. If you pour hot coffee into a thermos bottle and seal it, the coffee stays hot for hours because the insulation of the bottle allows the coffee to lose heat to the air outside only very slowly. Like a thermos bottle, the wall rock surrounding an intrusion acts as insulation, so magma in an intrusive environment cools relatively slowly. In contrast, if you spill coffee on a table, it cools quickly because it loses heat to the cooler air. Likewise, lava that erupts at the ground surface cools relatively quickly.

Not all magma trapped in the intrusive realm cools at the same rate. Three factors can control the cooling time of such magma:

- *The depth of intrusion:* Magma intruded deep in the crust, where hot wall rock surrounds it, cools more slowly than does magma intruded into cool wall rock near the ground surface.

- *The shape and size of a magma body:* Heat escapes from magma at an intrusion's surface, so the greater the surface area for a given volume of intrusion, the faster it cools. So a body of magma with the shape of a pancake cools faster than one with the shape of a melon. And because the ratio of surface area to volume increases as size decreases, a body of magma the size of a car cools faster than one the size of a ship (Fig. 6.9a).

FIGURE 6.9 Factors affecting the cooling rate of intrusions.

	Faster cooling	**Slower cooling**
Effect of size	●	⬤
	For a given shape, a smaller volume cools faster.	
Effect of shape	⬭	⬤
	For a given volume, a pancake shape—with its greater surface area—cools faster.	

(a) The shape and size of an intrusion affect the surface area-to-volume ratio.

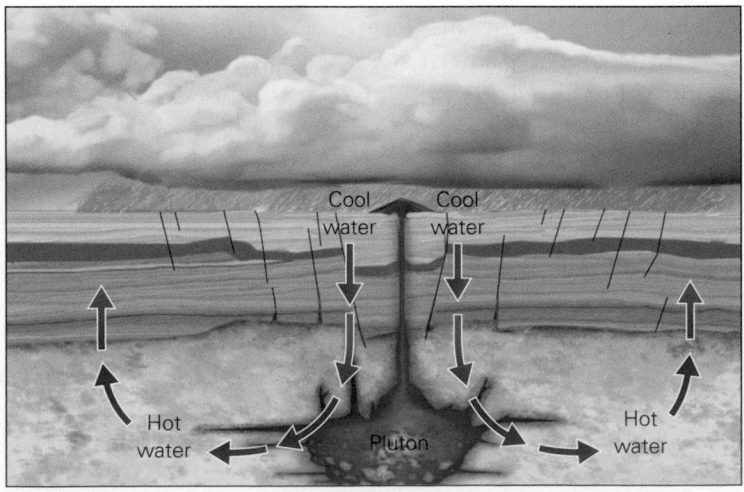

(b) Circulating groundwater can cause a pluton to cool more quickly.

FIGURE 6.10 Factors that affect the freezing of molten rock.

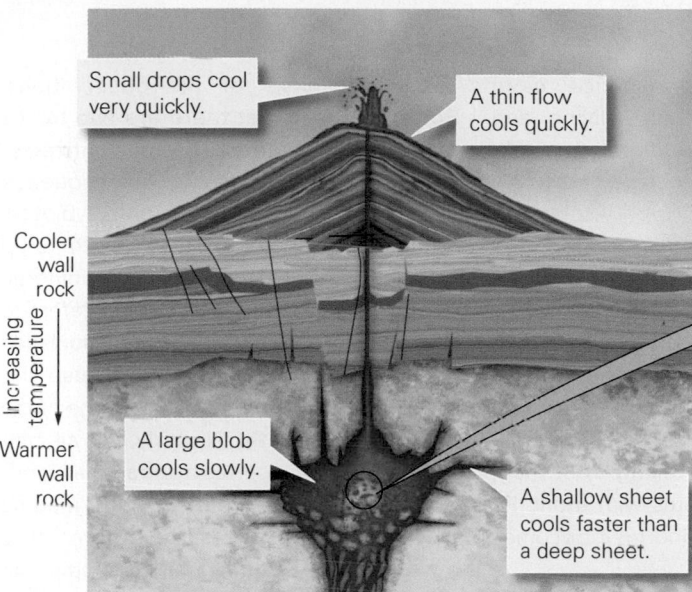

(a) The cooling rate of molten rock depends on the size and shape of the magma or lava body, and on its depth.

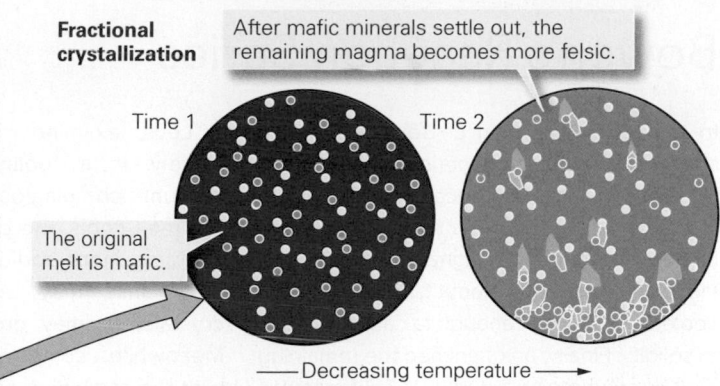

Fractional crystallization

After mafic minerals settle out, the remaining magma becomes more felsic.

The original melt is mafic.

Time 1 — Time 2

—— Decreasing temperature ——

(b) The process of fractional crystallization results in a progressive change in magma composition during freezing.

- *The presence of circulating groundwater:* Water passing through wall rock carries away heat, much like the coolant that flows around an automobile engine. Consequently, magma that interacts with circulating groundwater cools faster than does magma that intrudes dry rock (Fig. 6.9b).

The same factors that control cooling of magma also control cooling of lava. Specifically, a thin lava flow cools faster than a thick one because the proportion of surface area to volume for a thin flow exceeds that for a thick one. Tiny droplets of lava sprayed into the air in an explosive eruption freeze much faster than does a coherent lava flow, for the same reason. Lava immersed in water cools faster than lava immersed in air because water carries heat away from rock faster than air can.

Changes in Molten Rock during Cooling: Fractional Crystallization

Most people have observed the process of ice forming from liquid water—cool the water in a pan to a temperature of 0°C and crystals of ice start to form. Keep the temperature cold enough for long enough, and all of the water becomes solid, composed entirely of one type of mineral—water ice. Molten rock can freeze too, but during the freezing of molten rock, many different minerals form because the melt contains a variety of chemicals.

To gain insight into how magma freezes, let's look at an example. When a mafic magma cools to the freezing temperature, mafic minerals (meaning iron- and magnesium-rich

minerals), such as olivine and pyroxene, crystallize first. These solid crystals are denser than the remaining liquid magma, so they may start to sink (Fig. 6.10). Some of the crystals react chemically with the remaining magma, but some become isolated from the magma. This process of sequential crystal formation is called fractional crystallization. It extracts iron and magnesium from the magma so that the remaining magma becomes more felsic. If a magma freezes completely before much fractional crystallization has occurred, the magma becomes mafic igneous rock. As the process of fractional crystallization continues, more and more iron and magnesium separate from the magma, so the magma evolves to become progressively more felsic. Freezing of the magma that remains after lots of fractional crystallization has taken place can, therefore, yield a felsic igneous rock. In the early 20th century, an American geologist, N. L. Bowen, conducted a series of laboratory experiments that allowed him to work out the specific sequence in which minerals form during fractional crystallization (Box 6.1).

TAKE-HOME MESSAGE

Magma rises because it's buoyant and because pressure due to the weight of overlying rocks squeezes it upward. Viscosity, which depends primarily on composition and temperature, affects the rate of melt flow. When molten rock enters a cooler environment, it freezes. The rate of cooling depends on both the temperature of the surroundings and the shape of the magma body. Magma composition may evolve as the magma cools.

QUICK QUESTION: Which cools faster—a large blob of magma intruded at depth or a thin flow extruded at the surface? Why?

BOX 6.1 CONSIDER THIS . . .

Bowen's Reaction Series

In the 1920s, Norman L. Bowen began a series of laboratory experiments designed to determine the sequence in which silicate minerals crystallize from a melt. First, Bowen melted powdered mafic igneous rock by raising its temperature to about 1,280°C. Then he cooled the melt just enough to cause part of it to solidify. Finally, he quenched the remaining melt by submerging it quickly in cold mercury. *Quenching*, which means sudden cooling to form a solid, transformed any remaining liquid into glass, and the glass trapped the earlier-formed crystals within it. Bowen identified mineral crystals formed before quenching with a microscope, and he analyzed the chemical composition of the remaining glass to determine the composition of the magma.

After completing experiments at several different temperatures, Bowen found that as new crystals form, they extract certain chemicals preferentially from the melt **(Fig. Bx6.1a)**. As a result, the chemical composition of the remaining melt progressively changes as the melt cools. Bowen described the specific sequence of mineral-producing reactions that take place in a cooling, initially mafic, magma. This sequence has come to be known as **Bowen's reaction series** in his honor.

Let's examine the sequence more closely. In a cooling melt, olivine and calcium-rich plagioclase form first. As the melt cools, the plagioclase that forms contains more sodium (Na). This Na-rich plagioclase may encase earlier-formed crystals, or may grow as new crystals. Meanwhile, some olivine crystals react with the remaining melt to produce pyroxene, which may encase olivine crystals or even replace them. Some of the olivine and Ca-plagioclase crystals may become isolated from the melt, taking iron, magnesium, and calcium atoms with them. By this process, the remaining melt becomes enriched with silica.

As the melt continues to cool, pyroxene crystals react with the melt to form amphibole, and then some amphibole reacts with the remaining melt to form biotite. All the while, crystals continue to become isolated, so the remaining melt continues to become more felsic. At temperatures between 650°C and 850°C, only about 10% of the melt remains liquid, and this melt has a high silica content. At this stage, the final melt freezes, yielding quartz, potassium feldspar (also known as K-feldspar or orthoclase), and muscovite.

On the basis of his observations, Bowen realized that there are two tracks to the reaction series. The "discontinuous reaction series" refers to the sequence olivine, pyroxene, amphibole, biotite, K-feldspar/muscovite/quartz: each step yields a different class of silicate mineral. The "continuous reaction series" refers to the progressive change from calcium-rich to sodium-rich plagioclase because the steps yield different versions of the same mineral **(Fig. Bx6.1b)**. Notably, not all minerals listed in the series appear in all igneous rock. For example, a mafic magma may completely crystallize before felsic minerals such as quartz or K-feldspar can grow.

The succession of minerals in the discontinuous series is not random—it begins with minerals having isolated tetrahedra (olivine) and progresses to those having single chains of tetrahedra (pyroxene), then double chains (amphibole), and finally sheets (mica) or three-dimensional frameworks (quartz) (see Fig. 5.16b). Put another way, the minerals that crystallize later in the discontinuous series have more Si-O-Si bonds and smaller O/Si ratios than do those that crystallize earlier.

FIGURE Bx6.1 Bowen's reaction series indicates the succession of crystallization in cooling magma.

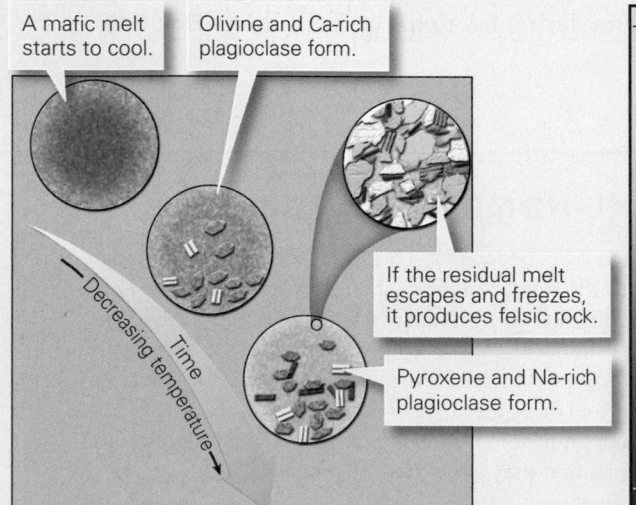

(a) With decreasing temperature, fractional crystallization begins, and the composition of the remaining magma becomes more felsic.

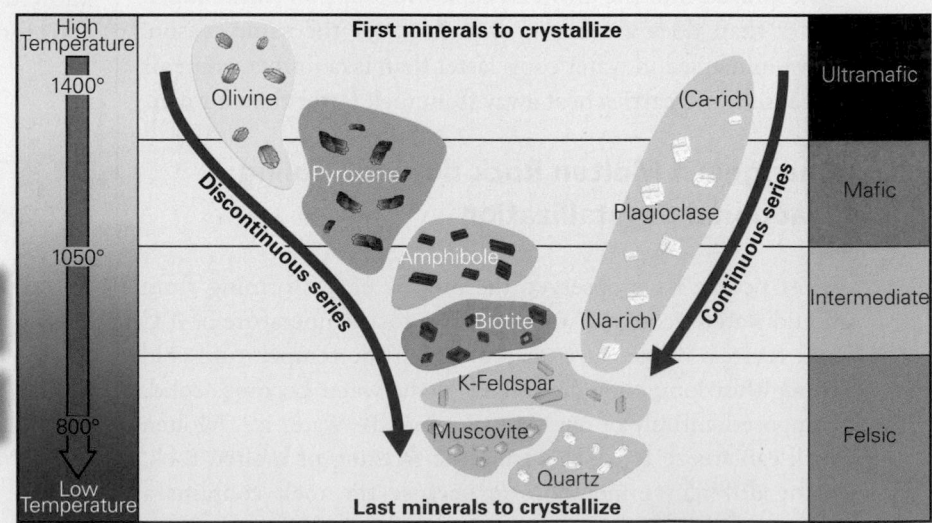

(b) This chart displays the discontinuous and continuous reaction series. Rocks formed from minerals at the top of the series are mafic, whereas rocks formed from the bottom of the series are felsic.

FIGURE 6.11 Examples of eruptions and extrusive volcanic materials.

(a) This volcano is producing lava flows and fountains.

(b) This volcanic explosion produced two styles of ash eruption.

(c) A stack of over 50 thin lava flows, capped by debris, is visible inside Mt. Vesuvius, Italy.

(d) Thick layers of ash deposited by explosive eruptions in New Mexico about 1.14 Ma. Note the highway for scale.

6.5 Comparing Extrusive and Intrusive Environments

Extrusive Igneous Settings

Not all volcanic eruptions are the same, so not all extrusive rocks are the same. Some volcanoes erupt streams of low-viscosity lava that cover broad swaths of the countryside with relatively thin lava flows **(Fig. 6.11a)**. Others erupt viscous masses of lava that pile into mounds of angular blocks. And still others erupt in cataclysmic explosions, which forcefully

(e) A close-up of a layer of ash containing lapilli.

eject turbulent clouds of pyroclastic debris that can rise several kilometers into the sky **(Fig. 6.11b)**. The debris includes pea- to golf ball-sized fragments, known as lapilli, that fall like hail on or near the volcano, as well as finer, dust-sized material, known as ash, which may be carried by the wind for great distances before it drifts down from the sky like snow. An explosive eruption may also produce a fast-moving pyroclastic flow, a scalding avalanche of ash and other debris that races down the surface of the volcano, destroying everything in its path.

Whether an eruption produces mostly sheets of lava, mounds of lava, or clouds of pyroclastic debris depends largely on a lava's viscosity and volatile content. Mafic lavas tend to have low viscosity and can spread in broad, thin flows **(Fig. 6.11c)**; gas-poor felsic lavas tend to form bulbous flows; and volatile-rich felsic lavas tend to erupt explosively, producing pyroclastic debris **(Fig. 6.11d, e)**. We'll discuss such products of extrusive settings in more detail in Chapter 9.

Intrusive Igneous Settings

Magma rises and intrudes into pre-existing rock (or *wall rock*) by slowly percolating upward between grains or by forcing open cracks. Magma that doesn't make it to the surface freezes solid underground. The boundary between this new intrusive igneous rock and the wall rock is called an **intrusive contact**.

We distinguish among different types of intrusions by their shape. *Tabular intrusions*, or *sheet intrusions*, are roughly planar and of fairly uniform thickness. Smaller ones may be only centimeters thick or meters long, but the largest reach sizes of meters to hundreds of meters wide and tens to hundreds of kilometers long. A **dike** is a tabular intrusion that cuts across pre-existing layering (bedding or foliation), whereas a **sill** is a tabular intrusion that injects between layers **(Fig. 6.12)**. Dikes may radiate out from a volcano, or they may develop

FIGURE 6.12 Igneous sills and dikes are examples of tabular intrusions.

Dike cuts across layers.

Sill pushes between layers.

Intrusive contact

Layers of sandstone

If all the sandstone were removed, the intrusions would look like this (before erosion).

(a) Dikes cut across pre-existing layering. Sills are parallel to pre-existing layering.

(b) A wall-like intrusion cutting into pre-existing igneous rock is also called a dike.

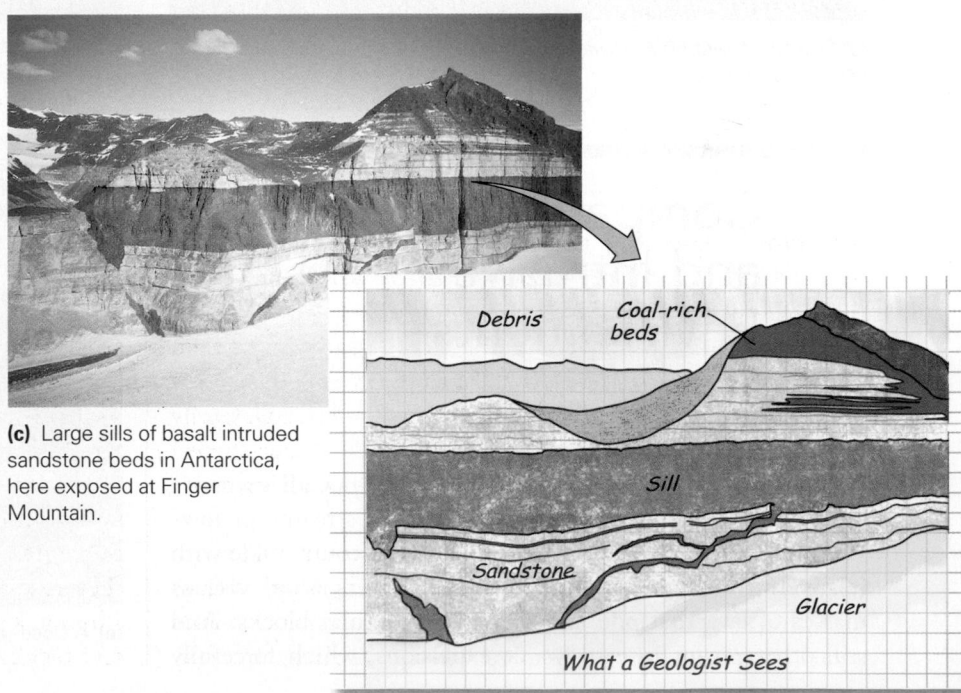

(c) Large sills of basalt intruded sandstone beds in Antarctica, here exposed at Finger Mountain.

Debris

Coal-rich beds

Sill

Sandstone

Glacier

What a Geologist Sees

FIGURE 6.13 Examples of igneous dikes.

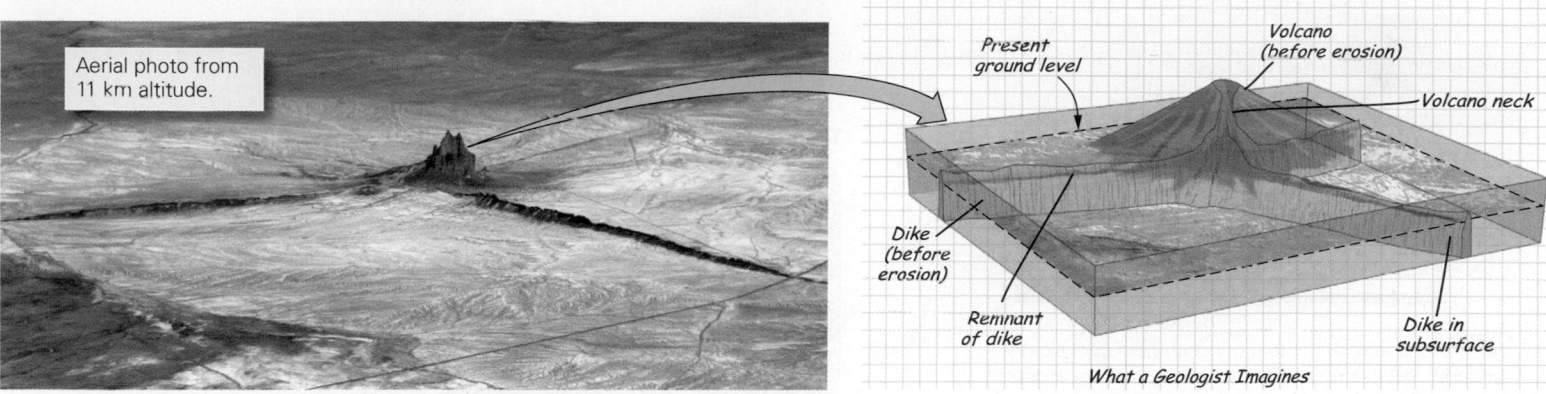

Aerial photo from 11 km altitude.

Present ground level

Volcano (before erosion)

Volcano neck

Dike (before erosion)

Remnant of dike

Dike in subsurface

What a Geologist Imagines

(a) Shiprock, in northwestern New Mexico, is the remnant of an eroded volcano. Large dikes radiate from the central hill. The central hill is the frozen magma chamber under what was once a volcano. The volcano has eroded away.

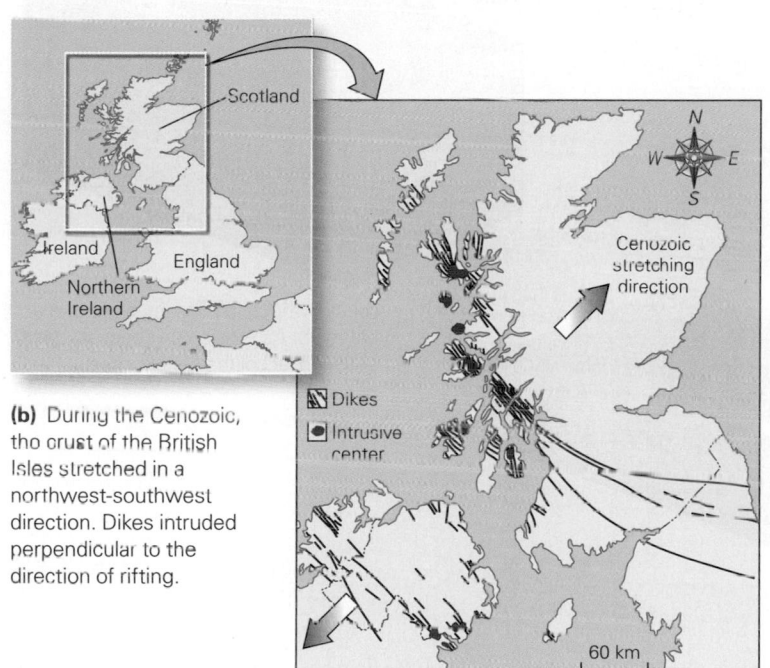

Scotland

Ireland

England

Northern Ireland

Cenozoic stretching direction

N
W   E
S

◪ Dikes
◙ Intrusive center

60 km

(b) During the Cenozoic, the crust of the British Isles stretched in a northwest-southwest direction. Dikes intruded perpendicular to the direction of rifting.

over broad regions of crust during the process of rifting **(Fig. 6.13a, b)**. In places where tabular intrusions lie within rock that does not have layering, we generally refer to a nearly vertical, wall-like example as a dike and a nearly horizontal, tabletop-shaped example as a sill. In some places, the magma injecting between layers gets blocked and can't spread laterally very far, so the magma accumulates to form a blister-shaped intrusion known as a **laccolith**, which pushes overlying strata upward into a dome.

Plutons are irregular or blob-shaped intrusions that range in size from tens of meters across to tens of kilometers across **(Fig. 6.14)**. Intrusion of numerous plutons in a region yields a vast composite igneous body, called a **batholith**, that may

be several hundred kilometers long and over 100 km wide. The rugged high peaks of the Sierra Nevada range in California expose the intrusive rocks of a batholith formed from plutons that intruded between 145 and 80 Ma **(Fig. 6.15)**.

Where does the space for intrusions come from? It's relatively easy to visualize how space for tabular intrusions can be produced. For example, igneous dikes form in regions where the crust has stretched horizontally in response to rifting or seafloor spreading, so as magma intrudes, vertical cracks open up **(Fig. 6.16a)**. In effect, the magma fills space being generated by stretching. Intrusion of sills occurs near the surface of the Earth, so the pressure of the magma simply pushes overlying rock and the Earth's surface upward to make room **(Fig. 6.16b)**.

Understanding how the space for plutons develops remains a subject of research. Some plutons may have originated as *diapirs*, meaning that they rose upward through the crust as buoyant, light-bulb-shaped blobs of magma that pierced overlying rock and pushed it aside as they rose **(Fig. 6.17a, b)**. Pluton intrusion may also involve **stoping**, a process during which magma assimilates wall rock, and blocks of wall rock break off and sink into the magma

FIGURE 6.14 Igneous plutons are blob-shaped intrusions.

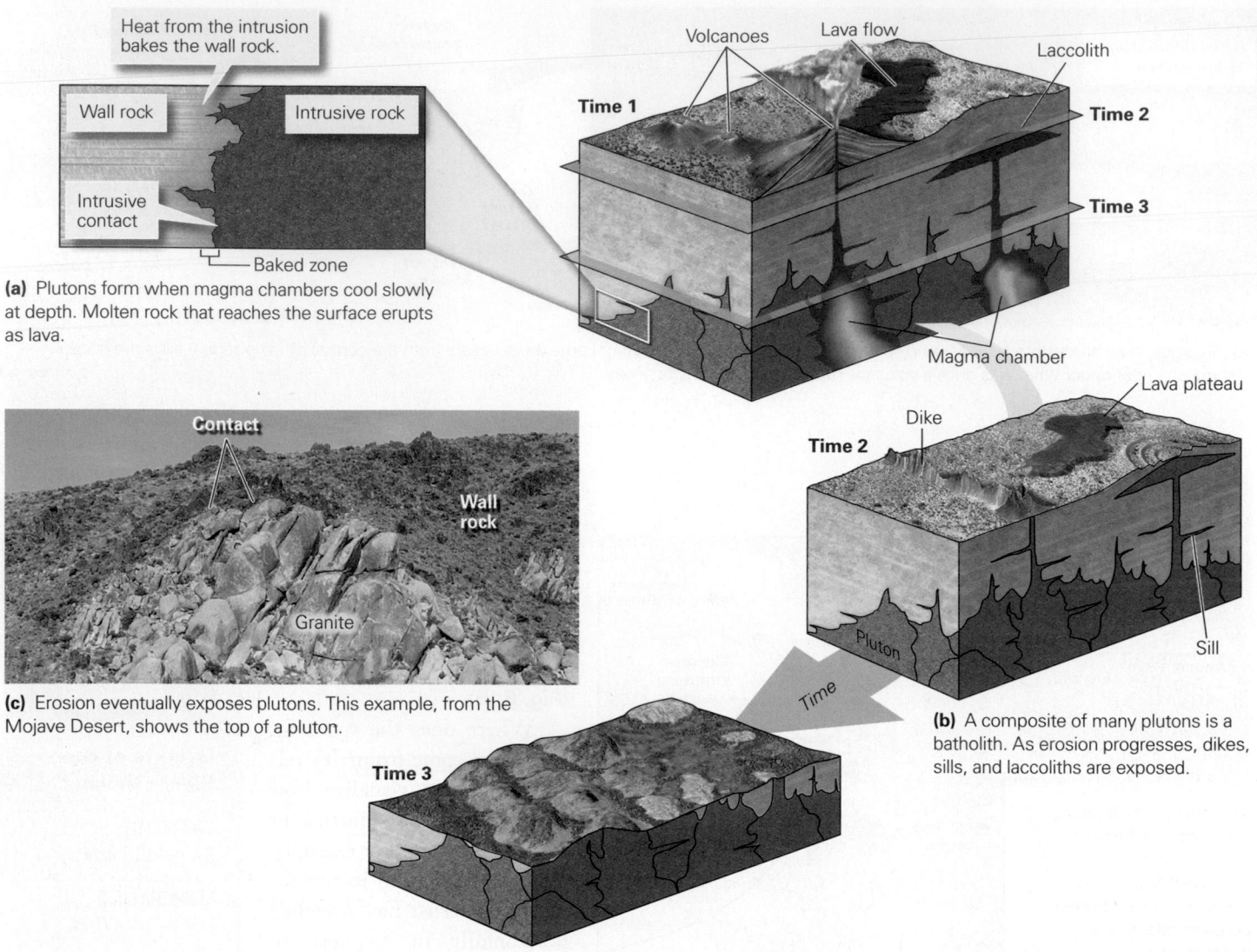

Heat from the intrusion bakes the wall rock.

Wall rock · Intrusive rock

Intrusive contact

Baked zone

(a) Plutons form when magma chambers cool slowly at depth. Molten rock that reaches the surface erupts as lava.

Time 1 · Volcanoes · Lava flow · Laccolith · Time 2 · Time 3 · Magma chamber

Contact · Wall rock · Granite

(c) Erosion eventually exposes plutons. This example, from the Mojave Desert, shows the top of a pluton.

Time 2 · Lava plateau · Dike · Pluton · Sill · Time · Time 3

(b) A composite of many plutons is a batholith. As erosion progresses, dikes, sills, and laccoliths are exposed.

(Fig. 6.17c). If a stoped block does not melt entirely, but rather becomes surrounded by new igneous rock, it becomes a **xenolith**, after the Greek word *xeno*, meaning foreign **(Fig. 6.17d)**.

Recent work has questioned the general applicability of diapiric and stoping models, leads to the alternative view that plutons form by intrusion of several superimposed dikes or sills, which coalesce to become a single, massive body. If high temperatures are sustained for a long time, diffusion can take place, and the rock may gradually recrystallize and its composition may evolve **(Fig. 6.17e)**. The overall space for plutons may form because of crustal stretching or because of uplift and erosion of overlying rock.

If intrusive igneous rocks form deep beneath the Earth's surface, why can we see them exposed today? Over long periods of geologic time, mountain building slowly uplifts belts of crust. Erosion by water, wind, and ice can gradually strip away the thick, overlying rock and expose the intrusive rock that has formed below. Some intrusive rocks exposed in mountain cliffs today solidified kilometers to tens of kilometers below the surface.

TAKE-HOME MESSAGE

Molten rock can extrude either as a lava flow or as pyroclastic debris. Intrusions form when magma injects into wall rock and then cools; intrusions can come in many different shapes. Tabular intrusions include dikes, which cut across layers, and sills, which intrude parallel to layers. Plutons have blob-like shapes. Huge batholiths consist of many plutons.

QUICK QUESTION: Intrusions occupy space—where does this space for plutons come from?

FIGURE 6.15 Mesozoic batholiths of western North America.

(b) These huge exposures of granite are part of the Sierra Nevada batholith in California.

(a) During the Mesozoic, subduction produced a huge volcanic arc. In the crust, beneath the arc, large granite batholiths formed. They are now exposed by erosion.

FIGURE 6.16 Making room for igneous dikes and sills.

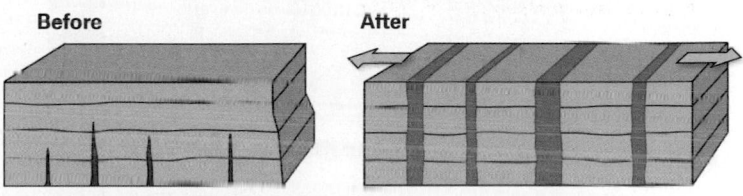

(a) Igneous dikes can form where the crust is stretched sideways.

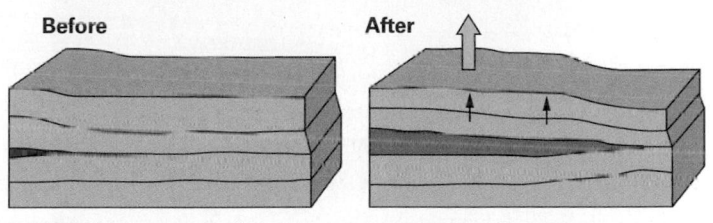

(b) The intrusion of igneous sills pushes the Earth's surface upward.

6.6 How Do You Describe an Igneous Rock?

Characterizing Color and Texture

Wander around a city and you'll find that the facades and lobbies of many buildings consist of slices of polished rock. In recent years, polished rock has also become a popular surfacing material for home kitchen countertops. Architects refer to this rock in a generic way as "granite" and like to use it because it's both beautiful and durable. Its durability comes from the minerals it contains—most architectural "granite" contains feldspar and other minerals with high numbers on the Mohs hardness scale—and from the way its grains interlock. But if you look at architectural granite closely, you'll discover that it doesn't all look the same. In fact, by the end of this section

> **Did you ever wonder . . .**
> why "granite" used in floors and countertops is so hard?

you'll discover that the word *granite*, as a geologist would use the term, has a relatively limited applicability, and that most architectural granite actually consists of different kinds of rock.

Imagine that you wanted to describe an example of the stone in a facade to a friend—what adjectives would you use? You would probably start by noting the rock's color. Overall, is the rock dark or light? More specifically, is it gray, pink, white, or black? Describing its color may not be easy because some igneous rocks contain many different minerals, each with a different color—but even so, you'll probably be able to

Crystalline	Fragmental	Glassy

(a) Granite is crystalline.   **(b)** Tuff is fragmental.   **(c)** Obsidian is glassy.

rates and at different times and, as the crystals grow, they interfere with each other. For example, a faster-growing crystal may surround a slower-growing crystal partly, or even entirely, and crystals that form later will then fill in the gaps between the earlier-formed crystals. Geologists distinguish subcategories of crystalline igneous rocks by the size of the crystals. Coarse-grained (*phaneritic*) rocks have crystals large enough to be identified with the naked eye. Fine-grained (*aphanitic*) rocks have crystals too small to be identified with the naked eye. *Porphyritic* rocks have larger crystals surrounded by a mass of fine crystals. In a porphyritic rock, the larger crystals are *phenocrysts*, while the mass of finer crystals is called *groundmass*. **Fragmental igneous rocks** form from pyroclastic debris and consist of igneous chunks, grains, or flakes that are packed together, welded together, or cemented together after they have solidified **(Fig. 6.18b)**. Geologists distinguish different types of fragmental igneous rocks from one another by fragment size. Rocks made of a solid mass of glass or of glass surrounding isolated small crystals are **glassy igneous rocks (Fig. 6.18c)**. Glassy rocks typically fracture conchoidally.

The texture of a nonfragmental igneous rock largely reflects its cooling rate. Why? Recall from Chapter 5 that mineral crystals grow when atoms diffuse (move) through melts and attach to crystal seeds. At high temperatures, seeds constantly form and dissolve because heat causes diffusion to occur

rapidly. Only some seeds are "successful" enough to grow into small crystals, and even most of these end up dissolving in the melt again. Relatively few crystals grow large enough to survive until the melt cools to a level at which crystals stop dissolving. If a melt cools very rapidly, it can become solid before seeds have grown into crystals, so the resulting rock has a glassy texture. If the melt cools rapidly, but not quickly enough to form glass, it will develop a fine-grained (aphanitic) texture. Many crystals in the resulting rocks will have grown from seeds, but none of those crystals will have had enough time to grow large. If a melt cools slowly, successful crystals have time to grow large before the rock solidifies completely. As the large crystals are growing, new seeds and small crystals do form, but because the melt remains hot, they tend to dissolve into the melt again before they can grow large.

Because of the relationship between cooling rate and texture, lava flows, dikes, and sills tend to be composed of fine-grained igneous rock. In contrast, plutons tend to be composed of coarse-grained rock. Plutons that intrude into hot wall rock at great depth cool particularly slowly and thus tend to have larger crystals than plutons that intrude into cool wall rock at shallow depth. Porphyritic rocks form when a melt cools in two stages. First, the melt cools at depth slowly enough that phenocrysts form. Then the melt erupts, and the remainder cools quickly, so groundmass forms around the phenocrysts.

Formation of Igneous Rocks

Igneous rocks are formed by the cooling of magma underground or of lava at the Earth's surface. Igneous rocks that solidify underground are intrusive, whereas those that solidify at the surface are extrusive. The type of igneous rock that forms depends on the composition of the melt and on the environment of cooling.

Stratified volcanic tuff

Increasing silica content

MAFIC	FELSIC
Scoria (glassy)	Obsidian (glassy)
Basalt (fine grained)	Rhyolite (fine grained)
Gabbro (coarse grained)	Granite (coarse grained)

Fast cooling

Slow cooling

In an extrusive environment, melt may cool quickly, so extrusive rocks tend to be fine-grained or may even have a glassy texture. Melt that explodes into the air forms ash and other pyroclastic debris. Rocks formed from this debris have a fragmental texture.

In an intrusive environment, magma can cool slowly, so larger crystals can grow.

Minerals in an igneous rock crystallize in succession as the melt cools. They interlock to produce a crystalline texture.

Cooler

Hotter

EXTRUSIVE ENVIRONMENT

Pyroclastic flow

Lava flow

Dike swarm

Sills

Lava dome

Ring dikes

Volcanic neck

Laccolith

Irregular stock

INTRUSIVE ENVIRONMENT

Pluton

Magma chamber

FIGURE 6.19 Crystalline igneous rocks are classified based on composition and texture.

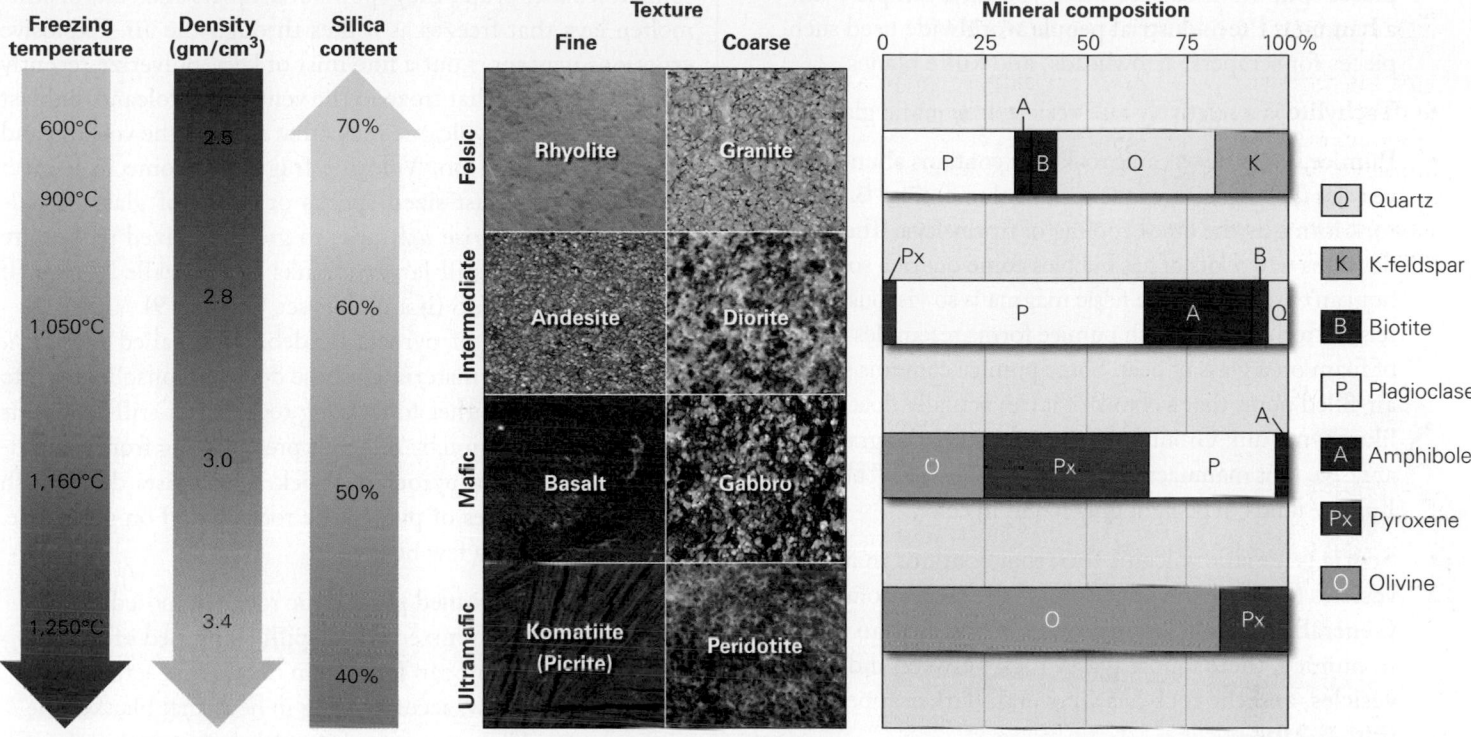

There is, however, an exception to the standard cooling rate-grain size relationship. A very coarse-grained igneous rock called **pegmatite** doesn't form by slow cooling. Pegmatite, which can contain crystals up to tens of centimeters across, typically occurs in dikes. Because dikes generally cool relatively quickly, the coarseness of this rock may seem surprising. Researchers have shown that pegmatite becomes coarse because it forms from water-rich melts in which atoms can diffuse so rapidly that large crystals can grow very quickly.

Classifying Igneous Rocks

Because melts can have a variety of compositions and can freeze to form igneous rocks in many different environments above and below the surface of the Earth, you can find a wide spectrum of distinct igneous rock types on the Earth. Geologists classify these types according to their texture and composition. A rock's texture provides clues to the rate at which it cooled, as we've seen, and therefore to the environment in which it formed (**Geology at a Glance**, p. 178). A rock's composition tells us about the original source of the magma and about the way in which the magma evolved before finally solidifying. Here, we introduce some key igneous rock types.

Crystalline Igneous Rocks To classify crystalline igneous rocks, we can make a simple chart with the different compositional groups (ultramafic, mafic, intermediate, and felsic)

listed on the vertical axis in sequence, from most felsic at the top to least felsic at the bottom. We show the different grain sizes (coarse and fine) on the horizontal axis. **Figure 6.19** provides such a chart for eight different types of crystalline igneous rocks. Note that the basalt, andesite, and rhyolite could have come from the same magmas as the gabbro, diorite, and granite, respectively. But the three fine-grained rocks probably cooled quickly in lava flows or near-surface dikes and sills, while the three coarse-grained rocks probably cooled more slowly in plutons. The color of an igneous rock provides a rough guide to its composition: mafic rocks tend to be black or dark gray, intermediate rocks tend to be lighter gray or greenish gray, and felsic rocks tend to be tan to pink or maroon (**Fig. 6.20**). Ultramafic rocks are relatively rare at the Earth's surface.

Glassy Igneous Rocks Glassy texture develops most commonly in felsic igneous rocks because the high concentration of silica inhibits diffusion and, therefore, the growth of crystals. But basaltic and intermediate lavas can form glass if they cool rapidly enough. In some cases, a rapidly cooling lava freezes while it still contains gas bubbles; these bubbles remain as open holes known as **vesicles**. Geologists distinguish among several different kinds of glassy rocks.

- **Obsidian** is a vesicle-free felsic glass that tends to be black or brown (see

> **Did you ever wonder . . .**
> how the black glass once used for arrowheads formed?

Fig. 6.18c). Because it breaks conchoidally, sharp-edged pieces split off its surface when you hit a sample with a hammer. Pre-industrial people worldwide used such pieces for scrapers, arrowheads, and knife blades.

- **Tachylite** is a relatively rare vesicle-free mafic glass.

- **Pumice** is a felsic volcanic rock that contains abundant vesicles (75% to 90% of the rock's volume) **(Fig. 6.21a)**. This rock forms by the quick cooling of frothy lava. The froth develops when lots of gas bubbles come out of a solution, but can't escape because felsic magma is so viscous. The felsic "froth" from which pumice forms resembles the head of foam on a glass of beer. Some pumice contains so many air-filled pores that a chunk of it can actually float on water, like Styrofoam. Ground-up pumice makes the grainy abrasive that manufacturers use to "stone-wash" blue jeans. Pumice tends to be light gray to tan in color.

- **Scoria** is a mafic volcanic rock that contains many vesicles (more than about 30% of the rock's volume). Generally, the vesicles in scoria are bigger than those in pumice, there's more glassy rock between individual vesicles, and the rock has an overall darker appearance **(Fig. 6.21b)**.

Fragmental Igneous Rocks As we have noted, when some volcanoes erupt, they spew out droplets and clots of still-molten lava that freezes as it flies through the air. Explosive eruptions may spray out a fine mist of lava, pulverize recently solidified pumice that froze in the vent of the volcano, or blast apart pre-existing volcanic rocks that formed the volcano and send them flying, too. Volcanic fragments come in a great range of sizes—dust-sized specks or flakes of glass or pulverized rock comprise *ash*; pea- to golf ball-sized pellets are called *lapilli*; and still larger chunks can be called *bombs* (if streamlined) or *blocks* (if angular; see Chapter 9).

Accumulations of pyroclastic debris are called *pyroclastic deposits*. When the material in these deposits consolidates into a solid mass, due either to welding together of still-hot clasts or to later cementation by minerals precipitating from groundwater, it becomes a **pyroclastic rock**. Geologists distinguish among several types of pyroclastic rocks based on grain size. We introduce just a few below.

- **Tuff** is a fine-grained pyroclastic rock composed of volcanic ash or of ash mixed with lapilli composed of pumice **(Fig. 6.21c)**. Tuff can form from debris that settles out of the air like snow, accumulating in beds that blanket the landscape. Or it may form from debris that rushes down

FIGURE 6.20 Examples of crystalline igneous rocks, arranged by grain size and composition.

Fine grained

Rhyolite

Coarse grained

Granite

Felsic

Increasing silica content

Andesite

Diorite

Basalt

Gabbro

Mafic

FIGURE 6.21 Glassy and fragmental igneous rocks.

(a) Pumice is a felsic glassy rock with abundant tiny vesicles.

(b) Scoria is a mafic glassy rock with many vesicles. Lens cap for scale.

(c) Tuff is a fragmental rock composed of ash and, locally, pumice fragments.

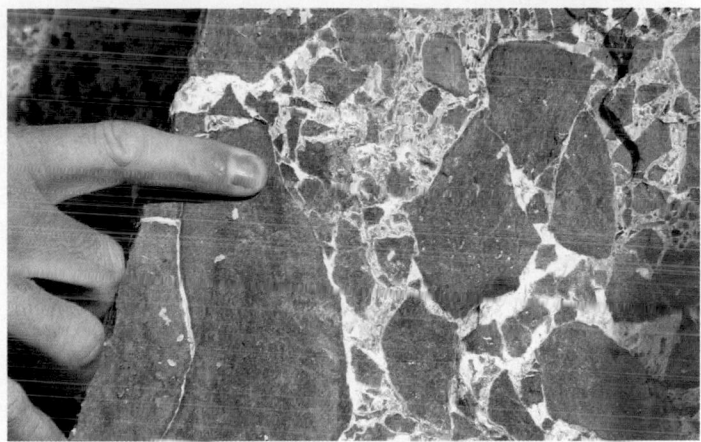

(d) Volcanic breccia consists of angular fragments.

the volcano's side in a pyroclastic flow. Ash in the interior of a pyroclastic flow may be so hot that it welds together, when the flow stops moving, to produce *welded tuff*.

- **Volcanic agglomerate** consists of accumulations of lapilli or bombs.

- **Volcanic breccia** consists of angular fragments of pyroclastic debris that have been cemented together **(Fig. 6.21d)**. A special type of breccia, called *hyaloclastite*, forms when lava erupts under water or ice and violently shatters into glassy fragments that react chemically with the surrounding water. Of note, not all volcanic breccias

form from debris falling from the sky or tumbling down a volcano. Some develop when the crust of a lava flow breaks up as the still-liquid interior continues to move.

When we investigate volcanic eruptions further in Chapter 9, we will see that pyroclastic debris may mix with water after deposition and move down the slope of a volcano before eventually being buried and turned to rock. Geologists use the term **volcaniclastic rock** to refer to any rock that contains a large proportion of volcanic fragments; this category includes both pyroclastic rocks and rocks formed from water-transported volcanic debris.

6.7 Plate Tectonic Context of Igneous Activity

Look at the map showing the distribution of *igneous activity*—the formation, movement, and in some cases, eruption of molten rock—around the world (Fig. 6.22), and compare that map with a map showing the locations of plate boundaries (see Fig. 4.4a). You'll realize that most igneous activity occurs along the volcanic arcs of convergent boundaries, along the mid-ocean ridge axes of divergent boundaries, or within continental rifts (where continents undergo stretching and

FIGURE 6.22 Settings of igneous activity.

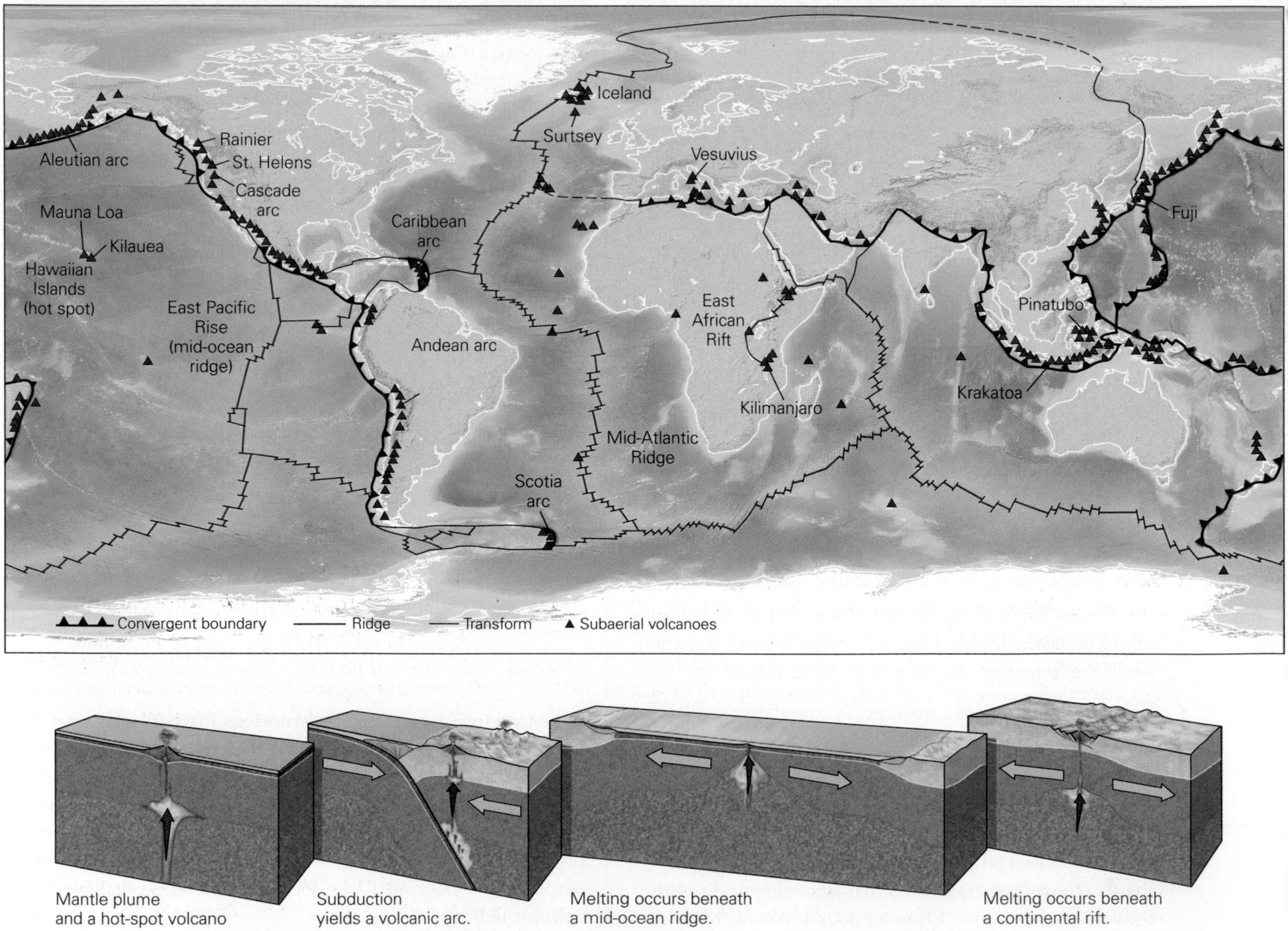

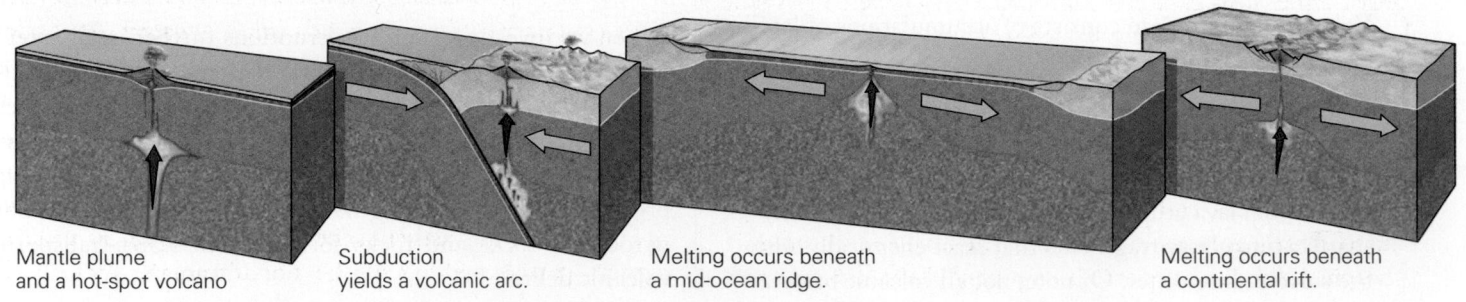

Mantle plume and a hot-spot volcano

Subduction yields a volcanic arc.

Melting occurs beneath a mid-ocean ridge.

Melting occurs beneath a continental rift.

FIGURE 6.23 The Aleutian arc of Alaska. The trench and volcanic islands are clearly visible. The enlargement shows Umnak Island, which hosts two volcanoes.

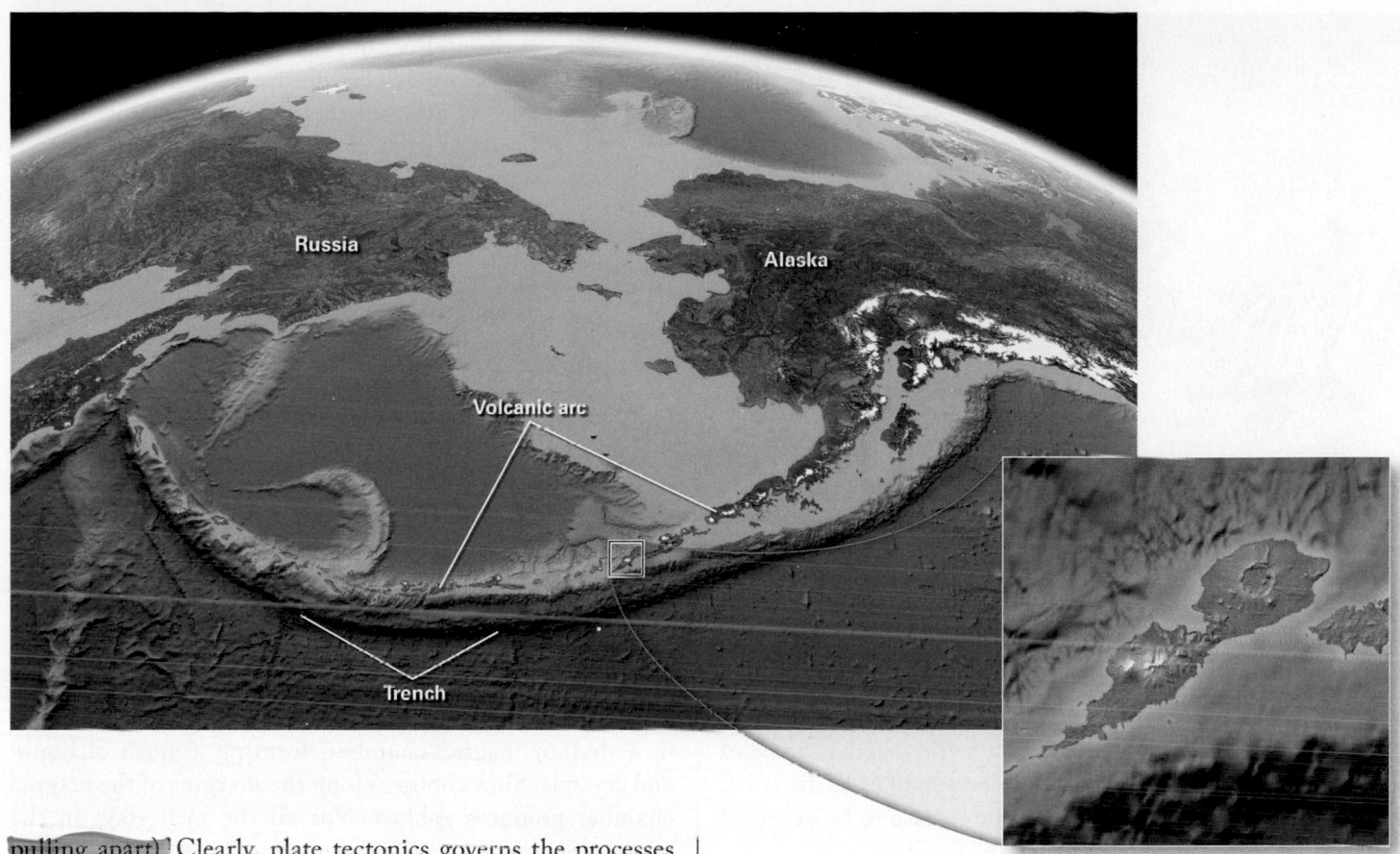

pulling apart). Clearly, plate tectonics governs the processes that lead to melt production in the Earth. But the map also shows that igneous activity happens at isolated hot spots, some of which occur on plate boundaries and some of which occur in plate interiors. Let's look at the settings where igneous activity occurs more closely, so that we can see why a particular setting leads to melt production and why different types of melts form in different settings.

Products of Subduction

A chain of volcanoes, called a **volcanic arc** (or, commonly, just an *arc*), develops along all convergent boundaries. Recall that at convergent boundaries, a downgoing plate of oceanic lithosphere subducts beneath an overriding plate, which can consist either of oceanic or continental lithosphere. You will find the volcanic arc along the edge of the overriding plate, generally at a distance of 150 to 300 km from the axis of the trench **(Fig. 6.23)**. Geologists use the word *arc* to emphasize that many of these chains of volcanoes define a curve on a map. *Continental volcanic arcs*, such as the Andean arc of South America and the Cascade arc of the northwestern United States, grew on continental crust, where the oceanic lithosphere subducts beneath a continent. *Volcanic island arcs*, such as the Aleutian arc of Alaska and the Mariana arc of the western Pacific, grow on oceanic crust at localities where one oceanic plate subducts beneath another. Igneous activity at volcanic arcs produces both extrusive rocks and intrusive rocks—in the crust beneath each volcano, dikes, sills, and plutons have developed or are developing. These intrusive rocks may eventually be exposed when the overriding rock has been eroded away.

How does subduction trigger melting? Most of the melt at convergent boundaries comes from flux melting of the asthenosphere in the region just above the downgoing plate. This melting happens because some of the minerals in oceanic-crust rocks contain volatile compounds (mostly water). At shallow depths, the volatiles chemically bond to the minerals in the crust. But when subduction brings oceanic crust down into the hot asthenosphere, the "wet" crustal rocks start to warm, and at a depth of about 150 km, they become so hot that the volatiles separate from the minerals and diffuse upward into the overlying asthenosphere. The addition of volatiles triggers partial melting of the hot ultramafic rock in the

FIGURE 6.24 The rounded boulders shown here are a consequence of in-place weathering of a homogeneous granite pluton in Joshua Tree National Monument, California.

asthenosphere, a process that yields mafic magma. Some of this magma rises to form basaltic sills and dikes in the crust, and some makes it all the way to the surface to be extruded as basaltic lava.

In continental volcanic arcs, not all the mantle-derived basaltic magma rises directly to the surface. Some gets trapped at the base of the continental crust and some in magma chambers deep in the crust. When this happens, fractional crystallization may occur, so part of the magma evolves into a felsic melt. Furthermore, heat can transfer from the magma into the continental crust and cause partial melting of this crust. Because much of the continental crust is mafic to intermediate in composition to start with, the resulting magmas are intermediate to felsic in composition. (Remember that partial melting always produces magma that is richer in silica than was the source rock.) This magma rises, leaving the basalt behind, and either cools higher in the crust to form plutons or rises to the surface and erupts. For this reason, granitic plutons, as well as felsic and andesitic lavas, form beneath continental arcs **(Fig. 6.24)**. The word *andesite,* in fact, comes from the name of the Andes Mountains, a continental volcanic arc.

The Formation of Igneous Rocks at Mid-Ocean Ridges

Most *subaerial volcanoes* on the the Earth—those that rise above sea level and erupt into the air, producing visible displays of igneous activity—occur in island arcs. So it may seem surprising that, in fact, most of the igneous activity at the Earth's surface happens at the mid-ocean ridges of divergent boundaries. Unless we descend to the seafloor in a submersible, we don't see this activity because it's hidden from our view by a 2-km-thick layer of ocean water. Think about it—the entire oceanic crust, a 7- to 10-km-thick layer of basalt and gabbro that covers 70% of the Earth's surface, forms at mid-ocean ridges. And this entire volume gets subducted and replaced by new crust about every 200 million years.

Magmas form at mid-ocean ridges because of decompression melting. As seafloor spreading takes place and oceanic lithosphere moves away from the ridge, hot asthenosphere rises beneath the ridge axis. As this asthenosphere rises, the weight of the overlying rock, and therefore the pressure on the asthenosphere, progressively decreases, which leads to partial melting and the generation of basaltic magma. As we noted in Chapter 4, some of this magma rises into the crust and collects in a shallow magma chamber, forming a mush of liquid and crystals. Slow cooling along the margins of the magma chamber produces gabbro. Not all the melt stays in the magma chamber—about 30% eventually rises still farther. Of this volume of melt, two-thirds intrudes and freezes within vertical cracks that propagate as newly formed crust splits apart; this magma forms basalt dikes (see Fig. 4.7a). The remaining third reaches the seafloor and extrudes as lava on its surface, commonly in the form of pillow basalt **(Fig. 6.25)**.

Igneous Rocks at Rifts

Rifts occur at places where continental lithosphere undergoes horizontal stretching and, therefore, vertical thinning. As this process takes place, pressure on the asthenosphere decreases and decompression melting takes place, producing basaltic magma. Some of this magma intrudes the crust to form sills or dikes, and some makes it to the Earth's surface and erupts as basaltic lava. Part of the magma produced in the asthenosphere beneath rifts may become trapped at the base of the continental crust, or even in the crust itself. This magma transfers enough heat to the continental crust to cause partial melting, which in turn, yields felsic magmas that erupt mostly as rhyolitic ash. As a result, extrusive rocks in a rift generally include both basaltic flows and sheets of rhyolitic tuff.

(a) The formation of pillow basalt. Lava squeezes through a conduit and emerges at the surface like toothpaste.

(b) This pillow basalt forms part of an ophiolite, a slice of seafloor that was pushed up onto the surface of a continent during mountain building.

Products of Hot Spots

Hawaii and numerous other South Pacific island volcanoes are *oceanic hot-spot volcanoes*, isolated volcanoes rising from the ocean floor that are not a consequence of plate-boundary interactions. Most oceanic hot-spot volcanoes erupt in the interior of an oceanic plate, away from convergent or divergent boundaries. But some, such as Iceland, sit astride a divergent boundary. (Geoscientists associate Iceland with a hot spot because its volcanoes generate far more lava than normal mid-ocean ridge volcanoes do.) Not all hot-spot volcanoes are oceanic, however—some erupt in the interior of a continent. Eruptions from such *continental hot-spot volcanoes* produced the colorful sulfur- and iron-stained layers of volcanic ash that form the "yellow stone" of Yellowstone National Park in northwestern Wyoming and adjacent states.

> **Did you ever wonder . . .**
> why volcanic islands like Iceland and Hawaii exist?

As we learned in Chapter 4, most researchers associate hot-spot igneous activity with mantle plumes—columns of hot rock rising from deeper in the mantle. According to the plume hypothesis, the plume itself does not consist of magma—it's composed of solid rock, but rock that is hot and soft enough to flow plastically at rates of a few centimeters a year. When the hot rock of a plume reaches the base of the lithosphere, decompression causes partial melting, a process that generates mafic magma. At oceanic hot spots, much of the mafic magma erupts as basalt. At continental hot spots, part of the mafic magma erupts as basalt, but some transfers heat to the continental crust, which itself partially melts, as happens beneath rifts, producing felsic magmas that erupt as rhyolite.

Large Igneous Provinces

In many places on the Earth, particularly voluminous quantities of mafic magma have erupted or intruded **(Fig. 6.26)**. Some of these regions occur along the margins of continents, some in the interiors of oceanic plates, and some in the interiors of continents. The largest of these, the Ontong Java Oceanic Plateau of the western Pacific, covers an area of about 5,000,000 km^2 of the seafloor and has a volume of about 50,000,000 km^3. It's no surprise that geologists refer to an occurrence of such a huge volume of igneous rock as a **large igneous province (LIP)**. More recently, this term has also been applied to areas covered by immense eruptions of felsic rocks, so Yellowstone National Park can also be called a LIP.

Mafic LIPs may form when the bulbous head of a mantle plume first reaches the base of the lithosphere. More partial

FIGURE 6.26 A map showing the distribution of LIPs on the Earth. The red areas are or once were underlain by immense volumes of basalt; not all of this basalt is exposed.

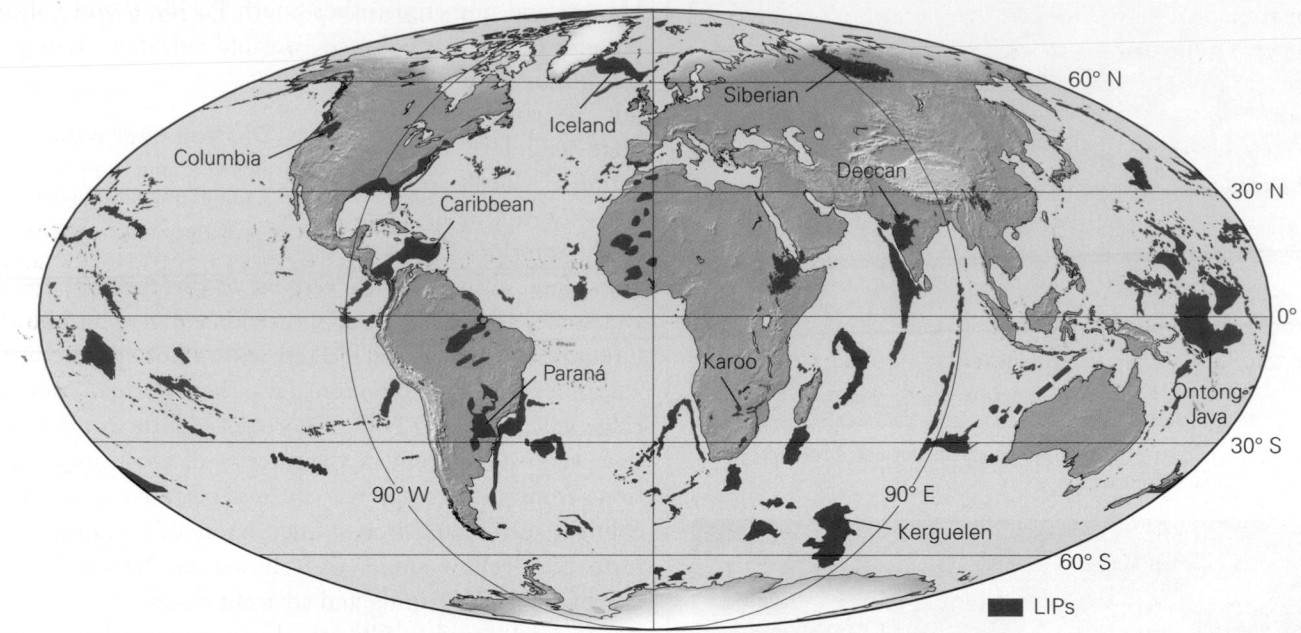

FIGURE 6.27 Flood basalts form when vast quantities of low-viscosity mafic lava "flood" the landscape and freeze into a thin sheet. Accumulation of successive flows builds a flat-topped plateau.

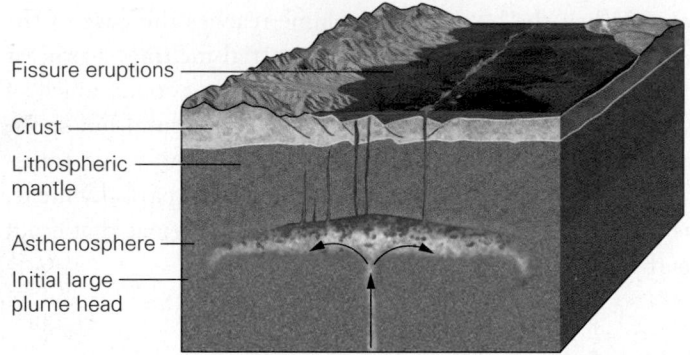

(a) The plume model for formation of flood basalts.

(b) The flood basalts of the Columbia Plateau form the layers exposed in Palouse Canyon, Washington.

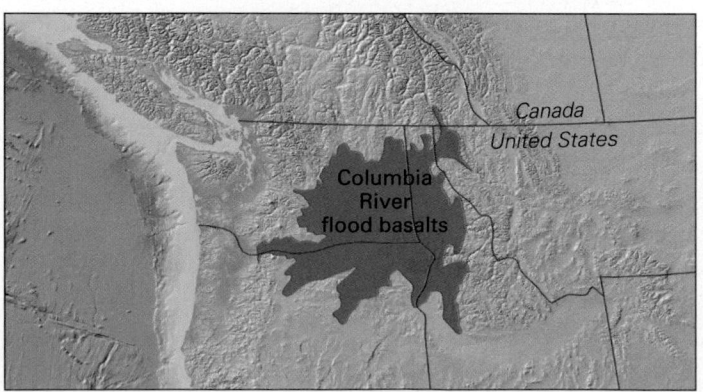

(c) Flood basalts underlie the Columbia Plateau in Washington and Oregon, the dark area of this map.

(d) Iguaçu Falls, on the Brazil-Argentina border. The falls flow over the huge flood basalt sheet (the black rock) of the Paraná Plateau. Flood basalt underlies all of the region in view.

melting can occur in a plume head than in asthenosphere rising from relatively shallow depths beneath normal mid-ocean ridges or rifts, because the rock in the plume head came from greater depth in the mantle, so it's hotter. Therefore, an unusually large quantity of unusually hot basaltic magma forms in the plume head. If the lithosphere over the plume head starts to stretch and rift, huge quantities of basaltic lava can rise and spew out of the ground. The particularly hot basaltic lava that erupts at such localities has such low viscosity that it can flow tens to hundreds of kilometers across the landscape. Geoscientists refer to such flows as **flood basalts (Fig. 6.27a)**. Flood basalts make up the bedrock of the Columbia Plateau in Oregon and Washington **(Fig. 6.27b, c)**, the Paraná Plateau in southeastern Brazil **(Fig. 6.27d)**, the Karoo region of southern Africa, and the Deccan region of southwestern India.

The volume of material that erupted at the biggest LIPs on the Earth may have exceeded the amount that erupted along the Earth's entire mid-ocean ridge system during the same time. So it seems that eruptions of such LIPs represent special events in Earth history. Some geologists attribute them to *superplumes* in the mantle—plumes that bring up vastly more hot asthenosphere than do normal plumes.

TAKE-HOME MESSAGE

Igneous activity occurs only in special locations where melting takes place. Flux melting takes place just above subducting plates, and decompression melting takes place beneath rifts, mid-ocean ridges, and hot spots. When igneous activity occurs beneath continental crust, heat-transfer melting may trigger partial melting of the crust itself.

QUICK QUESTION: Explain why both mafic and felsic igneous rocks form at continental rifts and at continental hot spots.

ANOTHER VIEW Convergent-margin tectonics has incorporated a slice of oceanic crust into the mountains of Cyprus, in the eastern Mediterranean. A canyon wall formed by stream erosion through this slice provides a cross sectional image of this crust. Here, we see vertical dikes of basalt, cutting through pillow basalts.

Pillow basalt

Dikes

Dikes

SUMMARY

- Magma is liquid rock (melt) under the Earth's surface. Lava is melt that has erupted from a volcano at the Earth's surface.

- Magma forms when hot rock partially melts. This process occurs when pressure decreases (decompression), when volatiles (such as water) diffuse into hot rock, and when heat transfers from hot magma into adjacent rock.

- Magma occurs in a range of compositions: felsic (silicic), intermediate, mafic, and ultramafic. Composition depends partly on the original composition of the source rock and partly on the way the magma evolves.

- During partial melting, only part of the source rock melts, so magma tends to be more felsic.

- Magma rises because of its buoyancy and because of pressure caused by the weight of overlying rock.

- Magma viscosity (resistance to flow) depends on its composition. Felsic magma is more viscous than mafic magma.

- Geologists distinguish between extrusive igneous rocks, formed from lava that erupts from a volcano, and intrusive igneous rocks that develop from magma that freezes inside the Earth.

- Lava may solidify to form flows or domes, or it may explode into the air to form ash and lapilli.

- Intrusive igneous rocks form when magma intrudes into pre-existing rock below the Earth's surface. Blob-shaped intrusions are called plutons. Sheet-like intrusions that cut across layering in wall rock are dikes, and sheet-like intrusions that form parallel to layering in wall rock are sills. Huge intrusions with many plutons are known as batholiths.

- The rate at which intrusive magma cools depends on the depth of intrusion, on the size and shape of the magma body, and on groundwater circulation. Cooling time controls igneous rock texture.

- Crystalline igneous rocks are classified according to grain size and composition. Glassy igneous rocks are classified according to texture and the presence of gas bubbles. Fragmental igneous rocks are classified by grain size.

- The origin of igneous rocks can be understood in the context of plate tectonics. Magma forms at continental or island volcanic arcs along convergent boundaries due to flux melting. Igneous rocks form at hot spots due to decompression melting. Igneous rocks form at rifts as a result of decompression melting or due to heat transfer into crustal rocks. Igneous rocks form along mid-ocean ridges because of decompression melting.

GUIDE TERMS

ash (p. 172)
assimilation (p. 166)
batholith (p. 173)
Bowen's reaction series (p. 170)
crystalline igneous rock (p. 176)
decompression melting (p. 164)
dike (p. 172)
extrusive igneous rock (p. 161)
flood basalt (p. 187)
flux melting (p. 164)
fractional crystallization (p. 169)
fragmental igneous rock (p. 177)

geotherm (p. 164)
glassy igneous rock (p. 177)
heat-transfer melting (p. 165)
igneous intrusion (p. 161)
igneous rock (p. 161)
intrusive contact (p. 172)
laccolith (p. 173)
lapilli (p. 172)
large igneous province (LIP) (p. 185)
lava (p. 161)
lava flow (p. 161)

magma (p. 161)
magma chamber (p. 162)
obsidian (p. 179)
partial melting (p. 166)
pegmatite (p. 179)
pluton (p. 173)
pumice (p. 180)
pyroclastic debris (p. 161)
pyroclastic flow (p. 172)
pyroclastic rock (p. 180)
scoria (p. 180)
sill (p. 172)

stoping (p. 173)
tachylite (p. 180)
tuff (p. 180)
vesicle (p. 179)
viscosity (p. 167)
volcanic agglomerate (p. 181)
volcanic arc (p. 183)
volcanic breccia (p. 181)
volcanic eruption (p. 161)
volcaniclastic rock (p. 181)
volcano (p. 161)
wall rock (p. 161)
xenolith (p. 174)

 **GEOTOURS** *THIS CHAPTER'S GEOTOURS WORKSHEET (D) FEATURES QUESTIONS AND GOOGLE EARTH SITES ON:*

- Batholiths and laccoliths
- Dikes

REVIEW QUESTIONS

The letters following each Review Question refer to the corresponding Learning Objective from the Chapter Opener.

1. How does the process of freezing magma resemble that of freezing water? How is it different? **(A)**

2. What sources produce heat inside the Earth? How did the first igneous rocks on the planet form? **(B)**

3. Describe the three processes responsible for the formation of magmas. **(C)**

4. Why can we find so many different types of magmas? **(B)**

5. Why do magmas rise from depth to the surface of the Earth? **(B)**

6. What factors control the viscosity of a melt? **(C)**

7. What factors control the cooling rate of a magma within the crust? **(C)**

8. How does grain size reflect the cooling rate of a magma? Did the pictured rock come from rapidly or slowly cooling magma? **(D)**

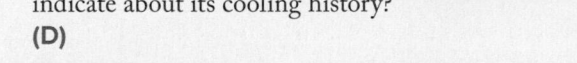

9. What does the mixture of grain sizes in a porphyritic igneous rock indicate about its cooling history? **(D)**

10. Why does magma form at a convergent boundary? **(E)**

11. What process in the mantle may be responsible for causing hot-spot volcanoes to form? **(E)**

12. Describe how magmas are produced at continental rifts. **(E)**

13. Why does melting take place beneath the axis of a mid-ocean ridge? **(E)**

14. What is a large igneous province (LIP)? **(E)**

ON FURTHER THOUGHT

15. The Cascade volcanic arc of the northwestern United States is only about 800 km long. The Andean arc of western South America is several thousand kilometers long. Look at a map showing the Earth's plate boundaries, and explain why the Andean arc is so much longer than the Cascade arc **(E)**.

16. This photograph shows a nearly horizontal layer of basalt that now crops out as a cliff along the Hudson River, across from New York City. It is parallel to layers of sandstone above and below, and is younger than all of those layers. The basalt intruded at 190 Ma, as Pangaea was breaking apart. What is this body of basalt, and in what geologic setting did it form? **(E)**

ONLINE RESOURCES

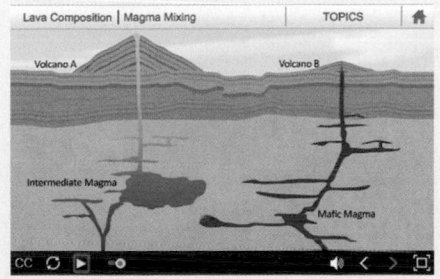

Animations
This chapter features animations on partial melting, fractional crystallization, assimilation, and magma mixing.

Videos
This chapter features a video on the topic of partial melting.

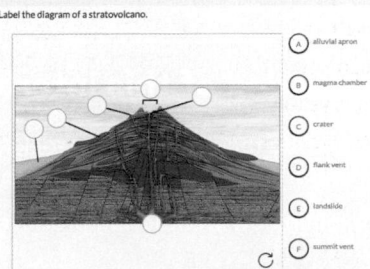

Smartwork5
Features include visual and labeling exercises on the processes involved in magma formation, the creation of various igneous rocks, and volcanic activity.

INTERLUDE B

A Surface Veneer:
Sediments and Soils

By the end of this interlude, you should be able to . . .

A. describe the changes that rocks undergo at and near the Earth's surface due to weathering.

B. explain how weathering produces sediment.

C. describe the difference between sediment and soil.

D. appraise factors that affect the character and thickness of soil.

B.1 Introduction

In the 1950s, the government of Egypt decided to build the Aswan High Dam to trap water of the Nile River in a huge reservoir before the water could reach the Mediterranean Sea. In the process of identifying a good site for the dam's foundation, geologists discovered that the present-day Nile flows on top of a 1,500-m-thick accumulation of gravel, sand, and mud that fills what was once a canyon as large as the Grand Canyon (Fig. B.1a). Evidently, at some time in the geologic past, the Nile had carved a canyon 1,500 m deep, and subsequently this canyon filled with debris. Geologists of the 1950s couldn't understand how such a sequence of events could have happened. As a general rule, the surface of a river where it empties into the sea can't be lower than sea level—otherwise, the river would have to flow upstream. The floor of the Nile's present-day channel lies only about 15 m below the river's surface, so it seemed impossible for the river to cut a canyon 100 times deeper.

The origin of the sub-Nile canyon remained a mystery until the summer of 1970, when geologists began to study the Mediterranean Sea's floor by drilling into it from a ship, as part of a 15-year-long research program called the *Deep Sea Drilling Project*. The geologists expected to find layers consisting of the shells of *plankton* (tiny floating organisms) that had settled out of the water and of mud that rivers had carried to the sea. The drill did penetrate such materials, but to everyone's surprise, the drill also penetrated a 2-km-thick layer of halite and gypsum (Fig. B.1b). Such minerals, known as salts, form only when seawater dries up. Since only about 3.5% of seawater consists of dissolved minerals, drying of the 1,500-m-deep Mediterranean would form a layer of salts less than 100 m thick. Therefore, formation of a 2-km-thick salts layer meant that the entire body of water must have dried up slowly over a long time, and probably dried up completely several times, with the sea refilling the Mediterranean basin, the low area beneath the water, after each drying event.

This explanation of the salt layer solved the mystery of the pre-Nile canyon. When the Mediterranean Sea dried up, the Nile River would have been flowing down into a basin whose floor was 1,500 m below present-day sea level. So, over time,

◀ (facing page) On the coast of Heimaey, an island off the south coast of Iceland, aprons of sediment collect at the base of lava cliffs. At the shore, waves break up this sediment to make the black sand. In the island's wet climate, soils quickly form on the sediment and grass roots in the soil. Sediment and soil form a veneer that hides bedrock.

FIGURE B.1 The Nile River flows on top of a sediment-filled canyon. This canyon formed when water evaporated from the Mediterranean, and the river cut down to the level of the basin floor.

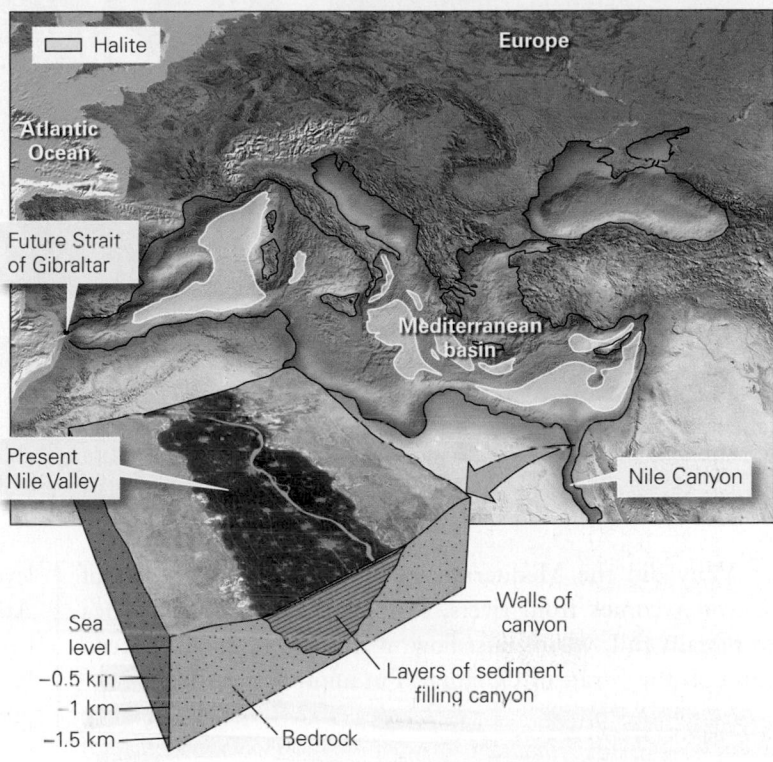

(a) The floor of the Mediterranean basin lies about 1.5 km below present sea level.

Not to scale

(b) Evidence for the past evaporation of the Mediterranean Sea was the result of drilling, which revealed a thick salt layer.

the river cut a canyon down to the low elevation of the dry sea floor. Later, when the sea refilled with water, the river could no longer cut down, and the flooded canyon filled with gravel, sand, and mud brought in from upstream.

FIGURE B.2 Various types of sediments form and accumulate at the surface of the Earth.

(a) Gravel, sand, and mud have washed or tumbled onto a valley floor in California.

(b) Shells accumulate on a beach in Florida and later break into smaller pieces. Lens cap for scale.

(c) Salt precipitates on fence posts protruding from the Dead Sea, which is a very salty lake.

Why did the Mediterranean Sea dry up? Only 10% of its water comes from rivers, so for the Mediterranean Sea to remain full, water must flow in from the Atlantic Ocean through the Strait of Gibraltar. Put another way, if the passage to the Atlantic were to be blocked, water would evaporate from the Mediterranean at a rate 10 times faster than it would flow into the Mediterranean from rivers. About 6 million years ago, the northward-drifting African Plate collided with the European Plate, forming a natural dam separating the Mediterranean from the Atlantic. At times when global sea level dropped, water stopped flowing over this dam from the Atlantic, and the Mediterranean evaporated. The salt that had been dissolved in its water accumulated as a solid deposit of halite and gypsum on the floor of the resulting basin, and the pre-Nile canyon formed. When sea level rose, water flooded in from the Atlantic, filling the Mediterranean basin again. This process was repeated many times. About 5.5 million years ago, the Mediterranean rose to its present level, and gravel, sand, and mud carried by the Nile River filled the canyon to its present level.

Geologists refer to the kinds of deposits just described—sand, mud, gravel, salt (halite and gypsum crystals), and shell fragments—as sediment. **Sediment**, broadly defined, consists of loose fragments of rocks or minerals broken off of bedrock, mineral crystals that precipitate directly out of water, and shells or shell fragments (Fig. B.2).

Sediments are produced by *weathering*, the physical and chemical breakdown of pre-existing rock at or near the Earth's surface. They form a surface veneer, or cover, on bedrock **(Fig. B.3)**. This cover ranges in thickness from 0 km, at places where bedrock crops out at the Earth's surface, to 20 km, where deep, sediment-filled depressions called *sedimentary basins* have developed. Interaction with water and organisms may modify sediment over time and transform it into *soil*, an essential material for life.

Geologists use the general term **regolith** for any loose material (sediment or soil) that covers bedrock at the Earth's surface. In Interlude B, we'll discover how weathering produces sediment and how soils form and evolve—in other words, how geologic processes produce regolith. This Interlude provides important background for Chapter 7, which considers how sediment can be transformed into sedimentary rock and why there are so many different kinds of sedimentary rock.

FIGURE B.3 A layer of unconsolidated sediment (sand, clay, and cobbles), topped by dark soil, overlies bedrock in this outcrop along the coast of western Ireland.

Soil

Unconsolidated sediment

Bedrock

B.2 Weathering: Forming Sediment

If you examine a new road cut excavated deep into bedrock, you'll find that the rock contains the minerals and textures that it has had since it first formed. Geologists informally refer to such rock as "fresh" **(Fig. B.4)**. But in the surface and near-surface realm of our planet, fresh rock doesn't last forever, both because the Earth System has an atmosphere and hydrosphere that contain water and other chemicals that react with minerals in rocks, and because the Earth hosts living organisms that can alter Earth materials. As a result, rocks in the surface or near-surface realm undergo weathering. In a general sense, the term **weathering** refers to the combination of processes that break up and corrode solid rock and may eventually transform it into loose debris or *detritus*. Because of weathering, rock at the Earth's surface will crumble away, sooner or later.

Just as a plumber can unclog a drain by using physical force (with a plumber's snake) or by causing a chemical reaction (with a dose of liquid drain opener), nature can attack rocks in two ways. So geologists distinguish between two types of weathering: physical and chemical. We'll discuss each of these separately.

Physical Weathering

Physical weathering, also referred to as *mechanical weathering*, breaks intact rock into unconnected **clasts** (grains or chunks). Each size range of clasts has a name **(Table B.1)**. Many different phenomena contribute to physical weathering, as we now describe.

FIGURE B.4 The contrast between fresh and weathered granite in an Arizona outcrop. Weathered granite can break apart. Grains fall off and collect as regolith at the base of the outcrop. The inset photos show how weathering visibly changes the rock.

TABLE B.1 Clasts Are Classified by Grain Diameter

Boulders	More than 256 mm
Cobbles	Between 65 mm and 256 mm
Pebbles	Between 3 mm and 64 mm
Sand	Between 1/16 mm and 2 mm
Silt	Between 1/256 mm and 1/16 mm
Mud	Less than 1/256 mm

Cobbles

Sand

Jointing Rocks buried deep in the Earth's crust endure enormous pressure due to the weight of *overburden*, the overlying rock. Deep rocks are also warmer than rocks nearer the surface because of the Earth's geothermal gradient. Over long periods, moving water, air, and ice grind away and remove overburden at the surface in a process called *erosion*. At the same time, tectonic activity may cause upward movement of the crust. As a consequence, with uplift and erosion acting simultaneously, rock that was deep in the Earth gradually moves closer to the surface. As this process of **exhumation** proceeds, both the pressure and the temperature acting on the rock decrease. The resulting changes in pressure and temperature can cause a rock's shape to change slightly. Specifically, as the weight acting on the rock decreases, the rock stretches a bit vertically and shortens slightly horizontally, and as the rock cools, it contracts. Such changes can generate forces sufficient to break the rock. Natural cracks that form in rocks due to removal of overburden or to cooling (and for other reasons as well; see Chapter 11) are known as **joints**.

Joints come in a variety of forms. Some have fairly smooth, planar surfaces, whereas others have irregular, jagged surfaces, and still others trace out broad curves. Intervals of sedimentary rock typically contain sets of planar vertical joints which, because they intersect beds, break rocks into rectangular blocks **(Fig. B.5a)**. Granite plutons, in contrast, commonly undergo *exfoliation*, meaning that they split along joints oriented nearly parallel to the rock face to create curving sheets that resemble the layers of an onion **(Fig. B.5b)**.

Regardless of their orientation, the formation of joints turns formerly intact bedrock into separate blocks. Gravity pulls on these blocks, which eventually fall from the outcrop in which they formed. In some cases, the blocks collect in an apron of **talus**, the rock rubble at the base of a slope **(Fig. B.5c)**. If the blocks fall into a stream or onto the shore, they may be carried away by water.

Frost Wedging Freezing water can burst pipes and shatter bottles because water expands when it freezes and pushes the walls of its container apart. The same phenomenon happens in rock. When water trapped in a joint freezes, it forces the joint open and may cause the joint to grow. Such **frost wedging** helps break blocks free from intact bedrock **(Fig. B.6a)**.

Salt Wedging In arid climates, dissolved salt in water precipitates and forms crystals in open pore spaces in rocks. The growing crystals push apart the surrounding grains and weaken the rock, so when exposed to wind and rain, the rock disintegrates into separate grains. This phenomenon, called **salt wedging**, also happens on outcrop faces in coastal areas, where salt spray percolates into rock and then dries; after a while, the rock develops a surface texture resembling a honeycomb **(Fig. B.6b)**.

Root Wedging Have you ever noticed how the roots of an old tree can break up a sidewalk? Even though the wood of roots doesn't seem very strong compared with rock or concrete, roots apply force to their surroundings as they grow and expand. Tree roots that grow into joints can push joints open in a process known as **root wedging (Fig. B.6c)**.

Thermal Expansion When the heat of an intense forest fire bakes a rock, the outer layer of the rock expands. On cooling, that layer contracts. This change produces forces in the rock sufficient to make the outer part of the rock break off in sheet-like scales. Heating does not need to be as intense as that of a fire in order to trigger cracking, however. Recent research suggests that the heat of the Sun's rays sweeping across dark rock in a desert day after day may, over time, cause the rock to fracture into thin slices.

Animal Attack Animal life also contributes to physical weathering. For example, burrowing creatures, from earthworms to gophers, move rock fragments. Recently, humans have become perhaps the most energetic agent of physical weathering on the planet. When we excavate quarries, foundations, mines, or roadbeds by digging and blasting, we shatter and displace rock that might otherwise have remained intact for millions of years more.

FIGURE 6.17 Making room for igneous plutons.

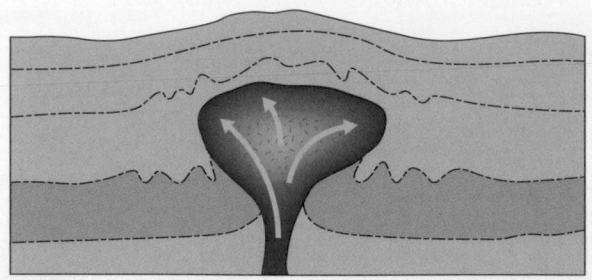

(a) During diapiric rise, a blob-shaped mass of magma forces its way up through wall rock, which plastically deforms to move out of the way.

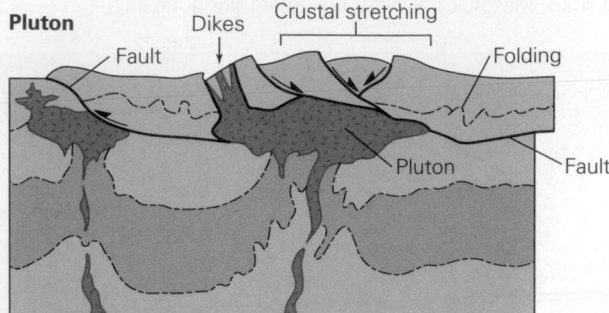

(b) Faulting associated with crustal stretching may accommodate the emplacement of a diapir.

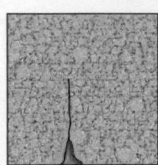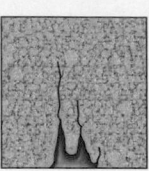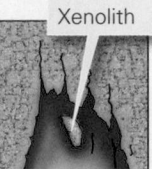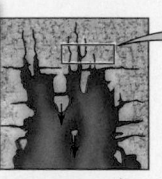

Xenolith

Time

(c) Magma pushes up into cracks and breaks off blocks of wall rock. The blocks may be incorporated in the melt by stoping.

The white rock (granite) is intruding into the dark wall rock.

Xenolith

(d) A xenolith of darker rock surrounded by light-colored granite.

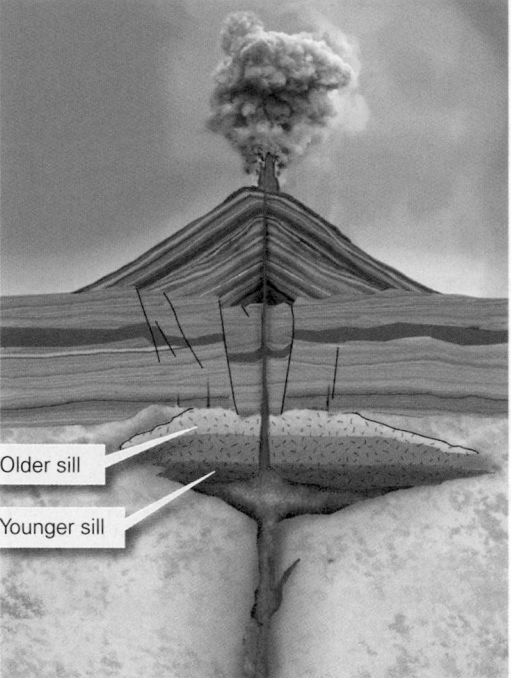

Older sill

Younger sill

(e) Plutons may intrude as a succession of sill-like sheets, whose composition evolves.

characterize the overall hue of the rock. What physical factors control the color of igneous rock? Generally, the color reflects the rock's composition, but it isn't always so simple, because color may also be influenced by grain size and by the presence of trace amounts of impurities. (For example, the presence of a small amount of iron oxide gives rock a reddish tint.)

Next, you would probably characterize the rock's texture. **Crystalline igneous rocks** consist of mineral crystals that intergrow when the melt solidifies, so that they fit together like pieces of a jigsaw puzzle **(Fig. 6.18a)**. The interlocking of crystals in these rocks happens because the rock does not solidify instantly. Rather, different crystals grow at different

FIGURE B.5 Joints (natural cracks) break bedrock into blocks and sheets, which can tumble down a slope.

(a) Vertical joints break beds of sedimentary rock into blocks that fall to the base of this 100-m-high cliff.

(b) Exfoliation joints on this granite mountain slope in the Sierra Nevada break the rock into onion-like sheets.

(c) Recently formed talus, collecting at the base of a cliff in the Canadian Rockies.

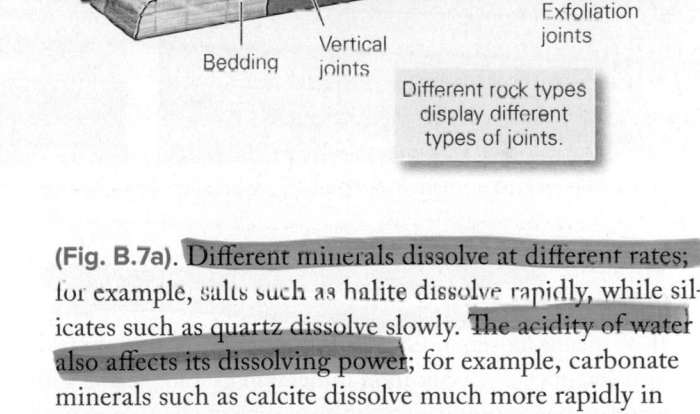

Sandstone

Granite

Exfoliation joints

Bedding

Vertical joints

Different rock types display different types of joints.

Chemical Weathering

Up to now we've taken the plumber's-snake approach to breaking up rock. Now let's look at the liquid-drain-opener approach. Chemical weathering refers to the chemical reactions that alter or destroy minerals when rock comes in contact with water solutions or air. Common reactions active during chemical weathering include the following:

• *Dissolution:* During **dissolution**, minerals in water separate into ions that become surrounded by water molecules

(Fig. B.7a). Different minerals dissolve at different rates; for example, salts such as halite dissolve rapidly, while silicates such as quartz dissolve slowly. The acidity of water also affects its dissolving power; for example, carbonate minerals such as calcite dissolve much more rapidly in acidic water (with an excess of hydrogen ions, or H^+), than in non-acidic water. As we saw in Chapter 5, the reaction of acidic water with calcite yields a solution of Ca^+ and CO_3^{2-} ions and releases CO_2 gas.

How does the water in rock near the surface of the Earth become acidic? As rainwater falls, it dissolves CO_2 gas in the atmosphere, and as the water sinks down through soil containing organic debris, it reacts with the debris. Both processes yield carbonic acid. Because of the solubility of calcite, limestone and marble—two types of rock composed of calcite—dissolve sufficiently to yield underground caverns (see Chapter 19).

FIGURE B.6 Wedging is one type of physical (mechanical) weathering.

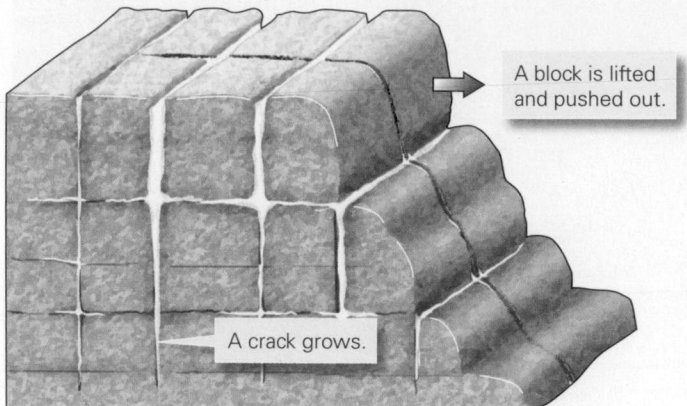

A block is lifted and pushed out.

A crack grows.

(a) Frost wedging occurs when the water that fills cracks freezes; it expands and wedges the cracks open.

(b) Salt wedging in rock on a beach cliff in Scotland yields "honeycomb weathering".

Tree growing in a joint.

Eventually, the blocks tumble to the base of the cliff.

(c) Root wedging pushes open a joint, slowly separating a block from the cliff.

- *Hydrolysis:* During *hydrolysis*, water reacts chemically with minerals and breaks them down to form other minerals (*lysis* means loosen in Greek). For example, potassium feldspar (orthoclase), a common mineral in granite, reacts with acidic water to produce kaolinite (a type of clay) along with a variety of dissolved ions. Hydrolysis reactions break down not only feldspars but many other silicate minerals as well—amphibole, pyroxene, mica, and olivine all react slowly with water and transform into various types of clay. Quartz also undergoes hydrolysis, but does so at such a slow rate that quartz grains survive such weathering in most climates.

- *Oxidation:* Chemists refer to a reaction during which an element loses electrons as an *oxidation reaction*, because such a loss commonly takes place when elements combine with oxygen. The oxidation, or rusting, of iron serves as an example. Oxidation reactions in rocks transform iron-bearing minerals (such as biotite and pyrite) into a weak, rusty-brown mixture of various iron-oxide and iron-hydroxide minerals **(Fig. B.7b)**.

- *Hydration:* *Hydration*, the absorption of water into the crystal structure of minerals, causes some minerals, such as certain types of clay, to expand. Such expansion weakens rock.

Not all minerals undergo chemical weathering at the same rates **(Table B.2)**. Some weather in a matter of months or years, whereas others remain unweathered for millions of years. In temperate climates, for example, calcite weathers faster than most silicate minerals. Of the silicate minerals, those that crystallize at the highest temperatures tend to be less stable under the cool temperatures at the Earth's surface than those that crystallize at lower temperatures. The difference depends partly on crystal structure and partly on chemical composition.

FIGURE B.7 Examples of chemical weathering.

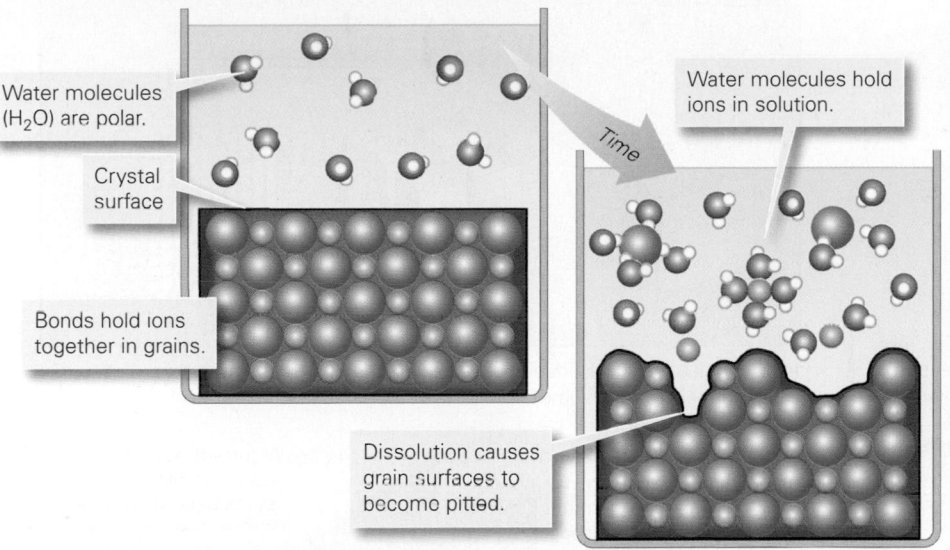

Water molecules (H₂O) are polar.

Crystal surface

Bonds hold ions together in grains.

Time

Water molecules hold ions in solution.

Dissolution causes grain surfaces to become pitted.

(a) Dissolution occurs when water molecules pluck ions off grain surfaces.

Water seeping into joints in limestone produced troughs.

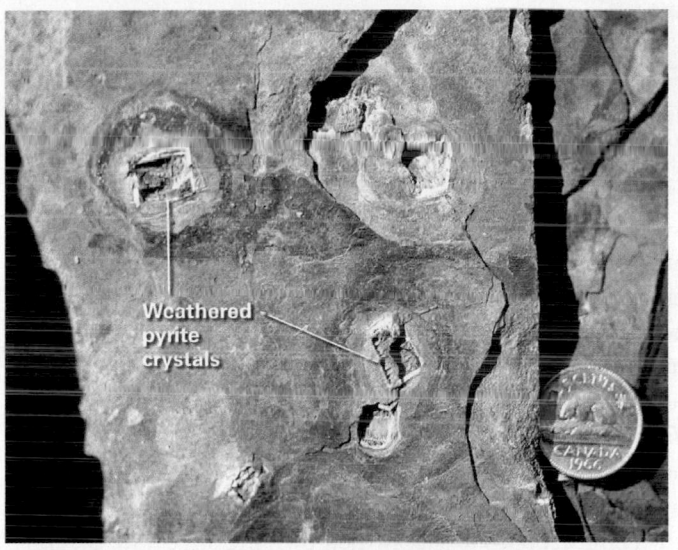

Weathered pyrite crystals

(b) Pyrite (FeS₂) is a sulfide mineral that reacts with air to form iron oxide.

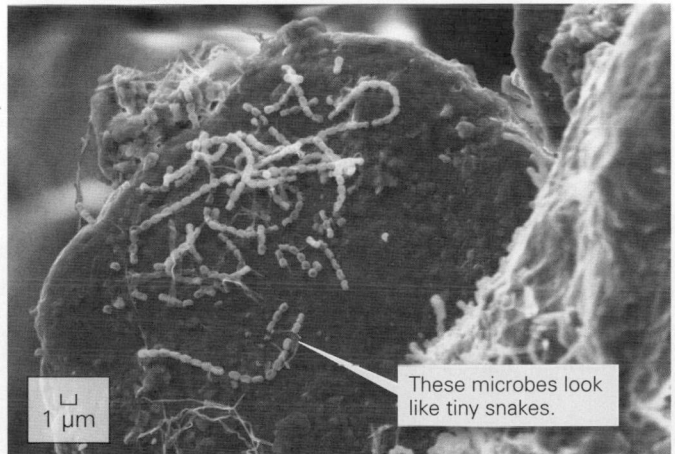

1 μm

These microbes look like tiny snakes.

(c) Certain types of microbes obtain their life energy from the chemical bonds in minerals.

Specifically, minerals with fewer linkages between silicon-oxygen tetrahedra tend to have weaker structures, so they weather faster than do minerals with more linkages. And minerals containing iron, magnesium, sodium, potassium, and aluminum tend to weather faster than minerals without

TABLE B.2 Relative Stability of Minerals at the Earth's Surface*

Fastest Weathering		Least Stable
	Halite	
	Calcite	
	Olivine	
	Ca-plagioclase	
	Pyroxene	
	Amphibole	
	Na-plagioclase	
	Biotite	
	Orthoclase (K-feldspar)	
	Muscovite	
	Clay (various types)	
	Quartz	
	Gibbsite (aluminum hydroxide)	
Slowest Weathering	Hematite (iron oxide)	Most Stable

*Note that minerals that form early in Bowen's reaction series (see Box 6.1) are among the least stable minerals at the Earth's surface. Minerals that are the products of weathering reactions (e.g., hematite) are among the most stable minerals at the Earth's surface. Mafic minerals weather by oxidation, felsic minerals by hydrolysis, and carbonates and salts by dissolution; oxide minerals don't weather at all.

FIGURE B.8 Physical and chemical weathering processes work together to break down rock.

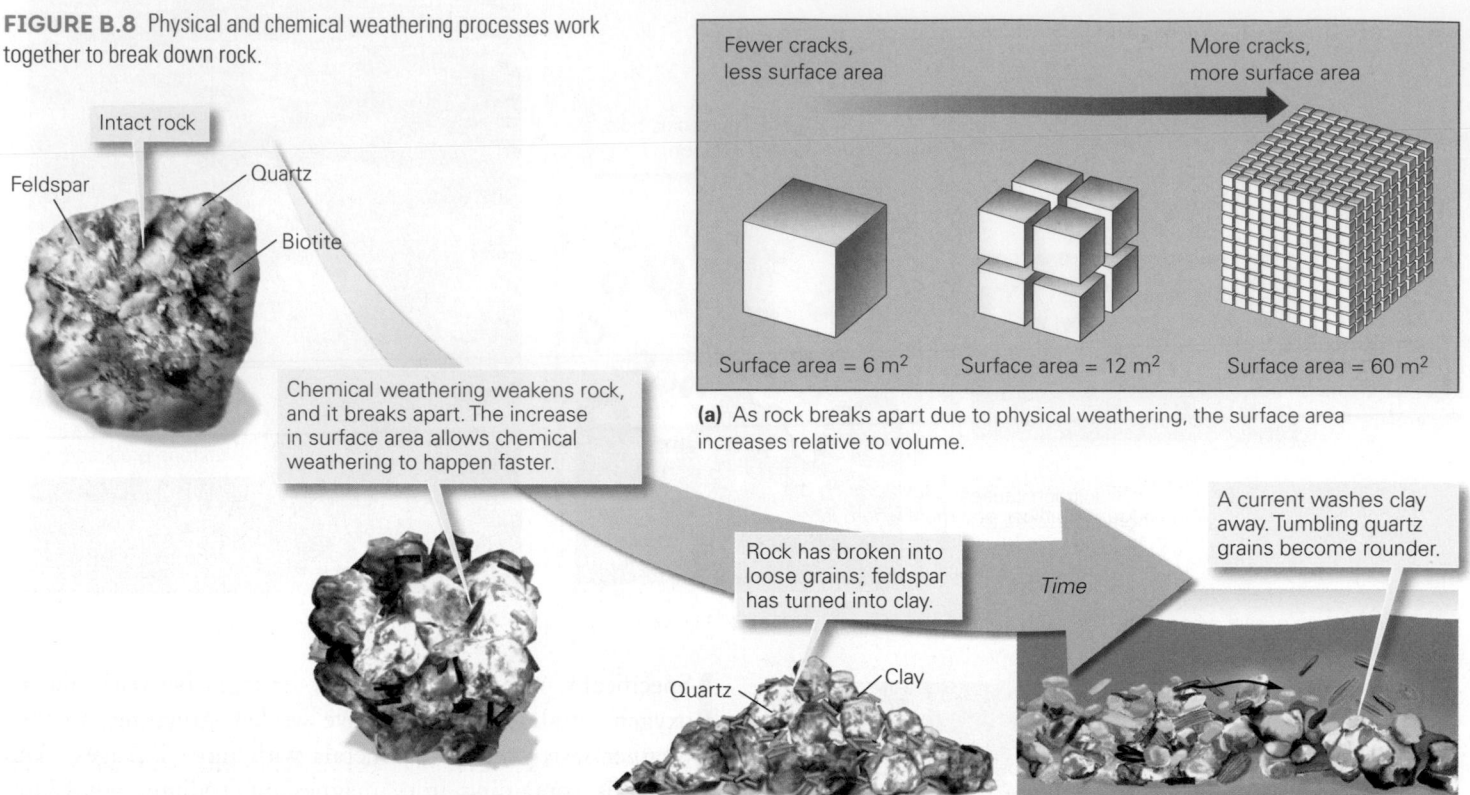

Intact rock

Feldspar

Quartz

Biotite

Chemical weathering weakens rock, and it breaks apart. The increase in surface area allows chemical weathering to happen faster.

Fewer cracks, less surface area

More cracks, more surface area

Surface area = 6 m²   Surface area = 12 m²   Surface area = 60 m²

(a) As rock breaks apart due to physical weathering, the surface area increases relative to volume.

Rock has broken into loose grains; feldspar has turned into clay.

A current washes clay away. Tumbling quartz grains become rounder.

Time

Quartz

Clay

(b) Chemical weathering weakens rock, so it breaks apart. As this happens, the surface area increases, so chemical weathering happens still faster. Eventually, the rock completely disaggregates to form sediment. Here, the weathering of granite produces quartz sand and clay.

these elements. Quartz (pure SiO_2), a three-dimensional network silicate mineral with strong bonds in all directions, tends to be very stable. When granite (a rock that contains quartz, mica, and feldspar) undergoes chemical weathering, everything but quartz transforms to clay. That's why beaches typically consist of quartz sand; quartz remains after the other minerals have turned to clay and washed away.

Until fairly recently, geologists tended to think of chemical weathering as a strictly inorganic chemical reaction, occurring entirely independently of biological activity. New research demonstrates, however, that organisms play a major role in the chemical-weathering process. For example, in order to extract nutrients from minerals, the roots of plants, fungi, and lichens secrete organic acids that help dissolve minerals in rocks. *Microbes*, microscopic single-celled organisms (including bacteria and archaea), literally eat minerals for lunch! Microbes can pluck molecules from a mineral's surface and use the energy from broken chemical bonds to supply their own life force (**Fig. B.7c**).

These slopes have smoothed surfaces because the rock they are made of has weathered.

FIGURE B.9 Spheroidal weathering.

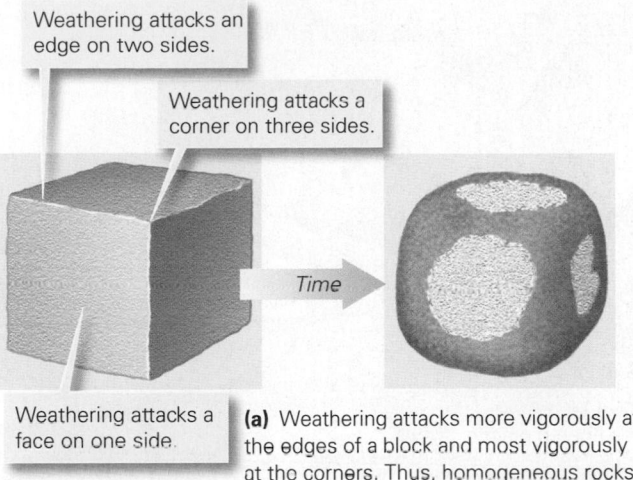

Weathering attacks an edge on two sides.

Weathering attacks a corner on three sides.

Weathering attacks a face on one side.

(a) Weathering attacks more vigorously at the edges of a block and most vigorously at the corners. Thus, homogeneous rocks tend to weather into rounded blocks.

(b) At this location in Nevada, we can spot the granite outcrops by looking for light-colored, spheroidally weathered rocks.

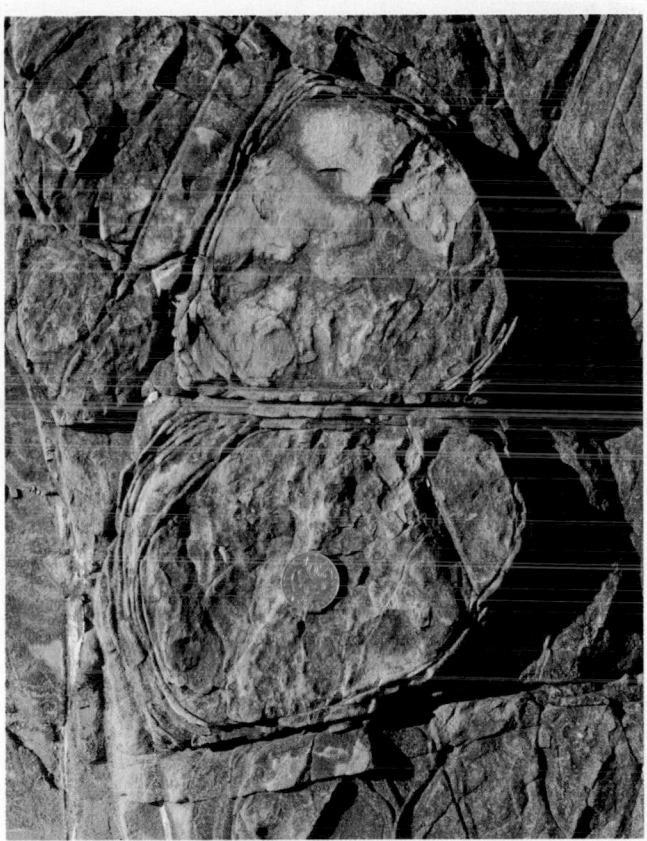

(c) Rock in New Caledonia is undergoing spheroidal weathering by the formation of small exfoliation cracks.

Physical and Chemical Weathering Working Together

So far, we've looked at the processes of chemical and physical weathering separately. But in the real world, they happen together, aiding each other in disintegrating rock to form sediment (**Geology at a Glance**, pp. 200–201).

Physical weathering speeds up chemical weathering. To understand why, keep in mind that chemical-weathering reactions take place at the surface of a material. Therefore, the overall rate at which chemical weathering occurs depends on the ratio of surface area to volume—the greater the surface area, the faster the volume as a whole can chemically weather. When jointing (physical weathering) breaks a large block of rock into smaller pieces, the surface area increases, so chemical weathering happens faster (**Fig. B.8a**).

Similarly, chemical weathering speeds up physical weathering by dissolving away grains or cements that hold a rock together, by transforming hard minerals (such as feldspar) into soft minerals (such as clay), and by causing minerals to absorb water and expand. These phenomena make rock weaker, so that it can disintegrate more easily (**Fig. B.8b**). If you drop a block of fresh granite on the ground, it will most likely stay intact, but if you drop a block of chemically weathered granite on the ground, it likely will crumble into a pile of debris.

Weathering happens faster at edges, and even faster at the corners of broken blocks, because weathering attacks a flat face from only one direction, an edge from two directions, and a corner from three directions. As a result, over time, edges of blocks become blunt and corners become rounded (**Fig. B.9a**). In fairly homogeneous rocks such as granite, which do not contain layering that can affect weathering rates, rectangular blocks transform into a rounded shape (**Fig. B.9b**). In some cases, exfoliation takes place at the scale of an individual rock block, so the block breaks into onion-like sheets that fall off to make the block somewhat spherical. Geologists refer to processes that produce rounded blocks as *spheroidal weathering* (**Fig. B.9c**).

GEOLOGY AT A GLANCE

Weathering, Sediment, and Soil Production

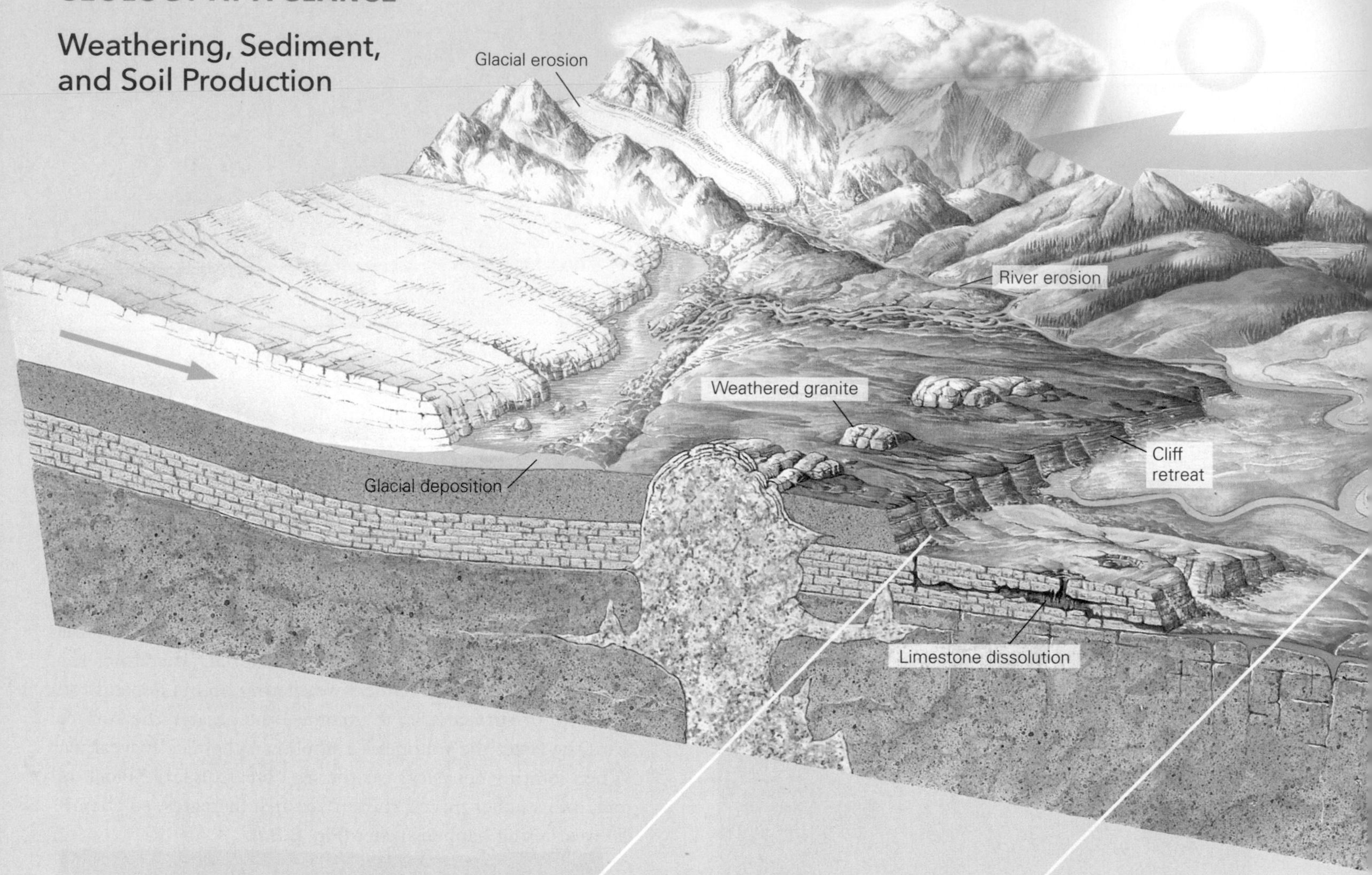
Glacial erosion
River erosion
Weathered granite
Cliff retreat
Glacial deposition
Limestone dissolution

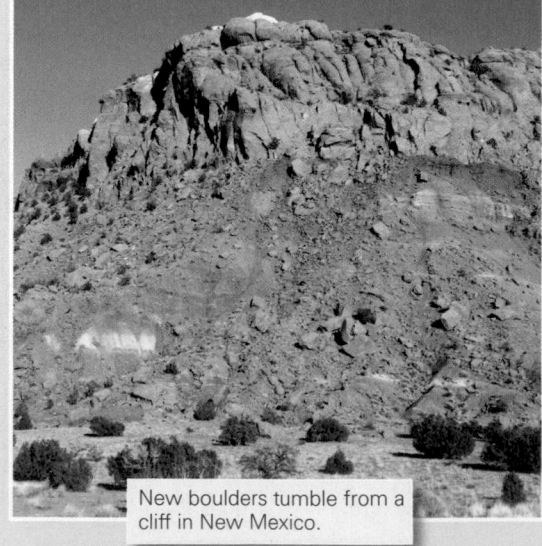

New boulders tumble from a cliff in New Mexico.

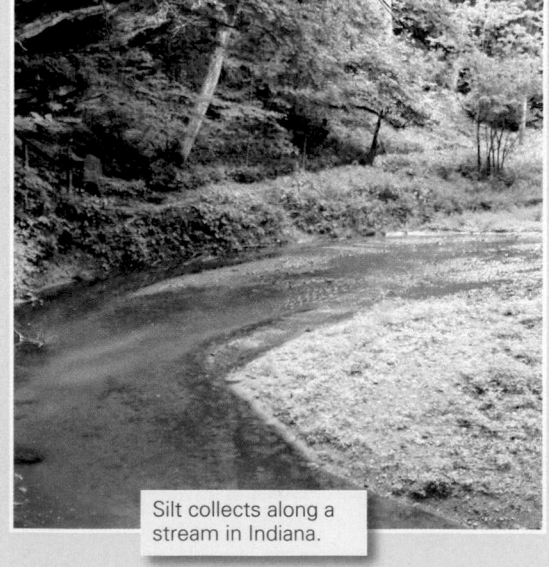

Silt collects along a stream in Indiana.

Wind

Tectonic processes raise the land surface above sea level. Once exposed, rock interacts with air and water and undergoes chemical and physical weathering, ultimately breaking down to produce sediment. Convection in the atmosphere generates wind, rain, and snow. Flowing water, ice, and air erode and transport sediment to sites of deposition. Leaching by downward-percolating rainwater, along with the addition of organic material, produces soil.

Coastal erosion

River deposition

Soil formation

Coastal deposition

Soil forms on bedrock of chalk in southern England.

Waves move sand on a beach in Brazil.

Erosion carves the coastal cliffs of Ireland.

FIGURE B.10 Differential weathering.

(a) Sawtooth weathering profiles develop in sequences of alternating strong and weak layers on this exposure in New Mexico. Weak layers are indented, whereas strong layers protrude.

(b) Inscriptions on a granite headstone (left) last for centuries, but those on a marble headstone (right) may weather away in decades. These gravestones are in the same cemetery and are about the same age.

When different rocks in an outcrop undergo weathering at different rates, we say that the outcrop has undergone *differential weathering*. As a result of differential weathering, cliffs composed of a variety of rock layers take on a stair-step or sawtooth shape **(Fig. B.10a)**. Weak layers weather away beneath a more resistant layer, so the resistant layer becomes an overhanging ledge. Similarly, the rate at which the land surface weathers depends on the rock type, so valleys tend to develop over weak rocks, while strong rocks hold up hills.

You can easily see the consequences of differential weathering if you walk through a graveyard. The inscriptions on some headstones remain sharp and clear, whereas those on other stones have become blunted or have even disappeared **(Fig. B.10b)**. That's because the minerals in these different stones have different resistances to weathering. Granite, an igneous rock with a high quartz content, retains inscriptions the longest. In contrast, marble, a metamorphic rock composed of calcite, dissolves away relatively rapidly in acidic rainwater.

B.3 Soil

If you've ever had the chance to dig in a garden, you've seen firsthand that the material in which flowers grow looks and feels different from beach sand or potter's clay. We call the material in a garden "dirt" or, more technically, soil. In a general sense, **soil** consists of rock or sediment that has been modified by physical and chemical interaction with organic material and rainwater, over time, to produce a substrate that can support the growth of plants. Soil serves as one of our planet's most valuable resources, for without it there could be no agriculture, forestry, ranching, or home gardening.

How Does Soil Form?

Three processes taking place at or just below the surface of the Earth contribute to soil formation.

- *Production of mineral debris:* Chemical and physical weathering produces loose debris, new mineral grains (such as clay), and ions in solution from sediment or bedrock.

- *Interaction with water:* When rain falls, some of the water percolates down through the debris and carries dissolved ions and clay flakes with it. We refer to the region in which this downward transport occurs as the **zone of leaching**, because *leaching* means extraction, absorption, and removal. Deeper down, new mineral crystals precipitate out of the downward-percolating water or form by reaction of the water with debris. In addition, the water leaves behind its load of fine clay. The region in which new minerals grow and clay collects is the **zone of accumulation (Fig. B.11a)**.

- *Interaction with organisms:* Soil serves as home for a remarkable number of organisms—it's a realm in which biological and physical components of the Earth System interact profoundly. For example, a single cubic centimeter of moist soil in a warm region hosts over 1 billion microbial cells, and over 1.5 million earthworms wriggle

FIGURE B.11 Formation of soil horizons.

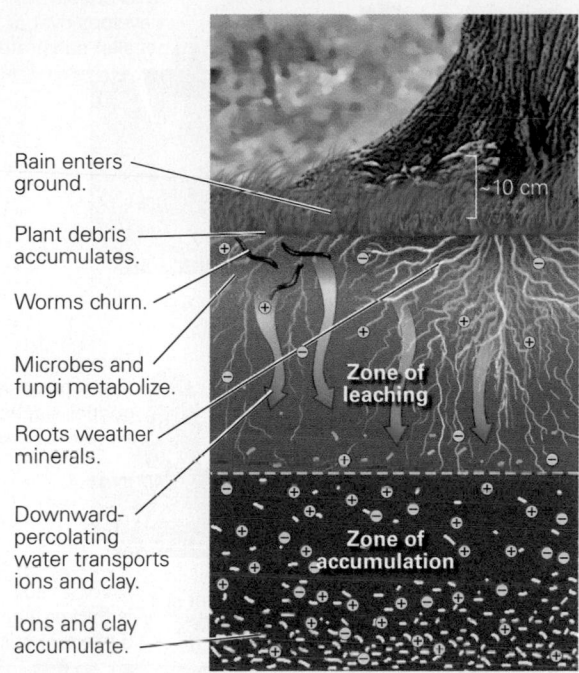

Rain enters ground.

Plant debris accumulates.

Worms churn.

Microbes and fungi metabolize.

Roots weather minerals.

Downward-percolating water transports ions and clay.

Ions and clay accumulate.

~10 cm

Zone of leaching

Zone of accumulation

(a) Soil horizons develop as water percolates downward, carrying ions and clay with it, and as organisms interact with the soil.

Horizon designation

O
A
E
B
C

Topsoil

Transition

Subsoil

Weathered bedrock

Solid bedrock

(b) Distinct soil horizons develop, each with a characteristic composition and texture.

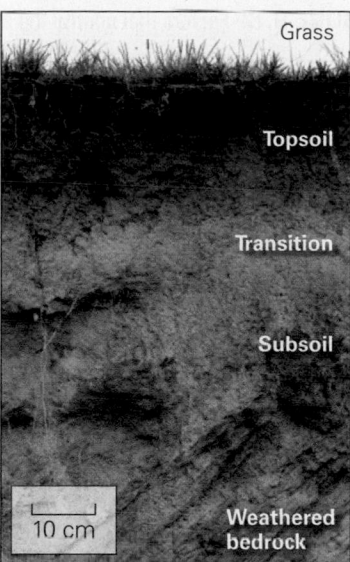

Grass

Topsoil

Transition

Subsoil

Weathered bedrock

10 cm

(c) Soil horizons exposed on the wall of a gully in eastern Brazil.

through each acre of such soil. The microbes, fungi, plants, and animals in soil absorb nutrient atoms, leave behind organic waste, produce acids that weather mineral grains, and also physically churn and break up the soil. When organisms die, they rot and transform into organic carbon that mixes with the mineral debris of soil. The accumulation of rotted organic debris makes up **humus**.

As a consequence of the above processes, sediment and bedrock evolve into soil, and the soil's "character"—its texture and composition—becomes very different from that of the starting material.

Because different soil-forming processes operate at different depths, soils typically develop distinct zones, known as **soil horizons**, arranged in a vertical sequence called a **soil profile (Fig. B.11b, c)**. Let's look at an idealized soil profile, from top to bottom, using a soil formed in a temperate forest as our example. The highest horizon, the *O-horizon* (the prefix stands for organic), consists almost entirely of organic matter and contains barely any mineral matter. Below the O-horizon we find the *A-horizon*, in which humus has decayed further and has mixed with mineral grains (clay, silt, and sand). Water percolating through the A-horizon causes chemical-weathering reactions to occur and produces ions in solution and new clay minerals. The downward-moving water eventually carries soluble chemicals and fine clay deeper into the subsurface. The O- and A-horizons constitute dark-gray to

blackish-brown **topsoil**, the fertile portion of soil that farmers till for planting crops. In some places, the A-horizon grades downward into the *E-horizon*, a soil level that has undergone substantial leaching but has not yet mixed with organic material. Ions and clay accumulate in the *B-horizon*, or **subsoil**. (Note from our description that the O-, A-, and E-horizons lie within the zone of leaching, whereas the B-horizon lies in the zone of accumulation.) Finally, at the base of a soil profile we find the *C-horizon*, which consists of detritus that's been chemically weathered and broken apart, but has not yet undergone leaching or accumulation. The C-horizon grades downward into unweathered bedrock or into unweathered sediment.

Farmers, foresters, and ranchers are aware that soil in one locality differs greatly from soil in another in terms of composition, thickness, and texture. Crops that grow well in one type of soil may wither and die in another. Such diversity exists because the makeup of a soil depends on several *soil-forming factors* **(Fig. B.12)**.

- *Climate:* The total rainfall, the distribution of rainfall over the year, and the temperature range and average over the year determine the rate and amount of chemical weathering and leaching at a given location. Large amounts of rainfall and warm temperatures accelerate chemical weathering and cause most soluble elements to be leached. In regions with small amounts of rainfall and cooler temperatures, soils take a long time to develop and may retain unweathered minerals and soluble components.

- *Substrate composition:* Some soils form on basalt, some on granite, some on volcanic ash, and some on recently deposited

FIGURE B.12 Factors that control the character of soil.

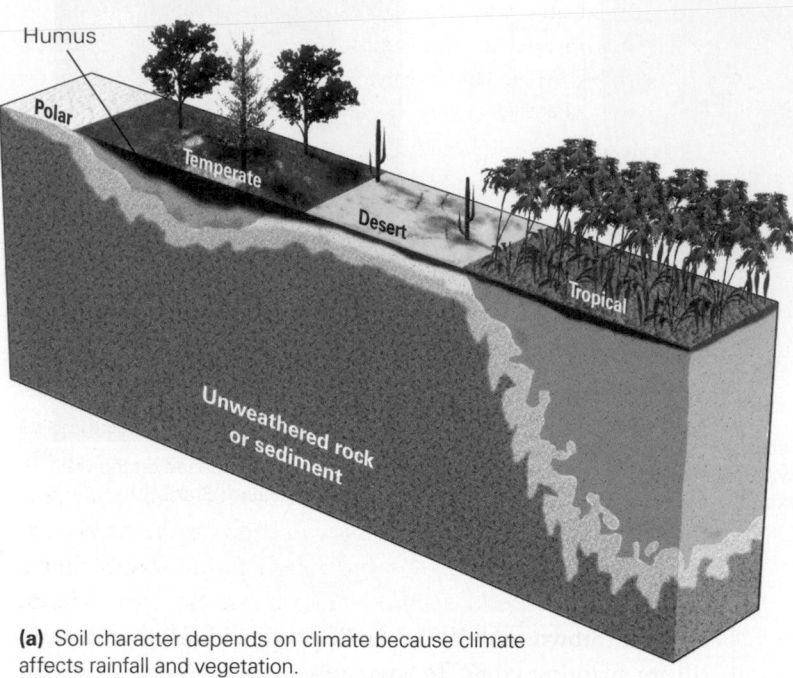

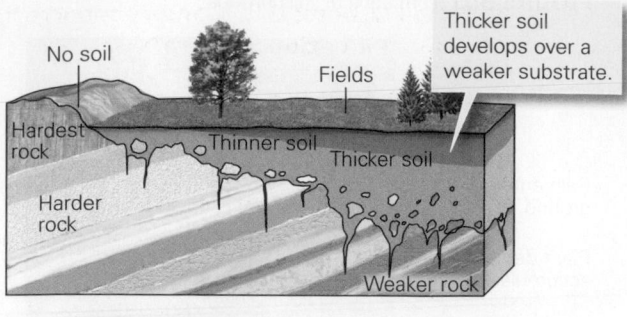

(a) Soil character depends on climate because climate affects rainfall and vegetation.

(b) Soil character also depends on the strength of the substrate, the steepness of the slope, and the length of time that soil has been forming.

quartz sand. These different substrates consist of different materials, so the soils formed on them end up with different chemical compositions. For example, a soil formed on basalt tends to be richer in iron than a soil formed on granite, because basalt contains more iron than does granite. Furthermore, soils tend to develop faster on *unconsolidated material* (loose ash or sediment) than on hard, solid bedrock.

• *Slope steepness:* A thick soil can accumulate under land that lies flat. But on a steep slope, sediment may wash away before it can evolve into soil. All other factors being equal, soil thickness increases as the slope angle decreases.

• *Wetness:* Depending on the details of local topography and on the depth below the surface at which groundwater occurs, some soil contains much more water than does other soil in the same region. Wetness affects soil weathering rates and organic content.

• *Time:* Because soil formation takes time, a younger soil tends to be thinner and less developed than does an older soil in the same location. The rate of soil formation varies greatly with environment. In a protected, moist, warm region, soil may develop over the course of a few years to a few decades. But in an exposed, cold, dry region, soil may take thousands of years or more to develop. In temperate regions, soil forms at a rate of 0.02 to 0.20 mm per year, thereby producing 1 meter of soil in about 10,000 years.

• *Vegetation type:* Different kinds of plants extract or add different nutrients and quantities of organic matter to a soil. Also, plants with deep root systems help prevent soil from washing away, so that it builds into a thicker layer.

Soil Classification

Depending on their evolution and composition, soils come in a variety of textures, structures, and colors. *Soil texture* reflects the relative proportions of sand-, silt-, and clay-sized grains in the soil. For many crops, farmers prefer to sow seeds in **loam,** a type of soil consisting of 10% to 30% clay and the rest silt and sand. In loam, pores (open spaces) remain between grains so that water and air can pass through and roots can easily penetrate. In soils with too much clay, the clay packs together and prevents water movement. *Soil structure* refers to the degree to which soil grains clump together to form lumps or clods, which soil scientists refer to as *peds* (from the Latin *pedo,* meaning soil). Soil structure evolves as the soil develops, because structure depends on clay content and organic content, both of which change over time. These materials give soil

TABLE B.3 Soil Orders: U.S. Comprehensive Soil Classification System*

Alfisol	Gray/brown, has subsurface clay accumulation and abundant plant nutrients. Forms in humid forests.
Andisol	Forms in volcanic ash.
Aridisol	Low in organic matter, has carbonate horizons. Forms in arid environments.
Entisol	Has no horizons. Formed very recently.
Gelisol	Underlain with permanently frozen ground.
Histosol	Very rich in organic debris. Forms in swamps and marshes.
Inceptisol	Moist, has poorly developed horizons. Formed recently.
Mollisol	Soft, black, and rich in nutrients. Forms in subhumid to subarid grasslands.
Oxisol	Very weathered, rich in aluminum oxide and iron oxide, low in plant nutrients. Forms in tropical regions.
Spodosol	Acidic, low in plant nutrients, ashy, has accumulations of iron and aluminum. Forms in humid forests.
Ultisol	Very mature, strongly weathered soils, low in plant nutrients.
Vertisol	Clay-rich soils capable of swelling when wet, and shrinking and cracking when dry.

*See also Fig. B.13.

its stickiness. Soil composition controls soil color: organic-rich soil tends to be gray or black, organic-poor and calcite-rich soil is whitish, and iron-rich soil is red or yellow.

Soil scientists worldwide have struggled mightily to develop a rational scheme for *soil classification*. Not all schemes use the same criteria, and even today, scientists do not agree on which classification system works best. In the United States, a country that includes many climates at mid-latitudes, soil scientists commonly use the U.S. Comprehensive Soil Classification System, which distinguishes among 12 orders of soil based on physical characteristics and the environment of soil formation (**Table B.3** and **Fig. B.13**). Canadians use a different scheme focusing only on soils that develop in cooler, high-latitude climates.

As we've seen, rainfall and vegetation play a key role in determining the type of soil that forms. For example, in deserts, which host very little rainfall and sparse vegetation, an *aridisol* forms (**Fig. B.14a**). (In older classifications, aridisols were known as pedocal soils.) Aridisols have no O-horizon (due to the lack of organic material), and only a

thin A-horizon. Soluble minerals, specifically calcite, that would be washed away entirely if there were more rainfall, instead accumulate in the B-horizon of an aridosol. In fact, capillary action may bring calcite up from deeper down as water evaporates at the ground surface. Calcite cements clasts together in the B-horizon to form a rock-like mass called *caliche* or *calcrete*.

In temperate environments, an *alfisol* forms. (In older classifications, alfisols were known as pedalfer soils.) This soil has an O-horizon, and because of moderate amounts of rainfall, materials leached from the A-horizon accumulate in the B-horizon (**Fig. B.14b**).

In a tropical climate, *oxisols* develop. Here, so much rainfall percolates down into the ground that all reactive minerals in the soil undergo chemical weathering, producing ions and clay that flush downward. This process leaves an A-horizon that contains substantial amounts of stable iron-oxide, aluminum-oxide, and aluminum-hydroxide residues (**Fig. B.14c**). The resulting soil tends to be brick red. So much water flushes down through some oxisols that a B-horizon cannot develop. Geologists use the term **laterite** for a type of oxisol that tends to be very rich in iron oxides and can be hard enough to be broken into blocks that can be used for construction—in fact, the name comes from the Latin word *later*, which means brick (**Fig. B.15**). Typically, laterite forms in regions that have a distinct dry season, during which capillary action brings additional oxide minerals back up into the A-horizon, where they precipitate.

Soil Destruction

As we have seen, soils take time to form, so soils capable of supporting crops or forests should be perceived as a natural resource worthy of protection (**Fig. B.16**). However, human activities such as agriculture, overgrazing, and clear-cutting have led to the destruction of soil. This destruction results from two processes: nutrient removal and soil erosion.

Nutrient Removal In a natural ecosystem, plants remove *nutrients* (specific chemicals, such as nitrogen and phosphorus, necessary for life) from soil, but when the plants die and transform into humus, the nutrients return to the soil. Agriculture, by its nature, involves harvesting (removal) of organic material. The nutrients incorporated in the crop, therefore, do not seep back into the soil. As a result, a few seasons of farming may decrease the concentration of nutrients in the soil sufficiently to inhibit the growth of successive crops. Farmers can overcome this problem by fertilizing the soil—fertilizer containing the missing nutrients can be mixed into the soil to add the nutrients back. But fertilizer can be expensive, so it adds significantly to the cost of food production.

FIGURE B.13 U.S. Department of Agriculture map of soil types around the world.

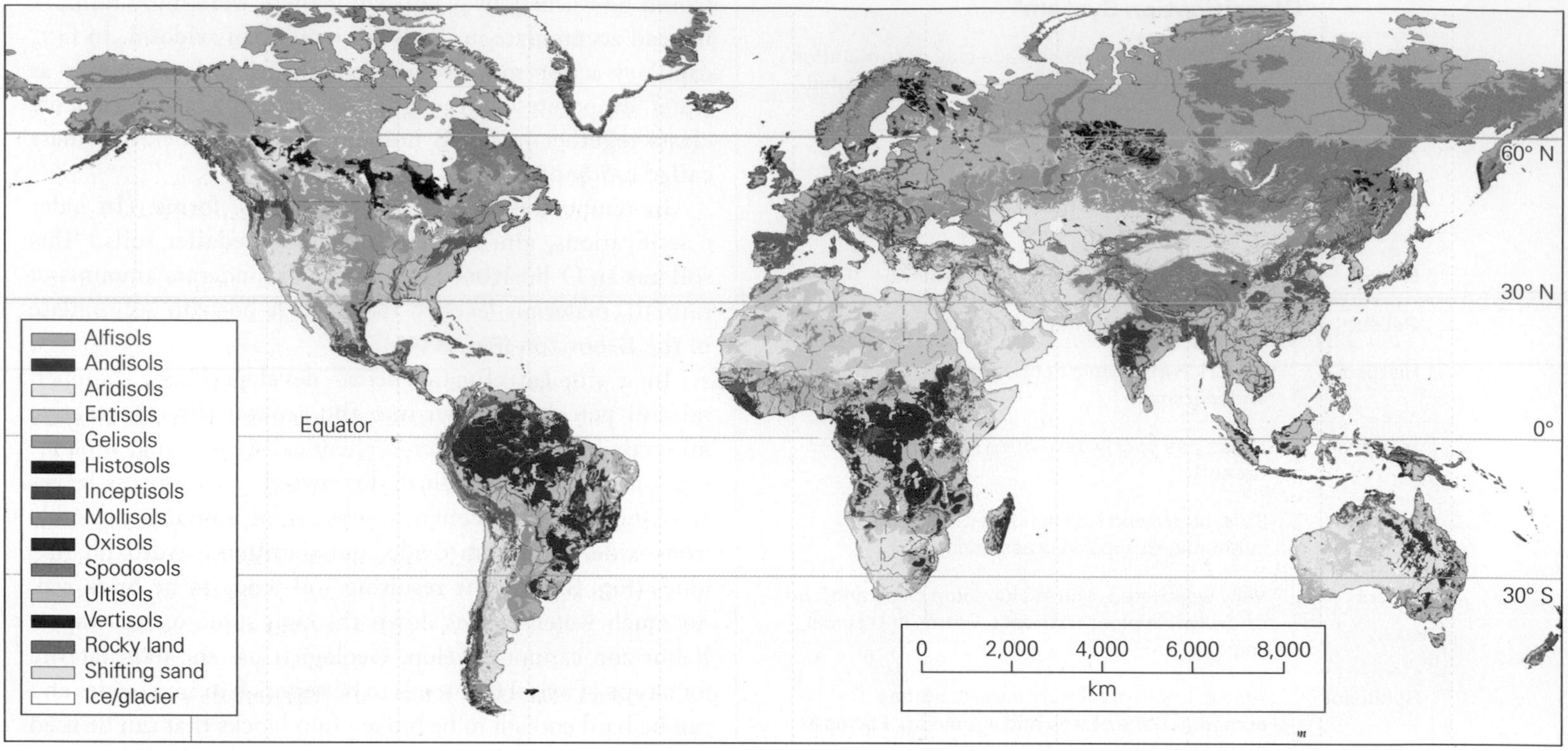

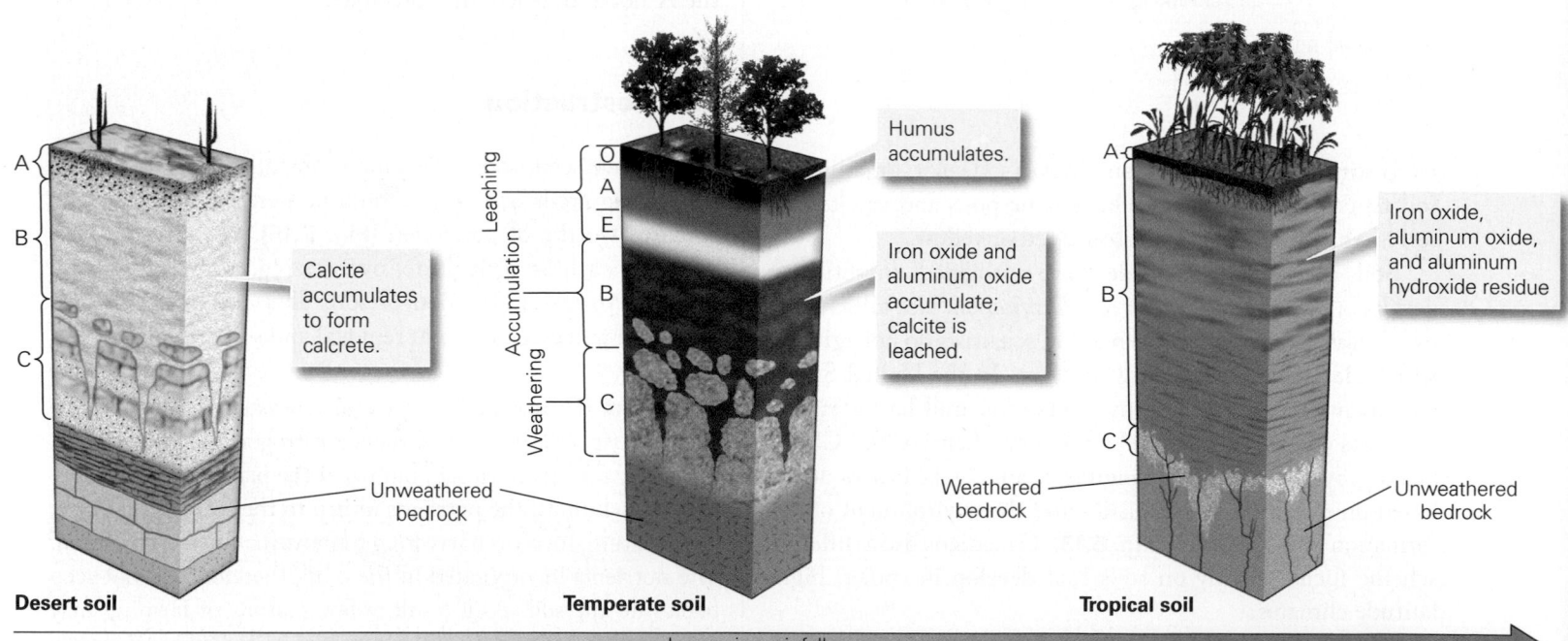

Alfisols
Andisols
Aridisols
Entisols
Gelisols
Histosols
Inceptisols
Mollisols
Oxisols
Spodosols
Ultisols
Vertisols
Rocky land
Shifting sand
Ice/glacier

60° N

30° N

Equator

0°

30° S

0   2,000   4,000   6,000   8,000
km

FIGURE B.14 Examples of soil classication.

Desert soil

Calcite accumulates to form calcrete.

Unweathered bedrock

Humus accumulates.

Leaching

Accumulation

Weathering

Iron oxide and aluminum oxide accumulate; calcite is leached.

Temperate soil

Unweathered bedrock

Tropical soil

Iron oxide, aluminum oxide, and aluminum hydroxide residue

Weathered bedrock

Unweathered bedrock

Increasing rainfall

(a) Aridisol forms in deserts. Rainfall is so low that no O-horizon forms, and soluble minerals accumulate in the B-horizon.

(b) Alfisol forms in temperate climates. An O-horizon forms, and less soluble materials accumulate in the B-horizon.

(c) Oxisol forms in tropical climates where percolating rainwater leaches all soluble minerals, leaving only iron- and aluminum-rich residues.

FIGURE B.15 A laterite in Brazil.

FIGURE B.16 Soils provide a substrate for the growth of vegetation.

The impact of rainforest destruction on soil nutrients can be particularly profound. In an established rainforest, lush growth provides sufficient organic debris so that trees can grow. But if the forest is logged or cleared for agriculture, the humus rapidly disappears, leaving laterite, which contains few nutrients. Crop plants consume the remaining nutrients so rapidly that the soil becomes infertile after only a year or two, useless for agriculture and unsuitable for immediate regrowth of rainforest trees.

Soil Erosion When natural plant cover disappears, as happens during deforestation or grassland destruction, the surface of the soil becomes exposed to wind and water. Forces such as the impact of falling raindrops or the rasping of a plow can break up the soil at the surface, which can then wash away in water or blow away as dust. When this happens, **soil erosion**, the removal of soil by wind or by running water, takes place **(Fig. B.17)**. Soil erosion can remove up to 6 tons of soil from an acre of land in a year. Human activities can increase rates of soil erosion by 10 to 100 times, so that the rate of erosion far exceeds the rate of soil formation. When soil erosion becomes severe, once-clear streams flowing through the affected area turn brown because of all the sediment that they carry.

(a) This soil in a temperate realm hosts a forest. It displays a good O-horizon.

(b) Recently tilled soil of an Illinois farm field.

FIGURE B.17 Examples of soil erosion.

(a) Farm fields, after harvest, are like deserts in that they have no plant cover. Plowing a dry field sends clouds of soil into the air as dust.

(b) When vegetation doesn't protect soil, water erosion carves a "badlands landscape" of closely spaced gullies and carries the soil away.

Droughts can cause rates of soil erosion to increase. For example, during the 1930s, a succession of droughts killed off so much vegetation in the American plains that wind stripped the land of soil and caused devastating dust storms. Large numbers of people were forced to migrate away from the Dust Bowl of Oklahoma and adjacent areas.

Nutrient removal and soil erosion are only two of the problems that face society. The overuse of fertilizers,

pesticides, and herbicides, as well as spills of a great variety of toxic chemicals, causes *soil contamination*. And too much irrigation in arid climates can make soils too saline for plant growth, for irrigation water contains trace amounts of salts that get left behind when irrigation water evaporates. Fortunately, people have begun to understand the fragility of soil and have been working on ways to promote *soil conservation*.

ANOTHER VIEW Over time, the rock forming this steep cliff in the Canadian Rockies breaks up and collects to form a pile of sediment downslope. If left undisturbed for long enough, the sediment starts to transform into soil, capable of supporting plant growth.

Interlude B Review

SUMMARY

- Sediment consists of loose fragments derived from pre-existing rock, precipitated from water, or formed by the breakup of shells.

- Rock at or near the Earth's surface weathers over time. Physical weathering breaks larger rock bodies into smaller pieces. Chemical weathering involves a variety of chemical reactions that dissolve and/or alter minerals.

- Soil develops over time when water percolates down through detritus or weathering bedrock. Soil can be modified by interactions with organisms. The character of soil depends on the composition of its source material, as well as on climate and time.

- Soil serves as an essential resource to society, for it provides the basis for agriculture. Various phenomena, some caused by humans, can lead to the loss of soil.

GUIDE TERMS

chemical weathering (p. 195)	joint (p. 194)	salt wedging (p. 194)	subsoil (p. 203)
clast (p. 193)	laterite (p. 205)	sediment (p. 192)	talus (p. 194)
dissolution (p. 195)	loam (p. 204)	soil (p. 202)	topsoil (p. 203)
exhumation (p. 194)	physical weathering (p. 193)	soil erosion (p. 207)	weathering (p. 193)
frost wedging (p. 194)	regolith (p. 192)	soil horizon (p. 203)	zone of accumulation (p. 202)
humus (p. 203)	root wedging (p. 194)	soil profile (p. 203)	zone of leaching (p. 202)

REVIEW QUESTIONS

The letters following each Review Question refer to the corresponding Learning Objective from the Chapter Opener.

1. Explain the difference between physical and chemical weathering. **(A)**

2. What processes can cause originally solid rock to break into pieces? **(B)**

3. What are the various reactions that can contribute to chemical weathering? **(A)**

4. Why doesn't weathering take place on the Moon? **(A)**

5. Explain the process of soil formation. **(C)**

6. Why do soils develop distinct horizons? Label the horizons on the figure. **(C)**

7. What factors determine the character (e.g., thickness, texture, types of horizons, etc.) of a soil? **(D)**

8. How does a soil that forms in a tropical climate differ from one that forms in an arid climate? **(D)**

9. Explain why soil erosion has been exacerbated by human activity. **(D)**

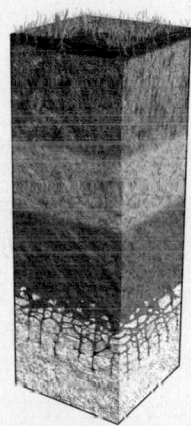

ONLINE RESOURCES

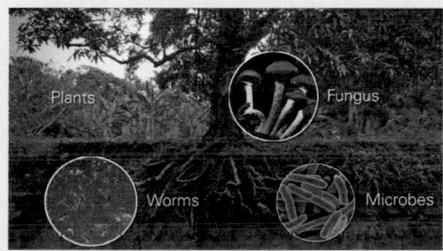

Videos
This interlude features a video on how soil is formed and the role of organisms in this process.

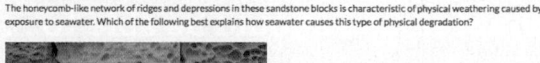
The honeycomb-like network of ridges and depressions in these sandstone blocks is characteristic of physical weathering caused by exposure to seawater. Which of the following best explains how seawater causes this type of physical degradation?

Smartwork5
This interlude features questions on topics including physical weathering, chemical weathering, and soil forming processes.

Pages of the Earth's Past: Sedimentary Rocks

By the end of this chapter, you should be able to . . .

A. distinguish among various classes of sedimentary rocks.

B. explain how clastic sedimentary rocks form, and recognize and name major types.

C. describe the role of life in the production of rocks such as limestone and coal.

D. produce a model to illustrate how the layering (bedding) in sedimentary rock forms.

E. recognize the shapes and textures preserved in sedimentary rocks that reflect depositional environments.

F. discuss why thick accumulations of sedimentary rock can be found only in certain locations.

> In every grain of sand there is a story of Earth.
>
> —Rachel Carson (American conservationist, 1907–1964)

7.1 Introduction

In this day when Google Earth can take you to every nook and cranny of our planet's surface at the touch of a phone screen or computer mouse, it's hard to imagine a world in which vast regions were blanks on a map. But only a little over a century ago, that was the state of affairs that members of the 1910–1913 British Antarctic Expedition were seeking to change. Led by Robert Falcon Scott, a team of explorers from the expedition set out to be the first to reach the South Pole. The early part of the journey took them over the Ross Ice Shelf, a broad plain of ice not far above sea level. But to reach the pole, they had to haul their heavy sledges up the Beardmore Glacier, a river of ice that had cut its way down through the rugged Transantarctic Mountains **(Fig. 7.1a)**, for the South Pole lies on the Polar Plateau, at an elevation of about 3 km, on the other side of the mountains. The cliffs overlooking the glacier, like those along much of the length of the Transantarctic Mountains, expose layer upon layer of a light-colored grainy rock **(Fig. 7.1b)**. One of the expedition's members, Edward Wilson, a physician and naturalist who served as the team's geologist, collected numerous specimens of these rocks.

Scott, Wilson, and the others succeeded in reaching the South Pole on January 17, 1912. But when they arrived, they found, to their profound disappointment, that the Norwegian explorer Roald Amundsen had beaten them there by 34 days. On their return journey, all of the British explorers perished in the blizzards and cold of the southernmost continent. When rescuers eventually came upon Scott's last campsite, they found Wilson's specimens, some of which contained fossils of *Glossopteris*, the plant whose distribution Alfred Wegener would use as evidence of continental drift.

What are the grainy, layered rocks that Wilson collected? They are a type of sedimentary rock called sandstone. Formally defined, **sedimentary rock** is rock that forms at or near the surface of the Earth in one of several ways: by the cementing together of loose *clasts* (fragments or grains) that were produced by physical or chemical weathering of pre-existing rock, by the growth of shell masses or the cementing together of shells and shell fragments, by the accumulation and subsequent alteration of organic matter derived from living

◄ (facing page) Sedimentary strata exposed in Utah. The cliff at the top consists of sandstone beds, and the slope below consists of shale beds.

organisms, or by the precipitation of minerals directly from surface-water solutions. Layers, or *beds*, of sedimentary rock are like the pages of a book, recording tales of ancient events and ancient environments on the ever-changing face of the Earth. They occur only in the upper part of the crust, where they form a "cover" that buries the underlying "basement" of igneous and/or metamorphic rock **(Fig. 7.2)**.

In Interlude B, we introduced the concept of weathering and showed how it breaks solid rock down into ions and loose sediment grains. Now, in this chapter, we see how these materials can be buried and transformed into sedimentary rock. We also introduce various specific types of sedimentary rock and show how geologists use the study of sedimentary rock to characterize Earth System history. Finally, we discuss the special settings, called sedimentary basins, in which particularly thick successions of sedimentary rock accumulate.

7.2 Classes of Sedimentary Rocks

Geologists divide sedimentary rocks into four major classes, based on their mode of origin:

1. *Clastic sedimentary rock* consists of cemented-together clasts, solid fragments and grains broken off of pre-existing rocks (the word *clastic* comes from the Greek *klastos*, meaning broken).

2. *Biochemical sedimentary rock* consists of shells grown by organisms.

3. *Organic sedimentary rock* consists of carbon-rich relicts of plants or other organisms.

4. *Chemical sedimentary rock* consists of minerals that precipitated directly from surface-water solutions.

Geologists sometimes also use various adjectives to characterize sedimentary-rock composition: *siliceous rocks* contain mostly quartz, *argillaceous rocks* contain mostly clay minerals, and *carbonate rocks* contain mostly calcite and/or dolomite. By some estimates, 70% to 85% of all the sedimentary rocks on the Earth are siliceous or argillaceous clastic rocks, and 15% to 25% are carbonate biochemical or chemical rocks. Other kinds of sedimentary rocks occur only in minor quantities.

Let's now look at the four major classes in more detail. For each class, we'll explain how the rocks form and introduce the names of common examples.

FIGURE 7.1 Robert Falcon Scott and his companions, including naturalist Edward Wilson, traversed the Transantarctic Mountains in 1912 to reach the Polar Plateau. Cliffs along the mountains expose sedimentary rocks.

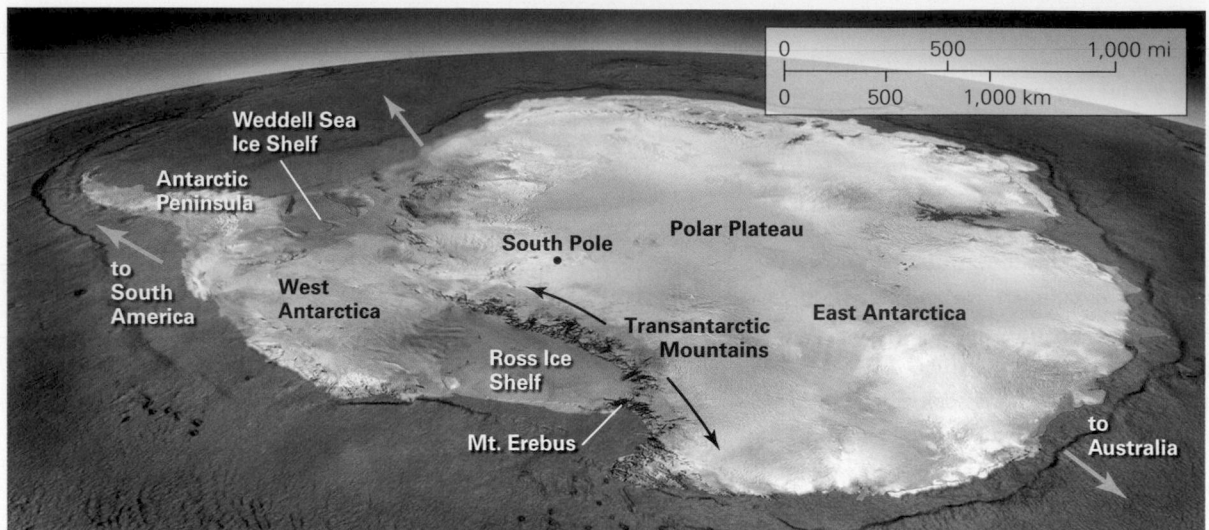

(a) Scott started near Mt. Erebus and crossed the Ross Ice Shelf and the Transantarctic Mountains to reach the South Pole. Roald Amundson beat him there.

(b) The white rock of these high cliffs in the Antarctic Dry Valleys is sedimentary. The black rock is a sill of basalt. White lines highlight the base and top of the sill.

Clastic Sedimentary Rocks

Formation Nine hundred years ago, a thriving community of Native Americans inhabited the high plateau of Mesa Verde, Colorado. In hollows beneath huge overhanging ledges, they built multistory stone-block buildings that have survived to this day **(Fig. 7.3)**. Clearly, the blocks are solid and durable—they are, after all, rock. But if you were to rub your thumb along one, it would feel gritty, and small grains of quartz would break free and roll under your thumb, because the block consists of quartz sand grains cemented together. Geologists call the rock in this block **sandstone**.

Sandstone serves as an example of a clastic sedimentary rock (also known as a *detrital sedimentary rock*) because it consists of loose clasts (detritus) that have been stuck together to form a solid mass. The clasts, or *grains*, can consist of individual mineral fragments, such as grains of quartz or flakes of clay minerals, or of chunks of rock, such as pebbles of granite. Production of a clastic sedimentary rock involves five steps **(Fig. 7.4a)**:

- *Weathering*: The clasts from which clastic rocks form come from the disintegration of pre-existing bedrock into separate grains due to physical and chemical weathering. The dissolved ions that eventually precipitate as new minerals to hold the grains together in sedimentary rocks come from chemical weathering.

- *Erosion*: Once formed, clasts produced by weathering do not stay in place forever. Gravity may cause them to fall off an outcrop, or they may be removed from the outcrop by **erosion**—the process by which moving water, ice, or air separates clasts from their substrate and carries them away.

- *Transportation*: Once produced by weathering, and removed by erosion, clasts and dissolved ions can be carried away in a *transporting medium* (wind, water, or ice). The ability of a medium to carry sediment depends on the medium's viscosity and velocity. Solid ice, for example, can transport clasts of any size, regardless of how slowly the ice moves. Very fast-moving, turbulent water can transport very coarse clasts (cobbles and boulders), moderately fast-moving water can carry only

FIGURE 7.2 Layers of sedimentary rock form a blanket that covers basement rock.

(a) During the formation of the Grand Canyon, erosion cut through the sedimentary cover into the basement.

(b) The contact between the sedimentary cover and the underlying basement lies at the top of the inner gorge.

sand and gravel, and slowly moving water carries only silt and mud. Strong winds can move sand and dust, but gentle breezes carry only dust.

- *Deposition:* A transporting medium does not carry sediment forever. Eventually, the sediment undergoes **deposition**, the process by which sediment settles out of the medium **(Fig. 7.4b)**. Sediment drops out of air or moving water when these fluids slow, because as their velocity decreases, the fluids no longer have the ability to carry sediment. Sediment carried by ice accumulates when the ice melts.

- *Lithification:* Geologists refer to the last stage of clastic sedimentary rock production, namely, the transformation of loose sediment into solid rock, as **lithification**.

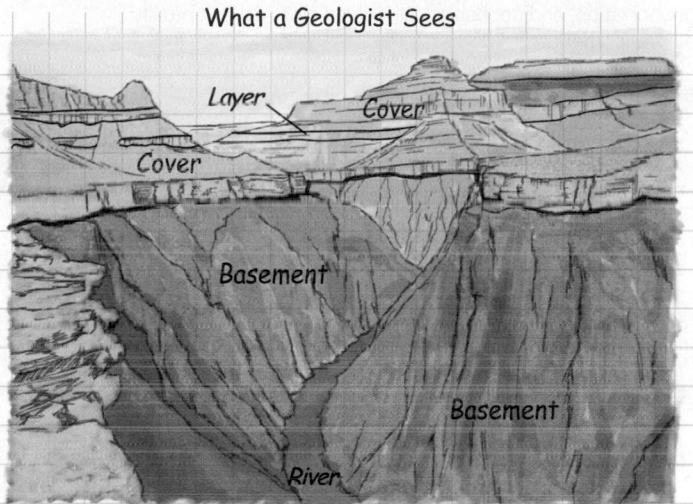

What a Geologist Sees

(c) A geologist's sketch emphasizes the contact. Here the basement consists of metamorphic and igneous rock.

FIGURE 7.3 The cliff dwellings nestled beneath a ledge at Mesa Verde, Colorado, are made of sandstone blocks. The inset shows the grainy character of the rock.

Sandstone layer

FIGURE 7.4 The five steps in clastic sedimentary rock formation.

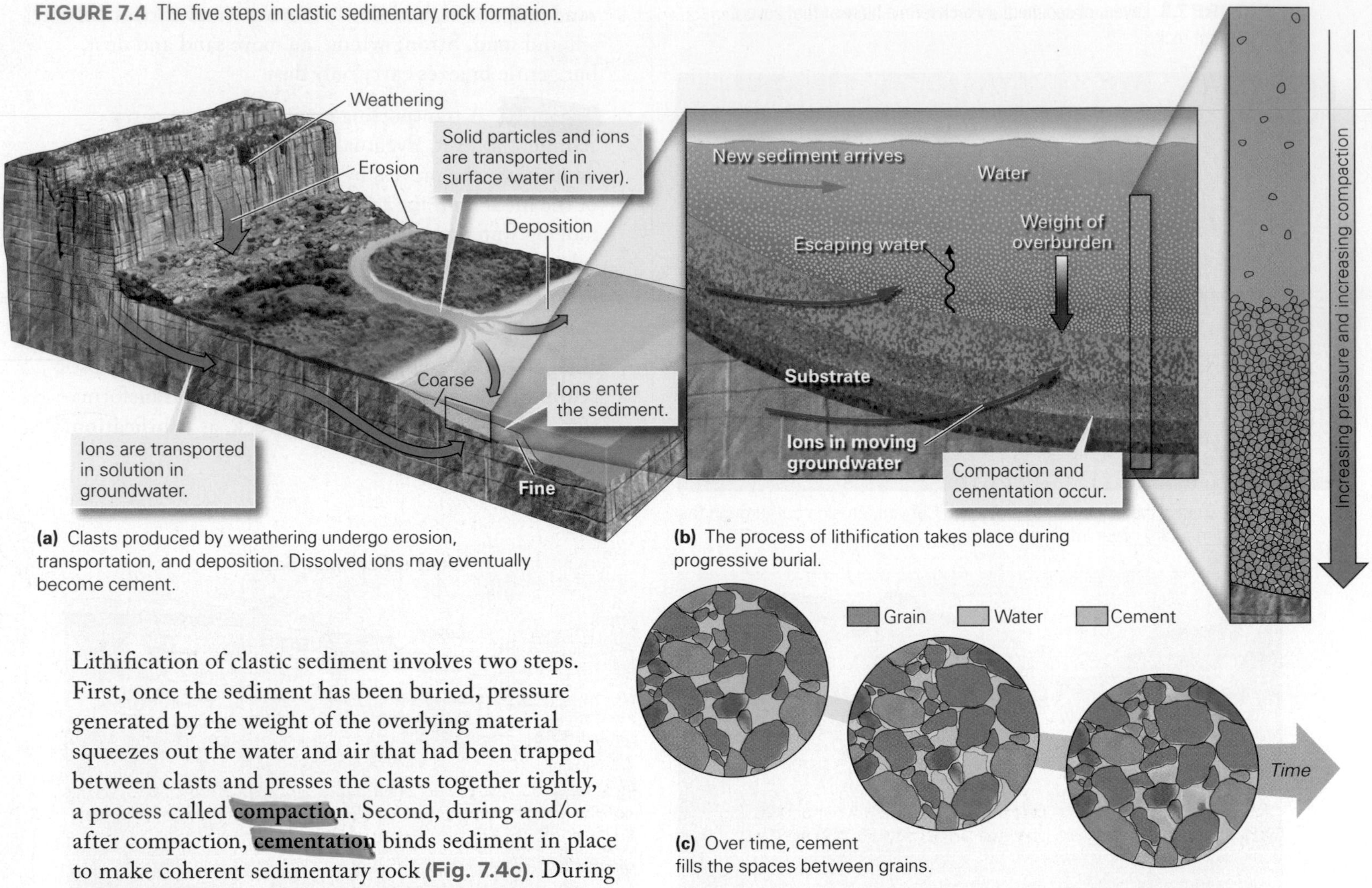

(a) Clasts produced by weathering undergo erosion, transportation, and deposition. Dissolved ions may eventually become cement.

(b) The process of lithification takes place during progressive burial.

(c) Over time, cement fills the spaces between grains.

Lithification of clastic sediment involves two steps. First, once the sediment has been buried, pressure generated by the weight of the overlying material squeezes out the water and air that had been trapped between clasts and presses the clasts together tightly, a process called **compaction**. Second, during and/or after compaction, **cementation** binds sediment in place to make coherent sedimentary rock **(Fig. 7.4c)**. During cementation, minerals (commonly quartz or calcite) precipitate from groundwater and fill spaces between clasts. The resulting **cement** acts like glue in that it holds grains together.

Classification Say that you pick up a clastic sedimentary rock and want to describe it sufficiently so that, from your words alone, another person can picture the rock. What characteristics should you mention? Geologists find the following characteristics most useful:

- *Clast size:* Geologists refer to the diameter of the grains making up a clastic sedimentary rock as the *clast size* or *grain size* **(Fig. 7.5a)**. Names used for clast size, listed in order from coarsest to finest, include boulder, cobble, pebble, sand, silt, and mud (see Table B.1 in Interlude B). Geologists informally use the term *gravel* for an accumulation of pebbles and cobbles and the term *mud* for an accumulation of wet clay and very fine silt. In this context, *clay* refers to all grains less than 0.004 mm in diameter—such grains consist mostly of clay minerals (several different types of sheet-silicate minerals) but also include tiny specks of quartz and other minerals.

- *Clast composition:* Not all clasts consist of the same type of mineral or rock. *Composition* refers to the makeup of clasts in sedimentary rock. Larger clasts (pebbles or larger) typically include rock fragments, meaning that the clasts themselves are aggregates of many mineral grains, whereas smaller clasts (sand, silt, or clay) typically consist of individual minerals. In some cases, *lithic clasts*, chips of fine-grained rock, may be mixed in with sand grains. Some sedimentary rocks contain clasts of only one composition, but others contain a variety of different kinds of clasts.

- *Angularity and sphericity: Angularity* indicates the degree to which clasts have smooth or angular corners and edges. *Sphericity*, in contrast, refers to the degree to which a clast has the same dimensions in all directions, so that it resembles the shape of a sphere **(Fig. 7.5b)**.

- *Sorting:* Geologists refer to the degree to which the clasts in a rock all have the same size as *sorting* **(Fig. 7.5c)**. Well-sorted sediment consists entirely of clasts of the same size, whereas poorly sorted sediment contains a mixture of more than one grain size. If a sedimentary

FIGURE 7.5 Grain characteristics and their evolution with increasing transport and weathering.

Grain size

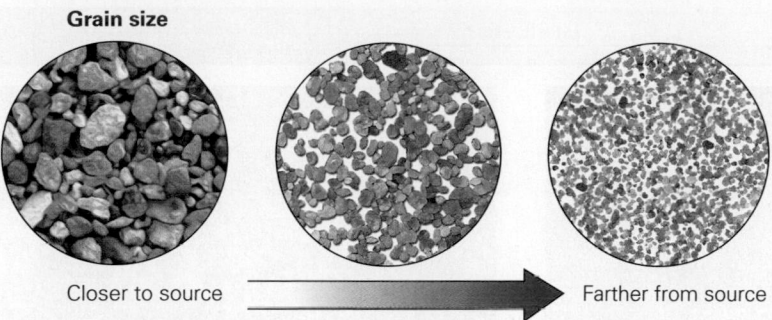

Closer to source →→→ Farther from source

(a) As the distance of transport of a sediment (by water or wind) increases, the sediment tends to become finer grained.

Sorting

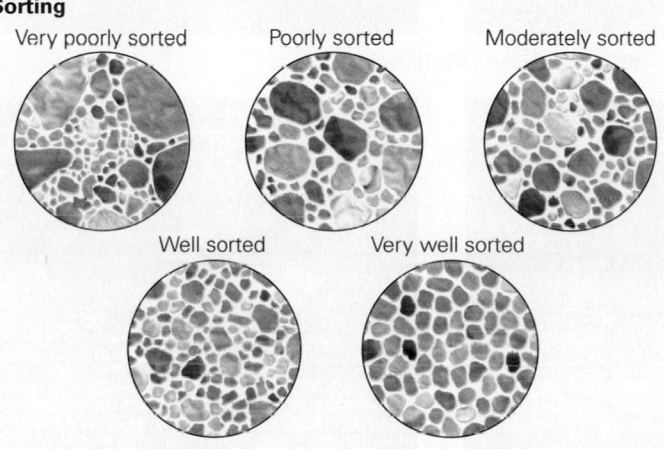

(c) If transport sifts grains, carrying smaller ones farther and leaving coarser ones behind, grains in a sediment tend to be the same size.

Angularity

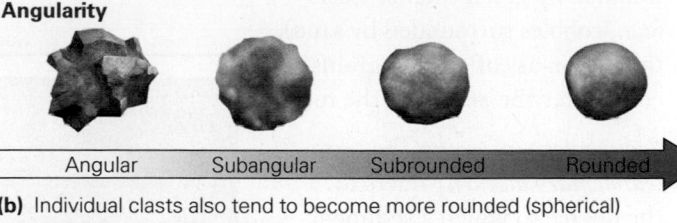

Angular   Subangular   Subrounded   Rounded

(b) Individual clasts also tend to become more rounded (spherical) and smoother.

Maturity

Feldspar weathers to clay; clay gets washed away.

Lithic clasts break into individual grains.

Lithic clast • Silt grain ⚬ Clay flakes

Quartz sand grain   Feldspar

Less mature →→→ More mature

(d) A less mature sediment consists of fragments of the original rock and contains both resistant and nonresistant minerals. A mature sediment contains only well-sorted resistant minerals.

TABLE 7.1 Classification of Clastic Sedimentary Rocks

Clast Size	Clast Character	Rock Name (Alternate Name)
Coarse to very coarse (> 2 mm)	Rounded pebbles and cobbles	**Conglomerate**
	Angular clasts	**Breccia**
	Large clasts in muddy matrix	**Diamictite**
Medium to coarse (0.07–2 mm)	Sand-sized grains	**Sandstone**
	• Quartz grains only	• Quartz sandstone (quartz arenite)
	• Quartz and feldspar sand	• Arkose
	• Sand-sized rock fragments	• Lithic sandstone
	• Sand and rock fragments in a clay-rich matrix	• Wacke (informally called graywacke)
Fine (0.004–0.06 mm)	Silt-sized clasts	**Siltstone**
Very fine (< 0.004 mm)	Clay and/or very fine silt	**Shale**, if it breaks into platy sheets
		Mudstone, if it doesn't break into platy sheets

rock contains larger clasts surrounded by much smaller clasts (e.g., cobbles surrounded by sand), then the mass of smaller grains constitutes the *matrix* of the rock.

- *Sedimentary maturity:* The term *sedimentary maturity* refers to the degree to which a sediment has evolved from being just a crushed-up version of its source rock into a well-sorted and well-rounded collection of clasts consisting only of the minerals that are most resistant to weathering **(Fig. 7.5d)**.

- *Character of cement:* Not all clastic sedimentary rocks contain the same kind of cement. In some, the cement consists predominantly of quartz, whereas in others it consists predominantly of calcite. Other kinds of mineral cements do occur, but they are rarer.

With these characteristics in mind, we can distinguish among several common types of clastic sedimentary rocks. **Table 7.1** provides common rock names—specialists sometimes use other, more precise names based on more complex classification schemes. Note that grain size serves as the primary basis for the classification of clastic rocks. Geologists further distinguish among kinds of sandstone (quartz sandstone, arkose, or wacke) based on clast composition or on sorting. We also distinguish between shale and mudstone based on the way in which the rock breaks (shale splits into thin sheets, whereas mudstone does not), and between breccia and conglomerate based on grain angularity.

Origins Characteristics of a clastic sedimentary rock provide clues to the source of the sediment and to the environment of deposition. To see how, let's follow the fate of rock fragments as they gradually move from a cliff face in the mountains via a river to the seashore. Different kinds of sediment develop along the route. Each kind, if buried and lithified, would yield a different type of sedimentary rock.

FIGURE 7.6 Different kinds of clasts lithify into different kinds of sedimentary rocks.

Sediment ——— Lithification ———▶ Sedimentary rock

(a) Lithification of an accumulation of angular clasts yields breccia.

(b) Layers of river gravel lithify into conglomerate.

Alluvial fan

(c) Sediment deposited in an alluvial fan, close to its source, can be feldspar rich. Lithification of this sediment yields arkose.

To start, imagine that large blocks of rock tumble off a cliff and slam into other blocks already at the bottom (see Fig. B.5c). The impact shatters the blocks, producing a pile of angular clasts with sharp edges. If these clasts were to be cemented together, the resulting rock would be **breccia (Fig. 7.6a)**. Later, a storm causes the clasts to slide downslope into a turbulent river. In the moving water, clasts bang into one another and into the riverbed, a process

Sediment ———— Lithification ————▶ Sedimentary rock

(d) Layers of beach or dune sand lithify into sandstone.

Sandstone

Shale

(e) Layers of mud, exposed beneath marsh grass, lithify to form shale. Here the thin-bedded shale is interbedded with sandstone.

(f) A rock formed from material containing large clasts surrounded by a fine-grained matrix is a diamictite.

that shatters them into still smaller pieces and breaks off their sharp edges. Angular clasts gradually become rounded clasts. When the river water slows, pebbles and cobbles stop moving and form a mound, or bar, of gravel. Burial and lithification of these rounded clasts produces **conglomerate (Fig. 7.6b)**.

If the gravel stays put for a long time, it undergoes chemical weathering. As a consequence, cobbles and

Did you ever wonder . . .
where beach sand comes from?

pebbles from the rock break apart into individual mineral grains. For example, disintegration of granite would yield a mixture of quartz, feldspar, and clay. Even slowly moving water can carry clay away, and leave behind sand containing a mixture of quartz and some feldspar grains. This sediment, if buried and lithified, becomes *arkose* **(Fig. 7.6c)**. Over time, feldspar grains in sand continue to weather into clay so that gradually, during successive events that wash the sediment downstream, the sand loses feldspar clasts and ends up being composed almost entirely of durable quartz sand grains. Some of this sand may make it to the sea, where waves carry it to beaches, and some may end up in desert dunes. Such sediment, when buried and lithified, becomes quartz sandstone **(Fig. 7.6d)**. Meanwhile, silt and clay may accumulate in the flat areas bordering streams, regions called *floodplains* that become submerged only during floods. And some silt and mud settles in a wedge, called a *delta*, at the mouth of the river, in *lagoons* (protected bodies of quiet water), or in *mudflats* (broad, quiet-water areas exposed at low tide) along the shore. The silt, when lithified, becomes **siltstone**, and the mud, when lithified, becomes **shale** or **mudstone (Fig. 7.6e)**. You may have sensed from our narrative that as sediment moves downstream, it becomes more mature.

Two of the rock names in Table 7.1 did not appear in the preceding narrative. *Diamicton* is a very poorly sorted sediment that contains cobbles or boulders surrounded by a matrix of sand, silt, and clay; when lithified, diamicton becomes *diamictite* (Fig. 7.6f). Research suggests that diamictites can form from the lithification of *debris flows* (viscous slurries consisting of mud mixed with larger clasts) both on land and underwater, or of *glacial till*, the debris left behind as glacial ice melts. *Wacke*, a poorly sorted sedimentary rock that consists of sand grains and lithic fragments suspended in a matrix of mud, forms from the deposits of submarine avalanches. Most wacke has a grayish color, so geologists informally refer to it as "graywacke."

Biochemical Sedimentary Rocks

Numerous organisms have evolved the ability to extract dissolved ions from seawater to make solid shells. Some of these organisms anchor to the seafloor, while others crawl or burrow on the seafloor or float in the water above. When these organisms die, the solid material in their shells survives, and this material, when lithified, makes up *biochemical sedimentary rock*. Geologists recognize several different types of biochemical sedimentary rocks, two of which we now describe.

Biochemical Limestone A snorkeler gliding above a coral reef sees an incredibly diverse community of corals and algae, around which creatures such as clams, oysters, snails (gastropods), and lamp shells (brachiopods) live, and above which plankton, including coccolithophores and forams, float (Fig. 7.7a). All of these organisms make solid shells out of calcium carbonate ($CaCO_3$), either as calcite or as its polymorph, aragonite. When the organisms die, the shells remain and may accumulate. Lithification of this sediment yields **biochemical limestone**, a type of *carbonate rock*.

Limestone comes in a variety of textures because the material that forms it has a range of grain sizes and accumulates in a variety of ways. For example, limestone can originate from reef builders (such as coral) that grew in place (Fig. 7.7b), from shell debris that was broken up and transported, or from carbonate mud consisting of plankton shells that settled like snow out of water. Because of this variety, we distinguish among *fossiliferous limestone*, containing visible fossil shells or shell fragments (Fig. 7.7c); *micrite*, consisting only of very fine carbonate mud (Fig. 7.7d); and *chalk*, consisting of plankton shells. (Experts recognize many other types as well and use a more precise, but complex, terminology for limestone classification.)

Typically, limestone occurs as a massive light-gray to dark-bluish-gray rock that breaks into chunky blocks—it doesn't look much like a pile of shell fragments (Fig. 7.7e). That's because several processes take place that change the texture of the rock over time. Prior to lithification, organisms may burrow into recently formed or deposited shells and break them up. Later, water passing through the rock not only precipitates cement but also dissolves some carbonate grains and causes new ones to grow. As a result, new crystals of calcite replace the original crystals. Typically, all the aragonite that was originally in shells transforms into calcite, a more stable mineral, and larger crystals of calcite replace smaller ones.

Biochemical Chert In several locations, along the west coast of California, you will find outcrops of reddish, almost porcelain-like rock occurring in 3- to 15-cm-thick layers (Fig. 7.8a). Hit it with a hammer, and the rock cracks, almost like glass, creating smooth, spoon-shaped (conchoidal) fractures. The rock consists of cryptocrystalline quartz (*crypto* is the Greek word for hidden), consisting of quartz grains that are too small to be seen without the extreme magnification of a scanning electron microscope. Such rock forms from plankton, such as radiolaria and diatoms, that produce shells composed of SiO_2 (silica), which accumulate along with clay to form a siliceous "ooze" on the seafloor. Gradually, the shells dissolve, forming silica solutions, which fill pores in the surrounding clay. Eventually, as tiny crystals grow from these solutions, the sediment lithifies to become rock. Geologists refer to this rock either as *biochemical chert*, to emphasize that it forms from the shells of organisms, or as *bedded chert*, to emphasize that it crops out as a succession of layers. Chert can come in a variety of colors, depending on the impurities that it contains. For example, chert containing traces of iron oxide tends to be red—such chert is also known as *jasper*.

Organic Sedimentary Rocks

We've seen how the mineral shells of organisms can accumulate and lithify to become biochemical sedimentary rocks. What happens to the "guts" of the organisms—the cellulose, fat, carbohydrate, protein, and other organic compounds that make up living tissue? Commonly, organic debris gets eaten by other organisms or decays at the Earth's surface. But in some environments, such as quiet, oxygen-poor water in swamps, lagoons, or lakes, the debris settles along with other sediment and eventually gets buried and preserved. At the elevated temperatures and pressures that exist at depth below the Earth's surface, organic matter undergoes chemical reactions that slowly transform it into *organic sedimentary rock*, which is distinct from other sedimentary rock in that it contains a high proportion of organic chemicals. Since the dawn of the industrial revolution in the early 19th century, organic sedimentary rock has provided the fuel for modern industry and transportation (see Chapter 14), because organic chemicals can burn to produce energy. We'll briefly consider two types of organic sedimentary rock: coal and oil shale.

FIGURE 7.7 The formation of carbonate rocks (limestone).

(a) In this modern coral reef, corals produce shells. If buried and preserved, these shells become limestone.

Relict of a small reef

(b) A Vermont quarry shows the gray color of 400-Ma limestone. The white mounds are relicts of small reefs.

(c) Fossil shells of ~415 Ma brachiopods protrude from an outcrop of limestone in New York.

(d) The grains of this micrite are almost too small to see. This outcrop contains very thin bedding.

(e) This road cut near Kingston, New York, exposes beds of limestone. The vertical stripes are drillholes.

Coal is a black, combustible rock containing between 40% and 90% carbon; the remainder consists of clay and quartz. The carbon of coal occurs in large, complicated molecules known as *macerals*—a typical maceral consists of about 85% carbon and 15% oxygen, nitrogen, and hydrogen. As discussed further in Chapter 14, coal forms from plant remains that have been buried deeply. Under the pressure and temperature conditions found at depth, the organic material of the plants becomes tightly compacted, and volatile molecules such as hydrogen, water, carbon dioxide, and ammonia break free and escape. As this happens, the carbon atoms reorganize into macerals **(Fig. 7.8b)**.

FIGURE 7.8 Examples of biochemical and organic sedimentary rocks.

Sandstone
and shale

Coal layer

(a) Chert is a biochemical sedimentary rock. This bedded chert developed on the deep seafloor by the deposition of plankton that secrete silica shells.

(b) Coal is an organic sedimentary rock. It is deposited in layers (beds), just like other kinds of sedimentary rocks.

Oil shale, like all shales, contains not only clay, but also between 15% and 30% organic material in a form called *kerogen*. The kerogen in oil shale comes from the fats and proteins that made up the living parts of plankton or algae. If, after dying, the tiny organisms settle in an environment where they do not immediately rot away or get eaten, they mix with clay minerals in mud. When the mud gets buried and lithified, the organic material undergoes chemical reactions that transform it into kerogen. The presence of organic material colors oil shale black.

Chemical Sedimentary Rocks

The colorful terraces, or mounds, that grow around the vents of hot-water springs; the immense layers of salt that underlie the floor of the Mediterranean Sea; the smooth, sharp point of an ancient arrowhead—these materials all have something in common. They all consist of rock formed primarily by the precipitation of minerals directly out of water solutions. We call such rocks *chemical sedimentary rocks*. They typically have a crystalline texture, formed partly during their original precipitation and partly when, at a later time, new crystals grow at the expense of old ones through the process of recrystallization. In some chemical sedimentary rocks, crystals are coarse, whereas in others, they are too small to see. Geologists distinguish among many types of chemical sedimentary rocks, primarily on the basis of composition.

Evaporites: Products of Saltwater Evaporation
In 1965, two daredevil drivers in jet-powered cars battled to be the first to surpass a speed of 600 mph on land.

On November 7, the *Green Monster* peaked at 576.127 mph. Eight days later, the *Spirit of America* reached 600.601 mph. Traveling at such speeds, a driver must maintain an absolutely straight line, for the slightest turn will catapult the vehicle out of control. Therefore, high-speed trials must take place on extremely long and flat racecourses. Not many places can provide such conditions—but the Bonneville Salt Flats of Utah do. The salt flats formed from the evaporation of an ancient salt lake. Under the heat of the Sun, the water turned to vapor and drifted up into the atmosphere, but the salt that had been dissolved in the water stayed behind. Such salt precipitation occurs wherever saltwater becomes saturated, as can happen in desert lakes with no outlet, or along the margins of restricted seas **(Fig. 7.9)**. For thick deposits of salt to form, large volumes of water must evaporate. This may happen when plate tectonic movements temporarily cut off arms of the sea (as we saw in the case of the Mediterranean Sea; see Interlude B), or during continental rifting, when seawater first begins to spill into the rift valley.

Because salt deposits form as a consequence of evaporation, geologists refer to them as **evaporites.** The specific type of salt mineral comprising an evaporite depends on the amount of evaporation. For example, when 80% of the seawater trapped in a basin evaporates, gypsum forms, and when 90% of the water evaporates, halite precipitates. If seawater were to evaporate entirely, the resulting evaporite would consist of 80% halite, 13% gypsum, and other salts and carbonates would make up the remainder.

Travertine (Chemical Limestone) Most limestone is
biochemical, in that it forms from the shells of organisms.

FIGURE 7.9 The formation of evaporite deposits.

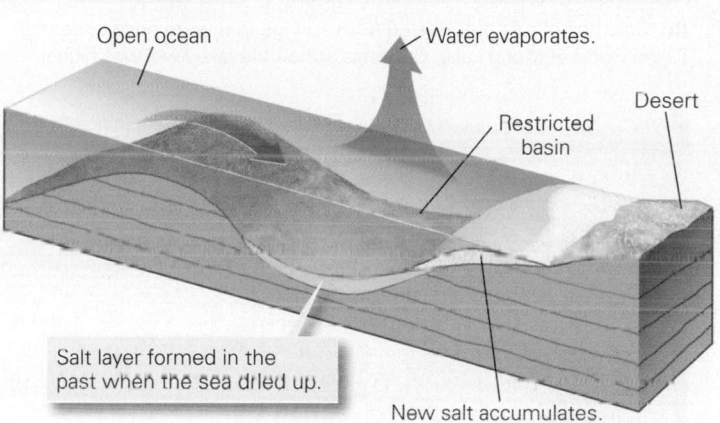

(a) In lakes with no outlet, tiny amounts of salt brought in by streams stay behind as the water evaporates. When the water evaporates entirely, a white crust of salt remains.

(b) Salt precipitation can also occur along the margins of a restricted marine basin if saltwater evaporates faster than it can be resupplied.

(c) Thick layers of salt may be buried deeply. Here salt is being mined deep underground.

But one type, called **travertine**, consists of crystalline calcium carbonate ($CaCO_3$) that precipitates directly from groundwater that has seeped out at the ground surface either in hot- or cold-water springs, or on the walls of caves. What causes this precipitation? It happens, in part, when the groundwater degasses, meaning that some of the carbon dioxide that had been dissolved in it bubbles out of solution, for removal of carbon dioxide decreases the ability of the water to hold dissolved carbonate. Precipitation also occurs when water evaporates, thereby increasing the concentration of carbonate. Recent research suggests that various kinds of microbes can accelerate the process of precipitation.

Travertine produced at springs forms terraces and mounds that can grow to be meters or even hundreds of meters thick **(Fig. 7.10a)**. Spectacular terraces of travertine occur at Mammoth Hot Springs in Yellowstone National Park. Amazing column-like mounds of travertine grew up from the floor of Mono Lake, California **(Fig. 7.10b)**, where hot springs seeped into the cold water of the lake. Travertine also grows on the walls of caves where groundwater seeps out **(Fig. 7.10c)**. In cave settings, travertine builds up beautiful and complex growth forms called *speleothems* (see Chapter 19).

Travertine has been quarried for millennia to make building stones and decorative stones **(Fig. 7.10d)**. The rock derives its beauty from its translucence (in thin slices) and from the colored growth bands it displays. The bands develop in response to changes in the composition of groundwater or in the environment into which the water drains. Some travertines (of a type called tufa) contain abundant large pores (open spaces).

Dolostone Not all carbonate rock consists of pure calcite. A variety of carbonate rock called *dolostone* differs from limestone in that it contains the mineral dolomite ($CaMg[CO_3]_2$). Where does the magnesium in dolomite come from? Dolomite is formed by a chemical reaction between solid calcite and magnesium-bearing groundwater. This change may take place beneath lagoons along a shore

FIGURE 7.10 Examples of travertine (chemical limestone) deposits.

Terrace of new travertine

2 m

(a) Travertine accumulates in terraces at Mammoth Hot Springs in Yellowstone National Park, Wyoming.

(b) Columns of tufa precipitated from springs that were once under the saline water of Mono Lake, California, when the lake level was higher.

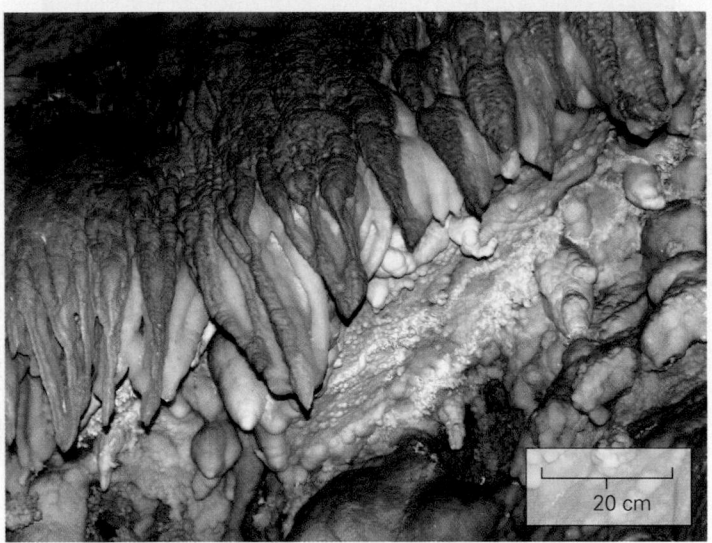

20 cm

(c) Travertine speleothems form as calcite-rich water drips from the ceiling of Timpanogos Cave in Utah.

(d) This slice of travertine forms a decorative panel for a building's interior wall.

soon after the limestone forms or, a long time later, after the limestone has been buried deeply.

Chemically Precipitated Chert A community of Native Americans, the Onondaga, once lived in eastern New York State. Here, outcrops of limestone contain *nodules* (small, rounded lumps or lenses) of black chert **(Fig. 7.11a, b)**. Because of the way it breaks, Onondaga artisans could fashion sharp-edged tools (arrowheads and scrapers) from this chert, or flints, so they collected it for their own toolmaking industry and for use in trade. Unlike the biochemical chert we described earlier, the chert collected

by the Onondaga didn't form from layers of siliceous plankton shells. Rather, the chert nodules grew when microscopic quartz crystals

Did you ever wonder . . .
how flint used for arrowheads first formed?

gradually precipitated and replaced calcite crystals within a bed of limestone, long after the limestone was originally deposited. Because of its mode of formation, geologists refer to this type of chert as *replacement chert*. Because replacement chert commonly occurs in nodules, it's also known as *nodular chert*.

FIGURE 7.11 Examples of chert that precipitated in place.

(a) Replacement chert forms as layers of nodules between tilted limestone beds in New York.

Labels: Layer of black chert; Tilted limestone beds

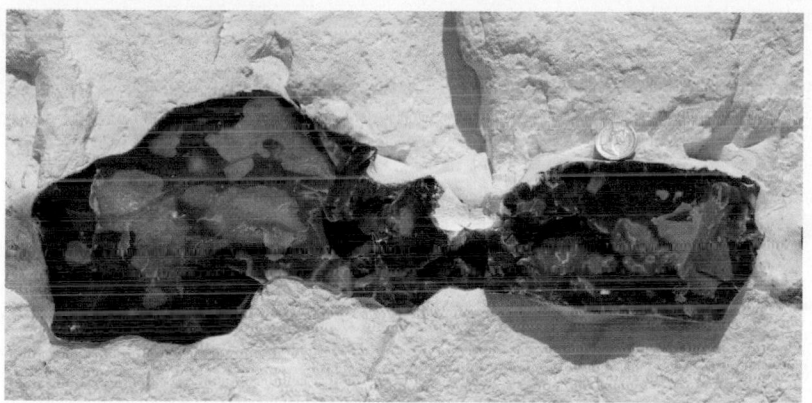

(b) Close-up of a black chert nodule surrounded by white chalk, exposed on a cliff in southern England. The coin provides scale.

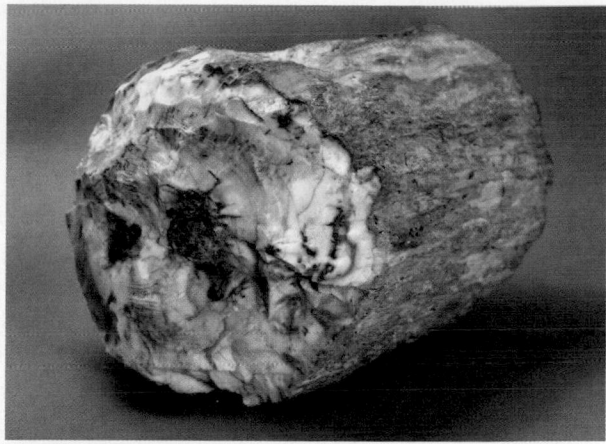

(c) This 14-cm-diameter log of petrified wood from Wyoming formed about 50 Ma.

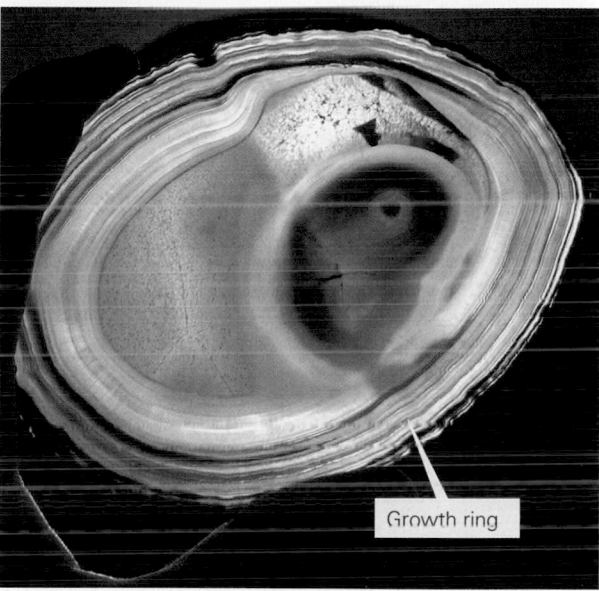

(d) A thin slice of Brazilian agate, lit from the back, shows growth rings.

Label: Growth ring

Replacement chert doesn't form only in limestone. Some grows in buried silica-rich volcanic ash beds when the silica in groundwater precipitates as microcrystalline quartz within wood that has been buried by the ash, forming *petrified wood* **(Fig. 7.11c)**. The quartz gradually replaces the interiors of the wood's cells, as the cell walls convert into carbon and other minerals. Because of this delicate replacement process, the chert retains the shape of the wood and the growth rings within it, so the resulting solid block of rock still looks like wood.

Not all chemically precipitated chert replaces other minerals. A type of chert, known as *agate*, precipitates in concentric rings inside open hollows in a rock. Commonly, the successive rings of chert have different colors, giving the rock a striped appearance **(Fig. 7.11d)**. These colors are caused by variations in the type and concentration of impurities.

TAKE-HOME MESSAGE

Geologists distinguish among many types of sedimentary rocks based on the mode of formation. Clastic sedimentary rocks consist of grains weathered and eroded from pre-existing rocks and transported by wind, water, or ice to a site of deposition where they were buried and lithified. Clastic rocks can be classified by grain size. Biochemical sedimentary rocks, such as limestone, consist of the shells of organisms, and organic sedimentary rocks, such as coal, form from the organic remains of organisms. Chemical sedimentary rocks precipitate from water solutions.

QUICK QUESTION: Do all sedimentary rocks have the same composition? Why or why not?

7.3 Sedimentary Structures

When you look at a large outcrop of sedimentary rock, you'll notice that the outcrop displays distinct layers. On the surfaces of layers you may see small ridges and/or indentations, and within the layers the grains may be sorted or oriented so as to define distinct textures. These features, which range from subtle to obvious, form during the process of deposition and, as we will see, provide key clues to the environment in which the sediments were deposited. Geologists use the term **sedimentary structure** for such features. Here we examine some of the more important types of sedimentary structures.

Bedding and Stratification

Let's start by introducing the terms geologists use for discussing sedimentary layers. A single layer of sediment or sedimentary rock with a recognizable top and bottom is called a **bed;** the boundary between two beds can be called a *bedding plane*. Several beds together constitute **strata** (singular *stratum*, from the Latin *stratum*, meaning pavement), and the overall division of sediment into a sequence of beds is **bedding**, or *stratification*.

When you examine strata in a region with good exposure, the bedding generally stands out clearly—it looks like bands or stripes across a cliff face **(Fig. 7.12a)**. Typically, contrasts in rock type distinguish one bed from adjacent beds. Each bed has a definable thickness (from a couple of centimeters to tens of meters) and may contrast with its neighbors in composition, color, or grain size. For example, if a sequence of strata contains a bed of sandstone overlain by a bed of shale and overlain in turn by a bed of siltstone, the surfaces separating one rock type from another define bedding. In many cases, however, adjacent beds all have the same overall composition, and bedding may be represented by subtle changes in grain size, by surfaces developed as a consequence of interruptions in deposition, or by cracks that formed parallel to bedding planes.

Why does bedding form? To find the answer, we need to think about how sediment accumulates. Changes in climate, water depth, current velocity, or the sediment source control the type of sediment deposited at a location at a given time. For example, on a normal day, a slow-moving river may carry only silt, which collects on the riverbed **(Fig. 7.12b, c)**. During a flood, the river flows faster and carries sand and pebbles, so a layer of sandy gravel forms over the silt layer. Then, when the flooding stops, more silt buries the gravel. If this succession of sediments becomes lithified and exposed for you to see, the layers appear as alternating beds of siltstone and sandy conglomerate. Bedding is not always well preserved, however, for in some environments, burrowing organisms disrupt the layering. Specifically, worms, clams, and other creatures can churn sediment and leave behind burrows; this process is called *bioturbation*.

Over geologic time, long-term changes in a depositional environment may take place, so a given sequence of strata may differ markedly from sequences of strata above or below. A sequence of strata distinctive enough to be traced across a fairly large region is called a **stratigraphic formation**, or simply a *formation* **(Fig. 7.13a)**. For example, a region may contain a succession of alternating sandstone and shale beds deposited by rivers, overlain by beds of marine limestone deposited later when the region was submerged by the sea. A geologist might identify the sequence of sandstone and shale beds as one formation and the sequence of limestone beds as another. Formations may be named after the locality where they were first found and studied. A geologic map portrays the distribution of stratigraphic formations in a region **(Fig. 7.13b)**.

Ripple Marks, Dunes, and Cross Bedding: Consequences of Deposition in a Current

Many clastic sediments accumulate in moving fluids (wind, rivers, or waves). Fascinating sedimentary structures develop at the interface between the sediment and the fluid. These structures, known as *bedforms*, reflect factors such as the velocity of the flow and the size of the clasts. Though there are many types of bedforms, we'll focus on only two—ripple marks and dunes. The growth of both produces cross bedding, a special type of lamination within beds.

Ripple marks are relatively small (generally no more than a few centimeters high) elongated ridges that form on a bed surface at right angles to the direction of current flow. If the current always flows in the same direction, the ripple marks tend to be asymmetric, with a steeper slope on

SEE FOR YOURSELF . . .

Grand Canyon, Arizona

LATITUDE
36°8'8.94" N

LONGITUDE
112°15'48.56" W

Look down from
15 km.

Here you can see spectacular color banding caused by the succession of stratigraphic formations exposed on the canyon's walls. (Gray is limestone, tan is sandstone, and red is shale.)

FIGURE 7.12 Bedding in sedimentary rocks.

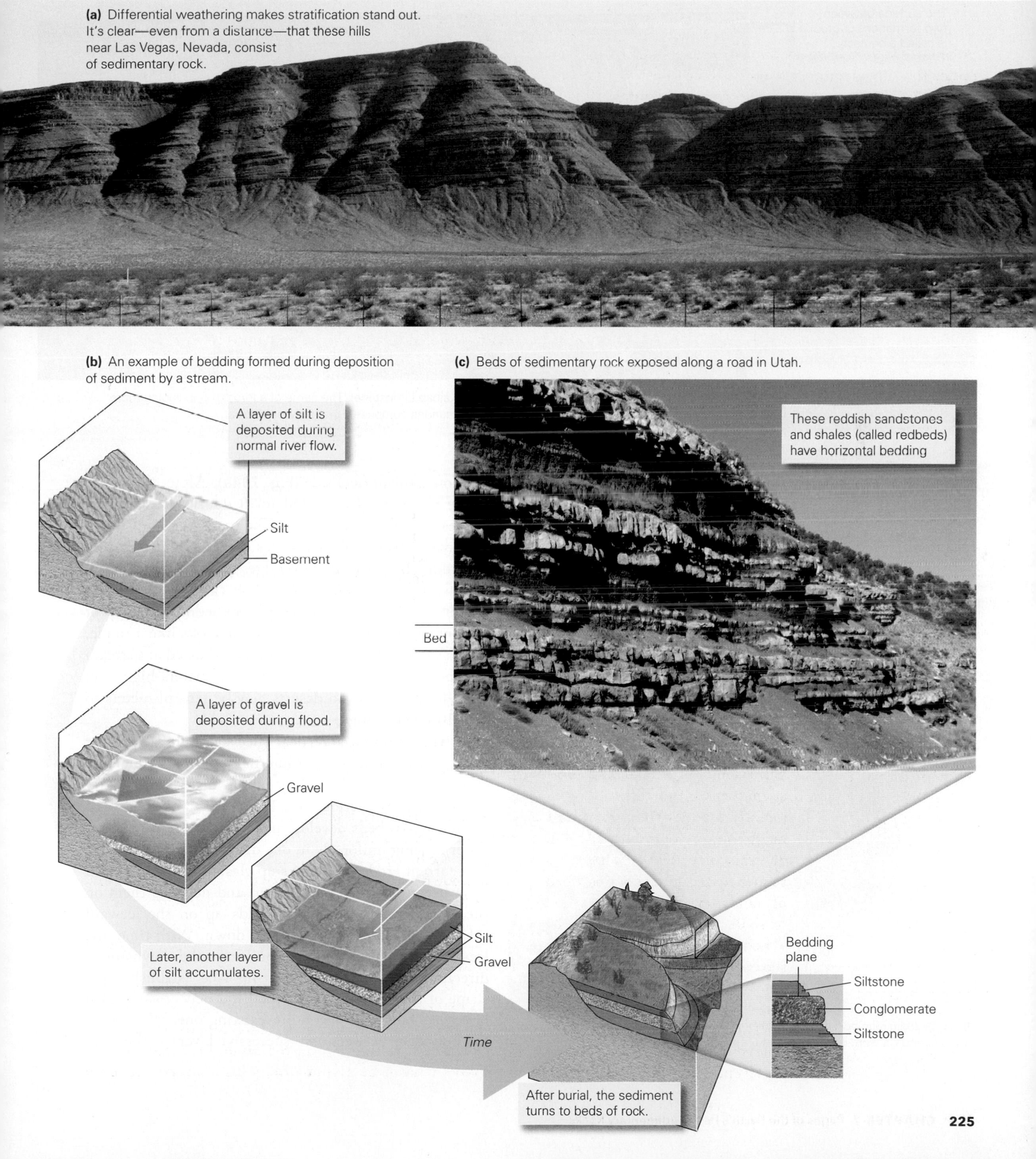

(a) Differential weathering makes stratification stand out. It's clear—even from a distance—that these hills near Las Vegas, Nevada, consist of sedimentary rock.

(b) An example of bedding formed during deposition of sediment by a stream.

A layer of silt is deposited during normal river flow.

Silt

Basement

A layer of gravel is deposited during flood.

Gravel

Later, another layer of silt accumulates.

Silt

Gravel

Time

After burial, the sediment turns to beds of rock.

(c) Beds of sedimentary rock exposed along a road in Utah.

These reddish sandstones and shales (called redbeds) have horizontal bedding

Bed

Bedding plane

Siltstone

Conglomerate

Siltstone

225

FIGURE 7.13 The concept of a stratigraphic formation as exemplified by the Grand Canyon.

The surface between two formations or groups is called a contact.

Kaibab Limestone
Toroweap Formation
Coconino Sandstone
Hermit Shale
Supai Group

Toroweap · Kaibab
Coconino
Hermit
Supai
Redwall
Muave
Bright Angel
Tapeats · Plateau on Tapeats
Vishnu

(a) The name of a formation consisting of one rock type may indicate the rock type (e.g., Kaibab Limestone). The name of a formation including more than one rock type includes the word *formation* (Toroweap Formation). Several related formations comprise a group (Supai Group).

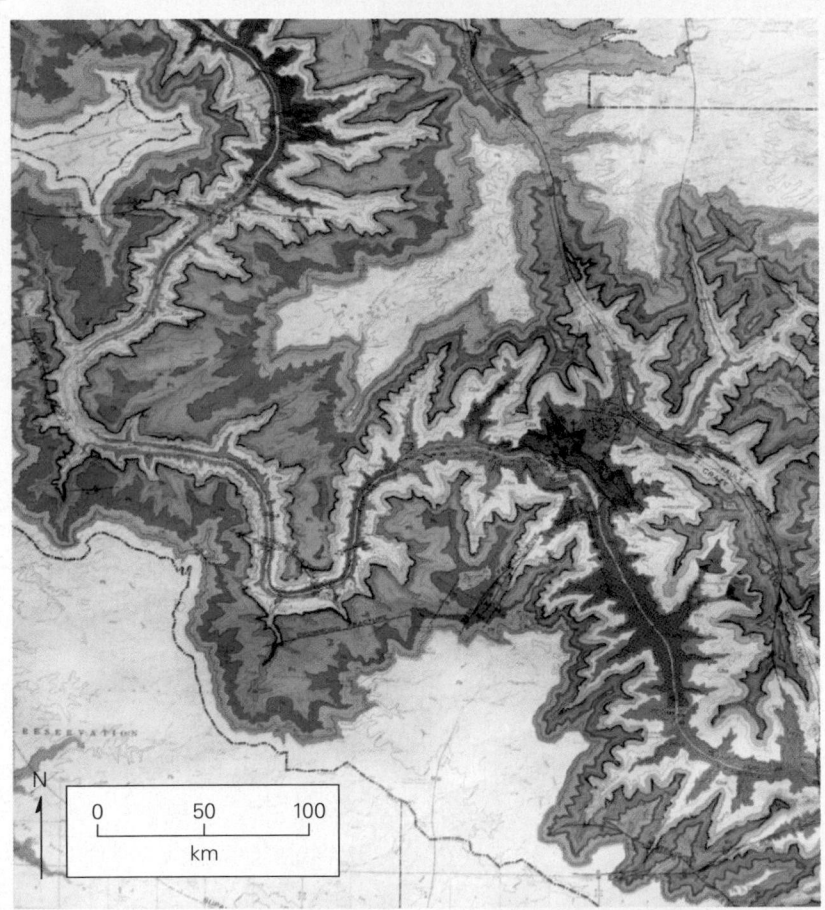

(b) A geologic map portrays the distribution of formations in a portion of the Grand Canyon. Each color band represents a specific formation.

N
0   50   100
km

the downstream (lee) side **(Fig. 7.14a)**. Along the shore, where water flows back and forth due to wave action, ripples tend to be symmetric. The crest (the high ridge) of a symmetric ripple delineates a sharp ridge, whereas the trough between adjacent ridges tends to be a smooth, concave-up curve **(Fig. 7.14b)**. You can find ripples on modern beaches and preserved on bedding planes of ancient rocks **(Fig. 7.14c, d)**. A **dune** looks like a ripple, only it's larger. For example, dunes on the bed of a stream may be tens of centimeters to several meters high, and wind-formed dunes in deserts may be tens of meters to over 100 meters high.

If you examine a vertical slice cut into a ripple or dune, you will find distinct internal laminations inclined at an angle to the boundary of the main sedimentary layer. Such laminations are known as **cross beds**. To see how cross beds develop, imagine a current of air or water moving uniformly in one direction **(Fig. 7.15a, b)**. The current erodes and picks up clasts from the upwind or upstream part of the bedform and deposits them on the leeward part. Sediment builds up on the leeward side until gravity causes it to slip down. With time, the leeward side of the bedform builds in the downstream direction. The surface of the slip face establishes the shape of the cross beds. Eventually, a new cross-bedded layer may build out over a pre-existing one. We refer to the boundary between two successive layers as the *main*

FIGURE 7.14 Ripple marks, a type of sedimentary structure, are visible on the surfaces of modern and ancient beds.

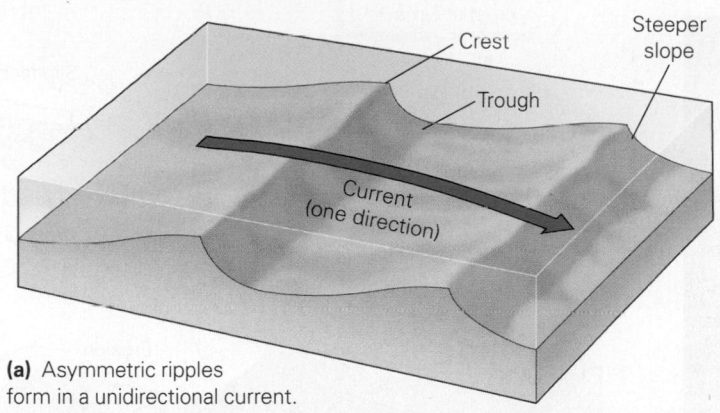

(a) Asymmetric ripples form in a unidirectional current.

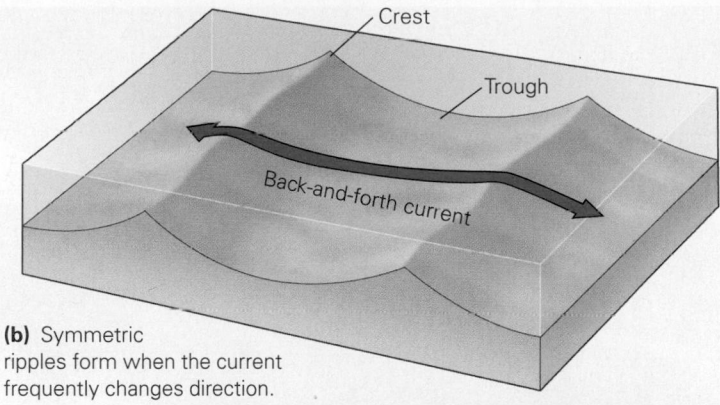

(b) Symmetric ripples form when the current frequently changes direction.

(c) Modern ripples exposed at low tide along a sandy beach on the shore of Cape Cod, Massachusetts.

(d) These 145-Ma ripples are preserved on a tilted bed of solid sandstone at Dinosaur Ridge, Colorado.

bedding, and the internal curving surfaces within the layer as *cross bedding* **(Fig. 7.15c, d)**.

Turbidity Currents and Graded Beds

Sediment deposited on a submarine slope might not stay in place forever. For example, an earthquake or storm might disturb this sediment and cause it to slip downslope, and mix with water to yield a murky, turbulent cloud. This cloud is denser than clear water, so it flows downslope like an underwater avalanche. Such a flow is known as a *turbidity current* **(Fig. 7.16)**. Downslope, the turbidity current slows, and the sediment that it carried starts to settle out. Larger grains sink faster through a fluid than do finer grains, so the coarsest sediment settles out first. Progressively finer grains accumulate on top, with the finest sediment (clay) settling out last. This process forms a *graded bed*—that is, a layer of sediment in which grain size varies from coarse at the bottom to fine at the top. Geologists refer to a deposit from a turbidity current as a *turbidite*.

Bed-Surface Markings

A number of features may appear on the surface of a bed as a consequence of events that happen during deposition or soon after. These bed-surface markings include the following:

- *Mud cracks:* If a mud layer dries up after deposition, it cracks into roughly hexagonal plates that typically curl up at their edges. We refer to the openings between the plates as **mud cracks**. If buried, mud cracks can be preserved **(Fig. 7.17a, b)**.

FIGURE 7.15 Cross bedding, a type of sedimentary structure within a bed.

(a) Large sand dunes formed in a strong wind.

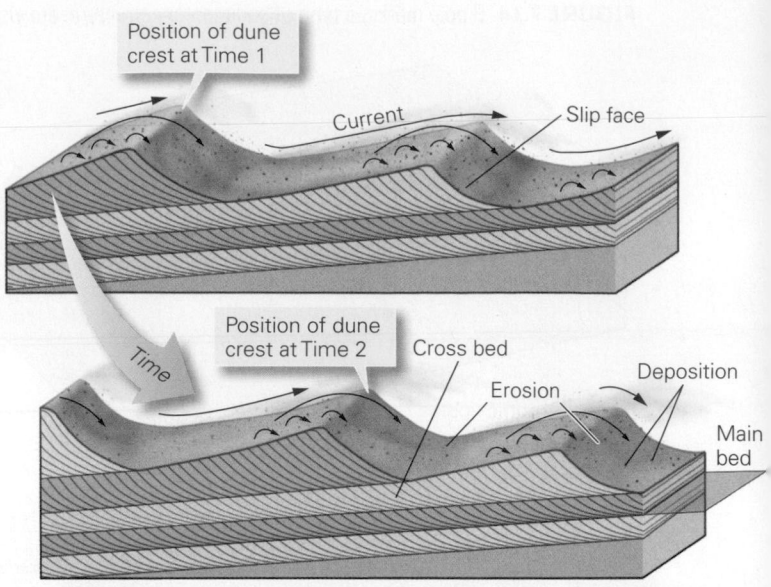

(b) Cross beds form as sand blows up the windward side of a dune or ripple and then accumulates on the slip face. With time, the dune crest moves.

(c) Slip face of a small sand dune, Death Valley. The edges of the slip face are highlighted.

(d) A cliff face in Zion National Park, Utah, displays large cross beds formed between 200 and 180 Ma, when the region was a desert with large sand dunes.

- *Scour marks:* As currents flow over a sediment surface, they may erode small troughs, called *scour marks*, parallel to the current flow. These indentations can be buried and preserved. **(Fig. 7.17c)**

- *Fossils:* Fossils are relics of past life. Some fossils are shell imprints, footprints, or feeding traces on a bedding surface (see Interlude E).

- *Paleosols:* If a soil that formed on the surface of a sedimentary bed becomes buried, it can be preserved as a **paleosol**. If lithified, a paleosol may delineate a bed surface because it has a different texture and,

in some cases, a different color than does the internal portion of the bed **(Fig. 7.17d)**.

Why Study Sedimentary Structures?

Sedimentary structures are not just a curiosity; they provide important clues that help geologists understand the environment in which clastic sedimentary beds were deposited (**Geology at a Glance**, pp. 232–233). For example, the presence of ripple marks and cross bedding indicates that layers were deposited in a current; the presence of mud cracks indicates that the sediment layer was exposed to the air and dried out on occasion; and

FIGURE 7.16 The deposition of graded bedding by turbidity currents.

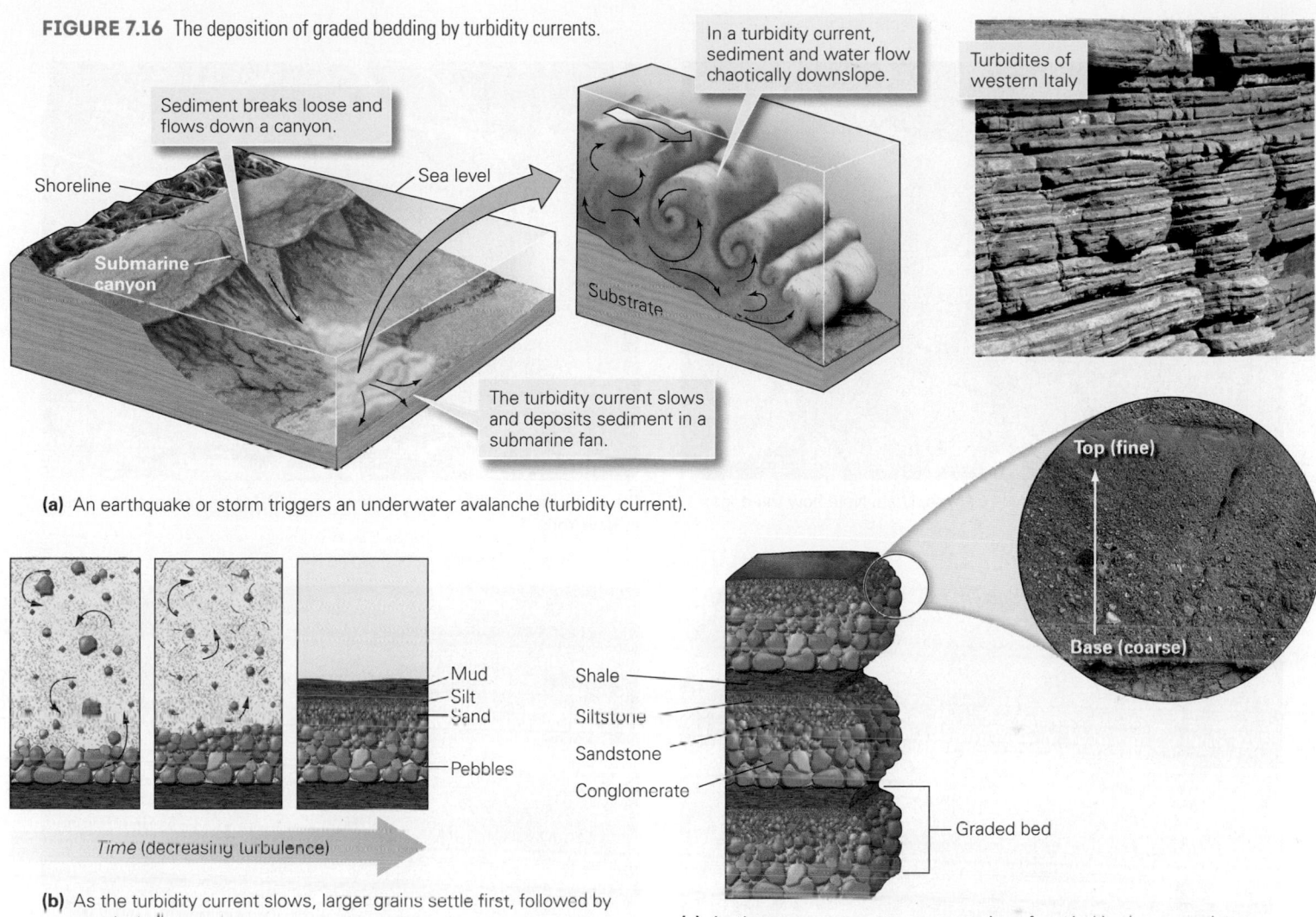

Sediment breaks loose and flows down a canyon.

Shoreline

Sea level

Submarine canyon

In a turbidity current, sediment and water flow chaotically downslope.

Substrate

Turbidites of western Italy

The turbidity current slows and deposits sediment in a submarine fan.

(a) An earthquake or storm triggers an underwater avalanche (turbidity current).

Mud
Silt
Sand
Pebbles

Time (decreasing turbulence)

(b) As the turbidity current slows, larger grains settle first, followed by progressively finer grains.

Top (fine)
Base (coarse)

Shale
Siltstone
Sandstone
Conglomerate

Graded bed

(c) As the process repeats, a succession of graded beds accumulates.

graded beds indicate deposition by turbidity currents. Fossil types can tell us whether sediment was deposited along a river or in the deep sea, for different species of organisms live in these different environments. In the next section of this chapter, we examine these environments in greater detail.

TAKE-HOME MESSAGE

Exposures of sedimentary rocks typically contain a variety of sedimentary structures, features or textures formed during deposition. Examples include bedding, ripple marks on bed surfaces, and cross bedding within beds. Features like these provide clues to the environment of deposition. For example, cross beds and dunes form only in a current.

QUICK QUESTION: What is the difference between a bed and a stratigraphic formation?

7.4 How Do We Recognize Depositional Environments?

Imagine that you could travel anywhere on the Earth and took the opportunity to visit a variety of natural settings where sediment accumulates. In each setting, if you looked closely, you'd see that the character of sediment being deposited was distinct. For example, the sediment carried by desert winds doesn't look like the sediment settling on the floor of a tropical lagoon. Geologists refer to the conditions in which sediment was deposited as a **depositional environment**. To identify the depositional environment in which ancient sedimentary strata accumulated, geologists, like detectives, look for clues. Detectives may seek fingerprints and bloodstains at

FIGURE 7.17 Bed-surface marking.

(a) Mud cracks in red mud at Bryce Canyon, Utah. Note how the edges of the mud plates curl up.

(b) Mud cracks visible on the surface of a 410-Ma bed exposed on a cliff in New York.

(c) The streaks and ridges on the base of this siltstone bed are casts of scour marks that had been carved into the underlying bed.

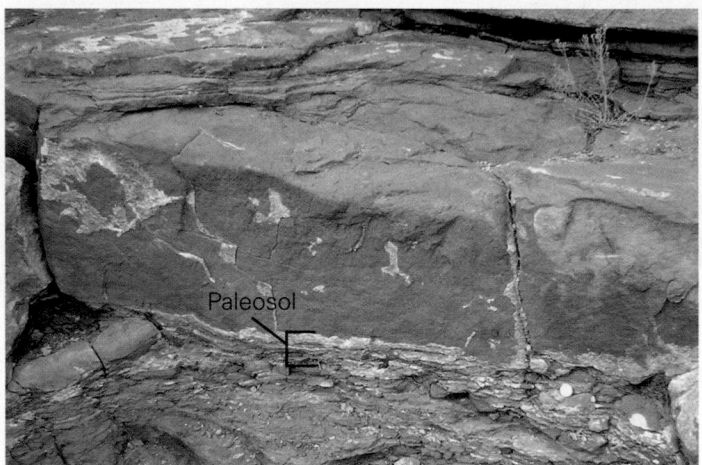

(d) The bleached layer between the red sandstone ledge and the red shale below represents a paleosol. It exists because a soil had time to form on the top of a mud layer before a flood buried the mud and soil with a new layer of sand.

a crime scene to identify a culprit. A geologist examines grain size, clast composition, sorting, bed-surface marks, cross bedding, and fossils to identify a depositional environment. With experience, a geologist can determine whether a succession of beds accumulated at the end of a glacier, in a desert, on a river floodplain, along a beach, in shallow seawater just offshore, or on the deep seafloor.

Let's now look at some examples of different depositional environments and the sediments deposited in them by taking an imaginary journey from the mountains to the sea, this time looking not only at clast size and composition, but also at sedimentary structures as we proceed. We will see that geologists distinguish between two basic categories of depositional environment: terrestrial and marine.

Terrestrial (Nonmarine) Depositional Environments

Terrestrial depositional environments are those that develop inland, far enough away from the ocean shoreline that ocean tides and waves do not affect them. The sediments of terrestrial deposits settle on dry land or under and adjacent to freshwater streams, glaciers, and lakes. In some settings, oxygen in surface water or groundwater reacts with the iron in terrestrial sediments to produce rust-like iron-oxide minerals, which give the sediment an overall reddish hue. Strata with this hue are informally called **redbeds** (see Fig. 7.12c).

FIGURE 7.18 Examples of nonmarine depositional environments.

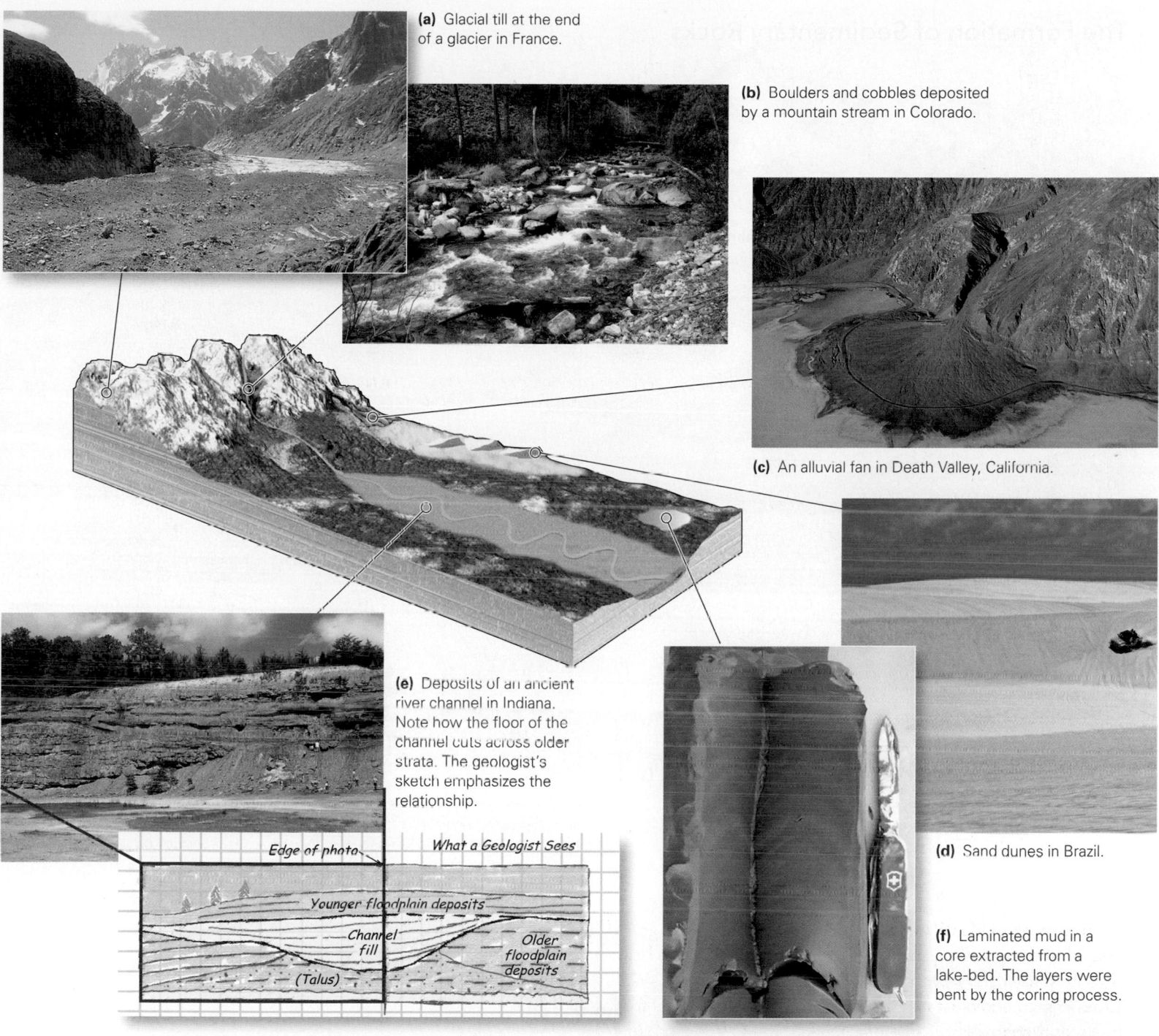

(a) Glacial till at the end of a glacier in France.

(b) Boulders and cobbles deposited by a mountain stream in Colorado.

(c) An alluvial fan in Death Valley, California.

(e) Deposits of an ancient river channel in Indiana. Note how the floor of the channel cuts across older strata. The geologist's sketch emphasizes the relationship.

Edge of photo

What a Geologist Sees

Younger floodplain deposits

Channel fill

Older floodplain deposits

(Talus)

(d) Sand dunes in Brazil.

(f) Laminated mud in a core extracted from a lake-bed. The layers were bent by the coring process.

Glacial Environments We begin high in the mountains, where it's so cold that more snow collects in the winter than melts away, so that glaciers—rivers or sheets of ice—develop and slowly flow. Solid ice can move sediment of any size. So as a glacier moves down a valley in the mountains, it carries along all the sediment that falls on its surface from adjacent cliffs or gets plucked from the ground at its base or sides. At the end of the glacier, where the ice finally melts away, it drops its sedimentary load and produces a pile of *glacial till* (Fig. 7.18a). Till is unsorted and unstratified—it contains clasts ranging from clay size to boulder size all mixed together—and therefore becomes a type of diamicton. We provide further details on glacial sediments in Chapter 22.

The Formation of Sedimentary Rocks

Glacial environment

Estuary

Beach

Bar

Continental shelf

Coastal
erosion

Turbidity
current

Submarine fan

Deep-sea current

Redbeds

Bedding

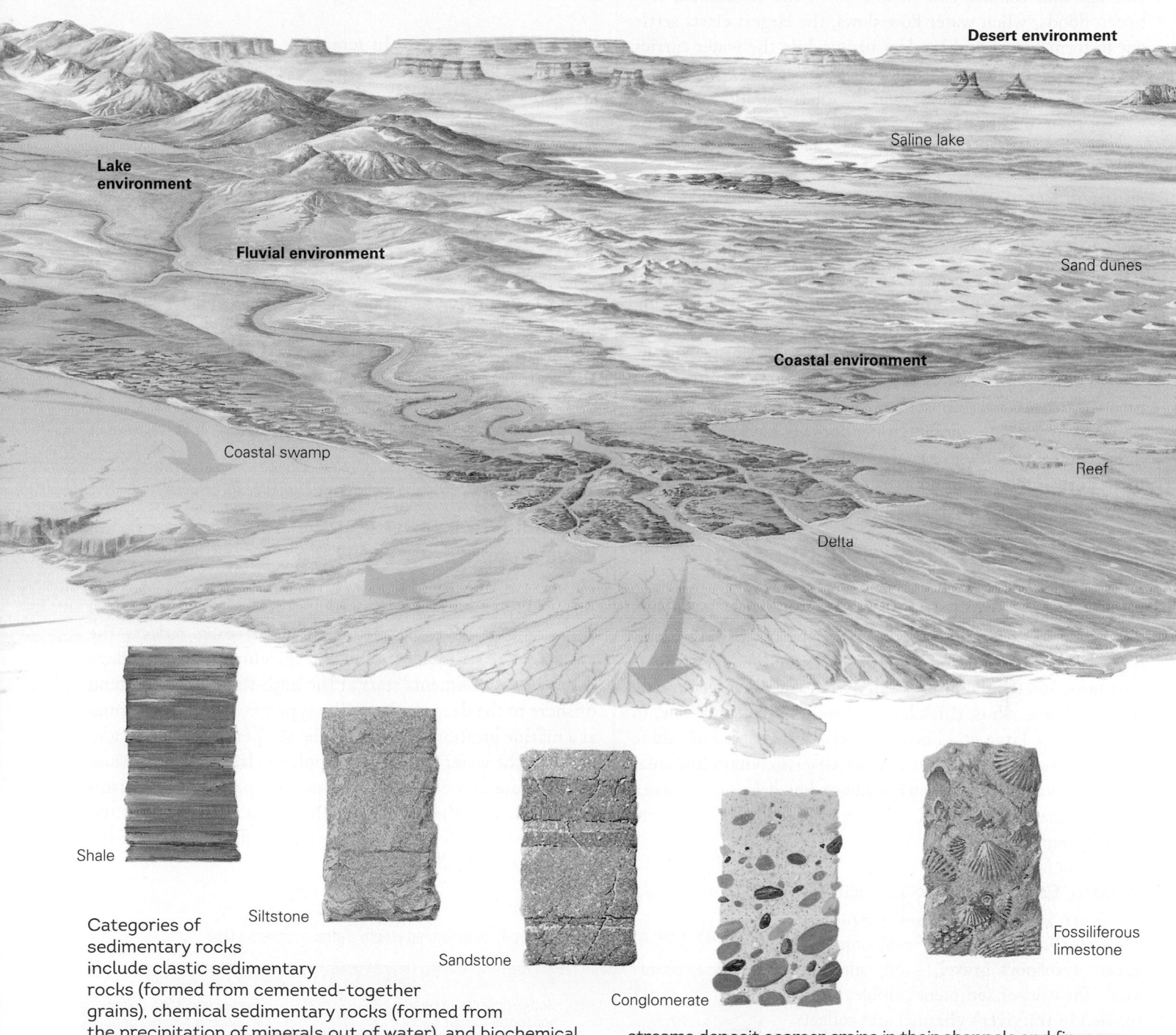

Desert environment

Saline lake

Sand dunes

Lake environment

Fluvial environment

Coastal environment

Coastal swamp

Reef

Delta

Shale

Siltstone

Sandstone

Conglomerate

Fossiliferous limestone

Categories of sedimentary rocks include clastic sedimentary rocks (formed from cemented-together grains), chemical sedimentary rocks (formed from the precipitation of minerals out of water), and biochemical sedimentary rocks (formed from the shells of organisms).

Clastic sedimentary rocks develop when grains (clasts) break off pre-existing rock by weathering and erosion and are transported to a new location by wind, water, or ice; the grains are deposited to create sediment layers, which are then lithified. We distinguish among types of clastic sedimentary rocks on the basis of grain size.

The character of a sedimentary rock depends on the composition of the sediment and on the environment in which it accumulated. For example, glaciers carry sediment of all sizes, so they leave deposits of poorly sorted (different-sized) till;

streams deposit coarser grains in their channels and finer ones on floodplains; a river slows down at its mouth and deposits an immense pile of silt in a delta. Fossiliferous limestone develops on coral reefs. In desert environments, sand accumulates into dunes and evaporites precipitate in saline lakes. Offshore, submarine canyons channel avalanches of sediment, or turbidity currents, out to the deep seafloor.

Sedimentary rocks tell the history of the Earth. For example, the layering, or bedding, of sedimentary rocks is initially horizontal. So where we see layers bent or folded, we can conclude that the layers were deformed during mountain building.

Mountain Stream Environments As we walk beyond the end of the glacier, we enter a realm where turbulent streams rush downslope in steep-sided valleys. This fast-moving water has the power to carry large clasts—in fact, during floods, boulders and cobbles can tumble down the streambed. Between floods, when water flow slows, the largest clasts settle out to form gravel and boulder beds, while the water carries finer sediments, such as sand and mud, farther downstream **(Fig. 7.18b)**. Sedimentary deposits of a mountain stream would, therefore, become breccia and conglomerate (depending on the degree of rounding).

Alluvial Fan Environments Our journey now takes us to the mountain front, where the fast-moving stream empties onto a plain. In arid regions, where there is not enough water for the stream to flow continuously, the stream deposits its load of sediment on the plain near the mountain front, producing a wedge-shaped apron of gravel and sand called an **alluvial fan (Fig. 7.18c)**. Deposition takes place here because when the stream emerges from a canyon mouth, it spreads out over a broader area, so friction with the ground causes the water to slow down, and slow-moving water does not have the power to move pebbles, cobbles, or coarse sediment. Notably, the sediment in the sand of an alluvial fan is derived from the erosion of granite, therefore it may still contain feldspar grains that have not yet weathered into clay. As a consequence, alluvial fan sediments become breccia, conglomerate, and arkose.

Desert Environments In very dry climates, few plants can grow, so the ground surface lies exposed. Strong winds can pick up dust and sand from the surface. The dust gets carried away, and the remaining well-sorted sand may accumulate in large dunes. Thus, thick layers of well-sorted sandstone, in which we see large cross beds, are relicts of desert sand-dune environments **(Fig. 7.18d)**. In places, deserts contain low areas in which water collects during floods but dries up between floods. Salts precipitate in such settings and can build into beds of evaporites.

River Environments In climates where streams flow, we find several distinctive depositional environments. Rivers transport gravel, sand, silt, and mud. The coarser sediment tumbles along the bed in the river's channel and collects in cross-bedded, rippled layers, while the finer sediment drifts along, suspended in the water. This fine sediment settles out along the floodplain. Mud layers on the floodplain dry out between floods, leading to the formation of mud cracks. River sediments lithify to form sandstone, siltstone, and shale. Typically, channels of coarser sediment are surrounded by layers of fine-grained floodplain deposits; in cross section, the channel has a lens-like shape **(Fig. 7.18e)**. Geologists commonly refer to river deposits as *fluvial sediments*, from the Latin word *fluvius*, meaning river.

Lake Environments In temperate climates, where water remains at the surface throughout the year, lakes form. In lakes, the relatively quiet water can't move coarse sediment; any coarse sediment brought into the lake by a stream settles out at the stream's outlet. Only fine clay makes it out into the center of the lake, where it settles to form mud on the lake bed. Not surprisingly, lake sediments typically consist of finely laminated shale **(Fig. 7.18f)**.

At the mouths of streams that empty into lakes, small deltas may form. A **delta** is a wedge of sediment that accumulates where moving water enters standing water. Deltas were so named because the map shape of some deltas resembles the Greek letter *delta* (Δ), as we discuss further in Chapter 17. In 1885, an American geologist named G. K. Gilbert showed that small deltas contain three components **(Fig. 7.19)**: topset beds of gravel, foreset beds of gravel and sand, and bottomset beds of silt.

Coastal and Marine Depositional Environments

Along the seashore, a variety of distinct coastal depositional environments develop; the character of each reflects the nature of the sediment supply and the climate. Marine depositional environments start at the high-tide line and extend offshore to the deep seafloor. The type of sediment deposited at a marine location depends on the temperature, clarity, and depth of the water and on the supply of clasts. That's because temperature and clarity determine the species of organisms that can live in the water, and the availability of clasts and the degree to which waves move water affect the size of the clasts that accumulate.

FIGURE 7.19 A simple "Gilbert-type" delta formed where a stream enters a lake.

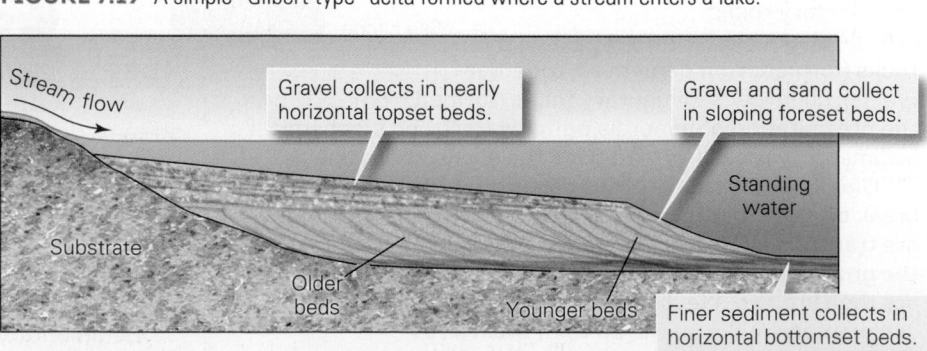

Stream flow

Gravel collects in nearly horizontal topset beds.

Gravel and sand collect in sloping foreset beds.

Standing water

Substrate

Older beds

Younger beds

Finer sediment collects in horizontal bottomset beds.

FIGURE 7.20 Examples of coastal depositional environments.

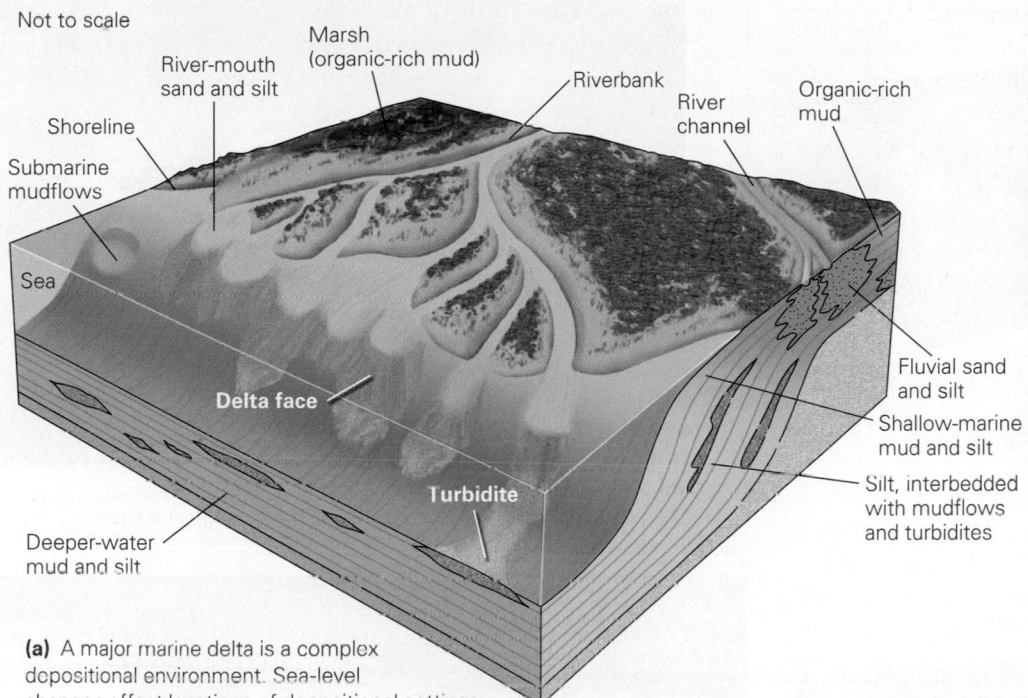

Not to scale

Submarine mudflows

Shoreline

River-mouth sand and silt

Marsh (organic-rich mud)

Riverbank

River channel

Organic-rich mud

Sea

Delta face

Deeper-water mud and silt

Turbidite

Fluvial sand and silt

Shallow-marine mud and silt

Silt, interbedded with mudflows and turbidites

(a) A major marine delta is a complex depositional environment. Sea-level changes affect locations of depositional settings.

(b) Waves along the coast sort beach sand.

Marine Delta Deposits After following the river downstream for a long distance, we reach its mouth, where it empties into the sea. Here the river water stops flowing, so water velocity slows, and sediment settles out to build a delta into the sea. Large marine deltas are much more complex than the small lake examples that Gilbert studied, for marine deltas host many different sedimentary environments, including swamps, channels, floodplains, and submarine slopes. Sea-level changes may cause the positions of the different environments to move inshore or offshore with time. As a result, the deposits of a marine delta produce a great variety of sedimentary rock types **(Fig. 7.20a)**.

Coastal Beach Sands Now we leave the delta and wander along the coast **(Fig. 7.20b)**. Ocean currents transport sand along the coastline. The sand washes back and forth in the surf, so it becomes well sorted (as waves winnow out mud and silt) and well rounded, and because of the back-and-forth movement of ocean water, the sand surface may become rippled. So, if you find well-sorted, medium-grained sandstone, perhaps with ripple marks, you may be looking at the remnants of a beach environment.

Shallow-Marine Clastic Deposits From the beach, we proceed offshore. In deeper water, where wave energy does not stir the seafloor, finer sediment can accumulate. Because the water here may be only meters to a few tens

of meters deep, geologists refer to this depositional setting as a *shallow marine environment*. Clastic sediments that accumulate in this environment tend to be fine-grained, well-sorted, and well-rounded silt. A great variety of organisms such as mollusks, gastropods, and worms live in or on this sediment. Therefore, if you see beds of siltstone and mudstone containing marine fossils, you may be looking at shallow-marine clastic deposits.

Shallow-Marine Carbonate Environments In shallow-marine settings where relatively little sand and mud enter the water, the warm, clear, nutrient-rich water can host an abundance of organisms with carbonate shells, which become carbonate sediment **(Fig. 7.21)**. Beaches collect sand composed of shell fragments; lagoons are sites where carbonate mud accumulates;

FIGURE 7.21 Reef environments for the deposition of carbonate sediments.

(a) Carbonate reefs form along shorelines in warm-water environments. In detail, reefs include many distinct depositional environments.

(b) A dramatic reef surrounds an island in the tropical Pacific. Deeper water is darker. Note the surf along the edge of the reef.

FIGURE 7.22 Examples of deep-marine sediment.

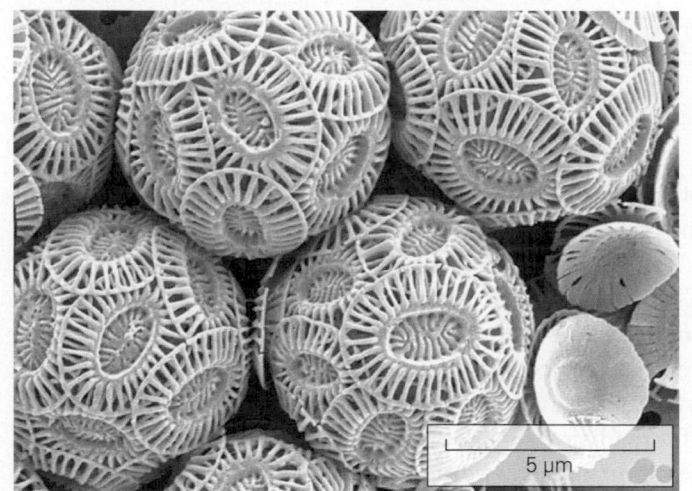

(a) These plankton shells, which make up some kinds of deep-marine sediment, are so small that they could pass through the eye of a needle.

(b) The chalk cliffs of southeastern England consist of plankton shells deposited on the seafloor tens of millions of years ago.

and reefs consist of coral and coral debris. Farther offshore from a reef, we can find a sloping apron of reef fragments. Shallow-marine carbonate sediments transform into various kinds of limestone.

Deep-Marine Deposits We conclude our journey by sailing far offshore. Along the transition between coastal regions and the deep ocean, turbidity currents deposit graded beds. In the deep-ocean realm, only fine clay and plankton provide sources of sediment. The clay eventually settles out onto the deep seafloor, forming deposits of finely laminated mudstone, and plankton shells settle to form chalk (from calcite shells; **Fig. 7.22**) or chert (from siliceous shells). Deposits of mudstone, chalk, or bedded chert, therefore, indicate a deep-marine origin.

TAKE-HOME MESSAGE

Different types of sediments accumulate in different depositional environments. Thus, sediments deposited along a river differ from sediments deposited by ocean waves, by glaciers, or in the deep sea. By studying sedimentary rocks at a location, geologists can distinguish among various marine and terrestrial deposits, and they can deduce environments that existed at the locality in the past.

QUICK QUESTION: How can you distinguish sediment deposited in an alluvial fan from sediment deposited in a shallow-marine environment?

7.5 Sedimentary Basins

The sedimentary veneer on the Earth's surface varies greatly in thickness. If you stand in central Siberia or Canada, you will find yourself on igneous and metamorphic basement that formed over a billion years ago—sedimentary rocks are nowhere in sight. Yet if you stand along the southern coast of Texas, you would have to drill through over 15 km of sedimentary beds before reaching igneous or metamorphic basement. Thick accumulations of sediment form only in special regions where the surface of the Earth's lithosphere sinks, providing space in which sediment can collect. Geologists use the term **subsidence** to refer to the process by which this sinking takes place, and the term **sedimentary basin** for the resulting sediment-filled depression. (Note that sedimentary basins fill with sediment as they subside, so they shouldn't be pictured as deep depressions.) In what geologic settings do sedimentary basins form? Plate tectonics theory provides a key.

Categories of Basins in the Context of Plate Tectonics Theory

Geologists distinguish among different kinds of sedimentary basins on the basis of the region of a lithospheric plate in which the basin forms, as defined in the context of plate tectonics theory (**Fig. 7.23**). Let's consider a few examples.

- *Rift basins:* Rift basins form in continental rifts, regions where the lithosphere has been stretched. During rifting, the surface of the Earth subsides simply because the crust becomes thinner as it stretches. (To picture this process, imagine pulling on either end of a block of clay with your hands—as the clay stretches, the central region of the block thins and sinks lower than the ends.) As the rift grows, slip on faults sends blocks of crust downward, producing low troughs bordered by narrow mountain ranges. Alluvial fan deposits form along the bases of the mountains, and salt flats or lakes develop in the interior portions of the troughs.

    Thinning is not the only reason why rifted lithosphere subsides. During rifting, warm asthenosphere rises beneath the rift and heats up the thin lithosphere. When rifting ceases, the rifted lithosphere cools, thickens, and becomes denser. This heavier lithosphere sinks, causing more subsidence, just as the deck of a tanker drops to a lower elevation when the ship is filled with ballast. Sinking due to cooling of the lithosphere is called *thermal subsidence.*

- *Passive-margin basins:* Passive-margin basins form along edges of continents that are not plate boundaries. They are underlain by stretched lithosphere, the remnants of a rift whose evolution successfully led to the formation of a mid-ocean ridge. Passive-margin basins form because thermal subsidence of stretched lithosphere continues long after rifting ceases and seafloor spreading begins. These basins fill both with sediment carried to the sea by rivers, and with carbonate sediment formed in coastal reefs. Sediment in a passive-margin basin can reach an astounding thickness of 15 to 20 km.

- *Intracontinental basins:* Intracontinental basins develop in the interiors of continents, initially because of subsidence over a rift. They may continue to subside in pulses, even hundreds of millions of years after they first formed, for reasons that are not yet well understood. Illinois and Michigan are each underlain by an intracontinental basin—the Illinois basin and the Michigan basin, respectively—in which up to 7 km of fluvial, deltaic, or shallow-marine sediment accumulated. At times, extensive swamps formed along the shorelines of these basins. The plant matter of these swamps transformed into coal when it was later buried deeply.

FIGURE 7.23 The geologic setting of sedimentary basins.

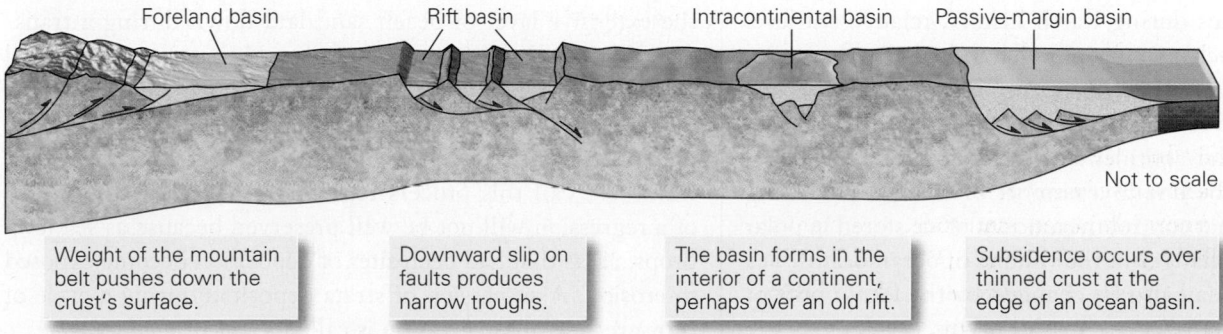

Foreland basin | Rift basin | Intracontinental basin | Passive-margin basin

Not to scale

| Weight of the mountain belt pushes down the crust's surface. | Downward slip on faults produces narrow troughs. | The basin forms in the interior of a continent, perhaps over an old rift. | Subsidence occurs over thinned crust at the edge of an ocean basin. |

FIGURE 7.24 The concept of transgression and regression during deposition of sedimentary sequence.

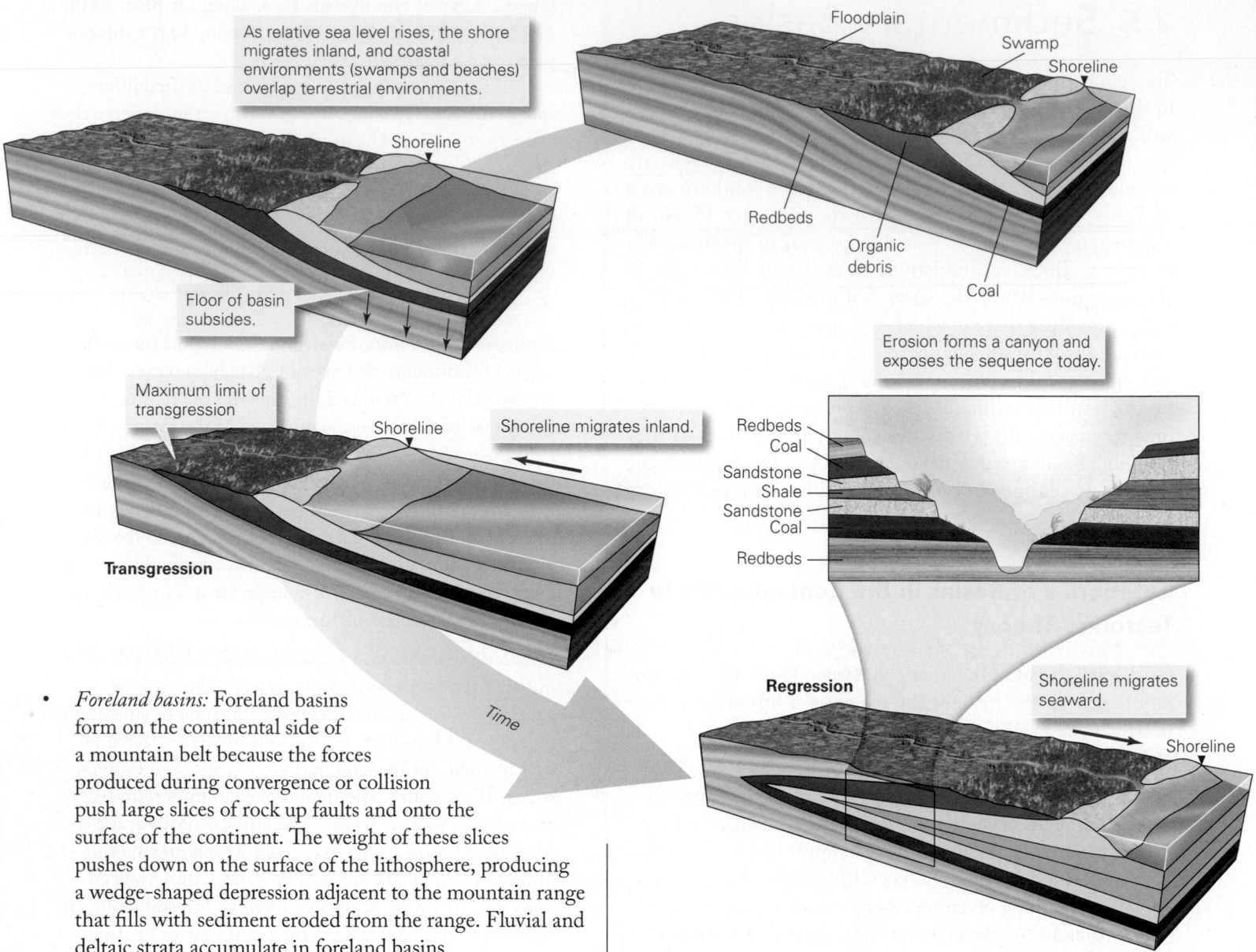

As relative sea level rises, the shore migrates inland, and coastal environments (swamps and beaches) overlap terrestrial environments.

Shoreline

Floodplain

Swamp

Shoreline

Redbeds

Organic debris

Coal

Floor of basin subsides.

Maximum limit of transgression

Shoreline

Shoreline migrates inland.

Transgression

Erosion forms a canyon and exposes the sequence today.

Redbeds
Coal
Sandstone
Shale
Sandstone
Coal
Redbeds

Regression

Shoreline migrates seaward.

Shoreline

Time

- *Foreland basins:* Foreland basins form on the continental side of a mountain belt because the forces produced during convergence or collision push large slices of rock up faults and onto the surface of the continent. The weight of these slices pushes down on the surface of the lithosphere, producing a wedge-shaped depression adjacent to the mountain range that fills with sediment eroded from the range. Fluvial and deltaic strata accumulate in foreland basins.

Transgression and Regression

Changes in sea level, relative to the land surface, can control the succession of sediments that we see in a sedimentary basin **(Fig. 7.24)**. At times during Earth history, relative sea level has risen by as much as a couple of hundred meters, creating shallow seas that submerge the interiors of continents. At other times, relative sea level has fallen by a couple of hundred meters, exposing the continental shelves to air. Global sea-level changes may be due to a number of factors—including climate change—that control the amount of ice stored in polar ice caps and cause changes in the volume of ocean basins. Sea level at a location may also be changed by the local uplift or sinking of the land surface.

When relative sea level rises, the shoreline migrates inland. During this process, known as **transgression**, coastal sediments progressively bury terrestrial sediments, and deeper-water sediments bury coastal sediments. Note that the extensive layer of beach sand laid down during a transgression may look like a blanket of sand that accumulated all at once, but in fact the sand deposited at one location differs in age from the sand deposited at another location.

When relative sea level falls, the coast migrates seaward. We call this process **regression**. Typically, the record of a regression will not be well preserved because as sea level drops, areas that had been sites of deposition become exposed to erosion. A succession of strata deposited during a cycle of transgression and regression is called a *depositional sequence*.

Diagenesis

Earlier in this chapter, we discussed the process of lithification, by which sediment hardens into rock. Lithification represents an aspect of a broader phenomenon called diagenesis. Geologists use the term **diagenesis** for all the physical, chemical, and biological processes that transform sediment into sedimentary rock and that alter characteristics of sedimentary rock after the rock has formed.

In sedimentary basins, sedimentary rocks may become buried very deeply. As a result, the rocks endure high pressures and temperatures and come in contact with warm groundwater. Diagenesis, under such conditions, can cause chemical reactions in the rocks that produce new minerals and can also cause cement to dissolve or precipitate. As temperatures and pressures increase with depth, the changes that take place in rocks become more profound. At sufficiently high temperatures and pressures, metamorphism begins: a new assemblage of minerals forms, and mineral grains may become aligned parallel to one another. The transition between diagenesis and metamorphism in sedimentary rocks occurs between temperatures of 150°C and 300°C. In the next chapter, we enter the realm of true metamorphism.

TAKE-HOME MESSAGE

In certain geologic settings, the Earth's surface sinks (subsides) to form a depression that fills with sediment. The depression with its thick fill of sediment is a sedimentary basin. As sea level rises or falls, the coast, and therefore depositional environments, can migrate inland or offshore, respectively. Once sediment has been deposited, it undergoes various changes, known as diagenesis, in response to rising temperature and pressure and to interaction with groundwater.

QUICK QUESTION: How thick can the fill of a sedimentary basin become?

ANOTHER VIEW These cliffs, near Bryce Canyon (Utah), expose beds of sedimentary rock deposited in lakes and by streams over 40 Ma. Present-day erosion has produced aprons of debris at the base of the cliffs.

Chapter 7 Review

SUMMARY

- Geologists recognize four major classes of sedimentary rocks. Clastic sedimentary rocks form from cemented-together grains that were first produced by weathering and were then transported, deposited, and lithified. Biochemical sedimentary rocks develop from the shells of organisms. Organic sedimentary rocks form from plant debris or the altered remains of plankton. Chemical sedimentary rocks precipitate directly from water.

- Formation of clastic rocks begins when grains erode from pre-existing rock. Moving water, air, or ice transport these grains to a site of deposition, where they accumulate. Lithification, involving compaction and cementation, converts loose sediment into rock.

- Clastic rocks are classified primarily by grain size. Other characteristics of grains (roundness, sorting, composition) help to distinguish sediment source and depositional environment.

- Sedimentary structures include bedding, cross bedding, graded bedding, ripple marks, dunes, and mud cracks. They serve as clues to depositional settings.

- Biochemical and organic rocks form from materials produced by living organisms. Limestone consists predominantly of calcite; chert forms from silica, and coal forms from carbon.

- Evaporites consist of minerals precipitated from saline water.

- Glaciers, streams, alluvial fans, deserts, rivers, lakes, deltas, beaches, shallow seas, and deep seas each accumulate a different, distinctive assemblage of sedimentary strata.

- Thick piles of sedimentary rocks collect in sedimentary basins, regions where the lithosphere subsided.

- Transgression occurs when sea level rises and the coastline migrates inland. Regression occurs when sea level falls and the coastline migrates seaward.

- Diagenesis involves processes leading to lithification and processes that alter sedimentary rock once it has formed.

GUIDE TERMS

alluvial fan (p. 234)
bed (p. 224)
bedding (p. 224)
biochemical limestone (p. 218)
breccia (p. 216)
cement (p. 214)
cementation (p. 214)
clast (p. 211)
coal (p. 219)
compaction (p. 214)

conglomerate (p. 217)
cross bed (p. 226)
delta (p. 234)
deposition (p. 213)
depositional environment (p. 229)
diagenesis (p. 239)
dune (p. 226)
erosion (p. 212)
evaporite (p. 220)
lithification (p. 213)

mud crack (p. 227)
mudstone (p. 217)
oil shale (p. 220)
paleosol (p. 228)
redbed (p. 230)
regression (p. 238)
ripple mark (p. 224)
sandstone (p. 212)
sedimentary basin (p. 237)
sedimentary rock (p. 211)

sedimentary structure (p. 224)
shale (p. 217)
siltstone (p. 217)
strata (p. 224)
stratigraphic formation (p. 224)
subsidence (p. 237)
transgression (p. 238)
travertine (p. 221)

GEOTOURS *THIS CHAPTER'S GEOTOURS WORKSHEET (F) FEATURES QUESTIONS AND GOOGLE EARTH SITES ON:*

- Sedimentary rocks exposed in the Grand Canyon, the Lewis Range, Death Valley, and Zion National Park

- Erosion and differential and physical weathering
- Regression and transgression

- Fluvial and arid depositional environments
- Modern and ancient depositional environments

REVIEW QUESTIONS

The letters following each Review Question refer to the corresponding Learning Objective from the Chapter Opener.

1. Describe how a clastic sedimentary rock forms from its unweathered parent rock. **(A, B)**

2. Explain how biochemical sedimentary rocks form. **(A)**

3. How do grain size, sorting, sphericity, and angularity change as sediments move downstream? **(B, E)**

4. Describe the two different kinds of chert. How are they similar? How are they different? **(A)**

5. What conditions lead to deposition of evaporites? **(E)**

6. How does dolostone differ from limestone, and how does dolostone form? What kinds of rock form in the environment shown here? **(C)**

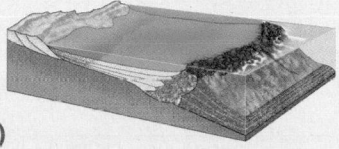

7. What are cross beds, and how do they form? How can you read the current direction from cross beds? **(D)**

8. Describe how a turbidity current forms and moves. How does graded bedding form? **(D)**

9. Compare the deposits of an alluvial fan with those of a deep-marine environment. **(E)**

10. What kinds of sediments accumulate in river and delta systems? How can they be distinguished from glacial sediments? **(E)**

11. Why don't sediments accumulate everywhere? What types of tectonic conditions are required to create sedimentary basins? **(F)**

12. Compare the consequences of a transgression with those of a regression. Which process is depicted in the diagram? **(F)**

13. What changes take place during diagenesis? **(F)**

ON FURTHER THOUGHT

14. Recent exploration of Mars by robotic vehicles suggests that layers of sedimentary rock cover portions of the planet's surface. The layers contain cross bedding and relics of gypsum crystals. At face value, what do these features suggest about depositional environments on Mars in the past? **(D, E)**

15. The Gulf Coast of the United States is a passive-margin basin that contains a very thick accumulation of sediment. Drilling reveals that the base of the sedimentary succession in this basin consists of redbeds. These are overlain by a thick layer of evaporite which, in turn, is overlain by deposits composed predominantly of sandstone and shale. In some strata, the sandstone occurs in channels and contains ripple marks, and the shale contains mud cracks. In other strata,

the sandstone and shale contain fossils of marine organisms. The sequence contains hardly any conglomerate or arkose. Be a sedimentary detective and explain the succession of sediment in the basin. **(D)**

16. Examine the Bahamas (Lat 23°58′40.98″ N, Long 77°30′20.37″ W) with Google Earth. Describe the depositional environments that you see. **(E)**

17. The bedrock of Florida consists mostly of shallow-marine limestone. What does this observation suggest about the nature of the Florida peninsula in the past? Keep in mind that Florida's land surface lies at less than 50 m above sea level. **(E)**

ONLINE RESOURCES

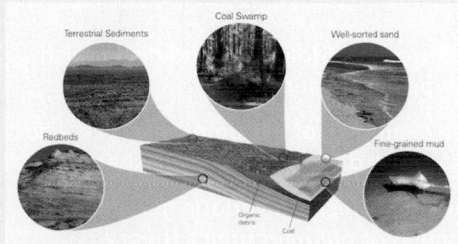

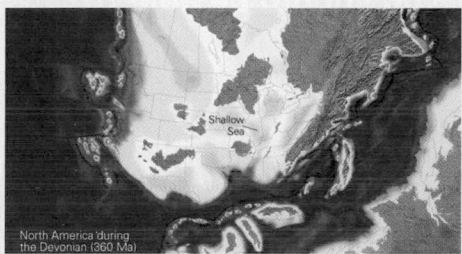

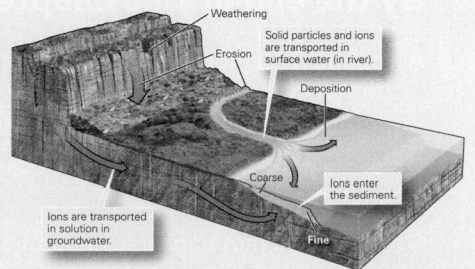

Videos
This chapter features videos on the formation of sedimentary rocks, transgression and regression, and how this appears in the stratigraphic record.

Smartwork5
This chapter features questions on recognizing types of sedimentary rocks, structures, and environments.

CHAPTER 8

Metamorphism: A Process of Change

By the end of this chapter, you should be able to . . .

A. define metamorphism and characterize the changes that a rock undergoes during the transformation of a protolith into a metamorphic rock.

B. explain the key processes that can take place during metamorphism.

C. contrast a metamorphic mineral assemblage with the assemblage of a protolith, and contrast a metamorphic texture with the texture of a protolith.

D. recognize the occurrence of foliation, and distinguish among different types of foliation.

E. identify and name examples of metamorphic rock, and explain how these rocks differ from one another.

F. relate kinds of metamorphism to various geologic settings in the context of plate tectonics theory.

G. describe how the character of a metamorphic rock reflects the grade of metamorphism and how metamorphic grade depends on temperature and pressure.

Nothing in the world lasts, save eternal change.

—Honorat de Bueil (French poet, 1589–1670)

8.1 Introduction

Cool winds sweep across Scotland for much of the year. In this blustery climate, vegetation has a hard time taking hold, so the landscape provides countless outcrops of barren rock. During the mid-18th century, James Hutton, the man who came to be known as the father of geology, examined these outcrops in earnest, hoping to learn how rock formed. Hutton discovered that many features in the outcrops resembled the products of present-day sediment deposition or volcanic activity, and from this observation he came to an understanding of how sedimentary and igneous rock forms. But Hutton also recognized that some rock contained minerals and textures quite different from those visible in sedimentary and igneous samples. He described this puzzling rock as "a mass of matter which had evidently formed originally in the ordinary manner . . . but which is now extremely distorted in its structure . . . and variously changed in its composition."

The rock that so puzzled Hutton is now known as metamorphic rock, from the Greek words *meta*, meaning change, and *morphe*, meaning form. In modern wording, a **metamorphic rock** forms when a pre-existing rock, or **protolith**, undergoes a solid-state change in response to the modification of its environment, generally at depth in the Earth. We use the word **metamorphism** for the overall process of such change.

Let's look at the components of this definition more closely. By *solid-state*, we mean that a metamorphic rock does not form by solidification of magma—remember that geologists consider rocks that solidified from magma to be igneous. By *change*, we mean that metamorphism produces new minerals that did not occur in the protolith, and produces a new *texture* (arrangement of mineral grains) that differs from that of the protolith. By *modification of environment*, we mean that metamorphism takes place when a protolith endures a rise in temperature or pressure, undergoes compression and shear, or reacts with *hydrothermal fluids* (solutions of very hot water). And by using the term *at depth*, we mean that metamorphism takes place at higher pressures and temperatures than those responsible for diagenesis (see Chapter 7).

◄ (facing page) A metamorphic rock exposed on a hillside in Scotland displays contorted foliation and contains minerals that formed at high pressure and temperature at depth in the crust.

Hutton did more than just note the existence of metamorphic rock—he also tried to understand why metamorphism takes place. Because he found metamorphic rock adjacent to igneous intrusions, he concluded that metamorphism can take place when heat from an intrusion "cooks" the rock into which it intrudes. And because he found that metamorphic rock can occur over broad regions in the absence of intrusions, he speculated that metamorphism can also take place when rock becomes deeply buried, as occurs during mountain building.

From Hutton's day to the present, geologists have undertaken field studies, laboratory experiments, and theoretical calculations to better characterize metamorphism. In this chapter, we introduce the results of their work. We begin by explaining the causes of metamorphism and the basis for classifying metamorphic rocks. We conclude by discussing the geologic settings in which these rocks form. As you will see, Hutton's speculations on the origin of metamorphic rock were basically correct, but they represented only part of the story—the rest of the story could not take shape until the theory of plate tectonics became established.

Did you ever wonder . . .
if, once formed, rocks ever change?

8.2 Consequences and Causes of Metamorphism

What Is a Metamorphic Rock?

If someone were to put a rock on a table in front of you, how would you know that it was metamorphic? First, metamorphic rocks can possess **metamorphic minerals**, new minerals that grow in place within solid rock during metamorphism. In fact, metamorphism can produce a group of minerals that together make up a *metamorphic mineral assemblage*. Second, metamorphic rocks have **metamorphic texture**, defined by the arrangement of their mineral grains. That texture can be manifested by **metamorphic foliation**, which is characterized by the parallel alignment of platy minerals (such as mica) or by the presence of alternating light-colored and dark-colored layers (Fig. 8.1).

Because of its metamorphic minerals and textures, a metamorphic rock can be as different from its protolith as a butterfly is from a caterpillar. For example, metamorphism of red shale can yield a metamorphic rock consisting of aligned mica flakes and brilliant garnet crystals (**Fig. 8.2a**); metamorphism of limestone composed of cemented-together fossil fragments can yield a metamorphic rock called marble, consisting of large interlocking crystals of calcite (**Fig. 8.2b**); and metamorphism of granite, a rock with randomly oriented crystals, can produce a rock with crystals aligned parallel to one another (**Fig. 8.2c**). As seen through a microscope, the change in the grains of a rock due to metamorphism looks every bit as dramatic as the change in the cells of a caterpillar becoming a butterfly through metamorphosis (**Fig. 8.2d**).

The formation of metamorphic minerals and textures takes place very slowly—it may take thousands to millions of years—and involves several processes, which sometimes occur alone and sometimes together. The most common metamorphic processes include the following:

- *Recrystallization* changes the shape and size of grains without changing the identity of the mineral making up the grains. For example, during recrystallization of sandstone, a tightly fitting mosaic of large, irregularly shaped quartz grains replaces a cluster of small, round, cemented-together quartz grains (**Fig. 8.3a**).

- *Phase change* transforms one mineral into another mineral with the same composition but a different crystal structure. For example, the transformation of quartz into a denser mineral called coesite represents a phase change, for these minerals have the same formula (SiO_2) but different crystal structures. On an atomic scale, phase change involves the rearrangement of atoms.

- *Metamorphic reaction,* or *neocrystallization* (from the Greek *neos*, for new) results in the growth of new mineral crystals that differ from those of the protolith (**Fig. 8.3b**). During neocrystallization, chemical reactions digest minerals of the protolith and grow new minerals of the metamorphic rock. For this process to take place, atoms must slowly *diffuse* (migrate) through solid crystals, or dissolve and reprecipitate at grain boundaries.

FIGURE 8.1 An outcrop of 2.7-billion-year-old metamorphic rock in Ontario, Canada, shows distinct foliation, in this case defined by alternating bands of light and dark minerals.

FIGURE 8.2 Metamorphism causes changes in mineral makeup and texture.

Protolith ⟶ Metamorphic rock

Red shale   Gneiss

Foliation plane

(a) A specimen of red shale (left) contains clay, quartz, and iron oxide. When this rock undergoes intense metamorphism, it may change into a gneiss containing different minerals. This gneiss sample (right) contains biotite, quartz, feldspar, and purple garnet.

(b) In a limestone (left), individual fossils are visible. During metamorphism, the texture changes profoundly to produce marble (right).

- *Pressure solution* happens when a wet rock is squeezed more strongly in one direction than in the other. Mineral grains dissolve where their surfaces press against other grains, producing ions that migrate through the water to precipitate elsewhere. Precipitation may take place on faces where the grains are squeezed together less strongly. Pressure solution, therefore, can cause grains to become shorter in one direction and longer in another **(Fig. 8.3c)**. For pressure solution to take place, liquid-water films must coat grain surfaces during compression. Such films can persist during diagenesis and during low-temperature metamorphism. At high temperatures, water molecules evaporate or break apart, so the films disappear.

- *Plastic deformation* happens when a rock undergoes squeezing or shearing at elevated temperatures and pressures, conditions during which minerals behave like soft plastic and can change shape without breaking **(Fig. 8.3d)**.

FIGURE 8.2 (*continued*)

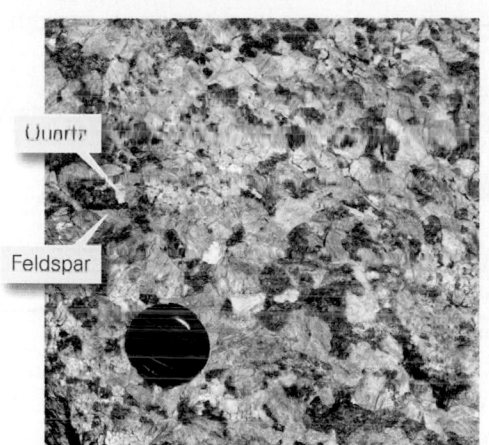

Feldspar crystals are about parallel to this line.

Quartz

Feldspar

(c) Metamorphism can transform a rock in which crystals are randomly oriented (left) into one in which they are aligned (right).

Caterpillars undergo metamorphosis because of hormonal changes in their bodies. Rocks undergo metamorphism when they are subjected to *metamorphic conditions*: heat, pressure, compression and shear, and/or very hot water. Let's see how these changes happen.

Metamorphism Due to Heating

When you heat cake batter sufficiently, the batter transforms into a new material—cake. Similarly, when you heat a rock sufficiently, its ingredients transform into a new material—metamorphic rock. Why? Think about what happens to atoms in a mineral grain as the grain warms. Heat causes the atoms to vibrate rapidly, so the chemical bonds that lock atoms to their neighbors stretch and bend. If bonds stretch and bend too far, they break, causing atoms to detach from their original neighbors, move slightly, and form new bonds with other atoms. Repetition of this process—trillions of times—leads to enough rearrangement of atoms within grains, or enough migration of atoms into and out of grains, that recrystallization or neocrystallization, or both, take place, and a new metamorphic mineral assemblage grows in the rock.

Metamorphism takes place at temperatures between those at which diagenesis (which modifies the rock without producing metamorphic minerals and textures) occurs and those that cause melting. Roughly speaking, this means that most of the metamorphic rocks you find in outcrops on continents formed at temperatures between 250°C and 850°C. Melting temperature, however, depends on the composition and water content of rock (see Chapter 6)—wet granite, for example, melts at temperatures below 650°C, and very dry peridotite melts at temperatures as high as 1,200°C. So the upper limit of the metamorphic realm actually ranges between 650°C and 1,200°C.

The depth in the Earth at which metamorphic temperatures exist depends on the *geothermal gradient*, which, in turn, reflects the geologic setting. For example, near a hot igneous intrusion, metamorphic

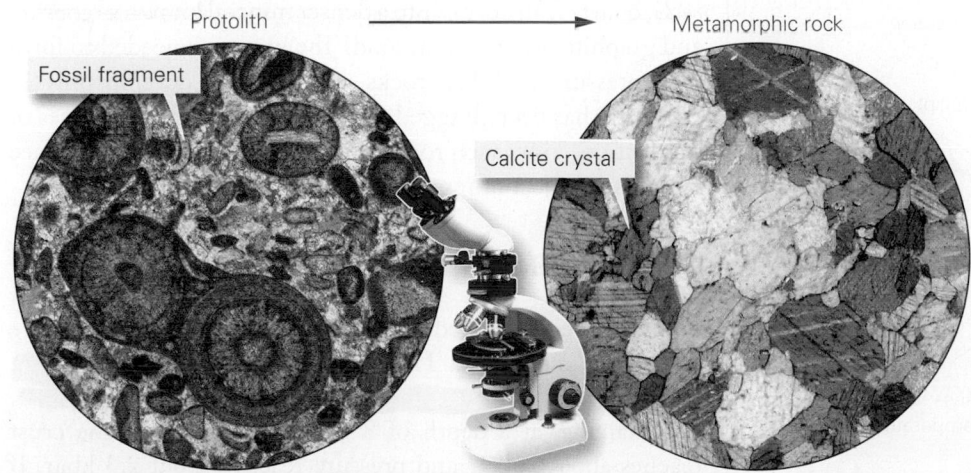

Protolith → Metamorphic rock

Fossil fragment

Calcite crystal

(d) The contrast in texture between a protolith of fossiliferous limestone and a marble formed by metamorphism is evident under a microscope.

FIGURE 8.3 Metamorphic processes as seen through a microscope.

Protolith ⟶ Metamorphic rock

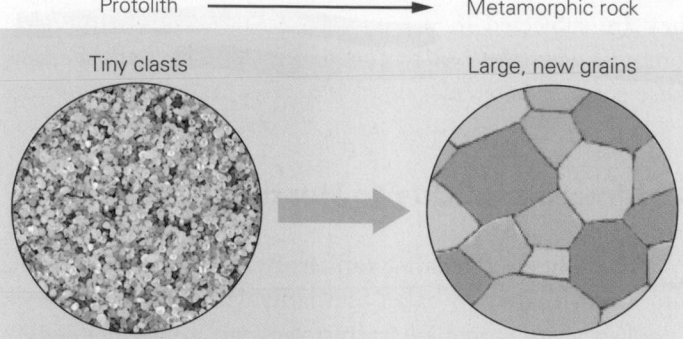

Tiny clasts   Large, new grains

(a) Mineral grains recrystallize to form new, interlocking grains of the same mineral. Typically, the new grains are larger.

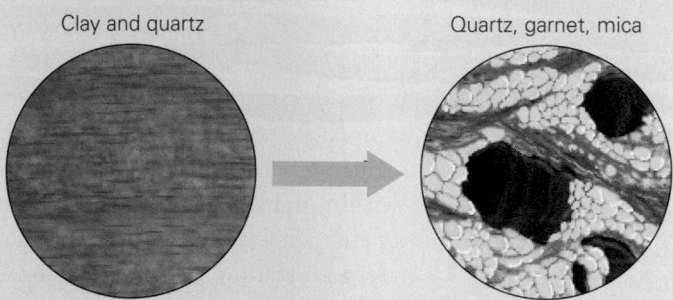

Clay and quartz   Quartz, garnet, mica

(b) Chemical reactions change the original assemblage of minerals into a new metamorphic assemblage of minerals.

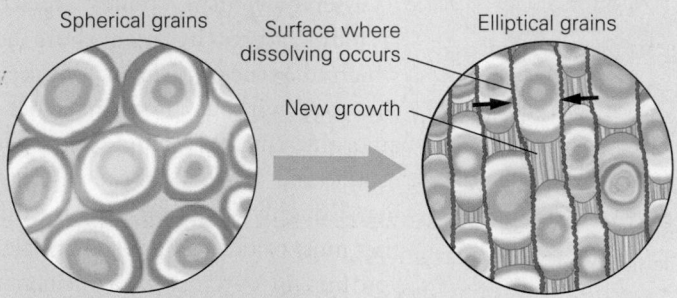

Spherical grains   Surface where   Elliptical grains
                    dissolving occurs
                    New growth

(c) Pressure solution dissolves grains on the sides undergoing the highest pressure and precipitates new mineral material where the pressure is lower. Arrows indicate the squeezing direction.

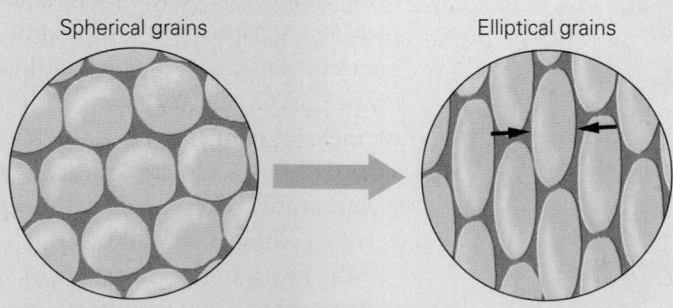

Spherical grains   Elliptical grains

(d) Plastic deformation changes the shape of grains—without breaking them—as a result of squeezing or shear at high temperatures.

temperatures affect rocks near the Earth's surface. But in the upper part of most continental crust, away from intrusions, the geothermal gradient is about 25°C per kilometer, and metamorphic temperatures occur only at depths greater than about 12 km.

Metamorphism Due to Pressure

As you swim underwater in a swimming pool, water squeezes against you equally from all sides—in other words, your body feels **pressure**. Pressure can cause a material to collapse inward. For example, if you pull an air-filled balloon down to a depth of 10 m in a lake, the balloon becomes significantly smaller. Pressure can have the same effect on minerals. Near the Earth's surface, minerals with relatively open crystal structures (meaning more space between atoms) can be stable. However, if you subject these minerals to extreme pressure, the atoms pack more closely together (resulting in less space between atoms) to yield denser minerals. Such transformations involve phase changes or neocrystallization.

Most metamorphic rocks that occur in outcrops on continents were metamorphosed at pressures of between 3 and 12 kbar (= 3,000 to 12,000 bars, where 1 bar is the air pressure at the Earth's surface). Pressure increases with depth at a rate of about 300 bars per kilometer due to the weight of overlying rock, so metamorphic pressures occur at depths between 10 and 40 km.

Some metamorphic rocks, however, form at pressures that indicate that they were once even deeper than 40 km. Metamorphic rocks formed under pressures between 12 and 27 kbar are known as *high-pressure metamorphic rocks*. Metamorphism under such conditions transforms mafic protoliths, such as basalt or gabbro, into *eclogite*, a rock composed of brilliant maroon garnets embedded in greenish pyroxene. If the pressure at the time of metamorphism exceeds 27 kbar, ultrahigh-pressure (UHP) metamorphic rocks form. In such *UHP rocks*, quartz transforms into a denser mineral known as coesite, and graphite becomes diamond. The pressures needed to form high-pressure or UHP rocks can probably develop only in crust that has been dragged down to depth by the processes of subduction. How such rocks can return to the Earth's surface remains a subject of research.

Changing Both Pressure and Temperature

So far, we've considered changes in pressure and temperature as separate phenomena. However, in reality, pressure and temperature in the Earth change together with increasing depth. For example, at a depth of 8 km, temperature in the crust reaches about 200°C and pressure reaches about 2.3 kbar. If a rock at 8 km later becomes buried to a depth of 20 km, as can happen during mountain building, temperature in the rock increases to more than 500°C, and pressure increases to 5.5 kbar. Experiments and calculations show that the *stability* of certain minerals (the ability of a mineral to form and survive) and of mineral assemblages depends on both pressure and temperature. Therefore, a metamorphic rock formed at a depth of 8 km does not contain the same minerals as one formed at 20 km.

We can illustrate the relationship of mineral stability to pressure and temperature by studying the behavior of the compound Al_2SiO_5 (aluminum silicate) as portrayed on a *phase diagram*, a graph with temperature indicated by one axis and pressure indicated by the other (Fig. 8.4). Al_2SiO_5 can exist as three different minerals: kyanite, andalusite, and sillimanite. Each of these minerals can survive only under a specific range of temperatures and pressures, indicated by an area called a *stability field* on the phase diagram. If geologic processes move a protolith containing the elements necessary to produce Al_2SiO_5 to a depth in the Earth where the pressure reaches 2 kbar and the temperature rises to 450°C (Point X), then andalusite grows. If the temperature stays at 450°C but pressure on the rock increases to 5 kbar (Point Y), then andalusite becomes unstable and kyanite grows. And if the pressure stays at 5 kbar but the temperature increases to 650°C (Point Z), then sillimanite grows. So the presence of one of these minerals provides a clue to the pressure and temperature at which metamorphism occurred.

Compression, Shear, and Development of Preferred Orientation

Put a ball of dough on the floor, lay a book on it, and step on the book. You'll see the ball flatten into a pancake, oriented

FIGURE 8.4 The stability fields for three metamorphic minerals (kyanite, andalusite, and sillimanite) that are polymorphs of Al_2SiO_5 (aluminum silicate) can be depicted on a phase diagram.

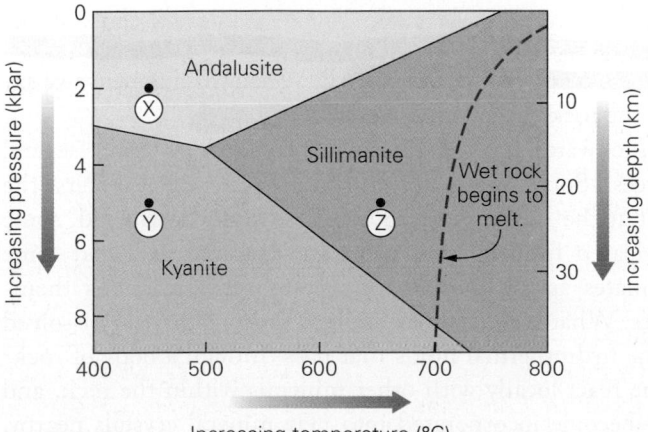

parallel to the floor, because the downward push you apply with your foot exceeds the push provided by air in other directions (Fig. 8.5a). If a material is squeezed (or stretched) unequally from different sides, we say that it has been subjected to **differential stress**. (Note that differential stress differs from pressure—in the latter, the push is the same in all directions.) In other words, under conditions of differential stress, the push or pull in one direction differs in magnitude from the push or pull in another direction. We distinguish between two kinds of differential stress:

- *Normal stress:* Normal stress pushes or pulls perpendicular to a surface. We call a push *compression* and a pull *tension*. Compression flattens a material (Fig. 8.5b), whereas tension stretches a material.

- *Shear stress:* Shear stress, or *shear*, moves one part of a material sideways relative to another part. If you place a ball of dough on a table, set your hand on top of it, and move your hand parallel to the floor, you're applying *shear* to the ball (Fig. 8.5c).

Compression and shear at metamorphic temperatures and pressures may cause a body of rock to change shape without cracking or breaking. During the process, a **preferred orientation** develops, in that pancake-shaped (platy) grains become aligned with one another, as do cigar-shaped (elongate) grains. Platy and elongate grains are examples of *inequant grains*, in that they have different dimensions in different directions (Fig. 8.5d, e); in contrast, *equant grains* have roughly the same dimensions in all directions.

Let's look more closely at how preferred orientation forms. In wet rocks at relatively low temperatures, pressure solution dissolves grains on the faces perpendicular to the direction of compression. So, as a result of pressure solution, grains become shorter in the direction of compression (Fig. 8.6a). (Sometimes, precipitation of new mineral crystals takes place on the faces where compression is less, so the grains may effectively get longer in the direction perpendicular to compression.) At relatively high temperatures, weaker grains flatten by means of plastic deformation due to differential stress (see Fig. 8.6a); the grains become narrower in one direction and become longer in the other. Preferred orientation also develops in response to shearing of a rock under elevated temperatures. Shearing either rotates inequant grains into alignment with the direction of shear, like logs in a river that align with a current, or smears and flattens grains (Fig. 8.6b). Growth of new metamorphic minerals during shear contributes to the development of preferred orientation—the new minerals grow faster in the direction in which a rock is stretching than in other directions.

FIGURE 8.5 Compression and shear change the shapes and sizes of mineral grains during metamorphism.

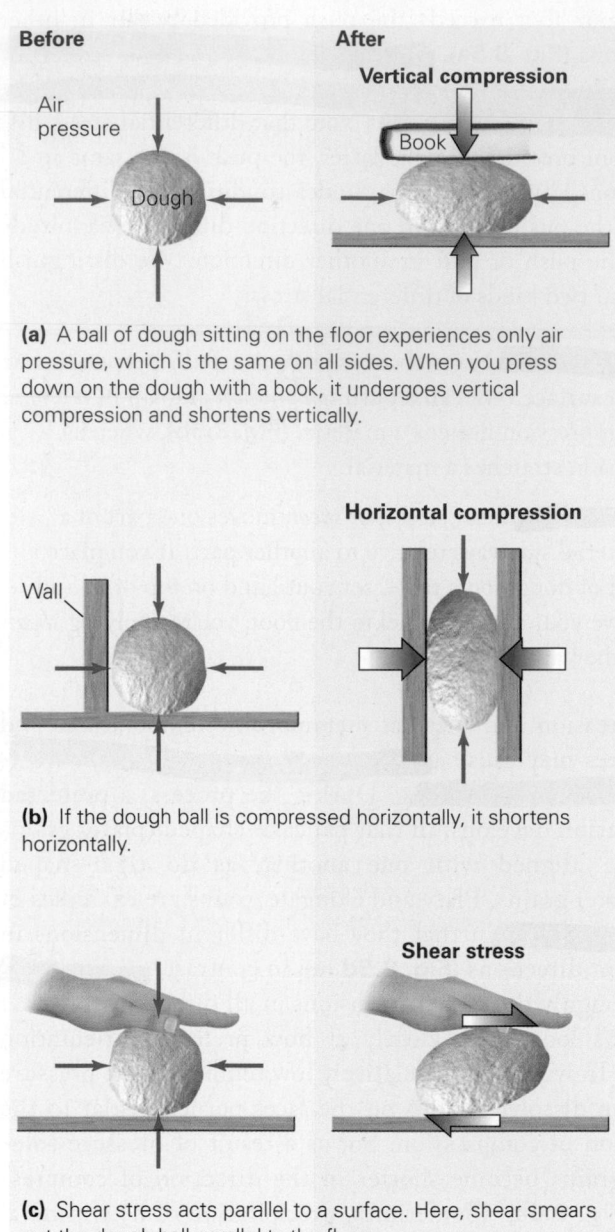

(a) A ball of dough sitting on the floor experiences only air pressure, which is the same on all sides. When you press down on the dough with a book, it undergoes vertical compression and shortens vertically.

(b) If the dough ball is compressed horizontally, it shortens horizontally.

(c) Shear stress acts parallel to a surface. Here, shear smears out the dough ball parallel to the floor.

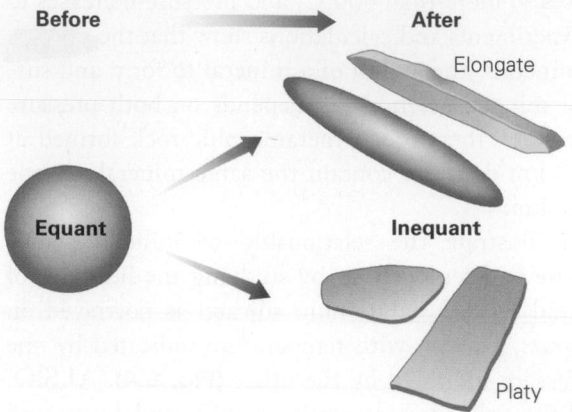

(d) Compression and shear can transform equant grains into inequant grains. Inequant grains can be elongate (cigar-shaped) or platy (pancake-shaped).

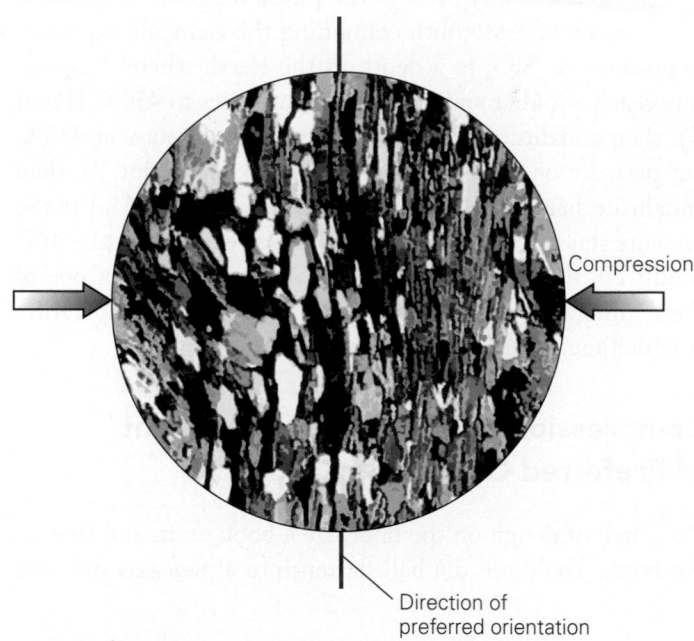

(e) In metamorphic rock, inequant grains may be aligned to form a preferred orientation. As seen through a microscope, the flat planes of grains are perpendicular to the compression direction.

The Role of Hydrothermal Fluids

Metamorphic reactions commonly take place in the presence of hydrothermal fluids. We initially defined hydrothermal fluids simply as solutions of very hot water. In fact, they actually can include hot water, steam, and so-called supercritical fluid. (A *supercritical fluid* is a substance that forms under high temperatures and pressures and has characteristics of both liquid and gas). Hydrothermal fluids react chemically with rock by dissolving, transporting, and providing ions. These fluids also provide water molecules that can become incorporated in minerals. In some cases, hydrothermal fluids passing through a rock during metamorphism pick up some ions of one element and drop off ions of another, thereby changing the overall chemical composition of the rock. When this happens, we say that the rock has undergone **metasomatism.**

The water in hydrothermal fluids comes from several sources. Some originates as groundwater that enters the crust at the Earth's surface and then sinks downward, some is released from magma when the magma rises, and some originates as the product of metamorphic reactions themselves. What is the final location of the ions that are dissolved in the hydrothermal fluids that pass through a body of rock? Some react locally with other minerals within the rock, and they become incorporated into new mineral crystals nearby. Others precipitate in spaces that open up where rock cracks,

FIGURE 8.6 Changes in grain shape and orientation lead to the development of preferred orientation.

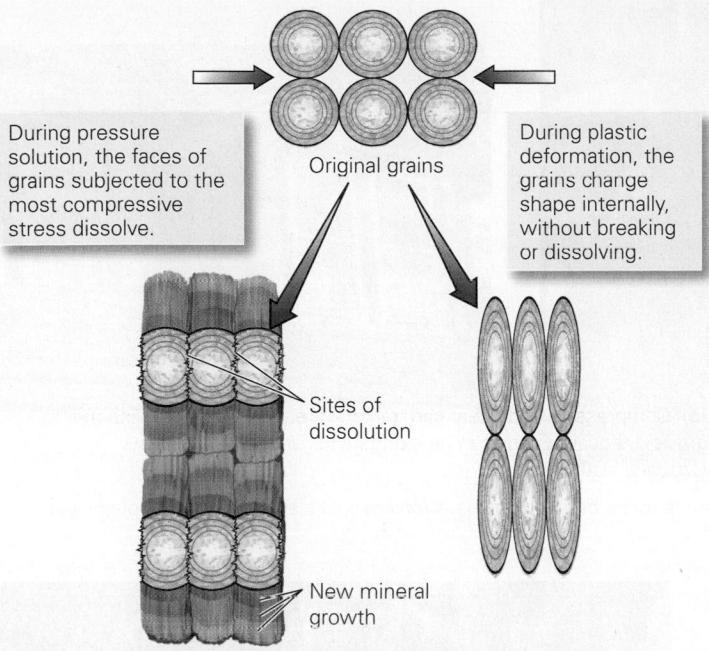

During pressure solution, the faces of grains subjected to the most compressive stress dissolve.

Original grains

During plastic deformation, the grains change shape internally, without breaking or dissolving.

Sites of dissolution

New mineral growth

(a) Application of differential stress during metamorphism can change the shapes of grains in two ways.

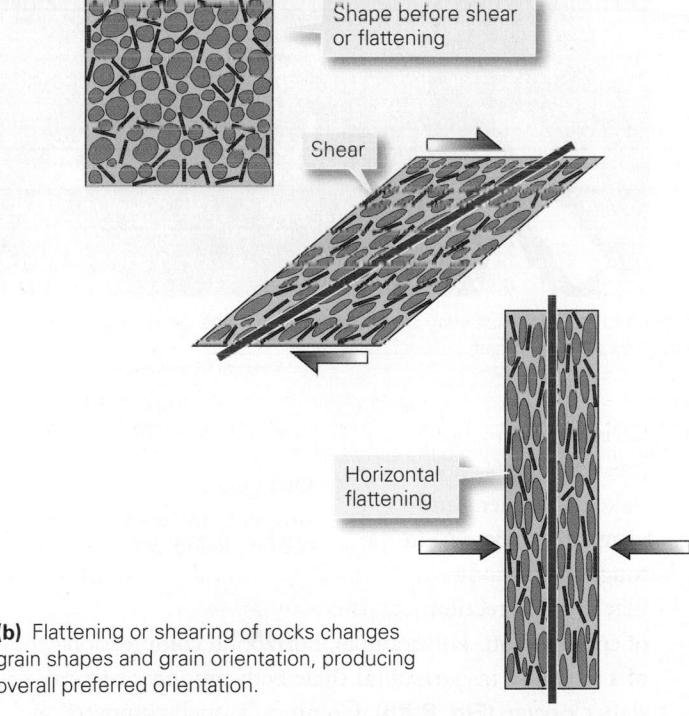

Shape before shear or flattening

Shear

Horizontal flattening

(b) Flattening or shearing of rocks changes grain shapes and grain orientation, producing overall preferred orientation.

producing a mineral filled crack known as a vein (Fig. 8.7). Veins commonly contain milky-white quartz, so they stand out against darker-colored rock surrounding them. In some cases, hydrothermal fluids carry ions hundreds of meters to tens of kilometers from where the ions originated.

FIGURE 8.7 Veins in metamorphic rock on the coast of Wales. Note that the veins themselves have been plastically deformed into complex shapes.

Vein

TAKE-HOME MESSAGE

Metamorphic rocks form in response to changes in temperature and pressure, application of compression and shear, and/or interaction with hydrothermal fluids. During metamorphism, new mineral grains grow as pre-existing ones disappear, and rock may develop a preferred orientation due to the alignment of inequant grains. A variety of processes, such as recrystallization, chemical reaction, and plastic deformation may be involved.

QUICK QUESTION: Can the overall composition of rock change during metamorphism?

8.3 Types of Metamorphic Rocks

Coming up with a way to classify and name the great variety of metamorphic rocks on the Earth hasn't been easy. After decades of debate, geologists have found it most convenient to divide metamorphic rocks into two fundamental classes: foliated rocks and nonfoliated rocks. Each class contains several rock types, as we now see.

Foliated Metamorphic Rocks

To understand foliated metamorphic rocks, we first need to discuss the nature of foliation in more detail. The word comes from the Latin *folium*, for leaf. Geologists use *foliation* to refer

FIGURE 8.8 Slate is a foliated metamorphic rock that forms at relatively low temperatures and pressures.

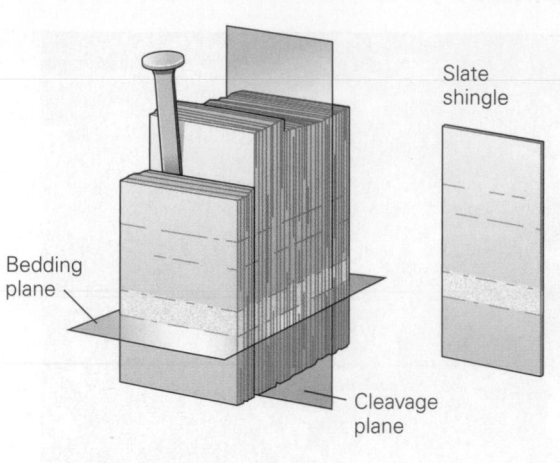

(a) A block of slate splits easily along cleavage planes, which may be at a high angle to the bedding planes. Workers split slate to produce roof shingles that, when overlapped, make a watertight surface.

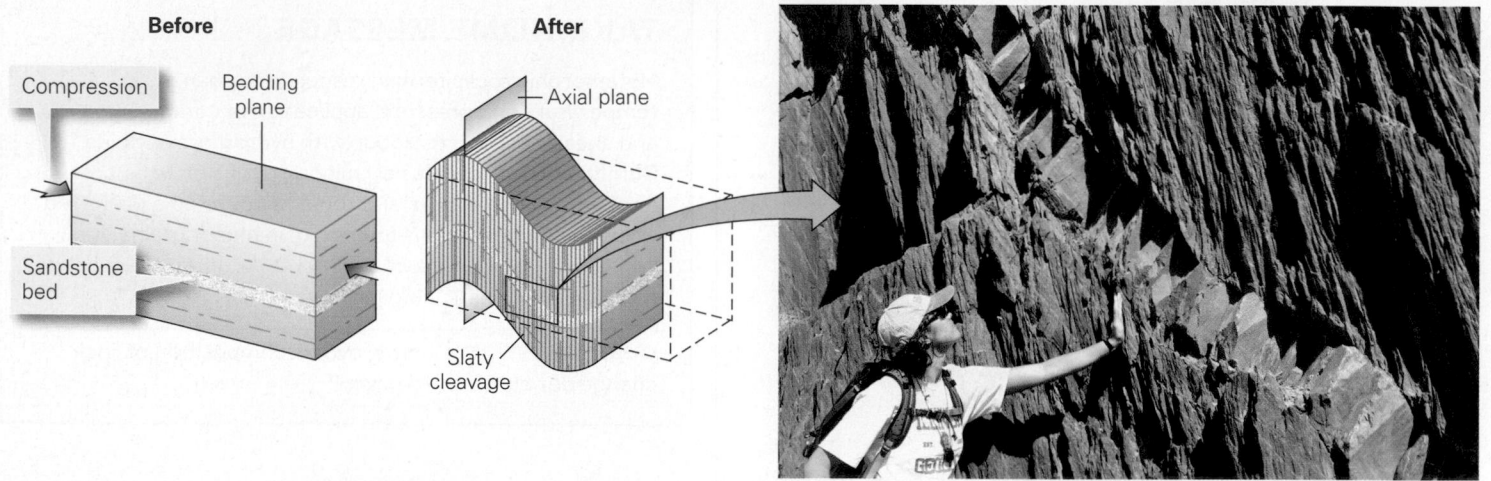

(b) Slaty cleavage forms in response to compression. In this example, beds also bend to form folds as slaty cleavage develops. The cleavage plane tends to be oriented parallel to the axial plane, an imaginary surface that, simplistically, divides the fold in half.

to an assemblage of parallel planar surfaces in a metamorphic rock. Foliation can give metamorphic rocks a striped or streaked appearance in an outcrop. A foliated metamorphic rock has foliation either because it has a preferred orientation (that is, it contains inequant mineral crystals that are aligned parallel to one another), or because it has alternating dark-colored and light-colored layers.

Foliated metamorphic rocks can be distinguished from one another by their composition, their grain size, and the nature of their foliation. Common examples include the following:

- *Slate:* The finest-grained foliated metamorphic rock, **slate**, forms by metamorphism of shale or mudstone—rocks composed of clay—under relatively low pressures and temperatures. Slate contains a type of foliation called *slaty cleavage*, which allows it to split into thin sheets that make excellent roofing shingles **(Fig. 8.8a)**. Slaty cleavage

develops when compression causes clay flakes to reorient and grow in an orientation roughly perpendicular to the direction of compression. For example, horizontal compression of a sequence of horizontal shale beds produces vertical slaty cleavage **(Fig. 8.8b)**. Commonly, such compression also causes the layers to bend into curves called *folds* (see Chapter 11). The preferred orientation of clay in slate develops in response to pressure solution and recrystallization—grains lying at an angle to the cleavage plane dissolve, whereas grains parallel to the cleavage plane grow. In addition, during this process, less-soluble inequant grains may rotate into the plane of cleavage.

> **Did you ever wonder . . .**
> why slate makes such nice roofing shingles?

- *Phyllite:* **Phyllite** is a fine-grained metamorphic rock with a foliation caused by the preferred orientation of very fine-grained white mica. The name comes from the Greek word *phyllon*, meaning leaf, as does the word *phyllo*, the flaky dough in Greek pastry. The alignment of translucent fine-grained mica gives phyllite a silky sheen known as *phyllitic luster* **(Fig. 8.9a)**. Phyllite forms when slate is heated to a temperature high enough to produce a new assemblage of metamorphic minerals (fine-grained white mica) out of clay.

- *Metaconglomerate:* Under the metamorphic conditions that produce slate or phyllite, a protolith of conglomerate becomes **metaconglomerate**. Typically, pressure solution and plastic deformation flatten pebbles and cobbles of the rock into pancake-like shapes. The alignment of inequant clasts defines a foliation **(Fig. 8.9b)**.

- *Schist:* **Schist**, a medium- to coarse-grained metamorphic rock type, possesses a type of foliation, called *schistosity*, defined by the preferred orientation of large mica flakes (muscovite or biotite; **Fig. 8.9c**). Schist forms at temperatures even higher than those needed to form phyllite, and it differs from phyllite in that it has larger mica grains. Typically, schists also contain other minerals, such as quartz, feldspar, garnet, and amphibole—the specific minerals that grow depend on the chemical composition of the protolith. Schist can form from a shale protolith, but also from a great variety of other protoliths, as long as the protolith contains the appropriate elements to make mica. In some cases, certain mineral grains in schists grow to be much larger than surrounding grains. For example, garnet crystals in schist may become many times larger than those of other minerals (see Fig. 8.3b). Geologists refer to especially large crystals that grow in a metamorphic rock as *porphyroblasts*. Smaller grains surrounding porphyroblasts constitute a rock's *matrix*.

- *Gneiss:* **Gneiss** generally forms at temperatures and pressures even higher than those that give rise to schist. This metamorphic rock type has layers of differing mineral composition, manifested by alternating darker- and lighter-colored bands. Typically, in rock with such *compositional layering*, the layers range in thickness from millimeters to meters, and their presence gives the rock a striped appearance, known as *gneissic banding* **(Fig. 8.10)**. The contrasting colors in gneiss represent contrasting mineral compositions. Light-colored layers contain predominantly felsic minerals, such as quartz and feldspar, whereas dark-colored layers contain predominantly mafic minerals, such as amphibole, pyroxene, and biotite. If gneiss contains mica, the mica-rich layers typically display schistosity. Gneiss can form from either sedimentary or igneous protoliths.

FIGURE 8.9 Examples of foliated metamorphic rocks formed at high temperatures and pressures.

(a) During formation of phyllite, clay recrystallizes to form tiny mica flakes that reflect light, giving the rock a sheen.

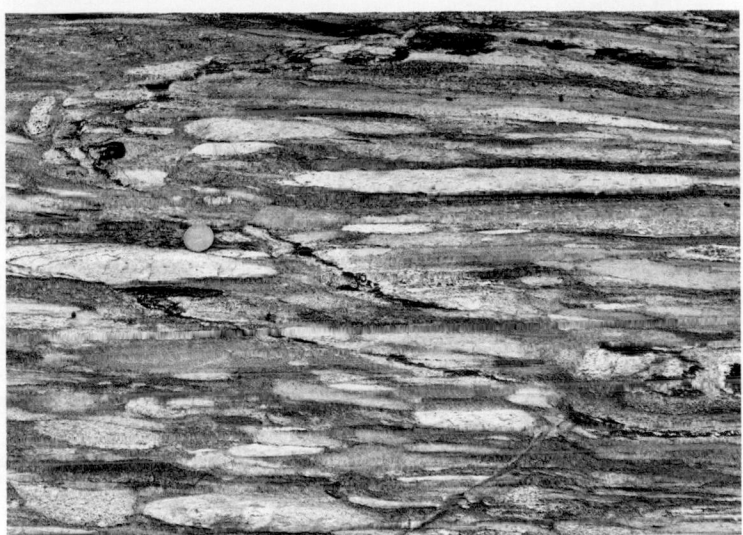

(b) In metaconglomerate, pebbles and cobbles flatten into a pancake shape without cracking.

(c) A schist contains coarse mica flakes, along with other metamorphic minerals.

(a) These huge (>100-m-high) cliffs in Greenland consist of gneiss. Some of the bands of light and dark are tens of meters across.

(b) A gneiss from northern Scotland. The light layers contain more felsic minerals, whereas the dark layers contain more mafic minerals.

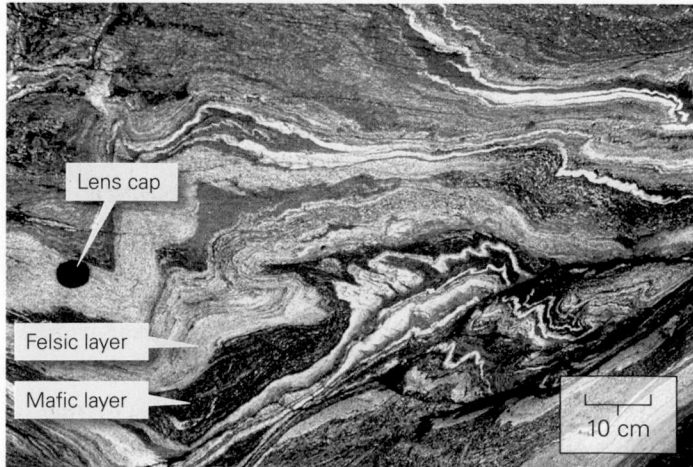

Lens cap

Felsic layer

Mafic layer

10 cm

(c) In this outcrop of gneiss from Brazil, the gneissic banding has been contorted by flow in the rock when it was at high temperature.

How does the banding in gneiss form? If the rock has a sedimentary protolith, the banding may evolve from bedding, because the composition of a metamorphic band reflects the composition of the bed from which it formed. Banding can also form by extreme shear of a nonhomogeneous rock, whether the protolith was sedimentary or igneous, if the rock accommodates the shear plastically, without breaking (Fig. 8.11a). Such plastic flow stretches, folds, and smears out pre-existing compositional contrasts in the rock and transforms them into aligned sheets. (To picture this process, take a rolling pin and flatten a ball consisting of red dough embedded in white dough. Then fold the layer in half and roll it flat again. If you cut through the end product, you'll see thin alternating layers of different-colored dough.) Finally, gneissic banding can develop by an incompletely understood process called *metamorphic differentiation*. This process may involve dissolution of minerals in some layers and migration (diffusion) of the chemical components of those minerals to other layers, where new minerals then grow (Fig. 8.11b). In effect, chemical reactions segregate different minerals into different layers.

- *Migmatite:* If the temperature rises high enough, gneiss may begin to melt, producing felsic magma and residual, still-solid mafic metamorphic rock. If the melt freezes again before flowing out of the source area, a mixture of igneous rock and relict metamorphic rock, called **migmatite**, forms (Fig. 8.12). Plastic flow during the process can contort the light-colored felsic rock and the dark-colored mafic rock into complex shapes.

Nonfoliated Metamorphic Rocks

Nonfoliated metamorphic rocks contain minerals that recrystallized or grew during metamorphism, but the rock overall has no foliation. The lack of foliation means either that metamorphism occurred in the absence of compression and shear, or that most of the new crystals are equant. We list below some of the metamorphic rock types that can occur without foliation.

- *Hornfels:* **Hornfels** is a fine-grained nonfoliated rock that contains a variety of metamorphic minerals (some equant and some inequant). The specific mineral assemblage in a hornfels depends on the composition of the protolith and on the temperature and pressure of metamorphism. Inequant minerals that grow in hornfels are randomly oriented—in other words, they do not display a preferred orientation.

- *Quartzite:* Most **quartzite** forms by the metamorphism of pure quartz sandstone. During metamorphism, pre-existing quartz grains recrystallize, yielding new, larger

FIGURE 8.11 The formation of gneiss, which takes place at very high temperatures and pressures.

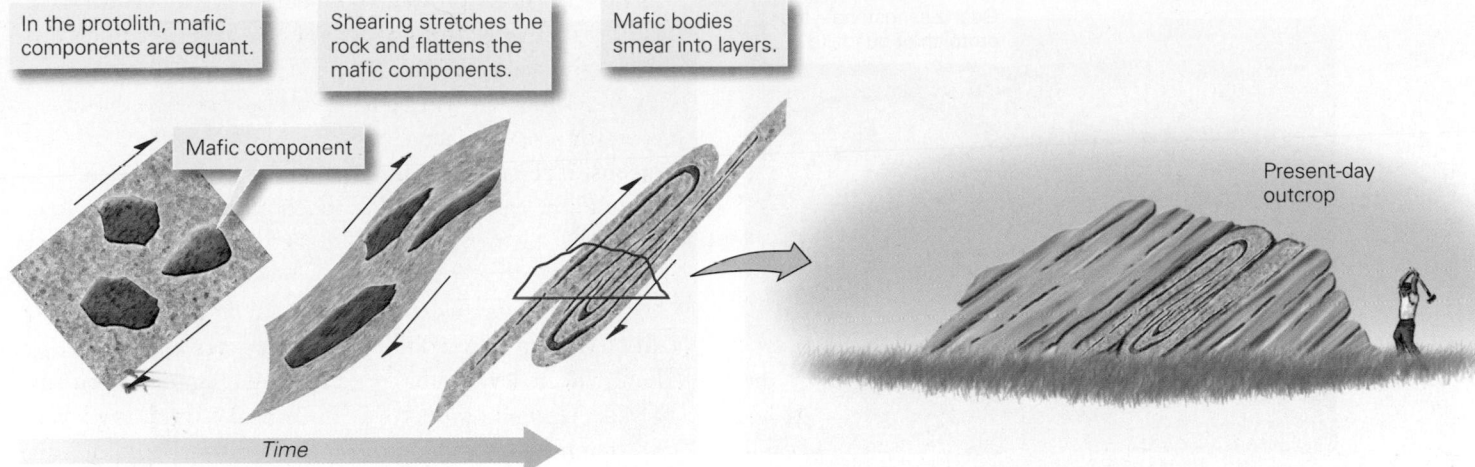

In the protolith, mafic components are equant.

Shearing stretches the rock and flattens the mafic components.

Mafic bodies smear into layers.

Mafic component

Present-day outcrop

Time

(a) Formation of gneiss, in some cases, involves extreme shear. Original contrasting rock types are smeared into parallel layers.

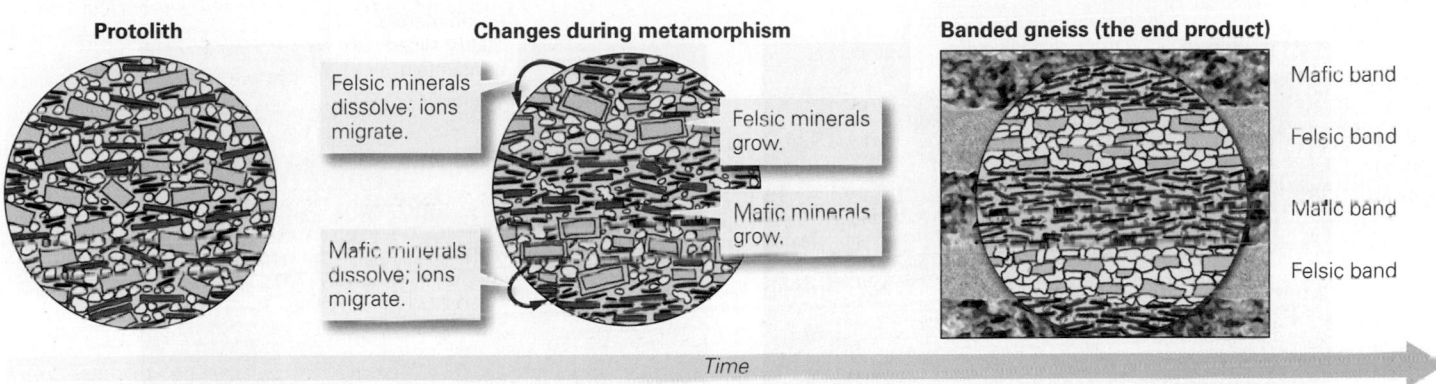

Protolith

Changes during metamorphism

Banded gneiss (the end product)

Felsic minerals dissolve; ions migrate.

Felsic minerals grow.

Mafic minerals grow.

Mafic minerals dissolve; ions migrate.

Mafic band

Felsic band

Mafic band

Felsic band

Time

(b) Gneiss may also form by metamorphic differentiation, during which chemical reactions cause felsic and mafic minerals to grow in distinct, separate layers.

FIGURE 8.12 An outcrop of migmatite in Ontario, Canada contains both light-colored (felsic) igneous rock and dark-colored (mafic) metamorphic rock. The mixture of the two rock types makes migmatite resemble marble cake.

Light (felsic) bands formed from melt.

Dark (mafic) bands remained solid.

grains. In the process, the distinction between cement and grains disappears, open pore space disappears, and the grains become interlocking. When quartzite cracks, the fracture cuts across grain boundaries—in contrast, fractures in sandstone curve around grains. Quartzite looks glassier than sandstone and does not have the grainy, sandpaper-like surface characteristic of sandstone **(Fig. 8.13a)**. Depending on the impurities it contains, quartzite can vary in color from white to gray, purple, or green. Most quartzite has no foliation because it does not contain aligned mica or compositional layering. In some cases, however, quartz grains deform plastically and become pancake shaped. The alignment of "pancakes" defines a foliation. (To avoid confusion, quartzite with this texture should be called *foliated quartzite*.)

- *Marble:* The metamorphism of limestone yields **marble** **(Fig. 8.13b)**. During the formation of marble, the calcite composing the protolith recrystallizes, so fossil shells, pore space, and the distinction between grains and

FIGURE 8.13 Examples of quartzite and marble, which are typically, but not always, nonfoliated metamorphic rocks.

Quartz sandstone—the protolith of quartzite.

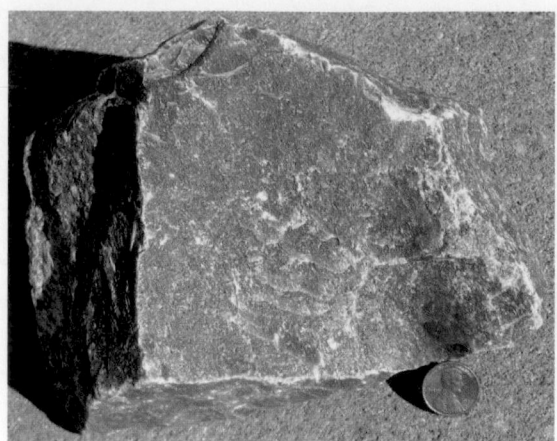

(a) In this sandstone from Kuwait (left), the sand grains stand out. In contrast, this nonfoliated maroon quartzite from Wisconsin (right) looks glassy and breaks on smooth fractures that cut across grains.

In this unmetamorphosed limestone, fossils and bedding are visible.

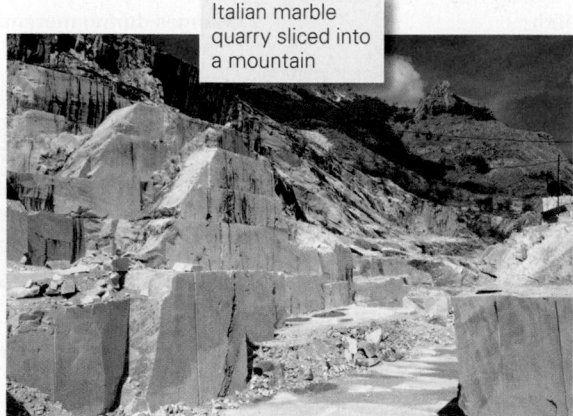

Italian marble quarry sliced into a mountain

(b) In this unmetamorphosed limestone from New York (left), you can see fossils and shell fragments. Such grains are not visible in the white marble (right) exposed in an Italian quarry.

(c) Sculptors, such as Michelangelo, like to work with nonfoliated white marble due to its softness and uniform texture.

(d) This marble floor has color banding. The layers became contorted during metamorphism.

cement disappears. Marble, therefore, typically consists of a fairly uniform mass of interlocking calcite crystals.

Sculptors love to work with marble because its relative softness and uniform texture give it the cohesiveness and homogeneity they need to fashion large, smooth, highly detailed sculptures **(Fig. 8.13c)**. Marble comes in a variety of colors—white, pink, green, and black—depending on the impurities it contains. (Michelangelo, one of the great Italian Renaissance artists, sought large, unbroken blocks of creamy white marble from quarries in western Italy for his masterpieces; see Fig. 8.13b). If the original protolith contained layers with different impurities, the resulting marble has color banding that makes it a prized decorative stone **(Fig. 8.13d)**. Marble can flow like soft plastic under metamorphic conditions, a process that smears out different-colored portions of marble into beautiful contorted, curving bands.

- *Amphibolite:* Metamorphism of mafic rocks (basalt or gabbro) can't produce quartz and muscovite, for these rocks don't contain the right mix of chemicals to yield such minerals. Rather, they transform into *amphibolite*, a dark-colored metamorphic rock containing predominantly hornblende (a type of amphibole) and plagioclase (a type of feldspar), and in some cases, a tiny bit of biotite. Where subjected to differential stress, amphibolites can develop a foliation, but the foliation tends to be poorly defined because the rock contains very little mica.

Chemical Composition of Metamorphic Rocks

Up to this point, we've emphasized the importance of temperature and pressure in determining the mineral assemblage in a metamorphic rock. Let's not forget that composition of the protolith plays a key role in determining which minerals form as well. For example, it's impossible to form a biotite-rich schist from a pure quartz sandstone because biotite contains elements, such as iron, that do not occur in quartz.

To distinguish among different compositions found in metamorphic rocks, geologists use the following terms: (1) *Pelitic* metamorphic rocks form from a sedimentary protolith composed of shale or mudstone. The clay in shale or mudstone contains a variety of elements, including aluminum. So pelitic metamorphic rocks contain aluminum-bearing minerals, such as muscovite [$KAl_2(AlSi_3O_{10})(OH)_2$]. (2) *Mafic* metamorphic rocks contain relatively little silica and an abundance of iron and magnesium. During metamorphism, iron- and magnesium-rich minerals such as biotite and hornblende grow. (3) *Calcareous* metamorphic rocks form from calcium-rich sedimentary rocks (limestone) and contain calcite ($CaCO_3$). (4) *Quartzo-feldspathic* metamorphic rocks form from protoliths (such as granite) that contain mostly quartz and feldspar.

8.4 Defining Metamorphic Intensity

Not all metamorphism takes place under the same physical conditions. For example, rocks carried to a great depth beneath a mountain range undergo more intense metamorphism than do rocks closer to the surface. Geologists use the term **metamorphic grade** in a somewhat informal way to indicate the intensity of metamorphism, meaning the amount or degree of metamorphic change **(Fig. 8.14a)**. Metamorphic grade depends primarily on temperature because temperature plays the dominant role in determining the extent of recrystallization and neocrystallization that takes place during metamorphism. Metamorphic rocks that form at relatively low temperatures (between about 250°C and 400°C) are *low-grade* rocks, and metamorphic rocks that form at relatively high temperatures (over 600°C) are *high-grade* rocks. *Intermediate-grade* rocks form at temperatures between these two extremes. To provide a more complete indication of the intensity of metamorphism, geologists use the related concept of *metamorphic facies* **(Box 8.1)**.

Different grades of metamorphism yield different metamorphic mineral assemblages **(Fig. 8.14b)**. As grade increases, recrystallization produces coarser grains, and neocrystallization produces mineral assemblages that are stable at higher temperatures and pressures. Geologists refer to metamorphism taking place while temperature and pressure progressively increase as *prograde metamorphism*. As grade increases, metamorphic reactions release water, so high-grade rocks tend to be "drier" than low-grade rocks. This means that minerals in high-grade rocks do not contain minerals with –OH in their chemical formula, whereas minerals in lower-grade rocks may.

To understand prograde metamorphism, consider the changes that a shale undergoes during mountain building as

FIGURE 8.14 Intensity of metamorphism is indicated by metamorphic grade.

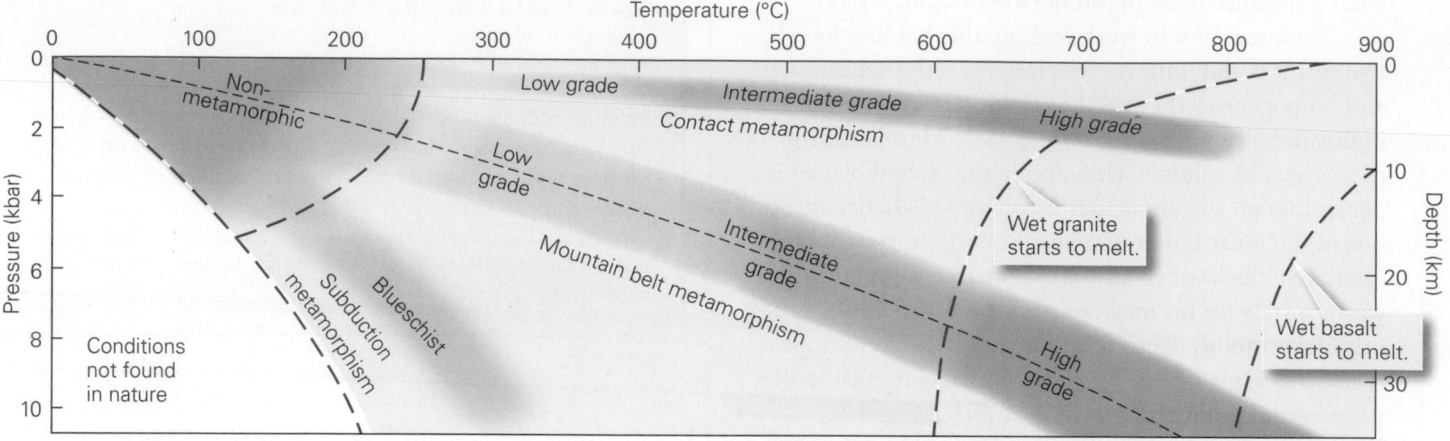

(a) This graph depicts the approximate temperatures and pressures of metamorphic grades. Different conditions occur in different geologic settings.

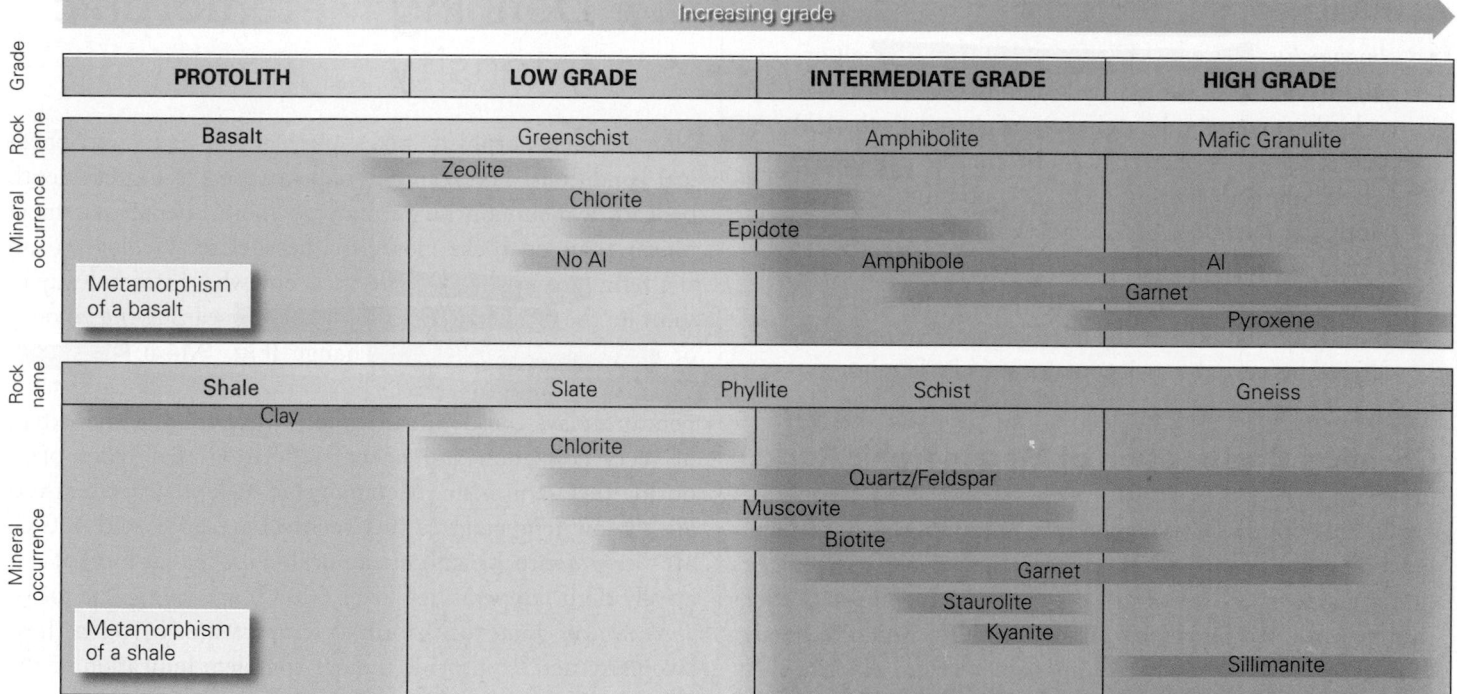

(b) The metamorphic minerals that form in a given rock depend on metamorphic grade and protolith composition. This chart contrasts important metamorphic minerals that form from a basalt protolith with those formed from a shale protolith.

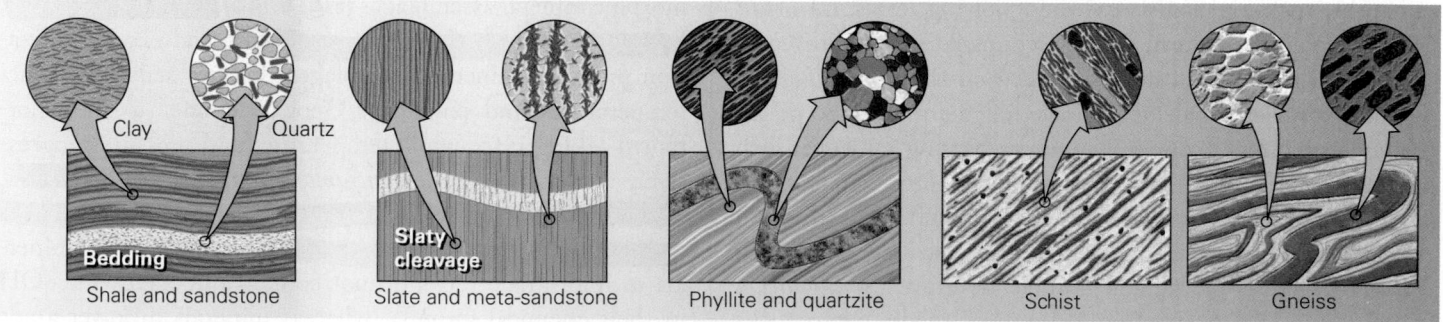

(c) Here we see the consequences of the progressive metamorphism of shale and sandstone during mountain building. The background colors indicate the range of grades in which key metamorphic minerals form. Note how textures change with increasing grade.

BOX 8.1 CONSIDER THIS . . .

Metamorphic Facies

In the early years of the 20th century, geologists working in Scandinavia, where erosion by glaciers has left beautiful, nearly unweathered outcrops, came to realize that metamorphic rocks, in general, do not consist of a hodgepodge of minerals formed at different times and in different places, but rather consist of a distinct assemblage of minerals that grew in association with one another at a certain pressure and temperature in a given time interval. It seemed that the mineral assemblage present in a rock more or less represents a condition of *chemical equilibrium*, meaning that the chemicals making up the rock organize into a group of mineral grains that are—to anthropomorphize a bit—comfortable with one another and their surroundings and thus do not feel the need to change further. These geologists concluded that the specific mineral assemblage in a rock depends both on pressure and temperature conditions and on the composition of the protolith.

This discovery led the geologists to propose the concept of metamorphic facies. A **metamorphic facies** is a set of metamorphic mineral assemblages indicative of a certain range of pressure and temperature. Each specific assemblage in a facies reflects a specific protolith composition. According to this definition, a given metamorphic facies includes several different kinds of rocks that differ from one another in terms of chemical composition and, therefore, mineral content—but all the rocks of a given facies formed under roughly the same temperature and pressure conditions. Geologists recognize several facies: zeolite, prehnite-pumpellyite, hornfels, greenschist, amphibolite, blueschist, eclogite, and granulite. The name of each facies comes from a distinctive feature or mineral found in some of the rocks of the facies.

We can represent the approximate conditions under which the rocks of a metamorphic facies formed by using a pressure-temperature graph **(Fig. Bx8.1)**. Each area on the graph, labeled with a facies name, represents the approximate range of temperatures and pressures in which mineral assemblages characteristic of that particular facies form. For example, a rock subjected to the pressure and temperature at Point A (4.5 kbar and 400°C) develops a mineral assemblage characteristic of the greenschist facies. This facies' mineral assemblages commonly contain chlorite, a green mica. As the graph implies, the boundaries between facies cannot be precisely defined because the transitions between facies are gradual.

We can also portray the geothermal gradients of different crustal regions on the graph. Beneath mountain ranges, for example, the geothermal gradient passes through the zeolite, greenschist, amphibolite, and granulite facies. In contrast, in the accretionary prism that forms at a subduction zone, temperature increases slowly with increasing depth, so blueschist assemblages can form. Note that rocks of the eclogite facies occur at great depth.

FIGURE Bx8.1 The common metamorphic facies. Boundaries between the facies are depicted as wide bands because they are gradational. Note that some amphibolite-facies rocks and all granulite-facies rocks form at pressure-temperature conditions to the right of the melting curve for wet granite. Such metamorphic rocks develop only if the protolith is dry.

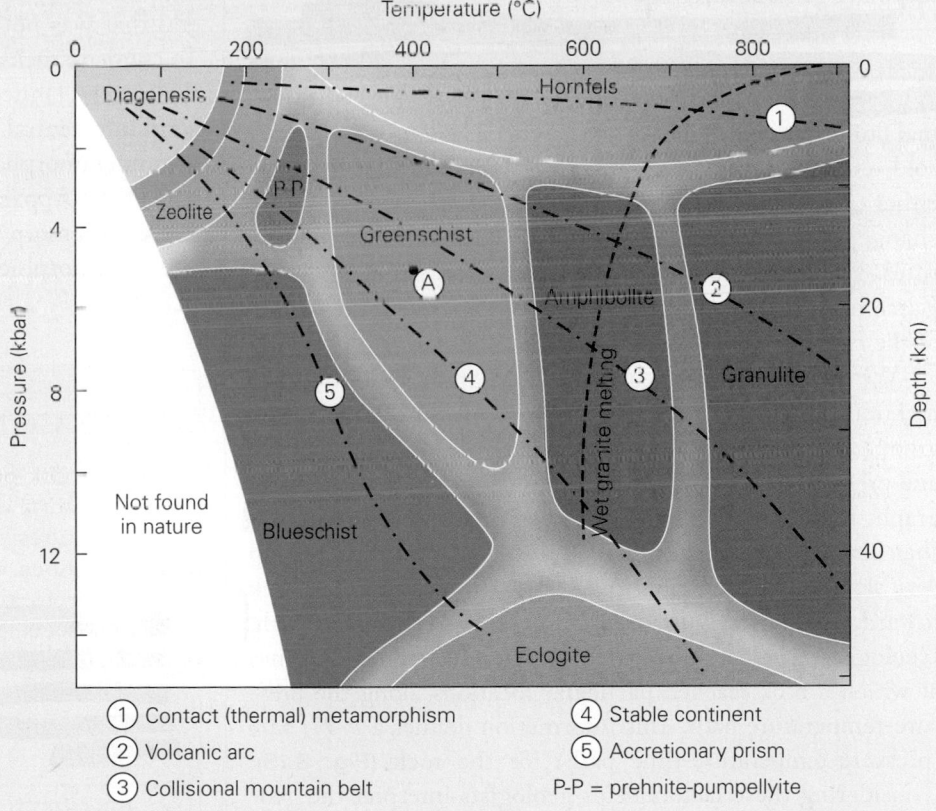

1. Contact (thermal) metamorphism
2. Volcanic arc
3. Collisional mountain belt
4. Stable continent
5. Accretionary prism

P-P = prehnite-pumpellyite

it travels from a location near the Earth's surface to a depth of 30 km below the surface **(Fig. 8.14c)**. The clay flakes in shale lie more or less parallel to the bedding plane. Under low-grade metamorphic conditions and differential stress, the shale transforms into slate. In slate, the clay flakes grow larger, become better formed, and align parallel to the cleavage plane. As metamorphic grade increases a little more, the clay flakes decompose, and new crystals of fine-grained white mica, as well as new crystals of chlorite and quartz, grow, transforming the rock into phyllite. Under intermediate-grade conditions, the minerals in phyllite react and decompose, yielding atoms that combine to produce large crystals of mica (such as muscovite and biotite) as well as other minerals (such as garnet). The reactions also release water, which escapes. In our example, metamorphism takes place under differential stress, so the micas grow with a preferred orientation and the rock becomes a schist. During metamorphic reactions under high-grade conditions, yet another assemblage of minerals forms. High-grade rocks do not contain much, if any, mica, because mica contains −OH and tends to decompose and release water at high temperatures. In fact, high-grade rock typically includes water-free minerals, such as feldspar, quartz, pyroxene, and garnet. As mica disappears, the rock loses its schistosity but may have gneissic layering.

Metamorphism that takes place while temperature and pressure progressively decrease is known as *retrograde metamorphism*. For retrograde metamorphism to occur, hydrothermal fluids must enter the rock to add back water. In fact, under cold and dry conditions, retrograde metamorphic reactions either cannot proceed or take place too slowly to cause much change, because atoms cannot diffuse easily at low temperatures. For this reason, high-grade rocks formed early during Earth history have survived and can be exposed at the surface of the Earth today.

We can represent the concepts of prograde and retrograde metamorphism as a path plotted on a pressure-temperature graph. As a rock is buried progressively deeper, temperature and pressure increase; the rock follows a prograde path on the graph. If the rock is buried quickly, pressure increases faster than temperature (rocks are good insulators and thus heat up very slowly). Later, as uplift and erosion move the rock back toward the Earth's surface, it follows the retrograde path. Geologists have developed techniques to determine the times at which a rock reached particular locations along the pressure-temperature path. This information defines a *P-T-t path* (pressure-temperature-time path) for the rock **(Fig. 8.15)**. Considering these factors helps geologists interpret the geologic history of the rock.

The presence of certain minerals, known as **index minerals,** in a rock indicates the approximate metamorphic grade of the rock. A line on a map along which an index

FIGURE 8.15 A P-T-t path for a hypothetical rock. The dates in parentheses indicate the time at which the specified pressure and temperature conditions were achieved. P_m is the peak pressure and T_m is the peak temperature.

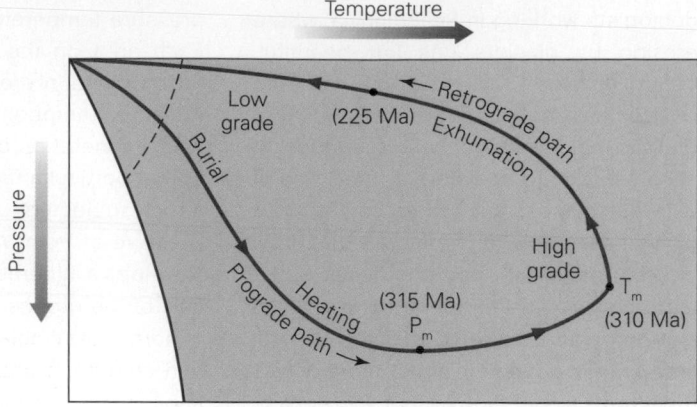

mineral first appears is an *isograd* (from the Greek *iso*, meaning equal). All points along an isograd have approximately the same metamorphic grade. **Metamorphic zones,** the regions between two isograds, are named after an index mineral that was not present in the adjacent, lower-grade zone. To compare rocks of different grades, you could take a hike in the eastern United States from central New York State eastward into central Massachusetts. Your path starts in a region of unmetamorphosed rocks, and takes you into the internal part of the Appalachian mountain belt, which contains high-grade metamorphic rocks. As a consequence, you cross several metamorphic zones **(Fig. 8.16)**.

TAKE-HOME MESSAGE

Metamorphic grade informally specifies the intensity of metamorphism; high-grade rocks form at higher temperatures, and low-grade rocks form at lower temperatures. The mineral assemblage that a metamorphic rock contains reflects its grade. A region in which a specific grade of rock occurs is a metamorphic zone, and the boundary between zones is an isograd. A metamorphic facies is a set of mineral assemblages indicative of certain pressure and temperature conditions.

QUICK QUESTION: The geothermal gradient on Mars (~8°C/km) is less than that on the Earth, and in some places the crust of Mars is only 30 km thick. Can high-grade metamorphic rocks form in this crust?

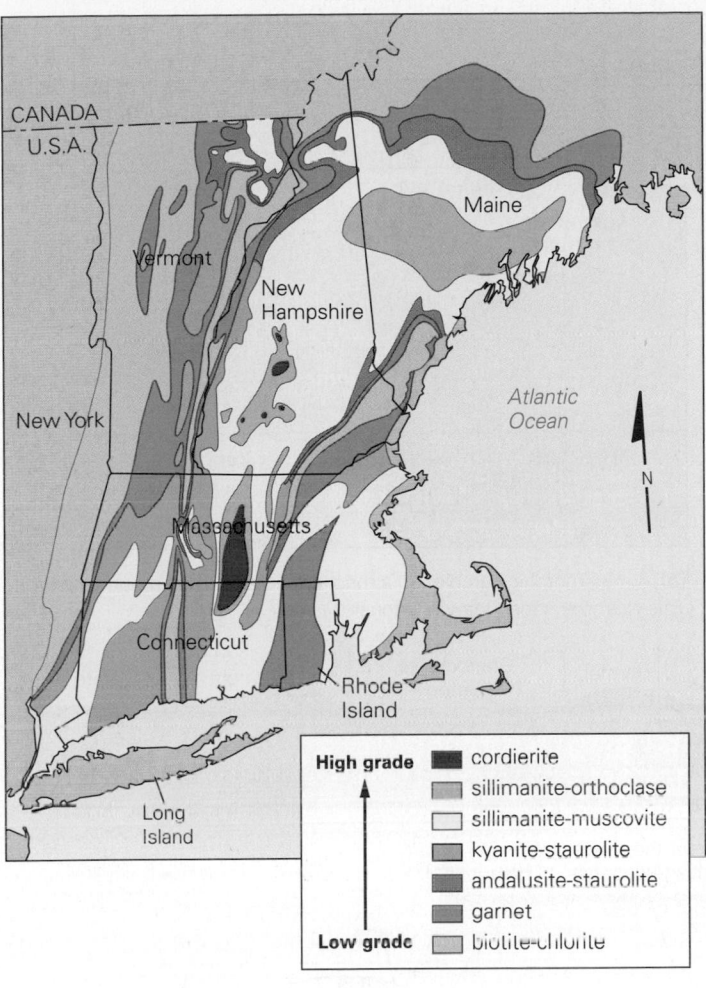

FIGURE 8.16 Metamorphic zones and isograds.

High grade

- cordierite
- sillimanite-orthoclase
- sillimanite-muscovite
- kyanite-staurolite
- andalusite-staurolite
- garnet

Low grade

- biotite-chlorite

extent to which rocks interact with hydrothermal fluids, varies with location.

Thermal or Contact Metamorphism

Imagine a hot magma that rises from greater depth beneath the Earth's surface and intrudes into cooler rock at a shallower depth. Heat flows from the magma into the wall rock, for heat always flows from hotter to colder materials. As a consequence, the magma cools and solidifies while the wall rock heats up. In addition, hydrothermal fluids circulate through both the intrusion and the wall rock. As a consequence of the heat and hydrothermal fluids, the wall rock undergoes metamorphism. The highest-grade rocks form immediately adjacent to the pluton, where the temperatures are highest, and progressively lower-grade rocks form farther away. The distinct belt of metamorphic rock that forms around an igneous intrusion is called a **metamorphic aureole**, or *contact aureole* **(Fig. 8.18a)**. The width of an aureole depends on the size and shape of the intrusion and on the amount of hydrothermal circulation—larger intrusions produce wider aureoles.

Geologists refer to the local metamorphism caused by an igneous intrusion either as **thermal metamorphism**, to emphasize that it develops in response to heat without a change in pressure and without differential stress, or as **contact metamorphism**, to emphasize that it develops adjacent to the contact of an intrusion with its wall rock **(Box 8.2)**. Because this metamorphism takes place without application of compression or shear, aureoles contain hornfels, a nonfoliated metamorphic rock. Contact metamorphism occurs anywhere that the intrusion of plutons occurs. In the context of plate tectonics theory, plutons intrude into the crust at convergent boundaries, in rifts, and during the mountain building that takes place when continents collide.

8.5 Where Does Metamorphism Occur?

So far, we've discussed the nature of changes that occur during metamorphism, the causes of metamorphism (heat, pressure, differential stress, and hydrothermal fluids), the rock types that form as a result of metamorphism, and the concepts of metamorphic grade and metamorphic facies. With this background, let's now examine the wide range of geologic settings on the Earth where metamorphism takes place, as viewed in the context of plate tectonics theory (**Geology at a Glance**, pp. 264–265). The conditions under which metamorphism happens differ among these settings because the geothermal gradient **(Fig. 8.17)**, the extent to which rocks endure differential stress during metamorphism, and the

FIGURE 8.17 The geothermal gradient varies among different locations in a continent. The solid lines are *isotherms* (in °C)—temperature in rock along an isotherm has the same value.

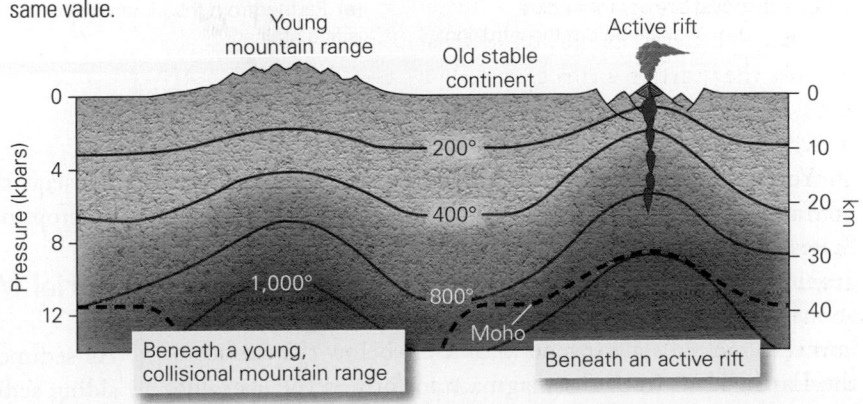

FIGURE 8.18 Contact (thermal) metamorphism occurs in an aureole adjacent to an igneous intrusion. This type of metamorphism produces hornfels. The grade of hornfels decreases away from the contact.

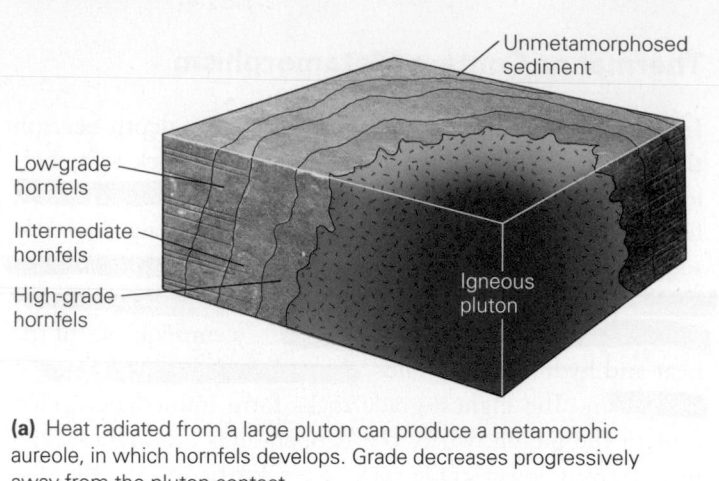

(a) Heat radiated from a large pluton can produce a metamorphic aureole, in which hornfels develops. Grade decreases progressively away from the pluton contact.

(b) A metamorphic aureole, on a map, appears as a ring around a pluton. This example is the Onawa pluton in Maine.

(c) A cross section shows the aureole in the subsurface. Grade decreases away from the contact with the pluton.

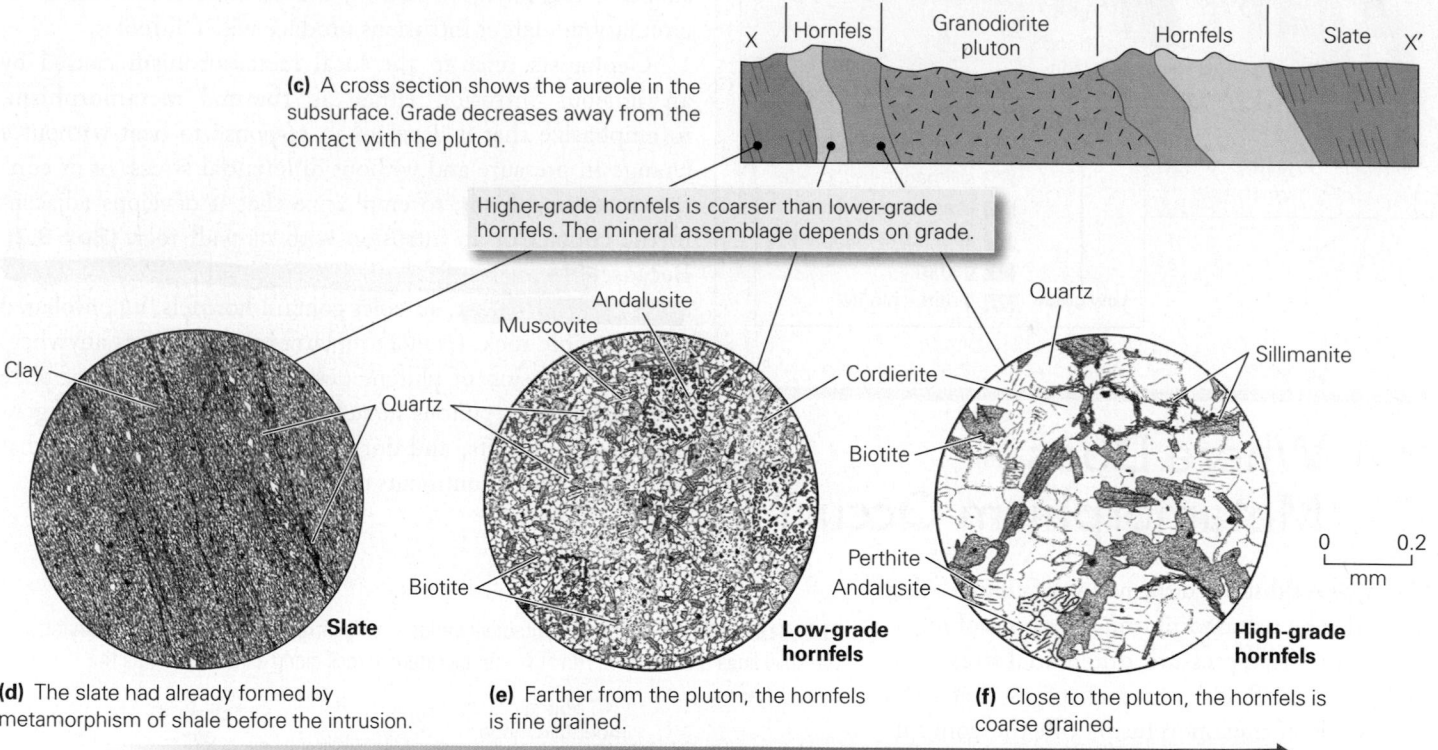

Higher-grade hornfels is coarser than lower-grade hornfels. The mineral assemblage depends on grade.

(d) The slate had already formed by metamorphism of shale before the intrusion.

(e) Farther from the pluton, the hornfels is fine grained.

(f) Close to the pluton, the hornfels is coarse grained.

Increasing grade

You can see a classic example of contact metamorphism by traveling to the state of Maine in the northeastern United States. Here you will find a 14-km-long by 4-km-wide granitic pluton, named the Onawa pluton, which formed about 400 million years ago when an 850°C magma intruded into wall rock of slate several kilometers below the surface of the Earth. Heat from the magma transformed the slate into hornfels in an aureole that reaches a maximum width of 2 km.

Subsequently, erosion stripped off overlying rock, exposing outcrops of the granodiorite and hornfels **(Fig. 8.18b–f)**.

Burial Metamorphism

As sediment becomes buried progressively deeper in a subsiding sedimentary basin, the pressure on it increases due to the increasing weight of overburden, and the temperature

BOX 8.2 CONSIDER THIS . . .

Pottery Making—An Analog for Thermal Metamorphism

A brick for the wall of an adobe house, an earthenware pot, a stoneware bowl, or a translucent porcelain teacup may all be formed from the same lump of soft clay, scooped from the surface of the Earth and shaped by human hands (Fig. Bx8.2). This pliable and slimy muck consists of water and very fine clay minerals and quartz grains left by the chemical weathering of rock. Fine potter's clay for making white china contains a particular clay mineral called *kaolinite*, named after the locality in China (Kauling, meaning high ridge) where it was originally discovered.

People in arid climates make adobe bricks by forming damp clay into blocks, which they then dry under the Sun. Such blocks can be used for construction only in arid climates because if it rains heavily, the blocks rehydrate and turn back into sticky muck—drying clay under the Sun does not change the structure of the clay minerals.

To make a more durable material, brick makers place clay blocks in a kiln and "fire," or bake, them at high temperatures. This process makes the blocks hard and impervious to water. Potters use the same process to make jugs. In fact, fired clay jugs that were used for storing wine and olive oil have been found intact in sunken Greek and Phoenician ships that have rested on the floor of the Mediterranean Sea for thousands of years! Clearly, the firing of a clay pot fundamentally and permanently changes clay in a way that makes it physically different. In other words, firing causes a thermal metamorphic change in the mineral assemblage that composes pottery. The extent of the transformation depends on the kiln temperature, just as the grade of metamorphic rock depends on temperature. Potters usually fire earthenware at about 1,100°C and stoneware (which is harder than a knife or fork) at about 1,250°C. To produce porcelain—fine china—the clay must partially melt at even higher temperatures. Just as it begins to melt, the potter cools it quickly. Such quenching of the melt creates glass, which gives porcelain its translucent, vitreous (glassy) appearance.

FIGURE Bx8.2 When metamorphosed, common mud becomes stronger and more durable.

(a) Mud can be shaped into blocks that, when dried, can be used to build houses like this one in Peru.

(b) Baking the mud turns it into much harder brick.

(c) At high temperatures, mud turns into porcelain, as in this plate from China.

increases due to the geothermal gradient. In the upper few kilometers, temperatures and pressures are low enough that the changes taking place represent diagenesis. But at depths greater than 8 to 15 km, depending on the geothermal gradient, temperatures may be high enough for metamorphic reactions to begin, and low-grade metamorphic rocks form. Geologists refer to metamorphism due only to the consequences of very deep burial as **burial metamorphism**. Of note, burial metamorphism destroys the organic molecules of oil. For this reason, oil drillers stop drilling when the bottom of the hole reaches depths at which burial metamorphism has begun.

Dynamic Metamorphism

Near the Earth's surface (in the upper 10 to 15 km) slip on faults can fracture rock, breaking it into angular fragments or even crushing it to a powder. But at greater depths, rock becomes warm enough to behave like soft plastic as shear along a fault takes place. During this process, the minerals in the rock recrystallize. We call this process **dynamic metamorphism** because it occurs solely as a consequence of shearing under metamorphic conditions, without requiring a change in temperature or pressure. The resulting rock, called a **mylonite**, is extremely fine

FIGURE 8.19 The formation of mylonite during dynamic metamorphism in a shear zone.

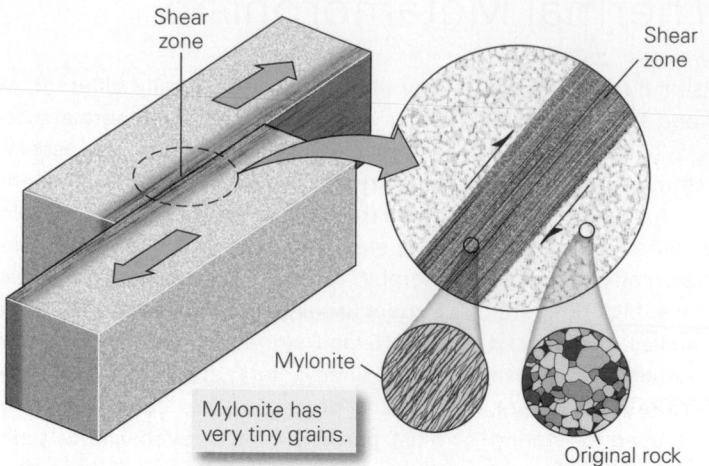

Shear zone

Mylonite

Mylonite has very tiny grains.

Original rock

(a) Shearing of a rock under plastic conditions causes its original crystals to divide into tiny crystals without breaking, forming a mylonite.

Shear zone

Foliation orientation

This shear zone has finer grains than the granitic rock from which it formed.

(b) Typically, a mylonite has a strong, very fine grain and a strong foliation, as in this example from Ontario.

grained and has a strong foliation that roughly parallels the fault **(Fig. 8.19)**. (More advanced geology books explain how the fine grains form from coarser ones.) Dynamic metamorphism can happen anywhere that faulting occurs at depth in the crust. So mylonites can be found at all plate boundaries, in rifts, and in collision zones.

Dynamothermal or Regional Metamorphism

During the development of large mountain ranges, in response to either convergent-boundary tectonics or continental collision, broad regions of crust undergo compression, and large slices of continental crust slip up and over other portions of the crust along faults. As a consequence, rock that was once near the Earth's surface along the margin of a continent may end up at great depth beneath the mountain range **(Fig. 8.20)**. In this environment, three changes happen: (1) the protolith heats up because of the geothermal gradient and because of igneous activity; (2) the protolith endures greater pressure because of the weight of overburden; and (3) the protolith undergoes compression and shearing. As a result of these changes, the protolith transforms into foliated metamorphic

rock. The type of foliated rock that forms depends on the grade of metamorphism—slate forms at shallower depths, whereas schist and gneiss form at greater depths. Since the metamorphism we've just described involves not only heat but also compression and shearing, we can call it **dynamothermal metamorphism**. Because such metamorphism tends to affect a broad region (tens to even hundreds of kilometers across and hundreds to thousands of kilometers long), so geologists also call it **regional metamorphism**. Erosion eventually removes the mountains, exposing a belt of metamorphic rock that once lay at depth.

Hydrothermal Metamorphism at Mid-Ocean Ridges

Hot magma rises beneath the axis of a mid-ocean ridge, so when cold seawater sinks through cracks down into the oceanic crust along the ridge, it heats up and transforms into hydrothermal fluid. This fluid then rises through the crust near the ridge, causing **hydrothermal metamorphism** of ocean-floor basalt **(Fig. 8.21)**. Eventually, the fluid escapes back into the sea at vents known as *black smokers* (see Chapter 4).

FIGURE 8.20 Dynamothermal metamorphism occurs during the development of mountain belts. The process is also called regional metamorphism.

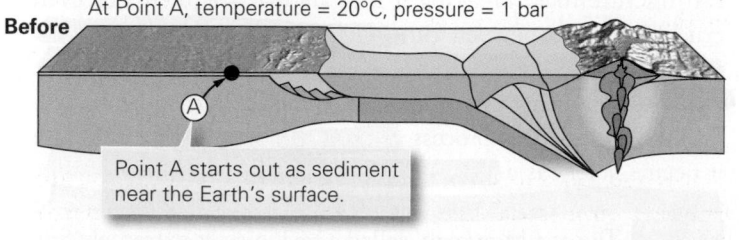

At Point A, temperature = 20°C, pressure = 1 bar

Before

Point A starts out as sediment near the Earth's surface.

Time

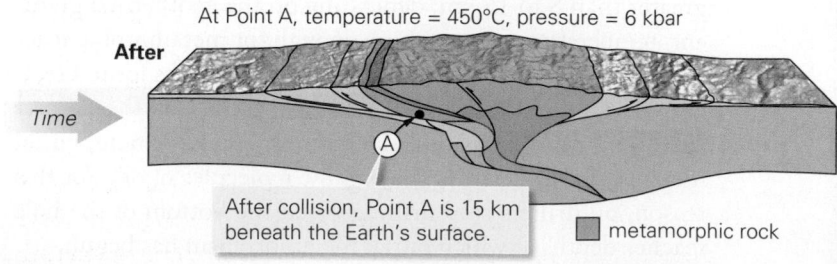

At Point A, temperature = 450°C, pressure = 6 kbar

After

After collision, Point A is 15 km beneath the Earth's surface.

metamorphic rock

FIGURE 8.21 Metamorphism due to hydrothermal circulation along mid-ocean ridges.

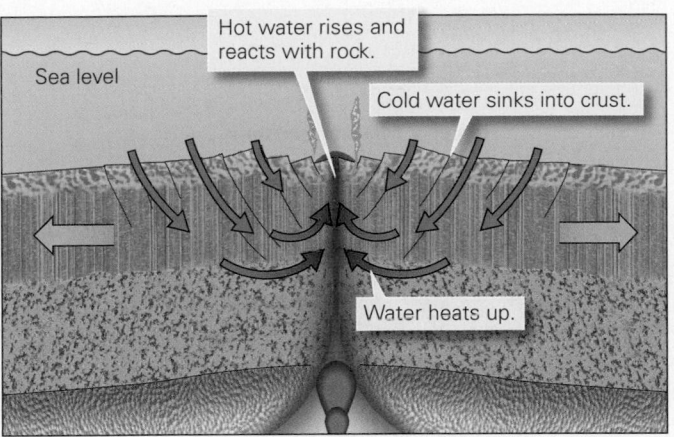

Sea level

Hot water rises and reacts with rock.

Cold water sinks into crust.

Water heats up.

(a) The rising magma at a ridge axis heats water, which then rises. The hot water reacts with the crust and forms metamorphic minerals.

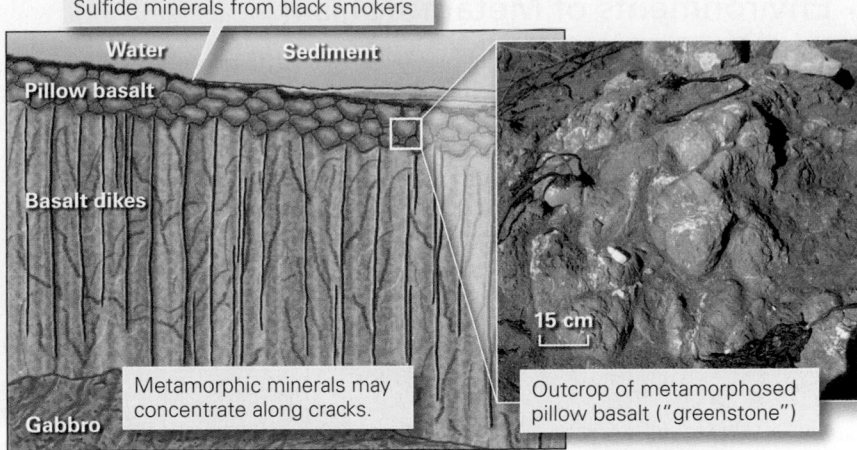

Sulfide minerals from black smokers

Water · Sediment

Pillow basalt

Basalt dikes

Metamorphic minerals may concentrate along cracks.

Gabbro

15 cm

Outcrop of metamorphosed pillow basalt ("greenstone")

(b) Hydrothermal metamorphism in oceanic crust produces greenish minerals (chlorite and epidote). Sulfide minerals typically precipitate from water, mostly at the surface.

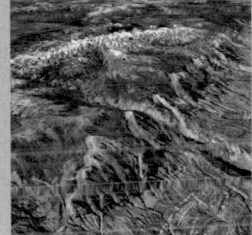
Metamorphism in Subduction Zones

Blueschist, a relatively rare metamorphic rock that contains an unusual blue-colored amphibole, forms only under conditions where there's very high pressure and relatively low temperature. Such conditions do not develop in continental crust, for at the high pressures needed to produce blue amphibole, temperatures are also high (see Box 8.1). So, to understand where blueschist forms, we must search for a geologic setting where high pressures can develop at relatively low temperatures.

Plate tectonics theory provides the answer to this puzzle. Researchers found that blueschist develops only in the accretionary prisms that form at subduction zones (see Geology at a Glance, pp. 264–265). They realized that because such prisms grow to be over 20 km thick, rock at the base of the prism is subjected to high pressures (due to the weight of overburden). But because the subducted oceanic lithosphere beneath the prism is relatively cool, temperatures at the base of the prism remain relatively low. Under these conditions, blue amphibole can form. Because of shear between the subducting plate and the overriding plate, blueschist develops a foliation.

Shock Metamorphism

When large meteorites slam into the Earth, a vast amount of kinetic energy instantly transforms into heat, and pulses of extreme compression (shock waves) propagate into the Earth. The heat may be sufficient to melt or even vaporize rock at the impact site, and the extreme compression of the shock wave causes the quartz grains in rocks below the impact site to suddenly undergo a phase change to a denser mineral called coesite. The phase changes that take place in rock due to the passage of a shock wave represent **shock metamorphism**. When astronauts sampled the Moon, they discovered that the regolith covering the lunar surface contains the products of shock metamorphism produced by countless impacts.

Where Do You Find Metamorphic Rocks?

When you stand on an outcrop of metamorphic rock, you are standing on material that once lay many kilometers beneath the surface of the Earth. How does metamorphic rock return

GEOLOGY AT A GLANCE

Environments of Metamorphism

Metamorphic rocks form when a pre-existing rock (a protolith) undergoes changes in texture and/or mineral content in the solid state in response to changes in temperature, changes in pressure, and/or differential stress. Metamorphism may also reflect interaction with hydrothermal fluids. Some metamorphic rocks are foliated (have metamorphic layering), whereas others are not.

Regional Metamorphism in an Orogenic Belt

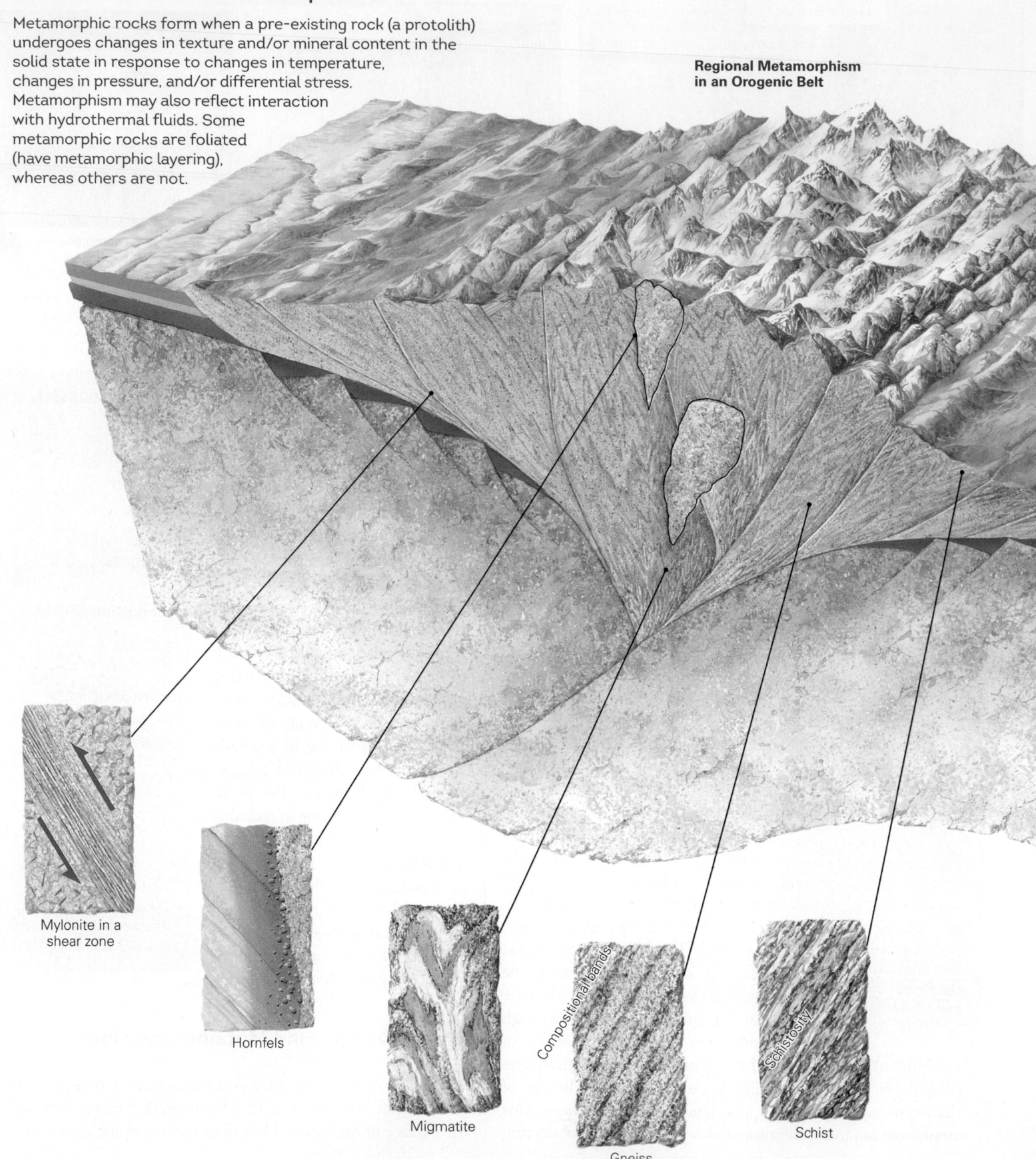

Mylonite in a shear zone

Hornfels

Migmatite

Compositional bands

Gneiss

Schistosity

Schist

Slate

Unmetamorphosed shale

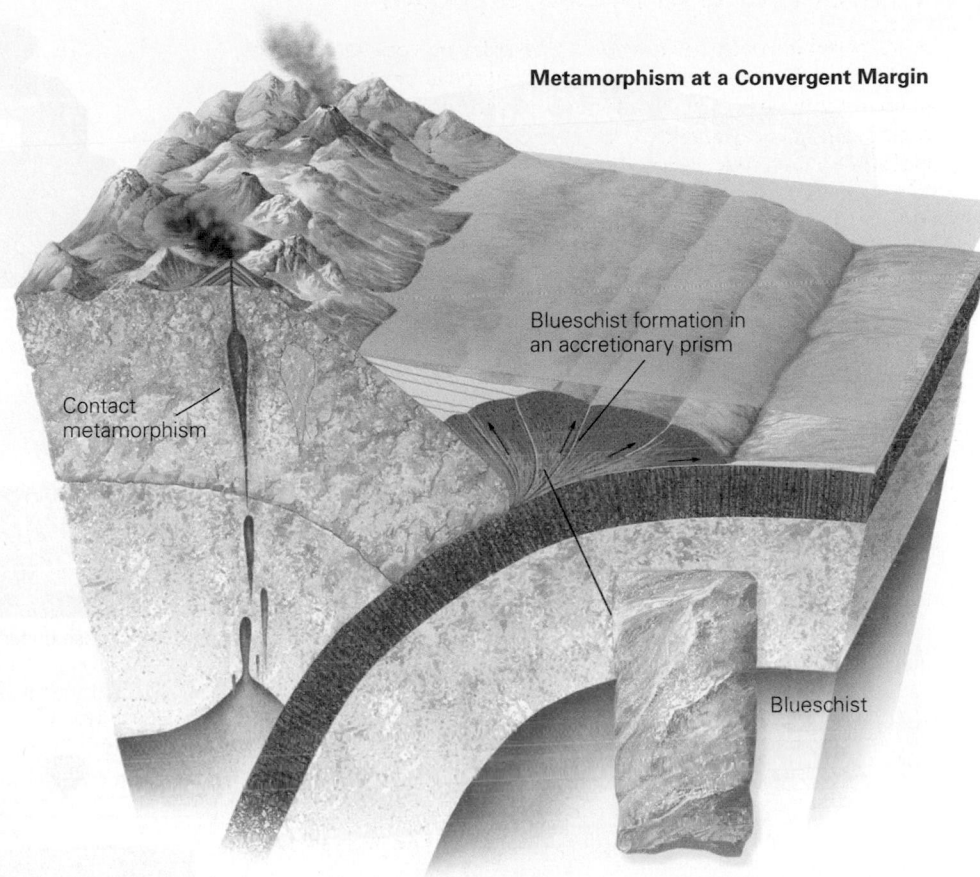

Contact metamorphism

Blueschist formation in an accretionary prism

Blueschist

Foliation results when rock is compressed or sheared during metamorphism, causing mineral crystals to grow or rotate into alignment with one another. Dynamothermal (regional) metamorphism occurs during mountain building. Contact metamorphism, which takes place around an igneous intrusion, or pluton, is caused by the heat released by the pluton.

Geologists distinguish among metamorphic rocks according to the type of foliation and the mineral assemblage a rock contains. Hornfels is nonfoliated and forms as a result of contact metamorphism. Mylonite develops when shearing creates a foliation but not necessarily a change in types of minerals. Slate, which forms from shale, contains slaty cleavage; clay flakes are typically aligned at an angle to the bedding plane. Schist contains coarse grains of mica (muscovite and/or biotite) aligned parallel to one another. Gneiss has compositional banding. (Migmatite forms when part of the rock melts, and thus it is a mixture of metamorphic and igneous rock.) Quartzite is composed predominantly of quartz (it is metamorphosed sandstone), whereas marble is composed predominantly of calcite or dolomite (it is metamorphosed limestone or dolostone). Quartzite and marble are usually unfoliated.

The types of minerals and foliation in a metamorphic rock indicate the rock's grade. High-grade rocks, such as gneiss, form at higher temperatures and pressures, whereas low-grade rocks, such as shale, form at lower pressures and temperatures. Blueschist is an unusual metamorphic rock that develops under relatively high pressures but relatively low temperatures—the environment of an accretionary prism.

Foliation resulting from differential stress

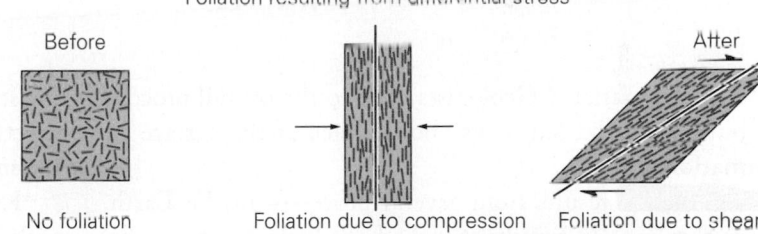

Before

No foliation

Foliation due to compression

After

Foliation due to shear

Increasing temperature

Increasing pressure

Increasing metamorphic grade

Shale

Low grade

Slate

Blueschist

Hornfels

Schist

Gneiss

Migmatite

High grade

FIGURE 8.22 Three processes contribute to exhumation of metamorphic rock formed at depth. Here the red dot (representing metamorphic rock formed at the base of a mountain range) gets progressively closer to the surface over time.

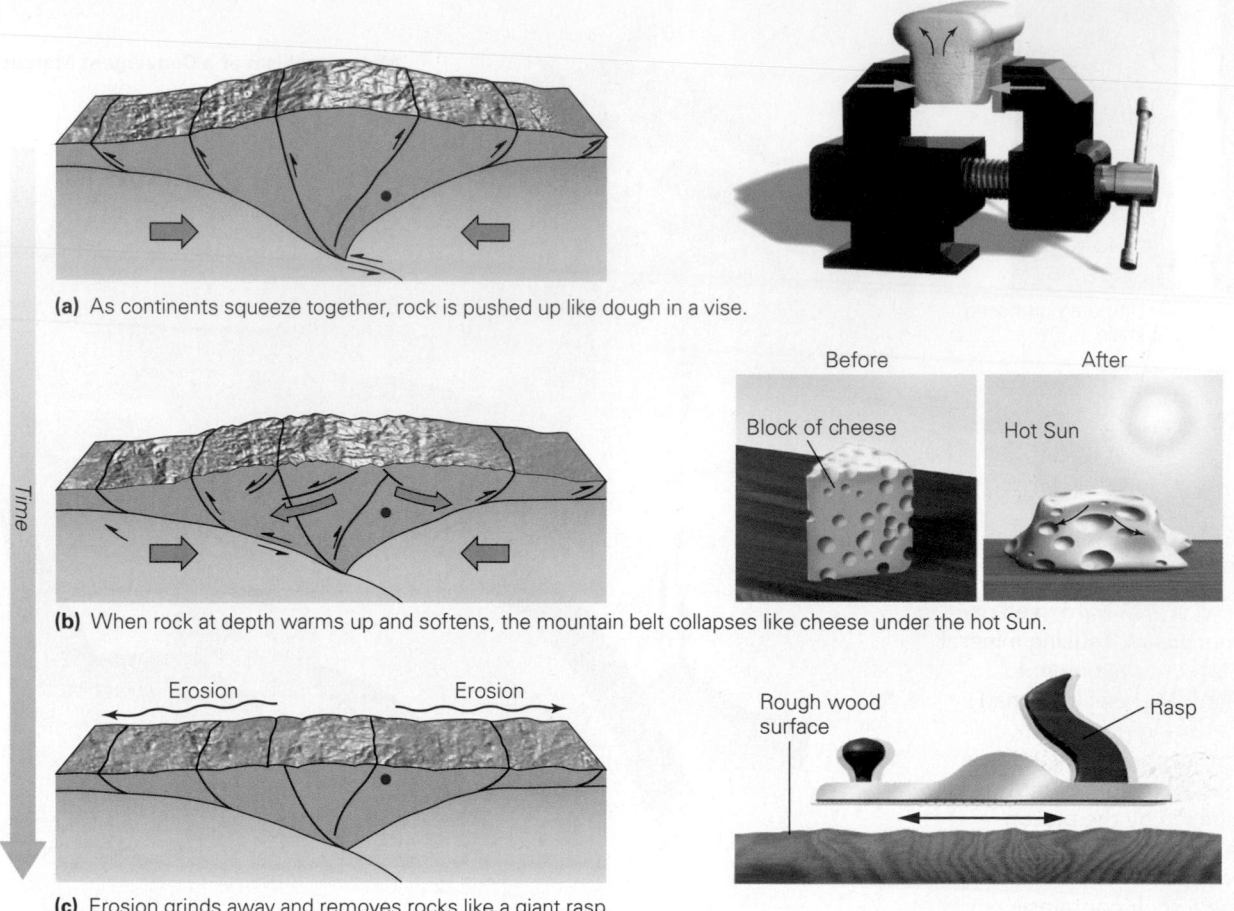

(a) As continents squeeze together, rock is pushed up like dough in a vise.

(b) When rock at depth warms up and softens, the mountain belt collapses like cheese under the hot Sun.

(c) Erosion grinds away and removes rocks like a giant rasp.

to the Earth's surface? Geologists refer to the overall process by which deeply buried rocks end up back at the surface as **exhumation**.

Exhumation results from several processes in the Earth System that happen simultaneously. Let's look at the specific processes that contribute to bringing high-grade metamorphic rocks from below a collisional mountain range back to the surface **(Fig. 8.22)**. First, as two continents progressively push together, the rock caught between them squeezes upward, much like dough pressed in a vise. The upward movement takes place by slip on faults and by plastic flow of rock. Second, as the mountain range grows, the crust at depth beneath it warms up and becomes softer and weaker. Eventually, the range starts to collapse under its own weight, much like a block of soft cheese placed under the hot Sun. As a result of this collapse, the upper crust spreads out laterally. Stretching of the upper part of the crust in the horizontal direction causes it to become thinner in the vertical direction, and as the upper part of the crust becomes thinner, the deeper crust ends up closer to the surface. Third, as erosion takes place at

the surface, weathering, landslides, river flow, and glacial flow together play the role of a giant rasp, stripping away rock at the surface and exposing rock that was once below the surface.

Keeping in mind the processes that form metamorphic rock and cause exhumation, let's ask the question, "Where can you find exposure of metamorphic rocks today?" You can start your quest by hiking into a mountain range. As we've seen, the process of mountain building produces and eventually exhumes metamorphic rocks. The towering cliffs in the interior of a mountain range typically reveal outcrops of schist, gneiss, and quartzite **(Fig. 8.23a)**. Even after the peaks have eroded away, the record of mountain building remains in the form of a belt of metamorphic rock at the ground surface. Vast expanses of metamorphic rock crop out in continental shields. A **shield** is a broad region of long-lived, stable continental crust where Phanerozoic sedimentary cover either was not deposited or has been eroded away **(Fig. 8.23b, c)**. As a result, the bedrock exposed at the Earth's surface today formed during the succession of Precambrian mountain-building events that led to the original growth of continents.

FIGURE 8.23 Examples of rock exposures consisting of Precambrian metamorphic rocks.

(a) This cliff, in the Wasatch Mountains of Utah, exposes gneiss, which has been intruded by granite. The gneiss has foliation, whereas the granite does not.

(b) A photograph from an airplane window of the flat landscape of the eastern Canadian Shield.

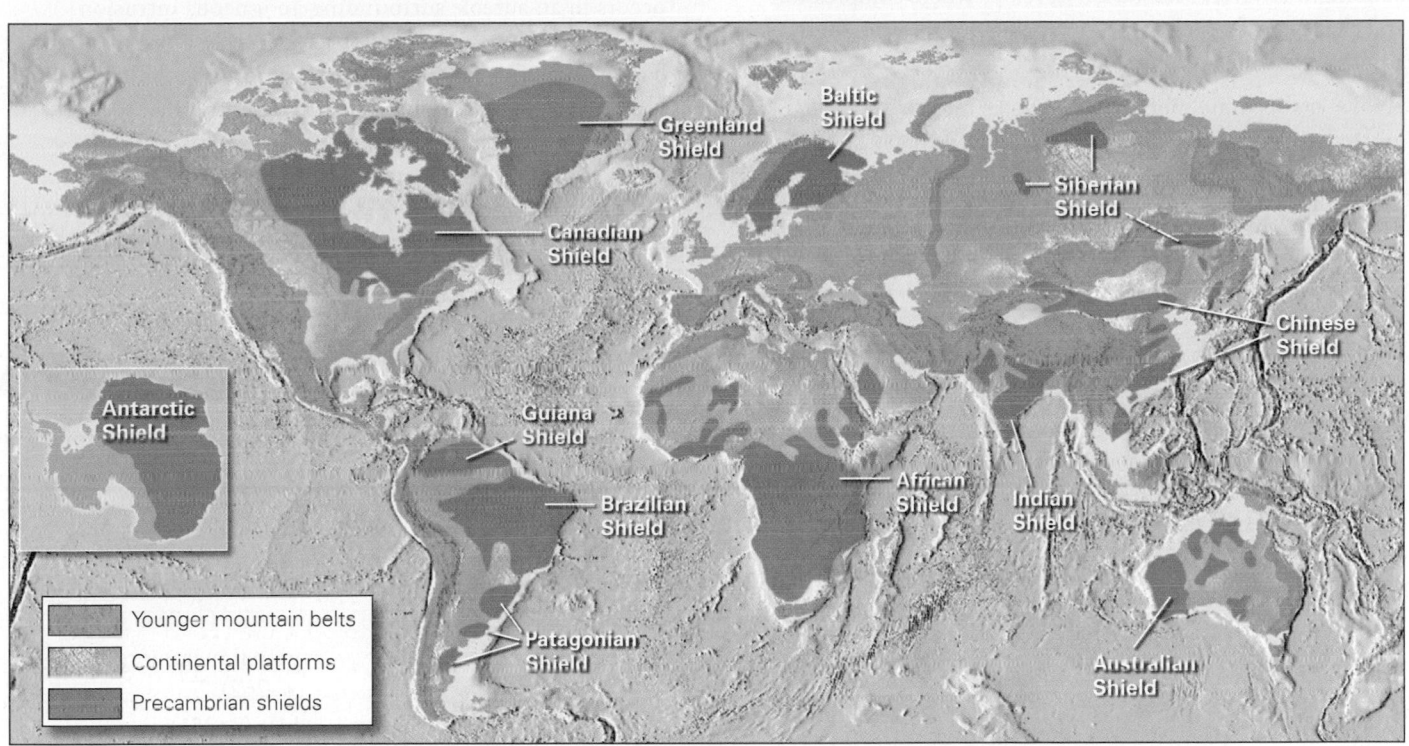

(c) A map showing the distribution of shields, areas where broad expanses of Precambrian crust, including Precambrian metamorphic rocks, crop out.

TAKE-HOME MESSAGE

Thermal (contact) metamorphism develops around an igneous intrusion due to heat from the pluton. Dynamothermal (regional) metamorphism develops beneath mountain ranges, where rock undergoes compression and shear at high temperatures and pressures. Metamorphism can also happen in response to deep burial, reaction with hydrothermal fluids, and meteorite impacts. Erosion and uplift eventually expose metamorphic rock in mountain ranges or continental shields.

QUICK QUESTION: Could you find a layer of metamorphic rock between the layers of sedimentary rock in a sedimentary basin? Why or why not?

SUMMARY

- Metamorphism refers to changes in a rock that result in the formation of a metamorphic mineral assemblage and metamorphic texture in response to change in temperature and/or pressure, application of differential stress, or interaction with hydrothermal fluids (hot-water solutions).

- Metamorphism involves recrystallization, phase changes, metamorphic reactions (neocrystallization), pressure solution, or plastic deformation. If hydrothermal fluids bring in or remove elements, we say that metasomatism has occurred.

- Metamorphic foliation can be defined either by preferred orientation (alignment of inequant crystals) or by compositional banding. Preferred orientation develops where compression and shearing of a rock causes its inequant grains to align parallel with one another.

- Geologists separate metamorphic rocks into two classes, foliated rocks and nonfoliated rocks, depending on whether the rocks contain foliation.

- The class of foliated rocks includes slate, phyllite, metaconglomerate, schist, and gneiss. The class of nonfoliated rocks includes hornfels, quartzite, and marble. Migmatite, a mixture of igneous and metamorphic rock, forms under conditions where melting begins.

- Rocks formed at relatively low temperatures are known as low-grade rocks, whereas those formed at high temperatures are known as high-grade rocks. Intermediate-grade rocks develop between these two extremes. Different mineral assemblages form at different metamorphic grades.

- Geologists track the distribution of different grades of rock by looking for index minerals. Isograds indicate the locations at which index minerals first appear. A metamorphic zone is the region between two isograds.

- A metamorphic facies consists of a group of metamorphic mineral assemblages that develop under a specified range of temperature and pressure conditions.

- Thermal metamorphism (also called contact metamorphism) occurs in an aureole surrounding an igneous intrusion. Burial metamorphism develops at depth in a sedimentary basin. Dynamic metamorphism occurs along faults, where rocks undergo plastic shearing. Dynamothermal metamorphism (also called regional metamorphism) results when rocks undergo heating and shearing during mountain building. Hydrothermal metamorphism takes place due to the circulation of hot water in oceanic crust at mid-ocean ridges. Shock metamorphism happens during the impact of a meteorite.

- We find belts of metamorphic rocks in mountain ranges. Blueschist forms in accretionary prisms. Shields expose broad areas of Precambrian metamorphic rocks.

GUIDE TERMS

burial metamorphism (p. 261)

contact metamorphism (p. 259)

differential stress (p. 247)

dynamic metamorphism (p. 261)

dynamothermal metamorphism (p. 262)

exhumation (p. 266)

gneiss (p. 251)

hornfels (p. 252)

hydrothermal metamorphism (p. 262)

index mineral (p. 258)

marble (p. 253)

metaconglomerate (p. 251)

metamorphic aureole (p. 259)

metamorphic facies (p. 257)

metamorphic foliation (p. 243)

metamorphic grade (p. 255)

metamorphic mineral (p. 243)

metamorphic rock (p. 243)

metamorphic texture (p. 243)

metamorphic zone (p. 258)

metamorphism (p. 243)

metasomatism (p. 248)

migmatite (p. 252)

mylonite (p. 261)

phyllite (p. 251)

preferred orientation (p. 247)

pressure (p. 246)

protolith (p. 243)

quartzite (p. 252)

regional metamorphism (p. 262)

schist (p. 251)

shield (p. 266)

shock metamorphism (p. 263)

slate (p. 250)

thermal metamorphism (p. 259)

vein (p. 249)

GEOTOURS THIS CHAPTER'S GEOTOURS WORKSHEET (G) FEATURES QUESTIONS AND GOOGLE EARTH SITES ON:

- Metamorphic zones
- Diverse metamorphic environments
- Protoliths
- Types of metamorphism
- Metamorphic foliation

REVIEW QUESTIONS

The letters following each Review Question refer to the corresponding Learning Objective from the Chapter Opener.

1. How do metamorphic rocks differ from igneous and sedimentary rocks? **(A)**

2. What two features can develop in a rock during metamorphism? **(C)**

3. What phenomena may be involved in transforming a protolith into metamorphic rock? **(B)**

4. What is metamorphic foliation, and how does it form? Identify the orientation of foliation on the thin section. **(D)**

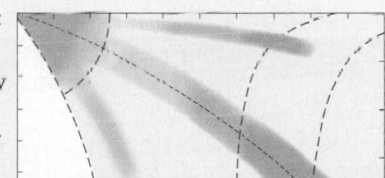

5. How does slate differ from a phyllite? How does phyllite differ from a schist? How does schist differ from a gneiss? **(E)**

6. Why doesn't hornfels contain foliation? Name two other nonfoliated metamorphic rocks. **(D)**

7. What does a metamorphic grade refer to, and how can it be determined? How does grade differ from facies? Identify the axes of the graph, and the areas representing conditions in which low-, intermediate-, and high-grade metamorphic rocks form. **(G)**

8. Describe the geologic settings where thermal, dynamic, and dynamothermal metamorphism take place. Identify the aureole of highest-grade rock on the diagram. **(F)**

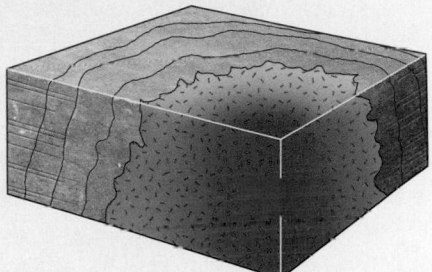

9. Why does metamorphism happen at the site of meteor impacts or along mid-ocean ridges? **(F)**

10. How does plate tectonics theory explain the combination of low-temperature but high-pressure minerals found in a blueschist? **(F)**

11. Where would you go if you wanted to find exposed metamorphic rocks, and how did such rocks return to the surface of the Earth after undergoing metamorphism at depth in the crust? **(F)**

ON FURTHER THOUGHT

12. Do you think that you would be likely to find a broad region (hundreds of kilometers across and hundreds of kilometers long) in which the outcrops consist of high-grade hornfels? Why or why not? (Hint: Think about the causes of metamorphism and the conditions under which a hornfels forms.) **(B, E)**

13. Would we be likely to find gneiss and schist on the Moon? Why or why not? **(E, F)**

ONLINE RESOURCES

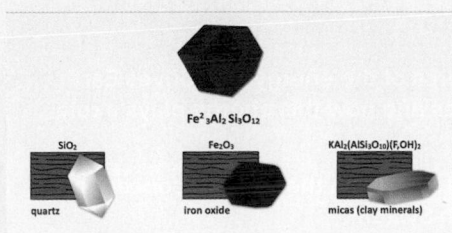

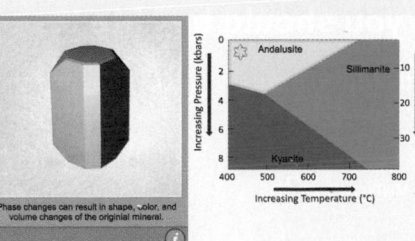

Animations
This chapter features animations on the six major types of metamorphic changes, focusing on each process close up.

Smartwork5
This chapter features visual exercises on metamorphism and its effects.

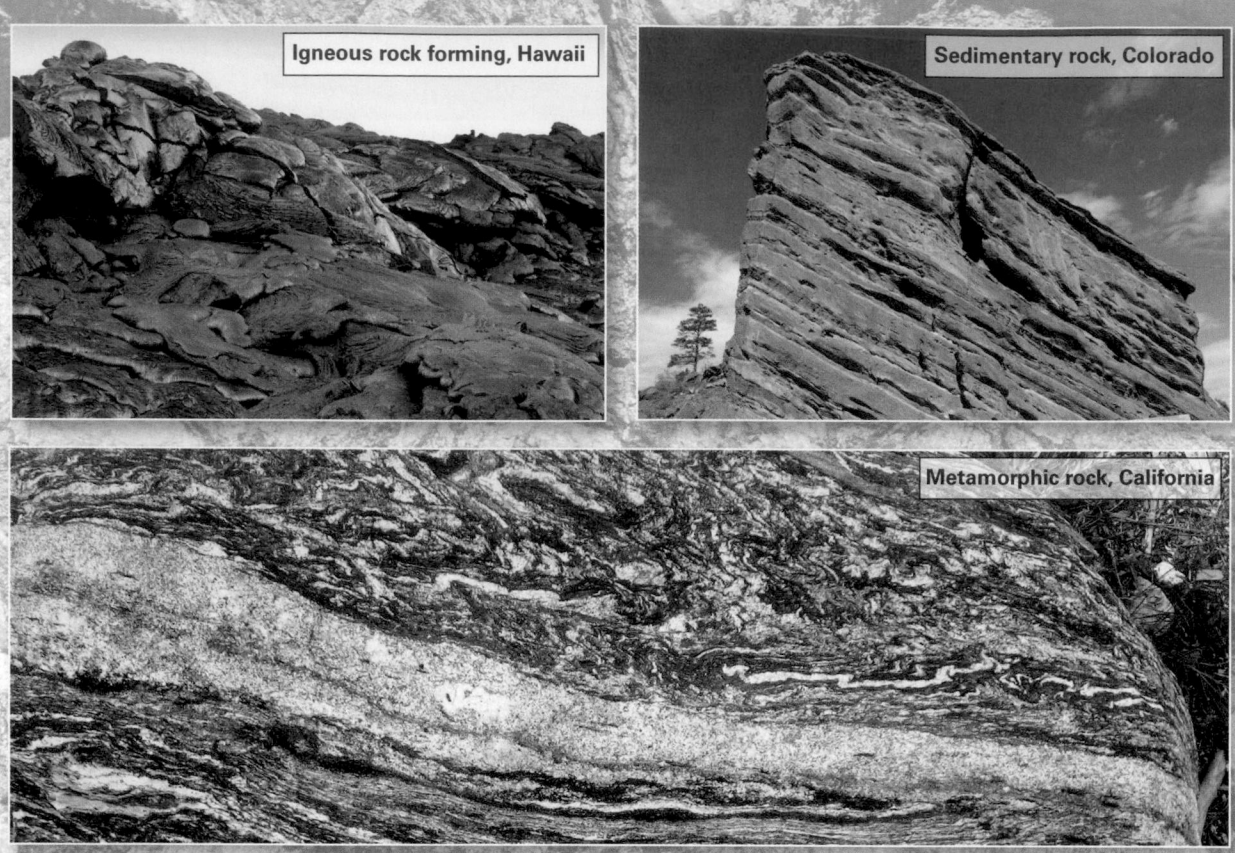

Igneous rock forming, Hawaii

Sedimentary rock, Colorado

Metamorphic rock, California

The Rock Cycle in the Earth System

By the end of this interlude, you should be able to . . .

A. explain why rocks don't last forever in the Earth's crust and why there are more younger rocks than older rocks at the surface of the Earth.

B. define the rock cycle and illustrate the various paths through it.

C. relate the paths through the rock cycle to geologic settings, in the context of plate tectonics theory.

D. describe the source of the energy that drives Earth System processes and how this energy plays a role in the rock cycle.

E. develop a model to explain the general concept of a cycle in the Earth System.

C.1 Introduction

"Stable as a rock." This familiar expression implies that a rock, once formed, remains a permanent entity in the Earth's crust. In fact, over the time frame of Earth history, a span of over 4.56 billion years, components making up a rock of one class may later be rearranged or moved elsewhere to form another rock of the same class or of a different rock class altogether. In some places, this process of change has happened many times. Geologists refer to the progressive transformation of Earth materials from one rock to another over time as the **rock cycle (Fig. C.1)**. Discussion of the rock cycle illustrates important relationships among the rock classes described in the previous three chapters. In this interlude, we illustrate how to apply the concept of the rock cycle, and we conclude by showing how it represents one of many important cycles in the Earth System.

C.2 Rock Cycle Paths

Paths Reflect Geologic History

Different rock types form in different environments (**Geology at a Glance**, pp. 272–273). During the rock cycle, materials can transfer among these environments. By following the arrows in Figure C.1, you can see that there are many paths around or through the rock cycle. For example, igneous rock formed by solidification of a melt that rose from the mantle may undergo weathering and erosion to produce sediment. Later, burial and lithification transforms that sediment into new sedimentary rock. This sedimentary rock may, in turn, become buried so deeply that it transforms into metamorphic rock. Extreme heating of the metamorphic rock might cause it to partially melt and produce new magma. This new magma

FIGURE C.1 The stages of the rock cycle, showing various alternative paths.

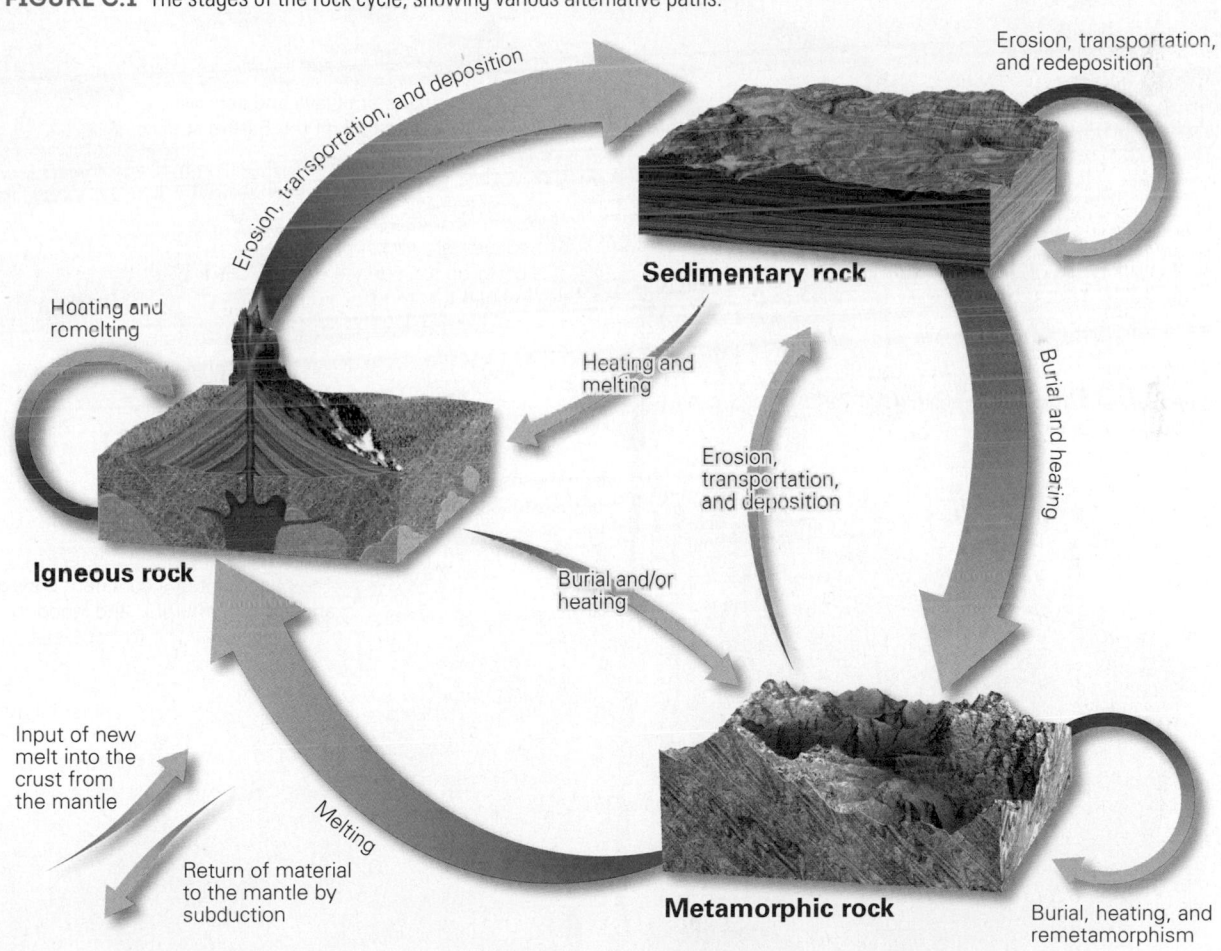

◀ (facing page) On our dynamic planet, the atoms in a rock of one class may, over time, be incorporated into rock of another class, and then another. Such transformations comprise the rock cycle.

Rock-Forming Environments and the Rock Cycle

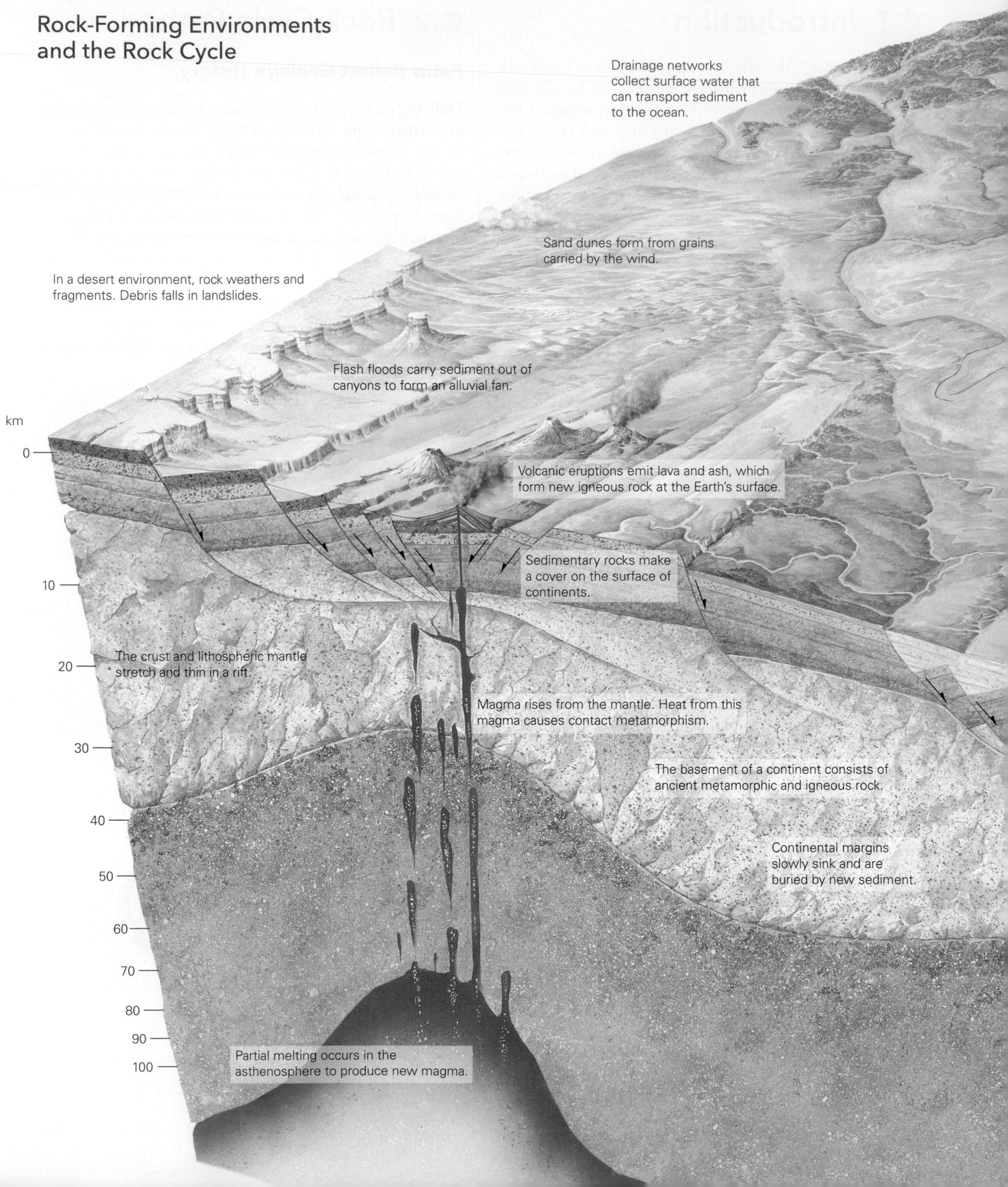

Drainage networks collect surface water that can transport sediment to the ocean.

Sand dunes form from grains carried by the wind.

In a desert environment, rock weathers and fragments. Debris falls in landslides.

Flash floods carry sediment out of canyons to form an alluvial fan.

Volcanic eruptions emit lava and ash, which form new igneous rock at the Earth's surface.

Sedimentary rocks make a cover on the surface of continents.

The crust and lithospheric mantle stretch and thin in a rift.

Magma rises from the mantle. Heat from this magma causes contact metamorphism.

The basement of a continent consists of ancient metamorphic and igneous rock.

Continental margins slowly sink and are buried by new sediment.

Partial melting occurs in the asthenosphere to produce new magma.

km

0

10

20

30

40

50

60

70

80

90

100

Glaciers erode rock and can transport sediment of all sizes.

In a region of continental collision, rocks that were near the surface are deeply buried and metamorphosed.

In humid climates, thick soils develop.

Magma that cools and solidifies underground forms igneous intrusions.

Along coastal plains, rivers meander. Sediment collects in the channel and floodplain.

Where a river enters the sea, sediment settles out to form a delta.

Reefs grow from calcite-secreting organisms. These will eventually turn into limestone.

Many different kinds of sediment accumulate along coastlines, building out a continental shelf.

Turbidity currents carry a cloud of sediment that settles to form a submarine fan.

Fine clay and plankton shells settle on the oceanic crust.

The oceanic crust consists of igneous rocks formed at a mid-ocean ridge.

Rocks form in many different environments. Igneous rocks develop where melt rises from depth and cools; sedimentary rocks form at or near the surface in many different environments; and metamorphic rocks form deep underground. Because the Earth is dynamic—because of plate motion and the circulation of water, ice, and air at the surface—environments at a given location change over time, so atoms don't necessarily stay within one rock type for all of geologic time. This progressive shift of material over time, from one rock to another, is the rock cycle. There are many paths through the rock cycle; which path (or succession of paths) an atom takes depends on the geologic setting.

might later solidify to form new igneous rock. We can express this path as follows:

$$Igneous \rightarrow Sedimentary \rightarrow Metamorphic \rightarrow Igneous$$

Given local geologic conditions, a rock at one stage in the cycle could follow another path. For example, the metamorphic rock, once formed, could itself be uplifted and eroded to form new sediment and, later, new sedimentary rock, without melting. This path takes a shortcut through the cycle that we can illustrate as follows:

$$Igneous \rightarrow Sedimentary \rightarrow Metamorphic \rightarrow Sedimentary$$

Likewise, the original igneous rock might be deeply buried before it could be eroded and could therefore undergo metamorphism directly, without first turning to sediment. The resulting metamorphic rock might then be uplifted and eroded to produce sediment that becomes sedimentary rock, defining another shortcut path:

$$Igneous \rightarrow Metamorphic \rightarrow Sedimentary$$

But not all steps have to yield a new rock type. For example, if a sedimentary rock, once formed, were uplifted and eroded to form new sediment that was then buried and lithified, its path would be

$$Sedimentary \rightarrow Sedimentary$$

To get a clearer sense of how the rock cycle works, let's look at a case study of the rock cycle, in the context of plate tectonics theory.

C.3 A Case Study of the Rock Cycle

Material may enter the rock cycle when magma rises from the mantle. Suppose the magma erupts and forms basalt (an igneous rock) at a continental hot-spot volcano **(Fig. C.2a)**. Interaction with wind, rain, and vegetation gradually weathers the basalt, physically breaking it into smaller fragments and chemically weathering it to yield clay. Water washes the newly formed clay away and transports it downstream—if you've ever seen a brown-colored river, you've seen clay traveling to a site of deposition. Eventually the river reaches the sea, where the water slows down and the clay settles out.

Let's imagine, in our case study, that the clay accumulates as a deposit of mud along the margin of Continent X.

Gradually, over time, the mud becomes buried beneath, say, 6 km of additional sediment, and the clay flakes pack tightly together to yield a new sedimentary rock, shale. The shale resides 6 km below the continental shelf for millions of years, until Continent X collides with Continent Y. As the two continents squeeze together, slip on faults transports the edge of Continent Y up and over the shale of Continent X. The shale that had once been only 6 km below the surface ends up at a depth of, say, 30 km, and under the pressure and temperature conditions existing at this depth, it metamorphoses into schist **(Fig. C.2b)**.

The story's not over. Once mountain building stops, erosion grinds away the mountain range, and exhumation brings some of the schist back up to the ground surface. Some of this schist erodes to form sediment that a river carries off and deposits elsewhere. Burial and lithification of this sediment produces new sedimentary rock (perhaps sandstone, siltstone, and shale). The rest remains preserved below the surface **(Fig. C.2c)**. Eventually, continental rifting takes place at the site of the former mountain range, and the crust containing the schist begins to split apart. Injection of new, hot magma may cause some of the schist to partially melt, producing new felsic magma. This felsic magma rises to the surface of the crust and freezes into rhyolite, a new igneous rock **(Fig. C.2d)**. In terms of the rock cycle, we're back at the beginning, having once again made igneous rock.

Note that atoms, as they pass through the rock cycle, do not always stay within the same mineral. In our case study, a silicon atom in a pyroxene crystal in the basalt may become part of a clay crystal in the shale, part of a mica crystal in the schist, or part of a quartz crystal in the rhyolite. Similarly, atoms that were adjacent in the starting rock don't necessary end up near one another at a later stage in the rock cycle. Moving water or wind, for example, may carry one atom of what was originally a single mineral crystal to a location hundreds or even thousands of kilometers away from what was originally an adjacent atom.

Rates of Transfer

We have seen that not all atoms pass through the rock cycle in the same way. Similarly, not all atoms pass through the rock cycle at the same rate, and for that reason, we find rocks of many different ages at the surface of the Earth. Some rocks have remained in one form for less than a few million years, while others have stayed unchanged for most of the Earth's history. For example, rocks exposed in the Appalachian Mountains today may have passed through stages of the rock cycle many times during the past several hundred million years, for the eastern margin of North America has been subjected to multiple occurrences of basin formation, mountain building, and rifting during this time.

FIGURE C.2 A case study of the rock cycle, in the context of plate tectonics theory.

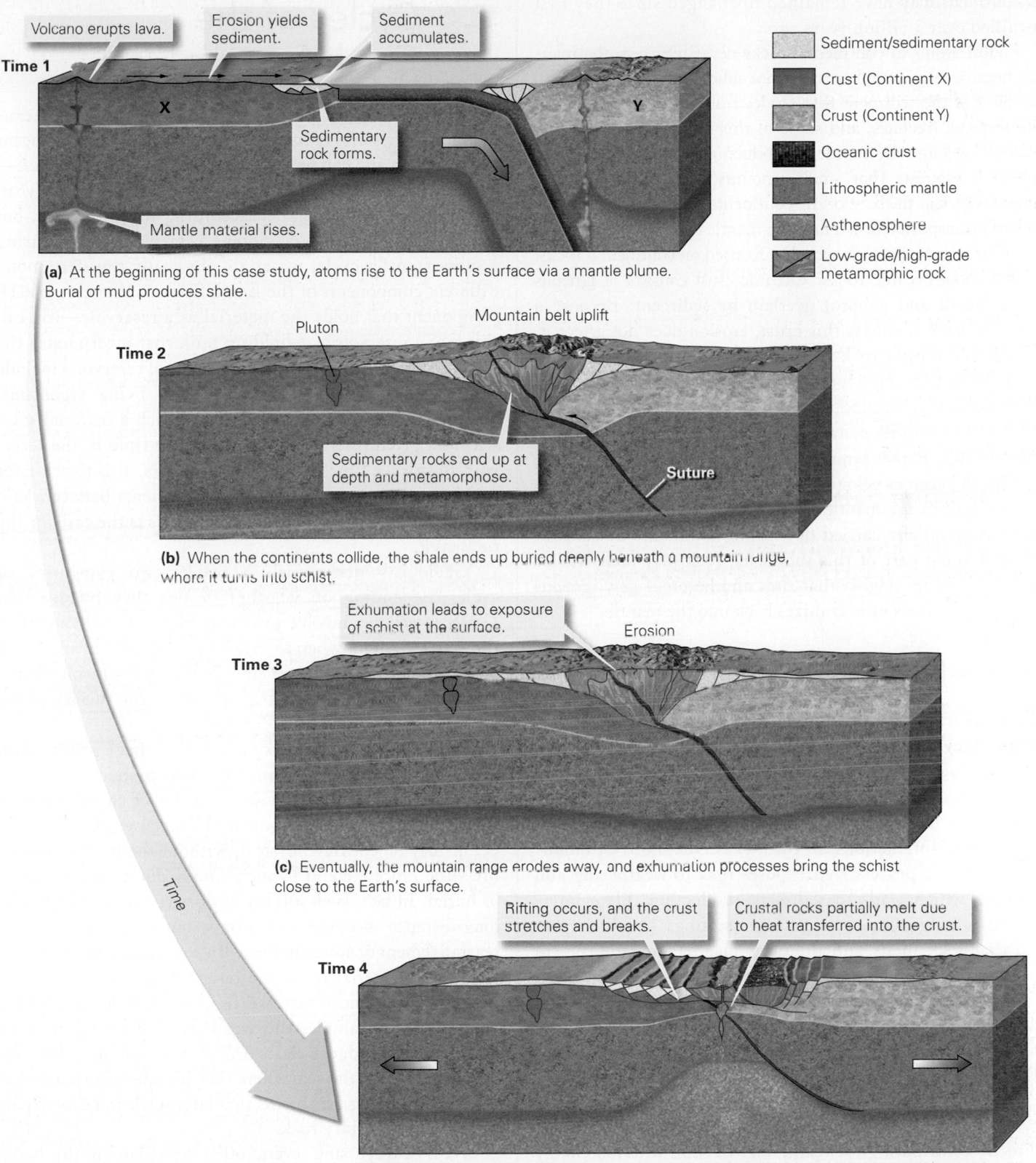

Time 1

Volcano erupts lava.

Erosion yields sediment.

Sediment accumulates.

Sedimentary rock forms.

Mantle material rises.

X Y

Sediment/sedimentary rock

Crust (Continent X)

Crust (Continent Y)

Oceanic crust

Lithospheric mantle

Asthenosphere

Low-grade/high-grade metamorphic rock

(a) At the beginning of this case study, atoms rise to the Earth's surface via a mantle plume. Burial of mud produces shale.

Time 2

Pluton

Mountain belt uplift

Sedimentary rocks end up at depth and metamorphose.

Suture

(b) When the continents collide, the shale ends up buried deeply beneath a mountain range, where it turns into schist.

Time 3

Exhumation leads to exposure of schist at the surface.

Erosion

(c) Eventually, the mountain range erodes away, and exhumation processes bring the schist close to the Earth's surface.

Time 4

Rifting occurs, and the crust stretches and breaks.

Crustal rocks partially melt due to heat transferred into the crust.

Time

(d) When rifting splits the continents apart, some of the schist melts, and its atoms become incorporated into magma again.

In contrast, Precambrian chert beds now in the interior of a continent may have remained unchanged since they first lithified over 3 billion years ago.

Most atoms in continental rocks never return to the mantle because continental crust cannot subduct. However, a small amount of the sediment that erodes from a continent ends up in deep-sea trenches, and some of this sediment is eventually carried back into the mantle by subduction. Furthermore, recent research suggests that small amounts of metamorphic and igneous rock at the base of the continental crust may be scraped off and transported down into the mantle by subduction.

Our tour of the rock cycle has focused on continental rocks. What about oceanic rocks? Oceanic crust consists of igneous rock (basalt and gabbro) overlain by sediment. Because a layer of water blankets this crust, erosion does not affect it, so oceanic crustal rock generally does not follow the path into the sedimentary loop of the rock cycle. But sooner or later, oceanic crust subducts. When this happens, the rock of the crust undergoes metamorphism, for as it sinks, it endures progressively higher temperatures and pressures. Specifically, the basalt becomes eclogite, a rock so dense that it can sink to great depth in the mantle. Most igneous rock of the oceanic crust is eventually carried down into the mantle by subduction. A small part of this subducted rock may melt and be incorporated in magma that rises and becomes new igneous rock, but most eventually mixes back into the mantle.

What Drives the Rock Cycle?

The rock cycle happens because the Earth is a dynamic planet that has many different rock-forming environments (see **Geology at a Glance**, pp. 272–273). Our planet's internal heat and gravitational field, acting together, ultimately drive plate movements and generate plume-associated hot spots. Plate interactions, in turn, cause the uplift of mountain ranges, a process that exposes rock to weathering and erosion, which leads to sediment production. Plate interactions also generate the geologic settings in which pre-existing rock melts and produces magma, metamorphism occurs, and sedimentary basins develop.

At the surface of the Earth, heat from the Sun, together with gravity, drives wind, rain, ice, and currents. These agents of weathering and erosion grind away at the surface of the Earth and send material into the sedimentary path of the cycle. In the Earth System, life also plays a key role in the rock cycle by adding corrosive oxygen to the atmosphere and by directly contributing to weathering. In sum, external energy (solar heat), internal energy (the Earth's internal heat), gravity, and life all play roles in driving the rock cycle by keeping the mantle, crust, atmosphere, and oceans in motion and by making portions of the system chemically reactive.

C.4 Cycles of the Earth System

In a general sense, a **cycle**, from the Latin word *cyclus*, meaning circle, is a series of interrelated events or steps that occur in succession and can be repeated. During a *temporal cycle*— such as the phases of the Moon, the seasons of the year, or the tides—events happen according to a timetable, but the materials involved do not necessarily change. During a **mass-transfer cycle**, materials physically transfer among different components of the Earth System. We refer to each component that holds the material as a **reservoir**—you can think of a reservoir as a holding tank that incorporates the material for a period of time. Examples of reservoirs include the ocean, rock, the atmosphere, and living organisms. Geologists refer to the time during which a material stays within a given reservoir as its **residence time** in the reservoir. Residence times can vary from hours, as is the case for cycles involving transfer of volatile elements between land and air, to millions or billions of years, as is the case for the rock cycle.

Geologists distinguish between two categories of cycles, depending on whether or not they involve life. *Geochemical cycles* involve primarily physical components in the Earth System, whereas *biogeochemical cycles* involve both physical and living components. Some cycles involve many different reservoirs, above, below, and on the surface of the Earth, whereas others involve just a few.

We can consider the rock cycle to be either a geochemical or a biogeochemical cycle in the Earth System, depending on circumstances. For example, transfer of chemicals from the sedimentary rock reservoir to the metamorphic rock reservoir may be entirely physical, without the involvement of life—the process could happen simply due to heating and/ or burial. In fact, such a transfer doesn't necessarily involve long-distance movement of atoms, but could just entail a rearrangement of atoms in place. In some cases, however, the transfer of material from one reservoir to another involves life, which may move material from one location to another far away. For example, the weathering and erosion of an igneous rock produces clasts and ions in solution. The ions may be carried to the sea, where they become incorporated in the shells of living organisms that later settle to become part of a new sedimentary rock.

We'll be discussing several other cycles later in this book. Examples include the *hydrological cycle* (movement of water from sea to air to rain to stream and/or to and from ice and living tissue) and the *carbon cycle* (movement of carbon among organisms, water solutions, air, and calcite or fossil fuel).

Interlude C Review

SUMMARY

- A given rock doesn't necessarily last forever. The atoms in a rock may, over time, be incorporated in different rock types. This transfer of atoms progressively from one rock type to another, over time, is the rock cycle.

- Not all atoms follow the same path through the rock cycle. For example, an igneous rock could later be eroded and turned into sediment, which could become a sedimentary rock, which might eventually be metamorphosed. Or the igneous rock could be metamorphosed directly.

- The rock cycle happens because the Earth is dynamic, and its internal and external sources of energy drive melting, uplift, faulting, weathering, erosion, and burial.

- The rock cycle is one of many cycles in the Earth System.

- Mass-transfer cycles involve transfer of material from one reservoir to another over time. Some of these cycles involve only inorganic materials, but some involve life.

GUIDE TERMS

cycle (p. 276)   reservoir (p. 276)   rock cycle (p. 271)
mass-transfer cycle (p. 276)   residence time (p. 276)

REVIEW QUESTIONS

The letters following each Review Question refer to the corresponding Learning Objective from the Chapter Opener.

1. Once formed, does a rock necessarily last for all of the Earth's history? **(A)**

2. Define the rock cycle, and give three examples of paths through it. Label the processes associated with the arrows shown in the diagram. **(B)**

3. Have all rocks on the Earth passed through the rock cycle the same number of times? Explain your answer. **(C)**

4. Is there a rock cycle on the Moon? Why or why not? **(D)**

5. Explain the concept of a cycle in the context of the Earth System. **(E)**

ONLINE RESOURCES

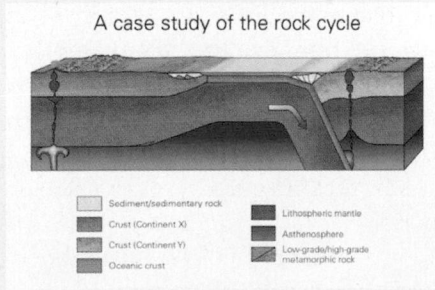

A case study of the rock cycle

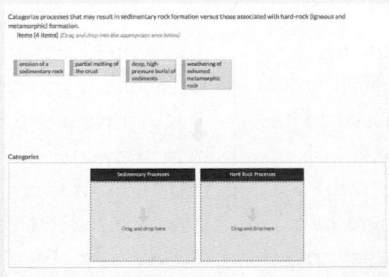

Videos
This interlude features a video that explains the rock cycle, and includes a case study.

Smartwork5
This interlude features video questions on the processes and outcomes of the rock cycle.

PART III

TECTONIC ACTIVITY OF A DYNAMIC PLANET

Cultures around the world have conjured a myriad of imaginative explanations for dynamic geologic events. The ancient Greeks and Romans attributed volcanic eruptions to sparks and smoke flying from the fire god's forges. The Maori of New Zealand believed that movements of the god Rūaumoko within the Earth Mother's womb caused earthquakes. In Norse folklore, Odin and his brothers killed the frost giant Ymir and fashioned the mountains from his bones. We now know that these phenomena are all, perhaps less dramatically, a result of plate movements and interactions. Why does molten rock rise like a fountain out of the ground? Why can the ground shake enough to topple a city? What processes cause the land surface to rise to form mountain belts? How do rocks bend, squash, stretch, and break? Read on, and you will be able to answer these questions and understand some of the deadly natural hazards that threaten society. In Part III, each chapter focuses on one of the dramatic consequences of tectonic activity in the Earth System: volcanoes in Chapter 9, earthquakes in Chapter 10, and mountain building in Chapter 11. Interlude D explores how seismic (earthquake) waves reveal the composition of the Earth's interior and how our planet generates the magnetic and gravity fields.

◀ These mountains, in Colorado, emphasize the continuing interplay between plate-tectonic forces displacing the Earth's surface and erosive forces grinding it away.

CHAPTER 9

The Wrath of Vulcan: Volcanic Eruptions

By the end of this chapter, you should be able to . . .

A. distinguish among the different types of volcanoes.

B. explain how volcanic eruptions produce such a variety of materials, including lava, pyroclastic debris, and gases.

C. explain why some eruptions yield streams of lava while others produce catastrophic explosions.

D. describe how the type of eruption reflects the character of lava and the geologic setting.

E. assess the many hazards that eruptions pose to life and environment.

F. interpret clues to impending eruptions.

9.1 Introduction

Every few hundred years, one of the hills on Vulcano, an island in the Mediterranean Sea off the western coast of Italy, rumbles and spews out molten rock, glassy cinders, and dense "smoke" (actually a mixture of various gases, fine ash, and very tiny liquid droplets). Ancient Romans thought that such eruptions happened when Vulcan, the god of fire, fueled his forges beneath the island to manufacture weapons for the other gods. Geologic study suggests instead that eruptions take place when hot magma, formed by melting inside the Earth, rises through the crust and emerges at the surface. No one believes the Roman myth anymore, but the island's name evolved into the English word **volcano**, which geologists use to designate either an erupting vent through which molten rock reaches the Earth's surface or the hill or mountain built from the products of eruption.

On the main peninsula of Italy, not far from Vulcano, another volcano, Mt. Vesuvius, towers over the Bay of Naples. Nearly 2,000 years ago, a prosperous Roman resort and trading town named Pompeii sprawled at the foot of Vesuvius. One morning in 79 c.e., earthquakes signaled the mountain's awakening. At 1:00 p.m. on August 24, a ferociously turbulent, mottled cloud boiled up above Mt. Vesuvius's summit to a height of 27 km. The cloud soon drifted over Pompeii, turning day into night. Blocks and pellets of rock fell like hail, while fine ash and choking fumes enveloped the town **(Fig. 9.1)**. People frantically rushed to escape, but for most it was too late. As the growing weight of the falling volcanic debris began to crush buildings, a scalding, turbulent avalanche of ash mixed with pumice fragments surged down the flank of the volcano and swept over Pompeii. When the next day dawned, the town, along with its neighbor, Herculaneum, had vanished beneath a 6-m-thick gray-black blanket of debris **(Fig. 9.2a–d)**. This covering protected the ruins of Pompeii and Herculaneum so well that when archaeologists started to excavate the towns 1,800 years later, they found artifacts and structures that gave an amazingly complete picture of Roman daily life. In addition, they discovered odd-shaped open spaces in the debris covering Pompeii. Out of curiosity, they filled the spaces with plaster and then dug away the surrounding ash. The spaces turned out to be fossil casts of Pompeii's unfortunate inhabitants, contorted in agony or huddled in despair **(Fig. 9.2e)**.

FIGURE 9.1 The eruption of Vesuvius buried Pompeii and nearby Herculaneum in 79 c.e.

(a) In this 1817 painting, the British artist J. M. W. Turner depicted the cataclysmic explosion.

(b) An artist's interpretation of the early phase of the eruption when roof-crushing debris rained on the town.

◄ (facing page) The drama of an eruption, in this case, at the peak of the Santiaguito Volcano, in Guatemala. Behind a cloud of ash (tiny glass shards mixed with fragmented rock), molten rock, which has risen from the mantle, spurts out explosively from the summit.

FIGURE 9.2 The burial of Pompeii.

(a) If we look southeast at Mt. Vesuvius from the air today, it's evident that the volcano was once much bigger. The red dot shows the location of Pompeii.

(b) Excavations exposed the ruins of Pompeii, shown here with Vesuvius in the distance. The dashed line shows the volcano's profile prior to its eruption.

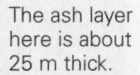

A modern suburb of Naples has been built on top of the ash.

The ash layer here is about 25 m thick.

Excavation exposes Roman columns.

(c) Volcanic debris still buries much of Herculaneum.

You can see ruts carved into streets by chariots.

(d) Streets and buildings of Pompeii are well preserved.

(e) Casts of people and animals convey the terror of that awful day.

Clearly, volcanoes are unpredictable and dangerous. Volcanic activity can build a towering mountain, or it can blast one apart. The materials a volcano produces can provide the fertile soil and mineral deposits that enable a civilization to thrive, or they can provide a rain of destruction that can snuff one out. To characterize volcanic activity, this chapter sets ambitious goals. We first build on Chapter 6 by looking more closely at the products of volcanic eruptions and at the basic components of volcanoes. Then we consider the different kinds of volcanic eruptions on the Earth and why they occur where they do. Finally, we review the hazards posed by volcanoes, the efforts made by geoscientists to predict eruptions and help minimize the damage they cause, and the possible influence of eruptions on climate and civilization.

9.2 The Products of Volcanic Eruptions

The drama of a volcanic eruption serves an important role in the Earth System because it transfers materials from inside the Earth to our planet's surface. Products of an eruption come in three forms—lava flows, pyroclastic debris, and gases. (Note that we use the term *lava flow* for both a molten, moving layer of lava and for the solid layer of rock that forms when the lava freezes.) We'll examine each of these products in turn.

Lava Flows

Sometimes lava races down the side of a volcano like a fast-moving, incandescent stream, sometimes it builds into a rubble-covered mound at a volcano's summit, and sometimes it oozes like a sticky but scalding paste into blob-like masses. Clearly, not all lava behaves in the same way when it rises out of a volcano, so not all lava flows look the same. Why?

The character of a lava primarily reflects its **viscosity** (resistance to flow), and not all lavas have the same viscosity. Differences in viscosity depend on a variety of factors, including chemical composition, temperature, gas content, and crystal content. Silica content (the proportion of SiO_2 in lava) plays a particularly key role in controlling viscosity. That's because silicon-oxygen tetrahedra link together to form long chains that can't move easily in a melt. Therefore, as the proportion of silica in lava increases, the lava's ability to flow decreases. Basaltic lava contains a low proportion of silica, so it has lower viscosity and can flow long distances. Rhyolitic lava contains a high proportion of silica, so it has higher viscosity and tends to build a mound at the vent **(Fig. 9.3)**. To emphasize the effects of viscosity on lava behavior, we'll now examine flows of different compositions more closely.

FIGURE 9.3 The characteristics of a lava flow depend on its viscosity.

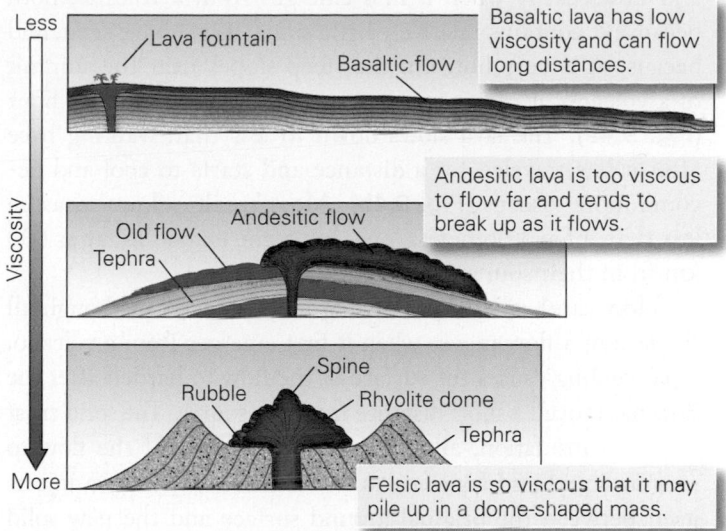

(a) Three different kinds of flows.

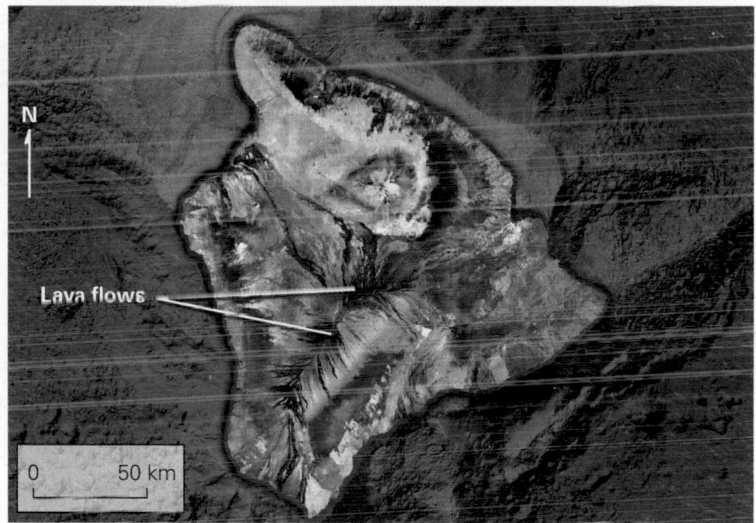

(b) A satellite view of Hawaii shows that numerous lava flows have come from fissures and vents on Kilauea.

(c) This rhyolite dome formed about 650 years ago in Panum Crater, California. Tephra (cinders) accumulated around the vent.

Basaltic Lava Flows Basaltic (mafic) lava has low viscosity and flows easily when it first emerges from a volcano, both because it contains relatively little silica, as we have seen, and because it is very hot. On the steep slopes near the summit of a volcano, it can move at speeds of up to 30 km per hour **(Fig. 9.4a)**. The lava slows down to less-than-walking pace after it has traveled for a distance and starts to cool and become more viscous **(Fig. 9.4b)**. Most basaltic flows measure less than a few kilometers long, but some extend as far as 600 km from their source.

How can basaltic lava travel great distances? Although all the lava of a flow moves when it first emerges from a volcano, rapid cooling causes the surface of the flow to harden after the flow has moved a short distance from the source. The solid crust serves as insulation, allowing the hot interior of the flow to remain liquid and continue to move. New molten lava injects itself between the original ground surface and the new solid crust—the addition of this lava effectively inflates the flow, jacking up the hardened crust and making the overall flow thicker. As time progresses, part of the flow's interior solidifies, so eventually molten lava moves only through a tunnel-like passageway, or **lava tube**, within the flow. The largest lava tubes may be tens of meters in diameter **(Fig 9.4c)**. In some cases, they eventually drain and become empty tunnels **(Fig. 9.4d)**.

The character of a basaltic lava flow's surface reflects the timing of freezing. Flows that have warm, pasty surfaces wrinkle into smooth, glassy, rope-like ridges **(Fig. 9.4e)**—geologists refer to such flows by their Hawaiian name, **pahoehoe** (pronounced pa-hoy-hoy). If the surface layer of the lava freezes, but then breaks up due to the continued movement of lava underneath, it becomes a jumble of sharp, angular fragments, creating a rubbly flow also known by its Hawaiian name, **a'a'** (pronounced ah-ah) **(Fig. 9.4f)**. Footpaths made by people living in basaltic volcanic regions follow the smooth surface of pahoehoe flows rather than the rough, foot-slashing surface of a'a' flows.

During the final stages of cooling, lava flows contract, because rock shrinks as it loses heat, and may fracture into polygonal columns. This type of fracturing, called **columnar jointing (Fig. 9.5a, b)**, typically terminates in the rubble that occurs at the top and bottom of a flow.

Basaltic flows that erupt underwater look different from those that erupt on land because lava cools so much more quickly in water. Because of rapid cooling, submarine basaltic lava can travel only a short distance before its surface freezes, producing a glass-encrusted blob, or *pillow* **(Fig. 9.5c)**. The glass rind of a pillow momentarily stops the flow's advance, but soon the pressure of the lava squeezing into the pillow breaks the rind, and a new blob of lava squirts out, freezes, and produces another pillow. In some cases, successive pillows add to the end of previous ones, forming worm-like chains. Geologists refer to a lava flow consisting of a pile of such blobs as **pillow lava**.

Andesitic and Rhyolitic Lava Flows Because of its greater viscosity, andesitic lava cannot flow as easily as basaltic lava. When erupted, andesitic lava forms a mound above the vent. This mound advances slowly down the volcano's flank at only 1 to 5 m a day, becoming a lumpy flow with a bulbous snout. Typically, andesitic flows are less than a few kilometers long. Because the lava moves so slowly, the outside of the flow has time to solidify, so as it moves, the surface breaks up into angular blocks, and the whole flow looks like a jumble of rubble known as *blocky lava*. On steep slopes, the blocks may tumble downhill, so the flow may evolve into a landslide of blocks.

Rhyolitic lava is the most viscous of all lavas because it has the highest silica concentration and the coolest temperature. Therefore, it tends to accumulate above the vent in a bulbous mass called a *lava dome* (see Fig. 9.3c). Sometimes rhyolitic lava freezes while still in the vent and then pushes upward to form a column-like *lava spire* or lava spine, which may rise up to 100 m above the vent. Rhyolitic flows, where they do form, rarely extend for more than 1 to 2 km from the vent and have broken and blocky surfaces.

Volcaniclastic Deposits

On a mild day in February 1943, as Dionisio Pulido prepared to sow the fertile soil of his field 330 km west of Mexico City, an earthquake jolted the ground, as it had dozens of times in the previous days. But this time, to Dionisio's amazement, the surface of his field visibly bulged upward by a few meters and then cracked. Ash and sulfurous fumes filled the air, and the farmer fled. When he returned the following morning, his rich land lay buried beneath a 40-m-high mound of gray cinders. Dionisio had witnessed the birth of a new volcano named Paricutín. During the next several months, Paricutín erupted continuously, at times blasting clots of lava into the sky like fireworks. By the following year, it had become a steep-sided cone over 300 m high. Nine years later, when the volcano stopped erupting, its lava and debris covered 25 km², and Dionisio's farm and those of his neighbors were gone.

> **Did you ever wonder . . .**
> if anyone has ever seen a brand-new volcano appear?

This description of Paricutín's eruption, and that of Vesuvius at the beginning of this chapter, emphasizes that volcanoes produce large quantities of fragmental material. Geologists use the term **pyroclastic debris** (from the Greek *pyro*, meaning fire) for fragmented igneous material forcefully ejected from a volcano. Pyroclastic debris includes both material solidified from clots or droplets of lava that freeze, either in the air or after they fall, and clasts formed by the breakup of already solid rock during an eruption. It accumulates on or

FIGURE 9.4 Features of basaltic lava flows. Basaltic lava has low viscosity and thus can flow long distances.

(a) A fast-moving flow coming from Mt. Etna, Sicily.

Lava flow

Highway

(b) A basaltic lava flow covers a highway on the Big Island of Hawaii.

A "skylight" into an active lava tube

(c) In a lava tube, still-molten lava flows beneath a crust of solid basalt.

(d) A drained lava tube exposed in a road cut on Hawaii.

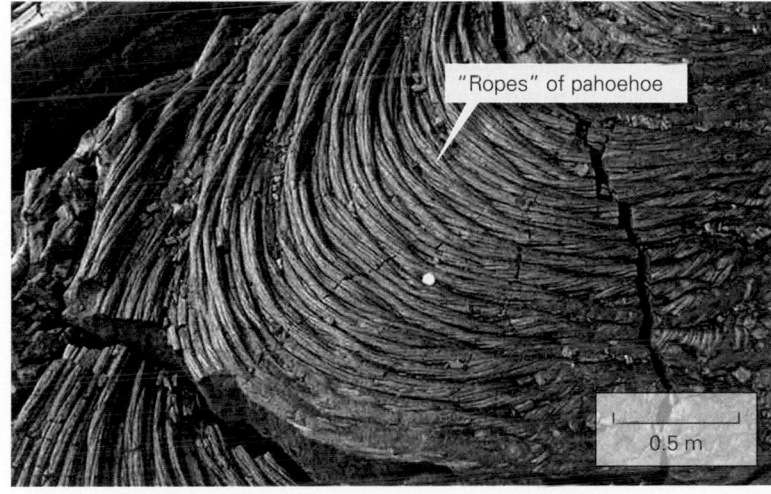

"Ropes" of pahoehoe

0.5 m

(e) Pahoehoe from a recent lava flow on Hawaii. Note the coin for scale.

(f) The rubbly surface of an a'a' flow, Sunset Crater, Arizona.

FIGURE 9.5 Examples of structures within lava flows.

(a) Columnar jointing develops when the interior of a flow cools and cracks. This example is Devils Postpile in California.

(c) Pillow basalt develops when lava erupts underwater. Later uplift may expose pillows above sea level, as in this outcrop in Cyprus.

(b) Columnar jointing can also be seen in this view from the side (top) and from above (bottom) at The Giant's Causeway in Northern Ireland.

around the volcano to form a layer of *tephra* **(Fig. 9.6a)**, which, if it becomes solid and coherent, becomes a layer of **pyroclastic rock**. In some cases, pyroclastic debris mixes with water (from rain, snow, or ice) and flows in a muddy slurry down the side of the volcano. Where there's enough water, streams may transport pyroclastic debris away from the volcano, eventually depositing it elsewhere in sorted sedimentary layers. We refer to any accumulation of fragmental volcanic material—pyroclastic debris, or the deposits of slurries and streams—as a *volcaniclastic deposit*. Let's now look more closely at the various components of volcaniclastic deposits. You'll see that different types form in association with different kinds of eruptions.

Pyroclastic Debris from Basaltic Eruptions Basaltic magma rising in a volcano may contain dissolved volatiles, such as water. As such magma approaches the surface, the volatiles form gas bubbles. In basaltic magma, the bubbles can rise faster than the magma itself, and when the bubbles reach the surface, they burst and eject clots and drops of molten lava upward to form dramatic fountains **(Fig. 9.6b)**. To picture this process, think of the droplets that spray from a just-opened bottle of soda.

Geologists recognize several different types of fragments formed from frozen clots or drops of lava. Pea- to golf-ball-sized fragments of glassy lava and scoria are known as *cinders*. Geologists consider cinders to be a type of **lapilli** (singular *lapillus*, from the Latin word for little stones). Lapilli accumulate in layers **(Fig. 9.6c)**. Occasionally, flying droplets of lava will develop long, thin tails as they move through the air. These tails freeze into hair-like filaments of glass known

FIGURE 9.6 Examples of pyroclastic debris from basaltic eruptions on Hawaii.

(a) Recent tephra on the flank of a volcano in Hawaii.

(b) A fountain of basaltic lapilli spouts from a vent on Hawaii.

(c) Blocks and lapilli on the flank of a Hawaiian volcano.

(d) A bomb has a smooth, streaked surface.

as *Pele's hair*, named after the Hawaiian goddess of volcanoes. The droplets themselves freeze into tiny, streamlined, glassy beads known as Pele's tears.

Basaltic eruptions also produce apple- to refrigerator sized fragments, called **blocks** (see Fig. 9.6c), which consist of already-solid volcanic rock, broken up during the eruption. Blocks tend to be angular and chunky. Lumps of erupted igneous rock that have streamlined, polished surfaces are known as **bombs (Fig. 9.6d)**. Bombs form when lava has already solidified, but remains hot and soft as it squirts out of the vent and then flies through the air.

Pyroclastic Debris from Andesitic and Rhyolitic Eruptions

Andesitic and rhyolitic lavas are more viscous than basalt and tend to be more gas rich. Eruptions of these lavas also tend to be explosive. Volcanic explosions can produce immense quantities of pyroclastic debris, much more than can come from a basaltic volcano **(Fig. 9.7a)**. Debris ejected from explosive eruptions includes **ash**, which consists of glassy particles less

than 2 mm in diameter formed when frothy lava or recently formed pumice breaks up explosively during an eruption, or when pre-existing volcanic rock gets pulverized by the force of an explosion; *pumice lapilli*, which consists of angular pumice fragments; and *accretionary lapilli*, which consists of snowball-like lumps of ash formed when ash mixes with water in the air and then sticks together to form small balls **(Fig. 9.7b–d)**.

Ash, or ash mixed with lapilli, becomes a type of pyroclastic rock called **tuff** when buried and lithified **(Fig. 9.8a, b)**. In some cases, the grains bond during deposition, because grains are so hot that they weld together. But more commonly, the coherence of tuffs occurs when the grains are cemented together either by minerals precipitated from groundwater or by minerals that grow in the ash as it reacts with groundwater.

Other Volcaniclastic Deposits

The nature and origin of fragmental material produced during and after eruptions continue to be the subject of active research because deposits of

FIGURE 9.7 Pyroclastic debris from andesitic and rhyolitic eruptions.

(a) Pyroclastic debris billowing from the 2008 eruption of Chaitén in Chile.

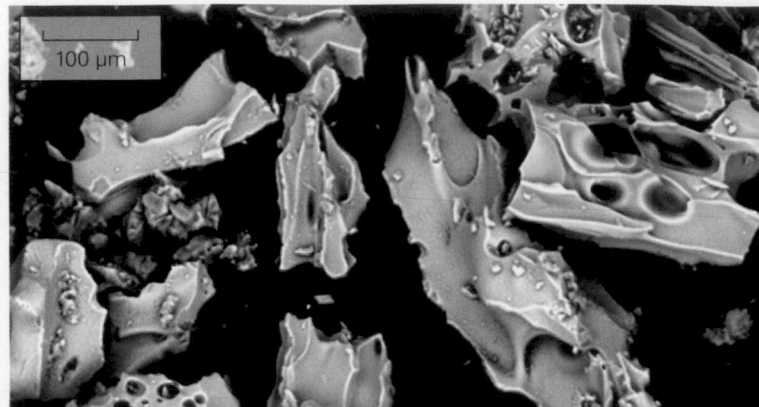

(b) Electron photomicrograph of ash.

(c) Pumice lapilli.

(d) Accretionary lapilli.

this material prove to be difficult to interpret. As we've noted, geologists use the term *volcaniclastic deposit* for any material that consists of volcanic igneous fragments, and they recognize three categories:

1. *Pyroclastic* deposits, as we have just seen, consist of fragments ejected during an eruption, and they accumulate directly from the clouds of debris ejected into the sky or sent in avalanches down the flank of the volcano. The fragments in such deposits have not moved subsequent to their original deposition.

2. *Volcanic-sedimentary* deposits consist of volcanic material (lava and pyroclastic debris) that later moved downslope and was redeposited elsewhere subsequent to accumulating after an eruption. Some of this material tumbles in landslides, breaking up to varying degrees as it moves. If volcanoes are covered with snow and ice or are drenched with rain, water mixes with debris to form a *volcanic debris flow*, which moves downslope like wet concrete (**Fig. 9.8c–f**). Very wet, ash-rich debris flows downslope in a fast-moving slurry called a **lahar**, which can reach speeds of 50 km per hour. Lahars tend to follow river channels and may travel for tens of kilometers away from the volcano. When debris flows and lahars stop moving, they yield a layer consisting of volcanic blocks suspended in ashy mud. Rivers may eventually sort and transport some of this volcanic

sediment. Where this material accumulates, perhaps far downstream, it forms deposits of *volcanic sandstone* and/or *volcanic conglomerate.*

3. *Fragmental lava* deposits consist of debris produced when lava breaks up into angular clasts while flowing, without ever being ejected into the air. As we've seen, fragmentation (also called brecciation) happens when the inside of a lava flow continues to move after its surface has frozen and the crust breaks up due to the movement. But fragmentation may also happen when lava freezes very quickly and shatters upon erupting into water or ice. The resulting material, *hyaloclastite,* consists of glassy fragments embedded in ash that has reacted with hot water.

FIGURE 9.8 Examples of volcaniclastic deposits.

(a) Tuff at Hole in the Wall, in the Mojave Desert. Lens cap for scale.

(b) A cliff of ~1-Ma tuff in New Mexico.

The landslide stripped away the forest.

(c) Water-soaked volcanic debris slid down the side of this volcano in Nicaragua.

Debris flow deposits

Stratified ash

Paleo-channel wall

(d) Deposits of a debris flow that accumulated at about 35 Ma in Utah. Note that the debris flow filled a channel cut into finer, stratified ash.

(e) A lahar fills a riverbed in New Zealand after an eruption in 2007.

(f) Deposits of a lahar from Mt. St. Helens, 20 years after its 1980 eruption, include logs ripped from hill slopes.

Volcanic Gases

Most magma contains dissolved volatiles, including water (H_2O), carbon dioxide (CO_2), sulfur dioxide (SO_2), and hydrogen sulfide (H_2S). Generally, felsic lavas can contain more volatiles than can mafic lavas—in fact, up to 9% by weight of a felsic magma consists of volatiles. As we've seen, these dissolved gases come out of solution when the magma approaches the Earth's surface. This process happens for two reasons. First, the ability of a liquid to hold dissolved gas decreases as the pressure acting on the liquid decreases. You see this phenomenon when you pop off the top of a carbonated beverage—the beverage was injected with CO_2 under pressure when it was bottled, and because popping the top decreases the pressure, bubbles form and give the beverage its sparkle. Second, gas comes out of solution as a side effect of crystallization. Gases can't fit easily into growing crystals, so they remain in the liquid magma, causing the concentration of gas in the liquid to increase until gas bubbles form.

The sulfurous gases emitted by some volcanoes smell like rotten eggs. These gases dissolve in tiny water droplets to produce corrosive sulfuric acid. This acid occurs in the form of an **aerosol**, meaning a haze of droplets or solid particles that are so small that they can remain suspended in air for a long time.

The fate of bubbles in magma depends on the viscosity of the magma. For example, in low-viscosity mafic magma, gas bubbles can rise faster than the magma moves, so most reach the surface of the magma and enter the atmosphere before the lava does. As a result, some volcanoes may, for a while, produce large quantities of steam, without much lava **(Fig. 9.9a)**. The last bubbles to form, however, remain as holes when the lava freezes around them. As we discussed in Chapter 6, these holes are called **vesicles (Fig. 9.9b)**. Mafic rock in which more than 50% of the volume consists of bubbles is called *scoria*. In high-viscosity felsic magmas, the gas has trouble escaping because bubbles can't push through the very sticky magma. When such magma reaches shallower depths, it effectively becomes a foam. As this foam approaches the Earth's surface, and the weight of overlying magma decreases, the gas expands, so that in some cases, bubbles may account for as much as 50% to 75% of the volume of the magma. As we've seen, pumice forms when this material freezes—the thin films of melt between bubbles turn into glass. In some cases, pumice has such low density that it can float.

9.3 Structure and Eruptive Style

Volcanic Architecture

As we saw in Chapter 6, melting in the upper mantle and lower crust produces magma, which rises into the upper crust. Typically, this magma accumulates underground in

FIGURE 9.9 The gas component of volcanic eruptions.

H_2O, CO_2, and SO_2

(a) A volcano in Alaska erupting large quantities of steam.

(b) Gas bubbles frozen in lava produce vesicles, as in this block from Sunset Crater, Arizona.

(a) During some eruptions, lava spouts from a chimney-shaped conduit. Such eruptions typically happen at the summit of a volcano.

(b) At a fissure eruption, lava comes out in a curtain, along the length of a crack.

a **magma chamber,** a zone of open spaces and/or fractured rock that can contain a large quantity of magma or a mush of magma mixed with crystals. Some of the magma may solidify in the magma chamber and transform into intrusive igneous rock, but the remainder rises along a pathway, or *conduit*, to the Earth's surface and erupts from an opening, or **vent.** The conduit may have the shape of a chimney or may be a long crack called a **fissure (Fig. 9.10)**. At times, the vent at the top of a chimney-shaped conduit erupts a tall fountain of lava, whereas a fissure erupts a long curtain of lava.

Over time, new igneous rock (lava and/or pyroclastic debris) builds up around the vent to form a *volcanic edifice.* *Summit eruptions* take place at the top of the edifice, whereas *flank eruptions* break through along the sides, or flanks, of the volcano. In some cases, the vent lies at the floor of a bowl-like depression called a **crater,** which can be up to 500 m across and 200 m deep **(Fig. 9.11a)**. Craters form either because lava and tephra accumulate around the vent during eruption or

because the top part of the edifice collapses into the drained conduit when the eruption ceases.

During some major eruptions, a large portion of the volcanic edifice collapses into the drained magma chamber below, producing a **caldera,** a large circular or elliptical depression up to thousands of meters across and up to several hundred meters deep **(Fig. 9.11b–e)**. Typically, a caldera has steep walls and a fairly flat floor and may be partially filled with new lava or pyroclastic debris. Some calderas fill with water and become lakes. Note that calderas differ from craters in terms of size, and mode of formation.

Geologists distinguish among several different shapes of *subaerial* (above sea level) volcanic edifices. **Shield volcanoes**—broad, gentle domes—have this name because they resemble a soldier's shield lying on the ground. They form when the products of eruption have low viscosity and cannot pile up around the vent, but rather spread out over large areas. The volcanoes of Hawaii, for example, which

FIGURE 9.11 Volcanic craters and calderas.

(a) The crater of Santa Ana Volcano in El Salvador.

produce layer upon layer of low-viscosity basaltic lava, are shield volcanoes **(Fig. 9.12a)**. **Cinder cones**, also known as *scoria cones*, consist of cone-shaped piles of basaltic lapilli and blocks, sometimes from a single eruption **(Fig. 9.12b)**. **Stratovolcanoes**, also known as *composite volcanoes*, grow to become large (up to 3 km high) cone-shaped mountains, generally with steeper slopes near the summit. They consist of interleaved contrasting layers (hence the prefix *strato-*) of lava and tephra **(Fig. 9.12c)**. Their shape, exemplified by Japan's Mt. Fuji, serves as the classic image that most people have of a volcano.

Let's look at the nature of stratovolcanoes a little more closely. They build from the products of many eruptions over an extended period of time. Not all eruptions produce the same kind of material. Specifically, eruptions producing

Time →

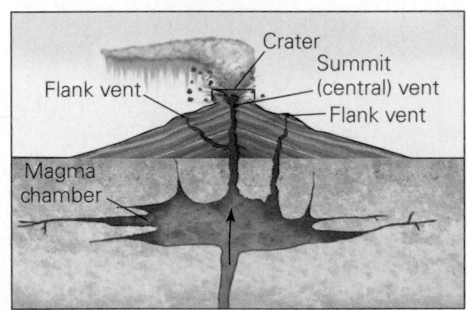

(b) As an eruption begins, the magma chamber inflates with magma. There can be a central vent and one or more flank vents.

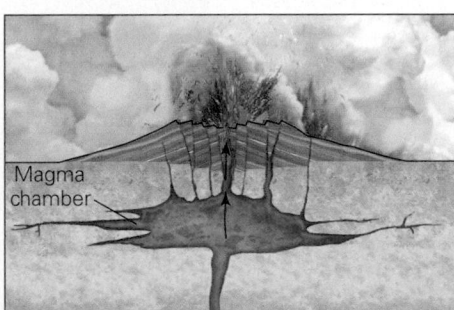

(c) During an eruption, the magma chamber drains, and the central portion of the volcano collapses downward.

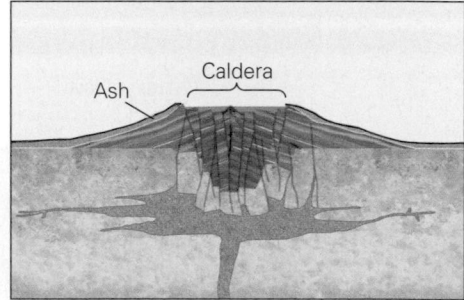

(d) The collapsed area becomes a caldera. Later, a new volcano may begin to grow within the caldera.

(e) This caldera in Oregon formed about 7,700 years ago. Afterward, it filled with water to become Crater Lake. Wizard Island, protruding from the lake, is a small volcano that grew on top of the caldera floor.

FIGURE 9.12 Volcanoes of different shapes.

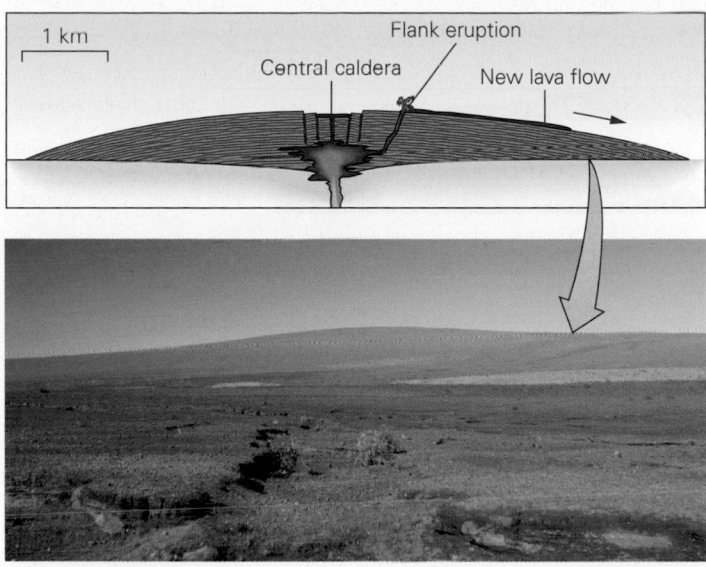

1 km

Flank eruption
Central caldera
New lava flow

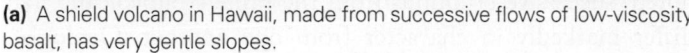

(a) A shield volcano in Hawaii, made from successive flows of low-viscosity basalt, has very gentle slopes.

100 m

(b) A cinder cone in Arizona, with a lava flow covering the land surface to its right.

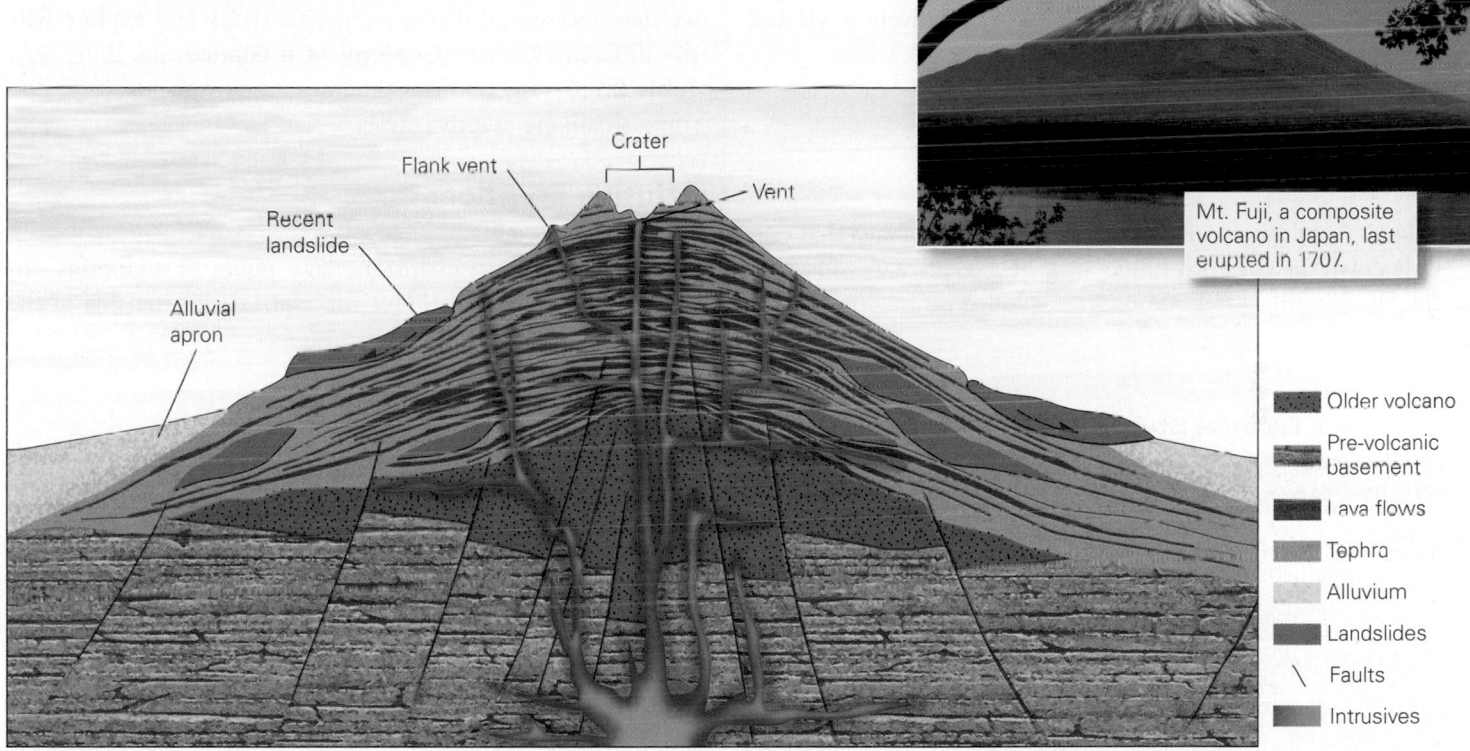

Crater
Flank vent
Vent
Recent landslide
Alluvial apron

Mt. Fuji, a composite volcano in Japan, last erupted in 1707.

Older volcano
Pre-volcanic basement
Lava flows
Tephra
Alluvium
Landslides
Faults
Intrusives

(c) A composite volcano consists of layers of tephra and lava. Volcanic debris flows and ash avalanches modify slopes and contribute to the development of a classic cone-like shape.

pyroclastic debris produce layers of tephra, eruptions of andesite produce blocky flows that break up into bouldery landslides on the volcano's slopes, and eruptions of low-viscosity lava produce flows that cascade down the flanks of the volcano. Lava flows resist erosion and can armor the underlying tephra, preventing it from being washed away. Over time, the volcano builds into a symmetric cone. For many reasons, this shape doesn't last indefinitely. For example, large landslides occasionally carry masses of rock down the slopes and build broad aprons around the base of the volcano. Heavy rains or spring snowmelt can trigger debris flows, which transport volcaniclastic sediment outward to form alluvial fans surrounding

FIGURE 9.13 These profiles emphasize that volcanoes come in different sizes. Large shield volcanoes, like those on Hawaii, are many times larger than cinder cones.

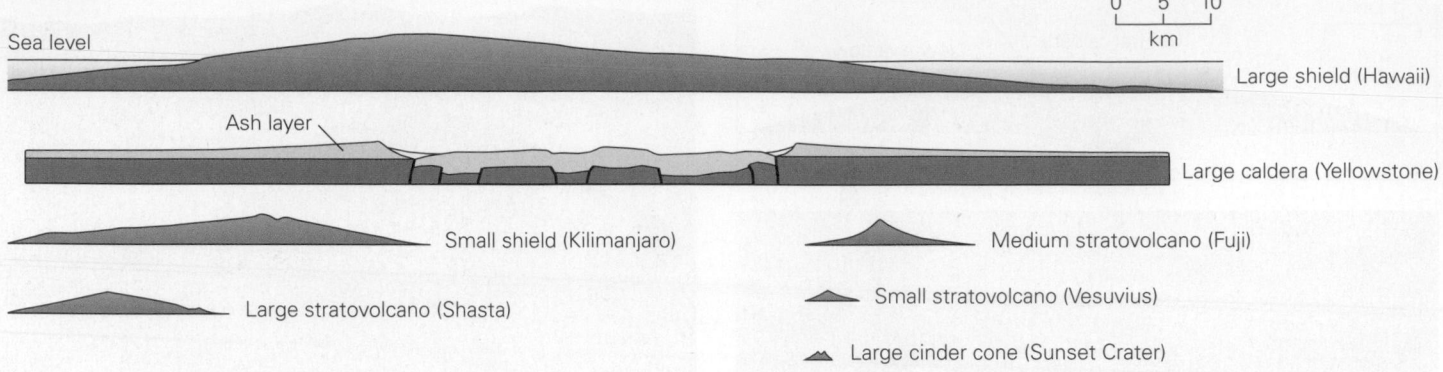

the volcano. Finally, explosive eruptions can blast away a large portion of the volcanic edifice.

The hills or mountains resulting from volcanism come in a great range of sizes **(Fig. 9.13)**. Shield volcanoes and volcanoes that produce large calderas tend to be the largest, followed by stratovolcanoes. Cinder cones tend to be relatively small and may occur in clusters on the flanks of larger volcanoes.

The Concept of Eruptive Style: Will It Flow or Will It Blow?

Kilauea, a volcano on Hawaii, produces rivers of lava that cascade down the volcano's flanks. Mt. St. Helens, a volcano near the Washington-Oregon border, exploded catastrophically in 1980 and blanketed the surrounding countryside with tephra. Clearly, different volcanoes erupt differently and, as we've noted, successive eruptions from the same stratovolcano may differ markedly in character from one another. Geologists refer to the character of an eruption as **eruptive style**. Below we describe several distinct eruptive styles and explore why the differences occur **(Geology at a Glance, pp. 296–297)**. **Table 9.1** lists the traditional names used by geologists to distinguish among eruptive styles.

Effusive Eruptions

During an **effusive eruption**, lava pours or fountains out of a vent or fissure. This lava may spill down the side of the

TABLE 9.1 Eruptive Styles of Volcanoes

Name	Source of name	Characteristics
Effusive	Latin word for pour out	Low-viscosity basaltic lava that fountains or flows easily
Hawaiian	Island of Hawaii	Predominantly effusive eruptions, producing flows with only minor scoria
Strombolian	Island of Stromboli (Italy)	Periodically erupts a fountain of scoria lapilli and ash
Vulcanian	Island of Vulcano (Italy)	Erupts both scoria lapilli and tall plumes of pyroclastic debris
Phreatic	Greek word for water well	Groundwater interacts with magma, producing a steam explosion and blocks of wall rock
Phreatomagmatic	Same root as *phreatic*	Eruption involves steam explosions, but also eruption of wet ash, lapilli, and blocks of lava
Surtseyan	Island of Surtsey (Iceland)	A phreatomagmatic eruption that happens when seawater enters the magma chamber, resulting in steam explosions
Plinean	Pliny (Roman scholar who saw Vesuvius erupt)	Huge explosion that sends a convective cloud to high altitudes, produces pyroclastic flows, and may destroy the volcanic edifice

FIGURE 9.14 The effects of effusive eruptions.

(a) A 1986 effusive eruption on Hawaii.

(b) A lava lake in the caldera of Kilauea, Hawaii.

(c) Layer upon layer of basaltic lava flows, exposed on the wall of a caldera in Hawaii.

volcano as a *lava flow* (Fig. 9.14a), or it may accumulate in a glowing *lava lake* around the crater (Fig. 9.14b). Successive eruptions can pile up lava flow over lava flow, eventually building a shield volcano (Fig. 9.14c; see Hawaiian eruptive style in Table 9.1).

Effusive eruptions take place when hot, mafic magma feeds the volcano, for such magma has low viscosity. The buoyancy of the magma, relative to the surrounding solid rock, along with pressure generated by injection of magma from below, pushes the magma up conduits within the volcano to a summit crater, flank crater, or fissure, where it extrudes as lava. The formation of gas bubbles in rising magma decreases its viscosity even more, and may increase pressure within it, causing the magma to rise faster. If a volcano's conduit narrows near the vent, rising magma accelerates, just as water coming out of a garden hose accelerates when you pinch the end of the hose. When this happens, the pressure may be great enough to drive

magma skyward in a lava fountain that spurts as high as a few hundred meters above the vent (see Fig. 9.10a).

Explosive Eruptions

Explosive eruptions forcefully emit significant quantities of pyroclastic debris. Geologists recognize a variety of different types based on the size and products of the explosion, as we now see.

Smaller Explosive Eruptions Ancient Romans referred to the island of Stromboli as "the lighthouse of the Mediterranean" because it has erupted about every 10 to 20 minutes throughout recorded history, and the red-hot clots that it ejects trace out glowing arcs of light in the night sky. Occasionally, Stromboli also produces basaltic lava flows, but most of the material it erupts comes out as a fountain that produces scoria lapilli and blocks, which build into a cone around the vent (Fig. 9.15a; see Table 9.1). Somewhat larger explosive eruptions emit both a fountain of lava and a dense plume of pyroclastic debris (Fig. 9.15b).

In some explosive eruptions, much of the pressure driving the eruption comes from the sudden heating of water by magma so that it flashes to steam and expands very rapidly. The expanding steam can rip pre-existing rock from the conduit of the volcano and send it skyward. In *phreatic eruptions*, groundwater (possibly from melting snow or heavy rain) interacts with the magma, and the resulting eruption blasts steam, ash, and coarser blocks skyward, but little if any lava appears at the surface. When the vent lies in relatively shallow seawater, heating of the water produces prodigious amounts of

GEOLOGY AT A GLANCE

Volcanoes

Beneath a volcano, magma rises to fill an open or cracked region of crust and forms a magma chamber. Some of the magma erupts at a surface vent. Once molten rock has erupted at the surface, it is called lava. Some lava spills down the side of the volcano in lava flows. Some may fountain out of a vent to form scoria fragments, which pile up in a cone around the vent. Eruptions may eject larger chunks as blocks or bombs.

The nature of eruptions depends on the viscosity of the magma, which in turn depends on magma composition. Explosive eruptions blast up a cloud of ash and lapilli,

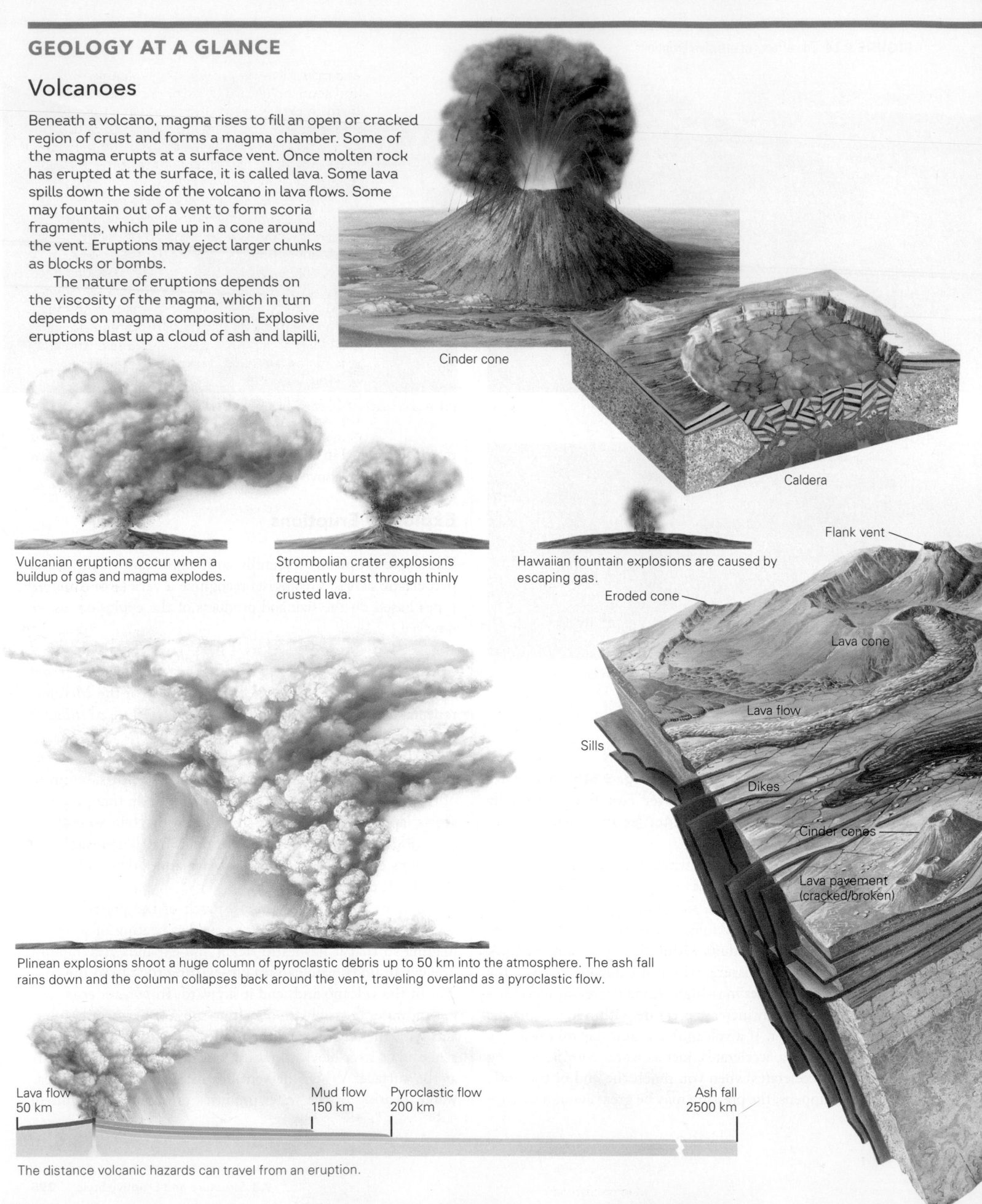

Cinder cone

Caldera

Vulcanian eruptions occur when a buildup of gas and magma explodes.

Strombolian crater explosions frequently burst through thinly crusted lava.

Hawaiian fountain explosions are caused by escaping gas.

Flank vent

Eroded cone

Lava cone

Lava flow

Sills

Dikes

Cinder cones

Lava pavement (cracked/broken)

Plinean explosions shoot a huge column of pyroclastic debris up to 50 km into the atmosphere. The ash fall rains down and the column collapses back around the vent, traveling overland as a pyroclastic flow.

Lava flow
50 km

Mud flow
150 km

Pyroclastic flow
200 km

Ash fall
2500 km

The distance volcanic hazards can travel from an eruption.

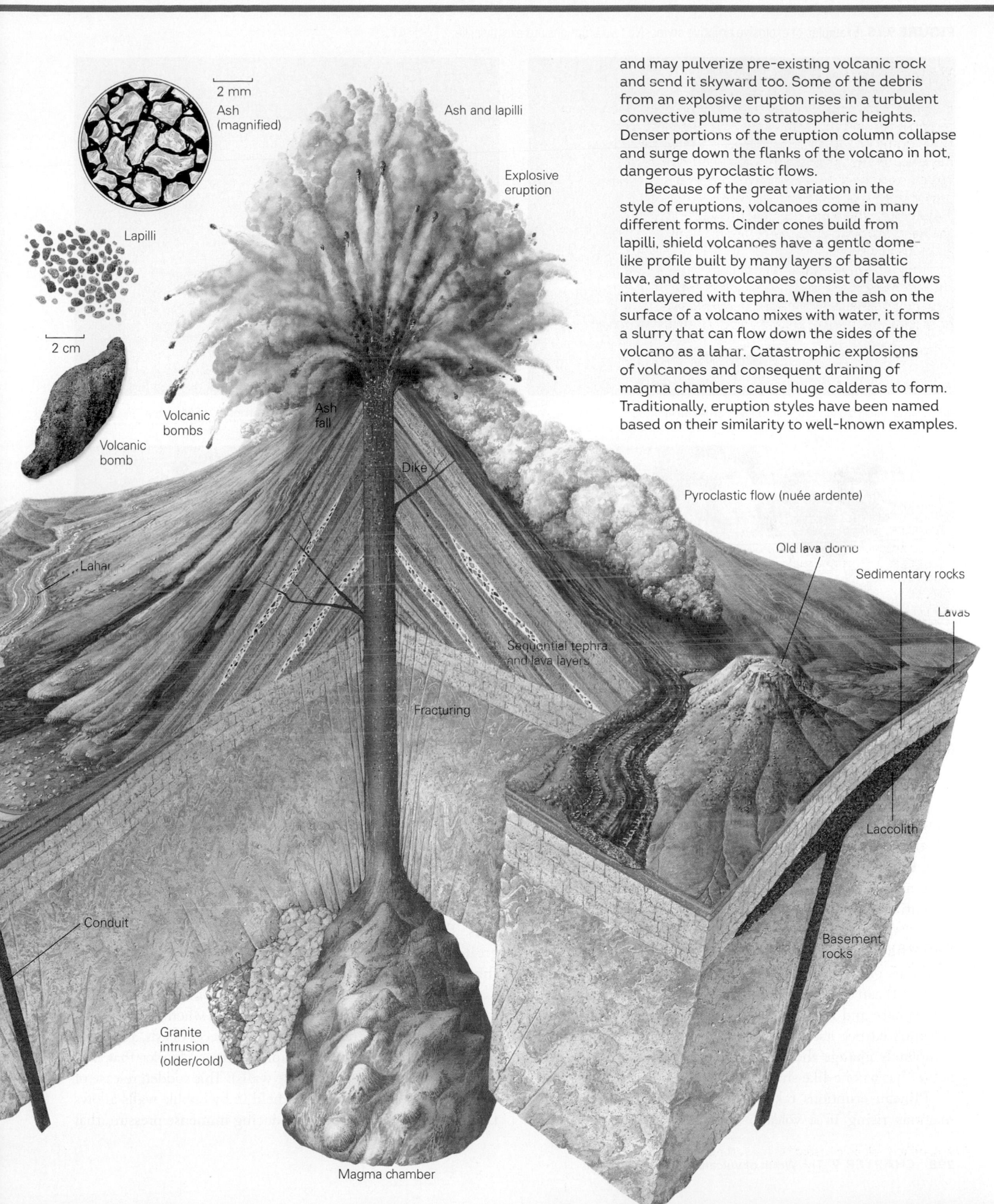

2 mm
Ash (magnified)

Lapilli

2 cm

Volcanic bomb

Ash and lapilli

Explosive eruption

Volcanic bombs

Ash fall

Dike

Lahar

Sequential tephra and lava layers

Fracturing

Conduit

Granite intrusion (older/cold)

Magma chamber

Pyroclastic flow (nuée ardente)

Old lava dome

Sedimentary rocks

Lavas

Laccolith

Basement rocks

and may pulverize pre-existing volcanic rock and send it skyward too. Some of the debris from an explosive eruption rises in a turbulent convective plume to stratospheric heights. Denser portions of the eruption column collapse and surge down the flanks of the volcano in hot, dangerous pyroclastic flows.

Because of the great variation in the style of eruptions, volcanoes come in many different forms. Cinder cones build from lapilli, shield volcanoes have a gentle dome-like profile built by many layers of basaltic lava, and stratovolcanoes consist of lava flows interlayered with tephra. When the ash on the surface of a volcano mixes with water, it forms a slurry that can flow down the sides of the volcano as a lahar. Catastrophic explosions of volcanoes and consequent draining of magma chambers cause huge calderas to form. Traditionally, eruption styles have been named based on their similarity to well-known examples.

FIGURE 9.15 Examples of explosive eruptive styles. No two eruptions are exactly alike.

(a) A large Strombolian eruption on Mt. Etna, Sicily.

(b) A lava fountain in the plume during an eruption of Mt. Etna.

Sea surface

(c) A Surtseyan eruption of a subsea volcano near Tonga.

Convective cloud

(d) The Plinean eruption of Mt. Pinatubo in the Philippines.

steam that billows out of the sea, along with fountains of wet ash and, on the seafloor, eruptions of volcanic-glass fragments **(Fig. 9.15c)**. If seawater suddenly gains access via cracks to a large magma chamber, the resulting flash to steam may blast the entire volcano apart in a huge explosion.

Immense Explosive Eruptions Really huge explosive eruptions of stratovolcanoes, known as *Plinean eruptions* (see Table 9.1) can eject many cubic kilometers of material into the atmosphere and may destroy a substantial part of the strato-volcano's edifice itself **(Fig. 9.15d)**. In fact, the explosion can completely change the profile of the volcano such that it no longer has a cone-like shape **(Box 9.1)**.

Plinean eruptions take place when andesitic or rhyolitic magmas rising in a volcano contains very large quantities

of gas. As we've seen, when this gas comes out of solution, it forms countless bubbles, which may comprise most of the magma's volume. Because of their high silica content, andesitic or rhyolitic magmas are so viscous that the gas bubbles cannot rise through the magma and escape while the gassy magma rises. The pressure within the trapped bubbles becomes much greater than that in the air above the volcano, and only the thin bubble walls keep the gas from bursting free. Eventually, as the rising froth shears against the walls of the conduit, the walls of the bubbles stretch and break, and when this happens, the bubble walls shatter into dust-sized pieces of ash, and these pieces surround the pumice lapilli (larger chunks that still contain both bubbles and bubble walls). This sudden release of the very hot gas that had been held in by bubble walls allows the gas to expand violently, producing immense pressure that

FIGURE 9.16 The components of a large explosive volcanic eruption.

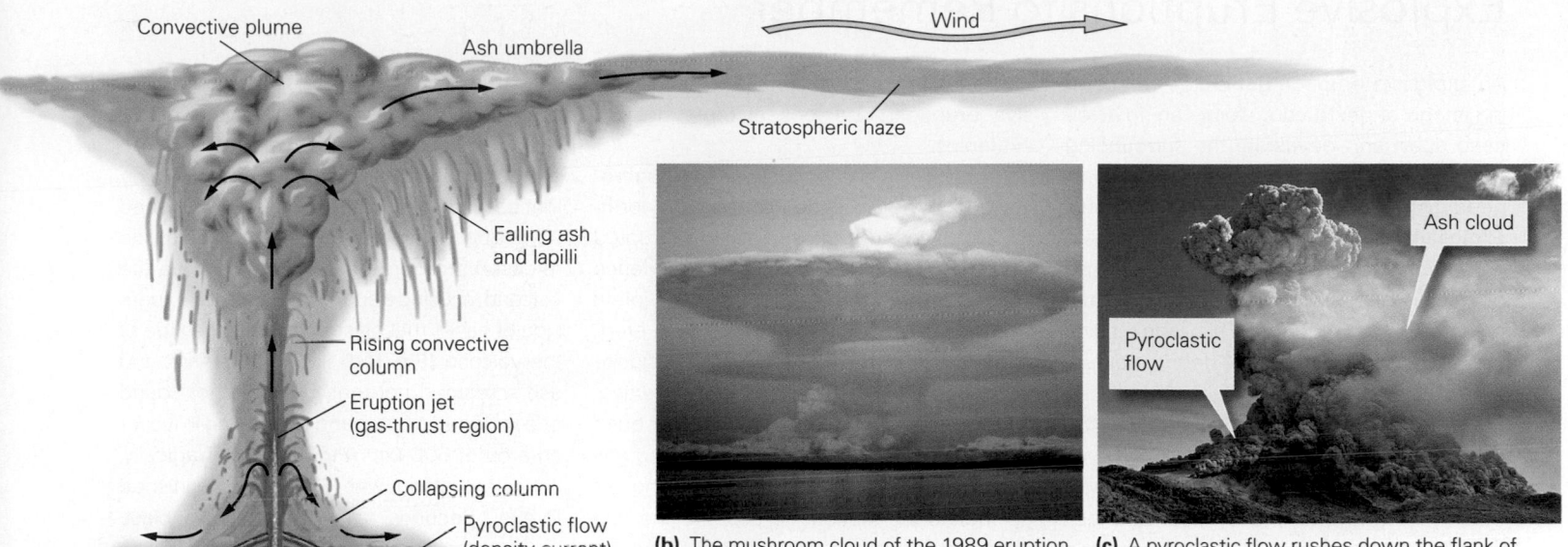

(a) A Plinean eruptive column contains several components.

(b) The mushroom cloud of the 1989 eruption of Redoubt Volcano, Alaska.

(c) A pyroclastic flow rushes down the flank of Mt. Merapi, Indonesia, in 2006.

pushes surrounding debris upward. If the pressure cracks the lava dome capping the vent, the mixture of fragments and gas bursts out of the volcano's vent at very high velocity (over 300 km per hour), like a giant shotgun blast. The eruption of this material decreases the pressure on magma deeper in the conduit, allowing bubbles in the deeper magma to expand and shatter, so more and more pyroclastic debris erupts until the magma chamber drains. The ejected debris rises above the volcano to produce a huge **eruption column**.

Geologists recognize several distinct regions within a Plinean eruptive column **(Fig. 9.16a)**. The blast from an explosive eruption can propel debris upward for hundreds of meters to a couple of kilometers and forms the lower part, or *gas-thrust region*, of the eruption column. Generally, the eruptive column of a huge eruption grows to become much taller than the gas-thrust region. That's because the density of the mixture of hot ash, hot volcanic gas, and hot air becomes less than that of the cooler air around it, so, like a hot-air balloon, the column starts to rise buoyantly as a turbulent, billowing cloud, or *convective plume*. At stratospheric heights of 10 to 20 km, the convective plume spreads out into a broad *ash umbrella*—the mushroom head of an overall mushroom cloud **(Fig. 9.16b)**. The formation of an umbrella occurs where the plume has cooled enough so that it no longer has buoyancy.

Debris from the convective plume and ash umbrella of a huge volcanic explosion falls from the air, like hail and snow, and blankets the countryside near the volcano. Strong winds of the jet stream may waft substantial amounts of ash in the umbrella great distances—hundreds or even thousands of kilometers—away from the volcano. When gravity causes some of the debris at the top of the gas-thrust region to collapse downward, a scalding avalanche composed of air, hot ash, and pumice lapilli rushes down the side of the volcano. This surge of air and debris stays relatively close to the ground because it's denser than the clear air above **(Fig. 9.16c)**. Such avalanches of hot ash are known as **pyroclastic flows**. In older literature, a pyroclastic flow was known as a *nuée ardente*, French for glowing cloud. Tuff formed from ash and/or pumice lapilli that fall from the sky is called *air-fall tuff*, whereas a sheet of tuff that formed from a pyroclastic flow is an **ignimbrite**. Ash and pumice lapilli in an ignimbrite are sometimes so hot that they weld together to form a hard mass.

Supervolcanoes In recent years, geologists have realized that some volcanic explosions of the geologic past dwarf any that have been observed during human history (see Box 9.1). Such incomprehensibly huge volcanoes have come to be known informally as **supervolcanoes**, and they can produce hundreds to a few thousand cubic kilometers of pyroclastic debris. After the eruption, a huge caldera forms as the land collapses into the drained magma chamber (see Fig. 9.11). Eruptions that occurred hundreds of thousands of years ago in the area that is now Yellowstone Park serve as an example—in the aftermath of one of these explosions, a caldera 72 km across formed! The largest supervolcano eruption yet recognized in the geologic record blasted 5,000 km³ of pyroclastic debris into the sky and left a huge caldera in southwestern Colorado.

BOX 9.1 CONSIDER THIS . . .

Explosive Eruptions to Remember

An explosive eruption generates an enduring image of destruction, for it can rip a volcano apart and devastate the surrounding region. To compare explosive eruptions, geologists sometimes use the **volcanic explosivity index** (**VEI**), a logarithmic scale on which the largest known eruption has been assigned a VEI of 8; larger ones could conceivably take place. This index takes into account the volume of debris ejected, the height of the eruptive column, and an estimate of the energy released during the explosion. The historic record shows that there have been about 100 eruptions with VEIs in the range of 4 to 6 since 1800 C.E., and the largest observed eruption in recorded history (Tambora, in 1815) ranks as a 7. The geologic record shows that "mega-colossal" explosions with a VEI of 8 have taken place during the past few million years. One of these formed the caldera of Yellowstone National Park, Wyoming, about 600,000 years ago **(Fig. Bx9.1a)**. To get a

sense of the consequences of an explosive eruption, let's look at three notable examples.

Mt. St. Helens, a snow-crested stratovolcano in the Cascade Range of the northwestern United States, had not erupted since 1857. However, geologic evidence suggested that the mountain had a violent past, punctuated by many explosive eruptions. On March 20, 1980, an earthquake announced that the volcano was awakening. A week later, a crater 80 m in diameter burst open at the summit and began emitting gas and pyroclastic debris. Geologists who set up monitoring stations to observe the volcano noted that its north side was beginning to bulge markedly, suggesting that the volcano was filling with magma and was expanding like a balloon. Their concern that an eruption was imminent led local authorities to evacuate people in the area.

The climactic eruption came suddenly. At 8:32 A.M. on May 18, a geologist, David

Johnston, monitoring the volcano from a distance of 10 km, shouted over his two-way radio, "Vancouver, Vancouver, this is it!" An earthquake had triggered a huge landslide that caused 3 km³ of the volcano's weakened north side to slide away. The sudden landslide released pressure on the magma inside the volcano, causing a sudden and violent expansion of gases that blasted through the side of the volcano **(Fig. Bx9.1b)**. Rock, steam, and ash screamed north at the speed of sound and flattened a forest and everything in it over an area of 600 km² **(Fig. Bx9.1c)**. Tragically, Johnston, along with 60 others, vanished forever. Seconds after the sideways blast, a vertical column carried about 540 million tons of ash (about 1 km³) 25 km into the sky, where the jet stream carried it away; the ash was able to circle the globe. In towns near the volcano, a blizzard of ash choked roads and buried fields. Water-saturated ash formed viscous slurries, or lahars, that flooded river valleys, carrying away everything in their path.

FIGURE Bx9.1 Examples of explosive eruptions.

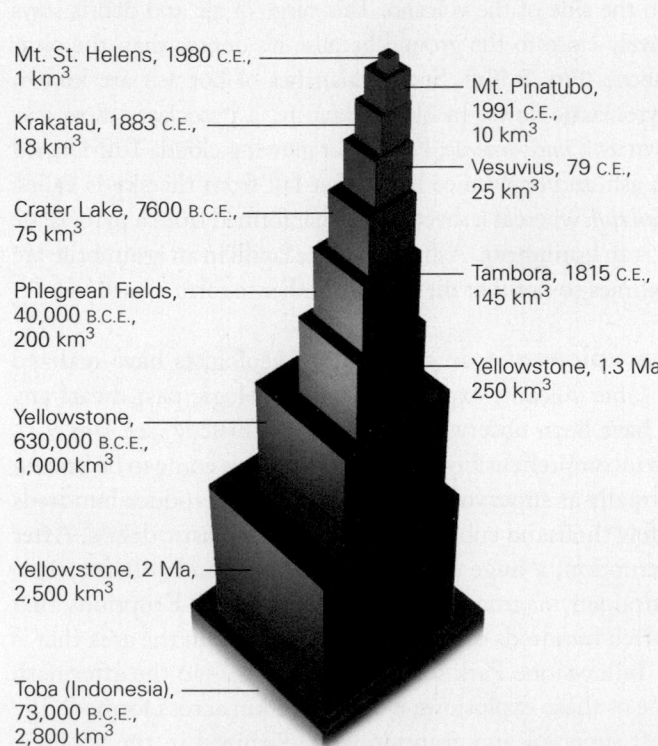

Mt. St. Helens, 1980 C.E., 1 km³
Krakatau, 1883 C.E., 18 km³
Crater Lake, 7600 B.C.E., 75 km³
Phlegrean Fields, 40,000 B.C.E., 200 km³
Yellowstone, 630,000 B.C.E., 1,000 km³
Yellowstone, 2 Ma, 2,500 km³
Toba (Indonesia), 73,000 B.C.E., 2,800 km³

Mt. Pinatubo, 1991 C.E., 10 km³
Vesuvius, 79 C.E., 25 km³
Tambora, 1815 C.E., 145 km³
Yellowstone, 1.3 Ma, 250 km³

(a) The relative amounts of pyroclastic debris (in km³) ejected during major explosive eruptions.

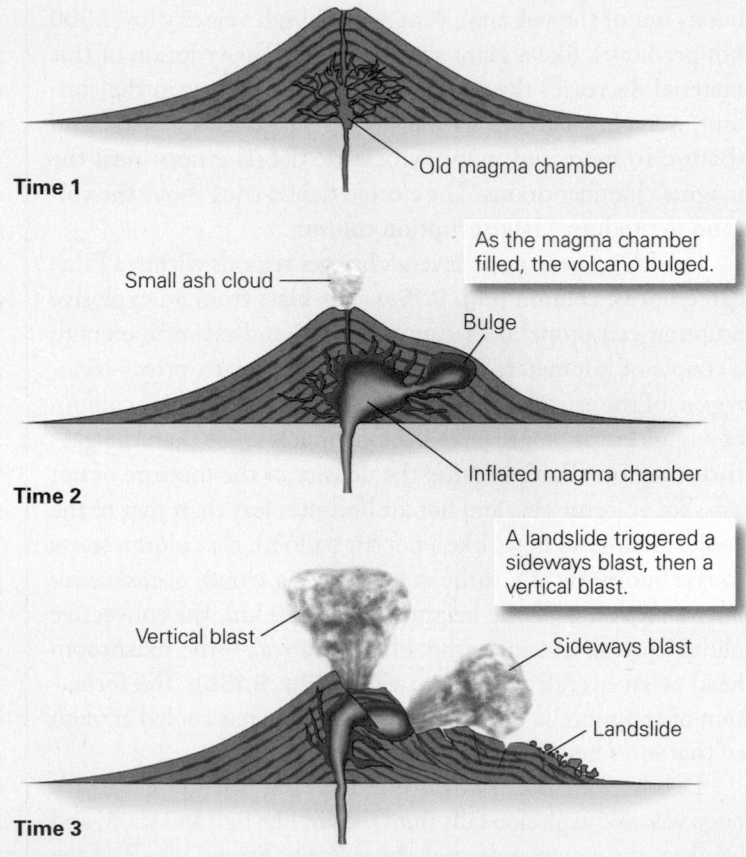

Old magma chamber

Time 1

Small ash cloud

As the magma chamber filled, the volcano bulged.

Bulge

Inflated magma chamber

Time 2

A landslide triggered a sideways blast, then a vertical blast.

Vertical blast

Sideways blast

Landslide

Time 3

(b) Stages during the eruption of Mt. St. Helens, 1980.

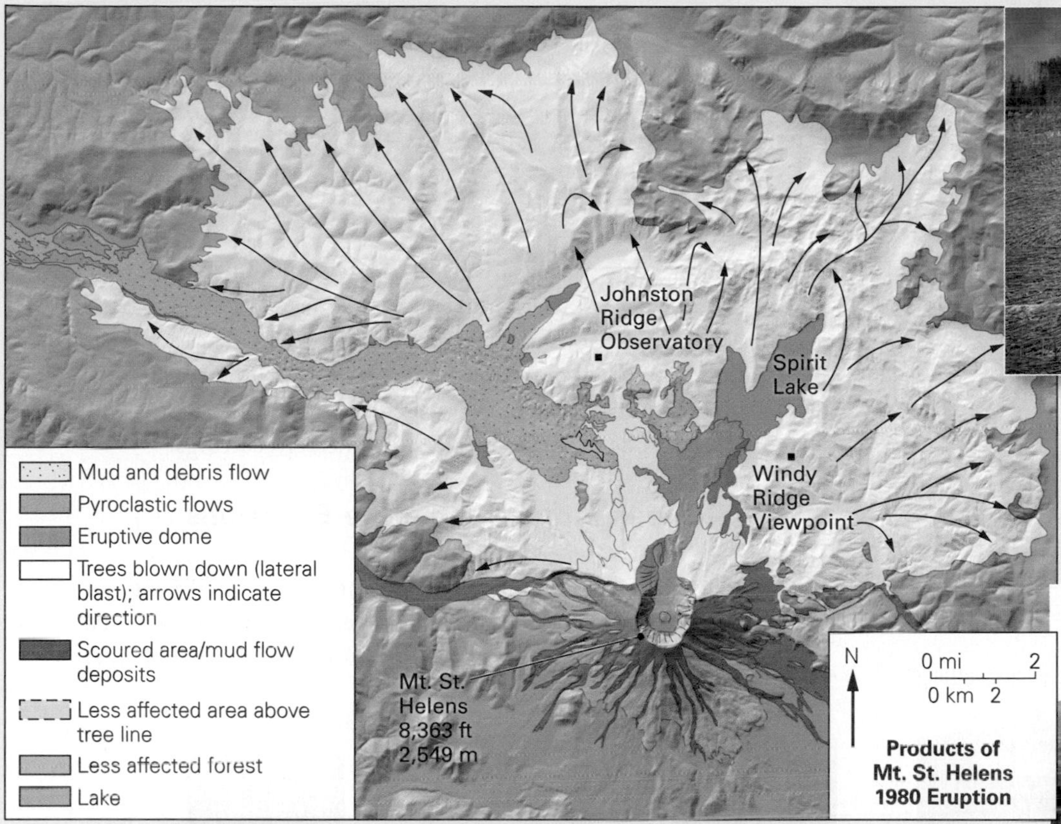

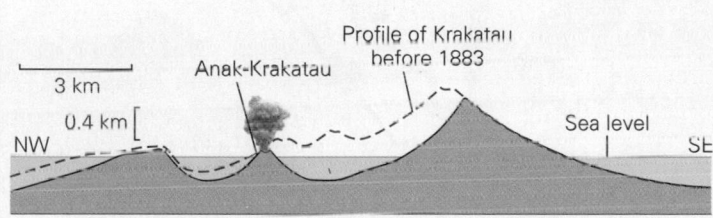

The blast knocked trees down as if they were toothpicks.

Thirty years later, the downed trees remained.

Mud and debris flow
Pyroclastic flows
Eruptive dome
Trees blown down (lateral blast); arrows indicate direction
Scoured area/mud flow deposits
Less affected area above tree line
Less affected forest
Lake

Johnston Ridge Observatory

Spirit Lake

Windy Ridge Viewpoint

Mt. St. Helens
8,363 ft
2,549 m

N

0 mi 2
0 km 2

Products of Mt. St. Helens 1980 Eruption

(c) A map shows the dimensions of the region destroyed by the eruption of Mt. St. Helens. The arrows indicate the blast direction. The neighboring forest was flattened by a blast of rock, steam, and ash.

Profile of Krakatau before 1883

Anak-Krakatau

3 km

0.4 km

NW

Sea level

SE

(d) Profile of Krakatau before and after the eruption. Note that a new volcano (Anak-Krakatau) has formed.

When the eruption was finally over, the once cone-shaped peak of Mt. St. Helens had disappeared—the summit now lay 440 m lower, and the once snow-covered mountain was a gray mound with a large gouge in one side. The volcano came alive again in 2004, but it did not explode.

The death toll due to the 1902 explosion of Mt. Pelée, on the Caribbean island of Martinique, was much higher. On the morning of May 8, the 100-m-high rhyolite spire that had plugged the conduit of the volcano suddenly disintegrated. The immense gas pressure that had been building beneath the obstruction was suddenly released, and in the same way that champagne bursts out of a bottle when the cork pops, a cloud of ash and pumice lapilli spewed out of Mt. Pelée. Collapse of the ash column formed a pyroclastic flow—at a temperature of 200°C to 450°C— that swept down Pelée's flank. This flow rode on a cushion of air and reached speeds of up to 300 km per hour before slamming into the busy port town of St. Pierre. Within moments, the town's buildings were flattened and its 28,000 inhabitants were dead of incineration or asphyxiation. Only two people survived—one was a prisoner who was protected by the stout walls of his underground cell.

An even greater explosive eruption happened in 1883. Krakatau, a volcano in Indonesia between Java and Sumatra, had grown to become a 9-km-long island rising 800 m above the sea. On May 20, the island began to erupt with a series of large explosions, yielding ash that settled as far as 500 km away. Smaller explosions continued through June and July, and steam and ash rose from the island, forming a huge black cloud that rained ash into the surrounding straits. Ships sailing by couldn't see where they were going, and their crews had to shovel ash off the decks. Krakatau's demise came at 10 A.M. on August 27, perhaps when the volcano cracked and the magma chamber suddenly flooded with seawater. The resulting blast, 5,000 times greater than the Hiroshima atomic bomb explosion, could be heard as far as 4,800 km away, and subaudible sound traveled around the globe seven times. Giant sea waves pushed out by the explosion slammed into nearby coastal towns, killing over 36,000 people. Near the volcano, a layer of ash up to 40 m thick accumulated. When the air finally cleared, Krakatau was gone, replaced by a submarine caldera some 300 m deep **(Fig. Bx9.1d)**. All told, the eruption shot 20 km³ of rock into the sky. Some ash reached altitudes of 27 km. Because of this ash, people around the world could view spectacular sunsets during the next several years.

9.4 Geologic Settings of Volcanism

Different styles of volcanism occur at different locations on the Earth. Most eruptions occur along plate boundaries, but major eruptions also take place at hot spots and in rifts (Fig. 9.17a). We'll now look at the settings in which eruptions occur, in the context of plate tectonics theory, and see why different kinds of volcanoes form in different settings.

Mid-Ocean Ridge Submarine Eruptions

Products of mid-ocean ridge volcanism cover 70% of our planet's surface. We don't generally see this volcanic activity, however, because the ocean hides most of it beneath a blanket

FIGURE 9.17 Volcanoes of the world and their geologic settings.

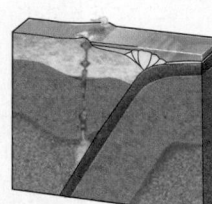 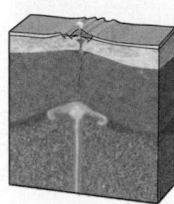  

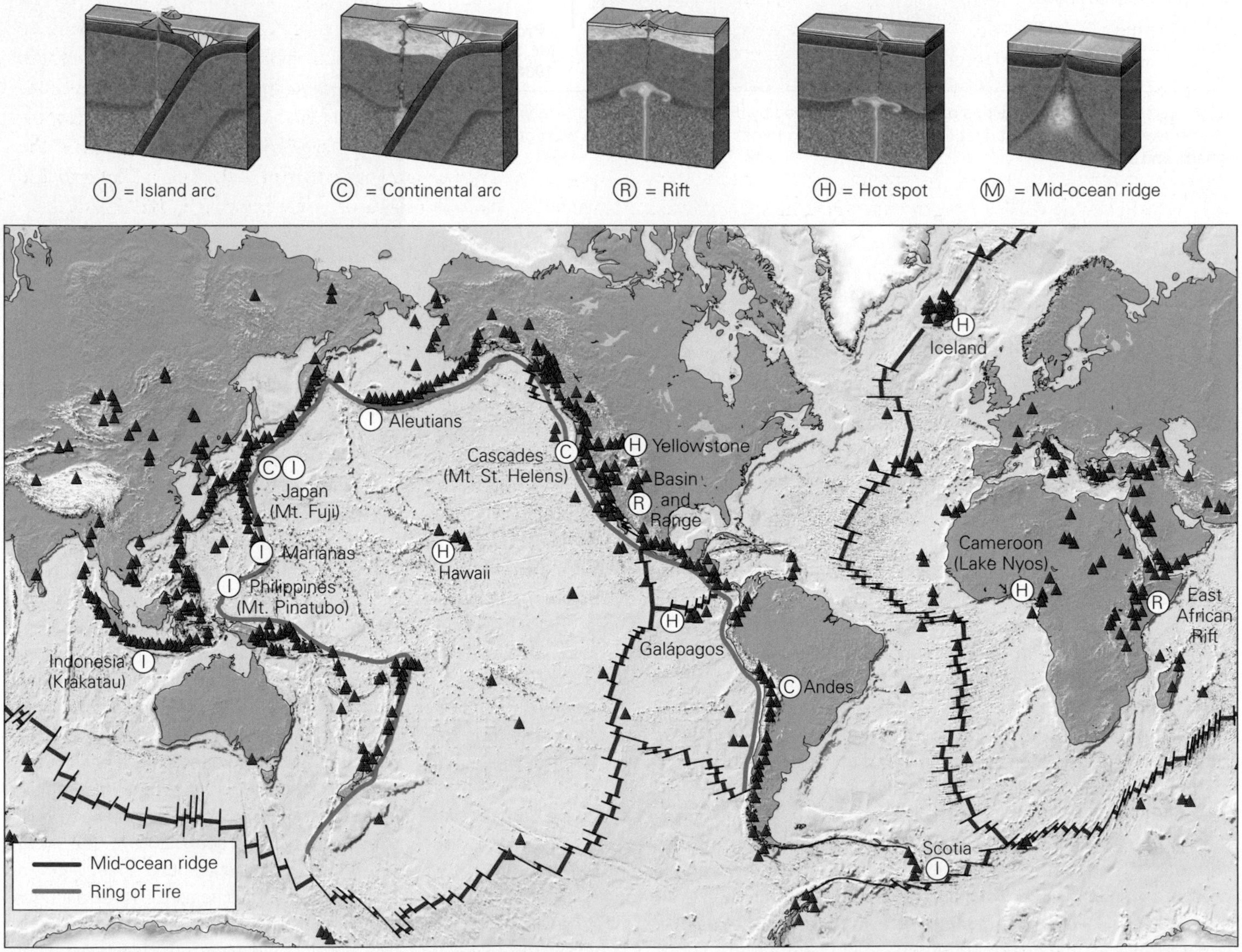

Ⓘ = Island arc   Ⓒ = Continental arc   Ⓡ = Rift   Ⓗ = Hot spot   Ⓜ = Mid-ocean ridge

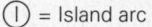

(a) A map showing the distribution of volcanoes around the world and the basic geologic settings in which volcanoes form, in the context of plate tectonics theory.

of water. Mid-ocean ridge volcanoes, which develop along fissures parallel to the ridge axis, are not all continuously active (Fig. 9.17b). Each one turns on and off on a time scale measured in tens to hundreds of years. These volcanoes erupt basalt, formed when hot mantle rock rises to shallow depths beneath the ridge and undergoes decompression melting (see Chapter 6). This basalt, because it cools so quickly underwater, forms pillow-lava mounds or aprons. The pillow basalts commonly occur in association with hyaloclastites. Water that heats up as it circulates through the crust near the magma chamber bursts out of hydrothermal (hot-water) vents along these mounds, producing black smokers (see Chapter 4).

Volcanic Arcs at Convergent Boundaries

Most subaerial volcanoes on the Earth lie along convergent boundaries (subduction zones). These volcanoes form when volatiles rise from the rock of the subducting plate into the overlying hot asthenosphere, causing flux melting in the asthenosphere. The resulting magma then rises and eventually erupts along the edge of the overriding plate. Some of these volcanoes grow on oceanic crust and become volcanic island arcs, such as the Marianas of the western Pacific and the Aleutians of the northern Pacific (Fig. 9.17c, d). Others grow on continental crust, building continental volcanic arcs such as the Cascade Range of Washington and Oregon or the Andes of South America. Typically, individual volcanoes in volcanic arcs lie about 50 to 100 km apart. Subduction zones border over 60% of the Pacific Ocean, creating a 20,000-km-long chain of volcanoes known as the *Ring of Fire*.

In island arcs, where magma rises from the mantle through oceanic crust, volcanoes initially produce primarily basaltic lava, formed by partial melting of the mantle. This lava builds an edifice of pillows and hyaloclastites that eventually rises above sea level. During its initial appearance above sea level, the eruption produces blasts of steam and ash. When the vent rises entirely above sea level, a shield of basalt starts to grow. Processes such as fractional crystallization and assimilation (see Chapter 6) may eventually yield andesitic lava, so the edifice may eventually evolve into a stratovolcano.

In continental arcs, some basalt rises to the surface, but andesitic and rhyolitic eruptions are more common because more of the magma stalls in the crust during its rise, where it undergoes fractional crystallization and assimilation. In addition, heat transfer partially melts some of the continental crust and produces felsic magma (see Chapter 6). Because many different kinds of magma form at continental arcs, these volcanoes sometimes have effusive eruptions and sometimes have pyroclastic eruptions, and they build large stratovolcanoes, which occasionally explode. Examples include the elegant symmetric cone of Mt. Fuji (see Fig. 9.12c) and the blasted-apart hulk of Mt. St. Helens (see Fig. Bx9.1b, c).

FIGURE 9.17 (*continued*)

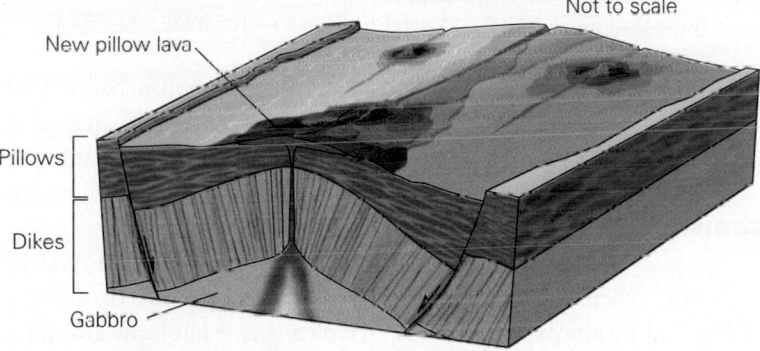

(b) Mounds of pillow basalt erupt along fissures.

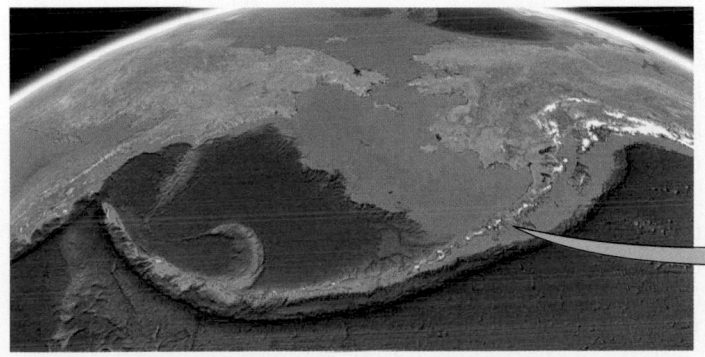

(c) The Aleutian arc forms the northern edge of the Pacific Plate. It displays a distinct curvature.

(d) An oblique view of the Aleutian arc, as seen looking northwest.

FIGURE 9.18 Oceanic hot-spot volcanism.

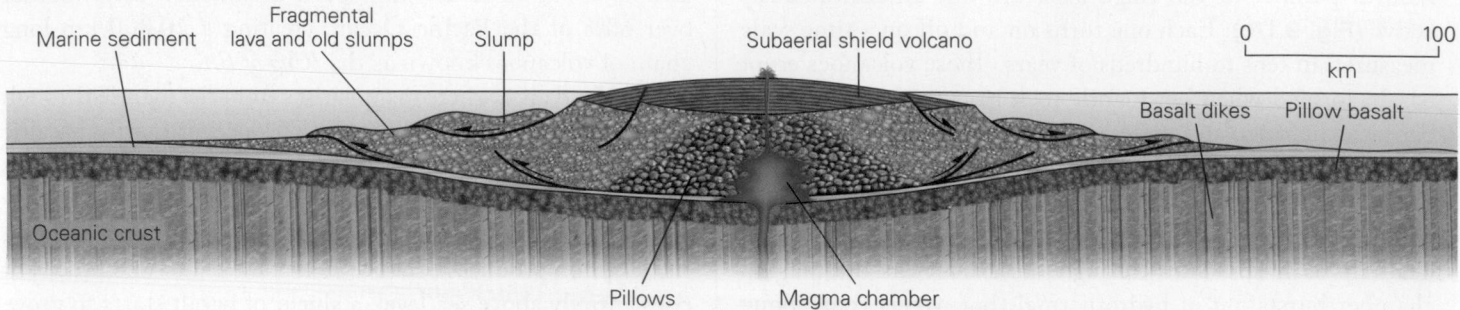

(a) The interior structure of an oceanic hot-spot volcano is complicated. Initially, eruption produces pillow basalts. When the volcano emerges above sea level, it becomes a shield volcano. The margins of the island frequently undergo slumping, and the weight of the volcano pushes down the surface of the lithosphere. The Hawaiian Islands exemplify this architecture.

Volcanism of Continental Rifts

The igneous activity at rifts happens because thinning of the continental lithosphere allows the underlying asthenosphere to rise to shallower depths, where it undergoes decompression and partially melts to produce basaltic magma. Some of this magma rises straight to the surface and erupts as basalt, but some gets trapped at the base of, or within, the continental crust, where it undergoes fractional crystallization or assimilates surrounding crust to produce intermediate or felsic magmas. Trapped magma can also partially melt continental crust through heat transfer, producing rhyolitic magma. As a result, rifts can host both basaltic fissure eruptions, in which curtains of lava fountain up or linear chains of cinder cones develop, and explosive rhyolitic eruptions. Therefore, you may find both large basalt flows and immense ignimbrites in rifts. In some locations, eruptions build stratovolcanoes, such as Mt. Kilimanjaro in Africa.

Oceanic Hot-Spot Volcanism

Oceanic hot-spot volcanoes form where asthenosphere undergoes decompression melting and produces voluminous amounts of basaltic magma. Most examples, such as the volcanoes that produced Hawaii, occur in the interior of plates, away from plate boundaries. A few, such as the ones that produced Iceland, sit astride a mid-ocean ridge. Most geologists favor the hypothesis that hot-spot volcanoes lie above mantle plumes, localized upwellings from the deep mantle, but some argue for alternative interpretations.

When a hot-spot volcano first forms on oceanic lithosphere, basaltic magma erupts at the surface of the seafloor. At first, such submarine eruptions yield a mound of pillow lava and hyaloclastite. With time, the volcano rises above the sea surface and becomes an island. After the volcano emerges from the sea, the erupting basaltic lava no longer freezes so quickly and thus flows as a thin sheet over a great distance.

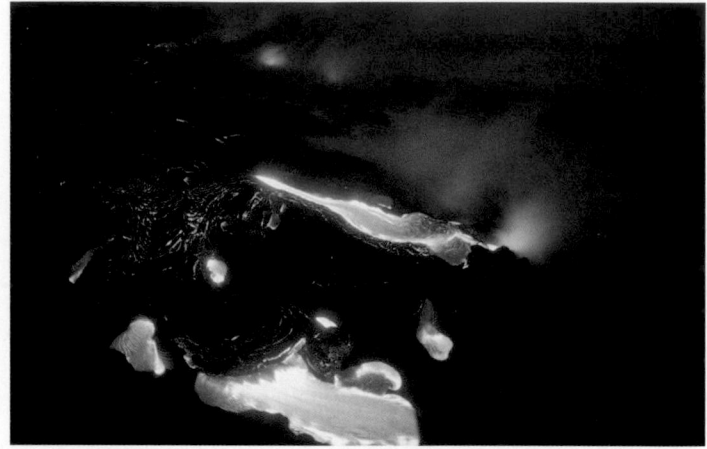

(b) Lava from a Hawaiian eruption spills into the sea at night. Its heat instantly turns water into steam.

Thousands of thin basalt flows pile up, layer upon layer, to build a broad, dome-shaped shield volcano with gentle slopes (Fig. 9.18). As the volcanic edifice grows, portions of it can no longer resist the pull of gravity and slip seaward, yielding large submarine slumps.

The Big Island of Hawaii, the tallest oceanic hot-spot volcano on the Earth today, currently consists of five shield volcanoes, each built around a different vent. The island now towers over 9 km above the adjacent ocean floor (about 4.2 km above sea level), the greatest relief from base to top of any mountain on the Earth—by comparison, Mt. Everest rises 8.85 km above the plains of India. Calderas up to 3 km wide have formed at the summit. Vast sheets of basaltic lava extrude from chimney-shaped conduits and fissures, both at the summits and on the flanks of the volcanoes (see Fig. 9.3b).

Iceland also formed over a hot spot, one that lies beneath the Mid-Atlantic Ridge—the presence of this hot spot means that far more lava erupted here than beneath other places along the ridge. As a result, Iceland sits on a broad oceanic plateau. Divergent-boundary plate motion constantly stretches Iceland in a northwest-southeast direction; this process produces numerous northeast-southwest-trending faults. The island's

FIGURE 9.19 Iceland, a hot spot on the Mid-Atlantic Ridge.

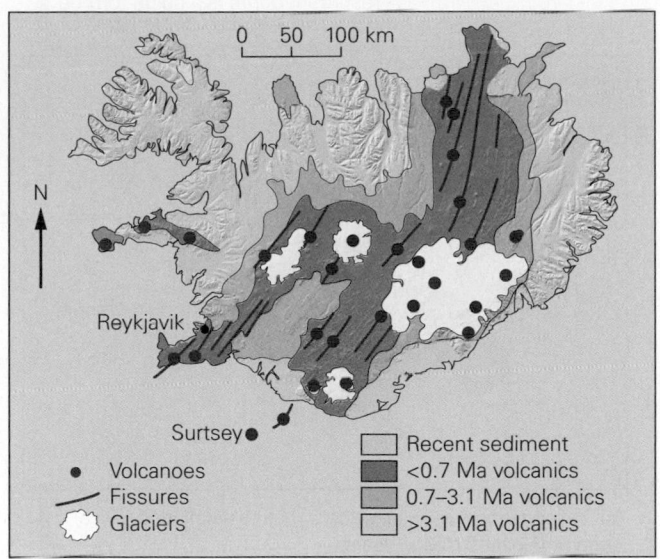

0   50   100 km

N

Reykjavik

Surtsey●

● Volcanoes
／ Fissures
⋯ Glaciers

☐	Recent sediment
☐	<0.7 Ma volcanics
☐	0.7–3.1 Ma volcanics
☐	>3.1 Ma volcanics

(a) A geologic map of Iceland shows that the youngest volcanoes occur in the central rift, effectively the on-land portion of the Mid-Atlantic Ridge.

Greenland

Iceland Plateau

Mid-Atlantic
Ridge

Ireland   UK

(b) A bathymetric map shows that Iceland sits atop a huge plateau straddling the Mid-Atlantic Ridge. Light blue is shallower water; dark blue is deeper.

youngest volcanic rocks erupt in a central rift valley that represents the trace of the Mid-Atlantic Ridge **(Fig. 9.19)**. Faulting cracks the crust and so provides a conduit to a magma chamber, so eruptions on Iceland tend to begin as fissure eruptions, spewing curtains of lava. Eventually, the curtains die out and are replaced by more localized eruptions that form linear chains of cinder cones.

Not all volcanic activity on Iceland occurs subaerially. Some eruptions take place under glaciers. During 1996, for example, an eruption at the base of a 600-m-thick glacier melted the ice and produced a column of steam that rose several kilometers into the air. Meltwater accumulated under the ice for 6 days, then burst through the edge of the glacier and became a flood (called a *jökulhlaup*, from the Icelandic for glacial burst) that lasted 2 days and destroyed roads, bridges, and telephone lines. The 2010 eruption of the volcano Eyjafjallajökull caused similar problems, and was also disruptive in other ways, as we will see.

Some of Iceland's volcanic activity occurs off the coast. Such activity produced the island of Surtsey. The birth of Surtsey was heralded by huge quantities of steam bubbling up from the

(c) A rift cutting through basalt.

ocean. Eventually, steam pressure explosively ejected ash as high as 5 km into the atmosphere. Surtsey finally emerged from the sea on November 14, 1963, building up a cone of ash and lapilli that rose almost 200 m above sea level in just 3 months. Waves could easily have eroded the cinder cone away, but the island has survived because lava erupted from the vent and flowed over the cinders, effectively encasing them in an armor-like blanket of solid rock.

Continental Hot-Spot Volcanism

Yellowstone National Park lies at the northeastern end of a string of calderas, known as the Yellowstone hot-spot track, whose remnants crop out in the Snake River Plain of Idaho **(Fig. 9.20a)**. The oldest of these calderas, at the southwestern end of the track, erupted 16 million years ago. Recent studies have found evidence of a mantle plume beneath Yellowstone, adding support to the hypothesis that the Yellowstone hot-spot track formed as the North American plate moved over a plume.

Ongoing activity beneath Yellowstone has yielded fascinating landforms, volcanic rock deposits, and geysers. Eruptions at the Yellowstone hot spot differ from those in Hawaii in an important way: the Yellowstone hot spot erupts both basaltic lava and rhyolitic pyroclastic debris. This happens because heat transfer from the rising basaltic magma partially melts the continental crust to produce felsic magma.

About 630,000 years ago (0.63 Ma), immense pyroclastic flows and convective clouds of ash and pumice lapilli blasted out of the Yellowstone region. Close to the eruption, numerous

FIGURE 9.20 Hot-spot volcanic activity in Yellowstone National Park.

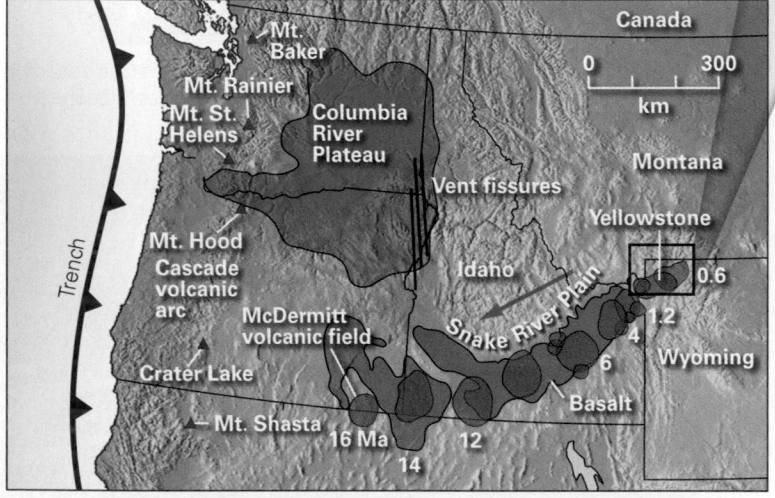

(a) Yellowstone lies at the end of a continental hot-spot track. Progressively older calderas follow the Snake River Plain. The blue arrow indicates plate motion.

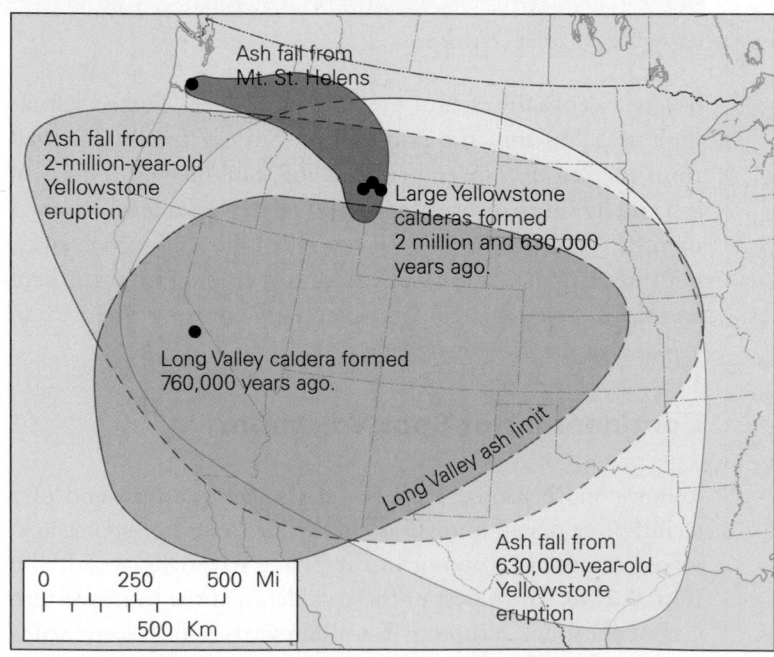

(b) The ash produced by explosions of the Yellowstone calderas covered vast areas—much more than did the Mt. St. Helens eruption.

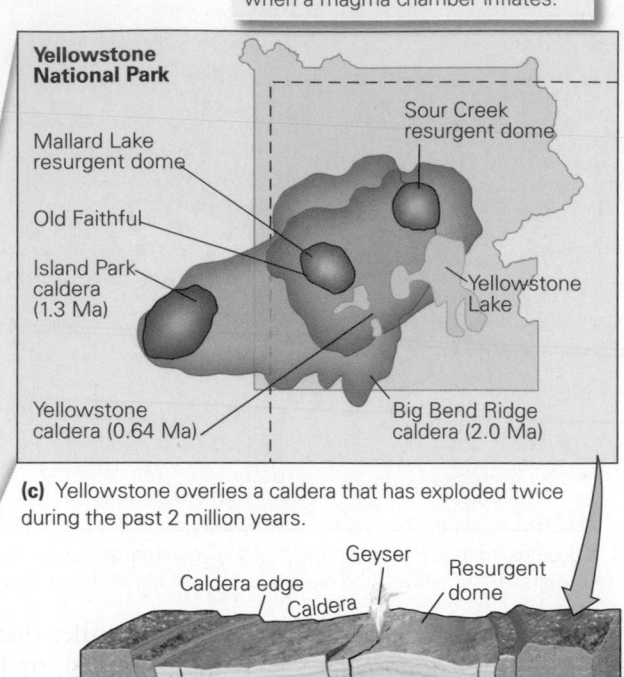

(c) Yellowstone overlies a caldera that has exploded twice during the past 2 million years.

(d) Felsic tuffs form the colorful walls of Yellowstone Canyon.

ignimbrites built up, and ash and lapilli from the giant cloud sifted down over the United States as far east as the Mississippi River **(Fig. 9.20b)**. The 0.63-Ma eruption produced an immense caldera that overlaps earlier calderas **(Fig. 9.20c)**. When the debris settled, it blanketed an area of 2,500 km² with tuffs that, in the park, reached a thickness of 400 m—Yellowstone was a supervolcano! The park's name reflects the brilliant color of volcaniclastic debris exposures in the park's canyons **(Fig. 9.20d)**. Volcanic activity, producing basaltic and rhyolitic lavas and more pyroclastic debris, continued until about 70,000 years ago. Magma remains in the crust beneath the park today. It's the energy radiating from this magma that heats the water filling hot springs and spurting out of geysers.

Flood-Basalt Eruptions

In several locations around the world, huge sheets of low-viscosity lava have erupted and spread out in vast sheets. Geologists refer to the lava of these sheets as **flood basalt** **(Fig. 9.21a)**. Over time, many successive eruptions of flood basalt can build up a broad *basalt plateau*. What causes flood-basalt eruptions? A popular hypothesis suggests that flood basalt forms when a mantle plume starts to rise beneath a region that is undergoing rifting (see Fig. 6.27a). As the plume reaches the base of the lithosphere, it develops a bulbous head containing a large amount of partially molten rock. Stretching and thinning of the overlying lithosphere results in further decompression of the plume head and causes even more melt to form. The melt intrudes along fissures that form in the rift and erupts spectacularly at the surface. Once the plume head no longer exists, the volume of eruption decreases, and "normal" hot-spot volcanism (with less magma production) takes place.

The aggregate volume of rock in a basalt plateau may be so great (over 175,000 km³) that geologists also refer to the region as a **large igneous province (LIP)** (see Fig. 6.26). (The term has also been used for regions of immense rhyolitic eruption, such as the Yellowstone region.) The Columbia River Plateau of Washington and Oregon serves as an example of a LIP

FIGURE 9.21 According to one hypothesis, flood basalts erupt when the head of a plume reaches the base of rifting lithosphere.

(a) Flood-basalt layers exposed on the wall of a canyon in Idaho.

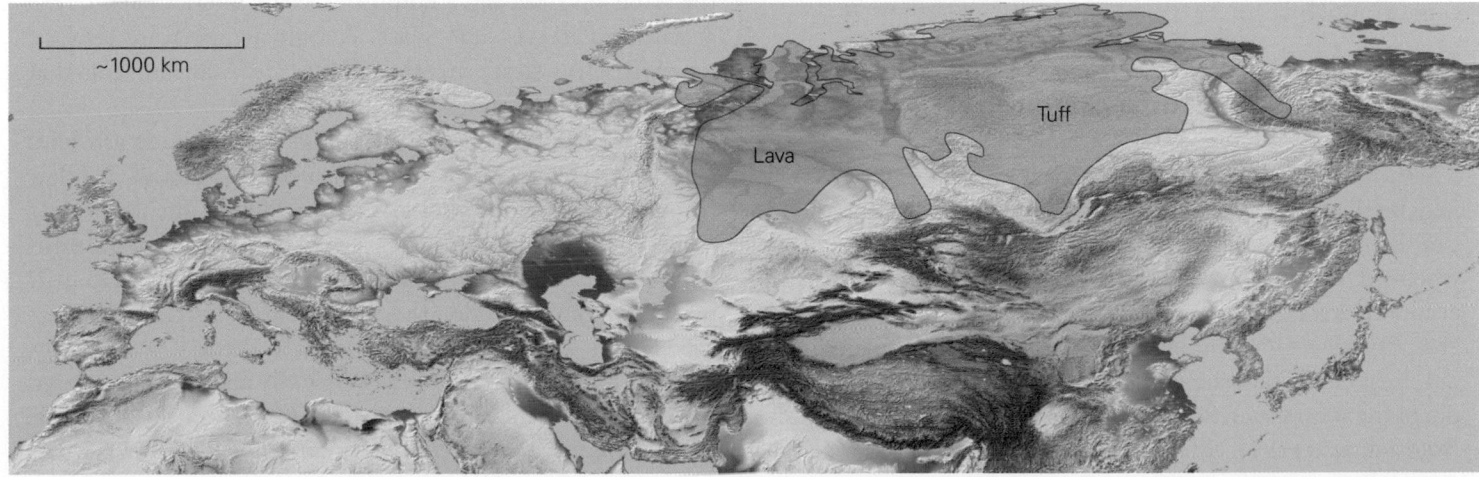

(b) In Siberia, an area the size of Europe was covered by flood basalt (the "Siberian Traps") and associated tuffs. The time of the eruption, about 250 Ma, coincides with the end of the Paleozoic, when there was mass extinction. The map shows the maximum extent of the volcanic rock before erosion.

(see Fig. 6.27b). The basalt here, which erupted about 15 Ma, reaches a thickness of 3.5 km. Geologists have identified about 300 individual flows in the Columbia River Plateau. Lava in some of these flows traveled great distances—up to 600 km—from its source. Eventually, basalt covered an area of 220,000 km². Even larger flood-basalt provinces include the Siberian Traps in eastern Siberia **(Fig. 9.21b)**, the Deccan Plateau of India, the Paraná region of Brazil, and the Karoo Plateau of South Africa.

> ## TAKE-HOME MESSAGE
>
> Most volcanic activity takes place on plate boundaries, but some occurs at hot spots. The style of eruption depends on the geologic setting. The sea hides divergent-boundary volcanoes, which erupt pillow basalt. Volcanic arcs and continental rifts may produce stratovolcanoes. Oceanic hot spots produce shield volcanoes. Continental hot spots and rifts produce both effusive and explosive eruptions.
>
> **QUICK QUESTION:** How does a volcano formed in a volcanic island arc differ from one formed at an oceanic hot spot?

9.5 Beware: Volcanoes Are Hazards!

Like earthquakes, volcanoes are natural hazards that have the potential to cause great destruction to humanity. According to one estimate, volcanic eruptions in the last 2,000 years have caused about a quarter of a million deaths. Considering the rapid expansion of cities, far more people live in dangerous proximity to volcanoes today than ever before, so if anything, the hazard posed by volcanoes has gotten worse—imagine if a large explosive eruption were to occur next to a major city today. Let's look at the different kinds of threats posed by volcanic eruptions.

Hazards Due to Eruptive Materials

Threat of Flows When you think of an eruption, lava flows may be the first threat that comes to mind. Indeed, lava flows can damage real estate, and on many occasions, lava has overwhelmed towns **(Fig. 9.22a–d)**. Basaltic lava from effusive eruptions is the greatest threat because it can spread over a broad area. On Hawaii, recent lava flows have covered roads, housing developments, and vehicles. Although people have time to get out of the way of such flows, they might have to watch helplessly from a distance as an advancing flow engulfs their homes. Even

before the lava touches it, a building will burst into flame from the intense heat. Kilauea, the most active volcano on the island of Hawaii, began its most intense eruption since 1955 in May of 2018. Immense volumes of lava erupted from fissures. At times, fountains carried the lava almost 100 m into the sky. Relatively fast-moving flows destroyed dozens of homes (displacing hundreds of people), buried highways, and shut down a geothermal power plant. Ash dusted large areas of forest and "vog," aerosols composed of volcanic gases dissolved in water, reduced visibility in areas downwind of the eruption.

Pyroclastic flows can move extremely fast (100 to 300 km per hour) and are so hot (500°C to 1,000°C) that they represent a profound hazard to humans and the environment **(Fig. 9.22e)**. Even relatively small examples, such as the flow that struck St. Pierre on Martinique, can flatten towns and devastate fields even though they leave only a few centimeters of ash and lapilli behind. People caught in the direct path of such flows may be incinerated, and even those protected from the ash itself may die from inhaling toxic, superhot gases.

Threat of Falling Ash and Lapilli During a large pyroclastic eruption, ash and lapilli erupt into the air, later to fall back to the ground **(Fig. 9.22f, g)**. Close to the volcano, pumice and lapilli tumble out of the sky and can accumulate to form a blanket up to several meters thick. The mass, especially when saturated with rainwater, causes roofs and power lines to collapse. Winds can carry fine ash over a broader region. In the Philippines, for example, typhoon winds during the 1991 eruption of Mt. Pinatubo covered a 4,000-km² area with ash. An ash fall buries crops, coats the leaves of trees, and may spread toxic chemicals that poison the soil. Ash also insidiously infiltrates machinery, causing moving parts to wear out.

Threat to Aircraft Fine ash from an eruption can be dangerous to airplanes. Like a sandblaster, sharp, angular shards of ash abrade turbine blades, greatly reducing engine efficiency. The ash, along with sulfuric acid aerosols, scores windows and damages the fuselage. Also, when heated inside a jet engine, the ash melts, yielding a liquid that coats interior parts of the engine and freezes into glass, coating temperature sensors, which falsely indicate that the engines are overheating, so that they automatically shut down.

Encounters between airliners and high-altitude ash have led to terrifying incidents. In 1982, a British Airways 747 flew through the ash cloud above a volcano in Java. The windshield turned opaque and all four engines shut down. For 13 minutes, the plane silently glided earthward, dropping from its initial 11.5-km altitude. The pilot frantically tried to restart the engines to no avail and prepared to ditch at sea. Finally, at 3.7 km, the engines cooled sufficiently and suddenly roared back to life, and the plane headed to Jakarta for an emergency landing. There, without functioning instruments, the

FIGURE 9.22 Hazards due to lava and ash from volcanic eruptions.

Lava Flows

(a) A lava flow reaches a house in Hawaii and sets it on fire.

(b) Lava from Mt. Etna threatens a town and olive grove in Sicily.

(c) Residents rescue household goods after a lava flow filled the streets of Goma, along the East African Rift.

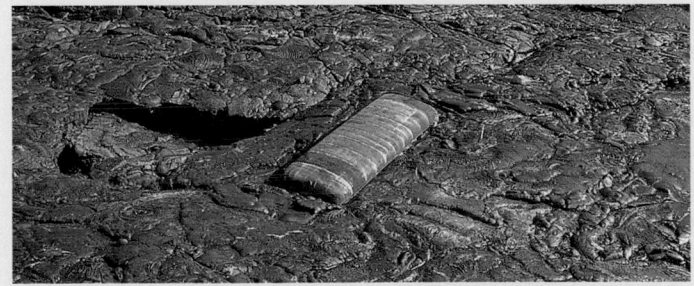

(d) This empty school bus was engulfed by lava in Hawaii.

Pyroclastic Debris

(e) A pyroclastic flow rushes down the slopes of the Soufrière Hills Volcano of Monserrate.

(f) A blizzard of ash fell from the cloud erupted by Mt. Pinatubo in the Philippines.

(g) Lapilli falls from an eruption in Iceland.

Lahar

(h) A lahar submerges farmland in Colombia.

pilot squinted out an open side window to see the runway and brought the plane safely to a halt with only his toes touching the pedals. A similar encounter took place between a KLM 747 and the ash cloud of Redoubt Volcano in Alaska in 1989.

Because of the lessons learned from such incidents, the 2010 eruption of the volcano Eyjafjallajökull in Iceland had a profound impact on air traffic. All told, the eruption sent about 0.25 km³ of pyroclastic material into the air. The jet stream, a high-altitude current of rapidly moving air, was passing over Iceland at the time of the eruption and dispersed the ash throughout European air space. Because of concern that this ash could damage planes, officials shut down almost all air traffic across Europe for 6 days. The closure directly cost airlines $200 million per day, disrupted travel plans for countless passengers, and halted shipment of everything from electronic goods to flowers, negatively impacting economies worldwide.

Other Hazards Related to Eruptions

Threat of the Blast Most exploding volcanoes direct their fury upward. But some, like Mt. St. Helens, explode sideways. The forcefully ejected gas and ash, like the blast of a bomb, flattens everything in its path. In the case of Mt. St. Helens, the region around the volcano had been a beautiful pine forest. But after the eruption, the once-towering trees, stripped of bark and needles, lay flattened onto the hill slopes, all pointing in the direction away from the blast (see Box 9.1).

Threat of Landslides Eruptions commonly trigger large landslides along a volcano's flanks. Volcanic debris, composed of ash and solidified lava that erupted earlier, can move downslope quite fast (250 km per hour) and far. During the eruption of Mt. St. Helens, 8 billion tons of debris took off down the mountainside, rocketed over a 360-m-high ridge, and tumbled down a river valley, until the last of it finally came to rest over 20 km from the volcano.

Threat of Lahars Mixing volcanic ash and other debris with water produces a *lahar*, an ashy slurry that resembles wet concrete. A lahar can move downslope at speeds of over 50 km per hour. Because lahars are denser and more viscous than clear water, they pack more force than clear water and can carry away everything in their path. Lahars that flow into stream valleys can travel far from the volcano. The lahars of Mt. St. Helens, for example, traveled along existing drainages for more than 40 km from the volcano. When they had passed, they left a gray and barren wake of mud, boulders, broken bridges, and crumpled houses, as if a giant knife had scraped across the landscape. The lahars generated during the 1991 eruption of Mt. Pinatubo engulfed whole villages.

Lahars may develop when heavy rains happen during an eruption, or in regions where snow and ice cover an erupting volcano, for the eruption melts the snow and ice, thereby generating a supply of water. Perhaps the most destructive lahar of recent times accompanied the eruption of the snow-crested Nevado del Ruiz in Colombia on the night of November 13, 1985. The lahar surged down a valley like a 40-m-high wave, hitting the sleeping town of Armero, 60 km from the volcano. Ninety percent of the buildings in the town vanished, replaced by a 5-m-thick layer of mud, which now entombs the remains of 25,000 people **(Fig. 9.22h)**.

Threat of Earthquakes Earthquakes accompany almost all major volcanic eruptions because the movement of magma breaks rocks underground. Such earthquakes may trigger landslides on the volcano's flanks and can cause nearby buildings to collapse and dams to rupture, even before the eruption itself begins.

Threat of Tsunamis Where explosive eruptions occur at an island arc, the blast and the underwater collapse of a caldera can generate huge sea waves, or *tsunamis*, tens of meters high. Most of the 36,000 deaths attributed to the 1883 eruption of Krakatau were not due to ash or lava, but rather to tsunamis that slammed into nearby coastal towns. Tsunamis may also be generated by huge submarine landslides that occur when part of a volcanic island suddenly slumps into the sea.

Threat of Gas We have already seen that volcanoes erupt not only solid material, but also large quantities of gases such as water vapor, carbon dioxide, sulfur dioxide, and hydrogen sulfide. Usually the eruption of gases accompanies the eruption of lava and ash, with the gas playing only a minor part in the calamity. For example, sulfur gases, mixing with moisture, produce sulfuric acid aerosols that can cause respiratory problems in people who live downwind. But occasionally volcanic gas alone snuffs out life in its path without causing any other damage. Such an event occurred in 1986 near Lake Nyos in western Africa.

Lake Nyos is a small but deep lake that fills the crater of an active volcano in Cameroon. Though only 1 km across, the lake reaches a depth of over 200 m. Because of its depth, the cool bottom water of the lake normally does not mix with warm surface water. Carbon dioxide gas slowly bubbles out of cracks in the floor of the crater and dissolves in the bottom water, eventually saturating the water with CO_2. On August 21, 1986, perhaps because a landslide or wind disturbed the water, the water within the lake overturned, and the saturated bottom water rose to the surface **(Fig. 9.23a, b)**. As it rose, the pressure acting on the saturated water decreased, the CO_2 came out of solution, and the lake forcefully expelled a froth of CO_2 bubbles. Because it's denser than air, this invisible CO_2 gas flowed down the flank of the volcano and spread out over the countryside for a distance of about 23 km before

FIGURE 9.23 The CO_2 gas disaster, Lake Nyos, Cameroon.

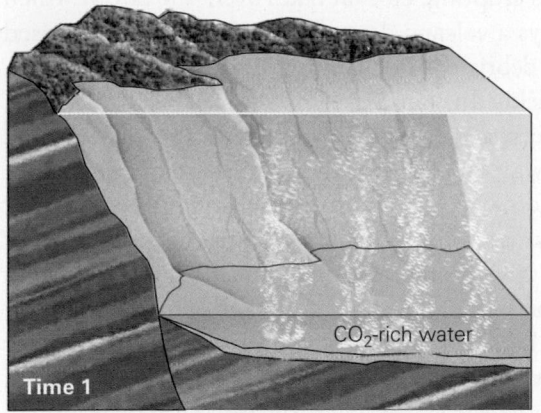

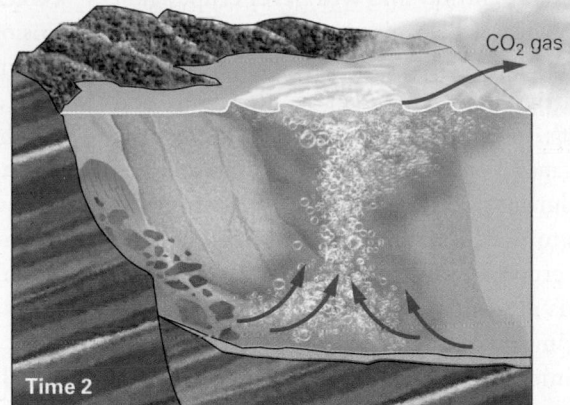

(a) Carbon dioxide dissolved in the colder bottom water. When a landslide or wind disturbed the lake, the saturated water rose. The CO_2 came out of solution and, in gas form, flowed out of the crater.

(b) Lake Nyos, after the disaster. The crater lake has been discolored by turbulence.

(c) The CO_2 suffocated cattle on the slopes below the volcano.

dispersing. Although it is not toxic, carbon dioxide cannot provide oxygen for metabolism or oxidation. When the gas cloud engulfed the village of Nyos, it quietly put out cooking fires and suffocated the sleeping inhabitants. The next morning, the landscape looked exactly as it had the day before, except for the lifeless bodies of 1,742 people and about 6,000 head of cattle **(Fig. 9.23c)**.

TAKE-HOME MESSAGE

Volcanoes can be dangerous! The lava and pyroclastic flows, pyroclastic debris, explosions, mudflows (lahars), landslides, earthquakes, and tsunamis that can be produced during eruptions can destroy cities and farmland. Ash that enters the air can be a hazard for air travel.

QUICK QUESTION: Why can a lahar do so much more damage than an equivalent-sized flood of clear water?

9.6 Protection from Vulcan's Wrath

Volcanic eruptions are a natural hazard of extreme danger. Can anything be done to protect lives and property from this danger? The answer is yes. In this section, we first examine the evidence that geologists use to determine whether a volcano has the potential to erupt, and then consider the suite of observations that may allow geologists to predict the timing of an impending eruption.

Active, Dormant, and Extinct Volcanoes

Geologists refer to volcanoes that are erupting, have erupted recently, or are likely to erupt soon as **active volcanoes** and distinguish them from **dormant volcanoes**, which have not erupted for hundreds to thousands of years but may erupt again in the future. Volcanoes that were active in the geologic past but have shut off entirely and will never erupt again because the geologic cause of volcanism no longer exists are called **extinct volcanoes**.

To clarify the distinction among active, dormant, and extinct volcanoes, let's look at a few examples. Geologists consider Hawaii's Kilauea Volcano to be active because it has erupted many times during recorded history and continues to erupt today. Washington's Mt. Rainier, in contrast, is a dormant volcano. It last erupted a millennium ago, and today it emits no visible clouds of ash or any flows of lava—but geologic conditions driving volcanism in western Washington persist, and Mt. Rainier will likely erupt at some time in the future. Devils Tower, in Wyoming, is an example of an extinct volcano **(Fig. 9.24)**. The rock exposed in this steep-sided hill solidified in a shallow intrusion beneath the volcano about 40.5 Ma. At that time, due to the subduction of Pacific Ocean floor, volcanoes were erupting in eastern Wyoming and central Colorado. So much time has passed since these eruptions took place that erosion has removed hundreds of meters to kilometers of rock that once lay above Devils Tower. Subduction no longer takes place there, so volcanism won't happen again at the site of Devils Tower (unless the geometry of plate interactions changes, which is not likely for tens of millions of years or more).

How do you determine whether a volcano is active, dormant, or extinct? One way is to examine the historical record. Another is to determine the age of erupted rocks and to search for evidence that the volcano still lies within a tectonically active area. Finally, you can examine the landscape character of the volcano. Specifically, the shape (shield, stratovolcano, or cinder cone) of an active volcano depends primarily on its eruptive style, because at an active volcano, the process of construction happens faster than the process of erosion. Once a volcano stops erupting, erosion takes over. The rate at which erosion destroys a volcano depends on whether it's composed of pyroclastic debris or lava. Cinder cones and ash piles can wash away quickly. In contrast, composite or shield volcanoes, which have been armor-plated by lava flows, can withstand the attack of water and ice for quite some time. In the end, however, erosion wins out, and you can distinguish a dormant volcano from an active volcano by the extent to which river or glacial valleys have been carved into its flanks **(Fig. 9.25)**. In some cases, the softer exterior of a volcano completely erodes away, leaving behind the plug of harder frozen magma that once lay within or beneath the volcano, as well as the network of dikes that radiate from this plug. You can see good examples of such landforms at Shiprock, New Mexico (see Fig. 6.13).

Predicting Eruptions

Predicting the time of an eruption at a given volcano decades or even years in advance is impossible. The only basis for constraining a long-term prediction comes from determining the *recurrence interval* (the average time between eruptions) for the volcano. Geologists calculate recurrence intervals by measuring the ages of erupted layers comprising the volcano. For example, Mt. Fuji has erupted about 65 times in the last 10,000 years. So it has an eruptive recurrence interval of about 150 years. Note that a recurrence interval does not indicate periodicity—some of Fuji's eruptions were decades apart while others were centuries apart. The recurrence interval just gives us a sense of the probability of an eruption during a time period.

> **Did you ever wonder...**
> if a volcano could erupt beneath London, England?

Some volcanoes send out distinct warning signals announcing that an eruption may take place within a time frame of days or months, for as magma squeezes into the magma chamber, it causes a number of changes that geologists can measure:

- *Earthquake activity:* Movement of magma generates vibrations in the Earth. When magma flows into a volcano, rocks surrounding the magma chamber crack, and blocks slip with respect to one another. Such cracking and shifting cause earthquakes at depths between 1 and 7 km beneath a volcano.
- *Changes in heat flow:* The presence of hot magma increases the local *heat flow*, the amount of heat passing upward through rock. In some cases, the increase in heat flow melts

FIGURE 9.24 Devils Tower, Wyoming, formed as an intrusion into sedimentary rocks beneath a volcano. It solidified hundreds of meters below the surface of the Earth and has been exposed by erosion. Cooling produced spectacular columnar joints.

The vertical lines are columnar joints.

FIGURE 9.25 The shape of a volcano changes as it is eroded.

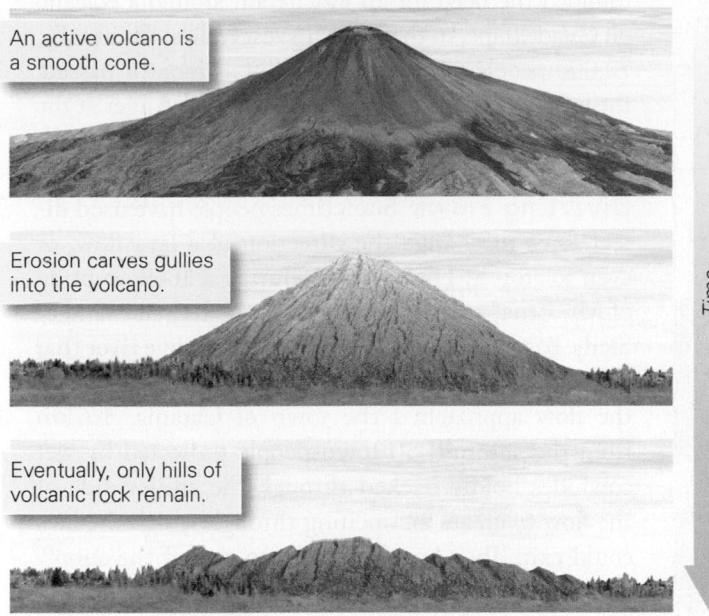

An active volcano is a smooth cone.

Erosion carves gullies into the volcano.

Eventually, only hills of volcanic rock remain.

Time

snow or ice on the volcano, triggering floods and lahars even before an eruption occurs.

• *Changes in shape:* As magma fills the magma chamber inside a volcano, it pushes outward and can cause the surface of the volcano to bulge; the same effect happens when you blow air into a balloon. Geologists now use laser sighting, accurate tiltmeters, surveys using global

FIGURE 9.26 Measuring the activity of volcanoes.

A satellite uses radar (InSAR) to measure the elevation of the same area at different times. Color bands indicate the amount of movement between the times.

positioning systems, and a technique known as InSAR (interferometric synthetic aperture radar), which uses radar beams emitted by satellites to measure elevation changes in order to detect any alteration in a volcano's shape due to the rise of magma **(Fig. 9.26a)**.

• *Increases in gas and steam emission:* Even though magma remains below the surface, gases bubbling out of the magma and steam formed as the magma heats groundwater percolate upward through cracks in the Earth and rise from the volcanic vent. So an increase in the volume and composition of gas emissions, or the birth of new hot springs, may indicate that magma has entered the ground below **(Fig. 9.26b)**.

Mitigating Volcanic Hazards

Danger Assessment Maps Let's say that a given active volcano has the potential to erupt in the near future. What can geologists do to prevent the loss of life and property? First, they compile a **volcanic-hazard assessment map (Fig. 9.27)**. Such maps delineate areas that lie in the path of potential lava flows, lahars, or pyroclastic flows and serve to guide evacuation plans.

For example, before the 1991 eruption of Mt. Pinatubo in the Philippines, geologists had defined areas potentially in the path of pyroclastic flows and had predicted which river valleys were likely hosts for lahars. When the eruption seemed imminent, many lives were saved by evacuating people in areas thought to be under threat.

Volcano Monitoring and Evacuation Because geologists can determine when magma has moved into the magma chamber of a volcano, government agencies now send monitoring teams to a volcano at the first sign of activity. These teams set up instruments to record earthquakes, measure the

Mount Longonot
9 cm UPLIFT
2004–2006

9 cm

6 cm

3 cm

(a) Monitoring ground movements that may reflect intrusion of magma below. Each color-spectrum band represents 3 cm of uplift. The measurements were made by satellite radar measurements.

(b) Geologists can detect changes in gas composition.

FIGURE 9.27 A volcanic-hazard assessment map for Mt. Rainier, Washington, showing the regions that might be affected by flows and lahars. Note that lahars may travel long distances down river valleys.

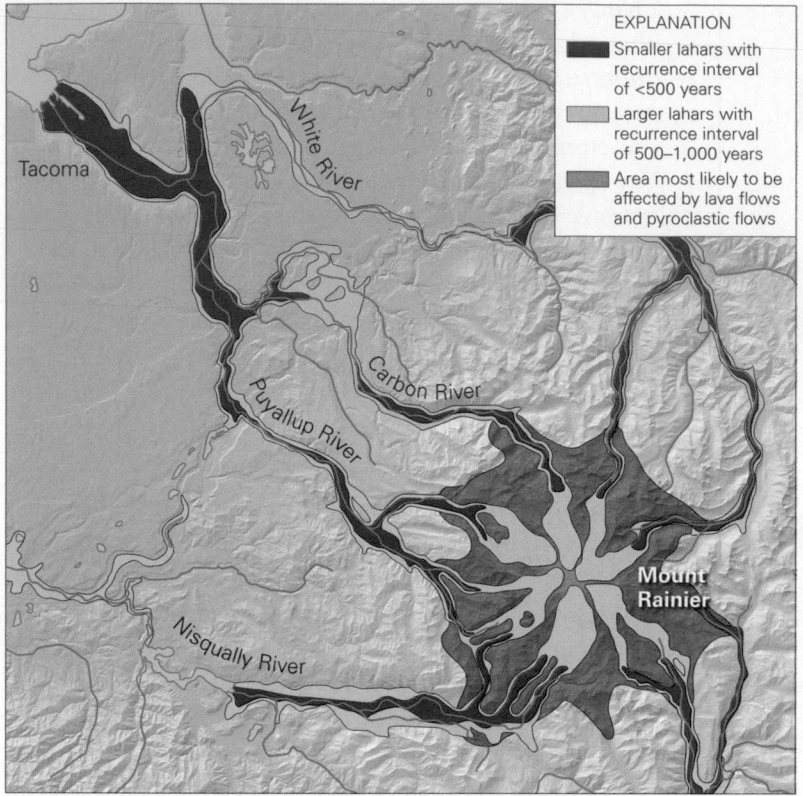

EXPLANATION

- Smaller lahars with recurrence interval of <500 years
- Larger lahars with recurrence interval of 500–1,000 years
- Area most likely to be affected by lava flows and pyroclastic flows

Tacoma

White River

Carbon River

Puyallup River

Nisqually River

Mount Rainier

heat flow, determine changes in the volcano's shape, and analyze emissions. If an eruption seems imminent, they may issue a warning and call for area residents to evacuate.

Unfortunately, because of the uncertainty of eruption predictions, the decision about whether or not to evacuate can't be easy. In the Philippines, as we've just seen, evacuation of people from the danger area around Mt. Pinatubo succeeded in saving thousands of lives. In 1976, however, fierce debate surrounded the need for an evacuation around a volcano on Guadeloupe, in the French West Indies. Eventually, 72,000 people were evacuated, but as months passed, the volcano only hosted a small eruption. Anger at the cost of the evacuation translated into lawsuits.

Diverting Flows Sometimes people have used direct force to change the direction of a lava flow, or even to stop it. For example, during a 1669 eruption of Mt. Etna, an active volcano on the Italian island of Sicily, basaltic lava formed a glowing orange river that began to spill down the side of the mountain. When the flow approached the town of Catania, 16 km from the summit, 50 townspeople protected by wet cowhides boldly hacked through the chilled side of the flow to create an opening through which the lava could exit. They hoped thereby to cut off the supply of lava feeding the end of the flow, near their homes. Their strategy worked, and the flow began to ooze through the new hole in its side. Unfortunately, the diverted flow began to move toward the neighboring town of Paterno. Five hundred men of Paterno then chased away the Catanians so that the hole would not be kept open, and eventually the flow once again headed toward Catania, covering part of the town.

More recently, people have used high explosives to blast breaches in the flanks of flows and use bulldozers to build dams and channels to divert them. Major efforts to divert flows from a 1983 eruption of Mt. Etna, and another in 1992, were successful **(Fig. 9.28a)**. Inhabitants of Iceland used a

Did you ever wonder...
whether people could redirect a lava flow?

FIGURE 9.28 Efforts to divert lava flows away from inhabited locations.

(a) Workers spraying a lava flow to solidify it and using a bulldozer to build an embankment to divert it on the flanks of Mt. Etna.

(b) Firefighters pumping 6 million cubic meters of water on a lava flow on the island of Heimaey, Iceland, in an effort to freeze it and stop it.

particularly creative approach in 1973 to stop a flow before it overran a town—they sprayed cold seawater onto the flow to freeze it in its tracks **(Fig. 9.28b)**. The flow did stop short of the town, but whether this was a consequence of the cold shower it received remains unknown.

TAKE-HOME MESSAGE

Volcanoes don't erupt continuously and don't last forever, so we can distinguish among active, dormant, and extinct volcanoes. Once a volcano ceases to erupt, erosion destroys its eruptive shape. Geologists can provide near-term predictions of eruptions so that people can take precautions.

QUICK QUESTION: Using the Web, discover how many times Vesuvius has erupted since 79 C.E. and calculate its recurrence interval. Is it wise to continue building housing on its flanks?

9.7 Effect of Volcanoes on Climate and Civilization

Can Eruptions Affect Climate?

In 1783, Benjamin Franklin was living in Europe, serving as the American ambassador to France. The summer of that year seemed to be unusually cool and hazy. Franklin, who was an accomplished scientist as well as a statesman, couldn't resist seeking an explanation for this phenomenon. He learned that in June of 1783, a huge volcanic eruption had taken place in Iceland. He wondered if the "smoke" from the eruption had prevented sunlight from reaching the Earth, thus causing the cooler temperatures. Franklin reported this idea at a meeting, and by doing so, he may well have become the first scientist ever to suggest a link between volcanic eruptions and climate.

Franklin's idea seemed to be confirmed in 1815, when Mt. Tambora in Indonesia exploded. Tambora's explosion ejected over 100 km³ of ash and pumice into the air (compared with 1 km³ from Mt. St. Helens). Ten thousand people were killed by the eruption and the associated tsunami. Another 82,000 died of starvation. The sky became so hazy that stars dimmed by a full magnitude. Temperatures dipped so low in the northern hemisphere that 1816 became known as "the year without a summer." The unusual weather of that year may have influenced artists and writers. For example, the hazy glow of the sky in the landscape paintings of the English artist

J. M. W. Turner brings to mind the fabulous sunsets caused by atmospheric aerosols from Tambora (see Fig. 9.1a). Byron's 1816 poem "Darkness" contains the gloomy lines, "The bright Sun was extinguish'd, and the stars / Did wander darkling in the eternal space . . . / Morn came and went—and came, and brought no day." Not long after, Mary Shelley, trapped indoors by bad weather, wrote *Frankenstein*, with its numerous scenes of gloom and doom.

How can a volcanic eruption cause the atmosphere to cool? As a result of a large explosive eruption, fine ash and sulfuric acid aerosols enter the stratosphere. It takes only about 2 weeks for the ash and aerosols to circle the planet **(Fig. 9.29a)**, and they stay suspended in the stratosphere for many months to years because they float above the weather and do not get washed away by rainfall. The resulting haze causes cooler average temperatures because it reflects or scatters incoming visible solar radiation during the day, but not the infrared radiation that rises from the Earth's surface at night. Therefore, the haze keeps energy from reaching the Earth, but unlike greenhouse gases (such as CO_2), it does not prevent heat from escaping. Such temperature dips seem to last up to a few years **(Fig. 9.29b)**. A Krakatau-scale eruption could lead to a drop in global average temperature of 0.3°C to 1°C. According to some calculations, a series of large eruptions over a short period of time could cause a global average temperature drop of 6°C.

To study the effect of volcanic activity on climate in the past, geologists have studied ice from the glaciers of Greenland and Antarctica. Glacial ice has layers, each of which represents the snow that fell in a single year. Some layers contain sulfuric acid, formed when sulfur dioxide from volcanic gas dissolves in the water from which snow forms. These layers indicate years in which major eruptions occurred. Years for which ice layers contain sulfuric acid correspond to years for which the thinness of tree rings elsewhere in the world indicates a cool growing season.

Some researchers are currently exploring the possibility that the eruption of LIPs may have had longer-lasting effects at times during the Earth's history, perhaps changing climate so much that species have gone extinct. The apparent coincidence of the eruption of the Siberian Traps with the extinctions that define the end of the Paleozoic may be an example.

The largest eruption to have happened in the last million years of the Earth's history took place at the Toba Volcano in Indonesia about 73,000 years ago. The explosion emitted huge quantities of ash, some of which was nearly white, covering much of southern Asia like a snowfall. In addition, the eruption injected a huge dose of aerosols into the atmosphere. The aerosols and ash diminished the solar radiation reaching the Earth, and the white ash on the ground reflected back into space some of the radiation that did reach the planet. The resulting cooling event may have

FIGURE 9.29 Effects of the eruption of Mt. Pinatubo in 1991.

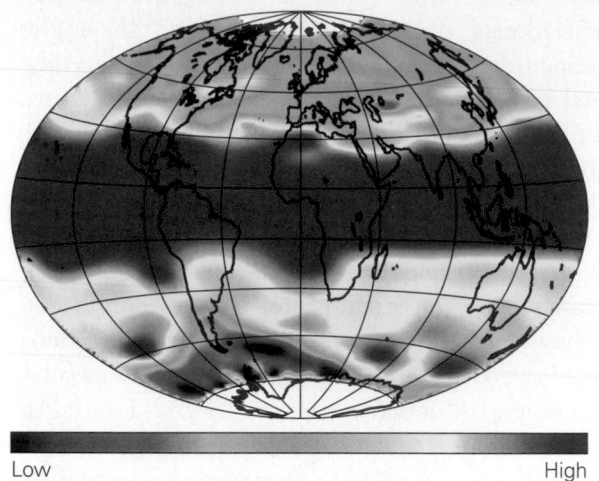

Low — Aerosol concentration — High

(a) Two months after the eruption, volcanic aerosols surrounded the Earth.

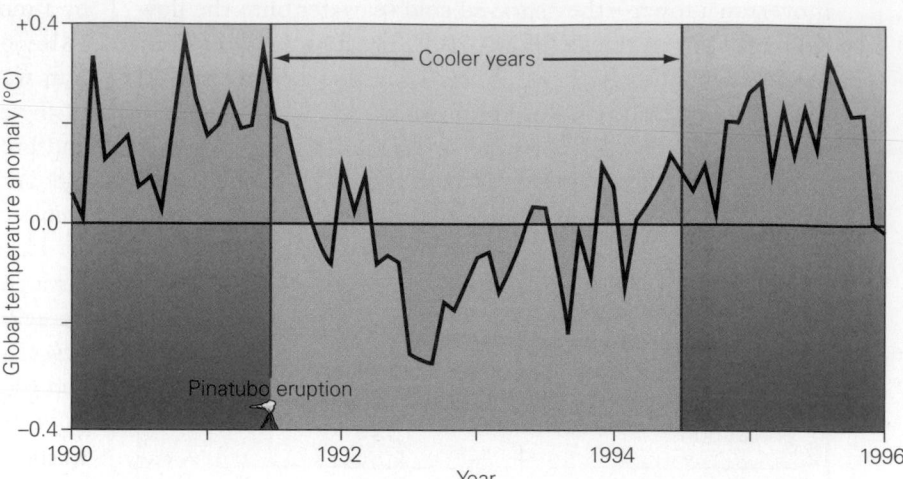

(b) The large injection of dust and asteroids into the atmosphere caused the average global temperature to drop for a few years after the eruption.

lasted for a decade—possibly for as long as 1,000 years. The duration of the cooling may have been sufficient to cause species to go extinct. Anthropologists have speculated that the event killed off all but 3,000 to 10,000 humans on the Earth. Genetic studies suggest that a bottleneck in human evolution occurred at about the same time as the Toba eruption, in that all modern humans are descended from a single ancestor who lived at about that time. If so, then the Toba eruption could have significantly affected human evolution.

Volcanoes and Civilization

Over time, volcanic activity has played a major role in making the Earth a habitable planet. Eruptions and underlying igneous intrusions produced the rock making up the Earth's crust, and gases emitted by volcanoes provided the raw materials from which the atmosphere and oceans formed. The black smokers surrounding hydrothermal vents along mid-ocean ridges may have served as a birthplace for life, and volcanic islands in the oceans have hosted isolated populations whose evolution adds to the diversity of life on the planet. Volcanic activity continues to bring nutrients (potassium, sulfur, calcium, and phosphorus) from the Earth's interior to the surface and to provide fertile soils that nurture plant growth. And in more recent times, people have exploited the mineral and energy resources generated by volcanic eruptions. In fact, volcanoes and people have lived in close association since the first human-like ancestors walked the Earth 3 million years ago. In fact, one of the earliest relics of human ancestors consists of footprints fossilized in a volcanic ash layer in East Africa.

But as we have seen, volcanic eruptions also pose a hazard. Eruptions may even lead to the demise of civilizations. The history of the Minoan people, who inhabited several islands in the eastern Mediterranean during the Bronze Age, illustrates this possibility. Beginning around 3000 B.C.E., the Minoans built elaborate cities and prospered. Then their civilization waned and disappeared (Fig. 9.30a). Geologists have discovered that the disappearance of the Minoans came within 150 years of a series of explosive eruptions of the Santorini Volcano in the first half of the 17th century B.C.E. Remnants of the volcano now constitute Thera, one of the islands of Greece (Fig. 9.30b). After a huge eruption, the center of the volcano collapsed into the sea, leaving only a steep-walled caldera (Fig. 9.30c). Archaeologists speculate that pyroclastic debris from the eruptions periodically darkened the sky, burying Minoan settlements and destroying crops. In addition, related earthquakes crumbled homes, and tsunamis generated by the eruptions damaged Minoan seaports. Perhaps the Minoans took these calamities as a sign of the gods' displeasure, became demoralized, and left the region. Or perhaps trade was disrupted, and bad times led to political unrest. Eventually the Mycenaeans moved in, bringing the culture that evolved into that of classical Greece.

Numerous cultures living along the Pacific Ring of Fire have evolved religious practices that are based on volcanic activity —no surprise, considering the awesome might of a volcanic eruption in comparison with the power of humans. In some cultures, this reverence took the form of sacrifice in hopes of preventing an eruption that could destroy villages and bury food supplies. In traditional Hawaiian culture, Pele, goddess of the volcano, created all the major landforms of the Hawaiian

(a) The ruins of palaces left by the Minoans.

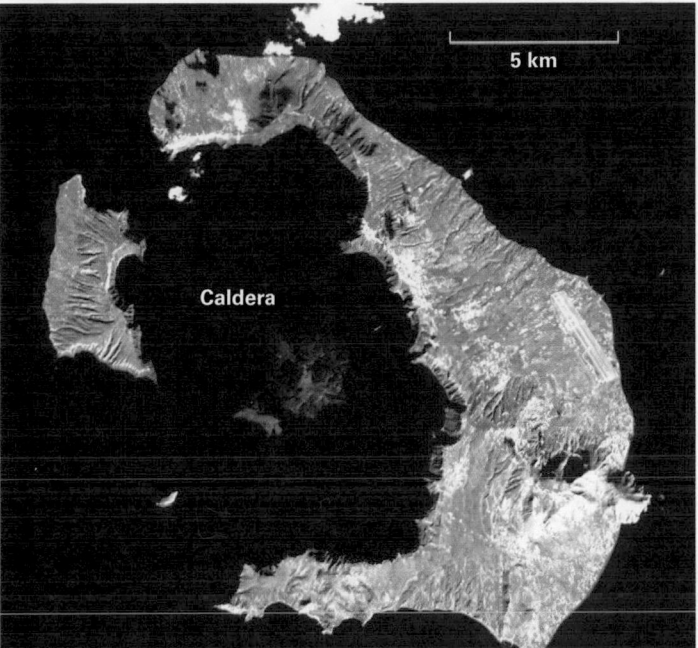

(b) From space, we can see the 11-km-wide caldera, all that is left of Santorini.

(c) The inner wall of the caldera reveals the typical geology of a stratovolcano; we see both lava and tuff layers.

Islands. She gouged out the craters that top the volcanoes, her fits and moods bring about the eruptions, and her tears are the smooth, glassy lapilli ejected from the lava fountains.

TAKE-HOME MESSAGE

The ash, gases, and aerosols produced by explosive eruptions can be blown around the globe. This material can cause significant global cooling. Climatic effects, as well as other consequences of eruptions, may have affected human evolution and civilization.

QUICK QUESTION: Not all eruptions of equivalent size (defined by the volume of material erupted) trigger equivalent amounts of global cooling. Why? (*Hint:* Think about eruptive style.)

9.8 Volcanoes on Other Planets

We conclude this chapter by looking beyond the Earth, for our planet is not the only object in the Solar System to host volcanic eruptions. We can see the effects of volcanic activity on our nearest neighbor, the Moon, just by looking up on a clear night. The broad darker areas of the Moon, the **maria** (singular *mare*, after the Latin word for sea), consist of flood basalts that erupted more than 3 billion years ago **(Fig. 9.31a)**. Geologists propose that the flood basalts formed after huge meteors collided with the Moon, blasting out giant craters and cracking the crust. Crater formation decreased the pressure in the Moon's mantle so that it underwent partial melting, producing basaltic magma. The magma eventually erupted through fissures about 100 million years after impact and filled the craters.

> **Did you ever wonder...**
> if the Earth is the only planet with volcanoes?

On Venus, about 22,000 volcanic edifices have been identified. Some of these have calderas at their crests **(Fig. 9.31b)**. Though no volcanoes currently erupt on Mars, the planet's surface displays a record of a spectacular volcanic past.

FIGURE 9.31 Volcanism on other planets and moons in the Solar System.

(a) The maria of the Moon were once seas of basaltic lava.

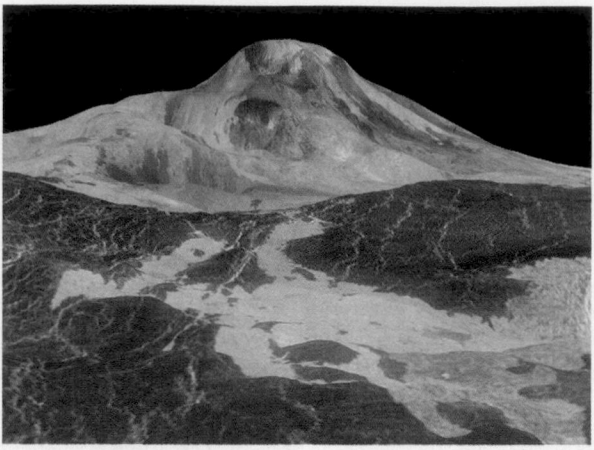

(b) A volcano rises above the plains of Venus. Lava flows cover part of the land in the foreground.

(c) Olympus Mons rises 27 km above the surface of Mars. It's the largest volcano in our Solar System.

(d) An active volcano erupts sulfur on Io, a moon of Jupiter.

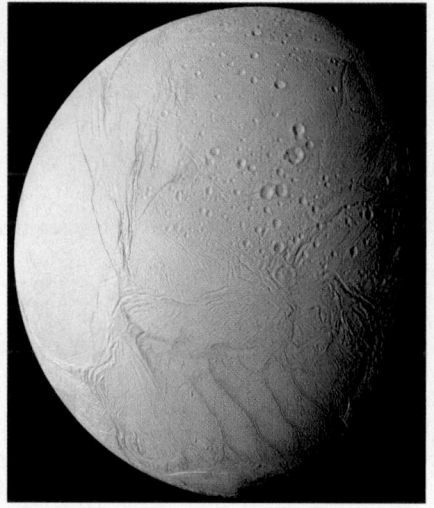

(e) Cracks at the southern end of Saturn's moon, Enceladus, erupt water vapor.

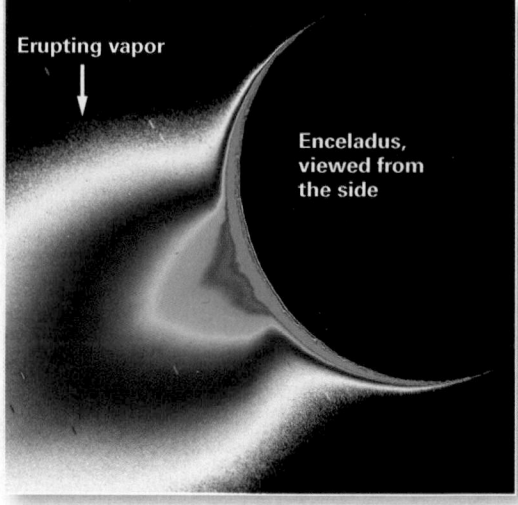

(f) The *Cassini* spacecraft detected gas erupting from Enceladus.

The largest known mountain in the Solar System, Olympus Mons **(Fig. 9.31c)**, is an extinct shield volcano on Mars. The base of Olympus Mons is 600 km across, and its peak rises 27 km above the surrounding plains.

Active volcanism currently occurs on Io, one of the many moons of Jupiter. Cameras in the *Galileo* spacecraft have recorded these volcanoes in the act of spraying plumes of sulfur gas into space **(Fig. 9.31d)** and have tracked immense, moving basaltic lava flows. Different colors of erupted material make the surface of this moon resemble a pizza. Researchers have proposed that the volcanic activity is due to tides: the gravitational pull exerted by Jupiter and by other moons alternately stretches and then squeezes Io, generating sufficient friction to keep Io's mantle hot. Geologists have also detected gas eruptions from the moons of Saturn **(Fig. 9.31e, f)**.

TAKE-HOME MESSAGE

Space exploration reveals that volcanism occurs not only on the Earth, but also on other terrestrial planets and on the moons of gas-giant planets. Satellites have detected active eruptions on the moons of Jupiter and Saturn.

QUICK QUESTION: Since most volcanic activity on the Earth is a result of plate tectonics, what does the lack of current volcanic activity on the Moon tell us about whether or not plate tectonics happens on the Moon?

ANOTHER VIEW The ash cloud from the 2011 eruption of a volcano in the Puyehue-Cordón Caulle chain of Chile. The ash circled the globe within 2 weeks, disrupting air traffic throughout the southern hemisphere.

CHAPTER SUMMARY

- Volcanoes are vents through which molten rock (lava), pyroclastic debris, gases, and aerosols erupt at the Earth's surface. A hill or mountain produced by volcanism is also called a volcano.

- The characteristics of a lava flow depend on the lava's viscosity, which in turn depends on its temperature and composition.

- Basaltic lava can flow great distances. Pahoehoe flows have smooth, ropy surfaces, whereas a'a' flows have rough, rubbly surfaces. Andesitic and rhyolitic lava flows tend to pile into mounds at the vent.

- Pyroclastic debris includes powder-sized ash, marble-sized lapilli, and apple- to refrigerator-sized blocks and bombs. Some falls from the air, whereas other debris forms incandescent pyroclastic flows that avalanche down the sides of the volcano.

- Eruptions may occur from a crater at a volcano's summit or flank or from fissures. Collapse of a volcano following an eruption may produce a large bowl-shaped depression called a caldera.

- A volcano's shape depends on its eruptive style. Shield volcanoes are broad, gentle domes. Cinder cones are steep-sided, symmetric hills composed of tephra. Composite volcanoes (stratovolcanoes) can become quite large and consist of alternating layers of pyroclastic debris and lava.

- A volcano's eruptive style depends on several factors, including the lava's viscosity and gas content. Effusive eruptions produce only flows of lava, whereas explosive eruptions produce clouds and flows of pyroclastic debris.

- Different kinds of volcanoes form in different geologic settings, as defined by plate tectonics theory.

- Volcanic eruptions pose many hazards: lava flows overrun roads and towns, ash falls blanket the landscape, pyroclastic flows incinerate towns and fields, landslides and lahars bury the land surface, earthquakes topple structures and rupture dams, tsunamis wash away coastal towns, and invisible gases suffocate people and animals.

- Eruptions can be predicted by earthquake activity, changes in heat flow, changes in the shape of the volcano, and the emission of gases or steam.

- We can minimize the consequences of an eruption by using volcanic-hazard assessment maps and by making evacuation plans.

- Immense flood basalts cover portions of the Moon. The largest known volcano, Olympus Mons, towers over the surface of Mars. Satellites have documented evidence for eruptions on moons of Jupiter and Saturn.

GUIDE TERMS

a'a' (p. 284)
active volcano (p. 311)
aerosol (p. 290)
ash (p. 287)
block (p. 287)
bomb (p. 287)
caldera (p. 291)
cinder cone (p. 292)
columnar jointing (p. 284)
crater (p. 291)
dormant volcano (p. 311)

effusive eruption (p. 294)
eruption column (p. 299)
eruptive style (p. 294)
explosive eruption (p. 295)
extinct volcano (p. 312)
fissure (p. 291)
flood basalt (p. 307)
ignimbrite (p. 299)
lahar (p. 288)
lapilli (p. 286)

large igneous province (LIP) (p. 307)
lava tube (p. 284)
magma chamber (p. 291)
mare (p. 317)
pahoehoe (p. 284)
pillow lava (p. 284)
pyroclastic debris (p. 284)
pyroclastic flow (p. 299)
pyroclastic rock (p. 286)
shield volcano (p. 291)

stratovolcano (p. 292)
supervolcano (p. 299)
tuff (p. 287)
vent (p. 291)
vesicle (p. 290)
viscosity (p. 283)
volcanic explosivity index (VEI) (p. 300)
volcanic-hazard assessment map (p. 313)
volcano (p. 281)

GEOTOURS *THIS CHAPTER'S GEOTOURS WORKSHEET (E) FEATURES QUESTIONS AND GOOGLE EARTH SITES ON:*

- Shield volcanoes   • Stratovolcanoes   • Cinder cone volcanoes

REVIEW QUESTIONS

The letters following each Review Question refer to the corresponding Learning Objective from the Chapter Opener.

1. Describe three different kinds of material that can erupt from a volcano. Identify them on the figure. **(B)**

2. How does pyroclastic flow differ from a lahar? **(D)**

3. Why do some volcanic eruptions consist mostly of lava flows, whereas others are explosive and do not produce lava flows? **(C)**

4. Distinguish among different types of explosive eruptions. **(D)**

5. Describe the activity in the mantle that leads to hot-spot eruptions. **(D)**

6. How do continental-rift eruptions form flood basalts? **(D)**

7. What is a large igneous province (LIP)? **(D)**

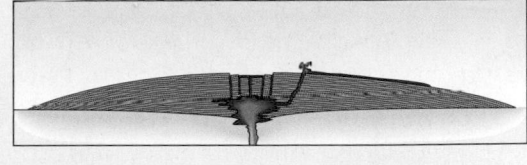

8. Identify some of the major volcanic hazards, and explain how they develop. **(E)**

9. Contrast shield volcanoes, stratovolcanoes, and cinder cones. Which is pictured in the diagram? How are these differences explained by the composition of their lavas and other factors? **(A)**

10. How do geologists predict impending volcanic eruptions? **(F)**

11. Distinguish among active, dormant, and extinct volcanoes. **(F)**

12. Explain how steps can be taken to protect people from the effects of eruptions. **(E)**

ON FURTHER THOUGHT

13. The Long Valley Caldera (see Fig. 9.20b), near the Sierra Nevada range, exploded about 700,000 years ago and produced a huge ignimbrite called the Bishop Tuff. About 30 km to the northwest lies Mono Lake, with an island in the middle and a string of craters extending south from its south shore. Hot springs can be found along the lake. You can see the lake on Google Earth at Latitude 37°59′56.58″ N, Longitude 119°2′18.20″ W. Explain the origin of Mono Lake. Is it a volcanic hazard? **(B, F)**

14. Mt. Fuji is a 3.6-km-high stratovolcano in Japan formed as a consequence of subduction. With Google Earth you can see the volcano at Latitude 35°21′46.72″ N, Longitude 138°43′49.38″ E. It contains volcanic rocks with a range of compositions, including some andesitic rocks. Why do andesites erupt at Mt. Fuji? Very little andesite occurs on the Mariana Islands, which are also subduction-related volcanoes. Why the difference? **(D)**

ONLINE RESOURCES

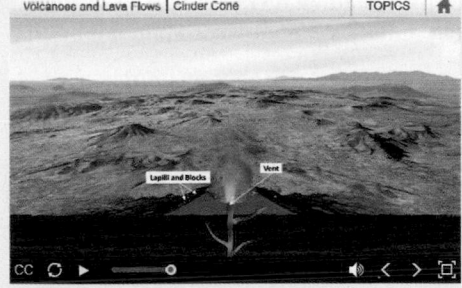

Animations
This chapter features animations covering the various types of volcanoes and lava flows, and the formation and activity of hot spots.

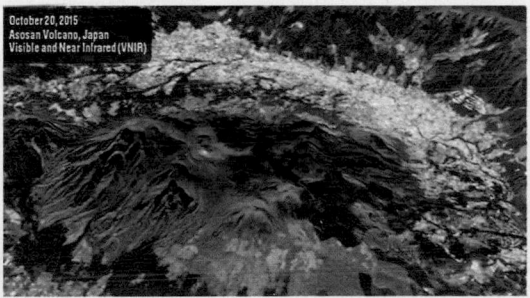

Videos
This chapter features videos on monitoring volcanoes and on lava activity, flows, and hazards.

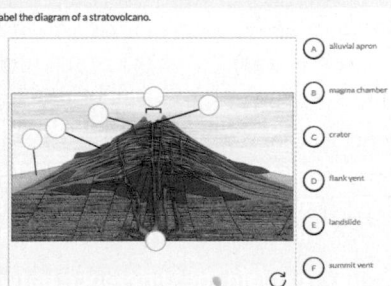

Smartwork5
Questions include visual and labeling exercises on the processes involved in the structure and eruptive styles of various types of volcanoes, and assessments of the hazards volcanoes present.

CHAPTER 10

A Violent Pulse: Earthquakes

By the end of this chapter, you should be able to . . .

A. describe an earthquake and explain where the energy released during an earthquake comes from.

B. relate earthquakes to specific geologic settings, in the context of plate tectonics theory.

C. draw a sketch illustrating how a seismometer operates, and explain what the squiggles on a seismogram mean.

D. distinguish among the different kinds of seismic waves, and show how the arrival times of seismic waves can indicate where an earthquake occurred.

E. explain the difference between the intensity and magnitude of an earthquake and how these indicators of earthquake size can be determined.

F. discuss the many ways in which earthquakes cause damage and injury.

G. distinguish between tsunamis and storm waves, and explain how large tsunamis can cause so much damage.

H. determine whether a prediction of an earthquake is worth listening to, and explain the difference between a prediction and an early warning.

I. identify steps that can help you and others prevent earthquake damage and avoid injury.

We learn geology the morning after the earthquake.

—Ralph Waldo Emerson (American poet, 1803–1882)

10.1 Introduction

It was mid-afternoon on March 11, 2011, and in the many seaside towns along the eastern coast of Honshu, the northern island of Japan, fishing fleets unloaded their catch, factories churned out goods, shoppers browsed the stores, and office workers tapped at their computers, unaware that their surroundings would suddenly be changing forever. This coast lies near the convergent boundary at which the Pacific Plate slips beneath the edge of Japan and sinks back into the mantle. Averaged over time, the relative movement across this boundary takes place at a rate of about 8 cm per year. But the motion doesn't take place smoothly. Rather, for a while, rocks adjacent to the boundary quietly flex, like a stick bent between your hands, to accommodate the motion, until suddenly, like a stick that snaps after you bend it too far, the rocks break, and a large amount of slip on a fracture takes place in a matter of seconds to minutes **(Fig. 10.1)**. On March 11, at 2:46 P.M., such a "snap" started about 130 km east of Japan's coast, at a depth of about 24 km below the seafloor. Japan lurched eastward by several meters, relative to the Pacific plate, and the seafloor moved upward by about 30 cm. The stage had been set for a disaster.

Geologists refer to a fracture on which such sliding takes place as a **fault**. The fault involved in the March 11 event delineates the boundary between Honshu and the Pacific Plate. The sudden breaking and sliding of rock along a fault produces energy that propagates through the crust in the form of vibrations. In essence, these vibrations resemble the shock waves that travel through a breaking stick to your hands. In the Earth, vibrations can race through

the crust at an average speed of 11,000 km per hour, 10 times the speed of sound in air. When the vibrations produced on March 11 reached Japan, the land surface lurched back and forth and bounced up and down for over a minute. People panicked, became disoriented, and even seasick—some lost their balance and crouched or fell, and some heard a dull rumbling or thumping. Bottles and plates flew off shelves and crashed to the floor, buildings twisted and swayed, ceilings and facades fell in a shower of debris, dust rose from the ground to create a fog-like mist, power lines stretched and sparked, and weaker buildings collapsed **(Fig. 10.2a)**. In addition, several natural gas tanks and pipes broke, sending flammable vapors into the air—in some places, the gas ignited in billows of flame that set fire to damaged buildings. A major **earthquake**—an episode of ground shaking—had occurred. Geologists now refer to this event as the Tohoku earthquake, named for the eastern province of Honshu.

FIGURE 10.1 What happens during an earthquake?

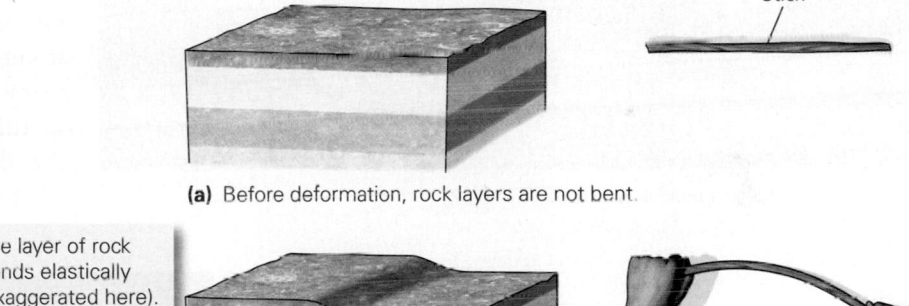

Stick

(a) Before deformation, rock layers are not bent.

The layer of rock bends elastically (exaggerated here).

(b) Before an earthquake, rock bends elastically, like a stick that you arch between your hands.

Due to elastic rebound, the rock layers return to their initial shape.

(c) Eventually, the rock breaks, and sliding suddenly occurs on a fault. This break generates vibrations. You feel such vibrations when you break a stick.

◄ (facing page) In 2017, a large earthquake shook Mexico City. Sadly, the vibrations caused some buildings to collapse, with the catastrophic consequences you can see here. Earthquakes, such as this one, emphasize that our planet remains a dynamic place.

In many cases, the societal calamity due to an earthquake results directly from ground shaking, which causes buildings to collapse, crushing their inhabitants, and sends debris tumbling down slopes. Building collapse and landslides triggered by the May 12, 2008, earthquake in Sichuan, in central China, leveled cities—70,000 people died, and millions were left homeless. But in Japan, a country struck by earthquakes fairly frequently, officials have established stringent building codes, which successfully prevented widespread building collapse in 2011. Unfortunately, even strong buildings could not resist what happened minutes after the earthquake shock. The sudden displacement of the seafloor off Japan's coast displaced a massive amount of ocean water and produced immense water waves known as *tsunamis*. When these waves washed over the land, they became a roaring jumble of water and debris that submerged low-lying areas as far as 10 km from the shoreline **(Fig. 10.2b, c)**. In the end, the immensity of the devastation due to the Tōhoku earthquake was almost beyond comprehension.

Earthquakes are a fact of life on planet Earth—almost 1 million detectable earthquakes happen every year. Most are a consequence of plate movement—they punctuate each step in the growth of mountains, the drift of continents, and the opening and closing of ocean basins. Fortunately, most cause no damage or casualties, either because they are too small or because they occur in unpopulated areas. But a few hundred earthquakes per year rattle the ground sufficiently to crack or topple buildings and injure their occupants, and every 5 to 20 years, on average, a great earthquake triggers a horrific calamity. In fact, during the past two millennia, ground shaking, tsunamis, landslides, fires, and other phenomena caused by earthquakes have killed over 3.5 million people **(Table 10.1)**.

What geologic phenomena trigger earthquakes? Why do earthquakes take place where they do? How do they cause damage? Can we predict when earthquakes will happen or even prevent them from happening? **Seismologists** (from the Greek word *seismos*, for shock or earthquake), geoscientists who study earthquakes, have addressed many of these questions during the past century. In this chapter, we present some of the answers that they have obtained.

FIGURE 10.2 The great Tōhoku earthquake and tsunami in Japan, 2011.

(a) Some buildings collapsed due to ground shaking.

(b) The tsunami filled harbors and spilled over sea walls.

(c) The tsunami washed over low coastal areas. In this photograph, the wave is starting to wash over a canal.

TABLE 10.1 Some Notable Earthquakes

Year	Location	Deaths
2017	Mexico City, Mexico	369
2011	Tōhoku, Japan (tsunami)	20,000
2011	Christchurch, New Zealand	180
2010	Haiti	230,000
2010	Concepción, Chile	1,000
2008	Sichuan, China	70,000
2005	Pakistan	80,000
2004	Sumatra (tsunami)	230,000
2003	Bam, Iran	41,000
2001	Bhuj, India	20,000
1999	Calaraca/Armenia, Colombia	2,000
1999	Izmit, Turkey	17,000
1995	Kobe, Japan	5,500
1994	Northridge, California	51
1990	Western Iran	50,000
1989	Loma Prieta, California	65
1988	Spitak, Armenia	24,000
1985	Mexico City	9,500
1983	Turkey	1,300
1978	Iran	15,000
1976	T'ang-shan, China	255,000
1976	Caldiran, Turkey	8,000
1976	Guatemala	23,000
1972	Nicaragua	12,000
1971	San Fernando, California	65
1970	Peru	66,000
1968	Iran	12,000
1964	Anchorage, Alaska	131
1963	Skopje, Yugoslavia	1,000
1962	Iran	12,000
1960	Agadir, Morocco	12,000
1960	Southern Chile	6,000
1948	Turkmenistan, USSR	110,000
1939	Erzincan, Turkey	40,000
1939	Chillán, Chile	30,000
1935	Quetta, Pakistan	60,000
1932	Gansu, China	70,000
1927	Tsinghai, China	200,000
1923	Tokyo, Japan	143,000
1920	Gansu, China	180,000
1915	Avezzano, Italy	30,000
1908	Messina, Italy	160,000
1906	San Francisco	500
1896	Japan	22,000
1886	Charleston, South Carolina	60
1866	Peru and Ecuador	25,000
1811-12	New Madrid, Missouri (3 events)	Few
1783	Calabria, Italy	50,000
1755	Lisbon, Portugal	70,000
1556	Shen-shu, China	830,000

10.2 What Causes Earthquakes?

What causes earthquake activity, or **seismicity**? Traditional cultures commonly attributed it to the action of supernatural beasts. In Japanese folklore, for example, seismicity happened when a giant catfish, Namazu, living in the mud below the surface of the ground, started to thrash about. Similarly, in Indian folklore, seismicity happened when one of eight elephants holding up the Earth shook its head. Scientific study instead associates seismicity with any of the following:

- the sudden formation of a new fault
- sudden slip on an existing fault
- a phase change, when the atoms in minerals suddenly rearrange
- movement of magma into, or explosive eruption of, a volcano
- a giant landslide
- a meteorite impact
- an underground nuclear-bomb test

In this chapter, we focus our attention on fault-generated earthquakes, as they are by far the most common.

When describing the location of an earthquake, seismologists use two terms. First, we define the point on a fault at which rock starts to rupture and slip as the *hypocenter*, or **focus**, of an earthquake (**Fig. 10.3a**). Simplistically, we can picture the focus as the place inside the Earth from which earthquake energy, in the form of vibrations, begins to propagate. (In reality, the area of a fault that slips during an earthquake extends beyond the focus, so not all of the energy produced during an earthquake actually originates right at the focus.) Second, we define the point on the surface of the Earth that lies directly above the focus as the earthquake's **epicenter**. We can portray the position of an epicenter as a spot on a map (**Fig. 10.3b**).

Faults in the Crust

What does a fault look like? At first glance, a fault may simply appear to be a break that cuts across rock or sediment. But on closer examination, you may be able to see evidence of the sliding that occurred on a fault. For example, the rock adjacent to the fault may be broken up into angular fragments, or may be pulverized into tiny grains, due to the crushing and grinding that can accompany slip. The surface of a fault may be polished and grooved as if scratched by a rasp. In some localities, a fault

FIGURE 10.3 Earthquake hypocenters and epicenters.

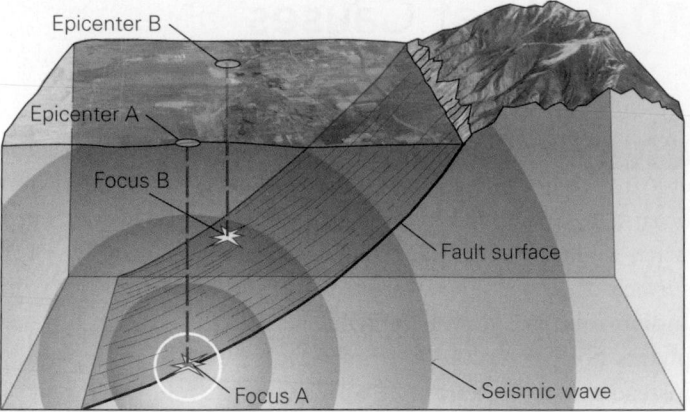

(a) The focus is the point on the fault where slip begins. Seismic energy starts radiating from it. The epicenter (circle diameter represents earthquake "size") is the point on the Earth's surface directly above the focus. Earthquake A just happened; earthquake B happened a while ago.

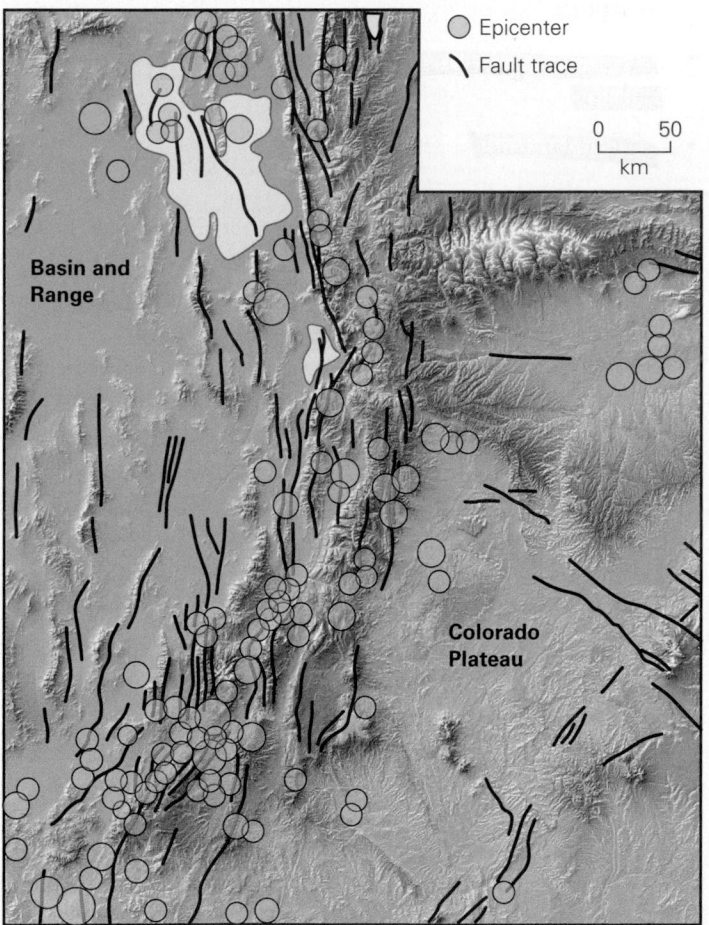

(b) A map of Utah showing the distribution of earthquake epicenters, recorded over several years. The map indicates that seismic activity happens mostly in a distinct belt following the boundary between the Basin and Range rift and the Colorado Plateau.

cuts through a *marker*, such as a distinct sedimentary bed, an igneous dike, or a fence. Where this happens, the marker on one side of the fault no longer lies adjacent to the marker on the other side after the fault slips. The distance between two ends of the marker, as measured along the fault surface in the direction of slip, is the fault's **displacement (Fig. 10.4)**.

Many faults do not cut the Earth's surface while active. Such *blind faults* are hidden completely underground while active, but they may later become visible if exposed by erosion of overlying rock. But some faults intersect and offset the ground surface, producing a step called a **fault scarp (Fig. 10.5)**. Geologists refer to the intersection between a fault and the Earth's surface as a *fault line* or **fault trace**.

In the 19th century, miners who encountered faults in mine tunnels referred to the rock mass above a sloping fault plane as the *hanging wall*, because it hung over their heads, and the rock mass below the fault plane as the *footwall*, because it lay beneath their feet. These miners described the direction in which rock masses slipped on a sloping fault by specifying the direction in which the hanging wall moved relative to the footwall, and we still use their terminology today (**Geology at a Glance**, pp. 330–331). Specifically, a fault whose hanging wall has slipped down the slope of the fault is a *normal fault*, and a fault whose hanging wall has slipped up the slope is a *reverse fault* (if steep) or a *thrust fault* (if shallowly sloping). A *strike-slip fault* is a near-vertical fracture on which slip moves rock parallel to an imaginary horizontal line, called a *strike line*, on the fault plane. No upward or downward motion takes place on a strike-slip fault. Note that although we depict faults as single surfaces in Figure 10.5, in reality, major faults commonly consist of a zone including several smaller faults.

The direction of slip on a fault reflects the nature of crustal movements causing the fault. For example, normal faults form in response to stretching or extension of the crust, whereas reverse faults and thrust faults develop in response to squeezing (compression) and shortening of the crust. Strike-slip faults form where one block of crust slides past another laterally.

Faults can be found in many locations—but don't panic! Not all faults are likely to be sources of earthquakes. Faults that have slipped recently or are likely to slip in the near future are called *active faults* (and if they generate earthquakes, news media sometimes refer to them as "earthquake faults"). Most active faults occur along plate boundaries or in currently developing collision zones and rifts. Faults that last slipped in the distant past and probably won't move again for at least millenia are called *inactive faults*. Some faults have been inactive for billions of years and probably will never slip again.

Generating Earthquake Energy: Stick-Slip Behavior

Imagine that you grip each side of a brick-shaped block of rock with a clamp. Apply an upward push on one of the clamps and a downward push on the other. By doing so, you have applied

FIGURE 10.4 Examples of fault displacement on the San Andreas fault in California.

What a Geologist Sees

Displacement

Fault trace

These ruptures formed where the fault broke the ground surface.

Dirt road

(a) A wooden fence built across the San Andreas fault was offset during the 1906 San Francisco earthquake. The displacement indicates strike-slip motion.

(b) An aerial photograph shows offset of a dirt road by slip on a fault in the Mojave Desert in 1999.

FIGURE 10.5 The basic types of faults. Fault types are distinguished from one another by the direction of slip relative to the fault surface.

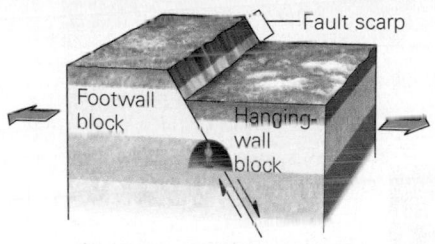

Fault scarp

Footwall block

Hanging-wall block

(a) Normal faults form during extension of the crust. The hanging wall moves down.

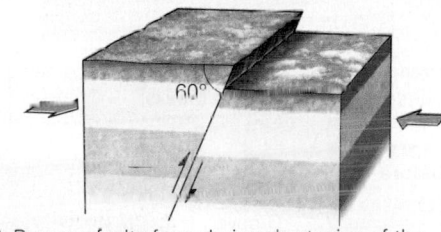

60°

(b) Reverse faults form during shortening of the crust. The hanging wall moves up and the fault is steep.

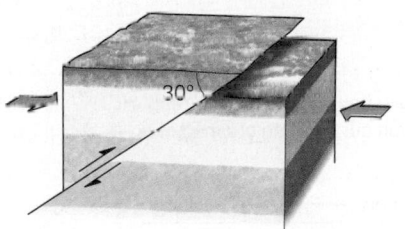

30°

(c) Thrust faults also form during shortening. The fault's slope is gentle (less than 30°).

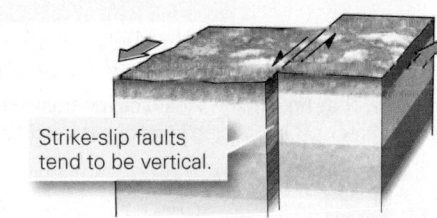

Strike-slip faults tend to be vertical.

(d) On a strike-slip fault, one block slides laterally past another, so no vertical displacement takes place.

let go. The change in shape due to elastic bending, stretching, or shortening is called *elastic strain*. Now repeat the experiment, but bend the rock even more.

If you bend the rock far enough, a number of small cracks start to form in the rock, typically in a diagonal zone. Eventually the cracks connect to one another to form a fracture (or rupture) that cuts across the entire block of rock **(Fig. 10.6b)**. The instant that such a throughgoing fracture forms, the block breaks in two, and the rock on one side of the fracture suddenly slides past the rock on the other side. When sliding occurs, the fracture has become a fault. Any elastic strain that had built up in the rock gets released, so the rock straightens out, or rebounds **(Fig. 10.6c)**. Such fault formation in a previously intact rock releases energy and generates earthquake vibrations.

A new fault can't slip forever, for fric-

a stress to the rock—simplistically, we can think of **stress** as a push, pull, or shear. At first, the rock bends slightly but doesn't break **(Fig. 10.6a)**. In fact, if you were to stop applying stress before the rock broke, the rock would return to its original shape. Geologists refer to such a phenomenon as **elastic behavior**—the same phenomenon happens when you stretch a spring and then

tion eventually slows and stops the movement, just as friction slows and stops a book sliding across a table. **Friction**, defined as the force that resists sliding on a surface, exists because fault surfaces are not perfectly smooth. The small protrusions, or *asperities*, on a fault surface act like tiny anchors and dig into the opposing surface **(Fig. 10.7a, b)**.

FIGURE 10.6 A model representing the development of a new fault. The process can generate earthquake-like vibrations.

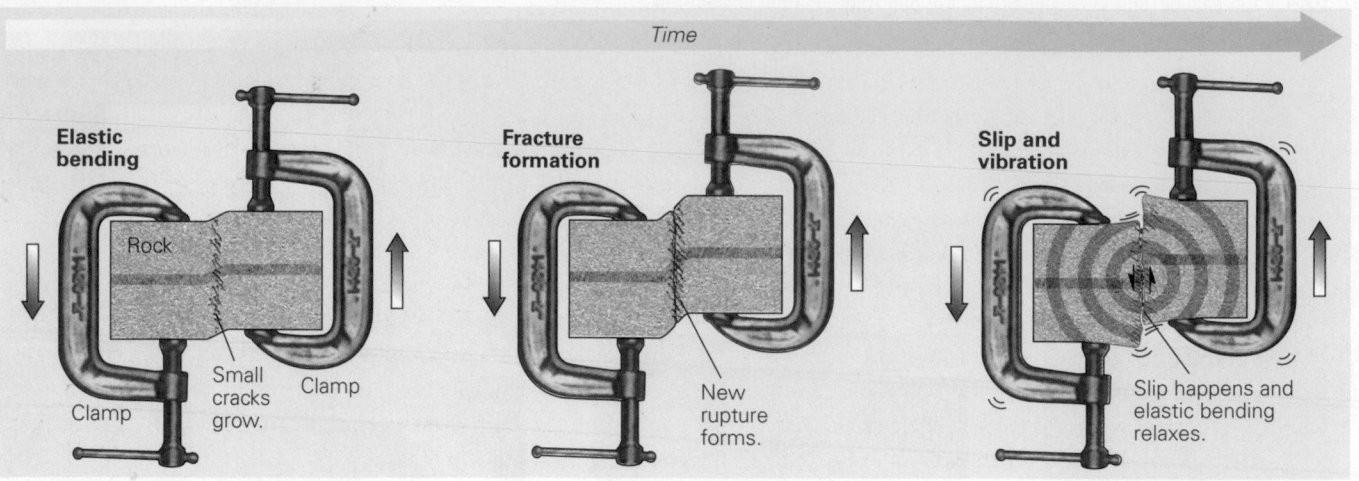

Time

Elastic bending

Rock

Small cracks grow.

Clamp

Clamp

Fracture formation

New rupture forms.

Slip and vibration

Slip happens and elastic bending relaxes.

(a) Imagine a block of rock gripped by two clamps. Move one clamp up, and the rock starts to bend. Small cracks develop along the bend.

(b) Eventually, the cracks link. When this happens, a throughgoing fracture forms.

(c) The instant that the fracture forms, the rock breaks into two pieces that slide past each other. The energy that is released generates vibrations.

FIGURE 10.7 The concept of friction and stick-slip behavior. On a microscopic scale, fault surfaces are rough and friction temporarily inhibits sliding.

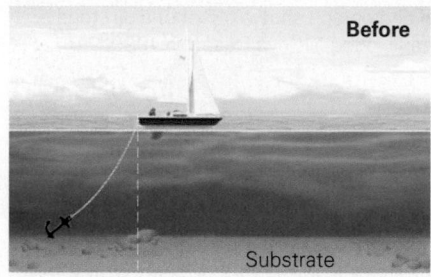
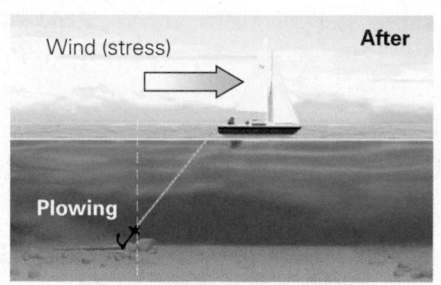
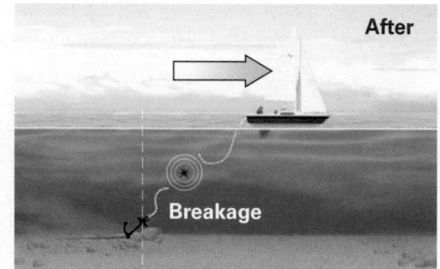

Before

Substrate

Wind (stress)

After

Plowing

or

After

Breakage

(a) An anchor analogy for frictional slip on a fault. The wind applies a stress. If the stress becomes great enough, the anchor "plows" by scratching a trough into the substrate, or the anchor line breaks.

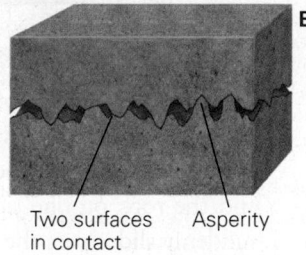
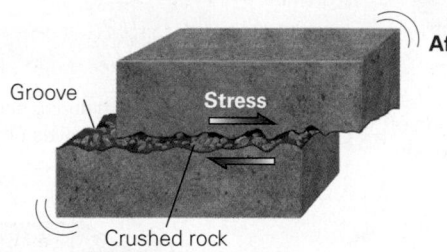

Before

Two surfaces in contact

Asperity

Groove

Stress

After

Crushed rock

(b) In the real world, fault surfaces are not perfectly smooth. They host many asperities (protrusions) that act like tiny anchors, preventing slip. If the stress becomes great enough, the asperities plow, or break off, producing striations (scratches) on the fault surface and crushed rock.

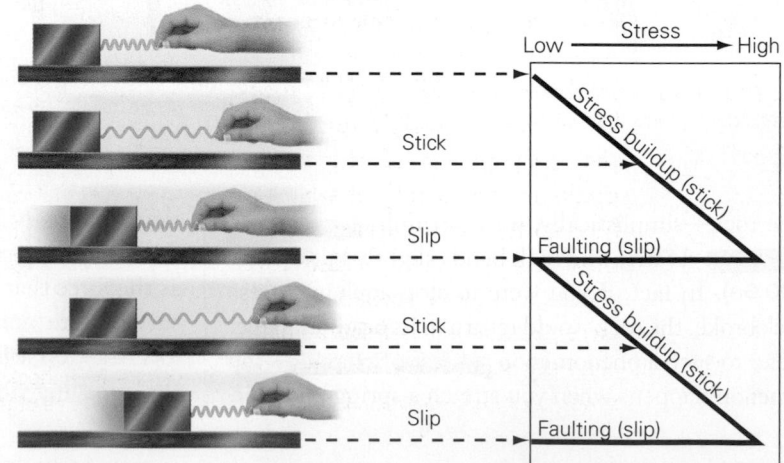

The block is at rest. Friction holds it in place.

Pulling on the spring stretches it and builds stress, but because of friction, it doesn't move the block.

When the stress exceeds friction, the block suddenly slips, and the spring relaxes, so the stress drops.

Friction causes sliding to stop, and the block sticks again, so stress builds.

Stress overcomes friction, and the block slips again.

Time

Stress

Low —————→ High

Stick

Slip

Stick

Slip

Stress buildup (stick)

Faulting (slip)

Stress buildup (stick)

Faulting (slip)

(c) Stick-slip behavior on a fault can be modeled by pulling on a heavy block with a spring. As the spring stretches, it applies stress to the block.

We've just seen how earthquakes can happen when intact rock ruptures. Earthquakes can also happen by the sudden reactivation of sliding on pre-existing faults. In this case, stress builds up in rock until it's sufficient to overcome friction, either by breaking off asperities or by causing them to plow a groove into the opposing fault surface. The breaking or plowing of asperities causes the release of energy and, like the fracture of intact rock, produces earthquakes. And, like a new fault, a *reactivated fault* can't slip forever. Friction stops the motion and prevents the fault from slipping until stress builds up sufficiently again. Geologists refer to such alternation between stress buildup and slip events as **stick-slip behavior**. You can model stick-slip behavior with a simple experiment. Attach a spring to a heavy block sitting on a rough table. By pulling on the spring, you elastically stretch it and apply a stress to the block. The more you stretch the spring, the greater the stress. Suddenly the block slips, and when this happens, the spring relaxes, so the stress decreases. You have to pull on the spring once again to increase stress enough to cause another sliding event **(Fig. 10.7c)**.

In sum, we see that earthquakes happen because stresses build up, causing rock adjacent to the fault to develop elastic strain until either intact rock breaks or a pre-existing fault reactivates. When slip takes place, the once-bent rocks adjacent to the fault rebound back to their original, unbent shape, thereby relieving the elastic strain. Geologists refer to this overall concept as the **elastic-rebound theory** of earthquake generation. Notably, the stress necessary to reactivate a fault tends to be less than the stress necessary to break intact rock, so most earthquakes probably represent slip on pre-existing faults.

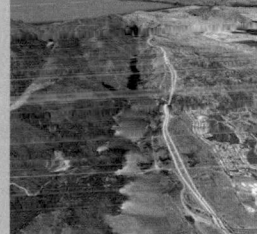
Foreshocks and Aftershocks

The strongest earthquake, or *mainshock*, along a fault in a given time period, may be preceded by smaller ones, called **foreshocks**. These shocks possibly result from the development of the smaller cracks in the vicinity of what will become the major surface of shear. Smaller earthquakes, called **aftershocks**, follow a mainshock—the largest of these are 10 times smaller than the mainshock, and most are much smaller. Most aftershocks, especially larger ones, take place soon after the mainshock. But they can occur sporadically for weeks or even years. Their distribution defines the overall area of the fault that slipped during the mainshock. Aftershocks of the Tōhoku earthquake indicate that the faults slipped over an area 600 km long, as measured parallel to the coast, by 200 km wide, as measured perpendicular to the coast **(Fig. 10.8)**. Aftershocks happen because slip during the mainshock does not leave the fault in a perfectly stable configuration. For example, after the mainshock, protrusions on one side of the fault surface, in their new position, may push into the opposing side and generate new stresses. Such stresses may become large enough either to cause a small portion of the fault around the protrusion to slip again, or to trigger slip on a nearby fault.

FIGURE 10.8 The distribution of aftershocks outlines the area of the main slip area of a major earthquake.

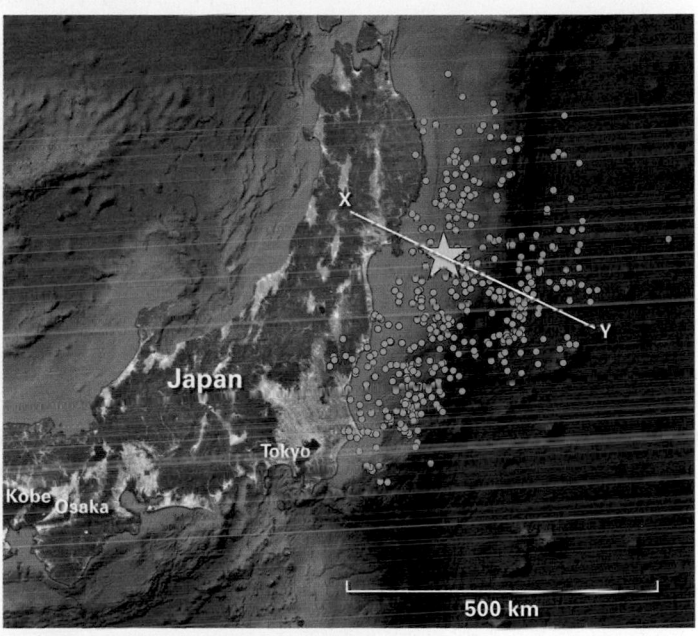

(a) The dots represent epicenters of aftershocks after the 2011 Tōhoku earthquake. The star is the epicenter of the main earthquake.
Source: Adapted from Lou Estey (UNAVCO's Jules Verne Voyager data).

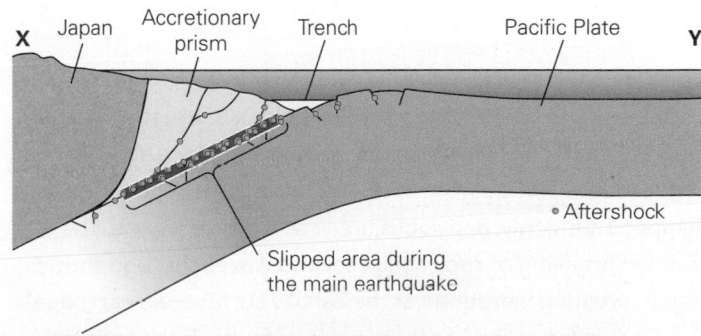

(b) A cross section, drawn along the section line in Fig. 10.8a, shows that the aftershocks occurred primarily along the main slip area, which lies at the base of the accretionary prism.

Faulting in the Crust

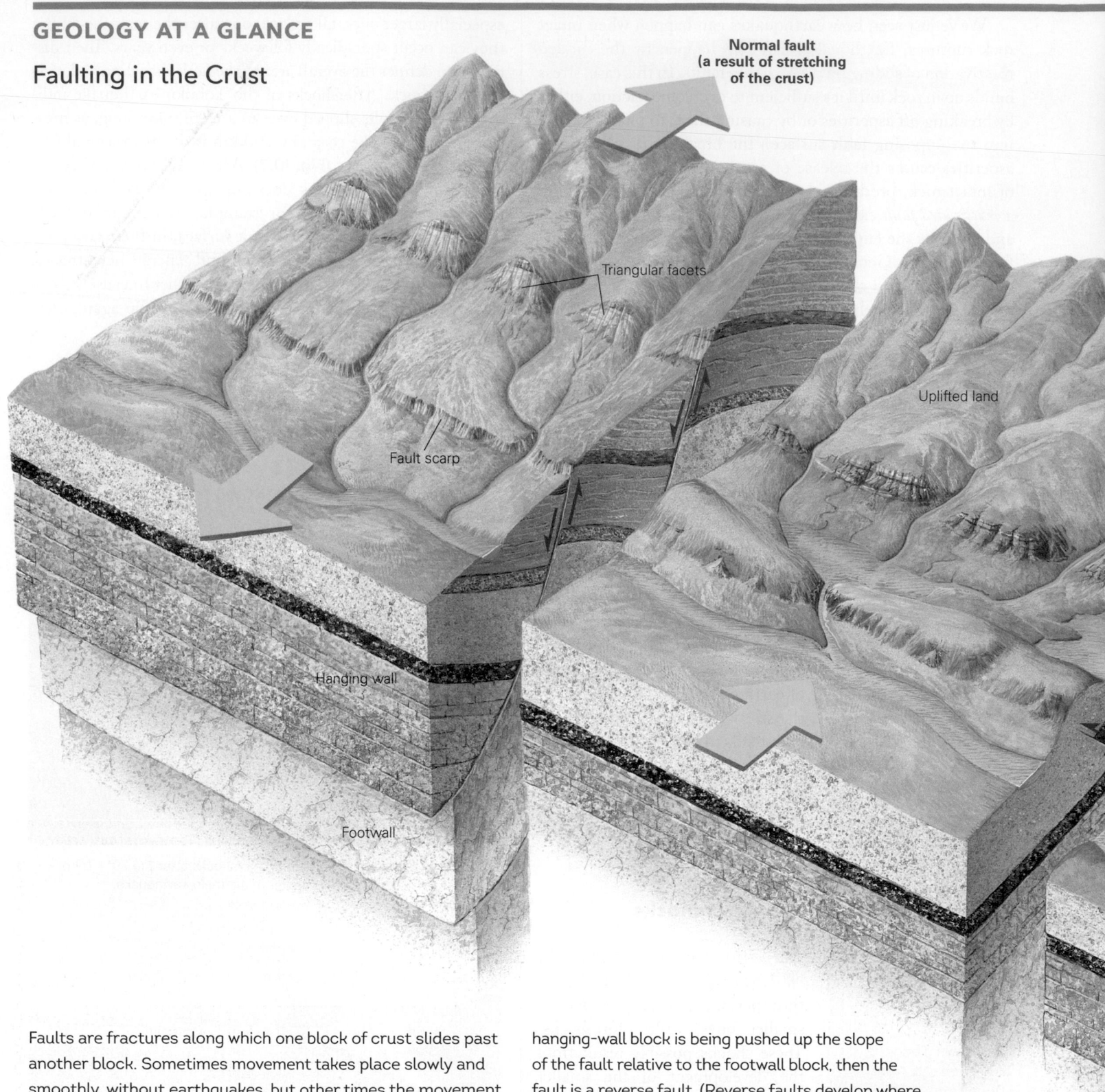

Normal fault
(a result of stretching of the crust)

Triangular facets

Uplifted land

Fault scarp

Hanging wall

Footwall

Faults are fractures along which one block of crust slides past another block. Sometimes movement takes place slowly and smoothly, without earthquakes, but other times the movement happens suddenly, and rocks break as a consequence. The sudden breaking of rock sends seismic waves through the crust, creating vibrations at the Earth's surface—an earthquake.

Geologists recognize three types of faults. If the hanging-wall block (the rock above a fault plane) slides down the fault's slope relative to the footwall block (the rock below the fault plane), the fault is a normal fault. (Normal faults form where the crust is being stretched apart, as in a continental rift.) If the

hanging-wall block is being pushed up the slope of the fault relative to the footwall block, then the fault is a reverse fault. (Reverse faults develop where the crust is being compressed or squashed, as in a collisional mountain belt.) If one block of rock slides past another and there is no up or down motion, the fault is a strike-slip fault. Strike-slip fault planes tend to be nearly vertical. If a fault displaces the ground surface, it creates a ledge called a fault scarp. Where a fault scarp cuts a system of rivers and valleys, it truncates ridges and produces triangular facets. Strike-slip faults may offset ridges and streams sideways.

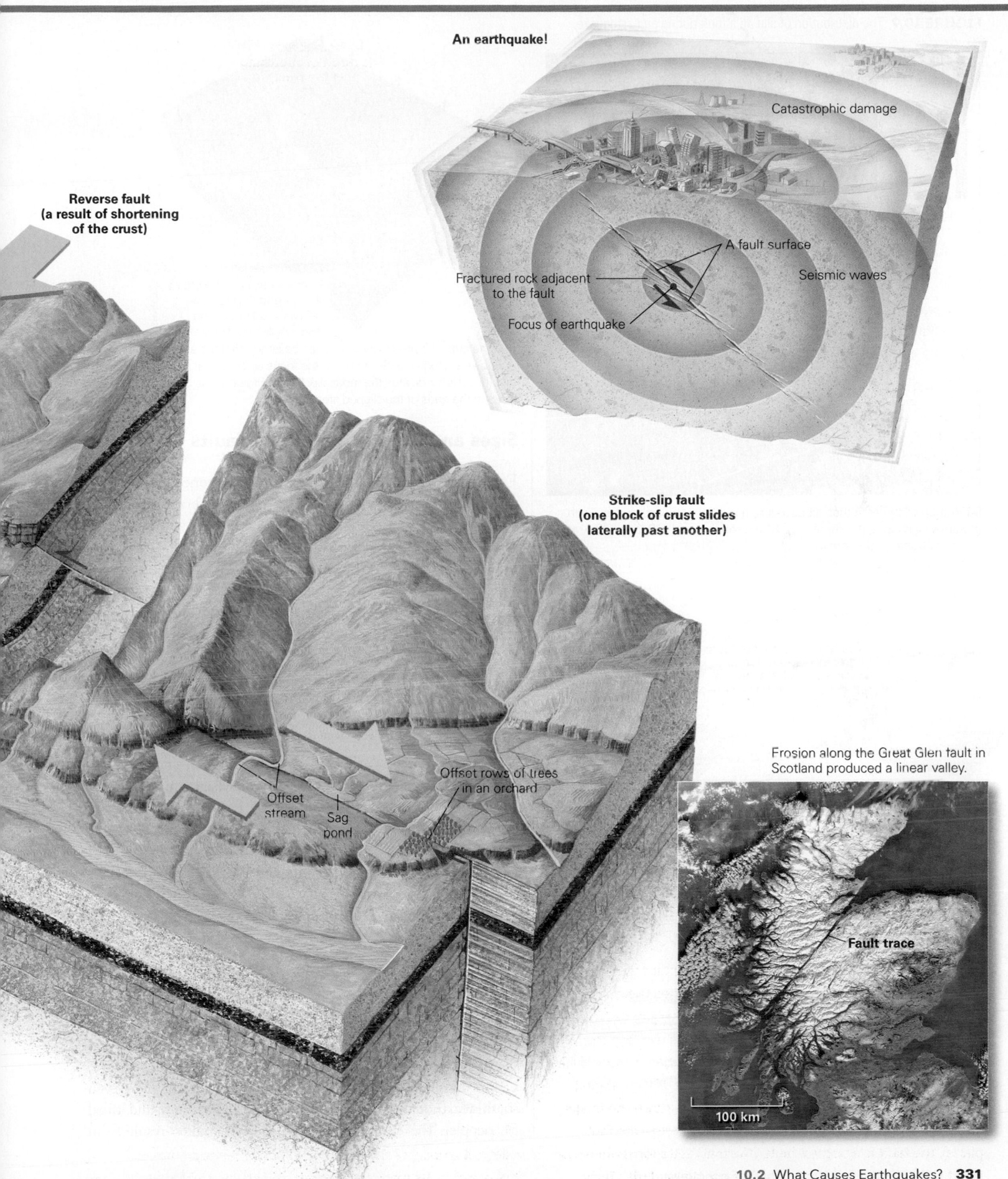

An earthquake!

Catastrophic damage

A fault surface

Fractured rock adjacent to the fault

Seismic waves

Focus of earthquake

**Reverse fault
(a result of shortening
of the crust)**

**Strike-slip fault
(one block of crust slides
laterally past another)**

Offset
stream

Sag
pond

Offset rows of trees
in an orchard

Erosion along the Great Glen fault in
Scotland produced a linear valley.

Fault trace

100 km

FIGURE 10.9 The distribution of slip on a fault during an earthquake.

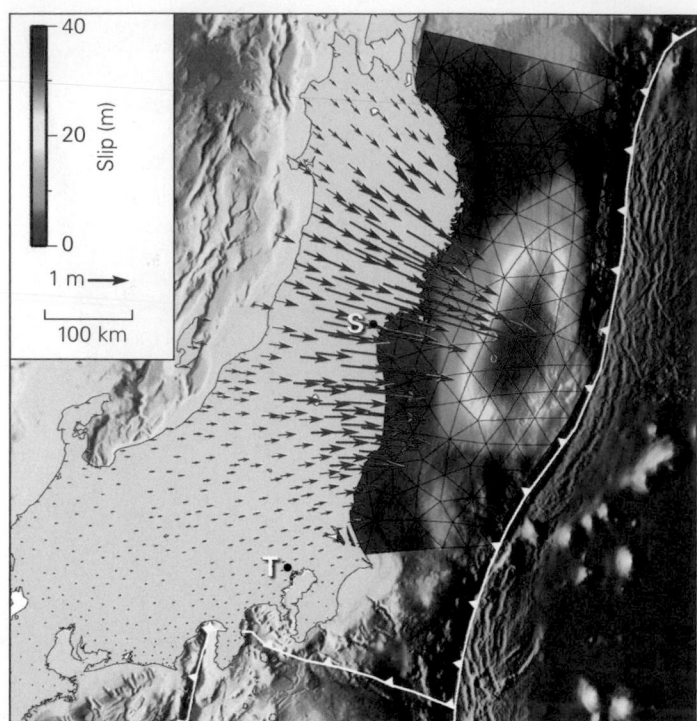

(a) A map of the area that slipped during the Tōhoku earthquake. Each red arrow indicates the motion of Japan, relative to the Pacific Plate, at the point beneath the arrow.

Source: Created by Lou Estey with UNAVCO's Jules Verne Voyager.

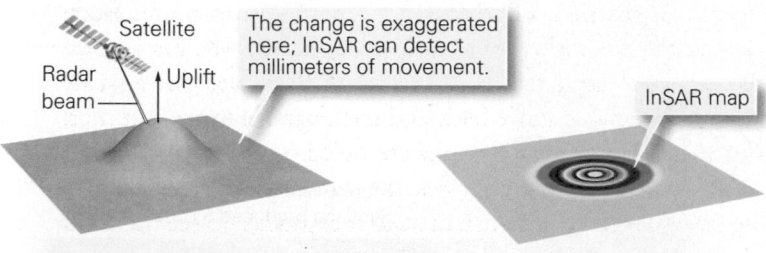

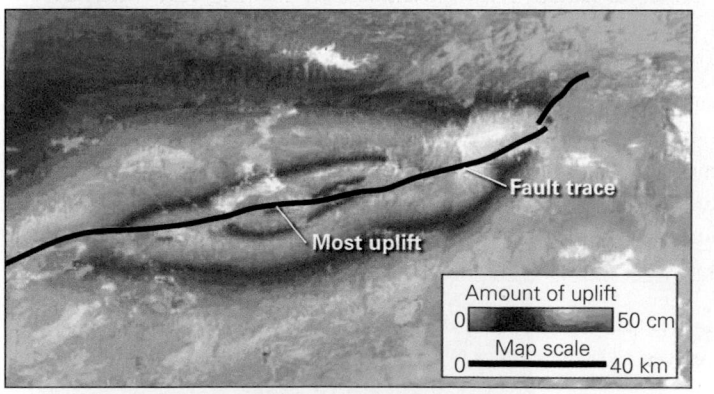

(c) Slip on a fault locally warps the Earth's surface adjacent to the fault. Measurements made using InSAR, a type of satellite-based radar, portray the warping of the Earth's surface with color bands on a map. The span of each rainbow indicates a specified amount of uplift. Note how displacement decreases toward the end of the fault.

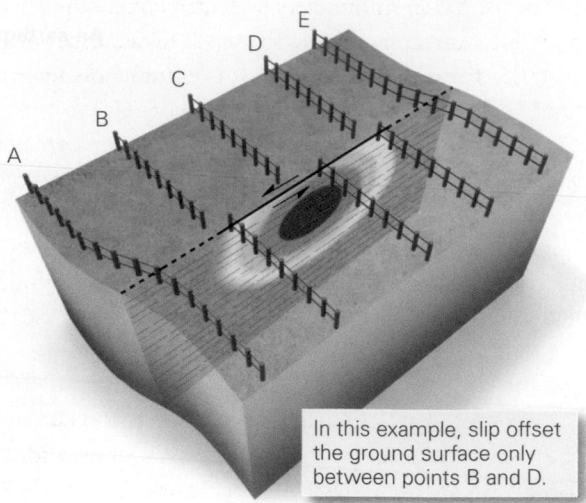

In this example, slip offset the ground surface only between points B and D.

(b) The area of a pre-existing fault can be larger than the area that slips during an earthquake. Slip starts at the focus and dies out with increasing distance. Red indicates the most slip. Rock of the crust is bent, not broken, beyond the ends of the slipped area.

Sizes and Amount of Slip on Faults

Faults come in a huge range of sizes. Small ones may be only a few centimeters or a few meters from top to bottom, as measured in cross section, so that it's possible for you to see the *fault tip*, the place where the displacement on the fault dies out. Beyond the fault tip, rock remains intact and unbroken, though it may be bent and distorted. Some faults are immense, and in cross section, extend entirely through the crust and even into the lithospheric mantle. In fact, such faults may be hundreds to thousands of kilometers long as measured on a map. For example, the San Andreas fault, a transform boundary, extends from the Earth's surface to the base of the lithosphere and has a trace that runs through California for a distance of about 1,400 km.

Generally, the "larger" the earthquake—meaning, as we will see, the greater the energy released and the more intense the vibrations generated—the larger the area of slip on a fault and the greater the displacement. The earthquake that destroyed San Francisco, California in 1906 ruptured a 430-km-long (measured parallel to the Earth's surface) by 15-km-deep (measured perpendicular to the Earth's surface) segment of the San Andreas fault and resulted in up to 7 m of strike-slip displacement. During the 1964 Good Friday earthquake in southern Alaska, slip on a thrust fault marking the boundary between the Pacific Plate and North America reached a maximum of 12 m. Slip on the thrust fault delineating the boundary between the Pacific Plate and Japan during the 2011 Tōhoku earthquake reached 30 m **(Fig. 10.9a)**.

Smaller earthquakes, such as the one that hit Northridge, California, in 1994, resulted in only about 0.5 m of slip—even so, this earthquake toppled homes, ruptured pipelines, and killed 51 people. The smallest perceptible earthquakes result from displacements measured in millimeters to centimeters. Slip on a fault varies with location along a fault. Slip tends to decrease with increasing distance from the focus, until it eventually decreases to zero, but not necessarily in a symmetrical pattern (Fig. 10.9b).

The amount of displacement during a slip event can be studied using a variety of techniques, ranging from field observation of offset markers to sophisticated satellite technologies involving GPS, which directly measure the movements of one point on the Earth's crust relative to another (see Fig. 10.9a), and InSAR, which uses radar beams to detect subtle vertical displacement of the crust (Fig. 10.9c).

Although the slip on an active fault during a human life span may not amount to much, over geologic time, the cumulative movement on the fault becomes significant. For example, if earthquakes occurring on a strike-slip fault cause 1 cm of displacement per year, on average, the fault's movement will yield 10 km of displacement after 1 million years. Thus, earthquakes mark the incremental movements that create mountains.

Can Faults Slip without Earthquakes?

When a material breaks along fractures, we say that it has undergone *brittle deformation* (see Chapter 11). In contrast, when a material changes shape by flowing slowly, while remaining solid, we say that it has undergone *plastic deformation*. To picture the contrast between these two types of behavior, consider what happens if you try to bend a wax candle that has been kept in a freezer, compared with what happens if you try to bend a candle that has been softened (but not melted) by placing it in a slightly warm oven. The frozen candle snaps, whereas the softened candle bends. Generation of energy in most earthquakes involves brittle deformation of rock, for if rock flows plastically, it can accommodate movements without causing earthquakes.

Rock, like wax, tends to be more brittle at cooler temperatures and more plastic at warmer temperatures. Because the temperature of continental crust increases with depth, continental crustal rocks typically become plastic at depths below 15 to 20 km. As a result, earthquakes on faults in continental crust occur mostly in the upper 15 to 20 km. If a fault extends below that depth, slip on that fault still takes place, but the movement occurs plastically without generating earthquakes. In crust (or lithospheric mantle) that has been subducted, earthquakes can happen at much deeper positions because cooler rock can be carried down to greater depth, as we'll see later in this chapter.

Even though most rock of the upper continental crust is brittle, geologists have found that some movement on faults in the upper crust appears to take place slowly and steadily, without generating earthquakes. When movement on a fault happens without generating earthquakes, we call the movement **fault creep**. Seismologists do not completely understand fault creep, but speculate that it occurs in particularly weak rock or when the fault surface has a coating of soft clay.

TAKE-HOME MESSAGE

Most earthquakes happen when enough stress builds to cause either sudden formation of a new fault or slip on a pre-existing fault. The focus of an earthquake is the point in the Earth where the slip begins, and the epicenter is the point on the surface of the Earth directly above the focus. Faults display stick-slip behavior in that stress builds until slip takes place. Then, stress builds again. Elastic rebound of rock when slip takes place generates earthquake energy. The size of faults and the amount of displacement during a fault vary greatly.

QUICK QUESTION: Why might foreshocks and aftershocks take place?

10.3 Seismic Waves and Their Measurement

How does the energy produced at the focus of an earthquake travel to the Earth's surface, or even pass through the entire Earth? As we've noted, earthquake energy moves through rock and sediment in the form of vibrations. We call these movements **seismic waves**, or *earthquake waves*. You feel such waves when you hold one end of a brick and strike the other end with a hammer.

Seismologists distinguish among different types of seismic waves on the basis of where and how the waves move. **Body waves** pass through the interior of the Earth, whereas **surface waves** travel along the Earth's surface. Waves that cause particles of material to move back and forth parallel to the direction in which the wave itself moves are called **compressional waves**. As a compressional wave passes, the material first compresses (or squeezes together) and then dilates (or expands). To see this kind of motion in action, push on the end of a spring and watch as the little pulse, in which the coils become more closely spaced, moves along the length of the spring. Waves that cause particles of material to move back and forth perpendicular to the direction in which the wave itself moves are called **shear waves**. To see shear-wave motion, jerk the end of a rope up and down and watch how the undulation travels along the rope.

With these concepts in mind, we can define four basic types of seismic waves (Fig. 10.10):

- *P-waves* (P stands for primary) are compressional body waves.

- *S-waves* (S stands for secondary) are shear body waves.

- *L-waves* (L stands for Love, the name of a seismologist) are surface waves that cause the ground to shift back and forth, producing a snake-like movement.

FIGURE 10.10 Different types of earthquake waves.

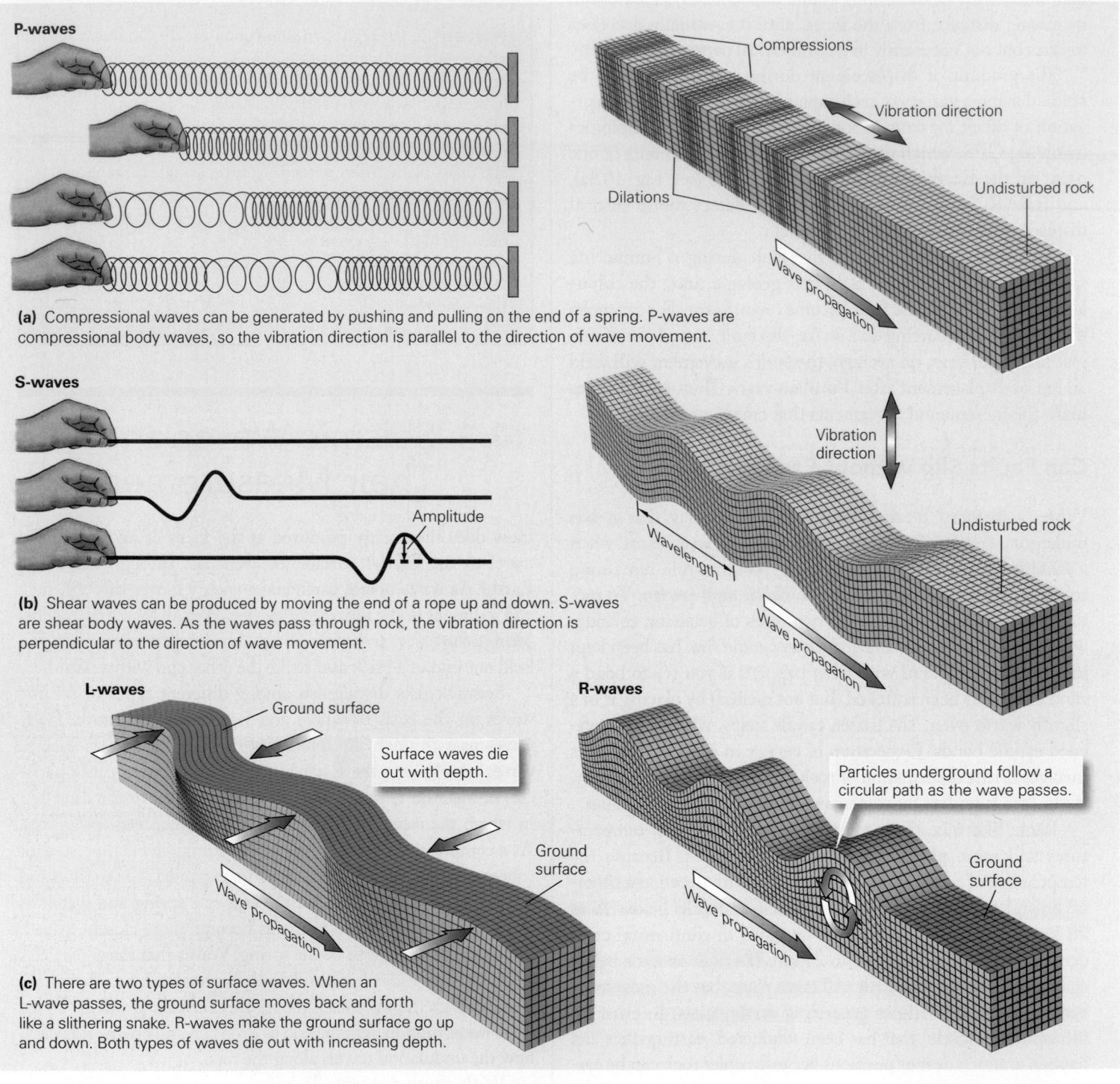

P-waves

(a) Compressional waves can be generated by pushing and pulling on the end of a spring. P-waves are compressional body waves, so the vibration direction is parallel to the direction of wave movement.

Compressions

Vibration direction

Dilations

Undisturbed rock

Wave propagation

S-waves

Amplitude

(b) Shear waves can be produced by moving the end of a rope up and down. S-waves are shear body waves. As the waves pass through rock, the vibration direction is perpendicular to the direction of wave movement.

Vibration direction

Wavelength

Undisturbed rock

Wave propagation

L-waves

Ground surface

Surface waves die out with depth.

Ground surface

Wave propagation

(c) There are two types of surface waves. When an L-wave passes, the ground surface moves back and forth like a slithering snake. R-waves make the ground surface go up and down. Both types of waves die out with increasing depth.

R-waves

Particles underground follow a circular path as the wave passes.

Ground surface

Wave propagation

- *R-waves* (R stands for Rayleigh, the name of a physicist) are surface waves that cause the ground to ripple up and down.

The different types of seismic waves travel at different velocities. P-waves travel the fastest. S-waves travel more slowly, at about 60% of the speed of P-waves. Surface waves are the slowest of all.

Friction absorbs energy as seismic waves pass through a material, and the waves bounce off layers and obstacles in the Earth, so the amount of energy carried by seismic waves decreases as they travel. In fact, even if people near the epicenter are thrown off their feet by a large earthquake, those 200 km away barely feel it. Similarly, an earthquake caused by slip on a fault deep in the crust causes less damage than one caused by slip on a fault near the surface.

FIGURE 10.11 Different types of seismometers.

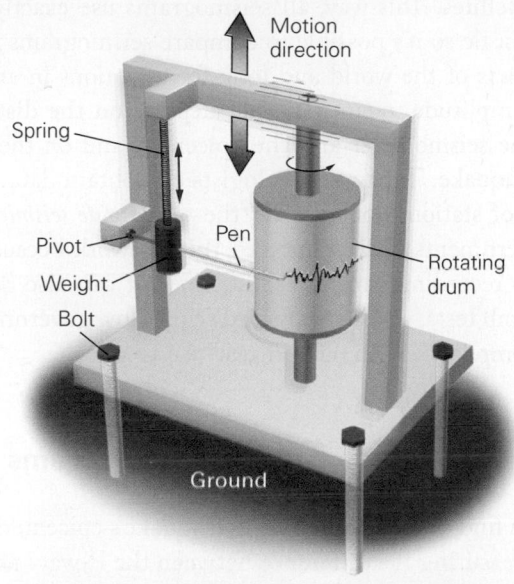

(a) A mechanical vertical-motion seismometer records up-and-down ground motion.

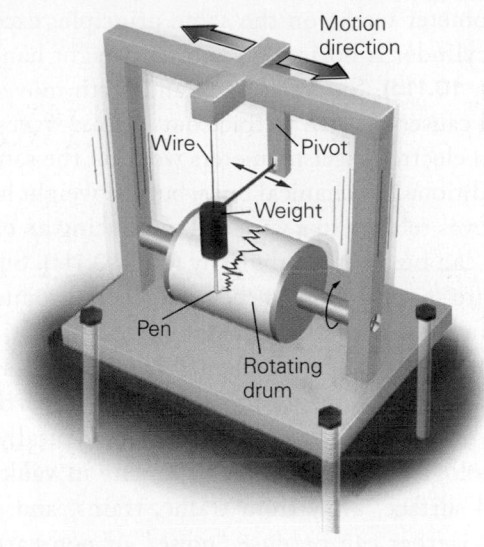

(b) A mechanical horizontal-motion seismometer records back-and-forth ground motion.

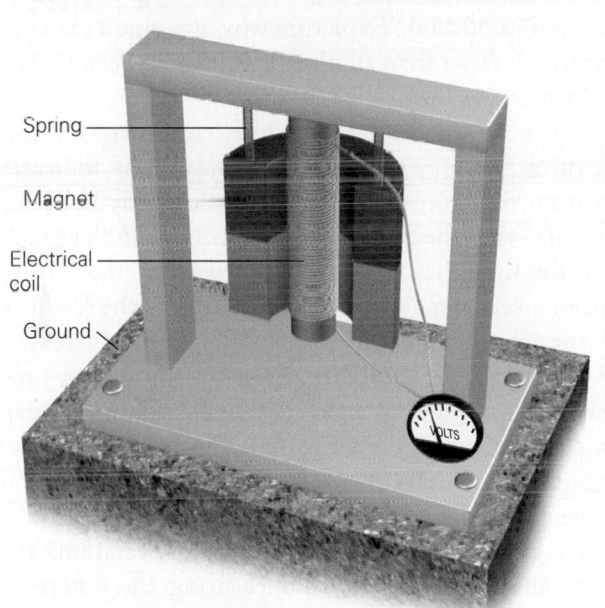

(c) In an electronic seismometer, a magnet moves relative to an electrical coil as shown in this simplified sketch (left). The voltage of the resulting electrical current represents the amount of motion. In reality, modern seismometers are complex inside (right).

Seismometers and the Record of an Earthquake

To detect and measure seismic waves, seismologists use a **seismometer**, an instrument that can detect even very slight movements of the Earth's surface. Seismologists use two basic configurations of seismometers, one for measuring vertical

(up-and-down) ground motion and the other for measuring horizontal (back-and-forth) ground motion.

A traditional, mechanical vertical-motion seismometer consists of a heavy weight (like a pendulum) suspended from a spring (**Fig. 10.11a**). The spring connects to a sturdy frame that has been bolted to the ground. A pen extends sideways from the weight and touches a vertical revolving cylinder of paper that has been connected to the seismometer frame. If the ground is steady, the pen traces out a straight reference line as the cylinder revolves. But when a seismic wave arrives and causes the ground surface to move up and down, it makes the seismometer frame and the attached paper cylinder move up and down as well. The weight, however, because of its *inertia* (the tendency of an object at rest to remain at rest), remains fixed in space. As the revolving paper roll moves up and down with respect to the fixed pen, the line traced by the pen on the paper deflects from the reference line. The deflection of the pen, therefore, represents the up-and-down movement of the ground. Note that if the paper cylinder did not revolve, the pen would simply move back and forth in the same place on the paper, but since the paper moves under the pen, the pen traces out curves that represent seismic waves. A mechanical horizontal-motion seismometer works on the same principle, except that the paper cylinder is horizontal and the weight hangs from a wire (Fig. 10.11b). Sideways back-and-forth movement of the ground causes the pen to trace out seismic waves.

Modern electronic seismometers work on the same principle as traditional mechanical ones, but the weight is a magnet that moves relative to a wire coil, producing an electrical signal that can be recorded digitally (Fig. 10.11c). Such seismometers are so sensitive that they can record ground movements of a millionth of a millimeter (only 2 to 10 times the diameter of an atom)—movements that people can't feel. This allows the instruments to measure large earthquakes happening on the other side of the planet. Typically, therefore, seismologists place these instruments in vaults below the ground surface, away from traffic, trains, and swaying trees that together can produce "noise," or non-earthquake vibrations (Fig. 10.12). The entire configuration comprises a *seismometer station*.

The deflections traced by a seismometer provide a record of an earthquake called a **seismogram** (Fig. 10.13). At first glance, a typical seismogram looks like a messy squiggle of lines, but to a seismologist it contains a wealth of information. The horizontal axis represents time, and the vertical axis represents the *amplitude* (the size) of the seismic waves. The instant at which a seismic wave appears at a seismometer station defines the *arrival time* of the wave. The first squiggles on the record represent P-waves because P-waves travel the fastest. Next come the S-waves, and finally the surface waves (R-waves and L-waves). Typically, the surface waves have the largest amplitude and arrive over a relatively long interval of time.

Seismologists the world over have agreed to certain standards for measuring earthquakes, and they calibrate time on their seismometers using very accurate time signals broadcast by GPS satellites. This way, all seismograms use exactly the same time scale so it's possible to compare seismograms from different parts of the world and look for variations in arrival time and amplitude, quantities that depend on the distance between the seismometer and the epicenter, and on the size of the earthquake. Today, seismologists can obtain data from thousands of stations constituting the *worldwide seismic network*. Governments have supported this network because in addition to recording natural earthquakes, it can also detect nuclear-bomb tests, so seismic records can allow governments to verify compliance with nuclear test-ban treaties.

Finding the Epicenter Using Seismograms

How do we find the location of an earthquake's epicenter? We begin by measuring the difference between the P-wave arrival time and the S-wave arrival time at seismometer stations. P-waves and S-waves pass through the interior of the Earth at different velocities, so the difference between P-wave and S-wave arrival times increases as the distance from the epicenter increases (Fig. 10.14a). To picture why, imagine a car race. If one car travels faster than another, the distance between the two cars increases as the race proceeds.

We can represent the time delay between P- and S-waves on a graph, where the horizontal axis indicates distance from the epicenter and the vertical axis indicates time (Fig. 10.14b). The red travel-time curves on the graph show that the time that it takes for the wave to move from its origin to a seismometer station increases as the distance between the epicenter and the seismometer station increases. Since P-waves travel faster than S-waves, the travel-time curve for an S-wave differs from that for a P-wave. The gap between the two curves (as measured along a vertical line) represents the time delay between the P-wave arrival and the S-wave arrival.

To use a graph of travel-time curves for determining the distance to an epicenter, start by measuring the difference between the P- and S-waves on your seismogram; this is called the *S – P time* (pronounced S minus P time). Then draw a line segment on a piece of tracing paper to represent this amount of time, at the scale used for the vertical axis of the graph. Orient the line segment parallel to the vertical time axis and move it back and forth until one end lies on the P-wave curve and the other end lies on the S-wave curve (this gives the S – P time). You have now identified the gap between the two travel-time curves for your seismic station. Extend the line down to the horizontal distance axis, and simply read off the distance to the epicenter.

The analysis of one seismogram tells you only the distance between the epicenter and the seismometer station—it does

FIGURE 10.12 Seismometers are bolted to bedrock in a protected shelter or vault.

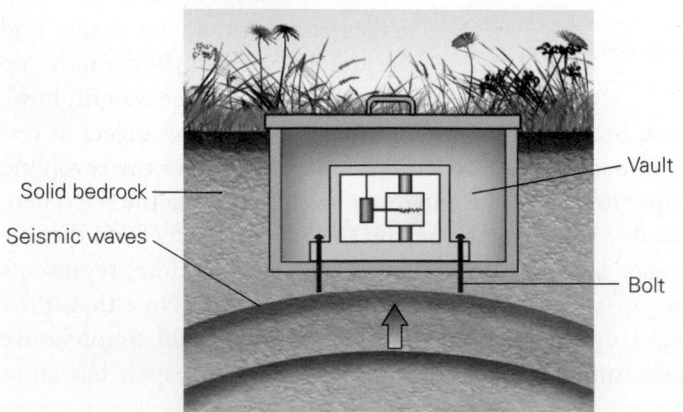

Solid bedrock

Seismic waves

Vault

Bolt

FIGURE 10.13 The nature of seismograms.

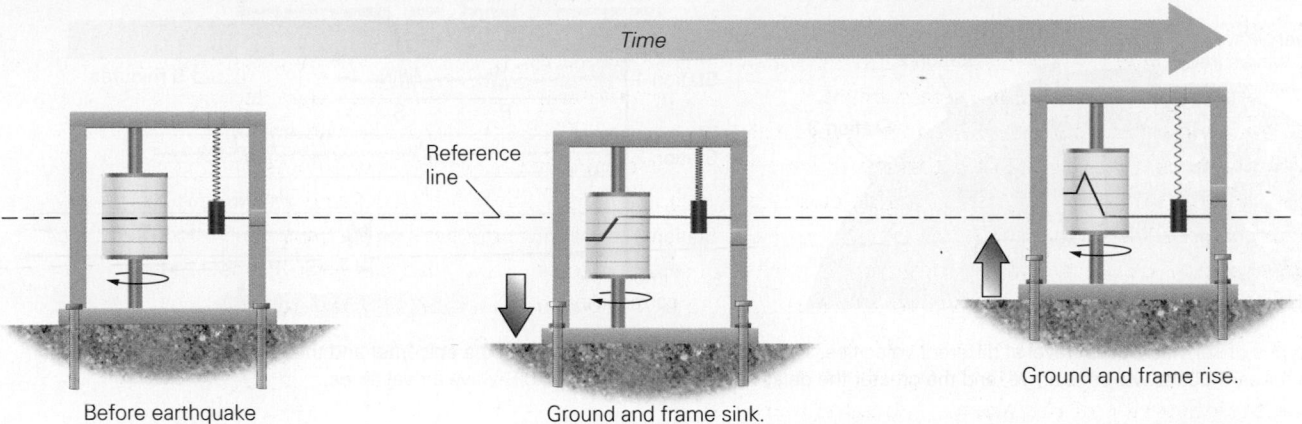

(a) Before an earthquake, the pen traces a straight line. During an earthquake, the paper cylinder moves up and down while the pen stays in place.

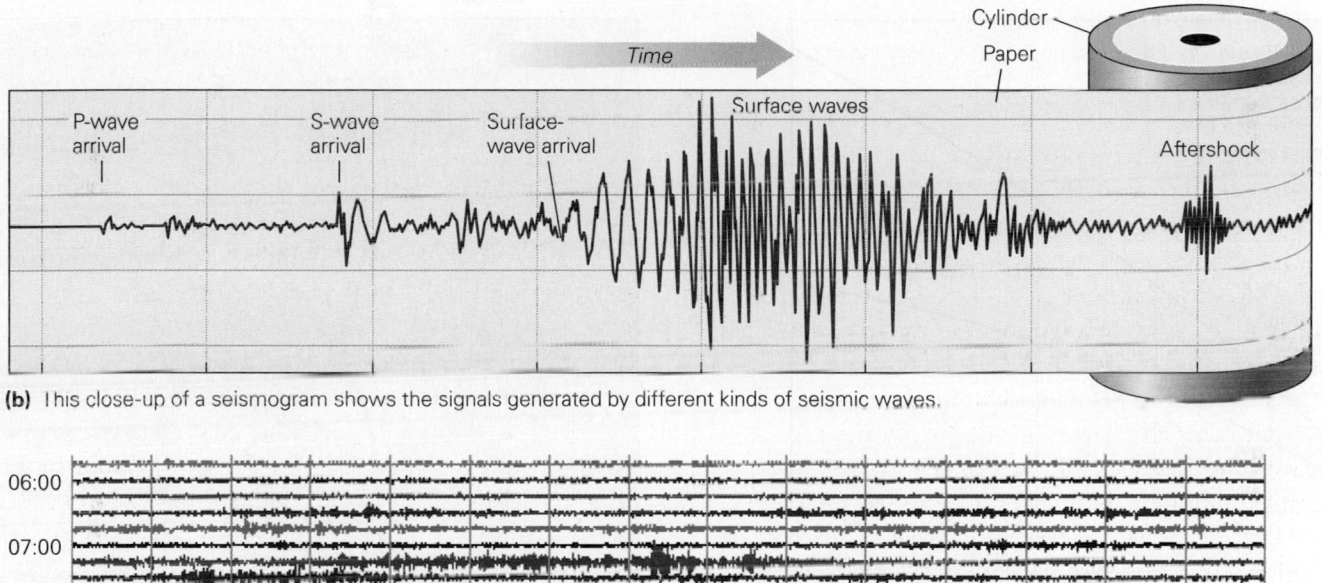

(b) This close-up of a seismogram shows the signals generated by different kinds of seismic waves.

(c) A digital seismic record from a seismometer station in Arkansas. The space between vertical lines represents 1 minute. Colors have no meaning but make the figure more readable. Each color band represents the record of 15 minutes.

not tell you in which direction from the station the epicenter lies. To determine the map position of the epicenter, we *triangulate*, meaning we plot the distance from the epicenter to three stations. For example, imagine that you determine that the epicenter lies 2,000 km from Station 1, 4,000 km from Station 2, and 6,000 km from Station 3. On a map, you can draw a circle around each station such that the radius of the circle represents the distance between the station and the epicenter, at the scale of the map. The epicenter lies at the intersection of the three circles, for this is the only point at which the epicenter has the appropriate measured distance from all three stations (**Fig. 10.14c**).

TAKE-HOME MESSAGE

Earthquake energy travels as seismic waves. Body waves travel through the interior of the Earth, whereas surface waves travel along its surface. Ground shaking due to seismic waves generally decreases with distance from the focus. We can detect an earthquake using a seismometer. A seismogram is the record of an earthquake.

QUICK QUESTION: How can you determine the location of an earthquake's epicenter?

FIGURE 10.14 The method for locating an earthquake epicenter.

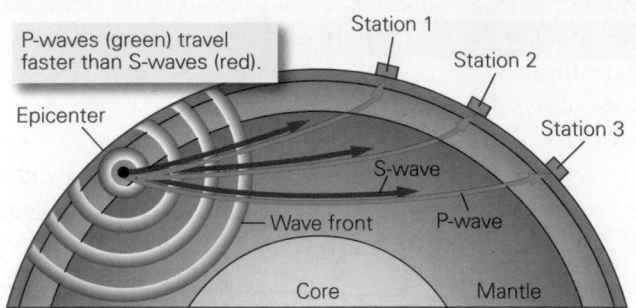

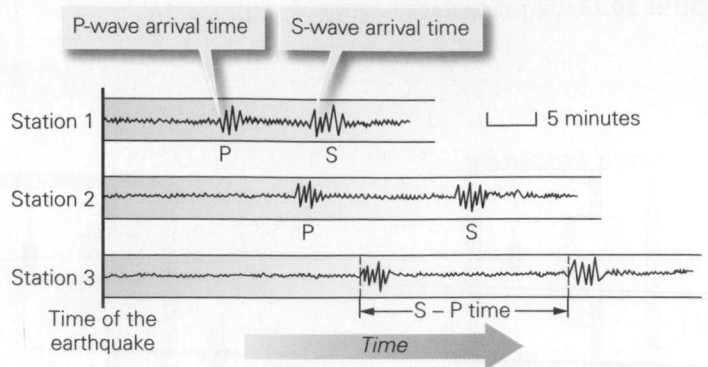

(a) Different types of seismic waves travel at different velocities. The greater the distance between the epicenter and the seismometer station, the longer it takes for earthquake waves to arrive, and the greater the delay between the P-wave and S-wave arrival times.

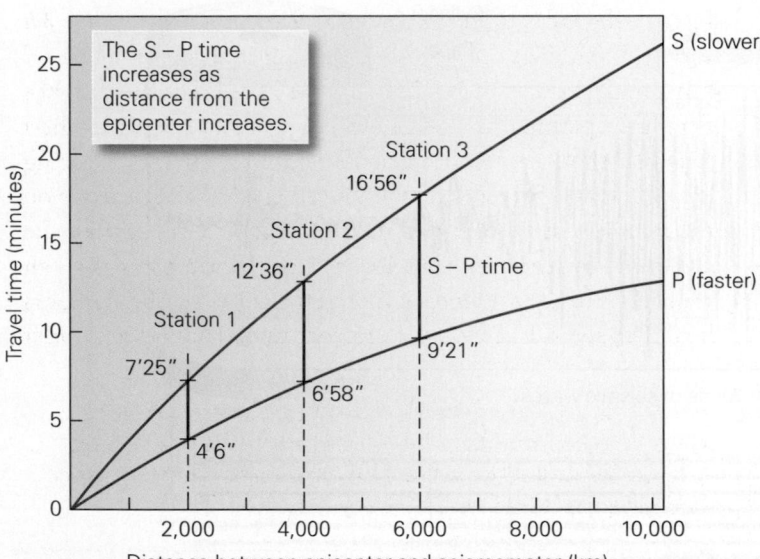

(b) We can represent the different arrival times of P-waves and S-waves on a graph of travel-time curves.

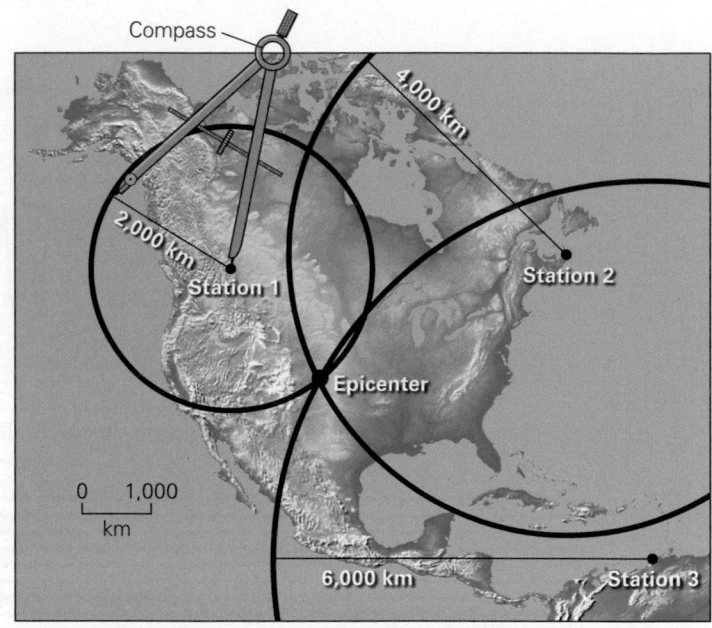

(c) If an earthquake epicenter lies 2,000 km from Station 1, we draw a circle with a radius of 2,000 km around the station at the scale of the map. We do the same for the other two stations. The intersection of the three circles is the epicenter.

10.4 Defining the "Size" of Earthquakes

Some earthquakes shake the ground violently, whereas others can barely be felt—in other words, some earthquakes are bigger than others. Seismologists have developed two scales to define "size" in a meaningful way so that they can systematically describe and compare earthquakes. The first scale, called the Mercalli Intensity Scale, depends on human perception of ground shaking and the damage resulting from it at a given locality. The second scale, called the magnitude scale, focuses on the measured amount of ground motion, as recorded by a seismometer at a specified distance from the epicenter.

Mercalli Intensity Scale

The **intensity** of a given earthquake refers to the effect or consequence of its ground shaking at a locality on the Earth's surface. In 1902, an Italian scientist named Giuseppe Mercalli devised a scale for defining intensity by systematically assessing human perception of shaking and of the damage that an earthquake caused. A version of this scale, called the

TABLE 10.2 Modified Mercalli Intensity Scale

MMI	Destructiveness (Perceptions of the Extent of Shaking and Damage)
I	Detected only by seismic instruments; causes no damage.
II	Felt by a few stationary people, especially in upper floors of buildings; suspended objects, such as lamps, may swing.
III	Felt indoors; standing automobiles sway on their suspensions; it seems as though a heavy truck is passing.
IV	Shaking awakens some sleepers; dishes and windows rattle.
V	Most people awaken; some dishes and windows break, unstable objects tip over; trees and poles sway.
VI	Shaking frightens some people; plaster walls crack, heavy furniture moves slightly, and a few chimneys crack, but overall little damage occurs.
VII	Most people are frightened and run outside; a lot of plaster cracks, windows break, some chimneys topple, and unstable furniture overturns; poorly built buildings sustain considerable damage.
VIII	Many chimneys and factory smokestacks topple; heavy furniture overturns; substantial buildings sustain some damage, and poorly built buildings suffer severe damage.
IX	Frame buildings separate from their foundations; most buildings sustain damage, and some buildings collapse; the ground cracks, underground pipes break, and rails bend; some landslides occur.
X	Most masonry structures and some well-built wooden structures are destroyed; the ground severely cracks in places; many landslides occur along steep slopes; some bridges collapse; some sediment liquefies; concrete dams may crack; facades on many buildings collapse; railways and roads suffer severe damage.
XI	Few masonry buildings remain standing; many bridges collapse; broad fissures form in the ground; most pipelines break; severe liquefaction of sediment occurs; some dams collapse; facades on most buildings collapse or are severely damaged.
XII	Earthquake waves cause visible undulations of the ground surface; objects are thrown up off the ground; there is complete destruction of buildings and bridges of all types.

Modified Mercalli Intensity (MMI) scale, continues to be used today **(Table 10.2).** The MMI scale uses Roman numerals ranging from I (low intensity) to XII (extremely high intensity). To use the scale, seismologists visit the affected region to interview residents and observe the damage. For example, if

an earthquake at a location causes everyone to notice shaking but does not damage buildings, then the earthquake had an intensity of III at that location. In contrast, if ground shaking feels significant and destroys poorly constructed buildings, but causes only minor damage to substantial buildings, then the earthquake had an intensity of VIII.

Note that the specification of earthquake intensity depends on a subjective assessment of perception and damage, not on a direct measurement with an instrument. Furthermore, the MMI number varies with location for a particular seismic event—we cannot assign a single MMI number to a given earthquake. Typically, the intensity decreases progressively away from the epicenter. To illustrate how intensity varies over a region for a particular earthquake, seismologists draw lines, called contours, which delimit zones where the earthquake had a given intensity **(Fig. 10.15).** Generally, seismologists refer to the area within the intensity II contour line as the *felt area*. Of note, not only is the maximum intensity of a "large" earthquake greater than that of a "small" one, but the distance from the epicenter to a given contour tends to be greater for a larger earthquake than for a smaller one. In detail, the spacing between intensity contours for a given earthquake depends on the strength of the crust in the region where the earthquake occurred. In regions underlain by strong crust, the crust can transmit earthquake energy more efficiently, so the contours lie father apart. In contrast, in regions underlain by weak crust

FIGURE 10.15 This map shows Modied Mercalli Intensity contours for the 1886 Charleston, South Carolina, earthquake. Note that near the epicenter, ground shaking reached MMI of X, but in New York City, ground shaking reached MMI of II to III.

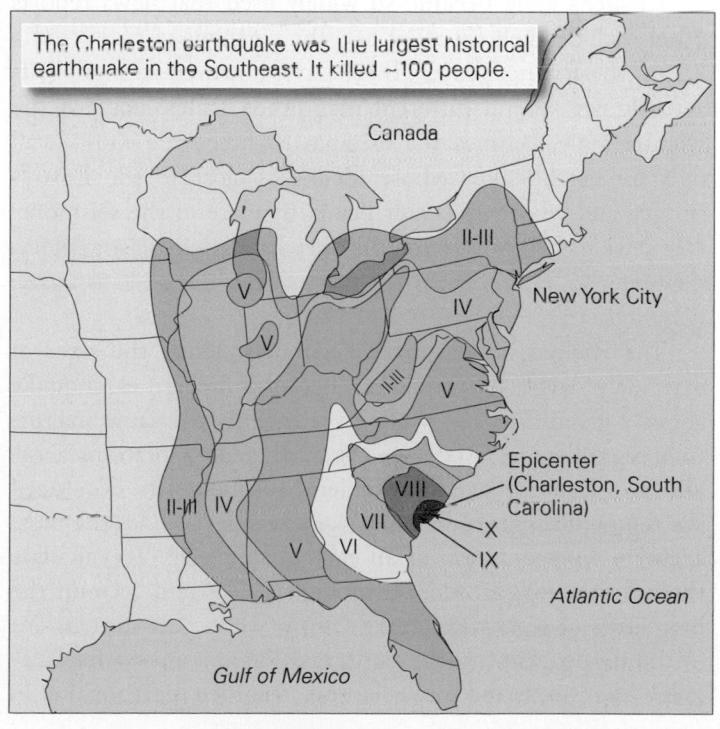

(where the crust contains lots of fractures and/or consists of soft rock), the contours lie closer together, so the earthquake affects a smaller area.

Earthquake Magnitude Scales

When you read a report of an earthquake disaster in the news, you will probably come across a phrase that states something like this: "An earthquake with a magnitude of 7.2 struck the city yesterday at noon." What does this mean? The **magnitude** of an earthquake indicates the maximum amplitude of ground motion recorded by a seismometer at a specific, standard distance from the focus. By "amplitude of ground motion," we mean the amount of up-and-down or back-and-forth motion of the ground. The larger the ground motion, the greater the deflection of a seismometer pen tracing out a seismogram. Because the magnitude does not depend on distance, use of the magnitude scale allows seismologists to define the size of an earthquake objectively.

An American seismologist, Charles Richter, developed a method for defining and measuring earthquake magnitude. The scale he proposed, in 1935, came to be known as the **Richter scale**. It indicates the maximum amplitude of motion as it would be recorded at a seismic station 100 km from the epicenter. Because the amount of deflection on a seismogram depends on the distance between the seismometer and the epicenter, and since most seismic stations do not happen to lie exactly 100 km from the epicenter, seismologists use a chart to adjust for distance of the station from the epicenter when calculating magnitude **(Fig. 10.16)**.

Richter's scale became so widely used that news reports often include such wording as "The earthquake registered a 7.2 on the Richter scale." These days, however, seismologists actually use several different magnitude scales, not just the Richter scale, because the original Richter scale works well only for earthquakes whose focus lies close to the Earth's surface and whose epicenter lies fairly close to the seismometer station. Because of the distance limitation, seismologists now refer to a number on the original Richter scale as a *local magnitude* (M_L).

The Richter scale cannot accurately define the sizes of extremely large earthquakes, because for an earthquake above a given size, the scale gives roughly the same magnitude regardless of how large the earthquake vibrations actually are. Because of this problem, seismologists developed the **moment magnitude scale** (M_W), which provides the most accurate representation of an earthquake's size. To calculate the moment magnitude, seismologists take into account the amplitude of several different seismic waves, the dimensions of the slipped area on the fault, and the amount of displacement that occurred. The largest recorded earthquake in

FIGURE 10.16 Using the Richter magnitude scale.

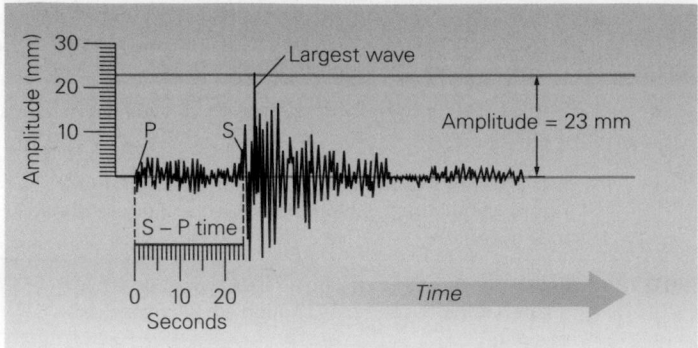

(a) To calculate the Richter magnitude from a seismogram, first measure the S – P time to determine the distance to the epicenter. Then measure the amplitude of the largest wave.

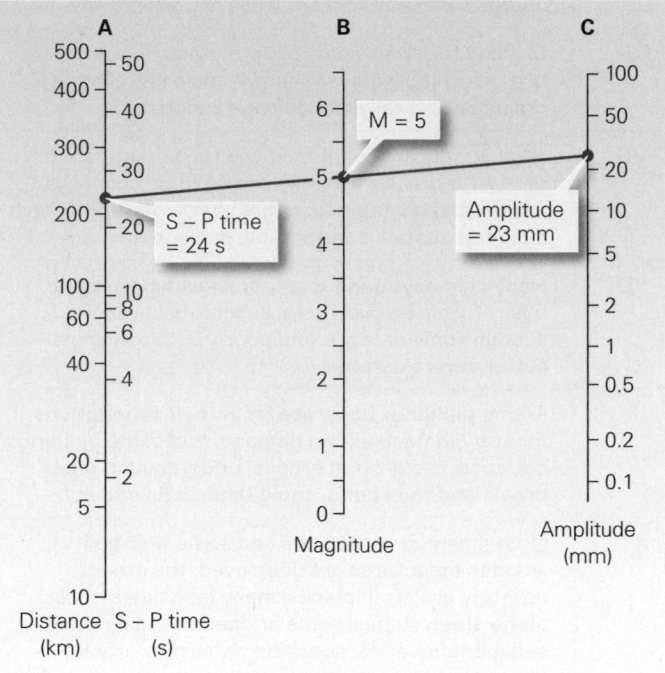

(b) Draw a line from the point on Column A representing the S – P time or the distance to the epicenter, to the point on Column C representing the wave amplitude. Read the Richter magnitude from Column B.

history, the great 1960 Chilean quake, registered as an 8.5 on the M_L scale, but as a 9.5 on the M_W scale. The larger number makes more sense because this earthquake was indeed much larger than other known events for which M_L = 8.5. Of note, the catastrophic 2011 Tōhoku earthquake had a magnitude of M_W = 9.0.

What magnitude do modern news reports of earthquakes provide? For early reports, seismologists report a preliminary magnitude, typically an M_L, which can be calculated quickly. Later on, after they have had the chance to collect the necessary data, seismologists report an M_W, which becomes the number that appears in archival records.

All magnitude scales are logarithmic, meaning that an increase of one integer of magnitude represents a 10-fold increase in the maximum ground motion. Therefore, an M_W 8 earthquake results in ground motion that is 10 times greater than that of an M_W 7 earthquake, and 1,000 times greater than that of an M_W 5 earthquake. To make discussion easier, seismologists use familiar adjectives to describe the size of an earthquake, as listed in **Table 10.3**. Note that we can roughly correlate magnitude with the intensity at the epicenter for earthquakes that take place in the upper continental crust.

Energy Release by Earthquakes

As we've pointed out, an earthquake releases energy. Seismologists have compared the energy released by earthquakes with that released by other events. According to some estimates, an M_W 5.3 earthquake releases about as much energy as the

FIGURE 10.17 Measuring the energy released by earthquakes.

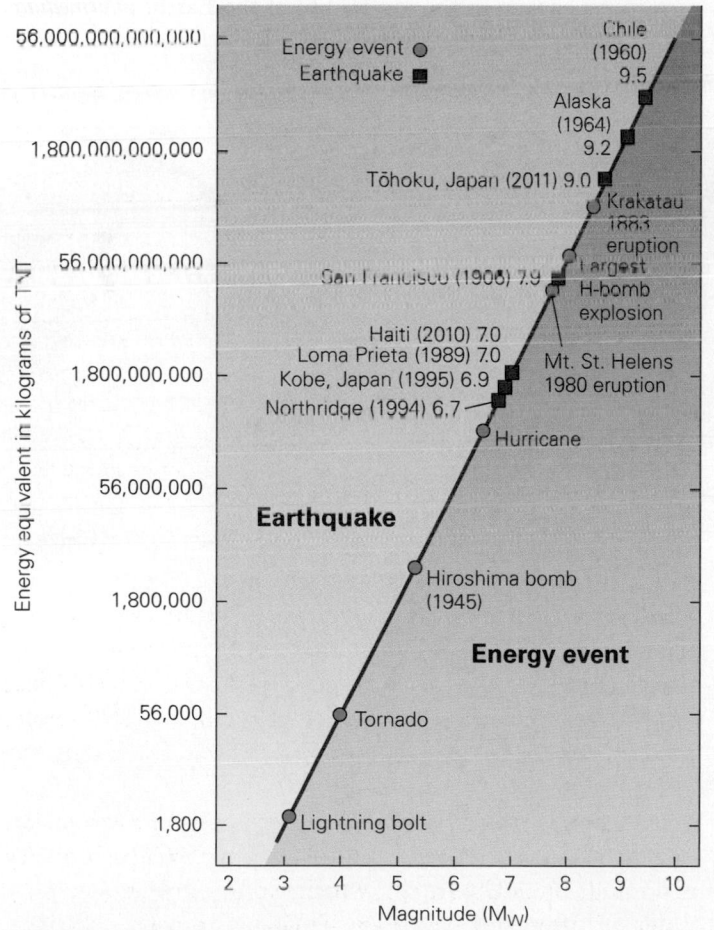

(a) Energy released by earthquakes increases dramatically with magnitude. Great earthquakes release vastly more energy than the largest bombs.

TABLE 10.3 Adjectives for Describing Earthquakes

Adjective	Magnitude	Approximate maximum intensity at epicenter*	Effects
Great	>8.0	X to XII	Major to total destruction
Major	7.0 to 7.9	IX to X	Great damage
Strong	6.0 to 6.9	VII to VIII	Moderate to serious damage
Moderate	5.0 to 5.9	VI to VII	Slight to moderate damage
Light	4.0 to 4.9	IV to V	Felt by most; slight damage
Minor	<3.9	III or smaller	Felt by some; hardly any damage

*For upper-crustal earthquakes in continents.

Hiroshima atomic bomb, and an M_W 9 earthquake releases significantly more energy than the largest hydrogen bomb ever detonated, and even more than the explosive eruption of Krakatau (see Box 9.1). Notably, an increase in magnitude by one integer represents approximately a 32-fold increase in energy. Therefore, an M_W 8 earthquake releases about 1 million times more energy than an M_W 4 earthquake **(Fig. 10.17a)**. In fact, a single M_W 8.9 earthquake releases as much

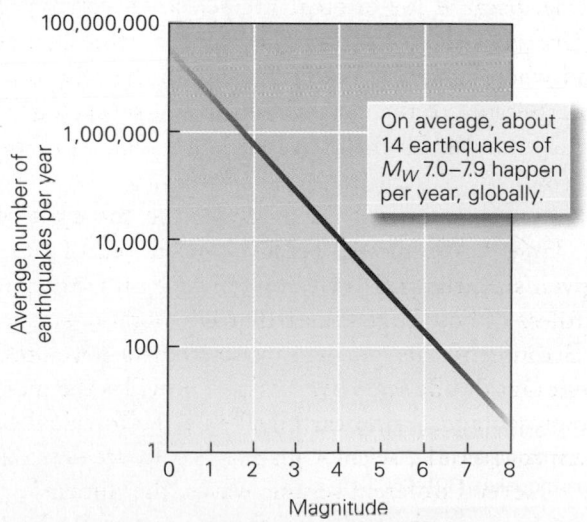

(b) The number of earthquakes per year of a given magnitude decreases with increasing magnitude.

energy as the entire average global annual release of seismic energy coming from all other earthquakes combined! Fortunately, such large earthquakes occur much less frequently than small earthquakes. About 100,000 M_W 3 earthquakes happen every year, but an M_W 8 earthquake happens only about once or twice a year **(Fig. 10.17b)**.

TAKE-HOME MESSAGE

We can specify earthquake size by intensity (based on the perception of shaking and damage caused) or by magnitude (based on the measurements of ground motion on a seismogram). An increase of one magnitude number represents a 10-fold increase in shaking and a 32-fold increase in energy release. A large earthquake releases more energy than the largest hydrogen bomb.

QUICK QUESTION: Can you specify the "size" of an earthquake by giving just one intensity number? How about a single magnitude number?

10.5 Where and Why Do Earthquakes Occur?

Earthquakes do not take place everywhere on the globe. By plotting the distribution of earthquake epicenters on a map, seismologists find that most, but not all, earthquakes

Did you ever wonder . . .
if an earthquake might happen near where you live?

occur in elongate **seismic belts**, or localized *seismic zones*. Most seismic belts coincide with plate boundaries, so we can refer to earthquakes within these belts as *plate-boundary earthquakes*. Earthquakes that happen away from plate boundaries are called *intraplate earthquakes*; the prefix *intra-* means within **(Fig. 10.18)**. Eighty percent of the earthquake energy released on the Earth comes from the plate-boundary earthquakes in the seismic belts surrounding the Pacific Ocean.

At what depths do we find earthquake foci? Seismologists distinguish three classes of earthquakes based on depth: *shallow earthquakes* happen in the top 60 km of the Earth, *intermediate*

FIGURE 10.18 A map of epicenters emphasizes that most earthquakes occur in distinct belts along plate boundaries.

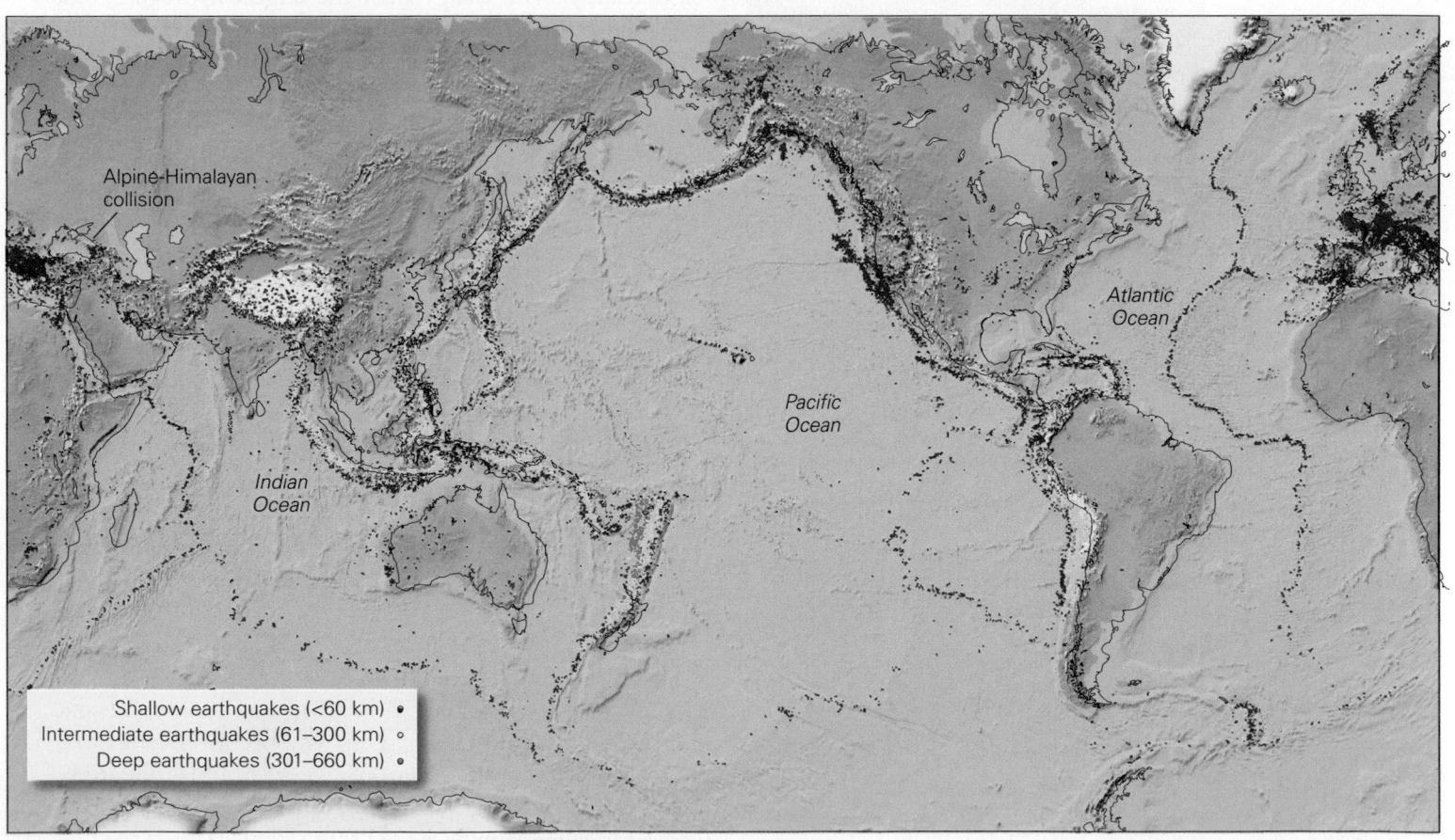

Alpine-Himalayan collision

Atlantic Ocean

Pacific Ocean

Indian Ocean

Shallow earthquakes (<60 km) •
Intermediate earthquakes (61–300 km) ∘
Deep earthquakes (301–660 km) ∘

earthquakes take place between 60 and 300 km, and *deep earthquakes* occur down to a depth of about 660 km. Earthquake foci do not lie below a depth of about 660 km. Let's look at the characteristics of earthquakes in various geologic settings and learn why earthquakes take place where they do.

Earthquakes at Plate Boundaries

The majority of earthquakes happen on faults along plate boundaries because the relative motion of plates triggers slip on faults. We find different kinds of faulting at different types of plate boundaries.

Divergent-Boundary Seismicity At divergent boundaries (mid-ocean ridges), two oceanic plates form and move apart. Divergent boundaries consist of spreading segments linked by transform faults. Therefore, two kinds of faults develop at divergent boundaries. Along spreading segments, stretching generates normal faults, whereas along transform faults, strike-slip displacement occurs **(Fig. 10.19)**. All earthquakes along mid-ocean ridges have foci at depths of less than 25 km, so we classify them as shallow earthquakes. Since most ridges lie out in the ocean, far away from settled areas, mid-ocean ridge earthquakes don't cause damage. Only a few populated localities (such as Iceland) lie astride divergent boundaries.

FIGURE 10.19 The distribution of earthquakes at a mid-ocean ridge. Note that normal faults occur along the ridge axis, and strike-slip faults occur along active transform faults. Earthquakes don't take place along inactive fracture zones.

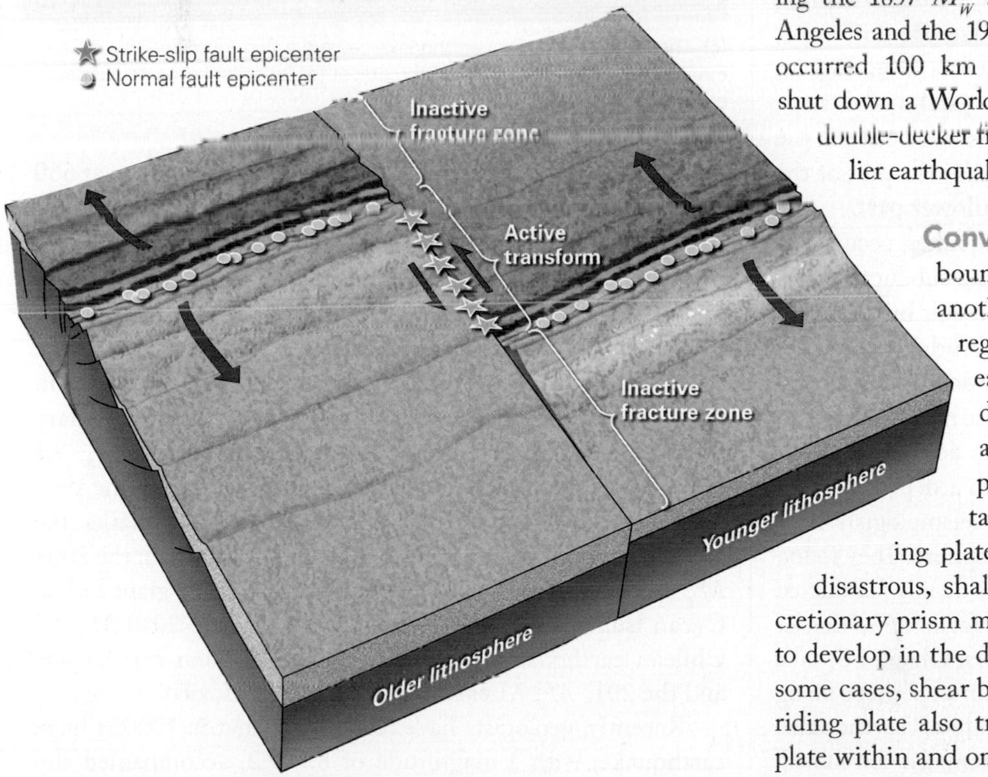

Strike-slip fault epicenter
Normal fault epicenter

Inactive fracture zone
Active transform
Inactive fracture zone
Younger lithosphere
Older lithosphere

Transform-Boundary Seismicity At transform boundaries, where one plate slides past another without the production or consumption of oceanic lithosphere, most faulting results in strike-slip motion. The majority of transform faults in the world link segments of mid-ocean ridges, as we've just discussed. But a few, such as the San Andreas fault of California, the Alpine fault of New Zealand, and the Anatolian fault in Turkey, cut across continental lithosphere. All transform-fault earthquakes have a shallow focus, so the larger ones on land can cause immense damage.

The San Francisco earthquake of 1906 serves as an example of a continental transform-fault earthquake **(Fig. 10.20a)**. In the wake of the gold rush, San Francisco was a booming city with broad streets and numerous large buildings. But it was built on the transform boundary along which the Pacific Plate moves north at an average rate of 6 cm per year relative to North America. Because of the stick-slip behavior of the fault, this movement tends to happen in sudden jerks, each of which causes an earthquake. At 5:12 A.M. on April 18, the fault near San Francisco slipped by as much as 7 m, and minutes later seismic waves struck the city. Witnesses watched in horror as the streets undulated, buildings swayed and banged together, laundry lines stretched and snapped, and weaker buildings collapsed. Fire followed soon after, consuming huge areas of the city **(Fig. 10.20b)**. Judging from the damage, seismologists estimate that the earthquake had an M_W of 7.9.

The San Francisco earthquake has not been the only one to strike along the San Andreas and nearby related faults during historic time. Over a dozen major earthquakes have happened on these faults during the past two centuries, including the 1857 M_W 7.7 Tejon Pass earthquake just east of Los Angeles and the 1989 M_W 7.1 Loma Prieta earthquake, which occurred 100 km south of San Francisco, but nevertheless shut down a World Series game and caused the collapse of a double-decker freeway in that city **(Fig. 10.20c)**. Even deadlier earthquakes have happened on other transform faults.

Convergent-Boundary Seismicity Such boundaries, where one plate subducts under another, tend to be geologically complicated regions at which several different kinds of earthquakes take place. Specifically, as the downgoing plate begins to subduct, it bends and scrapes along the base of the overriding plate. Large thrust faults define the contact between the downgoing and overriding plates, and shear on these faults can produce disastrous, shallow earthquakes. Thrust faults in the accretionary prism may also slip. Bending causes normal faults to develop in the downgoing plate, seaward of the trench. In some cases, shear between the downgoing plate and the overriding plate also triggers shallow faulting in the overriding plate within and on both sides of the volcanic arc.

FIGURE 10.20 The San Andreas fault system, a continental transform boundary.

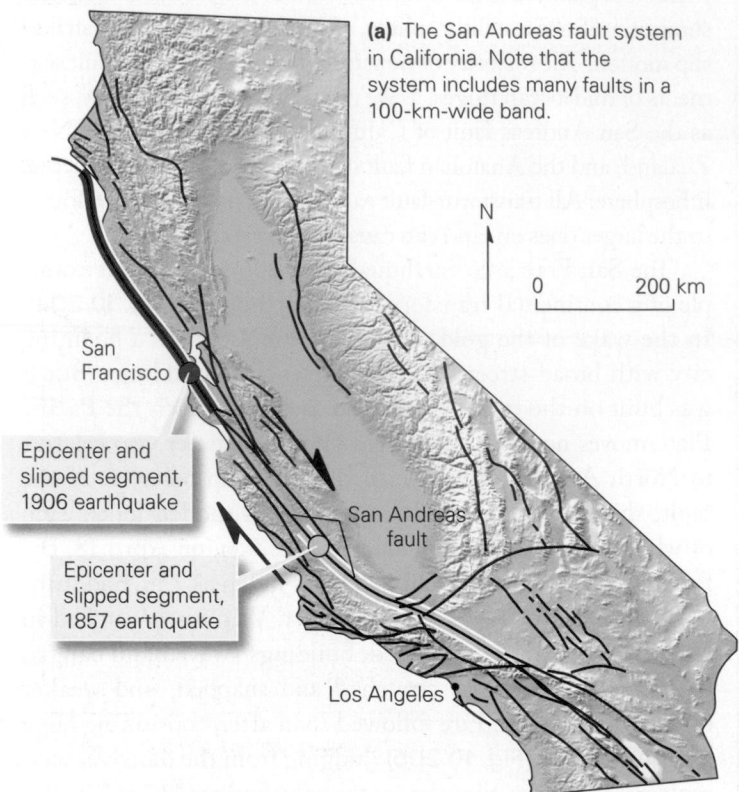

(a) The San Andreas fault system in California. Note that the system includes many faults in a 100-km-wide band.

San Francisco

Epicenter and slipped segment, 1906 earthquake

Epicenter and slipped segment, 1857 earthquake

San Andreas fault

Los Angeles

N

0   200 km

(b) A street in San Francisco after the 1906 earthquake. The event caused fires to sweep through the city, adding to the damage and devastation.

(c) The 1989 Loma Prieta earthquake caused a two-level freeway in San Francisco, 100 km northwest of the epicenter, to collapse, as its support columns gave way.

In contrast to other types of plate boundaries, where only shallow earthquakes occur, convergent boundaries also host intermediate and deep earthquakes. These earthquakes occur in the downgoing slab as it sinks into the mantle, defining a sloping band of seismicity called a **Wadati-Benioff zone**, after the seismologists who first recognized it (**Fig. 10.21a**; see also Fig. 4.11c). Intermediate and deep earthquakes happen in response to stresses caused partly by shear between the downgoing plate and the mantle and partly by the pull of the sinking, deeper part of the plate on the shallower part.

Why do intermediate and deep earthquakes occur in a Wadati-Benioff zone? Shouldn't the rock of a subducted plate at these depths be too warm and soft to break brittlely? To answer these questions, seismologists have studied the rate at which a subducting slab warms up as it sinks down through hot asthenosphere. They determined that because rock serves as an insulator, the interior of a downgoing plate actually remains cool enough to fracture seismically down to a depth of about 300 km. To explain deeper earthquakes, seismologists have studied the stability of minerals in the lithosphere. They found that at the extreme pressures developed in deeply subducted lithosphere, certain minerals collapse to form new, denser minerals. As such sudden phase changes (see Chapter 8) take place, the minerals abruptly decrease in volume, perhaps along a fault-like band. This process could generate an earthquake, but the exact process by which this occurs remains the subject of research. Some of a subducted plate may accumulate at 660 km, whereas some sinks still deeper. But at depths greater than 660 km, processes that generate earthquakes in a subducted plate can no longer take place.

Earthquakes in southern Mexico, in southern Alaska, in eastern Japan, on the western coast of South America, on the coast of Oregon and Washington, and along island arcs in the western Pacific serve as examples of convergent-boundary earthquakes. Notable examples include the 1960 M_W 9.5 earthquake in Chile, the largest earthquake on record; the 1964 M_W 9.2 Good Friday earthquake near Anchorage, Alaska; the 1995 M_W 6.9 earthquake in Kobe, Japan (**Fig. 10.21b**); the 2004 M_W 9.3 Sumatra earthquake, which triggered the giant Indian Ocean tsunami that killed 230,000 people; the 2010 M_W 8.8 Chilean earthquake; the 2011 M_W 9.0 Tōhoku earthquake; and the 2017 M_W 7.1 earthquake southern Mexico.

Recently, geologists have recognized that in 1700, a huge earthquake, with a magnitude of 8.7–9.2, accompanied slip

FIGURE 10.21 Convergent-boundary earthquakes.

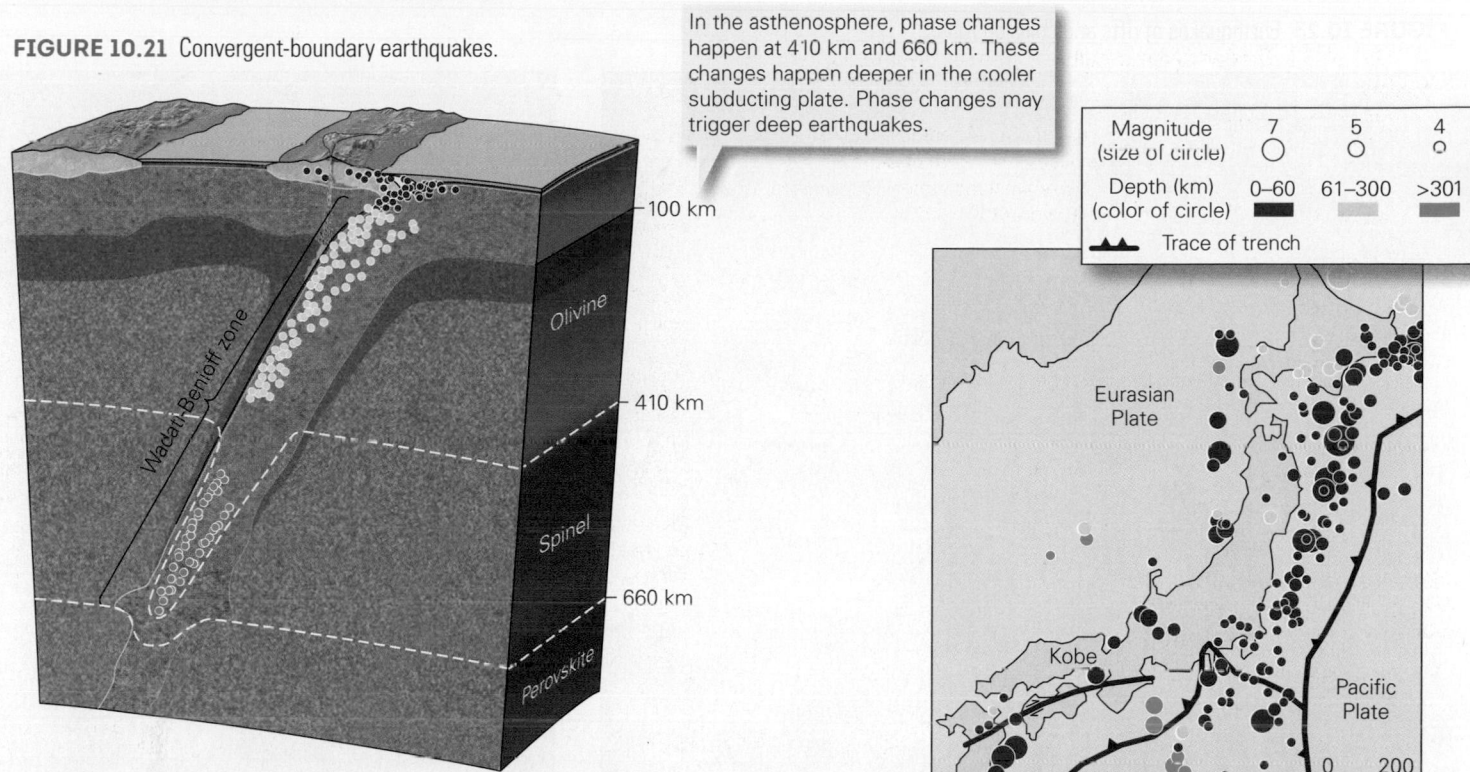

In the asthenosphere, phase changes happen at 410 km and 660 km. These changes happen deeper in the cooler subducting plate. Phase changes may trigger deep earthquakes.

Magnitude (size of circle)	7 ◯	5 ◯	4 ◦
Depth (km) (color of circle)	0–60 ■	61–300 ▨	>301 ▨
◣◣ Trace of trench			

(a) At a convergent boundary, earthquakes occur along the contact between the two plates, as well as in the downgoing plate. In the asthenosphere, phase changes happen at 410 km and at 660 km, but occur deeper in the cooler subducting plate. The changes may cause deep earthquakes.

(b) Map of earthquake epicenters and subduction zones in and near Japan.

on the *Cascadia subduction zone* off the now densely populated coast of Oregon and Washington. An earthquake of similar size today would be disastrous to the region. GPS measurements have detected movement of the crust of the region, indicating the buildup of stress **(Fig. 10.22)**.

Earthquakes Due to Continental Rifting and Collision

Continental Rifts The stretching of continental crust at continental rifts generates normal faults. Active rifts today

FIGURE 10.22 GPS measurements made as part of the EarthScope research program reveal that the crust is actively shortening in western Oregon and Washington due to subduction. The last earthquake to occur because of this deformation was in 1700; it was an M_w = 8.7 to 9.2 (i.e., a great earthquake) and caused a major tsunami. The observation of such movement leads to concern that another earthquake may happen again in the future.

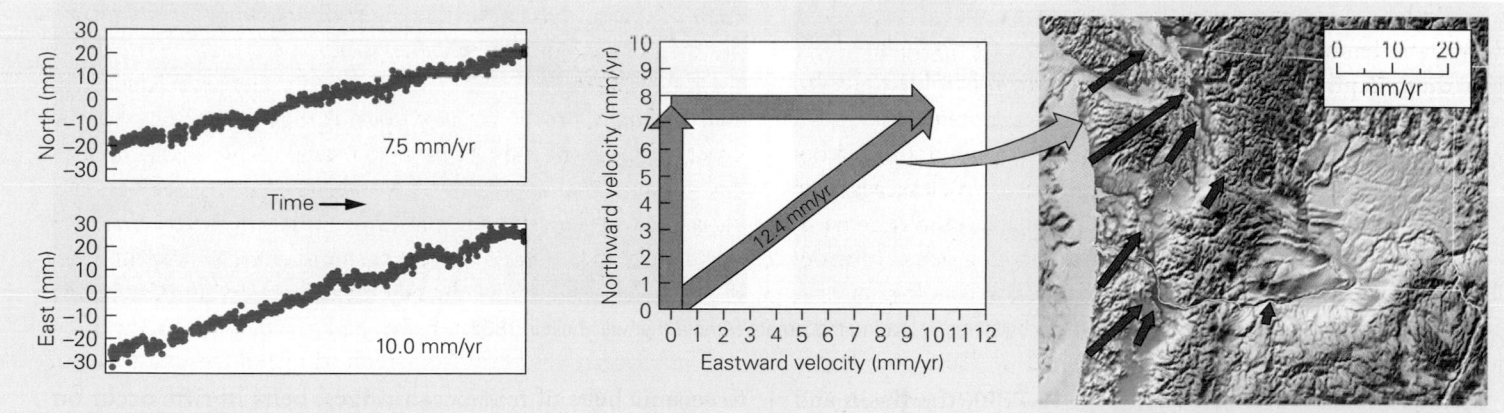

(a) The position of a survey point at Neah Bay, Washington, changes over time, as indicated by GPS measurements. The top left graph gives the northward component, and the bottom left graph gives the eastward component of its movement. Adding these components (graph on the right) indicates that the overall movement is 12.4 mm per year to the northeast.

(b) Each arrow indicates the velocity of the point at the rear end of the arrow. Note that the velocity decreases progressively eastward.

FIGURE 10.23 Earthquakes of rifts and collision zones.

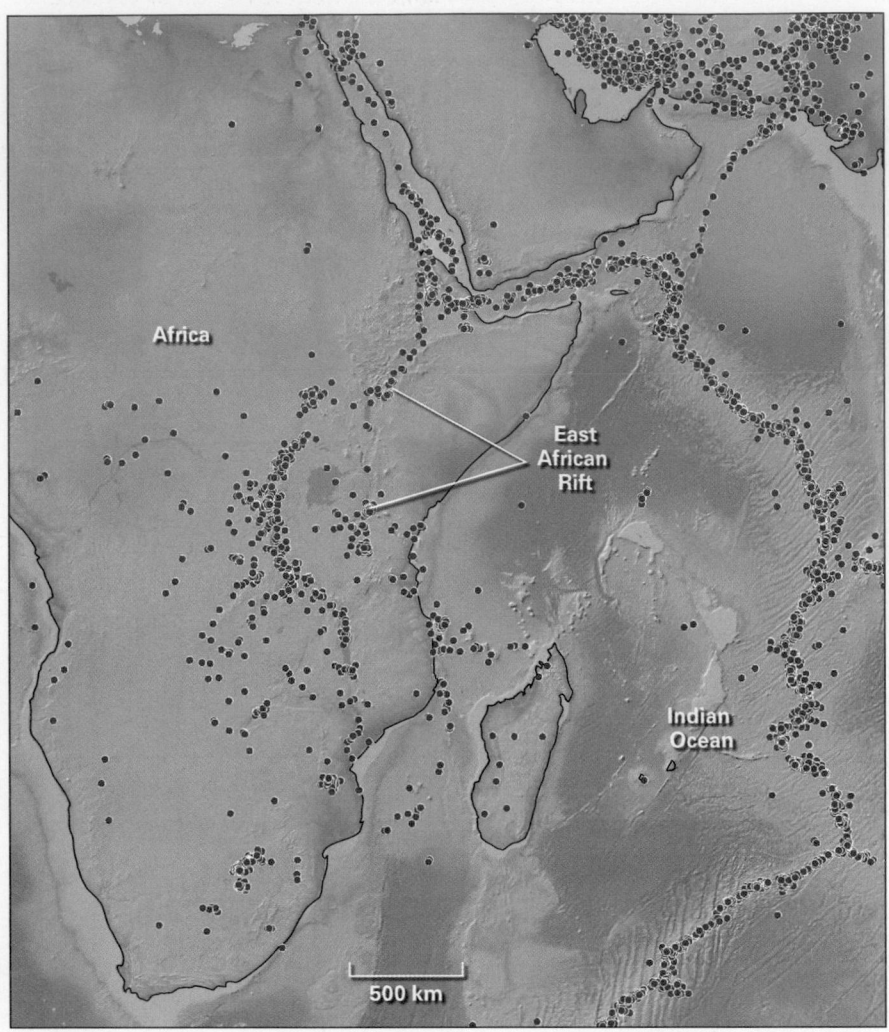

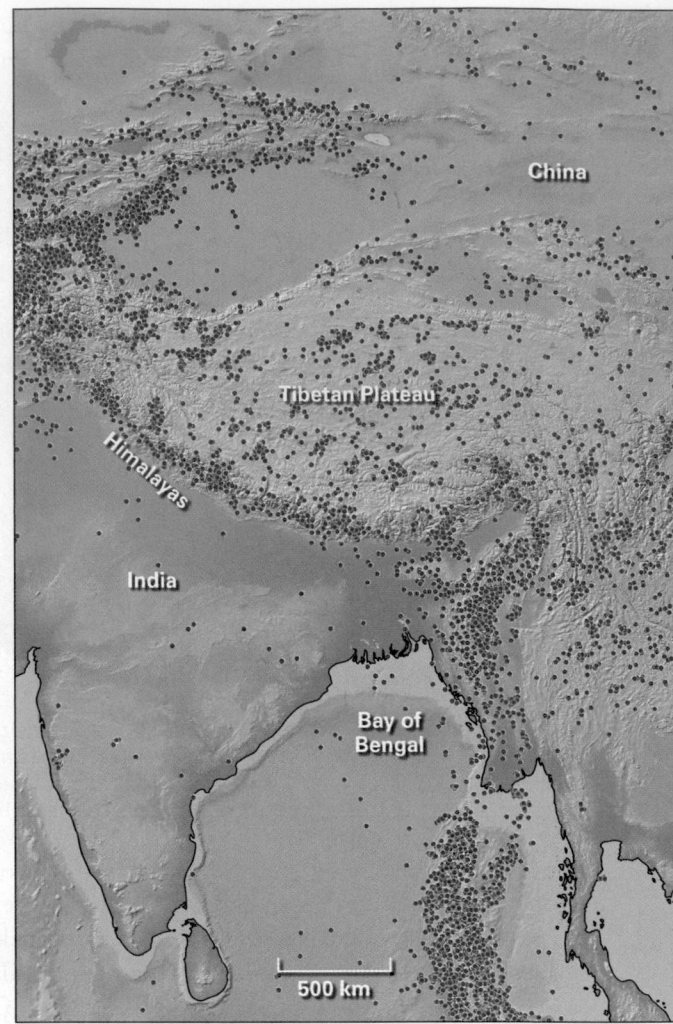

(a) Earthquakes in Africa occur mostly along the East African Rift.

(b) Earthquakes in southern Asia occur primarily in crust deforming due to the collision of Eurasia and India.

(c) The 2015 Nepal earthquake destroyed many buildings, including this ornate temple that was built in 1832.

include the East African Rift **(Fig. 10.23a)**, the Basin and Range Province (mostly in Nevada, Utah, and Arizona), and the Rio Grande Rift (in New Mexico). In all these places, shallow earthquakes, similar in nature to the earthquakes at mid-ocean ridges, rattle the landscape. In contrast to seismic belts of mid-ocean ridges, belts in rifts occur on land and can be located under or near populated areas.

Collision Zones Two continents collide when the oceanic lithosphere that once separated them has been completely subducted. Such collisions produce great mountain ranges such as the Alpine-Himalayan chain (Fig. 10.23b). Although a variety of earthquakes happen in collision zones, the most common earthquakes result from movement on thrust faults.

An example of a collision-zone earthquake took place in April 2015. Compression resulting from the northward push of the Indian subcontinent into Asia caused an M_W 7.8 earthquake in Nepal, a country that encompasses a portion of the Himalayas (Fig. 10.23c). The event occurred when a 7,000 km³ portion of a large thrust fault, 15 km below the surface, suddenly slipped by up to 3 m. Ground shaking destroyed whole towns. Tragically, thousands died and many more were injured.

Intraplate Earthquakes

Some earthquakes affect the interiors of plates and are not associated with plate boundaries, active rifts, or collision zones (Fig. 10.24). These **intraplate earthquakes**, almost all of which have a focus that lies at a depth of less than 25 km, account for only about 5% of the earthquake energy released in a year. Most intraplate earthquakes happen in continents, and because some can occur under populated areas, large ones have the potential to cause significant damage.

What causes intraplate earthquakes? Most seismologists favor the idea that stress applied to continental lithosphere triggers slip on pre-existing faults in the crust. Many of these faults may have first formed during Precambrian rifting events and represent long-lived weak "scars" in the crust. The source of the stress driving fault reactivation remains controversial. Some seismologists attribute the stress to the push acting on plate boundaries, whereas others suggest that it comes from the shear between the lithosphere and the underlying asthenosphere, or even from the unloading that happens when glaciers on the surface melt.

Seismologists have detected intraplate earthquakes on all continents. Most are concentrated in specific seismic zones.

In North America, for example, intraplate earthquakes occur in the vicinities of New Madrid, Missouri; Charleston, South Carolina; eastern Tennessee; Montreal, Quebec; and the Adirondack Mountains in New York. An M_W 7.3 earthquake occurred near Charleston in 1886, ringing church bells up and down the coast and vibrating buildings as far away as Chicago. In Charleston itself, over 90% of the buildings were damaged, and 60 people died. In 2011, an M_W 5.9 earthquake rattled central Virginia, abruptly reminding residents of the eastern United States that the region is not immune to seismicity. The tremor was felt from the Carolinas to New England, and people evacuated buildings in Washington, D.C., where some damage occurred to the Washington Monument and other structures.

The largest intraplate earthquakes to affect the United States took place in the early 19th century, near New Madrid, which lies near the Mississippi River in southernmost Missouri. At the time, the region was inhabited by a small population of Native Americans and an even smaller population of European descent. During the winter of 1811–1812, three M_W 7 to 7.4 earthquakes struck the region. The ground motion temporarily reversed the flow of the Mississippi River and toppled cabins (Fig. 10.25a). The earthquakes resulted from slip on thrust and strike-slip faults that underlie the Mississippi Valley (Fig. 10.25b). St. Louis, Missouri, and Memphis, Tennessee, lie close to the epicenter, so if large earthquakes were to happen in the New Madrid region again, they could cause significant damage.

Did you ever wonder . . .
if people could trigger earthquakes?

Induced Seismicity

Most earthquakes reflect geologic phenomena independent of human activity. But the timing of some earthquakes relative to human-caused events suggests that in certain cases, people can

FIGURE 10.24 The tectonic settings in which earthquakes occur in continental lithosphere. Subduction-related earthquakes in continental crust are not shown.

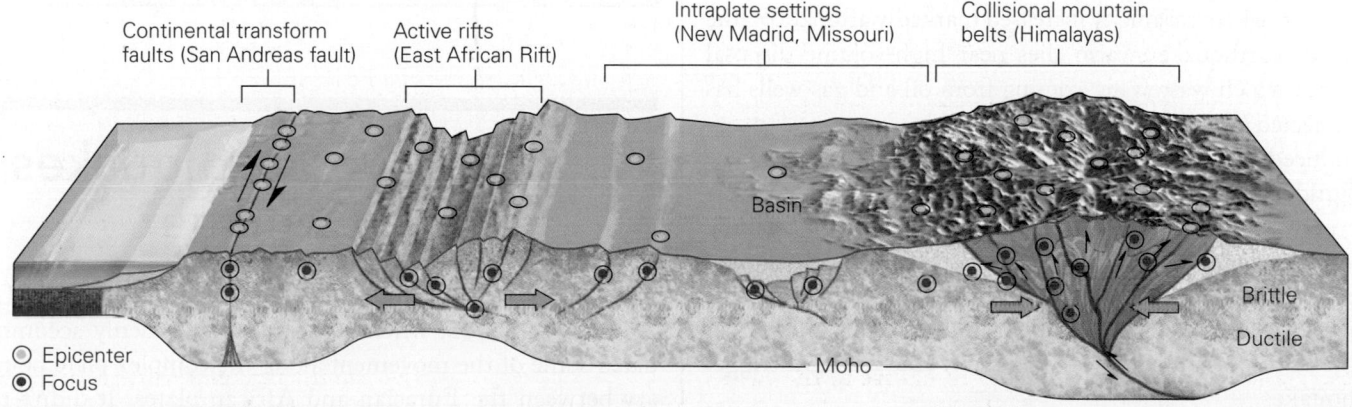

Continental transform faults (San Andreas fault)

Active rifts (East African Rift)

Intraplate settings (New Madrid, Missouri)

Collisional mountain belts (Himalayas)

Basin

Brittle

Ductile

Moho

◉ Epicenter
◉ Focus

FIGURE 10.25 New Madrid, Missouri, is a center of intraplate seismic activity.

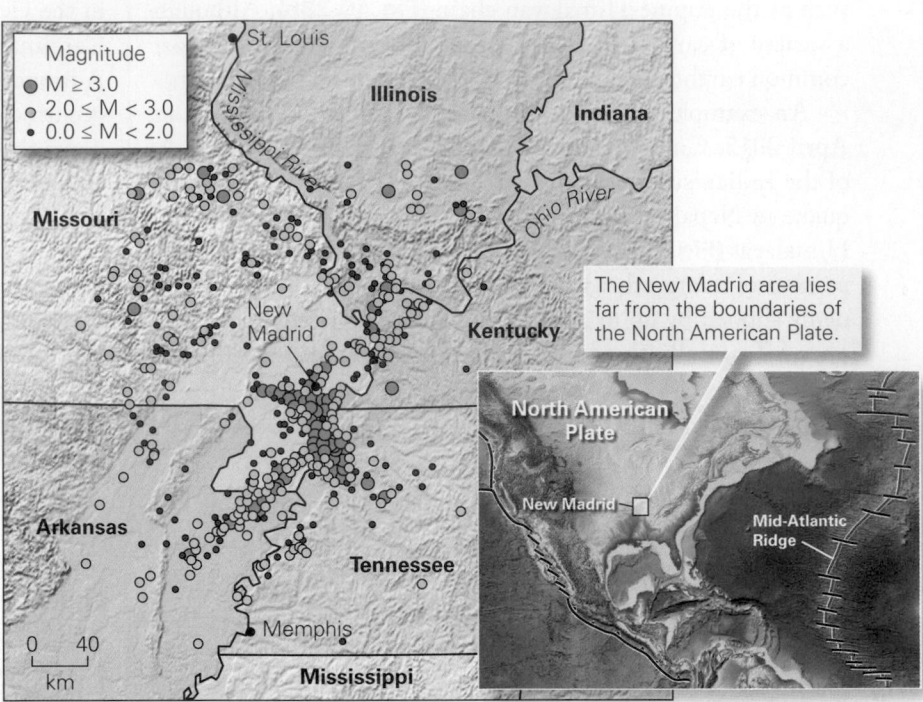

The New Madrid area lies far from the boundaries of the North American Plate.

(b) The epicenters of recent small earthquakes in the New Madrid area, as recorded by modern seismic instruments. The region remains active.

(a) The earthquakes of 1811–1812 destroyed cabins and disrupted the Mississippi River.

indeed influence seismicity. **Induced seismicity,** meaning seismic activity caused by actions of people, generally occurs in response to changes in groundwater pressure. Simplistically, the pressure of groundwater can slightly push apart the opposing surfaces of faults and, by doing so, effectively decrease the friction that resists slip on those faults. So when people increase groundwater pressure by pumping lots of water underground in a region containing an active fault, the fault may slip under regional stress conditions that might not have otherwise led to slip. Seismologists observed such a relationship near Denver, Colorado, when engineers pumped wastewater from a military installation down a deep well—as soon as the pumping began, small earthquakes started in the region. A similar phenomenon has happened in Oklahoma, where seismicity increased markedly after 2009. The observed "earthquake swarm" lies near high-volume disposal wells into which wastewater coming from oil and gas wells has been injected (see Chapter 14).

Induced seismicity can also become a danger where people build dams and create large reservoirs in valleys overlying active faults. Faulting generally breaks up rock, making it more erodible by rivers, so it's no surprise that deep river valleys form over large faults. When a reservoir fills the valley above a fault, water seeps down into the fault and, under the pressure caused by the water column above, might trigger earthquakes.

TAKE-HOME MESSAGE

Most, but not all, earthquakes happen along plate boundaries. Thrust faulting dominates at convergent boundaries, strike-slip faulting at transforms boundaries, and normal faulting at mid-ocean ridges. Normal faulting also occurs commonly in rifts, and thrust faulting occurs in collision zones. Earthquakes occasionally happen in plate interiors, probably along long-lived weak faults.

QUICK QUESTION: On a global basis, why are earthquakes in continental crust more dangerous to society than those along mid-ocean ridges?

10.6 How Do Earthquakes Cause Damage?

The great Lisbon earthquake happened on All Saints' Day, November 1, 1755, when a thrust fault suddenly accommodated some of the movement along the complex plate boundary between the Eurasian and African plates. It didn't take long for seismic waves from the M_w 8.5 to 9.0 earthquake to reach Lisbon, 440 km to the northeast. Lisbon, the capital of Portugal, was one of the great port cities and cultural centers of the day. The resulting ground shaking toppled 85% of the city's buildings, and fires set by overturned stoves then consumed much of the wreckage. Forty minutes later, a tsunami inundated the coast and washed away Lisbon's harbor. In the end more than 50,000 people lost their lives. Also, the library, which housed all the records of Portuguese exploration as well as countless Renaissance artworks, was destroyed. The event led philosophers such as Voltaire (1694–1778) to question the long-held belief that "this was the best of all possible worlds," and to suggest that bad things happen by chance, so that making a better world requires human effort. This concept set the stage for the Age of Enlightenment.

Visitors to Lisbon in the months after the 1755 earthquake attested that an area ravaged by a major earthquake is a heartbreaking sight. Let's consider the many factors that, sadly, can contribute to earthquake devastation. Understanding them, as we will see, may help prevent such devastation in the future.

Ground Shaking and Displacement

A large earthquake at a location on the Earth's surface may last from a few seconds to several minutes, not including the aftershocks. Its duration depends both on how long it took for slip on the earthquake-generating fault to take place, and on the distance between the focus and the surface location. The distance to the focus matters because different seismic waves travel at different velocities, so the difference between their arrival times increases as distance increases.

Did you ever wonder . . .
how long an earthquake lasts?

The intensity of an earthquake at a given location depends on four factors: (1) the magnitude of the earthquake, because larger-magnitude events release more energy; (2) the distance from the focus, because vibrations lose their energy as they pass through the Earth; (3) the nature of the substrate at the location (i.e., the character and thickness of materials beneath the ground surface), because earthquake waves tend to be amplified in weaker substrates; and (4) the frequency of the earthquake waves (where *frequency* equals the number of oscillations that pass a point in a specified interval of time)— high-frequency vibrations are not as dangerous as low-frequency vibrations because the former don't cause buildings to sway as much.

Different kinds of earthquake waves cause different kinds of ground motion (Fig. 10.26). For example, P-waves are almost perpendicular to the ground surface when they arrive and cause the ground to buck up and down. S-waves also reach the surface at a steep angle, but these waves are more complicated and tend to cause back-and-forth motion parallel to the ground surface. L-waves, the first surface waves to arrive, cause a snake-like side-to-side undulation. R-waves, the last waves to arrive, result in a rolling motion as particles near the surface of the ground follow elliptical paths, in cross section. Interference among the different kinds of waves causes motion to be anything but regular. In great earthquakes, the ground's movement can have an amplitude of as much as 1 m at the epicenter, but in moderate earthquakes, motions fall in the range of a few centimeters or less. Ground accelerations caused by moderate earthquakes lie in the range of 10%

FIGURE 10.26 Types of ground motion during earthquakes. The ground can shake in many ways at once, causing surface structures to move.

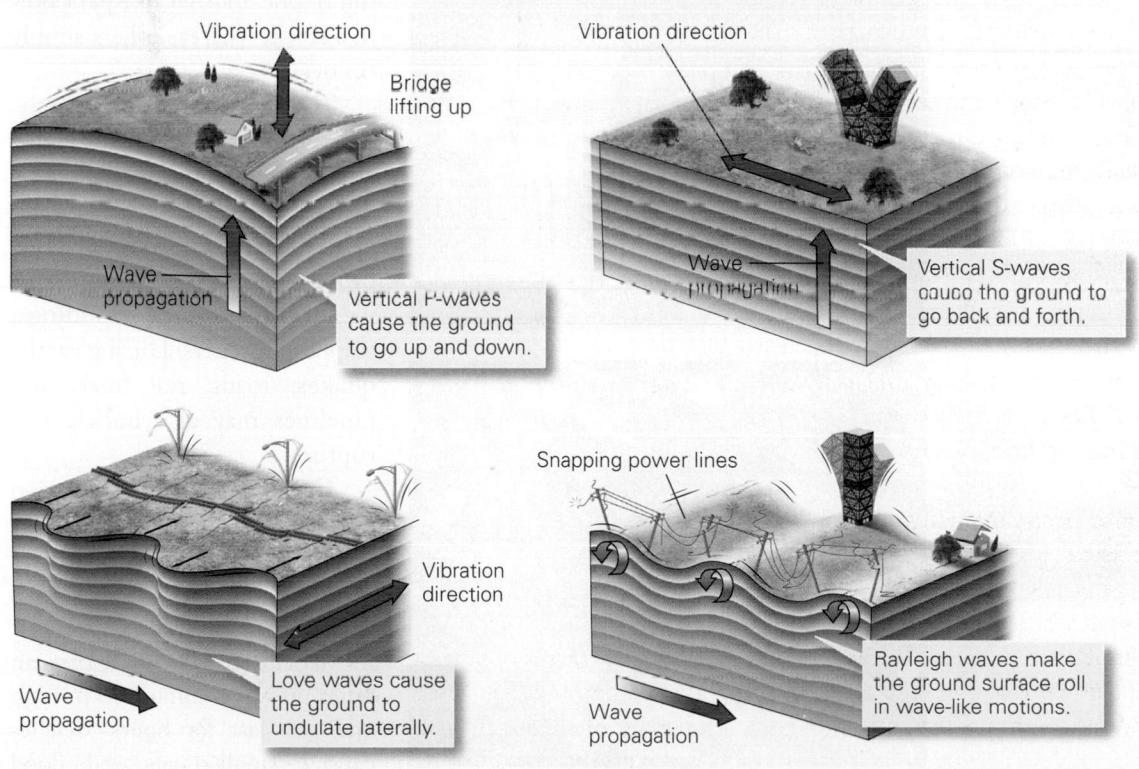

Vibration direction

Bridge lifting up

Wave propagation

Vertical P-waves cause the ground to go up and down.

Vibration direction

Wave propagation

Vertical S-waves cause the ground to go back and forth.

Vibration direction

Wave propagation

Love waves cause the ground to undulate laterally.

Snapping power lines

Wave propagation

Rayleigh waves make the ground surface roll in wave-like motions.

FIGURE 10.27 Examples of earthquake damage due to vibration.

Before	After

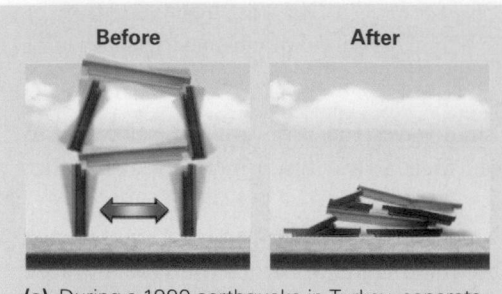

(a) During a 1999 earthquake in Turkey, concrete buildings collapsed when supports gave way and floors piled on one another like pancakes.

(b) An elevated bridge tipped over during the 1995 earthquake in Kobe, Japan.

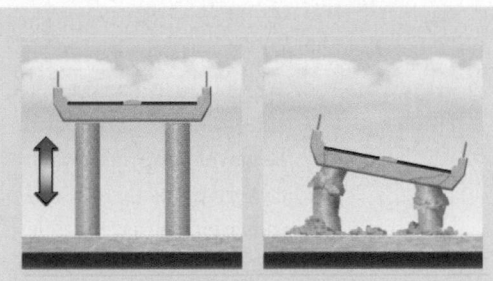

(c) Concrete bridge supports were crushed during the 1994 Northridge, California, earthquake when the overlying bridge bounced up and slammed back down.

(d) A neighborhood of masonry buildings in Armenia collapsed during a 1999 earthquake because the walls broke apart.

to 20% of g (where g represents the acceleration due to gravity), and during a great earthquake, accelerations may approach 1 g, so ground movements can toss you into the air.

If you're out in an open field during an earthquake, ground motion alone won't kill you, for your body flexes and bends. Buildings and bridges aren't so lucky **(Fig. 10.27)**. When earthquake waves pass, they sway, twist back and forth, or lurch up and down, depending on the type of wave motion. As a result, connectors between the frame and facade of a building may separate, so that the facade crashes to the ground. The flexing of walls shatters windows and wallboard and makes roofs collapse. Building floors or bridge decks may shift sideways and tip support columns over, or rise up and slam down on the support columns, thereby crushing them. Some buildings collapse with their floors piled on top of one another like pancakes in a stack, whereas others simply tip over.

The majority of earthquake-related deaths and injuries happen when falling debris or collapsing structures crush people. Aftershocks worsen the problem because they may topple already weakened buildings, trapping rescuers. During earthquakes, roads, rail lines, and pipelines may also buckle and rupture.

Occasionally, ground motion causes water in lakes, bays, reservoirs, and pools, in some cases far from the epicenter, to slosh back and forth. The water's rhythmic movement, known as a *seiche*, can build up waves almost 10 m high and can last for hours. Seiches capsize small boats and flood shoreline homes. And if they occur in reservoirs, seiches may wash over and weaken retaining dams.

Landslides

The shaking of an earthquake can cause ground on steep slopes or ground underlain by weak sediment to give way. This movement results in a *landslide*, the tumbling and flow of soil and rock downslope (see Chapter 16). Seismically triggered landslides occur all too frequently along the coast of California, for movement on faults has rapidly uplifted this coastline in the past few million years, resulting in the development of steep cliffs. When earthquakes take place, the cliffs collapse, often carrying expensive homes down to the beach below **(Fig. 10.28a, b)**. Such events lead to the misperception that "California will someday fall into the sea." Although small portions of the coastline do collapse, the state as a whole remains firmly attached to the continent, despite what Hollywood scriptwriters say.

> **Did you ever wonder...**
> will California fall into the sea?

Earthquake-triggered landslides that transport massive amounts of debris into reservoirs, lakes, or bays may cause huge waves. One of the biggest known examples happened in 1958, when an M_W 8.3 earthquake in southeastern Alaska triggered a landslide at the head of Lituya Bay. The splash from the displaced water washed the forest off the opposite wall of the bay up to an elevation of 516 m—this wall of water was 25% higher than the Empire State Building **(Fig. 10.28c)**!

Sediment Liquefaction

In 1964, an M_W 7.5 earthquake struck Niigata, Japan. A portion of the city had been built on land underlain by wet sand. During the ground shaking, foundations of over 15,000 buildings sank into their substrate, causing walls and roofs to crack. Several four-story buildings in a newly built apartment complex tipped over **(Fig. 10.29a)**. In 2011, an earthquake in Christchurch, New Zealand, not only caused sturdy stone buildings to crumble, but also caused sand to erupt and produce small, cone-shaped mounds, called *sand volcanoes* or *sand blows*, on the ground surface **(Fig. 10.29b)**. The transfer of sand from underground to the surface led to the formation of depressions large enough to swallow cars **(Fig. 10.29c)**.

These events serve as examples of **sediment liquefaction**. Liquefaction in beds of wet sand or silt happens because ground shaking causes the sediment grains to try and settle together. But because water fills the spaces (pores) between grains, water pressure in the pores increases and pushes the

FIGURE 10.28 Examples of landslide damage triggered by earthquakes.

(a) Shaking triggered a landslide that caused a steep slope along the coast of California to collapse, carrying part of a home with it.

(b) During the 1964 Alaska earthquake, slumping caused the land to give way beneath parts of Anchorage.

(c) The dark-gray cliff wall in the background was the failure surface of an earthquake-triggered landslide in Lituya Bay, Alaska. The resulting tsunami washed away the forest bordering the bay (the brown area). The wave was highest closest to the landslide.

FIGURE 10.29 Examples of liquefaction triggered by earthquakes.

(a) Liquefaction under their foundations caused these apartment buildings in Niigata, Japan, to tip over during a 1964 earthquake.

(b) A sand volcano (sand blow) formed during the 2011 Christchurch earthquake in New Zealand.

(c) During the 2011 Christchurch earthquake, liquefied sand spurted out and spread over the pavement. The process produced open space underground, so the pavement collapsed to form a sinkhole.

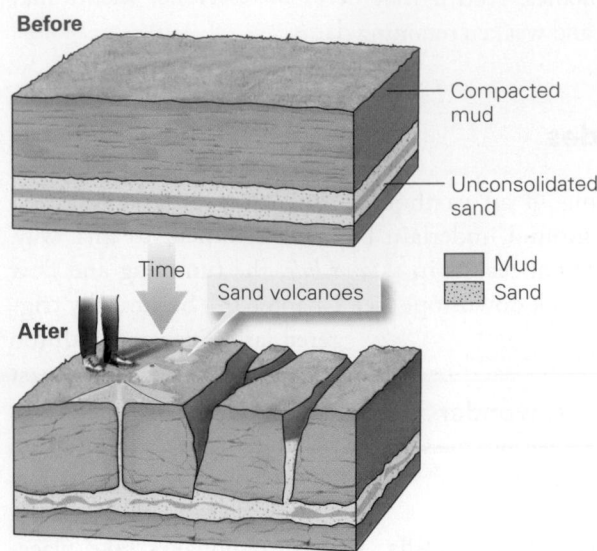

Before

Compacted mud

Unconsolidated sand

Mud

Sand

Time

Sand volcanoes

After

(d) Liquefaction of a sand layer causes the ground to crack and sand volcanoes to erupt.

(e) Liquefaction of a sand layer beneath the dry soil of this field caused the soil to crack and fissures to develop.

grains apart, and the wet silt or sand becomes a slurry called *quicksand*. As the material above the liquefied sediment settles downward, pressure squeezes the sand upward and out onto the ground surface—the result being sand volcanoes, as we've seen. The settling of sedimentary layers down into a liquefied

layer can also disrupt bedding and can lead to formation of open fissures of the land surface **(Fig. 10.29d, e)**.

The 1964 Good Friday earthquake in Alaska caused liquefaction beneath the Turnagain Heights neighborhood of Anchorage. Contractors had built the neighborhood on a small terrace of uplifted sediment. A 20-m-high escarpment that dropped down to Cook Inlet, a bay of the Pacific Ocean, formed the edge of the terrace. Here, as the ground shaking began, a layer of wet clay beneath the neighborhood liquefied. In this case, liquefaction allowed the overlying terrace, along with the houses built on top of it, to slide seaward. In the process, the terrace broke into separate blocks that tilted, turning the landscape into a chaotic jumble **(Fig. 10.30)**. Liquefaction beneath Turnagain Heights happened because in wet clay, the clay flakes stick together. The flakes stick together because there is surface tension in

FIGURE 10.30 The 1964 Turnagain Heights disaster.

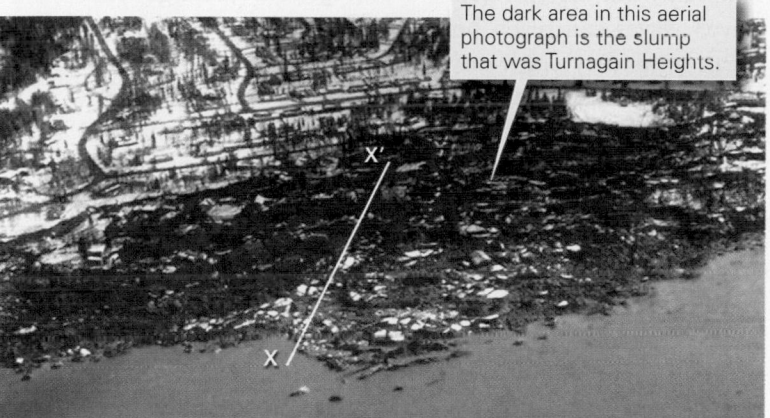

The dark area in this aerial photograph is the slump that was Turnagain Heights.

(a) A landslide carried a neighborhood into the sea.

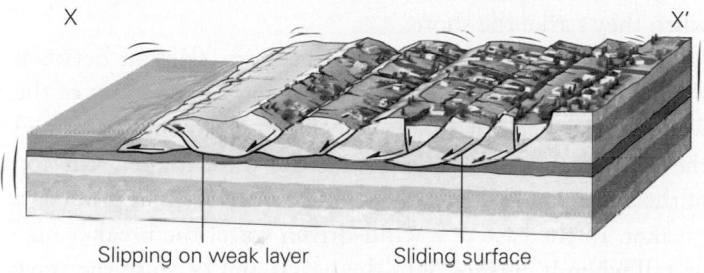

Slipping on weak layer   Sliding surface

(b) Slip occurred on a weak layer.

FIGURE 10.31 Fire sometimes follows an earthquake.

(a) Broken gas tanks erupt in fountains of flame after the 2011 Tōhoku, Japan, earthquake.

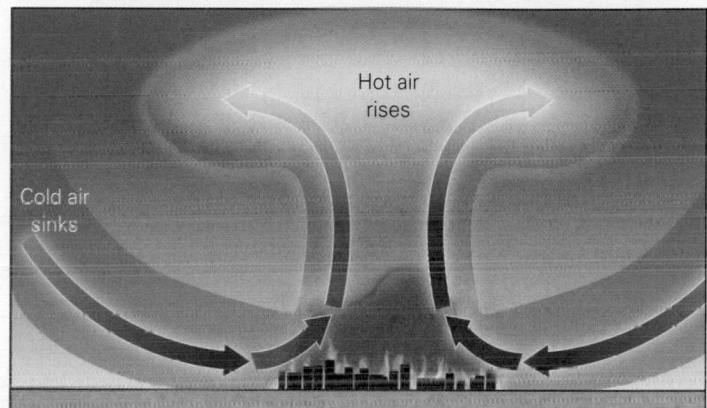

Hot air rises

Cold air sinks

(b) A firestorm develops when cool air rushes in to replace rising hot air above a huge fire. The cool air stokes the blaze, making it larger and hotter.

the water between the flakes, caused by the attraction of the water molecules to one another. When still, wet clay can behave like a solid gel, but when shaken, the weak bonds between water molecules break, and the clay transforms into a viscous liquid. Clay that displays this behavior is called thixotropic clay, or *quick clay*.

Fires Associated with Earthquakes

The shaking during an earthquake can tip over lamps, stoves, or candles with open flames, and it may break wires or topple power lines, generating sparks. As a consequence, areas already turned to rubble, and even areas not so badly damaged, may be consumed by fire. Ruptured gas pipelines and oil tanks feed the flames, sending columns of fire erupting skyward **(Fig. 10.31a)**. Firefighters may not even be able to reach the fires if the doors to the firehouse won't open or rubble blocks the streets. Moreover, firefighters may find themselves without water, for ground shaking and landslides damage water lines.

Once an earthquake-triggered fire starts to spread, it can become an unstoppable inferno. Most of the destruction caused by the 1906 San Francisco earthquake resulted from fire. For 3 days, the blaze spread through the city until firefighters contained it by blowing up buildings to form a firebreak. By then, 500 blocks of structures had turned to ash,

causing 20 times as much financial loss as the shaking itself. When a large earthquake hit Tokyo in September 1923, fires set by cooking stoves spread quickly through the wood-and-paper buildings, creating an inferno that heated the air above the city. The hot air rose like a balloon, and when cool air rushed in, wind gusts of over 100 mph stoked the blaze, which grew to engulf 120,000 people **(Fig. 10.31b)**.

Tsunamis

The azure waters and palm-fringed islands of the Indian Ocean's eastern coast hide one of the most seismically active plate boundaries on the Earth—the Sunda Trench. Along this convergent boundary, the Indian Ocean floor subducts at a rate of about 4 cm per year, leading to the accumulation of a large elastic strain during the "stick" phase of a stick-slip cycle. Just before 8:00 a.m. on December 26, 2004, the crust above a 1,300-km-long by 100-km-wide portion of one of these faults slipped and lurched westward by as much as 15 m.

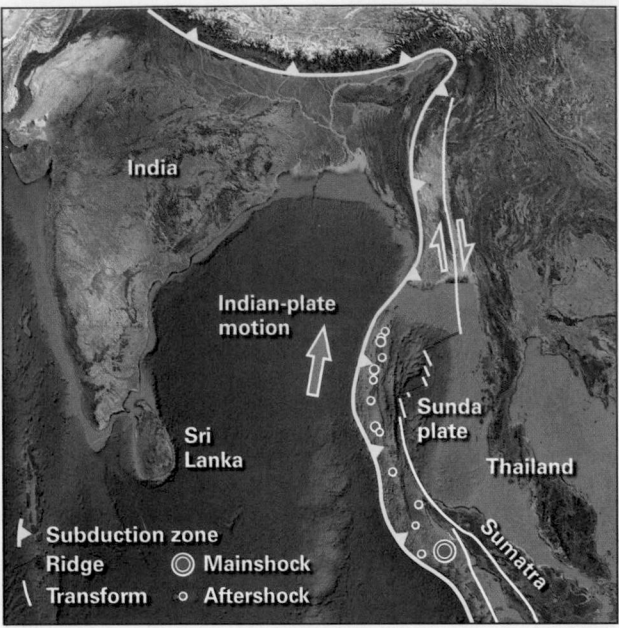

FIGURE 10.32 On December 26, 2004, a devastating tsunami was triggered by a subduction-related earthquake off Sumatra. Numerous aftershocks followed.

India

Indian-plate motion

Sri Lanka

Sunda plate

Thailand

Sumatra

▶ Subduction zone
\\ Ridge ◎ Mainshock
\\ Transform ○ Aftershock

The rupture started at the focus and then propagated north at 2.8 km per second, so that the rupturing process, overall, took about 9 minutes. This slip triggered a magnitude 9.3 earthquake—a great earthquake—and pushed the seafloor up by tens of centimeters **(Fig. 10.32)**. The rise of the seafloor, in turn, shoved up overlying ocean water. Because the area that rose was so broad, it displaced an immense volume of seawater. As a consequence, tragedy of an unimaginable extent began to unfold. Gravity caused the water that had been pushed up to sink and then spread outward in a wave moving at speeds of about 800 km per hour—almost the speed of a jet plane **(Fig. 10.33)**.

Geologists refer to such a wave, caused by the sudden displacement of a large volume of water, as a *tsunami*. This Japanese word translates literally as harbor wave, an apt name because tsunamis can be particularly damaging to harbor towns. Though we most often hear of tsunamis generated by displacement due to an earthquake, they can also be triggered by submarine landslides (see Chapter 16) or by explosive eruption of island volcanoes (see Chapter 9). In older literature, tsunamis were called "tidal waves" because when one arrives on shore, as we will see, water rises as if a huge tide were coming in. But the waves have nothing to do with the Earth's daily tidal cycles, so the name is misleading. Though in popular media the word *tsunami* tends to be associated with giant or very high waves, in fact tsunamis come in all sizes—some may be just centimeters high and are barely noticeable, whereas others reach colossal heights, causing the sea surface to rise by as much as 30 m near the shore.

Regardless of their cause, tsunamis differ greatly from familiar, wind-driven storm waves **(Fig. 10.34)**. Large wind-driven waves can reach heights of 10 to 30 m in the open ocean or as they crash on beaches. But even such monsters have wavelengths (the distance between adjacent wave crests) of only tens of meters, so they contain a relatively small volume of water. In contrast, although a tsunami in deep water may cause a rise in the sea surface of, at most, only a few tens of centimeters—a ship crossing one wouldn't even notice it—tsunamis have wavelengths of tens to hundreds of kilometers, and an individual crest can be several kilometers wide, as measured perpendicular to the wave front. Such waves contain an immense volume of water. In simpler terms, we can think of the width of a tsunami, in map view, as being more than 100 times the width of a wind-driven wave. Because of this difference, a storm wave and a tsunami have very different effects when they strike the shore.

When any wave approaches the shore, friction between the base of the wave and the seafloor slows the bottom of the wave, so the back of the wave catches up to the front, and the added volume of water builds the wave higher. The top of the wave may fall over the front of the wave and produce a breaker. In the case of a wind-driven wave, the breaker may be tall when it crashes onto the beach, but because the wave doesn't contain much water, it generally won't wash beyond the edge of the beach before it runs out of water. Friction slows the water running up the beach to a stop, and then gravity causes the water to recede back seaward. As is the case for a storm wave, when a tsunami approaches the shore, friction slows it down, so water farther offshore catches up to the water near the shore. Consequently, even though a tsunami rises only centimeters in the deep ocean, a large one can build into a monster that can be meters to tens of meters high near the shore. If a tsunami enters a bay or harbor that narrows landward, confinement of the water can build the wave even higher. In contrast to a storm wave, a single tsunami crest can be many kilometers wide and hundreds of kilometers long. If you picture a storm wave as a narrow ridge, you can picture a tsunami as a broad plateau. As a result, a large tsunami contains so much water that it crosses the beach and just keeps on going, eventually submerging all the low-lying land in a huge area. Notably, a single earthquake may generate several tsunamis that arrive on distant shores as much as an hour apart.

When the December 2004 wave struck Banda Aceh, a city at the north end of the island of Sumatra, the sea receded much farther than anyone had ever seen, exposing large areas of reefs that normally remained submerged even at low tide. People walked out onto the exposed reefs in wonder, not realizing that such a severe pullback of water provides an important warning of an impending tsunami. Then, with a rumble that grew to a roar, a wall of frothing water began to build in

FIGURE 10.33 Formation of a tsunami; an example from a convergent boundary.

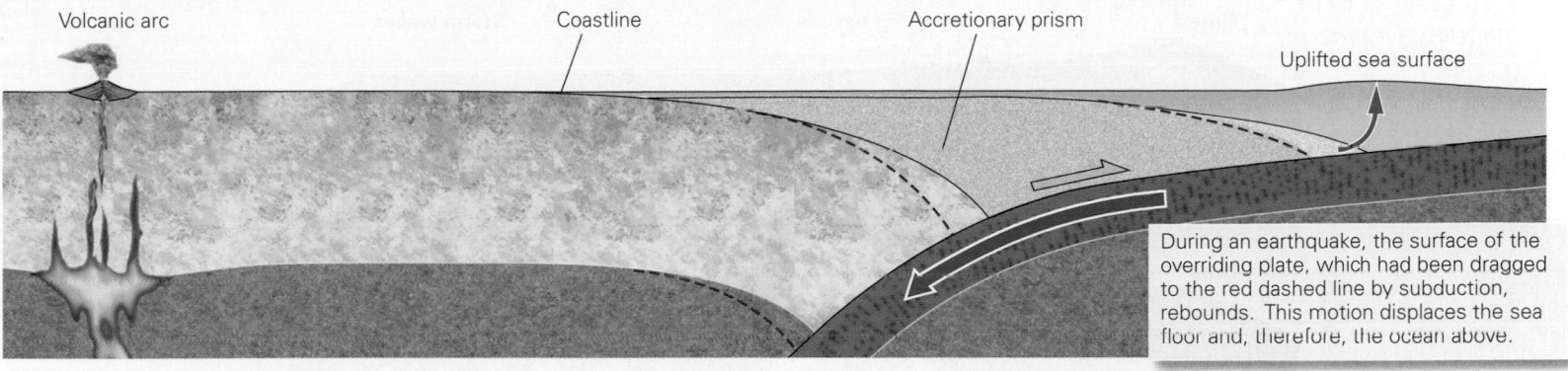

Volcanic arc   Coastline   Accretionary prism   Uplifted sea surface

During an earthquake, the surface of the overriding plate, which had been dragged to the red dashed line by subduction, rebounds. This motion displaces the sea floor and, therefore, the ocean above.

(a) A schematic showing how subduction leads to the buildup of elastic strain. Release of this strain by fault slip displaces the seafloor.

Time 1   Origin of the tsunami

Uplift of the seafloor pushes up a wide area of ocean water

Not to scale

Time 2   "Near-field tsunami"   "Far-field tsunami"

A near-field tsunami heads to nearby shores; a far-field tsunami heads across a wide ocean.

Not to scale

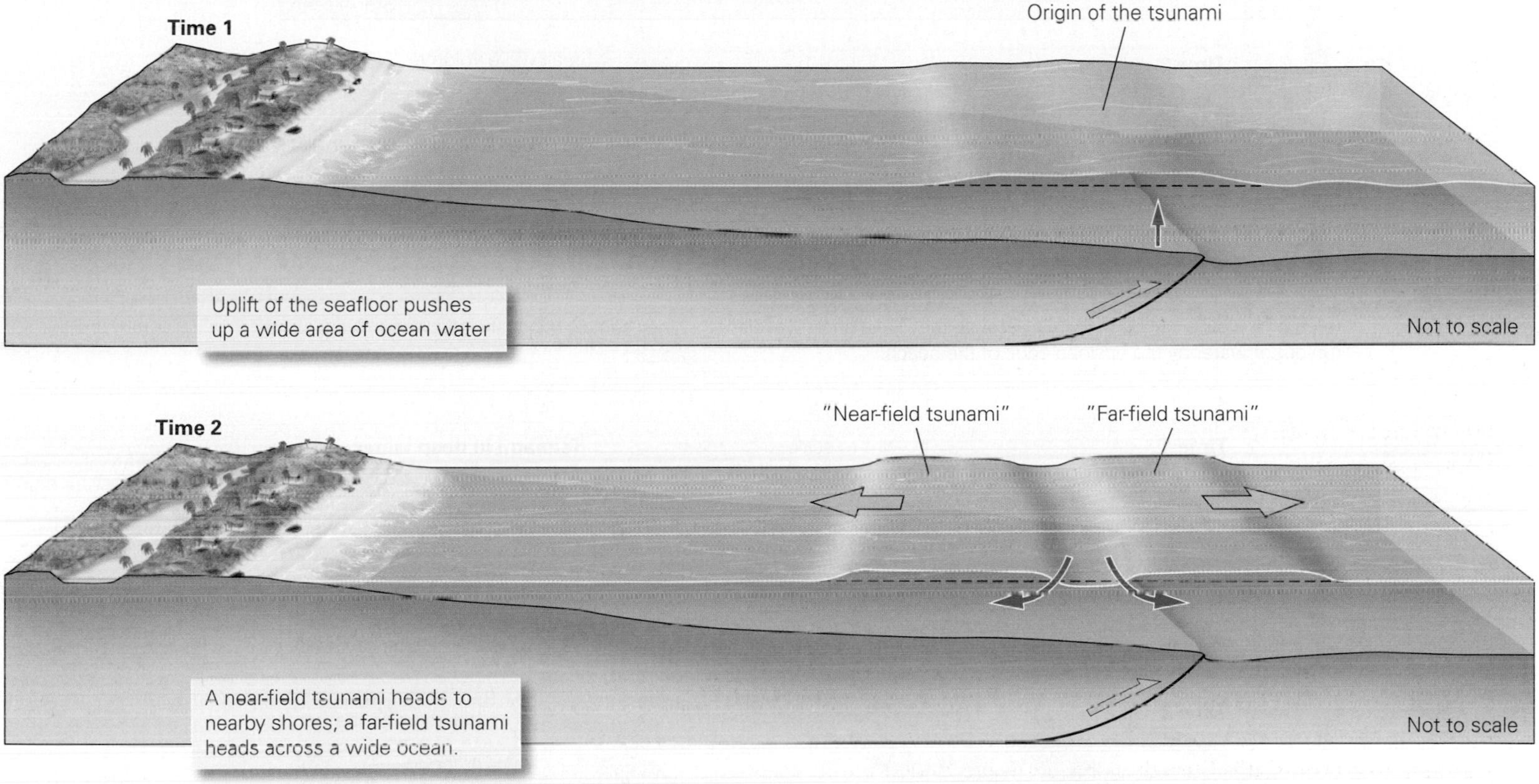

(b) After initial uplift of a mound of water, gravity causes the wave to spread out. The waves travel almost as fast as a jet plane.

the distance and approach land **(Fig. 10.35a)**. Puzzled bathers first watched, then ran inland in panic when the threat became clear. As the tsunami approached shore, friction with the seafloor had slowed it to less than 30 km per hour, but it still moved faster than people could run **(Fig. 10.35b)**. In places, the wave front reached heights of 15 to 30 m as it slammed into Banda Aceh **(Fig. 10.35c)**. The impact of the water ripped boats from their moorings, snapped trees, battered buildings into rubble, and tossed cars and trucks like toys. And the water just kept coming, eventually flooding low-lying land as far as

about 7 km inland. It drenched forests and fields with saltwater (deadly to plants) and buried fields and streets in up to a meter of sand and mud. Eventually the water slowed and then began to rush back to the shore. At Banda Aceh, at least two more tsunamis struck before the first one had entirely receded, so the water remained high for some time. The damage from the tsunamis at Banda Aceh was catastrophic **(Fig. 10.35d)**. When the water level finally returned to normal, a jumble of flotsam, as well as the bodies of unfortunate victims, floated out to sea and drifted away.

FIGURE 10.34 A comparison of a tsunami with storm waves.

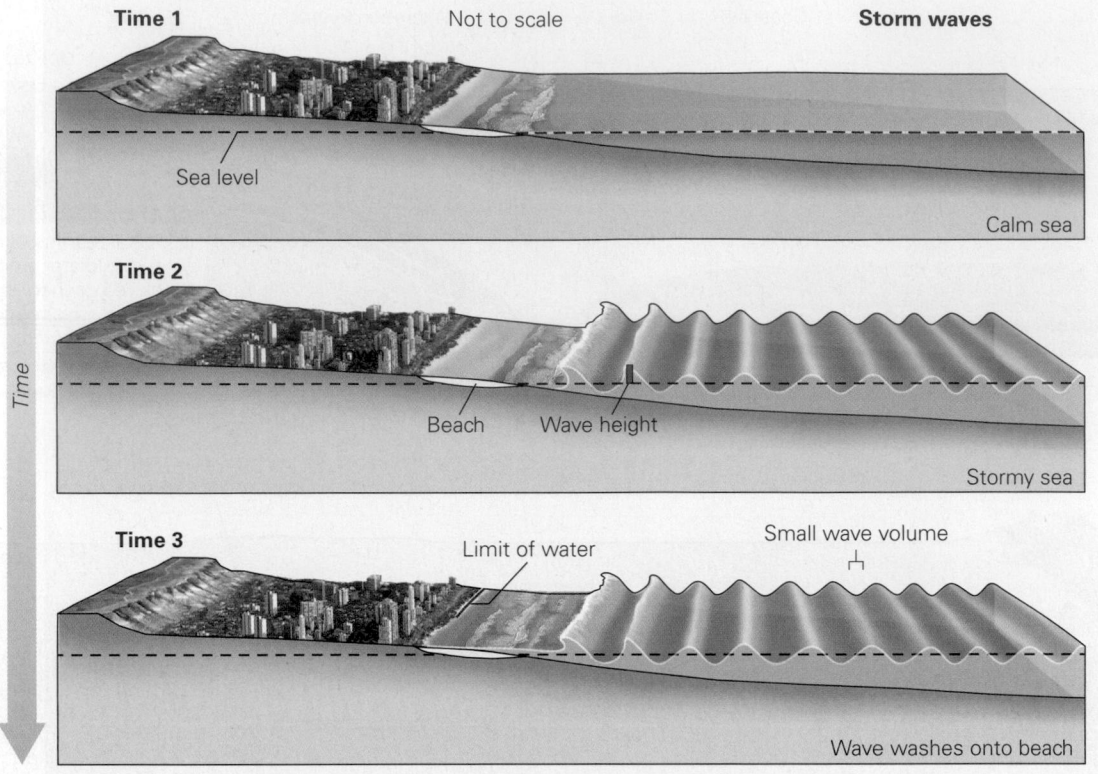

(a) Storm waves can be high, but because they have short wavelengths, they contain relatively little water. Most waves run out of water by the upslope edge of the beach.

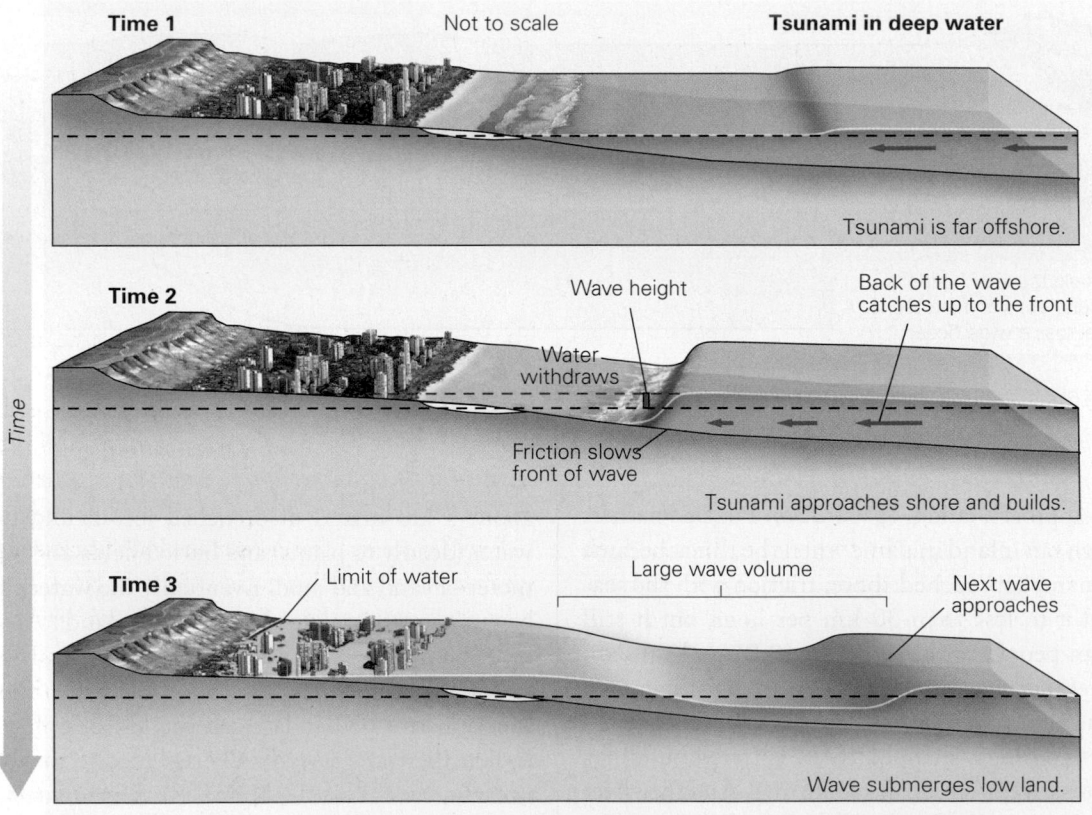

(b) A tsunami is not very high out in the open ocean, but as it approaches the land, friction slows the front of the wave, so the rear catches up and the wave grows. It is so wide that it can cover a wide area of low-lying land.

FIGURE 10.35 The great Indian Ocean tsunami of 2004.

(a) This snapshot shows the wave rushing toward the coast of Sumatra. Recession of water in advance of the wave exposed a reef.

(b) The wave blasts through a grove of palm trees as it strikes the coast of Thailand.

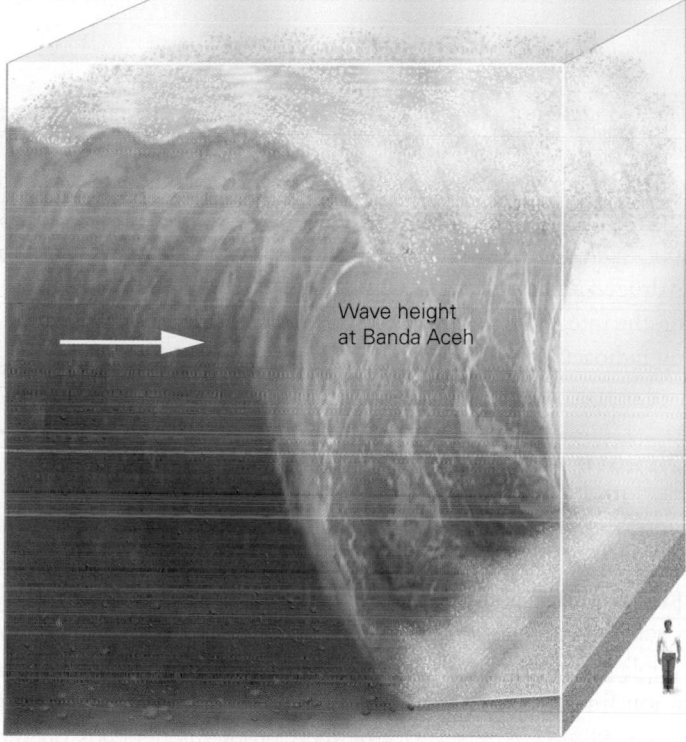

(c) At its highest, the tsunami's front was over 15 m high. Note the person for scale.

Wave height at Banda Aceh

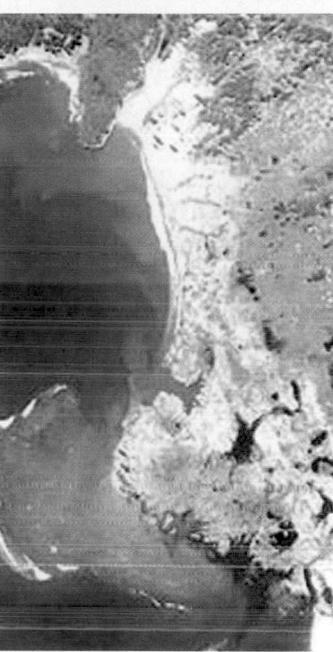

200 m

(d) Satellite photos of a beach near Banda Aceh before and after the 2004 tsunami struck. Note that the green fields were washed away and the beach vanished.

Geologists refer to the tsunami that struck Banda Aceh as a *near-field tsunami*, or local tsunami, because of its close proximity to the earthquake. A *far-field tsunami*, or distant tsunami, has crossed an entire ocean. A far-field tsunami struck Sri Lanka and the eastern coast of India two hours after the 2004 earthquake, and then the coast of Africa, on the west side of the Indian Ocean, eight hours after the earthquake **(Fig. 10.36)**. Wherever it struck, coastal towns vanished, fishing fleets sank, and beach resorts collapsed into rubble.

By the end of that horrible day, more than 230,000 people had died.

The 2004 Indian Ocean event remains etched in people's minds because of the immense death toll and the nonstop news coverage. But it is not unique. Tsunamis generated by the magnitude 9.5 Chilean earthquake in 1960 destroyed coastal towns in South America and crossed the Pacific, causing a 10.7-m-high wall of water to strike Hawaii 15 hours later. Twenty-one hours after the earthquake, when the tsunami reached Japan, it flattened coastal villages and left 50,000 people homeless. A tsunami following the 1964 Good Friday earthquake in Alaska destroyed ports at Valdez and Kodiak **(Fig. 10.37)**. And a devastating tsunami struck Japan in 2011, as we have seen.

Unlike earlier examples, the tsunami that struck Japan soon after the 2011 M_w 9.0 Tōhoku earthquake was captured

FIGURE 10.36 The 2004 Indian Ocean tsunami hit parts of Indonesia within minutes. It took two hours for the leading wave to reach Sri Lanka and India.

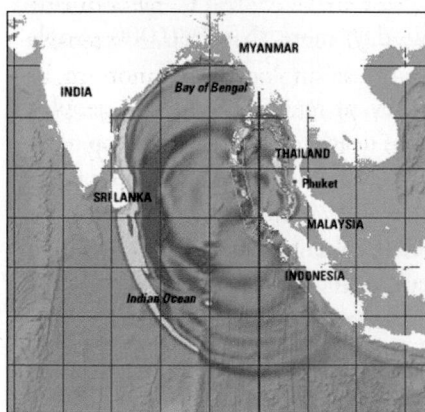

In this computer model, colors represent wave height—yellow is highest. There were several waves.

in high-definition video that was seen throughout the world, generating a new level of international awareness. Because the Tōhoku earthquake's epicenter was 130 km offshore, ground shaking on land during the event was not extremely intense. But, since tsunamis travel so fast, the first waves reached the shore only about 10 minutes after the earthquake struck, and there was not much time for residents of coastal towns to escape when the warning sirens went off. The 10-m-high seawalls that fringe the coast were not high enough to stop the advance of the wave that, in places, built to a height of 30 m when it reached shore. The rising sea picked up boats and ships in the harbor and flung them over the seawalls and, in some cases, onto the roofs of buildings. It crossed the beach at a speed of 30 km per hour, and once on land, it smashed through houses and tumbled cars as if they were pebbles. As the churning wave picked up dirt and debris, it became a viscous slurry resembling a volcanic lahar, moving with such force that nothing could withstand its impact (see Fig. 10.2c). When the wave finally ran out of water, it receded to the sea, carrying debris and victims with it and leaving behind a wasteland **(Fig. 10.38a)**.

But the catastrophe was not over. A portion of the wave struck the Fukushima nuclear power plant. Although the plant had withstood the ground shaking of the earthquake and had automatically shut down, its radioactive reactor cores still needed to be cooled by water in order to remain safe.

FIGURE 10.37 A tsunami inundated the shore of Valdez, Alaska, in 1964 and washed away the snow cover. The town's port was destroyed.

The tsunami not only destroyed power lines, cutting the plant off from the electrical grid, but also drowned the backup diesel generators, so the cooling pumps, now without power, stopped functioning. Eventually, the water surrounding the reactors cores boiled away. Some of the superheated water separated into hydrogen and oxygen gas, which then exploded, thereby breaching the integrity of three of the four reactor buildings and releasing radioactivity into the environment **(Fig. 10.38b)**.

Disease

Once the ground shaking and fires have stopped, disease may still threaten lives in an earthquake-damaged region. Earthquakes destroy housing, leaving victims exposed to the weather; sever water and sewer lines, there by contaminating clean-water supplies and exposing the public to bacteria; and cut transportation lines, preventing food and medicine from reaching the damaged area. The severity of such problems depends on the ability of emergency services to cope **(Box 10.1)**.

> ### TAKE-HOME MESSAGE
>
> Earthquakes cause devastation in many ways. Ground shaking, landslides, sediment liquefaction, and tsunamis can topple buildings and disrupt the land. Fire and disease may follow.
>
> **QUICK QUESTION:** Is ground shaking the major cause of loss of life in all earthquakes?

FIGURE 10.38 Damage due to the 2011 Tōhoku tsunami.

(a) This Japanese coastal town was completely destroyed by a tsunami that followed the Tōhoku earthquake of March 2011.

Before

The power plant was built next to the shore.

Reactor building 4

After

(b) Each cubic building houses a reactor of the Fukushima nuclear power plant. The tsunami washed over the seawalls and inundated the plant to a depth of 14 m, cutting off power to the cooling pumps. Hydrogen explosions destroyed three of the four reactor buildings.

10.7 Can We Predict the "Big One"?

We have seen that large earthquakes near population centers can cause catastrophe. Needless to say, many lives could be saved if we could determine which regions endure earthquakes more frequently so that building codes in these regions could require stronger structures. Many more lives could be saved if we knew exactly when and where a specific earthquake would happen, so people could evacuate dangerous buildings and potentially unstable slopes. Even if we had seconds to minutes of advance warning, it might provide enough time to shut off gas lines, electrical circuits, and water systems.

Can seismologists predict the next great earthquake, the proverbial "big one?" The answer depends on the time frame of the prediction. With their present understanding of the distribution of seismic belts and the frequencies at which earthquakes occur, seismologists can make *long-term predictions* (on a time scale of decades to centuries). For example, with some certainty, they can say that a major earthquake will probably rattle California during the next 100 years and that a major earthquake probably won't strike central Canada during the next 10 years. But despite extensive research, seismologists cannot make *short-term predictions* (on a time scale of hours to weeks or even years). They cannot say, for example, that an earthquake will happen in Montreal next month. But new technologies may permit warnings to be sent seconds in advance of the arrival of seismic waves if an earthquake does happen.

In this section, we look at the scientific basis of long-term and short-term predictions and consider the consequences of making a prediction. We also introduce earthquake early warning

BOX 10.1 CONSIDER THIS . . .

The 2010 Haiti Catastrophe

The history of Haiti changed forever on the sunny afternoon of January 12, 2010. At 4:53 P.M., a 70-km-long segment of a strike-slip fault suddenly slipped by an average amount of 1.8 m and, locally, by as much as 4 m. The motion began only 25 km west-south-west of Port-au-Prince, and 13 km beneath the ground surface, so the shock waves of the resulting M_w 7 earthquake reached the capital city in a matter of seconds, causing the ground to lurch violently for 35 seconds. The event released as much energy as a 32-megaton nuclear weapon, 1,000 times the energy released by the Nagasaki atomic bomb.

Ground shaking caused insufficiently reinforced structures to crack and crumble (Fig. Bx10.1a, b). Unable to hold up the shifting weight of the buildings above, columns gave way, bringing floors down into pancake-like stacks. Similarly, brick or block walls broke apart, roads buckled, and hillslopes slumped. And beneath the harbor, sediment liquefied, causing the wharfs to sink into the sea and the giant crane used to unload ships to topple sideways. When the shaking, which reached an intensity of IX on the MMI scale (Fig. Bx10.1c, d), finally stopped, most of Port-au-Prince had collapsed.

As a dense cloud of white dust slowly rose over the rubble, survivors began the frantic scramble to dig out victims, a task made more hazardous by aftershocks, of which there were over 50 with a magnitudes of between M_w 4.5 and 6.1—the renewed shaking caused still-standing but weakened walls to collapse on rescuers. Sadly, only about 150 people were eventually pulled from the rubble alive. No one will ever know exactly how many people died during the earthquake or of injuries in the weeks afterward, but some estimates place the death toll at 230,000, about 2.5% of the country's population. More than a million people, in a country of 9 million, lost their homes.

Why did the earthquake occur? Haiti sits astride the transform boundary along which the North American Plate moves westward at about 2 cm per year, relative to the Caribbean Plate. So earthquakes in Haiti are inevitable. Slip-accommodating plate motion in Haiti happens along a few large faults; the southernmost of these was responsible for the disaster. The last major earthquakes on this plate boundary happened about 240 years ago, so stress had been building on the fault for quite some time.

The impact of an earthquake on society depends not only on its size, but also on the nature of the substrate, on the steepness of the slopes, on construction practices, and on the quality of emergency services in the affected area. Much of Port-au-Prince was built on a basin of weak sediment, so as earthquake waves passed into this basin from regions of harder bedrock, they were amplified and therefore caused particularly large ground movements. In addition, many of the city's neighborhoods perch on steep slopes, which slumped downhill during the quake. Sadly, buildings in Haiti were not designed to withstand ground shaking, and local emergency services were overwhelmed. The country did not have enough rescue workers, doctors, or supplies to cope with the catastrophe, and debris made streets impassable. To make things worse, the country has only a small airport, slowing down delivery of supplies by air, and the destruction of the port prevented ships from bringing relief. Continued poor sanitation contributed to the spread of cholera over the next 3 years. The country will take many years to recover.

What does the future hold? Perhaps the earthquake released enough stress that another large event will not happen for another century or two. Alternatively, the 2010 earthquake might be the beginning of a sequence of earthquakes along the plate boundary. Because nothing can stop the movement of plates, the safety of the region's inhabitants will depend on the strength of the new buildings that rise from the rubble and on efforts to reinforce buildings in other regions along the fault.

---

systems. Seismologists refer to studies leading to predictions as seismic-risk assessment, or *seismic-hazard assessment*.

Long-Term Predictions

When making a prediction, we use the word *probability* because a prediction only gives the likelihood of an event. For example, a seismologist may say, "The probability of a major earthquake occurring in the next 20 years in this state is 20%." This sentence implies that there's a one-in-five chance that an earthquake will happen during the 20-year period. Urban planners and civil engineers use long-term predictions to create building codes for a region—codes requiring stronger, more expensive buildings make sense for regions with greater seismic risk. Planners may also use predictions to determine whether to build vulnerable structures such as nuclear power plants, hospitals, or dams in potentially seismic areas. Seismologists base long-term earthquake predictions on two kinds of information: the identification of seismic belts and the **recurrence interval** (the average time between successive events) of earthquakes along a given fault.

FIGURE Bx10.1 The disastrous January 2010 earthquake in Haiti and its geologic setting.

(a) Survivors salvage what they can in Port-au-Prince, the capital of Haiti, after the devastating earthquake of January 2010.

(b) Ground shaking during the 2010 Haiti earthquake caused most of the houses in this residential neighborhood to collapse.

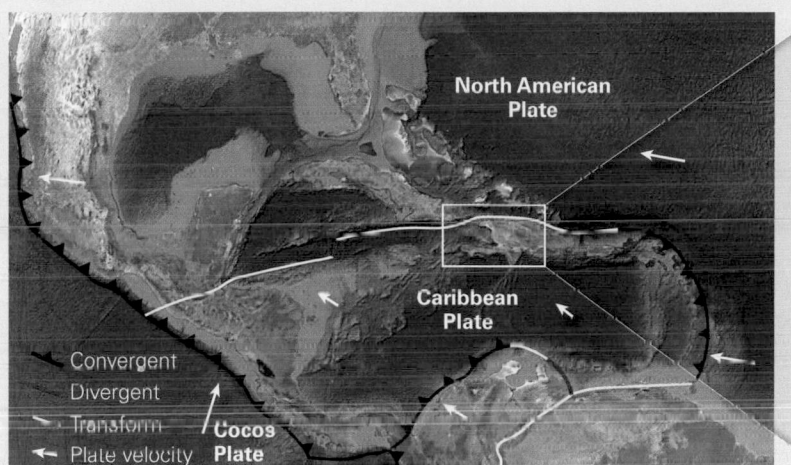

(c) The Caribbean Plate has complex boundaries, delineated by bathymetric features. The white rectangle shows the location of Haiti.

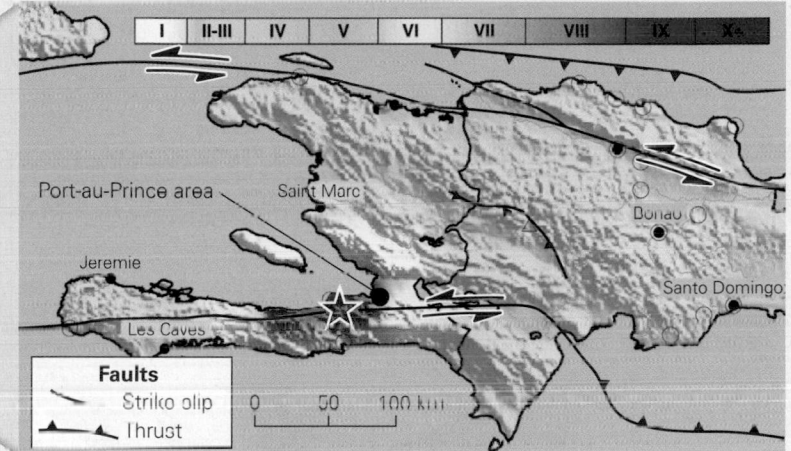

(d) A map showing the intensity of shaking in Haiti on the MMI scale. The star marks the epicenter of the quake.

To identify a seismic belt (or zone), seismologists produce a map showing the epicenters of earthquakes that have happened during a set period of time (say, 30 years). Clusters of epicenters define a seismic belt. The basic premise of long-term earthquake prediction can be stated as follows: a region in which there have been many earthquakes in the past will probably experience more earthquakes in the future. Seismic belts, therefore, are regions of greater seismic risk. This doesn't mean that disastrous earthquakes can't happen far from a seismic belt—they can and do—but the risk that an earthquake will happen in a given time window is less.

Seismologists can produce epicenter maps only with data from only the past 60 years or so, because before that time, they did not have enough seismometers to locate epicenters accurately. Fortunately, geologists can help provide insight into seismic risk by examining landforms for evidence of recent faulting. For example, the presence of a distinct fault scarp in a landscape indicates that faulting has happened so recently that erosion has not yet had time to grind away the evidence **(Fig. 10.39)**.

To determine the recurrence interval for large earthquakes at a location, geologists determine when large earthquakes

FIGURE 10.39 Identifying recent fault movement.

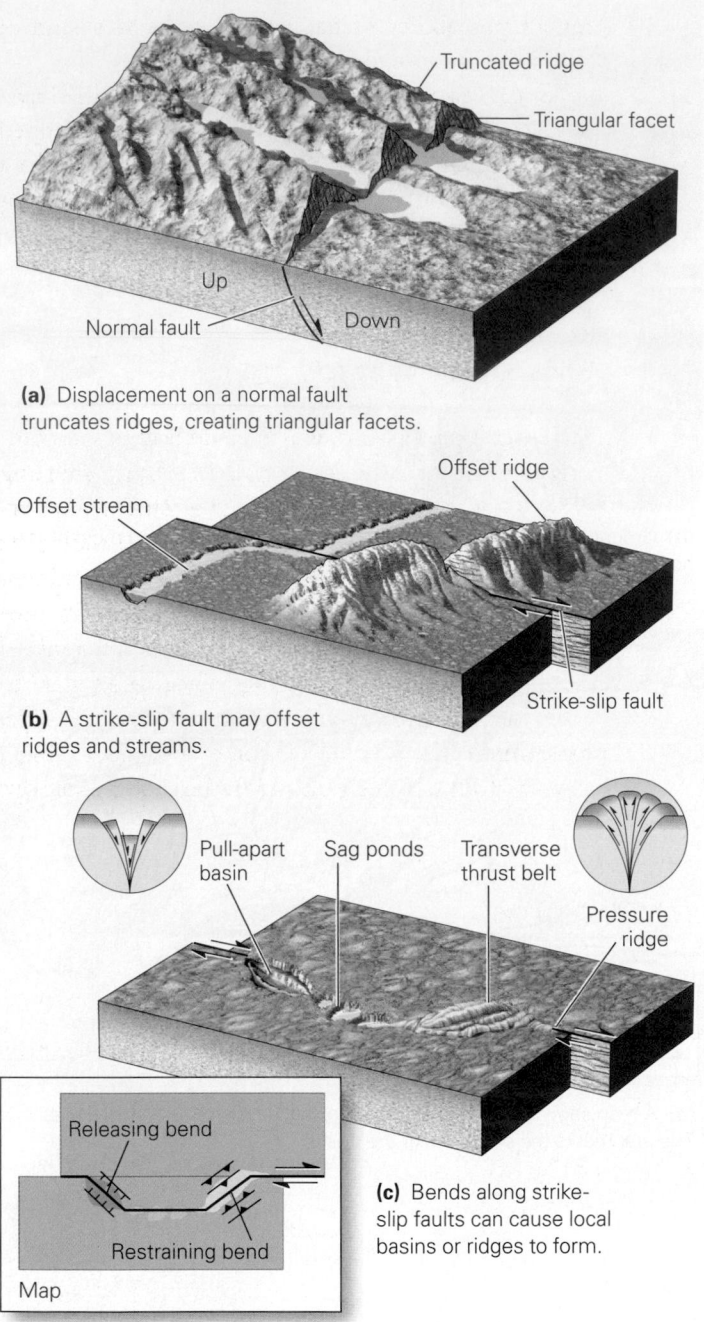

(a) Displacement on a normal fault truncates ridges, creating triangular facets.

(b) A strike-slip fault may offset ridges and streams.

(c) Bends along strike-slip faults can cause local basins or ridges to form.

FIGURE 10.40 Evidence of paleoseismicity can be used to determine recurrence interval. The block shows the walls of two trenches cut into the ground at a fault zone.

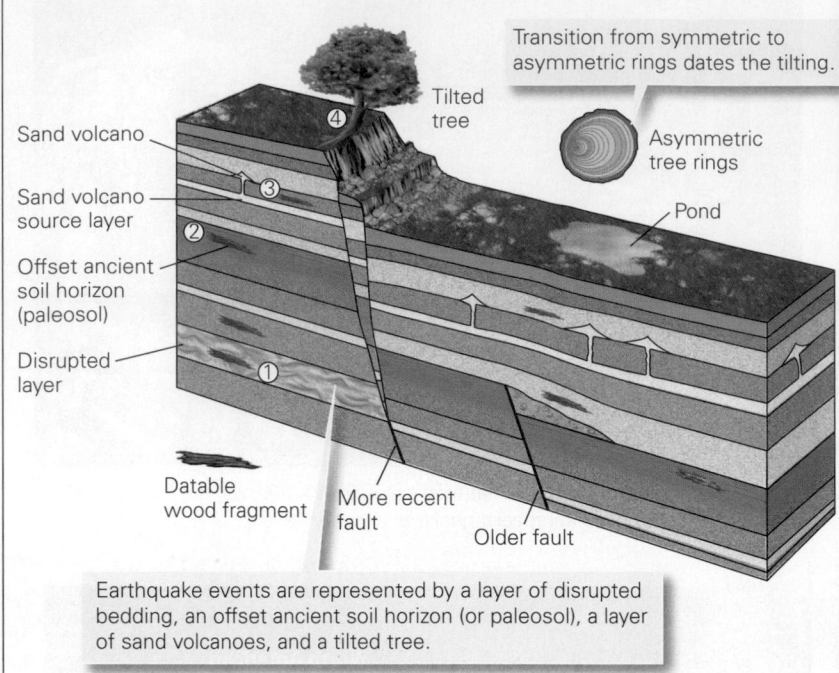

Earthquake events are represented by a layer of disrupted bedding, an offset ancient soil horizon (or paleosol), a layer of sand volcanoes, and a tilted tree.

happened at the location in the past. This type of research, called *paleoseismology*, uses the historical record to identify earthquakes that occurred during the past several centuries. To determine the ages of prehistoric earthquakes, they rely on geologic evidence. For example, a trench cut into sedimentary strata near a fault may reveal layers of sand volcanoes and disrupted bedding in the stratigraphic record. Each such layer, whose age can be determined by using radiocarbon dating of plant fragments, records the time of an earthquake **(Fig. 10.40)**. By calculating the number of years between

successive events and taking the average, seismologists obtain the recurrence interval. As an example, imagine that disrupted layers formed at 260, 820, 1,200, 2,100, and 2,300 years ago. We can say that the recurrence interval between events is about 510 years. Note again that a recurrence interval does not specify the exact number of years between events, only the average number. Since stress builds up over time on a fault, the probability that an earthquake will happen in a given year probably increases as time passes. Determining a recurrence interval allows seismologists to refine *seismic–hazard maps* representing risk **(Fig. 10.41)**.

To avoid confusion about the meaning of a recurrence interval, seismologists sometimes use another number, the *annual probability*, to represent seismic risk at a given location. Annual probability equals 1 divided by the recurrence interval. So, if the recurrence interval is 100 years, the annual probability is $1 \div 100 = 1\%$.

In some cases, patterns of seismicity along a fault may provide clues to future seismicity. For example, the North Anatolian fault, a large strike-slip fault along which Turkey slips westward **(Fig. 10.42a)**, has been the site of numerous earthquakes in historic time. Since 1939, 11 major earthquakes have occurred along the fault—each rupturing a different portion of the fault **(Fig. 10.42b)**. By mapping the extent of the area that slipped during each earthquake, seismologists recognized a general westward progression in the faulting

FIGURE 10.41 Examples of seismic-hazard maps.

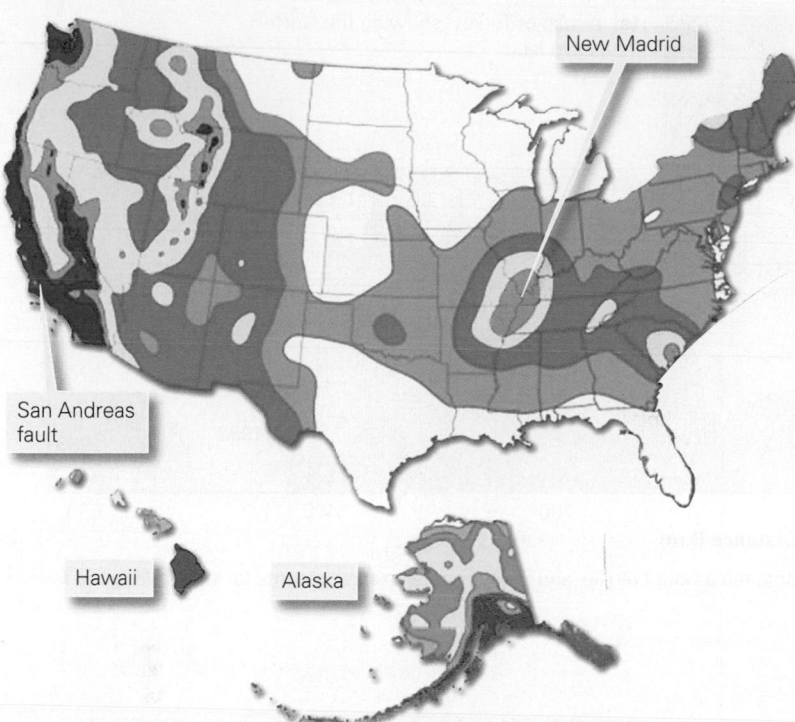

San Andreas fault

Hawaii

Alaska

New Madrid

(a) A seismic-hazard map of the United States. Red and pink regions have the greatest probability of experiencing large earthquakes. The San Andreas fault plate boundary is particularly hazardous. Risk in the New Madrid area remains subject to debate.

and therefore predicted that the next major earthquake on the North Anatolian fault would have a higher probability of happening at the west end of the fault than elsewhere.

Some seismologists suspect that places called *seismic gaps*, where a known active fault has not slipped for a long time, may be particularly dangerous. In a seismic gap, either the fault moves nonseismically or stress is building up to be released by a major earthquake at some time in the future.

Short-Term Predictions

Short-term predictions, which could lead to such precautions as evacuating dangerous buildings, shutting off gas and electricity, and readying emergency services, are not and may never be reliable. Seismologists have explored a number of possible clues to imminent earthquakes, but none have yet led to an accurate prediction, and often, possible clues can be recognized only in hindsight. For example, since rocks start to crack before a throughgoing rupture forms and slips, recognition of a "swarm" (a cluster of events during a short period) of foreshocks might be such a clue. But

Lowest hazard ⟶ Highest hazard

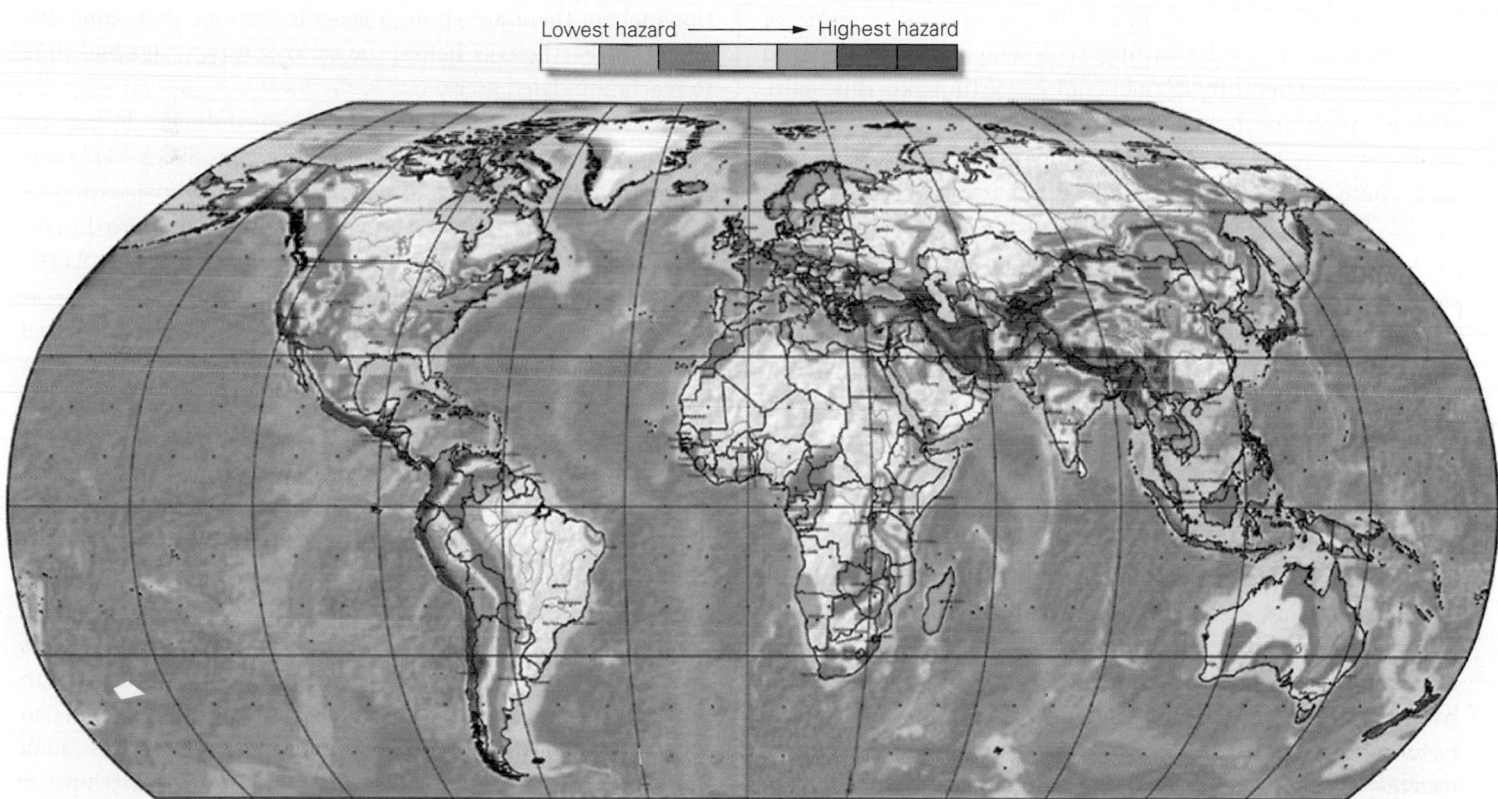

(b) A global seismic-hazard map. The redder regions have the greater probability of experiencing a large earthquake. Probability is high along plate boundaries.

FIGURE 10.42 Major earthquakes of the last century along the North Anatolian fault of Turkey have occurred roughly in sequence from east to west.

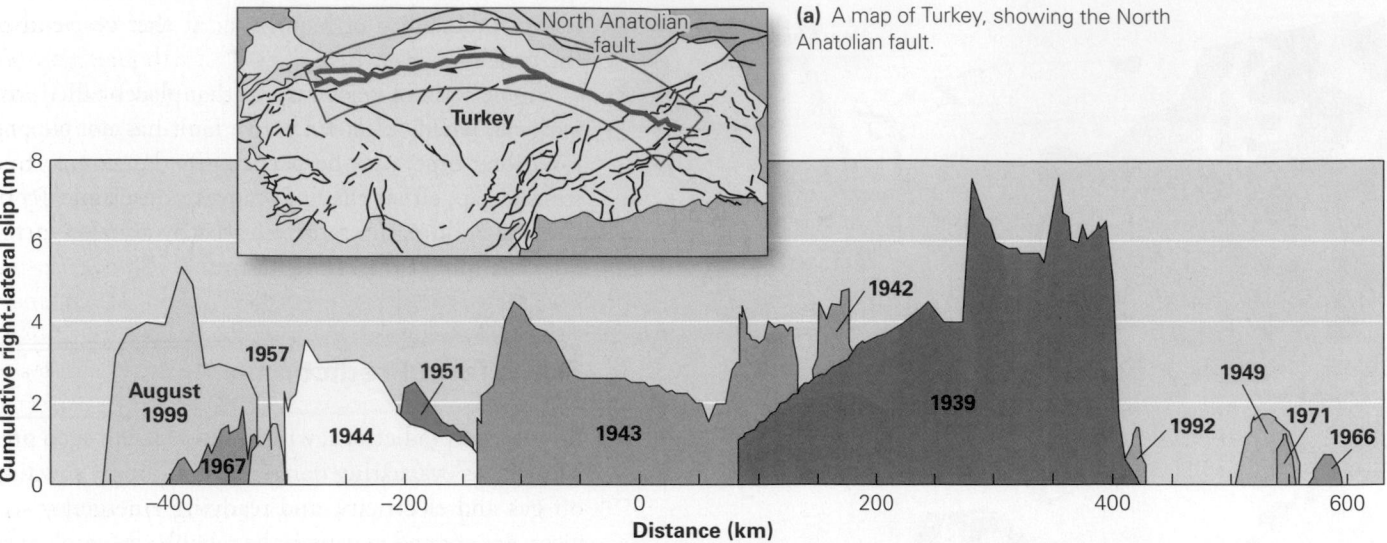

(a) A map of Turkey, showing the North Anatolian fault.

(b) A graph representing regions of the fault. The vertical axis represents the amount of slip, and the horizontal axis represents locations along the fault.

foreshocks do not always occur, and even if they do, they may be indistinguishable from other small earthquakes.

Similarly, since the crust deforms elastically prior to seismic slip (according to the elastic-rebound theory), and thus the local volume of rock may increase as open cracks start to develop, precise surveying of the ground (using lasers or InSAR) may detect upwarping or downwarping of the land surface, which may hint at an upcoming earthquake. But again, such warping may be recognized only in hindsight. Recently, geologists have begun to use computer models of stress to predict where stress buildups may lead to earthquakes, but these models have not yet proved to be a reliable predictor of events.

Other changes that have been explored, but have not been confirmed as precursors of earthquakes, include changes in the water level in wells; the appearance of gases, such as radon or helium, in wells; changes in the electrical conductivity of rock underground; and unusual animal behavior. Believers in these proposed clues suggest that they all reflect the occurrence of cracking in the crust prior to an earthquake, but most researchers remain very skeptical.

As long as short-term predictions remain questionable, emergency service planners must consider what should be done in the case of such a prediction. Should schools and offices be shut? Should millions of dollars be spent to evacuate people and leave cities open to looters? Should the public be notified, or should only officials be notified, creating a potential for rumor? If the prediction proves wrong, can seismologists be sued? Unfortunately, no one knows the answers to such questions.

Earthquake Early Warning Systems

Even though seismologists cannot provide weeks to years of advance notice that an earthquake will happen, they have successfully developed *earthquake early warning systems* in locations where there are enough seismic stations that some can detect the earthquake before the seismic waves have had time to reach populated areas.

An early warning system works as follows: When an earthquake happens, the seismic waves it produces start traveling through the Earth. The instant that multiple seismometer stations detect the earthquake, a computer approximates the epicenter location and then sends a signal to a control center, which automatically sends out emergency signals to areas that might be affected. Since broadcast warning signals travel at the speed of light, orders of magnitude faster than seismic waves, the signal arrives before the seismic waves. When the warning signal arrives, it activates electronic switches that automatically shut down gas pipelines, trains, nuclear reactors, power lines, and other vulnerable infrastructure. The signal also automatically activates sirens and alerts broadcasters to send out warnings on radio, TV, and cell phone networks to the public. Unless the focus lies directly under the city, the warning may precede the arrival of the first earthquake waves by several seconds—not a lot of time, but perhaps enough to prevent some infrastructure damage and to allow people to seek a safer location. Some earthquake-prone regions, such as Japan and California, have already installed earthquake warning systems.

10.8 Earthquake Engineering and Zoning

The destruction caused by an earthquake of a given size varies widely and depends on a number of factors. The most important of these factors include: the size of the earthquake; the proximity of the epicenter to a population center; the depth of the focus; the style of construction in the epicentral region; whether the earthquake occurred in a region of steep slopes or along the coast; whether building foundations are on solid bedrock or on weak substrate; whether the earthquake happened when people were outside or inside; and whether the government was able to provide emergency services promptly.

A comparison of two earthquakes illustrates the significance of these factors. The 1988 earthquake in Armenia and the 1971 earthquake in San Fernando, California, each had a magnitude of between M_W 6.6 and 6.8. But the former caused almost 500 times as many deaths as the latter (24,000 vs. 65). The difference in death toll primarily reflects differences in the style and quality of construction. The unreinforced concrete-slab buildings and masonry houses of Armenia collapsed, whereas structures in California had, by and large, been erected according to building codes that take into account stresses caused by earthquakes. Most flexed and twisted but did not fall down and crush people.

The terrible 1976 earthquake in T'ang-shan, China, illustrates the importance of substrate characteristics in determining damage. It killed over a quarter of a million people because the ground beneath the epicenter had been weakened by coal mining and collapsed, and because buildings were poorly constructed. During the 1989 Loma Prieta quake in California, portions of Route 880 in Oakland that were built on a weak substrate collapsed, whereas portions built on bedrock remained standing. Mexico City's 1985 earthquake proved disastrous because the city lies over a sedimentary basin whose composition and bowl-like shape focused seismic energy, and because the vibrations had a frequency that caused certain buildings to resonate (Box 10.2). Because of the same problem, Mexico City's 2017 earthquake also caused more damage than would have happened were the city not built on a basin.

Communities can mitigate, or diminish, the consequences of earthquakes by taking sensible precautions. Clearly, *earthquake engineering* (designing buildings that can withstand shaking) and *earthquake zoning* (determining where land is stable and restricting construction accordingly) can help save lives and property. In regions prone to large earthquakes, buildings and bridges should be constructed so that they are able to withstand vibrations without collapsing (Fig. 10.43a, b). They should be somewhat flexible so that ground motions cannot crack them, but they should have sufficient bracing so that movements don't become too severe. In addition, supports should be strong enough to maintain loads far in excess of their normal static (nonmoving) loads. Wrapping steel cables around bridge support columns makes them many times stronger. Bolting bridge spans to the tops of their columns prevents the spans from bouncing off. Bolting buildings to foundations keeps them in place, and adding diagonal braces to frames keeps them from twisting and shearing too much. Earthquake construction techniques have been improved greatly by testing scale models of designs on *shaking tables*, platforms that shake in response to computer-activated pistons.

Certain kinds of construction should be avoided in seismic zones. For example, concrete-block, unreinforced-concrete, and brick buildings crack and tumble under conditions in which wood-frame, steel-girder, and reinforced-concrete buildings remain standing. Traditional heavy, brittle tile roofs shatter and bury the inhabitants inside, whereas sheet-metal or asphalt-shingle roofs do not. Loose decorative stone and huge open-span roofs also do not fare well when shaken. Of note, inadequate structures can be made safer by *seismic retrofitting*, the process of strengthening existing buildings in potentially seismically hazardous areas. Examples of retrofitting include adding shock absorbers to foundations, adding braces, jacketing support columns, and coating masonry with resins.

Similarly, developers building in seismic zones should avoid construction on land underlain by weak sediment that could liquefy. They should not build on top of, on, or at the base of steep escarpments because the escarpments could fail and produce landslides, and they should avoid locating large population centers downstream of dams, which could crack

BOX 10.2 CONSIDER THIS . . .

When Earthquake Waves Resonate—Beware!

When you shine a flashlight, you produce white light. If you aim that beam at a prism, the light spreads into a spectrum of different colors, with each color representing light waves of a different frequency—the original beam of white light contained all those frequencies. Similarly, when an earthquake occurs, it produces seismic waves with a variety of frequencies. As the waves travel away from the focus, however, the Earth acts like a filter in that high-frequency waves (waves with short wavelengths) lose energy more rapidly than low-frequency waves (waves with long wavelengths). You've experienced this phenomenon if you've ever heard a car stereo playing loud music—when you stand near the car, you hear all the sound frequencies and can make out soprano voices and high-frequency guitar notes, but if you stand far away, all you can hear are the low-frequency thump-thump-thumps of the bass guitar and bass drum. Because of this effect, the frequency content (the variety of wave frequencies) of an earthquake changes with distance from the focus.

Why pay attention to frequency content? Different frequencies of seismic waves cause different amounts of ground acceleration. Also, waves of certain frequencies can cause resonance. **Resonance** happens when each new wave arrives at just the right time to add more energy to a system. To understand resonance, picture a boy on a swing. If the boy pumps his legs at just the right time, he swings higher, but if not, he slows down. The same phenomenon occurs if you slide a block of Jell-O that is resting on a plate back and forth on a table. If you move the plate too fast, the Jell-O merely trembles, but if you move at just the right frequency, resonance begins, and the block sways wildly.

Resonance often plays a major role in accentuating the damage during an earthquake. Even earthquakes that occur a few hundred kilometers away, on the convergent boundary along which the Cocos Plate grinds beneath North America can cause intense ground shaking in the city. That's because Mexico City sits on a thick sequence of unconsolidated lake-bed sediments, exposed when the Spanish conquistadors drained Lake Texcoco. The sedimentary basin, somewhat like a lens, focuses earthquake energy on the city. Because of the distance between the epicenter and the city, high-frequency waves weaken, but because of the nature of the sedimentary basin, low-frequency waves get amplified. These waves have just the right frequency to make the swaying motion of buildings between 8 and 18 stories high begin to resonate. In some cases, neighboring buildings slam together like clapping hands. Engineers did not design the buildings to accommodate such motion, so many buildings are susceptible to collapse. Between 8,000 and 30,000 people died, and another 250,000 were left homeless when an earthquake struck Mexico City in 1985, and hundreds of lives were again lost as the city suffered another earthquake in 2017.

and collapse, causing a flood. And they should also avoid constructing vulnerable buildings directly over active faults, because fault movement could crack and destroy the buildings. Cities in seismic zones need to draw up emergency plans to deal with disaster. Specifically, communication centers should be situated in safe localities, and strategies need to be implemented for providing supplies under circumstances in which roads may be impassable.

Because tsunamis are so dangerous, predicting their arrival can save thousands of lives. A tsunami warning center in Hawaii keeps track of earthquakes around the Pacific and uses data relayed via buoys from tide gauges and seafloor pressure gauges to determine whether a particular earthquake has generated a tsunami (**Fig. 10.43c**). If observers detect a tsunami, they flash warnings to authorities in communities around the Pacific. Warning signs posted along coasts direct people to safety on higher ground (**Fig. 10.43d**)

Finally, individuals should learn to protect themselves during an earthquake. In your home, keep emergency supplies accessible, bolt bookshelves to walls, strap the water heater in place, install locking latches on cabinets, know how to shut off the gas and electricity, know how to find the exit, have a fire extinguisher handy, and know where to go to find family members. Schools and offices should have earthquake-preparedness drills. When an earthquake strikes, stay away from buildings. If you are trapped inside, a heavy table or solid door frame may provide protection (**Fig. 10.43e**). As long as lithosphere plates continue to move, earthquakes will continue to shake. But we can learn to live with them.

TAKE-HOME MESSAGE

Earthquakes are a fact of life on this dynamic planet. People in regions facing high seismic risk should build on stable ground, avoid unstable slopes, and build structures that can survive shaking. Individuals should learn what to do in the event of an earthquake.

QUICK QUESTION: What factors influence the degree of devastation during an earthquake?

FIGURE 10.43 Preventing damage and injury during an earthquake.

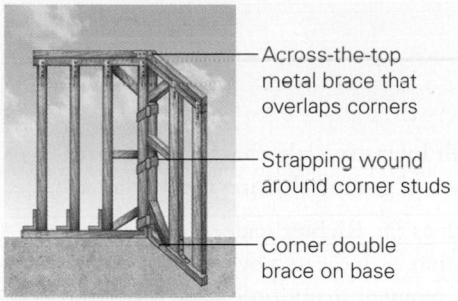

Across-the-top metal brace that overlaps corners

Strapping wound around corner studs

Corner double brace on base

Adding corner struts, braces, and connectors can substantially strengthen a wood-frame house.

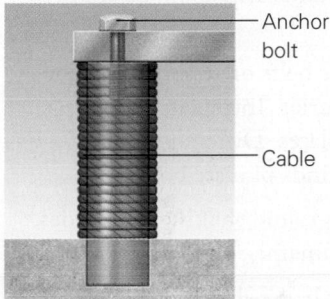

Anchor bolt

Cable

Wrapping a bridge's support columns in cable and bolting the span to the columns will prevent the bridge from collapsing so easily.

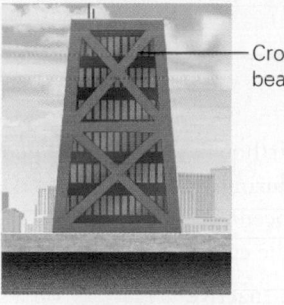

Cross-beam

Buildings are less likely to collapse if they are wider at the base and if crossbeams are added for strength.

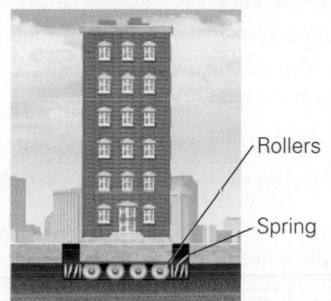

Rollers

Spring

Placing buildings on rollers or shock absorbers lessens the severity of the vibrations.

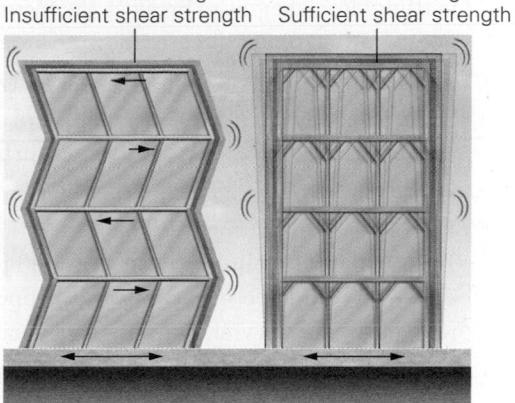

Unreinforced building: Insufficient shear strength

Reinforced building: Sufficient shear strength

(b) An unreinforced building will shear side to side in a way that causes floors to shift out of alignment.

(a) Damage can be prevented if buildings are designed to withstand vibration.

(c) Buoys can detect a tsunami in the open ocean so that people on land can be warned.

(d) Signs point out tsunami evacuation routes along the coast of Peru.

(e) If an earthquake strikes, take cover under a sturdy table near a wall.

Chapter 10 Review

CHAPTER SUMMARY

- Earthquakes are episodes of ground shaking. Earthquake activity, or seismicity, happens when rock slips during faulting. Rock begins to break at the focus (hypocenter). The point on the ground directly above the focus is the epicenter.

- Active faults are likely to host future movement. Inactive faults ceased being active long ago. Displacement on an active fault that intersects the ground surface may yield a fault scarp.

- During fault formation, rock bends elastically, then cracks. Eventually, cracks link to form a throughgoing fracture on which sliding occurs. When this happens, the elastic rebound generates vibrations (an earthquake).

- Once formed, faults exhibit stick-slip behavior. Most earthquakes happen when stress overcomes friction on a pre-existing fault.

- Earthquake energy travels in the form of seismic waves. Body waves (P and S) pass through the interior of the Earth, whereas surface waves (L and R) pass along the surface of the Earth.

- We can detect earthquake waves by using a seismometer. In principle, a seismometer consists of a weight whose inertia keeps it in place, while the Earth around it moves.

- Seismograms demonstrate that different types of earthquake waves arrive at different times. This difference allows seismologists to pinpoint the epicenter location.

- The Modified Mercalli Intensity scale characterizes earthquake size by perception of ground shaking and of damage.

- Magnitude scales, such as the Richter scale, are based on the amount of ground motion, as indicated by traces of waves on a seismogram. The moment magnitude scale takes into account the amplitudes of several seismic waves, the area of slip, and the amount of displacement.

- An M_W 8 earthquake yields about 10 times as much ground motion as an M_W 7 earthquake and releases about 32 times as much energy.

- Most earthquakes occur in seismic belts or zones, the majority of which lie along plate boundaries. Intraplate earthquakes happen in the interior of plates. Different kinds of earthquakes happen at different kinds of plate boundaries.

- Earthquake damage results from ground shaking, landslides, sediment liquefaction, fire, and tsunamis.

- Seismologists can identify seismic belts where earthquakes are most likely, and can determine the recurrence interval for great earthquakes. It may never be possible to pinpoint the exact time and place at which an earthquake will happen.

- Earthquake hazards can be reduced with better construction practices and zoning, and by educating people about what to do during an earthquake.

GUIDE TERMS

aftershocks (p. 329)
body wave (p. 333)
compressional wave (p. 333)
displacement (p. 326)
earthquake (p. 323)
elastic-rebound theory (p. 329)
elastic behavior (p. 327)
epicenter (p. 325)
fault (p. 323)

fault creep (p. 333)
fault scarp (p. 326)
fault trace (p. 326)
focus (p. 325)
foreshock (p. 329)
friction (p. 327)
induced seismicity (p. 348)
intensity (p. 338)
intraplate earthquake (p. 347)
magnitude (p. 340)

Modified Mercalli Intensity (MMI) Scale (pp. 338–339)
moment magnitude scale (p. 340)
recurrence interval (p. 360)
resonance (p. 366)
Richter scale (p. 340)
sediment liquefaction (p. 351)
seismic belt (p. 342)
seismicity (p. 325)

seismic wave (p. 333)
seismogram (p. 336)
seismologist (p. 324)
seismometer (p. 335)
shear wave (p. 333)
stick-slip behavior (p. 329)
stress (p. 327)
surface wave (p. 333)
Wadati-Benioff zone (p. 344)

 **GEOTOURS** *THIS CHAPTER'S GEOTOURS WORKSHEET (H) FEATURES QUESTIONS AND GOOGLE EARTH SITES ON:*

- Locating earthquakes
- Earthquake activity along a plate boundary
- Intraplate earthquake activity
- Earthquake prediction
- Tsunami devastation

REVIEW QUESTIONS

The letters following each Review Question refer to the corresponding Learning Objective from the Chapter Opener.

1. Compare normal, reverse, and strike-slip faults. **(A)**

2. Describe elastic-rebound theory and the concept of stick-slip behavior. **(A)**

3. Identify the focus and epicenter of the earthquake shown in the diagram. **(A)**

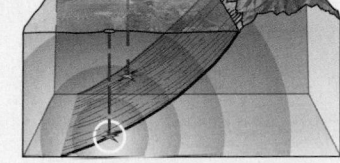

4. Explain how the vertical and horizontal components of earthquake motion are detected on a seismometer. **(C)**

5. Explain the differences among the scales used to describe the size of an earthquake. **(E)**

6. How does seismicity on mid-ocean ridges compare with seismicity at convergent or transform boundaries? Do all earthquakes occur at plate boundaries? **(B)**

7. What is a Wadati-Benioff zone? At what depth do the deepest earthquakes occur? **(B)**

8. Describe the types of damage caused by earthquakes. **(F)**

9. What are the four types of seismic waves? Which are body waves, and which are surface waves? Which type is depicted in the diagram? **(D)**

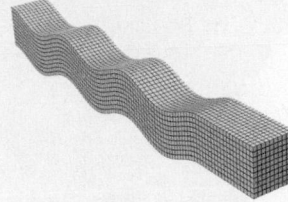

10. What is a tsunami, and why does it form? **(G)**

11. Explain how sediment liquefaction occurs in an earthquake and how it can cause damage. **(F)**

12. How are long-term and short-term earthquake predictions made? What is the basis for determining a recurrence interval, and what does a recurrence interval mean? What is an earthquake early warning system? **(H)**

13. What types of structures are most prone to collapse in an earthquake? What types are most resistant to collapse? **(I)**

ON FURTHER THOUGHT

14. Is seismic risk greater in a town on the west coast of South America or in one on the east coast? Explain your answer. **(B)**

15. The northeast-trending Ramapo fault crops out north of New York City. Precambrian gneiss forms the hills to the northwest of the fault, and Mesozoic sedimentary rock underlies the lowlands to the southeast. (You can see the fault on Google Earth by going to Latitude 41°10'21.12" N, Longitude 74°5'12.36" W. Where the fault crosses the Hudson River, there is an abrupt bend in the river. A nuclear power plant was built near this bend. Geologic studies suggest that the Ramapo fault first formed during the Precambrian, was reactivated during the

Paleozoic, and was active again during Mesozoic rifting. Imagine that you are a geologist with the task of determining the seismic risk of the fault today. What evidence of present-day or past seismic activity could you look for? **(B)**

16. On the seismogram of an earthquake recorded at a seismic station in Paris, France, the S-wave arrives 6 minutes after the P-wave. On the seismogram recorded at a station in Mumbai, India, for the same earthquake, the difference between the P-wave and S-wave arrival times is 4 minutes. Which station is closer to the epicenter? From the information provided, can you pinpoint the location of the epicenter? Explain. **(D)**

ONLINE RESOURCES

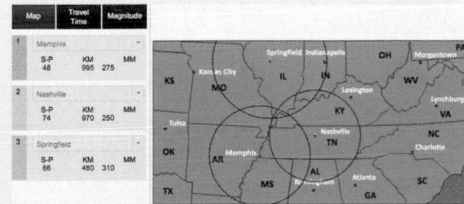

Animations
This chapter features animations that simulate and show the effects of tsunamis and an interactive activity on locating an earthquake's epicenter.

Videos
This chapter features videos on tsunami awareness and earthquake impacts.

Smartwork5
This chapter features analysis questions on earthquake magnitude, faults, and tsunamis.

Magnetic anomaly strength (nanoteslas)

+400
+100
+50
0
−50
−100
−400

INTERLUDE D

The Earth's Interior, Revisited: Seismic Layering, Gravity, and the Magnetic Field

By the end of this interlude, you should be able to . . .

A. explain how seismic waves behave as they pass through the Earth's interior.

B. describe how the study of seismic waves can tell us about layering inside the Earth.

C. provide a model illustrating how the surface elevation of lithosphere generally represents the consequences of Archimedes' principle.

D. explain how gravitational attraction varies with location and what these variations mean.

E. explain why the Earth's magnetic field exists and why its strength varies with location.

D.1 Introduction

Strange as it may seem, researchers had an understanding of our Solar System's configuration long before they had a clear concept of our own planet's internal structure. Why? We can see light-years into space just by looking up. But our eyes cannot see even a centimeter below the surface of the opaque soil and rock that make up the land. Tunnels and drillholes don't help much, for they literally only scratch the planet's surface. For example, even the world's deepest mine, a gold mine in South Africa that reaches a depth of 3.9 km, provides access to only the top 0.06% of the Earth; and the deepest drillhole, which penetrates 12.3 km into the crust beneath northwest Russia, samples only the top 0.19%. To study the Earth's interior, therefore, researchers seek insight from the study of **geophysics**, the subdiscipline of geoscience that focuses on understanding seismic waves, gravity, and magnetism. Geophysicists use mathematical calculations, instrumental measurements, and computer simulations to carry out their work.

This interlude builds on concepts introduced earlier in this book (in Chapters 2 through 6 and Chapter 10) to give you a sense of what geophysics can tell us about the Earth's interior. We begin by examining how seismic waves interact with the boundaries between the Earth's internal layers in a general sense, then discuss how an understanding of this interaction led first to the discovery of specific layers within the Earth,

and then later to an image of three-dimensional variations in the characteristics of each layer. We also discuss our planet's gravity field and examine why gravitational pull varies with location. The interlude concludes by revisiting the Earth's magnetic field, this time with a focus on understanding why the magnetic field exists.

D.2 The Basis for Seismic Study of the Earth's Interior

Setting the Stage

As discussed in Chapter 2, the first clues used to determine what's inside our planet came from measurements of the Earth's overall mass and shape. These measurements led geologists to conclude that the Earth consists of three concentric layers that differ from one another in terms of density **(Fig. D.1a)**. From the surface down, the layers are the crust, which has a low density; the mantle, which has an intermediate density; and the core, which has a high density.

To gain further insight into the nature of the crust, geologists examined samples of rocks exposed on land and samples dredged and drilled from the seafloor. To characterize the mantle, geologists analyzed rocks formed from magmas that originated in the mantle as well as chunks of the mantle brought up in those magmas. Further insight came from the study of meteorites thought to have come from planetesimals that had differentiated into a mantle and core before fragmenting. In fact, study of meteorites that came from the cores of planetesimals provides the only way for us to "see" core material firsthand.

Overall, this work led geologists to conclude that beneath a veneer of sediment, oceanic crust consists predominantly of mafic rock (basalt and gabbro), whereas continental crust consists of a variety of igneous and metamorphic rocks ranging from mafic to felsic in composition **(Fig. D.1b)**. The average chemical composition of the continental crust

FIGURE D.1 Simplified images of the Earth's interior.

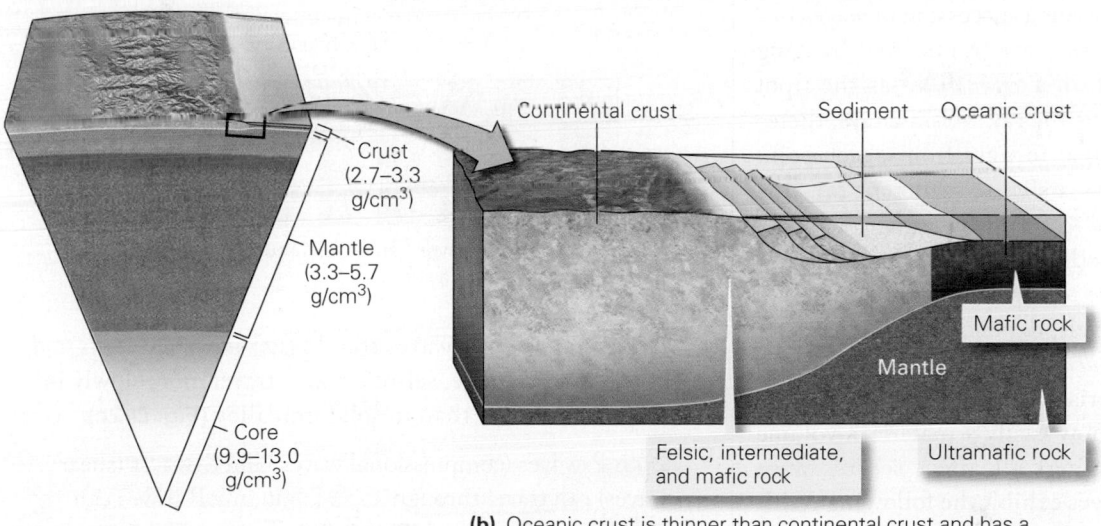

Crust
(2.7–3.3
g/cm³)

Mantle
(3.3–5.7
g/cm³)

Core
(9.9–13.0
g/cm³)

Continental crust   Sediment   Oceanic crust

Mafic rock

Mantle

Felsic, intermediate,
and mafic rock

Ultramafic rock

(a) The 19th-century three-layer image of the Earth.

(b) Oceanic crust is thinner than continental crust and has a different composition. The lower part of the continental crust tends to be more mafic than the upper.

◀ (facing page) A magnetic anomaly map of Missouri, Illinois, Indiana, and Kentucky shows variations in the magnetic field strength (purples and reds are positive; blues and greens are negative), and provides insight into the character of the crust below.

resembles that of granodiorite, a rock with a silica content part-way between that of granite and diorite. The mantle consists of ultramafic rock (peridotite), so its composition differs markedly from that of both kinds of crust. The core does not consist of rock at all, but rather of a metallic iron alloy.

To go beyond the basic understanding we've just reviewed, researchers searched for a tool that could provide an actual image of the Earth's interior. Study of seismic waves provides that tool. By measuring how fast seismic waves travel through the Earth and how they refract (bend) or reflect (bounce) as they travel, researchers could determine the exact depth to the crust-mantle boundary and to the mantle-core boundary, and they identified sublayers within the crust, mantle, and core, as we'll now see.

Seismic Wave Fronts and Travel Times

Energy travels from one location to another in the form of waves. For example, the impact of a pebble on the surface of a pond produces water waves that move outward from the point of impact and eventually cause a stick meters away to bob up and down. Similarly, the energy produced by a sudden rupture of intact rock or by the sudden slip of rock on a fault produces seismic waves that travel through rock or metal in the Earth. These waves transmit energy outward from the earthquake's focus (hypocenter) in all directions at once, and can eventually rattle distant seismometers.

The boundary between the rock through which a seismic wave has passed and the rock through which it has not yet passed is called a *wave front*. In three dimensions, a wave front expands outward from the earthquake focus like a growing bubble **(Fig. D.2a)**. We can represent a succession of waves in a drawing with a series of concentric wave fronts. The changing position of an imaginary point on a wave front as the front moves through rock defines a **seismic ray**. Seismic rays, there-fore, are lines drawn perpendicular to wave fronts; each point on a curving wave front follows a slightly different ray. The time it takes for a seismic wave to travel from the focus to a seismometer along a given ray is the **travel time** along that ray.

The ability of a seismic wave to travel through a material, and the velocity at which it travels, depends on the character of the material. Factors such as *density* (mass per unit volume), *rigidity* (the stiffness of a material, meaning its resistance to shearing), and *compressibility* (how easily a material's volume changes in response to squashing) all affect seismic-wave velocity. As a result, seismic waves exhibit the following traits:

- Seismic waves travel at different velocities in different rock types **(Fig. D.2b)**. For example, P-waves travel at 3.5 km per second in sandstone (a porous sedimentary rock) but at 8 km per second in peridotite (an ultramafic igneous rock).

- The velocity of seismic waves can change when the waves pass from one rock type into another.

FIGURE D.2 The propagation of earthquake waves.

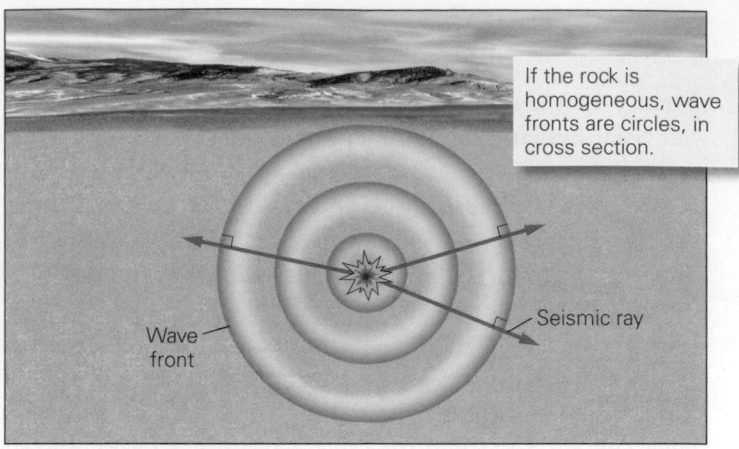

If the rock is homogeneous, wave fronts are circles, in cross section.

Wave front

Seismic ray

(a) An earthquake sends out waves in all directions. Seismic rays are perpendicular to wave fronts.

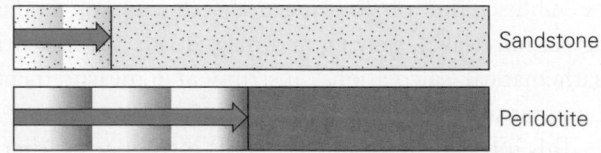

Sandstone

Peridotite

(b) Seismic waves travel at different velocities in different rock types. After a given time, the wave will have traveled farther in peridotite than in sandstone.

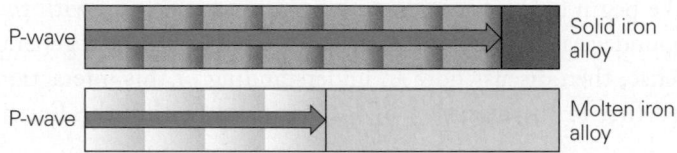

P-wave — Solid iron alloy

P-wave — Molten iron alloy

(c) P-waves travel faster in solid iron alloy than in liquid, such as molten iron alloy.

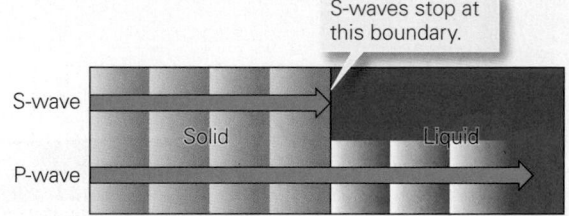

S-waves stop at this boundary.

S-wave

Solid   Liquid

P-wave

(d) Both P-waves and S-waves can travel through a solid, but only P-waves can travel through a liquid.

- In general, seismic waves travel faster in a solid than in a liquid. For example, seismic waves travel more slowly in molten iron alloy than in solid iron alloy **(Fig. D.2c)**.

- Both P-waves (compressional waves) and S-waves (shear waves) can travel through a solid, but only P-waves can travel through a liquid **(Fig. D.2d)**. To see why, picture what happens if you push down on the water surface in a pool—you send a pulse of compression to the bottom of the pool. Now move your hand sideways through the water. The water in front of your hand simply slides or flows past the water deeper down—your shearing motion has no effect on the water at the bottom of the pool **(Fig. D.3)**.

FIGURE D.3 Shear waves don't pass through a liquid.

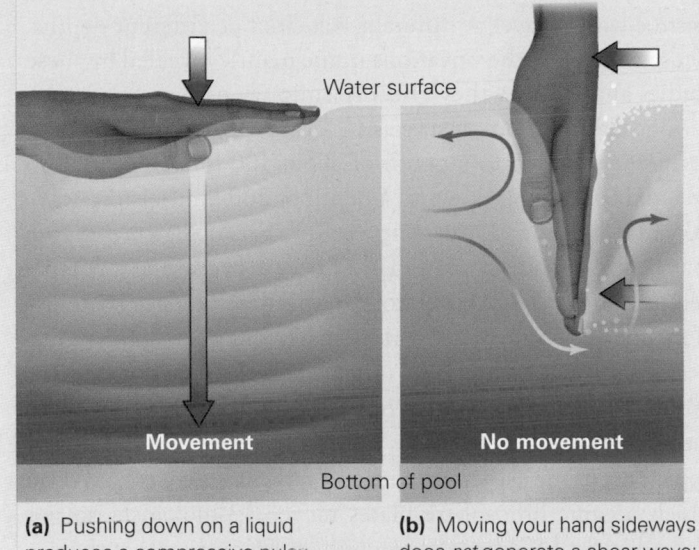

(a) Pushing down on a liquid produces a compressive pulse (P-wave) that can travel far through a liquid.

(b) Moving your hand sideways does *not* generate a shear wave; the moving water simply flows past deeper water.

Reflection and Refraction of Seismic Wave Energy

Shine a flashlight into a container of water so that the light ray hits the boundary between water and air at an angle. Some of the light bounces off the water surface and heads back up into the air, while some enters the water **(Fig. D.4a)**. The light ray that enters the water bends at the air-water boundary, so that the angle between the ray and the boundary in the air differs from the angle between the ray and the boundary in the water. Physicists refer to the light that bounces off the air-water boundary and heads back into the air as the *reflected light*; the light that bends at the boundary is the *refracted light*. The phenomenon of bouncing off is **reflection**, and the phenomenon of bending is **refraction**. Ray reflection and refraction take place at the interface between two materials if the wave travels at different velocities in the two materials.

The angle at which a reflected ray bounces off a boundary must be the same as the angle at which the incoming, or *incident*, ray strikes the surface. The angle by which a refracted ray bends at a boundary, however, depends on the contrast in wave velocity across the boundary and on the angle at which a ray hits the boundary. As a rule, if a ray enters a material through which it will travel more slowly, the ray bends down and away from the boundary **(Fig. D.4b)**. For example, the light ray on the left in Figure D.4b bends downward when hitting the air-water boundary because light waves travel more slowly in water. (To see why, picture a car driving from a paved surface diagonally onto a sandy beach—the wheel that rolls onto the sand first slows down relative to the wheel still on the pavement, causing the car to turn toward the sand.) Alternatively,

FIGURE D.4 Refraction and reflection of waves.

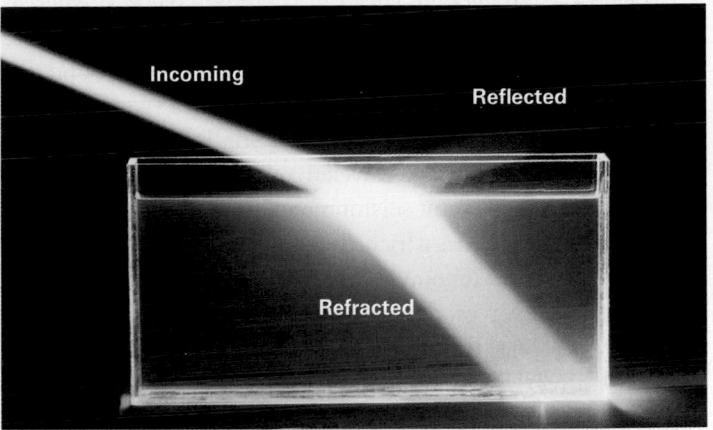

(a) A lab experiment showing how a ray of light partly reflect and partly refract when it crosses the boundary between two different materials.

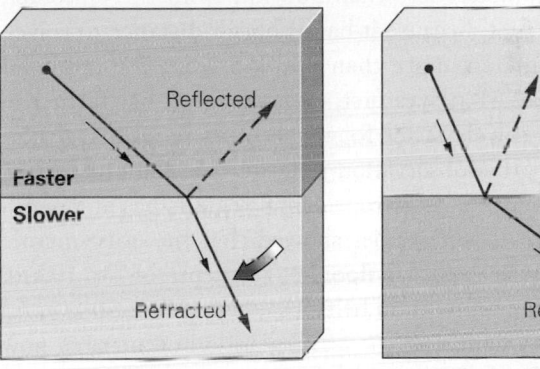

(b) Left: A ray that enters a material through which it travels more slowly bends away from the boundary. Right: A ray that enters a material through which it travels more rapidly bends toward the boundary.

if the ray were to pass from a layer in which it travels slowly into one in which it travels more rapidly, it would bend upward and toward the boundary.

D.3 Seismic Study of the Earth's Interior

Let's use our knowledge of seismic-wave velocity, refraction, and reflection to see how each of the major boundaries inside the Earth was discovered.

Discovering the Crust-Mantle Boundary

The knowledge that seismic waves refract at boundaries between the Earth's internal layers led to the identification of the crust-mantle boundary. In 1909, Andrija Mohorovičić, a Croatian seismologist, noted that P-waves arriving at seismometers less than 200 km from an earthquake's epicenter traveled at an average speed of 6 km per second, whereas P-waves arriving at seismometers more than 200 km from the epicenter traveled at an average speed of 8 km per second. To explain this observation, he suggested that P-waves reaching nearby seismometers followed a shallow path through the crust, in which they traveled relatively slowly, whereas P-waves reaching distant seismometers followed a deeper path through the mantle, in which they traveled relatively rapidly.

To understand Mohorovičić's proposal, examine **Figure D.5a**, which shows P-waves, depicted as rays, generated by an earthquake in the crust. Ray C, the shallower wave, travels through the crust directly to a seismometer. Ray M, the deeper wave, heads downward, refracts at the crust-mantle boundary, curves through the mantle, refracts again at the crust-mantle boundary, and then proceeds through the crust up to the seismometer. At seismometers less than 200 km from the epicenter, Ray C arrives first, because it has a shorter distance to travel. But at seismometers more than 200 km from the epicenter **(Fig. D.5b)**, Ray M arrives first, even though it has farther to go, because it travels faster for much of its length. Calculations based on this observation indicate that the crust-mantle boundary beneath most continental regions lies at a depth of 35 to 50 km. Later studies showed that the crust-mantle boundary beneath the ocean floor lies at a depth of 7 to 10 km, and beneath some mountain belts it's at a depth of 70 km. As we learned in Chapter 2, the crust-mantle boundary is now called the **Moho**, in honor of Mohorovičić.

Detailed analyses indicate that seismic waves travel faster in the lower continental crust than in the upper continental crust. To interpret these observations, geologists have measured wave velocities in rock samples of various compositions under laboratory conditions. Such experiments suggest that rocks in the upper crust are more mafic than those in the lower crust.

Defining the Structure of the Mantle

If the density, rigidity, and compressibility of mantle peridotite were exactly the same at all depths, seismic-wave velocities would be the same everywhere in the mantle, and seismic rays passing through the mantle would be straight lines. But, by studying travel times, seismologists have determined that seismic waves travel at different velocities at different depths. Let's now look at the variations in the mantle revealed by these studies and see how they affect seismic-ray paths.

Beneath oceanic crust, seismic-wave velocity in the mantle increases down to a depth of about 100 km. Beneath 100 km, and continuing down to a depth of 200 km, seismic-wave velocities become slower **(Fig. D.6a)**. Seismologists propose that this **low-velocity zone** (LVZ) occurs because at the temperature and pressure conditions in this depth range, 1% to 6% of mantle peridotite melts. The melt, a liquid, coats solid grains. Because seismic waves travel more slowly through liquids than through solids, the coatings of melt slow the waves down. In the context of plate tectonics theory, the LVZ represents the upper part of the asthenosphere and serves as the weak layer on which oceanic lithosphere plates move. Seismologists do not find a distinct LVZ beneath continents, suggesting that there is no partial melt in the upper asthenosphere beneath continents.

Below the LVZ under oceanic lithosphere, and throughout the mantle under continents, seismic-wave velocities increase progressively with depth in the mantle. Seismologists interpret this increase to mean that mantle peridotite becomes progressively less compressible, more rigid, and denser with depth. This proposal makes sense intuitively, considering that the weight of overlying rock increases with depth, and as pressure increases, the atoms making up rock squeeze together more tightly and are not so free to move. Because of refraction, the increase in seismic–wave velocity with depth causes seismic rays to curve in the mantle when traveling a long distance **(Fig. D.6b)**. So, at the scale of the mantle, the expanding "bubble" of a wave front actually has an elliptical shape.

To understand the shape of a curved ray, imagine representing a portion of the mantle with a series of imaginary layers, each of which has a slightly greater seismic-wave velocity than the layer above **(Fig. D.6c)**. Every time a seismic ray crosses the boundary between adjacent layers, it refracts a little toward the boundary. After the ray has crossed several layers, it has bent so much that it begins to head back up toward the top of the stack. If we replace the stack of distinct layers with

FIGURE D.5 Discovery of the Moho.

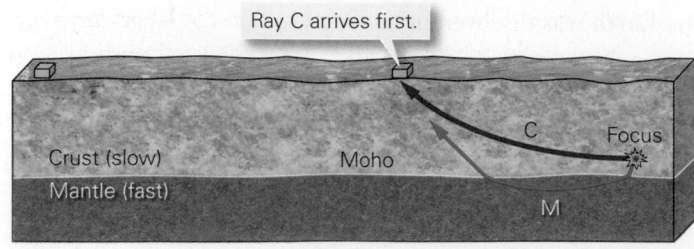

(a) Seismic waves traveling only in the crust reach a nearby seismometer first.

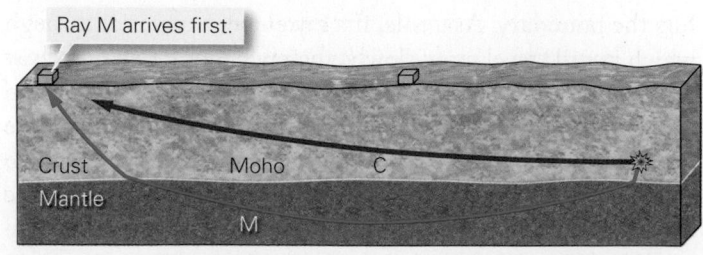

(b) Seismic waves traveling for most of their path in the mantle reach a distant seismometer first.

FIGURE D.6 The velocity of P-waves in the mantle changes because the physical properties of the mantle change with depth.

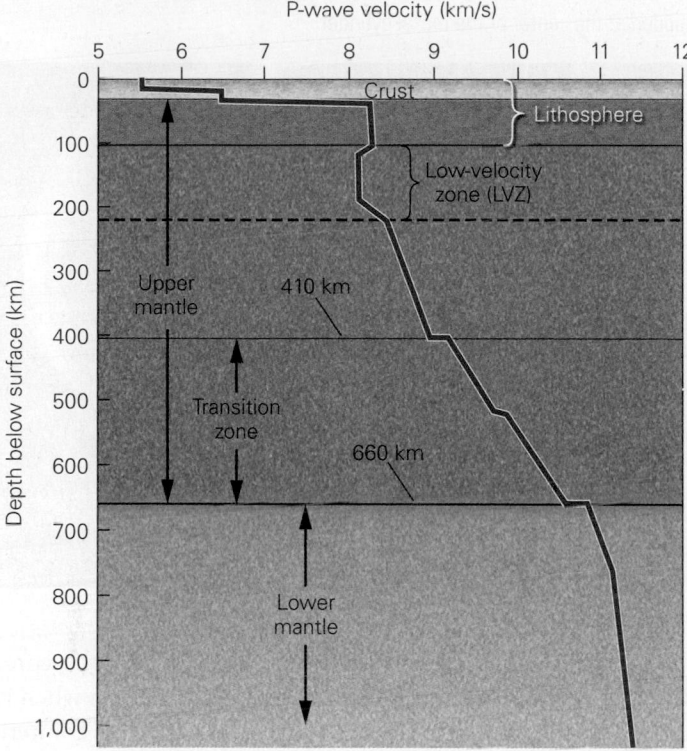

(a) The velocity of P-waves changes with depth in the mantle.

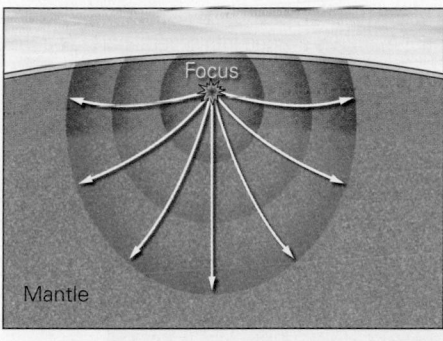

(b) The velocity of seismic waves increases with depth in the mantle, so rays curve and wave fronts are oblong.

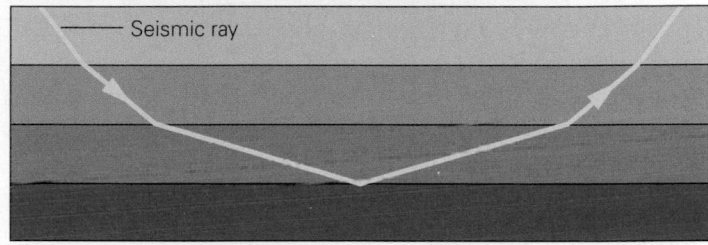

(c) In a stack of discrete layers, rays bend at each boundary. If the velocity is progressively faster in each lower layer, the ray eventually bends back and returns to the surface.

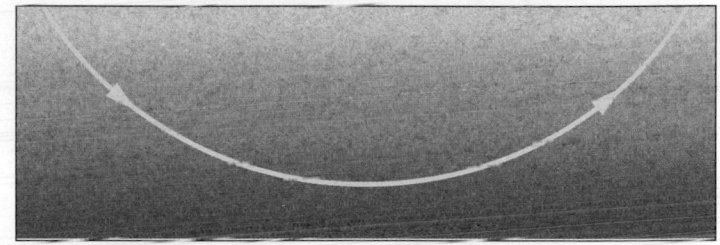

(d) In a material in which the velocity increases gradually with depth, rays curve smoothly and eventually return to the surface.

a single layer in which velocity increases with depth at a constant rate, the ray follows a smoothly curving path **(Fig. D.6d)**.

As shown in Figure D.6a, at depths between 410 km and 660 km, seismic velocity increases in a few abrupt steps, so in this interval, the stack of layers depicted in Figure D.6c actually provides a somewhat realistic image of how the mantle changes with depth. Experiments suggest that these **seismic-velocity discontinuities** occur at depths where pressure causes minerals in mantle rock to undergo a *phase change*, meaning that below the boundary, atoms crowd into a denser, more tightly packed configuration than they were above the boundary. This contrast means that, overall, the rock above the boundary has a different compressibility and rigidity than the rock below the boundary. Specifically, above the 410-km discontinuity, silicon, oxygen, iron, and magnesium, the most abundant elements of the mantle, bond together to form crystals of the mineral olivine, but below the discontinuity, the atoms bond together to form crystals of a mineral called magnesium spinel. Magnesium spinel can be stable down to a depth of about 660 km. Below this depth, atoms rearrange into crystals of an even denser mineral known as perovskite. Because of these seismic-velocity discontinuities, seismologists subdivide the mantle into the **upper mantle**, above 660 km, and the **lower mantle**, below 660 km **(Fig. D.7)**. The region between 410 km and 660 km, where successive seismic discontinuities occur, is the upper mantle's **transition zone**.

At this point, you may be wondering whether there is any way to test ideas about the identity of minerals that occur deep in the mantle. The answer is yes. Researchers use a laboratory device called a *diamond anvil* to simulate deep-mantle pressure and temperature conditions **(Fig. D.8)**. In a diamond-anvil experiment, tiny samples of minerals thought to occur in the mantle are squeezed between two diamonds. Due to the strength of diamonds, the pressure between the two diamonds can reach values comparable to those found in the deep mantle. Lasers can heat the sample to mantle temperatures during the experiment. Researchers then make measurements of the sample's physical characteristics. They use the results to calculate the sample's seismic velocity. By comparing the velocities measured in the laboratory with those observed for the real mantle, researchers gain insight into whether or not the laboratory sample could be a mantle mineral.

Discovering the Core-Mantle Boundary

During the first decade of the 20th century, seismologists installed seismometer stations around the world, expecting to

FIGURE D.7 The layering of the mantle, drawn to scale, beneath the ocean floor. Note that the upper mantle is divided into several parts and that, overall, it is much thinner than the lower mantle.

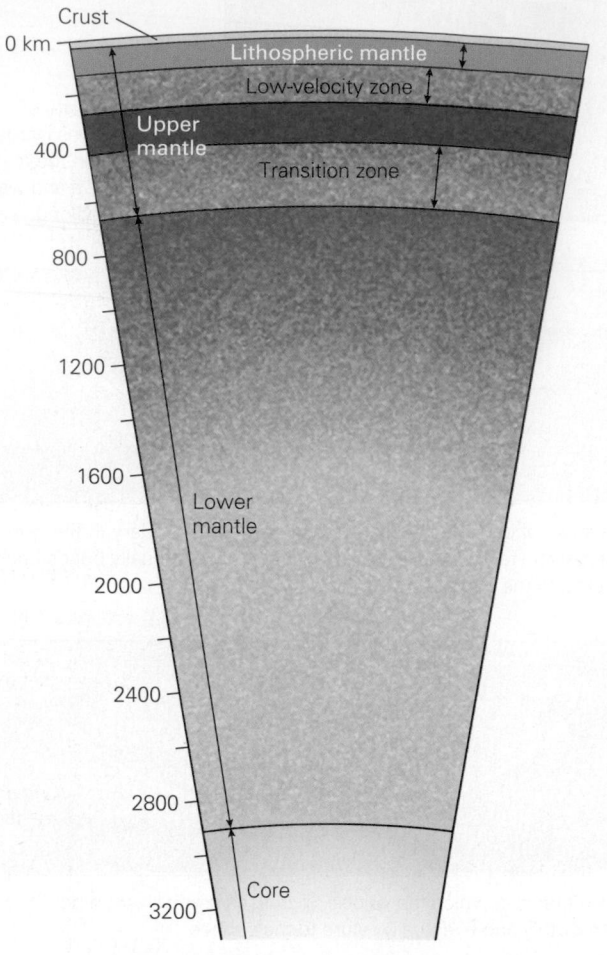

FIGURE D.8 A laboratory apparatus for studying the characteristics of minerals under very high pressures and temperatures. The green laser beam is heating up a microscopic sample being squeezed between two diamonds hidden at the center of the metal cylinder.

B penetrates the boundary and refracts down into the core. Ray B then curves through the core and refracts again when it crosses back into the mantle. As a consequence, Ray B intersects the surface at more than 143° from the epicenter.

Discovering the Nature of the Core

The iron alloy of the core contains about 85% iron, 5% nickel, and 10% of a lighter element (probably oxygen, silicon, and/or sulfur). Notably, the density of this iron alloy greatly exceeds that of mantle peridotite—in fact, the density contrast across the core-mantle boundary is greater than the density contrast between the Earth's crust and water! Seismological study provides insight into the physical state of the iron alloy in the core. Specifically, the downward bending of P-waves when they pass from the mantle into the core indicates that P-wave velocity, at least in the outer part of the core, is slower than in the mantle. So, even though the core is deeper and denser than the mantle, at least the outer part of the core must be less rigid than the mantle. How can this be? The answer to this question came from examining how S-waves pass through the Earth's interior. Seismologists found that S-waves do not arrive at any stations located beyond 103° from the epicenter. The band in which no S-waves arrive is called the **S-wave shadow zone**. The presence of this shadow zone indicates that S-waves cannot pass through the core at all. Keeping in mind that S-waves are shear waves, which by their nature can travel only through solids, the fact that S-waves do not pass through the core means that the core, or at least the outer part of it, consists of liquid **(Fig. D.9b)**.

be able to record waves produced by a large earthquake anywhere on the Earth. In 1914, one of these seismologists, Beno Gutenberg, noticed that P-waves from an earthquake do not arrive at seismometer stations lying in a band at a distance of between 103° and 143° from the epicenter, as measured along the surface of the Earth. This band is now called the **P-wave shadow zone (Fig. D.9a)**. If the density of the Earth increased gradually with depth all the way to the center, the shadow zone would not exist, because rays passing through the interior would reach every point on the surface. The presence of a shadow zone, therefore, means that deep in the Earth, a major interface exists at which seismic waves abruptly refract downward, implying that the velocity of the waves suddenly decreases. This interface, the **core-mantle boundary**, lies at a depth of about 2,900 km.

To see why the P-wave shadow zone exists, follow the two seismic rays labeled A and B in Figure D.9a. Ray A curves smoothly in the mantle (we are ignoring seismic-velocity discontinuities in the mantle, for simplicity) and passes just above the core-mantle boundary before returning to the surface. It reaches the surface at 103° from the epicenter. In contrast, Ray

FIGURE D.9 Shadow zones and the discovery of the Earth's core. (Note that the circumference of a circle is 360°, so it is 180° from a given point to a locality on the other side of the planet.)

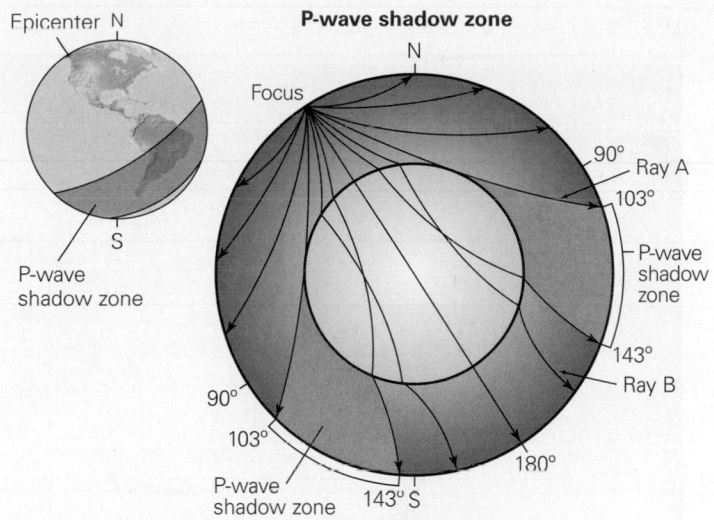

(a) P-waves do not arrive in the P-wave shadow zone because they refract at the core-mantle boundary.

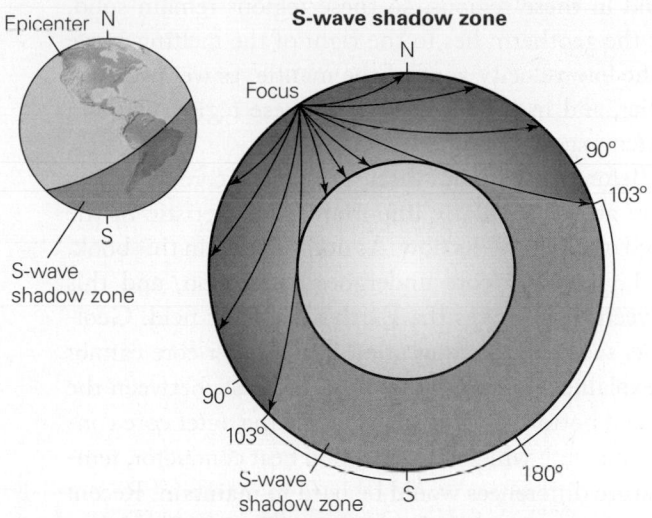

(b) S-waves do not arrive in the S-wave shadow zone because they cannot pass through the liquid outer core.

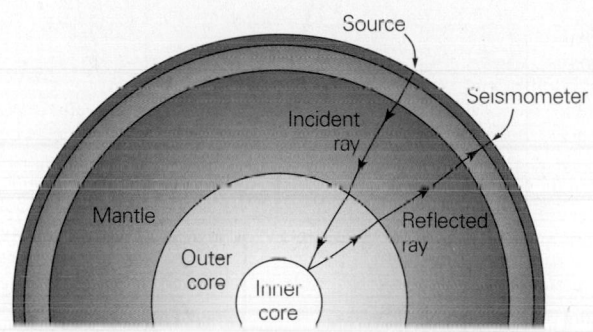

(c) Seismic waves reflect off the inner core–outer core boundary.

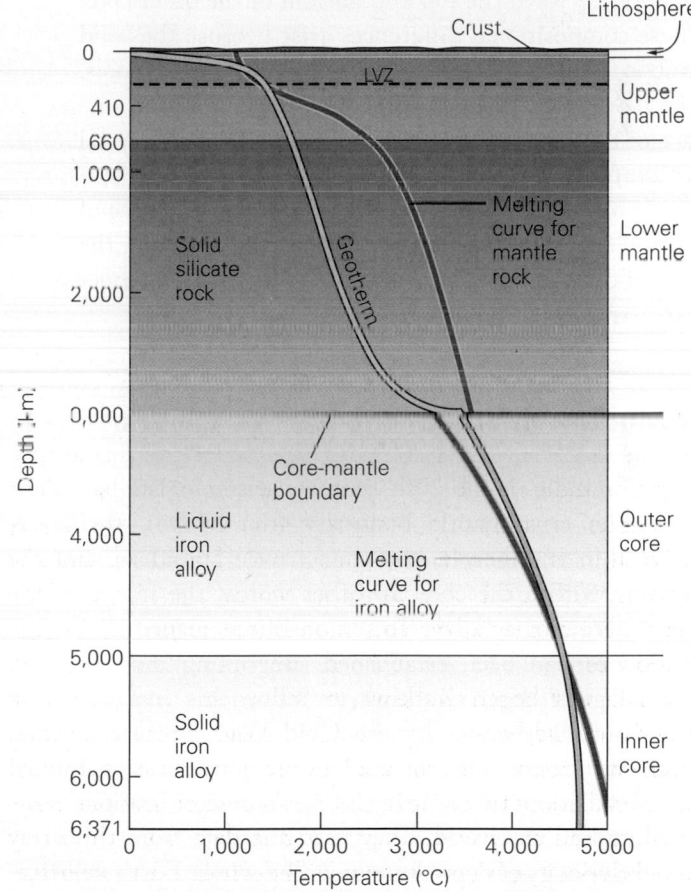

(d) A graph of the geotherm and melting curve for the Earth. Note that the melting temperature is less than the Earth's temperature in the outer core, so the outer core is molten.

At first, seismologists thought that the entire core might be liquid iron alloy. But in 1936, a Danish seismologist, Inge Lehmann, discovered that P-waves passing into the core reflect off a boundary within the core. Based on this observation, she proposed that the core consists of two parts: an **outer core**, made of liquid iron alloy, and an **inner core**, made of solid iron alloy. Lehmann's work defined the existence of the inner core but could not locate the depth of the inner core–outer core boundary. This depth was eventually located by measuring the exact time it took for seismic waves generated by nuclear explosions to penetrate the Earth, bounce off the inner core–outer core boundary, and return to the surface **(Fig. D.9c)**. The measurements showed that this boundary lies at a depth of about 5,155 km.

Why does the core have two layers—an outer liquid one and an inner solid one? An examination of **Figure D.9d** provides some insight. This graph shows two curves: the *geotherm*, which indicates how temperature changes with increasing depth in the Earth, and the *melting curve*, which indicates how the temperature at which materials start to melt changes with

increasing depth in the Earth. As the graph shows, the geotherm lies to the left of the melting curve through most of the mantle and in the inner core. This means that temperatures in most of the mantle and in the inner core do not get high enough to cause melting under the intense pressures found in these regions, so these regions remain solid. But the geotherm lies to the right of the melting curve in the low-velocity zone of the mantle, as we discussed earlier, and in the outer core, so these regions contain molten material.

Before we leave our discussion of the core, let's look again at a particularly important characteristic of the liquid outer core—its flow. As noted earlier in this book, the liquid outer core undergoes convection, and this convection generates the Earth's magnetic field. Geologists suggest that convection in the outer core cannot be explained entirely by thermal contrasts between the top and bottom of the core. Because the outer core consists of metal, which is a very good heat conductor, temperature differences would be hard to maintain. Recent research suggests that the density differences leading to convection may be due to differences in chemical composition between the top and bottom of the outer core. These compositional differences arise because the solid inner core is slowly growing as the Earth, overall, cools. The new crystals of solid iron that form along the surface of the inner core do not have room in their crystal structure for low-density elements such as silicon, sulfur, or oxygen. So these elements move into the base of the molten outer core. At any given time, therefore, the base of the outer core has a lower density than the upper parts, and it starts to rise convectively.

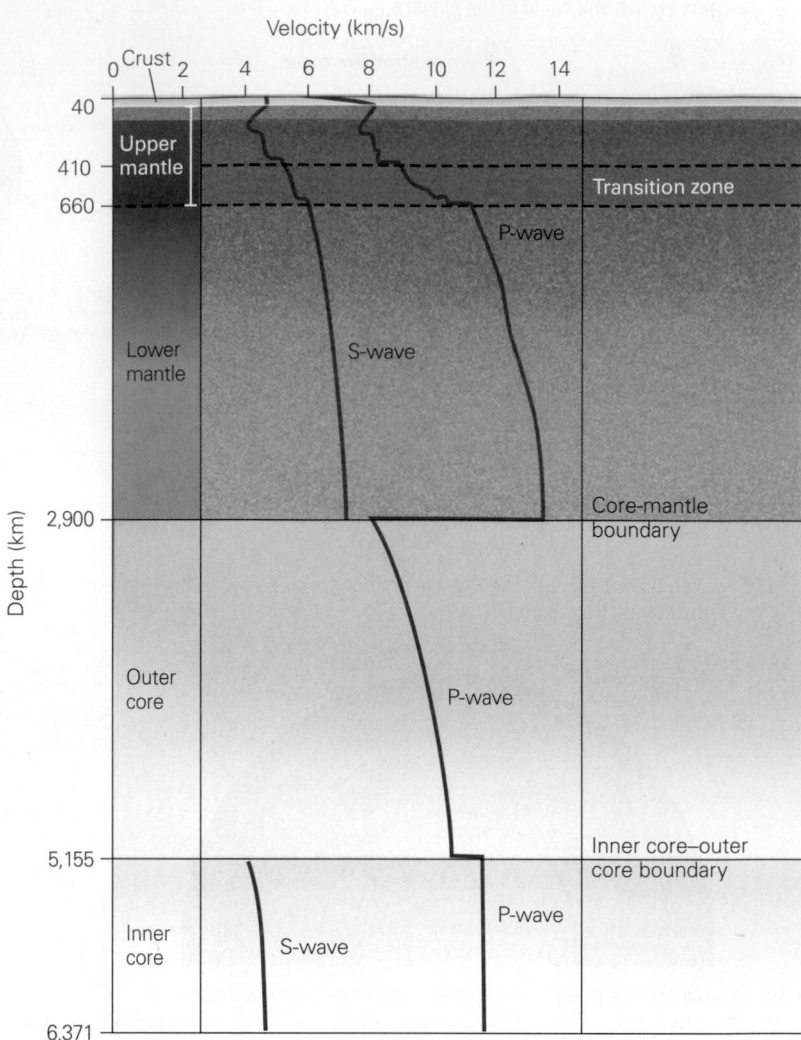

FIGURE D.10 The velocity-versus-depth profile of the whole Earth.

A Modern Image of the Earth's Layers

By the middle of the 20th century, seismologists had identified the crust-mantle boundary (the Moho), the layering within the mantle, the core-mantle boundary, and the layering within the core. In other words, the image of the Earth's interior as a layered, onion-like sequence of concentric zones had been established. Beginning in the 1950s, seismologists began working to refine this image, a task ironically made easier by the Cold War. Because of their desire to detect nuclear explosions, governments funded the installation of an array of seismometer stations scattered around the world. They used the data from this array to develop a graph now known as the whole Earth **velocity-versus-depth curve**. We've already presented the upper part of the curve (see Fig. D.6a)—**Figure D.10** presents the rest of it. This curve shows the depths at which seismic-wave velocity suddenly changes. These changes define the principal layers and sublayers in the Earth. Note that the graph does not show a velocity for S-waves in the outer core because S-waves cannot travel through a liquid.

Seismic Tomography

In recent decades, seismologists have developed a technique, called **seismic tomography**, to produce three-dimensional images of variation in seismic-wave velocities in the Earth's interior. This technique resembles the method used to produce three-dimensional CAT (computerized axial tomography) scans of the human body. In seismic tomographic studies, researchers compare the observed travel time of seismic waves following a specific ray path with the predicted travel time that waves following the same path would have if the average velocity-versus-depth model of Figures D.6 and D.10 were completely correct. They found that waves following some paths take more time than predicted, whereas waves following other paths take less time than predicted. By repeating the measurements for many different wave paths in many different directions, researchers can outline three-dimensional regions of the mantle in which travel times are unexpectedly fast or unexpectedly slow. Fast regions are probably cooler and more

FIGURE D.11 Tomographic images of the Earth's interior and their interpretation. Regions in which seismic-wave velocities are slower than expected are depicted by warm colors (yellow to red) and may be warmer than their surroundings. Regions in which velocities are faster than expected are depicted with cool colors (green to blue) and may be relatively cooler.

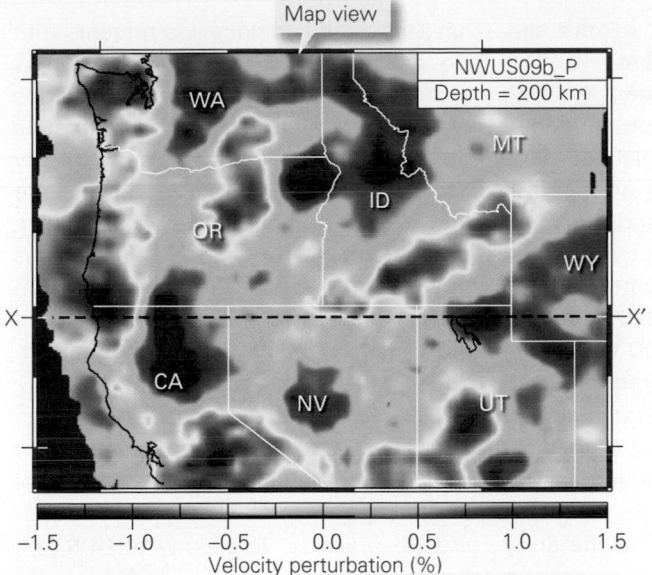

(a) A tomographic map, at a depth of 200 km, using data from EarthScope.

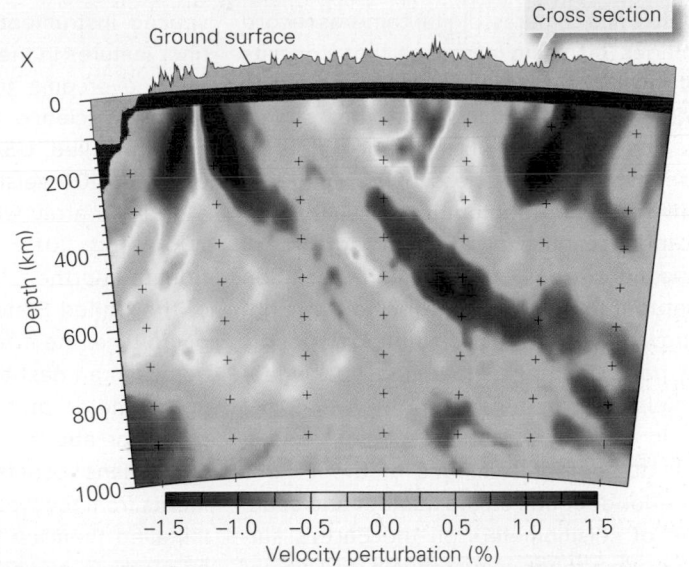

(b) A cross section of the Earth along line XX' beneath the western United States.

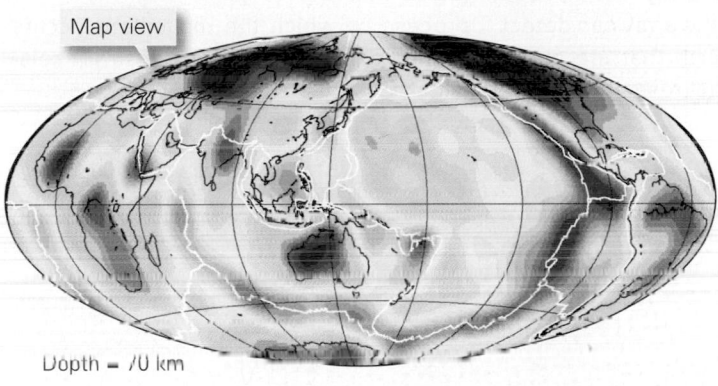

(c) A tomographic map image of the whole Earth showing relative seismic velocities at a depth of 70 km. Note that red (warmer) mantle underlies mid-ocean ridges.

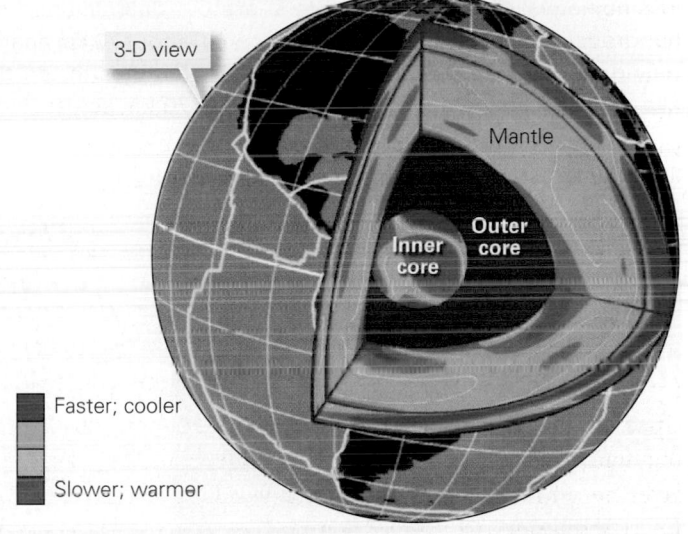

(d) A three-dimensional tomographic image of the Earth. Note that the outer core, a liquid, does not display velocity variations.

rigid than their surroundings, and slow regions are probably warmer and less rigid than their surroundings.

Tomographic studies emphasize that the simple onion-like image of the Earth, with velocities increasing with depth at the same rate everywhere, is an oversimplification. In reality, the velocities of seismic waves vary significantly with location at a given depth. Results of tomographic studies can be displayed by maps, cross sections, or three-dimensional models **(Fig. D.11)**. Generally, warmer colors (reds) on these images indicate slower regions, whereas cooler colors (blues and purples) indicate faster regions.

Seismic tomographic studies provide new insight into the Earth's interior. For example, the studies show that a region of cool mantle lies in the asthenosphere beneath North America. This region may represent the remnants of oceanic lithosphere that was subducted during the Mesozoic. The observation that the cool region extends into the lower mantle suggests that some subducted plates sink down into the lower mantle. In fact, some researchers suggest that there is a "plate graveyard" at the base of the mantle, where denser portions of subducted plates accumulate. Tomographic images of the uppermost mantle (70-km depth) show a region of warm mantle that occurs in the region

Resolving the Details of the Earth's Interior with EarthScope

Unlike film cameras, digital cameras record images by using a sensor that consists of an array of points, called pixels, each of which detects the light that strikes it. The *resolution* of an image refers to the detail or sharpness of an image—a low-resolution photograph of a highway billboard might show a blurry image of dark and light areas, while a high-resolution photograph of the same billboard might show distinct letters. Resolution depends, in part, on the number and spacing of pixels used on the sensor.

In a similar way, the resolution of seismic tomographic images of the Earth's interior depends on the number and spacing of seismometers on the Earth's surface. As a rough rule of thumb, the size of features visible on a tomographic image is roughly the same as the spacing between seismometers. So, if seismometers are hundreds of kilometers apart, they can only detect features that are hundreds of kilometers across. Data from such widely spaced instruments cannot resolve distinct features in the lithosphere.

To overcome this problem, the U.S. National Science Foundation funded a project called *USArray*, which consists of about 400 seismometers arranged in a grid-like array within a band. Between 2004 and 2013, the bands extended from the northern to the southern border of the United States **(Fig. BxD.1)**. At any given time, the array was 700 to 800 km across in an east-to-west direction. Each seismometer of the array remained in place for about 2 years, but a team of technicians constantly removed instruments from the west side of the array and installed them on the east, so the array as a whole moved eastward across the country like a giant bulldozer tread. After 2013, the array moved to Alaska.

Instruments in the array are spaced about 70 km apart, so the array can detect features inside the Earth that are about 70 km across. Individual research groups have installed additional instruments with closer spacing (10 km) in smaller areas to obtain even higher-resolution images of specific structures. The initial results of studies using records from USArray have revealed features in the mantle that researchers previously didn't know existed. Knowledge of these features has provided new insight into the convection of the mantle, the nature of hot spots, and the causes of uplift and sinking (subsidence) of the Earth's surface.

USArray is one component of an even larger project called *EarthScope*. Other projects operating under the banner of EarthScope include the Plate Boundary Observatory. This array of GPS stations and other instruments documents the exact distribution of relative motion between the North American and Pacific plates in order to gain insight into the process by which the movement occurs and to potentially better constrain seismic risk.

FIGURE BxD.1 USArray seismometers of EarthScope.

(a) A USArray seismometer after being placed in a vault, as viewed looking down into the vault before it was covered.

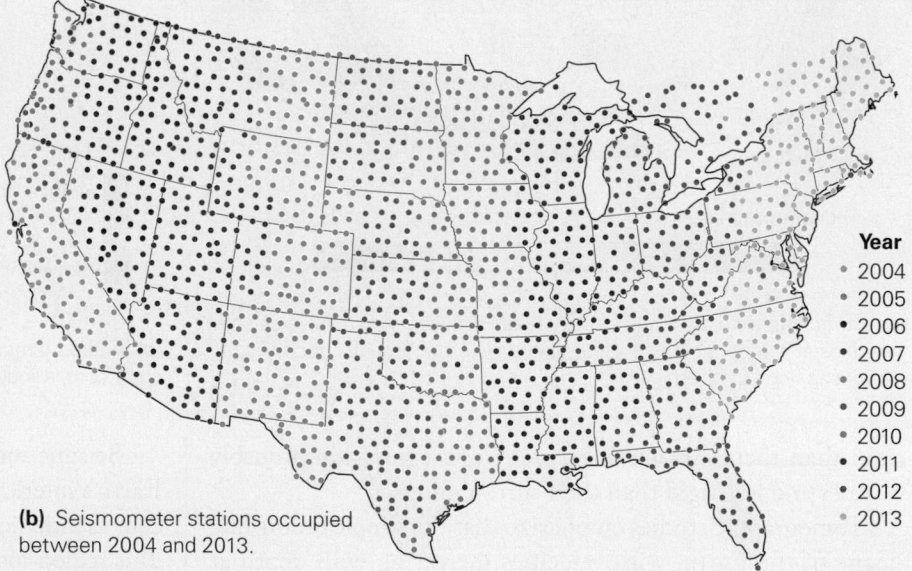

(b) Seismometer stations occupied between 2004 and 2013.

Year
- 2004
- 2005
- 2006
- 2007
- 2008
- 2009
- 2010
- 2011
- 2012
- 2013

---

beneath mid-ocean ridges (see Fig. D.11c). This result supports the concept that warm asthenosphere rises beneath ridges.

Tomographic studies focused on the very deep Earth have defined a 200-km-thick layer of warmer mantle just above the core-mantle boundary. This layer, called the D″ layer (pronounced dee double prime), may represent the region in which the mantle has absorbed heat radiating from the core. Some researchers speculate that it may be the source of mantle plumes, but this proposal remains a subject for debate.

On a global scale, seismologists have been able to outline broad areas of **mantle upwelling**, where warmer and less dense mantle is rising, and broad areas of **mantle downwelling**, where cooler and denser mantle is sinking. The existence of these zones characterizes the large-scale complex pattern of mantle convection. Recently, aided by the installation of arrays of more closely spaced seismometers **(Box D.1)**, seismologists may even be able to see localized mantle plumes associated with hot-spot volcanoes.

Tomographic studies of the heterogeneities in the mantle have inspired researchers to develop supercomputer models of what this convection looks like **(Fig. D.12a)**. By modeling the mantle, researchers have gradually changed their impressions of how subducted plates look. In earlier images of subduction, plates were pictured as rather rigid sheets of lithosphere that maintained their thickness and overall character down to depth. Now, however, researchers think of subducted plates as dense, very viscous fluids that sink like heavy cream poured into coffee, though much more slowly **(Fig D.12b)**. Heat transfer between subducting lithosphere and the surrounding asthenosphere causes the plates to warm as they descend, and the adjacent asthenosphere to cool. As a consequence, a subducting plate—defined now simply as a slab of mantle rock—does not maintain a constant thickness as it sinks, but may actually thicken as the adjacent asthenosphere becomes cool enough to start sinking along with the lithosphere. If you picture a subducting plate as a dense, viscous plastic material sinking downward, you may not be surprised that some plates fold into back-and-forth zigzags, like cake batter poured into a bowl, or that plates thin and detach into separate segments. If, as it descends, a subducting slab encounters mantle rock with increasing viscosity, it may reach depths where its density is no longer greater than that of the surrounding asthenosphere. In fact, the slab may pool at levels where the asthenosphere's viscosity increases abruptly (such as the seismic-velocity discontinuity at 660 km). Keep in mind that all this complex flow-like motion takes place because the mantle is so hot that rock, though solid, behaves plastically, meaning that it can slowly flow at rates of a few centimeters per year. Even at these slow rates, subducting plates can contort into many shapes over the course of tens of millions of years. Though many important questions remain, tomography has led geologists to picture the Earth's insides as a dynamic place, not just a region of static, concentric shells **(Fig. D.12c)**.

We've focused on the mantle so far. Can seismic tomography also see variability inside the core? Studies of the inner core have shown that it contains a distinct pattern of cooler and warmer regions (see Fig. D.11a). Significantly, the orientation of this pattern in the core changes over time relative to the

FIGURE D.12 Images of the Earth's mantle.

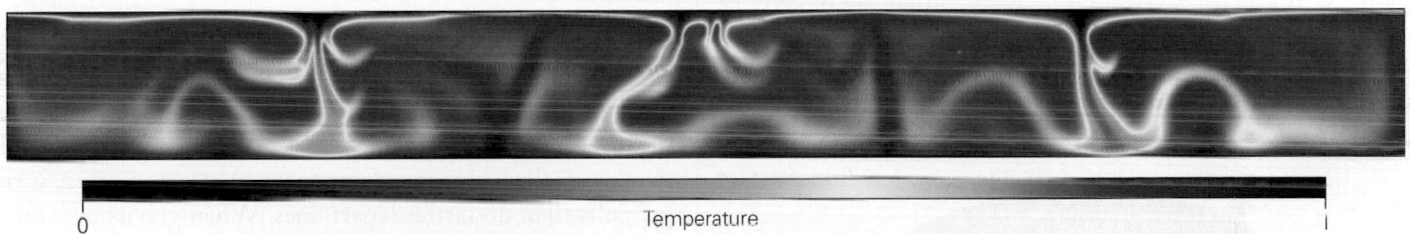

0   Temperature   1

(a) A computer model that simulates convective cells formed in part of the mantle. Blue areas are cooler, downwelling regions, whereas dark red areas are warmer, upwelling regions. Computer models provide insight into how the real Earth might work.

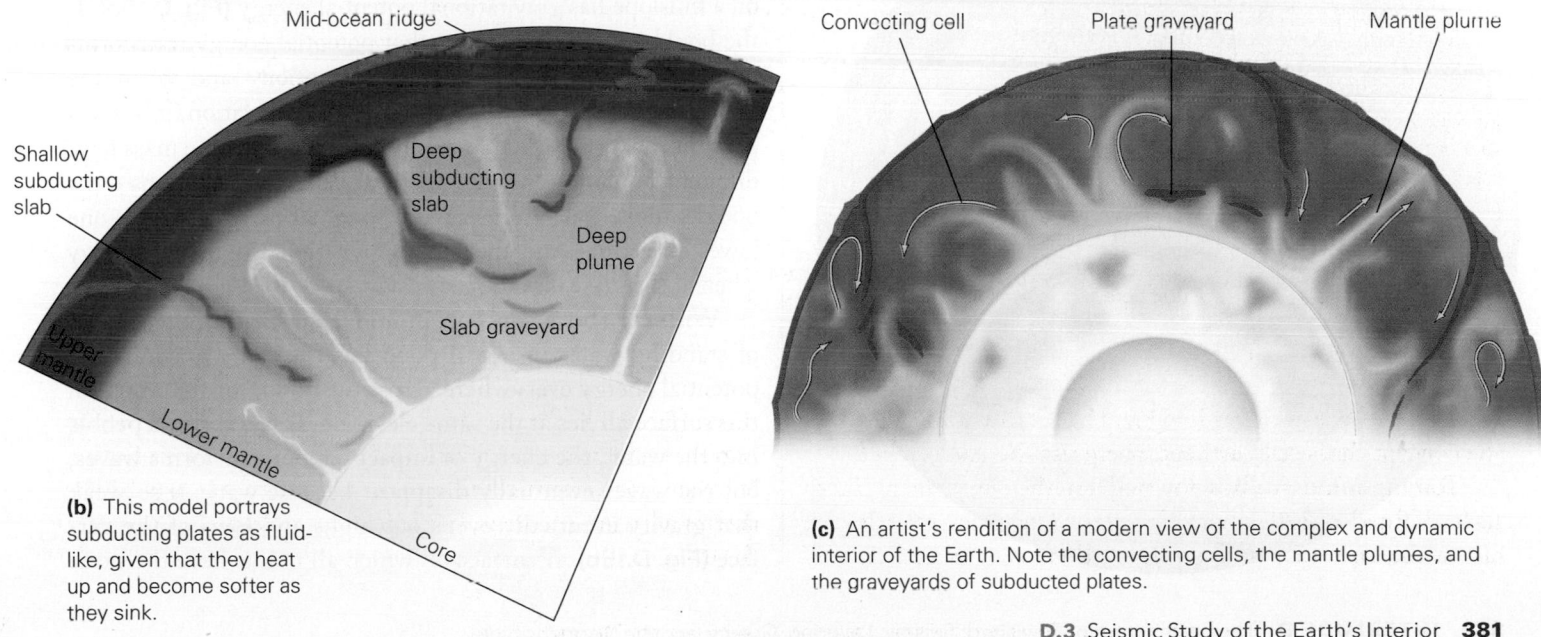

(b) This model portrays subducting plates as fluid-like, given that they heat up and become softer as they sink.

(c) An artist's rendition of a modern view of the complex and dynamic interior of the Earth. Note the convecting cells, the mantle plumes, and the graveyards of subducted plates.

orientation of the pattern in the mantle. Researchers interpret this to mean that the inner core rotates slightly faster than the rest of the Earth—it makes an extra rotation about once every several hundred years. This process has come to be known as **core superrotation**. Notably, this superrotation may be driven by interaction with the Earth's magnetic field **(Fig. D.13)**.

Seismic-Reflection Profiling

So far, we've focused our discussion in this interlude on how seismological research provides a picture of the deep interior of the Earth. Seismic techniques also help geologists fine-tune our image of the upper crust. During the past half century, geologists have found that by exploding dynamite, by banging large weights against the Earth's surface, or by releasing bursts of compressed air into the ocean, it's possible to create artificial seismic waves that propagate down into the Earth and reflect off the boundaries between different layers of rock in the crust. By recording the travel time of the reflected waves, geologists can determine the depth to these boundaries. With this information, they can produce a cross-sectional view of the crust called a **seismic-reflection profile (Fig. D.14a–c)**. Such an image can define subsurface contacts between beds and stratigraphic formations and can also reveal the presence of subsurface folds (bends in layers) and faults. Oil and gas companies purchase many seismic-reflection profiles, despite their high cost, to identify likely locations of energy resources underground. Research geologists have used these profiles to obtain images of the lower crust, the Moho, and even the upper mantle.

In the past decade, data-processing techniques have become so sophisticated, and computers have become so fast, that geologists can now produce three-dimensional seismic-reflection images of the crust **(Fig. D.14d)**. These images provide so much detail that they can even show an individual ribbon of sandstone, representing the channel of an ancient stream, buried kilometers below the Earth's surface.

D.4 The Earth's Gravity

We can develop additional understanding of the complexity of the Earth's interior from the study of gravity and magnetism, the two field forces that emanate from our planet. Recall that a *field force* can invisibly apply a push or pull across a distance. In other words, a field force exerted by one object can cause another object to *accelerate* (change speed and/or direction) without ever touching it. In this section, we discuss gravity, and in the next, we'll turn our attention to magnetism.

The Geoid

Gravity is an attractive field force that one mass exerts on another. As Isaac Newton first recognized, the magnitude of gravitational pull depends on the size of the masses and on the distance between them—larger masses exert stronger pulls than do smaller ones, and closer-together masses produce stronger pulls than do farther-apart ones. When gravity acts on an object but can't make the object move, then the object has *gravitational potential energy*. For example, a boulder sitting at rest on a hillslope has gravitational potential energy **(Fig. D.15a)**. If the boulder starts to tumble, that potential energy transforms into kinetic energy—the energy of motion—and when the boulder ultimately comes to rest at a lower elevation, it has less potential energy. Note that two boulders of the same mass have different potential energies when at different elevations—the boulder higher on a hill has more potential energy than the one lower on a hill—and the same gravitational potential energy when at the same elevation.

With the above concepts in mind, we see that the surface of standing water in a still pond has the same gravitational potential energy everywhere across the pond, for the water on this surface all lies at the same elevation. If you throw a pebble into the water, the energy of impact temporarily forms waves, but the waves eventually disappear because water is so weak that gravity eventually evens out highs and lows of the surface **(Fig. D.15b)**. A surface on which all points have the same

FIGURE D.13 The inner core of the Earth rotates about 0.5° per year faster than the rest of the Earth. The figure shows the changes in location of a reference line, anchored to the inner core, relative to the Earth's surface.

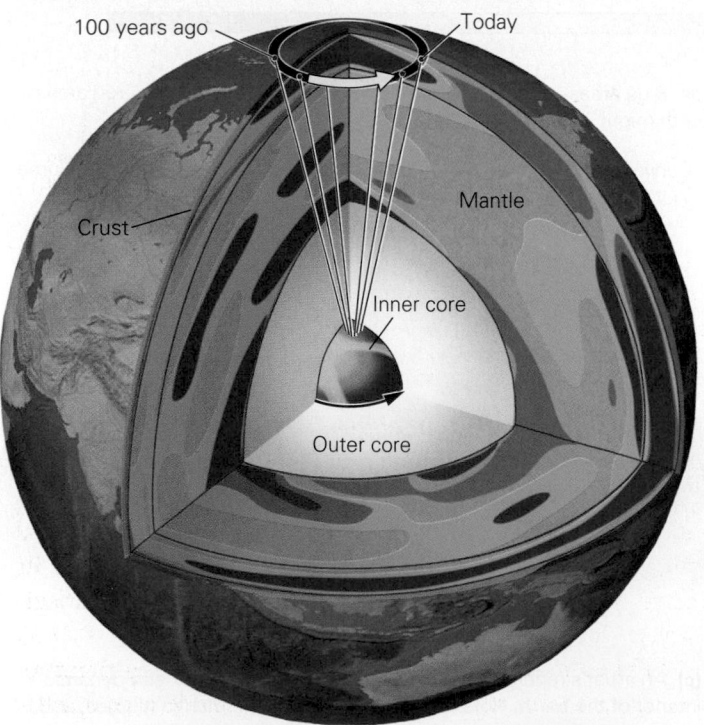

FIGURE D.14 Seismic-reflection proling.

(a) Trucks thumping on the ground to generate the signals needed for making a seismic-reflection profile.

(b) Geologists at Shell Oil Company process seismic data.

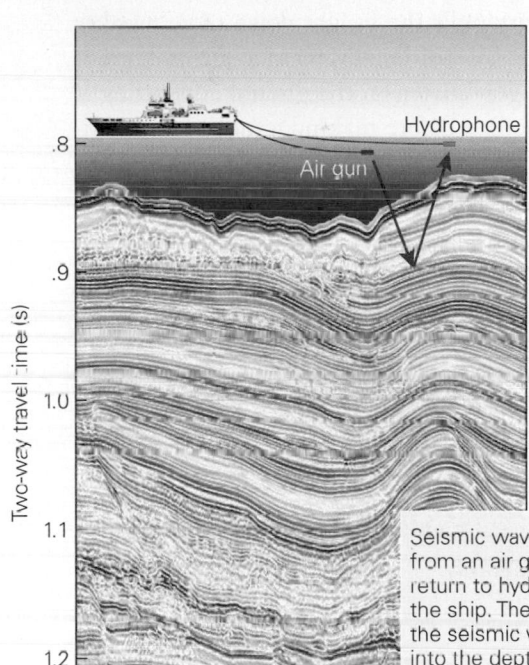

Seismic waves sent into the seafloor from an air gun bounce off layers and return to hydrophones also towed by the ship. The two-way travel time of the seismic waves can be translated into the depth of the layers below the surface.

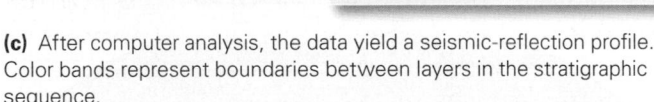

(c) After computer analysis, the data yield a seismic-reflection profile. Color bands represent boundaries between layers in the stratigraphic sequence.

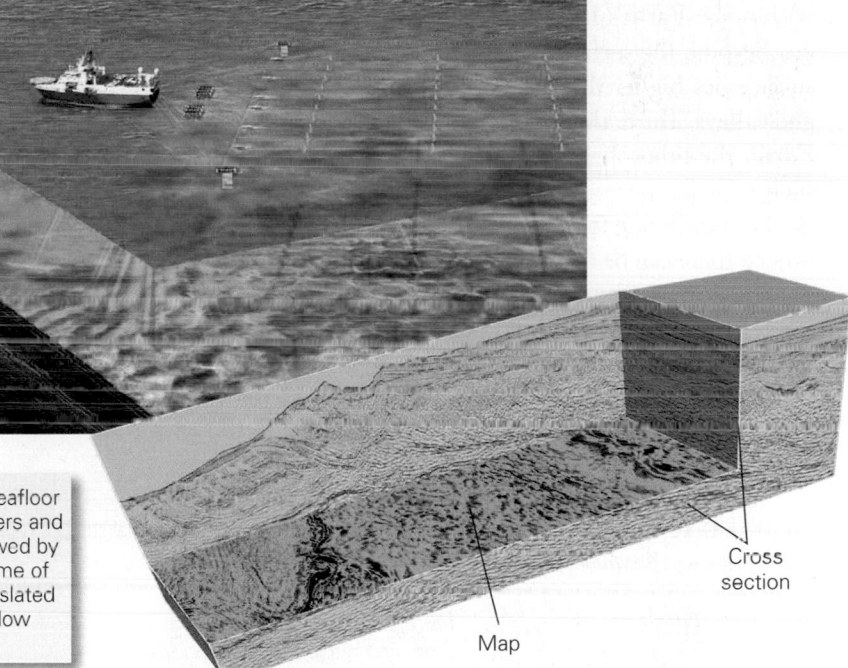

(d) Data can be used to produce a three-dimensional image of the subsurface. Geologists can study both map surfaces and cross sections through the block.

potential energy, such as the surface of standing water in a pond, is called an **equipotential surface**.

What does the gravitational equipotential surface of the Earth look like? Let's start to address this question by imagining what sea level would look like (not what it really is) if an ocean completely surrounded a stationary (nonrotating), perfectly smooth, spherical globe on which there were no winds or currents. This equipotential surface would itself be a sphere. In reality, however, the Earth's equipotential surface

is not a perfect sphere. Part of this deviation is due to our planet's spin on its axis. Rotation produces centrifugal force, which flattens the Earth into a *spheroid*. The imaginary spheroid that most closely has the shape of the Earth has a slight equatorial bulge, so that its radius at the equator is 6,378 km, while its radius at the poles is 6,357 km. Geologists refer to this imaginary equipotential surface as the *reference spheroid* (**Fig. D.16a**).

While the reference spheroid serves as a starting point for describing the Earth's gravity, it doesn't provide a completely

FIGURE D.15 The concept of gravitational potential energy.

(a) Potential energy increases as you go up the hill. Boulder 1 has the most potential energy. Boulders 2 and 3 have the same potential energy. Boulder 4 has less.

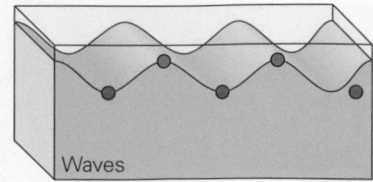

Waves

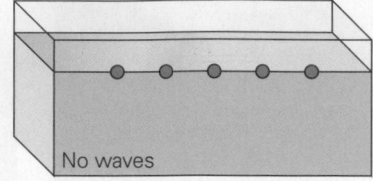

No waves

(b) Crests of waves have higher potential energy than do troughs. Without waves, water is an equipotential surface.

the geoid differs from that of the reference spheroid only slightly—the highest point on the geoid lies 85 m above the reference spheroid, and the lowest point lies 107 m below the reference spheroid. But even these small differences can noticeably influence the orbits of satellites or the accuracy of land surveys.

Gravity Anomalies

Geologists can measure the pull of gravity at a point on the Earth by using an instrument called a **gravimeter**, which consists of a weight hanging from a delicate spring—stronger gravity pulls the weight down more; weaker gravity pulls it down less. Traditionally, measurements of gravity required placing a gravimeter on the ground and letting it sit long enough to stabilize before making a measurement. In recent years, researchers have developed methods that allow satellites (such as the GRACE satellite, launched by NASA) to detect tiny variations in the Earth's gravitational pull at the surface over broad areas very rapidly. On-land and satellite-based measurements reveal that the pull of gravity at the surface varies with location and differs from the value that the geoid indicates.

How can we describe variations in the pull of gravity? Recall that gravity can cause an unrestrained object to accelerate, or increase its velocity. Researchers describe the acceleration

accurate picture for many reasons. First, the density of material within the Earth's layers is not uniform throughout those layers. Second, the surface of the Earth is not smooth, for the land surface lies higher than the seafloor, and both have mountains and valleys. Third, the rock making up the outermost layer of the Earth, the lithosphere, is strong enough to hold up heavy loads such as volcanoes or glaciers and to allow depressions such as trenches to persist for a long time. Because of these factors, and others, there can be more mass or less mass at a location than the reference spheroid predicts, so the real equipotential surface for the Earth has bumps and dimples. Geologists refer to this irregular surface, which provides the best representation of the Earth's equipotential surface, as the **geoid (Fig. D.16b)**. The surface of

FIGURE D.16 Representations of the geoid, the shape of an equipotential surface representing the Earth's gravity. Colors represent *elevations* of the geoid relative to the reference spheroid.

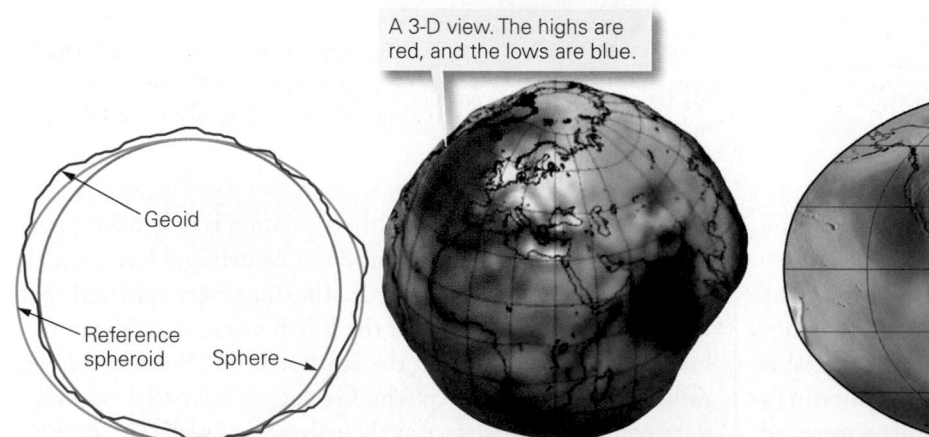

A 3-D view. The highs are red, and the lows are blue.

Geoid

Reference spheroid

Sphere

(a) The geoid is a spheroid with bumps and dimples (greatly exaggerated here).

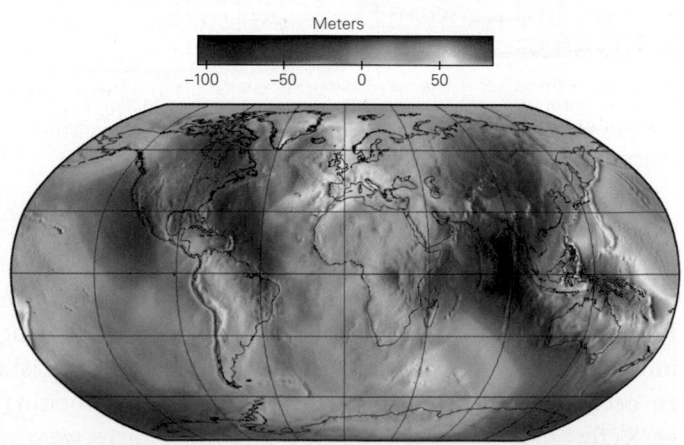

Meters

−100 −50 0 50

(b) A map of the geoid. Note the deep low in the Indian Ocean and the high in northeastern Atlantic.

caused by the Earth's gravity using a unit called the *Gal* (named for the Italian astronomer Galileo, 1564–1642), where 1 Gal = 1 cm/s². On average, the acceleration due to the Earth's gravity has a value of 981 Gal (≈ 9.8 m/s²), but it ranges from 976 Gal to 983 Gal. Differences in the gravitational pull between one location and an adjacent one may be so tiny that geologists must specify them in milliGals (mGal), where 1 mGal = 1/1,000 Gal. Geologists refer to a deviation of the geoid from the reference spheroid as a **gravity anomaly**. Gravity pulls more strongly over a *positive gravity anomaly* and pulls less strongly over a *negative gravity anomaly*. A positive anomaly indicates that extra mass lies below the site, perhaps due to a body of particularly dense rock underground, whereas a negative anomaly means that a deficit of mass lies below the site, perhaps due to the presence of less-dense rock **(Fig. D.17)**.

As a practical point, mapping of gravity anomalies helps geologists to find reserves of valuable metal ores (rocks that can be mined and processed to yield metal) underground, because metal ore tends to be denser than other rock. And from an academic perspective, gravity anomalies help geologists to detect upwelling and downwelling zones in the mantle.

The Concept of Isostasy

Archimedes (ca. 287–212 B.C.E.), a Greek mathematician, had been puzzling over the question of why some objects float and others sink, until one day when, according to legend, he noticed the water level rise in a tub as he stepped in to take a bath. He realized that this meant that an object sinking into water displaces the water, and that if the density of the object exceeds that of water, it sinks, but if it is less dense, it floats. Furthermore, a floating object sinks into water only until it has displaced an amount of water whose mass equals the mass of the whole object. This concept is now known as **Archimedes' principle (Fig. D.18)**.

According to Archimedes' principle, a cargo ship anchored in a harbor floats at just the right level so that the mass of the water displaced by the ship equals the mass of the whole ship. A ship, in effect, is just a steel-sheathed

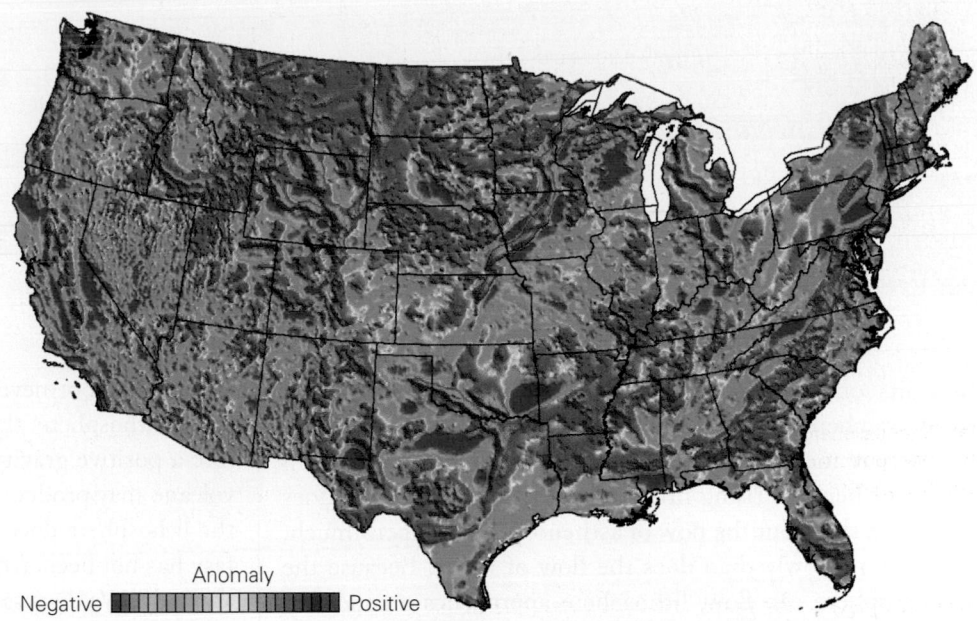

FIGURE D.17 A gravity anomaly map of the United States. Colors represent variations in the magnitude of gravitational pull: red is stronger, blue is weaker.

bubble—it floats because its overall average density, taking into account the fact that most of a ship's volume consists of air, is less than that of water, even though its hull consists of heavy steel. When the ship remains empty, the distance between its deck and the water surface—its so-called *freeboard*—can be large **(Fig. D.19)**. Addition of heavy cargo causes the ship's hull to settle down deeper into the water, so its freeboard decreases. Such movement can happen only if the water beneath the ship can flow out of the way. If the ship's keel rests on the harbor's solid floor, addition of cargo will not change its freeboard, because the floor would hold the ship up. When the freeboard of the ship has just the right value for a given cargo, so that ship's freeboard does not "want" to rise or sink, we say that the ship is in **isostatic equilibrium**, or that a condition of **isostasy** exists.

As we learned in Chapter 4, the outer layer of the Earth, the lithosphere, can be thought of as a relatively rigid shell

FIGURE D.18 According to Archimedes' principle of buoyancy, an iceberg sinks until the total mass of the water displaced equals the total mass of the whole iceberg. Since water is denser, the volume of the water displaced is less than the volume of the iceberg, so 13% of the iceberg protrudes above the water.

FIGURE D.19 A cargo ship illustrates the concept of isostasy. At isostatic equilibrium, the freeboard reflects Archimedes' principle, so the freeboard decreases as the mass of the ship increases. Water flows out of the way as the ship sinks downward (right image).

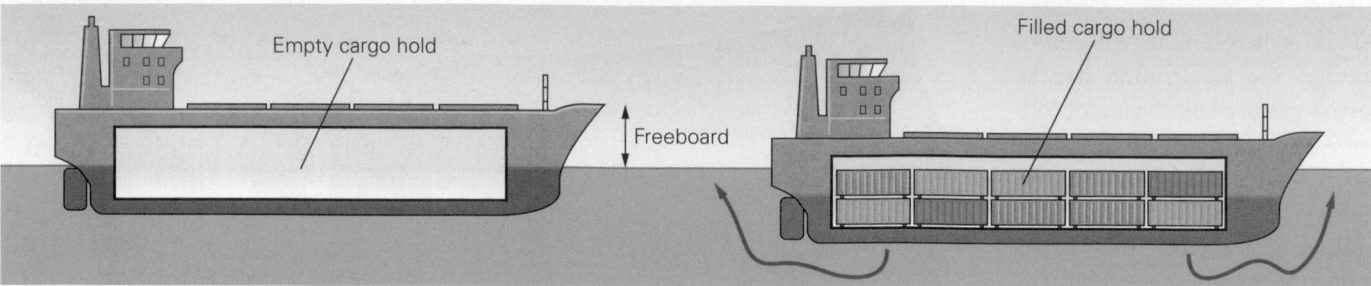

that sits on the underlying, relatively soft asthenosphere. Asthenosphere can flow out of the way when the base of the lithosphere moves downward, or it can flow back in to keep the space beneath rising lithosphere filled, just as water does beneath a ship. But the flow of asthenosphere happens much, much more slowly than does the flow of water. Because the asthenosphere can flow, lithosphere approaches a condition of isostatic equilibrium, over broad scales, in most places on the Earth. This means that the elevation of the lithosphere's surface reflects the density and thickness of the lithosphere. For this reason, the abyssal plains of oceans (underlain by relatively dense, cold, and thick lithosphere) lie lower than mid-ocean ridges (underlain by relatively warm, less-dense, and thinner lithosphere). Further, addition of a load, such as the growth of a large glacial ice sheet or the building of a large volcano, causes the lithosphere to sink, whereas removal of a load, say, by the melting of an ice sheet or erosion of a volcano, causes the lithosphere to rise.

At the local scale, however, lithosphere does not necessarily remain in isostatic equilibrium. First, deviations from isostasy develop because asthenosphere flows so slowly that the lithosphere hasn't had time to sink enough or rise enough to accommodate a change in loading in some places. As an example, regions that were covered by huge continental glaciers during the most recent ice age, which ended only about 10,000 years ago, are still in the process of rising and are not yet in full isostatic equilibrium. Second, since the lithosphere itself has strength, it can hold up relatively small loads without sinking. To understand why, imagine the surface of a trampoline. If a heavy person steps on it, the trampoline's surface bends down. But if a small block of wood sits on it, the surface supports the load and doesn't bend.

In locations where isostatic equilibrium has not been achieved, we observe gravity anomalies. Examples include deep-sea trenches, where the lithosphere has been bent and held down by the process of subduction—since the density of water filling the trench is much less than that of rock, there is a deficit of mass over a trench and, therefore, a negative gravity anomaly. In places where a mountain consists of particularly dense rock, but nevertheless can be held up by the strength of the lithosphere, there may be an excess of mass, and therefore a positive gravity anomaly. Similarly, rapid eruption of a volcano may produce a load on the lithosphere so quickly that the lithosphere does not have time to sink—here again, isostasy has not been achieved, and we'll detect a positive gravity anomaly. We discuss isostasy further in Chapter 11, where we explore controls on the height of mountain ranges.

D.5 The Earth's Magnetic Field, Revisited

As we learned in Chapters 2 and 3, the Earth produces a magnetic field. Here we look more closely at the magnetic field, focusing on why it exists and on why variations in its strength can develop. To set the stage, we review the fundamentals of magnetism.

A Brief Review of the Earth's Magnetism

If you hold a magnet over steel paper clips, it lifts the clips by exerting an attraction or pull on them that, at close range, exceeds the force of gravity. A magnet can also produce a repulsion, capable of pushing an object away. Recall that the push or pull exerted by a magnet represents *magnetic force*, and that the region around a magnet in which we can detect a magnetic force is a *magnetic field*. Magnetic fields can be represented by curving lines, called *magnetic field lines* **(Fig. D.20)**; a compass needle placed in the field aligns with these lines. All magnets have two *magnetic poles*, north at one end and south at the other. Opposite poles attract, and like poles repel, so the north end of one magnet attracts the south end of another, but it repels the north end of another. An imaginary arrow, called a **dipole**, indicates the direction from one pole to the other. Magnetic field lines form a continuous loop through the magnet's dipole.

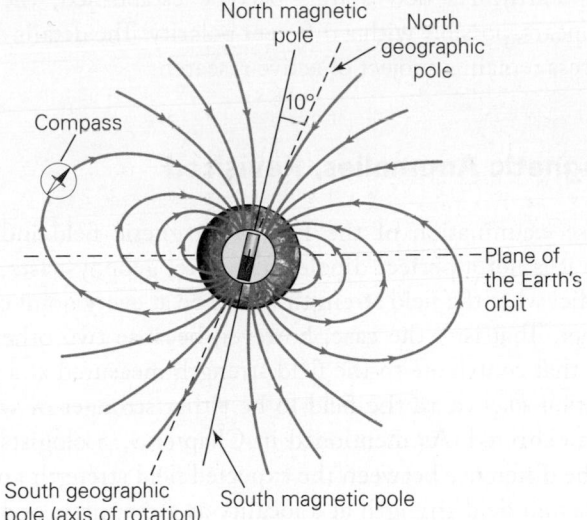

FIGURE D.20 The geometry of the Earth's dipole field.

In Chapter 2, we learned that our planet's magnetic field deflects solar wind and traps deadly cosmic rays. (Of note, not all planets have magnetic fields; in fact, neither Mars nor Venus, the two planets most like the Earth, has a measurable field.) The dipole representing most of the Earth's field currently lies about 10° from the Earth's axis of rotation—so, in the northern hemisphere, the *magnetic pole* (the point where the magnetic dipole intersects the planet's surface) currently lies about 500 km from the *geographic pole* (the point where the Earth's axis intersects the Earth's surface). Magnetic field lines are parallel to the planet's surface at the magnetic equator and are perpendicular to it at the magnetic poles. The angle between the magnetic dipole and the axis changes over time, so the location of the magnetic pole on the surface of the Earth changes. Currently, the magnetic pole moves at about 55 km per year. Measurements suggest that on a time scale of thousands of years, the Earth's magnetic poles follow looping paths, but don't stray more than about 15° of latitude from the geographic poles (see Fig. 2.3).

Geologists, by convention, refer to the tail end of the dipole, which lies close to the north geographic pole, as the **north magnetic pole**; the arrowhead end of the dipole, which lies near the south geographic pole, is the **south magnetic pole**. Based on this convention, the north-seeking end of a compass needle points toward geographic north. (This convention can be a bit confusing because, to a physicist, the north end of a magnet corresponds to the arrowhead end, and the arrowhead representing the Earth's dipole currently lies in the south.) As we've seen, the polarity of the Earth field suddenly changes every now and then—during times of reversed polarity, the dipole arrow points toward geographic north. The last reversal happened about 730,000 years ago.

Origin of the Earth's Magnetic Field

How can magnetic fields be produced? In an *electromagnet*, the field develops when a current runs through a wire. In a *permanent magnet*, the magnetism comes from the arrangement of atoms. Each atom can be pictured as a tiny dipole whose magnetism arises, simplistically, from the orbiting of its electrons. In nonmagnetic materials, these dipoles are randomly oriented, so the fields they produce cancel one another out **(Fig. D.21)**. But in a magnetic material, such as solid iron, the dipoles lock into parallelism with one another, so that the field produced by each adds to the fields produced by the others to create an overall measurable magnetic field.

Why does the Earth have a magnetic field? Though diagrams commonly portray the field as emanating from a giant permanent magnet in the interior, that's not the real situation, for at the temperatures that occur deep in the Earth, even iron cannot behave as a permanent magnet—the atoms in a hot material tumble chaotically, so their dipoles cannot align and lock into the same orientation. The path toward discovering the origin of the Earth's magnetic field started in 1926, when researchers learned that our planet's outer core consists of liquid iron alloy. Because flow of metal can produce an electrical current, it seemed that the flow of the outer core must somehow be responsible for the magnetic field. But how?

Clues to understanding the production of the Earth's magnetic field come from electrical power plants. In a power plant, water or wind power spins a wire coil (an electrical conductor) around an iron bar (a permanent magnet). Such an apparatus is called a **dynamo**. Because an electrical current in a wire produces a magnetic field, movement of a wire relative to a magnetic field produces an electrical current. The Earth's interior behaves, effectively, like a dynamo. In the case of the Earth, flow in the outer core serves the role of the spinning wire coil. But the inner core

FIGURE D.21 The origin of permanent magnetization in a material.

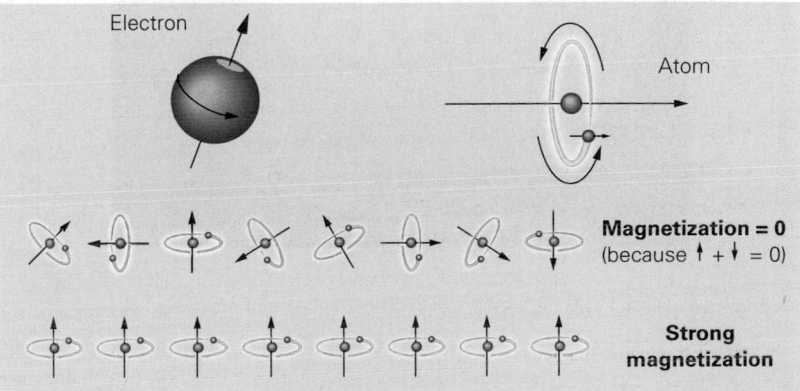

can't serve the role of the permanent magnet because it is so hot. Researchers suggest instead that the Earth is a **self-exciting dynamo**. Evidently, during the Earth's early history, flow in the outer core took place in the presence of a magnetic field. This flow generated an electrical current, and once the current existed, it generated a magnetic field. Continued flow in the presence of the generated magnetic field produced more electrical current which, in turn, maintains the magnetic field. Once started, the system perpetuates itself, as long as there's an input of energy to keep the outer core in motion.

The key to understanding the orientation of the Earth's magnetic dipole comes from the geometry of the convection cells in the outer core. Researchers suggest that the Coriolis force (see Chapter 18) causes convective cells in the outer core to become spirals that align roughly with the Earth's axis of rotation **(Fig. D.22)**. The spirals wobble, so at any given time the overall dipole axis they generate is not exactly parallel to the axis of rotation. Notably, the spirals apply a force to the inner core, causing it to rotate slightly faster than the rest of the Earth—the phenomenon we introduced earlier as core superrotation.

Why does the polarity of the Earth's magnetic field reverse every now and then? Computer models suggest that reversals happen because the convective spirals in the outer core don't remain stable. Over time, they slow and fade away as they interact, causing the magnetic field to weaken or disappear temporarily. As new spirals become established, the field reappears, possibly with a different polarity. The details of this process remain a subject of active research.

Magnetic Anomalies, Revisited

Close examination of the Earth's magnetic field indicates that it is not a perfect dipole. If it were, geophysicists could predict what the field strength would be at every point on the planet. That isn't the case, however, because two other factors that contribute to the field strength measured at a given location may cause the field to be either stronger or weaker than expected. As mentioned in Chapter 3, geologists refer to the difference between the expected field strength and the measured field strength at a locality as a **magnetic anomaly**. A positive anomaly occurs where the field is stronger than expected, and a negative anomaly occurs where the field is weaker than expected.

Why do magnetic anomalies occur? First, the Earth generates a *non-dipolar field*, perhaps as a consequence of turbulence in the flow of the outer core. The non-dipolar field may account for as much as 10% of the Earth's field strength measured at a given location. Second, some rocks contain minerals, such as magnetite and hematite, that behave as permanent magnets. These minerals make the rock, overall, behave like a weak magnet. The presence of magnetic rocks in the crust produces local magnetic anomalies that are independent of the Earth's internal magnetic field.

We noted in Chapter 3 that magnetic anomalies due to the magnetization of rocks take the form of parallel stripes on the seafloor, because the basalt comprising the upper layer of oceanic crust was produced at the mid-ocean ridge axis by seafloor spreading during successive polarity reversals. A positive anomaly stripe occurs over basalt with normal polarity, because the magnetization of the basalt adds to the Earth's magnetic field, whereas a negative anomaly stripe occurs over basalt with reversed polarity, because the magnetization of this basalt subtracts from the Earth's magnetic field. On continents, in contrast, the pattern of magnetic anomalies tends to be more complex because the distribution of rock types is more complex. We see anomalies due to igneous intrusions (which contain magnetic minerals), lava flows, concentrations of iron-rich sediments, and variations in the depth to basement. The shapes of the anomalies may reflect the shapes of intrusions or extrusions, or the shapes of iron-rich sedimentary layers.

FIGURE D.22 An artist's representation of the spiral convection cells in the Earth's outer core that may cause the magnetic field.

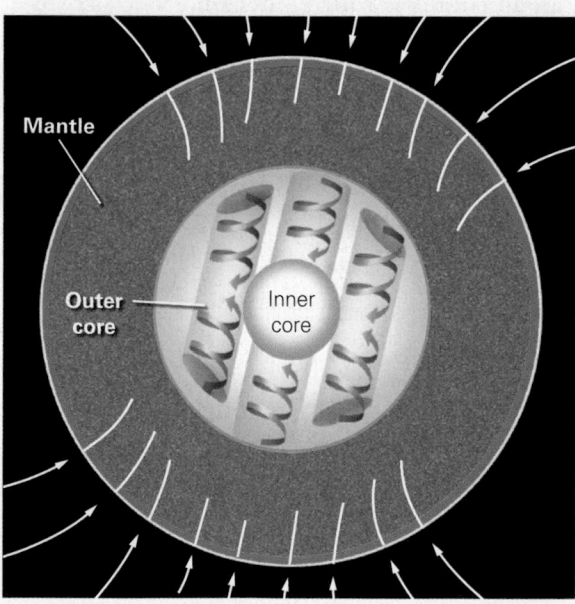

Interlude D Review

SUMMARY

- The depths of the Earth's internal layers, and of divisions within those layers, were discovered through the study of how seismic waves passing through the Earth refract and reflect.

- The Moho was discovered because seismic waves travel faster through the mantle than through the crust. Seismic-velocity discontinuities define boundaries within the mantle.

- Seismic shadow zones reveal the depth to the Earth's core. S-waves cannot pass through the core, so part of it must be liquid.

- Seismic tomographic studies indicate that the Earth's interior layers are not homogeneous.

- Seismic-reflection profiling allows geologists to identify strata structures in the upper crust.

- The geoid, the representation of the Earth's surface on which the pull of gravity is the same at all locations, is not a perfect sphere. Local gravitational anomalies occur.

- The elevation of the lithosphere's surface regionally reflects isostatic equilibrium.

- The Earth's magnetic dipole exists because of convective circulation in the outer core. Local anomalies may exist where rocks contain magnetic minerals.

GUIDE TERMS

Archimides' principle (p. 385)
core-mantle boundary (p. 376)
core superrotation (p. 382)
dipole (p. 386)
dynamo (p. 387)
equipotential surface (p. 383)
geoid (p. 384)
geophysics (p. 371)
gravimeter (p. 384)

gravity anomaly (p. 385)
inner core (p. 377)
isostasy (isostatic equilibrium) (p. 385)
lower mantle (p. 375)
low velocity zone (p. 374)
magnetic anomaly (p. 388)
mantle downwelling (p. 381)
mantle upwelling (p. 381)
Moho (p. 374)

north magnetic pole (p. 387)
outer core (p. 377)
P-wave shadow zone (p. 376)
reflection (p. 373)
refraction (p. 373)
seismic ray (p. 372)
seismic-reflection profile (p. 382)
seismic tomography (p. 378)

seismic-velocity discontinuity (p. 375)
self-exciting dynamo (p. 388)
south magnetic pole (p. 387)
S-wave shadow zone (p. 376)
transition zone (p. 375)
travel time (p. 372)
upper mantle (p. 375)
velocity-versus-depth curve (p. 378)

REVIEW QUESTIONS

The letters following each Review Question refer to the corresponding Learning Objective from the Chapter Opener.

1. What basic layers of the Earth were recognized before the use of seismic studies? **(B)**
2. Do seismic waves travel at the same velocities in all rock? **(A)**
3. What is the difference between refraction and reflection of seismic waves? **(A)**
4. What observation led to the discovery of the Moho? **(B)**
5. Why do seismic waves bend in the mantle? **(A)**
6. What are seismic-velocity discontinuities, and what do they tell us? **(B)**

7. What are P-wave and S-wave shadow zones, and what do they tell us? Which type does the figure show? **(B)**
8. Is seismic velocity constant at a given depth? Explain what the answer tells us about the mantle. **(B)**
9. What can we learn from seismic reflection profiling? **(B)**
10. What is the principle of isostasy? **(C)**
11. Explain what the geoid is. What is a gravity anomaly? **(D)**
12. What causes magnetic anomalies? **(E)**

ONLINE RESOURCES

Animations
This chapter features interactive animations on isostasy and the effects of different conditions on isostasy.

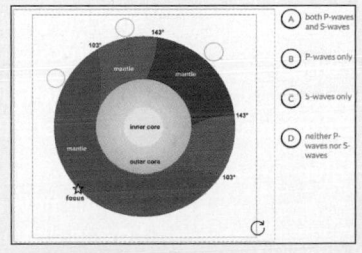

Smartwork5
This chapter features simulation and analysis questions on isostasy, seismic waves, and interpreting geophysical waves.

CHAPTER 11

Crags, Cracks, and Crumples: Crustal Deformation and Mountain Building

---

By the end of this chapter, you should be able to . . .

A. distinguish between mountain belts (orogens) and other regions of continental crust.

B. define what geologists mean by deformation when referring to rock, and explain how and why rocks undergo deformation and develop geologic structures.

C. contrast brittle with plastic deformation, and give examples of each.

D. sketch the various kinds of folds and faults that develop during deformation, and interpret these geologic structures.

E. correlate different kinds of mountain belts with specific geologic settings, in the context of plate tectonics.

F. explain why mountain uplift develops and why topography in mountain belts tends to become so rugged.

G. explain how a craton differs from an orogen, and show how to distinguish between basins and domes in cratonic platforms.

Innumerable peaks, black and sharp, rose grandly into the dark blue sky. . . . [Mountains] are nature's poems carved on tables of stone. . . . How quickly these old monuments excite and hold the imagination!

—John Muir (American naturalist, 1838–1914)

11.1 Introduction

Geographers call the peak of Mt. Everest "the top of the world," for this mountain, which lies in the Himalayas of southern Asia, rises higher than any other on the Earth. The cluster of flags on Mt. Everest's summit flap at 8.85 km above sea level—almost the cruising height of modern jets. No one can survive very long at the top, for the air there isn't dense enough to sustain human activity. So, when Sir Edmund Hillary, from New Zealand, and Tenzing Norgay, a Nepalese guide, became the first to reach the summit in 1953, they hauled oxygen bottles with them. About 5,000 other people have also succeeded, and now hundreds make the ascent every year—but more than 290 climbers and guides have died trying. Success depends not just on the skill of the climber but also on the path of the jet stream, a 200-km-per-hour current of air that flows at high elevations. If the jet stream crosses the summit, it engulfs everyone there in heat-robbing winds that can freeze a person's face, hands, and feet almost instantly, even if they're swaddled in high-tech clothing.

Mountains draw nonclimbers as well, for everyone loves a vista of snow-crested peaks. The stark cliffs, clear air, meadows, forests, streams, and glaciers of mountains provide a refuge from the mundane. Geologists have a particular fascination with mountains, for high peaks provide one of the most obvious indications of dynamic activity on the Earth. Think about it . . . to make a mountain, many cubic kilometers of rock must be pushed skyward, against the pull of gravity.

With the exception of the large volcanoes formed over hot spots, mountains do not occur in isolation, but rather as part of linear chains or ranges called **mountain belts**, or **orogens** (from the Greek words *oros*, meaning mountain, and *genesis*, meaning formation). Geographers define about a dozen major mountain belts and numerous smaller ones on our planet (**Fig. 11.1**).

Mountain building, or **orogeny**, the process of forming a mountain belt, not only raises the surface of the crust, a

◄ (facing page) The surface of this fault displays streak like slip lineations, formed when movement took place on the fault in the past. At the time these slip lineations formed, the fault was deep underground. Uplift brought the rock to the surface, and erosion removed the rock on one side of the fault, exposing the fault surface. Faults are a manifestation of deformation— the displacement, rotation, or distortion of rock bodies—that occurs during mountain building. Mountain building continues today on our dynamic planet.

process called *uplift*, but also causes rocks to undergo *deformation*, a process by which rocks bend, break, or flow in response to *stress* (compression, tension, or shearing). Deformation yields *joints* (cracks), *faults* (fractures on which one body of rock slides past another), *folds* (bends, curves, or wrinkles of rock layers), and *foliations* (fabric or layering in rock)—all of which are examples of *geologic structures*. Orogeny can also cause metamorphism and igneous activity.

Mountain building happens on the Earth for a variety of reasons. Some mountains develop along convergent boundaries, some rise during the collision of continents, and some result from continental rifting. A particular orogeny may last for tens of millions of years. Significantly, as the land surface rises and *relief*—the elevation difference between a higher area and a lower area—develops, erosion starts to grind away the land surface. This process produces immense volumes of sediment and sculpts awesome, jagged topography (**Fig. 11.2**). When uplift ceases, erosion can eventually bevel a mountain range down to near sea level over a period of tens of millions of years. But even after the high peaks are gone, a distinct belt of fractured, contorted, and metamorphosed rock remains. Such crustal scars serve as a permanent monument to what was once a region of high peaks.

Did you ever wonder . . . if mountain belts last forever?

In this chapter, we'll learn about deformation, uplift, and other phenomena that happen during mountain building and will discuss ways in which we can describe and interpret geologic structures. You'll find that, for better or worse, geologists have invented a lot of jargon for geologic structures, in order to streamline the description of these structures. This chapter concludes by addressing the broad question of why mountains form, in the context of plate tectonics theory.

11.2 Rock Deformation in the Earth's Crust

Deformation and Strain

As we've just noted, deformation, meaning the bending, breaking, shortening, stretching, or shearing of rock, produces a variety of geologic structures. To get a visual sense

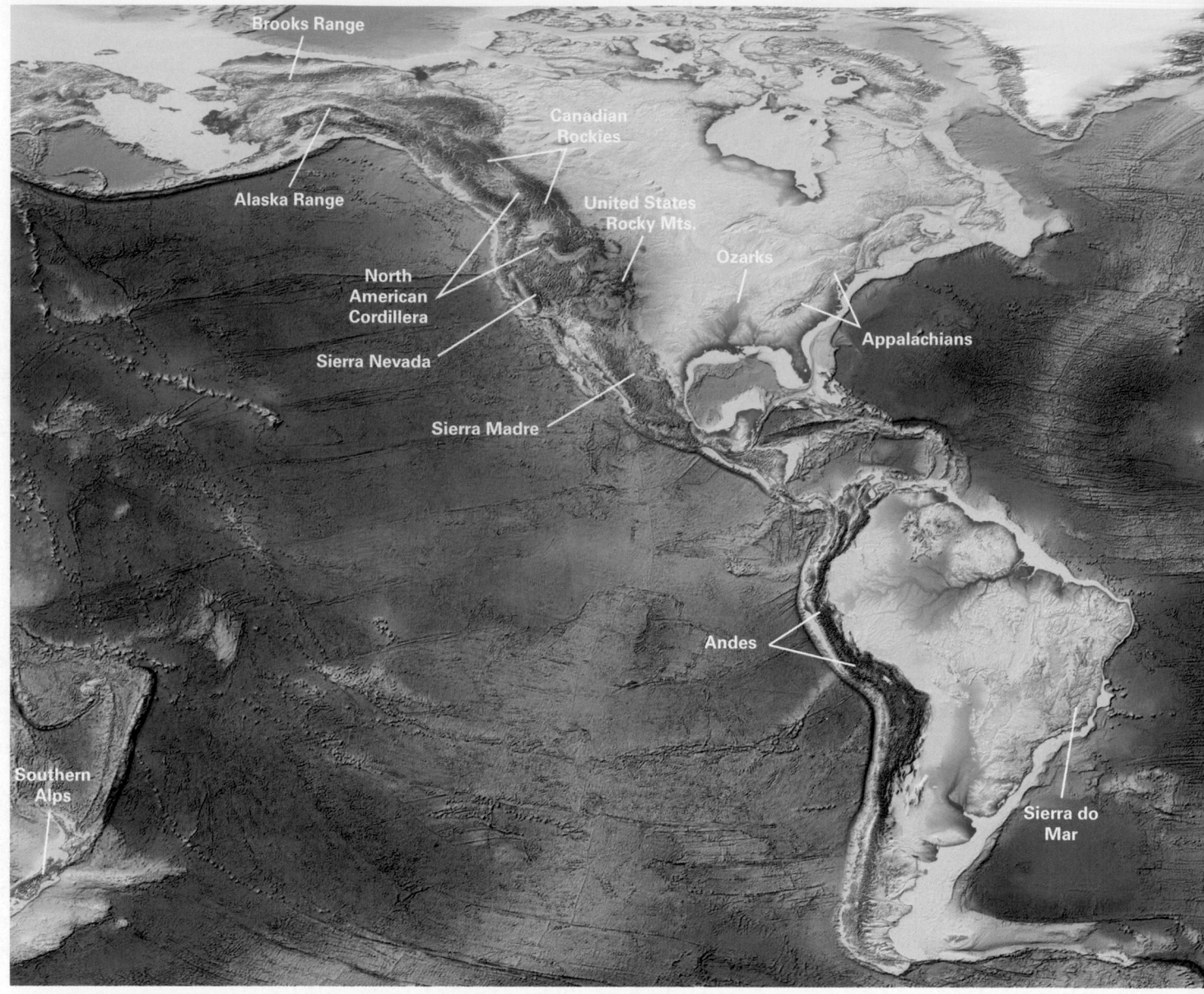

FIGURE 11.1 Digital map of world topography, showing the locations of major mountain ranges.

of deformation, let's compare a road cut along a highway in a region that has not undergone orogeny with a cliff face in a region that has **(Fig. 11.3a)**.

Our imaginary road cut lies at an elevation of only about 200 m above sea level. It exposes nearly horizontal beds of sandstone, shale, and limestone—these beds have the same orientation that they had when first deposited (see Fig. 11.3a). Sand grains in sandstone beds of this outcrop have a nearly spherical shape (the same shape they had when deposited), and clay flakes in the shale lie roughly parallel to bedding because of compaction. Geologists consider rock of such outcrops to be essentially *undeformed*, meaning that the rock contains no geologic structures other than a few joints **(Fig. 11.3b)**.

Rocks of our imaginary mountain cliff, exposed at an elevation of 3 km, look very different. Here we find layers of quartzite, slate, and marble, the metamorphic equivalents of sandstone, shale, and limestone, respectively. The layers

not only contain joints, but have also been contorted into folds **(Fig. 11.3c, d)**. On close examination, we also discover that the grains in the quartzite aren't spherical, like those of a sandstone, but have been flattened into ellipsoids. And in the slate, clay flakes align to define a type of foliation called slaty cleavage (see Chapter 8). Finally, if we try tracing the quartzite and slate layers along the outcrop face, we find that they abruptly terminate at a sloping surface along which rock appears to be broken into fragments. This surface is a fault. Marble, a different type of rock, lies below the fault, so we can infer that the quartzite and slate must have moved some distance along the fault from where they first formed to get to their present location, juxtaposed against the marble.

Clearly, the beds in the mountain cliff have been deformed, and as a result, the cliff exposes a variety of geologic structures. Note that because of deformation, beds no longer have the same shape and position that they had when first formed,

and even the shape and orientation of grains has changed. This example illustrates that deformation involves one or more of the following (Fig. 11.4): change in location (displacement); change in orientation (rotation); and change in shape (distortion).

Geologists refer to the change in shape that develops during deformation as **strain**. We distinguish among different kinds of strain according to the nature of the shape change. If a layer of rock becomes longer, it has undergone *stretching*, but if the layer becomes shorter, it has undergone *shortening* (Fig. 11.5a–c). Movement of one part of a rock body past another so that angles between features in the rock change results in *shear strain* (Fig. 11.5d, e).

Brittle versus Plastic Deformation

Imagine that a plate tumbles off a table and lands on a hard floor—the plate breaks and smashes into pieces. Similarly, if you strike a glass window with a ball, the window cracks and may even shatter. Such phenomena serve as familiar examples of **brittle deformation** (Fig. 11.6a, b). Now imagine that you squeeze a ball of soft dough between a book and a tabletop—the dough flattens into a pancake. Similarly, if you bend a stick of chewing gum, it changes from a plane into a curve. During such **plastic deformation** objects change shape without visibly breaking (Fig. 11.6c, d). Informally, geologists also refer to plastic deformation as *ductile deformation*, though the terms have slightly different meanings to specialists.

What actually happens within mineral grains during these two different kinds of deformation? Recall that in minerals, chemical bonds connect atoms to one another. During brittle deformation, large numbers of bonds break and stay broken, leading to the formation of a permanent crack across which material no longer connects. During plastic deformation, simplistically, some bonds break, but new ones quickly form.

FIGURE 11.2 The Matterhorn is located in the Alps along the Swiss-Italian border. The strata of the peak lay on the seafloor during the late Mesozoic. Mountain building thrust the strata upward so they now reach an elevation of 4.5 km. Erosion continuously sculpts uplifted land to yield jagged peaks.

In this way, the atoms within grains rearrange, and the grains can change shape without permanent cracks forming.

Why do rocks inside the Earth sometimes deform brittlely and sometimes plastically? The behavior of a rock depends on a number of variables:

- *Temperature:* Warmer rocks tend to deform plastically, whereas colder rocks tend to deform brittlely. To see this contrast, think again of our candle experiment (see Chapter 10). Chill a candle in a freezer, then press its middle against the edge of a table—the candle will brittlely snap in two. But if you first warm the candle in an oven, it will plastically bend without breaking when pressed against the table. Heat makes materials softer.

FIGURE 11.3 Deformation changes the character and configuration of rocks.

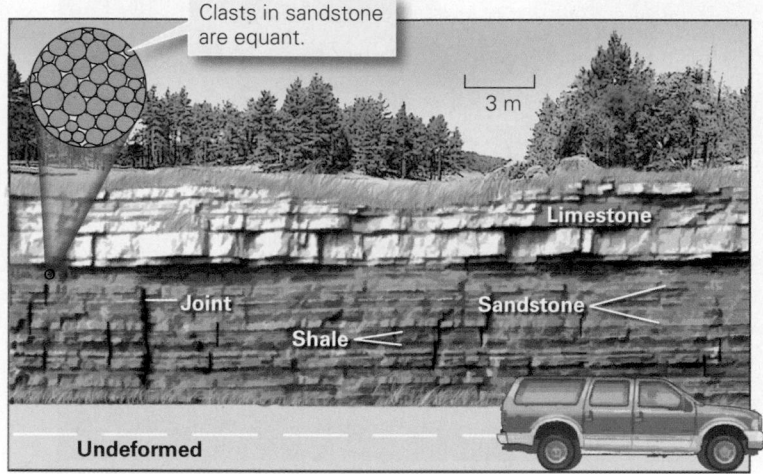

Clasts in sandstone are equant.

(a) These flat-lying beds along a highway are essentially undeformed. A few joints, formed when overlying rock eroded away, are visible.

(b) Another example of undeformed, horizontal beds. These beds form a 100-m-high sea cliff in western Ireland. The vertical lines are joints.

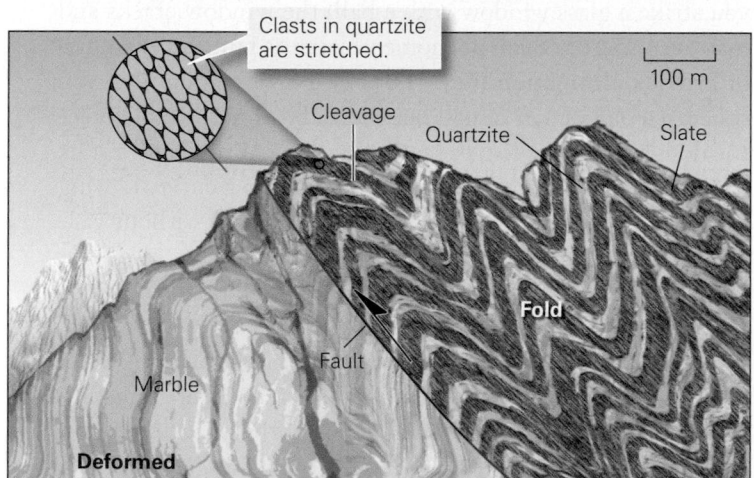

Clasts in quartzite are stretched.

(c) In a mountain belt, deformation may cause layers to undergo folding and faulting. In addition, foliation (such as slaty cleavage) and stretched clasts may develop.

(d) Another example of deformed beds. These beds along the coast of Wales have arched into a fold. Note how the beds look "wrinkled."

FIGURE 11.4 The components of deformation include displacement, rotation, and distortion.

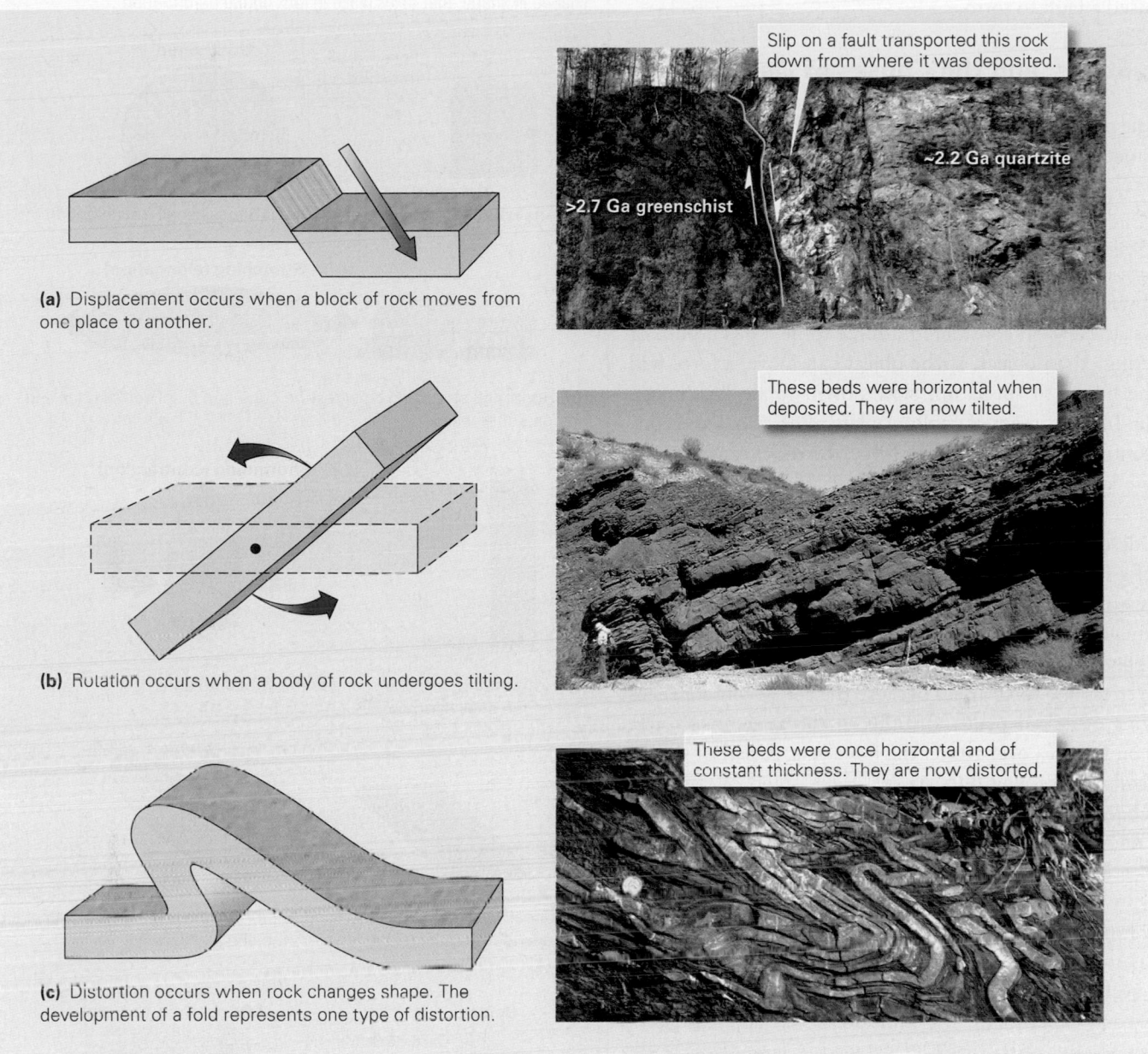

Slip on a fault transported this rock down from where it was deposited.

~2.2 Ga quartzite

>2.7 Ga greenschist

(a) Displacement occurs when a block of rock moves from one place to another.

These beds were horizontal when deposited. They are now tilted.

(b) Rotation occurs when a body of rock undergoes tilting.

These beds were once horizontal and of constant thickness. They are now distorted.

(c) Distortion occurs when rock changes shape. The development of a fold represents one type of distortion.

- *Pressure:* Under great pressures deep in the Earth, rock behaves more plastically than it does under low pressures near the surface. Pressure effectively prevents rock from separating into fragments.
- *Deformation rate:* A sudden change in shape causes brittle deformation, whereas a slow change in shape can cause plastic deformation. For example, if you hit a marble bench with a hammer, it shatters, but if you leave the bench alone for a century, gravity may cause it to gradually sag, without breaking.
- *Composition:* Some rock types are softer than others because bonds within their mineral grains are weaker. For example, halite (rock salt) deforms plastically under conditions in which granite deforms brittlely.

Considering that pressure and temperature both increase with depth in the Earth, geologists find that in typical continental crust, rocks generally behave brittlely above 10 to 20 km and plastically below this depth. We refer to the depth at which this change in behavior takes place as the *brittle-plastic transition.* Earthquakes in continental crust happen only above this depth because these earthquakes involve brittle breaking.

In some cases, you can see consequences of both brittle and plastic deformation in the same outcrop. For example, our mountain cliff (see Fig. 11.3c) displays both faulting (brittle deformation) and folding (plastic deformation). Such an occurrence may seem like a paradox at first. But a juxtaposition of different deformation styles can happen because of changes in the deformation rate during orogeny. Slow deformation yielded the folds, whereas a pulse of rapid deformation may have caused a fault to form.

Force, Stress, and the Causes of Deformation

Up to this point, we've focused on picturing the consequences of deformation. Describing the causes of deformation can be a bit more challenging in the context of an introductory geology book. In captions for displays about mountain building, museums and national parks typically dispense with the issue by using a broad-brush phrase such as, "The mountains were caused by forces deep within the Earth." But what does this mean? Isaac Newton considered a **force** to be a push, pull, or shear acting on an object. If the object can move, a force will cause the object to speed up, slow down, rotate, or change direction. In the context of geology, plate interactions and continent-continent collisions apply forces to rock, causing rock to change location, orientation, or shape. In other words, the application of forces in the Earth indeed causes deformation.

But describing the cause of deformation in terms of the action of a force doesn't provide the whole picture. Geologists use the word *stress* instead of *force* when talking about the cause of deformation. We define the **stress** acting on a plane as the force applied per unit area of the plane. The need to distinguish between stress and force arises because the actual consequences of applying a force depend not only on the amount of force, but also on the area over which the force acts. A pair of simple experiments shows why **(Fig. 11.7a)**. First, stand on a single, empty aluminum can. All of your weight—a force—focuses entirely on the can, and the can crushes. Second, place a board on top of 100 cans and stand on the board. In this case, your weight spreads across all 100 cans, and the cans don't crush. In both experiments, the force caused by the weight of your body was the same, but in the first experiment, the force was applied over a small area, so a large stress developed, whereas in the second experiment, the same force was applied over a large area, so only a small stress developed. A large stress could crush a can, but a small one could not. How does this concept apply to geology? During mountain building, the force of one plate interacting with another acts across the broad area of contact between the two plates, so the deformation resulting at any specific location actually reflects the stress developed at that location, not the total force produced by the plate interaction.

Several different kinds of stress occur in rock bodies **(Fig. 11.7b–e)**. **Compression** takes place when a rock is subjected to squeezing, *tension* exists when the ends of a rock body are pulled away from each other, and *shear stress* develops when one part of a rock body moves sideways past another. **Pressure** refers to a special stress condition that happens when the same push acts on all sides of an object.

Note that *stress* and *strain* have very different meanings to geologists, even though people tend to use these words

(a) An unstrained cube and an unstrained fossil shell (brachiopod).

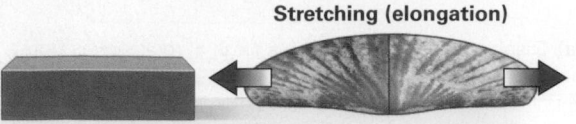

(b) Horizontal stretching changes the cube into a horizontal brick and elongates the shell.

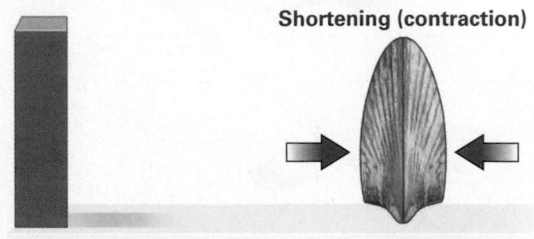

(c) Horizontal shortening changes the cube into a vertical brick and makes the shell narrower.

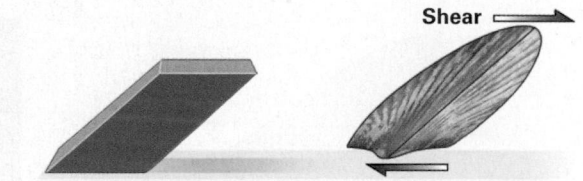

(d) Shear strain tilts the cube and transforms it into a parallelogram, and changes angular relationships in the shell.

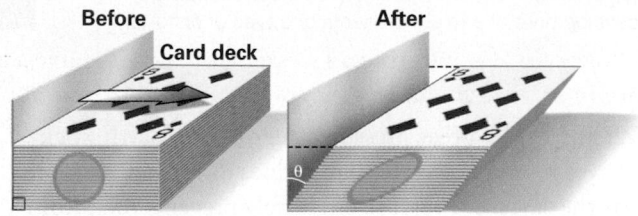

(e) You can produce shear strain by moving a deck of cards so that each card slides a little with respect to the one below.

interchangeably in everyday English: *stress* specifically refers to the amount of force applied per unit area of a rock, whereas *strain* specifically refers to a change in the shape of a rock. Stress causes strain—compression causes shortening, tension leads to stretching, and shear stress produces shear strain. Pressure can cause an object to become smaller, but will not cause it to move, rotate, or change shape. With this

FIGURE 11.6 Brittle versus plastic deformation.

Before

After

Cracks separate the plate into pieces.

(a) Brittle deformation occurs when you drop a plate and it shatters.

(b) Cracks in these quartzite beds in Utah are due to brittle deformation.

Before

GEOLOGY

Dough

After

The dough remains a single, coherent piece during deformation.

GEOLOGY

(c) Plastic deformation occurs when you squash a ball of dough.

(d) The quartzite cobbles in this conglomerate were flattened plastically.

knowledge of stress and strain, we can now look at the nature and origin of various classes of geologic structures and discuss their significance.

TAKE-HOME MESSAGE

Stress (compression, tension, or shear) can cause rocks to change shape, position, or orientation. Strain is change in the shape of a rock. During brittle deformation, rocks break, whereas during plastic deformation, rocks bend and distort without breaking. Which type of deformation takes place depends on factors such as temperature, pressure, and the rate of deformation.

QUICK QUESTION: What is the difference between stress and strain to a geologist?

11.3 Brittle Structures

Joints and Veins

If you look at the photographs of rock outcrops in this book, you'll notice thin dark lines that cross the rock faces. These lines represent traces of natural cracks along which the rock broke and separated into two pieces during brittle deformation. Geologists refer to such natural cracks as **joints** (Fig. 11.8a, b). Rock bodies do not slide past each other on joints. Since joints are roughly planar structures, we define their orientation by their *strike* and *dip* (Box 11.1).

Joints develop in response to tension in brittle rock: a rock splits open because it has been pulled slightly apart. Joints may form for a variety of geologic reasons. For example, some joints form when a rock cools and contracts, because contraction makes one part of a rock pull away from the adjacent part. Others develop when rock formerly at depth undergoes a decrease in pressure as overlying rock erodes away. Because of the change in pressure, the rock changes shape slightly by expanding vertically and contracting horizontally. Still other joints form when relatively brittle rock layers bend.

Rock bodies may contain two categories of joints. *Systematic joints* tend to be long and planar and occur fairly regularly through a rock body. *Nonsystematic joints* tend to be short and occur in a range of orientations; they are randomly spaced and may be irregular or curved. A group of systematic joints constitutes a *joint set*, a spectacular example of which stands out in sandstone beds of Arches National Park in Utah (see Fig. 11.8a). Here, erosion along the joints produced narrow gullies. In sedimentary rocks, systematic joints tend to be vertical planes (see Fig. 11.8b). If groundwater seeps through joints underground over a long period, minerals such as quartz or calcite may precipitate and fill the joint. Such mineral-filled joints are a type of **vein** and look like white stripes cutting across a body of rock (Fig. 11.8c).

FIGURE 11.7 There are several kinds of stress.

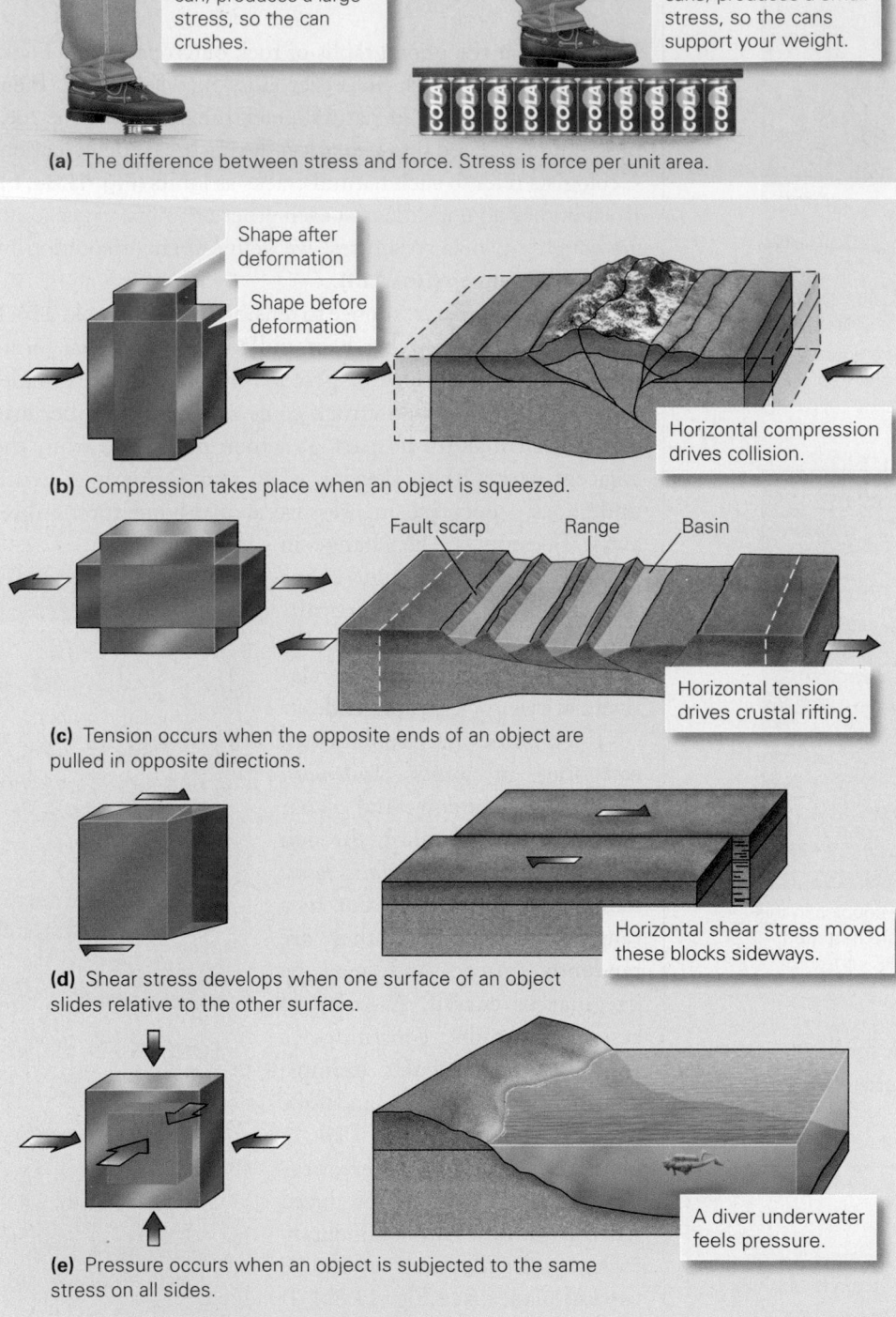

A force applied to a small area (the top of a can) produces a large stress, so the can crushes.

The same force applied to a large area (many cans) produces a small stress, so the cans support your weight.

(a) The difference between stress and force. Stress is force per unit area.

Shape after deformation

Shape before deformation

Horizontal compression drives collision.

(b) Compression takes place when an object is squeezed.

Fault scarp   Range   Basin

Horizontal tension drives crustal rifting.

(c) Tension occurs when the opposite ends of an object are pulled in opposite directions.

Horizontal shear stress moved these blocks sideways.

(d) Shear stress develops when one surface of an object slides relative to the other surface.

A diver underwater feels pressure.

(e) Pressure occurs when an object is subjected to the same stress on all sides.

Geotechnical engineers, people who study the geologic settings of construction sites, pay close attention to jointing when recommending where to put roads, dams, and buildings. Water flows much more easily through joints than it does through solid rock, so it would be a bad investment to situate a water reservoir over rock containing lots of joints—the water would leak down into the joints. Similarly, building a road on a steep cliff composed of jointed rock could be risky, for joint-bounded blocks separate easily from bedrock, and the cliff might collapse.

Faults: Surfaces of Slip

After the San Francisco earthquake of 1906, geologists found a rupture that seemingly had torn through the land surface near the city. Where this rupture crossed orchards, it offset rows of trees, and where it crossed a fence, it broke the fence in two—the western side of the fence moved northward by about 2 m (see Fig. 10.4a). The rupture represents the trace of the San Andreas fault. As we have seen, the term **fault** refers to a fracture on which sliding occurs, and slip events, or *faulting*, can generate earthquakes. Faults, like joints, tend to be more or less planar structures, so we represent their orientation by strike and dip (see Box 11.1).

Faults have formed throughout Earth history. Some are currently *active*, in that sliding has been occurring on them in recent geologic time, but most can be considered *inactive*, meaning that sliding on them ceased long ago and won't happen again any time soon. Some faults, such as the San Andreas, intersect the ground surface and displace the ground when they move. Others, known as *blind faults*, accommodate the sliding of rocks in the crust at depth and remain invisible at the surface unless they are later exposed by erosion. Geologists refer to the intersection of a fault with the land surface, regardless of whether the fault is active and has offset the surface or was blind and later exposed by erosion, as the **fault trace**, or *fault line*.

Geologists study faults not only because the movement on some faults causes earthquakes, but also because faults juxtapose bodies of rock that did not originally lie adjacent to each other and therefore complicate the arrangement of rocks in the crust. For example, in our mountain cliff (see Fig. 11.3c), movement on a fault placed quartzite and slate beds against marble beds. Geologists must understand these rearrangements in order to interpret orogenic events and to predict where resources lie underground.

Fault Classification

Not all faults result in the same kind of crustal deformation—some accommodate horizontal shear, some accommodate shortening, and some accommodate stretching. It's important for geologists to distinguish among different kinds of faults in order to interpret their geologic significance. Fault classification focuses on two characteristics of faults: (1) the dip, or slope, of the fault surface (see Box 11.1), which can be vertical, horizontal, or any angle in between; and (2) the *shear sense* across the fault, by which we mean the direction that material on one side of the fault moved relative to material on the other side. With this concept in mind, let's consider the principal kinds of faults. (This description builds on that of Chapter 10 and adds an additional category.)

- *Dip-slip faults:* On a dip-slip fault, movement parallels the dip line, a line on the fault surface that follows the steepest slope on the fault surface. (*Dip* refers to the angle that this line makes relative to horizontal; see Box 11.1.) We distinguish among different types of dip-slip faults based on whether the *hanging-wall* block, meaning the material above the fault surface, slides up or down relative to the *footwall* block, the material below the fault surface **(Fig. 11.9a, b)**. If the hanging-wall block

slides up, the fault is a **reverse fault**. Geologists refer to a reverse fault with a gentle dip (<30°) as a **thrust fault**. Reverse faults and thrust faults accommodate *shortening* of the crust, meaning that when they slip, the overall surface length of the crustal block that the fault cuts decreases in a direction perpendicular to the fault trace **(Fig. 11.9c)**. Not surprisingly, such faults develop during continental collision, when two crustal blocks squeeze together horizontally. Geologists refer to a fault on which the hanging-wall block slides down the dip of the fault surface as a **normal fault**. Normal faults accommodate *stretching* of the Earth's crust, meaning that the overall surface length of a crustal block cut by a normal fault becomes longer in a direction perpendicular to the fault trace (see Fig. 11.9c). Such faults commonly develop during crustal rifting.

- *Strike-slip faults:* When shear takes place on a **strike-slip fault**, material on one side of the fault moves horizontally along the fault plane. (The *strike* of a surface indicates the compass trend of a horizontal line on the surface; see Box 11.1). This means that the shear, ideally, does not result in any up or down movement relative to the ground surface. Most strike-slip faults have a steep to vertical dip. Geologists describe the shear sense on a strike-slip fault with reference to the view you have when you stand next to the fault and gaze across it **(Fig. 11.9d)**. If the block across the fault moved to your left during faulting, the fault displays left-lateral slip, whereas if the block moved to your right, the fault displays right-lateral slip.

FIGURE 11.8 Examples of joints and veins.

(a) Prominent vertical joints cut these red sandstone beds in Arches National Park, Utah.

(b) Vertical joints on a cliff face in shale near Ithaca, New York.

(c) Milky white quartz veins cut across gray limestone beds.

BOX 11.1 CONSIDER THIS . . .

Describing the Orientation of Geologic Structures

When discussing geologic structures, it's important to be able to communicate information about their orientation. For example, does a fault exposed in an outcrop at the edge of town continue beneath the nuclear power plant 3 km to the north, or does it go beneath the hospital 2 km to the east? If we knew the fault's orientation, we might be able to answer this question. To describe the orientation of a geologic structure, geologists picture the structure as a simple geometric shape, and then specify the angles that the shape makes with respect to a horizontal plane (a flat surface parallel to sea level), a vertical plane (a flat surface perpendicular to sea level), and the north direction (a line of longitude).

Let's start by considering planar structures such as faults, beds, and joints. We call these structures planar because they resemble a geometric plane. A planar structure's orientation can be specified by its strike and dip. The **strike** is the angle between an imaginary horizontal line (the strike line) on the structure and the direction to true north **(Fig. Bx11.1a, b)**. We measure the strike with a special type of compass that has a level bubble so that we can be sure that the compass surface is exactly horizontal **(Fig. Bx11.1c)**. The **dip** is the angle of the structure's slope. More precisely, it is the angle between a horizontal plane and the dip line (an imaginary line parallel to the steepest slope on the structure), as measured in a vertical plane perpendicular to the strike **(Fig. Bx11.1d)**. We measure the dip angle with a clinometer, a type of protractor. (A geologist's compass typically contains a built-in clinometer.) A horizontal plane has a dip of 0°, and a vertical plane has a dip of 90°. We represent strike and dip on a geologic map using the symbol shown in Figure Bx11.1b.

FIGURE Bx11.1 Specifying the orientation of planar and linear structures.

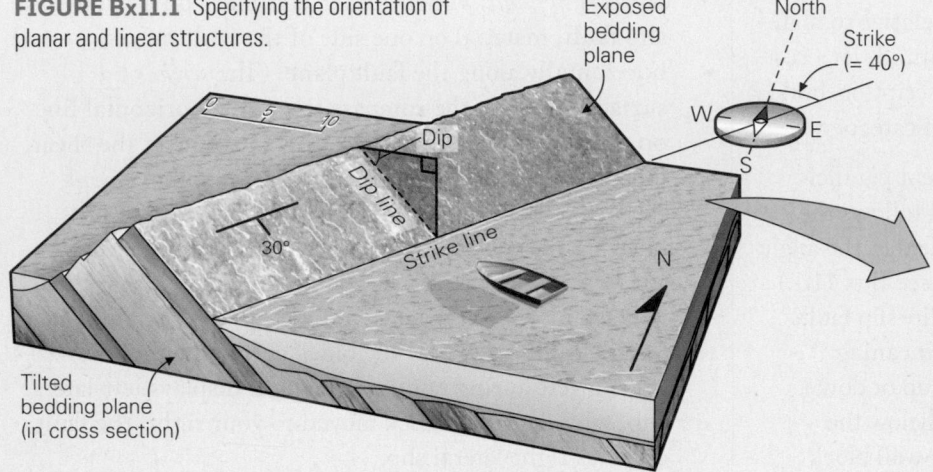

(a) We use strike and dip to indicate the orientation of planar structures, such as these tilted beds. The shaded plane is vertical.

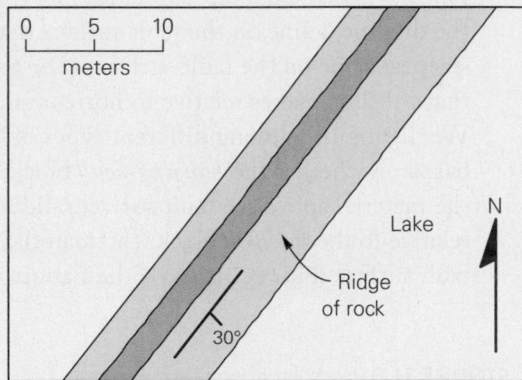

(b) On a map, the line segment represents the strike direction and the tick on the segment represents the dip direction. The number indicates the dip angle as measured in degrees.

• *Oblique-slip faults*: On an oblique-slip fault, movement takes place parallel to a diagonal line on the fault surface. These faults, therefore, display both a strike-slip and a dip-slip component of movement **(Fig. 11.9e)**.

Recognizing Faults

How do we recognize a fault when we see one? The most obvious indicator of a fault is an observation or measurement of displacement, or offset, meaning movement of one side of a fault relative to the other. Where displacement has occurred, layers on one side of the fault don't connect to the same layers on the other side. When layers have distinctive characteristics, you can easily see the displacement in an outcrop **(Fig. 11.10a, b)**. In some cases, faults juxtapose two different rock units (see Fig. 11.3c). Typically, thrust or reverse faults cutting sedimentary beds place older beds on younger ones, whereas normal faults place younger beds on older ones. In some cases, layers of rock cut by a fault undergo folding during or just before slip. Geologists informally refer to the resulting folds adjacent to a fault as *drag folds*.

Faults may also leave a mark on the landscape. Those that intersect the ground surface while active can displace natural landscape features, such as stream valleys or glacial moraines **(Fig. 11.10c)**, as well as human-made features, such as highways, fences, or rows of trees in orchards. Displacement of

A linear structure resembles a geometric line rather than a plane; examples of linear structures include scratches or grooves on a fault surface. Geologists specify the orientation of linear structures by giving their *plunge* and *bearing* (**Fig. Bx11.1e**). The **plunge** is the angle between a line and horizontal in the vertical plane that contains the line. A horizontal line has a plunge of 0°, and a vertical line has a plunge of 90°. The **bearing** is the compass heading of the line—more precisely, the angle between the projection of the line on the horizontal plane and the direction to true north.

The edge of the compass is parallel to the strike line.

The intersection of the water with the rock is horizontal.

(c) Geologists use a Brunton compass to measure strike and dip.

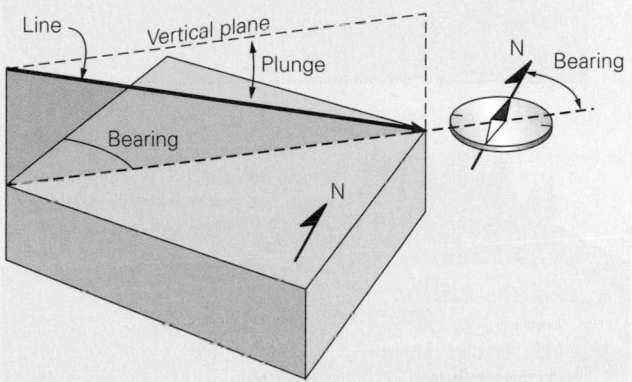

Line
Vertical plane
Plunge
N
Bearing
Bearing
N

(e) To specify the orientation of a line, we use plunge and bearing.

Dipping beds intersect a calm sea.

Bedding

Dip

(d) The intersection of dipping beds with the horizontal water surface represents a strike line. The slope of the bed surface represents the dip.

the land surface on a dip-slip or oblique-slip fault yields a small step called a **fault scarp** along the fault trace (**Fig. 11.11a**). Inactive faults may still be readily identifiable in the landscape. Because faulting tends to break up rock, fault traces tend to be more easily eroded. As a result, a linear valley may delineate the fault trace (see the inset photo of the Great Glen fault on p. 331). And, if a fault juxtaposes two different rock units with different resistances to erosion, the land surface may, over time, erode to be lower on the side of the weaker rock, so the fault trace will be marked by a noticeable *escarpment*, a steep slope or cliff.

Faulting under brittle conditions may crush or break adjacent rock. If this shattered rock consists of visible angular fragments, it's called *fault breccia* (**Fig. 11.11b**), but if it consists of a fine powder, it's called *fault gouge*. Because fault gouge contains such tiny grains, it may quickly undergo chemical weathering and turn into clay. Some fault surfaces have been polished and grooved by the movement of the hanging wall past the footwall. We refer to a polished fault surface as a **slickenside**, and to linear grooves on a fault surface as **slip lineations** (**Fig. 11.11c**).

The shear on some faults takes place under plastic conditions at depth in the crust. Where this happens, rock does not break up into breccia or gouge along the fault zone, but rather, shears plastically to form a band of fine-grained foliated rock called mylonite. The fine grain size of mylonite results not from brittle fracturing during shear, but instead from a type of metamorphic recrystallization that subdivides large grains

FIGURE 11.9 The different categories of faults.

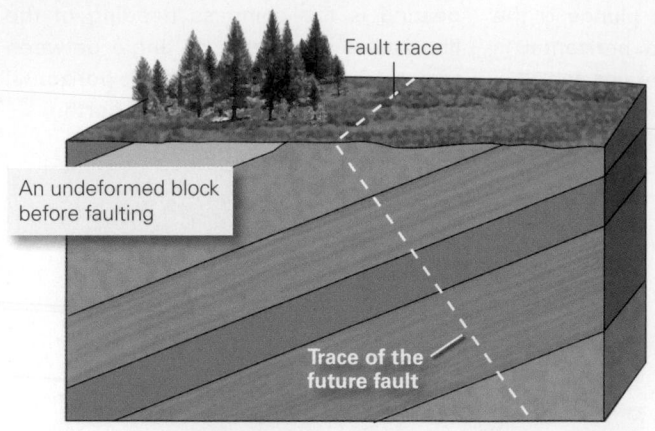

Fault trace

An undeformed block before faulting

Trace of the future fault

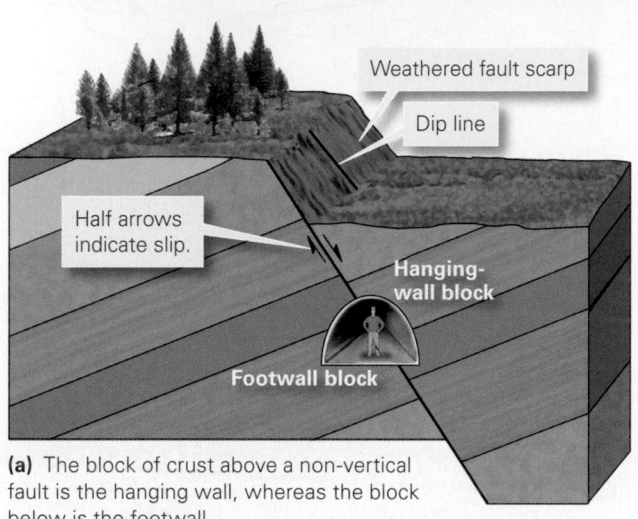

Weathered fault scarp

Dip line

Half arrows indicate slip.

Hanging-wall block

Footwall block

(a) The block of crust above a non-vertical fault is the hanging wall, whereas the block below is the footwall.

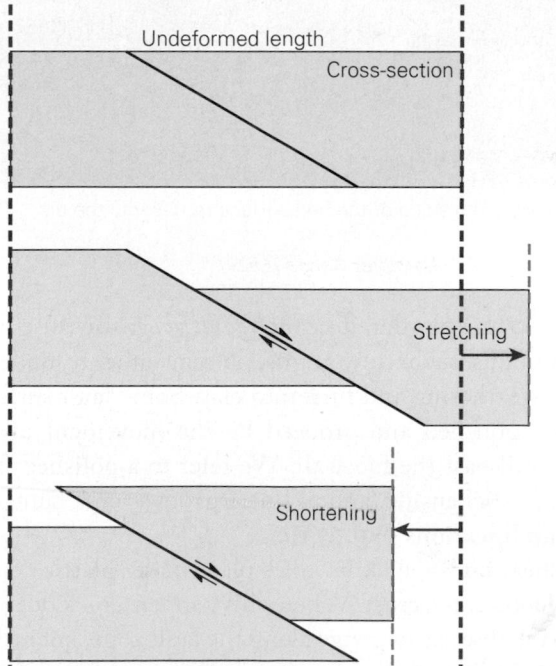

Undeformed length

Cross-section

Stretching

Shortening

(c) As seen in cross section, slip on a normal fault accommodates crustal stretching, whereas slip on a reverse fault or thrust fault accommodates crustal shortening.

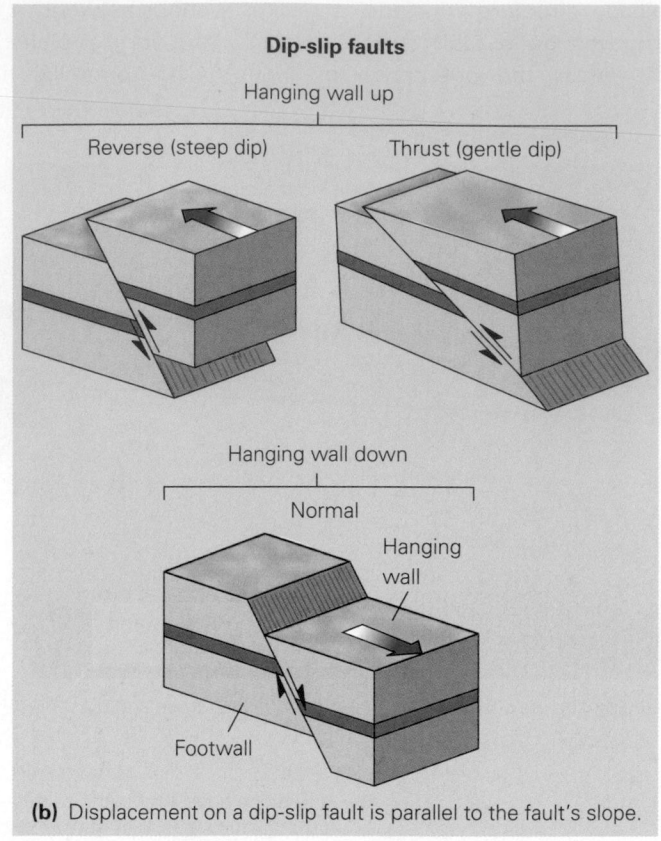

Dip-slip faults

Hanging wall up

Reverse (steep dip)

Thrust (gentle dip)

Hanging wall down

Normal

Hanging wall

Footwall

(b) Displacement on a dip-slip fault is parallel to the fault's slope.

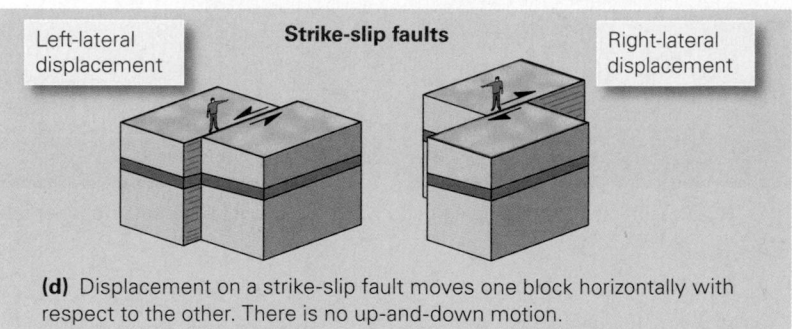

Left-lateral displacement

Strike-slip faults

Right-lateral displacement

(d) Displacement on a strike-slip fault moves one block horizontally with respect to the other. There is no up-and-down motion.

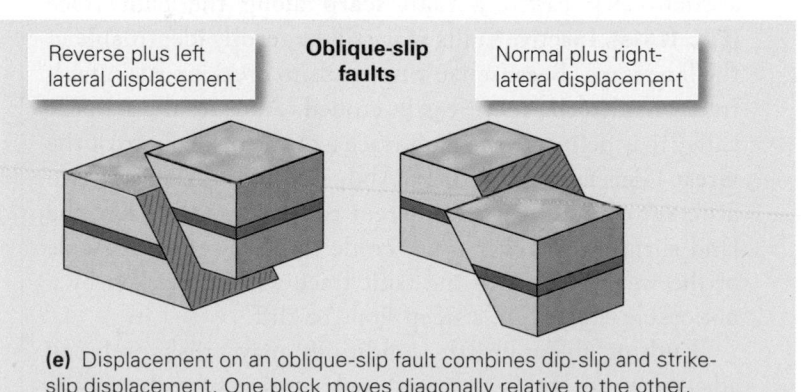

Reverse plus left lateral displacement

Oblique-slip faults

Normal plus right-lateral displacement

(e) Displacement on an oblique-slip fault combines dip-slip and strike-slip displacement. One block moves diagonally relative to the other.

FIGURE 11.10 Recognizing fault displacement in the field.

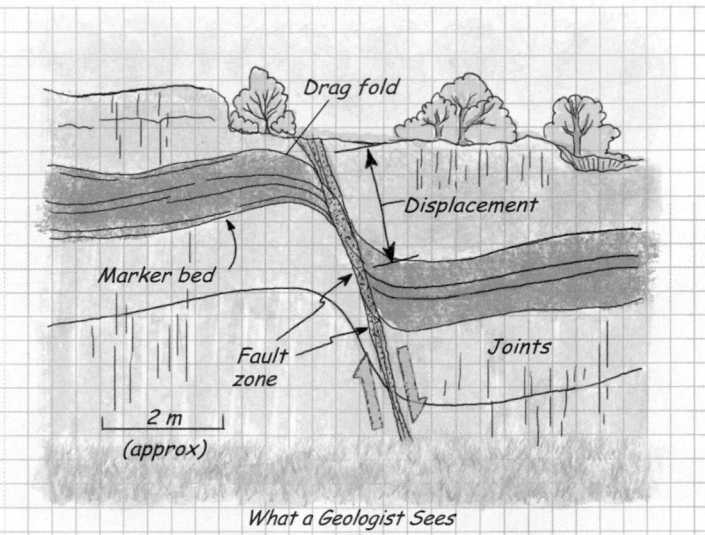

What a Geologist Sees

(a) A steep normal fault has displaced a distinctive red bed (a marker bed). Note that the displacement formed a 0.5-m-wide fault zone of broken rock. Drag folds developed adjacent to the fault.

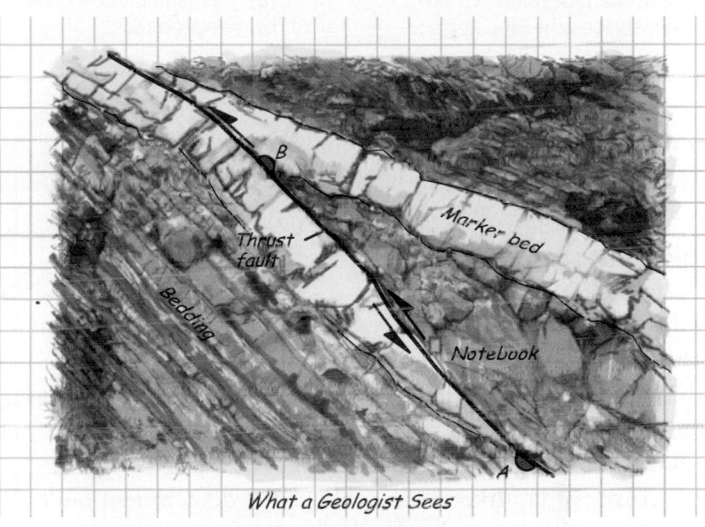

What a Geologist Sees

(b) Slip on a thrust fault caused one part of the light-colored marker bed to be shoved over another part, as emphasized in the drawing. Note that the beds are tilted to the right. The distance between A and B (the red dots) is the displacement.

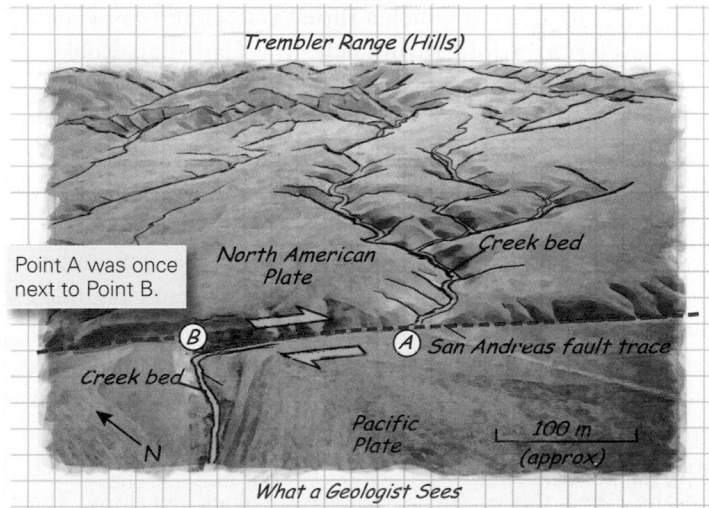

What a Geologist Sees

(c) An aerial photograph of a portion of the San Andreas fault. As emphasized by the sketch, the fault offsets a stream channel in a right-lateral sense.

FIGURE 11.11 Features of exposed fault surfaces.

A scarp due to slip on a normal fault in Nevada.

(a) A fault scarp develops where faulting displaces the land surface.

(b) This fault breccia consists of irregular fragments of light-colored rock.

Striation orientation

(c) Slip lineations or striations on the surface of a strike-slip fault may look like grooves or scratches.

into small ones. Geologists commonly refer to a fault whose movement occurred plastically as a **shear zone (Fig. 11.12)**.

Fault Systems

An array of several related faults comprises a **fault system**. In systems of dip-slip faults, individual faults in the system typically merge at depth with a nearly horizontal **detachment fault**. Displacements in a *thrust system* can accommodate substantial shortening of the crust because slip makes slices of the crust overlap like shingles above the detachment fault **(Fig. 11.13a, b)**. During the development of a thrust system, strata within the *thrust sheets*, the bodies of rock transported by a thrust fault, tend to become folded. Geologists commonly refer to areas in which a thrust system and associated folds develop as a **fold-thrust belt**. The Canadian Rockies serve as an example of a fold-thrust belt—the rugged cliffs and peaks of this mountain belt formed when compressive stress shoved rocks eastward **(Fig. 11.13c)**. The process resembles the building of a pile of snow or sand in front of a plow as the plow moves forward and pushes it.

Displacements in a *normal-fault system* accommodate stretching of the crust. During the process, fault-bounded slivers of crust drop down **(Fig. 11.13d)**. Many normal-fault systems consist of parallel faults that curve to shallower dips at depth, where they merge with the detachment fault. As slip progresses on these curved faults, the hanging-wall block rotates, and a wedge-shaped space between the fault surface and the tilted top of the hanging-wall block develops. The down-dropped block, bounded only on one side by a normal

FIGURE 11.12 An outcrop of a small shear zone in granite. Note that the boundaries of the shear zone are gradational, and that foliation has developed in the shear zone. A portion of the edge is highlighted.

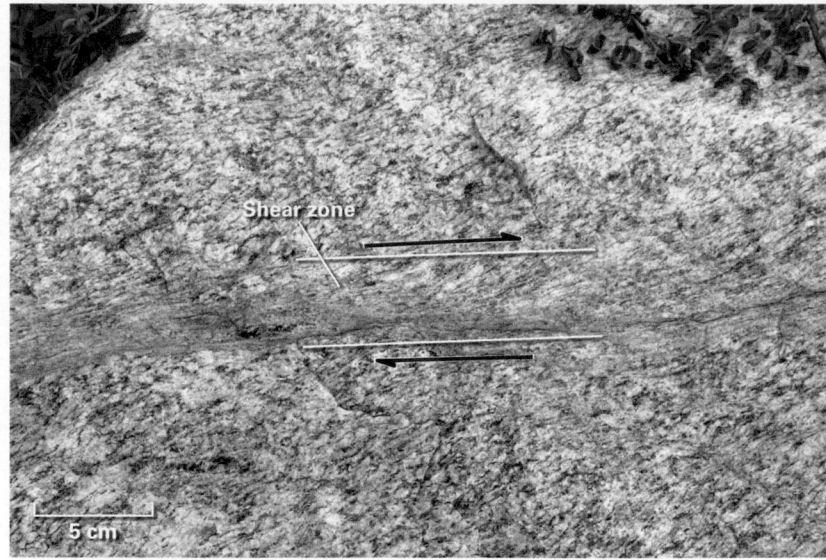

Shear zone

5 cm

fault, is a *half graben*. Eventually, the space above a half graben fills with a wedge of sediments and becomes a narrow sedimentary basin. Locally, two adjacent normal faults may dip toward each other—the down-dropped keystone-shaped block between the two faults is a *graben*, and the high block between two grabens is a *horst* **(Fig. 11.13e)**. Note that the two normal faults bounding a horst dip away from each other. A graben, like a half graben, tends to get buried by sediments eroded from the high areas on either side.

FIGURE 11.13 Examples of fault systems, arrays of related faults formed in sequence during a deformation event.

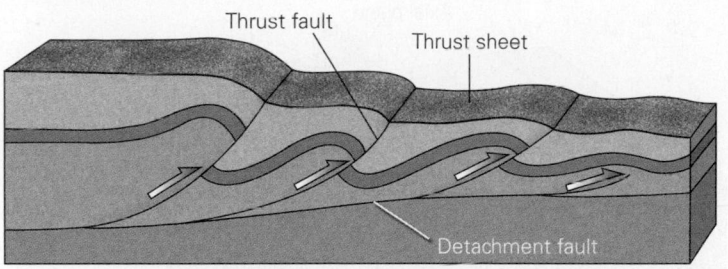

Thrust fault
Thrust sheet
Detachment fault

(a) In this thrust-fault system, several related thrust faults merge at depth with a detachment fault. Note that displacement on the thrust faults shortens the layers of rock above the detachment faults.

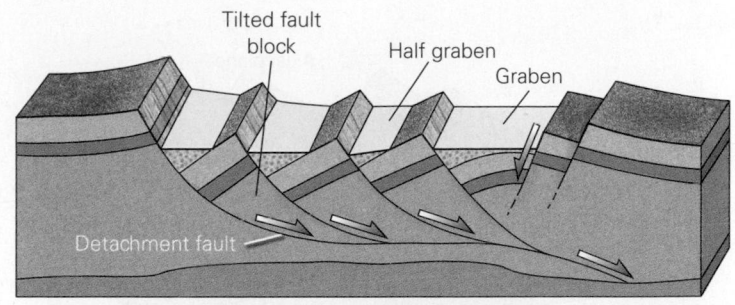

Tilted fault block
Half graben
Graben
Detachment fault

(d) A simplified diagram showing a normal-fault system. Here, several faults merge with a detachment at depth.

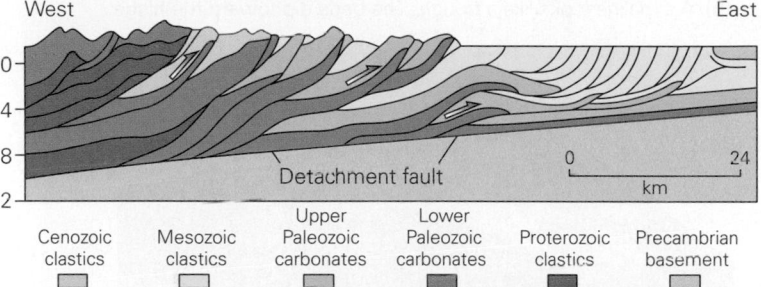

West   East

0
4
8
12

Detachment fault

0   24
km

| Cenozoic clastics | Mesozoic clastics | Upper Paleozoic carbonates | Lower Paleozoic carbonates | Proterozoic clastics | Precambrian basement |

(b) A cross-section of the Canadian Rockies, showing that many thrust sheets underlie the range. They were pushed toward the east.

(c) A close-up photo of the folding within one of the thrust sheets of the Canadian Rockies. Snow highlights curving layers.

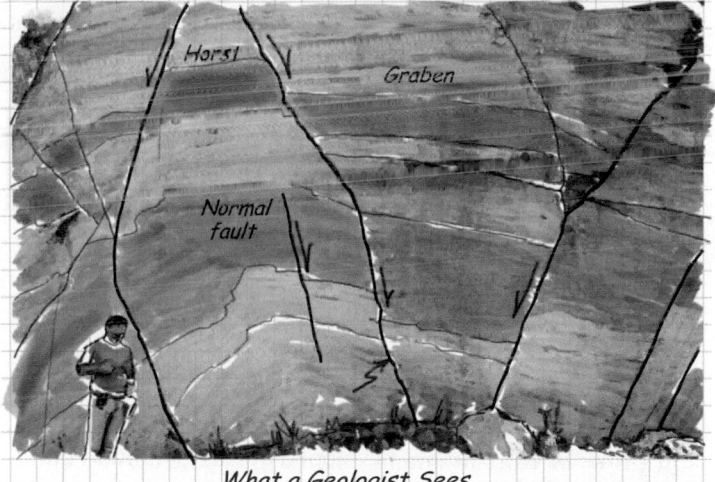

Horst
Graben
Normal fault

What a Geologist Sees

(e) An example of horsts and grabens exposed in the wall of a marble quarry in Brazil.

TAKE-HOME MESSAGE

Brittle structures include joints (cracks), veins (mineral-filled cracks), and faults (fractures on which sliding occurs). Geologists distinguish among different kinds of faults based on the relative displacement across the fault. Slip shatters or pulverizes rock and can polish and scratch a fault's surface. You can identify faults by searching for offset layers or by steps in the landscape. Slip on fault systems can accommodate significant crustal stretching or shortening.

QUICK QUESTION: How can you distinguish between a joint and a fault in the field?

11.4 Folds and Foliations

Geometry of Folds

Imagine a carpet lying flat on the floor. Push on one end of the carpet, and it will wrinkle or contort into a series of

FIGURE 11.14 Geometric characteristics of folds.

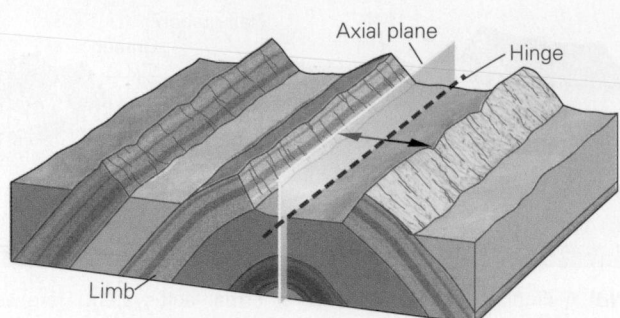

(a) An anticline looks like an arch. The beds dip away from the hinge.

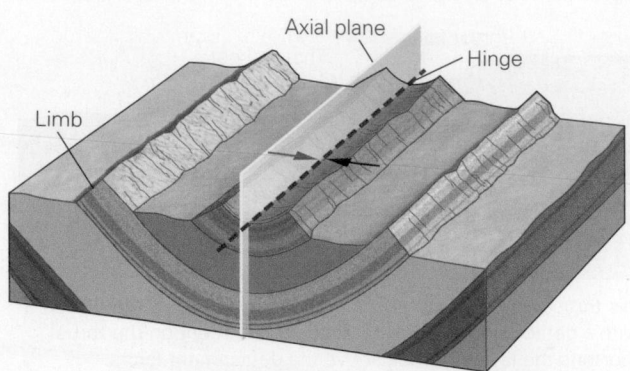

(b) A syncline looks like a trough. The beds dip toward the hinge.

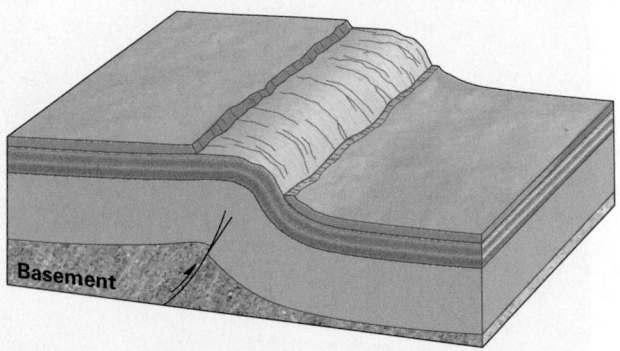

(c) A monocline looks like a stair step and is commonly draped over a fault block.

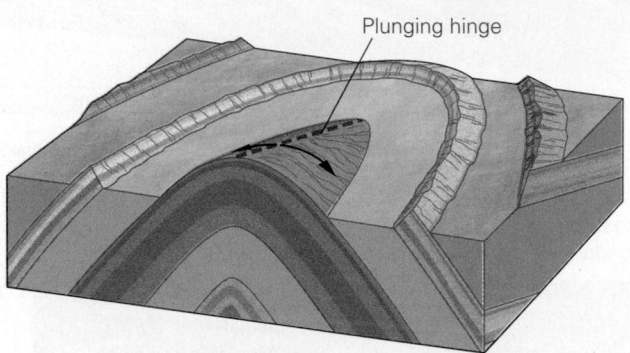

(d) A plunging anticline has a tilted hinge.

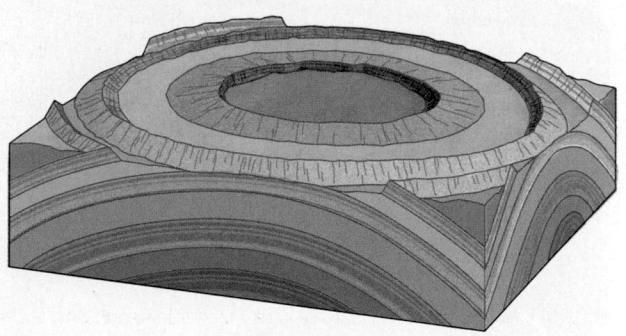

(e) A dome has the shape of an overturned bowl.

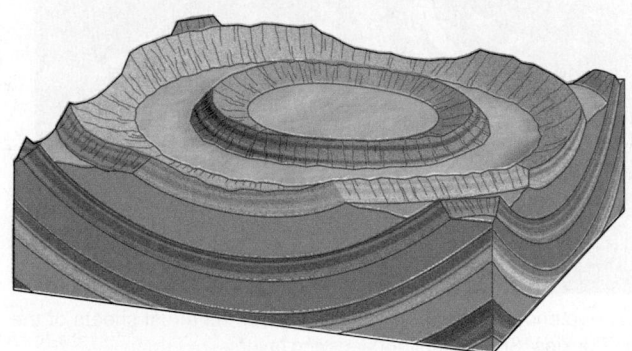

(f) A basin has the shape of an upright bowl.

wave-like curves. Stresses developed during mountain building can similarly warp or bend bedding, foliation, veins, or other planar features in rock. The result—a curving surface—is called a **fold**.

Not all folds look the same—some look like arches, some like troughs, and some have other shapes. To describe these shapes efficiently, we must first label the parts of a fold. The **hinge** refers to a line along which the greatest curvature develops. Commonly, the dip changes direction at the hinge. The **limbs** are the sides of the fold that display less curvature. Geologists refer to the imaginary plane that contains the hinges of successive layers as the **axial plane** of the fold. With these terms in hand, we can distinguish among the following characteristics:

- *Anticlines, synclines, and monoclines:* Folds that have an arch-like shape in which the limbs dip away from the hinge are called **anticlines (Fig. 11.14a)**, whereas folds with a trough-like shape in which the limbs dip toward the hinge are called **synclines (Fig. 11.14b)**. Where a fold intersects the ground surface and has undergone erosion, the oldest beds crop out near the hinge and the youngest away from the hinge in an anticline; the opposite is true in a syncline. A **monocline** has the shape of a carpet draped over a stair step **(Fig. 11.14c)**.

- *Nonplunging and plunging folds:* Geologists refer to a fold with a horizontal hinge as a *nonplunging fold*. If the hinge tilts, the fold is called a *plunging fold* **(Fig. 11.14d)**.

FIGURE 11.15 Examples of folds on outcrops and in the landscape.

(a) This anticline, exposed in a road cut near Kingston, New York, involves beds of Paleozoic limestone.

(b) This syncline, exposed in a road cut in Maryland, involves beds of Paleozoic sandstone and shale.

(d) A train of folds exposed in sea cliffs in eastern Ireland includes anticlines and synclines. The folds affect beds of Paleozoic sandstone and shale.

Fold hinge

Fold limb

(c) This fold, exposed along the coast of Brazil, occurs in Precambrian gneiss.

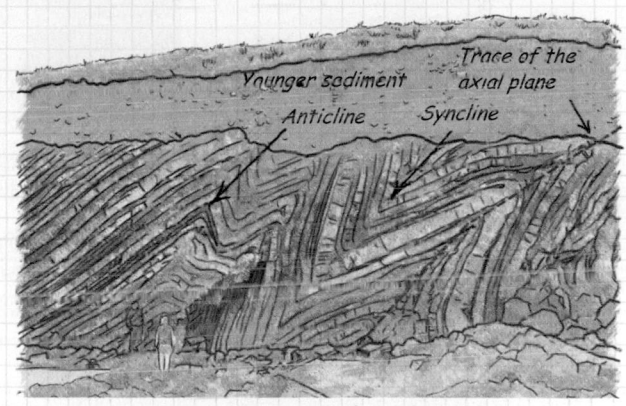

Younger sediment

Trace of the axial plane

Anticline   Syncline

What a Geologist Sees

Aerial view

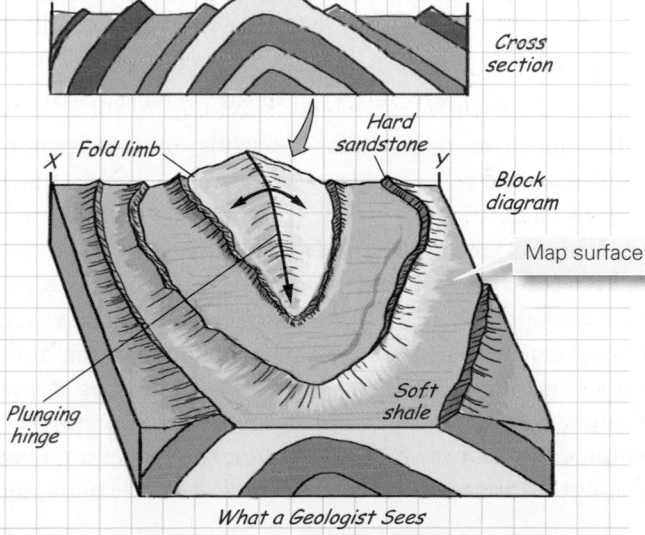

X   Y

Cross section

Fold limb

Hard sandstone

Block diagram

X   Y

Map surface

Plunging hinge

Soft shale

What a Geologist Sees

(e) The plunging anticline of Sheep Mountain, Wyoming, is easy to see because of the lack of vegetation. Resistant rock layers (sandstone) stand out as ridges, whereas weak rock layers (shale) erode away. A block diagram shows how surface exposures relate to underground structure.

FIGURE 11.16 Fold development in flexural-slip and passive-flow folding.

Before

After

Time

A flexural-slip fold in eastern Ireland

Beds protruding from a horizontal surface

Bed surface

Fold hinge

Cross-section plane

What a Geologist Imagines

(a) During the formation of flexural-slip folds, layers maintain constant thickness, so for the fold to form, layers must bend. To accommodate the bending, each bed slips relative to its neighbor.

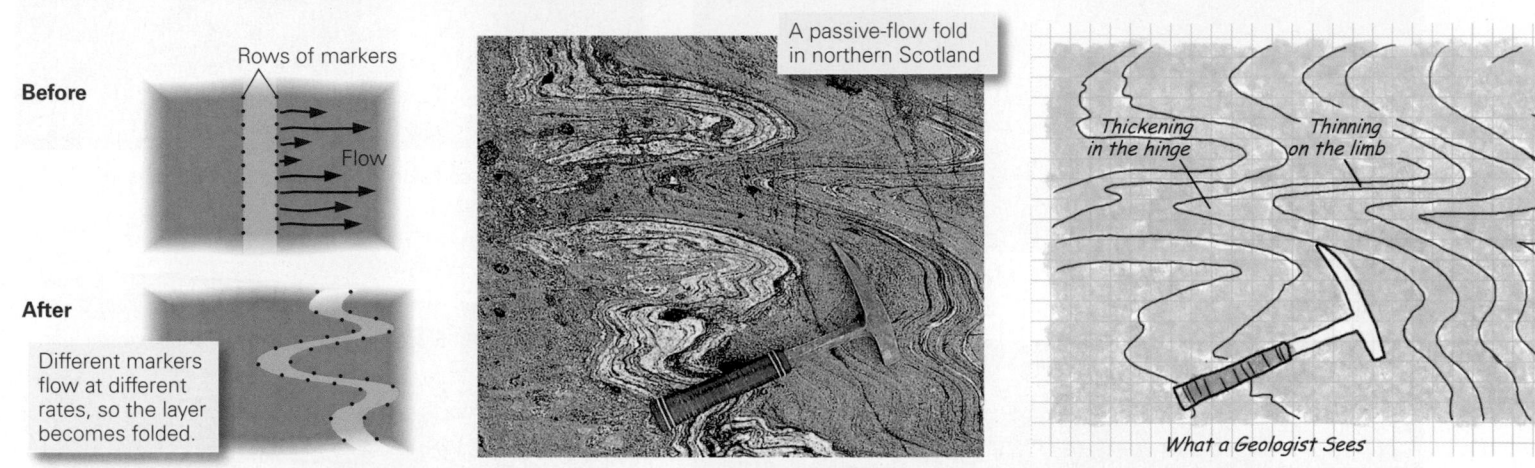

Rows of markers

Before

Flow

After

Different markers flow at different rates, so the layer becomes folded.

A passive-flow fold in northern Scotland

Thickening in the hinge

Thinning on the limb

What a Geologist Sees

(b) During the formation of passive-flow folds, the rock slowly flows. Different points along a marker line flow at different rates, causing the layer to become folded. Note that the layer's thickness doesn't stay constant during folding. In outcrops, the limb tends to thin and the hinge tends to thicken.

• *Domes and basins:* A fold with the shape of an overturned bowl is a **dome**, whereas a fold shaped like an upright bowl is a **basin (Fig. 11.14e, f)**. Domes and basins that intersect the ground surface and have been eroded both display circular outcrop patterns that look like bull's-eyes; the oldest layer crops out in the center of a dome, whereas the youngest layer crops out in the center of a basin.

Using the geologic terms for describing folds we've just presented, see if you can identify the various folds shown in **Figure 11.15**.

You can recognize folds not only in a cross section (a vertical slice through the crust), but also by the pattern of rock layers in map view (a horizontal plane representing the ground surface). For example, a nonplunging anticline involving sedimentary layers appears as a series of parallel stripes, with the oldest layer in the center and progressively younger layers away from the center; the stripes are symmetrically positioned around the hinge. Layers in a plunging fold

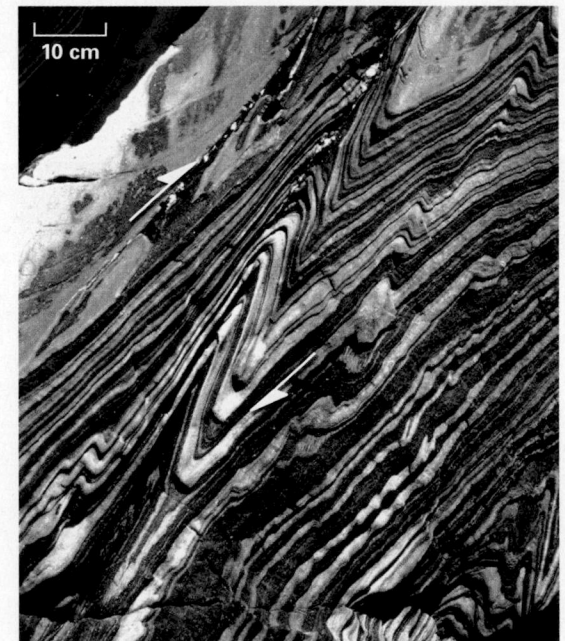

10 cm

(c) A flexural-slip fold resulting from shearing in South Australia.

FIGURE 11.17 Folding is caused by several different processes.

Before ➡ After

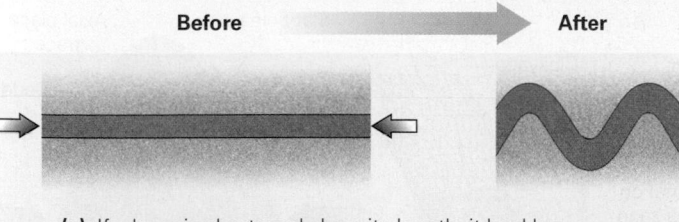

(a) If a layer is shortened along its length, it buckles.

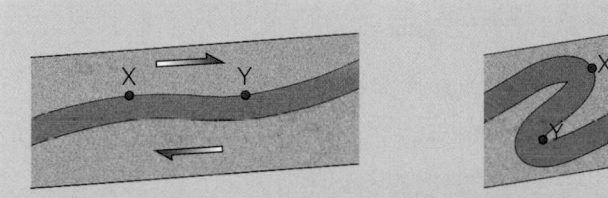

(b) If a layer is sheared, one part of the layer moves over another part to produce a fold.

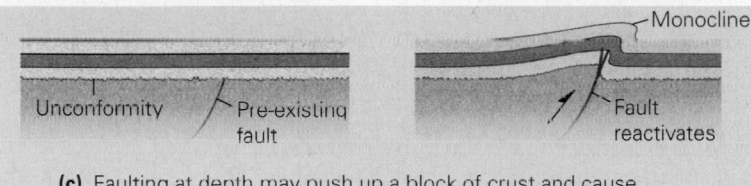

(c) Faulting at depth may push up a block of crust and cause overlying beds to bend into a monocline.

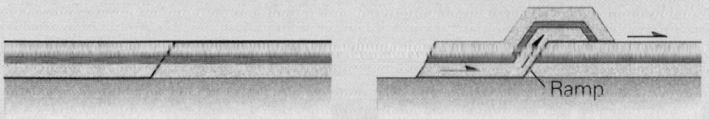

(d) When layers move up and over step-shaped faults, they must bend into folds. The portion of the fault that cuts across layering is a ramp.

What a Geologist Sees

(e) An anticline associated with faulting along a highway in New York State.

display a U-shape on the ground surface (see Fig. 11.15e). We can represent the hinge of the fold with a heavy line bordered by outward-pointing arrows for an anticline and inward-pointing arrows for a syncline. Note that because some layers erode more easily than others, a landscape underlain by folds may host ridges and valleys.

Formation of Folds

Folds develop in two principal ways. During formation of *flexural-slip folds*, a stack of layers bends, and slip takes place between the layers **(Fig. 11.16a, c)**. The same phenomenon happens when you bend a deck of cards—to accommodate the change in the shape of the deck, the cards slide with respect to one another. *Passive-flow folds*, in contrast, form when the rock, overall, behaves like weak plastic and slowly flows; these folds develop simply because different parts of the rock body flow at different rates **(Fig. 11.16b)**.

Why do folds form? Some layers wrinkle up, or buckle, in response to end-on compression **(Fig. 11.17a)**. Other folds develop where shear stress gradually shifts one part of a layer up and over another part **(Fig. 11.17b)**. Still others form when new slip on a fault causes a block of basement to move up so that the overlying sedimentary layers must warp **(Fig. 11.17c)**. And finally, some result from movement of rock layers up and over step-like bends in an underlying fault, for the layers must curve to conform to the fault's shape as they move **(Fig. 11.17d, e)**.

Tectonic Foliation in Rocks

In an undeformed sandstone, the grains of quartz resemble spheres, and in an undeformed shale, clay flakes are aligned with the bedding plane, so that shales tend to split parallel to the bedding. During deformation, internal changes take place in a rock that gradually modify the original shape and arrangement of grains. For example, plastic deformation of quartz grains may transform them into cigar shapes, elongate ribbons, or tiny pancakes,

FIGURE 11.18 The development of tectonic foliation in rock.

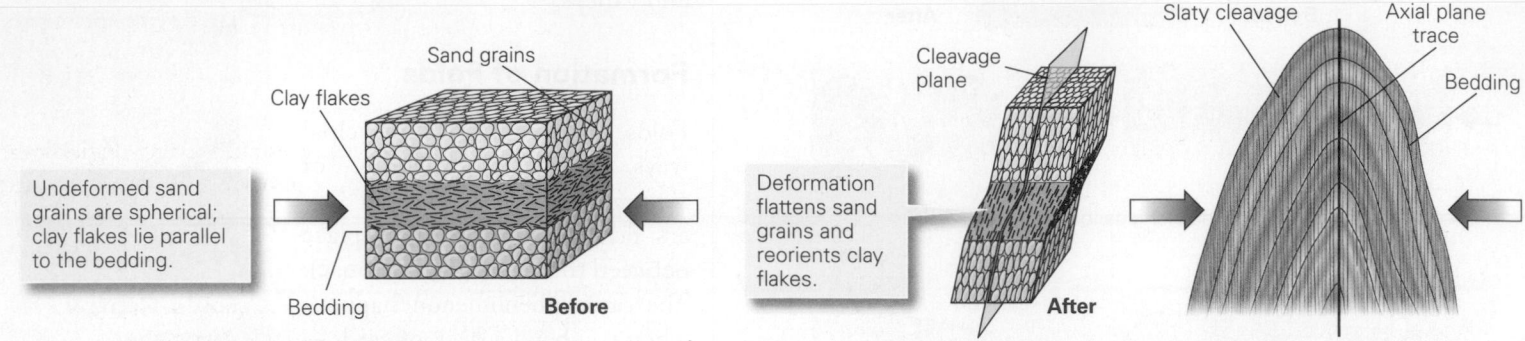

(a) Compression shortens beds, flattens sand grains, and reorients clay flakes. Clay flakes were originally parallel to bedding, but they become parallel to slaty cleavage during deformation. Folding may accompany cleavage formation.

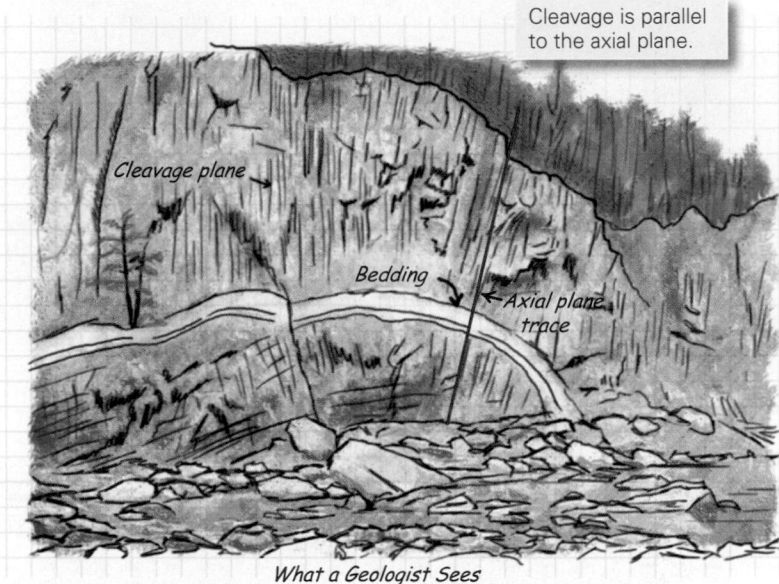

(b) An example of slaty cleavage developed in Paleozoic strata exposed in a stream cut in New York. Relict bedding is still visible, but note that the rock breaks more easily on the cleavage plane. This slaty cleavage formed in association with folding.

and clay flakes may recrystallize or reorient so that they lie at an angle to the bedding. Overall, deformation tends to produce *inequant grains*, meaning grains that have different orientations in different directions, and can cause such grains to align parallel to one another. We refer to layering developed by the alignment of grains in response to deformation as **tectonic foliation**.

We introduced foliation— such as slaty cleavage, schistosity, and gneissic layering—in Chapter 8 while discussing the effects of metamorphism. Here we add to the story by noting that such foliation forms in response to flattening and shearing in plastically deforming rocks—in other words, foliation indicates that the rock has developed a strain **(Fig. 11.18)**. For example, in rocks with slaty cleavage, the cleavage planes lie perpendicular to the direction of shortening, so the cleavage parallels the axial plane of folds. In schists and gneisses, the foliation commonly lies parallel to or at a slight angle

to planes on which shear took place, because shear smears grains out **(Fig. 11.19)**.

TAKE-HOME MESSAGE

Folds are bends or curves defined by the shape of rock layers. Arch-like folds are anticlines, and trough-like folds are synclines. Folds form for a variety of reasons; for example, buckling of layers due to end-on compression and crustal shortening can yield folds. Deformation can change the shape and orientation of grains, aligning them to produce a planar pattern called foliation. Foliation commonly develops when rocks simultaneously undergo metamorphism and deformation.

QUICK QUESTION: What mechanisms allow layers to undergo folding?

FIGURE 11.19 Development of foliation due to shearing. In this rock, quartz and mica-rich layers have become very fine grained and strongly foliated. The foliation is parallel to the shearing direction.

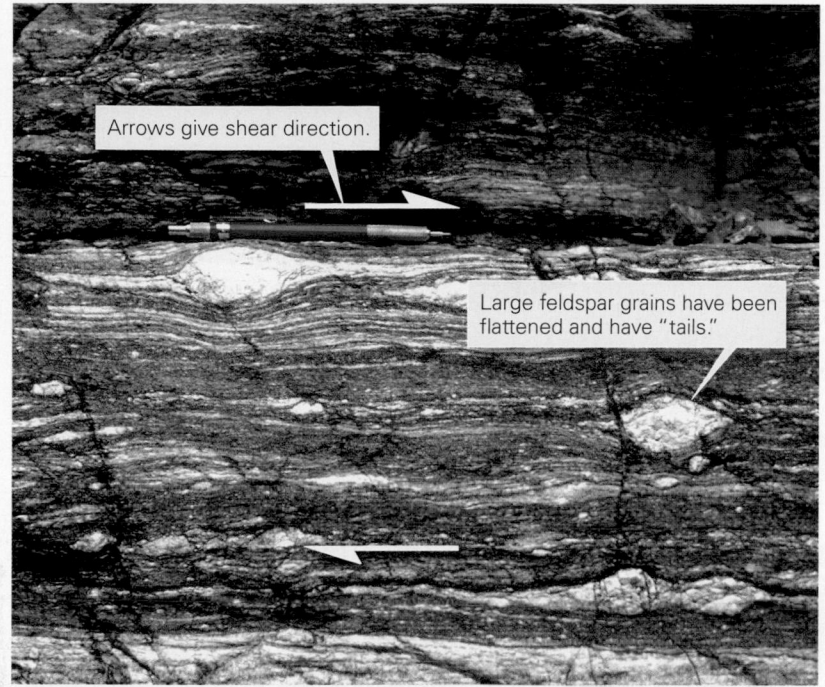

Arrows give shear direction.

Large feldspar grains have been flattened and have "tails."

11.5 Causes of Mountain Building

Before plate tectonics theory became established, geologists were just plain confused about why mountains formed. In the context of plate tectonics, however, the many processes driving mountain building became clear: mountains form in response to convergent-boundary deformation, continental collisions, and rifting. Mountain belts tend to be much longer than they are wide because collision zones, rifts, and plate boundaries tend to be elongate. In this section, we look at these different settings and the types of mountains and geologic structures that develop in each one.

> **Did you ever wonder . . .**
> why mountains occur in distinct belts?

Mountains Related to Subduction

As we saw in Chapter 9, subduction at a convergent boundary produces a volcanic arc. But that's not the only feature

FIGURE 11.20 The Andes exemplify the characteristics of convergent-boundary orogens.

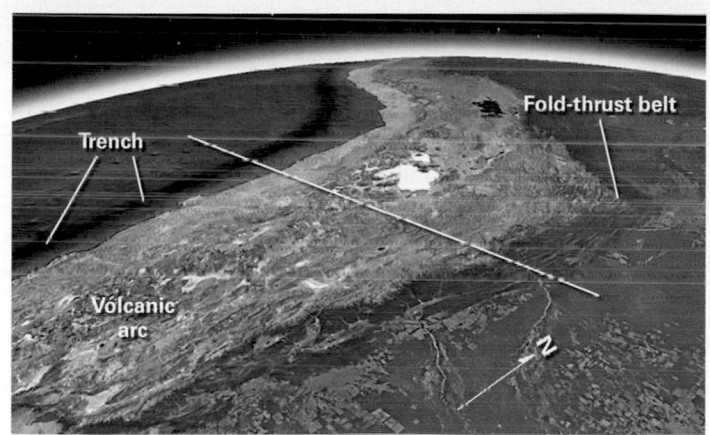

Trench

Fold-thrust belt

Volcanic arc

(a) An oblique view of the central Andes, showing the fold-thrust belt on the east side of the orogen and the trench on the west.

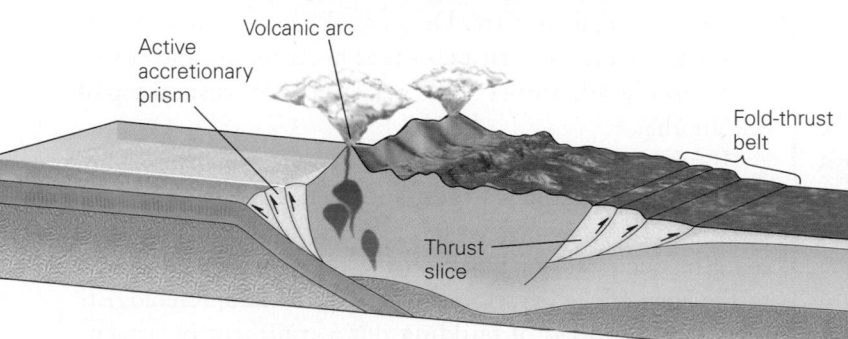

Active accretionary prism

Volcanic arc

Fold-thrust belt

Thrust slice

(b) A simplified cross section of the crustal structure beneath the Andes along the white line in (a). Compression uplifts the range and shortens the crust.

(c) Peaks of the Andes in Chile.

FIGURE 11.21 Characteristics of collisional orogens.

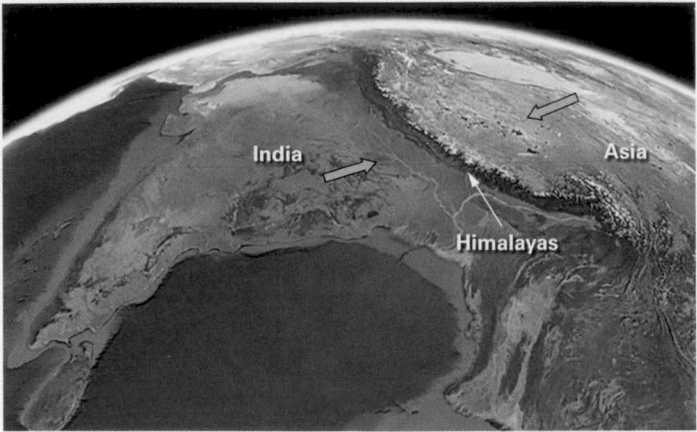

(a) The ongoing collision between India and Asia generated the Himalaya orogen.

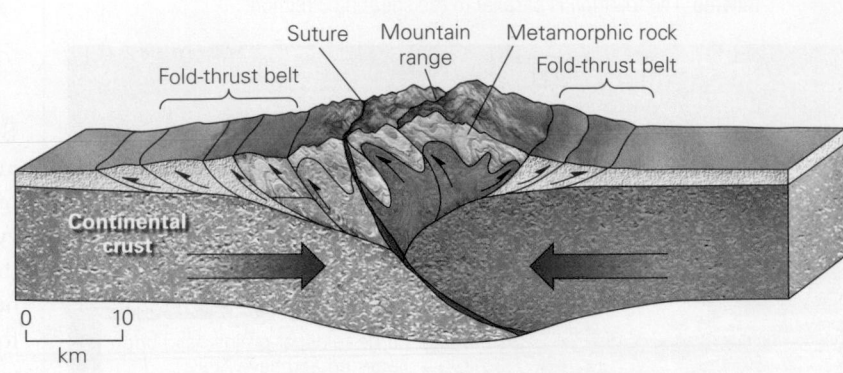

(b) A simplified cross section of a collisional orogeny.

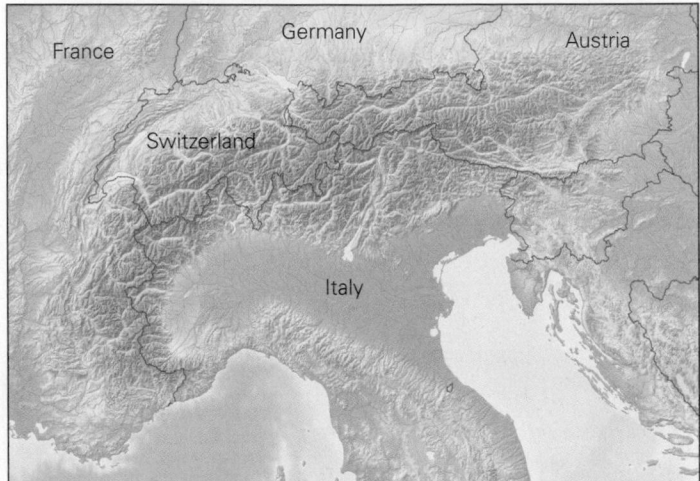

(c) The collision of the Italian Peninsula with continental Europe produced the Alps.

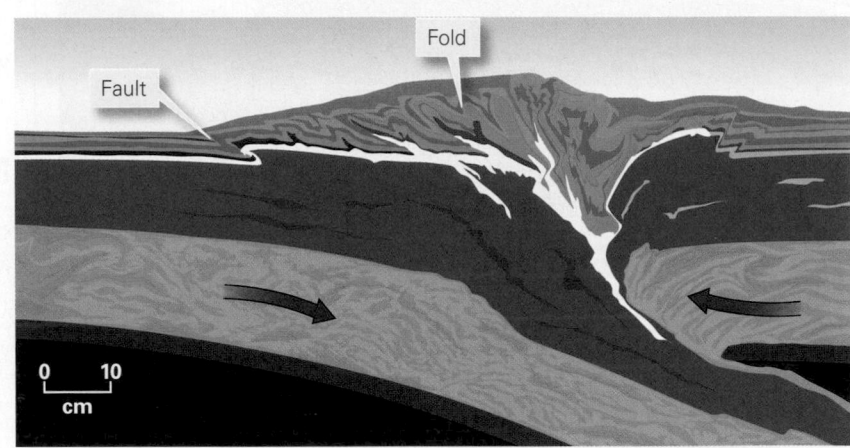

(d) Geologists simulate collision in the laboratory using layers of colored sand. Dragging the left side of the model under the right produces structures and uplift, as shown in this sketch of a model.

to develop in response to interactions at such boundaries. At some convergent boundaries, compressional stress develops and drives crustal shortening in the overriding plate. Such shortening produces a fold-thrust belt. The Andes orogen of western South America displays the rugged topography that can develop in a compressional convergent-margin orogen **(Fig. 11.20)**.

Mountains Related to Collision

Once the oceanic lithosphere between two relatively buoyant crustal blocks completely subducts, the blocks themselves collide with each other. The buoyant blocks may be large or small continents, island arcs, or large oceanic plateaus. Continental collision results in the formation of large mountain ranges, such as the present-day Himalayas of Asia, the Alps of Europe, and the Paleozoic Appalachian Mountains of eastern North America (**Geology at a Glance**, pp. 416–417).

The final stage in the growth of the Appalachians happened when Africa and North America collided.

During collision, intense compression generates fold-thrust belts on the margins of the orogen **(Fig. 11.21)**. In the interior of the orogen, where one block overrides the edge of the other, high-grade metamorphism occurs, accompanied by formation of passive-flow folds and tectonic foliation. The boundary between blocks that had been separate before the collision is called a **suture**. During collision, *crustal thickening* takes place as the crust shortens horizontally and thickens vertically, and thrust faults place slices of crust on top of one another.

If subduction continues over a long period, many offshore volcanic arcs, oceanic plateaus, and small fragments of continents (called *microcontinents*) may drift into a convergent margin. Such crustal blocks may collide with and attach to the edge of the overriding plate **(Fig. 11.22a)**. Geologists refer to the process of building out a continent by attaching new crustal fragments as **accretion**. The buoyant crustal block is called an *accreted terrane* once it has attached to the

FIGURE 11.22 Attachment of terranes.

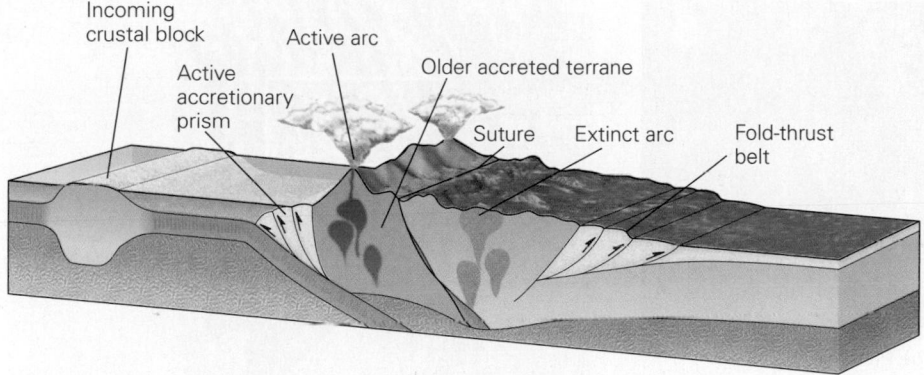

Incoming crustal block · Active arc · Older accreted terrane · Active accretionary prism · Suture · Extinct arc · Fold-thrust belt

(a) When subduction brings in terranes, they collide with and attach to the continent.

overriding plate. Once accretion occurs, the convergent boundary may jump to the seaward side of the accreted terrane so that subduction can continue and perhaps bring in additional crustal slivers. The process of accretion can add substantial new crust to a convergent-boundary orogen. For example, the western part of the North American Cordillera, the huge (7,000 km long and 600–1,500 km wide) orogen that extends from Alaska through Mexico, consists of accreted terranes that attached to the continent between 250 and 150 million years ago (Ma) **(Fig. 11.22b)**. At its widest, the belt of accreted crust reaches 500 km wide, as measured in an east-west direction.

Mountains Related to Continental Rifting

In a continental rift, tension stretches the crust horizontally. In deeper, warmer crust, rock can stretch plastically. But the upper, brittle layer of the crust breaks, and numerous normal faults form and slip **(Fig. 11.23a, b)**. Movement on the normal fault drops down blocks of crust, producing deep, sediment-filled basins (grabens and half grabens) separated by narrow, elongate mountain ranges that contain tilted crustal blocks. Traditionally, geologists refer to these ranges as *fault-block mountains*. Stretching during rifting thins the lithosphere, allowing hot asthenosphere to rise and undergo decompression melting (see Chapter 6). Heating may cause uplift of the rift and its borders. This process produces magmas that rise to form volcanoes within the rift. Today, the East African Rift clearly shows the configuration of rift-related mountains and volcanoes. In North America, rifting yielded the broad Basin and Range Province of Utah, Nevada, and Arizona **(Fig. 11.23c)**.

Rocks That Form during Orogeny

We've discussed how orogeny produces a variety of geologic structures and uplifts elongate belts of crust. The process of orogeny also establishes geologic conditions appropriate for the formation of a great variety of rocks. Here, we'll briefly consider examples from all three rock categories **(Fig. 11.24a)**:

- *Igneous activity during orogeny:* At convergent boundaries, melting takes place in the mantle above the downgoing plate and produces large volumes of magma that rise into the crust. Some of this magma erupts from volcanic arcs, but much of it solidifies underground to form numerous plutons, which together form batholiths

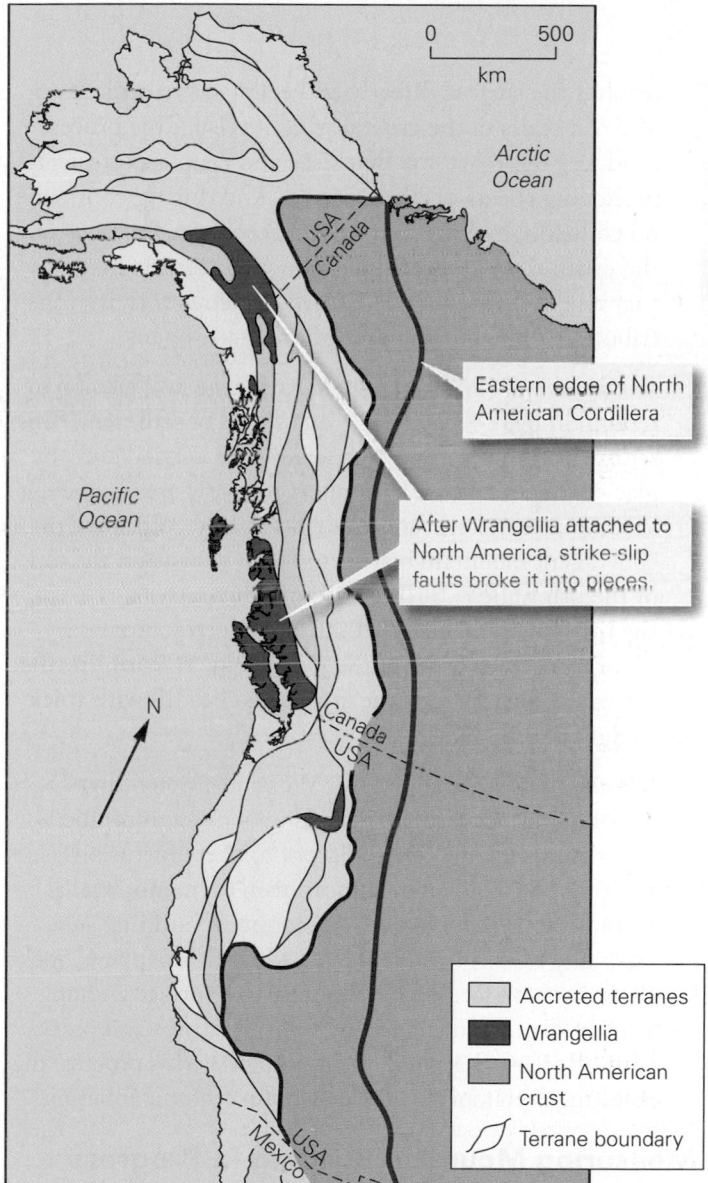

Eastern edge of North American Cordillera

After Wrangellia attached to North America, strike-slip faults broke it into pieces.

Arctic Ocean

Pacific Ocean

N

☐	Accreted terranes
■	Wrangellia
▨	North American crust
⬠	Terrane boundary

(b) The western portion of the North American Cordillera consists of accreted terranes that attached to the continent during the Mesozoic. During docking, a distinct terrane called Wrangellia (highlighted in red) was sliced into pieces that were displaced by strike-slip faults.

FIGURE 11.23 Rift-related orogens.

Range ▮ Basin ▮

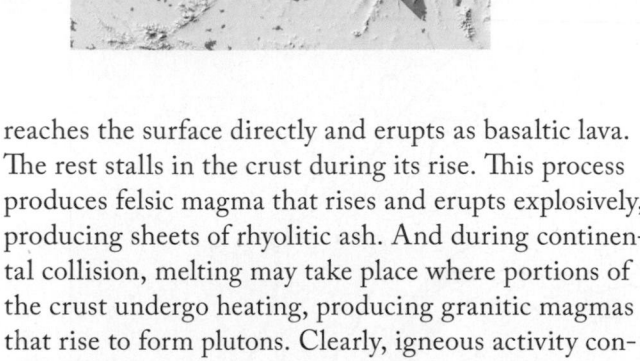

(a) A satellite image of a portion of the Basin and Range Province in Nevada. Note that the forested ranges trend north-south. The sediment-filled basins have unvegetated desert surfaces.

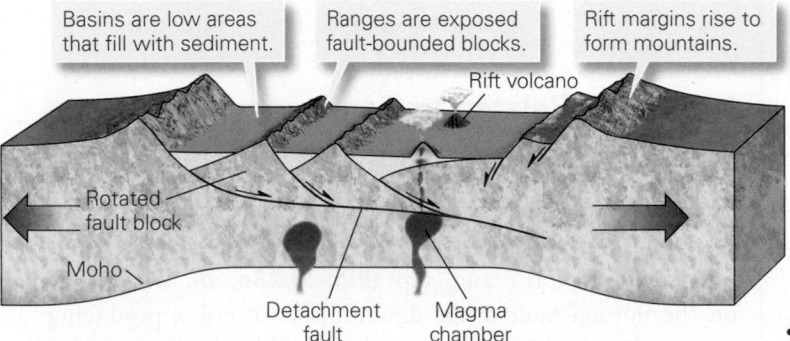

Basins are low areas that fill with sediment.

Ranges are exposed fault-bounded blocks.

Rift margins rise to form mountains.

Rift volcano

Rotated fault block

Moho

Detachment fault

Magma chamber

(b) Rifting leads to the development of numerous narrow mountain ranges. Rift-margin mountains may also form.

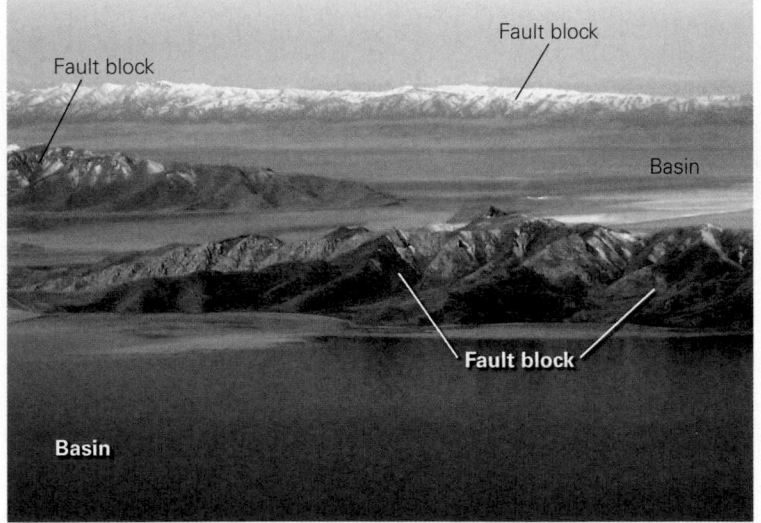

Fault block

Fault block

Basin

Fault block

Fault block

Basin

(c) The Basin and Range Province of the western United States formed during Cenozoic rifting.

(see Chapter 6). Therefore, orogens formed along convergent boundaries may include dramatic cliffs of granitic rocks. Stretching and thinning of lithosphere along rifts causes decompression melting of the underlying mantle, producing magmas. Some of the magma reaches the surface directly and erupts as basaltic lava. The rest stalls in the crust during its rise. This process produces felsic magma that rises and erupts explosively, producing sheets of rhyolitic ash. And during continental collision, melting may take place where portions of the crust undergo heating, producing granitic magmas that rise to form plutons. Clearly, igneous activity contributes significant volumes of rock to orogens.

- *Sedimentation during orogeny:* Weathering and erosion in mountain belts generate vast quantities of sediment. This sediment tumbles down slopes and gets carried away by glaciers or streams that transport it to low areas, where it accumulates in large fans or deltas. Large collisional or convergent mountain belts act as heavy loads on the top of the lithosphere. Such loads push down the surface of the lithosphere, thereby producing a deep sedimentary basin along the border of the range **(Fig. 11.24b)**. In rifts, the basins that form as the crust stretches fill with thick wedges of sediment.

- *Metamorphism during orogeny:* We've just noted that magma intrudes into orogens. Contact metamorphic aureoles commonly form adjacent to these intrusions (see Chapter 8). Regional metamorphism (dynamothermal metamorphism) occurs where mountain building substantially thickens the crust, for when this happens, rocks that may have formed near the Earth's surface end up at great depth and endure high temperatures and pressures. Because deformation accompanies this process, the resulting metamorphic rocks contain tectonic foliation.

Measuring Mountain Building in Progress

Not all mountains are just "old monuments," as John Muir mused. The rumblings of earthquakes and the eruptions of volcanoes attest to present-day, continuing movements in some

FIGURE 11.24 An example of the various rocks formed during orogeny.

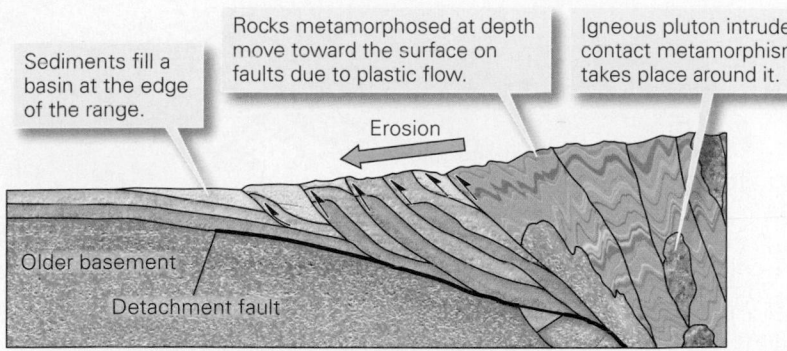

Sediments fill a basin at the edge of the range.

Rocks metamorphosed at depth move toward the surface on faults due to plastic flow.

Igneous pluton intrudes; contact metamorphism takes place around it.

Erosion

Older basement

Detachment fault

(a) In the internal zone of the range, metamorphic rocks and igneous rocks form. At the edge of the range, sedimentary rocks form.

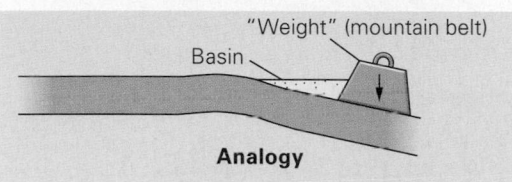

"Weight" (mountain belt)

Basin

Analogy

(b) The sedimentary basin develops because the mountain range acts as a weight that pushes down the surface of the lithosphere.

ranges. Geologists can measure the rates of these movements through field studies and satellite technology. For example, geologists can determine where coastal areas have been rising relative to sea level by locating ancient beaches that now lie high above the water. And they can tell where the land surface has risen relative to a river by identifying places where a river has recently carved a new valley. In addition, geologists now

FIGURE 11.25 GPS measurements of shortening in the Andes. The arrows indicate the velocity of locations in the Andes relative to locations in the interior of South America. The white arrow indicates relative plate motion.

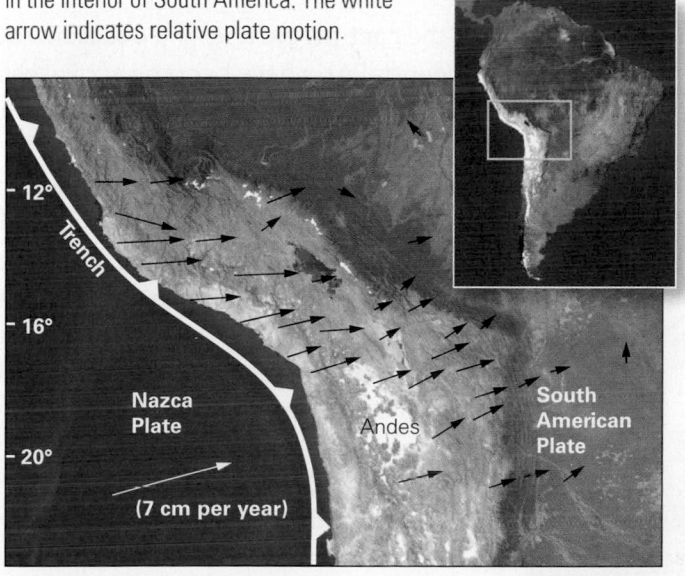

- 12°

Trench

- 16°

- 20°

Nazca Plate

Andes

South American Plate

(7 cm per year)

use the **global positioning system (GPS)** to measure rates of uplift and horizontal shortening in orogens. Research-quality GPS systems can specify locations to within 1 mm. By comparing the position of a location within an orogen with that of a location outside an orogen over a few years, geologists can detect crustal motion. Amazingly, we can "see" the Andes shorten horizontally at a rate of a couple of centimeters per year **(Fig. 11.25)**, and we can "watch" as mountains along this convergent boundary rise by a couple of millimeters per year.

TAKE-HOME MESSAGE

Mountain belts form in association with convergence, collision, and rifting. During continental collisions, and at some convergent margins, the crust thickens, large thrust faults and folds form, and regional metamorphism develops. Tilting of crustal blocks during rifting yields fault-block mountains.

QUICK QUESTION: Could fold-thrust belts develop in rifts? Why or why not?

11.6 Mountain Topography

Leonardo da Vinci, the Renaissance artist and scientist, enjoyed walking in the mountains, sketching ledges, and examining the rocks he found there. In the process, he discovered marine shells (fossils) in limestone beds cropping out a kilometer above sea level, and he suggested that the rock containing the fossils had risen from below sea level up to its present elevation. Modern geologists agree with Leonardo and now refer to the process by which the surface of the Earth moves vertically from a lower to a higher elevation as **uplift**. Mountain building requires substantial uplift of the Earth's surface **(Fig. 11.26)**. Think about it—to form a mountain range, on the order of 1 million km³ of rock rises up.

What kinds of distances are we talking about when referring to uplift in mountain ranges? As noted earlier, Mt. Everest rises 8.85 km above sea level. Although this distance may seem monumental—that's equal to 5,000 people standing one on top of another—it represents only about 0.06% of the Earth's diameter. In fact, if the Earth magically shrank to the size of a billiard ball, its surface (mountains and all) would feel smoother than that of an actual billiard ball. In general, the individual peaks that you see in a mountain range represent only a fraction of the range's total height, for

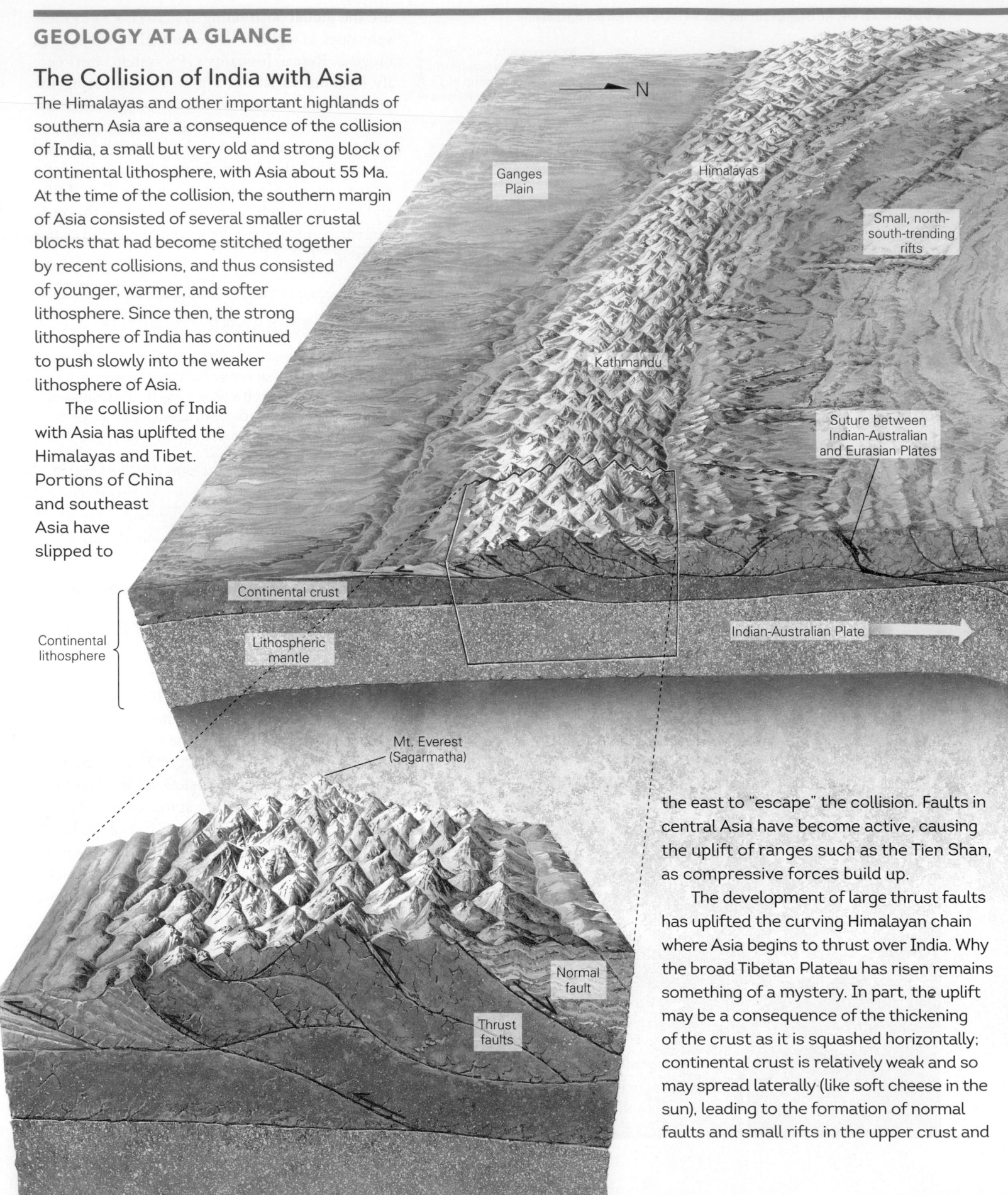

The Collision of India with Asia

The Himalayas and other important highlands of southern Asia are a consequence of the collision of India, a small but very old and strong block of continental lithosphere, with Asia about 55 Ma. At the time of the collision, the southern margin of Asia consisted of several smaller crustal blocks that had become stitched together by recent collisions, and thus consisted of younger, warmer, and softer lithosphere. Since then, the strong lithosphere of India has continued to push slowly into the weaker lithosphere of Asia.

The collision of India with Asia has uplifted the Himalayas and Tibet. Portions of China and southeast Asia have slipped to

N

Ganges Plain

Himalayas

Small, north-south-trending rifts

Kathmandu

Suture between Indian-Australian and Eurasian Plates

Continental crust

Lithospheric mantle

Continental lithosphere

Indian-Australian Plate

Mt. Everest (Sagarmatha)

Normal fault

Thrust faults

the east to "escape" the collision. Faults in central Asia have become active, causing the uplift of ranges such as the Tien Shan, as compressive forces build up.

The development of large thrust faults has uplifted the curving Himalayan chain where Asia begins to thrust over India. Why the broad Tibetan Plateau has risen remains something of a mystery. In part, the uplift may be a consequence of the thickening of the crust as it is squashed horizontally; continental crust is relatively weak and so may spread laterally (like soft cheese in the sun), leading to the formation of normal faults and small rifts in the upper crust and

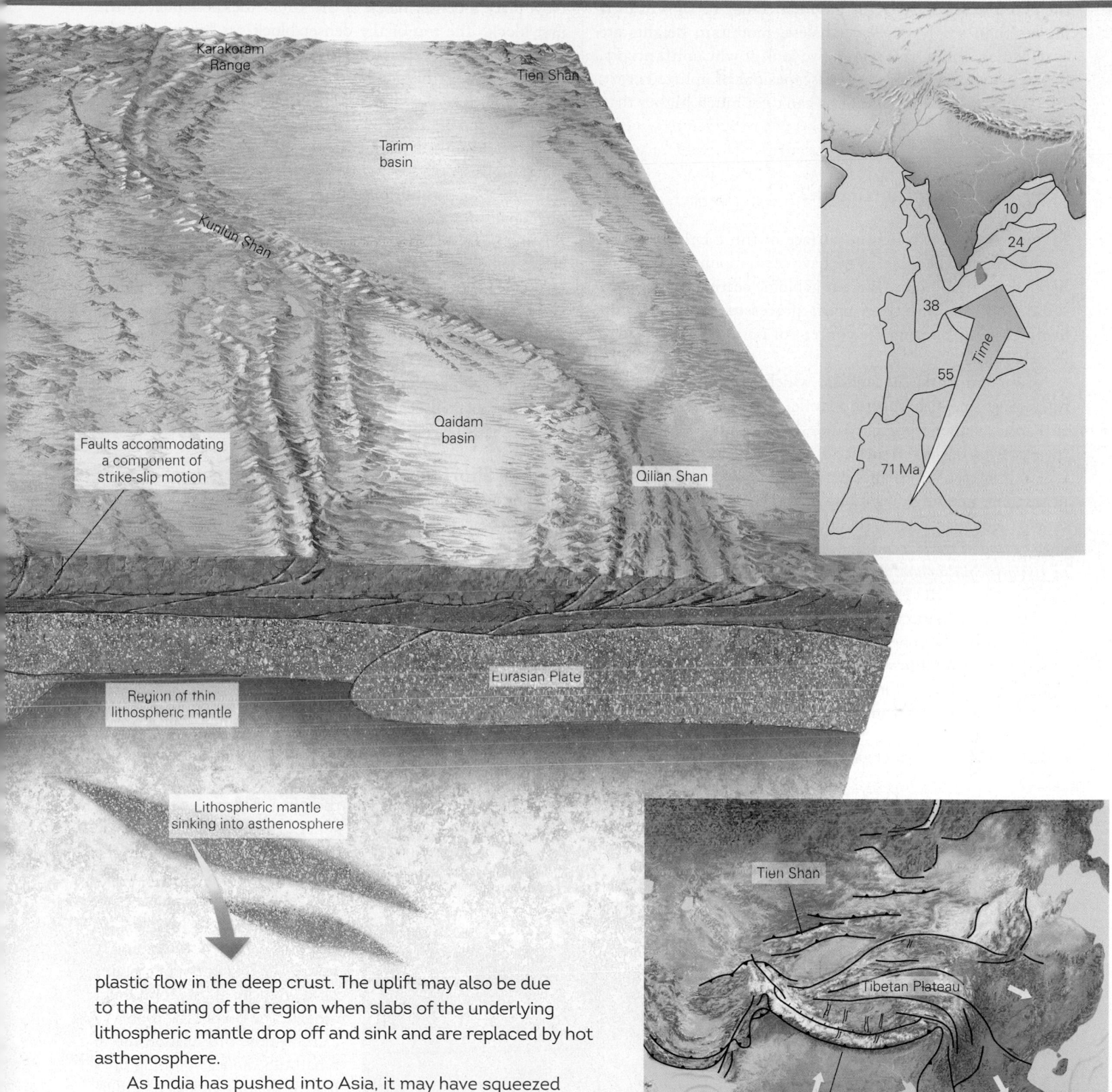

Karakoram Range

Tien Shan

Tarim basin

Kunlun Shan

Qaidam basin

Qilian Shan

Faults accommodating a component of strike-slip motion

Eurasian Plate

Region of thin lithospheric mantle

Lithospheric mantle sinking into asthenosphere

71 Ma

55

38

24

10

Time

Tien Shan

Tibetan Plateau

Himalayas

plastic flow in the deep crust. The uplift may also be due to the heating of the region when slabs of the underlying lithospheric mantle drop off and sink and are replaced by hot asthenosphere.

As India has pushed into Asia, it may have squeezed blocks of China and Southeast Asia sideways, toward the east; this motion is accommodated by slip on strike-slip faults. The collision may also have caused reverse faults in the interior of Asia to become active, uplifting a succession of small mountain ranges, such as the Tien Shan.

the plains at the base of the mountains may be significantly higher than sea level. Nevertheless, mountain heights are spectacular, and in this section we look at why uplift occurs, how erosion carves rugged landscapes out of uplifted crust, and why the Earth's mountains can't get much higher than Mt. Everest.

Why Are Mountains High?

Many processes can lift the surface of the Earth to mountainous heights because, as we have seen, mountain building happens in numerous different geologic settings. To understand how the variety of uplift processes work, we must begin by remembering the concept of *isostasy*, introduced in Interlude D.

The continental lithosphere, which consists of relatively rigid crust and lithospheric mantle, rests on the softer, but still solid, asthenospheric mantle below. As a consequence, the elevation of the top surface of the lithosphere, over a broad region, represents a balance between buoyancy force pushing lithosphere up and gravitational force pulling the lithosphere down. Geologists refer to the condition that exists when a balance between these two forces has been achieved as **isostasy**, or *isostatic equilibrium*. Put another way, isostasy exists where the elevation of the Earth's surface reflects the level at which the lithosphere naturally floats.

To picture the concept of isostasy, imagine placing a block of wood into a bathtub full of water. If the block is less dense than water, it floats, with part of the block remaining above the water surface and most of the block submerged below.

Now place a denser block of the same thickness next to the first block. The top of the denser block sits lower than the less-dense block of the same thickness. Similarly, the top surface of a thicker block sits higher than the top surface of a thinner block of the same density. If you were to add another block of wood on top of one that is already floating, the lower block would sink, adjusting for the additional weight so as to maintain isostasy.

From our bathtub experiment, we can deduce that any phenomenon that changes the thickness and/or density of a floating block will affect the elevation of the block's surface above the water surface. Similarly, because the lithosphere rests on the asthenosphere, the elevation of the lithosphere's surface at a location depends on the thickness and density of the lithosphere beneath. So, to answer the question of why mountain belts can rise, we must identify geologic processes that can change the thickness and/or density of layers in the lithosphere. Let's consider some examples of how these changes take place.

Crustal Shortening and Thickening During collisional orogeny or certain types of convergent-boundary orogeny, horizontal compression causes the crust to shorten horizontally and thicken vertically. In fact, the folding, faulting, and plastic flow that take place during such events can almost double the crust's thickness. For example, the crust beneath the Earth's tallest mountain range, the Himalayas, attains a thickness of 70 km, whereas crust beneath the plains of the central United States averages 40 km thick (**Fig. 11.27a**; see also Geology at a Glance, pp. 416–417). To compensate for the thickening of the crust (the geologic equivalent to adding another block of low-density wood to the top of a floating block) and maintain isostasy, the bases of the crust and underlying lithospheric mantle sink or subside (**Fig. 11.27b**). Indeed, since the Himalayas rise 8 km above sea level, most of the thickened crust extends downward beneath the range as a **crustal root**. Simply stated, the higher the mountains in a collisional or convergent orogen, the deeper the root. We can illustrate this relationship in a bathtub model by lining up a row of floating blocks of different thicknesses—if all the blocks have the same density, the thicker blocks rise farther above water and protrude deeper below the surface (**Fig. 11.27c**).

FIGURE 11.26 Mountain cliffs tower above a lake in the Canadian Rockies.

FIGURE 11.27 The concept of isostasy as applied to collisional mountain ranges.

(a) The Himalayas are the world's highest mountains. The crust beneath them is almost twice the normal thickness.

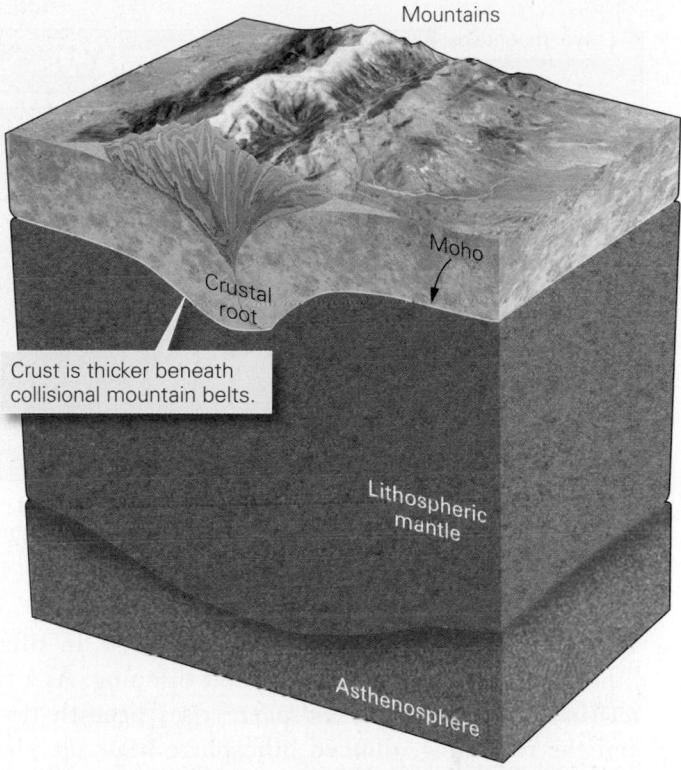

(b) Collision thickens the crust. Mountains form where the low-density crust becomes thicker.

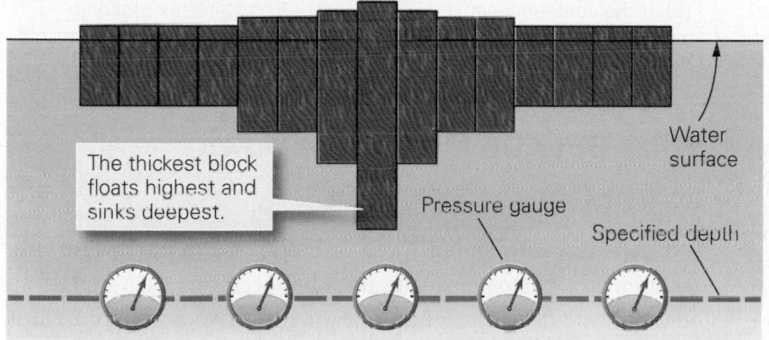

(c) Because of isostasy, blocks of wood floating in water sink to a depth such that the mass of the water displaced is the same as the mass of the block.

Where did the idea that mountains have roots come from? The discovery began with an observation by Sir George Everest, after whom the world's highest mountain was named. When surveying in India during the mid-1800s, Everest discovered that the gravitational attraction due to the mass of the Himalayas was large enough to deflect a plumb bob (a lead weight at the end of a string) away from vertical. Subsequent calculations demonstrated that the amount of deflection was actually less than expected, given the size of the range. A British scientist, George Airy, came up with an explanation. Realizing that crustal rocks aren't as dense as mantle rocks, Airy speculated that if the Himalayas had a low-density crustal root, the range would have less mass than expected and thus would not exert as much gravitational pull on a plumb bob. Note that Airy's proposal resembles the image shown in Figure 11.27c, so this image is known as the *Airy model* of isostasy.

Addition of Igneous Rock to the Crust When a volcano erupts, a thick pile of pyroclastic debris, lava, or both, may build up on the surface of the crust. Successive eruptions over millions of years, and granitic intrusions within and below the volcanic edifice, can produce a range whose peaks rise kilometers above sea level.

Geologists suspect that in some locations, basaltic magma formed by partial melting of the mantle gets trapped in sills at or near the base of the crust and accumulates. Addition of this hidden basalt thickens the crust from below, a process known as *underplating*. Even though basalt (a mafic rock) is denser than granite (a felsic rock), it is less dense than the peridotite (an ultramafic rock) that makes up the lithospheric mantle. Therefore, underplating causes the land surface to rise and the base of the crust to sink to maintain isostasy.

Removal of Lithospheric Mantle Lithospheric mantle consists of very dense rock (peridotite). The weight of this rock pulls the lithosphere down, just as heavy ballast makes a ship settle deeper into the water. Removal of some or all of the lithospheric mantle from the base of a plate, therefore, causes the surface of the remaining lithosphere to rise to maintain isostasy, even if the thickness of the crustal component remains unchanged **(Fig. 11.28)**. Such removal, a process known as *delamination*, resembles removal of ballast from the hold of a ship—as the weight of the ballast disappears, the deck of the ship rises. Several data sources suggest that the region of the Tibetan Plateau in Asia underwent an episode of uplift when some of the underlying lithospheric mantle dripped or peeled off the base of the plate and sank down into the asthenosphere.

FIGURE 11.28 Uplift due to delamination of the lithosphere root may happen after collision.

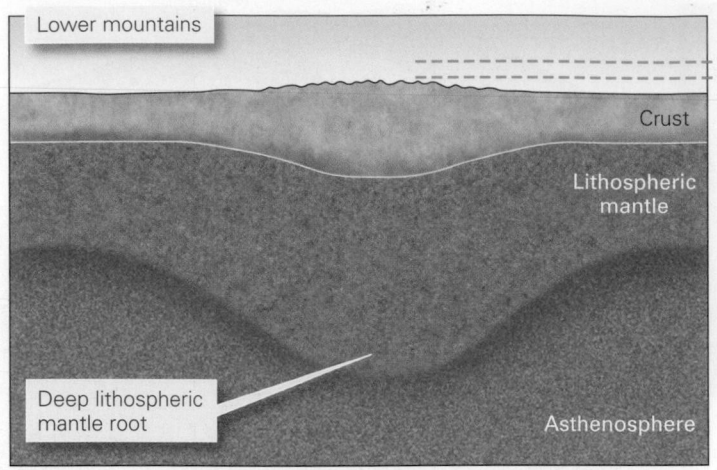

(a) Soon after collision, the crust and lithospheric mantle have thickened. The lithospheric mantle root is like ballast, holding the surface of the crust down.

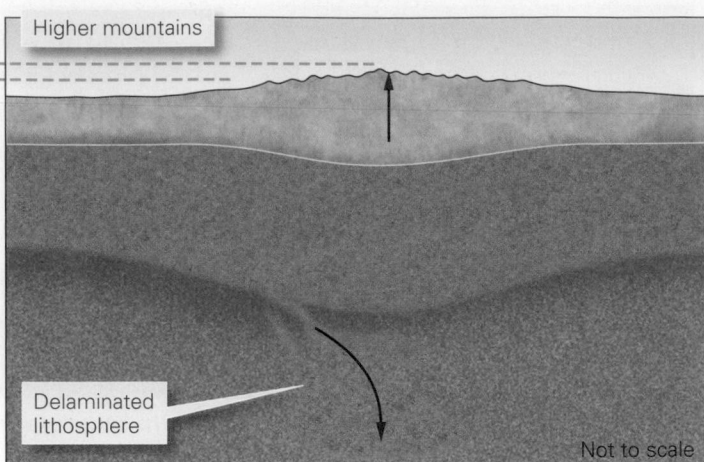

(b) If the lithospheric mantle root detaches and sinks, a process called delamination, the surface of the lithosphere may rise, like a ship dropping ballast.

Thinning and Heating of Lithosphere In rifts, the lithosphere undergoes stretching and thinning. As a result, relatively less-dense asthenosphere rises beneath the rift, and the remaining, thinned lithosphere heats up. Heating causes rocks to expand and therefore causes the density of the thinned lithosphere to decrease. Replacement of denser lithospheric mantle with less-dense asthenosphere, together with heating of the remaining thinned lithosphere, causes the overall region of the rift, as well as the margin of the rift, to rise in order to maintain isostasy. It's for this reason that the region of the East African Rift hosts dramatic cliffs.

What Goes Up Must Come Down

When the land surface rises significantly, for whatever reason, it doesn't remain a smooth welt or bulge on the Earth's surface. As soon as a difference in elevation between one location and an adjacent one develops, gravity begins to drive a variety of erosive processes. For example, as slopes steepen, landslides of various types cause rock and debris to tumble from higher to lower elevations (see Chapter 16); when rain falls, streams form and sculpt valleys and canyons (see Chapter 17); and if temperatures remain cold enough, glaciers grow and flow, carving sharp peaks

FIGURE 11.29 Manifestations of erosion in mountain ranges. When land rises, water and ice start cutting into it.

(a) Glaciers carved these rugged peaks in Switzerland.

(b) Streams cut valleys into weathered bedrock in Brazil.

at their origin and widening and deepening valleys downslope (see Chapter 22). Together, such processes grind away elevated areas and produce the jagged landscapes that we associate with mountain terrains (Fig. 11.29). Glaciers appear to be particularly effective in keeping mountains lower than they might otherwise grow—so efficient, in fact, that geologists informally refer to glacial erosion in mountains as the "glacial buzz saw."

It's important to keep in mind that uplift and erosion happen simultaneously in active mountain belts. So, if the elevation of a range increases over time, the rate of uplift must exceed the rate of erosion. However, if the erosion rate exceeds the uplift rate, then the range's elevation decreases over time, even if the tectonic processes driving uplift continue to operate. And, if the rate of erosion equals the rate of uplift, the elevation of the range stays the same.

When the tectonic processes driving orogeny eventually stop because convergence, collision, or rifting ceases, erosion continues and eventually bevels the mountain range back to near sea level. But the process takes a long time, at least 30 to 50 million years. It takes that long, in part, because of isostasy. Specifically, removal of rock from the top of a mountain range resembles removal of cargo from the deck of a ship—removing cargo causes the ship to float higher, and erosion of rock from the range causes isostatic rebound (uplift). Roughly speaking, for every 3 cm eroded, the range rises by 1 cm.

The highest point on our planet, the peak of Mt. Everest, lies 8.85 km above sea level. Can mountain ranges on the Earth get significantly higher? Probably not. Mountains as high as Olympus Mons on Mars, which rises 27 km above the plain at its base, couldn't form on the Earth because of the relatively high geothermal gradient (the rate of increase in temperature with depth) in our planet's crust. Due to the geotherm, quartz-rich crustal rocks at mid-crustal depths (15–30 km) become so warm and weak that they can flow plastically. When plastic flow begins, overlying mountains begin to collapse under their own weight, and spread laterally like soft cheese that has been left out in the summer sun (Fig. 11.30). Geologists call this process *orogenic collapse*. During orogenic collapse, the upper crust breaks, and a system of normal faults develops to accommodate the horizontal stretching.

Because erosion at the surface takes place as uplift continues, rock intruded or metamorphosed at depth eventually becomes exposed at the surface. This process of revealing deeper rocks by removal of the overlying crust is called unroofing, or **exhumation**. In large mountain ranges, rocks that were metamorphosed at depths of over 20 km are exhumed and can eventually crop out at the surface. Exhumation rates, averaged over time, range from less than 1 mm per year up to about 10 mm per year. These rates may seem slow, but even at a rate of just 1 mm per year, rock rises from depth by about 1 km every 1 million years, and therefore, rock from 10 km below the surface reaches the surface in 10 million years.

FIGURE 11.30 The concept of orogenic collapse.

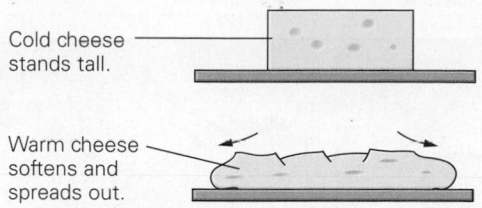

Cold cheese stands tall.

Warm cheese softens and spreads out.

(a) Cheese spreads out sideways as it warms up and softens.

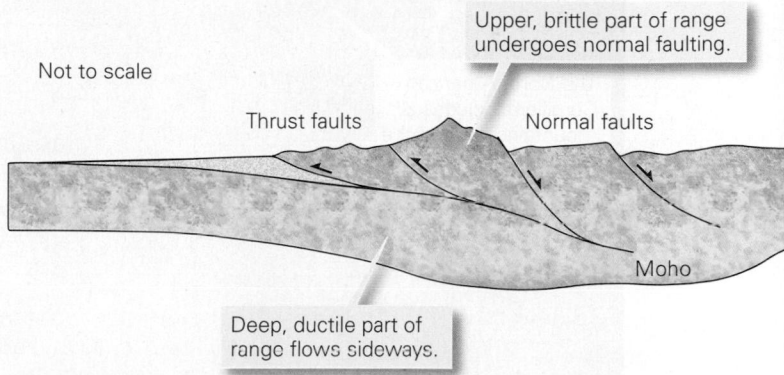

Not to scale

Upper, brittle part of range undergoes normal faulting.

Thrust faults

Normal faults

Moho

Deep, ductile part of range flows sideways.

(b) Similarly, mountain belts spread out sideways once they reach a certain thickness. The ductile crust at depth flows, whereas the upper (brittle) crust is broken by normal faults.

TAKE-HOME MESSAGE

Mountains exist where major uplift takes place. Uplift occurs for a variety of reasons but ultimately reflects the tendency of lithosphere to achieve isostasy (to "float" at the appropriate level). Crustal thickening causes uplift in collisional and convergent mountain ranges. Uplift of rifts happens because less-dense asthenosphere rises and replaces denser lithospheric mantle. Once uplift has occurred, erosion sculpts rugged topography. Mountain belts may eventually collapse under their own weight.

QUICK QUESTION: How do rocks metamorphosed at 20 km below the surface eventually crop out in a mountain range?

11.7 Basins and Domes in Cratons

A **craton** consists of crust that has not been affected by orogeny for at least the last 1 billion years and, as a result, has become relatively cool and therefore relatively strong and stable. Craton stability may also reflect the presence of underlying long-lived lithospheric mantle. Each continent

FIGURE 11.31 North America's craton consists of a shield, where Precambrian rock is exposed, and a cratonic platform, where Paleozoic sedimentary rock covers the Precambrian rock.

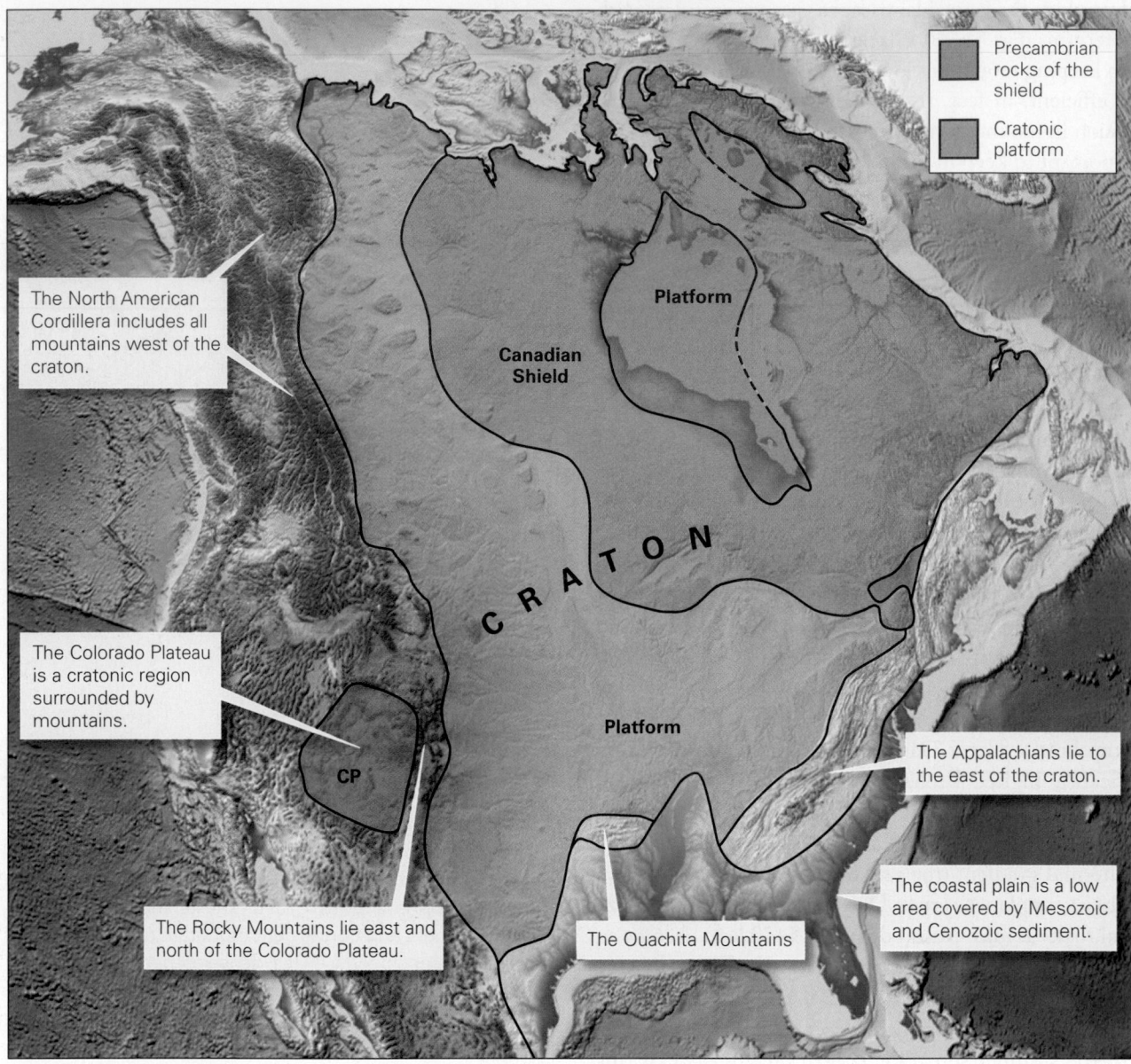

Precambrian rocks of the shield

Cratonic platform

The North American Cordillera includes all mountains west of the craton.

The Colorado Plateau is a cratonic region surrounded by mountains.

The Rocky Mountains lie east and north of the Colorado Plateau.

Platform

Canadian Shield

C R A T O N

CP

Platform

The Appalachians lie to the east of the craton.

The Ouachita Mountains

The coastal plain is a low area covered by Mesozoic and Cenozoic sediment.

contains one or more cratons. North America, for example, contains one large craton, bounded on the west by the North American Cordillera, an orogen that formed in the Mesozoic and Cenozoic, and bounded on the east and south by the Appalachian-Ouachita orogen, which developed during the Paleozoic **(Fig. 11.31)**. South America contains several smaller cratons, which were sutured together during late Precambrian and early Paleozoic time. Geologists divide cratons into two provinces: **shields**, where Precambrian metamorphic and igneous rocks crop out at the ground surface, and *cratonic platforms*, where a relatively thin layer of Paleozoic and locally younger sediments covers the Precambrian rocks (see Fig. 11.31).

Shields tend to be broad, low-lying regions in which outcrops expose metamorphic rocks displaying examples of passive-flow folds and tectonic foliation. These structures are relics of the tectonic processes that assembled the craton in the Precambrian.

Cratonic platforms tend to be broad plains. Outcrops in these regions expose nearly flat-lying undeformed strata. But at a regional scale, the pattern of contacts between stratigraphic formations on a geologic map of a platform defines regional-scale domes, basins, and arches. Cratonic basins are regions that, over geologic time, gradually subsided so that stratigraphic units are warped downward into the shape of a broad bowl, whereas domes and arches have stayed high or have risen, so that stratigraphic units have the shape of an overturned bowl (Fig. 11.32).

To see examples of basins and domes, let's look at the central Midwest of the United States. In the Illinois basin, strata warp downward into a huge bowl about 300 km across and up to 7 km deep. Significantly, strata thicken progressively toward the basin center, indicating that the floor of the basin was subsiding (sinking) as sediment was accumulating. There was more room for sediment to accumulate where the basin subsided the most. On a geologic map, the basin has a bull's-eye shape, with the youngest strata exposed in the center. Note that because the dome is so much wider than it is deep, bed dips are so gentle as to be hardly noticeable at an individual outcrop. The Ozark dome of Missouri also has a diameter of about 300 km. Individual sedimentary layers thin toward the crest of the dome. Why? Because the crest was relatively high during the time of deposition, less sediment could accumulate there than in adjacent basins. The dome also looks like a bull's-eye on a map, but in the case of a dome, the oldest rocks crop out at the center. Geologists refer to the vertical movements that generate broad mid-continent domes and basins as **epeirogeny**. The causes of epeirogeny remain poorly understood.

We noted earlier that strata in cratonic platforms are nearly undeformed, but they are not completely undeformed. Locally, faults cut the strata, and movement on faults, like the rise of a trapdoor beneath a carpet, causes overlying strata to bend into monoclines. Geologists refer to these faults as *intracratonic faults*. (The prefix *intra–* means within, so *intracratonic* means within the craton.) Most of these faults appear to have originated during the Precambrian, perhaps in response to episodes of rifting. They reactivated in pulses during the Paleozoic, generally at times when collisional orogens developed along the margins of the continent. Evidently, stress generated during the collisions was sufficient to cause slip on pre-existing faults in the cratons but was not sufficient to generate the pervasive folds and tectonic foliation found within orogens.

TAKE-HOME MESSAGE

Cratons are portions of continents that consist of very old and relatively stable crust. Parts of cratons, the cratonic platforms, may be covered by Paleozoic sedimentary strata. Variations in the dip and thickness of these strata define regional basins, arches, and domes.

QUICK QUESTION: Imagine that coal occurs in a particular stratigraphic unit. Will mines to reach the coal be deeper or shallower in the center of a basin?

FIGURE 11.32 Domes and basins of the North American cratonic platform.

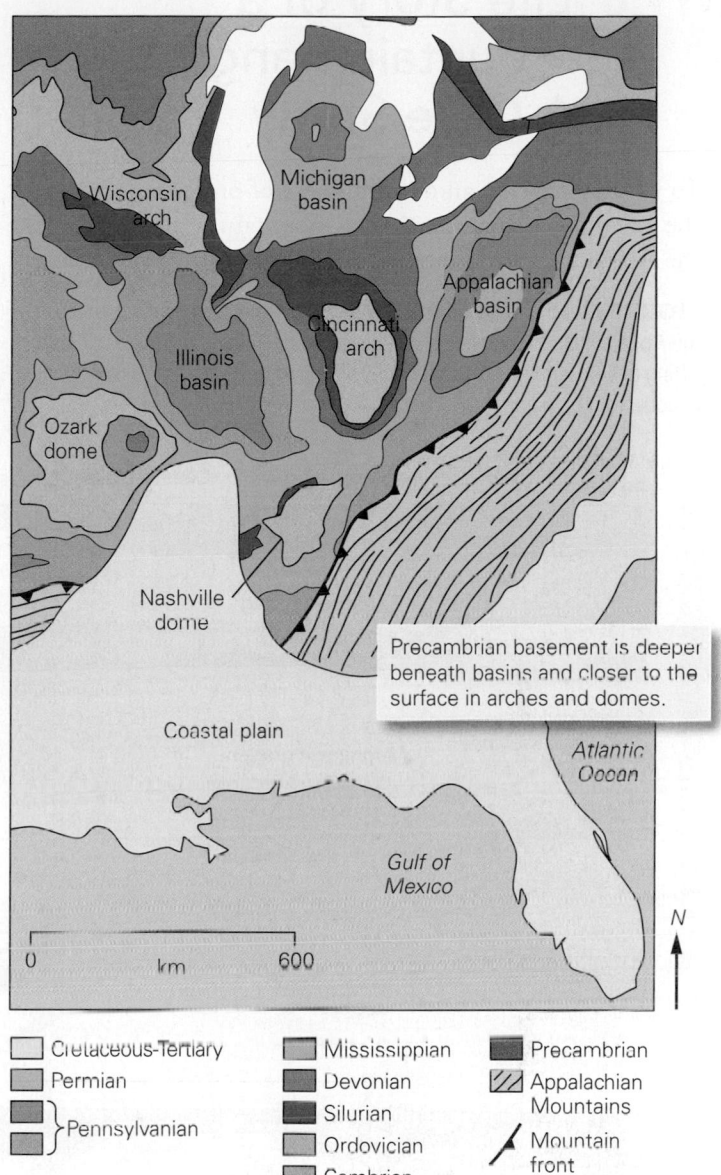

(a) A geologic map of the U.S. mid-continent platform region showing the locations of basins and domes.

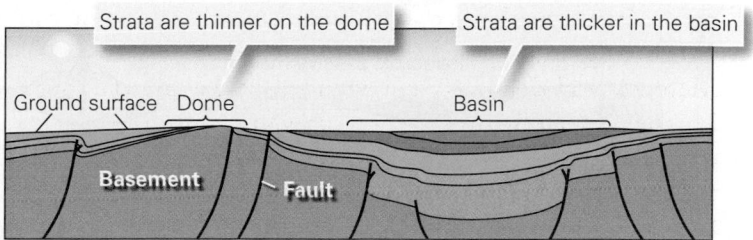

(b) A cross section showing how strata thin toward the crest of a dome and thicken toward the center of a basin. The cross section is vertically exaggerated. This cross section symbolically represents the Ozark dome and Illinois basin.

11.8 Life Story of a Mountain Range: A Case Study

To help us develop an overall image of orogeny, let's look at the life story of one particular mountain range. We'll take the Appalachian Mountains of eastern North America as our

FIGURE 11.33 These idealized stages show the tectonic evolution of the Appalachian Mountains. Note that although mountains do form during rifting events, geologists traditionally assign names only to the collisional or convergent events.

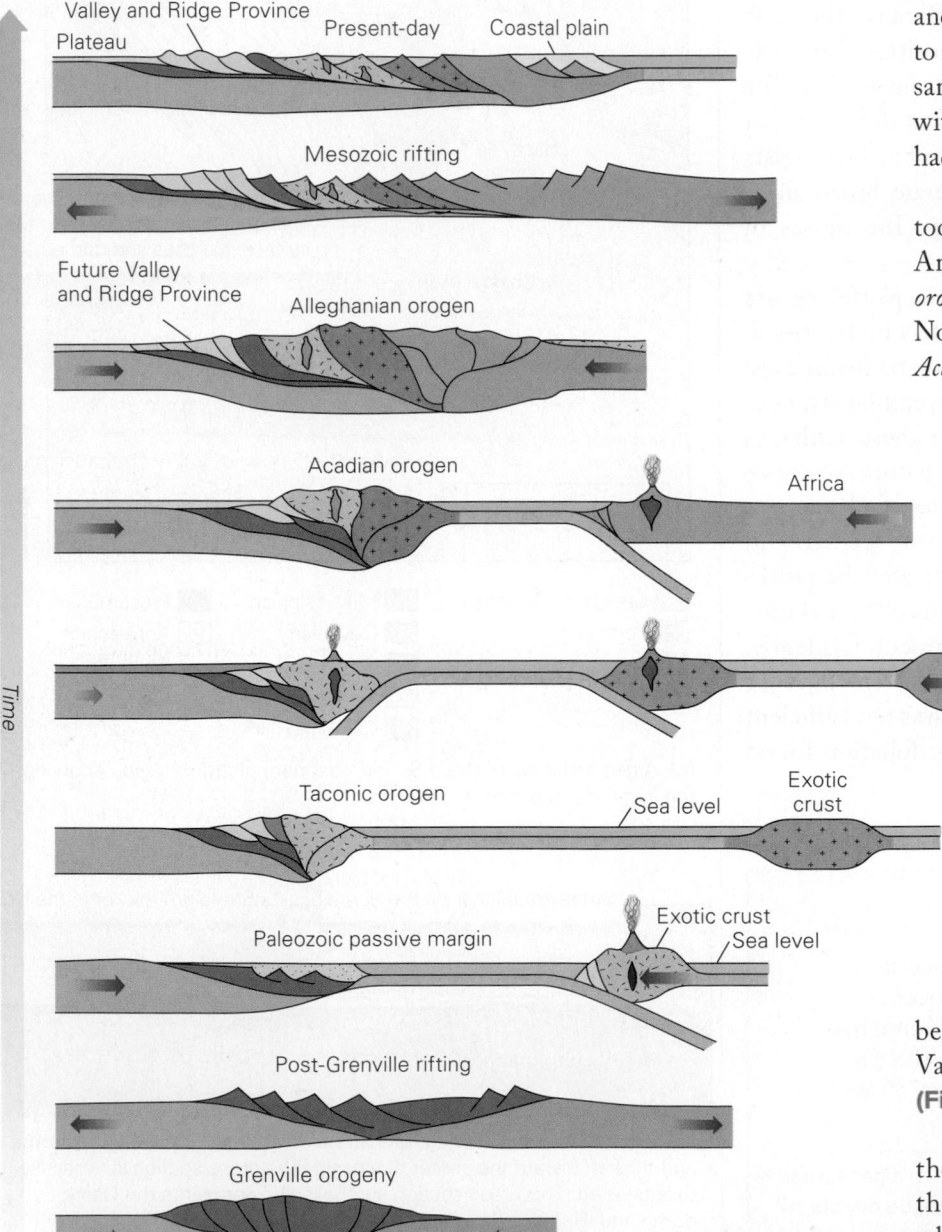

example. (We've simplified the story a bit, for ease of reading.) Geologists have constructed the range's life story by studying geologic structures, by examining the ages of igneous and metamorphic rocks, and by examining strata formed from sediment eroded from the range.

The story begins about 1.1 billion years ago, when the eastern margin of North America was involved in a massive collision with another continent **(Fig. 11.33)**. This event, called the *Grenville orogeny*, yielded a belt of deformed and dynamothermally metamorphosed rocks that underlie the eastern fifth of the continent. For a while after the Grenville orogeny, eastern North America lay in the middle of a supercontinent. But this supercontinent rifted apart around 600 Ma. Eventually, a new ocean formed to the east, and the rifted margin of eastern North America cooled, subsided, and evolved into a passive margin. From 600 to about 420 Ma, the passive margin slowly sank to become a basin, which gradually filled with a very thick sequence of sediment that had washed off the adjacent continent.

Between 420 and 370 Ma, two collisions took place along the eastern margin of North America. During the first, called the *Taconic orogeny*, a volcanic arc collided with eastern North America, and during the second, the *Acadian orogeny*, continental crustal slivers accreted to the continent. At times between these events, the margin of North America was a convergent boundary. The accretion of terranes during the Taconic and Acadian orogenies deformed and metamorphosed sediment and made the continent grow eastward. At 270 Ma, the Taconic and Acadian orogens were caught in a gigantic vise as the large continent of Africa collided with North America. This event, the *Alleghanian orogeny*, yielded a huge mountain range resembling the present-day Himalayas and generated a wide fold-thrust belt along the range's western margin. Eroded folds of this belt make up the topography of the present Valley and Ridge Province in Pennsylvania **(Fig. 11.34)**.

When the Alleghanian orogeny ceased, the Appalachian region once again lay in the interior of a supercontinent (Pangaea), where it remained until about 180 Ma (Mesozoic time). At that time, rifting split

FIGURE 11.34 The relationship between deformation and topography.

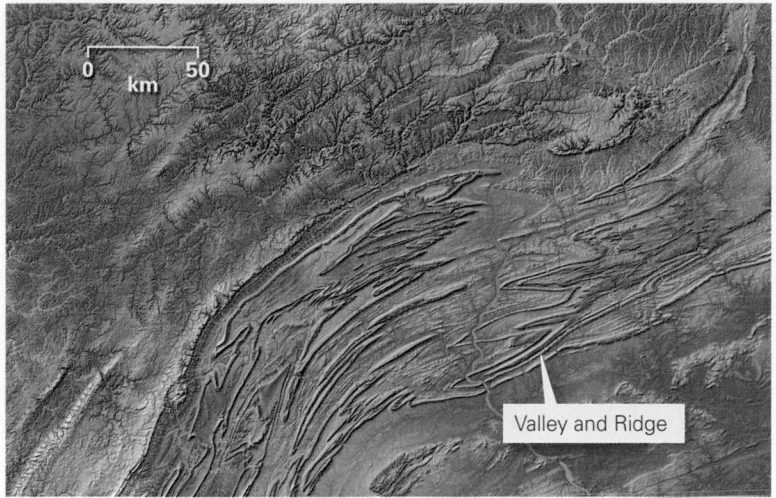

(a) The topography of eastern Pennsylvania, here shown in shaded relief, shows how erosional patterns reveal the folds that form the Appalachians, producing a region called the Valley and Ridge Province.

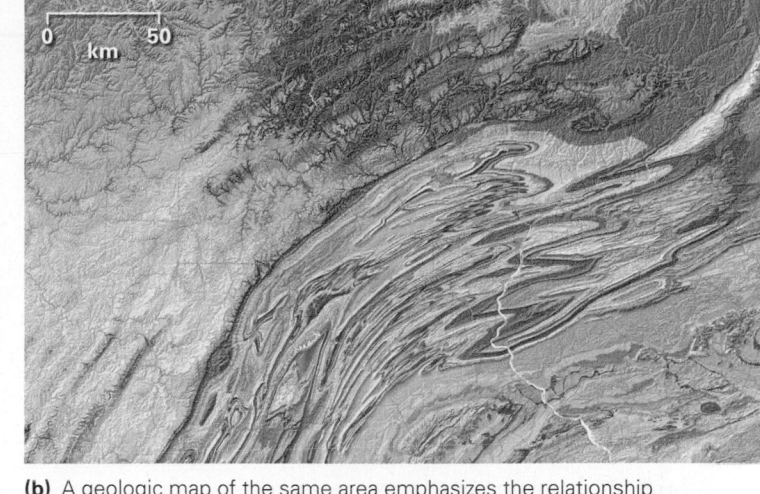

(b) A geologic map of the same area emphasizes the relationship between the distribution of folded stratigraphic formations (indicated by different colors) and the topographic features. A resistant sandstone formation forms the ridges.

the region open again, creating the Atlantic Ocean and a new passive margin that forms the east coast of North America today. As you can see from this example, major ranges such as the Appalachians incorporate the products of multiple orogenies and reflect the opening and closing of ocean basins. The succession of the opening of an ocean basin to form a passive margin, followed later by the closing of that ocean basin to form a collisional orogen, has come to be known as the *Wilson cycle*, after J. Tuzo Wilson, the Canadian geophysicist who first recognized the steps.

TAKE-HOME MESSAGE

The Appalachian Mountains expose remnants of structures and rocks formed during three distinct orogenies in the Paleozoic, each associated with a collision. The present eastern margin of the continent formed subsequently. Orogenic belts clearly can have complex histories.

QUICK QUESTION: How does the Appalachian orogen differ from the cratonic platform of the United States? See if you can find the boundary between the two on Google Earth.

ANOTHER VIEW An exposure in Painted Canyon, California, displays folded and faulted Cenozoic strata that has been deformed due to movement along the San Andreas fault.

CHAPTER SUMMARY

- Mountains occur in elongate ranges called mountain belts, or orogens. An orogen forms during an orogeny, or mountain-building event.

- Mountain building causes rock to undergo deformation during which the rock can change its location, orientation, and shape. Deformation produces geologic structures.

- During brittle deformation, rocks crack and break into two or more pieces. During plastic deformation, rocks change shape without breaking.

- Stress (compression, tension, and shear) causes deformation. Strain is distortion in response to a stress.

- Joints are natural cracks in rock, formed in response to tension under brittle conditions. Veins can form when minerals precipitate out of water passing through joints.

- Faults are fractures on which shear displacement takes place. Geologists distinguish among normal, reverse, strike-slip, and oblique faults, based on the sense of slip. Thrust faults are gently dipping reverse faults.

- Folds are curves defined by the shape of rock layers. Anticlines are arch-like, synclines are trough-like, monoclines resemble the shape of a carpet draped over a stair step, basins are shaped like a bowl, and domes are shaped like an overturned bowl.

- Fold-thrust belts form on the continental edge of collisional and convergent-boundary orogens.

- Tectonic foliation forms when grains flatten, rotate, or grow so that they align parallel with one another during deformation.

- Convergent-margin orogens may include terranes.

- Large collisional mountain ranges are underlain by buoyant roots. Folds, faults, and foliation's develop within these ranges, and crustal thickening takes place.

- With modern GPS technology, we can "see" the slow shortening and uplift of mountains.

- Uplift, the vertical rise of a mountain belt, can be controlled by isostasy. Uplift may develop due to crustal thickening, heating, or delamination.

- Erosion sculpts the rugged landscape of mountains. During mountain building, the deep crust may eventually become warm and weak, leading to orogenic collapse.

- Rifting produces tilted blocks of crust that become narrow, elongate mountain ranges. Heating of the lithosphere causes uplift of rifts.

- Cratons are the old, relatively stable parts of continental crust. They include shields and platforms. Broad regional domes and basins form in platform areas.

- Large orogens, such as the Appalachians, may record multiple events of collision and rifting.

GUIDE TERMS

accretion (p. 412)
anticline (p. 406)
axial plane (p. 406)
basin (p. 408)
bearing (p. 401)
brittle deformation (p. 393)
compression (p. 396)
craton (p. 421)
crustal root (p. 418)
detachment fault (p. 404)
dip (p. 400)
dome (p. 408)
epeirogeny (p. 423)

exhumation (p. 421)
fault (p. 398)
fault scarp (p. 401)
fault system (p. 404)
fault trace (p. 398)
fold (p. 406)
fold-thrust belt (p. 404)
force (p. 396)
global positioning system (GPS) (p. 415)
hinge (p. 406)
isostasy (p. 418)
joint (p. 397)

limb (of fold) (p. 406)
monocline (p. 406)
mountain belt (p. 391)
mountain building (p. 391)
normal fault (p. 399)
orogen (p. 391)
orogeny (p. 391)
plastic deformation (p. 393)
plunge (p. 401)
pressure (p. 396)
reverse fault (p. 399)
shear zone (p. 404)
shield (p. 422)

slickenside (p. 401)
slip lineation (p. 401)
strain (p. 393)
stress (p. 396)
strike (p. 400)
strike-slip fault (p. 399)
suture (p. 412)
syncline (p. 406)
tectonic foliation (p. 410)
thrust fault (p. 399)
uplift (p. 415)
vein (p. 397)

 **GEOTOURS** *THIS CHAPTER'S GEOTOURS WORKSHEET (I) FEATURES QUESTIONS AND GOOGLE EARTH SITES ON:*

- Calculating strike and dip from flatirons
- Fractures

- Normal, reverse, and strike-slip faults
- Folds, domes, and basins

REVIEW QUESTIONS

The letters following each Review Question refer to the corresponding Learning Objective from the Chapter Opener.

1. What changes do rocks undergo during formation of a mountain belt? **(A)**

2. What is the difference between brittle and plastic deformation? **(C)**

3. What factors determine whether a rock will behave brittlely or plastically? **(C)**

4. Discuss the differences between stress and strain. **(B)**

5. How is a fault different from a joint? **(D)**

6. Compare the motion of normal, reverse, and strike-slip faults. Which type of fault does the figure show? **(D)**

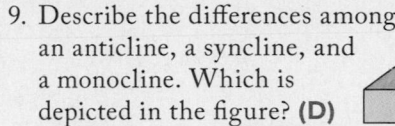

7. How do you recognize faults in the field? **(D)**

8. Explain the anatomy of a fold by defining hinge, limb, and axial plane. **(D)**

9. Describe the differences among an anticline, a syncline, and a monocline. Which is depicted in the figure? **(D)**

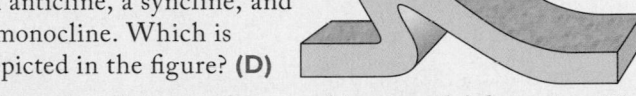

10. Discuss the relationship between foliation and deformation. **(B)**

11. Discuss the processes by which mountain belts form at convergent boundaries, in continental collisions, and in continental rifts. **(E)**

12. Describe the concept of isostasy and how it affects the elevation of a mountain range. **(E, F)**

13. What processes sculpt the rugged topography of mountain belts? How can rocks deep beneath a mountain range be exhumed? **(F)**

14. How does a craton differ from a mountain belt? What is the difference between a platform and a shield? **(G)**

15. Can we attribute all the structures of the Appalachian Mountains to a single orogeny? Explain your answer. **(E)**

ON FURTHER THOUGHT

16. Imagine that a geologist sees two outcrops of resistant sandstone, as depicted in the cross-section sketch to the right. The region between the outcrops is covered by soil. A distinctive bed of cross-bedded sandstone occurs in both outcrops, so the geologist has correlated the western outcrop (on the left) with the eastern outcrop. The curving lines in the cross-bedded layer indicate the shape of the cross beds.

   (a) Keeping in mind how cross beds form (see Chapter 7), sketch how the cross-bedded layer connected one outcrop to the other before erosion. What geologic structure have you drawn? **(D)**

   (b) Is the bedrock directly beneath the geologist older than or younger than the sandstone of the outcrops? **(D)**

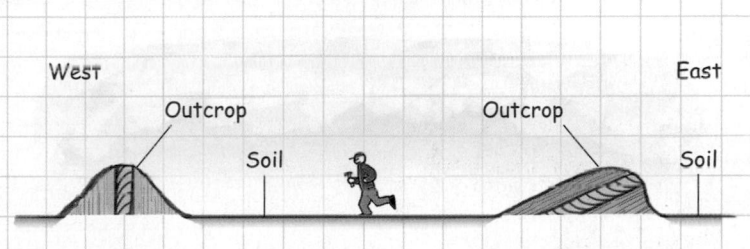

ONLINE RESOURCES

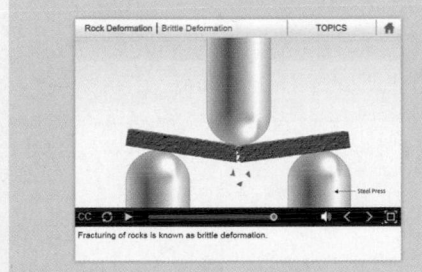

Animations
This chapter features animations on types of rock deformation, faults, and folds.

Videos
This chapter features a video on continental collision and the formation of mountains.

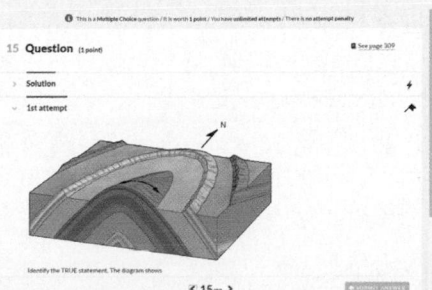

Smartwork5
This chapter features identification exercises on types of faults, folds, and deformations.

428

HISTORY BEFORE HISTORY

Geology has contributed greatly to humanity's understanding of its place in the Universe by demonstrating that our planet existed long, long before humans took their first steps. In Part IV, we peer back into this history. First, in Interlude E, we take an in-depth look at fossils, remnants of past life that allow geologists to fit life's evolution into the context of the Earth's history. Then, in Chapter 12, we learn how geologists gaze into deep time—geologic time, meaning the time since the Earth formed—first by determining the relative ages of geologic features (whether one feature is older or younger than another) and then by learning how to calculate numerical ages (in years) based on ratios of radioactive elements to their daughter products in minerals. With the background provided in Chapter 12, we're ready for Chapter 13's brief synopsis of Earth history, from the birth of the planet to the present. We see how plate tectonics has redistributed continents and built mountains, how sea level has risen and fallen, and how the Earth's climate has changed over time.

◄ A river has carved through layer upon layer of strata in southern Utah. Each layer tells a tale of a time in the Earth's very long history.

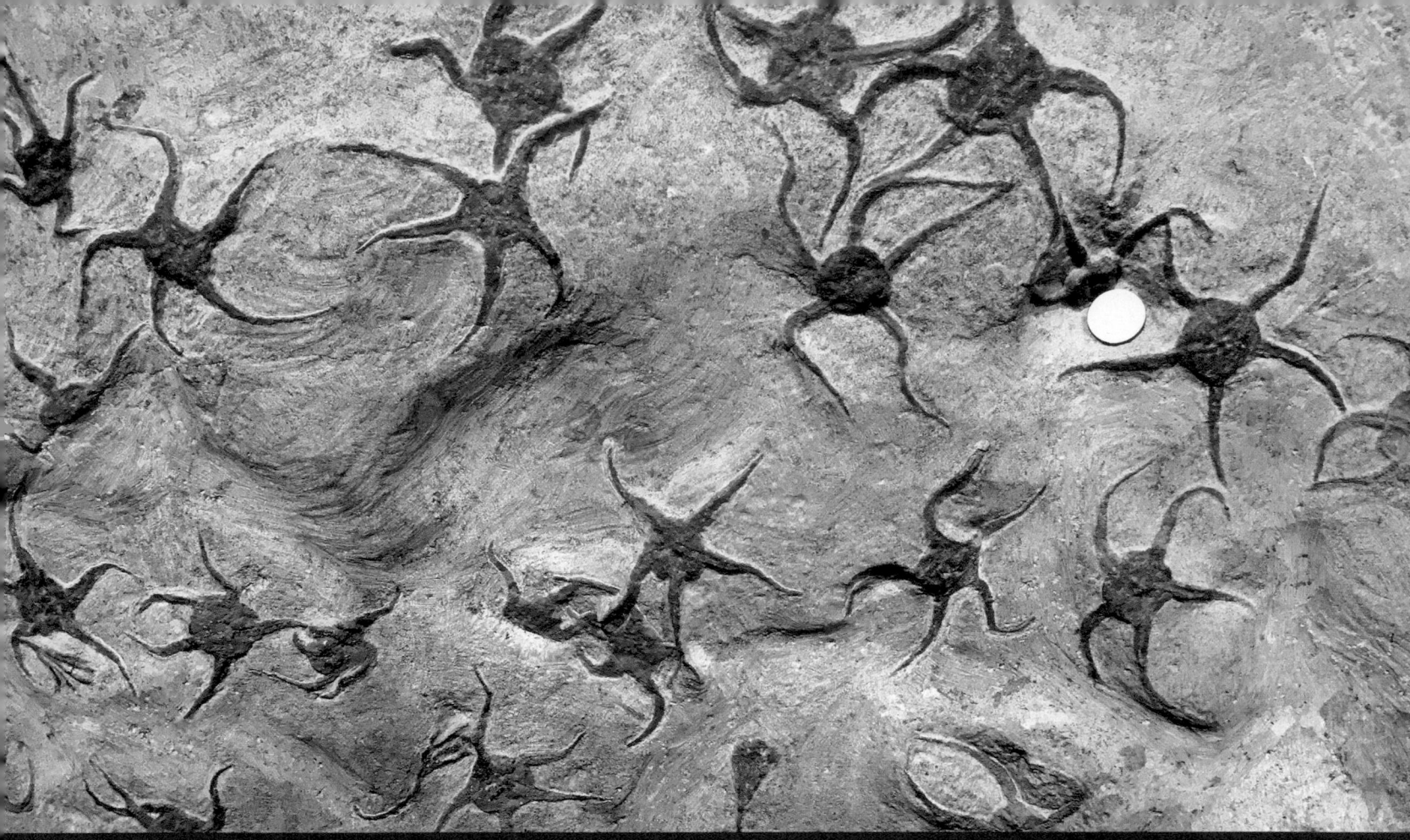

INTERLUDE E

Memories of Past Life: Fossils and Evolution

By the end of this interlude, you should be able to . . .

A. explain what a fossil is, how fossils form, and why relatively few organisms become fossils.

B. describe how fossils are classified, and recognize some of the more common fossils.

C. describe how the study of fossils contributes to an understanding of life's evolution.

E.1 The Discovery of Fossils

If you look closely at outcrops of sedimentary strata or air-fall tuff, you may occasionally find features that resemble shells, bones, leaves, or footprints, either on the surfaces of layers or within the rock itself **(Fig. E.1a–c)**. The origin of these features mystified early thinkers. Some thought that they had somehow grown underground, in already solid rock. Today geologists interpret such **fossils**—from the Latin word *fossilis*, which means dug up—to be the remnants or traces of ancient living organisms that were buried with the material from which the rock formed and were preserved after lithification. Surprisingly, this interpretation, though proposed by the Greek historian Herodotus in 450 B.C.E. and revived by Leonardo da Vinci in 1500 C.E., did not become widely accepted until the 1669 publication of a book in which a Danish physician named Nicolas Steno (1638–1686) argued that components of organisms could be incorporated in rock without losing their distinctive shape. A British contemporary of Steno, Robert Hooke (1635–1703), emphasized that fossils provide insight into the nature of life that existed earlier in Earth history because most fossils represent *extinct species*, meaning species that lived in the past but can no longer be found alive today. The concept of extinction remained controversial until Georges Cuvier (1769–1832), a French zoologist, carefully demonstrated that the skeletons and teeth of fossil organisms differ from those of any modern ones. Cuvier was also the first to classify fossils using the same systematic approach that biologists had developed for classifying modern organisms. Subsequently, **paleontologists**, researchers who specialize in the study of fossils, collected vast numbers of fossil specimens, which they named, classified, and cataloged. Museums accumulated extensive collections of fossils, which, in turn, could be used as a basis for comparison in the analysis of new discoveries **(Fig. E.1d)**. As a result, the 19th century saw **paleontology**, the study of fossils, ripen into a science.

FIGURE E.1 Examples of fossils and fossil collections.

(a) Fossil shells in 400-Ma sandstone. (Ma means million-year-old.)

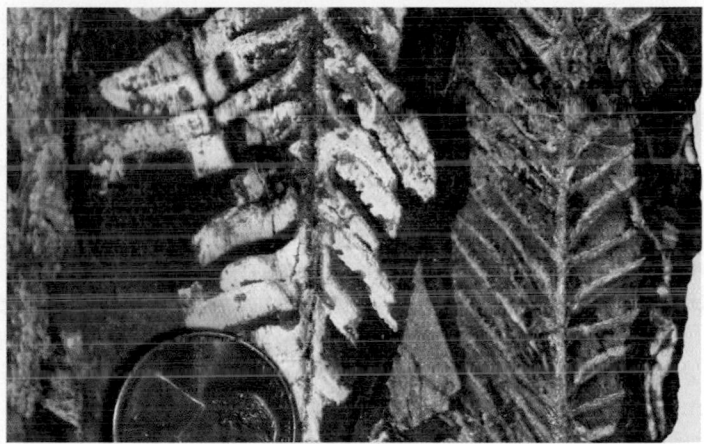

(b) Fossil leaves in 300-Ma shale.

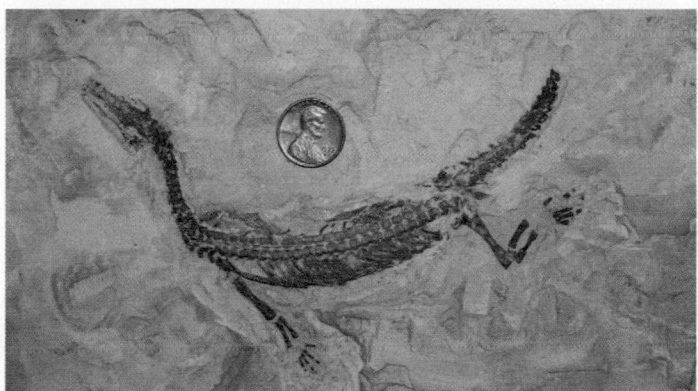

(c) Fossil skeleton in 200-Ma sandstone.

(d) A drawer of fossil specimens in a museum.

◄ (facing page) Fossil brittle stars of Ordovician age on a slab of rock from Morocco. The coin, for scale, is 2.5 cm across.

Paleontological research went beyond being merely an exercise in description when William Smith, a British engineer who supervised canal construction in England during the 1830s, noted that, in a sequence of sedimentary strata, fossil species in the lower beds differ from those in the higher beds. Smith eventually realized that sequences of strata, in fact, contain a predictable succession of fossils from base to top, and that a given fossil species can be found only in a specific interval of strata. His discovery made it possible for geologists to use fossils as a basis for determining the age of one sequence of sedimentary beds relative to another. Fossils, therefore, have become an indispensable tool for studying geologic history as well as for documenting the evolution of life.

How do fossils form? Where can we find them? What can they tell us about how life has changed over the Earth's history? To address these questions, this interlude provides a brief overview of fossils. We'll use the background provided in this interlude as the basis for the discussion of geologic time and Earth history presented in the next two chapters.

E.2 Fossilization

What Kinds of Rocks Contain Fossils?

Fossils form from buried remains of dead organisms or from tracks, burrows, or debris that organisms leave behind. The vast majority of fossils occur in sedimentary rocks, though some important examples have been found in volcanic tuffs (Fig. E.2). Fossils cannot survive the recrystallization, new mineral growth, or shearing that accompany all but the lowest grades of metamorphism, so they do not occur in metamorphic rocks. Similarly, fossils cannot be found in intrusive igneous rocks because organisms do not live in intrusive environments and could not survive the heat of igneous activity. And, with one exception, they don't exist in lava flows or hot pyroclastic flows because they would be incinerated before the material solidified into rock. The exception develops when very low-viscosity basaltic lava flows around tree trunks or other organisms and freezes into rock before the organisms have completely burned up.

Forming Fossils

Paleontologists refer to the process of forming a fossil as **fossilization**. To see how a fossil develops, let's consider an example (Fig. E.3a). Imagine an elderly dinosaur searching for food along the muddy shore of a lake on a scalding summer day in the geologic past. The hungry dinosaur succumbs to the heat and collapses dead into the mud. Soon after, scavengers strip the skeleton of meat and scatter the bones. But

FIGURE E.2 The famous fossil footprints at a site called Laetoli, in Olduvai Gorge, Tanzania. They were left when an adult and child of a human ancestor, *Australopithecus*, walked on two feet over ash that had recently been erupted by a nearby volcano. The ash had been dampened by rain before a second ash eruption buried it and thereby preserved the footprints.

before the bones have had time to weather away, it rains heavily, and streams draining nearby hills dump silty water into the lake, so the lake level rises and submerges the carcass. Silt settles out of the water and buries the bones along with the dinosaur's footprints. More sediment from succeeding floods buries the bones and prints still deeper, permanently protecting them from being destroyed by currents or by burrowing organisms. Much later, sea level rises and a thick sequence of marine sediment accumulates over the beds of mud and silt until, eventually, the sediment containing the bones and footprints becomes so deeply buried that it turns into sedimentary rock—the silt turns into siltstone and the mud turns into shale. The dinosaur's footprints remain outlined by the contact between the siltstone and shale, while its bones reside within the siltstone. Over time, minerals precipitating from groundwater replace some of the chemicals constituting the

FIGURE E.3 From a living organism to a museum display.

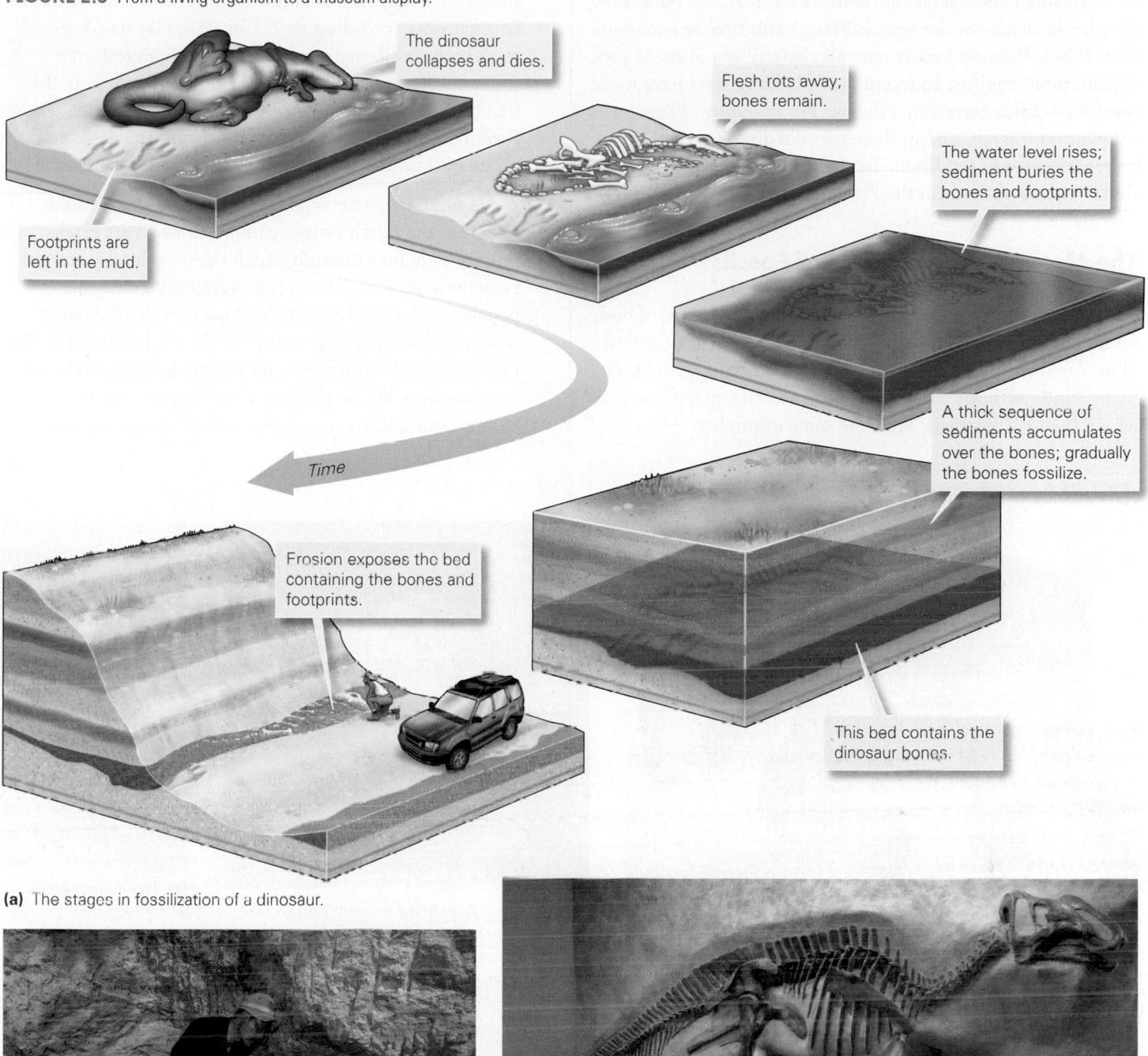

The dinosaur collapses and dies.

Footprints are left in the mud.

Flesh rots away; bones remain.

The water level rises; sediment buries the bones and footprints.

A thick sequence of sediments accumulates over the bones; gradually the bones fossilize.

Time

Erosion exposes the bed containing the bones and footprints.

This bed contains the dinosaur bones.

(a) The stages in fossilization of a dinosaur.

(b) A paleontologist collecting specimens.

(c) A display of dinosaur fossils (or plaster casts of fossils) in a museum.

bones, until the bones themselves become rock-like. The buried bones and footprints have become fossils.

How do fossils end up back at the Earth's surface, where they can be found? As time passes, the region containing the dinosaur-fossil-bearing beds stops subsiding and starts to undergo uplift. Erosion gradually strips away overlying strata until rocks containing the bones and footprints become exposed in an outcrop. If a team of paleontologists observes the fossils protruding from the outcrop, they may undertake an excavation, taking care to avoid breaking the specimens **(Fig. E.3b)**. If they're lucky, they can uncover enough bones to reconstruct a skeleton so that the dinosaur can rise again, although this time in a museum (**Fig. E.3c**). Paleontologists may also quarry out slabs of rock containing footprints. In recent years, bidding wars have made some fossil finds extremely valuable. For example, a skeleton of a *Tyrannosaurus rex*, a 67-million-year-old dinosaur, sold at auction in 1997 for $7.6 million. The specimen, named Sue after its discoverer, now stands in the Field Museum of Chicago.

The Many Different Kinds of Fossils

The dinosaur bones that we've just discussed are a type of **body fossil**, derived by fossilization of the body, or part of the body, of an organism. In detail, paleontologists distinguish among many kinds of body fossils, according to the specific way in which the fossil formed. Here are some examples:

- *Frozen or dried fossils:* In a few environments, whole bodies of organisms, including flesh and skin, may be preserved. Such body fossils can survive for only thousands, not millions, of years. Examples include woolly mammoths that became incorporated in the permafrost (permanently frozen ground) of Siberia (**Fig. E.4a**) and the desiccated (dried out) carcasses of creatures that died in desert caves.

- *Fossils preserved in tar or amber:* Small pools of oil accumulate at the Earth's surface in locations where cracks provide a conduit through which the oil can seep upward from underground. Once at the surface, volatile components of the oil evaporate, and bacteria degrade what remains, leaving behind sticky tar. At the La Brea Tar Pits in Los Angeles, California, tar accumulated in a swampy area long ago. While grazing, drinking, or hunting at the swamp, animals became mired in the tar and sank into it.

FIGURE E.4 Examples of different kinds of fossils.

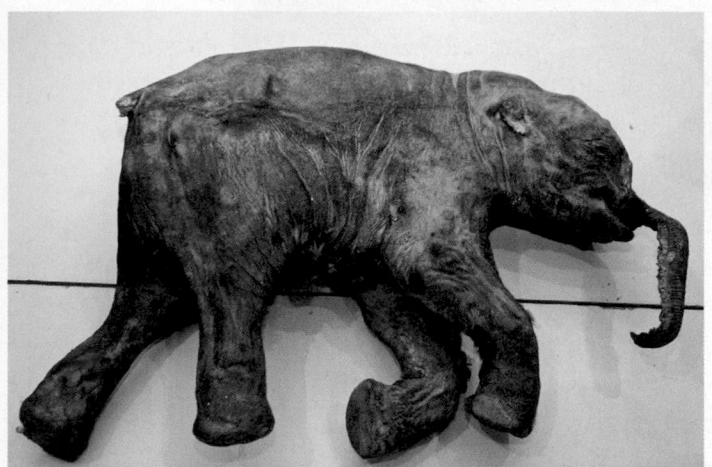

(a) This 1-m-long baby mammoth, found in Siberia, died 37,000 years ago.

(b) A fossil skeleton of a 2-m-high giant ground sloth from the La Brea Tar Pits in California.

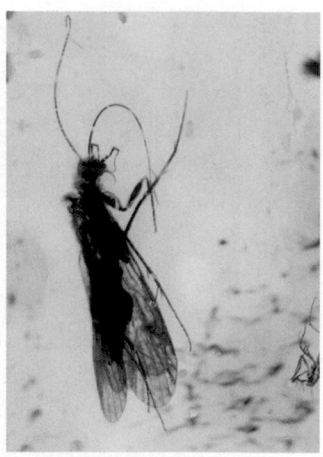

(c) This insect became embedded in amber about 200 million years ago.

(d) Fossil shells, the hard parts of invertebrates.

(e) The carbonized impressions of fern fronds in a shale.

(f) Petrified wood from Arizona. It is so hard that it remains after the rock that surrounded it has eroded away.

Tar acts as a preservative, and bones embedded in the La Brea Tar Pits have survived in great shape for over 40,000 years **(Fig. E.4b)**. Insects landing on the bark of trees may similarly become trapped in the sap that the trees produce. This golden syrup envelops the insects and, over time, hardens into **amber**, which, if buried, can last for at least 200 million years **(Fig. E.4c)**. Amber that contains insect fossils has become popular as jewelry.

- *Preserved hard parts:* Paleontologists refer to the bones (internal skeletons) of vertebrate animals and the shells (external skeletons) of invertebrate animals informally as *hard parts*. Bones and shells are hard because they consist of durable minerals, such as calcite or silica. In some cases, the original minerals of a bone or shell may survive in a rock long after the rock lithifies. More commonly, the minerals of the hard parts recrystallize during diagenesis of sedimentary rock (see Chapter 7). But even when this happens, the shape of the bone or shell can be preserved as a fossil **(Fig. E.4d)**.

- *Molds and casts:* As sediment compacts around the hard parts of an organism, it conforms to their shape. If the hard parts later dissolve away, the shape may still remain as an indentation called a **mold** (see Fig. E.1a)—sculptors use the same word to refer to the receptacle into which they pour bronze or plaster. The sediment that fills the mold forms a **cast** (see Fig. E.4d), which also depicts the shape of the hard part. Molds and casts can appear as indentations or protrusions from bedding surfaces, respectively, and since they are a relict of an organism's body, they are considered body fossils.

- *Carbonized impressions of bodies:* Impressions are simply flattened molds or casts created when soft or semi-soft organisms (such as leaves, insects, shell-less invertebrates, sponges, feathers, or jellyfish) get pressed between layers of sediment. Chemical reactions may eventually remove most of the organic material, leaving only a thin film of carbon on the surface of the impression **(Fig. E.4e)**.

- *Permineralized fossils:* **Permineralization** refers to the process by which minerals precipitate in porous material, such as wood or bone, underground. The ions from which the minerals grow come from groundwater solutions that slowly seep into the pores. **Petrified wood**, an example of a permineralized fossil, forms when volcanic ash buries a forest. Groundwater passing through the ash dissolves silica and carries it into the wood, where the silica replaces cell interiors. Eventually, the wood completely transforms into hard chert. (The word *petrified* literally means turned to stone.) During the process, the cell walls of the wood transform into organic films that survive permineralization, so you can see the fine detail of the wood's cell structure and bark in a petrified log **(Fig. E.4f)**. The colors of petrified logs come from impurities such as iron or carbon in the chert.

Not all visible fossils are body fossils. Paleontologists refer to a fossilized feature formed by the action of an organism as a **trace fossil**. In detail, we can distinguish among several types of trace fossils, including the following:

- *Footprints:* We've already noted that organisms may leave distinctive footprints in mud, which can be buried and preserved in rock as molds or casts **(Fig. E.5a, b)**. Paleontologists can study assemblages of fossil footprints to determine how organisms moved.

- *Burrows:* Sediment-filled holes can be left behind when an organism digs its way through sediment at or below the surface. Sediment within the burrow may have a slightly different texture or permeability than surrounding unburrowed sediment and can therefore remain visible in rock.

FIGURE E.5 Examples of trace fossils.

 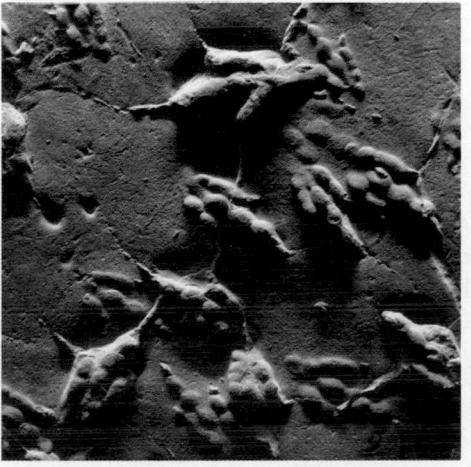 

(a) Molds of dinosaur footprints. These traces indent the top of a bed.

(b) Casts of dinosaur footprints. These traces protrude from the base of a bed.

(c) Feeding traces on the surface of a bed of sandstone in Ireland.

- *Feeding traces:* Marks formed when an organism disrupts the surface of a sedimentary bed in order to eat organisms that live in the sediment are called *feeding traces* (**Fig. E.5c**).

- *Coprolites:* Animals leave behind excrement that has passed through their digestive tract. This material, when buried and fossilized by the same processes that lead to fossilization of the organism itself, becomes a *coprolite*.

So far, we've described fossils that you can see. **Chemical fossils** are not visible—rather, they consist of distinct chemicals or isotope ratios that form during the life activity of organisms. Examples include the following:

- *Biomarkers:* The life activity of an organism produces organic chemicals. When the organism dies and becomes buried in sediment, these organic chemicals tend to break up. In some cases, specific segments of these chemicals may be durable enough to withstand lithification. Such distinctive, durable molecular segments are called molecular fossils, or **biomarkers**.

- *Isotopic signatures:* An *isotopic signature* is the relative proportion of a lighter isotope to a heavier isotope of the same element. (Recall that lighter and heavier isotopes have the same atomic number, but the lighter one has a smaller atomic weight than does the heavier one.) The life activity of organisms tends to use lighter isotopes slightly more than heavier ones, so the isotopic signature found in strata deposited at the time organisms lived differs from that found in strata from which they were absent. For example, photosynthesis preferentially uses ^{12}C instead of ^{13}C, so enrichment of ^{12}C in strata serves as fossil evidence for the appearance of photosynthetic organisms.

We've now seen how paleontologists describe a fossil based on how the fossil formed. They also distinguish among different fossils on the basis of the fossil's size. Specifically, **macrofossils** can be seen with the naked eye, and **microfossils** can be seen only with a microscope. Microfossils include the remnants of plankton, algae, bacteria, and pollen (**Fig. E.6**).

Fossil Preservation

Not all living organisms become fossils when they die. In fact, only a small percentage do, for it takes special circumstances to produce a fossil and allow it to survive. Examples of conditions that have an effect on the degree to which a fossil forms and can be preserved include the following:

- *How fast burial takes place:* If an organism dies in a depositional environment where sediment accumulates rapidly, it may be buried before it has time to rot, oxidize, or be eaten, and therefore has a better chance of being

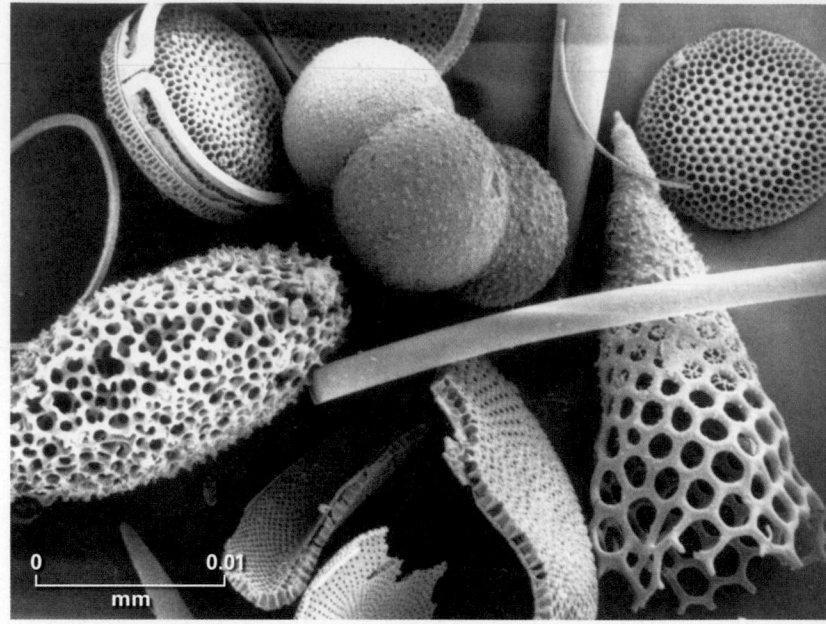

FIGURE E.6 Examples of fossil plankton shells. Because of their size, geologists refer to these fossils as microfossils or nanofossils.

0   0.01
mm

preserved. Organisms that lie exposed for a long time before burial most likely won't be preserved.

- *The energy of the depositional setting:* Organisms deposited in quiet water, a so-called *low-energy environment*, tend to be better preserved because the fossils remain intact as they are buried. In *high-energy environments*, in which strong currents or waves churn the water, organisms tumble about, break up, and tend to be dispersed as fragments before being buried.

- *The presence of hard parts:* Organisms without durable hard parts (shells or bones) usually won't be fossilized, for soft flesh decays long before hard parts do under most depositional conditions. For this reason, there are more examples of bivalves (a class of organisms, including clams and oysters, with strong shells) in the fossil record than there are of jellyfish (which have no shells) or spiders (which have very fragile shells).

- *Oxygen content of the depositional environment:* A dead squirrel by the side of the road won't become a fossil. As time passes, birds, dogs, or other scavengers may come along and eat the carcass. And if that doesn't happen, maggots, bacteria, and fungi infest the carcass and gradually digest it. Flesh that has not been eaten or does not rot can *oxidize* (react with oxygen), and that reaction breaks down organic chemicals. The remaining skeleton weathers in air and turns to dust. So, before roadkill can become incorporated in sediment, it has vanished. If, however, a carcass settles into an oxygen-poor (*anoxic*) environment, such as in a stagnant lagoon, oxidation reactions happen slowly, scavenging organisms aren't abundant, and bacterial metabolism takes place very slowly. In such environments, the organism won't rot away before it has a chance to be buried and preserved, so the likelihood that the organism will become fossilized increases.

By carefully studying modern organisms and by taking into account the concepts we've just described, paleontologists can provide rough estimates of the **preservation potential** of organisms, meaning the likelihood that an organism will be buried and eventually transformed into a fossil. For example, in a typical modern-day shallow-marine environment, such as the mud-and-sand seafloor close to a beach, about 30% of the organisms have sturdy shells and, therefore, a high preservation potential; 40% have fragile shells and a low preservation potential; and the remaining 30% have no hard parts at all and are not likely to be fossilized except in special circumstances. Of the 30% with sturdy shells, though, few happen to die in a depositional setting in which they undergo rapid burial and actually *do* become fossilized. As a result, fossilization is the exception rather than the rule.

Extraordinary Fossils: A Special Window to the Past

Although, as we've just seen, only hard parts survive in most fossilization environments, paleontologists have discovered a few special locations where relicts of soft parts remain as well; such fossils are known as **extraordinary fossils**. Extraordinary fossils include insects preserved in amber. The category also includes fossils that settled on the anoxic floor of a quiet-water lake or lagoon or the deep ocean, for in these settings, distinct carbonized impressions preserve the shape of soft parts.

A small quarry near Messel, Germany is one of the most famous sources of extraordinary fossils. By carefully prying apart thin beds in the quarry, paleontologists have been able to extract extraordinary fossils of 49-million-year-old mammals, birds, fish, and amphibians that died in a shallow-water lake **(Fig. E.7a)**. Bird fossils from the quarry include the delicate imprints of feathers, bat fossils retain impressions of ears and wings, and other mammal fossils have an aura of carbonized fur. In southern Germany, exposures of the Solnhofen Limestone, an approximately 150-million-year-old rock made of calcite mud deposited in a stagnant lagoon, contains extraordinary fossils of about 600 species, including *Archaeopteryx*, one of the earliest birds **(Fig. E.7b)**. And exposures of the Burgess Shale in the Canadian Rockies of British Columbia have yielded a plentitude of fossils, revealing what shell-less invertebrates that inhabited the seafloor about 510 million years ago looked like **(Fig. E.7c, d)**. Some of the organisms in this *Burgess Shale fauna* look so strange that it has been hard to determine how these organisms relate to present-day ones.

In a few cases, extraordinary fossils include actual tissue, a discovery that has led to a research race to find the oldest preserved DNA. Paleontologists have isolated small segments of DNA from amber-encased insects that are over 40 million years old. The amounts are not enough, however, to clone extinct species, as characters did in the popular 1993 film *Jurassic Park*.

E.3 Characterizing Life of the Past

Taxonomy Based on Morphology

Since the dawn of intellect, people have struggled to define a meaningful distinction between the living and nonliving components of the Earth System **(Box E.1)**. In recent centuries, biologists have also worked to classify living organisms in a meaningful and repeatable way. In the 18th century, a Swedish biologist named Carolus Linnaeus proposed a classification scheme based on the shared characteristics of organisms. He introduced the concept of defining a hierarchy of divisions, starting with *kingdoms* as the highest division and proceeding down through *phylum, class, order, family, genus,* and *species*. Linnaeus also introduced the idea of using the lowest two divisions, genus and species, to give each organism a two-part name. (Using this approach, we now refer to humans as *Homo sapiens*.) Researchers refer to the classification of organisms based on shared characteristics as **taxonomy**.

Before the discovery of microscopes and the development of modern biochemistry, biologists could use only studies of the *morphology* (visible shapes and forms) of organisms as a basis for taxonomic classification. Based on morphological comparisons, biologists conclude, for example, that dogs are more closely related to foxes than to bears because a dog resembles a fox more than it does a bear. By the same logic, dogs are more closely related to bears than to fish. Prior to the 1970s, classification began at the kingdom level, and biologists recognized the following kingdoms:

- *Bacteria:* Single-celled organisms composed of very tiny cells that lack a nucleus

- *Protista:* Various unicellular and simple multicellular organisms

- *Fungi:* Mushrooms and yeasts

- *Plantae:* Trees, grasses, and ferns

- *Animalia:* Sponges, corals, snails, dinosaurs, ants, and people

FIGURE E.7 Extraordinary fossils. These examples are particularly well preserved.

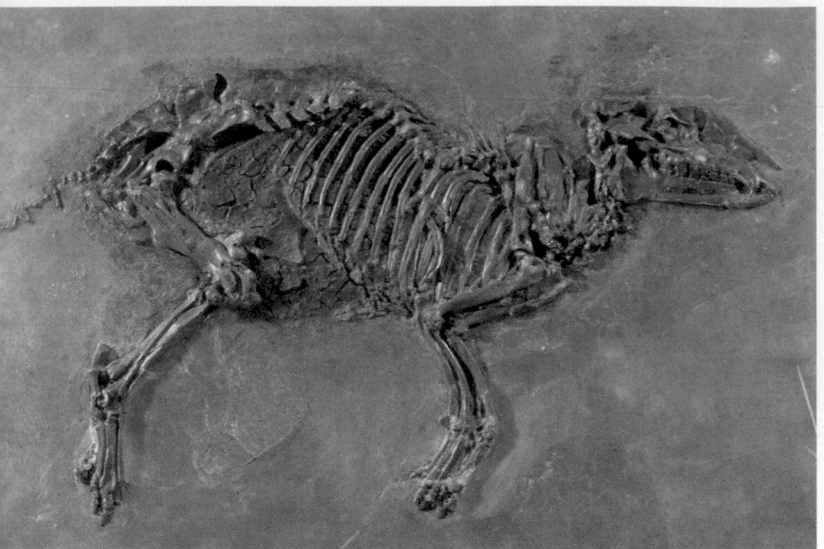

(a) A 50-million-year-old mammal fossil was chiseled from oil shale near Messel, Germany. It still contains the remains of skin.

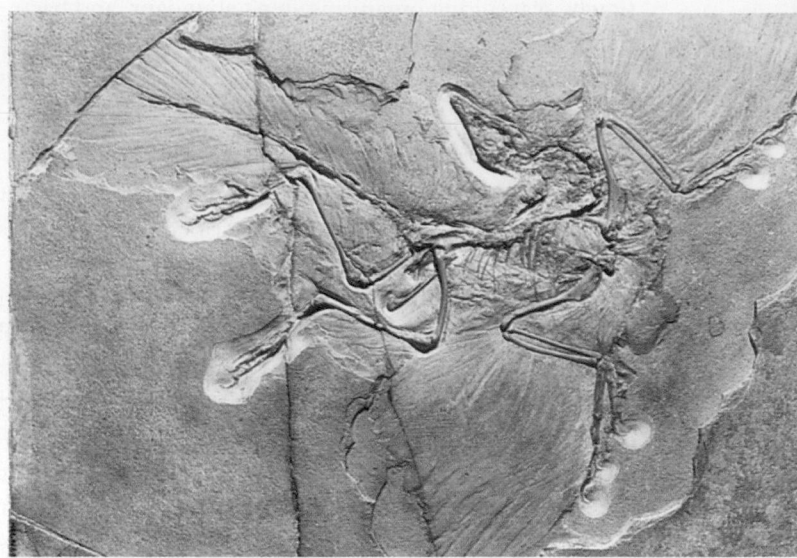

(b) *Archaeopteryx* from the 150-million-year-old Solnhofen Limestone of Germany. The imprints of feathers are clearly visible.

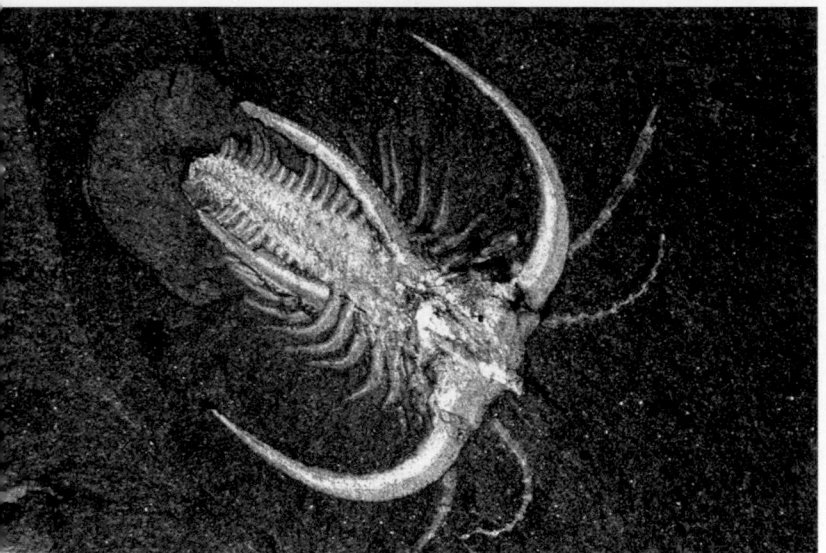

(c) The Burgess Shale of the Canadian Rockies contains unique Cambrian arthropods, such as *Marella*, shown here.

(d) An artist's reconstruction of a Cambrian ecosystem. The painting is based on Burgess Shale fossils.

Each kingdom consists of one or more phyla. A phylum, in turn, includes several classes, a class includes several orders, an order includes several families, a family includes several genera, and a genus includes one or more species **(Fig. E.8)**. Kingdoms, therefore, represent the broadest category, and species represent the narrowest. In the early 19th century, Georges Cuvier introduced Linnaeus's taxonomic approach to paleontology.

Keeping in mind the concepts of taxonomy, you'll find that there's nothing magical about identifying fossils. Typically, you can recognize common fossils in the field by examining their morphology. If the fossil has well-preserved, distinctive features, the process can be straightforward, so even beginners can distinguish the major groups of fossils from one another on sight. But if a fossil has been broken into fragments and parts have disappeared, or if a fossil shares many characteristics with other species, identification can be challenging. As a result, classification of some organisms has remained controversial.

Many fossil organisms resemble modern organisms, and this similarity provides a starting point for classification. To begin identifying a fossil, paleontologists look at

BOX E.1 CONSIDER THIS . . .

Defining Life

Biologists refer to the fundamental component or building block of a living entity as a *cell*. We consider a single cell, or a group of cells operating together, to be an **organism**. An organism is capable of doing some or all of the following tasks: it can (1) regulate its internal state; (2) interact with its environment to use energy or convert chemicals extracted from its environment into energy and new living materials; (3) grow; (4) adapt to its surroundings; (5) respond to stimuli; and (6) reproduce. Contrast a cell with a sand grain: while the sand grain can grow if atoms attach to its surface, it can't reorganize the atoms into new configurations, can't reproduce itself, can't move on its own, and doesn't produce or consume energy. This contrast emphasizes that life serves, in effect, as a reactor, processing materials extracted from its environment and converting them into new forms.

A cell consists of a *membrane*, an incredibly thin barrier capable of letting only certain elements in or out, surrounding a mass of organic molecules. Cells contain many different kinds of organic molecules, including proteins, fats, and carbohydrates. One of these molecules, **DNA** (deoxyribonucleic acid), is a large, complex molecule shaped like a double helix. DNA serves a particularly important role because it contains the genetic code of the organism, meaning that it provides the "instructions" for the organism's design, functions, and reproduction. This code comes in the form of *genes*, distinct segments of DNA that control a particular heritable characteristic.

Organisms vary greatly in their dimensions, complexity, and mobility. The smallest, known as *microorganisms* or **microbes**, consist of only one cell, a chain of cells, or a community of relatively few cells. The smallest microbes have a diameter of 0.5 to 5.0 µm—only 5% of that of a human hair—so researchers can observe them only through high-powered microscopes. *Macroorganisms*, those large enough to be visible to the naked eye, range in size from minuscule insects with a diameter of 500 µm (=0.5 mm) to blue whales with a length of 30 m.

Almost all macroorganisms are **multicellular organisms**, meaning that they consist of many cells—in fact, a blue whale's body may contain trillions of cells. In a *primitive multicellular organism*, many cells link together in a chain or some other shape, but all perform the same function. In a *complex multicellular organism*, cells tend to be specialized, and they associate to form distinct *organs* (such as a stomach or a lung) or *appendages* (such as an arm). Typically, muscle cells allow movement, stomach cells digest food, nerve cells transmit information about stimuli in the environment, and blood cells carry oxygen.

If someone asked you how to tell the difference between a live bear in a zoo and a stuffed bear in a museum, what would you say? Probably you'd point out that the live bear moves, eats, breathes, and emits heat, while the stuffed bear doesn't. In the language of biology, you have said that metabolism happens in the live bear but not in the stuffed one. Formally defined, **metabolism** refers to the set of chemical reactions that allow an organism to harvest energy and materials from its surroundings and then expend that energy or those materials to fuel life functions. Any element or compound that an organism uses to produce new materials and/or new energy is a *nutrient*. The life functions sustained through the metabolism of nutrients include constructing new components of cells and enabling those cells to operate, move, and, potentially, reproduce. Metabolic reactions resemble the chemical reactions that take place during an inorganic process: each reaction involves various reactants, yields various products, and, depending on the reaction, absorbs or releases energy. Through these reactions, organisms build molecules, such as proteins, and break these molecules down to obtain energy and the raw materials to produce other molecules.

---

the nature of the skeleton (is it internal or external?); the symmetry of the organism (is it bilaterally symmetric, like a mammal, so that one side is a mirror image of the other, or does it have fivefold symmetry, like a starfish?); the design of the shell (in the case of invertebrates); and the design of the jaws, teeth, or feet (in the case of vertebrates). A fossil mammal looks like a mammal and not a fish, and a fossil clam (class Bivalvia) looks like a clam and not a snail (class Gastropoda). Similarly, a fossil organism with a spiral shell that does not contain internal chambers fits in the class Gastropoda (the snails), whereas an organism with a chambered, spiral shell belongs in the class Cephalopoda **(Fig. E.9a)**. At taxonomic levels below class, identification may involve focusing on intricate details, such as the number of ridges on the surface of the shell.

Paleontologists have found countless species of fossils that resemble living organisms only at the level of orders or even higher. For example, a group of extinct organisms called trilobites have no close living relatives **(Fig. E.9b)**. But they were clearly segmented invertebrate animals, and as such, they resemble arthropods such as insects and crustaceans. Therefore, they represent a distinct class of the phylum Arthropoda.

Figure E.10 provides simplified sketches of some of the most common types of invertebrate fossils distinguished by their morphology. With this figure, you should be able

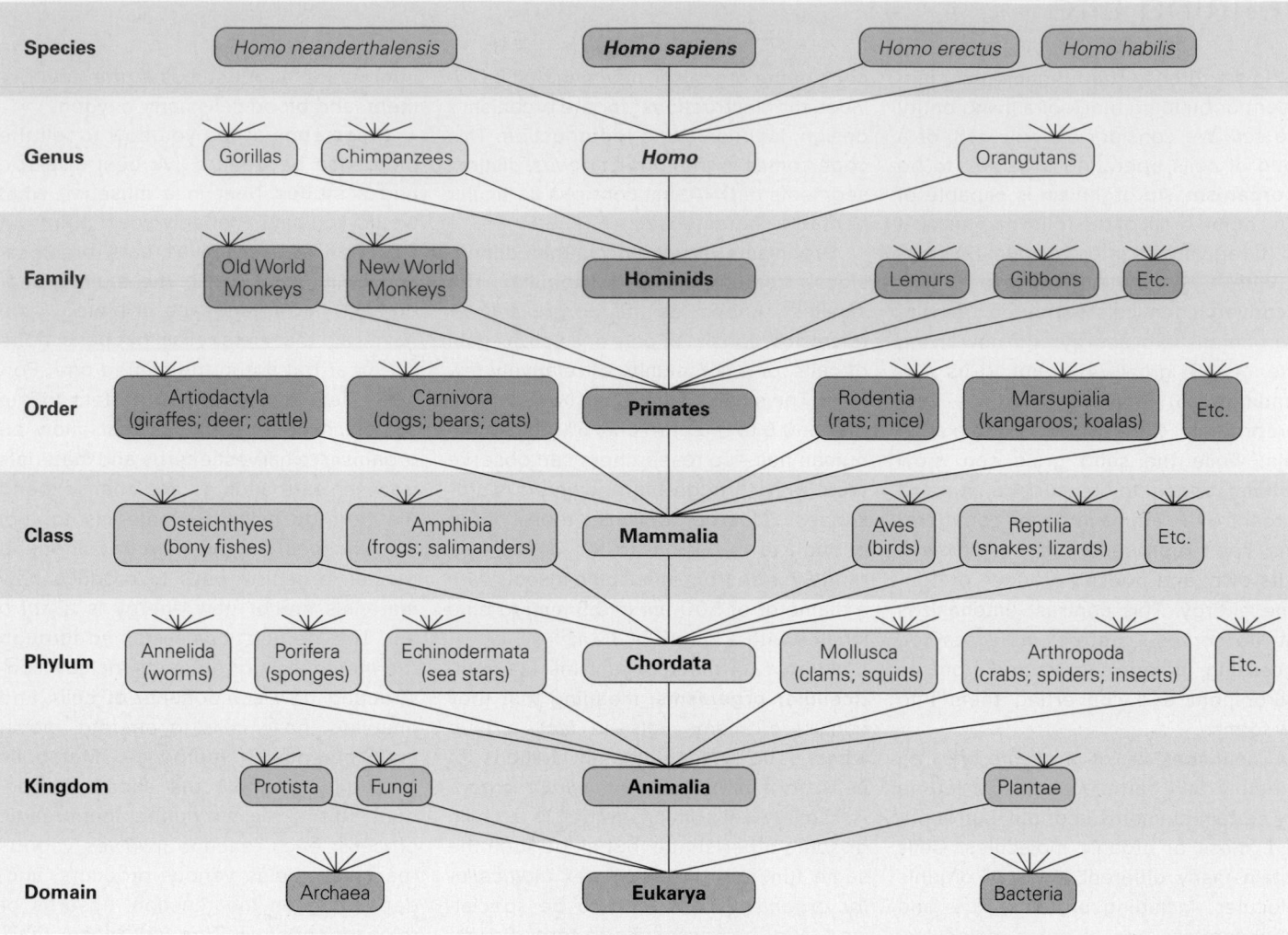

to identify many of the fossils you'll find in a typical bed of limestone. Some of the notable characteristics of these fossils include the following:

- *Trilobites:* These arthropods have a segmented shell that is divided lengthwise into three parts.

- *Gastropods (snails):* Most fossil specimens of gastropods have a spiral shell that does not contain internal chambers.

- *Bivalves (clams and oysters):* These invertebrates have a shell that can be divided into two similar halves. The plane of symmetry lies parallel to the plane of the shell.

- *Brachiopods (lamp shells):* The top and bottom parts of lamp shells have different shapes, and the plane of symmetry lies perpendicular to the plane of the shell. Fossils typically have ridges radiating out from the hinge.

- *Bryozoans:* These invertebrates are colonial animals. Their fossils resemble a screen-like grid. Each box in the grid represents the shell of a single animal.

- *Crinoids (sea lilies):* These organisms look like flowers but are actually animals. They have a stalk consisting of numerous circular plates stacked one on top of the other.

- *Graptolites:* These fossils look like tiny saw blades. They are remnants of colonial animals that floated in the sea.

- *Cephalopods:* These squid-like organisms include ammonites, with a spiral shell, and nautiloids, with a straight shell. Their shells contain internal chambers and have ridged surfaces.

- *Corals:* These invertebrates include colonial organisms that form distinctive mounds or columns in tropical reefs. Paleozoic examples include solitary individuals with a cone-like shell.

In some cases, using morphology to determine the relationships of organisms becomes difficult because, while specific features of different organisms may serve the same purpose, they may actually represent very different body parts. For example, both bats and birds have wings, but bats are mammals and birds

FIGURE E.9 Examples of fossil classification.

(a) Examples of ammonites, a type of cephalopod. The inset shows segmentation inside a shell.

(b) Examples of trilobites. Note that each specimen has three parts (hence the prefix *tri-*).

are not, so bats are actually more closely related to humans than they are to birds, despite this shared characteristic.

Ecosystems and Paleoecology

Organisms, with relatively few exceptions, do not live in isolation. Rather, they thrive in association with other organisms, forming a *community*. A community of coexisting organisms together with the nonliving components of the environment they live in comprise an **ecosystem**. Biologists refer to each component or role in an ecosystem as an *ecological niche*. Our

FIGURE E.10 Common types of invertebrate fossils.

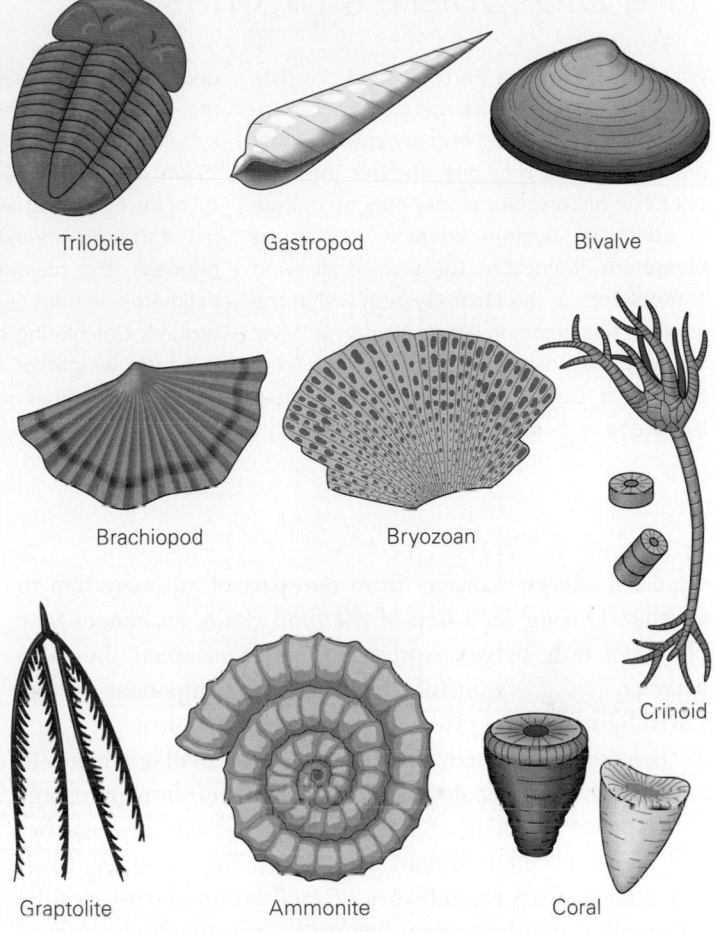

Trilobite   Gastropod   Bivalve

Brachiopod   Bryozoan

Crinoid

Graptolite   Ammonite   Coral

planet hosts an amazing number of different ecosystems, each of which contains many ecological niches. The ecosystem that develops at a given locality depends on climate, water availability, nutrient sources, substrate composition, land-surface slope, and many other parameters. Taken together, all the organisms living at a given time (and those just recently deceased), in all ecosystems, define the **biomass** of our planet **(Box E.2)**.

Most ecosystems include a variety of organisms. Organisms that extract energy directly from the Sun (through photosynthesis) or from chemical bonds in various inorganic minerals are called **primary producers** because these organisms do not use other organisms as an energy source. All the other organisms are **consumers**, in that they must consume other organisms to gain their nutrients, and they produce energy by breaking down molecules obtained from those organisms. We can divide consumers into several types: *herbivores* consume only plants; *carnivores* consume mostly animals; *omnivores* eat both plants and animals; *scavengers* consume already-dead creatures; and *decomposers* live by breaking down (causing the decay of) dead organisms. We refer to the succession of organisms consuming other organisms as the **food chain** or *food web*. In the context of the Earth System, steps in the food chain

The Biosphere's Biomass

We've seen that the Earth System consists of several realms—the geosphere, atmosphere, hydrosphere, and biosphere—that have interacted with one another throughout Earth history. Our planet may be unique in our Solar System because it hosts a **biosphere**, defined as the sum of all living components of the Earth System within the realm of this planet in which life can survive. Spatially, the biosphere extends from 2–5 km below the Earth's solid surface to about 10 km above it—it includes pores and cracks underground, soil, land, surface water, and the lower atmosphere. Note that the biosphere has a maximum thickness of about 15 km, only about 0.2% of the Earth's radius. All of the living and recently deceased material of the biosphere makes up our planet's **biomass**, the majority of which, by some estimates, resides in tiny organisms underground. Calculating biomass isn't easy, but the total weight of biomass on the Earth today may equal roughly the weight of 10 stratovolcanoes.

In recent decades, many researchers have focused their efforts on *geobiology*, the study of how the biosphere interacts with other components of the Earth System. They have paid particular attention to a spatial realm of the biosphere known as the **critical zone**, within which most interactions between living and nonliving components take place. Maintenance of resources and climate conditions in the critical zone is critically important to the sustainability of life.

---

represent energy transfers from one part of an ecosystem to another. During each step of the food chain, exchanges take place not only between producers and consumers, but also between living organisms and nonliving components of the Earth System; these processes tend to recycle nutrients.

Paleontologists examine assemblages of fossil organisms to work out the **paleoecology** of a particular environment, meaning the ways in which organisms interacted with one another and with their environment in the past. By comparing fossil assemblages with assemblages of modern organisms, and by interpreting the depositional environments in which the sedimentary rocks containing the fossils accumulated, paleontologists can classify a fossil assemblage as, say, members of a shallow-marine as opposed to a deep-marine community.

E.4 Evolution and Extinction

The Fossil Record

The fossil record demonstrates that life on the Earth evolves. Over time, some species disappear and new species appear, so that the assemblage of species inhabiting our planet changes progressively and continuously. Paleontologists have struggled to characterize the steps in this process of *evolution*. This work hasn't been easy because, although more than 250,000 species of fossil organisms have been collected and identified by thousands of collectors working on all continents, our planet's fossil record remains very incomplete. In fact, paleontologists estimate that between 5 billion and 50 billion distinct species of life have inhabited the Earth since life first appeared, so the tally of known fossils represents only a tiny fraction of these species. As a result, researchers cannot account for every step in the evolution of every organism, so in most cases we see only snapshots of brief moments during the course of evolution. Occasionally, however, paleontologists find *transitional fossils* that can fill gaps in the record of evolution. Such evidence of intermediate stages can help show how a species changed progressively into a new species. It can also demonstrate *divergence* (or separation) of distinct species from a *common ancestor*.

Why is the fossil record so incomplete? First, despite all the fossil-collecting efforts of the past two centuries, paleontologists have not even come close to sampling every cubic centimeter of exposed sedimentary rock on the Earth, much less the vast volumes of sedimentary rock that remain below the surface. Just as biologists have not yet identified every living species of insect, paleontologists have not yet identified every fossil species. New species and even genera of fossils continue to be discovered every year.

Second, not all organisms are represented in the fossil record because not all organisms have a high preservation potential. As noted earlier, fossilization occurs only under special conditions, so only a minuscule fraction of the organisms that have lived on the Earth have been fossilized. There may be few, if any, fossils of a vast number of extinct species, so we have no way of knowing what they looked like or even that they ever existed.

Finally, as we will learn in Chapter 12, the sequence of sedimentary strata that exists on the Earth does not account for every moment of time since the formation of our planet. Sediments accumulate only in environments where conditions allow deposition to take place. Sediments do not accumulate, for example, on dry plains or on mountain peaks, but they do accumulate in the sea and in the floodplains and deltas of rivers. Because the Earth's climate changes over time and because sea level rises and falls, certain locations on continents are sites of deposition at some times, but not at others, and on occasion they may become sites of erosion. Therefore, strata accumulate only episodically.

Darwin's Grand Idea

As a young man in England in the early 19th century, Charles Darwin had been unable to settle on a career, but he had developed a strong interest in natural history. Therefore, he jumped at the opportunity to serve as a naturalist aboard HMS *Beagle* on an around-the-world surveying cruise. During the five years of the cruise, from 1831 to 1836, Darwin made detailed observations of plants, animals, and geology in the field, and he amassed an immense specimen collection from South America, Australia, and Africa. Just before Darwin departed on the voyage, a friend gave him a copy of the first geology textbook, Charles Lyell's *Principles of Geology*, which argued in favor of James Hutton's proposal that the Earth had a long history and that geologic time extended much farther into the past than did human civilization.

A visit to the Galápagos Islands, off the coast of Ecuador, led to a turning point in Darwin's thinking. The naturalist was most impressed with the variability of Galápagos finches. He marveled not only at the fact that different varieties of the bird occurred on different islands, but also at how each variety had adapted to use a particular food supply. With Lyell's writings in mind, Darwin developed a hypothesis that the finches had begun as a single species that later evolved into several different species when populations of the birds became isolated on different islands. This process of diversifying species is now known as adaptive radiation. Darwin concluded that such evolution can happen because offspring can differ from their parents and new traits can be transferred to succeeding generations. If enough change accumulates over many generations, the living population ends up being so different from its distant ancestors that the population can be classified as a new species. During the course of this evolutionary process, old species vanish and new species appear. The accumulation of changes may eventually yield a population that is so different from its ancestors that taxonomists consider it to be a new genus. With this concept in mind, we can now refine our definition of evolution to mean a change in a population over successive generations due to the transfer of heritable characteristics.

Darwin and his contemporary, Alfred Russel Wallace, not only proposed that evolution took place, but also came up with an explanation for why it occurs. According to Darwin and Wallace, a population of organisms cannot increase in number forever because it is limited by competition for scarce resources in the environment. In nature, only organisms capable of survival can pass on their characteristics to the next generation. In each new generation, some individuals have characteristics that make them more fit, whereas some have characteristics that make them less fit. The fitter organisms are more likely to survive long enough to produce offspring, so the beneficial characteristics that they possess get passed on to the next

generation. Organisms that have traits that make them less fit either die or fail to produce offspring. Darwin called this process **natural selection** because it occurs on its own in nature. If environmental conditions change, or if competitors enter the environment, species that have not evolved and become better adapted to survive eventually die off and become extinct.

Darwin's view of evolution has been successfully supported by many observations and so far has not been definitively disproved by any observation or experiment. Furthermore, it can be used to make testable predictions. Therefore, scientists now refer to the concept as the **theory of evolution by natural selection** (often abbreviated "the theory of evolution"; see Box P.1 in the Prelude for the definition of a theory).

In the century and a half since Darwin published his work, *genetics* (the study of genes) has developed and provided insight into how evolution works. Progress began in the late 19th century when an Austrian monk, Gregor Mendel, who studied peas in the garden of his monastery, showed that genetic mutations led to new traits that could be passed on to offspring. After James Watson and Francis Crick discovered DNA in 1953, biologists began to understand the molecular nature of genes and mutations, and thus of evolution. And with the genome projects of the 21st century, which define the detailed architecture of DNA molecules for a given species, it has become possible to pinpoint the exact arrangement of genes responsible for specific traits.

The theory of evolution by natural selection provides a conceptual framework in which to understand paleontology. But because of the incompleteness of the fossil record, many questions remain as to the rates at which evolution takes place during the course of geologic time. As a result, different researchers have suggested different concepts of rates. For example, Cuvier, back in the early 19th century, thought that extinction happened primarily during catastrophes, such as giant floods, and that a whole new assemblage of life took over the Earth after each catastrophe—Cuvier's view came to be known as *catastrophism*. In contrast, James Hutton and his followers, such as Lyell and Darwin, assumed that evolution happened at a constant, slow rate—their view came to be called *gradualism*. (Gradualism reflects a strict interpretation of a broader idea, called uniformitarianism, which we'll discuss in Chapter 12.)

More recently, researchers have suggested that evolution takes place in fits and starts. According to this concept of *punctuated equilibrium*, evolution happens very slowly for quite a while, and then, during a relatively short period, it takes place very rapidly. Factors that could cause sudden accelerations in the rate of evolution include (1) a widespread extinction during which many organisms disappear, leaving environments open for new species to colonize; (2) a relatively rapid change in the Earth's climate that puts stress on organisms, so that new species that can survive the new climate survive, whereas those that can't become extinct;

(3) formation of new environments, as may happen when rifting splits apart a continent and generates a new ocean with new coastlines; and (4) the isolation of a breeding population. The punctuated equilibrium concept takes into account the observation that sudden events leading to widespread extinction do take place during Earth history, but that evolution can happen slowly or not at all during intervals between these events.

Regardless of the process of evolution, the longevity of different kinds of organisms is not all the same. Some populations survive as identifiable genera or even species for long intervals of geologic time (tens to hundreds of millions of years). But others appear and then disappear within a relatively short interval of geologic time (less than a few million years).

Extinction: When Species Vanish

Extinction occurs when the last members of a species die, so that there are no parents to pass on their genetic traits to off-spring. Some species become extinct as a population evolves into new species, whereas other species just vanish, leaving no hereditary offspring. These days, we take for granted that species become extinct because a great number have, unfortunately, vanished from the Earth during human history. Before the 1770s, however, few geologists believed that extinction occurred; they thought fossils that didn't resemble known species must have living relatives somewhere on the planet. Considering that large parts of the Earth remained unexplored, this idea wasn't so far-fetched. However, by the end of the 18th century, in light of Cuvier's work and the completion of the map of the Earth by European explorers, it became clear that many fossil organisms did not have modern-day counterparts. The bones of mastodons and woolly mammoths, for example, were too different from those of elephants to be of the same species, but the animals were too big to hide.

Twentieth-century studies concluded that many different phenomena can contribute to extinction. Some of the geologic factors that may cause extinction include the following:

- *Global climate change:* At times, the Earth's mean temperature has been significantly colder than today's, whereas at other times it has been much warmer. Because of a change in climate, an individual species may lose its habitat, and if it cannot adapt to new conditions or migrate to stay with its old habitat, the species will disappear.

- *Tectonic activity:* Tectonic activity causes both vertical movement of the crust over broad regions and changes in seafloor spreading rates. These phenomena can modify the distribution and area of habitats. Species that cannot adapt die off.

- *Asteroid or comet impacts:* Many geologists have concluded that impacts of large meteorites with the Earth have been catastrophic for life. A large impact would send dust and debris into the atmosphere that could blot out the Sun and plunge the Earth into darkness and cold (see Chapter 23). Such a change, though relatively short lived, could interrupt the food chain.

- *Voluminous volcanic eruptions:* Several times during Earth history, incredible quantities of lava have spilled out onto the surface, or incredible volumes of ash and gas have spewed into the air. These eruptions, perhaps due to the rise of superplumes in the mantle, were accompanied by the release of enough greenhouse gas into the atmosphere to alter the climate.

- *The appearance of a new predator or competitor:* Some extinctions may happen simply because a new predator appears on the scene and kills individuals of a given species at a faster rate than new individuals can be born. (For example, when humans first arrived in North America, they killed so many giant mammals that these mammals went extinct.) Similarly, if a more efficient competitor appears, that species steals an ecological niche from a less competitive species, whose members can't obtain enough food and thus die off.

Some extinctions happen over long intervals, during which when the replacement rate of a population simply becomes lower than the mortality rate, but others happen suddenly, when a cataclysmic event leads to the rapid extermination of many organisms. For example, in 1870s, the population of passenger pigeons in North America exceeded 3 billion. Due to widespread hunting by people, the population dropped rapidly during the next two decades, and the last representative of the species died in 1914.

Paleontologists have found that the number of different genera of fossils, a representation of **biodiversity** (the overall variation of life), changes over time and has abruptly decreased at specific times during Earth history. An abrupt worldwide decrease in the number of fossil genera is called a **mass-extinction event**. At least five major mass-extinction events have happened during the past half-billion years **(Fig. E.11)**. These events define the boundaries between some of the major intervals into which geologists divide time. For example, a mass-extinction event marks the end of the Cretaceous Period, at 66 Ma. During this event, all dinosaur species (with the exception of their modified descendants, the birds) vanished, along with most marine invertebrate species. A huge extinction event also brought the Permian to a close. Significantly, the rate at which species have been disappearing during the past few centuries, in association with the growth of human populations and the accompanying expansion of agriculture and industry, has been so rapid that some researchers and writers refer to our present time as the *sixth extinction*.

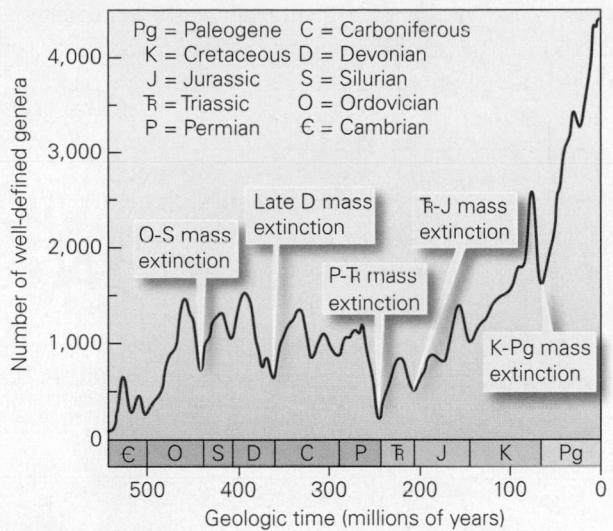

FIGURE E.11 This graph shows how the diversity of life has changed over time. Sudden drops indicate periods when mass extinctions occurred.

Evidence for Evolution

Biologists can test aspects of the theory of evolution in several ways: by examining how bacteria develop resistance to antibiotics; by studying embryology, because organisms can look very different at birth, yet look quite similar at early embryonic stages; by studying *vestigial organs,* because certain organs served a purpose in an ancestor but do not serve a purpose in its descendants; and by studying genetics. But because evolution takes place very slowly, researchers turn to the fossil record to find evidence of evolution that has taken place over a long time. As we've already noted, one approach to studying long-term evolution involves tracking a species back through ancestral forms and predicting the existence of transitional fossils. Evidence also comes from studying *homology,* evidence for shared ancestry in the form of certain structures or organs within an organism. For example, certain bones in the legs of different mammals now have very different forms, but share the same location in the overall architecture of mammalian legs, suggesting that all were derived from the same common ancestor.

Phylogenetics

The development of new techniques in recent years has allowed researchers to define the **genome** of an organism, the complete code held in the organism's DNA. The genome includes genes that define specific characteristics of an organism as well as segments of DNA molecules that do not provide codes for distinct characteristics and whose purpose remains somewhat unclear. Using their knowledge of genomes or parts of genomes, researchers have been able to define evolutionary relationships among organisms by comparing their genetic makeup. The discipline that studies such relationships is called *phylogeny.* In phylogenetic studies, researchers refer to a group of organisms that originated by adaptive radiation from a common ancestor as a *clade.*

Studies of phylogeny in recent decades have led to a fundamental change in how biologists and paleontologists classify organisms. Until 1977, kingdoms represented the highest division of life, as proposed by Linnaeus. Using DNA comparisons, Carl Woese, a microbiology professor at the University of Illinois, showed that life can be divided into three **domains,** named Archaea, Bacteria, and Eukarya. **Archaea** include a vast array of single-celled microbes that grow not only in the mild environments of oceans, soils, and wetlands, but also in the harsh environments of hot springs, black smokers, salt lakes, very acidic sediments, and saline groundwater **(Fig. E12a).** Organisms that survive in harsh environments are known as *extremophiles* (from the Latin, meaning extreme-loving). Many extremophiles do not need light or air, but rather live on the energy stored in the chemical bonds of minerals. **Bacteria,** which are also single-celled organisms, also inhabit almost all livable environments on the Earth **(Fig. E.12b).**

Though Archaea and Bacteria cells resemble each other morphologically, they differ profoundly at a genetic level. Nevertheless, both groups are **prokarya,** meaning that their cells do not contain a nucleus. In this regard, Archaea and Bacteria differ from **Eukarya,** organisms whose cells do contain a nucleus. Four of the traditional kingdoms that we introduced earlier in this interlude (Protista, Fungi, Plantae, and Animalia) have eukaryotic cells and therefore lie within the domain Eukarya.

The Tree of Life

Researchers have been working hard to understand phylogenetic relationships among organisms. Much of this work focuses on identifying common ancestors. The results can be represented on a diagram known as the *tree of life,* or more formally, as a **phylogenetic tree.** The designs of such diagrams have changed over time. Traditionally, they were portrayed in the form of a tree **(Fig. E.13a)** with its roots in *proto-life,* the molecules from which all life originated. More recently, researchers have portrayed phylogenetic relationships on a *cladogram,* a diagram that focuses on showing the branch points at which different organisms radiated from common ancestors **(Fig. E.13b).**

Advances in phylogeny have allowed researchers to make progress in answering the age-old question of when and how life began on the Earth and what form it took. Fossils of Archaea and Bacteria appear in rocks as old as 3.7 to 3.8 Ga, and biomarkers have been reported from even older rocks—in fact, some researchers argue that biomarkers may appear in

FIGURE E.12 The two prokaryotic domains of life.

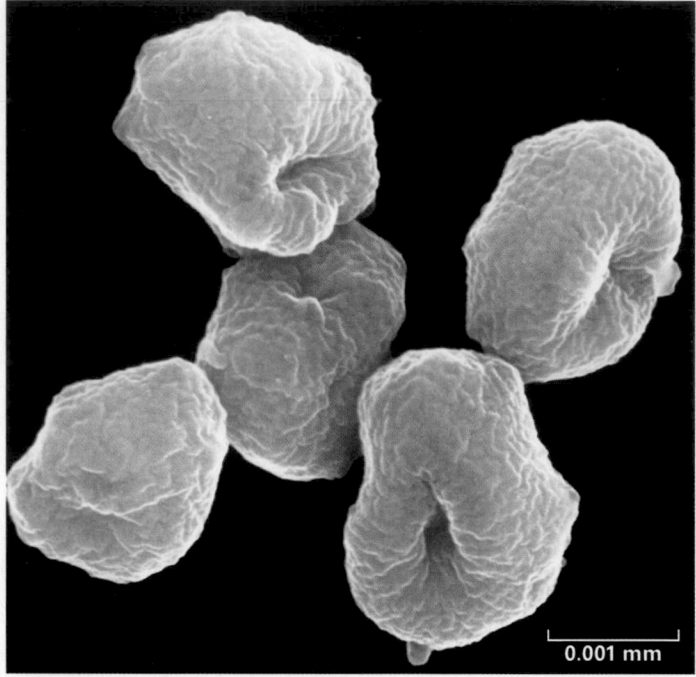

(a) Archaea cells.

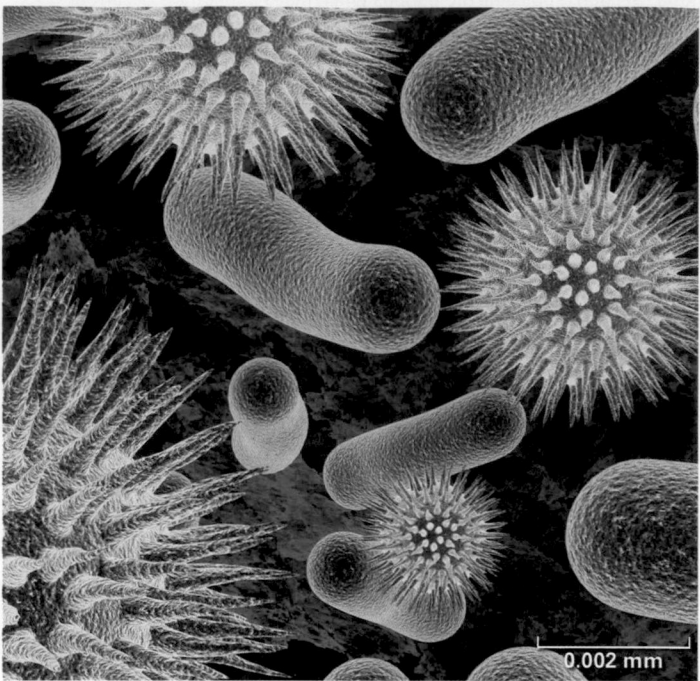

(b) Bacteria cells.

rocks as old as 4.3 Ga. Most biologists favor a model in which proto-life separated early into two distinct branches, one that became the Bacteria, and one that became the precursor of Archaea and Eukarya. Eukarya branched off to become a distinct domain by about 2.5 Ga, for fossils of Protista date back to this time. Before this date, the Earth's biomass consisted entirely of prokaryotic cells (Archaea and Bacteria). The date at which complex organisms (those with distinct organs) appeared remains a subject of research. Rocks dated between 0.6 and 0.7 Ga contain well-documented examples of organisms with distinct organs. Some researchers argue that complex organisms date back to 1.5 Ga, or even 2.1 Ga. The diversification of prokaryotic life into the major phyla that we see today took place during the Cambrian, which began 541 Ma.

FIGURE E.13 Two types of phylogenetic trees.

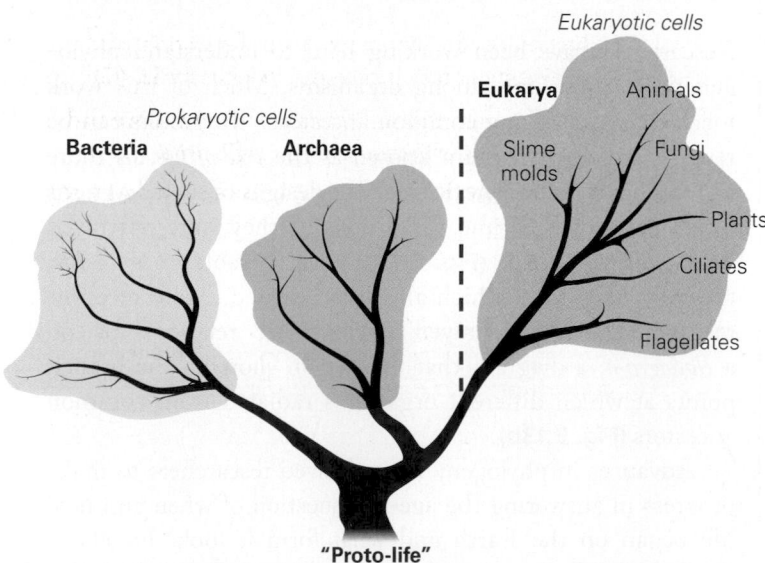

(a) This traditional "tree of life" symbolically illustrates the relationship among the three domains of organisms on the Earth. Archaea and Bacteria are both prokarya whose cells lack nuclei. All other life forms are Eukarya because they have cells with a nucleus.

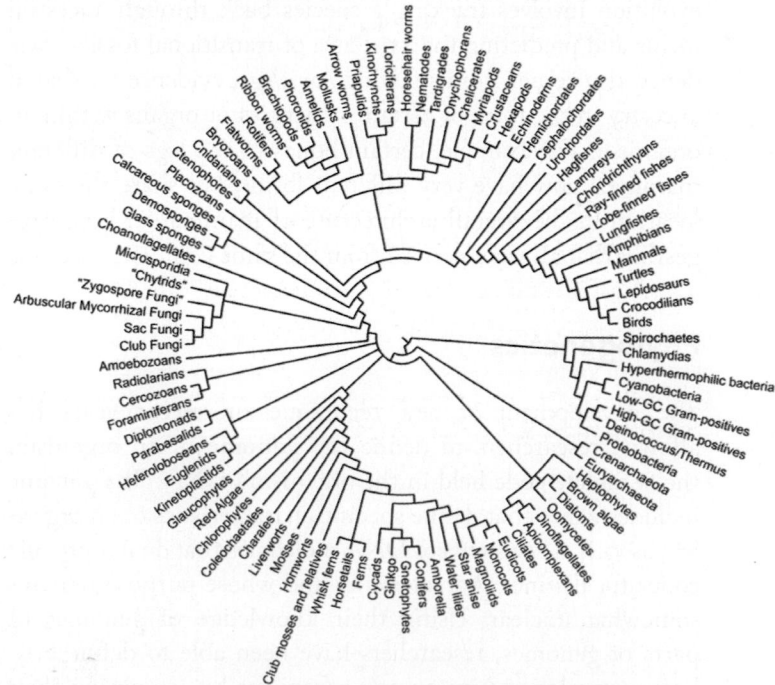

(b) In a cladogram, the position of each branch is based on DNA.

Interlude E Review

SUMMARY

- Fossils are records of past life on the Earth. There are two main categories: body fossils and trace fossils.

- Fossilization involves several steps, and not all organisms have the same likelihood of being preserved. Those with hard parts have a higher preservation potential.

- Some fossils are molds or casts of hard parts; others form when minerals replace organic materials. Chemical fossils are remnants of molecules formed by living organisms.

- Taxonomy classifies organisms into a hierarchy of divisions. Fossils may be classified by morphology.

- The fossil record is incomplete for geologic reasons. Whether evolution happens gradually remains unclear.

- Extinction can happen for many reasons. During mass-extinction events, huge numbers of species go extinct all at once and, therefore, biodiversity decreases.

- Using phylogenetics, based on the study of DNA, researchers recognize three domains (Archaea, Bacteria, and Eukarya). Animals and plants are Eukarya.

- Paleontologists can track the evolution of many organisms and can define the "tree of life."

GUIDE TERMS

amber (p. 435)
Archaea (p. 445)
Bacteria (p. 445)
biodiversity (p. 444)
biomarker (p. 436)
biomass (pp. 441, 442)
biosphere (p. 442)
body fossil (p. 434)
cast (p. 435)
chemical fossil (p. 436)
consumer (p. 441)

critical zone (p. 442)
DNA (p. 439)
domain (p. 445)
ecosystem (p. 441)
Eukarya (p. 445)
extraordinary fossil (p. 437)
food chain (p. 441)
fossil (p. 431)
fossilization (p. 432)
genome (p. 445)
macrofossil (p. 436)

mass-extinction event (p. 444)
metabolism (p. 439)
microbe (p. 439)
microfossil (p. 436)
mold (p. 435)
multicellular organism (p. 439)
natural selection (p. 443)
organism (p. 439)
paleoecology (p. 442)
paleontologist (p. 431)
paleontology (p. 431)

permineralization (p. 435)
petrified wood (p. 435)
phylogenetic tree (p. 445)
preservation potential (p. 437)
primary producer (p. 441)
prokarya (p. 445)
taxonomy (p. 437)
theory of evolution by natural selection (p. 443)
trace fossil (p. 435)

REVIEW QUESTIONS

The letters following each Review Question refer to the corresponding Learning Objective from the Chapter Opener.

1. What is a fossil, and how can fossils form? **(A)**

2. Do fossil bones and shells always have the same chemical composition as do modern ones? **(B)**

3. What is meant by "preservation potential?" **(A)**

4. Explain the basis for classifying fossils. **(B)**

5. What is the theory of evolution by natural selection, and how does the study of fossils provide evidence for it? **(C)**

6. What are the alternative ideas that paleontologists have had concerning the rate of evolution over geologic time? **(C)**

7. What is a mass-extinction event, and what causes them? **(C)**

8. What are the three domains of life? **(C)**

ONLINE RESOURCES

Animations
This chapter features animations on fossil formation and preservation.

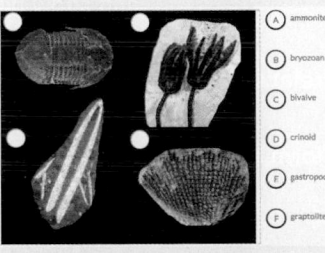

Smartwork5
This chapter features questions about fossils and life over geologic time.

CHAPTER 12

Deep Time: How Old Is Old?

By the end of this chapter, you should be able to . . .

A. explain the meaning of geologic time and the difference between relative and numerical ages.

B. use geologic principles (uniformitarianism, superposition, fossil succession) to determine relative ages.

C. make a model illustrating how unconformities form and what they represent.

D. explain the basis for correlating stratigraphic formations, and show how correlation led to the development of the geologic column.

E. illustrate the concept of a half-life, and use it to explain how geologists determine the numerical age of rocks.

F. describe the basis for determining dates on the geologic time scale and the age of the Earth.

> If the Eiffel Tower were now representing the world's age, the skin of paint on the pinnacle-knob
> at its summit would represent man's share of that age; and anybody would perceive that
> that skin was what the tower was built for. I reckon they would, I dunno.
>
> —Mark Twain (American author, 1835–1910)

12.1 Introduction

In May of 1869, a one-armed Civil War veteran named John Wesley Powell set out with a team of nine geologists and scouts to explore the previously unmapped expanse of the Grand Canyon, the greatest gorge on the Earth. Though Powell and his companions battled fearsome rapids and the pangs of starvation, most managed to emerge from the mouth of the canyon three months later (Fig. 12.1). During their voyage, seemingly insurmountable walls of rock both imprisoned and amazed the explorers and led them to pose important questions about the Earth and its history, questions that even casual tourists to the canyon ponder today: Did the Colorado River sculpt this marvel, and if so, how long did it take? When did the rocks making up the walls of the canyon form? Was there a time before the colorful layers accumulated? Such questions pertain to **geologic time**, the span of time since the Earth's formation.

In this chapter, we first learn the geologic principles that allowed geologists to provide a frame of reference for describing the *relative ages* (the age of one feature with respect to another) of rocks, fossils, structures, and landscapes. This information sets the stage for discussing the *geologic column*, the chart that geologists use to divide time into intervals. Then we describe how geologists can obtain a rock's *numerical age* (its age in years) by a procedure called *isotopic dating* or *radiometric dating*. Isotopic dating permitted the establishment of the *geologic time scale*, which provides numerical ages for intervals on the geologic column.

With the concept of geologic time in mind, a hike down a trail into the Grand Canyon becomes a trip into what some authors have called *deep time*. Humans are obsessed with time (Box 12.1), so the geologic discovery that our planet's history extends billions of years into the past changed humanity's perception of its place in the Universe as profoundly as did the astronomical discovery that the limit of space extends billions of light-years beyond the edge of our Solar System.

◀ (facing page) A satellite image shows kilometers of strata deposited over volcanic rocks and now exposed in the Dome Rock Mountains of western Arizona. These strata contain a record of the Earth's history at a time when dinosaurs still roamed.

FIGURE 12.1 Woodcut illustration of the "noonday rest in Marble Canyon," from J. W. Powell's *The Exploration of the Colorado River and Its Canyons* (1895). "We pass many side canyons today that are dark, gloomy passages back into the heart of the rocks."

12.2 The Concept of Geologic Time

Setting the Stage for Studying the Past

Until relatively recently, people in most cultures believed that geologic time began about the same time that human history began and that our planet has been virtually unchanged since its birth. With this concept in mind, an Irish archbishop named James Ussher (1581–1656) tallied successions of lineages and reigns described in the Old and New Testaments to determine the age of the Earth and, in 1654, stated his conclusion: the birth of the Earth took place on October 23, 4004 B.C.E.

BOX 12.1 CONSIDER THIS . . .

Time: A Human Obsession

When you plan your daily schedule, you have to know not only where you need to be, but when you need to be there. Because time assumes such significance in human consciousness today, we have developed elaborate tools to measure it and formal scales to record it. We use a *second* as the basic unit of time measurement. What exactly is a second? From 1900 to 1968, we defined the second as 1/31,556,925.9747 of the year 1900, but now we define it as the time that it takes for a cesium atom to change back and forth between two energy states 9,192,631,770 times. This change

can be measured with a device called an atomic clock, which has an expected error of only about 1 second per 30 million years. We sum 60 seconds into 1 minute, 60 minutes into 1 hour, and 24 hours into 1 day, about the time it takes for the Earth to spin once on its axis.

In the pre-industrial era, each locality kept its own time, setting noon as the moment when the Sun reached the highest point in the sky. But with the advent of train travel and telegraphs, people needed to calibrate schedules from place to place. So in 1883, countries around the globe agreed

to divide the world into 15°-wide bands of longitude called time zones—in each time zone, all clocks keep the same standard time. The times in each zone correlate to Greenwich mean time (GMT), the time at the astronomical observatory in Greenwich, England. Today, a group of about 200 atomic clocks together define Coordinated Universal Time (abbreviated as UTC, based on the French *Temps universel coordonné*). UTC serves as the basis for the global positioning system (GPS), used for precise navigation. When you look at the time on your cell phone, you're seeing UTC.

Not long after Ussher had implied that, in effect, the Earth has existed for only about 250 human generations (6,000 years), Nicolas Steno (1638–1686) proposed an idea that established the foundation for a very different approach to thinking about geologic time. Steno was serving as a physician in a nobleman's court in Florence, Italy, when he realized that unusual triangular-shaped rocks, known as *tongue stones* by the local people who had chiseled them out of outcrops in the nearby mountains, resembled the teeth of sharks (**Fig. 12.2**). He speculated that the tongue stones were not the tongues of dragons, as the locals thought, but rather were

FIGURE 12.2 A fossilized shark tooth. Before Steno explained the origin of such fossils, they were thought to be dragons' tongues.

actual shark teeth that had been buried with sediment on the seafloor, and that the sediment and teeth together had transformed into rocks that were later uplifted above sea level when the mountains formed. Steno eventually concluded that the presence of shark teeth, along with other **fossils** (remnants of ancient life preserved in rock; see Interlude E) of marine organisms, in rocks now exposed in mountains implied that the Earth could change over time. This realization set the stage for the founding of the science of geology by James Hutton a century later.

Hutton (1726–1797), a Scottish gentleman farmer and doctor, lived during the Age of Enlightenment when, sparked by the discovery of physical laws by Sir Isaac Newton, scientists began to seek natural rather than supernatural explanations for features of the world around them. While wandering among rock outcrops in the highlands of Scotland, Hutton noted that many features (such as ripple marks and cross beds) found in sedimentary rock types resembled features he could see forming in modern depositional environments. These observations led Hutton to propose that the formation of rocks and landscapes, in general, was a consequence of processes that he could see happening today.

Hutton's idea, discussed in his 1785 book called *The Theory of the Earth*, came to be known as the principle of **uniformitarianism**. According to this principle, physical processes we observe today also operated in the past at roughly the same rates, and these processes were responsible for the formation of geologic features that we now see in outcrops. More concisely, the principle can be stated as "the present is the key to the past." Because the rates of most geologic processes taking place today are so slow, Hutton deduced that the development of individual geologic features takes a very long time. Further, he

FIGURE 12.3 The difference between relative and numerical age.

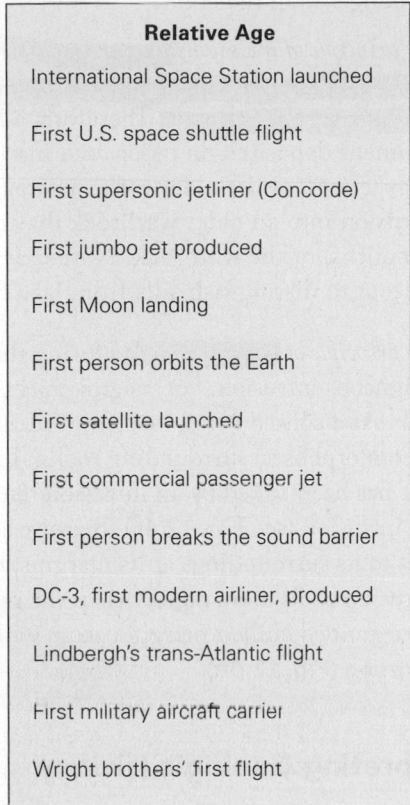

Relative Age

International Space Station launched

First U.S. space shuttle flight

First supersonic jetliner (Concorde)

First jumbo jet produced

First Moon landing

First person orbits the Earth

First satellite launched

First commercial passenger jet

First person breaks the sound barrier

DC-3, first modern airliner, produced

Lindbergh's trans-Atlantic flight

First military aircraft carrier

Wright brothers' first flight

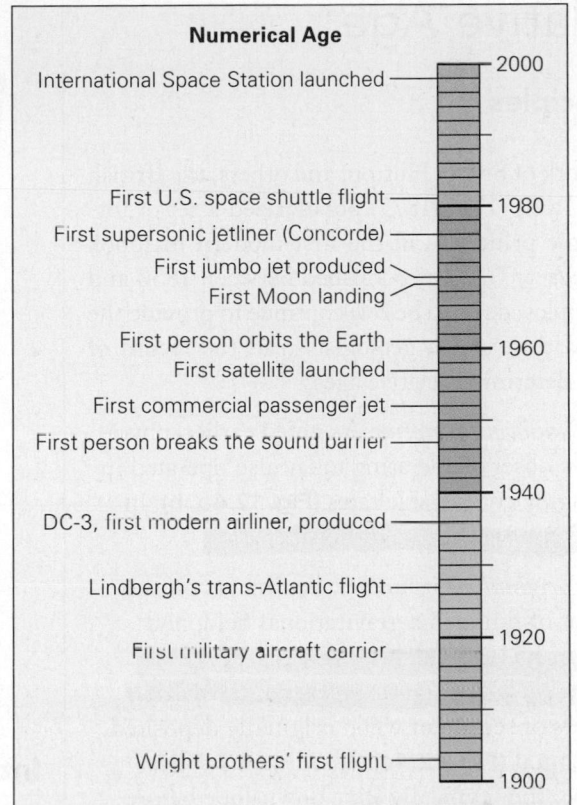

Numerical Age

International Space Station launched — 2000

First U.S. space shuttle flight — 1980
First supersonic jetliner (Concorde) —
First jumbo jet produced —
First Moon landing —

First person orbits the Earth — 1960
First satellite launched —
First commercial passenger jet —
First person breaks the sound barrier — 1940

DC-3, first modern airliner, produced —

Lindbergh's trans-Atlantic flight — 1920

First military aircraft carrier —

Wright brothers' first flight — 1900

(a) The relative ages of important moments in aviation and space flight.

(b) The numerical ages of these same events. Clearly this chart provides more information, for it displays the dates of the events and indicates the amount of time between the events.

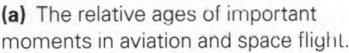

deduced that not all features formed at the same time, so the Earth must have a history that includes a succession of slow events, and therefore, the Earth existed for a long time before human history began. In fact, he speculated that we could see "no vestige of a beginning, nor prospect of an end."

Hutton was not a particularly clear writer, and it took the efforts of subsequent geologists to clarify the implications of the principle of uniformitarianism and to publicize them. Once this had been accomplished, geologists around the world began to apply their growing understanding of geologic processes to define and interpret geologic events of the Earth's past.

Relative versus Numerical Age

Like historians, geologists strive to establish both the sequence of events—those that created an array of geologic features such as rocks, structures, and landscapes—and, when possible, the date on which each event happened **(Fig. 12.3)**. We specify the age of one feature with respect to another in a sequence as its **relative age** and the age of a feature given in years as its **numerical age** (or its *absolute age*). Recall that

we can abbreviate numerical ages by using the units Ka, for thousands of years ago, Ma, for millions of years ago, and Ga, for billions of years ago. In these abbreviations, *K* stands for kilo-, *M* for mega-, *G* for giga- and *a* stands for *annum*, Latin for year. Geologists learned how to determine relative age long before they could determine numerical age, so we will look next at the principles leading to relative-age determination.

TAKE-HOME MESSAGE

The principle of uniformitarianism ("the present is the key to the past") implies that the Earth must be very old because geologic processes happen slowly, and that we can interpret events in the Earth's history to be a consequence of physical processes. Geologists distinguish between relative age (is one event older or younger than another?) and numerical age (how many years ago did an event happen?).

QUICK QUESTION: What observations led Hutton to propose uniformitarianism?

12.3 Relative Age

Geologic Principles

Building on the work of Steno, Hutton, and others, the British geologist Charles Lyell (1797–1875) popularized a set of formal, usable geologic principles in the first modern textbook of geology (*Principles of Geology*, published between 1830 and 1833). These principles, defined below, continue to provide the basic framework within which geologists read the record of Earth history and determine relative ages.

- *The principle of uniformitarianism*: As noted earlier, physical processes we observe operating today also operated in the past, at roughly comparable rates (Fig. 12.4a, b); in other words, the present is the key to the past.

- *The principle of original horizontality*: Sediments on the Earth settle out of fluids in a gravitational field, and the surfaces on which sediments accumulate (such as a floodplain or the bed of a lake or sea) are fairly flat. Therefore, layers of sediment when originally deposited are fairly horizontal (Fig. 12.4c). If sediments collect on a steep slope, they typically slide downslope before lithification and so will not be preserved as sedimentary rocks. With this principle in mind, we realize that when we see folds and tilted beds (see Chapter 11), we are seeing the consequences of deformation that postdates deposition.

- *The principle of superposition:* In a sequence of sedimentary rock layers, each layer must be younger than the one below, for a layer of sediment cannot accumulate unless there is already a substrate on which it can collect. Thus, the layer at the bottom of a sequence of strata is the oldest, and the layer at the top is the youngest (Fig. 12.4d).

- *The principle of lateral continuity:* Sediments generally accumulate in continuous sheets within a given region. So if you find a sedimentary layer cut by a canyon, you can assume that the layer once spanned the area that was later eroded by the river that formed the canyon (Fig. 12.4e).

- *The principle of cross-cutting relations:* If one geologic feature cuts across another, the feature that has been cut must be older. Applying this principle, we can conclude that if an igneous dike cuts across a sequence of sedimentary beds, the beds existed before the dike (Fig. 12.4f), and if a fault cuts across and displaces layers of sedimentary rock, then the fault must be younger than the layers.

In contrast, if a layer of sediment buries a fault, the layer is younger than the fault.

- *The principle of inclusions:* An *inclusion* (a fragment of one rock incorporated in another) must be older than the rock that contains it. Therefore, a layer of younger sediment deposited on older rock may contain inclusions (clasts) of the older rock, whereas a younger intrusion into an older wall rock may contain inclusions (xenoliths) of the wall rock. Note that we can use this concept to distinguish sills from lava flows (Fig. 12.4g).

- *The principle of baked contacts:* During the formation of an igneous intrusion, hot magma injects into cooler rock. As a consequence, heat from the intrusion "bakes" (metamorphoses) surrounding rocks. Therefore, rock that has been baked by an intrusion must be older than the intrusion (see Fig. 12.4f). Because an intrusion loses heat to its surroundings at its margins, the margin of an intrusion cools more rapidly than the rest. Therefore, a finer-grained chilled margin occurs within the younger intrusion (Fig. 12.4h).

Interpreting Geologic History

Geologists apply geologic principles to determine the relative ages of rocks, structures, and other geologic features at a given location. They interpret the development of each feature to be the consequence of a specific *geologic event*. Examples of geologic events include deposition of sedimentary beds, erosion of the land surface, intrusion or extrusion of igneous rocks, deformation (folding and/or faulting), and episodes of metamorphism. The succession of events, in order of relative age, that has produced the rocks, structure, and landscape of a region represents the *geologic history* of the region.

To understand how researchers figure out the geologic history of an area, let's work through an example (Fig. 12.5a). The principle of superposition requires that the oldest sedimentary layer of the figure is Bed 1 while the youngest is Bed 7, and the principle of original horizontality means that folding postdates deposition of the beds. The principles of cross-cutting relations, inclusions, and baked contacts allow us to determine the relative ages of the intrusions and faults relative to the beds and to one another. Using these interpretations together, we can propose the following geologic history for this region (Fig. 12.5b): deposition of the sedimentary sequence in order from Beds 1 to 7, intrusion of the sill, folding of the sedimentary beds and the sill, intrusion of the granite pluton, faulting, intrusion of the dike, and erosion to form the present-day land surface.

FIGURE 12.4 Major geologic principles used for determining relative ages.

Present-day mudcracks form in clay-rich sediment.

Ancient mud cracks in solid rock

(a) Uniformitarianism: The processes that formed cracks in the dried-up mud puddle on the left also formed the mudcracks preserved in the ancient, solid rock on the right. We can see these ancient mudcracks because erosion removed the adjacent bed.

Present-day volcanism produces molten lava.

Layers of basalt formed during volcanic activity.

(b) Uniformitarianism (cont.): We can observe lava flows forming today, so we can infer that solid lava flows represent the products of volcanic eruptions in the past.

Modern sediment, exposed at low tide, on the coast of France.

Horizontal sandstone beds in Wisconsin

— Youngest bed

— Bedding plane

— Cross beds

— Oldest bed

What a Geologist Sees

(c) Original horizontality: Gravity causes sediments to accumulate in horizontal sheets.

FIGURE 12.4 (continued)

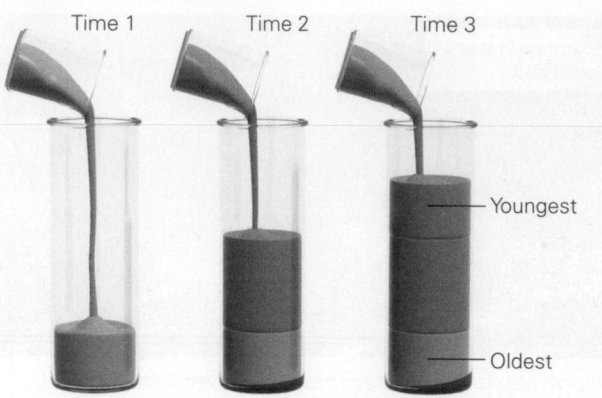

Time 1   Time 2   Time 3

Youngest

Oldest

(d) Superposition: In a sequence of strata, the oldest bed is on the bottom, and the youngest is on top. Pouring sand into a glass illustrates this point.

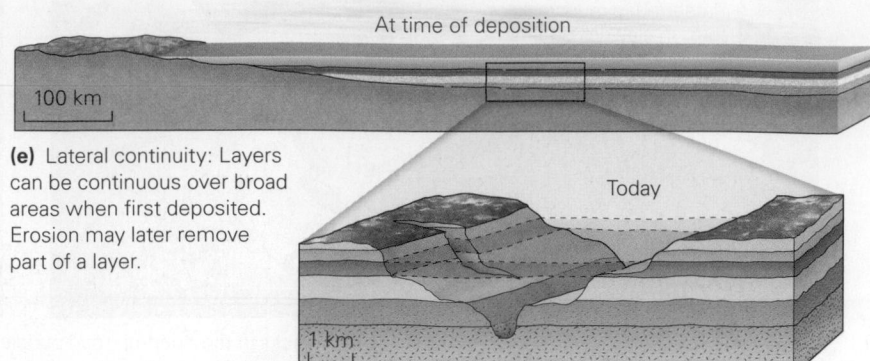

At time of deposition

100 km

Today

1 km

(e) Lateral continuity: Layers can be continuous over broad areas when first deposited. Erosion may later remove part of a layer.

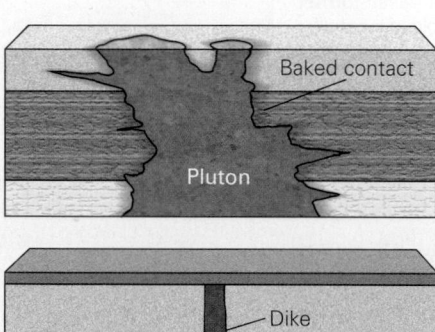

Baked contact

Pluton

Dike

(f) Top: By the principle of cross-cutting relations, the pluton is younger than the beds it cuts across. Top: A baked contact—metamorphic aureole—forms in the strata next to the pluton. Bottom: The dike is younger than the beds it cuts across. The sediment layer that buries the dike is younger than the dike.

Flow

Sill

(g) By the principle of inclusions, the pebbles of basalt in a conglomerate must be older than the conglomerate, and xenoliths of sandstone must be older than the basalt containing them.

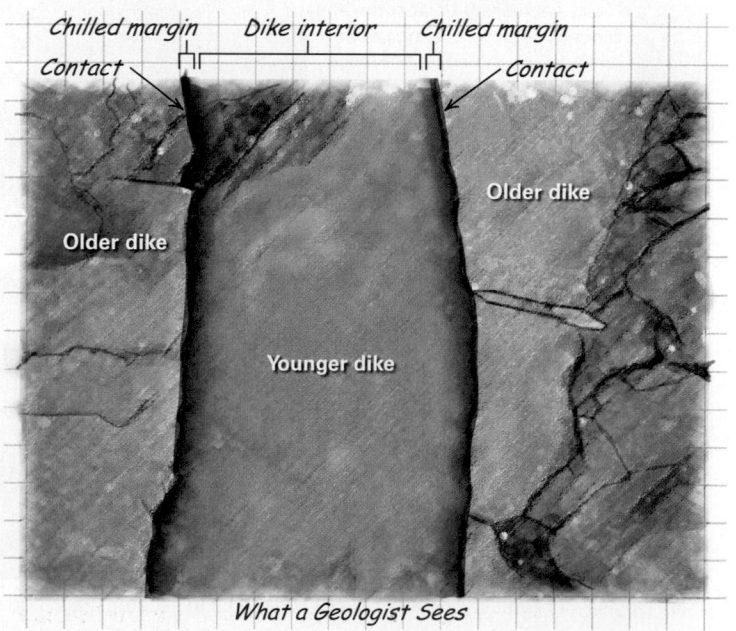

Chilled margin   Dike interior   Chilled margin

Contact   Contact

Older dike   Older dike

Younger dike

What a Geologist Sees

(h) Chilled margins can be used to determine age relationships. In this outcrop in California, a younger dike intrudes an older one. The younger dike is finer-grained and darker where it cooled faster at the contact with the older one.

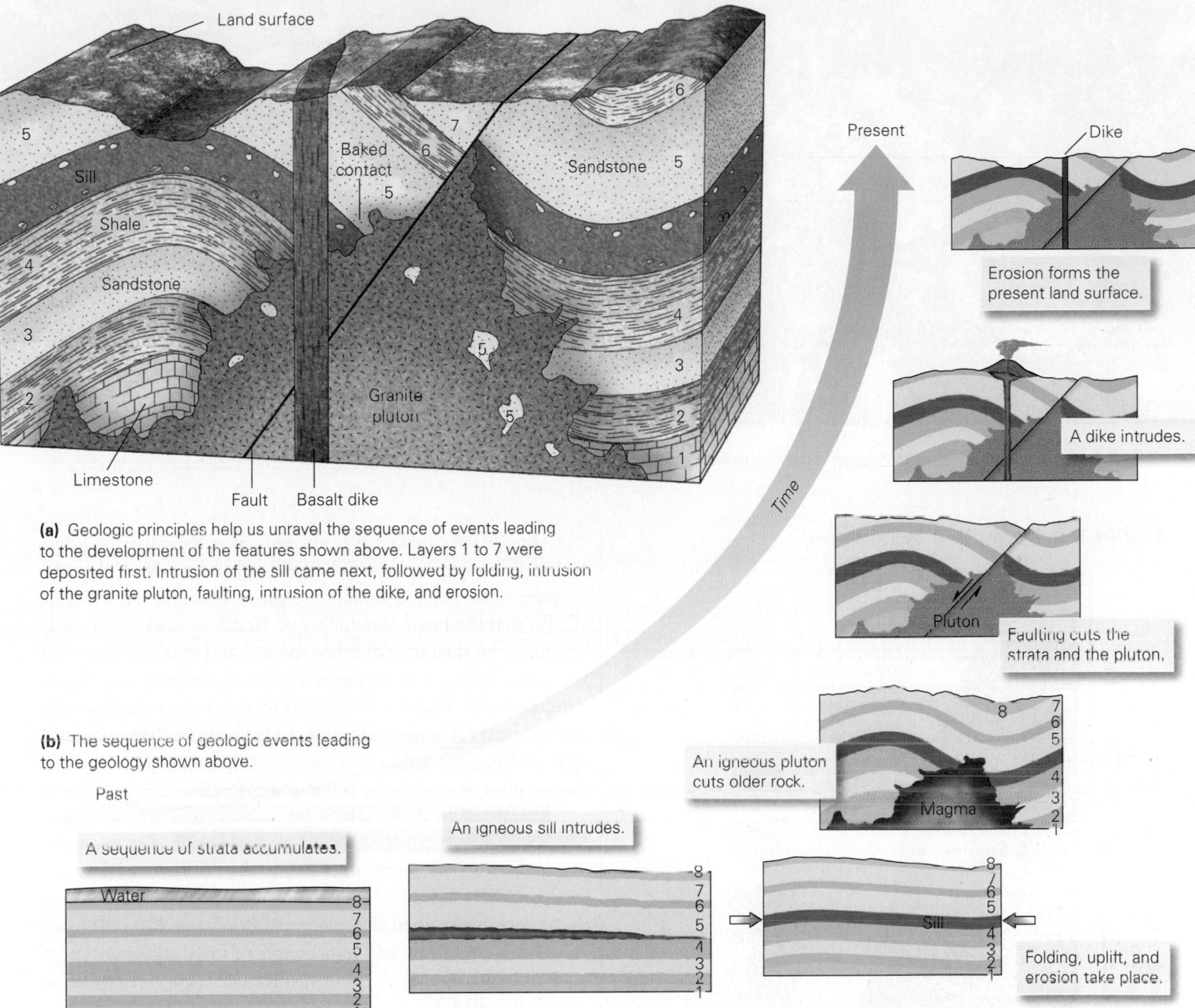

FIGURE 12.5 Interpreting the geologic history of a region, using geologic principles as a guide.

(a) Geologic principles help us unravel the sequence of events leading to the development of the features shown above. Layers 1 to 7 were deposited first. Intrusion of the sill came next, followed by folding, intrusion of the granite pluton, faulting, intrusion of the dike, and erosion.

(b) The sequence of geologic events leading to the geology shown above.

Past

A sequence of strata accumulates.

An igneous sill intrudes.

Folding, uplift, and erosion take place.

An igneous pluton cuts older rock.

Faulting cuts the strata and the pluton.

A dike intrudes.

Erosion forms the present land surface.

Present

Time

Fossil Succession

As Britain entered the industrial revolution in the late 18th and early 19th centuries, new factories demanded coal to fire their steam engines. Various companies decided to build a network of canals to transport coal and iron, and hired an engineer named William Smith (1769–1839) to survey the excavations. Canal digging provided fresh exposures of bedrock, which previously had been covered by vegetation. Smith learned to recognize distinctive layers of sedimentary rock and to identify the **fossil assemblage**—the group of fossil species—that each layer contained (**Fig. 12.6**). He also

realized, as we noted in Interlude E, that a particular fossil assemblage can be found in only a limited interval of strata and not in beds above or below that interval. Therefore, once a fossil species disappears at a horizon in a sequence of strata, it never reappears higher in the sequence. In other words, "extinction is forever." Smith's observation has been repeated at thousands of locations around the world and has been codified as the *principle of fossil succession*.

To see how this principle works, examine **Figure 12.7**, which depicts a sequence of strata. Bed 1 at the base contains fossil species A, Bed 2 contains A and B, Bed 3 contains B and C, Bed 4 contains C, and so on. From these data, we can define

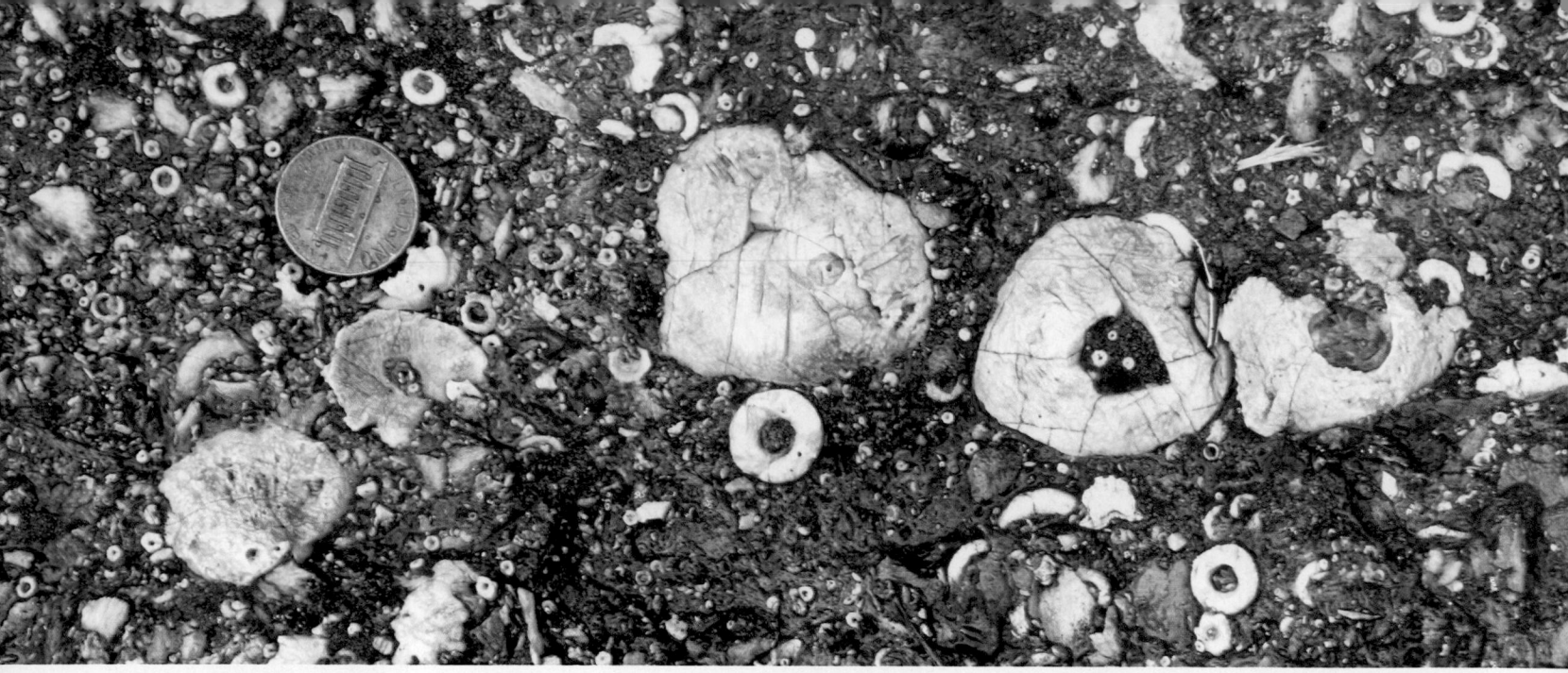

FIGURE 12.6 A close-up photo of a bedding surface in Devonian strata of New York, showing many fossils. The ring-shaped ones are pieces of crinoid stems.

FIGURE 12.7 The principle of fossil succession.

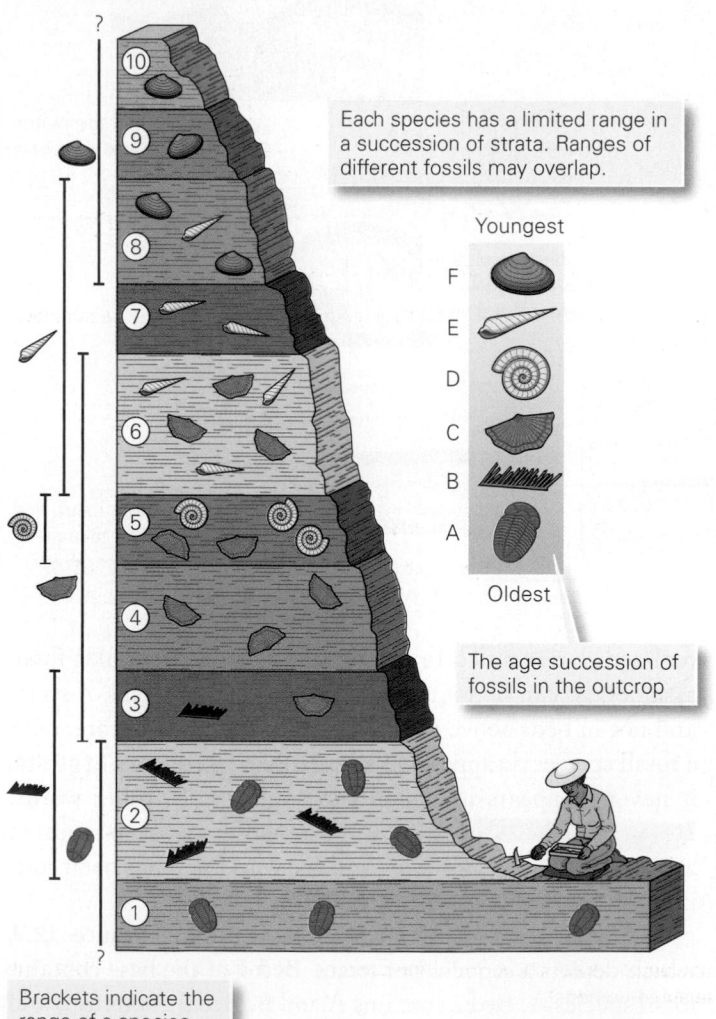

Each species has a limited range in a succession of strata. Ranges of different fossils may overlap.

Youngest

F
E
D
C
B
A

Oldest

The age succession of fossils in the outcrop

Brackets indicate the range of a species.

the *range* of specific fossils in the sequence, meaning the interval in the sequence in which the fossils occur. Note that the sequence contains a definable succession of fossils (A, B, C, D, E, F), that the range in which a particular species occurs may overlap with the range of other species, and that once a species vanishes, it does not reappear higher in the sequence. Some species can be found over a broad region, but existed for only a short interval of geologic time, so they can be diagnostic of a precise time interval in rocks at many different locations. The fossils of such species are called **index fossils.**

Because of the principle of fossil succession, we can define the relative ages of strata by looking at fossils. For example, if we find a bed containing Fossil A, we can say that the bed is older than a bed containing, say, Fossil F. Geologists have now determined the relative ages of over 200,000 fossil species. Recognition of the principle of fossil succession provides the geologic underpinnings for the theory of evolution (see Interlude E).

TAKE-HOME MESSAGE

Geologic principles (including uniformitarianism, superposition, cross-cutting relations, and fossil succession) provide the basis for determining the relative ages of rocks and other geologic features. By working out relative ages, we can reconstruct the geologic history of a region.

QUICK QUESTION: If a fault forms a fault scarp, what is the relative age of the fault and the landscape surface?

12.4 Unconformities: Gaps in the Record

James Hutton used a boat to explore the seaside of Scotland because its shore cliffs provided great exposures of rock, stripped of soil and shrubbery. He was particularly puzzled by an outcrop he found at Siccar Point on the east coast.

Here he saw a sequence of strata consisting of red sandstone and conglomerate that rested on a distinctly different sequence, one that consisted of gray sandstone and shale (Fig. 12.8a). Furthermore, the beds of gray sandstone and shale had a nearly vertical dip (slope), whereas the beds of red sandstone and conglomerate had a dip of less than 15°— the gently dipping layers seemed to lie across the truncated ends of the vertical layers, like a handkerchief resting on a row of books. (See Box 11.1 for the definition of *dip*.)

FIGURE 12.8 Examples of unconformities, as visible in outcrops.

What a Geologist Sees

(a) James Hutton found this unconformity along the east coast of Scotland. He deduced that the layers above were deposited long after the beds below had been tilted. Geologists have since determined that the strata above are about 80 million years younger than the strata below.

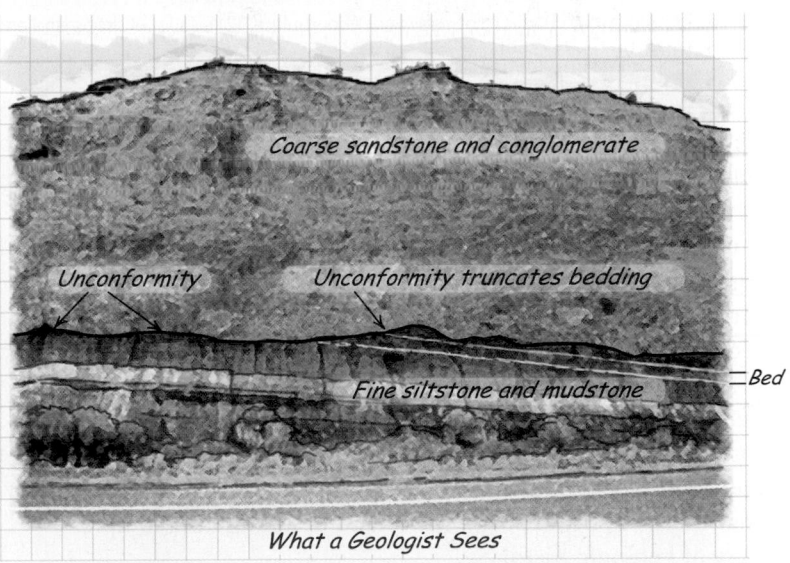

What a Geologist Sees

(b) A road cut in Utah reveals an unconformity between a coarse sandstone and conglomerate above and a fine siltstone and mudstone below. Note that the irregularities along this unconformity locally truncate bedding of the siltstone and mudstone.

FIGURE 12.9 The three kinds of unconformities and their formation.

Time →

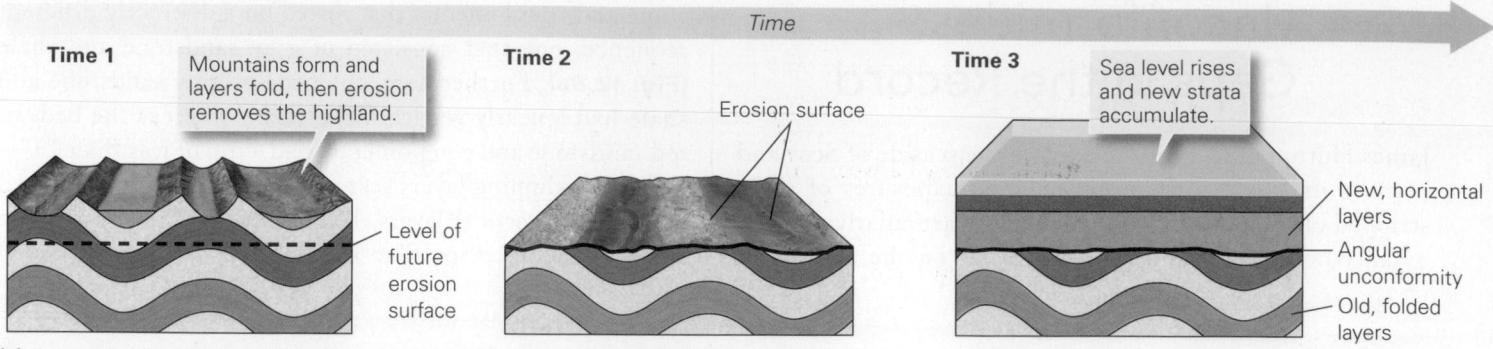

(a) An angular unconformity: (1) layers undergo folding; (2) erosion produces a flat surface; (3) sea level rises and new layers of sediment accumulate.

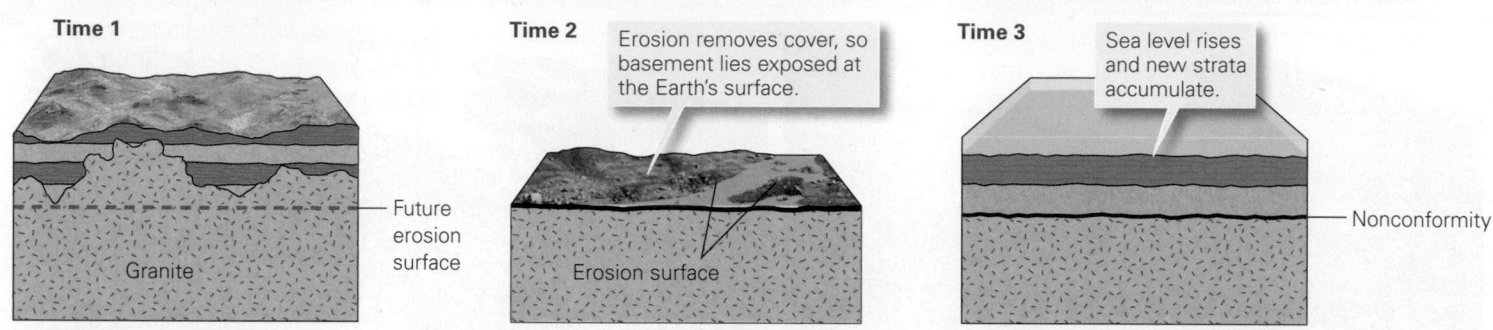

(b) A nonconformity: (1) a pluton intrudes; (2) erosion cuts down into the crystalline rock; (3) new sedimentary layers accumulate above the erosion surface.

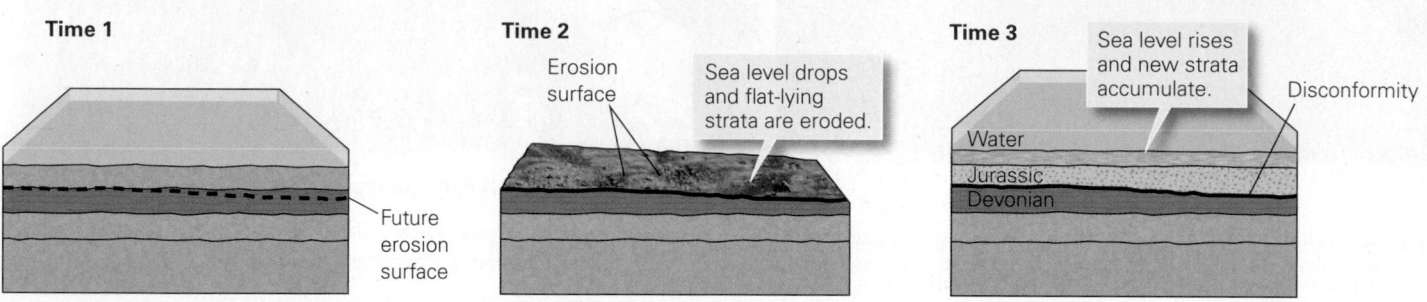

(c) A disconformity: (1) layers of sediment accumulate; (2) sea level drops and an erosion surface forms; (3) sea level rises and new sedimentary layers accumulate.

Perhaps as Hutton sat and stared at this odd geometric relationship, the tide came in and deposited a new layer of sand on top of the rocky shore. With the principle of uniformitarianism in mind, Hutton suddenly realized the significance of the outcrop—the gray sandstone and shale sequence had been deposited, turned into rock, tilted, and truncated by erosion before the red sandstone and conglomerate beds had been deposited. Therefore, the surface between the gray and red rock sequences represented a time interval during which the gray strata had been eroded away and new strata had not yet accumulated. Hutton realized that Siccar Point exposed the record of a long and complex saga of geologic history.

We now refer to a boundary surface between two units, which represents a period of nondeposition and possibly erosion, as an **unconformity (Fig. 12.8b)**. The gap in the geologic record

that an unconformity represents is called a *hiatus.* Geologists recognize three main types of unconformities:

- *Angular unconformity:* An **angular unconformity** represents an erosional surface that cuts across previously tilted or folded underlying layers. As a result, the orientation of layers below an unconformity differs from that of the layers above **(Fig. 12.9a)**, as is the case at Siccar Point. Angular unconformities form where rocks underwent deformation before being exposed at the Earth's surface.

- *Nonconformity:* At a **nonconformity**, sedimentary rocks overlie older intrusive igneous rocks and/or metamorphic rocks **(Fig. 12.9b)**. The igneous or metamorphic rocks underwent cooling, uplift, and erosion prior to becoming the substrate or "basement" on which the "cover" of new sediments accumulated.

- *Disconformity:* Imagine that a sequence of sedimentary beds has been deposited beneath a shallow sea. Then sea level drops, exposing the beds for some time. During this time, no new sediment accumulates, and some of the pre-existing sediment gets eroded away. Later, sea level rises, and a new sequence of sediment accumulates over the old. Geologists consider the boundary between the two sequences to be a **disconformity (Fig. 12.9c)**. Even though the beds above and below the disconformity are parallel, the boundary between them represents a hiatus. The unconformity in the outcrop shown in Figure 12.8b is a disconformity.

The succession of strata at a particular location provides a record of Earth history there. But because of unconformities, this record is incomplete. It's as if geologic history were being chronicled by a video recorder that turns on only intermittently—when it's on (= times of deposition), the rock record accumulates, but when it's off (= times of nondeposition and possibly erosion), an unconformity develops. Because of unconformities, no single location on the Earth contains a complete record of Earth history.

How can you recognize an unconformity in the field? At angular unconformities, the strata above have a different dip than the strata below, and at nonconformities, the juxtaposition of cover over basement serves as a clue. Disconformities may be hard to recognize because the beds above and below are parallel. A disconformity may be indicated by a gap in the fossil succession or by the presence of a surface of erosion and weathering. If the surface was exposed at the Earth's surface for a while, a pebbly layer of debris might occur just above it, or a *paleosol* (a remnant of a soil horizon that has been lithified) may be visible just below it **(Fig. 12.10)**.

FIGURE 12.10 This road cut in Utah shows a sand-filled channel cut down into floodplain mud. The mud was exposed between floods, and a soil formed on it. When later buried, all the sediment turned into rock; the channel floor is now an unconformity, and the ancient soil is now a paleosol. Note that the channel cut across the paleosol. The paleosol also represents a time during which deposition did not occur.

TAKE-HOME MESSAGE

Unconformities represent time intervals of nondeposition and possibly erosion. Where rock below the unconformity was folded or tilted before deposition of strata above, an angular unconformity occurs. Nonconformities juxtapose strata above with basement below. Strata above and below a disconformity are parallel, so a gap in the fossil record or evidence of erosion reveals the unconformity.

QUICK QUESTION: Is there any one place on the surface of the Earth where the exposed stratigraphic succession represents all of geologic time? Explain.

12.5 Stratigraphic Formations and Their Correlation

The Concept of a Formation

When William Smith first began to explore the strata exposed along the newly dug canals of England, he realized that distinctive sets of beds, with distinctive assemblages of fossils, crop out at many locations. Geologists now routinely divide thick successions of strata into recognizable units, called stratigraphic formations, which others can recognize and identify. Formally defined, a **stratigraphic formation** (or *formation*, for short) is an interval of strata composed of a specific rock type or group of rock types that together can be traced over a fairly broad region. A formation represents the products of deposition during a definable interval of time. The boundary surface between two formations is a type of **geologic contact** (or *contact*, for short). Fault surfaces, unconformities, and the boundary between an igneous intrusion and its wall rock are also types of contacts.

Geologists summarize information about the sequence of sedimentary strata at a location by drawing a **stratigraphic column**. Typically, we construct stratigraphic columns to scale so that the relative thicknesses of beds or formations portrayed on the column are in proportion to the thicknesses of these units in outcrop. Geologists may represent the relative resistance to erosion of beds (or formations) by making the right side of the column irregular to symbolize the way the units might erode on a cliff face—more resistant units stick

out farther. An unconformity can be represented on a stratigraphic column by a dark, wavy line.

Let's see how the concepts of a stratigraphic formation and a stratigraphic column apply to the Grand Canyon. The walls of the canyon look striped because they expose a variety of rock types that differ in color and in resistance to erosion. Geologists identify major contrasts distinguishing one interval of strata from another and use them as a basis for dividing the strata into formations, each of which may consist of many beds (**Fig. 12.11**). Some formations consist of beds of a single rock type, whereas others include interlayered beds of two or more rock types. Furthermore, not all formations have the same thickness. Commonly, geologists name a formation after a locality where it was first identified or first studied. For example, the Schoharie Formation was first defined in exposures in Schoharie Creek of eastern New York. If a formation consists of only one rock type, we may incorporate that rock type in the name (as in Kaibab Limestone), but if a formation contains more than one rock type, we use the word *formation* in the name (as in Toroweap Formation). Note that in the formal name of a formation, all words are capitalized. Several adjacent formations in a succession may be lumped together as a *stratigraphic group* (or simply a *group*).

Correlating Strata

How does the stratigraphy of a sedimentary succession exposed at one locality relate to one exposed at a different locality? Stated another way, how do we determine the age of strata at one location with respect to that of strata at another? Geologists determine such relations by a process called *stratigraphic correlation* (or **correlation**, for short). Geologists use two approaches for correlating intervals of strata.

When correlating formations among nearby regions, we can simply look for similarities in rock type. We call this method *lithologic correlation*. For example, the sequence of strata on the southern rim of the Grand Canyon clearly correlates with the sequence on the northern rim, because they contain the same

rock types in the same order. In some cases, a sequence contains a key bed, or marker bed, with such distinctive characteristics that its presence provides a definitive basis for correlation.

To correlate units over broad distances, lithologic correlation alone may not be sufficient. Uncertainty about correlation can arise for several reasons. In some cases, different strata of different ages may look more or less the same, so a geologist cannot decide which strata correlate based on rock types alone. Fortunately, **fossil correlation**—correlation of strata by comparing their fossil assemblages—provides another approach.

How does fossil correlation work? Let's consider a simple example. Imagine that you are studying strata in a particular area and find an outcrop of sandstone beds with a distinctive fossil assemblage. Then you walk somewhere else and find another outcrop, but this outcrop contains two units of sandstone beds, one at the base of a cliff and one at the top. The lower set of sandstone beds contains the same assemblage of fossils that you found at the first outcrop, whereas the upper set contains a different assemblage. Fossil correlation allows you to say that the sandstone beds you saw at the first outcrop correlate with the lower sandstone beds at the second outcrop.

Geologists also use fossil correlation to determine the relative ages of units that don't contain the same rock types. For example, to determine if an outcrop of sandstone correlates with (in this case, meaning "was deposited at the same time as") an outcrop of limestone 100 km away, you can't simply compare their fossil assemblages directly, because the organisms that lived in a depositional environment that produces limestone did not necessarily live in a depositional environment that produces sandstone. In such cases, geologists must use data collected from around the globe in order to assign specific fossil species to specific intervals of geologic time. Then, by comparing fossil ages, geologists can determine whether two units, even if composed of different rock types, are the same age. (We'll explore this concept further in Section 12.6.) Geologists also use this method to correlate strata that have been deposited on different continents or at different latitudes because, at any given time, different species live in different parts of the world.

Fossil correlation can reveal the existence of disconformities. For example, imagine that two of the formations shown in **Figure 12.12**, the Santuit Sandstone and the Oswaldo Sandstone, look the same. At Location A, the Milo Limestone lies between the two formations, but at Location C, the two formations are in direct contact, because between Location A and Location C, the Milo Limestone progressively thins and ultimately "pinches out." By studying fossils, a sharp-eyed geologist would be able to identify the disconformity separating the two sandstone formations, because the top and bottom parts of what might, at first glance, look like the same unit contain different fossil assemblages.

Let's now apply correlation principles to the challenge of determining the ages of formations exposed in the Grand

FIGURE 12.11 The concepts of a stratigraphic column and stratigraphic formations: examples from the Grand Canyon in Arizona.

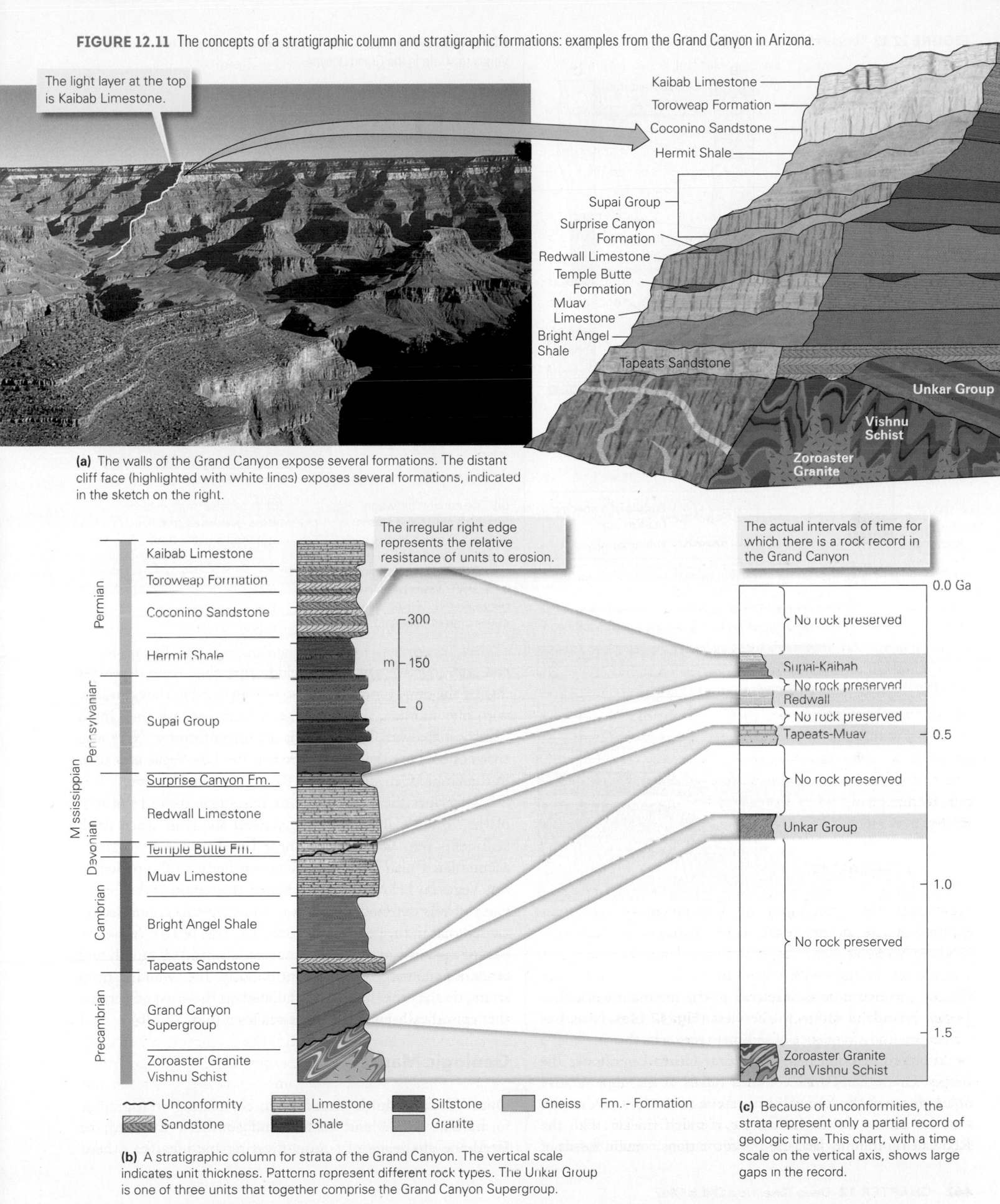

The light layer at the top is Kaibab Limestone.

Kaibab Limestone
Toroweap Formation
Coconino Sandstone
Hermit Shale
Supai Group
Surprise Canyon Formation
Redwall Limestone
Temple Butte Formation
Muav Limestone
Bright Angel Shale
Tapeats Sandstone
Unkar Group
Vishnu Schist
Zoroaster Granite

(a) The walls of the Grand Canyon expose several formations. The distant cliff face (highlighted with white lines) exposes several formations, indicated in the sketch on the right.

The irregular right edge represents the relative resistance of units to erosion.

The actual intervals of time for which there is a rock record in the Grand Canyon

Permian
Kaibab Limestone
Toroweap Formation
Coconino Sandstone
Hermit Shale

Pennsylvanian
Supai Group
Surprise Canyon Fm.

Mississippian
Redwall Limestone

Devonian
Temple Butte Fm.

Cambrian
Muav Limestone
Bright Angel Shale
Tapeats Sandstone

Precambrian
Grand Canyon Supergroup
Zoroaster Granite
Vishnu Schist

300
m 150
0

0.0 Ga
No rock preserved
Supai–Kaibab
No rock preserved
Redwall
No rock preserved
Tapeats–Muav
0.5
No rock preserved
Unkar Group
1.0
No rock preserved
1.5
Zoroaster Granite and Vishnu Schist

Unconformity   Limestone   Siltstone   Gneiss   Fm. - Formation
Sandstone   Shale   Granite

(b) A stratigraphic column for strata of the Grand Canyon. The vertical scale indicates unit thickness. Patterns represent different rock types. The Unkar Group is one of three units that together comprise the Grand Canyon Supergroup.

(c) Because of unconformities, the strata represent only a partial record of geologic time. This chart, with a time scale on the vertical axis, shows large gaps in the record.

FIGURE 12.12 The principles of correlation.

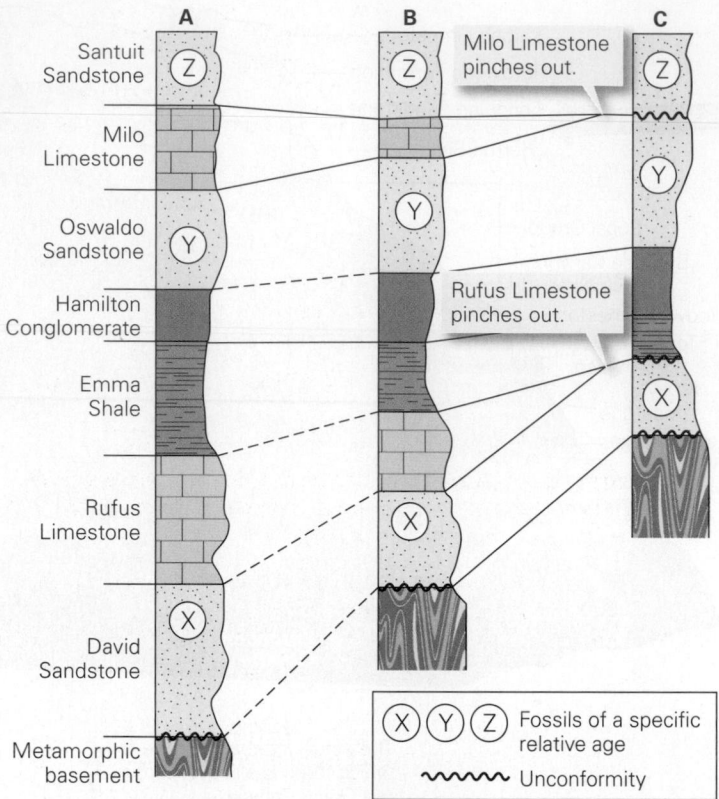

(a) Stratigraphic columns can be correlated by matching rock types (lithologic correlation). The Hamilton Conglomerate is a marker horizon. Because some strata pinch out, Column C contains unconformities. Fossil correlation indicates that the youngest beds in C are Santuit Sandstone.

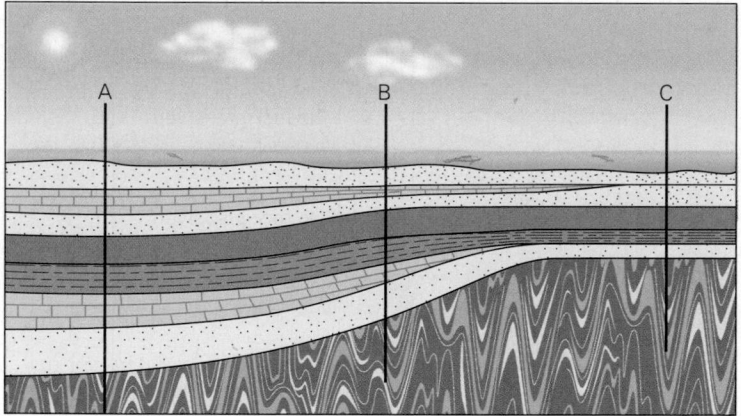

(b) At the time of deposition, locations A, B, and C (which correlate with the columns in part a) were in different parts of a basin. The basin floor was subsiding fastest at A.

Canyon relative to those exposed in the mountains near Las Vegas, Nevada, 150 km to the west (Fig. 12.13a). Near Las Vegas, we find a sequence of sedimentary rocks that includes a limestone formation called the Monte Cristo Limestone. The Monte Cristo Limestone contains fossils of the same relative age as those of the Redwall Limestone of the Grand Canyon, but the Monte Cristo Limestone is much thicker than the Redwall Limestone. Because the formations contain fossils of

FIGURE 12.13 An application of correlation to relate strata near Las Vegas to strata in the Grand Canyon.

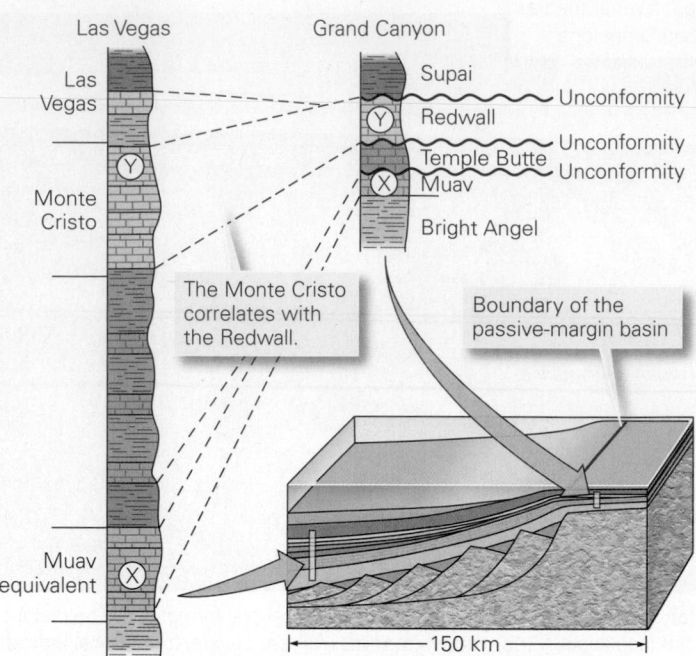

(a) The section between fossils of age X and fossils of age Y is much thicker near Las Vegas than in the Grand Canyon. The edge of a passive-margin basin lay between the two localities at the time of deposition.

(b) A passive-margin basin forms over crust that has stretched and thinned.

the same relative age, we conclude that they were deposited during the same time interval, so we can say that they correlate with one another. Therefore, the contact beneath the Grand Canyon's Redwall Limestone is an unconformity. Note also that not only are the units thicker in the Las Vegas area than in the Grand Canyon area, but there are more of them.

Why does the stratigraphy of Las Vegas differ from that of the Grand Canyon? During part of the time when thick sediments were accumulating near Las Vegas, no sediments accumulated near the Grand Canyon because the region of Las Vegas lay below sea level, whereas the region of the Grand Canyon was dry land. Geologists studying this contrast concluded that in the geologic past, the location of Las Vegas was part of a passive-margin basin that sank (subsided) rapidly and remained submerged almost continuously. The Grand Canyon strata, on the other hand, accumulated on the crust of a craton that episodically emerged above sea level (Fig. 12.13b).

Geologic Maps

Once William Smith succeeded in correlating stratigraphic formations throughout central and southern England, he faced the challenge of communicating his ideas to others.

FIGURE 12.14 A geologic map depicts the distribution of rock units and structures.

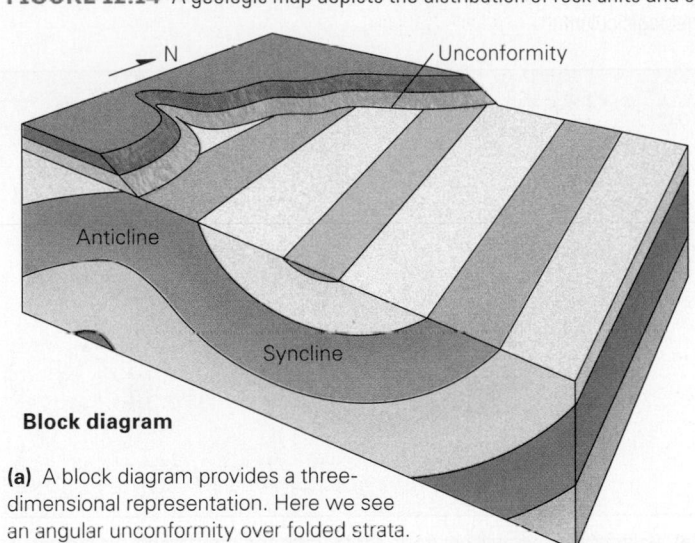

Block diagram

(a) A block diagram provides a three-dimensional representation. Here we see an angular unconformity over folded strata.

A geologic map shows the view looking straight down.

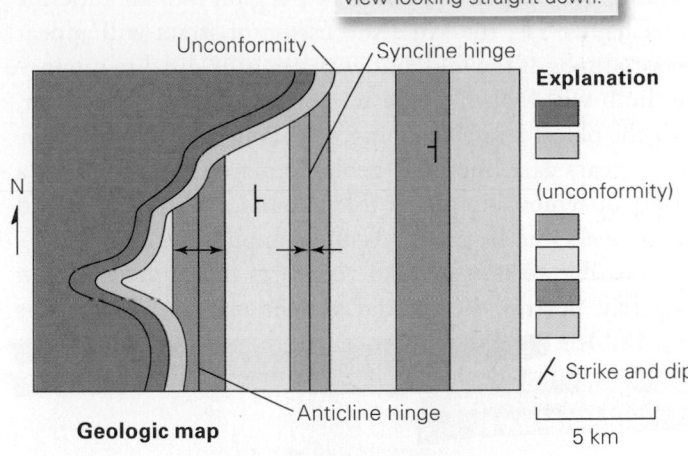

Geologic map

(b) A geologic map shows the distribution of units. Contacts between units are indicated by lines. Note that the map also shows geologic structures.

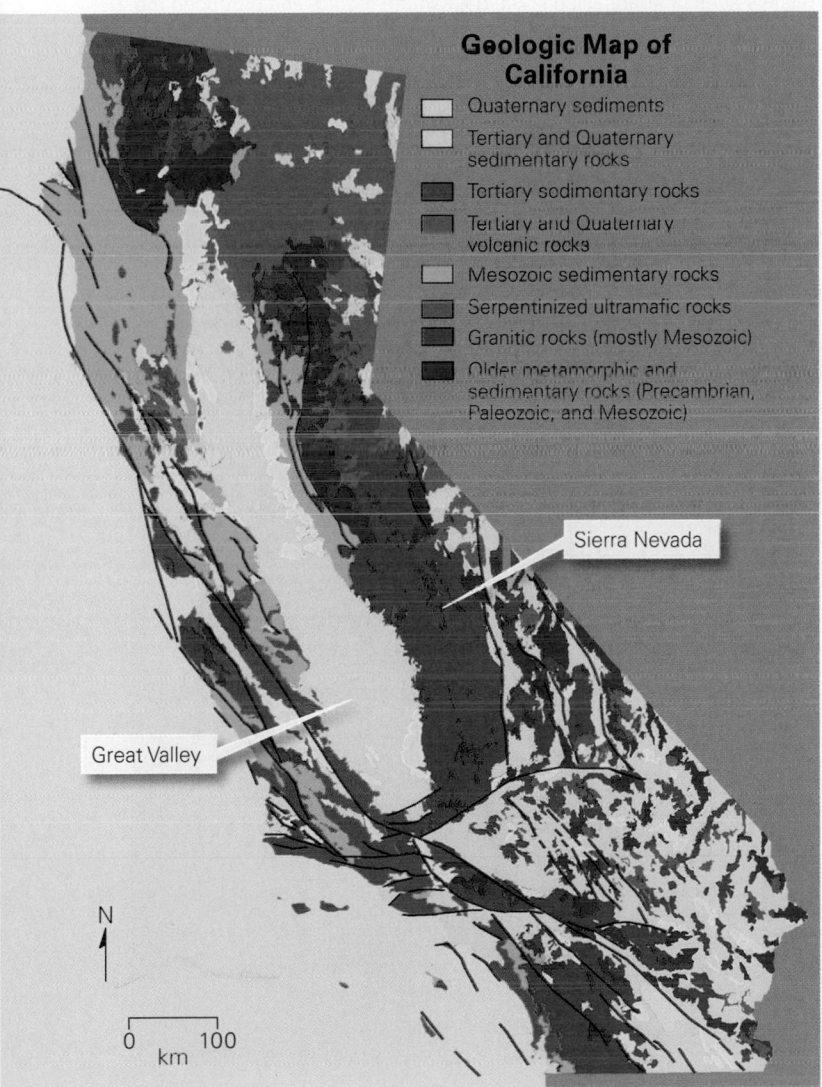

Geologic Map of California

- Quaternary sediments
- Tertiary and Quaternary sedimentary rocks
- Tertiary sedimentary rocks
- Tertiary and Quaternary volcanic rocks
- Mesozoic sedimentary rocks
- Serpentinized ultramafic rocks
- Granitic rocks (mostly Mesozoic)
- Older metamorphic and sedimentary rocks (Precambrian, Paleozoic, and Mesozoic)

Sierra Nevada

Great Valley

0   100
   km

N

(c) A geologic map of California indicates that the state is underlain by many different rock units. Granite underlies the Sierra Nevada, and Quaternary sediments underlie the Great Valley. The black lines are fault traces.

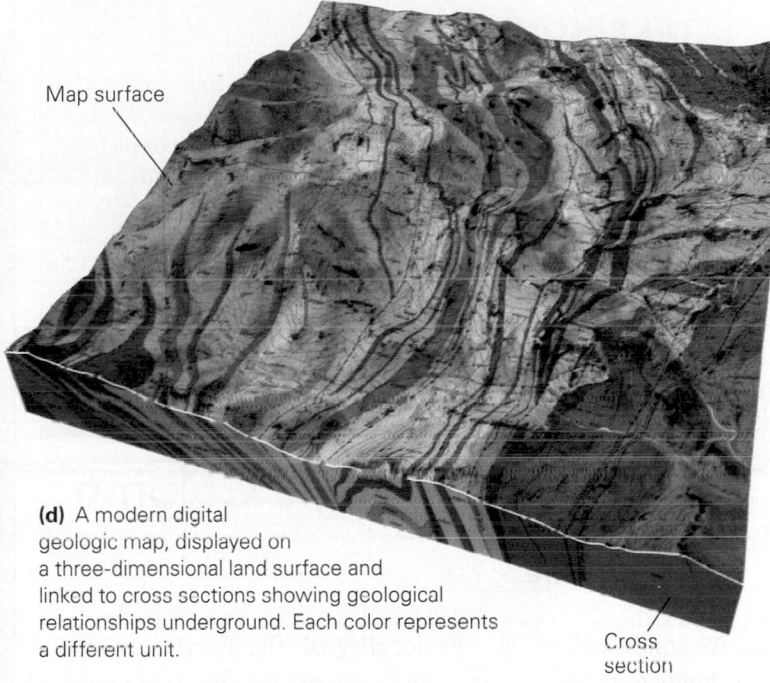

Map surface

Cross section

(d) A modern digital geologic map, displayed on a three-dimensional land surface and linked to cross sections showing geological relationships underground. Each color represents a different unit.

He could have produced a table that compared stratigraphic columns from different locations. But since Smith was a surveyor and worked with maps, it occurred to him that he could also outline and color the areas on a map that represent areas in which strata of a given relative age occurred. He did this using the data he had collected, and in 1815, he produced the first modern geologic map. In general, a **geologic map** portrays the spatial distribution of rock units at the Earth's surface.

Significantly, the pattern displayed on a geologic map can provide insight into the presence and orientation of geologic structures in the map area **(Fig. 12.14a–c)**. With experience, a geologist can interpret the pattern of contacts and distribution of formations on a map and can recognize folds, faults, and unconformities. For example, if mountain building has warped the strata in a region into an anticline (see Chapter 11), the same succession of strata will appear on both limbs of the fold, but with opposite dip directions—one limb will look like the mirror reflection of the other, with the oldest strata cropping out along the hinge. A contact appears as a line on a geologic map. Geologists use a variety of symbols to depict folds, faults, and the strike and dip of layers (see Box 11.1). Using computer technology, it's now possible to plot geologic contacts on *digital elevation maps* that portray the ground surface in three dimensions (Fig. 12.14d). *Geologic cross sections* indicate the relationships of underground geologic contacts and structures on the plane of a vertical slice.

TAKE-HOME MESSAGE

A stratigraphic formation is a recognizable sequence of beds that can be mapped across a broad region. Geologists correlate formations regionally on the basis of rock type and fossil content, and they portray the configuration of formations in a region on a geologic map.

QUICK QUESTION: What does a syncline look like on a geologic map?

12.6 The Geologic Column

Dividing Time

As stated earlier, no one locality on the Earth provides a complete record of our planet's history because stratigraphic successions can contain unconformities. But by correlating strata from locality to locality at millions of places around the world, geologists have pieced together a composite stratigraphic column, called the **geologic column**, that represents the entirety of Earth history (Fig. 12.15). They divide this column into segments, each of which represents a specific interval of time. The largest subdivisions break Earth history into the Hadean, Archean, Proterozoic, and Phanerozoic **Eons**—the first three of these, together, constitute the **Precambrian.** (Hadean rocks don't exist, so they do not appear in Figure 12.15.) In the names of the two youngest eons, the suffix –*zoic* means life, so Phanerozoic means visible life, and Proterozoic means first life. These names can be a bit confusing, though, because decades after the eons had been named, geologists discovered that the earliest living

FIGURE 12.15 Global correlation of strata led to the development of the geologic column.

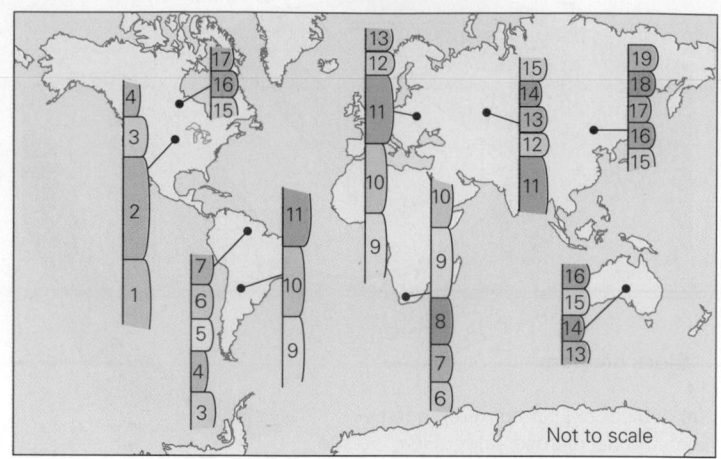

Not to scale

(a) Each of these small columns represents the stratigraphy at a given location. By correlating these columns, geologists determined the relative ages of strata, filled in the gaps in the record, and produced the geologic column.

	Eon	Era	Period	Epoch
19			Quaternary	Holocene
18				Pleistocene
17		Cenozoic	Neogene	Pliocene
16				Miocene
15			Tertiary	
14				Oligocene
13			Paleogene	Eocene
12				Paleocene
11	Phanerozoic		Cretaceous	
10		Mesozoic	Jurassic	
9			Triassic	
8			Permian / Pennsylvanian	
7			Carboniferous	
6		Paleozoic	Devonian / Mississippian	
5			Silurian	
4			Ordovician	
3			Cambrian	
2	Proterozoic			
1	Archean			
	Hadean			

(Precambrian)

(b) By correlation, strata from localities around the world were stacked in a chart representing geologic time to create the geologic column. Geologists assigned names to time intervals, but since the column was built without knowledge of numerical ages, it does not depict the duration of these intervals. Subdivisions of eons in the Precambrian are not shown. The Hadean is not shown because rocks do not preserve a record of it.

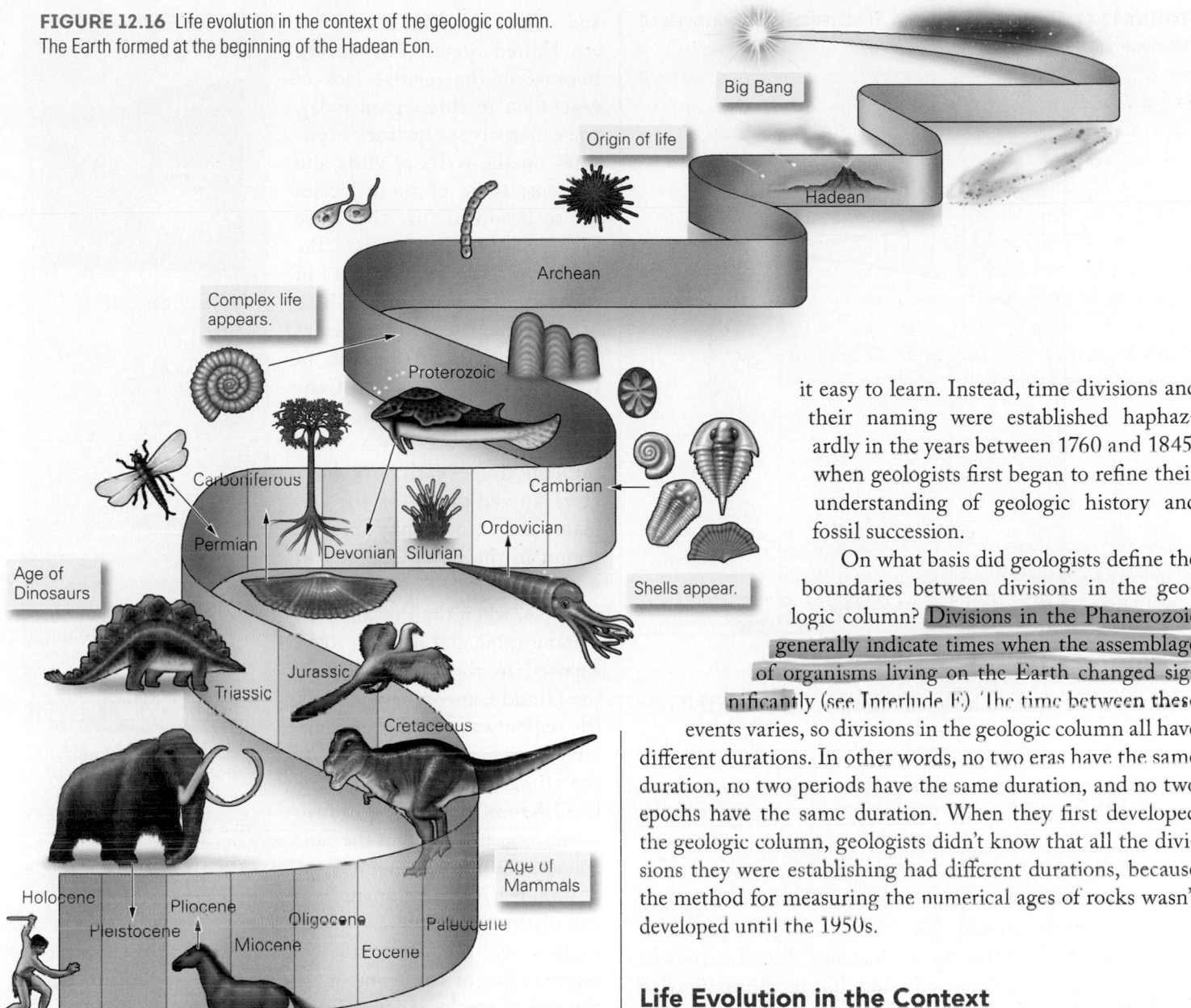

FIGURE 12.16 Life evolution in the context of the geologic column. The Earth formed at the beginning of the Hadean Eon.

Big Bang

Origin of life

Hadean

Archean

Complex life appears.

Proterozoic

Carboniferous

Cambrian

Permian

Ordovician

Devonian | Silurian

Shells appear.

Age of Dinosaurs

Jurassic

Triassic

Cretaceous

Holocene

Pliocene

Pleistocene

Oligocene

Paleocene

Miocene

Eocene

Age of Mammals

it easy to learn. Instead, time divisions and their naming were established haphazardly in the years between 1760 and 1845, when geologists first began to refine their understanding of geologic history and fossil succession.

On what basis did geologists define the boundaries between divisions in the geologic column? Divisions in the Phanerozoic generally indicate times when the assemblage of organisms living on the Earth changed significantly (see Interlude E). The time between these events varies, so divisions in the geologic column all have different durations. In other words, no two eras have the same duration, no two periods have the same duration, and no two epochs have the same duration. When they first developed the geologic column, geologists didn't know that all the divisions they were establishing had different durations, because the method for measuring the numerical ages of rocks wasn't developed until the 1950s.

Life Evolution in the Context of the Geologic Column

The succession of fossils preserved in strata of the geologic column defines the course of life's evolution throughout Earth history **(Fig. 12.16)**. Bacteria and Archaea appeared during the Archean Eon, but complex shell-less invertebrates did not evolve until the late Proterozoic.

The occurrence of beds containing trace fossils of a burrowing organism defines the base of the Cambrian. Invertebrates with shells appeared soon after. During the Cambrian, there was a sudden diversification of life, with many new families appearing over a relatively short interval—this event is called the **Cambrian explosion (Fig. 12.17)**.

Progressively more complex organisms populated the Earth during the Paleozoic. For example, the first fish swam in Ordovician seas, land plants started to spread

cells, of bacteria and archaea, actually appeared during the Archean Eon. Eons, in turn, can be subdivided into **eras**. The Phanerozoic Eon, for example includes, in order from oldest to youngest, the Paleozoic (ancient life), Mesozoic (middle life), and Cenozoic (recent life) Eras. We further divide each era into **periods** and each period into **epochs**.

Where do the names of the periods come from? They refer either to localities where a fairly complete stratigraphic column representing that time interval was first identified (e.g., rocks representing the Devonian Period crop out near Devon, England) or to a characteristic of the time (rocks from the Carboniferous Period contain a lot of coal). The terminology was not set up in a planned fashion that would make

FIGURE 12.17 The Cambrian explosion. The diversity of genera increased dramatically at about 530 Ma.

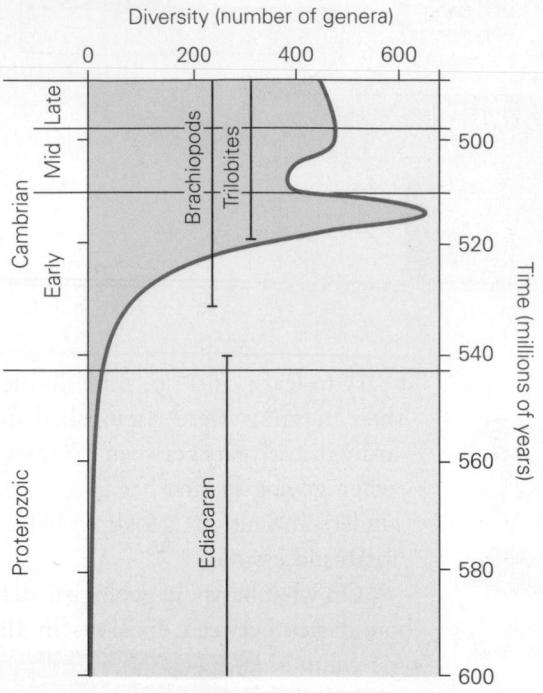

over the continents during the Silurian, and amphibians appeared during the Devonian. (Note that prior to the Silurian, the land surface was unvegetated and would have been a stark, dusty desert, even at the equator.) Though reptiles appeared during the Pennsylvanian Period, the first dinosaurs did not stomp across the land until the Triassic, at the start of the Mesozoic Era. Dinosaurs continued to inhabit the Earth until their sudden extinction at the end of the Cretaceous Period. For this reason, geologists refer to the Mesozoic as the *Age of Dinosaurs*. Small mammals appeared during the Triassic Period, but the diversification (development of many different species) of mammals to fill a wide range of ecological niches did not happen until the beginning of the Cenozoic Era, so geologists call the Cenozoic the *Age of Mammals*. Birds also appeared during the Mesozoic—specifically, at the beginning of the Cretaceous Period—but underwent great diversification in the Cenozoic Era.

Using the Geologic Column for Regional Correlation

To conclude our discussion of the geologic column, let's see how it comes into play when correlating strata across a region. Let's return to the Colorado Plateau of Arizona and Utah, in the southwestern United States **(Fig. 12.18)**. Because of the relative lack of vegetation in this region today, you can easily see bedrock exposures on the walls of cliffs and canyons; some of these locales are so beautiful that they have become national parks. The oldest sedimentary rocks of the region crop out at the base of the Grand Canyon, whereas the youngest form the cliffs of Cedar Breaks and Bryce Canyon. Walking through these parks takes you on a stroll through time—each rock layer gives an indication of the climate and topography of the region in the past (**Geology at a Glance**, pp. 468–469). For example, when the Precambrian metamorphic and igneous rocks exposed in the inner gorge of the Grand Canyon first formed, the region was a high mountain range, perhaps as dramatic as the Himalayas today. When the fossiliferous beds of the Kaibab Limestone at the rim of the canyon accumulated, the region was a Bahama-like carbonate reef and platform, bathed in a warm, shallow sea. And when the rocks making up the towering red cliffs of sandstone in Zion Canyon were deposited, the region was a Sahara-like desert, blanketed with huge sand dunes.

TAKE-HOME MESSAGE

Correlation of stratigraphic sequences from around the world led to the production of a chart, the geologic column, that represents the entirety of Earth history. The column, developed using only relative-age relationships, is subdivided into eons, eras, periods, and epochs.

QUICK QUESTION: What feature of living organisms appeared at the Precambrian/Cambrian boundary?

FIGURE 12.18 Correlation of strata among the national parks of Arizona and Utah.

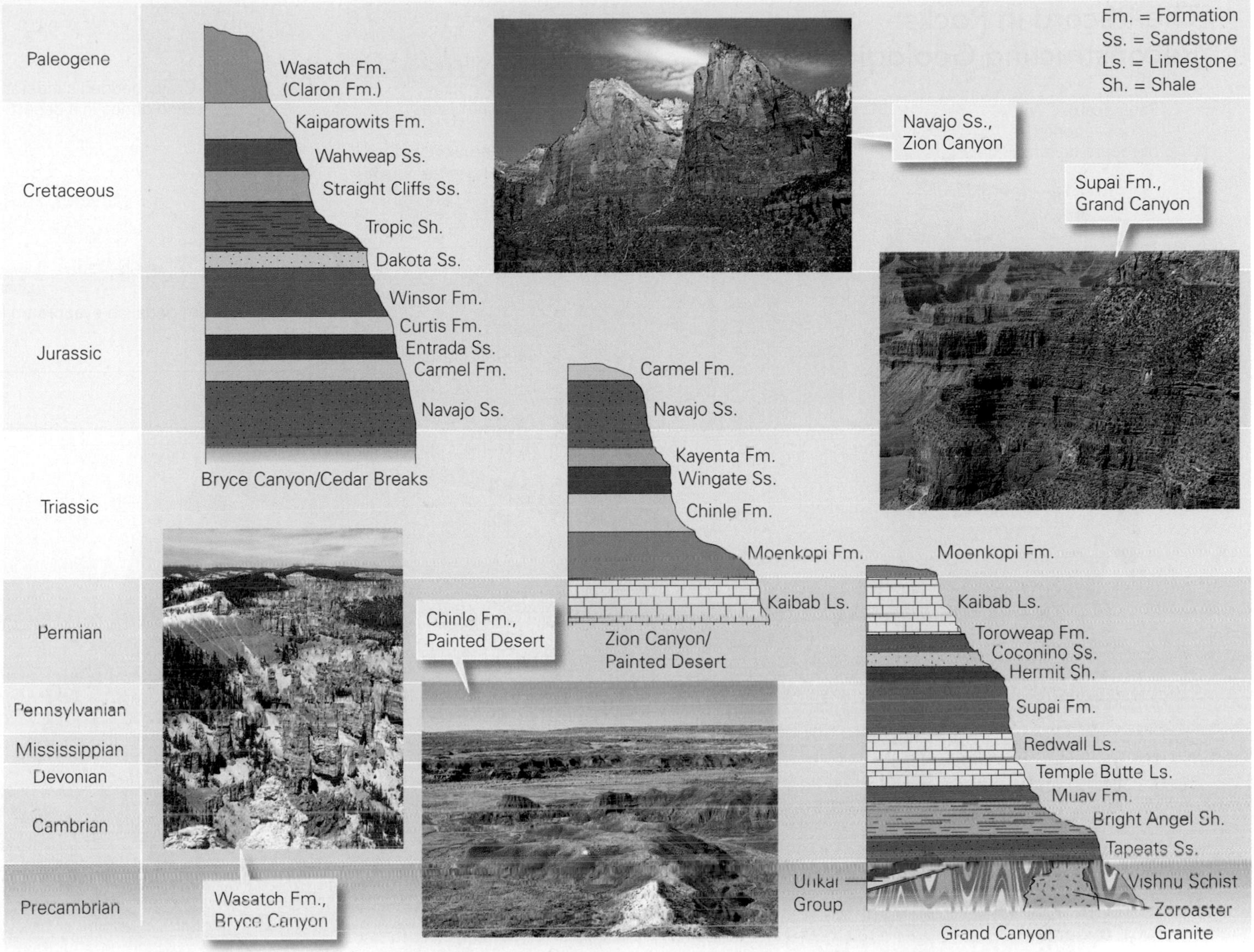

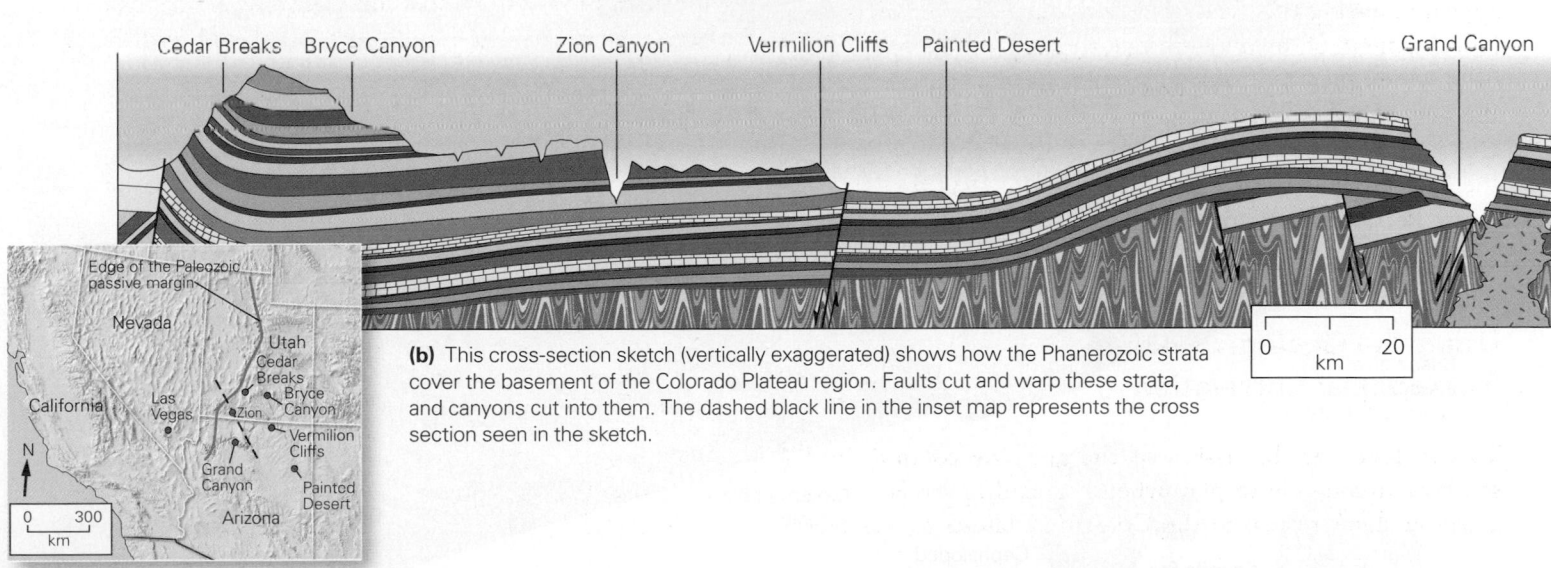

Fm. = Formation
Ss. = Sandstone
Ls. = Limestone
Sh. = Shale

Paleogene

Wasatch Fm.
(Claron Fm.)

Kaiparowits Fm.

Wahweap Ss.

Straight Cliffs Ss.

Cretaceous

Tropic Sh.

Dakota Ss.

Navajo Ss.,
Zion Canyon

Supai Fm.,
Grand Canyon

Winsor Fm.

Curtis Fm.
Entrada Ss.
Carmel Fm.

Jurassic

Carmel Fm.

Navajo Ss.

Navajo Ss.

Kayenta Fm.
Wingate Ss.

Chinle Fm.

Triassic

Bryce Canyon/Cedar Breaks

Moenkopi Fm.

Moenkopi Fm.

Kaibab Ls.

Kaibab Ls.

Chinle Fm.,
Painted Desert

Zion Canyon/
Painted Desert

Toroweap Fm.
Coconino Ss.
Hermit Sh.

Permian

Pennsylvanian

Supai Fm.

Mississippian

Redwall Ls.

Devonian

Temple Butte Ls.

Cambrian

Muav Fm.
Bright Angel Sh.

Tapeats Ss.

Precambrian

Wasatch Fm.,
Bryce Canyon

Unkar
Group

Grand Canyon

Vishnu Schist

Zoroaster
Granite

(a) Different intervals of geologic time are represented by the strata of different parks.

Cedar Breaks Bryce Canyon   Zion Canyon   Vermilion Cliffs   Painted Desert   Grand Canyon

Edge of the Paleozoic
passive margin

Nevada

Utah

Cedar
Breaks
Bryce
Canyon

California

Las
Vegas

Zion

Vermilion
Cliffs

Grand
Canyon

Painted
Desert

Arizona

N

0   300
km

(b) This cross-section sketch (vertically exaggerated) shows how the Phanerozoic strata cover the basement of the Colorado Plateau region. Faults cut and warp these strata, and canyons cut into them. The dashed black line in the inset map represents the cross section seen in the sketch.

0   20
km

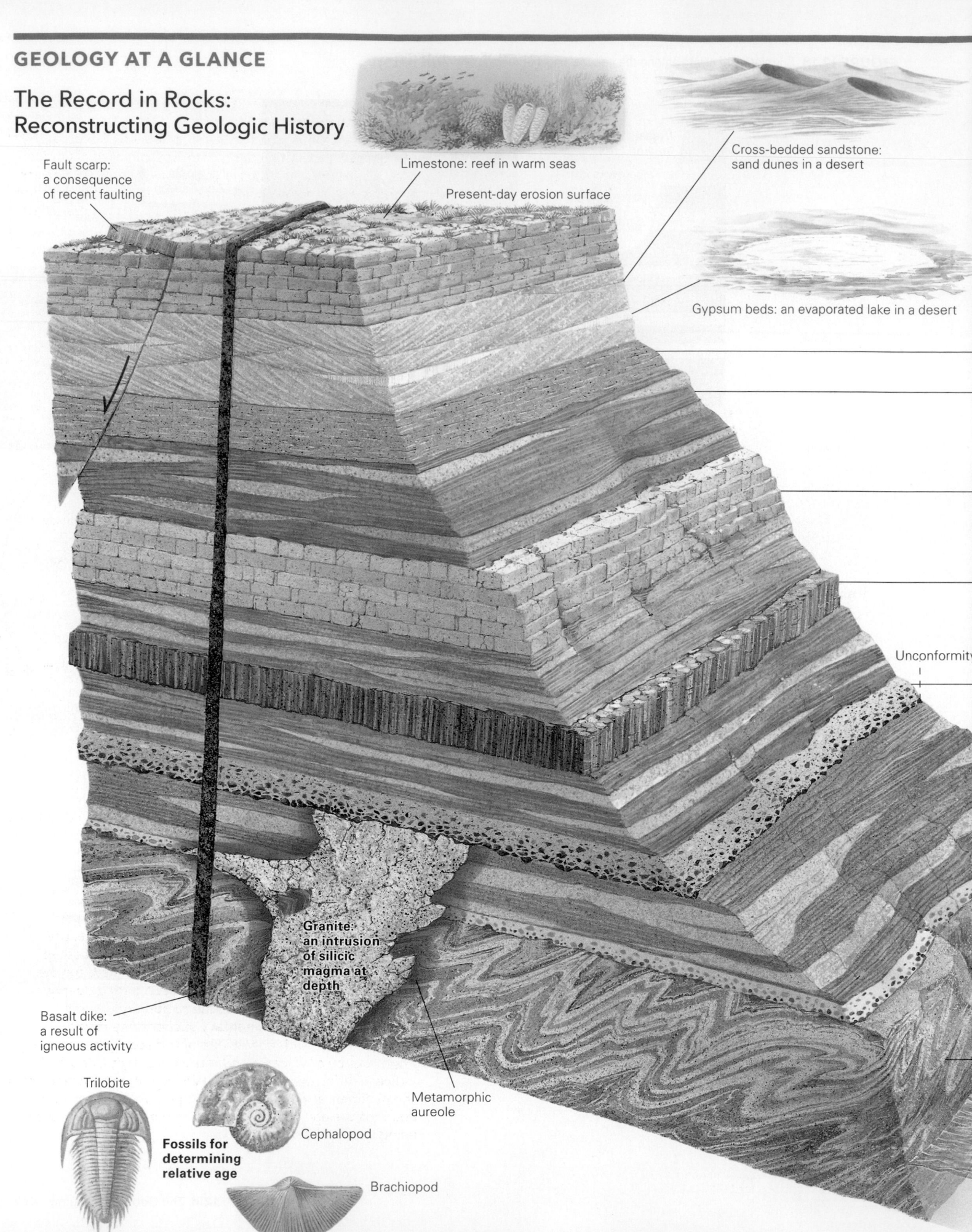

GEOLOGY AT A GLANCE

The Record in Rocks: Reconstructing Geologic History

Fault scarp:
a consequence
of recent faulting

Limestone: reef in warm seas

Present-day erosion surface

Cross-bedded sandstone:
sand dunes in a desert

Gypsum beds: an evaporated lake in a desert

Unconformity

Granite:
an intrusion
of silicic
magma at
depth

Basalt dike:
a result of
igneous activity

Metamorphic
aureole

Trilobite

Cephalopod

**Fossils for
determining
relative age**

Brachiopod

Ignimbrite (welded tuff): an explosive volcanic eruption

When geologists examine a sequence of rocks exposed on a cliff, they see a record of Earth history that can be interpreted by applying the basic principles of geology, by searching for fossils, and by using isotopic dating. On this cliff, we see evidence for many geologic events. The layers of sediment (and the sedimentary structures they contain), the igneous intrusions, and the geologic structures tell us about past climates and past tectonic activity.

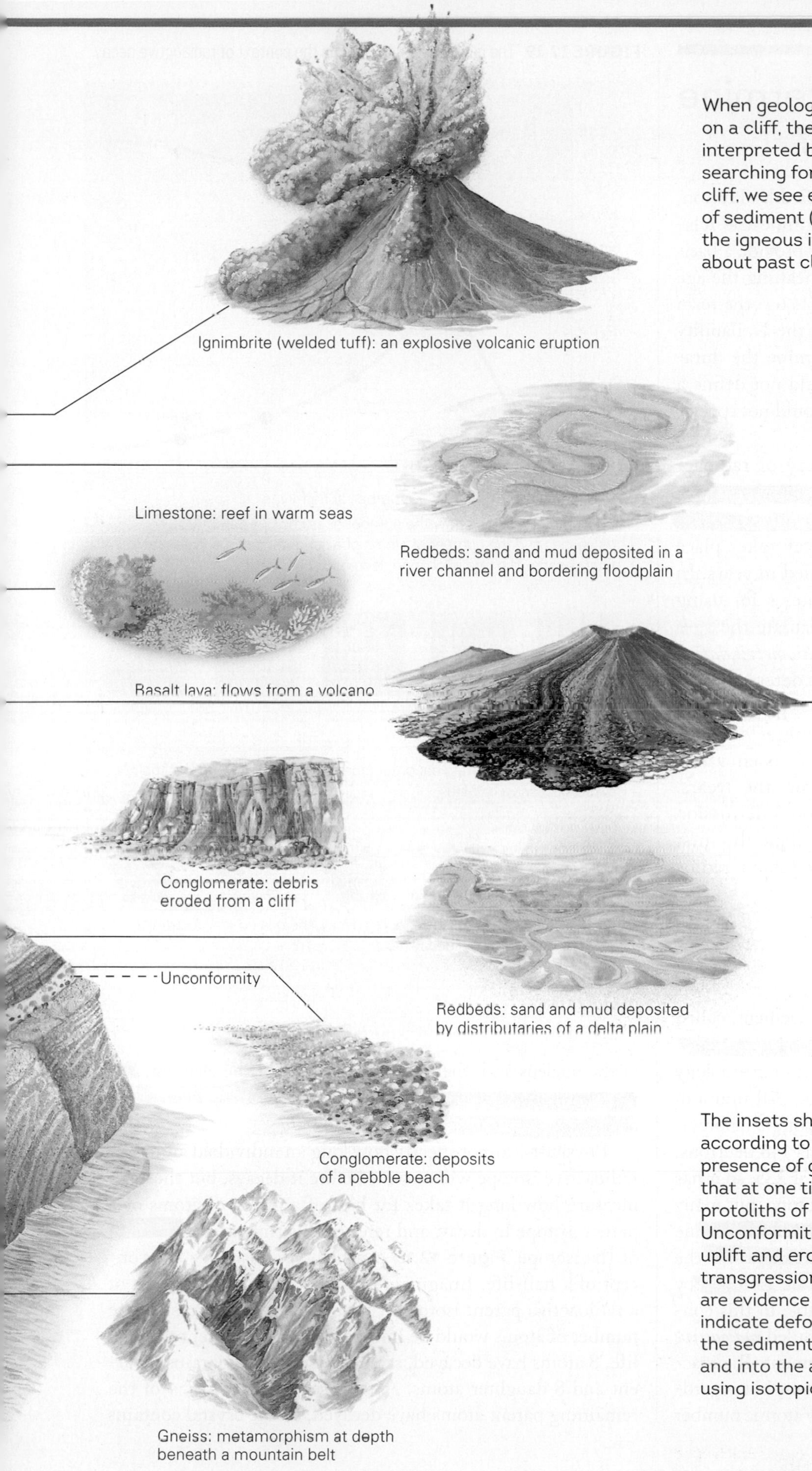

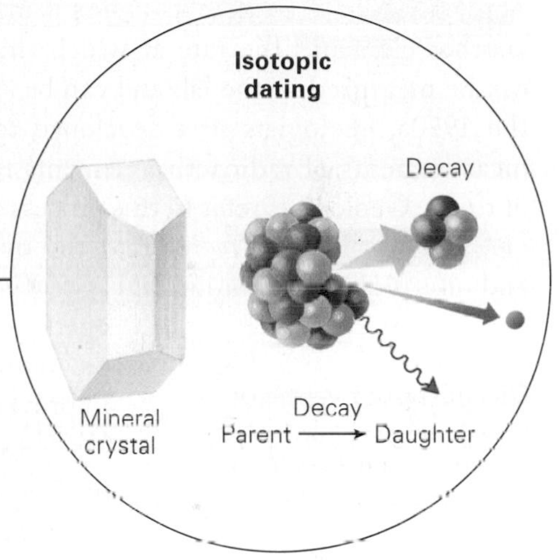

Limestone: reef in warm seas

Redbeds: sand and mud deposited in a river channel and bordering floodplain

Basalt lava: flows from a volcano

Isotopic dating

Decay

Mineral crystal

Decay

Parent ⟶ Daughter

Conglomerate: debris eroded from a cliff

- - - Unconformity

Redbeds: sand and mud deposited by distributaries of a delta plain

Conglomerate: deposits of a pebble beach

The insets show the way the region looked in the past according to the record in the rocks. For example, the presence of gneiss at the base of the canyon indicates that at one time the region was a mountain belt, for the protoliths of the gneiss must have been buried deeply. Unconformities indicate that the region underwent uplift and erosion. Sedimentary successions record transgressions and regressions of the sea, igneous rocks are evidence of volcanic and intrusive activity, and faults indicate deformation. We can gain insight into the ages of the sedimentary rocks by studying the fossils they contain, and into the age of the igneous and metamorphic rocks by using isotopic dating methods.

Gneiss: metamorphism at depth beneath a mountain belt

12.7 How Do We Determine Numerical Ages?

As we have seen, the geologic principles codified by Hutton, Steno, and others allowed geologists to determine the relative ages of rocks. But these principles did not provide a basis for determining the *numerical age* of a rock, meaning the age of the rock in years. (Note that some geologists use the term *absolute age* instead of numerical age.) Without the availability of numerical ages, geologists could not determine the durations of divisions in the geologic column, could not define a timeline of events in the Earth's history, and could not specify the age of our planet.

This situation changed with the discovery of radioactivity. Simply put, a **radioactive element** is one that spontaneously decays so that its atoms transform into atoms of another element. The rate at which this decay takes place can be measured in the lab and can be specified in years. In the 1950s, geologists first developed techniques for using measurements of radioactive elements to calculate the ages of rocks. Geologists refer to this process as *radiometric dating* or **isotopic dating**. We refer to the overall determination and interpretation of numerical ages of rocks as **geochronology**. The technique of isotopic dating has been vastly improved over the years. We begin our discussion of this technique by first learning more about radioactive decay.

> **Did you ever wonder ...**
> how geologists can specify the age of some rocks in years?

Radioactive Decay

As we've noted earlier, different versions of an element, called **isotopes** of the element, have the same atomic number but a different atomic weight (see Box 1.3). To see how this terminology applies, let's consider two isotopes of uranium. All uranium atoms have 92 protons, but the uranium-238 isotope (abbreviated ^{238}U) has an atomic weight of 238, so it has 146 neutrons, whereas the ^{235}U isotope has an atomic weight of 235, so it has 143 neutrons. Some elements have only one isotope, but many elements have two or more isotopes. Of these isotopes, some are stable, meaning that they will survive unchanged for the life of the Universe, but some are not. The unstable isotopes, by definition, are called *radioactive isotopes*. Unstable in this context means that the isotopes undergo a change called *radioactive decay*, which converts them into a different element. Radioactive decay can take place by a variety of reactions, but regardless of the details, all these reactions change the atomic number

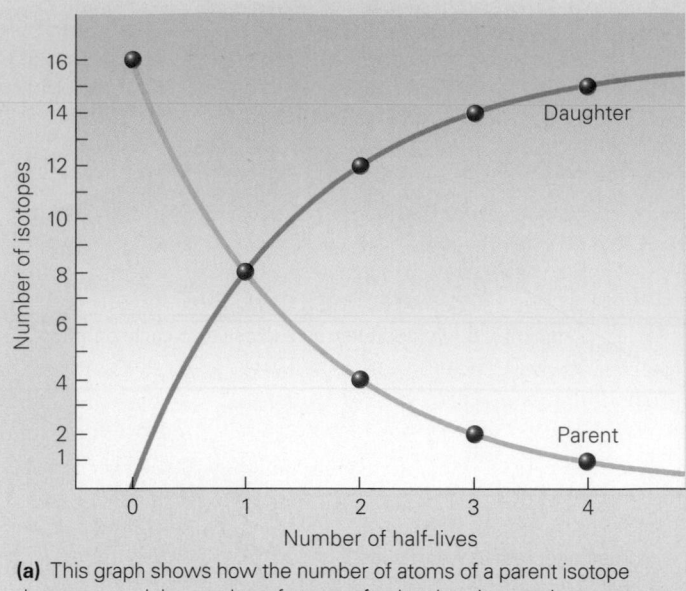

(a) This graph shows how the number of atoms of a parent isotope decreases and the number of atoms of a daughter isotope increases as time passes. The rate of change decreases over time.

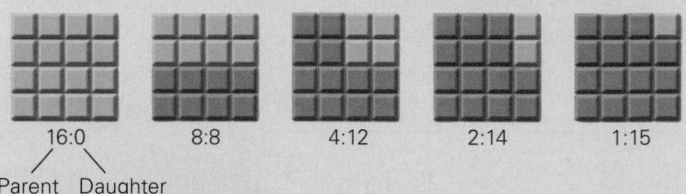

Parent Daughter

(b) The ratio of parent to daughter atoms changes with the passage of each successive half-life.

(c) In a cluster of atoms undergoing decay, there is no way to predict which parent atom will decay next.

of the nucleus and, therefore, the identity of the element. We refer to the isotope that undergoes decay as the *parent isotope* and the decay product as the *daughter isotope*.

Physicists cannot specify how long an individual atom of a radioactive isotope will survive before it decays, but they can measure how long it takes for half of a group of atoms of a parent isotope to decay, and refer to this time as the **half-life** of the isotope. **Figure 12.19** can help you visualize the concept of a half-life. Imagine a crystal containing 16 atoms of a radioactive parent isotope. (Note that in a real crystal, the number of atoms would be immensely larger.) After one half-life, 8 atoms have decayed, so the crystal now contains 8 parent and 8 daughter atoms. After a second half-life, 4 of the remaining parent atoms have decayed, so the crystal contains

BOX 12.2 CONSIDER THIS . . .

Carbon-14 Dating

Many people who have heard of carbon-14 (^{14}C) dating assume that it can be used to define the numerical ages of rocks. But this is not the case. Rather, ^{14}C dating tells us the ages of organic materials—such as wood, cotton fibers, charcoal, flesh, bones, and shells—that contain carbon originally extracted from the atmosphere by photosynthesis. ^{14}C, a radioactive isotope of carbon, forms naturally in the atmosphere when cosmic rays (charged particles from space) bombard atmospheric nitrogen-14 (^{14}N) atoms. When plants incorporate carbon dioxide during photosynthesis, or when animals consume plants, they ingest a tiny amount of ^{14}C along with ^{12}C, the more common isotope of carbon. After an organism dies and can no longer exchange carbon with the atmosphere, the ^{14}C in its body begins to decay back to ^{14}N. As a result, the ratio of ^{14}C to ^{12}C changes at a rate determined by the half-life of ^{14}C.

We can use ^{14}C dating to determine the age of prehistoric fire pits or of organic debris in sediment. Carbon-14 has a short half-life—only 5,730 years. So the method cannot be used to date anything older than about 70,000 years, for after that time essentially no ^{14}C remains in the material. But this range makes it a useful tool for geologists studying sediments of the last ice age and for archaeologists studying ancient cultures or prehistoric peoples. Again, since rocks do not contain organic carbon, and may be significantly older than 70,000 years, we cannot determine the age of rocks by using the ^{14}C dating method.

---

4 parent and 12 daughter atoms. And after a third half-life, 2 more parent atoms have decayed, so the crystal contains 2 parent and 14 daughter atoms. For a given decay reaction, the half-life is a constant, measured in years.

Isotopic Dating Technique

Since radioactive decay proceeds at a known rate, like the ticktock of a clock, it provides a basis for telling time. In other words, because an isotope's half-life stays constant, we can calculate the age of a mineral by measuring the ratio of parent to daughter atoms in the mineral. How do geologists actually obtain an isotopic date? First, we must find an appropriate radioactive isotope to work with. Only a few have long enough half-lives and occur in sufficient abundance in minerals to be useful for isotopic dating—we list some of these isotopes in **Table 12.1**. Each radioactive isotope has its own half-life. Note that carbon dating can't be used for dating rocks; the method only works for dating organic material that is less than 70 Ka **(Box 12.2)**. Second, we must identify the right kind of minerals to work with. Not all minerals contain radioactive isotopes, but fortunately, some common minerals do.

Once geologists have settled on which mineral to study, they use the following steps to obtain a date:

- *Collecting the rocks:* Geologists collect unweathered rocks for dating, for the chemical reactions that happen during weathering may remove parent or daughter atoms. If, for example, weathering allows daughter atoms to escape, then the isotopic clock becomes inaccurate.

- *Separating the minerals:* Once a good sample of fresh rock has been collected, geologists crush the sample and separate the appropriate minerals from the debris.

TABLE 12.1 Isotopes Used in the Isotopic Dating of Rocks

Parent → Daughter	Half-Life (years)	Minerals Containing the Isotopes
^{147}Sm → ^{143}Nd	106 billion	Garnets, micas
^{87}Rb → ^{87}Sr	48.8 billion	Potassium-bearing minerals (mica, feldspar, hornblende)
^{238}U → ^{206}Pb	4.5 billion	Uranium-bearing minerals (zircon)
^{40}K → ^{40}Ar	1.3 billion	Potassium-bearing minerals (mica, feldspar, hornblende)
^{235}U → ^{207}Pb	713 million	Uranium-bearing minerals (zircon)

Sm = samarium, Nd = neodymium, Rb = rubidium, Sr = strontium, U = uranium, Pb = lead, K = potassium, Ar = argon.

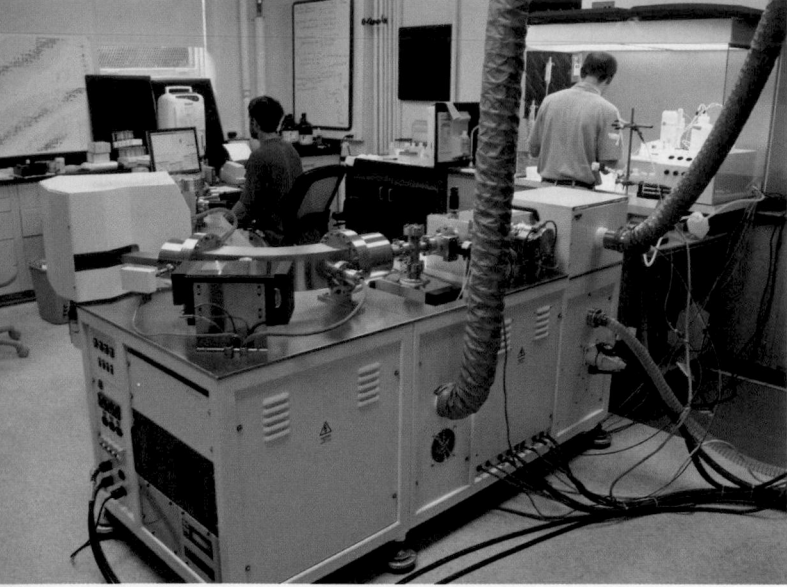

FIGURE 12.20 In an isotopic dating laboratory, samples are analyzed using a mass spectrometer. This instrument measures the ratio of parent to daughter isotopes.

- *Extracting parent and daughter isotopes:* To separate out the parent and daughter atoms from minerals, geologists use several techniques, including dissolving the minerals in acid or evaporating portions of them with a laser. This step must take place in a *clean lab*, a special facility with ultra-filtered air and water, to avoid contaminating the samples with stray atoms of a parent or daughter isotope.

- *Analyzing the parent-to-daughter ratio*: Geologists pass the atoms through a *mass spectrometer*, a sophisticated instrument that uses a strong magnet to separate isotopes from one another according to their respective weights **(Fig. 12.20)**. The instrument counts the number of atoms of specific isotopes separately.

At the end of the laboratory process, geologists can define the ratio of parent to daughter atoms in a mineral and from this ratio determine the age of the mineral. Needless to say, the description of the procedure here has been simplified—in reality, obtaining an isotopic date requires complex calculations and can be time-consuming and expensive.

What Does an Isotopic Date Mean?

At high temperatures, atoms in a crystal lattice vibrate so rapidly that chemical bonds break and reattach relatively easily. As a consequence, atoms escape from or move into crystals continually, so parent-to-daughter ratios do not represent the age of the mineral. Because isotopic dating is based on the parent-to-daughter ratio, the isotopic clock starts only when crystals become cool enough for atoms of both parent and daughter isotopes to be locked into the lattice. The temperature below which these atoms can no longer move freely is called the closure temperature of a mineral. When we specify an isotopic date for a rock, we are defining the time at which a specific mineral in the rock cooled below its closure temperature.

With the concept of closure temperature in mind, we can interpret the meaning of isotopic dates. In the case of igneous rocks, isotopic dating of minerals with relatively high closure temperatures (e.g., >650°C) tells you when a magma or lava cooled to form a solid, cool igneous rock. In the case of metamorphic rocks, isotopic dating of minerals with relatively high closure temperatures tells you when a rock cooled from a metamorphic temperature (e.g., >450°C) above the closure temperature to a temperature below it. Not all minerals have the same closure temperature, so different minerals in a rock that cools very slowly will yield different dates.

In recent years, geologists have started to date rocks using isotopic systems that have very low closure temperatures (as low as 75°C). This method, called *thermochronology*, constrains the time of *exhumation*, meaning the time at which uplift and erosion led to the return of a rock that was once, say, 20 km below the surface back to within a few kilometers of the surface.

Can we isotopically date a clastic sedimentary rock directly? No. If we date minerals in a sedimentary rock, we're determining only when those minerals first crystallized as part of an igneous or metamorphic rock, not the time when the minerals were deposited as sediment, nor the time when the sediment lithified to form a sedimentary rock. For example, if we date the feldspar grains contained within a granite pebble in a conglomerate, we're dating the time the granite cooled below feldspar's closure temperature, not the time the pebble was deposited by a stream.

Knowing the ages of mineral grains in sediment, however, can be useful. In recent years, geologists have used *detrital geochronology*, determination of the ages of detrital (clastic) grains, to learn the ages of the rocks in the sediment's source region and, from this information, to determine where the sediment came from. Such ages also put an upper limit on the age of the sedimentary rock, for according to the principle of inclusions, the sedimentary rock must be younger than the grains it contains.

Other Methods of Determining Numerical Age

Geologists have developed a variety of specialized methods for obtaining numerical ages of natural materials when isotopic measurements can't be used. Let's consider a few examples.

Counting Changes of Season The changes in seasons affect a wide variety of phenomena, including the following:

- *The growth rate of trees:* Trees grow seasonally, with different growth rates in different seasons. These variations produce alternating dense (dark) and less-dense (light) bands in wood, for growth rate affects cell size, which in turn affects wood density.

- *The organic productivity of lakes and seas:* During the winter, less light reaches the Earth, water temperatures decrease, and the supply of nutrients decreases. Therefore, more organic material will be deposited in summer than in winter. So alternations between organic-rich and organic-poor sediment layers indicate the seasons.

- *The growth rate of chemically precipitated sedimentary rocks:* Factors that control precipitation rates, such as the rate of groundwater flow and temperature, change seasonally in some locations. Such changes may form distinct bands in chemical sedimentary rocks such as travertine.

- *The growth rate of shell-secreting organisms:* Organisms grow and produce new shell material on a seasonal basis. Seasonal variations can produce ridges in the shells of mollusks and alternating dark and light bands in the heads of coral colonies.

- *Ice bands in glaciers:* During the snowy season, more snow falls relative to dust than during the dry season, so annual snowfall includes a dusty layer and a clean layer. These contrasts are preserved as visible banding in ice.

Note that because of seasonal changes in growth or deposition rates and styles, **growth bands** develop in trees, travertine deposits, and shelly organisms, and **rhythmic layering** develops in sedimentary accumulations and glacier ice **(Fig. 12.21a–c)**. By counting bands or layers, geologists can determine how long an organism survived and how long a sedimentary accumulation took to form. If rings or layers have developed right up to the present, we can count backward and determine how long ago they began to form. *Dendrochronologists*, scientists who study and date **tree rings**, the growth bands in trees, have found that such rings not only count time, but also preserve a record of the changing climate, for during warm, rainy years, trees grow faster than they do during drought years. In fact, climate variations over time yield distinctive patterns of tree rings, much like the bar codes on products in supermarkets. By correlating the older rings of a still-living tree with the youngest rings in a log from a dead tree, it's possible to extend the record of tree rings back before living trees started to grow. For example, the oldest living trees, bristlecone pines, are almost 4,000 years old. By correlating the older rings of these trees with rings in preserved logs, dendrochronologists have extended the tree-ring record back for many thousands of years more **(Fig. 12.21d)**.

Glacier ice also preserves a valuable record of past climate, for the ratio of different oxygen isotopes in the water molecules making up the ice reflects the global temperature at the time the snow fell to create the ice, as we'll discuss further in Chapter 22. Ice cores drilled through the thick Greenland ice cap contain a continuous record of climate in polar latitudes over the past 750,000 years. Geologists are racing to sample the record of past climate recorded in glaciers on mountain peaks in temperate and equatorial regions before those glaciers melt away entirely.

Magnetostratigraphy

As we discussed in Chapter 3, the polarity of the Earth's magnetic field flips every now and then over geologic time. Geologists have determined when the reversals took place and have constructed a reference column showing the succession of reversals over time **(Fig. 12.22)**. By comparing the pattern of the reversals in a sequence of strata with the pattern of reversals in the reference column—a method known as *magnetostratigraphy*—geologists can determine the age of the sequence.

Fission-Track Dating

In certain minerals, the ejection of an atomic particle during the decay of a radioactive isotope damages the nearby crystal, producing a line called a **fission track.** This track resembles the line of crushed grass left behind when you ride a bike across a lawn. As time passes, more atoms undergo fission, so the number of fission tracks in the crystal increases at a rate that can be calculated **(Fig. 12.23)**. If the crystal warms above a certain temperature, however, the tracks heal and disappear. Therefore, the number of fission tracks in a given volume of a crystal represents the time since the crystal was last warm enough for tracks to heal. Fission-track healing takes place at relatively low temperatures—for example, it takes place in the mineral apatite at 110°C. Therefore, geologists use fission-track dating primarily to determine the timing of exhumation.

TAKE-HOME MESSAGE

Isotopic dating specifies numerical ages in years. To obtain an isotopic date, we measure the ratio of parent radioactive atoms to stable daughter atoms in a mineral. An isotopic date for a mineral gives the time at which the mineral cooled below its closure temperature. Using this method, we can date the time an igneous rock solidifies, a metamorphic rock cools, or a body of rock undergoes exhumation. We cannot isotopically date sedimentary rock directly. Counting growth bands or seasonal layers can constrain ages of younger materials.

QUICK QUESTION: How can we date changes in climate during the past few hundred thousand years?

FIGURE 12.21 Using rhythmic layering and growth rings as a basis for dating.

(a) Dust settling during the summer highlights boundaries between snow layers, now turned to ice in this Oregon glacier.

(b) The ridges in this giant clamshell from the Red Sea formed during successive seasons of growth.

(c) Each ring in this slice of wood represents the growth of one year. The width of each ring depends on the temperature and rainfall during its growth. Patterns of rings are like bar codes—the pattern in a living tree can be correlated with the pattern in a dead tree, which can be correlated with the pattern in an ancient, buried tree. Such correlation extends the tree-ring record back in time.

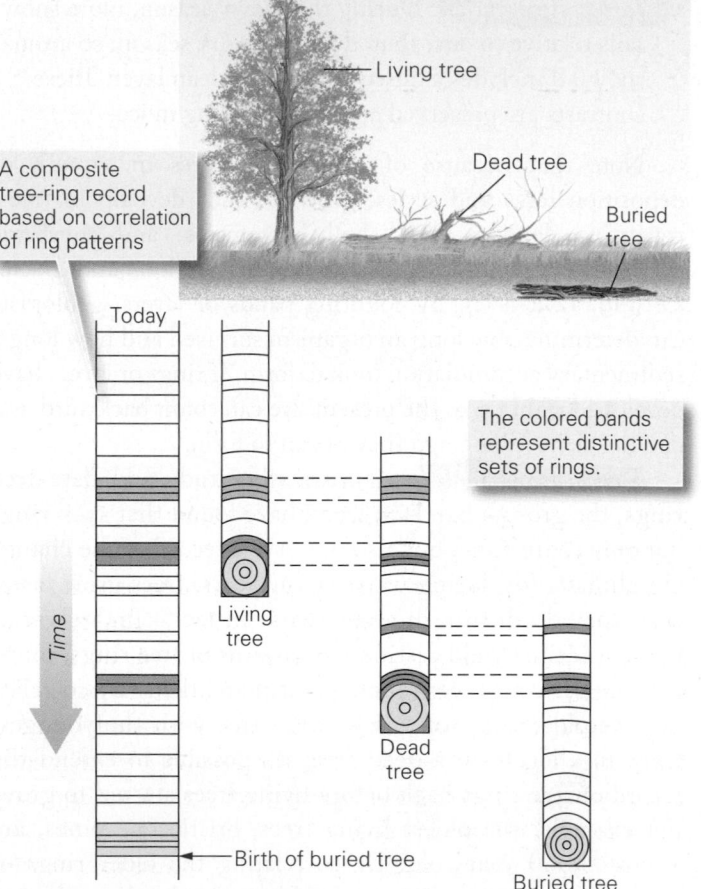

(d) Dendrochronology is based on the correlation of tree rings. Each of the columns in the diagram represents a core drilled out of a tree. Distinctive clusters of closely spaced rings indicate dry seasons. By correlation, researchers can extend the climate record back in time before the oldest living tree started to grow.

FIGURE 12.22 Magnetostratigraphy involves comparing the sequence of polarity reversals in strata with the sequence of polarity reversals in a global reference column to determine the age of the strata.

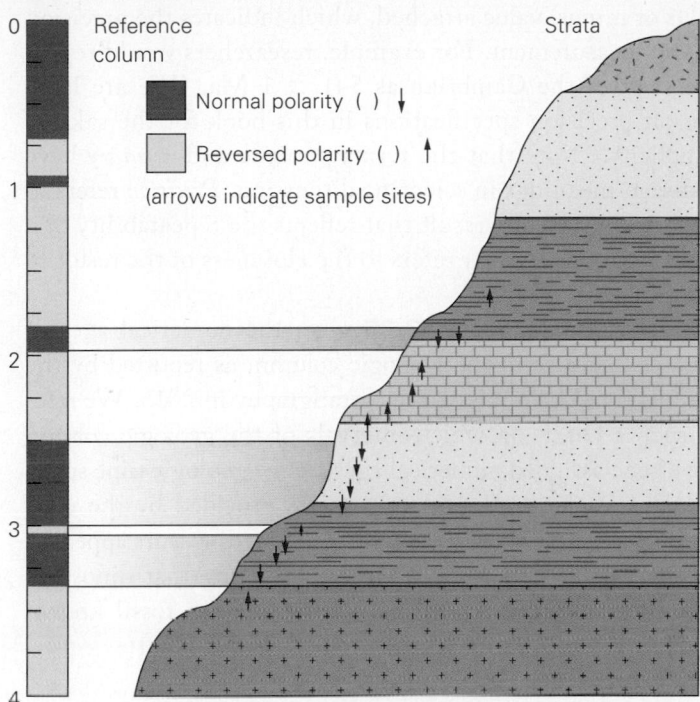

FIGURE 12.23 The fission-track method of dating.

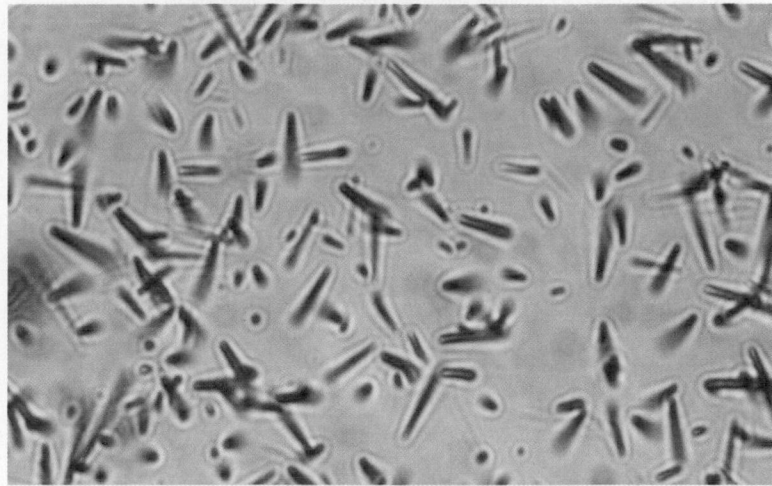

(a) This high-magnification photomicrograph shows fission tracks in a crystal. When formed, each track is only a few atoms wide. Treating the sample with acid enlarges the tracks so they are visible.

Time

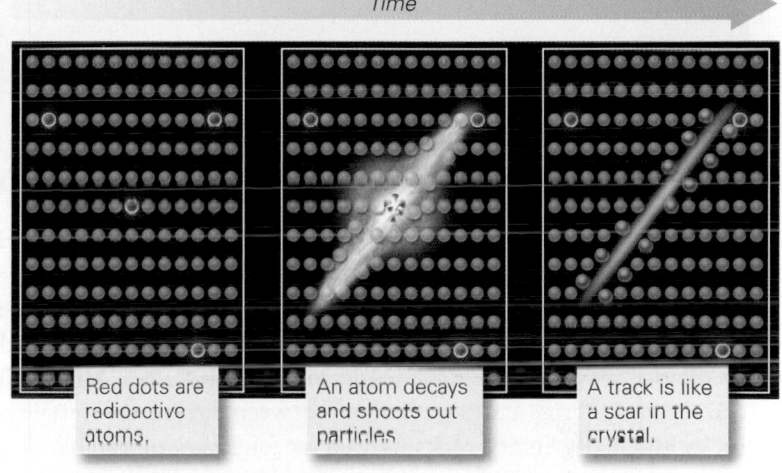

| Red dots are radioactive atoms. | An atom decays and shoots out particles | A track is like a scar in the crystal. |

(b) A fission track forms when a radioactive atom decays and blasts out particles that disrupt the crystal structure in a narrow band.

12.8 Numerical Ages and Geologic Time

Dating Sedimentary Rocks

The mind grows giddy gazing so far back into the abyss of time.
—John Playfair (British geologist, 1747–1819)

We have seen that isotopic dating can be used to date the time when igneous rocks formed and when rocks metamorphosed, but not when sedimentary rocks were deposited. So how do we determine the numerical age of a sedimentary rock? We must answer this question if we want to add numerical ages to the geologic column—remember, the column was originally constructed by studying only the relative ages of sedimentary rocks and did not specify dates.

Geologists obtain dates for sedimentary rocks by studying cross-cutting relationships between sedimentary rocks and datable igneous or metamorphic rocks. For example, if we find a sequence of sedimentary strata deposited unconformably on a datable granite, the strata must be younger than the granite. If a datable basalt dike cuts the strata, the strata must be older than the dike. And if datable volcanic ash buried the strata, then the strata must be older than the ash.

Let's see how cross-cutting relations can help us constrain the numerical age of an imaginary bed of sandstone that contains fossilized dinosaur bones **(Fig. 12.24)**. Geologists assign the beds to the Cretaceous Period by correlating the bones with fossils in known Cretaceous strata from elsewhere. Field study shows that the sandstone was deposited unconformably over an eroded pluton consisting of granite that has a numerical age, determined by uranium-lead (U-Pb) dating, of 125 Ma. Further study shows that a basalt dike whose numerical age, based on potassium-argon (K-Ar) dating, is 80 Ma, cuts across the bed. These measurements mean that this Cretaceous sandstone bed was deposited sometime between 125 Ma and 80 Ma. Note that the data provide

FIGURE 12.24 The Cretaceous sandstone bed was deposited on the granite, so it must be younger than 125 Ma. The dike cuts the bed, so the bed must be older than 80 Ma. Thus, the Cretaceous bed was deposited between 125 and 80 Ma. The Paleocene sandstone was unconformably deposited over the dike and lies beneath a 50-million-year-old layer of ash. Therefore, it must have been deposited between 80 and 50 Ma.

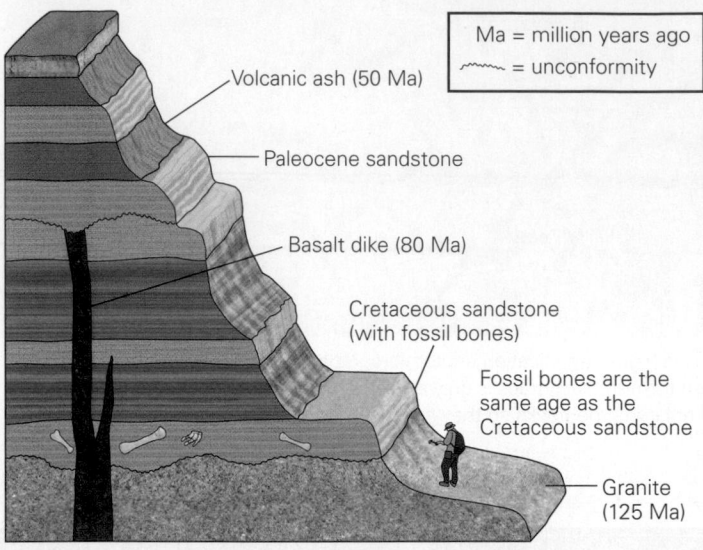

Ma = million years ago
~ = unconformity

Volcanic ash (50 Ma)

Paleocene sandstone

Basalt dike (80 Ma)

Cretaceous sandstone
(with fossil bones)

Fossil bones are the
same age as the
Cretaceous sandstone

Granite
(125 Ma)

an age range, not an exact age. Therefore, from this observation we can conclude only that the Cretaceous Period includes, but is not limited to, the time interval of 125 to 80 Ma.

The Geologic Time Scale

Geologists have searched the world for localities where they can recognize cross-cutting relations between datable igneous rocks and sedimentary rocks as well as for layers of datable volcanic rocks interbedded with sedimentary rocks. By isotopically dating the igneous rocks, they have provided numerical ages for the boundaries between all geologic periods. For example, a compilation of many studies from around the world has led to a refinement of the numerical ages assigned to the Cretaceous Period—currently, the beginning of the Cretaceous has been placed at 145 Ma and the end at 66 Ma. (With this information in hand, we can say that the bed in Figure 12.24 was deposited during the middle part of the Cretaceous, not at the beginning or end.)

Because numerical ages depend on the documentation of datable cross-cutting relations, discovery of new data can require changes in the numerical ages assigned to period boundaries. For example, around 1995, new dates on rhyolite ash layers above and below the Cambrian/Precambrian boundary showed that this boundary occurred closer to an age of 542 Ma than to 570 Ma, where previous, less definitive studies had placed it. More recently, that boundary has been

moved to 541 Ma. Furthermore, because repeated measurements of the same sample may yield slightly different results, a given numerical age is not perfectly precise. To be technically accurate, a reported age must have an uncertainty (plus or minus) value attached, which indicates the precision of the measurement. For example, researchers would report the start of the Cambrian as 541 ± 1 Ma. (We are leaving off precision specifications in this book for the sake of simplicity.) Note that the terms *precision* and *accuracy* have different meanings in scientific discussion. *Precision* refers to the uncertainty of a result that reflects the repeatability of a measurement. *Accuracy* refers to the closeness of the result to the real or true value.

The chart in **Figure 12.25** gives the numerical ages of periods and eras in the geologic column, as reported by the International Commission on Stratigraphy in 2013. We refer to such a chart, on which intervals of the geologic column have been assigned numerical ages, as the **geologic time scale**. Because of the numerical constraints provided by the geologic time scale, when we say that the first dinosaurs appeared during the Triassic Period, we are implying that dinosaurs appeared after 252 Ma. (The oldest dinosaur fossil known actually comes from strata that are 240 to 245 Ma.)

What Is the Age of the Earth?

During the 18th and 19th centuries, before the discovery of isotopic dating, scientists came up with a great variety of clever solutions to the question, "How old is the Earth?" All have since been proved wrong. Lord Kelvin, a 19th-century physicist renowned for his discoveries in thermodynamics, made the most influential scientific estimate of the Earth's age of his time. Kelvin calculated how long it would take for the Earth to cool down from a temperature as hot as the Sun's and concluded (in 1862) that our planet is about 20 million years old.

Kelvin's estimate contrasted with those being promoted by followers of Hutton, Lyell, and Darwin, who argued that if the concepts of uniformitarianism and evolution were correct, the Earth must be much older. They held that the physical processes that shape the Earth and form its rocks, as well as the process of natural selection that yields the diversity of species, all take a very long time. Geologists and physicists continued to debate the age issue for many years. The route to a solution appeared in 1896, when Henri Becquerel announced the discovery of radioactivity. Geologists immediately realized that the Earth's interior was producing some heat from the decay of radioactive material. This realization uncovered the key flaw in Kelvin's argument: Kelvin had assumed that no new heat was produced after the Earth first formed. Because radioactivity constantly generates new heat

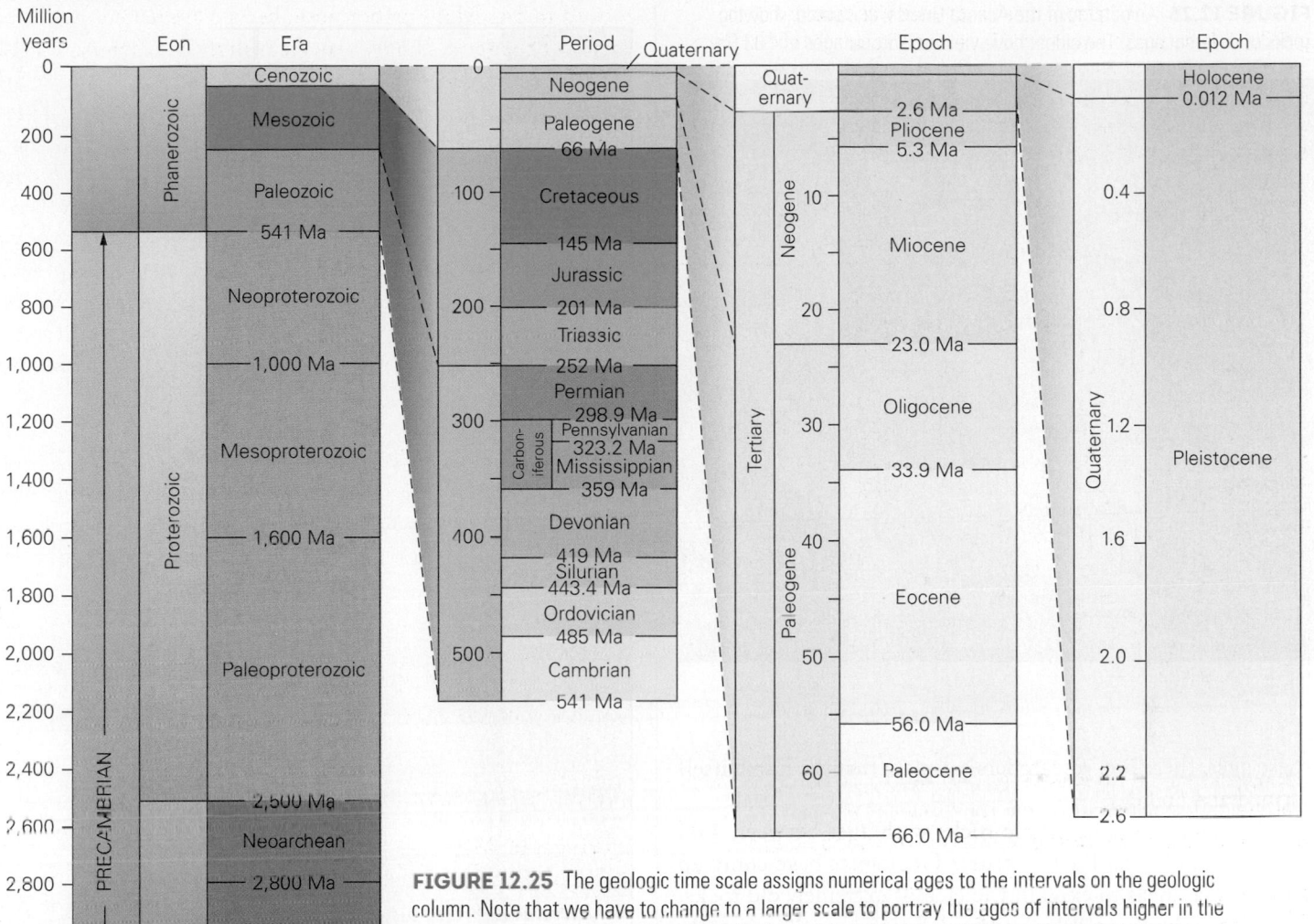

FIGURE 12.25 The geologic time scale assigns numerical ages to the intervals on the geologic column. Note that we have to change to a larger scale to portray the ages of intervals higher in the column, because these are shorter subdivisions. This time scale utilizes numbers favored by the International Commission on Stratigraphy.

Since the 1950s, geologists have scoured the planet to identify its oldest rocks. Samples from several localities (Wyoming, Canada, Greenland, and China) have yielded dates as old as 4.03 Ga **(Fig. 12.26)**. Individual clastic grains of the mineral zircon have yielded dates of up to 4.4 Ga, indicating that by 4.4 Ga solid crust existed, at least for a while. Isotopic dating of Moon rocks yields dates of up to 4.5 Ga, and dating of meteorites thought to reflect the most primitive solids of the Solar System has yielded ages as old as 4.57 Ga. Geologists consider these meteorites to be fragments of very early, undifferentiated planetesimals. Meteorites from the oldest differentiated planetesimals (ones that had separated into a core and mantle) are slightly younger, and their ages are taken to be the same as the age of the Earth itself. Based on

Did you ever wonder...
how old the Earth's oldest rock is?

in the Earth, the planet has cooled down much more slowly and could, therefore, be much older than Kelvin had calculated. The discovery of radioactivity not only invalidated Kelvin's estimate of the Earth's age, but also led to the development of isotopic dating.

FIGURE 12.26 An outcrop of the Acasta Gneiss, in Canada, showing rocks of different ages. The oldest have yielded isotopic ages of 4.03 Ga.

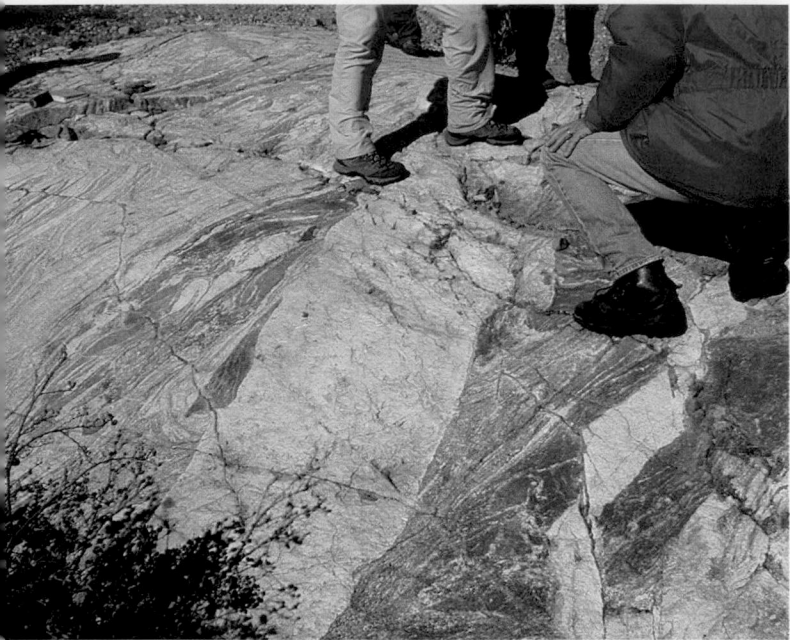

these ages, therefore, researchers estimate that the Earth itself formed at 4.56 Ga.

Why don't we find whole rocks with ages between 4.03 and 4.56 Ga in the Earth's crust? Geologists have come up with several ideas to explain the lack of extremely old rocks. One idea comes from calculations defining how the temperature of our planet's interior has changed over time. These calculations indicate that during the first half-billion years of its existence, the Earth might have been so hot that rocks in the crust remained above the closure temperature for minerals, so isotopic clocks could not start ticking. More recent studies, looking at isotope ratios in the oldest (4.4-Ga) zircons, suggest that the Earth had cooled sufficiently to host oceans of water within only a couple of hundred million years of its formation. So an alternative view is that intense bombardment of the Earth by meteorites just prior to 4.03 Ga destroyed or remelted any crust that existed and vaporized the earliest oceans. As noted earlier, geologists have named the time interval between the birth of the Earth and the approximate age of the oldest isotopically dated rock the Hadean Eon, to emphasize that conditions at the surface, at times, resembled literary images of Hades.

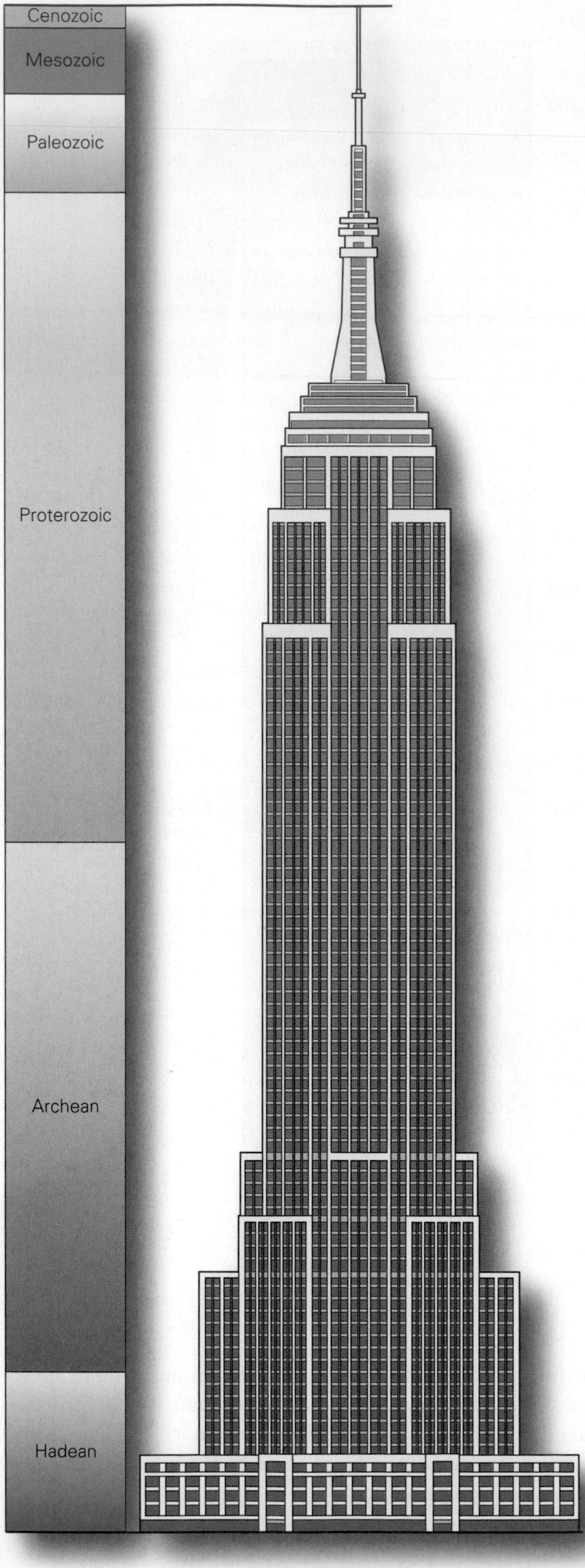
Cenozoic
Mesozoic
Paleozoic
Proterozoic
Archean
Hadean

FIGURE 12.27 We can use the analogy of distance (or height) to represent the duration of geologic time. Here we compare the durations of intervals of the geologic time scale to the height of the Empire State Building.

Picturing Geologic Time

How can we picture the relative durations of intervals on the geologic time scale? Some intervals last only several thousand years, while others last for millions or even billions of years. It helps to visualize these durations by comparing them either to distances along a familiar object or to durations of a familiar time interval.

Let's first compare the relative durations of geologic time intervals to the height of the Empire State Building, an iconic skyscraper in New York City. This building has 102 floors and a tall antenna, the top of which towers 443.2 m above the street. On average, each floor of the building has a height of 3.7 m. At this scale, 1 million years equals 9.7 cm, so the Hadean = 14.7 floors; the Archean = 39.0 floors; the Proterozoic = 51.1 floors; the Paleozoic = 7.5 floors; the Mesozoic = 4.8 floors; and the Cenozoic = 1.7 floors **(Fig. 12.27)**. Archaeological evidence suggests that the earliest *Homo sapiens* appeared about 150,000 years ago, a time equivalent to a height of only 1.5 cm. Clearly, Mark Twain had the right idea (see the chapter opening quote).

Another way to grasp the immensity of geologic time is to equate the entire 4.56 billion years to a single calendar year. On this scale, the oldest rocks preserved on the Earth date from early February, the first archaea appear in the sea on February 21, the first shelly invertebrates burrow through the mud on October 25, and the first amphibians crawl onto the land about November 20. On December 7, the continents coalesce into the supercontinent of Pangaea. The first mammals and birds appear about December 15, along with the dinosaurs, and the Age of Dinosaurs ends on December 25. The last week of December represents the last 66 million years of Earth history, including the entire Age of Mammals. The first humanlike ancestor appears on December 31 at 3 P.M., and our species, *Homo sapiens*, shows up an hour before midnight. The last ice age ends a minute before midnight, and all of recorded human history takes place in the last 30 seconds. To put it another way, human history occupies the last 0.0001% of Earth history. The Earth is so old that there has been more than enough time for the rocks and mountains to form and disappear and form again, and for life forms of the Earth to evolve.

TAKE-HOME MESSAGE

Numerical dates for sedimentary rocks come from isotopic dating of cross-cutting datable rocks. Such work led to the geologic time scale, which assigns dates to time intervals. The oldest rock of the Earth's crust is about 4.0 Ga. Dating of meteorites indicates that the Earth is 4.56 Ga.

QUICK QUESTION: Why don't all periods on the geologic time scale have the same duration in years?

ANOTHER VIEW The view from an airplane window of the Colorado Plateau region in Arizona and Utah reveals layer upon layer of strata deposited long ago and now exposed by erosion. These layers preserve a record of the Earth's very long history.

Chapter 12 Review

SUMMARY

- Geologic time refers to the time span since the Earth's formation.

- Relative age specifies whether one geologic feature is older or younger than another; numerical age provides the age of a geologic feature in years.

- Using such principles as uniformitarianism, original horizontality, superposition, and cross-cutting relations, we can construct the geologic history of a region.

- The principle of fossil succession states that the assemblage of fossils in strata changes from base to top of a sequence. Once a species becomes extinct, it never reappears.

- Strata are not necessarily deposited continuously at a location. An interval of nondeposition and/or erosion is called an unconformity. Geologists recognize three kinds: angular unconformities, nonconformities, and disconformities.

- A stratigraphic column shows the succession of strata in a region. Correlation allows geologists to determine the relationship between strata at one location and strata at another.

- A given succession of strata that can be traced over a fairly broad region is called a stratigraphic formation. A geologic map shows the distribution of formations.

- A composite chart called the geologic column represents the entirety of geologic time. The column's largest subdivisions, each of which represents a specific interval of time, are eons. Eons are subdivided into eras, eras into periods, and periods into epochs.

- The numerical ages of rocks can be determined by isotopic (radiometric) dating, because radioactive elements decay at a rate characterized by a known half-life.

- The isotopic date of a mineral specifies the time at which the mineral cooled below its closure temperature. We can use isotopic dating to determine when an igneous rock solidified and when a metamorphic rock cooled from high temperatures.

- To date sedimentary strata, we must examine cross-cutting relations among dated igneous or metamorphic rock and the strata.

- Other methods for dating materials include counting growth rings in trees and seasonal layers in glaciers.

- From the isotopic dating of meteorites and Moon rocks, geologists conclude that the Earth formed about 4.56 billion years ago.

GUIDE TERMS

angular unconformity (p. 458)

Cambrian explosion (p. 465)

closure temperature (p. 472)

correlation (p. 460)

disconformity (p. 459)

eon (p. 464)

epoch (p. 465)

era (p. 465)

fission track (p. 473)

fossil (p. 450)

fossil assemblage (p. 455)

fossil correlation (p. 460)

geochronology (p. 470)

geologic column (p. 464)

geologic contact (p. 459)

geologic map (p. 463)

geologic time (p. 449)

geologic time scale (p. 476)

growth band (p. 473)

half-life (p. 470)

index fossil (p. 456)

isotope (p. 470)

isotopic dating (p. 470)

marker bed (p. 460)

nonconformity (p. 458)

numerical age (p. 451)

period (p. 465)

Precambrian (p. 464)

radioactive element (p. 470)

relative age (p. 451)

rhythmic layering (p. 473)

stratigraphic column (p. 459)

stratigraphic formation (p. 459)

tree ring (p. 473)

unconformity (p. 458)

uniformitarianism (p. 450)

 **GEOTOURS** *THIS CHAPTER'S GEOTOURS WORKSHEET (J) FEATURES QUESTIONS AND GOOGLE EARTH SITES ON:*

- Relative-age dating and unconformities

- Stratigraphic formations in southern Utah

- Rock layers and monoclines, Circle Cliffs, Utah

REVIEW QUESTIONS

The letters following each Review Question refer to the corresponding Learning Objective from the Chapter Opener.

1. Contrast numerical age with relative age. **(A)**

2. Describe the principles that allow us to determine the relative ages of geologic events. **(B)**

3. How does the principle of fossil succession help determine relative ages? **(B)**

4. How does an unconformity develop? Distinguish among the three kinds of unconformities. Which type is pictured here? **(C)**

5. What is a stratigraphic formation? Describe two different methods of correlating formations. How was correlation used to develop the geologic column? **(D)**

6. What happens during radioactive decay and what does a half-life indicate? **(E)**

7. How do geologists obtain an isotopic date? What does the age of an igneous rock mean? What does the age of a metamorphic rock mean? **(E)**

8. Why can't we date sedimentary rocks directly? How do we assign numerical ages to intervals on the geologic column to produce a geologic time scale? **(F)**

9. How are growth rings and ice layering useful in determining the ages of geologic events? **(F)**

10. What is the age of the oldest rocks on the Earth? What is the current estimate of the numerical age of the Earth? Why is there a difference? **(F)**

ON FURTHER THOUGHT

11. Examine the photograph and the "What a Geologist Sees" interpretation of an outcrop in eastern New York State below. Write a brief geologic history that explains the relationships displayed in this outcrop. The strata directly above the unconformity are Late Silurian (Rondout Formation), whereas the strata below the unconformity are Middle Ordovician (Austin Glen Formation). **(D)**

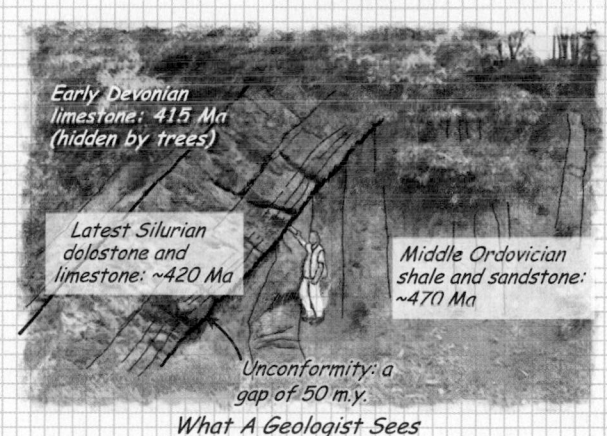

Early Devonian limestone: 415 Ma (hidden by trees)

Latest Silurian dolostone and limestone: ~420 Ma

Middle Ordovician shale and sandstone: ~470 Ma

Unconformity: a gap of 50 m.y.

What A Geologist Sees

ONLINE RESOURCES

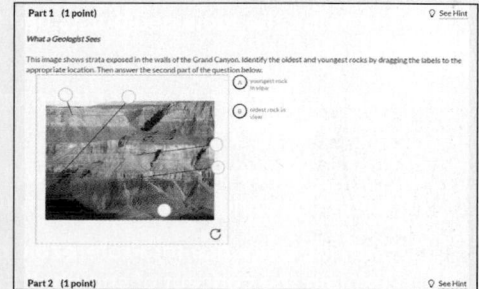

Animations
This chapter features a series of animations demonstrating relative-age dating and the different types of geologic unconformities.

Videos
This chapter features a video on a petrified forest and the clues it reveals to geologic history.

Smartwork5
This chapter covers the principles of defining relative age, unconformities, and identifying relative ages.

CHAPTER 13

A Biography of the Earth

By the end of this chapter, you should be able to . . .

A. describe the tools geologists use to study the Earth's history.

B. outline changes that our planet underwent during its first 2 billion years.

C. correlate key geologic events with the tectonic conditions and settings in which they formed.

D. provide evidence for Proterozoic glaciations, and explain the concept of snowball Earth.

E. outline major stages in life evolution, and describe how the land surface changed as life evolved.

F. identify supercontinents and mountain-building events (orogenies), and tie their occurrence to the geologic time scale.

G. place the Age of Dinosaurs in the context of Earth history, and indicate the favored hypothesis for why it came to an end.

> The man who should know the true history of the bit of chalk which every carpenter carries about in his breeches pocket, though ignorant of all other history, is likely, if he will think his knowledge out to its ultimate results, to have a truer and therefore a better conception of this wonderful universe and of man's relation to it than the most learned student who [has] deep-read the records of humanity [but is] ignorant of those of nature.
>
> —Thomas Henry Huxley (British biologist, 1825–1895) from *On a Piece of Chalk* (1868)

13.1 Introduction

In 1868, a well-known British scientist, Thomas Henry Huxley, presented a public lecture on geology to an audience in Norwich, England. Seeking a way to convey his fascination with Earth history to people with no geologic background, he focused his audience's attention on the piece of chalk he had been writing with (see the epigraph above). And what a tale the chalk has to tell! Chalk, a type of limestone, consists of microscopic marine algae shells and shrimp feces. The specific chalk that Huxley held came from beds deposited in Cretaceous time (the name *Cretaceous*, in fact, derives from the Latin word for chalk). These beds now form the white cliffs bordering the southeastern shore of England (Fig. 13.1). Geologists in Huxley's day knew of similar chalk beds in outcrops throughout much of Europe and had discovered that the chalk contains not only plankton shells, but also fossils of bizarre swimming reptiles, fish, and invertebrates—species absent in the seas of today. Clearly, when the chalk was deposited, warm seas holding unfamiliar creatures covered some of what is dry land today.

Clues in his humble piece of chalk allowed Huxley to demonstrate to his audience that the landscape and inhabitants of our planet in the past differed markedly from those of today, implying that the Earth has a history. Geologic research over the past few centuries has led to the conclusion that physical and biological components of the Earth System have interacted pervasively during this long history, in ways that have transformed a formerly barren, crater-pocked surface into countless environments and landscapes supporting a diversity of life.

In this chapter, we offer a concise geologic biography of our planet, from its birth 4.56 billion years ago to the present. We illustrate how continents came into existence and have waltzed across the globe ever since. We also describe mountain-building

◀ (facing page) The Pancake Rocks along the coast of New Zealand consist of thin layers of limestone deposited in the sea during the Oligocene, about 30 million years ago. Each layer represents a snapshot of Earth history, preserved for us to study today.

events, changes in the Earth's climate and sea level over time, and the evolution of life. To simplify our discussion, we use the following abbreviations: Ga for billion years ago, Ma for million years ago, and Ka for thousand years ago.

13.2 Methods for Studying the Past

When historians outline human history, they describe daily life, wars, economies, governments, leaders, inventions, and explorations. When geologists outline Earth history, they describe changes in depositional environments, mountain-building events (orogenics), past climates, the rise and fall of sea level, life evolution, past configurations of plate boundaries and continents, and changes in the composition of the atmosphere and oceans. Historians collect data by reading written accounts, examining relics and monuments, and for more recent events, by listening to recordings or watching videos. Geologists collect data by examining rocks, geologic structures, and fossils, and for more recent events, by studying sediments, ice cores, and tree rings.

Figuring out the Earth's past hasn't been an easy task for geologists. The record that we can see isn't complete because the materials that hold the record of the past don't form continuously through time and because many important clues have been eroded away or are covered by younger rocks. Futhermore, it can be very challenging to obtain and interpret accurate isotopic ages of old rocks. Nevertheless, enough of the record exists to outline major geologic events of the past. Geologists use several types of clues to study Earth history:

- *Ancient orogens:* We identify present-day orogens (mountain belts) by finding regions of high, rugged peaks. However, since it takes as little as 50 million years to erode a mountain range entirely away, we cannot identify orogens of the past simply by studying topography. Rather, we identify them based on the record that orogeny leaves behind. Orogeny causes igneous activity, deformation, and metamorphism. Therefore, a belt of crust containing these features represents an ancient orogen (Fig. 13.2). We can determine the age of an orogen

FIGURE 13.1 Horizontal chalk beds exposed along the coast of southeastern England, formed from layers of deep-sea sediment. The thin, dark beds between white chalk layers consist of chert. The chalk erodes easily, so the pebbles on the beach all consist of chert.

Bedding

Person

by isotopically dating the metamorphic and igneous rocks that crop out within it. Orogeny also leads to the development of unconformities, for uplift exposes rocks to erosion. And it leads to the formation of sedimentary basins in which the detritus produced by erosion of mountains accumulates. These basins, which border the orogen, develop because the weight of the orogen pushes the lithosphere's surface downward, forming a depression that traps sediment.

- *The growth of continents:* Not all continental crust formed at the same time. To learn how a continent grew, geologists determine the ages of different regions of the crust by using isotopic dating techniques. They can figure out not only when rocks of the region originally formed from magmas rising out of the mantle, but also when the rocks were metamorphosed during a subsequent orogeny. The identities of the rock types making up the crust indicate the tectonic environment in which the crust formed.

FIGURE 13.2 Evidence of mountain building in the past. Even after the topography of a mountain belt has eroded away, a record of mountain building remains; deformed and metamorphosed rock defines a distinct belt.

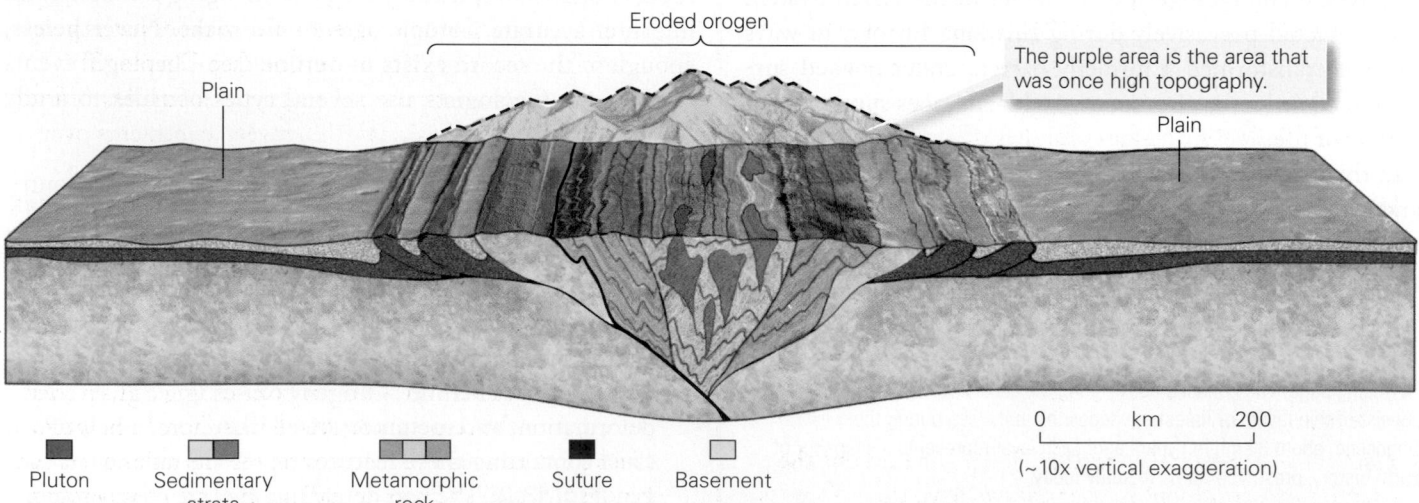

Eroded orogen

Plain

The purple area is the area that was once high topography.

Plain

Pluton Sedimentary strata Metamorphic rock Suture Basement

0   km   200

(~10x vertical exaggeration)

FIGURE 13.3 Paleomagnetic tools that geologists use to determine the past positions of the continents.

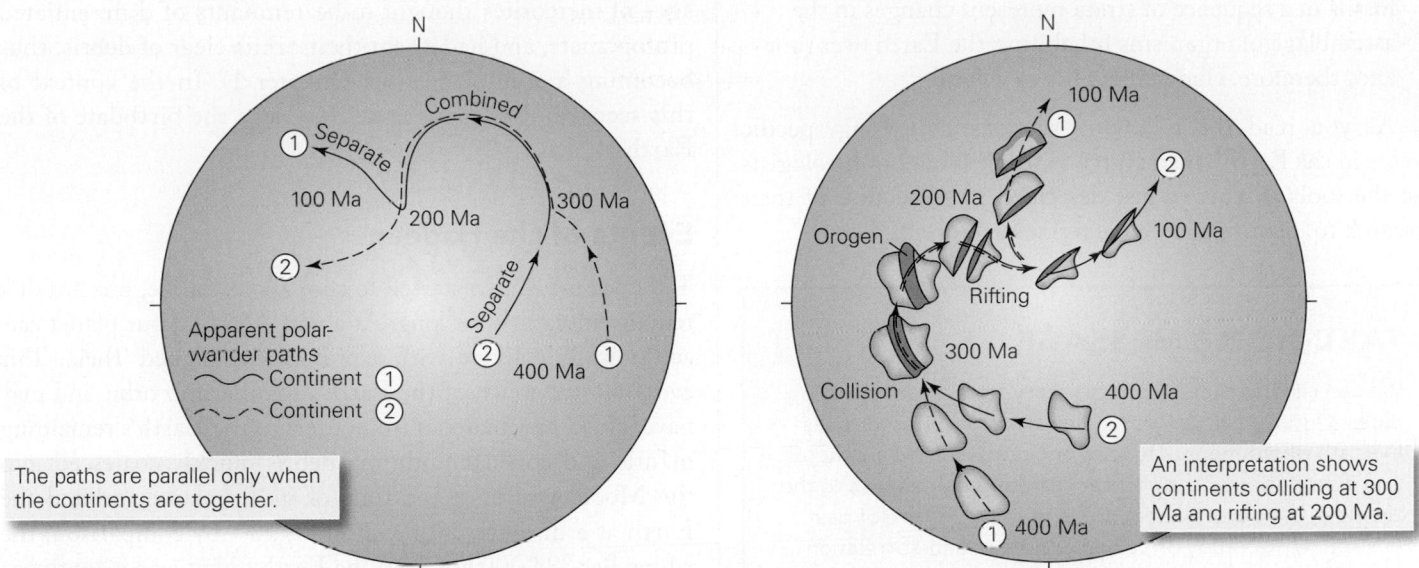

(a) Comparison of the apparent polar-wander paths for two continents indicates when they moved together and when they moved separately.

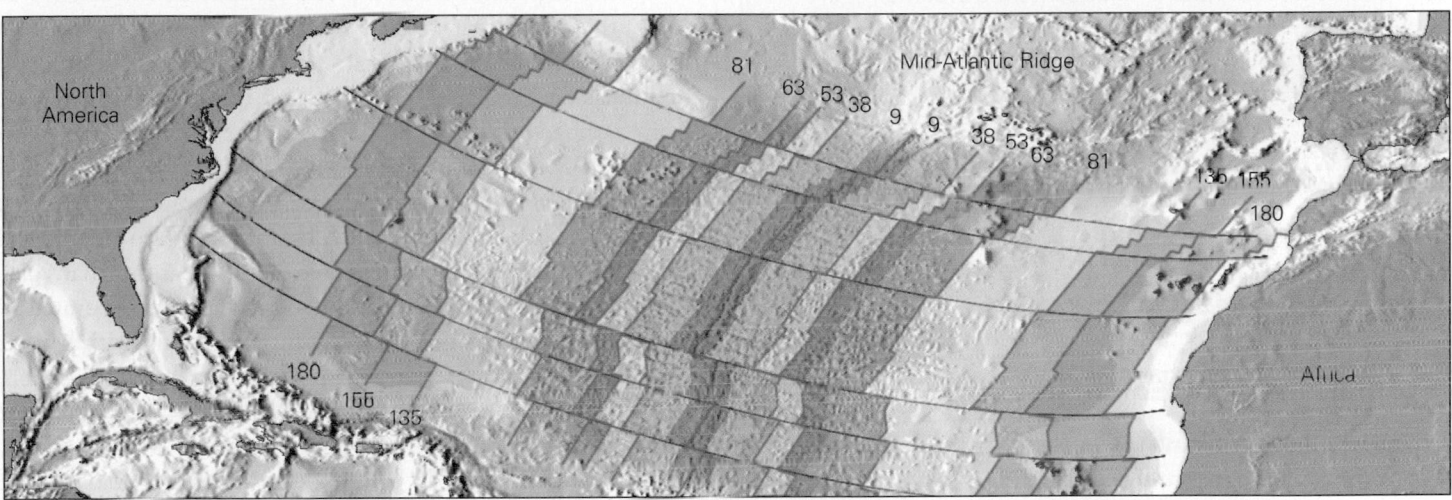

(b) Marine magnetic anomalies indicate how the distance between continents separated by a mid-ocean ridge changes over time. On this map, the ages of anomaly boundaries are indicated in millions of years.

• *Past depositional environments:* The environment at a particular location changes over time. To learn about these changes, we study successions of sedimentary rocks and the fossils they contain, for depositional environment controls both the type of sediment accumulating at a location and the types of organisms that live there.

• *Past changes in relative sea level:* We can determine when sea level has gone up or down relative to the land by looking for changes in the depositional environment. For example, a marine limestone above an alluvial fan conglomerate in a sedimentary sequence indicates a rise in relative sea level.

• *Positions of continents in the past:* To help us find out the position of a continent in the past, we have three sources

of information. First, apparent polar-wander paths give us a sense of continental movement (see Chapter 3) **(Fig. 13.3a)**. Second, marine magnetic anomalies define changes in ocean-basin width between continents over time **(Fig. 13.3b)**. Third, comparison of rocks and/or fossils from different continents permits correlations that indicate whether continents were adjacent at a given time.

• *Past climates:* We can gain insight into past climates by looking at fossils and rock types that formed at given latitudes. For example, if organisms now living in semi-tropical conditions lived near the poles during a given time period, then the atmosphere overall must have been warmer. Geologists have also learned how to use the ratios of certain isotopes in fossil shells as an indication of past temperatures.

- *Life evolution:* Progressive changes in the assemblage of fossils in a sequence of strata represent changes in the assemblage of organisms inhabiting the Earth over time and, therefore, characterize life's evolution.

As you read the following sections describing specific events in the Earth's history, try to think about how geologists use the tools that we've just described in the course of their research to identify and characterize each event.

TAKE-HOME MESSAGE

We can reconstruct geologic history from a variety of clues. Outcrops of deformed and metamorphosed rocks indicate where mountain belts once existed, the character and distribution of sedimentary strata provide clues to the timing of sea-level rise and fall and to the nature of past environments, study of paleomagnetism and correlation of rock units allows researchers to constrain the past positions of continents relative to one another, and the fossil record provides a window into the evolution of life.

QUICK QUESTION: How can you determine the numerical age of an ancient orogen?

13.3 The Hadean Eon and Before

Formation of the Earth, Revisited

The current scientific interpretation of the Solar System's formation places the beginning of the process at about 4.57 Ga, perhaps when a supernova explosion sent shock waves, as well as atoms of heavier elements, into a nebula within our region of the Milky Way Galaxy. Stimulated by the shock waves, this nebula collapsed rapidly to form the Sun, which ignited a few million years later.

According to the nebular theory of Solar System formation that we discussed in Chapter 1, during the next million years a protoplanetary disk formed, and dust and ice condensed in the disk. Based on the isotopic dating of meteorites thought to be remnants of the earliest planetesimals, the planetesimals soon began to grow, and less than 10 million years later, several had grown into sizable protoplanets by sweeping in material from their orbit. Once a protoplanet became large enough and hot enough, it underwent **differentiation**, during which iron within the body began to melt, and gravity pulled the iron into the body's center where it accumulated to form a core. Core formation left behind a mantle composed of ultramafic rock. By about 4.56 Ga, large

protoplanets had differentiated, as indicated by the isotopic ages of meteorites thought to be remnants of differentiated protoplanets, and had swept their orbits clear of debris, thus becoming a true planet (see Chapter 1). In the context of this scenario, geologists use 4.56 Ga as the birthdate of the Earth.

Events of the Hadean

The newborn Earth started to cool and stabilize, but it didn't remain unscathed for long. At about 4.53 Ga, our planet cataclysmically collided with a protoplanet named Theia. This event blasted much of the Earth's mantle into orbit and may have added new material from Theia to the Earth's remaining mantle and core. The orbiting debris quickly coalesced into the Moon, which, at the time of its formation, orbited the Earth at a distance of only 20,000 km—by comparison, the Moon lies 384,000 km from the Earth today, and it continues to move farther away at about 3.8 cm per year.

In the aftermath of the Moon's formation, the Earth was so hot that much of its surface was probably an ocean of seething magma **(Fig. 13.4a)**. But its surface temperature dropped rapidly as heat radiated into space and the supply of new heat from radioactive decay diminished (because elements with short half-lives quickly decayed). A skin of solid rock formed on the surface of the magma ocean, but much of this rock eventually sank and remelted to be recycled into new rocks. In addition, rapid outgassing took place, meaning that volatile (gassy) elements or compounds originally incorporated in mantle minerals escaped from the interior, along with lava, at volcanoes. These gases accumulated to form a toxic atmosphere consisting mostly of water (H_2O), methane (CH_4), ammonia (NH_3), hydrogen (H_2), nitrogen (N_2), carbon dioxide (CO_2), and sulfur dioxide (SO_2). Some researchers speculate that comets colliding with the Earth contributed additional gases to this early atmosphere.

By about 4.4 Ga, the Earth probably had become cool and stable enough for a solid crust and even liquid water to exist at its surface **(Fig. 13.4b)**. The evidence for this statement comes from analyzing grains of a durable mineral called zircon, extracted from sandstone beds exposed in Western Australia. This zircon, which yields isotopic ages of 4.4 Ga, formed originally in an igneous rock, indicating that the Earth was cool enough for rock to solidify at 4.4 Ga. Furthermore, the character of oxygen in the zircon indicates that the zircon underwent alteration by interaction with surface water, hinting at the presence of early oceans. What did the Earth's surface look like at this time? An observer from space probably would have found small, barren landmasses spotted with volcanoes poking up above pools of acidic water. But both land and sea would have been obscured by murky, dense (H_2O-, CO_2-, and SO_2-rich) air.

FIGURE 13.4 Visualizing the early Earth is a challenge, but by using geologic interpretations, artists have provided useful images.

(a) At a very early stage, during or after differentiation, the surface might have been largely molten. The loss of heat to space would have allowed patches of solid ultramafic crust to form. Meteorite impacts destroyed most of this crust.

(b) When Earth's surface temperature fell below the boiling point of water, water from the atmosphere rained onto the surface, submerging it beneath early oceans.

TAKE-HOME MESSAGE

The Earth formed not long after the birth of the Sun. It differentiated and swept its orbit clear of debris by about 4.56 Ga, a date taken as the birth of the Earth and the beginning of the Hadean. Soon after, the Moon formed. A rock record of this eon doesn't exist because our planet's surface may have undergone melting and recycling.

QUICK QUESTION: When might the earliest liquid water on the Earth have accumulated? What's the evidence?

13.4 The Archean Eon: Birth of Continents and Life

The Transition from Planet Formation to Planet Evolution

As we've noted, individual grains of zircon have yielded isotopic ages of 4.4 Ga. However, the oldest known whole rock (meaning a coherent aggregate of many minerals), a gneiss that crops out in northwestern Canada, dates to 4.03 Ga. Geologists take this age, rounded to 4.0 Ga, to define the end of the Hadean and the beginning of the **Archean Eon** (from the Greek words *arkhaios*, meaning ancient, and *arche*, meaning beginning). The Archean lasted for the next 1.5 billion years.

Hardly any rocks with ages between 4.0 Ga and 3.85 Ga remain, so the history of the Archean until about 3.85 Ga remains poorly known. What destroyed most of the pre-3.85-Ga rock (and any oceans, if they existed)? The answer may come from studies of cratering on the Moon. These studies suggest that the Moon—and, therefore, all inner planets of the Solar System—underwent a period of intense meteorite bombardment, called the *late heavy bombardment*, between about 4.1 and 3.8 Ga. Researchers speculate that this event pulverized or melted almost all crust that had existed on the Earth at the time and destroyed the existing atmosphere and ocean. Only after the bombardment ceased could long-lasting crust, atmosphere, and oceans begin to form. In addition to bombardment, convective movements involving mantle and crust on the still-hot Earth may have continuously recycled crust, so that soon after an area of crust formed, it might have been subducted and remelted.

Land Appears

By about 3.85 Ga, the Earth's crust had become cool and stable enough for isotopic clocks to start ticking; the rate of crustal recycling decreased, and marine strata started to be preserved. As a result, the Earth of the Archean clearly had land and sea, a situation that has persisted ever since. Geologists still argue about whether plate-tectonic processes in the form that exists today operated in the early part of the Archean. The Earth at this

FIGURE 13.5 A model for crust formation during the Archean Eon.

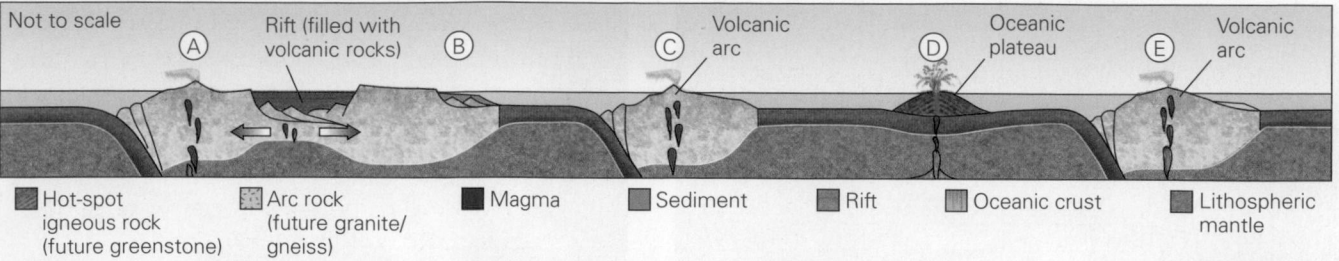

(a) In the Archean, island arcs and hot-spot volcanoes built small blocks of buoyant crust. Rifting of these blocks may have produced flood basalts, and erosion of the blocks produced sediment.

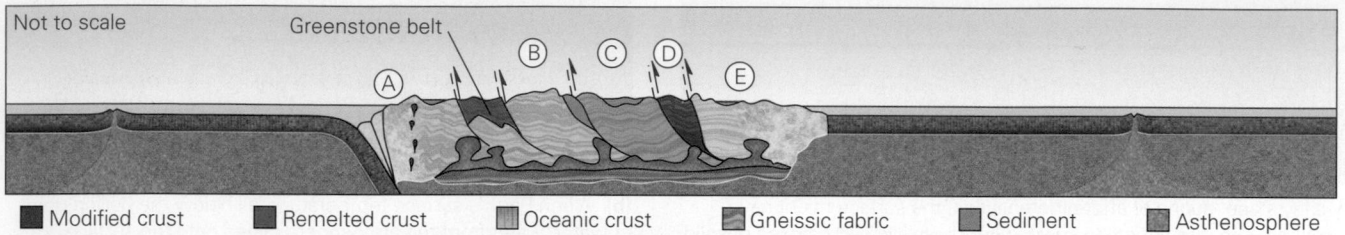

(b) Buoyant blocks collided and sutured together, forming protocontinents. Melting at depth produced granite. Eventually, regions of crust cooled, stabilized, and became cratons.

(c) An exposure of Archean rock in the Upper Peninsula of Michigan. This rock, migmatite, was once buried so deeply that it started to melt.

time was probably much hotter inside than it is today (due both to leftover heat from our planet's formation and to higher concentrations of radioactive elements with short half-lives). Some researchers suggest that, because of its higher temperature, the mantle convected more rapidly than it does today, and therefore that it underwent more partial melting (see Chapter 6) than it does today. If correct, this proposal implies that the early Archean Earth hosted many small, rapidly moving plates, numerous volcanic island arcs, and large hot-spot volcanoes. Others propose that early Archean lithosphere was too warm and buoyant to subduct, so plate tectonics could not have operated until the later part of

the Archean; these geologists argue that plume-related volcanism was the primary source of new crust until the late Archean. Regardless of which model ultimately proves to be more correct, the Archean was clearly a time during which significant volumes of new continental crust came into existence.

What processes produced continental crust? According to one model, the earliest crust formed from mafic igneous rocks that were originally extruded or intruded at convergent boundaries or at hot-spot volcanoes. When these small, relatively buoyant blocks of crust collided with one another, they were sutured together to form larger, more buoyant blocks that remained at the Earth's surface. The development of convergent boundaries along the margins of these blocks, and of rifts and hot-spot volcanoes within the blocks, led to production of new igneous rocks. Not only did flood basalts erupt, but partial melting of basaltic crust, perhaps at or near its base, yielded felsic and intermediate magmas, which rose and solidified in the upper crust or were extruded at the surface. As a result, continents differentiated over time into a more mafic lower crust and a more felsic upper crust. As collisions continued, the blocks coalesced into still larger *protocontinents* **(Fig. 13.5)**, which slowly cooled and became stronger. Between 3.2 and 2.7 Ga, these processes yielded the first long-lived blocks of durable continental crust, and by the end of the Archean Eon, about 80% of the Earth's continental area had formed **(Fig. 13.6)**. Since that time, relatively little "juvenile," (newly extracted from the mantle) rock has formed. Most rock younger than about 2.7 Ga is "recycled," in the sense that it either has gone through various stages of the rock cycle in the crust (see Interlude C) or has been carried back

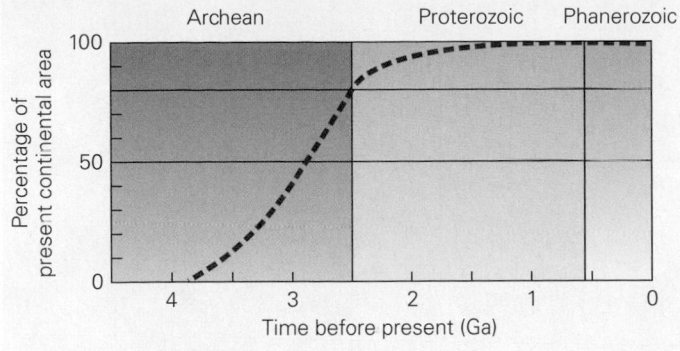

FIGURE 13.6 As time progressed, the area of the Earth covered by continental crust increased. Most crust had formed by the beginning of the Proterozoic.

into the mantle, where it subsequently became incorporated in new magma.

A clear stratigraphic record of marine sediment deposition has been preserved in remnants of Archean crust, indicating that oceans filled in the Archean and have existed ever since. Permanent oceans could survive only after the Earth's surface had cooled below the boiling point of water. Prior to that time, gaseous H_2O saturated the atmosphere—in fact, prior to ocean formation, the atmosphere consisted almost entirely of H_2O and CO_2. Once the oceans formed, however, the atmosphere lost most of its H_2O. And once liquid water existed, substantial amounts of atmospheric CO_2 dissolved into it. The Archean, therefore, saw the atmosphere transform from a foggy mixture of H_2O and CO_2 into a transparent gas dominantly composed of N_2, for N_2 doesn't chemically react with or dissolve in other materials, and it remained behind.

Archean crust, now found almost exclusively in cratons, consists of five principal rock types: *gneiss*, relics of Archean metamorphism in collisional zones; *greenstone*, metamorphosed relics of oceanic crust trapped between colliding blocks of continental crust, as well as of basalts that erupted in rifts, at hot spots, and from volcanoes of island arcs; *granite*, formed from magmas generated by the partial melting of the crust in continental volcanic arcs or above hot spots; *graywacke*, a mixture of sand and clay eroded from the volcanic areas and dumped into the ocean; and *chert*, formed by the precipitation of silica in the deep sea. Archean shallow-water sediments are rare, either because continents were so small that depositional environments in which such sediments could accumulate didn't exist, or because any that were present have eroded away.

Once land areas had formed, rivers flowed over their stark, unvegetated surfaces. Geologists reached this conclusion because sedimentary beds from this time contain clastic grains that were clearly rounded by transport in liquid water. Salts that weathered out of rock and were transported to the sea by rivers made the oceans salty.

The First Life

Clearly, the Archean Eon saw many firsts in Earth history. Not only did the first continents appear during the Archean, but probably also the first life. Geologists use three sources of evidence to identify early life:

- *Chemical fossils (biomarkers):* Chemical fossils are durable chemicals that represent pieces of larger molecules produced by the metabolism of living organisms.

- *Isotopic signatures:* By analyzing the ratio of ^{12}C to ^{13}C in carbon-rich sediment, geologists can determine if the sediment once contained the bodies of organisms because organisms preferentially incorporate ^{12}C.

- *Fossil forms:* Given appropriate depositional conditions, fossils of bacteria and archaea cells can be preserved in rock. However, identification of such fossil forms remains controversial because similar shapes may result from inorganic crystal growth.

The search for the earliest evidence of life continues to make headlines in the popular media **(Box 13.1)**. Most geologists currently conclude that life has existed on the Earth since at least 3.5 Ga, and perhaps since 3.8 Ga, for rocks of this age contain isotopic signatures of organisms. The oldest undisputed body fossils of bacteria and archaea occur in 3.2-Ga rocks **(Fig. 13.7a)**—shapes resembling such organisms occur in rocks at least as old as 3.5 Ga, but their identity remains less certain. Some rocks of this age contain **stromatolites**, distinctive mounds composed of very thin layers. Although some stromatolites may form simply by precipitation of minerals from water solutions, most develop by the growth of layer upon layer of *microbial mats* (sheets composed of archaea or bacteria). Microbial mats secrete a slimy mucus-like substance that traps silt or sand that settles from water after a storm has churned up nearshore sediment or flooding rivers have dumped sediment into the sea. The underlying cells die when silt and sand cover a microbial mat, but a new layer of cells soon colonizes the surface of the sediment layer and builds another microbial mat **(Fig. 13.7b)**. Burial and lithification of successive mats preserve the layering in the mound **(Fig. 13.7c)**. Modern examples of stromatolites occur locally in shallow, tropical waters **(Fig. 13.7d)**.

Did you ever wonder...
what the oldest relict of life is?

During the Archean, perhaps as early as 3.5 Ga, *cyanobacteria* appeared. These organisms, which may have been the principal stromatolite builders, are able to carry out photosynthesis, a process that produces free oxygen (O_2). Very little of this oxygen accumulated in the atmosphere, for it was either dissolved in the sea or absorbed by weathering reactions with rocks. Therefore, the composition of the Earth's atmosphere at the end of the

FIGURE 13.7 Archean life forms.

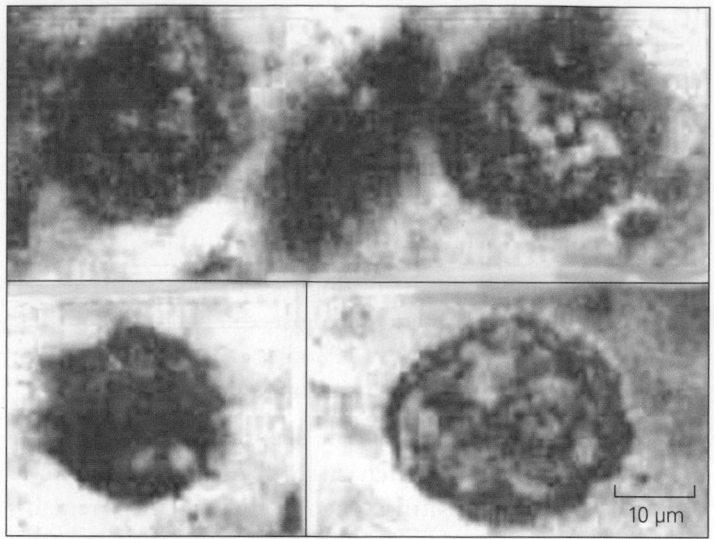

(a) These shapes in 3.2-Ga chert from South America are thought to be fossil bacteria or archaea.

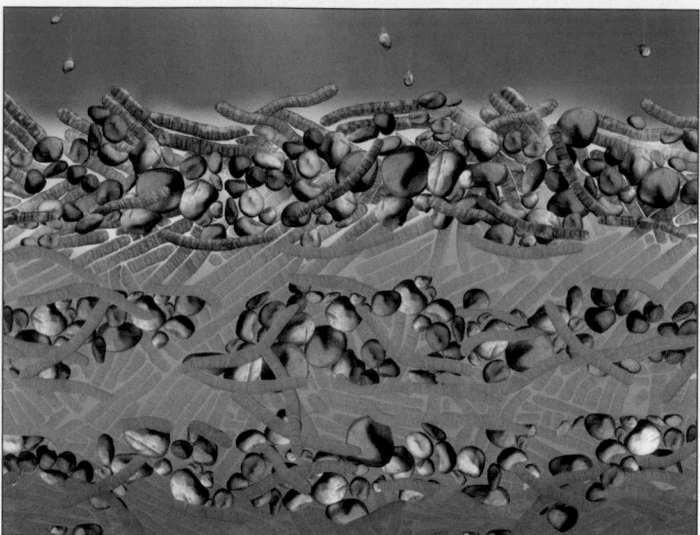

(b) Stromatolites form when sediment sticks to layers of cells. As one layer dies and compacts, a new one grows above.

(c) This weathered outcrop of 1.85-Ga dolostone near Marquette, Michigan, reveals the layered structure of stromatolites. The delicate ridges represent the fossilized remnants of bacterial mats. Similar stromatolites also occur in exposures of Archean rocks.

(d) Modern stromatolites in Shark Bay, Western Australia.

Archean was still unbreathable. It probably consisted of about 75% N_2 gas and 25% CO_2 gas, with only traces of oxygen.

As the Archean Eon came to a close, the first continents had formed, and life colonized not only the depths of the sea but also the shallow-marine realm. Plate-tectonic processes had commenced, continental drift was taking place, collisional mountain belts were forming, and erosion was occurring. Oxygen was beginning to enter the air but had not yet accumulated in any significant quantity. The stage was set for another major change in the Earth System.

Did you ever wonder...
if the atmosphere has always been breathable?

TAKE-HOME MESSAGE

The Archean (4.0–2.5 Ga) began with the late heavy bombardment, during which almost all crust was destroyed. A new, solid crust, re-formed by cooling of the mantle's skin, appeared soon after, and by 3.8 Ga, oceans formed, and have existed ever since. Plate-tectonic processes probably began in the Archean; there were also huge mantle plumes. During this eon, the first continental crust formed from colliding volcanic arcs and hot-spot volcanoes, the atmosphere changed, and life appeared. By the end of the eon, the first continents existed.

QUICK QUESTION: What kinds of rock form the blocks of Archean crust that remain as cratons today?

FIGURE 13.8 Major geologic provinces of the Earth. The black lines indicate the borders of regions underlain by Precambrian crust. Shields are regions where broad areas of Precambrian rocks are exposed.

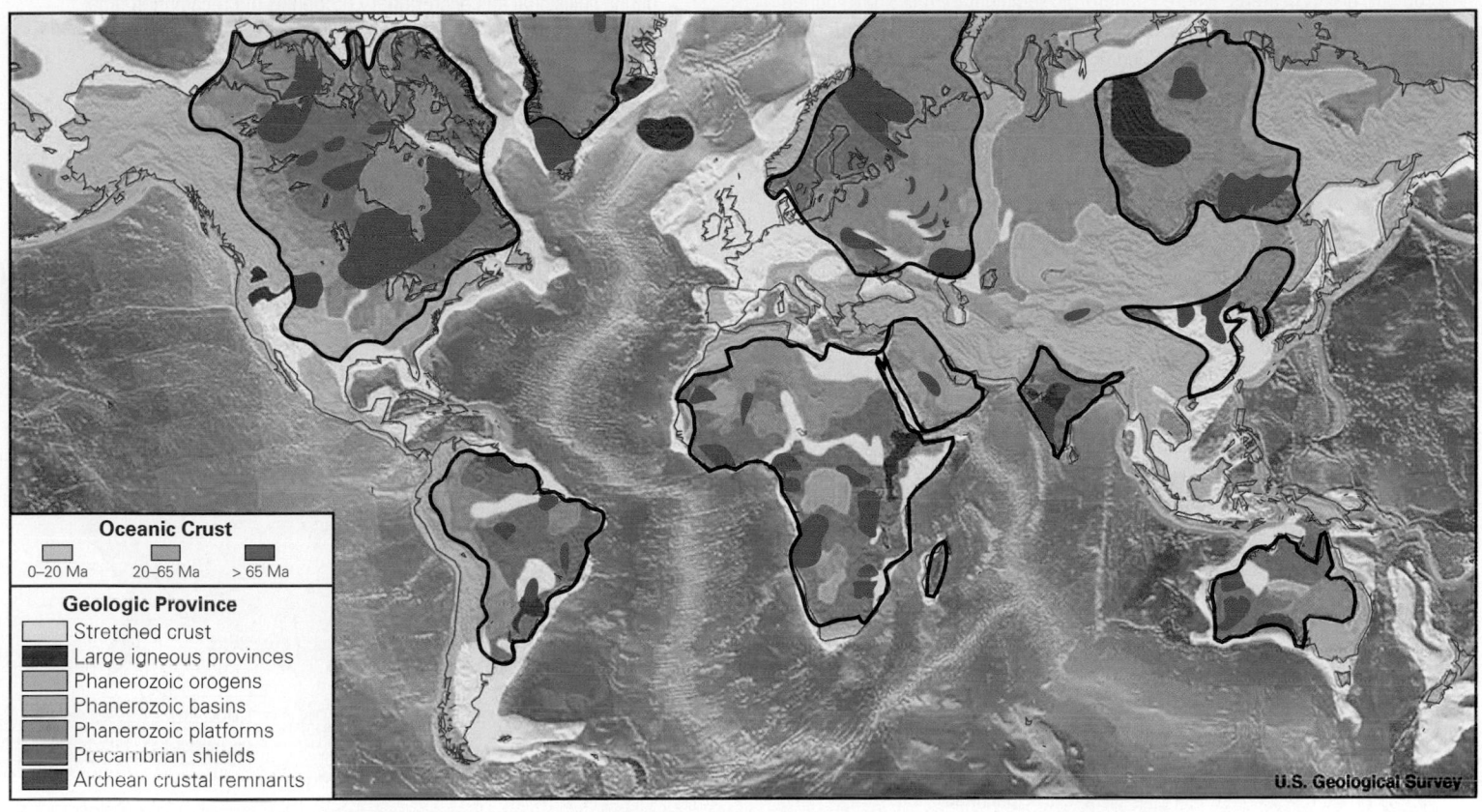

Oceanic Crust
0–20 Ma   20–65 Ma   > 65 Ma

Geologic Province
- Stretched crust
- Large igneous provinces
- Phanerozoic orogens
- Phanerozoic basins
- Phanerozoic platforms
- Precambrian shields
- Archean crustal remnants

U.S. Geological Survey

13.5 The Proterozoic Eon: The Earth in Transition

Continued Growth of Continents

The **Proterozoic Eon** (from the Greek meaning earlier life) spans roughly 2 billion years, from about 2.5 Ga to the beginning of the Cambrian Period at 541 Ma; therefore, it encompasses almost half of the Earth's history. During Proterozoic time, the Earth's surface environment changed from an unfamiliar world of small, fast-moving plates, small continents, and an oxygen-free atmosphere to the more familiar world of mostly large, slow-moving plates, large continents, and an oxygenated atmosphere.

First, let's look at changes to the continents. New continental crust continued to form during the Proterozoic Eon, but at progressively slower rates, and by the middle of the eon, over 90% of the Earth's continental crust had formed. Collisions between Archean continental blocks, and between these blocks and volcanic island arcs or hot-spot volcanoes, gradually assembled larger continents. Size matters when it comes to the geologic behavior of continents, for the interior of a larger continent can be isolated from heating by

subduction-related igneous activity that happens along its margins. Such interior regions, therefore, slowly cool and strengthen until they become rigid and durable. The resulting region of cold, relatively stable, long-lived continental crust, as we have seen, is called a **craton** (see Chapter 11). All cratons that exist today had formed by about 1 Ga **(Fig. 13.8)**, meaning that the crust of cratons ranges in age from 3.85 Ga to about 1 Ga.

To understand the character of a craton, let's examine North America's craton a bit more closely. We see that it consists of two regions **(Fig. 13.9)**. Throughout the **shield**, outcrops expose Precambrian basement, which consists of igneous and metamorphic rocks older than about 1 Ga. The landscape of the shield tends to have fairly low relief—there are small hills and valleys, but no dramatic mountain ranges. Most of North America's shield lies in Canada, so geologists refer to it as the *Canadian Shield*. Throughout the **cratonic platform**, which surrounds the shield and also underlies Hudson Bay, a blanket or *cover* of Paleozoic or Mesozoic strata overlies the Precambrian basement. In the eastern platform of North America, Paleozoic strata are the youngest bedrock, whereas in the western platform, most of the Paleozoic strata are covered by Mesozoic and Cenozoic strata derived from sediments eroded from mountains to the west.

FIGURE 13.9 On this map of North America, we see four different geologic provinces: shield areas, where Precambrian rocks of the craton crop out; platform areas, where Phanerozoic sedimentary rocks have buried Precambrian rocks; Phanerozoic orogenic belts composed of rocks deformed during the past half-billion years; and the coastal plain, which is underlain by Cretaceous and Cenozoic strata.

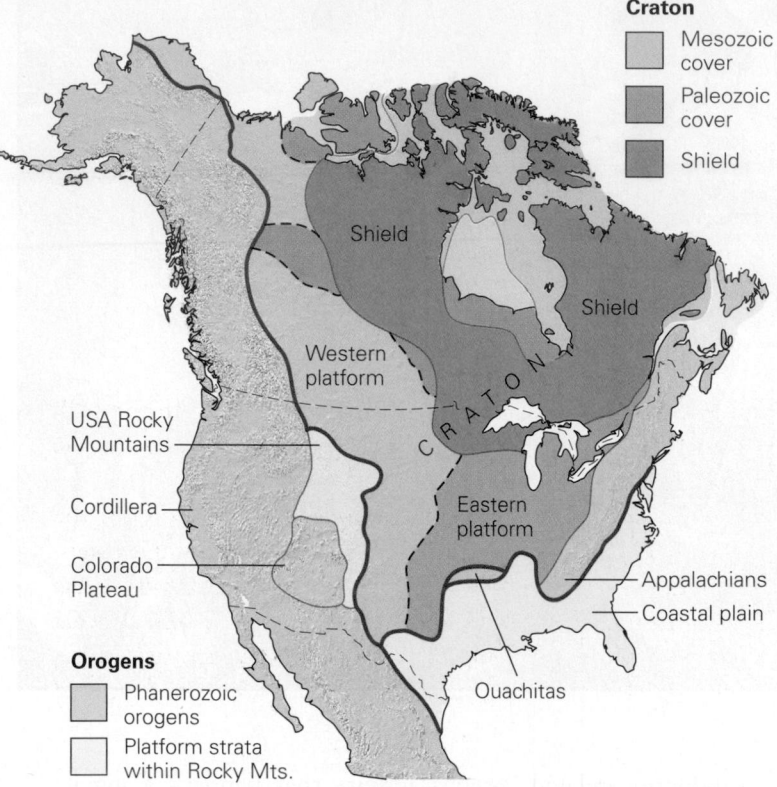

Craton
- Mesozoic cover
- Paleozoic cover
- Shield

Orogens
- Phanerozoic orogens
- Platform strata within Rocky Mts.

FIGURE 13.10 Geologic provinces of the North American craton consist of a collage of different belts and blocks stitched together during collisional and accretionary orogenies of Precambrian time.

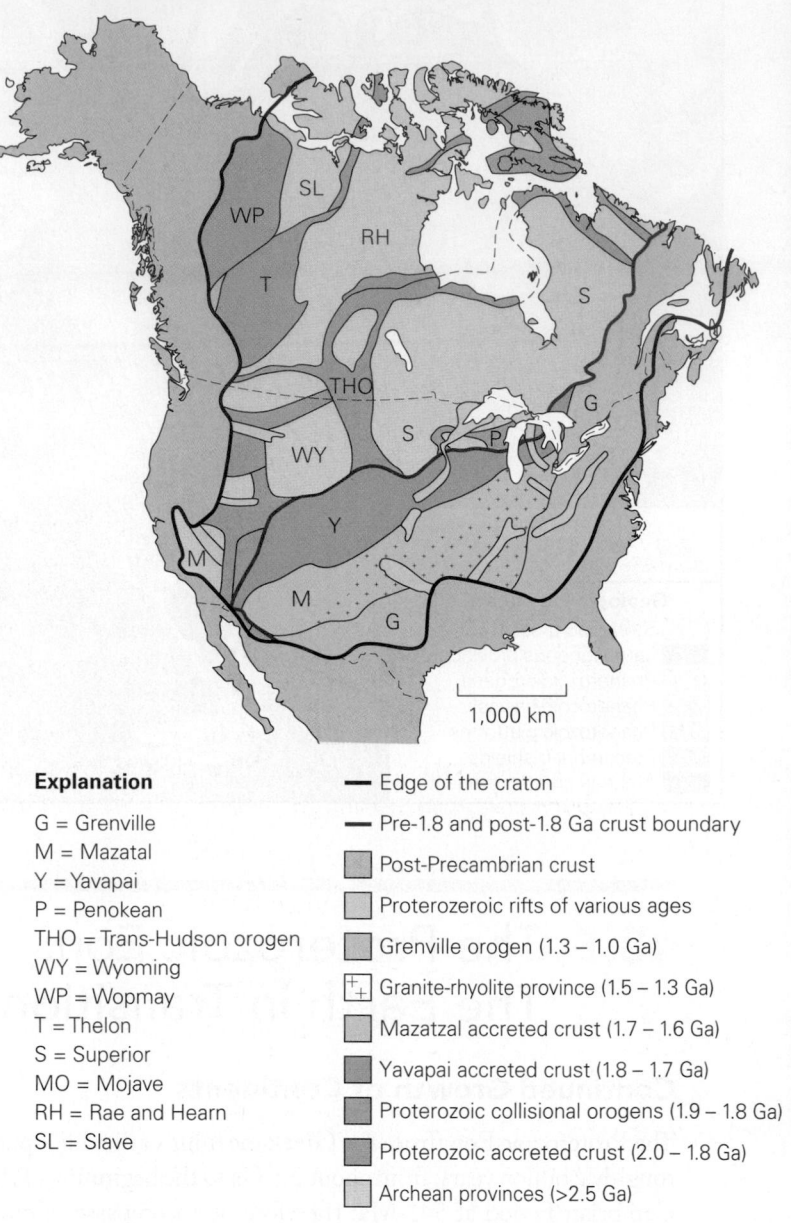

Explanation

G = Grenville
M = Mazatal
Y = Yavapai
P = Penokean
THO = Trans-Hudson orogen
WY = Wyoming
WP = Wopmay
T = Thelon
S = Superior
MO = Mojave
RH = Rae and Hearn
SL = Slave

- Edge of the craton
- Pre-1.8 and post-1.8 Ga crust boundary
- Post-Precambrian crust
- Proterozeroic rifts of various ages
- Grenville orogen (1.3 – 1.0 Ga)
- Granite-rhyolite province (1.5 – 1.3 Ga)
- Mazatzal accreted crust (1.7 – 1.6 Ga)
- Yavapai accreted crust (1.8 – 1.7 Ga)
- Proterozoic collisional orogens (1.9 – 1.8 Ga)
- Proterozoic accreted crust (2.0 – 1.8 Ga)
- Archean provinces (>2.5 Ga)

By using isotopic dating on samples from both outcrops and drill holes, geologists have been able to subdivide the Precambrian basement of North America's craton into distinct geologic provinces, each of which has been given a name **(Fig. 13.10)**. Provinces of the Canadian Shield consist of Archean crustal blocks sutured together by Proterozoic orogens. The province of the cratonic platform in the United States, in contrast, grew when a series of volcanic island arcs and continental slivers accreted, or attached, to the margin of the Canadian Shield between 1.8 and 1.6 Ga—these accreted provinces are known as the Yavapai and Mazatzal Provinces, respectively. In the Midwest, granite plutons intruded the accreted provinces, and rhyolite ash flows covered them, due to widespread felsic igneous activity between 1.5 and 1.3 Ga.

Successive collisions ultimately brought together most continental crust on the Earth into a single supercontinent, named **Rodinia**, by around 1 Ga. Geologists refer to the last major collision during the formation of Rodinia as the **Grenville orogeny**. The resulting Grenville orogen was probably as huge as the present-day Himalayas. If you look at a popular (though not universally accepted) reconstruction of Rodinia, you can identify the crustal areas that would eventually become the familiar continents of today **(Fig. 13.11a)**. Several studies suggest that sometime between 800 and 600 Ma, Rodinia effectively turned inside out, in that Antarctica, India, and Australia broke away from western North America and later collided with the future South America, possibly forming a short-lived supercontinent that some geologists refer to as *Pannotia* **(Fig. 13.11b)**.

Life Becomes More Complex

The map of the Earth clearly changed radically during the Proterozoic. But that's not all that changed—fossil evidence

BOX 13.1 CONSIDER THIS . . .

Where Was the Cradle of Life?

What specific environment on the Archean Earth served as the cradle of life? Laboratory experiments conducted in the 1950s led many researchers to think that life began in warm pools of surface water, beneath a methane- and ammonia-rich atmosphere streaked by bolts of lightning (see Interlude E). The only problem with this hypothesis is that more recent evidence suggests that the early atmosphere consisted mostly of CO_2 and N_2, with

relatively little methane and ammonia. More recent research suggests instead that submarine hydrothermal vents—so-called black smokers—along mid-ocean ridges served as the homes of the first organisms. These vents emit clouds of ion-charged solutions from which sulfide minerals precipitate and build chimneys. The chimneys are hollow, and their surfaces act like membranes in keeping two chemically distinct environments

separate. This difference creates a weak electric charge, perhaps providing energy for "proto-life molecules"—molecules that have structures that resemble proteins of living organisms—to start developing. The earliest life in the Archean Eon may well have been thermophilic (heat-loving) bacteria or archaea that originated at submarine hydrothermal vents and dined on pyrite at dark depths in the ocean alongside these vents.

suggests that this eon also saw important steps in the evolution of life. When the Proterozoic began, most life was *prokaryotic*, meaning that it consisted of single-celled organisms (archaea and bacteria) without a nucleus. Studies of chemical fossils (see Interlude E) hint that *eukaryotic* life, consisting of cells that have nuclei, originated as early as 2.7 Ga, but the first body fossils of eukaryotic organisms appear in rocks by about 1.9 Ga. Abundant body fossils of eukaryotic organisms can be found only in rocks younger than about 1.5 Ga. Clearly,

the proliferation of eukaryotic life, the foundation from which complex organisms eventually evolved, took place during the Proterozoic.

The last half billion years of the Proterozoic Eon saw the remarkable transition from simple organisms to complex ones. Ciliate protozoans (single-celled organisms coated with fibers that give them mobility) appeared at about 750 Ma. A great leap forward in the complexity of organisms occurred during the next 150 million years of the eon. Sediments deposited

FIGURE 13.11 Supercontinents in the late Precambrian.

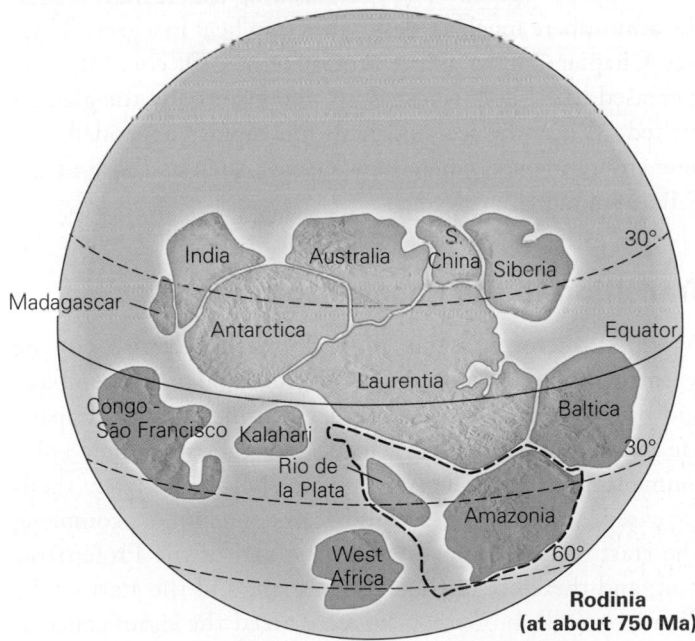

Rodinia
(at about 750 Ma)

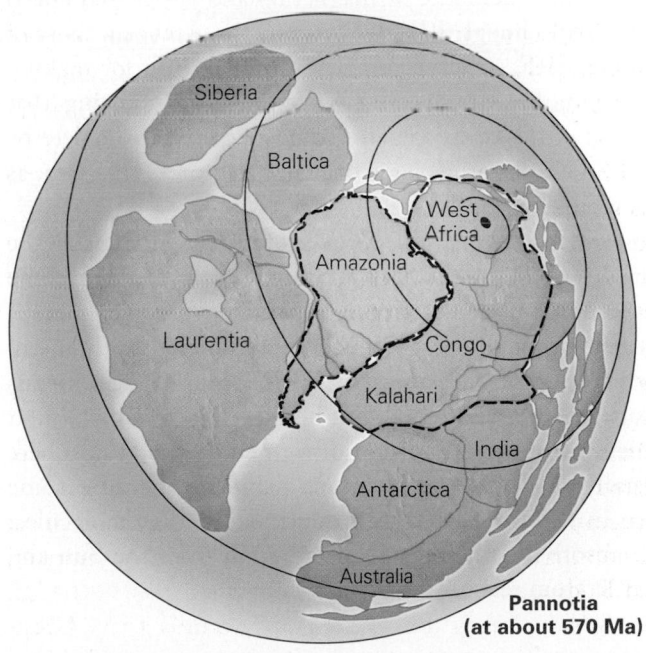

Pannotia
(at about 570 Ma)

(a) Rodinia formed around 1 Ga and lasted until about 700 Ma. North America and Greenland together comprise Laurentia.

(b) According to one model, by 570 Ma, Rodinia had broken apart; continents that once lay to the west of Laurentia ended up to the east of Africa. The resulting supercontinent, Pannotia, broke up soon after it formed.

perhaps as early as 620 Ma and certainly by 565 Ma contain several types of multicellular animals, probably with distinct organs. These organisms constitute the **Ediacaran fauna**, named for a region in southern Australia where their fossils were first found. Ediacaran species survived into the beginning of the Cambrian before becoming extinct. Their fossil forms suggest that some of these invertebrate organisms resembled jellyfish, while others resembled worms, and some looked like bags or quilts **(Fig. 13.12a)**.

The evolution of life played a key role in the evolution of the Earth's atmosphere. Before life appeared, there was hardly any free oxygen (O_2) in the atmosphere. With the appearance of photosynthetic organisms, oxygen began to enter the atmosphere. But it was not until between 2.4 Ga and 1.8 Ga that the concentration of oxygen in the atmosphere increased dramatically. This change, called the **great oxygenation event**, happened when the land and sea were no longer able to react with, absorb, or dissolve all the oxygen produced by organisms, so the oxygen began to accumulate as a gas in air.

One of the important consequences of this change is that the oceans became *oxidizing* environments (meaning that they contained atoms, such as oxygen, that can accept electrons from other atoms). Iron atoms can donate electrons to oxygen and become oxidized. Oxidized iron cannot dissolve in water, so when the oceans became oxidizing, they could no longer contain large quantities of dissolved iron. Between 2.4 Ga and 1.8 Ga, huge amounts of iron precipitated out of the ocean to form colorful sedimentary beds known as *banded iron formation* (*BIF*). BIF consists of alternating layers, or bands, of iron oxide minerals (hematite or magnetite) and jasper (red chert) **(Fig. 13.12b)**. Though BIF did form in the Archean, most of the world's BIF, society's primary source of iron for making steel, accumulated between 2.4 and 1.8 Ga, implying that the transition to an oxygenated atmosphere was complete by about 1.8 Ga. Other geologic evidence reinforces this idea, as described in **Box 13.2**.

The great oxygenation event profoundly influenced the evolution of life in the Earth System. Life could become more complex because oxygen-dependent (aerobic) metabolism can produce energy much more efficiently than can oxygen-free (anaerobic) metabolism—eating sulfide minerals may sustain an archaea cell, but it can't keep a multicellular organism alive. And addition of oxygen to the atmosphere eventually made the land surface habitable, because some oxygen in the air reacted to produce ozone (O_3) molecules, which absorb deadly ultraviolet radiation from the Sun and prevent it from reaching the surface.

Snowball Earth

Study of Proterozoic strata indicates that radical climate shifts took place near the end of the Proterozoic Eon. Specifically,

researchers have found accumulations of glacial sediments in late Proterozoic stratigraphic sequences, suggesting multiple episodes of glacial growth between about 720 and 660 Ma, and again from 650 to 635 Ma. What's strange about the occurrence of these sediments is that some occur even in regions that were located at the equator at the time they were deposited **(Fig. 13.12c)**.

This observation implies that, at times during the latter part of the Proterozoic, the entire planet was cold enough for glaciers to grow at all latitudes. Geologists are still debating the character and history of these global ice ages (see Chapter 22), but in one model, glaciers covered all land at times, and the entire ocean surface froze. The ice-covered globe that our planet may have become at the end of the Proterozoic has come to be known as **snowball Earth** **(Fig. 13.12d)**.

> **Did you ever wonder...**
> if the oceans have ever frozen over entirely?

The shell of ice covering snowball Earth would have cut off the oceans from the atmosphere, and indeed, many life forms died off at this time. Survival may have been possible only near submarine hydrothermal vents or land-based hot springs. The Earth might have remained a snowball forever were it not for volcanic CO_2. The icy sheath covering the oceans would have prevented atmospheric CO_2 from dissolving in seawater, but it could not have prevented volcanic activity from continuing to add CO_2 to the atmosphere. Carbon dioxide is a greenhouse gas, meaning that it traps heat in the atmosphere much as glass panes trap heat in a greenhouse (see Chapter 23), so as the atmospheric CO_2 concentration increased, the Earth warmed up, and eventually the glaciers melted. When the ice vanished, life rapidly expanded into new environments, where new species, such as those of the Ediacaran fauna, evolved.

Transition to the Phanerozoic Eon

As the Proterozoic came to a close, the Earth's climate warmed and supercontinents broke apart, reassembled, and broke apart again. As smaller continents drifted apart, life evolved and diversified to occupy the many new environments that formed. A relatively short time later, shells appeared, and the fossil record became much more complete. The start of this change defines the end of the Proterozoic Eon, and therefore of the Precambrian, and the start of the Phanerozoic Eon. Geologists recognized the significance of this event long before they could assign it a numerical age (currently 541 Ma).

The **Phanerozoic Eon** (Greek for visible life) encompasses the last 541 million years of Earth history. Its name reflects the appearance of diverse organisms with hard shells or skeletons that became the well-preserved fossils you can find easily in sedimentary rock outcrops. This eon consists of three eras—the Paleozoic (Greek for ancient life), the Mesozoic (middle life), and the Cenozoic (recent life). Geologists have divided the Mesozoic and Cenozoic into three periods each and the Paleozoic into six periods. In the sections that follow, we consider changes in the map of our planet's surface (its *paleogeography*) as manifested by the distribution of continents, seas, and mountain belts, as well as life evolution that happened during the three eras.

TAKE-HOME MESSAGE

During the Proterozoic (2.5–0.54 Ga), cratons formed and sutured together to form continents, and then, supercontinents. The atmosphere began to accumulate oxygen, and the chemical behavior of the seas changed. Near the end of the eon, the Earth may have been completely ice covered. As the eon ended, the climate warmed, supercontinents rifted apart, and multicellular organisms appeared.

QUICK QUESTION: What evidence suggests that oxygen began to accumulate in the air during the Proterozoic?

FIGURE 13.12 Major changes of the Earth System in the Proterozoic Eon.

(a) *Dicksonia*, a fossil of the Ediacaran fauna. These complex, soft-bodied marine organisms appeared in the late Proterozoic.

Iron oxide minerals (hematite or magnetite)
Red chert (jasper)

(b) An outcrop of BIF in the Iron Ranges of Michigan's Upper Peninsula. The red stripes are jasper (red chert) and the gray stripes are hematite. The rock was folded during a mountain-building event long after deposition. The hammer indicates scale.

Strata contain large clasts surrounded by mudstone, for glaciers can carry clasts of all sizes.

Bedding

(c) Layers of Proterozoic glacial till crop out in Africa, indicating that low-latitude landmasses were glaciated during the Proterozoic.

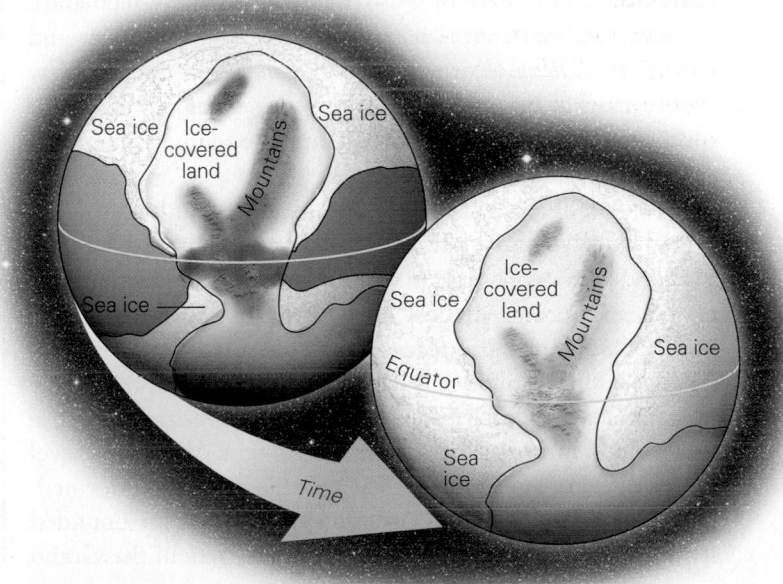

Sea ice Ice-covered land Sea ice Mountains Sea ice Sea ice Ice-covered land Mountains Sea ice Equator Sea ice Time

(d) This planet may have frozen over completely to form "snowball Earth." Glaciers first grew on land, and eventually the sea surface froze over.

FIGURE 13.13 Land and sea in the early Paleozoic Era.

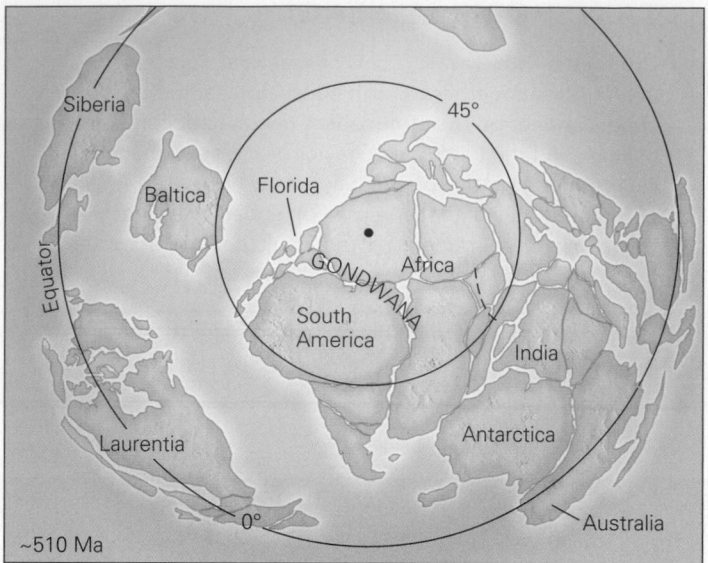

(a) The distribution of continents in the Cambrian Period (510 Ma), as viewed looking down on the South Pole.

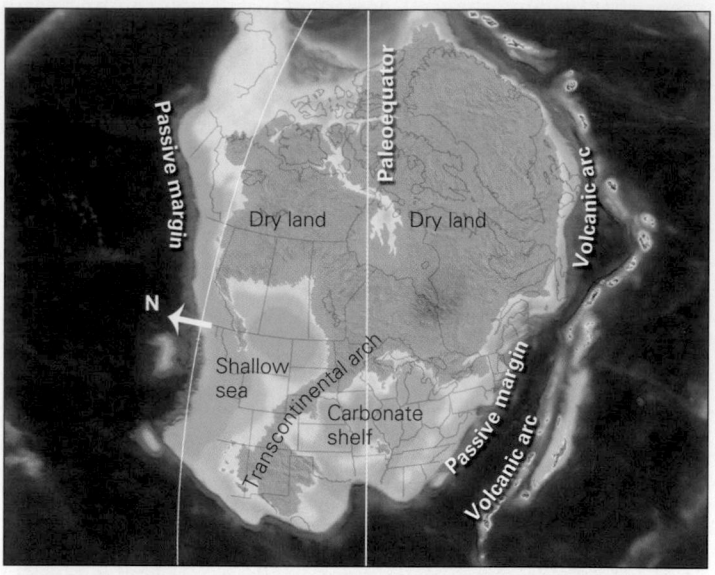

(b) A paleogeographic map of North America shows the regions of dry land and shallow seas in the Late Cambrian Period.

13.6 The Paleozoic Era: Continents Reassemble and Life Gets Complex

The Early Paleozoic Era (Cambrian–Ordovician Periods, 541–444 Ma)

Paleogeography At the beginning of the Paleozoic Era, Pannotia broke up, yielding smaller continents, including **Laurentia** (composed of North America and Greenland), **Gondwana** (South America, Africa, Antarctica, India, and Australia), *Baltica* (Europe), and *Siberia* **(Fig. 13.13a)**. New passive-margin basins formed along the edges of these new continents, and each of them accumulated large volumes of sediment.

Sea level rose and fell significantly multiple times during part of the early Paleozoic. At times of high sea level, transgression took place, and vast areas of continental interiors became shallow seas, known as *epicontinental seas* **(Fig. 13.13b)**. In many places, water depths in epicontinental seas reached only a few meters, creating a well-lit marine environment in which life abounded, so deposition in these seas yielded layers of fossiliferous sediment. When sea level dropped, regression took place, and unconformities formed. As a result, the craton was covered by unconformity-bounded sequences of sediment—the layer cake of strata in the Grand

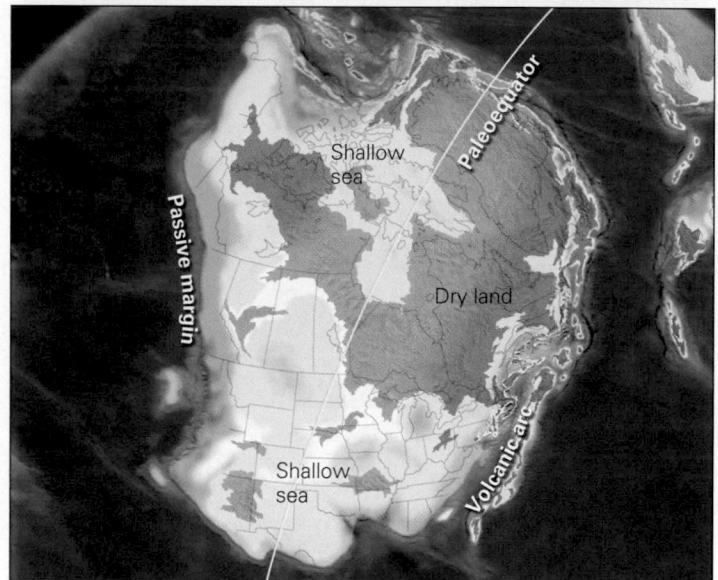

(c) During the Middle Ordovician Period, shallow seas covered much of North America. A volcanic arc formed off the east coast.

Canyon formed from such sediment **(Box 13.3)**. These strata comprise the *cover* of the cratonic platform.

The geologically peaceful world of the early Paleozoic Era in Laurentia abruptly came to a close in the Middle Ordovician Period, for at this time its eastern margin rammed into a volcanic island arc and other crustal fragments. The resulting collision, called the *Taconic orogeny*, deformed and metamorphosed strata of the continent's margin and produced a mountain range in what is now the eastern part of the Appalachians **(Fig. 13.13c)**.

BOX 13.2 CONSIDER THIS . . .

The Evolution of Atmospheric Oxygen

Without oxygen, the great variety of life that exists on the Earth today could not survive. Presently, the atmosphere contains 21% oxygen, but this has not always been the case. Throughout most of the Archean Eon and into the beginning of the Proterozoic Eon, the atmosphere contained less than 1% oxygen. Several lines of evidence lead geologists to conclude that a transformation from an oxygen-poor to an oxygen-rich atmosphere occurred during the *great oxygenation event* (GOE), which lasted from about 2.4 to about 1.8 Ga, in the early part of the Proterozoic.

As noted earlier, one line of evidence that led to the proposal of the GOE came from studying the deposition of BIFs, a process that could have happened only in an oxygenated environment. Another line of evidence came from examining clastic grains in sandstones. In sediments deposited before 2.4 Ga, well-formed pyrite (iron sulfide) occurs as clasts. The formation of

clasts would have been possible only if the atmosphere before 2.4 Ga contained very little oxygen, for in an oxygen-rich atmosphere, pyrite rapidly undergoes chemical weathering (oxidation) and doesn't survive long enough at the Earth's surface to become a sedimentary clast. A third line of evidence comes from studying the age of *redbeds*, clastic sedimentary rocks colored by the presence of bright red hematite (iron oxide). Redbeds form when oxygen-rich groundwater flows through sediment during lithification, and such rocks appear in the geologic record only after 1.8 Ga.

The oxygen concentration in the atmosphere has changed considerably over the Earth's history. As we have noted, oxygen constituted less than 1% of the atmosphere during the Archean. By 1.8 Ga, after the great oxygenation event, it had risen to about 3%, and it stayed at roughly that level until about 0.6 Ga (**Fig. Bx13.2**). Then, because all the materials on the

land and in the sea that could absorb oxygen had become *saturated* and could hold no more O_2, atmospheric concentrations gradually grew, reaching about 12% at the end of the Proterozoic. The proportion of oxygen grew substantially when photosynthetic organisms began to prosper on land, and it has oscillated between 35% and 15% during the past half-billion years. Oxygen has remained at 21% since about 25 Ma. Keeping in mind that atmospheric oxygen comes from photosynthetic organisms, geologists suggest that increases reflect proliferation of photosynthetic life (such as land-based plants) and that decreases reflect mass-extinction events. Notably, the concentration of O_2 can't get higher than about 35%, because if it did, land plants would become explosively combustible, since oxygen feeds fire, and so much vegetation would burn that the amount of photosynthesis would decrease until oxygen levels decreased.

FIGURE Bx13.2 The change in the proportion of oxygen in the atmosphere over time. Oxygen made up only a small percentage of the atmosphere until the end of the Proterozoic.

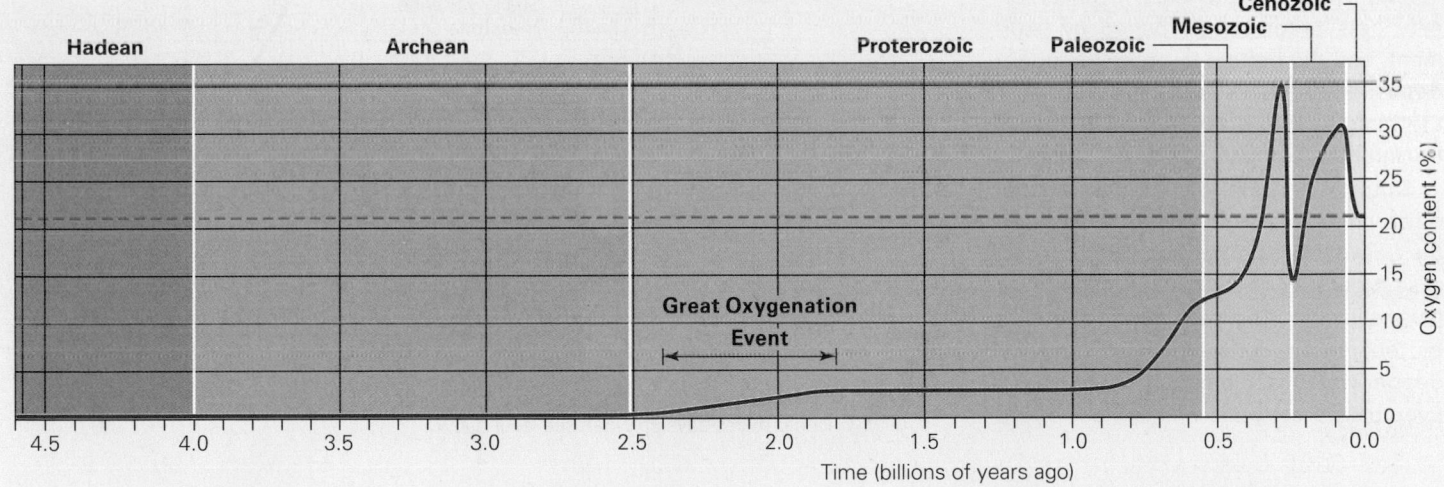

Life Evolution The first animals to appear in the Cambrian Period had simple tube- or cone-shaped shells. Shells may have evolved as a means of protection against predation by organisms such as conodonts, small, eel-like organisms with hard parts that resemble teeth. The fossil record indicates that soon after the Cambrian began, life underwent remarkable diversification. This event, which

paleontologists refer to as the **Cambrian explosion**, took several million years. What caused this event? No one can say for sure, but considering that it occurred roughly at the time a supercontinent broke up, it may have been triggered by the production of new ecological niches and by the isolation of populations that resulted when small continents formed and drifted apart.

BOX 13.3 CONSIDER THIS . . .

Stratigraphic Sequences and Sea-Level Change

As we have seen in this chapter, there have been intervals in geologic history when the interior of North America was submerged beneath a shallow sea and times when it was high and dry. This observation implies that sea level, relative to the surface of the continent, rises and falls through geologic time. When the continental surface was dry and exposed to the atmosphere, weathering and erosion ground away at previously deposited rock and, as a result, created a continent-wide unconformity. In the early 1960s, an American stratigrapher named Larry Sloss introduced the term **stratigraphic sequence** to refer to the strata deposited on a continent during periods when continents were submerged. Such sequences are bounded above and below by regional unconformities.

We can picture the deposition of an idealized sequence as follows. At our starting condition, much of the surface of a continent lies above sea level, and an unconformity develops. Then, as a transgression takes place, the shoreline migrates inland and the continent's interior progressively floods. Thus, the pre-existing unconformity gets progressively buried by sediment, and the bottom layers of the newly deposited sequence tend to be older near the margin of the continent than toward the interior.

During transgression, the depositional environment at a location changes as the sea deepens. Therefore, the base of a sequence consists of terrestrial strata (river alluvium). This sediment gets buried by nearshore sediment, and then by deeper-water sediment **(Fig. Bx13.3a)**. Eventually, nearly the entire width of the

continent, with the exception of highlands and mountain belts, lies below sea level.

As regression begins, the interior of the continent becomes exposed first, and then the margins. (In detail, the location and behavior of individual sedimentary basins influence the local positions of shorelines.) So a new unconformity forms first in the interior and then later along the margins. Many shorter-duration transgressions and regressions may happen during a single long-duration sea-level rise or fall.

Sloss recognized six major stratigraphic sequences in North America, and

FIGURE Bx13.3 The rise and fall of sea level and its manifestation in the stratigraphic record.

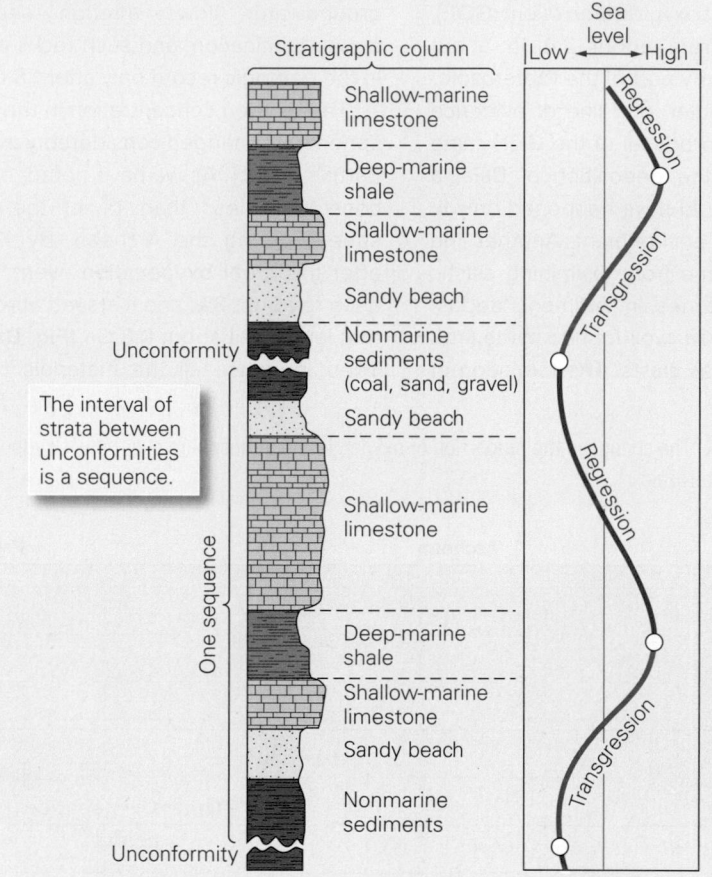

(a) Different types of strata are deposited as sea level rises and falls. Unconformities develop when sea level is low.

By the end of the Cambrian, trilobites were grazing the seafloor. Trilobites shared the environment with mollusks, brachiopods, nautiloids, gastropods, graptolites, and echinoderms **(Fig. 13.14**; see Interlude E). A complex food chain arose, which included plankton, bottom feeders, and predators. Many of the organisms crawled over or swam around reefs composed of mounds of sponges with mineral skeletons.

Such life forms continued into the Ordovician Period. But new organisms, such as crinoids (sea lilies) and jawless fish, the first vertebrate animals, appeared.

Although the sea teemed with organisms during the early Paleozoic Era, there were no land organisms for most of this time, so the land surface remained a stark landscape of rock and sediment, subjected to rapid erosion. As we mentioned he named each after a nation of Native Americans (Fig. Bx13.3b). Each sequence represents deposition during an interval of time lasting tens of millions of years. The transgressions and regressions could have been caused by global (eustatic) sea-level change, or they could reflect mountain-building processes, or they could reflect the subsidence or uplift of continents.

In more recent decades, studies of strata along continental passive margins have provided an even more detailed record of sequences, because less erosion takes place in these regions.

Taken together, this information may provide insight into the history of the global rise and fall of sea level, though this interpretation remains controversial (Fig. Bx13.3c). Some sea-level changes may reflect changes in seafloor-spreading rates which, in turn, affect the volume of mid-ocean ridges; some may reflect hot-spot activity; some may reflect variations in the Earth's climate that caused the formation or melting of ice sheets; others may represent changes in areas of continents accompanying continental collisions; and still others may reflect the warping of continental surfaces as a result of mountain building.

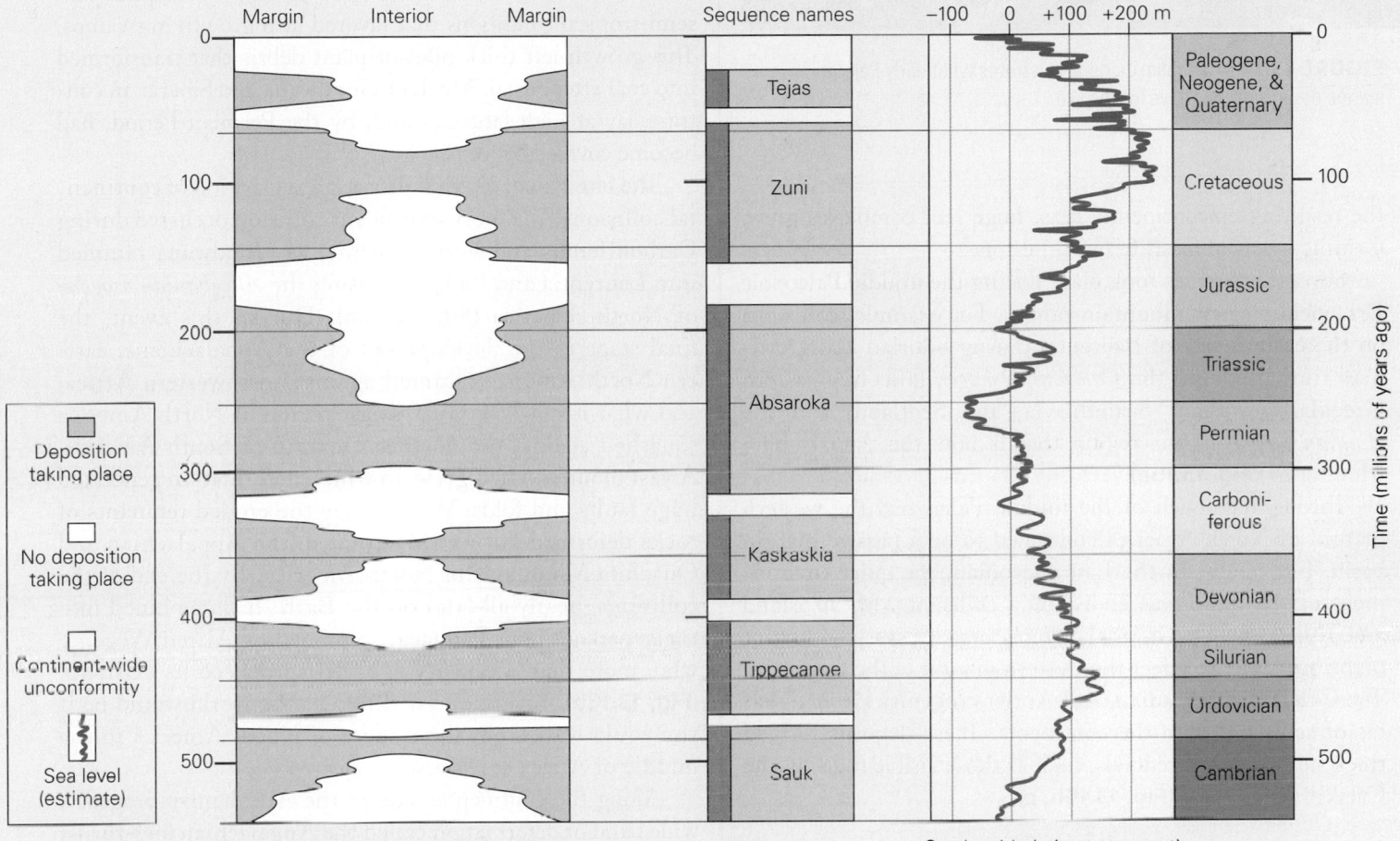

(b) Stratigraphic sequences of North America, as named by Sloss (1962). Symbolically, the chart on the left shows that, overall, continental margins were sites of deposition, while the interior was not. At times, the whole continent was a not a site of deposition—the resulting continent-wide unconformities separate the sequences, as shown by the chart in the middle.

(c) An interpretation of global sea-level change, based on analysis of stratigraphy in passive-margin basins. This interpretation remains controversial.

earlier, the invasion of the land could begin only when there was enough ozone in the atmosphere to protect the land surface from UV radiation. The earliest land plants appeared in the Late Ordovician Period. These plants were very small and occurred only along bodies of water. At the end of the Ordovician, a mass extinction took place, perhaps because of a brief ice age and associated sea-level lowering.

The Middle Paleozoic Era (Silurian–Devonian Periods, 444–359 Ma)

Paleogeography As the world entered the Silurian Period, the start of the middle Paleozoic, global climate warmed (leading to so-called greenhouse conditions), sea level rose, and the continents flooded once again. In the warm, clear waters of

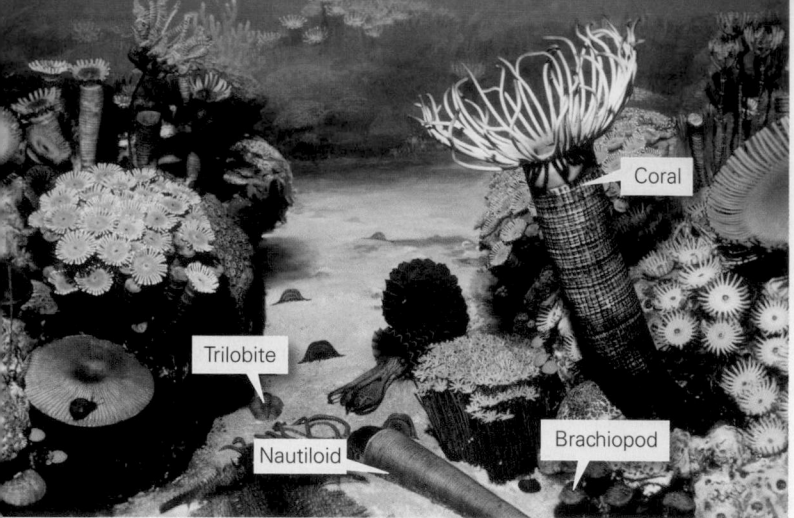

FIGURE 13.14 A museum diorama illustrates what early Paleozoic marine organisms may have looked like.

the resulting epicontinental seas, huge reef complexes grew, forming layers of fossiliferous limestone.

Several orogenies took place during the middle Paleozoic Era, yielding new mountain ranges. For example, collisions on the eastern side of Laurentia during Silurian and Devonian time produced the *Caledonian orogen* (affecting eastern Greenland, western Scandinavia, and Scotland) and the *Acadian orogen* in the region that is now the Appalachian Mountains **(Fig. 13.15a)**.

Throughout much of the middle Paleozoic, the western margin of North America continued to be a passive-margin basin. But finally, in the Late Devonian, the quiet environment of the basin was ended by a collision with an island arc. This event, known as the *Antler orogeny*, was the first of many orogenies to affect the western margin of the continent. The Caledonian, Acadian, and Antler orogenies all shed deltas of sediment onto the continents. These deposits formed thick successions of redbeds, such as those visible today in the Catskill Mountains **(Fig. 13.15b, c)**.

Life Evolution During the middle Paleozoic Era, new species of trilobites, gastropods, crinoids, and bivalves appeared in the sea, replacing species that had disappeared during the mass extinction at the end of the Ordovician Period. On land, vascular plants with woody tissues, seeds, and veins (for transporting water and food) rooted for the first time. With the evolution of veins and wood, plants could grow much larger, and by the Late Devonian Period the land surface hosted swampy forests with tree-sized relatives of club mosses and ferns. Also at this time, spiders, scorpions, insects, and crustaceans began to exploit both dry-land and freshwater habitats, and jawed fish, including sharks and bony fish, began to cruise the oceans. Finally, at the very end of the Devonian Period, the first amphibians crawled out onto land and inhaled air with lungs **(Fig. 13.15d)**.

The Late Paleozoic Era (Carboniferous–Permian Periods, 359–252 Ma)

Paleogeography The climate cooled significantly in the late Paleozoic (leading to so-called icehouse conditions). Seas gradually retreated from the continents, so that during the Carboniferous Period, regions that had hosted the limestone-forming reefs of epicontinental seas now became coastal areas and river deltas in which sand, shale, and organic debris accumulated. In fact, during the Carboniferous Period, Laurentia lay near the equator, so it enjoyed tropical and semi-tropical conditions that favored lush growth in swamps. This growth left thick piles of plant debris that transformed into coal after burial. Much of Gondwana and Siberia, in contrast, lay at high latitudes and, by the Permian Period, had become covered by ice sheets.

The late Paleozoic Era also saw a succession of continental collisions. The most significant collision occurred during Carboniferous and Permian time, as Gondwana rammed into Laurentia and Baltica, causing the *Alleghanian orogeny* of North America **(Fig. 13.16a)**. During this event, the final stage in the development of the Appalachians, eastern North America rammed against northwestern Africa, and what is now the Gulf Coast region of North America squashed against the northern margin of South America. A vast mountain belt grew, in which deformation generated huge faults and folds. We now see the eroded remnants of rocks deformed during this event in the Appalachian and Ouachita Mountains of North America. By the end of the collisions, nearly all land on the Earth had combined into the supercontinent **Pangaea**, so named by Alfred Wegener who, more than a century ago, first proposed its existence **(Fig. 13.16b)**. Imagine how different the world would be if you could walk from the middle of North America to the middle of Africa or Asia.

Along the continental side of the Alleghanian orogen, a wide band of deformation called the **Appalachian fold-thrust belt** formed. In this province, a distinctive style of deformation, called *thin-skinned deformation*, took place. In a thin-skinned fold-thrust belt, an array or system of thrust faults cuts across what was originally a layer cake of sedimentary strata **(Fig. 13.17)**. Movement on the faults displaces the strata and results in the formation of large folds. At depth, the thrust faults merge with a near-horizontal sliding surface, called a *detachment*, that lies just above the Precambrian basement. The adjective *thin-skinned* highlights the fact that the thrust faults do not cut into the basement, but occur only in the overlying cover or "skin" of sedimentary strata.

Stresses generated during the Alleghanian orogeny were so strong that pre-existing faults in the continental crust clear across North America became active again. The movement produced *basement uplifts* (local high areas, cored by

FIGURE 13.15 Paleogeography and fossils of the Silurian and Devonian Periods.

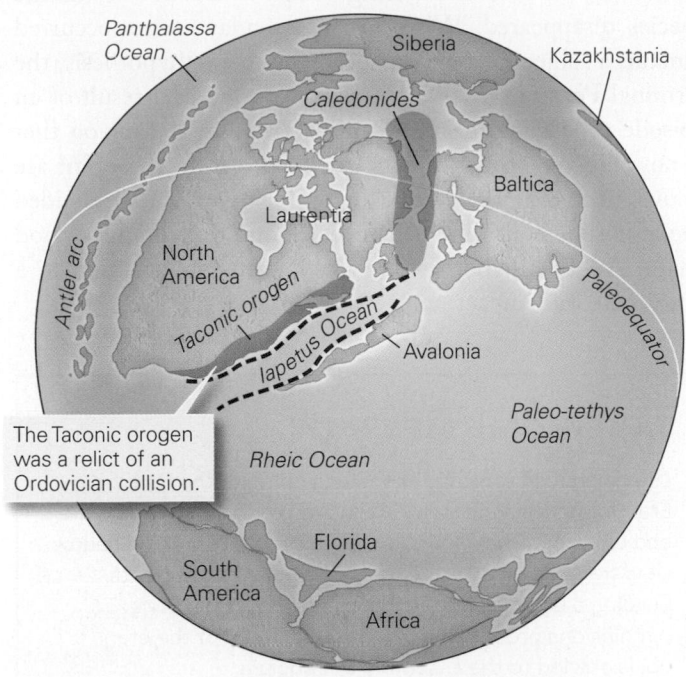

(a) During the Silurian and Devonian Periods, Laurentia collided with Baltica, Avalonia, and South America in succession, as oceans in between were consumed. The Antler arc formed off the west coast.

The Taconic orogen was a relict of an Ordovician collision.

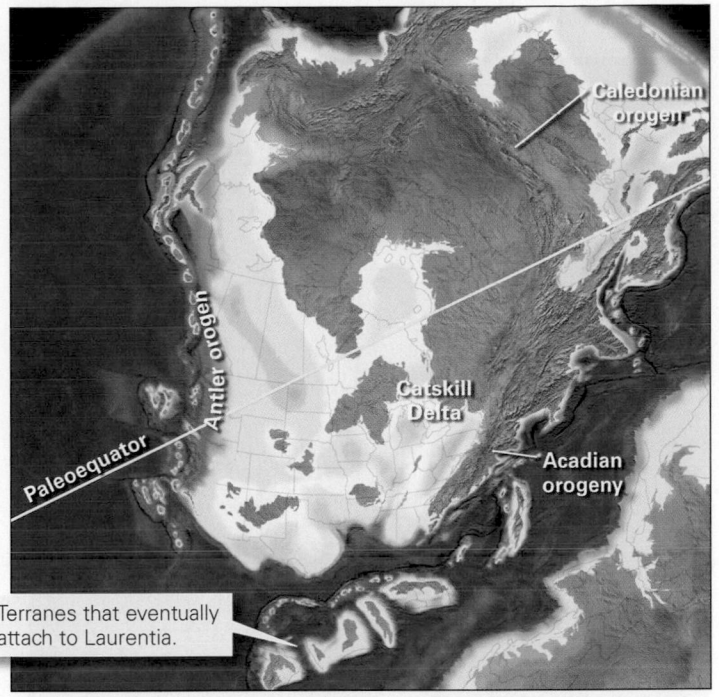

Terranes that eventually attach to Laurentia.

(b) During the Devonian Period, the Acadian orogeny shed sediments into a shallow sea to form the Catskill Delta on the east coast of Laurentia. The Antler orogeny shed sediments in the west.

(c) A road cut exposing Devonian redbeds (sandstone and shale) in New York.

Precambrian metamorphic and igneous rocks) and sediment-filled basins in the Midwest and in the region of the present-day Rocky Mountains (Colorado, New Mexico, and Wyoming). Geologists refer to the late Paleozoic basement uplifts of the Rocky Mountain region as the **Ancestral Rockies.**

The assembly of Pangaea involved a number of other collisions around the world as well. Notably, Africa collided with southern Europe during the *Hercynian orogeny* (also known as the *Variscan orogeny*). Also, a rift or small ocean in Russia closed, leading to the uplift of the Ural Mountains, and parts of China, along with other fragments of Asia, attached to southern Siberia.

Life Evolution The fossil record indicates that during the late Paleozoic Era, plants and animals continued to evolve toward forms more like those we see today. In coal swamps, fixed-wing insects such as huge dragonflies flew through a tangle of ferns, club mosses, and scouring rushes, and by the end of the Carboniferous Period, insects such as the cockroach, with foldable wings, appeared **(Fig. 13.18)**. Forests containing gymnosperms ("naked seed" plants, such as conifers) and cycads (trees with a palm-like stalk peaked by a fan of fern-like fronds) became widespread in the Permian Period.

~ 20 cm

(d) A Late Devonian fossil skeleton of *Tiktaalik*; this lobe-finned fish was one of the first animals to walk on land.

FIGURE 13.16 Paleogeography at the end of the Paleozoic Era.

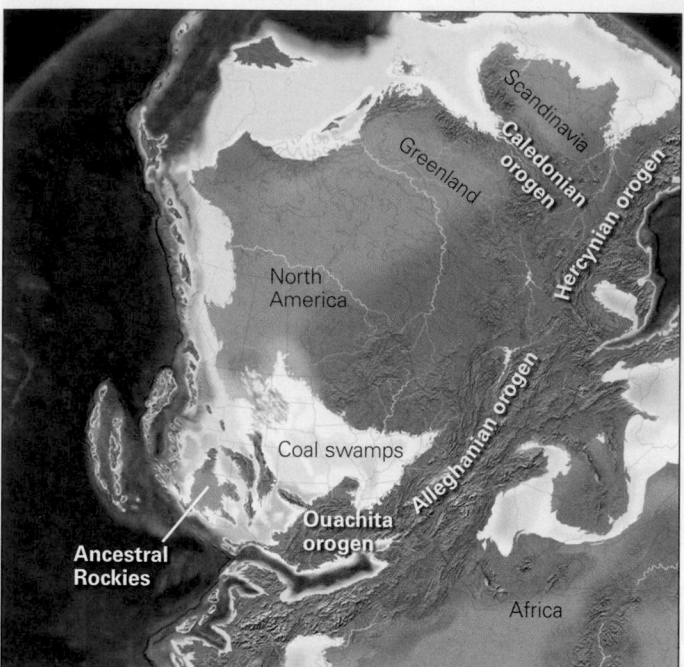

(a) During the Alleghanian and Hercynian orogenies, a huge mountain belt formed. The Caledonian orogen had formed earlier. Coal swamps bordered interior seas, and the Ancestral Rockies rose.

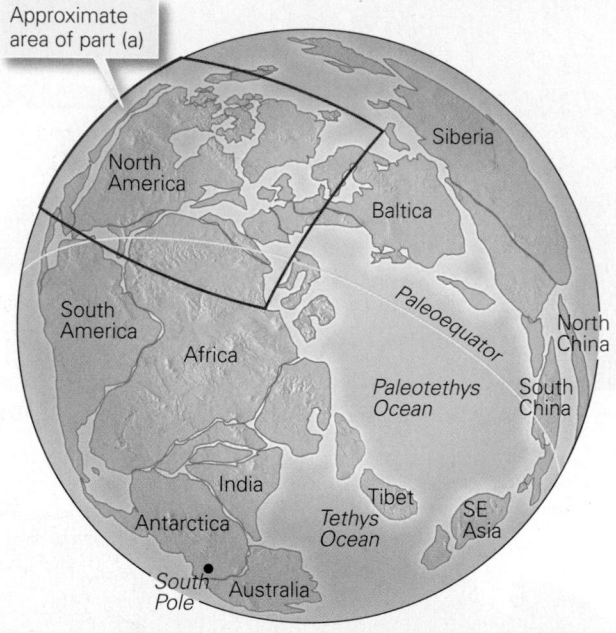

(b) At the end of the Paleozoic, almost all land had combined into a single supercontinent called Pangaea.

Amphibians and, later, reptiles populated the land. The appearance of reptiles marked the evolution of a radically new component in animal reproduction: eggs with a protective shell. By producing such eggs, reptiles could reproduce without returning to the water and thus could populate previously uninhabitable environments on land.

The late Paleozoic Era came to a close with two major mass-extinction events, during which over 95% of marine species disappeared. Why these particular events occurred remains a subject of debate. According to one hypothesis, the terminal **Permian mass extinction** occurred as a result of an episode of extraordinary volcanic activity in the region that is now Siberia—basalt sheets extruded during the event are known as the *Siberian traps*. Eruptions could have clouded the atmosphere, acidified the oceans, and disrupted the food chain. Another hypothesis relates the mass extinction to a huge meteorite impact.

> **TAKE-HOME MESSAGE**
>
> Life diversified radically at the beginning of the Paleozoic Era. During the Paleozoic, life moved onto land, and by the end of it, reptiles roamed through forests of trees. Shallow seas transgressed and regressed over the continents, building a layer cake of Paleozoic strata, and a series of collisions produced orogens. At the end of the era, collisions led to the assembly of Pangaea.
>
> **QUICK QUESTION:** What event, as indicated by the fossil record, marks the end of the Paleozoic?

13.7 The Mesozoic Era: When Dinosaurs Ruled

The Early and Middle Mesozoic Era (Triassic–Jurassic Periods, 252–145 Ma)

Paleogeography Pangaea existed for about 100 million years, until the Late Triassic. Then rifts developed and the supercontinent began to break apart. By the end of the Jurassic Period, the Mid-Atlantic Ridge had formed, the North Atlantic Ocean and Gulf of Mexico started to grow, and North America had split away from Europe and Africa **(Fig. 13.19a)**. During the early stages of its formation, the North Atlantic was narrow and shallow, and evaporation made its water so salty that thick evaporite deposits accumulated along its margins. These evaporites now lie buried beneath the younger strata in the passive-margin basins that rim the North Atlantic—they are particularly thick beneath the Gulf Coast region of the United States.

The Earth had a generally mild climate during the Triassic and Jurassic. At this time, large areas of North America's interior became nonmarine environments. In fact, during the

FIGURE 13.17 Features of the Appalachian Mountains in the eastern United States.

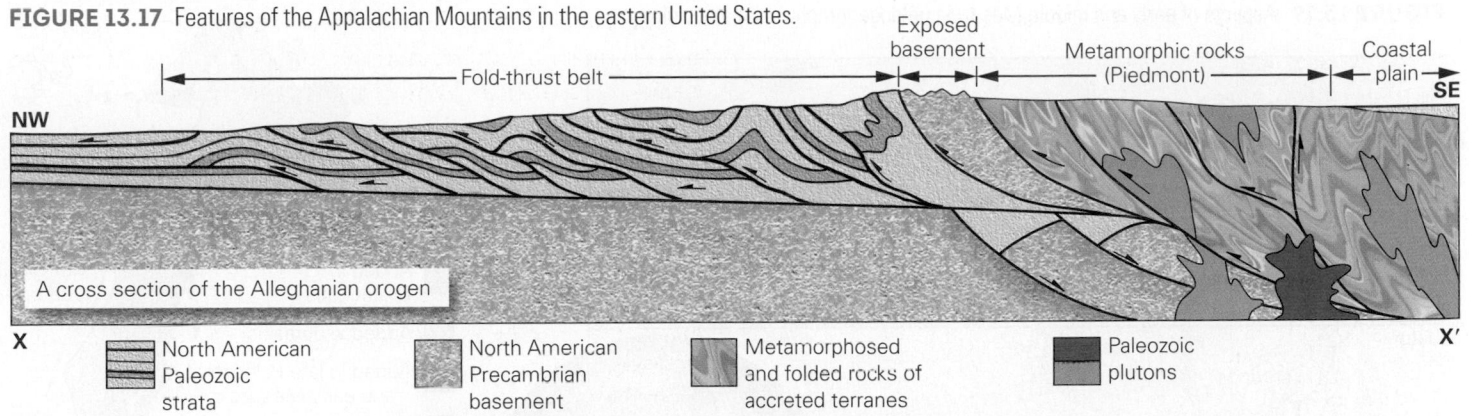

Exposed basement
Metamorphic rocks (Piedmont)
Coastal plain
Fold-thrust belt
NW
SE
A cross section of the Alleghanian orogen
X
X′

North American Paleozoic strata

North American Precambrian basement

Metamorphosed and folded rocks of accreted terranes

Paleozoic plutons

(a) In the fold-thrust belt, strata have been pushed westward and have been folded and faulted. The Blue Ridge exposes a slice of Precambrian basement. Metamorphic and plutonic rocks underlie the Piedmont. Cross-section location is shown in (b).

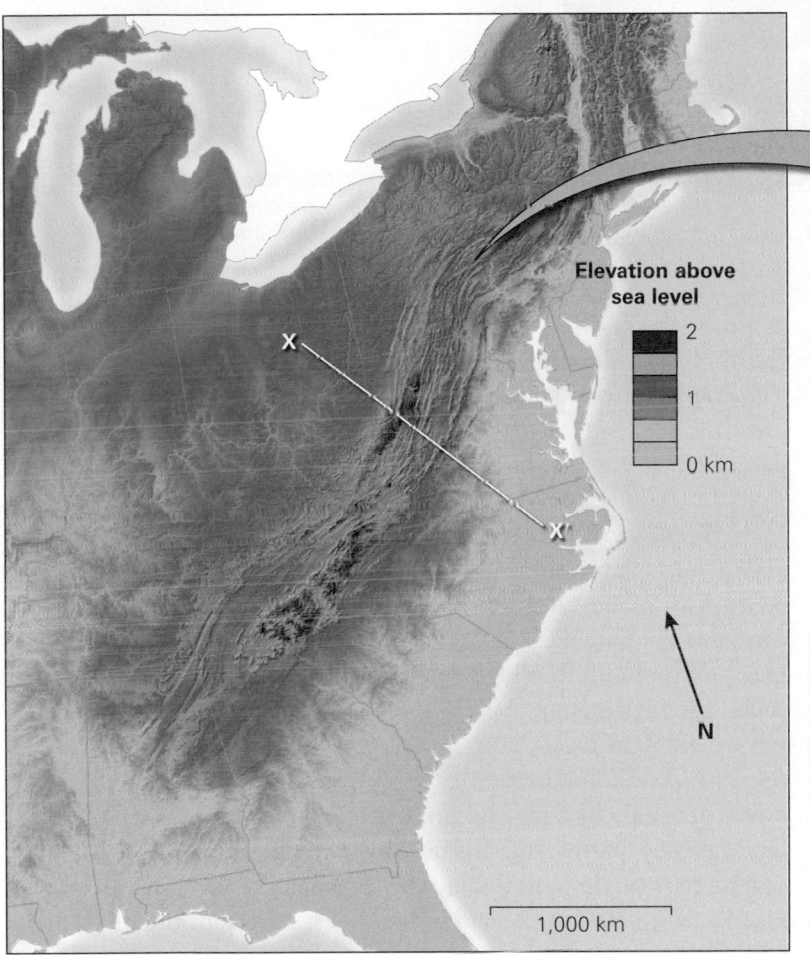

Elevation above sea level

2

1

0 km

N

1,000 km

(b) The eroded remnants of the Appalachian orogen stand out in the eastern United States. The white line shows the approximate position of the cross section in part (a).

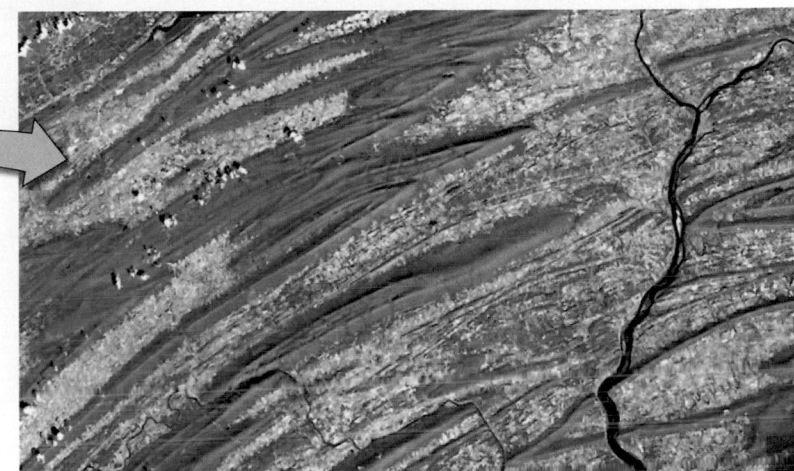

(c) Folds stand out in this satellite image of the Pennsylvania Valley and Ridge. Resistant sandstone layers form the ridges. The field of view is 80 km wide.

FIGURE 13.18 A museum diorama of a Carboniferous coal swamp. The inset image shows the size of a giant dragonfly (wingspan of about 1 m)—that inhabited such swamps during the Paleozoic—relative to the size of a human.

Early Jurassic, sand dunes covered parts of the southwestern United States. These dunes, when buried and lithified, became redbeds containing prominent cross beds that you can see today in Zion National Park **(Fig. 13.19b)**. Then, as the Middle Jurassic began, sea level rose, and parts of what are now the Rocky Mountains were submerged.

FIGURE 13.19 Aspects of early and middle Mesozoic paleogeography and paleobiology.

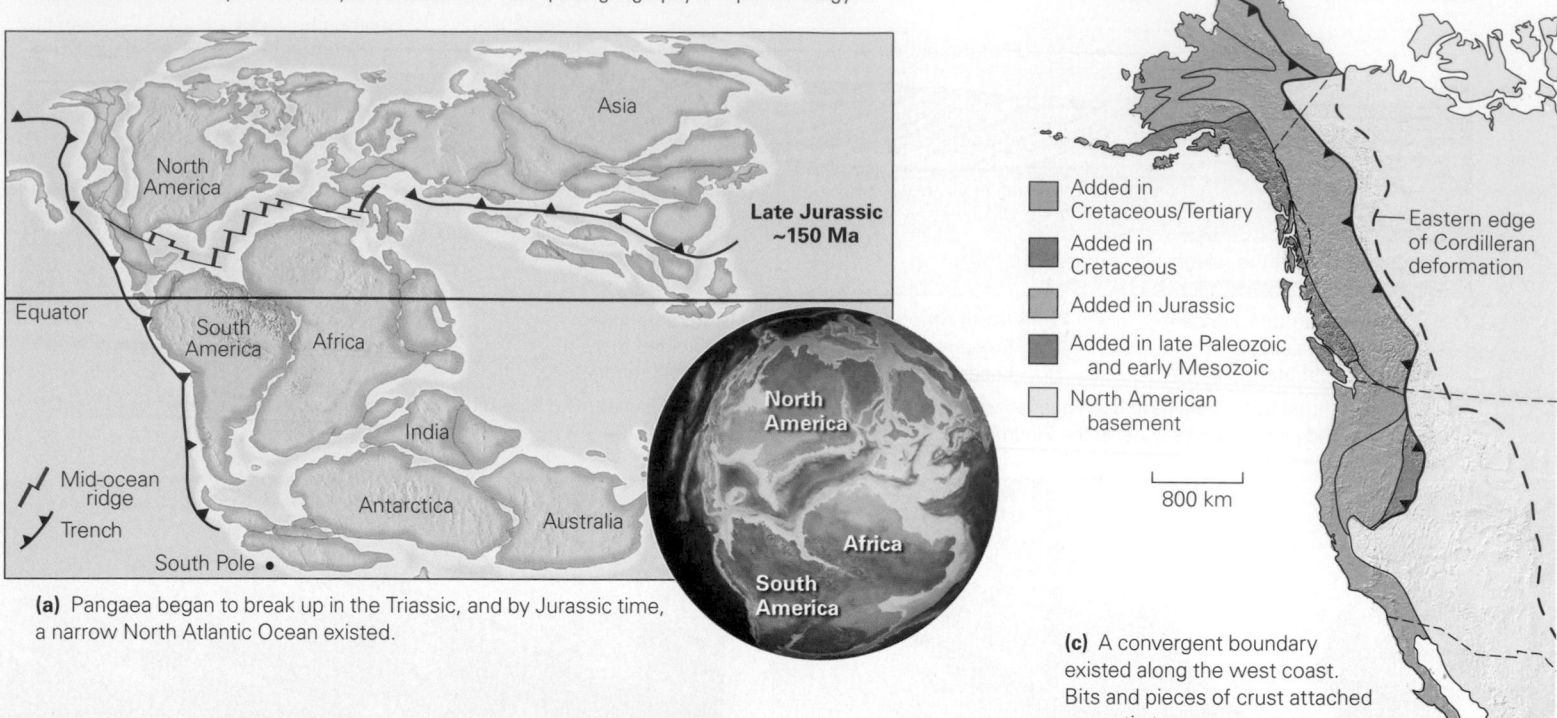

Late Jurassic ~150 Ma

Equator

Mid-ocean ridge

Trench

South Pole •

(a) Pangaea began to break up in the Triassic, and by Jurassic time, a narrow North Atlantic Ocean existed.

Added in Cretaceous/Tertiary

Added in Cretaceous

Added in Jurassic

Added in late Paleozoic and early Mesozoic

North American basement

Eastern edge of Cordilleran deformation

800 km

(c) A convergent boundary existed along the west coast. Bits and pieces of crust attached as exotic terranes.

(b) During the Jurassic, immense sand dunes blanketed the southwestern United States. Sandstone beds in Zion Park are the relicts of these dunes.

On the western margin of North America, convergent-boundary tectonics became the order of the day. Beginning with Late Permian and continuing through Mesozoic time, subduction generated volcanic island arcs and caused them, along with oceanic plateaus (the product of hot-spot volcanism), to collide with North America. As a result, North America grew in land area by the accretion (addition) of crustal fragments onto its western margin **(Fig. 13.19c)**. Because these fragments consist of crust that formed elsewhere, not originally on or adjacent to the continent, geologists refer to them

as **exotic terranes**. At the end of the Jurassic, subduction of Pacific Ocean floor beneath North America began, an event that produced a major continental volcanic arc, the *Sierran arc*. This arc continued to erupt through the Cretaceous, as described later in this section.

Life Evolution During the early Mesozoic Era, a variety of new plant and animal species appeared, filling the ecological niches left vacant by the Late Permian mass extinction. Reptiles, such as plesiosaurs, begin to swim in the oceans **(Fig. 13.20a)**, and new kinds of corals became the predominant reef builders. On land, gymnosperms and reptiles diversified, and the Earth saw its first turtles and flying reptiles (such as pterodactyls; **Fig. 13.20b**). And at the end of the Triassic Period, the first true dinosaurs evolved.

Dinosaurs differed from other reptiles in that their legs extended underneath their bodies rather than off to the sides. In addition, they may have been warm-blooded. By the end of the Jurassic Period, gigantic sauropod dinosaurs (weighing up to 91 metric tons), along with other beasts such as stegosaurs, thundered across the landscape **(Fig. 13.20c)**, and the first feathered birds, including the occasional archaeopteryx, took to the skies. The earliest ancestors of mammals appeared at the end of the Triassic Period in the form of small, rat-like creatures.

Did you ever wonder...
when the dinosaurs lived?

FIGURE 13.20 Reptiles take to the land and sea.

(a) Plesiosaurs were swimming reptiles that had flippers instead of legs.

(b) There were many species of pterodactyls, some of which had wing spans up to 11 m. The name means winged finger.

(c) During the Jurassic, giant dinosaurs roamed the land. This painting shows several species.

The Late Mesozoic Era (Cretaceous Period, 145–66 Ma)

Paleogeography During the Cretaceous Period, several more exotic island arcs accreted along the west coast of North America **(Fig. 13.21a)**. Also, the Earth's climate continued to shift to warmer greenhouse conditions, and sea level rose significantly, reaching levels that had not been attained for the previous 200 million years. Great shallow seas flooded most of the continents. In fact, during the latter part of the Cretaceous Period, a shark could have swum from the Gulf of Mexico to the Arctic Ocean in the *Western Interior Seaway* or, similarly, across much of western Europe **(Fig. 13.21b)**.

In western North America, the *Sierran arc*, a large continental volcanic arc that formed at the end of the Jurassic Period, continued to be active. This arc resembled the present-day Andean arc of western South America. Though the volcanoes of the Sierran arc have long since eroded away, we can see their roots in the form of the plutons that now constitute the granitic batholith of the Sierra Nevada range. A thick accretionary prism, formed from sediments and debris scraped

FIGURE 13.21 Cretaceous paleogeography.

Convergent boundary   Thrust fault   Volcanic arc

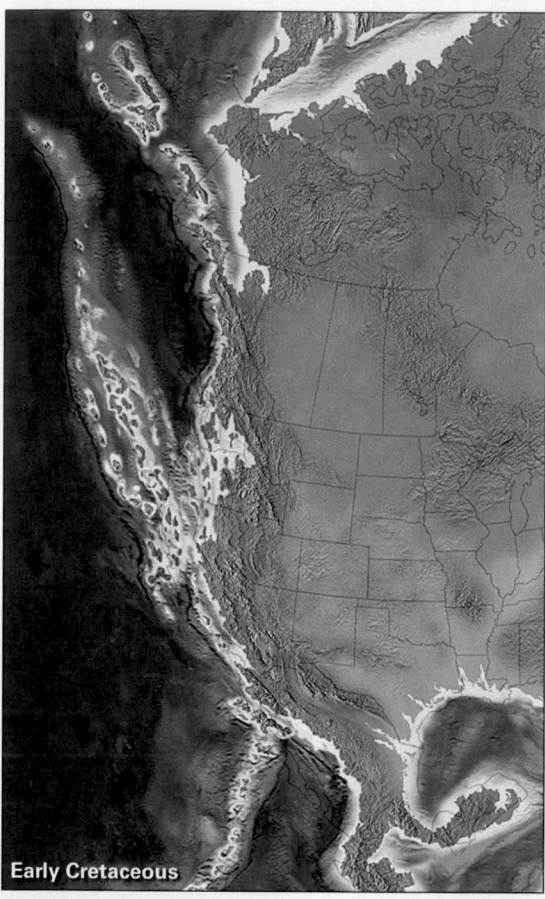

Early Cretaceous

(a) In the Early Cretaceous, several exotic island arcs collided with western North America.

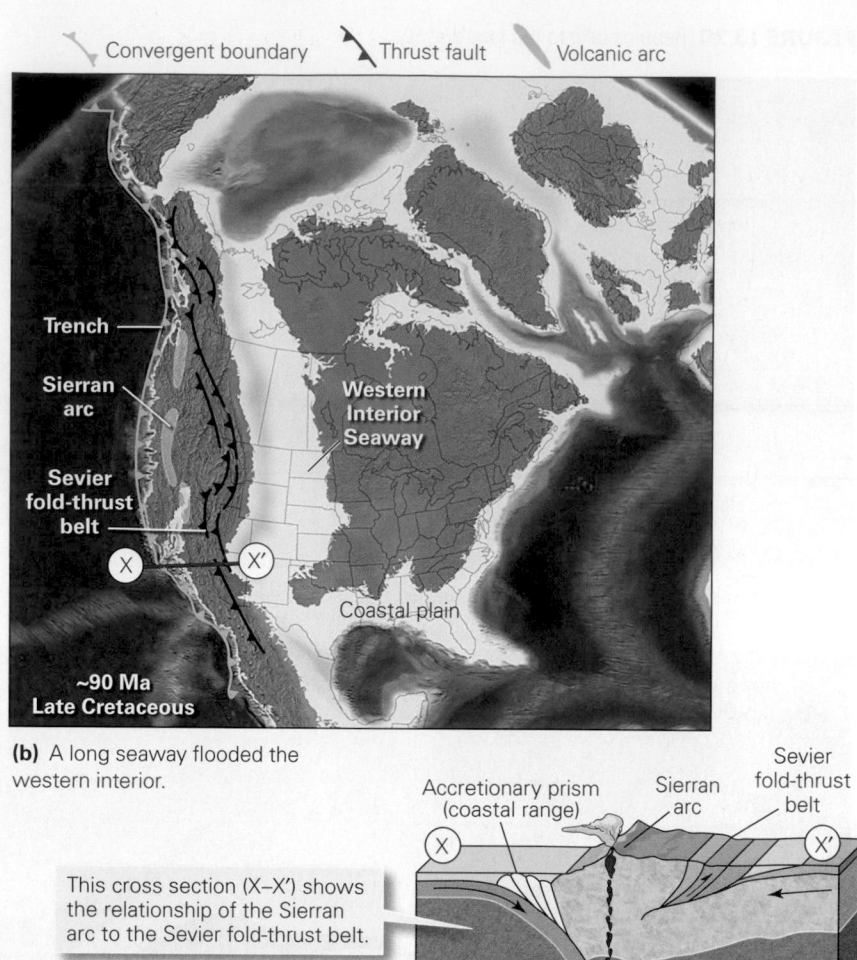

Trench

Sierran arc

Sevier fold-thrust belt

Western Interior Seaway

Coastal plain

X   X'

~90 Ma Late Cretaceous

(b) A long seaway flooded the western interior.

This cross section (X–X') shows the relationship of the Sierran arc to the Sevier fold-thrust belt.

Accretionary prism (coastal range)   Sierran arc   Sevier fold-thrust belt

X   X'

(c) The Sierran arc, formed in the Late Jurassic remained active through the Cretaceous. A fold-thrust belt formed to the east, as did a transcontinental seaway.

off the subducting oceanic plate, piled up to the west of the Sierran arc and now crops out in the Coast Ranges of California. Compressional stresses along the western North American convergent boundary activated faults east of the arc. This crustal-shortening event, which geologists refer to as the **Sevier orogeny**, produced a thin-skinned fold-thrust belt whose remnants can be seen today in the Canadian Rockies and in western Wyoming **(Fig. 13.21c)**. The weight of the fold-thrust belt helped push the surface of the continent down, forming a wide foreland basin that filled with sediment (see Chapter 7). The depth of the basin may have been enhanced by the downward pull of the subducting plate beneath it. This foreland basin lay beneath the western part of the Western Interior Seaway.

The breakup of Pangaea continued through the Cretaceous Period, with the opening of the South Atlantic Ocean and the separation of South America and Africa from Antarctica and Australia. India broke away from Gondwana and headed rapidly northward toward Asia **(Fig. 13.22a, b)**. Along the continental margins of the newly formed Mesozoic oceans,

large passive-margin basins developed, which filled with great thicknesses of sediments. The passive-margin basin along the Gulf Coast of the United States eventually accumulated a wedge of sediment that's over 15 km thick.

Subduction continued along the convergent plate boundary of North America's west coast as the Cretaceous came to a close. In the United States, however, the character of the deformation associated with compression across this plate boundary changed. Prior to 80 Ma, as we have seen, compression produced a thin-skinned fold-thrust belt. In contrast, starting at about 80 Ma and continuing until about 40 Ma, deformation swept eastward **(Fig. 13.22c)**, and slip began to take place on large reverse faults that penetrated deep into Precambrian basement, beneath the cover of Paleozoic and Mesozoic strata. Movement on these faults generated **basement uplifts** by bringing metamorphic and igneous rocks in the hanging-wall blocks up and over the Paleozoic and Mesozoic strata of the footwall blocks **(Fig. 13.22d)**. Geologists refer to these faults as *thick-skinned* reverse faults, in contrast to the *thin-skinned* thrust faults of fold-thrust belts. When slip

FIGURE 13.22 Paleogeography in Late Cretaceous through Eocene time.

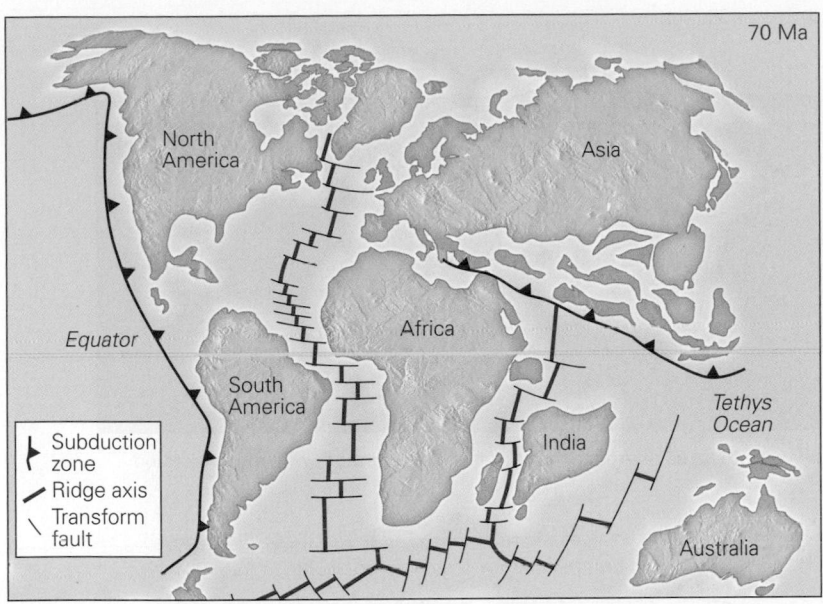

(a) By the Late Cretaceous Period, the South Atlantic Ocean had formed, and India was moving rapidly northward toward Asia.

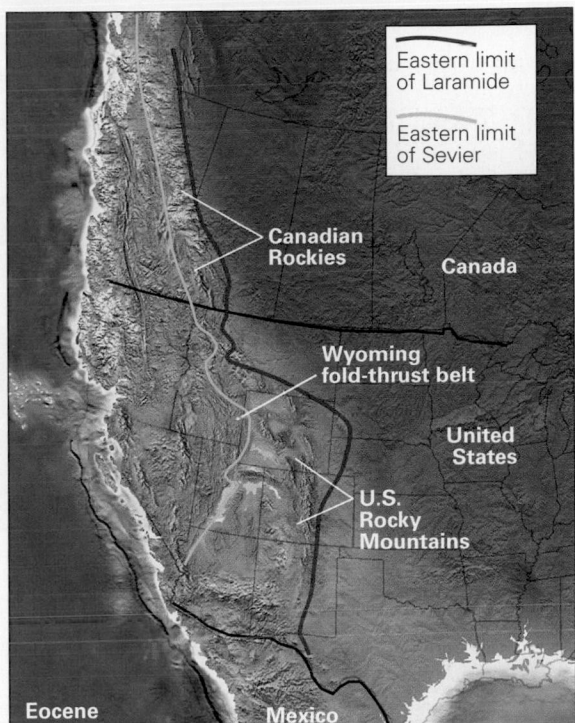

(c) During the Laramide orogeny, deformation shifted eastward in the United States, moving from the Sevier orogen to the Rocky Mountains, and the style of deformation changed.

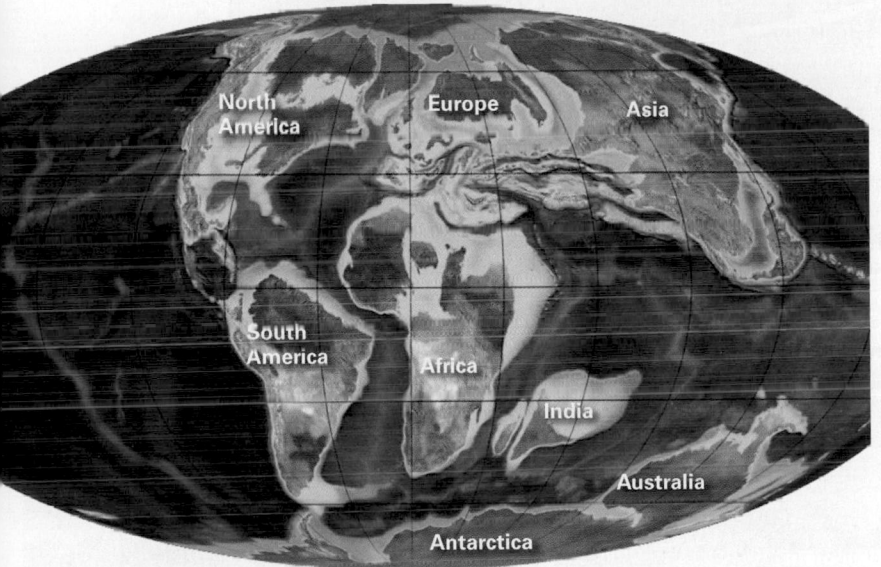

(b) In this Late Cretaceous paleogeographic reconstruction, southern Europe and Asia are beginning to form from a collage of many crustal blocks.

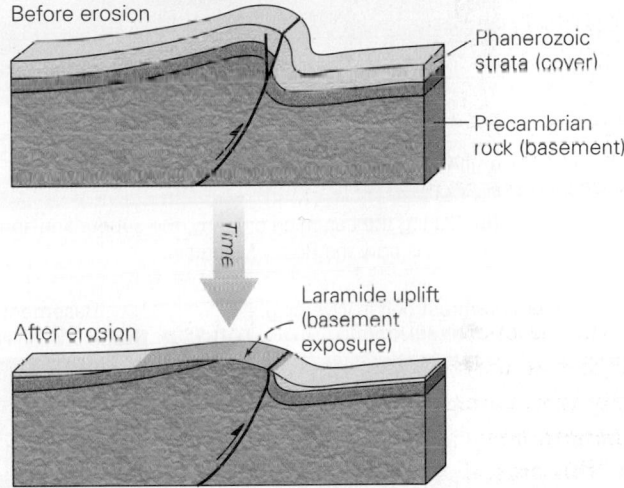

(d) The Laramide orogeny produced basement uplifts, in which faults lifted up blocks of basement, causing the overlying strata to bend into a stair-step-like fold.

takes place on a thick-skinned reverse fault, the strata overlying the fault bend into large *monoclines*, folds whose shape resembles the drape of a carpet over a step. The formation of basement uplifts due to thick-skinned reverse faulting produced the *Rocky Mountains* of the United States, a group of mountain ranges in Wyoming, Colorado, Montana, Utah, and New Mexico. (Note that the Rocky Mountains are part of the much wider North American Cordillera.)

Geologists refer to the thick-skinned deformation event that took place in the U.S. Rocky Mountains between 80 and 40 Ma as the **Laramide orogeny** to distinguish it from the thin-skinned deformation that took place earlier during the Sevier orogeny **(Fig. 13.23)**. Some geologists have suggested that the contrast in the location of faulting between the Sevier and Laramide orogenies in the United States may reflect a change in the dip of the subducting plate. During the Laramide orogeny, the subducting plate entered the mantle at a shallower angle and therefore scraped along and applied stress to the base of the continent farther inland. The change in subduction angle may have been due

FIGURE 13.23 Contrasting the Sevier orogeny with the Laramide orogeny.

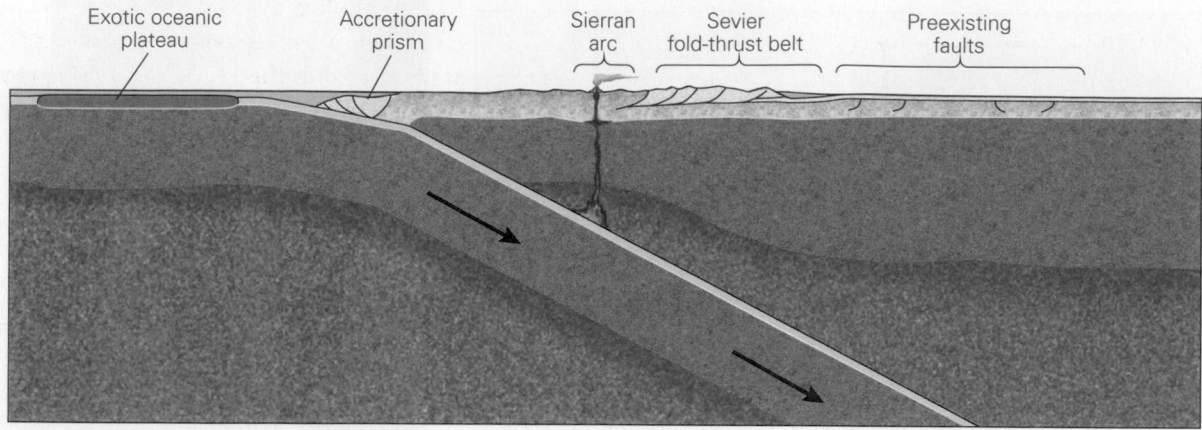

Exotic oceanic plateau | Accretionary prism | Sierran arc | Sevier fold-thrust belt | Preexisting faults

(a) During the Sevier orogeny, the subduction angle was steep, so an Andean-type volcanic arc formed, with a thin-skinned fold-thrust belt on the continental side.

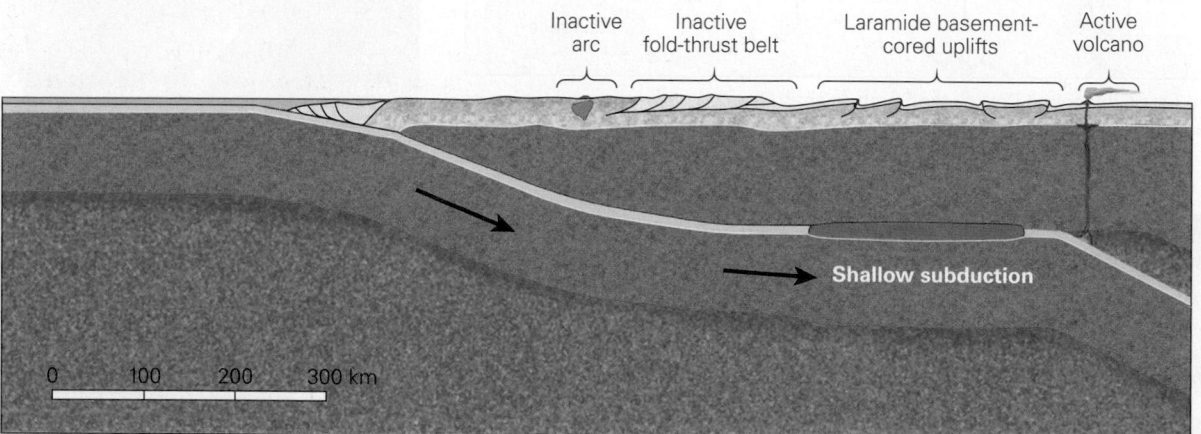

Inactive arc | Inactive fold-thrust belt | Laramide basement-cored uplifts | Active volcano

Shallow subduction

0   100   200   300 km

(b) During the Laramide orogeny, the subduction angle was shallow, and thick-skinned basement uplifts formed in the region of what is now the Rocky Mountains.

Part of the Sevier fold-thrust belt in Wyoming. Note the long ridges of tilted strata.

Basement uplift of the Wind River Mountains

Basement   Cover

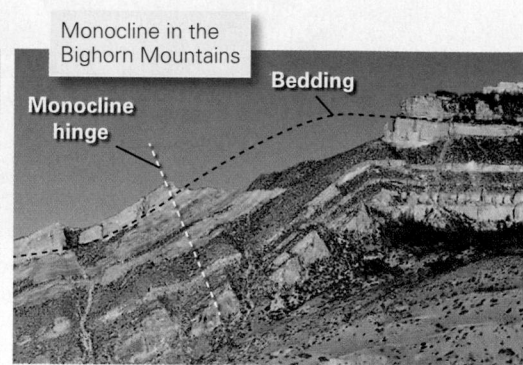

Monocline in the Bighorn Mountains

Bedding

Monocline hinge

(c) Photographs of Sevier and Laramide features.

to the presence of an oceanic plateau—a region of thicker oceanic crust—in the downgoing slab, or to a change in the rate of subduction. Significantly, this change in dip did not occur in Canada, so during the Laramide orogeny, the Canadian Rockies simply continued to grow eastward as a thin-skinned fold-thrust belt.

Geologists have determined that seafloor-spreading rates may have been as much as three times faster during the Cretaceous than they are today. As a result, more of the oceanic crust

was younger and warmer than it is today, and since young seafloor lies at a shallower depth than does older seafloor (due to isostasy; see Chapter 11), Cretaceous mid-ocean ridges occupied more volume than they do today. Also during the Cretaceous, huge oceanic plateaus formed from basalts erupted at hot-spot volcanoes. The existence of these plateaus implies that particularly active mantle plumes, or **superplumes**, reached the base of the lithosphere. Melting at the top of such plumes produced immense quantities of magma, which erupted and built up the plateaus. Growth of oceanic plateaus also displaced seawater. Broader mid-ocean ridges and large oceanic plateaus displaced seawater, causing sea level to rise during the Cretaceous. At this time, the sea submerged what is now the coastal plain of the eastern and southern United States as well as much of England and Europe. The chalk deposits we mentioned at the opening of this chapter accumulated in these shallow seas.

Notably, the volcanism associated with extra-rapid sea-floor spreading, as well as with oceanic plateau growth, probably released large quantities of CO_2, a greenhouse gas, into the atmosphere. Geologists hypothesize that the increased atmospheric CO_2 concentration led to a global rise in atmospheric temperature. Rising temperatures would cause seawater to expand and polar ice sheets to melt, both phenomena that would make sea level go up even more. Considering all the phenomena that caused sea level to rise during the Cretaceous, it's no surprise that the continents flooded and that large epicontinental seas formed during this period.

Life Evolution In the seas of the Cretaceous world, modern fish appeared and became dominant. In contrast to earlier fish, modern fish had short jaws, rounded scales, symmetric tails, and specialized fins. Huge swimming reptiles and gigantic turtles (with shells up to 4 m across) preyed on the fish. On land, cycads largely vanished, and *angiosperms* (flowering plants), including hardwood trees, began to compete successfully with conifers for dominance of the forest. Dinosaurs reached their peak of success at this time, inhabiting almost all environments on the Earth. Social herds of grazing dinosaurs roamed the plains, preyed on by the fearsome *Tyrannosaurus rex* (a Cretaceous, not a Jurassic, dinosaur, despite what Hollywood says!). Pterosaurs with wingspans of up to 11 m soared overhead, and birds began to diversify. Mammals also diversified and developed larger brains and more specialized teeth, but for the most part they remained small and rat-like.

The K-Pg Boundary Event Geologists first recognized the K-Pg boundary from 18th-century studies that identified an abrupt global change in fossil assemblages. (K stands for Cretaceous and Pg for Paleogene, the earliest period of the Cenozoic Era.) Until the 1980s, most geologists assumed that this faunal turnover took millions of years. But modern dating techniques indicate that the change happened almost instantaneously and that it signaled a sudden mass extinction of most species on the Earth. The dinosaurs, rulers of the planet for over 150 million years, simply vanished, along with 90% of plankton species and up to 75% of plant species, at 66 Ma. What catastrophe could cause such a sudden and extensive mass extinction? From data collected in the 1970s and 1980s, most geologists have concluded that the Cretaceous Period came to a close, at least in part, as a result of the impact of a 13-km-wide meteorite at the site of the present-day Yucatán Peninsula in Mexico **(Fig. 13.24a)**.

The discoveries leading up to this conclusion provide fascinating insight into how science works. The story began when Walter Alvarez, a geologist studying strata in Italy, noted that a thin layer of clay interrupted the deposition of deep-sea limestone precisely at the K-Pg boundary. Cretaceous plankton shells constituted the limestone below the clay layer, whereas Cenozoic plankton shells made up the limestone just above the clay. Apparently, for a short interval of time at the K-Pg boundary, all the plankton died, so only clay settled out of the sea. When Alvarez, his father, Luis (a physicist), and other colleagues analyzed the clay, they learned that it contained *iridium*, a very heavy element found only in meteorites. Soon geologists were finding similar iridium-bearing clay layers at the K-Pg boundary all over the world. Further study showed that the clay layer contained other unusual materials, such as tiny glass spheres (formed from the flash freezing of molten rock), wood ash, and shocked quartz (grains of quartz that had been subjected to intense pressure). Only an immense meteorite impact could explain all these features. The glass spheres formed when melt sprayed into the air from the impact site, the iridium came from fragments of the colliding object, and the shocked quartz grains were produced and scattered by the force of the impact. The wood ash resulted when forests were set ablaze, conceivably because the impact ejected super-heated debris at such high velocity that the debris almost went into orbit

A Tyrannosaurus rex (12.5 m from its nose to the tip of its tail)

Person for scale

FIGURE 13.24 The Cretaceous Paleogene (K-Pg) boundary event was a huge meteorite impact that caused a mass extinction.

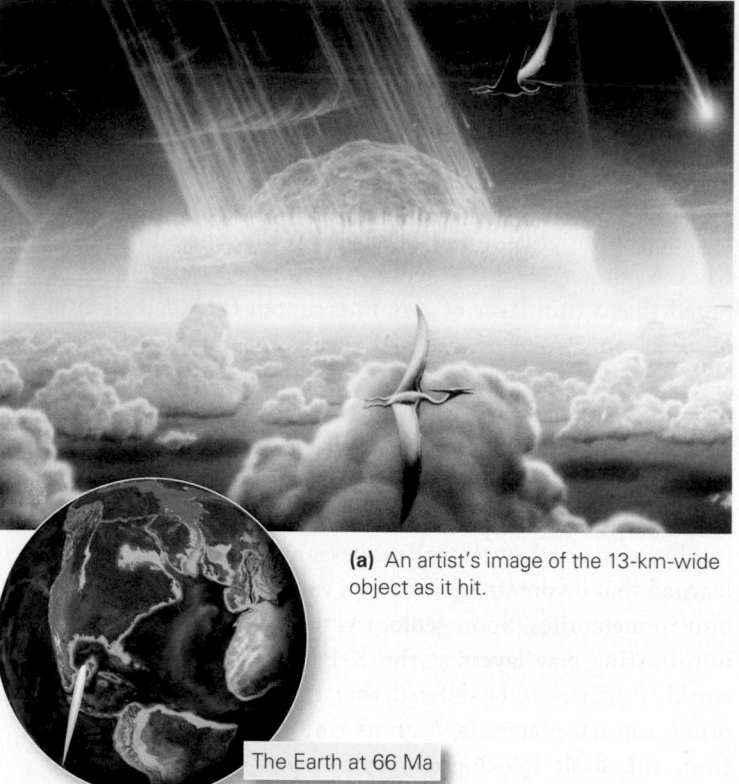

(a) An artist's image of the 13-km-wide object as it hit.

The Earth at 66 Ma

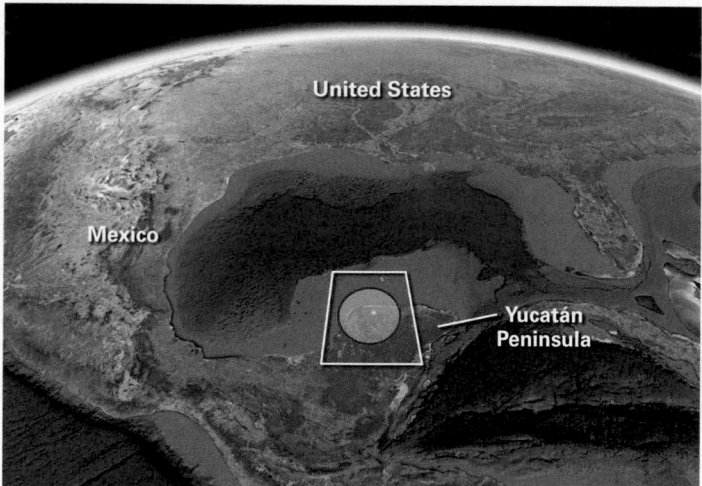

United States

Mexico

Yucatán Peninsula

(b) The location of the buried crater today. Gravity anomalies reveal the shape of the crater.

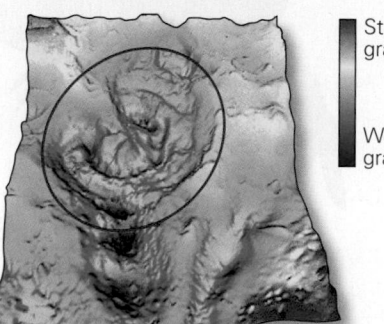

Stronger gravity

Weaker gravity

(c) The crater is now buried, but gravity anomalies outline its shape (see Interlude D).

and could reach forests worldwide. The impact also generated 2-km-high tsunamis that inundated the shores of continents, and it generated a blast of super-heated air.

Researchers suggest that the impact caused unfathomable destruction because the dust, ash, and aerosols that it produced lofted into the atmosphere and transformed the air into a murky haze that reflected incoming sunlight. As a result, photosynthesis became difficult, so all but the hardiest plants died, and winter-like cold might have lasted all year, perhaps for several years. Finally, sulfur-bearing aerosols could have combined with water to produce acid rain, acidifying the ocean and land. These conditions could have broken the food chain and, therefore, could have triggered extinctions.

Geologists suggest that the meteorite responsible for the K-Pg boundary event landed on the northwestern coast of the Yucatán Peninsula, because beneath the reefs and sediments of this tropical realm lies a 100-km-wide by 16-km-deep scar called the Chicxulub crater **(Fig. 13.24b, c)**. A layer of tiny glass spheres up to 1 m thick occurs at the K-Pg boundary in strata near the site. And radiometric dating indicates that igneous melts in the crater formed at the time of the K-Pg boundary event. The discovery of this event has led geologists to speculate that other such collisions may have punctuated the path of life evolution throughout Earth history.

TAKE-HOME MESSAGE

The Mesozoic Era began with the breakup of Pangaea and the formation of the Atlantic Ocean. A convergent boundary formed along the west coast of North America, yielding the Sierran arc and eventually leading to the uplift of the Rocky Mountains. Dinosaurs ruled the planet, and modern forests appeared. A huge meteorite impact marks the K-Pg boundary event—the end of the era—which may have caused the extinction of the dinosaurs.

QUICK QUESTION: Did all of today's continents break off Pangaea at the same time?

13.8 The Cenozoic Era: The Modern World Comes to Be

Paleogeography Plate tectonics doesn't stop, so during the last 66 million years the map of the Earth has continued to change, gradually producing the configuration of continents we see today. The final stages of the Pangaea breakup separated

Australia from Antarctica and Greenland from North America and formed the North Sea between Britain and continental Europe. The Atlantic Ocean continued to grow because of sea-floor spreading at the Mid-Atlantic Ridge, so the Americas moved relatively westward, away from Europe and Africa. Meanwhile, the continents that once made up Gondwana drifted northward as subduction consumed the intervening ocean. Several volcanic island arcs and microcontinents started colliding with Asia beginning around 50 to 60 Ma, and then about 40 Ma, India arrived and rammed into Asia. The overall result of this protracted period of collisional tectonics was the uplift of the Himalayas and the Tibetan Plateau. Meanwhile, Africa, along with some volcanic island arcs and microcontinents, collided with Europe to produce the Alps of southern Europe, the Caucasus Mountains between the Black Sea and Caspian Sea, and the Zagros Mountains of Iran. Finally, collision between Australia and New Guinea led to orogeny in Papua New Guinea. Thus, collisions of the former Gondwana continents with the southern margins of Europe and Asia resulted in the formation of the largest orogenic belt on the Earth today, the **Alpine-Himalayan chain (Fig. 13.25)**.

As the Americas moved westward, convergent boundaries evolved along their western margins. In South America, convergent-boundary activity built the Andes, which remains an active orogen to the present day. In North America, convergent-boundary activity continued without interruption until about 40 Ma (the Eocene Epoch), yielding, as we have seen, the Laramide orogeny. Then the Farallon-Pacific Ridge, beneath the waters of the Pacific Ocean, reached the continental margin, and the configuration of plates along the western shore of North America began to evolve. By 25 Ma, a transform boundary had replaced part of the convergent boundary in the western part of the continent **(Fig. 13.26)**. Where this happened, volcanism and compression ceased, and strike-slip faulting took over. This led to the formation of the San Andreas fault system in California and the Queen Charlotte fault system off the west coast of Canada. Along the San Andreas and Queen Charlotte faults today, the Pacific Plate moves northward with respect to North America at a rate of about 6 cm per year. In the western United States, convergent-boundary tectonics continues only in Washington, Oregon, and northern California, where subduction of the Juan de Fuca Plate generates the volcanism of the Cascade Range.

As compression associated with convergent-boundary tectonics ceased in the western United States south of the Cascades, the region began to undergo extension in a roughly east-west direction. The result was the formation of the **Basin and Range Province**, a broad continental rift. Continued extension in this rift over the past 20 Ma has caused the region

FIGURE 13.25 The two main active continental orogenic systems on the Earth today. The Alpine-Himalayan system formed when Africa, India, and Australia collided with Asia. The Cordilleran and Andean systems reflect the consequences of convergent-boundary tectonics along the shoreline of the eastern Pacific Ocean.

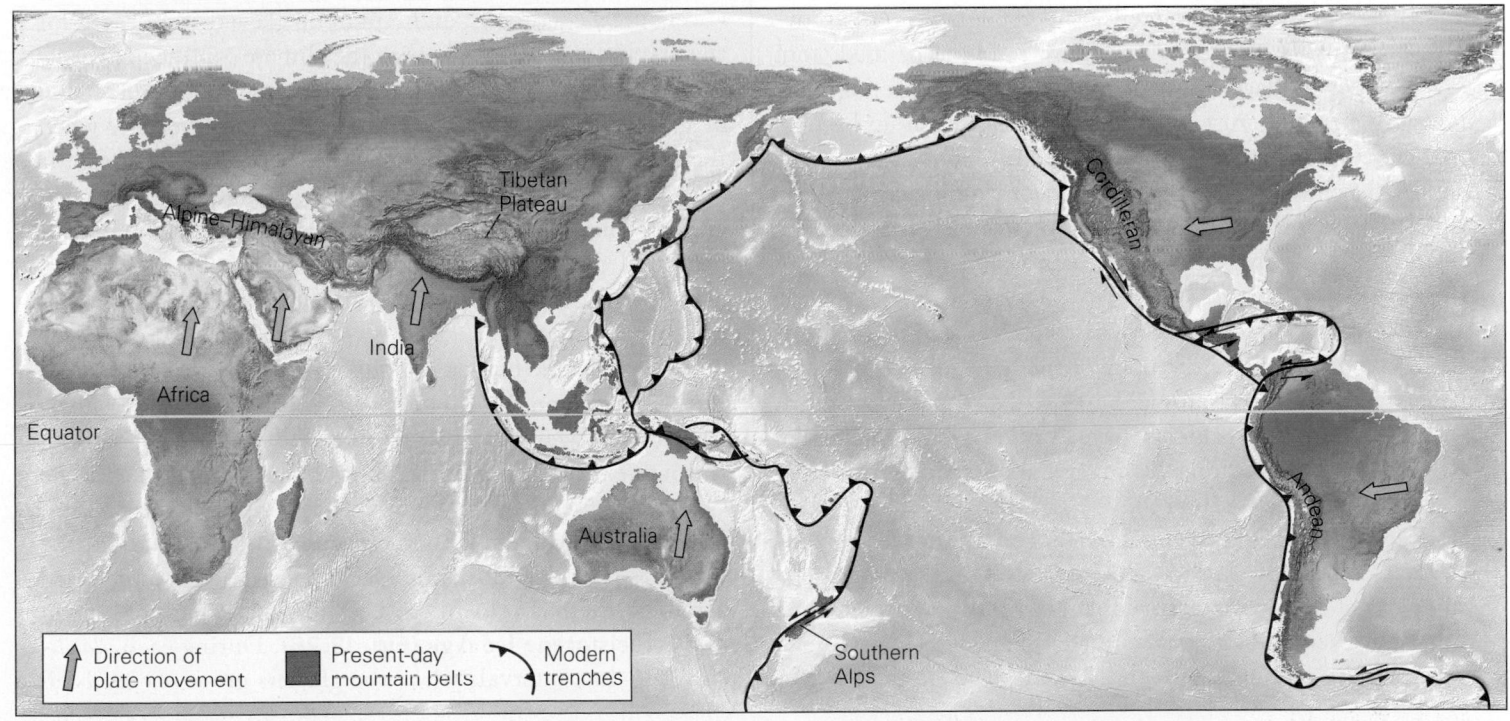

FIGURE 13.26 The western margin of North America changed from a convergent boundary into a transform boundary after subduction of the Pacific-Farallon Ridge. Then the San Andreas fault developed. To the east, the Basin and Range rift developed.

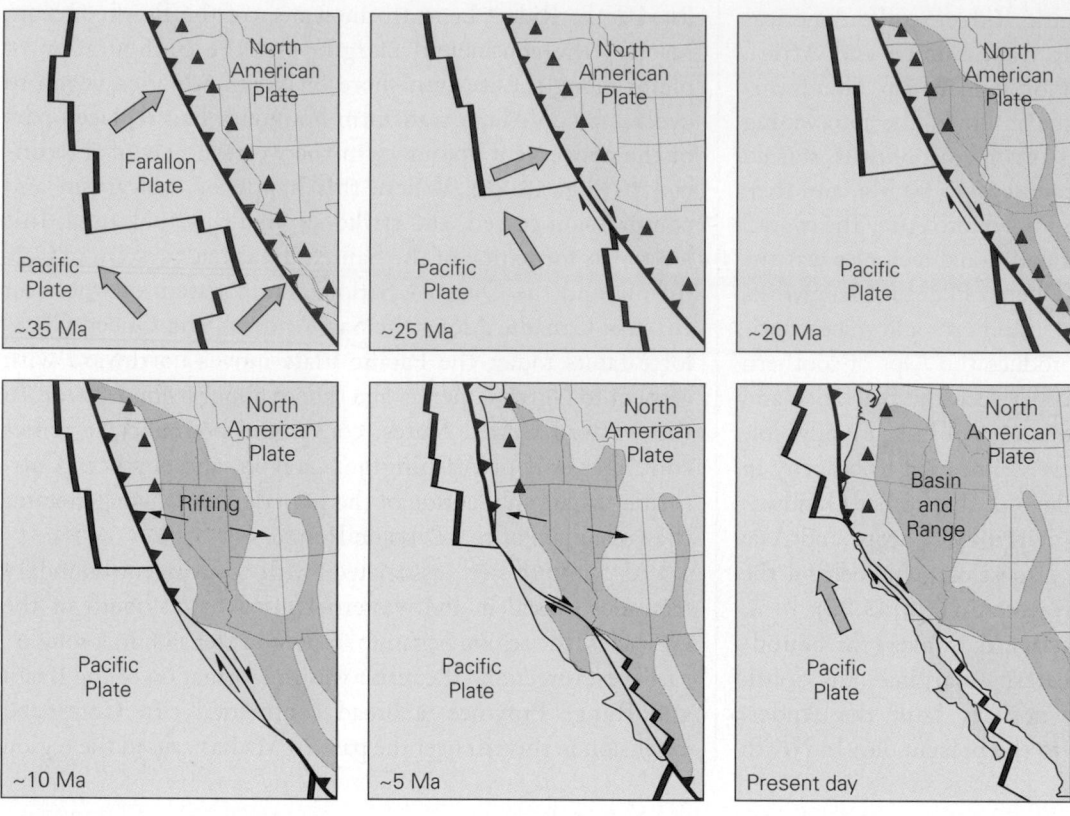

to stretch to twice its original width (**Fig. 13.27**). The Basin and Range Province's name reflects its topography—the province contains long, narrow mountain ranges separated from one another by flat, sediment-filled basins. This topography formed when the crust of the region was broken up by normal faults (see Chapter 11). Blocks of crust above these faults slipped downward and tilted. Upward-protruding crests of the tilted blocks form the ranges. The depressions between the ranges became basins that filled with sediment eroded from the ranges.

The Basin and Range Province terminates just north of the Snake River Plain, the track of the hot spot that now lies beneath Yellowstone National Park. As North America drifts westward, volcanic calderas have formed along the Snake River Plain; Yellowstone National Park straddles the most recent caldera (see Chapter 9).

Recall that during the Cretaceous Period, the world experienced greenhouse conditions and sea level rose so that extensive parts of continents were submerged. During the Cenozoic Era, however, the global climate changed. The climate became very warm during the Eocene Epoch, then began to cool. By the early Oligocene, the Earth returned to cooler ("icehouse") conditions, and Antarctic glaciers reappeared for the first time since the Triassic. The climate continued to grow colder through the Late Miocene, leading to the formation of grasslands in temperate climates. In the Pliocene, when the Isthmus of Panama formed, land separated the Atlantic completely from the Pacific, and the configuration of ocean currents changed. The change in currents, in turn, may have decreased the transport of oceanic heat to polar regions and may have triggered the formation of the sea ice that covered the Arctic Ocean.

The Cenozoic consists of three periods—the Paleogene, the Neogene, and the Quaternary. (Until recently, the Paleogene and Neogene together were called the Tertiary Period; this name is no longer official but remains widely used.) The Quaternary begins at 2.6 Ma and continues to the present. Between 2.6 Ma and 12 Ka, an interval of time called the *Pleistocene Epoch*, continental glaciers expanded and retreated across northern continents at least 20 times, an event known as the **Pleistocene Ice Age (Fig. 13.28)**. During each *glaciation*, the time intervals when the glaciers grew, sea level fell

FIGURE 13.27 The Basin and Range Province is a rift. Its opening caused rotation of the Sierra Nevada. The opening of the Rio Grande rift caused rotation of the Colorado Plateau, which is an unrifted block of cratonic crust. The inset shows a cross section along the red line.

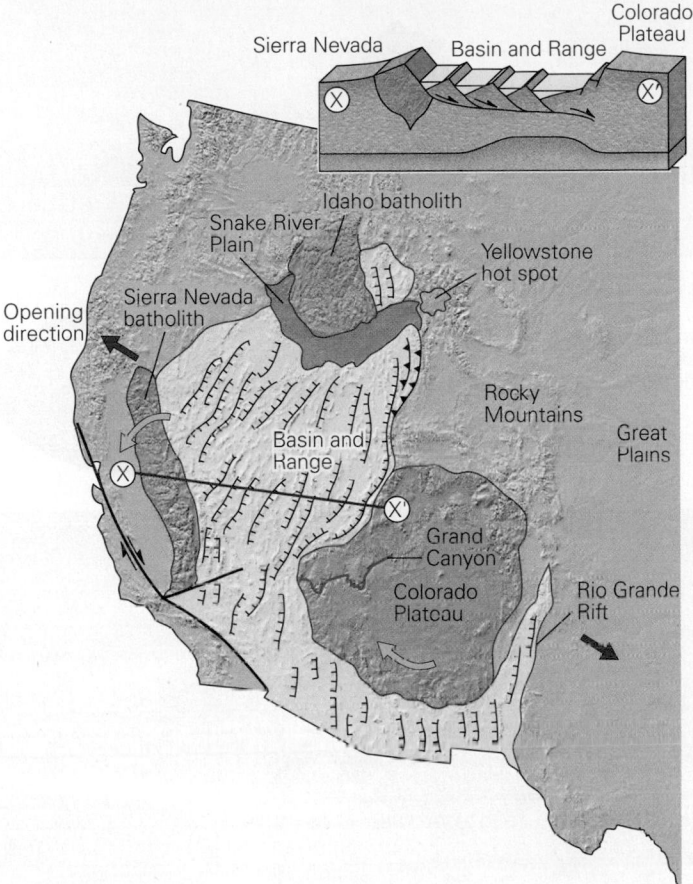

plains in temperate and subtropical climates by the middle of the Cenozoic Era, transforming them into vast grasslands. The dinosaurs, except for their distant relatives, the birds, were gone for good, and mammals rapidly diversified to fill the ecological niches that dinosaurs left vacant. In fact, most of the modern groups of mammals that exist today originated at the beginning of the Cenozoic Era, giving this time the nickname *Age of Mammals*. During the latter part of the era, particularly huge mammals appeared—including mammoths, mastodons, giant beavers, giant bears, and giant sloths—but these animals became extinct during the past 10,000 years, perhaps because of hunting by humans **(Fig. 13.29)**.

It was during the Cenozoic that our own ancestors first appeared. The fossil record indicates that ape-like primates diversified during the Miocene Epoch, at about 20 Ma, and the first human-like primates appeared at about 4 Ma **(Fig. 13.30)**. The first members of the human genus, *Homo*, have been found in 2.4-Ma strata. Evidence from studies in Africa indicates that *Homo erectus*, an ancestor capable of making stone axes, appeared at about 1.6 Ma, and the line leading to *Homo sapiens* (our species) diverged from *Homo neanderthalensis* (Neanderthal man) at about 500 Ka. The first

FIGURE 13.28 The maximum advance of the Pleistocene ice sheet in North America.

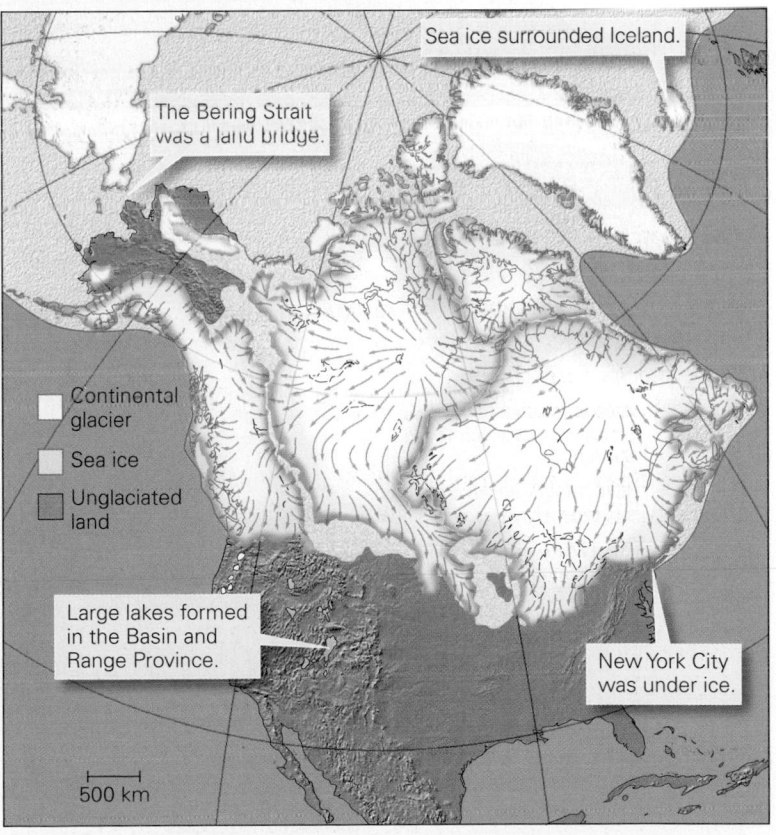

so much that the continental shelf became exposed to air, and at times, a *land bridge* formed across the Bering Strait, west of Alaska. Animals and people migrated across this bridge from Asia into North America. A partial land bridge also formed from Southeast Asia to Australia, making human migration to Australia easier. Erosion and deposition by the glaciers produced much of the landscape we see today in northern temperate regions. During *interglacials*, intervals of time when glaciers retreated, the land bridges and broad areas of continental shelves became submerged again. About 12,000 years ago, the climate warmed, and we entered the interglacial time interval we are still experiencing today (see Chapter 22). This most recent interval of time is the **Holocene Epoch**.

Life Evolution Within a few million years of the K-Pg mass extinction, plant life recovered, and forests of both angiosperms and gymnosperms reappeared. The grasses, which first appeared in the Cretaceous, spread across the

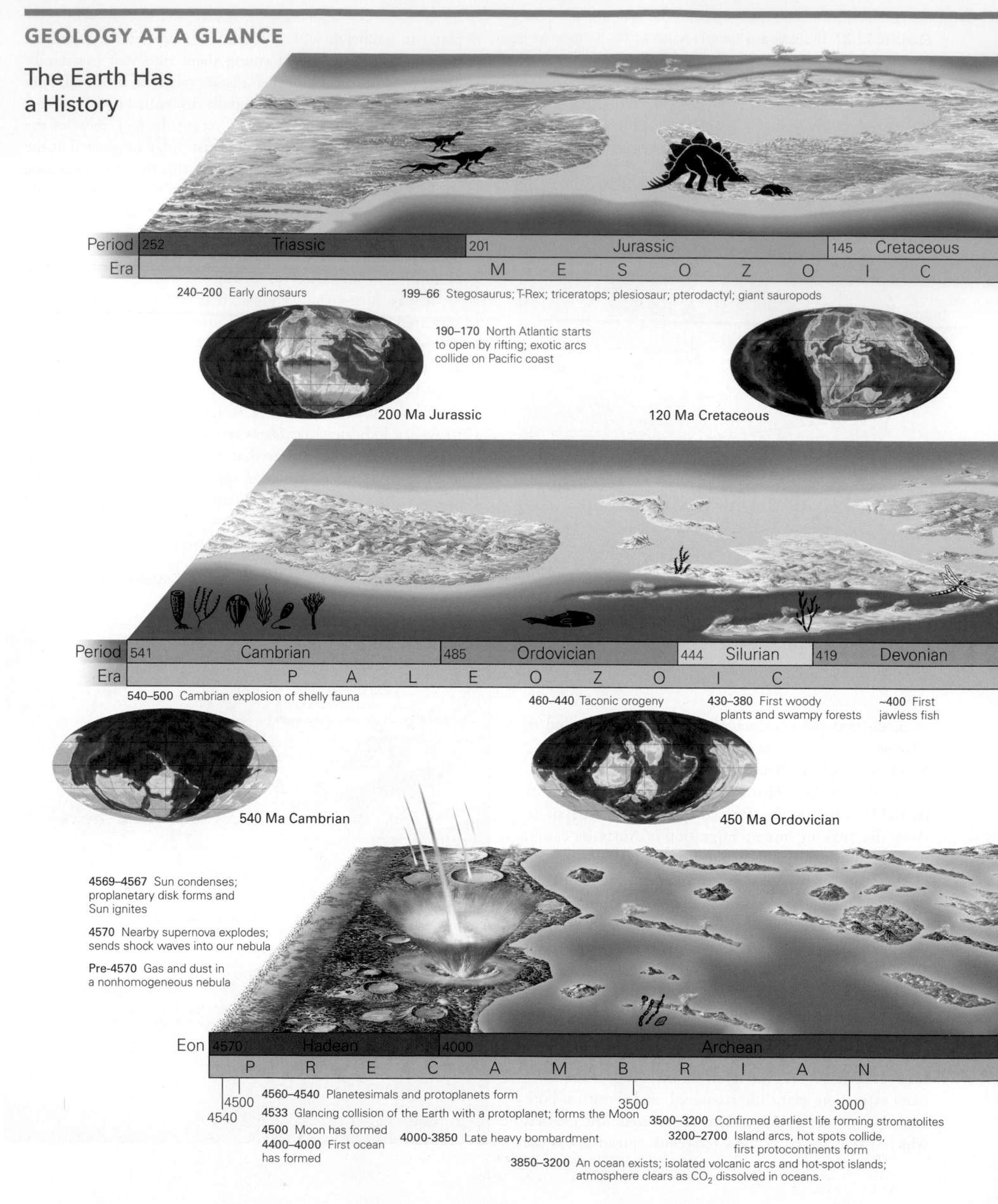

The Earth Has a History

Period	252	Triassic	201	Jurassic	145	Cretaceous
Era			M E S O Z O I C			

240–200 Early dinosaurs

199–66 Stegosaurus; T-Rex; triceratops; plesiosaur; pterodactyl; giant sauropods

190–170 North Atlantic starts to open by rifting; exotic arcs collide on Pacific coast

200 Ma Jurassic

120 Ma Cretaceous

Period	541	Cambrian	485	Ordovician	444	Silurian	419	Devonian
Era			P A L E O Z O I C					

540–500 Cambrian explosion of shelly fauna

460–440 Taconic orogeny

430–380 First woody plants and swampy forests

~400 First jawless fish

540 Ma Cambrian

450 Ma Ordovician

4569–4567 Sun condenses; proplanetary disk forms and Sun ignites

4570 Nearby supernova explodes; sends shock waves into our nebula

Pre-4570 Gas and dust in a nonhomogeneous nebula

Eon	4570	Hadean	4000	Archean
		P R E C A M B R I A N		

4500

4540

4560–4540 Planetesimals and protoplanets form

4533 Glancing collision of the Earth with a protoplanet; forms the Moon

4500 Moon has formed

4400–4000 First ocean has formed

4000-3850 Late heavy bombardment

3850–3200 An ocean exists; isolated volcanic arcs and hot-spot islands; atmosphere clears as CO_2 dissolved in oceans.

3500

3000

3500–3200 Confirmed earliest life forming stromatolites

3200–2700 Island arcs, hot spots collide, first protocontinents form

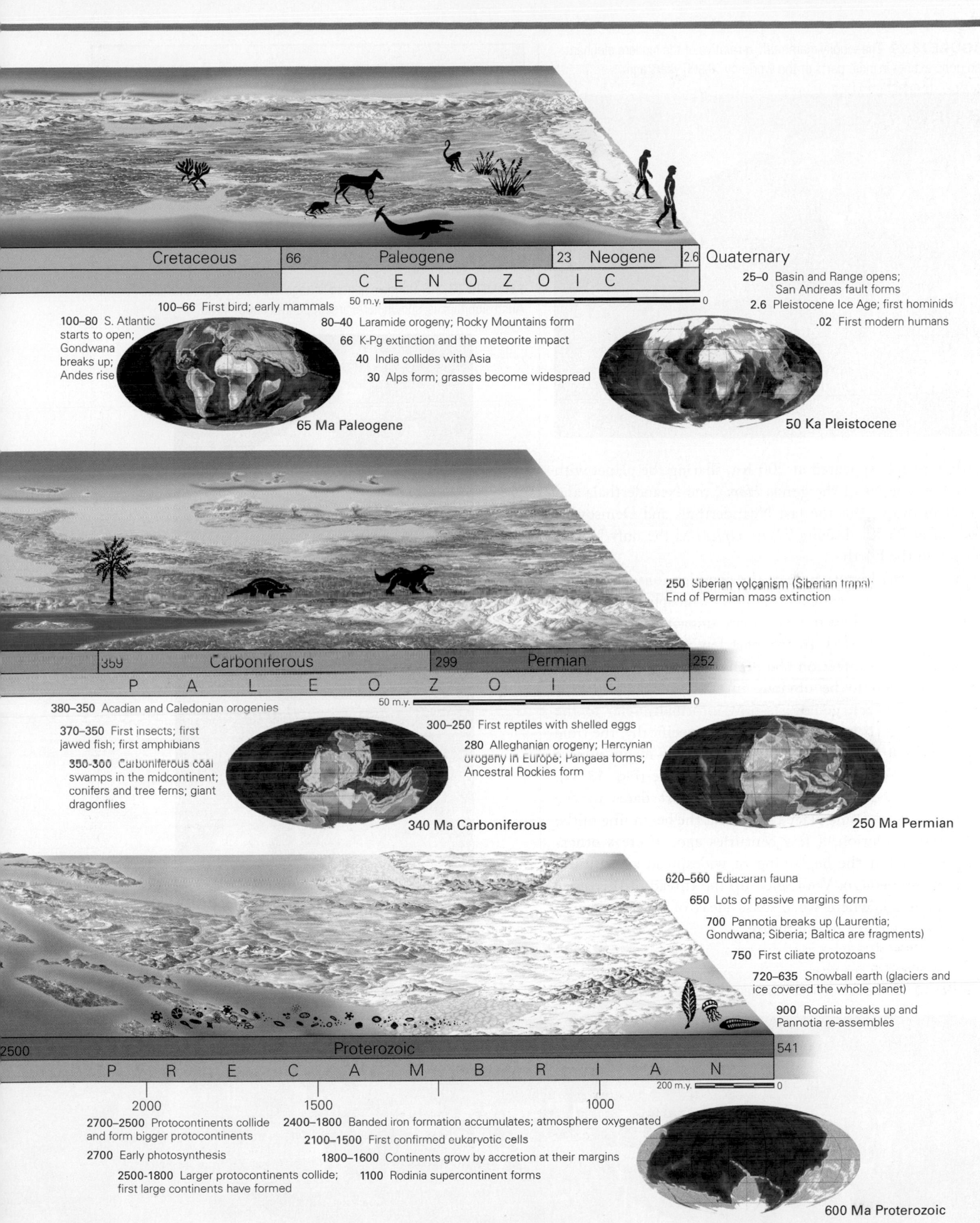

| Cretaceous | 66 | Paleogene | 23 | Neogene | 2.6 | Quaternary |

C E N O Z O I C

50 m.y. ━━━━━━━━━━ 0

100–66 First bird; early mammals

100–80 S. Atlantic
starts to open;
Gondwana
breaks up;
Andes rise

80–40 Laramide orogeny; Rocky Mountains form
66 K-Pg extinction and the meteorite impact
40 India collides with Asia
30 Alps form; grasses become widespread

25–0 Basin and Range opens;
San Andreas fault forms
2.6 Pleistocene Ice Age; first hominids
.02 First modern humans

65 Ma Paleogene

50 Ka Pleistocene

| 359 | Carboniferous | 299 | Permian | 252 |

P A L E O Z O I C

50 m.y. ━━━━━━━━━━ 0

250 Siberian volcanism (Siberian traps).
End of Permian mass extinction

380–350 Acadian and Caledonian orogenies

370–350 First insects; first
jawed fish; first amphibians

350–300 Carboniferous coal
swamps in the midcontinent;
conifers and tree ferns; giant
dragonflies

300–250 First reptiles with shelled eggs

280 Alleghanian orogeny; Hercynian
orogeny in Europe; Pangaea forms;
Ancestral Rockies form

340 Ma Carboniferous

250 Ma Permian

620–560 Ediacaran fauna

650 Lots of passive margins form

700 Pannotia breaks up (Laurentia;
Gondwana; Siberia; Baltica are fragments)

750 First ciliate protozoans

720–635 Snowball earth (glaciers and
ice covered the whole planet)

900 Rodinia breaks up and
Pannotia re-assembles

| 2500 | Proterozoic | 541 |

P R E C A M B R I A N

200 m.y. ━━━━━━━━━━ 0

2000   1500   1000

2700–2500 Protocontinents collide
and form bigger protocontinents

2700 Early photosynthesis

2500–1800 Larger protocontinents collide;
first large continents have formed

2400–1800 Banded iron formation accumulates; atmosphere oxygenated

2100–1500 First confirmed eukaryotic cells

1800–1600 Continents grow by accretion at their margins

1100 Rodinia supercontinent forms

600 Ma Proterozoic

FIGURE 13.29 The woolly mammoth, a relative of the modern elephant, had gone extinct in most parts of the world by 10,000 years ago.

TAKE-HOME MESSAGE

During the Cenozoic, the mountain belts of today rose, and modern plate boundaries became established. After the K-Pg mass extinction, mammals diversified. During the Pleistocene, glaciers covered large areas of continents, and humans appeared.

QUICK QUESTION: What interval of time does the Anthropocene refer to?

FIGURE 13.30 Our own human lineage evolved in the Cenozoic. Australopithecus, shown here, first appeared about 4 Ma.

modern people appeared at 200 Ka, sharing the planet with two other species of the genus *Homo*, the Neanderthals and the Denisovans. But the last Neanderthals and Denisovans died off at 25 Ka, leaving *Homo sapiens* as the only human species on the Earth.

The Earth's history reflects the complex consequences of plate interactions, sea-level changes, atmospheric changes, life evolution, and even a meteorite impact (**Geology at a Glance**, pp. 514–515). In the past few millennia, humans have had a huge effect on the planet, causing changes significant enough to be obvious in the geologic record of the future. In fact, geologists now informally refer to the more recent portion of the Holocene—specifically, the time during which human activities have had a major impact on the Earth System—as the **Anthropocene (Fig. 13.31)**. Different researchers assign different start dates to the Anthropocene—some place the start at the beginning of the industrial revolution, a few centuries ago, whereas others place its start at the beginning of widespread agriculture, a few millennia ago. We'll pick up the thread of this story in Chapter 23, where we discuss ideas of how the Earth System may change in the future.

FIGURE 13.31 The land surface has changed radically during the Anthropocene. What was a grassland has become a grid of streets and buildings in Chicago.

ANOTHER VIEW The present-day Bahamas serve as an example of what the interior of the United States might have looked like during intervals of the Paleozoic. Shallow land areas were submerged and became the site of shallow-marine sedimentation.

SUMMARY

- The Earth formed about 4.56 billion years ago. At times during the Hadean Eon, the planet was so hot that its surface was a magma ocean.

- The Archean Eon began at 4.0 Ga, about when the oldest rock that remains formed. Continental crust, assembled out of volcanic arcs and hot-spot volcanoes that were too buoyant to subduct, grew during the Archean, and the first life forms—bacteria and archaea—appeared.

- During the Proterozoic Eon, which began at 2.5 Ga, Archean cratons collided and sutured together along orogenic belts, forming large Proterozoic cratons. Photosynthesis added oxygen to the atmosphere.

- By the end of the Proterozoic, complex shell-less marine invertebrates populated the planet. Most continental crust assembled to form a supercontinent called Rodinia at about 1 Ga.

- At the beginning of the Paleozoic Era, rifting yielded several separate continents. During the Paleozoic, sea level rose and fell, yielding sequences of strata in continental interiors. Collisional orogenies produced another supercontinent, Pangaea, at the end of the era.

- Early Paleozoic evolution produced many invertebrates with shells, as well as jawless fish. Land plants and insects appeared in the middle Paleozoic. And, by the end of the era, there were land reptiles and gymnosperm trees.

- In the Mesozoic Era, Pangaea broke apart and the Atlantic Ocean formed. Convergent-boundary tectonics dominated along the western margin of North America. Dinosaurs appeared in the Late Triassic and became prominent land animals through the Mesozoic Era.

- During the Cretaceous Period, sea level was relatively high, and the continents flooded. Angiosperms appeared at this time, along with modern fish. A huge mass-extinction event, which wiped out the dinosaurs, occurred at the end of the Cretaceous Period, probably because of the impact of a large meteorite.

- In the Cenozoic Era, continental fragments of Pangaea collided again. The collision of Africa and India with Asia and Europe formed the Alpine-Himalayan orogen. Convergent tectonics has persisted along the margin of South America, creating the Andes, but ceased in North America when the San Andreas fault formed. Rifting in the western United States during the Cenozoic Era produced the Basin and Range Province.

- During the Cretaceous, various kinds of mammals filled niches left vacant by the disappearance of dinosaurs. The human genus, *Homo*, appeared and evolved throughout the radically shifting climate and ice ages of the Pleistocene Epoch.

GUIDE TERMS

Alpine-Himalayan chain (p. 511)
Ancestral Rockies (p. 501)
Anthropocene (p. 516)
Appalachian fold-thrust belt (p. 500)
Archean Eon (p. 487)
basement uplift (p. 506)
Basin and Range Province (p. 511)

Cambrian explosion (p. 497)
craton (p. 491)
cratonic platform (p. 491)
differentiation (p. 486)
Ediacaran fauna (p. 494)
exotic terrane (p. 504)
Gondwana (p. 496)
great oxygenation event (p. 494)

Grenville orogeny (p. 492)
Holocene Epoch (p. 513)
Laramide orogeny (p. 507)
Laurentia (p. 496)
Pangaea (p. 500)
Permian mass extinction (p. 502)
Phanerozoic Eon (p. 494)
Pleistocene Ice Age (p. 512)

Proterozoic Eon (p. 491)
Rodinia (p. 492)
Sevier orogeny (p. 506)
shield (p. 491)
snowball Earth (p. 494)
stratigraphic sequence (p. 498)
stromatolite (p. 489)
superplume (p. 508)

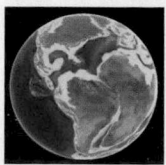

GEOTOURS *THIS CHAPTER'S GEOTOURS WORKSHEET (K) FEATURES QUESTIONS AND GOOGLE EARTH SITES ON:*

- Paleogeography of the Earth

REVIEW QUESTIONS

The letters following each Review Question refer to the corresponding Learning Objective from the Chapter Opener.

1. What do geologists date to determine when the Earth formed? **(A)**

2. Why are there no whole rocks on the Earth that yield isotopic dates older than about 4 billion years? **(B)**

3. Describe the formation of the crust, atmosphere, and oceans during the Hadean Eon. **(B)**

4. How did the atmosphere change during the first 2 billion years of Earth history? When did most continental crust form? **(B)**

5. Considering the map in the figure, what evidence do we have that the Earth froze during the Proterozoic Eon? **(D)**

6. Did supercontinents exist in the Proterozoic? **(F)**

7. How did the Cambrian explosion of life change the nature of the living world? **(E)**

8. What was the Earth's land surface like during the early Paleozoic? **(E)**

9. When in Earth history did vascular plants appear? When did trees appear? **(E)**

10. How did the Alleghanian and Ancestral Rockies orogenies affect North America? **(F)**

11. Describe the plate-tectonic conditions that led to the formation of the Sierran arc and the Sevier fold-thrust belt. How does the Laramide deformation differ from the Sevier deformation in the United States? **(C)**

12. What life forms appeared during the Mesozoic? **(E)**

13. What may have caused the Cretaceous flooding of the continents? **(C)**

14. What could have caused the K-Pg mass extinction? **(G)**

15. What continents formed as a result of the breakup of Pangaea? Which ocean is growing on this paleographic map? **(F)**

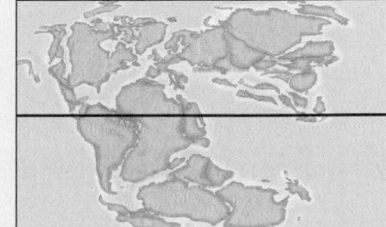

16. Why did the Himalayas and the Alps form? **(F)**

17. What major tectonic provinces formed in the western United States during the Cenozoic? **(C)**

18. What major climatic and biological events happened during the Pleistocene? **(E)**

ON FURTHER THOUGHT

19. During intervals of the Paleozoic, large areas of continents were submerged by shallow seas. Using Google Earth, tour North America from space. Do any present-day regions within North America consist of continental crust that was submerged by seawater? What about regions offshore? (Hint: Look at the region just east of Florida.) **(E)**

20. Geologists have concluded that 80% to 90% of the Earth's continental crust had formed by 2.5 Ga. But if you look at a geologic map of the world, you find that only about 10% of the Earth's continental crustal surface is labeled "Precambrian." Why? **(B)**

ONLINE RESOURCES

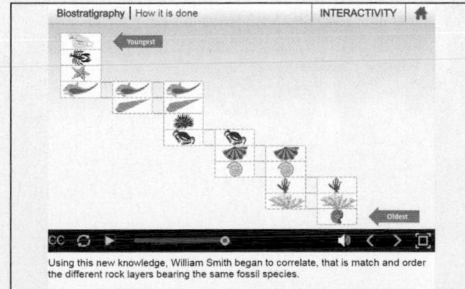

Animations
This chapter features interactive animations testing your knowledge of biostratigraphy.

Videos
This chapter features a video on how the study of Antarctic rocks reveals clues to the Earth's past.

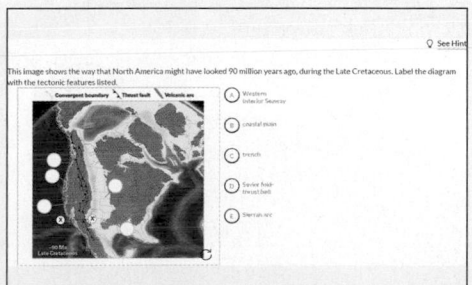

Smartwork5
This chapter features understanding and ranking exercises about the Earth throughout history, including how the planet's tectonic features have evolved over time.

PART V

EARTH RESOURCES

The earliest humans were hunter-gatherers and needed only food and water to survive. But then people discovered that fires made food easier to eat and campsites more comfortable, weapons made hunting more successful, shelters made daily life more pleasant, and farming made food supplies more reliable—and humanity's needs expanded. Specifically, people began to require *energy resources* (sources of heat and/or power) and *mineral resources* (materials from which metals and other chemicals can be derived). In a general sense, we use the term *resource* for any item that can be employed for a useful purpose. Modern society requires resources obtained directly from the Earth System in order to survive and prosper. Think about it . . . metals, oil, plastics, wallboard, brick, coal, pottery, cement, uranium, and more all come from the Earth's crust. To appreciate the value and cost of such Earth resources, it's important to know how they form, where they can be found, how they're extracted, whether or not they're sustainable, and how their use can impact the environment. In Chapter 14, we focus on the energy resources that come from the Earth. These resources include fossil fuels (oil and coal) as well as nuclear fuel and moving water. Chapter 15 describes nonenergy resources, particularly the mineral deposits from which we obtain metals.

> **14 Squeezing Power from a Stone: Energy Resources**
>
> **15 Riches in Rock: Mineral Resources**

◀ In the 15th century, the Incas built a stone city, Machu Picchu, on a mountaintop in the Andes in what is now Peru. They used stone from nearby to construct durable walls that have survived the centuries. Stone is one of many Earth materials that society uses for energy, construction, and much more.

CHAPTER 14

Squeezing Power from a Stone: Energy Resources

By the end of this chapter, you should be able to . . .

A. discuss the concept of an energy resource, and list the variety of energy resources available to society.

B. explain how oil and gas differ from each other and from coal, and how all these materials can be considered to be fossil fuels.

C. describe the steps that take place in geologic environments to form fossil fuel reserves.

D. distinguish between conventional and unconventional hydrocarbon reserves, and describe the technologies needed to access them.

E. sketch the basic layout of a nuclear power plant, describe where nuclear fuel comes from, and explain how nuclear fission differs from burning.

F. describe the challenges society faces regarding future reliance on fossil fuels and what options exist for alternative energy sources.

> To keep a lamp burning, we have to keep putting oil in it.
>
> —Mother Teresa (Nobel Peace Prize winner, 1910–1997)

14.1 Introduction

The extreme chill of an arctic midwinter doesn't stop a wolf from stalking its prey. The wolf's legs move through the snow, its heart pumps, its lungs inhale and exhale, and its body radiates heat. All of these activities require energy. **Energy**, as defined by a physicist, is the capacity to do work or, in other words, to cause a change in a physical or biological system. This means that energy can raise the temperature of a material, can drive chemical reactions, can cause an object to move or change shape, can generate light or magnetism, or can change the state of a material (from solid to liquid or liquid to gas). A wolf's energy comes from the metabolism of sugar, protein, and carbohydrates in its body. These chemicals, in turn, come from the food the wolf catches and eats. Therefore, we can think of mice, rabbits, and deer as an **energy resource**—a source of materials that yield energy — for a wolf.

Early humans, like wolves, could supply their energy needs entirely by eating food, so they could survive by hunting and gathering. But when people discovered how to use fire, tools, and weapons, their need for energy resources began to exceed that of other animals. Before the advent of civilization, wood and dried dung could supply humanity's energy-resource needs—these materials could serve as **fuel**, an energy resource stored in a usable and transportable form **(Fig. 14.1a)**. As people began to congregate in towns, however, they began to use energy for agriculture and transportation. At first, animal power, wind, and flowing water met society's additional energy demand, but as populations grew and new industries such as iron smelting emerged, energy-resource needs began to outpace the supplies available at the Earth's surface. In fact, to feed the smelting industry, 17th-century woodcutters devastated European forests **(Fig. 14.1b)**. The industrial revolution could not begin unless new fuel supplies could be found, so society turned to coal, the fossilized remains of woody plants, to meet the demand. Coal could be transported easily and contains about 1.7 times more energy per kilogram than wood, so coal kept 18th-century cooking and heating stoves hot, and it powered the steam engines of factories and trains around the industrializing world.

Since the industrial revolution, society's hunger for energy has increased unabated **(Fig. 14.2a)**. In the United States today,

FIGURE 14.1 Nongeologic energy resources dominated human society in the past.

(a) Women collecting dung to burn for cooking in India (ca. 1977).

(b) A painting of an ironworks in Norway from 1800 by John Edy (1760–1820).

◀ (facing page) Heavy equipment scrapes coal from the ground in an Indiana mine. The coal stores energy that came to the Earth from the Sun about 300 million years ago.

for example, an average city dweller uses more than 110 times the amount of energy that a prehistoric hunter did. And as countries in Asia continue to undergo industrialization, their energy needs have skyrocketed (**Fig. 14.2b**). Most energy for human consumption in the industrialized world now comes from oil, natural gas, and coal, but we still use wind and flowing water. In the last half century we've added nuclear energy, geothermal energy, and solar energy to the list of energy resources, and there has been growing interest in expanding the use of biofuels from plant crops. In 2007, the world's energy

FIGURE 14.2 Global energy usage by type and the increase of its consumption in Asia.

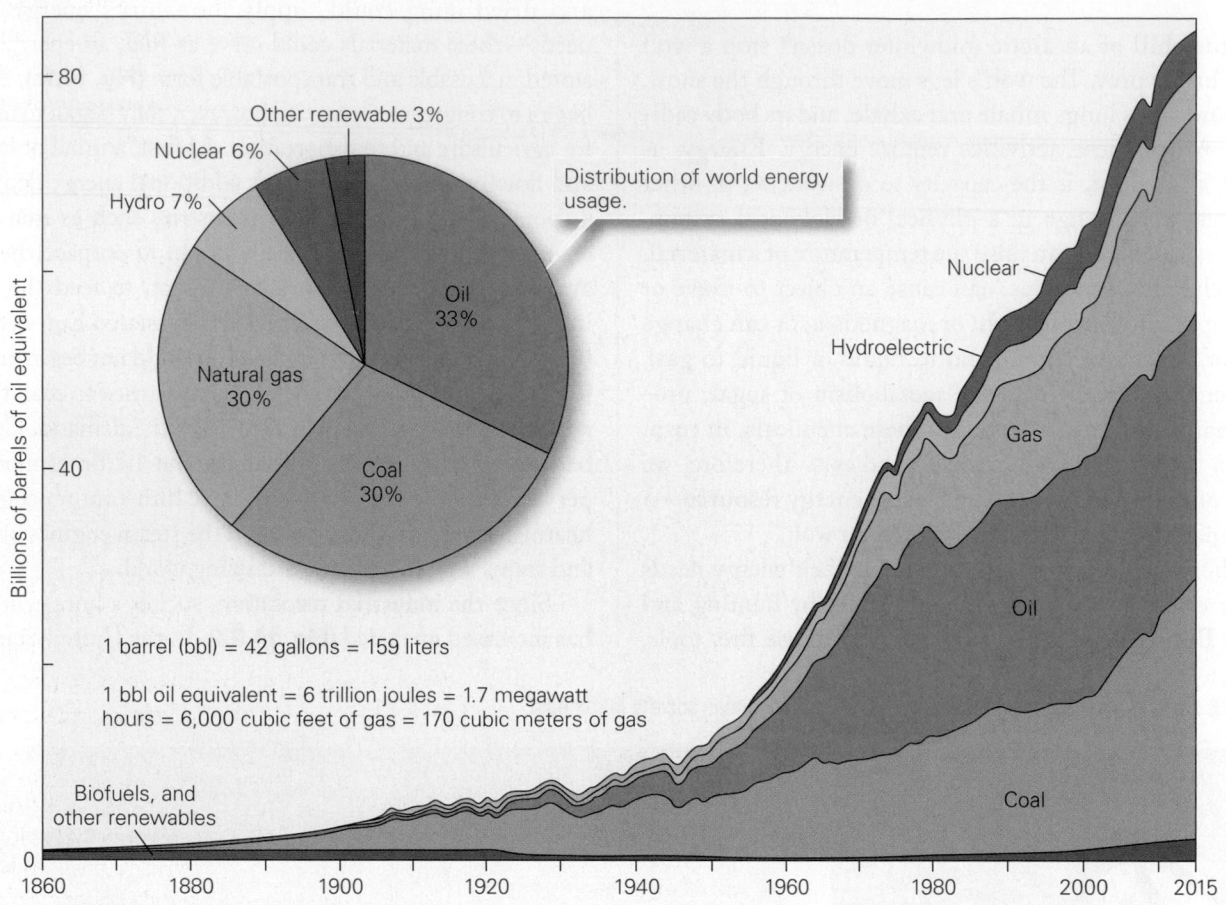

(a) The proportion of different sources of energy that people use has changed over time, and the amount of energy used almost continuously increases.

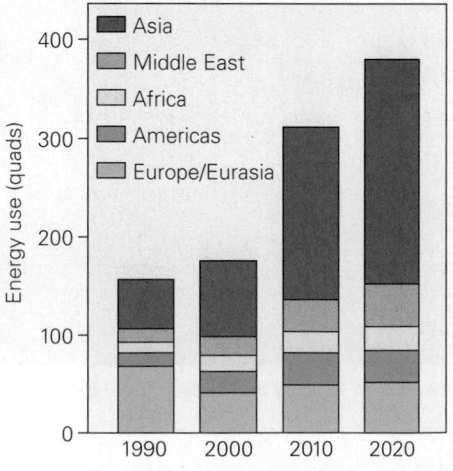

1 quad = 1 quadrillion British thermal units (BTU)
1 barrel (bbl) = 5.8 million BTU

(b) In the past few decades, Asia has become the largest consumer of energy. (Values for 2020 are estimates.)

picture changed significantly when technologies to extract oil and natural gas from shale improved, and natural gas has now started to substitute for other fuels. But use of energy brings with it myriad challenges to society: the distribution of supplies can spark discord among nations, production and consumption both can have undesirable consequences for the environment and climate, and questions concerning the long-term sustainability of energy resources remain incompletely answered.

Why does a geology book include a chapter devoted to energy resources? Because most of these resources originate in geologic materials or are the result of geologic processes. So, to understand the sources and limitations of energy resources, and to find new resources, we must understand their geologic context and geologic consequences. (That's why the energy industry employs tens of thousands of geologists.) To help you to understand the geology of energy, this chapter begins by surveying the various types of energy resources on the Earth.

It then focuses on fossil fuels (oil, gas, and coal), which are combustible materials derived from organisms that lived in the past. The chapter continues with a survey of other energy resources and concludes by outlining the dilemmas that society faces as conventional energy resources begin to run out and by-products of energy consumption impact our environment.

14.2 Sources of Energy in the Earth System

What comes to mind when someone asks you to name an energy resource? Perhaps you think about the gasoline that fills the tanks of cars, or the coal piled outside of a power plant, or the flowing stream that turns a waterwheel. Alternatively, you may think of windmills or arrays of solar panels, because they are appearing on the landscape with increasing frequency. Let's step back and consider where the energy in these energy resources comes from in the first place **(Fig. 14.3)**:

- *Energy directly from the Sun:* Solar energy, resulting from nuclear fusion reactions in the Sun, bathes the Earth's surface. It may be converted directly into electricity using solar panels, or it may be used to heat water in tanks.

- *Energy directly from gravity:* The gravitational attraction of the Moon and, to a lesser extent, the Sun, causes ocean *tides*, the daily up-and-down movement of the sea surface. The flow of water into and out of channels during tidal changes can drive turbines.

- *Energy from both solar energy and gravity:* Solar radiation heats air, which becomes buoyant and rises. As this happens, gravity causes cooler air to sink. The resulting air movement, *wind*, powers sails and windmills. Solar energy also evaporates water, which enters the atmosphere. When the water condenses, rain falls on the land, where it accumulates in streams that flow downhill in response to gravity. This moving water powers waterwheels and turbines.

- *Energy via photosynthesis:* Algae and green plants absorb some of the solar energy that reaches the Earth's surface. Their green color comes from a pigment called *chlorophyll*. With the aid of chlorophyll, plants produce sugar through a chemical reaction called *photosynthesis*. In chemist's shorthand, we can write the reaction as:

$$6CO_2 + 12H_2O + Light \rightarrow 6O_2 + C_6H_{12}O_6 + 6H_2O$$

carbon dioxide   water   oxygen   sugar   water

Plants use the sugar produced by photosynthesis to manufacture more complex chemicals, or they metabolize it directly to provide themselves with energy. We can refer to plant material, both living and dead—or even digested (such as cow dung)—as **biomass**. Burning biomass in a fire releases potential energy stored in the chemical bonds of organic chemicals.

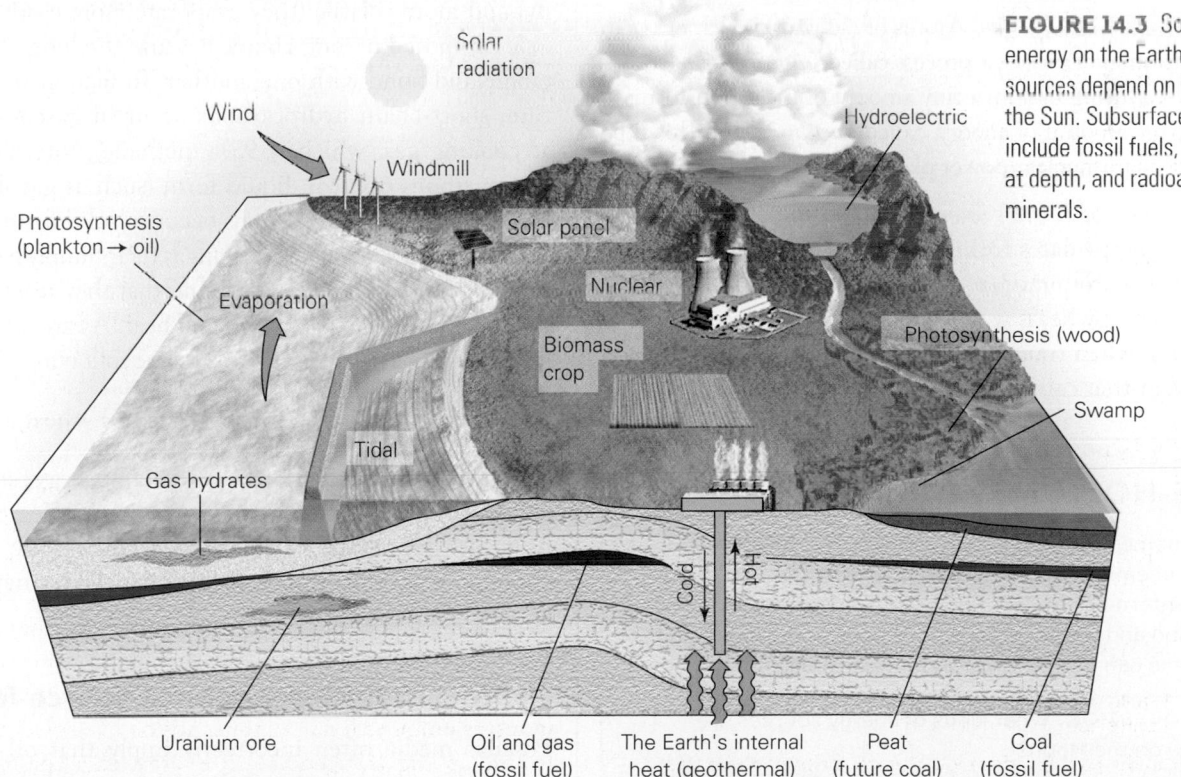

FIGURE 14.3 Sources of energy on the Earth. Surface sources depend on heat from the Sun. Subsurface sources include fossil fuels, hot rocks at depth, and radioactive minerals.

Solar radiation

Wind

Windmill

Photosynthesis (plankton → oil)

Solar panel

Evaporation

Nuclear

Biomass crop

Hydroelectric

Photosynthesis (wood)

Swamp

Tidal

Gas hydrates

Cold

Hot

Uranium ore   Oil and gas (fossil fuel)   The Earth's internal heat (geothermal)   Peat (future coal)   Coal (fossil fuel)

During burning, the organic molecules react with oxygen and break apart to produce carbon dioxide, water, and carbon (soot):

$$\underset{\text{molecules}}{\text{plant}} + O_2 \xrightarrow{\text{burning}} CO_2 + H_2O + C + \underset{\text{gases}}{\text{other}} + \underset{\text{energy}}{\text{heat}}$$

The flames you see in fire consist of glowing gases released and heated by this reaction.

For centuries, people have burned wood and cow dung to produce energy. More recently, corn, sugarcane, switchgrass, and algae are being used to produce *ethanol*, a flammable alcohol, and plant and animal oils are being converted into *biodiesel*. Ethanol and biodiesel are both *biofuels*.

- *Energy from fossil fuels*: Oil, natural gas, and coal come from organisms that lived long ago, so they store solar energy that reached the Earth long ago. We refer to these substances as **fossil fuels** to emphasize that they were derived from ancient organisms, the remains of which have been preserved in rocks over geologic time.

> **Did you ever wonder . . .**
> what the "fossils" are in fossil fuel?

- *Energy from chemical reactions:* A number of inorganic chemicals can burn to produce light and energy. The energy results from *exothermic* (heat-producing) chemical reactions. A dynamite explosion serves as an extreme example of such energy production. Recently, researchers have been studying *electrochemical devices*, such as hydrogen fuel cells, that produce electricity directly from chemical reactions.

- *Energy from nuclear fission:* Atoms of radioactive elements can split into smaller pieces, a process called *nuclear fission* (see Box 1.3). During fission, a tiny amount of mass transforms into a large amount of energy, called *nuclear energy*. This type of energy runs nuclear power plants and nuclear submarines.

- *Energy from the Earth's internal heat:* Some of the Earth's internal energy dates from the birth of the planet, while some comes from radioactive decay in minerals. This internal energy heats water underground. The resulting hot water, when transformed to steam, provides *geothermal energy* that can drive turbines or heat buildings.

TAKE-HOME MESSAGE

Energy comes from several sources: solar radiation, gravity, chemical reactions, radioactive decay, and the Earth's internal heat. Solar radiation and gravity together drive wind and water movement. Products of living organisms can be preserved in rocks as fossil fuels.

QUICK QUESTION: What kinds of energy sources are available on the Moon?

14.3 Introducing Hydrocarbon Resources

What Are Oil and Gas?

Industrialized societies today rely primarily on familiar products derived from oil (such as gasoline, jet fuel, kerosene, and diesel) and various kinds of natural gas (such as methane and propane) for their energy needs. Why are these fuels so popular? They have a relatively high **energy density**, meaning that they contain a relatively large amount of energy per unit of weight. For example, 1 g of oil provides twice as much energy as 1 g of coal and about 500 times more energy than a 1-g lead-acid battery provides. So an airplane can cross an ocean on a tank of jet fuel, but wouldn't even be able to get off the ground if it had to run on car batteries. Chemically, both oil and natural gas are **hydrocarbons**—organic chemicals that consist of chain-like or ring-like polymers made of carbon and hydrogen atoms. For example, bottled gas (propane) has the chemical formula C_3H_9. Hydrocarbons are so named because similar carbon-based chemicals make up living organisms.

Some hydrocarbons are gaseous and invisible, some resemble watery liquids, some appear syrupy, and some are solid **(Fig. 14.4)**. The *viscosity* (ability to flow) and the *volatility* (ability to evaporate) of a hydrocarbon product depend on the size of its component polymers. Products composed of short polymer chains tend to be less viscous (they can flow more easily) and more volatile (they evaporate more easily) than products composed of long chains because the long chains tend to tangle and bond with one another. In fact, at room temperature, short-chain hydrocarbons occur in gaseous form (such as cooking gas, which is 95% methane), intermediate-chain hydrocarbons occur in liquid form (such as gasoline and oil), and long-chain hydrocarbons occur in solid form (tar).

Why can we use hydrocarbons as fuel? Simply because hydrocarbons, like wood, *burn*, meaning that they react with oxygen to form carbon dioxide, water, and heat. As an example, we can describe the burning of gasoline by the following reaction:

$$2C_8H_{18} + 25O_2 \rightarrow 16CO_2 + 18H_2O + \text{heat and light}$$

During such reactions, potential energy stored in the chemical bonds of the hydrocarbon molecules converts into heat and light **(Fig. 14.5)**. This energy can be used to run engines or to transform water into the steam that drives generators in a power plant.

Hydrocarbon Generation in Source Rocks

Popular media often incorrectly imply that oil and gas are derived from buried trees or the carcasses of dinosaurs. In fact,

FIGURE 14.4 The diversity of hydrocarbon products in order of increasing viscosity. Note that the viscosity of a product reflects the length of its polymer chains.

	Product	Number of carbons in the polymer chain
Low viscosity		
	Natural gas	
	Bottled gas	C_1 to C_4
	Gasoline	C_5 to C_{10}
	Kerosene	C_{11} to C_{15}
	Heating oil	C_{16} to C_{20}
	Lubricating oil	C_{21} to C_{40}
High viscosity	Tar	$> C_{70}$

the hydrocarbon molecules in oil and gas form from organic chemicals, such as fatty molecules called lipids, that were once in plankton. *Plankton* consists of very tiny floating organisms that include single-celled and very small multicellular plants (algae) as well as protists and microscopic animals. Typically, most planktonic organisms range in size from 0.02 to 2.0 mm in diameter. Note that it's the organic cells of plankton that transform into hydrocarbons, not the mineral shells.

When plankton cells die, they sink to the floor of the lake or sea that they lived in, and if the water is relatively quiet (nonflowing), they accumulate **(Fig. 14.6)**. In most locations, relatively little of such organic matter settles out of the water column because most plankton gets consumed by organisms higher in the food chain. But if surface waters are nutrient-rich and receive a lot of sunlight, plankton blooms, and a significant amount can settle out. Commonly, the seafloor or lake floor hosts an oxygen-rich environment populated by scavengers

and microbes. As a result, dead plankton cells are consumed, decay, or oxidize before being buried, and the organic matter in the cells transforms into CH_4 or CO_2 that bubbles away. But in oxygen-poor waters, organic material can survive long enough to mix with clay and form an organic-rich, muddy ooze, which can then become buried by still more sediment so that it becomes preserved. Eventually, pressure from the weight of overlying sediment squeezes out the water, and the ooze becomes compacted and lithified to become black *organic shale*. (Shale that does not contain organic matter, in contrast, tends to be gray, tan, or red.) Organic shale contains the raw materials from which hydrocarbons form, so we refer to it as **source rock**.

If organic shale becomes buried deeply enough (2–4 km), it warms up, since temperature increases with depth in the Earth. Chemical reactions that take place in warm source rocks slowly transform the organic material in the shale into a variety of large waxy molecules that together comprise **kerogen**. Geologists refer to shale containing 25% to 75% kerogen as **oil shale**. If oil shale warms to temperatures greater than about 90°C, kerogen molecules break into smaller oil and natural gas molecules, a process known as **hydrocarbon generation**. At temperatures over about 160°C, any remaining oil breaks down to form natural gas. And at temperatures over 225°C, organic matter loses all its hydrogen and transforms into graphite (pure carbon). Note that oil forms only in a relatively narrow range of temperatures, called the **oil window (Fig. 14.7)**. For regions with a geothermal gradient of 25°C/km, the oil window lies at depths of 3.5 to 6.5 km. Since gas survives to a higher temperature, the gas window extends down to 9 km. If the geothermal gradient is low (15°C/km), oil can survive down to depths of about 11 km and gas down to 15 km. This means that, at most, hydrocarbons exist only in the topmost third of the crust.

Conventional versus Unconventional Hydrocarbon Reserves

Oil and gas do not occur in all rocks at all locations. Where they do occur, we refer to the volume of oil and gas underground as a **hydrocarbon reserve**—if the reserve consists predominantly

FIGURE 14.5 The Iraqi army set fire to 700 oil wells in Kuwait in 1991, a tragic display of the energy locked in fossil fuels underground.

FIGURE 14.6 The formation of oil. The process begins when organic debris settles with sediment. As burial depth increases, heat and pressure transform the sediment into black shale, in which organic matter becomes kerogen. At appropriate temperatures, kerogen becomes oil, which then seeps upward.

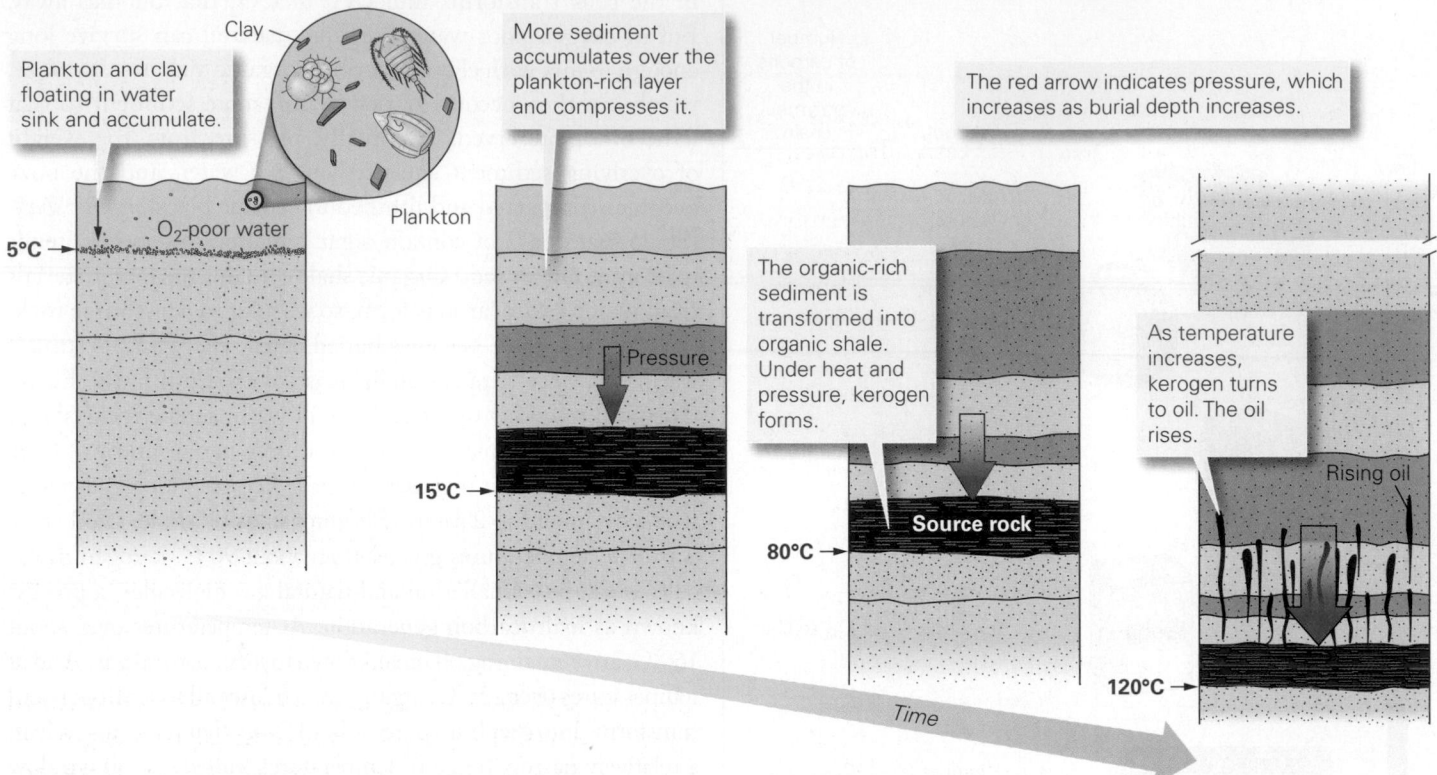

FIGURE 14.7 The oil window indicates subsurface conditions in which oil can form and survive. Deeper down, oil breaks down to form gas. At greater depth, metamorphism transforms organic material to graphite.

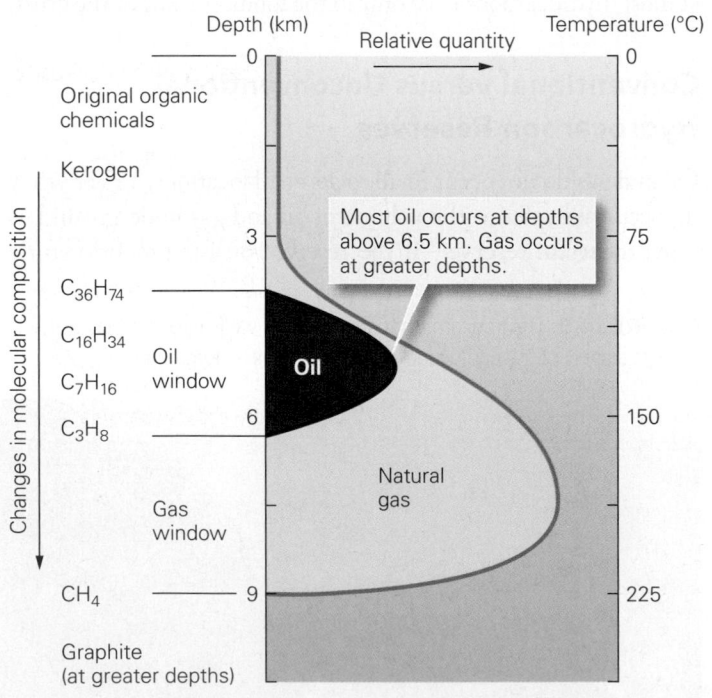

of oil, it's usually called an *oil reserve,* and if it consists predominantly of gas, it's a *gas reserve.* Some hydrocarbon reserves contain both oil and gas. Until relatively recently, oil companies could obtain hydrocarbons economically only from reserves that could be pumped from the ground relatively easily, meaning that the underground hydrocarbons could flow through the rock containing them to a drillhole. We refer to such hydrocarbon reserves as **conventional reserves.** In recent years, the rising price of hydrocarbons, as well as improvements in technology, have made it possible to obtain supplies that previously could not be pumped out easily. These hard-to-get supplies are known as **unconventional reserves.** The ability to access unconventional reserves has led to major changes in the energy industry.

TAKE-HOME MESSAGE

Oil and gas are hydrocarbons derived from the remains of plankton buried with clay in an organic ooze. Burial transforms ooze into source rock, a black shale containing kerogen. If the rock is subjected to appropriate temperatures, organic materials transform into oil and gas. Hydrocarbon reserves are accumulations of oil and/or gas underground.

QUICK QUESTION: What is the difference between a conventional and an unconventional reserve?

14.4 Conventional Hydrocarbon Systems

We've just noted that the hydrocarbons of a conventional reserve can be pumped out of the ground easily. The development of such a reserve requires a specific association of materials, conditions, and time. Geologists refer to this association as a **conventional hydrocarbon system (Fig. 14.8)**. We've already discussed the first two components of the system, namely, the formation of source rock and the existence of thermal conditions that transform the kerogen in the source rock into smaller hydrocarbon molecules. Let's now look at the remaining components: the migration of oil into a reservoir rock that lies within a configuration of rocks called an oil trap.

Reservoir Rocks and Hydrocarbon Migration

The clay flakes of a source rock fit together tightly and prevent kerogen or any hydrocarbons that form within the rock from moving easily through the rock. Therefore, you can't simply drill a hole into a source rock and pump out oil—the oil won't flow into the well fast enough to make the process cost efficient. To extract oil or gas from a conventional reserve, energy companies instead drill into **reservoir rocks**—rocks that contain, or could contain, accessible oil or gas. By "accessible," we mean that the oil or gas can flow through rock and be sucked into a well fairly easily by pumping.

> **Did you ever wonder . . .**
> whether there are actually lakes or pools of oil underground?

To be a reservoir rock, a body of rock must have space in which the oil or gas can reside and must have channels through which the oil or gas can move. The space can be in the form of openings, or **pores**, between grains (which exist because the grains didn't fit together tightly and because cement didn't fill all the spaces during cementation) or in the form of cracks and fractures that developed after the rock formed. In some cases, groundwater passing through rock dissolves minerals and creates new pores. **Porosity** refers to the overall amount of open space in a rock. Not all rocks have the same porosity **(Fig. 14.9)**. For example, shale typically has a low porosity (less than 10%), whereas poorly cemented sandstone has a high porosity (up to 35%, which means that about a third of a block of porous sandstone actually consists of open space). The oil or gas in a reservoir rock occurs in the pores and cracks

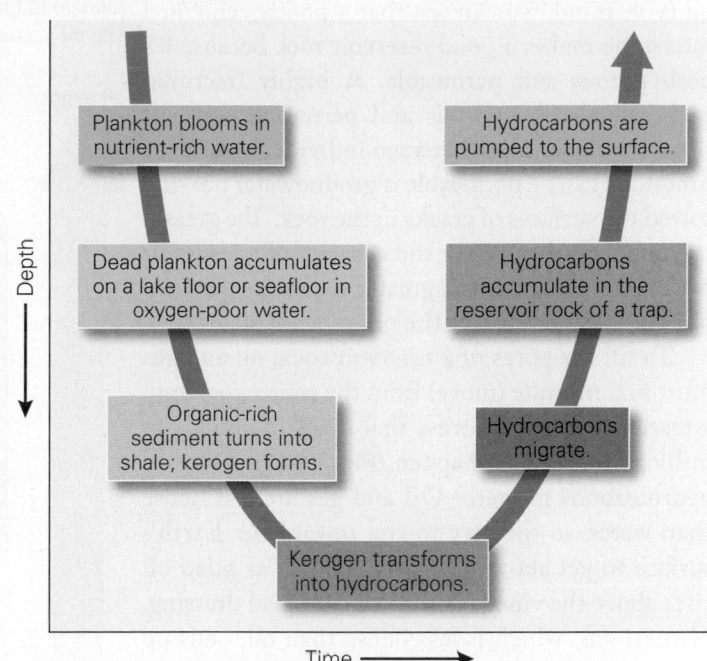

FIGURE 14.8 Stages of a conventional hydrocarbon system.

(just as water fills the holes in a sponge), so it is distributed through the rock—it does not occur in open pools underground. **Permeability** refers to the degree to which pores or cracks connect to one another. In a permeable rock, the pores and cracks are linked, so a fluid can flow slowly through the rock, following a tortuous pathway. Note that even if a rock has high porosity, it is not necessarily permeable (see Fig. 14.9c), for if the pores aren't connected, fluid can't move from one to another.

FIGURE 14.9 Porosity and permeability in sedimentary rocks. Rocks with high porosity and high permeability make the best reservoir rocks.

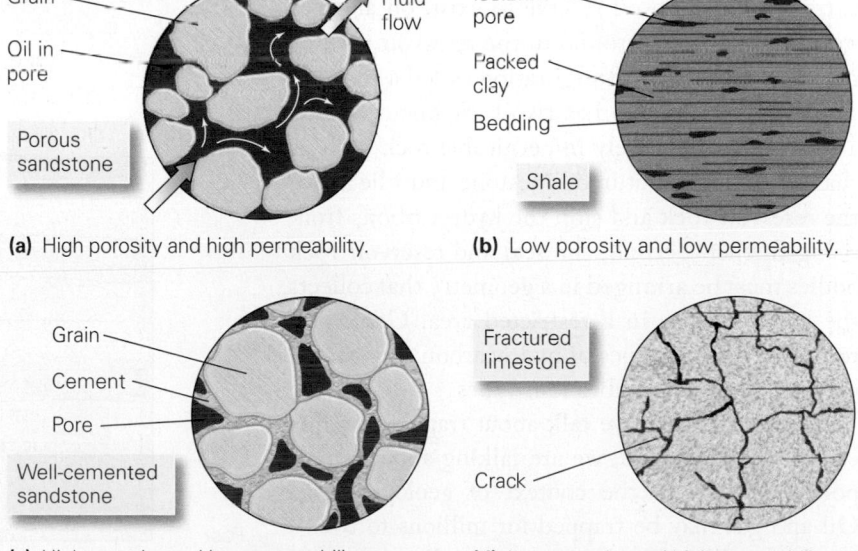

(a) High porosity and high permeability.

(b) Low porosity and low permeability.

(c) High porosity and low permeability.

(d) Low porosity and high permeability.

Keeping the concepts of porosity and permeability in mind, we can see that a poorly cemented sandstone makes a good reservoir rock because it's both porous and permeable. A highly fractured rock can also be porous and permeable, even if there is no pore space between individual grains. A limestone can be permeable if groundwater has dissolved the surfaces of cracks in the rock. The greater the porosity, the greater the capacity of a reservoir rock to hold oil, and the greater the rock's permeability, the easier it is for the oil to be extracted.

To fill the pores of a reservoir rock, oil and gas must first **migrate** (move) from the source rock into a reservoir rock, a process that takes thousands to millions of years to happen **(Fig. 14.10)**. Why do hydrocarbons migrate? Oil and gas are less dense than water, so they try to rise toward the Earth's surface to get above groundwater, just as salad oil rises above the vinegar in a bottle of salad dressing. Natural gas, which is less dense than oil, ends up floating above oil. In other words, buoyancy drives oil and gas upward. Typically, a hydrocarbon system must have a good *migration pathway*, such as a set of permeable fractures, in order for large volumes of hydrocarbons to move.

Traps and Seals

The existence of reservoir rock alone does not generate a conventional reserve, because if hydrocarbons can flow into a reservoir rock, they can also flow out. If oil or gas escapes from the reservoir rock and ultimately reaches the Earth's surface, where it can leak away at an **oil seep** or *gas seep* **(Fig. 14.11)**, there will be none left underground to extract. So, for an oil reserve to exist, oil and gas must be held underground in the reservoir rock by means of a geologic configuration called a **trap**.

An oil or gas trap has two components. First, a **seal rock**, a relatively impermeable rock such as shale, salt, or unfractured limestone, must lie above the reservoir rock and stop the hydrocarbons from rising further. Second, the seal and reservoir rock bodies must be arranged in a geometry that collects the hydrocarbons in a restricted area. Geologists recognize several types of hydrocarbon trap geometries; **Box 14.1** describes four types.

Note that when we talk about trapping hydrocarbons underground, we are talking about a temporary process in the context of geologic time. Oil and gas may be trapped for millions to over a hundred million years, but eventually they may

FIGURE 14.10 Initially, oil resides in the source rock. Because it is buoyant relative to groundwater, the oil migrates into the overlying reservoir rock. The oil accumulates beneath a seal rock in a trap.

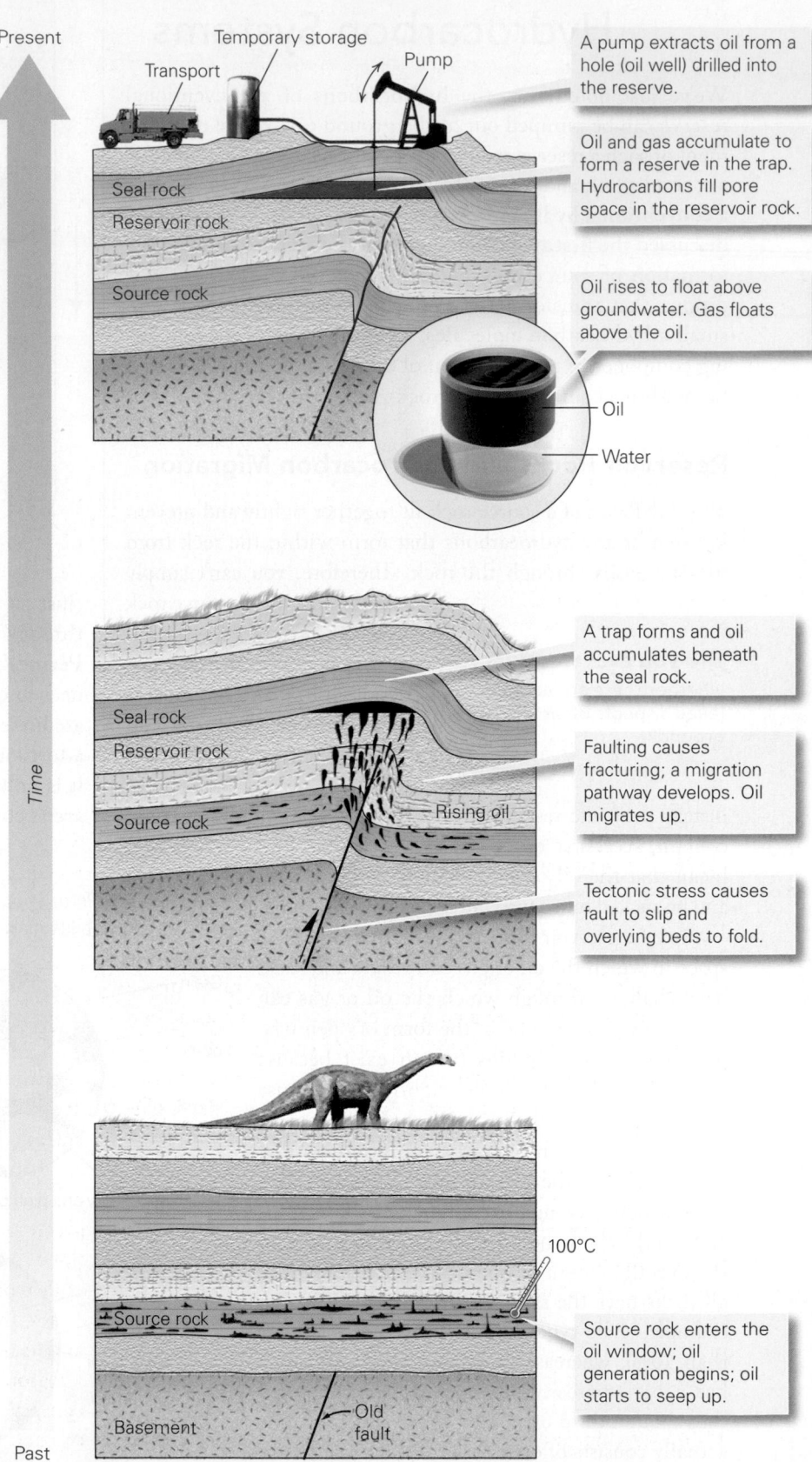

A pump extracts oil from a hole (oil well) drilled into the reserve.

Oil and gas accumulate to form a reserve in the trap. Hydrocarbons fill pore space in the reservoir rock.

Oil rises to float above groundwater. Gas floats above the oil.

A trap forms and oil accumulates beneath the seal rock.

Faulting causes fracturing; a migration pathway develops. Oil migrates up.

Tectonic stress causes fault to slip and overlying beds to fold.

Source rock enters the oil window; oil generation begins; oil starts to seep up.

FIGURE 14.11 Examples of hydrocarbon seeps at the Earth's surface.

(a) The La Brea tar pit is an oil seep that now forms the centerpiece of a small park in Los Angeles, California.

(b) The Darvasa gas crater (known as the "door to hell") in Turkmenistan formed in 1971 when a drilling rig tapped into a cavern filled with natural gas. Engineers lit a fire hoping to burn off the gas, but it's above a seep, so the fire has burned ever since.

manage to pass through a seal rock because no rock is absolutely impermeable—most rocks contain joints that can provide permeability. Also, in some cases, microbes eat hydrocarbons in the subsurface. As a result, innumerable oil reserves that existed in the past have vanished, and the oil fields we find today, if left alone, may disappear millions of years in the future.

Birth of the Conventional Oil Industry

People have used oil since the dawn of civilization—as a lubricant, as a waterproof sealant, and even as a preservative to embalm mummies. In the United States, during the first half of the 19th century, "rock oil" (later called petroleum, from the Latin words *petra*, meaning rock, and *oleum*, meaning oil), obtained at seeps, was used to grease wagon axles and to make patent medicines. But such oil was rare and expensive. In 1854, George Bissell, a New York lawyer, realized that oil might have broader uses, particularly as fuel for lamps, to replace increasingly scarce whale oil. Bissell and a group of investors hired Edwin Drake, a colorful character who had drifted among many professions, to find a way to drill for oil in rocks beneath a hill near Titusville, Pennsylvania, where oily films floated on the water of springs. Using the phony title "Colonel" to add respectability to his name, Drake hired drillers and obtained a steam-powered drill. Work was slow, and the investors became discouraged, but the very day that Drake received a letter ordering him to stop drilling, his drillers discovered that the hole, which had reached a depth of 21.2 m, had filled with oil. They set up a pump, and on August 27, 1859, for the first time in history, pumped oil out of the ground. No one had given much thought to the question of how to store the oil, so workers dumped it into empty whisky barrels. (These days, a **barrel of oil [bbl]** has become a standard unit of measurement: 1 bbl = 42 gallons = 159 liters. Oil may also be sold by weight: 1 metric ton [1 tonne] of oil = 6.5 barrels.)

Within a few years, thousands of oil wells had been drilled in many states, and by the turn of the 20th century, civilization had begun its addiction to oil. Initially, most oil went into the production of kerosene for lamps. Later, when electricity took over from kerosene as the primary source for illumination, gasoline derived from oil became the fuel of choice for the newly invented automobile. Oil also fueled electric power plants. In its early years, the oil industry was in perpetual chaos. When "wildcatters" discovered a new oil reserve, there would be a short-lived boom during which the price of oil could drop to pennies a barrel. In the midst of this chaos, John D. Rockefeller established the Standard Oil Company, which monopolized the production, transport, and marketing of oil. In 1911, the U.S. Supreme Court broke Standard Oil down into smaller companies (including Exxon, Chevron, Mobil, Sohio, Amoco, Arco, Conoco, and Marathon), some of which have recombined in recent decades. Oil became a global industry governed by the complex interplay of politics, profits, supply, and demand.

The Modern Search for Oil

Wildcatters discovered the earliest *oil fields* or *gas fields* (land areas above a reserve) either by blind luck or by searching for surface seeps. But in the 20th century, when most known seeps had been drilled and blind luck became too risky, oil companies realized that finding new fields would require systematic exploration. The modern-day search for hydrocarbons is a complex, sometimes dangerous, and often exciting endeavor with many steps.

Source rocks are always sedimentary, as are most reservoir and seal rocks, so geologists begin exploration by looking for a region containing appropriate sedimentary rocks. Then they compile a geologic map of the area, showing the distribution of rock units. From this information, it may be possible to construct a preliminary cross section depicting the geometry

BOX 14.1 CONSIDER THIS . . .

Types of Oil and Gas Traps

Geologists who work for oil companies spend much of their time trying to identify underground traps. No two traps are exactly alike, but we can classify many into the following four categories.

- *Anticline trap:* In some places, sedimentary beds are not horizontal, as they are when originally deposited, but have been bent by the forces involved in mountain building. These bends, as we have seen, are called folds. An anticline is a type of fold with an arch-like shape (**Fig. Bx14.1a**). If the layers in the anticline include a source rock overlain by a reservoir rock beneath a seal rock, then we have the recipe for a trap. The oil and gas rise from the source rock, enter the reservoir rock, and become trapped in the crest of the anticline.
- *Fault trap:* If the slip on a fault crushes and grinds the adjacent rock to make an impermeable layer along the fault, then oil and gas may migrate upward along bedding in the reservoir rock until they stop at the fault surface (**Fig. Bx14.1b**). A fault trap may also develop if the slip juxtaposes a seal rock against a reservoir rock.
- *Salt-dome trap:* In some sedimentary basins, the sequence of strata contains a thick layer of salt, deposited when the basin was first formed and seawater covering the basin was shallow and very salty. Sandstone, shale, and limestone overlie the salt. The salt layer is not as dense as sandstone, limestone, or shale, so it is buoyant and tends to rise up slowly through the overlying strata. Once the salt starts to rise, the weight of surrounding strata squeezes the salt out of the salt layer and up into a growing, bulbous *salt dome*. As the dome rises, it bends up the adjacent layers of sedimentary rock. Oil and gas in reservoir rock layers migrate upward until they are trapped against the boundary of the impermeable salt dome (**Fig. Bx14.1c**).
- *Stratigraphic trap:* In a stratigraphic trap, a tilted reservoir rock bed "pinches out" (thins and disappears) up-dip between two impermeable layers. Oil and gas migrating upward along the bed accumulate at the pinch-out (**Fig. Bx14.1d**).

FIGURE Bx14.1 Examples of oil and gas traps. A trap is a configuration of a seal rock over a reservoir rock in a geometry that keeps the oil underground.

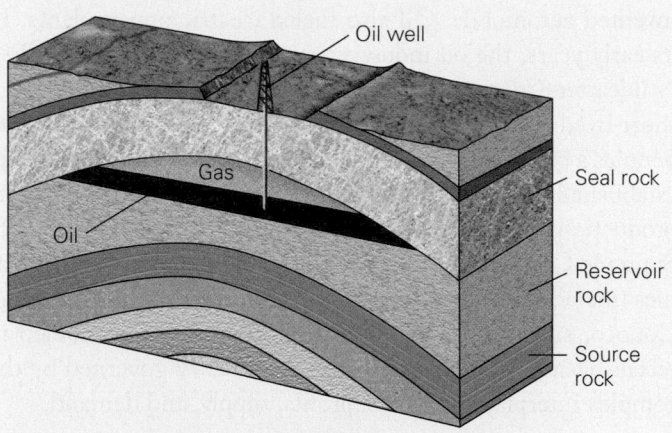

(a) Anticline trap. Oil and gas rise to the crest of the fold.

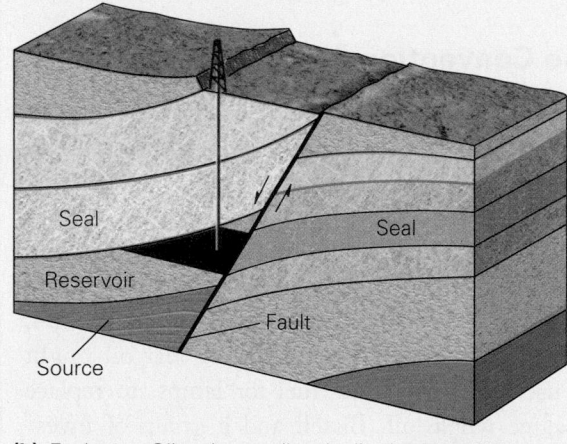

(b) Fault trap. Oil and gas collect in tilted strata adjacent to the fault.

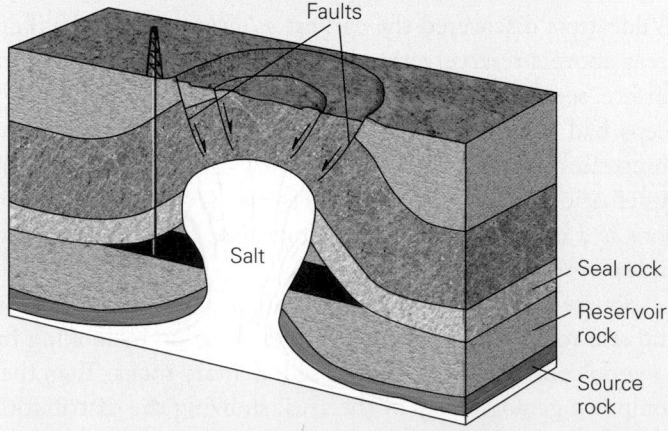

(c) Salt-dome trap. Oil and gas collect in strata on the flanks of the dome, beneath the salt.

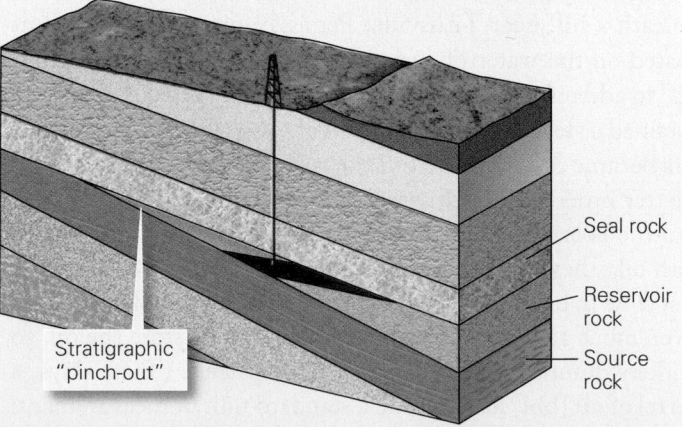

(d) Stratigraphic trap. Oil and gas collect where the reservoir layer pinches out.

of the sedimentary layers underground as they would appear on an imaginary vertical slice through the Earth.

To add detail to the cross section, geologists obtain a **seismic-reflection profile** of the region. On land, they do this by using a special vibrating truck or by setting off dynamite explosions that send seismic waves into the ground **(Fig. 14.12a; see Interlude D)**. The seismic waves reflect off boundaries between rock layers, just as sonar waves sent out by a submarine reflect off the bottom of the sea. Reflected seismic waves then return to the ground surface where sensitive seismometers (*geophones*) set into the ground record their arrival. A computer measures the time between the generation of a seismic wave and its return and, based on this information, defines the depth to the boundaries from which the wave reflected **(Fig. 14.12b)**. To explore for oil in sedimentary layers below the seafloor, geologists use an *air gun* dragged behind a ship to send pulses of compressed air into the water **(Fig. 14.12c)**. The pulses carry enough energy to produce seismic waves in the strata beneath the water. Reflected signals are received by *hydrophones* dragged behind the ship. Using seismic-reflection data, geologists work with computer programs to construct an image of the configuration of underground rock layers and, in some cases, to locate oil or gas reserves. Technological advances now enable geologists to create three-dimensional seismic-reflection data blocks of the subsurface, both underground and underwater **(Fig. 14.12d)**. Such blocks are expensive—just one may cost millions of dollars to create.

Drilling and Refining

If geologic studies identify a trap, as well as good source rocks and reservoir rocks, and a likelihood that strata have been heated into the oil or gas window, then geologists make a recommendation to drill. They do not make such recommendations lightly, for drilling a deep well may cost tens of millions of dollars. If the company management accepts the geologists' recommendation, drillers go to work.

These days, drillers use rotary drills to grind a hole down through rock. A rotary drill consists of a pipe tipped by a

FIGURE 14.12 Using seismic-reflection profiling to describe the character of underground beds and help locate reservoirs.

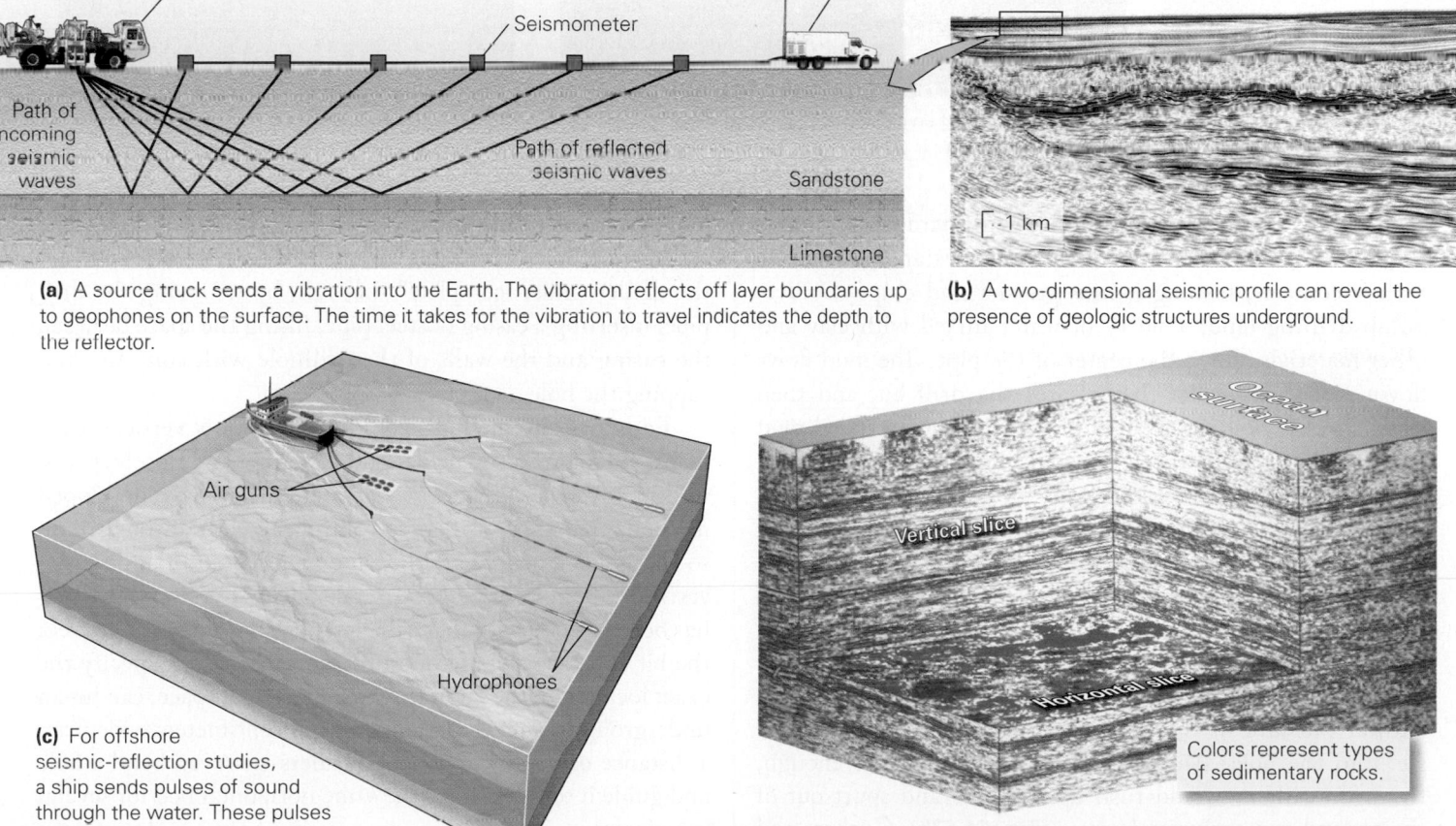

(a) A source truck sends a vibration into the Earth. The vibration reflects off layer boundaries up to geophones on the surface. The time it takes for the vibration to travel indicates the depth to the reflector.

(b) A two-dimensional seismic profile can reveal the presence of geologic structures underground.

(c) For offshore seismic-reflection studies, a ship sends pulses of sound through the water. These pulses penetrate the crust and reflect off boundaries back to hydrophones.

(d) Modern techniques produce three-dimensional blocks of data, so that geologists can study both vertical and horizontal slices through the subsurface.

FIGURE 14.13 Drilling rigs and oil pumps.

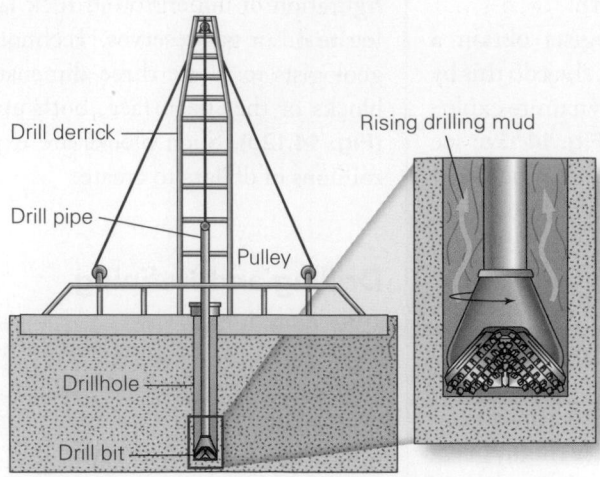

(a) An onshore drilling rig. The inset shows a close-up of the drill bit. Drilling mud carries the cuttings up and out of the hole.

(b) The Lakeview gusher, in California, spilled 9 million barrels of oil in 1910.

(c) An aerial view of an oil field with many closely spaced wells.

(d) An offshore drilling rig. The derrick is constructed on top of a platform. The flare burns off gas.

rotating bit, a bulb of metal studded with hard metal prongs **(Fig. 14.13a)**. As the bit rotates, it scratches and gouges the rock it contacts, turning it into powder and chips. Drillers pump **drilling mud**, a slurry of water mixed with clay and other materials, down the center of the pipe. The mud flows down, past a propeller that rotates the drill bit, and then squirts out of holes at the end of the bit. The extruded mud cools the bit, which otherwise would heat up due to friction with the rock, then flows back up the hole outside the drill pipe. As it rises, the mud carries *rock cuttings* (fragments of rock that were broken up by the drill bit) up and out of the hole. Drilling mud also serves another very important purpose—its weight counters the natural pressure of the oil and gas in underground reservoir rocks. By doing so, it prevents hydrocarbons from entering the hole. Without the mud, the natural pressure in the reservoir rock would drive oil and gas into the hole. And if the pressure were great enough, the hydrocarbons would rush up the hole and spurt out of the ground as a *gusher* or *blowout* **(Fig. 14.13b)**. Gushers and

blowouts can be disastrous, because they spill oil and, in some cases, ignite into an inferno. When drilling has been finished, the drillers must "complete" the well, by removing the drill pipe, inserting a casing of steel pipe, filling the space between the casing and the walls of the drillhole with concrete, and capping the hole.

Early drilling methods could produce only vertical drillholes. Therefore, an oil field might contain many closely spaced wells **(Fig. 14.13c)**. More recently, engineers have developed methods that allow drillers to control the path of the drill bit so the hole can curve and become inclined at an angle from vertical or can even be horizontal. Such **directional drilling** has become so precise that a driller, by using a joystick to steer the bit and by watching output from sensors that specify the exact location of the bit in three-dimensional space, can hit an underground target that is only several centimeters wide from a distance of a few kilometers. Drillers can even bend a hole and guide it to stay within the same horizontal bed for several kilometers.

Drillers use derricks to hoist the heavy drill pipe. To drill in an offshore hydrocarbon reserve, one that occurs in strata beneath the continental shelf, the derrick must be constructed on an *offshore-drilling platform* (Fig. 14.13d). These facilities may be built on huge towers rising from the seafloor or on giant submerged pontoons. Using directional drilling, it's possible to reach multiple targets from the same platform.

On completion of a well, workers remove the drilling rig and set up a pump. Some pumps resemble a bird pecking for grain; their heads move up and down to pull up oil that has seeped out of pores in the reservoir rock into the drillhole (Fig. 14.14a).

You may be surprised to learn that simple pumping gets only about 30% of the oil in a reservoir rock out of the ground. Oil companies use *secondary recovery techniques* to coax out more oil (as much as 20% more). For example, a company may drive oil toward a drillhole by forcing steam into holes in the ground nearby—the steam heats the oil in the ground, making it less viscous, and pushes it along. In some cases, drillers create artificial fractures in rock around the hole by pumping a high-pressure water, sand, and chemical mixture into a portion of the hole. This process, called **hydrofracturing** (also known as *hydraulic fracturing* or simply *fracking*), creates new fractures and opens up pre-existing ones—the fractures provide easy routes for the oil to follow from the rock to the well. We'll discuss hydrofracturing in more detail later in this chapter.

Crude oil, once extracted directly from the ground, flows first into storage tanks and then into a pipeline or tanker, which transports it to a refinery (Fig. 14.14b–d). At a refinery, workers distill crude oil into several separate components by first heating it to a temperature of about 400°C and then pumping it into a vertical pipe called a *distillation column* (Fig. 14.14e). Lighter molecules rise to the top of the column, while heavier molecules stay at the bottom. Outlets at different levels in the column allow removal of molecules of different sizes—tar comes out the bottom, oil above that, gasoline above that, and propane from the top. The heat may also "crack" larger molecules to make smaller ones. Chemical factories buy the largest molecules left at the bottom and transform them into plastics.

Natural gas burns more cleanly than oil, in that combustion of gas produces only carbon dioxide and water, while the burning of oil produces not only carbon dioxide and water, but also complex organic pollutants. Therefore, natural gas has long been a preferred fuel for home cooking and heating. And when the cost of natural gas decreased, many electricity-generating stations were converted to burn natural gas. It can also be used to fuel cars and trucks if the vehicles have been appropriately modified.

If gas is the main hydrocarbon of a conventional reserve, it may be economical to extract and transport gas. But in many conventional reserves, the volume of gas is too small to be produced economically because gas must be compressed for transport, which takes energy, and transportation requires high-pressure pipelines or special ships that are expensive to operate. So, oil-field operators commonly vent gas from a pipe and burn it as a flare where it enters the air (see Fig. 14.13d).

Where Do Conventional Hydrocarbon Reserves Occur?

Conventional hydrocarbon reserves are not randomly distributed around the world. Currently, countries in the Middle East bordering the Persian Gulf contain the world's largest reserves in 25 *supergiant fields*, meaning fields that contain more than 1 billion barrels of oil (Fig. 14.15a, b). In fact, this region has almost 60% of the world's conventional reserves, whereas the United States, the largest consumer of oil, has only a small percentage of the total.

What determines where conventional reserves occur? To start with, a region must have been in an environment where plankton could grow well, so that sediments deposited there were rich in organic content. Between the Jurassic (135 Ma) and the Late Cretaceous (66 Ma), the now oil-rich Middle East region was situated in tropical areas between latitude 20° S and 20° N, where biological activity was high (Fig. 14.15c). Sediments deposited at that time were rich in organic matter. Thick successions of porous sandstone buried the source rocks of the Middle East, and crustal compression due to the collision between Africa and Asia folded the strata to produce excellent fold traps.

In addition to the Middle East, conventional oil reserves occur in sedimentary basins formed along passive continental margins, such as the Gulf Coast of the United States and the Atlantic coasts of Africa and Brazil. They also occur in the intracratonic and foreland basins on continents (see Chapter 7).

> ## TAKE-HOME MESSAGE
>
> Conventional hydrocarbon reserves are ones in which oil has migrated from a source rock into a porous and permeable reservoir rock, situated within an oil trap, so that the oil or gas can be pumped fairly easily. The first oil drilling took place in 1859. Modern methods for finding, drilling, producing, and refining oil are very complex and expensive.
>
> **QUICK QUESTION:** Where do most of the conventional oil reserves in the world occur today?

FIGURE 14.14 Pumping, transporting, and reining oil.

The head of the pump goes up and down.

Storage tanks

Well

(a) When a well is complete, the derrick is replaced by a pump, which sucks oil out of the ground. This design is called a pumpjack.

(b) The Trans Alaska Pipeline transports oil from fields on the Arctic coast to a tanker port on the southern coast of Alaska.

(c) This oil tanker is capable of carrying 2.6 million barrels (enough to supply the entire United States for 7 hours).

(d) Distilling columns of an oil refinery transform crude oil into gasoline and other hydrocarbon products.

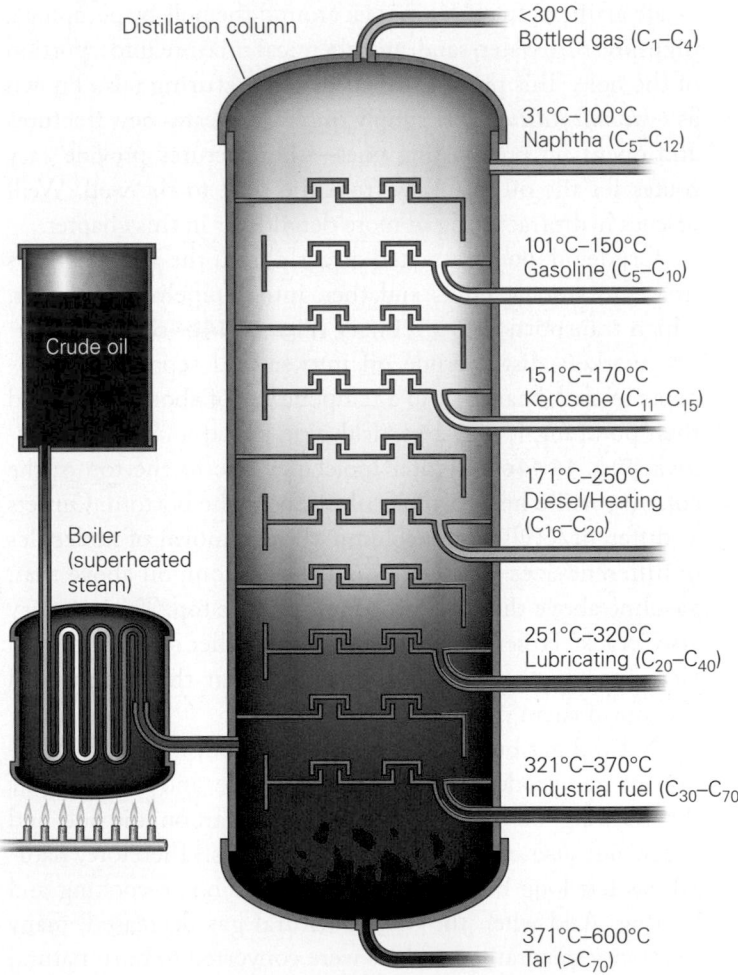

Distillation column

<30°C
Bottled gas (C_1–C_4)

31°C–100°C
Naphtha (C_5–C_{12})

101°C–150°C
Gasoline (C_5–C_{10})

151°C–170°C
Kerosene (C_{11}–C_{15})

171°C–250°C
Diesel/Heating
(C_{16}–C_{20})

251°C–320°C
Lubricating (C_{20}–C_{40})

321°C–370°C
Industrial fuel (C_{30}–C_{70})

Crude oil

Boiler
(superheated
steam)

371°C–600°C
Tar (>C_{70})

(e) A distillation column works by gravity. Heated oil separates into bubbles of lighter hydrocarbons and droplets of heavier ones. Heavier ones sink and light ones rise.

FIGURE 14.15 The distribution of conventional oil reserves around the world.

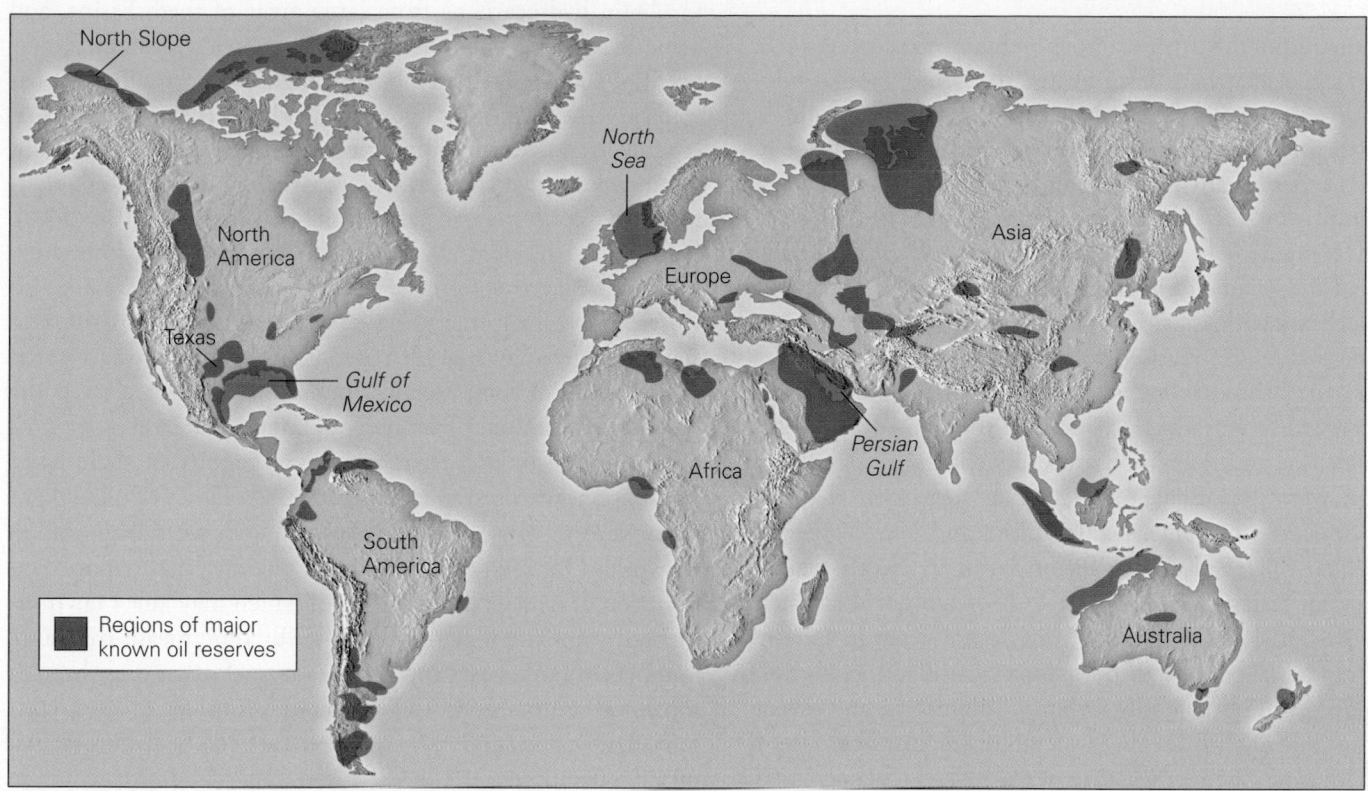

(a) Oil reserves are distributed on all continents — some onshore and some offshore.

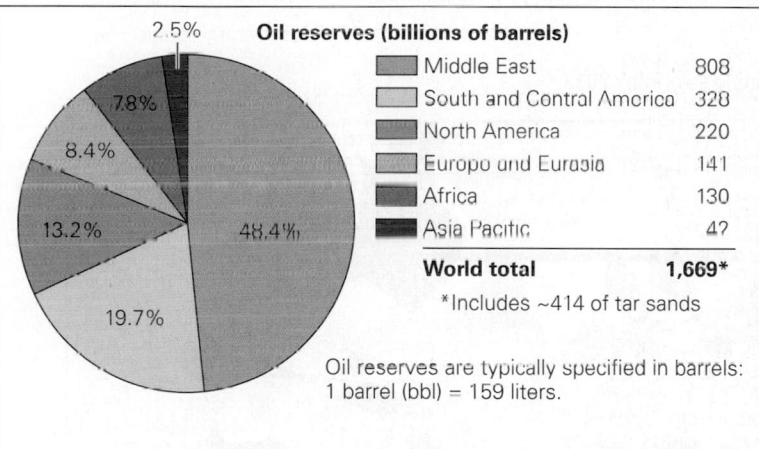

Oil reserves (billions of barrels)	
Middle East	808
South and Central America	328
North America	220
Europe and Eurasia	141
Africa	130
Asia Pacific	42
World total	**1,669***

*Includes ~414 of tar sands

Oil reserves are typically specified in barrels:
1 barrel (bbl) = 159 liters.

(b) Distribution of oil reserves among regions.

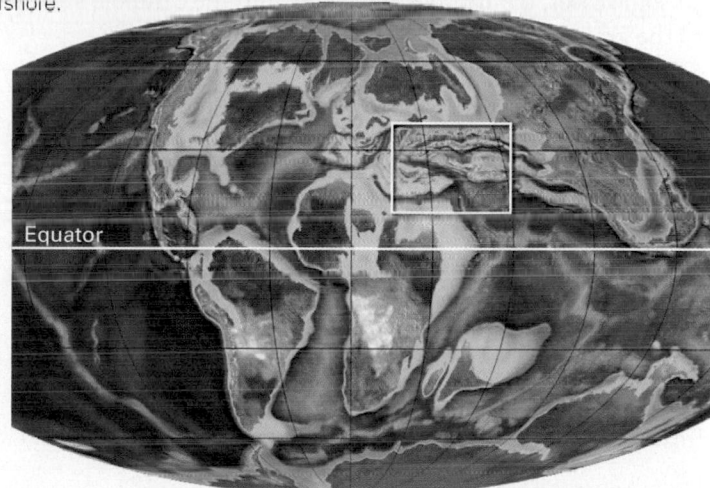

(c) A map of the Late Cretaceous world. The current source rocks of the Middle East were formed by sediments deposited in this remote time period. The box shows the region that would become the Middle East oil fields of today. Note that the region lay in the warm subtropics, in which marine plankton could thrive.

14.5 Unconventional Hydrocarbon Reserves

In an *unconventional hydrocarbon reserve*, rock or sediment contains significant quantities of hydrocarbons, but either the rock does not have adequate permeability or the hydrocarbons themselves are too viscous to flow. Therefore, the hydrocarbons cannot be extracted simply by drilling and pumping. Extraction of unconventional hydrocarbons was not possible until engineers developed new technologies and the price per barrel of hydrocarbons became high enough for the extraction to be profitable. These conditions were met 10 to 15 years ago. Since then, efforts to extract hydrocarbons from unconventional reserves have grown at a rapid rate, and now such reserves provide a significant portion of the energy used worldwide. Let's examine the geologic context of such reserves.

Shale Gas and Shale Oil

Huge quantities of hydrocarbons remain in source rocks (organic shale) that have been heated to the oil and gas windows. This resource is called **shale oil** or **shale gas** depending on its viscosity. Until recently, this oil and gas could not be extracted economically by pumping because of shale's very low permeability. In 2007, Terry Engelder of Penn State University and others pointed out that previous reports had greatly underestimated the volume of hydrocarbons available in source rock in sedimentary basins that lie near populated areas **(Fig. 14.16)**. The combination of increased volume estimates and the ability to access the hydrocarbons by directional drilling and hydrofracturing led to a production boom in the United States and worldwide. For example, thousands of wells have been drilled into the Marcellus Shale, a Devonian formation that lies beneath portions of Pennsylvania, Ohio, and New York, and the Bakken Shale, in the Williston basin in North Dakota and adjacent states. The shale source beds are not that thick, and in these basins, the hydrocarbons do not occur in conventional traps. But because of directional drilling, a single well can follow a bed horizontally for many kilometers, and several wells can be drilled from the same platform. And because of hydrofracturing, the permeability of the rock can be increased. In some cases, the hydrocarbons flow into the drillhole and up to the ground under their own pressure.

Energy companies spent billions of dollars to lease the right to obtain hydrocarbons from large areas of these basins, and some local landowners became millionaires overnight. Thousands of people obtained jobs working for energy companies as well as in the towns that house the workers. The boom has been controversial because of lingering questions about the amount of oil and gas that can really be recovered and because of environmental concerns associated with hydrofracturing **(Box 14.2)**. Not only are residents worried that chemicals in hydrofracturing fluids may contaminate water supplies, but there are also worries that hydrofracturing may release hydrocarbons into sources of groundwater. Of note, natural gas does occur in the groundwater of some regions, but it's not always clear if the gas entered the water from new or enlarged fractures, or if its presence reflects long-term natural seepage from shale beds. In addition, the need to transport large volumes of oil and gas from wells to consumers imposes demands on transportation networks. One method, transport by pipeline, may require construction of new pipelines, some of which may affect environmentally sensitive regions. Transport by train means that large numbers of tanker cars travel on train tracks that pass through inhabited communities. Oil extracted from shale tends to contain dissolved natural gas, so tanker cars can be explosive. An oil train derailment in the Canadian town of Lac-Mégantic, in 2013, triggered an explosion that leveled the center of the town. One outcome of the production boom has been a switch from

FIGURE 14.16 Map of basins with assessed shale oil and shale gas formations, as of May 2013.

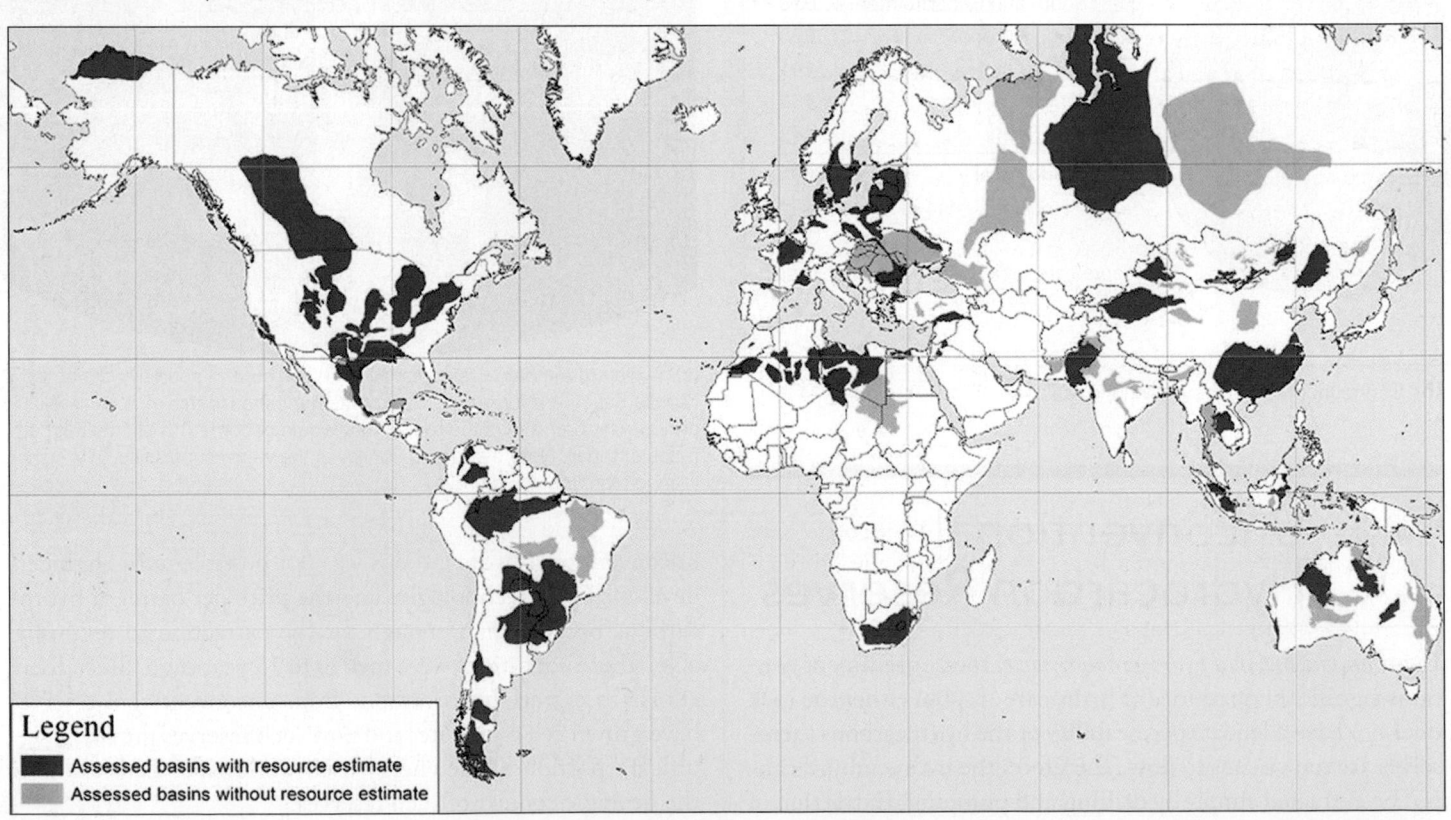

Legend

■ Assessed basins with resource estimate

▨ Assessed basins without resource estimate

coal and oil to natural gas for electricity generation because gas supplies near major cities have been more accessible. Notably, a drop in gas prices slowed the boom after 2015.

Tar Sand (Oil Sand)

In several locations around the world, most notably Alberta (in western Canada) and Venezuela, vast reserves of very viscous, tar-like *heavy oil* exist. This heavy oil, known also as *bitumen*, has the consistency of gooey molasses and thus cannot be pumped directly from the ground. It fills the pore spaces of sand or of poorly cemented sandstone, constituting up to 12% of the sediment or rock volume. Sand or sandstone containing high concentrations of bitumen is known as **tar sand** or *oil sand*.

The hydrocarbon system that leads to the generation of tar sand begins with the production and burial of a source rock in a large sedimentary basin. When subjected to temperatures in the oil window, the source rock yields oil and gas, which migrate into sandstone layers and then up the dip of tilted layers to the edge of the basin, where they become caught in stratigraphic traps (see Box 14.1). Initially, these hydrocarbons have relatively low viscosity, so in the geologic past they could have been pumped easily. But over time, microbes attack the oil reserve underground, digest lighter, smaller hydrocarbon molecules, and leave behind only the larger molecules, whose presence makes the remaining oil viscous. Geologists refer to such a transformation process as *biodegradation*. Generation of tar sand by biodegradation represents yet another example of the interaction between physical and biological components of the Earth System.

Production of usable oil from tar sand is difficult and expensive, but not impossible. It takes about two tons of tar sand, and a lot of energy, to produce one barrel of oil. Oil companies mine near-surface deposits in vast open-pit mines and then heat the tar sand in a furnace to extract the oil **(Fig. 14.17a)**. Producers then crack the heavy oil molecules to produce smaller, more usable molecules. Trucks dump the drained sand back into the mine pit. To extract oil from deeper deposits of tar sand, oil companies drill wells and pump steam or solvents down into the sand to liquefy the oil enough so that it can be pumped out.

Oil Shale

Some sedimentary basins host vast reserves of organic shale that have not been subjected to temperatures of the oil window, or if they have, they did not stay within the oil window long enough to complete the transformation to oil. Such rock still contains a high proportion of kerogen. Geologists refer to shale that contains at least 25% to 75% kerogen as **oil shale**. Oil shale differs from coal because the organic matter within it exists in the form of waxy hydrocarbon molecules, not as elemental carbon, and it differs from shale oil in that it

FIGURE 14.17 Examples of unconventional hydrocarbon sources.

(a) An open-pit tar-sand mine in Canada. Trucks haul the sand to a plant where it is heated so hydrocarbons can be extracted.

(b) Oil shale can be set ablaze.

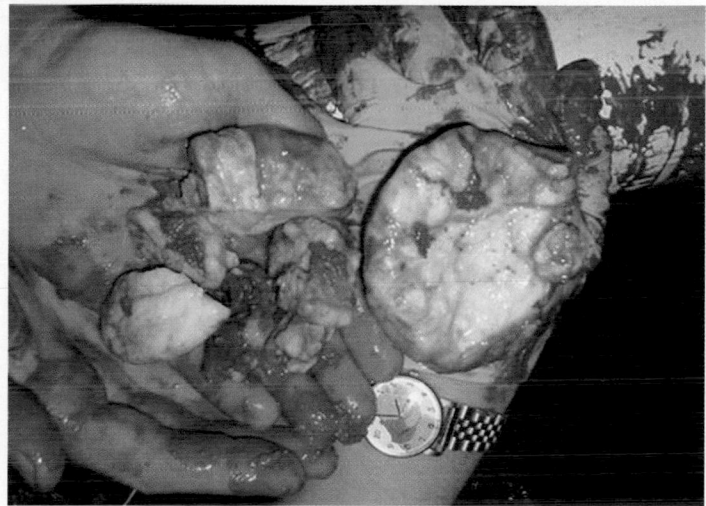

(c) Gas hydrate samples (white material) dug up from the muddy seafloor.

BOX 14.2 CONSIDER THIS . . .

Hydrofracturing (*Fracking*)

Hydrofracturing, a technology first used in the 1950s, has been in the news quite a bit in recent years, where it's commonly referred to as *fracking*. Drillers use the technique to open and propagate existing joints in rocks at depth as well as to generate new cracks in the rock. These fractures provide a permeability pathway through which hydrocarbons can flow to reach a drillhole and be extracted. Originally, hydrofracturing was used in conventional vertical wells as a means to enhance secondary recovery. Today it is also being widely used in horizontal wells following source beds **(Fig Bx14.2a, b)**. What is hydrofracturing, how does it work, and what risks are potentially involved in its use?

To hydraulically fracture a drillhole, drillers start by sealing off a length of a completed well with *packers*, which are simply inflatable balloons that, when filled with high-pressure fluid, press tightly against the wall of the hole **(Fig Bx14.2c)**. Then the drillers insert a pipe through one of the packers into the sealed-off section of the well and pump *fracking fluid* into this section under high pressure. When the pressure generated by the fluid within the sealed section becomes great enough, it cracks the well casing, forces open existing joints that intersect the hole, and may cause these joints to lengthen at their tips **(Fig Bx14.2d)**. The process may also generate new cracks in the rock adjacent to the hole. The area affected by hydrofracturing can extend tens of meters out from the drillhole. Once fracturing has been completed, drillers pump out the fluid. Hydrocarbons then flow along the fractures into the drillhole and up to the surface, where drillers capture it for transport **(Fig Bx14.2e)**. The volume of fluid used to hydrofracture a section of hole is roughly equivalent to the volume in an Olympic swimming pool—usually several sections of a hole may be subjected to hydrofracturing, so the process tends to be repeated several times in the hole.

What's in fracking fluid? A typical example consists of about 90% water, 9.5% quartz sand, and 0.5% other chemicals. The sand props the holes open once the fluid has been removed—without the sand, the cracks opened by hydrofracturing would close up tightly under the pressure applied by surrounding rock, and could not serve as permeability pathways. The 0.5% portion of fracking fluid that is not water or sand contains many chemicals, including oils that make the fluid more slippery so that it can penetrate farther into the rock; acid, which dissolves cement between grains and increases porosity; detergent, which lowers the surface tension of water so it doesn't stick to grains; guar gum, to make the fluid more viscous so that it can carry more sand; antifreeze, to prevent scale buildup; and biocides, which prevent the growth of bacteria that could clog pores.

Hydrofracturing requires a lot of equipment and materials, so there will be many trucks carrying water, sand, and other chemicals as well as trucks carrying portable pumps and giant mixing vats. The drill site may also have holding tanks or retaining ponds for storing fluid that has been removed from the ground once hydrofracturing has been finished.

Concerns about hydrofracturing have become the subject of intense public debate. Nearby residents worry that fracking fluid will contaminate drinking water underground. To understand the nature of this risk, it's necessary to understand how groundwater changes with depth. As we'll discuss further in Chapter 19, *groundwater* is water that fills or saturates pores and cracks in rock or sediment underground, beneath a surface called the *water table*—above the water table, pores and cracks contain some air. Typically, groundwater in the upper several hundred to a few thousand meters is fresh and drinkable, but below that depth groundwater tends to be saline **(Fig Bx14.2f)**. If the section of a hole subjected to hydrofracturing lies deeper than the saline boundary, the fluids from the section probably won't mix with drinkable groundwater, for they are denser than groundwater and won't rise. Leakage from the vertical portion of the hole above the saline boundary, however, can be problematic, so it is important that this portion of the hole be cased and sealed thoroughly before injecting fracking fluid into a hole. Fluid leakage at the surface, from tanks or holding ponds or from transporting trucks, is also of concern—to avoid contamination, handling of the fluids at the surface must be very carefully monitored.

---

contains kerogen, not oil. (The difference in meaning between oil shale and shale oil can be confusing.)

Lumps of oil shale can be burned directly and have been used as a fuel since ancient times **(Fig. 14.17b)**. Researchers have developed techniques to produce liquid oil from oil shale. The process involves heating the oil shale to a temperature of 500°C; at this temperature, the shale decomposes and the kerogen transforms into liquid hydrocarbon and gas. Large supplies of oil shale occur in Estonia, Scotland, China, and Russia, and in the Green River basin of Wyoming in the United States. As is the case with tar sand, production of oil from oil shale is possible, but expensive. In addition to the expense of mining and environmental reclamation, producers must pay for the energy needed to heat the shale. It takes about 40% of the energy yielded by a volume of oil shale to produce the oil itself.

Gas Hydrate

Gas hydrate is a chemical compound consisting of methane (CH_4) molecules surrounded by a cage-like arrangement of water molecules. An accumulation of gas hydrate occurs as a whitish solid that resembles ordinary water ice **(Fig. 14.17c)**. Gas hydrate forms when anaerobic bacteria (bacteria that live in the absence of oxygen) eat organic matter, such as dead plankton that have been incorporated into sediments of the seafloor. When the bacteria digest organic matter, they

FIGURE Bx14.2 Directional drilling and hydrofracturing. These new technologies have led to economical production of shale gas.

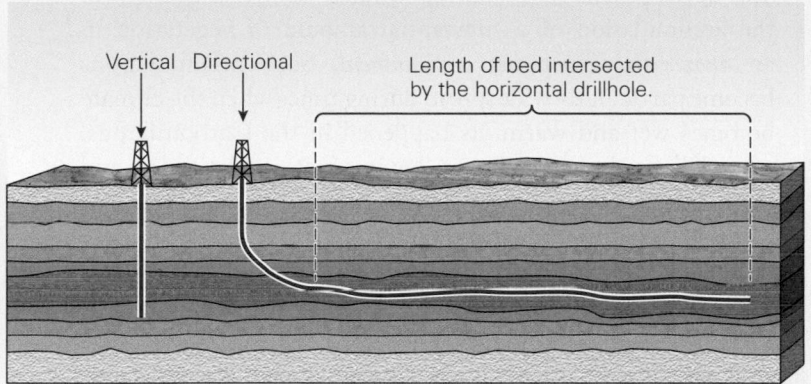

(a) Directional drilling permits the drillhole to follow a relatively thin bed for many kilometers, so the amount of shale oil or gas accessed can be large. A vertical hole intersects the shale for only a short distance.

(b) A drilling site. The trucks and the holding pond are used during hydrofracturing. Many holes can be drilled from this site, like spokes of a wheel.

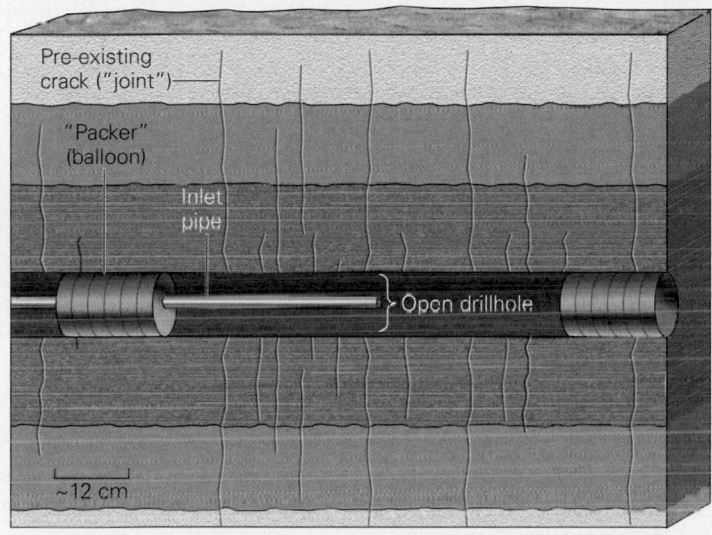

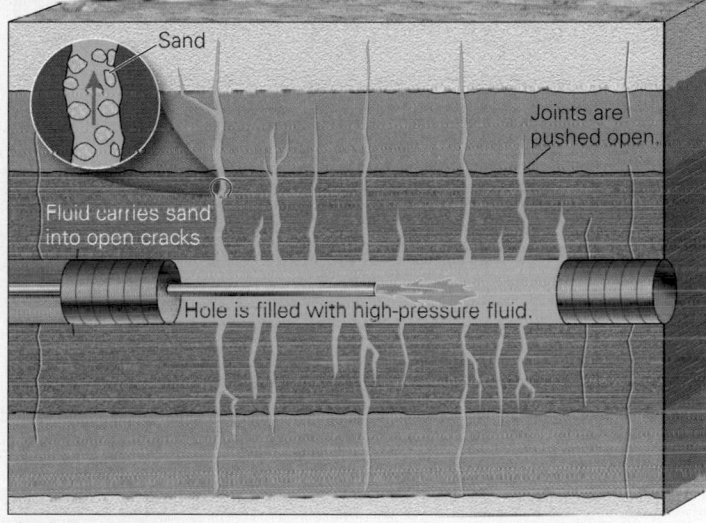

(c) The first step in hydrofracturing. After the hole has been drilled, packers seal off a portion, and a pipe is inserted through one of the packers.

(d) High-pressure fluid is pumped into the segment of hole. The pressure pushes open cracks and forms new ones. Sand injected with the fluid keeps the cracks from closing.

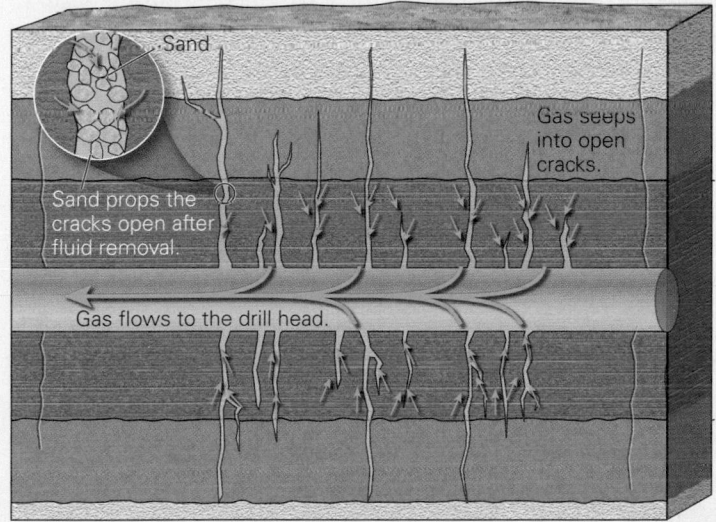

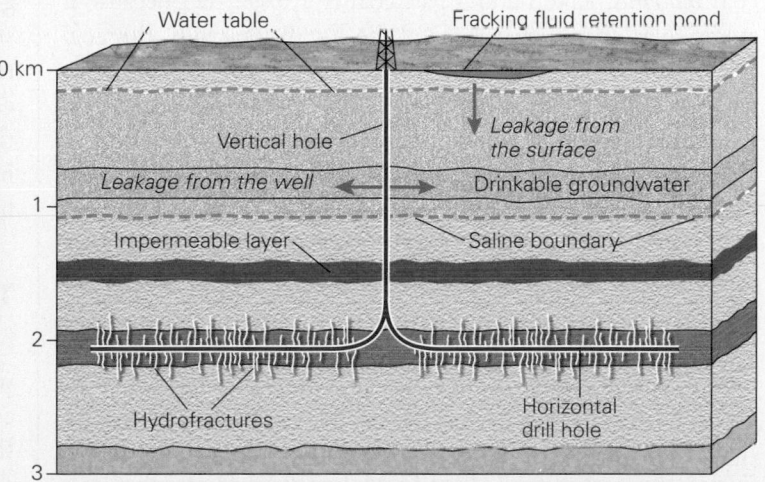

(e) After the packers and the fluid are removed, oil and gas can seep into the pipe and flow to the drill head.

(f) Potential for the contamination of groundwater by hydrofracturing.

produce methane as a by-product, and the methane bubbles into the cold seawater that fills pore spaces in sediments. Under the pressures found at depth in the ocean, the methane dissolves in the pore water and produces gas-hydrate molecules. The reaction can occur in water as shallow as 90 m in cold polar water, but not until depths of about 300 m in warm equatorial water. Exploration tests suggest that gas hydrate occurs as layers interbedded with sediment, or as a cement holding the sediment together, at depths of 90 to 900 m beneath the seafloor. Geologists estimate that an immense amount of methane lies trapped in gas-hydrate layers. In fact, there may be more organic carbon stored in gas hydrate than in all other hyrocarbon reservoirs combined! So far, however, techniques for safely recovering gas hydrate from the seafloor have not been devised.

TAKE-HOME MESSAGE

Shale gas, shale oil, tar sand, and oil shale are unconventional hydrocarbon reserves because it is difficult and expensive to extract hydrocarbons from them. Development of new technologies, however, has made extraction feasible. In particular, directional drilling and hydrofracturing have led to a huge increase in production of shale gas.

QUICK QUESTION: How can usable fuels be obtained from oil shale and tar sand?

14.6 Coal: Energy from the Swamps of the Past

Coal, a black, brittle sedimentary rock that burns, consists of elemental carbon mixed with minor amounts of organic chemicals, quartz, and clay. Typically, the carbon atoms in coal have bonded together, forming large, complicated molecules called coal *macerals*. Like oil and gas, coal is a fossil fuel because it stores solar energy that reached the Earth long ago. But coal does not have the same composition as oil or gas—coal contains carbon, not hydrocarbons. And, in contrast to oil, coal forms from plant material (wood, stems, leaves), not plankton. Coal commonly occurs in beds, called *coal seams*, that may be centimeters to meters thick and may be traceable over large regions.

Significant coal deposits could not form until vascular land plants appeared on the Earth in the late Silurian Period, about 420 million years ago. The most extensive deposits of coal in the world occur in Carboniferous (359 to 299 Ma) strata. In fact, geologists coined the name *Carboniferous* because strata representing this interval of the geologic column contain so much coal. Not all coal reserves, however, are Carboniferous—during the Cretaceous (145 to 66 Ma), organic matter that later became coal grew in Wyoming and adjacent states.

The Formation of Coal

The development of broad, continuous coal seams requires the accumulation of a substantial amount of vegetation in an *anoxic* (oxygen-poor) *environment*. Such accumulations become particularly widespread during times when the climate becomes wet and warm, as happened in the Carboniferous, when large areas of continents lay in equatorial latitudes, and in the Cretaceous, when volcanic eruptions dumped greenhouse gases into the atmosphere. Wet climates cause the water table to rise until broad plains become saturated with water and lower areas become submerged by shallow water. If it's warm, vegetation flourishes, so such wet regions become **coal swamps**, broad lowlands that resembled the wetlands and rainforests of modern tropical to semi-tropical regions **(Fig. 14.18a)**. In the stagnant swamps, little oxygen from the air mixes into the water, and microorganisms completely consume any oxygen that had been dissolved in the water, so the water becomes anoxic. Therefore, vegetation that dies and falls into the water doesn't rot entirely away, but rather gets buried by more dead vegetation until a compacted and partially decayed layer of organic matter, called **peat**, accumulates. The peat may be buried by fluvial (river) deposits or, if sea level rises and a transgression takes place (see Chapter 7), by layers of marine sediments such as sand, mud, or carbonate debris **(Fig. 14.18b)**. Burial causes the peat to undergo compaction, dewatering, and partial decay. Notably, peat consists of about 50% carbon and serves as a fuel in parts of the world where thick deposits formed during the last several thousand years from the moss and grass of bogs. This peat can easily be cut out of the ground and, once dried, it burns.

To transform peat into coal, the peat must be buried to a depth of 4 to 10 km by overlying sediment. Such deep burial can happen where the surface of the continent gradually sinks to produce a sedimentary basin (see Chapter 7). Because temperature increases with depth in the Earth, deeply buried peat gradually heats up. Heat accelerates chemical reactions that gradually destroy plant fiber and release volatile molecules such as H_2O, CO_2, CH_4, and N_2. These gases seep out of the reacting peat layer, leaving behind a residue of concentrated carbon. Once the proportion of carbon exceeds about 60%, the deposit formally becomes coal. With further burial and higher temperatures, chemical reactions yield progressively higher concentrations of carbon.

The Classification of Coal

Geologists classify coal according to its concentration of carbon, which in turn reflects the maximum temperature to which the coal was subjected while underground. At temperatures of less than 100°C, coal is a soft, dark brown material called *lignite*. At higher temperatures (100°C to 200°C), lignite transforms into dull black *bituminous coal*. At still higher temperatures

FIGURE 14.18 The formation of coal. Coal forms when plant debris becomes deeply buried.

(a) A museum diorama depicting a Carboniferous coal swamp.

(b) If there is a transgression, peat formed in the coal swamp can be buried and preserved.

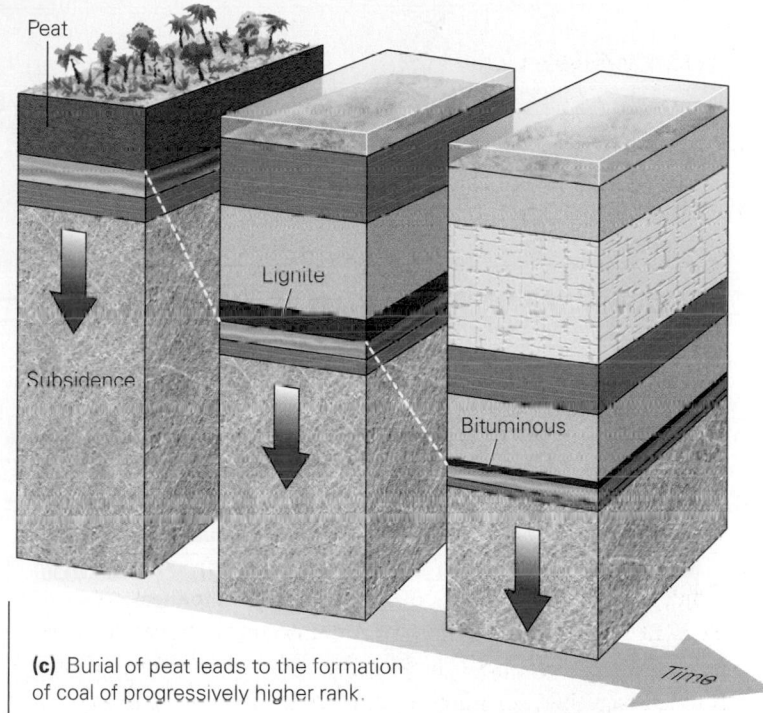

(c) Burial of peat leads to the formation of coal of progressively higher rank.

(201°C–300°C), bituminous coal transforms into shiny, black *anthracite* (also called "hard coal"). The progressive transformation of peat to anthracite, which occurs as the coal layer becomes more deeply buried and warmer, reflects the completeness of chemical reactions that remove hydrogen, oxygen, and nitrogen atoms from the organic chemicals of the peat and leave behind carbon **(Fig. 14.18c)**. As the carbon content of coal increases, we say the **coal rank** increases—lignite (60–70% carbon) is low-rank coal, bituminous (70–80% carbon) is intermediate-rank coal, and anthracite (90% carbon) is high-rank coal.

The burning of coal is a chemical reaction: $C + O_2 \rightarrow CO_2$. Therefore, the different ranks of coal produce different amounts of energy when burned. For example, anthracite contains more carbon per kilogram than lignite contains, so it produces more energy when burned. Specifically, burning a kilogram of anthracite yields about two to three times as much energy as does burning a kilogram.

Notably, the temperatures necessary to form anthracite develop only on the borders of mountain belts. Here mountain-building processes can push thick sheets of rock up along thrust faults and over the coal-bearing sediment, so the sediment ends up at depths of 8 to 10 km, where temperatures reach 300°C. In addition, mountain uplift drives very hot groundwater from great depth up through the coal, and the water's heat causes its rank to increase. Metamorphism in the interiors of mountain belts leads to the expulsion of all elements except carbon from organic layers, and the remaining carbon atoms rearrange to form graphite, the gray mineral used to make pencils. As a result, coal cannot be found in metamorphic rocks.

Finding, Mining, and Using Coal

Because the vegetation that eventually becomes coal was initially deposited in a sequence of sediment, coal occurs in sedimentary beds interlayered with other sedimentary rocks **(Fig. 14.19)**. To find coal, geologists search for sequences that

FIGURE 14.19 An example of coal beds interlayered with beds of sandstone and shale.

FIGURE 14.20 The distribution of coal, and its consumption.

(a) A map showing global distribution of coal reserves. Most coal accumulated in continental interior basins.

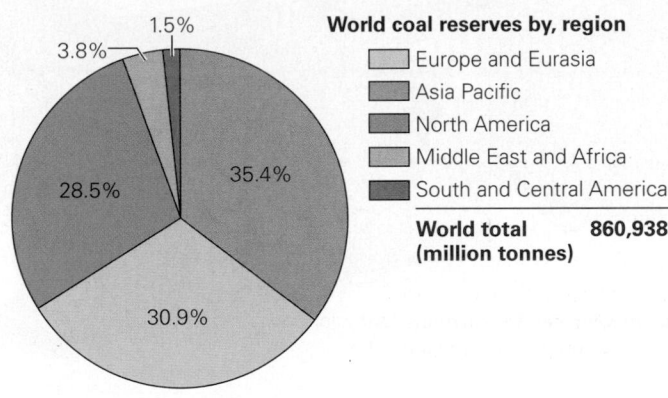

World coal reserves by, region

- Europe and Eurasia
- Asia Pacific
- North America
- Middle East and Africa
- South and Central America

World total (million tonnes) 860,938

(b) Distribution of coal reserves among regions.

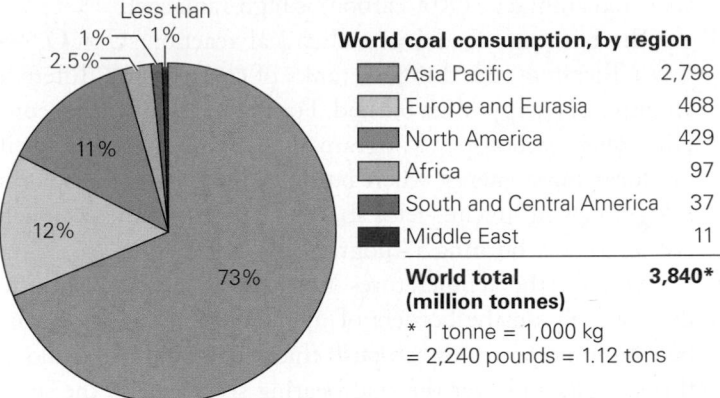

World coal consumption, by region

Asia Pacific	2,798
Europe and Eurasia	468
North America	429
Africa	97
South and Central America	37
Middle East	11

World total (million tonnes) 3,840*

* 1 tonne = 1,000 kg = 2,240 pounds = 1.12 tons

(c) Distribution of coal consumption among regions.

were deposited in tropical to semi-tropical shallow-marine to terrestrial environments—the environments in which a swamp could exist. The sedimentary strata of continents contain huge quantities of discovered coal, or **coal reserves**. For example, *economic seams* (beds of coal 1 to 3 m thick, thick enough to be worth mining) of Cretaceous age occur in the U.S. and Canadian Rocky Mountain region, while economic seams of Carboniferous age are found throughout the midwestern United States. Coal underlies large areas of Europe, Asia, and Australia **(Fig. 14.20a, b)**.

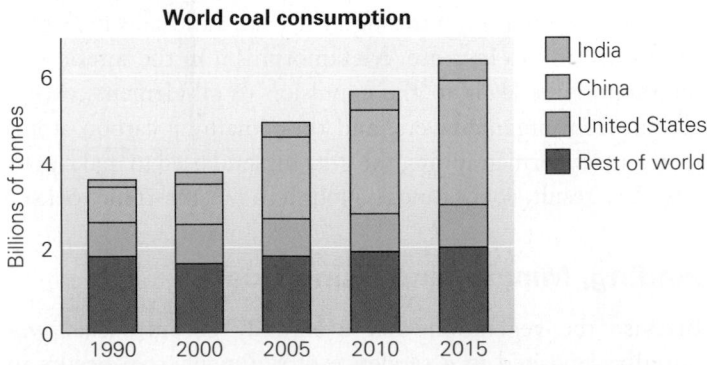

World coal consumption

- India
- China
- United States
- Rest of world

(d) Coal consumption varies greatly and is changing rapidly. Growth of consumption in China has more than tripled in the last 25 years.

The United States was the largest consumer of coal until about 1986, when China surpassed it. China's consumption tripled just between 2005 and 2015, due to the country's industrialization. In 2015, society worldwide burned almost 3.5 billion tons of coal (equivalent in volume to about 1,200 large football stadiums, or 34,000 Empire State Buildings), which provided about 29% of the world's energy supply. Of this total amount, countries in Asia consumed over 70% (Fig. 14.20c, d). Since 2015, coal consumption has leveled off and may have started to decrease. In the United States, this change reflects a shift from coal to natural gas as the preferred fossil fuel for electricity generation: between 2011 and 2016, coal's share of the electricity-generation market decreased from about 42% to 30%, while the natural gas share increased from about 25% to 34%.

The way in which energy companies mine coal depends on the depth of the coal seam. If the coal seam lies within about 100 m of the ground surface, *strip mining* proves to be most economical. In strip mines, miners use a giant shovel called a *dragline* to scrape off soil and layers of sedimentary rock above the coal seam (Fig. 14.21a). Draglines are so big that the shovel could swallow a two-car garage. Once the dragline has exposed the seam, smaller bulldozers scrape out the coal and dump it into trucks or onto a conveyor belt which carry it to storage piles (Fig. 14.21b, c). Before modern environmental awareness took hold, strip mining left huge scars on the landscape. Without topsoil, the rubble and exposed rock of the mining operation remained barren of vegetation. Coal beds that formed from coastal swamps tend to contain pyrite (FeS), which weathers when exposed to air and water to form sulfuric acid; sulfuric acid contributes to making the soil unsuitable for growth. In many contemporary mines, the dragline operator separates out and preserves soil. Then, when the coal has been scraped out, the operator fills the hole with the rock that was stripped to expose the coal and covers the rock back up with

FIGURE 14.21 Coal can be mined in strip mines or in underground mines.

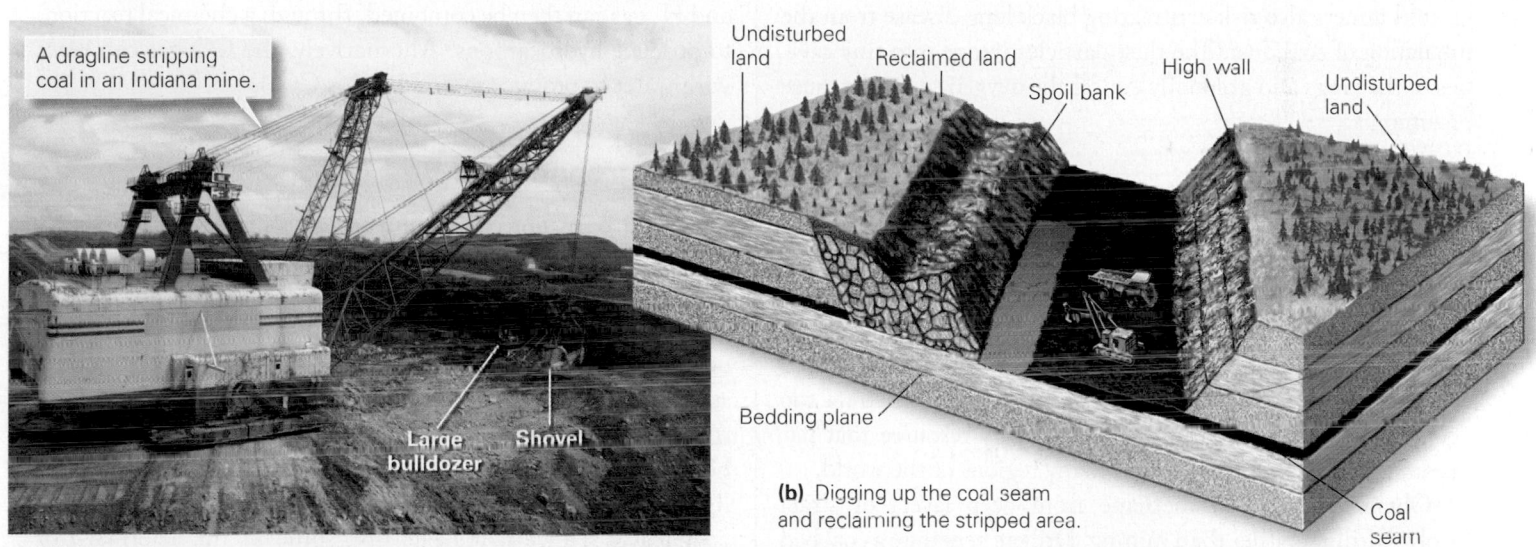

(b) Digging up the coal seam and reclaiming the stripped area.

(a) A dragline strips the overburden from a coal seam.

(c) Piles of recently mined coal.

(d) Underground coal mining.

the saved soil, on which grass or trees may eventually grow. Within years to decades, the former mine site can become a pasture or a forest. In hilly areas, however, miners may use a practice called *mountaintop removal*, during which they blast off the top of a mountain and dump the debris into adjacent valleys. This practice disrupts the landscape permanently.

Deep coal can be obtained only by *underground mining*. To develop an underground mine, miners dig a shaft down to the depth of the coal seam and then create a maze of tunnels, using huge grinding machines that chew their way into the coal (Fig. 14.21d). Depending on circumstances, miners either leave columns of coal behind to hold up the mine roof, or they let the mined area collapse, so that the ground level above sinks once mining is finished.

Underground coal mining can be very dangerous, not only because the sedimentary rocks forming the roof of the mine are weak and can collapse, but also because methane gas released by continuing chemical reactions in coal can accumulate in the mine, leading to the danger of a small spark triggering a deadly mine explosion. Unless they breathe through filters, underground miners also risk contracting black lung disease from the inhalation of coal dust. The dust particles wedge into tiny cavities of the lungs and gradually cut off the oxygen supply or cause pneumonia.

Gas from Coal

Coal-Bed Methane As we've noted, the natural process by which coal forms underground yields methane, a type of natural gas. Over time, some of the gas escapes to the atmosphere, but vast amounts remain within the coal in pores or bonded to coal molecules. Such **coal-bed methane**, trapped in strata too deep to be reached by mining, is an energy resource that has become a target for extraction in many regions of the world.

Obtaining coal-bed methane from deep layers of strata involves drilling rather than mining. Drillers penetrate a coal bed with a hole and then start pumping out groundwater. As a result of pumping out water, the pressure in the vicinity of the drillhole decreases relative to the surrounding bed. Methane bubbles into the hole and then up to the ground surface, where condensers compress it into tanks for storage. Disposal of the water produced by coal-bed methane extraction can be a major problem. If the water is pure, it can be used for irrigation, but in deep coal beds the water may be saline and thus cannot be used for crops. Producers either pump this water back underground or evaporate it in large ponds so that they can extract and collect the salt.

Coal Gasification. Traditional burning of coal produces clouds of smoke containing *fly ash* (solid residue left after the carbon in coal has been burned) and noxious gases (including sulfur dioxide, SO_2, which forms because coal contains sulfur-bearing minerals such as pyrite). Today, coal smoke can be partially cleaned by expensive scrubbers, but pollution remains a problem—the abundance of coal-burning plants in China has played a major role in producing the country's air pollution.

Alternatively, solid coal can be transformed into various gases, as well as solid by-products, before burning. The gases burn relatively cleanly. The process of producing gases from solid coal is called **coal gasification**; the process was invented in the late 18th and early 19th centuries and was used extensively to produce fuels during World War II. Coal gasification involves the following steps: First, workers place pulverized coal in a large container. Then a mixture of steam and oxygen passes through the coal at high pressure. As a result, the coal heats up to a high temperature but does not ignite. Under these conditions, chemical reactions break down and oxidize the molecules in coal to produce hydrogen (H_2) and other gases such as carbon monoxide (CO), H_2O, and CO_2. Solid ash, as well as sulfur and mercury, concentrate at the bottom of the container and can be removed before the gases are burned so that the contaminants do not go up the chimney and into the atmosphere. The CO gas and H_2 gas can then be combined, through a chemical reaction, to produce hydrocarbons. Alternatively, the CO gas can react with water to produce more H_2 (plus CO_2).

Underground Coal-Bed Fires

Coal will burn not only in furnaces but also in surface and subsurface mines as long as the fire has access to oxygen. Coal mining of the past two centuries has exposed large amounts of coal to the air and has provided many opportunities for fires to begin. Once started, a *coal-bed fire* progresses underground by sucking in oxygen from joints and pore spaces in surrounding rock. These fires may be difficult or impossible to extinguish because they are inaccessible. Some fires begin as a result of lightning strikes, some from spontaneous combustion (when coal reacts with air, it heats up), some in the aftermath of methane explosions, and some when people intentionally set trash fires in mines.

Major coal-bed fires can be truly disastrous. The most notorious of these fires in North America began as a result of trash burning in a mine near the town of Centralia, Pennsylvania (Fig. 14.22a). For the past 50 years, the fire has progressed underground, eventually burning coal seams beneath the town itself. The fire produces toxic fumes that rise through the ground and make the overlying landscape uninhabitable, and it also causes the land surface to collapse and sink. Inhabitants have had to abandon many neighborhoods of the town. Satellite imagery, by highlighting warm spots on the ground surface, indicates that thousands of coal-bed fires are currently burning around the world. Many of the fires occur in northern China—recent estimates suggest that 200 million tons of coal burn in China every year (Fig. 14.22b).

FIGURE 14.22 Coal-bed fires can be long-lived disasters.

(a) A coal-bed fire beneath Centralia, Pennsylvania, produces noxious gas and has forced the evacuation of many homes.

(b) A burning coal bed in China, exposed in a mine wall, glows red.

TAKE-HOME MESSAGE

Coal forms from the accumulation of plant material over time in anoxic coal swamps. When buried deeply, this organic material undergoes reactions that concentrate carbon. Higher-rank coal contains more carbon. Coal occurs as beds, called seams, in sedimentary sequences and must be mined underground or in open pits.

QUICK QUESTION: Why do underground coal fires start, and why are they so hard to stop?

14.7 Nuclear Power

How Does a Nuclear Power Plant Work?

So far we have looked at fuels, such as oil, gas, and coal, that release energy when they undergo *chemical burning*. During such burning, a chemical reaction between the fuel and oxygen releases the potential energy stored in the chemical bonds of the fuel. The energy that drives a nuclear power plant comes from a totally different process—*nuclear fission*, the breaking of the nuclear bonds that hold protons and neutrons together in the nucleus of an atom. Fission splits an atom into smaller pieces, and during this process, a small amount of mass transforms into thermal and electromagnetic energy. Einstein characterized this process in his famous equation, $E = mc^2$ (where E is energy, m is mass, and c is the speed of light).

A **nuclear reactor**, the heart of the power plant, commonly lies within a containment building made of reinforced concrete **(Fig. 14.23)**. The reactor contains *nuclear fuel*, consisting of pellets of concentrated uranium oxide or a comparable radioactive material, packed into metal tubes called *fuel rods*. Fission occurs when a speeding neutron strikes a radioactive isotope, causing it to split. For example, radioactive uranium-235 (^{235}U) splits into barium-141 (^{141}Ba) and krypton-92 (^{92}Kr) plus three neutrons. The neutrons released during the fission of one atom strike other atoms, thereby triggering more fission in a self-perpetuating process called a **chain reaction**, which overall produces lots of energy, including heat. Pipes carry water close to the heat-generating fuel rods, and the heat transforms the water into high-pressure steam. The pipes then carry this steam to a turbine, where it rotates fan blades—the rotation drives a dynamo that generates electricity. Eventually the steam moves into cooling towers where it condenses back into water, which can be reused in the plant or returned to the environment.

Nuclear fuel has an immense energy density 1 g of nuclear fuel contains almost 1,700 times as much energy as 1 g of gasoline and yields no air pollution or greenhouse gases. Nuclear power plants could produce a large proportion of global energy supply, but they don't, at present, because of society's concerns about handling nuclear fuel and nuclear waste. China has decreased its rate of building new coal-fired power plants, and has started to build more nuclear plants, in order to combat its air pollution problem.

The Geology of Uranium

Where does uranium come from? The Earth's radioactive elements, including uranium, probably developed during a supernova explosion that happened before the existence of our Solar System. Uranium atoms from this explosion became part of the nebula out of which the Earth formed and were eventually incorporated into the planet. Large atoms such as uranium don't fit well into the crystal structure of minerals, so when melts form, they preferentially enter a melt and rise with it. Eventually, uranium atoms were carried into the upper crust by rising granitic magma.

Even though granite contains uranium, it does not contain very much. But nature has a way of concentrating uranium. Hot water circulating through a pluton after intrusion dissolves the uranium and precipitates it as the mineral pitchblende (UO_2)

FIGURE 14.23 Producing electricity at a nuclear power plant.

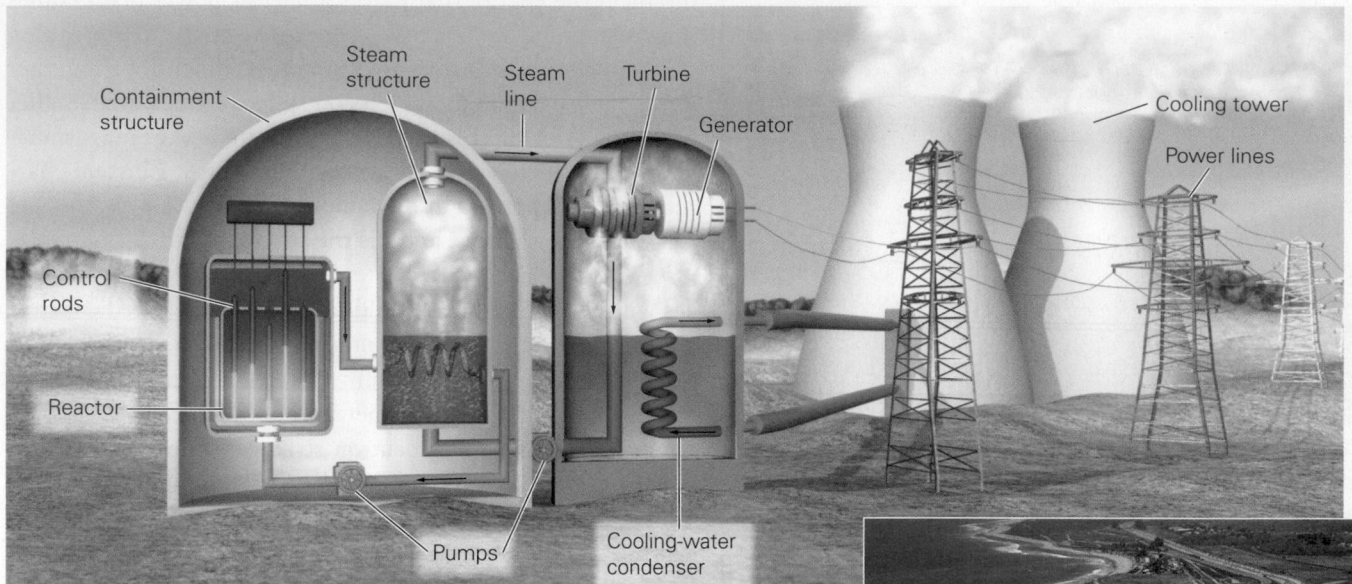

(a) In a nuclear power plant, a reactor heats water, which produces high-pressure steam. The steam drives a turbine that in turn drives a generator to produce electricity. A condenser transforms the steam back into water.

(b) This nuclear power plant in California has two reactors, each in its own containment building.

in veins. Uranium may be further concentrated when plutons, and the associated uranium-rich veins, weather and erode at the ground surface. Sand derived from a weathered pluton washes down a stream, and as it does so, uranium-rich grains stay behind because they are so heavy relative to quartz and feldspar grains. The world's richest uranium deposits, in fact, occur in ancient streambed gravels. Uranium deposits may also form when groundwater percolates through uranium-rich sedimentary rocks, for the uranium dissolves in the water and may move with it to another location where the chemical environment is different, causing new uranium-bearing minerals to precipitate out of solution and fill the pores of the host sedimentary rock.

You can't just put unprocessed uranium from a mine into a reactor. That's because ^{235}U, the isotope of uranium that serves as the most common fuel for conventional nuclear power plants, accounts for only about 0.7% of naturally occurring uranium—most uranium consists of ^{238}U. To make a reactor-grade fuel suitable for use in a power plant, the ^{235}U concentration in a mass of natural uranium must be increased to about 3% to 4%, an expensive process called *uranium enrichment*.

Challenges of Using Nuclear Power

The first nuclear power plant designed to generate electricity for the public became operational in 1954. Today, about 449 nuclear power plants are in operation, and about 60 are under construction. Maintaining safety at these plants requires hard work. Operators must ensure that circulating water constantly cools the nuclear fuel, and the rate of nuclear fission must be regulated by the insertion of *control rods* made of materials that absorb neutrons and therefore decrease the number of collisions between neutrons and radioactive atoms. Without control rods, the number of neutrons dashing around in the nuclear fuel would progressively increase, causing the rate of fission, and the accompanying heat production, to increase. Eventually, the fuel would become so hot that it would melt. Such a **meltdown** might cause a steam explosion, or it could generate such high temperatures that water molecules would break apart to form hydrogen gas and oxygen gas, a mixture that can explode. If the explosion were large enough, it could breach the containment building and scatter radioactive debris into the air. Note that a meltdown is not the same as an atomic bomb explosion. An atomic bomb explosion can only occur if there is a critical mass (sufficient quantity) of highly

Did you ever wonder...
whether a nuclear power plant could explode like an atomic bomb?

enriched ("weapons-grade") uranium (90% ^{235}U), in which fission reactions happen so quickly that the fuel itself explodes.

Over the past 60 years, there have been three significant accidents at nuclear power plants. The first occurred in 1979 at the Three Mile Island plant in Pennsylvania, when a stuck valve allowed coolant water to escape, causing the reactor to overheat. Eventually, some of that radioactive water leaked out into the environment, but contamination due to the leak was limited. The next, and much more serious, nuclear accident occurred at the power plant in Chernobyl, Ukraine, in April 1986, while engineers were conducting a test. The fuel became too hot, triggering a hydrogen explosion that ruptured the roof of the containment building and spread fragments of the reactor and its fuel around the plant grounds. Within 6 weeks, 20 people had died from radiation sickness. In addition, some of the radioactive material entered the atmosphere and dispersed over eastern Europe and Scandinavia, but no one yet knows whether this fallout affected the health of exposed populations. The reactor has been entombed in concrete, and the surrounding region has remained closed. In 2011, a disaster second only to Chernobyl, in terms of the amount of radiation released, occurred at the Fukushima power plant along the east coast of northern Japan. This disaster occurred when the catastrophic tsunami that followed the magnitude 9.0 Tōhoku earthquake knocked out the power lines providing electricity to the pumps that circulated cooling water and the plant's backup diesel generators (see Chapter 10). As a result, some of the reactors overheated, and suffered partial meltdowns and hydrogen gas explosions. The long-term health effects of this event are not known.

One of the biggest challenges to the nuclear power industry pertains to the storage of **nuclear waste**, the radioactive material produced in a nuclear plant. It includes spent fuel, which contains radioactive elements, as well as water and equipment that have come in contact with radioactive materials. Radioactive elements emit gamma rays and X-rays, which can damage living organisms and cause cancer. Some radioactive material decays quickly (in decades to centuries), but some remains dangerous for thousands of years or more. Nuclear waste cannot just be stashed in a warehouse or buried in a town landfill. Some of the waste stays hot enough that it needs to be cooled with water. Even relatively cool waste has the potential to leak radioactive elements into municipal water supplies or nearby lakes or streams. Ideally, waste should be sealed in containers that will last for thousands of years (the time needed for the short-lived radioactive atoms to undergo decay) and stored in a place where it will not come in contact with the environment. Finding an appropriate place is not easy, and so far, experts disagree about the best way to dispose of nuclear waste. For some years the U.S. government favored storing waste in tunnels under Yucca Mountain, in the Nevada desert, but the site has not been used and most nuclear waste remains on the site of the power plants that produced it.

14.8 Other Energy Sources

So far, we've discussed energy sources that use geologic materials (such as fossil fuels and uranium) extracted from the Earth. In this section, in order to complete our picture of energy resources, we discuss new sources of energy that society is turning to (**Geology at a Glance**, pp. 550–551).

Geothermal Energy

As the name suggests, **geothermal energy** comes from heat in the Earth's crust. We can distinguish between *high-temperature geothermal energy*, which can produce heat and electricity at a commercial scale, and low-temperature geothermal energy (also known as *ambient geothermal energy*), which can warm or cool the water of an individual household.

High-temperature geothermal energy exists because the crust becomes progressively hotter with increasing depth, at a rate defined by the *geothermal gradient*. In active volcanic areas, groundwater approaches its boiling point only several hundred meters below the Earth's surface. Power companies can use this geothermal energy in many ways. For example, in some places, they simply pump

GEOLOGY AT A GLANCE

Power from the Earth

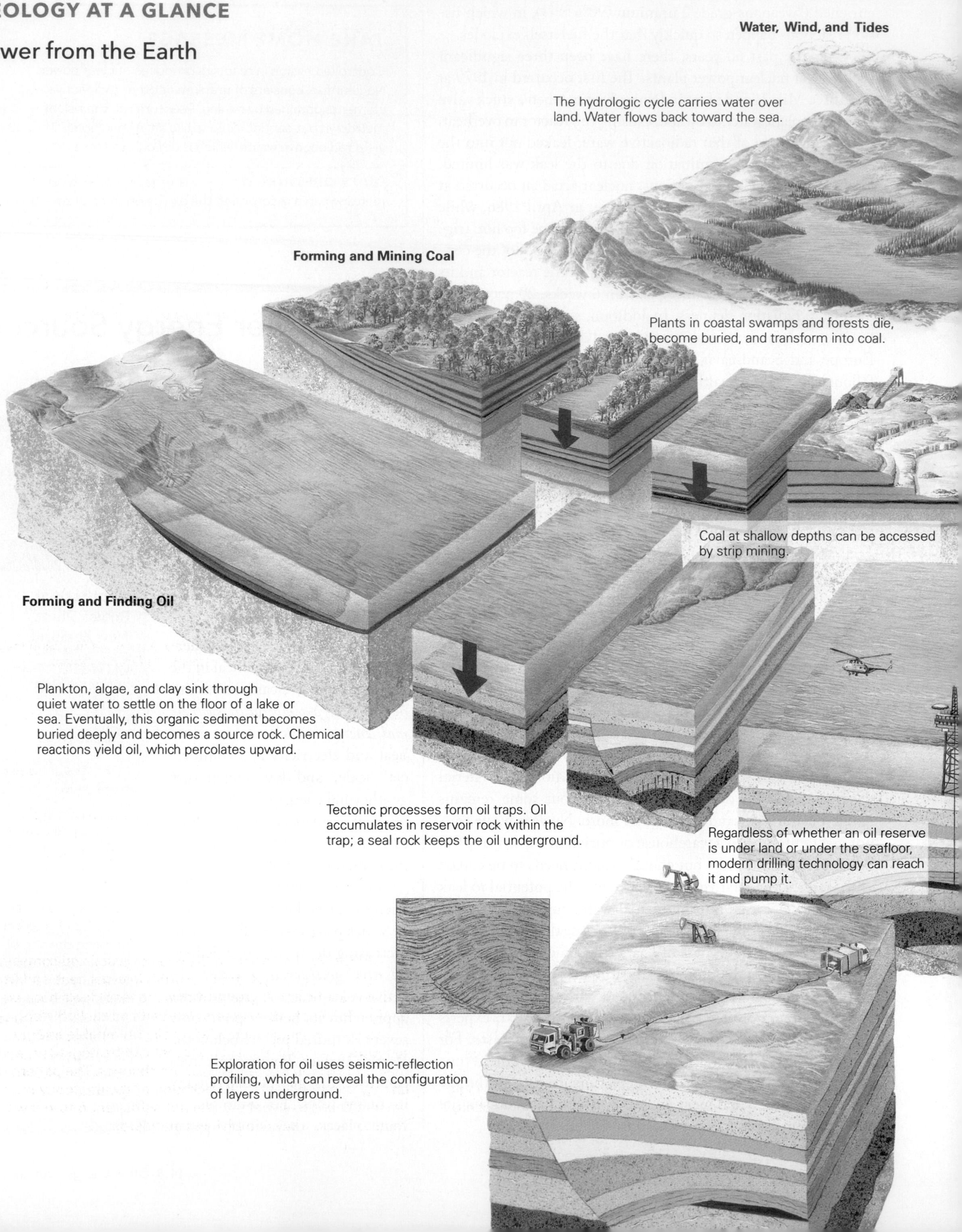

The hydrologic cycle carries water over land. Water flows back toward the sea.

Forming and Mining Coal

Plants in coastal swamps and forests die, become buried, and transform into coal.

Coal at shallow depths can be accessed by strip mining.

Forming and Finding Oil

Plankton, algae, and clay sink through quiet water to settle on the floor of a lake or sea. Eventually, this organic sediment becomes buried deeply and becomes a source rock. Chemical reactions yield oil, which percolates upward.

Tectonic processes form oil traps. Oil accumulates in reservoir rock within the trap; a seal rock keeps the oil underground.

Regardless of whether an oil reserve is under land or under the seafloor, modern drilling technology can reach it and pump it.

Exploration for oil uses seismic-reflection profiling, which can reveal the configuration of layers underground.

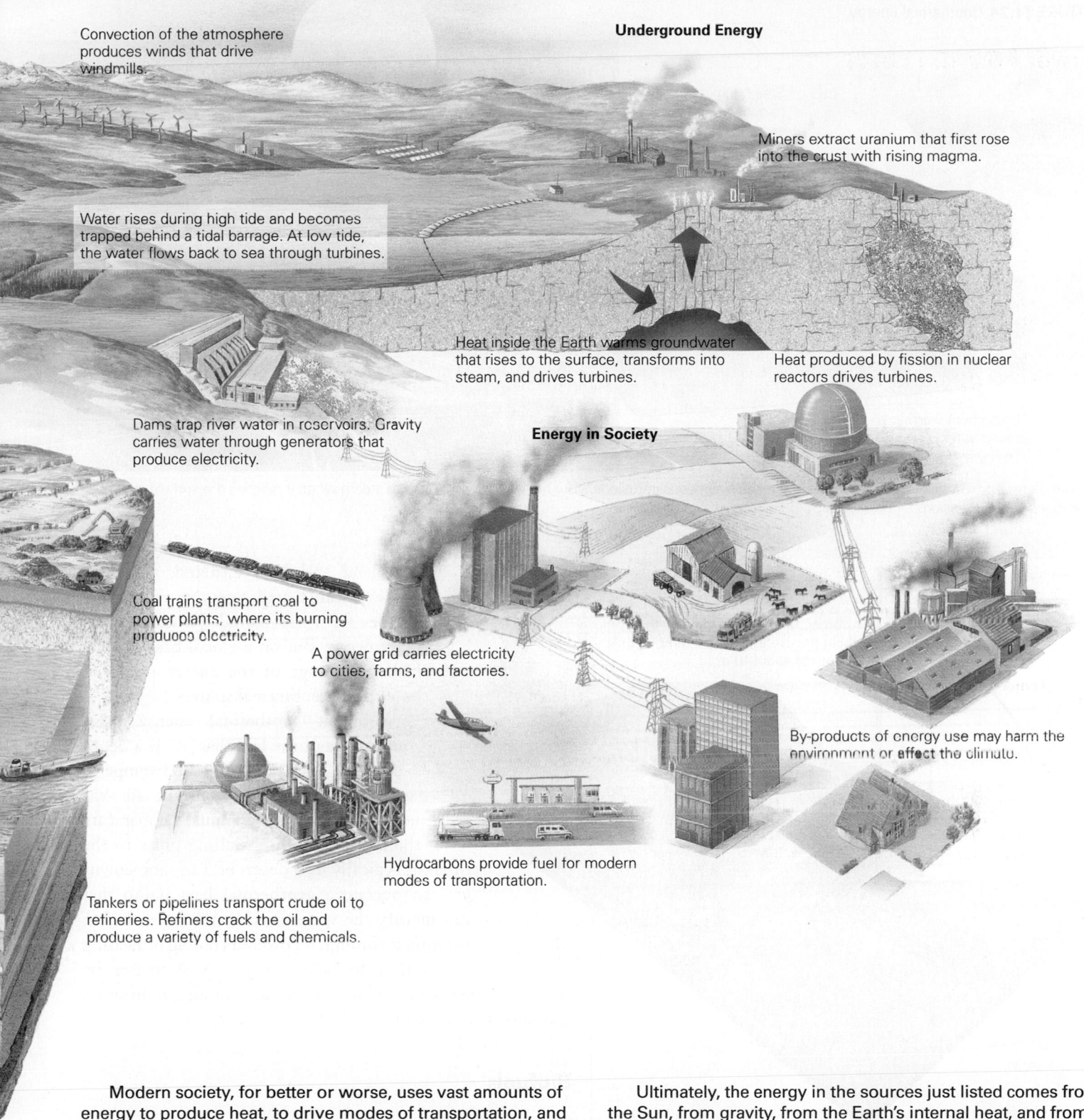

Convection of the atmosphere produces winds that drive windmills.

Underground Energy

Miners extract uranium that first rose into the crust with rising magma.

Water rises during high tide and becomes trapped behind a tidal barrage. At low tide, the water flows back to sea through turbines.

Heat inside the Earth warms groundwater that rises to the surface, transforms into steam, and drives turbines.

Heat produced by fission in nuclear reactors drives turbines.

Dams trap river water in reservoirs. Gravity carries water through generators that produce electricity.

Energy in Society

Coal trains transport coal to power plants, where its burning produces electricity.

A power grid carries electricity to cities, farms, and factories.

By-products of energy use may harm the environment or affect the climate.

Hydrocarbons provide fuel for modern modes of transportation.

Tankers or pipelines transport crude oil to refineries. Refiners crack the oil and produce a variety of fuels and chemicals.

Modern society, for better or worse, uses vast amounts of energy to produce heat, to drive modes of transportation, and to produce electricity. This energy comes either from geologic materials stored in the Earth or from geologic processes happening at our planet's surface. For example, oil and gas fill the pores of reservoir rocks at depth below the surface, coal occurs in sedimentary beds, and uranium concentrates in ore deposits. A hydroelectric power plant taps into the hydrologic cycle, windmills operate because of atmospheric convection, and geothermal energy comes from hot groundwater.

Ultimately, the energy in the sources just listed comes from the Sun, from gravity, from the Earth's internal heat, and from nuclear reactions. Oil, gas, and coal are fossil fuels because the energy they store first came to the Earth as sunlight, long ago.

As energy usage increases, easily obtainable energy resources dwindle, the environment can be degraded, and the composition of the atmosphere changes. The pattern of energy use that forms the backbone of society today may have to change radically in the not-so-distant future if we wish to avoid a decline in living standards.

FIGURE 14.24 Geothermal energy.

A geothermal power plant in New Zealand

Steam turns turbines to generate electricity.

Geothermal well

Rain

Hot water rises in a well and turns to steam.

Hot groundwater

Cold groundwater

Magma

Steam-filled fracture

Natural heat warms groundwater.

(a) Geothermal power plants use groundwater heated by igneous intrusions. Very hot groundwater turns to steam when it reaches the surface and undergoes decompression.

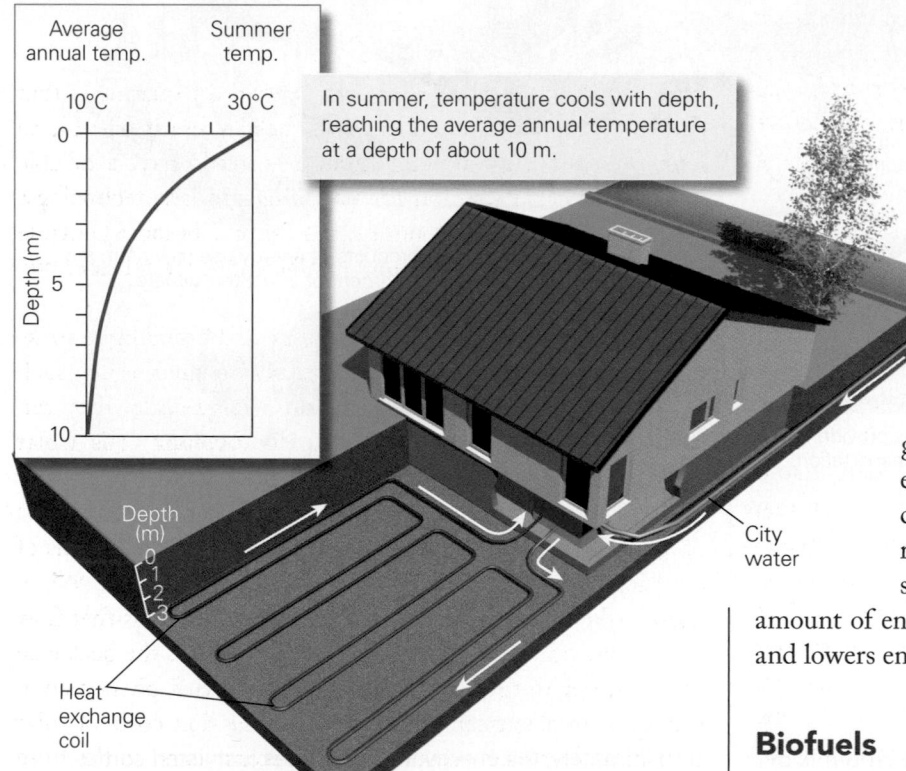

Average annual temp. 10°C

Summer temp. 30°C

In summer, temperature cools with depth, reaching the average annual temperature at a depth of about 10 m.

Depth (m)

0

5

10

City water

Depth (m)
0
1
2
3

Heat exchange coil

(b) Ambient geothermal energy, using a geothermal heat pump, can reduce home heating and cooling costs.

hot groundwater through pipes to heat houses or spas directly. Particularly hot groundwater turns to steam when it rises to the Earth's surface and undergoes decompression. This steam can drive turbines and generate electricity **(Fig. 14.24a)**.

In Iceland and New Zealand, which sit astride volcanic areas, geothermal energy provides for a substantial portion of electricity needs. But on a global basis, only a small percentage of the energy produced comes from geothermal sources.

Ambient geothermal energy takes advantage of the fact that below a depth of a few meters, the ground temperature remains nearly constant all year (at about the average annual temperature of the air above). By installing pipes in the ground, typically at a depth of 5 m, homeowners can operate a geothermal heat pump that can modify the water used in the home before running it through a powered cooling or heating system **(Fig. 14.24b)**. This process decreases the amount of energy needed to run furnaces or air conditioners and lowers energy costs.

Biofuels

In recent years, farmers have begun to produce crops specifically for the purpose of producing biomass for fuel production. The resulting liquids are called **biofuels**. The most commonly used biofuel is ethanol, a type of alcohol, which can substitute for gasoline in car engines. The process of producing ethanol from corn includes the following steps: First, producers grind corn into a fine powder, mix it with water, and cook it to produce a mash of starch. Then they add an enzyme to the mash, which converts the mash into sugar. The sugar, when mixed with yeast, ferments. Fermentation produces ethanol and CO_2. Finally, producers distill the fermented mash to concentrate the ethanol. While corn serves as the main source of ethanol in North America, ethanol in Brazil comes from fermentation of sugar extracted from sugarcane.

Researchers have been working to develop processes that yield ethanol from cellulose, which would permit perennial grasses and the stalks of other plants to become a source of liquid fuel. They have also experimented with the use of algae, which naturally produces fatty chemicals (lipids) from which hydrocarbons can be produced, and with the commercial production of biodiesel, a fuel produced by chemical modification of animal fats and vegetable oils.

Hydroelectric and Wind Power

For millennia, people have used flowing water and air to produce energy. The flow of streams can turn waterwheels that drive equipment in mills and factories, and the flow of air can drive windmills that pump water to irrigate farm fields. In the past century, engineers have begun to employ the same basic technology to drive generators that produce electricity.

In a modern hydroelectric power plant, the potential energy of water converts into kinetic energy as the water flows from a higher elevation to a lower elevation. The flowing water turns turbine blades placed in a pipe, and the turbine drives an electrical generator. Most hydroelectric plants rely on water from a reservoir held back by a dam. The largest of these, the Three Gorges Dam, blocks the Yangtze River in China (Fig. 14.25a). Dam construction increases the available potential energy of the water because the water level in a filled reservoir is higher than the level of the valley floor that the dam spans. Hydroelectric energy is clean, in that it does not release pollutants, and reservoirs may have the added benefit of providing flood control, irrigation water, and recreational opportunities. But the construction of dams and reservoirs may bring unwanted environmental changes to a region, so the benefits of their construction must be weighed against potential harm. Damming a river may flood a spectacular canyon, eliminate exciting rapids, or destroy an ecosystem. Furthermore, reservoirs trap sediment and nutrients, thereby disrupting the supply of these materials to downstream floodplains or deltas, a process that may adversely affect agriculture.

Not all hydroelectric power generation uses flowing river water—engineers have been developing new means to tap **tidal power** (the daily rise and fall of the sea; see Chapter 18). One approach involves building a dam, called a *tidal barrage,* across the entrance to a bay or *estuary* (the flooded mouth of a river) in which there are large tides. When the tide rises, water spills into the enclosed area through openings in the dam.

When the tide outside drops, water flows back to the sea via a pipe that carries it through a power-generating turbine (Fig. 14.25b). More recently, engineers have developed technologies to place huge fan blades underwater in nearshore regions where tidal currents naturally flow; the blades slowly turn in the current and run generators.

When you think of wind energy, you may picture a classic Dutch windmill driving a water pump. Modern efforts to harness the wind happen on a much larger scale. To produce wind-generated electricity, engineers identify regions, either onshore or just offshore, with steady breezes. In these regions, they build *wind farms* that consist of numerous towers, each of which holds a wind turbine, a giant fan blade that turns even in a gentle breeze (Fig. 14.25c). Some towers are on the order of 100 m tall, with fan blades that are over 40 m long (Fig. 14.25d). Large wind farms can host hundreds of towers. Wind power produces no pollutants, but like any type of energy source, it has some drawbacks. Cluttering the horizon with towers may spoil a beautiful view, the constant hum of the towers can disturb nearby residents, the moving blades may be a hazard to migrating birds, and offshore towers may interfere with marine life.

Solar Energy

The Sun drenches the Earth with energy in quantities that dwarf the amounts stored in fossil fuels. Were it possible to harness this energy directly, humanity would have a reliable and totally clean solution for powering modern technology. But using solar energy is not quite so simple because converting light into heat remains fairly inefficient. Let's consider two options for producing solar energy.

A *solar collector* consists of mirrors that focus light striking a broad area into a smaller area. On a small scale, such devices can be used for cooking; on a large scale, they can produce steam to drive turbines. **Photovoltaic cells** (solar cells), in contrast, convert light energy directly into electricity (Fig. 14.26a). Most photovoltaic cells consist of two wafers of silicon pressed together. One wafer also contains atoms of arsenic, and the other wafer includes atoms of boron. When light strikes the cell, arsenic atoms release electrons that flow over to the boron atoms. If a wire loop connects the back side of one wafer to the back side of the other, this phenomenon produces an electrical current. The production costs of solar cells have decreased substantially in recent years, so their use has increased exponentially. Between 2012 and 2016, global production of electricity by photovoltaic cells tripled.

Fuel Cells

In a hydrogen *fuel cell*, chemical reactions produce electricity directly. Within such a cell, hydrogen gas flows through a tube across an anode (made from a strip of platinum) that has

FIGURE 14.25 Hydroelectric and wind power.

(a) Gravity causes water held back by the Three Gorges Dam to flow through turbines and generate electricity.

(c) A wind farm in southwestern England. The towers are about 50 m high.

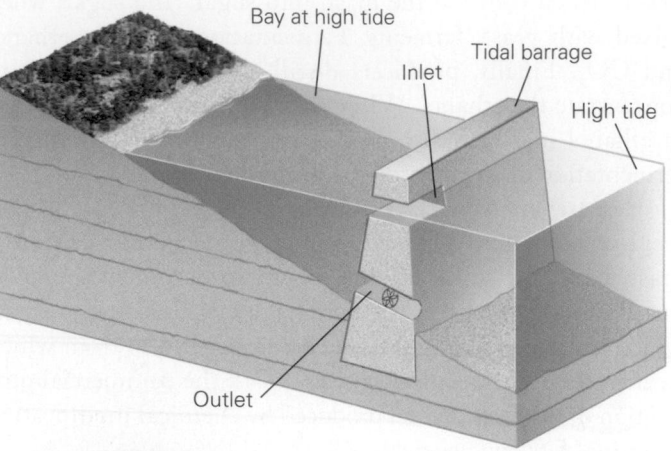

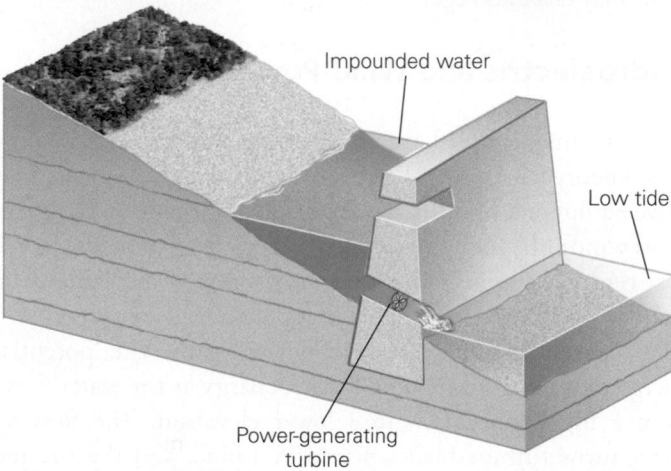

(b) To produce tidal power, a dam traps seawater at high tide; the water can flow through a turbine at low tide.

(d) A single giant blade being hauled down a highway, en route to a new windmill installation.

been placed in a water solution containing an electrolyte (a substance that enables the solution to conduct electricity). At the same time, a stream of oxygen gas flows onto a separate platinum cathode that has also been placed in the solution. A wire connects the anode and the cathode to provide an electrical circuit **(Fig. 14.26b)**. In this configuration, hydrogen reacts with oxygen to produce water and electricity. Fuel cells are efficient and clean. Their limitation lies in the need for a design that will protect the cells from damage by impact and that will enable them to store hydrogen, an explosive gas, in a safe way. Furthermore, it takes a significant amount of energy from other sources to produce the hydrogen used in fuel cells.

FIGURE 14.26 Photovoltaic cells and fuel cells.

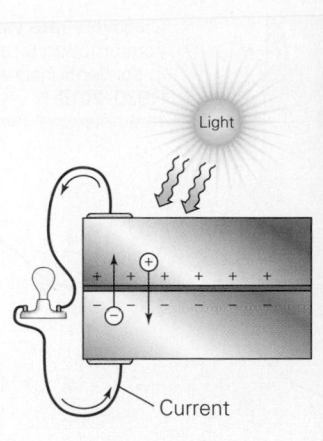

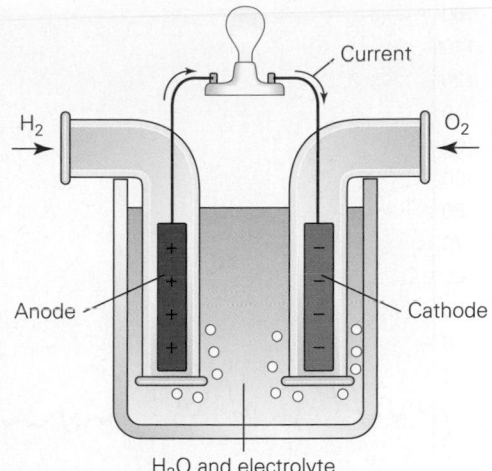

(a) Photovoltaic cells produce electricity directly from solar radiation.

(b) A hydrogen fuel cell.

> ### TAKE-HOME MESSAGE
>
> A variety of alternative energy resources are now under development. Geothermal energy uses groundwater that has been warmed by heat from the Earth's interior. Biomass can be transformed into burnable ethanol and biodiesel. Flowing water (in rivers or tides), flowing air, solar panels, and fuel cells can also produce energy.
>
> **QUICK QUESTION:** Is there any way that homeowners can use heat stored under their property to lower their energy costs?

14.9 Energy Choices, Energy Problems

The Age of Oil?

Energy use in industrialized countries grew with dizzying speed through the mid-20th century, and during this time people came to rely increasingly on oil (see Fig. 14.2). Oil remains the single largest source of energy globally, accounting for about 33% of global energy consumption. (Natural gas accounts for another 21%.) The United States, the European Union, and Japan together used more oil than the rest of the world combined until about 2006. Then consumption in these industrialized nations leveled off and started to decrease, while use in the rest of the world has increased as countries in Asia and South America industrialize. According to statistics published in 2016, the United States uses 19.6 million barrels per day of petroleum, or hydrocarbon, products (oil, gasoline, natural gas, tar, and plastic) and China uses 11.1 million barrels per day. Put another way, the daily volume of these products used in the United States would fill the equivalent of three Empire State Buildings.

During the 20th century, most of the petroleum products used by society were pumped from conventional oil and gas reserves. Up until the middle of the century, the United States could meet its needs using domestic supplies, but in the second half of the century, the United States, and many other industrialized countries, began to import more oil than they produced. Much of the imported oil came from the supergiant fields of the countries surrounding the Persian Gulf and was hauled across the ocean in supertankers, the largest of which could carry 2 million barrels (see Fig. 14.14b).

This oil remained fairly inexpensive through the 1960s **(Fig. 14.27a)**. In 1973, however, a complex tangle of politics led the *Organization of Petroleum Exporting Countries* (OPEC) to limit oil exports, and prices rose dramatically. In the United States, fear of an oil shortage turned to panic and newspaper headlines proclaimed, "Energy Crisis!" Gasoline rationing began, causing motorists to wait for hours in lines at gas stations, and governments encouraged energy conservation. Ever since the 1970s, oil prices have bounced up and down. Prices rose to $147 per barrel in 2008, just before the Great Recession. When the recession hit, prices suddenly dropped to $40 per barrel. In recent years, they have risen to $100 per barrel and then dropped again. The price per barrel, to some extent, controls the price consumers pay at gas stations, but that price can also be affected by the capacity of refineries to produce gasoline. Capacity tends to change over the course of a year because refineries reconfigure their distillation columns seasonally to produce different proportions of products at different times of the year.

FIGURE 14.27 Price, discovery, and consumption of oil.

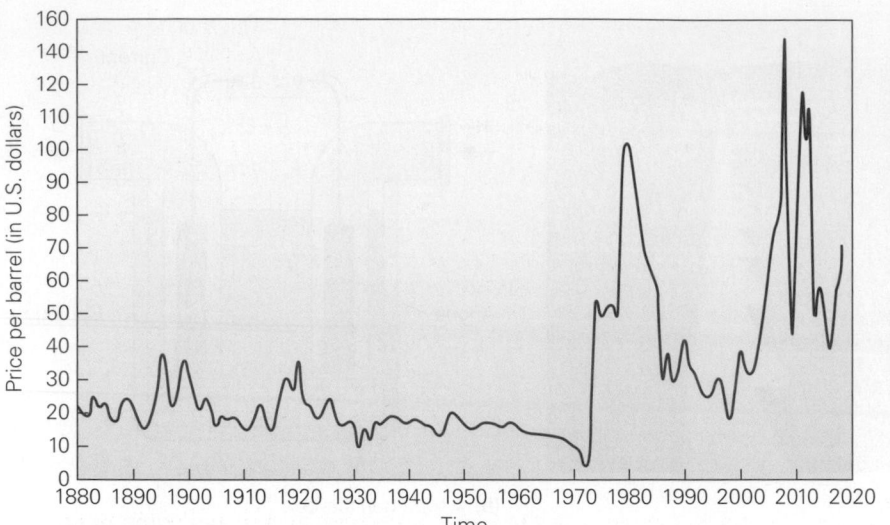

(a) The price of oil held fairly steady for almost 100 years. Starting in 1973, it has risen and fallen dramatically.

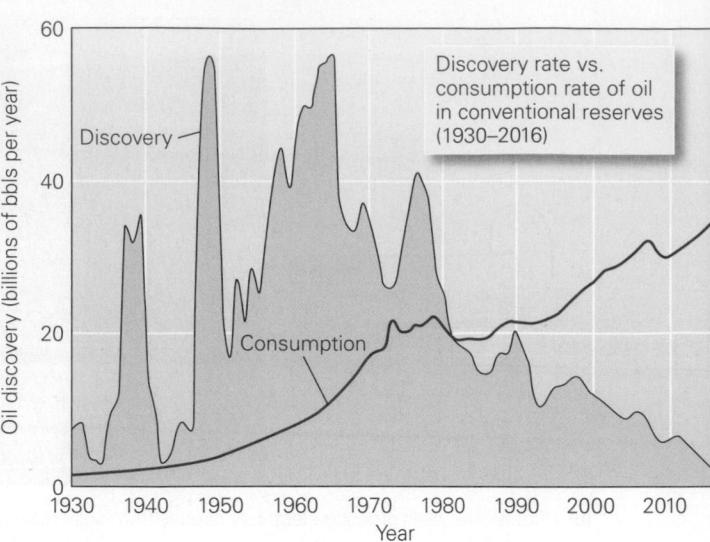

(b) Despite short-term dips, consumption of oil has continued to grow, but the rate of discovery of new conventional oil reserves has not.

Fluctuations in oil prices have reflected politics, market forces, the capacity of refineries, and the development of new technologies. When the price of oil rises beyond about $60 to $80 per barrel, it becomes economical to seek out and use unconventional reserves (particularly shale oil and shale gas). But because of the expense of the technologies needed to tap into these reserves, countries with conventional reserves can flood the market with more cheaply produced products, driving prices down, so that drilling into and producing from unconventional reserves becomes uneconomical.

Will the day come when shortages of conventional oil arise not because of politics, refining capacity, and market forces, but rather because there's just no more oil underground to extract?

> **Did you ever wonder...**
> how much longer the world's oil supply will last?

At current rates of consumption, the answer is probably yes, for the rate of new discoveries of conventional oil reserves no longer keeps up with the rate of consumption **(Fig. 14.27b)**. Unlike a **renewable resource**, such as vegetation used for biofuels, which can reproduce naturally in a short time relative to a human lifespan, oil is a **nonrenewable resource**: it takes thousands to millions of years or more for geologic processes to form oil.

When will conventional reserves run out? A clear answer to this question is hard to come by because resource experts do not all agree on the numbers that go into the calculation. In the 1950s, a geologist named M. King Hubbert considered the issue by looking at the production history of a particular oil field. He found that once the field had been discovered and drilled, production increased rapidly, reached a peak, and then decreased, so that overall production followed a bell curve—this pattern is unavoidable because oil is nonrenewable. The peak of such a curve has come to be known as **Hubbert's Peak**. In recent decades, resource experts have plotted production curves for countries and regions **(Fig. 14.28a)**. According to these curves, the United States passed Hubbert's Peak in the 1970s, and some experts suggest that the global peak may occur before 2020.

The conclusion that conventional oil reserves won't last forever also comes from simple arithmetic. Geologists estimate that we have already used a substantial proportion of the Earth's conventional reserves. Exploration completed to date indicates that sedimentary basins still hold about 1,250 billion barrels of proven conventional reserves. Optimistically, these experts believe that an additional 2,000 billion barrels may exist in conventional reserves that have not yet been discovered. If so, basins may still hold on the order of 3,250 billion barrels of conventional oil. At current global rates of consumption (about 35 billion barrels per year), such reserves will last for only about 100 years.

Facing the reality of a limited supply of conventional oil reserves, society faces three choices: we can decrease our rate of consumption; we can pay more and increase the use of unconventional oil supplies; or we can convert to other sources of energy. The near-term solution will probably involve using more unconventional reserves. Estimates of unconventional oil and natural gas reserves range up to 7,200 billion barrels, but experts disagree on these numbers because it's not clear which reserves can realistically be accessed economically using known technologies. Considering the most optimistic current

FIGURE 14.28 Hubbert's Peak and the future of energy.

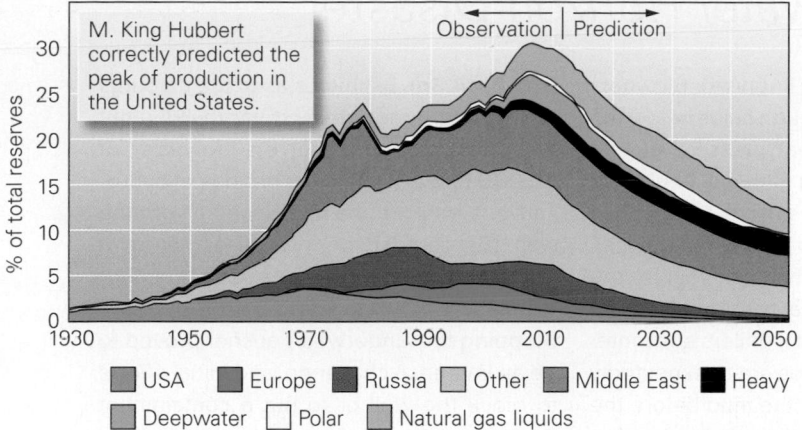

M. King Hubbert correctly predicted the peak of production in the United States.

Observation | Prediction

Legend: USA · Europe · Russia · Other · Middle East · Heavy · Deepwater · Polar · Natural gas liquids

(a) A plot of production versus time suggests that sometime in the early 21st century, the global production of oil will start to decrease.

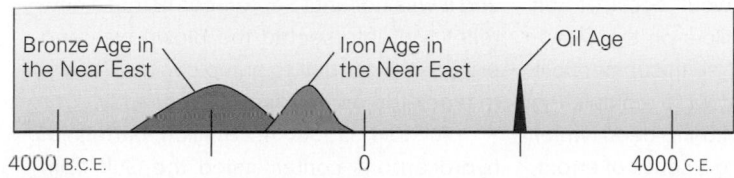

Bronze Age in the Near East · Iron Age in the Near East · Oil Age

4000 B.C.E. — 0 — 4000 C.E.

(b) The conventional oil age may be just a short blip in the timeline of human history.

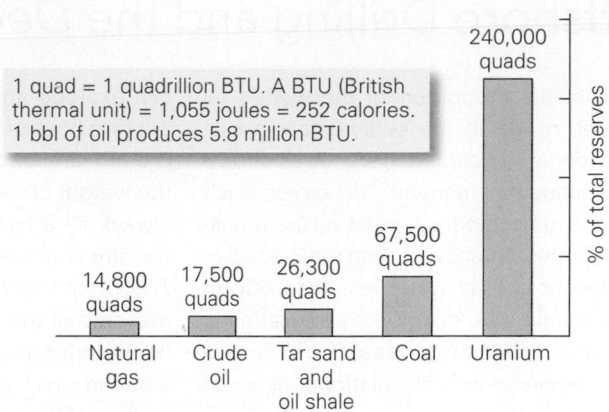

1 quad = 1 quadrillion BTU. A BTU (British thermal unit) = 1,055 joules = 252 calories. 1 bbl of oil produces 5.8 million BTU.

Natural gas: 14,800 quads · Crude oil: 17,500 quads · Tar sand and oil shale: 26,300 quads · Coal: 67,500 quads · Uranium: 240,000 quads

(c) Estimated nonrenewable energy resources on the Earth.

estimates leads to the conclusion that hydrocarbon reserves can last for at most another three centuries at current rates of use. Therefore, the **Oil Age**, the time when oil served as the primary source of energy for human society, can last no longer than about 400 years in all—just a short blip on a timeline of human history **(Fig. 14.28b)**. Most likely, because of cost and environmental concerns, society will have switched to other energy sources before oil supplies run out **(Fig. 14.28c)**.

Environmental Consequences of Fossil Fuel Use

Producing, transporting, and using fossil fuels of any type (oil, gas, or coal) has an impact on the environment. Let's first consider environmental problems associated with extraction of fossil fuels, then look at the consequences of using them.

Drilling oil or gas wells can cause localized damage to the landscape. Drillers must clear an area of vegetation and cover it with gravel in order to set up drilling equipment. In addition, networks of roads must be bulldozed across remote areas to provide access to drilling sites. In the past, drilling sites had to be closely spaced because a single well could access only a small volume of a resource. This problem has diminished somewhat in recent years because directional drilling allows a single site to be used for installing several wells, each aimed in a different direction.

Once drilling begins, there is a slight risk of a blowout, and if sparks ignite oil or gas spurting from a well, they may cause an *oil well fire*, an inferno spewing billows of sooty smoke. Needless to say, putting out such a fire takes special skill and equipment. Firefighters first haul hot metal away from the blaze. Then they set off a dynamite explosion, which snuffs out the fire, and finally they rush in to cap the well and prevent more oil or gas from escaping. Blowouts at an offshore drilling site present even greater challenges because the wellhead may lie beneath kilometers of water. As a consequence, offshore blowouts can be catastrophic, as was the case, sadly, for the *Deepwater Horizon*, a drilling platform in the Gulf of Mexico **(Box 14.3)**.

In addition to oil spillage that may happen at a drilling site or at storage and refining facilities, spills may come from leaking pipelines or from damaged tanker cars being hauled by trains or trucks. Any spillage of oil causes problems for the environment. On land, oil can contaminate and kill vegetation at the surface, can ruin the soil, or can enter streams and wetlands to harm organisms or make water undrinkable. Oil spilled on land sometimes seeps deep into the ground and contaminates groundwater. Because oil is an organic material, bacteria may eventually digest it, but this process can take a long time. Marine spills from tankers, pipelines, or offshore wells may produce vast slicks that spread over hundreds or even thousands of square kilometers **(Fig. 14.29a)**. When these slicks reach the shore, they can coat beaches and rocks and can kill shore life **(Fig. 14.29b)**.

Mining coal presents a different set of problems. Surface mining not only has the potential to scar the landscape—and, in the case of mountaintop removal, the damage can be impossible to repair—but also exposes coal to air. Most coal contains sulfur, typically in the form of small pyrite (FeS_2) crystals. Pyrite reacts with air and water to produce a solution of sulfuric acid, known as **acid-mine runoff**. This runoff, which typically develops a bright orange color due to the presence of iron oxide minerals as well as various archaea and bacteria, can flow into streams and kill fish and aquatic plants. Underground coal mines may also produce acid mine runoff that enters groundwater. The collapse of mines affects the

BOX 14.3 CONSIDER THIS . . .

Offshore Drilling and the *Deepwater Horizon* Disaster

A substantial proportion of the world's oil reserves reside in the sedimentary basins that underlie the continental shelves of passive continental margins. To access such reserves, oil companies build offshore drilling platforms. Offshore drilling requires complex technology. In water less than 600 m deep, companies position fixed platforms on towers resting on the seafloor. In deeper water, semi-submersible platforms float on huge submerged pontoons **(Fig. Bx14.3a)**. With these platforms, oil companies can now access reserves lying beneath 3 km of water.

North America's largest offshore oil fields occur in the passive-margin basin that fringes the coast of the Gulf of Mexico **(Fig. Bx14.3b)**. This basin started forming in Jurassic time, after the breakup of Pangaea, and its floor has been slowly sinking ever since. Up to 15 to 20 km of sediment (mud, salt, sand, marine shells) fills the basin. Some of the mud deposited was rich in organic matter which, when buried deeply, was converted to hydrocarbons. These now fill pores of reservoir rocks in salt-dome and stratigraphic traps. More than 3,500 platforms operate in the Gulf at present, collectively yielding up to 1.7 million barrels per day.

During both onshore and offshore drilling, workers worry about the possibility of a **blowout**. A blowout happens when the pressure within a hydrocarbon reserve penetrated by a well exceeds the pressure that drillers had planned for, causing the hydrocarbons (oil or gas) to rush up the well in an uncontrolled manner and burst out at the surface in an oil gusher or gas plume, which can explode and burn if ignited. Blowouts are rare because, although fluids below the ground are under great pressure due to the weight of overlying material, engineers always fill a hole with drilling mud whose density is greater than that of clear water. The weight of drilling mud can counter the pressure of the hydrocarbons and hold the fluids underground. But if drillers encounter a bed in which pressures are unexpectedly high, or if they remove the mud before the walls of the well have been sealed with a casing cemented in place by concrete, a blowout may happen.

A catastrophic blowout occurred on April 20, 2010, when drillers on the *Deepwater Horizon*, a huge semi-submersible platform leased by BP, were completing a 5.5-km-long hole in 1.5-km-deep water south of Louisiana. Due to a series of errors, the casing was not sufficiently strong when workers began to replace the drilling mud with clear water. Thus, the high-pressure, gassy oil in the reservoir that the well had punctured rushed up the drillhole. A backup safety device called a blowout preventer failed, so the oil reached the platform and sprayed 100 m into the sky. Sparks from electronic gear triggered an explosion, and the platform became a fountain of flame and smoke that killed 11 workers. An armada of fireboats could not douse the conflagration **(Fig. Bx14.3c)**, and after 36 hours, the still-burning platform tipped over and sank.

Robot submersibles sent to the seafloor to investigate found that the twisted mass of bent and ruptured pipes at the wellhead was billowing oil and gas **(Fig. Bx14.3d)**. Estimates as to the amounts of hydrocarbons released vary wildly, but most data suggest that on the order of 50,000 to 62,000 barrels per day of hydrocarbons entered the Gulf's water from the well. (By comparison, natural oil seeps at many localities around the Gulf together yield about 1,000 to 4,000 barrels per day.) Stopping this underwater gusher proved to be an immense challenge, and initial efforts to block the well or to put a containment dome over the wellhead failed. It was not until July 15, almost 3 months after the blowout, that the flow was finally stopped, and it was not until September 19 that a new relief well intersected the blown well and provided a conduit to pump concrete down to block the original well permanently.

All told, about 4.2 million barrels of hydrocarbons contaminated the Gulf from the *Deepwater Horizon* blowout. The more volatile components of the spill evaporated, but enough liquid hydrocarbons remained to form a slick that, at times, covered an area of over 100,000 km^2 **(Fig. Bx14.3e)**. Thousands of people labored to contain the damage—floating booms were set up to contain the oil, skimmer ships traversed the slick to suck up oil, sandbag barriers were placed along the coast from Louisiana to Florida to keep the oil out of wetlands and beaches, and planes spread chemical dispersants on the floating oil to break it into tiny particles. In the near term, the spill was devastating to wetlands, wildlife, and the fishing and tourism industries. Over the longer term, natural processes—microbes that eat the oil droplets—will conquer the spill.

---

environment because the ground surface above may then sink and break up.

Numerous types of air pollution can occur because of the use of fossil fuels. For example, leakage from well sites, underground storage reservoirs, or refineries releases fumes into the air that can be harmful to human or animal health. Burning fossil fuels sends soot, carbon monoxide, sulfur dioxide, nitrous oxide, unburned hydrocarbons, and in some cases, mercury into the air. Sulfur dioxide dissolves in atmospheric water to form dilute sulfuric acid, which precipitates as **acid rain**. Such rain can change soils downwind of the pollution source and can cause forests and fish populations to die.

But even if toxic air pollutants can be decreased, burning fossil fuels still releases carbon dioxide (CO_2) into the atmosphere. Carbon dioxide is a **greenhouse gas,** meaning that it traps heat in the Earth's atmosphere much as glass traps heat in a greenhouse (see Chapter 23). A global increase in atmospheric CO_2 will lead to a global increase in atmospheric temperature (global warming), which in turn may alter the distribution of climate belts and lead to a rise in sea level,

FIGURE Bx14.3 Offshore drilling and the *Deepwater Horizon* disaster.

(a) The *Deepwater Horizon* oil-drilling platform as it appeared in the Gulf of Mexico in July of 2009.

(c) Fireboats dousing the burning *Deepwater Horizon* rig before it sank.

(b) Drilling on the Gulf Coast margin. Most drilling sites are on the continental shelf, but more recently, deeper-water sites have been explored. The roughness of the slope to the abyssal plain is due to salt domes. Each yellow dot is a drilling platform. The location of the *Deepwater Horizon* platform is indicated by the larger, white dot.

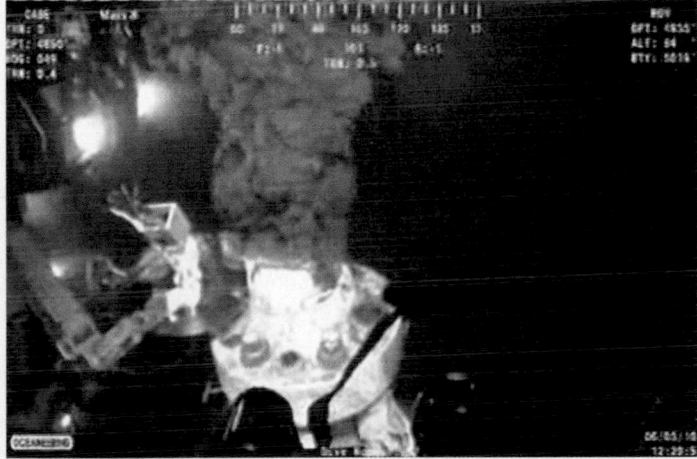

(d) A plume of oil billows into the water from the wellhead, as viewed by underwater cameras.

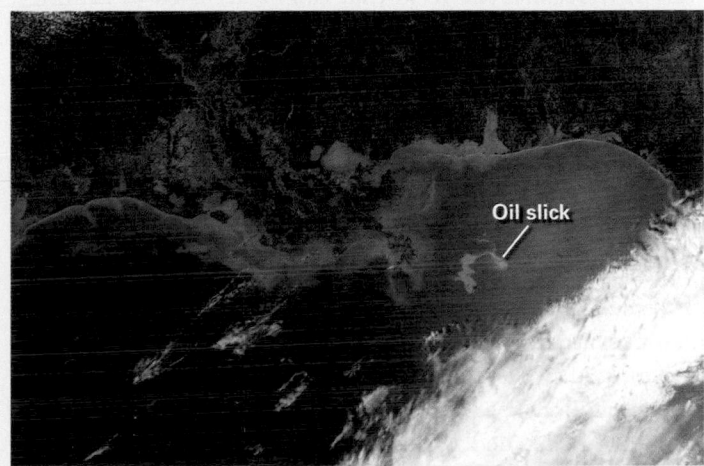

(e) A satellite image showing the oil slick in the Gulf of Mexico, southeast of the Mississippi Delta.

FIGURE 14.29 Marine oil spills can come from drilling rigs or from tankers.

(a) An oil tanker leaking oil on the sea surface.

(b) Oil spills can contaminate the shore and can be very difficult to clean up.

among many other effects. Because of concern about CO_2 production, efforts are under way to replace fossil fuels with biofuels derived from perennial plants, for these biofuels are "carbon neutral," in that, because plants absorb CO_2 as they grow, the CO_2 added to the air by burning these fuels will be absorbed by the growth of new plants. There are also efforts to develop techniques to capture CO_2 at power plants, liquefy it, and pump it into reservoir rocks deep underground. This process is called *carbon capture and sequestration*. We'll learn more about CO_2-related issues in Chapter 23.

Energy Options for the Future

In the early 21st century, hydrocarbons continue to be the dominant energy source for human society. Of course, technologies already exist to substitute coal for oil or natural gas in power plants, and the planet still hosts immense coal reserves—by some estimates, there are about 9,800 billion tons of *recoverable coal* (coal that can be accessed by mining), which contains as much energy as 4,700 billion barrels of oil. Large-scale substitution of coal for oil seems unlikely, though, because other energy sources have become cheaper, and because many people consider the environmental cost of mining and burning coal to be too high. Vast supplies of uranium, the fuel of nuclear power plants, remain untapped. Further, nuclear engineers have designed new types of plants, powered by breeder reactors that essentially produce new fuel. But many people view nuclear plants with concern because of issues pertaining to radiation, accidents, terrorism, and waste storage, and these concerns have slowed down the industry. Nevertheless, as we noted earlier, China has recently decided to increase its nuclear capacity in order

to combat its air pollution problem. The growth of geothermal energy also seems limited.

Because of the potential problems that might result from relying more on coal, hydroelectric, or nuclear energy, researchers have increasingly been exploring clean energy options. One possibility is solar power, whose cost has decreased exponentially in the past few decades. Similarly, we can turn to wind power for relatively small-scale energy production. Because sunlight reaching the Earth changes during the day, and with the amount of overcast, and because wind speed varies, solar power and wind power can't provide the steady supply of energy that the present-day *energy grid* (the interconnected network of power lines and transformers that distribute energy around the country) requires. This problem might someday be overcome with "smart grids," which can accommodate rapid fluctuations in energy input. Clearly, society will be facing difficult choices in the not-so-distant future about where to obtain energy, and we will need to invest in the research required to discover new alternatives.

> ### TAKE-HOME MESSAGE
>
> Oil is a nonrenewable resource, and conventional oil reserves may run out in less than a century—we live in the Oil Age, which may, when it ends, have lasted less than 350 years. Unconventional hydrocarbon reserves may last longer, but use of fossil fuels has environmental consequences.
>
> **QUICK QUESTION:** What is Hubbert's Peak, and when was it reached in the United States?

ANOTHER VIEW A line of electrical generators in a tunnel produces electricity from water flowing out of a reservoir that had been trapped by a large dam.

SUMMARY

- Oil and gas are hydrocarbons. Their viscosity and volatlity depend on the length of its molecules.

- Oil and gas develop from the bodies of plankton and algae, which settle in oxygen-poor quiet water, to form organic shale. Chemical reactions convert the organic matter into kerogen and then oil.

- To form a conventional hydrocarbon reserve, oil must migrate from a source rock into a reservoir rock overlain by a seal rock in a subsurface configuration called an oil trap.

- Unconventional reserves, from which hydrocarbons are difficult to extract, include shale, shale gas and shale oil, tar sand, and gas hydrate.

- For coal to form, abundant plant debris must be deposited in an oxygen-poor environment. Compaction creates peat. When buried deeply and heated, peat transforms to coal.

- Geologists rank coal based on the amount of carbon it contains, from lignite, to bituminous, to anthracite.

- Coal occurs interbedded with other sedimentary strata. Coal can be obtained by strip mining or underground mining.

- Nuclear power plants generate heat from the nuclear fission of radioactive elements.

- Some uranium deposits occur as veins in igneous rock bodies; others are found in sedimentary beds.

- Nuclear reactors must be carefully controlled to avoid overheating. Disposal of radioactive nuclear waste can create environmental problems.

- Geothermal power uses the Earth's internal heat; hydroelectric and wind power use the potential energy of water and wind respectively; and solar power converts sunlight to electricity.

- We now live in the Oil Age, but conventional oil reserves won't last forever.

- Production and use of fossil fuels have environmental consequences.

GUIDE TERMS

acid-mine runoff (p. 557)
acid rain (p. 558)
barrel of oil (bbl) (p. 531)
biofuel (p. 552)
biomass (p. 525)
blowout (p. 558)
chain reaction (p. 547)
coal (p. 542)
coal-bed methane (p. 546)
coal gasification (p. 546)
coal rank (p. 543)
coal reserve (p. 544)
coal swamp (p. 542)
conventional hydrocarbon
  system (p. 529)

conventional reserve (p. 528)
directional drilling (p. 534)
drilling mud (p. 534)
energy (p. 523)
energy density (p. 526)
energy resource (p. 523)
fossil fuel (p. 526)
fuel (p. 523)
gas hydrate (p. 540)
geothermal energy (p. 549)
greenhouse gas (p. 558)
Hubbert's Peak (p. 556)
hydrocarbon (p. 526)
hydrocarbon generation
  (p. 527)

hydrocarbon reserve (p. 527)
hydrofracturing (pp. 535)
kerogen (p. 527)
meltdown (p. 548)
migrate (p. 530)
nonrenewable resource (p. 556)
nuclear reactor (p. 547)
nuclear waste (p. 549)
Oil Age (p. 557)
oil seep (p. 530)
oil shale (pp. 527, 539)
oil window (p. 527)
peat (p. 542)
permeability (p. 529)
photovoltaic cell (p. 553)

pore (p. 529)
porosity (p. 529)
renewable resource (p. 556)
reservoir rock (p. 529)
seal rock (p. 530)
seismic-reflection profile
  (p. 533)
shale gas (p. 538)
shale oil (p. 538)
source rock (p. 527)
tar sand (p. 539)
tidal power (p. 553)
trap (p. 530)
unconventional
  reserve (p. 528)

 **GEOTOURS** *THIS CHAPTER'S GEOTOURS WORKSHEET (L) FEATURES QUESTIONS AND GOOGLE EARTH SITES ON:*

- Hydrocarbon reserves
- Coal resources
- Other energy resources

REVIEW QUESTIONS

The letters following each Review Question refer to the corresponding Learning Objective from the Chapter Opener.

1. What are the sources of energy in the Earth System? **(A)**

2. How does the length of a hydrocarbon's polymer chains affect its viscosity and volatility? **(C)**

3. What is the source of the organic material in oil? **(B)**

4. What is the oil window, and what happens to oil at temperatures higher than the oil window? **(C)**

5. How can organic matter be trapped and transformed into an oil reserve? **(D)**

6. Describe the different kinds of oil traps. Which type is pictured in the diagram? **(D)**

7. What are tar sand and oil shale, and how can oil be extracted from them? **(D)**

8. How do gas hydrates form, and where do they occur? **(B)**

9. How do porosity and permeability affect the oil-bearing potential of a rock? **(C)**

10. Where do most of the world's conventional oil reserves occur? At present rates of consumption, how long will those oil reserves last? **(D)**

11. How does coal form, and what class of rock is it? **(B)**

12. What conditions cause coal to transform in rank from peat to anthracite? **(B)**

13. How can the mining and burning of coal affect the environment? **(F)**

14. What is coal-bed methane and how is it extracted? **(C)**

15. Describe the reactions in a nuclear reactor and the means that engineers use to control reaction rates. **(E)**

16. Where does uranium form in the Earth's crust? Where does it usually accumulate in minable quantities? **(E)**

17. What are some of the drawbacks of nuclear energy? **(F)**

18. Discuss the pros and cons of alternative energy. **(F)**

19. What is geothermal energy? What limits its use? **(F)**

20. What is the difference between renewable and nonrenewable resources? **(A)**

21. What are the major environmental consequences of producing and burning fossil fuels? **(F)**

22. What is the likely future of hydrocarbon production and use in the 21st century? What trends are revealed in the graph? **(F)**

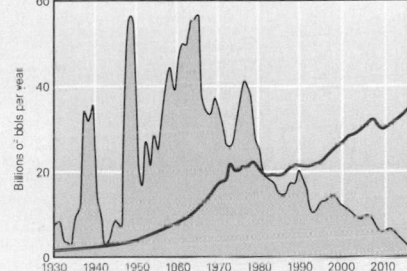

ON FURTHER THOUGHT

23. Much of the oil production in the United States takes place at offshore drilling platforms along the coast of the Gulf of Mexico. Consider the geologic setting of the Gulf Coast in the context of plate tectonics theory, and explain why an immensely thick sequence of sediment accumulated in this region and why so many salt-dome traps formed. **(C)**

24. Ethanol can be used as an alternative to petroleum as a liquid fuel. It can be produced by processing corn, sugarcane, or certain perennial grasses (e.g., switchgrass or *Miscanthus*). What factors should be considered in determining which of these crops would be the most appropriate for use as a source of ethanol in North America? **(F)**

ONLINE RESOURCES

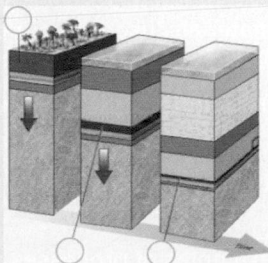

Videos
This chapter features videos on the basic process of hydraulic fracturing and on oil sands and plumes.

Smartwork5
This chapter features questions that check students' understanding of energy technologies and the formation of coal and oil.

CHAPTER 15

Riches in Rock: Mineral Resources

By the end of this chapter, you should be able to . . .

A. identify materials, used in your everyday life, that come from geologic resources.

B. discuss the nature of the ores from which we obtain metals.

C. explain the geologic settings in which ore deposits form.

D. provide examples of processes used to extract metals from ores.

E. distinguish between metallic and nonmetallic resources, and provide examples of nonmetallic resources and their origins.

F. discuss why reserves of mineral supplies may run out in the coming years.

G. describe the environmental consequences of extracting and using mineral resources.

> Truth, like gold, is to be obtained not by its growth,
> but by washing away from it all that is not gold.
>
> —Leo Tolstoy (Russian author, 1828–1910)

15.1 Introduction

In June 1845, James Marshall arrived by horse at Sutter's Fort in central California to start a new life. Having just finished a stint as a rancher and a few months as a soldier, Marshall decided to go into the lumber trade, and he convinced Captain John Sutter to finance the construction of a sawmill in the foothills of the Sierra Nevada range. Marshall's scruffy crew finished the mill by the beginning of 1848. As they stood admiring the new building, Marshall noticed a glimmer of metal in the gravel that littered the bed of the adjacent stream, picked it up, banged it between two rocks to test its hardness, and shouted, "Boys, by God, I believe I have found a gold mine!" For a short while, Marshall and Sutter managed to keep the discovery secret. But word of the gold soon spread, and within weeks the mill workers had disappeared into the mountains to seek their own fortunes.

Gold fever eventually reached the east coast, and as a result, 1849 saw 40,000 prospectors head to California. These "forty-niners," as they were called, had abandoned their friends and relatives on the gamble that they could strike it rich **(Fig. 15.1)**. Some traveled by land, some by sea, to reach San Francisco, a town of mud streets and plank buildings. In many cases, crews abandoned their ships in the harbor to join the scramble to the gold fields. Although miners collected $20 million worth of gold in 1849, few of the forty-niners actually became rich. Most lost whatever wealth they had found in the saloons, stores, and gambling halls that sprouted around the goldfields.

Gold is but one of many **mineral resources**, meaning minerals extracted from the Earth's crust for use by people. Without these resources, industrialized societies could not function. Geologists divide mineral resources into two categories: *metallic mineral resources* (materials containing a concentration of gold, copper, aluminum, iron, or other metals) and *nonmetallic mineral resources* (materials such as building stone, gravel, sand, gypsum, phosphate, and salt, used in construction or for chemical production). In this chapter, we look at the nature of such mineral resources, the geologic phenomena responsible for their formation, and the methods that people use to obtain them. We conclude by considering the sustainability of mineral reserves in the future.

15.2 Metals and Their Discovery

What Is Metal?

A **metal** is an opaque, shiny, smooth solid that can conduct electricity and can be bent, drawn into wire, or hammered into thin sheets. Metals look and behave quite differently from wood, plastic, meat, or rock because metallic bonds hold the

FIGURE 15.1 The Gold Rush of 1849.

(a) An 1849 poster promoting travel to the gold regions of California.

(b) A prospector, or "forty-niner," mining in the Sierra Nevada range.

◀ (facing page) Geology students explore an old, abandoned mine in Colorado. The colored materials on the wall are ore minerals containing valuable metals. When this mine operated, miners hauled the ore out in carts that they pushed along the steel tracks.

FIGURE 15.2 What makes a metal a metal?

(a) A copper crystal consists of wafer-like layers of copper atoms.

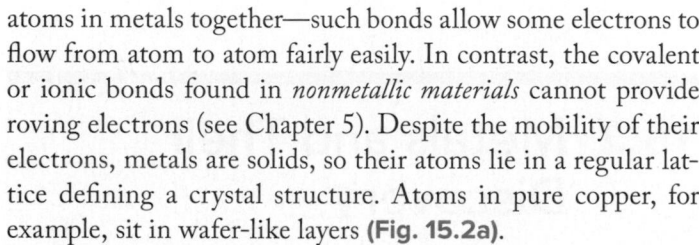

(b) The crystal can bend because the layers can easily slide past each other. Metals can be rolled into thin sheets without breaking.

Pennies

Sheets of rolled copper are stamped to make coins.

atoms in metals together—such bonds allow some electrons to flow from atom to atom fairly easily. In contrast, the covalent or ionic bonds found in *nonmetallic materials* cannot provide roving electrons (see Chapter 5). Despite the mobility of their electrons, metals are solids, so their atoms lie in a regular lattice defining a crystal structure. Atoms in pure copper, for example, sit in wafer-like layers **(Fig. 15.2a)**.

Not all metals behave the same way. Some have great strength, whereas others are particularly *malleable* (meaning they can be easily bent or molded). The behavior of a metal depends on the strength of the bonds between its atoms and on the architecture of its crystal structure. For example, the wafer-like layers of atoms in copper can slip past each other quite readily, so copper can be bent or stretched easily **(Fig. 15.2b)**. The shapes and dimensions of crystals in a piece of metal reflect the rate at which the metal cools.

The Discovery of Metals

Certain substances—namely, copper, silver, gold, and mercury—can occur in rock as a **native metal,** a material that consists only of metal atoms and thus looks and behaves like metal **(Fig. 15.3a)**. Prehistoric hunters collected nuggets of native metal from streambeds and pounded them with stone hammers to make tools, and because native metals are rare and durable, people began to use them to make coins. Gold attained particular popularity as a currency because of its unique warm yellow color and its resistance to tarnish or rust **(Fig. 15.3b)**. Pieces of native gold that have been eroded free of bedrock become gold flakes or *nuggets*. Though iron does not occur as a native metal in bedrock on the Earth, chunks of it fall to the Earth from space in the form of meteorites; such meteorite iron became highly valued for toolmaking.

If we had to rely on native metals as our only source of metal, we would have only a tiny fraction of our current metal supply. In fact, most of the metal atoms in materials we use today originated as ions bonded to nonmetallic elements in a great variety of minerals that themselves look nothing like metal. Only because of the chance discovery by some prehistoric genius that metals can be extracted from certain minerals do we now have the ability to produce sufficient metal for the needs of industrialized society. Workers extract metal from minerals by a process called **smelting**. The decomposition of minerals during smelting yields metal plus a nonmetallic residue known as *slag*.

Of the principal metals we rely on today (copper, iron, and aluminum), *copper* came into use first, because people can smelt copper from sulfide minerals relatively easily. Copper implements appeared as early as 4000 B.C.E. Pure copper has limited value for toolmaking or weapon making because it's too soft to retain a sharp edge. Around 2800 B.C.E., however, Sumerian craftsmen discovered that copper could be mixed with tin to produce bronze, an **alloy** (a compound containing two or more metals) whose strength exceeds that of either metal alone, and people came to rely on bronze for swords and ornaments **(Fig. 15.3c)**.

Iron is superior to copper or bronze for many purposes because of its strength, hardness, and abundance. Still, people didn't start using iron widely until 1,500 years after they had begun to use bronze. The delay happened, in part, because iron has such a high melting temperature that it can be difficult to work with. In addition, iron generally occurs in iron oxide minerals (such as hematite, Fe_2O_3, or magnetite, Fe_3O_4), and the liberation of the metal from oxide minerals requires a chemical reaction, not just heat. Therefore, the widespread use of iron became possible only after someone discovered that metallic iron can

Did you ever wonder . . .
where the iron in the steel bodies of cars comes from?

FIGURE 15.3 Native metals and bronze have been used for thousands of years.

(a) Native copper occurs in complex shapes, which remain when the surrounding rock weathers away.

Gold bracelets on display in a Kuwaiti marketplace

(b) Gold occurs as native metal within quartz veins. The quartz breaks up to form sand, leaving nuggets of gold.

(c) A bronze artifact made millennia ago in China.

be produced by heating iron oxide minerals in the presence of carbon monoxide (CO), a gas produced by burning charcoal **(Fig. 15.4)**. Chemists describe this reaction by the formula $Fe_2O_3 + 3CO \rightarrow 2Fe + 3CO_2$. The discovery of *steel*, an alloy of iron and carbon, came later, and industrial production of this particularly durable metal did not begin until the 1850s. *Stainless steel*, a corrosion-resistant alloy of iron and chromium, became available for widespread use only in the early 20th century.

Aluminum, a silvery metal, has a density only one-third that of iron, and it doesn't corrode. As a result, it has become the metal of choice for manufacturing objects, such as airplanes, that must be light in weight. But aluminum does not occur as a native metal, and extracting it from minerals requires complex methods that use lots of electricity, so its use only became economically practical beginning in the 1880s.

These days, in addition to iron, copper, tin, and aluminum, we use a vast array of different metals. Some are known as *precious metals* (gold, silver, and platinum) and others as *base metals* (copper, lead, zinc, and tin) because of the difference in their price. Surprisingly, of the 63 or so metals in use today, people knew of only 9 (gold, copper, silver, mercury, lead, tin, antimony, iron, and arsenic) before the year 1700.

TAKE-HOME MESSAGE

Metals are malleable materials in which metallic bonds hold atoms together. Though some occur in pure form as native metals, most occur in other minerals and have to be extracted by smelting. Copper was the first metal to be used by humans. Iron and aluminum are more difficult to extract.

QUICK QUESTION: Why did the Bronze Age come before the Iron Age in human history?

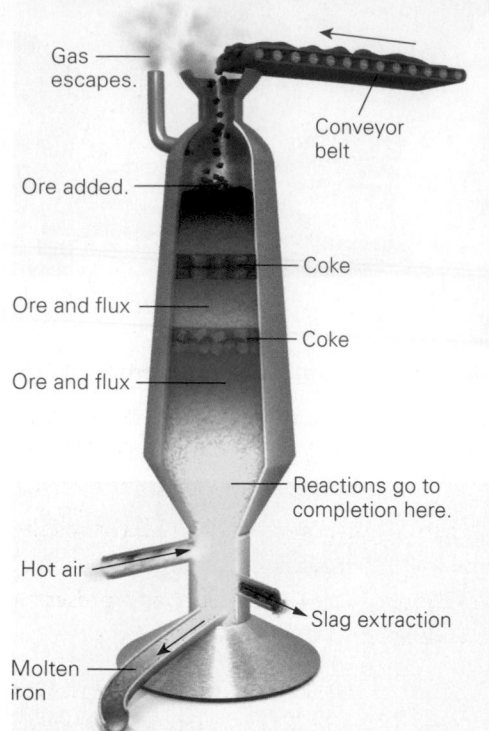

Gas escapes.

Conveyor belt

Ore added.

Coke

Ore and flux

Coke

Ore and flux

Reactions go to completion here.

Hot air

Slag extraction

Molten iron

(a) Molten iron sinks to the bottom and slag (silicates and oxides) floats.

(b) In a factory, huge tubs pour the end product, molten iron.

15.3 Ores, Ore Minerals, and Ore Deposits

What Is an Ore?

As we have seen, metals occur in two forms—as native metals or as ions bonded to nonmetallic elements in a great variety of minerals. Many metal-containing minerals can be found in common rocks. For example, if you pick up a chunk of common granite and analyze its mineral content, you will find that it consists mostly of quartz (SiO_2), plagioclase ($[Na,Ca]AlSi_3O_8$), and potassium feldspar ($KAlSi_3O_8$). Looking at these chemical formulas, you can see that feldspars contain about 8% aluminum, so common granite includes aluminum ions. But we don't mine granite to produce aluminum. Why? Simply because it's not economical—overall, granite contains relatively little aluminum, and it's difficult to separate aluminum atoms from feldspar, so processing costs would exceed the sale price.

To obtain the metals needed by industrialized society, we mine **ores**, rocks containing a significant concentration of native metals, or of ore minerals. **Ore minerals** (or *economic minerals*), in turn, are minerals that contain metals both in high

concentrations and in forms that can be easily extracted. For example, galena (PbS) contains about 50% lead, and lead can be separated from sulfur fairly easily, so we consider galena to be an ore mineral of lead **(Fig. 15.5a)**. Similarly, we obtain most of our iron from oxide minerals such as hematite and magnetite, so these minerals are ore minerals for iron. Geologists have identified many different kinds of ore minerals **(Table 15.1)**. As you can see from their chemical formulas, many ore minerals are sulfides, in which the metal occurs in combination with sulfur (S), or oxides, in which the metal occurs in combination with oxygen (O). Numerous ore minerals come in brilliant colors and interesting shapes **(Fig. 15.5b)**, and some have a metallic luster.

To be an ore, a rock must contain a sufficient amount of ore minerals to make the rock worth mining. For example, iron constitutes about 6.2% of the continental crust's weight, whereas it makes up 30% to 60% of the weight of iron ore **(Fig. 15.6)**. We refer to the concentration of a useful metal in an ore as the **grade** of the ore—the higher the concentration, the higher the grade. Whether or not ore of a certain grade can be worth mining depends on the price of metal in the market. For example, in 1880, copper-bearing rocks needed to contain at least 3% copper to be considered economic, but due to improvements in technology and a rise in the price of copper, rocks containing only about 0.4% copper can now be considered economic.

How Do Ore Deposits Form?

Ores do not occur uniformly through rocks of the crust. If they did, we would not be able to extract them economically. Fortunately for humanity, geologic processes concentrate these minerals in **ore deposits**, economically significant occurrences of ore. Geologists distinguish among many different kinds of ore deposits, which differ from one another in terms of which ore minerals they contain and which kind of rock they occur in. Here, we introduce a few examples.

Magmatic Deposits As a magma cools, sulfide ore minerals may crystallize early and then accumulate to form a

FIGURE 15.5 Examples of ore minerals.

(a) This lead ore from Missouri contains galena (PbS) crystals that grew in dolostone.

(b) Most copper comes from ore minerals that look nothing like metallic copper. This rock has a coating of malachite, a copper carbonate mineral.

TABLE 15.1 Some Common Ore Minerals

Metal	Mineral Name	Chemical Formula
Copper	Chalcocite	Cu_2S
	Chalcopyrite	$CuFeS_2$
	Bornite	Cu_5FeS_4
	Azurite	$Cu_3(CO_3)2(OH)_2$
	Malachite	$Cu_2(CO_3)(OH)_2$
Iron	Hematite	Fe_2O_3
	Magnetite	Fe_3O_4
Tin	Cassiterite	SnO_2
Lead	Galena	PbS
Mercury	Cinnabar	HgS
Zinc	Sphalerite	ZnS
Aluminum	Kaolinite	$Al_2Si_2O_5(OH)_4$
	Corundum	Al_2O_3
Chrome	Chromite	$(Fe,Mg)(Cr,Al,Fe)2O_4$
Nickel	Pentlandite	$(Ni,Fe)_9S_8$
Titanium	Rutile	TiO_2
	Ilmenite	$FeTiO_3$
Tungsten	Sheelite	$CaWO_4$
Molybdenum	Molybdenite	MoS_2
Magnesium	Magnesite	$MgCO_3$
	Dolomite	$CaMg(CO_3)_2$
Manganese	Pyrolusite	MnO_2
	Rhodochrosite	$MnCO_3$

magmatic deposit. When the magma freezes solid, the resulting igneous body contains concentrations of sulfide minerals. Because of their composition, these concentrations represent a type of **massive sulfide deposit (Fig. 15.7a).**

Hydrothermal Deposits *Hydrothermal activity* involves the circulation of hot-water solutions through a magma or through the rocks surrounding an igneous intrusion. These fluids dissolve metal ions. When a solution enters a region

FIGURE 15.6 The difference between ore and other rock. Iron in a block of granite is not worth extracting.

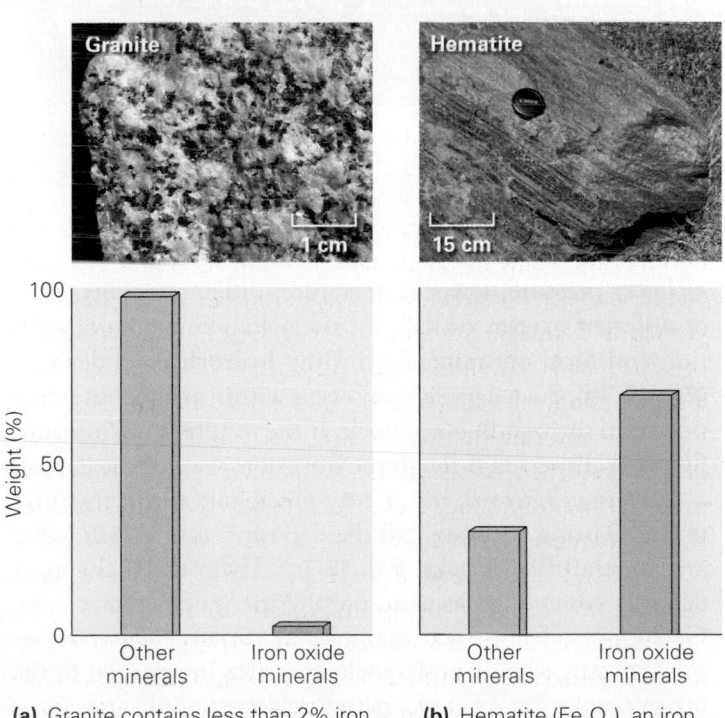

(a) Granite contains less than 2% iron.

(b) Hematite (Fe_2O_3), an iron ore, contains over 60% iron.

FIGURE 15.7 Various processes that form ore deposits.

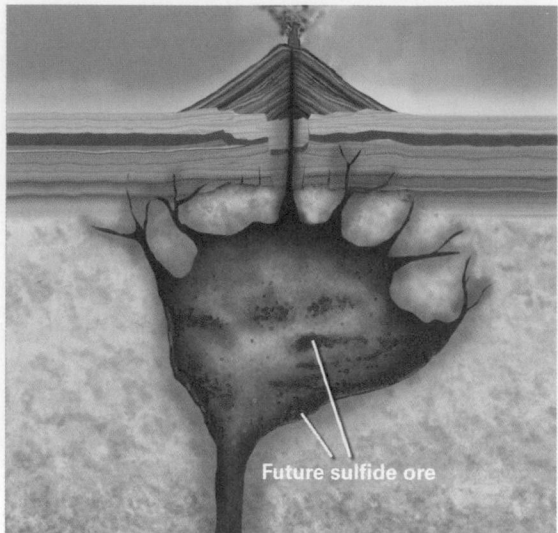

(a) Massive sulfide deposits can form when sulfide ore minerals become concentrated in a magma chamber.

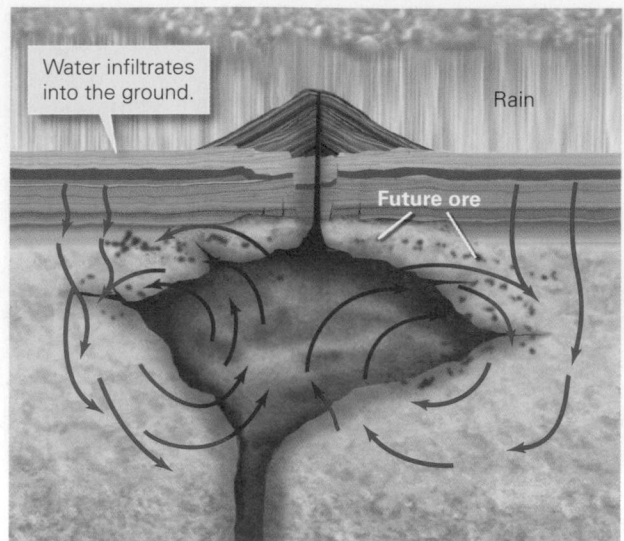

(b) Hydrothermal deposits form when water circulating around and through magma dissolves and redistributes metals (arrows indicate flowing water).

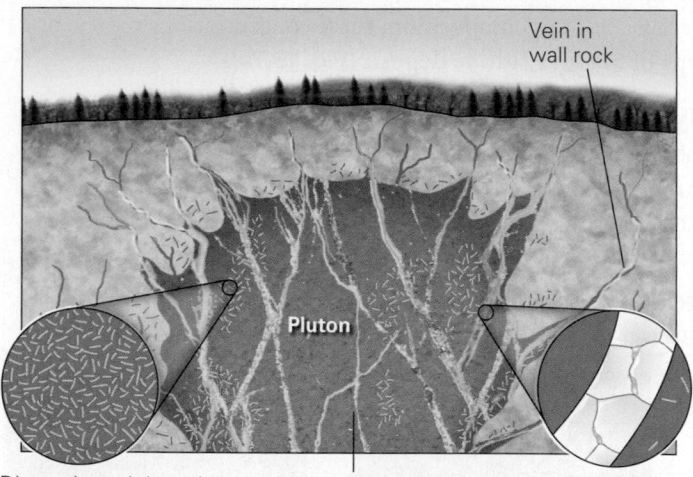

(c) Disseminated deposits consist of ore minerals that are dispersed. Vein deposits consist of ore minerals precipitated in cracks.

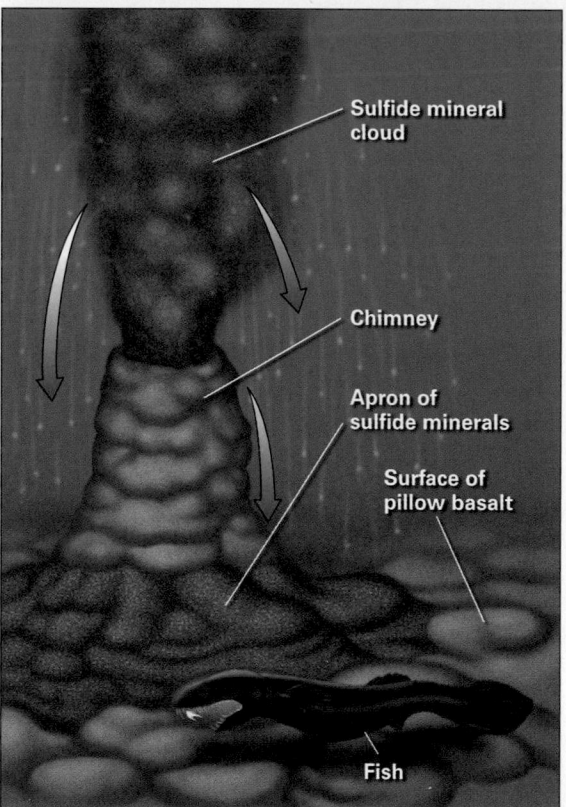

(d) Massive sulfide deposits also form when ore minerals precipitate around hydrothermal vents (black smokers) along a mid-ocean ridge.

of lower pressure, lower temperature, different acidity, and/or different oxygen availability, the metals come out of solution and form ore minerals yielding **hydrothermal deposits (Fig. 15.7b)**. Such deposits may occur within an igneous intrusion or in surrounding wall rock. If the resulting ore minerals fill pores dispersed through the intrusion, we call the deposit a *disseminated deposit*, but if they precipitate to fill fractures in pre-existing rock, we call the deposit a *vein deposit* (*veins* are mineral-filled cracks; **Fig. 15.7c**). Hydrothermal copper deposits commonly occur in porphyritic igneous intrusions. Geologists refer to such examples as *porphyry copper deposits*. Typically, vein deposits include quartz in addition to the ore minerals. For example, native gold commonly appears as flakes in milky-white quartz veins.

Hydrothermal Vent Deposits Hydrothermal activity at the submarine volcanoes along mid-ocean ridges leads to the eruption of hot water out of vents. This water contains high concentrations of dissolved metal and sulfur, and when the hot vent water comes in contact with cold seawater, the dissolved components precipitate as tiny crystals of sulfide minerals (Fig. 15.7d). The erupting water, therefore, looks like a black cloud, so the vents are called *black smokers* (see Chapter 4). The minerals in the cloud build a chimney of ore minerals around the vent and settle to form a layer of minerals on the surrounding seafloor. Because the ore minerals are typically sulfides, the resulting *hydrothermal vent deposits* constitute another type of massive sulfide deposit.

Secondary-Enrichment Deposits Sometimes groundwater passes through ore-bearing rock long after the ore formed. This water dissolves some of the ore minerals and carries away the ions. When the water eventually flows into a different chemical environment (for instance, one with a different amount of oxygen or acidity), it precipitates new ore minerals, commonly in concentrations exceeding that of the original ore. A new deposit containing metals that have been dissolved and carried away from a pre-existing ore deposit is a **secondary-enrichment deposit (Fig. 15.8a, b)**. Some of these deposits contain spectacularly beautiful copper-bearing carbonate minerals, such as azurite and malachite (see Fig. 15.5b).

MVT Deposits Rain falling along one margin of a large sedimentary basin may sink into the subsurface and then flow as groundwater along a curving path that takes it first down to the bottom of the basin, and then eventually back up to the opposite margin of the basin, hundreds of kilometers away. At the bottom of the basin, high temperatures warm the water enough for it to dissolve metals. When the solution returns to the surface and enters cooler rock, the metals that it carries precipitate as ore minerals. Ore deposits formed in this way commonly contain lead- and zinc-bearing minerals. Many such deposits appear in dolomite beds of the Mississippi Valley region in the central United States, so they have come to be known as **Mississippi Valley–type (MVT) ore deposits (Fig. 15.8c)**.

Sedimentary Deposits of Metals Some ore minerals accumulate in sedimentary environments under special circumstances. For example, between about 2.5 billion and 1.8 billion years ago, the oxygen concentration in the atmosphere began to increase, and more oxygen began to dissolve in seawater. This change affected the chemistry of seawater such that it could no longer hold large quantities of iron in solution. This iron precipitated as iron oxide minerals that settled as

FIGURE 15.8 Secondary enrichment occurs when groundwater leaches (extracts) ore minerals from a rock and moves them to another location where they precipitate.

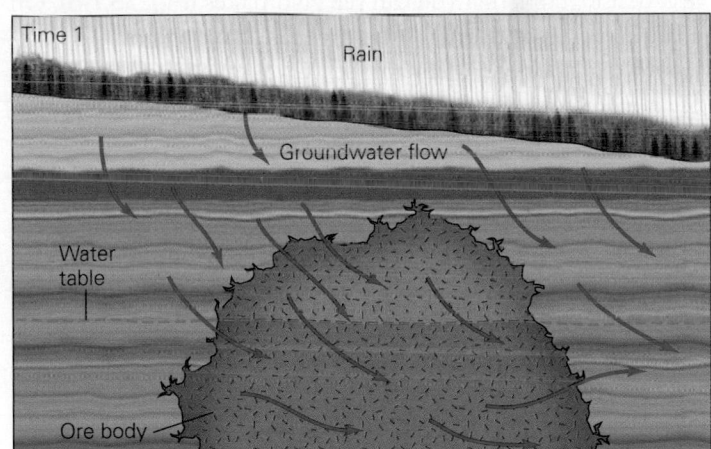

(a) Downward-flowing water dissolves ore minerals. The metal-rich solution moves down below the water table.

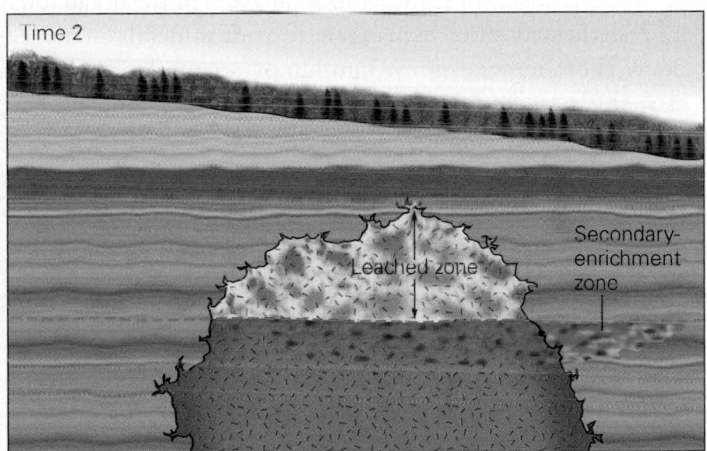

(b) Below the water table, the chemical environment changes, and new ore minerals precipitate.

(c) Mississippi Valley–type (MVT) deposits form when groundwater sinks into a deep sedimentary basin, warms, dissolves ore minerals, and then moves across the basin to a shallower, cooler, environment, where the metals precipitate.

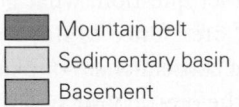

Mountain belt
Sedimentary basin
Basement

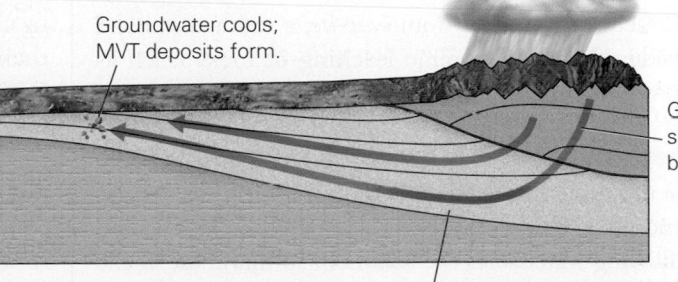

Groundwater cools; MVT deposits form.

Groundwater sinks into the basin.

Deep groundwater picks up heat and dissolves metals.

FIGURE 15.9 Examples of ore deposits that originated as sedimentary layers.

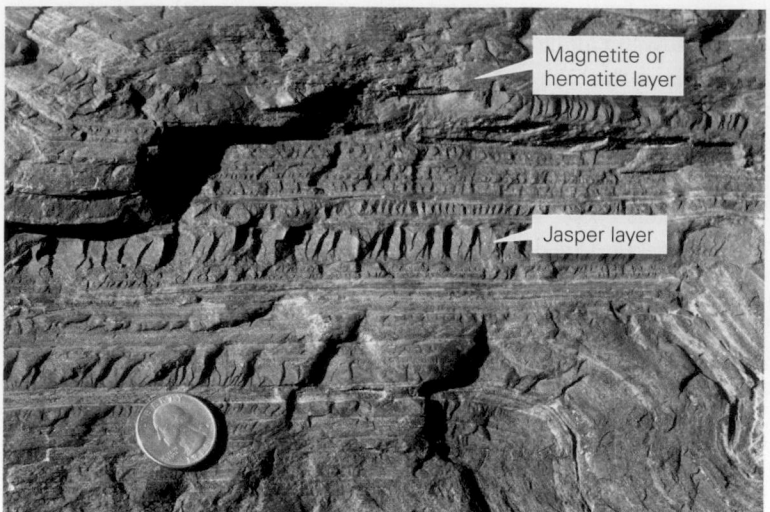

(a) Precambrian BIF from northern Michigan consists of hematite interbedded with jasper.

(b) A view from a submersible of seafloor on which polymetallic nodules have grown in the sediment. The inset provides a close-up of a 12-cm-long nodule.

sediment on the seafloor. As we learned in Chapter 13, the resulting iron-rich sedimentary deposits are known as **banded iron formation (BIF) (Fig. 15.9a)** because after lithification they consist of alternating layers of gray iron oxide (magnetite or hematite) and red layers of jasper (iron-bearing chert). Microbes may have participated in the precipitation process.

The chemistry of seawater in some parts of the ocean today leads to the deposition of manganese oxide minerals on the seafloor. These minerals grow into lumpy accumulations known as **manganese nodules (Fig. 15.9b)**. Mining companies have begun to explore technologies for vacuuming up these nodules because, according to some estimates, the worldwide supply of nodules contains 720 years' worth of copper and 60,000 years' worth of manganese, at current rates of consumption.

Residual Mineral Deposits Recall from Interlude B that as rainwater sinks into the Earth, it leaches (dissolves) certain elements and leaves behind others as part of the process of forming soil. In rainy tropical environments, the residuum left behind in soils after leaching includes iron and aluminum. Locally, these metals may become so concentrated that the soil itself becomes an ore deposit **(Fig. 15.10a, b)**. We refer to such deposits as **residual mineral deposits**. Most of the aluminum ore mined today comes from *bauxite*, a residual mineral deposit produced by the extreme leaching of rocks (such as granite) containing aluminum-bearing minerals **(Fig. 15.10c)**.

Placer Deposits Erosion of a rock outcrop containing gold ore yields a mixture of mineral grains, composed of quartz and feldspar, and metal flakes or nuggets. If the sediment falls into a stream, flowing water sorts the sediment so that more of the less-dense mineral grains get carried downstream, while more

of the denser gold clasts remain behind. This process produces a **placer deposit**, an ore formed where water flow concentrates valuable metals (such as gold) or valuable minerals (such as diamond) **(Fig. 15.11a)**. You can extract gold from a placer deposit by swirling sediment dug up from a streambed in a pan of water. As the water moves, it washes away enough of the less-dense grains that you can see gold flakes **(Fig. 15.11b)**. On a commercial scale, miners use dredges or high-pressure water jets to dig up modern streambed gravel or preserved deposits of ancient streambed gravel. They then spill the gravel into a gently sloping tank containing flowing water. Like natural stream flow, the water washes away less-dense minerals and leaves behind the gold. By tracking placer deposits upstream, prospectors sometimes find the *mother lode*, the bedrock source of the gold.

Where Are Ore Deposits Found?

The Inca Empire of 15th-century Peru built elaborate cities and temples, decorated with fantastic masks, jewelry, and sculptures made of gold. Then, around 1532, Spanish ships arrived, led by conquistadors who quipped, "We suffer from a disease that only gold can cure." The Incas, already weakened by civil war, were no match for the armor-clad Spaniards with their guns and horses. Within six years, the Inca Empire had vanished, and Spanish ships were transporting Inca treasure back to Spain. Why did the Incas possess so much gold? Or to ask the broader question, what geologic factors control the distribution of ore? Once again, we can find the answer by considering the consequences of plate tectonics.

Several of the ore-deposit types we've mentioned occur in association with igneous rocks. As we learned in Chapter 6,

FIGURE 15.10 The formation of residual mineral deposits by intense weathering and leaching.

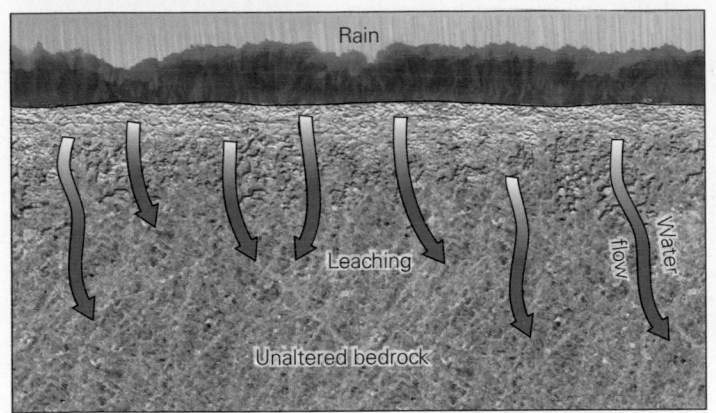

(a) When water sinks through the ground, bedrock weathers and slowly transforms into soil.

(b) If the amount of water is great, it leaches (dissolves and removes) many elements, leaving behind a residuum rich in iron or aluminum.

igneous activity does not happen randomly around the Earth. Rather, it develops along convergent boundaries (specifically, in the overriding plate of a subduction zone) or divergent boundaries (mid-ocean ridges), along continental rifts, or at hot spots. Therefore, magmatic and hydrothermal deposits (and secondary-enrichment deposits derived from them) form in these localities. Placer deposits accumulate in sediments eroded from magmatic or hydrothermal deposits, so we find them downstream of the igneous bodies. Inca gold, for example, rose into the crust with magmas that fed the Andean volcanic arc along the convergent boundary where the Pacific Ocean floor subducts beneath the South American Plate. As the mountains rose, erosion stripped away surface rocks to expose the hydrothermal deposits associated with plutons.

(c) Bauxite ore looks like reddish soil. It's typically obtained from near-surface open-pit mines.

FIGURE 15.11 Placer deposits.

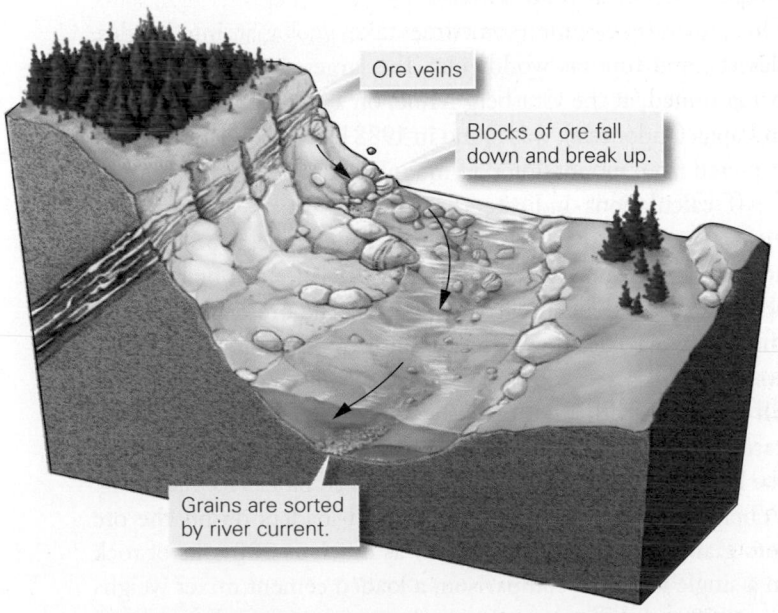

Ore veins

Blocks of ore fall down and break up.

Grains are sorted by river current.

(a) Placer deposits form where erosion produces clasts of native metals. Sorting by the stream concentrates the metals.

Inca miners quarried gold-bearing veins in the plutons and surrounding rock or panned for gold downstream.

As noted earlier, some massive sulfide deposits accumulate at black smokers along a mid-ocean ridge system and interleave with seafloor basalt. Miners can exploit such seafloor deposits only in places where continental collision slides a sliver

(b) A young woman panning for gold in the Mekong River, Laos.

of seafloor up along a fault and onto continental crust. (A slice of seafloor thrust up onto continents is known as an *ophiolite*.)

Not all ore deposits result directly from plate interactions, so not all occur in association with present-day or ancient plate boundaries. For example, BIF formed along passive continental margins during the Precambrian. We find BIF today only where Precambrian sedimentary rocks have been preserved, so most major exposures crop out in shield areas of continents. Bauxite forms where aluminum-rich bedrock undergoes extreme leaching during soil formation. As a result, bauxite deposits form by weathering of granite plutons (which contain aluminum-rich feldspars) in tropical regions.

TAKE-HOME MESSAGE

Ores (rocks or sediments that can be processed economically to produce useful metals) form in many ways, including crystallization in a melt, precipitation from hot water during igneous activity, accumulation in sediment, interaction with groundwater, and extreme weathering. The distribution of ore deposits can be explained by plate tectonics.

QUICK QUESTION: Why don't we use fresh granite as a source for aluminum?

15.4 Ore-Mineral Exploration and Production

Imagine an old prospector of days past clanking through the desert with a worn-out donkey. To find bedrock ore, such prospectors would eye hillsides for a "show," visible evidence on the ground surface that ore lies below. What does a show look like? It may be an outcropping of milky-white quartz veins, for veins could indicate the presence of hydrothermal deposits. It may be the sparkle of minerals with metallic luster disseminated through the outcrop. Or it may be the presence of orangish, yellowish, or bluish stains in outcrops, for these stains could result either from the presence of brightly colored ore minerals or from the oxidation (rusting) of oxide or sulfide minerals **(Fig. 15.12a)**. To find placer deposits, prospectors would pan a pile of sand or gravel from a streambed.

On finding a possible ore, a prospector would take samples back to town for an *assay*, a test to determine how much extractable metal the samples contain. If the assay indicated a significant concentration of metal, the prospector might "stake a claim" by literally marking off an area of ground with stakes. In some locations, a claim gave the prospector rights to all the mineral deposits on or under the land and the opportunity to develop mines. When one prospector would find ore, word would quickly spread, and others would rush to stake neighboring claims.

These days, large mining companies employ geologists to survey ore-bearing regions systematically. Once such a region has been identified, the geologists focus their studies on rocks that developed in settings appropriate for ore formation. They sometimes measure the local strength of the Earth's magnetic field and the local pull of gravity. These measurements may lead them to ore bodies because ore minerals tend to be denser and more magnetic than average rocks **(Fig. 15.12b)**. Geologists also sample rocks and soils to test for metal content and may even analyze plants in the area to detect traces of metals, for plants absorb metal ions through their roots. Once geologists have identified a possible ore deposit, they drill holes to sample subsurface rock and to determine the ore deposit's shape and extent **(Fig. 15.12c)**. Ore-mineral exploration sometimes takes geologists into jungles, deserts, and tundras worldwide. The largest gold deposit now being mined, at the Grasberg Mine, on a 4-km-high mountain in Papua (Indonesia), was found in 1988 by a geologist who helicoptered from mountaintop to mountaintop looking for shows.

If calculations indicate that mining an ore deposit will yield a profit, and if environmental concerns can be properly addressed, a company develops a mine. Mines can be below or above ground, depending on how close the ore deposit lies to the surface. To develop an **open-pit mine (Fig. 15.12d)**, workers first drill a series of holes into the solid bedrock and then fill the holes with high explosives. They must space the holes carefully and set off the charges in a precise sequence, so that the bedrock shatters into appropriate-sized blocks for handling. When the dust settles, large front-end loaders dump the ore into giant trucks, which can carry as much as 200 tons of rock in a single load. (In comparison, a loaded cement mixer weighs about 70 tons.) Tires on these mine trucks are so huge that a tall person comes up only to the base of the hub. The trucks

FIGURE 15.12 Finding and mining ore deposits.

The blue indicates the presence of copper ore.

20 cm

(a) Prospectors look for traces of ore minerals on rock outcrops.

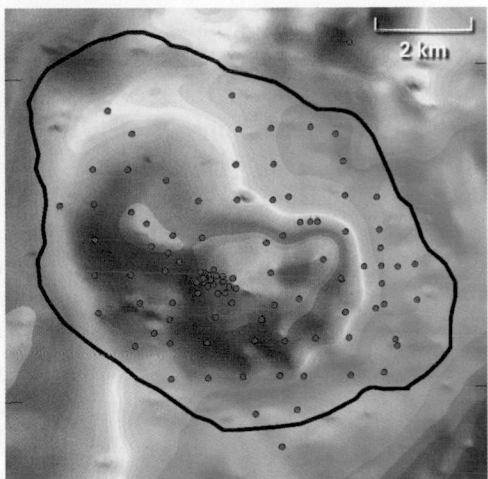

Stronger field

2 km

Weaker field

(b) A map of anomalies in the magnetic field may hint at ore bodies because metal may be magnetic. The black line outlines the ore body, and the red dots are drillhole locations for sampling.

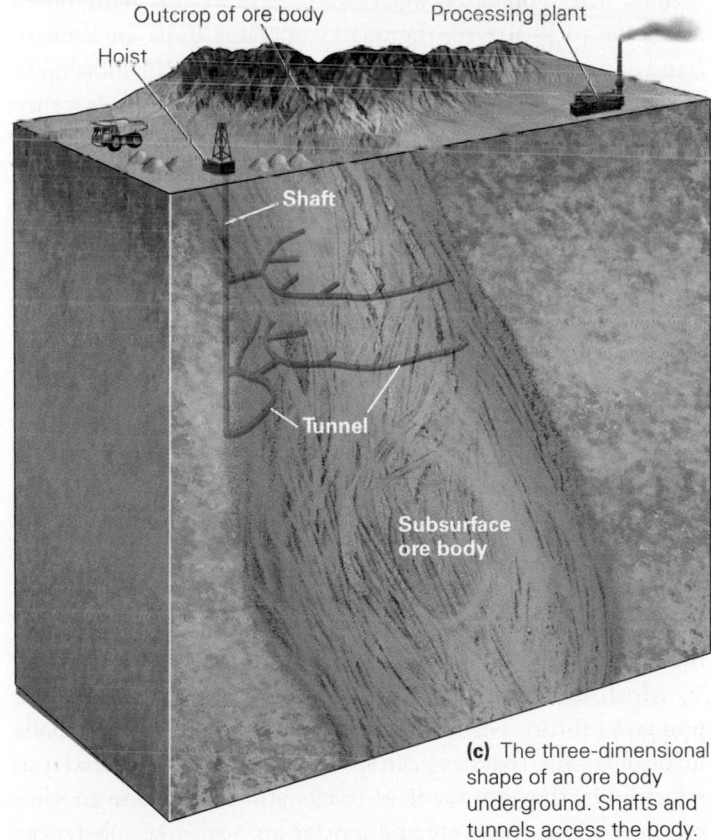

Hoist

Outcrop of ore body

Processing plant

Shaft

Tunnel

Subsurface ore body

(c) The three-dimensional shape of an ore body underground. Shafts and tunnels access the body.

(d) Miners employ open-pit mining when ore lies fairly close to the ground surface. Terracing the mine walls helps to stabilize them. Huge trucks haul the ore.

transport waste rock, or *tailings* (rock that doesn't contain ore), to a tailings pile and the ore to a crusher, a giant set of moving steel jaws that smash the ore into small fragments. Workers then separate ore minerals from other minerals and send the ore-mineral concentrate to a processing plant, where it undergoes smelting to separate metals from other elements. At the end of the process, workers melt the metal and then pour it into molds to make ingots (brick-shaped blocks) for transport to a manufacturing facility.

To develop an **underground mine**, miners either dig a horizontal tunnel into the side of a mountain (the entrance to such a tunnel is an *adit*), or they sink a vertical shaft in which they install an elevator. In some cases, they cut a spiral tunnel downward to provide a gentle ramp for trucks to carry ore up to the surface. At the level in the crust where the ore body appears, they build a maze of tunnels into the ore by drilling holes into the rock and then by blasting. The rock removed must be conveyed back to the surface. Rock columns between the tunnels hold up the ceiling of the mine.

The deepest mine on the planet, located in South Africa, currently reaches a depth of 3.5 km. Temperatures in the mine's deep tunnels exceed 55°C, making mining there a very uncomfortable occupation. Miners also face danger from mine collapse and rock falls **(Box 15.1)**. Some miners have been killed or injured by *rock bursts*, sudden explosions of rock off the ceiling or walls of a tunnel. These explosions happen because the rock surrounding the tunnel endures such great pressure that it sometimes spontaneously fractures.

TAKE-HOME MESSAGE

To find ore deposits, geologists study outcrops for ore shows, drill holes, and make maps of gravity and rock magnetism. Mining takes place either in open-pit mines or in shafts and tunnels underground. The process can be expensive and dangerous.

QUICK QUESTION: How can geologists confirm that an ore deposit lies deep beneath a given location?

15.5 Nonmetallic Mineral Resources

So far, this chapter has focused on resources that contain metal. But humans use many other geologic materials, commonly known as **industrial minerals**, as well. For example, the stone in roadbeds and buildings, the chemicals in fertilizers, the gypsum in drywall, the salt in saltshakers, and the clay in brick all come from the Earth. This section looks at a few of these nonmetallic materials and explains their origin **(Geology at a Glance**, pp. 580–581).

Dimension Stone

The Parthenon, a colossal stone temple rimmed by 46 carved columns, has stood atop a hill overlooking the city of Athens for almost 2,500 years. No wonder—*stone*, an architect's word for rock, outlasts nearly all other construction materials. We use stone to make facades, roofs, curbs, steps, countertops, and floors. We value stone for its visual appeal as well as its durability. The names that architects give to various types of stone may differ from the formal rock names that geologists use. For example, architects refer to any polished rock containing carbonate minerals as "marble," whether or not it has been metamorphosed. Likewise, they refer to any rock containing silicate minerals as "granite," regardless of whether the rock has an igneous or a metamorphic texture, or a felsic or mafic composition.

To extract intact slabs and blocks of rock—known as **dimension stone** for architectural purposes—workers must carefully cut rock out of the walls of quarries **(Fig. 15.13a, b)**. (Note that a *quarry* provides stone, whereas a *mine* supplies ore.) Quarry operators split rock blocks from bedrock either by hammering a series of wedges into the rock until a new crack propagates or by cutting directly through the bedrock with a wireline saw, a thermal lance, or a water jet. A *wireline saw* consists of a loop of braided wire moving between two pulleys. As the wire moves along the rock surface, the quarry operator spills an abrasive (sand or garnet grains) and water onto the wire. The movement of the wire drags the abrasive along the rock and grinds a slice into it. In some cases, a quarry operator uses a diamond-coated wire, cooled with water, to make cuts. A *thermal lance* looks like a long blowtorch—it produces an ultra-intense flame of burning diesel fuel stoked by high-pressure air that can cut a slot in rock. More recently, quarry operators have begun using *abrasive water jets*, which squirt out water and abrasive at extremely high pressure, to cut rock.

Crushed Stone and Concrete

Crushed stone forms the substrate of highways and railroads and serves as the raw material for manufacturing cement, concrete, and asphalt. In crushed-stone quarries **(Fig. 15.13c, d)**, operators use high explosives to break up bedrock into rubble that they then transport by truck to a crusher, which reduces the rubble into usable-sized chunks.

Most of the buildings and highways constructed in the past two centuries consist of walls, floors, columns, and roads made of *concrete* that has been spread into a layer or poured into a form. Or, they are made of bricks attached to one another by *mortar*. Both concrete and mortar are rock-like substances

BOX 15.1 CONSIDER THIS . . .

The Amazing Chilean Mine Rescue of 2010

The San José mine, near Copiapó in northern Chile, penetrates an ore-bearing body of diorite (an intermediate igneous rock) intruded during the Cenozoic. Like many other mines in the Andes, the mine produces copper and gold. The mine operators dug a long tunnel that spirals downward at a gentle slope to provide access into and out of the mine for miners, supplies, and ore. The spiral tunnel intersects the ore body at depths between 150 and 800 m below the ground surface. From this ramp, miners cut horizontal tunnels from which they removed ore. They also dug out "rooms" for repair shops and for shelters.

On August 5, 2010, at a depth of about 500 m below the surface, a catastrophic rock fall suddenly blocked a portion of the spiral tunnel with thousands of tons of debris. The fall isolated a group of 33 miners who were working near the bottom of the mine. The miners managed to retreat to a refuge room at a depth of over 688 m below the surface. Immediately, in an amazing example of self-reliance and strength, the miners organized into a functioning group under the leadership of their foreman, carefully rationed emergency food supplies, dug small wells to provide water, and established a daily routine in order to survive as they waited for rescue.

For 17 days, the miners sweltered in the humid, 35°C air of their refuge, while unbeknownst to them, rescuers were frantically drilling exploratory holes in hopes of reaching the miners to provide ventilation and communication. Finally, a drill broke through the ceiling of the refuge. The miners tapped on the pipe to indicate that they were alive and taped a paper message to the end of the drill bit, announcing to the world, "We are all right in the shelter—all 33." With contact established, rescuers drilled a wider hole that allowed them to send down supplies, letters, and even a video link.

Immediately, mining engineers at the surface set about the task of drilling a rescue hole wide enough for miners to fit through. Using a percussion drilling machine manufactured by a Pennsylvania supplier, drilling progressed at a rate of 40 m per day. It sent a cascade of rock debris down the ventilation hole into the refuge room. To keep the room open, the trapped miners had to haul away the debris—all 700 tons of it.

Once the rescue hole was complete, the engineers sent down a rescue capsule that could bring up one miner at a time. Before stepping into the cylinder, each miner donned sunglasses so the brightness of light at the Earth's surface would not harm his eyes. One by one, the miners were pulled to safety on a journey that took up to 18 minutes. As a huge crowd waited at the surface, and an audience of 1 billion people watched on TV worldwide, the miners emerged after having been underground for 69 days (Fig. Bx15.1).

FIGURE Bx15.1 The rescue capsule emerges at the surface.

consisting of **aggregate** (sand, gravel, or crushed rock) held together by **cement** (a complex group of minerals that precipitate from a water solution). In effect, we can think of concrete and mortar as human-made sedimentary rock **(Box 15.2)**.

Where does the cement in concrete or mortar come from? If you go to a building supply store, you'll find it in the form of a powder, usually packaged in tough paper bags. The powder consists of lime (CaO), quartz (SiO_2), aluminum oxide (Al_2O_3), and iron oxide (Fe_2O_3). Typically, lime accounts for 66% of the volume of cement powder, silica for 25%, and the other chemicals together for about 9%. When you mix this powder with water, the chemicals within it undergo reactions (mostly hydration), and a set of new minerals begins to grow. These minerals interlock as they grow, producing a hard solid. If you mix the water and cement solution with aggregate in a cement mixer and then let the resulting slurry set, the newly growing minerals bind grains of aggregate to one another.

The ancient Romans may have been the first people to use cement. They made it from a mixture of volcanic ash and limestone. In the 18th and early 19th centuries, workers produced the powder for making cement by heating specific types of

> **Did you ever wonder...**
> how concrete differs from rock?

FIGURE 15.13 Stone production in quarries.

(a) An active quarrying operation in Missouri that produces large blocks of cut dimension stone.

(b) Sheets of cut dimension stone being measured for cutting to become a kitchen countertop.

(c) A large crushed-stone quarry in Silurian limestone of Illinois. Drillers are working on the shelf in the distance.

(d) A large truck hauls debris from a quarry blast to the crusher.

limestone in a kiln up to a temperature of about 1,450°C. The heating releases CO_2 gas and produces "clinker," chunks consisting of lime and other oxide compounds. Workers then crushed the clinker into cement powder and packed it in bags for transport. The special limestone from which such *natural cement* can be produced is fairly rare, for it has to contain calcite, clay, and quartz in just the right proportions to yield the chemicals needed to make the minerals in cement. Therefore, since the mid-19th century we have generally used **Portland cement**, made by physically mixing crushed limestone, sandstone, and shale from different geologic formations in the correct ratios. Isaac Johnson, an English engineer, came up with the recipe for Portland cement in 1844, and named it after the town of Portland, England, because he thought that, when set, it resembled rock exposed there.

Nonmetallic Minerals for Homes and Farms

We use an astounding variety of nonmetallic geologic resources (Table 15.2) without ever realizing where they come from. Consider the materials in a typical house or apartment. We've already talked about how ore deposits provide metals, such as the copper in a house's electrical wiring or pipes, and the iron in household tools and appliances. We've also seen how concrete in the foundation, floors, and walls is made from limestone mixed with sand or gravel. Let's now look at some other components.

The *bricks* in exterior walls originated as clay, formed from the chemical weathering of silicate rocks and perhaps dug from the floodplain of a stream. To make bricks, workers mold wet clay into blocks and then bake it. Baking drives out water and causes metamorphic reactions that recrystallize the clay

BOX 15.2 CONSIDER THIS . . .

The Sidewalks of New York

Untold tons of concrete have gone into the construction of New York City. In fact, with the exception of a few city parks, most of the walking space in the city consists of concrete (Fig. Bx15.2). And concrete skyscrapers tower above the concrete plain. Where does all this concrete come from?

Much of the sand used in New York concrete was deposited during the last ice age. As vast glaciers moved southward over 14,000 years ago, they ground away the igneous and metamorphic rocks that constituted central and eastern Canada. These ancient rocks contained abundant quartz, and since quartz lasts a long time (it does not undergo chemical weathering easily), the sediment transported by the glaciers retained a large amount of quartz. Glaciers deposited this sediment in huge piles called moraines (see Chapter 22). As the glaciers melted, fast-moving rivers of meltwater washed the sediment, sorting sand from mud and pebbles. The sand was deposited in bars in the meltwater rivers, and these relict bars now provide thick lenses of sand that can be economically excavated.

What about the cement? Cement contains a mixture of lime, derived from limestone, and other elements (such as silica) derived from shale and sandstone. The bedrock of New York City, though, consists largely of schist and gneiss, not sedimentary rocks. Fortunately, a source of rocks appropriate for making cement lies up the Hudson River. A Silurian rock unit exposed in low hills just west of the river, called the Rosendale Formation, naturally contains exactly the right mixture of lime and silica needed to make durable cement. Beginning in the late 1820s, workers began quarrying the Rosendale Formation for cement. Quarry operators followed the Rosendale beds closely, making horizontal mine tunnels where the beds were horizontal, tilted mine tunnels where the beds tilted, and vertical mine tunnels where the beds were vertical. They then dumped the excavated rock into nearby kilns and roasted it to produce lime mixed with other oxides. The resulting powder was packed into barrels, loaded onto barges, and shipped downriver to New York City. As demand for cement increased, operators eventually dug open-pit quarries from which they excavated Devonian limestone and shale units, mixing them together in the correct proportion to make Portland cement.

The rocks making up the strata that provide the source materials for cement consist of shell fragments and small, reef-like colonies of organisms. In other words, the lime in the concrete of New York sidewalks was originally extracted from seawater by living organisms—brachiopods, crinoids, and bryozoans—over 400 million years ago.

FIGURE Bx15.2 The modern production of concrete. During this process, natural rock undergoes transformation into human-made stone.

(a) A limestone quarry. The trucks are carrying rock to a crusher.

(b) The heat of a kiln transforms limestone into lime (CaO).

(c) A concrete mixer pouring wet concrete.

(d) A sidewalk in New York City.

(Fig. 15.14a). Clay also serves as the raw material of pottery, porcelain, and other ceramic materials (see Chapter 8). The *glass* used to glaze windows consists largely of silica, formed by first melting and then freezing pure quartz sand from a beach deposit or a sandstone formation. Quartz may also be used in the construction of photovoltaic cells for solar panels. *Gypsum board* (drywall), used to construct interior walls, comes from a slurry of water and the mineral gypsum sandwiched between sheets of paper. Gypsum ($CaSO_4 \cdot 2H_2O$) occurs in evaporite strata precipitated from seawater or saline lake water. Evaporites provide

Forming and Processing the Earth's Mineral Resources

Mining and processing ore has environmental consequences, including acid runoff, acid rain, and groundwater contamination.

Ore deposits can be obtained either in strip mines or in underground mines.

Circulating, hot groundwater may extract and concentrate metals to form ore deposits.

Clay, when formed into blocks and baked, becomes brick.

Gravel itself may be quarried for construction purposes.

Mud, a mixture of clay minerals and water, accumulates in beds.

Ore minerals may collect on the bottom of a magma chamber.

Miners pan for gold in placer deposits where metal flakes and nuggets occur in sand and gravel.

Hydrothermal vents (black smokers) produce accumulations of sulfides.

From Mud to Brick

From Magma to Metal

Erosion tears down mountains and produces gravel and sand.

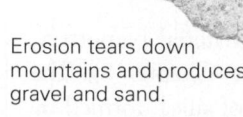

From Stream Channel to Roadbed

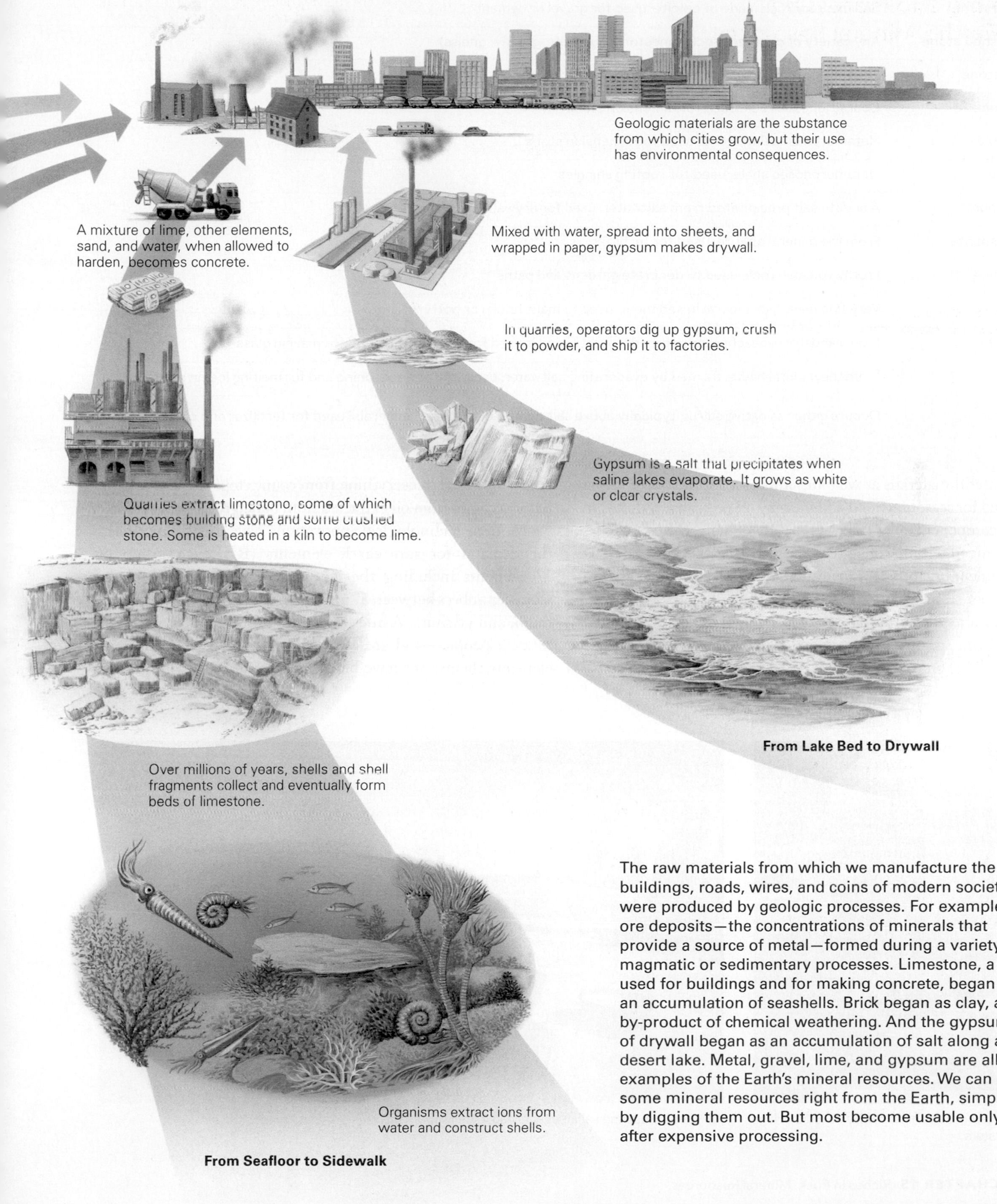

Geologic materials are the substance from which cities grow, but their use has environmental consequences.

A mixture of lime, other elements, sand, and water, when allowed to harden, becomes concrete.

Mixed with water, spread into sheets, and wrapped in paper, gypsum makes drywall.

In quarries, operators dig up gypsum, crush it to powder, and ship it to factories.

Quarries extract limestone, some of which becomes building stone and some crushed stone. Some is heated in a kiln to become lime.

Gypsum is a salt that precipitates when saline lakes evaporate. It grows as white or clear crystals.

Over millions of years, shells and shell fragments collect and eventually form beds of limestone.

From Lake Bed to Drywall

Organisms extract ions from water and construct shells.

From Seafloor to Sidewalk

The raw materials from which we manufacture the buildings, roads, wires, and coins of modern society were produced by geologic processes. For example, ore deposits—the concentrations of minerals that provide a source of metal—formed during a variety of magmatic or sedimentary processes. Limestone, a rock used for buildings and for making concrete, began as an accumulation of seashells. Brick began as clay, a by-product of chemical weathering. And the gypsum of drywall began as an accumulation of salt along a desert lake. Metal, gravel, lime, and gypsum are all examples of the Earth's mineral resources. We can use some mineral resources right from the Earth, simply by digging them out. But most become usable only after expensive processing.

TABLE 15.2 Common Nonmetallic Resources

Limestone	Sedimentary rock made of calcite; used for gravel or cement
Crushed stone	Any variety of coherent rock (limestone, quartzite, granite, gneiss)
Siltstone	Beds of sedimentary rock; used to make flagstones for patios
Granite	Coarse igneous rock; used for dimension stone
Marble	Metamorphosed limestone; used for dimension stone
Slate	Metamorphosed shale; used for roofing shingles
Gypsum	A sulfate salt precipitated from saltwater; used for drywall
Phosphate	From the mineral apatite; used for fertilizer
Pumice	Frothy volcanic rock; used to decorate gardens and paths
Clay	Very fine mica-like mineral in sediment; used to make bricks or pottery
Sand	From sandstone, beaches, or riverbeds; quartz sand is used for construction and for making glass
Salt	From the mineral halite, formed by evaporating saltwater; used for food seasoning and for melting ice on roads
Sulfur	Occurs either as native sulfur, typically above salt domes, or in sulfide minerals; used for fertilizer and chemicals

other useful materials as well, including *halite* (the source of the salt used for seasoning food and thawing ice on roads) and *lithium* (a key component of high-tech batteries). Of note, the largest currently mined resource of lithium occurs in dry lakebeds of Bolivia and Chile (**Fig. 15.14b**). Finally, recall that *asbestos*, once used to make roof shingles, floor tiles, and brake pads, comes from serpentine, a rock resulting from the reaction of olivine with water. (The olivine, in turn, occurs in ophiolites, slices of oceanic lithosphere thrust onto continental crust during continental collisions.)

And *plastic* used in everything from countertops to pipes to light fixtures comes from oil extracted from underground reserves.

Modern technological innovations have greatly increased the demand for **rare earth elements** (**REEs**), a group of 17 elements including the lanthanides (elements that have atomic numbers between 57 and 71 on the periodic table), scandium, and yttrium. While the names of REEs are unfamiliar to most people—and are also quite hard to pronounce—the elements themselves have become essential in the production

FIGURE 15.14 Clay and salt can be used for many products.

(a) These bricks consist of clay baked at a high temperature.

(b) Lithium-bearing salts being mined in Bolivia.

of lasers, magnets, X-ray tubes, night-vision goggles, camera lenses, high-tech lamps, and cell phones. REEs are not actually that rare in terms of their abundance relative to other elements in the Earth's crust. But localities where ores have a high enough concentration to be mined are rare. Most REEs used today come from strip mining either granitic plutons that contain REE-rich veins, or sediments and soils (residual deposits) derived from such plutons.

Chemicals employed as *fertilizer* also come from the ground. For example, potash (K_2CO_3) comes from the minerals in evaporite deposits. Phosphate (PO_4^{-3}) comes from the mineral apatite, which crystallizes in organic-rich muds that were deposited in shallow, oxygen-free (anoxic) seawater. Much of the phosphate now used in the United States has been extracted by strip mining 10-Ma shale beds that lie about 15 m below the land surface of Florida. Smaller quantities come from mines in a Permian stratigraphic unit called the Phosphoria Formation, which underlies portions of the western United States.

As you can see, geologic processes acting over thousands to billions of years provide many of the material goods used in homes, farms, and industry. Truly, without the geologic resources of the Earth, modern society would grind to a halt.

TAKE-HOME MESSAGE

Society uses a great variety of nonmetallic geologic materials. These materials include dimension stone, crushed stone, cement (made from roasted limestone), evaporites (including gypsum), and clay (to make bricks).

QUICK QUESTION: What is the difference between natural cement and Portland cement?

15.6 Global Mineral Needs

How Long Will Resources Last?

Each person in the United States uses about 600 kg of metallic resources, about 9,400 kg of nonmetallic resources each year (Fig. 15.15), and about 7,600 kg of energy resources (Table 15.3). The total of these numbers means that each person in the U.S. uses about 17,000 kg (17 metric tons) of energy and mineral resources each year. To obtain this material, workers must mine, quarry, or pump out about 20 billion metric tons of Earth materials each year. (By comparison, the Mississippi River carries about 0.2 billion metric tons of sediment in a given year.) Clearly, industrialized countries, such as the United States, churn through Earth materials at geologic rates. We've already discussed the sustainability of energy resources, given this rate of consumption (see Chapter 14). Now, let's consider the sustainability of mineral resources.

Mineral resources are nonrenewable—once mined, an ore deposit or a limestone hill disappears forever. Natural geologic processes do not happen fast enough to replace the deposits as quickly as we use them. Geologists have calculated **reserves** (known quantities of a commodity still in the ground) for various mineral deposits, just as they have for oil. Based on current definitions of reserves, which depend on today's prices and rates of consumption, supplies of some metals may run out in only decades to centuries (Table 15.4). But these estimates may change as supplies become depleted and prices rise, making previously uneconomical deposits worth mining. And supplies could increase if geologists discover new reserves or if new ways

---

FIGURE 15.15 Industrialized countries consume vast quantities of mineral resources in a year, as the diagram indicates. The numbers indicate the weight of the material used per person per year.

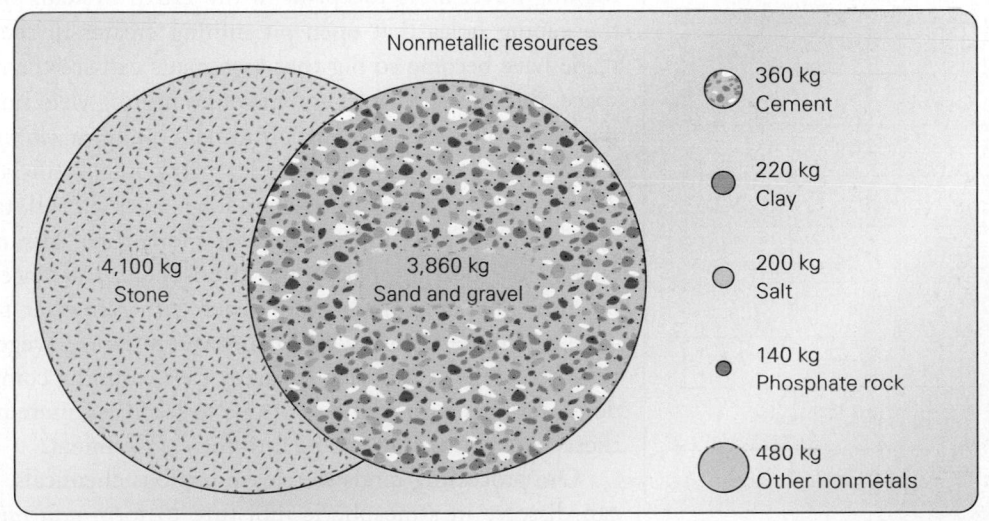
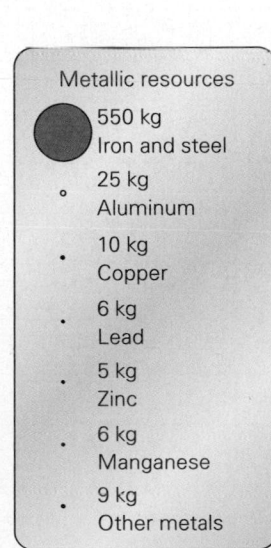

TABLE 15.3 Annual Per Capita Use of Earth Materials in the United States

Material	Weight Used (kg)
Stone	4,100
Sand and gravel	3,860
Petroleum	3,050
Coal	2,650
Natural gas	1,900
Iron and steel	550
Cement	360
Clay	220
Salt	200
Phosphate	140
Aluminum	25
Copper	10
Lead	6
Zinc	5

1 kg = 2.205 pounds; 1,000 kg = 1 metric ton.

TABLE 15.4 Expected Lifetimes of Currently Known Ore Reserves (in Years)

Metal	World Resources	U.S. Resources
Iron	120	40
Aluminum	330	2
Copper	65	40
Lead	20	40
Zinc	30	25
Gold	30	20
Platinum	45	1
Nickel	75	<1
Cobalt	50	<1
Manganese	70	0
Chromium	75	0

of mining become available. Further, increased efforts at conservation and recycling can cause a dramatic decrease in rates of consumption and thereby stretch the lifetime of existing reserves.

Ore deposits do not occur everywhere because, as we've seen, their formation requires specific geologic conditions. As a result, some countries possess vast supplies, whereas others have none. In fact, no single country owns all the mineral resources it needs, so nations must trade with each other to maintain supplies, and global politics inevitably affects prices. Many wars have their roots in competition for mineral reserves, and it is no surprise that the outcomes of some wars have hinged on who controls these reserves.

The United States worries in particular about supplies of **strategic minerals**, which include manganese, platinum, chromium, and cobalt—metals alloyed with iron to make the special-purpose steels needed in the aerospace industry. At present, the country must import 100% of its supply of many strategic minerals. Principal reserves of these metals lie in the crust of countries that have not always practiced open trade with the United States. As a defense precaution, the United States stockpiles these metals in case supplies are cut off.

We've already noted that many sources of lithium occur in nonindustrialized countries, and that industrialized countries are dependent on this material. The same is true of REE reserves, which have become a subject of international tension in recent years. Currently, about 90% of the mined supply occurs in China. But China has started to limit its exports, which has led other countries to reactivate mines. Mining of REEs is challenging because it often occurs in association with radioactive elements, which should require miners to use special precautions to prevent environmental contamination.

Mining and the Environment

Mining leaves a big footprint in the Earth System. Some of the gaping holes that open-pit mining creates in the landscape have become so big that astronauts can see them from space. Both open-pit and underground mining yield immense quantities of waste rock, which miners dump in *tailings piles* **(Fig. 15.16a)**. Some tailings piles grow into artificial hills 200 m high and many kilometers long. Lacking soil, tailings piles tend to remain unvegetated for a long time. Mining also exposes ore-bearing rock to the atmosphere, and since many ore minerals are sulfides, they react with rainwater to produce **acid mine runoff**, which can severely damage vegetation downstream **(Fig. 15.16b)**. In many cases, mining companies douse tailings with acidic solutions to leach out more metals; these acids sometimes escape into the environment.

Ore processing tends to release noxious chemicals, which can dissolve in atmospheric moisture to form *acid rain* that

FIGURE 15.16 Environmental consequences of producing metallic mineral resources.

(a) The tailings piles of the Bingham open-pit copper mine in Utah. The pit is 1.2 km deep and 4 km wide. During its century of operation, the mine has yielded about 17 million tons of copper, 700 tons of gold, and 6,000 tons of silver. The ore formed due to hydrothermal fluid circulation that accompanied igneous activity about 36 Ma.

Much of the color comes from bacteria and archaea living in the water.

(b) The orange color of this acid mine runoff is due to iron and sulfide in the water.

Smelter   Tailings pile

(c) Acidic smelter smoke killed off vegetation near Sudbury, Ontario, in the 1970s. A large tailings pile can be seen in the distance.

can harm vegetation. Before the installation of modern environmental controls, smoke from ore-processing plants caused severe air pollution. Plumes of smoke from the old smelters in Sudbury, Ontario, for example, created a wasteland for many kilometers downwind **(Fig. 15.16c)**.

Recent years have seen efforts to reclaim mining waste, and new technologies have been developed to extract metals in ways that are less deleterious to the environment and that treat waste more efficiently. Clearly, mining and ore processing can leave scars on the landscape, the size of which depends on the willingness of producers and consumers to minimize damage and the degree to which regulations are successfully designed and enforced.

TAKE-HOME MESSAGE

People in industrialized countries use a vast quantity of mineral resources over the course of a lifetime. Mineral resources are nonrenewable, and because reserves are not distributed uniformly around the planet, the trade in economic minerals is politically charged. Mineral extraction has significant environmental consequences that are challenging to address.

QUICK QUESTION: If one country restricts the export of a strategic mineral, would that change the minimum concentration of the mineral necessary to make an ore deposit elsewhere worth mining? Why?

SUMMARY

- Industrial societies use many types of minerals, all of which must be extracted from the upper crust. We distinguish metallic resources from nonmetallic resources.

- In metals, atoms are held together by metallic bonds. Metals are malleable and make good conductors.

- Metals come from ores, rocks containing native metals or ore minerals (minerals that contains a high proportion of metal) in sufficient quantities to be worth mining. An ore deposit is an accumulation of ore.

- Magmatic ore deposits form when sulfide ore minerals accumulate during solidification of magma. In hydrothermal deposits, ore minerals precipitate from hot-water solutions. Secondary-enrichment deposits form when groundwater carries metals away from a pre-existing deposit. MVT deposits precipitate from groundwater that has passed long distances through sedimentary basins. Sedimentary deposits

precipitate out of the ocean. Residual mineral deposits are the result of severe leaching in tropical soils. Placer deposits develop when heavy metal grains accumulate in sediment along a stream.

- Many ore deposits are associated with igneous activity in subduction zones, along mid-ocean ridges, along continental rifts, or at hot spots.

- Nonmetallic resources include dimension stone for architectural purposes, crushed stone for cement and asphalt production, clay for brick making, sand for glass production, and many other materials. A large proportion of materials in your home have a geologic ancestry.

- Mineral resources are nonrenewable. Many are now or may soon be in short supply.

- The production and processing of mineral resources can harm the environment if not done carefully.

GUIDE TERMS

acid mine runoff (p. 584)
aggregate (p. 577)
alloy (p. 566)
banded iron formation (BIF) (p. 572)
cement (p. 577)
dimension stone (p. 576)
grade (p. 568)
hydrothermal deposit (p. 570)

industrial mineral (p. 576)
magmatic deposit (p. 569)
manganese nodule (p. 572)
massive sulfide deposit (p. 569)
metal (p. 565)
mineral resource (p. 565)
Mississippi Valley–type (MVT) deposit (p. 571)
native metal (p. 566)

open-pit mine (p. 574)
ore (p. 568)
ore deposit (p. 568)
ore mineral (economic mineral) (p. 568)
placer deposit (p. 572)
Portland cement (p. 578)
rare earth element (REE) (p. 582)
reserve (p. 583)

residual mineral deposit (p. 572)
secondary-enrichment deposit (p. 571)
smelting (p. 566)
strategic mineral (p. 584)
underground mine (p. 576)

REVIEW QUESTIONS

The letters following each Review Question refer to the corresponding Learning Objective from the Chapter Opener.

1. Name some materials in your home that come from Earth materials. **(A)**

2. Why did people use stone weapons before using bronze weapons? **(B)**

3. What's the difference between an ore mineral and other minerals, and between an ore and other kinds of rock? **(B)**

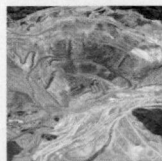

GEOTOURS *THIS CHAPTER'S GEOTOURS WORKSHEET (L) FEATURES QUESTIONS AND GOOGLE EARTH SITES ON:*

- Mineral resources
- Bingham Copper Mine, Utah

4. What types of ore deposits form in association with igneous activity? **(C)**

5. Which type of ore deposit does the figure show? **(C)**

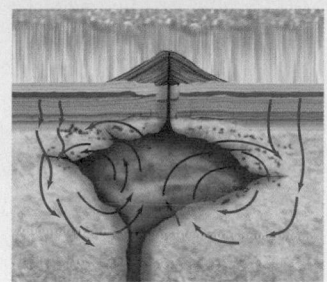

6. Explain how ores can occur or develop in sedimentary rocks. **(C)**

7. How do residual ore deposits form? **(C)**

8. What procedures are used to locate and mine mineral resources today? **(D)**

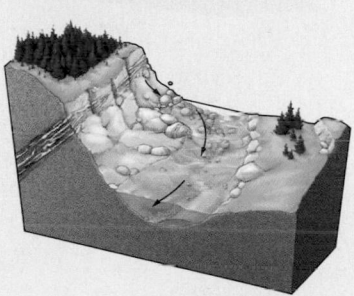

9. What type of ore deposit does this figure show? Describe the steps in the formation. **(C)**

10. How can dimension stone be obtained from a quarry? **(D)**

11. What are the ingredients of cement? How is Portland cement made? **(E)**

12. How many kilograms of Earth materials does the average person in an industrialized country use in a year? **(F)**

13. What are strategic minerals, and why have they become a political issue? **(F)**

14. What are some environmental hazards of large-scale mining? **(G)**

ON FURTHER THOUGHT

15. The costs of mining can be immense. To get a rough sense of this expense, imagine that an ore deposit of a certain metal contains 0.6% grade ore. This means that 0.6% by weight of a block of ore consists of the metal. On the open market, the pure metal sells for $8,000/ton. It costs $15/ton to mine the ore, $15/ton to transport the ore to the processing plant, and $15/ton to process the ore and produce pure metal. Start-up costs (building the mine and building the processing facility) are about $100 million. How much profit does the company make when it sells a ton of metal? How much ore (in tons) does the operation have to mine to pay back the start-up costs? Considering that a giant dump truck can carry 200 tons of ore at a time, how many dump-truck loads will have been transported at the break-even point? If the mine has eight trucks that can each transport six loads a day, about how many years will it take to break even? **(D)**

16. An ore deposit at a location in Arizona has the following characteristics: One portion of the ore deposit is an intrusive igneous rock in which tiny grains of copper sulfide minerals are dispersed among the other minerals of the rock. Another nearby portion of the ore deposit consists of limestone in which malachite fills cavities and pores in the rock. What types of ores are these? Describe the geologic history that led to the formation of these deposits. **(C)**

ONLINE RESOURCES

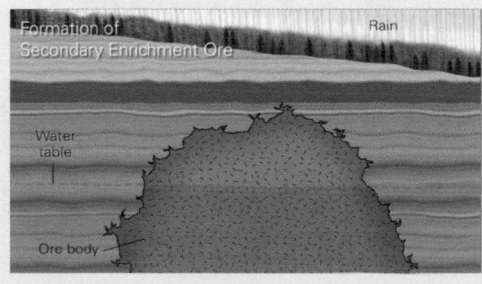

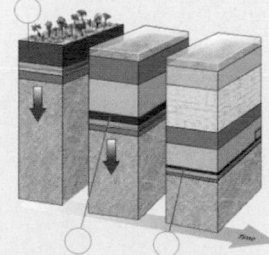

Videos
This chapter features videos on the formation of various ore deposits and the operation of a quarry.

Smartwork5
Questions include checks on a basic understanding of energy technologies and the formation of coal and oil.

588

PART VI

PROCESSES AND PROBLEMS AT THE EARTH'S SURFACE

In the last part of this book, we focus on the Earth's surface and near-surface realms, a portion of the Earth System that encompasses the interface among the lithosphere, hydrosphere, atmosphere, and biosphere. Here, the dynamic interplay between *internal processes* (driven by the Earth's internal heat) and *external processes* (driven by the warmth of the Sun) under the influence of the Earth's gravitational field has resulted in a diverse array of landscapes. In Chapters 16 through 22, we examine five of these landscapes, plus groundwater and the atmosphere. Finally, in Chapter 23, we see how forces at work in the Earth System—including human activities—cause the planet to change over time. These changes affect the *critical zone*, the realm that supports life.

◄ Landscapes can display dramatic and beautiful shapes. Here, at Zabriskie Point, near Death Valley, California, occasional heavy rains feed short-lived streams that carve intricate valleys.

Ever-Changing Landscapes and the Hydrologic Cycle

By the end of this interlude, you should be able to . . .

A. explain what a landscape is and what questions geologists ask about landscapes.

B. explain the difference between uplift and subsidence, and describe the forces that drive them.

C. contrast internal and external energy in the Earth System.

D. differentiate between erosional and depositional landscapes.

E. discuss the reservoirs and exchange processes of the hydrologic cycle.

Nothing that is can pause or stay–
The moon will wax, the moon will wane,
The mist and cloud will turn to rain,
The rain to mist and cloud again,
Tomorrow be today.

—Henry Wadsworth Longfellow (American poet, 1807–1882)

F.1 Introduction

The Earth's surface is a place of both endless variety and intricate detail. Observe the height of its mountains, the expanse of its seas, the desolation of its deserts, and you may be inspired, frightened, or calmed. It's no wonder that artists and writers from across the ages have sought inspiration from the **landscape**—the character and shape of the land surface in a region—for landscapes encompass the diversity of human emotion **(Fig. F.1)**. Geologists, like artists and writers, savor the impression of a dramatic landscape. But on seeing one, they can't help but ask, "How did it come to be, and how will it change in the future?"

This interlude introduces the driving forces behind landscape development and sets the stage for interpreting distinct **landforms**, the individual shapes (such as valleys, cliffs, fans, mesas, and beaches) that make up the landscapes which we describe later in Part VI. We also introduce the *hydrologic cycle*—the pathway water molecules follow as they move from ocean to air to land and back to ocean—later in Part VI you will see that water, in both its liquid and solid forms, serves many roles in surface and near-surface processes on the Earth. We conclude by taking a glimpse at landscapes found on other planets.

F.2 Shaping the Earth's Surface

Uplift and Subsidence; Erosion and Deposition

If the Earth's surface were totally flat, the great diversity of the landscapes that embellish our vistas would not exist. But the surface isn't flat, because a variety of geologic processes can cause one portion of the surface to move up or down relative to

an adjacent region. We refer to the relative upward movement of a region as **uplift** and the relative downward movement of a region as **subsidence**. Both uplift and subsidence occur for a variety of reasons, as outlined in **Table F.1**.

Cartographers and geologists refer to variations in land-surface elevation as **topography**. We can represent variations in elevation on a *topographic map* **(Box F.1)**, or on a *shaded-relief map*, which conveys the impression of three dimensions by shading appropriate slopes to appear as if they are in shadows cast when the Sun is low in the sky **(Fig. F.2a)**. In recent years, geologists have used radar beamed from satellites to produce highly detailed digital data on elevation variations on the Earth's surface. The resulting data sets, called **digital elevation models** (DEMs), can be analyzed with a computer to produce a shaded and colored image that gives the impression of three dimensions **(Fig. F.2b)**. The digital image can also be tilted to portray an oblique view **(Fig. F.2c)**.

TABLE F.1 Causes of Uplift and Subsidence

Causes of Uplift
• **Thickening of the crust.** At convergent and collisional boundaries, compression causes the crust to shorten horizontally (by development of folds, faults, and foliations) and thicken in the vertical direction. Because of isostasy (see Chapter 11), lithosphere with thickened crust rests relatively higher on the asthenosphere, with the result that the surface of the crust in mountain belts rises. Intrusion or extrusion of igneous rocks can also cause uplift, for these processes thicken the crust, or can build volcanoes on top of the crust.
• **Heating of the lithosphere.** Heating decreases the thickness and density of lithosphere, so in order to maintain isostatic equilibrium, lithosphere floats higher.
• **Rebound due to unloading.** Removal of a heavy load (such as a glacier or mountain) from the surface causes the Earth's surface to rise in a manner similar to the way a trampoline's surface rises when you step off it.
• **Delamination.** If dense lithospheric mantle separates from the base of a plate and sinks into the mantle, the surface of the lithosphere rises. The effect resembles the consequence of unloading ballast from a ship.

Causes of Subsidence
• **Thinning of the crust due to stretching.** In rifts, where the crust undergoes horizontal stretching, the axis of the rift drops down by slip on normal faults.
• **Cooling of the lithosphere.** Cooling thickens the lithospheric mantle and makes it denser, so to maintain isostatic equilibrium, the lithosphere sinks downward, and its surface lies at a lower elevation.
• **Sinking due to loading.** Where a heavy load (such as a glacier or volcano) forms on the Earth's surface, the lithosphere warps downward, in a manner similar to the way the surface of a trampoline bends down when you stand on it.

◀ (facing page) We see key stages of the hydrologic cycle along the coast of Skye, Scotland. Water evaporates from the sea to form clouds, which rain on the land. Some of the water sinks into the soil and becomes incorporated in plants; some flows down streams back to the sea.

FIGURE F.1 Examples of the great variety of landscapes on the Earth.

(a) Rounded "sugarloaf" mountains surround Rio de Janeiro, Brazil.

(b) Glaciated peaks of the Alps, France.

(c) Buttes of sandstone, Monument Valley, Arizona.

(d) Cliffs rise from the forest in the Blue Mountains, Australia.

(e) The Amazon River, Peru.

(f) Sandy beaches of Cape Cod, Massachusetts.

Topographic Maps and Profiles

We can distinguish one landform from another by its shape, as manifested by variations in elevation within a region. For example, as you will see in succeeding chapters, a river-carved valley simply does not look like a glacially carved valley. As we've noted, geologists use the term *topography* to refer to variations in elevation—effectively, the shape of the land surface.

How can we convey information about topography—a three-dimensional feature—on a two-dimensional sheet of paper? Cartographers and geologists do this by means of a **topographic map**, which uses contour lines to represent variations in elevation **(Fig. BxF.1a)**. A **contour line** is an imaginary line along which all points have the same elevation. For example, if you walked along the 200-m contour line on a hillslope, you would stay at exactly the same elevation (200 m). As another example, the shoreline on a flat, calm body of water is a contour line. In other words, you can picture a contour line as the intersection between the land surface and an imaginary horizontal plane. The elevation difference between two adjacent contour lines on a topographical map is called the **contour interval (Fig. BxF.1b)**. For a given topographic map, the contour interval stays constant, so the spacing between contour lines represents the steepness of a slope. Specifically, closely spaced contour lines represent a steep slope, whereas widely spaced contour lines represent a gentle slope.

We can also represent variations in elevation by means of a **topographic profile**, the trace of the ground surface as it would appear on a vertical plane that sliced into the ground. Put another way, a profile represents the shape of the ground surface as viewed from the side **(Fig. BxF.1c)**. If we add a representation of geologic features under the ground surface, then we have a **geologic cross section**. In some cases, geologists can gain insight into subsurface geology simply by looking at the shape of a landform **(Fig. BxF.1d)**. For example, a steep cliff in a region of dipping sedimentary strata may indicate the presence of a resistant (difficult to erode) layer; low areas may be underlain with nonresistant (easy to erode) layers.

FIGURE BxF.1 Topographic maps and profiles.

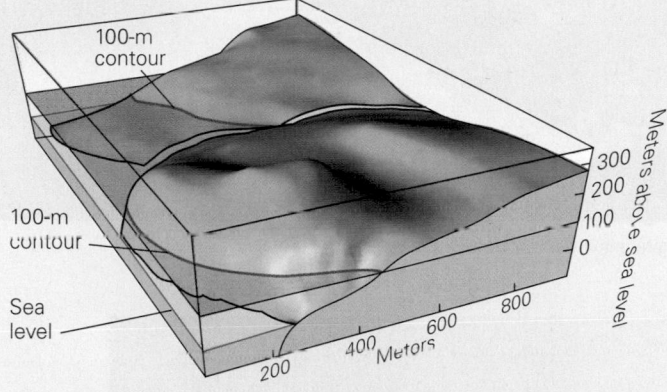

(a) A contour line is the intersection of a horizontal plane with the land surface. This block diagram shows the map area.

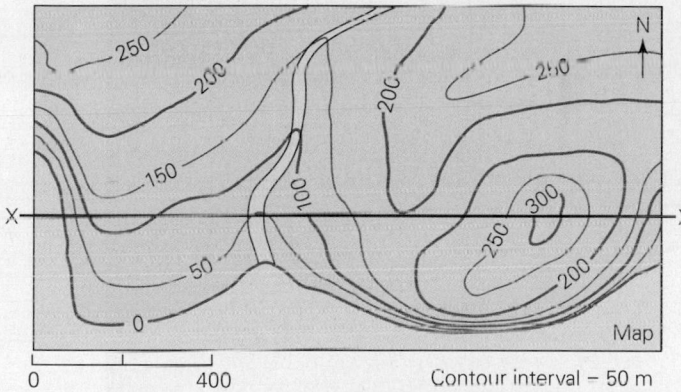

(b) A topographic map depicts the shape of the land surface through the use of contour lines. The difference in elevation between two adjacent lines is the contour interval.

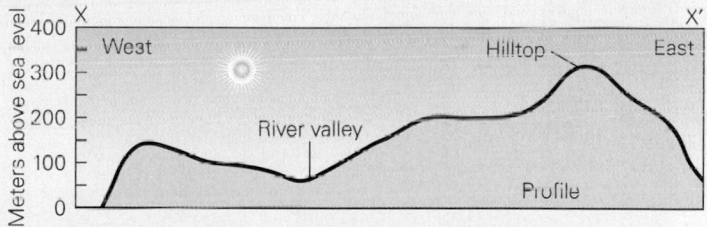

(c) A topographic profile (along section line X-X') shows the shape of the land surface as seen in a vertical slice.

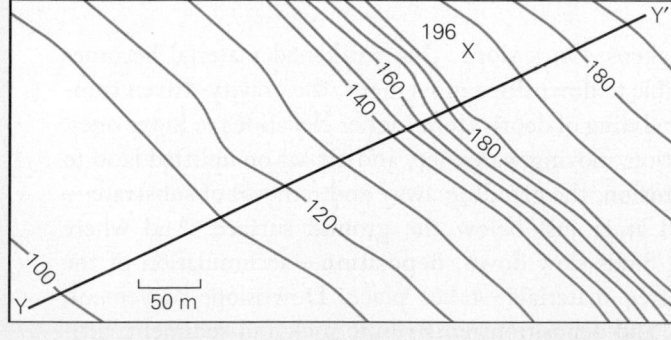

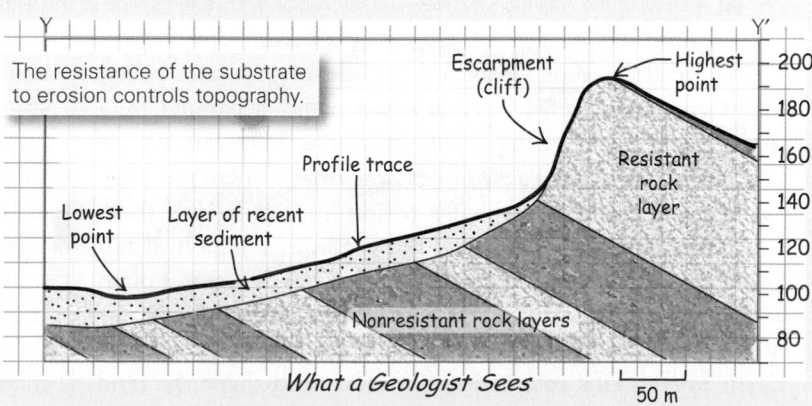

(d) This topographic map (left) shows a distinct cliff. A geologic cross section (right) depicts a geologist's interpretation of the subsurface along Y-Y'. The cliff is the edge of a resistant rock layer.

FIGURE F.2 Portraying the shape of the Earth's surface.

(a) A hand-painted shaded-relief map of Europe.

(b) An oblique digital elevation model (DEM) of Oahu, Hawaii.

(c) A DEM of the Atlantic City, New Jersey, region shows the shape of the land. Green areas are lower, orange areas are higher.

The occurrence of uplift and subsidence generates **relief**, the elevation difference between two points separated by a specified horizontal distance on a map **(Fig. F.3a)**. In discussion, we can say that a region has *high relief* if there are large elevation differences and that a region has *low relief* if there are small elevation differences. For example, a region of tall mountains cut by deep valleys has high relief, whereas a plain with a nearly flat land surface has low relief **(Fig. F.3b, c)**.

Wherever relief develops, various components of the Earth System kick into action to modify and shape the land surface. Rock at or near the land surface weathers, fractures, and weakens. On a slope, this weakened material becomes susceptible to **downslope movement**, the gravity-driven tumbling or sliding of debris from higher elevations to lower ones. In addition, moving water, ice, and air act on uplifted land to cause **erosion**, the grinding away and removal of **substrate**—material at or just below the ground surface. And where moving fluids slow down, **deposition**—accumulation of the transported materials—takes place. Downslope movement, erosion, and deposition redistribute rock and sediment, ultimately stripping it from higher areas and collecting it in low areas.

FIGURE F.3 The concept of relief.

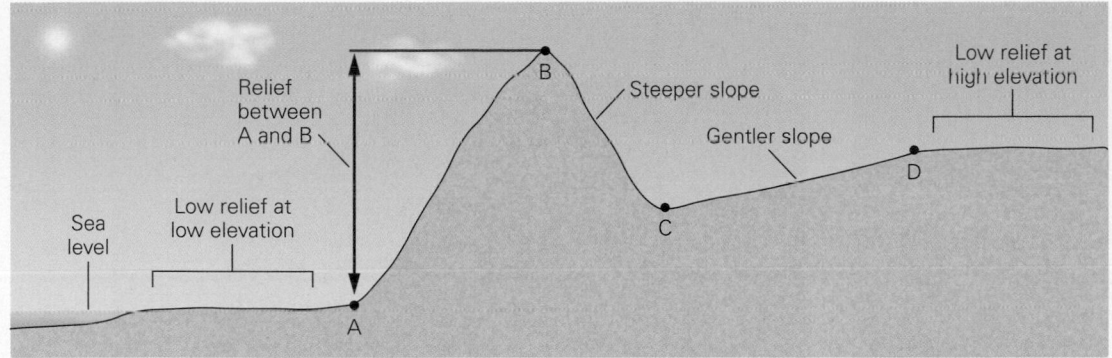

(a) The elevation difference between Points A and B on this profile is the relief of the region between those two points. The relief is steeper between B and C than between C and D.

(b) This mountainous area in Alaska has high relief.

(c) This plain in Ontario has low relief.

and depositional rates also vary between 0.10 and 10 mm per year. Although these rates seem small, a change in surface elevation of just 0.5 mm (the thickness of your fingernail) per year can yield a net change of 5 km in 10 million years. Uplift can build a mountain range, and erosion can whittle one down to near sea level—it just takes time!

What Drives Landscape Evolution?

The energy that drives landscape evolution comes from three sources: **internal energy**, the heat within the Earth, which ultimately keeps the asthenosphere hot and plastic enough that plates can move and interact and mantle plumes can rise; **external energy**, energy that comes to the Earth from the Sun, which warms the atmosphere and ocean; and **gravitational energy**, which exerts a downward pull on material and, along with external energy, causes the convection of water and air that yields winds and waves. In fact, we can think of landscape evolution as a "battle" between tectonic processes such as collision, convergence, rifting, and volcanism, caused by plate interactions and hot spots, which build relief by driving uplift or subsidence, and processes such as downslope movement, erosion, and deposition,

How rapidly do uplift and subsidence take place? The Earth's surface can rise or sink by as much as 3 m during a single major earthquake. But, averaged over time, rates of uplift and subsidence range between 0.01 and 10 mm per year **(Fig. F.4a)**. Similarly, erosion can carve out several meters of substrate during a single flood, storm, or landslide **(Fig. F.4b)**, and deposition in the aftermath of a single such event can produce a layer of debris several meters thick in a matter of minutes to days. But, averaged over time, erosional

which destroy relief by removing material from high areas and depositing it in low ones.

If, in a particular region, the rate of uplift exceeds the rate of erosion, the land surface rises, whereas if the rate of subsidence exceeds the rate of deposition, the land surface sinks. Without uplift and subsidence, erosion and deposition would have long ago transformed the Earth's surface into a flat plain. And without erosion and deposition, tectonically generated high and low areas would have lasted for the entirety of Earth history.

FIGURE F.4 The processes of uplift, subsidence, erosion, and deposition can be slow or rapid.

Uplifted terrace

New terrace forming

(a) Uplifted beach terraces form where the coast is rising relative to sea level. Present-day wave erosion is forming a new terrace and cutting a cliff on the edge of the old one.

(b) So much erosion can take place during a single hurricane that houses built along the beach become undermined.

F.3 Factors Controlling Landscape Development

Imagine traveling across a continent. On your journey, you pass plains, swamps, hills, valleys, mesas, and mountains. Some of these features are **erosional landforms**, which result from the breakdown and removal of rock or sediment and develop where **agents of erosion**, such as moving water, ice, or air, carve into the substrate. Other features are **depositional landforms**, which result from the deposition of sediment where the medium carrying the sediment evaporates, slows down, or melts. The specific landforms that develop at a given locality, which together make up the landscape, reflect several factors.

- *Eroding or transporting agents:* Moving water, ice, and wind all cause erosion and transport sediment. But the shapes of the landforms that each produces differ from one another, because of differences in the abilities of these agents to carve into the substrate and to carry debris. Of these three agents, water does the most work, on a global basis.

- *Relief:* The elevation difference, or relief, between adjacent places in a landscape determines the height and steepness of slopes (see Fig. F.3a). Steepness, in turn, controls the velocity of ice or water flow and determines whether rock or soil stays in place or tumbles downslope.

- *Climate:* The average mean temperature, the windiness, the volume of precipitation, and the distribution of precipitation through the year (in other words, the *climate*), determines whether moving water, flowing ice, or wind serves as the main agent of erosion or deposition

in a region. Climate also affects the processes by which substrate weathers.

- *Substrate composition:* The material that makes up a substrate determines how the substrate responds to erosion. For example, strong rocks can stand up to form steep cliffs, whereas soft sediment collapses to generate gentle slopes.

- *Life activity:* Some living organisms weaken the substrate (by burrowing, wedging, or digesting), while others hold it together (by binding it with roots).

- *Time:* Landscapes evolve over time in response to continued erosion and/or deposition. For instance, a gully that has just started to form in response to the flow of a stream does not look the same as the deep canyon that develops after the same stream has existed for a long time.

Although water, wind, and ice drive the development of most landscapes, human activities have had an increasingly important impact on the Earth's surface. We have dug pits (mines) where once there were mountains, have built hills (tailings piles and landfills) where once there were valleys, and have made steep slopes gentle and gentle slopes steep **(Fig. F.5)**. By constructing concrete walls, we modify the shapes of coastlines, change the courses of rivers, and fill new lakes (reservoirs). In cities, buildings and pavements completely seal the ground and cause water that might once have seeped down into the ground to spill into streams instead, increasing their flow. The area of land covered by pavement or buildings in the United States now exceeds the area of Ohio! And in rural areas, agriculture, grazing, water use, and deforestation substantially alter the rates at which natural erosion and deposition take place. For example, agriculture greatly increases the rate of erosion because for much of the year, farm fields have no vegetation cover.

FIGURE F.5 Human influence on a geologic scale.

(a) The pyramids of Egypt are human-made hills that rise above the desert sands. They have lasted for thousands of years.

(b) To construct highways through high ridges, workers effectively carve deep valleys. This example borders a highway near Denver.

(c) This stone dam holds back a reservoir in Colorado. Think about how long it would take a glacier to pile up so much sediment.

F.4 The Hydrologic Cycle

As the discussion above implies, water in its various states (liquid, gas, and solid) plays a major role in erosion and deposition on the Earth's surface (Fig. F.6). Our planet's water resides in certain distinct *reservoirs*, namely, oceans, glacial ice, groundwater, lakes, soil moisture, living organisms, the atmosphere, and rivers (Table F.2). Together, the water in these reservoirs constitutes Earth's **hydrosphere**. Water constantly flows from reservoir to reservoir in a never-ending process that geologists refer to as the **hydrologic cycle** (Geology at a Glance, pp. 598–599). Perhaps without realizing it, Longfellow, an American poet fascinated with reincarnation, provided an accurate if somewhat romantic image of the hydrologic cycle (see the epigraph at the start of this interlude). Without this cycle, the erosive force and transporting activity of running water in rivers and streams, or of flowing ice in glaciers, would not exist.

FIGURE F.6 Water in this pond seeped in from underlying springs. Evaporation is causing the water to enter the air.

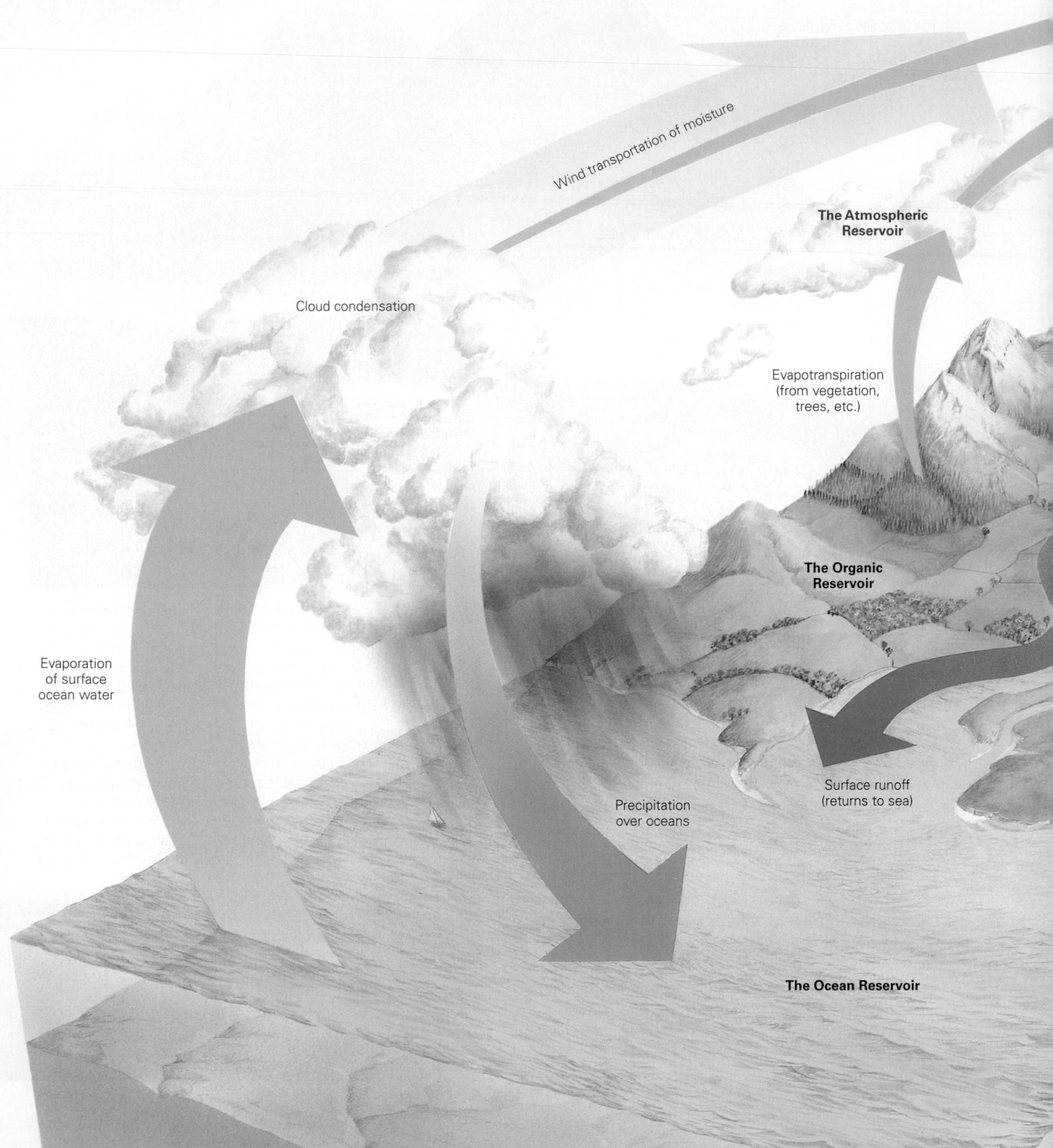

Wind transportation of moisture

Cloud condensation

The Atmospheric Reservoir

Evapotranspiration (from vegetation, trees, etc.)

The Organic Reservoir

Evaporation of surface ocean water

Precipitation over oceans

Surface runoff (returns to sea)

The Ocean Reservoir

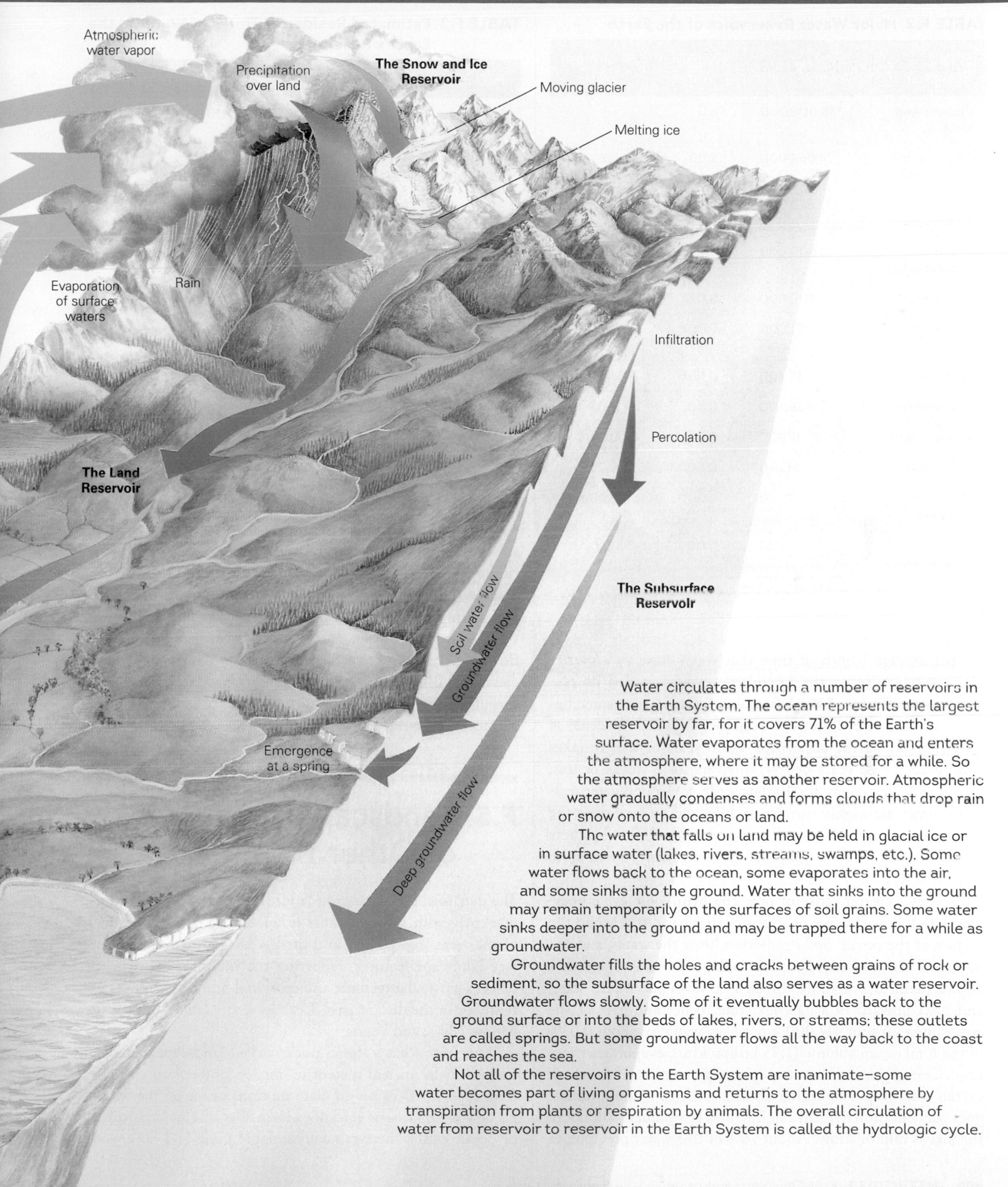

Atmospheric water vapor

Precipitation over land

The Snow and Ice Reservoir

Moving glacier

Melting ice

Evaporation of surface waters

Rain

Infiltration

Percolation

The Land Reservoir

Soil water flow

Groundwater flow

The Subsurface Reservoir

Emergence at a spring

Deep groundwater flow

Water circulates through a number of reservoirs in the Earth System. The ocean represents the largest reservoir by far, for it covers 71% of the Earth's surface. Water evaporates from the ocean and enters the atmosphere, where it may be stored for a while. So the atmosphere serves as another reservoir. Atmospheric water gradually condenses and forms clouds that drop rain or snow onto the oceans or land.

The water that falls on land may be held in glacial ice or in surface water (lakes, rivers, streams, swamps, etc.). Some water flows back to the ocean, some evaporates into the air, and some sinks into the ground. Water that sinks into the ground may remain temporarily on the surfaces of soil grains. Some water sinks deeper into the ground and may be trapped there for a while as groundwater.

Groundwater fills the holes and cracks between grains of rock or sediment, so the subsurface of the land also serves as a water reservoir. Groundwater flows slowly. Some of it eventually bubbles back to the ground surface or into the beds of lakes, rivers, or streams; these outlets are called springs. But some groundwater flows all the way back to the coast and reaches the sea.

Not all of the reservoirs in the Earth System are inanimate—some water becomes part of living organisms and returns to the atmosphere by transpiration from plants or respiration by animals. The overall circulation of water from reservoir to reservoir in the Earth System is called the hydrologic cycle.

TABLE F.2 Major Water Reservoirs of the Earth

H_2O Reservoir	Volume (km³)	% of Total Water	% of Fresh Water
Oceans and seas	1,338,000,000	96.5	–
Glaciers, ice caps, snow	24,064,000	2.05	68.7
Saline groundwater	12,870,000	0.76	–
Fresh groundwater	10,500,000	0.94	30.1
Permafrost	300,000	0.022	0.86
Freshwater lakes	91,000	0.007	0.26
Salt lakes	85,400	0.006	–
Soil moisture	16,500	0.001	0.05
Atmosphere	12,900	0.001	0.04
Swamps	11,470	0.0008	0.03
Rivers and streams	2,120	0.0002	00.006
Living organisms	1,120	0.0001	00.003

TABLE F.3 Estimated Residence Times of Water in the Earth's Reservoirs

H_2O Reservoir	Average Residence Time
Ice caps	10,000 to 200,000 years
Deep groundwater	3,000 to 10,000 years
Oceans and inland seas	3,000 to 3,500 years
Shallow groundwater	100 to 200 years
Valley glaciers	20 to 100 years
Freshwater lakes	50 to 100 years
Winter snow	2 to 6 months
Rivers and streams	2 to 6 months
Soil moisture	1 to 2 months
Atmosphere	5 to 15 days
Living organisms	Hours to days

The average length of time that water stays in a particular reservoir during the hydrologic cycle is called its **residence time (Table F.3)**. Water in different reservoirs has different residence times. For example, a typical molecule of water remains in the oceans for 4,000 years or less, in lakes and ponds for 10 years or less, in rivers for 2 weeks or less, and in the atmosphere for 10 days or less. Groundwater residence times are highly variable and depend on how deep the groundwater flows. Water can stay underground for anywhere from 2 weeks to 10,000 years before it inevitably moves on to another reservoir.

To get a clearer sense of how the hydrologic cycle operates, let's follow the fate of seawater that has just reached the surface of the ocean. Solar radiation heats the water, and the increased thermal energy of the vibrating water molecules allows them to evaporate (break free from the liquid state) and drift upward in a gaseous state to become part of the atmosphere. About 417,000 km³ of sea water, or about 0.03% of the total ocean volume (1.35 billion km³), evaporates every year. Convection of the atmosphere generates wind, which carries water vapor to higher altitudes, where it cools, undergoes condensation to form a liquid, and *precipitates* (falls out of the air) as rain or snow. About 76% of this water precipitates directly back into the ocean. The remainder precipitates on land, where most of it becomes trapped temporarily in the soil or in living organisms. It soon returns directly to the atmosphere by **evapotranspiration**, meaning the sum of evaporation from water bodies on land, evaporation from the ground surface, and release of water by the life activities of plants and animals. Rainwater that does not become trapped in the soil or in organisms enters lakes or rivers and ultimately flows back to the sea as surface water, becomes trapped in glaciers, or sinks deeper into the ground to become *groundwater*. Groundwater also flows and ultimately returns to the Earth's surface reservoirs.

F.5 Landscapes of Other Planets

The dynamic, ever-changing landscapes of the Earth contrast markedly with those of most other terrestrial planets. Each of the terrestrial planets and moons has its own unique surface landscape features, reflecting the interplay between the object's particular tectonic and erosional processes, either currently or in the distant past. Let's look at a few examples: the Moon, Mars, and Venus.

Our Moon has a static, pockmarked landscape generated exclusively by ancient meteorite impacts and volcanic activity **(Fig. F.7a, b)**. Because no plate motions occur on the Moon, no new mountains or volcanoes form. Because no atmosphere or ocean exists, there is no hydrologic cycle and no erosion by rivers, glaciers, or winds. Therefore, the lunar surface has remained largely unchanged for billions of years. This landscape can be divided into two general provinces. The *lunar highlands* are the heavily cratered, light-colored regions of the Moon, which expose rocks over 4.0 billion years old. The *maria* are the vast plains of flood basalt, possibly formed in response to impacts over 3.8 billion years ago—these impacts could have excavated such huge craters that that they caused decompression melting in the Moon's mantle and extrusion of flood basalts.

Landscapes on Mars differ from those of the Moon because Mars does have an atmosphere, though it's much less dense than that of the Earth. Martian winds generate huge dust storms that can obscure nearly the entire surface of the planet for months at a time. The landscapes of Mars also differ from the Moon's because Mars once had surface water (Box F.2).

FIGURE F.7 Landscapes of other planets.

(a) The heavily cratered surface of the Earth's Moon.

(b) A close-up view of the lunar landscape, with a lunar rover and an astronaut for scale.

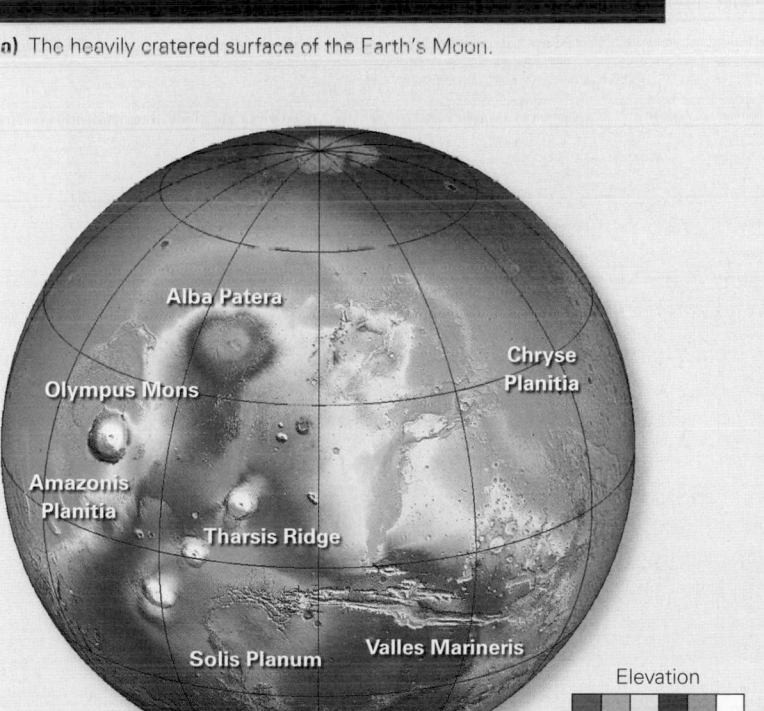

(c) A DEM depicting the surface of Mars. Note the huge bulge of Tharsis Ridge, the giant Olympus Mons volcano, and the deep Valles Marineris canyon.

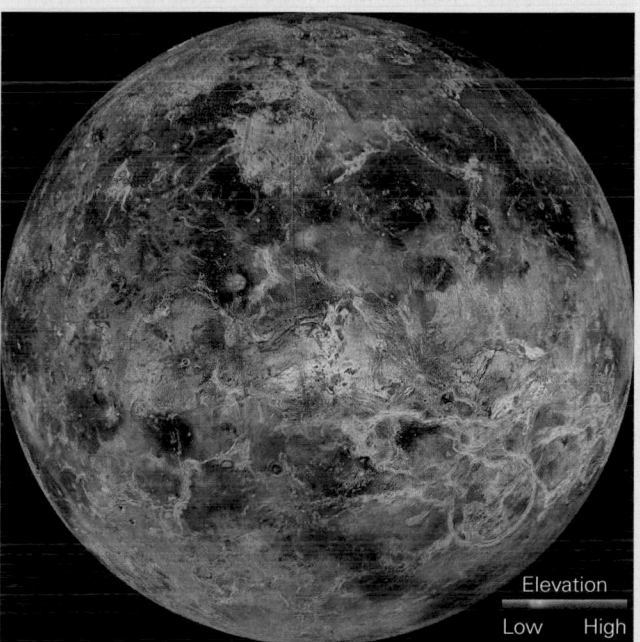

(d) A radar image of Venus. A thick blanket of clouds obscures the planet's surface, so it can't be seen through a telescope. Red areas are higher, blue areas lower.

Water on Mars?

In 1877, an Italian astronomer named Giovanni Schiaparelli studied the surface of Mars with a telescope and announced that long, straight *canali* crisscrossed the planet's surface. *Canali* should have been translated into the English word *channel*, but perhaps because of the recent construction of the Suez Canal, newspapers of the day translated the word into the English *canal*, with the implication that the features had been constructed by intelligent beings. An eminent American astronomer began to study the "canals" and suggested that they had been built to carry water from polar ice caps to Martian deserts.

Late-20th-century satellite mapping of Mars showed that the so-called canals do not exist—they were simply optical illusions. There are no lakes, oceans, or flowing rivers on the surface of Mars today. The atmosphere of Mars has such low density, and thus exerts so little pressure on the planet's surface, that any liquid water released at the surface would quickly evaporate. So Mars has no hydrologic cycle resembling the Earth's. But three crucial questions remain: Does liquid water ever form, even for short periods, on the Martian surface today? Did Mars ever have a significant amount of running water or standing water in the past? And, if the planet once had significant water, where is that water now? The question of the presence of water lies at the heart of an even more basic question: Given that the simplest life as we know it requires liquid water, is there, or was there ever, life on Mars?

The case for liquid water on Mars has become very strong. Much of the evidence comes from comparing landforms on the planet's surface with landforms of known origin on the Earth. High-resolution images of Mars reveal a number of landforms that appear to have been formed by flowing water. Examples include networks of channels resembling river networks on the Earth (**Fig. BxF.2**), scour features, deep gullies, and streamlined deposits of sediment. Studies by the *Odyssey* satellite in 2003, and by the Mars rovers (*Spirit* and *Opportunity*) that landed on the planet in 2004, added intriguing new data to the debate. *Odyssey* detected hints that hydrogen, an element in water, exists beneath the surface of the planet over broad regions, and the Mars rovers have documented the existence of hematite and gypsum, minerals that form in the presence of water. The rovers have also found sediment layers that appear to have been deposited in water. The *Phoenix* lander, in 2008, confirmed the existence of water ice by digging into the surface to expose some, and starting in 2012, the *Curiosity* rover has found further supporting evidence for ice. Researchers speculate that Mars was much wetter in its past, perhaps billions of years ago. But since the atmosphere became less dense, most of the water evaporated and what's left now lies hidden underground or trapped in polar ice caps.

FIGURE BxF.2 Evidence for water on Mars earlier in the planet's history.

(a) A DEM of the boundary between the Acidalia Plantia lowlands (blue, green, yellow) and the Tempe Terra Plateau (red and brown), based on data from the *Mars Express* satellite. Several stream channels are visible.

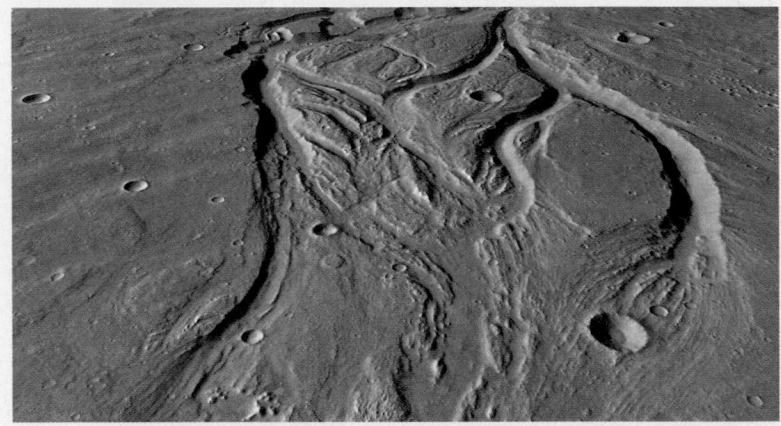

(b) An oblique view showing landforms on Mars resembling channels cut by rivers on the Earth.

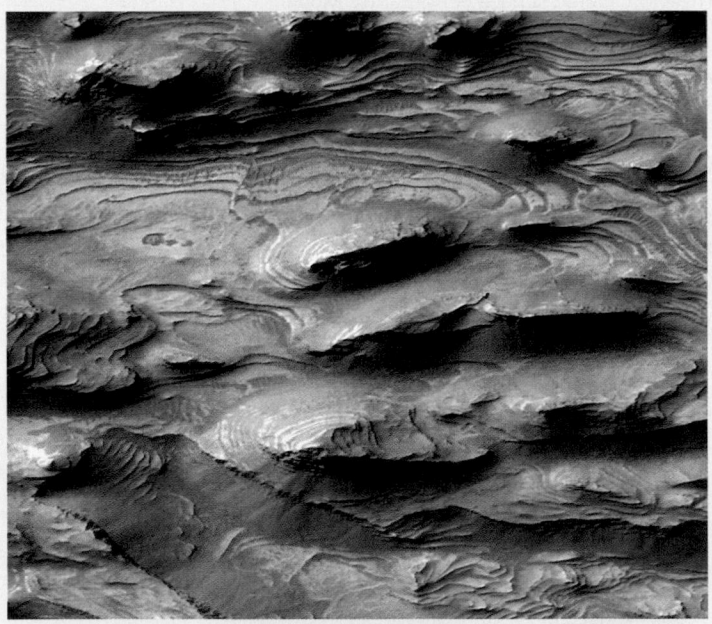

(c) Layers of strata, which appear to have been deposited in water, in Chasma Canyon, Mars. Note the elongate islands that resemble islands eroded by rivers on the Earth.

Therefore, the Martian surface appears to expose four kinds of materials: volcanic flows and deposits (primarily of basalt), debris from impacts, windblown sediment, and water-laid sediment. There is even evidence that soil-forming processes affected surface materials. Of note, Martian winds not only deposit sediment, but they also slowly erode impact craters and polish surface rocks.

Landscapes on Mars also differ from those on the Earth because, like the Moon, Mars lacks plate tectonics. So, unlike the Earth, Mars has no mountain belts or volcanic arcs. In fact, most landscape features on Mars, with the exception of wind-related ones, are over 3 billion years old. Significant relief, however, has developed on Mars (Fig. F.7c). Long ago, a huge mantle plume formed, causing the uplift of a 9-km-high bulge (the Tharsis Ridge) that covers an area comparable to that of North America. Rifting of the Tharsis Ridge produced an immense canyon, the Valles Marineris, that is 4,000 km long, 200 km wide, and 7 km deep. (By comparison, the Earth's Grand Canyon is about 230 km long, 30 km wide, and 1.8 km deep.) Martian igneous activity also led to the eruption of gargantuan hot-spot volcanoes, such as the 22-km-high Olympus Mons, the highest mountain in the Solar System. Because Mars has no vegetation and no longer has rain, its surface does not weather and erode like that of the Earth, so it still bears the scars of impacts by swarms of meteors earlier in the history of the Solar System. Mars does have a hydrologic cycle, of sorts, in that it has ice caps that grow and recede on a seasonal basis.

Venus resembles the Earth in size, and may still have operating mantle plumes. Virtually the entire surface of Venus was resurfaced by volcanic eruptions about 300 to 1,600 Ma, so the planet's surface is much younger than those of the Moon and Mars. Further, Venus has a dense atmosphere that protects it from impacts by smaller objects. Because there has been relatively little cratering since the resurfacing event, volcanic and tectonic features dominate the landscape of Venus (Fig. F.7d). Satellites have used radar to reveal a variety of volcanic landforms (such as shield volcanoes, lava flows, and calderas). Rifting on Venus has produced faults, some of which occur in association with volcanic features. Liquid water cannot survive the scalding temperatures of Venus's surface, so no hydrologic cycle operates there, and no life exists. Because of the density of the atmosphere, winds are too slow to cause much erosion or deposition, so volcanic landforms remain virtually unchanged.

During the past two decades, spacecraft visiting the moons and dwarf planets of the outer Solar System have sent home amazing images of surface features that differ markedly from any found on the Earth. As an example, consider Enceladus, a 500-km-diameter moon of Saturn (Fig. F.8a). Much of Enceladus's ice-covered surface is cracked and wrinkled and largely crater-free, suggesting that geologic processes rifted and folded this moon's crust after the intense meteorite bombardment episodes of the Solar System's early history. The still-cratered terrains may be older and stabler regions of the crust. Io, a 3,600-km-diameter moon of Jupiter, has significant ongoing volcanic activity, which has produced mafic and ultramafic flows as well as coatings of multicolored sulfur-rich ash (Fig. F.8b). Pluto has ragged mountains and strange icy plains (Fig. F.8c)

FIGURE F.8 Landscapes of distant objects in the Solar System.

(a) Enceladus is an icy moon of Saturn with very little cratering and many cracks.

(b) Io, a moon of Jupiter, is volcanically active.

(c) The strange surface of Pluto, seen for the first time by the *New Horizon* space probe in 2015.

ANOTHER VIEW The Mars 2020 rover will be launched in 2020. It will explore the harsh landscape of the red planet.

SUMMARY

- The character and shape of the land surface in a region is its landscape. Individual shapes are landforms. Topographic maps and DEMs can portray the shape of landscapes.

- Land can undergo uplift or subsidence to yield relief. Debris formed by the weathering of uplifted land moves downslope and collects in lower areas.

- The energy driving landscape development comes from three sources: the Earth's internal energy, gravitational energy, and energy radiating from the Sun.

- The nature of a landscape depends on climate, time, relief, slope angles, elevation, the activity of organisms, substrate composition, and the rate of uplift and subsidence.

- Water moves among various reservoirs during the hydrologic cycle. Landscape development involves erosion by flowing water or ice.

- Landscapes on other planets and moons differ markedly from those on the Earth.

GUIDE TERMS

agent of erosion (p. 596)
contour interval (p. 593)
contour line (p. 593)
deposition (p. 594)
depositional landform (p. 596)
digital elevation model (DEM) (p. 591)

downslope movement (p. 594)
erosion (p. 594)
erosional landform (p. 596)
evapotranspiration (p. 600)
external energy (p. 595)
geologic cross section (p. 593)

gravitational energy (p. 595)
hydrologic cycle (p. 597)
hydrosphere (p. 597)
internal energy (p. 595)
landform (p. 591)
landscape (p. 591)
relief (p. 594)

residence time (p. 600)
subsidence (p. 591)
substrate (p. 594)
topographic map (p. 593)
topographic profile (p. 593)
topography (p. 591)
uplift (p. 591)

REVIEW QUESTIONS

The letters following each Review Question refer to the corresponding Learning Objective from the Chapter Opener.

1. What is the difference between uplift and subsidence? **(B)**

2. Why do landscapes on the Earth change over geologic time, while landscapes on the Moon remain static? **(C)**

3. What is topography, and how can we portray it on a sheet of paper? **(A)**

4. Name the principal agents of erosion on the Earth. **(D)**

5. What factors affect the character of the erosional or depositional landforms that develop in a region? **(D)**

6. Explain the steps in the hydrologic cycle. **(E)**

7. How do landscapes of other planets differ from those of the Earth? **(A)**

ONLINE RESOURCES

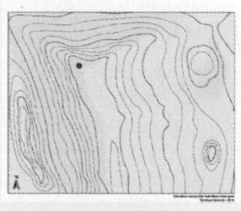

Videos
This interlude features videos on the Earth's water cycle and evapotranspiration.

Smartwork5
This interlude includes questions on features and processes that change the landscape and on the water cycle.

CHAPTER 16

Unsafe Ground: Landslides and Other Mass Movements

By the end of this chapter, you should be able to . . .

A. distinguish different types of mass movements, and explain the differences among them.

B. sketch a model illustrating the forces acting on the material of a slope, and discuss factors that determine whether a slope is stable or unstable.

C. describe the evidence that major mass movements have taken place under the sea and that some of these movements triggered tsunamis.

D. highlight events that may trigger mass movements, and explain why some regions are particularly susceptible to mass-movement events.

E. evaluate hazards related to mass movements, and discuss how such hazards can, in some cases, be prevented.

16.1 Introduction

It was Sunday, May 31, 1970, a market day, and thousands of people had crammed into the Andean town of Yungay, Peru, to shop. Suddenly they felt the jolt of an earthquake, which was strong enough to topple some masonry houses. But worse was yet to come. Shocks from the earthquake also caused an 800-m-wide ice slab to break off the end of a glacier at the top of Nevado Huascarán, a nearby 6.6-km-high mountain peak. Gravity instantly pulled the ice slab down the mountain's steep slopes. As it tumbled downward for a distance of over 3.7 km, the ice disintegrated and became a chaotic avalanche of chunks traveling at speeds of over 300 km per hour. Near the base of the mountain, most of the avalanche channeled into a valley and thickened into a turbulent mass as high as a 10-story building, ripping up rocks and soil along the way. Friction transformed the ice into water, which, when mixed with rock and dust, created 50 million m³ of a muddy slurry viscous enough to buoy along boulders larger than houses. This mass, sometimes riding on a compressed air cushion that allowed it to pass by without disturbing the grass below, traveled 14.5 km in less than 4 minutes.

On rounding a curve near the mouth of the valley, part of the mass shot up the sides and flew over the ridge between the valley and Yungay. As the town's inhabitants and visitors stumbled out of earthquake-damaged buildings, they heard a deafening roar and looked up to see a churning mud cloud descending on them. Moments later, the town was completely buried under several meters of mud and rock—only the top of the church and a few palm trees remained visible to show where Yungay once lay (Fig. 16.1). Over 18,000 people are forever entombed beneath the resulting debris layer. Today the site is a grassy meadow, spotted with memorials left by mourning relatives.

Could the Yungay tragedy have been prevented? Perhaps. A few years earlier, climbers had recognized the instability of glacial ice on Nevado Huascarán, and Peruvian newspapers had published a warning, but alas, no one took notice. In the aftermath of the event, geologists discovered that Yungay had been built on ancient layers of debris from past landslides. Peru subsequently prevented new construction in the danger zone.

People often assume that the ground beneath them is *terra firma*, a solid foundation on which they can build their lives. But the catastrophe at Yungay says otherwise. The substrate underlying sloping regions of the Earth's surface—regardless of whether it consists of rock or **regolith** (loose sediment, debris, or soil)—is inherently "unstable" in the sense that, under the relentless pull of gravity, it will eventually move downslope. Geologists refer to the downslope transport of rock, regolith, snow, or ice as **mass movement**, or *mass wasting*.

FIGURE 16.1 The May 1970 Yungay landslide disaster in Peru.

(a) Before the landslide, the town of Yungay perched on a hill near the ice-covered mountain Nevado Huascarán.

(b) The landslide completely buried the town beneath debris. A landslide scar is visible on the mountain in the distance.

◄ (facing page) During a rockfall in 2014, a huge boulder rolled right through a farmhouse, next to a vineyard in Italy. Clearly, uplifted land doesn't last forever, as gravity inevitably pulls it downslope. Fortunately, there were no injuries.

Like earthquakes, volcanic eruptions, storms, and floods, mass movements are a type of **natural hazard**, meaning a dangerous aspect of the environment that can cause damage to life and property. Unfortunately, mass movements become more of a threat every year because as the world's population grows, cities expand into areas of unsafe ground. In addition to representing a hazard that we must address, mass movement plays a critical role in the rock cycle, for it serves as the first step in the transportation of sediment. And it plays a critical role in the development of landscapes by modifying the shapes of slopes.

In this chapter, we look at the types, causes, and consequences of mass movement and the precautions society can take to protect people and property from its dangers. You might want to keep this information in mind when selecting a site for your home or when voting on land-use propositions for your community.

16.2 Types of Mass Movement

In general discussion, most people refer to any mass movement of rock or regolith down a slope as a **landslide**. Geologists and engineers, however, find it useful to distinguish among different kinds of mass movements based on four features: (1) the type of material involved (rock or regolith); (2) the velocity of movement (slow, intermediate, or fast); (3) the character of the moving mass (coherent or chaotic; wet or dry); and (4) the environment in which the movement takes place (subaerial or submarine). Similarly, most people think of an **avalanche** as a mass movement of snow. Geologists and engineers, however, apply the term more broadly to include any mass movement that moves like a turbulent cloud.

Why bother classifying mass movements? We make these distinctions because different types of mass movements have different consequences and therefore represent different kinds of natural hazards—by characterizing mass movements more completely, we can better prepare for them. In this section, we first examine mass movements that occur on land, roughly in order from slow to very fast. We then briefly introduce submarine mass movements.

Creep and Solifluction

Creep (also known as *soil creep*) refers to the slow, gradual downslope movement of regolith on a slope. Creep happens when regolith alternately expands and contracts in response to freezing and thawing or wetting and drying. During freezing or wetting, the regolith expands, and its particles move outward, perpendicular to the slope. During thawing or drying, the regolith contracts, and gravity makes the particles sink vertically. The vertical movement results in a slight downslope migration of the particles **(Fig. 16.2a, b)**. You can't see creep by staring at a hillslope because it occurs so slowly, but over a period of years, creep causes trees, fences, gravestones, walls, and foundations built on a hillside to tilt downslope. Notably, trees that continue to grow after they have been tilted display a pronounced curvature at their base **(Fig. 16.2c)**.

In arctic or high-elevation regions, regolith freezes solid to great depth during the winter. In the brief summer thaw, only the uppermost 1 to 3 m of the ground thaws. Since meltwater cannot sink into the underlying *permafrost* (permanently frozen ground), the melted layer becomes soggy and weak, and it slowly flows downslope in overlapping sheets. Geologists refer to this kind of creep, characteristic of cold, treeless tundra regions, as **solifluction (Fig. 16.2d)**.

Rock Glaciers

Slow mass movement also takes place in a **rock glacier**, a body made of rock fragments embedded in a matrix of ice **(Fig. 16.2e)**. Rock glaciers differ from more familiar ice-dominated glaciers in terms of the proportion of rock fragments to ice—in a rock glacier, most of the volume consists of rock, whereas in an ice-dominated glacier, most of the volume consists of ice. In effect, rock glaciers are breccias cemented by ice. Because of ice's weakness, the combined mass of rock and ice in a rock glacier can slowly flow downslope, as does the relatively pure ice of an ice-dominated glacier.

Rock glaciers form in two ways. First, they develop in cold regions where snow or rain percolates into a pile of rock debris that has accumulated above permafrost at the base of a cliff. This water can't infiltrate the frozen ground below the debris pile, so it freezes to form ice in the pores between clasts within the debris pile. Second, rock glaciers develop where an ice-dominated glacier already containing abundant rock debris begins to melt. As melting progresses, the proportion of rock to ice in the glacier increases, and if enough melting takes place, the glacier eventually contains more rock than ice.

Slumps

Near the Pacific Palisades, along the coast of southern California, Highway 1 runs between the beach and a 120-m-high cliff. On March 31, 1958, a gash developed about 200 m inland of the cliff's edge, and a semicoherent mass of sediment and rock began to move downslope. Four days later, when the movement finally stopped, a 1-km-long stretch of the coastal highway had been buried—it took weeks for bulldozers to make this stretch passable again. A similar event happened in northern New York State when, after weeks of drenching rains, a 1.5-km-wide portion of a slope began to move down and out into the floor of

FIGURE 16.2 The process and consequences of slow mass movements (creep and solifluction).

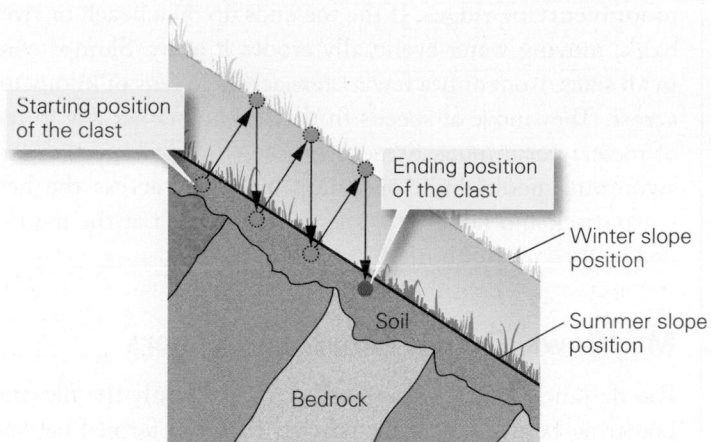

(a) Creep due to freezing and thawing: The clast rises perpendicular to the ground during freezing and sinks vertically during thawing. After 3 years, it migrates to the position shown.

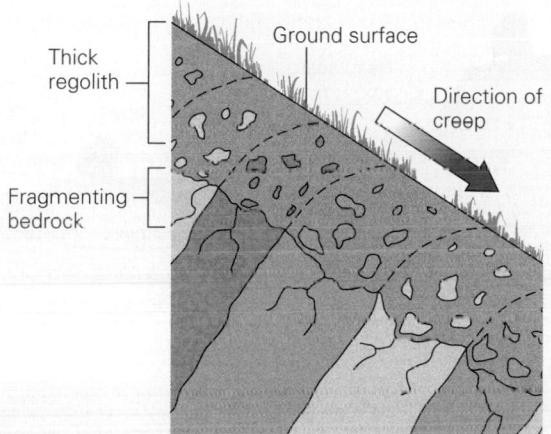

(b) As rock layers weather and break up, the resulting debris creeps downslope.

(d) Solifluction on a hillslope in the tundra.

(c) Soil creep causes walls to bend and crack, building foundations to sink, trees to bend, and power poles and gravestones to tilt.

(e) A rock glacier in Alaska. Note how the flow of the ice below wrinkles the rock layer on the glacier's surface.

Keene Valley. The mass moved at only centimeters to tens of centimeters per day, but even at this slow rate, the accumulated displacement destroyed several expensive homes. In places, the boundary between the moving mass and the unmoving land upslope developed into a 5-m-high escarpment.

Geologists refer to such relatively slow mass-movement events, during which moving rock or regolith does not disintegrate into a jumble of debris, but rather stays somewhat coherent, as a **slump**, and they refer to the moving mass itself as a *slump block* **(Fig. 16.3)**. A slump block slides down a **failure surface**—some failure surfaces are planar, but commonly they curve and resemble a spoon lying concave side up. Geologists refer to the exposed upslope edge of a failure surface as the **head scarp** and the downslope end of a slump block as the block's *toe*. The upslope and downslope ends of a slump block may break into a series of discrete slices, each separated from its neighbor by a small sliding surface. On a hillslope, slices of the toe can move up and over the pre-existing land surface to form curving ridges. If the toe ends up on a beach or riverbank, moving water eventually erodes it away. Slumps come in all sizes, from only a few meters across to tens of kilometers across. They move at speeds from millimeters per day to tens of meters per minute. Structures (such as houses, patios, and swimming pools) built on slump blocks or across the head scarp crack and fall apart, whereas those built at the toe may be knocked over or buried.

Mudflows, Debris Flows, and Lahars

Rio de Janeiro, Brazil, originally occupied only the flatlands bordering beautiful crescent beaches that had formed between steep hills. But in recent decades, the city's population

FIGURE 16.3 The process of slumping on a hillslope. Note the scarps that form at the head of the slump.

(a) A head scarp on a hillslope.

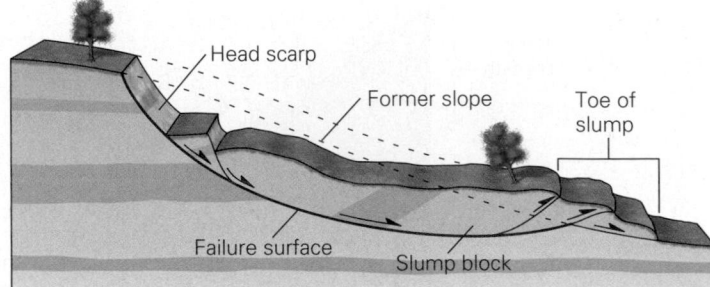

(b) Cross section of a slump.

(c) Slumping dumped sediment into this river in Costa Rica.

(d) A slump beginning to form along a highway in Utah.

has grown so much that in many places, densely populated communities of makeshift shacks have been built on steep slopes. These communities, many of which have no storm drains, have been built on the thick regolith that resulted from long-term weathering of bedrock in Brazil's tropical climate. Particularly heavy rains can saturate the regolith, transforming it into a viscous slurry of mud, resembling wet concrete, that flows downslope. The communities built on the regolith can disappear overnight, replaced by a muddle of mud and debris. And at the bases of the hills, the flowing mud may knock over and bury buildings of all sizes. In unpopulated areas of Brazil, similar mass movements rip away forests (**Fig. 16.4a, b**). Geologists refer to a moving slurry of mud as a **mudflow**, or *mudslide*, and to a slurry consisting of mud mixed with larger, pebble- to boulder-sized fragments as a **debris flow**, or *debris slide* (**Fig. 16.4c; Box 16.1**).

Mudflows are not just phenomena of tropical regions. Any slope underlain by poorly consolidated material can give way in a mudflow or debris flow during or following a heavy rain, and if people live nearby, the movement can have tragic consequences. In March 2014, near Oso, Washington, in the temperate Pacific Northwest, a forested hillslope bordering a river gave way, and within a few minutes it buried a 2.5-km² area of the river valley below with mud and wet sand, killing 44 people (**Fig. 16.5**).

The speed at which material moves in a debris flow or mudflow depends on its water content and on the slope angle. Drier mudflows have higher viscosity than do watery mudflows, so they move more slowly, and mudflows on gentler slopes move more slowly than do those on steeper slopes. In fact, on a steep slope, very wet mud may move at over 100 km per hour. Because mudflows and debris flows have greater viscosity than clear water does, they can carry large rock chunks as well as houses and cars. They typically follow channels downslope, and at the base of the slope they spread out into a broad lobe.

Particularly devastating mudflows spill down the river valleys bordering volcanoes. These mudflows, known as **lahars**, form when volcanic ash from an ongoing or previous eruption

FIGURE 16.4 Examples of mudflows and debris flows.

(a) This 2011 mudslide in Nova Friburgo, Brazil destroyed a high-rise building at the base of the hill.

(b) Mudslides in 2011 stripped away forests on hillslopes in Brazil.

(c) A recent debris flow in Utah. Note the chaotic mixture of rock chunks and mud.

BOX 16.1 CONSIDER THIS . . .

What Goes Up Must Come Down

Along the shore of California, waves slowly cut into the land to produce low, flat areas called *wave-cut benches* (see Chapter 18). Meanwhile, tectonic processes slowly push the land surface up, so that former wave-cut benches become small plateaus, or *terraces*. One such terrace lies at an elevation of 180 m above sea level, about 500 m east of the present-day beach at La Conchita. The western face of this terrace has become a cliff-like bluff **(Fig. Bx16.1a)**. Relatively little vegetation covers the bluff or the terrace above, so rain infiltrates into the ground, sinks down, and saturates clay and debris on the terrace and its bluff, making the material very weak. Every now and then, the weight of surface material causes the bluff to give way, and a mass of mud and debris flows downslope at rates of up to 10 m per second.

If the region of La Conchita were uninhabited, such mass wasting would just be part of the natural process of landscape evolution—as we've seen, gravity brings down land that has been raised by tectonic activity. But when downslope movements take place in La Conchita, they make headlines, because on the modern wave-cut bench between the shore and the base of the bluff, developers have built a community housing 350 people. In 1995, a flow of mud and debris overwhelmed 9 houses at the base of the bluff. An even more devastating flow happened in 2005, burying 13 houses, damaging 23, and killing 10 people **(Fig. Bx16.1b, c)**.

FIGURE Bx16.1 The 2005 La Conchita Mudslide along the coast of California.

(a) A housing development was built in a narrow strip between the beach and steep cliffs.

(c) Rescuers at the toe of the mudslide.

(b) During heavy rains, the slope gave way and heavy mud flowed down, burying houses and taking several lives.

FIGURE 16.5 The 2014 Oso Slide in Washington was a mudslide.

(a) The slide surged over the river, damming it temporarily, and tragically buried a small community on the opposite bank.

What a Geologist Sees

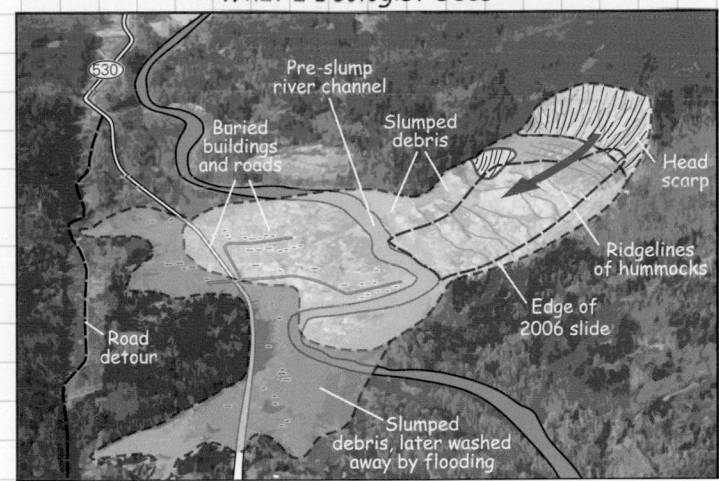

(b) The mudslide happened in an area where previous slides had taken place. The mountain slope was being undercut by the river.

mixes with water from snow and ice that melts in a volcano's heat or from heavy rains (**Fig. 16.6**; see Chapter 9). A lahar occurred on November 13, 1985, in the Andes of Colombia. That night, a major eruption melted a volcano's thick snow-cap, creating hot water that mixed with ash. A lahar rushed down river valleys and swept over the nearby town of Armero while most inhabitants were asleep. Of the 25,000 residents, 20,000 perished.

Rockslides and Debris Slides

In the early 1960s, engineers built a huge new dam across a river on the northern side of Monte Toc in the Italian Alps to create a reservoir for generating electricity. The Vaiont Dam was an engineering marvel, a concrete wall rising 260 m (as high as an 85-story skyscraper) above the valley floor (**Fig. 16.7a**). Unfortunately, the dam's builders did not appreciate the hazard posed by nearby Monte Toc. The side of the mountain facing the reservoir was underlain by a succession of limestone beds overlying a horizon of weak shale. These beds dipped parallel to the surface of the mountain and curved under the reservoir. As the reservoir filled, the flank of the mountain cracked, shook, and rumbled. Local residents began to call Monte Toc *la montagna che cammina* (the mountain that walks).

After several days of rain, Monte Toc began to rumble so much that on October 9, 1963, engineers lowered the water level in the reservoir. They thought the wet ground might slump a little into the reservoir, with minor consequences, so

FIGURE 16.6 Lahars develop when volcanic ash mixes with water from rain or melting snow and ice.

(a) A lahar that rushed down the side of Mt. St. Helens, Washington.

(b) Lahars can overrun populated areas far from the volcano.

no one ordered the evacuation of the town of Longarone, a few kilometers down the valley below the dam. Unfortunately, the engineers underestimated the problem. At 10:30 that evening, a huge chunk of Monte Toc—600 million tons of rock—detached from the mountain and slid downslope along the weak shale horizon into the reservoir. Some debris rocketed up the opposite wall of the valley to a height of 260 m above the original reservoir level **(Fig. 16.7b)**. The displaced water of the reservoir splashed over the top of the dam and rushed down into the valley below. When the flood had passed, nothing of Longarone and its 1,500 inhabitants remained. Though the dam itself still stands, it holds back only debris and has never provided any electricity.

Geologists refer to such a sudden movement of rock and debris down a nonvertical slope as a **rockslide** if the mass consists only of rock or as a **debris slide** if it consists mostly of regolith. Once a slide has taken place, it leaves a scar on the slope and forms a debris pile at the base of the slope. Slides happen when bedrock or regolith detaches from a slope, slips rapidly downhill on a failure surface, and breaks up into a chaotic jumble. Slides are most likely to occur where a weak layer of rock or sediment below the ground parallels the slope. (At the Vaiont Dam, the plane of weakness that would become the failure surface was a weak shale bed.) Slides may move at speeds of up to 300 km per hour—they are particularly fast when a cushion of air gets trapped beneath, in which case there is virtually no friction between the slide and its substrate, and the mass moves like a hovercraft. Rockslides and debris slides sometimes have enough momentum to climb the opposite side of the valley into which they fall. Slides, like slumps, come at a variety of scales. Most are small, involving blocks up to a few meters across. Some, such as the Vaiont slide, are large enough to cause a catastrophe.

Avalanches

In the winter of 1999, an unusual weather system passed over the Austrian Alps. First it snowed. Then the temperature warmed and the snow began to melt. But then the weather turned cold again, and the melted snow froze into a hard, icy crust. This cold snap ushered in a blizzard that blanketed the ice crust with tens of centimeters of new snow. With the frozen snow layer underneath acting as a failure surface, 200,000 tons of new snow began to slide down a mountain. As it accelerated, the mass transformed into a **snow avalanche**, a chaotic jumble of snow surging downslope. At the bottom of the slope, the avalanche overran a ski resort, crushing and carrying away buildings, cars, and trees and killing more than 30 people. It took searchers and their specially trained dogs many days to find buried survivors and victims under the 5- to 20-m-thick pile of snow that the avalanche deposited **(Fig. 16.8a)**.

Snow avalanches display a variety of behaviors, depending on both the temperature of the snow and on the steepness of the slope down which the snow moves. Specifically, *wet-snow avalanches*, which involve snow that has started to melt and contains some liquid water, behave like a viscous slurry in that they hug the slope as they move. Wet-snow avalanches generally travel at speeds of less than 30 km per hour and can pick up rock debris and vegetation along their path. *Dry-snow avalanches* contain cold, powdery snow that, upon tumbling down a steep slope, disintegrates into a turbulent, air-rich cloud that rushes along at speeds of up to 250 km per hour **(Fig. 16.8b)**. Both wet-snow and dry-snow avalanches can become powerful enough to flatten forests in their paths **(Fig. 16.8c)**.

What triggers snow avalanches? Some happen when a *cornice*, a large drift of snow that builds up on the lee side of a windy mountain summit, suddenly gives way and falls onto slopes below, where it knocks free additional snow. Others happen when a broad slab of snow on a moderate slope detaches from its substrate along an icy failure surface. Avalanches tend to affect the same localities year after year, because of the characteristics of snow buildup and because of the occurrence of *avalanche chutes*, shallow valleys running down the slope that confine the tumbling snow. To protect populated areas downslope of known avalanche chutes, experts may use special explosives to trigger small, controlled avalanches before the snow piles up enough to become a hazard.

As we've noted, geologists commonly use the word *avalanche* in a broader sense for any mass movement during which the solid fragments are suspended in so much fluid (air or water) that the flowing mixture behaves like a turbulent cloud. Thus, a *debris avalanche* contains rock fragments and regolith mixed with air, and a *submarine avalanche* consists of sediment suspended in water. Avalanches of all types flow downslope because the mixture of solid and fluid in an avalanche is denser than the surrounding pure fluid—for this reason, geologists refer to various kinds of avalanches as *density currents*.

FIGURE 16.7 The Vaiont Dam disaster—a catastrophic landslide that displaced the water in a reservoir with rock debris.

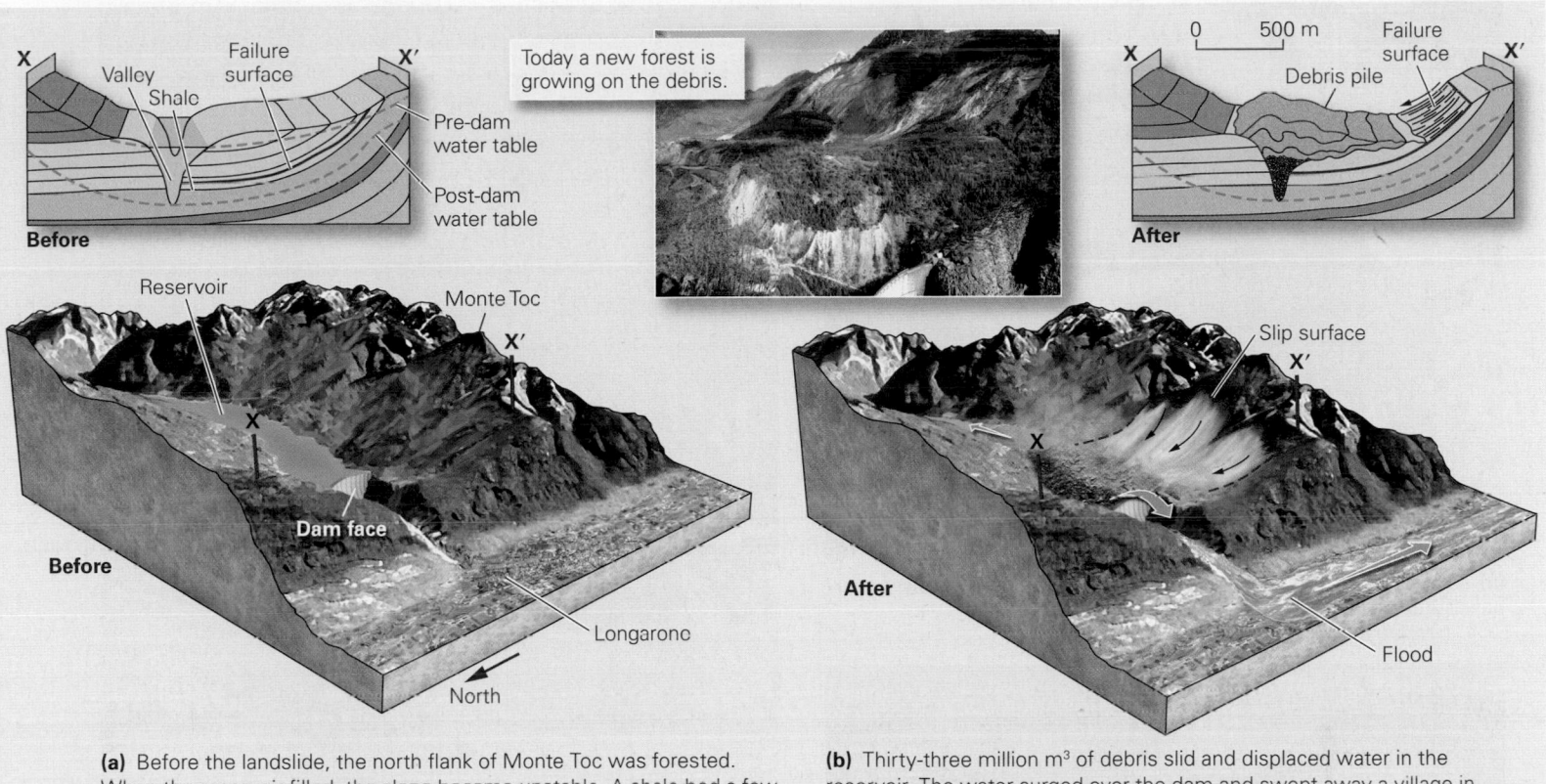

(a) Before the landslide, the north flank of Monte Toc was forested. When the reservoir filled, the slope became unstable. A shale bed a few hundred meters below the ground surface became a failure surface.

(b) Thirty-three million m³ of debris slid and displaced water in the reservoir. The water surged over the dam and swept away a village in the valley below.

Rockfalls and Debris Falls

Rockfalls and debris falls, as their names suggest, occur when a mass free-falls from a cliff **(Fig. 16.9a, b)**. Commonly, rockfalls happen when a body of rock separates from a cliff face along a *joint*, a natural crack in rock across which a block of rock no longer connects to bedrock (see Chapter 11). Most rockfalls involve only a few blocks, but some falls dislodge immense quantities of rock. In September 1881, a 600-m-high crag of slate, undermined by quarrying, suddenly collapsed onto the town of Elm in a valley of the Swiss Alps. Over 10 million m³ of rock fell to the valley floor, burying Elm and its 115 inhabitants to a depth of 10 to 20 m.

Friction and collision with other rocks may bring some blocks that have fallen to a halt before they reach the bottom of the slope; these blocks pile up to form an apron of loose rocks, known as a **talus** or *talus slope*, along the base of the cliff **(Fig. 16.9c)**. Debris that has fallen a long way can reach speeds of 300 km per hour and may have so much momentum that it keeps moving as an avalanche-like cloud of fragments mixed with air when it reaches the base of a cliff. Large, fast rockfalls push the air in front of them, creating a blast of hurricane-like wind. For example, the wind in front of a 1996 rockfall in Yosemite National Park flattened over 2,000 trees.

Rockfalls happen fairly frequently along steep highway road cuts, leading to the posting of "falling-rock zone" signs **(Fig. 16.9d)**. Such rockfalls take place because highway cuts are, effectively, new cliffs. In some cases, construction of the road weakens the material underlying the road cut, allowing frost wedging and root wedging to more easily break it free.

Did you ever wonder...
why highway engineers erect "Falling rock" signs?

Submarine Mass Movements

So far, we've focused on mass movements that occur sub-aerially, for these are the ones we can see and affect us the most. But mass wasting also happens underwater. In fact, the sedimentary record contains abundant evidence of submarine mass movements because after they take place, they tend to be buried by younger sediments and are preserved.

FIGURE 16.8 Examples of avalanches.

(a) Aftermath of a 1999 avalanche in the Austrian Alps. Masses of snow buried and destroyed several homes.

(b) Avalanche chutes down the side of a mountain in the Canadian Rockies. Recent avalanches have flattened trees.

(c) A dry-snow avalanche in Alaska is a turbulent cloud.

Geologists distinguish among three types of submarine mass movements according to whether the mass remains coherent or disintegrates as it moves **(Fig. 16.10)**. In **submarine slumps**, semicoherent blocks slip downslope on weak horizons, and the rock layers constituting the blocks become contorted as they move, like a tablecloth that slides off a table. In **submarine debris flows**, the moving mass breaks apart to form a slurry containing larger clasts (pebbles to boulders) suspended in a mud matrix. And in **turbidity currents**, sediment disperses in water to create a turbulent cloud of suspended sediment that rushes downslope as a submarine avalanche or density current. Turbidity currents commonly flow down submarine canyons—their movement, in fact, erodes the seafloor and contributes to the formation of the canyons. When a turbidity current starts to slow, the sediment it contains settles out in a sequence, with coarser grains at the base and finer grains at the top. The resulting deposits, therefore, consist of graded beds (see Chapter 7)—typically, the deposits accumulate in a fan at the mouth of a submarine canyon.

In recent years, geologists have used satellites as well as shipboard *side-scan sonar* (sonar that analyzes a 60-km-wide swath of the seafloor, instead of just a line, as it moves) to map the extent of submarine landslides **(Fig. 16.11a)**. The shapes of slumps and landslide deposits stand out on the resulting new generation of high-resolution seafloor maps. Geologists have found that submarine slopes bordering both hot-spot volcanoes and active plate boundaries are scalloped by immense slumps, because tectonic activity jars these areas with earthquakes that set masses of material in motion. For example, slumps up to

FIGURE 16.9 Examples of rockfalls.

(a) Successive rockfalls have littered the area below this sandstone cliff with boulders.

(b) This rockfall buried the forest bordering a lake in the Uinta Mountains, Utah. Fresh rock exposed by the rockfall has a lighter color.

(c) A talus pile in the Uinta Mountains, Utah.

(d) A rockfall along a highway in Vermont. Note that the blocks separated from the wall along joints.

200 km long and 100 km wide have substantially modified the flanks of the Hawaiian Islands (**Fig. 16.11b**). Some of the slumping events have even carried away large chunks of the subaerial parts of the islands—in fact, the steep portions of the islands' coasts are the head scarps of huge slumps. Studies suggest that huge slumping events off Hawaii happen, on average, about once every 100,000 years. Significantly, passive-margin coasts may also undergo slumping, and immense slumps have been mapped along the coasts of the Atlantic Ocean.

Since a submarine slump can develop fairly quickly, and since its movement can displace a large volume of sea water, it can trigger a tsunami. A submarine slump set in motion by a 1998 earthquake in Papua New Guinea generated a tsunami that devastated a 40-km-long stretch of coast and killed 2,100 people.

A prehistoric tsunami triggered by a slump off the coast of Norway left its trace all around the North Sea (**Box 16.2**).

TAKE-HOME MESSAGE

Mass movements differ from one another based on speed and character. Creep, slumping, and solifluction are slow. Mudflows and debris flows move faster, and avalanches and rockfalls move the fastest. Mass movements occur on land and underwater.

QUICK QUESTION: In what way is a snow avalanche like a turbidity current?

FIGURE 16.10 Submarine mass movements.

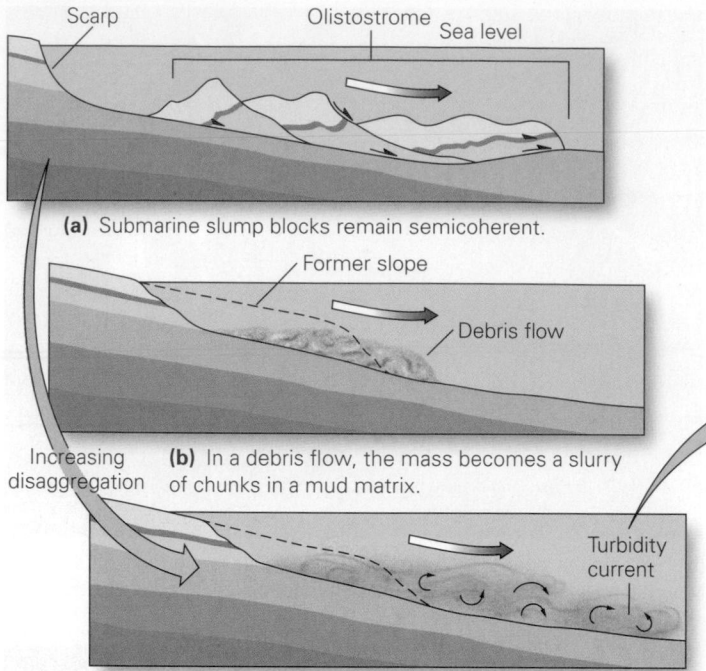

(a) Submarine slump blocks remain semicoherent.

(b) In a debris flow, the mass becomes a slurry of chunks in a mud matrix.

Increasing disaggregation

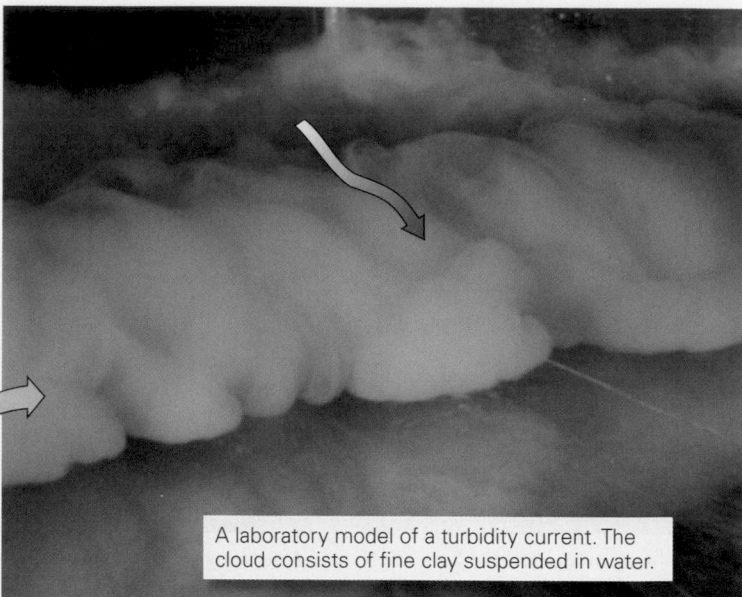

A laboratory model of a turbidity current. The cloud consists of fine clay suspended in water.

(c) A turbidity current is a cloud of sediment suspended in water; it flows near the seafloor because it is denser than clear water.

FIGURE 16.11 Examples of huge submarine slumps and debris flows.

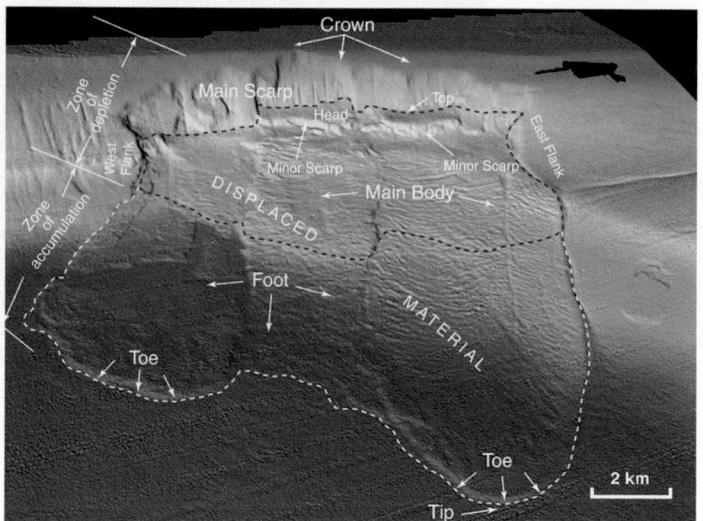

(a) A digital bathymetric map of a slump along the coast of California. The parts of the slump are labeled.

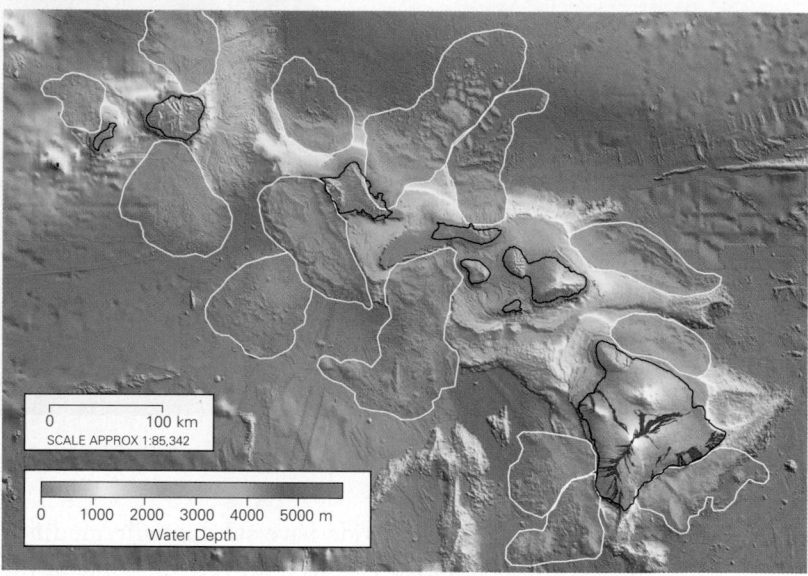

(b) A bathymetric map of the area around Hawaii shows several huge slumps, shaded in tan.

16.3 Why Do Mass Movements Occur?

We've seen that mass movements travel at a range of different velocities, from slow (creep) to faster (slumps, mud and debris flows, and rock and debris slides) to fastest (snow avalanches and rock- and debris falls; **Geology at a Glance,** pp. 622–623). The velocity depends on the steepness of the slope and the water or air content of the mass. Why do mass movements, regardless of their character, take place? The stage must be set by the following phenomena: (1) fracturing and weathering of the substrate, which weakens the substrate so it cannot hold up against the pull of gravity; (2) the development of relief, which provides slopes down which masses move; and (3) an event that sets a mass in motion. Let's look at these phenomena more closely.

BOX 16.2 CONSIDER THIS . . .

The Storegga Slide and North Sea Tsunamis

The Firth of Forth, a long inlet of the North Sea, forms the waterfront of Edinburgh, Scotland. At its western end, it merges with a broad plain in which mud and peat have been accumulating since the last ice age. Around 1865, geologists investigating this sediment discovered an unusual layer of sand containing smashed seashells, marine plankton, and torn-up fragments of marine substrate. This sand layer lies sandwiched between mud layers at an elevation of up to 4 m above the high-tide limit and 80 km inland from the shore. How could the sand layer have been deposited? The mystery baffled geologists for many decades. During this time, layers of shelly sand similar to the one found in Scotland were discovered at many other localities along both sides of the North Sea **(Fig. Bx16.2)**. In some cases, the layer occurred 20 m above the high-tide limit. Geologists studying the coast also found locations where coastal cliffs appeared to have been eroded by wave action at elevations well out of reach of normal storm waves.

While land-based geologists puzzled over these unusual sand layers and coastal erosional features, marine geologists investigating the continental shelf off the western coast of Norway discovered a region of very irregular seafloor underlain with a jumble of chaotic blocks, some of which were 10 km by 30 km across and 200 m thick. When mapped out, the blocks indicated that a 290-km-long sector of the continental shelf had collapsed in a series of submarine slides that together involved about 5,580 km³ of debris—the overall feature is known as the Storegga Slide.

Further studies show that the Storegga Slide formed during three movement events: one occurred 30,000 years ago, the second about 7,950 years ago, and the third about 6,000 years ago. Immense submarine slides displace enough ocean water to create tsunamis. The wave produced by the second Storegga Slide may have been responsible for the disappearance of Stone Age tribes that lived in villages bordering the North Sea—the tsunami washed the villages away.

If such a calamity happened in the past, could it happen again—could submarine slide-generated tsunamis inundate coastal cities that host large populations? Geologists now realize that submarine landslides can trigger tsunamis every bit as devastating as earthquake- and volcano-generated tsunamis. A slide in 1929 along the coast of Newfoundland, for example, not only created a turbidity current that broke the trans-Atlantic telephone cable, but also generated tsunamis that washed away houses and boats along the coast at elevations of up to 27 m above sea level. With a bit of looking, geologists have found huge boulders flung by tsunamis onto the land, layers of sand and gravel deposited well above the high-tide limit, and erosional features high up on shoreline cliffs along many coastal areas, even in regions (such as the Bahamas and southeastern Australia) far from seismic or volcanic regions. And submarine mapping reveals large slump scarps along all continental shelves.

FIGURE Bx16.2 Map of the North Sea region showing the locations of sites (red dots) where marine sand layers occur significantly above the high-tide limit. These layers were deposited by tsunamis generated by movement of the Storegga Slide. The map shows the estimated positions of the tsunamis at 2 hours, 4 hours, and 6 hours after the slide.

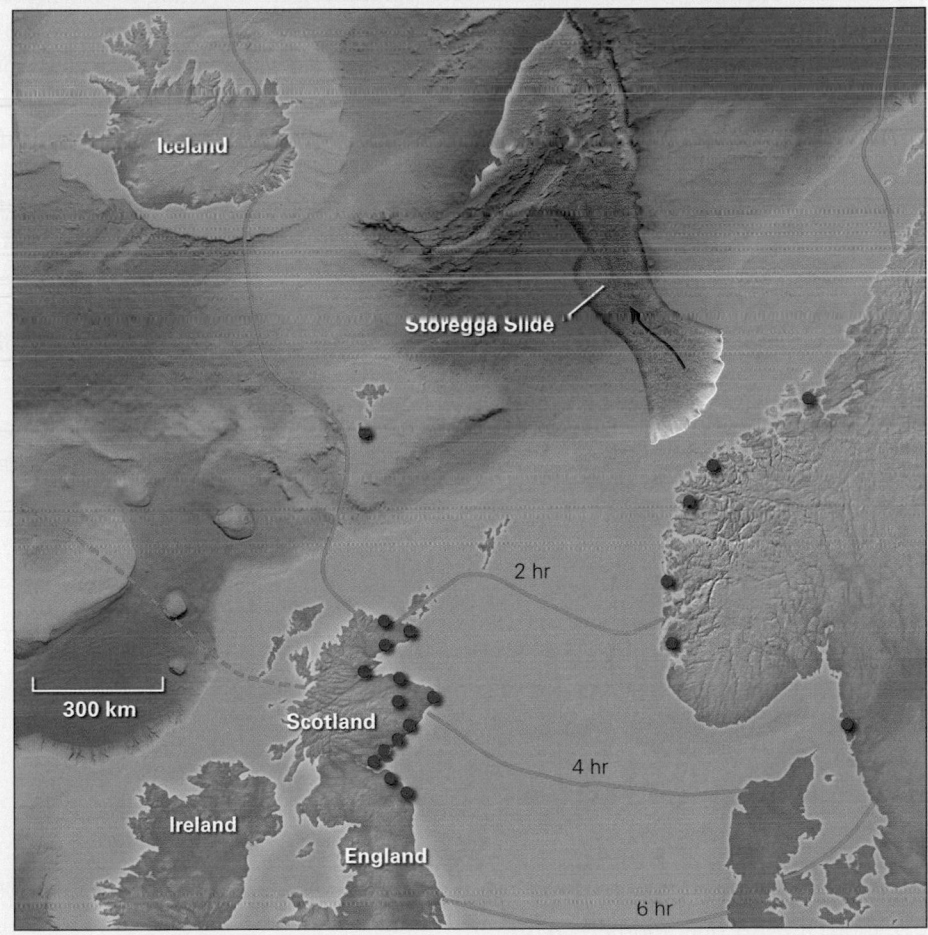

FIGURE 16.12 Jointing broke up this thick sandstone bed along a cliff in Utah. Blocks of sandstone break free along joints and tumble downslope.

Weakening the Substrate: Fragmentation and Weathering

If the Earth's surface consisted of unfractured and unweathered rock, mass movements would be of little concern, for intact fresh rock has great strength and could form stalwart mountain faces that would not tumble. But, in reality, the rock of the Earth's upper crust has been fractured by jointing and faulting and has undergone weathering. Furthermore, in many locations, regolith lies above bedrock. Regolith and fractured or weathered rock are much weaker than intact fresh rock and can collapse in response to gravitational pull **(Fig. 16.12)**, and ultimately make mass movements possible. To see why, let's consider the strength of the attachments holding materials together.

A mass of intact bedrock is relatively strong because the chemical bonds within its interlocking grains, or within the cement between grains, can't be broken easily. Weathered rock tends to be weaker because the strong bonds between the original grains of the rock have been replaced by weaker bonds between weathering products. Regolith is relatively weak because its grains are held together only by friction (caused by roughness along the contact between two grains), electrostatic attraction (the weak force that develops between clay flakes because the surfaces of the flakes are charged), or the surface tension of water (the weak force caused by the attraction of water molecules to one another). All of these forces combined are weaker than the chemical bonds holding together the atoms in the minerals of intact rock. To picture this contrast, think how much easier it is to break up a sand castle (whose strength comes primarily from the surface tension of water films on the sand grains) than it would be to break up a sculpture of a castle carved from rock.

Slope Stability

Mass movements do not take place on all slopes, and even on slopes where such movements are possible, they occur only occasionally. Geologists distinguish between *stable slopes*, on which sliding is unlikely, and *unstable slopes*, on which sliding will likely happen. When material starts moving on an unstable slope, we say that **slope failure** has occurred. Whether a slope fails or not depends on the balance between two forces—the *downslope force*, caused by gravity, and the *resistance force*, which inhibits sliding. If the downslope force exceeds the resistance force, the slope fails, and mass movement results.

Let's examine the battle between downslope force and resistance force more closely by imagining a block sitting on a slope. We can represent the gravitational attraction between the block and the Earth by an arrow (a vector) that points straight down, toward the Earth's center of gravity. This arrow can be separated into two components—the downslope force parallel to the slope and the normal force perpendicular to the slope. Note that for a given mass, the magnitude of the downslope force increases as the slope angle increases, so the downslope force is greater on steeper slopes. We can symbolize the resistance force by an arrow pointing uphill. (As we have seen, the resistance force comes from chemical bonds, electrostatic attraction, friction, and surface tension.) If the downslope force becomes larger than the resistance force, then the block moves—otherwise, it stays in place **(Fig. 16.13)**.

FIGURE 16.13 Forces that trigger downslope movement.

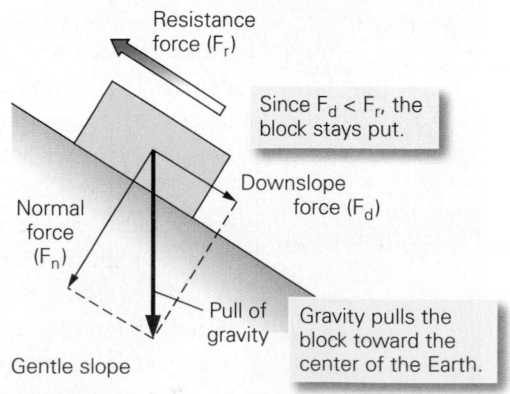

(a) Gravity can be divided into a normal force and a downslope force. If the resistance force, caused by friction, exceeds the downslope force, the block does not move.

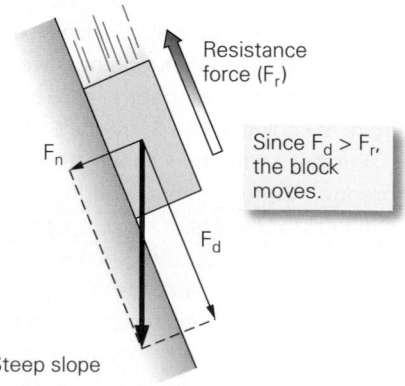

(b) If the slope angle increases, the downslope force due to gravity increases. If the downslope force becomes greater than the resistance force, the block starts to move.

Because of the resistance force, granular debris tends to pile up to produce the steepest slope it can without collapsing. This slope, called the **angle of repose**, generally has a value between 30° and 37° for dry, unconsolidated materials (such as dry sand). The angle of repose depends partly on the shape and size of grains, which determine the amount of friction across grain boundaries. For example, steeper angles of repose (up to 45°) characterize talus slopes composed of large, irregularly shaped grains **(Fig. 16.14)**.

In many locations, the resistance force is less than might be expected because a weak surface exists at some depth below ground level. If downslope movement begins on the weak surface, it becomes a *failure surface*. Geologists recognize several different kinds of weak surfaces that may become failure surfaces **(Fig. 16.15)**. Examples include wet clay layers; wet, unconsolidated sand layers; joints; weak bedding planes, such as shale beds or evaporite beds; and weak metamorphic foliation planes.

Weak surfaces that dip parallel to the land-surface slope are particularly likely to become failure surfaces. An example of failure on such a surface occurred in Madison Canyon, in Montana, on August 17, 1959. That day, a strong earthquake jarred the region. Metamorphic rock with weak foliation formed the bedrock of the canyon's southern wall. When the ground vibrated, rock detached along a foliation plane and tumbled downslope. Unfortunately, 28 campers lay sleeping on the valley floor. They were probably awakened by the hurricane-like winds blasting in front of the moving mass, but seconds later they were buried under 45 m of rubble.

Fingers on the Trigger: What Causes Slope Failure?

What triggers an individual mass-movement event? In other words, what causes the balance of forces to change so that the downslope force exceeds the resistance force and a slope suddenly fails? Here we look at various phenomena—natural and human-caused—that trigger slope failure.

Shocks, Vibrations, and Liquefaction Earthquake tremors, storms, the passing of large trucks, or blasting in construction sites may cause a mass that was on the verge of moving

FIGURE 16.14 The angle of repose is the steepest slope that a pile of unconsolidated sediment can have and remain stable. The angle depends on the shape and size of grains.

30°	45°
Well-rounded sand has a small angle.	Irregularly shaped gravel has a large angle.

FIGURE 16.15 Different kinds of weak surfaces can become failure surfaces.

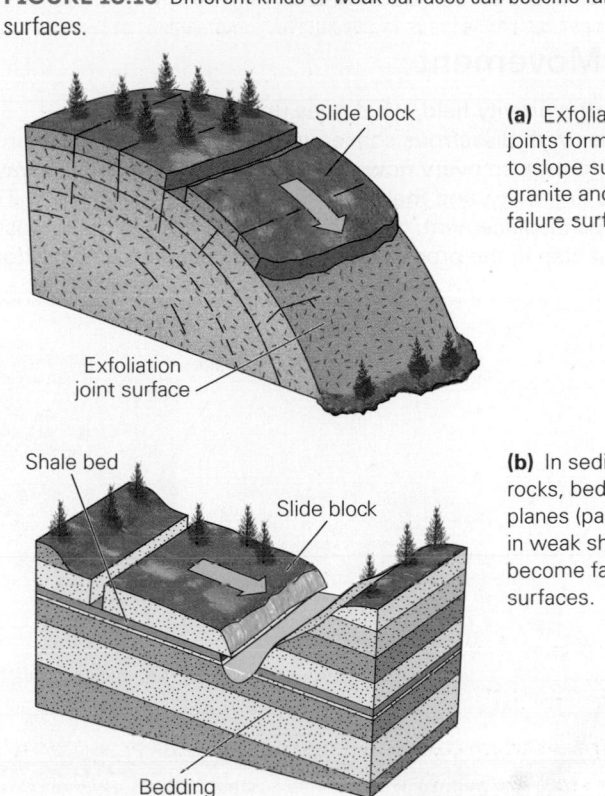

Slide block

Exfoliation joint surface

(a) Exfoliation joints form parallel to slope surfaces in granite and become failure surfaces.

Shale bed

Slide block

Bedding

(b) In sedimentary rocks, bedding planes (particularly in weak shale) become failure surfaces.

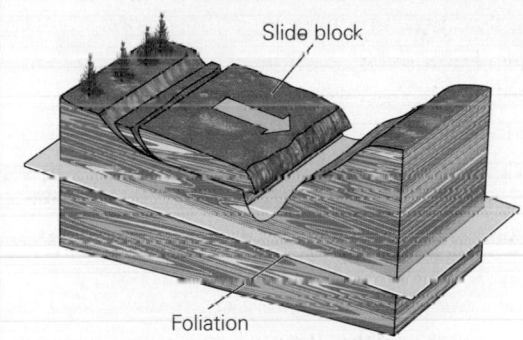

Slide block

Foliation

(c) In metamorphic rock, foliation planes (particularly in mica-rich schist) become failure surfaces.

to actually start moving. For example, an earthquake-triggered slide dumped debris into southeastern Alaska's Lituya Bay in 1958. The debris displaced the water in the bay, creating a 300-m-high splash that washed forests off the slopes bordering the bay and carried fishing boats anchored in the bay many kilometers out to sea. The vibrations of an earthquake break bonds that hold a mass in place and cause the mass and the slope to separate slightly, thereby decreasing friction. As a consequence, the resistance force decreases, and the downslope force sets the mass in motion.

In certain types of wet sediment, shaking can cause **liquefaction**, meaning that a solid sediment layer turns into a slurry. For example, **quick clay**, which consists of damp clay flakes, behaves like a solid when still, for surface tension holds the water-coated flakes together—but shaking separates the flakes from one another and suspends them in the water, thereby transforming the clay into wet mud that flows

Mass Movement

In the Earth's gravity field, what goes up must come down—
sometimes with disastrous consequences. Rock and regolith are not
infinitely strong, so every now and then slopes or cliffs give way in
response to gravity, and materials slide or tumble downslope. This
downslope displacement, called mass movement, or mass wasting,
is the first step in the process of erosion and sediment formation.

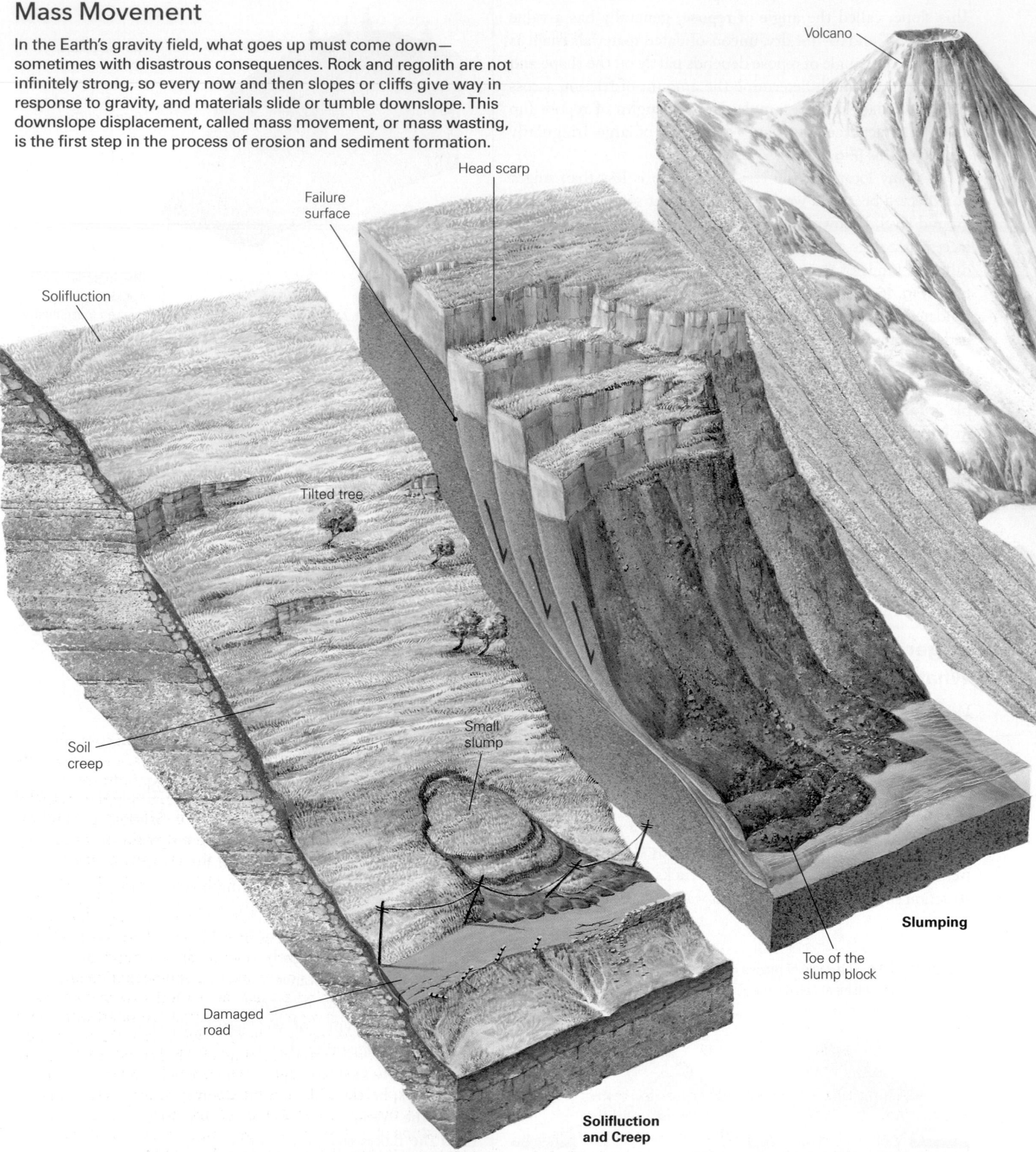

Solifluction

Failure
surface

Head scarp

Volcano

Tilted tree

Soil
creep

Small
slump

Damaged
road

Slumping

Toe of the
slump block

Solifluction
and Creep

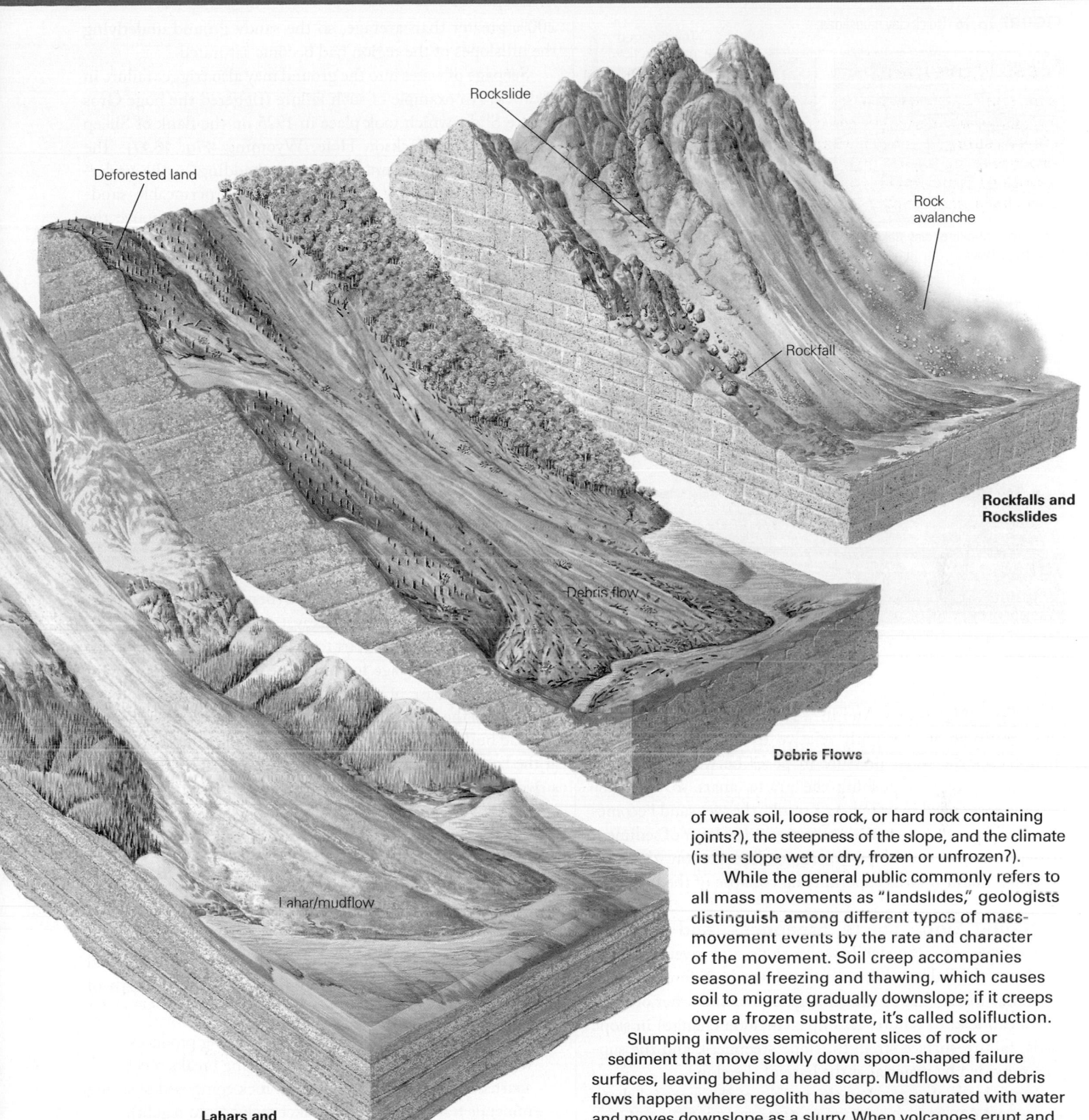

Rockslide

Rock
avalanche

Deforested land

Rockfall

**Rockfalls and
Rockslides**

Debris flow

Debris Flows

Lahar/mudflow

**Lahars and
Mudflows**

The resulting debris may eventually be carried away by water, ice, or wind.

The kind of mass movement that takes place at a given location reflects the composition of the slope (is it composed of weak soil, loose rock, or hard rock containing joints?), the steepness of the slope, and the climate (is the slope wet or dry, frozen or unfrozen?).

While the general public commonly refers to all mass movements as "landslides," geologists distinguish among different types of mass-movement events by the rate and character of the movement. Soil creep accompanies seasonal freezing and thawing, which causes soil to migrate gradually downslope; if it creeps over a frozen substrate, it's called solifluction. Slumping involves semicoherent slices of rock or sediment that move slowly down spoon-shaped failure surfaces, leaving behind a head scarp. Mudflows and debris flows happen where regolith has become saturated with water and moves downslope as a slurry. When volcanoes erupt and melt ice and snow at their summit, or if heavy rains fall during an eruption, water mixes with ash, producing a fast-moving lahar. Steep, rocky cliffs may suddenly give way in rockfalls. If the rock breaks up into a cloud of debris that rushes downslope at high velocity, it becomes a rock avalanche. In snow avalanches, the debris consists of snow.

FIGURE 16.16 Quick clay mudslides.

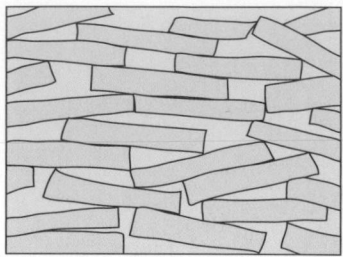

(a) Before shaking, clay flakes stick together.

The fluidized mud can flow.

(b) During shaking, clay flakes separate and become suspended in water.

(c) A mudslide near Namsos, Norway, destroyed ten houses. Movement may have been triggered by nearby blasting.

like a fluid **(Fig. 16.16)**. And in wet sand, shaking causes the sand grains to shift slightly relative to one another, which then causes the water pressure in pores between the grains to increase, thereby pushing the grains apart so that they become suspended in water. As a result, the wet sand becomes a sandy slurry with virtually no strength. If a layer of sediment at depth below a hillslope undergoes liquefaction, the layer becomes a failure surface, permitting collapse of the hillslope.

Changing Slope Loads, Steepness, and Support

As we have seen, the stability of a slope at a given time depends on the balance between the downslope force and the resistance force. Factors that change one or the other of these forces can lead to failure. Examples include changes in slope loads, failure-surface strength, slope steepness, and the support provided by material at the base of the slope.

Slope loads change when the weight of the material above a potential failure surface changes. If the load increases due to the addition of fill, the construction of buildings, or saturation of regolith with water due to heavy rains, the downslope force will increase and may exceed the resistance force. For example, the devastating Oso Slide of 2014 (see Fig. 16.5) happened after a six-week period during which rainfall was

200% greater than average, so the sandy ground underlying the hillslopes of the region had become saturated.

Seepage of water into the ground may also trigger failure in bedrock. An example of such failure triggered the huge Gros Ventre Slide, which took place in 1925 on the flank of Sheep Mountain, near Jackson Hole, Wyoming **(Fig. 16.17)**. The surface of Sheep Mountain parallels bedding in the underlying bedrock. Heavy rains saturated the permeable sandstone beds of the Tensleep Formation, which lay above the weak Amsden Shale, making the bedrock much heavier. On June 27, the downslope force exceeded the restraining force, and 40 million m³ of rock, as well as the overlying soil and forest, detached from the side of the mountain and slid 600 m downslope, with the Amsden Shale serving as the failure surface. The debris filled the valley below and built a 75-m-high natural dam across the Gros Ventre River. A new lake formed behind this dam.

Slope steepness may change over time, when rivers cut valleys or construction engineers pile up debris or cut into slopes **(Fig. 16.18)**. An increase in steepness causes an increase in the downslope force but does not change the resistance force. If the slope becomes too steep, it becomes unstable and may fail. Removing support at the base of a slope has a similar result. In effect, the material at the base of a slope acts like a retaining wall, holding back the material farther up the slope. When natural erosion, or bulldozer excavation, takes away this wall, the upslope material can start to move. Such a process contributed to setting the stage for the Gros Ventre Slide—the river flowing at the base of what would become the slide had been cutting into the base of the hill. Cutting terraces into hillslopes for road building can have the same effect, as can wave erosion at the base of a slope along a coast. In some cases, erosion by a river or by waves eats into the base of a cliff and produces an overhang. When such **undercutting** has occurred, rock making up the overhang eventually breaks away from the slope and falls **(Fig. 16.19)**.

Changing Slope Strength

The stability of a slope depends on the strength of the material constituting it. If the material weakens with time, the slope becomes weaker and eventually collapses. Three factors influence the strength of slopes: weathering, vegetation cover, and water.

- *Weathering:* With time, chemical weathering produces weaker minerals, and physical weathering breaks rocks apart. As a result, a formerly intact rock composed of strong minerals transforms into a weaker rock or into regolith.

- *Vegetation cover:* In the case of slopes underlain by regolith, vegetation tends to strengthen the slope because the roots hold otherwise unconsolidated grains together. Also, plants absorb water from the ground, preventing the substrate from turning into slippery mud. The removal

FIGURE 16.17 Stages leading to the 1925 Gros Ventre Slide in Wyoming.

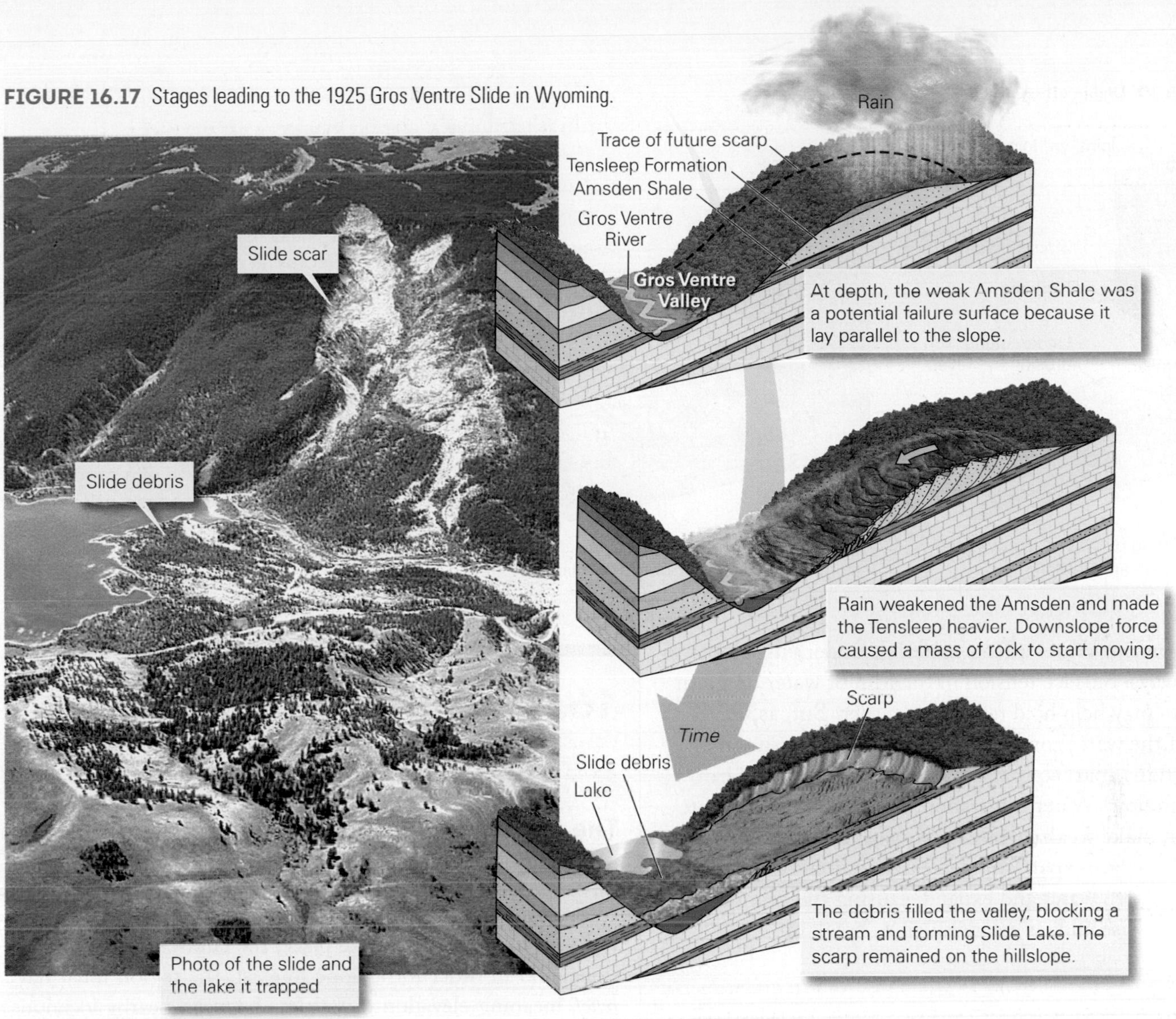

Slide scar

Slide debris

Photo of the slide and the lake it trapped

Rain

Trace of future scarp
Tensleep Formation
Amsden Shale
Gros Ventre River

Gros Ventre Valley

At depth, the weak Amsden Shale was a potential failure surface because it lay parallel to the slope.

Rain weakened the Amsden and made the Tensleep heavier. Downslope force caused a mass of rock to start moving.

Time

Scarp

Slide debris
Lake

The debris filled the valley, blocking a stream and forming Slide Lake. The scarp remained on the hillslope.

of vegetation therefore has the net result of making slopes more susceptible to downslope mass movement. Deforestation, for example, can lead to catastrophic mass movement of the forest's substrate (Fig. 16.20). And terrifying wildfires, stoked by strong winds, can destroy ground-covering vegetation. If heavy rains fall on burned regions, the unprotected soil can became saturated with water and turned into mud, which then flows downslope.

FIGURE 16.18 Processes that can steepen slopes and make them unstable.

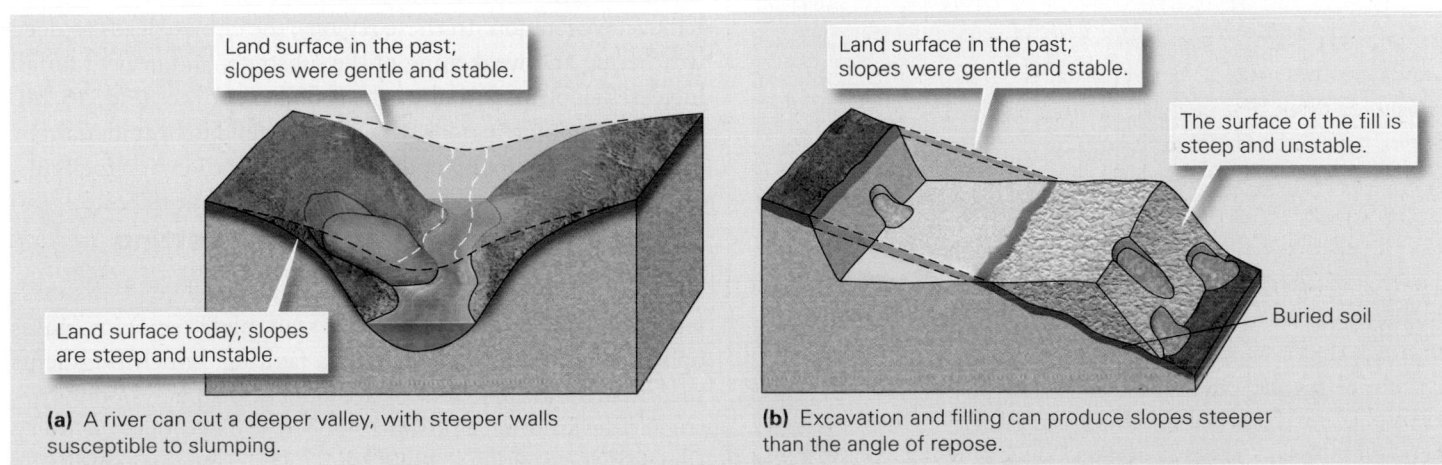

Land surface in the past; slopes were gentle and stable.

Land surface today; slopes are steep and unstable.

Land surface in the past; slopes were gentle and stable.

The surface of the fill is steep and unstable.

Buried soil

(a) A river can cut a deeper valley, with steeper walls susceptible to slumping.

(b) Excavation and filling can produce slopes steeper than the angle of repose.

FIGURE 16.19 Undercutting and collapse of a sea cliff.

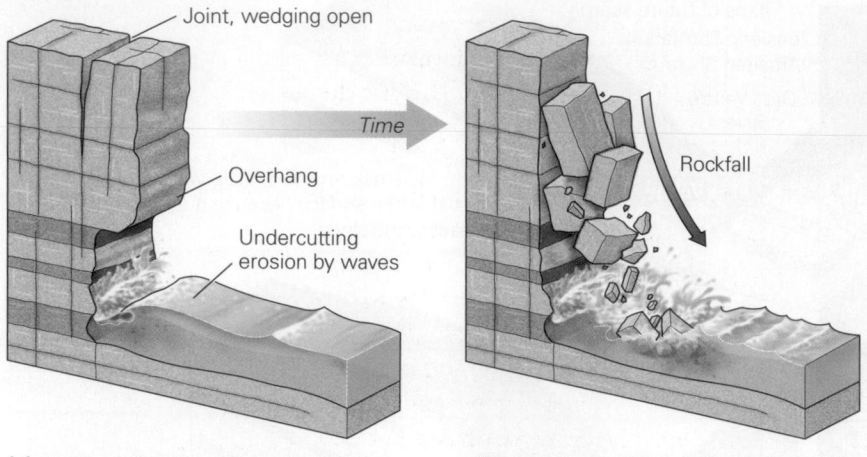

(a) Undercutting by waves removes the support beneath an overhang.

(b) Eventually, the overhang breaks off along joints, and a rockfall takes place.

• *Water content:* Water affects materials underlying slopes in many ways. Surface tension in the film of water on grain surfaces may help hold regolith together. But, as we've seen, if the water content increases, water pressure may push grains apart so that the regolith liquefies and can begin to flow. Water infiltration may make the substrate heavier, make weak surfaces underground more slippery, or push surfaces apart and decrease friction. Some types of clay absorb water and expand, causing the ground surface to rise and, as a consequence, break up.

FIGURE 16.20 Deforestation decreases slope stability. A large slump has formed on this deforested hill in Brazil.

Slump scar

16.4 Where Do Mass Movements Occur?

The Importance of Relief, Climate, and Substrate

Mass movements don't happen on a horizontal plain. In order for mass movements to take place, the landscape must have a slope, and in order for a slope to exist, the region must have *relief,* meaning elevation differences between nearby locations. Generally, the steeper the slope, the more susceptible the slope will be to mass movement. Therefore, mass movements happen most frequently where rivers or glaciers have cut deep valleys, where mountains have dramatic slopes, and where ocean waves have cut into an elevated coast.

Climate also affects where mass movements take place. For example, regions with heavy seasonal rainfall are subject to mudflows and landslides, especially if fire or deforestation has removed vegetation. In these regions, the climate leads to deep weathering and weakening of the substrate, and heavy rainfall increases slope loads and weakens failure surfaces (**Fig. 16.21**). Desert and alpine regions, where slopes tend to be underlain by unweathered bedrock, tend to host rare but dramatic rockfalls.

The Importance of the Tectonic Setting

Most unstable ground on the Earth ultimately owes its existence to plate tectonics. As we've seen, plate tectonics causes uplift, generates relief, and causes faulting, which fragments the crust. And, of course, earthquakes on plate boundaries trigger devastating landslides. Spend a day along the steep slopes of the Southern Alps, a mountain range now uplifting

FIGURE 16.21 An earthquake triggered the slumping of this rain-soaked hillside in Taiwan in 2010. The mass of the slump was so wet that part of it turned into a mudflow that spread over a highway.

along the plate boundary that transects New Zealand, and you can hear mass movement in progress—during heavy rains, rockfalls and landslides clatter with astounding frequency, as if the mountains were falling down around you.

A Case Study: Southern California

To see the interplay of plate tectonics and other factors, let's consider a case study in southern California. Contractors have built expensive homes on cliffs overlooking the Pacific so that homeowners can enjoy spectacular sunset views from their backyard patios. But the landscape isn't ideal for houses, from the standpoint of stability. Slumps and mudflows on coastal cliffs have consumed some of these homes over the years, with a cost to their owners (or insurance companies) of millions of dollars. What is special about southern California that makes it so susceptible to mass wasting?

First, California lies along an active plate boundary, the San Andreas fault. Repeated slip events along the fault over millions of years have shattered the rock of California's crust. Fractures not only act as planes of weakness but can provide paths for water to seep into bedrock and cause chemical weathering, producing slippery clay that weakens the rock. In addition, the rocks in many areas are weak to begin with because they formed as part of an accretionary prism, a chaotic mass of sediment and seafloor basalt that was scraped off subducting oceanic lithosphere during the Mesozoic Era.

> **Did you ever wonder...**
> why parts of the California coast slip into the sea?

Though most of the movement on the boundary between the North American and Pacific Plates involves strike-slip displacement, there is a component of compression across the boundary. This compression leads to uplift and slope formation in California (see Box 16.1). Since the uplifted region borders the coast, wave erosion steepens and, in some places, undercuts cliffs. And movement on the boundary causes numerous earthquakes to rock the region, shaking regolith loose.

The California coast also tends to be susceptible to mass movements because of its climate. In general, the region has a hot and dry climate that hosts only semidesert flora. Brush fires remove much of this cover, leaving large areas with no dense vegetation. But since the region lies adjacent to the Pacific Ocean, it endures occasional heavy winter "monsoonal" rains. The water sinks quickly into the sparsely vegetated ground, adds weight to the mass on the slope, and weakens failure surfaces.

Development of cities and suburbs locally contributes to triggering mass movements in southern California. Development has oversteepened and overloaded slopes and has caused the water content of regolith to change. The consequences can be seen in the Portuguese Bend Slide in the Palos Verdes area bordering the shore near Los Angeles **(Fig. 16.22)**. The Portuguese Bend region overlies a thick, seaward-dipping layer of weak volcanic ash (now altered to weak clay) resting on shale. The weak ash has served as a failure surface a few times during the past few thousand years. In 1956, developers deposited a 23-m-thick layer of fill over the ground surface and built homes on top. As residents watered their lawns and used septic tanks, water began to seep into the ground and decreased the strength of the ash layer. Because of the decrease in strength, the added weight, and the erosion of the toe of the hill by the sea, the upper 30 m of land began to move once again. Between 1956 and 1985, the Portuguese Bend Slide moved at rates of up to 2.5 cm per day. Eventually, portions of a 260-acre region

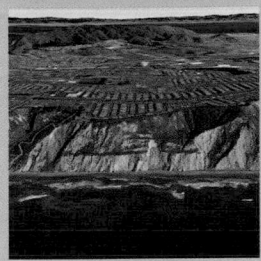

FIGURE 16.22 The Portuguese Bend Slide viewed from the air.

Head scarp

Approximate edge of slump

Hummocky slump surface

slid by over 200 m, and in the process more than 150 homes were destroyed.

16.5 How Can We Protect against Mass-Movement Disasters?

Identifying Regions at Risk

Clearly, landslides, mudflows, and slumps are natural hazards that we cannot ignore. Too many of us live in regions where mass movements have the potential to kill people and destroy property. In many cases, the best solution is avoidance: don't build, live, or work on slopes or below slopes where mass movement may take place. But avoidance will be possible only if we know where the hazards are greatest.

To pinpoint regions that are particularly susceptible to mass movements, and could be regions of elevated risk, geologists look for landforms known to result from mass movements, because where mass movements have happened in the past, they might happen again in the future. Features such as slump head scarps, swaths of forest in which trees have been tilted, piles of loose debris at the bases of hills, and hummocky (bumpy) land surfaces all indicate recent mass wasting.

Geologists may also be able to detect regions that are beginning to move **(Fig. 16.23)**. For example, roads, buildings, and pipes begin to crack over unstable ground. Power lines may be too tight or too loose because the poles to which they are attached move together or apart. Visible cracks may form on the ground at the potential head of a slump, while the ground may bulge up at the toe of the slump. In some cases, subsurface cracks may drain the water from an area and kill off vegetation, whereas in other areas land may sink and form a swamp. Even if mass wasting takes place too slowly to be perceptible to people, it can be documented with sensitive instruments that can detect a subtle tilt of the ground or changes in distance between nearby points. Specifically, measurements using satellite data, laser surveys, and tiltmeters permit geologists to identify movements of just a few millimeters that, while small, may indicate that a slump has reactivated.

If various clues indicate that a mass movement has commenced, and if conditions make accelerating movement likely

FIGURE 16.23 Surface features warn that a large slump is beginning to develop. Cracks that appear at the head scarp may drain water and kill trees. Power-line poles tilt and the lines become tight. Fences, roads, and houses on the slump begin to crack.

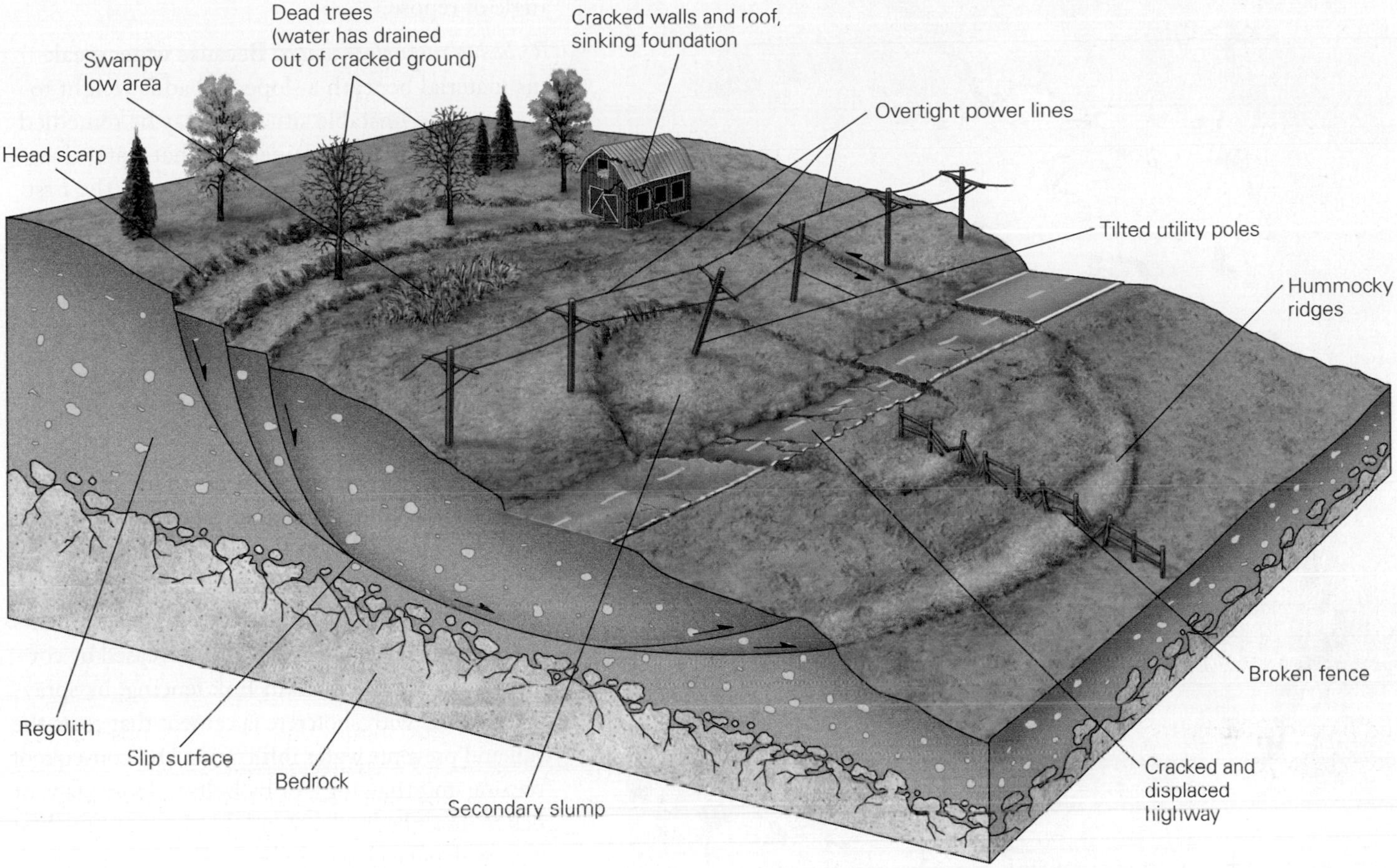

Dead trees
(water has drained
out of cracked ground)

Swampy
low area

Cracked walls and roof,
sinking foundation

Overtight power lines

Head scarp

Tilted utility poles

Hummocky
ridges

Regolith

Slip surface

Bedrock

Secondary slump

Broken fence

Cracked and
displaced
highway

(e.g., persistent rain, rising floodwaters, or continuing earthquake aftershocks), then officials may order an evacuation. Evacuations have saved lives, and ignored warnings have cost lives. But unfortunately, some mass movements happen without any warning, and some evacuations prove costly but unnecessary.

In places subject to monitoring, warnings can be quite precise. For example, an immense landslide that occurred within the Bingham Canyon Mine of Utah was predicted with enough advance warning to evacuate the miners before the landslide happened. The landslide, technically a debris flow, carried material almost 2 km at speeds of up to 100 km per hour **(Fig. 16.24).** It destroyed some buildings and equipment, but did not cause any injuries.

Even if a region doesn't display evidence of recent movement, a danger may still exist: just because a steep slope hasn't collapsed in the recent past doesn't mean it won't in the future. In recent years, geologists have begun to identify such potential hazards by using computer programs that evaluate factors that can trigger mass movement. With these data, they create maps that portray the degree of risk for a certain location. The factors that geologists consider include slope steepness; strength of substrate; degree of water saturation; orientation of bedding, joints, or foliation

relative to the slope; vegetation cover; potential for heavy rains; potential for undercutting to occur; and likelihood of earthquakes. From such hazard-assessment studies, geologists compile **landslide-potential maps**, which rank regions

FIGURE 16.24 A huge debris flow took away one of the walls of the Bingham Canyon Mine in Utah. No lives were lost in this open-pit copper mine because engineers detected the instability of the slope and evacuated the miners in time.

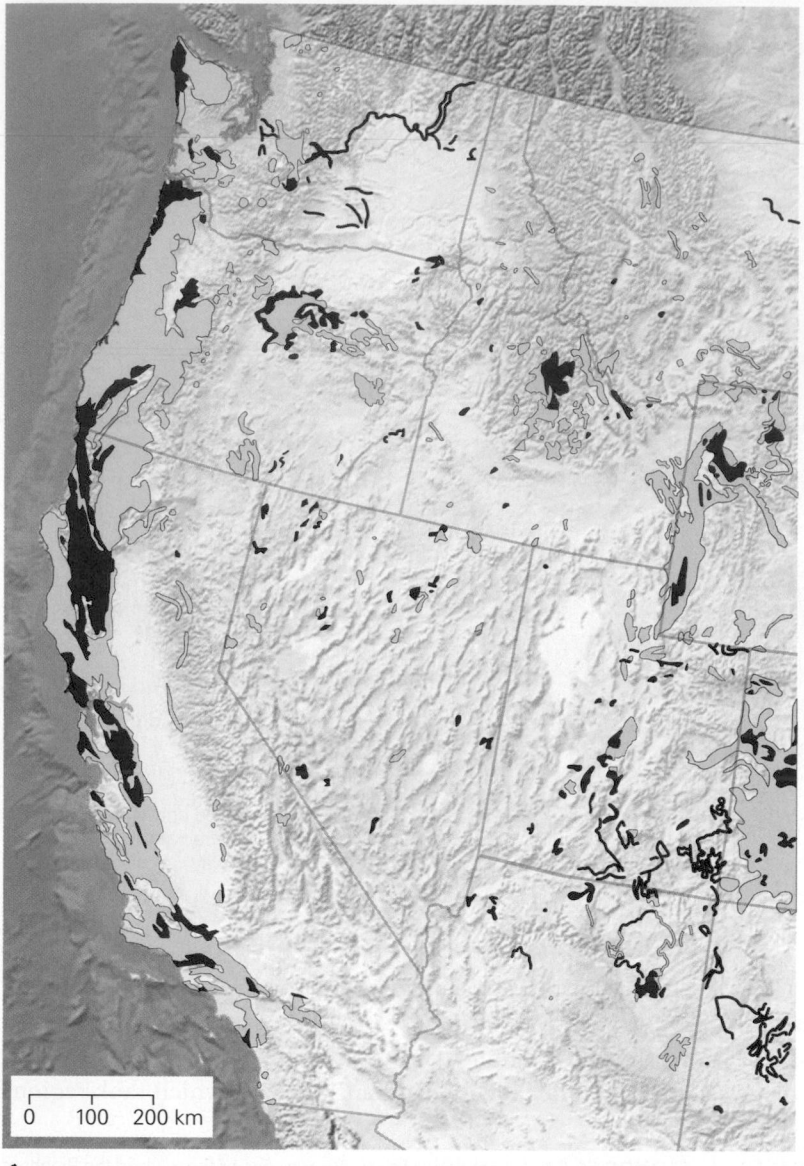

FIGURE 16.25 A landslide-potential map of the western United States.

0   100   200 km

High (More than 15% of area involved)

Moderate (1.5–15% of area involved)

Low (Less than 1.5% of area involved)

- *Regrading:* An oversteepened slope can be regraded or terraced so that it does not exceed the angle of repose.

- *Reducing subsurface water:* Because water weakens material beneath a slope and adds weight to the slope, an unstable situation may be remedied either by improving drainage so that water does not enter or remain in the subsurface in the first place or by pumping water out of the ground.

- *Preventing undercutting:* In places where a river undercuts a cliff face, engineers can divert the river. Similarly, in coastal regions, they may build an offshore breakwater or pile **riprap** (loose boulders or concrete) along the beach to absorb wave energy before it strikes the cliff face.

- *Building safety structures:* In some cases, engineers can build structures to stabilize a potentially unstable slope or to protect a region downslope from debris if a mass movement does occur. For example, retaining walls can stabilize highway embankments. The danger from rockfalls can be decreased by covering a road cut with chain-link fencing, by spraying road cuts with shotcrete (a cement that coats the wall and prevents water infiltration and consequent freezing and thawing), or by bolting loose slabs of rock to intact bedrock behind. Highways at the base of an avalanche chute can be covered by an avalanche shed, whose roof keeps debris off the road.

- *Controlled blasting of unstable slopes:* When it is clear that unstable ground or snow threatens a particular region, the best solution may be to blast the unstable ground or snow loose at a time when its movement can do no harm.

Clearly, preventing mass-movement calamities can be expensive, and people might not always be willing to pay the price. In such cases, they have a choice of avoiding the risky area, taking the chance that a calamity will not happen while they are around, buying appropriate insurance, or counting on relief agencies to help if disaster does strike. Once again, geology and society cross paths.

according to the likelihood that mass movement will occur **(Fig. 16.25)**. Officials may restrict construction in landslide-prone areas.

Preventing Mass Movements

In areas where landslide potential exists, people can take certain steps to remedy the problem and stabilize the slope **(Fig. 16.26)**:

- *Revegetation:* Since bare ground tends to be more vulnerable to downslope movement, slope stability in deforested areas will be greatly enhanced if landowners replant the region with vegetation that sends down deep roots.

TAKE-HOME MESSAGE

Various features of the landscape may help geologists to identify unstable slopes and estimate risk. Systematic study allows production of landslide-potential maps. Engineers can use a variety of techniques to stabilize slopes.

QUICK QUESTION: Why is mass movement of major concern during production of large road cuts?

FIGURE 16.26 A variety of remedial steps can stabilize unstable ground.

Roots stabilize the potential failure plane.

Potential failure plane

(a) Revegetating a slope results in the growth of roots that can hold a slope together.

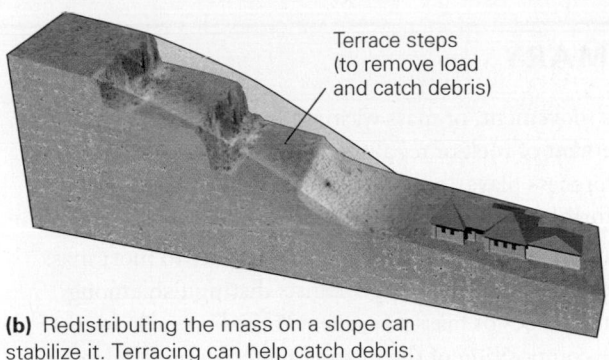

Terrace steps (to remove load and catch debris)

(b) Redistributing the mass on a slope can stabilize it. Terracing can help catch debris.

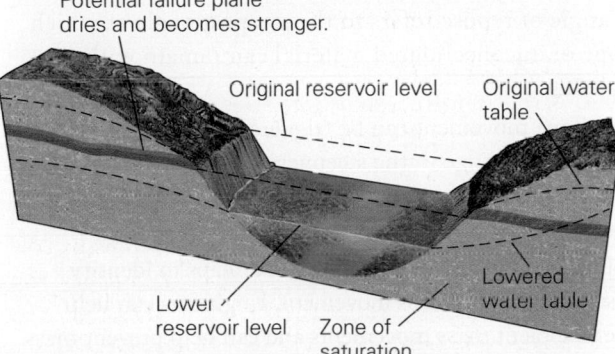

Potential failure plane dries and becomes stronger.

Original reservoir level

Original water table

Lower reservoir level

Zone of saturation

Lowered water table

(c) Lowering the level of the water table can strengthen a potential failure surface.

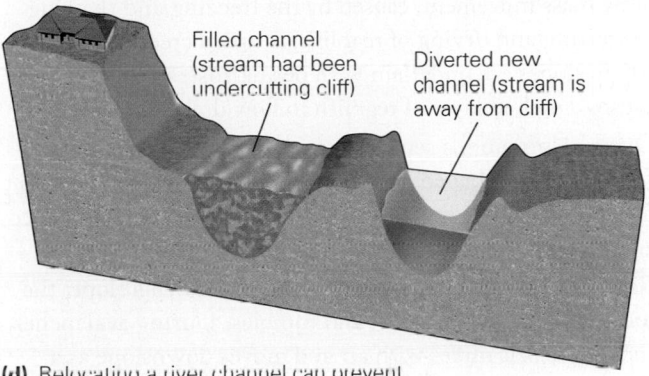

Filled channel (stream had been undercutting cliff)

Diverted new channel (stream is away from cliff)

(d) Relocating a river channel can prevent undercutting.

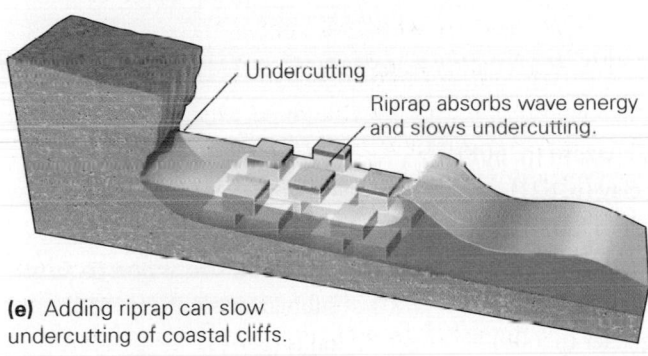

Undercutting

Riprap absorbs wave energy and slows undercutting.

(e) Adding riprap can slow undercutting of coastal cliffs.

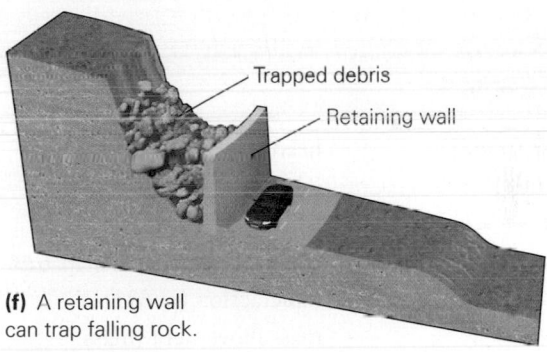

Trapped debris

Retaining wall

(f) A retaining wall can trap falling rock.

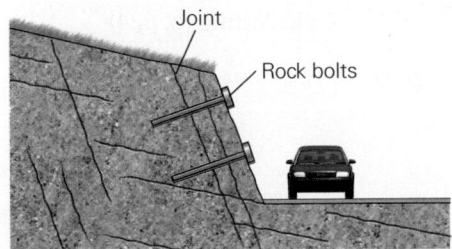

Joint

Rock bolts

(g) Bolting or screening a cliff face can hold loose rocks in place.

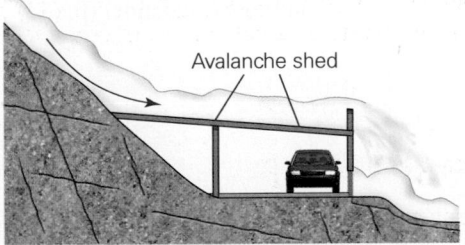

Avalanche shed

(h) An avalanche shed diverts debris or snow over a roadway.

SUMMARY

- Mass movement, or mass wasting, is the downslope movement of rock or regolith under the influence of gravity. This process plays an important role in the shaping of landforms and the transport of sediment.

- Although in everyday language people refer to most mass movements as landslides, geologists distinguish among different types of mass movements based on such factors as the composition of the moving materials and the rate of movement.

- Slow mass movement, caused by the freezing and thawing or wetting and drying of regolith, is called creep. In places where slopes are underlain with permafrost, solifluction causes a melted layer of regolith to flow downslope.

- During slumping, a semicoherent mass of material moves down a spoon-shaped failure surface. Mudflows and debris flows occur where regolith has become saturated with water and moves downslope as a slurry.

- Rock and debris slides move very rapidly down a slope; the rock or debris breaks apart and tumbles. During avalanches, snow or debris mixes with air and moves downslope as a turbulent cloud. And in a debris fall or rockfall, the material free-falls down a vertical cliff.

- Mass movements of various types can occur beneath the sea and, therefore, can trigger tsunamis.

- Intact, fresh rock is too strong to undergo mass movement. So, for mass movement to be possible, rock must be weakened by fracturing (joint formation) or weathering.

- Unstable slopes start to move when the downslope force exceeds the resistance force that holds material in place. The angle of repose refers to the steepest angle at which a slope of unconsolidated material can remain without collapsing.

- Downslope movement can be triggered by shocks and vibrations, a change in the steepness of a slope, removal of support from the base of the slope, a change in the strength of a slope, deforestation, weathering, or heavy rain.

- Geologists produce landslide-potential maps to identify areas susceptible to mass movement. Engineers can help detect incipient mass movements and can help prevent mass movements by using a variety of techniques to stabilize slopes.

GUIDE TERMS

angle of repose (p. 621)
avalanche (p. 608)
creep (p. 608)
debris fall (p. 615)
debris flow (debris slide) (p. 611)
debris slide (p. 614)
failure surface (p. 610)

head scarp (p. 610)
lahar (p. 611)
landslide (p. 608)
landslide-potential map (p. 629)
liquefaction (p. 621)
mass movement (mass wasting) (p. 607)
mudflow (mudslide) (p. 611)

natural hazard (p. 608)
quick clay (p. 621)
regolith (p. 607)
riprap (p. 630)
rockfall (p. 615)
rock glacier (p. 608)
rockslide (p. 614)
slope failure (p. 620)

slump (p. 610)
snow avalanche (p. 614)
solifluction (p. 608)
submarine debris flow (p. 616)
submarine slump (p. 616)
talus (p. 615)
turbidity current (p. 616)
undercutting (p. 624)

GEOTOURS *THIS CHAPTER'S GEOTOURS WORKSHEET (M) FEATURES QUESTIONS AND GOOGLE EARTH SITES ON:*

- Portuguese Bend Slide, CA
- La Conchita Mudslide, CA
- Gros Ventre Slide, WY

- Vaiont Dam Rockslide, Italy
- Frank Rockslide, Alberta, Canada

- Velarde Slumps, NM
- Landslide Remediation, Japan

REVIEW QUESTIONS

The letters following each Review Question refer to the corresponding Learning Objective from the Chapter Opener.

1. What factors do geologists use to distinguish among various types of mass movements? **(A)**

2. Identify the key differences among a slump, a debris flow, a lahar, an avalanche, a rockslide, and a rockfall. Which type of mass wasting is depicted here? Identify the parts of the sketch. **(A)**

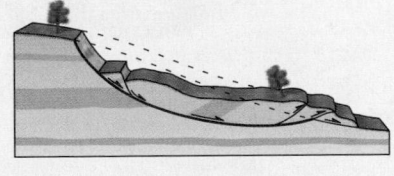

3. Explain the process of creep and discuss how it differs from solifluction. **(A)**

4. Distinguish among different types of submarine mass movements. Which of these types can trigger a major tsunami, and why? **(C)**

5. Why is intact bedrock stronger than fractured bedrock? Why is it stronger than regolith? **(B)**

6. Explain the difference between a stable and an unstable slope. What factors determine the angle of repose of a material? What features may serve as failure surfaces? **(B)**

7. Discuss the variety of phenomena that can cause a stable slope to become so unstable that it fails. **(D)**

8. How can ground shaking cause fairly solid layers of sand or mud to become weak slurries capable of flowing? **(D)**

9. Discuss the role of vegetation in slope stability. Why can fires and deforestation lead to slope failure? **(B)**

10. Identify the various factors that make the coast of southern California susceptible to mass movements. **(D)**

11. What factors do geologists take into account when producing a landslide-potential map, and how can they detect the beginning of mass movement in an area? **(E)**

12. What steps can people take to avoid landslide disasters? What is the purpose of the rock bolts shown in the sketch? **(E)**

ON FURTHER THOUGHT

13. Imagine that you have been asked by the World Bank to determine whether it makes sense to build a dam in a steep-sided, east-west-trending valley in a small central Asian nation. Initial investigation shows that the rock of the valley floor consists of schist containing a strong foliation that dips south. Outcrop studies reveal that abundant fractures occur in the schist along the valley floor; the surfaces of most fractures are coated with slickensides. Moderate earthquakes have rattled the region. Explain the hazards and what might happen if the reservoir behind the dam were filled. What would you advise? **(B, E)**

ONLINE RESOURCES

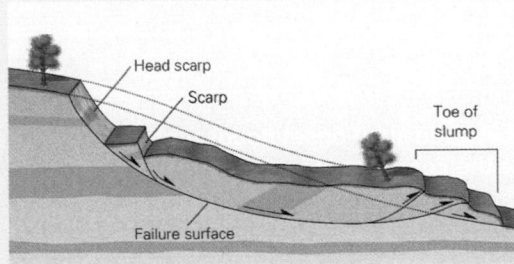

Head scarp
Scarp
Toe of slump
Failure surface

Videos
This chapter features videos on developing and detecting slumps, and landslide processes.

Smartwork5
This chapter includes questions on features and processes that change the landscape, and on mass wasting.

CHAPTER 17

Streams and Floods:
The Geology of Running Water

By the end of this chapter, you should be able to . . .

A. explain how streams and drainage networks form and evolve.

B. describe the processes that lead to erosion and deposition by streams.

C. characterize the changes that take place along the length of a stream from its headwaters to its mouth.

D. sketch the evolution of meanders in a stream, and explain the concept of a floodplain.

E. distinguish between slow-onset floods and flash floods, and describe the conditions that lead to each.

F. describe the various methods that engineers may use to protect areas from flooding, and interpret statements concerning flood frequency.

G. characterize environmental issues associated with streams.

Nothing in the world is more flexible and yielding than water.
Yet when it attacks the firm and the strong, none can withstand it.

—Laozi (legendary ancient Chinese philosopher)

17.1 Introduction

Wind swooping poleward across warm oceans can pick up vast quantities of water vapor, which it can transport into the interior of a continent. On a September day in 2013, such a mass of warm, moist air collided with a mass of cold air stalled along the eastern edge of the Rocky Mountains in Colorado. The warm air rose over the cold air, the moisture in it condensed, and drenching rains began to fall. In Boulder County, 430 mm of rain fell over a period of a few days—normally, the county receives 525 mm over the course of a whole year! Some of the water ended up in rivers that flow from the mountains toward the plains. One of these rivers, the Big Thompson, temporarily carried 30 times more water than normal. The water rushing down the river became much faster, more turbulent, and muddier than normal. Eventually, the roaring torrent washed away houses and trees, scoured away roads **(Fig. 17.1a)**, destroyed rail lines, inundated farmland, broke pipelines, overwhelmed sewage treatment plants, and undermined bridges. By the time the storms subsided and water levels returned to normal, the tragedy had claimed eight lives and destroyed $2 billion of property. During the same year, flooding in Alberta had a similar disastrous effect in the eastern Canadian Rockies **(Fig. 17.1b)**, where flood waters spread out onto the plains, submerging large portions of the city of Calgary **(Fig. 17.1c)**. In April of 2017, sudden floods and mudslides due to heavy rains dragged giant boulders and other debris through the streets of Moaca, Colombia, killing 262 people **(Fig. 17.1d)**.

The people of Colorado, Alberta, and Colombia had, unfortunately, experienced the immense power of *running water*, water that flows down the surface of sloping land in response to the pull of gravity. Geologists use the term **stream** for any body of running water that flows in a **channel**, an elongate depression or trough, and refer to the edges of the channel as the stream's *banks*, the floor of the channel as the *streambed* or *bed*, and a length of the channel as a *reach*. (In everyday English, we refer to large streams as *rivers* and medium-sized ones as *creeks*.) Water in a stream, overall, flows from the *upstream region*, closer to the source or **headwaters** of the stream, to the *downstream region*, closer to the end or **mouth** of the stream. Streams drain water from the landscape and carry it into lakes or to the sea, much as culverts drain water from a parking lot. In the process, streams relentlessly erode the landscape and transport sediment and debris to sites of deposition. Generally, a stream stays within the confines of its channel, but when the supply of water entering a stream exceeds the channel's capacity, water spills out and covers the surrounding land, thereby causing a **flood**, such as the ones that devastated parts of Colorado, Alberta, and Colombia.

Of all the planets in the Solar System, only the Earth currently hosts flowing streams. Streams are of great importance to humanity, not only because of how they modify the landscape, especially during floods, but also because they provide avenues for travel and commerce, nutrients and sediment for agriculture, water for irrigation and consumption, and energy to produce electricity and drive factories. In this chapter, we examine how streams operate in the Earth System. First, we learn about the origin of running water and about how the water draining a region organizes into networks of streams. Then we look at the process of stream erosion and deposition and at the landscapes that form as a consequence of these processes. Finally, we focus on the nature and consequences of flooding and address the question of how people can mitigate flooding risk.

17.2 Draining the Land

Forming Streams and Drainage Networks

Where does the water in a stream come from? Recall that water enters the hydrologic cycle by evaporating from the Earth's surface and rising into the atmosphere (see Interlude F). After a relatively short residence time, atmospheric water condenses and falls back to the Earth's surface as rain or snow, which accumulates in various reservoirs. Some rain or snow remains in relatively nonflowing accumulations on the land (puddles, swamps, lakes, snowfields, and glaciers), some flows down the land surface as a thin film called **sheetwash**, and some sinks into the ground, where it either becomes trapped in soil (as *soil moisture*) or descends below the water table to become *groundwater*. (As we will see Chapter 19, the *water table* defines the level below which groundwater fills all the pores and cracks in subsurface rock or sediment.) The flowing water in a stream can come from any or all of these reservoirs **(Fig. 17.2)**. Specifically,

◄ (facing page) A torrent of water rushes down the floor of a steep-sided canyon in the Andes of Peru. The silt and mud suspended in the water give it a brown color. During a major flood, this stream can move the boulders that litter its bed, and it would be a threat to the people who have built along its banks.

FIGURE 17.1 Examples of flooding damage illustrate the power of running water.

(a) A stream roaring down a Colorado canyon in 2013 undercut and carried away part of a highway.

(b) Heavy rains in the Canadian Rockies caused extensive flooding in Alberta, in 2013.

(c) The 2013 flood inundated parts of Calgary, including the stadium for the city's Stampede (rodeo).

(d) In April 2017, floods covered the streets of Moaca, Colombia with giant boulders and debris.

FIGURE 17.2 Excess surface water (runoff) comes from rain, melting ice or snow, and groundwater springs. On flat ground, water accumulates in puddles or swamps, but on slopes it flows downslope in streams.

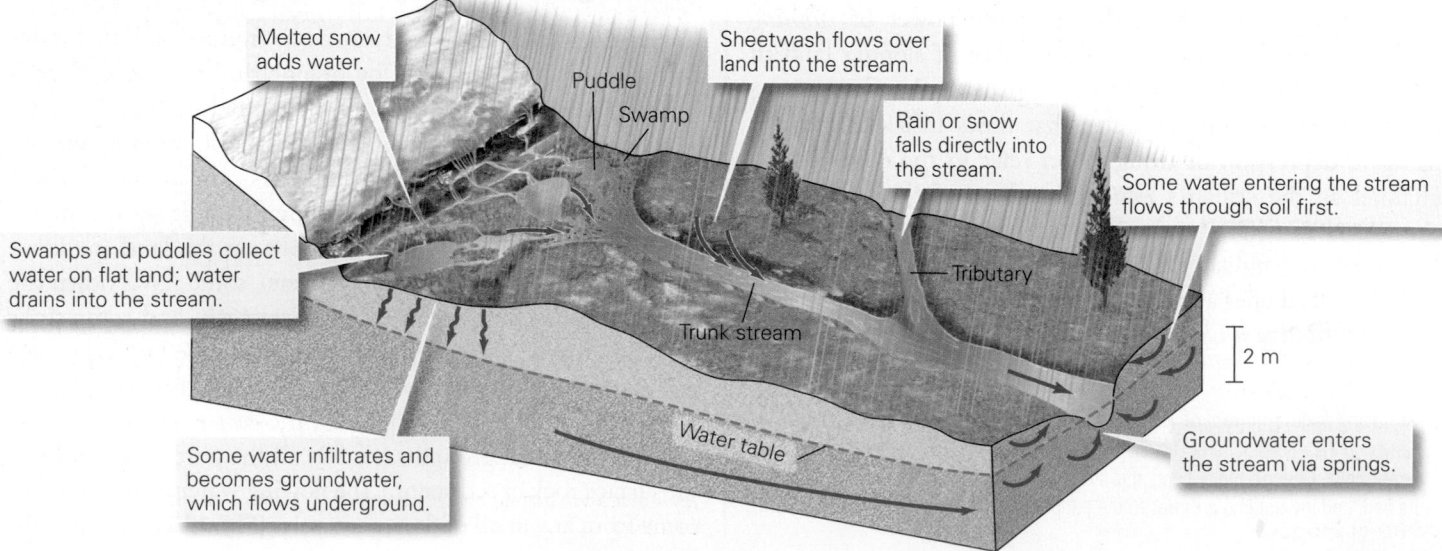

Melted snow adds water.

Puddle

Swamp

Sheetwash flows over land into the stream.

Rain or snow falls directly into the stream.

Some water entering the stream flows through soil first.

Swamps and puddles collect water on flat land; water drains into the stream.

Tributary

Trunk stream

2 m

Some water infiltrates and becomes groundwater, which flows underground.

Water table

Groundwater enters the stream via springs.

it can spill from the outlet of a lake or swamp, it can melt from the end of a glacier or snowfield, it can bubble up from springs that tap into groundwater, it can seep out of the soil, or it can spill into a stream as sheetwash.

Runoff, water flowing on the surface of the Earth, can collect in stream channels because the streambed lies at a lower elevation than the surrounding area, and gravity always moves material from a higher to a lower elevation. But how does a stream channel form in the first place? In some cases, the channel initiates at the mouth of a spring, where groundwater seeps out and starts to flow on the surface. The process of channel formation can also begin when sheetwash starts flowing downslope **(Fig. 17.3a)**, for like any flowing fluid, sheetwash erodes its *substrate*, the material it flows over. The efficiency of such erosion depends on the velocity of the flow (faster flows have more power, so they erode more rapidly) and on the resistance of the substrate to erosion. Where the flow happens to be faster, the substrate weaker, or protective groundcover more sparse, erosion scours (digs) a channel. Since the channel bed sits lower than the surrounding ground, sheetwash in adjacent areas starts to head toward it. Over time, the extra flow deepens the channel, a process called **downcutting**. More sheetwash diverts into the channel, in turn causing flow velocity to increase and erosion to take place even more rapidly. If this process continues for enough time, a distinct stream forms.

As time passes, water may dig into the land at the head of the channel, lengthening the stream by a process called **headward erosion (Fig. 17.3b).** Headward erosion occurs for two reasons. First, it happens when the surface flow converging at the head of a channel has sufficient erosive power to downcut. Second, it happens at locations where groundwater seeps out of the ground and enters the head of the stream channel. Such seepage, called *groundwater sapping,* gradually weakens and undermines the soil or rock just upstream of the channel's head until the material collapses into the channel—the collapsed debris eventually washes away during a flood. Each increment of collapse makes the channel longer.

FIGURE 17.3 The formation of stream channels and drainage networks.

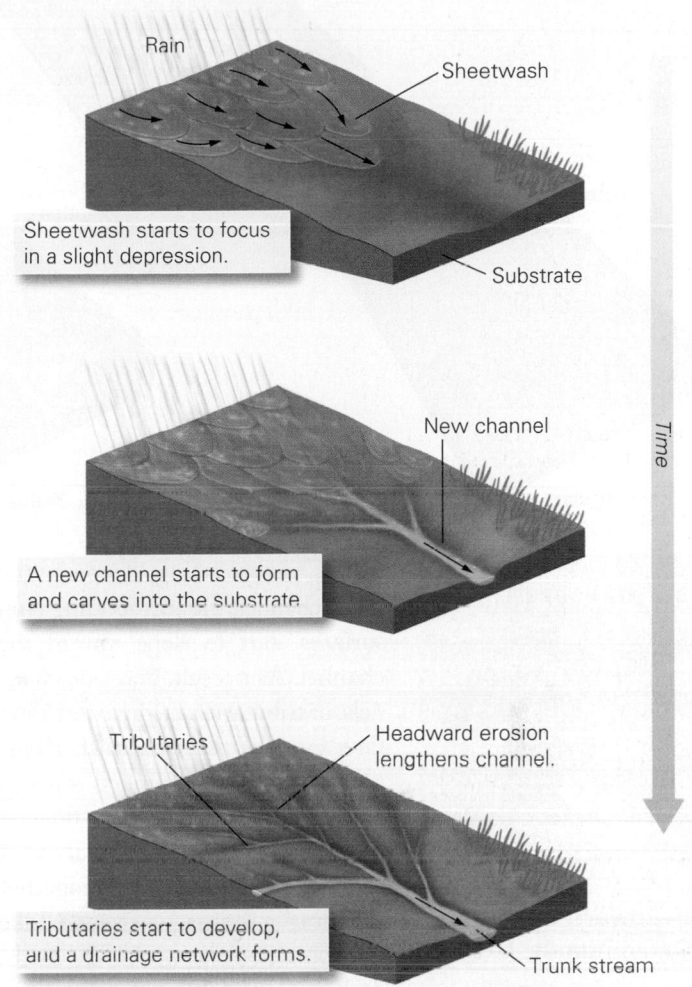

Rain

Sheetwash

Sheetwash starts to focus in a slight depression.

Substrate

New channel

A new channel starts to form and carves into the substrate

Time

Tributaries

Headward erosion lengthens channel.

Tributaries start to develop, and a drainage network forms.

Trunk stream

(a) Progressive stages during the development of a network of stream channels.

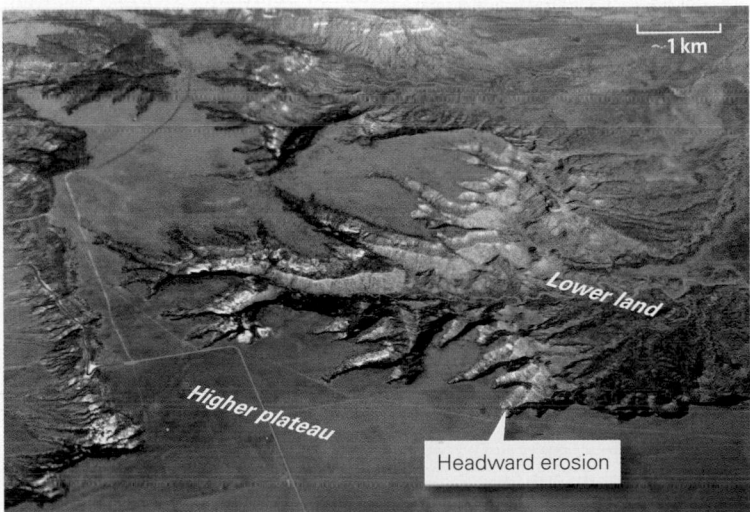

~1 km

Higher plateau

Lower land

Headward erosion

(b) As seen in this air photo, headward erosion is cutting into a plateau.

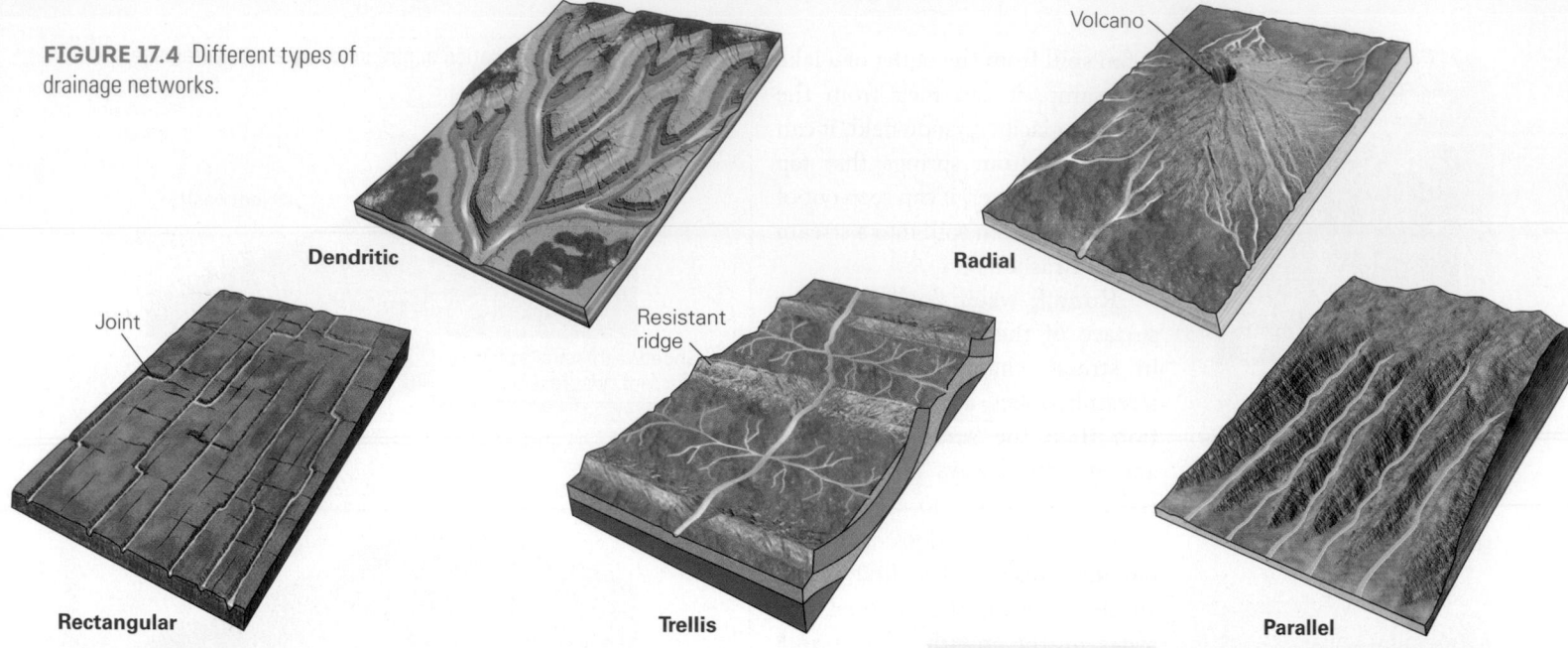

FIGURE 17.4 Different types of drainage networks.

Dendritic

Radial — Volcano

Joint — Rectangular

Resistant ridge — Trellis

Parallel

SEE FOR YOURSELF . . .

Headward Erosion, Canyonlands, Utah

LATITUDE
38°17'58.16" N

LONGITUDE
109°50'18.76" W

Look down from 8 km.

Tributaries flow down canyons into the Colorado River in Canyonlands National Park. The tributaries are downcutting into horizontal strata. Note the steep escarpments at the head of each canyon. Here, groundwater sapping causes collapse of debris into the canyon, resulting in the headward erosion.

As downcutting deepens the main channel, the surrounding land surfaces start to slope toward the channel. As a result, new side channels, or **tributaries**, begin to form and flow into the main channel. Eventually, an array of linked streams evolves, with tributaries flowing into a **trunk stream**. The array of interconnecting streams together constitute a **drainage network**. Like transportation networks of roads, drainage networks of streams reach into all corners of a region, providing conduits for the removal of runoff.

The configuration of tributaries and trunk streams defines the map pattern of a drainage network. The pattern that develops in a given region depends on the shape of the landscape and the character of the substrate. Geologists recognize several distinct geometries of drainage networks **(Fig. 17.4)**:

- *Dendritic:* When rivers flow over a fairly uniform substrate with a fairly gentle slope, a dendritic network develops. It looks like the pattern of branches connecting to the trunk of a deciduous tree. In fact, the word *dendritic* comes from the Greek *dendros*, meaning tree.

Badlands topography

- *Radial:* Drainage networks forming on the surface of a cone-shaped mountain, such as a volcano, flow outward from the mountain peak, like spokes on a wheel. Such a pattern defines a radial network.

- *Rectangular:* In places where a rectangular grid of fractures (vertical joints) breaks up the ground, channels form along the pre-existing fractures, and streams join one another at right angles, creating a rectangular network.

- *Trellis:* In places where a drainage network develops across a landscape of parallel valleys and ridges, major tributaries flow down a valley and join a trunk stream that cuts across the ridges. The place where a trunk stream cuts across a resistant ridge is a *water gap*. The resulting map pattern resembles a garden trellis, so the arrangement of streams constitutes a trellis network.

- *Parallel:* On a uniform, fairly steep slope, several streams with parallel courses develop simultaneously. The group of streams constitutes a parallel network. Downslope,

FIGURE 17.5 Drainage divides and basins.

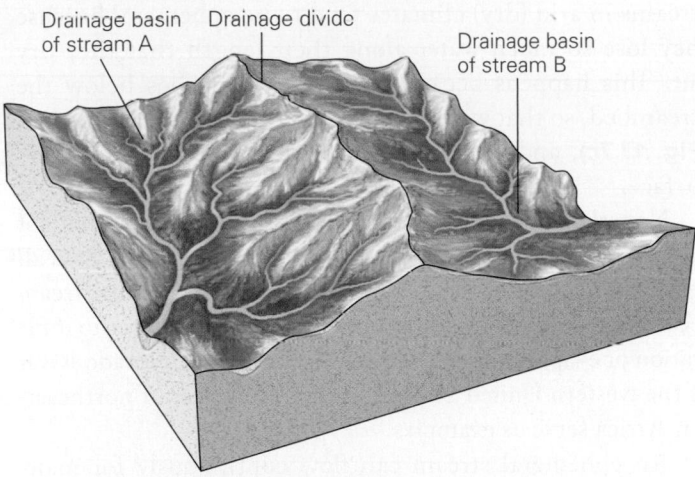

Drainage basin of stream A — Drainage divide — Drainage basin of stream B

(a) A drainage divide is a relatively high ridge that separates two drainage basins.

Main drainage divide

A trunk stream

Secondary drainage divide

A small drainage basin

What a Geologist Sees

(b) An air view shows numerous small drainage basins formed during erosion of this ridge in Oregon.

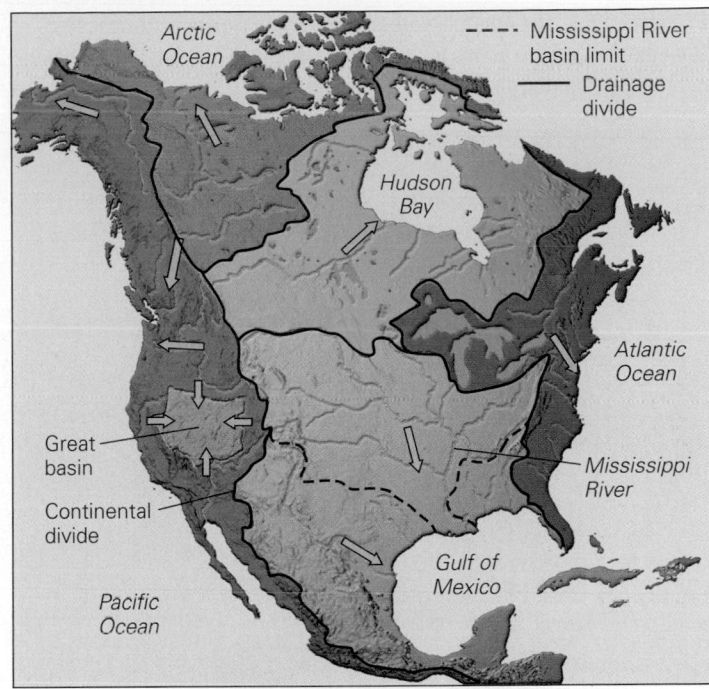

Arctic Ocean

- - - - Mississippi River basin limit

—— Drainage divide

Hudson Bay

Great basin

Continental divide

Pacific Ocean

Atlantic Ocean

Mississippi River

Gulf of Mexico

(c) The major drainage basins of North America.

the streams merge into fewer, but still parallel, channels. Parallel networks typically form on the sides of steep escarpments of weak substrate—if the surface of the substrate hosts no vegetation and many channels form, the resulting landscape can be called *badlands topography*.

Drainage Basins and Divides

A drainage network collects water from a region variously called a *drainage basin*, *catchment*, or **watershed**. A highland or ridge, known as a **drainage divide**, separates one watershed from another **(Fig. 17.5a)**. You can recognize a local divide on a topographic map or out an airplane window by looking for a change in the direction of slope **(Fig. 17.5b)**. *Regional divides* separate two large drainage basins—for example, a regional divide follows the crest of the Appalachians and separates Atlantic Ocean drainage from Gulf of Mexico drainage. A **continental divide** separates drainage that flows into one ocean from drainage that flows into another. In North America, one continental divide running the length of the North American Cordillera separates watersheds that ultimately drain into the Atlantic or Gulf of Mexico from those that drain into the Pacific **(Fig. 17.5c)**. Another continental divide separates watersheds that drain into Hudson Bay and the Arctic Ocean from those that drain into other oceans. The largest drainage basin in North America, the Mississippi, covers much of the interior of the United States, and the largest drainage basin in the world,

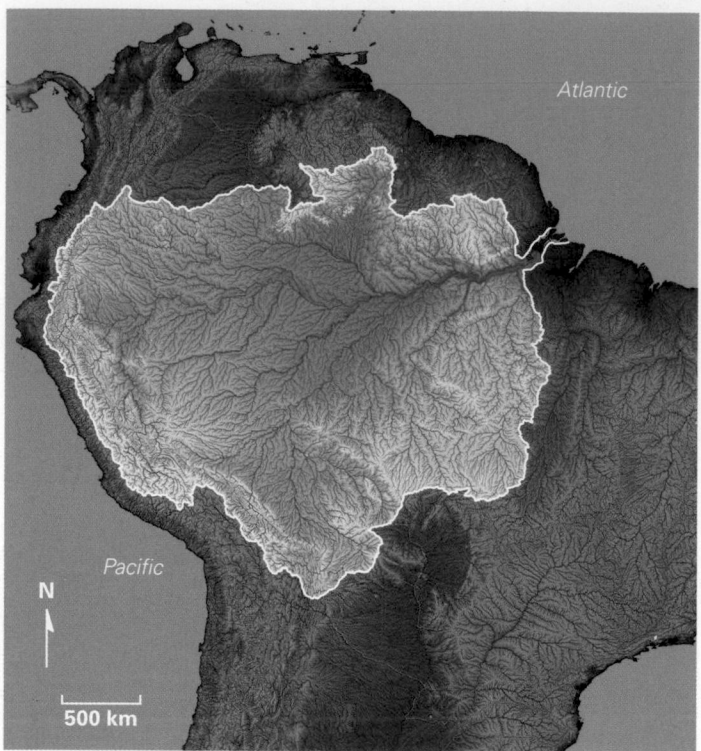

generally lies above the streambed (**Fig. 17.7a, b**). Most streams in arid (dry) climates tend to be ephemeral because they lose so much water along their length that they dry out. This happens because the water table lies below the streambed, so that water sinks down through the streambed (**Fig. 17.7c**), and because of evaporation from the stream's surface.

Note that all ephemeral streams are losing streams, but not all losing streams are ephemeral. A losing stream can flow all year if its headwaters lie in a wetter region, so that the stream receives more water from its upper reaches than it loses to infiltration or evaporation in its lower reaches. The Colorado River of the western United States and the Nile River of northeastern Africa serve as examples.

An ephemeral stream can flow continuously for many months during the wet season, when the water table rises and tributaries supply abundant water, but dry up during the dry season, when the water table sinks below the streambed and tributaries don't supply water. In contrast, some ephemeral streams flow only for a brief time after a heavy rain and remain dry most of the year. The dry channel of such an ephemeral stream is called a **dry wash**, *wadi*, or *arroyo* (**Fig 17.7d**).

Flow in the Ground around the Channel

When we look at a stream, we see only the flow taking place within the channel. But where a permanent stream flows over a permeable substrate, water also moves through the substrate in the direction of the stream flow, down to a depth of tens of centimeters to tens of meters below the streambed (**Fig. 17.8**). In this *hyporheic zone*, water flows more slowly than it does in the open channel, because the water has to find its way around clasts and along cracks. Notably, many organisms spawn within the hyporheic zone, where eggs can hatch without being washed away or eaten.

the Amazon, extends across almost the entire width of South America (**Fig. 17.6**). An *interior drainage basin*, such as the Great Basin of Utah and Nevada, has no outlet to the sea.

Streams That Last and Streams That Don't

Because streams can serve as water supplies for agriculture, industry, or municipalities, it's important to characterize how the volume of a stream changes along its length. In a *gaining stream*, the volume of water increases in the downstream direction, whereas in a *losing stream*, the volume of water decreases in the downstream direction. Whether a stream gains or loses depends on two factors: the amount of water entering the stream from tributaries, because water coming from tributaries can replace water lost to evaporation or to infiltration into the streambed; and the depth of the water table relative to the streambed, because if the streambed lies below the water table, then groundwater flows up into the stream through springs.

Geologists also find it useful to distinguish between types of streams based on the duration of flow—**permanent streams** flow all year long, whereas **ephemeral streams** flow for only part of the year. Streams in temperate or tropical climates tend to be permanent because the water table

---

TAKE-HOME MESSAGE

Streams drain water from the land. Their channels form by downcutting and lengthen by headward erosion, and they carry water from sheetwash, lakes, springs, and melting snow or ice. A drainage network of tributaries feeding a trunk stream carries water from a watershed to the sea. Divides separate watersheds. Permanent streams flow all year, while ephemeral streams flow only intermittently. The permanence of a stream depends on climate and on the supply of water from upstream.

QUICK QUESTION: Why do streams develop distinct channels?

FIGURE 17.7 The contrast between permanent and ephemeral streams.

(a) The channel of a permanent stream in a temperate climate lies below the water table. Springs add water from below, so the stream contains water even between rains.

An empty stream channel is called a dry wash.

(c) The channel of an ephemeral stream lies above the water table, so the stream flows only when water enters the stream faster than it can infiltrate into the ground.

(b) An example of a permanent stream in the Wind River Mountains, Wyoming

(d) An example of a dry wash in the Buckskin Mountains, Arizona.

FIGURE 17.8 Since streambeds are permeable, water from a permanent stream mixes with groundwater in a region beneath the streambed called the hyporheic zone. Water in this zone flows in the same direction as the stream, but not as fast.

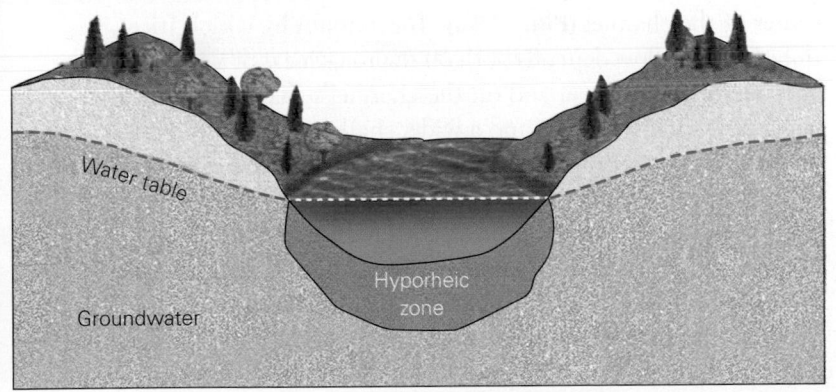

17.3 Describing Flow in Streams

Imagine two streams—a larger one in which water flows slowly and a smaller one in which water flows rapidly. Which stream carries more water? The answer is not obvious. To answer this question completely, geologists or engineers must calculate the streams' **discharge**, the volume of water passing a reference point in a given time. We can calculate discharge (D) by using a simple equation: $D = A_C \times v_A$. In this equation, A_C is the cross-sectional area of the stream as measured in a vertical plane oriented perpendicular to the stream bank, and v_A is the average velocity at which water moves in the downstream direction. If, for example, a stream has a cross-sectional area of 10 m² and its water flows at an average velocity of 5 m per second, then the discharge is 50 m³ per second. Note that we specify discharge by volume per unit time, using

FIGURE 17.9 Flow velocity and its measurement in streams.

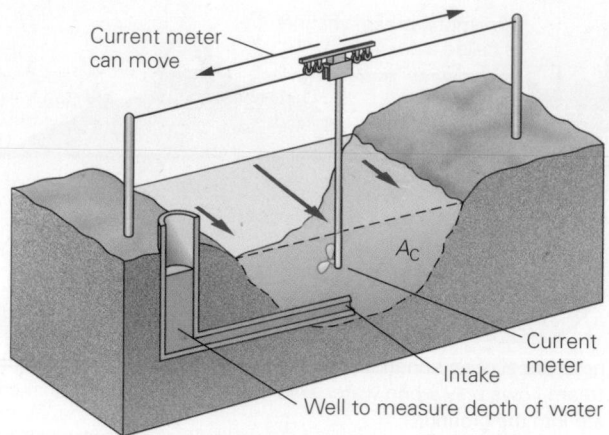

(a) At a stream-gauging station, geologists measure the cross-sectional area (A_c, the area within the dashed line), the depth, and the average velocity of the stream. Note that water flows more slowly near the banks.

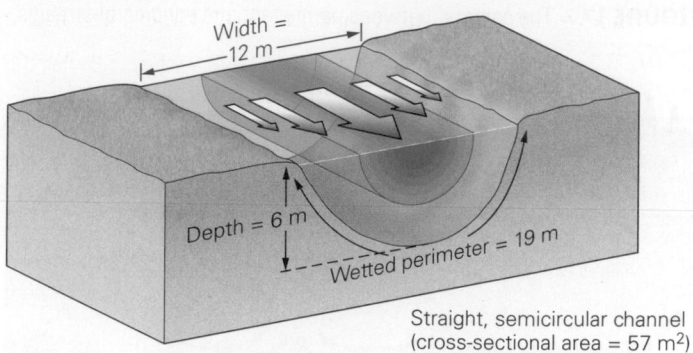

Straight, semicircular channel (cross-sectional area = 57 m²)

(b) In a straight channel, the fastest velocity occurs near the surface of the stream, equidistant from the banks.

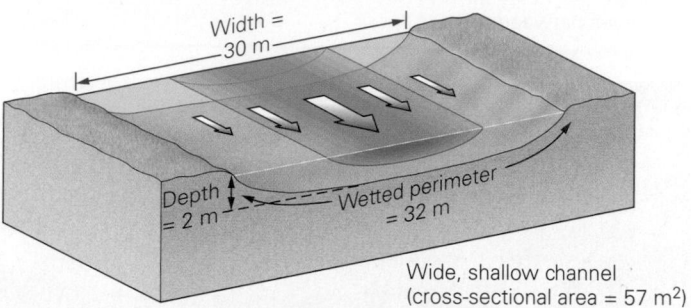

Wide, shallow channel (cross-sectional area = 57 m²)

(c) Velocity in a wide, shallow channel is less than that in a semicircular channel with the same cross-sectional area because the wetted perimeter is greater.

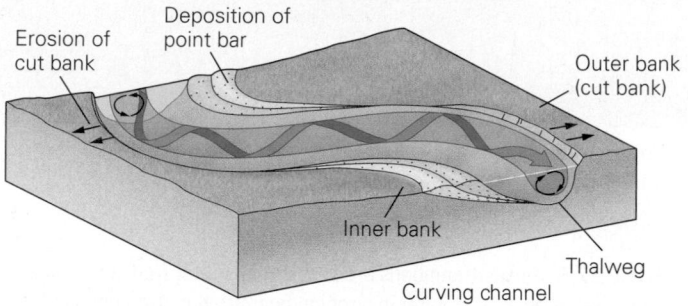

(d) In a curved channel, the fastest velocity shifts to the outer edge of the stream, and water follows a spiral-like path.

units such as cubic meters per second (m³/s), where 1 cubic meter per second = 35 cubic feet per second (ft³/s). We can measure discharge at a *stream-gauging station*, using instruments that analyze velocity and depth at various points across the stream **(Fig. 17.9a)**.

Discharge depends on the size of the watershed, on the amount of rain or snow falling in the watershed, and on whether the stream is gaining or losing. Discharge can also be affected by human activity. For example, if people divert a river's water to fields for irrigation, the river's discharge decreases. We can define an *average discharge* of a stream for a given location along the stream by taking into account daily measurements made over several years. The average discharge during the wet season may be double or triple that during the dry season, and during a flood, discharge may increase to more than 100 times normal.

Measurements of the average discharge at the mouth of a trunk stream allow us to compare different watersheds. The rivers that drain the Earth's two largest rainforests have the largest discharges. Specifically, the Amazon River has an average discharge of about 200,000 m³ per second, roughly 15% of global river discharge, and the Congo River has an average discharge of 40,000 m³ per second. In contrast, the "mighty" Mississippi, temperate North America's largest river, has a discharge of only 17,000 m³ per second.

It turns out that the average velocity of stream flow can be difficult to calculate because not all the water in a stream moves at the same velocity past a given point on the bank. Velocity variations happen, in part, because friction along the walls and bed of the stream slows the flow of water. In general, water near the stream banks or the streambed moves more slowly than water in the middle of the flow, and the fastest-moving part of the flow lies near the surface in the

center of the channel **(Fig. 17.9b)**. The amount by which friction slows the flow depends both on the roughness of the walls and bed of the channel and on the channel's shape. Rougher walls slow the flow more, and a wide, shallow stream channel has a larger *wetted perimeter* (the area in which water touches the channel) than does a semicircular channel, so water flows more slowly in the former than in the latter **(Fig. 17.9c)**.

In a curving stream channel, the fastest flow shifts toward the outside curve, similar to the way that a car swerves to the outer edge of a curve on a highway. Therefore, the deepest part of a channel, its *thalweg*, lies near the outside curve. In fact, as the water flows toward the outside wall of a curving channel, it starts to follow a spiral path, because as surface water moves toward the outer bank, water deeper down must flow toward the inner bank to replace the surface water **(Fig. 17.9d)**.

Velocity variations in a stream also happen because of **turbulence**, the twisting, swirling motion that, on a large scale, can generate *eddies* in which water curves and actually flows upstream or circles in place **(Fig. 17.10)**. Turbulence develops because the shearing motion of one water volume against its neighbor causes the neighbor to spin, and because obstacles such as boulders on the streambed deflect volumes, forcing them to change their direction of movement.

FIGURE 17.10 Turbulence in streams.

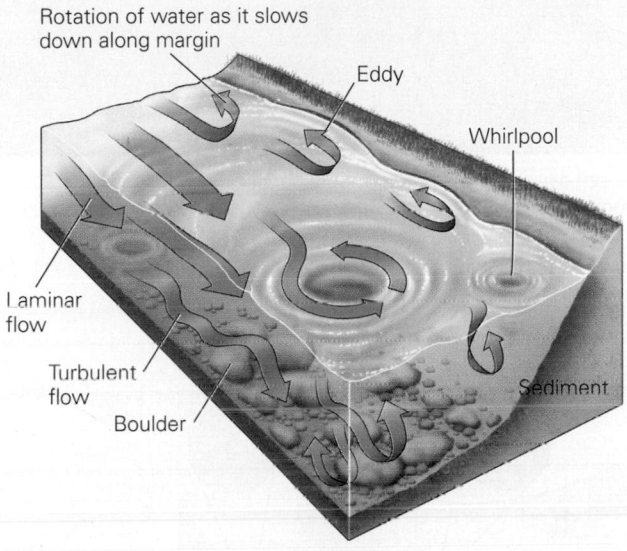

(a) Water in a stream doesn't usually follow a straight path. It swirls and twists, producing turbulence.

(b) The roughness of the water surface, and the chaos of the bubble trails, emphasize the turbulence of this river in Rochester, New York.

TAKE-HOME MESSAGE

Stream discharge, the amount of water passing through a cross section of the stream in a given time, depends on such factors as watershed area and climate. Discharge changes along the stream's length and changes seasonally. Water velocity varies across a stream due to turbulence and to friction with the channel.

QUICK QUESTION: Relative to a given point along the bank of a stream, where is the velocity of flow in a stream fastest? Why?

17.4 The Work of Running Water

How Do Streams Erode?

The energy that makes running water move comes from gravity. As water flows downslope from a higher to a lower elevation, the gravitational potential energy stored in water transforms into kinetic energy. About 3% of this energy goes into the work of eroding the walls and beds of stream channels. Running water causes erosion in four ways:

- *Scouring:* Running water can remove and carry away loose fragments of sediment, a process called *scouring*.

- *Breaking and lifting:* In some cases, the push of flowing water can break chunks of solid rock off the channel bed or walls. In addition, the flow of a current over a clast can cause the clast to rise or lift off the substrate.

- *Abrasion:* Clean water has little erosive effect, but sediment-laden water acts like sandpaper and grinds or rasps away at the channel bed and walls, a process called *abrasion*. In places where turbulence produces long-lived whirlpools, in which water swirls like a small tornado, abrasion by sand or gravel carves a bowl-shaped depression, called a **pothole**, into the bed of the stream **(Fig. 17.11a, b)**.

- *Dissolution:* Running water dissolves soluble minerals as it passes and carries the minerals away in solution.

The efficiency of erosion depends on the velocity and volume of water and on its sediment content. A large volume of fast-moving, turbulent, sandy water causes more erosion than does a trickle of quiet, clear water. Therefore, most erosion takes place during floods, which supply streams with large volumes of fast-moving, sediment-laden water.

FIGURE 17.11 Erosion and transportation in streams.

(a) A pothole in the bed of a stream near Ithaca, New York.

(b) This slot canyon in Arizona formed when many potholes were linked together.

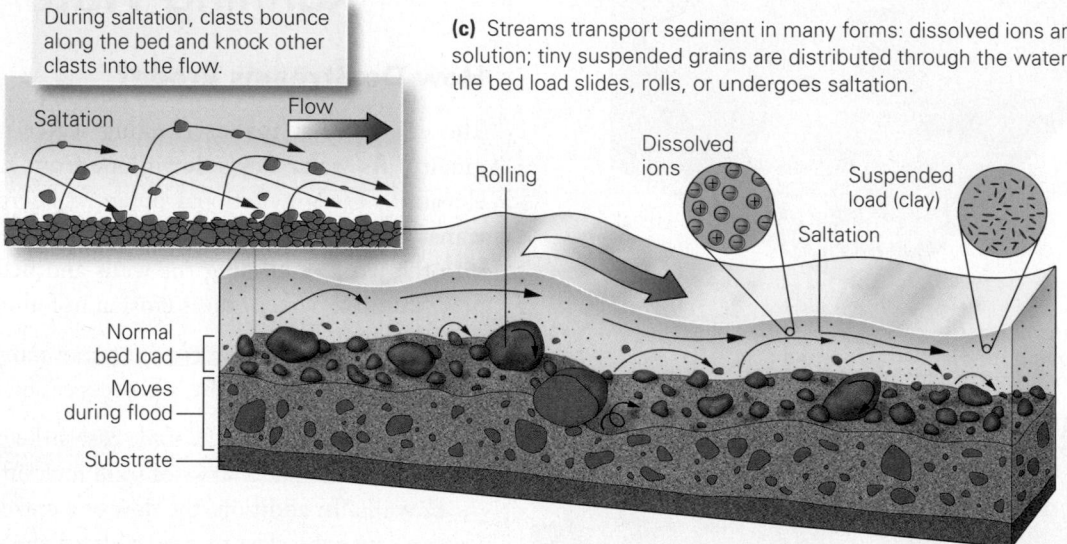

(c) Streams transport sediment in many forms: dissolved ions are in solution; tiny suspended grains are distributed through the water; and the bed load slides, rolls, or undergoes saltation.

How Do Streams Transport Sediment?

The Mississippi River received the nickname "Big Muddy" for a reason—its water can become chocolate brown because of all the clay and silt it carries. All streams carry sediment, though different streams carry different amounts at different times. Geologists refer to the total volume of sediment carried by a stream as its *sediment load*. The sediment load consists of three components **(Fig. 17.11c)**:

- *Dissolved load*: Running water dissolves soluble minerals in the sediment or rock of its substrate, and groundwater seeping into a stream through the channel walls and bed brings dissolved minerals with it. The ions of these dissolved minerals constitute a stream's **dissolved load**.

- *Suspended load*: The **suspended load** of a stream consists of tiny solid grains (silt or clay) that swirl along with the water without settling to the streambed; this sediment makes the water brown **(Fig. 17.12a–c)**.

- *Bed load*: The **bed load** of a stream consists of large particles, such as sand, pebbles, or cobbles, that bounce or roll along the streambed **(Fig. 17.12d)**. Bed-load movement commonly involves **saltation**, in which a multitude of grains bounce along in the direction of flow within a zone that extends up from the streambed for a distance of several centimeters to several tens of centimeters. Each saltating grain in this zone follows a curved trajectory up through the water and then back down to the bed. When it strikes the bed, it knocks other grains upward and supplies new grains to the saltation zone.

When describing a stream's ability to carry sediment, geologists specify its competence and capacity. The **competence** of a stream refers to the maximum particle size it carries—a

FIGURE 17.12 Sediment is carried and deposited by streams. The clast size depends on stream velocity.

(a) On a sunny day, this stream in Switzerland is fairly clear; you can see cobbles on the bed.

(b) On a rainy day, the stream's discharge increases, and the faster, more turbulent water becomes brown due to its load of sediment.

(c) An aerial photo shows a muddy river emptying into the Black Sea. Currents carry the sediment away.

(d) Gravel deposited within the channel of a stream in the Wasatch Mountains, Utah. This sediment moved as bed load.

(e) Gravel in the bed of a mountain stream in Denali National Park, Alaska. The large clasts were carried during floods. This stream has high competence.

(f) Mud deposited along a gentle, slowly moving stream in Brazil. This stream has low competence.

stream with high competence can carry large particles, whereas one with low competence can carry only small particles. A fast-moving, turbulent stream has greater competence than does a slow-moving stream, and a stream in flood has greater competence than does a stream with normal flow (**Fig. 17.12e, f**). In fact, the huge boulders that litter the bed of a mountain creek move only during floods. The **capacity** of a stream refers to the total quantity of sediment it can carry. A stream's capacity depends on both its competence and its discharge.

Depositional Processes

A raging torrent of water can carry coarse and fine sediment—the finer clasts rush along with the water as suspended load, whereas the coarser clasts may bounce and tumble as bed load. If the flow velocity decreases, either because the slope of the streambed decreases or because the channel broadens out and friction between the streambed and the water increases, then the competence of the stream decreases and sediment settles out. The size of the clasts that settle at a particular locality depends on how slow the flow has become. Therefore, coarser sediment tends to settle out upstream, where water flows faster, whereas finer grains settle out downstream, where water flows more slowly, and the finest sediment settles out when the stream flows into a standing body of water. Because of this process of **sediment sorting**, stream deposits tend to be segregated by size—gravel, sand, silt, and mud can collect in different locations.

Geologists refer to sediments transported by a stream as *fluvial deposits* (from the Latin *fluvius*, meaning river), or **alluvium**. Alluvium may accumulate along the streambed in elongate mounds called **bars** (see Fig. 17.12d). Some stream channels follow broad curves, which, as we'll see, are called *meanders*. Water slows along the inner edge of a meander, so crescent-shaped **point bars** bordering the shoreline develop along the inner edge (see Fig. 17.12f). During floods, a stream may overtop its banks and spread out over its **floodplain**, the broad, flat area bordering the stream. Friction slows the water on the floodplain, so a sheet of silt and mud settles out to form floodplain deposits. Where a stream empties at its mouth into a standing body of water, the water slows, and a wedge of sediment, called a **delta**, accumulates. We'll discuss meanders, floodplains, and deltas in more detail later in this chapter.

TAKE-HOME MESSAGE

Streams erode by scouring, breaking and lifting, abrasion, and dissolution. They carry sediment as dissolved, suspended, or bed loads. Competence, the ability to carry sediment, depends on flow velocity. Where the velocity decreases, sediment settles out.

QUICK QUESTION: Why do point bars form on the inner edge of a curve in a stream?

17.5 How Do Streams Change along Their Length?

Longitudinal Profiles

In 1803, under President Thomas Jefferson's leadership, the United States bought the Louisiana Territory, a vast tract of land encompassing the western half of the Mississippi drainage basin. At the time, the geography of the territory was a mystery. To fill the blank on the map, Jefferson asked Meriwether Lewis and William Clark to lead a voyage of exploration across the Louisiana Territory to the Pacific.

Lewis and Clark, along with a crew of 40, began their expedition at the mouth of the Missouri River, where it joins the Mississippi. At this juncture, the Missouri is a wide, languid stream of muddy water. The group found the Missouri's downstream reach, where the river has a deep channel and a smooth surface, to be easy going. But the farther upstream they went, the more difficult their voyage became, for the **stream gradient**, meaning the slope of the stream, became progressively steeper and the stream's discharge decreased. When Lewis and Clark reached the site of what is now Bismarck, North Dakota, they had to abandon their original boats and haul smaller vessels up *rapids*, where turbulent water plunges over a steep, bouldery bed, and occasionally they had to carry their boats around *waterfalls*, where water drops over an escarpment. Eventually, they had to abandon their boats and trudge by foot up steep stream valleys until they reached the continental divide.

If Lewis and Clark had been able to plot a graph showing their elevation above sea level relative to their distance along the Missouri, they would have created a **longitudinal profile** of the Missouri, an image showing the variation in the river's elevation along its length. The profile of the Missouri, like those of most rivers, roughly resembles a concave-up curve **(Fig. 17.13)**. A stream with this typical concave-up profile has a steep gradient near its headwaters, but flows over an almost horizontal plain near its mouth. As a result, streams typically cut deep valleys or canyons near their headwaters, but not near their mouths.

Geologists refer to the lowest elevation that a stream can attain at a locality as the **base level** of the stream. A *local base level* is one that occurs upstream of a stream's mouth, and the *ultimate base level* (the lowest possible elevation along the longitudinal profile) lies at sea level. The surface of a trunk stream where it enters the sea cannot be lower than sea level, for if it were, the stream would have to flow upslope.

Lakes or reservoirs act as local base levels along a stream, for where the stream enters such standing bodies of water, it slows almost to a halt and cannot downcut further **(Fig. 17.14a)**. A ledge of resistant rock can also act as a local base level, for the stream level cannot drop below the ledge until the ledge erodes away **(Fig. 17.14b)**. Finally, where a tributary joins a larger stream, the channel of the larger stream acts as the base level for the tributary. (As a result, the mouth of a tributary tends to lie at the same elevation as the stream that it joins at the point of intersection.) Local base levels do not last forever, because running water eventually removes the obstructions that control them.

Local base levels represent deviations from the "ideal" concave-up shape of a longitudinal profile. Over time, streams tend to erode away obstacles, or deposit sediments to fill in low areas, and therefore develop a smoother profile.

FIGURE 17.13 Change in the character of a stream along its longitudinal profile.

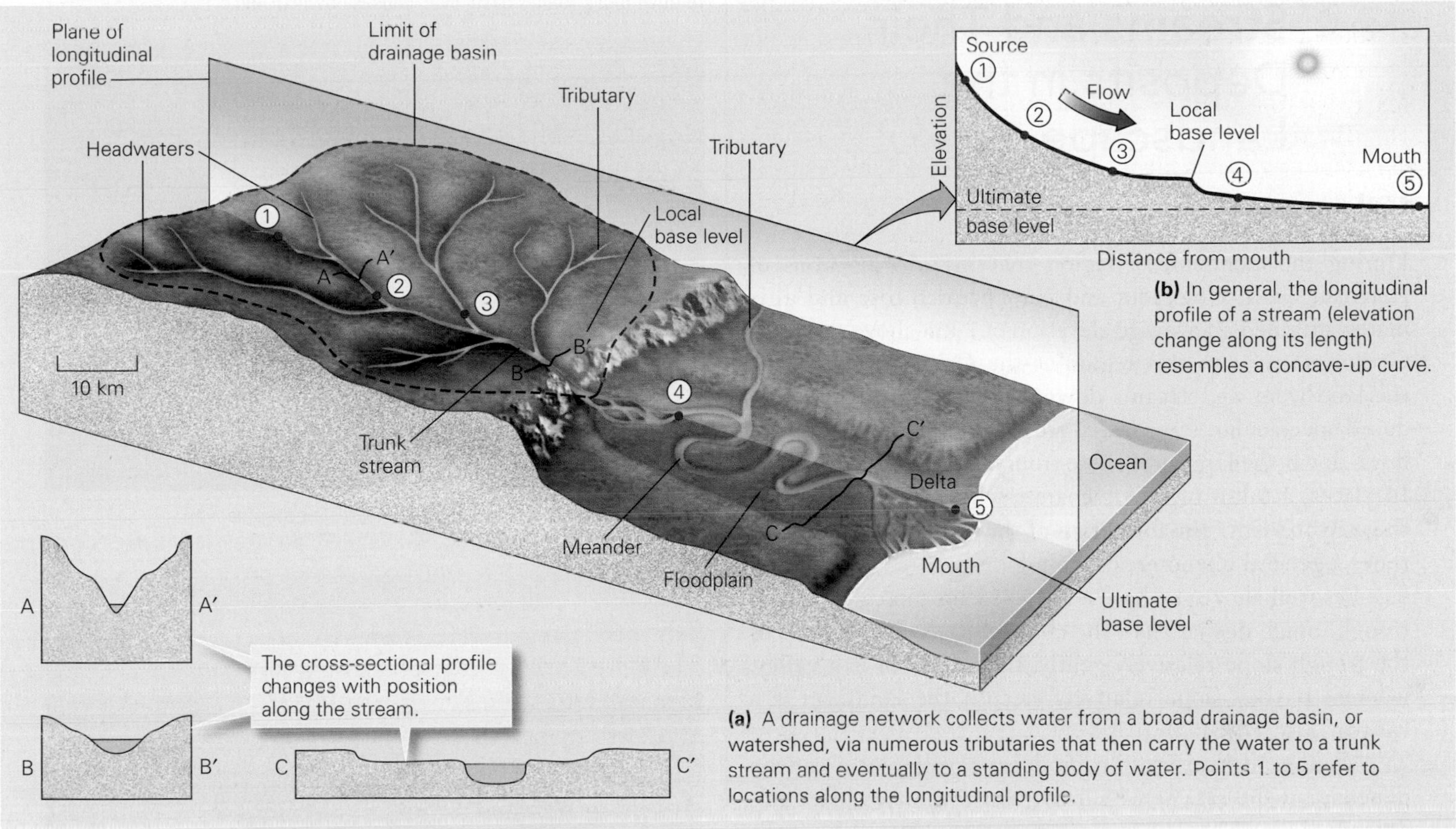

(b) In general, the longitudinal profile of a stream (elevation change along its length) resembles a concave-up curve.

The cross-sectional profile changes with position along the stream.

(a) A drainage network collects water from a broad drainage basin, or watershed, via numerous tributaries that then carry the water to a trunk stream and eventually to a standing body of water. Points 1 to 5 refer to locations along the longitudinal profile.

FIGURE 17.14 The concept of local base levels in a stream's longitudinal profile.

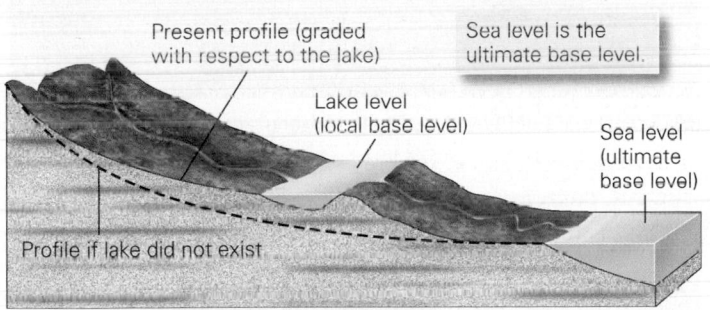

(a) A lake acts as a local base level. The stream profile uphill of the lake lies above the profile that would form if the lake didn't exist.

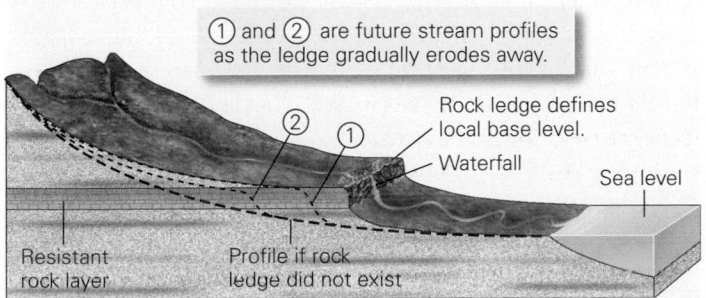

(b) A resistant rock ledge can form a local base level. Headward erosion gradually cuts back into the ledge.

Because the Earth remains dynamic, the ultimate base level of a stream can change over time. For example, if sea level rises, the ultimate base level rises, and if sea level falls, the ultimate base level falls. A local base level may rise or fall due to tectonic movements or to accumulation of sediment. For example, slip on a fault that cuts across a stream can cause the downstream reach of the stream to fall relative to the upstream reach; such movement produces a *nickpoint* (abrupt step) in the longitudinal profile. Where the base level falls, the stream will start to downcut and deepen the valley that it flows in.

TAKE-HOME MESSAGE

A stream typically has a steeper gradient toward its source and a gentler gradient near its mouth, so longitudinal profiles of streams tend to be concave-up curves. A stream's mouth cannot be lower than its ultimate base level, which for a trunk stream is sea level. Changes in the local base level disrupt this equilibrium and can cause renewed downcutting or deposition within a reach.

QUICK QUESTION: Why can't a stream downcut more deeply than its base level?

17.6 Streams and Their Deposits in the Landscape

Valleys and Canyons

During the Cenozoic, a region that includes portions of Arizona, Utah, Colorado, and New Mexico rose and ultimately attained an average elevation of 2 km above sea level. Geologists refer to this region as the *Colorado Plateau*. As the land went up, streams downcut into bedrock and produced spectacular, steep-sided troughs. The Colorado River flows down the largest of these troughs, the Grand Canyon. In places, 1.6 km of relief separates the river surface from the canyon's rim. The formation of the Grand Canyon illustrates a general phenomenon: in regions where the land surface lies well above the base level, a stream can carve a deep trough, much deeper than the channel itself. If the walls of the trough slope relatively gently, the landform is a **valley**, whereas if they slope relatively steeply, the landform is a **canyon (Fig. 17.15)**.

Whether stream erosion produces a valley or a canyon depends on the rate at which downcutting takes place relative to the rate at which mass movement causes the walls on either side of the stream to collapse. In places where the walls collapse as fast as the stream downcuts, landslides and slumps gradually cause the slope of the walls to approach the angle of repose (see Chapter 16). When this happens, the stream channel lies at the floor of a valley whose cross-sectional shape resembles the letter V (**Fig. 17.16a**; see Fig. 17.15c); this landform is called a *V-shaped valley*. In places where a stream downcuts through its substrate faster than the walls collapse, erosion creates a *slot canyon* (a relatively narrow, steep-walled canyon). Such canyons typically form in hard rock, which can hold up steep cliffs for a long time (**Fig. 17.16b**; see Fig. 17.15b). Where the walls consist of alternating layers of hard and soft rock, the walls develop a stair-step shape like that of the Grand Canyon (**Fig. 17.16c**; see Fig. 17.15a).

In places where active downcutting occurs, the valley floor remains relatively clear of sediment, because the stream—especially when it floods—carries away sediment that has fallen or slumped into the channel from the valley or canyon walls. But if the stream's base level rises, its discharge decreases, or its sediment load increases, the valley floor can fill with sediment to produce an *alluvium-filled valley* (**Fig. 17.17a, b**). The surface of the alluvium becomes a broad floodplain. If the stream's base level later drops again, or the discharge increases, the stream will start to cut down

FIGURE 17.15 Canyons and valleys form when uplift of the land, relative to a stream's base level, causes downcutting.

(a) Part of the Grand Canyon, Arizona, a stair-step canyon.

(b) Vertical walls of Canyon de Chelly, Arizona.

(c) V-shaped valleys in the Andes of Peru.

FIGURE 17.16 The shape of a canyon or valley depends on the resistance of its walls to mass movement.

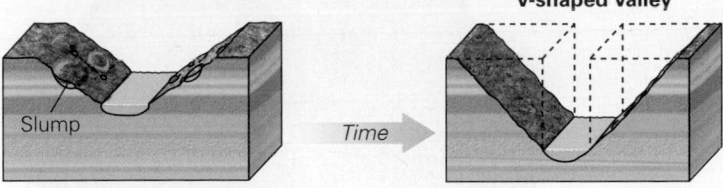

V-shaped valley

(a) If mass movement takes place as fast as downcutting occurs, a V-shaped valley develops.

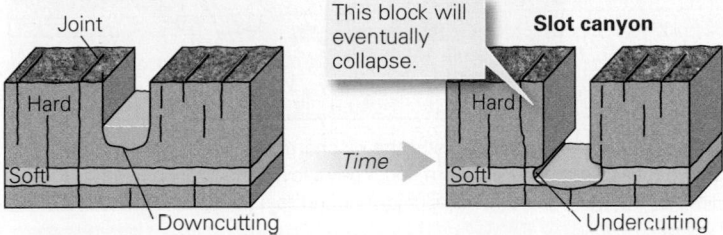

This block will eventually collapse.

Slot canyon

(b) If downcutting by the stream happens faster than mass movement on the walls, a slot canyon forms. The canyon widens as the stream undercuts the walls.

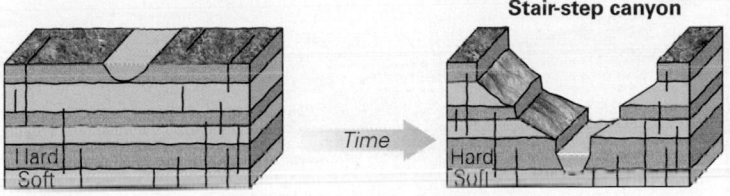

Stair-step canyon

(c) Downcutting through alternating hard and soft layers produces a stair-step canyon.

into the alluvium, a process that generates *stream terraces* bordering the present floodplain **(Fig. 17.17c)**.

Rapids and Waterfalls

When Lewis and Clark forged a path up the Missouri River, they came to reaches that could not be navigated by boat because of **rapids**, particularly turbulent water with a rough surface **(Fig. 17.18a)**. Rapids form where water flows over steps or large clasts on the streambed, where the channel abruptly narrows, or where its gradient abruptly changes. The turbulence in rapids produces eddies, waves, and whirlpools that roil and churn the water surface, in the process producing *whitewater*, a mixture of bubbles and water. Modern-day whitewater rafters or kayakers thrill to the unpredictable movement of rapids **(Fig. 17.18b)**.

A **waterfall** forms where the gradient of a stream becomes so steep that the water literally free-falls to

Did you ever wonder . . .
why a waterfall forms and whether it will always be there?

lower elevations **(Fig. 17.18c)**. The energy of falling water may scour a depression, called a *plunge pool*, at the base of the waterfall. Some waterfalls develop where a stream crosses a resistant ledge of rock, and some develop as a result of faulting because displacement produces an escarpment. Waterfalls also occur where glacial erosion has deepened a trunk valley relative to tributary valleys to form a *hanging valley* **(Fig. 17.18d)** whose mouth sits much higher than the floor of the trunk valley (see Chapter 22).

Although a waterfall may appear to be a permanent feature of the landscape, waterfalls eventually disappear because headward erosion slowly eats back the resistant ledge. We can see this process taking place at Niagara Falls, where water flowing from Lake Erie to Lake Ontario drops over a 55-m-high ledge of resistant Silurian dolostone that rests on a weak shale. Extreme turbulence of the water in the plunge pool erodes the shale and undercuts the dolostone. Gradually, the overhang of dolostone becomes unstable, breaks away at a joint, and collapses in a rockfall, with the result that the position of the waterfall migrates upstream. Before the industrial age, the edge of Niagara Falls cut upstream at an average rate of 1 m per year; but since then, the diversion of water from the Niagara River into a hydroelectric power station has decreased the discharge over the falls, cutting the rate of headward erosion in half. Nevertheless, geologists estimate that Niagara Falls will cut all the way back to Lake Erie in about 60,000 years **(Fig. 17.19)**.

Alluvial Fans and Braided Streams

Where a fast-moving ephemeral stream abruptly emerges from a mountain canyon onto an open plain, water that had been confined to a narrow channel spreads over a broader surface, slows, and deposits a lens of sediment. Eventually, the accumulation of sediment decreases the gradient at the mouth of the outlet that it came from, so the next time the stream flows, it follows a different, steeper path, to one side of the previous lens. Each new flow event deposits its load in a remaining lower area. Over time, therefore, deposition episodes build a broad, gently sloping, wedge-shaped apron of sediment called an **alluvial fan** **(Fig. 17.20a)**.

In some localities, streams carry abundant coarse sediment during floods, but cannot carry this sediment during normal flow. As a result, during normal flow, the sediment settles out and chokes the channel. Because the gravelly sediment can't stick together, the stream cannot cut a single deep channel with steep banks—the channel walls simply collapse. As a consequence, the stream divides into numerous strands weaving back and forth between elongate bars of gravel and sand. The result is a **braided stream**—the name emphasizes that the strands entwine like hair in a braid **(Fig. 17.20b)**.

FIGURE 17.17 The evolution of alluvium-filled stream valleys and the development of terraces.

Time 1

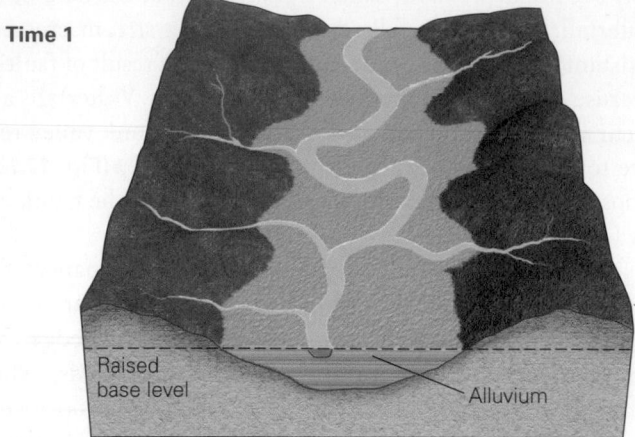

Time 2

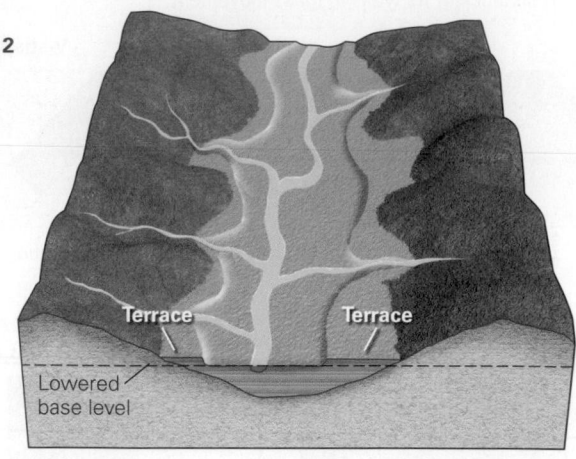

(a) A rise in the base level or a decrease in discharge causes the valley to fill with alluvium.

(b) Later, if the base level falls or the discharge increases, the stream downcuts through the alluvium and a new, lower floodplain develops. The remnants of the original alluvial plain remain as a pair of terraces.

(c) In this view of a valley in Utah, we can see a floodplain and two terrace levels.

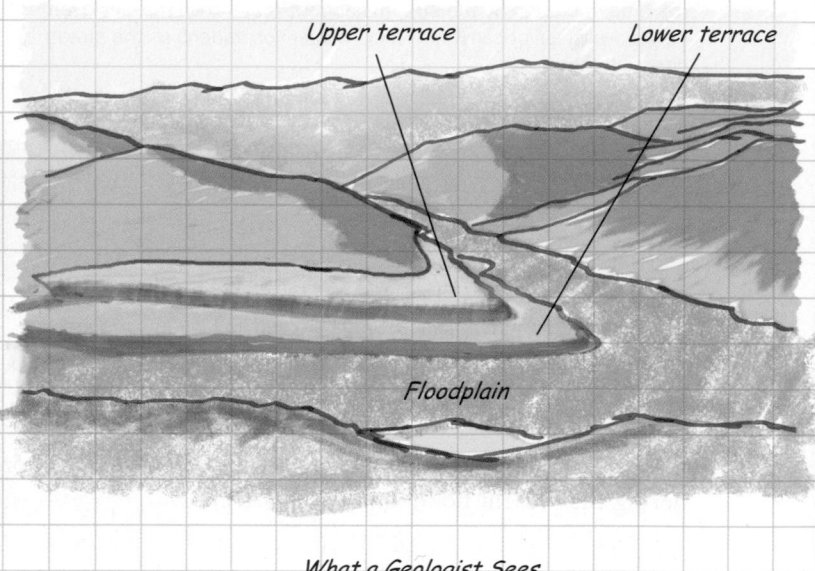

What a Geologist Sees

Meandering Streams and Their Floodplains

A riverboat cruising along the lower reaches of the Mississippi River cannot sail in a straight line, for the river channel winds back and forth in a series of snake-like curves, each of which is known as a **meander (Fig. 17.21)**. In fact, the boat has to go 500 km along the river channel to travel 100 km as the crow flies. A **meandering stream** curves around many meanders in succession. Effectively, the development of meanders increases the volume of water the stream can carry by increasing the stream's length in a region where the stream has a gentle gradient.

How do meanders evolve? Even if a stream starts out with a straight channel, natural variations in the water depth and associated friction cause the fastest-moving current to swing back and forth across the channel. The water erodes the wall

of the channel more effectively where it flows faster, so it begins to cut away faster on the outer edge of the curve. Each curve, therefore, begins to migrate sideways and grow more pronounced until it becomes a meander. On the outside edge of a meander, erosion continues to eat away at the channel wall, and a steep *cut bank* develops. Meanwhile, on the inside edge of the meander, water slows down so that its competence decreases and sediment accumulates, forming a wedge-shaped deposit called a *point bar*, as noted earlier. (Mark Twain, who worked as a riverboat pilot on the Mississippi River before writing such classic books as *Adventures of Huckleberry Finn*, took his pen name from the signals that the mate of a paddle-wheel riverboat called out to the skipper to indicate water depth, so that the boat wouldn't run aground on point bars; *mark twain* means that the water is 2 fathoms, or about 4 m,

FIGURE 17.18 Examples of rapids and waterfalls.

(a) These rapids in the Grand Canyon formed when a flood from a side canyon dumped debris into the channel of the Colorado River.

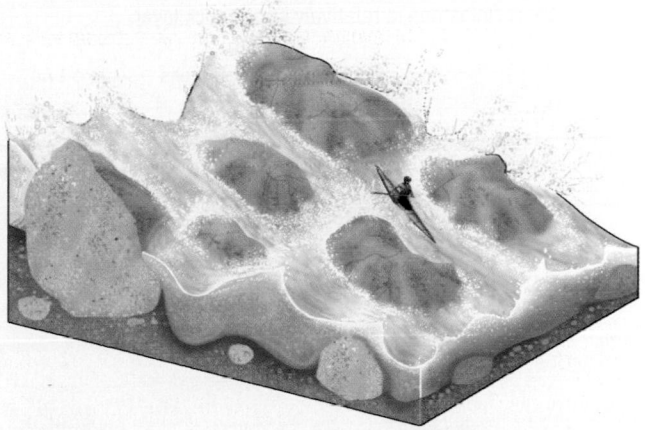

(b) Some rapids are extremely difficult to navigate.

(c) Iguaçu Falls, at the Brazil-Argentina border, spills across layers of basalt. The basalt acts as a resistant ledge.

(d) Waterfall spilling into Milford Sound, New Zealand, from a hanging valley.

deep.) With continued erosion, a meander may curve through more than 180°, so that the cut banks of adjacent meanders approach each other, leaving a *meander neck*, a narrow isthmus of land separating two meanders.

Meandering streams form where running water travels over a plain that has a gentle gradient. For the meanders to evolve, the substrate of a cut bank must be strong enough to resist collapsing constantly. In unconsolidated sediment, this strength comes from plant roots. Recent research suggests that meandering streams did not develop until abundant land plants appeared during the Silurian—before that, all streams on plains may have been braided.

People building communities along a riverbank may assume that the shape of a meander remains fixed for a long time. It doesn't. In a natural meandering river system, the river channel migrates back and forth across the floodplain. When erosion eats through a meander neck, a new straight reach, called a *cutoff*, develops. The abandonment of a portion of a river channel and the establishment of a new one, a process that geologists call an *avulsion*, can take place during a single

FIGURE 17.19 The formation of Niagara Falls, at the border between Ontario, Canada, and New York State. The falls tumble over the Lockport Dolostone, a relatively strong rock layer.

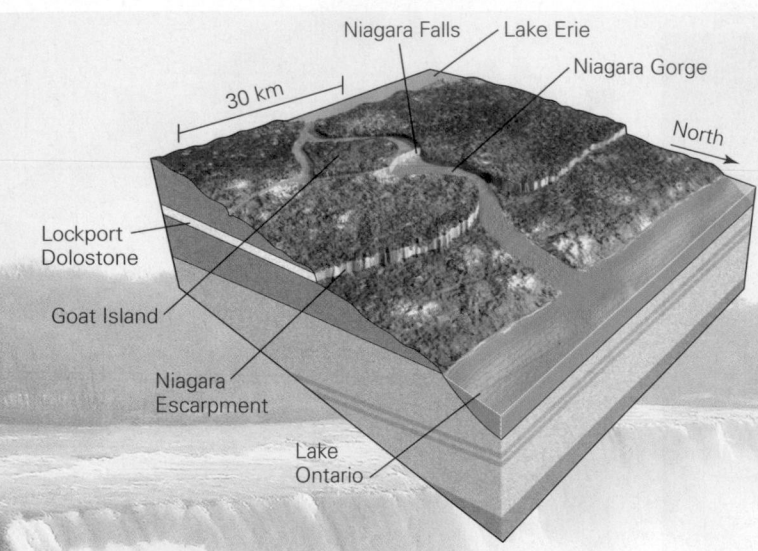

(a) Niagara Falls formed where the outlet of Lake Erie flowed over the Niagara escarpment.

Headward erosion causes the position of the falls to migrate upstream.

Rubble accumulates at the base of the falls.

(b) The American Falls, a part of Niagara Falls.

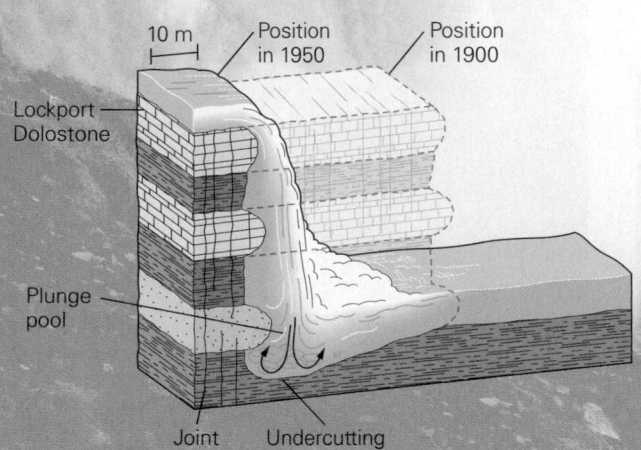

(c) The face of the falls retreats over time. The lower shale erodes, and stronger layers above break off at joints.

(d) Horseshoe Falls, on the Canadian side of the border.

FIGURE 17.20 Examples of depositional landforms produced from stream sediment.

(a) An alluvial fan in Death Valley, California, consists of sand, gravel, and debris flows. The curving black line is a road.

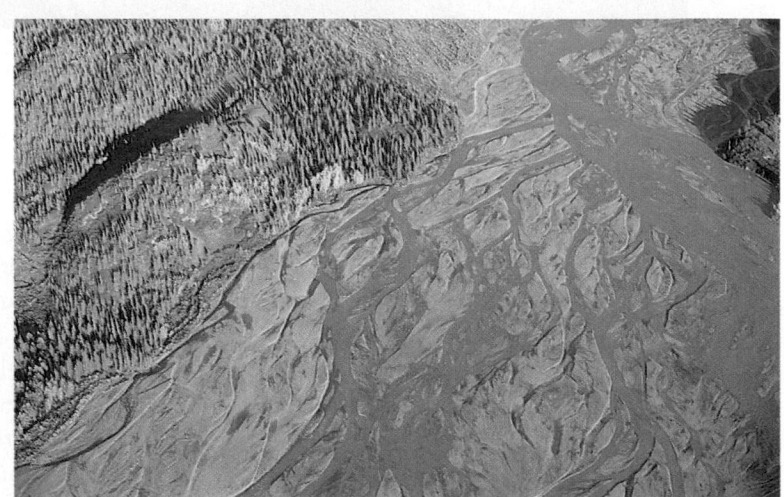

(b) A braided stream, carrying meltwater from a glacier near Denali, Alaska, deposits elongate bars of gravel.

FIGURE 17.21 The character and evolution of meandering streams and floodplains.

(a) A meandering stream wanders across a floodplain.

(b) The flat land of this floodplain hosts farm fields.

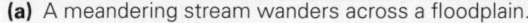
Time

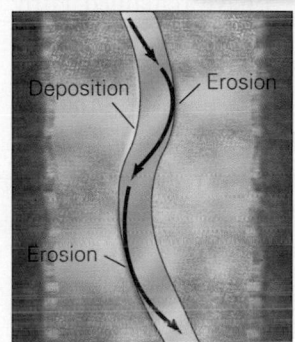
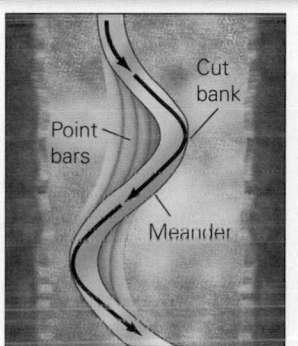
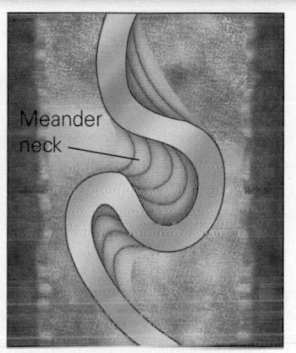
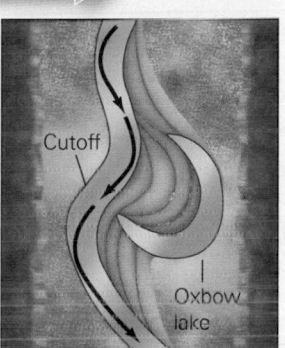

(c) Meanders evolve because erosion occurs faster on the outer edge of a curve, and deposition takes place on the inner curve. Eventually, a cutoff isolates an oxbow lake.

(d) Landforms along meandering streams include natural levees, point bars, and floodplains. Older deposits record the position of ancient channels and floodplains.

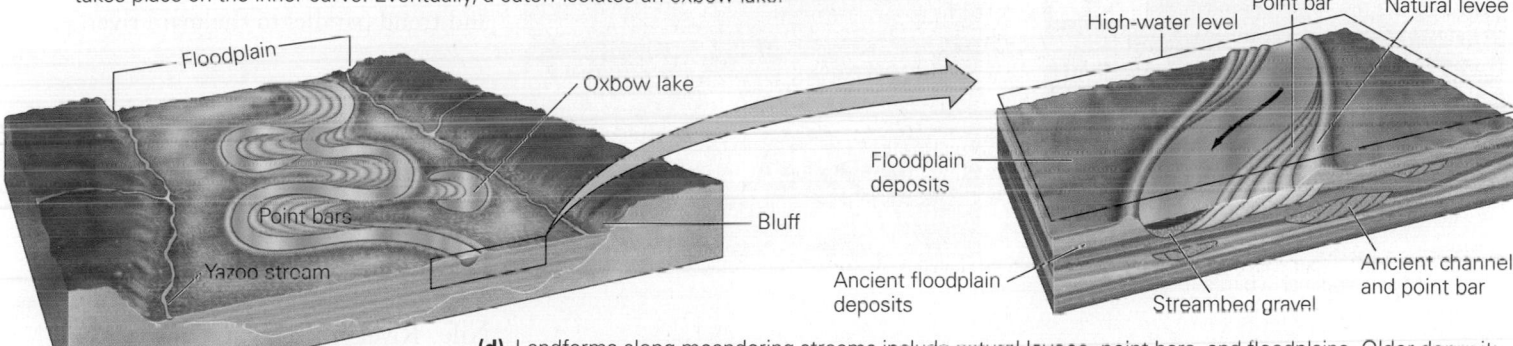

(e) A meandering stream in Brazil, as viewed from space. Note the oxbows, cutoffs, and abandoned meanders.

FIGURE 17.22 Formation of deltas and their variety of shapes.

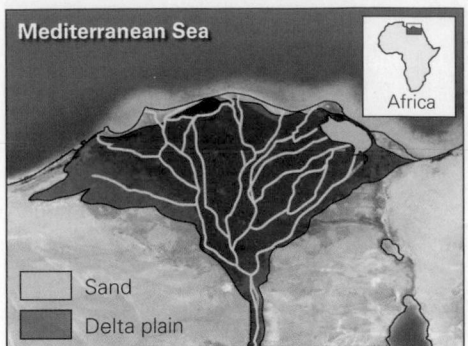

Mediterranean Sea

Africa

Sand

Delta plain

(a) The Nile has a Δ-shaped delta.

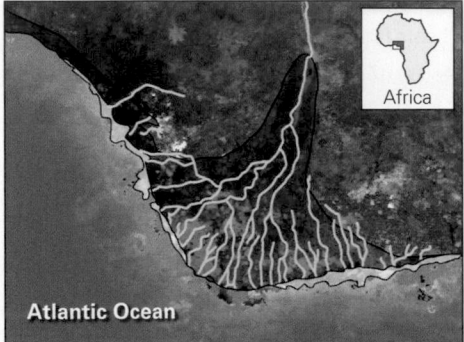

Africa

Atlantic Ocean

(c) The Niger has an arc-like delta.

USA

Swamp   Gulf of Mexico

(d) The Mississippi has a bird's-foot delta.

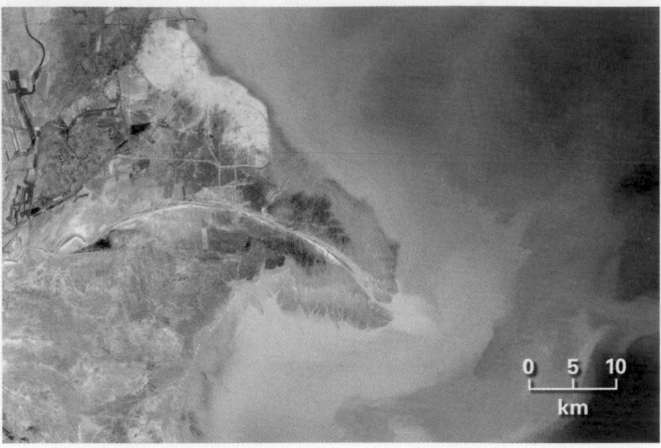

0 5 10
km

(b) The sediment of the Yellow River, China, settles out where it enters the sea, as seen from space.

Mouth of the Mississippi as seen from space.

Natural levee

out along the banks of the channel. Over time, the accumulation of this sediment builds a low ridge, called a **natural levee**, on either side of the stream. Natural levees may grow so large that the floor of the channel becomes higher than the surface of the floodplain. In fact, the higher neighborhoods of New Orleans, which have remained fairly dry during floods that submerged the rest of the city, were built on the Mississippi River's natural levees. In places where large natural levees exist, the region between the bluffs and the levees may become a low, marshy swamp. Also because of the levees, small tributaries may be blocked from joining the trunk stream—these tributaries, known as *yazoo streams,* flow in the floodplain and trend parallel to the main river.

Deltas: Deposition at the Mouth of a Stream

Along most of its length, only a narrow floodplain—covered by green, irrigated farm fields—borders the Nile River in Egypt. But at its mouth, the trunk stream of the Nile divides into a fan of smaller streams, called **distributaries**, and the area of green agricultural lands broadens into a triangular patch. The Greek historian Herodotus noted that this triangular patch resembles the shape of the Greek letter delta (Δ), so the region became known as the Nile Delta **(Fig. 17.22a)**. In general, a delta develops wherever sediment-laden stream water enters standing water, because as the current slows, the stream loses competence, so sediment settles out **(Fig. 17.22b)**. Small deltas form where small streams enter lakes. Huge ones develop where large river systems enter the sea.

Relatively few deltas display the simple triangular shape of the Nile Delta. Some curve out into the sea, whereas

flood. The meander that has been cut off becomes an **oxbow lake** if it remains filled with water or an *abandoned meander* if it dries out.

Most meandering stream channels, during normal flow, cover only a relatively small portion of their broad, gently sloping *floodplain.* As we noted earlier, water overtops the channel bank and submerges a floodplain during a flood. In many cases, a floodplain terminates at its sides along a *bluff,* or escarpment; large floods may cover the entire region from bluff to bluff. As the water leaves the channel, friction between the ground and the thin sheet of water moving over the floodplain slows down the flow. This slowdown decreases the competence of the running water, so sediment settles others, named *bird's-foot deltas*, consist of many elongate lobes (Fig. 17.22c, d). The shape of a delta depends on many factors. Deltas that form where the strength of the river current exceeds that of ocean currents have a bird's-foot shape, since the sediment can be carried far offshore. In contrast, deltas that form where the ocean currents are strong have a Δ shape, for ocean currents redistribute sediment in bars running parallel to the shore. And in places where waves and currents become strong enough to remove sediment as fast as it arrives, a river has no delta at all.

Why do rivers divide into distributaries at their mouths? When a river reaches standing water, its velocity slows. The sediment settles out at the mouth to form a midstream bar (Fig. 17.23a). The presence of the bar causes the stream to split into two channels. Similar bars produced at the mouths of these two subsidiary channels cause each of them to separate in turn, until eventually numerous distributary channels have formed (Fig. 17.23b).

With time, the sediment of a large marine delta compacts, and the lithosphere beneath the delta may subside (see Chapter 7). Where this happens, the surface of the delta slowly sinks. In a natural delta, distributaries provide sediment that fills the resulting space so that the delta's entire surface remains at or just above sea level, forming a broad flat area called a *delta plain*. But if people build levees or walls to restrict the river to its channel, sediment bypasses the interior part of a delta and gets carried directly to the seaward edge of the delta. Consequently, the delta's interior "starves" (does not receive sediment), and the land surface drops below sea level. Because of this process, much of New Orleans lies below sea level, so high waters from Hurricane Katrina in 2005 caused the city to flood extensively (see Chapter 20).

The position of a delta at the mouth of a major meandering river may change over time, because the main course of the river can shift over time. Sometimes these shifts occur as a result of an avulsion associated with a meander cutoff just upstream of the delta. And sometimes shifts happen when a toe of a bird's-foot delta builds so far out into the sea that the gradient of the stream becomes too gentle to allow the river to traverse the toe. At this point, the river overflows a natural levee upstream, an avulsion takes place, and the stream begins to flow along a new channel. The distinct lobes of the Mississippi Delta, a bird's-foot delta, reflect avulsions that have happened several times during the past 9,000 years (Fig. 17.24). New Orleans, built along one of the Mississippi's distributaries, may eventually lose its riverfront, for a break in a levee upstream of the city could divert the Mississippi into the Atchafalaya River channel.

TAKE-HOME MESSAGE

Erosion by running water carves valleys and canyons whose shapes depend on the balance between mass-movement and downcutting rates. Streams choked with sediment become braided; those following snake-like paths across floodplains are meandering. Meandering streams do not stay in the same place over time, due to the formation of cutoffs. Deltas form where a stream empties into standing water.

QUICK QUESTION: What factors cause the formation of rapids and waterfalls?

FIGURE 17.23 The formation of distributaries on a delta.

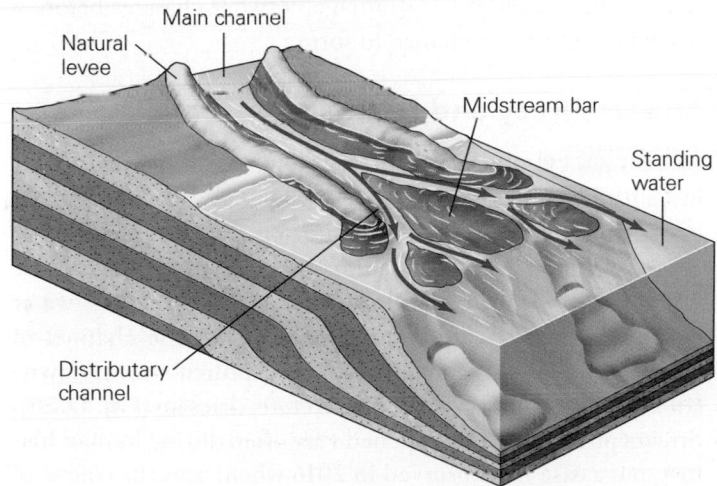

(a) Distributaries form because the river deposits sediment at its mouth when it reaches standing water.

(b) A few distributaries flow across this small delta in Costa Rica.

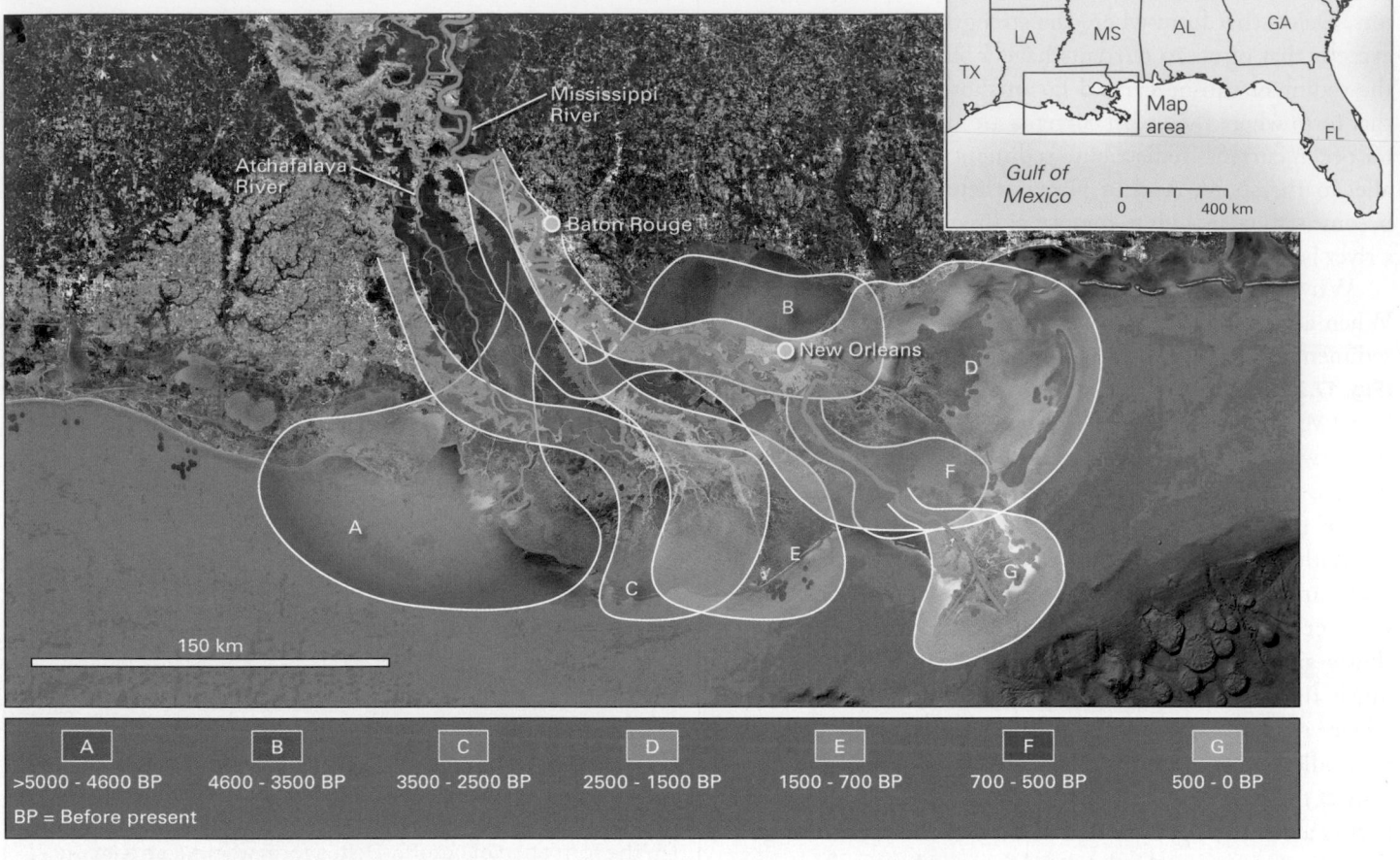

FIGURE 17.24 A map showing ancient lobes of the Mississippi Delta. A major flood could divert water from the Mississippi into the channel of the Atchafalaya.

A	B	C	D	E	F	G
>5000 - 4600 BP	4600 - 3500 BP	3500 - 2500 BP	2500 - 1500 BP	1500 - 700 BP	700 - 500 BP	500 - 0 BP

BP = Before present

17.7 The Evolution of Fluvial Landscapes

Beveling Topography

Over time, *fluvial landscapes*—those modified by erosion or deposition by streams—gradually evolve (**Fig. 17.25**). To picture this evolution, imagine that a region undergoes significant uplift. As soon as relief develops, running water gets to work, and a drainage network starts to form. In such a "young" fluvial landscape, streams of the network have steep gradients, and their water churns down rapids and tumbles over waterfalls as they carve deep valleys or canyons. But over time, mass movement along valley walls, combined with the removal of debris by running water, transforms rugged mountains into low, rounded hills, and formerly deep, narrow valleys broaden into wide floodplains with gentler gradients. Eventually, streams that had followed straight courses down steep gradients begin to meander over the surfaces of nearly horizontal floodplains. As more time passes, even low hills erode away, so that the landscape becomes a "mature" one with low relief, and its elevation lies close to the stream's base level. (Some geologists refer to such a low-relief landscape as a *peneplain*, from the Latin *paene*, which means almost.) Of course, this idealized model, like many models of components in the Earth System, is an oversimplification. Typically, tectonic processes cause the region to undergo uplift or subsidence again, or global sea level rises or falls over time, so that the base level of the drainage network changes before a peneplain ever has a chance to form.

Stream Piracy and Drainage Reversal

Stream piracy sounds like pretty violent stuff. In reality, it's just a natural process that happens when headward erosion by one stream causes the stream to intersect the course of another stream. When this happens, the pirate stream "captures" the water of the stream that it has intersected, so that the water of the captured stream starts to flow down the channel of the pirate stream. The channel of the captured stream downstream of the point of capture, therefore, dries up (**Fig. 17.26**). Stream piracy has not happened very often during human history, but a case was observed in 2016 when, over the course of just a few months, the Slims River in Canada's Yukon dried up due to piracy.

FIGURE 17.25 Evolution of a fluvial landscape when base level drops.

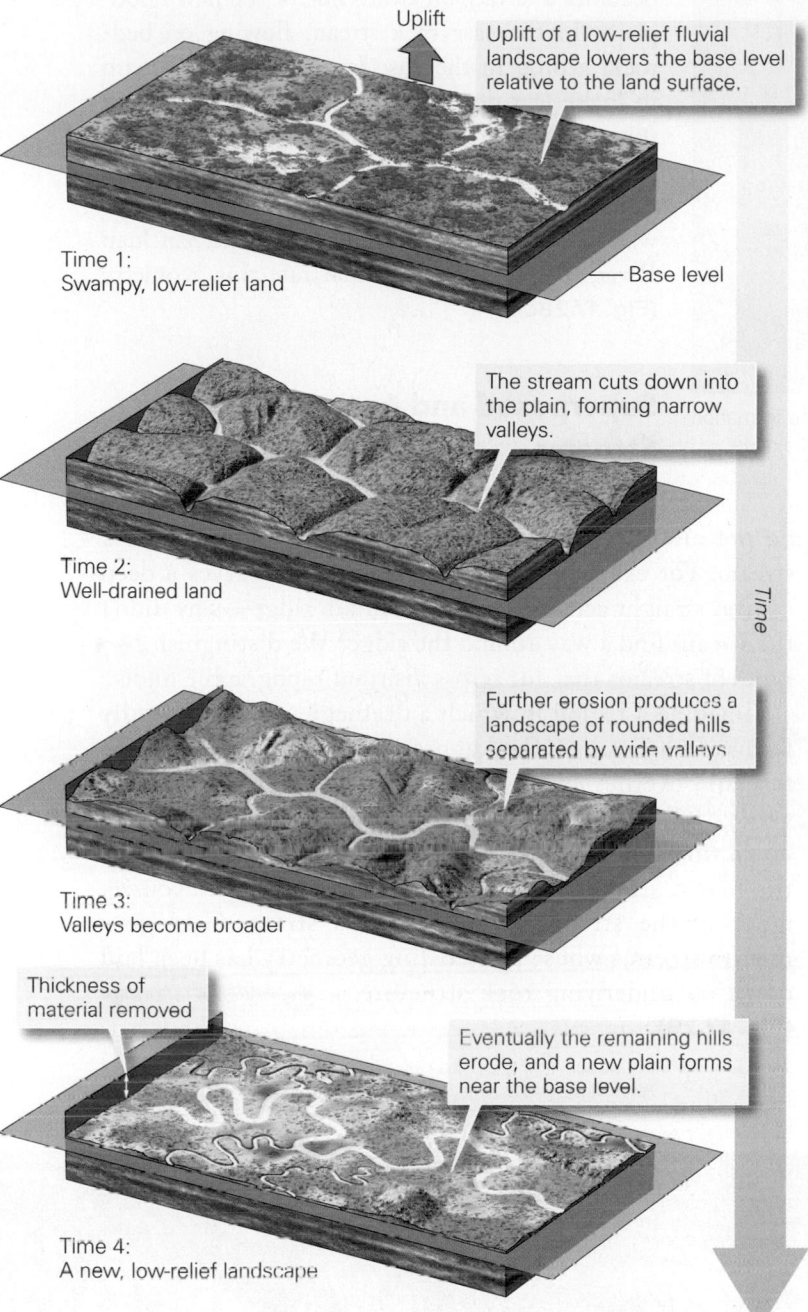

Uplift

Uplift of a low-relief fluvial landscape lowers the base level relative to the land surface.

Time 1:
Swampy, low-relief land

Base level

The stream cuts down into the plain, forming narrow valleys.

Time 2:
Well-drained land

Further erosion produces a landscape of rounded hills separated by wide valleys.

Time 3:
Valleys become broader

Thickness of material removed

Eventually the remaining hills erode, and a new plain forms near the base level.

Time 4:
A new, low-relief landscape

Time

We tend to think of a given drainage network as always flowing in the same direction because gravity always moves water from a higher to a lower elevation. But tectonic processes, acting over geologic time, can change the slope of the land on a continental scale. If the overall tilt of the land reverses, a **drainage reversal** will take place; that is, a drainage network reorganizes so that water flows, overall, in the opposite direction. As an example, consider the evolution of drainage in South America. Before about 100 million years ago, South America and Africa were adjacent parts of Pangaea, and a highland existed between what would become two continents. At this

time, the main drainage network of northern South America flowed westward to the Pacific. Later, when South America rifted away from Africa, the northeastern coast of the continent subsided, and the Andes Mountains rose along its western coast. As a consequence, westward flow became impossible, and the modern eastward-flowing Amazon drainage network developed **(Fig. 17.27)**.

Stream Rejuvenation

Where streams cut down into a landscape of low relief that was originally near the stream's base level, **stream rejuvenation** has occurred. Rejuvenation happens when the base level of a

FIGURE 17.26 The concept of stream piracy.

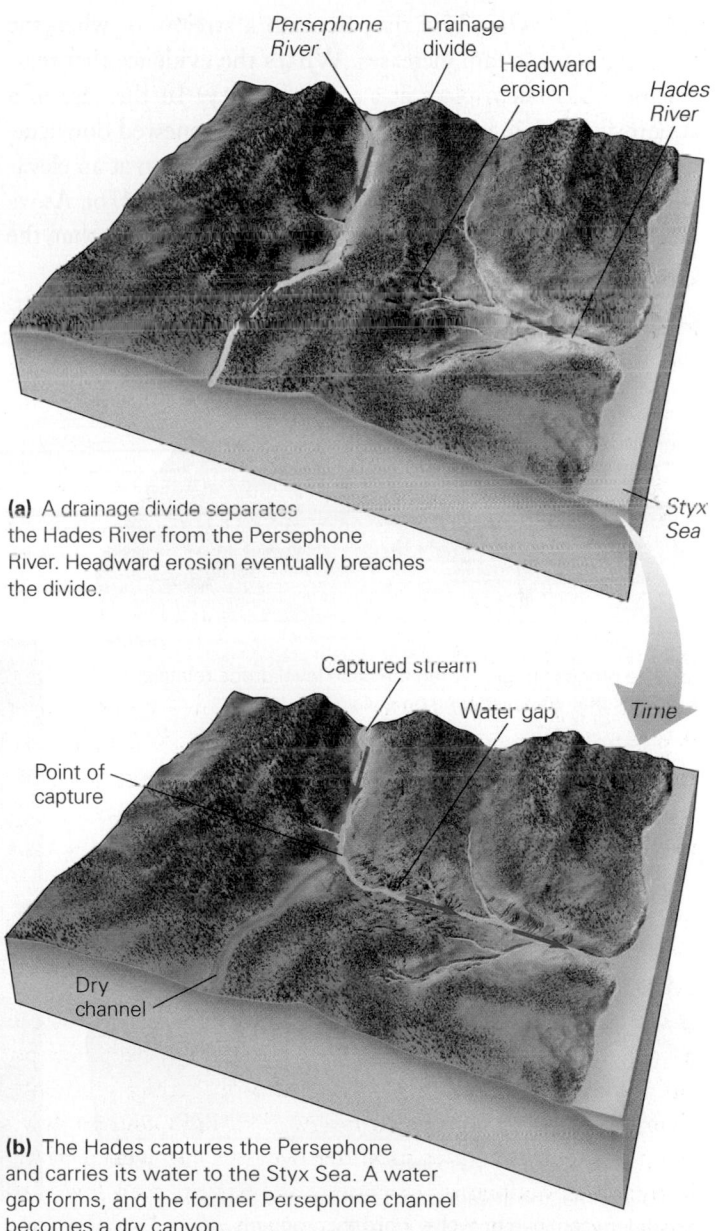

(a) A drainage divide separates the Hades River from the Persephone River. Headward erosion eventually breaches the divide.

Persephone River · Drainage divide · Headward erosion · Hades River · Styx Sea

Captured stream · Water gap · Time · Point of capture · Dry channel

(b) The Hades captures the Persephone and carries its water to the Styx Sea. A water gap forms, and the former Persephone channel becomes a dry canyon.

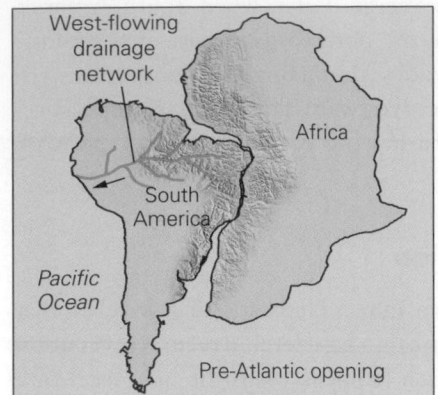

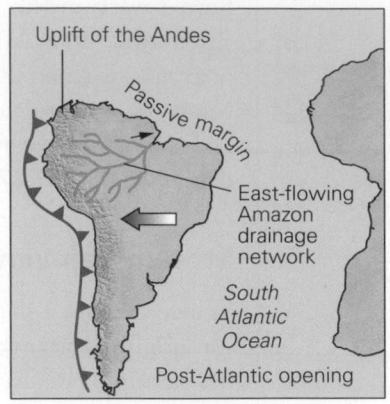

(a) In the early Mesozoic, rivers drained westward, from the interior of Pangaea.

(b) Later, after rifting and the formation of the Andes, the Amazon drainage flowed eastward.

older one, and the surface of the older floodplain becomes a terrace on either side of the new floodplain. In the case of a stream flowing on bedrock, a drop in the base level causes the stream to incise (cut down into) the underlying bedrock **(Fig. 17.28a, b)**. If the stream had a meandering course before rejuvenation, it will downcut to form *incised meanders* that lie at the bottom of a steep-walled canyon. The "goosenecks" of the San Juan River, in southern Utah, illustrate this geometry **(Fig. 17.28c)**.

Superposed and Antecedent Streams

The structure and topography of the landscape do not always appear to control the path, or course, of a stream. For example, consider a stream that carves a deep canyon straight across a strong mountain ridge—why didn't the stream find a way around the ridge? We distinguish two types of streams that cut across resistant topographic highs.

Imagine a region in which a drainage network initially forms on a layer of flat, nonresistant strata that unconformably overlies folded strata. Initially, the streams carve valleys into the flat strata. When they eventually erode down through the unconformity and start to downcut into the folded strata, they may maintain their earlier course, ignoring the structure of the folded strata. Geologists refer to streams whose pre-existing geometry has been laid down on underlying rock structure as *superposed streams* **(Fig. 17.29)**.

stream drops, when land rises beneath a stream, or when the discharge of a stream increases. What's the evidence that rejuvenation has taken place at a given locality? In the case of a stream flowing in an alluvium-filled valley, renewed downcutting allows the stream to produce a new floodplain at an elevation lower than that of the original one (see Fig. 17.17b). As we have seen, the younger floodplain tends to be narrower than the

FIGURE 17.28 The formation of incised meanders.

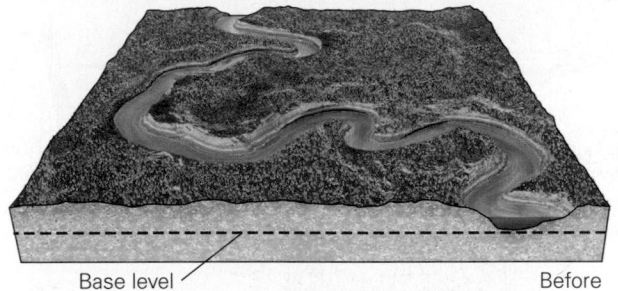

(a) The process begins when the base level drops relative to a stream that is meandering on a plain.

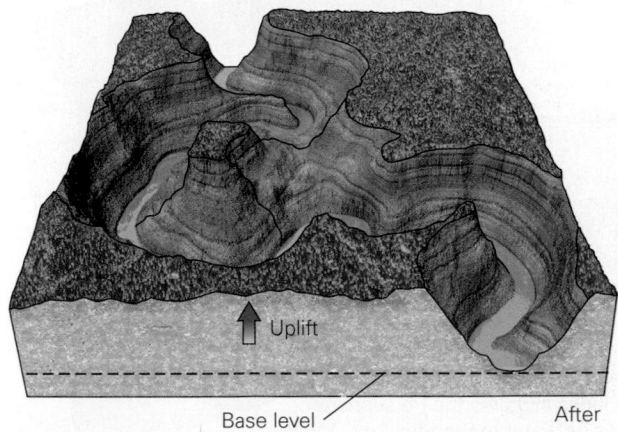

(b) Over time, the stream cuts down into the bedrock. Meanders continue to evolve while this happens.

(c) The "goosenecks" of the San Juan River, Utah, are incised meanders.

FIGURE 17.29 Formation of superposed drainage.

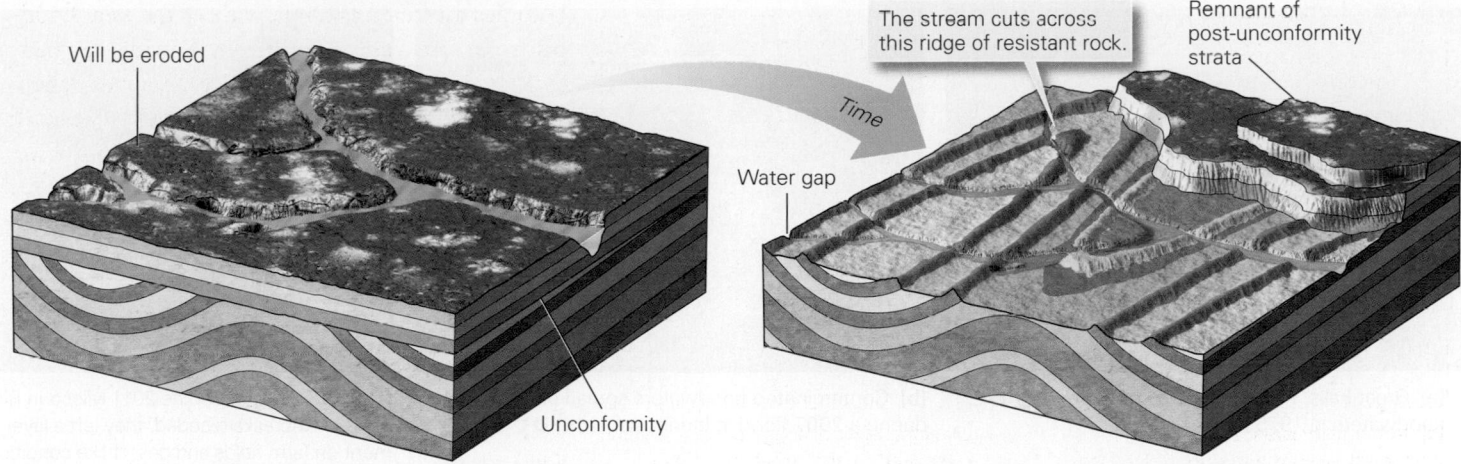

(a) A superposed stream establishes its geometry while flowing over a uniform substrate above an unconformity.

(b) When erosion exposes underlying rock with a different structure, the stream is superposed on the structure. As a result, it cuts across resistant ridges instead of flowing around them.

In some cases, tectonic activity, such as subduction or collision, causes localized uplift, so a mountain range rises beneath the course of an already established stream. If the stream downcuts as fast as the range rises, it can maintain its course and will cut right across the range. Geologists call such streams *antecedent streams*, from the Greek *ante*, meaning before, to emphasize that they existed before the range uplifted. Note that if the range rises faster than the stream can downcut, the new highlands divert the stream's course, and the *diverted stream* starts to flow parallel to the range face **(Fig. 17.30)**.

FIGURE 17.30 Development of antecedent and diverted streams.

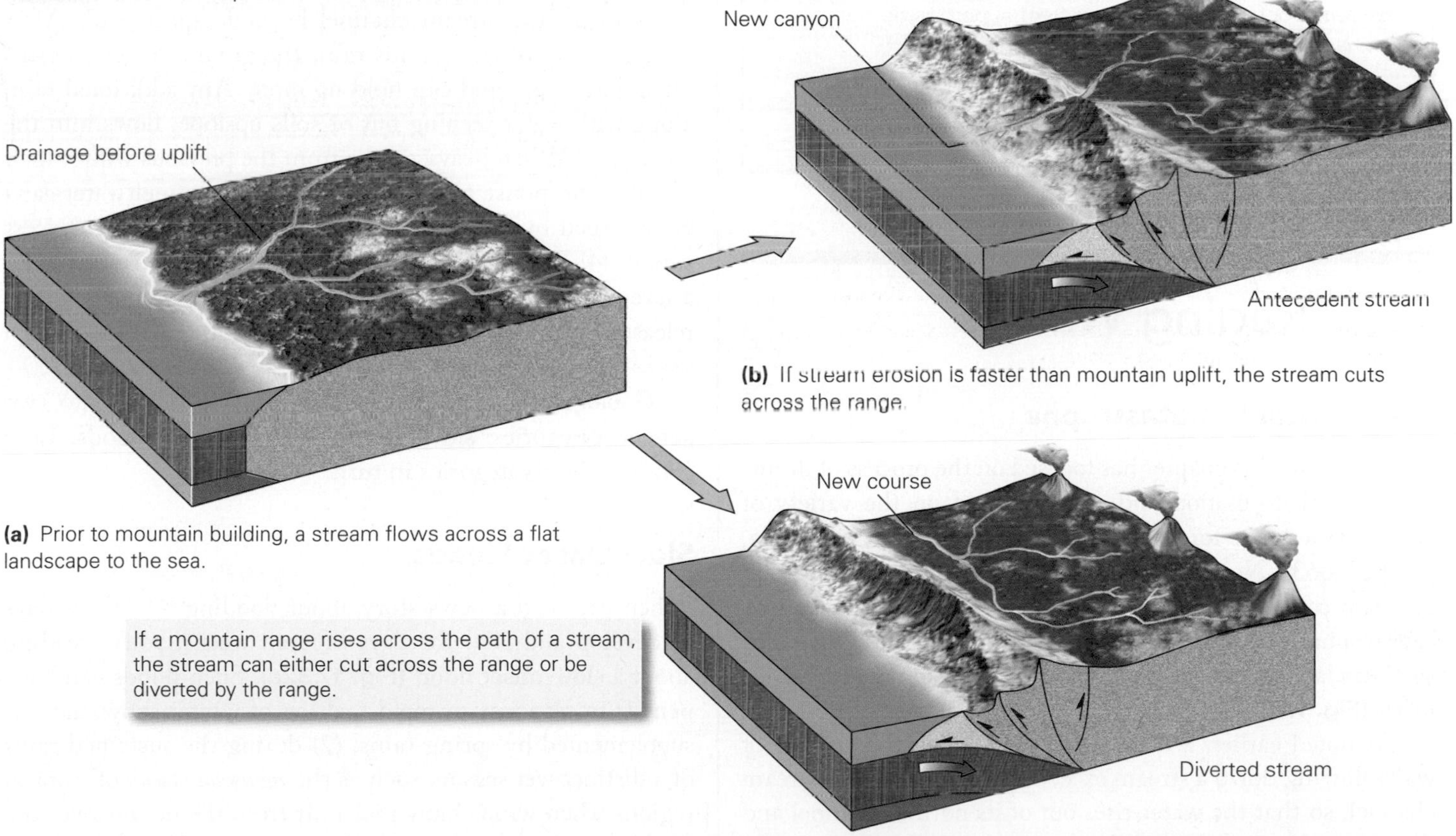

(a) Prior to mountain building, a stream flows across a flat landscape to the sea.

If a mountain range rises across the path of a stream, the stream can either cut across the range or be diverted by the range.

(b) If stream erosion is faster than mountain uplift, the stream cuts across the range.

(c) If uplift happens faster than erosion, the stream is diverted and flows along the edge of the range.

FIGURE 17.31 Flooding inundates fields and cities and deposits sediment.

(a) Great Falls, Montana, was submerged by floodwaters in 1975.

(b) Contaminated floodwaters spread disease during a 2007 flood in Indonesia.

(c) As floodwaters of the 2011 Missouri River flood in Nebraska receded, they left a layer of sediment on farm fields and desert-like conditions.

> ## TAKE-HOME MESSAGE
>
> Fluvial landscapes evolve over time as gradients diminish and ridges between valleys erode away. Eventually, the landscape may attain low relief. Subsequent changes in base level can trigger rejuvenation. Over time, some streams capture the flow of other streams. And if the regional tilt of the land surface changes, the direction of drainage reverses. Locations where streams cut across resistant rock structure may reflect superposition of drainage or downcutting through rising structure.
>
> **QUICK QUESTION:** What can cause a drainage-reversal event to take place?

17.8 Raging Waters

The Inevitable Catastrophe

Up until now, this chapter has focused on the process of drainage network formation and evolution and on the variety of landscape features formed by streams (**Geology at a Glance**, pp. 662–663). Now we turn our attention to the havoc that a stream can cause when flooding takes place. Floods can be catastrophic—they can strip land of forests and buildings, they can bury land in mud and silt, and they can submerge communities (**Fig. 17.31**).

As noted earlier, a flood takes place when the volume of water flowing down a stream exceeds the volume of the stream channel, so that the water rises out of its normal channel and fills the canyon in which it flows to a greater depth than normal, or spreads out over its floodplain or delta plain by breaking

through or overtopping levees. The news media may report that a stream "crested at 3 m above flood stage at 10 P.M." This means that the water surface in the stream rose to a level 3 m higher than the top of the normal channel at 10 P.M., and after that time it became lower—the *flood crest* is the highest level that the stream reaches.

Floods can happen for several reasons: (1) During abrupt, heavy rains, water falls on the ground faster than it can infiltrate deep into the subsurface. The excess becomes runoff that can fill the stream channel beyond capacity. (2) After a long period of continuous rain, the ground becomes saturated with water and can hold no more. Any additional rain, along with water seeping out of soils upslope, flows into the stream. (3) When heavy snows from the previous winter melt rapidly in response to a sudden hot spell, the meltwater can't be absorbed by the ground fast enough and becomes excess runoff. (4) When a dam holding back a lake or reservoir, or a levee holding back a river or canal, suddenly collapses and releases the water that it held back, sending a rush of high water downstream.

Geologists find it convenient to divide floods into two general categories: slow-onset floods and flash floods. Let's consider these categories in turn.

Slow-Onset Floods

When you read a news story about flooding that takes days to develop and lasts for days or even weeks, you're reading about a **slow-onset flood (Fig. 17.32a)**. Such floods can happen (1) in response to rapid melting of winter snow and ice, supplemented by spring rains; (2) during the sustained rains of a distinct wet season, such as the *monsoon season* of tropical regions when winds blow moist air from the oceans over the land for weeks on end; or (3) when a system of storms remains stationary over a broad region for a long time. In all these

FIGURE 17.32 Examples of slow-onset floods, in a floodplain.

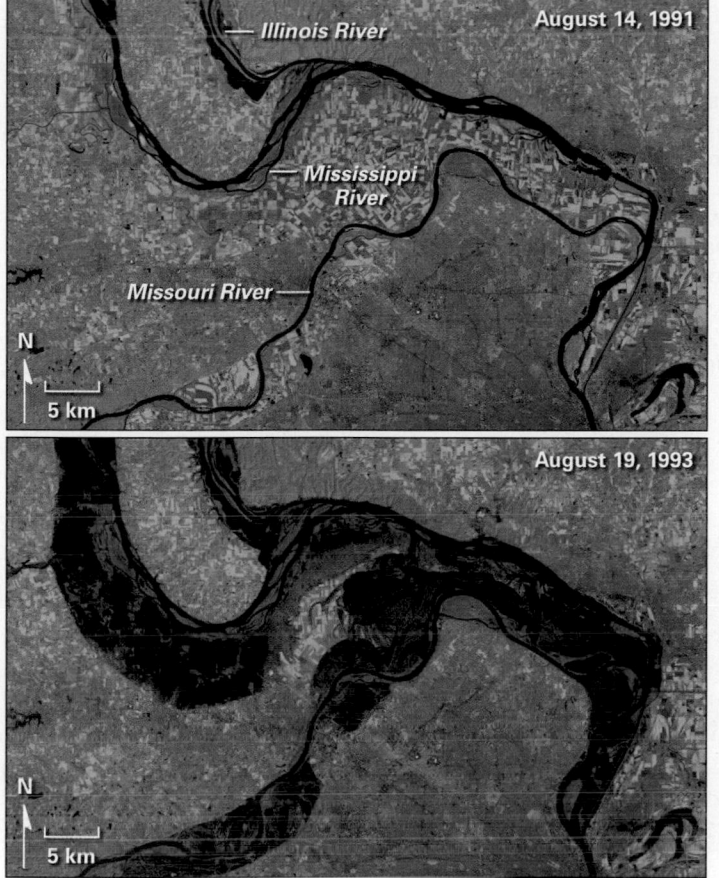

Illinois River — August 14, 1991

Mississippi River

Missouri River

N

5 km

August 19, 1993

N

5 km

(a) Satellite images show how the rivers in the midwestern United States can cover their floodplains. The top image shows the rivers at relatively normal water levels for comparison.

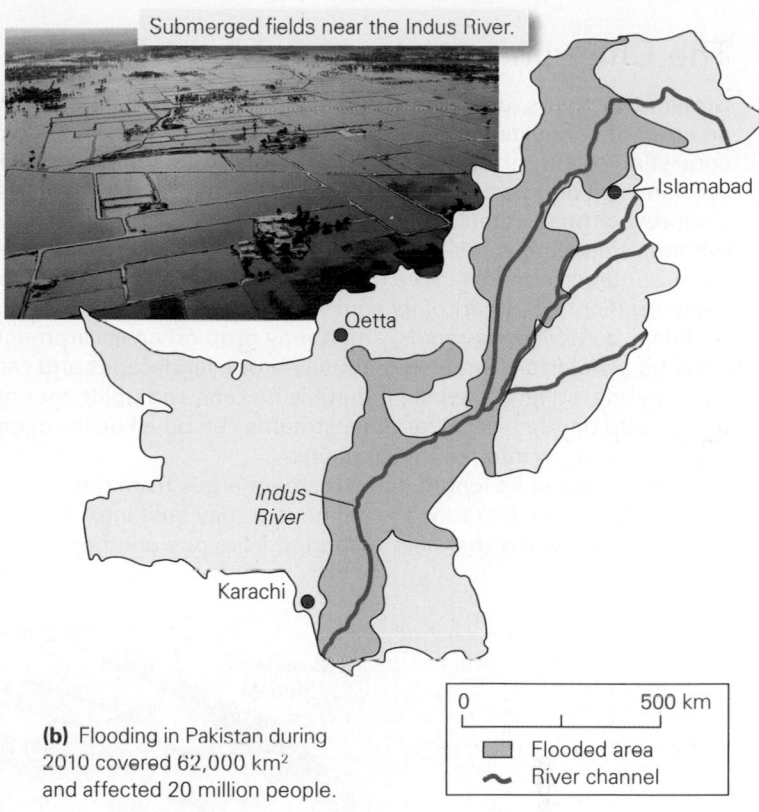

Submerged fields near the Indus River.

Islamabad

Qetta

Indus River

Karachi

0   500 km

▨ Flooded area
〰 River channel

(b) Flooding in Pakistan during 2010 covered 62,000 km² and affected 20 million people.

circumstances, the ground becomes saturated, so meltwater or rainfall becomes runoff that fills stream channels beyond their capacity.

Slow-onset floods that happen during specific times of the year are also known as **seasonal floods**. If the flood spreads out over a stream's floodplain, it's also a *floodplain flood*, whereas if the water spreads out over a delta plain, it's also a *delta-plain flood*.

Typically, slow-onset floods take time—hours or days—to develop, so authorities have time to evacuate potential victims and organize efforts to protect property. But because preparation or evacuation doesn't always take place, and because so many people live on floodplains and delta plains, these floods can cause a staggering loss of life and property.

Along many rivers, slow-onset flooding plays an important role in the sustainability of ecosystems and societies. Seasonal floods along the Nile River and its delta played a major role in replenishing nutrients to agricultural areas of Egypt. Unfortunately, some slow-onset floods cause so much damage and loss of life that they make a mark in history, either locally or globally.

For example, an 1887 flood of China's Hwang (Yellow) River—so named because of the yellow silt it carries—killed as many as 2.5 million people, and a 1931 flood of the Yangtze River in China led to a famine that killed 3.7 million people. During the 1990 monsoon season in Bangladesh, rain fell almost continuously for weeks; the Ganges Delta plain became inundated, and the resulting flood killed 100,000 people. Seasonal floods struck Indonesia in 2007, killing dozens of people and displacing almost half a million, nearly half of whom became sick from contact with the filthy water and mud that submerged 60% of the capital and hundreds of square kilometers of farmland. One of the most devastating floods of recent times began in July 2010, when a seasonal flood fed by intense monsoonal rains submerged floodplains of the Indus River drainage system in Pakistan **(Fig. 17.32b)**. On some days, it had rained up to 40 cm in 24 hours! The floods put almost 62,000 km² of the country under water and severely impacted the lives of over 20 million people (12% of the population)—many people lost all they owned. Crops growing in the fertile floodplain floated away or rotted, and over 200,000 cattle drowned. Due to the destruction of clean-water supplies and of road and rail networks, survivors were stranded for weeks or more, and sadly, disease spread despite relief efforts by organizations from around the world. Disastrous floods due to persistent storms affected

The Changing Landscape along a Stream

Streams, or rivers, drain the landscape of surface runoff. Typically, an array of connected streams, called a drainage network, develops, consisting of a trunk stream into which numerous tributaries flow. The land drained by the network is its watershed. A stream starts from a source, or headwaters. Some headwaters are in the mountains, perhaps collecting water from rainfall or from melting ice and snow. In the mountains, streams carve deep, V-shaped valleys and tend to have steep gradients. For part of its course, a stream may flow over a steep, bouldery bed, forming rapids, and it may drop off an escarpment, creating a waterfall. Streams gradually erode landscapes and carry away debris, so after a while, if there is no renewed uplift, mountains evolve into gentle hills. Over time, streams can bevel once-rugged mountain ranges into nearly flat plains.

Farther along its length, the stream emerges from the mountains. If it is choked with sediment, it may split into numerous entwined channels separated from one another

Developing drainage networks

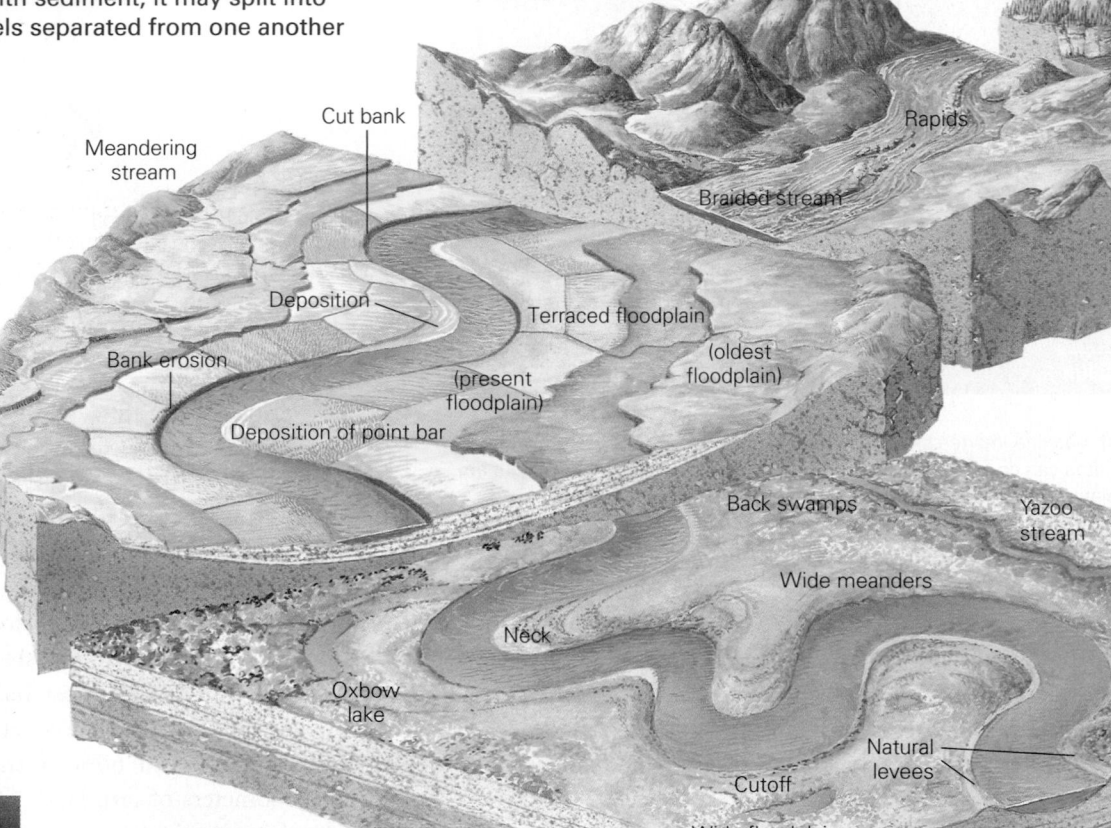

Transportation along the channel

Meandering stream

Cut bank

Rapids

Braided stream

Deposition

Bank erosion

Terraced floodplain

(oldest floodplain)

(present floodplain)

Deposition of point bar

Back swamps

Yazoo stream

Wide meanders

Neck

Oxbow lake

Cutoff

Natural levees

Wide floodplain

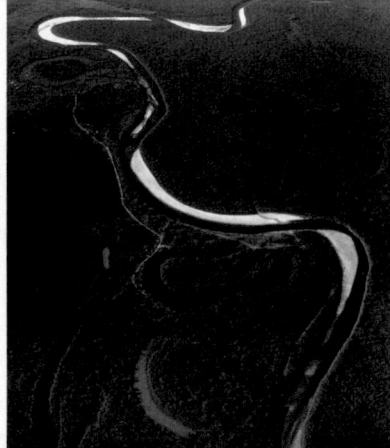

Point bars forming on inner curves.

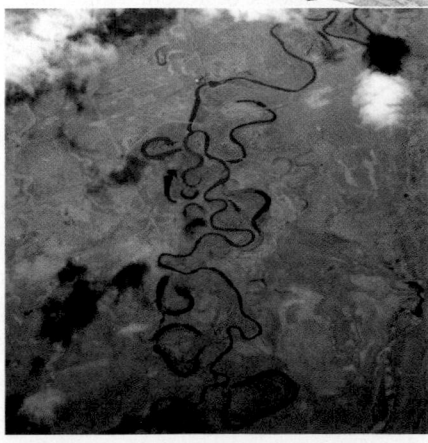

Meanders, abandoned meanders, and cutoffs.

A small delta in a mountain lake.

Headward erosion

Glaciers

Valleys with high relief

Melting ice

Lake

Dendritic drainage

Rapids

Waterfall

Collection of water in watershed

Streams contribute to carving mountains.

Waterfall in Hawaii spilling over a basalt ledge.

by gravel bars, creating a braided stream. Where a stream that has not been choked by sediment flows over flat ground, it becomes a meandering stream, winding back and forth in snake-like curves called meanders. The current flows faster on the outer edge of a curve, so erosion takes place there, whereas the current flows more slowly on the inner edge, where it drops sediment. Because of erosion and deposition, a meandering stream changes shape over time. Occasionally a meander may be cut off, leaving a curving lake called an oxbow lake. A broad floodplain, covered with water only during floods, may develop on either side of the stream. Natural levees build up between the channel and the floodplain from sediment dropped as a flooding river starts to spill out of its channel. Eventually, a stream reaches a standing body of water and slows down, and the sediment it carries is deposited to form a delta. On a delta, the trunk stream divides into many smaller channels called distributaries.

Deposition at mouth

Delta

Distributaries

Natural levees

Swamps and marsh

Tidal flats

Bar

Banks

FIGURE 17.33 Atmospheric rivers can bring torrential rainfalls, which cause severe flooding.

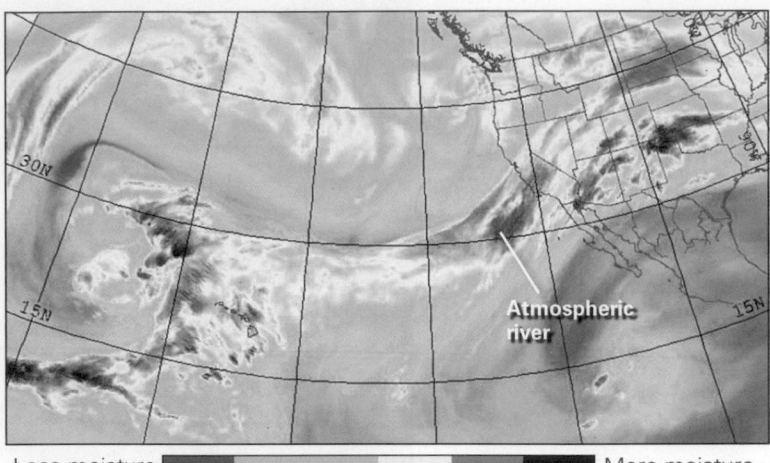

Less moisture ▮▮▮▮▮▮ More moisture

(a) Atmospheric rivers are streams of very moist air that flow rapidly in the lower layer of the atmosphere.

Moisture
less   more   Not to scale

(b) When the moist air is forced upward over coastal mountain ranges, the water precipitates as torrential rain.

(c) Flooding in the Sacramento Valley of California in February 2017 due to rain from an atmospheric river.

Texas and California in 2017. The floods were associated with *atmospheric rivers*, streams of very moist air that can drop torrential rains over land **(Fig. 17.33)**.

Slow-onset flooding happens frequently in the Mississippi Valley of the central United States, and in some years, the floods are disastrous. The 1993 flood can serve as a case study of such an event. High-altitude (10- to 15-km-high) winds drifted southward and for weeks formed an invisible wall that trapped warm, moist air from the Gulf of Mexico over the central United States. As this air rose to higher elevations and cooled, the water it held condensed and fell as rain, rain, and more rain. In fact, almost a whole year's supply of rain fell during the spring of 1993, and some regions received 400% more than usual. Eventually, the water in the Missouri and Mississippi Rivers rose above the height of their levees and spread out over the floodplain. By July, parts of nine states were underwater (see Fig. 17.32a).

The roiling, muddy flood of 1993 uprooted trees, swept cars away, and even unearthed coffins (which floated out of the inundated graveyards). All barge traffic along the Mississippi came to a halt, bridges and roads were washed away, and muddy water submerged towns along the river. For example, in Davenport, Iowa, streets in the riverfront district lay beneath 4 m of water. In Des Moines, Iowa, 250,000 residents lost their supply of drinking water when floodwaters contaminated the municipal water supply with raw sewage and chemical fertilizers. Rowboats replaced cars as the favored mode of transportation in towns where only the rooftops remained visible. In St. Louis, Missouri, the river crested 14 m above flood stage. For 79 days, the flooding continued. When the water finally subsided, it left behind a thick layer of silt and mud, filling living rooms and kitchens in floodplain towns and burying crops in floodplain fields. In the end, more than 40,000 km² of the floodplain had been submerged, 50 people died, at least 55,000 homes had been ruined, and countless acres of crops were buried. Officials estimated that the flood caused over $12 billion in damage. Comparable flooding happened again in the spring of 2011.

Flash Floods

Events during which the floodwaters rise so fast that it may be difficult to escape from their path are called **flash floods (Fig. 17.34a, b)**. A flash flood may be the harbinger of a seasonal flood in that the high waters may arrive rapidly but remain for a long time, or it may be a short-lived event in the wake of an intense rainstorm, dam collapse **(Box 17.1)**, or levee failure, in which the high waters subside in a matter of minutes to hours. Flash floods can be particularly unexpected in arid or semiarid regions, where water from an isolated rainstorm may suddenly fill an otherwise dry channel. Such a flood may even affect areas downstream that have not received a drop of rain. If a flash flood affects a drainage basin in which streams have cut steep-sided valleys or canyons without a floodplain, the flood may arrive as a wall of water, slamming downstream with great force, and water may quickly fill the canyon or valley to levels meters to tens of meters above normal in a matter of minutes.

Flash floods pose many dangers to human populations. Not only is there the threat of being trapped or washed away by the raging waters, but flash floods can also create devastating mudslides, flood or wash away homes and buildings, transport large boulders and other debris into populated areas (see Fig. 17.1d), and ruin farm land. In 2017, a furious flow of muddy water surged over the banks of the Roche River in northern Peru, blanketing miles of crops and fertile farmland along the river in a thick layer of suffocating mud **(Fig. 17.34c)**.

Another event, the historic Big Thompson River flood of July 31, 1976, illustrates the full power of a flash flood **(Fig. 17.34d)**. This river, which drains the Front Range of the Rocky Mountains, north of Denver, Colorado, normally seems quite harmless. It carries clear water from rains or from melting ice and snow. The stream has a steep gradient, so this water froths over and around boulders and cobbles on the streambed. The channel of the river does not occupy the whole valley floor, so a road follows the course of the river, and in places, vacation cabins and campgrounds dot the land between the river and the road. On July 31, the peaceful landscape of the valley turned to one of terror, as winds blew air

FIGURE 17.34 Flash floods can occur after torrential rains.

(a) A flash flood in a desert region of Israel has washed over a highway, forcing the evacuation of truckers.

(b) A 2013 flash flood in India washed away bridges and stranded 40,000 people.

(c) A 2017 flash flood in Peru sent this river over its banks, submerging farms in muddy water.

(d) During the 1976 Big Thompson River flash flood, this house was carried off its foundation and dropped on a bridge.

BOX 17.1 CONSIDER THIS . . .

The Johnstown Flood of 1889

By the 1880s, Johnstown, built along the Conemaugh River in scenic western Pennsylvania, had become a significant industrial town. Recognizing the attraction of the surrounding hills as a summer retreat, speculators built a mud-and-gravel dam across the river, upstream of Johnstown, to trap a pleasant reservoir of cool water. A group of industrialists and bankers bought the reservoir and established the exclusive South Fork Hunting and Fishing Club, a cluster of lavish 15-room "cottages" on the shore. Unfortunately, the dam had been poorly designed, and debris blocked its spillway (a passageway for surplus water), setting the stage for a monumental tragedy. On May 31, 1889, torrential rain drenched Pennsylvania, and the reservoir filled until water began to flow over the dam. Despite frantic attempts to strengthen the dam, the soggy structure abruptly collapsed, and the reservoir emptied into the Conemaugh River Valley. A 20-m-high wall of water roared downstream and slammed into Johnstown, transforming bridges and buildings into twisted wreckage (Fig. Bx17.1). When the water subsided, 2,300 people were dead, and Johnstown became the focus of national sympathy. The recently founded Red Cross set to work building dormitories, and citizens nationwide donated everything from clothes to beds. Nevertheless, it took years for the town to recover, and many residents simply picked up and left.

FIGURE Bx17.1 During the 1889 Johnstown flood, raging waters could tumble large houses.

north from the Gulf of Mexico. The warm, moist air rose over the mountains, towering thunderheads built up, and at 7:00 P.M., rain began to fall. It poured, in quantities that even old-timers couldn't recall. In a little over an hour, 19 cm of rain drenched the watershed of the Big Thompson River. The river's discharge grew to more than four times the maximum recorded during the previous century, and the water level rose by several meters. Turbulent currents swirled down the canyon at up to 8 m per second and churned up so much sand and mud that the once-clear river became a viscous slurry. Slides of rock and soil tumbled down the steep slopes bordering the river and fed the torrent with even more sediment. The raging water undercut foundations and washed the houses and bridges away. Parts of the road were eroded away, and other parts were buried by debris. Boulders that had stood like landmarks for generations bounced along in the torrent like beach balls, striking and shattering other rocks along the way; the largest rock known to be moved by the flood weighed 275 tons. Cars drifted downstream until they finally wrapped like foil around obstacles. When the flood subsided, the canyon had changed forever, and 144 people had lost their lives. As we described at the beginning of this chapter, flooding devastated the area again in 2013.

Glacial Torrents

Perhaps the greatest floods chronicled in the geologic record happen when natural ice dams burst. The Great Missoula Floods of about 11,000 years ago serve as an example of such **glacial torrents**. These floods occurred at the end of the last ice age, when a lobe of an ice sheet acted like a dam, holding back a large lake called Glacial Lake Missoula. When the ice sheet began to melt, the dam suddenly gave way, the lake abruptly drained, and water roared over what is now eastern Washington, eventually entering the Columbia River Valley and flowing on out to the Pacific Ocean. Subsequently, the ice sheet grew again, and the dam re-formed, trapping a new lake, which drained during a subsequent failure. The process repeated a few times. The resulting floods stripped off the regolith covering the dark basaltic bedrock of eastern Washington, leaving a barren, craggy landscape now known as the *channeled scablands* **(Fig. 17.35)**.

J. Harlan Bretz, a geologist studying landscape evolution in Washington, first made the association between scabland formation and catastrophic glacial torrents. When Bretz presented his interpretation in the 1920s, other geologists ridiculed him because the idea seemed to violate the well-accepted principle of uniformitarianism (see Chapter 12). Bretz fought back by demonstrating that the huge boulders that littered the landscape surface were moved by water, not ice; that the region's hills began as immense ripples (1,000 times bigger than those typically found on a streambed); and that the Grand Coulee, a large dry canyon, once hosted a waterfall hundreds of times larger than Niagara Falls. After years of debate, the geologic community concluded that Bretz was correct, and researchers have since identified several other examples of landscapes carved by glacial torrents.

Living with Floods

Flood Control Mark Twain once wrote of the Mississippi River that we "cannot tame that lawless stream, cannot curb it or confine it, cannot say to it, 'go here or go there,' and make it obey." Was Twain right? Since ancient times, people have attempted to confine rivers to set courses so as to prevent undesired flooding. In the 20th century, flood-control efforts intensified as the population living along rivers increased. For

FIGURE 17.35 The Great Missoula Floods. When ice dams broke, glacial torrents from Glacial Lake Missoula scoured portions of the Columbia River plateau.

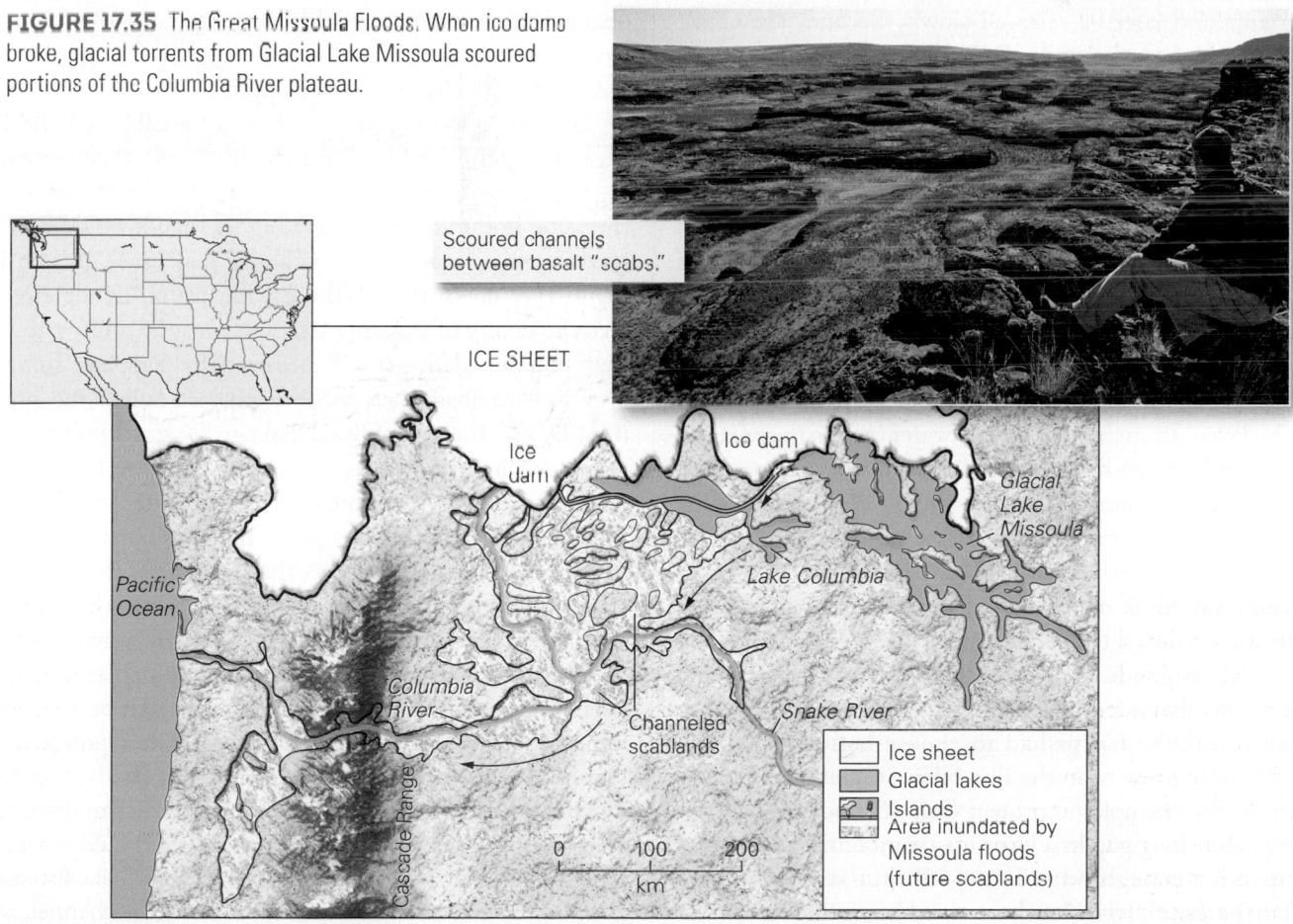

FIGURE 17.36 Flood control structures.

(a) Artificial levees built to protect the downtown of Galena, Illinois, which is built along a tributary of the Mississippi.

High-water marks of past floods.

(b) These floodwalls can be closed to protect Cape Girardeau, Missouri, from Mississippi River floods.

example, since the passage of the 1927 Mississippi River Flood Control Act, drafted after a cataclysmic flood took place that year, the U.S. Army Corps of Engineers has labored to control the Mississippi. First, engineers built about 300 dams along the river's tributaries so that excess runoff could be stored in reservoirs and later be released slowly. Second, they built *artificial levees* of sand and mud, and floodwalls of concrete, along the river **(Fig. 17.36)**. These structures effectively increase the channel's volume and isolate discrete areas of the floodplain to prevent them from being inundated.

Although the Corps' strategy worked for floods up to a certain size, it couldn't prevent the 1993 and 2011 floods, when the reservoirs filled to capacity and additional runoff headed downstream. The river rose until it spilled over the tops of some levees and undermined others **(Fig. 17.37)**. *Levee undermining* occurs when rising floodwaters increase the water pressure on the river side of the levee, forcing water down through sand under the levee. In susceptible areas, water begins to spurt out of the ground on the dry side of the levee, thereby washing away the levee's support. The levee finally becomes so weak that it collapses, and water rushes through the breach.

Using lessons learned from 1993, the Army Corps of Engineers undertook a number of proactive but controversial steps that reduced the impact of the 2011 Mississippi and Missouri River floods. The 2011 floodwaters simply didn't fit in the space bounded by existing levees and floodwalls, so at a few places, the Corps had to choose between flooding fields or flooding towns on the floodplain. Generally, it chose the former. For example, to protect Cairo, Illinois, the Corps blasted a 3-km-long gap in a levee upstream of the town. This action diverted enough water into a 530-km² area of farmed floodplain so that the river level remained lower than the levees

at Cairo, and water stayed out of the town's streets. Farther downstream, the Corps opened floodgates to divert water gradually into the Atchafalaya basin. Without this action, the Mississippi River might have broken through natural levees, and would then have flowed down the Atchafalaya River's channel. This avulsion would have prevented the Mississippi from reaching the port of New Orleans, a change that could have had huge economic implications.

Because of the expense, it's just not feasible to build levees high enough to handle all conceivable floods. And since building high levees confines the river to the channel, the levees cause the water in the channel to rise higher than it would have if no levees were present, and this change may cause greater flooding farther downstream. Those who live on floodplains must face the reality of flooding risk and must be willing to accept that flooding damage will occasionally happen. Temporary levees of sandbags can, at best, protect only some property **(Fig. 17.38)**. The cost of flood damage has quadrupled in recent years, despite the billions of dollars that have been spent on flood control, because more people have settled in floodplains. This has led to the challenge of figuring out ways to insure floodplain properties in ways that are fair both to floodplain inhabitants and to others who share in the cost of insurance.

Some communities are looking at new ways to mitigate flooding risk. For example, instead of building levees to isolate portions of the floodplain, these portions can be transformed into natural wetlands, which absorb water like a sponge, in order to diminish flooding hazards. Property may also be kept safe by moving levees farther away from the river so as to define *floodways*, regions that will flood, and therefore should not be used for building homes or businesses **(Fig. 17.39)**. The existence of a floodway effectively increases the volume of the channel, so that

FIGURE 17.37 How levees fail.

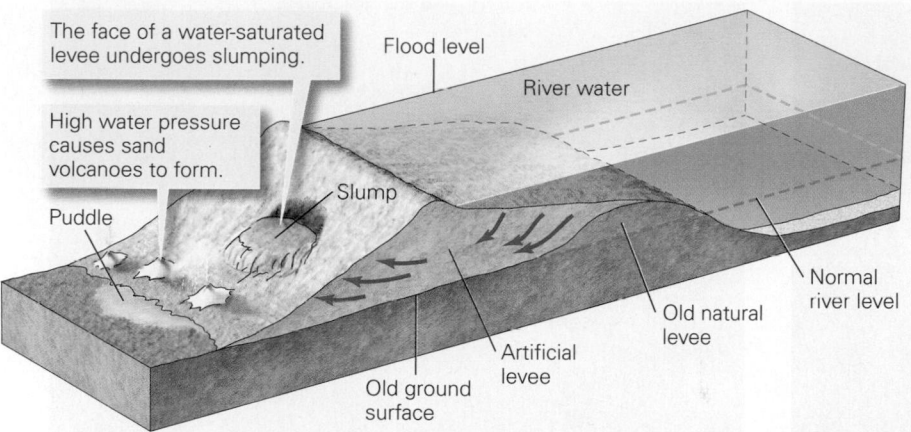

The face of a water-saturated levee undergoes slumping.

High water pressure causes sand volcanoes to form.

Flood level

River water

Slump

Puddle

Normal river level

Old natural levee

Artificial levee

Old ground surface

(b) Levees can be undermined if the river floods.

(b) Breaching of a levee lets water spill into the floodplain.

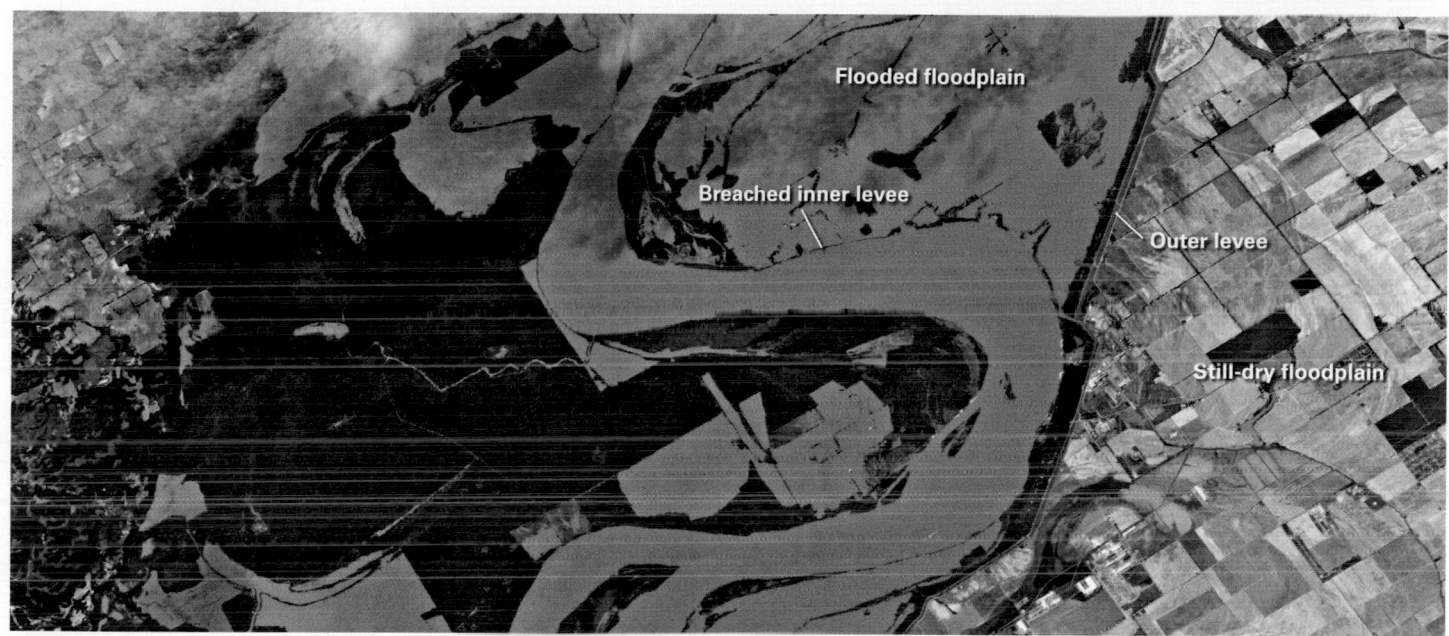

Flooded floodplain

Breached inner levee

Outer levee

Still-dry floodplain

(c) Astronauts in the International Space Station could see this flooding of the Mississippi River in 2011. Inner levees were breached, but outer levees held.

rising waters during a flood are less likely to overtop the levees. If there's no realistic way to prevent flooding, people must consider moving their communities to higher ground.

Evaluating Flooding Hazard When making decisions about investing in flood-control measures, mortgages, or insurance, planners need a basis for defining the hazard or risk posed by flooding. If floodwaters submerge a locality every year, a bank officer would be ill advised to approve a loan that would encourage building there. But if floodwaters submerge the locality only very rarely, then the loan might be worth the risk. Geologists characterize the risk of flooding in two ways. The **annual probability** of flooding—more formally known as the *annual exceedance probability*—is the likelihood that a flood of

a given size or larger will happen at a specified locality during any given year. For example, if we say that a flood of a given size has an annual probability of 1%, then we mean there is a 1 in 100 chance that a flood of at least that size will happen in any given year. The **recurrence interval** of a flood of a given size is the average number of years between successive floods of at least that size. For example, if a flood of a given size happens, on average, once in 100 years, then it has a recurrence interval of 100 years and can be called a *100-year flood*. Floods with shorter recurrence intervals happen more frequently than do floods with longer recurrence intervals

Did you ever wonder...
what newscasters mean by a "100-year flood"?

FIGURE 17.38 Emergency measures can protect local property.

(a) People increase the height of levees by adding sandbags.

(b) Some individual homeowners succeed in protecting their houses.

(Fig. 17. 40a). Note that we can relate annual probability to recurrence interval by the equation:

$$\text{Annual probability} = \frac{1}{\text{recurrence interval}}$$

For example, the annual probability of a 50-year flood is 1/50, which can also be written as 0.02 or 2%. To learn how to calculate annual probabilities and recurrence intervals of floods in more detail, see **Box 17.2**.

Unfortunately, some people may be misled by the meaning of *recurrence interval* and think that they do not face future flooding hazard if they buy a home within an area soon after a 100-year flood has occurred. Their false confidence comes from making the incorrect assumption that because such flooding just happened, it can't happen again until "long after I'm gone." They may regret their decision, because two 100-year floods can occur in consecutive years, or even in the same year (alternatively, the interval between such floods could be, say, 210 years). Because the term *recurrence interval* can lead to confusion, it may be better to report risk in terms of annual probability.

As an example of how to think about flood hazards, let's consider the case of Nashville, Tennessee. The "home of country

FIGURE 17.39 Building a floodway allows a stream to hold more water before flooding property.

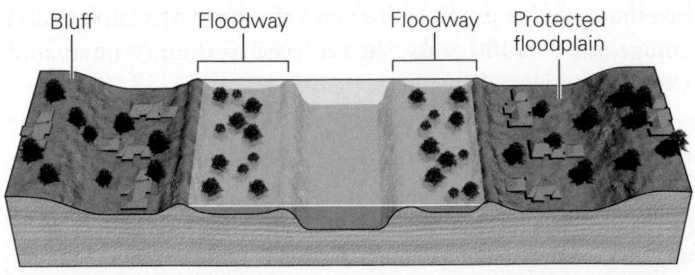

FIGURE 17.40 The conceptual relationship between flood size and probability.

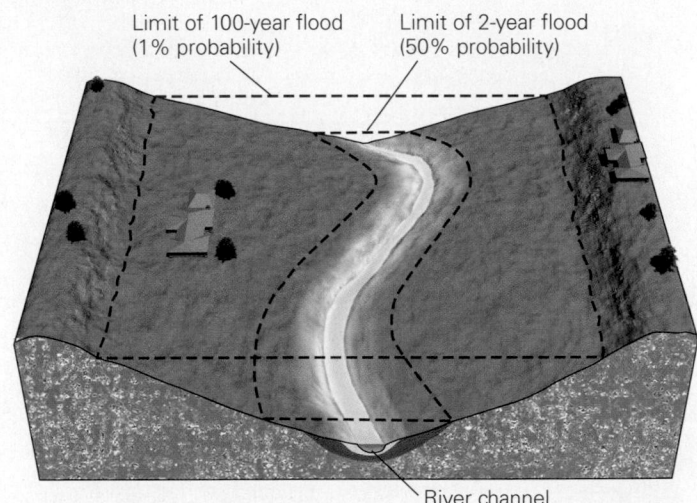

Limit of 100-year flood (1% probability) Limit of 2-year flood (50% probability)

River channel

(a) A 100-year flood covers a larger area than a 2-year flood and occurs less frequently.

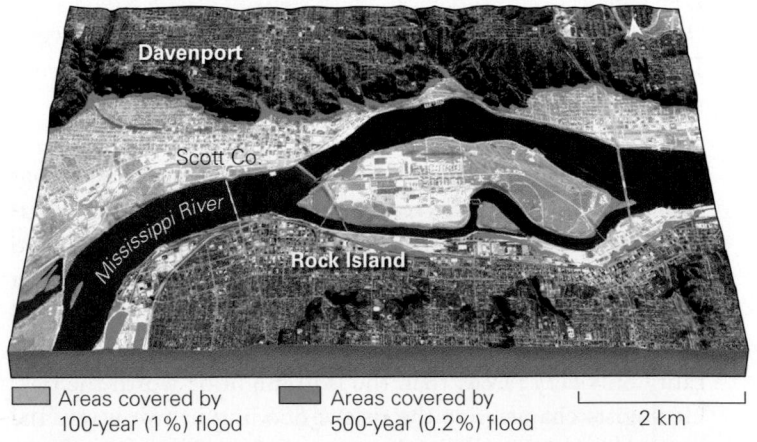

Davenport

Scott Co.

Mississippi River

Rock Island

☐ Areas covered by 100-year (1%) flood   ☐ Areas covered by 500-year (0.2%) flood   2 km

(b) A flood-hazard map shows areas likely to be flooded. Here, near Rock Island, Illinois, even large floods are confined to the floodplain.

BOX 17.2 CONSIDER THIS . . .

Calculating the Threat Posed by Flooding

How do we calculate the probability that a flood of a given size at a locality along a stream will happen in a given year? (Note that *size*, in this context, is the stream's discharge, as measured in cubic meters per second.) First, researchers collect data on the stream's discharge at the locality for at least 10 to 30 years to get a sense of how the discharge varies during a year and from year to year. Then, they pick the largest, or *peak*, *discharge* for each year and make a table listing the peak discharges. The largest peak discharge is given a rank of 1, the second-largest is given a rank of 2, and so on. Researchers can then calculate the recurrence

interval for each peak discharge by using a simple equation:

$$R = (n + 1) \div m$$

where R is the recurrence interval in years, n is the total number of years for which data are available, and m is the rank.

Once the recurrence interval for each peak discharge has been calculated, the researchers plot a graph: the vertical axis represents peak discharge, and the horizontal axis represents recurrence interval. In order for all the data to fit on a reasonably sized graph, the horizontal axis must be logarithmic. Typically, the data for a stream plot is roughly along a straight line (Fig. Bx17.2a). In the example shown in this figure, a flood with a peak discharge of about 460 ft³ per second has a

recurrence interval of 10 years (meaning an annual probability of 10%). We can extend the line beyond the data points (the dashed line on Fig. Bx17.2a) to make predictions about the recurrence interval and, therefore, the annual probability (= 1/R), of floods with discharges larger than the ones that have already been measured. As more data become available, the graph may need to be modified. In this example, note that the graph predicts that a flood with an annual probability of 1% (a 100-year flood) will have a discharge of about 650 ft³ per second.

The peak annual discharge of the Mississippi River at St. Louis has been measured almost continuously since 1850. A bar graph of these data shows that floods characterized as 100-year floods (meaning floods with a 1% annual probability) or larger happened in 1844, 1903, and 1993 (Fig. Bx17.2b). Note that the time between "100-year floods" does not exactly equal 100 years.

FIGURE Bx17.2 Flood frequency and peak discharge graphs.

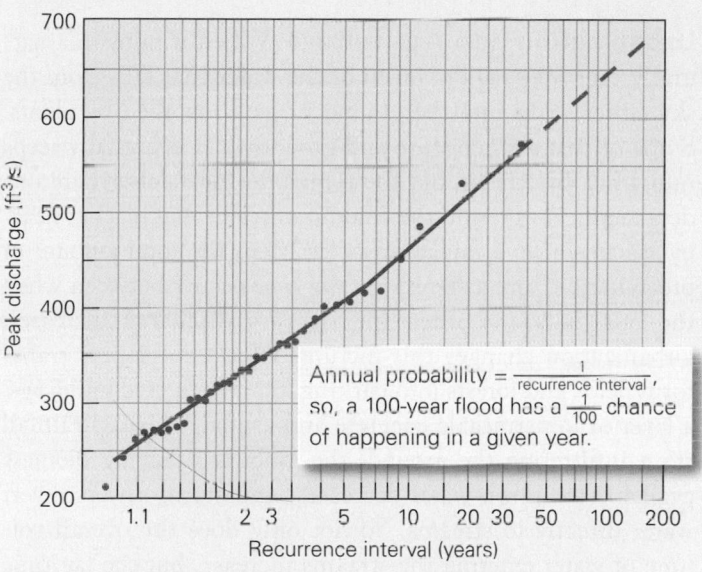

Annual probability = $\frac{1}{\text{recurrence interval}}$, so, a 100-year flood has a $\frac{1}{100}$ chance of happening in a given year.

(a) A flood-frequency graph shows the relationship between the recurrence interval and the peak discharge for an idealized river.

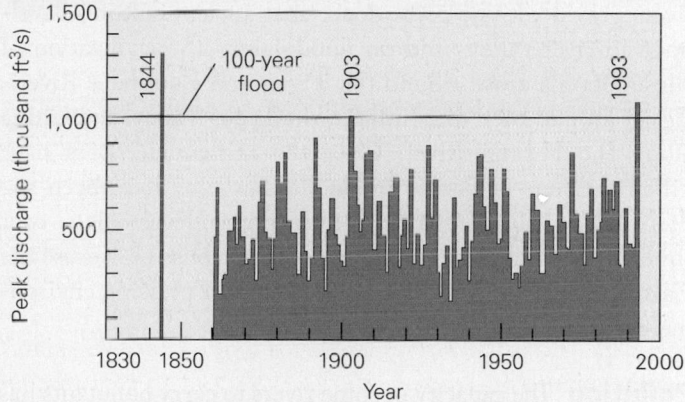

(b) A peak discharge graph for the Mississippi at St. Louis. Each bar represents the largest discharge for a given year.

music" endured a 500-year flood on May 1 and 2 of 2010. Put another way, the likelihood of such a flood, based on previous discharge records, is only 0.2% in any given year. But comparable floods had happened in 1927 and 1937. The 2010 disaster began when a storm system stalled, air from the Gulf of Mexico channeled over the region, and in a 2-day period, 0.34 m of rain fell over Nashville. The Cumberland River overtopped its banks and rose 3 m above flood stage, or 15 m above its normal height. Much of the downtown, as well as the Grand Ole Opry, a famous performance venue, lay beneath the floodwaters.

By knowing the discharge during a flood of a specified annual probability at a specified locality, and by knowing the shape of the river channel and the elevation of the land bordering the river, geologists can predict the extent of land that will be submerged by such a flood (see Fig. 17.40a). Such data, in turn, permit production of a **flood-hazard map**. In the United States, the Federal Emergency Management Agency (FEMA) produces maps that show the 1% annual probability (100-year) and the 0.2% annual probability (500-year) flood-risk zones **(Fig. 17.40b)**.

17.9 Vanishing Rivers

As *Homo sapiens* evolved from hunter-gatherers into farmers, areas along rivers became attractive places to settle. Rivers serve as sources of food, irrigation water, drinking water, power, recreation, and as avenues for transportation and (unfortunately) waste disposal. Further, their floodplains provide particularly fertile soil for fields, replenished annually by seasonal floods. Considering the multitudinous resources that rivers provide, it's no coincidence that ancient cultures developed in river valleys and on floodplains. The civilization of Mesopotamia arose around the Tigris and Euphrates Rivers, Egypt around the Nile, India in the Indus Valley, and China along the Hwang River. Over the millennia, rivers have killed millions of people in floods, but they have been the lifeblood for hundreds of millions more. Nevertheless, over time, humans have increasingly tended to abuse or overuse the Earth's rivers. In this section, we note four pressing environmental issues affecting rivers.

Pollution The capacity of some rivers to carry pollutants has long been exceeded, transforming them into deadly cesspools. Pollutants include raw sewage and storm drainage from urban areas, spilled oil, toxic chemicals from industrial sites, floating garbage, excess fertilizer, and animal waste **(Fig. 17.41a)**. Some pollutants poison aquatic life directly, some feed algal blooms that strip water of its oxygen, and some settle out to be buried along with sediments. River pollution has become overwhelming in countries with inadequate waste-treatment facilities.

Dam Construction In 1950, there were about 5,000 large dams (over 15 m high) worldwide, but today there are over 57,000. Damming rivers has both positive and negative results. Reservoirs provide irrigation water and hydroelectric power, and they trap some floodwaters and create popular recreation areas. But in some locations their construction destroys "wild rivers" (whitewater streams in hilly and mountainous areas) and alters the ecosystem of a drainage network by forming barriers to migrating fish, by decreasing the nutrient supply to organisms downstream, and by removing the source of sediment and nutrients for the floodplain and delta.

Overuse of Water Because of growing populations, our thirst for river water continues to increase, but the supply of water does not. The use of water has grown especially in response to the "green revolution" of the 1960s (an intense effort to increase food crops worldwide), during which huge new tracts of land came under irrigation. Today, 65% of the water taken from rivers is used for agriculture, 25% for industry, and 9% for drinking. In some places, human activity consumes the entire volume of a river's water, and as a result the channel contains little more than a saline trickle, if that, at its mouth. For example, except during unusually wet years, the Colorado River's channel contains almost no water where it crosses the Mexican border; pipes and canals carry the water instead to growing cities such as Phoenix and Los Angeles **(Fig. 17.41b)**.

Urbanization and Agriculture When it rains in a naturally vegetated region, much of the water that falls from the sky either soaks into the ground or gets absorbed by plants. Some of that soil moisture and groundwater eventually seeps into a nearby stream, but the remainder flows elsewhere underground. As a result, the amount of water that reaches nearby streams after a rainstorm is less than the total amount of precipitation, and a significant *lag time* occurs between when the rain falls and when the stream's discharge increases. Urbanization changes this picture. When developers transform fields and forests into parking lots, roads, and buildings, a layer of impermeable concrete and asphalt prevents rainfall from infiltrating the ground, the amount of living biomass available to absorb water decreases, and storm sewers divert water directly to streams. So not only does the overall volume of water entering the streams increase, but the lag time decreases. These changes can be illustrated by diagrams, called *hydrographs*, that show how discharge varies with time **(Fig. 17.42)**. Unfortunately, the changes in volume and lag time can lead to flash flooding.

Agriculture can have a similarly profound effect on streams. Though covered with green during the growing season, fields of many crops actually host relatively little vegetation because farmers keep the space between rows of the crop plants free of weeds. Further, if the crop consists of annual plants, such as corn or soybeans, the fields remain vegetation-free during the winter. As a consequence, the runoff from farm fields can be greater than that from forests or from naturally vegetated fields, and that runoff can carry significant amounts of soil into streams, thus increasing the sediment load that streams carry.

FIGURE 17.41 Environmental issues affecting rivers.

(b) The Central Arizona Project canal shunts water from the Colorado River to Phoenix.

TAKE-HOME MESSAGE

Society depends on streams for water supplies, irrigation, energy, and transport, but the growth of populations can negatively affect streams. Diversion of flow may decrease stream discharge to a trickle, dam construction can change flow, pollution can foul the water, and urbanization or agricultural runoff can change discharge and sediment load.

QUICK QUESTION: Why can urbanization decrease the discharge of a stream?

(a) Garbage along a polluted river.

FIGURE 17.42 Hydrographs, showing discharge as a function of time, are affected by urbanization.

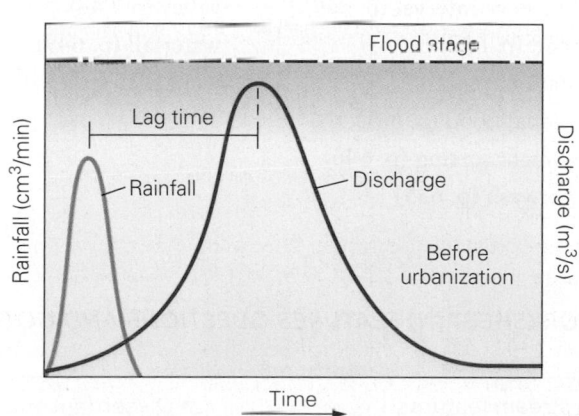

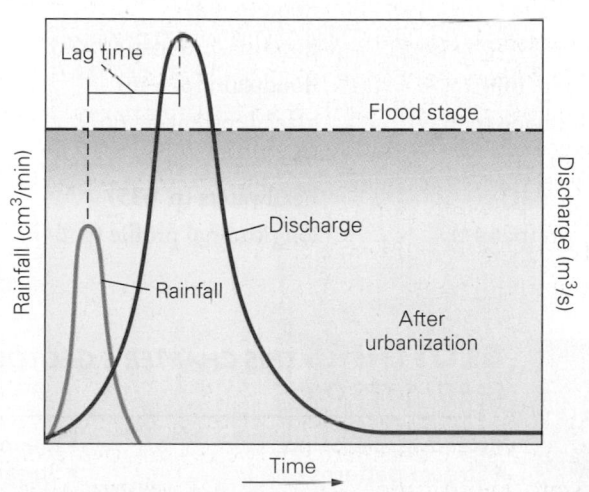

(a) Before urbanization, rain infiltrated the ground, so discharge was smaller and peak discharge occurred after a long lag time.

(b) After urbanization, water flows directly into streams, discharge is greater, and lag time is less.

Chapter 17 Review

SUMMARY

- Streams are bodies of water that flow down channels and drain the land surface. Channels grow by downcutting and headward erosion. Drainage networks consist of tributaries flowing into a trunk stream.

- Permanent streams last all year, while ephemeral streams dry up for part or most of the year.

- The discharge of a stream is the total volume of water passing a point along the bank in a second. Flow velocity slows where water comes in contact with the walls or bed of the channel.

- Streams erode the landscape and carry the resulting sediment. The total quantity of sediment carried by a stream is the stream's capacity. Capacity differs from competence, the maximum particle size a stream can carry.

- The longitudinal profile of a stream is typically a concave-up curve. Streams cannot cut below the base level.

- Whether streams cut valleys or canyons depends on the rate of downcutting relative to the rate at which the bordering slopes undergo mass movement. Where a stream flows down steep gradients or has an irregular bed, rapids develop.

- A meandering stream curves back and forth across a floodplain. Braided streams consist of many entwined channels.

- Where streams flow into standing water, they deposit deltas. Alluvial fans form at canyon mouths.

- With time, fluvial erosion can bevel landscapes to a nearly flat plain. If the base level drops or the land surface rises, stream rejuvenation takes place.

- If more water enters a stream than the channel can hold, a flood results. Slow-onset floods take time to develop and tend to happen seasonally. Flash floods happen very rapidly. Engineers try to prevent floods by building reservoirs and levees.

- Rivers are becoming a vanishing resource because of pollution, damming, and overuse.

GUIDE TERMS

alluvial fan (p. 649)	distributary (p. 654)	meander (p. 650)	slow-onset flood (p. 660)
alluvium (p. 646)	downcutting (p. 637)	meandering stream (p. 650)	stream (p. 635)
annual probability (p. 669)	drainage divide (p. 639)	mouth (p. 635)	stream gradient (p. 646)
bar (p. 646)	drainage network (p. 638)	natural levee (p. 654)	stream piracy (p. 656)
base level (p. 646)	drainage reversal (p. 657)	oxbow lake (p. 654)	stream rejuvenation (p. 657)
bed load (p. 644)	drywash (p. 640)	permanent stream (p. 640)	suspended load (p. 644)
braided stream (p. 649)	ephemeral stream (p. 640)	point bar (p. 646)	tributary (p. 638)
canyon (p. 648)	flash flood (p. 664)	pothole (p. 643)	trunk stream (p. 638)
capacity (p. 645)	flood (p. 635)	rapid (p. 649)	turbulence (p. 643)
channel (p. 635)	flood-hazard map (p. 671)	recurrence interval (p. 669)	valley (p. 648)
competence (p. 644)	floodplain (p. 646)	runoff (p. 637)	waterfall (p. 649)
continental divide (p. 639)	glacial torrent (p. 667)	saltation (p. 644)	watershed (p. 639)
delta (p. 646)	headward erosion (p. 637)	seasonal flood (p. 661)	
discharge (p. 641)	headwaters (p. 635)	sediment sorting (p. 646)	
dissolved load (p. 644)	longitudinal profile (p. 646)	sheetwash (p. 635)	

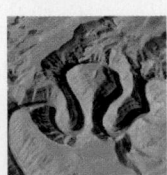 **GEOTOURS** *THIS CHAPTER'S GEOTOURS WORKSHEET (N) FEATURES QUESTIONS AND GOOGLE EARTH SITES ON:*

- Headward erosion
- Stream patterns
- Meandering stream features
- Stream piracy
- Discordant streams
- Catastrophic flooding

REVIEW QUESTIONS

The letters following each Review Question refer to the corresponding Learning Objective from the Chapter Opener.

1. What role do streams play in the hydrologic cycle? Indicate various sources of water in streams. **(A)**

2. Describe the five different types of drainage networks. What factors are responsible for the formation of each? Which does the figure show? **(A)**

3. What factors determine whether a stream is permanent or ephemeral, gaining or losing? **(A)**

4. How does discharge vary according to the stream's length, the climate, and position along the stream course? **(C)**

5. Why is average downstream velocity always less than maximum downstream velocity? **(C)**

6. Describe how streams and running water erode the Earth's surface. **(B)**

7. What are the three components of sediment load in a stream? **(B)**

8. Distinguish between stream competence and capacity. **(B)**

9. Describe how a drainage network changes along its length, from headwaters to mouth. **(C)**

10. What factors determine the position of a local base level? What is the ultimate base level, and why? **(C)**

11. What do lakes, rapids, waterfalls, and terraces indicate about the stream gradient and base level? Why do canyons form in some places and valleys in others? **(A)**

12. How does a braided stream differ from a meandering stream? **(D)**

13. Describe how meanders form, develop, are cut off, and then are abandoned. **(D)**

14. Describe how deltas grow and develop. How do they differ from alluvial fans? Which type of delta does the figure show? **(D)**

15. How does a fluvial landscape evolve over time pass? **(B)**

16. What is stream piracy? What causes a drainage reversal? **(A)**

17. How are superposed and antecedent streams similar? **(A)**

18. Explain the difference between a slow-onset flood and a flash flood, and describe the causes of each. **(E)**

19. What human activities tend to increase flood risk? **(E)**

20. What is the recurrence interval of a flood, and how is it related to the annual probability? **(F)**

21. How have humans abused and overused the resource of running water? **(A)**

ON FURTHER THOUGHT

22. Look closely at the graph from Box 17.2. What is the recurrence interval of a flood with a discharge of 650 ft^3 per second? In a given year, how much more likely is a flood with a discharge of 200 ft^3 per second than a flood of 400 ft^3 per second? For the sake of discussion, imagine that the floodplain of the river is completely covered when a flood with an annual probability of 1/300 occurs. Would you build a new home in the floodplain? Would it make a difference to your decision if the last flood with this probability happened 1 year ago? **(F)**

ONLINE RESOURCES

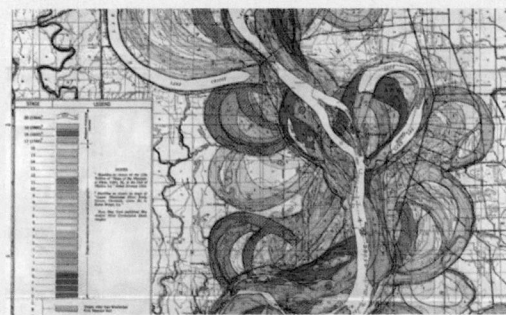

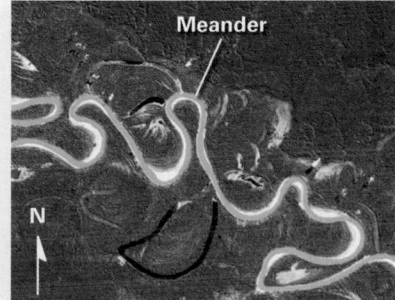

Videos
This chapter includes videos on river meanders, aquifer systems, and flash floods.

Smartwork5
This chapter features visual identification exercises on meandering stream systems, erosion, and lake formation.

CHAPTER 18

Restless Realm:
Oceans and Coasts

By the end of this chapter, you should be able to . . .

A. describe how ocean depth, temperature, and salinity correlate.

B. explain how tectonic processes produce bathymetric features of the seafloor.

C. describe the nature and causes of surface currents and deep currents.

D. relate the behavior of tides to the forces that cause tides.

E. discuss how waves form and how they behave as they approach the shore.

F. differentiate among the great variety of different coastal landforms, and explain how they develop and evolve.

G. explain how changes in sea level, wave erosion, and human activities affect the coast.

> The three great elemental sounds in nature are the sound of rain, the sound of wind
> in a primeval wood, and the sound of the outer ocean on a beach.
>
> —Henry Beston (American naturalist, 1888–1968)

18.1 Introduction

A thousand kilometers from the nearest shore, two scientists and a pilot wriggle through the entry hatch of the research submersible *Alvin*, ready for a cruise to the floor of the ocean and, hopefully, back. *Alvin* consists of a superstrong metal sphere embedded in a cigar-shaped tube **(Fig. 18.1a)**. The sphere protects its crew from the immense water pressures of the deep ocean, and the tube holds motors and oxygen tanks. When the hatch seals, *Alvin* sinks at a rate of 1.8 km per hour. Most of this journey takes place in utter darkness, for light penetrates only the top few hundred meters of ocean water.

On reaching the bottom, at a depth of 4.5 km, the cramped explorers turn on outside lights to reveal a stark vista of loose sediment, black rock, and the occasional sea creature. For the next five hours, they take photographs and use a robotic arm to collect samples. When finished, they release ballast and rise like a bubble, reaching the surface two hours later.

Alvin dives began in the 1970s, but humans have explored the ocean for tens of centuries. In fact, Phoenician traders had circumnavigated Africa by 590 B.C.E., Polynesian sailors in outrigger canoes traveled among South Pacific islands beginning around 700 C.E., and Chinese naval ships may have circled the globe in the 15th century. European mapmakers have known that the ocean spanned the entire globe since the tattered remnants of Ferdinand Magellan's crew completed a

FIGURE 18.1 Modern oceanographic research vessels explore the surface and subsurface realms.

(a) *Alvin* being launched from RV *Atlantis*.

(b) HMS *Challenger* explored the ocean in 1876.

(c) Submersibles use bright lights to see in the dark depths.

(d) *ABE* is a robotic submersible.

(e) Satellites can study the sea from space.

◄ (facing page) Waves from the Atlantic Ocean wash onto the sand of a beach on Puerto Rico. The movement of ocean water constantly modifies coasts, the boundary between land and sea.

round-the-world voyage in 1519–1522. But no one could systematically map the ocean until the late 18th century, when newly invented equipment allowed navigators to calculate longitude accurately. Subsequently, naval officers of many nations started to gather data on water depths in the ocean, and by 1839, data compilations demonstrated that the greatest ocean depths could swallow the highest mountains without a trace.

A converted British navy ship, HMS *Challenger*, made the first true ocean research cruise **(Fig. 18.1b)**. Beginning in 1872, onboard scientists spent four years dredging rocks from the seafloor, analyzing water composition, collecting specimens of marine organisms, and measuring water depths and currents. But still our knowledge of the ocean remained spotty. In fact, we knew less about the ocean floor than we did about the surface of the Moon, for at least we could see the Moon with a telescope.

Then, in the latter half of the 20th century, the fields of *oceanography* (the study of ocean water and its movements), *marine geology* (the study of the ocean floor and shorelines), and *marine biology* (the study of life forms in the sea) expanded rapidly. New technology became available and a fleet of oceanographic research ships began to crisscross the seas; submersibles roamed the depths, and satellites used remote sensing to characterize the sea from space **(Fig. 18.1c–e)**. It's now commonplace for ships to tow instruments, such as *side-scan sonar*, to generate detailed *bathymetric maps* (maps depicting the topography of the seafloor) along swaths of the ocean floor **(Fig. 18.2a)**. Some ships conduct surveys producing seismic-reflection profiles that reveal layering beneath the seafloor **(Fig. 18.2b; see Interlude D)**. Other ships, such as *JOIDES Resolution*, drill holes as deep as 4 km into the seafloor and bring up samples of the oceanic crust **(Fig. 18.2c)**. And while ships and satellites research the open ocean, land-based geologists continue to study the evolution of its margins.

When seen from space, the Earth glows blue, for ocean water covers 70.8% of its surface. The ocean provides the basis for life, tempers the Earth's climate, and spawns its largest storms. It serves as a vast reservoir for water and

FIGURE 18.2 Modern methods for surveying the seafloor.

(a) Side-scan sonar (also called multi-beam sonar) can map the detailed bathymetry of a swath of the seafloor rapidly.

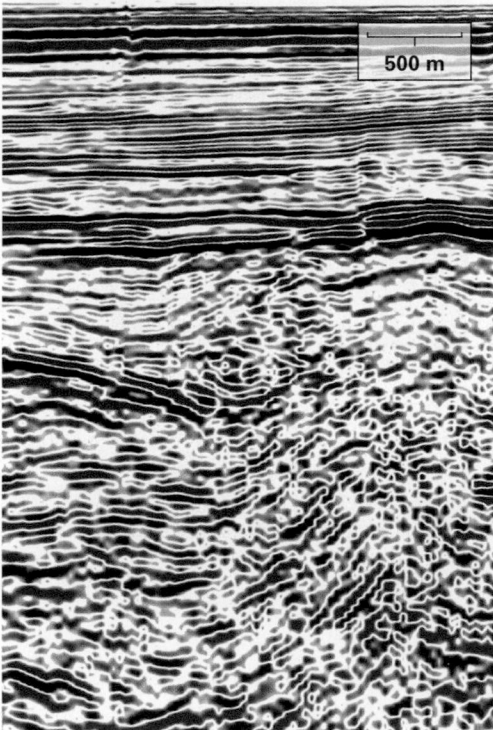

500 m

(b) Seismic-reflection profiles reveal layers and structures beneath the seafloor.

(c) *JOIDES Resolution* drills into the seafloor to extract cores.

chemicals that cycle into the atmosphere and crust and for sediment washed off the continents. In this chapter, we first learn about the fundamental characteristics of ocean basins and seawater and the role they play in the Earth System. Then we focus on the landforms that develop along the **coast**, the region where the land meets the sea and where over 60% of the global human population lives today. Finally, we consider the hazards of coastal areas and how people may confront them.

18.2 Landscapes beneath the Sea

If the top surface of the lithosphere were completely smooth, an ocean would surround the Earth as a uniform 2.5-km-deep layer. But in reality, dry land covers about 30% of this planet's surface, and most seawater resides in distinct *ocean basins*, which have an average depth of about 4.5 km **(Fig. 18.3a)**. These basins exist for a reason! Due to isostasy (see Interlude D), the surface of the denser, but thinner, oceanic lithosphere sinks deeper than the surface of the relatively buoyant, thicker continental lithosphere. It's the low areas over oceanic lithosphere that fill with water to become oceans—higher areas, underlain by continental lithosphere, remain as dry land

(Fig. 18.3b). You can model this contrast by floating wood blocks of different densities and thicknesses in a tub of water **(Fig. 18.3c)**. Cartographers delineate several distinct oceans and seas, which differ significantly in terms of their volumes **(Fig. 18.4a)**. Some of the boundaries between oceans are obvious—for example, the Americas separate the Pacific from the Atlantic—but some are rather arbitrary, defined by a line of latitude or longitude or by a chain of islands. Note that all oceans are, in effect, interconnected, for water can flow from one ocean to another.

Have you ever wondered what the ocean floor would look like if all the water evaporated? Marine geologists can now provide a clear image of the ocean's **bathymetry**, or variation in depth **(Fig. 18.4b)**. Depth measurements were first obtained by using a *plumb line*, a lead weight attached to a cable. Then, beginning in the mid-20th century, *sonar* became available, and today, satellites survey the ocean, measuring variations in the pull of gravity that can be translated into variations in water depth. Such studies indicate that the ocean contains broad *bathymetric provinces*, distinguished from one another by their water depth. Let's examine each of these provinces.

Continental Margins

Imagine you're in a submersible that is cruising just above the floor of the western half of the North Atlantic **(Fig. 18.5a)**.

FIGURE 18.3 Contrasts between continental lithosphere and oceanic lithosphere.

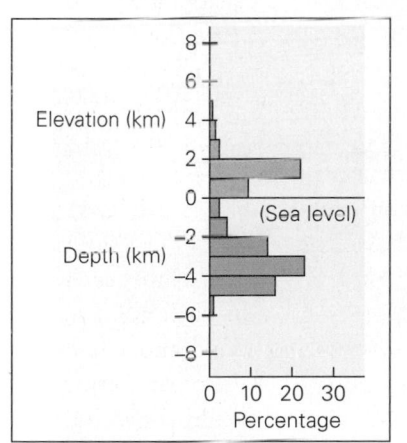

(a) Percentages of the Earth's surface at different elevations above or below sea level.

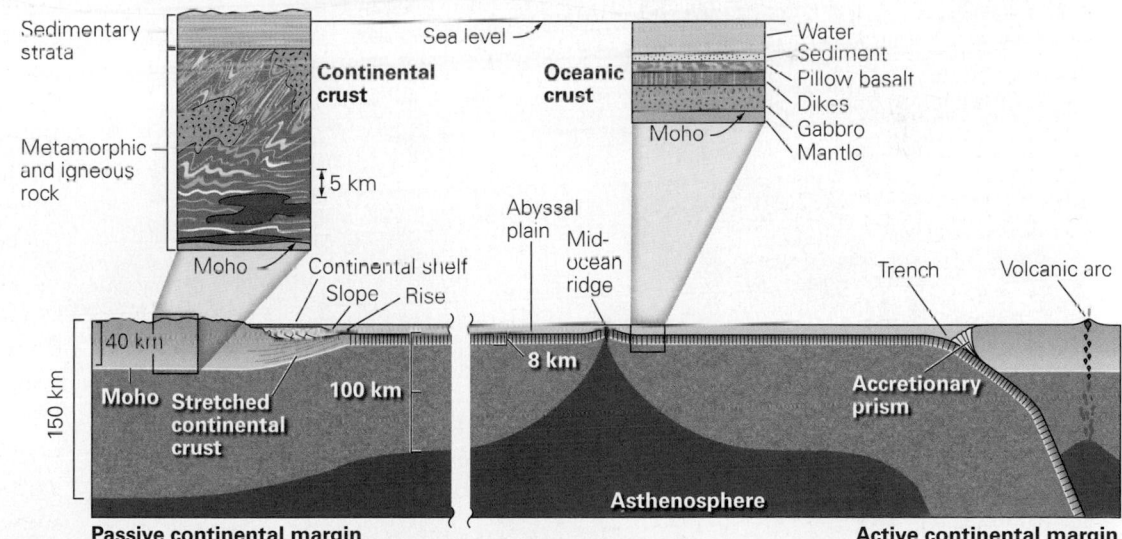

(b) The crustal portion of the continental lithosphere differs markedly from that of oceanic lithosphere.

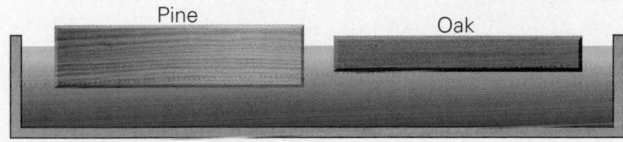

(c) A simple model, using wood blocks of differing densities and thicknesses, shows why the ocean floor sits at a lower elevation.

FIGURE 18.4 The oceans of the world.

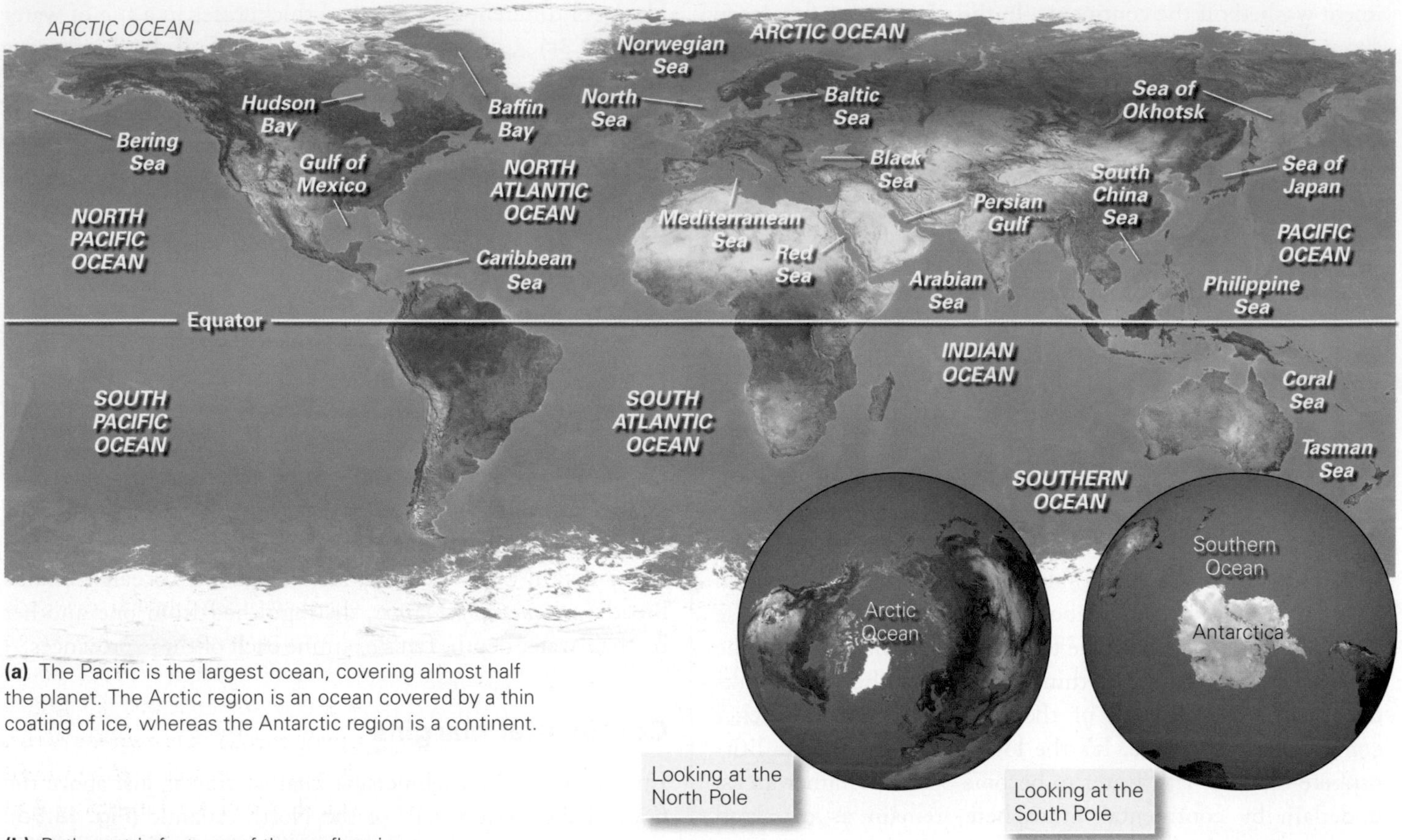

(a) The Pacific is the largest ocean, covering almost half the planet. The Arctic region is an ocean covered by a thin coating of ice, whereas the Antarctic region is a continent.

(b) Bathymetric features of the seafloor in the South Atlantic and southeastern Pacific.

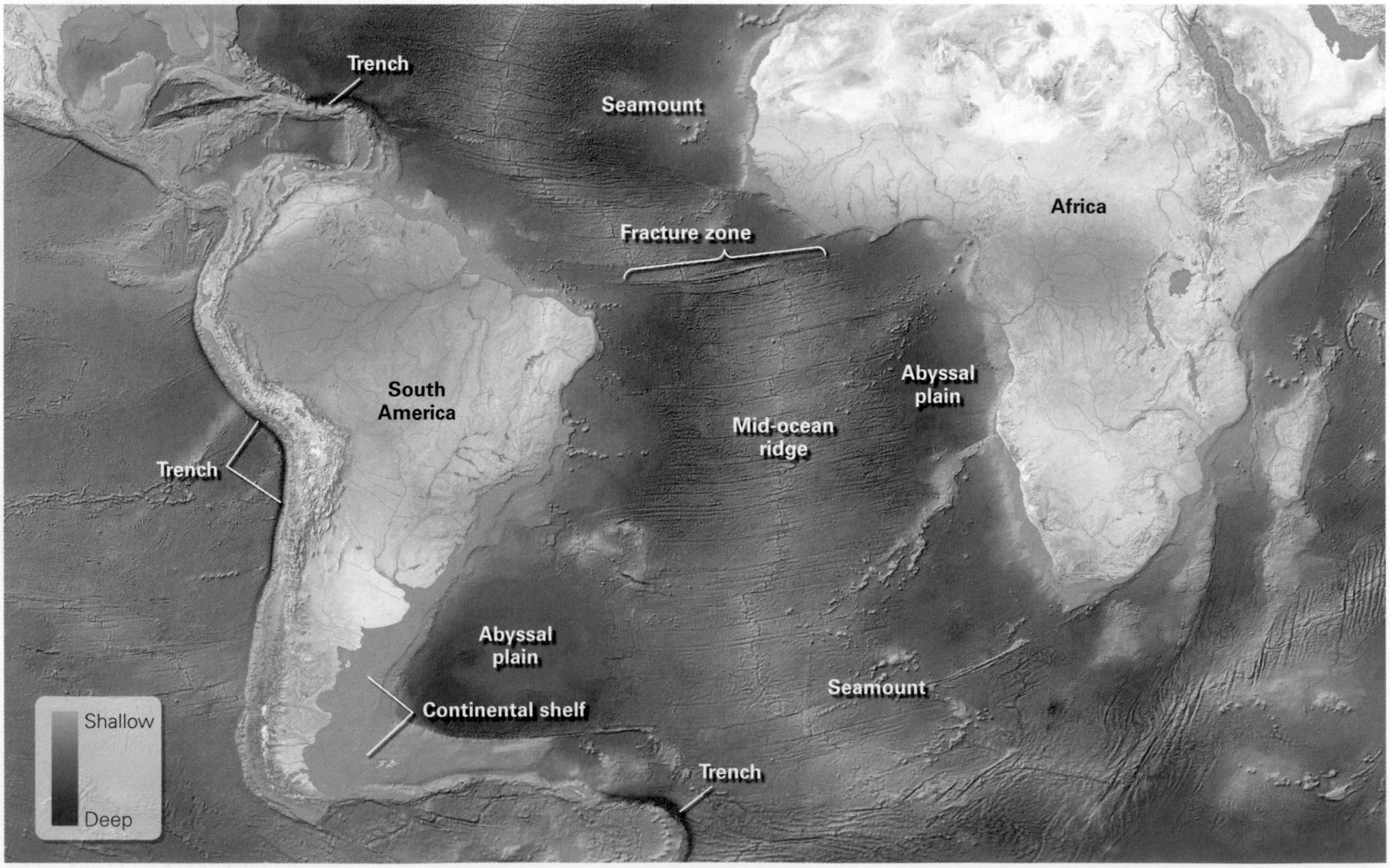

FIGURE 18.5 Bathymetry of passive continental margins.

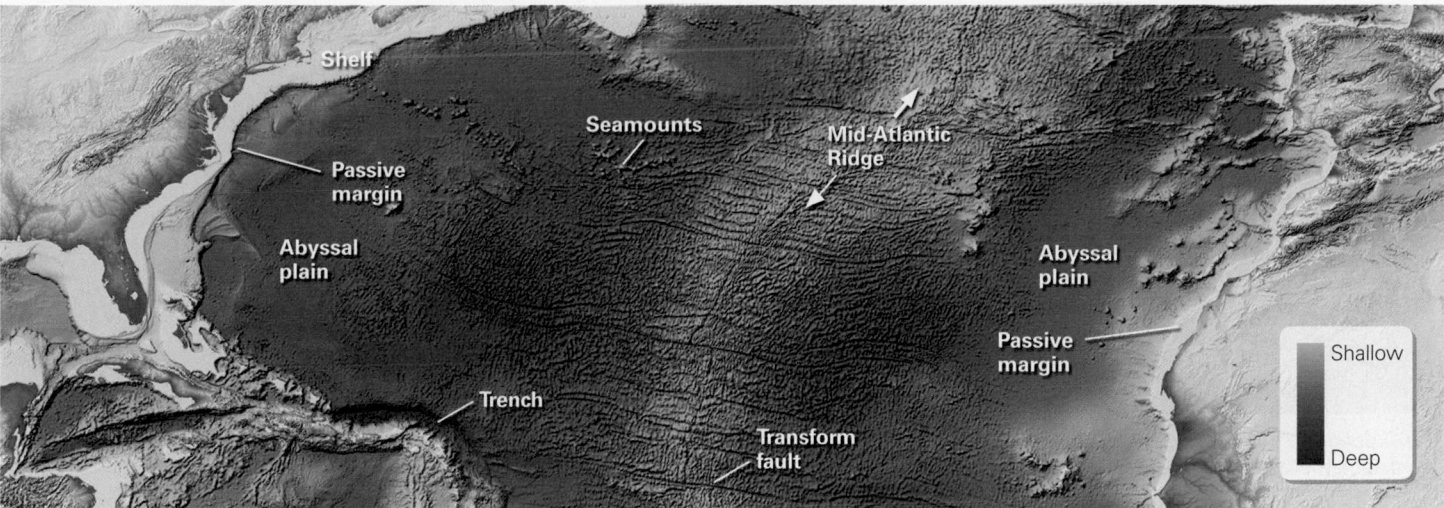

(a) Passive margins occur on both sides of the North Atlantic.

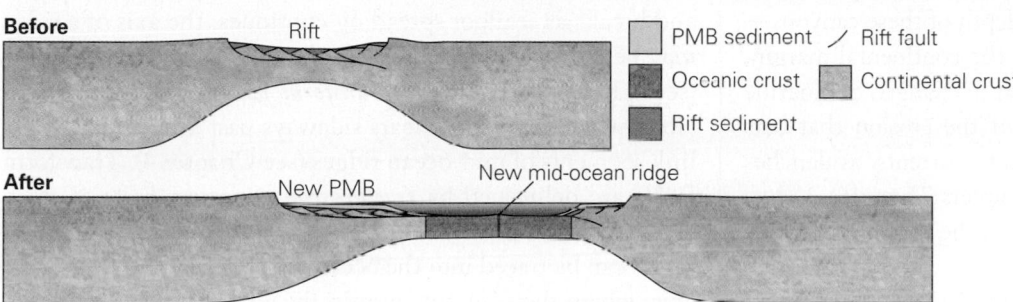

(b) The formation of a passive-margin basin (PMB). The top surface of the PMB is the continental shelf.

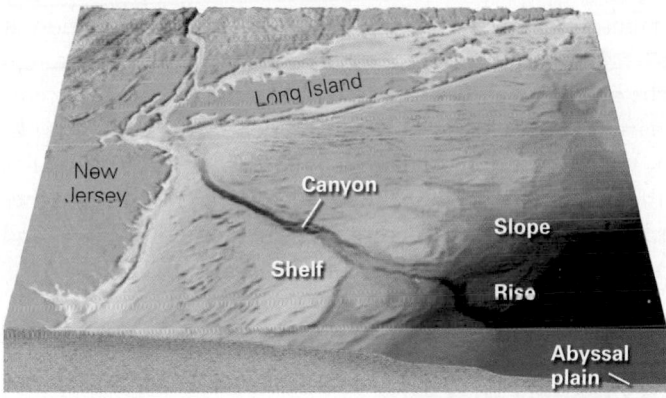

(c) A submarine canyon along the coast of New Jersey. Turbidity currents in submarine canyons carry sediment to the abyssal plain.

If you start at the shoreline of North America and head east, you will cross the 200- to 500-km-wide **continental shelf**, a relatively shallow portion of the ocean that fringes the continent. Water depth over the continental shelf does not exceed 500 m—most of the major marine fisheries occur in the relatively shallow waters of the shelf. Across the width of the shelf, the ocean floor slopes seaward at only about 0.3°, an almost imperceptible angle. At its eastern edge, the continental shelf merges with the

continental slope, which descends to depths of nearly 4 km at an angle of about 2°. From about 4 km down to about 4.5 km, a province called the *continental rise*, the angle decreases gradually until at a depth of 4.5 km, you'll find yourself above a vast, nearly horizontal plain—the **abyssal plain**.

Broad continental shelves, like the one bordering eastern North America, form along **passive continental margins**. Recall that such margins are not plate boundaries and thus lack seismicity (see Fig. 18.5a and Chapter 4). Passive margins originate after rifting has succeeded in breaking a continent in two, for when rifting ceases and seafloor spreading begins (along a newly formed mid-ocean ridge), the stretched lithosphere at the boundary between the ocean and continent gradually cools and sinks (**Fig. 18.5b**). Sand and mud that wash off the continent, along with the shells of marine creatures that grow on the seafloor or settle from the water above, bury the sinking crust, slowly producing a pile of sediment up to 15 or even 20 km thick. Geologists refer to the region along a passive margin that accumulates thick wedges of sediment as a *passive-margin basin*. The flat surface of a passive-margin basin constitutes the continental shelf. In some locations, a thick evaporite layer underlies that strata of a passive-margin basin. The weak salt comprising this layer can serve as a failure horizon or detachment, along which overlying strata slowly slump seaward; this movement generates faults, folds, and salt domes in the strata.

At many locations, relatively narrow and deep valleys called **submarine canyons** downcut into continental shelves and slopes (**Fig. 18.5c**). Some submarine canyons start offshore of major

FIGURE 18.6 The active continental margin along the west coast of South America.

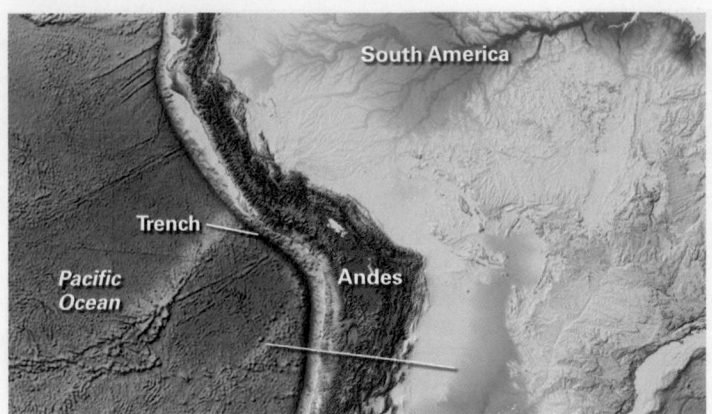

(a) A deep-sea trench lies along the west coast of South America, and a continental volcanic arc lines the coast.

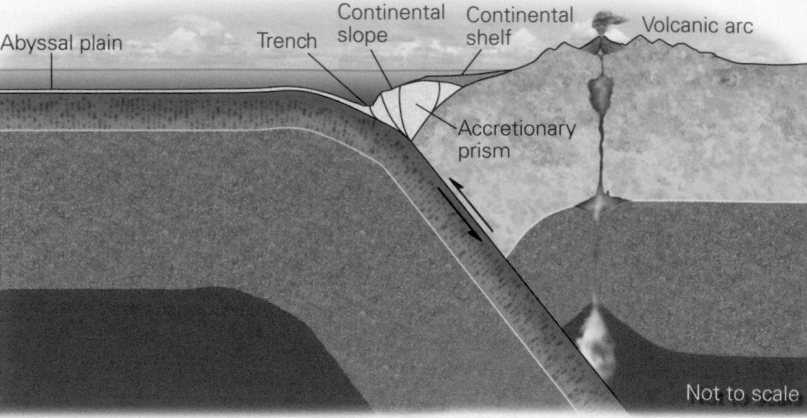

(b) A simplified cross section from the Pacific across the Andean volcanic arc, as represented by the yellow line in part (a).

rivers, and for good reason—rivers cut into the continental shelf at times when sea level was low and the shelf was exposed. But river erosion cannot explain the total depth of these canyons—some slice almost 1,000 m down into the continental margin, far deeper than the maximum sea-level change. Submarine exploration demonstrates that much of the erosion that cuts submarine canyons results from turbidity currents, avalanches of sediment mixed with water (see Chapters 7 and 16). When turbidity currents finally reach the base of the continental slope, the velocity of flow decreases, so sediments settle out, and turbidites, composed of graded beds, accumulate and build into a *submarine fan*.

If you were to take your submersible to the western coast of South America and cruise out into the Pacific, you would find a very different type of continental margin. This continental margin is an **active continental margin**, one that coincides with a plate boundary that hosts many earthquakes **(Fig. 18.6)**. The western South American coast is, in particular, a convergent boundary where subduction takes place. Because of subduction, a 100- to 200-km-wide *accretionary prism*, consisting of contorted and broken-up sediment and basalt that has been scraped off the subducting plate, lies along the edge of the coast. The upper part of the accretionary prism may be buried by sediment eroded from the land to form a narrow continental shelf. The shelf ends at the continental slope, which corresponds to the face of the accretionary prism—its surface slopes at an angle of 3.5° down to the axis of a deep-sea trench, which lies at a depth of 5 to 7.5 km. Submarine canyons locally cut into the accretionary prism and funnel turbidites into the trench.

Bathymetry of Oceanic Plate Boundaries

You can see all three types of plate boundaries by studying the bathymetry of the ocean floor (see Fig. 18.5a). Seafloor

spreading at a divergent boundary yields a *mid-ocean ridge*, a 2-km-high submarine mountain belt. Because crust stretches and breaks as seafloor spreading continues, the axis of a ridge may be bordered by fault scarps due to slip on normal faults (see Chapter 11). *Oceanic transform faults*, strike-slip faults along which one plate shears sideways past another, typically link segments of mid-ocean ridges (see Chapter 4). Transform faults are delineated by *fracture zones*, narrow belts of steep escarpments and broken-up rock **(Fig. 18.7a)**. These fracture zones can be traced into the oceanic plate away from the ridge axis, where they are not seismically active and, therefore, do not represent plate boundaries. Note, however, that the inactive portions of fracture zones are still distinct because they juxtapose portions of plates of different ages. Subduction at convergent boundaries yields a *trench*, a deep, elongate trough at the boundary between the subducting plate and the accretionary prism. Some trenches reach depths of over 8 km. In fact, the deepest point in the global ocean, at −11,035 m, lies in the Mariana Trench of the western Pacific. Some trenches border continents, as we described earlier; others border island arcs, which are curving chains of active volcanic islands.

Abyssal Plains, Seamounts, and Oceanic Plateaus

As oceanic crust ages and moves away from the axis of a mid-ocean ridge, two changes take place. First, the lithosphere cools and thickens, and as it does so, its surface sinks to maintain isostasy (see Interlude D). Second, a blanket of **pelagic sediment** gradually accumulates and covers the basalt of the oceanic crust. This sediment consists mostly of microscopic plankton shells and fine flakes of clay, which slowly fall like snow from the ocean water and settle on the seafloor. Because the oceanic crust gets progressively older away from the ridge axis, the thickness of the sediment blanket increases away from

FIGURE 18.7 Fracture zones and abyssal plains.

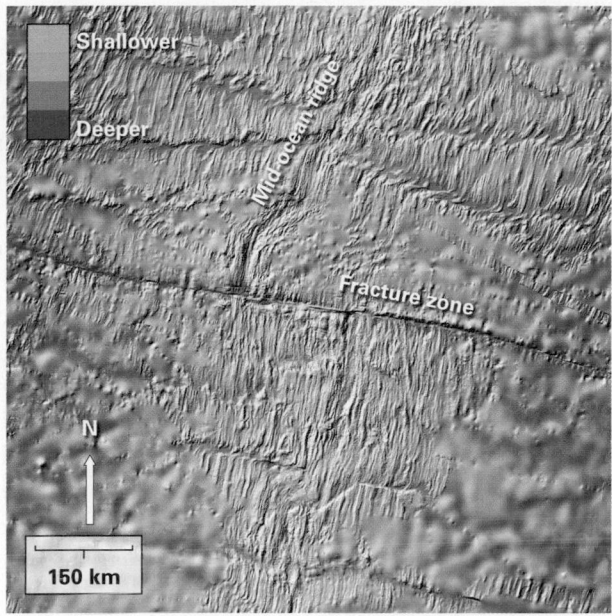

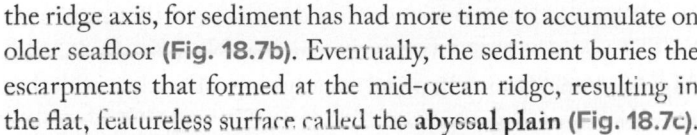

(a) Mapping of the seafloor shows a fracture zone linking two segments of a mid ocean ridge.

(b) Seafloor sinks as it moves away from a mid-ocean ridge and gets progressively covered with sediment.

(c) An abyssal plain has a cover of gray-brown pelagic sediment.

the ridge axis, for sediment has had more time to accumulate on older seafloor (Fig. 18.7b). Eventually, the sediment buries the escarpments that formed at the mid-ocean ridge, resulting in the flat, featureless surface called the **abyssal plain (Fig. 18.7c)**.

Hot-spot igneous activity has produced unusually thick accumulations of basalt on the seafloor. In places, the lava builds into a mound that protrudes above sea level to form an *oceanic island*. Oceanic islands that lie over hot spots host active volcanoes, whereas those that have moved off the hot spot are extinct. Notably, as the seafloor beneath an oceanic island gets older, it thickens and sinks due to isostasy. So, over time, the island gradually sinks. In addition, oceanic islands gradually undergo erosion, and every once in a while, chunks of the island slump and tumble to greater depths. Overall, these processes end up causing the island to sink beneath the waves and become a submerged **seamount (Fig. 18.8)**. Oceanic islands and seamounts that developed above the same hot spot line up in a chain (a hot-spot track) with the oldest seamount at one end and the youngest seamount or island at the other (see Chapter 4).

In several localities, broad areas of the seafloor have depths shallower than those that could be predicted based on the known relation between seafloor depth and lithospheric age. These areas of seafloor rise above the abyssal plain to form **oceanic plateaus** (see Fig. 18.8). Seismic-reflection profiles across oceanic plateaus demonstrate that they are underlain by unusually thick oceanic crust—whereas most oceanic crust has a thickness of about 7 km, the crust beneath oceanic plateaus is 10 to 12 km thick. Geologists speculate that oceanic plateaus develop where particularly voluminous hot-spot igneous activity once took place—in other words, they are large igneous provinces (LIPs;

FIGURE 18.8 Bathymetry of the southwestern Pacific, showing seamounts and an oceanic plateau.

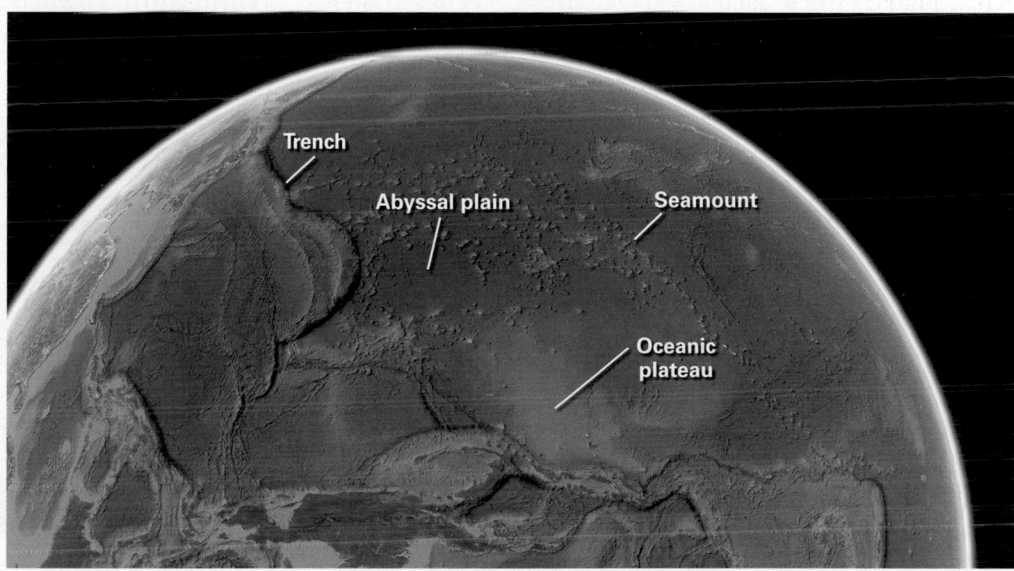

see Chapter 6), perhaps formed over superplumes, in which much more melting takes place than is typical of a plume.

18.3 Ocean Water Characteristics

Composition

If you've ever had a chance to swim in the ocean, you may have noticed that you float much more easily in ocean water than you do in freshwater. That's because ocean water contains an average of 3.5% dissolved salt **(Fig. 18.9a)**. Dissolved ions fit between water molecules without changing the volume of the water, so adding salt to water increases the water's density, and you float higher in a denser liquid. There's so much salt in the ocean that if all the water magically evaporated, a 60-m-thick layer of salt would coat the ocean floor. This layer would consist of about 75% halite (NaCl) with lesser amounts of gypsum (CaSO$_4$ · H$_2$O), anhydrite (CaSO$_4$), and other salts.

Oceanographers refer to the concentration of salt in water as the water's **salinity**.

Although ocean-surface salinity averages 3.5%, measurements from around the world demonstrate that salinity varies with location, ranging from about 1.0% to about 4.1%. Ocean-surface salinity reflects the balance between the addition of freshwater by rivers or rain and the removal of freshwater by evaporation. (When seawater evaporates, salt stays behind, and only H$_2$O goes into the vapor phase **(Fig. 18.9b)**.) Salinity also depends on water temperature, because warmer water can hold more salt in solution than can cold water. Therefore, salt concentration varies with latitude, proximity to the coast, and depth **(Fig. 18.10a, b)**. As shown in Figure 18.10a, variations in salinity generally occur only in the upper 1 km of seawater, where evaporation and addition of freshwater take place, and where temperature varies more. The salinity of deeper water tends to be more homogeneous. Oceanographers refer to the gradational boundary between shallow-water salinities and deep-water salinities as the *halocline*.

Where does the salt in seawater come from? Leonardo da Vinci, the famous Renaissance artist and scientist, speculated that sea salt came from buried salt layers, such as those found in salt mines. But modern studies demonstrate that most cations, or positive ions, in sea salt—such as Na$^+$, K$^+$, Ca^{2+}, and, Mg^{2+}—come from chemical weathering of rocks, and that the anions, or negative ions—such as Cl$^-$ and SO$_4^{2-}$—come from volcanic gases. Dissolved ions get carried to the sea by groundwater and river water because "freshwater," in fact, is actually a dilute solution typically containing 0.01% to 0.05% salt; rivers deliver over 2.5 billion tons of salt to the sea every year. Salt remains in the sea because, as we've noted, evaporation removes only H$_2$O molecules.

Temperature

When RMS *Titanic* sank after striking an iceberg in the North Atlantic, most of the unlucky passengers and crew

FIGURE 18.9 The salt in the sea.

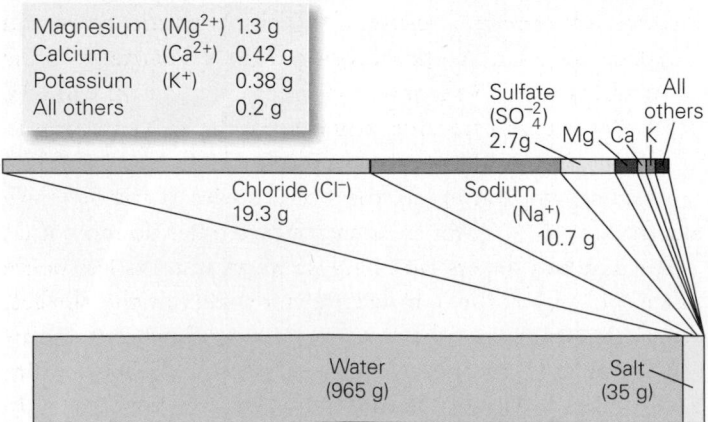

(a) The composition of average seawater. The expanded part of the graph shows the proportions of ions in the salt of seawater.

(b) Harvesting sea salt by evaporating seawater from shallow pools.

FIGURE 18.10 Physical characteristics of the ocean.

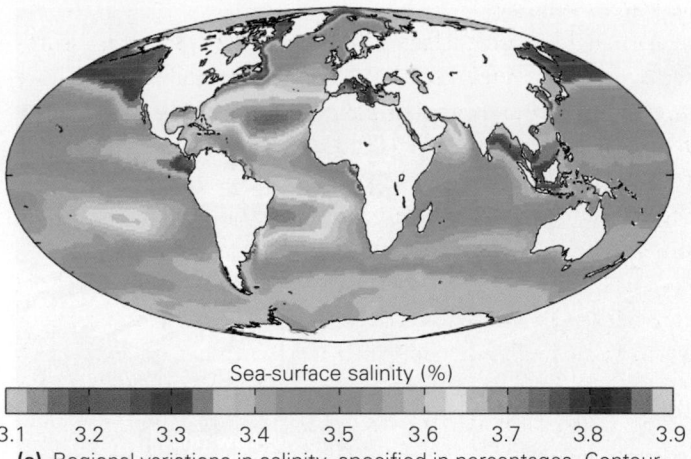

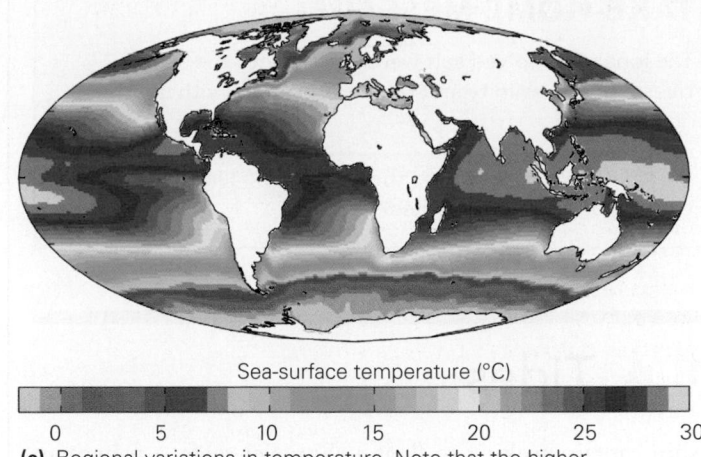

Sea-surface salinity (%)

3.1 3.2 3.3 3.4 3.5 3.6 3.7 3.8 3.9

(a) Regional variations in salinity, specified in percentages. Contour lines separate areas of different salinity.

Sea-surface temperature (°C)

0   5   10   15   20   25   30

(c) Regional variations in temperature. Note that the higher temperatures occur in equatorial regions.

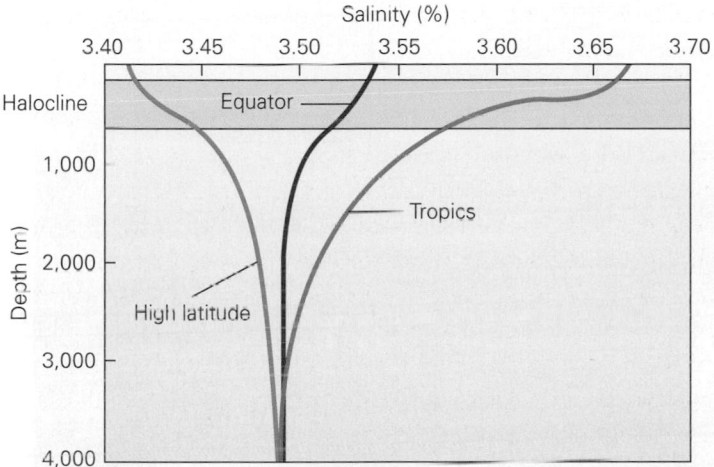

(b) Changes in salinity with depth. Note the halocline (blue band), an interval where salinity changes rapidly.

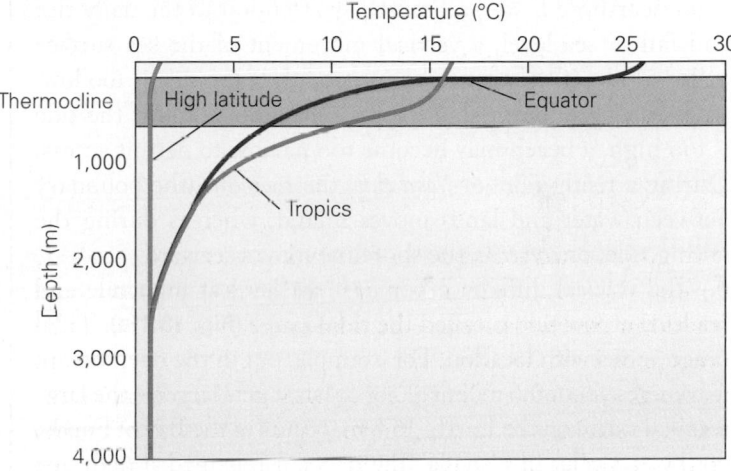

(d) Changes in temperature with depth. Note the thermocline (orange band), an interval where temperatures decrease rapidly.

who jumped or fell into the sea died within minutes because the seawater temperature at the site of the tragedy was about 1°C, and cold water removes heat from a body very rapidly. Yet swimmers can play for hours in the Caribbean, where sea-surface temperatures reach 29°C. Though the global average sea-surface temperature hovers around 17°C, it ranges from −2°C near the poles to almost 36°C in restricted tropical seas **(Fig. 18.10c)**. (Note that, because ocean water contains salt, it freezes at a temperature below that of freshwater.) The general correlation of average temperature with latitude exists because the intensity of solar radiation varies with latitude. As we'll see, ocean temperature at a given latitude actually varies significantly in detail because of currents.

Sea-surface temperature at a given locality varies with the season because the intensity of solar radiation varies with the season. But the difference is only about 2° in the tropics, 8° in the temperate latitudes, and 4° near the poles. Seasonal

seawater temperature changes remain in a narrow range because water has a large *heat capacity*, meaning that it can absorb or release large amounts of heat without changing temperature very much. Because of this characteristic, the ocean regulates the temperatures of coastal regions. For example, the air temperature in Vancouver, on the Pacific coast of Canada, rarely drops below freezing, while a cold day in Winnipeg, in the plains of central Canada, can reach −30°C.

Water temperature in the ocean varies markedly with depth **(Fig. 18.10d)**, for solar energy penetrates no more than a few hundred meters, and because warm water is less dense than cold water, the warmed water remains at the surface. Typically, a *thermocline* between warm water above and significantly colder water below appears at a depth of about 300 m in the tropics. A pronounced thermocline does not develop in polar seas because surface water has nearly the same temperature as deep water.

18.4 Tides

A ship captain seeking to float a ship over reefs, a fisherman hoping to set sail from a shallow port, marines planning to attack a beach from the sea, a tourist eager to harvest shellfish from nearshore mud—all must pay attention to the daily rise and fall of sea level, a vertical movement of the sea surface called a **tide**, if they are to be successful. If the tide is too low, ships run aground or stay trapped in harbors, and if the tide is too high, a beach may become too narrow to permit access. During a rising tide, or *flood tide*, the *shoreline* (the boundary between water and land) moves inland, whereas during the falling tide, or *ebb tide*, the shoreline moves seaward.

The vertical difference between sea level at high tide and sea level at low tide is called the **tidal range (Fig. 18.11a)**. Tidal range varies with location. For example, out in the open ocean, it averages about 0.6 m, but along coasts it gets larger—the largest tidal range on the Earth, 16.8 m, occurs in the Bay of Fundy, on the east coast of Canada. The region submerged at high tide and exposed at low tide, known as the *intertidal zone*, hosts fascinating ecological niches populated by marine organisms that must be able to survive out of water for hours at a time.

The horizontal distance over which the shoreline migrates between high and low tides depends on both the tidal range and the slope of the shore surface. At a location with a large tidal range and a gentle slope, the position of the shoreline can move a long way during a tidal cycle, leaving a broad *tidal flat* exposed to the air in the intertidal zone at low tide **(Fig. 18.11b, c)**. Some tidal flats extend out from the high-tide line for over a kilometer **(Fig. 18.12a)**. Wide tidal flats can be dangerous, for when the tide turns, the arrival of a flood tide can create a *tidal bore*, a visible wall of water ranging from a few centimeters to a couple of meters high, which moves inland at speeds of up to 35 km per hour—faster than a person can run **(Fig. 18.12b)**.

Tides develop in response to a **tide-generating force**, which is due partly to the gravitational attraction of the Sun and Moon and partly to centrifugal force caused by the revolution of the Earth-Moon system around its center of mass. (To understand the meaning of this complex statement, see **Box 18.1.**) Overall, the tide-generating force generates

FIGURE 18.11 Tidal range

(a) The tidal range is the vertical distance between low tide and high tide. This photo of a French harbor was taken at low tide.

(b) High tide at Perranporth in Cornwall, England (August 28, 2007).

(c) Low tide at Perranporth (August 29, 2007). Note that the rocks in the foreground are the same ones that the man was fishing from in part (b).

FIGURE 18.12 Tidal flats and tidal bores.

(a) A broad tidal flat is exposed at low tide around Mont-Saint-Michel, on the coast of France.

(b) This large tidal bore, entering the mouth of a river along the coast of China, is a tourist attraction.

two bulges in the global ocean **(Fig. 18.13a, b)**. One bulge, the *sublunar bulge*, lies on the side of the Earth closer to the Moon—it forms because the Moon's gravitational attraction attains its greatest magnitude at this point. The other bulge, the *secondary bulge*, lies on the opposite (far) side of the Earth (12,000 km—the diameter of the Earth—farther from the Moon); it forms because the Moon's gravitational attraction

is weakest at this point, so centrifugal force can push water outward (see Box 18.1). A depression in the global ocean surface separates the two bulges. Simplistically, when a location lies under a tidal bulge, it experiences a high tide, and when it passes under a depression, it experiences low tide.

If the Earth's solid surface were smooth and completely submerged beneath the ocean so that there were no continents

FIGURE 18.13 Tides on the Earth.

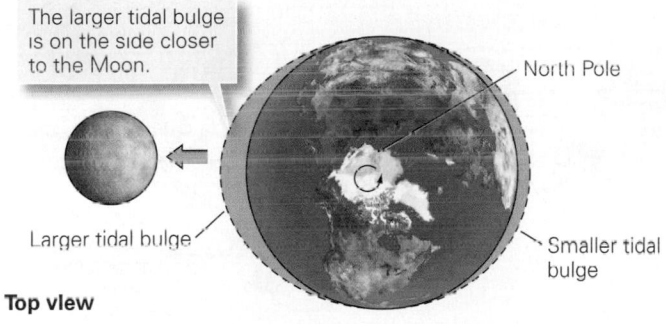

The larger tidal bulge is on the side closer to the Moon.

North Pole

Larger tidal bulge

Smaller tidal bulge

Top view

(a) Tides develop as the Earth spins relative to the two tidal bulges.

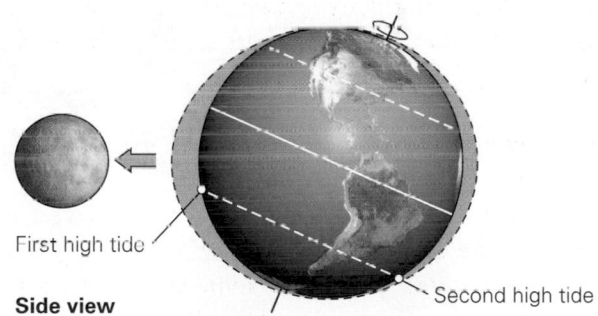

First high tide

Second high tide

Side view

(b) A side view shows that the tidal bulge does not align with the equator.

Spring tide

Solar tide
Lunar tide

Full moon

New moon

Sun

Extra-high tides are spring tides.

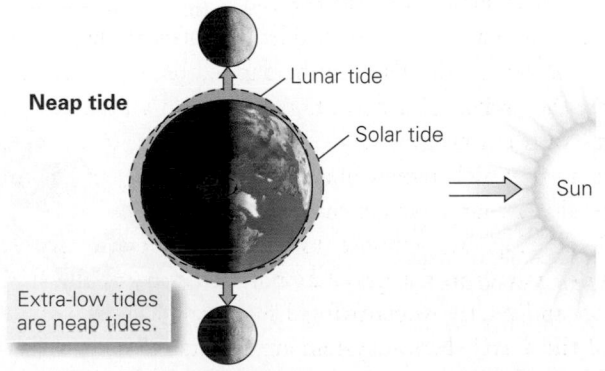

Neap tide

Lunar tide

Solar tide

Sun

Extra-low tides are neap tides.

(c) Gravitational pull of the Sun can add to that of the Moon to cause extra-high (spring) tides. If the Sun's pull is at right angles to that of the Moon, the tides are particularly low (neap tides).

BOX 18.1 CONSIDER THIS . . .

The Forces That Generate Tides

Fundamentally, tides result from interaction between two forces: gravitational attraction exerted by the Moon and the Sun on the Earth, and centrifugal force caused by the revolution of the Earth around the center of mass of the Earth-Moon system. To explain this statement, we must review some key terms from physics.

- *Gravitational pull* is the attractive force that one object exerts on another. The magnitude of gravitational pull depends on the amount of mass in each object and on the distance between the two objects.
- *Centrifugal force* is the apparent outward-directed ("center-fleeing") force that material on or in an object feels when the object spins or moves in orbit around a point. Note that centrifugal force differs from centripetal force, the "center-seeking" force; this distinction can be confusing. To experience

centripetal force, tie a ball to a string and swing it around your head. The string exerts an inward-directed centripetal force on the ball—if the string breaks, the centripetal force ceases to exist and the ball heads off in a straight-line path. To picture centrifugal force, imagine that the ball is hollow and that you've placed a marble inside. As you twirl the ball around your head, the marble moves to the outer edge of the ball. The apparent force pushing the marble outward is the centrifugal force. But as such, centrifugal force is not a real force—it is simply a manifestation of inertia, and it exists only from the perspective, or reference frame, of the orbiting object. (A physics book explains this contrast in greater detail.)

- The *Earth-Moon system* refers to this pair of objects viewed as a unit as they move together through space.

- *Center of mass* is the point within an object, or a group of objects, about which mass is evenly distributed; put another way, it is the location of the average position, or the balance point, of the total mass in a single object or a group of objects. Because the Earth has 81 times more mass than does the Moon, the center of mass of the Earth-Moon system actually lies 1,700 km below the surface of the Earth.

With these terms in mind, let's first consider the origin of centrifugal force in the Earth-Moon system. To do this, we must examine the way in which the Earth-Moon system moves. The center of the Earth itself does not follow a simple orbit around the Sun. Rather, it is the center of mass of the Earth-Moon system that follows this trajectory **(Fig. Bx18.1a)**; the Earth actually loops around this trajectory as it speeds around the Sun. To picture this

FIGURE Bx18.1 The concept of the center of mass of the Earth-Moon system and the tide-generating force.

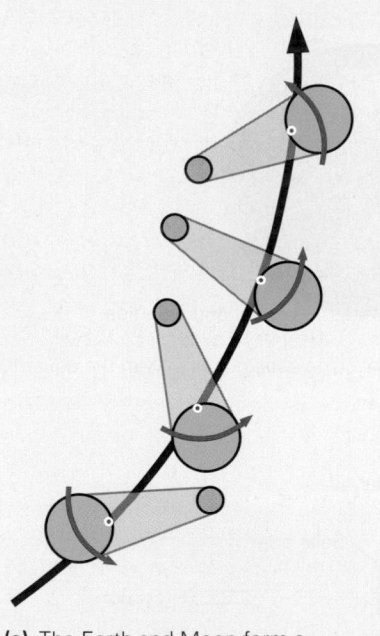

(a) The Earth and Moon form a system whose center of mass follows an orbit around the Sun.

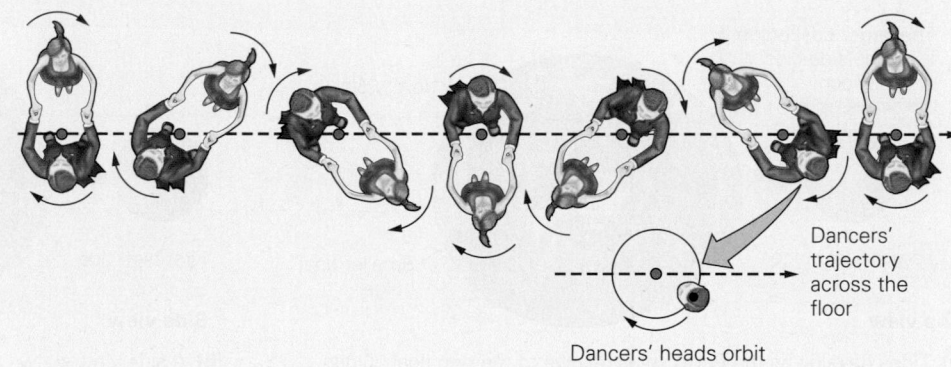

Dancers' trajectory across the floor

Dancers' heads orbit the center of mass.

(b) Two dancers as an analogy for the Earth-Moon system as it moves along its orbit. The dancers rotate around a center of mass that lies closer to the heavier dancer.

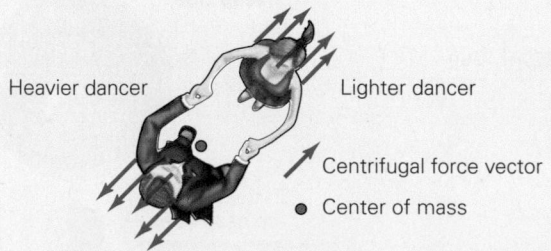

Heavier dancer

Lighter dancer

Recall that a vector is a quantity with magnitude and direction.

→ Centrifugal force vector

● Center of mass

Centrifugal force vectors point outward; they are the same magnitude for all points on a dancer.

(c) Each point on each dancer feels an outward-directed centrifugal force.

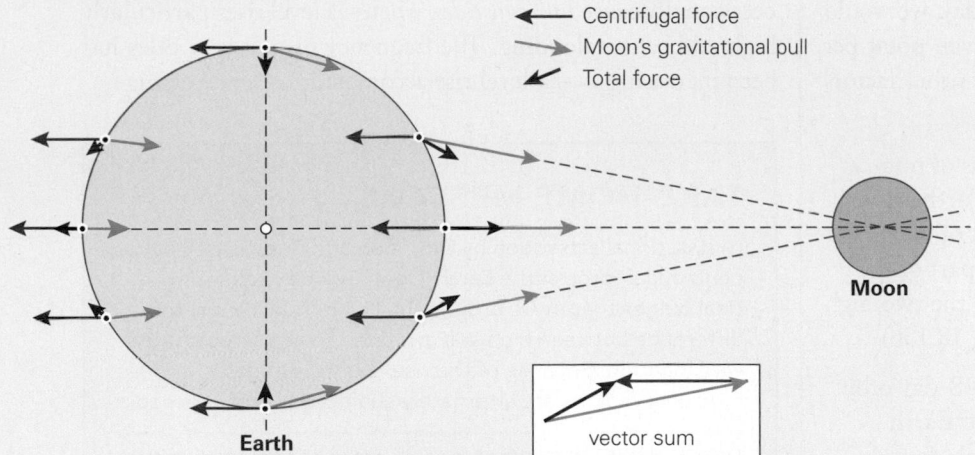

Legend:
— Centrifugal force
— Moon's gravitational pull
— Total force

Moon

Earth

vector sum

(d) Each point on the Earth's surface feels the same centrifugal force, but also feels a gravitational pull from the Moon. The total force can be represented by a vector sum. A vector is a quantity with both magnitude and direction.

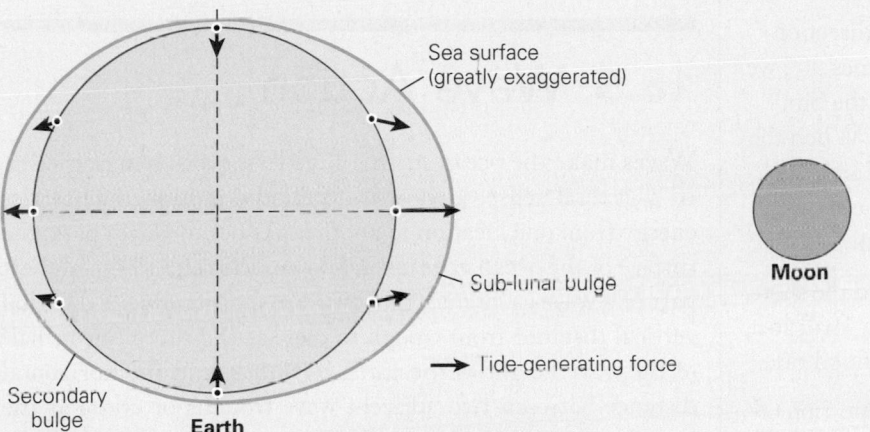

Sea surface (greatly exaggerated)

Sub-lunar bulge

Moon

→ Tide-generating force

Secondary bulge

Earth

(e) The tide-generating force is the sum of the centrifugal force vector and the gravitational force vectors. Note that the bulge of the sea surface is exaggerated.

motion, envision the Earth-Moon system as a pair of dancers, with one of the dancer much heavier than the other. The dancers face each other, hold hands, and whirl in a circle as they drift across the dance floor **(Fig. Bx18.1b)**. Each dancer's head orbits the center of mass.

Revolution of the Earth around the Earth-Moon system's center of mass generates centrifugal forces on both the Earth and the Moon that would cause the Earth and the Moon to fly away from each other were it not for the gravitational attraction holding them together. We can see this by looking again at our dancer analogy **(Fig. Bx18.1c)**—the centrifugal force acting on each dancer points outward, away from his or her partner, and is the same for all points on each dancer. We can represent the direction and

magnitude of centrifugal force by an arrow called a *vector*. (A vector is a quantity that has magnitude and direction.) In this case, the length of the arrow represents the magnitude of the force, and the orientation of the arrow indicates the direction of the force. If we think of the dancers as the Earth and the Moon, then centrifugal force vectors at all points on the surface of the Earth point away from the Moon **(Fig. Bx18.1d)**. On the Earth, therefore, centrifugal force causes the surface of the ocean to bulge outward, away from the center of mass of the Earth-Moon system, on the far side of the Earth.

Now let's consider how the force of gravity comes into play in causing tides. To simplify this discussion, we examine only the effect of the Moon's gravity on the Earth, a reasonable choice because

gravitational pull by the Moon contributes most of the tide-generating force. (The Sun, even though it is larger, is so far away that its contribution is only 46% that of the Moon.) Vectors representing the magnitude and direction of the Moon's gravitational pull at any point on the surface of the Earth all point toward the center of the Moon. Because the magnitude of gravity depends on distance, the Moon exerts more attraction on the near side of the Earth than at the Earth's center, and it exerts less attraction on the far side of the Earth than at the Earth's center. Gravity, therefore, causes the surface of the ocean on the near side of the Earth to bulge toward the Moon.

In the Earth-Moon system, both centrifugal force and gravitational pull operate at the same time. How do they interact? If we draw vectors representing both centrifugal force and gravitational force at various points on or in the Earth, we see that the vectors representing centrifugal force generally do not have the same length as those representing gravitational attraction. Moreover, the vectors representing centrifugal force do not point in the same direction as the vectors representing gravitational attraction. The sum of the two forces acting on the water represents force that affects the ocean water. You can determine the sum of two vectors by drawing the vectors so that they touch head to tail—the vector that completes the triangle indicates the sum. This vector sum is the *tide-generating force*.

The magnitude and direction of the two forces we've just described vary with location on the Earth. For example, on the side of the Earth closer to the Moon, gravitational vectors are larger than centrifugal force vectors, so adding the two gives a net tide-generating force that pulls the sea surface toward the Moon. On the side of the Earth farther from the Moon, the centrifugal force vectors are larger, so centrifugal force caused by the orbiting of the Earth-Moon system around the center of mass causes the surface of the sea to bulge outward, away from the Moon **(Fig. Bx18.1e)**. As a result, the ocean has two tidal bulges—one on the side close to the Moon and one on the opposite side of the Earth. The bigger bulge lies closer to the Moon.

or islands, the timing of tides would be fairly simple to understand. Because the Earth spins on its axis once a day, we would predict two high tides and two low tides at a given point per day. But the story isn't quite that simple—many other factors affect the timing and magnitude of tides.

- *Tilt of the Earth's axis:* Because the Earth's axis of rotation is not perpendicular to the plane of the Earth-Moon system, a given point passes under a high part of one bulge during one part of the day and under a lower part of the other bulge during another part of the day, so the two high tides at that point have different sizes (see Fig 18.13b).

- *The Moon's orbit:* The Moon progresses in its 28-day orbit around the Earth in the same direction as the Earth rotates. High tides arrive 50 minutes later each day because of the difference between the time it takes for the Earth to spin on its axis and the time it takes for the Moon to orbit the Earth.

- *The Sun's gravity:* When the angle between the direction to the Moon and the direction to the Sun becomes 90°, we experience particularly low tides, *neap tides*, for the Sun's gravitational attraction counteracts the Moon's. When the Sun and the Moon lie on the same side of the Earth, we experience particularly high tides, *spring tides*, because the Sun's attraction adds to the Moon's **(Fig. 18.13c)**.

- *Focusing effect of bays:* In a bay that narrows toward the shore, the flood tide brings a large volume of water into a small area, so the bay head experiences an especially high flood tide.

- *Basin shape:* The shape of the basin containing a portion of the sea influences the sloshing of water back and forth within the basin as tides rise and fall. Depending on the timing and magnitude of this sloshing, this effect can locally add to the global tidal bulge or subtract from it, so it can affect the rhythm and magnitude of tides. In some locations, the slosh cancels one of the daily tides entirely; as a result, the locality experiences only one high tide and one low tide in a day.

- *Air pressure:* The weight of the overlying atmosphere produces pressure that pushes down on the sea surface. Air pressure can be affected by temperature and circulation in the atmosphere, and can vary over time at a location (see Chapter 20). When air pressure decreases, sea level rises slightly. And when air pressure increases, sea level drops slightly. Normally, this effect has a minimal effect on tide height, but during a hurricane, air pressure can drop so much that sea level rise contributes to coastal flooding.

Because of the complexity of the factors contributing to tides, the timing and magnitude of tides vary significantly along a coast. Nevertheless, at a given location, the tides are periodic and can be predicted. Tides gave early civilizations a rudimentary way to tell time. In fact, some languages use the same word for *time* and *tide*. In some locations, combinations of factors occasionally cause *nuisance tides*, when sea level rises particularly high and causes flooding. The frequency of nuisance tides has been increasing as sea-level rise accompanies climate change.

TAKE-HOME MESSAGE

Gravitational attraction by the Moon and Sun, as well as centrifugal force in the Earth-Moon System, cause two tidal bulges that move around the Earth. Tidal range, the difference between high and low tides, varies significantly with location. Because of the rise and fall of tides, intertidal regions are alternately submerged and exposed.

QUICK QUESTION: What is the largest component of the tide-generating force?

18.5 Wave Action

Waves make the ocean surface a restless, ever-changing vista. In a general sense, a **wave** is a periodic motion that carries energy from one location to another. As ocean waves pass, the surface of the ocean goes up and down. We refer to the highest part of a wave as its *crest*, the lowest part as its *trough*, the total vertical distance from trough to crest as the *wave height*, half of the wave height as the *wave amplitude*, and the horizontal distance between two adjacent wave troughs or crests as the *wavelength* **(Fig. 18.14a)**.

How do ocean waves form? To picture the process, first imagine still water on a windless day. The horizontal surface represents an *equilibrium level* determined by the pull of the Earth's gravity. When air starts moving in a breeze, the air molecules apply a shear stress to the water surface (much as your hand applies a shear stress to a tabletop when you move your hand across the table). Because of *surface tension*, caused by the attraction of water molecules to one another, the water behaves somewhat like an elastic sheet, so the shear stress of a breeze stretches this sheet. Like a spring, the sheet can only stretch so far before it twangs back. This motion wrinkles the surface into small waves called *ripples*. Wind can now push against the sides of the ripples, causing them to build still higher. Eventually, ripples evolve into a wave that lifts the water well above the equilibrium level. When this happens, gravity acts on the elevated water in the wave, causing it to sink. Inertia causes the sinking water surface to sink below the equilibrium level. Like a weight suspended from a vertical spring, the water surface descends only so far before it bounces back up. Momentum carries it above the equilibrium level. The process repeats, so the surface of the water at a location effectively bounces up and down. Since gravity drives this

FIGURE 18.14 Ocean waves build in response to the shear of wind blowing over the water surface.

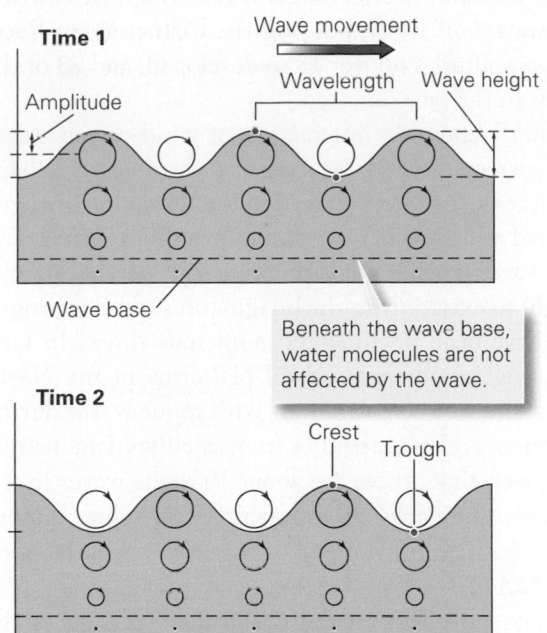

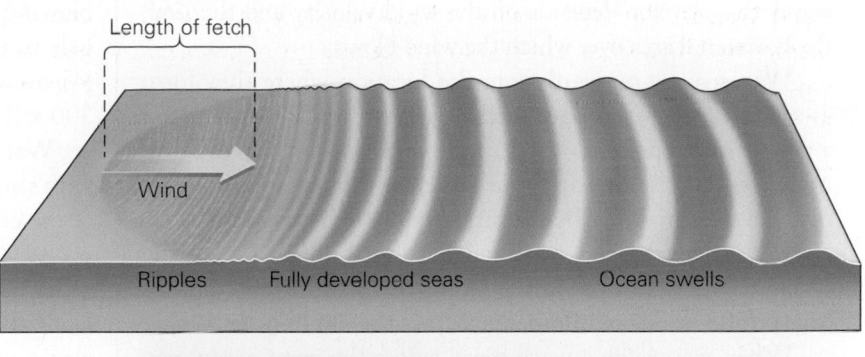

(a) Within a deep-ocean wave, water molecules follow a circular path. The diameter of the circle decreases with depth. Note that the wave height is twice the amplitude.

Beneath the wave base, water molecules are not affected by the wave.

(b) Waves form when wind blows along the sea surface. Waves can build into swells which can travel a long distance from where they were generated.

(c) This oil tanker was sailing through 7.5-m-high waves when, suddenly, a rogue wave that was 20 m high struck it broadside, completely submerging the deck, which normally lies 18 m above sea level.

Breakers along a California beach.

(d) As the wave approaches the shore, friction slows its base. Water motion in the wave becomes more elliptical, and the wave becomes a breaker.

process, researchers call such waves *gravity waves*. The size of the waves that develop depends on the wind velocity and the *fetch*, the horizontal area over which the wind blows.

Waves move outward from the location where they form as a *wave train*—you have seen a wave train form if you drop a pebble into a pool of water. In the region where waves form, numerous trains with different wavelengths and wave heights may develop, making the sea surface chaotic. Wave trains with large wavelengths and wave heights move out of the region where they formed, becoming **swells** that can travel hundreds or even thousands of kilometers across the sea **(Fig. 18.14b)**.

When you watch a wave travel across the open ocean, you may get the impression that the whole mass of water constituting the wave moves with the wave. But if you observe the motion of a cork floating on the water, you'll see that it mostly bobs up and down and shifts back and forth as a wave passes. It does not move horizontally at the same velocity as the wave—in fact, roughly 20 waves must pass before the cork moves horizontally by a distance equal to just one wavelength. Therefore, within a wave forming over deep water, a particle of water moves in a roughly circular motion, as viewed in cross section. The circle has the greatest diameter at the ocean's surface, where it equals the wave's amplitude. With increasing depth, the diameter of the circle decreases until, at a depth equal to about half the wavelength, no wave movement takes place (see Fig. 18.14a). Submarines traveling below this **wave base** cruise through smooth water, while ships on the surface may be tossing about.

How large can wind-driven waves in the open ocean get? A hurricane commonly produces 15- to 20-m-high waves. The largest storm waves recorded, in 2005 during Hurricane Ivan in the Gulf of Mexico, had a wave height of 30 m. Particularly large waves may also form where two sets of wind-driven waves coming from different directions constructively interfere (overlap) in such a way that wave crests add to each other.

This happened in 1979, when waves generated by an east-blowing gale collided with waves generated by a west-blowing gale in the waters off Ireland during the Fastnet Yacht Race. Waves with amplitudes of over 15 m developed, and 23 of the 300 sailboats in the race capsized.

Wave interference, the interaction of wind-driven waves with strong currents, and the focusing of waves due to the shape of the coastline or seafloor, can lead to the formation of **rogue waves**, defined as waves that rise two to five times higher than other large waves passing a location during a period of time. Long thought to exist only in the imagination of sailors, rogue waves now have been documented numerous times. In fact, wave-measuring instruments on oil platforms in the North Sea recorded almost 500 encounters with rogue waves during a 10-year period, and radar studies from satellites demonstrate that at any given time, there are about 10 rogue waves in the world's oceans. The decks of large ships—including famous cruise ships—have been swamped by rogue waves in the open ocean **(Fig. 18.14c)**.

Waves have no effect on the ocean floor as long as the floor lies below the wave base. However, near the shore, where the wave base just touches the seafloor, a wave causes a slight back-and-forth motion of sediment. Closer to shore, as the water gets shallower, friction between the wave and the seafloor slows the deeper part of the wave, and the motion of water in the wave becomes more elliptical. Eventually, water at the top of the wave curves over the base, and the wave becomes a *breaker*, ready for surfers to ride **(Fig. 18.14d)**.

Did you ever wonder...
why the breakers that surfers love form near the shore?

Waves may make a large angle with the shoreline as they're coming in, but they bend as they approach the shore, a phenomenon called **wave refraction**. When they reach the shore, their crests rarely trend at an angle of more than 5° with respect to the shoreline **(Fig. 18.15)**. To understand why wave refraction takes place, imagine a wave approaching the shore with a crest

FIGURE 18.15 Wave refraction.

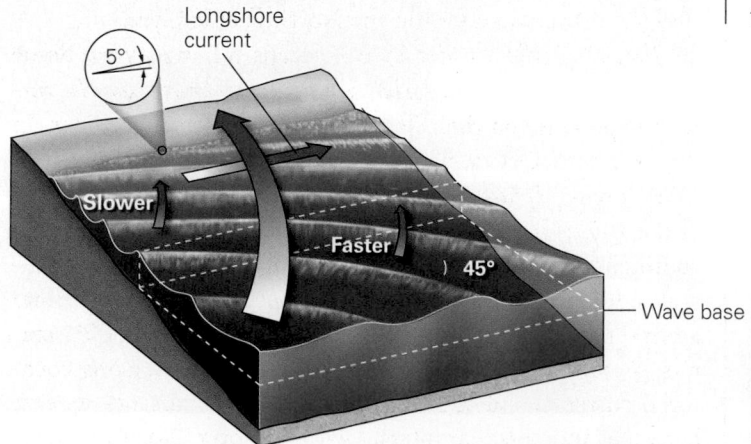

(a) Wave refraction occurs when waves approach the shore at an angle. Waves arriving at the shore obliquely can cause a longshore current.

(b) An example of waves refracting along a beach.

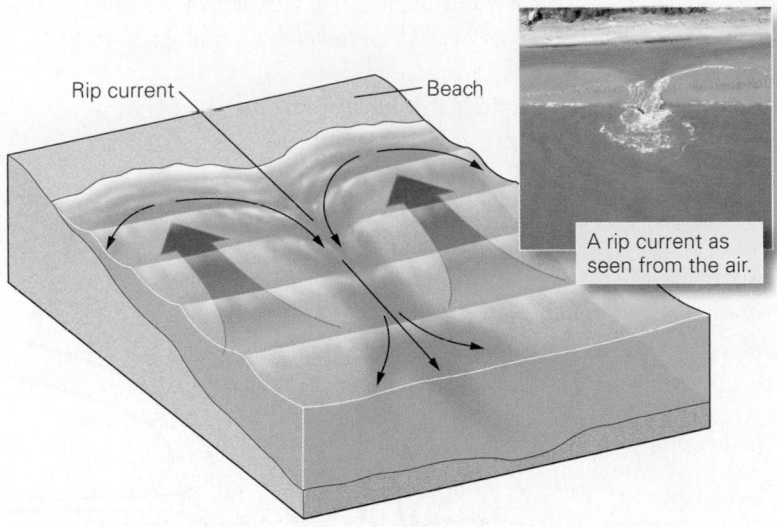

Rip current　　　　Beach

A rip current as seen from the air.

that initially makes an angle of 45° with the shoreline. The end of the wave closer to the shore touches bottom first and slows down because of friction, whereas the end farther offshore continues to move at its original velocity, swinging the whole wave around so that it is more parallel with the shoreline.

Although wave refraction decreases the angle at which waves move close to shore, it generally does not eliminate the angle. Where waves do arrive at the shore obliquely, water in the nearshore region has a component of motion that trends parallel to the shore. This **longshore current** causes swimmers floating in the water just offshore to drift gradually in a direction parallel to the beach.

Waves pile water up on the shore incessantly. As the excess water moves back to the sea, it may localize into a strong seaward flow perpendicular to the beach called a *rip current* (Fig. 18.16). Rip currents cause many drownings every year along beaches because they carry unsuspecting swimmers away from the beach and cause swimmers to panic. If caught in a rip current, your best strategy is to swim calmly parallel to the shore.

TAKE-HOME MESSAGE

The friction of the wind against the sea surface causes waves to start forming. Within a wave, water moves roughly in a circle; the amount of motion decreases with depth down to the wave base. Large storms generate large waves. Near the shore, friction between waves and the seafloor causes water to pile up into breakers. Waves refract when they approach the shore, but if waves wash ashore obliquely, longshore currents develop.

QUICK QUESTION: What are rogue waves, and why do they form?

18.6 Currents: Rivers in the Sea

Since first setting sail on the open ocean, people have known that its water does not stand still, but rather flows or circulates at velocities of up to several kilometers per hour in fairly well-defined streams called **currents**. Oceanographic studies made since the *Challenger* expedition demonstrate that circulation in the sea occurs at two levels: *surface currents* affect the upper hundred meters or so of water, while *deep currents* affect deeper levels, even down to the seafloor.

Surface Currents: A Consequence of the Wind

When ship captains plan their routes across the ocean, they pay close attention to the direction of surface currents, for sailing against one can slow a voyage substantially. In each ocean, surface currents flow at two scales. *Oceanic-scale currents* take water on journeys extending for thousands of kilometers and have long-established names (**Fig. 18.17a**). These currents travel at a range of velocities (**Fig 18.17b**). Note that some of these currents, known as **gyres**, define a huge loop around the margins of an ocean. As an example, the *North Atlantic Gyre* takes water completely around the North Atlantic Ocean. The western margin of this gyre flows along the east coast of North America as the familiar *Gulf Stream* (**Fig. 18.17c**), one of the world's fastest currents, locally traveling at speeds up to 6.4 km per hour. Most oceans contain one gyre—those in the northern hemisphere flow clockwise, whereas those in the southern hemisphere flow counterclockwise. Surface water may become trapped in the interior of a gyre, where currents barely exist, so these regions tend to accumulate floating seaweed or garbage. For example, the *Sargasso Sea*, named for a type of floating seaweed, lies within the North Atlantic Gyre, and the *Great Pacific Garbage Patch*, a mess of plastic and sludge, accumulated within the North Pacific Gyre.

Not all surface water in the oceans follows gyres. Some currents, known as *equatorial currents* or *equatorial counter currents* depending on their flow direction, follow the equator. In the Southern Ocean, which surrounds Antarctica, the clockwise-flowing *Antarctic Circumpolar Current* (traditionally known as the *West Wind Drift*) transports water completely around the continent. And, in addition to oceanic-scale currents, local-scale **eddies**, isolated swirls or ring-shaped currents, carry water around paths with diameters from tens to hundreds of kilometers. You'll find eddies along the margins of gyres, along equatorial currents, and at locations where larger currents squeeze between landmasses or interact with the coast (**Fig. 18.17d**).

Why do currents follow the paths that they do? Books focused on oceanography provide a complete explanation.

FIGURE 18.17 Surface currents of the world's oceans.

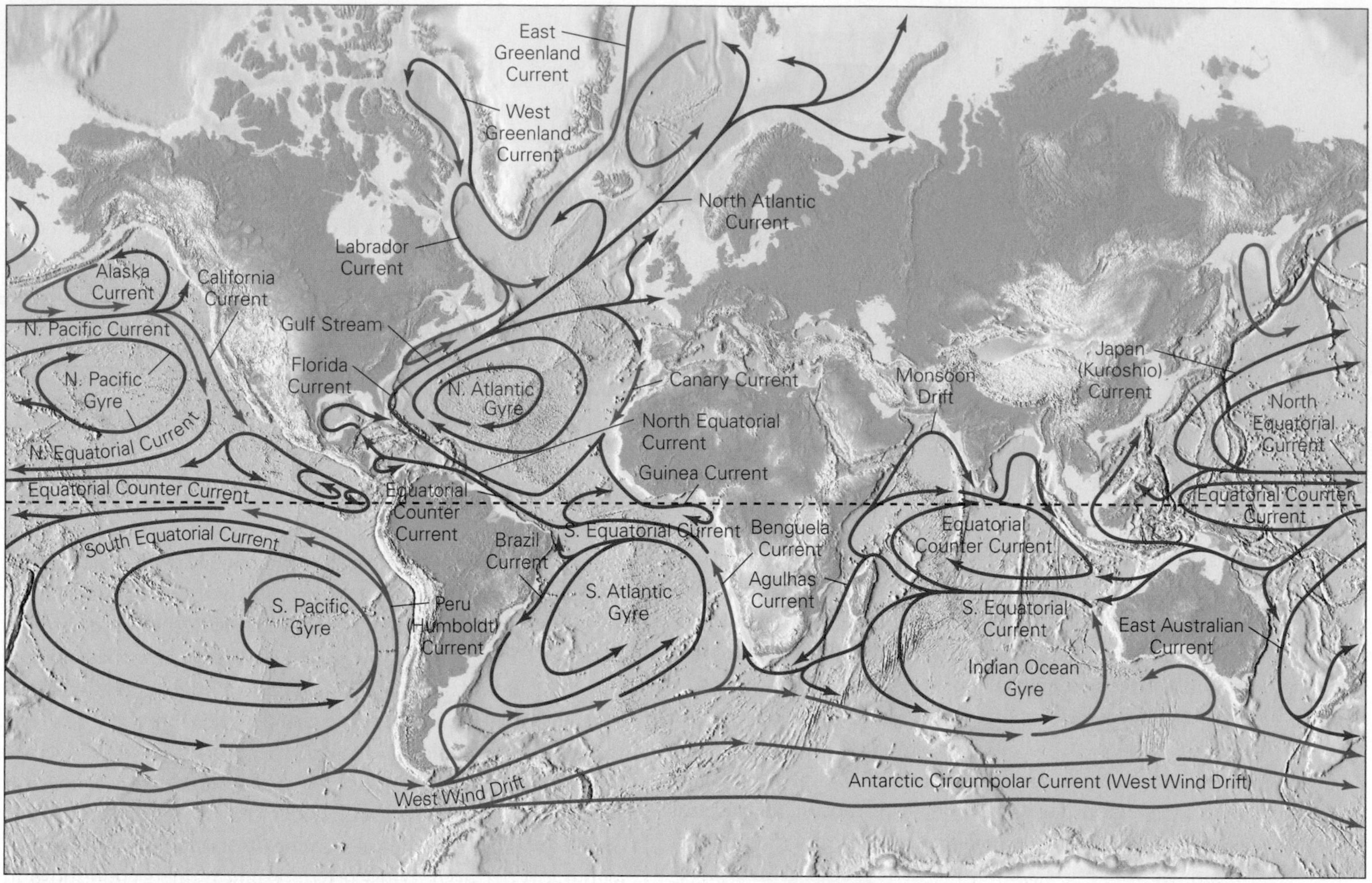

East Greenland Current

West Greenland Current

North Atlantic Current

Labrador Current

Alaska Current

California Current

N. Pacific Current

Gulf Stream

Japan (Kuroshio) Current

Monsoon Drift

N. Pacific Gyre

Florida Current

N. Atlantic Gyre

Canary Current

North Equatorial Current

North Equatorial Current

N. Equatorial Current

North Equatorial Current

Guinea Current

Equatorial Counter Current

Equatorial Counter Current

Equatorial Counter Current

Equatorial Counter Current

South Equatorial Current

Brazil Current

S. Equatorial Current

Benguela Current

Equatorial Counter Current

S. Pacific Gyre

Peru (Humboldt) Current

S. Atlantic Gyre

Agulhas Current

S. Equatorial Current

East Australian Current

Indian Ocean Gyre

West Wind Drift

Antarctic Circumpolar Current (West Wind Drift)

(a) A map of the major ocean-scale currents. Red indicates warm currents; blue indicates cold currents.

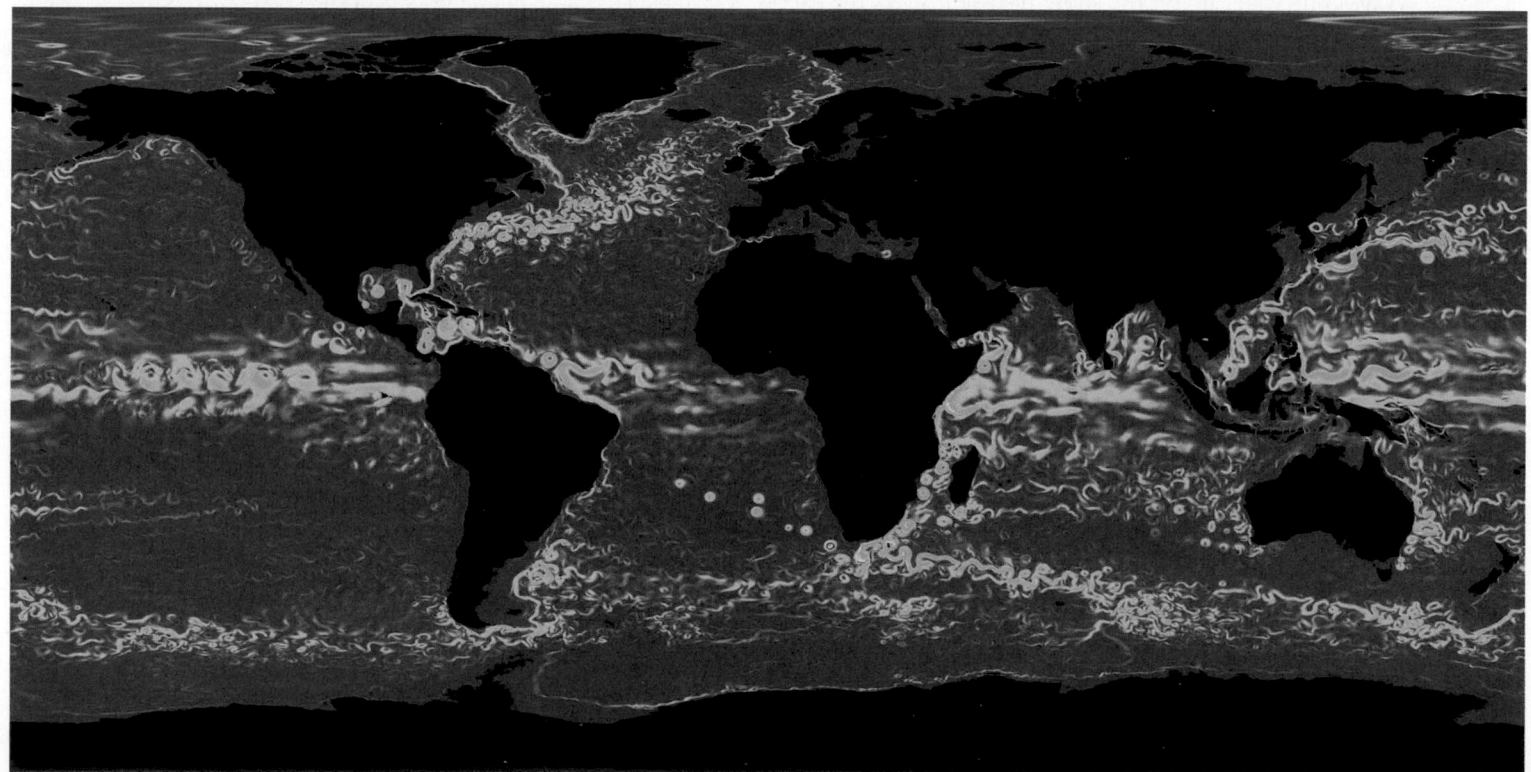

(b) On this map, the brightness of the flow lines indicates water velocity, and the orientation indicates direction. You can see a large number of eddies.

FIGURE 18.17 *Continued*

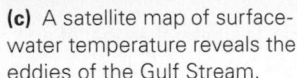

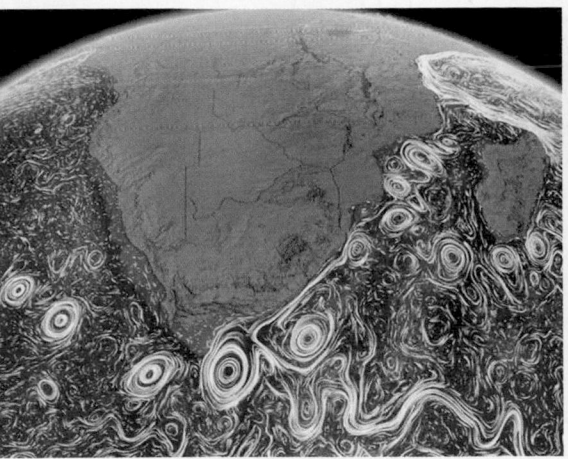

Southern Ocean, south of Africa

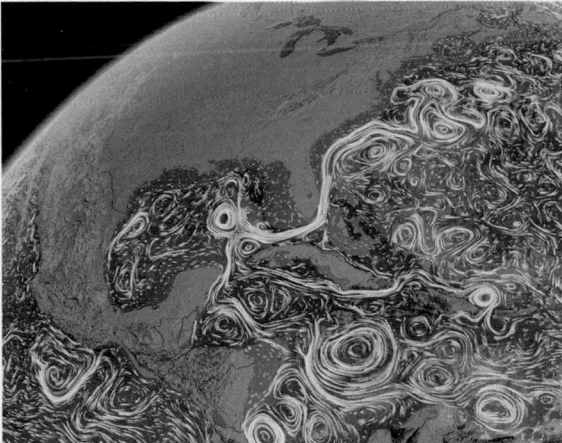

Western North Atlantic and Caribbean

(c) A satellite map of surface-water temperature reveals the eddies of the Gulf Stream.

(d) A computer animation of currents shows eddy formation.

Here, we note that global-scale surface current paths result from the action of three phenomena: (1) the shear of wind against the surface of water, which drags water molecules along; (2) the **Coriolis effect (Box 18.2)**, a consequence of the Earth's rotation, which causes moving water to veer to the right in the northern hemisphere and to the left in the southern hemisphere; and (3) pressure in water, directed outward from locations where the regional surface elevation of the water lies slightly higher to locations where the surface lies slightly lower. Let's consider an example of how these phenomena drive water flow.

Flow of water in the North Atlantic Gyre begins in response to the shear of *prevailing winds*, the most common surface winds of the region. Initially, surface water starts to flow in the direction of the prevailing wind, which tends to blow from west to east at higher latitudes and from east to west at lower latitudes (see Chapter 20). As soon as the water starts to move, however, the Coriolis effect causes it to veer to the right. The movement of the uppermost layer then shears the layer below, which in turn veers even farther to the right. In fact, each successive layer deflects farther to the right than the layer above, a pattern of flow called the *Ekman spiral*, named for the oceanographer who proposed it **(Fig. 18.18a)**. The net flow direction of the surface current, due to the Ekman spiral, trends roughly at right angles to the prevailing wind direction. Oceanographers refer to this net flow as **Ekman transport (Fig. 18.18b)**. Ekman transport, at the scale of the whole North Atlantic Ocean, drives water toward the interior of the ocean, forming a very subtle "mound" in which the ocean surface has a slightly higher than average elevation. The elevation difference (a maximum of only about 3 m) between water in the mound and water along its margins generates a pressure gradient, meaning that pressure decreases from the center of the mound toward its edge. The resulting *pressure-gradient force* (similar to the force produced when you squeeze on one end of

a balloon to push air out the opening at the other end) pushes outward, against the inward-directed force generated by the Coriolis effect. When theses forces equal each other—a condition known technically as *geostrophic balance*—water will flow in a path around the mound **(Fig. 18.18c)**. The resulting, somewhat circular, current, is the North Atlantic Gyre.

Eddies form along the margins of larger currents where faster-moving water shears against slower-moving water, and where a current shears against the slower-moving water of coastal areas. The shear between two volumes of water moving at different velocities occurs for the same reason that a pencil spins when you place it between your hand and a tabletop and then move your hand parallel to the tabletop.

Upwelling, Downwelling, and Deep Currents

Surface currents are not the only means by which water flows in the ocean; water also circulates in the vertical direction. Oceanographers have now identified *downwelling zones*, places where near-surface water sinks, and *upwelling zones*, places where deeper water rises. Let's consider some of the causes of upwelling and downwelling.

Local downwelling and upwelling take place near coasts where winds push surface water toward or away from the coast. If near-surface water moves toward the coast, then an oversupply of water develops along the shore, and excess water must sink, causing downwelling **(Fig. 18.19a)**. Alternatively, if near-surface water moves away from the coast, then a deficit of water develops near the coast and water rises to fill in the gap, causing upwelling **(Fig. 18.19b)**. Because of Ekman transport, flow of water toward or away from shore can develop where wind blows parallel to the shore. For example, southward-blowing winds along the west coast of the United States

BOX 18.2 CONSIDER THIS . . .

A Closer Look at the Coriolis Effect

The Earth revolves on its axis once a day, and because of this rotation, a point on the equator moves at a speed of 1,670 km per hour (1,037 mph) relative to an observer sitting out in space. The speed of a point on the Earth's surface decreases toward the poles, because a point at a higher latitude doesn't have as far to go in a day. In fact, at the poles, the velocity is 0 km per hour (Fig. Bx18.2a). Because of this change in velocity with latitude, the instant that a volume of seawater starts to move, it deflects from its initial

direction and starts to follow a curved path (Fig. Bx18.2b). If friction doesn't slow the water down, it will complete a circle (known technically as an *inertial circle*). In the northern hemisphere, it follows a clockwise path around the circle, and in the southern hemisphere, it follows a counterclockwise path (Fig. Bx18.2c). This means that in the northern hemisphere, an initially northward-traveling volume veers east, an eastward-traveling volume veers south, a southward-traveling volume veers west, and

a westward-traveling volume veers north. Gaspard-Gustave de Coriolis (1792–1843), who first described this phenomenon, inspired its name, the *Coriolis effect*.

Why does the Coriolis effect happen? We can visualize a simplistic answer by considering what happens when we anchor a cannon at a midlatitude point, aim due north, and fire. The *linear velocity* (the distance divided by time) of a point on the surface of the rotating Earth depends on latitude, even though the *angular*

FIGURE Bx18.2 The Coriolis effect occurs because the velocity of a point at the equator, in the direction of the Earth's spin, is greater than that of a point near the pole.

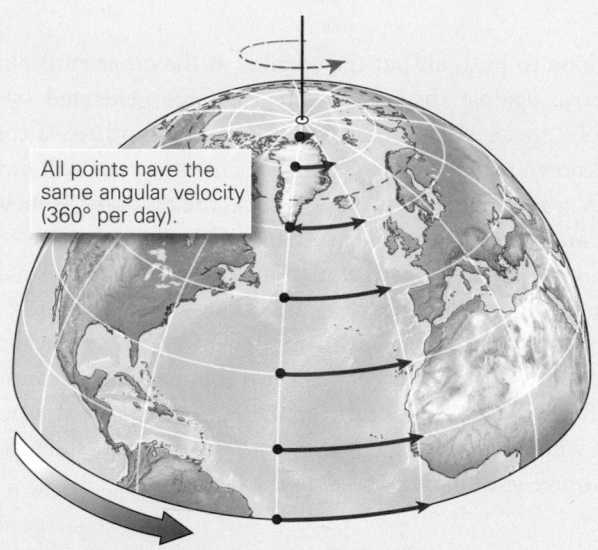

(a) The linear velocity of a point anchored to the Earth's surface decreases toward the poles, though the angular velocity does not change.

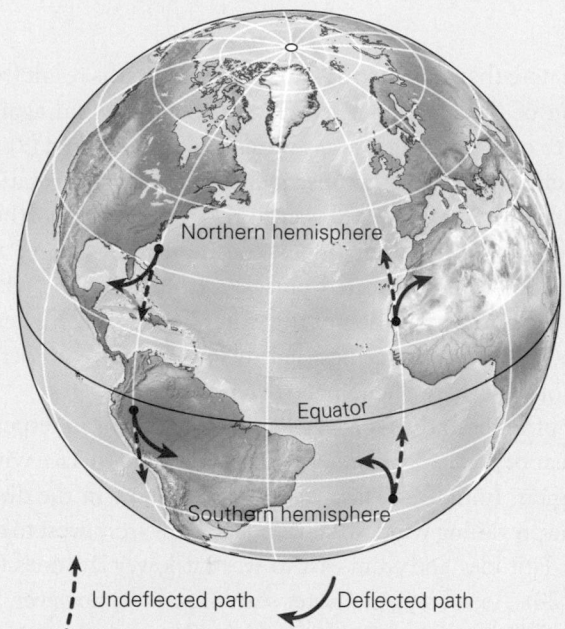

Undeflected path   Deflected path

(b) In the northern hemisphere, the Coriolis effect causes moving objects to veer right. In the southern hemisphere, they veer left.

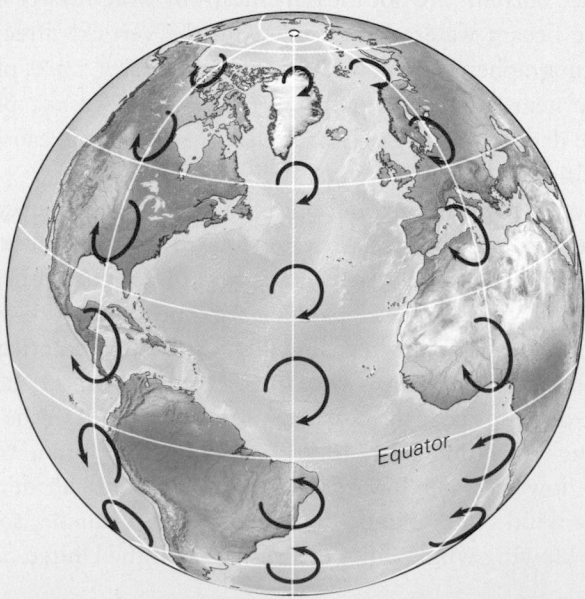

(c) If there were no friction, a moving object would follow an inertial circle due to the Coriolis effect. The diameter of the circle depends on the object's velocity and latitude.

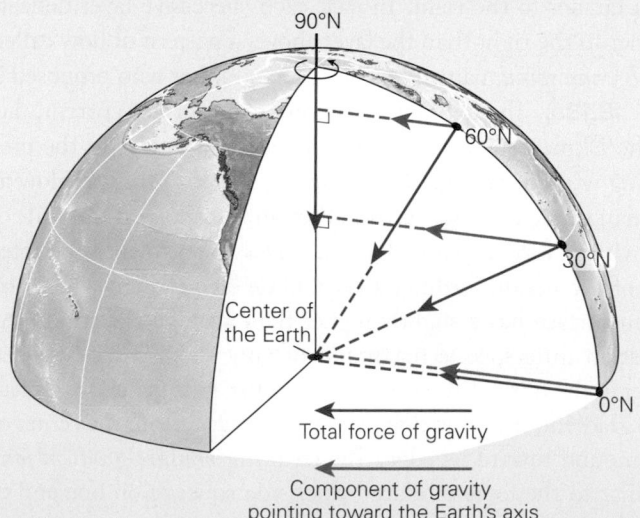

(d) The component of gravitational pull in the direction perpendicular to the Earth's axis decreases toward the poles.

FIGURE Bx18.2 *Continued*

Time 1: Object starts moving northward from its starting point.

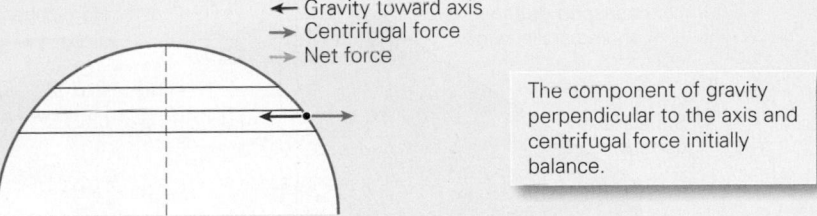

← Gravity toward axis
→ Centrifugal force
→ Net force

The component of gravity perpendicular to the axis and centrifugal force initially balance.

Time 2: Object moving northward turns east. At its highest latitude, it is moving eastward.

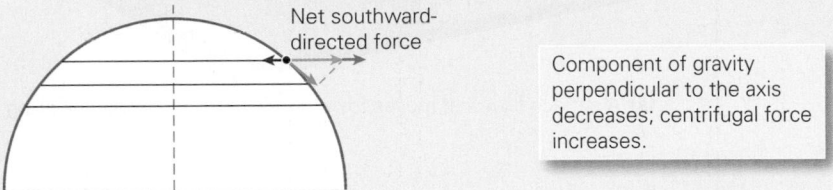

Net southward-directed force

Component of gravity perpendicular to the axis decreases; centrifugal force increases.

Time 3: Object moving eastward turns south, and passes across its starting latitude.

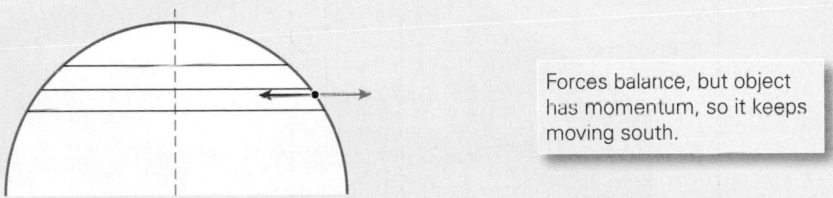

Forces balance, but object has momentum, so it keeps moving south.

Time 4: Object moving southward turns west. It reaches its lowest latitude.

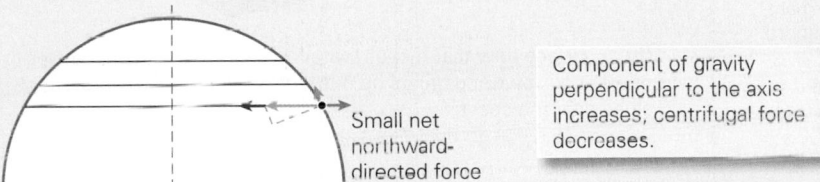

Small net northward-directed force

Component of gravity perpendicular to the axis increases; centrifugal force decreases.

Time 5: Object moving west starts moving north. It returns to its starting point.

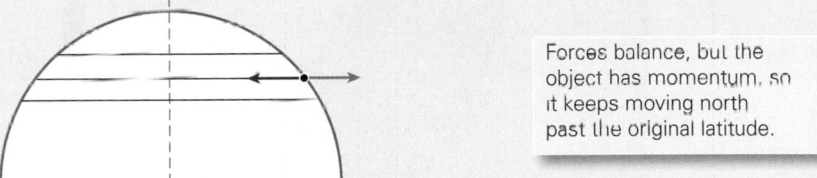

Forces balance, but the object has momentum, so it keeps moving north past the original latitude.

(e) The centrifugal force also changes with latitude, so the force acting on an object changes with latitude. As a result, a component of the net force parallel to the Earth's surface acts on an object, causing its direction to change.

velocity (the angle of a circle that the point covers divided by time) stays constant **(Fig. Bx18.2c)**. The cannonball's initial motion, relative to the cannon, tracks due north. But because the cannon is moving east at the linear velocity of the Earth's rotation at the cannon, the cannonball also has a component of motion to the east. As the ball travels north (and assuming it doesn't slow down due to friction), its eastward component of linear velocity becomes progressively greater than that of points on the Earth's surface over which it travels. Therefore, from the perspective of an observer sitting on the Earth, the ball veers to the east. It seems that another force must be acting on the ball to make it veer east, because *Newton's first law of motion* states that a moving body follows a straight line unless acted on by an

outside force. The imaginary force acting on the ball is called the **Coriolis force**.

This explanation, while fairly easy to visualize, doesn't actually explain the way the Coriolis effect operates in the real world. The physical explanation for the Coriolis effect is significantly more complicated and builds on the concepts of centrifugal force, gravitational force, and vectors (see Box 18.1), as well as on an understanding of angular momentum. An object moving around a circle has *angular momentum*, defined as the product of the object's mass, the radius of the circle, and the object's angular velocity. Newton showed that if no outside force acts on an object, angular momentum cannot change—physicists now refer to this concept as *conservation of angular momentum*. Note that, because of conservation of angular momentum, if the radius of the circle around which an object rotates decreases, then the object's angular velocity must increase.

With these definitions and concepts in mind, let's imagine the forces acting on an object that starts moving north over the surface of the Earth. As it moves north, its distance from the Earth's axis of rotation decreases. Therefore, conservation of angular momentum requires the object's angular velocity to increase, and as a result, the outward centrifugal force acting on it must increase. Because the Earth has mass, gravity pulls the object toward the Earth's center. We can picture gravity as a vector, and we can separate it into two components: one acting in the direction perpendicular to the Earth's axis of rotation and another acting parallel to the axis **(Fig. Bx18.2d)**. As the object moves north, the component of gravity perpendicular to the axis (the component that must balance the centrifugal force to keep the object from flying off into space) decreases. The *net force*, defined by the vector sum of centrifugal and gravitational forces in a direction perpendicular to the axis, can also be divided into two components: one perpendicular to the Earth's surface and one parallel to the Earth's surface (assuming the object or volume does not fly off the Earth) **(Fig. Bx18.2e)**. As the object starts moving north, it starts veering east, and the component of net force parallel to the Earth's surface pulls it southward. When the object reaches the top of its inertial circle, the force makes it continue to veer right and start heading south. When it reaches its original latitude, the surface-parallel component of the net force acting on it has decreased to zero, but since the object has momentum (like the swinging weight of a pendulum), it continues heading south and immediately starts veering right again. Now, the component of the net force parallel to the surface pushes it north, so it continues to swing north, eventually reaching its starting position and completing its inertial circle.

FIGURE 18.18 An explanation of Ekman transport and how it contributes to the formation of gyres.

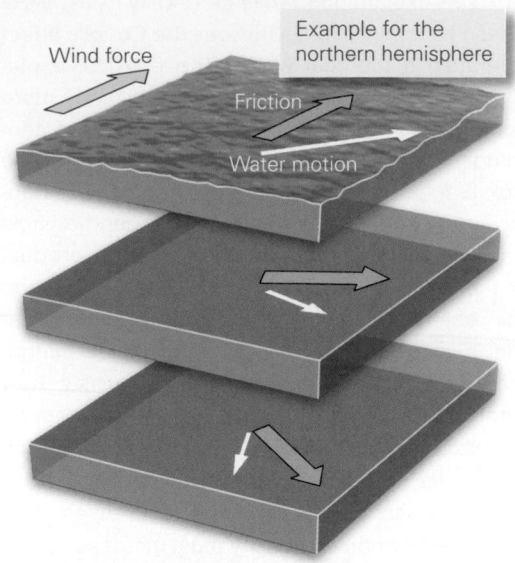

(a) Each layer of water veers to the right of the layer above (in the northern hemisphere) due to the Coriolis effect. The flow velocity decreases with depth due to friction.

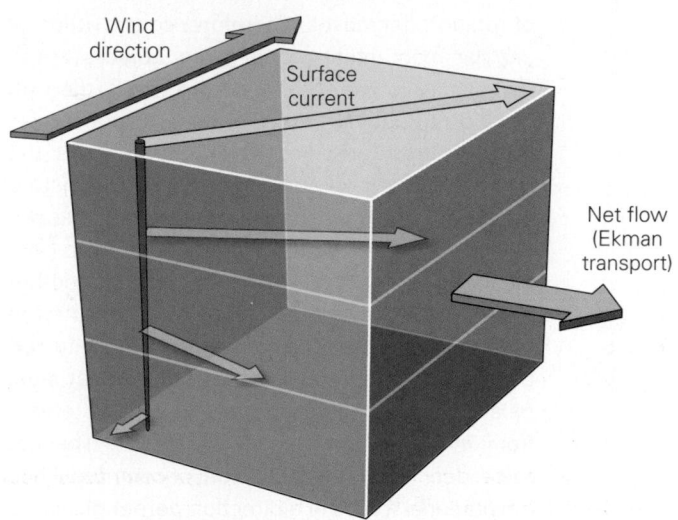

(b) Adding together the flow variations with depth leads to a total flow at right angles to the wind direction. This flow is Ekman transport.

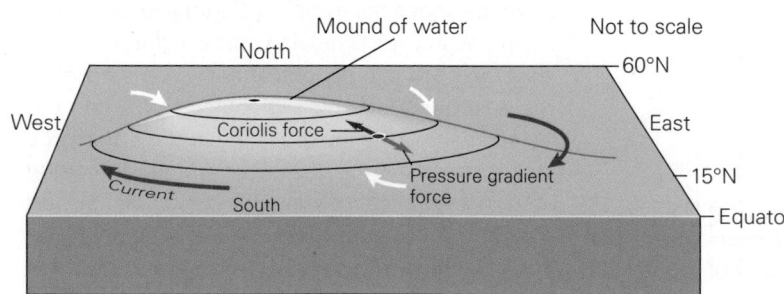

(c) Due to Ekman transport (indicated by white arrows), a mound of water builds in the center of an ocean. The resulting pressure-gradient force balances the Coriolis force and causes oceanic-scale gyres to flow around the mound.

FIGURE 18.19 Coastal upwelling and downwelling.

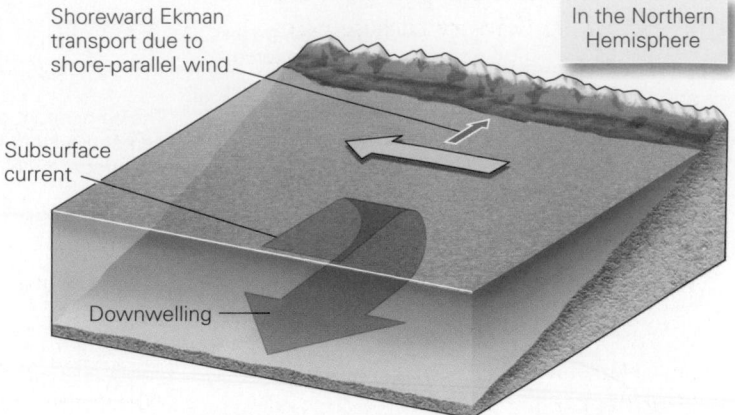

(a) A surface flow that moves toward the shore causes downwelling.

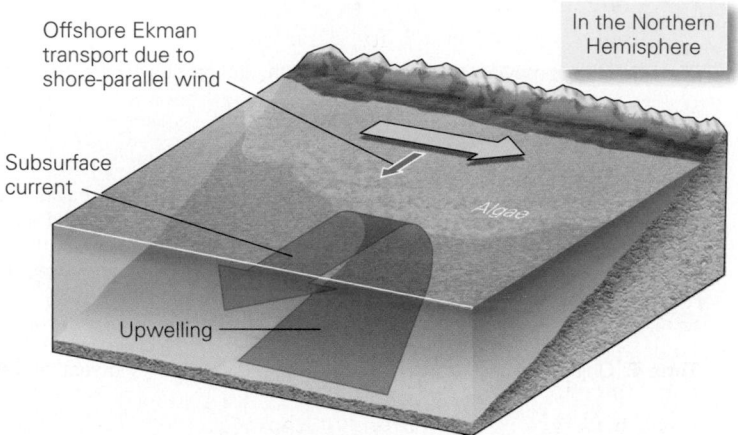

(b) A surface flow that moves water away from the shore causes upwelling. Upwelling brings up nutrients and fosters the growth of algae near the coast.

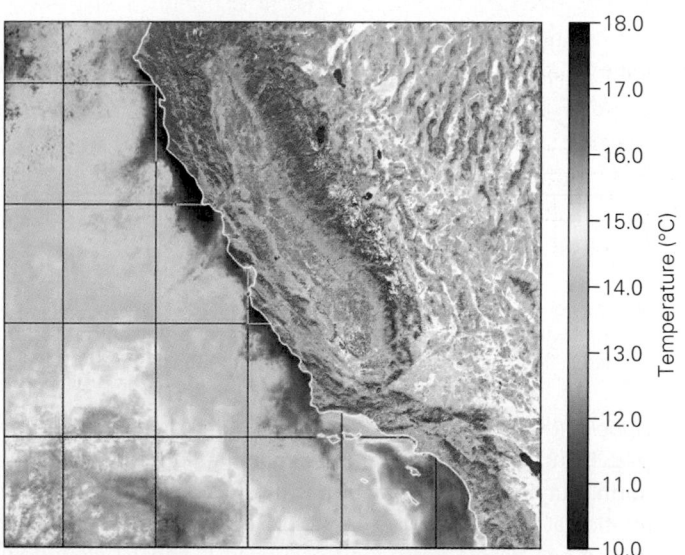

(c) The variations in surface temperature in the eastern Pacific Ocean show upwelling of cold water along the west coast of the United States.

cause upwelling adjacent to California **(Fig. 18.19c)**. Upwelling is important to the fishing industry because rising water brings up nutrients that encourage growth of the plankton on which fish feed. Similarly, upwelling occurs along the equator because the winds blow steadily from east to west—the Coriolis effect causes the surface water to deflect to the right in the northern hemisphere and to the left in the southern hemisphere, resulting in a net deficit of water along the equator. Upwelling replaces this deficit and causes the surface water at the equator to be cooler and rich in the nutrients that foster an abundance of life.

Contrasts in water density, caused by differences in temperature and salinity, can also drive upwelling and downwelling. We refer to the rising and sinking of water driven by such density contrasts as **thermohaline circulation**. During such circulation, denser (colder or saltier) water sinks, whereas less dense (warmer or less salty) water rises. As a result, the cold water in polar regions sinks and flows back along the bottom of the ocean toward the equator. This process divides the ocean vertically into a number of distinct *water masses*, which mix only very slowly with one another. In the Atlantic Ocean, for example, the Antarctic Bottom Water sinks along the coast of Antarctica, and the North Atlantic Deep Water sinks in the north polar region **(Fig. 18.20a)**. The combination of surface currents and thermohaline circulation, like a conveyor belt, moves water and heat among the various ocean basins **(Fig. 18.20b)**.

FIGURE 18.20 Global-scale upwelling and downwelling of ocean water.

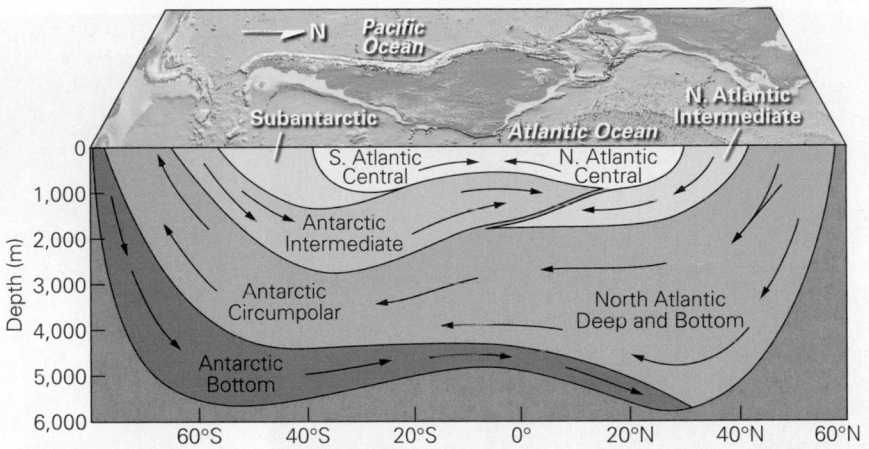

(a) Due to variations in density, the oceans are stratified into distinct water masses.

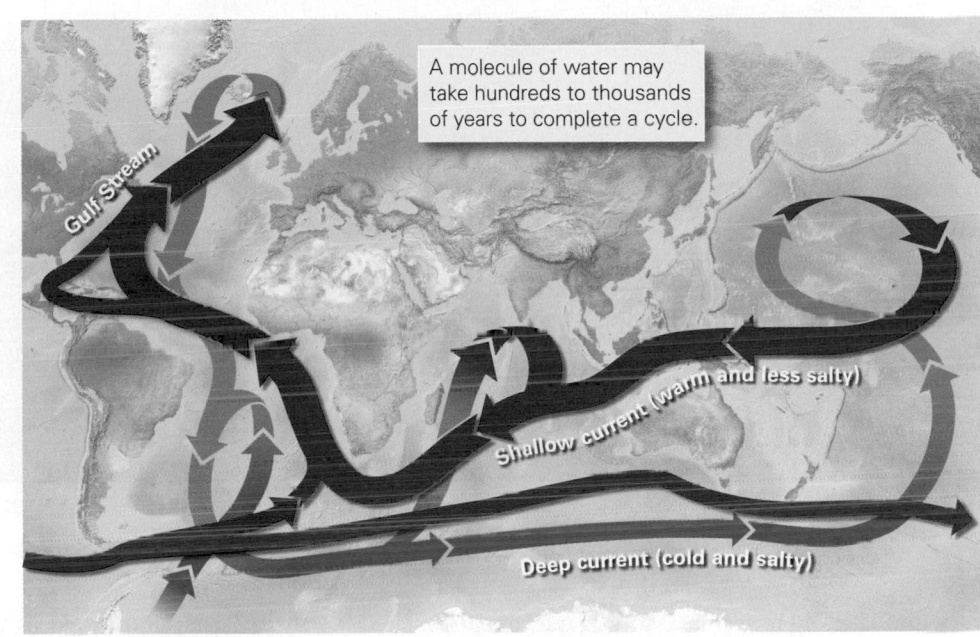

(b) Thermohaline circulation results in a global-scale "conveyor belt" that circulates water throughout the entire ocean system. The ocean mixes entirely in a 1,500-year period.

TAKE-HOME MESSAGE

The wind drives surface currents, forming large gyres. Upwelling and downwelling develop due to wind-driven water surplus or deficit, or due to density variations related to temperature and salinity.

QUICK QUESTION: Why does downwelling of ocean water occur at polar latitudes?

18.7 Where Land Meets Sea: Coastal Landforms

Tourists along the Amalfi Coast of Italy thrill to the sound of waves crashing on rugged rocky shores. But in the Virgin Islands, sunbathers can find seemingly endless white sand beaches. Large dome-like mountains rise directly from the sea in Rio de Janeiro, Brazil, but a 100-m-high vertical cliff marks the boundary between the Nullarbor Plain of southern Australia and the Southern Ocean **(Fig. 18.21)**. As these examples illustrate, *coasts*, the belts of land bordering the sea,

FIGURE 18.21 Spectacular coastal scenery from localities around the world.

(a) Rocky cliffs form the Amalfi Coast of western Italy.

(b) A sandy beach along the coast of St. John, U.S. Virgin Islands.

(c) Rounded "sugarloaf" mountains rise from the bays of Rio de Janeiro.

(d) The abrupt edge of the Nullarbor Plain forms the south coast of Australia.

vary dramatically in terms of topography and associated landforms **(Fig. 18.22)**.

Beaches and Tidal Flats

For many vacationers, the ideal holiday includes a trip to a **beach**, a gently sloping fringe of sediment along the shore. Some beaches consist of rounded pebbles or cobbles, but most are made up of a blanket of sand grains **(Fig. 18.23a, b)**. The character of the sediment on a beach is no accident, because beaches are subject to the constant action of waves. If the beach lies close to cliffs that shed coarse debris, or to the mouth of fast-moving streams that carry large clasts, wave action breaks up larger blocks, grinds away edges and corners, rounds the

> **Did you ever wonder...**
> why beautiful sandy beaches don't form along all coasts?

clasts, and produces a beach of cobbles. If the beach forms from fine- to medium- grained sediment deposited by rivers or glaciers, wave action winnows out finer sediment, such as silt and mud, and carries it offshore to quieter water, leaving sand behind. Wave action can't make sand grains smaller, because even when tossed in a current, the grains can't collide with enough energy to break up.

The composition of sand itself varies from beach to beach, because different sands come from different sources. Sands derived from the weathering and erosion of felsic-to-intermediate rocks consist mainly of quartz; other minerals in these rocks chemically weather to form clay, which washes away in waves. Beaches made from the erosion of coral reefs and shells consist of carbonate sand. And beaches derived from the erosion of basalt boast black sand, made of tiny basalt grains.

A **beach profile**, a cross section drawn perpendicular to the shore, illustrates the shape of a beach **(Fig. 18.23c)**. Starting from the sea and moving landward, a beach consists of a

FIGURE 18.22 Examples of different kinds of coasts.

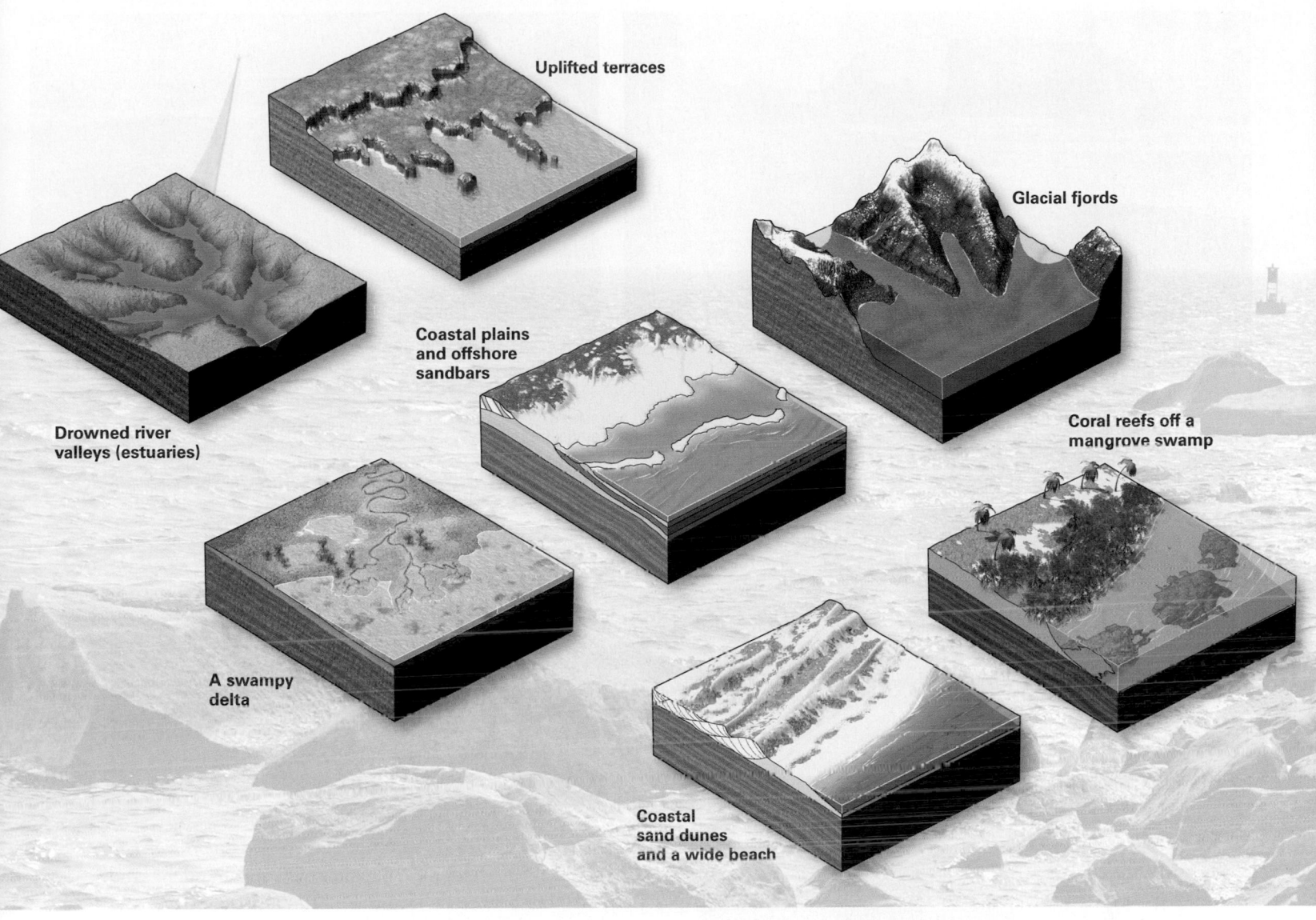

Uplifted terraces

Glacial fjords

Coastal plains and offshore sandbars

Drowned river valleys (estuaries)

Coral reefs off a mangrove swamp

A swampy delta

Coastal sand dunes and a wide beach

foreshore zone, which includes the intertidal zone across which the tide rises and falls. The *beach face*, a steeper, concave-up part of the foreshore zone, forms where the waves actively scour the sand. The *backshore zone* extends from a small step, cut by high-tide waves, to the front of the dunes or cliffs that lie farther inshore. The backshore zone includes one or more *berms*, horizontal to landward-sloping terraces that received sediment during a storm. A small *beach scarp* delineates the edge of a berm (**Fig. 18.23d**).

When breakers crash onto the shore, they send a surge of water up the beach. This upslope flow, or **swash**, continues until friction, together with the downslope pull of gravity, brings the water's motion to a halt. Then gravity takes over and draws the water back down the beach as **backwash**. If wave crests trend parallel to the beach face, beach sediment simply moves up and down the slope of the beach face along

with the swash and backwash, respectively. Geologists refer to the sediment that moves in response to waves on a calm to moderately windy day as *active sand*; a layer of *inactive sand* lies beneath the active sand. When storms build very large waves, even the inactive sand will start moving.

Where waves reach the shore at an angle, the active sand moves in a sawtooth pattern that results in the gradual net transport of sediment parallel to the beach, a phenomenon known as **longshore drift**. (Note that longshore drift on the beach moves in the same direction as the longshore current just offshore.) The sawtooth pattern happens because the swash of a wave moves perpendicular to the wave crest, so an oblique wave carries sediment diagonally up the beach, but the backwash flows straight down the beach slope due to gravity (**Fig. 18.24a**). Because of longshore drift, geologists sometimes refer to beaches as "rivers of sand" to emphasize that

FIGURE 18.23 Characteristics of beaches.

(a) A gravel beach along the Olympic Peninsula, Washington. The clasts were derived from erosion of adjacent cliffs.

(b) A sand beach in Puerto Rico. Wave action carries away finer sediment.

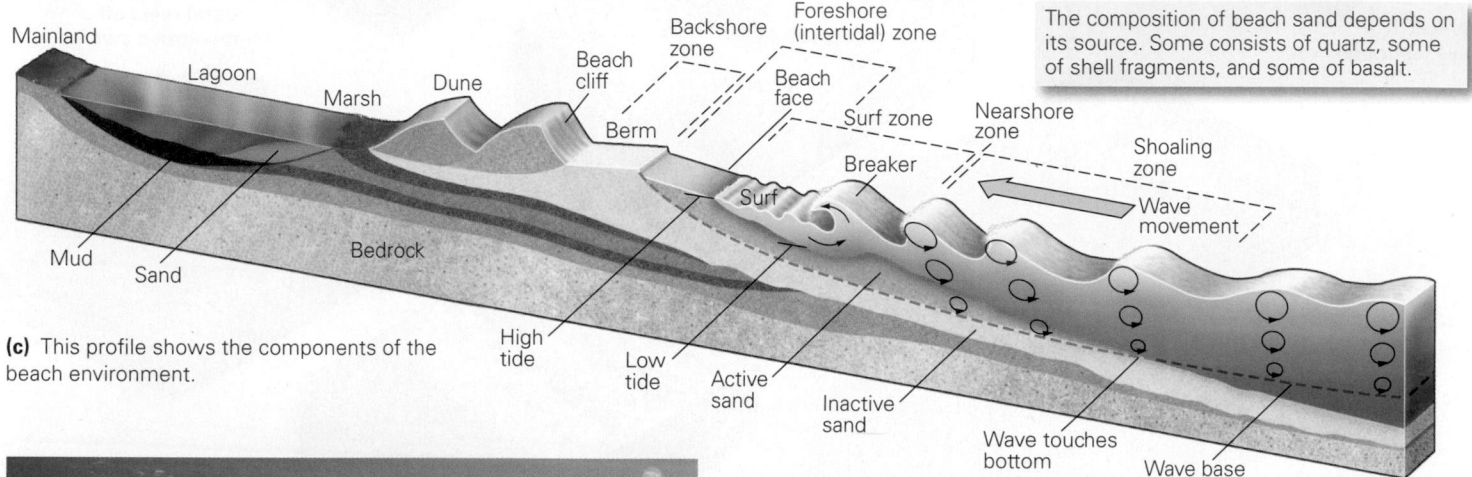

The composition of beach sand depends on its source. Some consists of quartz, some of shell fragments, and some of basalt.

(c) This profile shows the components of the beach environment.

(d) A berm on a Cape Cod beach.

beach sand moves along the coast over time. In fact, longshore drift can transport sand hundreds of kilometers along a coast.

Where the coastline indents landward, longshore drift stretches beaches out into open water to create a **sand spit**. Some sand spits grow across the opening of a bay, forming a *baymouth bar* (**Fig. 18.24b, c**). The scouring action of waves can pile sand

up in a narrow ridge away from the shore called an *offshore bar*, which parallels the shoreline. In regions with an abundant sand supply, offshore bars rise above the mean high-water level and become *barrier islands* (**Fig. 18.24d**). The water between a barrier island and the mainland becomes a quiet-water *lagoon*, a body of shallow seawater separated from the open ocean.

Though developers have covered some barrier islands with expensive resorts, in the time frame of centuries to millennia, barrier islands are temporary features. Wind and waves pick up sand from the ocean side of the barrier island and drop it on the lagoon side, causing the island to migrate landward. Storms may breach barrier islands and create an inlet (a narrow passage of water). Finally, longshore drift gradually transports the sand of barrier islands and modifies their shape.

Tidal flats, regions of mud and silt exposed or nearly exposed at low tide but totally submerged at high tide, develop in regions protected from strong wave action (**Fig. 18.25**; see Fig. 18.12a). They are typically found along the margins of lagoons or on shores protected by barrier islands. Here mud and silt accumulate to form thick, sticky layers. In tidal flats that provide a home for burrowing organisms such as clams

FIGURE 18.24 The concept and consequences of longshore drift.

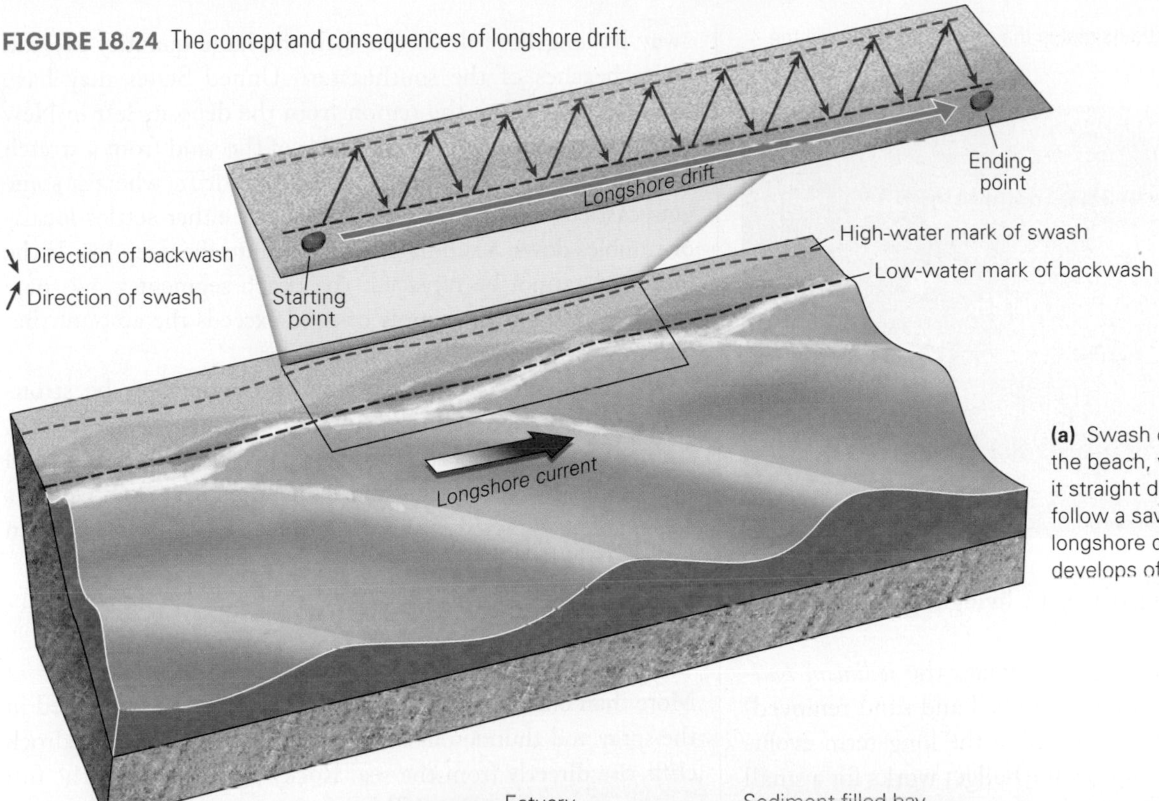

Longshore drift

Ending point

Direction of backwash
Direction of swash

Starting point

High-water mark of swash
Low-water mark of backwash

Longshore current

(a) Swash carries sand obliquely up the beach, whereas backwash carries it straight downslope. So sand grains follow a sawtooth pattern, yielding longshore drift. A longshore current develops offshore.

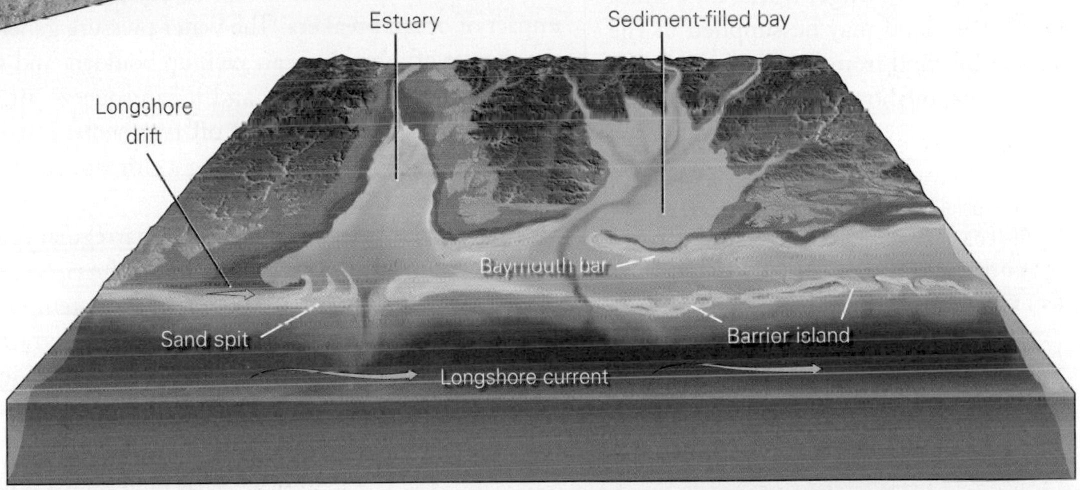

Estuary

Sediment-filled bay

Longshore drift

Baymouth bar

Sand spit

Barrier island

Longshore current

(b) Longshore drift on a beach can generate sand spits and baymouth bars. Sedimentation may fill in the region behind the bar.

(c) An example of a sand spit.

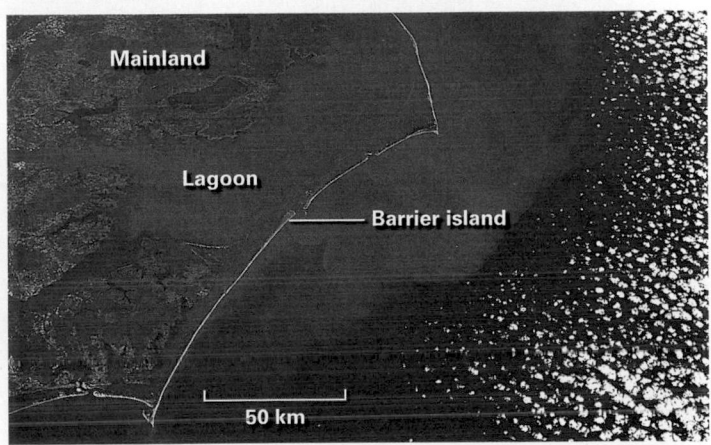

Mainland

Lagoon

Barrier island

50 km

(d) Barrier islands may be separated from the mainland by a lagoon.

FIGURE 18.25 At low tide, boats rest in the mud of a tidal flat on the coast of Brittany.

Tidal flat

and worms, **bioturbation** (stirring by living organisms) mixes sediments together.

Because of the movement of sediment, the *sediment budget* (the difference between sand supplied and sand removed) plays an important role in determining the long-term evolution of a beach. Let's look at how the budget works for a small segment of beach (Fig. 18.26). Sand may be supplied to the segment from local rivers or by wind from nearby dune fields. It may also be brought from just offshore by waves or from far

FIGURE 18.26 The sediment budget along a coast. Sediment is brought into the system by rivers, by the erosion of cliffs and moraines, and by wind. Sediment moves along the coast as a result of longshore drift. And sediment leaves the system by being blown off the beach, by sinking into deeper water, or by being carried out by the longshore current.

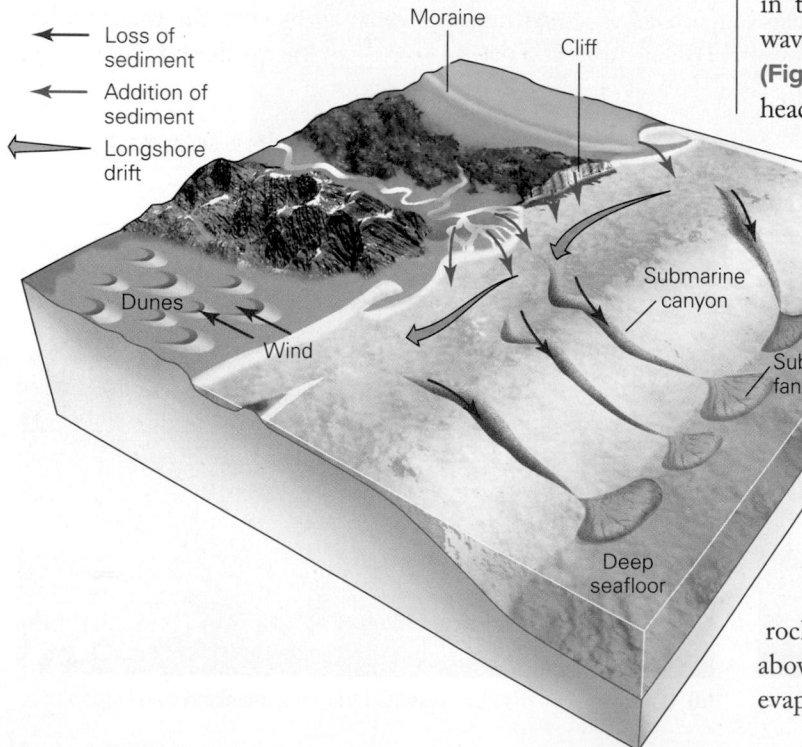

← Loss of sediment

← Addition of sediment

← Longshore drift

Moraine

Cliff

Dunes

Wind

Submarine canyon

Submarine fan

Deep seafloor

away by longshore drift. (In fact, the large quantity of sand along beaches of the southeastern United States may have been transported to the region from the deposits left in New England by ice-age glaciers.) Some of the sand from a stretch of beach may be removed by longshore drift, whereas some gets carried offshore by waves, where it either settles locally or tumbles down a submarine canyon into the deep sea. If the lost sand cannot be replaced, the beach segment grows narrower, whereas if the supply of sand exceeds the amount that washes away, the beach becomes wider.

In temperate climates, winter storms tend to be stronger and more frequent than summer ones. The larger, shorter-wavelength waves of winter storms wash beach sand into deeper water and thus make the beach narrower, whereas the smaller, longer-wavelength summer waves bring sand in from offshore and deposit it on the beach (Fig. 18.27).

Rocky Coasts

More than one ship has met its end smashed and splintered in the spray and thunderous surf of a rocky coast, where bedrock cliffs rise directly from the sea. Rocky coasts may feel the full impact of ocean breakers. The water pressure generated during the impact of a breaker can pick up boulders and smash them together until they shatter, and it can squeeze air into cracks, creating enough force to split off fragments. Further, because of its turbulence, the water hitting a cliff face carries suspended sand and thus can abrade the cliff.

Many rocky coasts start out with an irregular coastline, with *headlands* protruding into the sea and *embayments* set back from the sea. Such irregular coastlines tend to be temporary features in the context of geologic time, for wave refraction focuses wave energy on headlands and disperses it in embayments (Fig. 18.28). As a result, waves preferentially remove debris at headlands and deposit sediment in embayments.

The combined effects of shattering, wedging, and abrasion, together called *wave erosion*, gradually undercut a cliff face and make a *wave-cut notch* (Fig. 18.29a). Undercutting continues until the overhang becomes unstable, breaks away at a joint, and falls to produce a pile of rubble at the base of the cliff that waves immediately attack and break up. In this process, wave erosion cuts away at a rocky coast such that the cliff gradually migrates inland. Such cliff retreat leaves behind a *wave-cut platform* or *bench*, which becomes visible at low tide (Fig. 18.29b).

Other processes besides wave erosion also break up the rocks along coasts. For example, salt spray coats the cliff face above the waves and infiltrates into pores. When the water evaporates, salt crystals grow and push apart the grains,

FIGURE 18.27 Contrasts between winter and summer beaches.

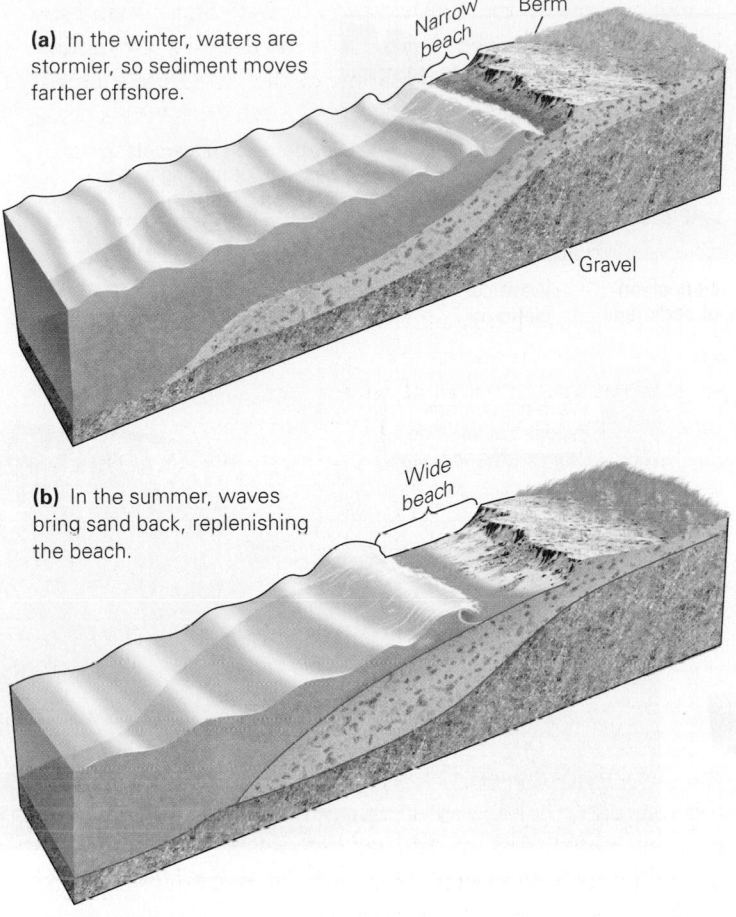

(a) In the winter, waters are stormier, so sediment moves farther offshore.

Narrow beach · Berm · Gravel

(b) In the summer, waves bring sand back, replenishing the beach.

Wide beach

thereby weakening the rock. Plants and animals in the intertidal zone also contribute to erosion; they weaken the rock by boring into it to create rootholds or burrows.

Many distinct landforms develop along a rocky coast **(Fig. 18.29c)**. A headland erodes in stages **(Fig. 18.29d-e)**.

Because of refraction, waves curve and attack the sides of a headland, slowly eating through it to create a *sea arch* connected to the mainland by a narrow bridge. Eventually the arch collapses, leaving isolated *sea stacks* just offshore. Once formed, a sea stack protects the adjacent shore from waves. Therefore, sand may collect in the lee of the stack, slowly building a *tombolo*, a narrow ridge of sand that links the sea stack to the mainland.

Estuaries

Along some coastlines, a relative rise in sea level causes the sea to flood river valleys that merge with the coast, resulting in **estuaries**, where seawater and river water mix **(Fig. 18.30)**. Oceanic and fluvial waters interact in two ways within an estuary. In quiet estuaries, protected from wave action or river turbulence, the water becomes stratified, with denser oceanic saltwater flowing upstream as a wedge beneath less-dense fluvial freshwater. Such a *saltwater wedge* migrates about 100 km up the Hudson River in New York, and about 40 km up the Columbia River in Oregon. In turbulent estuaries, such as the Chesapeake Bay, ocean and river water combine to create nutrient-rich *brackish water*, with a salinity between that of oceans and rivers. Estuaries are complex ecosystems inhabited by unique species of shrimp, clams, oysters, worms, and fish that can tolerate large changes in salinity.

Fjords

During the last ice age, glaciers carved deep valleys in coastal mountain ranges. When the ice age came to a close, the glaciers melted away, leaving deep, U-shaped valleys (see Chapter 22). The water stored in the glaciers, along with the water within the vast ice sheets that covered continents during the ice age,

FIGURE 18.28 Like a lens, wave refraction focuses wave energy on a headland, so erosion occurs there. Wave energy weakens in embayments, so deposition occurs there.

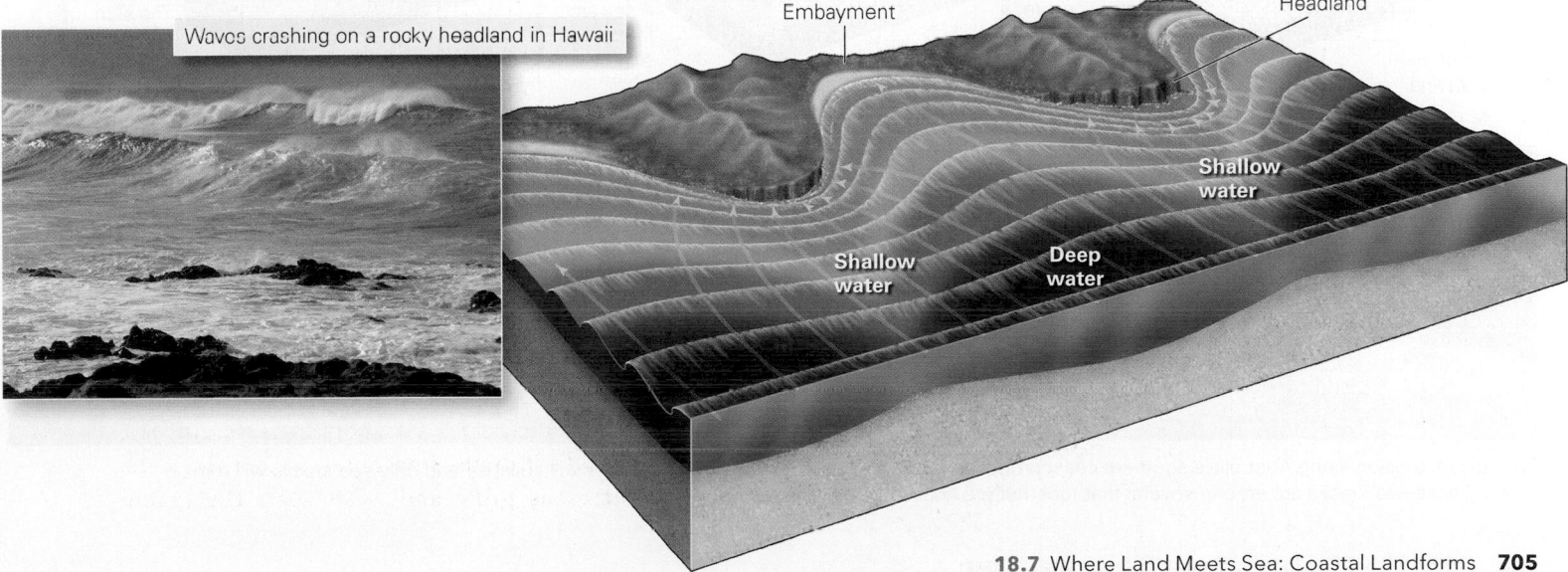

Waves crashing on a rocky headland in Hawaii

Embayment · Headland · Shallow water · Deep water · Shallow water

FIGURE 18.29 Erosion landforms of rocky shorelines.

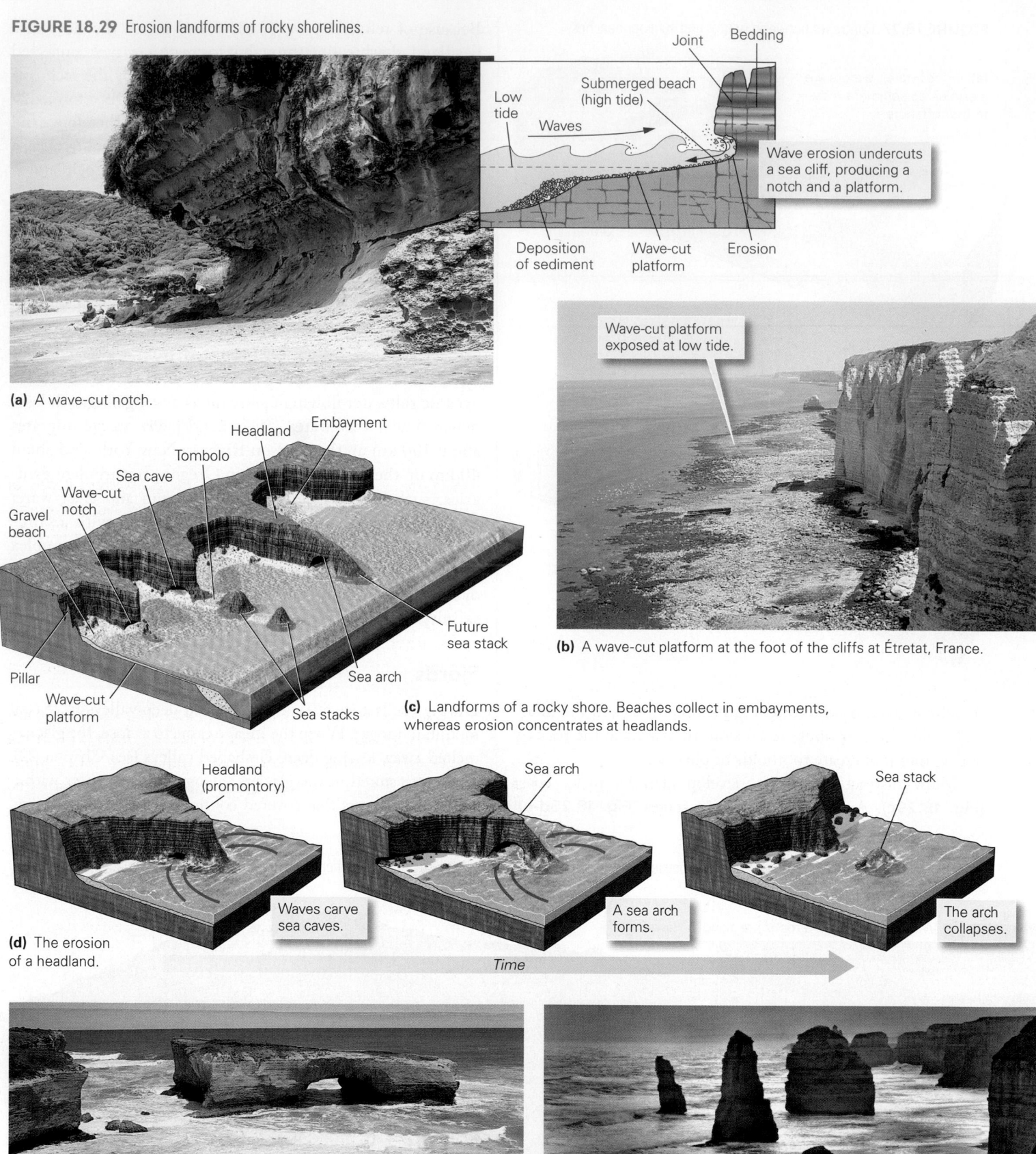

(a) A wave-cut notch.

(b) A wave-cut platform at the foot of the cliffs at Étretat, France.

(c) Landforms of a rocky shore. Beaches collect in embayments, whereas erosion concentrates at headlands.

(d) The erosion of a headland.

Waves carve sea caves.

A sea arch forms.

The arch collapses.

Time

(e) Coastal erosion along Australia's southern coast produced a sea arch (left). Eventually, the bridge will collapse and only sea stacks will remain (right). These sea stacks are among several that together are known locally as the Twelve Apostles.

FIGURE 18.30 Examples of estuaries.

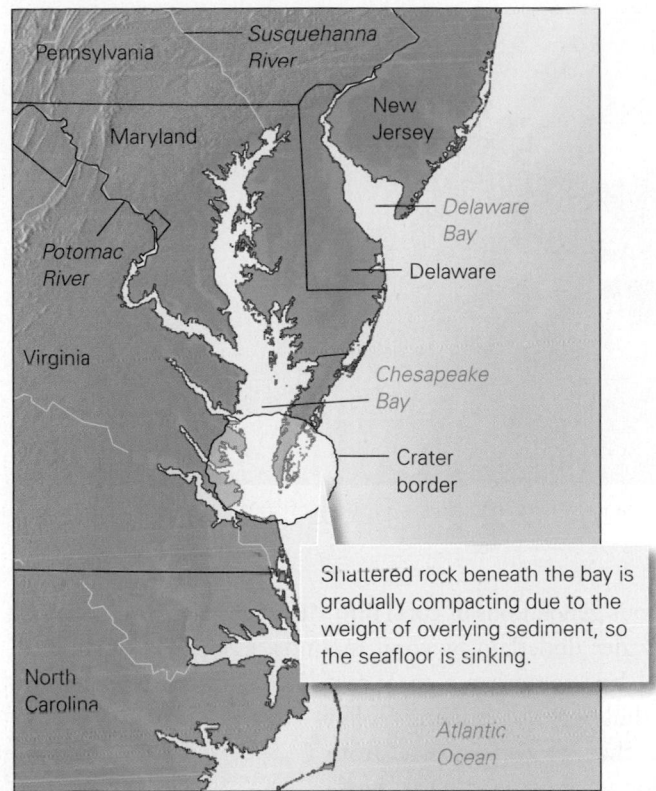

(a) Chesapeake Bay is an estuary at the mouth of the Susquehanna and Potomac Rivers. Recent research suggests that the mouth of these rivers has subsided because a meteorite crater lies underneath.

Shattered rock beneath the bay is gradually compacting due to the weight of overlying sediment, so the seafloor is sinking.

(b) The land surrounding the main channel of an estuary floods during high tide along the coast of the Isle of Wight, England.

flowed back into the sea, causing sea level to rise. The rising sea filled the deep valleys, yielding **fjords**, or flooded glacial valleys. Coastal fjords are fingers of the sea surrounded by mountains. Because of their deep blue water and steep walls of polished rock, they are distinctively beautiful **(Fig. 18.31)**. Some of the world's most spectacular fjords decorate the western coasts of Norway, British Columbia, and New Zealand. Smaller examples appear along the coast of Maine and southeastern Canada.

Organic Coasts

Coasts where living organisms control landforms along the shore are called **organic coasts**. The nature of an organic coast depends on the types of organisms that live there, which in turn depends on climate.

A **coastal wetland** is a vegetated, flat-lying stretch of coast that floods at high tide but does not feel the impact of strong waves. In temperate climates, coastal wetlands include *swamps* (wetlands dominated by trees), *marshes* (wetlands dominated

FIGURE 18.31 Fjord landscapes form where relative sea-level rise drowns glacially carved valleys.

Glacial valleys have a U shape, so fjords have steep sides.

Fjord

A fjord in Norway

FIGURE 18.32 Examples of coastal wetlands.

(a) A salt marsh along the coast of Cape Cod.

(b) A mangrove growing along the shore of southern Florida.

by grasses; **Fig. 18.32a**), and *bogs* (wetlands dominated by moss and shrubs). Many marine species spawn in wetlands, which despite their relatively small area when compared with the oceans as a whole, account for 10% to 30% of marine organic productivity.

In tropical or subtropical climates (between latitudes 30° N and 30° S), *mangrove swamps* thrive in wetlands **(Fig. 18.32b)**. Mangrove tree roots can filter salt out of water, so the trees have evolved to survive in either freshwater or saltwater. Some mangrove species form a broad network of roots above the water surface, making the plant look like an octopus standing on its tentacles, and some send up small protrusions from roots that rise above the water and allow the plant to breathe. Dense stands of mangroves counter the effects of stormy weather and thereby prevent coastal erosion.

Along the azure coasts of tropical regions, visitors may swim through colorful growths of living coral. Some corals look like brains, others like elk antlers, and still others like delicate fans **(Fig. 18.33a)**. Sea anemones, sponges, and clams grow on and around the coral. Though at first glance coral looks like a plant, it actually consists of a colony of tiny invertebrates. An individual coral animal, or polyp, has a tube-like body with a head of tentacles. Corals obtain part of their livelihood by filtering nutrients out of seawater; the remainder comes from algae that live on the corals' tissue. Corals have a symbiotic (mutually beneficial) relationship with the algae in that the algae photosynthesize and provide nutrients and oxygen to the corals, while the corals provide carbon dioxide and nutrients to the algae.

Coral polyps secrete calcite shells, which gradually build into a mound of solid limestone whose top surface may lie anywhere from just below the low-tide level down to a depth of about 60 m. At any given time, only the surface of the mound is alive—the mound's interior consists of shells from previous generations of coral **(Fig. 18.33b)**. The realm of shallow water underlain by coral mounds, associated organisms, and debris comprises a **coral reef**. Reefs absorb wave energy and thus serve as a living buffer zone that protects coasts from erosion. Corals need clear, well-lit, warm (18°C to 30°C) water with normal oceanic salinity, so coral reefs grow only along clean coasts at latitudes of less than about 30° **(Fig. 18.33c)**.

If an oceanic island rises above sea level in the warm waters of the tropics, it will be surrounded by a reef that extends almost right up to the beach. Geologists refer to such a reef as a *fringing reef* **(Fig. 18.33d)**. When Charles Darwin, later famous for his theory of evolution, was a young naturalist on HMS *Beagle*, he noticed that some islands of the tropics had fringing reefs, whereas others were surrounded by a *barrier reef* that lay offshore, separated from the island by a quiet lagoon. And in some locations, he would see a roughly circular coral reef, called an *atoll*, surrounding a lagoon, but without an island in the middle of the lagoon. Darwin correctly surmised that a fringing reef evolves into a barrier reef, which

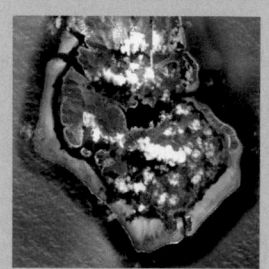

FIGURE 18.33 The character and evolution of coral reefs.

(a) The surface of this Hawaiian coral reef protects the shore from wave erosion. The underwater view emphasizes that a reef hosts a great variety of organisms. A close-up shows individual coral polyps.

(b) A close-up shows the internal skeleton of a long-dead coral.

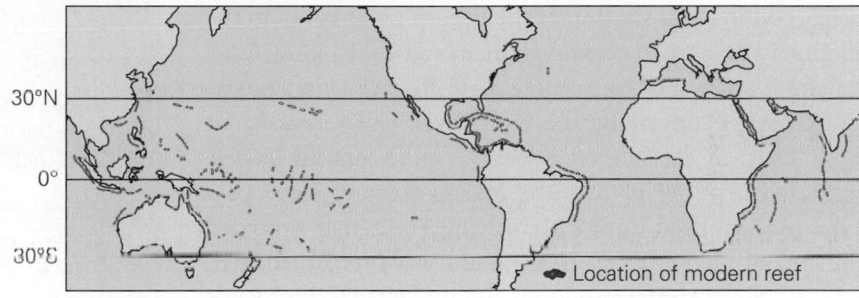

(c) The distribution of warm-water coral reefs on the Earth today.

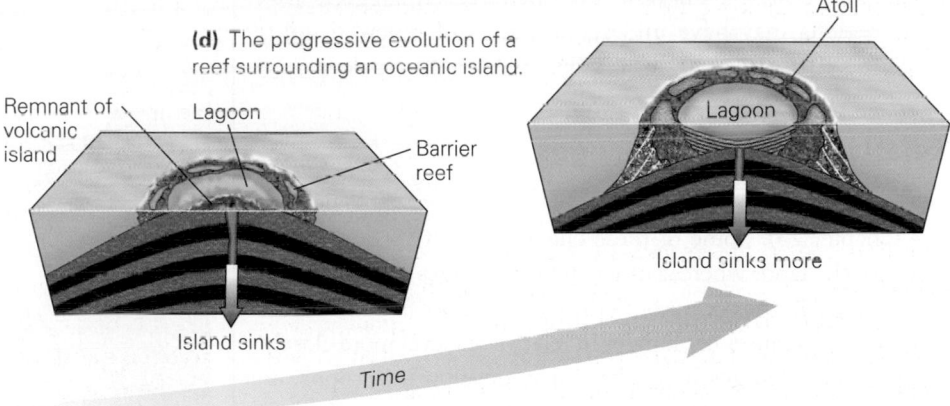

(d) The progressive evolution of a reef surrounding an oceanic island.

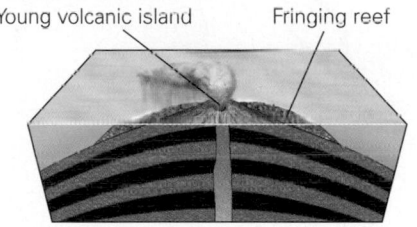

in turn evolves into an atoll as the island undergoes subsidence. As we've seen, this happens because the surface of the lithosphere sinks beneath the island as the lithosphere ages and because the island erodes and slumps away. Had Darwin been able to use sonar, invented more than a century after his historic voyage, he would have identified one more step in the evolution of a reef—eventually, even the reef itself sinks below sea level, for the rate of coral growth can't keep up with the rate of subsidence. When this happens, the resulting seamount has a flat top—geologists refer to a flat-topped seamount as a **guyot**.

TAKE-HOME MESSAGE

Beaches form from wave-washed and rounded sediment. Wave action may build bars, and longshore drift may produce sand spits. Rocky coasts evolve due to wave erosion. Submergence of coastal valleys produces fjords and estuaries. Some coasts host wetlands and reefs.

QUICK QUESTION: Does the sand of a beach stay in the same location for a long time? Why or why not?

18.8 Causes of Coastal Variability

Plate Tectonic Setting

Why does the character of the coast vary so much from place to place around the world? We can start to address this question from the perspective of plate tectonics theory. Specifically, the tectonic setting of a coast plays a role in determining whether the coast displays steep-sided mountain slopes or a broad lowland (**Geology at a Glance**, pp. 712–713). For example, along some active continental margins, compressive stresses have caused crustal shortening, squeezing the crust and pushing it up. Such stresses have contributed to the uplift of the Andes along the west coast of South America and the Coast Ranges of California. In contrast, along some passive continental margins, cooling and sinking of the lithosphere has led to the development of a broad **coastal plain**, a landscape of low relief that merges with the continental shelf. You can find examples of coastal plains along the Gulf Coast and southeastern Atlantic coast of the United States. Of note, not all passive margins have coastal plains. The coastal areas of some passive margins uplifted during the rifting event that preceded establishment of the passive margin and have remained high ever since. Such uplifts produced the present-day highlands bordering the Red Sea, southeastern Brazil, southern African coasts, and eastern Australia. Recent research suggests that these regions may have undergone pulses of renewed uplift long after rifting ceased, for reasons that remain unclear.

Relative Sea-Level Changes

Sea level, relative to the land surface, changes over geologic time (see Chapter 23). Some of these changes reflect vertical movements of the land, whereas others reflect changes in the amount of water held in the oceans or in the volume of ocean basins. Let's consider a few causes of changes in relative sea level more closely.

Regional changes in sea level may develop in response to the uplift or subsidence of the surface of the crust as a result of plate-tectonic processes. For example, compression across the coast of California has caused the coast to rise. They may also be due to flow in the asthenosphere, which can push the lithosphere up or pull it down from below. Addition or removal of large glaciers from the surface of a continent will cause the surface to sink or rise, respectively, in order to maintain isostasy (see Interlude D).

Global sea-level changes, resulting from global rise or fall of the ocean surface, may happen due to the advances and retreats of large glaciers during ice ages. Specifically, glaciers store water on land (**Fig. 18.34**), so as glaciers grow,

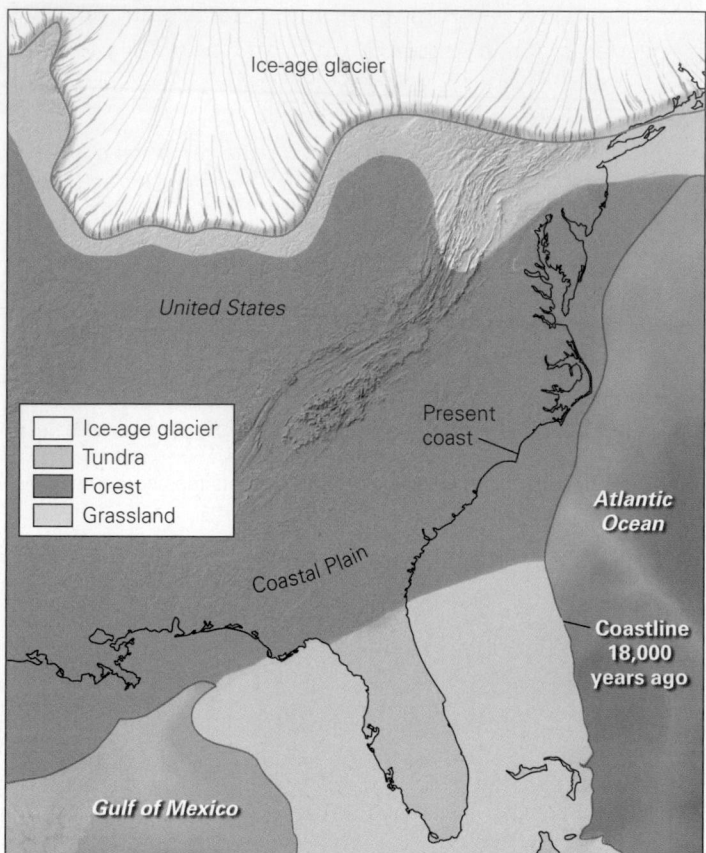

Ice-age glacier

United States

	Ice-age glacier
	Tundra
	Forest
	Grassland

Present coast

Atlantic Ocean

Coastal Plain

Coastline 18,000 years ago

Gulf of Mexico

(a) Exposed land along the coast of eastern North America.

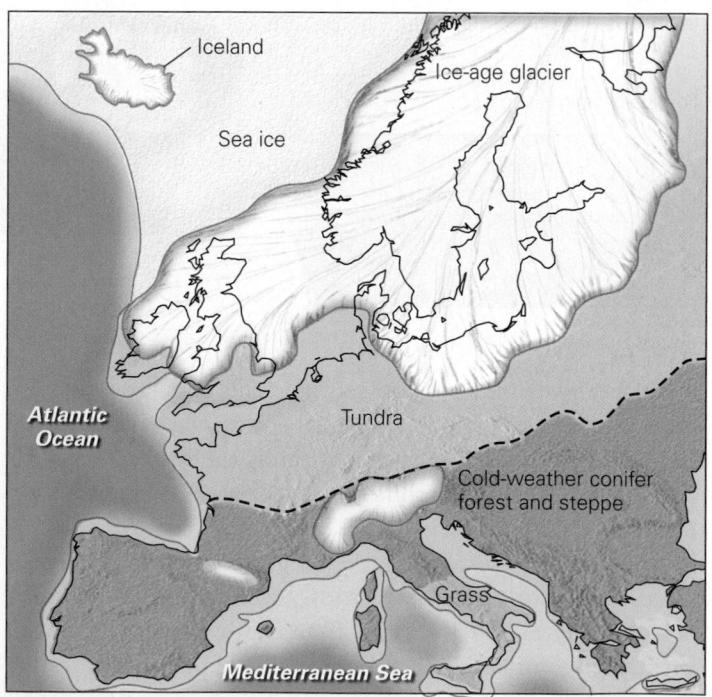

Iceland

Ice-age glacier

Sea ice

Atlantic Ocean

Tundra

Cold-weather conifer forest and steppe

Grass

Mediterranean Sea

(b) Exposed land along the coast of Europe.

sea level falls, and as the glaciers shrink, sea level rises. Sea-level changes may also reflect changes in the volume of mid-ocean ridges—an increase in the number or width of ridges, for example, displaces water and causes sea level to rise. Geologists use the term *eustatic sea-level change* for global sea-level fall or rise.

Sea-level changes, whether due to regional rise or fall of the land or to eustatic sea-level change, affect landforms along the coast. Geologists refer to coasts where the land is rising or has risen relative to sea level as **emergent coasts**. At emergent coasts, steep slopes typically border the shore. In some cases, a series of step-like terraces forms along emergent coasts **(Fig. 18.35a)**, reflecting pulses of uplift between times of erosion. Coasts at which the land sinks relative to sea level become **submergent coasts (Fig. 18.35b)**. Landforms of submergent coasts include estuaries and fjords that developed when the sea flooded coastal valleys. Many of the

coastal landforms of eastern North America are the consequences of submergence.

Sediment Supply and Climate

The amount of sediment supplied to a coast also affects its character. At *erosional coasts*, waves wash sediment away faster than it can be supplied. These coastlines recede landward and may become rocky. Coasts that receive more sediment than the sea erodes away, so-called *accretionary coasts*, grow seaward and develop broad beaches or tidal flats.

Regardless of whether we classify a coast as emergent or submergent, or as erosional or accretionary, what it looks like ultimately reflects climate. Shores that enjoy generally calm weather erode less rapidly than those constantly subjected to ravaging storms, so a sediment supply large enough to generate an accretionary coast in a calm environment may be insufficient to

FIGURE 18.35 Features of emergent coasts (where relative sea level is falling) and submergent coasts (where relative sea level is rising).

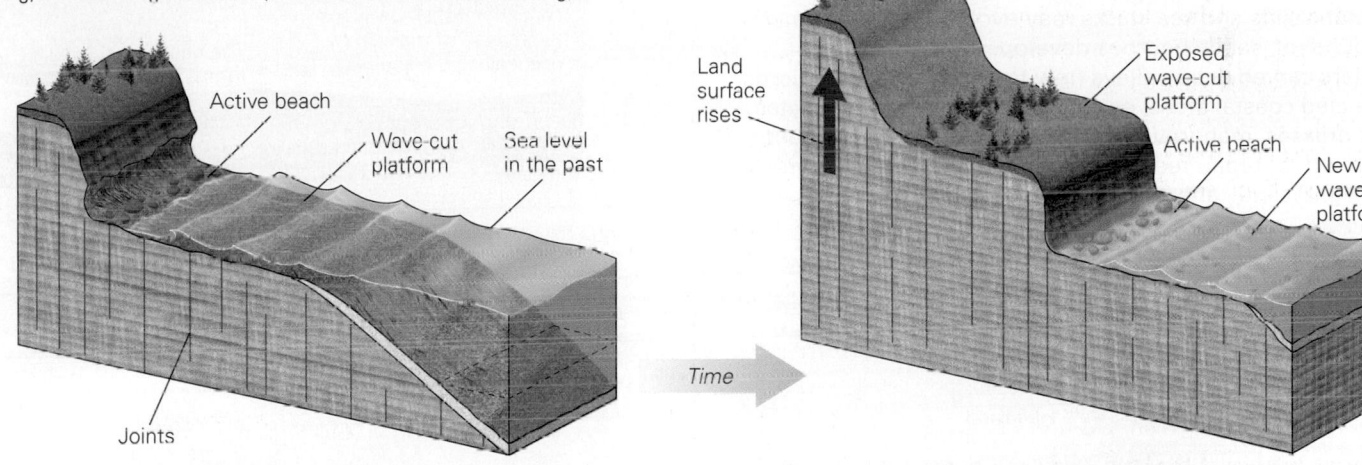

(a) *Emergent coasts*: Wave erosion produces a wave-cut platform along an emergent coast. As the land rises or relative sea level falls, the platform becomes a terrace, and a new wave-cut platform forms.

(b) *Submergent coasts*: Rivers drain valleys and deposit sediment on a coastal plain. As relative sea level rises, the valleys are flooded and waves erode the headlands.

Oceans and Coasts

The oceans of the world host a diverse array of environments and landscapes, illustrating the complexity of the Earth System. Tectonic processes and surface processes, working alone or in tandem, generate unique features beneath the sea and along its coasts.

The major structures of the ocean floor are the result of plate-tectonic activity. For example, mid-ocean ridges define divergent boundaries, fracture zones form along transform faults, and trenches mark subduction zones. Oceanic islands, seamounts, and oceanic plateaus build above hot spots. Along passive continental margins, broad continental shelves develop, locally incised by submarine canyons.

Within the ocean, currents circulate water due to friction between wind and the water surface as well as regional variations in water salinity and temperature. Tides cause sea level to rise and fall, and wind builds waves that churn the sea surface, erode shorelines, and transport sediment.

Coastal landscapes reflect variations in sediment supply, relative sea-level rise or fall, and climate. For example, where sediment is in short supply and the landscape is rising relative to sea level, rocky shores with dramatic cliffs and sea stacks may evolve. Where sediment is abundant, sandy beaches develop. Regions where glaciers carved deep valleys now feature spectacular fjords. Protected coastal areas, especially those in warm climates, host grasses, mangroves, or corals. Corals may contribute to growth of broad reefs along the shore. Human activities can significantly affect coastal landscapes.

Temperate and Tropical Coastal Landforms

Along sandy shores, sand builds beaches, sand spits, and bars.

Turbidity currents carve submarine canyons and produce submarine fans.

In tropical environments, mangroves live along the shore and coral reefs grow offshore.

Along rocky coasts, sea cliffs, sea arches, and sea stacks evolve.

Offshore, reefs grow

At a passive margin, a broad continental shelf develops. Submarine slumping may occur along the shelf.

The ocean teems with life.

Rocky Coast Evolution

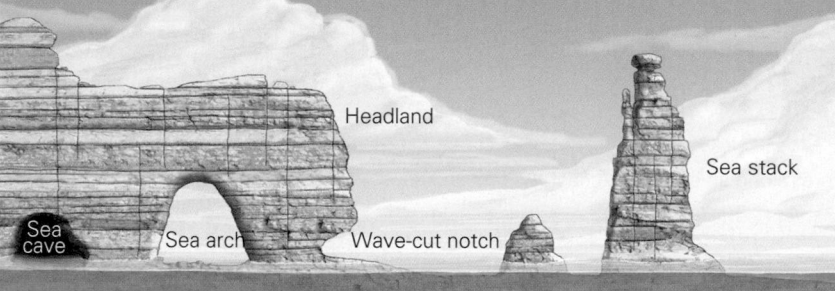

Sea cave

Sea arch

Wave-cut notch

Headland

Sea stack

As waves erode rocky coasts, bedrock breaks away at joints and along bedding planes. As a result, several distinct landforms evolve over time.

Water Masses

Water of the ocean is stratified, due to thermohaline circulation. The surface layer flows in ocean-spanning currents, and locally in eddies.

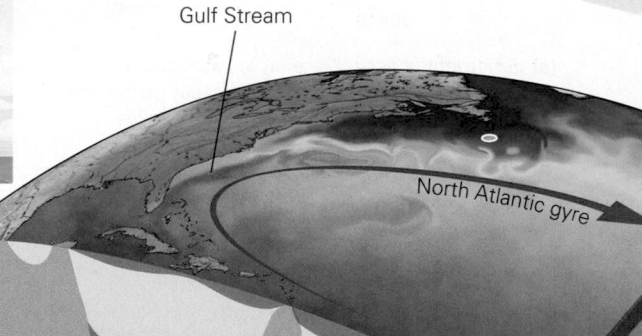

Gulf Stream

North Atlantic gyre

Surface water

Intermediate water

North Atlantic deep water

Antarctic bottom water

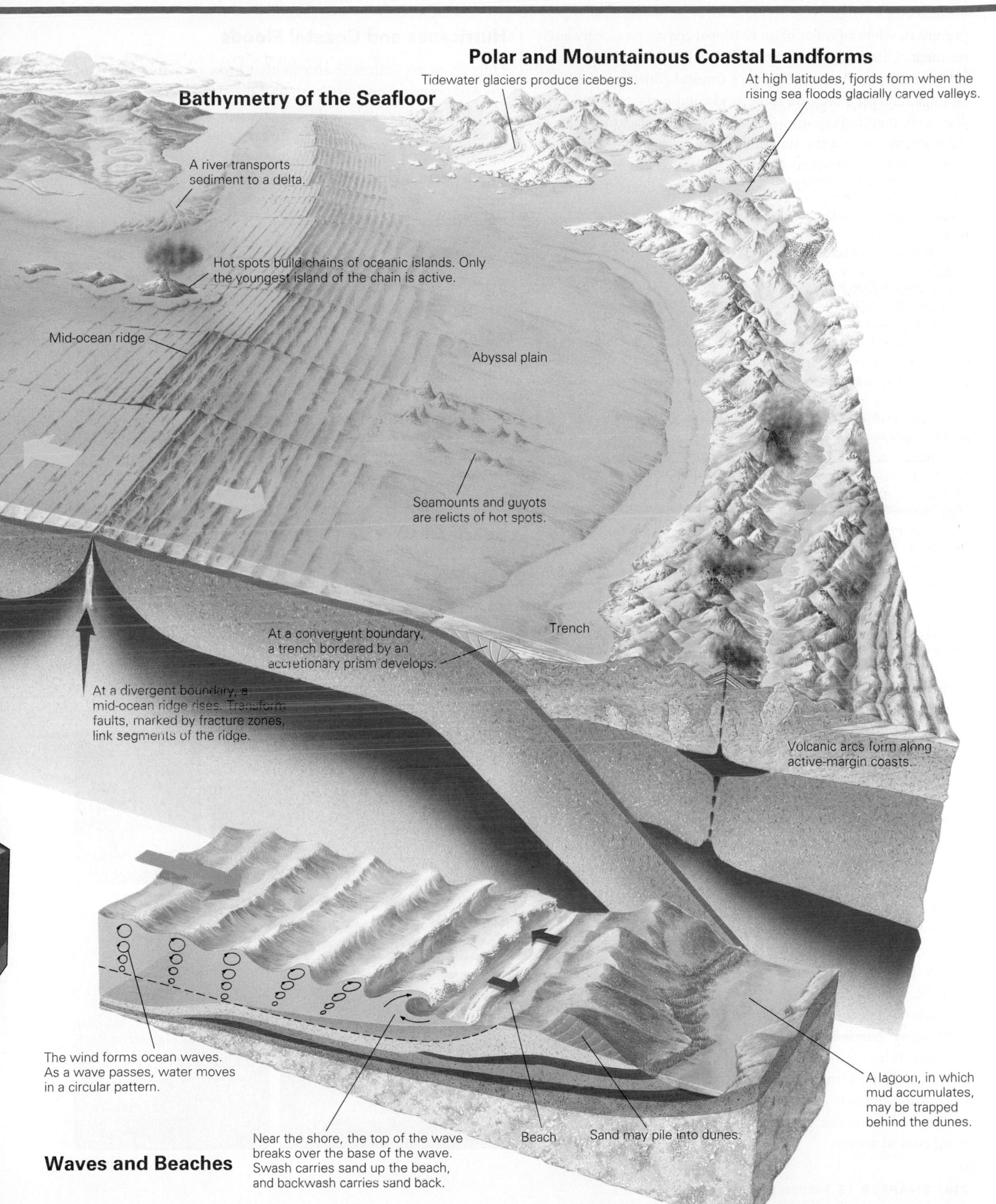

Bathymetry of the Seafloor

Polar and Mountainous Coastal Landforms

Tidewater glaciers produce icebergs.

At high latitudes, fjords form when the rising sea floods glacially carved valleys.

A river transports sediment to a delta.

Hot spots build chains of oceanic islands. Only the youngest island of the chain is active.

Mid-ocean ridge

Abyssal plain

Seamounts and guyots are relicts of hot spots.

Trench

At a convergent boundary, a trench bordered by an accretionary prism develops.

At a divergent boundary, a mid-ocean ridge rises. Transform faults, marked by fracture zones, link segments of the ridge.

Volcanic arcs form along active-margin coasts.

The wind forms ocean waves. As a wave passes, water moves in a circular pattern.

A lagoon, in which mud accumulates, may be trapped behind the dunes.

Near the shore, the top of the wave breaks over the base of the wave. Swash carries sand up the beach, and backwash carries sand back.

Beach

Sand may pile into dunes.

Waves and Beaches

prevent the development of an erosional coast in a stormy environment. Climate also affects biological activity along coasts. For example, in the warm water of tropical climates, mangrove swamps flourish along the shore, and coral reefs grow offshore. The reefs may build into a broad carbonate platform such as the one surrounding the Bahamas today. In cooler climates, salt marshes develop, whereas in arctic regions, the coast may be a stark environment of lichen-covered rock and barren sediment.

TAKE-HOME MESSAGE

Tectonic activity influences the character of a coastline. At an emergent coast, land is rising relative to sea level, whereas at a submergent coast, it is sinking. Changing mid-ocean ridge and glacial volumes affect global sea level. Factors such as sediment supply and climate determine whether a coast is a site of deposition or a site of erosion.

QUICK QUESTION: What landforms occur at submergent coasts?

18.9 Coastal Problems and Solutions

Contemporary Sea-Level Changes

People tend to view a shoreline as a rather permanent entity, whose position remains fixed over time. But in fact, shorelines are ephemeral geologic features, in that on a time scale of hundreds to thousands of years, a shoreline can move kilometers inland or seaward depending on whether relative sea level rises or falls or whether sediment supply increases or decreases. In places where relative sea level is rising today, shoreline towns may eventually be submerged. In fact, since global sea level rises at a rate of about 3.3 mm per year, it will likely rise by about 1 m over the next few centuries, and will have a significant effect on coastal cities (Fig. 18.36). About 15% of the islands in the Pacific Ocean may be submerged. Already, some vulnerable countries, such as the Netherlands, have been reinforcing barriers to hold out the sea, and nuisance tides are becoming frequent in parts of coastal Florida. Other regions may soon have to begin such projects or face the potential for loss of life and immense property damage in low-lying coastal areas.

Hurricanes and Coastal Floods

Hurricanes are immense storms that grow over the waters of tropical oceans. Some are born and die at sea, but some move over the coast. Winds in hurricanes can exceed 250 km per hour and can generate waves in excess of 15 m high. They can also push a bulge of water landward. Further, because air rises beneath a hurricane, atmospheric pressure (caused by the weight of the overlying atmosphere) decreases in the region beneath a hurricane, and without the downward push of the air, the sea surface rises still further. As a consequence, locally-high sea-surface elevation reaches the shore, producing a **storm surge** that inundates low areas along the shore. High waves add to the destruction by battering both the shore and offshore reefs. In regions where a low-lying delta plain borders the coast, a storm surge coupled with river flooding can submerge broad areas for days or more. Such catastrophic flooding has taken a dreadful toll on the Ganges Delta in Bangladesh and on the Gulf Coast of the United States. Flooding during Hurricane Sandy in 2012 caused tragic loss of life and tens of billions of dollars in damage in the northeastern United States. We discuss hurricanes and their consequences more fully in Chapter 20.

Beach Destruction—Beach Protection?

In a matter of hours, a storm—especially a hurricane—can radically alter a landscape that took centuries or millennia to form. The backwash of storm waves sweeps vast quantities of sand seaward, leaving the beach a skeleton of its former self.

FIGURE 18.36 U.S. coastal areas that may be submerged if global sea level rises by about 1.0 to 1.5 m.

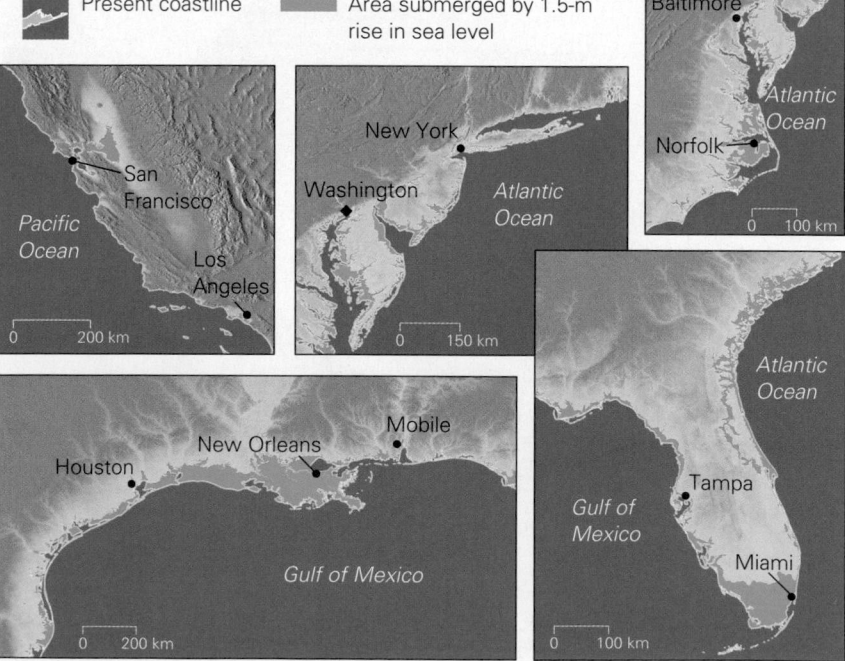

The surf submerges barrier islands and shifts them toward the lagoon. Waves and wind together rip out mangrove swamps and salt marshes and fragment coral reefs, thereby destroying the organic buffer that normally protects the coast and leaving it vulnerable to erosion for years to come. Major storms also destroy human constructions: erosion undermines shorefront buildings, causing them to collapse into the sea; wave impacts smash buildings to bits; and storm surge can float buildings off their foundations **(Fig. 18.37a)**.

But even less dramatic events, such as the loss of river sediment, a gradual rise in sea level, a change in the shape of a shoreline, or the destruction of coastal vegetation, can alter the balance between sediment accumulation and sediment removal on a beach, leading to **beach erosion (Fig. 18.37b)**. In some places, beaches retreat landward at rates of 1 to 2 m per year, forcing homeowners to pick up and move their houses. Even large lighthouses have been moved to keep them from washing away or tumbling down eroded headlands.

In many parts of the world, beachfront property has great value. But if a hotel loses its beach sand, it probably won't stay in business. As a consequence, property owners often construct artificial barriers to protect their stretch of coastline or to shelter the mouth of a harbor from waves. These barriers alter the natural movement of sand in the beach system and change the shape of the beach, sometimes with undesirable results. For example, people may build *groins*, concrete or stone walls protruding perpendicular to the shore, to prevent longshore drift from removing sand **(Fig. 18.38a)**. Sand accumulates on the updrift side of the groin, forming a long triangular wedge, but sand erodes away on the downdrift side. Needless to say, the property owner on the downdrift side doesn't appreciate this process. Harbor engineers may build a pair of walls called *jetties* to protect the entrance to a harbor **(Fig. 18.38b)**. But jetties erected at the mouth of a river channel effectively extend the river into deeper water and thus may lead to the deposition of an offshore sandbar. Engineers may also build an offshore wall called a *breakwater*, parallel or at an angle to the beach, to prevent the full force of waves from reaching a harbor. With time, however, sand builds up in the lee of the breakwater and the beach grows seaward, clogging the harbor **(Fig. 18.38c)**.

To protect expensive shorefront homes, people build *seawalls* out of riprap (large stone or concrete blocks) or reinforced

FIGURE 18.37 Examples of beach erosion.

(a) When Hurricane Ike hit Galveston, Texas, in September 2008, many beachfront homes were washed away.

(b) Wave erosion has completely removed the beach and has started to erode a beach cliff along the coast of Cape Cod.

FIGURE 18.38 Techniques used to preserve beaches.

Before | After

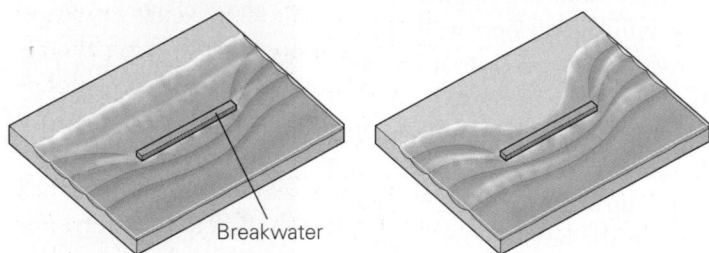

Groin

Longshore drift

(a) The construction of groins may produce a sawtooth beach.

Groins spaced along a beach in southern England are intended to prevent loss of sediment.

Jetty | Sandbar

(b) Jetties extend a river farther into the sea, but may cause a sandbar to form at the end of the channel.

Breakwater

(c) A beach may grow seaward behind a breakwater.

(d) Riprap will slow erosion of a parking lot along a California beach.

concrete **(Fig. 18.38d)**. But seawalls reflect wave energy back across the beach to the sea, so this process increases the rate of erosion at the foot of the seawall. During a large storm, the seawall may be undermined so that it collapses **(Fig. 18.39)**.

In some places, people have given up trying to decrease the rate of beach erosion and instead have worked to increase the rate of sediment supply. To do this, they pump sand from farther offshore or truck in sand from elsewhere to replenish a beach **(Fig. 18.40)**. This procedure, called *beach nourishment*, can be hugely expensive and at best provides only a temporary fix, for the backwash and longshore drift that removed the sand in the first place continue unabated as long as the wind blows and the waves break. Clearly, beach management remains a controversial issue, for beachfront properties are expensive, but protecting them can be even more expensive.

Coastal Pollution

Bad cases of beach pollution create headlines. Because of longshore drift, garbage dumped into the sea in an urban area may move along the shore and be deposited on a tourist beach far from its point of introduction. For example, hospital waste from New York City has washed up on beaches tens of kilometers to the south. Oil spills from ships that flush their bilges, tankers that have run aground or foundered in stormy seas, or blowouts on offshore oil rigs have contaminated shorelines at several places around the world.

The influx of nutrients from sewage and agricultural runoff into coastal waters can create dead zones along coasts. A *dead zone* is a region in which water contains so little oxygen that fish and other organisms within it die. Dead zones form when the concentration of nutrients rises enough to stimulate an algal bloom, for overnight respiration by algae depletes dissolved oxygen in the water, and the eventual death and decay of the algae depletes oxygen even more. Dead zones commonly develop at the mouths of rivers that drain agricultural areas. For example, drainage from the Mississippi River produced a large dead zone.

Wetland and Reef Destruction

Coastal wetlands and coral reefs are particularly susceptible to changes in the environment, and many of them have been

FIGURE 18.39 A seawall protects the cliff under most conditions, but during a severe storm, the wave energy reflected by the seawall helps scour the beach. As a result, the wall may be undermined and collapse.

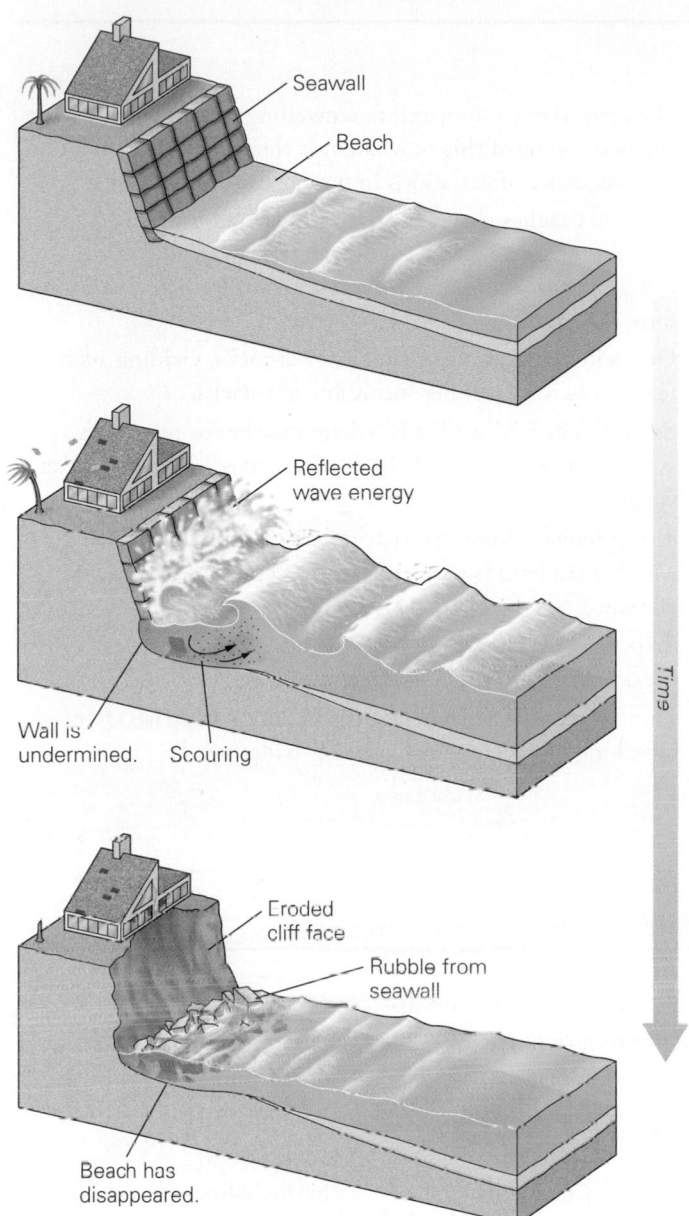

FIGURE 18.40 Beach nourishment involves bringing in vast amounts of sediment by truck, by pipe from offshore, or by barge.

quarrying operations intended to obtain construction materials. Chemicals and particulates entering coastal water from urban, industrial, and agricultural areas can cause havoc in wetlands and reefs, for these materials cloud water or trigger algal blooms, killing filter-feeding organisms. Toxic chemicals in runoff can also poison plankton and burrowing organisms and, therefore, other organisms progressively up the food chain.

Global climate change, which will be more fully described in Chapter 23, also impacts the health of organic coasts. For example, transformation of once-vegetated regions into deserts means that the amount of dust carried by winds from the land to the sea has increased. This dust can interfere with coral respiration and can bring dangerous viruses to reefs. A global increase in seawater temperature may be contributing to *reef bleaching*, the loss of coral color due to the death of the algae that live in coral polyps. Reef bleaching and other reef diseases have transformed once-living landscapes hosting diverse species of colorful coral into mounds and piles of whitish rubble.

destroyed in recent decades. Their loss increases a coast's vulnerability to erosion and, because these areas provide spawning grounds for marine organisms, their loss disrupts the global food chain of the ocean. The statistics of wetland and reef destruction worldwide are frightening—ecologists estimate that between 20% and 70% of wetlands have already been destroyed, and along some coasts 90% of reefs have died.

Destruction of wetlands and reefs happens for many reasons. Wetlands have been filled or drained to serve as farmland, housing developments, resorts, or garbage dumps. Reefs have been destroyed by boat anchors, dredging, the activities of divers, dynamite explosions intended to kill fish, and

TAKE-HOME MESSAGE

In a geologic context, the position and character of a coastal landscape is temporary. Sea-level rise and storms can flood coastal areas, and beaches and bars erode—especially during hurricanes—despite intense efforts to build protective structures or replenish sand. Pollution and a rise in ocean water temperature can destroy wetlands and coral reefs.

QUICK QUESTION: What problems can be caused by construction of groins or seawalls?

SUMMARY

- The landscape of the seafloor depends on the character of the underlying crust. Particularly wide continental shelves form over passive-margin basins. Continental shelves may be cut by submarine canyons. Abyssal plains develop on old, cool oceanic lithosphere. Seamounts and oceanic plateaus form above hot spots.

- Active continental margins host plate boundaries. Passive continental margins do not.

- The salinity, temperature, and density of seawater vary with location and depth.

- Tides—the daily rise and fall of sea level—are caused by a tide-generating force. The largest contribution to this force comes from the gravitational pull of the Moon.

- Friction where the wind shears across the surface of the ocean builds waves. Water particles approximately follow a circular motion in a vertical plane as a wave passes. Waves refract when they approach the shore because of friction with the seafloor.

- Water in the oceans circulates in currents. Wind drives surface currents. Because of the Coriolis force and associated Ekman transport, currents don't flow parallel to the wind. Large gyres circulate water around each ocean basin.

- The vertical upwelling and downwelling of water create deep currents. Some of this movement is thermohaline circulation, a consequence of variations in temperature and salinity.

- Sand on beaches moves with the swash and backwash of waves. During longshore drift, the sand gradually moves along the beach and may extend outward from headlands to form sand spits.

- On rocky coasts, waves grind away at rocks, yielding such features as wave-cut platforms and sea stacks.

- Some shores host wetlands, where marshes or mangrove swamps grow. Coral reefs build along coasts in warm, clear water.

- The differences among coasts reflect their tectonic setting, whether sea level is rising or falling, sediment supply, and climate.

- To protect beach property, people build groins, jetties, breakwaters, and seawalls. In some places, they add sand.

- Human activities have led to the pollution of coasts. Reef bleaching has become dangerously widespread.

GUIDE TERMS

abyssal plain (pp. 681, 683)

active continental margin (p. 682)

backwash (p. 701)

bathymetry (p. 679)

beach (p. 700)

beach erosion (p. 715)

beach profile (p. 700)

bioturbation (p. 704)

coast (p. 679)

coastal plain (p. 710)

coastal wetland (p. 707)

continental shelf (p. 681)

coral reef (p. 708)

Coriolis effect (p. 695)

Coriolis force (p. 697)

current (p. 693)

eddy (p. 693)

Ekman transport (p. 695)

emergent coast (p. 711)

estuary (p. 705)

fjord (p. 707)

guyot (p. 709)

gyre (p. 693)

longshore current (p. 693)

longshore drift (p. 701)

oceanic plateau (p. 683)

organic coast (p. 707)

passive continental margin (p. 681)

pelagic sediment (p. 682)

rogue wave (p. 692)

salinity (p. 684)

sand spit (p. 702)

seamount (p. 683)

storm surge (p. 714)

submarine canyon (p. 681)

submergent coast (p. 711)

swash (p. 701)

swell (p. 692)

thermohaline circulation (p. 699)

tidal range (p. 686)

tide (p. 686)

tide-generating force (p. 686)

wave base (p. 692)

wave refraction (p. 692)

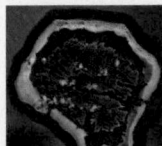

GEOTOURS *THIS CHAPTER'S GEOTOURS WORKSHEET (O) FEATURES QUESTIONS AND GOOGLE EARTH SITES ON:*

- Seafloor bathymetry
- Coral reefs
- Barrier islands and spits
- Beach preservation
- Sea-level rise

REVIEW QUESTIONS

The letters following each Review Question refer to the corresponding Learning Objective from the Chapter Opener.

1. How much of the Earth's surface do oceans cover? **(F)**

2. How does the lithosphere beneath a passive margin differ from that beneath an abyssal plain? **(B)**

3. How do the continental shelf and slope of an active continental margin differ from those of a passive margin? **(B)**

4. Where does the salt in the ocean come from? How do the salinity and temperature in the ocean vary? **(A)**

5. What causes the tides? Why does tidal range vary? **(D)**

6. Describe the motion of water molecules in a wave. How does wave refraction cause longshore currents? **(E)**

7. Describe the components of a beach profile. Identify the backshore zone, foreshore zone, and wave base in the profile. **(F)**

8. What factors control the direction of surface currents in the ocean? What is the Coriolis effect, and how does it affect oceanic circulation? Explain thermohaline circulation. **(C)**

9. How does beach sand migrate as a result of longshore drift? Explain the sediment budget of a coast. **(F)**

10. Describe how waves affect a rocky coast and how such coasts evolve. **(F)**

11. What is the difference between an estuary and a fjord? Which is pictured here? **(F)**

12. Discuss the different types of coastal wetlands. How does a reef surrounding an oceanic island change with time? **(F)**

13. How do plate tectonics, sea-level changes, sediment supply, and climate change affect the shape of a coastline? Explain the difference between emergent and submergent coasts, and between an erosional and depositional coast? **(G)**

14. In what ways do people try to modify or "stabilize" coasts? How do the actions of people threaten coastal areas? **(G)**

ON FURTHER THOUGHT

15. In 1789, the crew of HMS *Bounty* mutinied. Near Tonga, the crew, led by Fletcher Christian, forced the ship's commanding officer, Lieutenant Bligh, along with those crewmen who remained loyal to Bligh, into a rowboat and set them adrift in the Pacific Ocean. The castaways survived, and 47 days later, they landed at Timor, 6,700 km to the west. Why did they end up where they did? How fast were they moving? **(E)**

16. A hotel chain would like to build a new beachfront hotel along a north-south-trending stretch of beach where a strong longshore current flows from south to north. The neighbor to the south has constructed an east-west-trending groin on the property line. Will this groin pose a problem? If so, what solutions could the hotel try? **(G)**

ONLINE RESOURCES

Animations
This chapter features animations on living on the coasts and ocean wave motion and refraction.

Videos
This chapter includes videos on the science of currents, harnessing wave energy to power coastal communities, how sea-level rise impacts coasts, and the seafloor of central California.

Smartwork5
This chapter features questions on ocean composition and zones, currents and waves, coastal profiles, and sea-level rise.

CHAPTER 19

A Hidden Reserve: Groundwater

By the end of this chapter, you should be able to . . .

A. explain what groundwater is, where it resides in the Earth, and how its composition and flow varies.

B. discuss the difference between porosity and permeability and how these features form.

C. describe the difference between aquifers and aquitards and the nature of the water table.

D. differentiate among the various types of wells and springs that provide access to groundwater.

E. create a sketch showing how hot springs and geysers originate.

F. explain how groundwater supplies can be damaged or depleted and how to address these problems.

G. describe how caves and karst landscapes originate and evolve.

When the rain falls and enters the earth, when a pearl drops into the depth of the sea, you can dive in the sea and find the pearl, you can dig in the earth and find the water.

—Mei Yao-ch'en (Chinese poet, 1002–1060)

19.1 Introduction

Imagine Mae Rose Owens's surprise when, on May 8, 1981, she looked out her window and discovered that a large sycamore tree in the backyard of her Winter Park, Florida, home had suddenly disappeared. It wasn't a particularly windy day, so the tree hadn't blown over—it had just vanished! When Owens went outside to investigate, she found that more than the tree had disappeared. Her whole backyard had become a deep, gaping pit. The pit continued to grow for a few days until finally it swallowed Owens's house and six other buildings, as well as the municipal swimming pool, part of a road, and several expensive Porsches in a car dealer's lot **(Fig. 19.1a)**.

FIGURE 19.1 Development of sinkholes in central Florida.

(a) The Winter Park sinkhole, as seen from a helicopter.

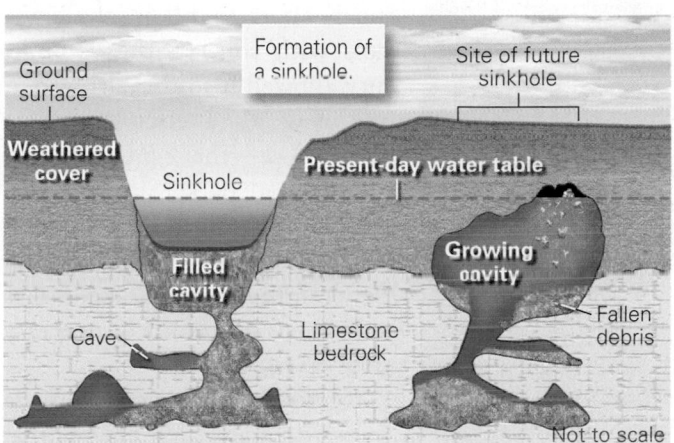

(b) As overburden slowly washes into underlying caves, a cavity forms. When the roof of this cavity collapses, a sinkhole forms.

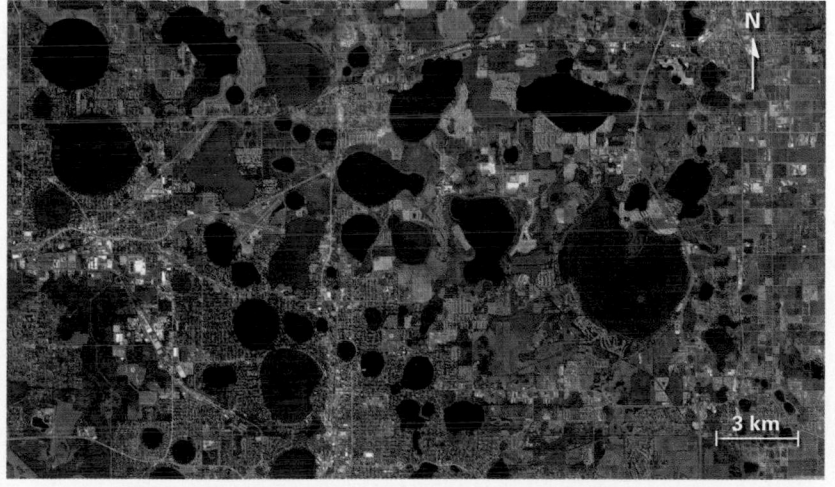

(c) An aerial view of Florida sinkholes that have become lakes.

(d) A new apartment building collapsed when a sinkhole formed under it.

◀ (facing page) This large cave in Vietnam (note the people, for scale) was hollowed out by flowing groundwater long ago. It now lies well above the water table and has filled with air. Calcite has precipitated from water dripping into the cave to form sculpture-like speleothems.

What had happened in Winter Park? The bedrock beneath the town consists of limestone. **Groundwater**, the liquid water that resides in sediment or rock under the surface of the Earth, had gradually dissolved the limestone over time, carving open rooms, or *caverns*, underground. On May 8, the roof of a cavern underneath Owens's backyard began to collapse, forming a circular depression called a **sinkhole (Fig. 19.1b)**. The sycamore tree and the rest of the neighborhood simply dropped down into the sinkhole. It would have taken too much effort to fill in the sinkhole with soil, so the community allowed it to fill with water, and now it's a circular lake—Lake Rose—the centerpiece of a municipal park. Similar lakes spot the landscape throughout central Florida **(Fig. 19.1c)**, and they continue to form, sometimes with tragic loss of life or property **(Fig. 19.1d)**.

Sinkholes serve as one of the more dramatic reminders that significant quantities of water reside underground. Though we can easily see the Earth's *surface water* (in lakes, rivers, streams, marshes, glaciers, and oceans) and some of its *atmospheric water* (in clouds and rain), *groundwater* lies hidden beneath the surface. This hidden reserve contains about 23,400,000 km³ of water—about 123 times as much water as all the visible lakes, rivers, and swamps combined. Clearly, groundwater serves as the largest reservoir of liquid freshwater in the hydrologic cycle. So it's not surprising that, globally, groundwater accounts for about two-thirds of the Earth's freshwater resources used by agriculture, industry, and homes.

In this chapter, we examine the nature of groundwater, discuss how it enters the subsurface, where it resides in the subsurface, and how it slowly moves underground until it returns to the surface through springs or wells. We also examine how human activities affect subsurface water. The chapter concludes with a brief survey of landscape features, such as hot springs and caves, that form as a consequence of the interaction between groundwater and its surroundings.

19.2 Where Does Groundwater Reside?

As we saw in Interlude F, water moves among various reservoirs (the ocean, the atmosphere, rivers and lakes, groundwater, living organisms, soil, and glaciers) during the hydrologic cycle. Of the water that falls on land, some evaporates directly back into the atmosphere, some gets trapped in glacial ice, and some becomes runoff that enters a network of streams and lakes that drains to the sea. The remainder sinks or percolates downward, by a process called *infiltration*, into the ground. In effect, the upper part of the crust behaves like a giant sponge that can soak up water. Infiltration can take place because most Earth materials (sediment and rocks) are not perfectly solid, but rather contain some interconnected open space. Let's examine such openings and their connections in more detail.

Porosity: The Open Space in Rock and Sediment

The existence of liquid water underground requires the existence of open space underground. What does this space look like? When asked this question, many people picture networks of large caves containing spooky lakes and inky streams. Indeed, as we see later in this chapter, caves do provide room for subsurface water to drip and flow, but only locally. Contrary to popular belief, only a small proportion of underground water actually occurs in caves. Most groundwater resides in tiny open spaces or voids within sediment and within seemingly solid rock. As we first noted in Chapter 14, geologists use the term **pore** for any open space within a volume of sediment or a body of rock, and they use the term **porosity** for the total amount of open space within a material, specified as a percentage. For example, if we say that a block of rock has 30% porosity, then 30% of the block consists of pores. What makes up pores? We distinguish between two basic kinds of porosity: primary and secondary.

Primary porosity develops during sediment deposition and during rock formation **(Fig. 19.2a–c)**. In clastic sedimentary rocks, it includes the gaps between grains. These gaps exist because the grains don't fit together tightly during deposition. Typically, the overall porosity of a poorly sorted sediment tends to be less than that of a well-sorted sediment because smaller clasts can fill spaces between larger grains. The primary porosity of a clastic sedimentary rock also depends on the amount of compaction and cementation, for these phenomena decrease porosity. Therefore, primary porosity tends to decrease with increasing burial depth. In chemical

FIGURE 19.2 Where water resides underground. Groundwater can fill microscopic pores (holes) and cracks between grains, or it can fill gaps formed along joints.

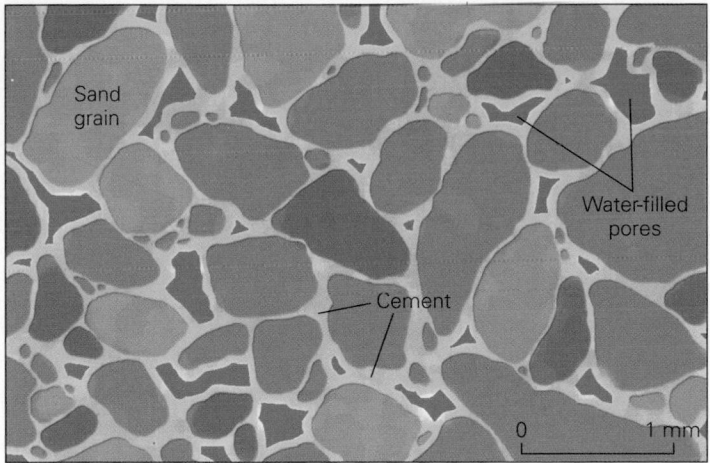

(a) Isolated pores in a sandstone occur in the spaces between grains. Water or air can fill pores.

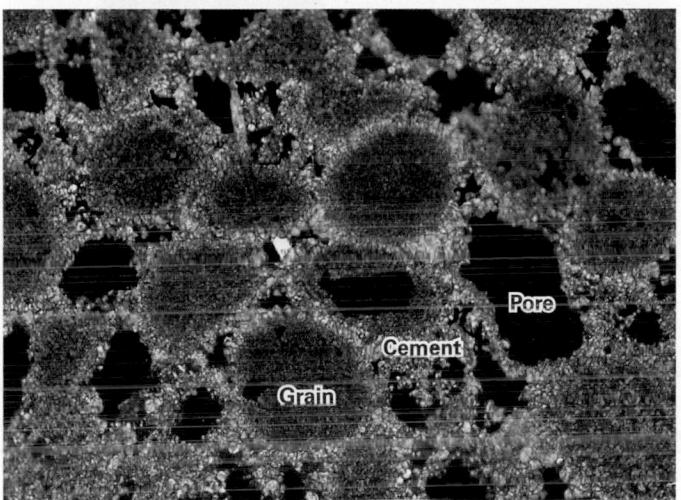

(b) This photo of a sedimentary rock, as seen through a microscope, shows grains, cement, and pores. The field of view is about 3 mm.

These fractures have been enlarged by dissolution.

(d) Limestone outcrop on the coast of Ireland contains abundant fractures that provide secondary porosity.

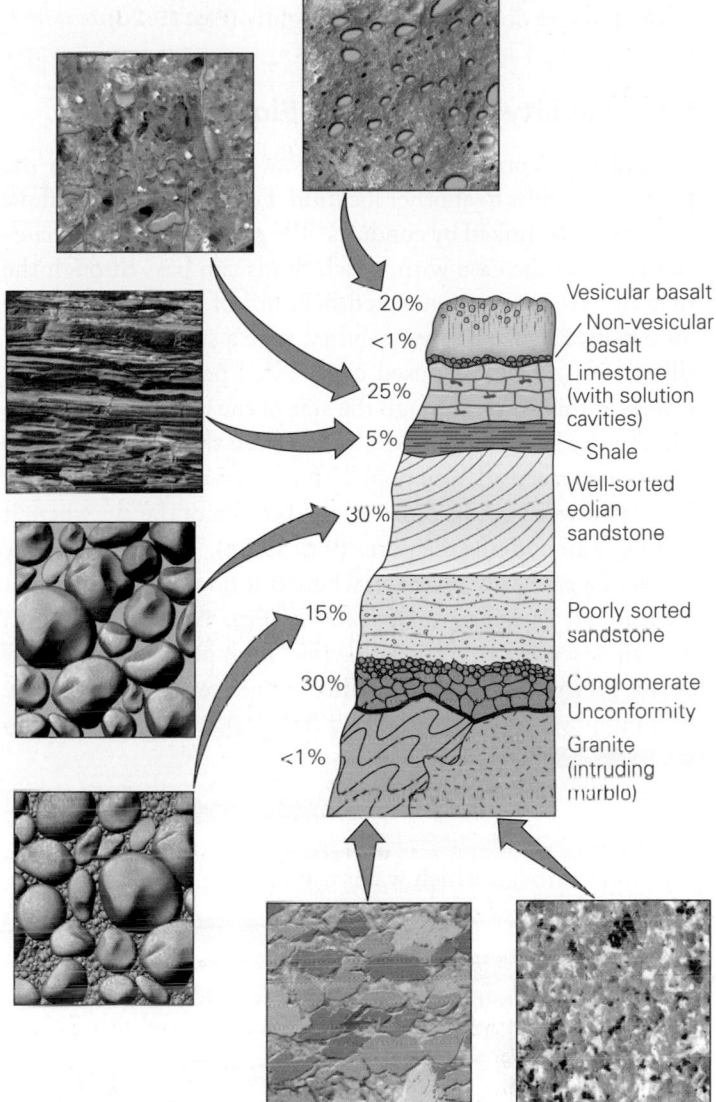

(c) Different kinds and amounts of porosity occur in different kinds of rocks. Well-sorted and poorly cemented sandstone contains high porosity, whereas granite contains low porosity.

sedimentary rocks or biochemical sedimentary rocks, primary porosity develops because mineral crystals do not all grow snugly against their neighbors during precipitation. In crystalline igneous or metamorphic rocks, primary porosity can persist if grains do not interlock perfectly. And in fine-grained or glassy igneous rocks, primary porosity may consist of vesicles, relics of air bubbles that were trapped during cooling. Primary porosity ranges from less than 1% in crystalline igneous rocks and metamorphic rocks to about 30% in well-sorted sands and poorly cemented sandstone.

Secondary porosity refers to new pore space produced in rocks some time after the rock first formed. Some secondary porosity forms when groundwater passes through rock and dissolves minerals or cement, yielding small solution cavities. Secondary porosity also forms when rocks fracture, for the opposing walls of the fracture do not fit together tightly **(Fig. 19.2d)**.

Permeability: The Ease of Flow

If solid rock completely surrounds a pore, the water in the pore cannot flow to another location. For groundwater to flow, pores must be linked by conduits. The **permeability** of a material refers to the ease with which fluids can pass through the material via an interconnected network of pores. To develop an intuitive image of permeability, take a sturdy glass jar and fill it with gravel composed of rounded pebbles. If you look closely at this gravel through the side of the jar, you will see air-filled pores between the pebbles, because the pebbles don't fit together perfectly. If you pour water into the jar, the water can trickle between grains down to the bottom of the jar, where it displaces air and fills the pores **(Fig. 19.3a)**. Water flows easily through a *permeable material*, whereas it flows slowly or not at all through an *impermeable material* **(Fig. 19.3b)**. Informally, we can describe materials as having "high permeability," "low permeability," or "no permeability" to convey the relative ease of fluid movement. The permeability of a material depends on several factors:

- *Number of available conduits:* As the number of conduits increases, permeability increases, for there are more routes through which water can move.

- *Size of the conduits:* Greater volumes of fluids can travel through wider conduits than through narrower ones because frictional resistance to flow develops along the conduits' walls.

- *Straightness of the conduits:* Water flows more rapidly through straight conduits than it does through crooked ones. In crooked conduits, the distance a water molecule actually travels may be many times the straight-line distance between the two end points.

Note that the factors that control permeability in rock or sediment resemble those that control the ease with which traffic moves through a city. Specifically, traffic can flow quickly through cities with many straight, multilane boulevards, whereas it flows slowly through cities with only a few narrow, crooked streets.

Porosity and permeability are not the same feature. A material containing isolated pores can have high porosity but low permeability. Cork exhibits this behavior—it has high porosity, but low permeability (so it can plug a bottle). In cork, a type of tree bark, woody cell walls isolate adjacent pores (empty cells) and prevent communication between them. Vesicular basalt, similarly, has high porosity but low permeability because rock surrounds each pore.

FIGURE 19.3 Permeability is the ability of water to flow through rock from pore to pore or along joints.

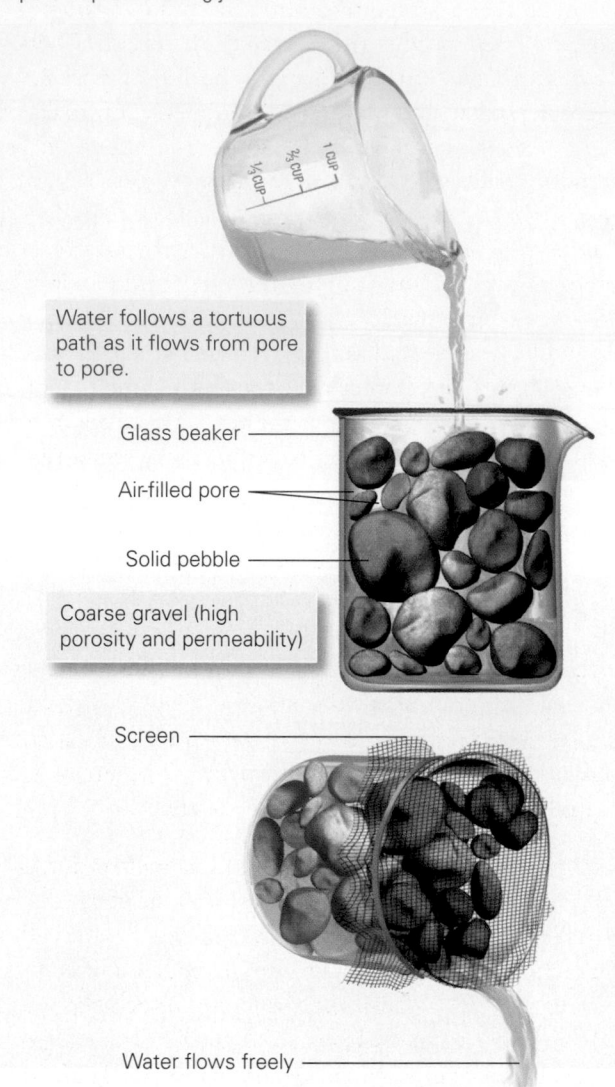

Water follows a tortuous path as it flows from pore to pore.

Glass beaker

Air-filled pore

Solid pebble

Coarse gravel (high porosity and permeability)

Screen

Water flows freely

(a) Gravel contains pore space because clasts don't fit together tightly. The connection of pores produces permeability, so water can flow through gravel.

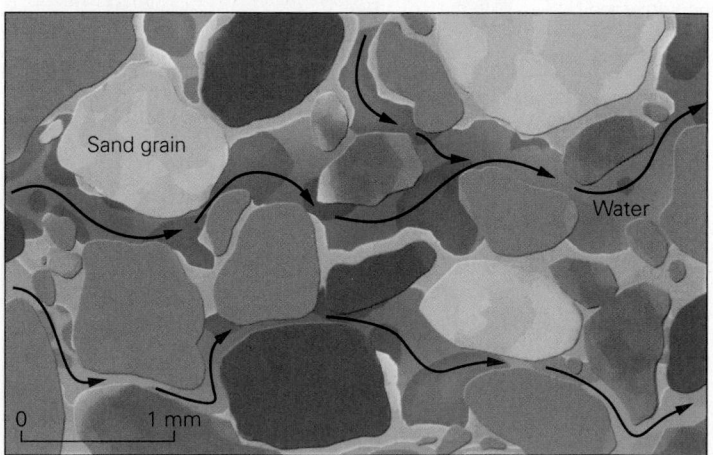

Sand grain

Water

0   1 mm

(b) Even at a microscopic scale, if passages connect pores, then water can flow through, and the rock has permeability.

Aquifers and Aquitards

With the concept of permeability in mind, *hydrogeologists* (researchers who study groundwater) distinguish between **aquifers**, sediments or rocks that have both high porosity and high permeability, and **aquitards**, sediments or rocks that have low permeability. Generally, aquitards do not hold much groundwater because they have low porosity, too. (A less used term, *aquiclude*, refers to a geologic material that does not accommodate groundwater flow at all.) Coarse gravels, poorly cemented sandstones, and highly fractured and partially dissolved limestones typically make good aquifers. Shales, evaporites, and well-cemented sandstones serve as aquitards.

As we learned in Chapter 7, the type of sediment deposited at a location can vary over time as the depositional environment changes. Therefore, successions of sedimentary strata typically contain interlayered aquifers and aquitards. Water can infiltrate directly into an *unconfined aquifer* from the Earth's surface. Water cannot infiltrate directly into a *confined aquifer,* which is isolated from the Earth's surface by an aquitard **(Fig. 19.4)**.

To get a sense of the diversity of characteristics that aquifers can have, and the diversity of ways in which they form, let's look at examples of specific aquifers that provide important groundwater supplies in the United States. Each aquifer has a name, based on the region in which it occurs or on the stratigraphic unit that hosts the groundwater.

- *Mahomet aquifer, Illinois:* When glaciers melted during the Pleistocene Ice Age, meltwater streams carried coarse gravel southward. This gravel filled stream valleys that had previously been cut into the bedrock of central Illinois. Younger glacial sediment later buried the filled valleys. The porous and permeable gravel of the largest buried valley comprises the Mahomet aquifer, named for a nearby town **(Fig. 19.5a)**.

- *Phoenix Basin aquifer, Arizona:* The city of Phoenix lies in the Basin and Range Province, a continental rift. During rifting, downward slip of crustal blocks on normal faults produced deep wedge-shaped basins that filled with gravels and sands eroded from adjacent mountain ranges **(Fig. 19.5b)**. These sediments serve as aquifers that provide groundwater for irrigation and drinking in Phoenix.

- *The Dakota Sandstone aquifer:* During the Cretaceous, rivers flowing from mountains in the western United States dumped sediment into an inland sea, forming a vast sheet of sand that was later buried by mud. The sand layer lithified to become the Dakota Sandstone, an aquifer, and the mud layer became a shale aquitard. Continued mountain building warped the strata into a syncline. Water infiltrates into the western limb of the fold and slowly flows eastward **(Fig. 19.5c)**. The Dakota Sandstone serves as a groundwater source where it lies close to the

FIGURE 19.4 An aquifer is a high-porosity, high-permeability rock or sediment. Some aquifers are unconfined, and some are confined.

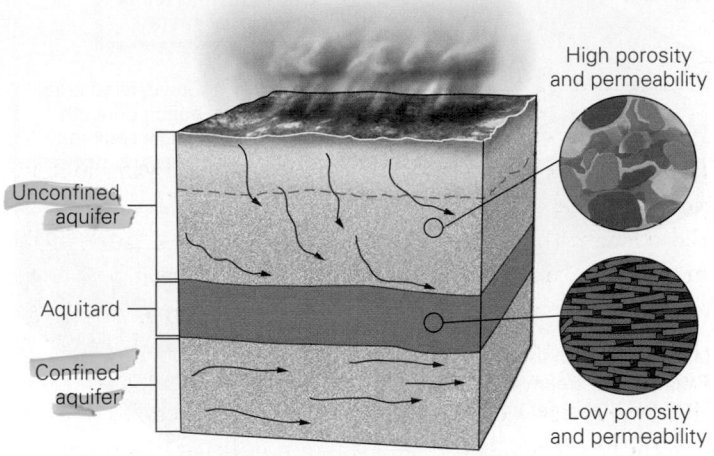

ground surface on the eastern limb of the syncline. Note that it represents a confined aquifer.

- *The High Plains (Ogallala) aquifer:* Erosion of the Rocky Mountains 3 million years ago led to the deposition of huge alluvial fans over what is now the High Plains region. The resulting deposits comprise the Ogallala Formation, a sheet of porous and permeable strata (fluvial sandstones and conglomerates and paleo-desert sands) that remains just below the ground surface **(Fig. 19.5d)**. Water can infiltrate this unconfined aquifer over a broad area. Note that the Ogallala Formation aquifer lies above the Dakota Sandstone aquifer over much of its extent.

- *Limestone aquifers of Florida:* Sheets of Cenozoic limestone, deposited when the region was submerged by a shallow sea, underlie the entire peninsula of Florida as well as adjacent areas of Georgia and Alabama. Fractures in the limestone have been widened by dissolution and now provide spaces that hold large quantities of groundwater.

TAKE-HOME MESSAGE

Most groundwater fills pores and cracks in rock or sediment. Porosity refers to the total amount of open space within a material, whereas permeability indicates the degree to which pores connect. Aquifers have high porosity and permeability whereas aquitards don't. An unconfined aquifer is one that connects to the ground surface, while a confined aquifer is one that lies beneath an aquitard. There are many different kinds of aquifers, each with distinct geologic characteristics.

QUICK QUESTION: Why do poorly cemented sandstones make good aquifers?

FIGURE 19.5 Examples of different types of aquifers. The aquifers are stippled.

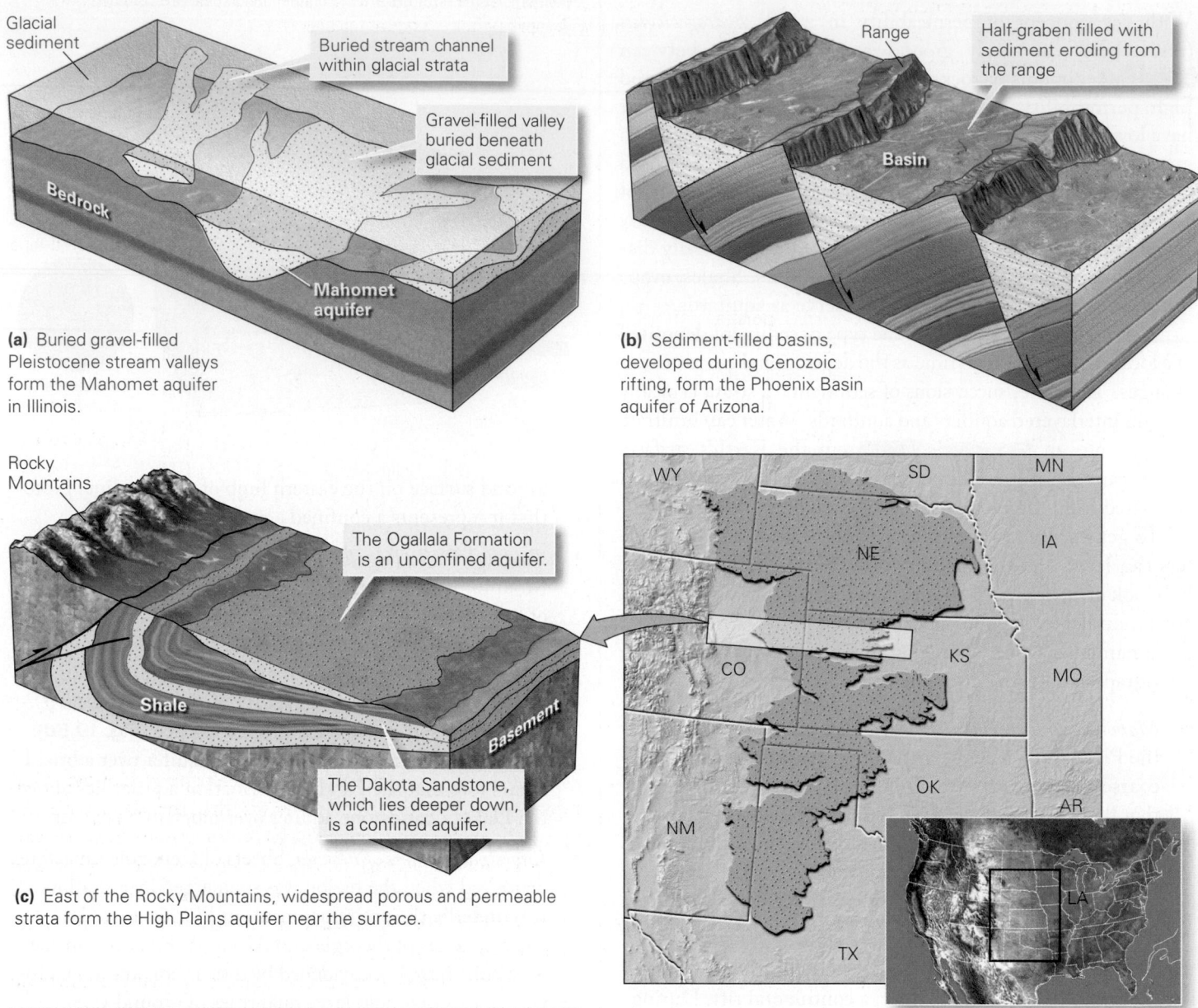

(a) Buried gravel-filled Pleistocene stream valleys form the Mahomet aquifer in Illinois.

(b) Sediment-filled basins, developed during Cenozoic rifting, form the Phoenix Basin aquifer of Arizona.

(c) East of the Rocky Mountains, widespread porous and permeable strata form the High Plains aquifer near the surface.

(d) Area of the High Plains aquifer (also called the Ogallala aquifer).

19.3 Characteristics of the Water Table

So far, we've used the term *groundwater* in a fairly general sense for all water underground. But now that we understand the concepts of porosity and permeability, and have been introduced to the various kinds of aquifers, we can look more closely at the distribution of subsurface water and see that we can define a boundary that marks the upper limit of what hydrogeologists formally define as groundwater.

Near the ground surface, water may only partially fill the pores, leaving some space that remains filled with air **(Fig. 19.6)**. The region of the subsurface in which water only partially fills pores is called the *unsaturated zone* (also known as the *vadose zone*). If the top of the unsaturated zone includes a soil, we can also refer to the water wetting the surfaces of the grains and organic material making up the soil as **soil moisture**—this water tends to evaporate back into the atmosphere, seep into streams, or get sucked up by plant roots and transpired back into the atmosphere. Below the unsaturated zone, water completely fills, or saturates, the pores, yielding the *saturated zone* (also called the *phreatic zone*). In a strict sense, hydrogeologists restrict use of the term *groundwater* to subsurface water in the saturated zone. The term **water table** specifies the horizon that separates the unsaturated zone above from the saturated zone below (see Fig. 19.6). We can picture the water table as the top boundary of groundwater in an unconfined aquifer. *Surface tension*, the electrostatic attraction of water molecules to one another and to mineral surfaces, causes water to seep upward from the saturated zone to form a layer called the **capillary fringe** just above the water table. Capillary fringes typically have a thickness of between 5 and 30 cm.

The depth of the water table in the subsurface varies greatly with location. In some places, the water table coincides with the surface of a permanent stream, lake, or marsh, so it effectively lies at or above the ground level (**Fig. 19.7a**). (In such locations, no unsaturated zone exists and the soil itself can be saturated, so we can't really distinguish soil moisture from groundwater.) Elsewhere, the water table lies hidden below the ground surface. In humid regions, it typically lies at depths of only meters to tens of meters, but in arid regions it may lie hundreds of meters below the surface. Note that permanent lakes or streams cannot exist in arid regions, unless fed by a water supply from upstream, because water can infiltrate down through the lake bed or streambed to the water table far below (**Fig. 19.7b**; see Chapter 17). Rainfall rates affect the water table depth in a given locality (**Fig. 19.7c**). Specifically, the water table drops during the dry season and rises during the wet season. In fact, streams or ponds that hold water during the wet season may dry up during the dry season.

We've defined the water table as the top of groundwater in the subsurface. Can we define the base of groundwater? Put another way, how deep down in the crust can we find groundwater? Recent research shows that liquid groundwater does circulate in basement igneous and metamorphic rock, perhaps to depths of over 15 km. Most of this deep circulation takes place along fractures. The ultimate base of groundwater in crust may be taken as the depth where water encounters such high temperatures and pressures that it becomes a hydrothermal fluid involved in metamorphic reactions, and rock becomes weak enough to flow plastically and close up pores. At high temperatures, water molecules split apart. The depth at which this transition takes place depends on the geothermal gradient, but roughly speaking, it lies at a depth of between 15 and 25 km. So groundwater occurs only in the upper crust.

Topography of the Water Table

In hilly regions, if the subsurface has low to moderate permeability, the water table won't be a planar surface. Rather, its

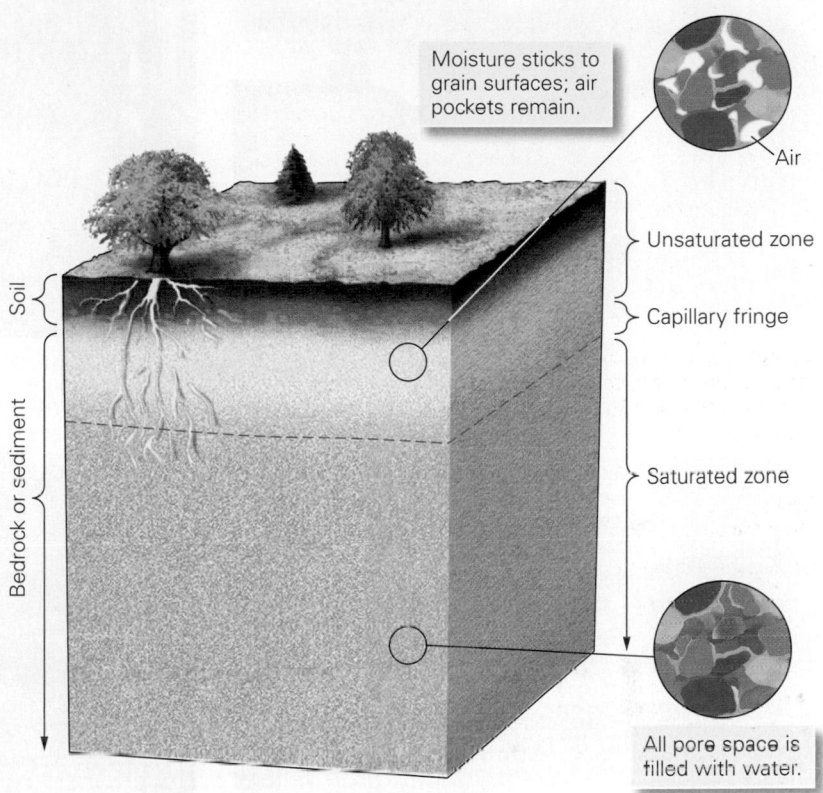

FIGURE 19.6 The water table is the top of the groundwater reservoir in the subsurface. It separates the unsaturated (vadose) zone above from the saturated zone below. A capillary fringe forms at the boundary.

Moisture sticks to grain surfaces; air pockets remain.

Air

Unsaturated zone

Capillary fringe

Saturated zone

Soil

Bedrock or sediment

All pore space is filled with water.

shape mimics, in a subdued way, the shape of the overlying topography (**Fig. 19.8a**). This means that the water table lies at a higher elevation beneath hills than it does beneath valleys. The relief (the vertical distance between the highest and lowest elevations) of the water table tends to be less than that of the overlying land, so the water table tends to be smoother than the ground surface.

At first thought, it may seem surprising that the elevation of the water table varies as a consequence of ground-surface topography. After all, when you pour a bucket of water into a pond, the surface of the pond immediately adjusts to remain horizontal. The elevation of the water table varies because groundwater moves so slowly through rock and sediment that it cannot quickly assume a horizontal surface. When it rains on a hill and rainwater infiltrates down to the water table, the water table rises a little. When it doesn't rain, the water table sinks slowly, but so slowly that rain will probably fall again, making the water table rise again before it has had time to sink very far.

Perched Water Tables

In some locations, layers of strata are *discontinuous*, meaning that they pinch out at their sides. As a result, a lens-shaped

FIGURE 19.7 The depth of the water table depends on the water supply from above.

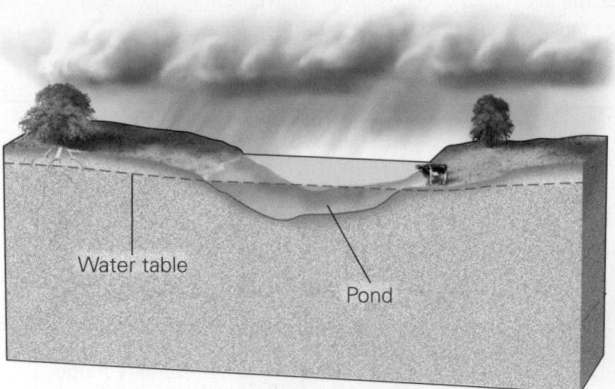

The water table lies above the ground in this swamp.

(a) Where the water table lies close to the ground surface, ponds remain filled. In fact, the water table coincides with the surface of the pond.

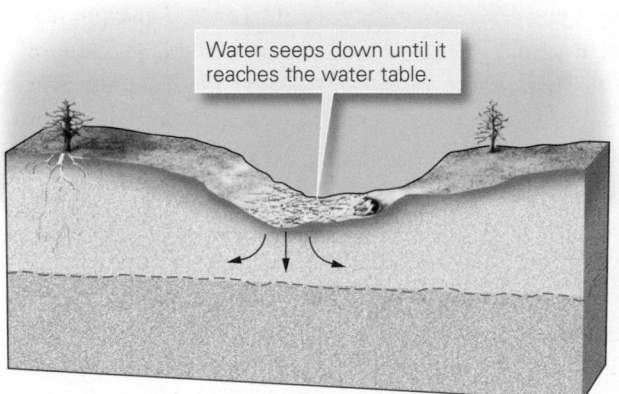

Water seeps down until it reaches the water table.

In a drought, the water table sank and this pond dried up.

(b) In dry regions, the water table sinks deep below the surface. Water that collects temporarily in low areas sinks into the subsurface.

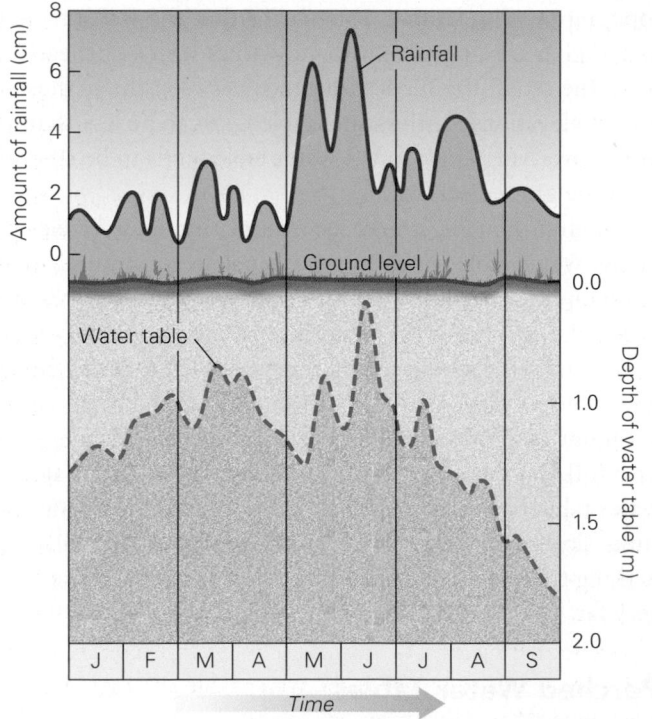

(c) This graph illustrates how the water table rises and falls depending on the amount of rainfall.

layer of impermeable rock (such as shale) may occur within a thick aquifer at depths above that of the regional water table. When this happens, a mound of groundwater accumulates above the aquitard lens because the lens prevents water from sinking down to the regional water table **(Fig. 19.8b)**. The top of the mound is called a **perched water table** because it lies above the regional water table.

TAKE-HOME MESSAGE

Groundwater is water that resides underground in the pores of rock or sediment. Below the water table, in the saturated zone, water fills available pore space; above the water table, pores are partially or entirely filled with air. The water table is higher in regions with wetter climates than in regions with drier climates; it can rise or fall depending on the amount of rainfall. In hilly areas, the water table itself has topography. Where the water table intersects valleys or depressions, streams or lakes form.

QUICK QUESTION: Why is the water table higher beneath hills than beneath valleys?

FIGURE 19.8 Factors that influence the position of the water table.

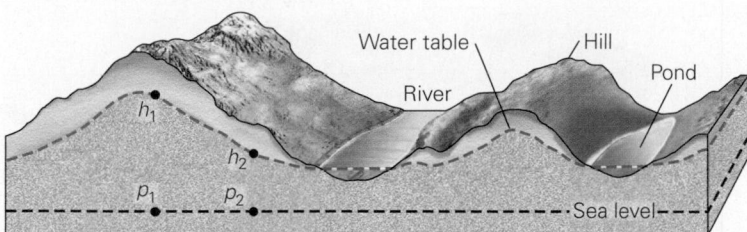

(a) The shape of a water table beneath hilly topography. Point h_1 on the water table is higher than point h_2, relative to a reference elevation (sea level). The pressure at p_1 is, therefore, more than the pressure at p_2.

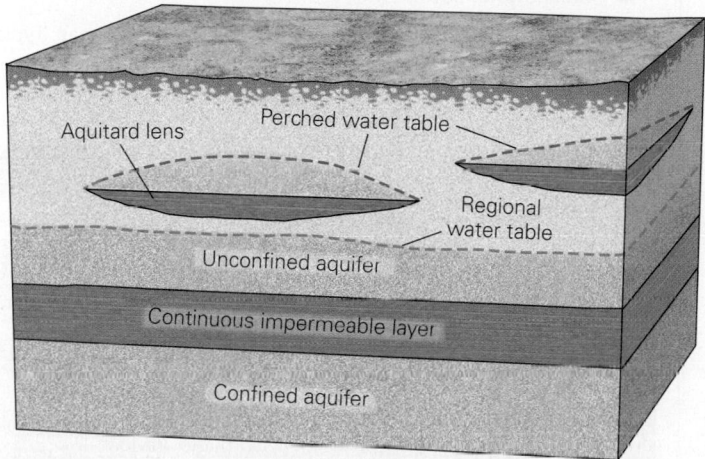

(b) A perched water table occurs where a mound of groundwater becomes trapped above a localized aquitard that lies above the regional water table.

19.4 Groundwater Flow

What happens to groundwater over time? Does it just sit, unmoving, like the water in a stagnant puddle, or does it flow so that some eventually finds its way back to the surface? Countless measurements confirm that groundwater not trapped in isolated pores enjoys the latter fate—most groundwater indeed flows, and in some cases it moves great distances underground, taking anywhere from a few weeks to tens of thousands of years before returning to the surface to pass once again into other reservoirs of the hydrologic cycle. We begin this section by examining factors that drive groundwater flow and determine the path that this flow takes. Then we examine the rate (velocity) at which groundwater moves.

Groundwater Flow Paths

In the unsaturated zone—the region between the ground surface and the water table—water percolates down, like the water passing through a drip coffeemaker, for this water moves only in response to the downward pull of gravity. But in the saturated zone—the region below the water table—water flow

tends to be more complex, for in addition to the downward pull of gravity, water responds to differences in pressure. Pressure can cause groundwater to flow sideways or even upward. (If you've ever watched water spray from a fountain, you've seen pressure pushing water upward.) So, to understand the nature of a *groundwater flow path*, the overall trajectory that a given volume of groundwater follows over time, we must first understand the origin of pressure in groundwater. For simplicity, we'll consider only the case of groundwater in an unconfined aquifer.

Pressure in groundwater at a specific point underground in an unconfined aquifer comes from the weight of all the overlying water from that point up to the water table. (The weight of overlying rock does not contribute to the pressure exerted on groundwater because the contact points between mineral grains bear the rock's weight.) As a consequence, a point at a greater depth below the water table feels more pressure than does a point at a lesser depth. At a location with a horizontal water table, the pressure acting on an imaginary horizontal reference plane at a specified depth below the water table is the same everywhere. But if the water table is not horizontal (see Fig. 19.8a), the pressure at points on a horizontal reference plane at depth differs with location. For example, point p_1, which lies below the hill in Figure 19.8a, feels greater pressure than does point p_2, which lies below the valley, even though both p_1 and p_2 lie at the same elevation (sea level, in this case).

Both the elevation of a volume of groundwater and the pressure within the water provide energy that, if given the chance, will cause the water to flow. Physicists refer to such stored energy as *potential energy*. To understand why elevation provides potential energy, imagine a bucket of water high on a hill; the water has potential energy due to the Earth's gravity, which would cause it to flow downslope if the bucket were suddenly to rupture. To understand why pressure provides potential energy, imagine a water-filled plastic bag sitting on a table; if you puncture the bag and then squeeze the bag to exert pressure, water spurts out. The potential energy available to drive the flow of a given volume of groundwater at a location is called the **hydraulic head**. To measure the hydraulic head at a point in an aquifer, hydrogeologists drill a vertical hole down to that point and then insert a pipe in the hole. The height above a reference elevation (such as sea level) to which water rises in the pipe represents the hydraulic head; water rises higher in the pipe where the head is higher. As a rule, groundwater flows from regions where it has a higher hydraulic head to regions where it has a lower hydraulic head. This statement generally implies that groundwater, regionally, flows from locations where the water table sits higher to locations where the water table sits lower.

Hydrogeologists have calculated how the hydraulic head changes with location underground by taking into account both the effect of gravity and the effect of pressure. These calculations reveal that groundwater flows along concave-up curved paths **(Fig. 19.9a)**. (Specialized books on hydrogeology provide the details of why flow paths have this shape.) These curved paths eventually take groundwater from regions where the water table sits high (under a hill) to regions where the water table sits low (below a valley), but because of flow-path shape, some groundwater may flow down into the crust along the first part of its path and then may flow back up, toward the ground surface, along the final part of its path. The location where water enters the ground, meaning the region where the flow has a downward trajectory, is the **recharge area**, and the location where groundwater flows back up to the surface is the **discharge area** (see Fig. 19.9a). Where landscapes have complex topography, not all water entering the ground in a given recharge zone ends up in the same discharge zone. Some flow follows longer paths to regional discharge areas, and some follows shorter paths to local discharge areas **(Fig. 19.9b)**. We can define a *groundwater divide* as the vertical plane separating flow that goes to different discharge areas, and these too can be local or regional.

Groundwater following short paths close to the Earth's surface travels tens of meters to a few kilometers before returning to the surface. Such *local flow* has a *residence time* (meaning the time that it stays underground) of only hours to weeks. Groundwater following paths of several kilometers to tens of kilometers constitutes *intermediate flow* and has a residence time of weeks to years. Groundwater following paths that carry it hundreds of kilometers across a large sedimentary basin constitutes *regional flow* and stays underground for centuries to millennia **(Fig. 19.9c)**.

Rates of Groundwater Flow

Flowing water in an ocean current moves at up to 3 km per hour (over 26,000 km per year), and water in a steep river channel can reach speeds of up to 30 km per hour (over 260,000 km per year). In contrast, groundwater moves at less than a snail's pace—hydrogeologists have found that typical flow rates range between 0.01 and 1.4 m per day (only about 4 to 500 m per year). They can measure the rate (velocity) of groundwater flow in a region by determining how long it takes a "tracer," such as a chemical dye or a radioactive isotope, injected into one well to arrive at another well a known distance away. Groundwater moves much more slowly than surface water for two reasons. First, groundwater must percolate through a complex, crooked network of tiny conduits, so it must travel a much greater distance than it would if it could follow a straight path. Second, the surface tension of water makes it stick to the solid material around it and therefore slows its escape from pores. Even in larger conduits, friction between groundwater and conduit walls slows the flow.

Simplistically, the velocity of groundwater flow depends on the slope of the water table and on the permeability

FIGURE 19.9 The flow of groundwater.

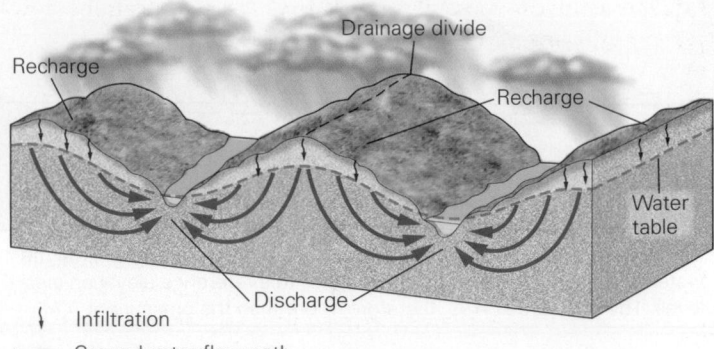

(a) Groundwater flows from recharge areas to discharge areas. Typically, the flow follows curving paths.

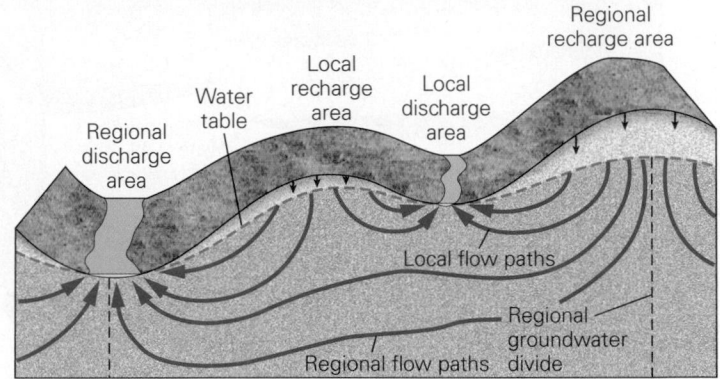

(b) In a region of complicated topography, local flow may be different from regional flow.

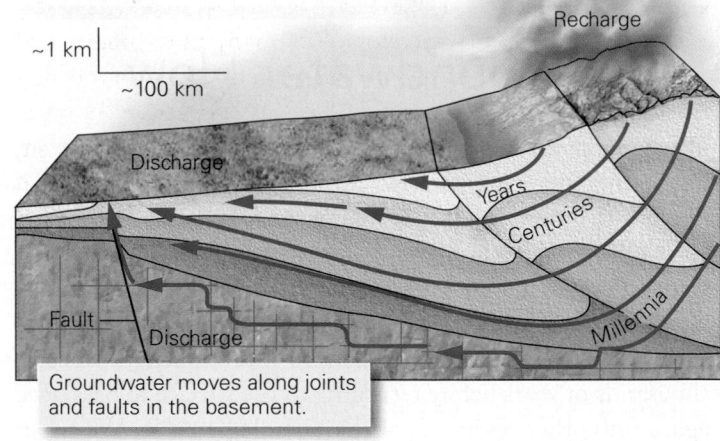

(c) The large hydraulic head resulting from uplift of a mountain belt may drive groundwater hundreds of kilometers across regional sedimentary basins. Deeper flow paths take longer.

of the material through which the groundwater moves. Groundwater flows faster through high-permeability rocks than it does through low-permeability rocks, and it flows faster in regions where the water table has a steep slope than it does in regions where the water table has a gentle slope. Therefore, groundwater flows relatively quickly through high-permeability gravel under a steep hillslope, but it flows relatively slowly through a low-permeability, well-cemented sandstone. Hydrogeologists use *Darcy's law* to determine flow rates at a location more precisely **(Box 19.1)**.

19.5 Tapping Groundwater Supplies

All living organisms rely on access to water. Of course, while marine life, coastal plants, and certain types of microbes survive in saline water, most terrestrial organisms, such as our own species, require freshwater. Although lakes and streams provide access to freshwater, they don't occur everywhere, so even our distant prehistoric ancestors, like many other plants and animals, needed access to **springs**, natural outlets from which groundwater flows or seeps onto the Earth's surface, in order to survive. Springs don't occur everywhere, either, so almost 10,000 years ago, people learned to dig *wells*, artificial holes that provide access to groundwater. The oldest known wells, found on the island of Cyprus, date from about 7500 B.C.E. In this section, we first examine the variation in natural groundwater quality, then focus on the nature of springs and wells.

Natural Groundwater Quality

Shallow, new groundwater has a composition similar to that of the rainwater or snowmelt that served as its source—pure H_2O. But as groundwater percolates downward, it reacts with the soil, sediment, or rock through which it passes, and these reactions transform it into a chemical solution. Deep in sedimentary basins, groundwater derived from rain or snow may mix with water left in pore spaces from the time of the original deposition of sediment millions of years ago. If the sediment was deposited in a marine environment, this ancient deep groundwater can be saline.

The concentration of dissolved ions in groundwater—meaning the quantity of ions dissolved per unit volume of water—depends on the temperature, pressure, acidity, availability of oxygen, and residence time of the groundwater. Warmer groundwater, for example, can hold more ions in solution than can cooler groundwater. And the longer that groundwater has to react with its surroundings, the greater the concentration can become. When groundwater attains *saturation*, the water contains as many dissolved ions as possible under the local environmental conditions.

Because groundwater flows, it eventually enters different environments. If groundwater enters a new environment where it has the capacity to contain more ions, it may dissolve surrounding rock or sediment and produce secondary porosity. Alternatively, if saturated groundwater enters an environment in which it cannot hold as many dissolved ions, some of the ions bond together to make solid mineral grains that precipitate to form cement or fill veins.

Groundwater in the upper few kilometers of the crust can be crystal clear and pure enough to drink right out of the ground, for rocks and sediment are natural filters capable of removing suspended solids—these solids get trapped in tiny pores or stick to the surfaces of clay flakes. In fact, the commercial distribution of bottled groundwater—so-called spring water—has become a major business worldwide. Small amounts of dissolved minerals may add flavor to water, so it can be sold as "mineral water." Dissolved CO_2 can give some groundwater a natural fizziness.

Did you ever wonder...
where bottled "spring water" comes from?

In too high a quantity, however, dissolved chemicals make natural groundwater undesirable. For example, deeper groundwater that has passed through salt-containing strata or has mixed with old saline pore water may become salty and unsuitable for irrigation or drinking. Groundwater that passed through limestone or dolostone may become saturated with calcium (Ca^{2+}) and magnesium (Mg^{2+}) ions—such water, called **hard water**, can be a problem for humans because carbonate minerals precipitate from it to form "scale," which clogs pipes and prevents soap from developing a lather. Groundwater that has passed through iron-bearing rocks may contain dissolved iron. When this groundwater rises from oxygen-poor environments at depth to oxygen-rich environments near the surface, the iron atoms bond to oxygen atoms to form tiny crystals of iron oxide ("rust"). Water containing these minerals develops an orange-brown color and may stain the surfaces of pipes and tanks. In some cases, bacteria living in oxygen-poor environments ingest sulfur from sulfur-bearing minerals to produce hydrogen sulfide (H_2S), a poisonous gas that smells like rotten eggs. Normally, the concentration of H_2S in groundwater doesn't reach dangerous levels, but even a small amount gives the water an unpleasant

BOX 19.1 CONSIDER THIS . . .

Darcy's Law for Groundwater Flow

The rate at which groundwater flows at a given location depends on the permeability of the material containing the groundwater—groundwater flows faster in a more permeable material than it does in a less permeable material. The rate also depends on the **hydraulic gradient**, meaning the change in hydraulic head per unit of distance between two locations, as measured along the flow path. To calculate the hydraulic gradient, we divide the difference in hydraulic head between two points by the distance between the two points as measured along the flow path. This calculation can be written as a formula:

$$\text{Hydraulic gradient} = \frac{h_1 - h_2}{j}$$

where $h_1 - h_2$ is the difference in head (given in meters or feet, because head is represented as an elevation) between two points along the water table, and j is the distance between the two points as measured along the flow path (**Fig. Bx19.1**). A hydraulic gradient exists anywhere that the water table is not horizontal. Typically, the slope of the water table is so small that the path length of groundwater following a shallow flow path is almost the same as the horizontal distance between two points. So, for shallow groundwater, the hydraulic gradient is roughly equivalent to the slope of the water table.

In 1856, a French engineer named Henry Darcy carried out a series of experiments designed to characterize factors that control the velocity of groundwater flow between two locations (1 and 2), each of which has a different hydraulic head (h_1 and h_2). Darcy represented the velocity of flow by a quantity called the *discharge* (Q), meaning the volume of water passing through an imaginary vertical plane perpendicular to the groundwater's flow path in a given time. He found that the discharge depends on the hydraulic head ($h_1 - h_2$); the area (A) of the imaginary plane through which the groundwater is passing; and a number called the hydraulic conductivity (K). The hydraulic conductivity represents the ease with which a fluid can flow through a material and depends on many factors, including the viscosity and density of the fluid, but mainly it reflects the permeability of the material. The relationship that Darcy discovered, now known as *Darcy's law*, can be written in the form of an equation as:

$$Q = \frac{KA(h_1 - h_2)}{j}$$

The equation states that if the hydraulic gradient increases, discharge increases, and that as conductivity increases, discharge increases. Put in simpler terms, the flow rate of groundwater increases as the permeability increases and as the slope of the water table gets steeper.

FIGURE Bx19.1 The level to which water rises in a drillhole is the hydraulic head (h). The hydraulic gradient (HG) is the difference in head divided by the length of the flow path.

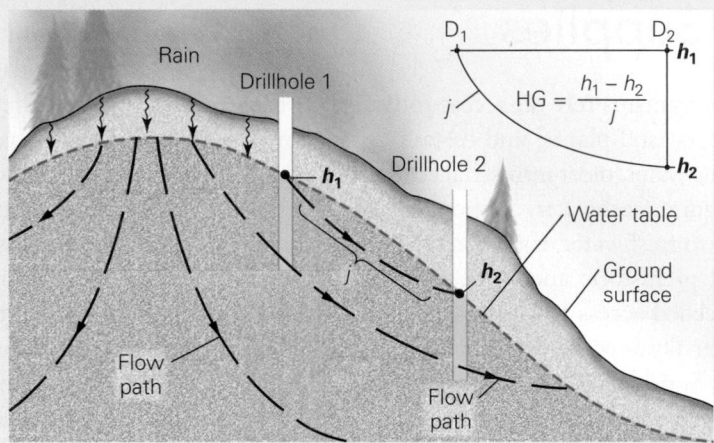

taste and odor. Natural gas (typically methane, CH_4) may enter groundwater that either passes through sediment containing decaying organic matter or comes from strata containing oil or gas. The gas dissolves in groundwater under the high pressures of the subsurface, but it bubbles out when the water rises in a well.

Arsenic, a highly toxic chemical, occurs in tiny amounts in many minerals. Over time, groundwater can dissolve some of this arsenic. Wells that fill with *arsenic-contaminated groundwater* can be dangerous. In the United States, for example, regulations require that drinking water contain less than 10 micrograms (µg) of arsenic per liter (L). Unfortunately, surveys suggest that about 5% of the groundwater coming out of U.S. wells contains 20µg/L or more. Arsenic-contaminated groundwater has become a major problem in parts of India and Bangladesh, where efforts to provide unpolluted water for drinking has resulted in a shift from using river water to using groundwater pumped up from deep wells. This groundwater has resided underground long enough to dissolve arsenic. As a consequence, over 25% of the wells contain 50µg/L or more, and some contain more than 3000µg/L. In villages using arsenic-contaminated groundwater, some inhabitants display skin lesions that develop due to arsenic poisoning.

The quality of groundwater can vary with depth in a given region because the age of groundwater typically increases with depth. Generally, groundwater in a deep sedimentary basin tends to be more saline (brackish) and harder with increasing depth. In fact, in deep basins, a boundary at a depth of a few hundred meters to a few thousand meters divides drinkable groundwater above from undrinkable groundwater below.

FIGURE 19.10 Geologic settings in which springs form.

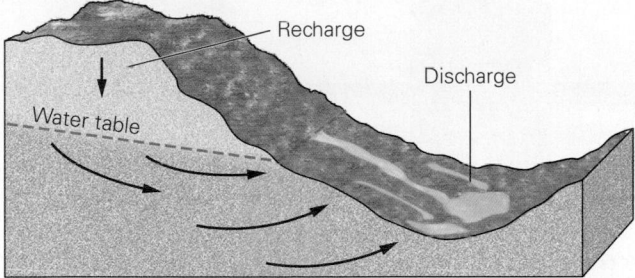

(a) Groundwater reaches the ground surface in a discharge area.

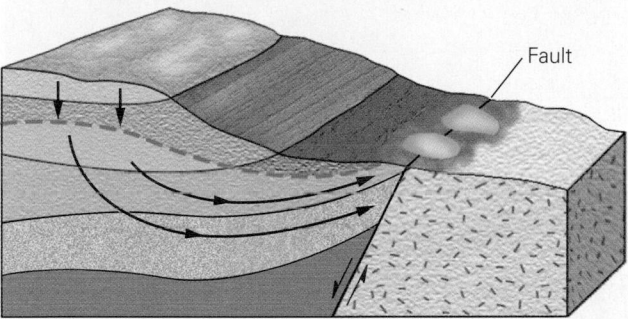

(b) Where groundwater reaches an impermeable barrier, it rises.

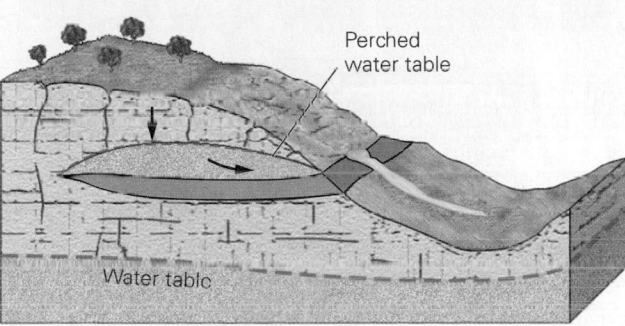

(c) Groundwater seeps out where a perched water table intersects a slope.

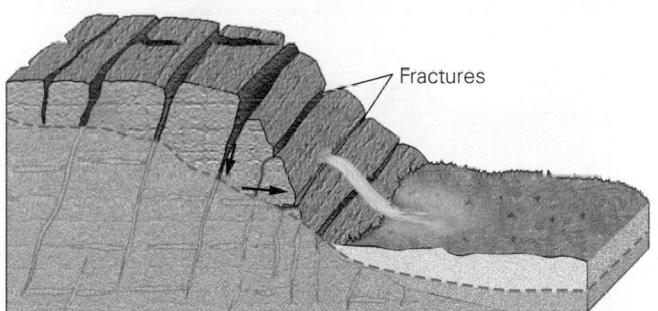

(d) A network of interconnected fractures channels water to the surface of a hill.

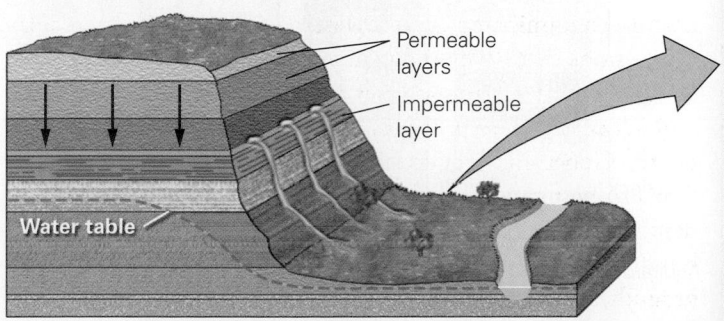

(e) Groundwater seeps out of a cliff face at the top of a relatively impermeable layer.

Springs

More than one town has grown up around a **spring**, a place where groundwater naturally flows or seeps onto the Earth's surface, for springs can provide fresh, clear water for drinking or irrigation without the expense of drilling or digging. Some springs spill water onto dry land. Others bubble up through the bed of a stream or lake. Springs form under a variety of conditions:

- where the ground surface intersects the water table in a discharge area **(Fig. 19.10a)**; such springs typically occur

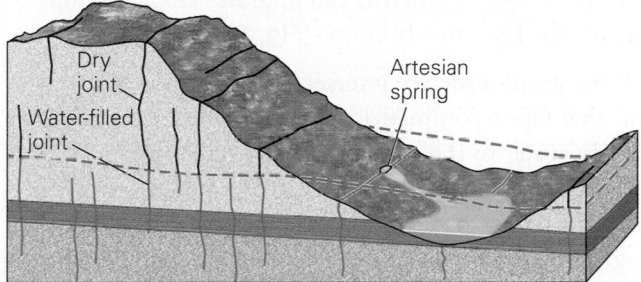

(f) In cases where water under pressure lies below an aquitard, a crack may provide a pathway for an artesian spring to form.

A spring on the wall of the Grand Canyon

on or near valley floors, where they may add water to lakes or streams.

- where flowing groundwater collides with a steeply dipping impermeable barrier, and pressure pushes it up the barrier to the ground **(Fig. 19.10b)**.

- where a perched water table intersects the surface of a hill **(Fig. 19.10c)**.

- where a network of interconnected fractures channels groundwater to the surface of a hill **(Fig. 19.10d)**.

- where downward-percolating water runs into a relatively impermeable layer (aquitard) and migrates along the top surface of the layer to a hillslope **(Fig. 19.10e)**.

- where the ground surface intersects a natural fracture (joint) that taps a confined aquifer in which the pressure drives the water to the surface. Geologists refer to such an occurrence as an **artesian spring (Fig. 19.10f)**—we'll explain the source of the pressure that drives these springs in our discussion of wells.

Springs can provide water in regions that would otherwise be uninhabitable. For example, a desert **oasis** forms where a spring permits plants to grow in an otherwise bone-dry region **(Box 19.2)**.

Wells

Where springs don't exist, people can tap into the groundwater by digging or drilling a well. In an *ordinary well*, the base of the well penetrates an aquifer below the water table **(Fig. 19.11)**. Water from the pore space in the aquifer seeps into the well and fills it to the level of the water table. Drilling into rock that lies above the water table, or into an aquitard, will not supply water and thus yields a *dry well*. Seasonal ordinary wells function only during the rainy season, when the water table rises; during the dry season, the water table lies below the base of the well, so the well goes dry.

Ideally, we would like an ordinary well to be as shallow as possible (to decrease the cost of digging or drilling). So hydrogeologists search for particularly porous and permeable aquifers in which the water table lies near the surface. Contrary to legend, a "dowser" cannot find water simply by using a forked stick—when dowsers do strike water, either they have enjoyed dumb luck or they have prior knowledge of the water table in the area of their search.

To obtain water from an ordinary well, people can either pull water up in a bucket or pump the water out. As long as the rate at which groundwater fills the well exceeds the rate at which water is removed, the level of the water table near the well remains about the same. However, if people pump water out of the well too fast, then the water table sinks down around the well, in a process called *drawdown*, and becomes

FIGURE 19.11 The nature of ordinary wells.

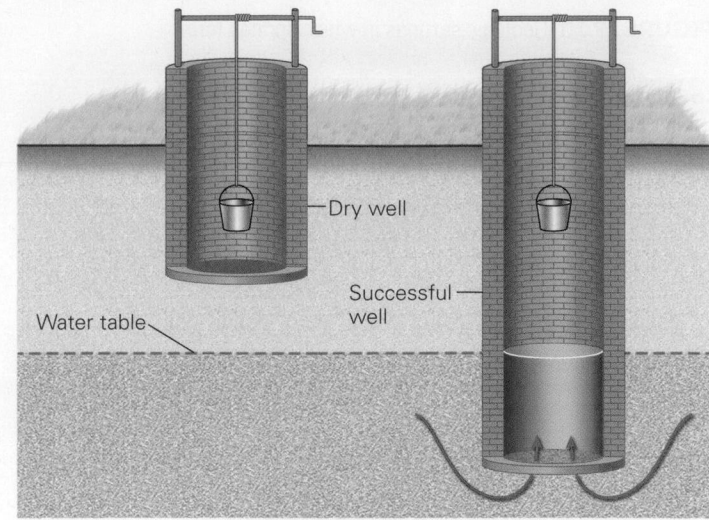

(a) The basic concept of an ordinary well. It's a vertical cylindrical hole that extends below the water table.

(b) At a traditional ordinary well, villagers obtain water by lifting it with a bucket.

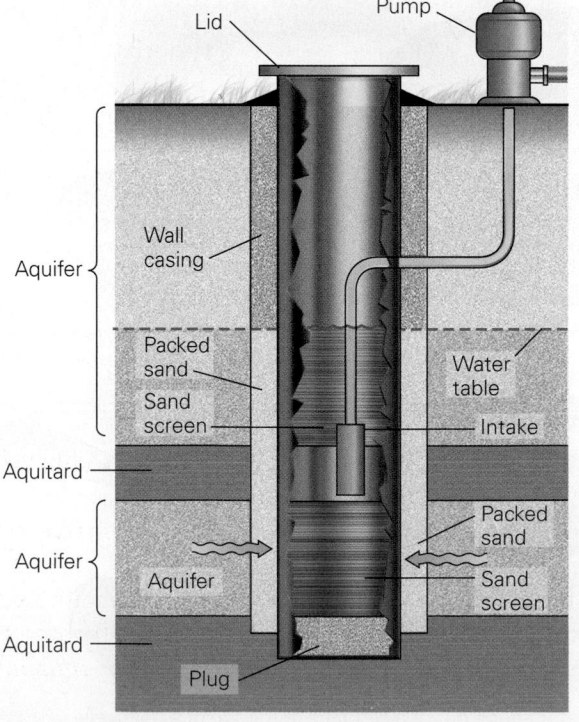

(c) A modern ordinary well sucks up water with an electric pump. The packed sand filters the water.

BOX 19.2 CONSIDER THIS . . .

Oases

The Sahara Desert of northern Africa is now one of the most barren and desolate places on the Earth, for it lies in a climate belt where rain seldom falls. But it wasn't always that way. During the last ice age, when glaciers covered parts of northern Europe on the other side of the Mediterranean, the Sahara enjoyed a more temperate climate, and the water table was nearer to the surface, so permanent streams dissected the landscape. These streams have vanished, victims of the drier, warmer climate of recent millennia, but vast reserves of groundwater fill a huge underground aquifer composed of porous sandstone. In general, the water of this aquifer can be obtained only by drilling deep wells, but at certain locations, water spills out at the surface—either because tectonic folding brings the aquifer particularly close to the ground, so that valley floors intersect the water table, or because pressure pushes groundwater up along joints or faults (**Fig. Bx19.2a**). In either case, the aquifer feeds springs that quench the thirst of plants and produce an *oasis*, an island of green in a sea of sand (**Fig. Bx19.2b**). Oases became important stopping points along caravan routes, allowing both people and camels to replenish water supplies.

In some oases, people settled and used the groundwater to irrigate date palms and other crops. For example, the Bahariya Oasis, about 400 km southwest of Cairo, Egypt, hosted a town of perhaps 30,000 people between 300 B.C.E. and 300 C.E. During that time, the water table lay only 5 m below the ground and could easily be reached by shallow wells. Unfortunately, as a result of changing climates and centuries of use, the water table now lies 1,500 m below the ground, almost out of reach. Bahariya's glorious past came to light in 1996, quite by accident. A man was riding his donkey in the desert near the oasis when the ground beneath the donkey suddenly caved in. The man had inadvertently opened the roof of a tomb filled with over 150 mummies, along with thousands of well-preserved artifacts. The site has since come to be known as the Valley of the Mummies.

FIGURE Bx19.2 The formation of oases in deserts.

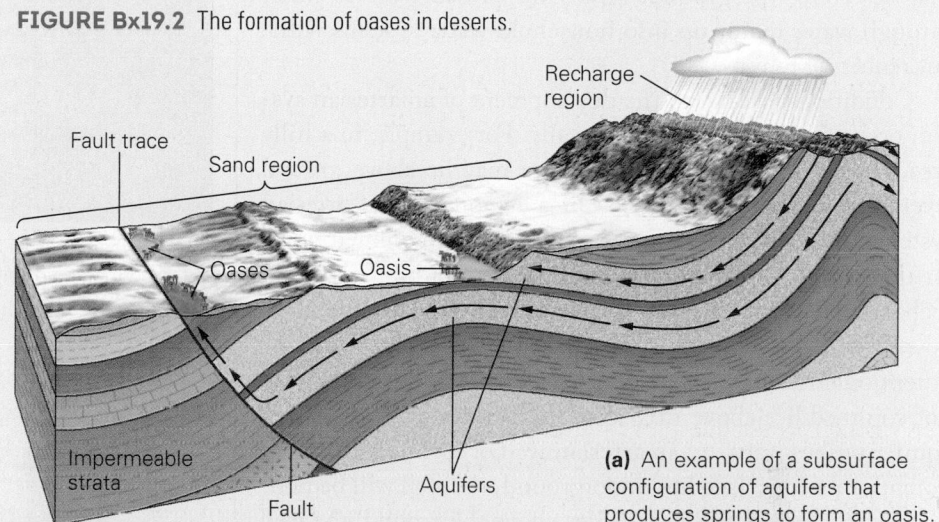

(a) An example of a subsurface configuration of aquifers that produces springs to form an oasis.

(b) A small oasis in the Sahara Desert.

a downward-pointing, cone-shaped surface called a **cone of depression (Fig. 19.12a)**. Drawdown in a deep well can lower the water table over a broad region causing nearby shallower wells to go dry **(Fig. 19.12b)**.

Artesian wells, named for the province of Artois in France, penetrate confined aquifers in which pressure causes the water to rise on its own to a level above the top surface of the aquifer. If this level lies below the ground surface, we call the well a *nonflowing artesian well*. But if the level lies above the ground surface, the well is a *flowing artesian well*, and its water actively fountains out of the ground **(Fig. 19.13a)**. (We've seen that the same phenomenon happens at an artesian spring, where a natural fracture cuts across an aquitard to provide access to the pressurized water.)

We can understand why artesian wells exist if we look first at the configuration of a city water supply (Fig. 19.13b). City water companies pump water into an elevated tank that has a significant hydraulic head relative to the surrounding areas. If the tank were connected by a water main to a series of vertical pipes, pressure caused by the elevation of the water in the tank would make the water rise in the pipes until it reached an imaginary surface, called a *potentiometric surface*, that lies above the ground. This same pressure drives water through water mains up into household water systems without requiring pumps.

Conditions leading to the development of an artesian system occur both locally and regionally. For example, in a hilly area, the local potentiometric surface may lie above ground level in valleys (Fig. 19.13c). On a regional scale, artesian systems develop where water enters a tilted, confined aquifer that intersects the ground in the hills of a high-elevation recharge area (Fig. 19.13d). The confined groundwater flows down to the adjacent plains, which lie at a lower elevation. The potentiometric surface to which the water would rise were it not confined lies above this aquifer. Pressure in the confined aquifer pushes water up an artesian well or spring. Where the potentiometric surface lies underground, the well will be nonflowing, but where the surface lies above the ground, the well will be flowing.

FIGURE 19.12 Pumping groundwater from an ordinary well can affect the water table.

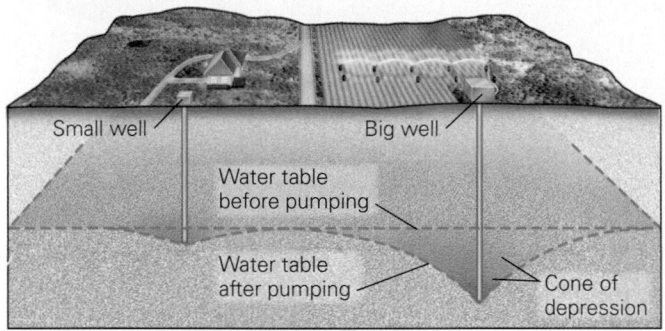

(a) Pumping forms a cone of depression in the water table.

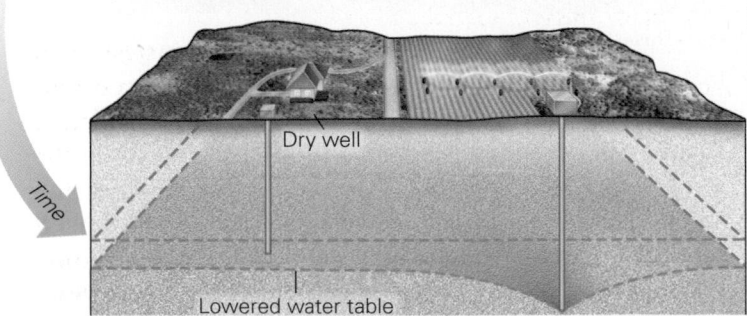

(b) Pumping by the big well may be enough to make the small well run dry.

FIGURE 19.13 Artesian wells, where water rises from the aquifer without pumping.

(a) A flowing artesian well in Wisconsin; the water rises from underground in the corrugated pipe without the need of pumping.

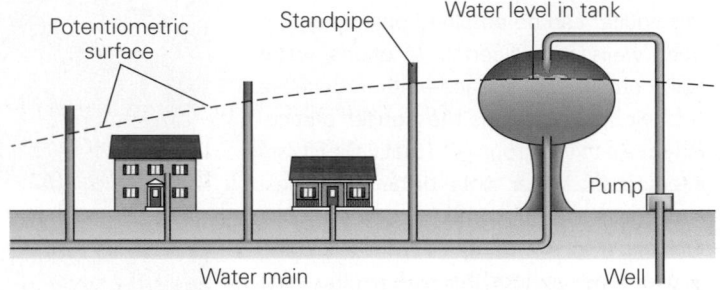

(b) The configuration of a city water supply. Water rises in vertical pipes up to the level of the potentiometric surface.

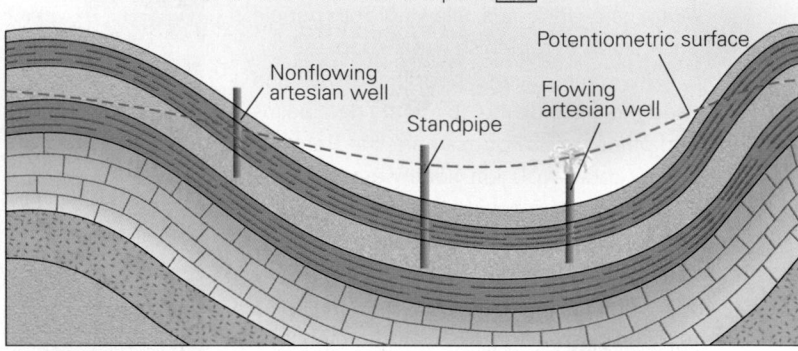

(c) The configuration of a local artesian system. Water rises above the confined aquifer because the water is under pressure.

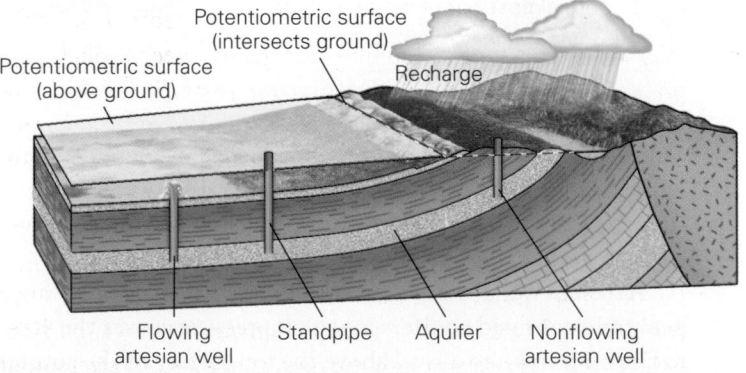

(d) The configuration of a regional artesian system.

19.6 Hot Springs and Geysers

Hot springs, springs that emit water ranging in temperature from 30°C to 104°C, develop in two geologic settings. First, they occur where groundwater, as it slowly flows from recharge area to discharge area, follows a flow path that takes it many kilometers down into the crust, where bedrock, naturally warm due to the geothermal gradient, heats the groundwater. The curving path of flow eventually carries the deep groundwater back up to the surface. Second, hot springs develop in **geothermal regions**, places where igneous activity increases the geothermal gradient substantially, so that hot rock lies close to the ground surface (**Fig. 19.14**). In geothermal regions, even shallow groundwater can become very hot. Since it has to rise only a short distance before returning to the surface at a discharge area, it comes out of the ground while still at a high temperature.

Notably, hot groundwater efficiently dissolves minerals from rock that it passes through because water becomes a more effective solvent when hot. So water emitted at hot springs can contain a high concentration of dissolved minerals. People use the water emitted at hot springs to fill relaxing mineral baths (**Fig. 19.15a**). Furthermore, the minerals in hot water feed microbes, so natural pools of geothermal water may become brightly colored—the gaudy greens, blues, and oranges of these pools come from *thermophilic* (heat-loving) bacteria and archaea that thrive in hot water and metabolize the sulfur-containing minerals dissolved in the groundwater (**Fig. 19.15b**).

Numerous distinctive geologic features form in geothermal regions as a result of the eruption of hot water. In places where the hot water rises into soils rich in volcanic ash and clay, a viscous slurry forms and fills goopy *mudpots* (**Fig. 19.15c**). Bubbles of steam rising through the mud cause it to splatter about.

FIGURE 19.14 Hot springs in geothermal regions.

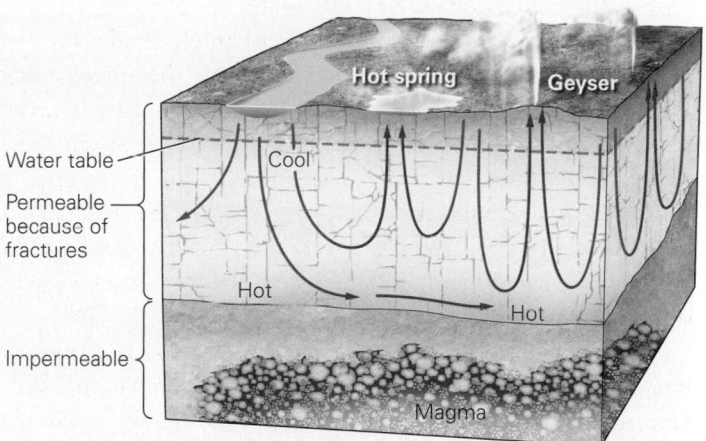

(a) Geothermal springs form where very hot rock in a region that hosts igneous activity lies relatively close to the Earth's surface and heats groundwater.

(b) The town of Rotorua, New Zealand, was built in a geothermal region associated with a volcanic arc. Steam rises at many locations in the town.

Where geothermal waters that have passed through limestone bedrock spill out of natural springs and then cool, dissolved carbonate minerals in the water precipitate, forming travertine mounds or *travertine terraces* (**Fig. 19.15d**).

Under special circumstances, a fountain of steam and boiling hot water erupts every now and then from a vent in a geothermal region (**Fig. 19.16**). An episodic fountain of hot, steamy water is a **geyser**—the name comes from the Icelandic *geysir*, a word that means gusher. To understand why a geyser erupts, we first need to picture the underground plumbing beneath one. A geyser overlies a network of irregular fractures in very hot rock; one or more of these fractures connects to the ground surface. Groundwater sinks and fills these fractures and absorbs heat from the rock. Because

FIGURE 19.15 Various features of hot springs.

(a) Hot springs in Iceland, warmed by magma below, attract tourists from around the world.

(b) Colorful bacteria- and archaea-laden pools, Yellowstone National Park, Wyoming.

(c) Mudpots in New Zealand form where boiling water mixes with ash. The ash changes into clay and forms a muddy soup.

(d) Terraces of travertine grew at Pamukkale, Turkey.

the boiling point of water—the temperature at which it vaporizes—increases with growing pressure, hot groundwater at depth can remain in liquid form even if its temperature has become greater than the boiling point of water at the Earth's surface (100°C). When such "superheated" groundwater begins to rise through a conduit toward the surface, the pressure on it decreases until eventually some of the water boils and transforms into steam. This process drives water upward in the fracture so that it spills out of the conduit at the ground surface. When this spill happens, the pressure deeper in the conduit exerted by weight of overlying water suddenly decreases. This sudden drop in pressure causes the superheated water at depth to "flash" into steam instantly, and this steam quickly rises, ejecting all the water and steam above it from the conduit in a geyser eruption. Once the conduit empties, the eruption ceases, and the conduit fills once again with water that gradually heats up, starting the eruptive cycle all over again.

Hot springs can be found in many localities around the world. Examples include Hot Springs, Arkansas, where deep groundwater rises to the surface; Yellowstone National Park, above a continental hot spot; around the Salton Sea in southern California, and in the Geysers geothermal field of northern California, both places where rifting has triggered local volcanism; in Iceland, which has grown on top of an oceanic hot spot along the Mid-Atlantic Ridge; and in Rotorua, New Zealand, which lies in an active volcanic field above a subduction zone.

People do live in some geothermal regions, despite the fact that the hot water can be a natural hazard. In Rotorua, signs along the road warn, "STEAM!" Steam indeed spills out of holes in backyards and parking lots, and can obscure visibility along roads. But all this hot water does offer a benefit as well. In geothermal regions worldwide, steam provides a clean means of generating electricity, and hot water can circulate through pipes to provide home heating.

FIGURE 19.16 Geysers form where steam erupts from the ground.

(a) A vent in Iceland begins to bubble.

(b) Water spurts out, releasing pressure on superheated water below.

(c) Water at depth transforms into steam, which rises and pushes overlying water out of the vent.

Time

(d) The Old Faithful geyser in Yellowstone National Park erupts predictably.

TAKE-HOME MESSAGE

At hot springs, warm or even boiling water emerges at the ground surface. Some hot springs form where groundwater rises from warm rock several kilometers down; others form where igneous activity heats water near the surface. At a geyser, a fountain of steam and scalding water spurts from the ground episodically.

QUICK QUESTION: Why do geysers generally erupt episodically instead of continuously?

19.7 Groundwater Problems

Since prehistoric times, groundwater has been a resource that people have relied on for drinking, irrigation, and industry. Groundwater feeds the lushness of desert oases in the Sahara, the amber grain in the North American High Plains, and the growing cities of the sunbelt. The proportion of a region's water supply that comes from groundwater depends on the local climate, the availability of other water supplies (such as lakes or rivers), local demand, and politics. While many people assume that groundwater is primarily consumed for drinking, on a global basis, agricultural and industrial applications actually account for most groundwater use—roughly 80% to 95% of the groundwater pumped by human society goes into crop irrigation or into factories. So, as once-empty land comes under cultivation and countries become increasingly industrialized, demands on the groundwater supply soar. Currently, groundwater provides only 20% to 30% of the water we use worldwide, but this percentage has been increasing as surface-water resources decrease. There are problems inherent in our use of groundwater that human society must address. In this section, we examine some of these problems.

Groundwater Depletion

Is groundwater a renewable resource? In the context of geologic time, the answer is yes, for the hydrologic cycle will eventually resupply depleted reserves, and new reserves will form as old ones disappear. But in a time frame of decades to millennia—the span of a human lifetime or a civilization—groundwater in many regions of the Earth must be viewed as a nonrenewable resource. The renewability of a particular groundwater reservoir depends on the balance between the rate of recharge and the rate of depletion.

Before the advent of civilization, a region's water table rose or sank primarily in response to seasonal variations in rainfall.

FIGURE 19.17 Irrigation can turn dry areas green.

(a) Irrigating a hayfield in Utah consumes vast quantities of water.

(b) A view from an airplane shows irrigated crop circles in an otherwise dry landscape.

The exponential growth of human populations in the past few centuries, along with improvements in living standards, have altered this natural balance. Why? Feeding ever larger numbers of people would not be possible without increasing supplies of freshwater, especially since the *green revolution* of the 1960s, when broader use of irrigation and fertilization, as well as the introduction of new crop hybrids, greatly increased not only the amount of cultivated land, but also the water consumption of existing farmland **(Fig. 19.17a, b)**. Global industrial growth also accelerates water use. For example, it takes about 10 liters of water to make a sheet of paper. And improvements in living standards increase per capita consumption of water: people use water for bathing, keeping lawns green, washing cars, and filling swimming pools.

Because of this increased demand, groundwater withdrawal significantly exceeds groundwater recharge in many regions. In fact, the sum of surface flow and recharge—cannot keep pace with the amount used **(Fig. 19.18)**. As a consequence, according to some estimates, we are depleting over 20% of the world's aquifers, and in some places we use groundwater at a rate that is 50 times the rate of recharge. Groundwater depletion problems have become particularly severe in arid regions, such as the southwestern United States, where recharge of groundwater supplies takes place very slowly and rivers and lakes are few. Researchers also

FIGURE 19.18 This map shows the combination of surface water and recharge, as measured in cubic meters per capita per year. Given that, globally, per capita water consumption averages 1,000 m³ per year, the map emphasizes that freshwater supplies are becoming insufficient in many parts of the world.

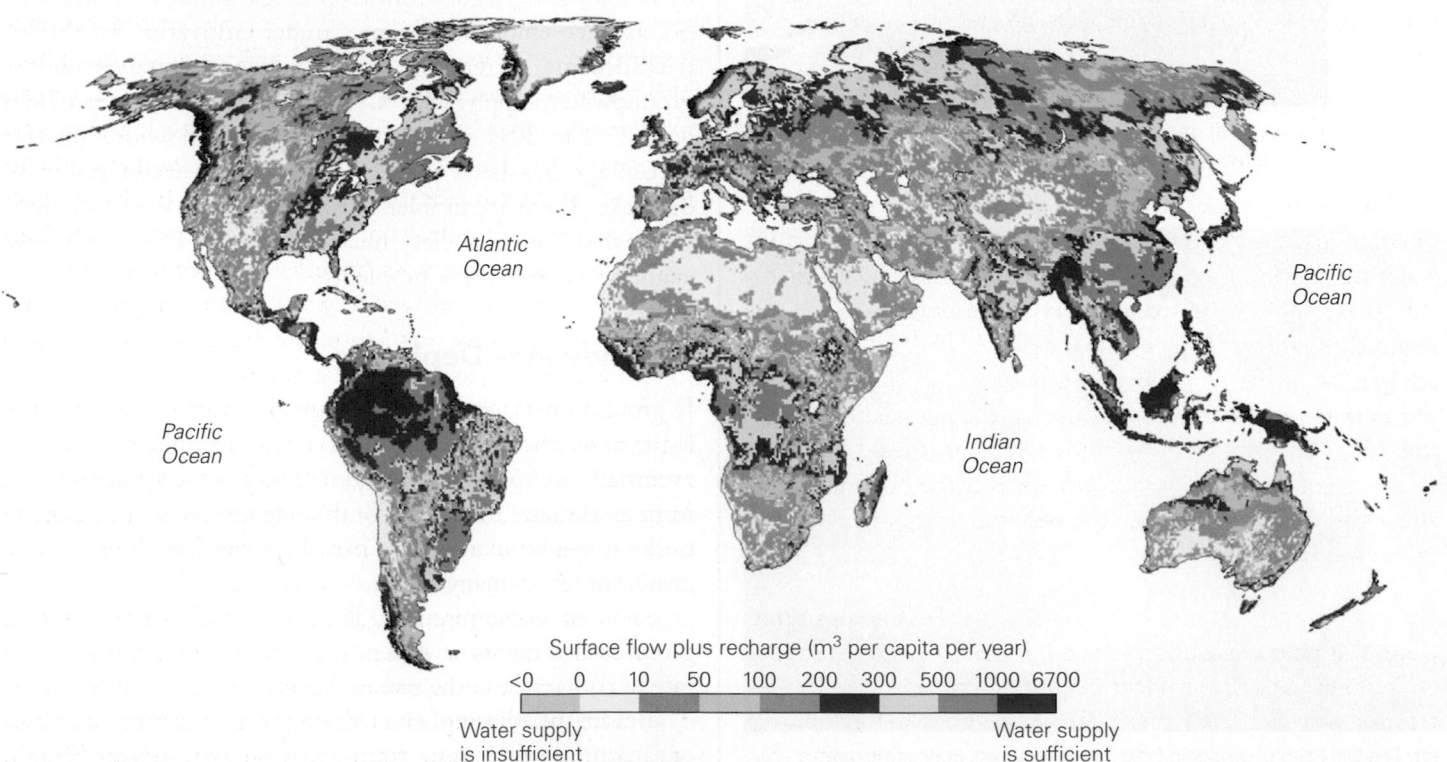

worry that changing climates, which tend to affect the distribution of rainfall, may decrease rates of groundwater recharge the most in places where it is needed the most (see Chapter 23).

When we extract groundwater at a rate faster than it can be resupplied by nature, the water table gradually becomes lower over a broad region. As a consequence, existing wells, springs, and rivers dry up (Fig. 19.19a, b), and to continue tapping into the water supply, we must drill progressively deeper. During the past decade, researchers have used data from GRACE satellites to study such *groundwater depletion*. These satellites provide precise measurements of how the Earth's gravitational pull in a given region varies over time. In some regions, a very small but detectable decrease in gravitational pull reflects groundwater depletion, because when air replaces water in subsurface pores, the mass of the near-surface realm becomes smaller. By calibrating the magnitude of gravity decrease in a region to the amount of groundwater depletion, researchers can produce maps depicting the amount by which the water table has dropped in a region. Such maps emphasize the amount of groundwater depletion during a drought in California and due to intense farming in India (Fig. 19.19c, d).

Notably, the water table can also drop when people divert surface water from the recharge area. Such a problem has developed in the Everglades of southern Florida, a huge swamp where, before the expansion of Miami and the development of agriculture in the region, the water table lay at the ground surface (Fig. 19.19e, f). Diversion of water from the Everglades' recharge area into canals has significantly lowered the water table, causing parts of the Everglades to dry up.

Other Consequences of Groundwater Use

Saline Intrusion In many regions, fresh groundwater lies in a layer above saline (salty) groundwater because of their differences in density. The boundary between fresh and saline water may be hundreds or even a few thousand meters below the surface of a large sedimentary basin, but it may lie at a relatively shallow depth along the coast, where water enters aquifers from the adjacent ocean (Fig. 19.20a, b). If people pump water out of a well too quickly, the boundary between the saline and the fresh groundwater rises. Geologists refer to this phenomenon as *saline intrusion*. If the boundary rises above the base of a well, then the well will start to yield useless saline water.

Pore Collapse and Land Subsidence When groundwater fills pore space underground, it holds the grains of rock or sediment apart, for water cannot be compressed. The extraction of water from a pore eliminates the support holding the pore open because the air that replaces the water can be compressed. As a result, the grains pack more closely together. Such *pore collapse* permanently decreases the porosity and permeability of the rock or sediment and thus lessens its value as an aquifer (Fig. 19.20c, d).

Pore collapse also decreases the volume of an aquifer, with the result that the ground above the aquifer sinks. Such *land subsidence* may cause fissures to develop at the surface and the ground to tilt (Fig. 19.21a). Buildings constructed over regions undergoing land subsidence may themselves tilt, or their foundations may crack. In the San Joaquin Valley of California, the land surface subsided by 9 m between 1925 and 1975 because so much water was removed to irrigate farm fields (Fig. 19.21b). In coastal areas, land subsidence due to groundwater removal may even make the land surface sink below sea level. The flooding of Venice, Italy, for example, accompanies land subsidence due to the withdrawal of groundwater beneath the city (Fig. 19.21c).

To avoid such problems, communities have sought to prevent groundwater depletion either by directing surface water into recharge areas or by pumping surface water back into the ground. For example, some communities excavate to lower the land surface of a park and then configure storm sewers so that they drain onto its grassy surface. The park then acts as a catchment for stormwater (Fig. 19.22).

Human-Caused Groundwater Contamination

As we've noted, some contaminants in groundwater occur naturally—sulfur, iron, calcium carbonate, methane, and salt can all be introduced to groundwater directly from the rock through which it is flowing. But in recent decades, **groundwater contamination** has increased as human activities have introduced contaminants of many kinds. These contaminants include agricultural waste (pesticides, fertilizers, and animal sewage); industrial waste (both organic and inorganic chemicals); effluent from landfills and septic tanks (including bacteria and viruses); petroleum products and other chemicals that do not dissolve in water (together referred to as nonaqueous-phase liquids); radioactive waste (from weapons manufacture, power plants, and hospitals); and acids leached from sulfide minerals in coal and metal mines.

A cloud of contaminated groundwater, known as a **contaminant plume**, moves away from a source of contamination (Fig. 19.23a, b). The plume flows in the same direction as groundwater. Note that the development of a large cone of depression around a big well can cause the flow direction of a nearby plume to change (Fig. 19.23c, d).

How do contaminants get into groundwaters? Some of these contaminants seep into the ground from leaks in surface or subsurface tanks; some infiltrate from the surface when downward-percolating water dissolves and carries chemicals with it; and some come from spills on the surface. Finally,

FIGURE 19.19 Effects of human modification of the water table.

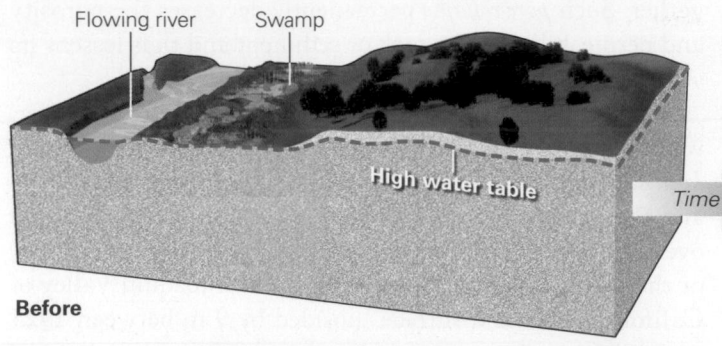

Flowing river   Swamp

High water table

Before

(a) Before humans start pumping groundwater, the water table is high. A swamp and permanent stream exist.

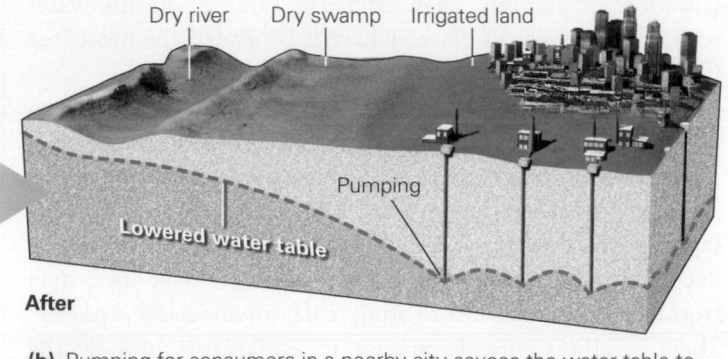

Dry river   Dry swamp   Irrigated land

Time

Lowered water table   Pumping

After

(b) Pumping for consumers in a nearby city causes the water table to sink, so the swamp dries up.

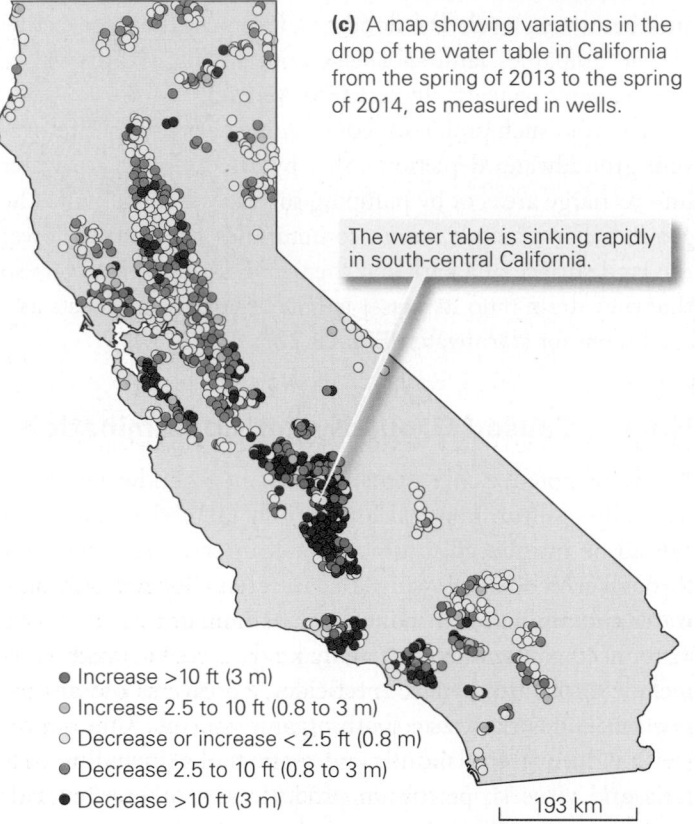

(c) A map showing variations in the drop of the water table in California from the spring of 2013 to the spring of 2014, as measured in wells.

The water table is sinking rapidly in south-central California.

- Increase >10 ft (3 m)
- Increase 2.5 to 10 ft (0.8 to 3 m)
- Decrease or increase < 2.5 ft (0.8 m)
- Decrease 2.5 to 10 ft (0.8 to 3 m)
- Decrease >10 ft (3 m)

193 km

Equivalent Height Anomaly (cm)
−12 −8 −4 0 4 8 12

India

350 km

(d) Changes in water table depth in northwestern India, based on GRACE satellite data.

(e) The Florida Everglades before the advent of urban growth and intensive agriculture.

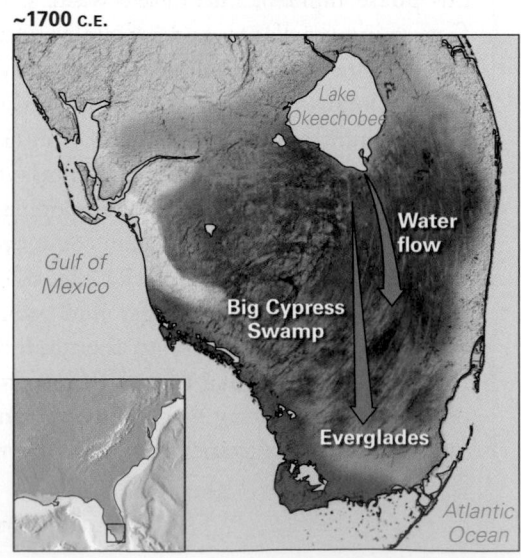

~1700 C.E.

Lake Okeechobee

Gulf of Mexico

Water flow

Big Cypress Swamp

Everglades

Atlantic Ocean

(f) Channelization and urbanization have removed water from recharge areas, disrupting groundwater flow paths.

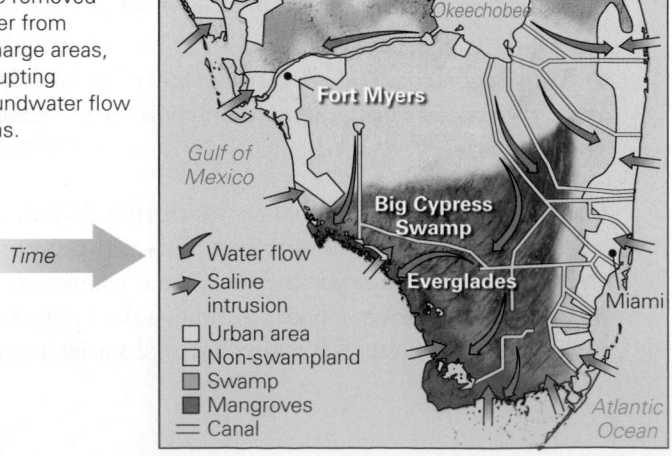

Today

Lake Okeechobee

Fort Myers

Gulf of Mexico

Big Cypress Swamp

Everglades

Miami

Time

→ Water flow
→ Saline intrusion
☐ Urban area
☐ Non-swampland
☐ Swamp
■ Mangroves
— Canal

Atlantic Ocean

FIGURE 19.20 Some problems caused by groundwater use.

Before

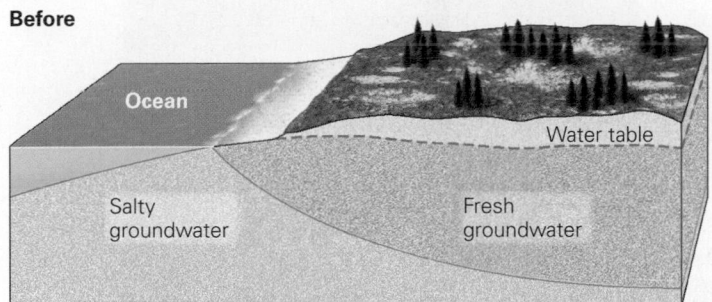

(a) Before pumping, fresh groundwater forms a lens below the ground.

After

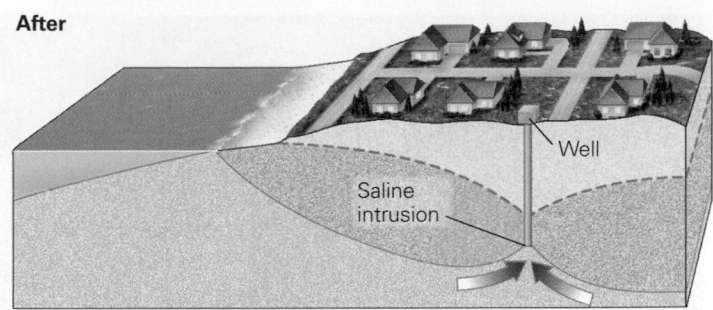

(b) If the freshwater is pumped too fast, saltwater from below is sucked up into the well. This is saltwater intrusion.

Before

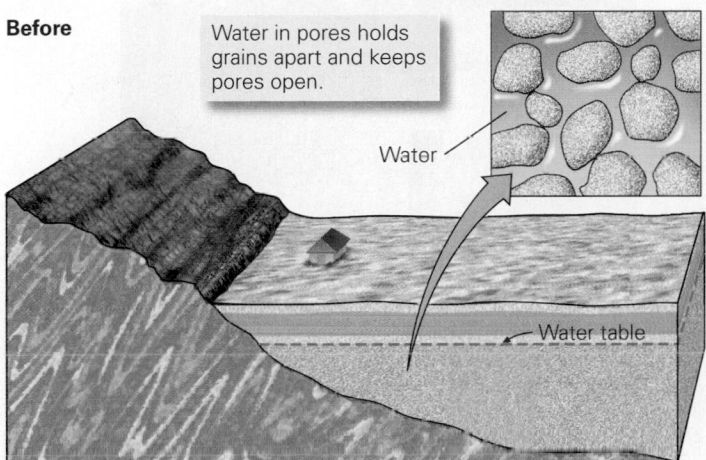

(c) When intensive irrigation removes groundwater, pore space in an aquifer collapses.

After

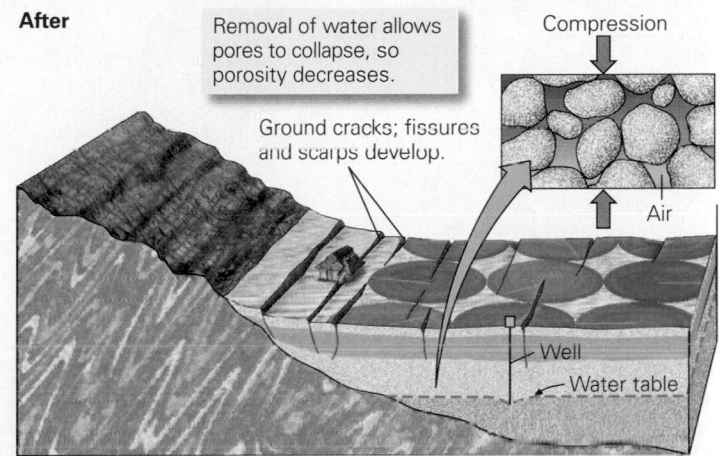

(d) As a result, the land surface sinks, leading to the formation of ground fissures and causing houses to crack.

some are intentionally forced into aquifers through *injection wells*, meaning wells in which a liquid is pumped down into the ground under pressure so that it passes from the well back into the pore space of the rock or sediment.

Sadly, staggering quantities of contaminating liquids (trillions of gallons in the United States alone) enter the groundwater system every year. Contaminants can be carried with local or regional groundwater flow up to tens of kilometers from the source. In some cases, local groundwater flow carries contaminants into a nearby stream that may serve as a municipal water supply. Such a situation developed in West Virginia in 2014, when an organic chemical used to wash coal spilled from a storage tank onto the ground and then migrated underground to the nearby Elk River, which provides the water supply for the city of Charleston and surrounding regions. Three hundred thousand residents temporarily lost their water supply.

To avoid groundwater contamination, people must prevent contaminants from entering groundwater in the first place. This can be done by locating potential sources of contamination on impermeable bedrock, isolated from aquifers. If such a site is not available, the storage area should be lined with a thick layer of clay or impermeable plastic. Clay not only acts as an aquitard, but can hold onto contaminants, and plastic liners keep contaminants from coming in contact with substrate. Government agencies have studied various options for safely storing containers containing contaminants. One option involves stockpiling the containers in tunnels cut into salt domes, because salt is impermeable. Another option involves placing the containers in tunnels above the water table. For example, officials considered a proposal to store American nuclear waste in a network of tunnels 300 m beneath Yucca Mountain, Nevada, a hill that consists of dry, fairly impermeable tuff high above the water table.

In some cases, natural processes can clean up groundwater contamination. For example, chemicals may be absorbed by clay, oxygen in the water may oxidize them, and bacteria in the water may metabolize them, thereby turning them into harmless substances. If natural processes don't solve the problem, what can geologists do? First, *environmental geologists*, who work with engineers to remediate contaminated sites, drill test wells to determine which way and how fast the contaminant plume is flowing. Once they know the flow path, they can close wells in the path to prevent consumption of

FIGURE 19.21 Consequences of groundwater removal and subsurface collapse.

(a) Ground fissures in Arizona.

The dates indicate ground elevation in the past.

(b) Sinking ground surface, California.

(c) Sinking and submergence continue to cause damage in Venice, Italy.

To carry out bioremediation, geologists inject oxygen and nutrients into a contaminated aquifer to foster the growth of bacteria that can consume and break down contaminant molecules. The second technique involves the insertion of *permeable reactive barriers* underground. These barriers are, in effect, subsurface walls of materials such as iron filings that react with contaminants chemically to transform them into safer, less soluble materials. To produce a permeable reactive barrier, contractors dig a trench in the path of the contaminant plume, fill it with the reactive material, and bury it. When the plume flows through the barrier, the reactions take place. Needless to say, cleaning groundwater can be expensive and may be only partially effective.

contaminated water. Engineers may also attempt to clean the groundwater by drilling a series of *extraction wells* to pump it out of the ground. If the contaminated water does not rise fast enough, engineers drill injection wells to force clean water or steam into the ground beneath the contaminant plume **(Fig. 19.24)**. The injected fluids then push the contaminated water up into the extraction wells.

More recently, scientists have begun exploring two techniques that may help remediate groundwater contamination. The first technique, *bioremediation*, involves microbes.

FIGURE 19.22 Sketch of an enhanced recharge catchment in a city.

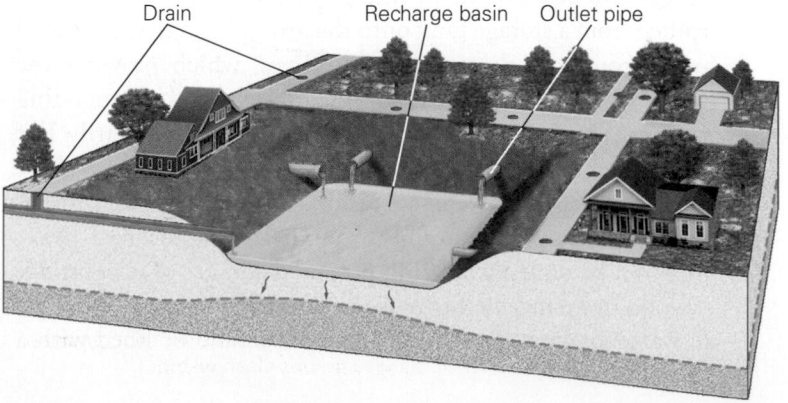

Drain   Recharge basin   Outlet pipe

Unwanted Effects of Rising Water Tables

We've seen the negative consequences of sinking water tables, but what happens when the water table rises? Is that necessarily good? Sometimes, but not always. If the water table rises above the level of a house's basement, water seeps through the foundation and floods the basement floor. As we noted in Chapter 16, catastrophic damage may occur when a rising water table weakens the base of a hillslope or an underground failure surface triggers landslides and slumps **(Fig. 19.25)**.

FIGURE 19.23 Contamination plumes in groundwater.

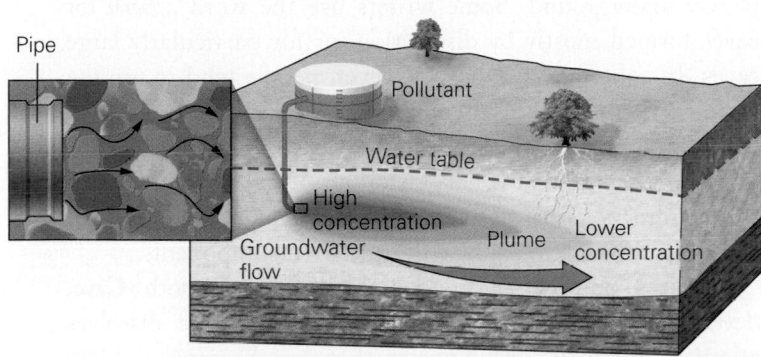

(a) A contaminant plume as seen in cross section. The darker the color, the greater the concentration of contaminant.

FIGURE 19.24 Steam injected beneath the contaminant plume drives the contaminated water upward in the aquifer, where extraction wells remove it.

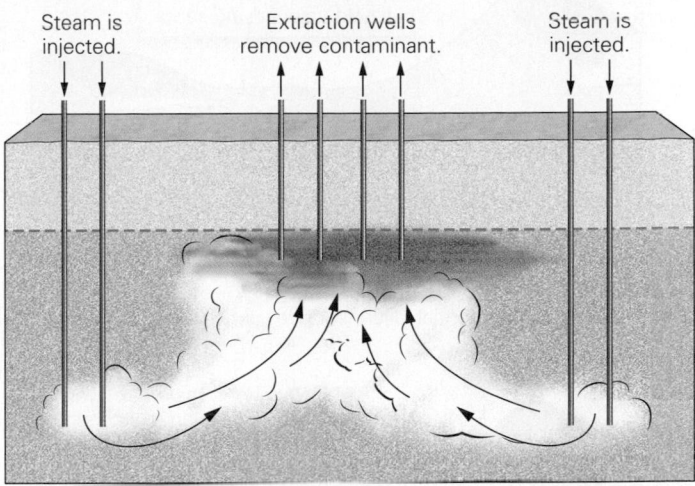

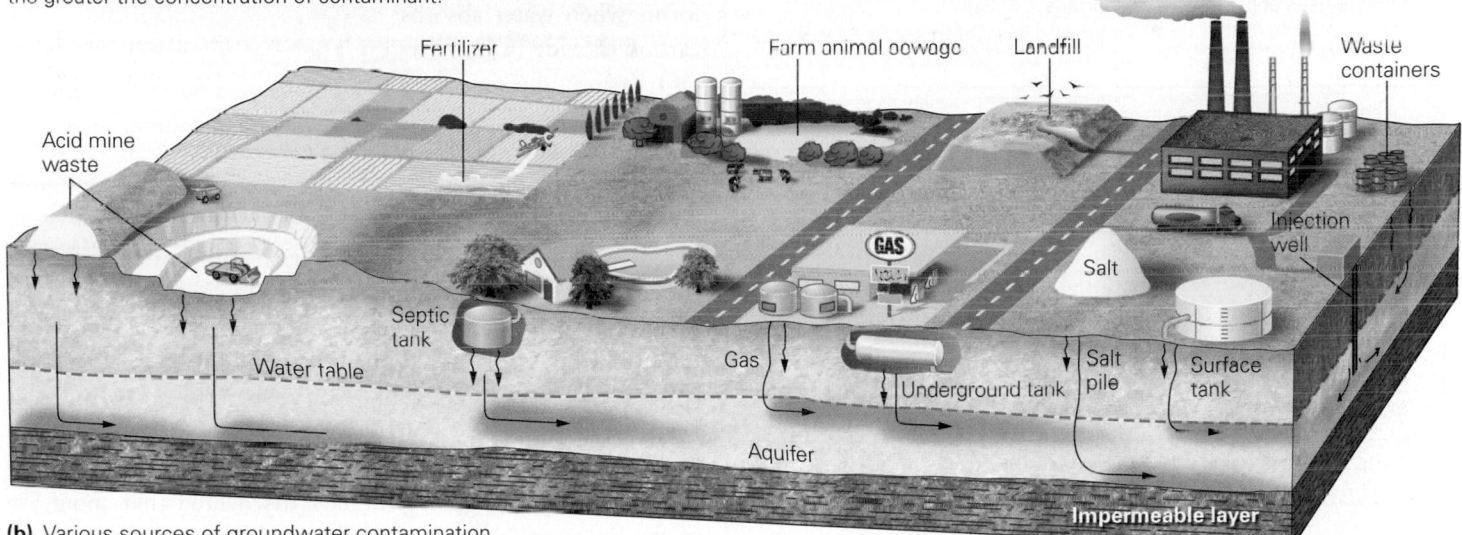

(b) Various sources of groundwater contamination.

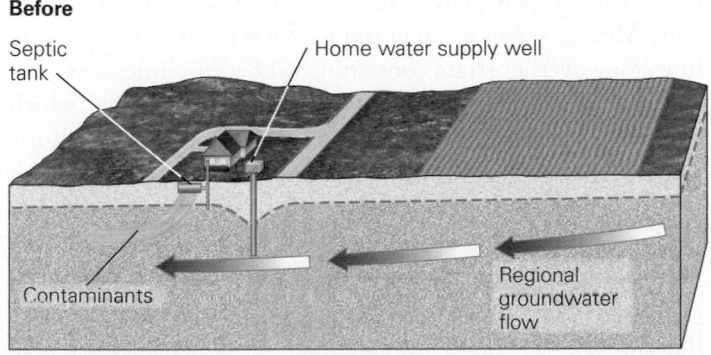

(c) Before pumping, effluent from a septic tank drifts with the regional groundwater flow, and the home well pumps clean water.

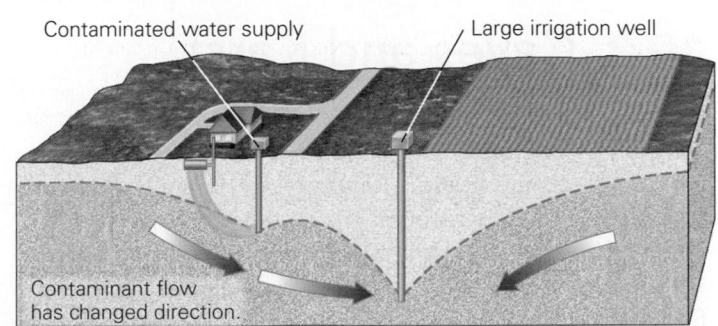

(d) After pumping by a nearby irrigation well, effluent flows into the home well in response to the new local slope of the water table.

FIGURE 19.25 When the water table rises, material above a weak layer of rock begins to slump, and a landslide may result.

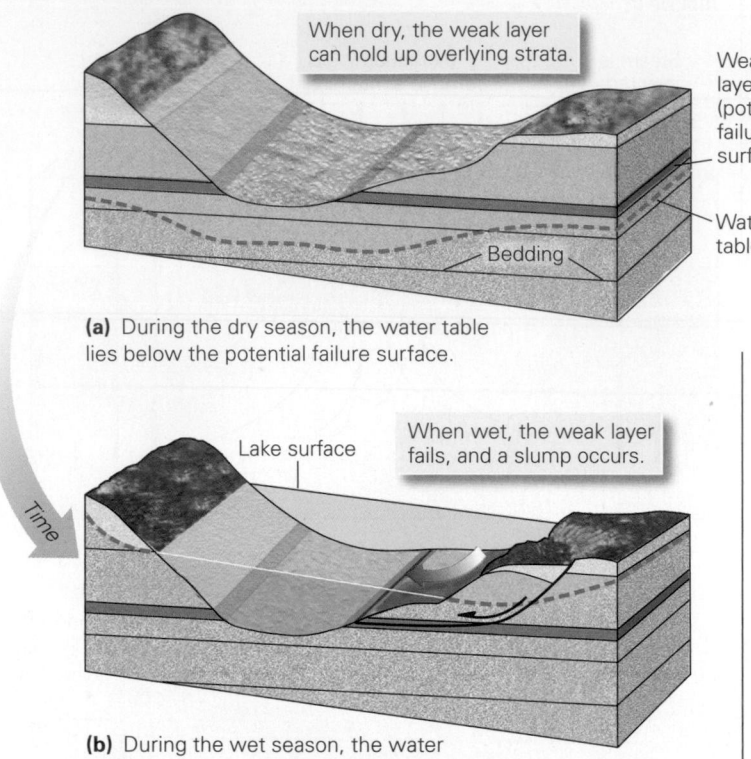

When dry, the weak layer can hold up overlying strata.

Weak layer (potential failure surface)

Water table

Bedding

(a) During the dry season, the water table lies below the potential failure surface.

When wet, the weak layer fails, and a slump occurs.

Lake surface

Time

(b) During the wet season, the water table rises above the failure surface.

TAKE-HOME MESSAGE

Groundwater use can cause problems, and growing evidence shows that in many locations around the globe, people are extracting groundwater at rates far in excess of natural recharge. Too much pumping lowers the water table, causing land subsidence or saltwater intrusion. Contamination can ruin a groundwater supply. Remediation of a contaminant plume can be extremely expensive.

QUICK QUESTION: What is bioremediation?

19.8 Caves and Karst

The Development of Caves

In 1799, as legend has it, a hunter by the name of Houchins was tracking a bear through the wooded hills of Kentucky when the bear suddenly disappeared. Baffled, Houchins plunged through the brambles trying to sight his prey. Suddenly he felt a draft of surprisingly cool air flowing downslope from uphill. Now curious, Houchins climbed up the hill and found a dark portal into the hillslope beneath a ledge of rocks. Bear tracks were all around—was the creature inside? He returned later with a lantern and cautiously stepped into the passageway. After walking a short distance, he found himself in a large, underground room. Houchins was the first person of European descent to enter Mammoth Cave.

In a general sense, a **cave** is any underground open space, most or all of which does not receive direct sunlight. A cave may grow due to the dissolution of rock, or by the removal of rock by rockfalls or erosion. Many caves open up along a cliff face, whereas others lie completely underground. Some writers use the word *cavern* for caves formed mostly by dissolution, or for particularly large caves. In colloquial English, however, people tend to use the words cave and cavern interchangeably. Commonly, a region may contain many interconnected caves which together comprise a *cave network*. The name Mammoth Cave actually refers to an immense cave network—you would have to walk (or crawl) for 630 km to explore all its known components.

Most large cave networks, including Mammoth Cave, develop in limestone bedrock because limestone dissolves relatively easily in groundwater that has become a dilute solution of carbonic acid (H_2CO_3). Such acidic groundwater forms when water absorbs carbon dioxide (CO_2). The CO_2 in groundwater comes from two sources—a small amount dissolves in rain as it falls through the sky, but most dissolves in water percolating down through organic-rich soil on its way to the water table. When carbonic acid comes in contact with calcite ($CaCO_3$) in limestone, the two chemicals react to produce HCO_3^{1-} and Ca^{2+} ions, which then dissolve. The reaction releases CO_2 back into the air—you've seen this reaction if you used the acid test to identify calcite when studying minerals in Chapter 5.

In recent years, geologists have discovered that about 5% of the limestone caves around the world formed due to reactions with sulfuric-acid-bearing water—Carlsbad Caverns in New Mexico serves as an example. Such caves develop where limestone overlies strata containing oil because microbes convert the sulfur in the oil to hydrogen sulfide (H_2S) gas, which rises and reacts with oxygen and water to produce sulfuric acid. This acid, in turn, eats into limestone and reacts with it to produce gypsum and CO_2 gas.

Geologists debate about the depth at which limestone cave networks form. Some limestone dissolves above the water table, particularly along joints, which act as conduits for water flowing into the subsurface, and some limestone may dissolve deep below the water table. But it appears that most cave growth,

> **Did you ever wonder...**
> why huge underground caverns form?

FIGURE 19.26 Development of caves and speleothems.

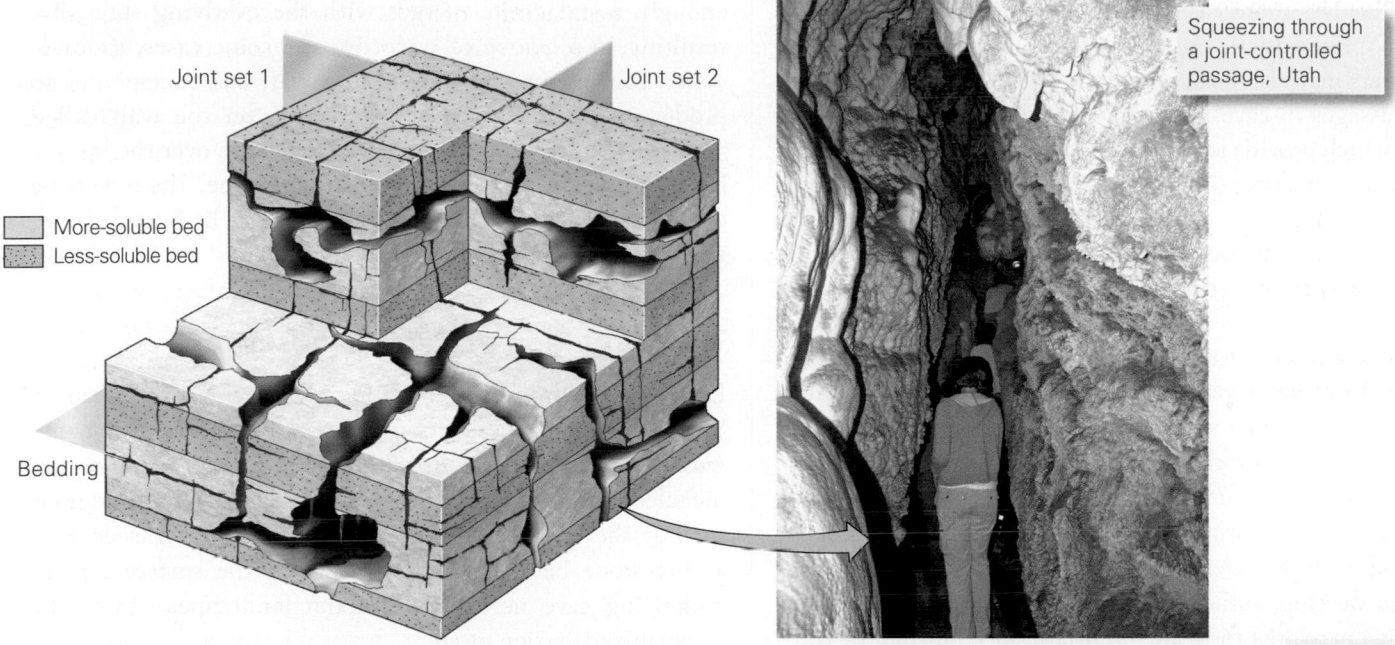

Joint set 1   Joint set 2

More-soluble bed
Less-soluble bed

Bedding

Squeezing through
a joint-controlled
passage, Utah

(a) Joints act as conduits for water in cave networks. Passages follow joints and preferentially form in more soluble beds.

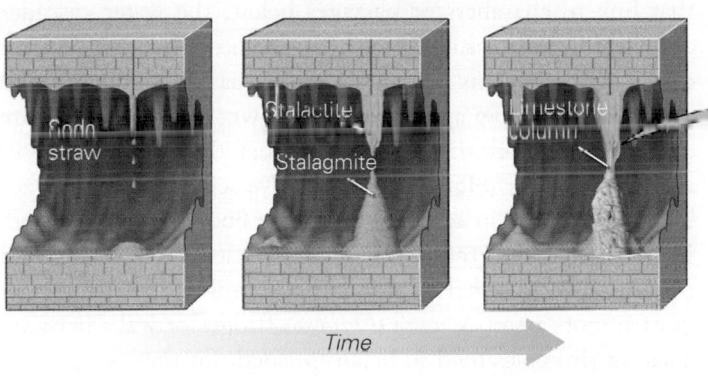

Soda
straw

Stalactite

Stalagmite

Limestone
column

Time

(b) The evolution of a soda straw stalactite into a limestone column.

Stalactites and
stalagmites in
a cave in Italy

or *speleogenesis*, takes place in limestone that lies just below the water table. Here groundwater acidity remains high, the mixture of groundwater and newly added rainwater is undersaturated (meaning that it has the capacity to dissolve more ions), and groundwater flows the fastest. If the water table goes up or down, the depth at which speleogenesis takes place goes up or down, so a region may host several levels of caves.

The Character of Cave Networks

Cave networks include *rooms*, or *chambers*, which are large, open spaces, sometimes with cathedral-like ceilings, and *passages*, which are tunnel- or slot-shaped corridors (**Geology at a Glance**, pp. 750–751). Some chambers may host underground lakes, and some passages may serve as conduits for underground streams. The shape of a cave network reflects variations in the permeability and the composition of the rock from

1m

(c) Flowstone on the wall of a cave in Vietnam.

which the caves formed. Chambers develop where the limestone dissolves more easily, so in a sequence of strata, caves develop preferentially in the most soluble limestone beds. Chambers may also form where groundwater flows most rapidly. Passages in cave networks typically follow pre-existing joints, which provide secondary porosity along which groundwater can flow faster **(Fig. 19.26a)**. Because joints commonly occur in orthogonal systems (consisting of two sets of joints oriented at right angles to each other; see Chapter 11), cave passages may form a grid.

Why do extensive cave networks, with large chambers and abundant passages, develop in some locations but not in others? There are several reasons. First, most caves form in limestone, so without a thick layer of limestone in the subsurface, extensive networks can't form. Second, the dissolution that forms caves occurs primarily in freshwater at the water table, so unless the water table lies above sea level and below the land surface, extensive networks can't form. Third, for caves to develop, sufficient liquid water must be present, so extensive networks form preferentially in temperate or tropical regions drenched by rain—they do not develop in polar regions, where ice covers the ground and subsurface water remains permanently frozen, nor do they develop in desert regions, which lack sufficient water. Finally, since percolation through organic matter provides the acidity that groundwater must have in order to dissolve limestone, most caves form faster beneath regions that have organic-rich soil.

Precipitation and the Formation of Speleothems

When the water table drops below the level of a cave that has developed, speleogenesis slows down or ceases, and the cave becomes an open space filled with air. In places where downward-percolating groundwater containing dissolved calcite emerges from the rock above the cave and drips from the ceiling, the surface of the cave gradually changes. As soon as this water re-enters the air, it evaporates a little and releases some of its dissolved carbon dioxide. As a result, calcite precipitates out of the water and produces a type of travertine called *dripstone*. The various intricately shaped formations that grow in caves by the accumulation of dripstone are called **speleothems**.

Cave explorers (spelunkers) and geologists have developed a detailed nomenclature for different kinds of speleothems **(Fig. 19.26b)**. Where water drips from the ceiling of the cave, calcite initially precipitates around the outside of the drip, forming a delicate, hollow *soda straw*. But eventually, the soda straw fills up and water migrates down the margin to form a massive, icicle-like cone called a **stalactite**, which grows downward from the ceiling. Where the drips hit the floor, the resulting precipitate builds an upward-pointing cone called a **stalagmite**, which grows upward from the floor.

If the process of dripstone formation in a cave continues long enough, a stalagmite merges with the overlying stalactite, resulting in a *column* of travertine. In some cases, groundwater flows along the surface of a wall and precipitates to produce cloth-like sheets of travertine on the wall called *flowstone* **(Fig. 19.26c)**. If groundwater flows over the lip of a ledge, it may precipitate a *curtain* of travertine. The travertine of caves tends to be translucent and, when lit from behind, glows with an eerie amber light.

The Formation of Karst Landscapes

Limestone bedrock underlies most of the Kras Plateau in Slovenia, along the east coast of the Adriatic Sea. The name *kras*, which means rocky ground, is apt because this region includes abundant rock exposures. Geologists refer to regions such as the Kras Plateau, whose surface landforms develop as limestone bedrock dissolves both at the surface and in underlying cave networks, as **karst landscapes**—from the Germanized version of *kras*.

Karst landscapes typically display a number of distinct landforms. Where surface streams intersect cracks (joints) or holes that link to chambers or passages below, the water cascades downward into the subsurface and vanishes **(Fig. 19.27)**. Such **disappearing streams** may flow through passages underground and re-emerge from a cave entrance downstream. *Sinkholes* are circular depressions that form either when the ground collapses into a cave below **(Fig. 19.28a)**, as we've seen, or when surface bedrock dissolves in acidic water on the floor of a bog or pond. In cases where the ground collapses over a long, joint-controlled passage, sinkholes may be elongate and canyon-like. Remnants of cave roofs stand as *natural bridges*. Ridges or walls between adjacent sinkholes tend to be steep-sided, for they were originally joint controlled. Over time, the walls erode, leaving only

FIGURE 19.27 A small disappearing stream in the Hudson Valley region of New York; the water is dropping into a subsurface cave.

jagged, isolated limestone *spires*—a karst landscape dominated by such spires is called *tower karst*. The surreal collection of pinnacles constituting the tower karst landscape in the Guilin region of China has inspired generations of artists who portray them in scroll paintings (Fig. 19.28b).

Karst landscapes typically take a long time to form. Their development involves a series of stages (Fig. 19.29):

- *The establishment of a water table in limestone:* The story of a karst landscape begins after the deposition and lithification of a thick interval of limestone. If, later in geologic time, erosion and uplift cause exhumation of the limestone, a water table can develop in the limestone below the ground surface.

- *The formation of a cave network:* Once the water table has been established, dissolution begins, and a cave network develops.

- *A drop in the water table:* If the water table later becomes lower, either because of a decrease in rainfall or because nearby rivers cut down through the landscape and drain the region, newly formed caves dry out. Downward-percolating groundwater emerges from the roofs of the caves; speleothems precipitate.

- *Roof collapse:* If rocks fall off the roof of a cave for a long time, the roof eventually collapses. Such collapse yields sinkholes and troughs, leaving behind hills, ridges, and natural bridges of limestone.

Life in Caves

Despite their lack of light, caves are not sterile, lifeless environments. Caves that are open to the air provide a refuge for bats as well as for various insects and spiders. Similarly, fish and crustaceans enter caves where streams flow in or out. Species living in caves have developed some unusual characteristics. For example, some of the fish species that evolved in caves lost their pigment and, in some cases, their eyes

FIGURE 19.28 Features of karst landscapes.

(a) Karst terrains typically have a rough, rocky surface. Here, we see sinkholes of the Kras Plateau.

(b) The tower karst landscape of Guilin, China, today (photo), and in the past (painting).

Tower karst of Halong Bay, Vietnam, have been submerged by the sea.

Caves and Karst Landscapes

Limestone is soluble in acidic water. Much of the water that falls to the ground as rain or seeps through the ground as groundwater is acidic, so in regions of the Earth where bedrock consists of limestone, we find signs of dissolution. Networks of underground openings that develop by dissolution are called caves or caverns. Some parts of these may be large, open chambers, whereas others are

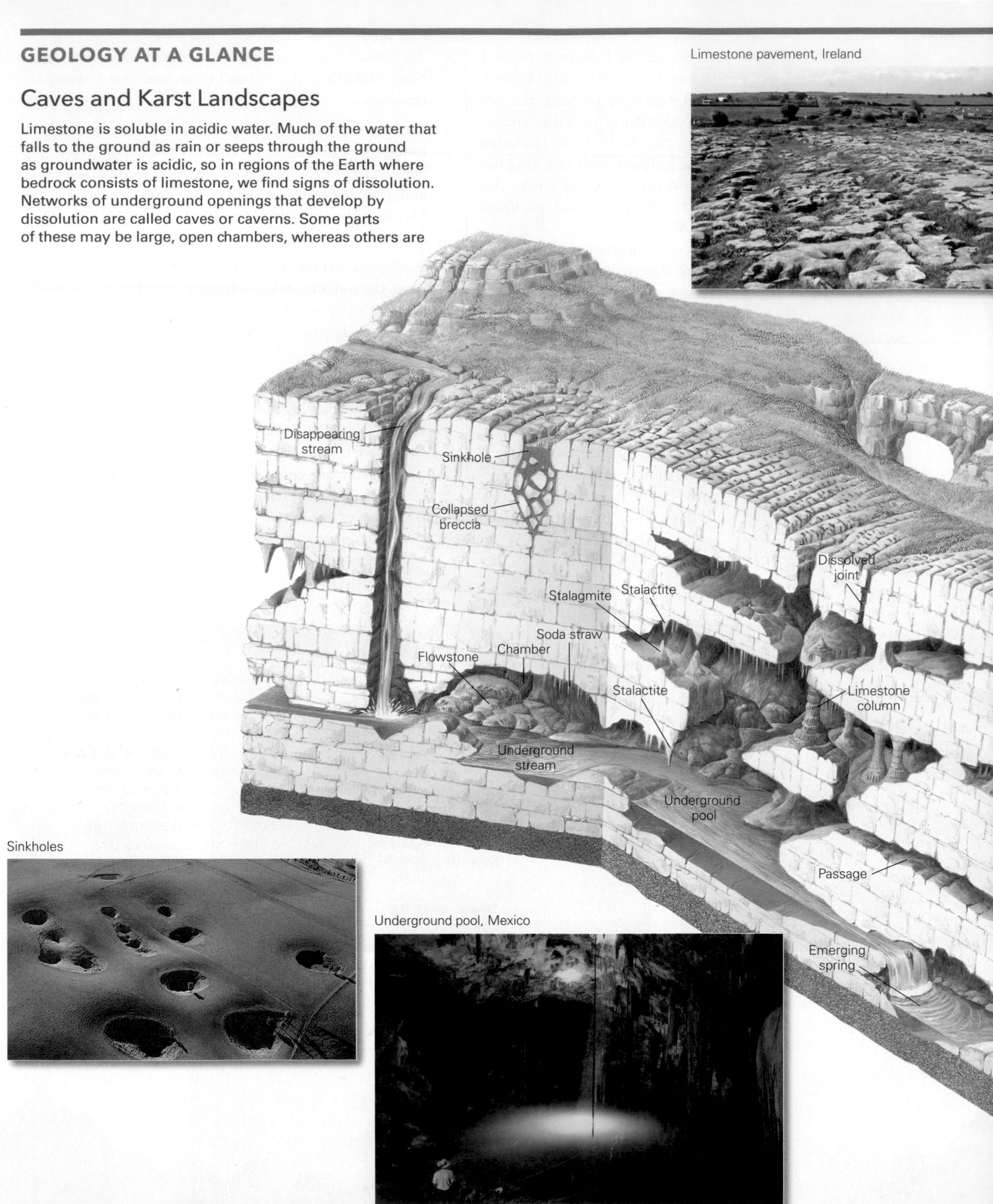

Limestone pavement, Ireland

Disappearing stream

Sinkhole

Collapsed breccia

Dissolved joint

Stalagmite

Stalactite

Soda straw

Chamber

Flowstone

Stalactite

Limestone column

Underground stream

Underground pool

Passage

Sinkholes

Underground pool, Mexico

Emerging spring

Natural Bridge, Virginia

Spelunker crawling in a cave

long, narrow passages. Underground lakes and streams may cover the floor. A cave's location depends on the orientation of bedding and joints, for these features localize the flow of groundwater.

In many locations, groundwater drips from the ceiling of a cave or flows along its walls. As the water evaporates and loses its acidity, calcite precipitates. Over time, this calcite builds into cave formations, or speleothems, such as stalactites, stalagmites, columns, and flowstone. Distinctive landscapes, called karst landscapes, develop at the Earth's surface over limestone bedrock that has undergone dissolution.

Soda-straw stalactites, Utah

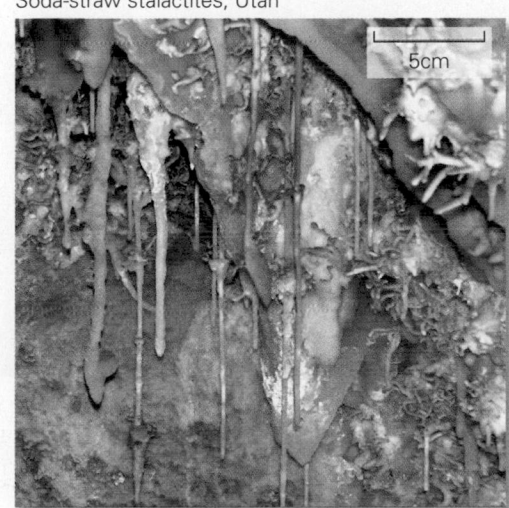
5cm

FIGURE 19.29 The progressive formation of caves and a karst landscape.

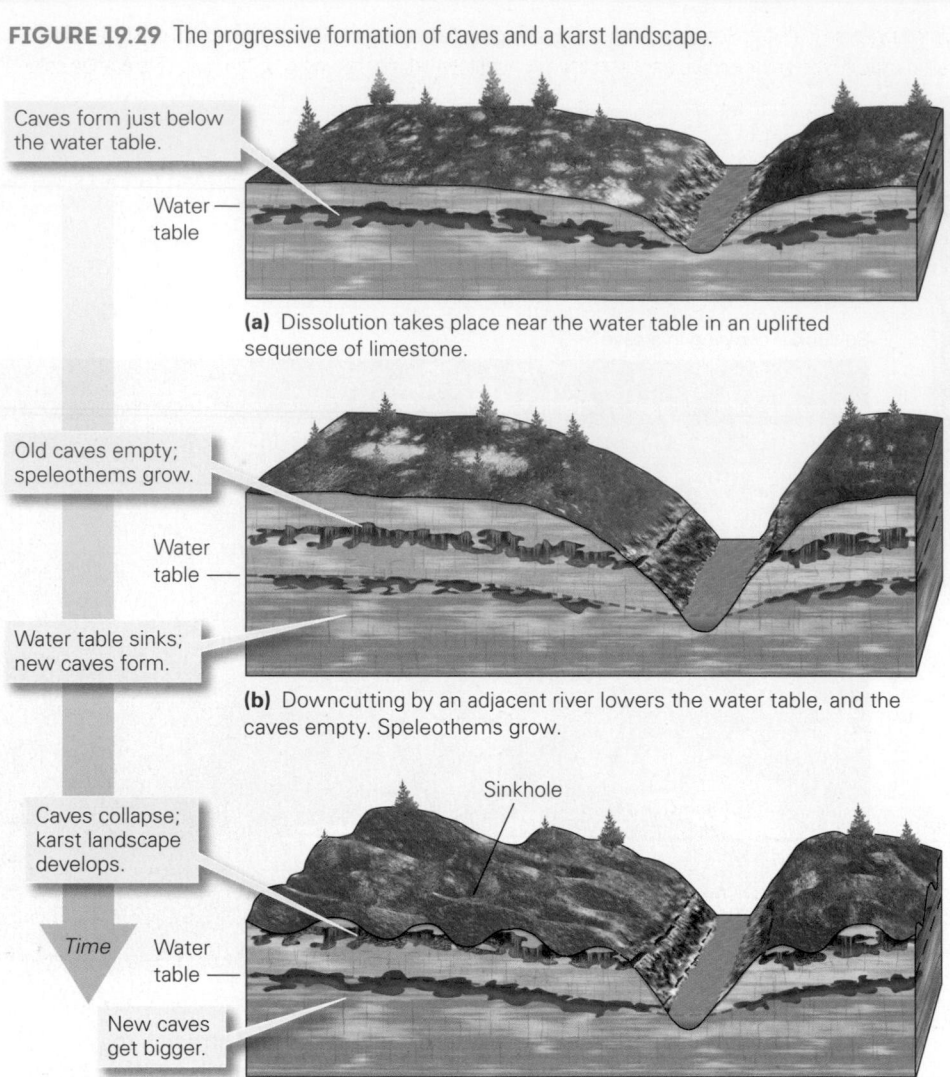

Caves form just below the water table.

Water table

(a) Dissolution takes place near the water table in an uplifted sequence of limestone.

Old caves empty; speleothems grow.

Water table

Water table sinks; new caves form.

(b) Downcutting by an adjacent river lowers the water table, and the caves empty. Speleothems grow.

Caves collapse; karst landscape develops.

Sinkhole

Time   Water table

New caves get bigger.

(c) After roof collapse, the landscape becomes pockmarked with sinkholes.

(Fig. 19.30a). Explorers have discovered caves in Mexico in which warm, mineral-rich groundwater currently flows. Colonies of bacteria metabolize sulfur-containing minerals in this water and create thick mats of living ooze in the complete darkness of the cave. Long gobs of these bacteria slowly drip from the ceiling. Because of the mucus-like texture of these drips, they have come to be known as "snottites" (Fig. 19.30b).

TAKE-HOME MESSAGE

Reaction with natural acids in groundwater dissolves limestone underground to form cave networks. Most dissolution takes place near the water table. If the water table sinks relative to a cave, dripping water in the cave can produce speleothems. Collapse of a cave network produces karst terrain.

QUICK QUESTION: Can organisms live in the pitch black of caves? If so, how?

FIGURE 19.30 Unusual organisms have evolved in the darkness of caves.

(a) A blind fish—there's no need for eyes in a cave.

(b) Gobs of bacteria drip from the ceiling of a cave to form snottites.

ANOTHER VIEW Not all hot-spring deposits consist of travertine (calcite). Some, such as these at Pagosa Hot Springs in Colorado, contain dissolved silica (SiO_2), which precipitates to form tufa. In this photo, tufa deposits coat a small escarpment and have accumulated along the edges of terraces. The orange color comes from thermophilic microbes.

SUMMARY

- During the hydrologic cycle, water infiltrates the ground and fills the pores and cracks in rock and sediment. This subsurface water is called groundwater. The amount of open space in rock or sediment determines its porosity; the degree to which pores are interconnected so that water can flow through defines its permeability.

- Geologists classify rock and sediment according to their permeability. Aquifers are relatively porous and permeable, and aquitards are relatively impermeable.

- The water table is the surface in the ground above which pores contain some air and below which pores are filled with water. The shape of a water table tends to be a subdued imitation of the shape of the overlying land surface.

- Groundwater flows wherever a hydraulic gradient exists, meaning that it flows from where it's under more pressure to where it's under less pressure. This flow carries it slowly from recharge areas to discharge areas. Darcy's law shows that flow rate depends on permeability and on the hydraulic gradient.

- Groundwater contains dissolved ions. These ions may come out of solution to form cement or veins.

- At a spring, groundwater exits the ground on its own. There are many geologic configurations that lead to spring formation.

- An ordinary well fills with water because it penetrates ground below the water table. In an artesian well, water rises on its own. Pumping water out of a well too fast yields a cone of depression.

- Hot springs and geysers release hot water to the Earth's surface. This water may have been heated by passing deep in the crust or by the proximity of a magma chamber or of recently formed and still hot igneous rock.

- Groundwater is a precious resource used for municipal water supplies, industry, and agriculture. In recent years, some regions have lost their groundwater supply because of overuse or contamination.

- Pore collapse, lowering of the water table, and saline infiltration can result from groundwater overuse.

- When limestone dissolves just below the water table, underground caves are the result. Soluble beds and joints determine the location and orientation of caves.

- If the water table drops, caves empty out. Limestone precipitates out of water dripping from cave roofs and creates speleothems, such as stalagmites and stalactites.

- Regions where abundant caves have collapsed to form sinkholes are karst landscapes. These terrains contain sinkholes, natural bridges, and disappearing streams.

GUIDE TERMS

aquifer (p. 725)
aquitard (p. 725)
artesian spring (p. 734)
artesian well (p. 735)
capillary fringe (p. 727)
cave (p. 746)
cone of depression (p. 735)
contaminant plume (p. 741)
disappearing stream (p. 748)

discharge area (p. 730)
geothermal region (p. 737)
geyser (p. 737)
groundwater (p. 722)
groundwater contamination (p. 741)
hard water (p. 731)
hot spring (p. 737)
hydraulic gradient (p. 732)

hydraulic head (p. 729)
karst landscape (p. 748)
oasis (p. 734)
perched water table (p. 728)
permeability (p. 724)
pore (p. 722)
porosity (p. 722)
recharge area (p. 730)
sinkhole (p. 722)

soil moisture (p. 726)
speleothem (p. 748)
spring (pp. 731, 733)
stalactite (p. 748)
stalagmite (p. 748)
water table (p. 726)

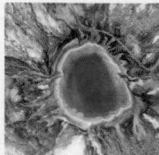

GEOTOURS *THIS CHAPTER'S GEOTOUR WORKSHEET (P) FEATURES QUESTIONS AND GOOGLE EARTH SITES ON:*

- Groundwater reserves and irrigation in the Arabian Desert

- Surface and groundwater flow in the Everglades

- Hot springs in Yellowstone National Park
- Karst features around the world

REVIEW QUESTIONS

The letters following each Review Question refer to the corresponding Learning Objective from the Chapter Opener.

1. What is groundwater, and where does it reside on the Earth? **(A)**

2. How do porosity and permeability differ? Where would porosity and permeability be higher in this figure? Give examples of substances with high porosity but low permeability. **(B)**

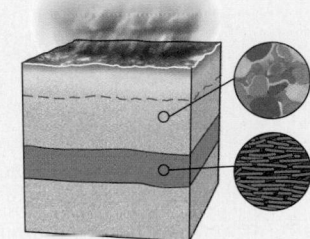

3. What is a water table, and what factors affect its level? What factors affect the flow direction of the water below the water table? **(C)**

4. How does the rate of groundwater flow compare with that of moving ocean water or river currents? **(A)**

5. What does Darcy's law tell us about rates of discharge in groundwater? **(A)**

6. How does the chemical composition of groundwater change over time? Why is "hard water" hard, and saline water saline? **(A)**

7. How does excessive pumping affect the local water table? **(F)**

8. Why do natural springs form? **(D)**

9. How is an artesian well different from an ordinary well? **(D)**

10. Explain why hot springs form and what makes a geyser erupt. **(E)**

11. Is groundwater a renewable or a nonrenewable resource? Explain your answer. **(F)**

12. Describe some of the ways in which human activities adversely affect the water table. What impact has water pumping on a large scale had on the nearby environment in this diagram? **(F)**

13. What are some sources of groundwater contamination? How can it be prevented? **(F)**

14. Describe the process leading to the formation of caves and the speleothems within caves. **(G)**

15. Describe the various features of a karst landscape, and explain how such landscapes evolve. **(G)**

ON FURTHER THOUGHT

16. The population of Desert Paradise (DP; a fictitious town in the southwestern United States) has been doubling every seven years. Most new inhabitants are "snowbirds," people escaping the cold winters of more northerly latitudes. There are no permanent streams or lakes near DP. In fact, the only standing water in town is in the ponds of the many golf courses that have been built recently. The water in these ponds needs to be replenished almost constantly, for without supplementation, the water evaporates or seeps into the ground quickly, and the ponds dry up. The golf courses and yards of the suburban-style developments of DP all have lawns of green grass. DP has been growing on a flat, sediment-filled basin between two small mountain ranges. Much of the water supply of DP comes from wells. What do you predict will happen to the water table of the area in coming years, and how might the land surface change as a consequence? Is there a policy that you might suggest to the residents of DP that could slow the process of change? **(F)**

ONLINE RESOURCES

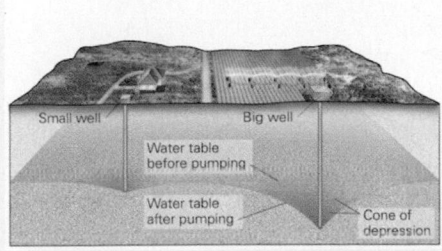

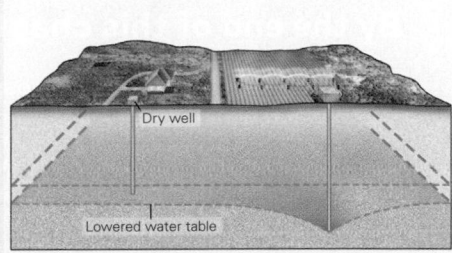

Videos
This chapter includes videos on groundwater removal, groundwater as a source to streams, and the recent drought in California.

Smartwork5
This chapters questions include visual labeling exercises on water tables, wells, and porosity and permeability.

CHAPTER 20

An Envelope of Gas: The Earth's Atmosphere and Climate

By the end of this chapter you should be able to . . .

A. describe how the Earth's atmosphere has evolved over time and where its gases come from.

B. distinguish among the atmospheric layers and explain how temperature and pressure vary with elevation.

C. sketch atmospheric circulation, and show how it causes the basic pattern of prevailing winds.

D. explain phenomena that cause fronts and clouds to form.

E. describe the evolution of mid-latitude cyclones, thunderstorms, tornadoes, and hurricanes, and explain how they cause damage.

F. distinguish between weather and climate, and explain why climate varies with location.

Who has seen the wind?
Neither you nor I:
But when the trees bow down their heads,
The wind is passing by.

—Christina Rossetti (British poet, 1830–1894)

20.1 Introduction

On March 1, 1999, Bertrand Piccard, a Swiss psychiatrist, and Brian Jones, a British ballooning instructor, silently rose from a launch site in Switzerland aboard the *Breitling Orbiter 3* **(Fig. 20.1)**, an airtight gondola suspended from a giant silvery helium-filled balloon. Their gondola provided life support and the balloon provided lift. Nineteen days later, more than two centuries after the first documented manned balloon flight, they became the first people to circle the globe nonstop by balloon.

This voyage could take place only because an **atmosphere**, a layer consisting of a unique mixture of gases, called **air**, surrounds the Earth. A balloon in the atmosphere feels a *buoyancy force*, because air in the balloon is less dense than the surrounding air, and because air can flow out of the balloon's way. Note that if the Earth were surrounded by a vacuum, balloon travel would be impossible, because no buoyancy force can develop in a vacuum.

Balloonists control their vertical movements by changing either the buoyancy of the balloon or the weight of the payload, but they cannot directly control their horizontal motions. They can travel horizontally only because the atmosphere circulates, resulting in the horizontal flow of air that we know as *wind*—balloons float with the wind, as sticks float with the current in a stream. To reach their destination before running out of supplies, long-distance balloonists must change their elevation to find wind flowing rapidly in the correct direction—which can be challenging, because atmospheric circulation is complex. To stay safe, Piccard and Jones also had to pay close attention to the *weather*—the overall physical conditions (temperature, pressure, moisture content, wind velocity, and wind direction) of the atmosphere at any given time and location. They had to avoid *storms*—episodes of extreme or turbulent wind, accompanied by rain, ice, snow, or lightning. And to stay alive, Piccard and Jones had to carry a heater and bottled oxygen, for conditions of pressure and temperature at high elevations cannot sustain life.

◀ (facing page) As the Sun sets over Lake Superior, the clouds turn orange. Cloud formation represents one consequence of the dynamic movement of air in the lower layer of the atmosphere.

FIGURE 20.1 The balloon and gondola used by Piccard and Jones during their successful attempt to circle the globe in March 1999.

In this chapter, we explore the envelope of air—the atmosphere—through which Piccard and Jones traveled. We begin by exploring the evolution of the atmosphere, and learn about the sources of its gases. Then we look at the structure of the atmosphere and the global-scale and local-scale circulation of its lowermost layer. This circulation controls the weather and leads to the growth of storms. We conclude by introducing the climate, the average of weather conditions for a region over many years. We discuss the atmosphere in a book about geology because interactions among air, water, and land not only shape landscapes, but they also control conditions in the *critical zone*—the life-sustaining part of the Earth System.

20.2 The Formation of the Atmosphere

The First and Second Atmospheres

When the Earth formed at 4.56 Ga, its gravitational pull held onto hydrogen and helium molecules that had been swirling in the protoplanetary disk (see Chapter 1). This *first atmosphere* survived only a short time, for heat from the Sun caused lightweight atoms to zip about so rapidly that they eventually achieved escape velocity and zoomed into space, and atoms that didn't escape on their own were carried away by intense solar winds.

While the Earth's primary atmosphere was disappearing, volcanic activity yielded new gases. Once the Earth had differentiated and had developed a magnetic field capable of deflecting the solar wind, these gases began to accumulate to form the *second atmosphere*. Where did the volcanic gases come from? Large quantities of volatile elements reside inside the Earth, bonded to solid minerals. During melting, the volatile elements separate from the minerals and dissolve in magma. Near the Earth's surface, where pressure decreases, the gases come out of solution and form bubbles that burst out of lava and into the air. Volcanic gas consists of 70% to 90% water (H_2O), with smaller amounts of carbon dioxide (CO_2) and sulfur dioxide (SO_2), along with traces of other gases, including nitrogen (N_2) and ammonia (NH_3). So the second atmosphere consisted of these gases plus other gases, such as methane (CH_4) and carbon monoxide (CO). Not all gases came from volcanoes—some were brought to the Earth by comets.

The Earth's second atmosphere evolved over geologic time. When the Earth's surface cooled sufficiently, water vapor could condense into rain that filled the oceans, lakes, and streams or sank into the shallow crust to become groundwater. This transfer of water from a gaseous state in the atmosphere to a liquid state on or near the surface caused the proportion of water in the atmosphere to decrease radically. As we noted in Chapter 13, the timing of the first liquid oceans remains a subject of debate—growing evidence suggests that ephemeral oceans had formed as early as 4.4 Ga, but permanent oceans appear to have existed only since 3.85 Ga.

The accumulation of liquid water on or near the surface also led to a radical drop in the concentration of CO_2 in the atmosphere, for two reasons. First, CO_2 dissolves in water to form carbonate ions, which in turn can combine with calcium ions to form solid carbonate minerals. Second, CO_2 reacts with certain silicate rocks during chemical weathering, in the presence of water, to produce carbonate ions. Formation of carbonate minerals effectively traps CO_2 in rock and keeps it out of the atmosphere.

With the removal of large quantities of H_2O and CO_2 from the air, what was left? The second atmosphere also initially contained NH_3. Ultraviolet radiation from the Sun split molecules of NH_3 into nitrogen and hydrogen atoms. The lightweight hydrogen atoms eventually escaped into space, but the nitrogen atoms combined to form molecular nitrogen (N_2) molecules. Molecular nitrogen doesn't react with other elements, so it can remain in the air for a long time. As the proportion of H_2O and CO_2 decreased, the proportion of nitrogen in the atmosphere increased.

Notably, the small amount of CO_2 that remained in the air after the oceans formed serves an essential role in keeping the Earth habitable. CO_2 is a **greenhouse gas**, meaning that it lets solar radiation (light) from space pass through but prevents infrared radiation (heat) rising from the Earth's surface from escaping back into space. Like other greenhouse gases (such as CH_4 and H_2O), CO_2 effectively traps heat in the atmosphere. The tiny amounts of CO_2 and other greenhouse gases in the air regulate atmospheric temperature and have kept the surface from becoming cold enough to freeze over, except perhaps for a relatively brief time in the Proterozoic. What would have happened if the Earth's surface had been too hot for liquid water to accumulate on it? Our atmosphere would have had a totally different history—CO_2 would not have been removed from the atmosphere, so instead of having about 0.04% CO_2, the Earth's atmosphere today would resemble the atmosphere of Venus, which currently contains 96.5% CO_2. Because of its high concentration of CO_2, Venus's atmosphere stays so hot that lead can melt at the planet's surface.

The Third and Modern Atmospheres

If you were suddenly to travel back through time and appear on the Earth at 3.8 Ga, you would instantly suffocate, for the atmosphere back then contained virtually no molecular oxygen (O_2). It wasn't until the first photosynthetic organisms—cyanobacteria—appeared on the Earth between 3.8 and 3.5 Ga that significant O_2 began to be produced, ushering in the Earth's *third atmosphere*. At first, virtually all of this O_2 was absorbed by dissolution in water

Did you ever wonder...
if our atmosphere has always been breathable?

FIGURE 20.2 Stages in the evolution of the Earth's atmosphere over time.

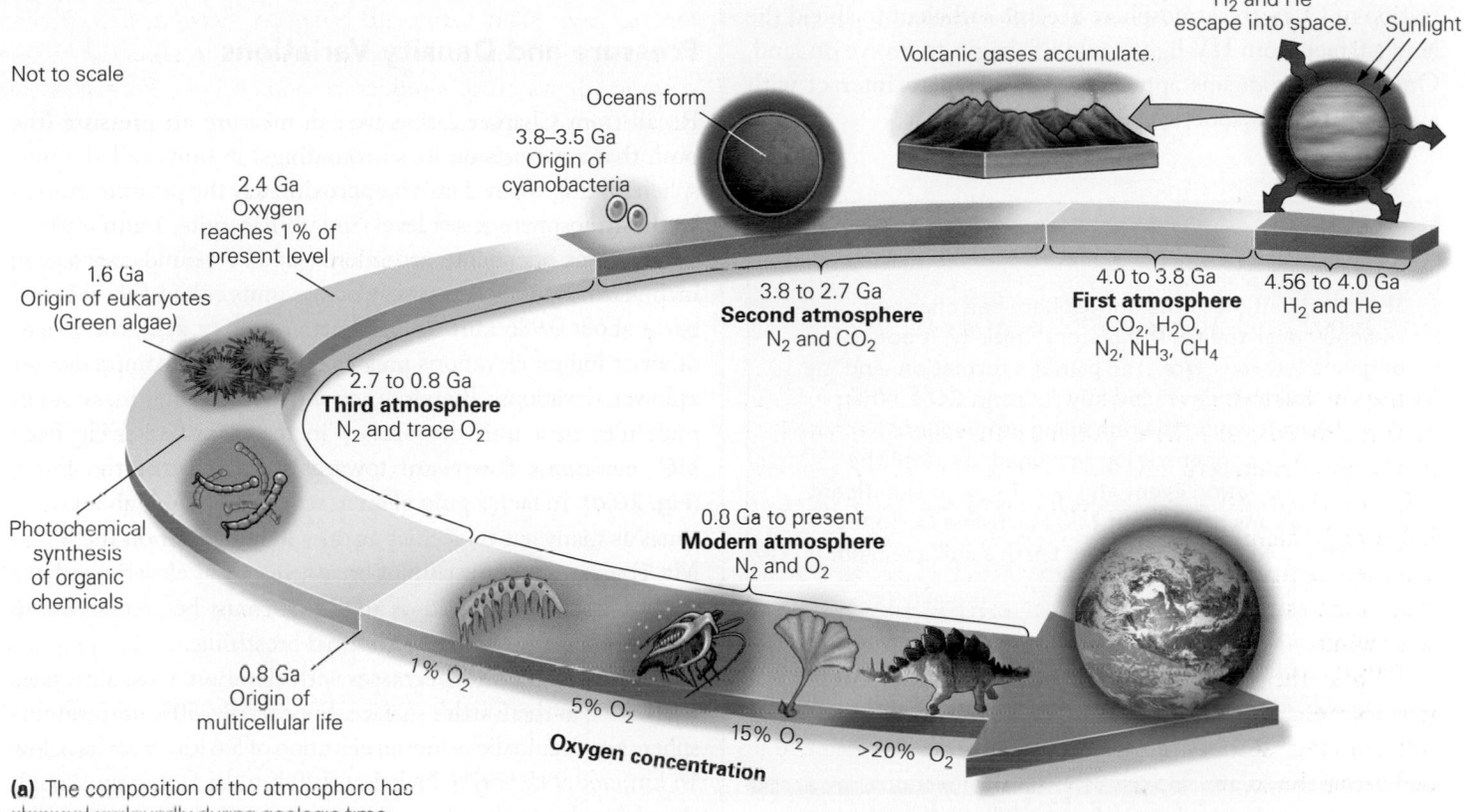

Not to scale

2.4 Ga
Oxygen
reaches 1% of
present level

1.6 Ga
Origin of eukaryotes
(Green algae)

3.8–3.5 Ga
Origin of
cyanobacteria

Oceans form

Volcanic gases accumulate.

H_2 and He
escape into space. Sunlight

3.8 to 2.7 Ga
Second atmosphere
N_2 and CO_2

4.0 to 3.8 Ga
First atmosphere
CO_2, H_2O,
N_2, NH_3, CH_4

4.56 to 4.0 Ga
H_2 and He

2.7 to 0.8 Ga
Third atmosphere
N_2 and trace O_2

Photochemical
synthesis
of organic
chemicals

0.8 Ga to present
Modern atmosphere
N_2 and O_2

0.8 Ga
Origin of
multicellular life

1% O_2

5% O_2

15% O_2

>20% O_2

Oxygen concentration

(a) The composition of the atmosphere has changed profoundly during geologic time.

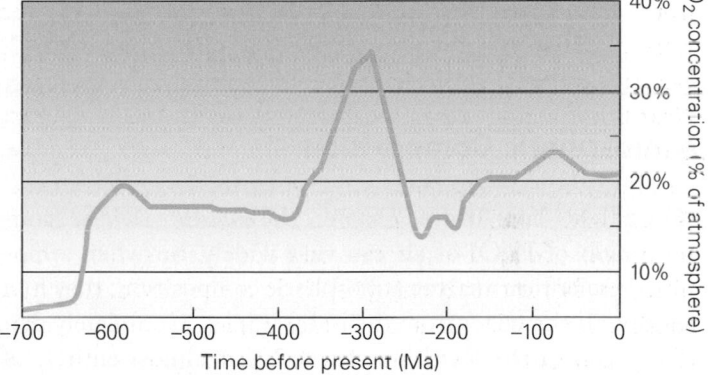

(b) During the Phanerozoic, oxygen composition has varied. It was particularly high during Carboniferous and Permian time.

or by reactions with rocks. But by about 2.5 Ga, both the land and oceanic oxygen reservoirs became saturated, and O_2 began to accumulate in the air. As a result, by about 2.4 Ga, the Earth's atmosphere contained 1% of its present oxygen level. Geologists refer to this transition—from an oxygen free atmosphere to an oxygen-containing atmosphere—as the *Great Oxygenation Event*. Oxygen concentrations increased slowly during the Proterozoic. At about 1.2 Ga, there may have been a boost in the production of O_2 with the appearance of photosynthetic algae, but only by about 0.6 Ga (600 Ma) did oxygen levels in the air become substantial, ushering the Earth's *modern atmosphere*—dominated by a breathable mixture of N_2 and O_2—into existence.

How oxygen levels varied during the Phanerozoic remains the subject of research. The occurrence of charcoal in Silurian strata indicates that there must have been at least 13% O_2 in the air by about 425 Ma, when vascular plants appeared, for vegetation cannot burn in air with less than 13% O_2. Oxygen concentrations appear to have reached a peak of about 35% during the late Paleozoic and have fluctuated since then, declining since the end of the Mesozoic to the present value of 21% **(Fig. 20.2)**. Oxygen concentrations probably cannot exceed 35%. If they did, plants would

become so combustible that forests would burn worldwide, and then, without land plants to produce O_2, the gas's concentration would drop dramatically.

Oxygen not only allows complex multicellular organisms to breathe, but also supplies the raw components for the production of *ozone* (O_3), a gas that absorbs harmful ultraviolet (short-wavelength) radiation from the Sun. Ozone, which accumulates primarily at an elevation of about 30 km, forms by a two-step reaction:

(1) O_2 + energy (from the Sun) $\rightarrow$ 2O
(2) O_2 + O $\rightarrow$ O_3

Note that, because ocean water absorbs UV light, life began and could survive in the sea. Only when the concentration of O_3 in the atmosphere became sufficient to shield the land surface from UV light could life begin to evolve on land. Once land organisms appeared, they began to interact with and modify atmospheric composition.

> **TAKE-HOME MESSAGE**
>
> The composition of the atmosphere has changed radically over the Earth's history. Initially, it consisted of gases left over from the planet's formation, and then of volcanic gases (mostly H_2O and CO_5). When the oceans formed, the remaining atmosphere became rich in N_2. O_2 concentration remained low until the Great Oxygenation Event and didn't rise to significant concentrations until about 600 Ma. Some O_2 bonds to form O_3, which protects the Earth's surface from ultraviolet rays.
>
> ---
>
> **QUICK QUESTION:** What caused oxygen concentrations in the atmosphere to increase?

20.3 General Atmospheric Characteristics

Atmospheric Composition

As we'll see later in this chapter, the *moisture content* (concentration of H_2O) of air can vary widely. So when atmospheric scientists analyze atmospheric composition, they first remove all the water from their air samples. Completely dry, clean air near the Earth's surface consists almost entirely of two gases, 78% N_2 and 21% O_2. The remaining 1%—the *trace gases* of air—includes argon (Ar), CO_2, neon (Ne), CH_4, He, H_2, and O_3. Some of these trace gases serve an important function in the Earth System—as we've seen, CO_2 and CH_4 are greenhouse gases that regulate the Earth's atmospheric temperature, and O_3 protects the surface from ultraviolet radiation.

Because air moves and has mass, it can keep aloft very tiny liquid droplets and solid particles, known collectively as **aerosols**. Aerosols, which range between 0.000003 and 0.003 mm in diameter, are so small that they remain suspended in the air, just as fine mud remains suspended in river water. Aerosols include sulfuric acid, sea salt, volcanic ash, clay flakes, mineral dust, soot, viruses and bacteria, specks of decayed organic material, and pollen **(Fig. 20.3)**. Some of these aerosols, along with

some types of trace gases, are produced by human activities and comprise *air pollution* **(Box 20.1)**.

Pressure and Density Variations

Recall from Chapter 2 that we can measure **air pressure** (the push that air exerts on its surroundings) in units called atmospheres (atm), where 1 atm is approximately the pressure exerted by the atmosphere at sea level. In familiar units, 1 atm is about 1,035 grams per square centimeter (or 14.7 pounds per square inch). Atmospheric scientists also use units called *bars*, where 1 bar is about 0.986 atm. In the Earth's gravity field, the weight of air at higher elevations presses down on and compresses air at lower elevations. Therefore, *air density* (the total mass of gas molecules in a unit of volume) increases progressively from high elevations downward toward the surface of the Earth **(Fig. 20.4)**. In fact, a gulp of air at sea level contains about three times as many gas molecules as does a gulp of air on the top of Mt. Everest. Humans cannot breathe easily at elevations above about 5 km, so the cabin of an airliner must be pressurized to provide adequate oxygen for normal breathing.

Because air density decreases with elevation, most air resides fairly close to the Earth's surface. Specifically, 50% of the atmosphere's molecules lie below an elevation of 5.6 km, 90% lie below 16 km, and 99.99997% lie below 100 km. In fact, even though the outer edge of the atmosphere, a gradual boundary where the gas density becomes the same as that of interplanetary space, technically lies as far as 10,000 km from the Earth's surface, most of the atmosphere's molecules lie within a shell only 0.5% as wide as the solid Earth. Though thin, the atmospheric shell contains sufficient gas to turn the sky blue when viewed during the day **(Box 20.2)**.

FIGURE 20.3 This 2017 forest fire in southern Georgia added aerosols (particularly soot) to the air.

BOX 20.1 CONSIDER THIS . . .

Air Pollution

Human activities (especially energy production, resource processing, manufacturing, and transportation) have changed atmospheric concentrations of trace gases and have added new gases and aerosols to the air. We refer to materials that would not be in air—or would not be there at the concentrations observed—due to natural phenomena alone as *pollutants* (Fig. Bx20.1). Pollutants include SO_2, nitrous oxide (N_2O), nitrogen dioxide (NO_2), CO, O_3, organic chemicals, chlorofluorocarbons (CFCs), metals (lead, uranium, mercury), CH_4, and CO_2, among many other chemicals as well as a great variety of solid or liquid aerosols (also known as *particulates*).

The addition of some pollutants can have major effects on *air quality*—the clarity and healthiness of air. For example, in the 19th century, the burning of coal for home heating in large cities created clouds of *soot* (carbon aerosols), which mixed with fog to produce *smog*. Today, automobile exhaust creates chemicals that react in sunlight to produce a dangerous mix of gases and aerosols, called *photochemical smog*, over cities. Sulfur emitted into the air can have effects that extend far beyond the source. Sulfur, in its various forms, dissolves in atmospheric moisture to produce sulfuric acid (H_2SO_4) aerosols. These aerosols can become incorporated in raindrops to produce **acid rain**. Where acid rain falls, lakes, streams, and the ground become more acidic and thus toxic to fish and vegetation (particularly coniferous trees). Similarly, the emission of CFCs into the air can have global consequences. In 1985, researchers discovered that CFCs react with ultraviolet light from the Sun to release chlorine atoms, which, in turn, react with ozone and break it down. These reactions appear to happen mainly in high clouds above polar regions during certain times of the year, thus preferentially removing ozone from these regions to create an *ozone hole*. Fortunately, global regulation of CFCs seems to be diminishing the problem. Finally, as we discuss in Chapter 23, the burning of fossil fuels has significantly increased the amount of CO_2 in the atmosphere.

To give a sense of the severity of air pollution in a region, governments now routinely issue regional measurements of air quality. The U.S. Environmental Protection Agency defines air quality on a scale, the *air quality index* (AQI), from 0 to 500. If the AQI value exceeds 100, the air is potentially unhealthy, and if it is over 300, it is hazardous. Air quality in rapidly industrializing cities has become a major problem.

FIGURE Bx20.1 Air pollution makes Beijing air murky.

FIGURE 20.4 Air pressure versus elevation. Climbers on top of Mt. Everest breathe an atmosphere that contains only about 33% of the gas molecules at sea level.

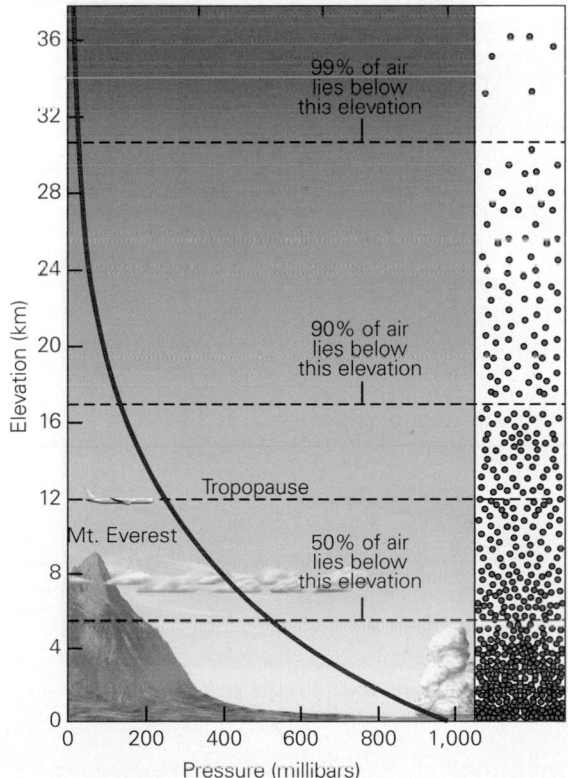

Heat and Temperature in Air

The molecules that constitute the atmosphere, or any gas, do not stand still, but remain in constant motion. We refer to the total kinetic energy (energy of motion) resulting from the movement of molecules in a gas as its *thermal energy*, or *heat*. Note that heat and temperature are not the same. A gas's *temperature* serves as a measure of the average kinetic energy of its molecules (see Chapter 2). Consequently, a volume of gas with a small number of very rapidly moving molecules has a higher temperature, but may contain less heat, than a volume with a larger number of slowly moving molecules. If we add heat to a gas in a given volume, its molecules move faster and its temperature rises.

Where does the heat in the atmosphere come from? Contrary to popular belief, only a small amount (12%) of this energy comes directly from the Sun, because incoming solar radiation consists predominantly of short-wavelength energy that air molecules cannot absorb. (In other words, air is transparent to visible light.) Most of the heat in air actually comes from long-wavelength (infrared) radiation that was absorbed by solids or liquids on the Earth's surface, then re-radiated and absorbed by greenhouse gases in the air. In fact, about 70% of the heat in the atmosphere comes from rising infrared energy, meaning, in effect, that the atmosphere is "baked from below" rather than

BOX 20.2 CONSIDER THIS . . .

Why Is the Sky Blue?

When astronauts standing on the Moon looked up, even during the day, they saw a black sky filled with stars and a nearly white Sun. On the Earth, when we look up during the day, we see a blue sky when there are no clouds and a white sky when there are. The color we see in the sky is the result of the dispersal of energy that occurs when light interacts with air molecules and other particles in the atmosphere, a process called *scattering*. When scattered, a single light ray divides into countless beams, each heading off in a different direction. You see a similar process when you shine a spotlight on a mirror ball over a dance floor.

Sunlight consists of a broad spectrum of electromagnetic radiation. Different components of the atmosphere scatter different wavelengths of this light because the ability of a particle to scatter light depends on the particle's size relative to the light's wavelength. Aerosols are larger than all wavelengths of light, so all wavelengths scatter off these particles. As a result, light scattered by these particles appears white—that's why clouds are white.

Light interacts differently with tiny gas molecules in the air. These molecules absorb some wavelengths and, shortly afterward, re-emit them. Blue light (with shorter wavelengths) is absorbed and re-emitted more than red light (with longer wavelengths). Put another way, blue light scatters more effectively in air, while red light passes through it more easily. Since scattered light heads in all directions, some of it is returned to space, or *backscattered*. Because of backscattering, the Earth's surface receives less intense light than it would if our planet had no atmosphere. In addition, because scattering sends light into regions that are blocked from the Sun, shadows on the Earth aren't completely dark.

When you look at an object, you see the color of light either emitted from the object or reflected by it. A plum appears violet because it reflects violet light and absorbs other wavelengths, and a traffic light appears green because it emits green light. On a clear day, with the Sun high in the sky, the gas in the atmosphere scatters

primarily blue light. When we look up, we are seeing the blue light scattered by the gas molecules (**Fig. Bx20.2a**). Because blue light is scattered, not all of it reaches the Earth (some has been backscattered to space), so the Sun appears yellow rather than white—if you subtract a little blue light from white light, you get yellow light. On a cloudy or hazy day, there are more aerosols (particularly water droplets) in the atmosphere, which scatter all wavelengths of light. Therefore, the sky appears white, unless the clouds are so dense that they do not let light through, in which case they appear gray.

At the beginning or end of the day, when the Sun lies close to the horizon, light passes through a thicker layer of atmosphere. As a result, so much of the blue light scatters back to space that the sunlight reaching the Earth contains mostly red wavelengths—if you subtract a lot of blue light from white light, you are left with red light. So, the Sun at sunrise or sunset appears red, as does the light that reflects off clouds (**Fig. Bx20.2b**).

FIGURE Bx20.2 Atmospheric color depends on the thickness of the atmosphere that light passes through.

(a) Scattering of light by gas molecules leads to the brilliant blue of a clear sky at noon on top of a volcano in Hawaii.

(b) So much scattering happens when sunlight enters the atmosphere at a low angle during sunrise that all that's left is red light.

"broiled from above." Of the remaining heat, 4% conducts into the air from the Earth's surface, and 14% comes from the condensation of water in the air, for when a gas changes state to become a liquid, it releases heat.

Relationships between Pressure and Temperature

Adding heat to a given volume of air will cause the air to expand, for as the molecules in the air start to move faster, they try to move outward. Similarly, removing heat from a given volume of air causes the air to contract, since its molecules slow down. Because of this behavior, a sealed balloon filled with air expands when you place it outside on a hot day, as the molecules within it exert outward force on the balloon's skin. Similarly, the balloon contracts if you place it in the refrigerator, where the molecules do not exert as much outward force.

You can see the same relationship between pressure and temperature when you change air pressure without adding or subtracting thermal energy. For example, when a given volume of air rises from a region of greater pressure to a region of lower pressure, without adding or subtracting thermal energy, it expands, and when this happens, its temperature decreases (the total kinetic energy per unit volume decreases). Researchers describe such a process as *adiabatic cooling*, from the Greek *adiabatos*, meaning impassable. Air cools adiabatically at a rate of 6°C to 10°C per kilometer that it rises, depending on its moisture content. (That's why it's cooler on a mountain peak than at its base.) Similarly, if air moves from a region of lower pressure to a region of greater pressure, without adding or subtracting heat, *adiabatic heating* takes place as the air contracts and its temperature increases. Therefore, when air rises (in *updrafts*) and sinks (in *downdrafts*), adiabatic cooling and heating, respectively, take place.

Water in the Air

Air contains varying concentrations of water vapor—from 0.3% above a hot desert to 4% in a rainforest before a heavy downpour. The moisture content of air changes over time at a given location, and with location at a given time. These changes affect your comfort level as well as the likelihood of rain or snow. How do we quantitatively distinguish wet air from dry air?

Meteorologists, scientists who study weather, specify the water content of air by a number called the **relative humidity**—the ratio between the measured water vapor content and the maximum possible amount of water vapor that the air could hold, expressed as a percentage. (Note that relative humidity differs from *absolute humidity*. The latter, which refers to the mass of water in a volume of air, is given in grams per cubic meter.) The maximum vapor content varies with temperature because warmer air can hold more water vapor than can colder air. Air that contains as much water vapor as possible is *saturated*, whereas air that doesn't is *unsaturated*. If we say that air at a given temperature has a relative humidity of 20%, we mean that the air contains only 20% of the water that it could hold at that temperature when saturated. Such air feels dry. Air with a relative humidity of 100% is saturated and feels very humid, or damp. The higher the relative humidity, the more difficult it is for your body to cool itself by perspiring, so a hot, humid day feels less comfortable than a hot, dry day.

Because cold air can't hold as much water vapor as warm air, unsaturated air may become saturated when cooled, without the addition of any new water. We refer to the temperature at which the air becomes saturated as the *dewpoint*; when unsaturated air cools at night and becomes saturated, liquid water condenses on surfaces to produce *dew*. (When the temperature drops below freezing, the moisture turns into *frost*.) When saturated air rises and adiabatically cools, its moisture condenses to form a **cloud**, a mist of tiny water droplets or tiny ice crystals **(Fig. 20.5)**. Clouds contain about 50% vapor and 50% liquid or solid water in droplets.

As we noted earlier, when water in the air changes in state from liquid to gas, or vice versa, the temperature of the air changes. That's because when water evaporates, it absorbs heat, allowing molecules to break free from the liquid state, and when it condenses, it releases heat as molecules become locked into position and slow down. The heat released during condensation is "hidden" in the sense that it comes only from the change of state and does not require an external energy source, so it is known as the **latent heat of condensation**.

FIGURE 20.5 As air rises and enters regions of lower pressure, it expands and adiabatically cools. Here we see moist air rising and becoming less dense; its moisture condenses at elevations above 3 km and produces a cloud.

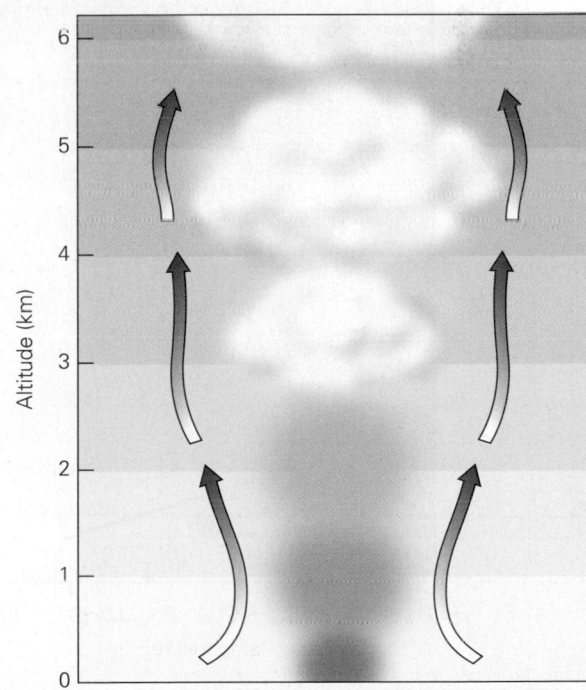

FIGURE 20.6 The principal layers of the atmosphere.

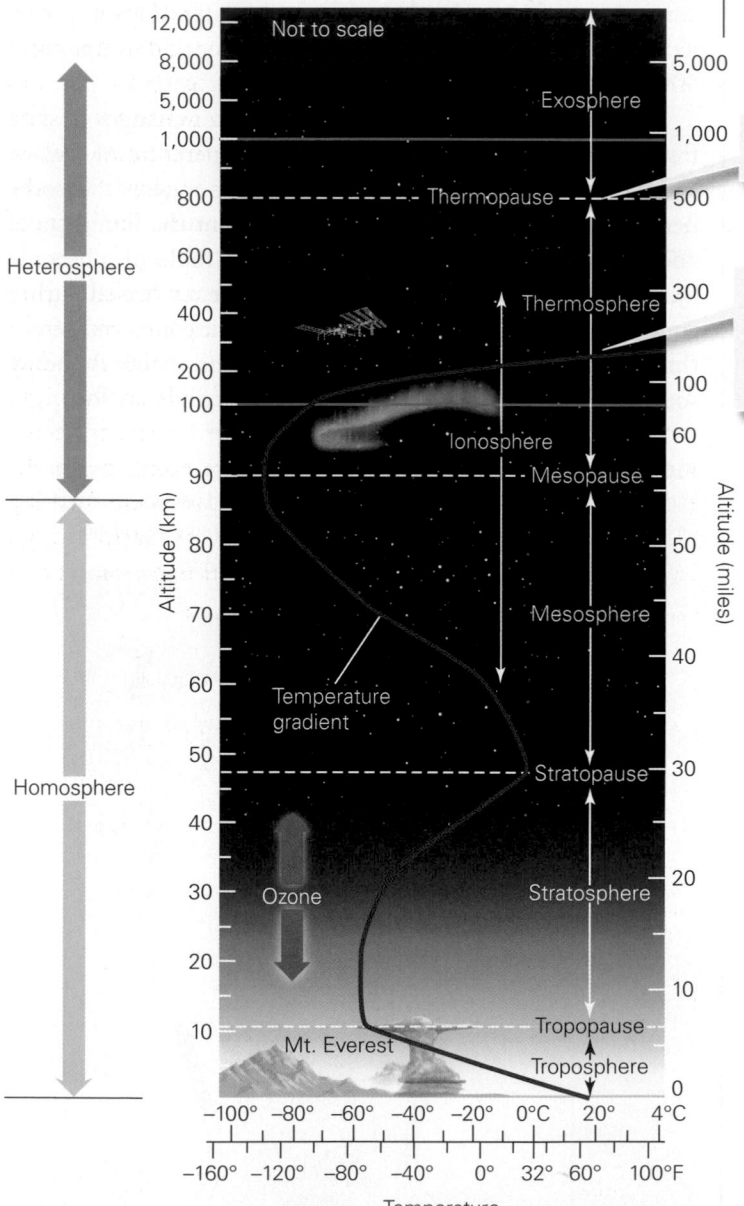

The base of the exosphere varies from 500 to 1,000 km.

Temperature reaches 500°C at about 300 km, and stays about the same upwards, though it varies over time due to variations in solar radiation.

(a) The atmosphere includes several layers. Temperature varies with elevation.

20.4 Atmospheric Layers

Layers Based on Temperature Variation

While air pressure in the atmosphere decreases uniformly from sea level upward, air temperature does not. In fact, starting from the Earth's surface and going up, temperature decreases, then increases, then decreases, then increases. Elevations where temperatures stop decreasing and start increasing, or vice versa, divide the Earth's atmosphere into four main layers and the exosphere. Let's consider these layers, starting from the base **(Fig. 20.6):**

- *Troposphere:* The layer of air that lies between the surface of the Earth and an elevation of 5 km at the poles and 18 km at the equator is called the **troposphere**. Within this layer, the temperature decreases gradually from an average of 18°C at the surface to about −55°C at the top, a boundary called the *tropopause*. The name *troposphere* comes from the Greek *tropos*, which means turning, because air in the troposphere constantly undergoes *convection* as warm air rises in updrafts and cold air sinks in downdrafts.

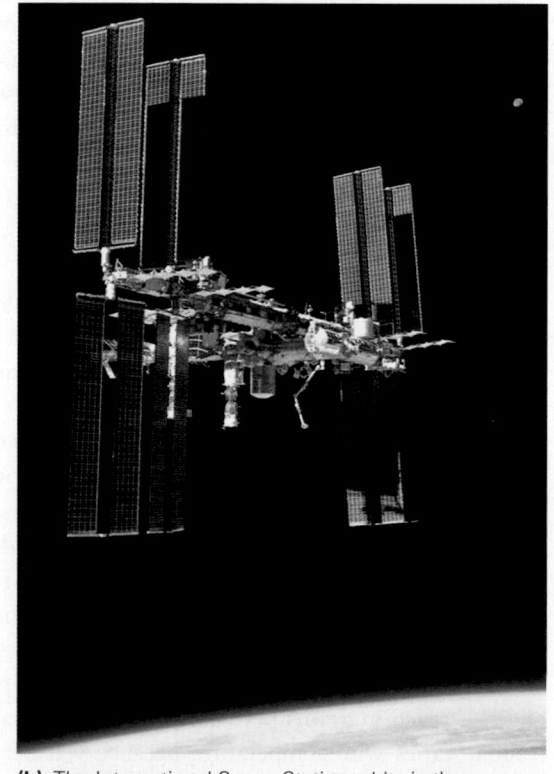

(b) The International Space Station orbits in the thermosphere.

Why does this convection take place? As we noted, most of the heat in the air comes from infrared radiation rising from the Earth's surface, meaning that air is heated from below. As soon as it warms, the near-surface air becomes buoyant, so it rises. Cold air must sink to take the place of the rising air. As we will see, this air movement in the troposphere drives most weather phenomena, so the troposphere can also be thought of as the "weather layer."

- *Stratosphere*: Beginning at the tropopause and continuing up for about 10 km, the temperature stays about the same. Then it slowly rises, reaching a maximum of about 0°C at an elevation of about 47 km, a boundary called the *stratopause*. The layer between the tropopause and the stratopause is the **stratosphere**, so named because it doesn't convect much and therefore remains relatively stable and stratified. The stratosphere generally doesn't mix with the underlying troposphere because at the tropopause, hotter (less-dense) air already lies on top of cooler (denser) air, so convection doesn't take place. Most of the ozone in the Earth's atmosphere resides in the stratosphere; heating in the stratosphere happens because ozone absorbs ultraviolet radiation directly from the Sun.

- *Mesosphere:* Temperature decreases in the layer called the **mesosphere**, which lies between 47 and 82 km. At the *mesopause*, the top of the mesosphere, the temperature has dropped to −85°C. The mesosphere does not absorb much solar energy, so it cools with increasing distance from the hotter stratosphere below. Most *meteors* (so-called shooting stars) begin burning in the mesosphere and have vaporized by the time they descend to an elevation of 25 km.

- *Thermosphere:* The outermost layer of the atmosphere, which lies between 82 and 700 km, is called the **thermosphere**. It holds less than 1% of the atmosphere's gas molecules. Temperature increases with elevation in this layer because its gases absorb short-wavelength solar energy. In fact, the temperature at the top becomes very high. But because the thermosphere has so little gas, it contains very little heat, so an astronaut walking in space at an elevation of 200 km doesn't feel hot.

- *Exosphere:* The layer between 700 and 10,000 km, known as the **exosphere**, represents a gradual transition between the atmosphere and beyond. At 10,000 km, the gas concentration becomes the same as interplanetary space.

Layering Based on Composition

In the relatively dense gas in the lower three layers of the atmosphere, moving atoms and molecules frequently collide. Like billiard balls, they bounce off one another and shoot off in different directions. This constant, chaotic motion stirs the gas sufficiently to make a homogeneous mixture, meaning that air in the lower three layers has essentially the same proportion of different gases regardless of location. For this reason, atmospheric scientists refer to the troposphere, stratosphere, and mesosphere together as the *homosphere*. In contrast, atoms and molecules in the low-density thermosphere collide so infrequently that this layer does not homogenize. Rather, the air of the homosphere separates into distinct layers, with the heaviest gas (nitrogen) on the bottom, followed in succession by oxygen, helium, and, at the top, hydrogen. To emphasize this composition, atmospheric scientists refer to the thermosphere as the *heterosphere*.

So far, we've distinguished atmospheric layers according to their thermal structure (troposphere, stratosphere, mesosphere, and thermosphere) and according to the degree their gases mix (homosphere and heterosphere). We need to add one more "sphere" to our discussion. The term **ionosphere** refers to the interval between 60 and 500 km, so it includes most of the mesosphere and the lower part of the thermosphere. It received its name because in this layer, short-wavelength solar energy strips nitrogen molecules and oxygen molecules of their electrons and transforms them into positive ions. The ionosphere plays an important role in modern communication in that, like a mirror, it reflects radio transmissions from the Earth back down so that they can travel around the curve of the surface.

The ionosphere also hosts a spectacular atmospheric phenomenon, the *aurorae* (*aurora borealis* in the northern hemisphere and *aurora australis* in the southern), which look like undulating, ghostly curtains of varicolored light in the night sky **(Fig. 20.7)**. Aurorae appear when charged particles ejected from the Sun, especially when solar flares erupt, reach the Earth and interact with the ions in the ionosphere, making them release energy. This interaction occurs primarily at high latitudes because the Earth's magnetic field traps solar particles and carries them to the poles.

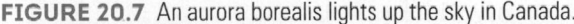

FIGURE 20.7 An aurora borealis lights up the sky in Canada.

20.5 Wind and Global Circulation in the Atmosphere

A gentle breeze on a summer day, the steady currents of air that once blew clipper ships across the oceans, and the fierce gusts of a hurricane are all examples of **wind**, the horizontal component of air movement. We can feel wind because of the impacts of air molecules as they strike us. The existence of wind emphasizes that the lower part of the atmosphere constantly moves, swirling and overturning at rates of between a fraction of a kilometer and a few hundred kilometers per hour. This circulation happens on two scales, local and global. *Local circulation* refers to the movement of air over a distance of tens to hundreds of kilometers. *Global circulation* refers to the movement of volumes of air in paths that ultimately carry it around the entire planet. To understand both kinds of circulation, we must first see what drives air from one place to another and examine how energy inputs into the atmosphere vary with location.

Lateral Pressure Changes and the Cause of Wind

The air pressure of the atmosphere not only changes vertically, but also changes horizontally at a given elevation. The rate of pressure change over a given horizontal distance, called a *pressure gradient*, can be represented by the slope of a line on a graph plotting pressure on the vertical axis and map distance on the horizontal axis **(Fig. 20.8a)**. We can use a map to illustrate how air pressure at a given elevation varies with location. A contour line on a map along which the air has a specified pressure is called an **isobar (Fig. 20.8b)**; pressure has the same value at all points on an isobar.

Winds form wherever a pressure gradient exists. In a general sense, air accelerates from a high-pressure region to a low-pressure region; in other words, it starts flowing down a pressure gradient. To see why, step on one end of a balloon filled with air; you momentarily increase the pressure at that end, so the air flows toward the other end **(Fig. 20.8c)**. Similarly, a difference in pressure exists between one isobar and the next, so air initially moves perpendicular to these lines. The Coriolis effect, however, modifies wind direction, so in general, air flow is not perpendicular to isobars—in fact, in many locations, air flow becomes almost parallel to isobars, following the path that represents the *geostrophic balance* (see Chapter 18) between the force caused by the flow of air from higher-pressure to lower-pressure regions and the force caused by the Coriolis effect.

Regional Circulation in the Atmosphere

Solar energy constantly bathes the Earth. Of this energy, 30% reflects back to space (off clouds, water, and land); air or clouds absorb 19%; and land and water absorb 51%. As we've seen, the energy absorbed by land and water re-radiates as infrared

FIGURE 20.8 The concept of a pressure gradient.

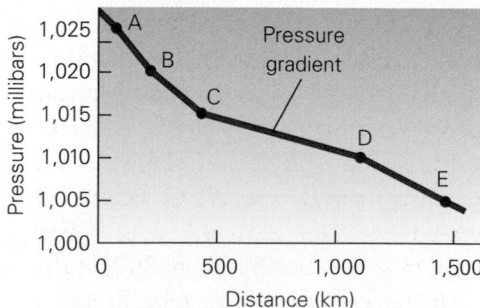

(a) A graph of pressure changes from Point A to Point E. The gradient is greater where the slope is steeper.

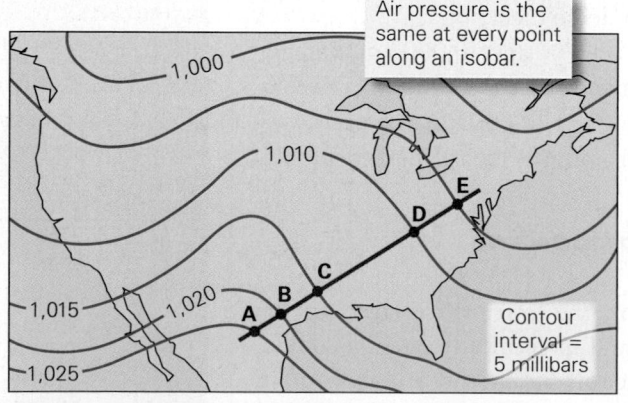

(b) The air pressure decreases between Points A and E. Note that the rate of change, the pressure gradient, is greater between Points A and C than between Points C and D.

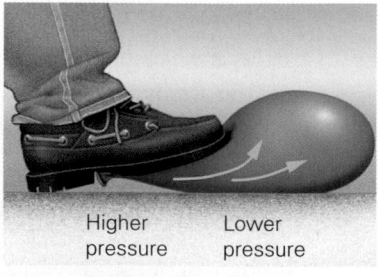

(c) Air flows from a region of high pressure to a region of low pressure.

radiation and essentially bakes the atmosphere from below. Because the Earth is a sphere, not all latitudes receive the same amount of incoming solar energy, or **insolation**. Specifically, portions of the Earth's surface hit by direct rays of the Sun receive more energy per square meter than portions hit by oblique rays. We can illustrate this contrast with a flashlight. If you point a flashlight beam straight down, you get a small but bright spot of light on the ground, but if you aim the beam so that it hits the ground at an angle of 45°, the spot covers a broader area but does not appear as bright **(Fig. 20.9)**. Higher latitudes, therefore, receive less energy than lower latitudes.

FIGURE 20.9 The amount of insolation depends on the angle at which sunbeams strike the Earth, so the poles are colder.

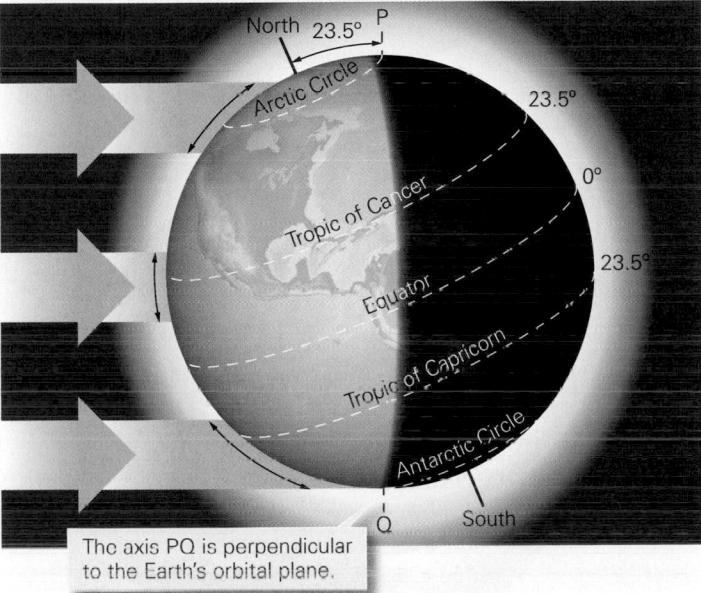

(a) The Earth's surface at high latitudes receives less solar energy per square meter than does the surface at low latitudes, so the poles are colder than the equatorial regions.

(b) The beam from a flashlight serves as an analogy to a beam from the Sun. A square centimeter of the table receives less energy from an oblique beam than it does from a direct beam.

Because of the tilt of the Earth's axis, the amount of solar radiation that any point on the surface receives changes during the year, which is why we have seasons **(Box 20.3)**. The contrast in the amount of solar radiation received at different latitudes means that polar regions are cooler than equatorial regions.

In 1735, George Hadley, a British mathematician, realized that latitudinal contrasts in temperature could drive air circulation over broad areas. Specifically, he suggested that very warm air at the equator would rise, and then, at high elevations, flow toward the poles to be replaced by cool polar air, which would flow toward the equator at lower elevations **(Fig. 20.10a)**. Hadley's proposal of a *hemispheric-scale circulation cell*, however, did not take into account the Earth's rotation and the resulting Coriolis effect. The Coriolis effect, as we learned in Chapter 18, causes an object moving from the equator to the pole above a rotating sphere to deflect relative to the ground. Because of the Coriolis effect, northward-moving high-altitude air in the northern hemisphere deflects to the east, so by the latitude of 30° N, it is essentially moving due east and cannot flow the rest of the way to the North Pole. Further, by the time it reaches this latitude, the air has cooled significantly and starts to sink. When the sinking air reaches low elevations, it divides, some moving back toward the equator near the surface and some moving north near the surface. The southern flow heads back toward the equator, gradually deflecting to the southwest until it flows due west at the equator, where it warms and rises. As a result, convection cells, now known as *Hadley cells*, develop in the troposphere between the equator and 30° N (in the northern hemisphere), and between the equator and 30° S (in the southern hemisphere) **(Fig. 20.10b)**.

In the early 20th century, atmospheric scientists thought that three circulation cells develop within each hemisphere. The low-latitude cells, extending from the equator to a latitude of about 30°, were the *Hadley cells*; the mid-latitude cells were named *Ferrel cells*, in honor of the American meteorologist William Ferrel, who proposed them; and the high-latitude cells were called *polar cells* **(Fig. 20.10c)**. In the context of this model, where the air of one circulation cell approaches that of another at the Earth's surface, a surface **convergence zone** develops—at this zone, the air has nowhere to go but up. Alternatively, where the air of one cell moves away from that of another near the surface, a surface **divergence zone** develops. In the three-cell model (which actually consists of six cells—three in each hemisphere), *an intertropical convergence zone (ITCZ)* occurs at the equator, the southern edge of the Hadley cell; a *subtropical divergence zone* forms at subtropical latitudes (about 30°), at the boundary between the Hadley cell and the Ferrel cell; and a *polar convergence zone* forms at the southern edge of the polar cell (about 60°), a boundary also known as the **polar front**.

Where moist air rises above the ITCZ at the equatorial limit of the Hadley cells, masses of clouds develop and

FIGURE 20.10 Global circulation patterns in the atmosphere.

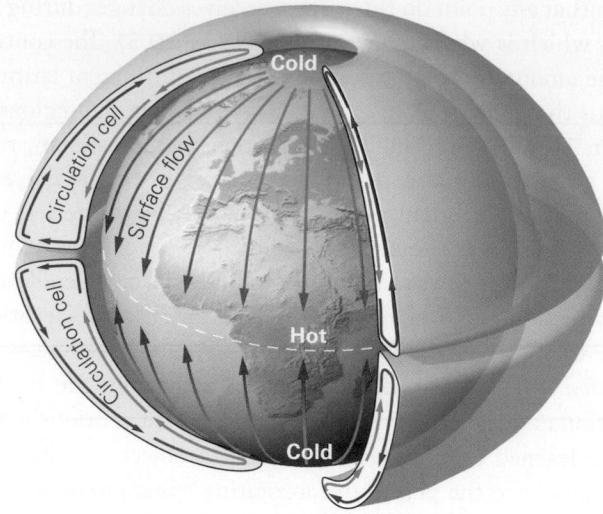

(a) If the Earth did not rotate, one simple circulation cell would exist in each hemisphere, each stretching from the equator to the pole.

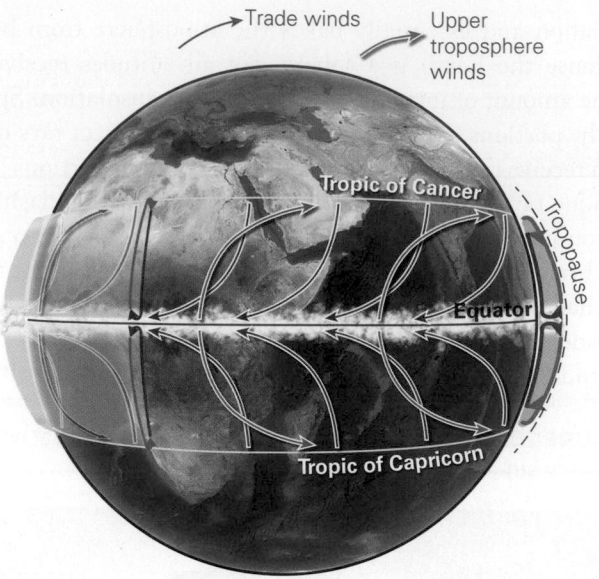

(b) Large circulation cells form between the equator and a latitude of 30°, both north and south of the equator.

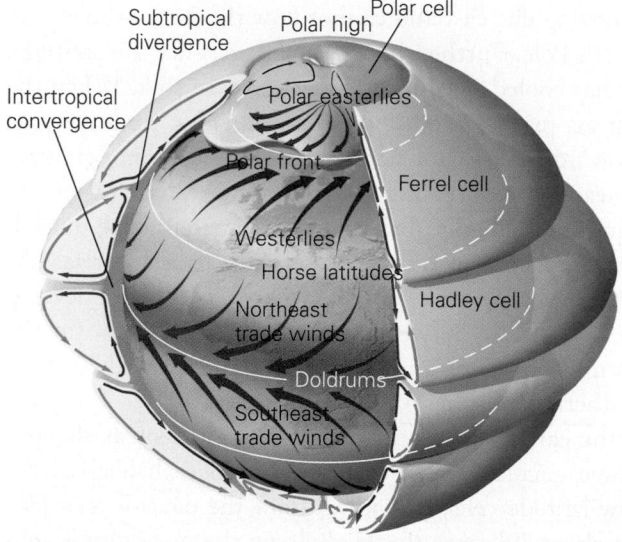

(c) The three-cell per hemisphere circulation model accommodates the Coriolis effect, but this model is an oversimplification.

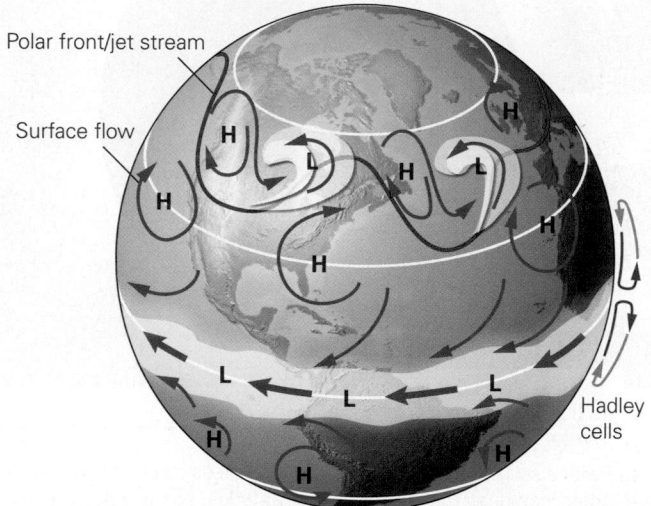

(d) The modern image of global circulation is far more complex. The Hadley cells exist. But farther north, there are several large, moving eddies, and the polar front (red line) follows a wavy path.

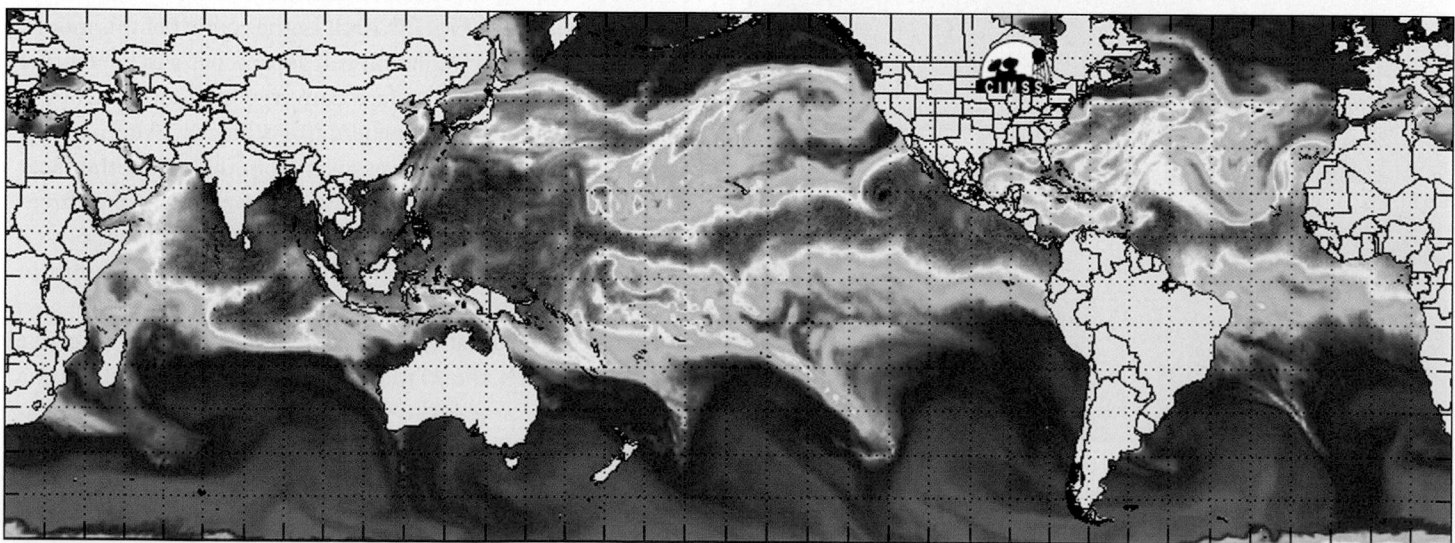

(e) A satellite map showing the amount of "precipitable moisture" (water that can fall as rain or snow) in the air (warm colors represent more moisture; cool colors represent less moisture). The pattern of colors gives a sense of true complexity of air flow in the near-surface atmosphere.

BOX 20.3 CONSIDER THIS . . .

The Earth's Tilt: The Cause of Seasons

A satellite view of the Earth emphasizes that snow cover distribution changes over the course of a year (Fig. Bx20.3a). This change occurs because the Earth has seasons. Seasons exist because the Earth's axis of rotation tilts (currently at 23.5°) relative to the plane of its orbit. As our planet moves around the Sun, the direction of the tilt relative to the Sun varies, so the insolation at a given latitude, and therefore, the average monthly temperature, changes during the course of a year (Fig. Bx20.3b). For example,

when the northern hemisphere tilts toward the Sun, it receives more radiation, warms up, and enjoys summer, but when the northern hemisphere tilts away from the Sun, it receives less radiation, cools down, and endures winter.

The boundary between the day side and the night side of the Earth is called the *terminator*. Because of the Earth's tilt, the terminator does not always pass through the North and South Poles (see Fig. Bx20.3b). On June 21, a special day called a *solstice*, the terminator lies 23.5° away from the

poles, as measured along the Earth's surface. We refer to the line of latitude at this position as the *Arctic Circle* in the northern hemisphere and the *Antarctic Circle* in the southern hemisphere. On June 21, a person standing on the Arctic Circle can see the midnight sun, meaning it's light all day. Anyone south of the Antarctic Circle sees night for a full 24 hours. During the northern summer, regions north of the Arctic Circle see the midnight sun for more than a day, and the North Pole itself experiences

FIGURE Bx20.3 Because of the tilt of the Earth's axis, we have seasons.

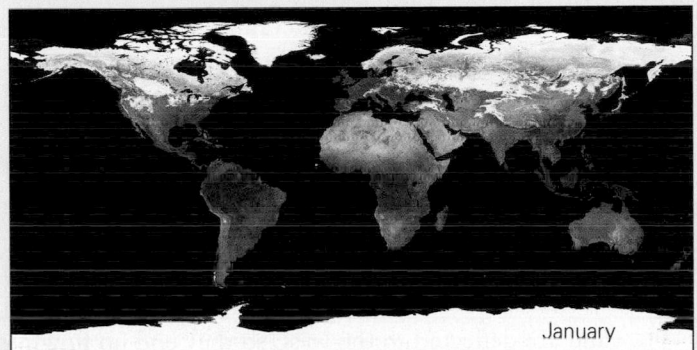 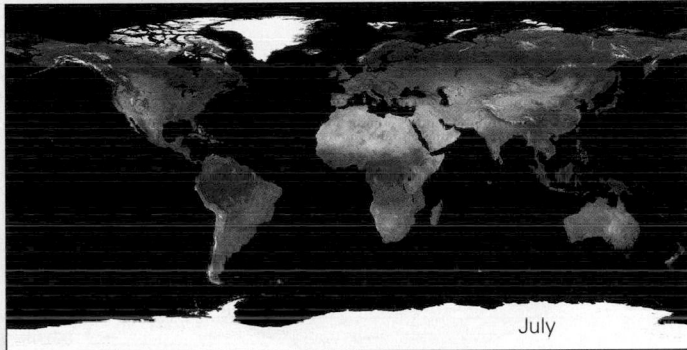

(a) Satellite photos emphasize the contrast in snow cover of the northern hemisphere as seasons change.

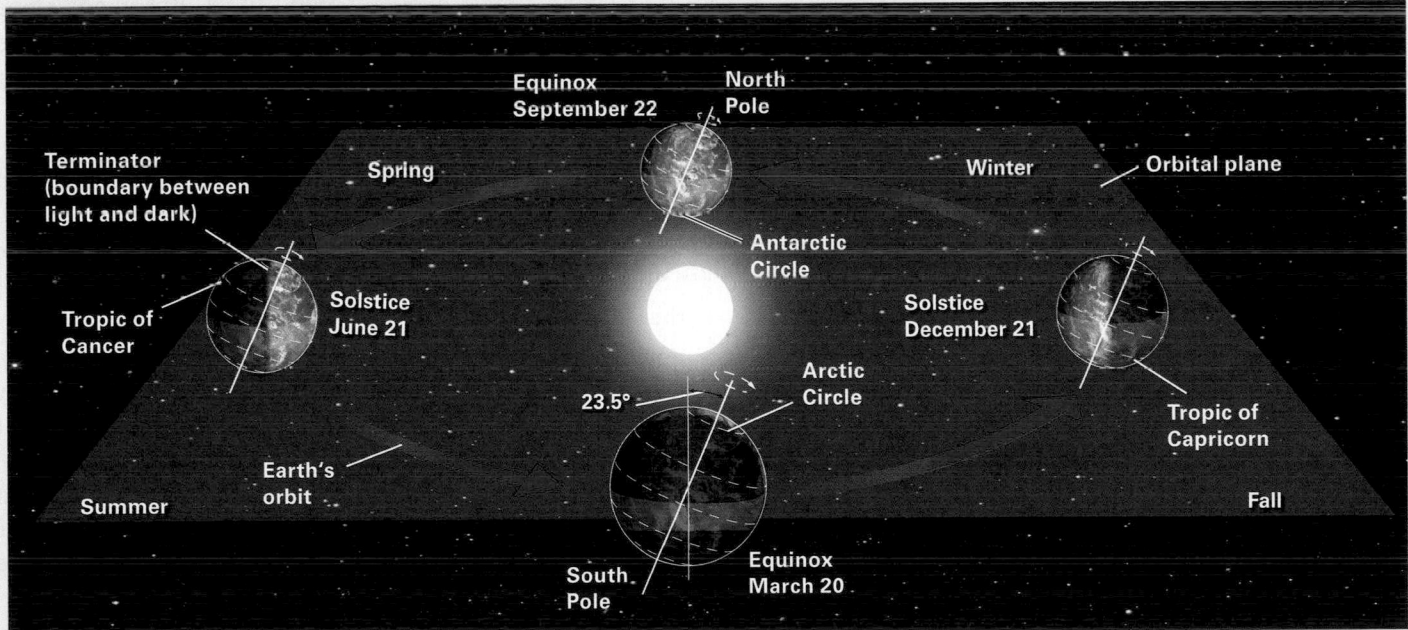

(b) In the northern hemisphere's summer solstice, the noonday Sun appears directly overhead at the Tropic of Cancer, and the region above the Arctic Circle sees the midnight sun, whereas at the winter solstice, the noonday Sun appears directly overhead at the Tropic of Capricorn, and the region south of the Antarctic Circle sees the midnight sun. At the equinoxes, the noonday sun appears directly overhead at the equator.

(Continued)

the midnight sun for a full 6 months. Similarly, regions south of the Antarctic Circle experience 24-hour nights for more than a day, and the South Pole itself sees night for a full 6 months. On December 21, the other solstice, a person at the Antarctic Circle sees the midnight sun, whereas all regions north of the Arctic Circle have perpetual night. During the southern summer, the South Pole sees daylight for 6 months.

On the June 21 solstice, the Sun's rays are exactly perpendicular to the Earth (the Sun is directly overhead) at latitude 23.5° N, the Tropic of Cancer, and on December 21, the Sun's rays are exactly perpendicular to the Earth at 23.5° S, the Tropic of Capricorn. On September 22 and March 20, each known as an *equinox,* the Sun is directly overhead at the equator. The two solstices and two equinoxes divide the year into four astronomical seasons.

FIGURE 20.11 A simplified map of average surface-wind patterns

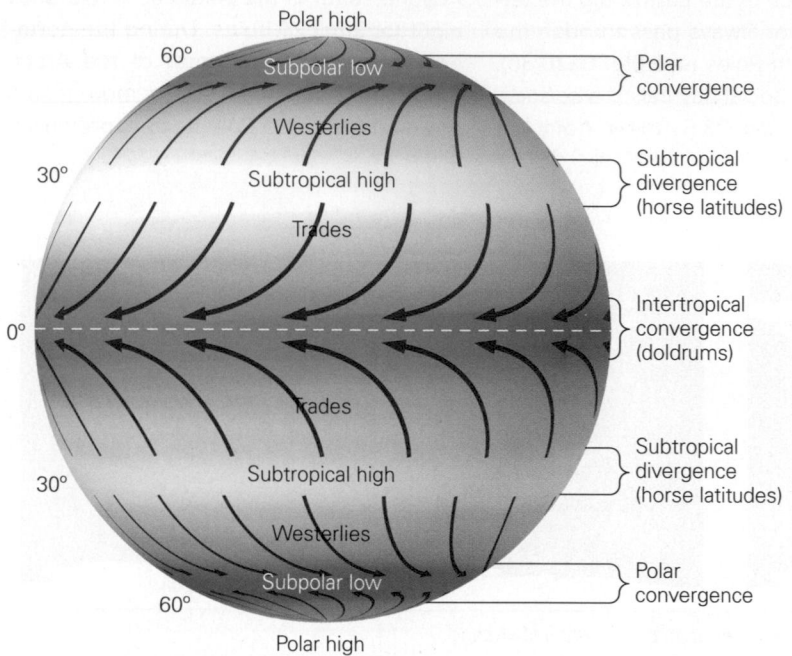

Prevailing Surface Winds

Despite its complexities, the overall circulation of the atmosphere yields belts in which surface air, on average, moves in a consistent direction. Such air flows are called **prevailing winds**. Note that when describing these winds, meteorologists label them according to the direction the air comes from. Therefore, a "westerly wind" blows from west to east.

Let's start our tour of prevailing winds at the subtropical divergence zone in the northern hemisphere. Near-surface winds start to flow to the south and are deflected west. Between the equator and a latitude of 30° N, therefore, surface winds come out of the northeast and are known as the **trade winds**, so named because they once carried trading ships westward from Europe to the Americas. Similar winds in the southern hemisphere start flowing northward and are deflected to the west, so they end up flowing from southeast to northwest. Where the southeast and northeast trade winds merge at the equator, they flow almost due west **(Fig. 20.11)**. But winds along the equator tend to be very slow because the air is also rising. Ships tended to be becalmed in this belt, which came to be known as the **doldrums**.

In mid-latitudes, the direction of the wind at a given time and location depends on the position of that location relative to high-pressure or low-pressure zones. Averaged over the course of a year, however, mid-latitude winds come primarily from the west, so they have come to be known as the *surface westerlies.* These winds, of course, shift as waves in the polar front pass by and as eddies pass over regions. In subtropical divergence zones, where air flows primarily downward, winds are weak and tend to shift in different directions. In the past, these conditions inhibited the progress of sailing ships. Perhaps because so many horses being transported by ship died of heat exhaustion while the vessels traversed the subtropics, the region came to be known as the *horse latitudes.*

Finally, in polar latitudes, surface air starts by flowing from the pole toward the equator but deflects to the west. The resulting prevailing winds are known as the *polar easterlies.* These winds converge with the westerlies of mid-latitudes at the polar front.

drench the tropics with rain (see Fig. 20.10b). Where dry air descends at the mid-latitude limit of the Hadley cells, desert realms exist.

In reality, atmospheric circulation is far more complex than the three-cell-per-hemisphere flow model of Figure 20.10c implies. Because of many factors—local geostrophic balance; the distribution of land and sea, which absorb and release heat and moisture at different rates; the distribution of topography, which deflects the wind; seasonal variations in the amount of insolation; and the transport of heat by ocean currents—atmospheric flow displays turbulence at many scales. As a result, while the Hadley cells tend to be recognizable, the Ferrel and polar cells actually represent only long-term averages of air flow. At any given time, air flow follows regional-scale eddies, and the position of the polar front on a map outlines huge, wave-like curves **(Fig. 20.10d, e)**. Zones of high pressure (H) or zones of low pressure (L) develop at the Earth's surface in the interiors of the regional-scale eddies. In the northern hemisphere, air flows clockwise around highs and counterclockwise around lows (see Figure 20.10d).

High-Altitude Winds in the Troposphere: The Jet Streams

In two special places, over the polar front and over the horse latitudes, air masses of very different temperatures come in contact. This temperature difference causes a step in *isobaric surfaces*. In isobaric surfaces (meaning imaginary surfaces of equal pressure), especially near the top of the troposphere. Therefore, a steep gradient develops at high elevation **(Fig. 20.12a)**. Due to this gradient, high-altitude westerly winds flow particularly fast—at speeds of between 200 and 400 km per hour—producing the **jet streams** **(Fig. 20.12b)**. The *polar jet stream*, which tends to be the stronger of the two, typically flows at 7 to 11 km above the Earth's surface, whereas the subtropical jet stream flows at 9 to 14 km above the surface. Airliners commonly encounter the jet streams. Planes flying east have shorter flying times than those flying west because the former benefit from strong tailwinds that help to push them along, whereas the latter battle headwinds that slow them down.

As we've noted, the polar front tends to develop large, wavelike undulations. As viewed on a map, therefore, the positions of the polar front and the associated jet stream follow large, curving trajectories that sometimes bring the jet stream down to southern latitudes of North America and sometimes to northern latitudes of Canada **(Fig. 20.12c)**. Furthermore, the average position of the polar jet stream varies with the seasons.

TAKE-HOME MESSAGE

Where air pressure changes horizontally, a pressure gradient exists. Air starts to move from high-pressure to low-pressure regions, producing a wind. The Coriolis force modifies the direction of air flow. Due to latitudinal variation in insolation, regional-scale air circulation cells develop in each hemisphere. In detail, air circulation is much more complicated, and only the Hadley cell is well defined. Regional variations in pressure drive prevailing surface winds. Jet streams form at high elevation in association with elevation steps in the tropopause.

QUICK QUESTION: What are doldrums, and why do they form?

FIGURE 20.12 Jet streams form just above the tropopause and follow wavy paths that go around the Earth.

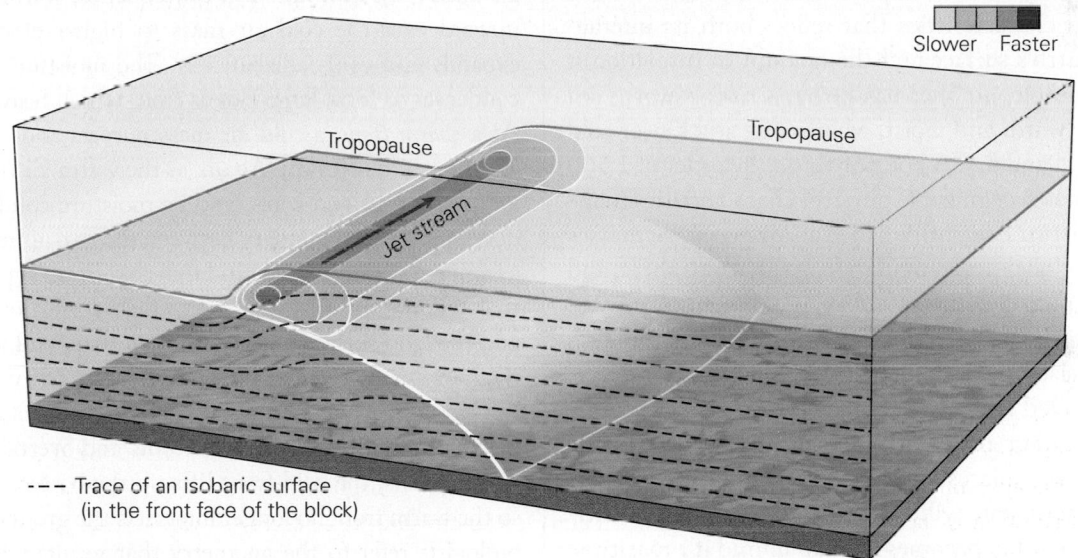

(a) The tropopause is an isobaric surface. At steps in this surface, high-altitude winds are stronger and jet streams form.

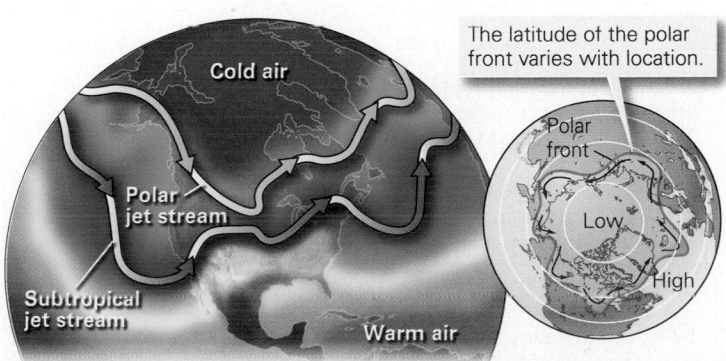

(b) Jet streams form at the boundary between regions of the troposphere with different temperatures. The boundaries are wavy, so the jet streams are wavy. The polar jet stream brings cold air southward.

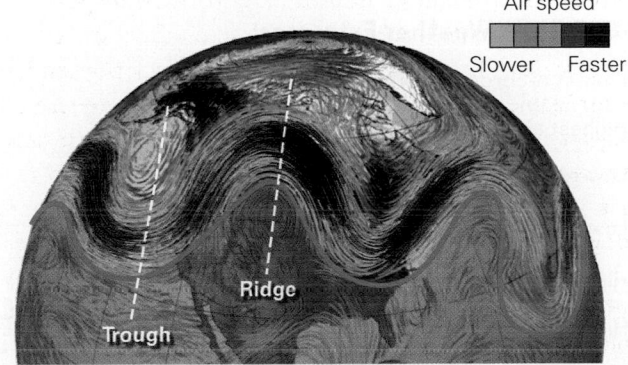

(c) Winds of the jet stream. Note the large curves. (Troughs bow southward; ridges bow northward.)

20.5 Wind and Global Circulation in the Atmosphere **771**

20.6 Weather and Its Causes

Weather refers to the local-scale conditions as defined by temperature, air pressure, relative humidity, rainfall or snowfall, cloud cover, and wind speed. If the surface of the Earth were a perfectly uniform sphere, air flow might follow the simple global circulation patterns, and weather patterns, therefore, might be fairly simple as well. But because of the wave-like undulations that develop along convergence and divergence zones; the local turbulence of atmospheric flow; interactions between air flow and landscape features; and exchanges of heat among land, sea, and air, weather can vary dramatically over time at a given location and with location at a given time **(Table 20.1)**. To understand what controls weather, we must begin by characterizing regions of the lower atmosphere called air masses and the boundaries, or *fronts*, that form between them.

Air Masses and Fronts

Air that remains or passes over a particular region for a length of time takes on characteristics that reflect both its interaction with the Earth's surface and the amount of insolation it receives. For example, air that has hovered over a warm sea tends to become warm and moist, whereas air stuck over cold land becomes cold and dry. A body of air on the order of 1,500 km across, that has recognizable physical characteristics (temperature and moisture content), is called an **air mass**. Air masses move within the global-scale circulation of the atmosphere, and their paths are controlled by prevailing winds. Simplistically, distinct air masses, which have been named by meteorologists **(Fig. 20.13)**, persistently develop in given regions; their development is controlled by the locations of land and sea and by latitude. The weather of a location can change drastically when one air mass replaces another. For example, a summer day in a midwestern state will be cool and dry under a continental polar air mass but becomes hot and humid if a maritime tropical *air mass* moves in.

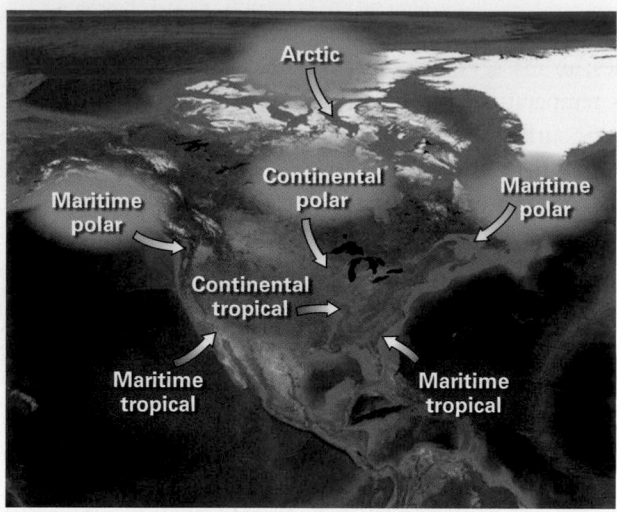

FIGURE 20.13 Air masses that form in particular places have been assigned different names. The arrows indicate the average directions in which the air masses move.

The boundary between two air masses is called a **front**. Meteorologists recognize several kinds of fronts—we introduce three. At a *cold front*, a cold air mass pushes underneath a warm air mass **(Fig. 20.14a)**. As a consequence, warm, moist air flows upward over the cold air mass to higher elevations, where it expands and cools adiabatically. The moisture it contains then condenses to form large clouds from which heavy rains may fall. At a *warm front*, a cold air mass retreats and a warm air mass advances **(Fig. 20.14b)**. Again, as the warm air rises over the cold air, it expands, and cools, and its moisture condenses, so clouds develop over the boundary between the two air masses. Note that a front is a sloping surface and that cold fronts typically are steeper than warm fronts. We can portray the intersection between a front and the land surface with a line on a map, and we can represent the movement of a front by the movement of this line.

Not all air masses move at the same velocity. Typically, cold fronts move faster than warm fronts and overtake them. Where this happens, the cold front lifts up the base of the warm front, so the warm front no longer intersects the ground surface. Meteorologists refer to the geometry that results when a cold front pushes underneath a warm front as an *occluded front* **(Fig. 20.14c)**.

TABLE 20.1 Weather Extremes

Lowest recorded air pressure at sea level	870 millibars; during Typhoon Tip (1979), Pacific Ocean
Highest air pressure at sea level	1,084 millibars; Agata, Siberia
Lowest air temperature (world)	−89°C; Vostok, Antarctica
Lowest air temperature (North America)	−63°C; Yukon, Canada
Highest air temperature (world)	58°C; Libya
Highest air temperature (North America)	57°C; Death Valley, California
Highest recorded wind speed (world), not in a tornado	372 km per hour; Mt. Washington, New Hampshire

FIGURE 20.14 The formation of fronts between air masses.

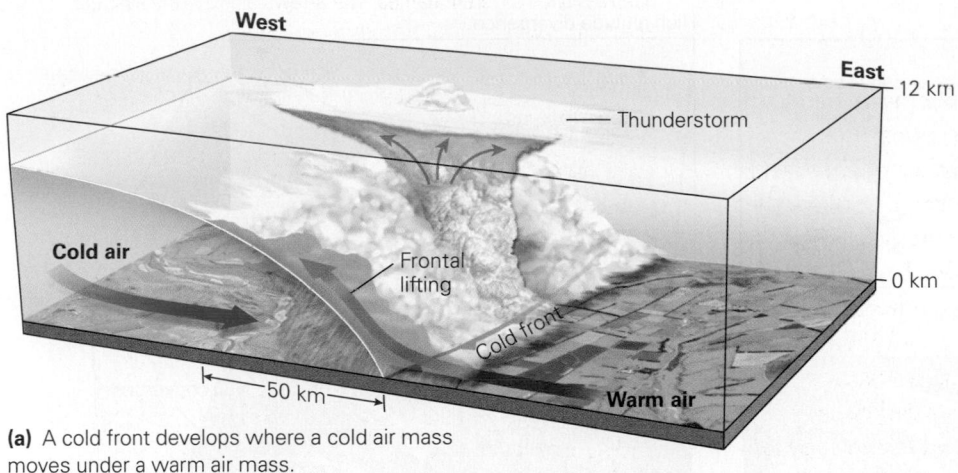

(a) A cold front develops where a cold air mass moves under a warm air mass.

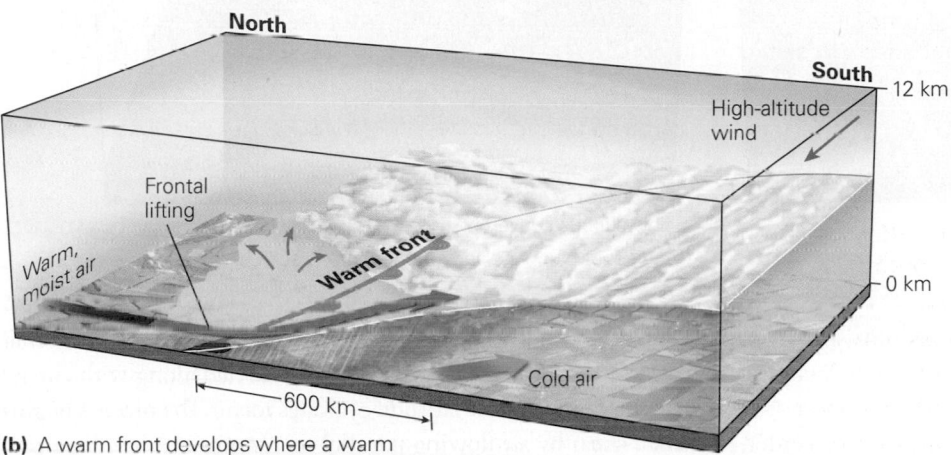

(b) A warm front develops where a warm air mass moves over a cold air mass.

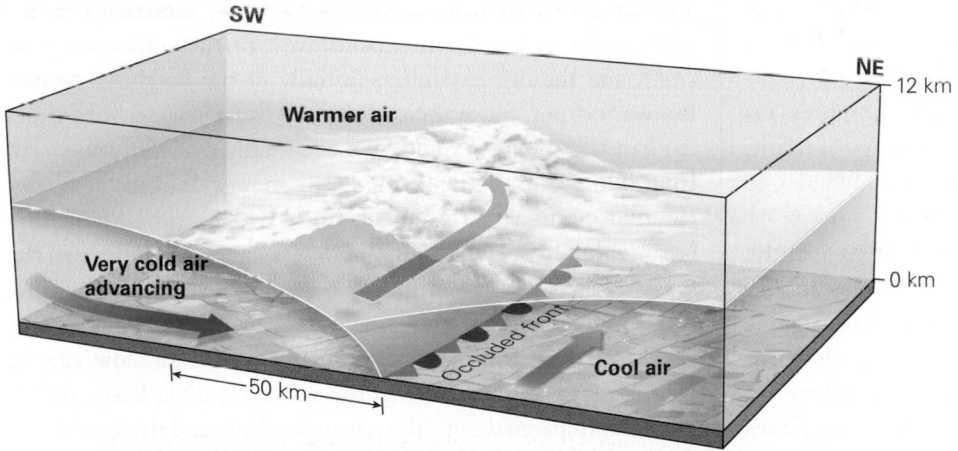

(c) An occluded front develops where a fast-moving cold front overtakes a warm front and lifts the base of the warm front off the ground. Note the symbols used to represent fronts.

Cyclonic and Anticyclonic Flow

Huge swirling currents of air occasionally develop at latitudes between 30° and 60° north or south of the equator when tropical air masses and polar air masses interact. These currents, known as **mid-latitude cyclones** if they rotate counterclockwise and **mid-latitude anticyclones** if they rotate clockwise (in the northern hemisphere), play a major role in controlling the weather across the United States, Europe, and other regions occupying the Earth's mid-latitudes. (Note that *cyclone*, in this context, does not mean a tornado or hurricane.)

Why do cyclones and anticyclones form? Simplistically, they result from variations in the velocity of air flow within the polar jet stream. At a locality where air flow in the jet stream slows down, air undergoes high-altitude *convergence*, meaning that horizontal flow carries more air into the locality (in this case, a segment of the jet stream) than gets carried away. This extra air at the top of the troposphere causes the mass of the underlying column of air to increase, so a *high-pressure center* develops at the base of the troposphere **(Fig. 20.15a)**. In contrast, at a locality where air flow in the jet stream speeds up, *divergence* takes place. This means that horizontal flow carries more air out of a locality (again, a segment of the jet stream) than has entered that locality **(Fig. 20.15b)**. The resulting air deficit at the top of the troposphere causes the mass of the underlying column of air to decrease, and as a result, air pressure decreases at the base of the troposphere. The position of the lowest pressure defines a *low-pressure center*.

The development of high-pressure and low-pressure centers at the base of the troposphere affects the pattern of air flow near the Earth's surface. Specifically, when a high-pressure center develops at the Earth's surface, it generates a pressure-gradient force that causes air to flow outward from the center into surrounding regions. Recall that when air starts to move horizontally on the spinning Earth, the Coriolis force deflects it to the right (in the northern hemisphere). If the pressure-gradient force and the Coriolis force were the only forces affecting air movement, air would circle the high-pressure center in a clockwise direction. But because friction slows the air, and because speed determines the magnitude of the Coriolis force, the pressure-gradient

FIGURE 20.15 Formation of high-pressure and low-pressure systems.

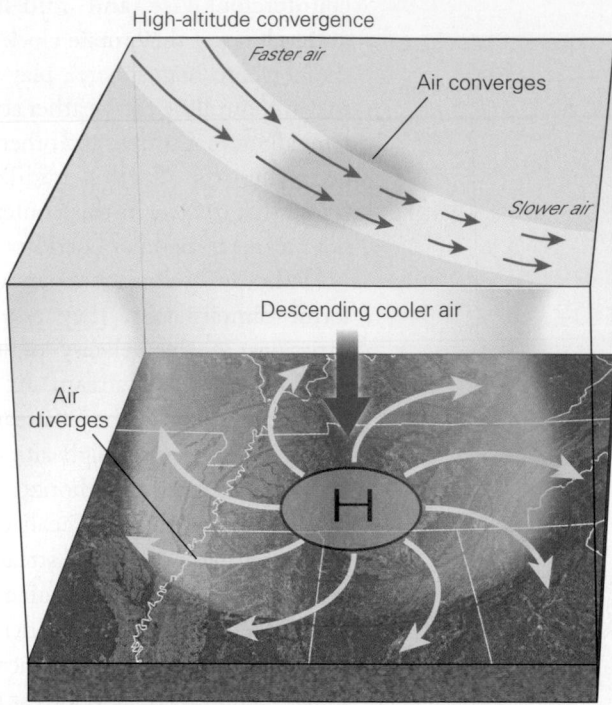

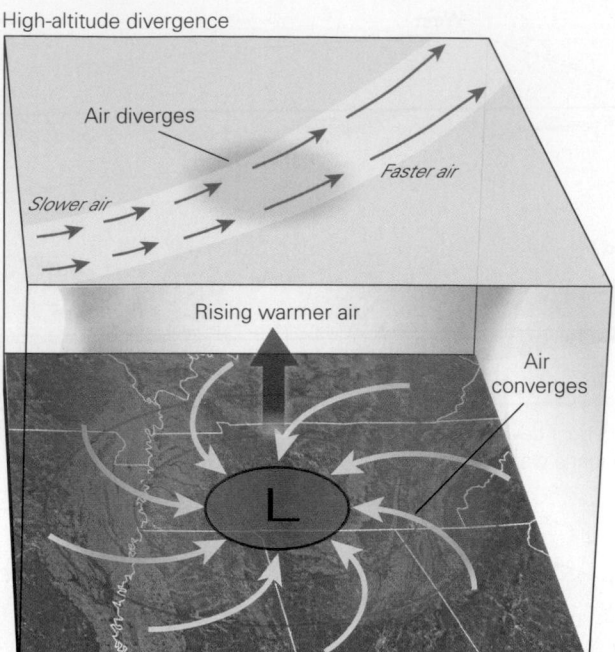

(a) Air spirals downward and clockwise at a high-pressure center.

(b) Air spirals upward and counterclockwise at a low-pressure center.

force exceeds the Coriolis force, so air spirals outward from the high-pressure center. This outward-spiraling clockwise flow generates divergence at the base of the troposphere (see Fig. 20.15a). Meteorologists refer to a high-pressure center, together with the associated three-dimensional pattern of air circulation, as a **high-pressure system**, and its presence results in the formation of an anticyclone. In a high-pressure system, air from high altitudes spirals downward to replace air spiraling outward. As this air sinks, it undergoes compression and warms, so its relative humidity decreases. As a result, high-pressure systems tend to host relatively dry, cloud-free air—in other words, they produce fair weather.

The development of a low-pressure center at the base of the troposphere also leads to the formation of a pressure gradient. Alone, this gradient would cause air from surrounding regions to flow inward toward the center. However, the Coriolis force deflects moving air to the right (in the northern hemisphere). If the pressure-gradient and Coriolis forces were the only forces acting on the moving air, it would circle the low-pressure center in a counterclockwise direction. However, because friction slows the air, the pressure-gradient force exceeds the Coriolis force, so air spirals inward toward the low-pressure center (see Figure 20.15b). This inward-spiraling, counterclockwise flow causes convergence at the base of the troposphere. The low-pressure center together with the fronts, clouds, and air circulation pattern within it is called a **low-pressure system**, and its presence produces a cyclone.

Mid-latitude cyclones tend to build into large storms that traverse the United States as they are carried along with the jet stream. How do mid-latitude cyclones form? The process begins when shear by air flowing parallel to a *stationary front* (one that doesn't move) causes the front to warp into a curve, with the apex of the curve at a low-pressure center. The front extending initially to the south or southwest of the center is a cold front, and the one extending initially to the northeast or east is a warm front. Since the cold front advances faster with time, it gradually swings around and eventually occludes the warm front **(Fig. 20.16a)**. The lifting of air above the fronts generates clouds, some of which coalesce to form strong storms, drenching areas with rain or blanketing them with snow, depending on the season. Once the mid-latitude cyclone has formed, the swirling of air around the low-pressure system gives the cloud pattern a characteristic comma shape, within which air flow can be complex **(Fig. 20.16b)**. When the warm front has been entirely occluded (separated out), the storm weakens and dissipates.

Forming Clouds and Precipitation

Let's look at clouds a little more closely to understand how to describe them and to see why they yield rain and snow **(Fig. 20.17)**. A **cloud** is a region of the atmosphere where about half of the atmospheric moisture exists as tiny water droplets (about 20 μm across, less than a third of the diameter of a human hair; 1 μm = 0.001 mm) or as tiny ice crystals. (Most

FIGURE 20.16 Formation of a mid-latitude cyclone.

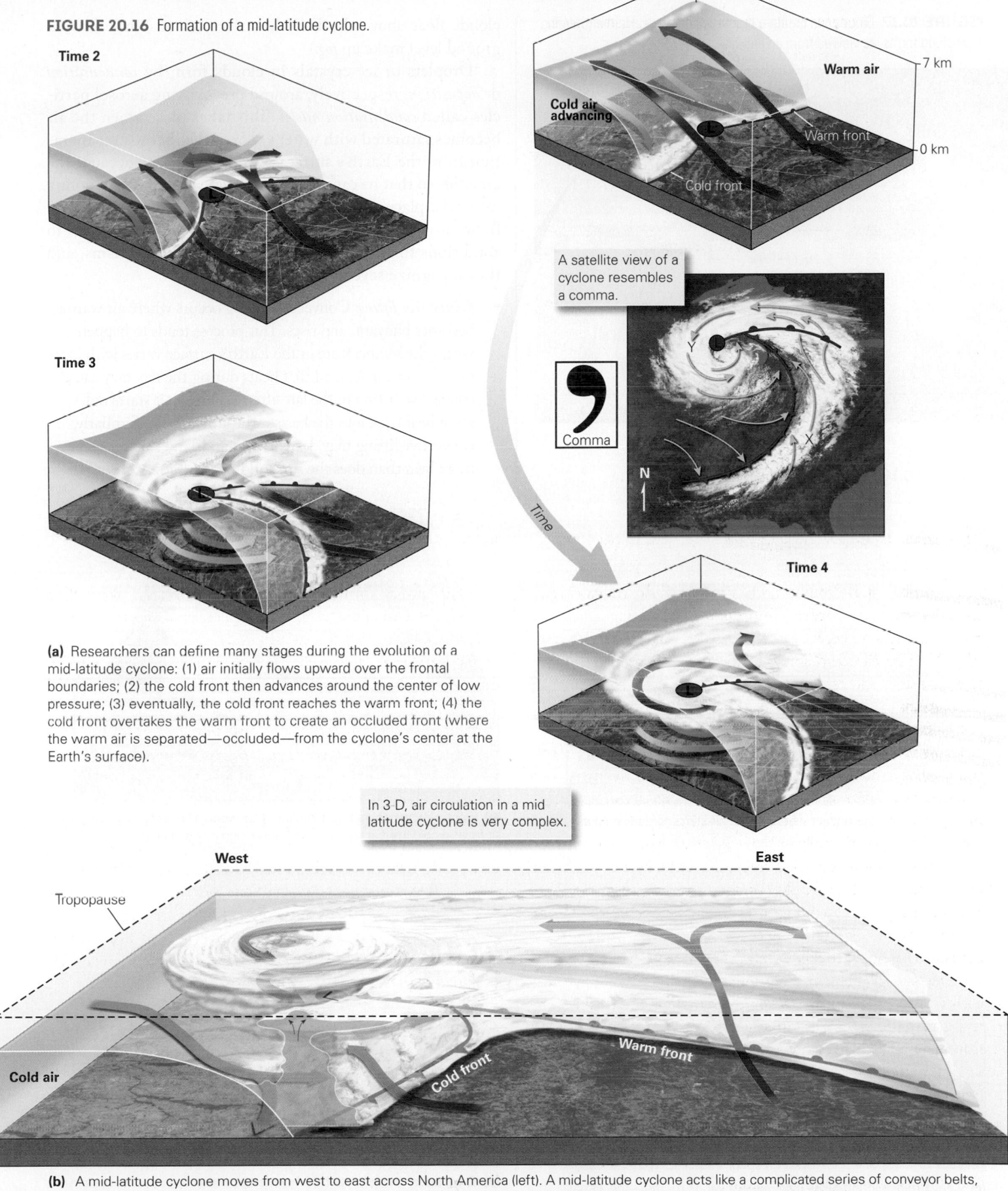

Time 1

Warm air

Cold air advancing

Warm front

Cold front

7 km

0 km

A satellite view of a cyclone resembles a comma.

Comma

Y

X

N

Time

Time 2

Time 3

(a) Researchers can define many stages during the evolution of a mid-latitude cyclone: (1) air initially flows upward over the frontal boundaries; (2) the cold front then advances around the center of low pressure; (3) eventually, the cold front reaches the warm front; (4) the cold front overtakes the warm front to create an occluded front (where the warm air is separated—occluded—from the cyclone's center at the Earth's surface).

Time 4

In 3-D, air circulation in a mid latitude cyclone is very complex.

West

East

Tropopause

Cold air

Cold front

Warm front

(b) A mid-latitude cyclone moves from west to east across North America (left). A mid-latitude cyclone acts like a complicated series of conveyor belts, bringing warm air northward, cold air southward, high-altitude air downward, and low-altitude air upward as the fronts evolve (right).

FIGURE 20.17 Clouds can create a dramatic spectacle. Here a towering anvil cloud forms, as viewed from an airplane.

clouds float above the Earth's surface; clouds that form at ground level make up *fog*.)

Droplets or ice crystals in clouds form by *condensation* or *deposition*, respectively, around pre-existing aerosol particles called *condensation nuclei*. This takes place when the air becomes saturated with water vapor, either because evaporation from the Earth's surface provides water or because the air cools so that its capacity to hold water decreases. Cooling may take place at night, simply because of the loss of sunlight, or when air rises and expands. Meteorologists refer to conditions that cause air to rise as **lifting mechanisms**, and they recognize several different types:

- *Convective lifting:* Convective lifting occurs where air warms, becomes buoyant, and rises. This process tends to happen where the temperature of the Earth's surface varies with location. For example, land that heats during the day may cause convective lifting in the late afternoon when it starts radiating heat back into the base of the atmosphere. Similarly, convective lifting may develop over an island that releases more heat than does the surrounding sea **(Fig. 20.18a)**.

FIGURE 20.18 Lifting mechanisms.

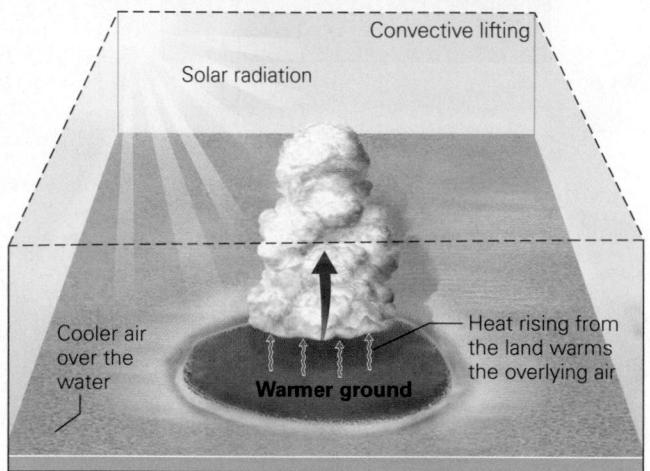

(a) Convective lifting occurs where warm air starts to rise, such as over the warmer ground of an Island.

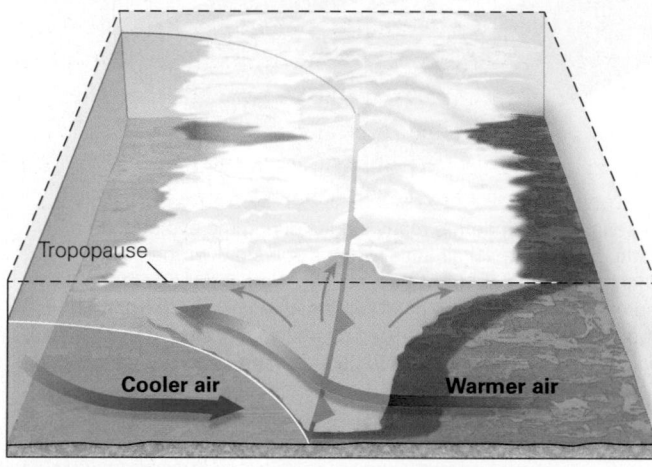

(b) Frontal lifting occurs along the fronts between air masses. As warm air faces a cold front, it rises over the advancing mass of cold air.

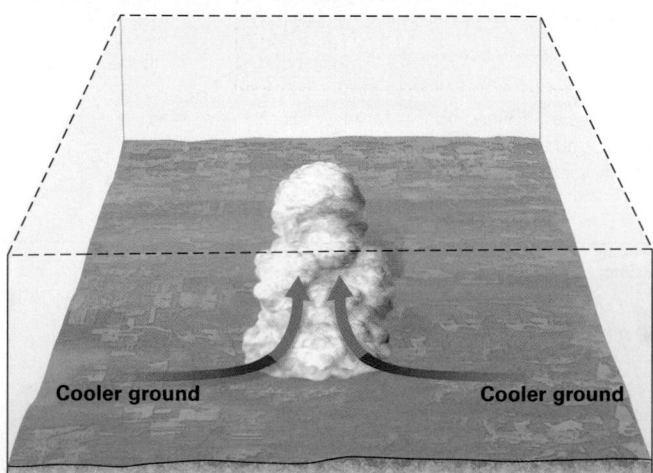

(c) Convergence lifting takes place where winds merge—the air has nowhere to go but up.

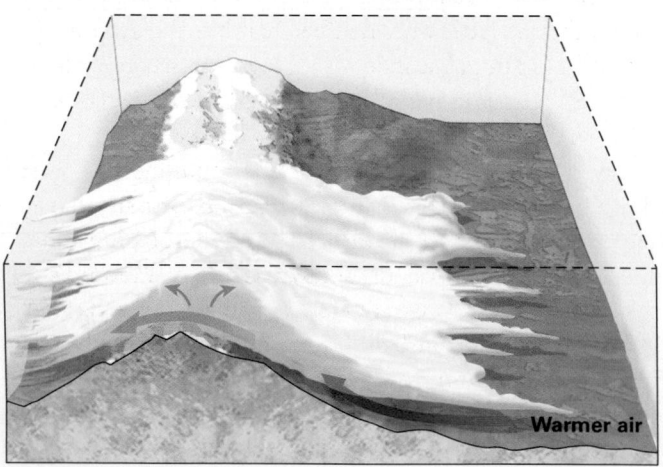

(d) Orographic lifting happens where moist winds run into a mountain range and are forced up.

- *Frontal lifting:* Frontal lifting takes place along the fronts between air masses, as we've seen **(Fig. 20.18b)**. At cold fronts, warm air pushes up and over a steep wall of cold air, rising rapidly to form large clouds. At warm fronts, the advancing warm air rides up the gentle slope of the front and condenses.
- *Convergence lifting:* Where air converges near the ground surface, as happens when two opposing winds collide, the air has nowhere to go but up, so it rises and cools, forming clouds **(Fig. 20.18c)**.
- *Orographic lifting*: This type of lifting happens when moisture-laden wind blows toward a mountain range and on reaching the mountain range, can go no farther and must rise. As a result, clouds form above the mountain range **(Fig. 20.18d)**.

Precipitation, in the form of rain, snow, hail, and sleet, falls from clouds in two ways, depending on the temperature of the cloud. In warm clouds, rain develops by **collision and coalescence**, during which the tiny droplets that compose the cloud collide and stick together to create a larger drop **(Fig. 20.19)**. Eventually, some water drops become too big to be held in suspension by circulating air and start to fall, incorporating more droplets as they descend. The sizes of the drops that hit the ground depend on how far the drops have fallen and on how much moisture is available. Typical raindrops have a diameter of 2 mm and fall at a velocity of about 20 km per hour. Any drops larger than 5 mm tend to break into smaller ones before reaching the ground. If rain falls through colder air near the ground, it freezes to become *sleet*.

Cold clouds contain a mixture of very cold water droplets and tiny ice crystals. The water droplets evaporate faster than the ice (because water molecules are less tightly bound to liquid than to solid) and provide moisture that condenses onto pre-existing ice crystals, leading to the growth of hexagonal snowflakes. If the air below the cloud is very cold, the snow falls as powder-like flakes, but if the air is close to the melting temperature, large, wet clumps of snowflakes fall. And, if the lower air is warmer than 0°C, the snow becomes rain before it hits the ground. This kind of precipitation, involving the growth of ice crystals in a cloud at the expense of water droplets, is called the *Bergeron process*, after the Swedish meteorologist who discovered it.

Many kinds of clouds develop in the troposphere. It wasn't until 1803, however, that Luke Howard, a British pharmacist, proposed a simple terminology for describing clouds **(Fig. 20.20)**. We still use a modified version of Howard's terminology today. First, we divide clouds into types based on their shape: Puffy, cotton-ball- or cauliflower-shaped clouds are *cumulus* (from the Latin word for stacking). Clouds that occur in relatively thin, stable layers and thus have a sheet-like or layered shape are called *stratus*. High clouds that have a wispy shape and taper into delicate, feather-like curls are called *cirrus*. We can then add a prefix to the name of a cloud to indicate its elevation:

high-altitude clouds (above about 7 km) take the prefix *cirro-*, mid-altitude clouds take the prefix *alto-*, and low-altitude clouds (below 2 km) do not have a prefix. Finally, we add the suffix *-nimbus* or the prefix *nimbo-* if the cloud produces rain.

Applying Howard's scheme, we see that the name nimbostratus refers to a layered, sheet-like rain cloud, and the name cumulonimbus refers to a rain-producing puffy cloud. Cumulonimbus clouds can be truly immense, with their bases lying at less than 1 km elevation and their tops butting up against the tropopause at an elevation of over 14 km. Large cumulonimbus clouds spread laterally at the tropopause to form broad, flat-topped clouds called *anvil clouds* (see Fig. 20.17).

The differences in cloud types depend on whether the clouds develop in stable or unstable air. *Stable air* does not have a tendency to rise because it is colder than its surroundings—stratus clouds may form in such stable air. *Unstable air* has a tendency to rise in updrafts because it remains warmer and less

FIGURE 20.19 Mechanisms of raindrop formation.

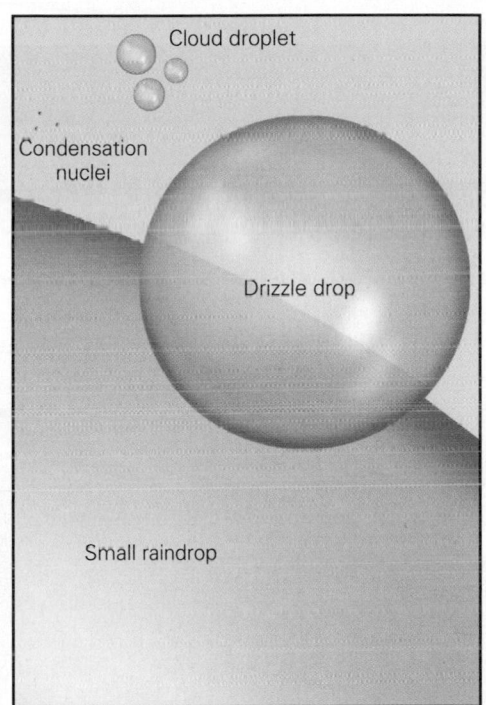

(a) Cloud droplets form when water vapor condenses around condensation nuclei. Rain forms as droplets collide with one another and grow.

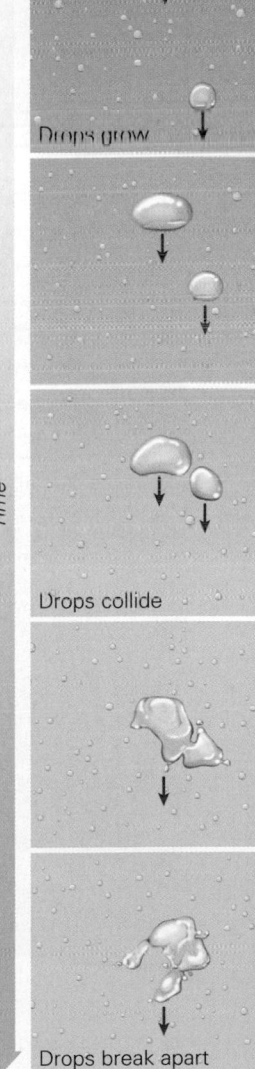

(b) Once a drop becomes large enough to fall, it incorporates more drops on the way down. Really large drops can break apart.

FIGURE 20.20 The various types of clouds on the Earth.

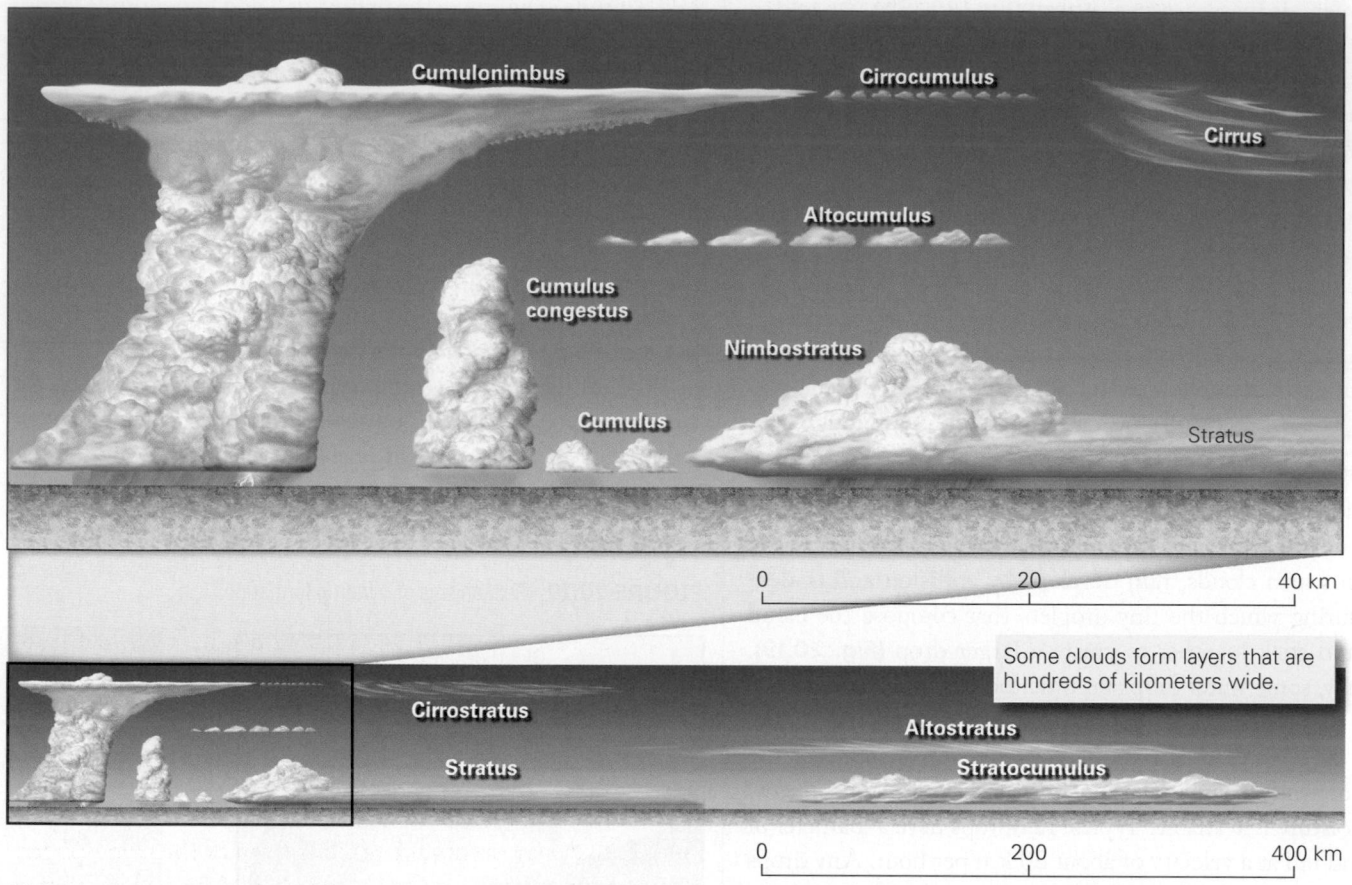

(a) The type of cloud that forms depends on the stability of the air, the temperature at which moisture condenses, and the wind speed.

(b) A satellite photo of the Earth displays the distribution of clouds. At times, more than half of the surface has cloud cover.

dense than its surroundings, even after it has risen substantially. Some of the heat that drives its rise comes from the condensation of water droplets, for the process releases the latent heat of condensation. Cumulus clouds, which in time-lapse photography look as if they're boiling, form in unstable air and contain intense updrafts, bordered by downdrafts. Plane flights through the unstable air of cumulus clouds can be intensely bumpy.

TAKE-HOME MESSAGE

An air mass is a body of air whose temperatures and moisture content distinguish it from adjacent masses. Air masses interact along a front. Air spirals toward a low-pressure center to cause cyclonic flow and away from a high-pressure center to cause anticyclonic flow. Large mid-latitude cyclones may result in strong storms. Where air rises and cools, clouds form. Water droplets in clouds can coalesce to form raindrops, or the water may crystallize onto tiny ice particles.

QUICK QUESTION: What names do meteorologists use for different kinds of clouds?

20.7 Storms: Nature's Fury

An overcast sky may be inconvenient for a picnic, but it won't threaten life or property—but a storm can! A *storm* is an episode of weather during which winds, rainfall, snowfall, and in some cases, lightning become strong enough to be bothersome and even dangerous **(Fig. 20.21)**. Storms are commonly associated with large pressure gradients, which may exist across a steep front, for pressure gradients produce strong winds. Storms also form where local conditions cause atmospheric instability, which can cause clouds to grow. Once the clouds become thick enough to start producing heavy rain, or the wind becomes strong enough to be troublesome, we can say that a storm has been born. We'll now look at various types of storms.

Thunderstorms

If you live in a tropical or temperate environment, you've probably experienced the spectacle—and terror—of a **thunderstorm**, an episode during which a mass of towering clouds produces strong winds and heavy rains, accompanied by flashes of lightning and claps of thunder. Some thunderstorms develop in isolation, but commonly multiple thunderstorms develop in a row along a front, producing a *squall line*. At any moment, our planet's surface hosts about 2,000 thunderstorms. In fact, over the course of a year, about 100,000 develop over the United States. What conditions produce thunderstorms?

FIGURE 20.21 A thunderstorm can drench an area with rain and strike it with lightning.

The formation of a thunderstorm begins when a lifting mechanism forces warm, moist air from near the Earth's surface up to higher elevation. As the moisture in the rising air condenses, it releases latent heat of condensation. If this process keeps the rising air warmer than its surroundings, the air retains its buoyancy. Such *unstable air* continues to rise as an updraft that may eventually reach the top of the troposphere, resulting in a towering cumulonimbus cloud. Note that a supply of warm, moist air must "feed" the storm to keep it growing, so most thunderstorms happen in tropical regions, or during the summer in temperate regions. Meteorologists distinguish between two general categories of thunderstorms: ordinary thunderstorms and supercells.

Ordinary Thunderstorms An ordinary thunderstorm has a relatively short lifespan, lasting from under an hour to a few hours, and has a diameter of a few kilometers to 10 km **(Fig. 20.22)**. These storms evolve through distinct stages. In a *developing thunderstorm*, cumulus clouds build and a strong vertical updraft develops. When the billowing clouds reach the top of the troposphere, they spread laterally to form a flat-topped *anvil cloud*. Eventually, air at high elevation undergoes enough cooling for the moisture in it to condense into rain or to form ice. Precipitation begins, pulling air down as it falls, to generate a downdraft. Falling ice turns to rain below the freezing level. As the rain descends and warms, it evaporates. The resulting evaporative cooling makes the air in the downdraft denser than its surroundings, so the air continues sinking down to the ground, where it spreads laterally as a strong wind. In this now *mature thunderstorm*, rain pelts the ground and, as we'll see, lightning flashes. Once formed, the downdraft competes with the remaining updraft, making the air within the storm very turbulent. Eventually, the sinking cool air of the downdraft cuts off the

FIGURE 20.22 Evolution of an ordinary a thunderstorm takes place in three stages.

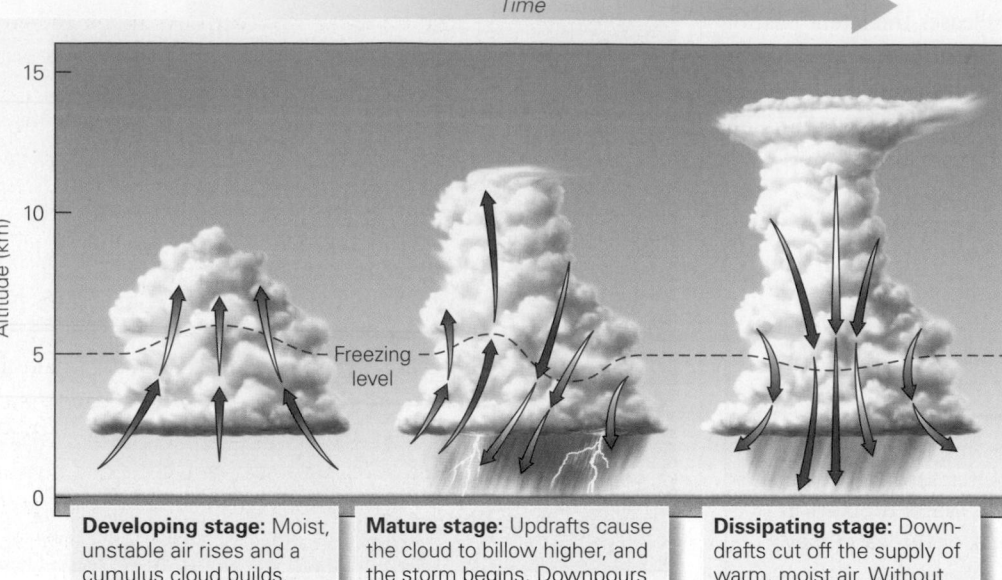

Time

Altitude (km)

- - - Freezing level

Developing stage: Moist, unstable air rises and a cumulus cloud builds.

Mature stage: Updrafts cause the cloud to billow higher, and the storm begins. Downpours start downdrafts.

Dissipating stage: Downdrafts cut off the supply of warm, moist air. Without fuel, the storm dissipates.

supply of rising warm, moist air, and starved of fuel, the *dissipating thunderstorm* becomes inactive. Notably, ordinary thunderstorms can be carried along by prevailing winds, traveling horizontally relative to the ground at rates of up to 100 km per hour.

Supercell Thunderstorms

In places where wind velocities at higher elevations are greater than those at lower elevations, shear between higher and lower air forms a cylinder-like roll of air near the ground **(Fig. 20.23a)**. When this cylinder moves under an ordinary thunderstorm, updrafts pull it up into a tall arch; in one arm of the arch, the air spirals upward, and in the other arm, it spirals downward. Downdrafts push the center of the arch down, so that the initial storm separates into two storms

FIGURE 20.23 Evolution of a supercell thunderstorm.

Time 2

Time 1

W N S E

(a) Wind shear causes a horizontal roll of air. When this roll moves beneath a thunderstorm, the updraft arches it upward.

W N S E

(b) The downdraft produces a double arch, and two storms form.

W N S E

(c) One storm dies off, leaving a single rotating supercell.

(d) A view of the anvil and overshooting top, where updrafts push the cloud into the stratosphere.

FIGURE 20.24 Formation of a lightning bolt happens because a charge separation forms in a cloud.

(a) Lightning can shoot between two parts of a cloud, between two adjacent clouds, or from a cloud to the ground.

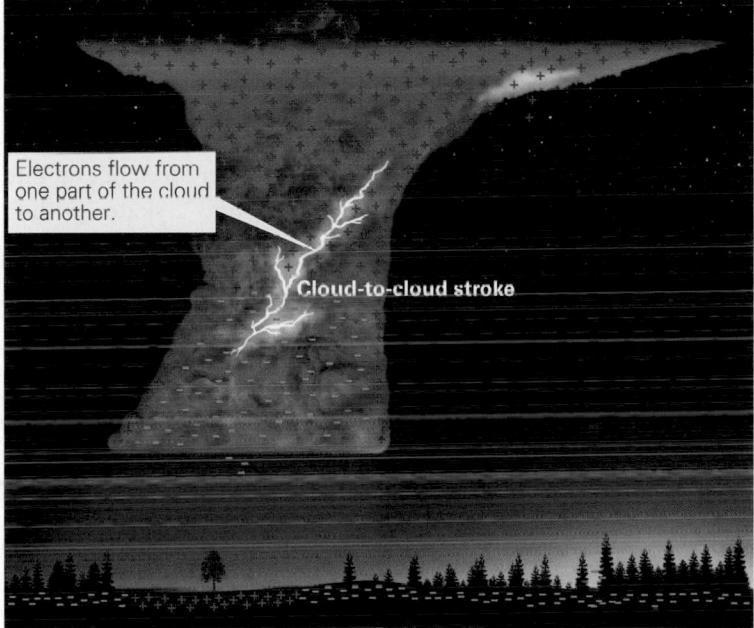

(b) Lightning discharges can occur in three ways: as negative polarity cloud-to-ground strokes, as positive polarity cloud-to-ground strokes, or as cloud-to-cloud strokes.

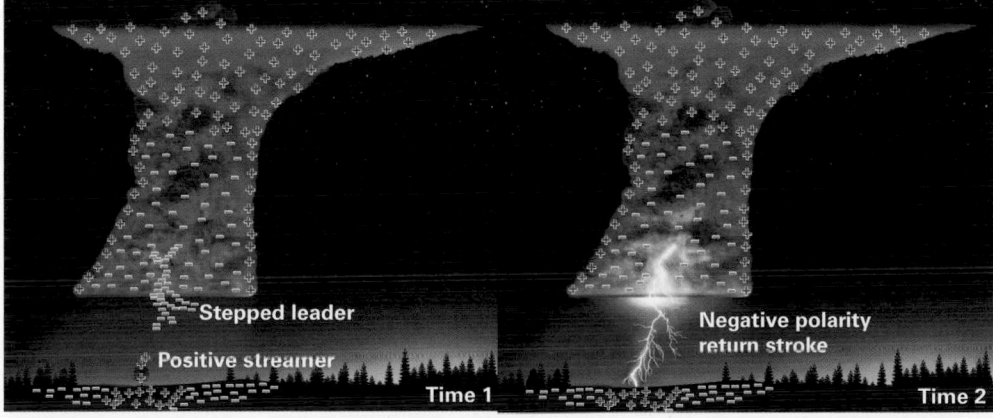

(c) The formation of a cloud-to-ground stroke takes place in stages, over the course of microseconds.

(**Fig. 20.23b**). Typically, one of the storms blocks the supply of warm moist air to the other, so one storm weakens and disappears. The remaining thunderstorm, which can now be called a **supercell**, remains (**Fig. 20.23c**). A supercell differs from an ordinary thunderstorm in two fundamental ways. First, air in the updraft of a supercell spirals, so the whole storm rotates, whereas in an ordinary thunderstorm, the updraft is vertical and the storm does not rotate (**Fig. 20.23d**). Second, strong winds at high elevation tilt a supercell, so its rains and downdrafts do not compete with, and defeat, the updraft. Therefore, updrafts can continue to carry moist air into the storm, so a supercell can last longer, and become stronger and larger, than an ordinary thunderstorm. Some supercells have a diameter as wide as 50 km.

Hail Due to updrafts in thunderstorms, water droplets can be carried to high elevation where their temperatures cool below freezing. When such "supercooled" droplets collide with ice particles, which also exist at high elevations, the water of the droplets instantly freezes, so the ice particle grow. With continued collisions, an ice particle quickly grows into a ball or clump of ice called a *hailstone*. If updrafts remain strong, as can happen in a supercell, hailstones remain aloft for enough time to become pea-sized or larger. The largest hailstone ever recorded was the size of a softball and weighed 0.8 kg. Eventually, the weight of hailstones carries them down, producing a type of precipitation called **hail**. If hailstones survive their transit through warmer air at lower elevations, they reach the ground while still solid. Conditions favoring hailstone formation can last for a few minutes to over an hour, so as a hail-producing storm moves, it leaves a *hail swath*, an area of hail-covered land, typically 2 to 5 km wide by 10 to 30 km long, elongated in the direction that the storm is moving.

Lightning and Thunder Thunderstorms are so named because of the crashes and rumbles of thunder that typically accompany them. The thunder is a consequence of lightning. A **lightning stroke** (or *lightning bolt*) is a giant spark that forms between two parts of a cloud, between two clouds, or between a cloud and the ground (**Fig. 20.24a**).

It resembles the spark that shocks you when you shuffle along a rug and then touch a door handle. But the spark from the door handle is only about a millimeter long and a fraction of a millimeter in diameter, whereas a single lightning bolt may be up to 10 km long, up to 3 cm in diameter, and can contain enough energy to power a house for many months.

Lightning develops, in part, because of the movement of various kinds of particles within storm clouds. Simplistically, the particles transfer electrons to one another when they come in contact, and different types of particles with different charges move to different parts of the storm cloud. Overall, positive charges accumulate toward the top of the cloud, and negative charges accumulate toward the bottom. Air serves as an insulator, so the charge separation can become very large before a giant spark or pulse of current, the lightning stroke, jumps across the gap.

Essentially, lightning is like a giant short circuit across which a huge (30-million-volt) pulse of electricity flows. A cloud-to-cloud lightning bolt represents a sudden flow of electrons from the negatively charged part of the cloud to the positively charged part (Fig. 20.24b). Most cloud-to-ground lightning develops in two stages. First, a charge separation develops between the cloud and the ground because the negative charges at the base of the cloud repel negative charges on the ground below, leaving a zone of positive charge on the ground (Fig. 20.24c). Once this charge separation exists, electrons start to leak incrementally downward from the base of the cloud, producing a *stepped leader*. Streams of electrons head in different directions, so the stepped leader has many branches. Meanwhile, positive charges flow upward to the cloud, through conducting materials such as trees and buildings, producing a *positive streamer*. The instant that the positive streamer connects to a branch of the stepped leader, a very strong current of electrons flows to the ground, yielding the main lightning bolt (known technically as the *return stroke*).

We hear **thunder**, the cracking or rumbling noise that accompanies lightning, because the immense energy of a flash heats the surrounding air to a temperature of 8,000° to 33,000°C, causing it to expand and then collapse almost instantly. This expansion and contraction, like a giant pair of hands clapping, creates sound waves that travel through the air to our ears. Sound travels much more slowly than light, so we hear thunder after we see lightning. A five-second time delay between the two means that the lightning flashed about 1.6 km away.

Over 80 people a year die from lightning strikes in the United States alone, and many more are seriously burned or shocked.

Did you ever wonder...
why you hear claps of thunder during a storm?

Lightning that strikes trees heats the sap so quickly that the trees can literally explode. Lightning can spark devastating forest fires and set buildings on fire. People can reduce the hazard to buildings by installing lightning rods, upward-pointing iron spikes that conduct electricity directly to the ground so that it doesn't pass through the building.

Tornadoes

A **tornado** is a near-vertical, funnel-shaped cloud in which air rotates extremely rapidly around the axis of the funnel. In other words, a tornado is an intense vortex. The word probably comes either from the Spanish *tonar*, meaning to turn, or *tronar*, meaning thunder (perhaps referring to the loud noise generated by a tornado).

Most tornadoes form when a sudden, intense downdraft rushes down the back side of a supercell. When this downdraft reaches the ground, it forms an intense rotating cylinder of air whose axis is parallel to the ground (Fig. 20.25a). The updraft of the supercell then pulls this cylinder up so that the spinning air starts to spiral upward around a vertical axis (Fig. 20.25b). As the cylinder stretches up into the storm, it becomes narrower, and like a skater pulling her arms inward while spinning, the cylinder spins faster (Fig. 20.25c).

In the mid-latitudes of the northern hemisphere, where most tornadoes form, air in the funnel normally rotates counterclockwise around the center and spirals upward. Air in the fiercest tornadoes may move at speeds in excess of 322 km per hour. The diameter of the base of the funnel in a very small tornado may be only 5 m across, while in the largest tornadoes it may be as wide as 1,500 m across (Fig. 20.26a, b). Because of the centrifugal force caused by rotating air, air diverges at the top of a tornado, so the land beneath the tornado becomes an intense low-pressure zone.

Small tornadoes may cut a swath less than a kilometer long, but large tornadoes rake the ground for tens of kilometers, and the largest have left a path of destruction up to 500 km long (Fig. 20.26c–f). In 1925, the Tri-State Tornado, one of the most enormous tornadoes on record, ripped across Missouri, Illinois, and Indiana, killing 689 people before it dissipated. In some cases, two or three tornadoes may erupt from a single thunderstorm. Squall lines may produce a *tornado swarm*, or *outbreak*, generating dozens of tornadoes along the same front. Between April 25 and April 28, 2011, a "super outbreak" of tornadoes occurred in the midwestern and southeastern United States. At least 336 tornadoes were recorded. Particularly huge tornadoes struck Joplin, Missouri, in 2013 and Moore, Oklahoma, in 2014.

FIGURE 20.25 Formation of a tornado.

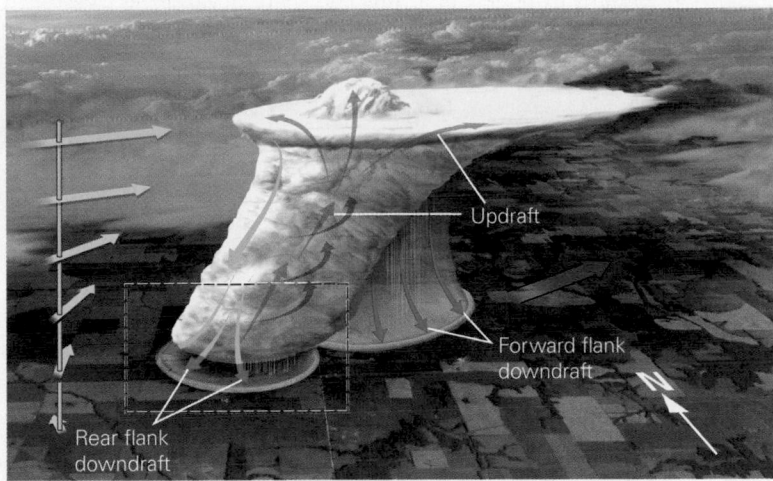

(a) A strong rear-flank downdraft develops on the back side of a supercell thunderstorm. Where it reaches the ground, a horizontal roll of air forms.

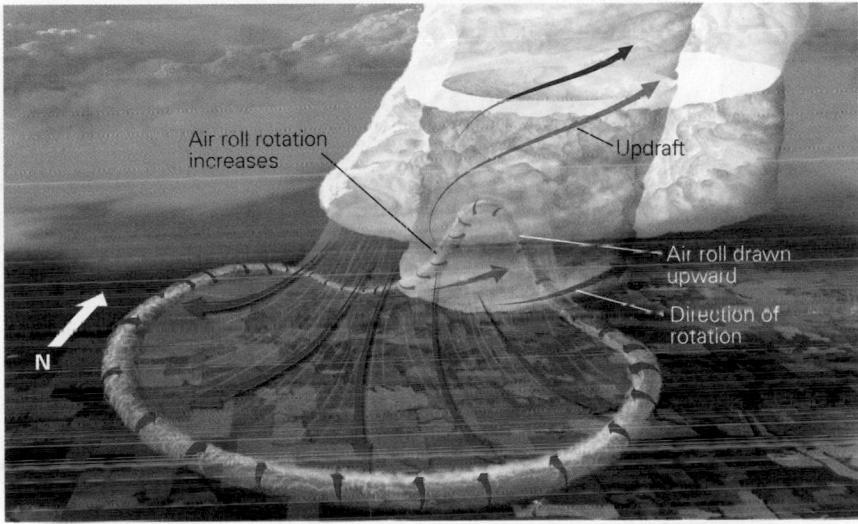

(b) Part of the roll at the base of the rear-flank downdraft gets pulled upward by the updraft of the supercell. As it stretches and becomes narrower, it speeds up.

(c) A rapidly spinning vortex—a tornado—stretches upward within the supercell's updraft. The wall cloud defines the edge of the updraft.

In North America, tornadoes drift with a thunderstorm from southwest to northeast because of the prevailing wind direction. They tend to hopscotch across the landscape, touching down for a stretch, then rising up into the air for a while before touching down again. This characteristic leads to a bizarre incidence of damage—one house may be blasted off its foundation while its next-door neighbor remains virtually unscathed.

Tornadoes cause damage because of the force of their rapidly moving wind. They lift trucks and tumble them for hundreds of meters, uproot trees, and flatten buildings. Particularly large tornadoes can even rip asphalt off a highway. In some cases, tornadoes cause strange kinds of damage: they have been known to drive straw through wood, lift cows and carry them unharmed for hundreds of meters. Because of the range of damage tornadoes can cause, T. T. Fujita of the University of Chicago proposed a scale that distinguishes among tornadoes on the basis of damage **(Table 20.2)**. The wind speeds in the **Enhanced Fujita scale** are estimates based on the damage assessment **(Geology at a Glance**, pp. 788–789).

The special weather conditions that spawn tornadoes in the midwestern United States and Florida develop when cold polar air from Canada collides with warm tropical air from the Gulf of Mexico. These conditions happen most frequently during the months of March through September but sometimes occur at other times of the year. So many tornadoes occur during the summer in a belt from Texas to Indiana that this region has the unwelcome nickname *Tornado Alley* **(Fig. 20.27a)**. Over a 30-year span, an average of 770 tornadoes have raked across Tornado Alley each year. In contrast, fewer than 20 strike Canada annually. About 80 people a year die in tornadoes, on average, but in some years the toll may be in the hundreds. A 2011 super outbreak alone killed about 350 people because some of its tornadoes rampaged through cities **(Fig. 20.27b)**.

Because of the threat tornadoes pose to life and property, meteorologists have worked hard to be able to forecast them. First they search for appropriate weather conditions. If these conditions exist, meteorologists issue a *tornado watch*. If observers spot an actual tornado or see one forming, they issue a **tornado warning** for the region in its general path (the exact path can't be predicted). If you hear a warning, it's best to take cover immediately in a basement or at least in an interior room

FIGURE 20.26 Tornadoes and the damage they cause.

(a) A moderate-sized tornado touches down near Mulvane, Kansas.

(b) A huge tornado near Eureka, Illinois.

(c) The swath of destruction left by a 1999 tornado in Oklahoma.

(d) Even downtowns are not immune—this tornado struck Miami in 1997.

(e) Close-up of the damage due to a 2007 tornado in Kansas.

(f) A helicopter view of the devastation track left in Tuscaloosa, Alabama, by a tornado from the April 2011 super outbreak.

FIGURE 20.27 North American tornadoes are most common in Tornado Alley, a band extending from Texas to Indiana, where the continental polar air mass collides with the Gulf Coast maritime tropical air mass. The storm systems move eastward.

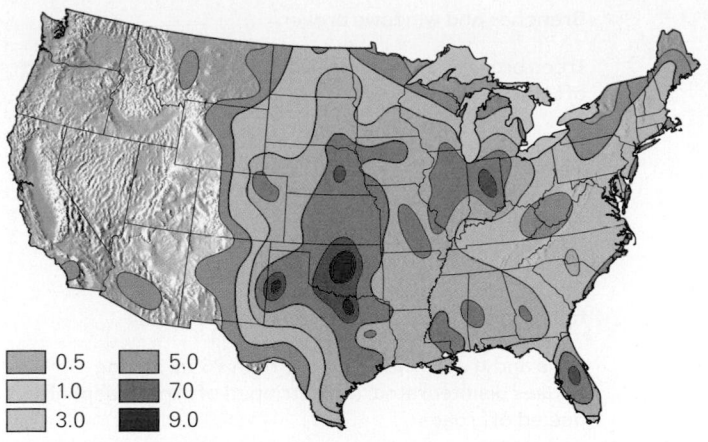

0.5   5.0
1.0   7.0
3.0   9.0

(a) Number of tornadoes per year (per 26,000 km², for a 27-year period).

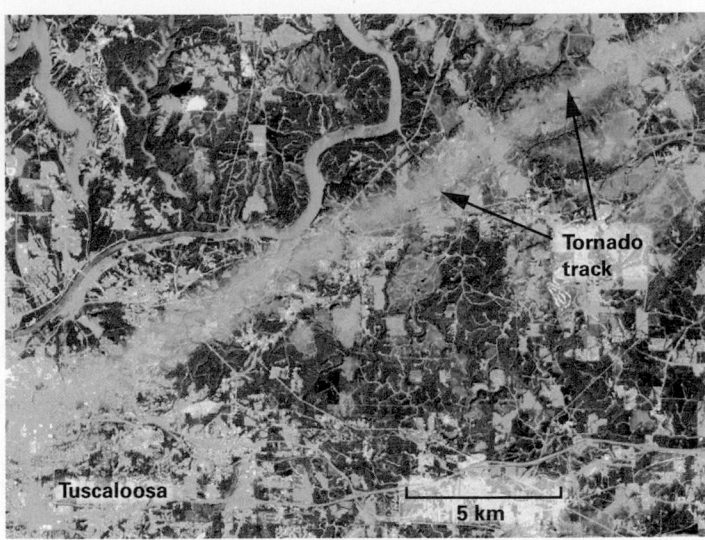

(b) This false-color satellite image shows the track of a tornado near Tuscaloosa, Alabama. Red areas are forested—the winds ripped out the trees.

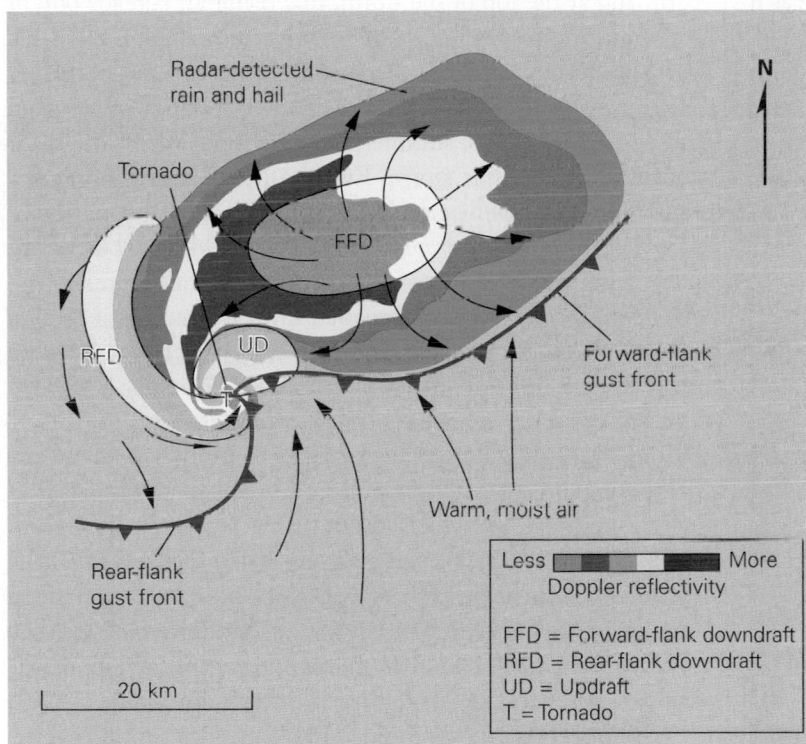

(c) A radar image shows the characteristic "hook" of a tornado, as viewed by satellite, as it passes near Vilonia, Arkansas, in 2011.

away from windows. With the invention of *Doppler radar*, which uses the Doppler effect (see Chapter 1) to identify rain moving in strong winds, meteorologists may detect tornadoes without even going outside. Probable tornadoes appear in Doppler images as a distinct hook-like shape at the edge of a thunderstorm **(Fig. 20.27c)**. The ball of debris carried by a tornado causes intense reflectivity on a Doppler image.

Mid-latitude Cyclones and Nor'easters

Mid-latitude cyclones (also known as extra-tropical cyclones) can produce intense thunderstorms and associated flash floods. In colder weather they can produce blizzards (snowstorms of immense proportions). A blizzard associated with a mid-latitude cyclone in 1888 dumped up to 1.5 m of snow, and a blizzard in 1996 buried New York under 1.2 m of snow. Large mid-latitude cyclones of North America that affect the Atlantic coast are called *nor'easters* because the cold, counterclockwise winds of these cyclones come out of the northeast. Some nor'easters are truly phenomenal storms **(Fig. 20.28)**. Their winds are not as strong as those of a hurricane, but they cover such a large area that waves in the open ocean build to a height of over 11 m, a disaster for ships. When the waves reach shore, they erode huge stretches of beach.

Hurricanes: A Coastal Calamity

What Is a Hurricane? As we saw in Section 20.5, global-scale convection of the atmosphere, influenced by the Coriolis effect, causes currents of warm air to flow steadily from east to

TABLE 20.2 Enhanced Fujita Scale for Tornadoes

Scale	Category	Wind Speed (km per hour)	Average Path Length; Average Path Width	Typical Damage
EF0	Weak	104–137	0–1.6 km; 0–17 m	Branches and windows broken
EF1	Moderate	138–177	1.7–4.9 km; 18–55 m	Trees broken; shingles peeled off; mobile homes moved off their foundations
EF2	Strong	178–217	5–16 km; 56–175 m	Large trees broken; mobile homes destroyed; roofs torn off
EF3	Severe	218–266	17–50 km; 176–556 m	Trees uprooted; cars overturned; well-constructed roofs and walls removed
EF4	Devastating	267–322	51–160 km; 0.56–1.5 km	Strong houses destroyed; buildings torn off foundations; cars thrown; trees carried away
EF5	Incredible	Over 322	161–500 km; 1.51–5.0 km	Cars and trucks carried more than 90 m; strong houses disintegrated; bark stripped off trees; asphalt peeled off roads

west in tropical latitudes (**Fig. 20.29a**). As the air flows over the ocean, it absorbs moisture. Because air becomes less dense as it gets warmer, the tropical air eventually begins to rise, producing a cluster of large thunderstorms, which consolidate to form a single, very large storm. Because of the Coriolis effect, this storm cluster starts rotating around a vertical axis and evolves into a broad swirl of thick clouds called a *tropical disturbance.*

FIGURE 20.28 A large nor'easter moves up the east coast of the United States.

At the center of the swirl, air rises, following an upward spiral path, and at the top of the storm, this rising air spreads out, or diverges. The resulting loss of air molecules at high elevation over the storm creates low pressure at the base of the storm.

If a tropical disturbance remains over warm ocean water, as can happen in late summer and early fall, warm, moist air continues to feed the storm. Eventually, the storm organizes into a spiral of rapidly circulating clouds, and the tropical disturbance becomes a *tropical depression.* Additional nourishment from moist, warm air causes the tropical depression to spin even faster and grow broader, until it becomes a *tropical storm* and receives a name. If a tropical storm becomes powerful enough, it becomes a *tropical cyclone.* Formally defined, a **tropical cyclone** is a huge rotating storm that forms in tropical latitudes and in which winds exceed 119 km per hour. In the northern hemisphere, it resembles a giant counterclockwise spiral of clouds—300 to 1,500 km wide—when viewed from space (**Fig. 20.29b**). Such a storm is called a *hurricane* in the Atlantic, Caribbean, Gulf of Mexico, and eastern Pacific; a *typhoon* in the northwestern Pacific, and simply a *cyclone* in the southwestern Pacific and the Indian Ocean. To avoid confusion, we'll generally refer to these storms as hurricanes.

Note that the process that forms a tropical cyclone (hurricane) differs from the process that forms a mid-latitude cyclone. Though both types of storms can produce devastating winds and torrential rains, they have different structures and evolve in different ways.

Atlantic hurricanes generally form in the open ocean to the east of the Caribbean Sea, though some form in the Caribbean itself. They first drift westward at speeds of up to 60 km per hour, then may eventually turn north and head into the North Atlantic or into the interior of North America. Over the cooler waters of high latitudes, or over land, they dissipate when their supply of warm water is cut

FIGURE 20.29 The structure and distribution of hurricanes.

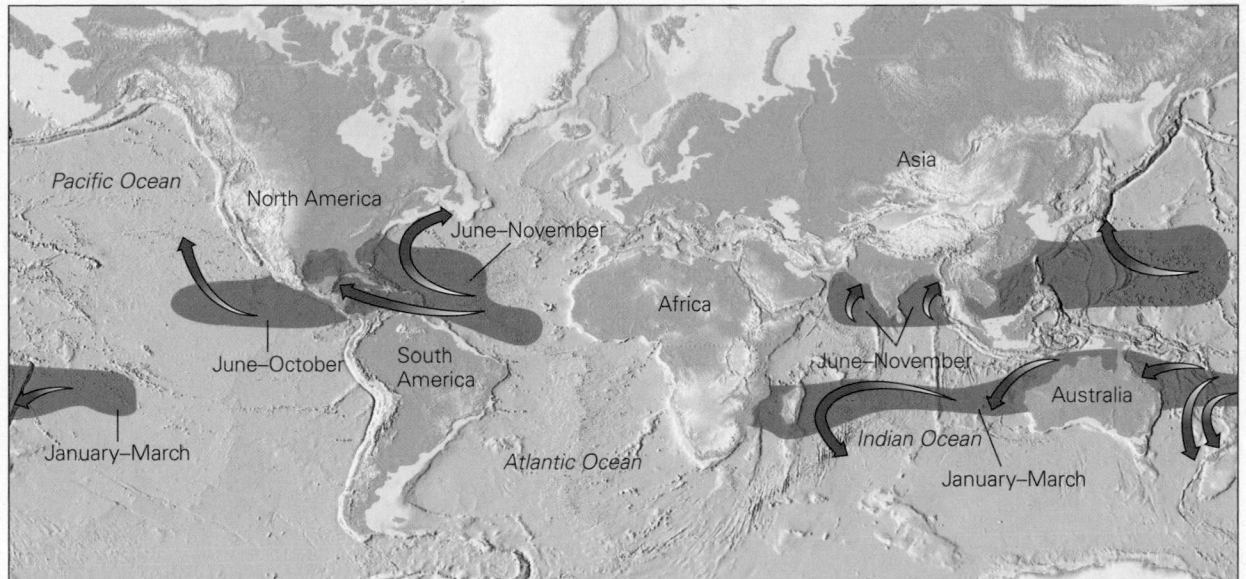

(a) Hurricanes form only in certain regions of the ocean, where water temperatures are high. They generally follow regional tracks and occur during certain times of the year.

(b) Hurricane Sandy approaches the east coast of the United States in 2012, as seen from space.

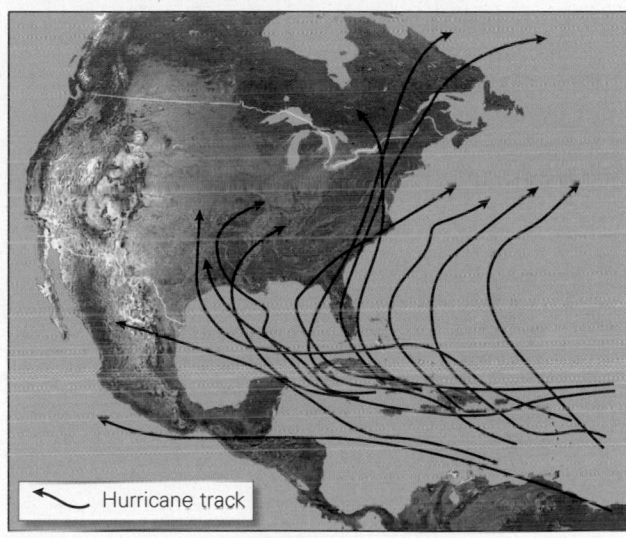

(c) Tracks of several Atlantic hurricanes show how most drift westward, then northward.

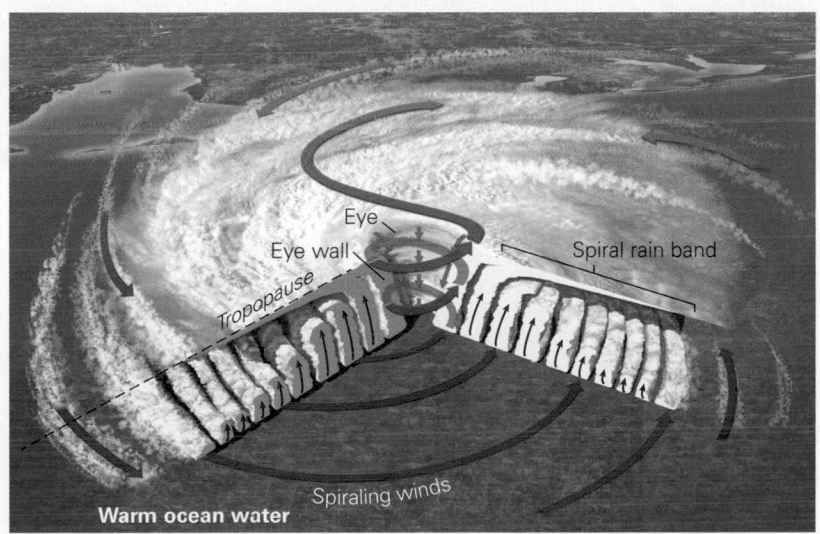

(d) In this cutaway drawing, we can see the rain bands, the eye, and the eye wall of a hurricane. Dry air descends in the eye.

off **(Fig. 20.29c)**. Weather researchers classify the strength of hurricanes using the **Saffir-Simpson scale**, which runs from 1 to 5 **(Table 20.3)**; they use somewhat different scales for typhoons and cyclones. On the Saffir-Simpson scale, a Category 5 hurricane hosts sustained winds of greater than 250 km per hour. The highest wind speed ever recorded during a hurricane was in excess of 300 km per hour.

A typical hurricane (or typhoon or cyclone) consists of several spiral arms, called *rain bands*, extending inward to a central zone of relative calm known as the *eye* **(Fig. 20.29d)**. A rotating vertical cylinder of clouds, the *eye wall*, surrounds the eye. Winds spiral toward the eye and, like those of a tornado, accelerate near the interior of the storm. Air spirals up

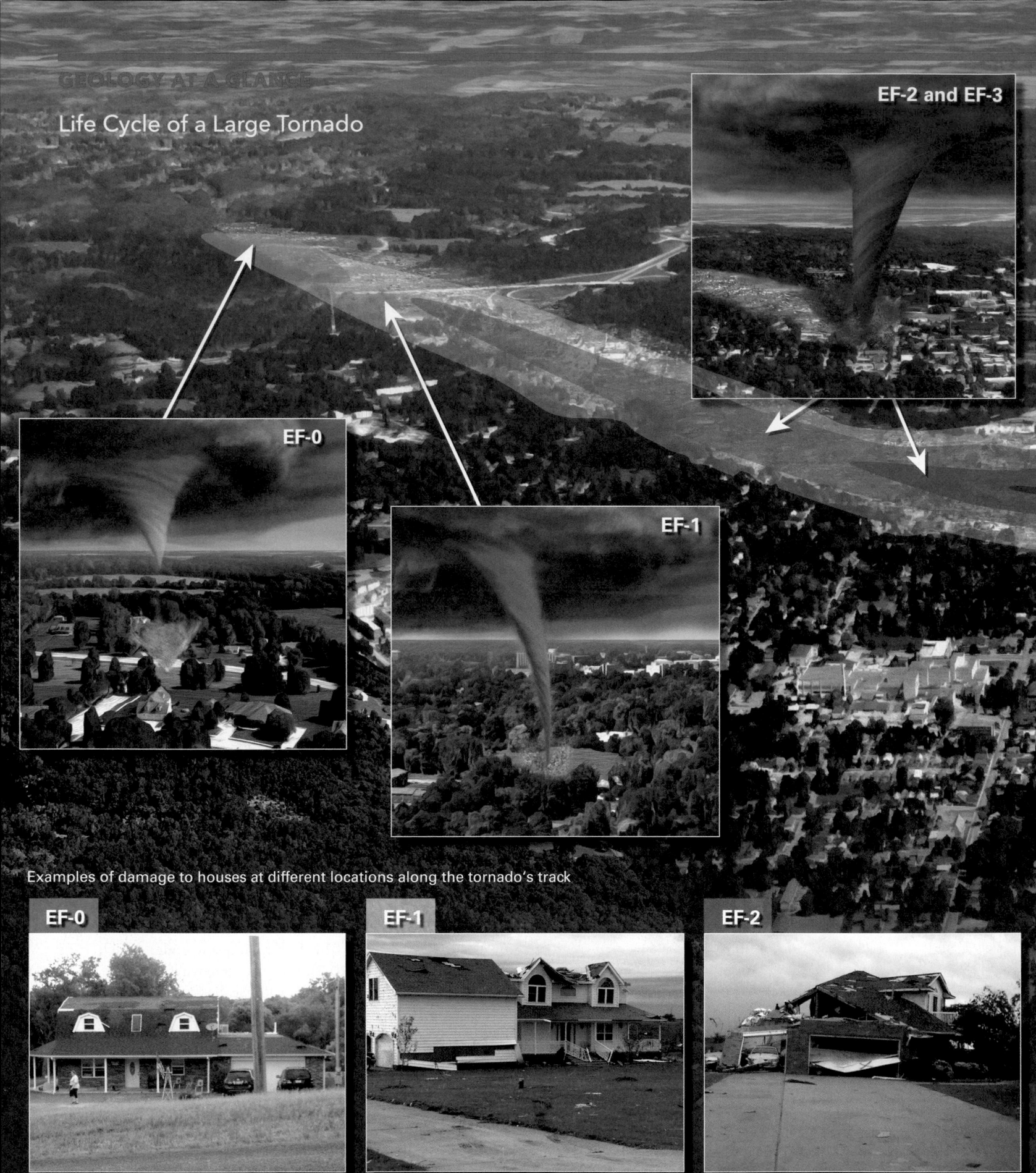

GEOLOGY AT A GLANCE

Life Cycle of a Large Tornado

EF-2 and EF-3

EF-0

EF-1

Examples of damage to houses at different locations along the tornado's track

EF-0

EF-1

EF-2

EF-4 and EF-5

EF-2 and EF-3

EF-1

A tornado evolves over time. It starts small, grows larger, and then eventually dissipates. In the record books, the overall rating of a tornado using the Enhanced Fujita (EF) scale reflects the maximum damage that the tornado causes somewhere along a portion of its path. Here, we see the swath of destruction that a tornado produced as it passed through a town, and a representation of the shape and size of the tornado at various times during its life cycle. This example was an EF-5 tornado along only a small part of its track.

EF-3

EF-4

EF-5

TABLE 20.3 Saffir-Simpson Scale for Hurricanes

Scale	Category	Wind Speed (km per hour)	Air Pressure in Eye (millibars)	Damage
1	Minimal	119–153	980 or more	Branches broken; unanchored mobile homes damaged; some flooding of coastal areas; no damage to buildings; storm surge of 1.2 to 1.5 m
2	Moderate	154–177	965–979	Some roofs, doors, and windows damaged; mobile homes seriously damaged; some trees blown down; small-boat moorings broken; storm surge of 1.6 to 2.4 m
3	Extensive	178–209	945–964	Some structural damage to small buildings; large trees blown down; mobile homes destroyed; structures along coastal areas destroyed by flooding and battering; storm surge of 2.5 to 3.6 m
4	Extreme	210–250	920–944	Some roofs completely destroyed; extensive window and door damage; major damage and flooding along coast; storm surge of 3.7–5.4 m. Widespread evacuation of regions within up to 10 km of the coast may be necessary.
5	Catastrophic	Over 250	Less than 920	Many roofs and buildings completely destroyed; extensive flooding; storm surge greater than 5.4 m. Widespread evacuation of regions within up to 16 km of the coast may be necessary.

the eye wall, then diverges near the tropopause. In a hurricane, the fastest winds occur along the eye wall, so the hurricane-force winds of a given storm generally affect a belt that is only 15% to 35% as wide as the whole storm. On the side of the eye where winds blow in the same direction that the whole storm is moving, the winds become the strongest, because the storm's overall speed adds to the rotational motion.

The Damage Due to Hurricanes Hurricanes pose extreme danger. Their extensive damage happens for several reasons.

- *Wind:* Winds of weaker hurricanes tear off branches and smash windows. Stronger hurricanes uproot trees, rip off roofs, and collapse walls. Extreme and catastrophic hurricanes can carry away or flatten whole towns, causing wind damage as intense as that of a tornado but over a much broader area **(Fig. 20.30a)**.
- *Waves:* Winds shearing across the sea surface during a hurricane generate huge waves. In the open ocean, these waves can capsize ships. Near the shore, waves batter and erode beaches, rip boats from moorings, and destroy coastal property **(Fig. 20.30b)**.
- *Storm surge:* The sustained winds of a hurricane drive water forward and can build a substantial bulge that raises the sea surface by up to several meters. This bulge, known as **storm surge,** can cover an area over 100 km in diameter. In addition, extremely low air pressure develops beneath a hurricane—in fact, the lowest sea-level air pressure ever recorded (0.87 bars) occurred during such a storm. This decrease in pressure causes the surface of the sea to bulge upward even more. When a hurricane

reaches the coast, the storm surge swamps the land **(Fig. 20.30c, d)**. If the bulge hits the land at high tide, the sea surface will be especially high, and storm surge will affect a broader area. Storm waves develop on top of storm surge, so the heights of the wave crests become the sum of the surge height plus the tide height plus wave height.

- *Rain, stream flooding, and landslides:* Rain drenches the Earth's surface beneath a hurricane. In places, half a meter or more of rain falls in a single day. Rain causes streams to flood, even far inland, and can trigger deadly landslides **(Fig. 20.30e)**.
- *Disruption of social structure:* When the storm passes, the hazard is not over. By disrupting transportation and communication networks, breaking water mains and power lines, and washing away sewage-treatment plants, hurricane damage creates severe obstacles to search and rescue, and can lead to the spread of disease, fire, and looting.

Nearly all hurricanes that reach the coast cause death and destruction, but some are truly catastrophic. Hurricane Katrina, Hurricane Sandy, Typhoon Haiyan, and 2017's trio of Atlantic hurricanes serve as examples of particularly devastating storms, as we now see.

Case Studies of Hurricanes

Katrina, 2005 Tropical Storm Katrina came into existence over the Bahamas and headed west. Just before landfall in southeastern Florida, its winds strengthened, and the storm became Hurricane Katrina. This hurricane sliced across the

FIGURE 20.30 Examples of hurricane damage.

(a) Wind damage due to Hurricane Andrew, 1992.

(b) Waves pummel the shore during Hurricane Sandy, 2012.

(c) Wind and low atmospheric pressure produce a bulge of water called storm surge.

(d) Damage due to storm surge from Typhoon Haiyan in the Philippines, 2013.

(e) Flooding caused by rains from Hurricane Irene in 2011 affected Vermont, far inland from the coast.

southern tip of Florida, causing several deaths and millions of dollars in damage. It then entered the Gulf of Mexico and passed a region of particularly warm water (32°C). Moisture rising from this warm water stoked the storm, injecting it with a burst of energy sufficient for the storm to morph into a Category 5 monster whose swath of hurricane-force winds reached a width of 325 km **(Fig. 20.31a, b)**. Katrina then turned north and began to bear down on the Louisiana-Mississippi coast. The eye of the storm passed just east of New Orleans and then across the coast of Mississippi **(Fig. 20.31c)**. Storm surges broke records, in

September 21, 2005

February 6, 2013

The 17th Street Canal, New Orleans

LATITUDE
30° 1'7.42" N

LONGITUDE
90° 7'17.53" W

Move the Historical Images Slider to the dates listed above and look down from 1.5 km. You're looking at the 17th Street Canal, in New Orleans; the failure of the eastern wall caused the city to flood after Hurricane Katrina.

The top view shows water from New Orleans draining back into the canal, but streets remain flooded. The bottom image shows the extent of reconstruction. A flood gate now blocks the entrance to the canal.

places rising 7.5 m above sea level, and they washed coastal communities off the map along a broad swath of the Gulf Coast **(Fig. 20.32a, b)**. In addition to the devastating wind and surge damage, Katrina led to the drowning of New Orleans.

To understand what happened to New Orleans, we must consider the city's geologic history. New Orleans grew on the Mississippi Delta between the banks of the Mississippi River on the south and Lake Pontchartrain (actually a bay of the Gulf of Mexico) on the north. The older parts of the town grew up on the relatively high land of the Mississippi's natural levee. Younger parts of the city, however, spread out over the topographically lower delta plain. As decades passed, people modified the surrounding delta landscape by draining wetlands, constructing artificial levees that confined the Mississippi River, and extracting groundwater. Sediment beneath the delta compacted, and the delta's surface, now starved of new sediment, sank below sea level. Today most of New Orleans lies in a bowl-shaped depression as much as 2 m below sea level—the hazard implicit in this situation had been recognized for years **(Fig. 20.32c)**.

The winds of Hurricane Katrina ripped off roofs, toppled trees, smashed windows, and triggered the collapse of weaker buildings, but their direct consequences were not catastrophic. However, when the winds blew storm surge into Lake Pontchartrain, its water level rose beyond most expectations and pressed against the system of artificial levees and floodwalls that had been built to protect New Orleans. Hours after the hurricane's eye had passed, the high water found a weakness along the floodwall bordering a drainage canal and pushed out a section. Breaks eventually formed in a few other locations as well. So, a day after the hurricane was over, New Orleans began to flood. As

FIGURE 20.31 Hurricane Katrina's path through the warm waters of the Gulf of Mexico.

Warmer   Cooler
Ocean water temperature

(a) A satellite photo of Hurricane Katrina over the warm water of the Gulf of Mexico.

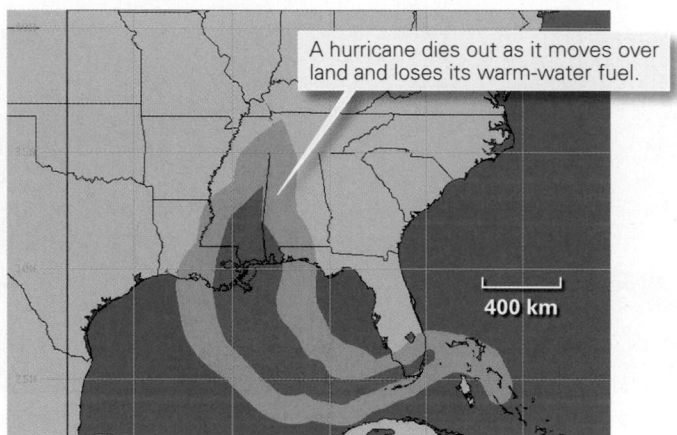

A hurricane dies out as it moves over land and loses its warm-water fuel.

400 km

(b) A wind-swath map of Hurricane Katrina. Red areas represent hurricane winds; orange areas represent tropical-storm winds.

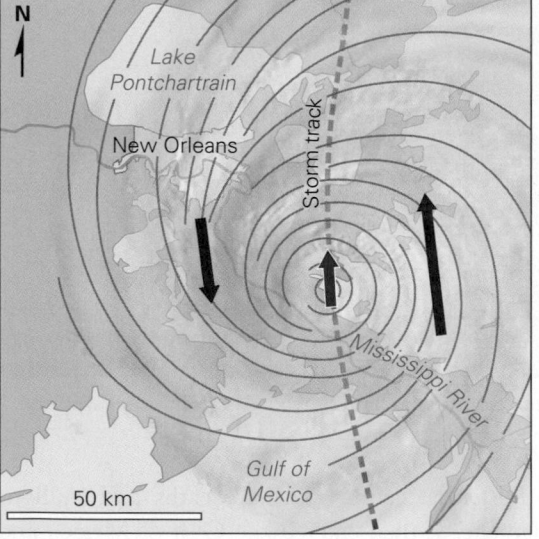

N

Lake Pontchartrain

New Orleans

Storm track

Mississippi River

Gulf of Mexico

50 km

(c) The eye of Katrina passed east of New Orleans. Because the storm was moving northward overall (red arrow) and spinning counterclockwise, winds (blue arrows) to the east of the eye were much stronger than those to the west of the eye.

FIGURE 20.32 The devastation of coastal areas by Hurricane Katrina.

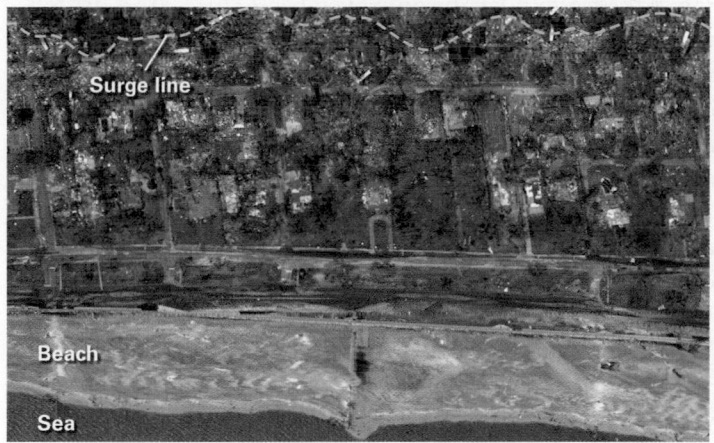

(a) Storm surge destroyed homes along the Alabama coast.

(b) Officials survey the storm damage.

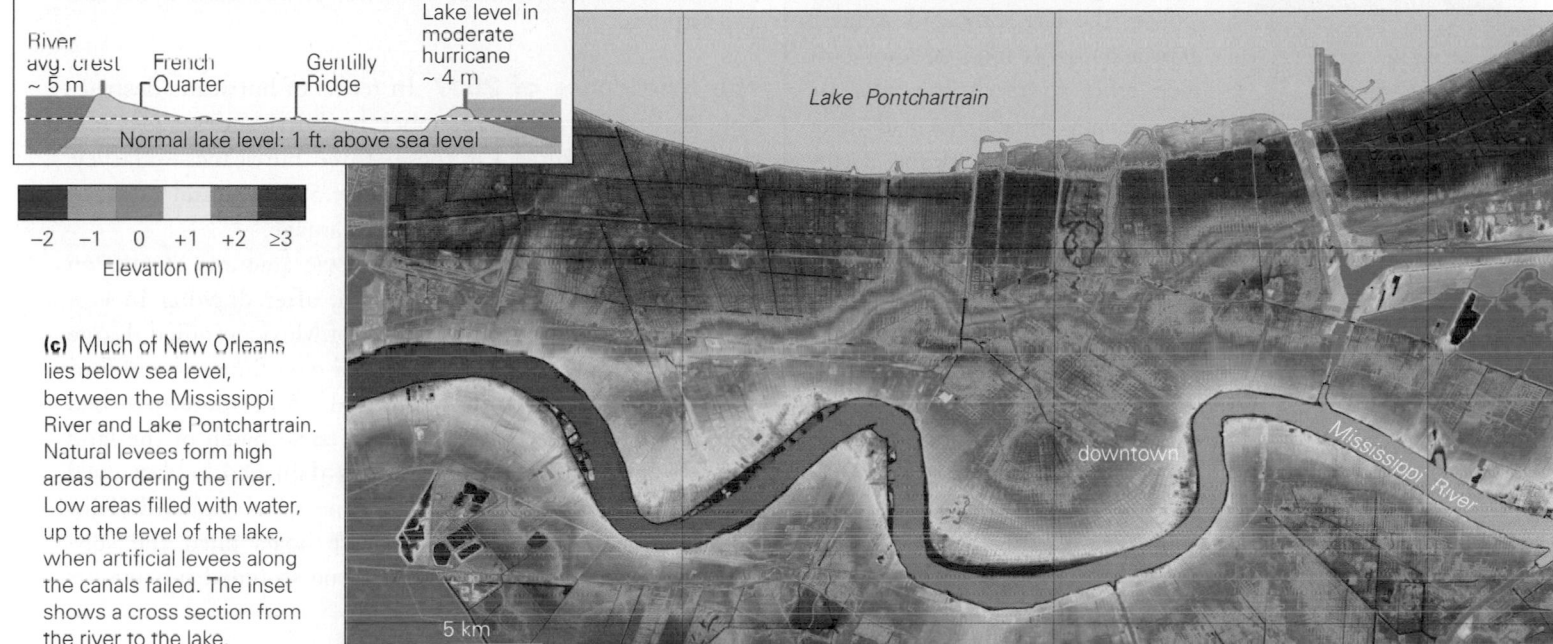

(c) Much of New Orleans lies below sea level, between the Mississippi River and Lake Pontchartrain. Natural levees form high areas bordering the river. Low areas filled with water, up to the level of the lake, when artificial levees along the canals failed. The inset shows a cross section from the river to the lake.

(d) Water flowing across the levees bordering the 17th Street Canal after the hurricane had passed.

(e) Damage due to flooding in a New Orleans home.

Seaside Heights, New Jersey: Hurricane Sandy Damage

LATITUDE
39°56'33.51" N

LONGITUDE
74° 4'8.64" W

Look straight down from 500m.

You are seeing the boardwalk amusement park of a coastal resort. Using the time travel option, set the clock to see the image of November 11, 2012, about two weeks after Hurricane Sandy struck (the top photo). Turn the clock back to September 21, 2010 (the bottom photo), and you can see the park before the hurricane. The damage reveals the power of both storm winds and the waves they drive.

the waterline climbed the walls of houses, residents fled first upstairs, then to their attics, and finally to their roofs. Water spread across the city until the bowl of New Orleans filled to the same level as Lake Pontchartrain, submerging 80% of the city **(Fig. 20.32d)**.

Floodwaters washed some houses away and filled others with debris **(Fig. 20.32e)**. The disaster took on national significance as the trapped population sweltered without food, drinking water, or adequate shelter. With no communications, no hospitals, and few police, the city almost descended into anarchy. It took days for outside relief to reach the city. By then, many inhabitants had died, and parts of New Orleans, a cultural landmark and major port, had become uninhabitable. It has taken years for the city to rebuild, and parts remain devastated.

Sandy, 2012 Sandy began as a tropical depression on October 22, 2012, and became a hurricane just before hitting Jamaica. Then it curved north and struck New Jersey on October 28. When it hit, its maximum sustained winds reached 185 km per hour. Sandy then collided and merged with a large mid-latitude cyclone to form the widest Atlantic hurricane ever documented, now known as "Superstorm Sandy." At its peak, it spanned a diameter of 1,800 km. Storm surge washed away landmarks of the New Jersey shore and devastated expensive beachfront property. The surge also inundated parts of New York City and even caused extensive flooding of the subway system. The $68 billion of damage that Sandy caused was a shock to the northeastern metropolitan areas and has led city governments to begin thinking about how to plan for the consequences of rising sea level.

Haiyan, 2013 Typhoon Haiyan began as a tropical depression in the western Pacific on November 2, 2013. As it drifted westward over very warm ocean water, it grew to typhoon status within two days. By November 6, it had become a Category 5 behemoth and took aim at the Philippines. Measurements indicated that it had sustained winds of 315 km per hour, the highest wind speeds ever recorded in a hurricane. In fact, at times, the wind gusted to 378 km per hour. Indeed, Haiyan may be the strongest storm ever to reach land in recorded history. In addition to intense wind and a record-breaking storm surge, the storm dumped as much as 282 mm of rain over the course of a few hours. By the time it had passed over the Philippines, more than 6,300 residents had lost their lives, whole towns had been flattened, and over a million homes sustained damage. Relief efforts struggled for weeks to reach people who were cut off by the destruction of roads, harbors, and airports.

The Hurricanes of 2017 In terms of hurricane destruction, 2017 was a year to remember: ten hurricanes raged across the western Atlantic! Of these, three hurricanes—Harvey, Irma, and Maria—reached Category 5 status, and together they caused close to $300 billion of damage.

Hurricane Harvey dumped historic amounts of rain on portions of Texas, where the storm, after drawing in vast amounts of moisture from the Gulf of Mexico, stalled. Large areas received over a meter of rainfall over the course of four days, and in places, rainfall exceeded 1.5 m, more than had ever been recorded. In the Houston area, much of the land lies at low elevation, so floodwaters damaged or destroyed tens of thousands of homes and businesses and caused major oil refineries to shut down. Private boats plied the submerged streets and highways to rescue stranded inhabitants **(Fig. 20.33)**.

Before Harvey had dissipated, Irma was already cutting its path of destruction across the Caribbean. The strength of its peak winds (295 km per hour) made it the strongest hurricane ever in the Atlantic. Irma slammed into islands of the eastern Caribbean, and on Barbuda and Saint Martin it severely damaged or destroyed about 95% of the buildings. Maria then skirted the north coast of Cuba, cutting a swath of damage across the Florida Keys, before turning north and damaging the entire west coast of Florida.

Maria followed a similar path to that of Irma, so some of the islands devastated by Irma were raked again by the winds of Maria. But while Irma's eye passed just to the north of Puerto Rico, Maria's eye made a direct hit on the island before turning north and heading into the Atlantic. The storm was a catastrophe for Puerto Rico, where the winds wiped out the electrical grid. In parts of the island, people have remained without power for many months, and food and clean water became scarce.

FIGURE 20.33 Houston's streets turned into rivers during Hurricane Harvey.

TAKE-HOME MESSAGE

Thunderstorms develop where convection causes warm, moist air to rise, resulting in billowing clouds and rain. Charge separation develops in clouds, resulting in lightning. The spiral of air in a tornado produces wind speeds of up to 500 km per hour. Mid-latitude cyclones can produce large storms. Hurricanes begin over warm seawater. Energy from latent heat of condensation enlarges these storms, producing winds ranging between 119 and 315 km per hour.

QUICK QUESTION: Why do hurricanes dissipate after they cross onto land or head to higher latitudes?

20.8 Global Climate

Climate and weather are related, but the words are not synonymous. *Weather* specifically refers to the atmospheric conditions (temperature, humidity, winds, and precipitation) at a certain location at a specified time. A description of **climate**, in contrast, characterizes the average weather conditions of a region, the typical range of weather conditions for the region, the character of the region's storms, and the nature of the region's seasons, as observed over many decades. To clarify this difference, let's consider an example. On a given summer day in Winnipeg in central Canada, the weather may be sunny, hot, and humid, and on a given winter day in Miami in the southeastern United States, it may be freezing and overcast. But averaged over time, the overall temperature of Winnipeg is lower than that of Miami, the range of temperature change between winter and summer in Winnipeg is much greater than that in Miami, and Miami endures hurricanes while Winnipeg endures mid-latitude cyclones. Therefore, we can say that Winnipeg and Miami have different climates, and for this reason, the native flora and fauna of Winnipeg differ significantly from those of Miami.

Controls on Climate

Climatologists, scientists who study the Earth's climate, suggest that several distinct factors control the climate of a region:

- *Latitude:* Latitude determines the amount of solar energy a region receives, as well as the contrasts among seasons, so it serves as the single most important factor in controlling climate. Polar regions, which receive much less solar radiation over the year, have colder climates than do equatorial regions. And the contrast between winter and summer is greater in mid-latitudes than at the poles or the equator. We can easily see the influence of latitude by examining a map showing the global distribution of temperature, as represented by **isotherms**, lines connecting locations that have the same temperature **(Fig. 20.34)**.
- *Elevation:* Because temperature decreases with elevation, colder climates exist at high elevations, even at the equator. Hiking from the base of a high mountain in the

FIGURE 20.34 The temperature of the atmosphere changes with the seasons, as depicted by isotherms. In the North Atlantic, the Gulf Stream deflects the isotherms northward, so the United Kingdom and Ireland are warmer than lands at equivalent latitudes in North America.

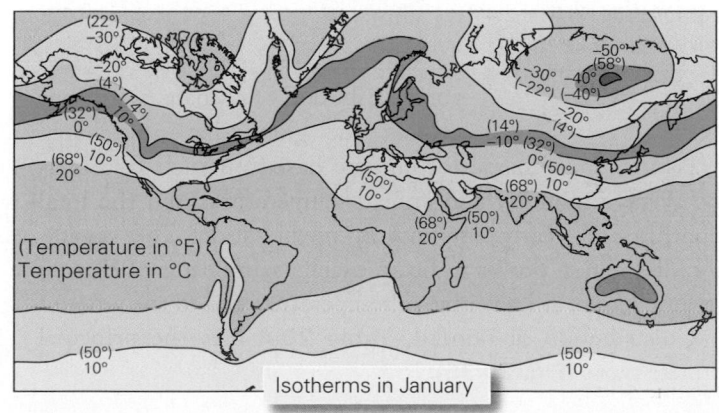

Isotherms in January

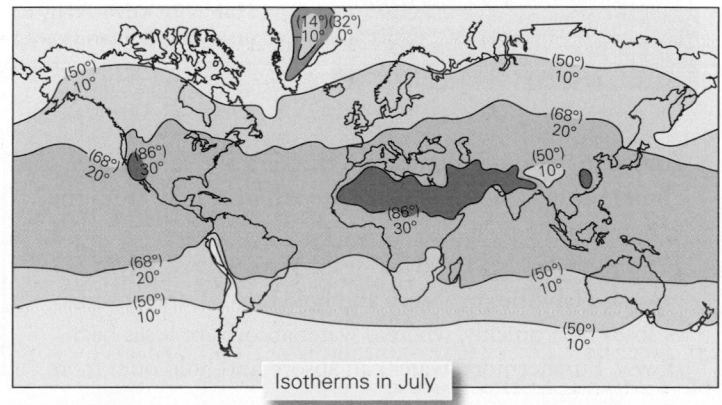

Isotherms in July

TABLE 20.4 Climate Types of the Earth

Climate Type	Regions and Characteristics
	**Tropical:** Tropical rainforests lie at equatorial latitudes and experience rain throughout the year. Rain commonly falls during afternoon thunderstorms. Tropical savannas (grasslands with brush), which lie on either side of a rainforest, have a rainy season and a dry season.
	**Arid:** Arid regions include deserts (regions with very little vegetation cover) and steppes. Steppe regions (vast grassy plains with no forest) border deserts.
	**Temperate:** This category includes regions that have moist air and warm temperatures for much of the year, and in which mixed deciduous-coniferous forest thrives. The category includes Mediterranean climates, coastal climates with most rainfall in the winter, very hot summers, and scrub forests; and marine west-coast climates, where the sea tempers the climate.
	**Cold:** These higher-latitude temperate climates, which occur only in the northern hemisphere, include humid continental regions with long summers, in which deciduous forest thrives; regions with short summers, characterized by coniferous forest; and subarctic climates.
	**Polar:** Polar climates include tundras and ice caps. Tundras are regions with no summer and an extremely cold winter, in which only low, cold-resistant plants can survive. Much of the ground in tundra is permanently frozen. In ice-cap regions, the climate is subfreezing year-round, and land not covered by ice has essentially no vegetation. Highlands are regions that lie at lower latitudes but have such a high elevation that they have polar-like climates.

Andes to its summit takes you through the same range of climate belts you would pass through on a hike from the equator to the pole.

- *Proximity of water:* Land and water have very different *heat capacities* (abilities to absorb and hold heat). Land absorbs or loses heat quickly, whereas water absorbs or loses heat slowly. Furthermore, water can absorb and hold onto more heat than land can because water is semitransparent. Consequently, the proximity of the sea tempers the climate of a region. In fact, as a rule, locations in the interior of a continent experience a much greater range of weather conditions than do locations along the coast.

- *Proximity to ocean currents:* Where a warm current flows, it may heat the overlying air, and where a cold current flows, it may absorb heat from the overlying air. For example, the Gulf Stream brings warm water into the North Atlantic from the Gulf of Mexico and keeps Ireland, the United Kingdom, and Scandinavia warmer than they would be otherwise.

- *Proximity to orographic barriers:* An *orographic barrier* is a landform (such as a mountain range) that diverts air flow upward or laterally. This diversion affects the amount of precipitation and wind a region receives.

- *Position in the Hadley cell*: Convergence at the ITCZ causes moist air to rise to build thunderstorms that drench the Earth below. At 30° N and S, sinking dry air produces deserts (see Fig. 20.10)

- *Proximity to distinct air masses:* Large, distinct air masses remain over specific regions of the Earth during much of the year (see Fig. 20.13). These influence the weather within the regions for much of the year, and they can therefore control the regions' climates.

Climatologists, who have studied the distribution of climate conditions around the globe, have developed a classification scheme for climate based on such factors as the average monthly and annual temperatures and the total monthly and yearly amounts of precipitation. The vegetation of a region proves to be an excellent indicator of climate because plants are sensitive to temperature and to the amount and distribution of rainfall. **Table 20.4** lists the principal climate types **(Fig. 20.35)**.

FIGURE 20.35 Climate belts and their effect on vegetation.

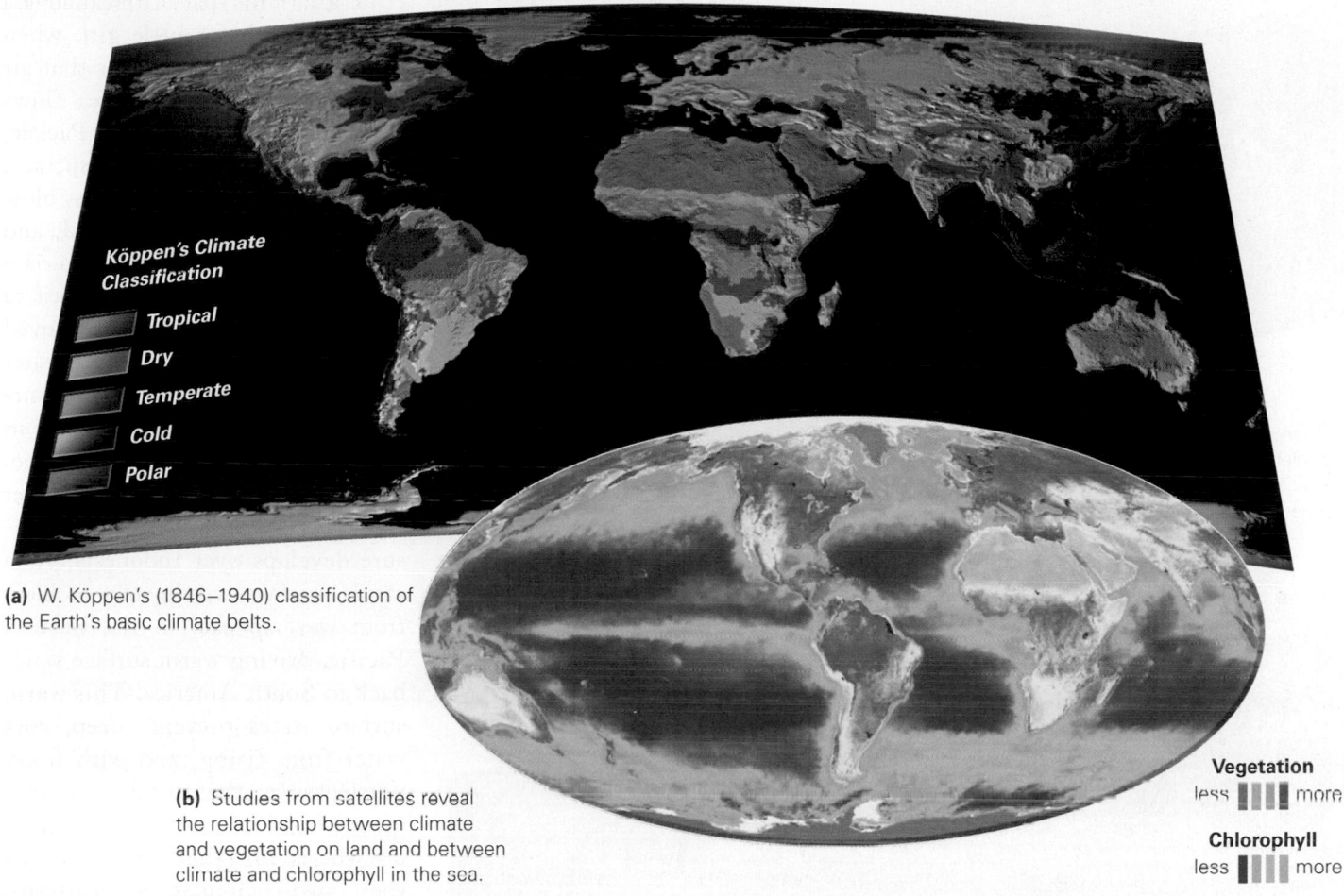

Köppen's Climate Classification

Tropical
Dry
Temperate
Cold
Polar

(a) W. Köppen's (1846–1940) classification of the Earth's basic climate belts.

(b) Studies from satellites reveal the relationship between climate and vegetation on land and between climate and chlorophyll in the sea.

Vegetation
less ▮▮▮▮ more
Chlorophyll
less ▮▮▮▮ more

Aspects of Climate Variability: Monsoons and El Niño

Tropical and subtropical regions of the world are subject to seasonal or multi-year climatic variations that are distinct from those of other regions. Monsoons and El Niño are examples of climate variability that cause significant impacts on human society.

Monsoons A **monsoon** is a major reversal in the wind direction that causes a shift from a dry season to a rainy season. In southern Asia, home to about half the world's population, people depend on monsoonal rains to bring moisture for their crops. The Asian monsoon develops primarily because Asia includes vast tracts of land far from the sea. Further, a substantial part of this land, the Tibetan Plateau, lies at a high elevation. During the winter, central Asia becomes very cold, much colder than the coastal regions to the south. This coldness yields a region of high pressure over central Asia. Dry air sinks and spreads outward from this region

and flows southward over southern Asia, pushing the ITCZ out over the Indian Ocean, south of Asia **(Fig. 20.36a)**. Consequently, during the winter, southern Asia experiences a dry season. During the summer, central Asia warms up dramatically. As warm air rises and expands over central Asia, a low-pressure region develops, and the ITCZ moves north. When this happens, warm air flows northward from the Indian Ocean, bringing with it substantial moisture, and the summer rains begin. Especially heavy rain falls on the southern slope of the Himalayas because of orographic lifting **(Fig. 20.36b)**.

El Niño Long before the modern science of meteorology became established, fishermen from Peru and Ecuador who ventured into the coastal waters west of South America knew that some years, in late December, the fish population that provided their livelihood diminished. Because of the timing of this event, it came to be known as **El Niño**, Spanish for little boy, or the Christ child. Why did the fish vanish? Fish rely on a food chain that begins with plankton, which live on nutrients in the water. Along coastal South America, the nutrients

FIGURE 20.36 Monsoons cause southern Asia to have a distinct wet season and dry season.

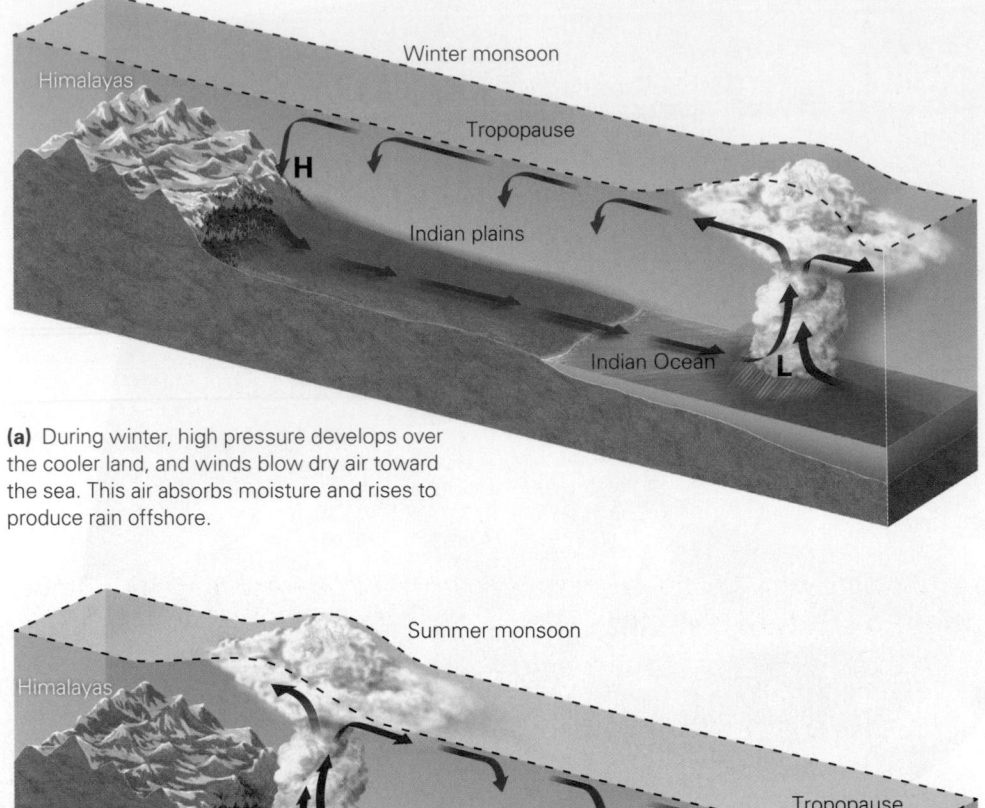

(a) During winter, high pressure develops over the cooler land, and winds blow dry air toward the sea. This air absorbs moisture and rises to produce rain offshore.

(b) During summer, low pressure develops over the warmer land, and winds blow moist air from the oceans toward the land. Where this air rises, large rainstorms form.

along the coast of equatorial South America. This pattern (called La Niña, Spanish for little girl, when particularly intense) means that air rises in the western Pacific, flows east, sinks in the eastern Pacific, and then flows west at the surface. The easterly surface winds blow warm surface water westward, and cold water from the deep ocean rises along the South American coast to replace the warm water that moved west. It is this rising cold water that brings nutrients to the surface, feeding the plankton, which in turn feed the fish. During El Niño, low pressure moves eastward over the central Pacific, and high pressure develops over Indonesia. As a result, surface winds start to blow from west to east in the western Pacific, driving warm surface water back to South America. This warm surface water prevents deep, cold water from rising, and with fewer nutrients in the water, plankton quantities decrease and fish move elsewhere. The alternation between the two circulating patterns, back and forth across the Pacific, is called the *southern oscillation*. Roughly speaking, El Niño events happen on a 4-year cycle.

that feed plankton come from upwelling deep water. During El Niño, warm-water currents flow eastward from the central Pacific, and the cold, nutrient-rich water that supports the marine food chain can't upwell. With fewer nutrients, there are fewer plankton, and without the plankton, the fish migrate elsewhere.

To understand why El Niño occurs, we need to look at atmospheric flow and related ocean surface currents in the equatorial Pacific **(Fig. 20.37)**. When El Niño is not in progress, a low-pressure region exists in the western Pacific over Indonesia and Papua New Guinea, while a high-pressure region forms over the eastern Pacific,

TAKE-HOME MESSAGE

Climate refers to the long-term overall range of weather conditions and the nature of seasons in a region. Generally, climate varies with latitude. In detail, factors such as proximity to the ocean or to orographic barriers affect local climate. Monsoons form because of changes in the position of the intertropical convergence zone. El Niño develops because of shifts in the pattern of air circulation at low latitudes.

QUICK QUESTION: How does El Niño affect fish supplies near the coast of South America?

FIGURE 20.37 El Niño exists because of a change in winds and currents in the central Pacific.

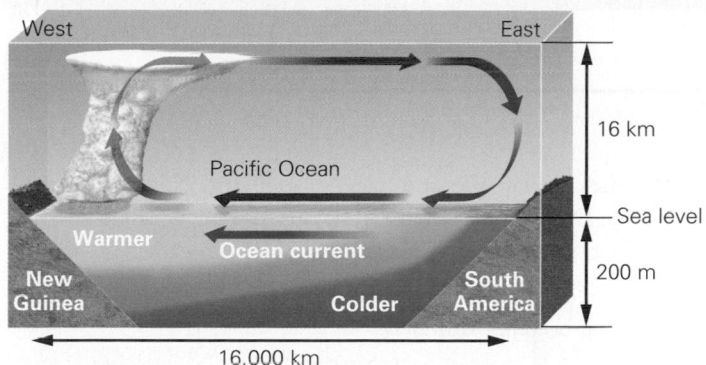

(a) During times between El Niño events, low pressure lies over the western Pacific and surface winds blow west. Cold water rises along the coast of South America.

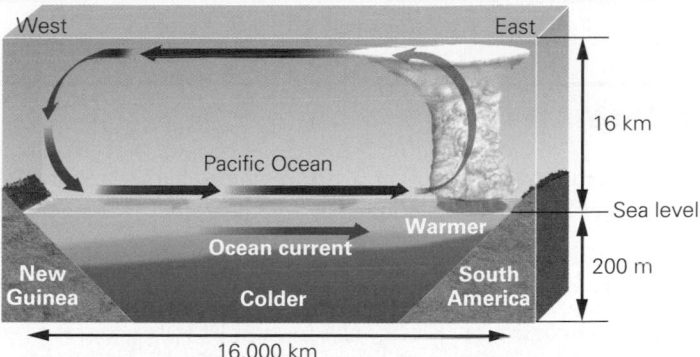

(b) During El Niño, the low-pressure cells move eastward and the westward flow stops, so upwelling ceases.

ANOTHER VIEW The atmosphere is an integral part of the Earth System. From an airplane, we can see cumulus clouds growing over the Great Plains of the United States. These contain moisture evaporated from the Earth's surface into the air. They may grow into cumulonimbus clouds that could drop much-needed rain back onto the surface.

Chapter 20 Review

SUMMARY

- The early atmosphere of the Earth contained high concentrations of H_2O and CO_2 erupted by volcanoes.

- When oceans formed, most H_2O and CO_2 left the atmosphere, and N_2 remained. Later, photosynthetic organisms added O_2.

- Air now consists of 78% N_2, 21% O_2, and 1% trace gases. Air pressure decreases with elevation, so 90% of air lies below 16 km.

- Air expands and cools as it rises, and contracts and warms as it sinks. Relative humidity of air changes, therefore, as air moves vertically.

- We divide the atmosphere into layers based on the trend of temperature change with increasing elevation. Weather occurs in the troposphere.

- High latitudes receive less solar energy, initiating convective flow, which the Coriolis force modifies. Hadley cells develop at low latitudes.

- Prevailing surface winds develop due to atmospheric convection. Where strong pressure gradients develop at the top of the troposphere, jet streams develop.

- Weather refers to atmospheric conditions at a given time in a given location. It can reflect interactions between air masses along fronts.

- Air rises in a low-pressure center and sinks in a high-pressure center. Due to the Coriolis effect, this movement produces mid-latitude cyclones beneath areas of low pressure.

- Clouds consist of tiny water droplets or ice crystals, which form when air becomes saturated.

- Thunderstorms develop when lifted air becomes unstable. Ordinary thunderstorms dissipate quickly, but supercells can survive. Lightning flashes when a giant spark jumps across charge separations.

- Tornadoes, rapidly rotating funnels, can develop in supercells. They can cause severe damage, ranked by the Fujita scale.

- Hurricanes are large tropical cyclones that host intense wind and drop immense amounts of rain. Category designation depends on wind speed.

- Climate refers to a region's typical range of weather conditions, the nature of seasons, and the character of weather extremes, averaged over a long time. It depends on many factors, including latitude, and is classified into types.

- Monsoonal climates of tropical or semi-tropical regions host seasonal contrasts in rainfall. El Niño reflects shifts in equatorial Pacific atmospheric circulation.

GUIDE TERMS

acid rain (p. 761)
aerosol (p. 760)
air (p. 757)
air mass (p. 772)
air pressure (p. 760)
atmosphere (p. 757)
climate (p. 795)
cloud (pp. 763, 774)
collision and coalescence (p. 777)
convergence zone (p. 767)
divergence zone (p. 767)
doldrums (p. 770)
El Niño (p. 797)

Enhanced Fujita scale (p. 783)
exosphere (p. 765)
front (p. 772)
greenhouse gas (p. 758)
hail (p. 781)
high-pressure system (p. 774)
insolation (p. 767)
ionosphere (p. 765)
isobar (p. 766)
isotherms (p. 795)
jet streams (p. 771)
latent heat of condensation (p. 763)
lifting mechanisms (p. 776)

lightning stroke (p. 781)
low-pressure system, (p. 774)
mesosphere (p. 765)
mid-latitude anticyclones (p. 773)
mid-latitude cyclones (p. 773)
monsoon (p. 797)
polar front (p. 767)
prevailing winds (p. 770)
relative humidity (p. 763)
Saffir-Simpson scale (p. 787)
storm surge, (p. 790)
stratosphere (p. 765)

supercell (p. 781)
thermosphere (p. 765)
thunder (p. 782)
thunderstorm (p. 779)
tornado (p. 782)
tornado warning (p. 783)
trade winds (p. 770)
tropical cyclone (p. 786)
troposphere (p. 764)
wind (p. 766)

REVIEW QUESTIONS

The letters following each Review Question refer to the corresponding Learning Objective from the Chapter Opener.

1. Describe the stages in the formation and evolution of the Earth's atmosphere. Where does the ozone in the atmosphere come from, and why is it important? **(A)**

2. Describe the composition of air (considering both its gases and its aerosols). Why are trace gases important? **(A)**

3. How does air pressure change with elevation? Does the density of the atmosphere also change with elevation? Explain why or why not. **(B)**

4. Describe the atmosphere's structure from base to top. What characteristics define the boundaries between layers? **(B)**

5. What is the relative humidity of the atmosphere? What is the latent heat of condensation, and what is its relevance to the evolution of a thunderstorm or hurricane? **(B)**

6. Explain the relation between the wind and variations in air pressure. **(C)**

7. Why do changes in atmospheric temperature depend on latitude and the seasons? **(C)**

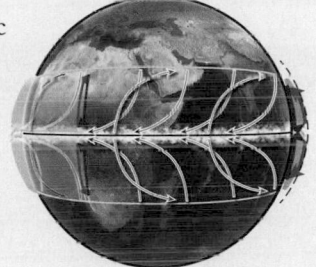

8. Identify the Hadley cells on the diagram. Explain their relationship to the locations of the world's tropical rainforests and subtropical deserts. **(C)**

9. Why do prevailing winds develop at the Earth's surface? Why do the jet streams form? **(C)**

10. Explain the nature of cyclones and anticyclones and note their relationship to high-pressure and low-pressure systems. What is a mid-latitude cyclone? **(D)**

11. How does a cold front differ from a warm front and from an occluded front? Which is depicted in the figure? **(D)**

12. Why do clouds form? (Include a discussion of lifting mechanisms.) What are the basic categories of clouds? **(D)**

13. Under what conditions do thunderstorms develop? What provides the energy that drives clouds to the top of the troposphere? How do meteorologists explain lightning? **(E)**

14. What conditions lead to the formation of a tornado? Where do most tornadoes appear? **(E)**

15. Describe the stages in the development of a hurricane. Describe a hurricane's basic geometry. **(E)**

16. What factors control the climate of a region? What special conditions cause monsoons and El Niño events? **(F)**

ON FURTHER THOUGHT

17. Explain why large rainforests occur in equatorial Africa (the Congo) and in equatorial South America (the Amazon). **(F)**

18. A typhoon is moving due west. Weather forecasters have predicted that the eye of the storm will make landfall in the middle of a narrow, north-south-trending island. To escape the highest winds, should they move to the north or to the south of the eye? **(E)**

ONLINE RESOURCES

Animations
This chapter features animations on how fronts form and operate and cloud features and formation.

Videos
This chapter features videos on rising ozone concentrations and air pollution, major hurricanes, tornado and superstorm tracking, and monsoons.

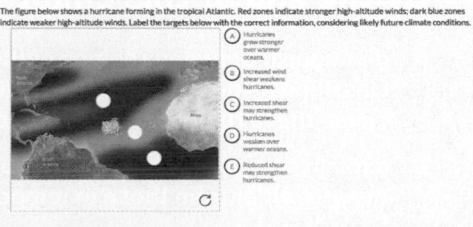

Smartwork5
This chapter features questions on atmospheric composition and layers, climate and weather systems, and hurricane and storm formation.

CHAPTER 21

Dry Regions:
The Geology of Deserts

By the end of this chapter you should be able to . . .

A. distinguish between regions of the land that have been classified as deserts, and characterize factors that cause these regions to have arid climates.

B. explain how weathering and erosional processes in deserts differ from those in temperate lands.

C. describe distinctive landforms and landscapes of deserts, and explain how they form.

D. characterize the species of plants and animals that can survive in desert climates.

E. discuss how human activity may transform vegetated regions into deserts.

> The bare hills are cut out with sharp gorges, and over their stone skeletons scanty
> earth clings. . . . A white light beat down, dispelling the last tract of shadow, and
> above hung the burnished shield of hard, pitiless sky.
>
> —Clarence King (first director of the U.S. Geological Survey, 1842–1901, describing a desert)

21.1 Introduction

For generations, nomadic traders have saddled camels to traverse the Sahara in northern Africa **(Fig. 21.1a)**. The Sahara, the world's largest desert, receives so little rainfall that it has hardly any surface water or vegetation. So camels must be able to walk for up to three weeks without drinking or eating. They can survive these journeys because they sweat relatively little, thereby conserving their internal water supply, and they have the ability to metabolize their own body fat to produce new water. Also, camels can withstand severe dehydration—most mammals die after losing only 10% to 15% of their body fluid, but camels can survive 30% dehydration with no ill effects. Camels do get thirsty, though. After a marathon trek across the desert, a camel may guzzle up to 100 liters of water in less than ten minutes.

The survival challenges faced by a camel emphasize that deserts are lands of extremes—extreme dryness, extreme heat, and extreme cold. But they can also be places of extreme beauty. Desert vistas include everything from sand seas with sculpture-like dunes, to stony pavements spotted with flowers, to cactus-covered hills, to dramatic cliffs of colorful rock. Although less populated than other regions of the Earth, deserts cover a significant proportion of the land surface—about 25%—so they constitute an important component of the Earth System **(Fig. 21.1b)**. In this chapter, we introduce the desert landscape. We learn why deserts occur where they do and how erosion and deposition shape their surfaces. We conclude by exploring life in the desert and by examining the problem of desertification, the gradual transformation of temperate lands into desert.

21.2 The Nature and Location of Deserts

What Is a Desert?

Formally defined, a **desert** is a region that is so *arid* (dry) that it contains no permanent streams, except for rivers that bring

◀ (facing page) In arid landscapes such as this one in central Utah, vegetation covers only a small part of the landscape, streambeds stay dry for most of the year, and erosion produces steep cliffs and debris-covered slopes.

water in from temperate regions elsewhere. In deserts, vegetation covers no more than 15% of the land surface. In general, desert conditions exist where, on average, less than 25 cm of rain falls per year. But rainfall amounts alone do not determine the aridity of a region. Aridity also depends on rates of evaporation and on whether rainfall occurs only sporadically or more continuously during the year. If all the rain in a region drenches the land during isolated downpours only once every few years, the region becomes a desert because the intervals of *drought* (lack of rainfall) last so long that plants cannot survive. Similarly, if high temperatures and dry air cause the rate of evaporation from the ground to exceed the rate of rainfall, then the region becomes a desert even if it receives more than 25 cm per year of rain.

Note that the definition of a desert depends on a region's aridity, not on its temperature. Geologists distinguish between *cold deserts*, where temperatures generally stay below about 20°C all year, and *hot deserts*, where summer daytime temperatures exceed 35°C. Cold deserts exist at high latitudes, where the Sun's rays strike the Earth obliquely and don't provide much energy, and at high elevations, where the thin air can't hold much heat. Hot deserts develop at low latitudes, where the Sun's rays strike the desert at a high angle, and at low elevations, where dense air can hold a lot of heat. The hottest recorded temperatures on the Earth occurred in low-latitude, low-elevation deserts—58°C in Libya and 57°C in Death Valley, California.

> **Did you ever wonder . . .**
> how hot it can become in a desert?

Heat contributes to aridity by increasing the rate of evaporation. In fact, evaporation rates in hot deserts may be so great that even when it rains, the ground stays dry because falling raindrops evaporate in midair. In low latitudes, the bare ground of the desert absorbs so much energy from sunlight that a thin layer of very hot air (up to 77°C) forms just above the ground. This layer refracts sunlight, creating a *mirage*, an optical illusion that makes the dry sand of a desert wasteland look like a shimmering lake and make distant mountains look like islands **(Fig. 21.2)**. But even the hottest of hot deserts becomes cold at night because the dry air doesn't hold heat and there are no clouds or vegetation to trap heat. In fact, the air temperature at the ground surface in a desert may change by as much as 80°C in a single day.

FIGURE 21.1 Deserts and their hardy inhabitants.

(a) Camels can survive the harsh conditions of the Arabian desert.

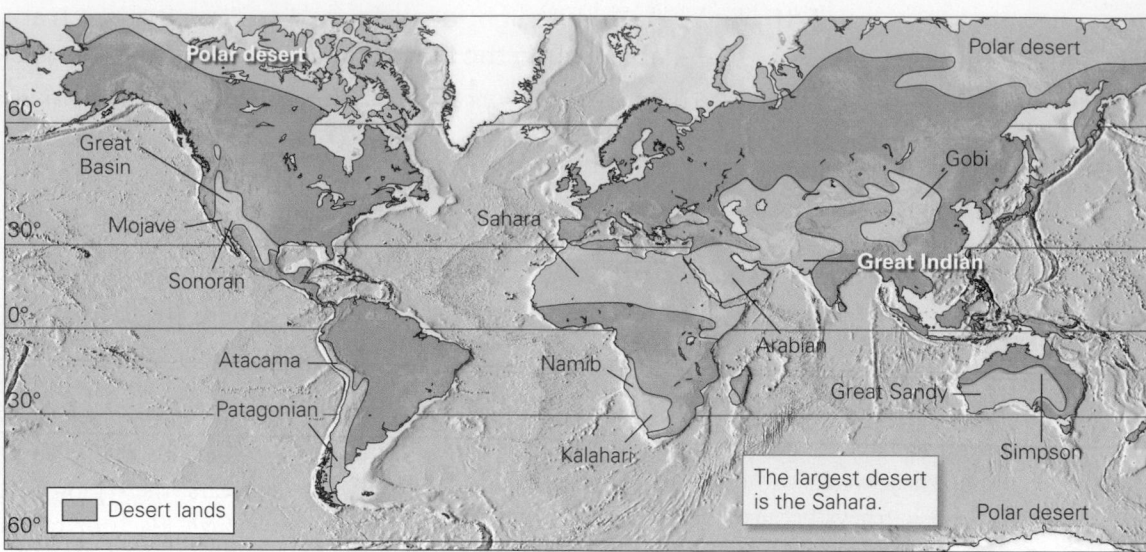

(b) The global distribution of deserts. Arid regions cover 25% of the land surface.

The largest desert is the Sahara.

Weathering, erosion, and depositional processes in deserts differ significantly from those in temperate or tropical regions. Without plant cover, rain and wind batter and scour the ground, and during particularly heavy rains, water rushes down stream channels in flash floods of immense power. In deserts, rocks and sediment do not undergo rapid chemical weathering, and humus (organic matter) does not collect on the ground surface. Therefore, a desert land surface may host one or more of the following: exposed bedrock, accumulations of clasts, relatively unweathered sediment, precipitated salt, or windblown sand. As we'll see, soils do develop in deserts, but they tend to be thinner and more mineralized than those of temperate or tropical regions.

FIGURE 21.2 A desert mirage forms when light from the sky refracts upward as it passes from cooler air into a thin layer of hot air near the ground. Your eyes superimpose the image of the sky on that of the ground. Air movement makes the mirage shimmer.

Types of Deserts

Each desert on the Earth has unique characteristics of landscape and vegetation that distinguish it from others. Nevertheless, geologists can group deserts into five general classes, based on the setting in which the desert forms:

- *Subtropical deserts:* Subtropical deserts (such as the Sahara, Arabian, Kalahari, and Australian) form because of the pattern of air circulation in the atmosphere (see the discussion of the Hadley cell in Chapter 20). At the equator, hot, moisture-laden air rises to great heights **(Fig. 21.3)**. As this air rises, it expands and cools, so water condenses and falls in downpours that feed the lushness of the equatorial rainforest. The now-dry air high in the troposphere spreads laterally north or south. The air sinks at latitudes of about 30°, a region called the *subtropics*. Because of the air's dryness, no clouds form, and intense solar radiation strikes the Earth's surface. As the air sinks, pressure within it increases, so it contracts and heats up, soaking up any moisture present. At the Earth's surface, this hot, dry air sweeps back toward the equator. In the regions that the air passes over, evaporation rates greatly exceed rainfall rates, so land becomes parched.

- *Rain-shadow deserts:* When a coastal mountain range blocks moist air flowing landward from the sea, the air must rise **(Fig. 21.4)**. As the air rises, it expands and cools. The water it contains condenses and falls as rain on the seaward flank of the mountains, nourishing a coastal rainforest. When this air finally reaches the inland side of the mountains, it has lost all its moisture

FIGURE 21.3 Subtropical deserts form because the air that convectively flows downward in the subtropics warms and absorbs water as it sinks.

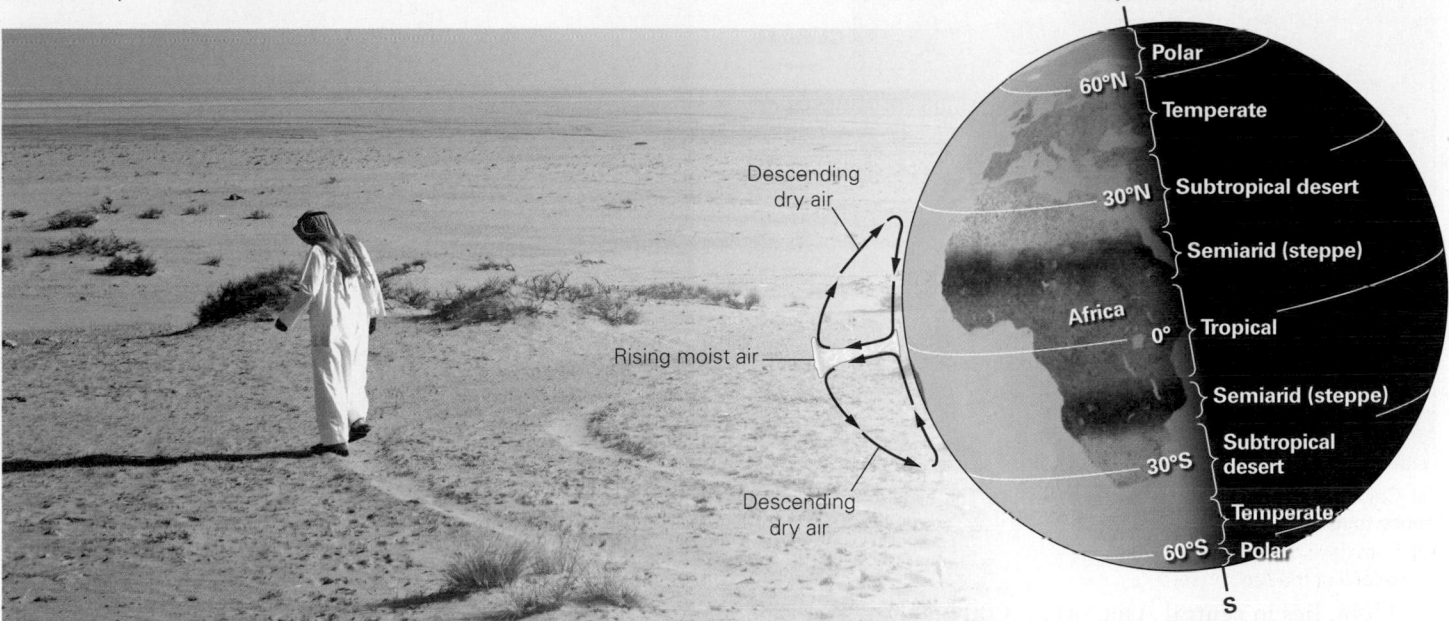

and can no longer provide rain. As a consequence, a **rain shadow** forms on the leeward (downwind) side of the range, and the land beneath becomes a *rain-shadow desert*. The landscapes flanking the Cascade Range of western Washington illustrate this pattern—the west side of the range hosts a dense temperate rainforest, whereas the east side has become a rain-shadow desert.

FIGURE 21.4 The formation of a rain-shadow desert.

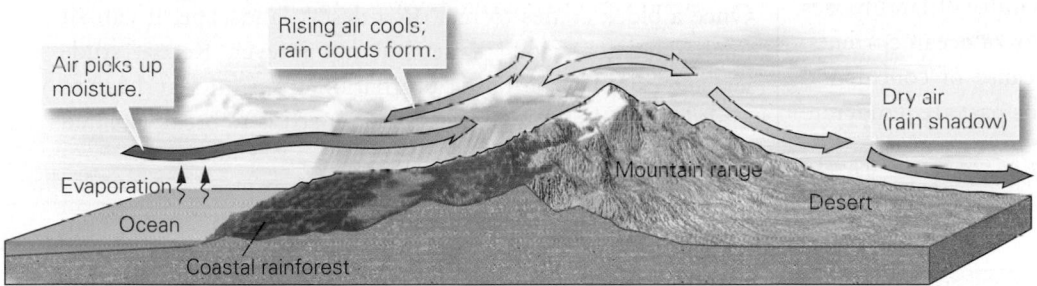

(a) Moist air rises and drops rain on the coastal side of a mountain range. By the time the air has crossed the mountains, it is dry.

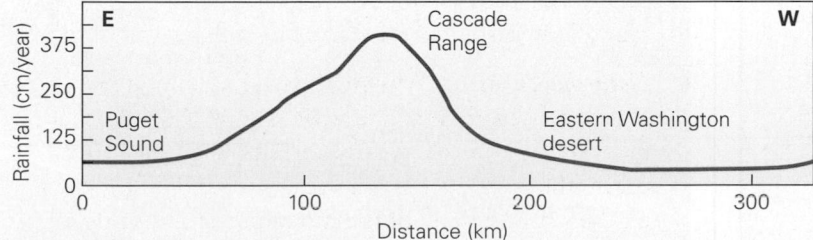

(b) Measured rainfall is much greater on the western side of the Cascade Range in Washington than it is on the eastern side.

- *Coastal deserts formed along cold ocean currents:* In places where cold ocean currents flow along the margin of a continent, a desert can develop right along the coast. This happens because the cold water of the current cools the overlying air by absorbing heat. The resulting cooler and denser air layer flows over the land, but it underlies a warmer layer and can't rise to form clouds, so the land beneath it receives no rain and becomes a *coastal desert*. Such a desert has formed along the western coast of South America, where the Humboldt Current, which carries icy water northward from Antarctica, cools the air that blows east over the coast of Chile and Peru. As a result, this region hosts one of the driest landscapes in the world, the Atacama Desert **(Fig. 21.5)**. Portions of this narrow (less than 200-km-wide) desert received no measurable rain at all between 1570 and 1971.

- *Continental-interior deserts:* As an air mass moves long distances across a continent, it gradually loses moisture by dropping rain. If this air mass reaches the interior of a particularly large continent, such as Asia, it will have grown so dry that the land beneath becomes arid. The largest present-day example of such a *continental-interior desert*, the

FIGURE 21.5 The formation of a coastal desert.

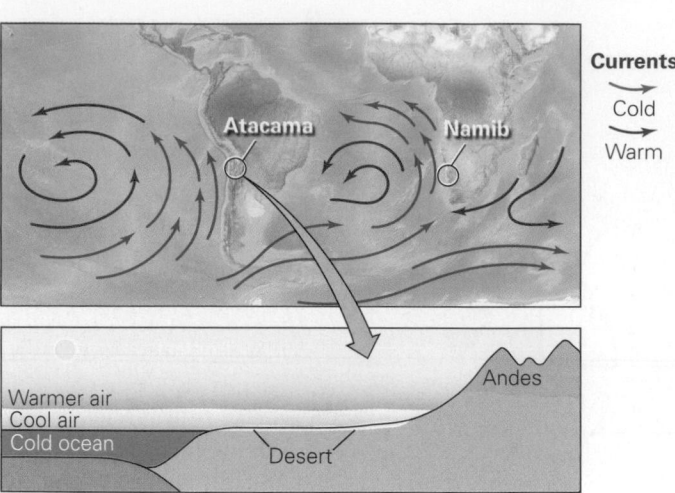

(a) Cold ocean currents cool air along the coast. This air is too dense to rise and produce rain clouds.

(b) The Atacama Desert is one of the driest places on the Earth.

Gobi, lies in central Asia, over 2,000 km away from the nearest ocean.

- *Polar deserts:* So little precipitation falls in the Earth's polar regions (north of the Arctic Circle, at 66°30′ N, and south of the Antarctic Circle, at 66°30′ S) that these areas are, in fact, arid. Polar regions stay dry for two reasons: First, the general pattern of global-scale air circulation causes the air flowing over these regions to be dry. Second, cold air holds relatively little moisture.

The distribution of deserts around the world, over geologic time, reflects the process of plate tectonics, for plate movements determine the latitude of landmasses, the position of landmasses relative to the coast, the proximity of landmasses to a mountain range, and the configuration of ocean currents. The stratigraphic record shows that, because of continental drift, some regions that, were deserts in the past are temperate or tropical regions now, and vice versa.

TAKE-HOME MESSAGE

Deserts are so arid that they host only very sparse vegetation. Temperature extremes happen in deserts, but not all deserts are hot. Deserts form in several settings: subtropical dry climates, rain shadows, coasts bordered by cold currents, continental interiors, and polar regions. Because of plate movements, regions that are now deserts were not deserts in the past, and vice versa.

QUICK QUESTION: Why does the world's largest desert, the Sahara, exist?

21.3 Producing Desert Landscapes

Weathering and Soil Formation in Deserts

If you stand in the Mojave Desert of California and look around, you'll see barren cliffs exposing fractured rock, and slopes or plains littered with gravel **(Fig. 21.6)**. Clearly, physical weathering happens in deserts—over time, blocks break off along joints and tumble downslope, perhaps shattering on the way down because of collisions with other blocks. Once a block comes to rest on a desert landscape, it can sit unchanged for a long time, but not forever. Recent studies suggest that the daily alternation from midday heat to midnight cold can generate sufficient stress to break isolated blocks apart, in place.

FIGURE 21.6 Dark gravel, formed by physical weathering of rock exposed on this cliff in the Mojave Desert, litters the gentle slope at the cliff's base.

FIGURE 21.7 Soils in deserts.

(a) Calcrete forms when calcite precipitates and binds rock fragments together.

Stream erosion exposed this calcrete layer.

(b) The red hues of the Painted Desert in Arizona are due to the oxidation of iron in the rock.

Do chemical weathering reactions take place in deserts? Yes. Over time, the moisture from dew and occasional rain allows oxidation, hydrolosis, and dissolution reactions to destroy cements and to transform silicate minerals into clay, thereby causing rocks to disaggregate into pebble- or sand-sized debris. But without the presence of acidic organic matter, and without a steady supply of water to seep into rock or infiltrate into the ground, these reactions take place relatively slowly.

Does soil form in deserts? Yes, but not everywhere. On steep slopes and in locations where vegetation doesn't protect and stabilize regolith, wind and water remove regolith before a soil has time to evolve. Where sediment does stay in place, soil can form, for infiltration of water after heavy rains leaches ions and transports fine clay downward (see Interlude D). In deserts, however, rain falls so infrequently that not enough water infiltrates to flush leached ions away entirely. Therefore, the ions precipitate to form new mineral cement not far below the ground surface. If the new cement consists of calcite, it can bind the regolith into a solid, rock-like material known as *caliche* or *calcrete* (Fig. 21.7a). Calcrete deposits can grow rapidly—some encase tools left behind by prospectors just a few decades ago.

Due to the lack of organic content, the black or brown colors of temperate soils don't develop in deserts, and variations in bedrock color tend to control the color of soil. Variations in the concentration of iron or in the degree of iron oxidation in adjacent beds result in spectacular color bands in rock layers and in the thin soils derived from them. For example, iron-rich, well-oxidized strata will be dark red or even maroon. Rocks with less iron will be lighter red or orange, and rocks in which iron concentration is low or has been chemically reduced can be tan, gray, or even greenish. The Painted Desert of northern Arizona earned its name from the brilliant and varied hues of oxidized iron in the region's shale bedrock (Fig. 21.7b).

Desert Varnish

Shiny **desert varnish**, a dark, rusty-brown coating of iron oxide, manganese oxide, and clay, locally coats the surfaces of rocks in deserts (Fig. 21.8a). Desert varnish was once thought to form when water from rain or dew seeped into a rock, dissolved iron and magnesium ions, and carried the ions by capillary action back to the surface of the rock, where the ions precipitated. More recent studies, however, suggest that the minerals in desert varnish are not necessarily derived from the rocks they coat. Rather, desert varnish may form from wind-borne dust that settles on the surface of a rock, for in the presence of moisture, microbes (bacteria and archaea) extract elements from the dust and transform them into iron or manganese oxide precipitates. Desert varnish doesn't develop in humid climates because rain washes away dust before ions can be extracted from it.

Desert varnish takes a long time to form. In fact, measuring the thickness of a desert varnish layer can provide an estimate of how long a rock has been exposed at the ground surface. In past centuries, people have used desert-varnished rock as a medium for art—by chipping away the varnish to reveal the underlying lighter-colored rock, they were able to create figures or symbols on a dark background. The resulting drawings are known as **petroglyphs** (Fig. 21.8b).

Water Erosion

Although rain rarely falls in deserts, when it does come it can radically alter a landscape in a matter of minutes. Because deserts lack plant cover, rainfall, sheetwash, and stream flow are all extremely effective agents of erosion. It may seem surprising, but in most deserts, water causes more erosion than wind does (Fig. 21.9a). If the substrate eroded by water

FIGURE 21.8 Desert varnish.

(a) A coating of desert varnish turns the surface of light tan sandstone beds to a dark, rusty brown.

(b) By chipping away desert varnish to reveal the lighter rock beneath, Native Americans created art and symbols.

consists of soft, homogeneous material, then a network of closely spaced parallel drainages develops—geologists refer to bare landscapes containing such drainages as *badlands* **(Fig. 21.9b)**.

Water erosion begins with the impact of raindrops, which eject sediment from the ground into the air. On a hill, the ejected sediment lands downslope, so sediment gradually migrates to lower elevations. The ground quickly becomes saturated with water during a heavy rain, so water starts flowing across the surface, carrying loose sediment with it. Within minutes after a heavy downpour begins, dry stream channels fill with a turbulent mixture of water and sediment, which rushes downstream as a flash flood. When the rain stops, the water sinks into the streambed's gravel and disappears. As we noted in Chapter 17, such drainages are called *ephemeral streams*; the channels of these streams are known as **dry washes** or **arroyos** in North America and **wadis** in the Middle East and North Africa. The water in a flash flood can cause intense erosion because it moves so fast and can carry so much sediment. As a result, flash floods can polish bedrock that borders streams **(Fig. 21.9c)**, cut steep-walled channels, and transport huge boulders downstream. As rocks roll and tumble along, they strike one another and shatter, yielding smaller pieces that can be carried still farther. Between floods, the floor of a dry wash consists of gravel littered with boulders **(Fig. 21.9d)**.

Wind Erosion

In temperate and humid regions, plant cover protects the ground surface from the wind, but in deserts, the wind has direct access to the ground. Wind, just like flowing water, can carry sediment both as suspended load and as surface load.

Suspended load (fine-grained sediment such as dust and silt held in suspension) stays in the air for a long time and moves with it. The suspended sediment can be carried so high into the atmosphere (up to several kilometers above the Earth's surface) and so far downwind (tens to thousands of kilometers) that it may move completely out of its source region. Of note, while a gentle breeze may be able to transport very fine grains as suspended sediment, it may not be able to break the grains free of the ground, for electrostatic charges on sediment grains keep the grains "stuck" to the ground. To break grains free of the ground and put them into suspension in the first place requires stronger breezes or turbulence. For example, on an otherwise calm day, tiny vortices that develop locally due to the instability of the hot desert air can churn up dust. In some cases, they become *dust devils* up to 100 m high that look like miniature tornadoes **(Fig. 21.10)**. Cars driving down dirt roads in deserts can also break dust free from the ground—that's why a huge plume of suspended sediment develops in the wake of a car.

Stronger winds can carry **surface loads**, sand and silt that rolls and bounces along the ground **(Fig. 21.11a, b)**. This process, known as *saltation*, begins when turbulence caused by wind shearing along the ground surface lifts sand grains. The grains move downwind, following an asymmetric arch-like trajectory. Eventually, gravity causes them to return to the ground, where they strike other sand grains, causing the new grains to bounce up and drift or roll downwind. The collisions between sand grains make the grains rounded and frosted. Saltating grains generally rise no more than 0.5 m above the ground surface, but where sand bounces on bedrock, the grains may rise 2 m.

The size of clasts that wind can transport depends on the wind velocity. Wind, therefore, does an effective job of sorting

FIGURE 21.9 Evidence of erosion by running water in deserts.

(a) These hills in the desert near Las Vegas, Nevada, are bone dry, but their shape indicates erosion by water. Note the numerous stream channels.

(b) Badlands topography develops where flowing water erodes a soft substrate in a desert.

(c) Flash floods carved this steep-walled channel bordered by polished rock in Mosaic Canyon, California.

(d) Gravel and sand are left behind on the floor of a dry wash after a flash flood in Death Valley. Erosion by the water is cutting a channel.

sediment, sending dust-sized particles skyward and sand-sized particles bouncing along the ground, while pebbles and larger grains remain behind. In some cases, wind carries away so much fine sediment that a **lag deposit**, a concentration of pebbles and cobbles, remains at the ground surface **(Fig. 21.11c)**. Over time, in regions where the substrate consists of soft sediment, wind picks up and removes so much sediment that the land surface becomes lower, a process known as **deflation**. Shrubs can stabilize a small patch of sediment with their roots, so after deflation, forlorn shrubs with residual pedestals of soil stand isolated above a lowered ground surface **(Fig. 21.11d)**. In some places, the shape of the ground surface twists the wind into a local vortex that causes enough deflation to scour a deep, bowl-like depression called a *blowout*.

Particularly strong winds, such as the downdrafts in front of a thunderstorm, can generate a dramatic **dust storm** (known as a *haboob* in the Middle East) that can be 100 km

FIGURE 21.10 A dust devil carries dust skyward in Death Valley.

FIGURE 21.11 Movement of the surface load during wind erosion.

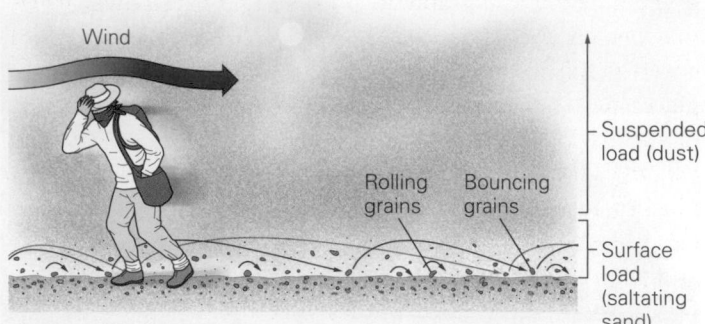

(a) Wind transports desert sediment as suspended load and surface load.

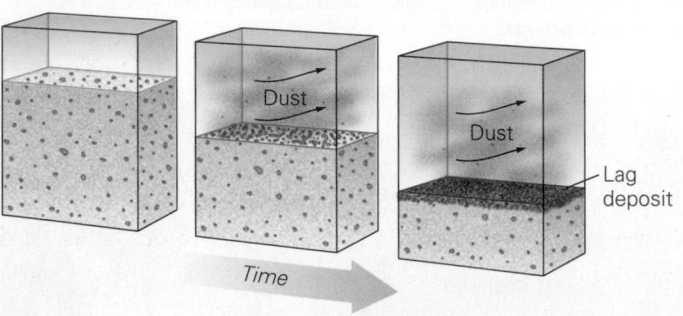

(c) A lag deposit develops when wind blows away finer sediment, leaving behind a layer of coarser grains.

(b) An example of wind-transported sediment.

(d) Deflation has removed the sediment between these shrubs in Death Valley, California.

long and up to 1.5 km high **(Fig. 21.12a)**. Dust storms can be dangerous, for they can cut visibility to almost nothing. They can shut down airports and bring highway traffic to a halt. They can also damage structures and infiltrate machinery with sand and grit, fouling the mechanisms. A particularly large dust storm rolled over Phoenix, Arizona, in July 2011—in photographs, it dwarfs the city below **(Fig. 21.12b)**.

Just as sandblasting cleans the grime off the surface of a building, windblown sand and dust grind away at surfaces in the desert. Over long periods, such wind abrasion produces smooth faces, or facets, on pebbles, cobbles, and boulders. If a rock rolls or tips relative to the prevailing wind direction after it has been faceted on one side, or if the wind shifts direction, a new facet with a different orientation forms, and the two facets

FIGURE 21.12 Examples of dust storms.

(a) A 2005 dust storm (haboob) approaches an army base in Iraq.

(b) A huge dust storm approaching Phoenix, Arizona.

FIGURE 21.13 The consequences of wind erosion.

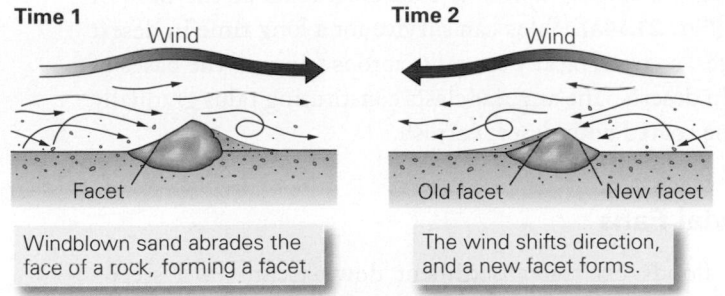

Time 1
Wind

Time 2
Wind

Facet

Old facet   New facet

Windblown sand abrades the face of a rock, forming a facet.

The wind shifts direction, and a new facet forms.

(a) Ventifacts form when windblown sediment erodes the surface of a rock.

2 cm

(b) An example of a multifaceted ventifact from the Dry Valleys of Antarctica.

(c) Yardangs develop when wind erodes a weaker layer of rock beneath a stronger layer.

(d) This NASA photo shows wind erosion on Mars.

join at a sharp edge. Geologists refer to rocks whose surface has been faceted by the wind as **ventifacts (Fig. 21.13a, b)**. Wind abrasion also gradually polishes and bevels away irregularities on the ground surface and polishes the surfaces of desert-varnished outcrops, giving them a reflective sheen. In places where a resistant layer of rock overlies a softer layer of rock, wind abrasion may create a formation, known as a *yardang*, consisting of a resistant block perched on an eroding mushroom-like column of softer rock **(Fig. 21.13c)**. If a strong wind blows in only one direction, yardangs become elongate and align with the wind direction.

Exploration by space probes over the last 20 years has shown that wind affects the desert-like surface of Mars just as it does the deserts on the Earth. During the Martian fall and spring, when differences in temperature between the ice-covered poles and the ice-free equator increase, strong winds blow from the poles to the equator, transporting so much dust that the planet's surface, as seen from the Earth, visibly changes. At times, the entire planet becomes enveloped in a cloud of dust. Close-up images taken by spacecraft that have landed on Mars show that rocks have been abraded by saltating sand and that sand has accumulated on the leeward side of the rocks **(Fig. 21.13d)**.

TAKE-HOME MESSAGE

Chemical weathering occurs slowly in deserts. In the dry climate, unique soils form. Colors of rocks and soils tend to be controlled by the amount and degree of oxidation of iron, and desert varnish may coat rock surfaces. Water causes most erosion and sediment transport, but rarely flows, so stream channels are usually dry. Wind transports sediment and can produce lag deposits and carve ventifacts.

QUICK QUESTION: Why do the unusual shapes of ventifacts and yardangs develop?

21.4 Deposition in Deserts

We've seen that erosion relentlessly eats away at bedrock and sediment in deserts. Where does the debris go? In this section, we examine the various desert settings in which sediment accumulates.

Talus

Over time, joint-bounded blocks of rock break off ledges and cliffs on the sides of hills. Under the influence of gravity, the

(a) This talus apron along the base of a desert cliff formed from rocks that broke off and tumbled down the cliff.

(b) An alluvial fan has accumulated at the mouth of a small canyon along the edge of Death Valley.

(c) Distributaries carry water and sediment to different parts of an alluvial fan at different times, thereby maintaining the fan shape.

resulting debris tumbles downslope and accumulates as a **talus** pile, a sloping apron of rock fragments at the base of a hill **(Fig. 21.14a)**. Talus can survive for a long time in desert climates, so we typically see talus aprons fringing the bases of cliffs in deserts. The angular clasts constituting talus gradually become coated with desert varnish.

Alluvial Fans

Flash floods can carry sediment downstream in a steep-walled canyon. When the turbulent, sediment-laden water flows out onto a plain at the mouth of a canyon, it spreads out over a broader surface with a gentler slope. When this happens, the water slows, and its sediment load drops out and builds into an **alluvial fan**, a conical wedge of sediment that builds outward from the canyon mouth. Once the fan's conical shape has been established, water spilling out of the canyon during a flash flood divides into numerous braided streams, called *distributaries*, that spread water and clasts radially outward over the fan's surface **(Fig. 21.14b)**. For a while, one distributary may carry most of a flood's water flow. Eventually, however, sediment accumulates at the end of this distributary, so the slope of the distributary's channel decreases. When this happens, the flowing water abandons that distributary and finds another one with a steeper channel. As a result, over time, different parts of the fan build out in succession **(Fig. 21.14c)**. This process maintains the overall cone-like shape of the fan as it grows. Notably, because the sediment in an alluvial fan has high permeability, water running down a distributary during a flood gradually infiltrates into the fan.

During some flash floods, debris flows (slurries of mud mixed with larger clasts) spill out onto the fan, depositing sediment in the form of lobes, generally closer to the canyon's mouth. At other times, water-dominated flows emerge from the canyon's mouth. These faster-moving flows carve channels into the fan and carry sediment farther downslope. Such flows sort sediment, leaving behind gravel bars as they carry sand and mud toward the fan's toe. Over time, alluvial fans emerging from adjacent valleys may merge and overlap along the front of a mountain range, producing an elongate wedge of sediment called a *bajada* **(Fig. 21.14d)**.

(d) A bajada accumulating at the base of a mountain range in the Mojave Desert. Note that a bajada consists of overlapping fans.

Playas and Salt Lakes

During a particularly large rainstorm or an unusually wet spring, a temporary lake may develop over the low part of a basin in a desert. In drier times, such desert lakes evaporate entirely, leaving behind a dry, flat, exposed lake bed known as a **playa (Fig. 21.15a, b)**. Over time, a crust of clay and various salts (halite, gypsum, borax, and other minerals) accumulate on the surface of a playa. Some of these minerals have industrial uses and thus have been mined.

Locally, playa surfaces dry out enough for deep mud cracks to develop **(Fig. 21.15c)**. And in a few localities, isolated cobbles sit out on the surface of a playa, far from any slope down which they could have rolled. At the best-known example, Racetrack Playa of Death Valley, each cobble lies at the end of a groove formed when the cobble slid across the soft surface of the playa **(Fig. 21.15d)**. Until recently, no one had seen the cobbles move, and geologists simply assumed that movement somehow occurred when the playa was wet and so slippery that strong winds could blow the cobbles along. In 2014, GPS-triggered cameras showed that, in fact, the cobbles slide only when very shallow water on the surface of the laya freezes into a thin sheet of ice during the night. When the air warms the next day, the ice breaks into broad plates, which move in response to strong winds and push the cobbles.

Where sufficient water flows from surrounding regions into a desert basin, a permanent lake will fill. If the lake occurs in an *interior basin*, meaning a basin that has no outlet to allow water to flow out, the lake becomes very salty. The salt accumulates because when water evaporates, only H_2O goes into the air—salt stays behind. The Great Salt Lake in Utah exemplifies this process **(Fig. 21.15e)**. Even though the streams feeding the lake are fresh enough to drink, their water contains trace amounts of dissolved ions. Because the lake has no outlet, these ions have become concentrated in the lake over time, making it even saltier than the ocean.

In subtropical deserts that border the sea, high tides may cause warm seawater to flood coastal tidal flats. Under the hot sun, the shallow seawater evaporates and becomes saturated. As a result, salts precipitate onto the underlying organic-rich mud, producing a salty crust overlying the mud. Geologists refer to a salt-encrusted muddy tidal flat as a *sabkha*.

Deposition by Wind

As mentioned earlier, wind carries two kinds of sediment loads—a suspended load of dust-sized particles and a surface load of sand. Much of the dust blows out of the desert and accumulates elsewhere. Locally, it builds into **loess**, layers of fine-grained sediment. Sand, however, cannot travel far, and it accumulates within the desert in piles called **dunes**, ranging in size from less than a meter to over 300 m high. In favorable locations, dunes accumulate to form vast sand seas hundreds of meters thick. We'll look at dunes in more detail in the following section.

> ### TAKE-HOME MESSAGE
>
> Sediment carried by water and wind in deserts accumulates in a variety of landforms. Alluvial fans form at the mouths of canyons; playas form where water temporarily collects in basins, and dunes form where large amounts of sand are available. Lakes in interior basins become salty.
>
> **QUICK QUESTION:** How do bajadas develop?

21.5 Desert Landforms and Life

The popular media commonly portray deserts as endless seas of sand piled into dunes, which hide the occasional palm-studded oasis. In reality, immense sand seas are only one type of desert landscape. Deserts may also host vast, rocky plains, a stubble of cacti and other hardy desert plants, or intricate rock formations that look like medieval castles. Nomads living in the Sahara emphasized these differences by distinguishing among *hamada* (barren, rocky highlands), *reg* (vast, stony plains), and *erg* (sand seas in which large dunes form). As we've noted, desert landscapes, overall, tend to be harsher and more rugged than temperate or tropical ones. For example, if eastern North America, a temperate region, were instead a desert, the Appalachians would not be gentle, forested hills, but rather would consist of stark, rocky ridges **(Fig. 21.16)**. In this section, we examine various desert landscapes.

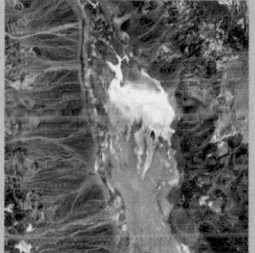

SEE FOR YOURSELF...

Death Valley, California

LATITUDE
36°12'42.34" N

LONGITUDE
116°48'25.43" W

Look down from 35 km.

You can see many desert landforms in Death Valley, a narrow basin whose floor lies below sea level. The white patch is a playa, a bajada forms the slope between the playa and the mountains to the west, and small alluvial fans spill into the basin on the east.

> **Did you ever wonder...**
> whether all deserts are completely covered by sand?

FIGURE 21.15 Playas form where a shallow, salty lake dries up.

An oblique air photo

Playa

Bajada

Range

(a) This playa in California formed at the base of a bajada.

A close-up of salt crystals.

(b) White salt crystals encrust the floor of a playa in Death Valley.

(c) The salt and mud layer on the bed of this playa in Death Valley has dried out and cracked.

(d) Wind-driven ice pushed these cobbles, leaving tracks along the slippery clay surface of Racetrack Playa in Death Valley, California.

(e) The Great Salt Lake in Utah has no outlet. Sediments deposited along its shores are quite salty.

Rocky Cliffs and Mesas

In hilly desert regions, the lack of soil exposes rocky ridges and cliffs. As cliffs erode when rocks split away along joints, the cliff face's position moves, but the face retains roughly the same shape. This process, known as **cliff retreat**, or *scarp retreat*, takes place in fits and starts **(Fig. 21.17a)**. A cliff may remain unchanged for decades or centuries, and then suddenly a wall of rock may fall off and crumble into rubble. Cliff height typically reflects bed thickness—in places where particularly thick, resistant beds crop out, tall cliffs develop, for joints tend to be widely spaced in thick beds, so the collapse of a portion of the wall generates a huge new face.

FIGURE 21.16 Contrasts between desert and temperate hillslopes.

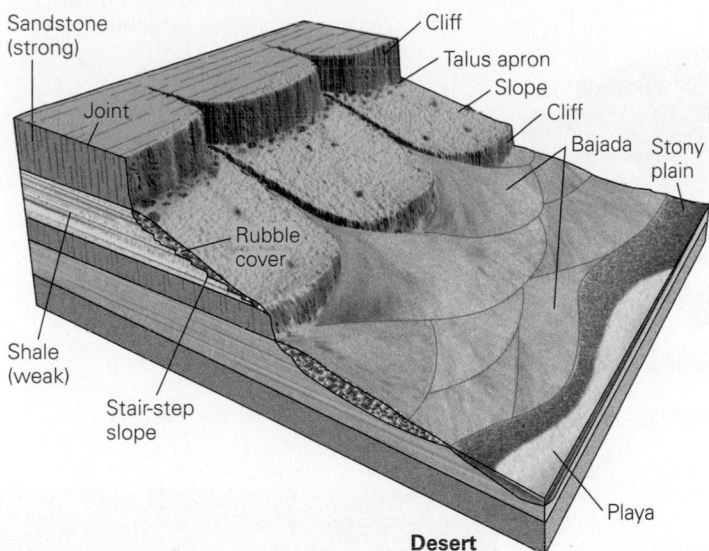

Sandstone (strong)
Cliff
Talus apron
Slope
Joint
Cliff
Bajada
Stony plain
Rubble cover
Shale (weak)
Stair-step slope
Playa
Desert

(a) In a desert, steep cliffs form out of resistant rock, fans of debris form a bajada at the base of the cliffs, and playas form on the plain.

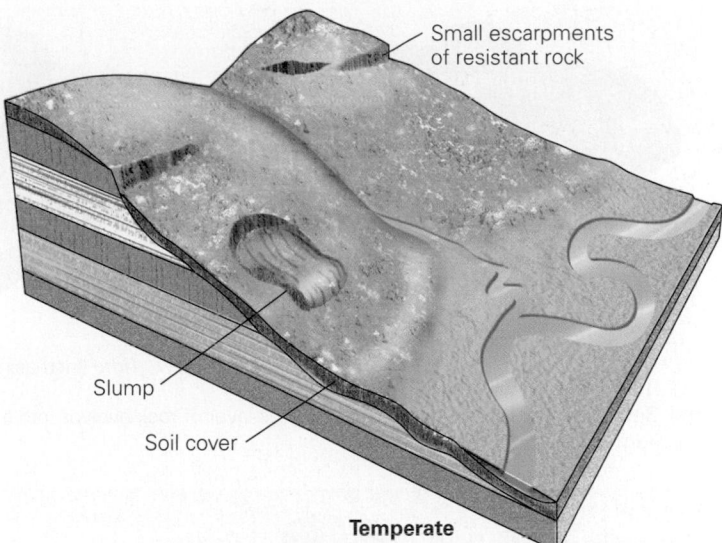

Small escarpments of resistant rock
Slump
Soil cover
Temperate

(b) In a temperate region, thick soils form, and slumping prevents steep slopes from developing. Vegetation covers the surface.

In thinly bedded shale, joints tend to be more closely spaced, so shale beds erode to make an overall gradual slope consisting of many tiny stair steps. Therefore, cliffs formed from strata of contrasting strength develop a stair-step-like shape—strong layers (sandstone or limestone) become cliffs, whereas weak layers (shale) become slopes **(Fig. 21.17b)**.

With continued erosion and cliff retreat, a plateau of rock slowly evolves into a cluster of isolated hills, ridges, or

FIGURE 21.17 Cliff retreat in a desert environment.

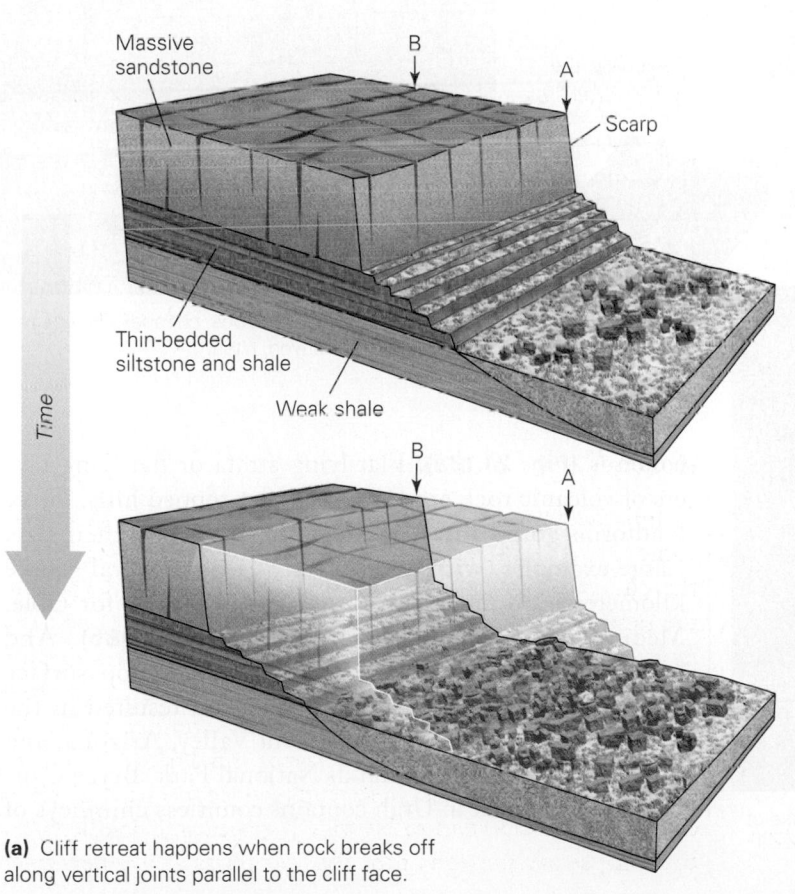

Massive sandstone
B
A
Scarp
Thin-bedded siltstone and shale
Weak shale
Time
B
A

(a) Cliff retreat happens when rock breaks off along vertical joints parallel to the cliff face.

Stronger sandstone
Weak shale
Debris from a recent rockfall

(b) On a hill in Utah, the strong sandstone holds up a cliff face, whereas weak shale forms a slope.

FIGURE 21.18 Mesas and buttes form in deserts as cliffs retreat over time.

Butte
Chimney

Mesa

Time (increasing amount of erosion)

(a) Because of cliff retreat, a once-continuous layer of rock evolves into a series of isolated remnants. If the bedding is horizontal, the resulting landforms have flat tops.

(b) Buttes and mesas tower above the floor of Monument Valley, Arizona.

Joint

Preferential erosion along joints leaves walls ("fins").

Erosion through a fin produces an arch.

Eventually, the arches collapse.

(c) Erosion produced hoodoos, chimney-like columns of rock, in Bryce Canyon, Utah.

(d) Erosion along joints in sandstone produces fins. Erosion at the base of these fins then produces arches.

columns **(Fig. 21.18a)**. Flat-lying strata or flat-lying layers of volcanic rock erode to make flat-topped hills. These landforms go by different names depending on their size. Large examples, with a top surface area of several square kilometers, are **mesas**, from the Spanish word for table. Medium-sized examples are **buttes (Fig. 21.18b)**. And small examples, whose height exceeds their top surface area, are **chimneys**. Erosion of strata has resulted in the skyscraper-like buttes of Monument Valley, Arizona, and the stark cliffs of Canyonlands National Park. Bryce Canyon National Park in Utah contains countless chimneys of brightly colored shale and sandstone—locally, these chimneys are known as *hoodoos* **(Fig. 21.18c)**. *Natural arches*, such as those of Arches National Monument, form when erosion removes the lower part of a wall of rock, while the upper part remains. Such walls of rock are bounded by joints and became walls because erosion preferentially removed rock along the joints **(Fig. 21.18d)**

In places where bedding dips at an angle, an asymmetric ridge called a *cuesta* develops. A joint-controlled cliff forms the steep front side of a cuesta, and the tilted top surface of a resistant bed forms the gradual slope on the backside

(Fig. 21.19a). Because this slope tilts at the same angle as the bed dips (see Chapter 11), it is known as a *dip slope*. Where beds have a near-vertical dip, a narrow symmetrical ridge, called a *hogback*, forms. If desert hills consist of homogeneous rock such as granite, rather than stratified rock, they typically erode to make a pile of rounded blocks **(Fig. 21.19b)**.

Over time, erosion may grind away all sides of a desert mountain range so that it becomes an *inselberg* (from the German, meaning island rock). In some cases, alluvium-filled basins surround an inselberg and cover bedrock right up to the base of the inselberg's slopes **(Fig. 21.19c)**. In other cases, a relatively flat shelf or ramp of bedrock separates the steep slope of the inselberg from the basins. These shelves, known as **pediments**, can develop when sheetwash during heavy rains carries sediment and weathered bedrock a distance away from the range. The moving sediment abrades the bedrock over which it travels.

Depending on the rock type, the orientation of stratification in the rock, and rates of erosion, inselbergs may be sharp-crested, plateau-like, or loaf-shaped with steep sides and a rounded crest. Geologists refer to inselbergs that have a loaf geometry, as exemplified by Uluru (Ayers Rock) in central Australia **(Box 21.1)**, as *bornhardts*.

FIGURE 21.19 Examples of erosional landscapes in deserts.

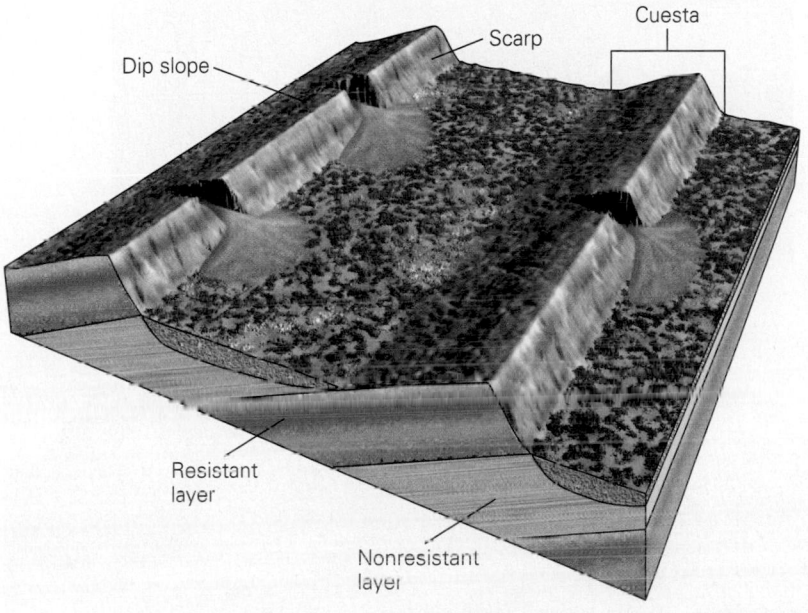

(a) Asymmetric ridges called cuestas develop where strata in a region are not horizontal.

(b) The homogeneous granite of a pluton breaks along joints into blocks, which erode into rounded boulders. This example crops out in southern Nevada.

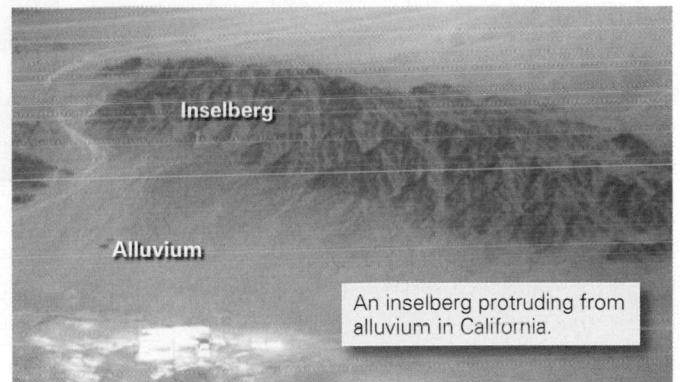

An inselberg protruding from alluvium in California.

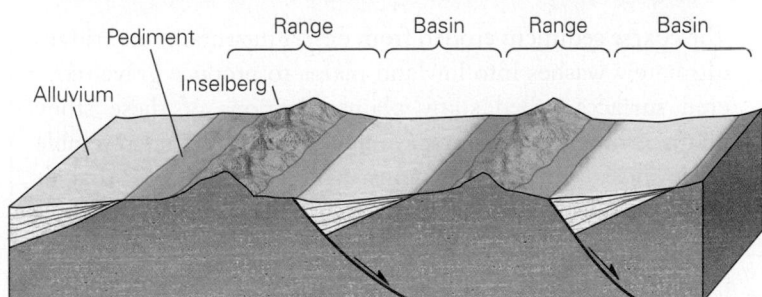

(c) In the Basin and Range Province of the southwestern United States, tilted fault-block ranges evolve into inselbergs, bordered by sediment-filled basins.

BOX 21.1 CONSIDER THIS . . .

Uluru (Ayers Rock)

In the immense deserts of central Australia, Uluru, also known by its English name, Ayers Rock, towers 360 m above a scrub-covered plain **(Fig. Bx21.1)**. This rock mass, 3.6 km long and 2 km wide, consists of nearly vertical dipping sandstone beds. It makes up one limb of a huge regional syncline. The other limb is also a bornhardt, known locally as the Olgas.

Alluvium buries the entire area in between. Uluru formed because the sandstone that comprises it resisted erosion, whereas adjacent rock formations did not. Thus, over geologic time, alluvium buried the surrounding landscape, but Uluru remained high. Because its strata dip vertically, it has not developed the stair-step shape of mesas and buttes.

According to Aboriginal Australian traditions, erosional features on the surface of the rock are scars from a fierce battle between ancient clans. In recent years, the rock has attracted tourists from around the world. Plaques at the base of the rock record the names of those who slipped and fell from its steep sides while climbing to the top.

FIGURE Bx21.1 The formation of Uluru (Ayers Rock) in central Australia.

Mountain building folded layers of sedimentary rock.

Time 1

After erosion, only resistant beds protrude from alluvium.

The Olgas   Uluru   Alluvium

Time 2

Uluru viewed from space

1 km

Strike of bedding

Stony Plains and Desert Pavement

The coarse sediment eroded from desert mountains and ridges ultimately washes into lowland plains to produce gravel-covered surfaces called stony plains. Portions of these stony plains evolve into **desert pavements**, surfaces that resemble a tile mosaic in that they consist of separate stones that fit together tightly, forming a fairly smooth surface layer above a soil composed of silt and clay **(Fig. 21.20a, b)**. Typically, desert varnish coats the top surfaces of the stones forming desert pavement. Geologists have proposed several explanations for the origin of desert pavements. Traditionally, pavements were thought to be lag deposits, formed when wind blows away the fine sediment between clasts, so that the clasts settle down and fit together. Recently, researchers have suggested instead that pavements form when windblown dust slowly sifts down onto the stones and then washes down between them. In this model, the pavement is "born at the surface," meaning that the stones forming the pavement have been progressively lifted up as sediment collects and builds up beneath them **(Fig. 21.20c)**. Over time, the rocks at the surface crack, perhaps due to differential heating by the Sun over time. During downpours, sheetwash washes away fine sediment between fragments, and when soils dry and shrink between storms, the clasts settle together, locking into a stable, jigsaw-puzzle-like arrangement.

FIGURE 21.20 Desert pavement and a hypothesis for how it forms by building up a soil from below.

(a) A well-developed desert pavement in the Sonoran Desert, Arizona. The inset shows a close-up of the pavement.

(b) Students standing at the edge of a trench cut into desert pavement. Soil lies between the pavement and the underlying alluvium. The inset shows the top surface of the pavement (lens cap diameter = 8 cm).

Desert pavements are remarkably durable and can last for hundreds or thousands of years when left alone. But like many features of the desert, they can be disrupted in a moment by human activity. For example, people driving vehicles across the pavement indent and crack its surface, making it susceptible to erosion. In parts of Arizona, vast desert pavements have become parking lots for campers who migrate to the desert in motor homes for the winter season.

Seas of Sand: The Nature of Dunes

Walk out onto the sand dunes of Mesquite Flats, in Death Valley, and you enter a different world **(Fig. 21.21)**. **Sand dunes**—asymmetric ridges of unconsolidated sand deposited by a flowing current—rise around you, defining a natural sculpture of dramatic curves, edges, and slopes. When the wind blows, and saltation can occur, the sand comes to life as it tumbles and bounces up the windward (upwind) side of each dune and slides down the leeward (downwind) side.

Dunes can be formed by currents of water on the beds of rivers as well as by currents of air on the land surfaces of deserts. Dunes in deserts start to

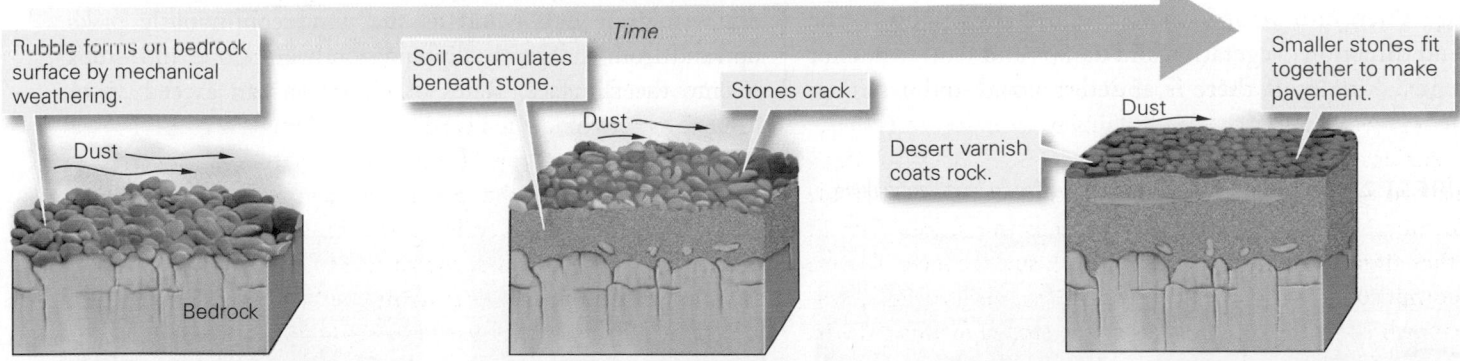

(c) Desert pavement forms in stages. First, loose pebbles and cobbles collect at the surface. Dust settles among the stones and builds up a soil layer below. The stones eventually crack into smaller pieces and settle to form a mosaic-like pavement.

FIGURE 21.21 Sand dunes of Mesquite Flats, in Death Valley, California.

form where sand becomes trapped on the windward side of an obstacle, such as a rock or a shrub **(Fig. 21.22)**. Gradually, the sand builds downwind into the lee of the obstacle. Once initiated, the dune itself affects the wind flow, and sand accumulates on the leeward side of the dune's crest. Here the sand eventually slides down the leeward surface of the dune.

In places where abundant sand accumulates, *sand seas* (ergs) bury the landscape. The wind builds the sand in these ergs into dunes that display a variety of shapes and sizes, depending on the character of the wind and the sand supply **(Fig. 21.23a)**. Where the sand is relatively scarce and the wind blows steadily in one direction, beautiful crescents called *barchan dunes* develop, with the tips of the crescents pointing downwind. If the wind shifts direction frequently, crescents pointing in different directions overlap one another, creating a constantly changing *star dune*. Where enough sand accumulates to bury the ground surface completely and only moderate winds blow, sand piles into simple, wave-like shapes called *transverse dunes*; the crests of these dunes lie perpendicular to the wind direction. Strong winds may break through transverse dunes and change them into *parabolic dunes*, whose ends point in the upwind direction. Vegetation and damp sand tend to anchor the ends. Finally, if there is abundant sand and a strong,

steady wind, the sand streams into *longitudinal dunes* (also called *seif dunes* after the Arabic word for sword) whose axis lies parallel to the wind direction. In the southern third of the Arabian Peninsula, a region called the Empty Quarter because of its lack of population, a vast erg called the Rub al Khali includes seif dunes that stretch for almost 200 km and reach heights of over 300 m.

As we've noted, sand saltates up the windward side of the dune, blows over the crest of the dune, and then settles on the steeper, leeward face of the dune, where the air slows down. The slope of this face attains the angle of repose, the slope angle of a freestanding pile of sand. As sand collects on this surface, it eventually becomes unstable and slides down the slope, so geologists refer to the leeward side of a dune as the **slip face**. As more and more sand accumulates on the slip face, the crest of the dune migrates downwind, and former slip faces become preserved inside the dune. In cross section, these slip faces appear as cross beds **(Fig. 21.23b–d)**. The surfaces of dunes are not, in general, smooth, but rather host ripples, whose orientation depends on local wind direction.

With the exception of star and longitudinal dunes, sand dunes migrate downwind as the wind continuously picks up sand from the gently dipping windward slope and drops it onto the slip face. Rates of migration can exceed 25 m

FIGURE 21.22 Progressive stages in the growth of a small sand dune.

Blowing sand
Obstacle

Time 1

Trapped sand

Time 2

Sand grows around obstacle.

Time 3

Obstacle is buried, and dune grows.

Time 4

FIGURE 21.23 The types of sand dunes and the cross beds within them.

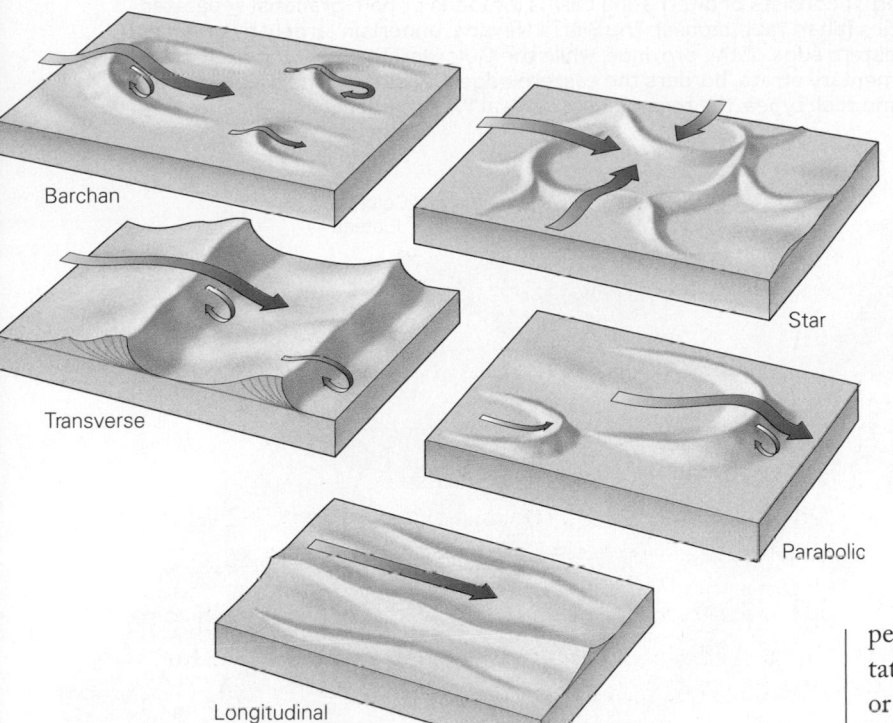

Barchan

Transverse

Star

Parabolic

Longitudinal

(a) The various types of sand dunes.

(b) A sand dune with surface ripples.

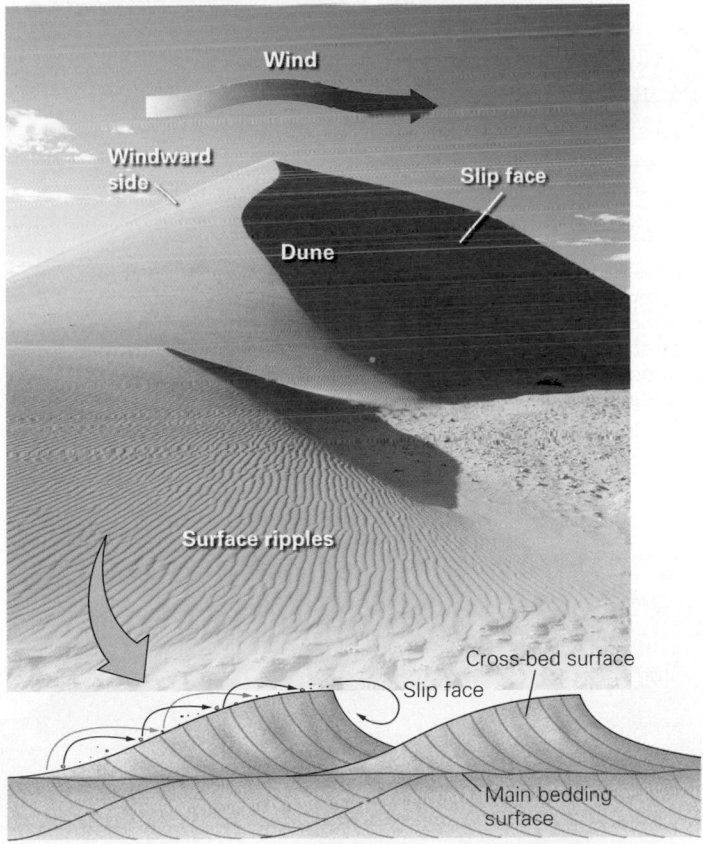

Wind

Windward side

Slip face

Dune

Surface ripples

Cross-bed surface

Slip face

Main bedding surface

(c) Cross bedding inside a dune.

Main bed

Cross bed

(d) Cross beds preserved in Mesozoic sandstone beds of Zion National Park.

per year. Because of the moving sand on an active dune, vegetation can't grow there. If a change in climate brings more rain or less wind, plant cover may grow and stabilize the dunes. At the end of the last ice age, for example, the Sand Hills region of western Nebraska was a vast active dune field, but in the past 11,000 years it has been covered by grasslands, and the dunes have become stabilized.

Life in the Desert

We've seen that deserts host a variety of landscapes (**Geology at a Glance**, pp. 822–823). In the midst of a large erg, there may be nothing growing or moving at all. But most desert landscapes do include plants and animals. These organisms must possess special characteristics to enable them to survive in the desert: they must be able to withstand extremes of temperature—oppressive heat during the day and chilling cold at night—and to survive without abundant water.

Plants have evolved a number of different means to survive desert conditions. Some produce thick-skinned seeds that last until a heavy rainfall, then quickly germinate, grow, and generate new seeds while water remains. The new generation of seeds then waits until the next rainfall to start the cycle over again. Other plants have evolved the ability to send long roots downward to find deep groundwater. Still others have shallow root systems that spread over a broad area so they can efficiently soak up water when it does rain.

Many desert plants have thick, fleshy stems and leaves. These plants, known as *succulents*, can store water for long periods (**Fig. 21.24**). Because succulents may be the only source of water during a time of drought, they have developed threatening thorns or needles to keep away thirsty

GEOLOGY AT A GLANCE

The Desert Realm

The desert of the Basin and Range Province in Utah, Nevada, and Arizona formed due to Cenozoic rifting. It consists of alternating basins (grabens or half-grabens) separated by narrow ranges (tilted fault blocks). The Sierra Nevada, underlain largely by granite, borders the western edge of the province, while the Colorado Plateau, underlain by flat-lying sedimentary strata, borders the eastern edge. Because of the great variety of elevations and rock types, the region hosts several types of desert landscapes.

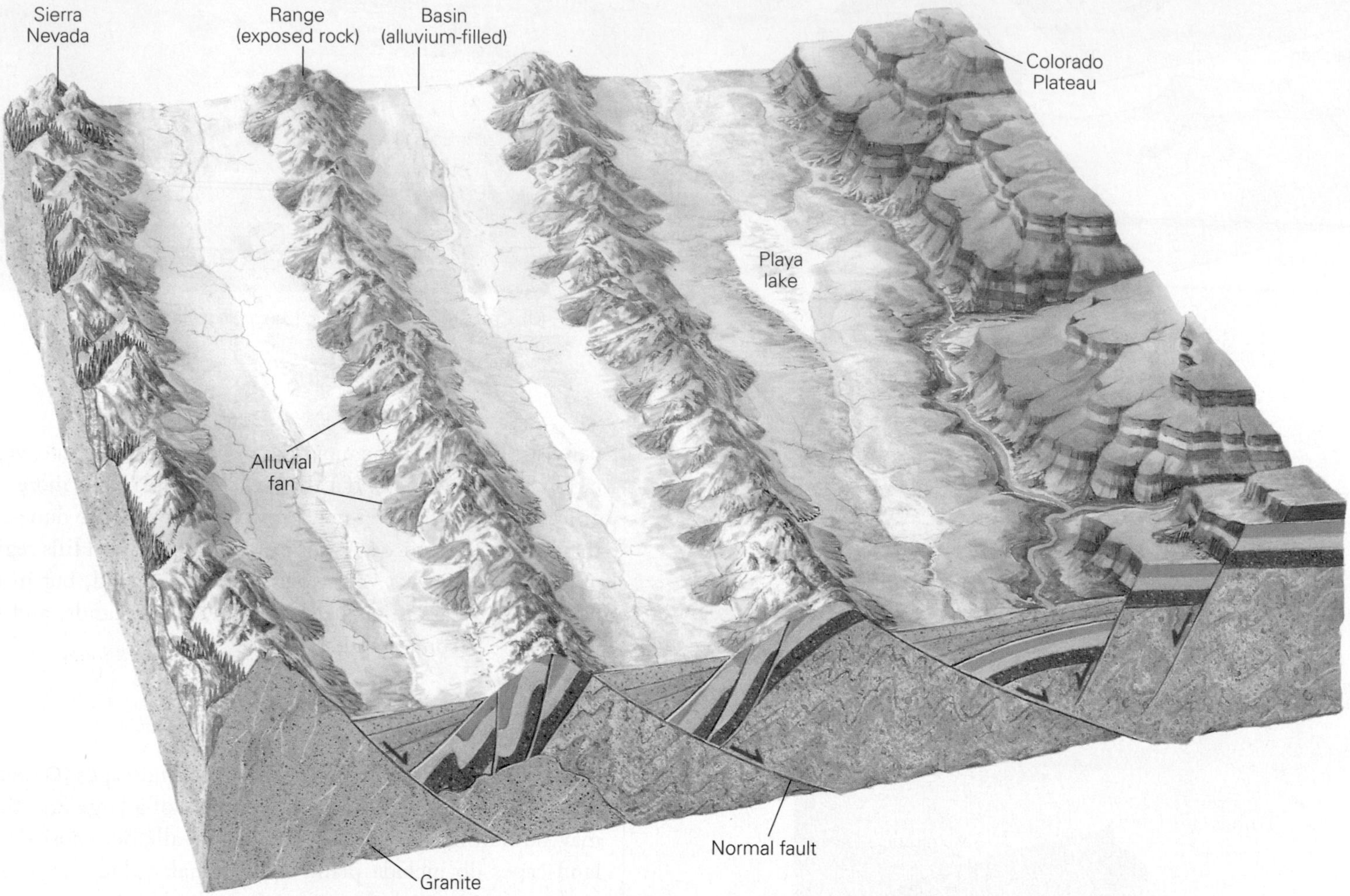

Sierra Nevada

Range (exposed rock)

Basin (alluvium-filled)

Colorado Plateau

Playa lake

Alluvial fan

Normal fault

Granite

Barchan dune

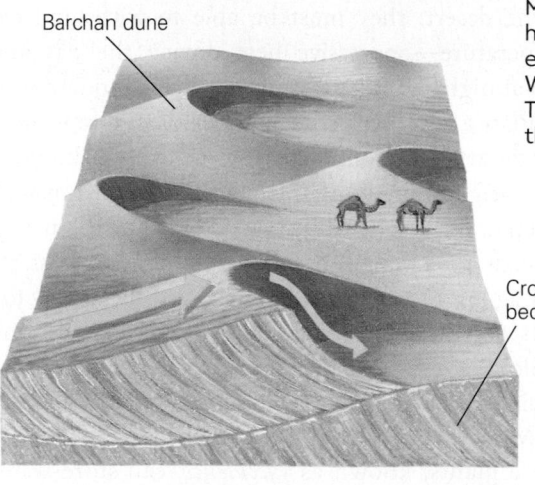

Cross beds

Most streams in deserts fill with water only during flash floods after heavy rains. The turbulent, muddy water of a flash flood can transport even large boulders. At other times, the stream channels are dry washes. Where there is a large supply of sand, a variety of sand dunes develop. The geometry of a particular sand dune depends on the sand supply and the wind. Inside sand dunes, we find cross beds.

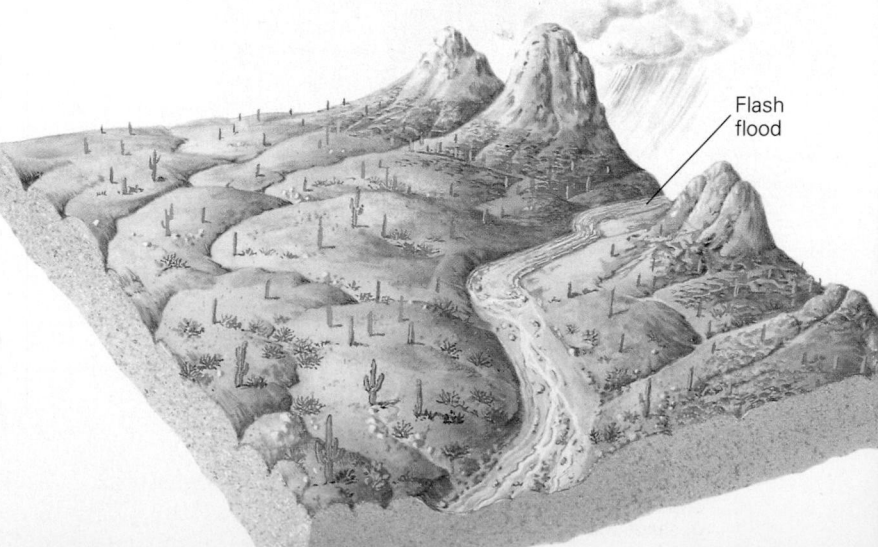

Flash flood

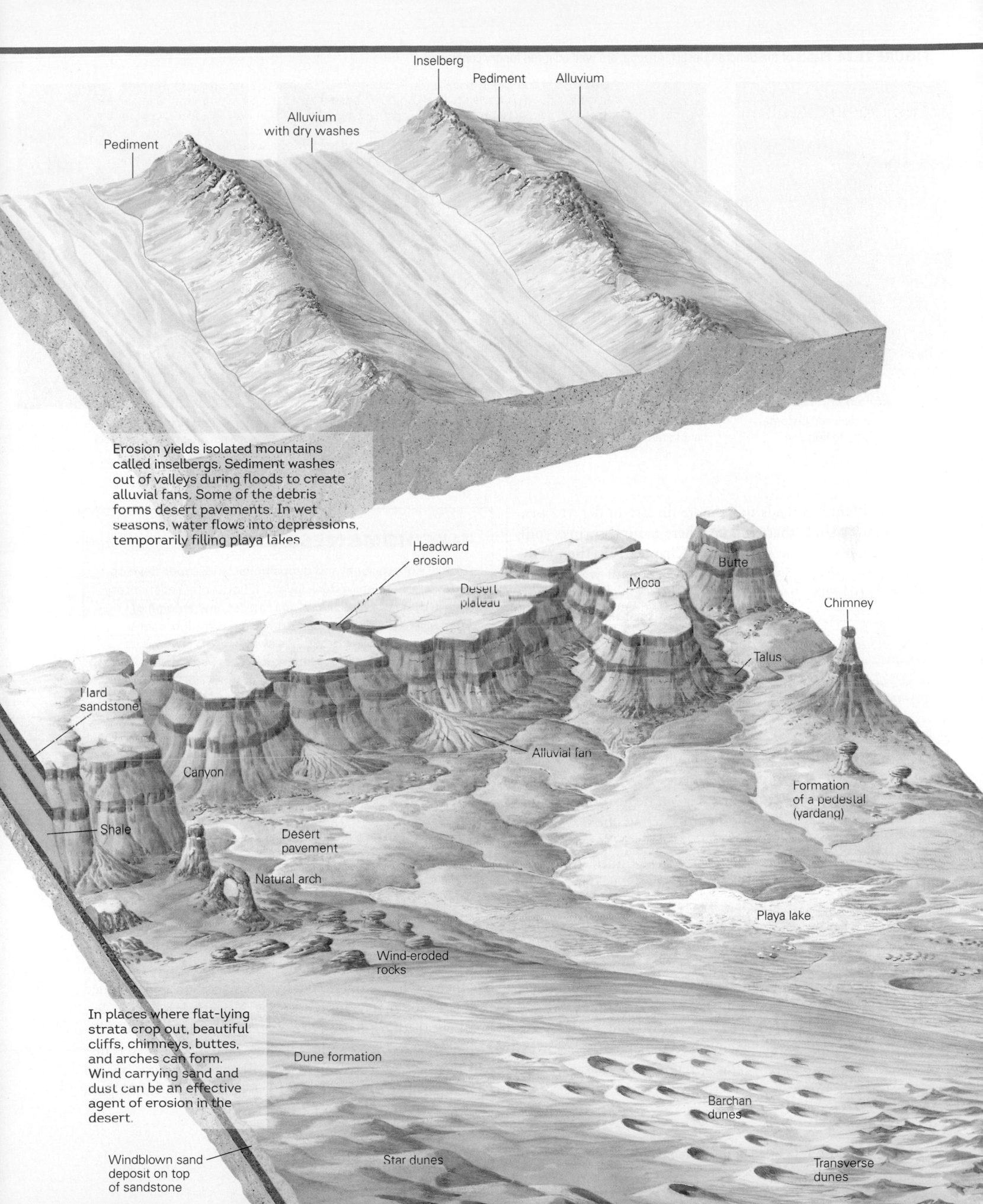

Pediment

Inselberg

Pediment

Alluvium

Alluvium
with dry washes

Erosion yields isolated mountains
called inselbergs. Sediment washes
out of valleys during floods to create
alluvial fans. Some of the debris
forms desert pavements. In wet
seasons, water flows into depressions,
temporarily filling playa lakes

Headward
erosion

Desert
plateau

Mesa

Butte

Chimney

Hard
sandstone

Talus

Canyon

Alluvial fan

Formation
of a pedestal
(yardang)

Shale

Desert
pavement

Natural arch

Playa lake

Wind-eroded
rocks

In places where flat-lying
strata crop out, beautiful
cliffs, chimneys, buttes,
and arches can form.
Wind carrying sand and
dust can be an effective
agent of erosion in the
desert.

Dune formation

Barchan
dunes

Windblown sand
deposit on top
of sandstone

Star dunes

Transverse
dunes

(a) Saguaro cacti can become huge. Don't try to hug one.

(b) "Teddy bear" cholla look soft, but the spines are extremely sharp and hard to remove.

(c) Cactus flowers stand out against brown sand and rock.

animals. Plant life tends to be more diverse in desert *oases*, the verdant islands that crop up where natural springs spill groundwater onto the surface (see Chapter 19). The nearly year-round supply of water in an oasis nourishes a variety of palms and other nonsucculent plants.

Animal life in the desert includes plant eaters, hunters, and scavengers. Animals face the same challenge as plants—they must be able to retain water and survive extreme temperatures. To accomplish these goals, desert animals have also evolved numerous strategies. Frogs, for example, burrow beneath the ground and remain dormant for months, waiting until the next rain. Reptiles escape the midday heat by crawling into dark cracks between rocks. Rodents forage for food only during the cool night. And kit foxes, jackrabbits, and mule deer have disproportionately large ears, through which they efficiently lose body heat. Many desert mammals, such as camels, retain body water by not sweating, as we noted earlier.

TAKE-HOME MESSAGE

A variety of erosional and depositional landscapes develop in deserts. Erosion of thick layers of horizontal sedimentary rock in deserts yields buttes and mesas, and erosion of tilted layers forms cuestas. Stony plains develop where finer sediment blows or washes away. Desert pavements form when fine sediment filters down between larger clasts, which settle together to form a mosaic. Regions with abundant sand contain many types of dunes. Particularly hardy organisms have evolved to survive in these harsh realms.

QUICK QUESTION: What factors control the shape, dimensions, and orientation of sand dunes?

21.6 Desert Problems

Humans can't live in the desert easily. The loss of body moisture in extreme heat can be so rapid that a person will die in less than 24 hours unless shaded from the Sun and supplied with at least 8 liters of water per day. Nevertheless, people live in all but the most barren deserts **(Fig. 21.25a)**. Before technology provided water wells, pipelines, and mechanized transportation, desert peoples lived in small nomadic groups, spaced far enough apart that they could live off the land. In the past,

FIGURE 21.25 Living in the desert is a challenge because of the need for water.

(a) A West African village struggles with blowing sand and the lack of vegetation.

(b) Suburbs of Las Vegas, Nevada, push into the desert. Residents use groundwater and water brought from the Colorado River to keep grass green.

nomadic desert dwellers either built temporary shelters out of local materials or traveled with tents. Locally, people carved underground dwellings in sediment or soft rock, for rock serves as such a good insulator that a few meters below the ground surface, temperatures remain close to the region's mean temperature all year.

In recent times, more and more people in industrialized countries have moved into desert regions. In fact, the population of the desert in the southwestern United States has been growing faster than that of any other part of the country. As desert cities grow, environmental problems soon follow **(Fig. 21.25b)**. As noted in Chapter 19, growing cities must either pump water out of the ground or bring in water via canals from rivers or reservoirs to meet their needs. As a result, water tables in deserts are dropping, rivers are drying up, the land surface is cracking, and vegetation is dying **(Fig. 21.26a)**. People also have imported exotic plants and animals, some of which have become *invasive species* that upset the desert's ecological balance.

The modern era has seen a remarkable change in desert margins. Natural droughts (periods of unusually low rainfall), aggravated by overpopulation, overgrazing, careless agriculture, and diversion of water supplies, have transformed semi-arid grasslands into true deserts, leading to tragic famines that have killed millions of people. **Desertification**, the transformation of nondesert areas to desert, has accelerated.

The consequences of desertification have devastated portions of the Sahel, the belt of semi-arid land that fringes the

southern margin of the Sahara **(Fig. 21.26b)**. In the past, the Sahel provided sufficient vegetation to support a small population of nomadic people and animals. But during the second half of the 20th century, large numbers of people migrated into the Sahel to escape overcrowding in central Africa. The immigrants began farming and maintained herds of cattle and goats. Plowing and overgrazing removed soil-preserving grass and caused the soil to dry out. In addition, trampling by animals compacted the ground so it could no longer soak up water. In the 1960s and again in the 1980s, a series of natural droughts hit the region, bringing catastrophe **(Fig. 21.26c)**. Wind erosion stripped off the remaining topsoil. Without vegetation, the air grew drier, and the semi-arid grassland of the Sahel became desert, with mass starvation as the result. In some cases, desertification happens when winds blow in new dunes, which bury fields or forests.

Other arid regions on the Earth have endured similar problems. The Aral Sea in Kazakhstan, for example, has almost entirely dried up. Rivers that once carried water into the sea have been diverted to provide water for irrigation, and now the rate at which the sea loses water to evaporation greatly exceeds the supply of new water. Consequently, the area of the sea has shrunk to a small fraction of what it once was. Boats that once plied the waters of the Aral Sea now lie as rusting hulks in the salty dust **(Fig. 21.27)**.

Desertification does not happen only in less-industrialized nations. People in the western Great Plains of the United

FIGURE 21.26 Desertification is happening in parts of Africa.

(a) A woodland has dried out and died as huge dunes encroach.

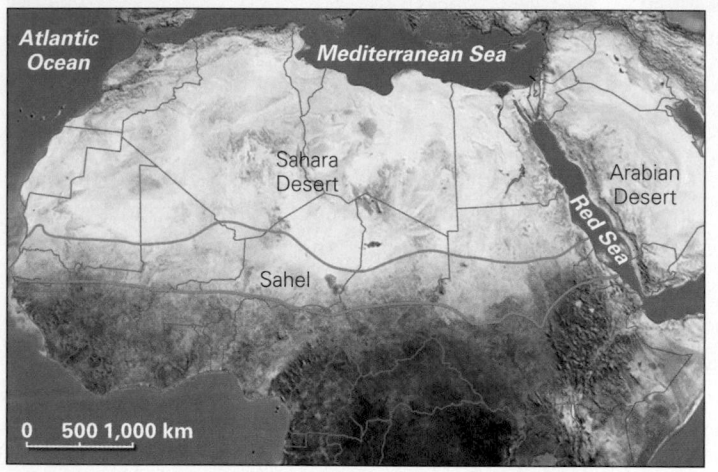

(b) The Sahel is the semi-arid land along the southern edge of the Sahara. Large parts have undergone desertification.

(c) Drought in the Sahel has brought deadly consequences. Here residents seek water from a dwindling well.

States and Canada suffered from the problem beginning in 1933, the fourth year of the Great Depression. Banks had failed, workers had lost their jobs, the stock market had crashed, and hardship burdened all. No one needed yet another disaster—but that year, even nature turned hostile. All through the fall, so little rain fell in the plains of Texas and Oklahoma that the region's grasslands and croplands browned and withered, and the topsoil turned to powdery dust. Then, in November, strong storms blew eastward across the plains. Without vegetation to protect the ground, the wind lapped at it, stripped off the topsoil, and sent it skyward to form rolling black clouds that literally blotted out the Sun **(Fig. 21.28)**. People caught in the resulting dust storms found themselves choking and gasping for breath. When the dust finally settled, it had buried houses and roads under huge drifts and had dirtied every nook and cranny. The dust blew east as far as New England, where it turned the snow brown. What had once been rich farmland in the southwestern plains turned into a wasteland, known as the Dust Bowl.

For several more years the drought persisted, leading to starvation and bankruptcy. Many of the region's residents were forced to move to wherever they could find work. People from Texas and Oklahoma piled into jalopies and drove on old Route 66 out to California, looking for jobs in the state's still-green agricultural regions. Many of these people were subjected to exploitation once they arrived in California. John Steinbeck dramatized this staggering human tragedy, which came to symbolize the Depression, in his novel *The Grapes of Wrath*.

FIGURE 21.27 Desertification of the Aral Sea in central Asia occurred when the Soviet Union diverted the rivers that flowed into the sea for irrigation projects. Once home to a fishing fleet, the sea has been reduced to a few ultra-salty ponds.

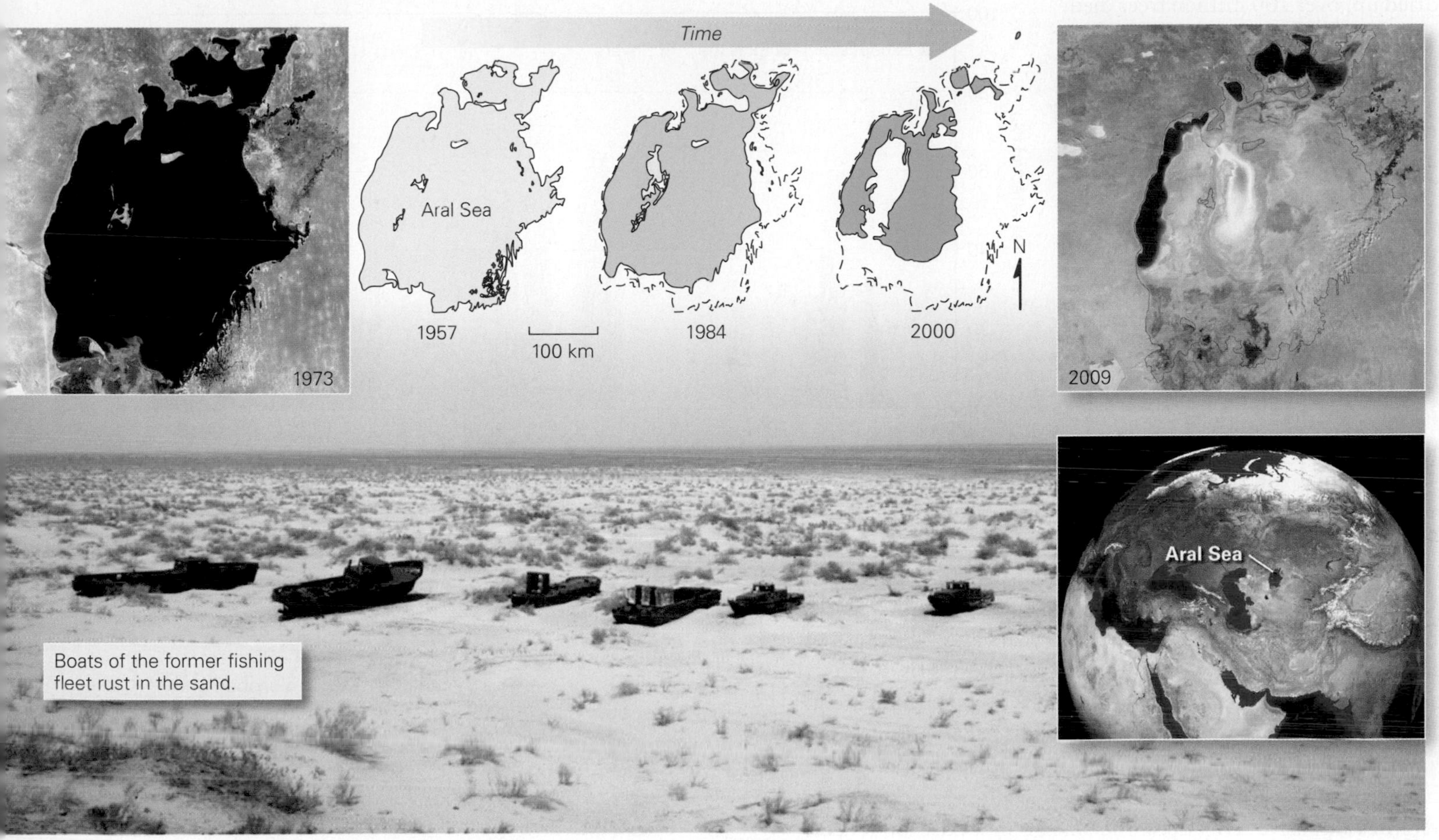

Time

1973

Aral Sea

1957

100 km

1984

2000

N

2009

Boats of the former fishing fleet rust in the sand.

Aral Sea

FIGURE 21.28 An iconic photo of a farmer walking in a 1930s dust storm in Oklahoma. Much of the topsoil of the region blew away.

Why did the fertile soils of the southern Great Plains suddenly dry up? The causes were complex—some were natural and some were human-induced. The region, which has a semiarid climate, was settled in the 1880s and 1890s, which were unusually wet decades. Far more people moved into the region than it could sustain, and they farmed the land too intensively. Plowing destroyed the fragile grassland root systems that held the thin soil in place, so when the drought of the 1930s came, it brought catastrophe. Clearly, the Dust Bowl of the 1930s reminds us of how fragile the Earth's green blanket of vegetation can be. And while such desertification can be reversed by irrigation and planting, water to nourish the plants has to come from somewhere, and when people obtain it by diverting rivers or by pumping groundwater, they can generate a new set of problems.

Notably, not all droughts happen in desert regions or along the margins of deserts. For example, between 2011 and 2017, rains avoided California, so that its verdant agricultural areas turned brown and its high mountains did not receive a blanket of snow. Eventually, nearly all of the state was enduring extreme drought conditions (Fig. 21.29). As a result, streams, lakes, and reservoirs dried up, over 100 million trees died, and the water table dropped so much that the land surface sank. Farmers lost their crops, resorts lost their recreational attractions, city dwellers had to drain swimming pools and let lawns die, and salmon lost access to their spawning grounds. To address the emergency, the state government instituted mandatory water use restrictions. Unlike a region that has undergone desertification, however, California eventually saw its rains return. In fact, *atmospheric rivers* (high-altitude flows of very moist air) drenched the state in 2017, and some of the temporarily parched lands were submerged with floodwaters.

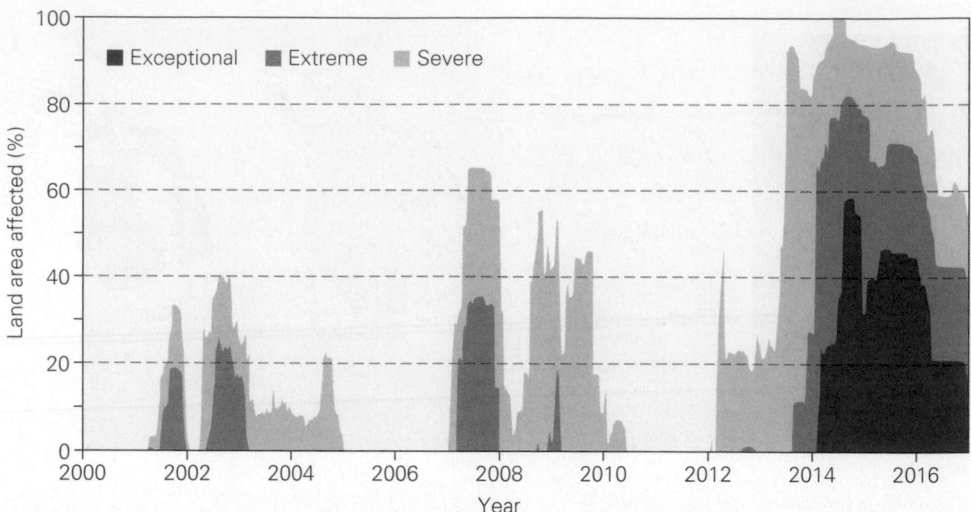

FIGURE 21.29 The percentage of California affected by drought since 2000. The 2011–2017 drought affected more area and lasted longer than any other.

On a time scale of millennia, global climate change can shift climate belts sufficiently to transform agricultural regions into deserts. For example, some 5,000 years ago, the swath of land between the Nile Valley in Egypt and the Tigris-Euphrates Valley of Mesopotamia was known as the Fertile Crescent. Here people first abandoned their nomadic ways and settled in agricultural communities. Now the original "land of milk and honey" hosts desert landscapes, and agriculture in the region requires intensive irrigation. The change in landscape reflects a change in climate—the beginning of Western civilization occurred during the warmest and wettest period since the last ice age. So much water drenched the Middle East and North Africa that rivers flowed where Saharan sands now blow. In fact, geologists using ground-penetrating radar have mapped abandoned river channels buried beneath the sand. Unfortunately, as we'll see in Chapter 23, if current trends in climate change continue, our present agricultural belts could someday become new Saharas.

Desertification has an additional dangerous side effect—global transmission of chemicals and pathogens by blowing dust. As desert areas expand in response to desertification, and as desert pavement gets disrupted, windblown dust becomes more of a problem. Not only do winds have larger areas of dry, dusty land to churn, but the dust generated from lands that were once agricultural and are now desert may contain harmful chemicals (such as residues of herbicides and pesticides), fungi, and microbes that can themselves become windborne. In recent years, satellite images have revealed that windblown dust from deserts can travel across oceans and affect regions on the other side. For example, dust blown off the Sahara can traverse the Atlantic and settle over the Caribbean (Fig. 21.30).

> ### TAKE-HOME MESSAGE
>
> Human populations in arid regions have been burgeoning. Droughts and stresses caused by agriculture, grazing, and water diversion have led to desertification. Dust from desertified regions can carry pollutants and microbes and may potentially cause problems elsewhere in the world.
>
> **QUICK QUESTION:** What factors transformed Oklahoma and Texas into the Dust Bowl during the 1930s?

FIGURE 21.30 In this satellite image, a huge dust cloud that originated in the Sahara blows across the Atlantic.

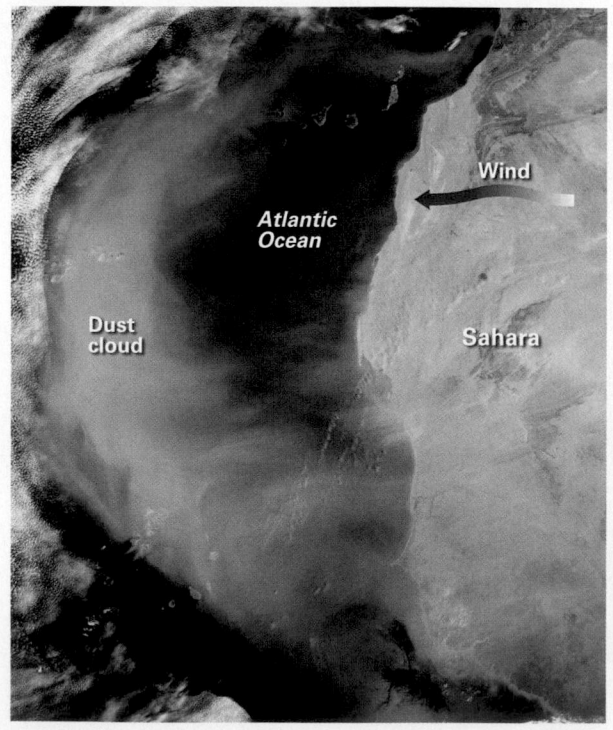

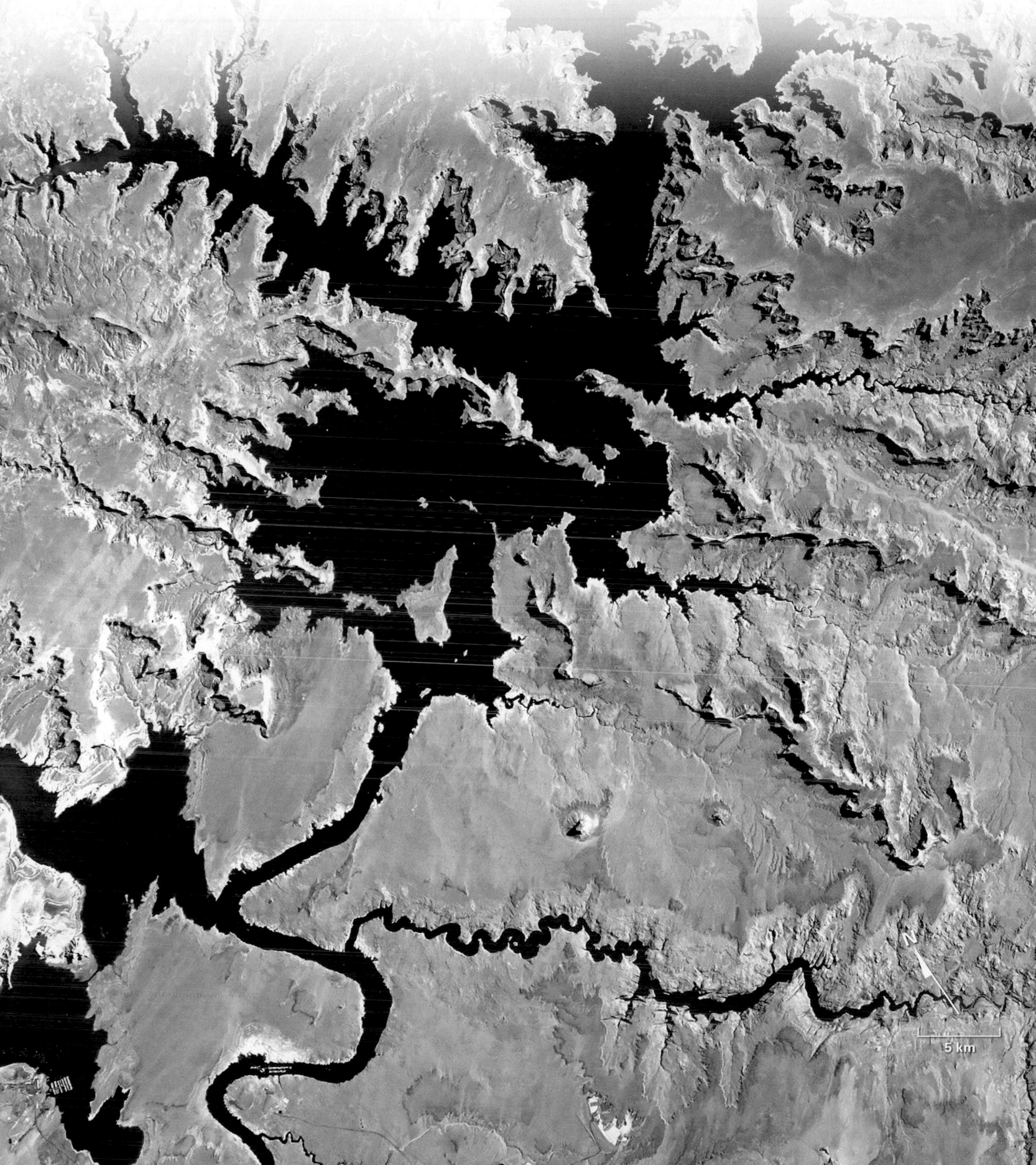

ANOTHER VIEW Construction of the Glen Canyon Dam across the Colorado River produced Lake Powell, whose clear blue water contrasts with the iron-oxide-stained barren rocks of the Colorado Plateau's desert in northern Arizona and southern Utah.

5 km

Chapter 21 Review

SUMMARY

- Deserts generally receive less than 25 cm of rain per year. Vegetation covers no more than 15% of their surface.

- Subtropical deserts form at latitudes of about 30°, rain-shadow deserts occur on the inland side of a mountain range, coastal deserts form on the land adjacent to cold ocean currents, continental-interior deserts exist in landlocked regions far from the ocean, and polar deserts develop at high latitudes.

- In deserts, chemical weathering happens slowly, so rock bodies tend to erode primarily by physical weathering. Desert varnish forms on rock surfaces, and soils tend to accumulate soluble minerals.

- Water causes significant erosion in deserts, mostly during heavy downpours. Flash floods carry large quantities of sediments down ephemeral streams. When the rain stops, these streams dry up, leaving dry washes.

- Wind can pick up dust and silt as suspended load and can cause sand to undergo saltation. Where wind blows away finer sediment, a lag deposit remains. Windblown sediment abrades the ground and sculpts ventifacts and yardangs.

- Talus piles form when rock fragments accumulate at the base of a slope. Alluvial fans form at a mountain front where water in ephemeral streams deposits sediment. When temporary desert lakes dry up, they leave playas.

- In some desert landscapes, erosion causes cliff retreat, eventually resulting in the formation of mesas, buttes, and inselbergs. Pediments of nearly flat or gently sloping bedrock surround some inselbergs.

- Desert pavements are mosaics of varnished stones armoring the surface of the ground.

- In places hosting abundant sand, the wind builds it into dunes. Common types include barchan, star, transverse, parabolic, and longitudinal (seif) dunes.

- Deserts contain a great variety of plant and animal species. All have adapted to survive extremes in temperature and without abundant water.

- Changing climates and land abuse may cause desertification, the transformation of semi-arid land into deserts. Wind-blown dust, sometimes carrying microbes and toxins, may waft from deserts across oceans.

GUIDE TERMS

alluvial fan (p. 812)
arroyo (p. 808)
butte (p. 816)
chimney (p. 816)
cliff retreat (p. 814)
deflation (p. 809)
desert (p. 803)

desertification (p. 823)
desert pavement (p. 818)
desert varnish (p. 807)
dry wash (p. 808)
dune (p. 813)
dust storm (p. 809)
lag deposit (p. 809)

loess (p. 813)
mesa (p. 816)
pediment (p. 817)
petroglyph (p. 807)
playa (p. 813)
rain shadow (p. 805)
sand dune (p. 819)

slip face (p. 820)
surface load (p. 808)
suspended load (p. 808)
talus (p. 812)
ventifact (p. 811)
wadi (p. 808)

GEOTOURS *THIS CHAPTER'S GEOTOURS WORKSHEET (Q) FEATURES QUESTIONS AND GOOGLE EARTH SITES ON:*

- Deserts and desert features from around the world

- Water use in arid regions

REVIEW QUESTIONS

The letters following each Review Question refer to the corresponding Learning Objective from the Chapter Opener.

1. What factors determine whether a region can be classified as a desert? **(A)**

2. Explain several settings that can cause deserts to form. What type of desert is depicted in the figure? Explain how it formed. **(A)**

3. Have today's deserts always been deserts? **(A)**

4. How do weathering processes in deserts differ from those in temperate or humid climates? **(B)**

5. Describe how water modifies the landscape of a desert. Be sure to discuss both erosional and depositional landforms. **(B)**

6. Explain the ways in which desert winds transport sediment. **(B)**

7. Explain how the following features form: (a) desert varnish, (b) desert pavement, (c) ventifacts, and (d) yardangs. **(C)**

8. Describe the process of formation of alluvial fans, pediments, bajadas, and playas. **(C)**

9. Describe the process of cliff (scarp) retreat and the landforms that result from it. **(C)**

10. What are the various types of sand dunes? What factors determine which type of dune develops at a particular location? Which type is pictured here? **(C)**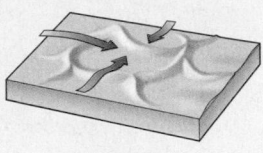

11. Discuss various adaptations that life forms have evolved in order to survive in desert climates. **(D)**

12. What is the process of desertification, and what causes it? How can desertification in Africa affect the Caribbean? **(E)**

ON FURTHER THOUGHT

13. Death Valley, California, lies to the east of a high mountain range, and its floor lies below sea level. During the summer, Death Valley becomes very hot and dry. Explain why it has such weather. **(A)**

14. You are working for an international nongovernmental organization and have been charged with the task of providing recommendations to an African nation that wishes to slow or halt the process of desertification within its borders. What are your recommendations? **(E)**

15. The Namib Desert lies to the north and west of the Kalahari Desert, in southern Africa. The two deserts formed for different reasons. Explain this statement. **(A)**

ONLINE RESOURCES

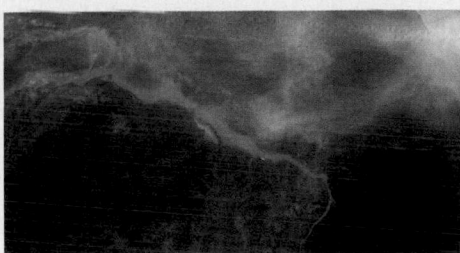

Videos
This chapter includes videos on the evolution of deserts and the transport of Saharan dust to the Amazon.

Smartwork5
This chapter includes labeling exercises about desert landscapes and the different types of deserts.

CHAPTER 22

Amazing Ice: Glaciers and Ice Ages

By the end of this chapter, you should be able to . . .

A. describe how glacial ice forms and flows, and how to categorize various kinds of glaciers.

B. explain why glaciers advance and retreat, and how their flow modifies the landscape.

C. recognize sedimentary deposits and associated landforms left by glaciers.

D. describe evidence showing that glaciers covered large areas of continents during the Pleistocene Ice Age, and that there have been four or five earlier ice ages during Earth history.

E. discuss why ice ages happen and why glaciations during an ice age occur periodically.

22.1 Introduction

There's nothing like a good mystery, and one of the most puzzling in the annals of geology came to light in northern Europe early in the 19th century. When farmers of the region prepared their land for spring planting, they occasionally broke their plows on large boulders that were buried randomly throughout otherwise fine-grained sediment. Many of these boulders did not consist of local bedrock, but rather came from outcrops hundreds of kilometers away. Because the boulders had apparently traveled so far, they came to be known as **erratics** (from the Latin *errare*, to wander).

The mystery of the wandering boulders became a subject of great interest to early 19th-century geologists, who realized that deposits of extremely *unsorted sediment*, meaning sediment that contains a great variety of different clast sizes, could not be examples of typical stream alluvium, for running water sorts clasts by size. Most attributed the deposits to a vast flood, during which a slurry of boulders, sand, and mud supposedly spread across the continent. In 1837, however, a young Swiss geologist named Louis Agassiz proposed a radically different interpretation. Agassiz often hiked in the Alps near his home, where he could study **glaciers**, rivers or sheets of ice that last all year long and flow slowly under the influence of gravity. He observed that glacial ice could carry enormous boulders as well as sand and mud because solid ice has enough strength to support the weight of rock. Agassiz realized that, because ice does not sort sediment as it flows, glaciers leave behind extremely unsorted sediment when they melt. On the basis of this observation, he proposed that the mysterious erratic-laden sediments of Europe were deposits left by *continental ice sheets*, vast glaciers that cover large areas of a continent **(Fig. 22.1)**. In Agassiz's view, Europe had once been in the grip of an **ice age**, a time when the climate was significantly colder and glaciers grew to be immensely larger than they are today.

Agassiz's radical proposal faced intense criticism for the next two decades. But he didn't back down, and instead challenged opponents to visit the Alps and examine the

sedimentary deposits that Alpine glaciers had left behind. By the late 1850s, most doubters had changed their minds, and the geological community concluded that the notion that Europe once had Arctic-like climates was correct. Later in life, Agassiz traveled to the United States and documented many glacier-related features in North America's landscape, proving that an ice age had affected vast areas of the planet.

Glaciers cover only about 10% of the land on the Earth today, but during the most recent ice age, which ended less than 12,000 years ago, as much as 30% of the continental land surface had a coating of ice. New York City, Montreal, and many of the great cities of Europe occupy land that once lay beneath hundreds of meters to a few kilometers of ice. The work of Louis Agassiz brought the subject of glaciers and ice ages into the realm of geologic study and led people

FIGURE 22.1 Agassiz's thoughts about the ice age.

(a) Agassiz found boulders protruding from the ground in places that are not currently glaciated. He proposed that the boulders are erratics left by now-vanished glaciers.

(b) Agassiz envisoned that vast areas of the northern hemisphere were once covered by vast ice sheets comparable to the one covering Antarctica today.

◀ (facing page) Glaciers imperceptibly flow down valleys in the mountains of southern Alaska. The rugged topography of many high mountain ranges has been sculpted by glaciers.

to recognize that major climate changes have happened during the Earth's history. In this chapter, after considering the nature of ice, we see how glaciers form, why they move, and how they modify landscapes by erosion and deposition. A substantial portion of our discussion concerns the most recent ice age, known as the **Pleistocene Ice Age**, whose impact on the landscape can still be seen today. We briefly introduce ice ages that happened earlier in Earth history, too. We conclude by considering hypotheses to explain why ice ages happen.

22.2 Ice and the Nature of Glaciers

What Is Ice?

Ice consists of solid water, formed when liquid water cools below its freezing point. We can consider a single ice crystal to be analogous to a mineral specimen: it's a naturally occurring, inorganic solid with a definite chemical composition (H_2O) and a regular crystal structure. Ice crystals have a hexagonal form, so snowflakes grow into six-pointed stars (**Fig. 22.2a**). We can think of a layer of fresh snow as a layer of sediment, and of a layer of snow that has been compacted so that its grains stick together as a bed of sedimentary rock (**Fig. 22.2b**). We can view a coating of ice that appears on the surface of a pond in winter as an igneous rock, for it forms when molten ice—liquid water—solidifies. Glacial ice, in this context, is a metamorphic rock. It develops when pre-existing ice recrystallizes, meaning that the molecules in solid water rearrange to form new crystals (**Fig. 22.2c**).

Pure new ice has the transparency of glass, but if ice contains tiny air bubbles or cracks that disperse light, it becomes milky white. Glacial ice tends to have a bluish color because it absorbs red light. Like glass, ice has a high **albedo**, meaning that it reflects light well—so well, in fact, that if you walk on ice without eye protection, you risk blindness from the glare. Ice differs from most other familiar materials in that its solid form is not as dense as its liquid form, for the architecture of an ice crystal holds water molecules apart. Ice, therefore, floats on water. This unusual characteristic prevents the oceans from freezing solid when it gets cold. If ice didn't float, ice in oceans would sink, leaving room for new ice to form above.

Ice also has the unusual property of being slippery—that's why skaters can skate! Surprisingly, researchers still don't completely understand this property. An older explanation—that skaters can glide on ice because a film of liquid water forms on the surface of the ice beneath their skates in response to frictional heating or pressure-induced melting—can't explain how ice remains slippery even when it's so cold that water can't exist as liquid. Modern studies suggest that ice remains slippery at very low temperatures because the surface of ice consists of a layer of water molecules that are not completely fixed within a crystal lattice. The existence of unattached bonds permits the surface molecules to behave somewhat like a liquid, even while attached to the solid.

How Does a Glacier Form?

In order for a glacier to form, four conditions must be met. First, the local climate must be sufficiently cold that winter snow does not melt away entirely during the summer; second, there must be or must have been sufficient snowfall for a large amount of snow to accumulate; third, the surface on which the snow accumulates must have a gentle enough slope so that snow falling on it does not slide away in avalanches; and fourth, the area where snow falls must be protected enough so that snow doesn't blow away.

Glaciers develop in polar regions because even though relatively little snow falls there today, temperatures remain so low that most ice and snow survive all year. Glaciers develop in mountains, even at low latitudes, because temperature decreases with elevation. So, at high elevations, the mean temperature stays cold enough for ice and snow to survive all year. Since the temperature of a region depends on latitude, the specific elevation at which glaciers form in mountains depends on latitude. In the Earth's present-day climate, glaciers form only at elevations above 5 km at latitudes between 0° and 30°, but they can flow down to sea level at latitudes of between 60° and 90°. Therefore, you can see high-latitude glaciers from a cruise ship, but at the equator you have to climb high into the mountains to find glaciers. Mountain glaciers tend to develop on the side of a mountain that receives less wind and on the side that receives less sunlight. Glaciers do not form on slopes greater than about 30° because avalanches clear such slopes.

The transformation of snow to glacial ice takes place slowly as younger snow progressively buries older snow. Freshly fallen snow consists of delicate hexagonal crystals with sharp points. The crystals do not fit together tightly, so this snow contains about 90% air. Over time, the points of the snowflakes become blunt because they either *sublimate* (evaporate directly into vapor) or melt, and the snow packs more tightly. As the snow becomes buried, the weight of the overlying snow increases pressure on it, which causes any remaining points of contact between snowflakes to melt. This process of melting at points of contact, where the pressure is greatest, is an example of *pressure solution* (see Chapter 8). Gradually, the snow transforms into a packed granular material called **firn**, which contains only about 25% air (**Fig. 22.2d**). Melting of firn grains at contact points produces water that crystallizes

FIGURE 22.2 The nature of ice and the formation of glaciers. Snow falls like sediment and metamorphoses to ice when buried.

(a) The hexagonal shape of snowflakes. No two are alike.

A boundary between layers

The layers in the photo at left are part of this glacier in the Alps.

(b) Layers of snow accumulate. They recrystallize to become ice.

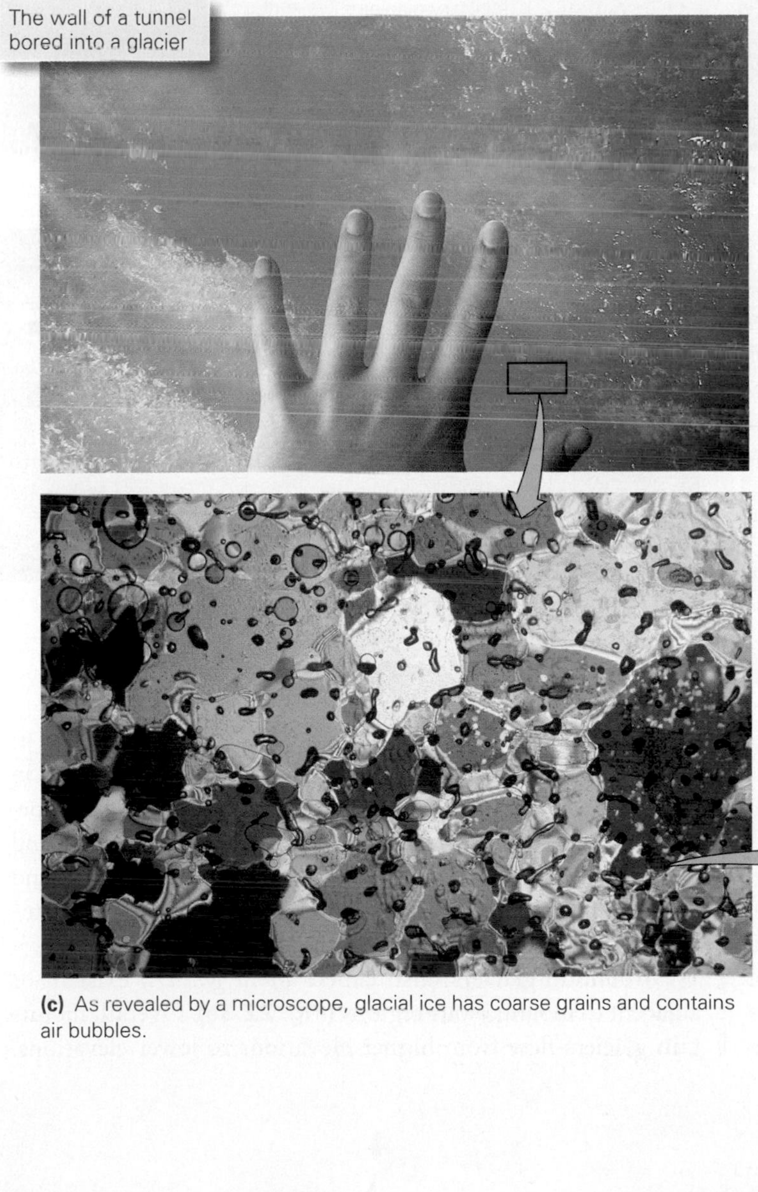

The wall of a tunnel bored into a glacier

Loose snow (90% air)

Granular snow (50% air)

Firn (25% air)

Fine-grained ice (<20% air, in bubbles)

Coarse-grained ice (<20% air, in bubbles)

10,000 years (250 m)

130,000 years (2,000 m)

Not to scale

(c) As revealed by a microscope, glacial ice has coarse grains and contains air bubbles.

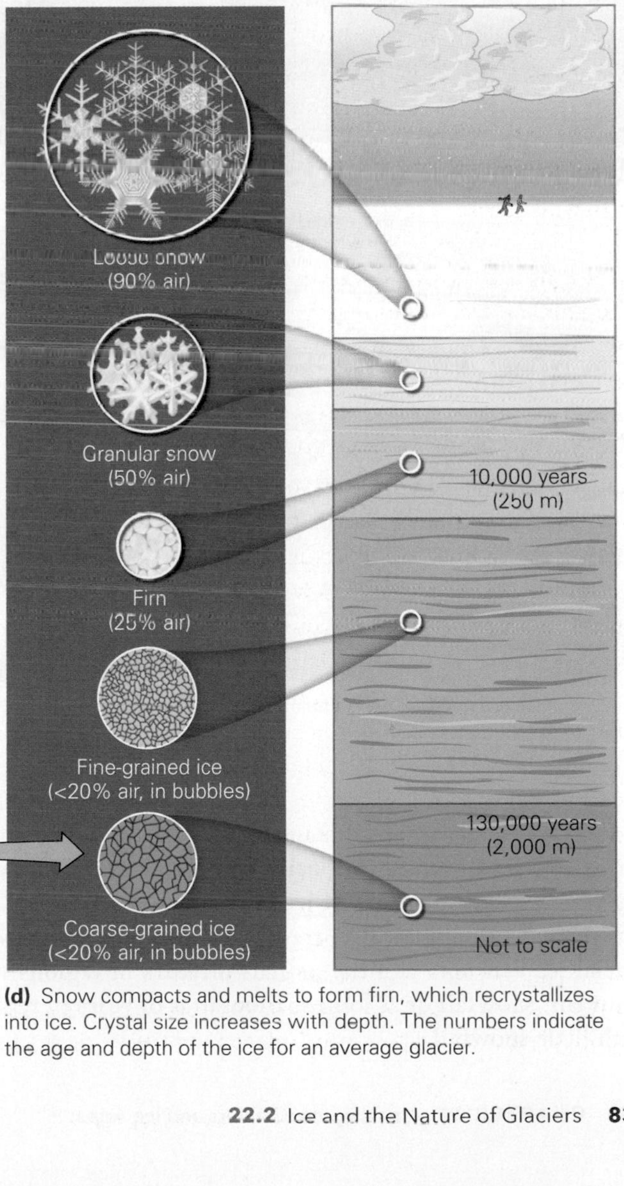

(d) Snow compacts and melts to form firn, which recrystallizes into ice. Crystal size increases with depth. The numbers indicate the age and depth of the ice for an average glacier.

FIGURE 22.3 Several types of glaciers form in mountainous areas.

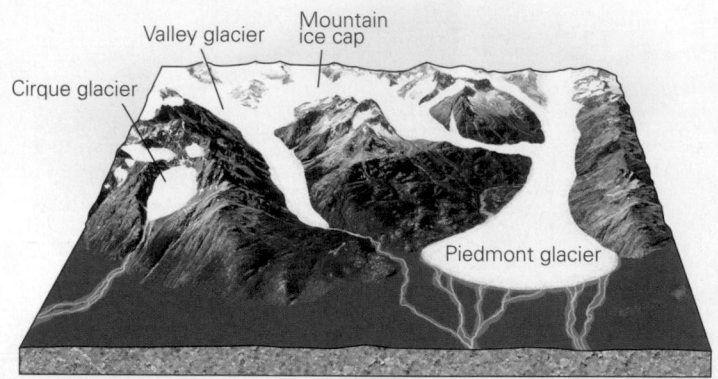

(a) Mountain glaciers are classified based on shape and position.

(d) A valley glacier and cirque glaciers in Switzerland.

(b) A mountain ice cap in Alaska.

(e) A large trunk valley glacier and tributary glaciers in Pakistan.

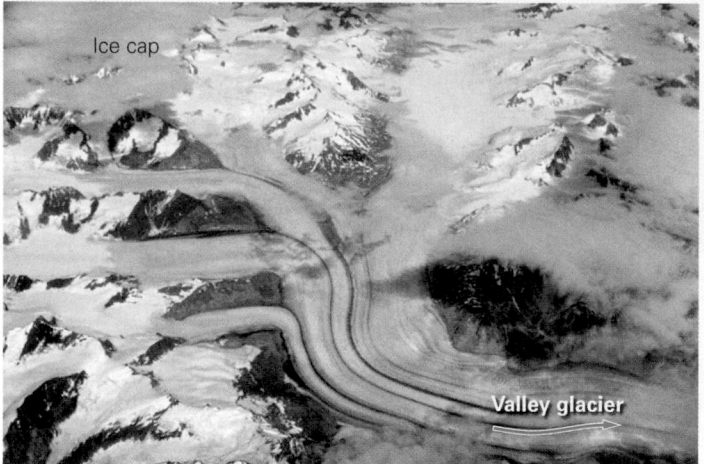

(c) Valley glaciers draining a mountain ice cap in Alaska.

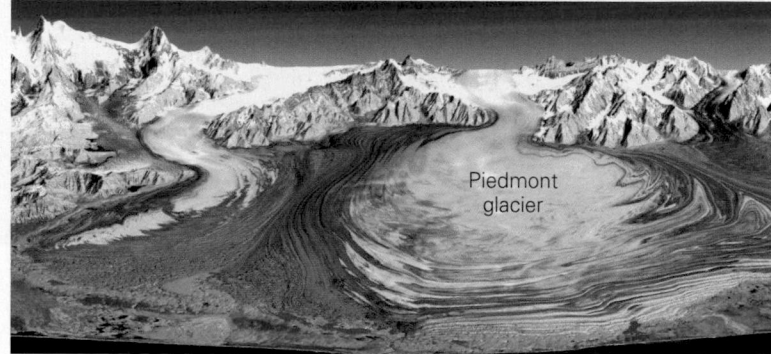

(f) A piedmont glacier near the coast of Alaska.

in the spaces between grains until eventually, the firn transforms into a solid mass of glacier ice composed of interlocking ice crystals. Such glacial ice may still contain up to 20% air trapped in bubbles. The transformation of fresh snow to glacial ice can take as little as tens of years in regions with abundant snowfall or as long as thousands of years in regions with little snowfall.

Categories of Glaciers

Today, glaciers highlight coastal and mountain scenery in Alaska, the North American Cordillera, the Alps of Europe, the Southern Alps of New Zealand, the Himalayas of Asia, and the Andes of South America, and they cover most of Greenland and Antarctica. Geologists distinguish between two main categories: mountain glaciers and continental glaciers.

Mountain glaciers (also called *alpine glaciers*) exist in or adjacent to mountainous regions **(Fig. 22.3a)**. Overall, mountain glaciers flow from higher elevations to lower elevations.

FIGURE 22.4 Two major continental glaciers exist today—one in Antarctica and one in Greenland.

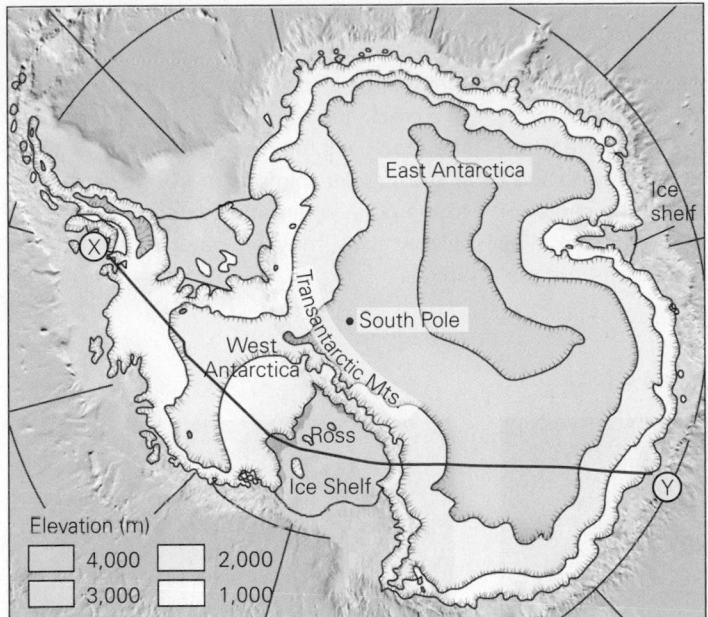

(a) A contour map of the Antarctic ice sheet. Valley glaciers carry ice from the ice sheet of East Antarctica down to the Ross Ice Shelf.

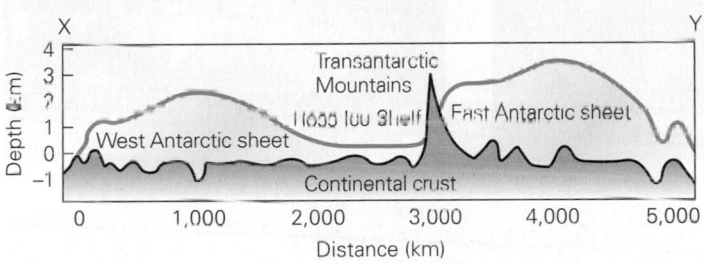

(b) A cross section X to Y of the Antarctic ice sheet. The Transantarctic Mountains separate East Antarctica from West Antarctica.

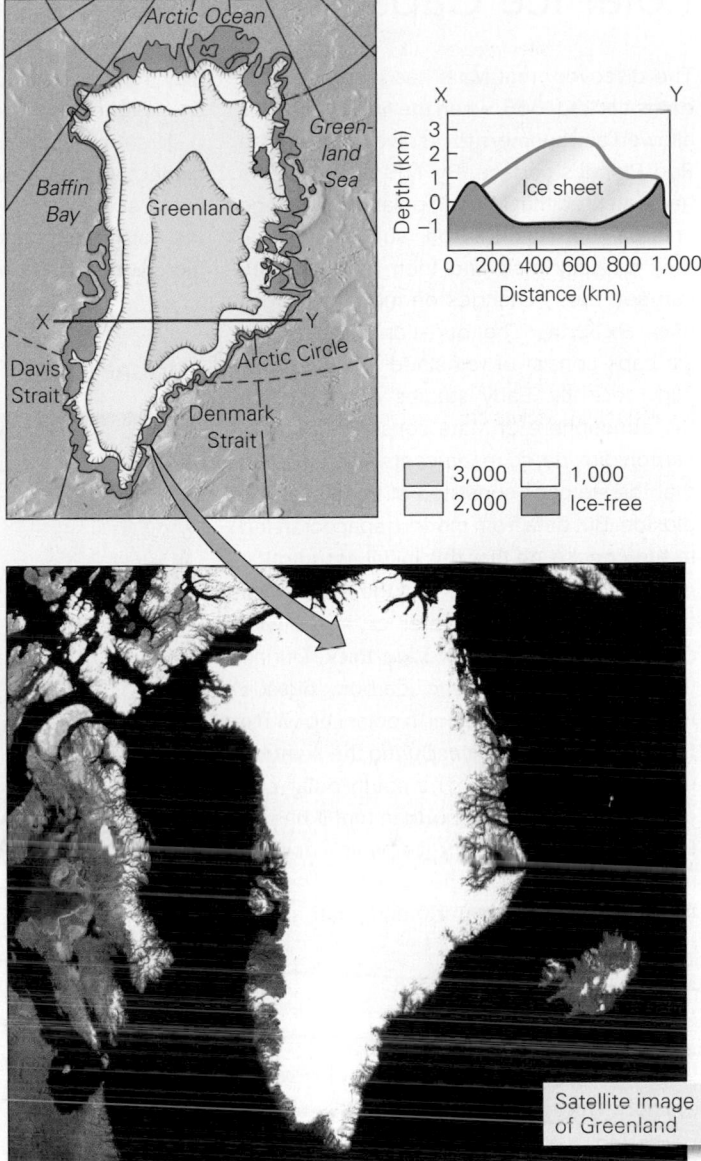

Satellite image of Greenland

(c) Greenland is also covered by an ice sheet that at its thickest is 1 km thinner than Antarctica's.

Mountain glaciers include *cirque glaciers*, which fill bowl-shaped depressions, or *cirques*, on the flank of a mountain; *mountain ice caps*, mounds of ice that submerge peaks and ridges at the crest of a mountain range; *valley glaciers*, rivers of ice that flow down valleys; and *piedmont glaciers*, fans or lobes of ice that form where a valley glacier emerges from a valley and spreads out into the adjacent plain **(Fig. 22.3b–f)**. Mountain glaciers range in size from a few hundred meters to a few hundred kilometers long.

Continental glaciers are vast ice sheets that spread over thousands of square kilometers of continental crust. Today, they exist only in Antarctica and Greenland **(Fig. 22.4)**, but during ice ages they have covered other continental areas. Keep in mind that Antarctica is a continent, so the ice beneath the South Pole rests mostly on solid ground. (We say *mostly* because new research reveals that at least three lakes underlie the glacier—the largest of these, Lake Vostok, has an area of 5,400 km².) The ice beneath the North Pole, in contrast, forms part of a thin sheet of sea ice floating on the Arctic Ocean.

Continental glaciers flow outward from their thickest point (up to 3.5 km thick) and thin toward their margins, where they may be only a few hundred meters thick. The front edge of the glacier may divide into several tongue-shaped lobes because not all of the glacier flows at the same speed. Notably, the Earth is not alone in hosting polar ice sheets—Mars has them too **(Box 22.1)**.

Geologists also find it valuable to distinguish between types of glaciers based on thermal conditions in the glacier. **Temperate glaciers** exist where atmospheric temperatures become warm enough for the glacial ice to be at or near its melting temperature during part or all of the year. In such glaciers, liquid water forms films between ice grains in the

BOX 22.1 CONSIDER THIS . . .

Polar Ice Caps on Mars

The discovery that Mars has polar ice caps dates back to 1666, when the first telescopes allowed astronomers to resolve details of the Red Planet's surface. By 1719, astronomers had detected that Martian polar caps change in area with the season, suggesting that they partially melt and then refreeze. You can see these changes on modern images **(Fig. Bx22.1a)**. The question of what the ice caps consist of remained a puzzle until fairly recently. Early studies revealed that the atmosphere of Mars consists mostly of carbon dioxide, so researchers first assumed that the ice caps consisted of frozen carbon dioxide. But data from modern spacecraft led to the conclusion that this initial assumption is wrong. It now appears that the Martian ice caps consist mostly of water ice, mixed with dust, in layers from 1 to 3 km thick. During the winter, atmospheric carbon dioxide freezes and covers the north polar cap with a 1-m-thick layer of dry ice. During the summer this layer melts away. The south polar cap differs from that of the north in that it has an 8-m-thick blanket of dry ice, which doesn't melt away entirely in the summer. The difference between the north and south poles may reflect elevation—the south pole sits 6 km higher and therefore remains colder.

High-resolution photographs reveal that distinctive canyons, up to 10 km wide and 1 km deep, spiral outward from the center of the north polar ice cap **(Fig. Bx22.1b)**. Why did this pattern form? Calculations suggest that if the ice sublimates on the sunny side of a crack and refreezes on the shady side, the crack will migrate sideways over time. If the cracks migrate more slowly closer to the pole, where it's colder, than they do farther away, they will naturally evolve into spirals.

FIGURE Bx22.1 The ice caps of Mars.

 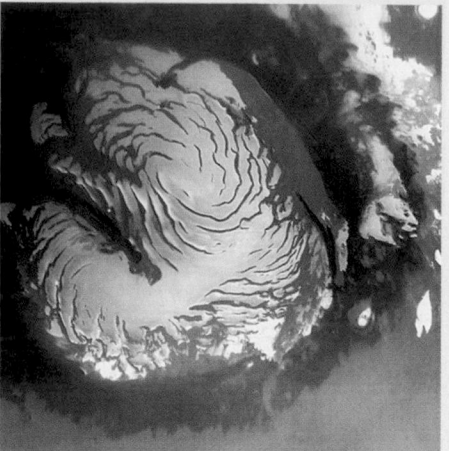

(a) During the winter, the ice caps expand to lower latitudes.

(b) A close-up of the north polar cap in summer.

glacier or collects in lenses and streams at the base of the glacier. Because of this water, a temperate glacier can also be called a *wet-based glacier*. **Polar glaciers** occur in regions where atmospheric temperatures stay so cold all year long that the glacial ice remains below melting temperature throughout the year—they are solid ice through and through. Geologists may also refer to a polar glacier as a *dry-based glacier*, because no liquid water collects at the base of the glacier.

The Movement of Glacial Ice

When Louis Agassiz became fascinated by glaciers, he decided to find out how fast the ice in them moved, so he hammered stakes into an Alpine glacier and measured how the stakes change position during the course of a year. More recently, researchers have observed glacial movement with the aid of time-lapse photography, which shows the evolution of a glacier over several years in a movie that lasts a few minutes. In such movies, the glacial ice seems to flow across the screen. How does this movement occur? Geologists have found that glacial flow involves two mechanisms—plastic deformation and basal sliding.

At conditions found below depths of about 60 m in a glacier, ice deforms by *plastic deformation*, meaning that the grains within it change shape very slowly or that new grains grow while old ones disappear **(Fig. 22.5a, b)**. Simplistically, we can picture such changes to be a consequence of the rearrangement of water molecules within a crystal lattice as some chemical bonds break and new ones form. If ice becomes warm enough for thin water films to form along grain boundaries, plastic deformation may also involve the microscopic slip of ice grains past their neighbors along water films.

When liquid water collects at the base of a glacier, it can occur as a lens of liquid between ice and bedrock. Commonly, however, it mixes with subglacial sediment to form a slurry. The presence of liquid water or a wet slurry beneath a glacier allows the glacier to move by **basal sliding**. During this process, friction or bonding between the ice and its substrate has

Did you ever wonder . . .
how a glacier moves?

FIGURE 22.5 Mechanisms of glacial movement.

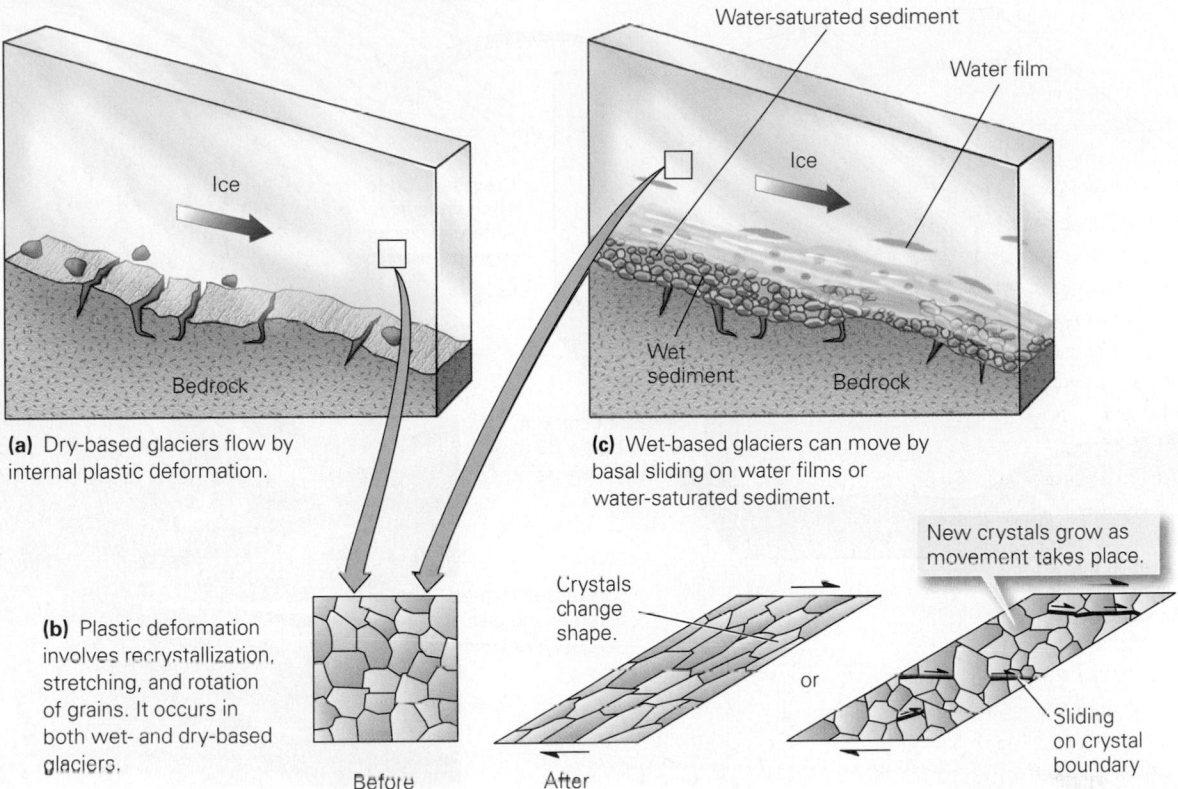

(a) Dry-based glaciers flow by internal plastic deformation.

(c) Wet-based glaciers can move by basal sliding on water films or water-saturated sediment.

(b) Plastic deformation involves recrystallization, stretching, and rotation of grains. It occurs in both wet- and dry-based glaciers.

diminished, and the glacier effectively glides along on a wet cushion (Fig. 22.5c).

The liquid water under a wet-based glacier forms when sunlight and atmospheric warming heat the glacier sufficiently to produce meltwater on the surface of the glacier. Recent studies show that surface meltwater ponds may drain in a matter of minutes to hours into cracks or tunnels that provide a conduit between the surface and the base of the glacier. Melting may also occur inside the glacier due to the trapping of heat rising from the ground beneath the glacier (for ice is an insulator) or due to the weight of overlying ice (for at elevated pressures, ice can melt even if its temperature remains below 0°C).

In the case of polar glaciers, which are so cold that they have no internal water films and have a dry base, flow takes place generally by plastic deformation alone. In the case of temperate glaciers, which contain some intergranular water or have a wet base, flow can involve both plastic deformation and basal sliding. Not all parts of a given glacier necessarily flow in the same way. For example, imagine a continental glacier that originates in a very cold polar realm but eventually flows into temperate realms at lower latitudes. Near its cold origin, the glacier has a dry base and moves only by plastic deformation, but near its warmer margin, it becomes wet based and moves by both plastic deformation and basal sliding.

As we noted earlier, plastic deformation takes place only at depths greater than about 60 m in a glacier—above this depth, known as the *brittle–plastic transition*, ice remains too brittle to

flow. (Note that, by comparison, plastic deformation in silicate rocks of the Earth's crust occurs primarily under metamorphic temperatures greater than about 300°C, so the brittle–plastic transition in continental crust occurs at depths of about 10 to 15 km; see Chapter 11.) As a glacier overall undergoes movement, its upper 60 m of ice deforms predominantly by cracking. An open crack that develops by brittle deformation of a glacier is called a **crevasse** (Fig. 22.6). In large glaciers, crevasses can be hundreds of meters long and tens of meters deep, and they may open into gashes with a width of several meters. Tragically, explorers, hikers, and skiers have died by falling into crevasses, whose openings sometimes become covered by a bridge of weak, windblown snow. Crevasse formation typically localizes in regions where a glacier flows over steps or hills in the underlying bedrock surface, for the ice of the glacier must bend to accommodate the surface shape of the substrate.

Why do glaciers move? Ultimately, they move because the pull of gravity exceeds the strength of ice and can cause the ice to flow (Fig. 22.7a). A glacier flows in the direction in which its top surface slopes. Therefore, valley glaciers flow down their valleys, and continental ice sheets spread outward from their thickest point. Note that it's the slope of the top surface that matters in driving ice forward—at its base, ice can flow up and over hills or ridges in the substrate.

To picture the movement of an ice sheet, imagine that a thick pile of ice builds up. Gravity causes the top of the pile to push down on the ice at the base. Eventually, the basal ice can

FIGURE 22.6 Crevasses form in the upper layer of a glacier, in which the ice is brittle. Commonly, cracking takes place where the glacier bends while flowing over steps or ridges in its substrate.

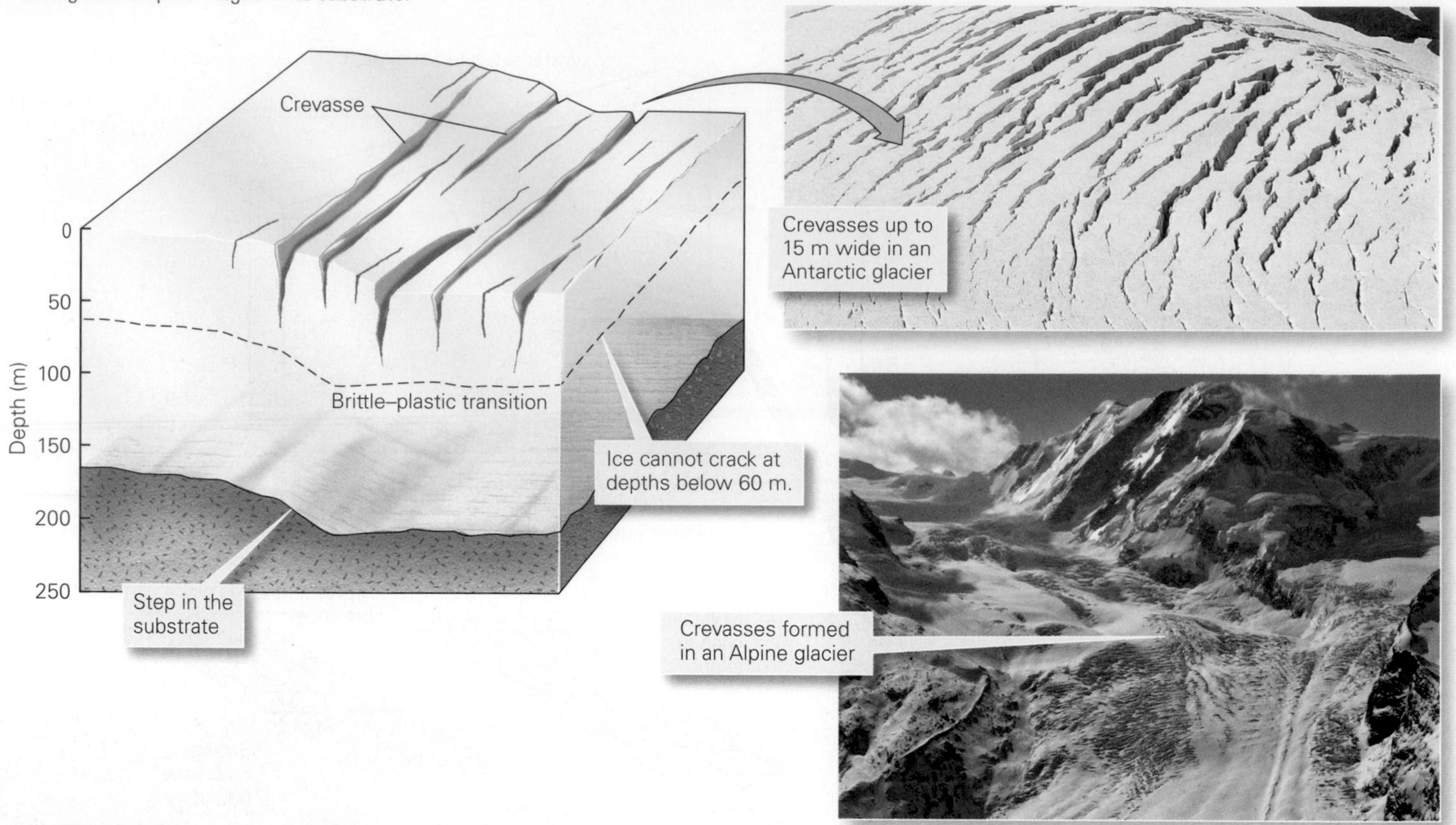

Crevasse
0
50
100
150
200
250
Depth (m)
Brittle–plastic transition
Ice cannot crack at depths below 60 m.
Step in the substrate
Crevasses up to 15 m wide in an Antarctic glacier
Crevasses formed in an Alpine glacier

FIGURE 22.7 Forces that drive the movement of glaciers.

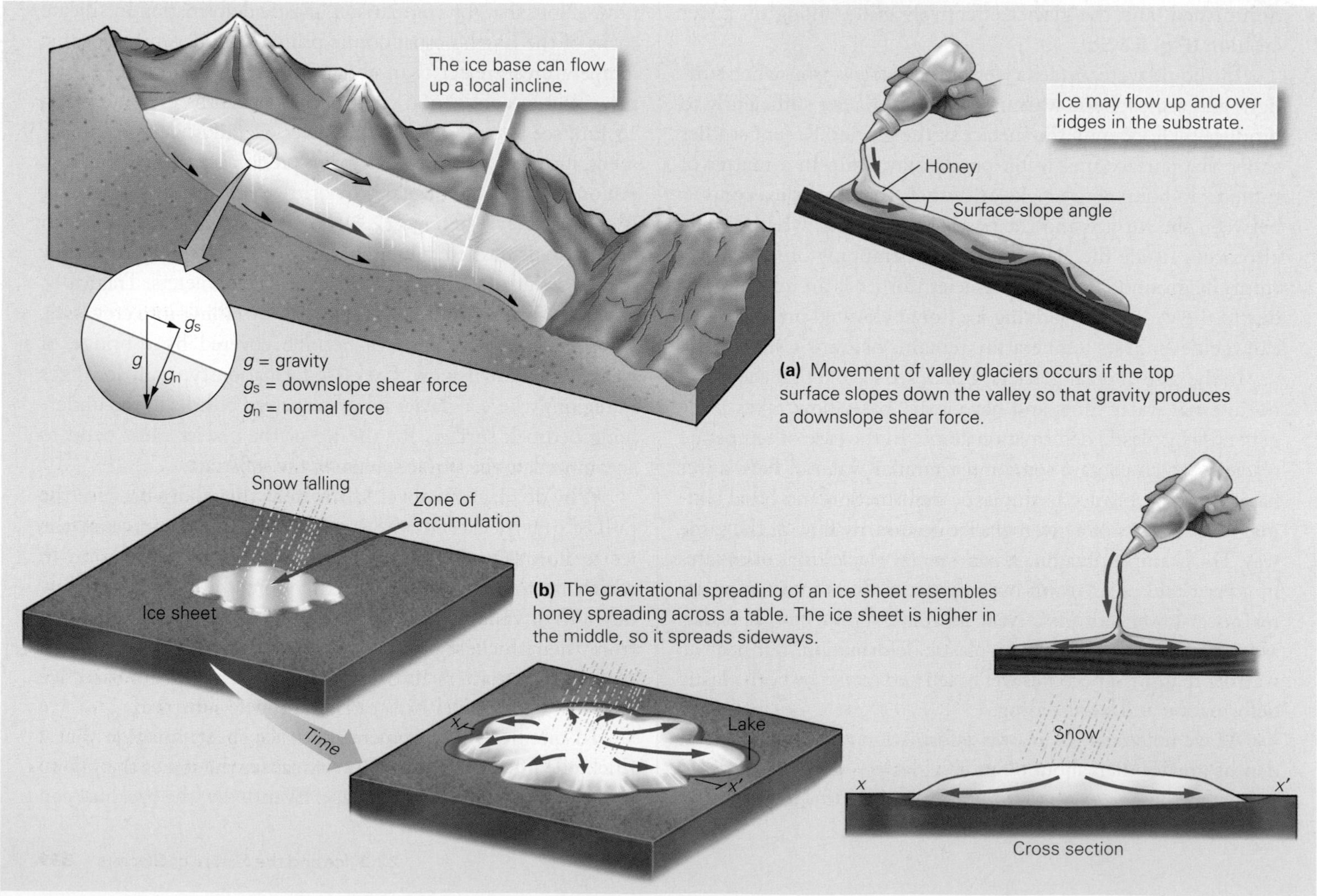

The ice base can flow up a local incline.
Ice may flow up and over ridges in the substrate.
Honey
Surface-slope angle
g_s
g
g_n
g = gravity
g_s = downslope shear force
g_n = normal force
(a) Movement of valley glaciers occurs if the top surface slopes down the valley so that gravity produces a downslope shear force.
Snow falling
Zone of accumulation
Ice sheet
(b) The gravitational spreading of an ice sheet resembles honey spreading across a table. The ice sheet is higher in the middle, so it spreads sideways.
Time
Lake
x
x′
Snow
Cross section

FIGURE 22.8 Glacial flow, accumulation, and ablation.

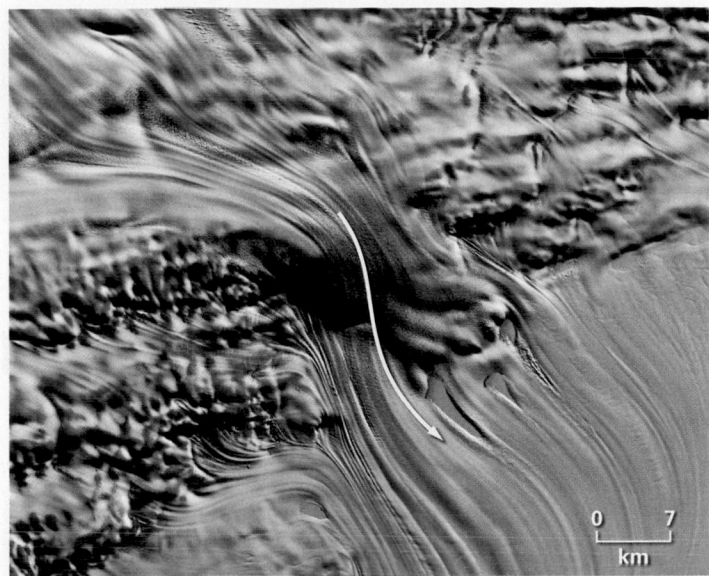

(a) A satellite image of ice flowing from the Polar Plateau of Antarctica down a 400-m-high ice fall to the Lambert Glacier. The curving line indicates the flow direction.

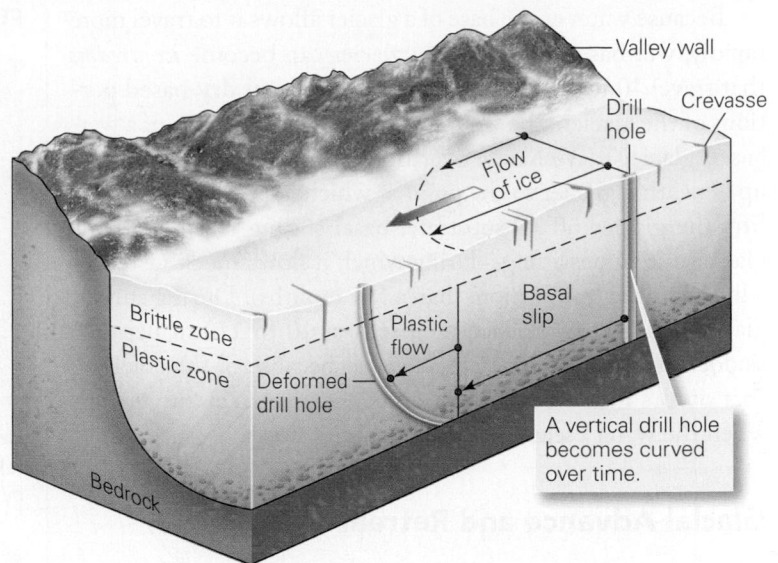

(b) Different parts of a glacier flow at different velocities due to friction with the substrate. The top and center regions flow fastest.

(c) Blocks of blue glacial ice, which calved off a glacier in the Alps.

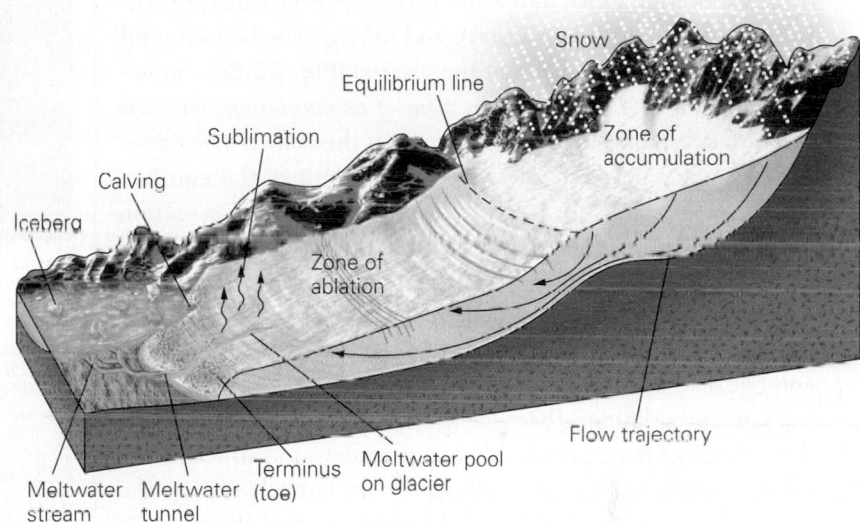

(d) The equilibrium line separates the zone of accumulation from the zone of ablation. Arrows indicate the flow path of ice within the glacier.

no longer support the weight of the overlying ice and begins to deform plastically or slide on its substrate. When this happens, the basal ice starts squeezing out to the side, carrying the overlying ice with it. The greater the volume of ice that builds up, the wider the sheet of ice can become. You've seen a similar process of *gravitational spreading* if you've ever poured honey onto a plate. The honey can't build up into a narrow column because it's too weak; rather, it flows laterally away from the point where it lands to form a wide, thin layer (Fig. 22.7b). The more honey you add, the farther it spreads.

Glaciers generally flow at rates between 10 and 300 m per year—far slower than a river but far faster than a silicate rock, even under high-grade metamorphic conditions. The velocity of a particular glacier depends, in part, on the magnitude of the force driving its motion. For example, a glacier whose surface slopes steeply moves faster than one with a gently sloping surface (Fig. 22.8a). Flow velocity also depends on the condition of the ice—temperate glaciers tend to move faster than polar glaciers because wet ice has less strength. Not all parts of a glacier move at the same rate. For example, friction or bonding between rock and ice slows a glacier, so the center of a valley glacier moves faster than its margins, and the top of a glacier moves faster than its base (Fig. 22.8b).

Because water at the base of a glacier allows it to travel more rapidly, wet-based portions of a glacier can become *ice streams* that travel 10 to 100 times faster than adjacent dry-based portions of the glacier. The volume of water at the bottom of a wet-based glacier may change over time. If water suddenly builds up beneath a glacier to the point at which a large area of water lifts the glacier off its substrate, basal sliding starts, and the glacier undergoes a *surge* during which it flows much faster for a limited time (rarely more than a few months). During surges, glaciers have been clocked at speeds of 10 to 110 m per day! Sudden surges may generate *ice quakes* because of the cracking that occurs in the brittle portion of the glacier. A surge stops when the water escapes and basal sliding slows.

Glacial Advance and Retreat

Glaciers resemble bank accounts. Snowfall adds to the account, while **ablation**—the removal of ice—subtracts from the account. Ablation involves three processes: *sublimation* (the evaporation of ice into water vapor); *melting* (the transformation of ice into liquid water); and *calving* (the breaking off of chunks of ice at the end of the glacier) **(Fig. 22.8c)**. Snowfall adds ice to a glacier in the *zone of accumulation*, whereas ablation subtracts ice from the glacier in the *zone of ablation*—the boundary between these two zones defines the **equilibrium line (Fig. 22.8d)**. Note that, as viewed in a cross-section plane parallel to the length of the glacier, ice follows an overall concave-up curving path inside the glacier. Why? Ice always has a component of movement toward the toe. But because of burial, ice moves down beneath the zone of accumulation, and because of sublimation or melting, ice moves up beneath the zone of ablation. The zone of accumulation occurs where the temperature remains cold enough year-round so that winter snow does not melt or sublimate away entirely during the summer. Therefore, elevation and latitude control the position of the equilibrium line.

The position of a glacier's *toe*, its leading edge or margin, remains fixed when the rate of accumulation equals the rate of ablation **(Fig. 22.9a)**. If the rate at which ice builds up in the zone of accumulation exceeds the rate at which ablation occurs below the equilibrium line, then the toe moves forward into previously unglaciated regions, a change called a **glacial advance (Fig. 22.9b)**. In mountain glaciers, the position of a toe moves downslope during an advance, and in continental glaciers, the toe moves outward, away from the glacier's origin. But if the rate of ablation exceeds the rate of accumulation, then the position of the toe moves back toward the origin of the glacier—such a change is called a **glacial retreat (Fig. 22.9c)**. During a mountain glacier's retreat, the position of the toe moves upslope. Note that when a glacier retreats, it's only the position of the toe that moves back toward the origin. Even during glacial retreat, ice continues to flow toward

FIGURE 22.9 Glacial advance and retreat.

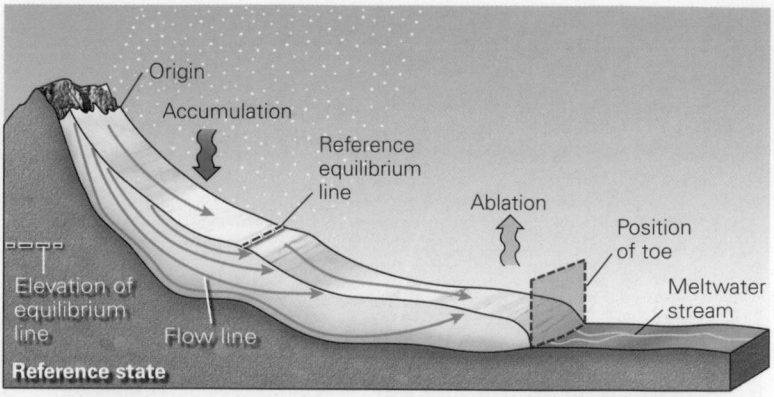

(a) The position of the toe represents a balance between addition of ice by accumulation and loss of ice by ablation.

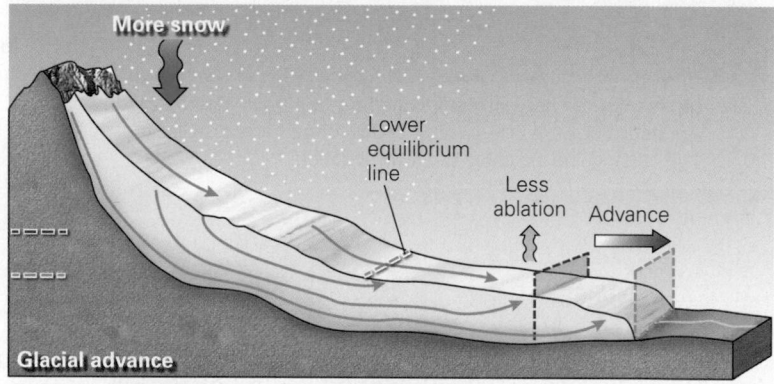

(b) If accumulation exceeds ablation, the glacier advances, the toe moves farther from the origin, and the ice thickens.

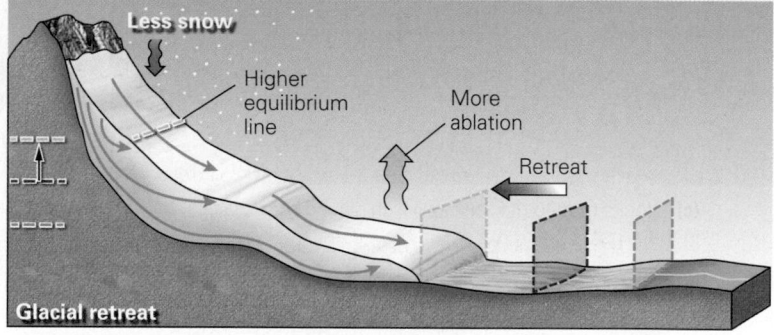

(c) If ablation exceeds accumulation, the glacier retreats and thins. The toe moves back, even though ice continues to flow toward the toe.

the toe as long as the surface of the glacier slopes toward the toe—glacial ice cannot flow back toward the glacier's origin.

One final point before we leave the subject of glacial flow: Because ice overall follows curved trajectories inside a glacier (see Fig. 22.9), rocks picked up by ice can slowly move down to depth beneath one part of the glacier, and can return to the surface near the toe. The upward flow of ice where the Antarctic ice sheet collides with the Transantarctic Mountains, for example, brings up meteorites long buried in the ice **(Fig. 22.10)**.

FIGURE 22.10 Meteorites accumulate along the Transantarctic Mountains.

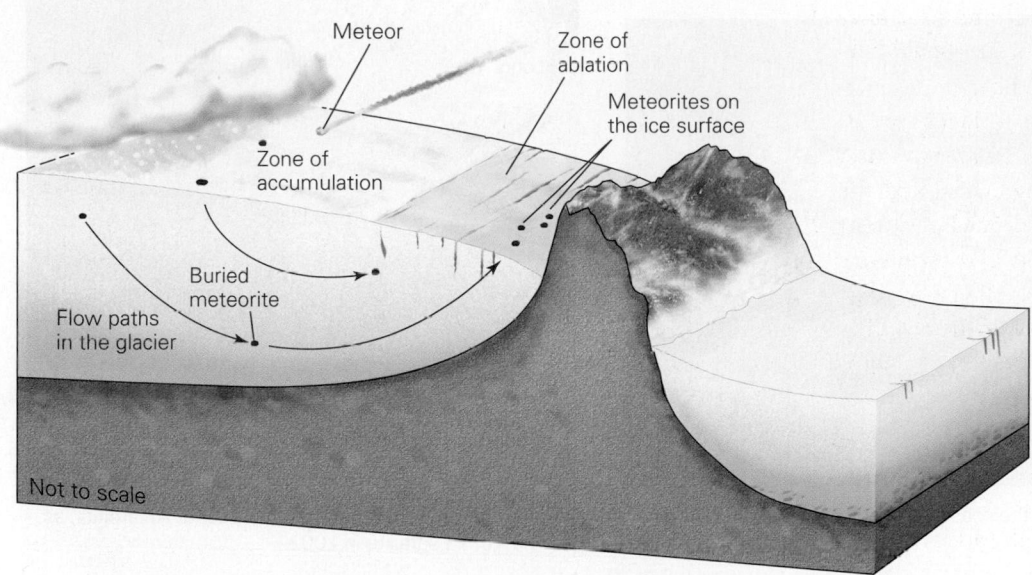

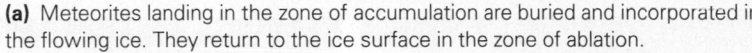

(a) Meteorites landing in the zone of accumulation are buried and incorporated in the flowing ice. They return to the ice surface in the zone of ablation.

(b) Researchers document a new meteorite discovery.

Ice in the Sea

On the moonless night of April 14, 1912, the great ocean liner *Titanic* plowed through the calm but frigid waters of the North Atlantic on her maiden voyage from Southampton, England, to New York. Although radio broadcasts from other ships warned that **icebergs**, large blocks of ice floating in the water, had been sighted in the area and might pose a hazard, the ship sailed on, its crew convinced that they could see and avoid the biggest bergs and that smaller ones would not be a problem for the steel hull of this "unsinkable" vessel. But in a story now retold countless times, their confidence was fatally wrong. At 11:40 P.M., while first-class passengers danced, the *Titanic* struck an iceberg. Lookouts had seen the ghostly mass only minutes earlier and had alerted the ship's pilot, but the ship had been unable to turn fast enough to avoid disaster. The force of the blow split the steel hull spanning 5 of the ship's 16 watertight compartments. The ship could stay afloat if 4 compartments flooded, but the flooding of 5 sealed its fate. At about 2:15 A.M., the bow disappeared below the water, and the stern rose until the ship protruded nearly vertically from the water. The ship then drifted downward through over 3.5 km of water to the silent seafloor below. Because of an inadequate number of lifeboats, only 705 passengers survived; 1,500 expired in the frigid waters of the Atlantic. The *Titanic* remained lost until 1985, when a team of oceanographers located the sunken hull and photographed its eerie form.

Where do icebergs, like the one responsible for the *Titanic*'s demise, originate? At high latitudes, as we've seen, glaciers flow down to sea level. Glaciers that flow into the sea along the coast become **tidewater glaciers**. Large valley glaciers may protrude several kilometers out into the ocean as elongate *ice tongues* (**Fig. 22.11a**). Continental glaciers entering the sea become broad, flat sheets called **ice shelves** (**Fig. 22.11b**). In shallow water, glacial ice remains grounded in that the base of the glacier rests on the seafloor (**Fig. 22.11c**). But in deeper water, the ice floats with four-fifths of its volume below the water's surface. At the boundary between glacier and ocean, blocks of ice calve off and tumble into the water with an impressive splash, producing large waves. A free-floating chunk of ice that rises 6 m above the water and has a length of at least 15 m can be called an *iceberg*. Smaller pieces, formed when ice blocks fragment before entering the water or after icebergs have had time to melt, include *bergy bits*, rising 1 to 5 m above the water and covering an area of 100 to 300 m², and *growlers* (**Fig. 22.11d**), rising less than 1 m above the water and covering an area of about 20 m². Growlers get their name because of the sound they make as they bob in the sea and grind together.

Most large icebergs form along the western coast of Greenland or along the coast of Antarctica. Icebergs that calve off valley glaciers tend to be irregularly shaped, with pointed peaks rising upward. Such icebergs are called *castle bergs* or *pinnacle bergs*—one of the largest on record protruded about 180 m above the sea. Since four-fifths of the ice lies below the surface of the sea, the base of a large iceberg may actually be a few hundred meters below the surface

FIGURE 22.11 Ice along the edges of continents—shelves, tongues, and bergs.

Land

Ice shelf

Water

(a) An ice tongue protruding into the Ross Sea along the coast of Antarctica.

(b) The Larsen Ice Shelf along the coast of Antarctica, as viewed from a satellite in 2002.

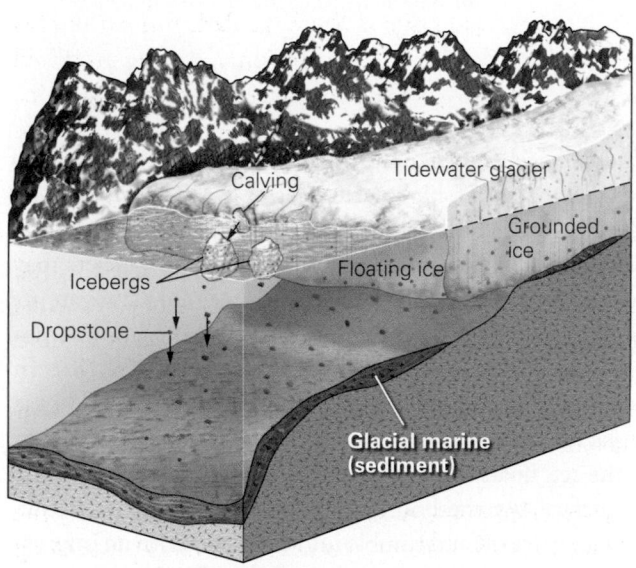

(c) Ice is grounded in shallow water but floats in deep water.

(e) This artist's rendition of an iceberg emphasizes that most of the ice is underwater.

(f) In summer, some of the sea ice of Antarctica breaks up to form tabular icebergs.

(d) A "growler" of floating ice off Alaska. Note the layering in the ice.

(g) In winter, sea ice covers most of the Arctic Ocean (left) and surrounds Antarctica (right).

(Fig. 22.11e). Icebergs that originate in Greenland float into the Iceberg Alley region of the North Atlantic. These are the bergs that threaten ships, although the danger has diminished in modern times because there's less ice and because ice patrols report the locations of floating ice. Blocks that calve off the vast ice shelves of Antarctica tend to have flat tops and nearly vertical sides—such blocks are called *tabular bergs*. Some of the tabular bergs in the Antarctic are truly immense. In 2017, a tabular berg measuring 200 km long by 65 km wide broke free of the Larsen Ice Shelf and began to drift into the Southern Ocean.

Not all ice floating in the sea originates in glaciers on land. In polar climates, the surface of the sea itself freezes, forming **sea ice (Fig. 22.11f, g).** Some sea ice, such as that covering the interior of the Arctic Ocean, floats freely, but some protrudes outward from the shore. Ships called *icebreakers* can crunch through sea ice up to 2.5 m thick; the icebreaker rides up on the ice, and its weight crushes the ice below. Vast areas of sea ice have been melting in recent years in association with global warming (see Chapter 23). For example, open regions develop in the Arctic Ocean during the summer, and the ice shelf in Antarctica has been decreasing rapidly in area. In some locations, large openings, known as *polynyas*, have developed in the sea ice of Antarctica. Some sea ice forms in winter and melts away in summer, but at high latitudes sea ice may last for several years. For example, in the Arctic Ocean, sea ice may last long enough to make the 7- to 10-year voyage around the Arctic Ocean, in response to currents, at least once.

The existence of icebergs leaves a record in the stratigraphy of the seafloor, for icebergs carry *ice-rafted sediment*. This sediment includes clay, sand, gravel, and boulders, and it collects on the seafloor to form a type of sediment known as *glacial marine*. Sometimes, isolated rocks called **dropstones** sink from the ice into otherwise fine-grained sediment of the seafloor. In ancient glacial deposits that underwent lithification, dropstones appear as isolated blocks surrounded by shale.

TAKE-HOME MESSAGE

Glaciers form where snow lasts all year, turns to ice, and gradually recrystallizes. Glacial ice flows by plastic deformation or by basal sliding. Mountain glaciers form at high elevations and flow to lower elevations. Continental ice sheets form at high latitudes and spread over continents. The balance of accumulation and ablation controls glacial advance or retreat. Where glaciers reach the sea, they may calve off icebergs. In polar regions, sea ice covers large areas.

QUICK QUESTION: Does ice actually flow uphill during a glacial retreat?

22.3 Carving and Carrying by Ice

The Process of Glacial Erosion

During the Mesozoic, a large continental volcanic arc lay just to the west of what is now the California-Nevada border. Beneath the volcanoes, so many granite plutons intruded that a batholith developed (see Chapter 6). Subsequent uplift produced the Sierra Nevada. During this process, erosion stripped off overlying rock and exposed the batholith at the Earth's surface. Weathering and exfoliation eventually transformed the top surface of this batholith into many rounded domes. Then, during the last ice age, the landscape of the Sierra Nevada changed radically as mountain glaciers cut deep, steep-sided valleys into the range. In a few locations, glacial erosion cut away one side of a dome to produce a steep cliff. Half Dome in Yosemite National Park formed in this way **(Fig. 22.12a).** Such glacial erosion has also produced the knife-edge ridges and pointed spires of high mountains **(Fig. 22.12b)** and broad expanses of land where rock outcrops have been stripped of overlying sediment. Glacial erosion can keep pace with tectonic uplift—in fact, glacial erosion grinds away mountain peaks so efficiently that geologists sometimes refer to the phenomenon as the "glacial buzz saw." Continental glaciers similarly strip material from the surface of continents. In fact, glacial erosion stripped off at least 30 m of rock from the Canadian Shield during the last ice age.

How does glacial erosion take place? In mountainous areas, landslides or rockfalls drop debris onto glaciers, which then carry it away. Freezing and thawing in glacial environments may accelerate the mechanical weathering that sets the stage for such mass wasting. But erosion also takes place along the contact between glacial ice and the land surface, for as ice moves, clasts embedded in the ice act like the teeth of a giant rasp and grind away the substrate. This process, *glacial abrasion*, produces very fine sediment called *rock flour*, just as sanding wood produces sawdust.

Rasping by embedded sand can smooth rock faces and produce **glacially polished surfaces (Fig. 22.12c).** Individual hard clasts protruding from moving ice yield grooves or scratches called **glacial striations** in the bedrock below **(Fig. 22.12d).** These striations trend parallel to the flow direction of the glacier. Glacial striations typically look like scratches in bedrock; they range from 0.1 to 10 cm across, depending on the size of the clasts that generated them. In places where a long train of clasts grinds along the bedrock surface, *glacial grooves* that range from 0.5 to 10 m in width may develop **(Fig. 22.12e).** Locally, when boulders entrained in the base of the ice strike bedrock below as the ice moves, asymmetric wedges of bedrock

FIGURE 22.12 Products of glacial erosion. Ice is a very aggressive agent of erosion.

(a) Half Dome in Yosemite National Park, California.

(b) Rugged, glacially carved peaks in the Swiss Alps.

(c) Glacially polished outcrop in Central Park, New York City.

(d) Small striations on an outcrop in Scotland.

(e) Glacial grooves in Victoria, British Columbia.

(f) Close-up of striations and chatter marks, Switzerland.

break off, leaving behind indentations called *chatter marks* **(Fig. 22.12f)**. In regions of wet-based glaciers, sediment-laden water rushing through tunnels at the base of the glaciers can carve substantial subglacial channels.

Glaciers pick up fragments of their substrate in several ways. During *glacial incorporation*, ice surrounds loose debris so that the debris starts to move with the ice **(Fig. 22.13a)**. During *glacial plucking* (or *glacial quarrying*), a glacier breaks off fragments of bedrock. Plucking occurs when ice freezes around rock that has just started to separate from its substrate; movement of the ice lifts off pieces of the rock. At the toe of an advancing glacier, ice may actually bulldoze sediment slightly before flowing over it **(Fig. 22.13b)**.

Landforms Produced by Glacial Erosion

Let's now look more closely at the erosional features associated with mountain glaciers. If the glacier builds into a mountain ice cap that completely covers the mountain, it smooths and rounds peaks **(Fig. 22.14a)**. But, if the glacier's *head* (the edge of the ice at the origin of the glacier) lies below the peak of the mountain, then the ice carves rugged topography. Freezing

FIGURE 22.13 The processes of incorporation, plucking, and plowing.

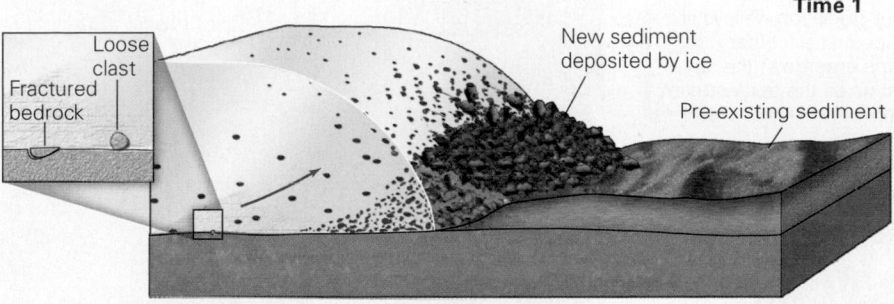

Time 1

Loose clast

Fractured bedrock

New sediment deposited by ice

Pre-existing sediment

(a) Glacial ice can pluck, pick up, and incorporate chunks of rock that it flows over. The chunks then move with the ice, following the ice's flow trajectories, until they are deposited at the toe of the glacier.

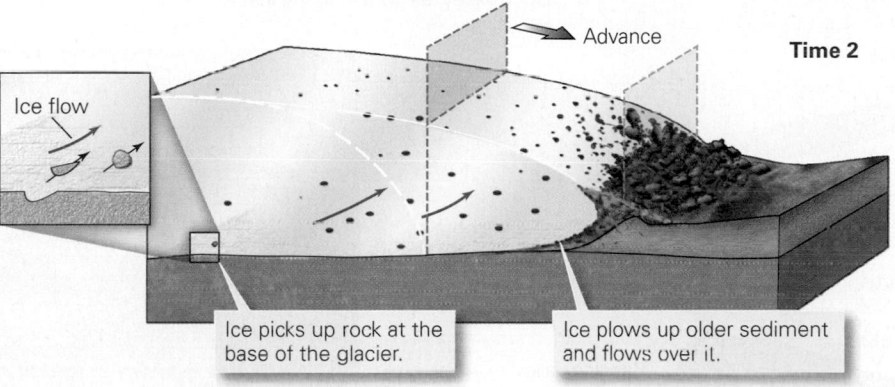

Advance

Time 2

Ice flow

Ice picks up rock at the base of the glacier.

Ice plows up older sediment and flows over it.

(b) At the toe of the glacier, ice can flow up and over pre-existing sediment. Locally, it bulldozes sediment, pushing it up into a ridge.

FIGURE 22.14 Landforms of glacially carved peaks.

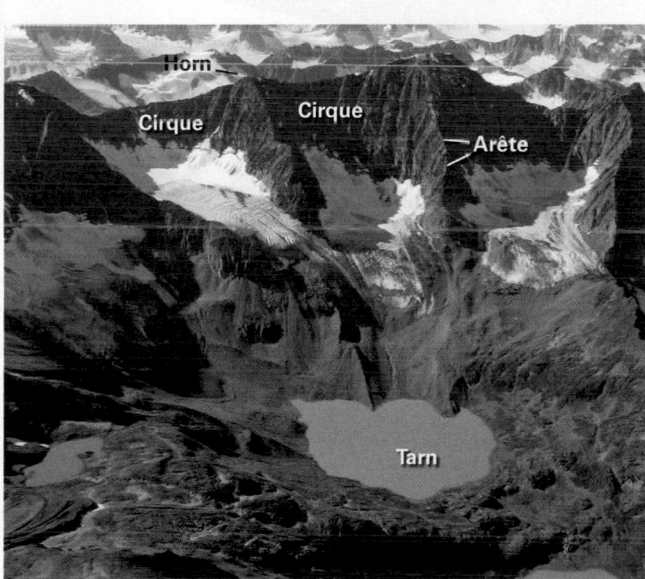

Horn

Cirque

Cirque

Arête

Tarn

(a) This peak in Wyoming was rounded when it lay beneath a mountain ice cap.

(b) This aerial photo of glacially carved mountains in Alaska illustrates cirques, aretes, horns, and tarns.

(c) The Matterhorn, in the Swiss Alps, challenges climbers.

FIGURE 22.15 Landscape features formed by the glacial erosion of a mountainous landscape.

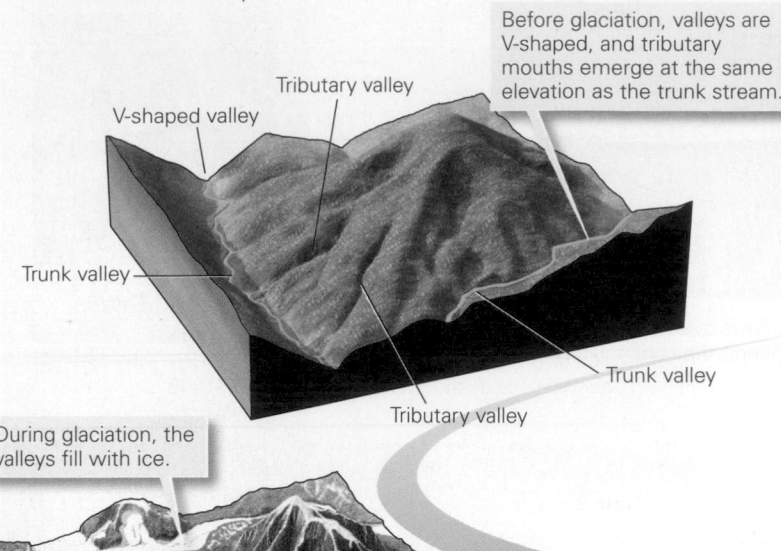

Tributary valley

V-shaped valley

Trunk valley

Before glaciation, valleys are V-shaped, and tributary mouths emerge at the same elevation as the trunk stream.

Trunk valley

Tributary valley

Profile

Profile

(b) A U-shaped valley in the Swiss Alps. The glacier that carved it has melted away. In contrast, a river carves a V-shaped valley, as shown in the inset.

During glaciation, the valleys fill with ice.

Time

After glaciation, the region contains U-shaped valleys, hanging valleys, truncated spurs, and horns.

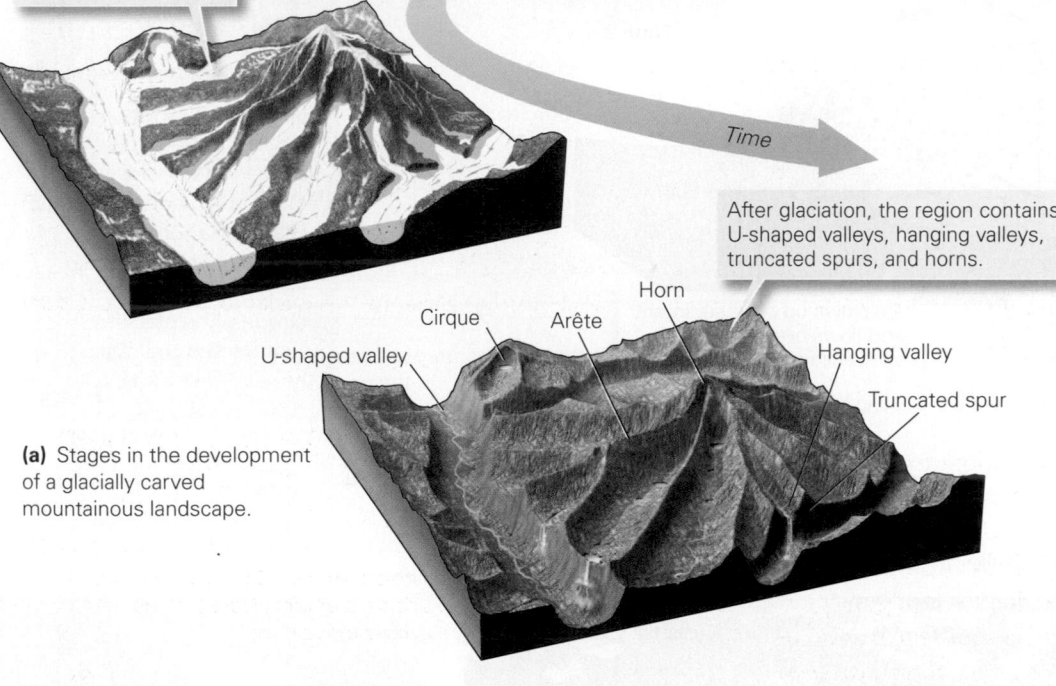

Horn

Cirque   Arête

U-shaped valley

Hanging valley

Truncated spur

(a) Stages in the development of a glacially carved mountainous landscape.

(c) A waterfall spilling out of a U-shaped hanging valley in the Sierra Nevada.

and thawing during the fall and spring help fracture the rock bordering the head of the glacier. This rock falls on the ice or gets picked up at the base of the ice and moves downslope with the glacier. As a consequence, a bowl-shaped depression, or **cirque**, develops on the side of the mountain at the head of a glacier **(Fig. 22.14b)**. If the ice later melts, a lake called a *tarn* may remain at the base of the cirque, filling the base of the depression. An **arête** (French for ridge), a residual knife-edged ridge of rock, separates two adjacent cirques, and a pointed mountain peak surrounded by at least three cirques is a **horn**. The Matterhorn, a famous peak in Switzerland, serves as a particularly beautiful example of a horn; each of its four faces originated as a cirque **(Fig. 22.14c)**.

Glacial erosion radically modifies the shape of valleys **(Fig. 22.15a)**. To see how, compare a river-eroded valley with a glacially eroded valley. If you look along the length of a river in unglaciated mountains, you'll see that it flows down a V-shaped valley, with the river channel forming the point of the V. The V develops because river erosion occurs only in the channel, and mass wasting causes the valley slopes to approach the angle of repose. But if you look down the length of a glacially eroded valley, you'll see that it resembles a U, with steep walls. A **U-shaped valley (Fig. 22.15b)** forms because the combined processes of glacial abrasion and plucking not only lower the floor of the valley, but also erode its sides. Remember that mountain faces above the ice level of a valley glacier erode as mechanical weathering breaks rock apart, and landslides carry debris onto the surface of the glacier below.

FIGURE 22.16 The sculpting of hills by glacial erosion.

(a) Polished and striated bedrock in Ontario.

(b) Rounded hills in the highlands of Scotland formed when the entire region was covered by an ice sheet.

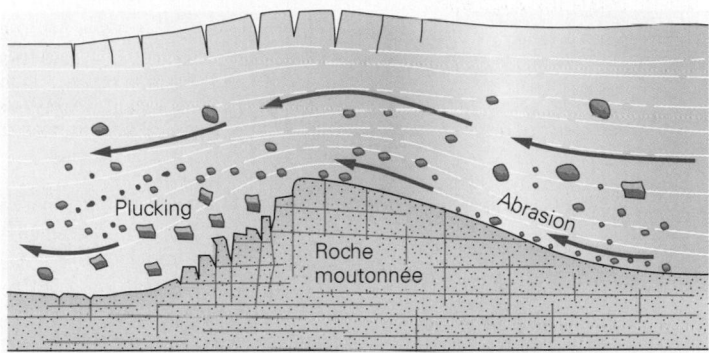

(c) Abrasion rasps the upstream side, and plucking carries away joint-bounded blocks on the downstream side.

(d) An example of a roche moutonnée. The glacier flowed from right to left.

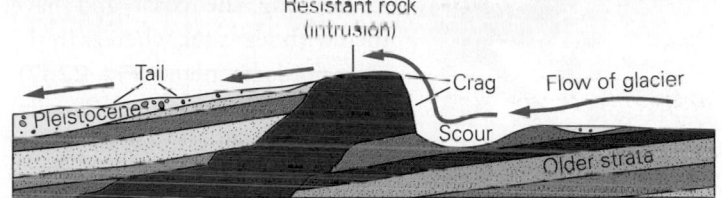

(e) A crag and tail forms when a resistant rock holds up a crag that protects rock on the downflow side from erosion. Note that the asymmetry differs from that of a roche moutonnée.

Glacial erosion in mountains also modifies the intersections between tributaries and the trunk valley. In a river system, the trunk stream serves as the local base level for tributaries (see Chapter 17), so the mouths of the tributary valleys lie at the same elevation as the trunk valley. The ridges (spurs) between valleys taper to a point where they join the trunk valley floor. During glaciation, tributary glaciers flow down side valleys into a trunk glacier. But the trunk glacier cuts the floor of its valley down to a depth that far exceeds the depth cut by the tributary glaciers. Thus, when the glaciers melt away, the mouths of the tributary valleys perch at a higher elevation than the floor of the trunk valley. Such side valleys are called **hanging valleys**. The water in post-glacial tributary streams that flow down a hanging valley cascades over a spectacular waterfall to reach the post-glacial trunk stream **(Fig. 22.15c)**. As they erode, trunk glaciers also remove the ends of spurs between valleys, producing *truncated spurs*.

Now let's look at the erosional features produced by continental ice sheets. To a large extent, these features depend on the nature of the pre-glacial landscape. Where an ice sheet spreads over a region of low relief, such as parts of Canada **(Fig. 22.16a)**, glacial erosion creates a vast region of polished, flat, striated surfaces. Where an ice sheet spreads completely over a hilly area, it smooths hills **(Fig. 22.16b)**. Glacially

eroded hills may become elongate in the direction of glacial flow, and because glacial rasping smooths and bevels the upstream part of the hill, creating a gentle slope—whereas glacial plucking eats away at the downstream part, making a steep slope—the hills may become asymmetric. Ultimately, the hill becomes a **roche moutonnée** (French for sheep rock), so named because its profile resembles that of a sheep lying in a meadow **(Fig. 22.16c, d)**. In some cases, a glacier erodes three sides of a hill, but deposits debris on the downstream or wake side of the hill. This process produces a *crag and tail*, with a steep cliff on the upstream side and a long ramp on the downstream side **(Fig. 22.16e)**. The castle of Edinburgh, Scotland, was built on a crag. The steep scarps on three sides were easy to defend, while the tail provided a site for the growth of a town.

FIGURE 22.17 Examples of fjords.

(a) The Finger Lakes of central New York State are freshwater fjords.

(b) A fjord in Norway reflects the cloudy sky.

(c) Tourists standing on Pulpit Rock (Preikestolen) look down on a cruise ship.

Fjords: Submerged Glacial Valleys

If the floor of a glacially carved valley lies below sea level along the coast, or beneath the water table inland, the floor of the valley becomes submerged with water. Geologists refer to any glacially-carved valley that has filled partially or entirely with water as a **fjord**. Marine fjords occur along the coast and have filled with seawater, whereas freshwater fjords lie inland (**Fig. 22.17**).

Spectacular examples of marine fjords can be found along the coasts of Norway, New Zealand, Chile, Alaska, and Greenland—in some cases, the walls of submerged U-shaped valleys rise straight from the sea as vertical cliffs up to 1,000 m high, and the water depth just offshore may exceed a few hundred meters. How do such dramatic fjords develop? As noted earlier, where a valley glacier meets the sea, the glacier's base remains in contact with the ground until the water depth exceeds about four-fifths of the glacier's thickness. Further, during an ice age, water extracted from the sea becomes locked in the ice sheets on land, so sea level drops significantly. Therefore, the floors of valleys cut by coastal glaciers during the Pleistocene Ice Age could be cut much deeper than present sea level.

TAKE-HOME MESSAGE

A glacier scrapes up and plucks rock from its substrate and carries debris that falls on its surface. Glacial erosion polishes and scratches rock and carves distinctive landforms, such as U-shaped valleys, cirques, and striations. An elongate bay or lake formed when water partially fills a glacial valley is a fjord.

QUICK QUESTION: Why do we find hanging valleys spilling waterfalls into trunk valleys in regions that have been eroded by mountain glaciers?

22.4 Deposition Associated with Glaciation

The Glacial Conveyor and Glacial Moraines

Glaciers can carry sediment of any size and, like a conveyor belt, transport it in the direction of flow (Fig. 22.18a). Remember that ice flows toward the toe regardless of whether the glacier is advancing or retreating, so sediment always moves toward the toe. Where does the sediment come from? It either falls onto the surface of the glacier from bordering cliffs (in mountainous regions) or gets plucked and lifted from the substrate and incorporated into the moving ice.

Sediment dropped on the glacier's surface from its margins becomes a stripe of debris, known as a **lateral moraine**, along the side edge of the glacier. When a glacier melts, its lateral moraines remain, stranded along the side of the glacially carved valley, like bathtub rings. In places where two valley glaciers merge, the debris constituting two lateral moraines merges to become a **medial moraine**, running as a stripe down the interior of the composite glacier (Fig. 22.18b, c). Trunk glaciers created by the merging of many tributary glaciers contain several medial moraines. Sediment transported to a glacier's toe by the glacial conveyor accumulates in a pile at the toe and builds up to form an **end moraine**. If the glacier recedes, the end moraine will remain as a low ridge, outlining the former position of the toe. The name *moraine* originated from a term used by Alpine farmers and shepherds for

FIGURE 22.18 The glacial conveyor and the formation of moraines.

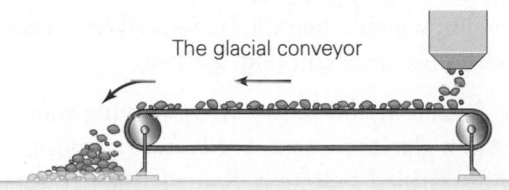

The glacial conveyor

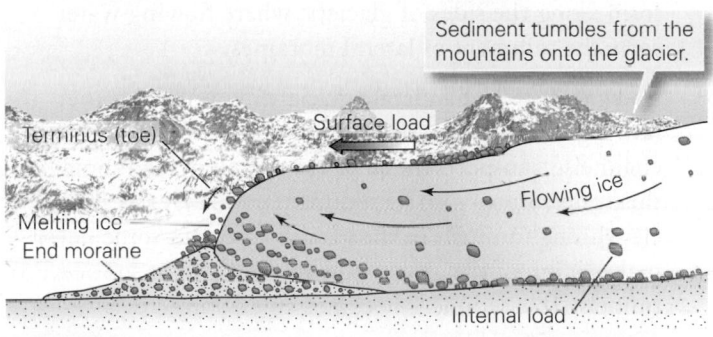

Sediment tumbles from the mountains onto the glacier.

Surface load

Terminus (toe)

Flowing ice

Melting ice

End moraine

Internal load

(a) Sediment falls on a glacier from bordering mountains and gets plucked up from below. Glaciers are like conveyor belts, moving sediment toward the toe of the glacier.

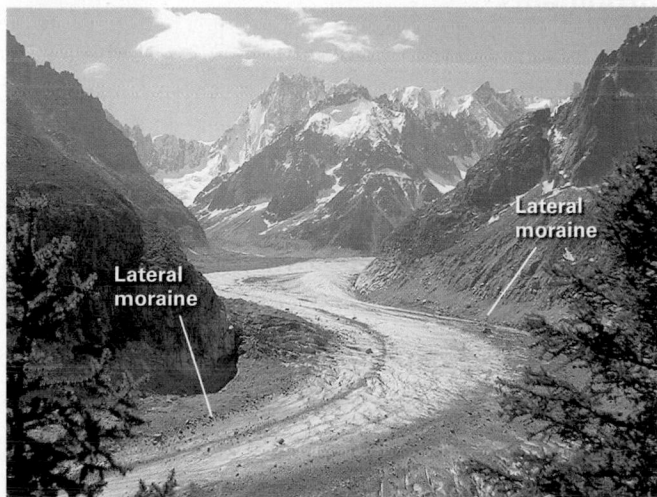

Lateral moraine

Lateral moraine

(b) This glacier in the French Alps carries lots of sediment.

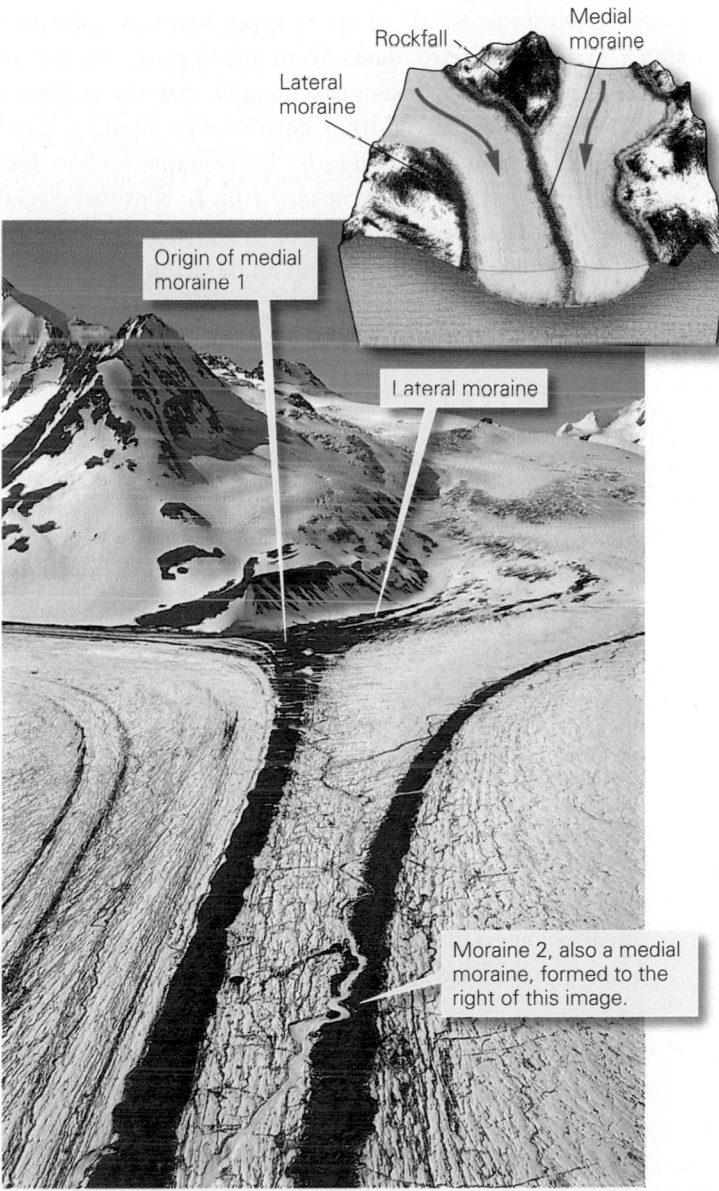

Rockfall

Medial moraine

Lateral moraine

Origin of medial moraine 1

Lateral moraine

Moraine 2, also a medial moraine, formed to the right of this image.

(c) A medial moraine forms where lateral moraines of two valley glaciers merge.

any pile of rock and dirt. The word now applies exclusively to debris piles carried by or left by glaciers.

Types of Glacial Sedimentary Deposits

If you drill into the soil throughout much of the upper midwestern and northeastern United States and adjacent parts of Canada, the drill penetrates a layer of sediment deposited during the Pleistocene Ice Age. A similar story holds true for much of northern Europe. As a consequence, many of the world's richest agricultural regions rely on soil derived from sediment deposited by glaciers during the ice age. In some locations, this sediment buries a pre–ice age landscape, like frosting that fills the irregularities on a cake. In fact, some pre-glacial valleys may be completely filled with sediment.

Several different types of sediment can be deposited in glacial environments; all of these types together constitute *glacial drift*. (The term dates from pre-Agassiz studies of glacial deposits, when geologists thought that the sediment had "drifted" into place during an immense flood.) Glacial sediment carried and deposited by ice contains no layering, so geologists refer to it as *unstratified drift*. In contrast, glacial sediment that has been redistributed by flowing water, or has settled through water, tends to contain layering and is called *stratified drift*. In detail, glacial drift includes the following features:

- *Till:* Sediment transported by ice and deposited beneath, at the side of, or at the toe of a glacier is called **glacial till**. Glacial till has not undergone sorting, so geologists consider it to be a type of diamicton (see Chapter 7) **(Fig. 22.19a)**.

- *Erratics:* Boulders or cobbles that have been dropped by a glacier are called *glacial erratics* **(Fig. 22.19b)**. Some erratics protrude from till piles, and others rest on glacially polished surfaces.

- *Glacial marine:* Where a sediment-laden glacier flows into the sea, icebergs calve off the toe and raft clasts out to sea. As the icebergs melt, they drop the clasts, which settle into the muddy sediment on the seafloor. Dropstones may be incorporated in these deposits. As we have seen, sediment consisting of ice-rafted clasts mixed with marine sediment makes up *glacial marine*. Glacial marine can also consist of sediment carried into the sea by water flowing at the base of a glacier.

- *Glacial outwash:* Till deposited by a glacier at its toe may be picked up and transported by meltwater streams that sort the sediment. Braided streams deposit these clasts in a broad area of gravel and sandbars called an *outwash plain*. This sediment becomes *glacial outwash* **(Fig. 22.19c)**.

- *Loess:* When the warmer air above ice-free land beyond the toe of a glacier rises, the colder, denser air from above the glacier rushes in to take its place; a strong wind, called *katabatic wind*, therefore blows at the margin of a glacier. This wind picks up fine silt and clay and transports it away from the glacier's toe. Where the winds die down, the sediment settles and forms a thick layer. This sediment, called **loess**, tends to stick together, so steep escarpments can develop by erosion of loess deposits **(Fig. 22.19d)**.

- *Glacial lake-bed sediment:* Streams transport fine clasts, including rock flour, away from the glacier's toe. This sediment eventually settles in meltwater lakes, forming a thick layer of glacial lake-bed sediment. This sediment commonly contains varves. A **varve** consists of a pair of thin layers deposited during a single year. One layer consists of silt brought in during spring floods and the other of clay deposited in winter when the lake's surface freezes over and the water becomes still **(Fig. 22.19e)**.

- *Kame deposits:* A *kame deposit* forms where flowing water on the surface of a glacier transports till and redeposits it as sorted and stratified sediment. Some kame deposits form along the sides of glaciers, where flowing water sorts the sediment of lateral moraines.

- *Esker deposits:* In temperate glacial environments, the flowing water at the base of the glacier, moving through channels, transports much of the sediment load. Some of this water, along with its sedimentary load, exits the glacier through tunnels in the glacier's toe. But some never makes it out of the glacier and accumulates in sub-ice tunnels. This sediment becomes an *esker deposit*.

Depositional Landforms of Glacial Environments

Picture a hunter, dressed in deerskin, standing at the toe of a continental ice sheet in what is now southern Canada. It's about 12,000 years ago, and the glacier has been receding for at least a millennium. Milky, sediment-laden streams gush from tunnels and channels at the base of the glacier and pour off the top as the ice melts. No mammoths venture by today, so the bored hunter climbs to the top of the glacier for a view. The climb isn't easy, partly because of the incessant katabatic wind and partly because of deep crevasses. Reaching the top of the ice sheet, the hunter looks northward, and squinting, sees the white of snow, and where the snow has blown away, a rippled, glassy surface of bluish ice (see Fig. 22.1b). Here and there, a rock protrudes from the ice. Now looking southward, the hunter surveys a stark landscape of low, sinuous

FIGURE 22.19 Sedimentation processes and deposits associated with glaciation. Glacial sediment is distinctive.

(a) This glacial till in Ireland is unsorted because ice can carry sediment of all sizes.

(b) Glacial erratics resting on a glacially polished surface in Wyoming.

(c) Braided streams choked with glacial outwash in Alaska. The streams carry away finer sediment and leave the gravel behind.

(d) Thick loess deposits underlie parts of the prairie in Illinois.

(e) In the quiet water of an Alaskan glacial lake, fine-grained sediments accumulate. Varves in lake-bed sediment, now exposed in an outcrop near Puget Sound, Washington, reflect seasonal changes.

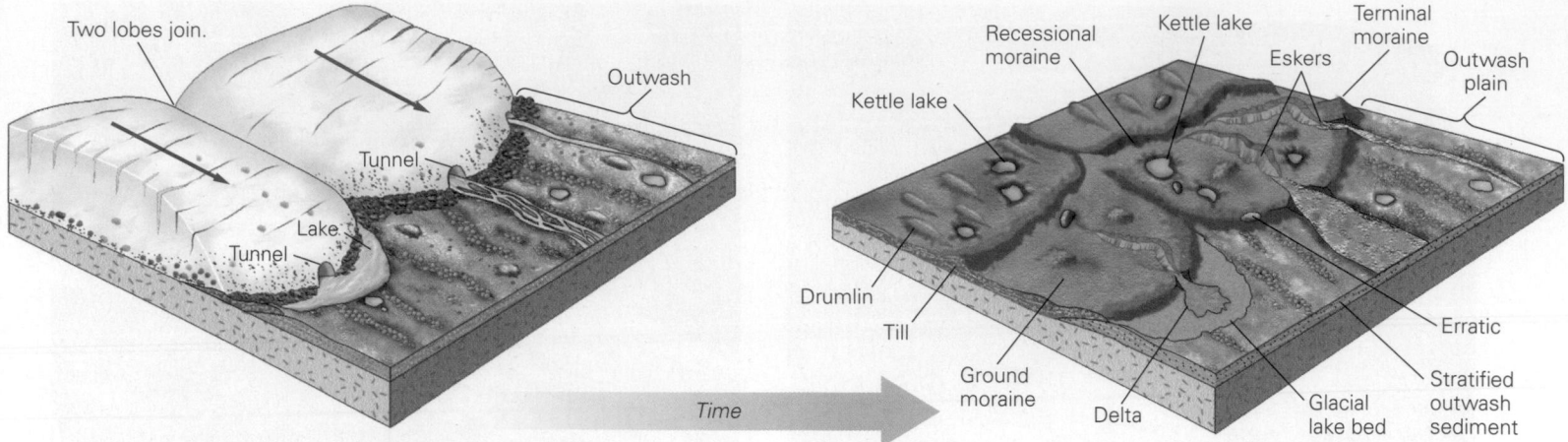

(a) The ice in continental glaciers flows toward the toe; sediment accumulates at the base and at the toe of the ice sheet.

(b) Several distinct depositional landforms form during glaciation; some developed under the ice and some at the toe.

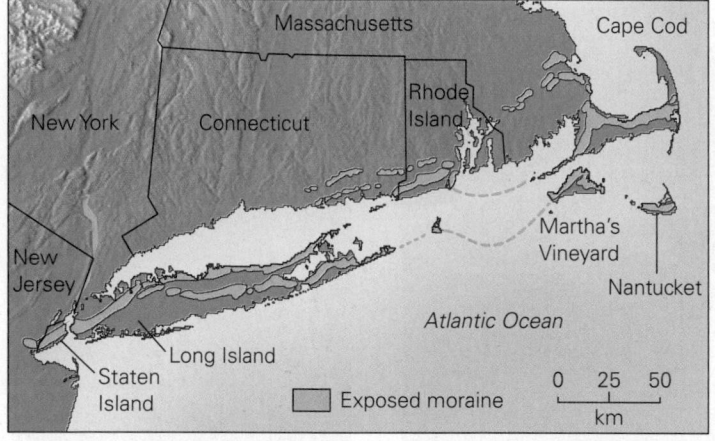

(c) Cape Cod, Long Island, and other landforms in the northeastern United States formed at the end of the continental ice sheet.

(d) This glacial moraine, in Wyoming, formed when glaciers covered the region between the moraine and the mountains in the distance.

ridges separated by hummocky (bumpy) plains (**Geology at a Glance**, pp. 858–859). Braided streams, which carry meltwater out across this landscape, flow through the hummocky plains and supply a number of lakes. The wind fills the air with dust.

All of the landscape features that the hunter observes were formed by glacial deposition (**Fig. 22.20a, b**). The low, sinuous ridges, which outline the former edge of the ice, are end moraines, developed when the toe of a glacier stalls in one position for a while. Geologists refer to the specific end moraine that lies at the farthest limit of glaciation as the **terminal moraine**. The ridge of sediment that makes up Long Island, New York, and continues east-northeast into Cape Cod, Massachusetts, forms part of the terminal moraine of the ice sheet that covered New England and eastern Canada during the Pleistocene Ice Age (**Fig. 22.20c**). End moraines that form when a glacier stalls temporarily as it recedes overall are *recessional moraines*.

Till that has been released from the ice at the base of a flowing glacier and remains after the glacier has melted away makes up *lodgment till* (**Fig. 22.20d**). Clasts in lodgment till may be aligned and scratched during glacial flow. The till left behind during rapid recession forms a thin, hummocky layer on the land surface. This till, together with lodgment till, forms a landscape feature known as *ground moraine*. The flow of the glacier may mold till and other subglacial sediment into a streamlined, elongate hill called a **drumlin** (from the Gaelic word for hills). Drumlins are aligned with the flow direction, and have a gentle downstream slope and a steeper upstream slope (**Fig. 22.21a, b**).

The hummocky surface of moraines results partly from variations in the amount of sediment supplied by the glacier and partly from the formation of *kettle holes*, circular depressions made when blocks of ice from a sediment-laden glacier separates from the toe as the glacier recedes; when the ice blocks melt, the overlying and adjacent sediment collapses

FIGURE 22.21 Drumlins, knob-and-kettle topography, and ice-margin lakes.

(a) The formation of a drumlin beneath a glacier.

Crevasse
Drumlin — Glacier
Ground moraine

View looking southeast

(b) Drumlins dominate this landscape near Rochester, New York.

Shaded relief map of the drumlins in central New York. Their SSE angle gives the direction of glacial flow.

Lake Ontario
N   0   4 km
☐ Sodus
Flow

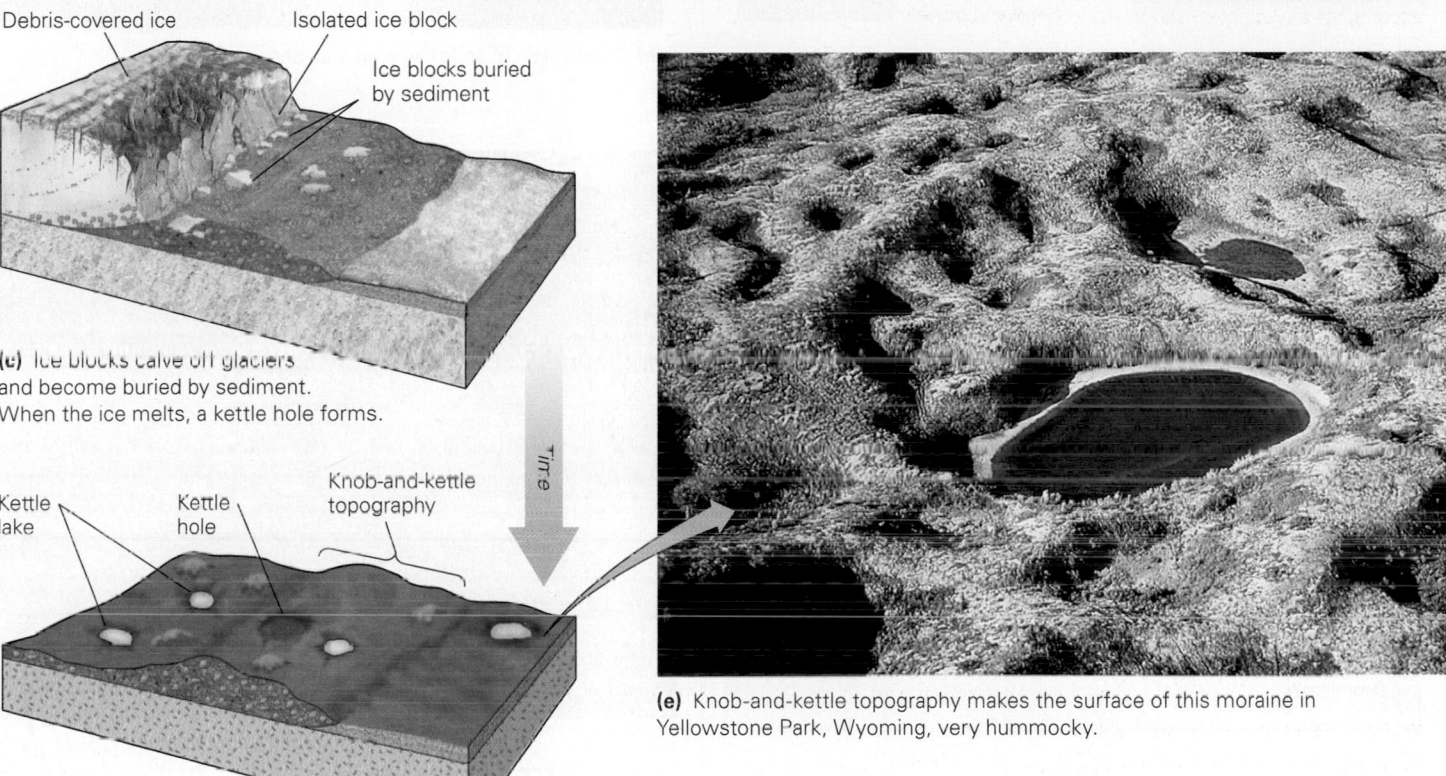

Debris-covered ice   Isolated ice block
Ice blocks buried by sediment

(c) Ice blocks calve off glaciers and become buried by sediment. When the ice melts, a kettle hole forms.

Kettle lake   Kettle hole   Knob-and-kettle topography

Time

(d) If the water table is high, kettle holes fill with water and turn into roughly circular lakes.

(e) Knob-and-kettle topography makes the surface of this moraine in Yellowstone Park, Wyoming, very hummocky.

downward **(Fig. 22.21c,d)**. Geologists refer to land surfaces with many kettle holes separated by round hills of till as *knob-and-kettle topography* **(Fig. 22.21e)**.

As we've noted, ice does not directly deposit all of the sediment associated with glacial landscapes, for meltwater also carries and deposits sediment. Water-transported sediment, in contrast to till, tends to be sorted and stratified. When the underlying ice melts away, a kame deposit becomes a small hill or ridge known as a *kame*. Esker deposits, filling meltwater tunnels beneath a glacier, may remain as narrow, sinuous ridges, called **eskers**, when the glacier melts away; eskers tend to trend at a high angle to the toe of the glacier **(Fig. 22.22a, b)**. Meltwater collecting adjacent to the glacier's toe, trapped in some cases by an end moraine, forms an ice-margin lake **(Fig. 22.22c)**. Additional lakes and swamps may form in low areas on the ground moraine. Braided meltwater streams that flow beyond the end of a glacier deposit layers of sand and gravel over a broad area, yielding a **glacial outwash plain**. Sediments deposited in eskers, kames, and glacial outwash plains serve as important sources of sand and gravel for construction, and the fine sediment of former glacial lake beds evolves into fertile soil for agriculture.

FIGURE 22.22 Landscape features resulting from glacial meltwater.

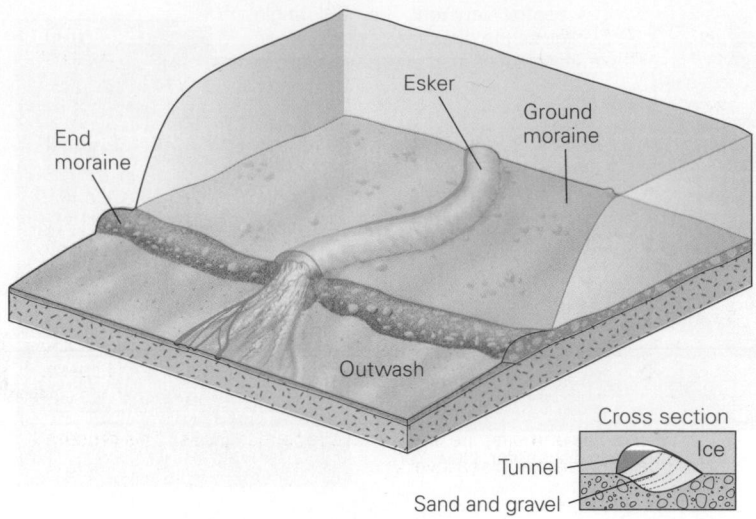

(a) Eskers are snake-like ridges of sand and gravel that form when sediment fills meltwater tunnels at the base of a glacier. In cross section (inset), wedges of sand accumulate in the tunnel.

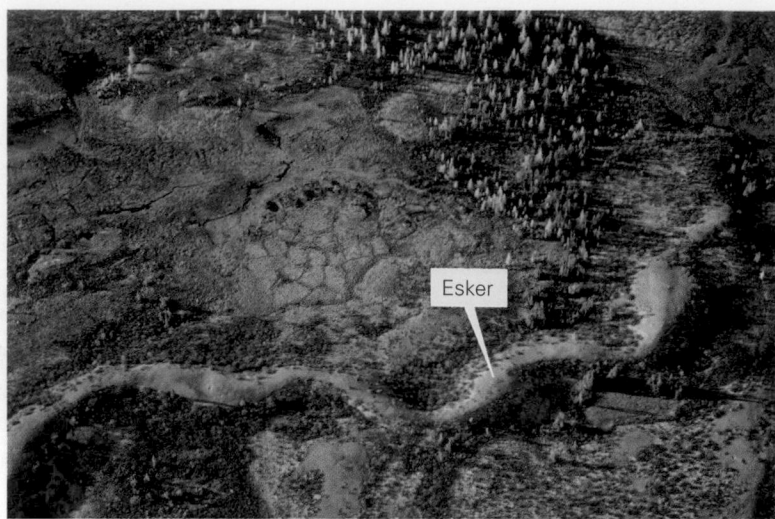

(b) An example of an esker in an area once glaciated.

(c) An ice-margin lake formed between the toe of the Bear Glacier and its terminal moraine in Alaska.

TAKE-HOME MESSAGE

Glaciers carry sediment of all sizes toward the toe. Along the surface of the glacier, this sediment becomes lateral or medial moraines. When ice melts, it deposits unsorted till in an end moraine or ground moraine. Meltwater streams and wind transport and sort the sediment to form outwash-plain gravels and loess deposits, respectively. Sediment carried to the sea by glaciers settles on the seafloor. Deposition by glaciers produces distinctive landforms, such as moraines, eskers, kames, and kettle holes.

QUICK QUESTION: How does knob-and-kettle topography develop?

22.5 Consequences of Continental Glaciation

Ice Loading and Glacial Rebound

When a large ice sheet (more than 50 km in diameter) grows on a continent, its weight causes the surface of the lithosphere to sink. In other words, ice loading causes **glacial subsidence**. Lithosphere, the relatively rigid outer shell of the Earth, can sink because the underlying asthenosphere can flow slowly out of the way **(Fig. 22.23a)**. As an analogy, imagine this simple experiment: Fill a bowl with honey and then place a thin rubber sheet over the honey. The rubber represents the lithosphere, and the honey represents the asthenosphere. If you place an ice cube on the rubber sheet, the sheet sinks because the weight of the ice pushes it down; the honey flows out of the way to make room. Because of ice loading, the rock surfaces underlying large areas of Antarctica's and Greenland's ice sheets now lie below sea level (see Fig. 22.4), so if the ice were instantly to melt away, these continents would be flooded by a shallow sea.

FIGURE 22.23 The concept of glacial subsidence and post-glacial rebound due to continental glaciation and deglaciation.

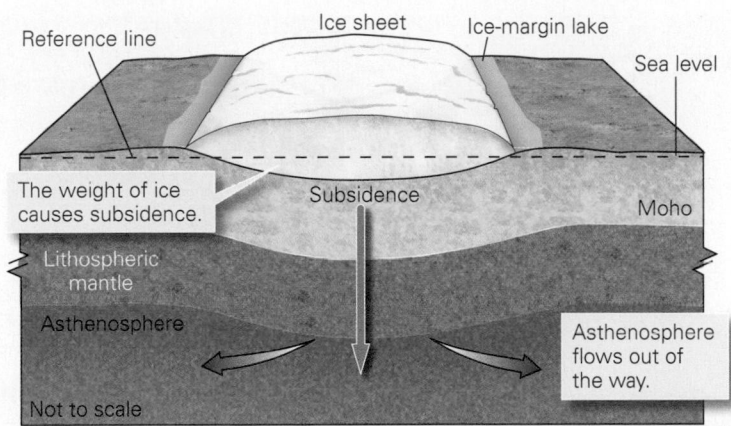

(a) The weight of the ice sheet causes the surface of the lithosphere to sink (subside).

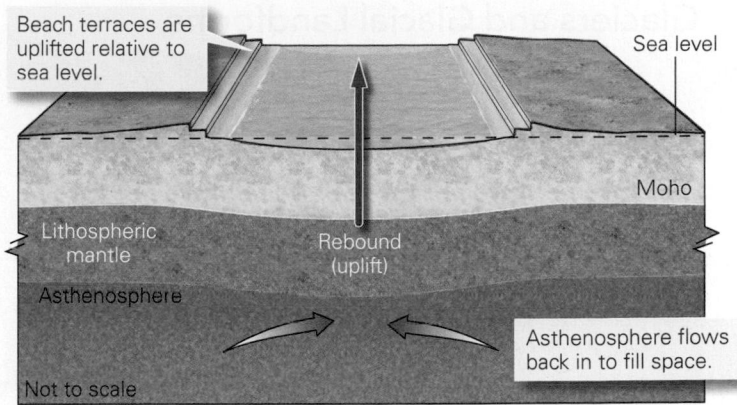

(b) After the glacier melts, the land surface rebounds (rises). This process uplifts beaches relative to sea level.

(c) Uplifted beaches along the coast of Arctic Canada form terraces as the land undergoes post-glacial rebound.

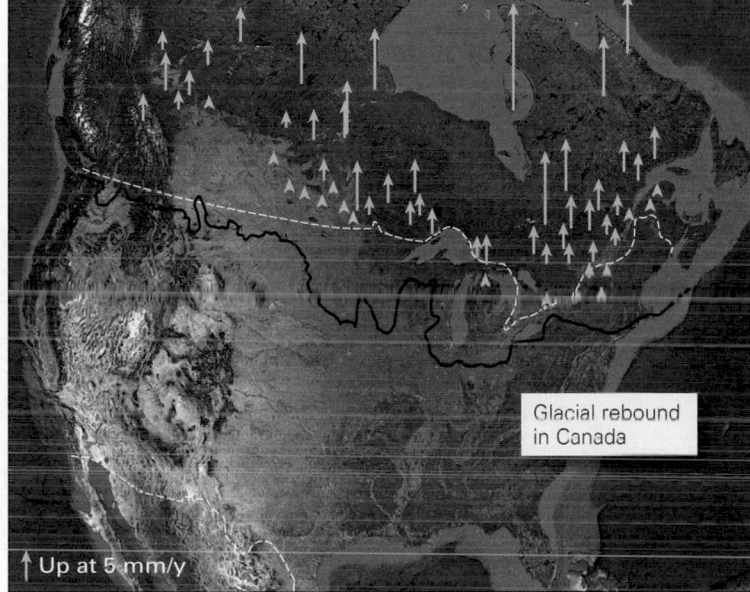

(d) GPS measurements show that the region north of the dark line is rebounding. Different rates of uplift occur at different locations.

What happens when continental ice sheets do melt away? Gradually, the surface of the underlying continent rises back up to re-achieve isostasy (see Interlude D) by a process called **post-glacial rebound.** As this happens, the asthenosphere flows back underneath so no space develops **(Fig. 22.23b).** Where rebound affects coastal areas, beaches along the shoreline rise several meters above sea level and become terraces **(Fig. 22.23c).** In the honey and rubber analogy, when you remove the ice cube, the rubber sheet slowly returns to its original shape. This process doesn't take place instantly because the honey can only flow slowly. Similarly, because the asthenosphere flows so slowly, it takes thousands of years for ice-depressed continents to rebound. Consequently, glacial rebound continues in some regions that were burdened by ice during the last glacial advance of the Pleistocene Ice Age. Recently, researchers in North America have documented

this movement by using GPS measurements **(Fig. 22.23d),** and have found that parts of Canada and the northern United States are now rising relative to sea level.

Sea-Level Changes: The Glacial Reservoir

More of the Earth's surface and near-surface freshwater resides in glacial ice than in any other reservoir. In fact, glacial ice accounts for 2.15% of the Earth's total water supply, while lakes, rivers, soil, and the atmosphere together contain only 0.03%. The melting of glacial ice transfers this water back into the ocean, causing sea level to rise. In fact, if today's ice sheets

Continental ice sheet

Crevasses

Ice shelf

Higher sea level

Lower sea level

Dropstones

Iceberg

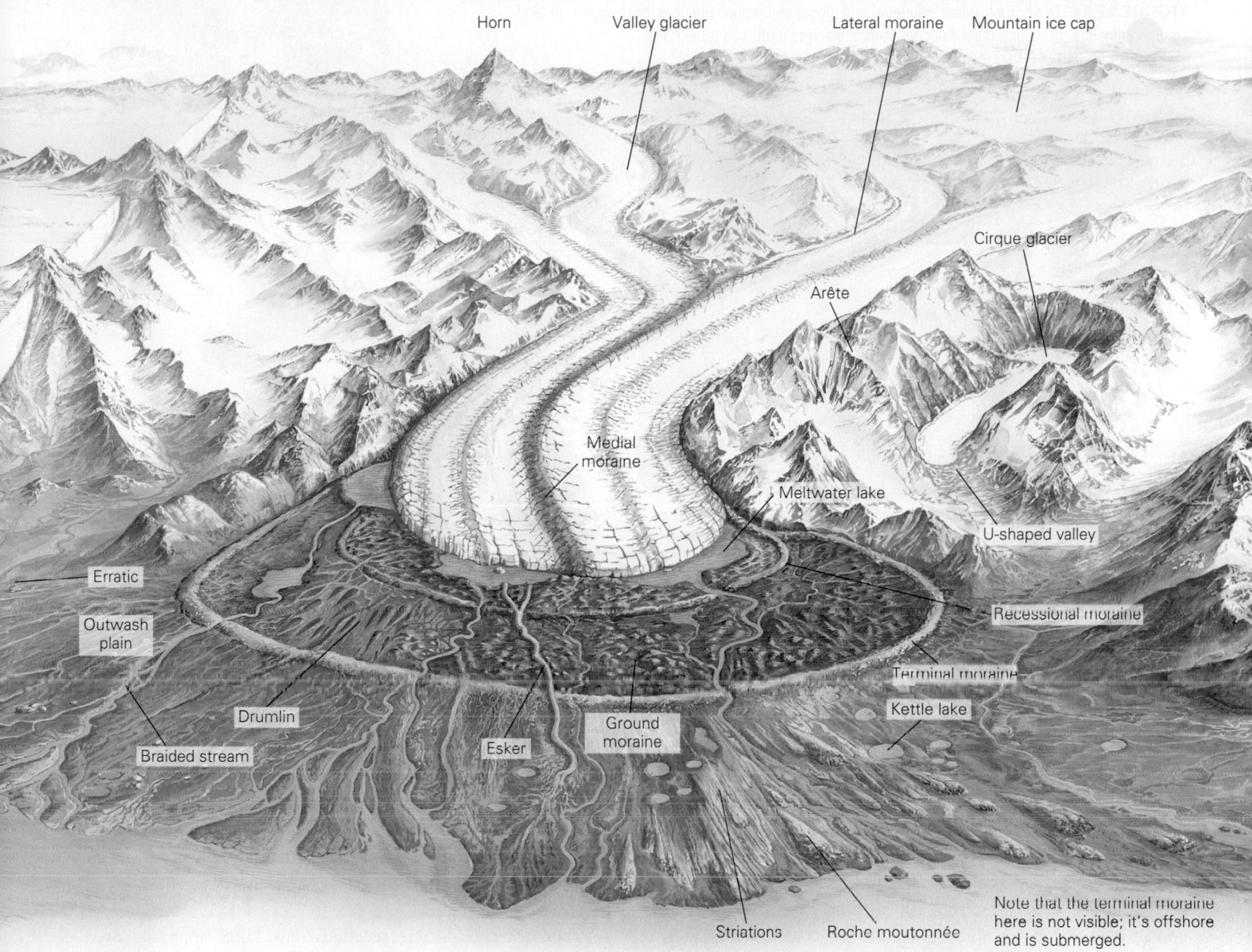

Horn   Valley glacier   Lateral moraine   Mountain ice cap

Cirque glacier

Arête

Medial moraine

Meltwater lake

U-shaped valley

Erratic

Outwash plain

Recessional moraine

Drumlin

Terminal moraine

Braided stream

Kettle lake

Esker   Ground moraine

Striations   Roche moutonnée

Note that the terminal moraine here is not visible; it's offshore and is submerged.

Glaciers are rivers or sheets of ice that last all year and slowly flow. Continental ice sheets, vast glaciers up to a few kilometers thick, covered extensive areas of land during times when the Earth had a colder climate. At the peak of the last ice age, ice sheets covered almost all of Canada, much of the United States, northern Europe, and parts of Russia. Because ice sheets store so much of the Earth's water, sea level becomes lower during an ice age.

The upper part of a glacier is brittle and may crack to form crevasses. When an ice sheet reaches the sea, it becomes an ice shelf. Rocks that the glacier has plucked up along the way can be carried out to sea with the ice; when the ice melts, these rocks fall to the seafloor as dropstones. At the edge of the shelf, icebergs calve off and float away.

Glacial recession may happen when the climate warms, so ice melts away faster at the toe (terminus) of the glacier

than it can be added at the source. Consequences of glacial erosion and deposition remain when a glacier melts away. Erosion features include striations on bedrock and roches moutonnées. Deposition features include glacial moraines, glacial outwash, and esker deposits. Even when the toe remains fixed in position for a while, the ice continues to flow and molds underlying sediment into drumlins. Ice blocks buried in till melt to form kettle holes.

Mountain or alpine glaciers grow in mountainous areas because snow can last all year at high elevations. In the mountains, glaciers fill valleys or form ice caps. Sediment falling from the mountains creates lateral and medial moraines. Glaciers carve distinct landforms in the mountains, such as cirques, arêtes, horns, and U-shaped valleys.

FIGURE 22.24 The link between sea level and global glaciation. Glaciers store water on land, so when glaciers grow, sea level falls, and when glaciers melt, sea level rises.

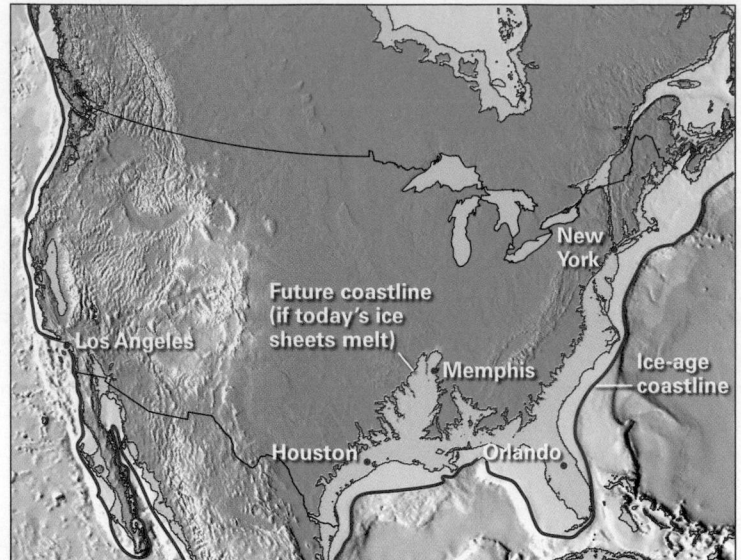

(a) The red line shows the coastline during the last ice age; much of the continental shelf was dry. If present-day ice sheets melt, coastal lands will flood.

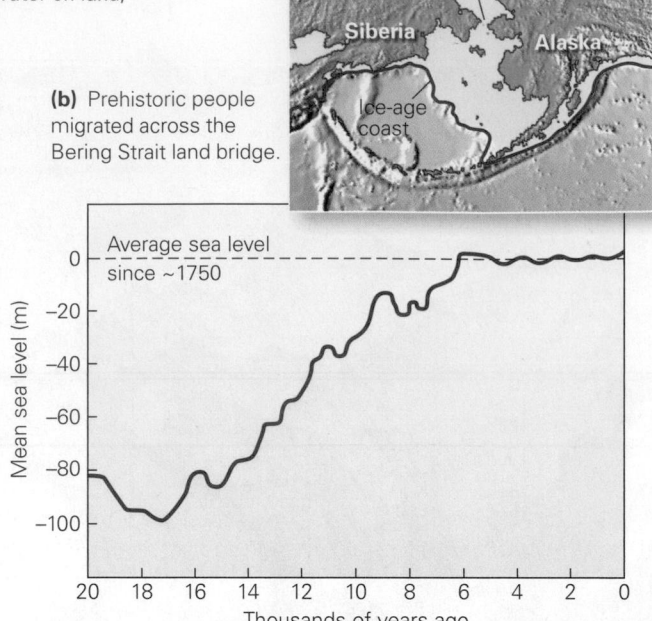

(b) Prehistoric people migrated across the Bering Strait land bridge.

(c) Sea-level rise between 17,000 and 7,000 B.C.E. was due to the melting of ice-age glaciers.

in Antarctica and Greenland were to melt, large areas of the coastal plain along the east coast and Gulf Coast of North America would become submerged, as would much of the Ganges Delta of Bangladesh.

During the last ice age, when glaciers covered almost three times as much land area as they do today, they held almost three times more water (70 million km³, as opposed to 25 million km³ today). In effect, during the ice age, water from the ocean reservoir was transferred to the glacial reservoir and remained trapped on land. As a consequence, sea level dropped by as much as 100 m, and extensive areas of continental shelves became exposed as the coastline migrated seaward, in places by more than 100 km (**Fig. 22.24**). People and animals migrated into the newly exposed ice-age coastal plains. In fact, fishermen dragging their nets along the Atlantic Ocean floor off New England today occasionally recover human artifacts. The drop in sea level also produced dry-land access, known as land bridges, across the Bering Strait between North America and northeastern Asia and between Australia and Indonesia, providing migration routes for early people.

Effects on Drainage and Lakes

Continental glaciation can significantly modify the locations and character of rivers and streams draining the land. Locally, the growth of a glacier or the deposition of a moraine can block

an individual stream. Diverted flow finds a new route and can carve new channels. By the time the glacier melts away, these new streams have become so well established that pre-glacial channels remain abandoned.

At a regional scale, glaciation during the Pleistocene Ice Age profoundly modified North America's interior drainage. Before this ice age, several major rivers drained much of the interior of the continent to the north, into the Arctic Ocean (**Fig. 22.25a, b**). The ice sheet buried this drainage network and diverted the flow into the Mississippi–Missouri network, which became larger.

When ice-age glaciers receded, new lakes appeared on the land. For example, kettles in regions covered by knob-and-kettle topography turned into small lakes, as now occur in central and southern Minnesota by the thousands. In the Canadian Shield, scouring left innumerable depressions that have now become lakes.

Erosion by glacial meltwater can carve valleys. Especially dramatic examples of this process have resulted when ice dams that held back large ice-margin lakes or moraine-dammed lakes melted and broke. In a matter of hours to days, the contents of the lakes could drain, yielding an immense flood called an outburst flood or *glacial torrent*. Torrents can carve huge valleys and steep cliffs, strip the land of soil, and leave behind immense ripple marks (**Fig. 22.25c**). For example, when the ice dam holding back Glacial Lake Missoula in Montana broke, it released an immense torrent—known as

FIGURE 22.25 Ice-age glaciation changed the position of the divide between north-draining and south-draining river networks.

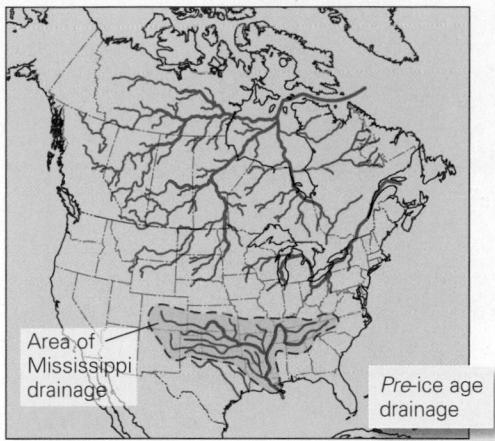

Area of Mississippi drainage

Pre-ice age drainage

(a) Before the last ice age, more rivers flowed north; the Mississippi network was smaller.

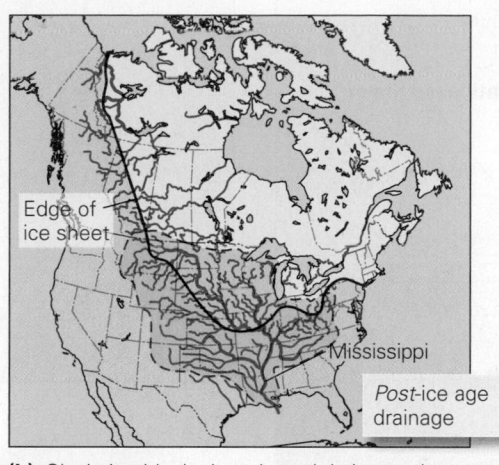

Edge of ice sheet

Mississippi

Post-ice age drainage

(b) Glaciation blocked northward drainage; the Mississippi network grew larger.

(c) Giant ripple marks formed during the Great Missoula Flood.

the *Great Missoula Flood*—that scoured eastern Washington, yielding a barren, soil-free landscape known as the *channeled scablands* (see Chapter 17). Recent evidence suggests that this process repeated several times. Another torrent flowed down the channel of what is now the Illinois River and, in northern Illinois, carved a broad, steep-sided valley that is much too large to have been cut by the present-day Illinois River.

Large ice-margin lakes formed when the weight of continental ice sheets depressed the surface of the lithosphere sufficiently to generate depressions that filled with water. The largest known ice-margin lake covered portions of Manitoba and Ontario in south-central Canada and North Dakota and Minnesota in the United States **(Fig. 22.26a)**. This body of water, known as Glacial Lake Agassiz, existed between 11,700 and 9,000 years ago, a time during which the last ice age came to a close and the continental ice sheet retreated north. At its largest, the lake covered over 250,000 km², an area greater than that of all the present Great Lakes

combined. Eventually, the ice sheet receded from the north shore of Glacial Lake Agassiz, so near the end of its life, the lake was surrounded by ice-free land. Field evidence suggests that the lake's demise came when it drained catastrophically, sending a torrent down what is now the St. Lawrence Seaway.

Pluvial Features

During the Pleistocene Ice Age, regions to the south of continental ice sheets were wetter than they are today. Fed by enhanced rainfall, lakes accumulated in low-lying areas even at a great distance from the ice front. The largest of these **pluvial lakes** (from the Latin *pluvia*, meaning rain) in North America flooded interior basins of the Basin and Range Province in Utah and Nevada **(Fig. 22.26b)**. Examples include Lake Bonneville, which covered almost a third of western Utah. When this lake suddenly drained after a natural dam holding it back broke, it left a bathtub ring of shoreline rimming the mountains near Salt Lake City (see Fig. 22.26b). Today's Great Salt Lake itself represents a small remnant of Lake Bonneville.

Periglacial Environments

In regions adjacent to the fronts of continental ice sheets during the last ice age, the mean annual temperature stayed cold enough (below −5°C) that soil moisture and groundwater froze and remained solid all year. Such permanently frozen ground, or **permafrost**, may extend to depths of 1,500 m below the ground surface. Regions with widespread permafrost that do not have a cover of snow or ice are called *periglacial environments* (the Greek *peri* means around, or encircling; periglacial environments appear around the edges of glacial environments) **(Fig. 22.27a)**. Today, such environments exist only at high latitudes and high elevations.

The upper few meters of permafrost may melt during the summer months, only to refreeze again when winter comes. As a consequence of the freeze-thaw process, the ground of some permafrost areas splits into pentagonal or hexagonal shapes, producing a landscape called **patterned ground (Fig. 22.27b)**. Water fills the gaps between the cracks and freezes to create wedge-shaped walls of ice. In some places, freeze-thaw cycles in permafrost gradually push cobbles and pebbles up from the

FIGURE 22.26 Ice-age lakes in North America.

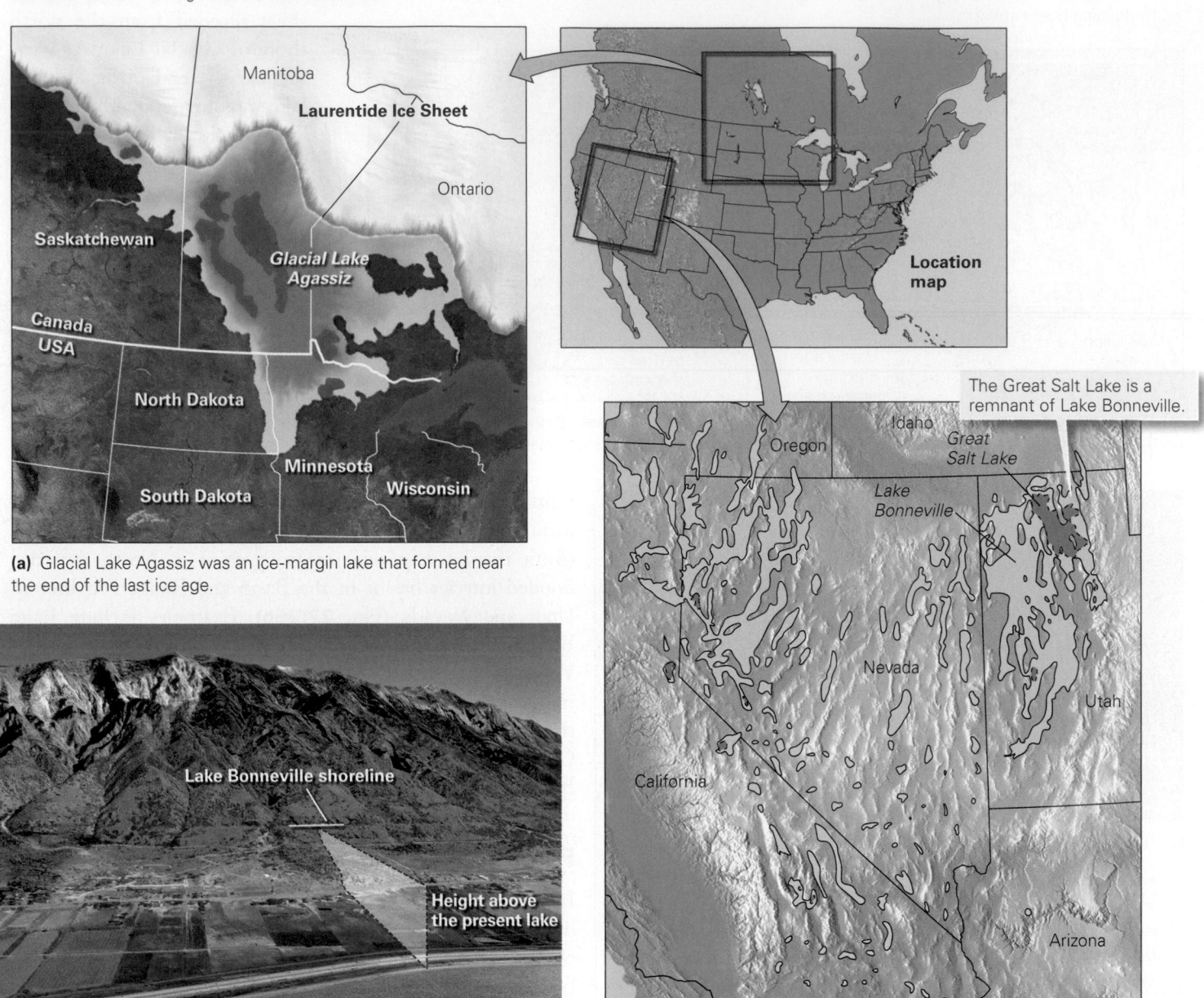

(a) Glacial Lake Agassiz was an ice-margin lake that formed near the end of the last ice age.

The Great Salt Lake is a remnant of Lake Bonneville.

(b) Pluvial lakes occurred throughout the Basin and Range Province during the last ice age due to the wetter climate. The largest of these was Lake Bonneville. Subtle horizontal terraces define the remnants of beaches, now over 100 m above the present level of the Great Salt Lake.

subsurface. These rocks, for reasons that remain poorly understood, gradually collect between adjacent bulges to form *stone circles* **(Fig. 22.27c)**.

Permafrost presents a unique challenge to people who live in polar regions or who work to extract resources from these regions. For example, heat from a building may warm and melt underlying permafrost, yielding mud into which the building settles. For this reason, buildings in permafrost regions must be placed on stilts so that cold air can circulate beneath them to keep the underlying ground frozen. When

geologists discovered oil on the northern coast of Alaska, oil companies faced the challenge of shipping the oil to markets outside of Alaska. After much debate over the environmental impact, they built the Trans Alaska Pipeline, which now carries oil 1,000 km to a seaport in southern Alaska (see Chapter 14). The oil must be warmed to lower its viscosity so that it flows more easily during transport; to prevent the warm pipeline from melting underlying permafrost, it had to be built on a frame that holds it above the ground for most of its length.

FIGURE 22.27 Periglacial regions are not ice covered but do include substantial areas of permafrost.

(a) The distribution of periglacial environments in North America.

(b) An example of patterned ground near a pond in Manitoba, Canada.

TAKE-HOME MESSAGE

In addition to erosion and deposition, the growth of continental glaciers has many consequences. For example, the weight of a large glacier can cause the land surface to subside, and melting of the glacier allows the surface to rebound. Glaciation can affect drainage patterns, and release of water from glacial lakes yields torrents. Continental ice sheets store water, so their growth or melting affects sea level. The land beyond an ice sheet may be covered with permafrost or pluvial lakes.

QUICK QUESTION: Why does the process of glacial rebound take thousands of years?

(c) These stone circles near Spitsbergen, Norway, were formed by repeated freezing and thawing, which separates gravel from silt.

22.6 The Pleistocene Ice Age

The Pleistocene Glaciers

Today, most of the land surface in New York City lies hidden beneath concrete and steel, but in Central Park, it's still possible to see land in a seminatural state. If you stroll through the park and study the rock outcrops, you'll find that their top surfaces are smooth and polished (see Fig. 22.12c) and in places have been grooved and scratched. You can also find numerous erratics. You're seeing evidence that an ice sheet once scraped along this now-urban ground. Geologists estimate that the ice sheet that overrode the New York City area may have been 250 m thick, enough to engulf the Empire State Building up to the 75th floor.

Glacial features such as those on display in Central Park first led Louis Agassiz to propose the idea that vast continental ice sheets advanced over substantial portions of North America, Europe, and Asia during a great ice age. Since Agassiz's day, geologists, by mapping out the distribution of glacial deposits and landforms, have gradually defined the extent of ice-age glaciers and a history of their movement **(Box 22.2)**.

The fact that glacially modified landscapes decorate the surface of the Earth today means that the most recent ice age took place fairly recently during the Earth's history. This ice age, responsible for the glacial landforms of North America and Eurasia, happened mostly during the Pleistocene Epoch, which began about 2.6 million years ago (Ma) (see Chapter 13), so, as we've noted earlier, geologists refer to it as the *Pleistocene Ice Age*. They use the term *Holocene* for about the

BOX 22.2 CONSIDER THIS . . .

So You Want to See Glaciation?

Though the Pleistocene continental ice sheet that once covered much of North America vanished about 6,000 years ago, you can find evidence of its presence quite easily. The Great Lakes, along the U.S.–Canada border, the Finger Lakes and drumlins of New York, the low-lying moraines and outwash plains of Illinois, and the polished outcrops of the southern Canadian Shield all formed in response to the existence of this glacier. But if you want to see continental glaciers in action today, you must trek to Greenland or Antarctica.

Mountain glaciers are easier to reach. A trip to the mountains of western North America (including Alaska), the Alps of France or Switzerland, the Andes of South America, or the mountains of southern New Zealand will bring you in contact with active glaciers. You can even spot glaciers from the comfort of a cruise ship. Some of the most spectacular glacial landscapes in North America formed during the Pleistocene Epoch, when mountain glaciers were more widespread. These landscapes are now on display in national parks.

- **Glacier National Park (Montana):** This park, which borders Waterton Lakes National Park in Canada, displays giant cirques, U-shaped valleys, hanging valleys, and terminal moraines. In 1850, there were about 150 glaciers in the park, and some of these were quite large. Now only 25 active relics of formerly larger glaciers remain, all in a mountainous terrain that reaches elevations of over 3 km. Unfortunately, these glaciers are melting away quickly and may vanish entirely by 2030.
- **Yosemite National Park (California):** A huge U-shaped valley carved into the Sierra Nevada granite batholith makes up the centerpiece of this park. Waterfalls spill out of hanging valleys bordering the valley.
- **Voyageurs National Park (Minnesota):** This park lacks the high peaks of mountainous parks but nevertheless reveals the dramatic consequences of glacial scouring and deposition on the Canadian Shield. The low-lying landscape,

dotted with lakes, contains abundant polished surfaces, glacial striations, and erratics, along with moraines, glacial lake beds, and outwash plains.
- **Acadia National Park (Maine):** During the last glacial advance, the continental ice sheet overrode low bedrock hills and flowed into the sea along the coast of Maine. This park provides some of the best examples of the consequences. Its hills were scoured and shaped into large roches moutonnées by glacial flow. Some of the deeper valleys have now become small fjords.
- **Glacier Bay National Park (Alaska):** In Glacier Bay, huge tidewater glaciers fringe the sea, creating immense ice cliffs from which icebergs calve off. Cruise ships bring tourists up to the toes of these glaciers. More adventurous visitors can climb the coastal peaks and observe lateral and medial moraines, crevasses, and the erosional and depositional consequences of glaciers that have already retreated up the valley.

---

last 12,000 years, the time since the last Pleistocene ice sheet melted away.

By mapping glacial striations and deposits, geologists have determined where the great Pleistocene ice sheets originated and flowed. In North America, the *Labrador ice sheet* started to grow over northeastern Canada and the *Keewatin ice sheet* began to form over northwestern Canada. These ice sheets eventually merged to form the *Laurentide ice sheet*, which covered all of Canada east of the Rocky Mountains and extended southward across the U.S. border as far as southern Illinois (**Fig. 22.28a**). At their maximum, the ice sheets attained a thickness of 2 to 3 km; each thinned toward its toe. In northeastern Canada, the ice sheet eroded the land surface, but farther south and west, it deposited sediment (**Fig. 22.28b, c**). It eventually merged with the Greenland ice sheet to the northeast and the Cordilleran ice sheet to the west. The *Cordilleran ice sheet* covered the mountains of western Canada as well as the southern third of Alaska.

During the Pleistocene Ice Age, mountain ice caps and valley glaciers also grew in the southern Rocky Mountains, the Sierra Nevada, and the Cascade Mountains, regions to

the south of the continental ice sheet. In Eurasia, a large ice sheet formed in northernmost Europe and adjacent Asia, and it gradually covered all of Scandinavia and northern Russia. This ice sheet flowed southward across France until it reached the Alps and merged with the Alpine mountain ice cap. Ice also covered almost all of Ireland and the United Kingdom. A smaller ice sheet grew in eastern Siberia, and glaciers expanded in the mountains of central Asia. In the southern hemisphere, Antarctica remained ice covered, and mountain ice caps expanded in the Andes, but there were no continental glaciers in South America, Africa, or Australia.

In addition to continental ice sheets, sea ice in the northern hemisphere covered all of the Arctic Ocean and parts of the North Atlantic during the Pleistocene. Sea ice surrounded Iceland, approached Scotland, and also fringed most of western Canada and southeastern Alaska.

Life and Climate in the Pleistocene World

During the Pleistocene Ice Age, all climatic belts of the northern hemisphere shifted southward (**Fig. 22.29a, b**).

FIGURE 22.28 Pleistocene ice sheets and their consequences.

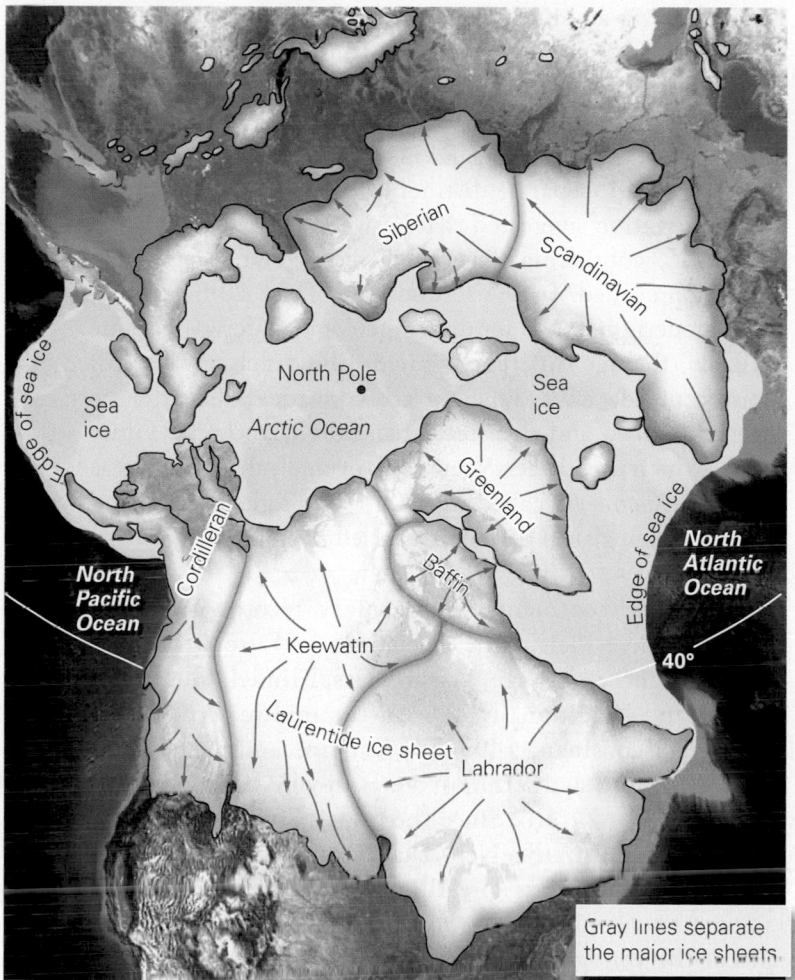

(a) During the Pleistocene, several distinct ice sheets formed. In several places, neighboring sheets merged.

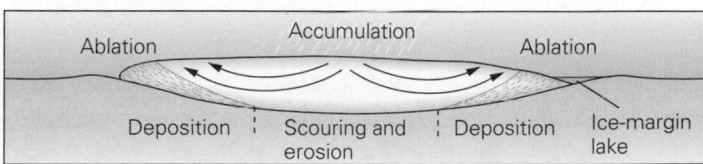

(b) Erosion dominates beneath the interior of a continental ice sheet, and deposition dominates along its margins.

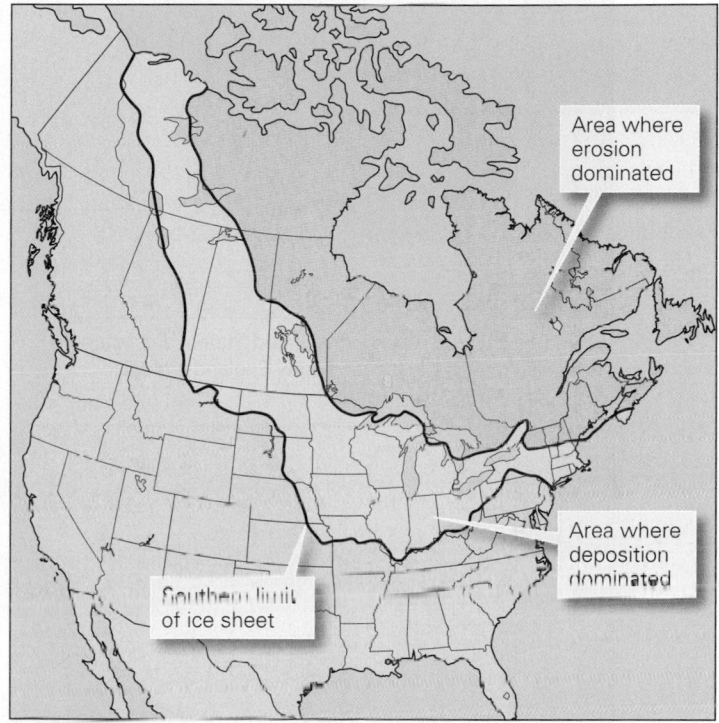

(c) Erosion by the Laurentide ice sheet dominated in northern and eastern Canada; deposition dominated in the Great Plains.

Geologists can document this shift by examining fossil pollen, which can survive for thousands of years if preserved in sediment. Presently, the southern boundary of North America's **tundra**, a treeless region supporting only low shrubs, moss, and lichens capable of living on permafrost, lies at a latitude of 68° N. During the Pleistocene Ice Age, its southern limit moved down to 48° N. Much of the interior of the United States, which now has temperate deciduous forest, harbored cold-weather spruce and pine forest. Ice-age climates also changed the distribution of rainfall on the planet. As we noted earlier, rainfall increased in North America south of the ice sheet, leading to the filling of pluvial lakes in Utah and Nevada. In contrast, rainfall decreased in equatorial regions, leading to shrinkage of the rainforest. Overall, the contrast between colder, glaciated regions and warmer, unglaciated regions caused windier conditions worldwide. These winds sent glacial rock flour skyward,

making the atmosphere hazy, presumably causing spectacular sunsets. The dust settled to form extensive deposits of loess. And because glaciers trapped so much water, as we have seen, sea level dropped.

Numerous species of now-extinct large mammals inhabited the Pleistocene world **(Fig. 22.29c)**. Giant mammoths and mastodons, relatives of the elephant, along with woolly rhinos, musk oxen, reindeer, giant ground sloths, bison, lions, saber-toothed cats, and giant cave bears wandered forests and tundra in North America. Ancestral human species already walked on the Earth by the beginning of the Pleistocene Epoch, and by its end, modern *Homo sapiens* lived on every continent except Antarctica and had discovered fire and invented tools. Rapidly changing climates may have triggered a global migration of early humans, who gained access to the Americas, Indonesia, and Australia via land bridges that became exposed when sea level dropped.

FIGURE 22.29 Climate belts during the Pleistocene.

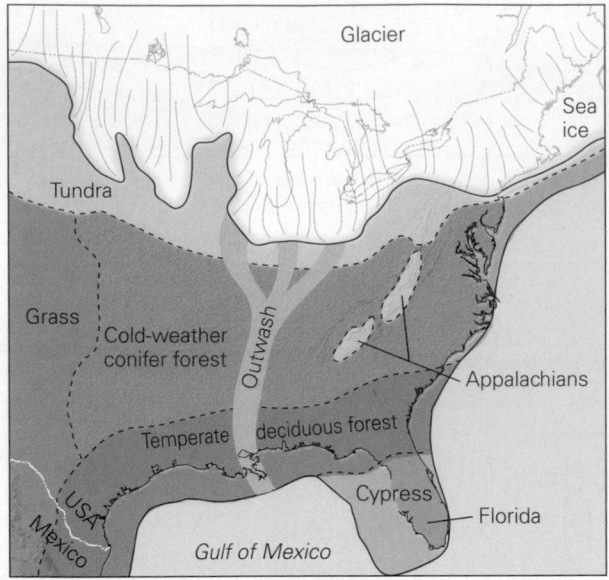

(a) Tundra covered parts of the United States, and southern states had forests like those of New England's today.

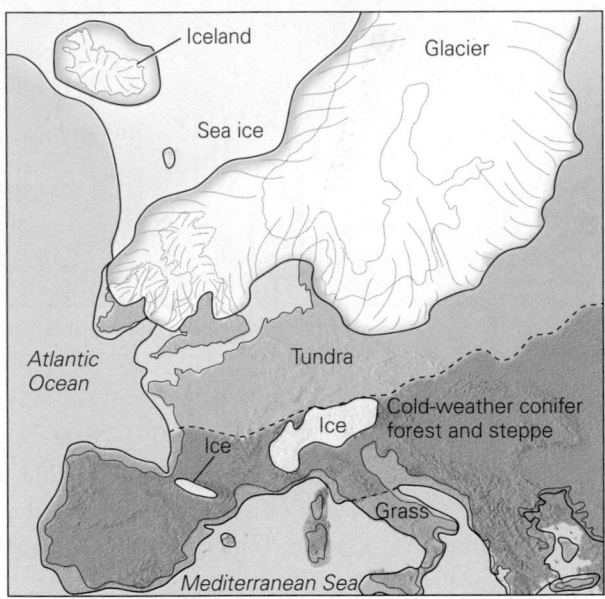

(b) Regions of Europe that support large populations today would have been barren tundra during the Pleistocene.

(c) Cold-adapted, now-extinct, large mammals roamed regions that are now temperate.

Timing of the Pleistocene Ice Age

Louis Agassiz assumed that only one ice age had affected the planet during the Pleistocene. But close examination of the stratigraphy of glacial deposits on land revealed that *paleosols* (ancient soils preserved in the stratigraphic record), as well as beds containing fossils of warmer-weather animals and plants, separate distinct layers of glacial sediment. This observation suggested that between episodes of glacial deposition, continental ice sheets receded and temperate climates prevailed. In the second half of the 20th century, when modern methods for dating geologic materials became available, the age differences among the different layers of Pleistocene glacial sediment were confirmed. Clearly, ice sheets advanced and then retreated more than once. Times during which the glaciers grew and covered substantial areas of the continents are called *glacial periods*, or **glaciations**, and times between glacial periods are called *interglacial periods*, or **interglacials**.

Using the continental sedimentary record, geologists recognized five Pleistocene glaciations in Europe (named, in order of increasing age, Würm, Riss, Mindel, Gunz, and Donau) and, traditionally, four in the midwestern United States (Wisconsinan, Illinoian, Kansan, and Nebraskan, named after the southernmost states in which their till was deposited; **Fig. 22.30**). Since the mid-1980s, geologists no longer distinguish the Nebraskan from the Kansan, but lump them together as "pre-Illinoian." With the advent of radiometric (isotopic) dating in the mid-20th century, the ages of the younger glaciations were determined by dating

FIGURE 22.30 Pleistocene glacial deposits in the north-central United States. Curving moraines reflect the shape of glacial lobes.

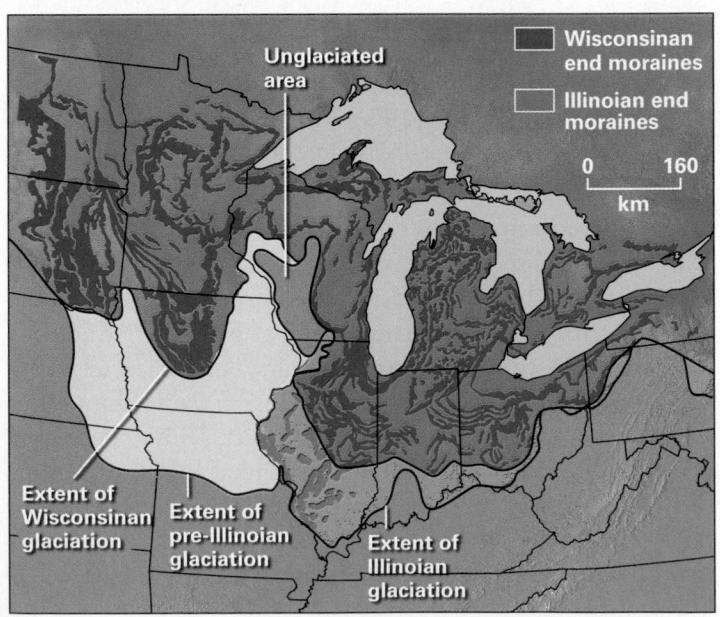

FIGURE 22.31 The timing of glaciations. Ice ages have occurred at several times in the geologic past.

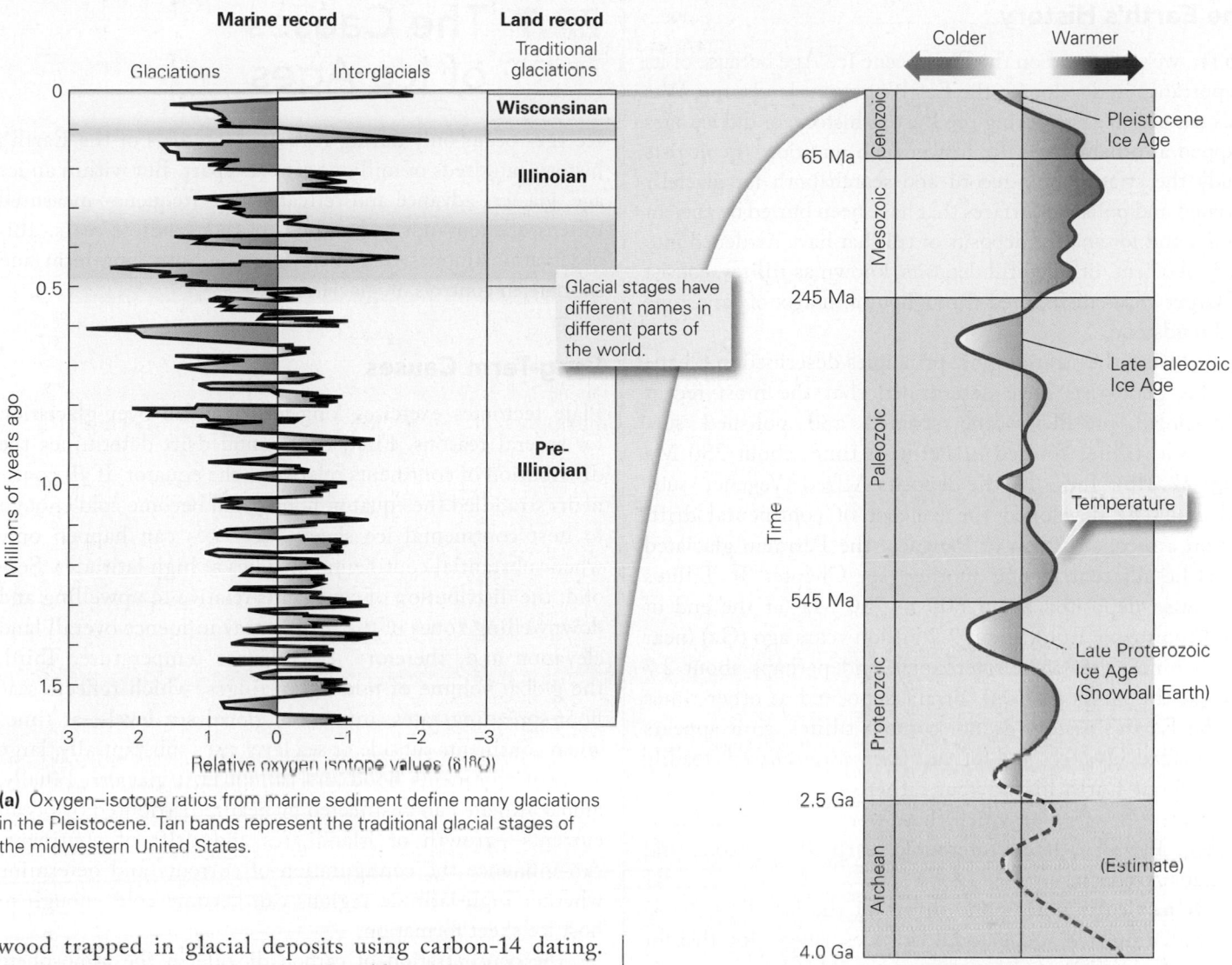

(a) Oxygen–isotope ratios from marine sediment define many glaciations in the Pleistocene. Tan bands represent the traditional glacial stages of the midwestern United States.

(b) The Pleistocene is not the only ice age in the Earth's history. Glacial advances happened in colder intervals of earlier eras, too.

wood trapped in glacial deposits using carbon-14 dating. Geologists estimate the ages of the older continental glaciations by identifying fossils in the deposits.

The four- or five-stage chronology of glaciations was turned on its head in the 1960s, when geologists began to study marine sediment containing the fossilized shells of microscopic marine plankton. Because the assemblage of plankton species living in warm water differs from the assemblage living in cold water, geologists can detect changes in the temperature of the ocean by studying plankton fossils. Researchers found that post-2.6-Ma sediment, assuming that cold water indicates a glacial period and warm water an interglacial period, records of 20 to 30 different glacial advances during the Pleistocene Epoch. The four or five traditionally recognized glaciations may represent only the largest of these. Sediments deposited on land by other glaciations were eroded and redistributed during subsequent glaciations or were eroded away by streams and wind during interglacials.

Geologists refined their conclusions about the frequency of Pleistocene glaciations by examining the isotopic composition of fossil shells. Shells of many plankton species consist of calcite ($CaCO_3$). The oxygen in the shells includes two isotopes, a heavier one (^{18}O) and a lighter one (^{16}O). The ratio of these isotopes tells us the relative water temperature in which the plankton grew, for as water gets colder, plankton incorporate a higher proportion of ^{18}O into their shells (see Chapter 23). Therefore, intervals in the stratigraphic record during which plankton shells have a large ratio of ^{18}O to ^{16}O define times when the Earth had a colder, glacial climate. The isotope record confirms that 20 to 30 glaciations occurred during the last 2.6 Ma **(Fig. 22.31a).**

Did you ever wonder...
how many ice ages have happened during Earth history?

Older Ice Ages during the Earth's History

So far, we've focused on the Pleistocene Ice Age because of its importance in developing the Earth's present landscape. Was this the only ice age during the Earth's history, or did ice ages happen at other times? To answer such questions, geologists study the stratigraphic record and search both for glacially striated and polished surfaces that have been buried by ancient strata, and for ancient deposits of till that have hardened into rock. Ancient, lithified till deposits, known as **tillites**, consist of larger clasts distributed throughout a matrix of sandstone and mudstone.

By using the stratigraphic principles described in Chapter 12, geologists have determined that the most recent widespread pre-Pleistocene striated and polished surfaces and tillites formed in Permian time, about 280 Ma (Fig. 22.31b). These are the deposits Alfred Wegener studied when he developed the concept of continental drift, for on a reconstruction of Pangaea, the Permian glaciated areas lie adjacent to one another (see Chapter 3). Tillites were also deposited about 600 to 700 Ma (at the end of the Proterozoic Eon), about 2.2 billion years ago (Ga) (near the beginning of the Proterozoic), and perhaps about 2.7 Ga (in the Archean Eon). Strata deposited at other times in the Earth's history do not contain tillites, so it appears that glacial advances and retreats have not occurred steadily throughout Earth history, but rather only in specific time intervals—*ice ages*—of which there were four or five: Pleistocene, Permian, late Proterozoic, early Proterozoic, and perhaps Archean.

Of particular note, some tillites of the late Proterozoic event accumulated at equatorial latitudes, suggesting that for at least a short time, the continents worldwide were largely glaciated, and that the sea may have been covered worldwide by ice. Geologists refer to the ice-encrusted planet of that time as **snowball Earth** (see Chapter 13).

TAKE-HOME MESSAGE

The Pleistocene Ice Age, responsible for continental glacial landscapes of today, occurred during the past 2.6 Ma. The land record shows four to five discrete glaciations, but the marine record reveals that, in fact, ice sheets advanced and retreated about 20 to 30 times. Ice ages also happened earlier in Earth history. Proterozoic glaciers and sea ice may have covered all of "snowball Earth."

QUICK QUESTION: What's the evidence for multiple Pleistocene glaciations?

22.7 The Causes of Ice Ages

Ice ages occur only during restricted intervals of the Earth's history, hundreds of millions of years apart. But within an ice age, glaciers advance and retreat with a frequency measured in tens of thousands to hundreds of thousands of years. This observation implies that there must be both long-term and short-term controls on glaciation.

Long-Term Causes

Plate tectonics exercises long-term control over glaciation for several reasons. First, continental drift determines the distribution of continents relative to the equator. If all continents straddled the equator, none could become cold enough to host continental ice sheets—ice ages can happen only when substantial continental area lies at high latitudes. Second, the distribution of continents relative to upwelling and downwelling zones of the mantle may influence overall land elevation and, therefore, land-surface temperature. Third, the global volume of mid-ocean ridges, which reflects sea-floor-spreading rates, influences global sea level—at times when continents subside or sea level rises substantially, large areas of continents flood and cannot host glaciers. Finally, global climate can be affected by heat redistributed by ocean currents—growth of island arcs and drift of continents can influence the configuration of currents and determine whether high-latitude regions can become cold enough to host ice sheet formation.

The concentration of carbon dioxide in the atmosphere may also play a key role in determining whether an ice age can or can't occur. As we've noted, carbon dioxide (CO_2) is a greenhouse gas—it traps infrared radiation rising from the Earth—so if the concentration of CO_2 increases, the atmosphere becomes warmer. Ice sheets cannot form during periods when the atmosphere has a relatively high concentration of CO_2, even if other factors favor glaciation. But what might cause long-term changes in CO_2 concentration? Possibilities include changes in the number of marine organisms that extract CO_2 from seawater to make shells; changes in the amount of chemical weathering taking place on land caused by growth of mountain ranges (weathering absorbs CO_2); changes in the amount of volcanic activity; and changes in the distribution and volume of photosynthetic organisms (these organisms remove CO_2 from the air). The widespread appearance of coal swamps may have triggered Permian glaciations of Pangaea.

Short-Term Causes

Now we've seen how the stage could be set for an ice age to occur, but why do glaciers advance and retreat periodically during an ice age? In 1920, Milutin Milanković, a Serbian astronomer and geophysicist, came up with an explanation. Milanković studied how the Earth's orbit changes shape and how its axis changes orientation over time, and he calculated the frequency of these changes. In particular, he evaluated three aspects of the Earth's movement around the Sun:

- *Orbital eccentricity:* The Earth's orbit gradually changes from a more circular shape to a more elliptical shape and back again. The degree to which an orbit deviates from a perfect circle is *orbital eccentricity*. The Earth's orbital eccentricity cycle takes about 100,000 years **(Fig. 22.32a)**.

- *Tilt of the Earth's axis:* We have seasons because the Earth's axis of rotation tilts relative to the plane of its orbit. Over time, the tilt angle varies between 22.5° and 24.5°, over a time period of 41,000 years **(Fig. 22.32b)**.

- *Precession of the Earth's axis:* If you've ever set a top spinning, you've probably noticed that its axis gradually traces a conical path. This motion, or wobble, is called *precession* **(Fig. 22.32c)**. The Earth's axis precesses over the course of about 23,000 years. Right now, the Earth's axis points toward Polaris, making Polaris the North Star, but 12,000 years ago, the axis pointed to Vega, a different star. Precession determines the relationship between the timing of the seasons and the position of the Earth along its orbit around the Sun.

Milanković showed that precession combines with variations in orbital eccentricity and tilt to affect the total annual amount of *insolation* (exposure to the Sun's rays) and the seasonal distribution of insolation that the Earth receives at the mid- to high latitudes (such as 65° N) by as much as 25%. For example, such regions receive more insolation when the Earth's axis lies almost perpendicular to its orbital plane than when its axis tilts significantly. According to Milanković, glaciers tend to advance during times of cool summers at 65° N, which occur periodically **(Fig. 22.32d)**. When geologists began to study the climate record, they found climate cycles with the frequency predicted by Milanković. These climate cycles are now called **Milankovitch cycles.**

The discovery of Milankovitch cycles in the geologic record strongly supports the contention that changes in the Earth's orbit and tilt help trigger short-term glacial advances and retreats during an ice age. But orbit and tilt changes cannot be the whole story because they could cause only about

FIGURE 22.32 Milankovitch cycles influence the amount of insolation received at high latitudes.

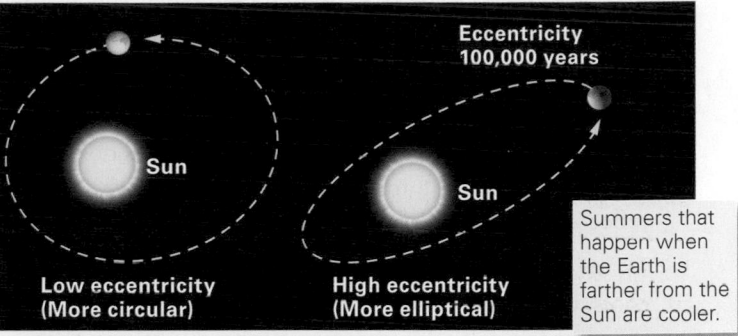

Summers that happen when the Earth is farther from the Sun are cooler.

(a) Variations caused by changes in orbital shape.

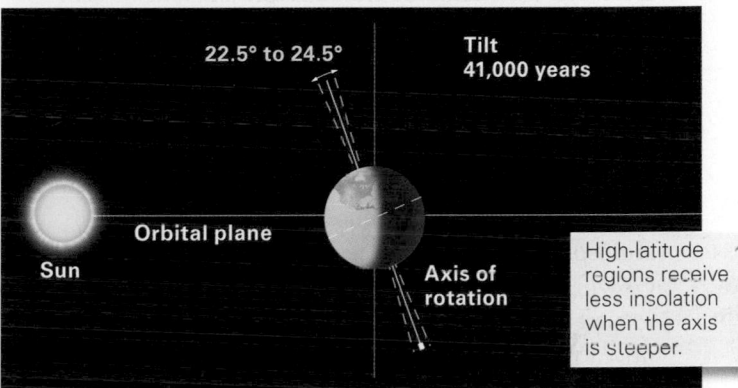

High-latitude regions receive less insolation when the axis is steeper.

(b) Variations caused by changes in axial tilt.

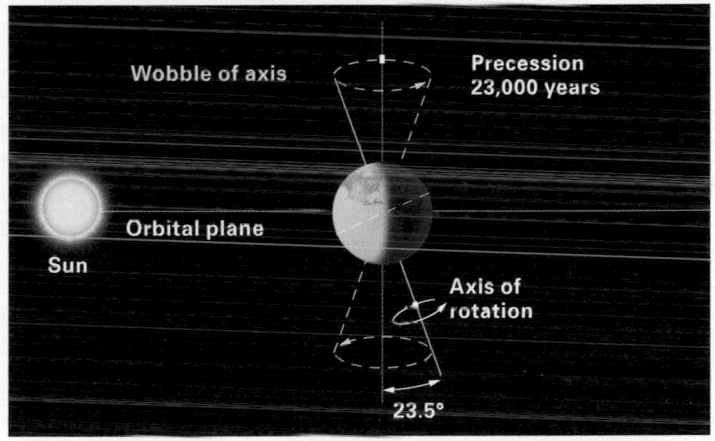

(c) Variations caused by the precession of the Earth's axis.

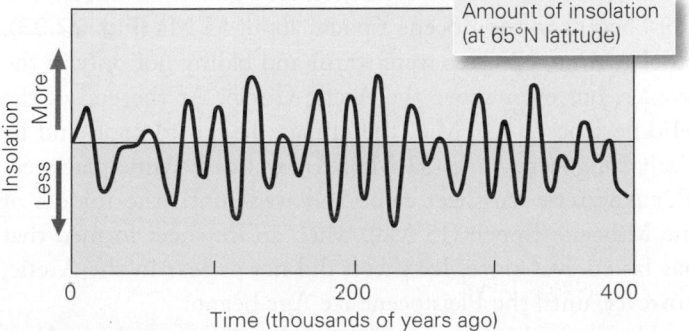

(d) Combining the effects of eccentricity, tilt, and precession produces distinct periods of more or less insolation.

a 4°C temperature decrease (relative to today's temperature), and during glaciations, the temperature decreased 5°C to 7°C along coasts and 10°C to 13°C inland. Geologists suggest that several other factors may come into play in order to trigger a glacial advance:

- *A changing albedo:* When snow remains on land throughout the year, or clouds form in the sky, the albedo (reflectivity) of the Earth increases, so the Earth's surface reflects incoming sunlight and thus becomes even cooler.

- *Interruption of thermohaline circulation:* As the climate cools, evaporation rates from the sea decrease, so seawater becomes less salty. Decreasing salinity might stop the system of thermohaline circulation that brings warm water to high latitudes (see Chapter 18). Consequently, the high latitudes might become even colder than they would otherwise.

- *Biological processes that change CO_2 concentration:* Several kinds of biological processes may have amplified climate changes by altering the concentration of CO_2 in the atmosphere. For example, a greater amount of plankton growing in the oceans could absorb more CO_2 and thus remove it from the atmosphere. A bloom of plankton might reflect an addition of nutrients to the oceans, perhaps because of changing patterns of upwelling or changing amounts of runoff.

The above processes represent *positive feedback* in that they enhance the effects of the phenomenon that causes them. Because of positive feedback, the Earth could cool more than it would otherwise during the cooler stage of a Milankovitch cycle, and this change could trigger a glacial advance. Variations in the radiation output of the Sun might also affect the amount of energy the Earth receives, but the long-term periodicity of such variation remains unclear.

A Model for Pleistocene Ice Age History

Long-Term Cooling in the Cenozoic Era Taking all of the above causes into account, we can propose a scenario for the events that led to the Pleistocene glacial advances. Our story begins in the Eocene Epoch, about 55 Ma **(Fig. 22.33)**. At that time, climates were warm and balmy not only in the tropics, but even above the Arctic Circle. At the end of the Middle Eocene (37 Ma), the climate began to cool, and by Early Oligocene time (33 Ma), Antarctica became glaciated. The Antarctic ice sheet came and went until the middle of the Miocene Epoch (15 Ma), when an ice sheet formed that has lasted ever since. Ice sheets did not appear in the Arctic, however, until the Pleistocene Ice Age began.

Cenozoic long-term climate changes may have been caused in part by changes in the pattern of ocean currents,

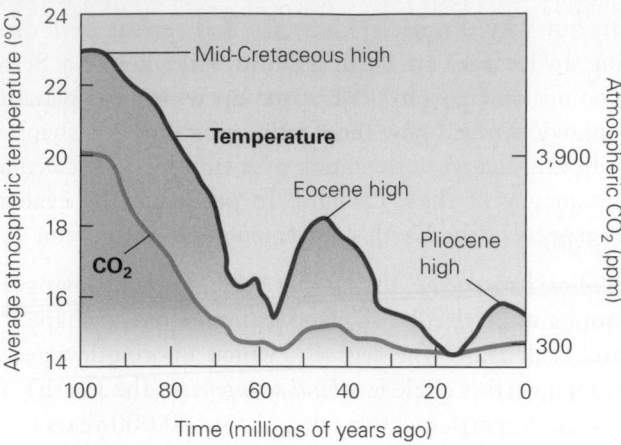

FIGURE 22.33 Until recently, the Earth's atmosphere has been gradually cooling, overall, since the Cretaceous.

which happened, in turn, because of plate tectonics. For example, in the Eocene, the collision of India with Asia cut off warm equatorial currents that had been flowing in the Tethys Sea. And in the Miocene and Oligocene, Australia and South America drifted away from Antarctica, allowing the cold Antarctic Circumpolar Current to develop. This new current prevented warm southward-flowing currents from reaching Antarctica, allowing ice to form and survive in the region. Without the warm currents, the climate of Antarctica overall underwent cooling, and this change could have cooled the global ocean. Changes to atmospheric circulation and temperature may also have happened at this time. Models suggest that the uplift of the Himalayas and the Tibetan Plateau diverted winds in a way that cooled the climate. Further, this uplift exposed more rock to chemical weathering, perhaps leading to extraction of CO_2 from the atmosphere (as noted, chemical weathering reactions absorb CO_2). A decrease in the concentration of this greenhouse gas would contribute to atmospheric cooling.

So far, we've examined hypotheses that explain long-term cooling since about 40 Ma, but what caused the sudden appearance of the Laurentide ice sheet at about 2.6 Ma? This event may coincide with other plate-tectonic events. For example, the gap between North and South America closed when the Isthmus of Panama grew and separated the waters of the Caribbean from those of the tropical Pacific for the first time. When this happened, warm currents that previously flowed out of the Caribbean into the Pacific were blocked and diverted northward to merge with the Gulf Stream. This current transfers warm water from the Caribbean up the Atlantic coast of North America and ultimately to the British Isles. As the warm water moves up the Atlantic coast, it generates warm, moisture-laden air that provides a source for the snow that falls over New England, eastern Canada, and Greenland. In other words, the Arctic has long been cold enough for ice caps, but until the Gulf Stream was diverted northward by the growth of Panama, there was no source of moisture to make the abundant snow and ice needed for glacial growth.

Short-Term Advances and Retreats in the Pleistocene Epoch

Once the Earth's climate had cooled overall, the Milankovitch cycles controlled periodic advances and retreats of ice sheets. To understand how, let's look at a possible case history of a single advance and retreat of the Laurentide ice sheet. (Such models remain the subject of vigorous debate.)

- *Stage 1:* During the overall cooler climates of the late Cenozoic Era, the Earth reached a point in the Milankovitch cycle when the average temperature in temperate latitudes dropped. Because of glacial rebound, the ice-free surface of northern Canada rose to an elevation of several hundred meters above sea level. With lower temperatures and higher elevations, not all of the winter's snow melted away during the summer. Eventually, snow covered the entire region of northern Canada, even during the summer. Because of the snow's high albedo, it reflected sunlight, so the region grew still colder (a positive feedback), and even more snow accumulated. Precipitation rates increased because evaporation from the Gulf Stream provided moisture. Finally, the snow at the base of the pile turned to ice, and the ice began to spread outward under its own weight. A new continental ice sheet had been born.

- *Stage 2:* The ice sheet continued to grow as more snow piled up in the zone of accumulation. And as the ice sheet grew, the atmosphere continued to cool because of the albedo effect. The weight of the ice loaded the continent and made it sink, so the elevation of the glacier decreased and its surface approached the elevation of the equilibrium line. Also, the temperature became cold enough that in high latitudes, the Atlantic Ocean began to freeze. As sea ice covered the ocean, the amount of evaporation decreased; removing the source of moisture diminished the amount of snowfall. The glacial advance basically choked on its own success. The decrease in the glacier's elevation (leading to warmer summer temperatures on the ice surface), as well as the decrease in snowfall, caused ablation to occur faster than accumulation, and the glacier began to retreat.

- *Stage 3:* As the glacier retreated, temperatures gradually increased and the sea ice began to melt. The supply of water to the atmosphere from evaporation increased once again, but with the warmer temperatures and lower elevations, this water precipitated as rain during the summer. The rain drastically accelerated the rate of ice melting, and the retreat progressed quite rapidly.

Will There Be Another Glacial Advance?

What does the future hold? Considering the periodicity of glacial advances and retreats during the Pleistocene Epoch, we may be living in an interglacial period. Pleistocene interglacials lasted about 10,000 years on average, and since the present interglacial began about 12,000 years ago, the time seems ripe for a new glaciation. If a glacier on the scale of the Laurentide ice sheet were to develop, major cities and agricultural belts would, over thousands of years, be overrun by ice, and their populations would have to migrate southward. Long before the ice front arrived, though, the climate would become so hostile that northern cities would already be abandoned.

The Earth actually had a brush with ice-age conditions between the 1300s and the mid-1800s, when average annual temperatures in the northern hemisphere fell sufficiently for mountain glaciers to advance significantly. During this period, now known as the *Little Ice Age*, sea ice surrounded Iceland and canals froze in the Netherlands, leading to that country's tradition of skating **(Fig. 22.34a, b)**. Some researchers speculate that the depopulation of the western hemisphere in the wake of European conquest, which brought devastating epidemics, caused temporary reforestation, for without inhabitants, farmlands went untended and new forests, which absorbed CO_2, grew. This process caused atmospheric concentrations of CO_2 to decrease, leading to the cooler conditions that triggered the Little Ice Age. Others speculate that the change reflects increased cloud cover, not a change in CO_2 concentration. Researchers will probably propose additional ideas as work on this problem continues.

During the past 150 years, temperatures have warmed, and most mountain glaciers have retreated significantly **(Fig. 22.34c)**. Large slabs have been calving off Antarctic ice shelves. In fact, most of the Larsen Ice Shelf of Antarctica, an area larger than Rhode Island, disintegrated in 2002, over a period of only one month, and as we noted earlier, a 200-km long iceberg calved off in 2017. Greenland's continental glacier, in particular, is showing signs of accelerating retreat **(Fig. 22.35)**. Large meltwater ponds are forming on the surface of the ice sheet, and some of these ponds drain abruptly through cracks to the base of the ice a kilometer below. The vast majority of researchers suggest that this global warming trend is due to increased CO_2 concentrations in the atmosphere from the burning of fossil fuels (see Chapter 23). Global warming could conceivably cause a "super-interglacial," meaning that the next glaciation could be substantially delayed or might never happen.

Current trends of global warming have led to concern that the ice sheet of West Antarctica might begin to float and then break up rapidly. If all of today's ice caps melted, global

FIGURE 22.34 The Little Ice Age and its demise. Glaciers that advanced between 1550 and 1850 have since retreated.

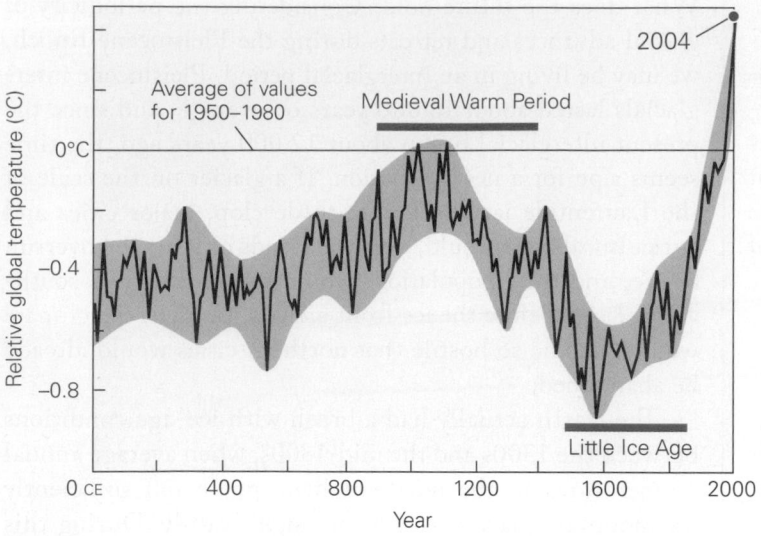

(a) A model of global temperature for the past 2,000 years. Overall trends display the Medieval Warm Period followed by the Little Ice Age. Since 1850, temperatures have warmed.

(b) Skaters (ca. 1600) on the frozen canals of the Netherlands during the Little Ice Age.

(c) During the Little Ice Age, a glacier filled this valley. In this 2003 photo, most of the glacier has vanished. Most of the retreat has happened in the last century.

sea level would rise by 70 m, extensive areas of coastal plains would be flooded, and major coastal cities such as New York, Miami, and London would be submerged (see Fig. 22.24). Instead of protruding from ice, the tip of the Empire State Building would protrude from the sea. We may not know which scenario—icehouse or greenhouse—will play out in the future until it happens. However, researchers have voiced concern that, at least in the near term, glacial retreat will be the order of the day, as global temperatures seem to be rising. The next chapter addresses such change.

TAKE-HOME MESSAGE

Ice ages occur when the distribution of continents, ocean currents, and the concentration of atmospheric CO_2 are appropriate. Advances and retreats during an ice age are controlled by Milankovitch cycles that take into account variations in the Earth's orbit and axis of rotation. Because of global warming, we may be living in a super-interglacial period.

QUICK QUESTION: How does positive feedback contribute to a glaciation during an ice age, and why does the glaciation eventually cease?

FIGURE 22.35 Greenland's melting glaciers. Melting has accelerated in the last few decades.

(a) Lakes of meltwater accumulate on the surface of the ice sheet during the summer.

(c) Where glaciers meet the sea, huge masses calve off and crash into the water. This is happening so fast that the ice front is retreating.

(b) Lakes suddenly drain through cracks that carry the water to the base of the glacier, 1 km down. Addition of liquid water to the base allows the glacier to move faster, causing a "surge."

☐ Melting ice

(d) The melting area has increased dramatically in recent years.

SUMMARY

- Glaciers are streams or sheets of recrystallized ice that survive for the entire year and flow in response to gravity. Mountain glaciers fill cirques and valleys. Continental glaciers (ice sheets) spread over large areas of continents.

- Glaciers form when snow accumulates over a long period. The snow turns first to firn and then to ice.

- Temperate glaciers undergo melting during part of the year; polar glaciers do not. Glaciers move by basal sliding or by plastic deformation of ice grains.

- Glaciers move because of gravitational pull; they flow in the direction of their surface slope.

- Whether the toe of a glacier stays fixed in position, advances, or retreats depends on the balance between the rate at which snow builds up in the zone of accumulation and the rate at which glaciers melt, sublimate, or calve in the zone of ablation.

- Continental glaciers that flow out into the sea make ice shelves. Sea ice forms where the ocean's surface freezes. Icebergs break off glaciers that flow into the sea.

- As glacial ice flows over sediment, it incorporates clasts. These clasts act like a rasp that abrades the substrate.

- Mountain glaciers carve numerous landforms, including cirques, arêtes, horns, U-shaped valleys, hanging valleys, and truncated spurs. Fjords are glacially carved valleys that filled with water.

- Glaciers transport sediment of all sizes, some of which may later be sorted by water or wind. Lateral moraines accumulate along the sides of valley glaciers, and medial moraines form down the middle. End moraines accumulate at a glacier's toe.

- Glacial depositional landforms include moraines, knob-and-kettle topography, drumlins, kames, eskers, lake beds, and outwash plains.

- Continental lithosphere subsides as a result of ice loading. When the glacier melts away, the crust rebounds.

- When water is stored in continental glaciers, sea level drops. When glaciers melt, sea level rises.

- During past ice ages, the climate in regions south of the continental ice sheets was wetter, and pluvial lakes formed. Permafrost exists in periglacial environments.

- During the Pleistocene Ice Age, large continental ice sheets covered much of North America, Europe, and Asia.

- The stratigraphy of Pleistocene glacial deposits on land records 5 European and 4 North American glaciations. The record in marine sediments records 20 to 30 such events.

- Long-term causes of ice ages include plate tectonics and changes in the concentration of CO_2 in the atmosphere. Short-term causes include the Milankovitch cycles (caused by periodic changes in the Earth's orbit and tilt).

GUIDE TERMS

ablation (p. 842)
albedo (p. 834)
arête (p. 848)
basal sliding (p. 838)
continental glacier (p. 837)
crevasse (p. 839)
dropstone (p. 845)
drumlin (p. 854)
end moraine (p. 851)
equilibrium line (p. 842)
erratic (p. 833)
esker (p. 855)
firn (p. 834)

fjord (p. 850)
glacial advance (p. 842)
glacially polished surface (p. 845)
glacial outwash plain (p. 855)
glacial retreat (p. 842)
glacial striation (p. 845)
glacial subsidence (p. 856)
glacial till (p. 852)
glaciation (p. 866)
glacier (p. 833)
hanging valley (p. 849)
horn (p. 848)

ice age (p. 833)
iceberg (p. 843)
ice shelf (p. 843)
interglacial (p. 866)
lateral moraine (p. 851)
loess (p. 852)
medial moraine (p. 851)
Milankovitch cycle (p. 869)
mountain (alpine) glacier (p. 836)
patterned ground (p. 861)
permafrost (p. 861)
Pleistocene Ice Age (p. 834)

pluvial lake (p. 861)
polar glacier (p. 838)
post-glacial rebound (p. 857)
roche moutonnée (p. 849)
sea ice (p. 845)
snowball Earth (p. 868)
temperate glacier (p. 837)
terminal moraine (p. 854)
tidewater glacier (p. 843)
tillite (p. 868)
tundra (p. 865)
U-shaped valley (p. 848)
varve (p. 852)

 **GEOTOURS** *THIS CHAPTER'S GEOTOURS WORKSHEET (R) FEATURES QUESTIONS AND GOOGLE EARTH SITES ON:*

- Continental glacier features in the northeastern United States
- Alpine valley glacial features around the world
- Piedmont glacial features in Alaska

REVIEW QUESTIONS

The letters following each Review Question refer to the corresponding Learning Objective from the Chapter Opener.

1. What evidence did Louis Agassiz offer to support the idea of an ice age? **(D)**

2. How do mountain glaciers and continental glaciers differ in terms of dimensions, thickness, and movement? **(A)**

3. Describe the transformation from snow to glacier ice. **(A)**

4. Explain how arêtes, cirques, and horns form. **(B)**

5. Describe the mechanisms that enable glaciers to move and explain why they move. **(B)**

6. How fast do glaciers normally move? How fast can they move during a surge? **(B)**

7. Explain how the balance between accumulation and ablation controls glacial advance and retreat. Identify the zones of accumulation and ablation in the figure. **(B)**

8. How can a glacier continue to flow toward its toe even though its toe is retreating? **(D)**

9. How does a glacier transform a V-shaped river valley into a U-shaped valley? Discuss how hanging valleys develop. **(B)**

10. Describe the various kinds of glacial deposits and the materials from which they are made. Identify the various depositional landforms on the figure. **(C)**

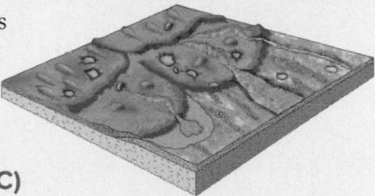

11. How does lithosphere respond to the weight of glacial ice? **(B)**

12. How was the world different during the glacial advances of the Pleistocene Ice Age? Be sure to mention the relationship between glaciations and sea level. **(D)**

13. How was the standard four-stage chronology of North American glaciations developed? Why is it so incomplete? How was it modified with the study of marine sediment? **(D)**

14. Were there ice ages before the Pleistocene? If so, when? **(D)**

15. What are some of the long-term causes that lead to ice ages? What are the short-term causes that trigger glaciations and interglacials? **(E)**

ON FURTHER THOUGHT

16. If you fly over the fields of central Illinois during the early spring, you will see slight differences in soil color due to variations in moisture content—wetter soil is darker. These variations outline the shapes of polygons that are tens of meters across. What do these patterns represent? **(E)**

17. An unusual late Precambrian rock unit crops out in the Flinders Range, a small mountain belt in South Australia. This unit consists of clasts of granite and gneiss, in a wide range of sizes, suspended throughout a matrix of slate. What is this unusual rock? **(C)**

ONLINE RESOURCES

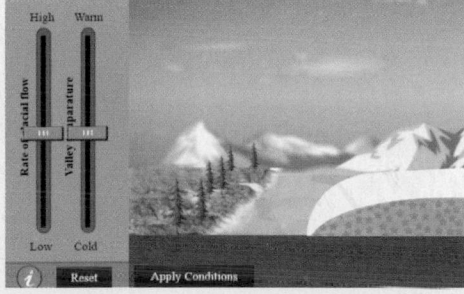

Animations
This chapter features interactive animations that simulate glacial bodies under different conditions.

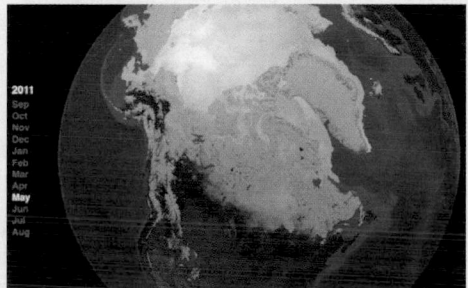

Videos
This chapter features videos on rapid glacier change and studies of Greenland's retreating glaciers.

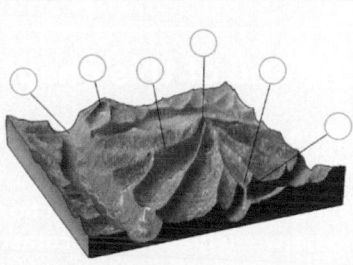

Smartwork5
This chapter features visual identification and labeling exercises on the effects of glacial systems on the landscape.

CHAPTER 23

Global Change in the Earth System

By the end of this chapter, you should be able to . . .

A. explain why the Earth has been able to change in many ways over its history.

B. distinguish between unidirectional and cyclic changes and between gradual and catastrophic changes, and give examples of each.

C. describe how elements or compounds flow among various reservoirs during biogeochemical cycles.

D. characterize changes in the Earth's climate over geologic history, and discuss phenomena that play a role in regulating climate at different time scales.

E. discuss ways in which human society has significantly modified the Earth System, particularly since the industrial revolution.

F. evaluate evidence that climate has changed during the past few centuries, and describe the causes and consequences of this change, as well as possible methods of investigating it.

G. provide scenarios for changes in the Earth System in the near and very distant future.

23.1 Introduction

Did the Earth's surface look the same in the Jurassic Period as it does today? Definitely not! Two hundred million years ago, the North Atlantic Ocean was just a narrow sea, and the South Atlantic Ocean didn't exist at all, so most dry land connected to form a single vast supercontinent, Pangaea **(Fig. 23.1)**. Today, the Atlantic is a wide ocean, and the Earth has seven separate continents. Moreover, during the Jurassic, the call of the wild rumbled from the throats of dinosaurs, whereas today mammals dominate the land. In essence, what we see of the Earth today is just a snapshot, an instant in the life story of a constantly changing planet. This idea arguably stands as geology's greatest philosophical contribution to humanity's understanding of our Universe.

Why has the Earth changed so much over geologic time, and why does it continue to change? Ultimately, change happens because the Earth's internal heat keeps the asthenosphere weak enough to flow and because the Sun's heat keeps most of the Earth's surface at temperatures above the freezing point of water. Flow in the asthenosphere permits plate tectonics, which leads to continental drift, volcanism, and mountain building. Solar heat, together with gravity, keeps streams, glaciers, waves, and wind in motion, thereby causing erosion and deposition. Solar heat also makes the Earth's surface and near-surface regions hospitable to life. If the Earth did not have just the right mix of tectonic activity and solar heat, it would be a frozen dust bowl like Mars, a crater-pocked wasteland like the Moon, or a cloud-choked oven like Venus.

Many of the changes that take place on the Earth reflect complex interactions among geologic and biological phenomena. For example, photosynthetic organisms affect the composition of the atmosphere by providing oxygen, and atmospheric composition, in turn, determines the nature of the chemical weathering reactions that take place in rocks. We've referred to the global interconnecting web of physical and biological

phenomena on the Earth as the **Earth System**. We can now define **global change**, in a general sense, as transformations or modifications of physical and biological components of the Earth System over time.

Geologists distinguish among different types of global change on the basis of the rate of a change or the way in which the change progresses over time. Specifically, *gradual change* takes place over long periods of geologic time (millions to billions of years); *catastrophic change* takes place relatively rapidly in the context of geologic time (seconds to millennia); *unidirectional change* involves transformations that never repeat; and *cyclic change* repeats the same steps over and over, though not necessarily with the same results or at the same rate.

In this chapter, we begin by reviewing examples of global change involving phenomena discussed earlier in the book. Then we introduce the concept of a *biogeochemical cycle*, the exchange of chemicals among living and nonliving reservoirs, for certain kinds of global change reflect an alteration in the proportions of chemicals held in different reservoirs. Finally, we focus on *global climate change*, transformations or modifications in the Earth's climate over time, some of which have

FIGURE 23.1 The map of the Earth's surface changes over time because of plate motions.

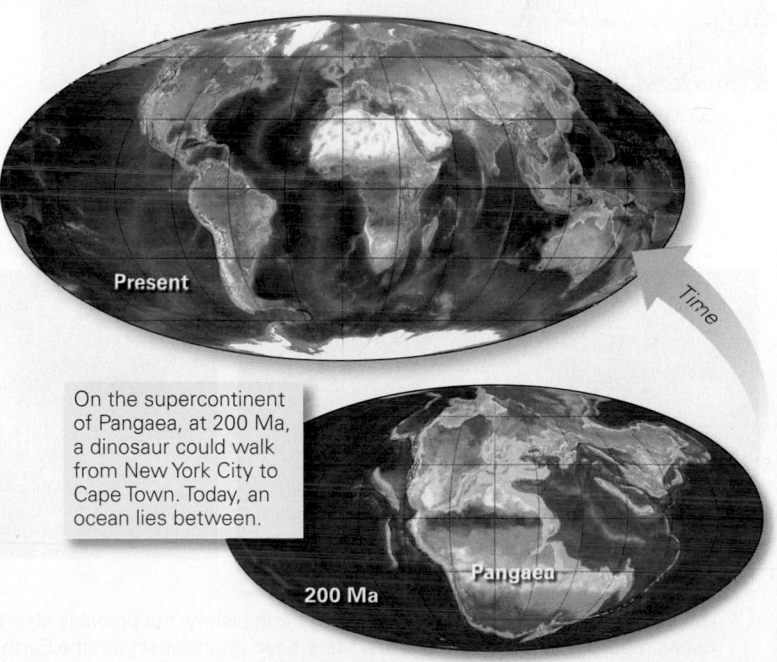

Present

Time

On the supercontinent of Pangaea, at 200 Ma, a dinosaur could walk from New York City to Cape Town. Today, an ocean lies between.

Pangaea

200 Ma

been attributed to the actions of human society. We conclude this chapter, and this book, by considering hypotheses that describe the ultimate global change—the end of the Earth—in the very distant future.

23.2 Unidirectional Changes

The Evolution of the Solid Earth

Recall from Chapter 1 that the Earth began as a fairly homogeneous mass formed by the coalescence of planetesimals. The homogeneous proto-Earth did not last long, for within about 10 million to 100 million years of its birth, the planet began to melt, yielding liquid iron alloy that sank to the center to form the core **(Fig. 23.2a)**. This process of **differentiation** represents a major unidirectional change in the Earth System—it produced a layered, onion-like planet with an iron alloy core surrounded by a rocky mantle.

FIGURE 23.2 Examples of major unidirectional change in Earth history.

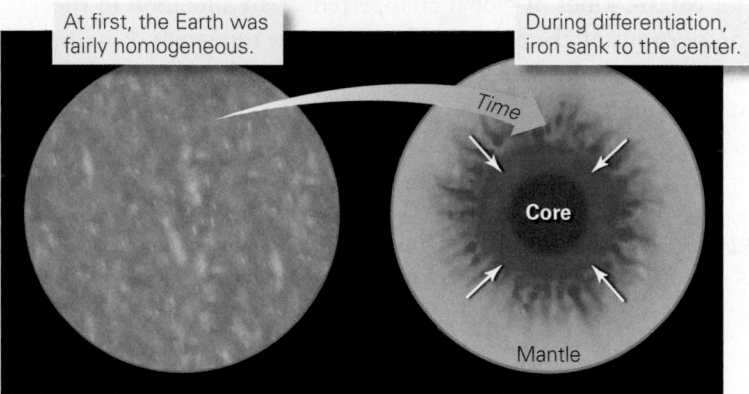

At first, the Earth was fairly homogeneous.

During differentiation, iron sank to the center.

Time

Core

Mantle

(a) When the Earth became hot enough inside, the rock within it started to melt, and droplets of iron sank to the center, accumulating to form a core. Once differentiation happened, it could not happen again.

According to a widely held model, a large protoplanet collided with the newborn Earth soon after its differentiation. This collision caused a catastrophic change in that much of both the Earth and the colliding object fragmented and vaporized, yielding a ring of debris that quickly coalesced to form the Moon **(Fig. 23.2b)**. Immediately after this collision, the Earth's mantle remained largely molten, and the planet's surface was a sea of magma. But cooling probably happened fairly quickly so that, according to recent research, the surface of the Earth had solidified and may even have hosted liquid water before 4.0 billion years ago (Ga).

The Earth underwent another catastrophic change between about 4.0 and 3.9 Ga, when it endured pummeling by asteroids and comets, an event known as the *late heavy bombardment*. Almost all the crust that had formed prior to 3.9 Ga was pulverized or melted—in fact, hardly any rock older than 4.0 Ga has yet been found. Geologists refer to the half-billion years between the birth of the Earth and the beginning of the rock record as the Hadean Eon. When bombardment slowed, our planet changed again, cooling enough for a new crust to form, for new, permanent seas to accumulate, and probably for an early form of plate tectonics to begin operating. These changes allowed the rock record to start accumulating, and the Archean Eon began.

During the Archean, another type of unidirectional change began when partial melting, perhaps associated with subduction or the upwelling of mantle plumes, started to produce relatively low-density rocks, such as granite. Low-density rocks could not be subducted, so they remained at or near the Earth's surface to constitute relatively buoyant blocks of crust. Eventually, these blocks sutured together to form the first continents. Effectively, partial melting "distilled" the crust out of the mantle. The amount of continental crust increased progressively as the process continued until, by the end of the Archean, continental crust covered about 25% of the Earth's surface. Subsequently, continental area has continued to grow,

Time

A protoplanet collided with the Earth.

Debris sprayed into space.

The debris coalesced to form the Moon.

(b) Moon formation happened early during Earth history, but probably after differentiation. According to a popular theory, the Moon coalesced from debris resulting from the collision of a Mars-sized protoplanet with the Earth.

but at a slower pace—today, continental crust underlies about 30% of the surface.

The Evolution of the Atmosphere and Oceans

The Earth's atmosphere has also undergone major unidirectional change over time. Changes in atmospheric composition have been so profound that researchers distinguish among first, second, and third atmospheres. As discussed in Chapter 20, the *first atmosphere* consisted of gases from the protoplanetary disk that had been trapped by our planet's gravity. These gases eventually escaped into space and were replaced by gases belched from volcanoes, perhaps mixed with gases brought to the Earth by comets. In this way, the Earth accumulated a *second atmosphere* composed mainly of carbon dioxide (CO_2) and water (H_2O). Other gases, such as nitrogen (N_2), composed only a minor proportion of the second atmosphere. When the Earth's surface cooled, however, water condensed and fell as rain, collecting in low areas to form oceans. Gradually, CO_2 dissolved in the oceans and was absorbed by chemical-weathering reactions on land, so its concentration in the atmosphere decreased. Molecular nitrogen, which doesn't react with other chemicals, was left behind. Consequently, the atmosphere's composition changed again to form the *third atmosphere*, dominated by N_2. Photosynthetic organisms appeared early in the Archean. But it probably wasn't until between 2.5 and 2.0 Ga, in the early Proterozoic, that oxygen (O_2) became a significant proportion of the atmosphere, and it didn't reach breathable concentrations for another billion years.

Did you ever wonder...
whether the Earth's atmosphere has always been breathable?

The Evolution of Life

During its earliest stages, the Earth's surface was probably lifeless, for carbon-based organisms could not survive the high temperatures of the time. The fossil record indicates that life had appeared at least by 3.8 Ga, and that it has undergone evolution, a unidirectional change, in fits and starts ever since (see Interlude E). Though simple organisms such as archaea and bacteria still exist, life evolution during the late Proterozoic and early Phanerozoic yielded multicellular organisms **(Fig. 23.3)**. Complex life could develop only once the air and sea contained sufficient oxygen, for oxygen-based metabolism is more efficient than methane- or sulfide-based metabolism. Life now inhabits regions from a few kilometers below the Earth's surface to a few kilometers above, yielding a diverse biosphere.

TAKE-HOME MESSAGE

Since it first formed, the Earth has undergone major unidirectional changes, which will never repeat. Examples include internal differentiation (core formation), Moon formation, growth of continental crust, atmospheric evolution, ocean formation, and life evolution.

QUICK QUESTION: How are life evolution and atmospheric evolution linked?

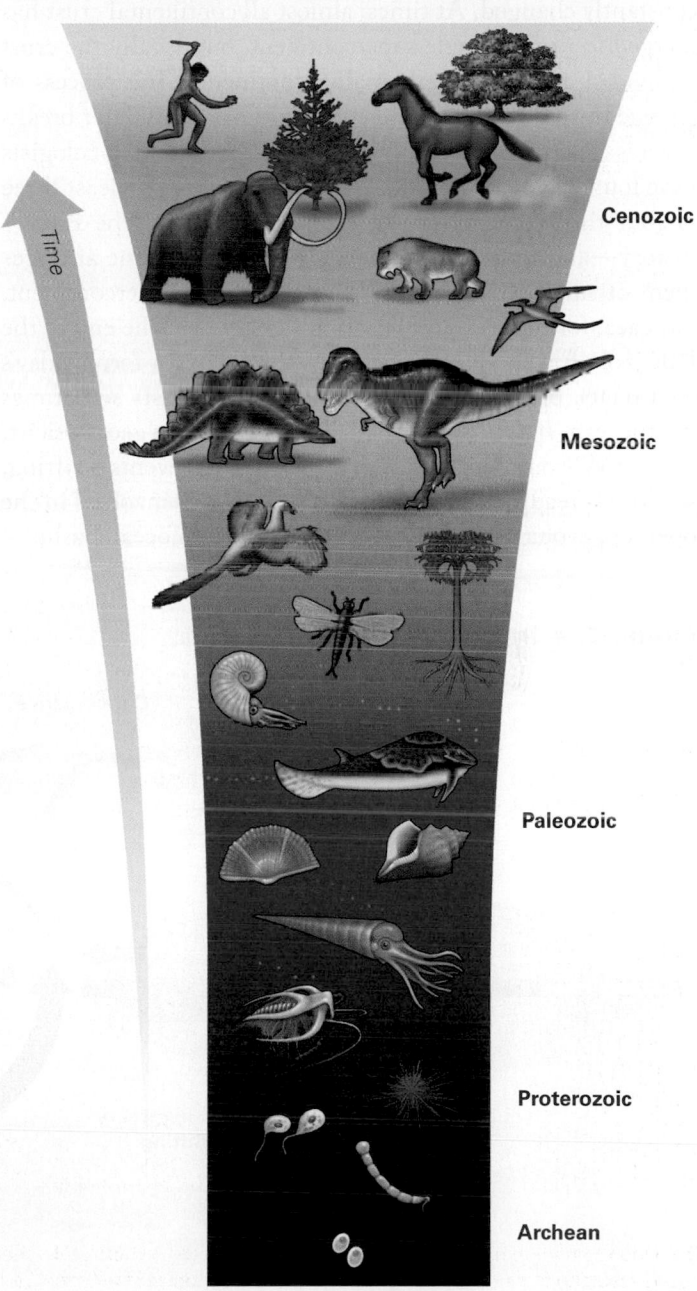

FIGURE 23.3 New species of life have evolved over geologic time. Though some of the simplest still exist, more complex organisms have appeared more recently.

Time

Cenozoic

Mesozoic

Paleozoic

Proterozoic

Archean

23.3 Cyclic Changes

During *cyclic changes*, a sequence of stages may be repeated over time. Some cyclic changes are periodic in that the cycles happen with a definable frequency, but others are not. Below, we look at several examples of cyclic change—you'll see that some involve movements of physical components of the Earth, whereas others involve transfer of chemicals among both living and nonliving reservoirs.

The Supercontinent Cycle

During the Earth's history, the map of the planet's surface has constantly changed. At times, almost all continental crust has merged to form a single supercontinent, but usually the crust is divided among several smaller continents. The process of change during which a supercontinent forms and later breaks apart is called the **supercontinent cycle (Fig. 23.4)**. Geologists have found evidence that supercontinents existed at least three or four times during the past 3 billion years of the Earth's history—and that no two included exactly the same arrangement of smaller continents. The most recent supercontinent, Pangaea, formed 300 million years ago (Ma) at the end of the Paleozoic Era and survived until it broke up to form today's continents, beginning about 200 Ma. Geologists sometimes use the term *Wilson cycle* (named for the Canadian geophysicist, J. Tuzo Wilson) for the succession of geologic events—rifting, seafloor spreading, subduction, and collision—involved in the opening, evolution, and eventual closing of an ocean basin.

The Sea-Level Change Cycle

Global sea level rose and fell by as much as 300 m during the Phanerozoic and probably did the same in the Precambrian. When sea level rose, the shoreline migrated inland, and low-lying plains became submerged. In fact, during periods of particularly high sea level, shallow seas covered more than half of the Earth's continental area **(Fig. 23.5a, b)**. At such times, shallow marine sediment buried continental deposits. When sea level fell, the continents became dry again, and regional unconformities developed. We can see the record of this sea-level change cycle preserved in the sedimentary beds of the midwestern United States. Here a succession of strata record at least six continent-wide advances and retreats of the sea, each of which left behind a blanket of sediment called a **sedimentary sequence**. Unconformities define the boundaries between the sequences **(Fig. 23.5c)**. Of note, the sequence deposited during the Pennsylvanian contains at least 30 shorter repeated intervals, called *cyclothems*, each of which contains a layer of coal. Cyclothems represent short-term cycles of sea-level rise and fall.

After studying sedimentary sequences around the world, geologists pieced together a chart defining the succession of global transgressions and regressions during the Phanerozoic Eon. This global sedimentary cycle chart may largely reflect the cycles of *eustatic* (worldwide) *sea-level change*. However, the chart probably does not give us an exact image of sea-level change because the sedimentary record reflects other factors as well, such as changes in sediment supply. Eustatic sea-level changes may be due to advances and retreats of continental glaciers, changes in the volume of mid-ocean ridge systems,

FIGURE 23.4 The stages of the supercontinent cycle.

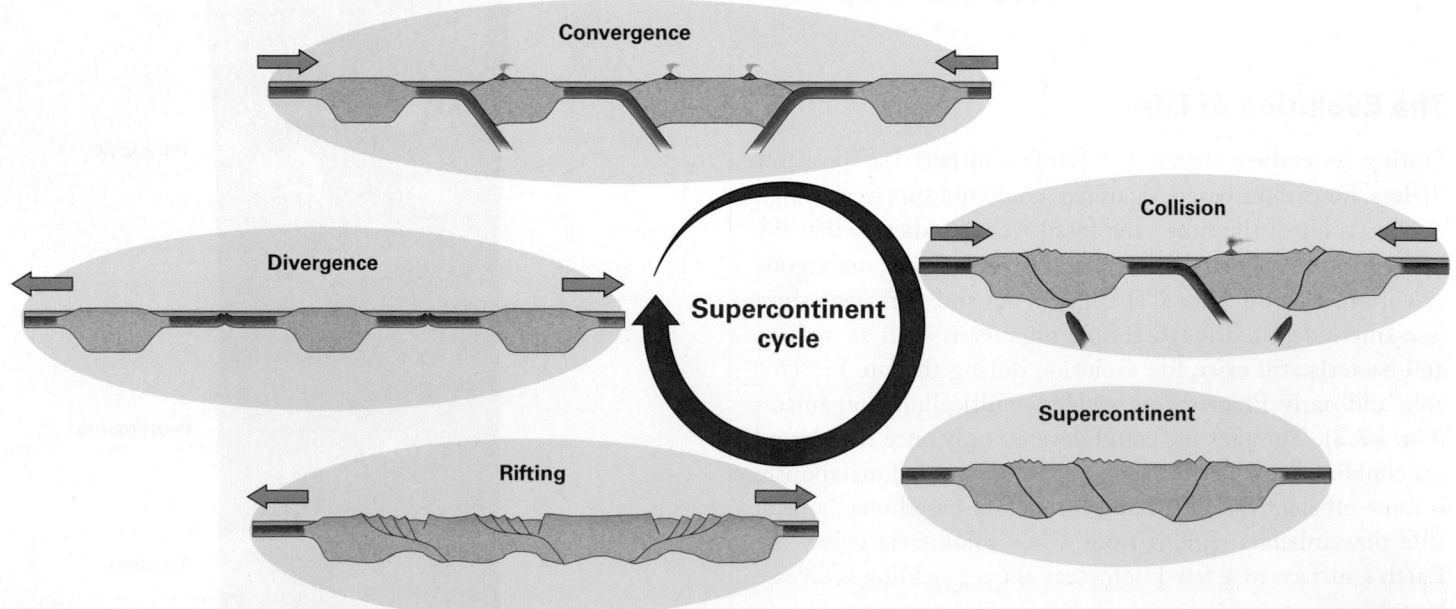

FIGURE 23.5 Sea-level change, and its manifestations, over geologic time.

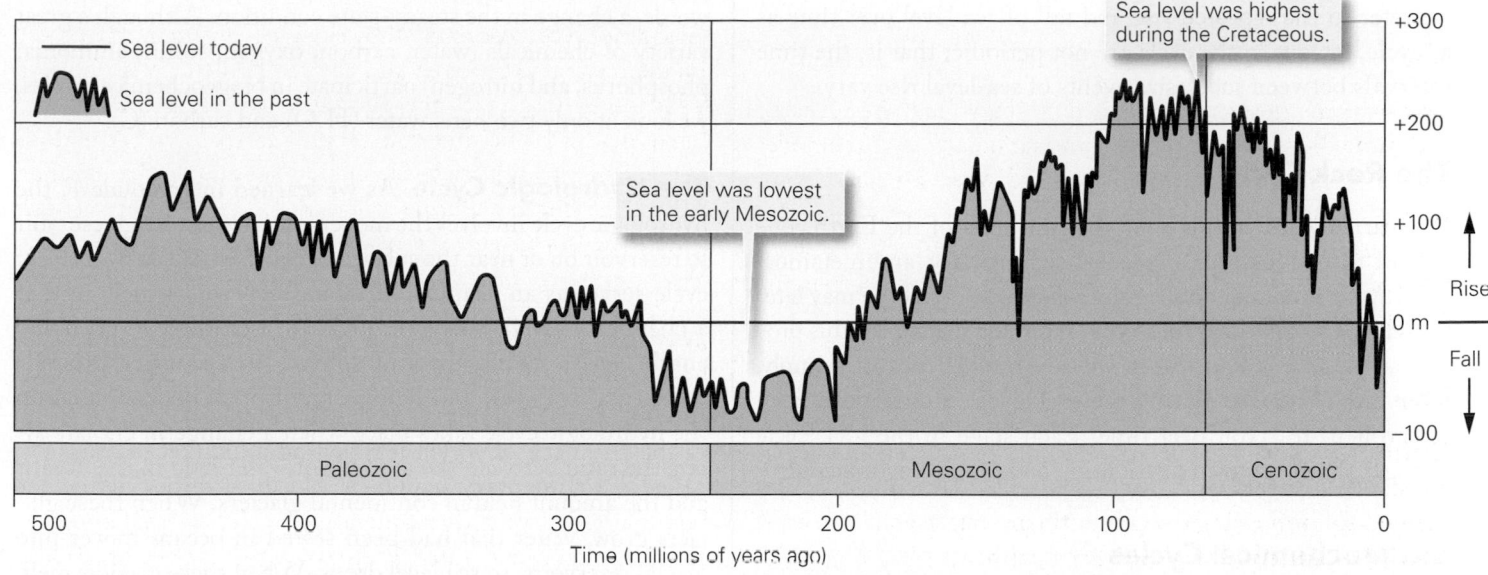

Sea level was highest during the Cretaceous.

Sea level was lowest in the early Mesozoic.

Sea level today

Sea level in the past

Rise

Fall

Paleozoic | Mesozoic | Cenozoic

Time (millions of years ago)

(a) This chart provides one interpretation of sea-level change during the past half-billion years, based on the stratigraphic record. There is not full agreement about this interpretation; it remains the subject of research.

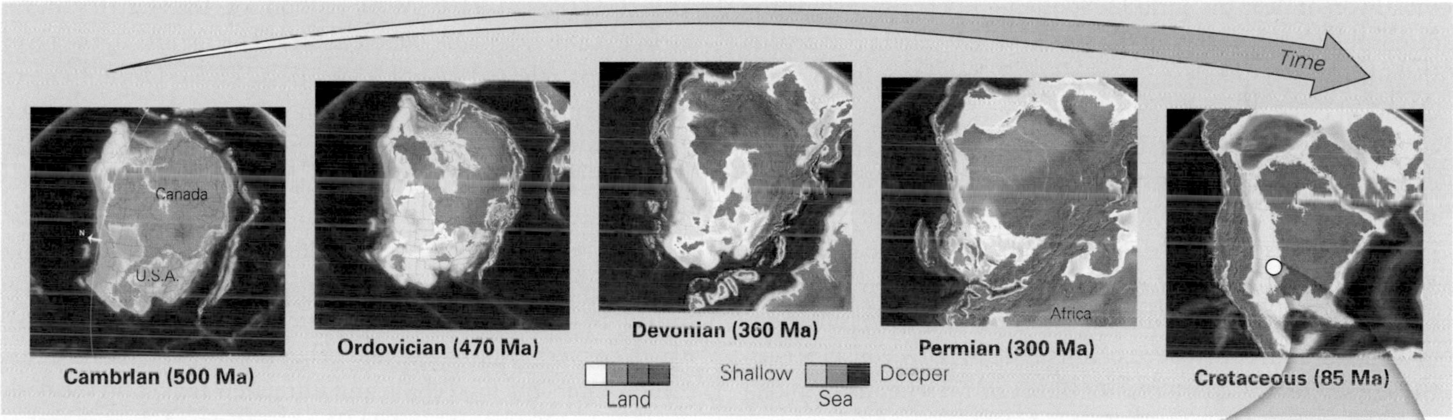

Canada

U.S.A.

Cambrian (500 Ma)

Ordovician (470 Ma)

Devonian (360 Ma)

Africa

Permian (300 Ma)

Cretaceous (85 Ma)

Land — Shallow — Deeper — Sea

Time

(b) When sea level was high, large parts of North America's interior were submerged by shallow seas.

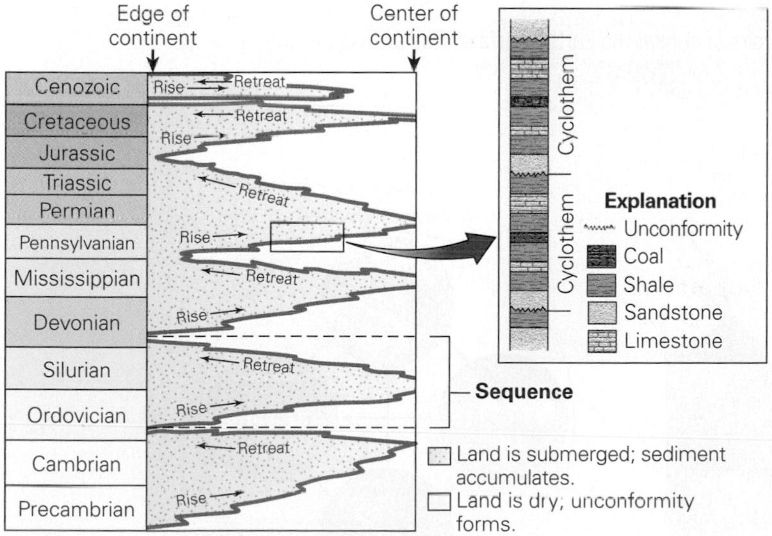

Edge of continent — Center of continent

Cenozoic	Rise ← Retreat
Cretaceous	← Retreat
Jurassic	Rise ←
Triassic	
Permian	← Retreat
Pennsylvanian	Rise →
Mississippian	← Retreat
Devonian	Rise →
Silurian	← Retreat
Ordovician	Rise →
Cambrian	← Retreat
Precambrian	Rise →

Cyclothem

Cyclothem

Cyclothem

Explanation
- ∿ Unconformity
- Coal
- Shale
- Sandstone
- Limestone

Sequence

☐ Land is submerged; sediment accumulates.

☐ Land is dry; unconformity forms.

(c) Sea-level rise and fall left sedimentary sequences separated by unconformities. Pennsylvanian sequences contain cyclothems.

A plesiosaur searches for food in the Cretaceous seaway of North America.

and changes in continental elevation and area. Note that while we refer to the repeated rise and fall of sea level over time as a "cycle," stages in the cycle are not periodic; that is, the time intervals between successive events of sea-level rise vary.

The Rock Cycle

We learned early in this book that the crust of the Earth consists of three rock types: igneous, sedimentary, and metamorphic. Atoms making up the minerals of one rock type may later become part of another rock type. Interlude C refers to this process as the *rock cycle*. In effect, we can think of rocks as, simply, reservoirs of atoms—during the rock cycle, atoms move from reservoir to reservoir over time. Each stage in the rock cycle changes the Earth by redistributing and modifying material.

Biogeochemical Cycles

A **biogeochemical cycle** involves the passage of a chemical among nonliving and living reservoirs in the Earth System, mostly on or near the surface. Nonliving reservoirs include the atmosphere, the crust, and the ocean, whereas living reservoirs include plants, animals, and microbes. Some stages in a biogeochemical cycle may take only hours, some may take thousands of years, and others may take millions of years. The transfer of a chemical from reservoir to reservoir during these cycles doesn't really seem like a change in the Earth in the way that the movement of continents or the metamorphism of rock seems like a change, because for intervals of time, biogeochemical cycles attain a **steady-state condition**, meaning that the proportions of a chemical in different reservoirs remain fairly constant even though there is a constant *flux* (flow) of the chemical among reservoirs. When we speak of global change in a biogeochemical cycle, we mean a change in the relative proportions of a chemical held in different reservoirs at a given time—in other words, a change in the steady-state condition. Although a great variety of chemicals (water, carbon, oxygen, sulfur, ammonia, phosphorus, and nitrogen) participate in biogeochemical cycles, we look at only two here: water (H_2O) and carbon (C).

The Hydrologic Cycle As we learned in Interlude F, the hydrologic cycle involves the movement of water from reservoir to reservoir on or near the surface of the Earth. The hydrologic cycle serves as an example of a biogeochemical cycle in that a chemical (H_2O) passes through both nonliving and living entities—the oceans, the atmosphere, surface water, groundwater, glaciers, soil, and living organisms. Global change in the hydrologic cycle takes place when a change in climate alters the ratio between the amount of water held in the oceans and the amount held in continental glaciers. When these glaciers grow, water that had been stored in oceans moves into glacial reservoirs, so sea level drops. When these glaciers melt, water returns to the oceans, so sea level rises.

The Carbon Cycle Most carbon in the Earth's near-surface realm originally bubbled out of the mantle in the form of CO_2 gas released by volcanoes **(Fig. 23.6)**. Once it enters the atmosphere, it moves through various reservoirs of the Earth System in the **carbon cycle**. Some dissolves in seawater to form bicarbonate (HCO_3^-) ions, which may later become incorporated in the shells of organisms that settle onto the seafloor. Some reacts with rock during chemical weathering and becomes incorporated in minerals. And some gets absorbed by photosynthetic organisms (microbes and plants), which convert it into sugar and other organic chemicals—this carbon enters the food chain and ultimately makes up the flesh of animals. Of note, about 63 billion tons of carbon move from the atmosphere into life forms every year.

FIGURE 23.6 In the carbon cycle, carbon moves among various reservoirs at or near the Earth's surface. Red arrows indicate release to the air, and green arrows indicate absorption from air.

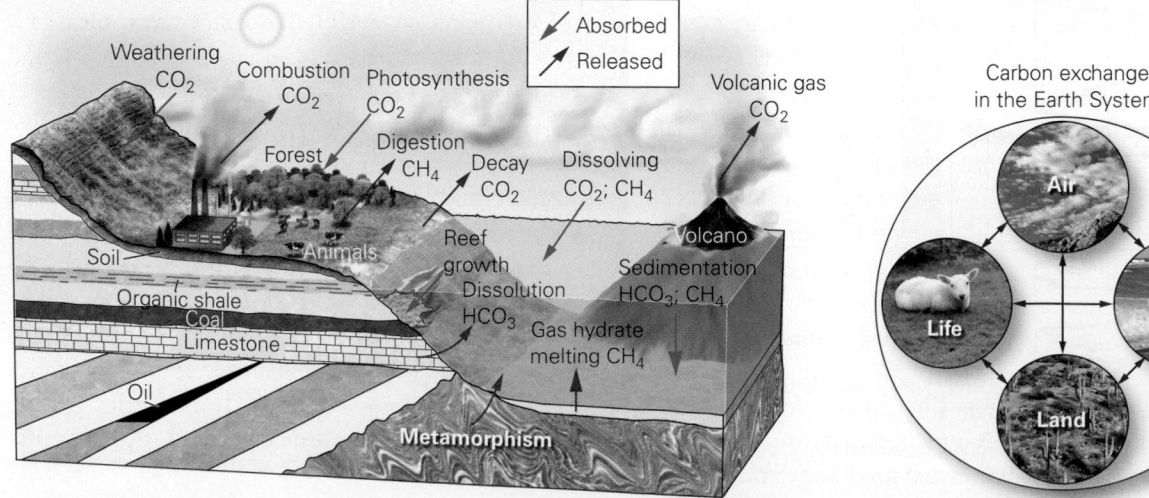

Of the carbon incorporated in organisms, some returns directly to the atmosphere (as CO_2) through the respiration of animals, the flatulence of animals (as methane, CH_4), or the decay of dead organisms. But some can be stored for long periods in fossil fuels (oil, gas, and coal), in organic shale, in methane hydrates (ice containing CH_4), or in limestone ($CaCO_3$). This carbon can return to the atmosphere (as CO_2) through the burning of fossil fuels, the melting of methane hydrates, production of cement, or the metamorphism of limestone. Or it can return to the sea after undergoing chemical weathering followed by dissolution in river water or groundwater.

TAKE-HOME MESSAGE

Some changes in the Earth System are cyclic in that they have stages that may be repeated. Examples include the supercontinent cycle, the sea-level cycle, the rock cycle, the hydrologic cycle, and biogeochemical cycles. During the carbon cycle, carbon transfers among living and nonliving reservoirs.

QUICK QUESTION: What evidence indicates that sea level rises and falls over time?

23.4 Global Climate Change

What Is Climate Change?

How often have you seen a newspaper proclaim "Record High Temperatures!" when thermometers register temperatures several degrees above "normal" for days on end. Do such headlines mean that "the climate" is changing? To start addressing this question, we first need to distinguish between two common terms: weather and climate.

As discussed in Chapter 20, atmospheric conditions during a specific time interval in a given region define the region's **weather** for that time interval. The weather may be "windy, rainy, and cool" in the morning and "calm, sunny, and dry" in the afternoon. The term **climate**, in contrast, refers to the overall range of weather conditions, as well as the typical daily to seasonal variability of weather conditions, as observed over a period of decades for a region. We can contrast, for example, a "tropical climate" with a "temperate climate"—the former tends to have hot, humid days all year, whereas the latter tends to have contrasting seasons and a wide range of temperature and humidity. So a newspaper's headline about a single hot spell or cold snap does not mean that the climate is changing. But if a new set of conditions—say, an overall increase in average temperature, a rising snow line, a longer growing season, or a change in storm frequency or intensity—becomes the new norm for a region, then climate change has occurred. And if such changes happen worldwide, then **global climate change** has occurred.

The stratigraphic record clearly shows that global climate change has taken place repeatedly throughout the Earth's long history. As we described in Chapter 13, for example, the Cretaceous was a time of relatively warm temperatures during which no polar ice caps existed, while the late Proterozoic may have seen "snowball Earth," when all oceans were frozen over and glaciers covered all continents. This well-established principle has been recognized since geologists first learned how to interpret the stratigraphic record, so it's not news. But global climate change has become a staple of public interest in recent years, and that's because research over the past few decades leads to the conclusion that global climate change is now taking place, not at the slow pace typical of most geologic phenomena, but at rates fast enough to have potential impacts on society in the next few decades. In other words, readers of this book may see the effects of *contemporary global climate change* in their lifetimes.

In this section, we begin our study of global climate change by reviewing the fundamental role that greenhouse gases play in regulating atmospheric temperature. Then, we describe how researchers study past climates. Finally, we discuss the different rates at which climate has changed over geologic time. For the purpose of our discussion, we distinguish between *long-term climate change*, which takes place over millions to tens of millions of years, and *short-term climate change*, which takes place over tens to hundreds of thousands of years. If average atmospheric and sea-surface temperatures rise, we have **global warming**, and if they fall, we have **global cooling**.

The Role of Greenhouse Gases

The Sun constantly bathes the Earth in visible and ultraviolet light. Some of this energy reflects off the atmosphere, off clouds, or off the Earth's surface, so that our planet shines when viewed from space. The Earth's surface absorbs the remainder of the incoming visible light and then releases it in the form of infrared radiation that radiates upward and provides thermal energy to warm the base of the atmosphere. If the Earth had no atmosphere, all of this thermal energy would escape back into space. But our planet does have an atmosphere, and certain gases in the atmosphere (H_2O, CO_2, CH_4, N_2O, and O_3) absorb infrared radiation and re-radiate it. Some of the re-radiated energy continues upward into space, but some heads downward and warms the lower atmosphere (**Fig. 23.7**; see Chapter 20). In effect, these gases trap thermal energy and keep the lower atmosphere warm, somewhat as glass traps heat in a greenhouse. The overall trapping process is known as the **greenhouse effect**, and the gases that cause it are **greenhouse gases**.

FIGURE 23.7 The greenhouse effect traps thermal energy in the atmosphere.

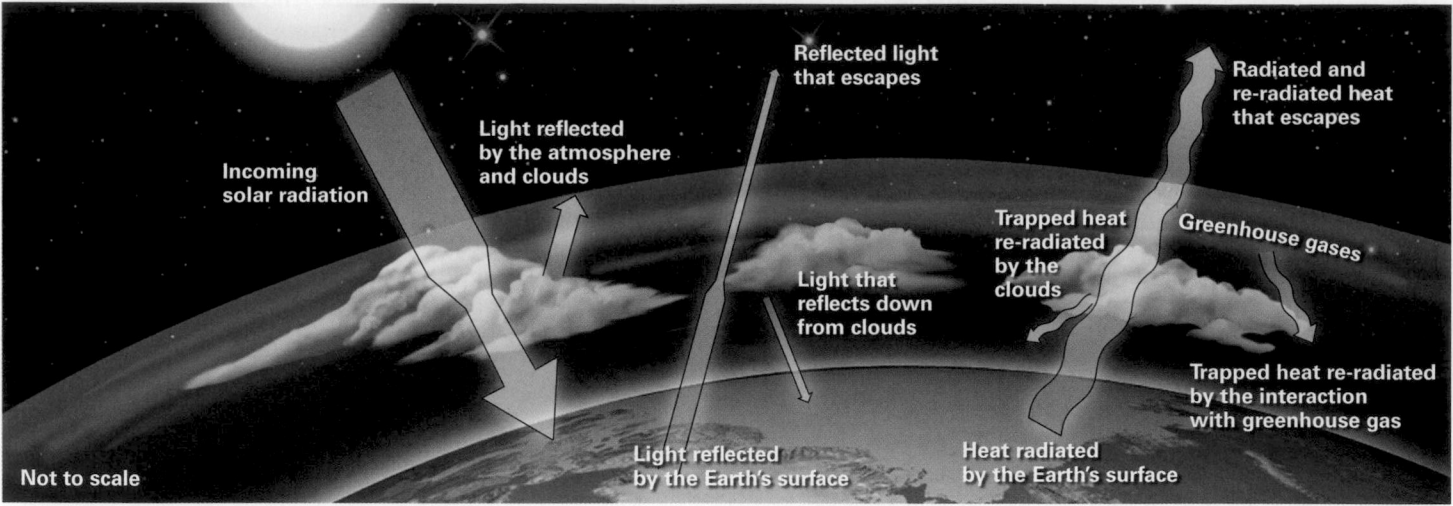

Researchers estimate that were it not for the greenhouse effect, the global average surface temperature of the Earth would be about 33°C lower than it is today. Put another way, an Earth without greenhouse gases would have an average global temperature of about –19°C, and our planet's surface would be frozen—it is the presence of greenhouse gases that makes the Earth habitable! Because greenhouse gases trap heat, any process that transfers these gases from underground, oceanic, or biomass reservoirs to the atmospheric reservoir will cause the climate to warm. Similarly, any process that removes greenhouse gases from the atmospheric reservoir and transfers them to biomass, or to oceanic or underground reservoirs, will cause the climate to cool.

Of the greenhouse gases, CO_2 and CH_4 play the most significant roles in influencing changes in global atmospheric temperature. Because these compounds are only trace elements in the atmosphere, one might wonder how they can be so important. Researchers suggest two reasons. First, these gases are much more efficient at trapping heat than other gases are. For example, CO_2 molecules trap 20 times as much heat as do H_2O molecules; CH_4 molecule are even more efficient. As a result, despite its low concentration in the atmosphere, CO_2 (with a concentration of about 400 parts per million, or *ppm*) contributes about 9% to 30% of the total greenhouse effect. And CH_4, with a concentration (~1.85 ppm) much less than that of CO_2, contributes about 4% to 9% of the greenhouse effect. Second, CO_2 and CH_4 mix into the atmosphere and can remain there for a long time—in other words, they have a very long *residence time* in air. As a result, any new CO_2 and CH_4 from land or sea added to the atmosphere changes the overall concentration of these gases for centuries to millennia. The greenhouse effect due to these gases warms the air, and therefore warms the global climate.

Notably, of the various greenhouse gases in air, H_2O occurs in the greatest concentration. In fact, this gas causes between 30% and 70% of the greenhouse effect in a given region—generally more than CO_2 and CH_4 combined. Researchers emphasize, however, that decadal to millennial changes in global temperature are not driven by progressive increases in H_2O concentration in air. That's because water has a very short residence time (9 days) in the atmosphere. When a volume of atmosphere becomes saturated with water, it rains or snows, and the water returns to the Earth's surface. Overall atmospheric warming on a time frame of decades to millennia depends on the concentration of greenhouse gases (CO_2 and CH_4) with long residence times.

Warming or cooling due to changes in the concentrations of greenhouse gases can be amplified by **feedback mechanisms**. *Negative* feedback slows a process down or even reverses it, whereas *positive* feedback enhances a process and amplifies its consequences. Let's consider an example of how positive feedback affects climate. Imagine that global average atmospheric temperature has increased due to an increase in CO_2. This increase will, in turn, cause the oceans to warm and evaporate more, so more water vapor moves into the atmospheric reservoir globally and increases the greenhouse effect due to water vapor. The increase also causes CH_4 to enter the atmosphere through decay of organic matter in melting permafrost and possibly through the melting of gas hydrates (ice containing dissolved CH_4). As a result, the warming due to the initial addition of CO_2 can cause the concentration of other greenhouse gases to increase overall, forcing the atmosphere to warm even more than it would have in the first place.

Later in this chapter, we will examine the causes of changes in greenhouse-gas concentrations in more detail. But first, let's consider how geologists study the record of climate change.

Methods of Studying Climate Change

Geologists and climatologists are working hard to understand the nature of climate change, the rates at which change can take place, and the effects that change may have on our planet. They use three basic approaches to study global climate change: (1) they measure past climate change, as recorded by sedimentary strata, to document the magnitude of changes that can happen and the rate at which such changes can take place; (2) they conduct experiments and calculations to see how changes in the concentration of atmospheric components, such as CO_2 or dust, might affect climate; and (3) they develop computer programs (called *general circulation models*, or *GCMs*) to simulate how factors such as atmospheric composition, topography, ocean currents, and the Earth's orbit affect the circulation of the atmosphere and, therefore, the global distribution of climate belts. Researchers can use GCMs to develop broader **climate-change models** that seek to provide insight into when and why changes took place in the past and how changes might affect the future. They use climate-change models to try to predict changes in rainfall, sea level, ice cover, and other physical features that may be influenced by the warming or cooling of the atmosphere.

> **Did you ever wonder ...**
> how researchers study past climates?

Let's look more closely at how geologists study the **paleoclimate** (past climate) so as to document climate changes throughout the Earth's history. Any feature of the Earth System whose character depends on the climate, and whose age can be determined, provides a clue to defining paleoclimate:

- *Stratigraphic record:* The nature of sedimentary strata deposited at a certain location reflects the climate at that location. For example, an outcrop exposing sandstone with huge cross beds, overlain successively by coal and glacial till, indicates that the site of the outcrop has endured different climates (desert, then tropical, then glacial) over time.

- *Paleontological evidence:* Different assemblages of species survive in different climates, so the succession of fossils in a sedimentary sequence provides clues to the changes in climate at that site. For example, a record of short-term climate change can be obtained by studying the succession of plankton fossils in seafloor sediments, because cold-water species of plankton differ from warm-water species. Fossil plant pollen preserved in the mud of bogs also provides information about paleoclimate because different plant species live in different climates **(Fig. 23.8a)**. Pollen studies show, for example, that spruce forests, indicative of cool climates, have slowly migrated north since the last ice age **(Fig. 23.8b)**.

FIGURE 23.8 Change in the assemblage of pollen in sediment indicates a shift in climate.

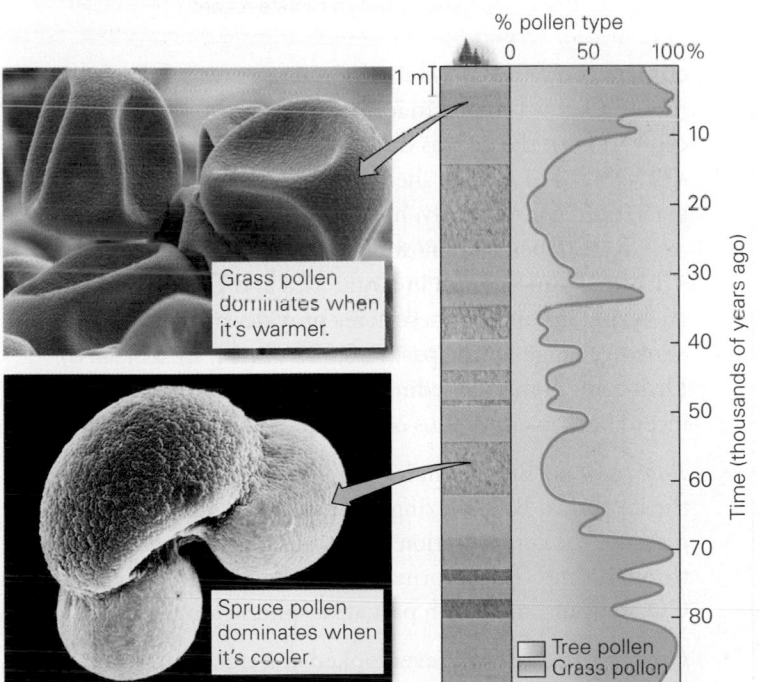

Grass pollen dominates when it's warmer.

Spruce pollen dominates when it's cooler.

(a) The proportion of tree pollen relative to grass pollen can change in a sedimentary sequence over time. Researchers plot changes in the proportion of pollen types over time by examining samples from a column of sediment.

Example of a lake sediment core.

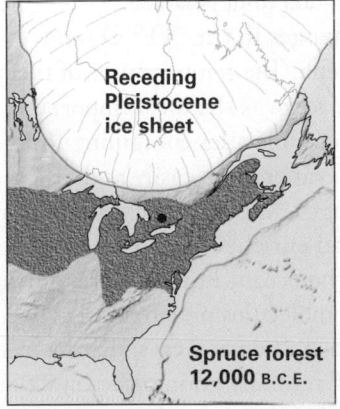

Receding Pleistocene ice sheet

Spruce forest 12,000 B.C.E.

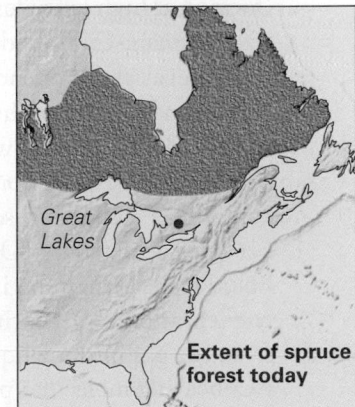

Great Lakes

Extent of spruce forest today

(b) Based on studying the pollen record, geologists discovered that spruce forests (green) grew farther south 12,000 years ago than they do today.

FIGURE 23.9 The proportion of oxygen isotopes transferred between reservoirs during evaporation or precipitation depends on temperature. The $^{18}O/^{16}O$ ratio can be studied in glacial ice (H_2O) and fossil shells ($CaCO_3$).

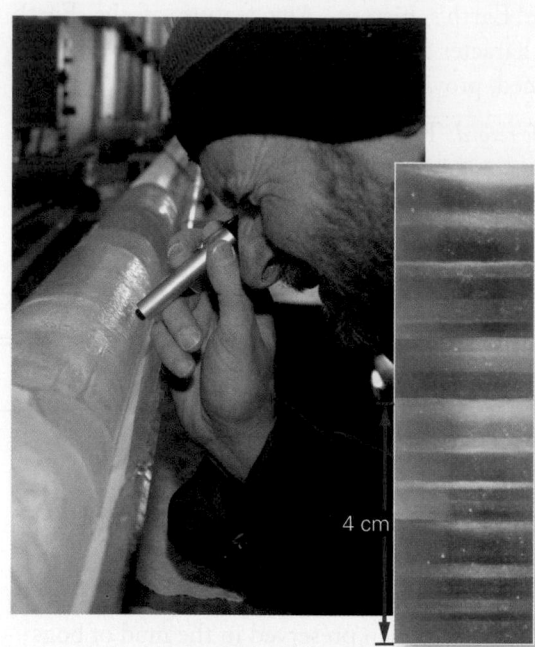

(a) A researcher examines an ice core in the field. Lab photos reveal annual layers.

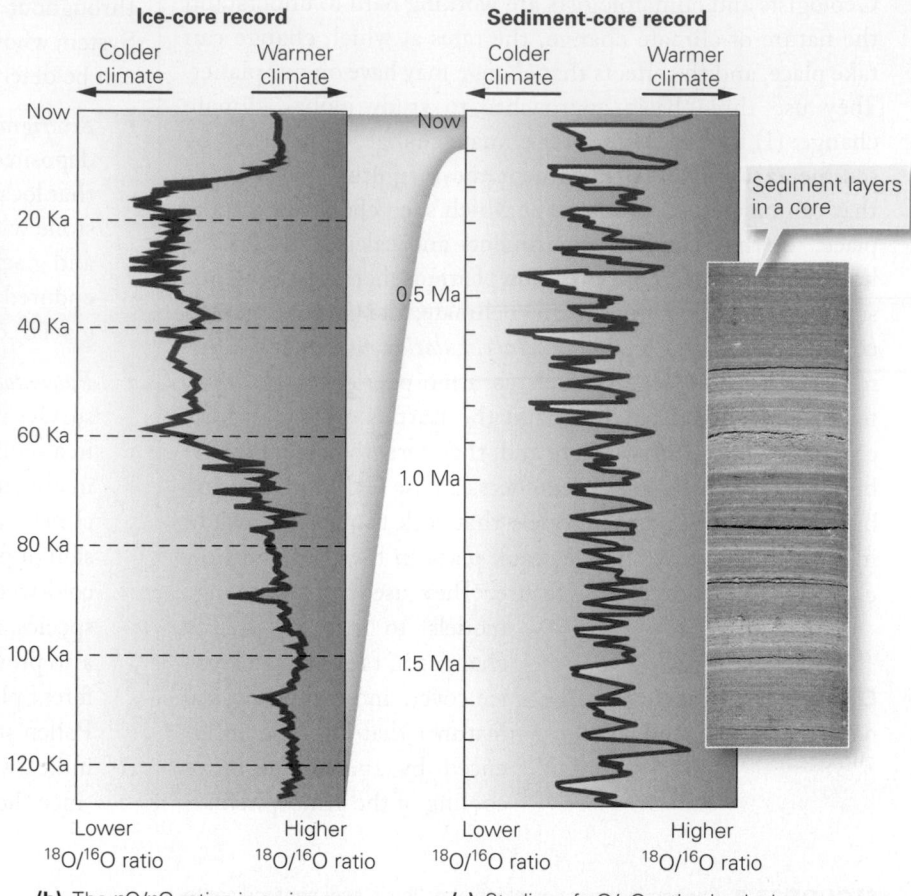

(b) The $^{18}O/^{16}O$ ratios in an ice core represent temperature changes.

(c) Studies of $^{18}O/^{16}O$ ratios in plankton shells in cores of deep-marine sediment also provide a climate record.

- Oxygen-isotope ratios: Researchers have found that the ratio of ^{18}O to ^{16}O, two isotopes of oxygen, preserved in the calcite ($CaCO_3$) or silica (SiO_2) molecules of shells, or in the water (H_2O) molecules of glacial ice, depends on global climate at the time the shells or ice formed. Therefore, by measuring the $^{18}O/^{16}O$ ratio in fossil shells in a succession of layers in marine sediment, or in a succession of layers in glacial ice, researchers can quantify how global climate has changed over time. Specifically, during times of cooler climates, when continental glaciers grow, the $^{18}O/^{16}O$ ratio in shells tends to be higher, and the $^{18}O/^{16}O$ ratio in glacial ice (formed at high latitudes) tends to be lower.

Climate-controlled changes in the $^{18}O/^{16}O$ ratio reflect a variety of phenomena. The temperatures of the oceans and the atmosphere serve as the most important control. That's because water molecules containing the heavier isotope (^{18}O) don't evaporate as easily as do those containing the lighter isotope (^{16}O), so water vapor in air becomes depleted in ^{18}O and enriched in ^{16}O, relative to liquid water remaining in the ocean. Furthermore, during condensation, ^{18}O-bearing molecules preferentially go from the gas into the liquid, and during crystallization, ^{18}O-bearing molecules preferentially go into the crystal. The extent of this fractionation depends on temperature, as does the relative distribution of isotopes between glacial and oceanic reservoirs. For example, when the

climate cools and continental glaciers grow, relatively more ^{16}O gets trapped in ice on land, so the ^{18}O concentration in the oceans decreases. Calcite ($CaCO_3$) and silica (SiO_2) in marine shells formed during the time of a glaciation, consequently, have measurably more ^{18}O.

Researchers have now obtained ice cores down to a depth of almost 3 km in Antarctica and Greenland; analyzing isotopes in these cores provides a record of global climate change for the past 800,000 years **(Fig. 23.9a, b)**. Drill cores of marine sediments extend the paleoclimate record back over millions of years **(Fig. 23.9c)**.

- *Bubbles in ice:* Bubbles in ice trap the air present at the time the ice forms. By analyzing these bubbles, geologists can measure the concentration of CO_2 in the atmosphere back through time. This information can be used to correlate CO_2 concentration with past atmospheric temperature.

- *Growth rings:* If you've ever looked at a tree stump, you will have noticed the concentric rings visible in the wood. Each ring represents one year of growth, and the thickness of the ring indicates the rate of growth in a given year.

BOX 23.1 CONSIDER THIS . . .

Global Climate Change and the Birth of Legends

Some of the myths passed down from the early days of civilization may have their roots in global climate change. For example, recent evidence suggests that before 7,600 years ago, the region now occupied by the Black Sea contained a much smaller freshwater lake surrounded by settlements. When the most recent ice-age glaciers retreated, sea level rose, and the Mediterranean Sea eventually broke through a natural dam at the site of the present Bosporus Strait. Researchers suggest that seawater from the Mediterranean then spilled into the Black Sea basin via a waterfall 200 times larger than Niagara Falls. This influx of water caused the lake level to rise by as much as 10 cm per day, and within a year, 155,000 km² of populated land had become submerged beneath hundreds of meters of water. This traumatic flooding presumably forced many people to migrate, and its timing has led some researchers to speculate that it may have inspired the Epic of Gilgamesh (written in Babylonia ca. 2000 B.C.E.) and, later, the biblical epic of Noah's Ark.

FIGURE 23.10 Records of recent climate change.

Researchers extract cores to see rings in living trees without damaging the tree.

(a) Tree rings provide a record of climate; more growth happens in wet years than in dry years, leading to wider rings.

(b) Historical archives provide records of floods and droughts.

Trees grow faster during warmer, wetter years and more slowly during cold, dry years **(Fig. 23.10a)**. Therefore, the succession of ring widths provides a calibrated record of climate during the lifetime of the tree. Bristlecone pines supply a record back through 4,000 years. To go further into the past, *dendrochronologists* (scientists who study tree rings) look at the record of rings in logs dated by the radiocarbon technique or in logs whose ages overlap with that of the oldest living tree. Growth rings in corals and shells can provide similar information.

• *Human history:* Researchers have been able to make careful direct measurements of climate change only for the past few decades. But history, both written and archaeological, contains important clues to climates centuries or millennia in the past. Events such as floods, droughts, or periods of unusual cold leave an impression on people, who record the impact of such events in paintings, stories, and records of crop success or failure **(Fig. 23.10b; Box 23.1)**.

Long-Term Climate Change

Using the variety of techniques described above, geologists have reconstructed an approximate record of global climate, represented by the Earth's average atmospheric temperature, over geologic time. The record shows that, with the exception of snowball Earth intervals (see Chapter 13), temperature has stayed above the freezing point of water and well below the boiling point of water since the beginning of the Archean. But the temperature has not always been the same—at some times in the past, the Earth's atmosphere was significantly warmer than it is today, and at other times, it was significantly cooler. The warmer periods have come to be known as **greenhouse** (or **hothouse**) **periods** and the colder ones as **icehouse periods**. (The more familiar term *ice age* refers to portions of an icehouse period when the Earth became cold enough for ice sheets to advance and cover substantial areas of the continents.) As the chart in **Figure 23.11a** shows, there have been at least five major icehouse periods during geologic time.

FIGURE 23.11 Estimates of global temperature change over geologic time, relative to a reference value.

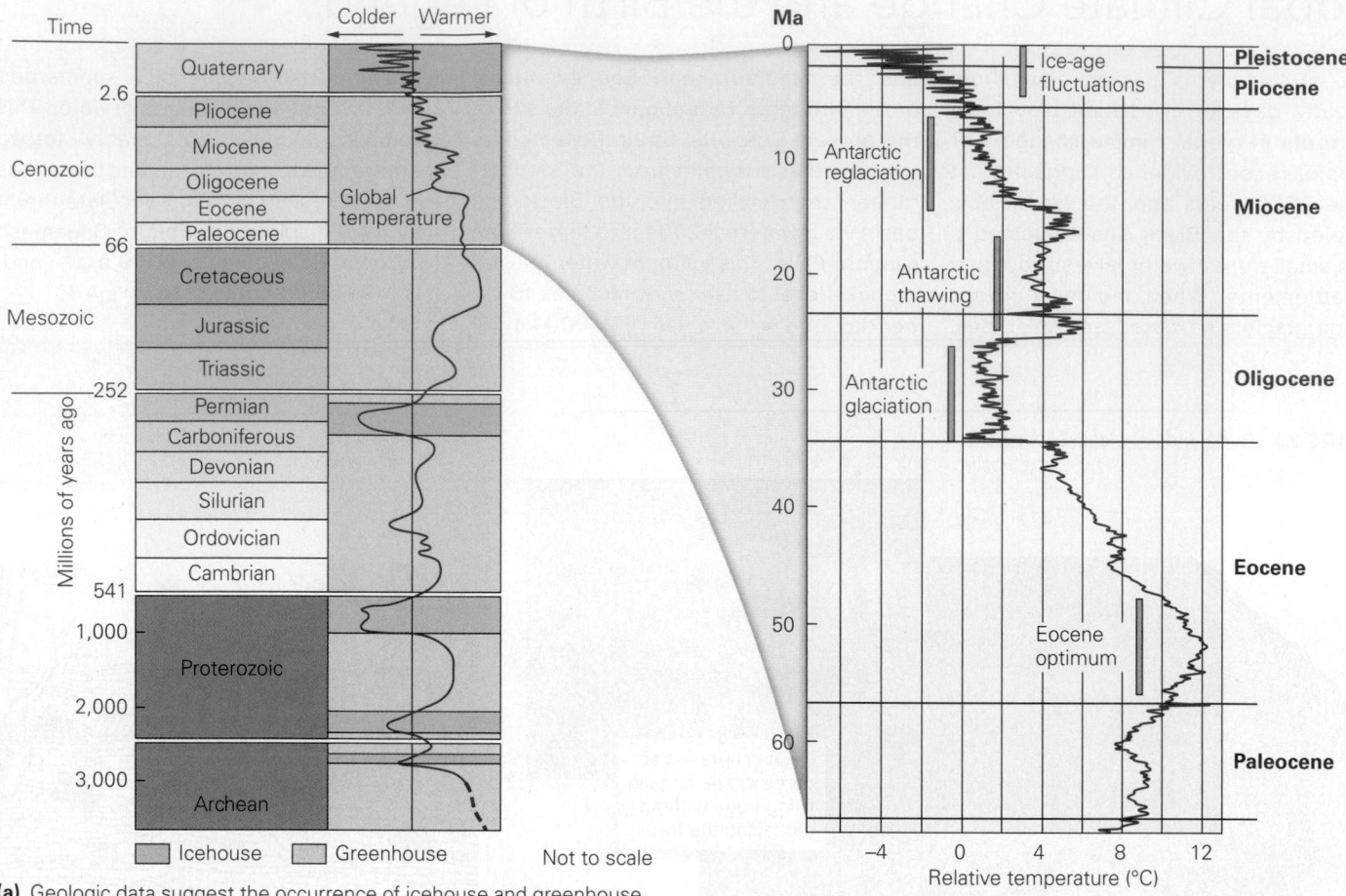

(a) Geologic data suggest the occurrence of icehouse and greenhouse periods at various times during Earth history.

(b) During the Cenozoic, there was an overall cooling trend; climate alternately warmed and cooled dramatically during the Pleistocene Ice Age.

Let's look a little more closely at the climate record of the last 100 million years, for this time interval includes a transition between a greenhouse and an icehouse period. Paleontological and other data suggest that the climate of the Mesozoic Era, the Age of Dinosaurs, was much warmer than the climate of today. At the equator, average annual temperatures may have been 2°C to 6°C warmer, while at the poles, temperatures may have been 20°C to 60°C warmer. In fact, during the Cretaceous Period, dinosaurs could live at high latitudes, and there were no polar ice caps on the Earth. But starting at about 80 Ma, the Earth's atmosphere began to cool. The cooling trend continued—except for an interval of about 10 million years, a time now called the *Eocene climatic optimum* **(Fig. 23.11b)**—and at about 33 Ma, our planet entered an icehouse period. Temperatures in polar regions dropped below freezing, the Antarctic ice sheet formed, and at about 2.6 Ma, the Pleistocene Ice Age began.

What caused these long-term global climate changes? The answer may lie in the complex relationships among the various solar, geologic, and biogeochemical cycles of the Earth System **(Box 23.2)**. Factors that may have played a role include the following:

- *Positions of continents:* Continental drift influences the climate by controlling the pattern of ocean currents, which redistribute heat around the planet's surface **(Fig. 23.12a)**. Drift also determines whether land lies at high or low latitudes (and, therefore, how much solar radiation strikes it); whether or not large continental interior regions exist where extremely cold winter temperatures can develop; and whether there is a lot of rainfall, which could cause weathering reactions that absorb CO_2.

- *Volcanic activity:* A long-term global increase in volcanic activity may contribute to long-term global warming if it increases the concentration of CO_2 in the atmosphere. For example, when Pangaea broke up during the Cretaceous Period, numerous rifts formed, and seafloor-spreading rates were particularly high, so volcanoes were more abundant than they are today. Such voluminous volcanic activity may have emitted enough CO_2 to trigger Cretaceous greenhouse conditions.

- *Uplift of land surfaces:* Tectonic events that lead to the long-term uplift of land affect atmospheric CO_2 concentrations

FIGURE 23.12 Changes in the distribution and elevation of landmasses can affect climate.

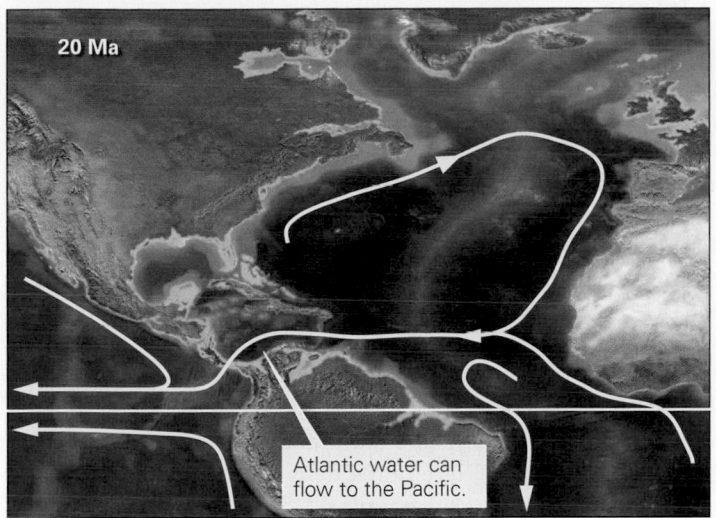

Atlantic water can flow to the Pacific.

Closure of Panama diverts currents.

(a) When the Isthmus of Panama, a volcanic arc, formed during the Miocene, the patterns of ocean currents in the North Atlantic changed. The white line is the equator.

☐ Uplifted area

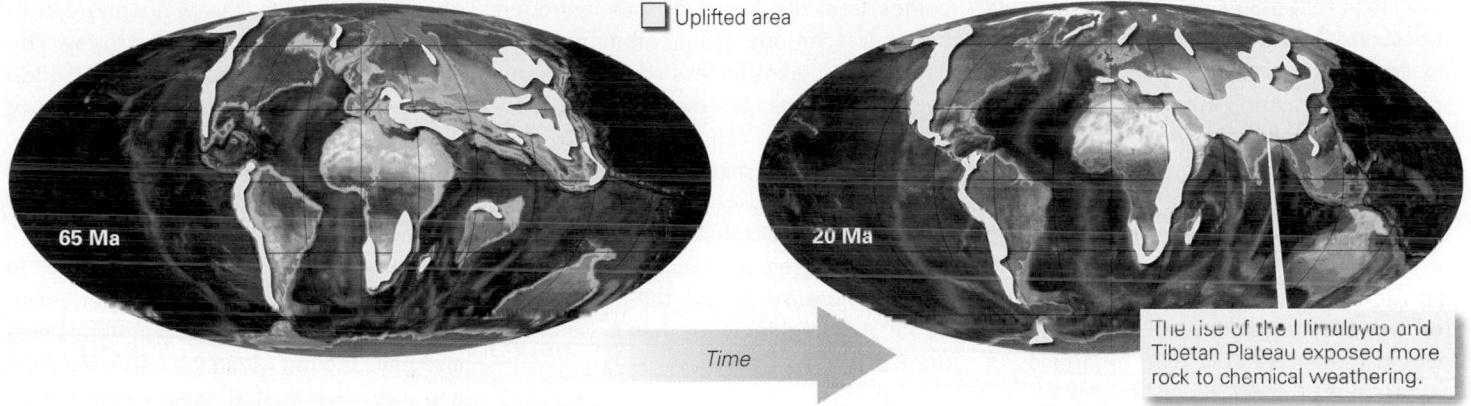

Time

The rise of the Himalayas and Tibetan Plateau exposed more rock to chemical weathering.

(b) Orogenies increased the percentage of elevated land between 65 Ma and 20 Ma. The exposure of more rock leads to more weathering, potentially affecting the concentration of greenhouse gases in the atmosphere.

because such events expose land to weathering, and chemical-weathering reactions absorb CO_2. Consequently, uplift potentially decreases the greenhouse effect and causes global cooling. For example, uplift of the Himalayas and the Tibetan Plateau may have triggered Cenozoic icehouse conditions **(Fig. 23.12b)**. Such uplift also affects atmospheric circulation and rainfall rates (see Chapter 20). Uplift of broad continental areas at the end of the Precambrian may have resulted in so much rock weathering that CO_2 concentrations dropped low enough to trigger the snowball Earth.

- *Formation of fossil fuels:* At various times during the Earth's history, environments suitable for the growth, accumulation, and burial of abundant organic material become particularly widespread. Once buried, the organic material transforms into coal or oil and can remain trapped underground. This overall process removes CO_2 from the atmosphere and, as a consequence, may result in global cooling. The cooling that occurred in the late

Paleozoic, a time when coal swamps were widespread, may be a manifestation of this process.

- *Life evolution:* The appearance or extinction of certain life forms may have affected climate significantly. For example, some researchers speculate that the appearance of lichens in the late Proterozoic may have decreased atmospheric CO_2 concentrations and may have contributed to triggering icehouse conditions. Similarly, the appearance of grasses at about 30 to 35 Ma may have triggered Cenozoic icehouse conditions.

Natural Short-Term Climate Change

So far, we've focused on climate change that takes place on a time scale of millions to tens or even hundreds of millions of years. The geologic record of the past few million years provides enough detail to allow geologists to detect cycles of climate change that have durations of centuries to hundreds of

BOX 23.2 CONSIDER THIS . . .

Goldilocks and the Faint Young Sun

Like Baby Bear's porridge in the tale of *Goldilocks and the Three Bears*, the Earth is not too hot, and it's not too cold . . . it's just right for liquid water and, therefore, for life, to exist **(Fig. Bx23.2)**. Researchers informally refer to the two related factors that make the Earth "just right" as the **Goldilocks effect**.

What factors keep the Earth's surface habitable? First, our planet orbits the Sun at a distance where the intensity of the radiation that reaches the surface can keep water in liquid form, given an appropriate atmosphere. Second, the concentrations of CO_2 and other greenhouse gases in the atmosphere provide just the right amount of greenhouse effect. If the Earth orbited too close to the Sun, solar radiation would be so intense that water could not exist in liquid form, regardless of atmospheric composition. And without liquid water, most CO_2 emitted by volcanoes would not have dissolved in the oceans, and the atmosphere would contain so much water and CO_2 that the greenhouse effect would make surface temperatures on the Earth way too hot for life. In contrast, if the Earth orbited too far from the Sun, temperatures would remain so cold that, regardless of atmospheric composition, any water present would freeze solid and life could not have evolved.

Astronomers define the distance from the Sun at which the Goldilocks effect is possible as the **habitable zone** of the Solar System. The habitable zone currently extends roughly from 0.8 AU to 1.3 AU. (An AU, or *astronomical unit*, represents the mean distance between the Earth's center and the Sun's center.) Notably, the distance of the habitable zone from the Sun has increased over the history of the Solar System. That's because the intensity of the radiation emitted by the Sun has increased over time. This change has taken place because the Sun's energy comes from the fusion of four hydrogen atoms to form one helium atom, and one helium atom takes up less space than four hydrogen atoms. So, over time, production of helium has caused the Sun to contract. As a result, material inside the Sun has been compressed into a smaller area, causing its temperature to rise, which in turn, has caused the rate of fusion reactions to increase, In fact, the Sun may be about 30% brighter today than it was when the Earth first formed.

If the Sun's intensity were the only factor controlling the Earth's temperature, our planet should have been over 20°C cooler during the Archean than it is today, and all water should have been frozen. But that wasn't the case. Stratigraphic and fossil records indicate that water has existed in liquid form on our planet's surface since at least the early Archean (~3.8 Ga). Researchers refer to this apparent contradiction between the calculated temperature and the observed temperature of the early Earth as the **faint young Sun paradox**. Most researchers agree that the paradox can be resolved by keeping in mind that earlier in the Earth's history, before the widespread appearance of photosynthetic life, the atmosphere contained more CO_2 than it does today. The greenhouse effect caused by the additional CO_2 increased the temperature of the Earth's atmosphere enough to counteract the lack of radiation from the faint young Sun, and this kept surface temperatures above freezing.

With the faint young Sun paradox in mind, astronomers speculate that when the Solar System was younger, Venus may also have orbited within the habitable zone and may have hosted liquid water. But as the Sun became brighter, Venus warmed up until its surface water evaporated and CO_2 dissolved in the water returned to the atmosphere. Addition of water to Venus's atmosphere caused the planet's surface to warm even more, leading to a drastic positive feedback that could not be stopped. Such a situation is called the *runaway greenhouse effect*. As a consequence, Venus's atmosphere eventually became so hot that water molecules broke apart, forming hydrogen atoms that escaped to space and oxygen atoms that reacted with rocks on the planet's surface to produce iron-oxide minerals. Volcanic CO_2 became the dominant gas of Venus's atmosphere, forming a dense blanket that now keeps the surface temperature of the planet at about 460°C, hot enough to melt lead.

FIGURE Bx23.2 The "Goldilocks effect," as applied to the Earth and its neighbors.

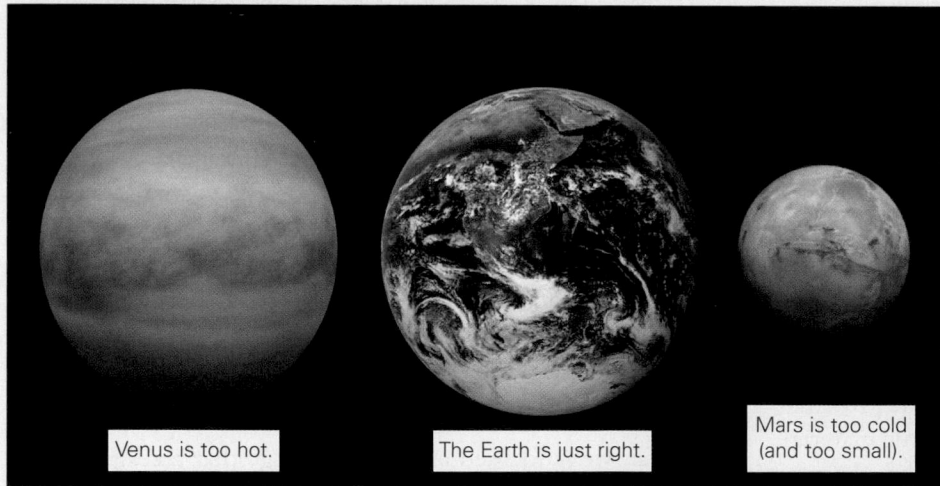

Venus is too hot.

The Earth is just right.

Mars is too cold (and too small).

FIGURE 23.13 Climate change during the Holocene. Measurements suggest that temperature has varied significantly over the past 18,000 years.

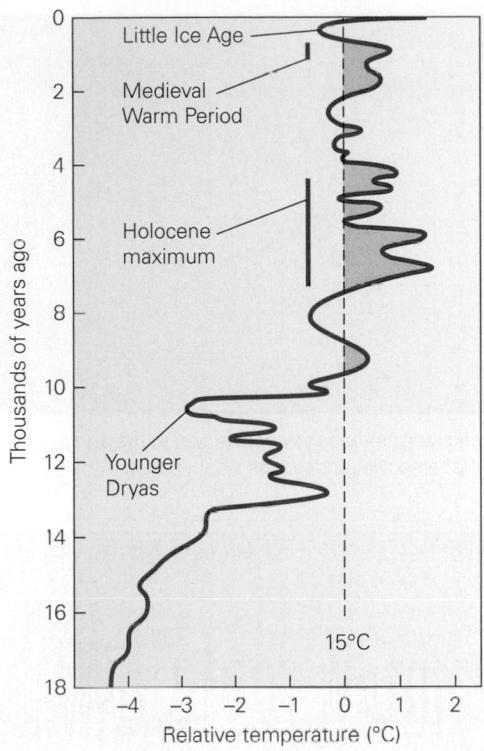

(a) There have been several temperature highs and lows during the Holocene.

At the end of the Little Ice Age, this tributary glacier in France reached the main valley floor.

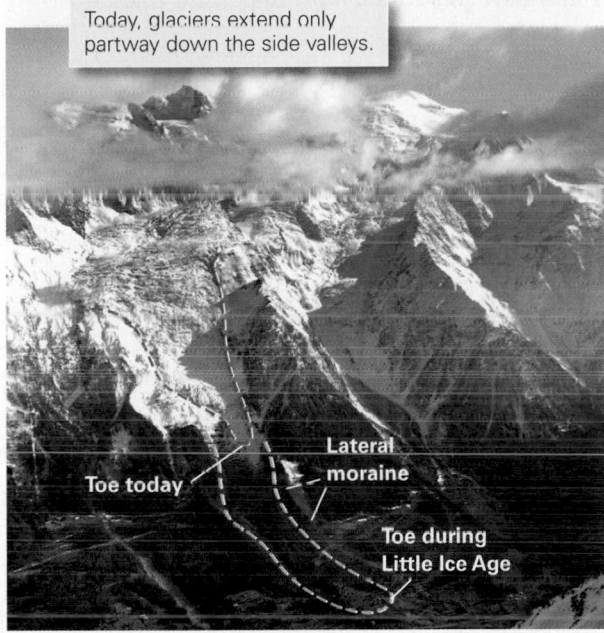

Today, glaciers extend only partway down the side valleys.

(c) Glaciers advanced in Europe during the Little Ice Age.

(b) Vikings settled in Greenland during the Medieval Warm Period, when the climate was warm enough for agriculture.

thousands of years. Such *short-term climate change* must be a consequence of factors that can operate quickly in the context of geologic time. We can get a sense of short-term climate change by examining the Pleistocene stratigraphic record. Changes in fossil assemblages, sediment composition, and isotope ratios indicate that continental glaciers advanced and retreated about 20 to 30 times in the northern hemisphere during the past 2.5 million years (see Chapter 22). Each advance (glaciation) represents an interval of global cooling, and each retreat (interglacial) represents an interval of global warming.

If we focus on the last 15,000 years, a period that includes all of the Holocene, we see trends of cooling or warming that last thousands of years, within which there are climate-change events whose duration lasts centuries or less **(Fig. 23.13a)**. For example, the time between about 15,000 and 10,500 B.C.E. was a warming period during which the last ice-age glaciers retreated. At 10,500 B.C.E., the Earth entered the *Younger Dryas*, an interval of cooler temperatures named for an Arctic flower that became widespread at the time. Then temperatures increased again, reaching a peak about 5,000 to 6,000 years ago, during a period called the *Holocene maximum*, in which average temperatures peaked at about 2°C above the temperatures of today. This warming led to increased evaporation and therefore to increased precipitation, making the Middle East region unusually wet and fertile, conditions that may partially account for the rise of civilization in Mesopotamia. The temperature dipped to a low about 3,000 years ago, before returning to a high during the Middle Ages, a time called the *Medieval Warm Period*. During this time, Vikings landed on the coast of Greenland and established settlements that were able to be self-supporting because of the relatively mild climate (**Fig. 23.13b**). The temperature dropped again from 1500 c.e. to about 1800 c.e., a period known as the *Little Ice Age*, when Alpine glaciers advanced and the canals of the Netherlands froze over in winter (**Fig. 23.13c**; see Fig. 22.34b). Overall, the climate has warmed since the end of the Little Ice Age, and today's global temperature is comparable to that of the Medieval Warm Period.

What factors might control short-term climate change? There may be many, but geologists speculate that the following phenomena may have played the most important roles:

- *Changes in the Earth's orbit and tilt:* As Milanković first recognized in 1920, the tilt of the Earth's axis changes over a period of 41,000 years, the Earth's axis undergoes precession over a period of 23,000 years, and the eccentricity of the Earth's orbit changes over a period of 100,000 years. Together, these phenomena cause the amount of summer insolation, and therefore the temperature in high latitudes, to vary (see Chapter 22).

- *Changes in ocean currents:* Recent studies suggest that the configuration of currents can change quickly and that such changes could affect the climate. The Younger Dryas, for example, may have resulted when a layer of freshwater from melting glaciers spread out over the North Atlantic and inhibited thermohaline circulation in the ocean, thereby shutting down the Gulf Stream (see Chapter 18).

- *Large eruptions of volcanic aerosols:* Not all of the sunlight that reaches the Earth penetrates its atmosphere and warms the ground. Some gets reflected by the atmosphere. The degree of reflectivity, or **albedo**, of the atmosphere increases not only when cloud cover increases, as we have noted, but also when the concentration of volcanic aerosols in the atmosphere increases. Very large eruptions can emit enough aerosols (particularly SO_2) to affect global temperature for months to years. As we noted in Chapter 9, the year following the 1815 eruption of Mt. Tambora in the western Pacific became known as the "year without a summer," for the sulfur aerosols it erupted encircled the Earth and blocked the Sun. Note that the effect of volcanic aerosol emission is global cooling, the

FIGURE 23.14 The abundance of sunspots varies over time and affects the radiation emitted by the Sun.

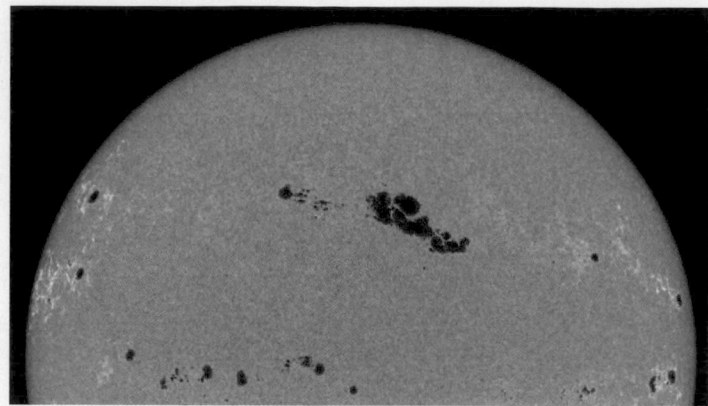

(a) Sunspots are magnetic storms that slow convection at the Sun's surface, producing a cooler area that appears as a dark patch.

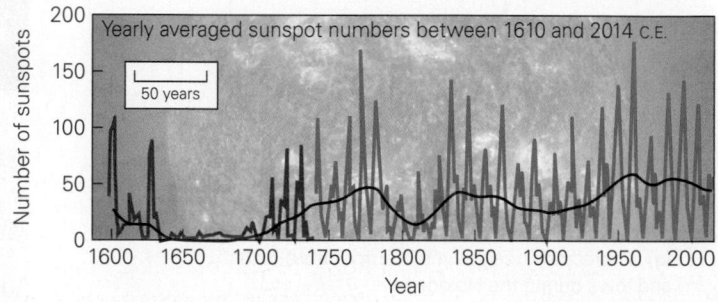

(b) The abundance of sunspots varies cyclically.

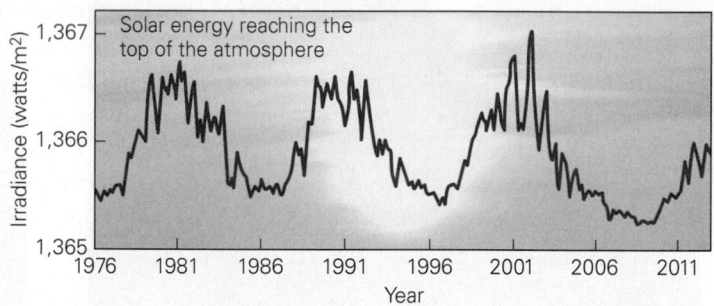

(c) Measurements since 1978 suggest that the solar energy reaching the atmosphere fluctuates periodically over a relatively short time frame.

opposite of the effect of volcanic CO_2 emission, which causes global warming.

- *Fluctuations in solar radiation:* The amount of energy produced by the Sun varies with the **sunspot cycle**. At peaks during the cycle, there are many sunspots (black spots on the Sun's surface) (**Fig. 23.14a, b**). Researchers have determined that the more sunspots there are, the more solar energy reaches the Earth (**Fig. 23.14c**). A reliable record of sunspot activity can be taken back for about 400 years and reveals that sunspot cycles last 9 to 11.5 years, a period too short to correlate with the rates of climate change observed

FIGURE 23.15 Life evolution has proceeded in fits and starts. During geologic time, there have been several catastrophic extinction events in which a large percentage of the genera on the Earth went extinct and biodiversity abruptly decreased.

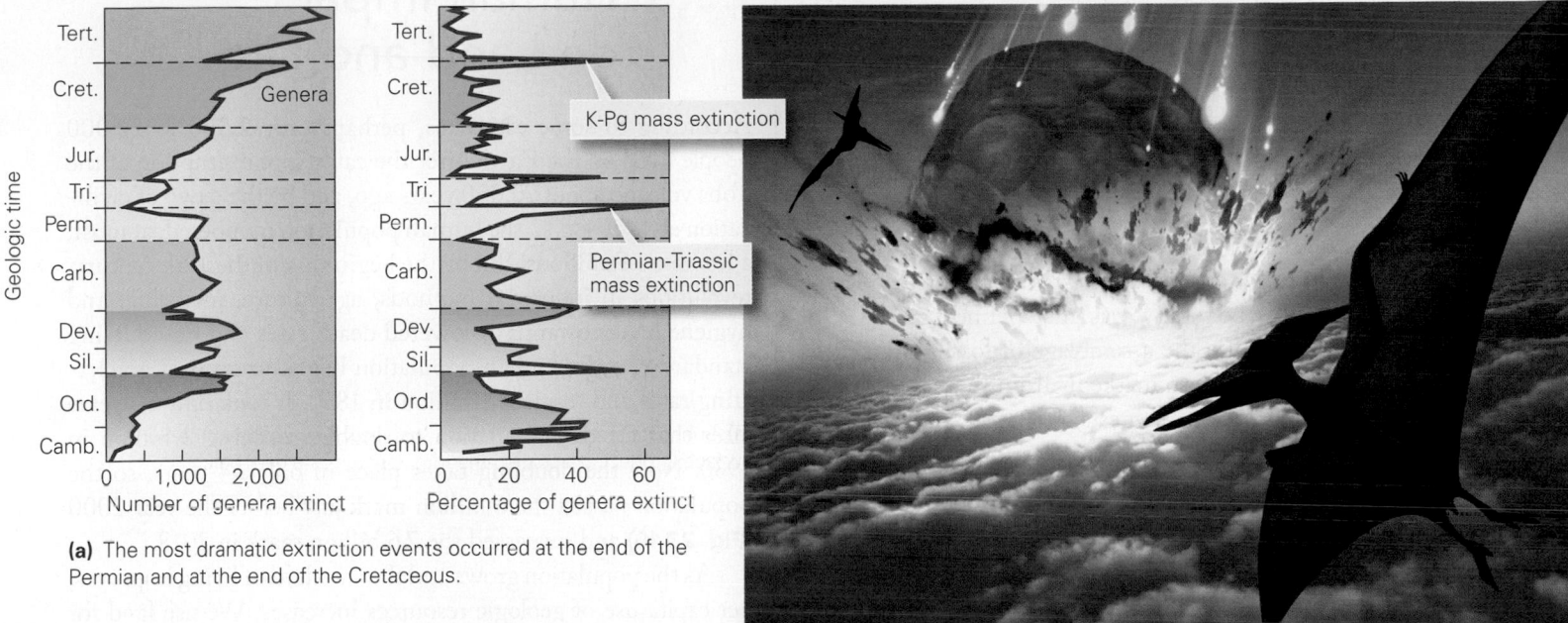

(a) The most dramatic extinction events occurred at the end of the Permian and at the end of the Cretaceous.

(b) The mass extinction at the end of the Cretaceous killed all dinosaur species. It is attributed to an asteroid impact.

in the geologic record. There may be, however, longer-term cycles that have not yet been identified.

- *Fluctuations in cosmic rays:* Some researchers have speculated that changes in the rate of influx of cosmic rays may affect climate, perhaps by generating clouds. Specifically, recent research suggests that cosmic rays striking the atmosphere produce clusters of ions that become condensation nuclei around which water molecules congregate, thus forming the mist droplets making up clouds. But how cloud formation changes climate remains uncertain. High-elevation clouds could reflect incoming solar radiation and, therefore, would cool the planet. Low-elevation clouds could absorb infrared rays rising from the Earth's surface and, therefore, would warm the planet.

- *Changes in surface albedo:* Regional-scale changes in the nature of continental vegetation cover, or the proportion of snow and ice on the Earth's surface, or the sudden deposition of reflective volcanic ash, could affect our planet's albedo. Increasing albedo causes cooling, whereas decreasing albedo causes warming.

- *Abrupt changes in concentrations of greenhouse gases:* A relatively sudden change in greenhouse gas concentrations in the atmosphere could affect climate. One such change might happen if the oceans warmed or sea level dropped, causing some of the methane hydrate that has crystallized in sediment on the seafloor to melt suddenly. Such melting would release CH_4 to the atmosphere. Similarly, algal blooms and reforestation (or deforestation) could conceivably change CO_2 concentrations.

Catastrophic Climate Change and Mass-Extinction Events

Changes on the Earth that happen almost instantaneously are called *catastrophic changes.* For example, a volcanic explosion, an earthquake, a tsunami, or a landslide can change a local landscape in seconds or minutes. But such events affect only relatively small areas. Can such catastrophes happen on a global scale? In recent decades, geoscientists have come to the conclusion that the answer is yes. The stratigraphic record shows that the Earth's history so far includes five **mass-extinction events** (see Interlude E and Chapter 13), during which large numbers of species abruptly vanished **(Fig. 23.15a)**. Some of these events define boundaries between geologic periods. A mass-extinction event decreases the *biodiversity*—the number of different species that exist at a given time—of life on the Earth. It takes millions of years after a mass-extinction event for biodiversity to recover, and the new species that appear differ from those that vanished, for evolution is unidirectional.

Geologists speculate that some mass-extinction events reflect a catastrophic change in the planet's climate, brought about by incredibly voluminous volcanic eruptions or by the impact of a comet or an asteroid with the Earth **(Fig. 23.15b)**. Either of these events could eject enough debris into the atmosphere to block sunlight. Without the warmth of the Sun, winter-like or night-like conditions would last for weeks to years, long enough to disrupt the food chain. In addition, either event could eject aerosols that would turn into global acid rain, scatter hot debris that would ignite forest fires, or give off chemicals that, when dissolved in the ocean, would make the ocean either toxic, killing marine life, or so nutritious that oxygen-consuming algae would thrive.

Let's examine possible causes for two of the more profound mass-extinction events in the Earth's history. The first event marks the boundary between the Permian and Triassic Periods and, therefore, defines the boundary between the Paleozoic and Mesozoic Eras. During the Permian-Triassic mass-extinction event, over two-thirds of the species on the Earth became extinct. In fact, this boundary was first defined in the 19th century, precisely because the assemblage of fossils from rocks below the boundary differs so markedly from the assemblage in rocks above. Isotopic dating suggests that the extinction event roughly coincided with the eruption of vast quantities of basalt in Siberia. So much basalt erupted that geologists have proposed that its source was a *superplume*, a mantle plume many times larger than the one currently beneath Hawaii. Because of the correlation between the time of the basalt eruptions and the time of the mass extinction, geologists suggest that the former caused the latter. Other, still controversial evidence suggests that, alternatively, a large asteroid collided with the Earth at the Permian-Triassic boundary and caused the mass extinction.

The second event, called the K-Pg boundary event, caused the mass extinction that marks the boundary between the Cretaceous and Paleogene Periods and, therefore, the boundary between the Mesozoic and Cenozoic Eras. (This boundary was once called the K-T boundary; T stands for Tertiary.) The time of this event correlates well with the time at which an asteroid collided with the Earth at a site now called the Chicxulub crater in Yucatán, Mexico. Consequently, most geologists suggest that this collision caused the mass extinction. A minority of researchers emphasize that the extinction occurred at roughly the same time as the eruption of extensive flood basalts in India and suggest that volcanic activity caused, or at least contributed to causing, the mass extinction.

TAKE-HOME MESSAGE

Geologic studies show that the Earth's climate has alternated between greenhouse and icehouse conditions over geologic time. Factors including life evolution, uplift of mountain belts, and continental drift may cause long-term change. Orbital cycles, eruption of volcanic aerosols, and possibly changes in solar radiation contribute to short-term change. Superplume eruptions or asteroid impacts can cause catastrophic change.

QUICK QUESTION: How do researchers document past climate change?

23.5 Human Impact on Land and Life

According to some estimates, perhaps only 2,000 to 20,000 people lived on the Earth after the catastrophic eruption of the Toba volcano about 70,000 years ago, and by the dawn of civilization at 4000 B.C.E., the human population numbered, at most, a few tens of millions. But by the beginning of the 19th century, revolutions in industrial methods, agriculture, medicine, and hygiene had substantially lowered death rates and raised living standards, so the human population began to grow at accelerating rates and reached 1 billion in 1850. It took only 80 years after that for the population to double, reaching 2 billion in 1930. Now the doubling takes place in only 44 years, so the population passed the 6 billion mark just before the year 2000 **(Fig. 23.16)** and surpassed the 7.6 billion mark in 2018.

As the population grows and the standard of living improves, per capita use of geologic resources increases. We use land for agriculture and grazing, forests for wood, rock and sand for construction, oil and coal for energy and plastics, and ores for metals (see Chapter 12). Without a doubt, our use of resources has affected the Earth System profoundly, and humanity has become a major agent of global change. In this section, we examine some of these *anthropogenic* (human-induced) *impacts* on the land surface, on the environment, and finally on the climate.

The Modification of Landscapes

Every time we pick up a shovel and move a pile of rock or soil, we redistribute a portion of the Earth's crust, an activity

FIGURE 23.16 The human population now doubles about every 44 years. The Black Death pandemic in the Middle Ages caused an abrupt drop that lasted for a few decades.

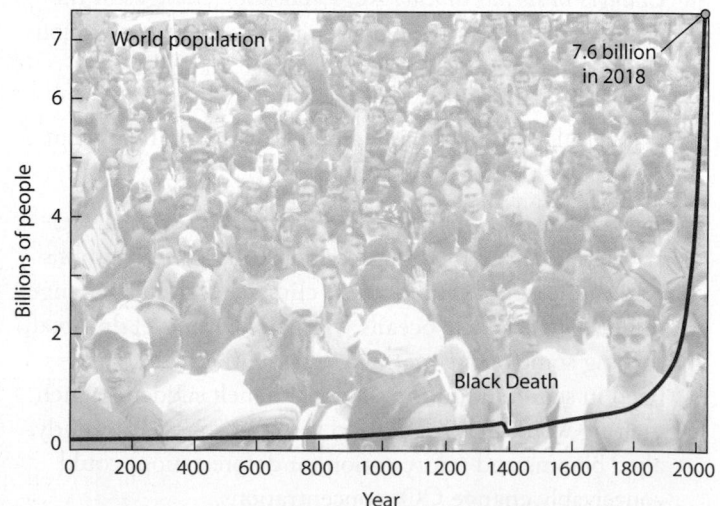

that prior to humanity was accomplished only by rivers, the wind, rodents, and worms. In the last century, the pace of human-driven Earth movement has accelerated, for now we have shovels in coal mines that can move 300 m³ of coal in a single scoop, trucks that can carry 200 tons of ore in a single load, and tankers that can transport 500 million liters of oil during a single journey. In North America, human activity now moves more sediment each year than rivers do. The extraction of rock during mining, the building of levees and dams along rivers or seawalls along the coast, and the construction of highways and cities all involve the redistribution of Earth materials (**Fig. 23.17**). In addition, people clear and plow fields, drain and fill wetlands, and pave over the land surface. All these activities change the landscape, the water table, and the supply of sediment.

Landscape modification has side effects. For example, it may make the ground unstable and susceptible to landslides. It may expose the land to erosion, thereby changing the volume of sediment transported by natural agents, such as running water and wind. Locally, flood-control projects may diminish the sediment supply downstream, with unfortunate consequences. For example, the damming of the Nile by the Aswan High Dam cut off the sediment supply to the Nile Delta, so ocean waves along the Mediterranean coast of the delta now eat into the coastline by more than 1 m per year.

The Modification of Ecosystems

In an undisturbed area, the **ecosystem**—the interconnected network of organisms, together with the physical environment in which they live—represents the product of evolution over an extended period. The ecosystem's flora (plant life) include species that have adapted to living together in that particular climate and on the substrate available, and its fauna (animal life) can survive local climate conditions and use local food supplies. Human-caused deforestation, overgrazing, agriculture, and urbanization disrupt ecosystems and lead to a decrease in biodiversity.

Archaeological studies have found that the earliest major example of human modification of an ecosystem took place in the Stone Age (pre-6000 B.C.E.). Spear-bearing hunters, over a relatively short time, caused the mass extinction of mammoths and other large mammals. With the advent of agriculture and deforestation and the growth of towns and cities, modification of the land accelerated, and today less than 5% of the landscapes in Europe and the United States retain their original ecosystems. In the developing world, landscape modification has claimed more than half of the area once occupied by tropical

FIGURE 23.17 Excavation, agriculture, and construction modify topography, drainage, infiltration, and ecology.

Humans move vast amounts of rock.

Agriculture eliminates diverse ecosystems.

Urbanization changes the water table.

FIGURE 23.18 The area of forests, particularly in the tropics, has been shrinking rapidly.

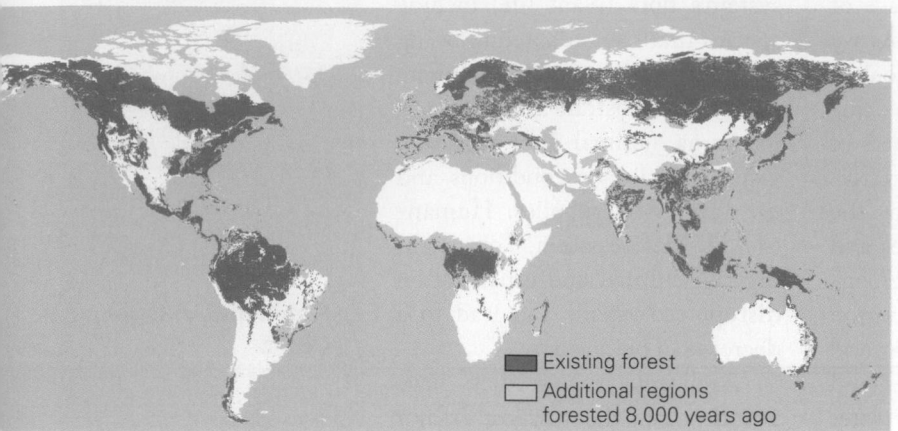

Existing forest
Additional regions
forested 8,000 years ago

(a) Remaining forests today are much smaller than forests of 8,000 years ago.

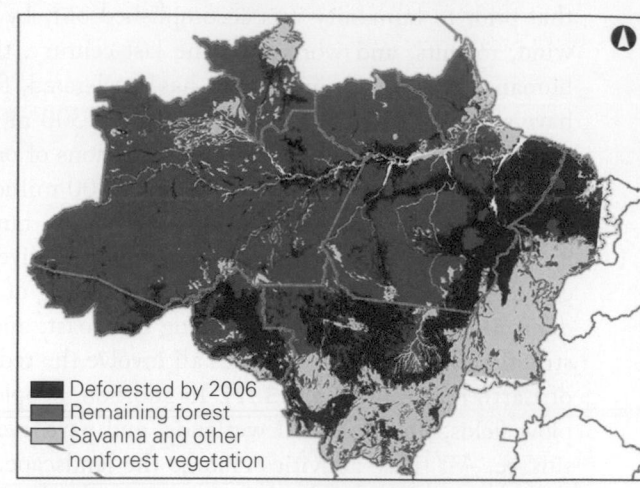

Deforested by 2006
Remaining forest
Savanna and other
nonforest vegetation

(b) A significant percentage of the Amazon rainforest has been lost in the past few decades.

(c) Part of the Amazon rainforest's destruction is due to slash-and-burn agriculture.

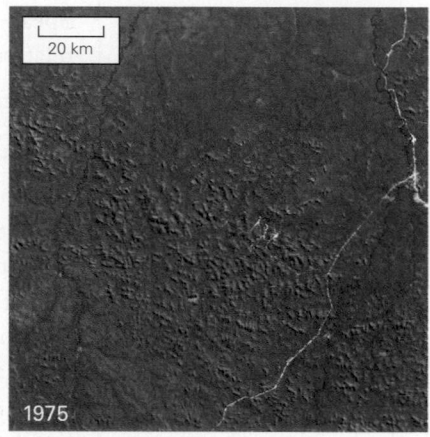

20 km

1975

20 km

2014

(d) Comparison of satellite imagery highlights areas that have undergone deforestation, such as this one near Ariquemes, Brazil.

rainforests, and such forests now disappear at a rate of about 1.8% per year **(Fig. 23.18a, b)**. Much of this loss comes from slash-and-burn agriculture, the destruction of forest to make open land for farming and grazing **(Fig. 23.18c, d)**.

Some of the changes we make to the land have permanent consequences in a human time frame. For example, the heavy rainfall of tropical regions removes nutrients from the soil, making the soil useless in just a few years, so forests cannot regrow quickly even if farming or grazing of the land stops. Overgrazing by domesticated animals can remove vegetation so completely that some grasslands have undergone desertification. And urbanization replaces the natural land surface with concrete or asphalt, a process that not only completely destroys ecosystems but also radically changes the amount of rain that infiltrates the land surface to become groundwater.

Anthropogenic changes to ecosystems affect the broader Earth System because they modify biogeochemical cycles and the Earth's albedo. For example, deforestation increases the CO_2 concentration in the atmosphere, for the carbon stored in

trees returns to the atmosphere when the trees burn. Further, the replacement of forest cover with concrete or fields increases the Earth's albedo.

Pollution

The environment has always contained various natural contaminants such as soot, dust, chemical runoff from sulfide minerals, and waste produced by organisms. In general, ecosystems could manage such contaminants naturally by absorbing them, by breaking them down, or by locking them into accumulating strata so that the contaminants didn't damage the environment. The nearly exponential growth of human population and its consequences (urbanization, industrialization, mining, expansion of framing and ranching, and a switch to engine-driven transportation) has changed this situation. We've greatly increased both the quantity and diversity of contaminants that enter the air, surface water, and groundwater. **Pollution**, the addition of these contaminants to the environment, can harm

FIGURE 23.19 Acid rain forms when the sulfur dioxide in industrial smoke dissolves in water and produces sulfuric acid. Acid rain is a problem in all industrialized countries.

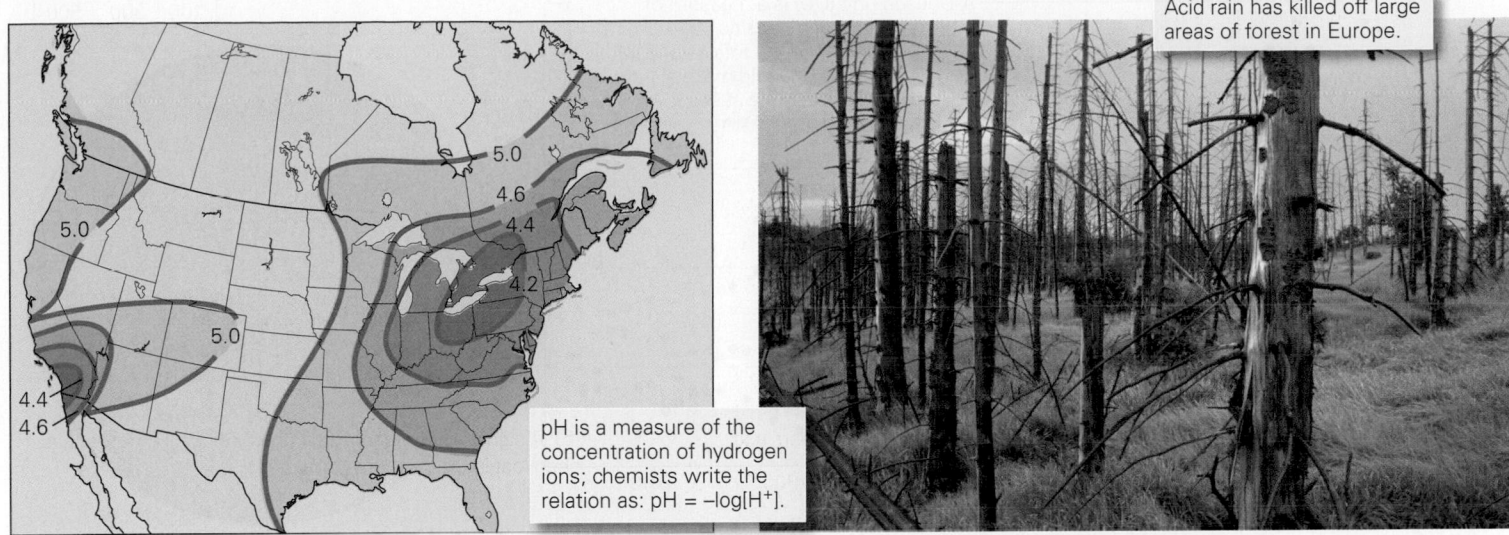

Acid rain has killed off large areas of forest in Europe.

pH is a measure of the concentration of hydrogen ions; chemists write the relation as: pH = −log[H⁺].

(a) Acid rain in North America occurs downwind of major industrial cities. We can specify the acidity of rainwater by stating its pH.

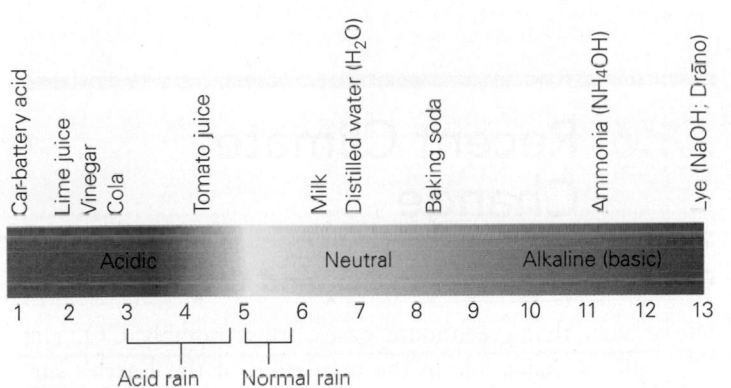

(b) The pH scale. Acid rain has a pH between 3 and 5.

ecosystems and society. Pollution, which includes both natural and synthetic materials in liquid, solid, or gaseous form, has become a major problem because there is too much of it for the Earth System to accommodate. Such pollution can be considered a type of global change because it represents a redistribution and reformulation of materials. Problems associated with this change include the following:

- *Smog:* The term *smog* was originally coined to refer to the dank, dark air that resulted when smoke from the burning of coal mixed with fog in industrial cities. It became a deadly problem between the early 19th and mid-20th centuries. Beginning in the mid-20th century, another kind of smog, *photochemical smog*, has plagued cities. It forms when exhaust from cars and trucks reacts with air in the presence of sunlight to produce an ozone-rich brown haze. When this haze gets trapped in the lower kilometer or so of the atmosphere, it can make the air dangerous to breathe. In cities where photochemical smog

has become a common problem, authorities publicize the *air-quality index*, a number representing the degree to which the air has become dangerous to breathe.

- *Water contamination:* Society dumps vast quantities of chemicals into surface water and groundwater. Examples include gasoline, other organic chemicals, radioactive waste, acids, fertilizers—the list could go on for pages. These chemicals can make groundwater supplies undrinkable and in some cases can kill off species in ecosystems where springs return the groundwater to the surface.

- *Acid runoff and acid rain:* When sulfide minerals remain buried in bedrock, water and air can't reach them. But in mines and tailings piles, where rocks are exposed and broken up, these minerals do come into contact with oxygen and water. The minerals may then react in the water to produce sulfuric acid. The resulting *acid runoff* that may flow out of mined areas can be toxic to life. Similarly, when rain passes through air that contains sulfur-containing aerosols (emitted from power plants or factories), the water dissolves the sulfur and yields *acid rain*. Wind can carry aerosols far from their source, so acid rain can damage a broad region **(Fig. 23.19)**.

- *Radioactive materials:* The by-products of nuclear weapons production, nuclear power plants, and certain medical procedures can yield radioactive materials, including new radioactive isotopes, some of which have relatively short half-lives (see Chapter 14). In effect, society has changed the distribution and composition of radioactive material worldwide. Radioactive pollution arises when radioactive material escapes into the air or into rivers or groundwater.

FIGURE 23.20 The ozone hole over Antarctica.

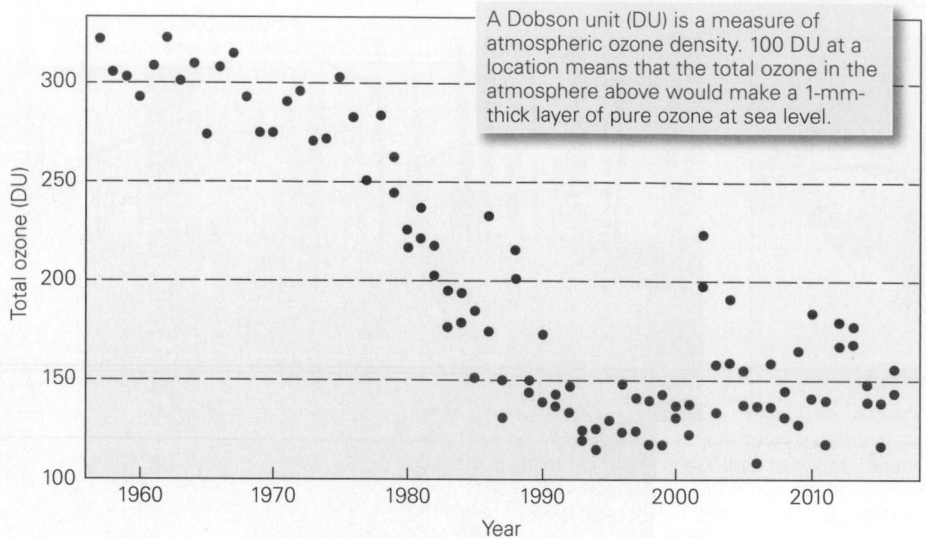

A Dobson unit (DU) is a measure of atmospheric ozone density. 100 DU at a location means that the total ozone in the atmosphere above would make a 1-mm-thick layer of pure ozone at sea level.

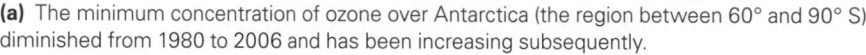

(a) The minimum concentration of ozone over Antarctica (the region between 60° and 90° S) diminished from 1980 to 2006 and has been increasing subsequently.

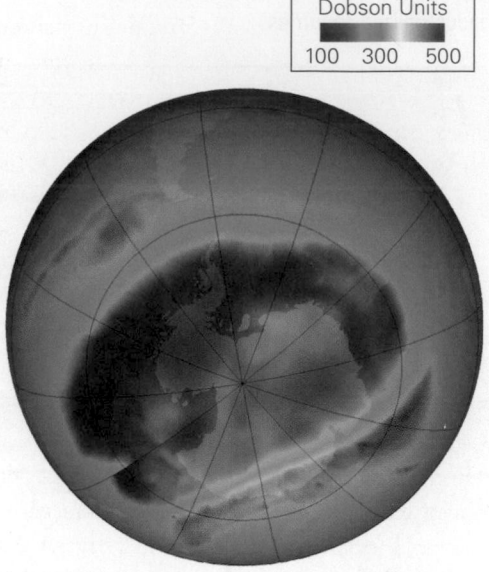

Dobson Units
100 300 500

(b) A map showing the dimensions of the ozone hole in 2006—the largest hole ever recorded.

- *Ozone depletion:* When emitted into the atmosphere, human-produced chemicals, most notably chlorofluorocarbons (CFCs), react with ozone in the stratosphere. This reaction, which happens most rapidly on the surfaces of tiny ice crystals in polar stratospheric clouds, destroys ozone molecules and produces an **ozone hole** over high-latitude regions, particularly Antarctica **(Fig. 23.20)**. Note that the "hole" refers to a region in which atmospheric ozone has been reduced substantially but not totally eliminated. Ozone holes have dangerous consequences, for they affect the ability of the atmosphere to shield the Earth's surface from harmful ultraviolet radiation. In 1987, a summit conference in Montreal proposed a global reduction of ozone-destroying CFC emissions. As a consequence, such emissions have decreased, and correspondingly, the amount of ozone depletion has decreased.

TAKE-HOME MESSAGE

Human activities impact the landscape and ecosystems significantly by moving rock and soil, paving the surface, and changing the character of vegetation cover. Society also introduces contaminants into the environment at rates that cannot be accommodated by the Earth System, leading to pollution of land, air, surface water, and groundwater.

QUICK QUESTION: What is the ozone hole, and why did it form?

23.6 Recent Climate Change

Observed Changes in CO_2 and CH_4

We've seen that greenhouse gases, most notably CO_2 and CH_4, play a major role in the regulation of the Earth's surface temperatures—without these gases, the Earth could not be a home for life. Both CO_2 and CH_4 cycle through various biogeochemical reservoirs of the Earth System during the carbon cycle, and the rate of movement between reservoirs influences the amount in any given reservoir at any given time. For most of geologic time, the concentration of greenhouse gases in the atmosphere was governed by natural processes—volcanic eruptions, life-evolution events, forest fires, weathering of mountain belts, warming and cooling due to the Milankovitch cycles, changes in solar activity or cosmic-ray flux, or meteorite impacts. But beginning around 8,000 years ago, human activities began to modify the environment significantly, first with the invention of agriculture and then, during the past two centuries, with the advent of industrialization.

Both industry and agriculture produce greenhouse gases and, in effect, transfer carbon that had been stored in underground or biomass reservoirs into the atmospheric reservoir. For example, burning fossil fuels oxidizes vast quantities of carbon, which had previously been locked for millions of years in fossil fuel underground, to yield CO_2, which mixes into the atmosphere. Heating calcite ($CaCO_3$) to produce the lime (CaO) used in cement takes carbon that had been locked in limestone underground for millions of years and produces CO_2, which also mixes into the air. Clear-cutting forests to make way for grazing land or fields replaces high-biomass vegetation (trees) with low-biomass vegetation (grasses or crops), thereby leaving CO_2 in the atmosphere. Decay of organic material in soggy rice paddies or melting permafrost, as well as the flatulence of cattle herds, produce significant quantities of CH_4 that would otherwise remain locked in biomass.

A sense of the relative volume of human society's input of greenhouse gases into the atmosphere comes from comparing human and volcanic production of these gases. Specifically, researchers estimate that all volcanic eruptions together in a given year—including both submarine and subaerial eruptions—emit about 0.15 to 0.26 gigatons of CO_2. By comparison, human activities emit about 35 gigatons of CO_2 every year (about 135 times as much). Put another way, human activities now produce more CO_2 in three days than all of the volcanoes on the Earth produce in a typical year. About 85% of the CO_2 that we send into the atmosphere comes from burning fossil fuels and producing cement, while about 15% is a consequence of deforestation.

Significantly, not all of the greenhouse gases that society sends into the atmosphere stay there. In the case of CO_2, some dissolves in the ocean, some reacts with minerals in rocks during chemical weathering, and some becomes incorporated into plants during photosynthesis. Before about 1900, natural *sinks* (the ocean, rock weathering, plants) could absorb most *anthropogenic* CO_2. But since then, the amount of anthropogenic CO_2 emitted has exceeded the ability of natural sinks to absorb it **(Fig. 23.21)**. In fact, calculations and isotopic studies suggest that only about 40% to 50% gets reabsorbed by oceans, organisms, or land during biogeochemical cycles. The remainder stays in the atmosphere. Put another way, human activities have pushed the Earth System's carbon cycle far beyond the natural range of steady-state conditions that existed during the Pleistocene and Holocene. This happened by transferring vast amounts of carbon from the underground reservoirs of hydrocarbons and limestone—where it had been locked away for millions to hundreds of millions of years or more—into the atmospheric reservoir, at rates significantly faster than it can be removed.

Can we detect increases in atmospheric CO_2 concentration? In the early 1960s, a chemist named Charles Keeling set out to measure the concentration of CO_2 in the atmosphere using the most accurate methods available. To avoid areas with urban pollution, he collected air samples every month

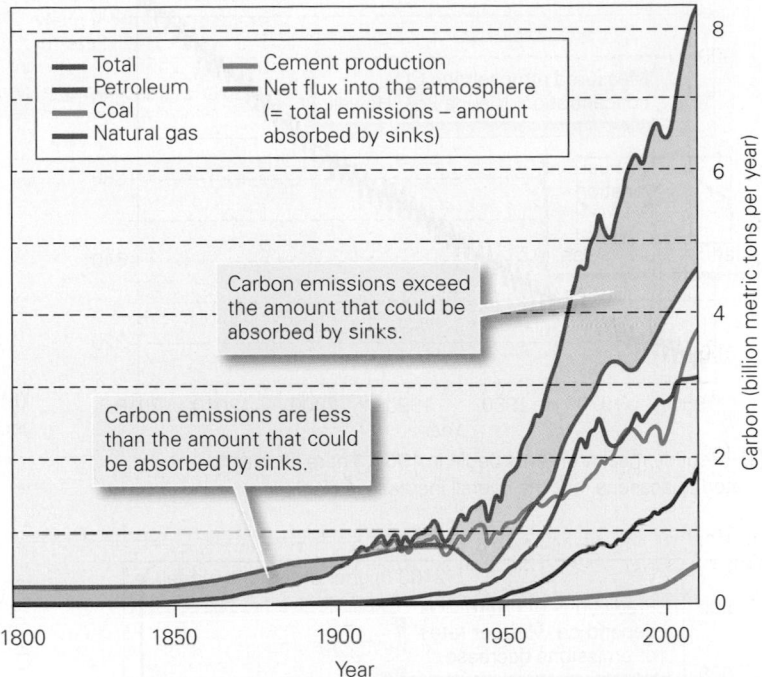

FIGURE 23.21 Until about 1900, anthropogenic CO_2 could be absorbed by natural sinks. Since then, about 50% of anthropogenic CO_2 remains in the atmosphere.

at the summit of the Mauna Kea volcano in Hawaii. After completing many years of measurements, Keeling showed not only that distinct seasonal variations occur (CO_2 concentration goes down in the warm summer when rates of photosynthesis increase, and it goes up in the winter when organic matter dies and decays) but also that the average annual concentration of CO_2 steadily rises **(Fig. 23.22a)**. Specifically, he demonstrated that the average annual CO_2 concentration went from 320 ppm in 1965 to 360 ppm in 1995. Keeling died in 2005, but measurements at Mauna Kea have continued. On May 9, 2013, the daily CO_2 concentration surpassed 400 ppm, probably for the first time in over 3 million years. During 2018, CO_2 concentration exceeded 412 ppm. Using records in ice cores, researchers have extended the record of CO_2 concentration further back in time and have found that in 1750, the average annual CO_2 concentration was only about 280 ppm **(Fig. 23.22b)**. Overall therefore, the atmospheric CO_2 concentration has increased by over 130 ppm (45%) since the beginning of the industrial revolution.

Studies of gas bubbles trapped in the glaciers of Antarctica allow researchers to extend the record of atmospheric CO_2 concentrations back through almost the last 800,000 years. This record demonstrates that during the alternating glacial and interglacial periods of the Late Pleistocene, CO_2 concentrations varied between about 180 and 300 ppm **(Fig. 23.22c)**, meaning that the increases since the beginning of the industrial revolution are beyond the range of natural fluctuations that occurred during the last 800,000 years.

FIGURE 23.22 Changes in carbon dioxide (CO_2) and methane (CH_4) concentrations in the atmosphere over time.

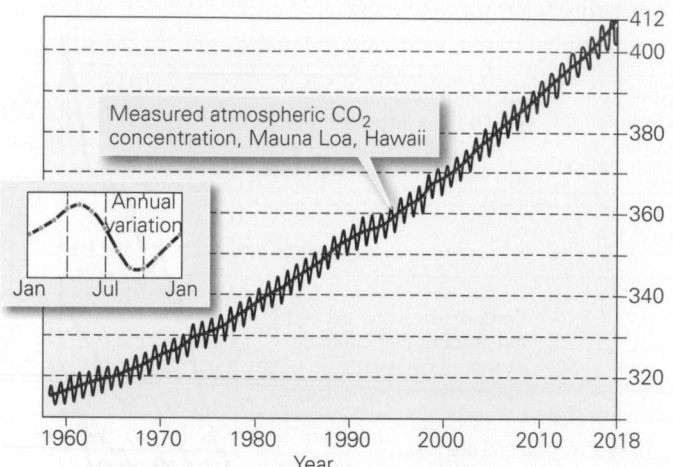

(a) Monthly measurements begin in 1960. There is an annual cycle, related to seasons, but the overall increase is clear.

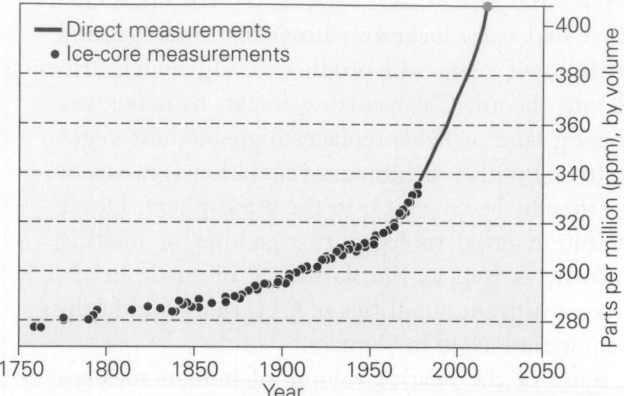

(b) Since the industrial revolution, CO_2 concentrations have steadily increased.

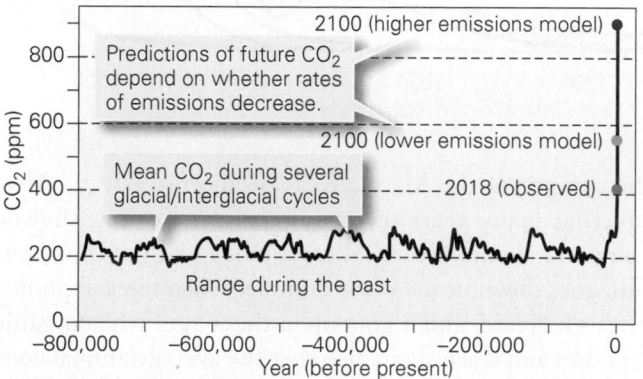

(c) In studies of ice cores from glaciers, researchers find that CO_2 concentrations varied between 180 and 300 ppm throughout glacial advances and retreats. The current value is far above this range.

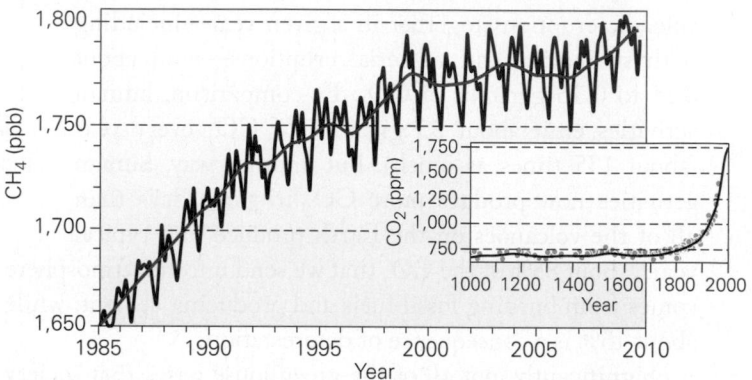

(d) Concentrations of methane, another greenhouse gas, have been increasing, too.

Have CO_2 concentrations ever been higher than they are today? Yes. For example, the CO_2 concentration during the Eocene climatic optimum (49 to 56 Ma) was over 1,000 ppm, about 2.5 times what it is today. But whether rates of change have ever been comparable to what researchers have measured in the past 200 years remains uncertain.

CO_2 is not the only greenhouse gas whose concentration has increased in the past two centuries. CH_4 (methane) concentrations have also risen **(Fig. 23.22d)**. Some of this gas comes from combustion of fossil fuels, some from venting or leakage of gas from oil and gas fields, some from decay of organic matter in rice paddies and in melting tundra, and some from the melting of gas hydrates.

Observations of Climate Change

The fundamental principle of the greenhouse effect requires that the increase in CO_2 concentrations during the past 200 years has caused atmospheric warming. Has such warming taken place during the past 200 years? Researchers have published thousands of observations suggesting that it has:

- Large ice shelves, such as the Larsen Ice Shelf, along the Antarctic Peninsula, and the Ayles Ice Shelf, along Ellesmere Island in northernmost Canada, are breaking up rapidly **(Fig. 23.23a)**. An iceberg of more than 11,000 km² (the size of Delaware) broke off from the Larsen Ice Shelf in 2017.

- The area covered by sea ice in the Arctic Ocean has decreased substantially **(Fig. 23.23b)**. Some estimates place the rate of ice-cover loss at about 3% per decade. But the trends are not simple—a graph of ice area from 1979 to 2013 shows many ups and downs lasting a few years **(Fig. 23.23c)**. Still, the overall average appears to trend in the direction of decreasing ice, and if this trend continues, it may be possible to sail across the Arctic Ocean within decades. Support for this prediction comes from studies of the age of the sea ice—the area of "old ice" (ice over 5 years old) has decreased by about 50% over the last 25 years **(Fig. 23.23d)**.

- The Greenland ice sheet is melting at an accelerating pace. Studies suggest that the rate of ice loss has increased from 90 to 220 km³ per year in the last 10 years and that in places, the ice sheet thins by about 1 m per year. Warming has caused the number of days during the year when melting takes place to increase **(Fig. 23.24a)**. In addition, the annual melt zone along the margins of the ice sheet has widened dramatically because the elevation of the equilibrium line (see Chapter 22) has risen **(Fig. 23.24b)**. Further indication of melting comes from observing the flow rates of valley glaciers draining the ice sheet.

- Valley glaciers worldwide have been retreating rapidly, so that areas that were once ice covered are now bare. The change is truly dramatic in many locations **(Fig. 23.25a)**.

- Worldwide, glacial volumes have been diminishing. About 400 km³ of ice disappears every year **(Fig. 23.25b)**.

FIGURE 23.23 The melting of sea ice.

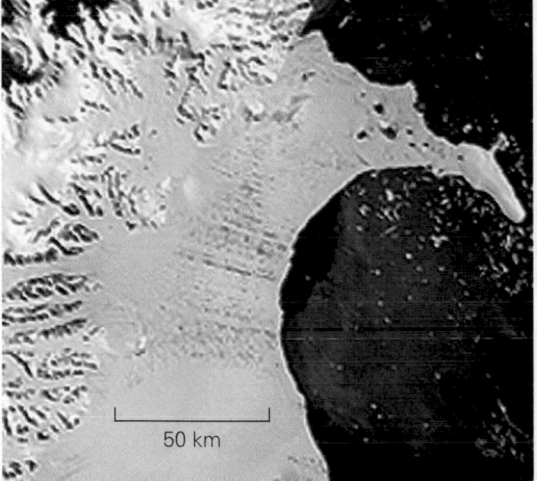

50 km

January 31, 2002

March 7, 2002

(a) In 2002, a large portion of Larsen Ice Shelf of Antarctica disintegrated over the course of a month.

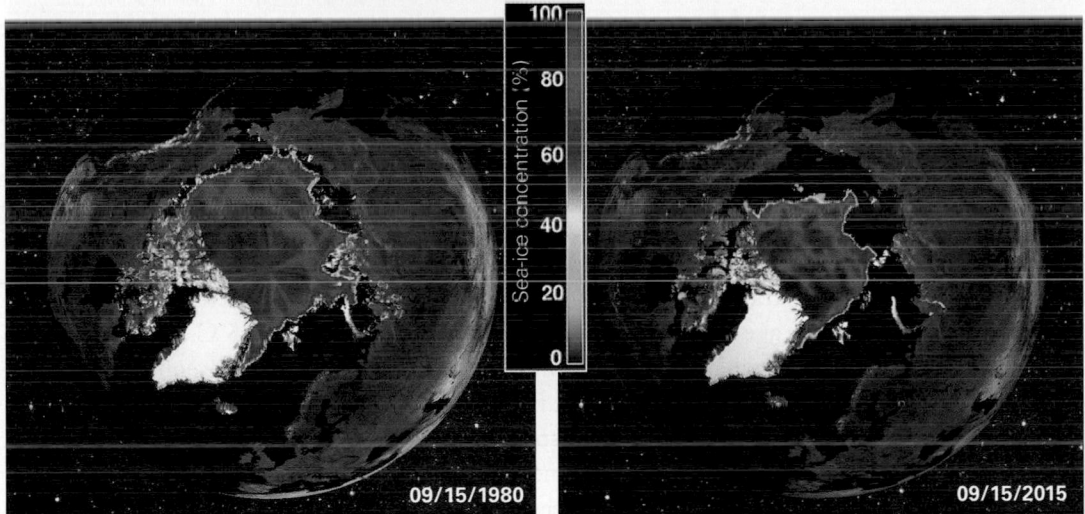

Sea-ice concentration (%)

09/15/1980

09/15/2015

(b) Between 1980 and 2015, the coverage of sea ice in the Arctic decreased.

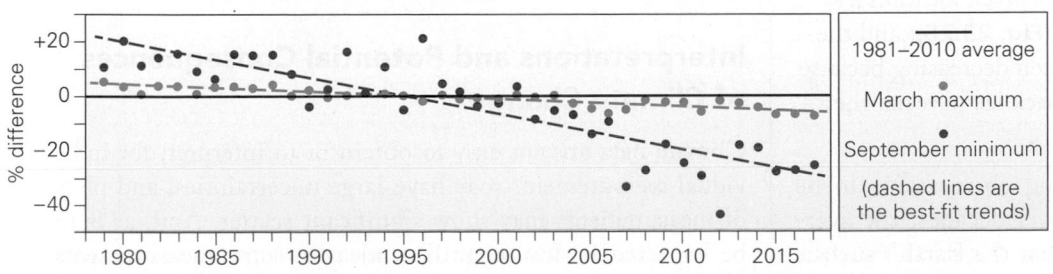

1981–2010 average

March maximum

September minimum

(dashed lines are the best-fit trends)

% difference

(c) The area of sea ice in the Arctic has gone up and down yearly, but on average, it is decreasing.

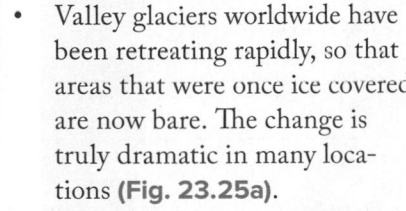

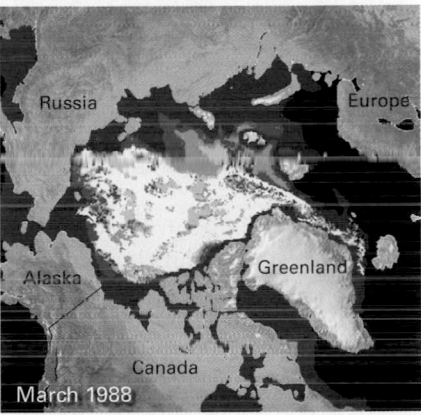

Russia | Europe

Alaska | Greenland

Canada

March 1988

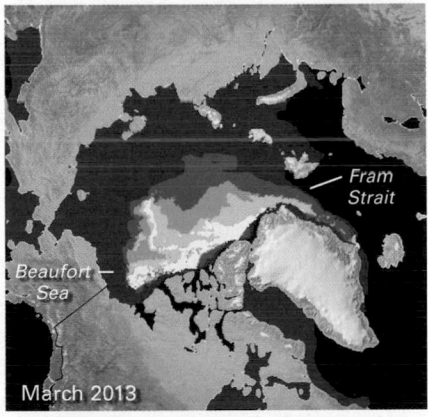

Fram Strait

Beaufort Sea

March 2013

Sea ice age (years)

Open water | 1 | 2 | 3 | 4 | 5+

(d) The age of sea ice in the Arctic Ocean has decreased substantially during the past 25 years.

FIGURE 23.24 Melting of the Greenland ice sheet.

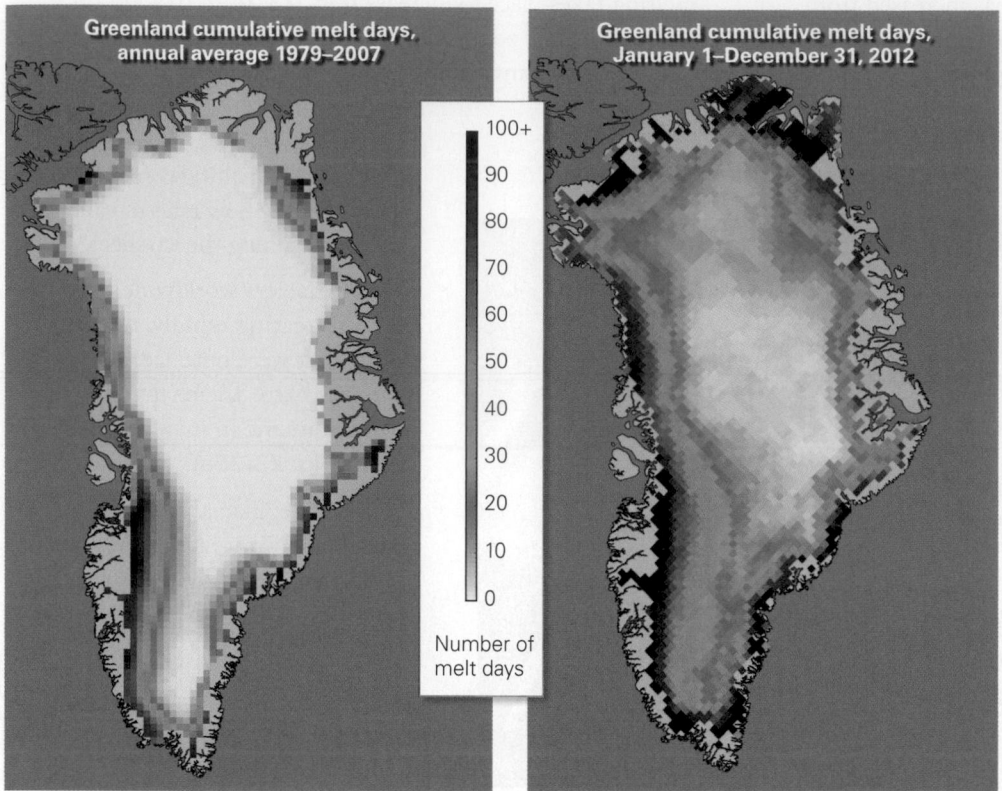

The gray area, spotted with puddles, is the melt zone.

June 13, 2002

June 17, 2003

(a) The number of days in a year during which melting of the Greenland ice sheet takes place has increased.

(b) The summer melt line has risen to higher elevations. (Each photo shows the same area.)

• The area of permafrost in high latitudes has substantially decreased, and melt ponds have formed on the surface of once-frozen land **(Fig. 23.25c)**. In fact, large regions that once stayed frozen all year are now melting in the summer.

• Average annual water vapor concentration in the atmosphere has been increasing due to evaporation of warmer seas. This trend is hard to characterize, however, because humidity can vary significantly around the world at a given time.

• Biological phenomena that are sensitive to climate have been disrupted. For example, the time at which sap in the maple trees of the northeastern United States starts to flow has changed; the mosquito line (the elevation at which mosquitoes can survive) has risen substantially; plant hardiness zones have migrated northward in the United States, which means that at a given location it's possible to plant earlier in the year **(Fig. 23.26)**; and the average weight of polar bears has been decreasing because the bears can no longer walk over pack ice to reach their hunting grounds in the sea.

Researchers consider the preceding observations to be substantive evidence that, averaged over decades, the average global atmospheric temperature near the Earth's surface is rising, a phenomenon often referred to as *global warming*.

Direct measurements of temperatures, collected at recording stations around the world since about 1880, support this proposal. Specifically, global average atmospheric temperature has risen by almost 1°C during the last century and is higher now than it has been at any time during the past 2,000 years **(Fig. 23.27a, b)**. Although such a change may seem small, the magnitude of temperature change between the last ice age and now, by comparison, was only 3°C to 5°C. A small change in average temperature may have major consequences. Significantly, the amount of temperature change varies around the world—some regions appear to be warming more than others, and some areas have been cooling **(Fig. 23.27c)**. Because of the increase in average atmospheric temperature, average measured values of near-surface ocean temperatures have also been rising **(Fig. 23.28)**.

Interpretations and Potential Consequences of Climate Change

Climate data are not easy to obtain or to interpret, for individual measurements may have large uncertainties, and plots of measurements may show significant scatter. And, as is to be expected in any scientific endeavor, some measurements and conclusions based on them turn out to be incorrect and

FIGURE 23.25 Global change in glaciers and permafrost.

(a) The Muir Glacier in Alaska retreated 12 km between 1941 and 2004.

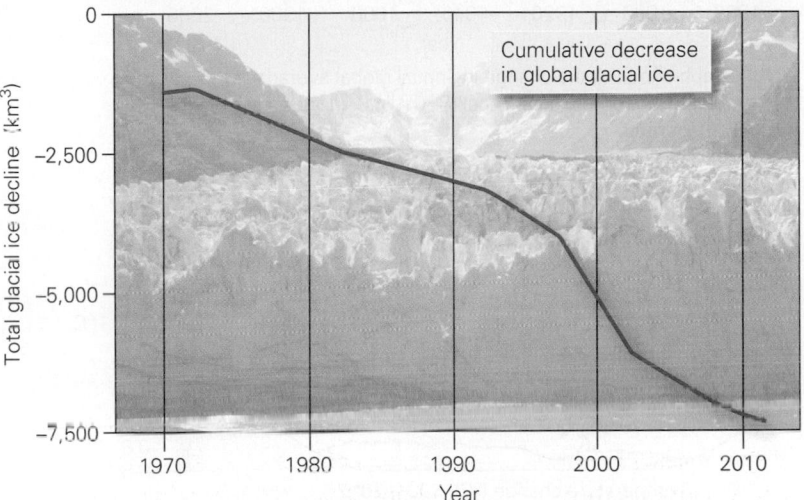

Cumulative decrease in global glacial ice.

(b) Global glacial ice volume has been decreasing by 400 km³ per year.

(c) Melt ponds and puddles forming as permafrost melts along the coast of the Arctic Ocean.

FIGURE 23.26 Plant hardiness zones have been migrating northward.

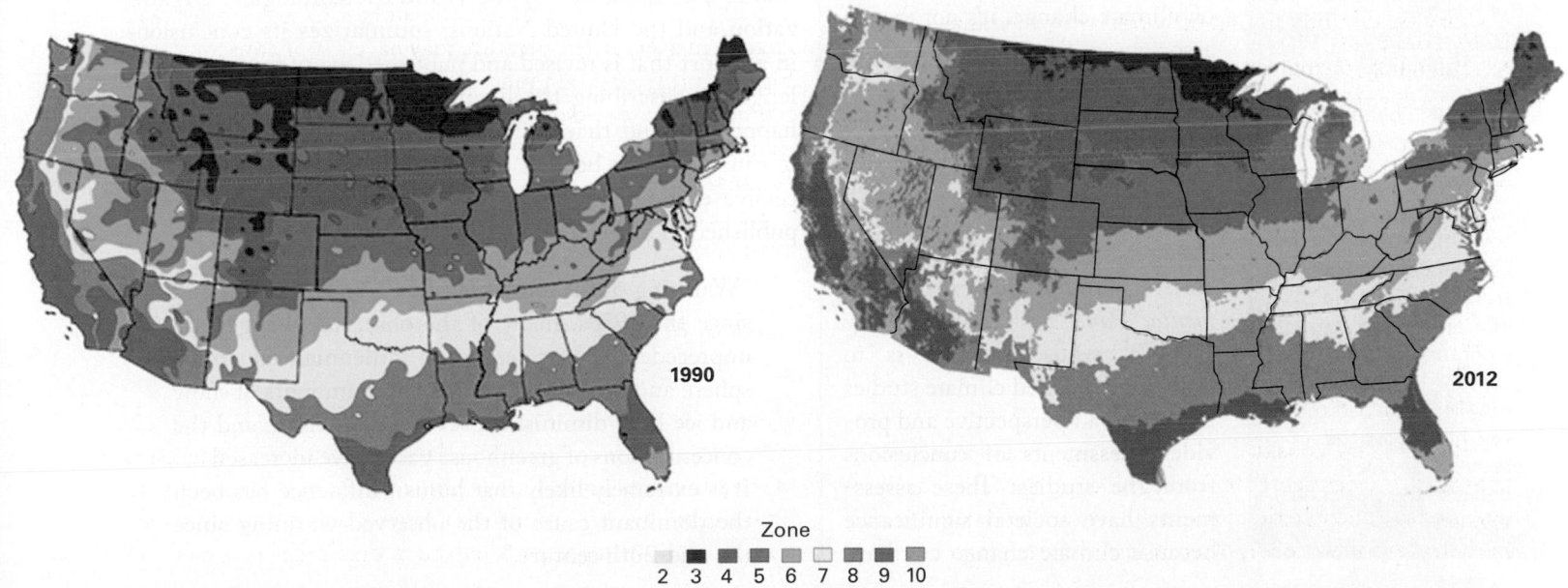

Zone
2 3 4 5 6 7 8 9 10

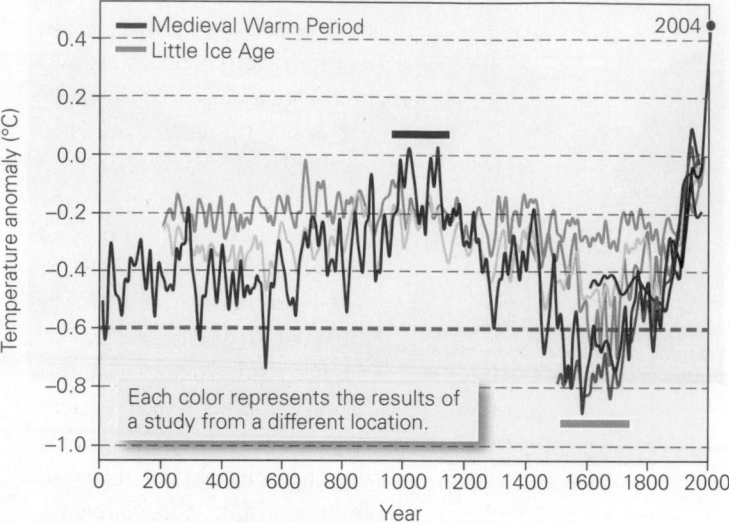

(a) Reconstructions of temperature during the past 2,000 years. Each color represents a published estimate from a research team. The black dashed line is the average of direct measurements.

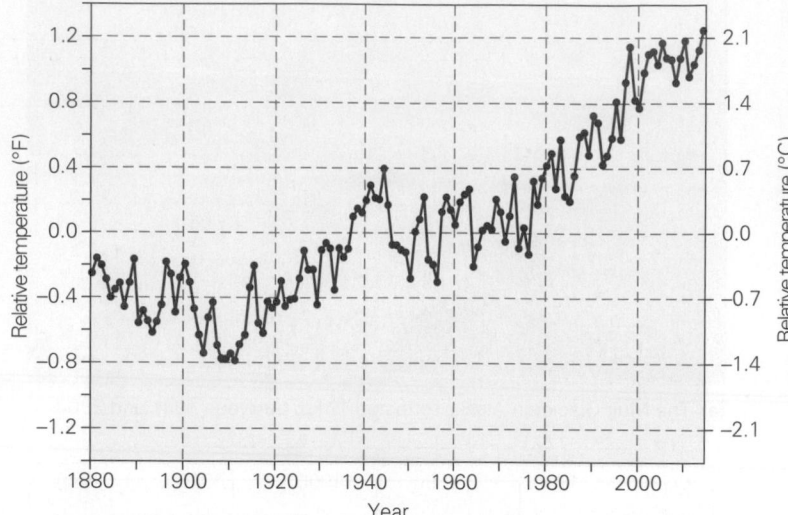

(b) Graph showing the change in annual global average temperature (combined land and ocean) since 1880 relative to the average for the 20th century.

Receding Glacier, Switzerland

LATITUDE
46°34'48.75" N

LONGITUDE
8°22'58.53" E

Look NNE obliquely from 4 km.

Above, you will see a valley glacier near the town of Gletsch. The bare land downslope from the toe of the glacier was ice covered a century ago. Note that the lateral moraine now lies like a bathtub ring high above the present-day ice because the glacier has thinned.

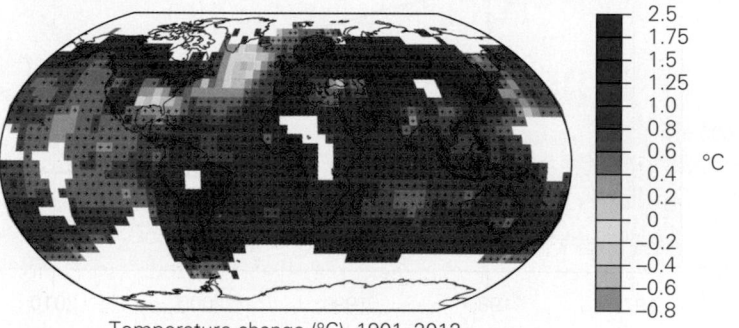

(c) Colors represent the change in average atmospheric temperature at the Earth's surface between 1901 and 2012. Redder areas have become warmer; bluer areas have become cooler.

must be replaced by the results of newer studies. In the case of climate-change studies, the data show ups and downs in temperature from year to year and in some cases from decade to decade. This fact can lead to confusion in interpreting the data because a trend that lasts for, say, 5 years does not necessarily represent a trend that lasts for 100 years. Further, because so many variables control climate change, and because so many feedbacks have an impact on climate change, it's not easy to associate specific individual effects with specific individual causes.

In an attempt to make sense of overwhelming volumes of sometimes contradictory data and interpretations relevant to climate change, a group of leading scientists founded the *Intergovernmental Panel on Climate Change* (IPCC), whose purpose is to evaluate published climate studies from a broad perspective and provide assessments of conclusions from the studies. These assessments have societal significance because climate change can have

major implications for society and for global policy decisions. The IPCC, sponsored by the World Meteorological Organization and the United Nations, summarizes its conclusions in a report that is revised and published every five years. The language describing the likelihood that global warming is happening, and that humans have contributed significantly to causing it, has become progressively less equivocal in successive editions of the report. The *Fifth Assessment Report*, published in 2014, states,

> "Warming of the climate system is unequivocal, and since the 1950s, many of the observed changes are unprecedented over decades to millennia. The atmosphere and ocean have warmed, the amounts of snow and ice have diminished, sea level has risen, and the concentrations of greenhouse gases have increased . . . It is extremely likely that human influence has been the dominant cause of the observed warming since the mid-20th century."

FIGURE 23.28 The global annual average temperature of shallow ocean water relative to the average temperature between 1961 and 2006.

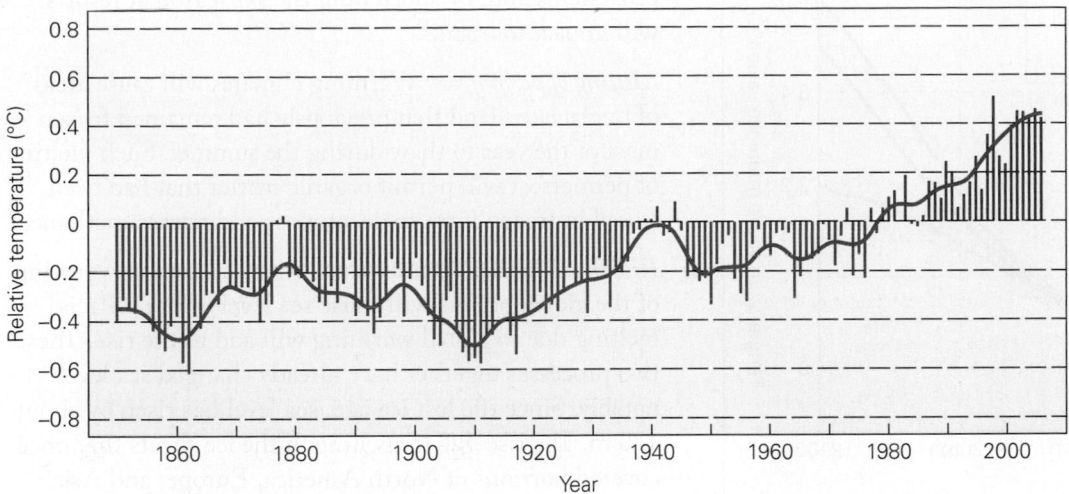

This wording means that the vast majority of climate researchers have concluded that global warming is real, and that the influence of human activities—referred to as *anthropogenic forcing*—has played a significant role in the change (see Fig. 23.21). Natural forcing, such as changes in solar radiation or cosmic-ray flux, does not appear to be of sufficient magnitude to have caused all observed warming. Calculations suggest that the rate of temperature change during the past 50 years is greater than the rate for the previous 50 years.

Some researchers suggest that human effects on climate became noticeable as far back as 8,000 years ago and that climate has been trending toward warmer conditions ever since. Deviations from the warming trend have been attributed, speculatively, to times when the human population abruptly decreased (due to pandemics). During these times, the production of greenhouse gases slowed and forests returned. But such deviations have been relatively short-lived, so overall,

researchers conclude that without anthropogenic forcing, the Earth's climate would be significantly cooler, and we might even have been heading toward another ice age **(Fig. 23.29)**. Natural forcing has changed climate in the geologic past, but does not appear to be responsible for post-industrialization trends in global temperature.

The *Fifth Assessment* of the IPCC implies that society faces the challenge of either slowing climate change or dealing with its consequences. The effects of global warming over the coming decades to centuries remain the subject of intense debate because predictions depend on computer models, and not all researchers agree on how to construct or interpret these models. The role of clouds in climate change, for example, remains poorly understood and inadequately addressed by such models. But newer models, running on faster computers, provide increasingly reliable constraints on future trends.

In a worst-case scenario, global warming will continue into the future at the present rate, so that by 2050—within the lifetimes of many readers of this book—the average annual temperature will have increased in some parts of the world by 1.5°C to 2.0°C. At these rates, by the end of the century, temperatures could be over 4°C warmer, depending on the model used **(Fig. 23.30a)**, and by 2150, global temperatures could be 5°C to 11°C warmer than at present—the warmest since the Eocene Epoch, 40 million years ago. Models predict that warming will not be the same everywhere—the greatest impact will be in the Arctic **(Fig. 23.30b)**. According to many climate models, the following events will happen as a result:

- *Shift in climate belts:* As the climate overall warms, temperate and desert regions will occur at higher latitudes, so regions that are now agricultural areas may dry out and become unfarmable **(Fig. 23.31a)**. The change in climate will also affect the amount and distribution of precipitation (rain and snow) that falls. For example, in North America, summers will become drier and winters will become wetter **(Fig. 23.31b)**. One way to picture this change is to think of how the "climate latitude" of a state in the United States might change over time. Models suggest that if current warming rates continue, the future climate of Illinois will be like the present climate of Texas, and the future climate of New Hampshire will be

FIGURE 23.29 Model comparing the consequences of natural forcing to natural plus anthropogenic forcing and to observed changes in temperature.

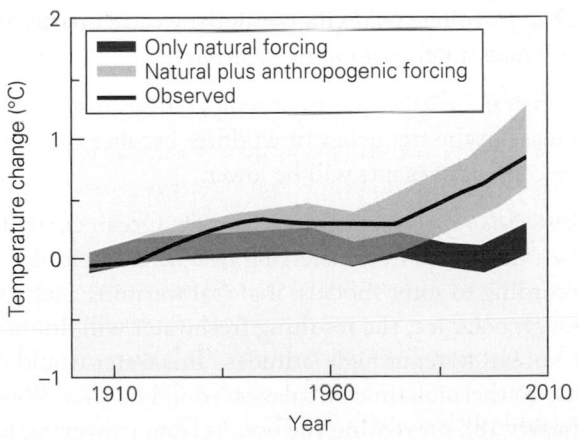

FIGURE 23.30 Model predictions of future global warming.

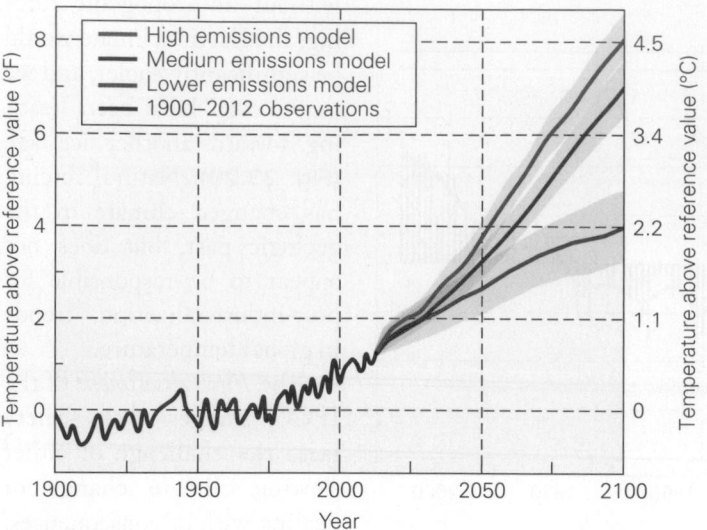

(a) Model calculations of global warming. Different models assume different rates of CO_2 emissions. All suggest a significant global temperature increase by 2100.

2070–2100 prediction
vs. 1960–1990 average

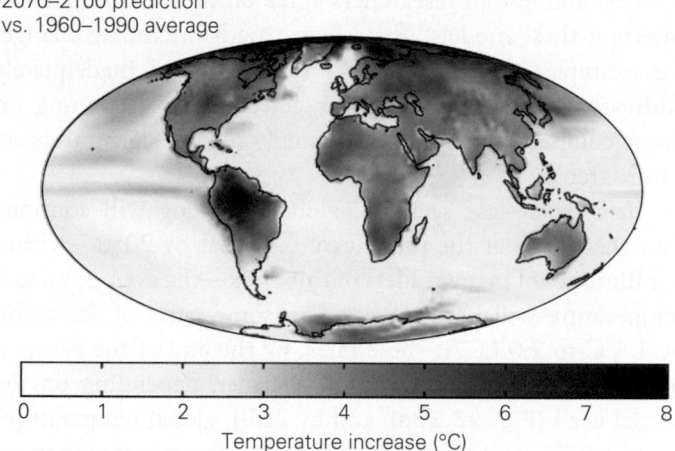

(b) Global warming does not mean that all locations warm by the same amount. This map shows a model prediction of how temperature increases may vary with location for the time period 2070–2100.

like the present climate of North Carolina **(Fig. 23.31c)**. Models also suggest that global warming will dramatically increase the number of health-threatening *heat waves* (loosely defined as prolonged periods of excessively hot weather, above 32°C) that affect particular regions **(Fig. 23.31d)**.

- *Ice retreat and snow-line rise:* We've already seen evidence that the volume of glacial ice worldwide has decreased as manifested by the shrinkage and retreat of most glaciers. (Some glaciers, however, are advancing. Current glacial advance could be a consequence either of surging due to addition of liquid water to the base of the glacier or of local increases in snowfall due to the addition of moisture to the atmosphere as the oceans warm.) As temperatures

increase, the snow line in mountains will rise to higher latitudes and higher elevations. This change will affect ecosystems and, by shortening the ski season at resorts, will impact tourism.

- *Melting of permafrost:* Warming climates will cause areas of unglaciated land that previously had remained frozen most of the year to thaw during the summer. Such melting of permafrost will permit organic matter that had been stored in frozen form to decompose and release methane.

- *Rise in sea level:* Water expands when heated, so warming of the global oceans will cause sea level to rise. Glacial melting due to global warming will add to the rise. These two processes together have already changed sea level notably. Since the last ice age, sea level has risen by about 120 m. The rise due to melting of the ice sheets that once covered portions of North America, Europe, and Asia tapered off about 8,000 years ago **(Fig. 23.32a)**. But a closer look at sea level for the past 130 years shows that it is continuing to rise **(Fig. 23.32b)**. In fact, measurements indicate that there has been a rise of almost 12 cm in the past century. This rise correlates with warming of ocean water **(Fig. 23.32c)** and with the melting of glaciers (see Fig. 23.25b). Sea-level rise is already causing flooding of coastal areas and has submerged some islands, and it has led some nations to begin investing in new coastal flood-control measures.

The amount of sea-level rise in the future depends on the rate of global warming. Models suggest that by 2100, sea level may rise by an additional 20 to 60 cm **(Fig. 23.33)**. A map showing the elevations of coastal areas emphasizes that a meter or two of sea-level rise could inundate regions of the world where 20% of the human population currently lives **(Geology at a Glance,** pp. 910–911).

- *Stronger storms:* An increase in average ocean and atmospheric temperatures may lead to increased evaporation from the oceans. The additional moisture might nourish stronger hurricanes. Some climate models predict that global warming could change global weather patterns or cause more intense flooding or drought.

- *Increases in wildfires:* Warmer temperatures may lead to an increase in the frequency of wildfires because the moisture content of plants will be lower.

- *Interruption of thermohaline circulation:* Ocean currents play a major role in transferring heat across latitudes. According to some models, if global warming melts enough polar ice, the resulting freshwater will dilute surface ocean water at high latitudes. This water would not sink, so thermohaline circulation would be shut off (see Chapter 18), preventing the oceans from conveying heat.

FIGURE 23.31 Model predictions of changes that may happen if current levels of global warming continues.

Tundra ☐   Deciduous forest ▨   Evergreen forest ■   Boreal forest ☐   Shrub and grassland ☐   Sparse vegetation ▨

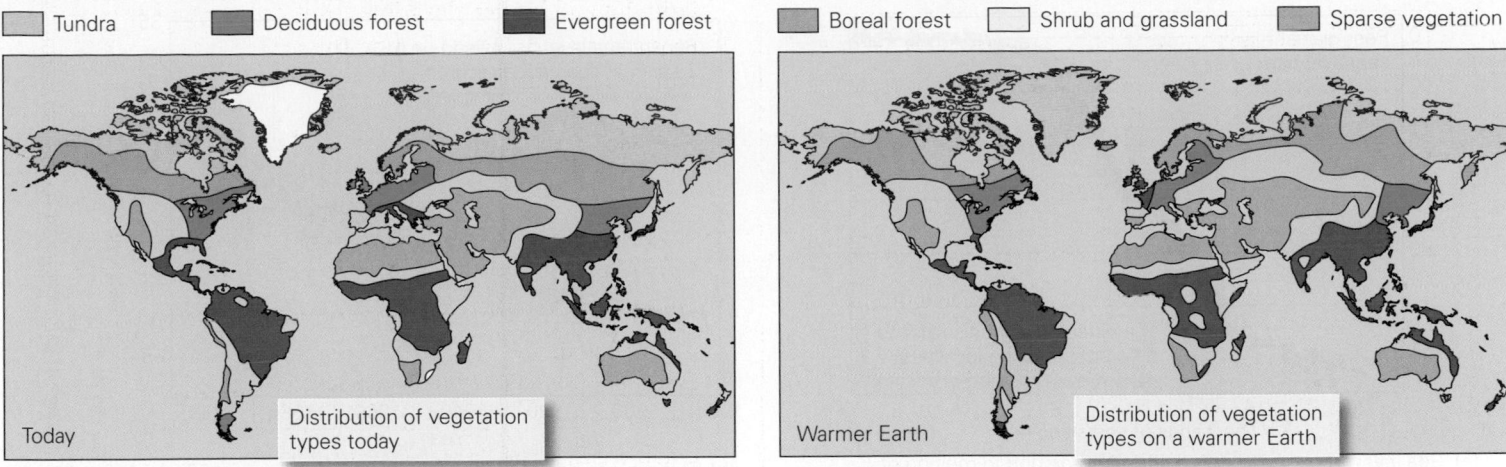

Today   Distribution of vegetation types today

Warmer Earth   Distribution of vegetation types on a warmer Earth

(a) Models predict that climate zones will shift to higher latitudes.

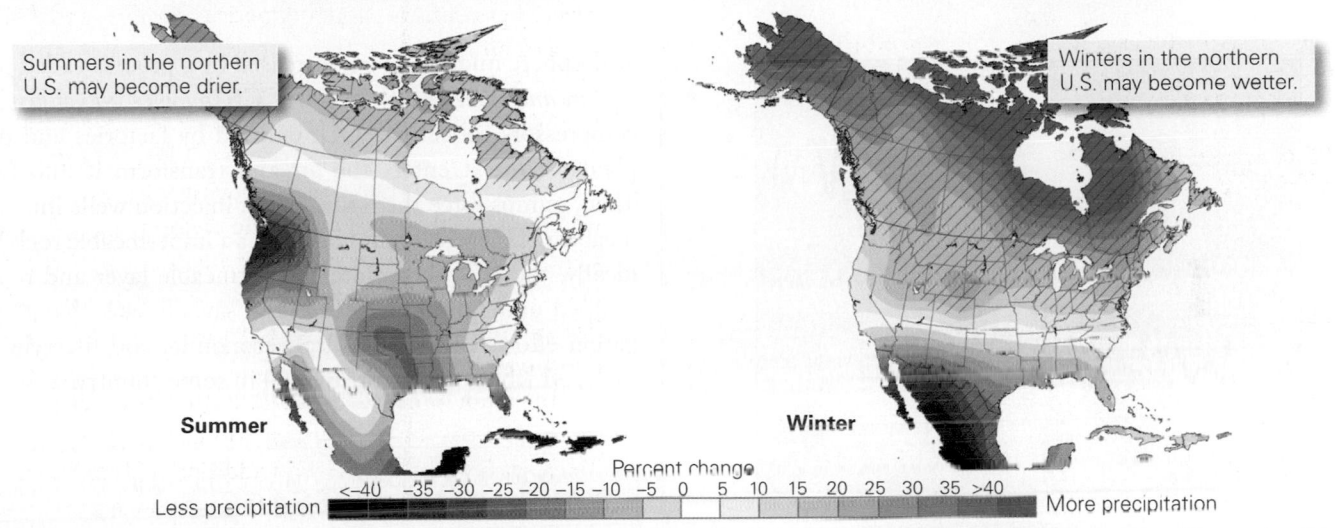

Summers in the northern U.S. may become drier.

Winters in the northern U.S. may become wetter.

Summer

Winter

Percent change

Less precipitation   <–40   –35   –30   –25   –20   –15   –10   –5   0   5   10   15   20   25   30   35   >40   More precipitation

(b) Models suggest that the amount of precipitation (rain and snow) at locations in North America about 100 years from now will be different than it is today.

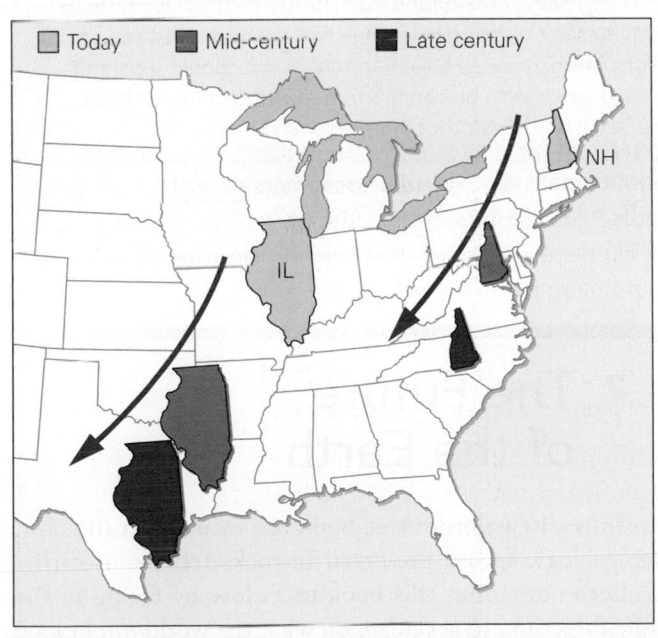

Today ☐   Mid-century ▨   Late century ■

NH

IL

(c) This map represents how the climate of a given state may change if the global climate warms. At the end of the 21st century, northern states may have climates that are like those of southern states today.

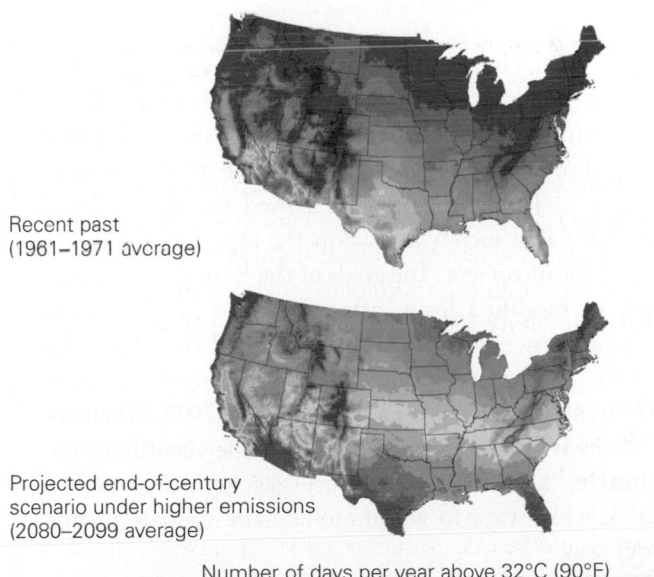

Recent past (1961–1971 average)

Projected end-of-century scenario under higher emissions (2080–2099 average)

Number of days per year above 32°C (90°F)

0   15   30   45   60   75   90   105   120   135   150   165   180   >180

(d) Recent (1961–1971) and predicted (2080–2099) numbers of days above 32°C (90°F) per year in the United States.

FIGURE 23.32 Sea-level rise of the recent past and possible sea-level rise of the future.

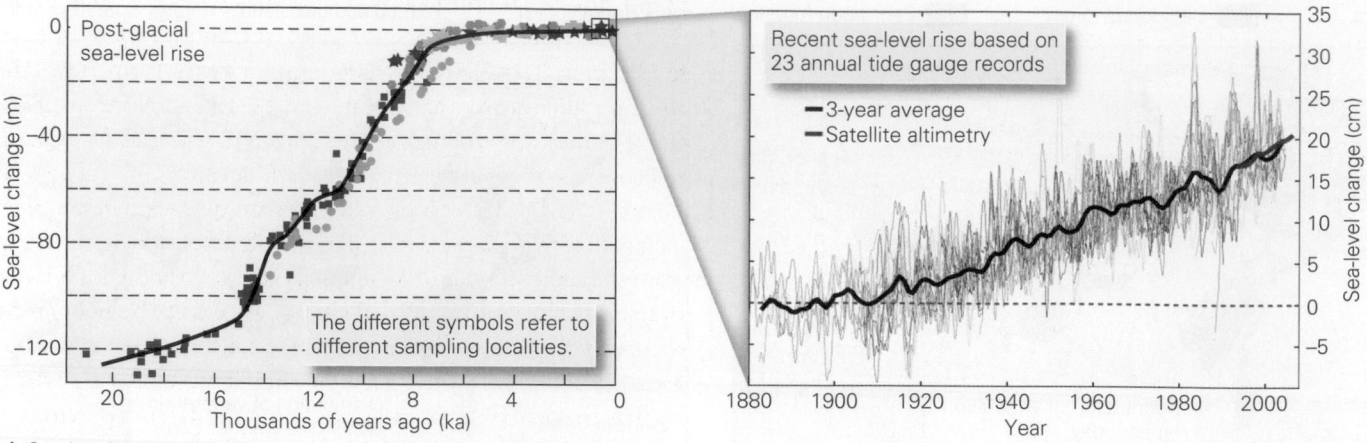

(a) Sea level rose rapidly after the last ice age due to melting of continental glaciers.

(b) Tide gauges document sea-level rise over the past 130 years.

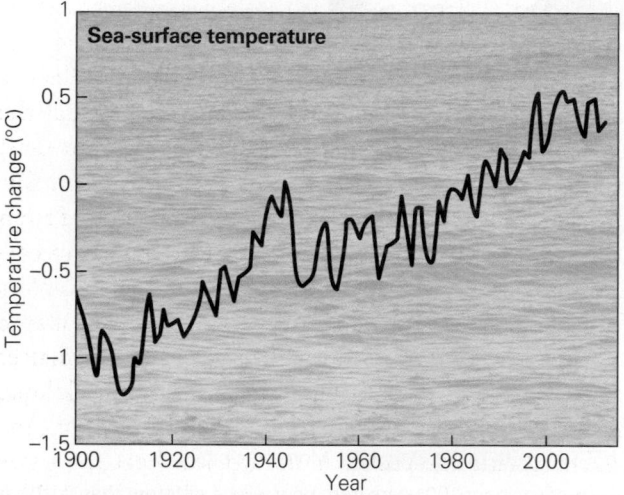

(c) Average sea-surface temperature change since 1900.

The potential changes described above, along with the results of other studies that estimate the large economic cost of global warming, imply that the issue needs to be addressed seriously and soon. But what can be done? In 1997, at a summit meeting in Japan, 160 nations signed the *Kyoto Accord*, which declared that society must slow the input of greenhouse gases into the atmosphere. The goals of the Kyoto Accord were amplified, in modified form, after the 2015 U.N. Climate Change Conference, when 195 countries (all but three that attended the conference) signed the *Paris Climate Accord* in 2016. The roster of signatories decreased in 2017, when the United States withdrew from the accord. The accord encourages countries to reduce greenhouse-gas emissions and to begin to develop ways to accommodate the climate change now under way.

Society can decrease greenhouse-gas emissions by switching to power-generation systems that do not burn fossil fuels, and by increasing energy conservation. Emissions to the atmosphere might also be decreased by a process called *carbon capture and sequestration* (*CCS*). CCS involves two steps: First, compressors collect the CO_2 emitted by factories and power plants before it enters the air and transform it into liquid. Then, pumps force the liquid down injection wells into a permeable rock layer that lies beneath an impermeable rock layer; ideally, the CO_2 seeps into the permeable layer and remains trapped underground. Needless to say, climate-change mitigation efforts involve political, economic, and lifestyle decisions, which remain controversial in some countries.

TAKE-HOME MESSAGE

Growing evidence indicates that greenhouse gases (CO_2 and CH_4) being introduced into the atmosphere by human activities are causing warming of both the air and sea that might not otherwise have happened. This global warming is associated with glacial melting, shifting climate zones, sea-level rise, and other phenomena.

QUICK QUESTION: How do researchers develop predictions of future climate change?

23.7 The Future of the Earth

Most of the discussion in this book has focused on the past, for the geologic record preserved in rocks tells us of earlier times. Let's now bring this book to a close by facing in the opposite direction, to speculate on what the world might look like in geologic time to come.

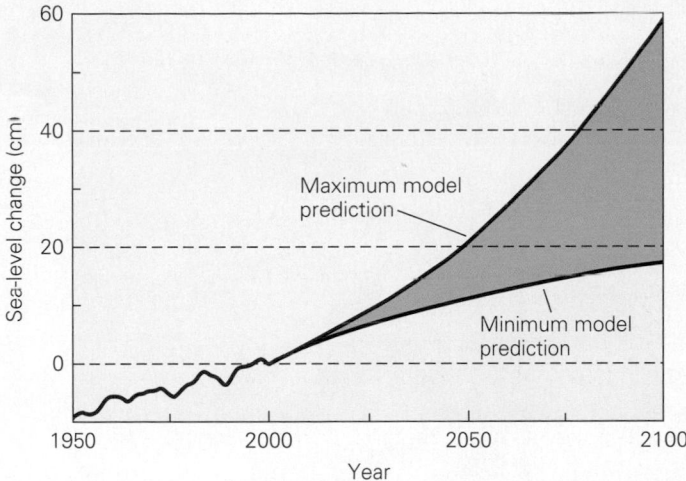

FIGURE 23.33 Estimates of future sea-level rise if global warming continues.

In the geologic near term, the future of the Earth probably depends largely on human activities. Whether the Earth System shifts to a new equilibrium, whether a mass-extinction event takes place, or whether society achieves **sustainable growth** (an ability to prosper within the constraints of the Earth System) will depend on our own foresight and ingenuity. Projecting thousands of years into the future, we might well wonder if the Earth will return to ice-age conditions, with glaciers growing over major cities and the continental shelf becoming dry land, or if the ice age is over for good because of anthropogenic global warming. No one really knows for sure.

If we project millions of years into the future, it is clear that the map of the planet will change significantly because of the continuing activity of plate tectonics. For example, during the next 50 million years or so, the Atlantic Ocean will probably become bigger, the Pacific Ocean will shrink, and the western part of California will migrate northward. Eventually, Australia may crush against the southern margin of Asia, and the islands of Indonesia will be flattened in-between.

Predicting the map of the Earth beyond that point is hard because we can't predict where new subduction zones will develop. Perhaps subduction of the Pacific Plate will lead to the collision of the Americas with Asia to produce a new supercontinent ("Amasia"). A subduction zone will eventually form on one side (or both sides) of the Atlantic Ocean, and the ocean will be consumed. As a consequence, the eastern margin of the Americas will collide with the western margin of Europe and/or Africa. The sites of major cities—New York, Miami, Rio de Janeiro, Buenos Aires, and London—will be incorporated in a collisional mountain belt and probably will be subjected to metamorphism and igneous intrusion before being uplifted and eroded. Shallow seas may once again cover the interiors of continents and then later retreat, and glaciers may once again cover the continents—it happened in the past, so it could happen again!

And if the past is the key to the future, we *Homo sapiens* might not be around to watch our cities enter the rock cycle, for biological evolution may have introduced new species to the biosphere. Perhaps, 100 million years from now, the stratigraphic record of our time will be several centimeters of strata containing anomalous isotopes and unusual chemicals, trace fossils of concrete structures, and a record of widespread extinctions. In fact, some researchers suggest that we are now facing the *sixth extinction*, a mass extinction that is resulting from the rapid destruction of global ecosystems now underway. Because anthropogenic global change appears to be happening so rapidly now, some researchers use the informal term *Anthropocene* in reference to the geologic time interval during which humans have had a profound impact on the Earth System.

And what of the end of the Earth? Geologic catastrophes resulting from asteroid and comet collisions will undoubtedly occur in the future as they have in the past. We can't predict when the next strike will come, but unless the object can be diverted, the Earth is in for another radical readjustment of surface conditions. It's not likely, however, that such collisions will destroy our planet. Rather, astronomers predict that the end of the Earth will occur some 5 billion years from now, when the Sun begins to run out of nuclear fuel. When this happens, outward-directed thermal pressure caused by fusion reactions will no longer be able to prevent the Sun from collapsing inward because of the immense gravitational pull of its mass. Were the Sun a few times larger than it is, the collapse would trigger a supernova explosion that would blast matter out into space to form a new nebula surrounding a black hole. But since the Sun is not that large, the thermal energy generated when its interior collapses inward will heat the gases of its outer layers sufficiently to cause them to expand. As a result, the Sun will become a red giant, a huge star whose radius will grow beyond the orbit of the Earth. Our planet will then vaporize, and its atoms will join an expanding ring of gas—the ultimate global change. If this happens, the atoms that once formed the Earth and all its inhabitants through geologic time may eventually be incorporated in a future solar system, where the cycle of planetary formation and evolution will begin anew.

TAKE-HOME MESSAGE

In the near term, the Earth's surface will be affected by decisions of human society. Over longer time scales, the map of the Earth will change due to plate interactions and sea-level change.

QUICK QUESTION: What might happen to the atoms that make up the Earth long after the red-giant stage of Solar System evolution?

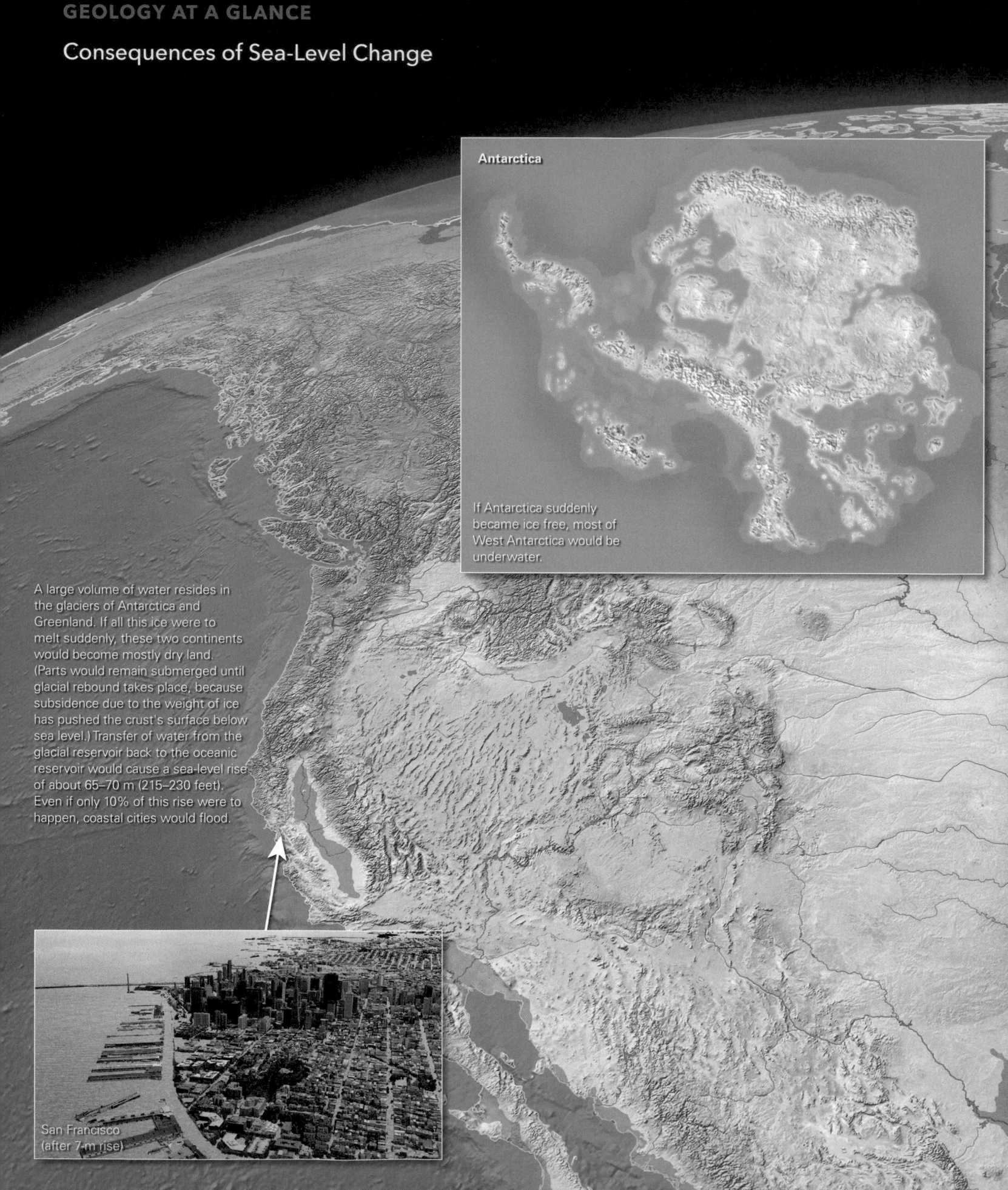

Consequences of Sea-Level Change

Antarctica

If Antarctica suddenly became ice free, most of West Antarctica would be underwater.

A large volume of water resides in the glaciers of Antarctica and Greenland. If all this ice were to melt suddenly, these two continents would become mostly dry land. (Parts would remain submerged until glacial rebound takes place, because subsidence due to the weight of ice has pushed the crust's surface below sea level.) Transfer of water from the glacial reservoir back to the oceanic reservoir would cause a sea-level rise of about 65–70 m (215–230 feet). Even if only 10% of this rise were to happen, coastal cities would flood.

San Francisco (after 7-m rise)

Greenland

A suddenly ice-free Greenland would host a central lake.

New Orleans (after 7-m rise)

Measured sea-level rise varies with location. The maximum current observed rate is about 1 m/century. At this rate, the 7-m rise depicted here in the city images will not happen for several centuries. But even a 1 m rise can lead to a costly increase in nuisance flooding and storm-surge damage.

A sea-level rise of 70 m would flood large areas of coastal land worldwide (areas shown in purple). In fact, most of Florida would be underwater, and New York City buildings would be submerged up to about the 20th floor. Even a 7-m rise would turn streets in many major cities into canals, as shown in the insets.

New York (after 7-m rise)

SUMMARY

- Global change involves transformations or modifications of physical and biological components of the Earth System over time. Unidirectional change results in transformations that never repeat, whereas cyclic change involves repetition of the same steps over and over.

- Examples of unidirectional change include differentiation of the Earth, growth of continents, formation of oceans, change in atmospheric composition, and life evolution.

- Examples of physical cycles that take place on the Earth include the supercontinent cycle, the sea-level cycle, and the rock cycle.

- A biogeochemical cycle involves the passage of a chemical among nonliving and living reservoirs. Examples include the hydrologic cycle and the carbon cycle. Global change occurs when the relative proportions of the chemicals in different reservoirs change.

- Tools for documenting global climate change include the stratigraphic record, paleontology, oxygen-isotope ratios, bubbles in ice, growth rings in trees, and human history.

- Studies of long-term climate change show that at some times in the past the Earth experienced greenhouse periods, while at other times there were icehouse periods. Factors leading to long-term climate change include the positions of continents, volcanic activity, the uplift of land, and life evolution.

- Short-term climate change can be seen in the record of the last million years. In fact, the climate has warmed and cooled several times during the past 15,000 years. Causes of short-term climate change include changes in the Earth's orbit and tilt, changes in surface albedo, changes in ocean currents, and perhaps fluctuations in solar radiation and cosmic rays.

- Mass extinction, a catastrophic change in biodiversity, may be caused by the impact of an asteroid or by intense volcanic activity.

- During the last two centuries, humans have changed landscapes and modified ecosystems and have added pollutants to the land, air, and water at rates faster than the Earth System can process them.

- The addition of greenhouse gases (CO_2 and CH_4) to the atmosphere causes global warming, which could shift climate belts and lead to a rise in sea level. In the industrial era, sources for the added CO_2 include fossil-fuel burning, cement production, and deforestation.

- In the future, in addition to climate change, the Earth will witness a continued rearrangement of continents resulting from plate tectonics and will probably suffer the impacts of asteroids and comets. The end of the Earth may come in about 5 billion years when the Sun runs out of fuel and becomes a red giant.

GUIDE TERMS

albedo (p. 892)

biogeochemical cycle (p. 882)

carbon cycle (p. 882)

climate (p. 883)

climate-change model (p. 885)

differentiation (p. 878)

Earth System (p. 877)

ecosystem (p. 895)

faint young Sun paradox (p. 890)

feedback mechanism (p. 884)

global change (p. 877)

global climate change (p. 883)

global cooling (p. 883)

global warming (p. 883)

Goldilocks effect (p. 890)

greenhouse effect (p. 883)

greenhouse gas (p. 883)

greenhouse (hothouse) period (p. 887)

habitable zone (p. 890)

icehouse period (p. 887)

mass-extinction event (p. 893)

ozone hole (p. 898)

paleoclimate (p. 885)

pollution (p. 896)

sedimentary sequence (p. 880)

steady-state condition (p. 882)

sunspot cycle (p. 892)

supercontinent cycle (p. 880)

sustainable growth (p. 909)

weather (p. 883)

GEOTOURS *THIS CHAPTER'S GEOTOURS WORKSHEET (S) FEATURES QUESTIONS AND GOOGLE EARTH SITES ON:*

- Effects of global warming
- Effects of deforestation
- Water use in arid regions
- Preservation of natural habitats
- Sea-level change

REVIEW QUESTIONS

The letters following each Review Question refer to the corresponding Learning Objective from the Chapter Opener.

1. What does the term *Earth System* refer to? **(C)**

2. How have the Earth's interior, crust, and atmosphere changed since the planet first formed? **(A)**

3. Does the map of the Earth stay the same over time? Why? **(A)**

4. What processes control the rise and fall of sea level on the Earth? **(B)**

5. Describe the reservoirs that play a role in the carbon cycle and how carbon transfers among reservoirs. **(C)**

6. Explain the role of greenhouse gases in regulating climate. **(D)**

7. How do paleoclimatologists study ancient climate change? From what geologic source were the data in this figure collected and what does the chart tell us? **(D)**

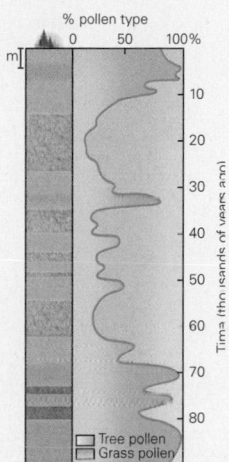

8. Contrast icehouse and greenhouse conditions. Did icehouse conditions happen prior to the most recent ice age? **(D)**

9. What are the possible causes of long-term climate change? What factors explain short-term climate change? **(D)**

10. What events can cause catastrophic change in the Earth System? **(B)**

11. How have humans changed the solid Earth? **(E)**

12. What are pollutants, and why do they pose a problem? What is the ozone hole, and why did it form? **(E)**

13. What does this graph tell us about the change in CO_2 content since 1960? **(E)**

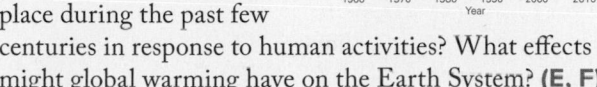
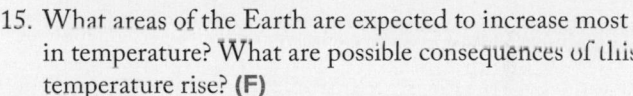

14. What evidence do researchers use to argue that global warming has been taking place during the past few centuries in response to human activities? What effects might global warming have on the Earth System? **(E, F)**

15. What areas of the Earth are expected to increase most in temperature? What are possible consequences of this temperature rise? **(F)**

16. What are some likely scenarios for the long-term future of the Earth? **(G)**

ON FURTHER THOUGHT

17. If global warming continues, how will the distribution of grain crops change? Might this change affect national economies, and if so, how? **(F)**

18. Currently, tropical rainforests are being cut down at a rate of 1.8% per year. At this rate, how many more years will the forests survive? **(F)**

ONLINE RESOURCES

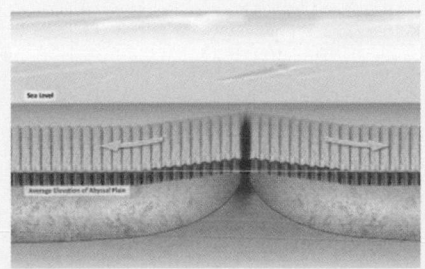

Animations
This chapter features animations on global sea-level change, including glacial ice volume, thermal expansion, seafloor-spreading rate, and oceanic volcanic plateaus.

Videos
This chapter features real-world videos on rising CO_2 levels, deforestation, ice sheet stratigraphy, and more.

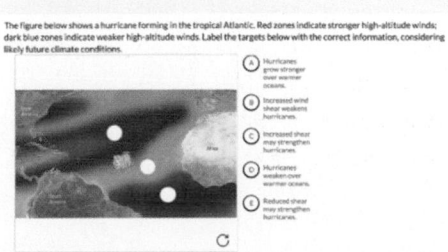

Smartwork5
This chapter features questions on the causes and impacts of long- and short-term climate change and the factors that control and impact regional climate conditions.

Metric Conversion Chart

Length

1 kilometer (km) = 0.6214 mile (mi)
1 meter (m) = 1.094 yards = 3.281 feet
1 centimeter (cm) = 0.3937 inch
1 millimeter (mm) = 0.0394 inch
1 mile (mi) = 1.609 kilometers (km)
1 yard = 0.9144 meter (m)
1 foot = 0.3048 meter (m)
1 inch = 2.54 centimeters (cm)

Area

1 square kilometer (km^2) = 0.386 square mile (mi^2)
1 square meter (m^2) = 1.196 square yards (yd^2)
= 10.764 square feet (ft^2)
1 square centimeter (cm^2) = 0.155 square inch (in^2)
1 square mile (mi^2) = 2.59 square kilometers (km^2)
1 square yard (yd^2) = 0.836 square meter (m^2)
1 square foot (ft^2) = 0.0929 square meter (m^2)
1 square inch (in^2) = 6.4516 square centimeters (cm^2)

Volume

1 cubic kilometer (km^3) = 0.24 cubic mile (mi^3)
1 cubic meter (m^3) = 264.2 gallons
= 35.314 cubic feet (ft^3)
1 liter (l) = 1.057 quarts
= 33.815 fluid ounces
1 cubic centimeter (cm^3) = 0.0610 cubic inch (in^3)
1 cubic mile (mi^3) = 4.168 cubic kilometers (km^3)
1 cubic yard (yd^3) = 0.7646 cubic meter (m^3)
1 cubic foot (ft^3) = 0.0283 cubic meter (m^3)
1 cubic inch (in^3) = 16.39 cubic centimeters (cm^3)

Mass

1 metric ton = 2,205 pounds
1 kilogram (kg) = 2.205 pounds
1 gram (g) = 0.03527 ounce
1 pound (lb) = 0.4536 kilogram (kg)
1 ounce (oz) = 28.35 grams (g)

Pressure

1 kilogram per
square centimeter (kg/cm^2)* = 0.96784 atmosphere (atm)
= 0.98066 bar
= 9.8067×10^4 pascals (Pa)
1 bar = 0.1 megapascals (Mpa)
= 1.0×10^5 pascals (Pa)
= 29.53 inches of mercury (in a barometer)
= 0.98692 atmosphere (atm)
= 1.02 kilograms per square centimeter (kg/cm^2)
1 pascal (Pa) = 1 $kg/m/s^2$
1 pound per square inch = 0.06895 bars
= 6.895×10^3 pascals (Pa)
= 0.0703 kilogram per square
centimeter

Temperature

To change from Fahrenheit (F) to Celsius (C):
$$°C = \frac{(°F - 32°)}{1.8}$$
To change from Celsius (C) to Fahrenheit (F):
$$°F = (°C \times 1.8) + 32°$$
To change from Celsius (C) to Kelvin (K):
$$K = °C + 273.15$$
To change from Fahrenheit (F) to Kelvin (K):
$$K = \frac{(°F - 32°)}{1.8} + 273.15$$

---

*Note: Because kilograms are a measure of mass whereas pounds are a unit of weight, pressure units incorporating kilograms assume a given gravitational constant (g) for the Earth. In reality, the gravitational for the Earth varies slightly with location.

FIGURE A.1 The modern periodic table of the elements. Each column groups elements with related properties. For example, inert gases are listed in the column on the right. Metals are found in the central and left parts of the chart.

Glossary

a'a A lava flow with a rubbly surface.

abandoned meander A meander that dries out after it was cut off.

ablation The removal of ice at the toe of a glacier by melting, sublimation (the evaporation of ice into water vapor), and/or calving.

abrasion The process in which one material (such as sand-laden water) grinds away at another (such as a stream channel's floor and walls).

absolute age Numerical age (the age specified in years).

absolute plate velocity The movement of a plate relative to a fixed point in the mantle.

absolute zero The lowest temperature possible ($-273.15°C$); at absolute zero, vibrations and movements of atoms in a material cease.

abyssal plain A broad, relatively flat region of the ocean that lies at least 4.5 km below sea level.

Acadian orogeny A convergent mountain-building event that occurred around 400 million years ago, during which continental slivers accreted to the eastern edge of the North American continent.

accreted terrane A block of crust that collided with a continent at a convergent margin and stayed attached to the continent.

accretion The addition of new crust to a continent, by collision of island arcs and oceanic plateaus.

accretionary coast A coastline that receives more sediment than erodes away.

accretionary lapilli Hailstone-like clumps of wet ash that fall from a volcanic eruptive cloud.

accretionary orogen An orogen formed by the attachment of numerous buoyant slivers of crust to an older, larger continental block.

accretionary prism A wedge-shaped mass of sediment and rock scraped off the top of a downgoing plate and accreted onto the overriding plate at a convergent plate margin.

acid-mine runoff A dilute solution of sulfuric acid, produced when sulfur-bearing minerals in mines react with rainwater, that flows out of a mine.

acid rain Precipitation in which air pollutants react with water to make a weak acid that then falls from the sky.

active continental margin A continental margin that coincides with a plate boundary.

active fault A fault that has moved recently or is likely to move in the future.

active margin A continental margin that is also a plate boundary.

active sand The top layer of beach sand, which moves daily because of wave action.

active volcano A volcano that has erupted within the past few centuries and will likely erupt again.

adiabatic cooling The cooling of a body of air or matter without the addition or subtraction of thermal energy (heat).

adiabatic heating The warming of a body of air or matter without the addition or subtraction of heat.

advection A process of heat transfer in which heat is carried into a solid by a liquid or gas moving through fractures or pores in the solid.

aerosols Tiny solid particles or liquid droplets that remain suspended in the atmosphere for a long time.

aftershocks The series of smaller earthquakes that follow a major earthquake.

agent of erosion Natural entities that remove material from the Earth's surface, and transport it elsewhere; examples include rivers, glaciers, and the wind.

aggregate (n) In the context of discussing concrete, aggregate refers to the solid chunks (such as gravel) that the cement of concrete holds together.

air The mixture of gases that make up the Earth's atmosphere.

air-fall tuff Tuff formed when ash settles gently from the air.

air mass A body of air, about 1,500 km across, that has recognizable physical characteristics.

air pressure The push that air exerts on its surroundings.

albedo The reflectivity of a surface.

Alleghenian orogeny The orogenic event that occurred about 270 million years ago when Africa collided with North America.

alloy A metal containing more than one type of metal atom.

alluvial fan A gently sloping apron of sediment dropped by an ephemeral stream at the base of a mountain in arid or semiarid regions.

alluvium Sorted sediment deposited by a stream.

alluvium-filled valley A valley whose floor fills with sediment.

Alpine-Himalayan chain The largest orogenic belt on the Earth today, formed by collisions of the former Gondwana continents with the southern margins of Europe and Asia.

amber Hardened (fossilized) ancient sap or resin.

amphibolite facies A set of metamorphic mineral assemblages formed under intermediate to high pressures and temperatures.

amplitude The height of a wave from crest to trough.

Ancestral Rockies The late Paleozoic uplifts of the Rocky Mountain region; they eroded away long before the present Rocky Mountains formed.

angiosperm A flowering plant.

angle of repose The angle of the steepest slope that a pile of uncemented material can attain without collapsing from the pull of gravity.

angularity The degree to which grains have sharp or rounded edges or corners.

angular unconformity An unconformity in which the strata below were tilted or folded before the unconformity developed; strata below the unconformity therefore have a different tilt than strata above.

anhedral grains Crystalline mineral grains without well-formed crystal faces.

anion A negatively charged ion.

annual probability The likelihood that a flood of a given size or larger will happen at a specified locality during any given year.

Antarctic bottom water mass The volume of cold, dense water that sank along the coast of Antarctica and spreads northward at the base of the ocean.

antecedent stream A stream that cuts across an uplifted mountain range; the stream must have existed before the range uplifted and must then have been able to downcut as fast as the land was rising.

anthracite coal Shiny black coal formed at temperatures between 200°C and 300°C. A high-rank coal.

Anthropocene A term used informally in reference to the past few thousand years, to emphasize that during this time, human society has modified the Earth System significantly.

anticline A fold with an arch-like shape in which the limbs dip away from the hinge.

anticyclone   A region in which a broad volume of air undergoes clockwise flow (in the northern hemisphere) around a high-pressure area.

anticyclonic flow   A circulation of air around a high-pressure region in the atmosphere; it rotates clockwise in the northern hemisphere.

Antler orogeny   The Late Devonian mountain-building event in which slices of deep-marine strata were pushed eastward, up and over the shallow-water strata on the western coast of North America.

anvil cloud   A large cumulonimbus cloud that spreads laterally at the tropopause to form a broad, flat top.

aphanitic   A textural term for fine-grained igneous rock.

Appalachian fold-thrust belt   The western portion of the Appalachian Orogen in which compression during the Alleghanian orogeny produced an array of thrust faults and associated folds above a regional detachment fault.

apparent polar-wander path   A path on the globe along which a magnetic pole appears to have wandered over time; in fact, the continents drift, while the magnetic pole stays fairly-fixed.

aquiclude   Sediment or rock that transmits no water.

aquifer   Sediment or rock that transmits water easily.

aquitard   Sediment or rock that does not transmit water easily and therefore retards the motion of the water.

Archaea   A kingdom of "old bacteria," now commonly found in extreme environments like hot springs. (Also called archaeobacteria.)

Archean Eon   The middle Precambrian eon (4.0–2.5 Ga).

Archimedes' principle   The mass of the water displaced by a block of material equals the mass of the whole block of material.

arête   A residual knife-edge ridge of rock that separates two adjacent cirques.

argillaceous sedimentary rock   Sedimentary rock that contains abundant clay.

arkose   A clastic sedimentary rock containing both quartz and feldspar grains.

arroyo   The channel of an ephemeral stream; dry wash; wadi.

artesian spring   A location where the ground surface intersects a natural fracture (joint) that taps a confined aquifer in which the pressure can drive the water to the surface.

artesian well   A well in which water rises on its own.

artificial levee   A man-made retaining wall to hold back a river from flooding.

ash   *See* Volcanic ash.

ash fall   Ash that settles to the ground, like snow, out of an ash cloud.

ash flow   An avalanche of ash that tumbles down the side of an explosively erupting volcano.

assimilation   The process of magma contamination in which blocks of wall rock fall into a magma chamber and dissolve.

asteroid   One of the fragments of solid material, left over from planet formation or produced by collision of planetesimals, that resides between the orbits of Mars and Jupiter.

asthenosphere   The layer of the mantle that lies between 100–150 km and 350 km deep; the asthenosphere is relatively soft and can flow when acted on by force.

astronomer   A scientist who studies the Universe outside of the Earth System.

astronomical unit   The distance between the Sun and the Earth, used as a reference frame for describing distances among objects in our Solar System.

atm   A unit of air pressure that approximates the pressure exerted by the atmosphere at sea level.

atmosphere   A layer of gases that surrounds a planet.

atoll   A roughly circular coral reef surrounding a lagoon.

atom   The smallest piece of an element that has the properties of the element; it consists of a nucleus surrounded by an electron cloud.

atomic mass   The amount of matter in an atom; roughly, it is the sum of the number of protons plus the number of neutrons in the nucleus.

atomic number   The number of protons in the nucleus of a given element.

atomic weight   In a general sense, the number of protons plus the number of neutrons in the nucleus of a given element. (The term is commonly used interchangeably with atomic mass, though in detail they have slightly different meanings.)

aurora australis   The same phenomenon as the aurora borealis, but in the southern hemisphere.

aurora borealis   A ghostly curtain of varicolored light that appears across the night sky in the northern hemisphere when charged particles from the Sun interact with the ions in the ionosphere.

aurorae   The currents of color, visible in polar regions, due to the influx of charged particles from space along magnetic field lines.

avalanche   A turbulent cloud of debris mixed with air that rushes down a steep hill slope at high velocity; the debris can be rock and/or snow.

avalanche chute   A downslope hillside pathway along which avalanches repeatedly fall, consequently clearing the pathway of mature trees.

avulsion   The process in which a river overflows a natural levee and begins to flow in a new direction.

axial plane   The imaginary surface that encompasses the hinges of successive layers of a fold.

axial trough   A narrow depression that runs along a mid-ocean ridge axis.

axis   An imaginary line around which an object spins.

backscattered light   Atmospheric scattered sunlight that returns to space.

backshore zone   The zone of beach that extends from a small step cut by high-tide swash to the front of the dunes or cliffs that lie farther inshore.

backswamp   The low marshy region between the bluffs and the natural levees of a floodplain.

backwash   The gravity-driven flow of water back down the slope of a beach.

Bacteria   A type of tiny prokaryotic single-celled organism.

bajada   An elongate wedge of sediment formed by the overlap of several alluvial fans emerging from adjacent valleys.

Baltica   A Paleozoic continent that included crust that is now part of today's Europe.

banded-iron formation (BIF)   Iron-rich sedimentary layers consisting of alternating gray beds of iron oxide and red beds of iron-rich chert.

bar   (1) A sheet or elongate lens or mound of alluvium; (2) a unit of air-pressure measurement approximately equal to 1 atm.

barchan dune   A crescent-shaped dune whose tips point downwind.

barrel of oil (bbl)   A unit of measurement for describing a volume of oil; a bbl equals 159 liters or 42 gallons.

barrier island   An offshore sand bar that rises above the mean high-water level, forming an island.

barrier reef   A coral reef that develops offshore, separated from the coast by a lagoon.

basal sliding   The phenomenon in which meltwater accumulates at the base of a glacier, so that the mass of the glacier slides on a layer of water or on a slurry of water and sediment.

basalt   A fine-grained, mafic, igneous rock.

base level   The lowest elevation a stream channel's floor can reach at a given locality.

basement   Older igneous and metamorphic rocks making up the Earth's crust beneath sedimentary cover.

basement uplift   Uplift of basement rock by faults that penetrate deep into the continental crust.

base metals   Metals that are mined but not considered precious. Examples include copper, lead, zinc, and tin.

basin   A fold or depression shaped like a right-side-up bowl.

Basin and Range Province   A broad, Cenozoic continental rift that has affected a portion of the western United States in Nevada, Utah, and Arizona; in this province, tilted fault blocks form ranges, and alluvium-filled valleys are basins.

batholith A vast composite, intrusive, igneous rock body up to several hundred km long and 100 km wide, formed by the intrusion of numerous plutons in the same region.

bathymetric map A map illustrating the shape of the ocean floor.

bathymetric profile A cross section showing ocean depth plotted against location.

bathymetry Variation in depth.

bauxite A residual mineral deposit rich in aluminum.

baymouth bar A sandspit that grows across the opening of a bay.

beach A gently sloping fringe of sediment along the shore.

beach drift The gradual migration of sand along a beach, due to the swash and backwash of waves that strike the shore at an angle.

beach erosion The removal of beach sand caused by wave action and longshore currents.

beach face A steeply concave part of the foreshore zone formed where the swash of the waves actively scours the sand.

beach profile A cross section illustrating the shape of a beach's surface.

bearing The compass heading of a line.

bed An individual layer of sediment or sedimentary rock in a deposit.

bedding Layering or stratification in sedimentary rocks.

bed load Large particles, such as sand, pebbles, or cobbles, that bounce or roll along a streambed.

bedrock Rock still attached to the Earth's crust.

Bergeron process Precipitation involving the growth of ice crystals in a cloud at the expense of water droplets.

berm A horizontal or landward-sloping terrace in the backshore zone of a beach that receives sediment during a storm.

Big Bang theory A theory suggested by scientists in which a cataclysmic explosion represents the formation of the Universe; before this event, all matter and all energy were packed into one volume-less point.

Big Bang nucleosynthesis The formation of new atomic nuclei (mostly hydrogen and helium) during the Big Bang.

biochemical limestone Limestone incorporating shells produced by organisms.

biochemical sedimentary rock Sedimentary rock formed from material (such as shells) produced by living organisms.

biodiversity The number of different species that exist at a given time.

biofuel Gas or liquid fuel made from plant material (biomass). Examples of biofuel include alcohol (from fermented sugar), biodiesel from vegetable oil, and wood.

biogenic minerals Substances that meet the definition of a mineral and are produced naturally by organisms (e.g., calcite in shells).

biogeochemical cycle The exchange of chemicals between living and nonliving reservoirs in the Earth System.

biomarker A molecule or set of molecules that could only have been produced by living organisms; their presence can be used for correlation and for determining the time at which life on the Earth began.

biomass The amount of organic material in a specified volume.

bioremediation The injection of oxygen and nutrients into a contaminated aquifer to foster the growth of bacteria that will ingest or break down contaminants.

biosphere The region of the Earth and atmosphere inhabited by life; this region stretches from a few km below the Earth's surface to a few km above.

bioturbation The mixing of sediment by burrowing animals such as clams and worms.

bituminous coal Dull, black intermediate-rank coal formed at temperatures between 100°c and 200°C.

black-lung disease Lung disease contracted by miners from the inhalation of too much coal dust.

black smoker The cloud of suspended minerals formed where hot water spews out of a vent along a mid-ocean ridge; the dissolved sulfide components of the hot water instantly precipitate when the water mixes with seawater and cools.

blind fault A fault that does not intersect the ground surface.

block Large, angular pyroclastic fragments consisting of volcanic rock, broken up during the eruption.

blocky lava Lava that is so viscous that it breaks into boulder-like blocks as it moves; typically, such lavas are andesitic or rhyolitic.

blowout A deep, bowl-like depression scoured out of desert terrain by a turbulent vortex of wind.

blue shift The phenomenon in which a source of light moving toward you appears to have a higher frequency.

body fossil A relict of an organism's body, preserved in rock.

body waves Seismic waves that pass through the interior of the Earth.

bog A wetland dominated by moss and shrubs.

bolide A solid extraterrestrial object such as a meteorite, comet, or asteroid that explodes in the atmosphere.

bomb A stream-lined block of rock ejected by a volcano while still hot; it gets shaped as it flies through the air.

bornhardt An inselberg with a loaf geometry, like that of Uluru (Ayers Rock) in central Australia.

Bowen's reaction series The sequence in which different silicate minerals crystallize during the progressive cooling of a melt.

braided stream A sediment-choked stream consisting of entwined subchannels.

breaker A water wave in which water at the top of the wave curves over the base of the wave.

breakwater An offshore wall, built parallel or at an angle to the beach, that prevents the full force of waves from reaching a harbor.

breccia Coarse sedimentary rock consisting of angular fragments; or rock broken into angular fragments by faulting.

breeder reactor A nuclear reactor that produces its own fuel.

brine Water that is not fresh but is less salty than seawater; brine may be found in estuaries.

brittle deformation The cracking and fracturing of a material subjected to stress.

brittle-ductile transition (brittle-plastic transition) The depth above which materials behave brittlely and below which materials behave ductilely (plastically); this transition typically lies between a depth of 10 and 15 km in continental crustal rock, and 60 m deep in glacial ice.

buoyancy The upward force acting on an object immersed or floating in fluid; the tendency of an object to float when placed in a fluid.

burial metamorphism Metamorphism due only to the consequences of very deep burial.

butte A medium-sized, flat-topped hill in an arid region.

caldera A large circular depression with steep walls and a fairly-flat floor, formed after an eruption as the center of the volcano collapses into the drained magma chamber below.

caliche A solid mass created where calcite cements the soil together. (Also called calcrete.)

calorie A unit of energy approximately equal to 4.2 joules; 1 calorie can raise the temperature of 1 gm of water by 1°C.

calving The breaking off of chunks of ice at the edge of a glacier.

Cambrian explosion The remarkable diversification of life, indicated by the fossil record, that occurred at the beginning of the Cambrian Period.

Canadian Shield A broad, low-lying region of exposed Precambrian rock in the Canadian interior.

canyon A trough or valley with steeply sloping walls, cut into the land by a stream.

capacity (of a stream) The total quantity of sediment a stream can carry.

capillary fringe The thin subsurface layer in which water molecules seep up from the water table by capillary action to fill pores.

carbonate minerals Minerals that contain the CO_3 ion. Examples include calcite and dolomite.

carbonate rocks Rocks containing calcite and/or dolomite.

carbon cycle the progressive transfer of carbon, from reservoir to reservoir in the Earth System.

carbon-14 A radioactive isotope of the element carbon; the ratio of C14 to C12 can provide an isotopic date of organic carbon.

carbon-14 dating A radiometric dating process that can tell us the age of organic material containing carbon originally extracted from the atmosphere.

carbon sequestration The process of extracting carbon dioxide from sources (e.g., power plants) and sending it back underground to keep it out of the atmosphere and diminish the greenhouse effect.

cast Sediment that preserves the shape of a shell it once filled before the shell dissolved or mechanically weathered away.

catabatic winds Strong winds that form at the margin of a glacier where the warmer air above ice-free land rises and the cold, denser air from above the glaciers rushes in to take its place.

catastrophic change Change that takes place either instantaneously or rapidly in geologic time.

catchment *See* Drainage network.

cation A positively charged ion.

celestial object An object (star, moon, planet, or galaxy) that lies outside of the Earth.

Celsius scale A metric-system measure of temperature in which the difference between the freezing point (0°C) and boiling point (100°C) of water is divided into 100 units; equivalent to centigrade scale.

cement Mineral material that precipitates from water and fills the spaces between grains, holding the grains together.

cementation The phase of lithification in which cement, consisting of minerals that precipitate from groundwater, partially or completely fills the spaces between clasts and attaches each grain to its neighbor.

Cenozoic Era The most recent era of the Phanerozoic Eon, lasting from 66 Ma up until the present.

chain reaction A self-perpetuating process in a nuclear reaction, whereby neutrons released during the fission trigger more fission.

chalk Very fine-grained limestone consisting of weakly cemented plankton shells.

change of state The process in which a material changes from one phase (liquid, gas, or solid) to another.

channel A trough dug into the ground surface by flowing water.

channeled scablands A barren, soil-free landscape in eastern Washington, scoured clean by a flood unleashed when a large glacial lake drained.

chatter marks Wedge-shaped indentations left on rock surfaces by glacial plucking.

chemical A material consisting of a distinct element or compound.

chemical bond The invisible link that holds together atoms in a molecule and/or in a crystal.

chemical formula The "recipe" that specifies the elements and their proportions in a compound.

chemical fossil (biomarkers) Distinctive molecules or molecular fragments, formed from the remains of living organisms, that can be preserved in rock.

chemical reaction Interactions among atoms and/or molecules involving breaking or forming chemical bonds.

chemical sedimentary rocks Sedimentary rocks made up of minerals that precipitate directly from water solution.

chemical weathering The process in which chemical reactions alter or destroy minerals when rock comes in contact with water solutions and/or air.

chert A sedimentary rock composed of very fine-grained silica (cryptocrystalline quartz).

Chicxulub crater A circular depression buried beneath younger sediment on the Yucatán Peninsula; geologists suggest that it formed due to impact of a meteorite there at 66 Ma.

chimney (1) A conduit in a magma chamber in the shape of a long vertical pipe through which magma rises and erupts at the surface; (2) an isolated column of rock in an arid region.

chron The time interval between successive magnetic reversals.

cinder cone A subaerial volcano consisting of a cone-shaped pile of tephra whose slope approaches the angle of repose for tephra.

cinders Fragments of glassy rock ejected from a volcano.

cirque A bowl-shaped depression carved by a glacier on the side of a mountain.

cirrus cloud A wispy cloud that tapers into delicate, feather-like curls.

clast A fragment or grain produced by the physical or chemical weathering of a pre-existing rock.

clastic (detrital) sedimentary rock Sedimentary rock consisting of cemented-together detritus derived from the weathering of preexisting rock.

cleavage (1) The tendency of a mineral to break along preferred planes; (2) a type of foliation in low-grade metamorphic rock.

cleavage planes A series of surfaces on a crystal that form parallel to the weakest bonds holding the atoms of the crystal together.

cliff (scarp) retreat The change in the position of a cliff face caused by erosion.

climate The average weather conditions, along with the range of conditions, of a region over a year.

climate-change model A computer-generated model designed to provide insight into how climate changed in the past and may change in the future, and what the consequences of climate change may be.

closure temperature The temperature at which ions can no longer move in or out of a mineral quickly; when a rock drops below the closure temperature for a mineral, the mineral can be used for isotopic dating.

cloud A mist of tiny water droplets in the sky.

coal A black, organic rock consisting of greater than 50% carbon; it forms from the buried and altered remains of plant material.

coal-bed methane Natural gas created during the formation of coal, that gets trapped within the coal.

coal gasification The process of producing relatively clean-burning gases from solid coal.

coal rank A measurement of the carbon content of coal; higher-rank coal forms at higher temperatures.

coal reserve The quantities of discovered, but not yet mined, coal in sedimentary rock of the continents.

coal swamp A swamp whose oxygen-poor water allows thick piles of woody debris to accumulate; this debris transforms into coal upon deep burial.

coast The belt of land bordering the sea.

coastal plain Low-relief regions of land adjacent to the coast.

coastal wetland A flat-lying coastal area that floods during high tide and drains during low tide, and hosts salt-resistant plants.

cold front The boundary at which a cold air mass pushes underneath a warm air mass.

collision The process of two buoyant pieces of lithosphere converging and squashing together.

collision and coalescence The process of forming raindrops, when tiny droplets merge together to form larger drops.

color The characteristic of a material due to the spectrum of light emitted or reflected by the material, as perceived by eyes or instruments.

columnar jointing A type of fracturing that yields roughly hexagonal columns of basalt; columnar joints form when a dike, sill, or lava flow cools.

comet A ball of ice and dust, probably remaining from the formation of the Solar System, that orbits the Sun.

compaction The phase of lithification in which the pressure of the overburden on the buried rock squeezes out water and air that was trapped between clasts, and the clasts press tightly together.

competence (of a stream) The maximum particle size a stream can carry.

composite volcano *See* Stratovolcano.

compositional banding A type of metamorphic foliation, found in gneiss, defined by alternating bands of light and dark minerals.

compound A material composed of two or more elements that cannot be separated mechanically; the smallest piece is a molecule.

compressibility The degree to which a material's volume changes in response to squashing.

compression A push or squeezing felt by a body.

compressional waves Waves in which particles of material move back and forth parallel to the direction in which the wave itself moves.

concentration The proportion of one substance (the solute) dissolved within another (the solvent).

conchoidal fractures Smoothly curving, clamshell-shaped surfaces along which materials with no cleavage planes tend to break.

condensation The process of gas molecules linking together to form a liquid.

condensation nuclei Preexisting solid or liquid particles, such as aerosols, onto which water condenses during cloud formation.

conduction A process of heat transfer involving progressive migration of thermal energy from cooler to warmer regions in a material, without the physical flow of the material itself.

cone of depression The downward-pointing, cone-shaped surface of the water table in a location where the water table is experiencing drawdown because of pumping at a well.

confined aquifer An aquifer that is separated from the Earth's surface by an overlying aquitard.

conglomerate Very coarse-grained sedimentary rock consisting of rounded clasts.

consumer An organism in a food web that ingests material around it to obtain nutrients.

consuming boundary *See* Convergent plate boundary.

contact The boundary surface between two rock bodies (as between two stratigraphic formations, between an igneous intrusion and adjacent rock, between two igneous rock bodies, or between rocks juxtaposed by a fault).

contact metamorphism *See* Thermal metamorphism.

contaminant plume A cloud of contaminated groundwater that moves away from the source of the contamination.

continental crust The crust beneath the continents.

continental divide A highland separating drainage that flows into one ocean from drainage that flows into another.

continental drift The hypothesis that continents have moved and are still moving slowly across the Earth's surface.

continental glacier A vast sheet of ice that spreads over thousands of square km of continental crust.

continental-interior desert An inland desert that develops because by the time air masses reach the continental interior, they have lost all of their moisture.

continental lithosphere Lithosphere topped by continental crust; this lithosphere reaches a thickness of 150 km.

continental margin A continent's coastline.

continental rift A linear belt along which continental lithosphere stretches and pulls apart.

continental rifting The process by which a continent stretches and splits along a belt; if it is successful, rifting separates a larger continent into two smaller continents separated by a divergent boundary.

continental rise The sloping sea floor that extends from the lower part of the continental slope to the abyssal plain.

continental shelf A broad, shallowly submerged fringe of a continent; ocean-water depth over the continental shelf is generally less than 200 meters; the widest continental shelves occur over passive margins.

continental slope The slope at the edge of a continental shelf, leading down to the deep-sea floor.

continental volcanic arc A long, curving chain of subaerial volcanoes on the margin of a continent adjacent to a convergent plate boundary.

control rod Rods that absorb neutrons in a nuclear reactor and thus decrease the number of collisions between neutrons and radioactive atoms.

contour interval The difference in value between adjacent contour lines; on a topographic map the contour interval is the elevation difference between two contour lines.

contour line A line on a map along which a parameter has the same value; for example, on a topographic map, all points on a contour line lie at the same elevation.

convection Heat transfer that results when warmer, less dense material rises while cooler, denser material sinks.

convective cell A distinct flow configuration for a volume of material that is moving during convective heat transport; simplistically, the material rises when warm and sinks when cool, and thus follows a loop-like path.

conventional hydrocarbon system The development and occurrence of oil or gas, that involves formation in a source rock, migration into a reservoir rock, and trapping in the reservoir rock.

conventional reserve A volume of oil or gas in a reservoir rock within a trap; it can be pumped relatively easily from the reservoir rock.

convergence zone A place where two surface air flows meet, forcing air to rise.

convergent margin *See* Convergent plate boundary.

convergent (consuming) boundary A boundary at which two plates move toward each other so that one plate sinks (subducts) beneath the other; only oceanic lithosphere can subduct.

coral reef A mound of coral and coral debris forming a region of shallow water.

core The dense, iron-rich center of the Earth.

core-mantle boundary An interface 2,900 km below the Earth's surface separating the mantle and core.

Coriolis effect The deflection of objects, winds, and currents on the surface of the Earth owing to the planet's rotation.

Coriolis force The apparent force that causes an object moving on the spinning Earth to undergo a deflection.

cornice A huge, overhanging drift of snow built up by strong winds at the crest of a mountain ridge.

correlation The process of defining the age relations between the strata at one locality and the strata at another.

cosmic rays Nuclei of hydrogen and other elements that bombard the Earth from deep space.

cosmology The study of the overall structure of the Universe.

country rock (wall rock) The preexisting rock into which magma intrudes.

covalent bonding The attachment of one atom to another that develops when the atoms share electrons; one type of chemical bond.

crater (1) A circular depression at the top of a volcanic mound; (2) a depression formed by the impact of a meteorite.

craton A long-lived block of durable continental crust commonly found in the stable interior of a continent.

cratonic platform A province in the interior of a continent in which Phanerozoic strata bury most of the underlying Precambrian rock.

creep The gradual downslope movement of regolith.

crevasse A large crack that develops by brittle deformation in the top 60 m of a glacier.

critical mass A sufficiently dense and large mass of radioactive atoms in which a chain reaction happens so quickly that the mass explodes.

critical zone The portion of the Earth System, including the surface, the near-surface including soil, liquid water bodies, and the lower atmosphere, in which resources and conditions are important for sustaining life.

cross bed Internal laminations in a bed, inclined at an angle to the main bedding; cross beds are a relict of the slip face of dunes or ripples.

cross section A diagram depicting the geometry of materials underground as they would appear on an imaginary vertical slice through the Earth.

crude oil Oil extracted directly from the ground.

crust The rock that makes up the outermost layer of the Earth.

crustal root Low-density crustal rock that protrudes downward beneath a mountain range.

crustal thickening The process by which the continental crust increases in thickness, becoming up to 70 km thick (vs. normal thickness of about 35–40 km); it can occur during continental collision.

crystal A single, continuous piece of a mineral bounded by flat surfaces that formed naturally as the mineral grew.

crystal face The flat surfaces of a crystal, formed during the crystal's growth.

crystal form The geometric shape of a crystal, defined by the arrangement of crystal faces.

crystal habit The general shape of a crystal or cluster of crystals that grew unimpeded.

crystal lattice The orderly framework within which the atoms or ions of a mineral are fixed.

crystal structure The arrangement of atoms in a crystal.

crystalline Containing a crystal lattice.

crystalline igneous rock An igneous rock that consists of minerals that grew when a melt solidified, and eventually interlock like pieces of a jigsaw puzzle.

crystalline material A substance in which atoms are arranged in a crystalline lattice.

cuesta An asymmetric ridge formed by tilted layers of rock, with a steep cliff on one side cutting across the layers and a gentle slope on the other side; the gentle slope is parallel to the layering.

cumulonimbus cloud A rain-producing, puffy cloud.

cumulus cloud A puffy, cotton-ball-shaped cloud.

current (1) A well-defined stream of ocean water; (2) the moving flow of water in a stream.

cut bank The outside bank of the channel wall of a meander, which is continually undergoing erosion.

cutoff A straight reach in a stream that develops when erosion eats through a meander neck.

cyanobacteria Blue-green algae; a type of archaea.

cycle A series of interrelated events or steps that occur in succession and can be repeated, perhaps indefinitely.

cyclone (1) The counterclockwise flow of air around a low-pressure mass; (2) the equivalent of a hurricane in the Indian Ocean.

cyclonic flow A circulation of air around a low-pressure region in the atmosphere; it rotates counterclockwise in the northern hemisphere.

cyclothem A repeated interval within a sedimentary sequence that contains a specific succession of sedimentary beds.

Darcy's law A mathematical equation stating that a volume of water, passing through a specified area of material at a given time, depends on the material's permeability and hydraulic gradient.

daughter isotope The decay product of radioactive decay.

day The time it takes for the Earth to spin once on its axis.

debris avalanche An avalanche in which the falling debris consists of rock fragments and dust.

debris fall A mass movement event in which fragments of various sizes, including large chunks and fine sediment, free fall down a slope.

debris flow (debris slide) A downslope movement of mud mixed with larger rock fragments.

debris slide A sudden downslope movement of material consisting only of regolith.

decompression melting The kind of melting that occurs when hot mantle rock rises to shallower depths in the Earth so that pressure decreases while the temperature remains unchanged.

deep current An ocean current at a depth greater than 100 m.

deep-focus earthquake An earthquake that occurs at a depth between 300 and 670 km; below 670 km, earthquakes do not happen.

deep-sea trench The deep trough on the ocean floor that forms at a subduction zone where one plate slides beneath another.

deflation The process of lowering the land surface by wind abrasion.

deformation A change in the shape, position, or orientation of a material, by bending, breaking, or flowing.

dehydration Loss of water.

delamination (plate tectonics) The process by which dense lithospheric mantle separates from the base of a plate and sinks into the mantle.

delta A wedge of sediment formed at a river mouth when the running water of the stream enters standing water, the current slows, the stream loses competence, and sediment settles out.

delta plain The low, swampy land on the surface of a delta.

delta-plain flood A flood in which water submerges a delta plain.

dendritic drainage network A drainage network whose interconnecting streams resemble the pattern of branches connecting to a deciduous tree.

dendrochronologist A scientist who analyzes tree rings to determine the geologic age of features.

density Mass per unit volume.

denudation The removal of rock and regolith from the Earth's surface.

deposition The process by which sediment settles out of a transporting medium.

depositional environment A setting in which sediments accumulate; its character (fluvial, deltaic, reef, glacial, etc.) reflects local conditions.

depositional landform A landform resulting from the deposition of sediment where the medium carrying the sediment evaporates, slows down, or melts.

desert A region so arid that it contains no permanent streams, except for those that bring water in from elsewhere, and has very sparse vegetation cover.

desertification The process of transforming nondesert areas into desert.

desert pavement A mosaic-like stone surface forming the ground in a desert.

desert varnish A dark, rusty-brown coating of iron oxide and magnesium oxide that accumulates on the surface of the rock.

detachment fault A nearly horizontal fault at the base of a fault system.

detritus The chunks and smaller grains of rock broken off outcrops by physical weathering.

dewpoint temperature The temperature at which air becomes saturated so that dew can form.

diagenesis All of the physical, chemical, and biological processes that transform sediment into sedimentary rock and that alter the rock after the rock has formed.

differential stress A condition causing a material to experience a push or pull in one direction of a greater magnitude than the push or pull in another direction; in some cases, differential stress can result in shearing.

differential weathering What happens when different rocks in an outcrop undergo weathering at different rates.

differentiation (of a planet) A process early in a planet's history during which dense iron alloy melted and sank downward to form the core, leaving less-dense mantle behind.

diffraction The splitting of light into many tiny beams that interfere with one another.

digital elevation map (DEM) A computer-produced portrayal of elevation differences commonly using shading to simulate shadows; the data used to produce the map assigns elevations to each point on the map.

dike A tabular (wall-shaped) intrusion of rock that cuts across the layering of country rock.

dimension stone An intact block of granite or marble to be used for architectural purposes.

dip The angle of a plane's slope as measured in a vertical plane perpendicular to the strike.

dipole A magnetic field with a north and south pole, like that of a bar magnet.

dipole field (for the Earth) The part of the Earth's magnetic field, caused by the flow of liquid iron alloy in the outer core, that can be represented by an imaginary bar magnet with a north and south pole.

dip-slip fault A fault in which sliding occurs up or down the slope (dip) of the fault.

dip slope A hill slope underlain by bedding parallel to the slope.

directional drilling The process of controlling the trajectory of a drill bit to make sure that the drill hole goes exactly where desired.

disappearing stream A stream that intersects a crack or sinkhole leading to an underground cavern, so that the water disappears into the subsurface and becomes an underground stream.

discharge The volume of water in a conduit or channel passing a point in 1 second.

discharge area A location where groundwater flows back up to the surface and may emerge at springs.

disconformity An unconformity parallel to the two sedimentary sequences it separates.

displacement (offset) The amount of movement or slip across a fault plane.

disseminated deposit A hydrothermal ore deposit in which ore minerals are dispersed throughout a body of rock.

dissolution A process during which materials dissolve in water.

dissolved load Ions dissolved in a stream's water.

distillation column A vertical pipe in which crude oil is separated into several components.

distributaries The fan of small streams formed where a river spreads out over its delta.

divergence zone A place where sinking air separates into two flows that move in opposite directions.

divergent boundary A boundary at which two lithosphere plates move apart from each other; they are marked by mid-ocean ridges.

diversification The development of many different species from a common ancestor.

DNA (deoxyribonucleic acid) The complex molecule, shaped like a double helix, containing the code that guides the growth and development of an organism.

doldrums A belt with very slow winds along the equator.

dome Folded or arched layers with the shape of an overturned bowl.

Doppler effect The phenomenon in which the frequency of wave energy appears to change when a moving source of wave energy passes an observer.

dormant volcano A volcano that has not erupted for hundreds to thousands of years but does have the potential to erupt again in the future.

downcutting The process in which water flowing through a channel cuts into the substrate and deepens the channel relative to its surroundings.

downdraft Downward-moving air.

downgoing plate (slab) A lithosphere plate that has been subducted at a convergent margin.

downslope force The component of the force of gravity acting in the downslope direction.

downslope movement The tumbling or sliding of rock and sediment from higher elevations to lower ones.

downwelling The downward movement of a volume of moving material; downwelling in the mantle carries cooler, denser mantle to greater depth; downwelling in the sea carries water from shallower depths to deeper depths.

downwelling zone A place where near-surface water sinks.

drag fold A fold that develops in layers of rock adjacent to a fault during or just before slip.

drainage divide A highland or ridge that separates one watershed from another.

drainage network (basin) An array of interconnecting streams that together drain an area.

drainage reversal When the overall direction of flow in a drainage network becomes the opposite of what it once had been.

drawdown The phenomenon in which the water table around a well drops because the users are pumping water out of the well faster than it flows in from the surrounding aquifer.

drilling mud A slurry of water mixed with clay that oil drillers use to cool a drill bit and flush rock cuttings up and out of the hole.

dripstone Limestone (travertine in a cave) formed by the precipitation of calcium carbonate out of groundwater.

dropstone A rock that drops to the sea floor once the iceberg that was carrying the rock melts.

drumlin A streamlined, elongate hill formed when a glacier overrides glacial till.

dry-based (polar) glacier A glacier so cold that its base remains frozen to the substrate.

dry wash The channel of an ephemeral stream when empty of water.

dry well (1) A well that does not supply water because the well has been drilled into an aquitard or into rock that lies above the water table; (2) a well that does not yield oil, even though it has been drilled into an anticipated reservoir.

ductile (plastic) deformation The bending and flowing of a material (without cracking and breaking) subjected to stress.

dune A pile of sand generally formed by deposition from the wind.

dust storm An event in which strong winds hit unvegetated land, strip off the topsoil, and send it skyward to form rolling dark clouds that block out the Sun.

dwarf planet A celestial object that orbits the Sun, but has not cleared its orbit of debris; Pluto is a dwarf planet.

dynamic metamorphism Metamorphism that occurs as a consequence of shearing alone, with no change in temperature or pressure.

dynamo A power plant generator in which water or wind power spins an electrical conductor around a permanent magnet.

dynamothermal metamorphism Metamorphism that involves heat, pressure, and shearing.

Earth materials A general term for the great variety of substances that make up this planet.

earthquake A vibration caused by the sudden breaking or frictional sliding of rock in the Earth.

earthquake belt A relatively narrow, and long band of earthquakes; earthquake belts define the position of plate boundaries.

earthquake engineering The design of buildings that can withstand shaking.

earthquake early warning system A communications network that provides an alert within microseconds after the first earthquake waves arrive at a seismograph near the epicenter, but before damaging vibrations reach population centers.

earthquake zoning The determination of where land is relatively stable and where it might collapse because of seismicity.

Earth System The global interconnecting web of physical and biological phenomena involving the solid Earth, the hydrosphere, and the atmosphere.

ebb tide   The falling tide.

eccentricity cycle   The cycle of the gradual change of the Earth's orbit from a more circular to a more elliptical shape; the cycle takes around 100,000 years.

ecliptic   The plane defined by a planet's orbit.

ecosystem   An environment and its inhabitants.

eddy   An isolated, ring-shaped current of water.

Ediacaran fauna   Multicellular invertebrate organisms that lived perhaps as early as 620 Ma and certainly by 565 Ma. They were named for a region in southern Australia.

effusive eruption   An eruption that yields mostly lava, not ash.

Ekman spiral   The change in flow direction of water with depth, caused by the Coriolis effect.

Ekman transport   The overall movement of a mass of water, resulting from the Eckman spiral, in a direction 90° to the wind direction.

elastic behavior   A response of a material to stress, during which the material changes shape. The amount of change depends on the magnitude of stress, and the change disappears when stress is removed. The behavior occurs when chemical bonds bend or stretch, but do not break.

elastic-rebound theory   The concept that earthquakes happen because stress builds up, causing rock adjacent to a fault to bend elastically until breaking and slip on a fault occurs; the slip relaxes the elastic bending and decreases stress.

elastic strain   A change in shape of a material; the change disappears instantly when stress is removed.

electromagnet   An electrical device that produces a magnetic field.

electromagnetic radiation   A type of energy emitted into space by a source; it travels in the form of waves that can pass through a vacuum.

electron   A negatively charged subatomic particle that orbits the nucleus of an atom; electrons are about 0.0005 × the size of a proton.

electron microprobe   A laboratory instrument that can focus a beam of electrons on a small part of a mineral grain to create a signal that defines its chemical composition.

electron shell   The distinct space around an atomic nucleus in which electrons move.

element   A material consisting entirely of one kind of atom; elements cannot be subdivided or changed by chemical reactions.

El Niño   The flow of warm water eastward from the Pacific Ocean that reverses the upwelling of cold water along the western coast of South America and causes significant global changes in weather patterns.

embayment   A low area of coastal land.

emergent coast   A coast where the land is rising relative to sea level or sea level is falling relative to the land.

end moraine (terminal moraine)   A low, sinuous ridge of till that develops when the terminus (toe) of a glacier stalls in one position for a while.

energy   The capacity to do work.

energy density   The amount of energy contained by a unit volume of material.

energy resource   Something that can be used to produce work; in a geologic context, a material (such as oil, coal, wind, flowing water) that can be used to produce energy.

Enhanced Fujita scale   A scale that represents the intensity of a tornado based on study of the damage that it caused.

eon   The largest subdivision of geologic time.

epeirogenic movement   The gradual uplift or subsidence of a broad region of the Earth's surface.

epeirogeny   An event of epeirogenic movement; the term is usually used in reference to the formation of broad mid-continent domes and basins.

ephemeral (intermittent) stream   A stream whose bed lies above the water table, so that the stream flows only when the rate at which water enters the stream from rainfall or meltwater exceeds the rate at which water infiltrates the ground below.

epicenter   The point on the surface of the Earth directly above the focus of an earthquake.

epicontinental sea   A shallow sea overlying a continent.

epoch   An interval of geologic time representing the largest subdivision of a period.

equant   A term for a grain that has the same dimensions in all directions.

equator   The great circle around the Earth that defines the boundary between the northern and southern hemisphere; it is halfway between the poles.

equatorial low   The area of low atmospheric pressure that develops over the equator because of the intertropical convergence zone.

equilibrium line (of a glacier)   The boundary between the zone of accumulation and the zone of ablation.

equinox   One of two days out of the year (September 22 and March 21) in which the Sun is directly overhead at noon at the equator.

equipotential surface   In the context of discussing groundwater, it is an imaginary surface that groundwater will rise to in a stand pipe.

era   An interval of geologic time representing the largest subdivision of the Phanerozoic Eon.

erg   Sand seas formed by the accumulation of dunes in a desert.

erosion   The grinding away and removal of the Earth's surface materials by moving water, air, or ice.

erosional coast   A coastline where sediment is not accumulating and wave action grinds away at the shore.

erosional landform   A landform that results from the breakdown and removal of rock or sediment.

erratic   A boulder or cobble that was picked up by a glacier and deposited hundreds of kilometers away from the outcrop from which it detached.

eruption column

eruptive style   The character of a particular volcanic eruption; geologists name styles based on typical examples (e.g., Hawaiian, Strombolian).

esker   A ridge of sorted sand and gravel that snakes across a ground moraine; the sediment of an esker was deposited in subglacial meltwater tunnels.

estuary   An inlet in which seawater and river water mix; created when a coastal valley is flooded because of either rising sea level or land subsidence.

Eubacteria   The kingdom of "true bacteria."

euhedral crystal   A crystal whose faces are well formed and whose shape reflects crystal form.

Eukaryote (Eukarya)   An organism whose cells contain a nucleus; all plants and animals consist of eukaryotic cells.

eukaryotic cell   A cell with a complex internal structure; multicellular organisms consist of such cells.

eustatic sea-level change   A global rising or falling of the ocean surface.

evaporate   To change state from liquid to vapor.

evaporite   Thick salt deposits that form as a consequence of precipitation from saline water.

evapotranspiration   The sum of evaporation from bodies of water and the ground surface and transpiration from plants and animals.

exfoliation   The process by which an outcrop of rock splits apart into onion-like sheets along joints that lie parallel to the ground surface.

exhumation   The process (involving uplift and erosion) that returns deeply buried rocks to the surface.

exoplanet   A planet that orbits a star outside of our Solar System.

exosphere   The outermost layer of the atmosphere.

exotic terrane   A block of land that collided with a continent along a convergent margin and attached to the continent; the term exotic implies that the land was not originally part of the continent to which it is now attached.

expanding Universe theory   The theory that the whole Universe must be expanding because galaxies in every direction seem to be moving away from us.

explosive eruptions Violent volcanic eruptions that produce clouds and avalanches of pyroclastic debris.

external energy In the context of the Earth System, this is the energy that comes to the Earth from the Sun.

external process A geomorphologic process—such as downslope movement, erosion, or deposition—that is the consequence of gravity or of the interaction between the solid Earth and its fluid envelope (air and water). Energy for these processes comes from gravity and sunlight.

extinction The death of the last members of a species so that there are no parents to pass on their genetic traits to offspring.

extinct volcano A volcano that was active in the past but has now shut off entirely and will not erupt in the future.

extraordinary fossil A rare fossilized relict, or trace, of the soft part of an organism.

extratropical cyclone (wave cyclone, mid-latitude cyclone) A large, rotating storm system, in mid-latitudes, associated with a regional-scale low-pressure zone.

extrusive igneous rock Rock that forms by the freezing of lava above ground, after it flows or explodes out (extrudes) onto the surface and comes into contact with the atmosphere or ocean.

eye The relative calm in the center of a hurricane.

eye wall A rotating vertical cylinder of clouds surrounding the eye of a hurricane.

facet (of a gem) The ground and polished surface of a gem, produced by a gem cutter using a grinding lap.

facies (1) Sedimentary: a group of rocks and primary structures indicative of a given depositional environment; (2) metamorphic: a set of metamorphic mineral assemblages formed under a given range of pressures and temperatures.

Fahrenheit scale An English-system measure of temperature in which the difference between the freezing point (32°F) and the boiling point (212°F) is divided into 180 units.

failure surface A weak surface that forms the base of a landslide.

faint young Sun paradox The apparent contradiction implied by the fact that much of the Earth's surface temperature has remained above the melting point of water during the past 4 Ga, even though calculations indicate that the Sun produced much less energy when young.

fault A fracture on which one body of rock slides past another.

fault-block mountains An outdated term for a narrow, elongate range of mountains that develops in a continental-rift setting as normal faulting drops down blocks of crust or tilts blocks.

fault breccia Fragmented rock in which angular fragments were formed by brittle fault movement; fault breccia occurs along a fault.

fault creep Gradual movement along a fault that occurs in the absence of an earthquake.

fault gouge Pulverized rock consisting of fine powder that lies along fault surfaces; gouge forms by crushing and grinding.

faulting Slip events along a fault.

fault scarp A small step on the ground surface where one side of a fault has moved vertically with respect to the other.

fault system A grouping of numerous related faults.

fault trace (fault line) The intersection between a fault and the ground surface.

feedback mechanism A condition that arises when the consequence of a phenomenon influences the phenomenon itself.

felsic An adjective used in reference to igneous rocks that are rich in elements forming feldspar and quartz.

Ferrel cells The name given to the middle-latitude convection cells in the atmosphere.

fetch The distance across a body of water along which a wind blows to build waves.

field force A push or pull that applies across a distance (i.e., without contact between objects); examples are gravity and magnetism.

fine-grained A textural term used to describe rock in which crystals or clasts are so small that they are hard to see without a microscope.

firn Compacted granular ice (derived from snow) that forms where snow is deeply buried; if buried more deeply, firn turns into glacial ice.

fission reaction A nuclear reaction during which the nucleus of a large atom splits to form two nuclei of smaller atoms; the process also releases neutrons and energy.

fission track A line of damage formed in the crystal lattice of a mineral by the impact of an atomic particle ejected during the decay of a radioactive isotope.

fissure A conduit in a magma chamber in the shape of a long crack through which magma rises and erupts at the surface.

fjord A deep, glacially carved, U-shaped valley flooded by rising sea level.

flank eruption An eruption that occurs when a secondary chimney, or fissure, breaks through the flank of a volcano.

flash flood A flood that occurs during unusually intense rainfall or as the result of a dam collapse, during which the floodwaters rise very fast.

flexural slip The process of folding in slip occurs between layers to accommodate bending of the layers.

flocculation The clumping together of clay suspended in river water into bunches that are large enough to settle out.

flood An event during which the volume of water in a stream becomes so great that it covers areas outside the stream's normal channel.

flood basalt Vast sheets of basalt that spread from a volcanic vent over an extensive surface of land; they may form where a rift develops above a continental hot spot, and where lava is particularly hot and has low viscosity.

flood-hazard map A representation of a portion of the Earth's surface that is designed to show how the danger of flooding varies with location.

floodplain The flat land on either side of a stream that becomes covered with water during a flood.

floodplain flood A flood during which a floodplain is submerged.

flood stage The stage when water reaches the top of a stream channel.

flood tide The rising tide.

floodway A mapped region likely to be flooded, in which people avoid constructing buildings.

flow fold A fold that forms when the rock behaves like weak plastic, and layering does not influence the folding process.

flowstone A sheet of limestone that forms along the wall of a cave when groundwater flows along the surface of the wall.

fluvial deposit Sediment deposited in a stream channel, along a stream bank, or on a floodplain.

flux Flow.

flux melting The transformation of hot solid to liquid that occurs when a volatile material injects into the solid.

focus The location where a fault slips during an earthquake (hypocenter).

fog A cloud that forms at ground level.

fold A bend or wrinkle of rock layers or foliation; folds form as a consequence of ductile deformation.

fold axis An imaginary line that, when moved parallel to itself, can trace out the shape of a folded surface.

fold-thrust belt An assemblage of folds and related thrust faults that develop above a detachment fault.

foliation Layering formed as a consequence of the alignment of mineral grains, or of compositional banding in a metamorphic rock.

food chain The succession of organisms, defined so that each organism in the chain eats organisms in levels below.

foraminifera Microscopic plankton with calcitic shells, components of some limestones.

force A push, pull, or shear that can change the velocity or direction of movement of an object.

foreland sedimentary basin A basin located under the plains adjacent to a mountain front, which develops as the weight of the mountains pushes the crust down, creating a depression that traps sediment.

foreshocks The series of smaller earthquakes that precede a major earthquake.

foreshore zone The zone of beach regularly covered and uncovered by rising and falling tides.

formation *See* Stratigraphic formation.

fossil The remnant, or trace, of an ancient living organism that has been preserved in rock or sediment.

fossil assemblage A group of fossil species found in a specific sequence of sedimentary rock.

fossil correlation A determination of the stratigraphic relation between two sedimentary rock units, reached by studying fossils.

fossil fuel An energy resource such as oil or coal that comes from organisms that lived long ago and thus stores solar energy that reached the Earth then.

fossiliferous limestone Limestone consisting of abundant fossil shells and shell fragments.

fossilization The process of forming a fossil.

fractional crystallization The process by which a magma becomes progressively more silicic as it cools, because early-formed crystals settle out.

fracture zone A narrow band of vertical fractures in the ocean floor; fracture zones lie roughly at right angles to a mid-ocean ridge, and the actively slipping part of a fracture zone is a transform fault.

fragmental igneous rock A rock consisting of igneous chunks and/or shards that are packed together, welded together, or cemented together after having solidified.

freezing The process by which a material in a liquid state transforms into a solid state.

frequency The number of waves that pass a point in a given time interval.

fresh rock Rock whose mineral grains have their original composition and shape.

friction Resistance to sliding on a surface.

fringing reef A coral reef that forms directly along the coast.

front The boundary between two air masses.

frost wedging The process in which water trapped in a joint freezes, forces the joint open, and may cause the joint to grow.

fuel A substance that can be used to produce energy.

fuel rod A metal tube that holds the nuclear fuel in a nuclear reactor.

Fujita scale A scale that distinguishes among tornadoes on the basis of wind speed, path dimensions, and possible damage.

fusion reaction A type of nuclear reaction during which nuclei collide and bond; fusion occurs in stars and hydrogen bombs.

Ga Billions of years ago (abbreviation).

gabbro A coarse-grained, intrusive, mafic igneous rock.

Gaia The term used for the Earth System, with the implication that it resembles a complex living entity.

galaxy An immense system of hundreds of billions of stars.

gas Matter that consists of atoms or molecules that are not attached to each other; a gas fills a container that contains it; air is an example of a gas.

gas hydrate A substance consisting of water ice in which methane or other organic substances, have dissolved.

gem A finished (cut and polished) gemstone ready to be set in jewelry.

gemstone A mineral that has special value because it is rare and people consider it beautiful.

gene An individual component of the DNA code that guides the growth and development of an organism.

general circulation model (GCM) A numerical calculation that simulates the flow of the atmosphere and resulting phenomena, due to changes in atmospheric temperature and other parameters.

genetics The study of genes and how they transmit information.

genome The succession of genes in strands of DNA.

geocentric model An ancient Greek idea suggesting that the Earth sat motionless in the center of the Universe while stars and other planets and the Sun orbited around it.

geochronology The science of dating geologic events in years.

geode A cavity in which euhedral crystals precipitate out of water solutions passing through a rock.

geographic pole The locations (north and south) where the Earth's rotational axis intersects the planet's surface.

geoid A reference surface representing the elevations, worldwide, at which gravitational potential energy is the same.

geologic column A composite stratigraphic chart that represents the entirety of the Earth's history.

geologic contact The surface between two distinct geologic units.

geologic cross section A depiction of contacts in the subsurface as represented by their traces on an imaginary vertical slice into the Earth.

geologic history The sequence of geologic events that has taken place in a region.

geologic map A map showing the distribution of rock units and structures across a region.

geologic time The span of time since the formation of the Earth.

geologic time scale A scale that describes the intervals of geologic time.

geologist A scientist who specializes in studying the Earth.

geology (geoscience) The study of the Earth, including our planet's composition, behavior, and history.

geophysics The subdiscipline of geology focused on the quantitative analysis and modeling of physical characteristics of the Earth; it includes the study of earthquakes, gravity, and magnetism.

geotherm The change in temperature with depth in the Earth.

geothermal energy Heat and electricity produced by using the internal heat of the Earth.

geothermal gradient The rate of change in temperature with depth.

geothermal region A region of current or recent volcanism in which magma or very hot rock heats up groundwater, which may discharge at the surface in the form of hot springs and/or geysers.

geyser A fountain of steam and hot water that erupts periodically from a vent in the ground in a geothermal region.

giant planets The four outer, or Jovian, planets of our Solar System, which are significantly larger than the rest of the planets and consist largely of gas and/or ice.

glacial abrasion The process by which clasts embedded in the base of a glacier grind away at the substrate as the glacier flows.

glacial advance The forward movement of a glacier's toe when the supply of snow exceeds the rate of ablation.

glacial drift Sediment deposited in glacial environments.

glacial incorporation The process by which flowing ice surrounds and incorporates debris.

glacial marine Sediment consisting of ice-rafted clasts mixed with marine sediment.

glacial outwash Coarse sediment deposited on a glacial outwash plain by meltwater streams.

glacial outwash plain The region in front of a glacier where the land surface has been covered by gravel transported by meltwater streams.

glacially polished surface A polished rock surface created by the glacial abrasion of the underlying substrate.

glacial plucking (glacial quarrying) The process by which a glacier breaks off and carries away fragments of bedrock.

glacial rebound The process by which the surface of a continent rises back up after an overlying continental ice sheet melts away and the weight of the ice is removed.

glacial retreat The movement of a glacier's toe back toward the glacier's origin; glacial retreat occurs if the rate of ablation exceeds the rate of supply.

glacial striation Grooves or scratches cut into bedrock when clasts embedded in the moving glacier act like the teeth of a giant rasp.

glacial subsidence The sinking of the surface of a continent caused by the weight of an overlying glacial ice sheet.

glacial till Sediment transported by flowing ice and deposited beneath a glacier or at its toe.

glacial torrent A catastrophic flood caused by the sudden release of water when an ice dam, or a moraine dam, fails.

glaciation (glacial period) A portion of an ice age during which huge glaciers grew and covered substantial areas of the continents.

glacier A river or sheet of ice that slowly flows across the land surface and lasts all year long.

glass A solid in which atoms are not arranged in an orderly pattern.

glassy igneous rock Igneous rock consisting entirely of glass, or of tiny crystals surrounded by a glass matrix.

glide horizon The surface along which a slump slips.

global change The transformations or modifications of both physical and biological components of the Earth System through time.

global circulation The movement of volumes of air in paths that ultimately take it around the planet.

global climate change Transformations or modifications in the Earth's climate over time.

global cooling A fall in the average atmospheric temperature.

global positioning system (GPS) A satellite system people can use to measure rates of movement of the Earth's crust relative to one another, or simply to locate their position on the Earth's surface.

global warming A rise in the average atmospheric temperature.

gneiss A compositionally banded metamorphic rock typically composed of alternating dark- and light-colored layers.

Goldilocks effect A phrase used to emphasize that the Earth is not too hot, and not too cold, but is just right for liquid water to exist.

Gondwana A supercontinent that consisted of today's South America, Africa, Antarctica, India, and Australia. (Also called Gondwanaland.)

graben A down-dropped crustal block bounded on either side by a normal fault dipping toward the basin.

grade (of an ore) The concentration of a useful metal in an ore—the higher the concentration, the higher the grade.

graded bed A layer of sediment, deposited by a turbidity current, in which grain size varies from coarse at the bottom to fine at the top.

graded stream A stream that has attained an equilibrium longitudinal profile in which the sediment input into an area equals sediment removal.

gradualism The theory that evolution happens at a constant, slow rate.

grain A fragment of a mineral crystal or of a rock.

grain rotation The process by which rigid, inequant mineral grains distributed through a soft matrix may rotate into parallelism as the rock changes shape owing to differential stress.

granite A coarse-grained, intrusive, silicic igneous rock.

granulite facies A set of metamorphic mineral assemblages formed at very high pressures and temperatures.

gravimeter A device used to measure the strength of gravitational pull at a location.

gravitational potential energy The energy stored in an object at rest in a gravitational field.

gravitational spreading A process of lateral spreading that occurs in a material because of the weakness of the material; gravitational spreading causes continental glaciers to grow and mountain belts to undergo orogenic collapse.

gravity The attractive force that one mass exerts on another; the magnitude depends on the size of the objects and the distance between them.

gravity anomaly A value of gravitational pull that is greater than or lesser than the pull predicted by the geoid.

graywacke An informal term used for sedimentary rock consisting of sand-sized up to small-pebble-sized grains of quartz and rock fragments all mixed together in a muddy matrix; typically, graywacke occurs at the base of a graded bed.

great oxygenation event (GOE) The time in Earth's history, about 2.4 Ga, when the concentration of oxygen in the atmosphere increased dramatically.

greenhouse (hothouse) period Relatively warm global climate leading to the rising of sea level for an interval of geologic time.

greenhouse effect The trapping of heat in the Earth's atmosphere by carbon dioxide and other greenhouse gases, which absorb infrared radiation; somewhat analogous to the effect of glass in a greenhouse.

greenhouse gases Atmospheric gases, such as carbon dioxide and methane, that regulate the Earth's atmospheric temperature by absorbing infrared radiation.

greenschist facies A set of metamorphic mineral assemblages formed under relatively low pressures and temperatures.

greenstone An informal term for a low-grade metamorphic rock formed from basalt; if foliated, the rock is called greenschist.

Greenwich mean time (GMT) The time at the astronomical observatory in Greenwich, England; time in all other time zones is set in relation to GMT.

Grenville orogeny The orogeny that occurred about 1 billion years ago and yielded the belt of deformed and metamorphosed rocks that underlie the eastern fifth of the North American continent.

groin A concrete or stone wall built perpendicular to a shoreline in order to prevent beach drift from removing sand.

ground moraine A thin, hummocky layer of till left behind on the land surface during a rapid glacial recession.

groundwater Water that resides under the surface of the Earth, mostly in pores or cracks of rock or sediment.

groundwater contamination Addition of chemicals or microbes (e.g., from agricultural and industrial activities, and landfills or septic tanks) to the groundwater supply.

group A succession of adjacent stratigraphic formations that together comprise a single, thicker stratigraphic entity.

growth band A layer of shell that forms in given year.

growth ring A rhythmic layering that develops in trees, travertine deposits, and shelly organisms as a consequence of seasonal changes.

gusher (blowout) A fountain of oil formed when underground pressure causes the oil to rise on its own out of a drilled hole.

guyot A seamount that had a coral reef growing on top of it, so that it is now flat-crested.

gymnosperm A plant whose seeds are "naked," not surrounded by a fruit.

gyre A large, circular flow pattern of ocean surface currents.

habitable zone (astronomy) The region in the Solar System where the intensity of radiation is sufficient to allow water to exist in liquid form on the surface of a planet.

Hadean Eon The oldest of the Precambrian eons; the time between the Earth's origin and the formation of the first rocks that have been preserved.

Hadley cells The name given to the low-latitude convection cells in the atmosphere.

hail Falling ice balls from the sky, formed when ice crystallizes in turbulent storm clouds.

hail streak An approximately 2-by-10-km area of ground, elongate in the direction of a storm, onto which hail has fallen.

half-graben A wedge-shaped basin in cross section that develops as the hanging-wall block above a normal fault slides down and rotates; the basin develops between the fault surface and the top surface of the rotated block.

half-life The time it takes for half of a group of a radioactive element's isotopes to decay.

halocline The boundary in the ocean between surface-water and deep-water salinities.

hamada Barren, rocky highlands in a desert.

hand specimen A piece of rock, about the size of a fist, that can be collected for study.

hanging valley A glacially carved tributary valley whose floor lies at a higher elevation than the floor of the trunk valley.

hanging wall The rock or sediment above an inclined fault plane.

hardness (of a mineral) A measure of the relative ability of a mineral to resist scratching; it represents the resistance of bonds in the crystal structure from being broken.

hard water Groundwater that contains dissolved calcium and magnesium, usually after passing through limestone or dolomite.

head (1) The elevation of the water table above a reference horizon; (2) the edge of ice at the origin of a glacier.

headland A place where a hill or cliff protrudes into the sea.

head scarp The distinct step along the upslope edge of a slump where the regolith detached.

headward erosion The process by which a stream channel lengthens up its slope as the flow of water increases.

headwaters The beginning point of a stream.

heat Thermal energy resulting from the movement of molecules.

heat capacity A measure of the amount of heat that must be added to a material to change its temperature.

heat flow The rate at which heat rises from the Earth's interior up to the surface.

heat-transfer melting Melting that results from the transfer of heat from a hotter magma to a cooler rock.

heliocentric model An idea proposed by Greek philosophers around 250 B.C.E. suggesting that all heavenly objects including the Earth orbited the Sun.

heliosphere A bubble-like region in space in which solar wind has blown away most interstellar atoms.

Hercynian orogen The late Paleozoic orogen that affected parts of Europe; a continuation of the Alleghenian orogen.

heterosphere A term for the upper portion of the atmosphere, in which gases separate into distinct layers on the basis of composition.

hiatus The interval of time between deposition of the youngest rock below an unconformity and deposition of the oldest rock above the unconformity.

high-altitude westerlies Westerly winds at the top of the troposphere.

high-grade metamorphic rocks Rocks that metamorphose under relatively high temperatures.

high-level waste Nuclear waste containing greater than 1 million times the safe level of radioactivity.

high-pressure system A region in which air pressure at the ground level is high, and an anticyclone develops.

hinge The portion of a fold where curvature is greatest.

hogback A steep-sided ridge of steeply dipping strata.

Holocene Epoch The period of geologic time since the last glaciation.

Holocene climatic maximum The period from 5,000 to 6,000 years ago, when Holocene temperatures reached a peak.

homosphere The lower part of the atmosphere, in which the gases have stirred into a homogenous mixture.

hoodoo The local name for the brightly colored shale and sandstone chimneys found in Bryce Canyon National Park in Utah.

horn A pointed mountain peak surrounded by at least three cirques.

hornfels Rock that undergoes metamorphism simply because of a change in temperature, without being subjected to differential stress.

horse latitudes The region of the subtropical high in which winds are weak.

horst The high block between two grabens.

hot spot A location at the base of the lithosphere, at the top of a mantle plume, where temperatures can cause melting.

hot-spot track A chain of now-dead volcanoes transported off the hot spot by the movement of a lithosphere plate.

hot-spot volcano An isolated volcano not caused by movement at a plate boundary, but rather by melting at a point locality, probably at the top of a mantle plume.

hot spring A spring that emits water ranging in temperature from about 30°C to 104°C.

Hubbert's Peak The high point on a graph of production vs. time; the concept that we can define Hubbert's Peak for a resource emphasizes that supplies of resources are limited.

hummocky surface An irregular and lumpy ground surface.

humus The organic material within a soil.

hurricane A huge rotating storm, resembling a giant spiral in map view, in which sustained winds blow over 119 km per hour.

hurricane track The path a hurricane follows.

hyaloclastite A rubbly extrusive rock consisting of glassy debris formed in a submarine or sub-ice eruption.

hydration The absorption of water into the crystal structure of minerals; a type of chemical weathering.

hydraulic conductivity The coefficient K in Darcy's law; hydraulic conductivity takes into account the permeability of the sediment or rock as well as the fluid's viscosity.

hydraulic gradient The slope of the water table.

hydraulic head The potential energy available to drive the flow of a given volume of groundwater at a location; it can be measured as an elevation above a reference.

hydrocarbon A chain-like or ring-like molecule made of hydrogen and carbon atoms; petroleum and natural gas are hydrocarbons.

hydrocarbon generation A process in which oil shale warms to temperatures of greater than about 90°C so kerogen molecules transform into oil and natural gas molecules.

hydrocarbon reserve A known supply of oil and gas held underground.

hydrocarbon system The association of source rock, migration pathway, reservoir rock, seal, and trap geometry that leads to the occurrence of a hydrocarbon reserve.

hydrofracturing (fracking) A process by which drillers generate new fractures or open preexisting ones underground, by pumping a high-pressure fluid into a portion of the drill hole, in order to increase the permeability of surrounding hydrocarbon-bearing rocks.

hydrogen bond The attraction of a hydrogen atom to a negatively charged atom or molecule (e.g., hydrogen bonds attach water molecules to each other).

hydrologic cycle The continual passage of water from reservoir to reservoir in the Earth System.

hydrolysis The process in which water chemically reacts with minerals and breaks them down.

hydrosphere The Earth's water, including surface water (lakes, rivers, and oceans), groundwater, and liquid water in the atmosphere.

hydrothermal deposit An accumulation of ore minerals precipitated from hot-water solutions circulating through a magma or through the rocks surrounding an igneous intrusion.

hydrothermal metamorphism When very hot water passes through the crust and causes metamorphism of rock.

hypocenter (focus) The place within the Earth where earthquake energy originates; commonly, the hypocenter is the place on a fault where slip took place.

hypothesis An idea that has the potential to explain a phenomenon; a hypothesis must be rigorously tested if it is to eventually become a theory.

hypsometric curve A graph that plots surface elevation on the vertical axis and the percentage of the Earth's surface on the horizontal axis.

ice age An interval of time in which the climate was colder than it is today, glaciers occasionally advanced to cover large areas of the continents, and mountain glaciers grew; an ice age can include many glacials and interglacials.

iceberg A large block of ice that calves off the front of a glacier and drops into the sea.

icehouse period A period of time when the Earth's temperature was cooler than it is today and ice ages could occur.

ice-margin lake A meltwater lake formed along the edge of a glacier.

ice-rafted sediment Sediment carried out to sea by icebergs.

ice sheet A vast glacier that covers the landscape.

ice shelf A broad, flat region of ice along the edge of a continent formed where a continental glacier flowed into the sea.

ice stream A portion of a glacier that travels much more quickly than adjacent portions of the glacier.

ice tongue The portion of a valley glacier that has flowed out into the sea.

igneous rock Rock that forms when hot molten rock (magma or lava) cools and freezes solid.

ignimbrite Rock formed when deposits of pyroclastic flows solidify.

inactive fault A fault that last moved in the distant past and probably won't move again in the near future, yet is still recognizable because of displacement across the fault plane.

inactive sand The sand along a coast that is buried beneath a layer of active sand and moves only during severe storms or not at all.

incised meander A meander that lies at the bottom of a steep-walled canyon.

index fossil A fossil of an organism that lived during a relatively short period of time over a relatively large area of the Earth, and can be used for stratigraphic correlation.

index minerals Minerals that serve as good indicators of metamorphic grade.

induced seismicity Seismic events caused by the actions of people (e.g., filling a reservoir that lies over a fault with water).

industrial minerals Minerals that serve as the raw materials for manufacturing chemicals, concrete, and wallboard, among other products.

inequant grains A term for mineral grains whose length and width are not the same.

inertia The tendency of an object at rest to remain at rest, and an object in motion to remain in motion, unless acted on by an outside force.

infiltrate Seep down into.

injection well A well in which a liquid is pumped down into the ground under pressure so that it passes from the well back into the pore space of the rock or regolith.

inner core The inner section of the core, extending from 5,155 km deep to the Earth's center at 6,371 km and consisting of solid iron alloy.

inselberg An isolated mountain or hill in a desert landscape created by progressive cliff retreat, so that the hill is surrounded by a pediment or an alluvial fan.

insolation Exposure to the Sun's rays.

intensity (seismology) A measure of the relative size of an earthquake (the severity of ground shaking) at a location, as determined by examining the amount of damage caused.

interglacial A period of time between two glaciations.

interior basin A basin with no outlet to the sea.

interlocking texture The texture of crystalline rocks in which mineral grains fit together like pieces of a jigsaw puzzle.

internal energy Energy in the Earth System that comes from within the planet, due to residual heat from the Earth's formation, or radioactive decay.

internal process A process in the Earth System, such as plate motion, mountain building, or volcanism, ultimately caused by the Earth's internal heat.

interplanetary space The region of vacuum between the orbits of planets.

interstellar space The region of vacuum between stars.

intertidal zone The area of coastal land across which the tide rises and falls.

intertropical convergence zone (ITCZ) The equatorial convergence zone in the atmosphere.

intraplate earthquakes Earthquakes that occur away from plate boundaries.

intrusive contact The boundary between country rock and an intrusive igneous rock.

intrusive igneous rock Rock formed by the freezing of magma underground.

ion A version of an atom that has lost or gained electrons, relative to an electrically neutral version, so that it has a net electrical charge.

ionic bond The attachment of one atom to another that happens when one atom transfers electrons to another; one type of chemical bond.

ionosphere The interval of the Earth's atmosphere, at an elevation between 50 and 400 km, containing abundant positive ions.

iron catastrophe The proposed event very early in Earth history when the Earth partly melted and molten iron sank to the center to form the core.

isobar A line on a map along which the air has a specified pressure.

isograd (1) A line on a pressure-temperature graph along which all points are taken to be at the same metamorphic grade; (2) a line on a map making the first appearance of a metamorphic index mineral.

isostasy (isostatic equilibrium) The condition that exists when the buoyancy force pushing lithosphere up equals the gravitational force pulling lithosphere down.

isostatic compensation The process in which the surface of the crust slowly rises or falls to reestablish isostatic equilibrium after a geologic event changes the density or thickness of the lithosphere.

isotherm Lines on a map or cross section along which the temperature is constant.

isotopes Different versions of a given element that have the same atomic number but different atomic weights.

isotopic dating Another term for radiometric dating, meaning the determination of the numerical age of rocks and minerals.

jet stream A fast-moving current of air that flows at high elevations.

jetty A man-made wall that protects the entrance to a harbor.

joints Naturally formed cracks in rocks.

joint set A group of systematic joints.

Jovian A term used to describe the outer gassy, Jupiter-like planets (gas-giant planets).

kame A stratified sequence of lateral-moraine sediment that's sorted by water flowing along the edge of a glacier.

karst landscape A region underlain by caves in limestone bedrock; the collapse of the caves creates a landscape of sinkholes separated by higher topography, or of limestone spires separated by low areas.

Kelvin (K) scale A measure of temperature in which 0 K is absolute zero and the freezing point of water is 273.15 K; divisions in the Kelvin scale have the same value as those in the Celsius scale.

kerogen The waxy molecules into which the organic material in shale transforms on reaching about 100°C. At higher temperatures, kerogen transforms into oil.

kettle hole A circular depression in the ground made when a block of ice calves off the toe of a glacier, becomes buried by till, and later melts.

knob-and-kettle topography A land surface with many kettle holes separated by round hills of glacial till.

K-Pg boundary event The mass extinction that happened at the end of the Cretaceous Period, 66 million years ago, possibly due to the collision of an asteroid with the Earth. This event used to be called the K-T boundary event.

Kuiper Belt A diffuse ring of icy objects, remnants of Solar System formation, that orbit our Sun outside the orbit of Neptune.

laccolith A blister-shaped igneous intrusion that forms when magma injects between layers underground in a manner that pushes overlying layers upward to form a dome.

lag deposit The coarse sediment left behind in a desert after wind erosion removes the finer sediment.

lagoon A body of shallow seawater separated from the open ocean by a barrier island.

lahar A thick slurry formed when volcanic ash and debris mix with water, either in rivers or from rain or melting snow and ice on the flank of a volcano.

landform A particular land-surface shape at a location.

landscape The overall shape and character of the land surface in a region.

landslide A sudden movement of rock and debris down a nonvertical slope.

landslide-potential map A map on which regions are ranked according to the likelihood that a mass movement will occur.

land subsidence Sinking elevation of the ground surface; the process may occur over an aquifer that is slowly draining and decreasing in volume because of pore collapse.

La Niña Years in which the El Niño event is not strong.

lapilli Any pyroclastic particle that is 2 to 64 mm in diameter (i.e., marble-sized); the particles can consist of frozen lava clots, pumice fragments, or ash clumps.

Laramide orogeny The mountain-building event that lasted from about 80 Ma to 40 Ma, in western North America; in the United States, it formed the Rocky Mountains as a result of basement uplift and the warping of the younger overlying strata into large monoclines.

large igneous province (LIP) A region in which huge volumes of lava and/or ash erupted over a relatively short interval of geologic time.

latent heat of condensation The heat released during condensation, which comes only from a change in state.

lateral moraine A strip of debris along the side margins of a glacier.

laterite A hard, brick-red, soil formed from iron-rich rock in a tropical environment; it consists primarily of insoluble iron and aluminum oxide and hydroxide and forms due to extreme leaching.

Laurentia A continent in the early Paleozoic Era composed of today's North America and Greenland.

Laurentide ice sheet An ice sheet that spread over northeastern Canada during the Pleistocene ice age(s).

lava Molten rock that has flowed out onto the Earth's surface.

lava dome A dome-like mass of rhyolitic lava that accumulates above the eruption vent.

lava flows Sheets or mounds of lava that flow onto the ground surface or sea floor in molten form and then solidify.

lava lake A large pool of lava produced around a vent when lava fountains spew forth large amounts of lava in a short period of time.

lava tube The empty space left when a lava tunnel drains; this happens when the surface of a lava flow solidifies while the inner part of the flow continues to stream downslope.

leach To dissolve and carry away.

leader A conductive path stretching from a cloud toward the ground, along which electrons leak from the base of the cloud, and which provides the start for a lightning flash to the ground.

lifting mechanisms Phenomena that cause a volume of air to start rising.

lightning stroke (lightning flash, lightning bolt) A giant spark or pulse of current that jumps across a gap of charge separation.

light-year The distance that light travels in one Earth year (about 6 trillion miles or 9.5 trillion km).

lignite Low-rank coal that consists of 50% carbon.

limb (of fold) The side of a fold, showing less curvature than at the hinge.

limestone Sedimentary rock composed of calcite.

liquefaction The process by which saturated, unconsolidated sediments are transformed into a substance that acts like a liquid as a result of ground shaking.

liquid Matter that can flow to conform to the shape of the container that holds it.

liquidus The lowest temperature at which all the components of a material have melted and transformed into liquid.

liquification The process by which wet sediment becomes a slurry; liquification may be triggered by earthquake vibrations.

lithification The transformation of loose sediment into solid rock through compaction and cementation.

lithologic correlation A correlation based on similarities in rock type.

lithosphere The relatively rigid, nonflowable, outer 100- to 150-km-thick layer of the Earth, constituting the crust and the top part of the mantle.

lithosphere plate One of many distinct pieces of the lithosphere (the Earth's relatively rigid shell) that are separated from one another by breaks (plate boundaries).

lithospheric mantle The part of a plate, below the crust, in which mantle is cool enough to behave rigidly.

Little Ice Age A period of cooler temperatures, between 1500 and 1800 C.E., during which many glaciers advanced.

loam A type of soil consisting of roughly equal parts of sand, silt, and clay; it tends to be good for growth of crops.

local base level A base level upstream from a drainage network's mouth.

lodgment till A flat layer of till smeared out over the ground when a glacier overrides an end moraine as it advances.

loess Layers of fine-grained sediments deposited from the wind; large deposits of loess formed from fine-grained glacial sediment blown off out-wash plains.

longitudinal dunes (seif dunes) A dune formed when there is abundant sand and a strong, steady wind, and whose axis lies parallel to the wind direction.

longitudinal profile A cross-sectional image showing the variation in elevation along the length of a river.

longshore current The flow of water parallel to the shore just off a coast, because of the diagonal movement of waves toward the shore.

longshore drift The movement of sediment laterally along a beach; it occurs when waves wash up a beach diagonally.

lower mantle The deepest section of the mantle, stretching from 670 km down to the core-mantle boundary.

low-grade metamorphic rocks Rocks that underwent metamorphism at relatively low temperatures.

low-pressure system A region in which air pressure at the land surface is relatively low, so that a cyclonic flow can begin.

low-velocity zone The asthenosphere underlying oceanic lithosphere in which seismic waves travel more slowly, probably because rock has partially melted.

luster The way a mineral surface scatters light.

L-waves (Love waves) Surface seismic waves that cause the ground to shimmy back and forth, creating a snake-like movement.

Ma Millions of years ago (abbreviation).

macrofossil A fossil large enough to be seen with the naked eye.

mafic A term used in reference to magmas or igneous rocks that are relatively poor in silica and rich in iron and magnesium.

magma Molten rock beneath the Earth's surface.

magma chamber A space below ground filled with magma.

magma contamination The process in which flowing magma incorporates components of the country rock through which it passes.

magmatic deposit   An ore deposit formed when sulfide ore minerals accumulate at the bottom of a magma chamber.

magnetic anomaly   The difference between the expected strength of the Earth's magnetic field at a certain location and the actual measured strength of the field at that location.

magnetic declination   The angle between the direction a compass needle points at a given location and the direction of true north.

magnetic dipole   An imaginary vector that points from the north magnetic pole to the south magnetic pole of a magnetic field.

magnetic field   The region affected by the force emanating from a magnet.

magnetic field lines   The trajectories along which magnetic particles would align, or charged particles would flow, if placed in a magnetic field.

magnetic force   The push or pull exerted by a magnet.

magnetic inclination   The angle between a magnetic needle free to pivot on a horizontal axis and a horizontal plane parallel to the Earth's surface.

magnetic poles   The ends of a magnetic dipole; all magnetic dipoles have a north pole and a south pole.

magnetic reversal   The change of the Earth's magnetic polarity; when a reversal occurs, the field flips from normal to reversed polarity, or vice versa.

magnetic-reversal chronology   The history of magnetic reversals through geologic time.

magnetism   An attractive or repulsive field force generated by permanent magnets or by an electrical current.

magnetization   The degree to which a material can exert a magnetic force.

magnetometer   An instrument that measures the strength of the Earth's magnetic field.

magnetosphere   The region protected from the electrically charged particles of the solar winds by the Earth's magnetic field.

magnetostratigraphy   The comparison of the pattern of magnetic reversals in a sequence of strata, with a reference column showing the succession of reversals through time.

magnitude (of an earthquake)   The number that represents the maximum amplitude of ground motion that would be measured by a seismometer placed at a specified distance from the epicenter.

manganese nodules   Lumpy accumulations of manganese-oxide minerals precipitated onto the sea floor.

mantle   The thick layer of rock below the Earth's crust and above the core.

mantle downwelling   The sinking of denser mantle in a region of the mantle.

mantle plume   A column of very hot rock that rises up through the mantle.

mantle upwelling   The rise of less-dense material in a region of the mantle.

marble   A metamorphic rock composed of calcite and transformed from a protolith of limestone.

mare   The broad, darker areas on the Moon's surface; they consist of flood basalts that erupted over 3 billion years ago and spread out across the Moon's lowlands.

marginal sea   A small ocean basin created when sea-floor spreading occurs behind an island arc.

marine magnetic anomaly   The difference between the *expected* strength of the Earth's main dipole field at a certain location on the sea floor and the *actual* measured strength of the magnetic field at that location.

maritime tropical air mass   A mass of air that originates over tropical or subtropical oceanic regions.

marker bed   A particularly unique layer that provides a definitive basis for correlation.

marsh   A wetland dominated by grasses.

mass   The amount of matter in an object; mass differs from weight in that its value does not depend on the strength of gravity.

mass-extinction event   A time when vast numbers of species abruptly vanish.

massive sulfide deposit   A type of ore deposit containing a high-concentration of sulfide minerals.

mass movement (mass wasting)   The gravitationally caused downslope transport of rock, regolith, snow, or ice.

mass-transfer cycle   The progressive movement of material from one reservoir in the Earth System to another.

matter   The material substance of the Universe; it consists of atoms and has mass.

matrix   Finer-grained material surrounding larger grains in a rock.

meander   A snake-like curve along a stream's course.

meandering stream   A reach of stream containing many meanders (snake-like curves).

meander neck   A narrow isthmus of land separating two adjacent meanders.

mean sea level   The average level between the high and low tide over a year at a given point.

mechanical force   A push, pull, or shear applied by one object on another; it can be applied only if the objects are in contact.

mechanical weathering   *See* Physical weathering.

medial moraine   A strip of sediment in the interior of a glacier, parallel to the flow direction of the glacier, formed by the lateral moraines of two merging glaciers.

Medieval Warm Period   A period of high temperatures in the Middle Ages.

melt   Molten (liquid) rock.

meltdown   The melting of the fuel rods in a nuclear reactor that occurs if the rate of fission becomes too fast and the fuel rods become too hot.

melting curve   The line defining the range of temperatures and pressures at which a rock melts.

melting temperature   The temperature at which the thermal vibration of the atoms or ions in the lattice of a mineral is sufficient to break the chemical bonds holding them to the lattice, so a material transforms into a liquid.

meltwater lake   A lake fed by glacial meltwater.

mesa   A large, flat-topped hill (with a surface area of several square km) in an arid region.

mesopause   The boundary that marks the top of the mesosphere of the Earth's atmosphere.

mesosphere   The cooler layer of atmosphere overlying the stratosphere.

Mesozoic Era   The middle of the three Phanerozoic eras; it lasted from 252 Ma to 66 Ma.

metaconglomerate   A metamorphic rock produced by metamorphism of a conglomerate; typically, it contains flattened pebbles and cobbles.

metal   A solid composed almost entirely of atoms of metallic elements; it is generally opaque, shiny, smooth, malleable, and can conduct electricity.

metallic bond   A chemical bond in which the outer atoms are attached to each other in such a way that electrons flow easily from atom to atom.

metamorphic aureole   The region around a pluton, stretching tens to hundreds of meters out, in which heat transferred into the country rock and metamorphosed the country rock.

metamorphic facies   A set of metamorphic mineral assemblages indicative of metamorphism under a specific range of pressures and temperatures.

metamorphic foliation   A fabric defined by parallel surfaces or layers that develop in a rock as a result of metamorphism; schistocity and gneissic layering are examples.

metamorphic grade   A representation of the intensity of metamorphism, meaning the amount or degree of metamorphic change.

metamorphic mineral   New minerals that grow in place within a solid rock under metamorphic temperatures and pressures.

metamorphic mineral assemblage   A group of minerals that form in a rock as a result of metamorphism.

metamorphic rock   Rock that forms when preexisting rock changes into new rock as a result of an increase in pressure and temperature and/or shearing under elevated temperatures; metamorphism occurs without the rock first becoming a melt or a sediment.

metamorphic texture A distinctive arrangement of mineral grains produced by metamorphism.

metamorphic zone The region between two metamorphic isograds, typically named after an index mineral found within the region.

metamorphism The process by which one kind of rock transforms into a different kind of rock.

metasomatism The process by which a rock's overall chemical composition changes during metamorphism because of reactions with hot water that bring in or remove elements.

meteor A streak of bright, glowing gas created as a meteoroid vaporizes in the atmosphere due to friction.

meteoric water Water that falls to the Earth from the atmosphere as either rain or snow.

meteorite A piece of rock or metal alloy that fell from space and landed on the Earth.

micrite Limestone consisting of lime mud (i.e., very fine-grained limestone).

microfossil A fossil that can be seen only with a microscope or an electron microscope.

mid-latitude anticyclone Clockwise (in the northern hemisphere) motion of air around a high-pressure zone.

mid-latitude cyclone Counterclockwise (in the northern hemisphere) motion of air around a low-pressure zone; these can build into huge storm systems.

mid-ocean ridge A 2-km-high submarine mountain belt that forms along a divergent oceanic plate boundary.

migmatite A rock formed when gneiss is heated high enough so that it begins to partially melt, creating layers, or lenses, of new igneous rock that mix with layers of the relict gneiss.

Milankovitch cycles Climate cycles that occur over tens to hundreds of thousands of years because of changes in the Earth's orbit and tilt.

Milky Way Galaxy The group of 300 billion stars, including our Sun, displaying the geometry of a huge spiral.

mine A site at which ore is extracted from the ground.

mineral A homogenous, naturally occurring, solid inorganic substance with a definable chemical composition and an internal structure characterized by an orderly arrangement of atoms, ions, or molecules in a lattice. Most minerals are inorganic.

mineral classes Groups of minerals distinguished from each other on the basis of chemical composition.

mineral resource An accumulation of a useful ore, in which valuable elements are sufficiently concentrated to be worth mining.

mineralogist A geoscientist specializing in the study of minerals.

mineralogy The study of minerals and their characteristics.

mineral resources The minerals extracted from the Earth's upper crust for practical purposes.

Mississippi Valley–type (MVT) deposit An ore deposit, typically in dolostone, containing lead- and zinc-bearing minerals that precipitated from groundwater that had moved up from several km depth in the upper crust; such deposits occur in the upper Mississippi Valley.

mixture A material consisting of two or more substances that can be separated mechanically (i.e., without chemical reactions).

Modified Mercalli Intensity (MMI) Scale An earthquake characterization scale based on the amount of damage that the earthquake causes.

Moho The seismic-velocity discontinuity that defines the boundary between the Earth's crust and mantle. Named for Andrija Mohorovičić.

Mohs hardness scale A list of ten minerals in a sequence of relative hardness, with which other minerals can be compared.

mold A cavity in sedimentary rock left behind when a shell that once filled the space weathers out.

molecule The smallest piece of a compound that has the properties of the compound; it consists of two or more atoms attached by chemical bonds.

moment magnitude scale A modern scale for measuring the relative size of earthquakes that involves studying the amplitude of waves on a seismograph, along with other parameters.

monocline A fold in the land surface whose shape resembles that of a carpet draped over a stair step.

monsoon A seasonal reversal in wind direction that causes a shift from a very dry season to a very rainy season in some regions of the world.

moon A sizable solid body locked in orbit around a planet.

moraine A sediment pile composed of till deposited by a glacier.

mountain belt An elongate band of mountains, formed as the result of an orogeny.

mountain building The process of generating a mountain range.

mountain front The boundary between a mountain range and adjacent plains.

mountain (alpine) glacier A glacier that exists in or adjacent to a mountainous region.

mountain ice cap A mound of ice that submerges peaks and ridges at the crest of a mountain range.

mouth The outlet of a stream where it discharges into another stream, a lake, or a sea.

mud crack A small fissure in a fine-grained sediment that forms when the sediment dries; mud cracks are arranged to outline the shape of polygons.

mudflow (mudslide) A downslope movement of mud at slow to moderate speed.

mud pot A viscous slurry that forms in a geothermal region when hot water or steam rises into soils rich in volcanic ash and clay.

mudstone Very fine-grained sedimentary rock that will not easily split into sheets.

multicellular organism An organism containing many different cells.

mylonite Rock formed during dynamic metamorphism and characterized by foliation that lies roughly parallel to the fault (shear zone) involved in the shearing process; mylonites have very fine grains formed by the nonbrittle subdivision of larger grains.

native metal A naturally occurring pure mass of a single metal in an ore deposit.

natural arch An arch that forms when erosion along joints leaves narrow walls of rock; when the lower part of the wall erodes while the upper part remains, an arch results.

natural hazard A natural feature of the environment that can cause injury to living organisms and/or damage to buildings and the landscape.

natural levees A pair of low ridges that appear on either side of a stream and develop as a result of the accumulation of sediment deposited naturally during flooding.

natural selection The process by which the fittest organisms survive to pass on their characteristics to the next generation.

neap tide An especially low tide that occurs when the angle between the direction of the Moon and the direction of the Sun is 90°.

near-Earth object A meteoroid whose path takes it close to the Earth, so that there is a slight possibility that it could collide with the Earth.

nebula A cloud of gas or dust in space.

nebular theory of planet formation The concept that planets grow out of rings of gas, dust, and ice surrounding a newborn star.

negative anomaly An area where the magnetic field strength is less than expected.

negative feedback Feedback that slows a process down or reverses it.

neocrystallization The growth of new crystals during metamorphism.

neutron A subatomic particle, in the nucleus of an atom, that has a neutral charge.

Nevadan orogeny A convergent-margin mountain-building event that took place in western North America during the Late Jurassic Period.

nonconformity   A type of unconformity at which sedimentary rocks overlie basement (older intrusive igneous rocks and/or metamorphic rocks).

nonflowing artesian well   An artesian well in which water rises on its own up to a level that lies below the ground surface.

nonfoliated metamorphic rock   Rock that has undergone metamorphism, and therefore contains metamorphic minerals, but has no foliation.

nonmetallic mineral resources   Mineral resources that do not contain metals; examples include building stone, gravel, sand, gypsum, phosphate, and salt.

nonplunging fold   A fold with a horizontal hinge.

nonrenewable resource   A resource that nature will take a long time (hundreds to millions of years) to replenish or may never replenish.

nonsystematic joints   Short cracks in rocks that occur in a range of orientations and are randomly placed and oriented.

nor'easter   A large, midlatitude North American cyclone; when it reaches the east coast, it produces strong winds that come out of the northeast.

normal fault   A fault in which the hanging-wall block moves down the slope of the fault.

normal force   In the context of mass movement, it is the component of gravitational force acting perpendicular to a slope.

normal polarity   Polarity in which the paleomagnetic dipole has the same orientation as it does today.

normal stress   The push or pull that is perpendicular to a surface.

North Atlantic deep-water mass   The volume of cold, dense water that sinks to depth in the north polar regions and spreads southward over the Antarctic bottom-water mass.

northeast trade winds   Surface winds that come out of the northeast and occur in the region between the equator and 30°N.

nuclear bond   The force that attaches subatomic particles to each other within the nucleus of an atom.

nuclear fuel   Pellets of concentrated uranium oxide or a comparable radioactive material that can provide energy in a nuclear reactor.

nuclear fusion   The process by which the nuclei of atoms fuse together, thereby creating new, larger atoms.

nuclear reaction   A process that results in changing the nucleus of an atom by breaking or forming nuclear bonds.

nuclear reactor   The part of a nuclear power plant where the fission reactions occur.

nuclear waste   The radioactive material produced as a byproduct in a nuclear plant that must be disposed of carefully due to its dangerous radioactivity.

nucleus   The central ball of an atom that consists of protons and neutrons (except for hydrogen, whose nuclei contains only a proton).

nuée ardente   *See* Pyroclastic flow.

numerical age   (in older literature, "absolute age") The age of a geologic feature given in years.

oasis   A verdant region surrounded by desert, occurring at a place where natural springs provide water at the surface.

oblique-slip fault   A fault in which sliding occurs diagonally along the fault plane.

obsidian   An igneous rock consisting of a solid mass of volcanic glass.

occluded front   A front that no longer intersects the ground surface.

oceanic crust   The crust beneath the oceans; composed of gabbro and basalt, overlain by sediment.

oceanic plateau   A region of oceanic floor that is higher than surrounding areas; such regions have particularly thick oceanic crust and are relicts of submarine large igneous provinces.

oceanic lithosphere   Lithosphere topped by oceanic crust; it reaches a thickness of 100 km.

offshore bar   A narrow ridge of sand that forms off the shore of a coast; some offshore bars rise above sea level, and separate a lagoon on one side from the open ocean on the other.

oil   (geology) A liquid hydrocarbon that can be burned as a fuel.

Oil Age   The period of human history, including our own, so named because the economy depends on oil.

oil field   A region containing a significant amount of accessible oil underground.

oil reserve   The known supply of oil held underground.

oil seep   A location where oil bubbles out of the ground on its own, without pumping.

oil shale   An organic shale containing abundant kerogen.

oil trap   A geologic configuration that keeps oil underground in the reservoir rock and prevents it from rising to the surface.

oil window   The narrow range of temperatures under which oil can form in a source rock.

olistotrome   A large, submarine slump block, buried and preserved.

Oort Cloud   A cloud of icy objects, left over from Solar System formation, that orbit the Sun in a region outside of the heliosphere.

open-pit mine   A mine at which ore is obtained by digging a large excavation at the Earth's surface.

ophiolite   A slice of oceanic crust that has been thrust onto continental crust.

ordinary well   A well whose base penetrates below the water table and can thus provide water.

ore   Rock containing native metals or a concentrated accumulation of ore minerals.

ore deposit   An economically significant accumulation of ore.

ore minerals   Minerals that have metal in high concentrations and in a form that can be easily extracted.

organic carbon   Carbon that has been incorporated in an organism.

organic chemical   A carbon-containing compound that occurs in living organisms, or that resembles such compounds; it consists of carbon atoms bonded to hydrogen atoms along with varying amounts of oxygen, nitrogen, and other chemicals.

organic coast   A coast along which living organisms control landforms along the shore.

organic sedimentary rock   Sedimentary rock (such as coal) formed from carbon-rich relicts of organisms.

organic shale   Lithified, muddy, organic-rich ooze that contains the raw materials from which hydrocarbons eventually form.

organism   A self-contained living entity.

orogen (orogenic belt)   A linear range of mountains.

orogenic collapse   The process in which mountains begin to collapse under their own weight and spread out laterally.

orogeny   A mountain-building event.

orographic barrier   A landform that diverts air flow upward or laterally.

outcrop   An exposure of bedrock.

outer core   The section of the core, between 2,900 and 5,150 km deep, that consists of liquid iron alloy.

outwash plain   A broad area of gravel and sandbars deposited by a braided stream network, fed by the melt water of a glacier.

overburden   The weight of overlying rock on rock buried deeper in the Earth's crust.

overriding plate (slab)   The plate at a subduction zone that overrides the downgoing plate.

oversaturated solution   A solution that contains so much solute (dissolved ions) that precipitation can begin.

oversized stream valley   A large valley with a small stream running through it; the valley formed earlier when the flow was greater.

oxbow lake   A meander that has been cut off yet remains filled with water.

oxidation reaction   A reaction in which an element loses electrons; an example is the reaction of iron with air to form rust.

ozone O3, an atmospheric gas that absorbs harmful ultraviolet radiation from the Sun.

ozone hole An area of the atmosphere, over polar regions, from which ozone has been depleted.

pahoehoe A lava flow with a surface texture of smooth, glassy, ropelike ridges.

paleoclimate The past climate of the Earth.

paleoecology The study of how assemblages of organisms live together at times in the past.

paleomagnetism The record of ancient magnetism preserved in rock.

paleontologist A scientist who specializes in studying and interpreting fossils.

paleontology The study of ancient life and its evolution as recorded by fossils.

paleopole The supposed position of the Earth's magnetic pole in the past, with respect to a particular continent.

paleosol Ancient soil preserved in the stratigraphic record.

Paleozoic Era The oldest era of the Phanerozoic Eon (541–252 Ma).

Pangaea A supercontinent that assembled at the end of the Paleozoic Era.

Pannotia A supercontinent that may have existed sometime between 800 Ma and 600 Ma.

parabolic dunes Dunes formed when strong winds break through transverse dunes to make new dunes whose ends point upwind.

parallax The apparent movement of an object seen from two different points not on a straight line from the object (e.g., from your two different eyes).

parallax method A trigonometric method used to determine the distance from the Earth to a nearby star.

parent isotope A radioactive isotope that undergoes decay.

partial melt The magma formed when the lower-melting-temperature component of a rock has melted.

partial melting The melting in a rock of the minerals with the lowest melting temperatures, while other minerals remain solid.

passive continental margin A continental margin that does not coincide with a plate boundary, and therefore does not display seismicity.

passive margin (see passive continental margin)

passive-margin basin A thick accumulation of sediment along a tectonically inactive coast, formed over crust that stretched and thinned when the margin initially formed by rifting.

patterned ground A polar landscape in which the ground splits into pentagonal or hexagonal shapes.

pause In the context of atmospheric science, it is an elevation in the atmosphere where temperature stops decreasing and starts increasing, or vice versa, as elevation changes.

peat Compacted and partially decayed vegetation accumulating beneath a swamp.

pedalfer soil An older term for a temperate-climate soil characterized by well-defined soil horizons and an organic A-horizon.

pediment The broad, nearly horizontal bedrock surface at the base of a retreating desert cliff.

pedocal soil An older term for thin soil, formed in arid climates. It contains very little organic matter, but significant precipitated calcite.

pegmatite A coarse-grained igneous rock containing crystals of up to tens of centimeters across and occurring in dike-shaped intrusions.

pelagic sediment Microscopic plankton shells and fine flakes of clay that settle out and accumulate on the deep-ocean floor.

Pelé's hair Droplets of basaltic lava that mold into long, glassy strands as they fall.

Pelé's tears Droplets of basaltic lava that mold into tear-shaped, glassy beads as they fall.

peneplain A nearly flat surface that lies at an elevation close to sea level; thought to be the product of long-term erosion.

perched water table A quantity of groundwater that lies above the regional water table because an underlying lens of impermeable rock or sediment prevents the water from sinking down to the regional water table.

percolation The process by which groundwater meanders through tiny, crooked channels in the surrounding material.

peridotite A coarse-grained ultramafic rock.

periglacial environment A region with widespread permafrost but without a blanket of snow or ice.

period An interval of geologic time representing a subdivision of a geologic era.

permafrost Permanently frozen ground.

permanent magnet A special material that behaves magnetically for a long time all by itself.

permanent stream A stream that flows year-round because its bed lies below the water table, or because more water is supplied from upstream than can infiltrate the ground.

permeability The degree to which a material allows fluids to pass through it via an interconnected network of pores and cracks.

Permian mass extinction The radical drop in biodiversity that took place at the Permian-Triassic boundary; it is thought to be due to widespread volcanism or to a meteorite impact.

permineralization The fossilization process in which plant material becomes transformed into rock by the precipitation of silica from groundwater.

petrified A term used by geologists to describe organic material that has transformed into rock by permineralization.

petrified wood Wood that has undergone permineralization and has turned into agate; growth rings and cell walls may remain visible in samples.

petroglyph Drawings formed by chipping into the desert varnish of rocks to reveal the lighter rock beneath.

petrographic microscope A specialized microscope designed to view thin sections of rock by using transmitted polarized light.

petroleum *See* Oil.

phaneritic A textural term used to describe coarse-grained igneous rock.

Phanerozoic Eon The most recent eon, an interval of time from 542 Ma to the present.

phenocryst A large crystal surrounded by a finer-grained matrix in an igneous rock.

photochemical smog Brown haze that blankets a city when exhaust from cars and trucks reacts in the presence of sunlight.

photomicrograph A photograph of a thin section taken through a microscope.

photosynthesis The process during which chlorophyll-containing plants remove carbon dioxide from the atmosphere, form tissues, and expel oxygen back to the atmosphere.

photovoltaic cell A devise capable of transforming solar energy directly into electricity.

phreatomagmatic eruption An explosive eruption that occurs when water enters the magma chamber and turns into steam.

phyllite A fine-grained metamorphic rock with a foliation caused by the preferred orientation of very fine-grained mica.

phyllitic luster A silk-like sheen characteristic of phyllite, a result of the rock's fine-grained mica.

phylogenetic tree A chart representing the ideas of paleontologists showing which groups of organisms radiated from which ancestors.

physical weathering The process in which intact rock breaks into smaller grains or chunks.

piedmont glacier A fan or lobe of ice that forms where a valley glacier emerges from a valley and spreads out into the adjacent plain.

pillow basalt Glass-encrusted basalt blobs that form when magma extrudes on the sea floor and cools very quickly.

pillow lava Mafic lava that extruded underwater to form blob-like shapes, or pillows, which typically have a glassy rind.

placer deposit Concentrations of metal grains in stream sediment that develop when rocks containing native metals erode and create a mixture of sand grains and metal fragments; the moving water of the stream carries away lighter mineral grains.

planet An object that orbits a star, is roughly spherical, and has cleared its neighborhood of other objects.

planetesimal Tiny, solid pieces of rock and metal that collect in a planetary nebula and eventually accumulate to form a planet.

plankton Tiny plants and animals that float in sea or lake water.

plastic In geology, the word is used as an adjective to describe when a material responds to stress by slowly flowing, without loss of cohesion, rather than by breaking.

plastic deformation The deformational process in which mineral grains behave like plastic and, when compressed or sheared, become flattened or elongate without cracking or breaking.

plate One of about 20 distinct pieces of the relatively rigid lithosphere.

plate boundary The border between two adjacent lithosphere plates.

plate-boundary earthquakes The earthquakes that occur along and define plate boundaries.

plate-boundary volcano A volcanic arc or mid-ocean ridge volcano, formed as a consequence of movement along a plate boundary.

plate interior A region away from the plate boundaries that consequently experiences few earthquakes.

plate tectonics *See* Theory of plate tectonics.

playa The flat, typically salty lake bed that remains when all the water evaporates in drier times; forms in desert regions.

Pleistocene Epoch The period of time from about 2.6 Ma to 14,000 years ago, during which the Earth experienced an ice age.

Pleistocene Ice Age The ice age that began about 2.6 Ma, and involves many advances and retreats of continental glaciers.

plunge In the context of defining the orientation of a line, plunge refers to the angle between the line and horizontal, as measured in a vertical plane.

plunge pool A depression at the base of a waterfall scoured by the energy of the falling water.

plunging fold A fold with a tilted hinge.

pluton An irregular or blob-shaped intrusion; can range in size from tens of m across to tens of km across.

pluvial lake A lake formed to the south of a continental glacier as a result of enhanced rainfall during an ice age.

point bar A wedge-shaped deposit of sediment on the inside bank of a meander.

polar cell A high-latitude convection cell in the atmosphere.

polar easterlies Prevailing winds that come from the east and flow from the polar high to the subpolar low.

polar front The convergence zone in the atmosphere at latitude 60°.

polar glacier *See* Dry-bottom glacier.

polar high The zone of high pressure in polar regions created by the sinking of air.

polarity The orientation of a magnetic dipole.

polarity chron The time interval between longer-duration polarity reversals of the Earth's magnetic field.

polarity subchron The time interval between magnetic reversals if the interval is of short duration (less than 200,000 years long).

polarized light A beam of filtered light waves that all vibrate in the same plane.

polar wander The phenomenon of the progressive changing through time of the position of the Earth's magnetic poles relative to a location on a continent; significant polar wander probably doesn't occur—in fact, poles seem to remain fairly fixed, while continents move.

polar-wander path The curving line representing the apparent progressive change in the position of the Earth's magnetic pole, relative to a locality X, assuming that the position of X on the Earth has been fixed through time (in fact, poles stay fixed while continents move).

pollen Tiny grains involved in plant reproduction.

pollution Natural and synthetic contaminant materials introduced to the Earth's environment by the activities of humans.

polymorphs Two minerals that have the same chemical composition but a different crystal lattice structure.

pore A small, open space within sediment or rock.

pore collapse The closer packing of grains that occurs when groundwater is extracted from pores, thus eliminating the support holding the grains apart.

porosity The total volume of empty space (pore space) in a material, usually expressed as a percentage.

porphyritic A textural term for igneous rock that has phenocrysts distributed throughout a finer matrix.

Portland cement Cement made by mechanically mixing limestone, sandstone, and shale in just the right proportions, before heating in a kiln, to provide the correct chemical makeup of cement.

positive anomaly An area where the magnetic field strength is stronger than expected.

positive-feedback mechanism A mechanism that enhances the process that caused the mechanism in the first place.

post-glacial rebound The rise of the surface of a continent after an ice sheet has melted away, so that isostasy is reestablished.

potentiometric surface The elevation to which water in an artesian system would rise if unimpeded; where there are flowing artesian wells, the potentiometric surface lies above ground.

pothole A bowl-shaped depression carved into the floor of a stream by a long-lived whirlpool carrying sand or gravel.

Precambrian The interval of geologic time between the Earth's formation about 4.57 Ga and the beginning of the Phanerozoic Eon 542 Ma.

precession The gradual conical path traced out by the Earth's spinning axis; simply put, it is the "wobble" of the axis.

precious metals Metals (like gold, silver, and platinum) that have high value.

precipitate (chemistry, n.) A solid substance formed when atoms attach and settle out of a solution, or attach to the walls of the container holding the solution; (chemistry, v.) the action of forming a solid substance from a solution; (meteorology, v.) the dropping of snow or rain from the sky.

precipitation (1) The process by which atoms dissolved in a solution come together and form a solid; (2) rainfall or snow.

preferred mineral orientation The metamorphic texture that exists where platy grains lie parallel to one another and/or elongate grains align in the same direction.

preferred orientation The parallelism of inequant grains in a metamorphic rock.

preservation potential The likelihood that an organism will be preserved as a fossil.

pressure Force per unit area, or the "push" acting on a material in cases where the push (compressional stretch) is the same in all directions.

pressure gradient The rate of pressure change over a given horizontal distance.

pressure solution The process of dissolution at points of contact, between grains, where compression is greatest, producing ions that then precipitate elsewhere, where compression is less.

prevailing winds Surface winds that generally flow in the same direction for long time periods.

primary porosity The space that remains between solid grains or crystals immediately after sediment accumulates or rock forms.

primary producer An organism that obtains its energy from photosynthesis, or by metabolizing minerals, and therefore does not consume other organisms.

principal aquifer The geologic unit that serves as the primary source of groundwater in a region.

principle of baked contacts When an igneous intrusion "bakes" (metamorphoses) surrounding rock, the rock that has been baked must be older than the intrusion.

principle of cross-cutting relations If one geologic feature cuts across another, the feature that has been cut is older.

principle of fossil succession In a stratigraphic sequence, different species of fossil organisms appear in a definite order; once a fossil species disappears in a sequence of strata, it never reappears higher in the sequence.

principle of inclusions If a rock contains fragments of another rock, the fragments must be older than the rock containing them.

principle of original continuity Sedimentary layers, before erosion, formed fairly continuous sheets over a region.

principle of original horizontality Layers of sediment, when originally deposited, are fairly horizontal.

principle of superposition In a sequence of sedimentary rock layers, each layer must be younger than the one below, for a layer of sediment cannot accumulate unless there is already a substrate on which it can collect.

principle of uniformitariansim The physical processes we observe today also operated in the past in the same way, and at comparable rates.

product (chemistry, n.) materials produced or formed by a chemical reaction.

prograde metamorphism Metamorphism that occurs as temperatures and pressures are increasing.

prokaryote (prokarya) An organism whose cells do not contain a nucleus; archaea and bacteria consist of prokaryotic cells.

Proterozoic Eon The most recent of the Precambrian eons (2,500–541 Ma).

protocontinent A block of crust composed of volcanic arcs and hotspot volcanoes sutured together.

protolith The original rock from which a metamorphic rock formed.

proton A positively charged subatomic particle in the nucleus of an atom.

protoplanet A body that grows by the accumulation of planetesimals but has not yet become big enough to be called a planet.

protoplanetary disk The plate-shaped region of gas and dust, surrounding the newborn Sun, from which the planets formed.

protostar A dense body of gas that is collapsing inward because of gravitational forces and that may eventually become a star.

pumice A glassy igneous rock that forms from felsic frothy lava and contains abundant (over 50%) pore space.

pumice lapilli Marble-sized chunks consisting of frothy, siliceous igneous rock that fall from a volcanic eruptive cloud.

punctuated equilibrium The hypothesis that evolution takes place in fits and starts; evolution occurs very slowly for quite a while and then, during a relatively short period, takes place very rapidly.

P-waves Compressional seismic waves that move through the body of the Earth.

P-wave shadow zone A band between 103° and 143° from an earthquake epicenter, as measured along the circumference of the Earth, inside which P-waves do not arrive at seismograph stations.

pycnocline The boundary between layers of water of different densities.

pyroclastic debris Fragmented material that sprayed out of a volcano and landed on the ground or sea floor in solid form.

pyroclastic flow A fast-moving avalanche that occurs when hot volcanic ash and debris mix with air and flow down the side of a volcano.

pyroclastic rock Rock made from fragments that were blown out of a volcano during an explosion and were then packed or welded together.

quarry A site at which stone is extracted from the ground.

quartzite A metamorphic rock composed of quartz and transformed from a protolith of quartz sandstone.

quenching A sudden cooling of molten material to form a solid.

quick clay Clay that behaves like a solid when still (because of surface tension holding the water-coated clay flakes together) but that flows like a liquid when shaken.

radial drainage network A drainage network in which the streams flow outward from a cone-shaped mountain and define a pattern resembling spokes on a wheel.

radiation (physics) Electromagnetic energy traveling away from a source through a medium or space.

radioactive decay The process by which a radioactive atom undergoes fission or releases particles, thereby being transformed into a new element.

radioactive element An element that that includes isotopes that can undergo decay.

radioactive isotope An unstable isotope of a given element.

radiometric dating (also called numerical dating) The science of determining the age of materials in years by measuring the ratio of parent radioactive atoms to daughter product atoms in the material.

rain band A spiral arm of a hurricane from which intense rain can fall.

rain shadow The inland side of a mountain range, which is arid because the mountains block rain clouds from reaching the area.

range (for fossils) The interval of a sequence of strata in which a specific fossil species appears.

rapids A reach of a stream in which water becomes particularly turbulent; as a consequence, waves develop on the surface of the stream.

rare earth element (REE) One of a group of 17 elements including the lanthanides, scandium, and yttrium; they are essential in the production of high-tech devices.

reach A specified interval of a stream's path.

reactant (chemistry) The starting materials of a chemical reaction.

recessional moraine The end moraine that forms when a glacier stalls for a while as it recedes.

recharge area A location where water enters the ground and infiltrates down to the water table.

recurrence interval The average time between events of a given size or magnitude; the term is commonly used to give a sense of the frequency of earthquakes or of flooding.

recrystallization The process in which ions or atoms in minerals rearrange to form new minerals.

rectangular drainage network A drainage network in which the streams join each other at right angles because of a rectangular grid of fractures that breaks up the ground and localizes channels.

recurrence interval The average time between successive geologic events.

Redbed A sedimentary rock layer displaying a reddish color due to the occurrence of iron oxide minerals; such rocks typically form in terrestrial deposits.

red giant A huge red star that forms when Sun-sized stars start to die and expand.

red shift The phenomenon in which a source of light moving away from you very rapidly shifts to a lower frequency; that is, toward the red end of the spectrum.

reef bleaching The death and loss of color of a coral reef.

reflected ray A ray that bounces off a boundary between two different materials.

reflection What happens when energy bounces off a boundary; when this happens, the incoming angle and outgoing angle are equal.

refracted ray A ray that bends as it passes through a boundary between two different materials.

refraction The bending of a ray as it passes through a boundary between two different materials.

refractory materials Substances that have a relatively high melting point and tend to exist in solid form.

reg A vast stony plain in a desert.

regional metamorphism *See also* Dynamothermal metamorphism; metamorphism of a broad region, usually the result of deep burial during an orogeny.

regolith Any kind of unconsolidated debris that covers bedrock.

regression The seaward migration of a shoreline caused by a lowering of sea level.

relative age The age of one geologic feature with respect to another.

relative humidity The ratio between the measured water content of air and the maximum possible amount of water the air can hold at a given condition.

relative plate velocity The movement of one lithosphere plate with respect to another.

relief The difference in elevation between adjacent high and low regions on the land surface.

renewable resource A resource that can be replaced by nature within a short time span relative to a human life span.

reserve A known occurrence of a resource in sufficient quantities and concentration to make it potentially worth extracting.

reservoir rock Rock with high porosity and permeability, so it can contain an abundant amount of easily accessible oil.

residence time The average length of time that a substance stays in a particular reservoir.

residual mineral deposit Soils in which the residuum left behind after leaching by rainwater is so concentrated in metals that the soil itself becomes an ore deposit.

resonance (seismology) A situation that arises when earthquake waves of a particular frequency cause particularly large-amplitude movements because energy input happens at just the right time.

resurgent dome The new mound, or cone, of igneous rock that grows within a caldera as an eruption begins anew.

retrograde metamorphism Metamorphism that occurs as pressures and temperatures are decreasing; for retrograde metamorphism to occur, water must be added.

return stroke An upward-flowing electric current from the ground that carries positive charges up to a cloud during a lightning flash.

reversed polarity Polarity in which the paleomagnetic dipole points north.

reverse fault A steeply dipping fault on which the hanging-wall block slides up.

rhythmic layering Banding in sediments, shells, trees, corals, or ice that repeats periodically; it may be correlated to annual cycles.

Richter scale A scale that defines earthquakes on the basis of the amplitude of the largest ground motion recorded on a seismogram.

ridge axis The crest of a mid-ocean ridge; the ridge axis defines the position of a divergent plate boundary.

ridge-push force A process in which gravity causes the elevated lithosphere at a mid-ocean ridge axis to push on the lithosphere that lies farther from the axis, making it move away.

rifting The process by which continental lithosphere stretches horizontally and thins vertically.

right-lateral strike-slip fault A strike-slip fault in which the block on the opposite fault plane from a fixed spot moves to the right of that spot.

rip current A strong, localized seaward flow of water perpendicular to a beach.

ripple mark Relatively small elongated ridges that form on a sedimentary bed surface at right angles to the direction of current flow.

riprap Loose boulders or concrete piled together along a beach to absorb wave energy before it strikes a cliff face.

roche moutonnée A glacially eroded hill that becomes elongate in the direction of flow and asymmetric; glacial rasping smooths the upstream part of the hill into a gentle slope, while glacial plucking erodes the downstream edge into a steep slope.

rock A coherent, naturally occurring solid, consisting of an aggregate of minerals or a mass of glass.

rock burst A sudden explosion of rock off the ceiling or wall of an underground mine.

rock composition The chemical makeup of a rock, as represented by the proportions of different minerals that it contains.

rock cycle The succession of events that results in the transformation of Earth materials from one rock type to another, then another, and so on.

rockfall A mass of rock that separates from a cliff, typically along a joint, and then free-falls downslope.

rock flour Fine-grained sediment produced by glacial abrasion of the substrate over which a glacier flows.

rock glacier A slow-moving mixture of rock fragments and ice.

rockslide A sudden downslope movement of rock.

rocky coast An area of coast where bedrock rises directly from the sea, so sandy beaches are largely absent.

Rodinia A proposed Precambrian supercontinent that existed around 1 billion years ago.

rogue wave Waves that are two to five times the size of most of the large waves passing a locality in a given time interval.

root wedging The phenomenon that happens when the roots of a tree or shrub grow in a joint and eventually push on the walls of the joint to make it wider.

rotational axis The imaginary line through the center of the Earth around which the Earth spins.

running water Water that flows down the surface of sloping land in response to the pull of gravity.

Runoff The water the flows on the surface of the Earth to drain the land; it includes streamflow and sheetflow.

R-waves (Rayleigh waves) Surface seismic waves that cause the ground to ripple up and down, like water waves in a pond.

sabkah A region of formerly flooded coastal desert in which stranded seawater has left a salt crust over a mire of mud that is rich in organic material.

Saffir-Simpson scale A scale used to measure the intensity of a hurricane, based on sustained wind speed; Category 5 hurricanes are the strongest.

salinity The degree of concentration of salt in water.

saltation The movement of a sediment in which grains bounce along their substrate, knocking other grains into the water column (or air) in the process.

salt dome A rising bulbous dome of salt that bends up the adjacent layers of sedimentary rock.

salt wedging The process in arid climates by which dissolved salt in groundwater crystallizes and grows in open pore spaces in rocks and pushes apart the surrounding grains.

sand dune A relatively large ridge of sand built up by a current of wind (or water); cross bedding typically occurs within the dune.

sand spit An area where the beach stretches out into open water across the mouth of a bay or estuary.

sandstone Coarse-grained sedimentary rock consisting almost entirely of quartz.

sand volcano (sand blow) A small mound of sand produced when sand layers below the ground surface liquify as a result of seismic shaking, causing the sand to erupt onto the Earth's surface through cracks or holes in overlying clay layers.

saprolite A layer of rotten rock created by chemical weathering in warm, wet climates.

Sargasso Sea The center of North Atlantic Gyre, named for the tropical seaweed sargassum, which accumulates in its relatively noncirculating waters.

saturated solution Water that carries as many dissolved ions as possible under given environmental conditions.

saturated zone (phreatic zone) The region below the water table where pore space is filled with water.

scattering The dispersal of energy that occurs when light interacts with particles in the atmosphere.

schist A medium-to-coarse-grained metamorphic rock that possesses schistosity.

schistosity Foliation caused by the preferred orientation of large mica flakes.

science The systematic study of natural phenomena via observation, computation, experiment, and modeling.

scientific cosmology The study of the overall structure and evolution of the Universe, based on the application of the laws of physics and chemistry, and based on astronomical observations and measurements.

scientific law A concise statement that completely describes a natural relationship or phenomenon; it does not, however, explain the phenomenon.

scientific method A sequence of steps for systematically analyzing scientific problems in a way that leads to verifiable results.

scientific revolution

scoria A glassy, mafic, igneous rock containing abundant air-filled holes.

scoria cone An accumulation of lapilli-sized or larger fragments formed from a volcanic eruption that spatters clots of basaltic lava. (Also called cinder cone.)

scouring A process by which running water removes loose fragments of sediment from a streambed.

sea arch An arch of land protruding into the sea and connected to the mainland by a narrow bridge.

seafloor spreading The gradual widening of an ocean basin as new oceanic crust forms at a mid-ocean ridge axis and then moves away from the axis.

sea ice Ice formed by the freezing of the surface of the sea.

seal rock A relatively impermeable rock, such as shale, salt, or unfractured limestone, that lies above a reservoir rock and stops the oil from rising further.

seam A sedimentary bed of coal interlayered with other sedimentary rocks.

seamount An isolated submarine mountain.

seasonal floods Floods that appear almost every year during seasons when rainfall is heavy or when winter snows start to melt.

seasonal ordinary well A well that provides water only during the rainy season when the water table rises below the base of the well.

sea stack An isolated tower of land just offshore, disconnected from the mainland by the collapse of a sea arch.

seawall A wall of riprap built on the landward side of a backshore zone in order to protect shore cliffs from erosion.

second The basic unit of time measurement, now defined as the time it takes for the magnetic field of a cesium atom to flip polarity 9,192,631,770 times, as measured by an atomic clock.

secondary-enrichment The process by which a new ore deposit forms from metals that were dissolved and carried away from preexisting ore minerals; the resulting ore body is a secondary-enrichment deposit

secondary porosity New pore space in rocks, created some time after a rock first forms.

secondary recovery technique A process used to extract the quantities of oil that will not come out of a reservoir rock with just simple pumping.

sediment An accumulation of loose mineral grains, such as boulders, pebbles, sand, silt, or mud, that are not cemented together.

sediment liquefaction When pressure in the water in the pores push sediment grains apart so that they become surrounded by water and no longer rest against each other, and the sediment becomes able to flow like a liquid.

sedimentary basin A depression, created as a consequence of subsidence, that fills with sediment.

sedimentary rock Rock that forms either by the cementing together of fragments broken off preexisting rock or by the precipitation of mineral crystals out of water solutions at or near the Earth's surface.

sedimentary sequence A grouping of sedimentary units bounded on top and bottom by regional unconformities.

sedimentary structure A geometry or arrangement of material in sediment or sedimentary rock that formed during or shortly after deposition, not in response to later tectonic stress; examples include cross beds and mudcracks.

sediment budget The proportion of sand supplied to sand removed from a depositional setting.

sediment load The total volume of sediment carried by a stream.

sediment maturity The degree to which a sediment has evolved from a crushed-up version of the original rock into a sediment that has lost its easily weathered minerals and become well sorted and rounded.

sediment sorting The segregation of sediment by size.

seiche Rhythmic movement in a body of water caused by ground motion.

seismic belts (seismic zones) The relatively narrow strips of crust on the Earth under which most earthquakes occur.

seismicity Earthquake activity.

seismic ray The changing position of an imaginary point on a wave front as the front moves through rock.

seismic-reflection profile A cross-sectional view of the crust made by measuring the reflection of artificial seismic waves off boundaries between different layers of rock in the crust.

seismic retrofitting The strengthening of an already existing structure (building, bridge, etc.) so that it can withstand earthquake vibrations.

seismic tomography Analysis by sophisticated computers of global seismic data in order to create a three-dimensional image of variations in seismic-wave velocities within the Earth.

seismic velocity The speed at which seismic waves travel.

seismic-velocity discontinuity A boundary in the Earth at which seismic velocity changes abruptly.

seismic (earthquake) waves Waves of energy emitted at the focus of an earthquake.

seismogram The record of an earthquake produced by a seismometer.

seismologist A scientist who specializes in the study of earthquakes, or in the study of how seismic waves characterize the interior of the Earth.

seismometer (seismograph) An instrument that can record the ground motion from an earthquake.

self-exciting dynamo A configuration that perpetually generates a magnetic field due to the movement of an electrical conductor in the presence of the magnetic field.

semipermanent pressure cell A somewhat elliptical zone of high or low atmospheric pressure that lasts much of the year; it forms because high-pressure zones tend to be narrower over land than over sea.

Sevier orogeny A mountain-building event that affected western North America between about 150 Ma and 80 Ma, a result of convergent margin tectonism; a fold-thrust belt formed during this event.

shale Very fine-grained sedimentary rock that breaks into thin sheets.

shale gas Gas extracted directly from a source rock (organic shale).

shale oil Oil extracted directly from a source rock.

shatter cones Small, cone-shaped fractures formed by the shock of a meteorite impact.

shear When one part of a material moves sideways, relative to another.

shear strain A change in shape of an object that involves the movement of one part of a rock body sideways past another part so that angular relationships within the body change.

shear stress A stress that moves one part of a material sideways past another part.

shear waves Seismic waves in which particles of material move back and forth perpendicular to the direction in which the wave itself moves.

shear zone A fault in which movement has occurred ductilely.

sheetwash A film of water less than a few mm thick that covers the ground surface during heavy rains.

shell (biology) A relatively hard, protective structure formed of minerals and surrounding the soft part of an invertebrate organism.

shield An older, interior region of a continent.

shield volcano A subaerial volcano with a broad, gentle dome, formed either from low-viscosity basaltic lava or from large pyroclastic sheets.

shocked quartz Grains of quartz that have been subjected to intense pressure such as occurs during a meteorite impact.

shock metamorphism The changes that can occur in a rock due to the passage of a shock wave, generally resulting from a meteorite impact.

shoreline The boundary between the water and land.

shortening The process during which a body of rock or a region of crust becomes shorter.

short-term climate change Climate change that takes place over hundreds to thousands of years.

Sierran arc A large continental volcanic arc along western North America that was initiated at the end of the Jurassic Period and lasted until about 80 million years ago.

silica The chemical name for a substance composed of SiO2.

silicate rock Rock composed of silicate minerals.

silicates (silicate minerals) Minerals built from silicon-oxygen tetrahedra arranged in chains, sheets, or 3-D networks; they make up most of the Earth's crust and mantle.

siliceous sedimentary rock Sedimentary rock that contains abundant quartz.

silicon-oxygen tetrahedron The SiO_4^{4-} anionic group, in which four oxygen atoms surround a single silicon atom, thereby defining the corners of a tetrahedron.

silicic Rich in silica with relatively little iron and magnesium.

sill A nearly horizontal tabletop-shaped tabular intrusion that occurs between the layers of country rock.

siltstone Fine-grained sedimentary rock generally composed of very small quartz grains.

sinkhole A circular depression in the land that forms when an underground cavern collapses.

slab-pull force The force that downgoing plates (or slabs) apply to oceanic lithosphere at a convergent margin.

slate Fine-grained, low-grade metamorphic rock, formed by the metamorphism of shale.

slaty cleavage The foliation typical of slate, and reflective of the preferred orientation of slate's clay minerals, that allows slate to be split into thin sheets.

slickensides The polished surface of a fault caused by slip on the fault; lineated slickensides also have grooves that indicate the direction of fault movement.

slip face The leeward slope of a dune; sand that builds up at the crest of the dune slides down this face; slip faces are preserved as cross beds within sandstone layers.

slip lineations Linear marks on a fault surface created during movement on the fault; some slip lineations are defined by grooves, some by aligned mineral fibers.

slope failure The downslope movement of material on an unstable slope.

slow-onset flood A flood that takes days to weeks to develop; these include seasonal floods that cover flood plains and delta plains.

slump A semi-coherent volume of regolith that slipped down a slope above a spoon-shaped failure surface at slow to moderate speed.

slumping Downslope movement in which a mass of regolith detaches from its substrate along a spoon-shaped, sliding surface and slips downward semi-coherently.

smelting The heating of a metal-containing rock to high temperatures in a fire so that the rock will decompose to yield metal plus a nonmetallic residue (slag).

snottite A long gob of bacteria that slowly drips from the ceiling of a cave.

snow avalanche Rapid downslope movement of a mass of snow; typically, the movement transforms the snow into a turbulent cloud.

snowball Earth A model proposing that, at times during Earth history, glaciers covered all land, and the entire ocean surface froze.

snow line The boundary above which snow remains all year.

soda straw A hollow stalactite in which calcite precipitates around the outside of a drip.

soil Sediment that has undergone changes at the surface of the Earth, including reaction with rainwater and the addition of organic material.

soil erosion The removal of soil by wind and runoff.

soil horizon Distinct zones within a soil, distinguished from each other by factors such as chemical composition and organic content.

soil moisture Underground water that wets the surface of the mineral grains and organic material making up soil, but lies above the water table.

soil profile A vertical sequence of distinct zones of soil.

Solar System Our Sun and all the materials that orbit it (including planets, moons, asteroids, Kuiper Belt objects, and Oort Cloud objects).

solar wind A stream of particles with enough energy to escape from the Sun's gravity and flow outward into space.

solid A material that can maintain its shape indefinitely.

solid-state diffusion The slow movement of atoms or ions through a solid.

solidus The highest temperature at which all the components of a material are solid; at the solidus temperature, the material begins to melt.

solifluction The type of creep characteristic of tundra regions; during the summer, the uppermost layer of permafrost melts, and the soggy, weak layer of ground then flows slowly downslope in overlapping sheets.

solstice A day on which the polar ends of the terminator (the boundary between the day hemisphere and the night hemisphere) lie 23.5° away from the associated geographic poles.

solution A material containing dissolved ions.

Sonoman orogeny A convergent-margin mountain-building event that took place on the western coast of North America in the Late Permian and Early Triassic periods.

sorting (1) The range of clast sizes in a collection of sediment; (2) the degree to which sediment has been separated by flowing currents into different-sized fractions.

source rock A rock (organic-rich shale) containing the raw materials from which hydrocarbons eventually form.

southeast trade winds Trade winds in the southern hemisphere, which start flowing northward, deflect to the west, and end up flowing from southeast to northwest.

southern oscillation The movement of atmospheric pressure cells back and forth across the Pacific Ocean, in association with El Niño.

specific gravity A number representing the density of a mineral, as specified by the ratio between the weight of a volume of the mineral and the weight of an equal volume of water.

speleothem A formation that grows in a limestone cave by the accumulation of travertine precipitated from water solutions dripping in a cave or flowing down the wall of a cave.

sphericity The measure of the degree to which a clast approaches the shape of a sphere.

spreading boundary *See* Divergent plate boundary.

spreading rate The rate at which sea floor moves away from a mid-ocean ridge axis, as measured with respect to the sea floor on the opposite side of the axis.

spring A natural outlet from which groundwater flows up onto the ground surface.

spring tide An especially high tide that occurs when the Sun is on the same side of the Earth as the Moon.

stable air Air that does not have a tendency to rise rapidly.

stable slope A slope on which downward sliding is unlikely.

stalactite An icicle-like cone that grows from the ceiling of a cave as dripping water precipitates limestone.

stalagmite An upward-pointing cone of limestone that grows when drips of water hit the floor of a cave.

standing wave A wave whose crest and trough remain in place as water moves through the wave.

star An object in the Universe in which fusion reactions occur pervasively, producing vast amounts of energy; our Sun is a star.

star dune A constantly changing dune formed by frequent shifts in wind direction; it consists of overlapping crescent dunes pointing in many different directions.

states of matter Versions of a substance that differ from each other in the degree to which atoms or molecules in the substance are bonded to each other; physicists recognize four states: solid, liquid, gas, and plasma.

steady-state condition The condition when proportions of a chemical in different reservoirs remain fairly constant even though there is a constant flux (flow) of the chemical among the reservoirs.

stellar nucleosynthesis The production of new, larger atoms by fusion reactions in stars; the process generates more massive elements that were not produced by the Big Bang.

stellar wind The stream of atoms emitted from a star into space.

stick-slip behavior Stop-start movement along a fault plane caused by friction, which prevents movement until stress builds up sufficiently.

stone rings Ridges of cobbles between adjacent bulges of permafrost ground.

stoping A process by which magma intrudes; blocks of wall rock break off and then sink into the magma.

storm An episode of severe weather in which winds, precipitation, and in some cases lightning become strong enough to be bothersome and even dangerous.

storm-center velocity A storm's (hurricane's) velocity along its track.

storm surge Excess seawater driven landward by wind during a storm; the low atmospheric pressure beneath the storm allows sea level to rise locally, increasing the surge.

strain The change in shape of an object in response to deformation (i.e., as a result of the application of a stress).

strata A succession of several layers or beds together.

strategic mineral A mineral containing elements, typically metals, of strategic importance to technology.

stratified drift Glacial sediment that has been redistributed and stratified by flowing water.

stratigraphic column A cross-section diagram of a sequence of strata summarizing information about the sequence.

stratigraphic formation A recognizable layer of a specific sedimentary rock type or set of rock types, deposited during a certain time interval, that can be traced over a broad region.

stratigraphic group (group) Several adjacent stratigraphic formations in a succession.

stratigraphic sequence An interval of strata deposited during periods of relatively high sea level, and bounded above and below by regional unconformities.

stratopause The temperature pause that marks the top of the stratosphere.

stratosphere The stable, stratified layer of atmosphere directly above the troposphere.

stratovolcano A large, cone-shaped subaerial volcano consisting of alternating layers of lava and tephra.

stratus cloud A thin, sheet-like, stable cloud.

streak The color of the powder produced by pulverizing a mineral on an unglazed ceramic plate.

stream A ribbon of water that flows in a channel.

streambed The floor of a stream.

stream capacity The total quantity of sediment a stream carries.

stream capture (stream piracy) The situation in which headward erosion causes one stream to intersect the course of another, previously independent stream, so that the intersected stream starts to flow down the channel of the first stream.

stream competence The maximum particle size that a stream can carry.

stream gradient The slope of a stream's channel in the downstream direction.

stream piracy A process that happens when headward erosion by one stream causes the stream to intersect the course of another stream and capture its flow.

stream rejuvenation The renewed downcutting of a stream into a floodplain or peneplain, caused by a relative drop of the base level.

stream terrace When a stream downcuts through the alluvium of a floodplain so that a new, lower floodplain develops and the original floodplain becomes a step-like platform.

stress The push, pull, or shear that a material feels when subjected to a force; formally, the force applied per unit area over which the force acts.

stretching The process during which a layer of rock or a region of crust becomes longer.

striations Linear scratches in rock.

strike The compass orientation of a horizontal line on a plane.

strike-slip fault A fault in which one block slides horizontally past another (and therefore parallel to the strike line), so there is no relative vertical motion.

strip mining The scraping off of all soil and sedimentary rock above a coal seam in order to gain access to the seam.

stromatolite Layered mounds of sediment formed by cyanobacteria; cyanobacteria secrete a mucous-like substance to which sediment sticks, and as each layer of cyanobacteria gets buried by sediment, it colonizes the surface of the new sediment, building a mound upward.

structural control The condition in which geologic structures, such as faults, affect the distribution and drainage of water or the shape of the land surface.

subaerial Pertaining to land regions above sea level (i.e., under air).

subduction The process by which one oceanic plate bends and sinks down into the asthenosphere beneath another plate.

subduction zone The region along a convergent boundary where one plate sinks beneath another.

sublimation The evaporation of ice directly into vapor without first forming a liquid.

submarine canyon A narrow, steep canyon that dissects a continental shelf and slope.

submarine fan A wedge-shaped accumulation of sediment at the base of a submarine slope; fans usually accumulate at the mouth of a submarine canyon.

submarine slump The underwater downslope movement of a semi-coherent block of sediment along a weak mud detachment.

submergent coast A coast at which the land is sinking relative to sea level.

subpolar low The rise of air where the surface flow of a polar cell converges with the surface flow of a Ferrel cell, creating a low-pressure zone in the atmosphere.

subsidence   The vertical sinking of the Earth's surface in a region, relative to a reference plane.

subsoil   The B-horizon, or zone of accumulation, in a soil; it underlies the topsoil.

substrate   A general term for material just below the ground surface.

subtropical high (subtropical divergence zone)   A belt of high pressure in the atmosphere at 30° latitude formed where the Hadley cell converges with the Ferrel cell, causing cool, dense air to sink.

subtropics   Desert climate regions that lie on either side of the equatorial tropics between the lines of 20° and 30° north or south of the equator.

summit eruption   An eruption that occurs in the summit crater of a volcano.

sunspot cycle   The cyclic appearance of large numbers of sunspots (black spots thought to be magnetic storms on the Sun's surface) every 9 to 11.5 years.

supercell   A large thunderstorm in which there is rotation around a vertical axis.

supercontinent   A very large continent formed by the suturing together of smaller continents.

supercontinent cycle   The process of change during which supercontinents develop and later break apart, forming pieces that may merge once again in geologic time to make yet another supercontinent.

supernova   A short-lived, very bright object in space that results from the cataclysmic explosion marking the death of a very large star; the explosion ejects large quantities of matter into space to form new nebulae.

superplume   A huge mantle plume.

superposed stream   A stream whose geometry has been laid down on a rock structure and is not controlled by the structure.

superrotation (the principle of)   The faster rotation of the core, relative to the rest of the Earth.

supervolcano   A volcano that erupts a vast amount (more than 1,000 cubic km) of volcanic material during a single event; none have erupted during recorded human history.

surface current   An ocean current in the top 100 m of water.

surface load   (bed load)   Sediment that rolls and bounce along the ground (under the air) or along a stream bed (under water).

surface water   Liquid or seasonally frozen water that resides at the surface of the Earth in oceans, lakes, streams, and marshes.

surface waves   Seismic waves that travel along the Earth's surface.

surface westerlies   The prevailing surface winds in North America and Europe, which come out of the west or southwest.

surf zone   A region of the shore in which breakers crash onto the shore.

surge (glacial)   A pulse of rapid flow in a glacier.

suspended load   Tiny solid grains carried along by a stream without settling to the floor of the channel.

sustainable growth   The ability of society to prosper without depleting the supply of natural resources, and without destroying the environment.

suture   The contact defining the boundary of what were two separate crustal blocks, prior to collision.

swamp   A wetland dominated by trees.

swash   The upward surge of water that flows up a beach slope when breakers crash onto the shore.

S-waves   Seismic shear waves that pass through the body of the Earth.

S-wave shadow zone   A band between 103° and 180° from the epicenter of an earthquake inside of which S-waves do not arrive at seismograph stations.

swell   A broad, long-wavelength, ocean wave that has traveled a long distance from its source.

swelling clay   Clay possessing a mineral structure that allows it to absorb water between its layers and thus swell to several times its original size.

symmetry   The condition in which the shape of one part of an object is a mirror image of the other part.

syncline   A trough-shaped fold whose limbs dip toward the hinge.

systematic joints   Long planar cracks that occur fairly regularly throughout a rock body.

tabular intrusions   Sheet intrusions that are planar and of roughly uniform thickness.

tachylite   Mafic (basaltic) volcanic glass.

Taconic orogeny   A convergent mountain-building event that took place around 400 million years ago, in which a volcanic island arc collided with eastern North America.

tailings pile   A pile of waste rock from a mine.

talus   A sloping apron of fallen rock along the base of a cliff.

tar   Hydrocarbons that exist in solid form at room temperature.

tarn   A lake that forms at the base of a cirque on a glacially eroded mountain.

tar sand   Sandstone reservoir rock in which less viscous oil and gas molecules have either escaped or been eaten by microbes, so that only tar remains.

taxonomy   The study and classification of the relationships among different forms of life.

tectonic foliation   A planar fabric, such as cleavage, schistocity, or gneissic banding, that develops in rocks; caused by compression or shearing during deformation (e.g., during mountain building).

temperate glacier   A glacier that exists under climatic conditions that allow it to partly melt during part of the year.

temperature   A measure of the hotness or coldness of a material.

tension   A stress that pulls on a material and could lead to stretching.

tephra   Unconsolidated accumulations of pyroclastic grains.

terminal moraine   The end moraine at the farthest limit of glaciation.

terminator   The boundary between the half of the Earth that has daylight and the half experiencing night.

terrace   The elevated surface of an older floodplain into which a younger floodplain had cut down.

terrestrial planets   Planets that are of comparable size and character to the Earth and consist of a metallic core surrounded by a rock mantle.

thalweg   The deepest part of a stream's channel.

theory   A scientific idea supported by an abundance of evidence that has passed many tests and failed none.

theory of evolution by natural selection   The idea that species change over time, new species appear, and old species disappear, due to the survival of the fittest.

theory of plate tectonics   The theory that the outer layer of the Earth (the lithosphere) consists of separate plates that move with respect to one another.

thermal energy   The total kinetic energy in a material due to the vibration and movement of atoms in the material.

thermal metamorphism   Metamorphism caused by heat conducted into country rock from an igneous intrusion.

thermocline   A boundary between layers of water with differing temperatures.

thermohaline circulation   The rising and sinking of water driven by contrasts in water density, which is due in turn to differences in temperature and salinity; this circulation involves both surface and deep-water currents in the ocean.

thermosphere   The outermost layer of the atmosphere, containing very little gas.

thin section   A 3/100-mm-thick slice of rock that can be examined with a petrographic microscope.

thin-skinned deformation   A distinctive style of deformation characterized by displacement on faults that terminate at depth along a subhorizontal detachment fault.

thrust fault   A gently dipping reverse fault; the hanging-wall block moves up the slope of the fault.

thunder   The audible clap or rumble produced due to the sudden heating of air by the passage of a lightning bolt.

thunderstorm A period of intense wind and rainfall, accompanied by lightning and thunder.

tidal bore A visible wall of water that moves toward shore with the rising tide in quiet waters.

tidal flat A broad, nearly horizontal plain of mud and silt, exposed or nearly exposed at low tide but totally submerged at high tide.

tidal power Energy produced by the daily rise and fall of the tides; people can utilize this energy, for example, by damming a bay or estuary, so that water passes through turbines when the tide changes.

tidal range The difference in sea level between high tide and low tide at a given point.

tide The daily rising or falling of sea level at a given point on the Earth.

tide-generating force The force, caused in part by the gravitational attraction of the Sun and Moon and in part by the centrifugal force created by the Earth's spin, that generates tides.

tidewater glacier A glacier that has entered the sea along a coast.

till A mixture of unsorted mud, sand, pebbles, and larger rocks deposited by glaciers.

tillite A rock formed from hardened ancient glacial deposits and consisting of larger clasts distributed through a matrix of sandstone and mudstone.

toe (terminus) The leading edge or margin of a glacier.

tombolo A narrow ridge of sand that links a sea stack to the mainland.

topographic map A map that uses contour lines to represent variations in elevation.

topographic profile A line representing the intersection of the land surface with an imaginary vertical plane at a locality.

topography Variations in elevation.

topsoil The top soil horizons, which are typically dark and nutrient-rich.

tornado A near-vertical, funnel-shaped cloud in which air rotates extremely rapidly around the axis of the funnel.

tornado swarm Dozens of tornadoes produced by the same storm.

tornado warning A statement that conditions are appropriate for tornado formation, and that a tornado has been sighted in the vicinity.

tower karst A karst landscape in which steep-sided residual bedrock towers remain between sinkholes.

trace fossil Fossilized imprints or debris that an organism leaves behind while moving on or through sediment; examples include footprints, burrows, and fecal matter.

trade winds Fairly steady breezes on the surface of the Earth representing the flow of air at the base of the Hadley cells.

transform fault A fault marking a transform plate boundary; along mid-ocean ridges, transform faults are the actively slipping segment of a fracture zone between two ridge segments.

transform boundary A boundary at which one lithosphere plate slips laterally past another.

transgression The inland migration of shoreline resulting from a rise in sea level.

transition zone The middle portion of the mantle, from 400 to 670 km deep, in which there are several jumps in seismic velocity.

transpiration The release of moisture as a metabolic by-product.

transverse dune A simple, wave-like dune that appears when enough sand accumulates for the ground surface to be completely buried, but only moderate winds blow.

trap A subsurface configuration of seal rocks and structures that keep oil and/or gas underground, so it doesn't seep out at the surface.

travel time The time that it takes for a seismic wave to travel from the focus of an earthquake to a seismometer along a given ray path.

travel-time curve A graph that plots the time since an earthquake began on the vertical axis and the distance to the epicenter on the horizontal axis.

travertine A rock composed of crystalline calcium carbonate (CaCO3) formed by chemical precipitation from groundwater that has seeped out at the ground surface.

tree ring A growth ring visible on a cross section of a tree.

trellis drainage network A drainage system that develops across a landscape of parallel valleys and ridges so that major tributaries flow down the valleys and join a trunk stream that cuts through the ridge; the resulting map pattern resembles a garden trellis.

trench A deep, elongate trough bordering a volcanic arc; a trench defines the trace of a convergent plate boundary.

triangulation The method for determining the map location of a point from knowing the distance between that point and three other points; this method is used to locate earthquake epicenters.

tributary A smaller stream that flows into a larger stream.

triple junction A point where three lithosphere plate boundaries intersect.

tropical cyclone A large spiral-shaped rotating storm that forms over the ocean in tropical latitudes; the categories includes hurricanes, typhoons, and cyclones.

tropical depression A tropical storm with winds reaching up to 61 km per hour; such storms develop from tropical disturbances, and may grow to become hurricanes.

tropical disturbance Cyclonic winds that develop in the tropics.

tropopause The temperature pause marking the top of the troposphere.

troposphere The lowest layer of the atmosphere, where air undergoes convection and where most wind and clouds develop.

truncated spur A spur (elongate ridge between two valleys) whose end was eroded off by a glacier.

trunk stream The single larger stream into which an array of tributaries flow.

tsunami A large wave along the sea surface triggered by an earthquake or large submarine slump.

tuff A pyroclastic igneous rock composed of volcanic ash and fragmented pumice, formed when accumulations of the debris cement together.

tundra A cold, treeless region of land at high latitudes, supporting only species of shrubs, moss, and lichen capable of living on permafrost.

turbidite A graded bed of sediment built up at the base of a submarine slope and deposited by turbidity currents.

turbidity current A submarine avalanche of sediment and water that speeds down a submarine slope.

turbulence The chaotic twisting, swirling motion in flowing fluid.

typhoon The equivalent of a hurricane in the western Pacific Ocean.

ultimate base level Sea level; the level below which a trunk stream cannot cut.

ultramafic A term used to describe igneous rocks or magmas that are rich in iron and magnesium and very poor in silica.

unconfined aquifer An aquifer that intersects the surface of the Earth.

unconformity A boundary between two different rock sequences representing an interval of time during which new strata were not deposited and/or were eroded.

unconsolidated Consisting of unattached grains.

unconventional reserve A supply of oil or gas that cannot be easily pumped; it includes forms of hydrocarbons that are too viscous to pump, or occur in impermeable rock; examples include tar sand, oil shale, shale oil, and shale gas.

undercutting Excavation at the base of a slope that results in the formation of an overhang.

underground mine A mine that consists of tunnels and shafts underground, rather than of an open pit.

undersaturated A term used to describe a solution capable of holding more dissolved ions.

uniformitarianism (the principle of) (see principle of uniformitarianism)

Universe   All of space and all the matter and energy within it.

unsaturated zone (vadose zone)   The region of the subsurface above the water table.

unstable air   Air that is significantly warmer than air above and has a tendency to rise quickly.

unstable ground   Land capable of slumping or slipping downslope in the near future.

unstable slope   A slope on which sliding will likely happen.

updraft   Upward-moving air.

uplift   (n. geology) The upward vertical movement of the Earth's surface.

upper mantle   The uppermost section of the mantle, reaching down to a depth of 400 km.

upwelling   The upward flow of air or water.

upwelling zone   A place where deep water rises in the ocean, or where hot magma rises in the asthenosphere.

U-shaped valley   A steep-walled valley shaped by glacial erosion into the form of a U.

vacuum   Space that contains very little matter in a given volume (e.g., a region in which air has been removed).

valley   A trough with sloping walls, cut into the land by a stream.

valley glacier   A river of ice that flows down a mountain valley.

Van Allen radiation belts   Belts of solar wind particles and cosmic rays that surround the Earth, trapped by the Earth's magnetic field.

van der Waals bonding   The relatively weak attachment of two elements or molecules due to their polarity and not due to covalent or ionic bonding.

varve   A pair of thin layers of glacial lake-bed sediment, one consisting of silt brought in during the spring floods and the other of clay deposited during the winter when the lake's surface freezes over and the water is still.

vascular plant   A plant with woody tissue and seeds and veins for transporting water and food.

vein   A seam of minerals that forms when dissolved ions carried by water solutions precipitate in cracks.

vein deposit   A hydrothermal deposit in which the ore minerals occur in veins that fill cracks in preexisting rocks.

velocity-versus-depth curve   A graph that shows the variation in the velocity of seismic waves with increasing depth in the Earth.

vent   In the context of discussing volcanoes, it is the location where lava, ash, or gas exit the ground.

ventifact (faceted rock)   A desert rock whose surface has been faceted by the wind.

vesicles   Open holes in igneous rock formed by the preservation of bubbles in magma as the magma cools into solid rock.

viscosity   The resistance of material to flow.

volatiles (volatile materials)   Elements or compounds such as H_2O and CO_2 that evaporate at relatively low temperatures and can exist in gaseous forms at the Earth's surface.

volatility   A specification of the ease with which a material evaporates.

volcanic agglomerate   An accumulation consisting dominantly of volcanic bombs and other relatively large chunks of igneous material.

volcanic arc   A curving chain of active volcanoes formed adjacent to a convergent plate boundary.

volcanic island arc   A volcanic arc formed on oceanic lithosphere.

volcanic ash   Tiny glass shards formed when a fine spray of exploded lava freezes instantly upon contact with the atmosphere.

volcanic bomb   A large piece of pyroclastic debris thrown into the atmosphere during a volcanic eruption.

volcanic breccia   A pyroclastic igneous rock that consists of fragments of volcanic debris, which either fall through the air and accumulate, or form when solidifying lava breaks up during flow.

volcanic debris flow   A mixture of water and pyroclastic debris that moves downslope like wet concrete.

volcanic eruption   An event during which lava or pyroclastic debris comes out of a volcanic vent.

volcanic gas   Elements or compounds that bubble out of magma or lava in gaseous form.

volcanic hazard-assessment map   A map delineating areas that lie in the path of potential lava flows, lahars, debris flows, or pyroclastic flows of an active volcano.

volcanic island arc   The volcanic island chain that forms on the edge of the overriding plate where one oceanic plate subducts beneath another oceanic plate.

volcaniclastic deposit   An accumulation of large quantities of fragmental igneous material (including both pyroclastic debris, and water-transported debris).

volcaniclastic rock   A material composed of cemented-together grains of volcanic material; it includes both pyroclastic rocks and rocks formed from accumulations of water-transported volcanic debris.

volcano   (1) A vent from which melt from inside the Earth spews out onto the planet's surface; (2) a mountain formed by the accumulation of extrusive volcanic rock.

V-shaped valley   A valley whose cross-sectional shape resembles the shape of a V; the valley probably has a river running down the point of the V.

Wadati-Benioff zone   A sloping band of seismicity defined by intermediate- and deep-focus earthquakes that occur in the downgoing slab of a convergent plate boundary.

wadi   The name used in the Middle East and North Africa for a dry wash.

wall rock   The rock surrounding an igneous intrusion.

warm front   A front in which warm air rises slowly over cooler air in the atmosphere.

waste rock   Rock dislodged by mining activity yet containing a non-economic concentration of ore minerals.

waterfall   A place where water drops over an escarpment.

water gap   An opening in a resistant ridge where a trunk river has cut through the ridge.

watershed   The region that collects water that feeds into a given drainage network.

water table   The boundary, approximately parallel to the Earth's surface, that separates substrate in which groundwater fills the pores from substrate in which air fills the pores.

wave   A disturbance that transmits energy from one point to another in the form of periodic motions.

wave base   The depth, approximately equal in distance to half a wavelength in a body of water, beneath which there is no wave movement.

wave-cut notch   A notch in a coastal cliff cut out by wave erosion.

wave-cut platform (wave-cut bench)   A platform of rock, cut by wave erosion, at the low-tide line that was left behind a retreating cliff.

wave erosion   The combined effects of the shattering, wedging, and abrading of a cliff face by waves and the sediment they carry.

wave front   The boundary between the region through which a wave has passed and the region through which it has not yet passed.

wavelength   The horizontal difference between two adjacent wave troughs or two adjacent crests.

wave refraction (ocean)   The bending of waves as they approach a shore so that their crests make no more than a 5° angle with the shoreline.

weather   Local-scale conditions as defined by temperature, air pressure, relative humidity, and wind speed.

weathered rock   Rock that has reacted with air and/or water, or has undergone mechanical disaggregation, at or near the Earth's surface.

weathering   The processes that break up and corrode solid rock, eventually transforming it into sediment.

weather system A specific set of weather conditions, reflecting the configuration of air movement in the atmosphere, that affects a region for a period of time.

welded tuff Tuff formed by the welding together of hot volcanic glass shards at the base of pyroclastic flows.

well A hole in the ground dug or drilled in order to obtain water.

Western Interior Seaway A north-south-trending seaway that ran down the middle of North America during the Late Cretaceous Period.

wet-bottom (temperate) glacier A glacier with a thin layer of water at its base, over which the glacier slides.

wetted perimeter The area in which water touches a stream channel's walls.

wind The horizontal component of air flow.

wind abrasion The grinding away at surfaces in a desert by windblown sand and dust.

wind gap An opening through a high ridge that developed earlier in geologic history by stream erosion, but that is now dry.

xenolith A relict of wall rock surrounded by intrusive rock when the intrusive rock freezes.

yardang A mushroom-like column with a resistant rock perched on an eroding column of softer rock; created by wind abrasion in deserts where a resistant rock overlies softer layers of rock.

yazoo stream A small tributary that runs parallel to the main river in a floodplain because the tributary is blocked from entering the main river by levees.

Younger Dryas An interval of cooler temperatures that took place 4,500 years ago during a general warming/glacier-retreat period.

zeolite facies The metamorphic facies just above diagenetic conditions, under which zeolite minerals form.

zone of ablation The area of a glacier in which ablation (melting, sublimation, calving) subtracts from the glacier.

zone of accumulation (1) The layer of regolith in which new minerals precipitate out of water passing through, thus leaving behind a load of fine clay; (2) the area of a glacier in which snowfall adds to the glacier.

zone of aeration *See* Unsaturated zone.

zone of leaching The layer of regolith in which water dissolves ions and picks up very fine clay; these materials are then carried downward by infiltrating water.

Photo Credits

by-sa/3.0/deed.en; *p. 130 (bottom right):* Stephen Marshak; *p. 132 (bottom left):* © 1996 Jeff Scovil; *p. 134 (a):* Richard P. Jacobs/ JLM Visuals; *p. 134 (b):* Richard P. Jacobs/ JLM Visuals; *p. 134 (c):* Colorado School of Mines Geology Museum. Photo: Stephen Marshak; *p. 134 (d):* Richard P. Jacobs/ JLM Visuals; *p. 134 (e):* Colorado School of Mines Geology Museum. Photo: Stephen Marshak; *p. 134 (f):* Stephen Marshak; *p. 134 (g):* Richard P. Jacobs/ JLM Visuals; *p. 134 (h):* GIPhotoStock/Science Source; *p. 135 (top left):* Richard P. Jacobs/ JLM Visuals; *p. 135 (top center):* © 1996 Jeff Scovil; *p. 135 (top right):* Marli Miller/Visuals Unlimited, Inc.; *p. 135 (bottom left):* Richard P. Jacobs/ JLM Visuals; *p. 135 (bottom center):* Richard P. Jacobs/ JLM Visuals; *p. 135 (bottom right):* Thomas Hunn/ Visuals Unlimited, Inc.; *p. 136 (top left):* On display at the Harvard Museum of Natural History, courtesy Mineralogical and Geological Museum, Photo by Stephen Marshak © President & Fellows, Harvard College. Photo by Stephen Marshak; *p. 136 (top right):* Stephen Marshak; *p. 136 (bottom left):* Ann Bryant www.Geology.com; *p. 136 (bottom right):* Arco Images GmbH/Alamy; *p. 137:* Javier Trueba/MSF/Science Source; *p. 138 (top):* Farbled/Dreamstime.com; *p. 138 (inset):* Dennis Kunkel/Science Source; *p. 138 (bottom left):* Huguette Roe/Dreamstime.com; *p. 138 (bottom right):* Construction Photography/Alamy; *p. 140 (left):* Stephen Marshak; *p. 140 (right):* Stephen Marshak; *p. 141 (left):* Jason Pineau/Getty Images; *p. 141 (right):* Ken Lucas/Visuals Unlimited; *p. 142 (bottom):* The Hope Diamond/Smithsonian Institution, Washington DC, USA/ Bridgeman Images; *p. 142 (both):* Images provided by Google Earth mapping services/NASA, © DigitalGlobe, © Terra Metrics, © GeoEye, © Europa Technologies, Copyright 2014; *p. 142 (both):* Images provided by Google Earth mapping services/NASA, © DigitalGlobe, © Terra Metrics, © GeoEye, © Europa Technologies, Copyright 2014; *p. 144 (left):* Albert Copley/Visuals Unlimited; *p. 144 (inset):* Stephen J. Krasemann/Science Source; *p. 144 (right):* Miljko/Getty Images; *p. 145:* Stephen Marshak; *p. 146:* Image provided by Google Earth mapping services/NASA, © DigitalGlobe, © Terra Metrics, © GeoEye, © Europa Technologies, Copyright 2018; *p. 147:* © 1996 Jeff Scovil.

INTERLUDE A

Page 148: Stephen Marshak; *p. 149 (top):* Stephen Marshak; *p. 149 (bottom):* Stephen Marshak; *p. 150 (top left):* Stephen Marshak; *p. 150 (top center):* Courtesy David W. Houseknecht, USGS; *p. 150 (bottom left):* sciencephotos/ Alamy; *p. 150 (bottom center):* Courtesy of Kent Ratajeski, Dept. of Geology and Geophysics, U of Wisconsin, Madison; *p. 152 (all):* Stephen Marshak *153 (all):* Stephen Marshak; *p. 155 (both):* Stephen Marshak; *p. 156 (all):* Stephen Marshak *157 (top right):* Stephen Marshak; *p. 157 (inset):* Scenics & Science/ Alamy *158 (top left):* Stephen Marshak; *p. 158 (top right):* Courtesy of Joseph H. Reibenspies, Texas A&M University; *p. 158 (bottom):* Stephen Marshak; *p. 159:* Stephen Marshak.

CHAPTER 6

Page 160: Stephen Marshak; *p. 162 (top left):* Liysa/Pacific Stock/Agefotostock; *p. 162 (top right):* J.D. Griggs/ U.S. Geological Survey; *p. 162 (center):* Stephen Marshak; *p. 163 (top right):* Stephen Marshak; *p. 163 (center right):* Stephen Marshak; *p. 163 (bottom left):* Images provided by Google Earth mapping services/NASA, © DigitalGlobe, © Terra Metrics, © GeoEye, © Europa Technologies, Copyright 2014; *p. 166:* USGS; *p. 171 (top left):* USGS; *p. 171 (top right):* Stephen Marshak; *p. 171 (center left):* Stephen Marshak; *p. 171 (center right):* Stephen Marshak; *p. 171 (bottom right):* Stephen Marshak; *p. 172 (both):* Stephen Marshak; *p. 172:* Stephen Marshak; *p. 173 (top left):* Stephen Marshak; *p. 173 (bottom right):* Images provided by Google Earth mapping services/ NASA, © DigitalGlobe, © Terra Metrics, © GeoEye, © Europa Technologies, Copyright 2014; *p. 174:* Stephen Marshak; *p. 175 (top right):* Stephen Marshak; *p. 176 (both):* Stephen Marshak; *p. 176:* Stephen Marshak; *p. 177 (bottom left):* Stephen Marshak; *p. 177 (top left):* Dr. Kent Ratajeski; *p. 177 (bottom center):* Mark A. Schneider/Science Source; *p. 177 (top center):* Omphacite. 2006. Wikimedia, http://en.wikipedia.org/wiki/Public_domain; *p. 177 (bottom right):* Doug Sokell/Visuals Unlimited; *p. 177 (top right):* Dr. Matthew Genge; *p. 180 (top left):* geoz/ Alamy; *p. 180 (top right):* Stephen Marshak; *p. 180 (center left):* Mark A. Schneider/Science Source; *p. 180 (center right):* Siim Sepp/ Alamy; *p. 180 (bottom left):* Joyce Photographics/Science Source; *p. 180 (bottom right):* Wally Eberhart/Getty Images; *p. 181 (top left):* Stephen Marshak; *p. 181 (top right):* Stephen Marshak; *p. 181 (bottom left):* Stephen Marshak; *p. 181 (bottom right):* Stephen Marshak; *p. 183:* Images provided by Google Earth mapping services/NASA, © DigitalGlobe, © Terra Metrics, © GeoEye, © Europa Technologies, Copyright 2014; *p. 183 (inset):* Images provided by Google Earth mapping services/NASA, © DigitalGlobe, © Terra Metrics, © GeoEye, © Europa Technologies, Copyright 2014; *p. 184:* Stephen Marshak;

p. 185: Stephen Marshak; *p. 186 (bottom left):* Stephen Marshak; *p. 186 (bottom right):* Stephen Marshak; *p. 187:* Stephen Marshak; *p. 188:* Images provided by Google Earth mapping services/NASA, © DigitalGlobe, © Terra Metrics, © GeoEye, © Europa Technologies, Copyright 2018; *p. 189:* Images provided by Google Earth mapping services/NASA, © DigitalGlobe, © Terra Metrics, © GeoEye, © Europa Technologies, Copyright 2018.

INTERLUDE B

Page 190: Stephen Marshak; *p. 192 (top left):* Stephen Marshak; *p. 192 (top center):* Stephen Marshak; *p. 192 (top right):* Emma Marshak; *p. 192 (bottom):* Stephen Marshak; *p. 193 (all):* Stephen Marshak; *p. 194 (both):* Stephen Marshak; *p. 195 (all):* Stephen Marshak; *p. 196 (all):* Stephen Marshak; *p. 197 (top):* Stephen Marshak; *p. 197 (center):* Carlo Giovanella, UBC, 1997-2005; *p. 197 (bottom):* British Geology Survey; *p. 198 (bottom):* Stephen Marshak; *p. 199 (both):* Stephen Marshak; *p. 200 (both):* Stephen Marshak; *p. 201 (all):* Stephen Marshak; *p. 202 (both):* Stephen Marshak; *p. 203:* Stephen Marshak; *p. 206:* U.S. Department of Agriculture; *p. 207 (top):* Stephen Marshak; *p. 207 (bottom):* Stephen Marshak; *p. 208 (all):* Stephen Marshak.

CHAPTER 7

Page 210: Stephen Marshak; *p. 212:* Stephen Marshak; *p. 213 (top):* Stephen Marshak; *p. 213 (center left):* Stephen Marshak; *p. 213 (bottom):* Stephen Marshak; *p. 213 (inset):* Stephen Marshak; *p. 216 (all):* Stephen Marshak; *p. 217 (top left):* Stephen Marshak; *p. 217 (top right):* Stephen Marshak; *p. 217 (center left):* Stephen Marshak; *p. 217 (center right):* Stephen Marshak; *p. 217 (bottom left):* Stephen Marshak; *p. 217 (bottom right):* D. G. F. Long; *p. 219 (all):* Stephen Marshak; *p. 220 (both):* Stephen Marshak; *p. 221 (top center):* Visuals Unlimited; *p. 221 (top right):* Stephen Marshak; *p. 221 (bottom right):* Jason Bye/ Alamy; *p. 222 (top left):* Stephen Marshak; *p. 222 (top right):* Photo by Yukinobu Zengame, 2005. http://creative-commons.org/licenses/ by/2.0/deed.en; *p. 222 (bottom left):* Stephen Marshak; *p. 222 (bottom right):* Stephen Marshak; *p. 223 (all):* Stephen Marshak; *p. 224:* Images provided by Google Earth mapping services/NASA, © DigitalGlobe, © Terra Metrics, © GeoEye, © Europa Technologies, Copyright 2014; *p. 225 (both):* Stephen Marshak; *p. 225:* Stephen Marshak; *p. 226 (top left):* Stephen Marshak; *p. 226 (top right):* All Canada Photos/ Alamy; *p. 226 (bottom):* 1980 Grand Canyon Natural History Association; *p. 227 (both):* Stephen Marshak; *p. 228 (top):* Imagina Photography/ Alamy; *p. 228 (bottom left):* Stephen Marshak; *p. 228 (bottom right):* Stephen Marshak; *p. 229 (inset):* Marli Miller/Visuals Unlimited; *p. 229 (top right):* Stephen Marshak; *p. 230 (all):* Stephen Marshak; *p. 231 (top left):* Emma Marshak; *p. 231 (top center):* Stephen Marshak; *p. 231 (top right):* Marli Miller/Visuals Unlimited; *p. 231 (bottom right):* Stephen Marshak; *p. 231 (bottom left):* Stephen Marshak; *p. 231 (bottom center):* PolarTREC/ Arctic Research Consortium of the United States (ARCUS); *p. 232:* Stephen Marshak; *p. 235 (top):* Stephen Marshak; *p. 235 (bottom)* Images provided by Google Earth mapping services/NASA, © DigitalGlobe, © Terra Metrics, © GeoEye, © Europa Technologies, Copyright 2014; *p. 236 (top right):* David Wall/Alamy Stock Photo; *p. 236 (bottom left):* The Natural History Museum/ Alamy Stock Photo; *p. 236 (bottom right):* G. R. 'Dick' Roberts © Natural Sciences Image Library; *p. 239:* Stephen Marshak; *p. 240:* Images provided by Google Earth mapping services/NASA, © DigitalGlobe, © Terra Metrics, © GeoEye, © Europa Technologies, Copyright 2018.

CHAPTER 8

Page 242: Stephen Marshak; *p.244 (top):* Stephen Marshak; *p. 244 (center left):* Stephen Marshak; *p. 244 (center right):* Visuals Unlimited; *p. 244 (bottom left):* Stephen Marshak; *p. 244 (bottom right):* Stephen Marshak; *p. 245 (top left):* Stephen Marshak; *p. 245 (top right):* Stephen Marshak; *p. 245 (bottom left):* Visuals Unlimited; *p. 245 (bottom center):* Stephen Marshak; *p. 245 (bottom right):* Kurt Freihauf; *p. 249:* Stephen Marshak; *p. 250 (top left):* Emma Marshak; *p. 250 (top right):* Stefano Clemente/Alamy Stock Photo; *p. 250 (bottom):* Stephen Marshak; *p. 251 (top):* Stephen Marshak; *p. 251 (center):* Stephen Marshak *251 (bottom):* Stephen Marshak *52 (top):* Dr. Jane A. Gilotti *252 (center):* Stephen Marshak; *p. 252 (bottom):* Stephen Marshak; *p. 253:* Stephen Marshak; *p. 254 (all):* Stephen Marshak; *p. 261 (all):* Stephen Marshak; *p. 263:* Dr. Terry Wright; *p. 263 (left):* Images provided by Google Earth mapping services/NASA, © DigitalGlobe, © Terra Metrics, © GeoEye, © Europa Technologies, Copyright 2014; *p. 263 (right):* Images provided by Google Earth mapping services/NASA, © DigitalGlobe, © Terra Metrics, © GeoEye, © Europa Technologies, Copyright 2014; *p. 267 (left):* Stephen Marshak; *p. 267 (right):* Stephen Marshak; *p. 268:* Images provided by Google Earth mapping services/NASA, © DigitalGlobe, © Terra Metrics, © GeoEye, © Europa Technologies, Copyright 2018.

INTERLUDE C

Page 270 (all): Stephen Marshak.

PART OPENER 3

Pages 278-279: Stephen Marshak.

CHAPTER 9

Page 280: Westend61 GmbH/Alamy Stock Photo; *p. 281 (top):* Bridgeman Art Library; *p. 281 (bottom):* Peter V. Bianchi/National Geographic/Getty Images; *p. 282 (top left):* Images provided by Google Earth mapping services/NASA, © DigitalGlobe, © Terra Metrics, © GeoEye, © Europa Technologies, Copyright 2014; *p. 282 (top right):* Stephen Marshak; *p. 282 (center left):* Stephen Marshak; *p. 282 (center right):* Jack Repcheck; *p. 282 (bottom left):* Stephen Marshak; *p. 282 (bottom center):* Jack Repcheck; *p. 282 (bottom right):* Jack Repcheck; *p. 283 (top):* Images provided by Google Earth mapping services/NASA, © DigitalGlobe, © Terra Metrics, © GeoEye, © Europa Technologies, Copyright 2014; *p. 283 (bottom):* Marli Miller/Visuals Unlimited; *p. 285 (top left):* Robert Francis/Agefotostock; *p. 285 (top right):* Stephen Marshak; *p. 285 (center left):* USGS; *p. 285 (center right):* Stephen Marshak; *p. 285 (bottom left):* Stephen Marshak; *p. 285 (bottom right):* Stephen Marshak; *p. 286 (top left):* Stephen Marshak; *p. 286 (center left):* Thomas Hallstein/Alamy; *p. 286 (bottom left):* Stephen Marshak; *p. 286 (top right):* Stephen Marshak; *p. 287 (top left):* Stephen Marshak; *p. 287 (top right):* AP Photo; *p. 287 (bottom left):* Stephen Marshak; *p. 287 (bottom right):* Stephen Marshak; *p. 288 (left):* Alvaro Vidal/AFP/Getty Images; *p. 288 (top right):* Photo by Suzanne MacLachlan, British Ocean Sediment Core Research Facility, National Oceanography Centre, Southampton; *p. 288 (center left):* Stephen Marshak; *p. 288 (center right):* Stephen Marshak; *p. 288:* Images provided by Google Earth mapping services/NASA, © DigitalGlobe, © Terra Metrics, © GeoEye, © Europa Technologies, Copyright 2014; *p. 289 (top left):* Stephen Marshak; *p. 289 (top right):* Stephen Marshak; *p. 289 (center left):* USGS; *p. 289 (center right):* Stephen Marshak; *p. 289 (bottom left):* Anthony Phelps/Reuters/Newscom; *p. 289 (bottom right):* Stephen Marshak; *p. 290 (both):* Stephen Marshak; *p. 291 (top):* Sunshine Pics/ Alamy; *p. 291 (bottom):* USGS 292 (top): John J. Bangma/Getty Images 292 (bottom):* Marli Miller/Visuals Unlimited; *p. 293 (top left):* Marli Miller/Visuals Unlimited; *p. 293 (top right):* Google EarthImage Landsat/Copernicus/Data LDEO-Columbia, NSF, NOAA/Data SIO, NOAA, U.S. Navey, NGA, GEBCO; *p. 293 (inset):* Robert Harding World Imagery/Alamy; *p. 295 (top left):* USGS; *p. 295 (top right):* George Dimijian/Science Source; *p. 295 (bottom):* Stephen Marshak; *p. 298 (top left):* Westend61 GmbH/Alamy; *p. 298 (top right):* © Tom Pfeiffer/ www.volcanodiscovery.com; *p. 298 (bottom left):* AFP/Getty Images; *p. 298 (bottom right):* USGS; *p. 299 (left):* USGS; *p. 299 (right):* AFP/Getty Images; *p. 301 (top):* Stephen Marshak; *p. 301 (bottom):* Lyn (topinka/USGS; *p. 303 (both):* Images provided by Google Earth mapping services/NASA, © DigitalGlobe, © Terra Metrics, © GeoEye, © Europa Technologies, Copyright 2014; *p. 304:* Chris Johns/Getty Images; *p. 305 (center right):* Arctic Images/Alamy; *p. 305:* Images provided by Google Earth mapping services/NASA, © DigitalGlobe, © Terra Metrics, © GeoEye, © Europa Technologies, Copyright 2014; *p. 306* Stephen Marshak; *p. 307:* Bob Rauber; *p. 309 (a):* USGS; *p. 309 (b):* Vittoriano Rastelli/CORBIS/Corbis via Getty Images; *p. 309 (c):* AP Photo; *p. 309 (d):* Roy Whiddon; *p. 309 (e):* Stocktrek Images, Inc./Alamy Stock Photo; *p. 309 (f):* REUTERS/Cristobal Saavedra/Newscom; *p. 309 (g)* Magnus T. Gudmundson, University of Iceland; *p. 309 (h):* USGS; *p. 311 (left):* THIERRY ORBAN/Sygma via Getty Images; *p. 311 (right):* Photo by Peter Turnley/Corbis/VCG via Getty Images; *p. 312:* Stephen Marshak; *p. 313:* Jennifer L. Lewicki, USGS; *p. 314 (left):* Vittoriano Rastelli/CORBIS/Corbis via Getty Images ; *p.314 (right):* Sigurgeir Jonasson/Frank Lane Picture Agency/Corbis/Getty Images; *p. 316:* NASA; *p. 317 (top left):* Gail Mooney/Corbis/VCG/Getty Images; *p. 317 (right):* Santorini Caldera. Photo by Steve Jurvetson. 2012. http://creativecommons.org/licenses/by/2.0/deed.en; *p. 317 (bottom left):* Earth Sciences and Image Analysis Laboratory, NASA; *p. 318 (all):* NASA; *p. 319:* Francisco Negroni/Alamy Stock Photo; *p. 320:* Images provided by Google Earth mapping services/NASA, © DigitalGlobe, © Terra Metrics, © GeoEye, © Europa Technologies, Copyright 2018.

CHAPTER 10

Page 322: Kyodo News via Getty Images; *p. 324 (top):* AFP/Getty Images; *p. 324: (bottom left):* JIJI Press/AFP/Getty Images, *p. 324 (bottom right):* AP Photo/Kyodo News; *p. 327 (left):* Photo courtesy of Paul "Kip" Otis-Diehl, USMC, 29 Palms CA; *p. 327 (right):* Photo courtesy of Paul "Kip" Otis-Diehl, USMC, 29 Palms CA; *p. 329 (right):* UNAVCO/NSF; *p. 329 (left):* Images provided by Google Earth mapping services/NASA © DigitalGlobe, © Terra Metrics, © GeoEye, © Europa Technologies, Copyright 2014; *p. 331* Peltzer et al. (1999), Evidence of nonlinear elasticity of the Crust. Science, v. 286. Copyright © 1999, AAAS; *p. 332:* UNAVCO/NSF; *p. 335:* Inga Spence/Visuals Unlimited; *p. 344 (top)* Library of Congress/Getty Images; *p. 344 (bottom):* AP Photo/Paul Sakuma; *p. 345 (right):* Omar Havana/Getty Images; *p. 345 (left):* Anna Kompanek/CIPE; *p. 348:* New Madrid earthquake woodcut from Deven's Our First Century (1877)/Wikimedia Commons; *p. 350 (top):* AP Photo; *p. 350; (top center)* Pacific Press Service/Alamy; *p. 350 (bottom center):* M. celebi, U.S. Geographical Survey; *p. 350 (bottom):* Piero Pomponi/Getty Images; *p. 351 (top):* Barry Lewis/Alamy; *p. 351 (center):* Bettmann/Getty Images; *p. 351 (bottom):* Charles and Emma Jean Mader, Mader Consulting Grange/Getty Images; *p. 352 (bottom left):* AP Photo/New Zealand Herald, Geoff Sloan; *p. 352 (right):* Courtesy of the National Information Service for Earthquake Engineering, PEER-NISEE, University of California, Berkeley; *p. 353 (left):* Karl V. Steinbrugge Collection, University of California, Berkeley; *p. 353 (right):* National Geophysical Data Center (NGDC) ; *p. 357 (top left):* AFP/Getty Images; *p. 357 (top right):* David Rydevik. 2004. Wikimedia http://en.wikipedia.org/wiki/Public_domain; *p. 357 (bottom right):* Ikonos images copyright Centre for Remote Imaging, Sensing and Processing, National University of Singapore and Space Imaging; *p. 358:* National Geophysical Data Center (NGDC) ; *p. 358 (left):* Images provided by Google Earth mapping services/NASA, © DigitalGlobe, © Terra Metrics, © GeoEye, © Europa Technologies, Copyright 2014; *p. 359 (top)* AP Photo/Kyodo News; *p. 359 (center):* EPA; *p. 359 (bottom):* GAMMA/Gamma-Rapho via Getty Images; *p. 361 (top left):* Reuters/Eduardo Munoz/Newscom; *p. 361 (top right)* Reuters/Eduardo Munoz/Newscom; *p. 361 (bottom left):* Images provided by Google Earth mapping services/NASA, © DigitalGlobe, © Terra Metrics, © GeoEye, © Europa Technologies, Copyright 2014; *p. 361 (bottom right):* USGS; *p. 367 (center left):* NOAA/NOA Center for Tsunami Research; *p. 367 (bottom right)* Stephen Marshak; *p. 368:* Images provided by Google Earth mapping services/NASA, © DigitalGlobe, © Terra Metrics, © GeoEye, © Europa Technologies, Copyright 2018.

INTERLUDE D

Page 370: Illinois State Geological Survey; *p. 373:* George Resch/Fundamental Photographs, NYC; *p. 376:* Jennifer Jackson, Caltech and Jay Bass; *p. 380:* Inga Spence/Visuals Unlimited; *p. 381:* Adapted/Reproduced from Naliboff and Kellogg, "(2005) Dynamic effects of a step-wise increase in thermal conductivity and viscosity in the lowermost mantle," Geophysical Research Letters, 33. Copyright 2006 by the American Geophysical Union; *p. 383 (top left):* John Q. Thompson, Courtesy of Dawson Geophysical Co. ; *p. 383 (top right):* Shell Oil Company; *p. 383 (bottom left):* Courtesy of Greg Moore, University of Hawaii and Nathan Bangs, University of Texas; *p. 383:* Courtesy of Sercel and CGG Veritas/Shell Oil Company; *p. 384 (top):* Stephen Marshak; *p. 384 (bottom left):* Figure provided courtesy F. Lemoine & J. Frawley, NASA Goddard Space Flight Center; *p. 384 (bottom right):* USGS; *p. 385:* USGS.

CHAPTER 11

Page 390: Stephen Marshak; *pp. 392-393:* NOAA/E(TOPO1382; *p. 394 (all):* Stephen Marshak; *p. 395 (all):* Stephen Marshak; *p. 397 (top):* Stephen Marshak; *p. 397 (center):* Stephen Marshak; *p. 397 (right):* Images provided by Google Earth mapping services/NASA, © DigitalGlobe, © Terra Metrics, © GeoEye, © Europa Technologies, Copyright 2014; *p. 399 (left):* Russ Bishop/Alamy Stock Photo; *p. 399 (center):* Stephen Marshak; *p. 399 (right):* Stephen Marshak; *p. 401 (both):* Stephen Marshak; *p. 403 (top):* Stephen Marshak; *p. 403 (center):* USGS; *p. 403 (bottom):* Lloyd Cluff/Getty Images; *p. 404 (all):* Stephen Marshak; *p. 405 (center left):* © Doug Sherman; *p. 405 (bottom left):* Stephen Marshak; *p. 407 (all):* Stephen Marshak; *p. 408 (all):* Stephen Marshak; *p. 409:* Stephen Marshak; *p. 409 (right):* Images provided by Google Earth mapping services/NASA, © DigitalGlobe, © Terra Metrics, © GeoEye, © Europa Technologies, Copyright 2014; *p. 410* Stephen Marshak; *p. 411 (top)* Stephen Marshak; *p. 411 (bottom left):* Images provided by Google Earth mapping services/NASA, © DigitalGlobe, © Terra Metrics, © GeoEye, © Europa Technologies, Copyright 2014; *p. 411 (bottom right):* Stephen Marshak; *p. 412 (top left):* Images provided by Google Earth mapping services/NASA, © DigitalGlobe, © Terra Metrics, © GeoEye, © Europa Technologies, Copyright 2014; *p. 412 (bottom left):* Wikimedia Commons; *p. 412 (bottom right):* https://readtiger.com/wkp/en/Mont_Blanc;

Stephen Marshak; *p. 567 (inset):* Layne Kennedy/Getty Images; *p. 567 (bottom right):* Stephen Marshak; *p. 568 (right):* Science VU-ASIS/Visual Unlimited; *p. 569 (top left):* Richard P. Jacobs/JLM Visuals; *p. 569 (top right):* Stephen Marshak; *p. 569 (bottom left):* Stephen Marshak; *p. 569 (bottom right):* Stephen Marshak; *p. 572 (top left):* Stephen Marshak; *p. 572 (top right):* K. L. Smith Jr. (MBARI) and S. E. Beaulieu (WHOI) ; *p. 573 (center):* Art Directors & TRIP/Alamy; *p. 573 (bottom right):* Hemis/Alamy; *p. 574:* Images provided by Google Earth mapping services/NASA, © DigitalGlobe, © Terra Metrics, © GeoEye, © Europa Technologies, Copyright 2014; *p. 575 (top left):* Robert W. Gerling/Visuals Unlimited; *p. 575 (top right):* USGS; *p. 575 (center right):* Stephen Marshak; *p. 575 (bottom right):* Stephen Marshak; *p. 577:* Rodrigo Arangua/AFP/Getty Images; *p. 578 (top left):* Richard P. Jacobs/JLM Visuals; *p. 578 (top right):* Stephen Marshak; *p. 578 (bottom left):* Stephen Marshak; *p. 578 (bottom right):* Stephen Marshak; *p. 579 (left):* Stephen Marshak; *p. 579 (center left):* Stephen Marshak; *p. 579 (center right):* Stephen Marshak; *p. 579 (right):* Stephen Marshak; *p. 582 (left):* Stephen Marshak; *p. 582 (right):* Bloomberg/Getty Images; *p. 585 (top):* Stephen Marshak; *p. 585 (bottom left):* Doug Sokell/Visuals Unlimited; *p. 585 (bottom right):* Bill Brooks/Alamy Stock Photo; *p. 586:* Images provided by Google Earth mapping services/NASA, © DigitalGlobe, © Terra Metrics, © GeoEye, © Europa Technologies, Copyright 2018.

PART OPENER 6

Pages 588-589: Stephen Marshak.

INTERLUDE F

Page 590: Stephen Marshak; *p. 592 (all):* Stephen Marshak; *p. 594 (top left):* KennethTownsend (artist); http://www.shadedreliefarchive.com/Europe_townsend.html; *p. 594 (bottom):* NOAA; *p. 594 (top right):* JLP/NASA; *p. 595 (both):* Stephen Marshak; *p. 596 (left):* G. R. 'Dick' Roberts © Natural Sciences Image Library; *p. 596: (right):* Julie Dermansky/Corbis via Getty Images; *p. 597 (all):* Stephen Marshak; *p. 600 (left table):* Data from P.H. Gleick, Encyclopedia of climate and weather (New York: Oxford University Press, 1996); *p. 601 (top left):* NASA; *p. 601 (top right):* JSC/NASA; *p. 601 (bottom left):* Dr. David Smith, NASA Goddard Space Flight Center/MOLA Science Team; *p. 601 (bottom right):* NASA/USGS/Flagstaff; *p. 602 (top):* ESA/DLR/FU Berlin (G. Neukum); *p. 602 (center):* ESA/DLR/FU Berlin; *p. 602 (bottom):* NASA, *p. 603 (left):* NASA/JPL/Space Science Institute; *p. 603 (center):* NASA/JPL/University of Arizona; *p. 603 (right):* NASA/JHUAPL; *p. 604:* NASA/JPL/Caltech.

CHAPTER 16

Page 606: Markus Hell/Tareom Aerials; *p. 607 (top):* Lloyd Cluff/Getty Images; *p. 607 (bottom):* Lloyd Cluff/Getty Images; *p. 609 (left) (inset):* Stephen Marshak; *p. 609 (top right):* Marli Miller/Visuals Unlimited, Inc.; *p. 609 (bottom right):* George Herben Photo/Visuals Unlimited; *p. 610 (top left):* Stephen Marshak; *p. 610 (bottom left):* Stephen Marshak; *p. 610 (bottom right):* Stephen Marshak; *p. 611 (top right):* Shana Reis/EPA; *p. 611 (center right):* Cascades Volcano Observatory/USGS; *p. 611 (bottom right):* Stephen Marshak; *p. 611 (left):* Images provided by Google Earth mapping services/NASA, © DigitalGlobe, © Terra Metrics, © GeoEye, © Europa Technologies, Copyright 2014; *p. 612 (top right):* Bob Schuster, USGS; *p. 612 (right):* National Geographic Image Collection/Alamy; *p. 612 (bottom left):* Ron Varela/Ventura County Star; *p. 613 (top left):* AP Photo/Ted S. Warren; *p. 613 (bottom left):* Stephen Marshak; *p. 613 (bottom right):* Guido Alberto Rossi/Age Fotostock; *p. 614:* Images provided by Google Earth mapping services/NASA, © DigitalGlobe, © Terra Metrics, © GeoEye, © Europa Technologies, Copyright 2014; *p. 615 (top center):* Stephen Marshak; *p. 616 (top left):* Mediacolor's/Alamy Stock Photo; *p. 616 (bottom left):* Stephen Marshak; *p. 616 (right):* Alaska Stock/Alamy; *p. 617 (top left):* Stephen Marshak; *p. 617 (top right):* Stephen Marshak; *p. 617 (bottom left):* Stephen Marshak; *p. 617 (bottom right):* AP Photo; *p. 618 (top right):* USGS/Barry W. Eakins; *p. 618 (bottom left):* USGS, Geologic Investigations Series I-2809 by Barry W. Eakins, Joel E. Robinson, Toshiya Kanamatsu, Jiro Naka, John R. Smith, Eiichi Takahashi, and David A. Clague; *p. 618 (bottom right):* Jerome Neufeld and Stephen Morris, Nonlinear Physics, University of Toronto; *p. 620:* Stephen Marshak; *p. 624:* AP Photo/Scanpix, Norway; *p. 625 (top left):* Breck P. Kent/JLM Visuals; *p. 626:* Stephen Marshak; *p. 627 (left):* National Airborne Service Corps; *p. 627 (bottom):* Images provided by Google Earth mapping services/NASA, © DigitalGlobe, © Terra Metrics, © GeoEye, © Europa Technologies, Copyright 2014; *p. 627 (bottom):* Images provided by Google Earth mapping services/NASA, © DigitalGlobe, © Terra Metrics, © GeoEye, © Europa Technologies, Copyright 2014; *p. 628:* Stephen Marshak; *p. 629:* AP Photo/The Deseret News, Ravell Call, File; *p. 630:* © 2008, All Rights Reserved, City of Seattle; *p. 632:* Images provided by Google Earth mapping services/NASA, © DigitalGlobe, © Terra Metrics, © GeoEye, © Europa Technologies, Copyright 2018.

CHAPTER 17

Page 634: Stephen Marshak; *p. 636 (top left):* Helen H. Richardson/The Denver Post/Getty Images; *p. 636 (top right):* John Gibson/Getty Images; *p. 636 (bottom left):* Andy Clark/Reuters/Newscom; *p. 636 (bottom right):* LUIS ROBAYO/AFP/Getty Images; *p. 637:* Stephen Marshak; *p. 637:* Images provided by Google Earth mapping services/NASA, © DigitalGlobe, © Terra Metrics, © GeoEye, © Europa Technologies, Copyright 2014; *p. 638 (right):* Stephen Marshak; *p. 638 (left):* Images provided by Google Earth mapping services/NASA, © DigitalGlobe, © Terra Metrics, © GeoEye, © Europa Technologies, Copyright 2014; *p. 639 (left center):* Stephen Marshak; *p. 640:* NASA; *p. 641 (center left):* Stephen Marshak; *p. 641 (center right):* Stephen Marshak; *p. 643:* Stephen Marshak; *p. 644 (top left):* Stephen Marshak; *p. 644 (top right):* Ron Niebrugge/Alamy Stock Photo; *p. 645 (all):* Stephen Marshak; *p. 648 (top):* Stephen Marshak; *p. 648 (center):* Amar and Isabelle Guillen- Guillen Photography/Alamy; *p. 648 (bottom):* Stephen Marshak; *p. 650:* Stephen Marshak; *p. 651 (all):* Stephen Marshak; *p. 652 (top right):* Stephen Marshak; *p. 652 (center right):* Stephen Marshak; *p. 652 (bottom left):* Marti Miller/Visuals Unlimited; *p. 652 (bottom right):* Courtesy of Jenny Jackson, Caltech; *p. 653 (top left):* Cedric Favero/Getty Images; *p. 653 (top right):* Stephen Marshak; *p. 653 (bottom):* Stephen Marshak; *p. 654 (top right):* NASA Earth Observatory; *p. 654 (inset):* NASA/GSFC/meti/ersdac/jaros, and US/Japan ASTER Science Team; *p. 655 (right):* Rob Crandall/Alamy Stock Photo; *p. 656:* NASA/GSFC/meti/ersdac/jaros, and US/Japan ASTER Science Team; *p. 658 (bottom right):* Images provided by Google Earth mapping services/NASA, © DigitalGlobe, © Terra Metrics, © GeoEye, © Europa Technologies, Copyright 2014; *p. 660 (left):* Stephen Marshak; *p. 660 (center):* Sigit Pamungkas/Newscom; *p. 660 (right):* Mike Hollingshead/Getty Images; *p. 661 (left) (both):* NASA images created by Jesse Allen, Earth Observatory, using data provided courtesy of the Landsat Project Science Office- copyright 2008; *p. 661 (right):* United Nations, Evan Schneider; *p. 662 (top):* Stephen Marshak; *p. 662 (bottom left):* Stephen Marshak; *p. 662 (bottom center):* Stephen Marshak; *p. 662 (bottom right):* Stephen Marshak; *p. 663 (right):* Stephen Marshak; *p. 663 (left):* Stephen Marshak; *p. 664 (top left):* U.S. Office of Naval Research; *p. 664 (bottom left):* Randy Pench/The Sacramento Bee via AP; *p. 665 (top left):* Reuters/Yoray Cohen/Eilat Rescue Unit; *p. 665 (top right):* Reuters/Danish Siddiqui/Newscom; *p. 665 (bottom left):* HANDOUT/Reuters; *p. 665 (bottom right):* USGS; *p. 666 (left):* Courtesy Johnstown Area Heritage Association; *p. 666 (right):* Mary Evans Picture Library/The Image Works; *p. 667:* © Tom Foster; *p. 668 (left):* Stephen Marshak; *p. 668 (right):* Stephen Marshak; *p. 669 (top right):* AP Photo/The News-Star, Margaret Croft; *p. 669 (bottom):* NASA; *p. 670 (left):* Scott Olson/Getty Images); *p. 670 (top right):* Scott Olson/Getty Images; *p. 673 (left):* Chinch Gryniewicz/UIG/Science Source; *p. 673 (right):* Photo courtesy of the Bureau of Reclamation; *p. 674:* Images provided by Google Earth mapping services/NASA, © DigitalGlobe, © Terra Metrics, © GeoEye, © Europa Technologies, Copyright 2018.

CHAPTER 18

Page 676: Stephen Marshak; *p. 677 (background):* Stephen Marshak; *p. 677 (top left):* Mountains in the Sea Research Team/IFE Crew/NOAA/OAR/OER; *p. 677 (center):* Topham/The Image Works; *p. 677 (top right):* Harbor Branch Oceanographic Institution/NOAA; *p. 677 (bottom right):* NOAA; *p. 677 (bottom right):* NASA; *p. 678 (top):* NOAA; *p. 678 (bottom left):* USGS; *p. 678 (bottom right):* William Crawford, Integrated Ocean Drilling Program/TAMU; *p. 681 (top):* NOAA; *p. 681 (bottom):* NOAA; *p. 682 (left):* NOAA; *p. 683 (top left):* http://www.geomapapp.org. Global Multi-Resolution (topography (GMRT) Synthesis by Ryan et al., 2009; *p. 683 (center right):* NOAA; *p. 683 (bottom):* Images provided by Google Earth mapping services/NASA, © DigitalGlobe, © Terra Metrics, © GeoEye, © Europa Technologies, Copyright 2014; *p. 684:* Premraj K.P/Alamy Stock Photo; *p. 686 (top):* Stephen Marshak; *p. 686 (center):* www.michaelmarten.com; *p. 686 (bottom):* www.michaelmarten.com; *p. 687 (left):* Wikimedia Commons; *p. 687 (right):* Imaginechina via AP Photos; *p. 691 (top right):* Photo by Phillip Capper/Wikimedia Commons; *p. 691 (left):* NOAA; *p. 691 (bottom right):* Stephen Marshak ; *p. 692 (right):* Stephen Marshak; *p. 693 (inset):* NOAA; *p. 694 (top):* Los Alamos National Laboratory; *p. 694 (bottom):* NASA; *p. 698:* NOAA; *p. 700 (top left):* Stephen Marshak; *p. 700 (top right):* Stephen Marshak; *p. 700 (bottom left):* Stephen Marshak; *p. 700 (bottom right):* Manfred Gottschalk/Alamy; *p. 702 (all):* Stephen Marshak; *p. 703 (bottom left):* Panther Media GmbH/Alamy Stock Photo; *p. 703 (bottom right):* NASA; *p. 704:* Stephen

Marshak; *p. 705 (inset):* Stephen Marshak; *p. 706 (top left):* G. R. 'Dick' Roberts © Natural Sciences Image Library; *p. 706 (center right):* Stephen Marshak; *p. 706 (bottom left):* Stephen Marshak; *p. 706 (bottom right):* Emma Marshak; *p. 707 (top right):* The Natiuonal Trust Photolibrary/Alamy Stock Phot; *p. 707 (bottom right):* Cody Duncan/Alamy; *p. 708 (left):* Stephen Marshak; *p. 708 (right):* Stephen Marshak; *p. 708 (bottom):* Images provided by Google Earth mapping Services/NASA, © DigitalGlobe, © Terra Metrics, © GeoEye, © Europa Technologies, Copyright 2017; *p. 709 (inset):* Steve Bloom Images/Alamy; *p. 709 (top left):* Stephen Marshak; *p. 709 (top right):* Stephen Marshak; *p. 709 (bottom left):* Stephen Marshak; *p. 715 (top left):* USGS; *p. 715 (bottom left):* USGS; *p. 715 (right):* Stephen Marshak; *p. 715:* Images provided by Google Earth mapping Services/NASA, © DigitalGlobe, © Terra Metrics, © GeoEye, © Europa Technologies, Copyright 2017; *p. 716 (top):* Stephen Marshak; *p. 716 (bottom):* Stephen Marshak; *p. 717:* imageBROKER/Alamy Stock Photos; *p. 718:* Images provided by Google Earth mapping services/NASA, © Digital-Globe, © Terra Metrics, © GeoEye, © Europa Technologies, Copyright 2018.

CHAPTER 19

Page 720: Stephen Marshak; *p. 721 (top left):* USGS; *p. 721 (bottom left):* Images provided by Google Earth mapping Services/NASA, © DigitalGlobe, © Terra Metrics, © GeoEye, © Europa Technologies, Copyright 2017; *p. 721 (bottom right):* Red Huber/Orlando Sentinel/MCT via Getty Images; *p. 722:* Images provided by Google Earth mapping Services/NASA, © DigitalGlobe, © Terra Metrics, © GeoEye, © Europa Technologies, Copyright 2017; *p. 723 (center left):* Photo courtesy of Eric Prokacki and Jim Best, University of Illinois; *p. 723 (bottom left):* Stephen Marshak; *p. 728 (top right):* Stephen Marshak; *p. 728 (bottom right):* Larry W. Smith/EPA; *p. 733 (bottom):* Stephen Marshak; *p. 734:* Vince Streano/Getty Images; *p. 735:* Vladimir Melnik | Dreamstime; *p. 736:* Stephen Marshak; *p. 737:* Emma Marshak; *p. 738 (top left):* Allan Tuchman; *p. 738 (top right):* Stephen Marshak; *p. 738 (bottom left):* Emma Marshak; *p. 738 (bottom right):* Stephen Marshak; *p. 739 (top):* Allan Tuchman; *p. 739 (top center):* Allan Tuchman; *p. 739 (bottom center):* Allan Tuchman; *p. 739 (bottom):* Stephen Marshak; *p. 740 (top):* Stephen Marshak; *p. 740 (center):* Stephen Marshak; *p. 740 (bottom):* Doll, P., Fiedler, K. (2008): Global-scale modeling of groundwater recharge. *Hydrol. Earth Syst. Sci.*, 12, 863-885; *p. 744 (top left):* Arizona Geological Survey; *p. 744 (top right):* Stephen Marshak; *p. 744 (bottom):* Stephen Marshak; *p. 747 (top):* Stephen Marshak; *p. 747 (center):* Francesco Tomasinelli/Science Source; *p. 747 (bottom):* Stephen Marshak; *p. 748:* Stephen Marshak; *p. 749 (center left):* Stephen Marshak; *p. 749 (center):* Karst landscapes of Guillin, China; *p. 749 (center right):* Stephen Marshak; *p. 749 (bottom):* Stephen Marshak; *p. 749 (top):* Images provided by Google Earth mapping Services/NASA, © DigitalGlobe, © Terra Metrics, © GeoEye, © Europa Technologies, Copyright 2017; *p. 750 (bottom right):* ML Sinibaldi/Media Bakery; *p. 750 (bottom left):* Paul F. Hudson, University of Texas; *p. 750 (top right):* Lois Kent; *p. 751 (bottom):* Stephen Marshak; *p. 751 (top):* Lois Kent; *p. 751 (center):* Ashley Cooper/Getty Image; *p. 752 (left):* NHPA/Photoshot/Photoshot/Superstock 752 (right):* Photo by Jim Pisarowicz/NPS; *p. 753:* Stephen Marshak; *p. 754:* Images provided by Google Earth mapping services/NASA, © DigitalGlobe, © Terra Metrics, © GeoEye, © Europa Technologies, Copyright 2018.

CHAPTER 20

Page 756: Stephen Marshak; *p. 757:* Nicolas LE CORRE/Gamma-Rapho via Getty Images; *p. 760:* InciWeb.gov; *p. 761:* Stephen Marshak; *p. 762 (left):* Stephen Marshak; *p. 762 (right):* Kathryn Marshak; *p. 764:* NASA; *p. 765:* Suranga Weeratuna/Alamy Stock Photo; *p. 768:* CIMSS; *p. 769 (left):* NASA; *p. 769 (right):* NASA; *p. 776:* Stephen Marshak; *p. 778:* NASA; *p. 779:* Stephen Marshak; *p. 780:* NASA; *p. 781:* Mike Hollingshead/Science Source; *p. 784 (top left):* Eric Nguyen/Getty Images; *p. 784 (top right):* NOAA, photo by Brian Bill; *p. 784 (center left):* AFP/Getty images; *p. 784 (center right):* Photo by Miami Herald/Getty Images; *p. 784 (bottom left):* FEMA Photo by Greg Henshall; *p. 784 (bottom right):* NOAA, photo by Brian Bill; *p. 785:* NASA; *p. 786:* NOAA; *p. 787:* NASA; *pp. 788-789 (background):* Google Earth mapping services/NASA, © DigitalGlobe, © Terra Metrics, © GeoEye, © Europa Technologies, Copyright 2017; *p. 788 (all):* Marco Antonio 2345, https://creativecommons.org/licenses/org/licenses/by-sa/3.0/deed.en; *p. 789:* NOAA; *p. 789:* Volkan Yuksel/Wikimedia Commons; *p. 789:* National Weather Service, Birmingham, AL; *p. 791 (top left):* NOAA; *p. 791 (top right):* AP Photo; *p. 791 (bottom left):* Images & Stories/Alamy; *p. 791 (bottom right):* Universal Images Group/Getty Images; *p. 792 (top right):* NASA; *p. 792 (top left):* Images provided by Google Earth mapping services/NASA, © DigitalGlobe, © Terra Metrics, © GeoEye,

© Europa Technologies, Copyright 2014; *p. 792 (bottom left):* Images provided by Google Earth mapping services/NASA, © DigitalGlobe, © Terra Metrics, © GeoEye, © Europa Technologies, Copyright 2014; *p. 793 (top left):* NOAA; *p. 793 (top right):* AP Photo/Susan Walsh; *p. 793 (center):* U.S. Geological Survey (2007) ; *p. 793 (bottom left):* POOL/AFP/Getty Images; *p. 793 (bottom right):* Stephen Marshak; *p. 794 (top):* Images provided by Google Earth mapping services/NASA, © DigitalGlobe, © Terra Metrics, © GeoEye, © Europa Technologies, Copyright 2014; *p. 794 (bottom):* Images provided by Google Earth mapping services/NASA, © DigitalGlobe, © Terra Metrics, © GeoEye, © Europa Technologies, Copyright 2014; *p. 795 (top):* AP Photo/David J. Phillip; *p. 795 (bottom left):* FAO-SDRN Agrometeorology Group; *p. 795 (bottom right):* NASA/GSFC; *p. 796 (top):* Stephen Marshak; *p. 796 (top center):* Stephen Marshak; *p. 796 (center):* Stephen Marshak; *p. 796 (bottom center):* Stephen Marshak; *p. 796 (bottom):* Stephen Marshak; *p. 797 (top):* FAO-SDRN Agrometeorology Group; *p. 797 (bottom):* NASA/GSFC; *p. 799:* Stephen Marshak.

CHAPTER 21

Page 802: Stephen Marshak; *p. 804:* Stephen Marshak; *p. 805:* Stephen Marshak; *p. 806 (top):* Professor Andre Danderfer; *p. 806 (bottom):* Stephen Marshak; *p. 807 (both):* Stephen Marshak; *p. 808 (both):* Stephen Marshak; *p. 809 (all):* Stephen Marshak; *p. 810 (top):* Liba Taylor/Getty Images; *p. 810 (center):* Stephen Marshak; *p. 810 (bottom left):* Shannon Arledge/USMC/Getty Images; *p. 810:* Stephen Marshak; *p. 810 (bottom right):* Daniel J Bryant/Getty Images; *p. 811 (bottom):* NASA/JPL-CALTECH/MSSS; *p. 811 (center):* O. Alamany & E. Vicens/Getty Image; *p. 811 (top):* Stephen Marshak; *p. 812 (all):* Stephen Marshak; *p. 813:* Images provided by Google Earth mapping services/NASA, © DigitalGlobe, © Terra Metrics, © GeoEye, © Europa Technologies, Copyright 2014; *p. 814 (top left):* Stephen Marshak; *p. 814 (top right):* Stephen Marshak; *p. 814 (inset):* Stephen Marshak; *p. 814 (center left):* Stephen Marshak; *p. 814 (center right):* Amar and Isabelle Guillen-Guillen Photography/Alamy; *p. 814 (bottom):* Stephen Marshak; *p. 815:* Stephen Marshak; *p. 816 (top left):* Stephen Marshak; *p. 816 (top right):* Stephen Marshak; *p. 816 (bottom):* Photo by Flicka. Sep. 2007. Wikimedia Commons; *p. 817 (both):* Stephen Marshak; *p. 818 (top):* Stephen Marshak; *p. 818 (bottom):* NASA; *p. 819 (all):* Stephen Marshak; *p. 820:* Stephen Marshak; *p. 821:* Stephen Marshak; *p. 822 (left):* Stephen Marshak; *p. 822 (center):* Stephen Marshak; *p. 822 (right):* Stephen Marshak; *p. 822 (bottom):* Images provided by Google Earth mapping services/NASA, © DigitalGlobe, © Terra Metrics, © GeoEye, © Europa Technologies, Copyright 2014; *p. 823 (left):* Eitan Simanor/Alamy; *p. 823 (right):* Stephen Marshak; *p. 824 (top):* Mark Phillips/Alamy; *p. 824 (bottom right):* JB Russell/Panos Pictures; *p. 825 (center):* BDR/Alamy; *p. 825 (inset):* Stocktrek Images, Inc./Alamy; *p. 825 (top left):* USGS; *p. 825 (top right):* USGS; *p. 825 (bottom):* Library of Congress; *p. 828:* Image courtesy Jacques Descloitres MODIS Rapid Response Team; *p. 829:* NASA/EarthKAM.org; *p. 830:* Images provided by Google Earth mapping services/NASA, © DigitalGlobe, © Terra Metrics, © GeoEye, © Europa Technologies, Copyright 2018.

CHAPTER 22

Page 832: Stephen Marshak; *p. 833 (both):* Stephen Marshak; *p. 835 (top left):* Shutterstock; *p. 835 (top center):* Stephen Marshak; *p. 835 (top right):* Stephen Marshak; *p. 835 (center):* Emma Marshak; *p. 835 (bottom):* Ted Spiegel/National Geographic Creative; *p. 836 (center left):* Stephen Marshak; *p. 836 (bottom left):* Stephen Marshak; *p. 836 (top right):* Stephen Marshak; *p. 836 (center right):* Galen Rowell/Getty Images; *p. 837:* NASA; *p. 838 (both):* NASA; *p. 840 (both):* Stephen Marshak; *p. 841 (top):* NASA/USGS; *p. 841 (bottom):* Emma Marshak; *p. 843 (right):* Antarctic Search for Meteorites Program, Linda Martel; *p. 844 (top left):* Stephen Marshak; *p. 844 (top right):* Ted Scambos, National Snow and Ice Data Center, Clevenger/Getty Images; *p. 844 (bottom left):* Stephen Marshak; *p. 844 (center left):* Ralph A. Clevenger/Getty Images; *p. 844 (center right):* Stephen Marshak; *p. 844 (bottom/center, right):* ESA; *p. 846 (top left):* Shutterstock; *p. 846 (top right):* Stephen Marshak; *p. 846 (center left):* Stephen Marshak; *p. 846 (center right):* Stephen Marshak; *p. 846 (bottom left):* Stephen Marshak; *p. 846 (bottom left):* Stephen Marshak; *p. 847 (bottom left):* Stephen Marshak; *p. 847 (bottom center):* Photo by Bruce F. Molnia, USGS; *p. 847 (bottom right):* Stephen Marshak; *p. 847:* Images provided by Google Earth mapping services/NASA, © DigitalGlobe, © Terra Metrics, © GeoEye, © Europa Technologies, Copyright 2014; *p. 848 (top):* Stephen Marshak; *p. 848 (inset):* Stephen Marshak; *p. 848 (bottom):* 986 Keith S. Walklet/Quietworks; *p. 849 (top left):* Stephen Marshak; *p. 849 (bottom left):* Stephen Marshak; *p. 849 (bottom right):* Marli Miller/Visuals Unlimited, Inc.;

p. 850 (top left): Stephen Marshak; p. 850 (top right): Max Gollin; p. 850 (bottom): Wolfgang Meier/zefa/Getty Images; p. 850 (bottom left): Images provided by Google Earth mapping services/NASA, © DigitalGlobe, © Terra Metrics, © GeoEye, © Europa Technologies, Copyright 2014; p. 851 (both): Stephen Marshak; p. 853 (top left): Stephen Marshak; p. 853 (top right): Stephen Marshak; p. 853 (inset): Stephen Marshak; p. 853 (center left): Stephen Marshak; p. 853 (center right): Stephen Marshak; p. 853 (bottom left): Stephen Marshak; p. 853 (bottom right): Kevin Schafer/Alamy; p. 854: Stephen Marshak; p. 855 (both): Stephen Marshak; p. 856 (top): All Canada Photos/Alamy Stock Photo; p. 856 (bottom): USGS photo by Bruce F. Molnia, U.S. Geological Survey; p. 857: Michael Beauregard; p. 861: © Tom Foster; p. 862: Images provided by Google Earth mapping services/NASA, © DigitalGlobe, © Terra Metrics, © GeoEye, © Europa Technologies, Copyright 2014; p. 863 (top): Lynda Dredge/Geological Survey of Canada; p. 863 (bottom): Shutterstock; p. 872 (top): PRISMA ARCHIVO/AlamyStock Photo; p. 872 (bottom): Stephen Marshak; p. 873 (top left): NaturePL/Superstock; p. 873 (right): Roger J. Braithwaite/JPL/NASA; p. 873 (center left): Michael Mclford/Getty Images; p. 873 (bottom/both): NASA; p. 874: Images provided by Google Earth mapping services/NASA, © DigitalGlobe, © Terra Metrics, © GeoEye, © Europa Technologies, Copyright 2018.

CHAPTER 23

Page 876: Stephen Marshak; p. 877: Ron Blakey, Colorado Plateau Geosystems; p. 881: Ron Blakey, Colorado Plateau Geosystems, Inc.; p. 881: Bournemouth News and Picture Service; p. 882 (all): Stephen Marshak; p. 885 (top left): Bob Sacha/Getty Images; p. 885 (bottom left): ISM/medicalimages.com; p. 885 (top right): Nick Krug/Lawrence Journal-World; p. 886 (left): Karim Agabi/Science Source; p. 886 insert: NOAA; p. 886 (right): Core Repository Lab at Lamont-Doherty Geological Observatory of Columbia University; p. 887 (left): Stephen Marshak; p. 887 (inset): Provided courtesy of JRTC & Fort Polk. Photography by Bruce Martin, Natural Resources Management Branch, ENRMD; p. 887 (right): Courtesy of the Climatic Research Unit, University of East Anglia; p. 889 (both): Ron Blakey, Colorado Plateau Geosystems, Inc.; p. 890 (left): NASA; p. 890 (both): Lunar and Planetary Institute; p. 891 (bottom left): Wikimedia Commons; p. 891 (top right): Stephen Marshak; p. 891 (bottom right): Stephen Marshak; p. 892 (top): NASA; p. 892 (center): Stephen Marshak; p. 892 (right): Stephen Marshak; p. 893: Mark Garlick/Science Source; p. 894: Stephen Marshak; p. 895 (bottom left): Courtesy P&H Mining Equipment; p. 895 (bottom center): Richard Hamilton Smith/Getty Images; p. 895 (bottom right): Stephen Marshak; p. 895 (top): Images provided by Google Earth mapping services/NASA, © DigitalGlobe, © Terra Metrics, © GeoEye, © Europa Technologies, Copyright 2014; p. 896 (top right): National Institute for Space Research (INPE) ; p. 896 (bottom left): Nigel Dickinson/Alamy; p. 896 (bottom right) (both): Images provided by Google Earth mapping services/NASA, © DigitalGlobe, © Terra Metrics, © GeoEye, © Europa Technologies, Copyright 2014; p. 897: Oliver Strewe/Getty Images; p. 898 (left): NASA Ozone Watch; p. 898 (right): NASA; p. 901 (top) (both): NASA/Goddard Space Flight Center Scientific Visualization Studio; p. 901 (bottom) (both): The Cryosphere Today, University of Illinois, Urbana-Champaign; p. 901 (both): National Snow and Ice Data Center; p. 902 (right): Jacques Descloitres, MODIS Rapid Response Team, NASA/GSFC; p. 903 (top left): USGS; p. 903 (top right): USGS photograph by Bruce Molnia; p. 903 (bottom left): Stephen Marshak; p. 903 (bottom right): Steven Kaziowski/Alamy; p. 904: Images provided by Google Earth mapping services/NASA, © DigitalGlobe, © Terra Metrics, © GeoEye, © Europa Technologies, Copyright 2014; pp. 910-911 (all): Images provided by Google Earth mapping services/NASA, © DigitalGlobe, © Terra Metrics, © GeoEye, © Europa Technologies, Copyright 2017; p. 912: Images provided by Google Earth mapping services/NASA, © DigitalGlobe, © Terra Metrics, © GeoEye, © Europa Technologies, Copyright 2018.

Index

ductile deformation, 393
dunes, **226**, **813**
   see also sand dunes
dung, as fuel, 523, *523*, 526
dust, wind-blown, 828, *828*
   see also dust storms
Dust Bowl in 1930s, 208, 825–827,
   827
dust (cosmic), *16–17*, 32, 33, *33*,
   34–35, 36
dust devils, 808, *809*
dust particles, 32, *34–35*
dust storms, 208, 601, **809**, 809–810,
   810, 811, 826, *827*
dwarf planets, **22**, 41
dynamic metamorphism, **261**, 261–
   262, *262*
dynamo, **387**
dynamothermal (regional)
   metamorphism, 262, **262**,
   264–265, 414, 424

Earth
   age of, 4, 9, 271, 476–479, 483,
      486, 758
   atmosphere of, 37, 43, 46–47
   axis of, 24, *78*
   biography of, 482–519
   changes in from plate motions,
      877
   circumference of, 23, *23*
   compared to other terrestrial
      planets, *43*
   density, 52
   differentiation of, *36*, 36–37,
      162–163
   elemental composition of, 49, *50*
   formation of, *34–35*
   future of, 908–909
   Goldilocks effect, 890
   map of, *47*, *49*
   orbital eccentricity, 869, *869*, 892
   orbital speed, 25
   orbit of, 25
   place in Solar System, 18, *21*
   as planet, 4–5, 9, 19
   precession of, 24, 869, *869*, 892
   pressure in, 50–51, 52–53
   rotation of, 24–25, 696
   seismic tomography image of, *379*
   as self-exciting dynamo, 77, 388
   shape of, 37
   surface of, 48–49, 60 (*see also*
      landforms; landscapes)
   surface temperature, *888*, 890
   temperature in, 51, *51*, 163
   tilt of axis of, 24, 43, 690, 767,
      769–770, 869, *869*, 892
   topographic map of, *49*, *392–393*
   topography of, 48, *49*
   viewed from space, *14–15*
   see also Earth history; Earth's
      interior; *specific topics*
Earth history, 482–519
   age of Earth, 4, 9, 271, 476–479,
      483, 486
   atmosphere formation, 35, 37
   collision with planetesimal and
      Moon formation, *34*, 36,
      163, 486, 878, *878*

differentiation, *36*, 36–37, *162–
   163*, 878, *878*
end of, 909
in exposed strata, *428–429*, 468–
   469, *469*
fossils and, 430–447
in geologic column, 464–467
geology's role in understanding,
   429
and human history, 479
meteorite bombardment, 36–37,
   54, 163, 479, 487, *487*, 878
methods for studying, 483–486
ocean formation, 35, 37
overview, *514–515*
proto-Earth, *34*, 878
in sedimentary rocks, *233*
   see also geologic time; *specific topics
   and time periods*
Earth materials, **49**, 49–50, 119
Earth-Moon system, 686, 688–689,
   690
earthquake engineering, 365–366,
   367
earthquake magnitude scales,
   340–341
earthquakes, 322–369
   adjectives for describing, *341*
   causes of, 325–333, 342–348
   in continental crust, 333, 341
   damage from, 5, 348–358
   deaths from, 324, *325*
   deep earthquakes, 343, 344, 345
   defined, **53**, **323**, *330*
   Earth's interior information from,
      53, 55
   energy released from, 53, 327–329,
      341, 341–342
   engineering and zoning for,
      365–366, *367*
   epicenter, defined, 95, 325, *326*
   epicenter, finding, 336–337, *337*,
      338
   epicenter, maps of, *326*, *342*, *345*,
      348
   factors affecting damage from,
      365
   faulting and, 53, *53*, 323, 325–
      329, *330–331*
   folklore of, 325
   and fracture zones, 75, 102
   household precautions, 366, *367*
   and induced seismicity, 347–348
   intensity, 338
   intermediate earthquakes, 342–
      343, 344
   introduction, *323*, 323–324
   location, *342*, 342–348, *347*
   magnitude (size) of, 338–342
   mass movement triggered by, 607,
      616–617, 621, 626–627
   measuring and locating, 333–337,
      338
   at mid-ocean ridges, 75, 97, 343,
      343
   notable earthquakes, *325*
   number per year, 324, *341*
   and plate boundaries, *94*, 95, 342,
      342, 343–345
   precursors of, 363–364
   predicting of, 359–364

preventing damage and injury
   from, 365–366, *367*
recurrence intervals, 360, 361–
   362, *362*
resonance in waves of, 366
and seafloor spreading, 75–76
seismic belts, 75, *76*
shallow earthquakes, 342, 343–
   344, 345
and subducted plates, 100
tectonic settings of, *342*, 342–348,
   347
and volcano eruption prediction,
   312, 313
as volcano threat, 310
   see also faults; seismic waves;
      specific locations
earthquake swarms, 348, 363
earthquake warning systems, 359, 364
earthquake waves. *see* seismic waves
earthquake zoning, 365
"Earthrise," *17*
EarthScope, *345*, 379, 380, *380*
Earth's interior, 51–62, 370–389
   basalt in modern view of Earth's
      interior layers, 56
   early images of, *52*, *53*
   earthquakes, information from,
      53, 55
   layers of, modern view, 55, 59, *56*,
      60–61, 62
   layers of, simplistic 19th-century
      view, 8, *8*, 51–55, *53*, 371,
      371–372
   meteorites, information from,
      55, *371*
   modern view of Earth's interior
      layers, 375, 378, *378*, 381
   seismic study of, 55, 371–382
   see also specific layers
Earth System
   anthropogenic impact on,
      894–898
   defined, **8**, 47, 877
   and global change, 876–913
   life processes and, 9–10
   and mass-transfer cycle, 276
   overview, 6, *6–7*, 8–9, 47–48
   rocks as insight into, 149
   sources of energy in, 6, 9, 525,
      525–526
   view from orbit, *63*
   see also specific topics
East African Rift
   cliffs, 420
   and continental rifting, 107, *108*,
      413
   and earthquakes, 346, *346*
   lava flows, 308
   rift-related mountains, 413
   triple junction, *108*
   viewed by astronauts, 107, *108*
   and volcanoes, 75, 308, 413
ebb tide, 686
echinoderms, 498
eclogite, 246, 257, *257*, 276
eclogite facies, 257, *257*
ecological niches, 441, 442, 446, 497,
   504, 513
economic minerals, 568
   see also ore minerals

ecosystems
   estuaries as, 705
   human modification of, 672, *895*,
      895–896, 897
   overview, **441**, 441–442, **895**
   Precambrian, *438*
eddies, 643, **693**, 695
Ediacaran fauna, **494**, *495*
Edinburgh, Scotland, 849
Edy, John, *523*
effusive eruptions, 294, **294**, 294–295,
   295, 308
Egypt
   Alexandria, 23
   Nile Delta, 654, *654*, 661, 895
   Nile River, 191, 640, 661, 672
   pyramids of, *597*
   Syene, 23
   Valley of the Mummies, 735
   see also Sahara Desert
E-horizon, 203, *203*
Einstein, Albert, 29, 91, 547
Ekati Diamond Mine, *142*
Ekman spiral, 695
Ekman transport, **695**, *698*
elastic behavior, **327**
elastic-rebound theory, **329**, 364
elastic strain, 327
electric power plants
   dynamos, 387
   hydroelectric power plants, *551*,
      553, *554*, *561*
   switch to natural gas for electricity
      generation, 539, 545
   see also nuclear power plants
electrochemical devices, 526
electromagnetic radiation, 25
electromagnets, 387
electron clouds, 28
electronic seismometers, 335, 336
electron microprobes, 121, 157, *158*
electron microscopes, 121, 127, *127*,
   157, 218
electrons, 28, 124
electron shells, *28*, **28**, 124
electrostatic attraction, 620
elements
   in crust, 57
   defined, **28**, **124**
   formation in Big Bang, 27, 31, 32
   formation in stars, 31–32, 49
   top ten elements in Milky Way, *32*
   in whole Earth, 49, *50*
elevation and climate, 795–796
elevation differences. *see* topographic
   maps; topographic profiles
Elk River contamination, 743
Ellesmere Island, 900
Elm, Switzerland, 615
El Niño, **797**, 797–798, *799*
elongate (cigar-shaped) grains, 247,
   248
El Salvador, 163, 292
embayments, 704, *705*, 706
emerald, 142, 143, *143*, 144
emergent coasts, 711, **711**
Emperor seamount chain, 105, *106*, 113
Empire State Building–geologic time
   analogy, *478*, 479
Enceladus (moon of Saturn), *22*, 318,
   603, *603*

and growth of continental crust, *489, 491–492*
ice ages in, 494, *495, 867,* 868, 883
life forms in, 465, 492–494, *495,* 879
mountain belts of, *71*
tillites from, 868
Protista (kingdom), 437, *440,* 445, 446
protocontinents, 488, *488*
proto-Earth, *34,* 878
proto-life, 445, 446, *446,* 493
protolith, defined, *243, 264*
protons, **28**
protoplanetary disk, **32,** *32–33, 34,* 36, 41, 486, 758
protoplanets, *33, 34,* 36–37, 486
protostars, **30**
proto-Sun, *34*
protozoans, ciliate, 493
pterodactyls, 504, *505*
pterosaurs, 509
Ptolemy, 18, *19*
P-T-t (pressure-temperature-time) path, 258, *258*
Puerto Rico, *676–677, 702, 749,* 794
Puerto Rico Trench, *73*
Pulido, Dionisio, 284
Pulpit rock, *850*
pumice, **180,** *181,* 290
pumice lapilli, 287, *288*
pumps, oil, *530,* 531, *534, 535, 536*
punctuated equilibrium, 443–444
Puyehue-Cordón Caulle volcano, *319*
P-waves (primary waves)
defined, 333, *334, 336, 349, 349*
discovery of crust-mantle boundary and, 374, *374*
and ground motion, 349, *349*
propagation through solids and liquids, 372, *372, 373*
refraction at core-mantle boundary, 376
S – P (S minus P) time, 336, *338*
velocity, 334, 372, 374–375, *375*
see also seismic waves
P-wave shadow zone, **376,** *377*
pyramids of Egypt, *597*
pyrite
chemical formula, *134,* 137, *197,* 557
in coal, 545, 546, 557
metallic luster, *134*
oxidation, 196, *197,* 497
sulfuric acid formation from, 545, 557
pyroclastic debris
from andesitic and rhyolitic eruptions, 287, *288*
from basaltic eruptions, *287*
defined, **161,** 284
from explosive volcanic eruptions, 161, *162, 172, 178,* 287, *288,* 300
fragment sizes, 172
from Plinean eruptions, 299
rock formed from, 161, 177, 180–181, 286
tephra, *283,* 286, *287,* 291–293, *293, 294, 297*

threat from, 308, *309*
at Yellowstone, 306–307
see also ash; fragmental igneous rocks
pyroclastic deposits, 180, 284, 286–288, *289*
pyroclastic flows
defined, **172,** *297,* **299**
Mt. Pelée, Martinique, 301, 308
prediction of, 313
threat from, *296, 297,* 308, *309*
from volcanic explosions, *171, 172, 297, 299*
in volcanic-hazard assessment map for Mt. Rainier, *314*
see also nuée ardente
pyroclastic rocks, **180,** *180–181, 181,* **286**
pyroxene
crystallization from magma, 169, 170, *170*
in eclogite, 246
in gneiss, 251
planes of cleavage, *135*
single chains of silica tetrahedra, *139,* 140, 170
stability of, *197*

Qaidam Basin, *417*
quarries
crushed-stone quarries, 576, *578, 581*
defined, 576
dimension stone extraction, 576, *578*
human-made outcrops, *152*
limestone quarries, 10, *12, 122, 219,* 578, *579, 581*
marble quarries, *254, 405*
travertine quarries, 221
quartz
as amethyst, 121, *132*
and Bowen's reaction series, *170, 170*
as cement, 214, 216
in cement, 577, 578
chemical formula, *123,* 137, 140
and coesite, 244, 246, 263
colors, 133, *134*
compression of, 409–410, *410*
conchoidal fractures, *136*
cryptocrystalline quartz, 218
crystallization from magma, 170, *170*
crystals of, *123,* 126, *126, 140*
crystal structure, *123*
framework silicate, *139,* 140, 170
in gneiss, *244,* 251
in granite, 568
hardness of, 133, *133*
and hydrolysis, 196
in joints, 397, *399*
in metamorphic rocks, 244, 249, 251, 253
and metamorphism, 244, 246, 251, 252–253
in New York cement, 579
in photovoltaic cells, 579
and recrystallization, 252
in sand, 198, 217, 700
and sandstone, 212

in schists, 251
in sediment transported by glaciers, 579
in shale, *244*
shocked quartz, 509
as silicate, 137, 198
solubility of, 195
specific gravity, 133, 135
stability of, *197*
and weathering, 195, 196, 198, *198*
quartzite
deformation, 392, *394, 395, 397,* 399
foliated quartzite, 253
metamorphism of quartz sandstone, 252–253, *254, 256,* 265, 392
outcrops, 266
overview, **252–253**
quartz grains in, 392, *394*
quartzo-feldspathic metamorphic rocks, 255
quartz sandstone (quartz arenite), *215,* 216, 217, 252, *254,* 255
quartz veins, gold in, *567,* 570
Quaternary Period beginning date, 512
Queen Charlotte fault system, 511
quenching, 170, 261
quick clay, 353, **621,** *624*
quicksand, 352

Racetrack Playa, 813, *814*
radial drainage networks, 638, *638*
radiant energy, 20
radiation sickness, 549
radiative heating, 58, *58*
radioactive decay, 163, *469, 470,* 470–471, 476, 486
radioactive elements, **470**
radioactive isotopes, 470, 471, *471,* 473
radioactive materials, pollution from, 741, 897
radioactivity
discovery of, 476–477
energy from, 526, 547
and heating of Earth, 163, 476, 486
radiocarbon dating, 362, 471
radiometric dating, 449, 470, 510
see also isotopic dating
radio transmissions, and ionosphere, 765
railroad building, through Sierra Nevada, 149
railroad cuts, *149, 152*
rain bands of hurricanes, 787, *787*
raindrops, formation of, 777, *777*
rainfall
atmospheric water extracted by, 758
at cold fronts, 772
in deserts, 803, 805, 807–808
in hurricane, 790
from low-pressure system, 774
during Pleistocene Ice Age, 865
and soil formation, 202, 203, 205, *206*
and water table, 727, *728*

rainforests, *2,* 207, *796,* 895–896, *896, 896*
see also deforestation
rain shadow, **805**
rain-shadow deserts, 804–805, *805*
rapids, 449, 646, *649, 651, 662*
rare earth elements (REE), **582,** *582–583,* 584
reach, of stream, 635, 646, 647, 651
reactants, 125
reactivated faults, 329, 347, *409,* 423
recessional moraines, 854, *854, 859*
recharge areas, *730,* **730,** *733, 735,* 736, 737, 741, 742
recharge (groundwater), *730,* 732, *733,* 739–741, *740, 744*
recrystallization
of clay, 250, *251*
defined, 244
during dynamic metamorphism, 261
and fossils, 432
during marble formation, 253
of mylonite, 401
of quartzite, 252
of sandstone, 244, *246*
rectangular drainage networks, 638, *638*
recurrence intervals
defined, 312, **360, 669**
earthquakes, 360, 361–362, *362*
floods, 669–670, 671
volcanoes, 312
redbeds
defined, **230**
depositional environments, 230, *469, 497,* 503
formation in Late Devonian, 500, *501*
horizontal bedding, *225, 232*
Red Cross, 666
red giants, 31, 909
Redoubt Volcano, Alaska, *299,* 310
Red Sea, 107, *108*
red shift, **26,** *26–27*
Redwall Limestone, *461,* 462, *462, 467*
reef bleaching, 717
reefs
carbonate reefs, 235–236, *236*
continental drift and, 70
see also coral reefs
REEs (rare earth elements), 582–583, 584
reference spheroid, 383–384, *384,* 385
refineries, 535, *536,* 555–556, 557, 558
reflection, *373,* **373**
refraction
defined, **373,** **692**
light waves, 373, *373*
ocean waves, *692,* 692–693
seismic waves, 373, 374, 376, *377*
refractory materials, **33**
reg, 852
regional basins, 422–423
regional correlation, 466, *467*
regional divides, 639
regional domes, 422–423
regional (dynamothermal) metamorphism, **262,** *262, 264–265,* 414, 424

unidirectional change, 877, 878–879
uniformitarianism, principle of, 443,
    450, 450–451, 452, *453,*
    458, 476, 667
Union Pacific railroad, 149
United Kingdom, *173,* 864
    see also England; Scotland; Wales
United States
    aquifers in, 725, *726*
    coal consumption, *544,* 545
    gravity anomalies, *385*
    during ice age, *866,* 866–867, *867*
    landslide-potential map of
        western United States, *630*
    Pleistocene deposits in, *867*
    seismic-hazard map of, *363*
    stockpiling of strategic resources
        by, 584
    yearly per capita usage of geologic
        materials in, *584*
    see also Midwestern United States;
        North America; *individual
        states*
universal ocean, 151
Universe
    about, 17–25
    age of, 27
    Big Bang theory and, 27, *27*
    defined, **17**
    expanding Universe theory,
        26–27, *27*
    formation of, 25–30
    size of observable Universe, 24
    structure of, 17–19
unloading, *591*
unsaturated air, 763
unsaturated zone (vadose zone),
    726–727, *727,* 729
unstable slopes, 620, 630
unstratified drift, 852
updrafts, 763
uplift
    during Alleghanian orogeny,
        500–501
    and antecedent streams, 659, *659*
    basement uplifts, 500–501,
        506–507, *507, 508*
    causes of, 591
    and chemical weathering
        reactions, 889, *889*
    defined, 391, **415,** 591
    and erosion, 421, 594
    and global climate change,
        888–889, *889*
    rates of, 595
    relief generated by, 594, 595
    in Southern California, 627
upper mantle
    defined, **57,** 375
    depth, 57
    layers in, *376*
    melting in, 59, 164, *164,* 290, 374
    in modern view of Earth's interior
        layers, 56
    olivine in, *345,* 375
    seismic wave velocities in, 375
    transition zone, 56, 57, 375
    see also asthenosphere; lithospheric
        mantle
upstream region, 635

upwelling, 59, **112,** *112,* 304, 381,
    381, 385
Ural Mountains, 501
uranium
    atomic mass, 124–125
    atomic number, 28, 124
    geology of, 547–548, *551*
    isotopes, 470, *471,* 547
    nuclear fission, 29, 547
    uranium-lead dating, *471,* 475
    uranium reserves, 560
    weapons-grade uranium, 548–549
uranium enrichment, 548–549
Uranus, 19, *21,* 42
urbanization
    ecosystem destruction from, 895,
        895, 896
    effect on streams, 672, *673*
    and Everglades, 741, *742*
U.S. Comprehensive Soil
        Classification System,
        205, *205*
U.S. Environmental Protection
        Agency, 761
USArray, 380, *380*
U-shaped valleys, *848,* **848,** 850, *859,*
    864
Ussher, James, 449–450
Utah
    Arches National Park in, 397, *397,*
        399, 817
    arid landscape in, *802–803*
    Basin and Range Province in, 413
    bedding in sedimentary rock,
        155, 225
    Bingham Canyon Copper Mine,
        574, 585, 629, *629*
    Bonneville Salt Flats in, 220
    Bryce Canyon in, *230, 239,* 466,
        467, 816, *816*
    Canyonlands National Park in,
        638, 816
    cave in, *747, 751*
    Cedar Breaks National
        Monument, 466, *467*
    debris flow in, *289,* 611, 629, *629*
    earthquake epicenters in, *326*
    exposed sedimentary rock in, *2,*
        210–211
    floodplain and terraces in, *650*
    Great Salt Lake, 813, *814*
    pluvial lakes, 864
    quartzite beds in, *397*
    rockfall and talus pile in Uinta
        Mountains, *617*
    sandstone outcrop in, *158,* 620
    San Juan River in, 658, *658*
    slumping in, *610*
    strata exposed by river, *428–429*
    unconformities in road cut, *457,*
        459
    volcanic debris flow, *289*
    Wasatch Mountains in, *267,* 645
    Zion National Park in, *228,* 503,
        504, 821

vacuum, **41**
vadose zone, 726, *727*
Vaiont Dam, rockslide disaster at,
    613–614, *615*

Valdez, Alaska, tsunami damage of,
    357, *358*
Valles Marineris, *601,* 603
Valley and Ridge Province,
    Pennsylvania, 424, *425,*
    503
valley glaciers, *836,* 837, *837,* 839,
    841, 850, *859,* 864
valley glaciers, shrinking of, 901, *903*
Valley of the Mummies, 735
valleys
    alluvium-filled valleys, 648, *650,*
        658
    creation by rivers, 420, 648–649
    defined, **648**
    drowned valleys, *701, 711*
    and glacial erosion, 847–848, *859*
    hanging valleys, 649, *651,* 864
    shape of, 648, *649,* 848
    U-shaped valleys, 848, *848,* 850,
        859, 864
    V-shaped valleys, 648, *648, 649,*
        662, 848
Van Allen radiation belts, 43, *45*
Vancouver, air temperature in, 685
varve, **852,** *853*
vascular plants, 500
vectors, 620, *688,* **689,** *689,* 697
vegetation, and slope stability,
    624–625
vegetation types and global warming
    effect, *907*
vegetation types and soil formation,
    204, 205
vein deposits, 570, *570*
veins, *249,* **249,** *398, 399,* 570
velocity-*versus*-depth curve, *375, 378,*
    378
Venezuela, tar sand in, 539
Venice, flooding of, 741, *744*
ventifacts, *811,* **811**
vents, defined, **291**
Venus
    atmosphere of, 46, 48, 603, 758,
        890
    atmospheric pressure, 47
    Earth contrasted with, 10, 877
    inner (terrestrial) planet, 19, *21,*
        42, *43*
    lack of magnetic field, 387
    lack of moons, 22
    landscape of, 317, *318, 601,* 603
    layers of, 60
    radar image, *601*
    runaway greenhouse effect, 890
    temperature of, 47, 603, 890, *890*
    volcanic edifices on, 317, *318,* 419
Vermilion Cliffs, 466, *467*
Vermont, 219, 617
Verne, Jules, 52
vertical-motion seismometers, 335,
    335
vertisols, 205, *206*
vesicles, **179,** 179–180, *181,* **290,** 723
vestigial organs, 445
Victoria, British Columbia, glacial
        grooves in, *846*
Vietnam, cave in, *720–721, 747*
Vikings, *891,* 892
Virginia, 347, 743, *751*

Virgin Islands, 699, *700*
viscosity
    defined, **167, 283,** 526
    of hydrocarbons, 526, *527,* 535,
        537, 538, *539*
    of lavas or magmas, 167, *168,*
        171–172, *283,* 283–284
    and temperature, 167
vog, 347
volatiles (volatile materials)
    defined, **33,** 50, 164
    in magma, 165, 167, 290, 758
    melting triggered by, 164, *165,*
        168, 183
    and viscosity, 167, 172
volatility, 526
volcanic agglomerate, **181**
volcanic arcs
    Aleutian Arc, *182,* 183, *183,* 303,
        303
    continental volcanic arcs, 101,
        183–184, 303
    convergent boundaries and, *95,*
        101, *101,* 104, 183, 303, 411
    defined, **74, 101,** *111,* **183**
    magma variation in, 303
    subduction and, *100,* 101, *101,*
        175, 411
    trenches and, 74, *75, 111, 183*
    volcanic island arcs, 101, *101,*
        183, 303
volcanic ash. *see* ash
volcanic blast, as volcano threat, 310
volcanic breccia, *181,* **181**
volcanic conglomerate, 288
volcanic debris flow, 288, *289,* 293,
    310
volcanic edifice, 291, 294, 304
volcanic eruptions
    along convergent boundaries, 303
    along mid-ocean ridges, 302–303
    carbon dioxide produced by, 165,
        290, 310, 758, 868, 882,
        888, 899
    cool climate due to, 315–316, 892
    crater eruptions, *291*
    defined, **161**
    explosive eruptions, *171,* 172, *294,*
        295, 296–297, 298–301
    and extinction, 444
    fissure eruptions, 291, *291,*
        304–305
    flank eruptions, 292
    geologic settings of, 302–308
    as hypothesis for anomalies at
        Midwest site, 10
    and long-term climate change, 888
    and mass extinction, 444, 502,
        893–894
    memorable examples of, *300,*
        300–301, *301*
    plate tectonics and, 302–308
    prediction of, 312–313
    products of, 283–290
    protection from, 311–315
    pyroclastic debris from, 161, *162*
    summit eruptions, 291, *291*
    through fissures *vs.* circular vents,
        291, *291*
    see also lava flows; *specific topics*